MICRO BIOLOGY

THE HUMAN EXPERIENCE

John W. Foster
University of South Alabama

Zarrintaj Aliabadi
University of South Alabama

Joan L. Slonczewski
Kenyon College

W·W·NORTON

NEW YORK · LONDON

W. W. Norton & Company has been independent since its founding in 1923, when William Warder Norton and Mary D. Herter Norton first published lectures delivered at the People's Institute, the adult education division of New York City's Cooper Union. The firm soon expanded its program beyond the Institute, publishing books by celebrated academics from America and abroad. By midcentury, the two major pillars of Norton's publishing program—trade books and college texts—were firmly established. In the 1950s, the Norton family transferred control of the company to its employees, and today—with a staff of four hundred and a comparable number of trade, college, and professional titles published each year—W. W. Norton & Company stands as the largest and oldest publishing house owned wholly by its employees.

Editor: Betsy Twitchell
Developmental Editors: Carol Pritchard-Martinez and Michael Zierler
Associate Managing Editor, College: Carla L. Talmadge
Copyeditors: Janet Greenblatt and Gabe Waggoner
Assistant Editors: Katie Callahan and Taylere Peterson
Managing Editor, College: Marian Johnson
Managing Editor, College Digital Media: Kim Yi
Production Manager: Sean Mintus
Media Editor: Kate Brayton
Associate Media Editor: Cailin Barrett-Bressack
Media Project Editor: Jesse Newkirk
Assistant Media Editor: Victoria Reuter
Media Editorial Assistant: Gina Forsyth
Digital Production: Mateus Teixeira
Marketing Manager, Biology: Todd Pearson
Design Director: Rubina Yeh
Design, Chapter Openers, and Life Cycle Figures: DeMarinis Design LLC
Photo Editor: Nelson Colon
Photo Researcher: Dena Digilio Betz
Permissions Manager: Megan Schindel
Permissions Associate: Elizabeth Trammell
Composition and Illustrations: Precision Graphics, a Lachina Company
Manufacturing: Transcontinental

Permission to use copyrighted material is included at the back of this book.

ISBN: 978-0-393-60257-9

W. W. Norton & Company, Inc., 500 Fifth Avenue, New York, NY 10110
wwnorton.com
W. W. Norton & Company Ltd., 15 Carlisle Street, London W1D 3BS

1 2 3 4 5 6 7 8 9 0

We dedicate this book to the thousands of health care workers who voluntarily risked, and in some cases gave, their lives to confront the Ebola epidemic that ravaged East Africa in 2014. Their selfless commitment to ease human suffering and conquer an infection that threatened the world was nothing short of heroic. Today (May 2017) the world calls on these dedicated professionals once more as a new cluster of Ebola cases erupts in the Democratic Republic of Congo.

[Contents]

Part II Essential Biology and Control of Infectious Agents

8
Bacterial Genetics and Biotechnology 222

Dental Biofilm Inflames the Gums 223

9
Bacterial Genomes and Evolution 258

Diphtheria, Toxins, and Natural Disasters 259

10
Bacterial Diversity 288

Poor Dental Care Comes Back to Bite 289

11

Eukaryotic Microbes and Invertebrate Infectious Agents 318

A Desert Sand Fly Parasite 319

12

Viruses 350

Virus from a Needle Stick 351

13
Sterilization, Disinfection, and Antibiotic Therapy 386

A Needless Death 387

Part III The Immune System

14
Normal Human Microbiota: A Delicate Balance of Power 428

Case of the Unsatisfactory Stool 429

Part V Epidemiology and Biotechnology

25
Diagnostic Clinical Microbiology 842

26
Epidemiology: Tracking Infectious Diseases 878

27

Environmental and Food Microbiology 906

Explosive Toxic Soil 907

[About the Authors]

John W. Foster received his BS from the Philadelphia College of Pharmacy and Science (now the University of the Sciences in Philadelphia) and his PhD from Hahnemann University (now Drexel University School of Medicine), also in Philadelphia, where he worked with Albert G. Moat. After postdoctoral work at Georgetown University, he joined the Marshall University School of Medicine in West Virginia. He is currently teaching in the Department of Microbiology and Immunology at the University of South Alabama College of Medicine in Mobile, Alabama. Dr. Foster has coauthored three editions of the textbook *Microbial Physiology* and has published more than 100 journal articles describing the physiology and genetics of microbial stress responses. He has served as Chair of the Microbial Physiology and Metabolism division of the American Society for Microbiology and as a member of the editorial advisory board of the journal *Molecular Microbiology*.

Zarrintaj (Zari) Aliabadi is a physician assistant and a microbiologist. She received her Pharm D from the University of Tehran College of Pharmacy, in Iran, and her PhD in biomedical sciences from Marshall University in Huntington, West Virginia, where she worked with John W. Foster. After her postdoctoral work at the University of South Alabama (USA), in Mobile, Alabama, she joined the Department of Biochemistry at the USA College of Medicine, where she taught biochemistry and conducted research on sickle-cell anemia. Dr. Aliabadi then earned a master's in health sciences from the USA Physician Assistant Studies Program, practiced medicine as a PA in endocrinology, served as director of the USA Diabetic Foot Clinic, and became Chair of the USA Physician Assistant Studies Program. Recently she was named professor emeritis for her contributions. Dr. Aliabadi has taught extensively on infectious disease, pathophysiology, and clinical medicine to undergraduate pre–health profession students, graduate physician assistant students, and medical students. Her publications span the realms of microbiology and medicine.

Joan L. Slonczewski received her BA from Bryn Mawr College and her PhD in molecular biophysics and biochemistry from Yale University, where she studied bacterial motility with Robert M. Macnab. After postdoctoral work at the University of Pennsylvania, she has since taught undergraduate microbiology in the Department of Biology at Kenyon College, where she earned a Silver Medal in the National Professor of the Year program of the Council for the Advancement and Support of Education. She has published numerous research articles with undergraduate coauthors on bacterial pH regulation and has published six science fiction novels, including *A Door into Ocean* and *The Highest Frontier,* both of which earned the John W. Campbell Memorial Award. She served as At-Large Member representing Divisions on the Council Policy Committee of the American Society for Microbiology and as a member of the editorial board of the journal *Applied and Environmental Microbiology*.

When we began writing this textbook, we asked ourselves whether a new introductory microbiology textbook was needed at all. Could we create a book with a fresh look and more vibrant appeal for the allied health and non–science major students taking the course? We studied the approaches taken by other textbooks and reflected on how we, ourselves, teach microbiology to health career students. Our conclusion was that the currently available textbooks do not fully realize a truly human-centric approach. We believe that this is a missed opportunity, for what could be more captivating than learning how tiny, unseen, living things can so greatly influence human life and death? As a result, we decided to write our book so that human experiences drive all discussions, from basic microbiological concepts to infectious diseases.

Learning the basic concepts of microbiology from a human health perspective is something we find allied health students really enjoy. After all, microbiology as a science was forged largely from a desire to ease human suffering, so why not embrace that theme while teaching microbiology to students who share that desire? From the death of a small child with measles to the pneumonia that killed an elderly grandparent, infectious diseases have been a driving force in our biological and intellectual evolution for millions of years. Even today, infectious diseases kill two-thirds of the nearly 9 million children who die each year. Understanding how microbes live, grow, and die is a good way to gain some control over our own mortality.

We find that students interested in pursuing health careers become energized when aspects of medicine are injected early and often. In Chapter 2, for instance, we quickly introduce the basics of infectious disease and then employ multiple case histories to illustrate those concepts. This strategy, pairing the explanation of core microbiology concepts with applications to medicine and human health, is maintained throughout the book. More than 100 clinical case histories propel coverage of basic microbiological concepts such as microbial growth, metabolism, genetics, differentiation, ecology, and immunology. The chapters on immunology, for instance, take the novel approach of using patient case histories to highlight what happens when one part of the immune system fails and then build a discussion of the immune system around those failures. Cases in every chapter provide a window through which students can unravel the mystery of infections: how they happen, how they are diagnosed, and how they are treated. Sections of several chapters also reveal important connections between the environment, climate change, and emerging infections, a growing concern among national and world health organizations.

Many features of our book lend themselves to new as well as traditional teaching paradigms. The writing style is engaging and conversational and includes a touch of humor to drive learning. Clearly written explanations and a lavish art program support the "flipped" classroom approach in which students learn independently through reading before attending interactive classroom sessions. The case histories themselves can be used to design innovative, team-based, active-learning exercises. We also know that health career students learn microbiology most easily when it relates to their own infectious disease experiences. Toward that end, frequent thought exercises embedded in each chapter revolve around infections that students may have had or may wonder about. These exercises link concepts from earlier chapters while developing a student's critical thinking skills. For these and many other reasons, we think you will find the pedagogical tools included in this book superior to those of other undergraduate allied health textbooks.

Over the years, we have taught microbiology to a wide variety of undergraduate, graduate, and health career students. But beyond that, we've listened to dozens of colleagues and thousands of students over the past decade while we wrote. Then, in 2015 we published a preliminary version of our textbook, which has now been adopted at more than 75 colleges and universities across North America. We made this decision because we wanted to be certain the content and tone of our book was appropriate for introductory microbiology students. We are honored to have so many early fans of our book. But more importantly, we are thankful for the feedback offered by those adopters, which we were able to use in preparing this edition.

John W. Foster
Zarrintaj (Zari) Aliabadi
Joan L. Slonczewski

Major Features

CHAPTER-OPENING CASE HISTORIES Each chapter opens with an elegantly illustrated case history that sets the theme. For example, the chapter on biochemistry (Chapter 4) begins with a small boy stricken with cholera; our chapter on bacterial growth (Chapter 6) starts with a newborn suffering from meningitis; and the bacterial metabolism chapter (Chapter 7) begins with a college student vacationing in Mexico who contracts shigellosis. Later, we explain how these cases connect to the basic concepts presented in the chapter.

CHAPTER AND SECTION OBJECTIVES Every chapter begins with a set of objectives that outline the major concepts that will be covered in the following pages. The chapter objectives are followed by individual section objectives and section summaries that alert students about what they should be able to explain, discuss, and compare after reading each section.

CONCEPTUAL LEARNING THROUGH EMBEDDED CASE HISTORIES In addition to the chapter-opening cases, there are more than 100 patient cases integrated into this textbook. Some are real, some are embellished, but all of them convey the human toll of infectious disease and why we as a species chose to explore the hidden world of microbes in the first place. Each story provides a focal point for discussing:

- Basic concepts of microbiology (Chapters 1–13) such as microbe structure, genetics, biochemistry, biotechnology, antibiotics, and disinfectants
- The human microbiome and our immune responses to infection (Chapters 14–17). Topics include innate immunity, adaptive immunity, immunological diseases, and immunological tools
- Microbial pathogenesis and infectious diseases by organ system (Chapters 18–24), including skin, respiratory, systemic, gastrointestinal, urinary, reproductive, and central nervous systems
- Clinical microbiology and epidemiology (Chapters 25 and 26)
- Environmental and food microbiology (Chapter 27)

The cases not only provide a framework for concept building, but repeatedly outline how diseases are diagnosed and how they are tracked and treated. Some cases explore the intimate connections between evolution and emerging diseases; between climate change and epidemiology; and between the immune system and the severity of disease.

BACKWARD AND FORWARD LINKS An important part of learning is to connect new concepts to earlier ones. Toward that end we have placed a series of links within each chapter that recap relevant concepts presented in earlier chapters and sometimes foretell important concepts. The links redirect students to relevant sections of the book if they need to refresh their knowledge.

END-OF-CHAPTER ASSESSMENT QUESTIONS Questions at the end of each chapter are tied to learning outcomes for each section and allow students to quickly evaluate their mastery of the material.

THOUGHT QUESTIONS Opportunities for critical thinking are central to student learning. Scattered throughout every chapter are eight to ten thought questions that help students think about the concepts they just read and connect them to concepts in other chapters. Answers to these questions are sometimes difficult but always insightful; and if students get stumped, the answers can be found at the end of the chapter.

CLINICAL CORRELATION QUESTIONS Because this textbook was designed in large part for students interested in health science careers, each chapter ends with clinical correlation questions. These questions are designed around clinical scenarios that students must evaluate by explaining how they would diagnose, treat, or track a pathogen.

THE ORGAN SYSTEM APPROACH TOWARD TEACHING INFECTIOUS DISEASE There are two basic ways infectious diseases are taught: the taxonomic approach, in which all the diseases a single microbe can cause are presented together, and the organ system approach, which focuses on an individual organ system and the array of pathogens that infect it. From the health care provider's standpoint, the organ system approach is the most useful way in which information on infectious disease is compiled because it reflects how a clinician interacts with a patient. Sick patients visiting a clinician will describe their symptoms ("I have a fever, cough, and chest pain"). This tells the clinician which organ system is principally involved (the respiratory system in this case). Then the clinician, knowing what pathogens can infect that organ system, will collect appropriate clinical samples to scientifically confirm the microbiological cause and, while waiting for that result, prescribe an antibiotic, if called for, that can stop the growth of the most likely pathogens. In this textbook, we use the organ system approach in a way that does not require the instructor to learn medicine, yet hooks the student on learning microbiology through real-world medical examples.

Art Program

The vast majority of concepts introduced in our book take place at a scale not visible to the naked eye. Yet, one of the most important skills introductory students can master is the ability to visualize key microbial processes and structures. As a result, we focused considerable effort on making sure that the images selected and developed for our book achieve the highest level of visual engagement and pedagogical value.

- Figures depicting processes and structures are colorful and engaging and include helpful bubble captions that provide essential information students need to learn from the figure.
- Meticulously designed life cycle figures combine drawn art, photographs, and a system of helpful schematic arrows that convey the key steps in the progression of important diseases.
- Carefully chosen micrographs include scale bars with size information and an acronym indicating the kind of microscope used to capture the image.
- Ample images of the physical manifestations of disease on the human body draw students in and help future clinicians become comfortable recognizing the outward signs of important diseases.

Case Histories are drawn from diverse sources and include both cases clinicians will encounter every day and those focused on global health issues.

3

Observing Microbes

[Pathogen in a Can]

SCENARIO Marianna and Julio, two children in Oklahoma aged 8 and 11, fell ill with food poisoning after consuming commercially canned hot dog chili sauce. The children complained of double vision and an inability to move their facial muscles.

SIGNS AND SYMPTOMS
Examination showed normal vital signs, symmetrical weakness (equal on right and left sides), and fixed pupils due to cranial nerve palsy (paralysis). Upper body paralysis gradually spread downward, and breathing became labored.

TESTING Botulinum toxin was identified in their blood by fluorescent monoclonal antibodies (antibodies produced by a genetically uniform cell culture).

TREATMENT The children were placed on mechanical ventilation (a machine that assists breathing) and treated with botulinum antitoxin. After several days, the children were removed from mechanical ventilation. They underwent physical rehabilitation for a full recovery.

Case Histories throughout each chapter drive the narrative, showing students how what they're learning applies in a clinical context.

FOLLOW-UP The chili sauce was traced to a canning facility where six swollen cans tested positive for botulinum toxin A. From the cans, inspectors cultured a bacterium that grew only without oxygen, as in a closed can. A microscope showed that the bacterium had a distinctive club-shaped appearance. The bacterium was *Clostridium botulinum*, which produces botulinum toxin, the cause of botulism. One end of the club-shaped cell contains an endospore, an inert form of the cell that can germinate and grow in a closed container of food. Growing cells produce botulinum toxin, leading to botulism, a life-threatening form of paralysis. Microscopy of the pathogen's unique form confirms the identification.

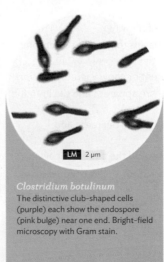

LM 2 μm

Clostridium botulinum
The distinctive club-shaped cells (purple) each show the endospore (pink bulge) near one end. Bright-field microscopy with Gram stain.

LEARNING OBJECTIVES GUIDE STUDENT LEARNING

CHAPTER OBJECTIVES

After reading this chapter, you will be able to:

- Describe the relationships among a host, its microbiome, and pathogens.
- Explain the basic concepts of infection and infectious disease.
- Discuss how infectious diseases impact communities and how communities shape emerging pathogens.

Each chapter opens with a set of **big-picture objectives**, outlining the most important concepts.

To further ensure student comprehension, there is also a set of objectives in each section.

SECTION OBJECTIVES

- Distinguish between the signs and symptoms of a disease.
- Explain the role of immunopathogenesis in infectious disease.
- Describe the five basic stages of an infectious disease.

SECTION SUMMARY

- **Diseases** are recognized by their signs and symptoms.
- **Symptoms** are caused by bacterial products and by the host immune response (immunopathology).
- **Stages of an infectious disease** include the following:
  —**Incubation period**, where the organism begins to grow but symptoms have not developed
  —**Prodromal phase**, which can be unapparent or show vague symptoms
  —**Illness phase**, when signs and symptoms are apparent and the immune system is fighting the disease
  —**Decline phase**, when the numbers of pathogens decrease and symptoms abate
  —**Convalescence**, when symptoms are gone and the patient recovers
- **Morbidity** is a measure of how many are sick from an infectious disease. **Mortality** is a measure of how many died.

An extensive **Section Summary** at the end of each section reiterates key concepts.

End-of-chapter review questions are explicitly tied to each section's learning objectives in a visual road map of the chapter. "Clinical Correlation" questions on the following page help students to effectively apply what they have learned to real-world situations.

LEARNING OUTCOMES AND ASSESSMENT

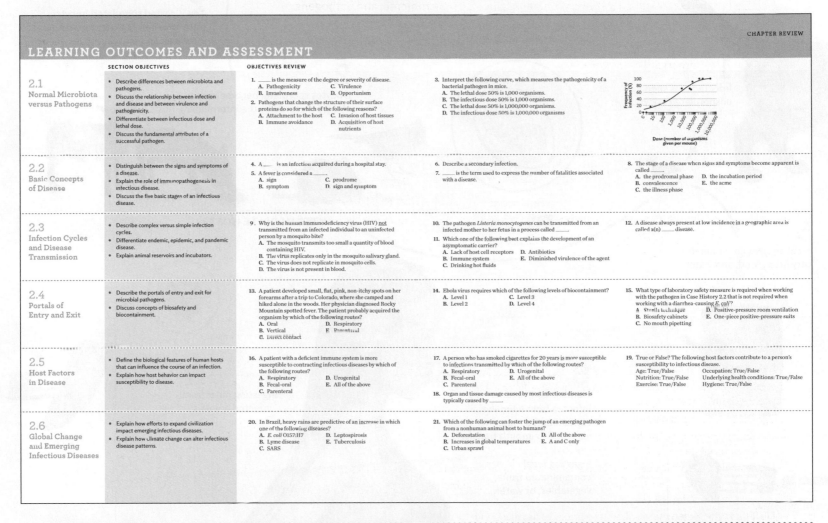

	SECTION OBJECTIVES	OBJECTIVES REVIEW
2.1 Normal Microbiota versus Pathogens	• Describe differences between microbiota and pathogens. • Discuss the relationship between infection and disease and between virulence and pathogenicity. • Differentiate between infectious dose and lethal dose. • Discuss the fundamental attributes of a successful pathogen.	1. _____ is the measure of the degree or severity of disease. A. Pathogenicity C. Virulence B. Invasiveness D. Opportunism 2. Pathogens that change the structure of their surface proteins do so for which of the following reasons? A. Attachment to the host C. Invasion of host tissues B. Immune avoidance D. Acquisition of host nutrients 3. Interpret the following curve, which measures the pathogenicity of a bacterial pathogen in mice. A. The lethal dose 50% is 1,000 organisms. B. The infectious dose 50% is 1,000 organisms. C. The lethal dose 50% is 1,000,000 organisms. D. The infectious dose 50% is 1,000,000 organisms

2.2 Basic Concepts of Disease	• Distinguish between the signs and symptoms of a disease. • Explain the role of immunopathogenesis in infectious disease. • Discuss the five basic stages of an infectious disease.	4. A _____ is an infection acquired during a hospital stay. 5. A fever is considered a _____ A. sign C. prodrome B. symptom D. sign and symptom 6. Describe a secondary infection. 7. _____ is the term used to express the number of fatalities associated with a disease. 8. The stage of a disease when signs and symptoms become apparent is called _____ A. the prodromal phase D. the incubation period B. convalescence E. the acme C. the illness phase

2.3 Infection Cycles and Disease Transmission	• Describe complex versus simple infection cycles. • Differentiate endemic, epidemic, and pandemic disease. • Explain animal reservoirs and incubators.	9. Why is the human immunodeficiency virus (HIV) *not* transmitted from an infected individual to an uninfected person by a mosquito bite? A. The mosquito transmits too small a quantity of blood containing HIV. B. The virus replicates only in the mosquito salivary gland. C. The virus does not replicate in mosquito cells. D. The virus is not present in blood. 10. The pathogen *Listeria monocytogenes* can be transmitted from an infected mother to her fetus in a process called _____. 11. Which one of the following best explains the development of an asymptomatic carrier? A. Lack of host cell receptors D. Antibiotics B. Immune system E. Diminished virulence of the agent C. Drinking hot fluids 12. A disease always present at low incidence in a geographic area is called a(n) _____ disease.

2.4 Portals of Entry and Exit	• Describe the portals of entry and exit for microbial pathogens. • Discuss concepts of biosafety and biocontainment.	13. A patient developed small, flat, pink, non-itchy spots on her forearms after a trip to Colorado, where she camped and hiked alone in the woods. Her physician diagnosed Rocky Mountain spotted fever. The patient probably acquired the organism by which of the following routes? A. Oral D. Respiratory B. Vertical E. Parenteral C. Direct contact 14. Ebola virus requires which of the following levels of biocontainment? A. Level 1 C. Level 3 B. Level 2 D. Level 4 15. What type of laboratory safety measure is required when working with the pathogen in Case History 2.2 that is not required when working with a diarrhea-causing *E. coli*? A. Sterile technique D. Positive-pressure room ventilation B. Biosafety cabinets E. One-piece positive-pressure suits C. No mouth pipetting

2.5 Host Factors in Disease	• Define the biological features of human hosts that can influence the course of an infection. • Explain how host behavior can impact susceptibility to disease.	16. A patient with a deficient immune system is more susceptible to contracting infectious diseases by which of the following routes? A. Respiratory D. Urogenital B. Fecal-oral E. All of the above C. Parenteral 17. A person who has smoked cigarettes for 20 years is more susceptible to infections transmitted by which of the following routes? A. Respiratory D. Urogenital B. Fecal-oral E. All of the above C. Parenteral 18. Organ and tissue damage caused by most infectious diseases is typically caused by _____. 19. True or False? The following host factors contribute to a person's susceptibility to infectious disease. Age: True/False Occupation: True/False Nutrition: True/False Underlying health conditions: True/False Exercise: True/False Hygiene: True/False

2.6 Global Change and Emerging Infectious Diseases	• Explain how efforts to expand civilization impact emerging infectious diseases. • Explain how climate change can alter infectious disease patterns.	20. In Brazil, heavy rains are predictive of an increase in which one of the following diseases? A. *E. coli* O157:H7 D. Leptospirosis B. Lyme disease E. Tuberculosis C. SARS 21. Which of the following can foster the jump of an emerging pathogen from a nonhuman animal host to humans? A. Deforestation D. All of the above B. Increases in global temperatures E. A and C only C. Urban sprawl

smartw⊛rk5

Smartwork5 is an interactive assessment tool that improves student problem-solving skills. Students receive rich answer feedback and hints that help them understand key concepts.

INQUIZITIVE

Norton's new formative adaptive quizzing program, InQuizitive, reinforces core concepts in the text through interactive quiz questions.

Figure 21.26   Life Cycle of *Toxoplasma gondii*

The only known definitive hosts for *Toxoplasma gondii* are domestic cats and their relatives. Infected cats can release up to 100 million gametes (unsporulated oocytes) every day for 7–21 days. The oocytes will sporulate in the environment and become infective, which will complete their growth cycle in the intermediate host.

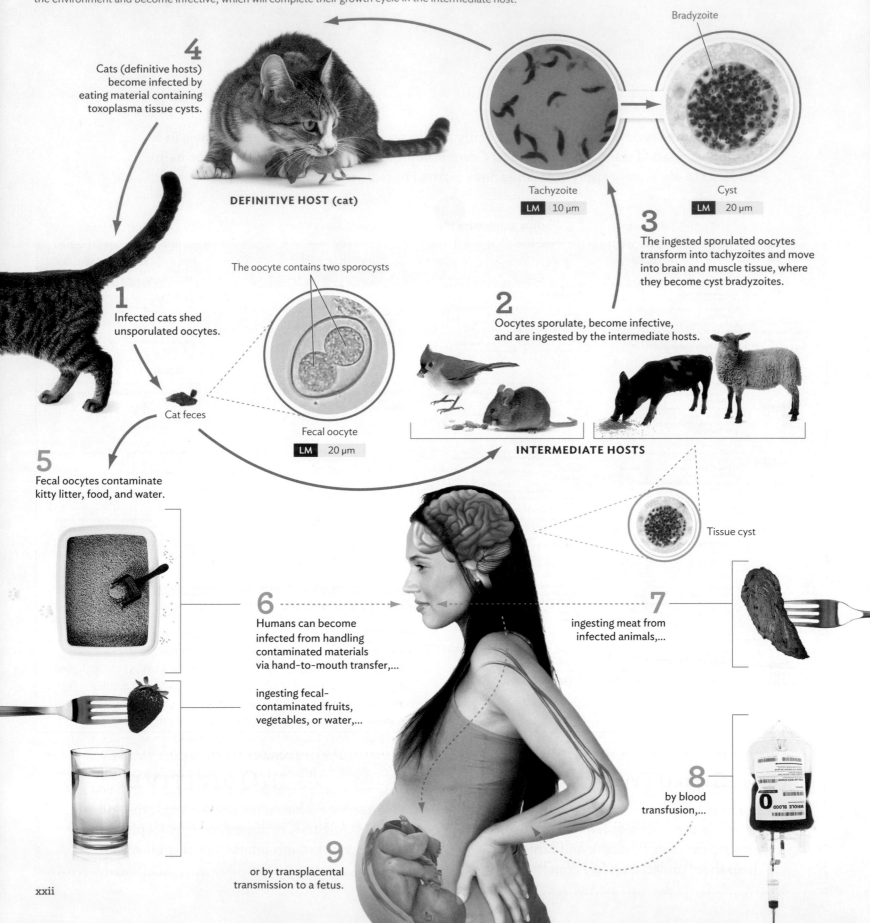

4
Cats (definitive hosts) become infected by eating material containing toxoplasma tissue cysts.

Bradyzoite

Tachyzoite
LM   10 μm

Cyst
LM   20 μm

DEFINITIVE HOST (cat)

3
The ingested sporulated oocytes transform into tachyzoites and move into brain and muscle tissue, where they become cyst bradyzoites.

The oocyte contains two sporocysts

1
Infected cats shed unsporulated oocytes.

Cat feces

Fecal oocyte
LM   20 μm

2
Oocytes sporulate, become infective, and are ingested by the intermediate hosts.

INTERMEDIATE HOSTS

Tissue cyst

5
Fecal oocytes contaminate kitty litter, food, and water.

6
Humans can become infected from handling contaminated materials via hand-to-mouth transfer,...

ingesting fecal-contaminated fruits, vegetables, or water,...

7
ingesting meat from infected animals,...

8
by blood transfusion,...

9
or by transplacental transmission to a fetus.

VISUALLY DYNAMIC ART BRINGS MICROBIAL PROCESSES AND STRUCTURES TO LIFE

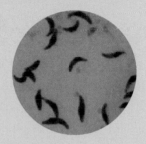

4
Cats (definitive hosts) become infected by eating material containing toxoplasma tissue cysts.

Bubble captions help students interpret what they see without having to toggle between the figure and the caption.

Photos emphasize the real-life physical manifestations of human disease.

Meticulously developed and visually stunning life-cycle drawings highlight the stages of important diseases.

Micrographs improve student understanding of and appreciation for the microbes underlying human diseases.

Chapter features help students get the most out of the text

Link The role of microbial siderophores in **iron uptake** is described in Section 5.2.

Links throughout each chapter encourage students to use the text in a more fluid way—referencing other chapters where a concept is covered.

Note Bacterial names include designations for genus (*Salmonella*) and species (*enterica*). Both are written in italics, with the *Genus* capitalized and *species* lowercase. Serovar subclass names of a species are not italicized but are capitalized (for instance, Typhimurium).

Notes provide tips students may need to develop a fuller understanding of a concept or to navigate the text more effectively.

Thought Questions designed for discussion both inside and outside the classroom encourage students to consider the big picture.

Thought Question 2.1 Is a microbe with an LD_{50} of 5×10^4 more or less virulent than a microbe with an LD_{50} of 5×10^7? Why?

Thought Question 2.2 The genus of bacteria called *Salmonella* causes a wide range of illnesses in humans. *Salmonella enterica* serovar Typhimurium (*S.* Typhimurium) is usually transmitted by contaminated food and causes a diarrhea that typically resolves without medical intervention. The organism remains localized to the intestine of humans. *Salmonella enterica* serovar Typhi (*S.* Typhi) is also transmitted via contaminated food, but can penetrate the intestinal wall and enter the lymphatic system and bloodstream to cause high fever and in some cases death. Clearly, *S.* Typhi is more virulent than *S.* Typhimurium to humans. However, when tested for lethal dose (LD_{50}) in mice, *S.* Typhi fails to kill any mice, whereas *S.* Typhimurium has a very low lethal dose (it takes very few organisms to kill mice). How might you explain this apparent contradiction?

smartw✷rk5

DIGITAL.WWNORTON.COM/MICHUM

Smartwork5 is an interactive assessment tool that improves student problem-solving skills. Assign visual, animation, case-history questions, and more, all with rich answer-specific feedback that builds knowledge and skills. The first edition course includes new image-based questions, critical-thinking questions, and interactive questions. All questions are tagged to the ASM Curriculum Guidelines as well as the Section Objectives in the book. Questions are book specific, matching the terminology and conventions students see in the textbook, include ebook links, and have been developed to address the learning objectives in the book. Pre-built activities and intuitive settings make getting started easy and instructors are able to customize and create assignments to best fit their classrooms. Smartwork5 integrates with the most popular campus learning-management systems, and intuitive performance reports for both individual students and entire classes help instructors gauge student comprehension and adjust their teaching accordingly.

INQUIZITIVE

DIGITAL.WWNORTON.COM/MICHUM

Norton's new formative adaptive quizzing platform, easily accessible via any mobile device, preserves valuable lecture and lab time by personalizing quiz questions for each student. A variety of question types test students' knowledge in different ways across the learning objectives in each chapter. The first edition course includes additional image-based questions to promote visual understanding of microbiology concepts. All questions are tagged to the ASM Curriculum Guidelines as well as the Section Objectives in the book. Students are motivated by the engaging, gamelike elements that allow them to set their confidence level on each question to reflect their knowledge, track their own progress, earn bonus points for high performance, and review learning objectives they might not have mastered. Instructors can assign InQuizitive activities out of the box, or use simple tools to customize the learning objectives they want students to work on or how much time they want students to spend on each module.

Coursepack

The free coursepack offers a variety of activities and assessment materials for instructors who use Blackboard and other learning-management systems. Three assignments for each chapter reflect and reinforce the emphases in the textbook: an assignment based on section-level learning objectives, a supplementary case-history exercise, and a visual quiz based on art from the text. Access to streaming animations, flashcards, and more is also available.

Ebook

The Norton Ebook for *Microbiology: The Human Experience* includes dynamic features that engage students, including process animations and drop-down answers to the Thought Questions. Instructors can focus student reading by sharing notes with their classes, including embedded images and video. Art expands, pop-up key terms are linked to definitions, and students can search, highlight, and take notes with ease. The ebook can be viewed on any device and will sync across devices.

The Ultimate Guide to Teaching *Microbiology: The Human Experience*

This all-in-one print resource is for instructors who want to integrate active learning into their course. For every chapter of the book, instructors will find two multimedia suggestions (videos or podcasts) with suggested classroom uses and discussion questions, two active-learning activities with premade, student-facing handouts, as well as a step-by-step guide to using the supplemental case-history PowerPoints supplied for that chapter. Activity handouts will be featured online as PDFs for easy printing and distribution. Each supplied asset is tied closely to text learning objectives. For the first edition, the guide has been thoroughly reviewed by current users of the materials. The new edition features several updated activities and multimedia descriptions as well as all NEW animation descriptions with discussion questions. The guide is now available in print format and the resources are incorporated into the searchable Interactive Instructor's Guide.

Interactive Instructor's Guide

Searchable by chapter, phrase, topic, or learning objective, the Interactive Instructor's Guide compiles the many valuable teaching resources available with *Microbiology: The Human Experience*. In this database can be found activities with downloadable handouts, streaming video with discussion questions, animations with discussion questions, supplemental case history PowerPoints, and more. This repository of lecture and teaching materials functions both as a course-prep tool and as a means of tracking the latest ideas in teaching the allied health microbiology course.

Supplemental Case History PowerPoints

For every chapter in the text, a new case study is introduced in PowerPoint form, including the symptoms of a fictional patient, the various tests that would be administered by a doctor, the "results" of these tests, and finally, the diagnosis and treatment. Bolded discussion questions appear throughout, allowing students to apply their knowledge of the subject to predict outcomes or consider the "doctor's" reasoning. Instructor notes throughout the file answer discussion questions and elaborate on key topics.

Lecture PowerPoints

Visually dynamic, chapter-specific lecture PowerPoints include outlines, notes, and clicker questions, and can be used in the classroom or for student self study.

Art Slides

All of the art in *Microbiology: The Human Experience* is available for instructor use and presentation, in both labeled and unlabeled versions, as JPEGs and in PowerPoint.

Test Bank

Every chapter of the test bank includes a minimum of 50 multiple-choice, fill-in-the-blank, and short-answer questions, all designed to evenly cover text learning objectives. Up to 10 questions per chapter are based on modified art from the text, and three questions tie in case histories from the book. Every test question is classified by Bloom's taxonomy, text section, learning objective, and difficulty level, making it easy to construct meaningful and diagnostic tests and quizzes.

Animations

Based on feedback from reviewers regarding the topics that students find most challenging, 37 high-quality, ADA-compliant animations bring the course's most complex topics to life. Questions based on the animations are embedded in Smartwork5 and the coursepack. All of the animations are available for students in tablet- and mobile-compatible formats in the coursepack and the ebook. Animation topics include:

Microscopy and Staining
Oxidation-Reduction Reactions
Osmosis and Water Balance
Endocytosis in a Eukaryotic Cell
Active Transport
Bacterial Cell Division
Dilution Streaking Technique
Biofilm Formation
Endospore Formation
Glycolysis and the TCA (Krebs) Cycle
The Electron Transport System
Photosynthesis
DNA Replication
Mutations and Base Excision Repair
Polymerase Chain Reaction
Operons
Transcription
Translation
Conjugation

Transduction
Influenza Virus Replication
HIV Replicative Cycle
Phagocytosis
Inflammation
Clonal Selection Theory
Antigen Processing and Presentation
Humoral Immunity/Antibody-Mediated Immunity
Cell-Mediated Immunity
Allergic Response: Type I Hypersensitivity
Type II Hypersensitivity
Type III Hypersensitivity
Type IV Hypersensitivity
Intracellular Pathogens
Varicella-Zoster Virus (VZV)
Neisseria meningitidis
Enzyme-Linked Immunosorbent Assay (ELISA)
Fermentation

PACKAGE OPTIONS

Contact your Norton representative for information on pricing, packaging, and customizing.

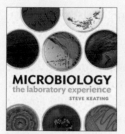

Microbiology: The Laboratory Experience, by Steve Keating

(Pennsylvania State University)

Students get more out of their microbiology lab experience because the manual has thorough introductions that emphasize important concepts and applications—written in a uniquely engaging authorial voice—and is accompanied by an unpatrolled visual program.

ISBN: 978-0-393-92364-3

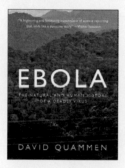

Ebola, by David Quammen

David Quammen writes about the past, present, and potential future of the Ebola virus. Rather than focus on the sensational stories in the popular press, Quammen analyzes the disease's origins, epidemiology, and possible reservoirs and poses questions yet to be answered by scientists. Microbiology students asking questions about Ebola will start to find answers in this short, readable book.

ISBN: 978-0-393-35155-2

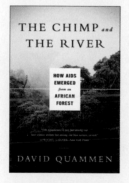

The Chimp and the River: How AIDS Emerged from an African Forest, by David Quammen

In this "frightening and fascinating masterpiece" (Walter Isaacson), David Quammen explores the true origins of HIV/AIDS—how it originated with a virus in a chimpanzee, jumped to one human, and then infected more than 60 million people—including how recent research has revealed dark surprises and yielded a radically new scenario of how AIDS began and spread.

ISBN: 978-0-393-35084-5

Infectious Disease Flashcards, by Julie Harless

(Lone Star College)

This deck of cards includes information on 48 diseases that affect different body systems. One side of each card includes two photographs: one showing the disease agent itself and one showing the effect of the disease on the human body. The other side includes an organized summary of information about the disease. The cards can be used for individual study and review or for several students to play with together.

ISBN: 978-0-393-26323-7

[Acknowledgments]

We thank the dozens of people whose incredible care, dedication, and nurturing brought this project to fruition. Special thanks go to our amazing editor, Betsy Twitchell, whose skill and personality helped focus everyone's efforts and who went the extra mile on many occasions to carry this first edition from its inception to the printed page; and to Carla Talmadge, our tireless project editor who kept the countless moving parts of this project from seizing and whose attention to detail was, well . . . detailed. We thank Kathy Gillen for her expert biochemistry content incorporated into Chapter 4. Our developmental editors, Carol Pritchard-Martinez and Michael Zierler, used their impressive skills and vision to polish our writing and clarify our logic, as did the outstanding pair of copyeditors, Janet Greenblatt and Gabe Waggoner. A huge debt of gratitude goes to Dena Betz, who miraculously tracked down elusive images from sources all around the world (we sure hope she enjoyed the trip), and to photo editor Nelson Colon for being with her every step of the way. We are lucky to have worked with an incredibly talented artist, Anne DeMarinis, whose captivating art and design grace the cover of this book and every chapter within it. We must also thank the unsung hero of the project, Katie Callahan, who relentlessly tracked the countless emails, chapter drafts, and Excel files that constantly zipped back and forth between editors, authors, and reviewers and whose electronic savvy helped guide us to and through the Norton website. Assistant editor Taylere Peterson has stepped in to fill Katie's enormous shoes; she is doing a fantastic job. Production manager Sean Mintus adeptly managed the process of translating our raw material into the polished final product; for that he has our deepest thanks. The amazing folks at Precision Graphics (a Lachina company) deserve medals for the excellent work they did. Thank you to Megan Calderwood, Eric Bramer, Rebecca Marshall, and Terri Hamer.

We have an absolutely tireless team at Norton creating the print and digital supplementary resources for our book. Media editor Kate Brayton, associate editor Cailin Barrett-Bressack, assistant media editor Victoria Reuter, and media editorial assistant Gina Forsyth worked on every element of the package as a team, and the content meets our very high standards as a result. Thank you also to Kim Yi's media project editorial group for the invaluable work they do shepherding content through many stages of development. We thank everyone involved in Norton's sales and marketing team for their unflagging support of our book as well as Drake McFeely, Roby Harrington, Julia Reidhead, and Marian Johnson for believing in us all these years.

We also thank the many colleagues who encouraged us and the dozens of reviewers across the country who lent their keen eyes and deep experience to critique early drafts of our work. We are grateful, too, to the countless scientists who graciously contributed the micrographs that appear in this textbook and to our students, who remain for us a constant source of wonder and unbridled energy. Finally, we thank our families—the parents, spouses, children, and grandchildren who fill our lives with joy and just the right amount of chaos.

Reviewers

Mari Aanenson, Western Illinois University

Lawrence R. Aaronson, Utica College

Sherrice V. Allen, Fayetteville State University

Cindy B. Anderson, Mt. San Antonio College

Daniel Aruscavage, Kutztown University of Pennsylvania

Dave Bachoon, Georgia College

Glenn Barnett, Central College

Jennifer Bess, Hillsborough Community College

R. Clark Billinghurst, St. Lawrence College

Will Blackburn, Tarrant County College

Lisa Ann Blankinship, University of North Alabama

Lanh Bloodworth, Florida State College at Jacksonville

Shaun Bowman, Clarke University

Jacqueline Brown, Angelo State University

Brad A. Bryan, Worcester State College

Sybil K. Burgess, Brunswick Community College

Evan Burkala, Oklahoma State University—Oklahoma City

Kristin Burkholder, University of New England

Kari Cargill, Montana State University

Bradley W. Christian, McLennan Community College

Tin-Chun Chu, Seton Hall University

Jenny H. Clark, Saddleback College

Bela Dadhich, Delaware County Community College

H. Kathleen Dannelly, Indiana State University

Deborah Dardis, Southeastern Louisiana University

Margaret Das, Southern Connecticut State University

Joyce Davis, Carroll Community College

Sondra Dubowsky, McLennan Community College

Elizabeth A. B. Emmert, Salisbury University

Tracey Emmons, Sandhills Community College

Clifton Franklund, Ferris State University

Ashley D. Frazier, Walters State Community College

Monica Friedrich, Saddleback College

David E. Fulford, Edinboro University of Pennsylvania

Ellen F. Fynan, Worcester State University

Julie Galvin, Halifax Community College

Ronald Girmus, New Mexico State University—Carlsbad

Carla J. Guthridge, Cameron University

Chris Guyer, Henderson State University

Kimberly Hale, Quincy University

Georgia A. Hammond, Radford University

Leanne Hanson, Clarke University

Steven Harris, University of Nebraska—Lincoln

Anne Hemsley, Antelope Valley College

Diane Hilker, Mercer County Community College

Dale Holen, Penn State Worthington Scranton Campus

Sara Reed Houser, Jefferson College of Health Sciences

Jane E. Huffman, East Stroudsburg University

Karen Huffman-Kelly, Genesee Community College

Julie Huggins, Arkansas State University

Abdallah M. Isa, Tennessee State University

Sayna A. Jahangiri, Folsom Lake College

V. Karunakaran, St. Francis Xavier University

D. Sue Katz, Rogers State University

Judy Kaufman, Monroe Community College

Christine A. Kirvan, California State University, Sacramento

Laurieann Klockow, Marquette University

Richard Knapp, University of Houston

Malda Kocache, George Mason University

Hari Kotturi, University of Central Oklahoma

Anne Kruchten, Linfield College

Ashwini Kucknoor, Lamar University

Melissa Lail-Trecker, Western New England University

Maia Larios-Sanz, University of St. Thomas

Carol Lauzon, California State University, East Bay

Leo G. Leduc, Laurentian University

Kate LePore, Monroe Community College

Heather Lofton-Garcia, Troy University

Suzanne Long, Monroe Community College

Aaron Lynne, Sam Houston State University

Shannon R. Mackey, St. Ambrose University

Bernard MacLennan, Cape Breton University

Barry Margulies, Towson University

Nancy Marthakis, Purdue University North Central

Carolyn F. Mathur, York College of Pennsylvania

Meghan May, Towson University

Sherry L. Meeks, University of Central Oklahoma

Karin Melkonian, LIU Post

Brian J. Merkel, University of Wisconsin—Green Bay

Stephanie Leah Molloy, California State University, East Bay

Ellyn Mulcahy, Johnson County Community College

Jacqueline Nesbit, University of New Orleans

Tanya Noel, University of Windsor

Lourdes Norman, Florida State College at Jacksonville

Maura Pavao, Worcester State University

Nancy A. Perigo, William Paterson University of New Jersey

Wendy L. Picking, Oklahoma State University

Laraine Powers, East Tennessee State University

Madhura Pradhan, The Ohio State University

Gregory Pryor, Francis Marion University

Mark Randa, Cumberland County College

Syed Raziuddin, Richard J. Daley College

Sarah M. Richart, Azusa Pacific University

Lori Rink, Penn State Abington Campus

Frances Sailer, University of North Dakota

Tony Schountz, University of Northern Colorado

Melissa Schreiber, Valencia College

Larry J. Scott, Central Virginia Community College

Heather Seitz, Johnson County Community College

Debbie Sesselmann, Fox Valley Technical College

Josh Sharp, Northern Michigan University

Michael E. Shea, Hudson Valley Community College

Shana Shields, Curry College

Jean M. Shingle, Immaculata University

Uma Singh, Valencia College

Jennifer Staiger, Mount St. Mary's University

Paula Steiert, Southwest Baptist University

David Straus, Virginia Commonwealth University

Anne O. Summers, University of Georgia

Sonia Suri, Valencia Community College

Sundeep Talwar, Richard J. Daley College

Renato V. Tameta, Schenectady County Community College

Steven J. Thurlow, Jackson College

Michael Troyan, Penn State University

Jorge Vasquez-Kool, Wake Technical Community College

Charles Vo, University of Arkansas—Fort Smith

Jeremiah Wagner, Mid Michigan Community College

Michael H. Walter, University of Northern Iowa

Mark Watson, University of Charleston

John E. Whitlock, Hillsborough Community College

Michael L. Womack, Gordon State College

Accuracy Reviewers

Bradley W. Christian, McLennan Community College

Elizabeth A. B. Emmert, Salisbury University

Sherry L. Meeks, University of Central Oklahoma

Madhura Pradhan, The Ohio State University

Focus Group Participants

Dale Amos, University of Arkansas—Fort Smith

Lance Bowen, Truckee Meadows Community College

Donald Breakwell, Brigham Young University

Amy Warenda Czura, Suffolk County Community College
Sharon Gusky, Northwestern Connecticut Community College
Michael Ibba, The Ohio State University
Nastassia N. Jones, Philander Smith College
D. Sue Katz, Rogers State University
Suzanne Long, Monroe Community College
Aaron Lynne, Sam Houston State University
Stephanie Leah Molloy, California State University, East Bay
Francine Norflus, Clayton State University
Kathleen A. Page, Southern Oregon University
Christopher Parker, Texas Wesleyan University
Todd P. Primm, Sam Houston State University
Sarah Richart, Azusa Pacific University
Melissa Schreiber, Valencia College
Heather Seitz, Johnson County Community College
Eric Yager, Albany College of Pharmacy and Health Sciences
Sonja B. Yung, Sam Houston State University

Authors of the Supplementary Resources

Holly Ahern, SUNY Adirondack
Alexandra Armstrong, University of Arizona
Jennifer Bess, Hillsborough Community College
Carrie Bottoms, Collin College
Nancy Boury, Iowa State
Bradley Christian, McLennan
Elizabeth Collins, Iowa Central Community College
Jamie Cunningham, Johnson County Community College
Bela Dadhich, Delaware County Community College
Jenny Gernhart, Iowa Central CC
Kathy Gillen, Kenyon College
Julie Harless, LSC-Montgomery
Amy Helms, Collin College
Meg Howard, University of Alaska, Anchorage
Erin Lentz, Galen College of Nursing
Jacqueline Nesbit, UNO
Tiffany Randall, John Tyler Community College

Deb Scheiwe, Tarrant County Community College
Heather Seitz, Johnson County Community College
Michael Shea, Hudson Valley Community College
Shana Shields, Curry College
Jennifer Smith, Tarrant County Community College
Suzanne Wakim, Butte College
Eric Warrick, State College of Florida

Preliminary Edition Survey Respondents

Lynn Bedard, DePauw University
Kristen Butela, Seton Hill University
Shauni Calhoun, Mt. San Jacinto College
Kathleen Dannelly, Indiana State University
Larry Feinstein, University of Maine at Presque Isle
Tom Firak, Oakton Community College, Des Plaines
David Fulford, Edinboro University of Pennsylvania
Eileen Gregory, Rollins College
Tray Hamil, University of South Alabama
Katina Harris-Carter, Paul D. Camp Community College
Amber Huber, Madison Area Technical College
Cristi Hunnes, Rocky Mountain College—Montana
Dave Johnson, Genesee Community College
George Keller, Samford University
Rachel Larsen, University of Southern Maine
Antje Lauer, California State University, Bakersfield
Shannon Mackey, St. Ambrose University
Matt Ruddy, Onondaga Community College
Zenda Rushing, Brunswick Community College
Josh Sharp, Northern Michigan University
John L. Sloyer, Jr., Cape Cod Community College
Mary Ann Smith, Marywood University
Larry Weiskirch, Onondaga Community College
Matthew Wood, Lake Sumter State College
Ruth Wrightsman, Flathead Valley Community College
Daniel Zamzow, University of Wisconsin, Rock County Campus

MICRO BIOLOGY

THE HUMAN EXPERIENCE

1

Microbes Shape Our History

Half a Lung Is Better than None

SCENARIO Debi was an ordinary teenager attending an affluent American public high school when she contracted tuberculosis (TB). She did not know the person who infected her. Infection requires inhalation of the causative bacteria.

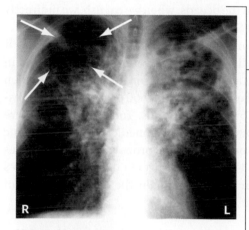

X-Ray of Lung
The X-ray reveals infiltration into the lungs (white wisps) caused by *Mycobacterium tuberculosis,* and a hole where dead tissue was semiliquefied.

SIGNS AND SYMPTOMS
Debi coughed all the time, felt tired, and was losing weight. Her coughing brought up blood. An X-ray revealed the signs of infection in her lung, including a large hole eaten away by the bacteria.

DIAGNOSIS From Debi's sputum sample, a DNA sequence was amplified by PCR (polymerase chain reaction). The DNA sequence revealed *Mycobacterium tuberculosis,* the cause of TB. Doctors prescribed isoniazid and rifampin, antibiotics that kill most strains of *M. tuberculosis.* But Debi's TB strain proved resistant to nearly all known drugs (MDR-TB).

TREATMENT Because drugs failed to eliminate the MDR strain, surgeons removed nearly half of her right lung to help the antibiotics overcome the infection. Debi recovered and returned to high school. She would have to continue taking antibiotics for years afterward. All the teachers and students in Debi's school were screened, and over 200 were found to have been infected by a student with tuberculosis misdiagnosed for two years. All required treatment to prevent disease.

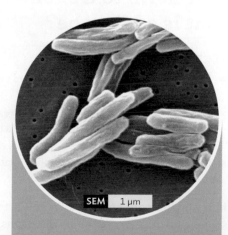

Mycobacterium tuberculosis
Bacteria revealed by scanning electron microscopy.

The disease that threatened Debi, a healthy American teenager, was tuberculosis, an ancient killer. Today, we fight this ancient disease by using modern antibiotics. But the causative bacterium, *Mycobacterium tuberculosis*, evolves to resist every antibiotic we use (discussed in Section 20.3). The emergence of multidrug-resistant (MDR) strains of disease-causing microbes is just one of the microbial challenges faced by health professionals. Despite all the advances of modern medicine and public health, tuberculosis and other microbial infections remain the world's leading cause of death in humans.

Yet bacteria that cause disease are surprisingly related to other kinds of bacteria that inhabit our environment without causing us harm—and actually enable us to live. All the plants in natural environments require nitrogen fixed by bacteria. Much of the oxygen we breathe is released by algae. And soil bacteria such as streptomycetes produce most of the antibiotics that we use to kill disease-causing pathogens.

In this chapter, we will introduce the concept of a microbe. We will then describe how humans discovered the roles of microbes in disease and in our environment. Finally, we will see how advances in microbial science, especially the discovery of the structure of DNA, have transformed medicine and society.

1.1
From Germ to Genome: What Is a Microbe?

Where did life come from? Life on Earth began early in our planet's history with microscopic organisms, or microbes. Microbial life has since shaped our atmosphere, our geology, and the energy cycles of all ecosystems. Some early microbes eventually evolved into multicellular plants and animals, including ourselves. Today, microbes generate the very air we breathe, including nitrogen gas and much of the oxygen and carbon dioxide. They fix nitrogen into forms used by plants, and they make essential vitamins that we consume, such as vitamin B_{12}. Microbes are the primary producers of food webs, particularly in the oceans; when we eat fish, we indirectly consume tons of algae at the base of the food chain.

The human body contains ten times as many microbes as it does human cells, including numerous bacteria on the skin and in the digestive tract. Throughout history, humans have had a hidden partnership with microbes ranging from food production and preservation to mining for precious minerals. Today, microbes serve as tools for biotechnology in fields from medicine to microscopic robots. Nevertheless, a small but critical proportion of all microbes are **pathogens**—that is, the causative agents of disease. Diseases caused by pathogens, or "germs," remain the principal cause of human mortality.

A Microbe Is a Microscopic Organism

A **microbe** is commonly defined as a living organism that requires a microscope to be seen. Microbial cells range in size from millimeters (mm) down to 0.2 micrometer (μm), and viruses may be tenfold smaller (Figure 1.1). Some microbes consist of a single cell, the smallest unit of life, a membrane-enclosed compartment of water solution containing molecules that carry out metabolism. Each microbe contains in its genome the capacity to reproduce its own kind. Microbes are found throughout our biosphere, from the superheated black smoker vents at the depths of the ocean floor to the subzero ice fields of Antarctica. Bacteria such as *Escherichia coli* live in our intestinal tract, whereas algae and cyanobacteria turn ponds green.

Our simple definition of a microbe, however, leaves us with contradictions. Most single-celled organisms require a microscope to render them visible and thus fit the definition of "microbe." Nevertheless, some protists and algae are large enough to be seen with the naked eye. An example is the ameba *Pelomyxa*, which can span several millimeters (Figure 1.1A). Some amebas can cause meningitis; others can harbor thousands of *Legionella* bacteria, the cause of legionellosis, a severe form of pneumonia.

Other kinds of microbes form complex multicellular assemblages, such as mycelia (multicellular filaments) and biofilms. In a

Figure 1.1 Representative Microbes

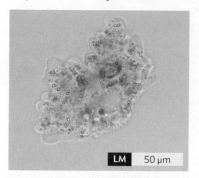

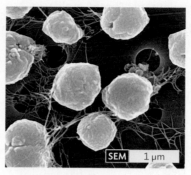

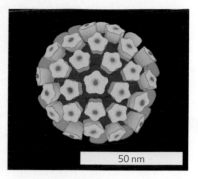

A. *Pelomyxa* species, a large ameba.

B. *Escherichia coli* bacteria colonizing the stomata of a lettuce leaf cell (colorized scanning electron micrograph).

C. *Methanocaldococcus jannaschii*, an archaeon that produces methane.

D. Human papillomavirus, the cause of genital warts and possibly cervical or penile cancer (model based on electron microscopy).

biofilm, cells are differentiated into distinct types that complement one another's function, as in multicellular organisms. On the other hand, some complex multicellular animals, such as mites and roundworms, require a microscope for us to see, but they are *not* considered microbes. Allied health courses may cover parasitic invertebrates under "microbiology" because these infectious agents are transmitted in a manner similar to that of disease-causing microbes.

Different Kinds of Microbes

Microbes, like other organisms, are classified as members of a **species** according to a shared set of genes and traits. The scientific name of the species, such as that of *Staphylococcus epidermidis*, a common skin bacterium, consists of a capitalized genus name (*Staphylococcus*) and a lowercase species name (*epidermidis*), both italicized. In addition, members of a genus are often referred to informally by a romanized vernacular term such as "staphylococci." The names of some microbial species are sometimes changed to reflect our new understanding of genetic relationships. For example, the causative agent of bubonic plague has formerly been called *Bacterium pestis* (1896), *Bacillus pestis* (1900), and *Pasteurella pestis* (1923), but it is now called *Yersinia pestis* (1944). The older names, however, still appear in the literature.

Microbes are classified according to their genetic relatedness. The more closely related two organisms are, the more recently they diverged from a common ancestor. Relatedness is important for understanding how microbes respond to treatment. For example, an antibiotic used against an intestinal pathogen will also kill many beneficial bacteria that normally live in the intestine; thus, the antibiotic may cause digestive problems. The degree of genetic relatedness between microbes is calculated by comparing their DNA sequences in the **genome**, the total DNA sequence content of each organism. Genome comparison is now the fundamental basis for classifying all life-forms.

A major trait distinguishing microbes is possession or lack of a membrane-enclosed nucleus. Microbes that lack a nuclear membrane are called **prokaryotes**, which include bacteria and archaea. Microbial **eukaryotes** (cells with a nucleus) include fungi, protozoa, and algae.

- **Bacteria** are prokaryotic cells, usually 0.2–20 μm in size; one-tenth to one-hundredth the size of a sentence period. Different species may grow as single cells, as filaments (chains), or as communities with simple differentiated forms. An example is *Escherichia coli* (Figure 1.1B), a bacterial species that grows in the human intestine. Most strains of *E. coli* are harmless commensals, aiding human digestion, but some strains cause acute gastroenteritis that may lead to kidney failure. Bacteria are found in every habitat of our biosphere, even several kilometers underground.

- **Archaea** are a genetically distinct group of prokaryotes that evolved by diverging from bacteria and eukaryotes more than 3 billion years ago. Archaea include "extremophiles" that live in seemingly hostile environments, such as the boiling sulfur springs of Yellowstone. Other archaea are methanogens, whose metabolism releases methane (natural gas) (Figure 1.1C). Methanogens are common in the gut of humans and animals, the source of the "gas" passed by one's intestinal tract. Their metabolism increases the efficiency of digestion. A remarkable feature of archaea is that none cause disease. The absence of pathogenesis (disease causation) in archaea is of great interest to medical researchers studying the cause and prevention of disease.

- Eukaryotic microbes include **protozoa**, which are motile heterotrophs (consuming organic food), usually single-celled. A protozoan such as an ameba (see Figure 1.1A) may be free-living or parasitic. **Algae** are eukaryotic microbes containing chloroplasts that conduct photosynthesis. Algae form an essential base of the food web, although overgrowth causes "algal blooms" that poison fish. Protozoa and algae together are classed as **protists**. Distinct from protists are **fungi**, heterotrophic organisms that are usually nonmotile and grow by absorbing nutrients from their surroundings. Fungi may grow as single cells (yeast) or as filaments (bread mold), or they may form complex structures such as mushrooms. Some fungi cause infections, especially in people with a depressed immune system.

- **Viruses** are noncellular microbes. A virus particle contains genetic material (DNA or RNA) that takes over the metabolism of a cell to generate more virus particles. Some viruses, such as papillomaviruses (Figure 1.1D), consist of only a few molecular parts. Other viruses, such as herpes viruses, show complexity approaching that of a cell, although none are fully functional cells. Engineered viruses are used as tools for gene therapy. For example, a nonpathogenic derivative of the human immunodeficiency virus (HIV) was used to deliver a gene to the white blood cells of a child, enabling the child's immune system to overcome leukemia.

SECTION SUMMARY

- **Microbes are microscopic**; that is, they are organisms too small to be seen without a microscope. Different species of microbes grow as single cells, in filaments, in biofilms, or in simple differentiated structures.
- **Bacteria** are cells lacking a nucleus (prokaryotes). Bacteria grow in all habitats. Most human-associated species are harmless, but some cause disease.
- **Archaea** are cells lacking a nucleus (prokaryotes) and are distantly related to other microbes. Methanogens live in the human intestine (among other places), where their metabolism releases methane. Archaea never cause disease.
- **Eukaryotic microbes** include protists (protozoa and algae) and fungi. Parasitic protozoa and fungi may infect humans.
- **Viruses** are noncellular microbes that must infect a host cell.

1.2
Microbes Shape Human History

SECTION OBJECTIVES

- Explain how microbial diseases have changed human history.
- Describe how microbes participate in human cultural practices such as production of food and drink.

Have microbes changed the course of human history? Until microscopes were developed in the seventeenth century, we humans were unaware of the unseen living organisms that surround us, that float in the air we breathe and the water we drink, and that inhabit our own bodies. Yet microbes have molded human culture since our earliest civilizations. Yeasts and bacteria made foods such as bread and cheese, as well as alcoholic beverages (**Figure 1.2A**). "Rock-eating" bacteria known as lithotrophs leached copper and other metals from ores exposed by mining, enabling ancient human miners to obtain these metals. Today, bacterial leaching produces about 20% of the world's copper. Unfortunately, microbial acid consumes the stone of ancient monuments (**Figure 1.2B**), a process intensified by airborne acidic pollution.

As we humans became aware of microbes, our relationship with the microbial world changed in important ways (**Table 1.1**). Early microscopists in the seventeenth and eighteenth centuries formulated key concepts of microbial existence, including microbes' means of reproduction and death. In the nineteenth century, the "golden age" of microbiology, key principles of disease pathology and microbial ecology were established that scientists still use today. This period laid the foundation for modern science, in which genetics and molecular biology provide powerful tools for scientists to manipulate microorganisms for medicine and industry.

Microbial Diseases Devastate Human Populations

Historians traditionally emphasize the role of warfare in shaping human history. Yet throughout history, more soldiers have died of microbial infections than of wounds in battle. Microbes often determined the fate of human societies. For example, smallpox, carried by European invaders, exterminated much of the native population of North America.

Throughout history, microbial diseases such as tuberculosis and leprosy have profoundly affected human demographics and cultural practices (**Figure 1.3**). The bubonic plague, which wiped out a third of Europe's population in the fourteenth century, was caused by *Yersinia pestis*, a bacterium spread by rat fleas.

Figure 1.2   Production and Destruction by Microbes

A. Roquefort cheeses ripening in France.

B. Statue at the Cathedral of Cologne, in Germany, undergoing decay from the action of lithotrophic microbes. Acid rain accelerates the process.

Table 1.1

Microbes and Human History

Date	Microbial Discovery	Discoverer(s)
Microbes Impact Human Culture without Detection		
10,000 BCE	Food and drink are produced by microbial fermentation.	Egyptians, Chinese, and others
1500 BCE	Tuberculosis, polio, leprosy, and smallpox are evident in mummies and tomb art.	Egyptians
50 BCE	Copper is recovered from mine water acidified by sulfur-oxidizing bacteria.	Roman metal workers under Julius Caesar
1000 CE	Smallpox immunization is accomplished by transfer of secreted material.	Chinese and Africans
1025 CE	Diseases are observed to be contagious. The basis of hygiene and quarantine is proposed.	Avicenna, or Ibn Sina (Persia)
1300–1400 CE	The Black Death (bubonic plague) killed 17 million people in Europe and Asia.	Catherine of Siena nursed plague victims; canonized as patron saint of nurses
1546 CE	Syphilis and other diseases are observed to be contagious.	Girolamo Fracastoro (Padua)
Early Microscopy and the Origin of Microbes		
1676	Microbes are observed under a microscope.	Antonie van Leeuwenhoek (Netherlands)
1717	Smallpox is prevented by inoculation of pox material, a rudimentary form of immunization.	Africa and Asia; Turkish women taught Lady Montagu, who brought the practice to England
1765	Microbes fail to grow after boiling in a sealed flask: evidence against spontaneous generation.	Lazzaro Spallanzani (Padua)
1798	Cowpox vaccination prevents smallpox.	Edward Jenner (England)
1835	Fungus causes disease in silkworms (first pathogen to be demonstrated in animals).	Agostino Bassi de Lodi (Italy)
1847	Chlorine as antiseptic wash for doctors' hands decreases pathogens.	Ignaz Semmelweis (Hungary)
1881	Bacterial spores survive boiling but are killed by cyclic boiling and cooling.	John Tyndall (Ireland)
"Golden Age" of Microbiology		
1855	Poor sanitation leads to mortality (Crimean War).	Florence Nightingale (England)
1866	Microbes are defined as a class distinct from animals and plants.	Ernst Haeckel (Germany)
1867	Antisepsis during surgery prevents patient death.	Joseph Lister (England)
1857–1881	Microbial fermentation produces lactic acid or alcohol. Microbes fail to appear spontaneously, even in the presence of oxygen. The first artificial vaccine is developed (against anthrax).	Louis Pasteur (France)
1877–1884	Bacteria are a causative agent in developing anthrax. First pure culture of colonies on solid medium, *Mycobacterium tuberculosis*. Koch's postulates demonstrate the microbial cause of a disease (anthrax and tuberculosis).	Robert Koch (Germany)
1884	Gram stain is devised to distinguish bacteria from human cells.	Hans Christian Gram (Denmark)
1886	Intestinal bacteria include *Escherichia coli*, the future model organism.	Theodor Escherich (Austria)
1889	Bacteria oxidize iron and sulfur (lithotrophy).	Sergei Winogradsky (Russia)
1889–1899	Bacteria isolated from root nodules fix nitrogen. The concept of a virus is proposed to explain tobacco mosaic disease.	Martinus Beijerinck (Netherlands)

(continued on next page)

Table 1.1

Microbes and Human History (continued)

Date	Microbial Discovery	Discoverer(s)
Biochemistry, Genetics, and Medicine		
1908	Antibiotic chemicals are synthesized and identified (chemotherapy).	Paul Ehrlich (USA)
1911	Cancer in chickens can be caused by a virus.	Francis Peyton Rous (USA)
1917	Bacteriophages are recognized as viruses that infect bacteria.	Frederick Twort (England) and Félix d'Herelle (France)
1918	Influenza A pandemic kills 50 million people worldwide.	
1928	*Streptococcus pneumoniae* bacteria are transformed by a genetic material from dead cells.	Frederick Griffith (England)
1929	Penicillin, the first widely successful antibiotic, is made by a fungus. The molecule is isolated in 1941.	Alexander Fleming (Scotland), Howard Florey (Australia), and Ernst Chain (Germany)
1937	The tricarboxylic acid cycle is discovered.	Hans Krebs (England)
1938	The microbial "kingdom" is subdivided into eukaryotes and prokaryotes (Monera).	Herbert Copeland (USA)
1938	*Bacillus thuringiensis* spray is produced as the first bacterial insecticide.	Insecticide manufacturers (France)
1941	One gene encodes one enzyme in *Neurospora*.	George Beadle and Edward Tatum (USA)
1941	Poliovirus is grown in human tissue culture.	John Enders, Thomas Weller, and Frederick Robbins (USA)
1944	The genetic material responsible for transformation of *S. pneumoniae* is DNA.	Oswald Avery, Colin MacLeod, and Maclyn McCarty (USA)
1946	Bacteria transfer DNA by conjugation.	Edward Tatum and Joshua Lederberg (USA)
1946–1956	X-ray diffraction crystal structures are obtained for the first complex biological molecules, penicillin and vitamin B_{12}.	Dorothy Hodgkin, John Bernal, and co-workers (England)
1951	Transposable elements are discovered in maize and later shown in bacteria, where they play key roles in evolution.	Barbara McClintock (USA)
1952	DNA is injected into a cell by a bacteriophage.	Martha Chase and Alfred Hershey (USA)
1953–1971	Oral rehydration therapy (ORT) is developed for dehydration due to diarrhea and cholera, saving millions of lives in developing countries.	Hemendra Chatterjee (India), Robert Phillips (USA), and Dilip Mahalanabis (India)
Molecular Biology and Medicine		
1953	The overall structure of DNA is a double helix, based on X-ray diffraction analysis.	Rosalind Franklin and Maurice Wilkins (England)
1953	Double-helical DNA consists of antiparallel chains connected by the hydrogen bonding of AT and GC base pairs.	James Watson (USA) and Francis Crick (England)
1959	Expression of the messenger RNA for the *E. coli lac* operon is regulated by a repressor protein.	Arthur Pardee (England) and François Jacob and Jacques Monod (France)
1960	Radioimmunoassay for detection of biomolecules is developed.	Rosalyn Yalow and Solomon Berson (USA)
1961	The chemiosmotic hypothesis, which states that biochemical energy is stored in a transmembrane proton gradient, is proposed and tested.	Peter Mitchell and Jennifer Moyle (England)
1966	The genetic code by which DNA information specifies protein sequence is deciphered.	Marshall Nirenberg, H. Gobind Khorana, and others (USA)

(continued on next page)

Table 1.1

Microbes and Human History (*continued*)

Date	Microbial Discovery	Discoverer(s)
1967	Bacteria can grow at temperatures above 80°C (176°F) in hot springs at Yellowstone National Park.	Thomas Brock (USA)
1968	Serial endosymbiosis explains the evolution of mitochondria and chloroplasts.	Lynn Margulis (USA)
1969	Retroviruses contain reverse transcriptase, which copies RNA to make DNA.	Howard Temin, David Baltimore, and Renato Dulbecco (USA)
1972	Artemisinin, the most effective antimalarial agent, is discovered from an ancient Chinese herb.	Tu Youyou and co-workers (China)
1973	A recombinant DNA molecule is created in vitro (in a test tube).	Annie Chang, Stanley Cohen, Robert Helling, and Herbert Boyer (USA)
1974	The bacterial flagellum is driven by a rotary motor.	Howard Berg, Michael Silverman, and Melvin Simon (USA)
1975	The dangers of recombinant DNA are assessed at the Asilomar Conference.	Paul Berg, Maxine Singer, and others (USA)
1975	Monoclonal antibodies are produced indefinitely in tissue culture by hybridomas, antibody-producing cells fused to cancer cells.	Georges Köhler and César Milstein (USA)
1977	A DNA sequencing method is invented and used to sequence the first genome of a virus.	Frederick Sanger, Walter Gilbert, and Allan Maxam (USA)
1977	Archaea are a third domain of life, the others being eukaryotes and bacteria.	Carl Woese (USA)
1978	Biofilms are a major form of existence of microbes.	J. William Costerton and others (Canada)
1979	Smallpox is declared eliminated, the culmination of worldwide efforts of immunology, molecular biology, and public health.	World Health Organization
Genomics and Medicine		
1981	Invention of the polymerase chain reaction (PCR) makes available large quantities of DNA.	Kary Mullis (USA)
1982	Viable but nonculturable bacteria contribute to ecology and pathology.	Rita Colwell, Norman Pace, and others (USA)
1982	Prions, infectious agents consisting solely of protein, are characterized.	Stanley Prusiner (USA)
1981–1983	AIDS pandemic begins (continues into present). Human immunodeficiency virus (HIV) is discovered as the cause of AIDS.	Françoise Barré-Sinoussi and Luc Montagnier (France), Robert Gallo (USA), and others
1984	*Helicobacter pylori*, an acid-resistant bacterium, is discovered in the stomach, where it causes gastritis.	Barry Marshall and Robin Warren (Australia)
1988	Molecular Koch's postulates are devised to show that a gene found in a pathogenic microorganism contributes to the disease caused by the pathogen.	Stanley Falkow (USA)
1993	Gene therapy using a retroviral vector succeeds in treating severe combined immunodeficiency disorder (SCID).	Donald Kohn and others (USA)
1995	The first bacterial genome is sequenced, *Haemophilus influenzae*.	Craig Venter, Hamilton Smith, Claire Fraser, and others (USA)
2008	Over 1,000 genome sequences of bacteria and archaea are publicly available.	National Center for Biotechnology Information (USA)
2014	Human Microbiome Project releases first compilation of microbes associated with healthy human bodies.	National Institutes of Health (USA)

Figure 1.3   Microbial Disease in History and Culture

A. Medieval church procession to ward off the Black Death (bubonic plague).

B. The AIDS Memorial Quilt spread before the Washington Monument. Each panel of the quilt memorializes an individual who died of AIDS.

Ironically, the plague-induced population decline enabled the social transformation that led to the Renaissance, a period of unprecedented cultural advancement. In the nineteenth century, the bacterium *Mycobacterium tuberculosis* stalked overcrowded cities, and tuberculosis became so common that the pallid appearance of tubercular patients became a symbol of tragic youth in European literature. Today, societies throughout the world have been profoundly shaped by the epidemic of acquired immunodeficiency syndrome (AIDS), caused by the human immunodeficiency virus (HIV).

Before the nineteenth century, the role of microbes as infectious agents was unknown. But people had a sense that those suffering from diseases such as plague and leprosy were to be avoided. The rare individuals who chose to nurse such people were considered spiritual heroes. A prominent example is Catherine of Siena (1347–1380), who served God through nursing the sick (Figure 1.4). When a wave of bubonic plague swept the Italian city of Siena, Catherine and her followers stayed to care for the ill and bury the dead. After her death, Catherine of Siena was canonized and became known as the patron saint of nursing. In modern times, nursing continues to require personal courage—for instance, during epidemics of influenza and AIDS, when ill individuals have often been abandoned by their communities.

Figure 1.4   Catherine of Siena

Catherine of Siena nursed victims of plague and leprosy. She became known as the patron saint of nurses.

Microscopes Reveal the Microbial World

The seventeenth century was a time of growing inquiry and excitement about the "natural magic" of science and patterns of our world, such as the laws of physics and chemistry formulated by Isaac Newton (1642–1727) and Robert Boyle (1627–1691). Physicians attempted new treatments for disease involving the application of "stone and minerals" (that is, chemicals), what today we would call chemotherapy. Minds were open to consider the astounding possibility that our surroundings, indeed our very bodies, were inhabited by tiny living beings.

ROBERT HOOKE OBSERVES THE MICROSCOPIC WORLD

The first microscopist to publish a systematic study of the world as seen under a microscope was Robert Hooke (1635–1703). As Curator of Experiments to the Royal Society of London, Hooke built the first compound microscope—a magnifying instrument containing two or more lenses that multiply their magnification in series. With his microscope, Hooke observed biological materials such as nematodes ("vinegar eels"), mites, and mold filaments, illustrations of which he published in *Micrographia* (1665), the first publication that illustrated objects observed under a microscope (Figure 1.5).

Hooke was the first to observe distinct units of living material, which he called "cells." Hooke first named the units cells because the shape of hollow cell walls in a slice of cork reminded him of the shape of monks' cells in a monastery. But his crude lenses achieved at best 30-fold power (30×), so he never observed single-celled microbes such as bacteria.

ANTONIE VAN LEEUWENHOEK OBSERVES BACTERIA WITH A SINGLE LENS   Hooke's *Micrographia* inspired other microscopists, including Antonie van Leeuwenhoek (1632–1723). Leeuwenhoek became the first individual to observe single-celled microbes (Figure 1.6A). As a young man, Leeuwenhoek lived in the

Figure 1.5
Robert Hooke's *Micrographia*

An illustration of mold sporangia, drawn by Hooke in 1665, from his observations of objects with a compound microscope.

Dutch city of Delft, where he worked as a cloth draper, a profession that introduced him to magnifying glasses. (The magnifying glasses were used to inspect the quality of the cloth, enabling the worker to count threads.) Later in life, he took up the hobby of grinding ever-stronger lenses to see into the world of the unseen.

Leeuwenhoek ground lenses stronger than Hooke's, which he used to build single-lens magnifiers, complete with sample holder and focus adjustment (Figure 1.6B). First Leeuwenhoek observed insects, including lice and fleas, then the relatively large single cells of protists and algae, and finally bacteria. One day he applied his microscope to observe matter extracted from between his teeth. He wrote, "to my great surprise [I] perceived that the aforesaid matter contained very many small living Animals, which moved themselves very extravagantly."

Leeuwenhoek recorded page after page on the movement of microbes, reporting their size and shape so accurately that we can often determine the species he observed (Figure 1.6C). He performed experiments, comparing, for example, the appearance of "small animals" from his teeth before and after drinking hot coffee. The disappearance of microbes from his teeth after drinking a hot beverage suggested that heat kills microbes—a profoundly important principle for the study and control of microbes ever since.

Historians have often wondered why it took so many centuries for Leeuwenhoek and his successors to determine the link between microbes and disease. The very ubiquity of microbes—most of them harmless—may have obscured the more deadly roles of pathogens (disease-causing microbes). Also, it was hard for the microscopist to distinguish between microbes and the single-celled components of the human body, such as blood cells and sperm. It was not until the nineteenth century that human tissues could be distinguished from microbial cells by the application of differential chemical stains. Microscopy is discussed further in Chapter 3.

Spontaneous Generation: Do Microbes Have Parents?

The observation of microscopic organisms led priests and philosophers to wonder where they came from. In the eighteenth century,

scientists and church leaders intensely debated the question of **spontaneous generation**, the theory that living microbes can arise spontaneously, without parental organisms. Some chemists supported spontaneous generation, arguing that microbes appear the same way chemicals precipitate from a solution. Christian church leaders, however, argued the biblical view that all organisms have "parents" going back to the first week of creation.

The Italian priest Lazzaro Spallanzani (1729–1799) sought to disprove the spontaneous generation of microbes. Spallanzani showed that a sealed flask of meat broth sterilized by boiling failed to grow microbes. The priest also noticed that microbes often appeared in pairs. Were these two parental microbes coupling to produce offspring, or did one microbe become two? Through long observation, Spallanzani watched a single microbe grow until it

Figure 1.6   Antonie van Leeuwenhoek

A. A portrait of Leeuwenhoek, the first person to observe individual microbes.

Leeuwenhoek microscope (circa late 1600s)

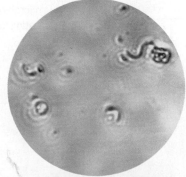

Lens
Sample holder
Focus knob
Sample mover

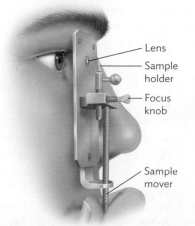

C. Spiral bacteria viewed through a replica of Leeuwenhoek's instrument.

B. "Microscope" (magnifying glass) used by Leeuwenhoek.

split in two. Thus, he demonstrated cell fission, the process by which cells arise by the splitting of preexisting cells.

Spallanzani's experiments, however, did not put the matter to rest. Supporters of spontaneous generation argued that the microbes in the priest's flask lacked access to oxygen and therefore could not grow. The pursuit of this question was left to future microbiologists, including the famous French microbiologist Louis Pasteur (1822–1895) (Figure 1.7A). In addressing spontaneous generation and related questions, Pasteur and his contemporaries laid the foundations for modern microbiology.

Pasteur began his career as a chemist. As a chemist, Pasteur was asked to help with a problem encountered by French manufacturers of wine and beer. Alcoholic beverages are made by **fermentation**, a process by which microbes gain energy by converting sugars to alcohol. In the time of Pasteur, however, the role of microbes was unknown. The conversion of grapes or grain to alcohol was believed to be a spontaneous chemical process. No one could explain why some fermentation mixtures produced vinegar (acetic acid) instead of alcohol. Pasteur discovered that fermentation is actually caused by living yeast, a single-celled fungus. In the absence of oxygen, yeast produces alcohol as a terminal waste product. But when the yeast culture is contaminated with bacteria, the bacteria outgrow the yeast and produce acetic acid instead of alcohol.

Pasteur's work on fermentation led him to test a key claim made by proponents of spontaneous generation. The proponents claimed that Spallanzani's failure to find spontaneous appearance of microbes was due to lack of oxygen. From his studies of yeast fermentation, Pasteur knew that some microbial species do not require oxygen for growth. So he devised an unsealed flask with a long, bent "swan neck" that admitted air but kept the boiled contents free of microbes (Figure 1.7B). The famous swan-necked flasks remained free of microbial growth for many years; but when a flask was tilted to enable contact of broth with microbe-containing dust, microbes

immediately grew. Thus, Pasteur disproved that lack of oxygen was the reason for failure of spontaneous generation in Spallanzani's flasks.

But even Pasteur's work did not prove that microbial growth requires preexisting microbes. The Irish scientist John Tyndall (1820–1893) attempted the same experiment as Pasteur but sometimes found the opposite result. Tyndall found that the broth sometimes gave rise to microbes, no matter how long it was sterilized by boiling. The microbes appeared because some kinds of organic matter, particularly hay infusion, are contaminated with a heat-resistant form of bacteria called endospores (or spores). The spore form can be eliminated only by repeated cycles of boiling and resting, in which the spores germinate to the growing, vegetative form that is killed at 100°C (212°F).

It was later discovered that endospores could be killed by boiling under pressure, as in a pressure cooker, which generates higher temperatures than can be obtained at atmospheric pressure. The steam pressure device called the **autoclave** became a standard method to sterilize materials required for the controlled study of microbes and for medical therapy. (Microbial control and antisepsis are discussed further in Chapter 13.)

The Origin of Life

If all life on Earth shares descent from a microbial ancestor, how did the first microbe arise? Although spontaneous generation has been discredited as a continual source of microbes, at some point in the past the first living organisms must have originated from nonliving materials. The earliest fossil evidence of cells in the geological record appears in sedimentary rock that formed over 2 billion years ago.

The components of the first living cells may have formed from spontaneous reactions sparked by ultraviolet absorption or electrical discharge. Such "early-Earth" conditions were simulated in 1953 during famous experiments by chemist Stanley Miller (1930–2007). Miller combined hydrogen gas, methane, and ammonia and applied an electrical discharge (comparable to a lightning strike), which generated simple amino acids such as glycine and alanine. Similar experiments conducted in 1961 by Spanish-American researcher Juan Oró (1923–2004) combined hydrogen cyanide and ammonia under electrical discharge to obtain adenine, a fundamental component of DNA and of the energy carrier adenosine triphosphate (ATP). These small organic molecules are also found in meteorites. Thus, chemical reactions both on Earth and in outer space could have generated the fundamental components of life; but how they assembled into living cells remains a mystery.

SECTION SUMMARY

- **Microbes affected human civilization** for centuries, long before humans guessed at their existence, through their contributions to our environment, food and drink production, and infectious diseases.
- **Robert Hooke and Antonie van Leeuwenhoek** were the first to record observations of microbes through simple microscopes.

Figure 1.7 **Louis Pasteur, Founder of Medical Microbiology and Immunology**

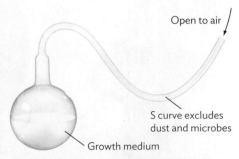

A. Pasteur's contributions to the science of microbiology and immunology earned him lasting fame.

Open to air

S curve excludes dust and microbes

Growth medium

B. Pasteur's swan-necked flask. After boiling, the contents in the flask remain free of microbial growth, despite access to air.

- **Spontaneous generation** is the theory that microbes arise spontaneously, without parental organisms.
- **Lazzaro Spallanzani** showed that microbes arise from preexisting microbes and demonstrated that heat sterilization can prevent microbial growth.
- **Louis Pasteur** discovered the microbial basis of fermentation. He also showed that supplying oxygen does not enable spontaneous generation.
- **John Tyndall** showed that repeated cycles of heat were necessary to eliminate spores formed by certain kinds of bacteria.
- **All life evolved from microbial cells.** Simple organic molecules can form out of inorganic chemicals, but how they assembled into the first cells is unknown.

1.3
Medical Microbiology and Immunology

SECTION OBJECTIVES

- Define the germ theory of disease.
- Explain how Florence Nightingale first drew a statistical correlation between infectious disease and human mortality.
- Explain how Koch's postulates can show that a specific kind of microbe causes a disease. Explain the problems in interpreting Koch's postulates in practice.

How did medical workers first figure out the connection between microbes and disease? As early as the eleventh century, the Persian physician and philosopher Avicenna (Ibn Sina) discovered the role of contagion in sexually transmitted diseases and established the principle of quarantine. Over the centuries, thoughtful observers, such as the Venetian physician Girolamo Fracastoro in the sixteenth century, noted a connection between disease and some kind of transmissible entity. Early physicians, however, could never see the actual agent of transmission. In the eighteenth century, researchers combined the tools of microscopy, microbial culture, and statistical analysis to develop the **germ theory of disease**. The germ theory holds that specific diseases are caused by specific kinds of microbes.

Linking Infectious Disease with Mortality

The significance of disease in warfare and other conditions of overcrowding, such as cities, was first recognized by the British nurse and statistician Florence Nightingale (1820–1910) (Figure 1.8A).

Born into a wealthy British family, Nightingale rejected her family's expectations for her future and chose a career as a nurse. At the time, nursing was not well codified as a professional discipline. In 1859, Nightingale published *Notes on Nursing*, the first major textbook for nursing instruction. She raised funds to establish the Nightingale Training School for nurses at St. Thomas' Hospital, which is now called the Florence Nightingale School of Nursing and Midwifery and is part of King's College London.

Nightingale also founded the science of medical statistics. Before Nightingale, health professionals lacked the analytical tools to calculate the proportions of populations that succumbed to infectious diseases. Nightingale used methods invented by French statisticians to quantify the role of disease in population mortality. She first applied her methods to demonstrate the high mortality rate due to disease among British soldiers during the Crimean War. In 1854, Nightingale arrived at the war hospital with a group of nurses seeking to improve the soldiers' conditions. Initially, their main aim was to improve nutrition and decrease overwork of the soldiers.

To Nightingale's surprise, however, the death rates continued to rise. To represent the deaths of soldiers due to various causes, Nightingale devised the "polar area chart" (Figure 1.8B). The area of each wedge, measured from the center, represents the proportion of deaths due to a particular cause: blue wedges represent deaths due to infectious diseases such as typhus and cholera; red wedges represent deaths due to wounds; and black wedges represent all other causes of death. After Nightingale compiled the data and made such charts, she discovered that infectious disease, rather than poor nutrition, accounted for more than half of all mortality. The death rates due to disease were highest in the summer months, when the

Figure 1.8 **Florence Nightingale, Founder of Medical Statistics**

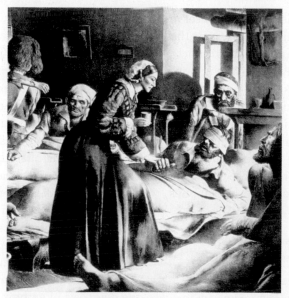

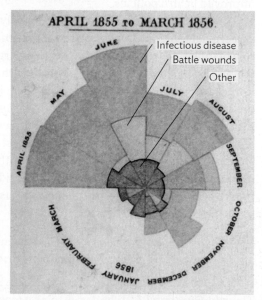

A. Florence Nightingale was the first to use medical statistics to demonstrate the significance of mortality due to disease.

B. Nightingale's polar area chart of mortality data during the Crimean War.

Figure 1.9   Epidemiological Chart of the H1N1 Influenza Epidemic
The number of people in the United States hospitalized for complications related to H1N1 influenza, and deaths reported, from August 30, 2009 to April 3, 2010.

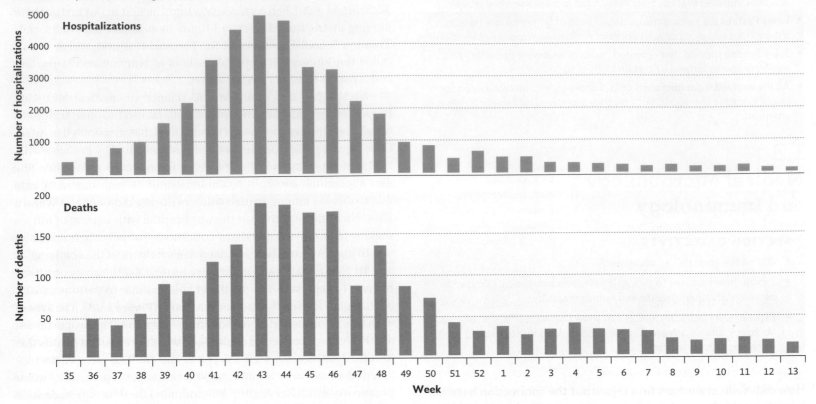

pathogens multiply fastest. Her statistics persuaded the British government to improve army hygiene and to upgrade the standards of army hospitals. Thus, for the first time, a statistical analysis led to improved health policies.

In **epidemiology** today, statistical analysis continues to serve as a crucial tool in determining the causes of disease. Assessing the role of infectious diseases in the health of large populations is now a major field of service known as **public health**. Public health is monitored and managed by government agencies, most prominently the Centers for Disease Control and Prevention (CDC), based in Atlanta. For example, the CDC tracks epidemics of seasonal influenza and tries to predict which influenza strains will require vaccination in a given year. **Figure 1.9** plots the course of the H1N1 influenza epidemic in 2009. During the epidemic, these data were tracked daily across the country and were used to allocate supplies of vaccine and antiviral drugs.

Link   Chapter 26 provides a detailed discussion of **epidemiology** and its critical role in tracking infectious diseases.

Thought Question 1.1   Why do you think it took so long for humans to connect microbes with infectious disease?

Growth of Microbes in Pure Culture

Although statistics reveal important correlations, diagnosis for a patient requires more direct evidence that a given microbe causes a given disease.

CASE HISTORY 1.1

Sickened by Dead Cattle

In 2000, on a farm in North Dakota, 67-year-old Caleb helped bury five cows that had died of anthrax. Wearing heavy leather gloves, Caleb placed chains around the heads and hooves of the carcasses and moved them to the burial site. Four days later, he noticed a small lump on his left cheek. Over 2 days, the lump enlarged, and a lesion opened. Caleb then sought medical attention. The physician reported a firm, superficial nodule surrounded by a purple ring, with an overlying black eschar (piece of dead tissue sloughed from the skin; **Figure 1.10A**). The physician prescribed ciprofloxacin, the standard antibiotic for cutaneous (affecting skin) anthrax. Testing the patient's serum with a bacterial antigen revealed the presence of antibodies, confirming the diagnosis of anthrax. The ciprofloxacin was continued, and the patient slowly improved over several weeks.

Figure 1.10 Anthrax Infection

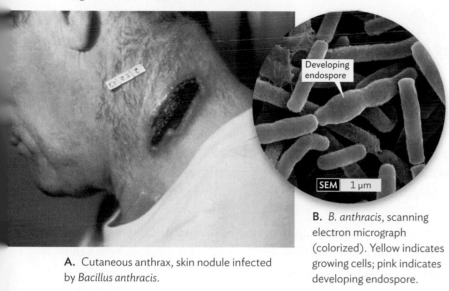

B. *B. anthracis*, scanning electron micrograph (colorized). Yellow indicates growing cells; pink indicates developing endospore.

A. Cutaneous anthrax, skin nodule infected by *Bacillus anthracis*.

The patient in this case was infected by the bacterium *Bacillus anthracis* (**Figure 1.10B**), a well-known veterinary pathogen common in soil, causing anthrax outbreaks among livestock every year. Anthrax bacteria produce tough, drought-resistant forms called endospores that can remain viable for decades in soil. Human infection is rare, but it does occur, as Caleb found out. In Caleb's case, he was fortunate that the infection remained limited to the skin (cutaneous anthrax) and was not a respiratory infection, which becomes systemic and progresses rapidly with high mortality. Respiratory anthrax is of concern to the government for use in bioterrorism because specially treated endospores can survive dry shipment through the mail. In 2001, mailed anthrax spores contaminated post offices throughout the northeastern United States as well as an office building of the United States Senate, causing several deaths.

In the nineteenth century, without antibiotics, anthrax was a scary business for farmers. The disease seemed to appear out of nowhere when the spores came up in plowed soil; the animal deaths threatened the livelihood of communities, and farmers became infected and died. Thanks to the German physician Robert Koch (1843–1910), anthrax became the first infectious disease for which scientific method established the microbial cause of a disease.

As a college student (**Figure 1.11**), Koch had conducted biochemical experiments on his own digestive system. Koch's curiosity about the natural world led him to develop principles and techniques crucial to modern microbial investigation, including the pure-culture technique and the famous Koch's postulates for identifying the causative agent of a disease. He applied his methods to many lethal diseases around the world, including anthrax and tuberculosis in Europe, malaria in Africa and the East Indies, and bubonic plague in India.

Unlike Pasteur, who was a university professor, Koch took up a medical practice in a small Polish-German town. To make space in his home for a laboratory to study anthrax and other deadly diseases, his wife curtained off part of his patients' examining room. Anthrax interested Koch because its epidemics in sheep and cattle caused economic hardship. To investigate whether anthrax was a transmissible disease, Koch used blood from an anthrax-infected carcass to inoculate a rabbit. When the rabbit died, he used the rabbit's blood to inoculate a second rabbit, which then died in turn. The blood of the unfortunate animal had turned black with long, rod-shaped cells of the bacterium *Bacillus anthracis*. Upon introduction of these bacteria into healthy animals, the animals became ill with anthrax. Thus, Koch demonstrated an important principle of epidemiology—the **chain of infection**, or transmission of a disease. In retrospect, his choice of anthrax was fortunate, for the microbes generate disease very quickly, multiply in the blood to an extraordinary concentration, and remain infective outside the body for long periods.

Figure 1.11 Robert Koch, Founder of the Scientific Method of Microbiology

A. Robert Koch as a university student.

B. Koch's sketch of anthrax bacilli in mouse blood.

C. Koch (second from left) during his visit to New Guinea to investigate malaria.

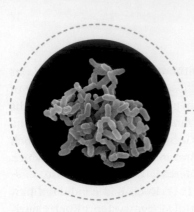

inSight

Tuberculosis: From Mummies to Multidrug Resistance

Tuberculosis is an ancient human disease. Traces of the causative bacterium, *Mycobacterium tuberculosis*, are found in Egyptian mummies 4,000 years old. The mummified lungs show lesions characteristic of the disease. During the seventeenth and eighteenth centuries, tuberculosis caused up to 30% of the deaths in Europe. Famous individuals who died of tuberculosis include Anton Chekhov and George Orwell.

As nutrition and housing standards improved during the twentieth century, people had healthier bodies to resist infection, and the rate of tuberculosis began to decline. In 1944, the first antibiotic, streptomycin, was used to cure a tuberculosis patient who was near death. Antibiotics largely eliminated tuberculosis from developed countries. But in poorer countries throughout Asia and Africa, tuberculosis grew with growing populations. New strains of the bacteria developed resistance to antibiotics. And the rise of AIDS accelerated the disease's spread through patients with impaired immune systems. Today, all countries are at risk for the epidemic spread of tuberculosis—including strains resistant to all known antibiotics.

How do bacteria infect a person and cause tuberculosis? The *M. tuberculosis* bacteria have one of the thickest cell walls of any bacterium, enabling their survival for days in air or water. When a person with tuberculosis talks or sneezes, he or she spreads droplets of mucus and saliva into the air. Another person can inhale these droplets, even days later. Fewer than ten bacteria may be enough to establish infection. The inhaled bacteria reach the alveoli, tiny sacs of tissue where oxygen and carbon dioxide are exchanged. There, macrophages, white blood cells that protect the body (Figure 1A), engulf the bacteria. But the bacterium's thick cell wall helps it resist the white blood cell's lysozomal enzymes that degrade the bacterial cell wall. Instead of getting digested by the macrophage, some of the bacteria grow and multiply within the white blood cell's cytoplasm. The cell dies, releasing bacteria that can infect other macrophages. The body's immune response may then seal the bacteria and infected macrophages within a lump called a "tubercle," where they can remain for many years. As the immune system weakens, however, the bacteria may break out and multiply, causing an active infection. An active case of tuberculosis can then spread bacteria to other people.

How is tuberculosis treated? A standard course of treatment for tuberculosis requires 6 months of antibiotic therapy (Figure 1B). A combination of antibiotics must be taken, because certain drugs (such as isoniazid and pyraminazide) are most effective against the rapidly multiplying bacteria, whereas other drugs (such as ethambutol) are more effective against long-term persistent bacteria. Simultaneous use of multiple drugs also decreases the chance of survival of mutant strains

Figure 1　Tuberculosis

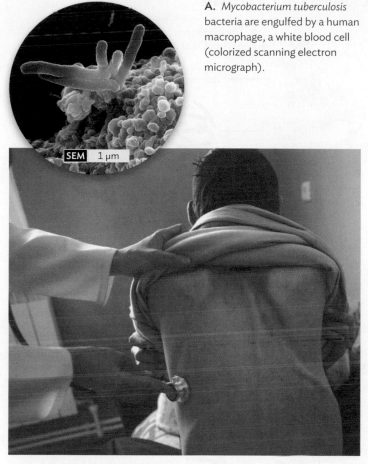

A. *Mycobacterium tuberculosis* bacteria are engulfed by a human macrophage, a white blood cell (colorized scanning electron micrograph).

SEM　1 µm

B. Patient being treated for tuberculosis.

resistant to any one drug. Unfortunately, in low-income communities, patients often stop taking the drugs once they feel well, failing to complete the full course of treatment. The result is that bacteria resume growing after a period of selection for drug resistance.

To avoid spread of drug-resistant strains, many countries have adopted the program of Direct Observed Treatment, Short-Course (DOTS). The DOTS program requires a nurse to visit the patient and observe the patient swallowing every pill of the medication. DOTS requires a substantial commitment to public health, but it is the only effective way to curb the spread of tuberculosis. Today, the World Health Organization recommends DOTS for all countries. The United States requires all immigrants to be tested for tuberculosis in their country of origin; if they test positive, they must undertake DOTS before entering the United States.

Koch and his colleagues then applied their experimental logic and culture methods to a more challenging disease: tuberculosis. In Koch's day, tuberculosis caused one-seventh of reported deaths in Europe; today, tuberculosis bacteria continue to infect millions of people worldwide (see **inSight**). Koch's approach to anthrax, however, was less applicable to tuberculosis, a disease that develops slowly after many years of dormancy. Furthermore, the causative bacterium, *Mycobacterium tuberculosis*, is small and difficult to detect in human tissue or distinguish from different bacteria of similar appearance associated with the human body. How could Koch prove that a particular bacterium causes a particular disease?

What was needed was to isolate a **pure culture** of microorganisms, a culture grown from a single "parental" cell. This had been done by previous researchers using the laborious process of serial dilution of suspended bacteria until a culture tube contained only a single cell. Alternatively, inoculation of a solid surface such as a sliced potato could produce isolated **colonies**, distinct populations of bacteria, each grown from a single cell. For *M. tuberculosis*, Koch inoculated serum, which then formed a solid gel after heating. Later, he refined the solid-substrate technique by adding gelatin to a defined liquid medium, which could then be chilled to form a solid medium in a glass dish. A covered version of the dish called a **petri dish** (also called a petri plate) was invented by a colleague, Richard J. Petri (1852–1921). The petri dish consists of a round dish with vertical walls covered by an inverted dish of slightly larger diameter. Today, the petri dish, generally made of disposable plastic, remains an indispensable part of the microbiological laboratory.

Another improvement in solid-substrate culture was the replacement of gelatin with materials that remain solid at higher temperatures, such as the gelling agent **agar** (a sugar polymer). The use of agar was recommended by Angelina Hesse (1850–1934), a microscopist and illustrator, to her husband, Walther Hesse (1846–1911), a young medical colleague of Koch's (**Figure 1.12**). Agar comes

Figure 1.12　Angelina and Walther Hesse

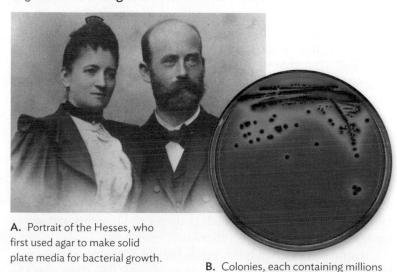

A. Portrait of the Hesses, who first used agar to make solid plate media for bacterial growth.

B. Colonies, each containing millions of bacteria, growing on an agar plate.

from red algae (seaweed), which East Indian birds use to build nests; it is the main ingredient in the delicacy "bird's nest soup." Dutch colonists used agar to make jellies and preserves, and a Dutch colonist from Java introduced it to Angelina Hesse. The Hesses used agar to develop the first effective growth medium for tuberculosis bacteria. (Pure culture is discussed further in Chapter 6.)

Some kinds of microbes cannot be grown in pure culture without other organisms. Bacteria such as *Chlamydia* species, which cause trachoma and sexually transmitted diseases, grow only within living, eukaryotic host cells. All viruses must be cultured with their host cells. Viruses are discussed in Chapter 12.

Koch's Postulates Link a Pathogen with a Disease

For his successful discovery of the bacterium responsible for tuberculosis, *M. tuberculosis*, Koch was awarded the Nobel Prize in Physiology or Medicine in 1905. Koch formulated his famous set of criteria for establishing a causative link between an infectious agent and a disease (Figure 1.13). These four criteria are known as **Koch's postulates**:

1. The microbe is found in all cases of the disease but is absent from healthy individuals.

2. The microbe is isolated from the diseased host and grown in pure culture.

3. When the microbe is introduced into a healthy, susceptible host (or animal model), the same disease occurs.

4. The same strain of microbe is obtained from the newly diseased host. When cultured, the strain shows the same characteristics as before.

Koch's postulates continue to be used to determine whether a given strain of microbe causes a disease. Modern examples include Lyme disease, a tick-borne infection that has become widespread in New England and the Mid-Atlantic states, and hantaviral pneumonia, an emerging disease particularly prevalent in the southwestern United States. Nevertheless, the postulates remain only a guide; individual diseases and pathogens may confound one or more of the criteria. For example, tuberculosis bacteria are now known to cause symptoms in only 10% of the people infected. If Koch had been able to detect these silent bacilli, they would not have fulfilled his first criterion.

Another disease for which Koch's postulates were difficult to establish is acquired immunodeficiency syndrome (AIDS). The first patients identified with AIDS had blood concentrations of HIV, the causative virus, so low that initially no virus could be detected. It took the invention of the polymerase chain reaction (PCR), a method of specifically amplifying the number of copies of any DNA or RNA sequence, to detect the presence of HIV. Today, PCR test kits are used routinely to identify pathogens in patients with many kinds of illnesses, including influenza (Figure 1.14).

A difficulty with studying certain human diseases is the absence of an animal host that exhibits the same disease. For AIDS, even the

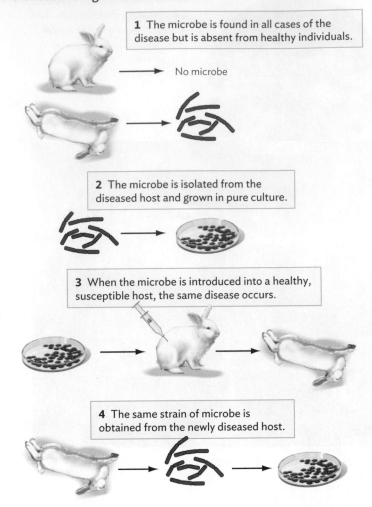

Figure 1.13   Koch's Postulates Defining the Causative Agent of a Disease

1 The microbe is found in all cases of the disease but is absent from healthy individuals.

No microbe

2 The microbe is isolated from the diseased host and grown in pure culture.

3 When the microbe is introduced into a healthy, susceptible host, the same disease occurs.

4 The same strain of microbe is obtained from the newly diseased host.

chimpanzees, our closest relatives, are not susceptible, although they exhibit a similar disease from a related pathogen, simian immunodeficiency virus (SIV). Experimentation on humans is prohibited, although in rare instances researchers have voluntarily exposed themselves to a proposed pathogen. For example, Australian researcher Barry Marshall ingested *Helicobacter pylori* to convince skeptical colleagues that this organism could colonize the extremely acidic stomach. *H. pylori* turned out to be the causative agent of gastritis and stomach ulcers, conditions that had long been thought to be caused by stress rather than infection. For the discovery of *H. pylori*, Marshall and colleague Robin Warren won the 2005 Nobel Prize in Physiology or Medicine.

Immunization Prevents Disease

Identifying the cause of a disease is, of course, only the first step to developing an effective therapy and preventing further transmission. Early microbiologists achieved some remarkable insights on how to control pathogens (see Table 1.1).

Figure 1.14
H1N1 Influenza Detection Kit

Patient samples are used for PCR amplification of DNA to provide sufficient quantity for sequence analysis.

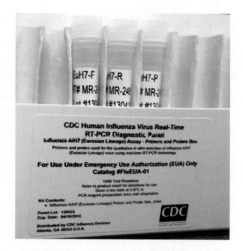

The first clue as to how to protect an individual from a deadly disease came from the dreaded smallpox. Smallpox was an ancient disease that originated in Africa about 10,000 BCE and spread throughout Europe and Asia. By 1000 BCE, in India and China it was known that individuals could be made immune to the disease by transferring secretions from a diseased individual to healthy ones. Presumably, if some time elapsed between removal of the secretions and transfer to the next patient, the virus became naturally "attenuated"; that is, its form deteriorated, rendering it less infectious but still capable of inducing an immune response.

In the eighteenth century, smallpox infected a large fraction of the European population, killing or disfiguring many people. In Turkey, however, the incidence of smallpox was decreased by deliberately inoculating children with material from smallpox pustules. Inoculated children usually developed a mild case of the disease and were protected from smallpox thereafter. The practice of smallpox inoculation was introduced from Turkey to other parts of Europe in 1717 by Lady Mary Montagu, a smallpox survivor (Figure 1.15A). Stationed in Turkey with her husband, the British ambassador, Lady Montagu learned that many elderly women there had perfected the art of inoculation: "The old woman comes with a nut-shell full of the matter of the best sort of small-pox, and asks what vein you please to have opened." Lady Montagu arranged for the procedure on her own son and then brought the practice back to England, where it became widespread.

A similar practice of smallpox inoculation was introduced to the American colonies by a slave, Onesimus, from the Coromantee people of Africa. Onesimus persuaded his master, Reverend Dr. Cotton Mather, to promote smallpox inoculation while an epidemic was devastating Boston.

Preventive inoculation with smallpox was dangerous, however, as some infected individuals still contracted serious disease and were contagious. Thus, doctors continued to seek a better method of prevention. In England, milkmaids claimed that they were protected from smallpox after they contracted cowpox, a related but much milder disease. This claim was confirmed by English physician Edward Jenner (1749–1823), who deliberately infected patients with matter from cowpox lesions (Figure 1.15B). The practice of cowpox inoculation was called vaccination, after the Latin word *vacca* for "cow." To this day, cowpox, or vaccinia virus, remains the basis of the modern smallpox vaccine.

Pasteur was aware of vaccination as he studied the course of various diseases in experimental animals. In the spring of 1879, he was studying fowl cholera, a transmissible disease of chickens with a high death rate. He had isolated and cultured the bacteria responsible, but he left his work during the summer for a long vacation. No refrigeration was available to preserve cultures, and when he returned to work, the aged bacteria failed to cause disease in his chickens. Pasteur then obtained fresh bacteria from an outbreak of

Figure 1.15   Smallpox Vaccination

A. Lady Mary Wortley Montagu, shown in Turkish dress. The artist avoided showing Montagu's facial disfigurement from smallpox.

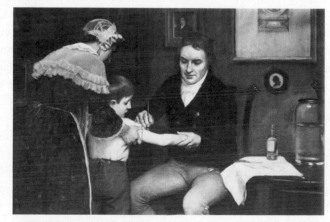

B. Dr. Edward Jenner, depicted vaccinating 8-year-old James Phipps with cowpox matter from the hand of milkmaid Sarah Nelmes, who had caught the disease from a cow.

C. Newspaper cartoon depicting public reaction to cowpox vaccination.

Figure 1.16

Pasteur Cures Rabies

Cartoon in French newspaper depicts Pasteur protecting children from rabid dogs.

LE BON PASTEUR

Laissez venir à moi les petits enragés.

disease elsewhere, as well as some new chickens. But the fresh bacteria failed to make the original chickens sick (those that had been exposed to the aged bacteria). Yet all the new chickens, exposed only to the fresh bacteria, contracted the disease. Grasping the clue from his mistake, Pasteur had the insight to recognize that an attenuated ("weakened") strain of microbe, altered somehow to eliminate its potency to cause disease, could still confer immunity to the virulent disease-causing form.

Pasteur was the first to recognize the significance of attenuation and extend the principle to other pathogens. We now know that the molecular components of pathogens generate **immunity**, the resistance to a specific disease, by stimulating the **immune system**, an organism's exceedingly complex cellular mechanisms of defense (discussed in Chapters 15 and 16). Understanding the immune system awaited the techniques of molecular biology a century later, but nineteenth-century physicians developed several effective examples of **immunization**, the stimulation of an immune response by deliberate inoculation with an attenuated pathogen.

The way to attenuate a strain for vaccination varies greatly among pathogens. Heat treatment or aging for various periods often turned out to be the most effective approach. A far more elaborate treatment was required for the most famous disease for which Pasteur devised a vaccine: rabies. The rabid dog loomed large in folklore, and the disease was dreaded for its particularly horrible and inevitable course of death. Pasteur's vaccine for rabies required a highly complex series of heat treatments and repeated inoculations. Its success led to his instant fame (**Figure 1.16**). Grateful survivors of rabies founded the Pasteur Institute, one of the world's greatest medical research institutions, whose scientists in the twentieth century discovered HIV, the virus that causes AIDS.

Antiseptics and Antibiotics Control Pathogens

Before the work of Koch and Pasteur, many patients died of infections transmitted unwittingly by their own doctors. In 1847, Hungarian physician Ignaz Semmelweis (1818–1865) noticed that the death rate of women in childbirth due to puerperal fever was much higher in his own hospital than in a birthing center run by midwives. He guessed that the doctors in his hospital were transmitting pathogens from cadavers that they had dissected. So he ordered the doctors to wash their hands in chlorine, an **antiseptic** agent (a chemical that kills microbes). The mortality rate fell; but this revelation displeased other doctors, who refused to accept Semmelweis's findings.

In 1865, the British surgeon Joseph Lister (1827–1912) noted that half his amputee patients died of sepsis. Lister knew from Pasteur that microbial contamination might be the cause. So he began experiments to develop the use of antiseptic agents, most successfully carbolic acid, to treat wounds and surgical instruments. After initial resistance, Lister's work, with the support of Pasteur and Koch, drew widespread recognition. In the twentieth century, surgeons developed fully **aseptic** environments for surgery—that is, environments completely free of microbes. Nevertheless, even in hospitals today, ensuring that physicians wash their hands between patient visits remains a challenge.

Although the use of antiseptic chemicals was a major advance, most such substances could not be taken internally, as they would kill the human patient. Researchers sought a "magic bullet," an **antibiotic** molecule that would kill only the microbes, leaving their host unharmed.

An important step in the search for antibiotics was the realization that microbes themselves produce antibiotic compounds with highly selective effects. This finding followed from the famous accidental discovery of penicillin by the English medical researcher Alexander Fleming (1881–1955). In 1929, Fleming was culturing *Staphylococcus*, which infects wounds. He found that one of his plates of *Staphylococcus* was contaminated with a mold, *Penicillium notatum*, which he noticed was surrounded by a clear region, free of *Staphylococcus* colonies (**Figure 1.17**). Following up on this observation, Fleming showed that the mold produced a substance that killed bacteria. We now know this substance as penicillin.

In 1941, British biochemists Howard Florey (1898–1968) and Ernst Chain (1906–1979) purified the penicillin molecule, which we now know inhibits formation of the bacterial cell wall. Penicillin saved the lives of many Allied troops during World War II, the first war in which an antibiotic became available to soldiers.

The second half of the twentieth century saw the discovery of many powerful antibiotics. Most of the new antibiotics, however, were produced by obscure strains of bacteria and fungi from dwindling ecosystems—a circumstance that focused attention on wilderness preservation worldwide. Furthermore, the widespread and often indiscriminate use of antibiotics has selected for pathogens that are antibiotic resistant. As a result, antibiotics have lost their

Figure 1.17   Alexander Fleming, Discoverer of Penicillin

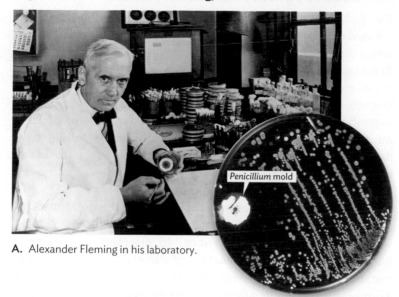

A. Alexander Fleming in his laboratory.

B. Fleming's original plate of bacteria with *Penicillium* mold inhibiting the growth of bacterial colonies.

effectiveness against certain strains of major pathogens. For example, multidrug-resistant *Mycobacterium tuberculosis* is now a serious threat to public health.

Fortunately, biotechnology provides new approaches to antibiotic development, including genetic engineering of microbial producers and artificial design of antimicrobial chemicals. This industry has become ever more critical because the indiscriminate use of antibiotics has led to a "molecular arms race" in which our only hope is to succeed faster than the pathogens develop resistance. (Antibiotics are described in Chapter 13 and their medical uses are discussed in Chapters 19–24.)

The Discovery of Viruses

The discovery of ever-smaller and more elusive microorganisms continues to this day. From the nineteenth century on, researchers were puzzled to find contagious diseases whose agent of transmission could pass through a filter with tiny pores that blocked known microbial cells. One of these researchers was the Dutch plant microbiologist Martinus Beijerinck (1851–1931), who studied tobacco mosaic disease, a condition in which the tobacco leaves become mottled and the crop yield

is decreased or destroyed altogether. Beijerinck concluded that because the agent of disease passed through a filter that retained bacteria, it could not be a bacterial cell.

The filterable agent was ultimately purified by the American scientist Wendell Stanley (1904–1971), who processed 4,000 kilograms (kg) of infected tobacco leaves and crystallized the infective particle. What he had crystallized was the tobacco mosaic virus (TMV), the causative agent of tobacco mosaic disease. TMV infects many kinds of plants; the virus is so infectious that plants can be infected by people smoking TMV-contaminated cigarettes. The crystallization of this virus particle earned Stanley the 1946 Nobel Prize in Chemistry. The fact that an entity capable of biological reproduction could be inert enough to be crystallized amazed scientists and ultimately led to a new, more mechanical view of living organisms.

The individual TMV particle consists of a helical tube of protein subunits containing its genetic material coiled within (**Figure 1.18A**). Stanley thought that the virus was a catalytic protein, but colleagues later determined that it contained RNA as its genetic material. The structure of the coiled RNA was solved through X-ray diffraction crystallography by the British scientist Rosalind Franklin (1920–1958). Since then, X-ray analysis has solved the structure of other viruses, including herpes virus (**Figure 1.18B**). The herpes virus contains a DNA chromosome coiled within an icosahedral (20-sided) protein capsid.

Toward the end of the twentieth century, even smaller infective particles were discovered consisting of a single molecule of RNA (viroids) or of protein (prions). Prions are suspected as a factor in the development of Alzheimer's disease. (The infectious processes of viruses, viroids, and prions are discussed in Chapter 12.)

Figure 1.18   Viruses

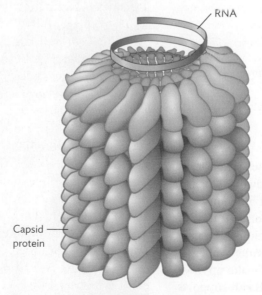

RNA

Capsid protein

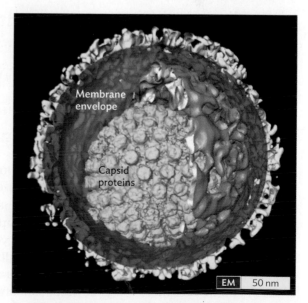

Membrane envelope

Capsid proteins

EM   50 nm

A. Tobacco mosaic virus (TMV). The RNA chromosome is surrounded by protein subunits.

B. Herpes virus structure. 3D structure of herpes simplex virus type 1.

SECTION SUMMARY

- **Florence Nightingale** quantified statistically the impact of infectious disease on human populations.

- **Robert Koch** devised techniques of pure culture to study a single species of microbe in isolation. A key technique is to culture the microbe on a solid medium by using agar, as developed by Angelina and Walther Hesse, in a double-dish container devised by Richard Petri.

- **Koch's postulates** provide a set of criteria to establish a causative link between an infectious agent and a disease.

- **Edward Jenner** established the practice of vaccination, inoculation of cowpox to prevent smallpox. Jenner's discovery was based on earlier observations by Lady Mary Montagu and others that a mild case of smallpox could prevent future infection.

- **Louis Pasteur** developed the first vaccines based on attenuated strains; one such vaccine is the rabies vaccine.

- **Ignaz Semmelweis** and Joseph Lister showed that antiseptics could prevent transmission of pathogens from doctor to patient.

- **Alexander Fleming** discovered that the *Penicillium* mold generates a substance that kills bacteria. **Howard Florey** and **Ernst Chain** purified the substance, penicillin, the first commercial antibiotic to save human lives.

Thought Question 1.2   How could you use Koch's postulates to demonstrate the causative agent of influenza? What problems would you need to overcome that were not encountered with anthrax?

Thought Question 1.3   Why do you think some pathogens generate immunity readily, whereas others evade the immune system?

1.4
Microbes in Our Environment

SECTION OBJECTIVES

- Describe examples of how microbes contribute to natural ecosystems.
- Explain how mitochondria and chloroplasts evolved by endosymbiosis.

How do microbes shape Earth's environment?   Koch's growth of microbes in pure culture was a major technological advance that led to amazing revelations in microbial physiology and biochemistry. In hindsight, this discovery eclipsed the equally important study of microbial ecology. Microbes are responsible for cycling the many minerals essential for all life. Yet barely 0.1% of all microbial species can be cultured in the laboratory—and the rest make up most of Earth's entire biosphere. Only the outer skin of Earth supports complex multicellular organisms. The depths of Earth's crust, to at least 2 miles down, as well as the atmosphere 10 miles out into the stratosphere, remain the domain of microbes. So, for the most part, Earth's ecology *is* microbial ecology.

Microbes Support Natural Ecosystems

The first microbiologists to culture microbes in the laboratory selected the kinds of nutrients that feed humans, such as beef broth or potatoes. Some of Koch's contemporaries, however, suspected that other kinds of microbes living in soil or wetlands existed on more exotic fare. Soil samples were known to oxidize hydrogen gas, and this activity was eliminated by treatment with heat or acid, suggesting microbial origin. Ammonia in sewage was oxidized to nitrate, and this process was eliminated by antibacterial treatment. These findings suggested the existence of microbes that "ate" hydrogen gas or ammonia instead of beef or potatoes, but no one could isolate these microbes in culture.

Among the first to study microbes in natural habitats was the Russian scientist Sergei Winogradsky (1856–1953). Winogradsky waded through marshes (wetlands) to discover new forms of microbes. In wetlands, Winogradsky discovered microbes with metabolism alien from human digestion. For example, species of the bacterium *Beggiatoa* oxidize hydrogen sulfide (H_2S) to sulfuric acid (H_2SO_4). *Beggiatoa* fixes carbon dioxide into biomass without consuming any organic food. Organisms that feed solely on inorganic minerals are known as lithotrophs. Today, wetland microbes are known for their critical roles in environmental quality, particularly filtering our groundwater (**Figure 1.19**).

The lithotrophs Winogradsky studied could not be grown on Koch's plate media containing agar or gelatin. The bacteria that Winogradsky isolated could grow only on inorganic minerals. For example, nitrifiers convert ammonia to nitrate, forming a crucial part of the nitrogen cycle in natural ecosystems. Winogradsky

Figure 1.19   Wetland Habitat for Microbes

The marshes of the Everglades act as natural microbial filters for the aquifers that supply drinking water for southern Florida.

cultured nitrifiers on a completely inorganic solution containing ammonia and silica gel, which supported no other kind of organism. This experiment was an early example of **enrichment culture**, the use of selective growth media that support certain classes of microbial metabolism while excluding others. Enrichment culture is important in clinical microbiology labs to help identify and properly treat the specific microbes that cause disease. A kind of enrichment culture used today in hospitals is MacConkey agar, a formula that permits growth of Gram-negative bacteria such as *E. coli* but excludes Gram-positive soil bacteria such as *Bacillus* species.

Winogradsky and later microbial ecologists showed that bacteria perform unique roles in **geochemical cycling**, the global interconversion of inorganic and organic forms of nitrogen, sulfur, phosphorus, and other minerals. Without these essential conversions (nutrient cycles), no plants or animals could live. Bacteria and archaea fix nitrogen (N_2) by reducing it to ammonia (NH_3), the form of nitrogen assimilated by plants. This process is called **nitrogen fixation**. Other bacterial species oxidize ammonium ions (NH_4^+) in several stages back to nitrogen gas. (Geochemical cycles are presented in Chapter 27.)

Within plant cells, certain bacteria fix nitrogen as **endosymbionts**, organisms living symbiotically inside a larger organism, like a guest that never leaves (but eventually becomes your cook). Endosymbiotic bacteria known as rhizobia induce the roots of legumes (beans and lentils) to form special nodules to facilitate bacterial nitrogen fixation. Rhizobial endosymbiosis was first observed by Martinus Beijerinck. Other bacteria serve animals as digestive endosymbionts (**Figure 1.20**). Animals such as cattle and termites require digestive bacteria to break down cellulose and other plant polymers. Even we humans obtain as much as 15% of our nutrition from bacteria growing in our colon. Research on obesity is investigating the role of intestinal bacteria in determining human weight gain.

Animals and Plants Evolved through Endosymbiosis

Microbial endosymbiosis, in many diverse forms, is widespread in all ecosystems. Many interesting cases involve animal or human hosts. The endosymbiotic origin of eukaryotic cells was proposed by Lynn Margulis (1938–2011), of the University of Massachusetts, Amherst (**Figure 1.21**). Margulis tried to explain how it is that eukaryotic cells

Figure 1.20 Biofilm of Digestive Bacteria in the Human Intestine

Bacteria growing on the surface of residual food particles aid our digestion.

Figure 1.21 Lynn Margulis and the Serial Endosymbiosis Theory

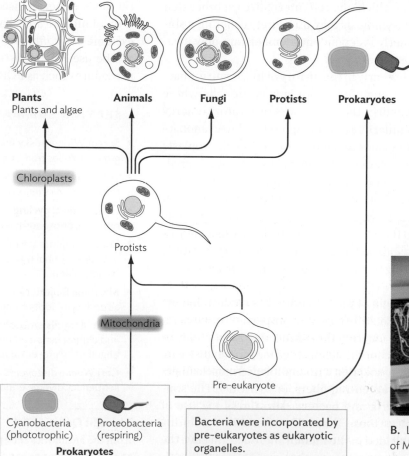

A. Five-kingdom scheme, modified by the endosymbiosis theory.

Bacteria were incorporated by pre-eukaryotes as eukaryotic organelles.

B. Lynn Margulis (University of Massachusetts, Amherst) proposed that organelles evolve through endosymbiosis.

Figure 1.22   Archaea, a Newly Discovered Form of Life

A. Yellowstone National Park hotsprings contain archaea growing above 80°C (176°F) in water containing sulfuric acid.

B. Carl Woese (University of Illinois at Urbana-Champaign) proposed that archaea constitute a third domain of life.

respiratory proteins in their cell membrane, whereas human mitochondria have similar proteins in the mitochondrial inner membrane.

New Microbes Continue to Emerge

We continue to discover surprising new kinds of microbes deep underground and in places previously thought uninhabitable, such as the hot springs of Yellowstone National Park (Figure 1.22A). Microbes shape our biosphere and provide new tools that impact human society. For example, a bacterial DNA polymerase (a DNA-replicating enzyme) from a Yellowstone hot spring is used for the PCR technology that identifies pathogens in ill patients. In 1977, Carl Woese (1928–2012), of the University of Illinois (Figure 1.22B), discovered that some of the microbes from Yellowstone hot springs have genomes very different from those of all known life-forms. The genomes of these microbes had diverged so far from that of any known bacteria that the newly discovered prokaryotes were seen as a distinct form of life, the archaea (discussed in Section 1.1). Archaea living in extreme environments produce exceptionally sturdy enzymes that can be used for industrial processes and for clinical identification procedures such as PCR amplification of DNA.

contain mitochondria and chloroplasts, membranous organelles that possess their own chromosomes. She proposed that eukaryotes evolved by merging with bacteria to form composite cells by intracellular endosymbiosis, in which one cell internalizes another that grows within it. The endosymbiosis may ultimately generate a single organism whose formerly independent members are now incapable of independent existence.

Margulis proposed that early in the history of life, respiring bacteria similar to *E. coli* were engulfed by pre-eukaryotic cells, where they evolved into mitochondria, the eukaryote's respiratory (energy generating) organelle. Similarly, a phototroph related to cyanobacteria was taken up by a eukaryote, giving rise to the chloroplasts of phototrophic algae and plants. Ultimately, DNA sequence analysis produced compelling evidence of the bacterial origin of mitochondria and chloroplasts. Both these classes of organelles contain circular molecules of DNA, whose sequences show unmistakable homology (similarity) to those of modern bacteria. DNA sequences and other evidence established the common ancestry between mitochondria and respiring bacteria and between chloroplasts and cyanobacteria.

The endosymbiotic origin of mitochondria has medical importance because mitochondria still share key properties with bacteria, including the structure and function of the respiratory (energy generating) complex. Mitochondrial defects cause human diseases, such as a form of epilepsy associated with malformed muscle fibers (myoclonic epilepsy with ragged-red fibers, or MERRF). The similarity between mitochondria and bacteria also limits the use of certain antibiotics, such as those that kill bacteria by inhibiting respiratory proteins embedded in the membrane. Bacteria have the

SECTION SUMMARY

- **Sergei Winogradsky** first developed a system of enrichment culture to grow microbes from natural environments.
- **Lithotrophs** metabolize inorganic minerals, such as ammonia, instead of the organic nutrients used by the microbes isolated by Koch.
- **Geochemical cycling** depends on bacteria and archaea that cycle nitrogen, phosphorus, and other minerals throughout the biosphere.
- **Endosymbionts** are microbes that live within multicellular organisms and provide essential functions for their hosts, such as nitrogen fixation for legume plants.
- **Martinus Beijerinck** first demonstrated that nitrogen-fixing rhizobia grow as endosymbionts within leguminous plants.
- **Lynn Margulis** proposed that eukaryotic organelles such as mitochondria and chloroplasts evolved by endosymbiosis from prokaryotic cells engulfed by pre-eukaryotes.
- **Carl Woese** discovered a domain of life, Archaea, whose genetic sequences diverge equally from those of bacteria and those of eukaryotes.

Thought Question 1.4   How do you think microbes protect themselves from the antibiotics they produce?

1.5
The DNA Revolution

How did science change medicine during the twentieth century?
Amid world wars and societal transformations, the field of microbiology exploded with new knowledge (see Table 1.1). More than 99% of what we know about microbes today was discovered since 1900 by scientists too numerous to cite in this book. Advances in biochemistry and microscopy revealed the fundamental structure and function of cell membranes and proteins and the cell's central information molecule, deoxyribonucleic acid, better known as DNA (Figure 1.23). Beyond microbiology, these advances produced the technology of "recombinant DNA," or genetic engineering, the creation of molecules that combine DNA sequences from unrelated species. These microbial tools offered unprecedented applications to human medicine and industry.

The Discovery of the Structure of DNA

The famous double helix of DNA was determined by X-ray crystallography, a method developed by British physicists in the early 1900s. The field of X-ray analysis included an unusual number of women, including Dorothy Hodgkin (1910–1994), who won a Nobel Prize for determining the structures of penicillin and vitamin B_{12}. In 1953, crystallographer Rosalind Franklin (1920–1958) joined a laboratory at King's College to study the structure of DNA (Figure 1.24A). As a woman and as a Jew who supported relief work in Palestine, Franklin felt socially isolated at the male-dominated Protestant university. Nevertheless, the exceptional quality of her X-ray micrographs impressed her colleagues. The X-shaped pattern in her micrograph (Figure 1.24B) arises from the double-helical form of the DNA molecule. Without Franklin's knowledge, her colleague Maurice Wilkins showed her data to a competitor, James Watson (born 1928). The X-shaped pattern

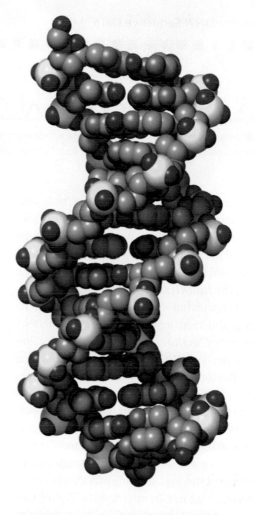

Figure 1.23
DNA: The Central Information Molecule
The sequence of base pairs in DNA encodes all the genetic information of an organism.

led Watson and Francis Crick (1916–2004) to guess that the four bases of the DNA "alphabet" are paired in the interior of Franklin's double helix (Figure 1.24C). They published their model in the journal *Nature*, without acknowledging their use of Franklin's data.

Figure 1.24   Discovering the DNA Double Helix

A. Rosalind Franklin discovered that DNA forms a double helix.

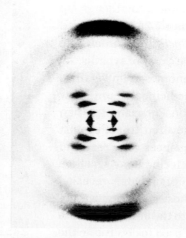

B. X-ray diffraction pattern of DNA, obtained by Rosalind Franklin.

C. James Watson and Francis Crick proposed a structure for the complementary pairing between bases of DNA.

Figure 1.25   DNA Sequence Data

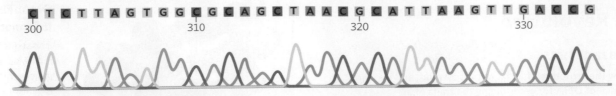

Sequence obtained from genomic DNA of a microbe isolated from a discarded beverage container. Each colored trace represents the concentration of one of the four bases terminating an interrupted chain of DNA.

The discovery of the double helix earned Watson, Crick, and Wilkins the 1962 Nobel Prize in Physiology or Medicine. Franklin died of ovarian cancer before the Nobel Prize was awarded. Before her death, however, she turned her efforts to the structure of ribonucleic acid (RNA). She determined the form of the RNA chromosome within tobacco mosaic virus, the first viral RNA to be characterized.

The structure of DNA base pairs led to the development of techniques for **DNA sequencing**, the reading of the sequence of DNA base pairs (the DNA sequencing technique is outlined in Section 8.4). Figure 1.25 shows an example of DNA sequence data, in which each color represents one of the four bases and each peak represents a DNA fragment terminating in that particular base. The order of fragment lengths yields the sequence of bases in one strand. Reading the sequence enabled microbiologists to determine the beginning and endpoint of genes and ultimately entire genomes.

A method of DNA sequencing was devised by Frederick Sanger (Figure 1.26A) and used to reveal the first genome of a virus in 1977. Sanger shared the 1980 Nobel Prize in Chemistry with Walter Gilbert and Paul Berg. The first genome sequence of a cellular microbe was obtained in 1995 for *Haemophilus influenzae*, a bacterium that causes ear infections and meningitis in children. The *H. influenzae* genome has nearly 2 million base pairs that specify about 1,700 genes. It was sequenced by a team of scientists led by Hamilton Smith and Craig Venter, who devised a special computational strategy for assembling large amounts of sequence data. This strategy was later applied to sequencing the human genome. Another group of scientists led by Claire Fraser-Liggett (Figure 1.26B) sequenced the genomes of many microbes, including *Bacillus anthracis*, the cause of anthrax, and *Treponema pallidum*, the cause of syphilis.

The growing availability of sequenced genomes at universities and in industry makes possible advances in medicine that until now could only have been imagined in science fiction. For example, when the new antibiotic diarylquinoline was discovered to act against *Mycobacterium tuberculosis*, the genome sequences were read and compared for two different strains of the bacterium, one sensitive to the antibiotic and one resistant. The result immediately revealed the one gene in the *M. tuberculosis* genome whose gene product (the proton pump of ATP synthase) was sensitive to the antibiotic. More recently, human genome sequences are being used to determine the gene variants of given patients and thus optimize their treatment according to their genes.

Microbial Discoveries Transform Medicine and Industry

Twentieth-century microbiology transformed the practice of medicine and generated entire industries of biotechnology and bioremediation. After the discovery of penicillin, Americans poured millions of dollars of private and public funds into medical research. The March of Dimes campaign for private donations to prevent polio led to a vaccine that has nearly eliminated the disease. With the end of World War II, research on microbes and other aspects of biology drew increasing financial support from U.S. government agencies, such as the National Institutes of Health and the National Science Foundation, as well as from governments of other countries, particularly the European nations and Japan. Further support has come from private foundations, such as the Pasteur Institute, the Wellcome Trust, and the Howard Hughes Medical Institute.

Careers in diverse fields recruit microbiologists (Table 1.2; Figure 1.27). Nurses, physician assistants, pharmacists, and other

Figure 1.26   Microbial Genome Sequencers

A. Frederick Sanger, who shared the 1980 Nobel Prize in Chemistry for devising the method of DNA sequence analysis that is the basis of modern genome sequencing. He is shown reading sequence data from bands of DNA separated by electrophoresis.

B. Claire Fraser-Liggett led a team that completed the sequences of many microbial genomes.

Table 1.2
Careers Involving Microbiology

Nursing	Providing comprehensive care for patients with illness
Physician assistant	Practicing medicine with supervision of a licensed physician
Medical microbiology	Diagnosis, treatment, and developing therapies for human diseases
Veterinary microbiology	Diagnosis, treatment, and developing therapies for animal diseases
Public health	Surveillance, assessment, and promotion of policies influencing the health of populations
Food and industrial microbiology	Microbial and industrial food products, food contamination, and bioremediation
Agricultural microbiology	Managing plant pathogens and plant-associated microbes
Environmental microbiology	Environmental remediation through microbial processes
Forensic microbiology	Analysis of microbial strains as evidence in criminal investigations
Experimental microbiology	Investigating fundamental questions about microbial form and function, genetics, and ecology

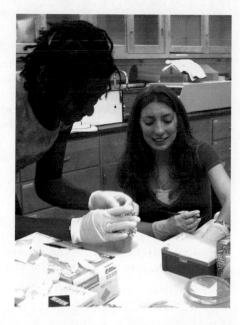

Figure 1.27
Microbiology Students at Kenyon College Prepare for Careers in Nursing and Medical Research

health care professionals use microbiological science to keep patients healthy and diagnose their illnesses. Public health workers apply statistical models to manage the health of communities and nations. Forensic technologists use DNA and microbial analysis to solve crimes. In environmental science, newly discovered microbes provide new ways to bioremediate wastes and control insect pests. On a global level, the challenges of pollution and climate change increasingly involve contributions of environmental microbiology to human health.

Perspective

Microbiology: The Human Experience is your guide to microbes and the enormous impact they have on human life. Our focus is medical by design, but we also explore intriguing connections between microbes, Earth's ecology, and human existence. Part 1 of the book lays the foundation for studying microorganisms and for understanding infectious disease. The chapter just completed provided the human timeline of discovery. We next introduce the basic concepts of microbial disease in Chapter 2 and then in Chapter 3 describe the tools of microscopy that allow us to see cells and their inner workings. Chapter 4 explains the fundamental biochemistry needed to understand microbial life (and infectious disease), while Chapter 5 reveals the intricate structures of the microbes themselves. Finally, Chapter 6 introduces microbial nutrition and growth within the human body and human-related environments. Throughout, we invite readers to share with us the excitement of discovery in microbiology and the potential applications of that knowledge to improving human health.

SECTION SUMMARY

- **The structure and function of DNA** was revealed by a series of experiments in the twentieth century.
- **Genome sequence determination** has shaped the study of biology in the twenty-first century.
- **Microbial discoveries transformed medicine and industry.** Biotechnology enables the production of new kinds of pharmaceuticals and industrial products.
- **Microbiology today offers many different careers** in the health field and in industry.

LEARNING OUTCOMES AND ASSESSMENT

1.1
From Germ to Genome: What Is a Microbe?

- Describe how we define a microbe, and explain why the definition is a challenge.
- Describe the three major domains of life: Archaea, Bacteria, and Eukarya. Explain what the three domains have in common and how they differ.
- Define viruses, and explain how they relate to living cells.

1. Which trait is absent from some kinds of cellular microbes?
   A. A nucleus with a nuclear membrane
   B. A cell membrane (plasma membrane)
   C. One or more chromosomes of DNA
   D. Enzymes that conduct metabolism

1.2
Microbes Shape Human History

- Explain how microbial diseases have changed human history.
- Describe how microbes participate in human cultural practices such as production of food and drink.

4. Which person was admired for tending victims of a deadly disease?
   A. Robert Hooke
   B. Antonie van Leeuwenhoek
   C. Catherine of Siena
   D. Lazzaro Spallanzani

1.3
Medical Microbiology and Immunology

- Define the germ theory of disease.
- Explain how Florence Nightingale first drew a statistical correlation between infectious disease and human mortality.
- Explain how Koch's postulates can show that a specific kind of microbe causes a disease. Explain the problems in interpreting Koch's postulates in practice.

7. Florence Nightingale's polar area chart (see Figure 1.8) and the CDC chart of influenza (see Figure 1.9) have what in common?
   A. Both charts graphically represent patterns of public nutrition.
   B. Both charts represent the rise and fall of an infectious disease.
   C. Both charts depict the role of disease in an army campaign.
   D. Both charts include data from culturing the causative agent of disease.

1.4
Microbes in Our Environment

- Describe examples of how microbes contribute to natural ecosystems.
- Explain how mitochondria and chloroplasts evolved by endosymbiosis.

10. Which class of microbes gains energy from oxidizing inorganic minerals?
    A. Enteric bacteria
    B. Lithotrophs
    C. Rhizobia
    D. Algae

1.5
The DNA Revolution

- Describe how the structure of DNA was discovered, and explain the significance of DNA for determining the traits of life.
- Describe how the manipulation of DNA information has transformed the practice of medicine.

13. The person whose X-ray crystallography data first revealed the fundamental structures of DNA and RNA was _____.
    A. James Watson
    B. Francis Crick
    C. Dorothy Hodgkin
    D. Rosalind Franklin

2. Which of the following applies <u>only</u> to Archaea?
  A. These microbes have no nucleus or nuclear membrane.
  B. These microbes possess mitochondria.
  C. These microbes are not true cells.
  D. These microbes have never been shown to cause disease.

3. Which trait of cells do viruses share?
  A. A genome that undergoes replication
  B. Ribosomes that synthesize protein
  C. Active metabolism
  D. Motility

5. Which person first demonstrated that the failure of spontaneous generation of microbes was not due to lack of oxygen?
  A. Catherine of Siena
  B. Lazzaro Spallanzani
  C. Robert Hooke
  D. Louis Pasteur

6. Which epidemic disease inspired a large demonstration at the Washington Monument?
  A. Bubonic plague
  B. Influenza
  C. AIDS
  D. Smallpox

8. Which of Koch's postulates requires modification for study of a human disease?
  A. The microbe is found in all cases of the disease but is absent from healthy individuals.
  B. The microbe is isolated from the diseased host and grown in pure culture.
  C. When the microbe is introduced into a healthy, susceptible host, the same disease occurs.
  D. The same strain of microbe is obtained from the newly diseased organism. When cultured, the strain shows the same traits as before.

9. Which statement about antibiotics is incorrect?
  A. Antibiotics reproduce themselves to form more antibiotics.
  B. Antibiotics are chemicals produced by specific microbes.
  C. A specific antibiotic can kill certain classes of microbes.
  D. Fewer antibiotics are known for viruses than for bacterial pathogens.

11. Which class of microbes most closely resembles the ancestor of chloroplasts?
  A. Enteric bacteria
  B. Mycobacteria
  C. Cyanobacteria
  D. Archaea

12. Which important cellular process requires bacteria or archaea?
  A. Fermentation
  B. Nitrogen fixation
  C. DNA replication
  D. Membrane formation

14. Who led the team of scientists that sequenced the genomes of numerous microbes, including *Bacillus anthracis* and *Treponema pallidum*?
  A. Frederick Sanger
  B. Walter Gilbert
  C. Claire Fraser-Liggett
  D. Hamilton Smith

15. Which molecular technique is now used to identify a very specific strain of a disease-causing microbe?
  A. Centrifugation
  B. Plate culture
  C. Polymerase chain reaction (PCR) amplification of DNA
  D. Gram stain

Key Terms

agar (p. 17)
alga (p. 5)
antibiotic (p. 20)
antiseptic (p. 20)
archaeon (p. 5)
aseptic (p. 20)
autoclave (p. 12)
bacterium (p. 5)
chain of infection (p. 15)

colony (p. 17)
DNA sequencing (p. 26)
endosymbiont (p. 23)
enrichment culture (p. 23)
eukaryote (p. 5)
fermentation (p. 12)
fungus (p. 5)
genome (p. 5)
geochemical cycling (p. 23)

germ theory of disease (p. 13)
immune system (p. 20)
immunity (p. 20)
immunization (p. 20)
Koch's postulates (p. 18)
microbe (p. 4)
nitrogen fixation (p. 23)
pathogen (p. 4)
petri dish (p. 17)

prokaryote (p. 5)
protist (p. 5)
protozoan (p. 5)
public health (p. 14)
pure culture (p. 17)
species (p. 5)
spontaneous generation (p. 11)
virus (p. 6)

Review Questions

1. Explain the apparent contradictions in defining microbiology as the study of microscopic organisms or the study of single-celled organisms.

2. What is the genome of an organism? How do genomes of viruses differ from those of cellular microbes?

3. Under what conditions might microbial life have originated? What evidence supports current views of microbial origin?

4. List the ways in which microbes have affected human life throughout history.

5. Summarize the key experiments and insights that shaped the controversy over spontaneous generation. What key questions were raised, and how were they answered?

6. Explain how microbes are cultured on liquid and solid media. Compare the culture methods of Koch and Winogradsky. How did their different approaches to microbial culture address different questions in microbiology?

7. Explain how a series of observations of disease transmission led to development of immunization to prevent disease.

Clinical Correlation Questions

1. A patient comes to the emergency ward of a hospital complaining of shortness of breath. The patient also has fever, coughing, and shaking chills. The patient demands antibiotics, but after testing, the physician declines to supply them. Instead, the hospital provides corticosteroids, fluids, and oxygen. Why does the physician not prescribe antibiotics?

2. A patient visits the doctor with symptoms of gastrointestinal illness, including diarrhea and vomiting. A stool culture is performed, revealing *Escherichia coli* bacteria. Does this test prove that *E. coli* caused the illness? Which of Koch's postulates might not be fulfilled?

Thought Questions: CHECK YOUR ANSWERS

Thought Question 1.1   Why do you think it took so long for humans to connect microbes with infectious disease?

ANSWER: Human eyes cannot see microbes. Furthermore, microbes are so ubiquitous in the environment that even if they had been seen, it would have been hard to tell which ones cause disease and which do not. By the late nineteenth century, microbiologists were discovering ways to distinguish microbial traits, such as the Gram-negative and Gram-positive cell wall structures. The sequencing of microbial DNA has vastly improved our ability to define the precise causative agents of disease.

Thought Question 1.2   How could you use Koch's postulates to demonstrate the causative agent of influenza? What problems would you need to overcome that were not encountered with anthrax?

ANSWER: To demonstrate that the virus causes influenza, it is necessary to culture the virus. However, the culture must be a "coculture" together with a source of cells to infect, such as human cells in tissue culture. Moreover, aspects of the influenza replicative cycle and transmission must be studied in nonhuman animals such as guinea pigs. Using guinea pigs, it is possible to obtain influenza virus from the sick animals only (postulate 1); to replicate the virus in tissue culture (postulate 2); then show that the reisolated virus can infect guinea pigs again (postulate 3), producing influenza virus again (postulate 4).

Thought Question 1.3   Why do you think some pathogens generate immunity readily, whereas others evade the immune system?

ANSWER: We do not fully understand why some pathogens are more immunogenic than others. Pathogens that readily evade the immune system usually have genomes that can change rapidly, allowing the microbe to display different surface proteins that the immune system does not yet recognize. Pathogens whose genomes mutate the most rapidly (such as influenza virus and HIV) are the most challenging to develop effective vaccines against.

Thought Question 1.4   How do you think microbes protect themselves from the antibiotics they produce?

ANSWER: Microbes can protect themselves from the antibiotics they produce by expressing enzymes that disable those antibiotics within the producer cell. Alternatively, the microbe may form an altered version of the antibiotic target, such as a modified ribosome subunit. Another possibility is that the producer microbe may export the antibiotic from the cell as soon as it is produced.

Basic Concepts of Infectious Disease

SCENARIO Brandon, a 30-year-old stockbroker living in Chicago, visited his physician's office. When the nurse asked Brandon why he was there, he blushed and said that he wanted to talk only to the physician about his problem.

SIGNS AND SYMPTOMS Once the doctor entered the room, Brandon explained he had a small round lesion on his penis. When asked about his sexual partners, Brandon initially said he was dating only one woman, but when pressed, he admitted he had been intimate with two women over the past month and one man. Upon examination, the lesion appeared to cause no pain but exuded a clear fluid.

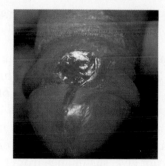

Chancre of Syphilis
Note single ulcer on the foreskin, usually painless and indurated (hardened) with a well-defined margin and a clear base.

TESTING The physician quickly sent a sample of the fluid to the clinical laboratory. There the sample was found to contain highly motile, corkscrew-shaped bacteria.

DIAGNOSIS The diagnosis was syphilis, caused by the bacterium *Treponema pallidum*. Left untreated, the disease could eventually cause horrible disfiguration and death.

TREATMENT Confident that he knew the cause, the physician gave Brandon a shot of long-acting penicillin.

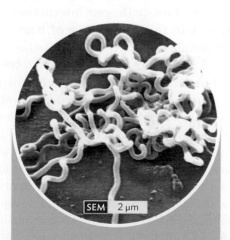

SEM 2 μm

Treponema pallidum
A sexually transmitted, spiral-shaped bacterium that can grow only in a host.

CHAPTER OBJECTIVES

After reading this chapter, you will be able to:

- Describe the relationships among a host, its microbiome, and pathogens.
- Explain the basic concepts of infection and infectious disease.
- Discuss how infectious diseases impact communities and how communities shape emerging pathogens.

Early in the sixteenth century, Ulrich von Hutten wrote: "In the year of Chryst 1493 or there aboute, this most foul and most grievous disease beganne to spreade amonge the people." Symptoms included disfiguring rashes, ulcers, violent fevers, and painful bone aches. It could do irreparable damage to the internal organs and frequently caused death. It appeared first in Spain (coinciding with Columbus's return from "Española"), reached epidemic proportions in Italy, and by 1500 had spread throughout Europe. Physicians, surgeons, and laypeople believed it was a new disease and realized it was transmitted by venereal (sexual) contact. The French called it the disease of Naples; the Italians called it the French disease; the English blamed it equally on the French and Spanish or simply referred to it as the pox. By the 1520s, it was called syphilis, the same disease Brandon had in the opening case. The significance of this tale is that the relationship between a horrible disease and its transmission was recognized even before bacteria were known to exist.

Wanting to understand the nature of infectious disease is one good reason for taking a course in microbiology. We begin this text with basic concepts of infectious disease, knowing that many readers probably intend to pursue a career in one of the health sciences. Even if you envision a career outside of health science, our guess is that you experience infections yourself and would like to learn more. **You do not need to know much about bacterial structure, genetics, or metabolism to grasp the basics of infectious disease. However, understanding the basics of infectious disease and host-pathogen interactions early on will help you appreciate the importance of bacterial structure, genetics, and metabolism.** You will learn in this chapter, for example, that bacterial structure dictates how microbes interact with their hosts and how hosts recognize invading organisms.

Perhaps the following question, more than any other, exemplifies the mystery of microbiology. How is it possible that an organism too small to be seen with the naked eye can kill a human who is 1 million times larger? In exploring this and other aspects of infectious disease, we'll introduce some basic types of pathogens and describe how they manage to colonize parts of our bodies. We'll outline the stages of an infectious disease and discuss where microbes that cause disease reside when not causing disease and how these infectious agents are transmitted to humans. Some familiar and not-so-familiar microbial diseases will be mentioned along the way, including some examples of new, emerging infectious diseases.

2.1
Normal Microbiota versus Pathogens

SECTION OBJECTIVES

- Describe differences between microbiota and pathogens.
- Discuss the relationship between infection and disease and between virulence and pathogenicity.
- Differentiate between infectious dose and lethal dose.
- Discuss the fundamental attributes of a successful pathogen.

What is a pathogen? You may not want to think about it, but we humans harbor at least 2–3 times more microbial hitchhikers than we have human cells in our bodies. The collection of bacteria, archaea, and eukaryotic microbes that call us home is called our normal **microbiota** and will be discussed in more depth in Chapter 14. Although they are commonly called "commensal organisms" (organisms that take, but do not give, sustenance from their host), most members of our microbiota both derive and give benefit to their hosts, which defines **mutualism**, or a mutualistic relationship. To stay with us, members of our normal microbiota have proteins on their surface (called **adhesins**) that allow them to attach to and colonize epithelial cells lining mucous membranes (intestine, urinary tract, mouth, nose). **Colonization** refers to the ability of a microbe to stay affixed to a body surface and replicate (**Figure 2.1**). Think about your own intestine. Without an ability to attach, some very useful bacteria would too easily be swept out of your body with your feces. However, membership in an individual's microbiota (or microbiome) is not forever. Different bacteria and strains of microbial species can enter or leave our bodies, depending on what we eat, how we exercise, and whom we meet. A changing microbiota can alter our susceptibility to pathogens, the development of our immune system, and, believe it or not, our weight.

Defining a member of our normal microbiota can be tricky. Take, for example, the bacterium *Escherichia coli*. *E. coli* is a normal part of our intestinal microbiota. We all harbor this organism. But news outlets carry alarming stories about *E. coli* causing death and disease. So is *E. coli* a disease-causing organism (pathogen) or isn't it? The answer is yes—to both questions. Every species is composed of different strains that display unique characteristics. Some strains of *E. coli* are relatively tame and actually contribute to our

Figure 2.1 Bacterial Attachment to Human Cells

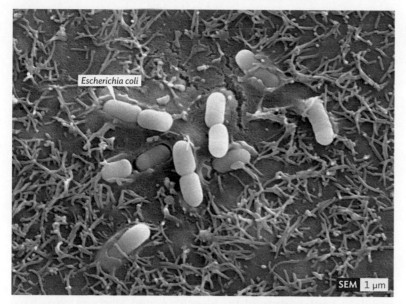

Escherichia coli

SEM | 1 μm

3D micrograph showing a small cluster of bacteria (*Escherichia coli*) adhering tightly to human cells. Note that these bacteria are rod-shaped.

health as residents of the colon. However, other strains of *E. coli* can cause diarrhea, septicemia (blood infection), meningitis, and urinary tract infections. Wherein lies the difference? It lies in the pathogen's genes.

Although defining a **pathogen** is easy—it is any bacterium, virus, fungus, protozoan, or worm (helminth) that causes disease—defining what actually makes a pathogen a pathogen is not so easy. The **pathogenicity** of an organism—that is, its ability to cause disease—depends on a combination of the organism's genetic makeup, its location on the host's body, and the effectiveness of the host's immune response. For example, an *E. coli* strain that is innocuous in you could possibly kill your immune-compromised neighbor.

The term "parasite" is often used in casual conversation but requires some definition and context. **Parasites**, in the broadest sense, include any organism that colonizes and harms its host. In medical usage, the word "parasite" is reserved for disease-causing protozoa (single-celled eukaryotic organisms) and worms. Insects such as fleas or lice that live on the body surface are also called parasites—ectoparasites. The term "pathogen" is typically used by health professionals to refer to bacterial, viral, and fungal agents of disease.

Pathogens and parasites infect their animal and plant hosts in a variety of ways and enter into different host-pathogen relationships, depending on the site of colonization and the capabilities of the organism. For example, some organisms can live only on the surface of a host; an example is the fungus *Trichophyton rubrum*, one cause of athlete's foot (**Figure 2.2**). *T. rubrum* does not have the molecular tools that would allow it to grow inside a host. In contrast, *Wuchereria bancrofti*, the parasitic worm that causes elephantiasis, cannot grow on the skin but causes disease by living inside the body (**Figure 2.3**). Parasites such as *W. bancrofti* that live inside the body are called endoparasites.

Note The various types of microscopy used to see the microbes in this chapter will be described in Chapter 4. Right now it is important only to appreciate the appearance and size of these organisms, not the techniques used to capture their images.

Terminology of Pathogenesis

Before examining the mechanisms microbes use to cause disease, it is helpful to know the terminology of pathogenesis. An **infection** occurs when a pathogen or parasite enters or begins to grow on a host. Be aware that the term "infection" does not necessarily imply overt disease. Any potential pathogen growing in or on a host is said to cause an infection, but that infection may be temporary if immune defenses kill the pathogen before noticeable disease results. Indeed, most infections go unnoticed. For example, every time you have your teeth cleaned by a dentist, your gums bleed and

Figure 2.2 A Skin Fungus

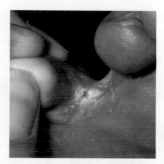

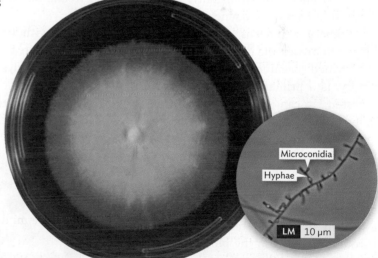

Microconidia

Hyphae

LM | 10 μm

A. Athlete's foot is caused by the fungus *Tricophyton rubrum*.

B. *T. rubrum* colony shape (approx. 3 inches, or 75 mm, in diameter). Inset is a magnification of *T. rubrum* showing the branching filaments (called **hyphae**) with small spores (called microconidia) that form the colony.

Link As will be described in Section 11.2, **hyphae** are branching strands of fungal cells lined end to end. Some cells at the tips of hyphae can transform into dormant, fungal spores called conidia (or microconidia if small) that can germinate to produce new fungal colonies at a later time.

Figure 2.3   An Endoparasite

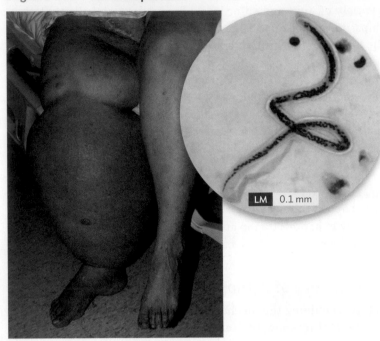

LM 0.1 mm

The disease filariasis, commonly known as "elephantiasis" for obvious reasons, is caused by the worm *Wuchereria bancrofti* (inset), which enters the lymphatics (vessels connecting lymph nodes) and blocks lymphatic circulation. Adult worms are threadlike and measure 4–10 cm (1.5–4 inches) in length. The young microfilaria (shown) are approximately 0.5 mm in length. Although not a problem in the United States, *W. bancrofti* is found throughout middle Africa, Asia, and New Zealand.

your oral microbiota transiently enter the bloodstream, but you rarely suffer any consequences.

Primary pathogens are disease-causing microbes with the means to breach the defenses of a healthy host. For example, the rod-shaped (bacillus) *Shigella flexneri*, the cause of bacillary dysentery, is a primary pathogen. When ingested, it can survive the natural defensive barrier of an acidic stomach, enter the intestine, and begin to replicate. The organism in the chapter-opening case, *Treponema pallidum*, is an example of a primary pathogen because it can survive initial attack by a healthy immune system and spread through the body in the bloodstream. **Opportunistic pathogens**, on the other hand, cause disease only in a compromised host. The fungus *Pneumocystis jirovecii* (previously *P. carinii*, Figure 2.4A) easily infects people but rarely causes disease. It is, however, an opportunistic pathogen that causes life-threatening infections in AIDS

patients, whose immune systems have been eroded by HIV. Some microbes even enter into a **latent state** during infection, where the organism cannot be found by culture. Herpes virus, for instance, can enter the peripheral nerves and remain dormant for years and then suddenly emerge to cause cold sores (Figure 2.4B). The bacterium *Rickettsia prowazekii* causes epidemic typhus, but it can also enter a mysterious latent phase and emerge months or years later to cause a disease relapse called recrudescent typhus. Where in the body does this bacterium hide? With all our technology, evidence of a latent form of *R. prowazekii* has never been found . . . yet it must be there.

Pathogenicity is defined in terms of how easily an organism causes disease (infectivity) and how severe that disease is (virulence). Pathogenicity, overall, is shaped by the genetic makeup of the pathogen. In other words, an organism is more—or less—pathogenic depending on the tools at its disposal (such as toxins) and their effectiveness.

Virulence is a measure of the degree or severity of disease. For instance, Ebola virus and the closely related Marburg virus have case fatality rates near 70%. This means that they are highly virulent (Figure 2.5). On the other hand, rhinovirus, the cause of the common cold, is very effective at causing disease but never kills its victims. So it is highly infective but has a low virulence. Both organisms are pathogenic, but with one you live and with the other you probably die.

One way to measure virulence is to calculate the **lethal dose 50% (LD$_{50}$)**: the number of bacteria or virus particles (virions) required to kill 50% of an experimental group of animal hosts (usually mice or guinea pigs). An organism with a low LD$_{50}$, in which very few organisms are required to kill 50% of the hosts, is more virulent than one with a high LD$_{50}$ (Figure 2.6). For organisms that colonize but do not kill the host, the infectious dose needed to colonize 50% of the experimental hosts, called the **infectious dose 50% (ID$_{50}$)**,

Figure 2.4   Opportunistic and Latent Infections

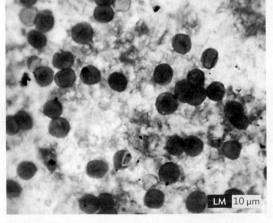

LM 10 μm

A. *Pneumocystis jirovecii* cysts in broncheoalveolar material. Notice how the fungi look like crushed Ping-Pong balls.

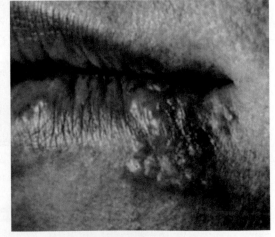

B. Cold sore produced by a reactivated herpes virus hiding latent in nerve cells.

Figure 2.5   Highly Virulent Ebola and Marburg Viruses

A. Ebola virus.

B. The body of a victim of Marburg virus is placed in a coffin for safe burial in Angola. Marburg and Ebola viruses cause hemorrhagic infections in which patients bleed from the mouth, nose, eyes, and other orifices. They have a 70%–80% mortality rate.

can be measured. Infectious dose is measured by determining how many microbes are required to cause disease symptoms in half of an experimental group of hosts.

Although one might be able to determine the infectious dose rather than the lethal dose for a lethal pathogen, the infectious dose is not typically measured because the end point (illness) can be difficult to assess in lab animals. Instead, the LD_{50} gives a clear end point (death) and is much easier to use when trying to determine the effectiveness of a given treatment (an antibiotic, for example) or when quantifying the role of a given gene in pathogenesis.

Two other aspects of pathogenesis are invasion and invasiveness. Although they sound similar, the terms mean very different things. **Invasion** describes the ability of some pathogens to actually enter (invade) and live inside cells of a human or nonhuman animal host. Some live inside small membrane-enclosed cavities within the host cell called vacuoles (**Figure 2.7A**). Examples are *Salmonella*, a cause of diarrhea; *Coxiella*, the cause of Q fever; and *Legionella*, the cause of Legionnaires' disease. Other bacteria prefer to live directly in the cytoplasm of the host cell, not in a vacuole (**Figure 2.7B**). Examples of bacteria that grow in cytoplasm are *Shigella*, a cause of bloody diarrhea, and *Listeria*, which can cause diarrhea and neonatal meningitis.

Note The disease processes and pathogenic mechanisms of the many microbes mentioned in this chapter will be explained more completely in later chapters. The basic concepts described here, however, can be applied to all the case histories that anchor the chapters of this book.

Invasiveness, in contrast to invasion, refers to the ability of a bacterial pathogen to rapidly spread through tissue. For instance, some strains of *Streptococcus pyogenes* are said to be highly invasive because they secrete enzymes that degrade host tissues. These strains are called "flesh-eating bacteria" because of the rapidity with which they destroy tissue as the organism spreads.

It is also important to remember that different microbes have different **host ranges** in terms of which animals they can infect and produce disease. Pathogens can have a very narrow host range; for example, *Salmonella enterica* serovar Typhi, the bacterium that causes typhoid fever, can infect only humans. (Serovar is a subclass within a species.) But a close relative, *S. enterica* serovar Typhimurium, which produces diarrhea in humans, can infect many different animals. *S. enterica* serovar Typhimurium, then, has a broad host range. *T. pallidum*, in the chapter-opening syphilis case, has a very narrow host range. It infects only humans.

Note Bacterial names include designations for genus (*Salmonella*) and species (*enterica*). Both are written in italics, with the *Genus* capitalized and *species* lowercase. Serovar subclass names of a species are not italicized but are capitalized (for instance, Typhimurium).

Figure 2.6   Measurement of Virulence

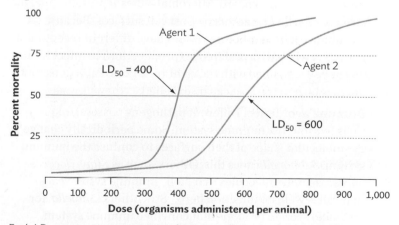

Each LD_{50} measurement requires infecting small groups of animals with increasing numbers of infectious agent and observing how many animals die. The number of microbes that kill half the animals is called the LD_{50}. In this example, agent 1, which requires fewer organisms to kill half the animals than does agent 2, is more virulent than agent 2.

Figure 2.7   Bacterial Life in a Host Cell

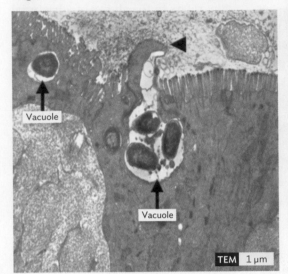

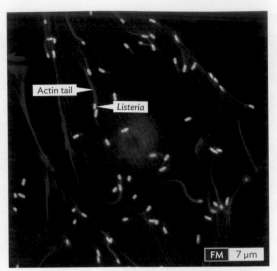

A. Invasion of intestinal cells (enterocytes) by *Salmonella enterica* serovar Typhimurium. Intracellular bacteria are located within membrane-enclosed vacuoles (arrows). Bacteria are first captured by cytoplasmic projections coming from the enterocyte surface (arrowhead).

B. The bacterium *Listeria* similarly invades eukaryotic host cells by being engulfed in a vacuole but then escapes the vacuole to live in the cytoplasm. *Listeria* (green) can propel itself in the cytoplasm by linking together (polymerizing) many copies of a host protein (actin) at one end of the bacterium. The structure is called an actin tail (red). Regions where the red and green fluorescence overlap appear yellow.

bacteria need in order to grow. One example involves iron uptake. To keep iron away from invading bacteria, the human body makes compounds (lactoferrin and transferrin) that tightly bind this metal. Consequently, most bacteria that enter the bloodstream cannot access the iron in the blood and cannot grow. Pathogens, however, have ways to steal the iron from host tissues to use for their own purposes.

Link The role of microbial siderophores in iron uptake is described in Section 5.2.

SECTION SUMMARY

- **Infection** with a microbe does not always lead to disease.
- **Primary pathogens** have mechanisms that help the organism circumvent host defenses.
- **Opportunistic pathogens** cause disease only in a compromised host.
- **Pathogenicity** refers to the mechanisms a pathogen uses to produce disease and how efficient the organism is at causing disease, whereas **virulence** is a measure of disease severity.
- **Invasion** refers to the ability of a pathogen to enter and live in host cells.
- **Invasiveness** describes the ability of an organism to spread through tissues.
- **Host range** refers to how readily a pathogen can infect different animals.
- **Pathogens** must be able to attach to host cells, evade the immune system (at least temporarily), and obtain nutrients from the host.

Fundamental Host-Pathogen Interactions

Regardless of the pathogen—viral, bacterial, or eukaryotic—all must interact with a host in several ways to be successful. The pathogen must attach to a host tissue, avoid the host's immune system, and steal nutrients from the host. We will delve more deeply into these interactions later (especially in Chapters 18–24) but discuss the basic concepts here.

Attachment. Recall that specific proteins (adhesins) on the surfaces of microbes help them adhere to host cells, like a thumbtack to a corkboard. Microbial adhesin proteins bind to structures called **receptors** on host cell surfaces. Because different host cells in different tissues have different receptors, a pathogen will have a preference (predilection) to infect tissues whose cells are lined with the right receptors. Pathogens that fail to attach to the host can be expelled.

Immune avoidance. Different pathogens possess unique tools to avoid the immune system. Some periodically change the molecular shape of their surfaces to confuse the immune system; *Salmonella* does this. Others, such as *Staphylococcus aureus*, secrete molecules that "tell" immune system cells that "all is well, no infection here." Still others (*Shigella*, for instance) secrete proteins that convince immune system cells to kill themselves in a process known as apoptosis. Many examples of the ingenuity displayed by microbes in avoiding the immune system will be described in Chapters 15 and 16.

Obtaining nutrients from the host. Animals, including humans, have evolved ways to tightly hold nutrients that

Thought Question 2.1   Is a microbe with an LD_{50} of 5×10^4 more or less virulent than a microbe with an LD_{50} of 5×10^7? Why?

Thought Question 2.2   The genus of bacteria called *Salmonella* causes a wide range of illnesses in humans. *Salmonella enterica* serovar Typhimurium (*S.* Typhimurium) is usually transmitted by contaminated food and causes a diarrhea that typically resolves without medical intervention. The organism remains localized to the intestine of humans. *Salmonella enterica* serovar Typhi (*S.* Typhi) is also transmitted via contaminated food, but can penetrate the intestinal wall and enter the lymphatic system and bloodstream to cause high fever and in some cases death. Clearly, *S.* Typhi is more virulent than *S.* Typhimurium to humans. However, when tested for lethal dose (LD_{50}) in mice, *S.* Typhi fails to kill any mice, whereas *S.* Typhimurium has a very low lethal dose (it takes very few organisms to kill mice). How might you explain this apparent contradiction?

2.2
Basic Concepts of Disease

How do we define disease? A **disease** is a disruption of the normal structure or function of any body part, organ, or system that can be recognized by a characteristic set of symptoms and signs. An **infectious disease**, then, is a disease caused by a microorganism (bacterial, viral, or parasitic) that can be transferred from one host to another. Table 2.1 presents the terms commonly used to describe the features of infectious disease. It is important for you to learn these terms because they will be used repeatedly throughout this book.

Table 2.1
Terms Used to Describe Infections and Infectious Diseases

Term	Definition	Example
Acute infection	Infection in which symptoms develop rapidly; its course can be rapid or protracted	Strep throat (*Streptococcus pyogenes*)
Chronic infection	Infection in which symptoms develop gradually, over weeks or months, and are slow to resolve (heal), taking 3 months or more	Tuberculosis (*Mycobacterium tuberculosis*)
Subacute disease	Infection in which symptoms take longer to develop than in an acute (rapid) infection but arise more quickly than for a chronic infection	Subacute bacterial endocarditis (*Enterococcus faecalis*)
Latent infection	A type of infection that may occur after an acute episode; the organism is present but symptoms are not; after time, the disease can reappear	Cold sores due to herpes virus
Focal infection	Initial site of infection from which organisms can travel via the bloodstream to another area of the body	Boils (*Staphylococcus aureus*)
Disseminated infection	Infection caused by organisms traveling from a focal infection; when affecting several organ systems, it is called a "systemic" infection	Tularemia (*Francisella tularensis*)
Metastatic lesion	A site of infection resulting from dissemination	Blastomycosis, fungal infection of the lung; can disseminate to form abscesses in the extremities (arms/legs)
Bacteremia	Presence of bacteria in blood. Usually transient, little, or no replication	May occur during dental procedures (*Streptococcus mutans*)
Septicemia	Presence and replication of bacteria in the blood (blood infection)	Bubonic plague (*Yersinia pestis*)
Viremia	Presence of viruses in the blood	HIV
Toxemia	Presence of toxin in the blood	Diphtheria, toxic shock syndrome
Primary infection	Infection in a previously healthy individual	Syphilis (*Treponema pallidum*)
Secondary infection	Infection that follows a primary infection; damaged tissue (e.g., lung) is more susceptible to infection by a different organism	Infection following viral influenza (*Haemophilus influenzae*)
Mixed infection	Infection caused by two or more pathogens	Appendicitis (*Bacteroides fragilis* and *Escherichia coli*)
Iatrogenic infection	Infection transmitted from a health care worker to a patient	Some septicemias (*Staphylococcus aureus*)
Nosocomial infection	Infection acquired during a hospital stay (hospital-acquired infection)	MRSA (methicillin-resistant *Staphylococcus aureus*)
Community-acquired infection	Infection acquired in the community, not in a hospital	Some MRSA strains are community acquired

Signs versus Symptoms

Most diseases are recognized by their signs and symptoms. A **sign** is something that can be observed by a person examining the patient—for instance, a runny nose or a rash. Brandon's lesion in the chapter-opening case was an obvious disease sign. A **symptom** is something that can be felt only by the patient, such as pain or general discomfort (malaise). A **syndrome** is a collection of signs and symptoms that occur together and signify a particular disease.

Even after an infectious disease has resolved, pathological consequences called **sequelae** may develop. For instance, a strep throat can result in heart or kidney damage, two sequelae known as rheumatic fever and glomerulonephritis, respectively.

It is important to realize that many of the signs and symptoms of an infectious disease, including some sequelae, are actually caused by the host's response to the infection (called **immunopathology**). Cells given the task of killing the microbe can also damage nearby host tissue, causing signs and symptoms such as runny nose, rash, and headaches.

Stages of an Infectious Disease

The stages (or phases) of a disease reflect how the contest between an infectious agent and a host progresses (Figure 2.8). At first, the microbe has the "upper hand" and the patient's signs and symptoms get worse. Eventually, the host's adaptive immune response kicks in and begins to kill the invader. So as the host recovers, the symptoms decline in severity. Even for diseases treated with antibiotics, most infections go through these stages.

The **incubation period** is the time after a microbe first infects a person through a portal of entry (discussed shortly) but before the first signs of disease. Depending on the disease, patients may be

contagious during the incubation period—that is, able to transmit the organism to someone else—even though they themselves appear healthy. During the incubation period, the microbe is trying to replicate to higher and higher numbers, possibly even using devices such as a thick capsule coat to hide from the host's immune system. Most infections never progress to active disease because the invader succumbs to early host defenses. This explains why people who never had symptoms of Lyme disease, for example, can have antibodies against the bacterial agent, *Borrelia burgdorferi*, in their blood. How long the incubation period lasts before signs of disease appear depends on many factors, including the infectious dose, the host's health status, and the tools of pathogenicity available to the microbe.

The **prodromal phase** is short and may not even be apparent. It involves vague symptoms, such as headache or a general feeling of **malaise** (fatigue and mild body discomfort), that serve as a warning of more serious symptoms to come. For example, a child infected with measles virus will experience a prodromal phase (or **prodrome**) marked by fever and a runny nose before the characteristic red, spotty rash develops. As with the incubation period, people in the prodromal phase can spread the microbe to others.

The **illness phase** begins when typical symptoms and signs of the disease appear. The point at which disease symptoms are most severe is called the acme. The battle between microbe and host is at its peak during this time. Fever is a symptom often seen during the illness phase.

Fever, though uncomfortable, is actually a protective mechanism to fight infection. High body temperatures can help our immune system function better and can inhibit microbial growth. Children can generally tolerate a fever of 41°C (106°F), but for adults, 39.4°C (103°F) is about the safe limit. Above that temperature brain damage can occur.

Figure 2.8   Stages of an Infectious Disease

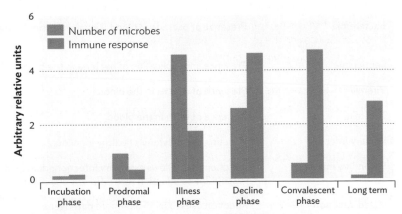

After initial infection, the numbers of organisms (or virus particles) rise in the body until the immune system recognizes that an invader is present; this is the prodromal phase. Symptoms start to develop at the same time the immune system begins attacking the organism or virus. During the illness phase, the pathogen reaches peak numbers. Once the immune system becomes more efficient at killing the pathogen, its numbers decrease (the decline phase). The immune system remains at the ready during convalescence, but its ability to attack the same invader will wane over the years.

Link Fever, how it is triggered and why it is important, is discussed further in Section 15.6 in the context of inflammation and the immune response.

The **decline phase** of a disease begins as the symptoms subside. Host defenses have won. As the infection recedes, fewer pyrogens (the compounds that trigger fever) are made and the body's thermostat is reset to the lower temperature. To restore body temperature, blood vessels will dilate to lose heat (vasodilation), and the patient will start to sweat. These are the signs of "breaking a fever."

Convalescence is the period after symptoms have disappeared and the patient begins to recover normal health.

When discussing infectious diseases, we often use the terms "morbidity" and "mortality." **Morbidity** refers to the existence of a disease state and the rate of incidence of the disease. You do not have to die to be included in a morbidity statistic, only sick. **Mortality**, on the other hand, is a measure of how many patients have died from a disease. The Centers for Disease Control and Prevention (CDC) publishes a weekly report called *Morbidity and Mortality Weekly Report* (*MMWR*) that discusses current outbreaks, statistics, and other health topics. Table 2.1 lists several other terms often used by medical professionals that you should master before proceeding.

 Centers for Disease Control and Prevention *Morbidity and Mortality Weekly Report*: cdc.gov

SECTION SUMMARY

- **Diseases** are recognized by their signs and symptoms.
- **Symptoms** are caused by bacterial products and by the host immune response (immunopathology).
- **Stages of an infectious disease** include the following:
  —**Incubation period**, where the organism begins to grow but symptoms have not developed
  —**Prodromal phase**, which can be unapparent or show vague symptoms
  —**Illness phase**, when signs and symptoms are apparent and the immune system is fighting the disease
  —**Decline phase**, when the numbers of pathogens decrease and symptoms abate
  —**Convalescence**, when symptoms are gone and the patient recovers
- **Morbidity** is a measure of how many are sick from an infectious disease. **Mortality** is a measure of how many died.

Thought Question 2.3 Why would quickly killing a host be a bad strategy for a pathogen?

Thought Question 2.4 Chapter 1 explained Koch's postulates and how they are used to determine the etiology of a disease (see Figure 1.13). The postulates say that you must find the same organism in each case of a disease in order to say that the organism causes the disease. Say three individuals with septicemia have been admitted to the hospital. The clinical laboratory identifies *E. coli* in the blood of one, *Enterococcus* in the blood of another, and *Yersinia pestis* in the third. Do these findings defy Koch's postulates? After all, the same organism was not identified in each case.

2.3
Infection Cycles and Disease Transmission

SECTION OBJECTIVES

- Describe complex versus simple infection cycles.
- Differentiate endemic, epidemic, and pandemic disease.
- Explain animal reservoirs and incubators.

How are diseases spread? Somehow, pathogens must pass from one person, or nonhuman animal, to another if a disease is to spread. The route of transmission an organism takes to infect additional hosts is called the infection cycle. These routes can be direct or roundabout, as seen in the case history presented next.

CASE HISTORY 2.1

A Hike, a Tick, and a Telltale Rash

 Emma Katherine, a 21-year-old college student attending school in Massachusetts, was in good health until one day when she developed a fever, nausea, headache, and muscle pain. A few days later, she developed a rash on her forearms—small, flat, pink, non-itchy spots (**Figure 2.9**). She went to her physician, who asked, among other things, what she had been doing over the past month. (**On pp. 42–43, we discuss the role of patient histories in diagnosing infectious diseases.**) She told him about a trip she had taken 3 weeks earlier to North Carolina, where she hiked through the woods for a couple of days. The expedition was incident-free until one morning when she discovered an ugly tick on her back, which was removed quickly. She encountered no other people or animals during her trek. About 1 week later, she started exhibiting the symptoms that drove her to the doctor's office.

Figure 2.9 **Rash Associated with Rocky Mountain Spotted Fever**

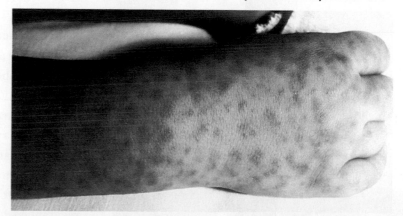

The rash is caused by the immune system's reaction to the presence of the organisms. Rocky Mountain spotted fever will be described in more detail in Chapter 21.

The crucial art of diagnosing infectious disease

How Do Health Care Providers Determine What Ails You?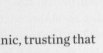

You don't feel well. You're worried about it. You go to the clinic, trusting that the clinician who greets you can diagnose your illness and prescribe a treatment. You really, really want to believe the treatment will work. But how does the clinician make the right diagnosis?

This book describes the many infections that befall humans and the pathogenic mechanisms by which microbes cause them. You will learn about infectious diseases in the numerous case histories interwoven into every chapter of this book as well as in Chapters 19–24, which are devoted to infectious disease. Here we describe a skill that is critical for diagnosing disease and at the heart of each of the case histories presented: the taking of a patient history.

Patient Histories: A "Gold Mine" of Diagnostic Clues

When you go to the clinic, what is the first thing that happens when the physician, physician assistant, or nurse practitioner walks in the room? You are asked questions such as "What brings you here today?" or "How long have you had these symptoms?" and perhaps "Where have you traveled recently?" This is not idle conversation; the clinician is taking a patient history (Figure 1).

What is a patient history, and why is it important? Because many infectious diseases display similar symptoms, a patient history can provide clues about the possible culprit. *Vibrio cholerae* and enterotoxigenic *E. coli,* for instance, both produce diarrheal diseases characterized by cramps, lethargy, and liters of watery stool each day. So when a patient presents these symptoms, how does the clinician determine which microbe is the cause? Here is where a good patient history can make all the difference. Although cholera is not commonly seen in the United States, a clinician might suspect cholera if the patient recently traveled to or emigrated from an endemic area where the disease is regularly observed. This travel information is not gained by examining the patient but comes from talking with the patient and taking a patient history.

A **patient history**, or medical history, is the written summary of a carefully choreographed dialogue between clinician and patient. It is composed of several parts:

Figure 1
Clinician Taking a Patient History

- **Chief complaint:** a statement of what prompted the patient's visit
- **History of present illness:** a chronological summary of the current illness from its inception to the present (often prompted by the question, "When did your symptoms start?")
- **Medical history:** a list of prior illness
- **Social history:** marital status, occupation, travel history, hobbies, stresses, diet, alcohol or drug use (legal or otherwise)
- **Family history:** present health or cause of death of parents and any sibling(s) (very useful for determining genetic diseases)

Many questions asked during the course of taking a patient's social history can seem irrelevant or intrusive to the patient, who only wants relief from his or her symptoms. For instance:

- Do you have any hobbies?
- What is your occupation?
- What foods have you eaten recently?
- Does your child attend day care?
- How long have you been in your current sexual relationship?

Although these questions may sometimes seem trivial, the answers can provide important diagnostic clues to the astute clinician. For example, learning that a man suffering from enlarged glands, fever, and headaches also likes to hunt and recently killed rabbits can be an important clue (**Figure 2**). The patient may have been exposed to the Gram-negative rod-shaped bacterium *Francisella tularensis,* a facultative intracellular pathogen that infects various wild animals and is the cause of tularemia, an illness also known as "rabbit fever." The patient could have accidentally infected a cut while cleaning his "kill."

Or consider the case of a woman with acute pneumonia. The clinician discovers from taking a patient history that the woman is a sheep farmer. Knowing this, the clinician considers *Coxiella burnetii* as one of the possible causes of the infection (the list of possible causes is called the **differential**). *Coxiella burnetii* is a pathogenic bacterial species that infects sheep and is shed in large quantities in the animal's amniotic fluid or placenta. Dried soil contaminated with *C. burnetii* can become aerosolized whenever the dirt is disturbed, and any human (such as a farmer) who inhales the dried particles can develop the lung infection Q fever. *C. burnetii* would be lower on the differential if the woman were an accountant.

Keep the idea of patient history in mind as you read the numerous case histories to come. Take note of how often a patient history can tilt a diagnosis away from one suspected pathogen and toward another. In the end you will have a greater understanding of infectious diseases and will better appreciate your clinician's diagnostic skills. Perhaps after reading the stories in this book and learning some of the secrets of infectious diseases, you may even decide to become a health care provider or clinical laboratory scientist yourself.

Figure 2   Patient with Tularemia

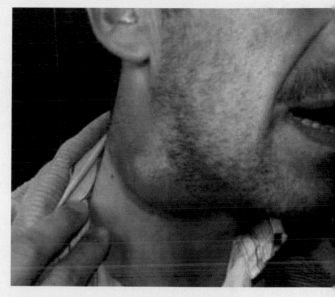

Francisella tularensis is a Gram-negative bacillus that causes tularemia, a zoonotic bacterial disease typically transmitted by direct contact with rodents such as rabbits. The man in this photo had a swollen lymph node in his neck (called cervical lymphadenitis), a fever of 38.8°C (101°F), fatigue, and weakness.

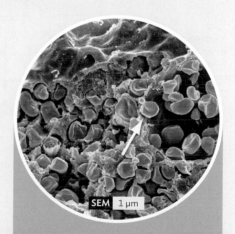

SEM   1 μm

Francisella tularensis
A murine macrophage infected with *Francisella tularensis* strain LVS (arrow).

Emma's signs, symptoms, and travel history led the doctor to suspect that Emma might have Rocky Mountain spotted fever. How could Emma have contracted this disease if she was near no people who might carry this disease?

Rocky Mountain spotted fever is caused by the bacterium *Rickettsia rickettsii*. Although initially recognized in the Rocky Mountain states, most cases of the disease occur on the East Coast. The organism lives in infected ticks without harming them. However, the bite of an infected tick can transmit the organism (present in salivary secretions) to unsuspecting humans and cause serious, life-threatening disease. This case illustrates several concepts of disease transmission and **infection cycles**.

Link This chapter is an introductory chapter. In-depth discussions of the **infection cycles** of specific organisms are presented in Chapters 19–24. Chapter 26 includes advanced coverage of epidemiology, including hospital- and community-acquired infections.

Direct and Indirect Modes of Transmission

A cycle of infection can be simple or complex (Figure 2.10). (See Table 2.2 for a list of various terms we will use to discuss infection cycles.) Organisms that spread directly from person to person, such as the rhinovirus (which causes the common cold) or *Shigella* (which causes bacillary dysentery), have simple infection cycles. Sneezing, for instance, produces an **aerosol** of secretion particles called droplet nuclei laden with bacterial or viral pathogens from the respiratory tract. Droplet nuclei can be inhaled, thereby transmitting the organism directly from person to person, a process called **direct contact transmission** (Figure 2.11). Indirect transmission is somewhat more complex. **Indirect transmission** involves transmitting the agent of disease through an intermediary, which may be living or nonliving. Not washing hands after defecating can spread pathogens such as *Shigella* directly by touch or indirectly via inanimate objects. Inanimate objects on which pathogens can be passed from one host to another are called **fomites**—a doorknob

Figure 2.10   Infection Cycles

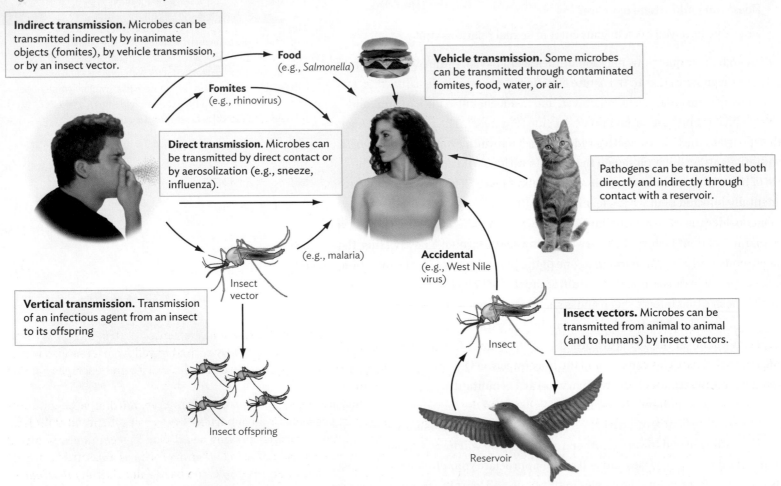

Indirect transmission. Microbes can be transmitted indirectly by inanimate objects (fomites), by vehicle transmission, or by an insect vector.

Food (e.g., *Salmonella*)

Fomites (e.g., rhinovirus)

Vehicle transmission. Some microbes can be transmitted through contaminated fomites, food, water, or air.

Direct transmission. Microbes can be transmitted by direct contact or by aerosolization (e.g., sneeze, influenza).

Pathogens can be transmitted both directly and indirectly through contact with a reservoir.

(e.g., malaria)

Accidental (e.g., West Nile virus)

Insect vector

Vertical transmission. Transmission of an infectious agent from an insect to its offspring

Insect

Insect vectors. Microbes can be transmitted from animal to animal (and to humans) by insect vectors.

Insect offspring

Reservoir

Infectious agents can be transmitted horizontally from one member of a species to another by a variety of means: direct contact between people; indirect transmission by fomites (inanimate objects) or insect vectors; or vehicle transmission by air, food, or water. Transmission can also occur by contact with a reservoir, such as cats harboring *Bartonella henselae* (cat scratch fever). Vertical transmission is passage from parent to offspring during birth, while accidental transmission happens when a host that is not part of the normal infectious cycle unintentionally encounters that cycle.

Table 2.2

Terms to Describe Disease Transmission and Frequency

Term	Definition	Examples
Direct contact	Intimate interaction between two people	Touching, kissing, sexual intercourse (mononucleosis or gonorrhea)
Indirect contact or transmission	Transmission of an infectious agent from one person to another by an insect vector or an inanimate object	Mosquito (malaria); sharing a spoon or fork (strep infection)
Vehicle transmission (fomites)	A form of indirect transmission whereby an infectious agent is transferred to an inanimate object (fomite) by the touch of one person and then transferred to another person touching the same object; or by ingesting contaminated food or water; or by inhaling the agent in air	Doorknobs, shared utensils (influenza, strep infection)
Reservoir	Nonhuman animal, plant, human, or environment that can harbor the organism; a reservoir may or may not exhibit disease	Cattle (*E. coli* O157:H7), horses (West Nile virus), alfalfa sprouts (*Salmonella*)
Aerosol	Organisms in air suspension	Sneezing (rhinovirus), air conditioning or heating (*Legionella*), water from a shower (*Mycobacterium avium*)
Vehicle	Means of pathogen transmission, as by air, food, or liquid	Air (anthrax), food (*Salmonella*), water (*Leptospira*)
Vector	A living carrier of an infectious organism	Mosquitoes (West Nile), fleas (*Yersinia pestis*), body lice (*Rickettsia*)
Fecal-oral route	Pathogen exits the body in feces, which contaminates food, water, or fomite; pathogen is introduced into a new host by ingestion	*Shigella, E. coli, Salmonella*, rotavirus
Respiratory route	Pathogen enters the body through breathing	Influenza, *Streptococcus pneumoniae*
Urogenital route	Pathogen enters the body through urethra or vagina	Urinary tract infections (nonsexual transmission), syphilis (sexual contact)
Parenteral route	Pathogen enters body through insect bite or needle injection	Malaria, HIV (AIDS)
Endemic	Disease (humans) is present in a specific geographic location; pathogen is usually harbored in an animal or human reservoir	Lyme disease, common cold
Epidemic	When the number of disease cases rises above endemic level	Influenza, plague
Pandemic	Worldwide epidemic	Influenza, plague, acquired immunodeficiency disease (AIDS)

or a shared fork, for example. Indirect transmission via inanimate objects is also called **vehicle transmission**. In vehicle transmission, the disease agent is acquired by contact with fomites or through a medium such as water, food, or air.

More complex infection cycles often involve **vectors**, usually insects, as intermediaries. The tick in our case history served as the vector that passed *Rickettsia rickettsii* to Emma Katherine. Vectors can carry infectious agents from one animal to another. A mosquito vector, for example, transfers the virus causing yellow fever (**Figure 2.12**) from infected to uninfected individuals in what is called **horizontal transmission**. The mosquito can also bequeath this particular virus to its offspring via infected eggs in a form of **vertical transmission** called **transovarial transmission**. Note that there are no examples of human transovarial transmission of infectious diseases.

Yellow fever is not a problem in the United States today. However, West Nile virus, a virus closely related to yellow fever, is also transmitted by the mosquito and claims several victims each year in this country.

Because insects are instrumental in transmitting pathogens, killing an insect vector is an important way to halt the spread of disease. Interventions include spraying insecticide in a community during egg-hatching season or using other microbes as assassins "trained" to kill the vector. For example, the bacterium *Bacillus thuringiensis* will kill many types of insects that carry infectious agents. More recently, an insect virus called baculovirus has been developed that kills the *Culex* mosquito vectors carrying West Nile virus. The advantage of these vector-targeting microbes is that they do not kill other insects or animals, unlike many chemical insecticides.

Figure 2.11   Spreading Respiratory Pathogens by Sneezing or Coughing

Sneezing or coughing will aerosolize respiratory pathogens, such as *Mycobacterium tuberculosis* (the cause of tuberculosis) and rhinoviruses (which cause common colds). Droplets from a sneeze can travel far in some circumstances (when carried by the wind, for example); but generally you can limit your exposure to aerosols by standing 10–15 feet away from someone who is coughing or sneezing. Contaminated aerosolized droplets tend to arc toward the ground over that distance.

Although transovarial transmission of a pathogen is limited, for the most part, to arthropods (insects or ticks), other forms of vertical transmission are applicable to humans. A pregnant woman with syphilis caused by *Treponema pallidum* can transmit the organism to her fetus though the placenta (a process called **transplacental transmission**). As a result, the infected child can be delivered with serious congenital deformities or may even be stillborn. Likewise, HIV, Rubella virus (German measles), and *Listeria* bacteria can be transplacentally transmitted from mother to fetus. All these cases represent congenital infections. *Neisseria gonorrhoeae*, the cause of gonorrhea, can also be transferred from mother to child but is not transplacentally transmitted. The organism resides in the vagina of the infected mother and will infect a newborn during delivery (parturition). With any of these pathogens, it is important to identify and treat infected mothers early in their pregnancies to prevent transmission to their offspring.

Reservoirs of Infection

Another critical factor in an infection cycle is the **reservoir** of infection. A reservoir is an animal (such as a bird, rat, horse, or insect) or an environment (such as soil or water) that normally harbors the pathogen. For yellow fever, the mosquito is not only the vector but also the reservoir, because the insect can pass the virus to future generations of mosquitoes through vertical transmission. Likewise, the tick in the Rocky Mountain spotted fever case history served as both reservoir and vector.

In contrast to the yellow fever virus, the virus causing eastern equine encephalitis (EEE, a potentially lethal brain infection) uses birds as a reservoir. The microbe is normally a bird pathogen transmitted from bird to bird via a mosquito vector. The virus does not persist in the insect, but transmission by the insect vector keeps the virus alive by passing it to new avian hosts. Humans or horses entering geographic areas harboring the disease (called endemic areas) can also be bitten by the mosquito. When this happens, they become accidental hosts and contract the disease. The virus does not replicate to high numbers (titers) in mammals, which means that horses and humans are poor reservoirs for the virus. The virus does, however, replicate to high numbers in the avian host. Reservoirs are crucial for the survival of a pathogen and as a source of infection. If the eastern equine encephalitis virus had to rely on humans to survive, the virus would cease to exist because of limited replication potential and limited access to mosquitoes. Remember: the reservoir of a given pathogen might not exhibit disease.

A special type of reservoir is the asymptomatic **carrier**, a person who harbors a potential disease agent but does not have disease. This is how *Neisseria meningitidis*, an important

Figure 2.12   Insect Vector and Yellow Fever

A. Walter Reed, a member of the University of Virginia Medical School class of 1869, proved in 1901 that the mosquito *Aedes aegypti* is the vector of transmission for yellow fever, a disease named for the jaundice produced by liver damage.

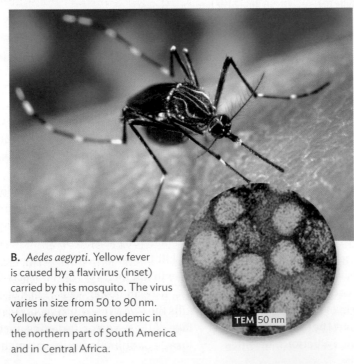

B. *Aedes aegypti*. Yellow fever is caused by a flavivirus (inset) carried by this mosquito. The virus varies in size from 50 to 90 nm. Yellow fever remains endemic in the northern part of South America and in Central Africa.

TEM 50 nm

cause of meningitis, remains in a population. The bacterium has no animal reservoir other than humans, so where does it go between outbreaks? It colonizes the nasopharynx (the area behind the nose down to the throat) of unsuspecting hosts whose immune systems keep the bacterium out of the bloodstream. A simple sneeze from the carrier, however, can aerosolize the pathogen and transfer it to a susceptible host who then contracts the disease and passes the organism to others.

Even a simple infection cycle can become complex. For example, rhinovirus can be spread person-to-person by a sneeze (airborne) or through the sharing of inanimate objects (fomites) such as contaminated utensils (fork, pen), towels, cloth handkerchiefs, and doorknobs. Handshaking is also an efficient means of transferring some pathogens. Imagine that one woman in a city of 100,000 people has a cold and sneezes on her hands. Then, without washing her hands, she goes through the day shaking hands with ten people, and each of those people shakes hands with another ten people per day, and so on. If no repeat handshakes take place and if none of the contacts washes his or her hands, the virus could theoretically spread throughout the population in only 4 days.

In this contrived example, the entire populace of the city comes in contact with the virus, but not everyone would actually contract disease. Additional factors, such as the effectiveness of the host's immune system, influence whether the virus successfully replicates in a given individual. Sometimes the infected host can become a carrier.

Endemic, Epidemic, and Pandemic

CASE HISTORY 2.2

The Third Pandemic

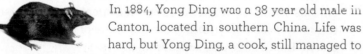

In 1884, Yong Ding was a 38-year-old male in Canton, located in southern China. Life was hard, but Yong Ding, a cook, still managed to support his family. As he walked to his restaurant each day, he barely noticed the small bands of rats scurrying through the streets. Disease was rampant that year; victims of the Shuyi (rat epidemic) were stacked like firewood in the streets, five bodies high in places, waiting to be taken for burial (Figure 2.13). Yong knew that the disease started as a swollen gland in the armpit and often had a black appearance (now called a bubo). He checked himself daily for these swellings and, seeing none, always felt relief. Then, one morning, he found one. Within days, Yong Ding began coughing blood as the agent (a mystery at the time) spread though his bloodstream to his lungs. Once that happened, Yong Ding knew death was not far behind. More than 60,000 died this way in what was to be the start of the Third Pandemic of bubonic plague. Yong Ding's body was one of many lining the street that year.

Yersinia pestis, a bacterium, is the cause (**etiologic agent**) of bubonic plague. The organism is transmitted from infected rats to humans by the rat flea, whose bite delivers bacterium-laden saliva.

Figure 2.13 Chinese Plague Victims, 1910

To stem the plague pandemic, the bodies of plague victims were piled into "plague pits" and burned. The homes of these people were also burned to the ground.

Several pandemics of plague have occurred over the centuries, including the Black Death that killed more than a third of Europe's population in the fourteenth century.

We have all heard the terms "endemic," "epidemic," and "pandemic" in reference to disease, but what do they mean? An **endemic disease** is one that is always present in a community and seen at a low rate. Often an animal reservoir harbors the organism. An **epidemic** occurs when many cases develop in a community over a short time. An organism need not be endemic to an area to cause an epidemic. A **pandemic** is essentially a worldwide epidemic. The Third Pandemic of plague (1855–1959) that started in China spread to many other areas of the globe, carried by people (and rats) as they traveled. In India alone, an estimated 12 million people died of plague over 30 years spanning the late nineteenth and early twentieth centuries.

Although many microbes can cause epidemics, surprisingly few have proved capable of causing pandemics. The major organisms known to have caused pandemics are influenza virus, the bacterium *Y. pestis* (bubonic plague), and HIV (Table 2.3). One might wonder why more organisms do not cause pandemics. There are several reasons. The organism must be easily transmitted from person to person. This is most readily accomplished if the microbe causes a respiratory disease and becomes airborne as a result of coughing. The agent must also require a relatively low infectious dose to cause disease.

HIV is not an airborne infectious agent, so why is AIDS a pandemic? HIV does not rapidly kill its host, and because the virus tampers with the immune system, the host does not kill the virus. Thus, over a person's life, ample opportunity exists to transmit the virus to a partner, especially considering that the symptoms of AIDS can take years to appear. During this extended prodromal period, the virus is present in an assortment of body fluids and is easily transmitted by various sexual practices. Having multiple sexual partners,

Table 2.3
Examples of Deadly Pandemics

Disease	Organism	Years	Link to Disease Coverage
Influenza	Influenza virus	1918–1919 (40 million–50 million dead) 1957 (2 million dead) 1968 (1 million dead) 2009 (150,000–500,000 dead)	Chapter 20
Bubonic plague	*Yersinia pestis*	541–542 CE (Plague of Justinian[a]; the number of dead is unclear but is in the millions) 1347–1351 (Black Death, 25 million dead) 1855–1959 (Third Pandemic, 137 million dead total)	Chapter 21
AIDS	Human immunodeficiency virus	1981– (>25 million dead)	Chapter 23
Cholera	*Vibrio cholerae*	Seven pandemics: the first, in 1816, began in India (over 15 million dead); the third, in 1852, began in Russia (over 1 million dead); the seventh, in 1961, began in Peru	Chapter 22

[a]Justinian was a Byzantine emperor (519–602 CE).

straight or gay, combined with rapid air travel and an inability to cure victims has led to the rapid spread of the virus around the globe.

Note that in some parts of the World, such as sub-Saharan Africa, HIV has actually become endemic. An infection like HIV that relies on person-to-person transmission becomes endemic when each person who becomes infected with the disease, on average, passes it to one other person. In this way, the infection doesn't die out, nor does the number of infected people increase exponentially—the infection is now said to be in an endemic steady state. An infection that starts as an epidemic will eventually either die out or reach the endemic steady state, depending on disease virulence and mode of transmission.

You might now ask, if AIDS is considered a pandemic, why isn't the common cold considered a pandemic? Every country around the globe deals with seasonal epidemics of the common cold. Why not call it a pandemic? The reason here is technical. Pandemics are caused by a single strain or closely related strains of a single microbe—for example, H1N1 influenza. Many different types of viruses and strains of viruses cause cold symptoms. Viral replication, pathogenesis, and the diseases of influenza and the common cold will be described in more detail in Chapters 12, 18, and 20.

Zoonotic Diseases

You might be surprised to learn that some infectious agents cross species barriers to infect humans. Infections that normally afflict animals but can be transmitted to humans are called **zoonotic diseases**. Humans typically contract a zoonotic disease after accidentally encountering the animal reservoir. Bubonic plague is a case in point. The bacterium usually infects rodents but can be transferred to humans by a flea bite. Other examples include Lyme disease (*Borrelia burgdorferi*), Eastern equine encephalitis (EEE virus), and Rocky Mountain spotted fever (*Rickettsia rickettsii*), whose

agents naturally infect deer mice, birds, and ticks, respectively, but can infect humans if they get close enough.

Animal and insect reservoirs can also function as "incubators" for new infectious diseases yet to emerge in humans. For example, two different strains of influenza virus can infect the same animal (usually pigs) at the same time. When both virus strains infect the same cell, the viruses can exchange discrete chunks of their RNA genomes, and a new virus more infectious and deadly than either of the original forms can emerge. Here are other examples of how infectious agents are thought to have crossed between species.

- We gave lions ulcers. It is thought that a large cat dining on the entrails of one of our ancestors became infected with the bacterium *Helicobacter pylori,* the major cause of human stomach ulcers. Today, large cats suffer from ulcers.

- Cats gave us *Toxoplasma gondii*, a parasite that infects the brains of more than half the human population. It is thought that this organism may even contribute to schizophrenia.

- HIV, the virus that causes acquired immunodeficiency syndrome (AIDS), may have originated from chimpanzees and crossed to humans a century ago.

- Rabies, a virus that causes a deadly and painful neurological disease, is transmitted from dogs to humans by dog bite. (However, there is no report of a human transferring the virus back to dogs by human bite.)

Clearly, the animal kingdom can serve as an incubator of new infectious diseases that can cross species.

SECTION SUMMARY

- **Infectious agents can be spread** by aerosols, inanimate objects (fomites), direct contact, or vectors.

- **Insect vectors can horizontally transmit disease** from animals to humans (zoonotic disease) or vertically from insect parent to offspring (vertical transmission).

- **Transplacental transmission** is the transfer of a pathogen across the placenta from human mother to fetus.

- **A disease reservoir** is a site where an infectious agent survives and thus serves as a source of infection. Animals (including humans), plants, soil, and water can serve as reservoirs of infectious agents.

- **Endemic disease** is always present in an area, usually in low numbers, compared with **epidemic disease**, in which the numbers of victims suddenly rise above endemic levels. **Pandemics** are worldwide epidemics.

Thought Question 2.5   Yellow fever is a viral disease found primarily in South America and Africa. The symptoms include a high fever, headache, and muscle aches, and the disease can lead to kidney and liver failure. Liver damage produces a yellowing of skin (jaundice), giving the disease its name. The virus can be transmitted person-to-person by way of a mosquito whose bite can transmit the virus from an infected person to an uninfected person. However, HIV, another virus that is found in blood, cannot be transmitted by a mosquito bite. How might that be?

2.4
Portals of Entry and Exit

SECTION OBJECTIVES

- Describe the various portals of entry and exit for microbial pathogens.
- Discuss concepts of biosafety and biocontainment.

CASE HISTORY 2.3
Risky Business in the Lab

In 2004, three people in Boston came down with a virulent form of pneumonia. An investigation by public health officials discovered that all three worked at the same laboratory studying *Francisella tularensis*, a bacterium that is highly infectious (although not usually spread by person-to-person contact). Under specific conditions, the organism can be aerosolized and inhaled and cause deadly pneumonia—making it a possible bioterrorism agent. Its handling is highly restricted by U.S. Homeland Security. Scientists studying this bacterium must use extreme precautionary measures to ensure that it cannot escape the laboratory. The investigation determined that the Boston researchers had indeed contracted tularemia. The scientists appear to have handled the organism in several instances without wearing or using proper protective gear—for example, examining agar plates containing the organism outside a biosafety containment hood. From the type of disease and the laboratory procedures performed, it seemed that the victims, who all fully recovered, inhaled the organism while working with it.

How do infectious agents gain access to the body? Each organism is adapted to enter the body in different ways (**Figure 2.14**). Food-borne pathogens (for example, *Salmonella*, *E. coli*, *Shigella*, and rotavirus) are ingested by mouth and ultimately colonize the intestine. They have an oral portal of entry (**oral-fecal route**). Airborne organisms, in contrast, infect through the respiratory tract (**respiratory route**). Some microbes enter through the conjunctiva of the eye, others through the mucosal surfaces of the genital and urinary tracts (**urogenital route**). Agents that are transmitted by mosquitoes or other insects enter their human hosts via the **parenteral route**, meaning injection into the bloodstream. Wounds and needle punctures can also serve as portals of entry for many microbes. Shared needle use between drug addicts has been a major factor in the spread of HIV.

Denying a pathogen access to its portal of entry is an effective way to halt the spread of disease. For example, condoms greatly limit the spread of sexually transmitted disease. And face masks were used at the height of the deadly 1918 influenza pandemic in an attempt to cloak the respiratory tract (**Figure 2.15**). Simple processes like covering a sneeze and hand washing can limit the spread of respiratory diseases.

"Picking" a Portal

Why does a pathogen use one portal of entry and not another? Several factors impact why a pathogen "prefers" one portal over another. One factor is the reservoir for the organism. Is the agent found in contaminated food (as for *Salmonella* bacteria) or can it be aerosolized by a sneeze (like influenza virus)? The first favors ingestion, the second inhalation. Is its natural reservoir an animal (as with *Francisella*, the subject of Case History 2.3)? If so, the organism can be transmitted to hunters if they cut themselves while cleaning a dead animal (the parenteral route) or by insect vectors that can feed on the animal and then on humans.

Portal of entry preference is also related to the pathogen's attachment capability. Some pathogens have access to many different entry points but do not succeed in establishing infection unless they enter by the appropriate route. Different body tissues have different receptor structures on their surface. The bacterium or virus must have a matching receptor in order to bind. If it can't bind, the disease-causing organism can be expelled by the host.

Once they are in, how do they get out? Hosts have a limited life span. When the host dies, the pathogen can die with it owing to the lack of food. Even in a live host, microbial growth can be limited by the space in which the pathogen can grow. So, to continue propagating, most pathogens need to leave and find new hosts.

Link You will learn more about microbial growth in Section 6.3. For now, suffice it to say that growth of a microbe within a constricted space cannot go on indefinitely because nutrients become depleted and harmful metabolic by-products accumulate to toxic levels as bacterial cell numbers increase. These factors conspire to eventually slow the pathogen's growth rate until it halts altogether.

Figure 2.14   Portals of Entry/Modes of Transmission

Entry via the eye. The conjuctiva (a membrane lining the eye) is subject to infection by some organisms carried there by fingers or air.

Oral route. Pathogens can be ingested (food-borne, waterborne, fecal-oral).

Entry through skin. Skin can be colonized by some pathogens and penetrated by others. Wounds, too, offer pathogens access to the body's interior.

Respiratory route. Airborne pathogens can be inhaled.

Parenteral route. Pathogens can be injected into the bloodstream (e.g., by insect vector or contaminated needle).

Genital or sexual transmission. Pathogens can enter via direct contact with genital mucosa or sexual transmission.

Do pathogens always leave the same way they came in? Not always. Obvious examples are diarrhea-causing bacteria and viruses, which enter by ingestion but leave by defecation, where they can again be ingested when food, water, or hands become contaminated (this is the **fecal-oral route** of transmission). Some organisms gain entrance through one route but disseminate to other organs and exit that way. A good example is *Yersinia pestis*, the agent in Case History 2.2. The bacterium enters the body via the parenteral route (via a flea bite) but spreads to the lung and is expelled by coughing (the respiratory route). Some microbes use the same portal to enter and exit the body; for example, influenza virus comes and goes via the respiratory tract. Other microbes can enter and exit through multiple portals. For instance, human immunodeficiency virus can enter and leave the body by the urogenital route or parentally through injection and shared needles.

Note In Chapters 19–24, which describe infectious diseases in more detail, you will learn more about the portals of entry and exit for specific pathogens. These chapters conclude with a figure that diagrams the portals of entry and exit of the key pathogens discussed in each chapter.

Biosafety Procedures

Medical and laboratory personnel are exposed to extremely dangerous pathogens on a daily basis. The microbiologists in Case History 2.3 did not protect themselves properly and ended up with a laboratory-acquired infection leading to tularemia. The Centers for Disease Control and Prevention (CDC) has published a series of regulations designed to protect workers at risk of infection by human pathogens. Infectious agents are ranked by the severity of disease and ease of transmission. Based on this ranking, four levels of containment are employed, as indicated in Table 2.4.

Biosafety level 1 organisms have little to no pathogenic potential and require the lowest level of containment. Standard sterile techniques and laboratory practices are sufficient.

Biosafety level 2 agents have greater pathogenic potential, but vaccines and/or therapeutic treatments (for example, antibiotics) are readily available. Biosafety level 2 agents require somewhat rigorous containment procedures, such as limiting laboratory access

Figure 2.15   Guarding a Portal of Entry

Seattle policemen wearing protective gauze face masks during the influenza pandemic of 1918, which claimed millions of lives worldwide.

Table 2.4

Biological Safety Levels (BSL) and Select Agents[a]

	Biosafety Levels (BSLs)			
	BSL 1	**BSL 2**	**BSL 3**	**BSL 4**
Class of Disease Agent	Agents not known to cause disease	Agents of moderate potential hazard; also required if personnel may have contact with human blood or tissues	Agents may cause disease by inhalation route	Dangerous and exotic pathogens with high risk of aerosol transmission; only 11 labs in the United States handle these
Recommended Safety Measures	Basic sterile technique; no mouth pipetting	Level 1 procedures plus limited access to lab; biohazard safety cabinets used; hepatitis vaccination recommended	Level 2 procedures plus ventilation providing directional airflow into room, exhaust air directed outdoors; restricted access to lab (no unauthorized persons)	Level 3 procedures plus one-piece positive-pressure suits; lab is completely isolated from other areas present in the same building or is in a separate building
Representative Organisms in Class	*Bacillus subtilis* *E. coli* K-12 *Saccharomyces* spp.	*Bordetella pertussis* *Campylobacter jejuni* *Chlamydia* spp. *Clostridium* spp. *Corynebacterium diphtheriae* *Cryptococcus neoformans* *Cryptosporidium parvum* Dengue virus Diarrheagenic *E. coli* *Entamoeba histolytica* *Giardia lamblia* *Haemophilus influenzae* *Helicobacter pylori* Hepatitis virus *Legionella pneumophila* *Listeria monocytogenes* *Mycoplasma pneumoniae* *Neisseria* spp. *Salmonella* spp. *Shigella* spp. *Staphylococcus aureus* *Toxoplasma* Pathogenic *Vibrio* spp. *Yersinia enterocolitica*	*Bacillus anthracis* (anthrax) *Brucella* spp. (brucellosis) *Burkholderia mallei* (glanders) California encephalitis virus *Coxiella burnetii* (Q fever) EEE (eastern equine encephalitis) virus *Francisella tularensis* (tularemia) Japanese encephalitis virus La Crosse encephalitis virus LCM (lymphocytic choriomeningitis) virus *Mycobacterium tuberculosis* Rabies virus *Rickettsia prowazekii* (typhus fever) Rift Valley fever virus SARS (severe acute respiratory syndrome) virus Variola major (smallpox) and other poxviruses VEE (Venezuelan equine encephalitis) virus West Nile virus Yellow fever virus *Yersinia pestis*	Ebola virus Guanarito virus Hantavirus Junin virus Kyasanur Forest disease virus Lassa fever virus Machupo virus Marburg virus Tick borne encephalitis viruses

[a]Organisms in blue are on the list of CDC select agents that are considered possible agents of bioterrorism.

when experiments are in progress and using biological laminar flow cabinets if aerosolization is possible (Figure 2.16).

Biosafety level 3 pathogens produce a serious or lethal human disease, but vaccines or therapeutic agents may be available. For safe handling of these organisms, level 2 procedures are supplemented with a lab design that ensures that ventilation air flows only *into* the room (negative pressure) and that exhaust air vents directly to the outside. Negative pressure will keep any organism that may aerosolize from escaping into hallways. In addition, access to the lab is strictly regulated and includes double-door air locks at the entrance. This was the containment level needed to prevent accidental infection by *F. tularensis* in Case History 2.3.

Figure 2.16   Biological Safety Cabinet

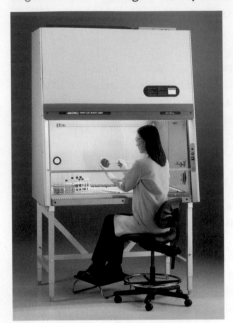

Side view

→ Blue = room air
→ Red = contaminated air
→ Green = filtered air

A. A clinical laboratory technician is transferring an infectious pathogen to a growth medium inside a biosafety cabinet.

B. Airflow in a biosafety cabinet. Air from the room enters the cabinet through the cabinet opening (1) or is pumped in (2) through a HEPA filter (3). It then passes behind the negative-pressure exhaust plenum (4) and is passed from the cabinet through another HEPA filter (5). The HEPA filter traps infectious microbes, preventing the escape of aerosolized infectious agents.

Microbes

SEM   2 μm

C. Microbes trapped in a HEPA filter.

Link Much more will be said about **standard sterile techniques** and other antimicrobial measures in Chapter 13.

By law, extremely dangerous pathogens for which no treatment or vaccine is available, such as the Ebola virus, are biosafety level 4 pathogens and may only be studied at a biosafety level 4 containment facility. Practices here dictate that lab personnel wear positive-pressure lab suits connected to a separate air supply (Figure 2.17). The positive pressure ensures that if the suit is penetrated, organisms will be blown *away* from the breach and not sucked into the suit. As of this writing, 11 biosafety level 4 laboratories are operating in the United States.

Although these regulations may seem reasonable or even obvious, they were not always in effect. Prior to 1970, scientists had an almost cavalier approach toward handling pathogens. For instance, culture material was routinely transferred from one vessel to another by mouth pipetting (essentially using a glass or plastic pipette as a straw). This practice is now forbidden for obvious reasons and has been replaced with mechanical pipettors with disposable tips.

SECTION SUMMARY

- **Diseases can be spread** by direct or indirect contact between infected and uninfected persons/animals or by insect vectors.
- **Pathogens use portals of entry** best suited to their mechanisms of pathogenesis.
- **Various levels of protective measures** are used when handling potentially infectious biological materials.

Figure 2.17   Biosafety Level 4 Containment

Dr. Kevin Karem, at the CDC, performs viral plaque assays to determine the neutralization potential of serum from smallpox vaccination trials. Working in a biosafety level 4 laboratory, he is protected by a positive-pressure suit. The airflow into his suit is so loud he must wear earplugs to protect his hearing.

- **Biosafety level 1 agents** are generally not pathogenic and require the lowest level of containment.
- **Biosafety level 2 agents** are pathogenic but not typically transmitted via the respiratory tract. Laminar flow hoods are required.
- **Biosafety level 3 agents** are virulent and transmitted by respiratory route. They require laboratories with special ventilation and air-lock doors.
- **Biosafety level 4 agents** are highly virulent and require the use of positive-pressure suits.

2.5
Host Factors in Disease

SECTION OBJECTIVES
- Define the biological features of human hosts that influence the course of an infection.
- Explain how host behavior can impact susceptibility to disease.

Does underlying health or lifestyle factor into whether hosts will contract an infectious disease? Until now we have focused on the pathogen as the main culprit in infectious disease. Although the pathogen is surely the cause of infectious disease, the host is not blameless in the process. Many host factors can influence a person's susceptibility to a disease, and more often than not, host factors actually contribute to the disease process itself. What host factors influence susceptibility?

Age. As a general rule, the people most susceptible to an infection are either very young (under 3 years old) or very old (over 60 years old). The very young and very old are more susceptible largely because of the natural development of our immune system. Immune systems are still developing in newborns and babies and are waning in the elderly. As a result, these two populations have a harder time fending off infections than do healthy children or adults under 60.

Host genetic makeup (genotype). Genetics not only determines whether we have blue or brown eyes but can also influence our susceptibility to infectious diseases. The host receptors to which bacteria and viruses bind are all encoded by host genes. Losing or altering a host receptor will affect susceptibility to a given pathogen. As one example, people with blood group O, rather than A, B, or AB, are at higher risk of cholera caused by the bacterium *Vibrio cholerae*, although the reason is not clear. This relationship was evident during the cholera epidemic striking Peru in 1991, where the indigenous people were virtually all blood type O and suffered with severe cholera. The good news, however, is that people with blood type O are more resistant to malaria. Malarial parasites release a protein that can make red blood cells sticky, causing them to clump and form rosettes that clog blood vessels leading to the brain. The infected type O red blood cells are less likely to form rosettes. Other examples of

genetics influencing disease susceptibility will be discussed in our chapters on immunity (see Chapters 15 and 16).

Host hygiene and behavior. Hygiene plays an important role in preventing disease despite exposure. For instance, among a population of people exposed to influenza virus, those who frequently wash their hands or use hand sanitizers are less likely to contract the flu. The simple act of shaking a flu victim's hand and then scratching your nose can transmit the virus, but using a hand sanitizer after the handshake and before scratching your nose can block that transfer. This measure also works if the flu victim sanitizes his or her hands before shaking. Hand washing or sanitation blocks the transfer of many respiratory diseases and even many gastrointestinal ones. Washing hands after a bowel movement will prevent you from transmitting diseases you might have, such as hepatitis virus or shigellosis, onto someone else. Cooking food properly can also interrupt the fecal-oral cycle of many food-borne pathogens. In addition to hygiene, behavior obviously plays a big role in the transfer of sexually transmitted diseases such as syphilis and gonorrhea.

Nutrition and exercise. Several reports suggest that good nutrition and moderate exercise can enhance a person's immune system. In an extreme example, people who are starving produce less acid in their stomach (a condition called achlorhydria). An acidic stomach is an important first line of defense against food-borne pathogens such as *Salmonella* and *Vibrio cholerae*. Thus, a decrease in stomach acidity translates into the organism needing a lower infectious dose to survive passage through the stomach. Recent studies suggest that moderate exercise can boost the immune system and decrease the number and severity of colds. The mechanism by which exercise boosts immunity is unknown.

Underlying noninfectious diseases or conditions. Genetic defects in the immune system, chronic infections (such as HIV), cancer, as well as drugs designed to prevent transplant rejection will immunocompromise a person and make them extremely susceptible to infectious diseases. Substance abuse can also place an individual at higher risk of infections. For example, alcoholics are very susceptible to infections because, along with decreasing libido, alcohol use can depress facets of the immune system. Diabetics with elevated blood glucose (hyperglycemia) are at increased risk for many infections because of hyperglycemia-related impairment of immune responses, vascular insufficiency (decreased blood flow), increased skin and mucosal colonization, as well as other factors. Serious infections on the soles of the feet are common in diabetic patients because sensory nerve damage (neuropathy) caused by hyperglycemia prevents them from feeling the beginning of an infection. Smoking not only increases the risk of lung cancer but also increases the number of respiratory infections because of smoke's paralyzing effect on the mechanisms designed to sweep bacteria up and out of the lung. Even sleep deprivation can dampen immune responses.

Occupation. The job, or hobby, a person pursues can also influence the exposure and thus the susceptibility to certain infectious agents. Health care workers are the obvious example. Not only are these individuals in almost constant contact with sick people spewing germs, but they place themselves at risk of exposure to blood-borne pathogens (for example, HIV) every time they use a hypodermic needle to administer a drug or draw a blood sample. Other, lesser known, examples of at-risk endeavors also exist. Agricultural workers may be exposed to anthrax spores, nearly ubiquitous in soil, as well as the agent of Q fever (*Coxiella burnetii*), present in the placenta of infected pregnant cattle. Hunters can contract zoonotic diseases such as tularemia and plague. People who work with animal hides, which may carry anthrax spores, can occasionally contract anthrax, although this is rare. And, of course, prostitutes (male or female) expose themselves to a host of sexually transmitted agents.

Immune status and immunopathogenesis. Although the public typically thinks that bacterial, viral, and fungal pathogens directly cause the signs and symptoms of a disease, this often is not the case. Most of the damage caused by an infection is actually due to the immune response meant to rid us of the invader. Various white blood cells can become quite aggressive when attacking an infection and can cause a considerable amount of collateral damage to healthy host tissues (immunopathology). So the better a host's immune system is, the faster it can rid the host of infection, but the more damage can occur in the process. Ebola virus is an extreme example of a pathogen that can disable some features of the immune system and overstimulate others. The virus is transmitted by direct contact with secretions of an infected patient and causes a hemorrhagic fever in which the patient's blood pressure drops and bleeding occurs from multiple orifices (Figure 2.18). The immune system reacts to the presence of any pathogen by sending specific types of white blood cells to the infected area. In the case of ebola, these white cells also become infected and carry the virus to more body sites. The infected white cells release a number of potent signal molecules that activate more white cells, trigger a high fever, and cause the increased permeability of blood vessels throughout the body. As part of the immunopathology of ebola, blood escapes from these leaky blood vessels to produce internal and external hemorrhaging. The consequence is 60–70% of those infected will die. Fortunately, few infectious agents trigger that degree of immunopathology.

Link Section 15.1 describes the array of white blood cells (WBCs) the body uses to fight infection and illustrates how WBCs can be identified and counted in a blood sample to help diagnose illness. Immunopathogenesis and the strategies that pathogens use for immune avoidance are further described in Chapter 18.

Figure 2.18   Ebola Hemorrhagic Fever

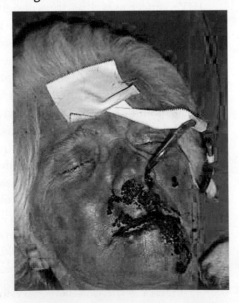

Ebola virus infection causes a deadly hemorrhagic fever during which chemical factors released by the immune system cause bleeding from various body orifices.

SECTION SUMMARY

- **The people most susceptible to infectious disease** are the very old, the very young, and the immunocompromised.
- **An individual's genetic makeup** can influence the person's susceptibility to pathogens.
- **Hygiene** (washing hands, covering your mouth when sneezing, properly cooking and storing foods) is an important factor in preventing the spread of disease.
- **Proper nutrition and exercise can boost the immune system** and lessen susceptibility to infections. In contrast, smoking, diabetes, and alcohol can increase susceptibility by weakening the immune response.
- **Exposure to infectious agents increases with certain occupations** (health care workers) and hobbies (hunting).
- **Tissue damage caused by an infection is mostly due to the immune response** trying to combat the infection.

Thought Question 2.6   Unbeknownst to you, you have contracted the flu (influenza), a viral disease whose prodromal period is followed by fever, muscle aches, coughing, and lethargy. However, before you develop symptoms, you attend a party with 100 guests. During the evening, you greet, hug, and speak to each person for an equal amount of time. A week later, you find yourself quite ill from flu and are worried that you transmitted the disease to all the party guests. After another 10 days, you are feeling better and begin to ask about the health of the other guests. Much to your surprise, only 20 of them became ill. No one had been vaccinated with this year's flu shot. How could 80% of the guests have escaped disease?

2.6
Global Change and Emerging Infectious Diseases

SECTION OBJECTIVES

- Explain how efforts to expand civilization impact emerging infectious diseases.
- Explain how climate change can alter infectious disease patterns.

Are all infectious diseases known to us? With all that we know about infectious diseases, we will never know enough because new diseases are continually emerging or reemerging. Reemerging diseases are diseases that have been around but are now rapidly increasing in incidence, geographic range, or both. Since 1976, the World Health Organization (WHO) has recorded more than 40 emerging and reemerging infectious diseases. Lyme disease, *E. coli* O157:H7 (one cause of bloody diarrhea), drug-resistant tuberculosis, and hantavirus (the cause of a lethal respiratory infection) are but a handful of diseases and disease-causing microbes that have emerged (or reemerged) over the past several decades. Where do these new diseases come from?

Global increases in death and disease are often connected to human activities that bring humans closer to disease reservoirs and vectors. Deforestation, human settlement sprawl, industrial development, road construction, large water control projects (such as dams, canals, irrigation systems, and reservoirs), and climate change alter the environment in ways that can expose humans to agents of infectious disease (**inSight**). Towns built on the margins of woodlands harboring deer populations, for example, make it easier for the deer tick to transmit the Lyme disease agent from deer to humans. Here are other recent examples:

- Malaria has spread in South America because deforestation in the Amazon altered mosquito breeding patterns.

- In Malaysia, pig farms built where the fruit bat lives led to the transfer of the Nipah virus from bats to pigs and then to humans. Nipah viruses can cause severe respiratory and brain infections. Fruit bats are also natural reservoirs for Australian rhabdovirus and Hendra virus.

- The hunting of civet cats, an Asian delicacy, is thought to have introduced the virus causing severe acute respiratory syndrome (SARS) to humans, leading to the SARS outbreak in Asia in 2005.

- Leptospirosis, a zoonotic bacterial disease transmitted by water contaminated with animal urine, is a globally reemerging infectious disease now spreading from its rural base to cause urban epidemics in industrialized and developing nations. In urban Brazil, for instance, outbreaks of leptospirosis can be predicted by heavy rain and flooding, which brings increased risk of human contact with waterborne *Leptospira*.

The global rate of tropical deforestation, one factor contributing to increased malaria infections, continues at staggering levels, with nearly 2%–3% of forest habitats lost per year. One recent study found that a mere 4.3% increase in deforestation correlated to a 48% increase in malaria incidence (**Figure 2.19**). Urban sprawl, especially in developing countries, is also on the rise. These practices

Figure 2.19   **A Link between Deforestation and Malaria**

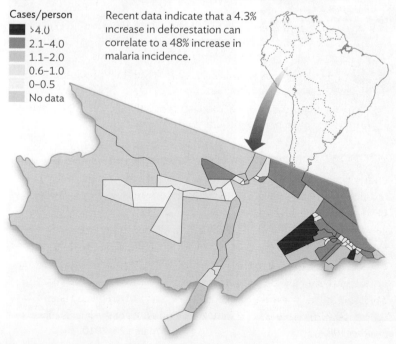

Cases/person

- >4.0
- 2.1–4.0
- 1.1–2.0
- 0.6–1.0
- 0–0.5
- No data

Recent data indicate that a 4.3% increase in deforestation can correlate to a 48% increase in malaria incidence.

A. Malaria incidence in Mancio, Brazil, in 2006. Health districts are outlined in black.

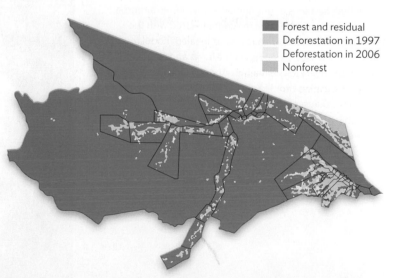

- Forest and residual
- Deforestation in 1997
- Deforestation in 2006
- Nonforest

B. Deforestation trends in the same location. The highest incidence of malaria (locations marked red and orange in panel A) corresponds to areas suffering the greatest increase in deforestation between 1999 and 2006 (light tan spots in panel B).

inSight

Unintended Consequences:
Rebirth of Tuberculosis Follows Death of the Aral Sea

Figure 1 Tuberculosis

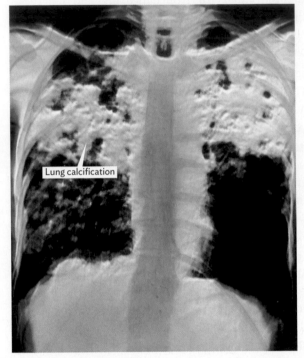

Lung calcification

Lung tissue calcification, resulting from pulmonary tuberculosis, appears as artificially colored yellow patches within the chest area of this human X-ray. When airborne sputum (expectorated phlegm) contaminated with the bacillus *Mycobacterium tuberculosis* is inhaled, nodular lesions, called tubercles, form in the lungs and spread through the nearest lymph node. About a quarter of the world's population (not U.S. population) is infected with the bacillus that causes tuberculosis, but most of these people show no signs of the disease as long as their immune systems can keep it in check.

A coughing woman stands outside a tuberculosis sanatorium in Nukus, in Uzbekistan, near the Aral Sea in central Asia. She is one of many waiting for treatment. The gray hospital building, now supervised by Doctors Without Borders (an international medical humanitarian organization), is a remnant of the Soviet era. Tuberculosis, a lung disease caused by the bacterium *Mycobacterium tuberculosis* (Figure 1), was once uncommon to this area, but its incidence has risen to alarming levels in recent decades as a result of a preventable environmental disaster.

The Aral Sea was once the world's fourth largest inland body of water. Now it is barely a lake. The people who lived near the Aral Sea of old now live in desert, many of them dying of tuberculosis (Figure 2). The World Health Organization (WHO) defines the lower limit for a tuberculosis epidemic at 50–70 infected people per 100,000. In cities near the shrinking Aral Sea, the number is now about 220 per 100,000. What led to this environmental and infectious disease disaster?

In the middle of the last century, the Soviet Union built an elaborate series of irrigation channels taking water from the Amu Darya river, which once fed the Aral Sea with freshwater. Soon after the project was completed, the river no longer reached the

Figure 2 The Aral Sea in Uzbekistan

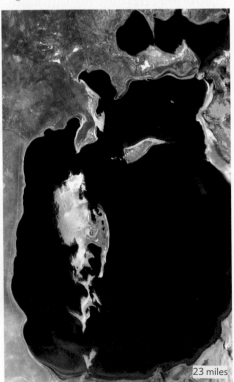
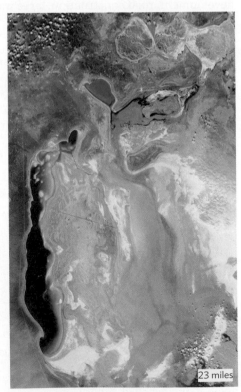

23 miles

23 miles

July–September 1989   March 26, 2010

Soviet Union irrigation projects starting in the 1960s have reduced what was the fourth largest body of inland water in the world to 10% of its former size.

Figure 3   The Aral Sea Ship Cemetery

Old rusting ships lie in the sand where the Aral sea used to be.

sea. Unreplenished with freshwater, the sea began to dry up. As the sea began to shrink, fishing communities were left miles away from the water that traditionally provided a livelihood, some as far as 30 miles (Figure 3). The collapse of the fishing industry in the early 1980s was the first in a chain of ecological reactions, including a change in the climate. Once noted for its mild temperatures, the area now has shorter, hotter, and rainless summers, while winters tend to be longer, colder, and without much snow. Arid conditions combined with wind produce windblown sand that damages lungs and makes them more susceptible to infection.

Deeply affected by the closing of the fishing industry, traditional communities have been slow to adapt, leaving economies once reliant on the sea in deep recession. Massive unemployment has led to extreme poverty. Most families have five or six children on average, and nutritional standards have fallen sharply. As a result of the arid conditions and deepening poverty, many infections and diseases traditionally found in desert or semidesert areas and among malnourished people have entered the area. This includes tuberculosis.

Adding to the nightmare, many of the new strains of *M. tuberculosis* isolated from patients in this region of Uzbekistan are resistant to almost every anti-tuberculosis antibiotic available. And treatment itself is arduous. Patients must stay in the sanatorium for 3 months—until they are no longer contagious—and then return 3 days a week for 6 months to receive medication under direct observation by hospital staff.

Worldwide, there are many examples of climate change caused by ill-advised human endeavors. Many of these have led to new emerging infectious diseases or the reemergence of ancient ones, such as tuberculosis.

 Recent news suggesting an association between climate change and emerging infectious diseases: sciencedaily.com

will continue to bring humans closer to disease reservoirs and vectors. We can also expect global warming to change worldwide disease patterns. For instance, as a result of rising global temperatures, insect vectors now limited to tropical areas of the world will be able to migrate to higher (historically colder) latitudes, bringing infectious microbes with them. All this stirring up of the global environment not only fosters the spread of existing diseases but also builds breeding grounds in which new pathogens will evolve.

SECTION SUMMARY

- **Emerging infectious diseases** are diseases that have not previously been recognized and are the result of either newly evolved organisms or organisms that have made the "leap" from animal to human host.
- **Reemerging diseases** are diseases that have existed but are exhibiting a recent increase in incidence or geographic location.
- **Deforestation and urban sprawl** bring insect vectors of disease into closer proximity to humans.
- **Temperature changes and/or drought or excessive rain** affect the geographic distribution of insect and other animal vectors.

Perspective

The interaction between host and pathogen can be likened to a war with each side evolving new weapons to outwit the opponent—a biological arms race. Of course, pathogens do not engage in this warfare out of ill intent, but out of the need to survive. Even though infectious agents continue to evolve, the stages of an infectious disease remain the same: entry, attachment, immune avoidance, recognition by the host, immunopathology, exit from the host, and resolution of disease (recovery or death). To understand the strategies different pathogens use to invade a host and cause disease, you must first appreciate their physical tools and the rich diversity of their biochemical and genetic capabilities. The next several chapters will discuss the basic structure, biochemistry, and growth of microbes in their natural environments, all in the context of infectious disease.

LEARNING OUTCOMES AND ASSESSMENT

2.1
Normal Microbiota versus Pathogens

- Describe differences between microbiota and pathogens.
- Discuss the relationship between infection and disease and between virulence and pathogenicity.
- Differentiate between infectious dose and lethal dose.
- Discuss the fundamental attributes of a successful pathogen.

1. _____ is the measure of the degree or severity of disease.
   - A. Pathogenicity
   - B. Invasiveness
   - C. Virulence
   - D. Opportunism
2. Pathogens that change the structure of their surface proteins do so for which of the following reasons?
   - A. Attachment to the host
   - B. Immune avoidance
   - C. Invasion of host tissues
   - D. Acquisition of host nutrients

2.2
Basic Concepts of Disease

- Distinguish between the signs and symptoms of a disease.
- Explain the role of immunopathogenesis in infectious disease.
- Discuss the five basic stages of an infectious disease.

4. A _____ is an infection acquired during a hospital stay.
5. A fever is considered a _____.
   - A. sign
   - B. symptom
   - C. prodrome
   - D. sign and symptom

2.3
Infection Cycles and Disease Transmission

- Describe complex versus simple infection cycles.
- Differentiate endemic, epidemic, and pandemic disease.
- Explain animal reservoirs and incubators.

9. Why is the human immunodeficiency virus (HIV) not transmitted from an infected individual to an uninfected person by a mosquito bite?
   - A. The mosquito transmits too small a quantity of blood containing HIV.
   - B. The virus replicates only in the mosquito salivary gland.
   - C. The virus does not replicate in mosquito cells.
   - D. The virus is not present in blood.

2.4
Portals of Entry and Exit

- Describe the portals of entry and exit for microbial pathogens.
- Discuss concepts of biosafety and biocontainment.

13. A patient developed small, flat, pink, non-itchy spots on her forearms after a trip to Colorado, where she camped and hiked alone in the woods. Her physician diagnosed Rocky Mountain spotted fever. The patient probably acquired the organism by which of the following routes?
   - A. Oral
   - B. Vertical
   - C. Direct contact
   - D. Respiratory
   - E. Parenteral

2.5
Host Factors in Disease

- Define the biological features of human hosts that can influence the course of an infection.
- Explain how host behavior can impact susceptibility to disease.

16. A patient with a deficient immune system is more susceptible to contracting infectious diseases by which of the following routes?
   - A. Respiratory
   - B. Fecal-oral
   - C. Parenteral
   - D. Urogenital
   - E. All of the above

2.6
Global Change and Emerging Infectious Diseases

- Explain how efforts to expand civilization impact emerging infectious diseases.
- Explain how climate change can alter infectious disease patterns.

20. In Brazil, heavy rains are predictive of an increase in which one of the following diseases?
   - A. *E. coli* O157:H7
   - B. Lyme disease
   - C. SARS
   - D. Leptospirosis
   - E. Tuberculosis

3. Interpret the following curve, which measures the pathogenicity of a
   bacterial pathogen in mice.
   A. The lethal dose 50% is 1,000 organisms.
   B. The infectious dose 50% is 1,000 organisms.
   C. The lethal dose 50% is 1,000,000 organisms.
   D. The infectious dose 50% is 1,000,000 organisms

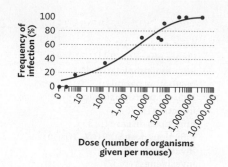

6. Describe a secondary infection.

7. _____ is the term used to express the number of fatalities associated
   with a disease.

8. The stage of a disease when signs and symptoms become apparent is
   called _____.
   A. the prodromal phase   D. the incubation period
   B. convalescence   E. the acme
   C. the illness phase

10. The pathogen *Listeria monocytogenes* can be transmitted from an
    infected mother to her fetus in a process called _____.

11. Which one of the following best explains the development of an
    asymptomatic carrier?
    A. Lack of host cell receptors   D. Antibiotics
    B. Immune system   E. Diminished virulence of the agent
    C. Drinking hot fluids

12. A disease always present at low incidence in a geographic area is
    called a(n) _____ disease.

14. Ebola virus requires which of the following levels of biocontainment?
    A. Level 1   C. Level 3
    B. Level 2   D. Level 4

15. What type of laboratory safety measure is required when working
    with the pathogen in Case History 2.2 that is not required when
    working with a diarrhea-causing *E. coli*?
    A. Sterile technique   D. Positive-pressure room ventilation
    B. Biosafety cabinets   E. One-piece positive-pressure suits
    C. No mouth pipetting

17. A person who has smoked cigarettes for 20 years is more susceptible
    to infections transmitted by which of the following routes?
    A. Respiratory   D. Urogenital
    B. Fecal-oral   E. All of the above
    C. Parenteral

18. Organ and tissue damage caused by most infectious diseases is
    typically caused by _____.

19. True or False? The following host factors contribute to a person's
    susceptibility to infectious disease.
    Age: True/False   Occupation: True/False
    Nutrition: True/False   Underlying health conditions: True/False
    Exercise: True/False   Hygiene: True/False

21. Which of the following can foster the jump of an emerging pathogen
    from a nonhuman animal host to humans?
    A. Deforestation   D. All of the above
    B. Increases in global temperatures E. A and C only
    C. Urban sprawl

Key Terms

adhesin (p. 34)
aerosol (p. 44)
carrier (p. 46)
colonization (p. 34)
convalescence (p. 41)
decline phase (p. 41)
differential (p. 43)
direct contact transmission (p. 44)
disease (p. 39)
endemic disease (p. 47)
epidemic (p. 47)
etiologic agent (p. 47)
fecal-oral route (p. 50)
fomite (p. 44)
horizontal transmission (p. 45)

host range (p. 37)
illness phase (p. 40)
immunopathology (p. 40)
incubation period (p. 40)
indirect transmission (p. 44)
infection (p. 35)
infectious disease (p. 39)
infectious dose 50% (ID$_{50}$) (p. 36)
invasion (p. 37)
invasiveness (p. 37)
latent state (p. 36)
lethal dose 50% (LD$_{50}$) (p. 36)
malaise (p. 40)
microbiota (p. 34)
morbidity (p. 41)

mortality (p. 41)
mutualism (p. 34)
opportunistic pathogen (p. 36)
oral-fecal route (p. 49)
pandemic (p. 47)
parasite (p. 35)
parenteral route (p. 49)
pathogen (p. 35)
pathogenicity (p. 35)
patient history (p. 42)
primary pathogen (p. 36)
prodromal phase (p. 40)
prodrome (p. 40)
receptor (p. 38)
reservoir (p. 46)

respiratory route (p. 49)
sequela (p. 40)
sign (p. 40)
symptom (p. 40)
syndrome (p. 40)
transovarial transmission (p. 45)
transplacental transmission (p. 46)
urogenital route (p. 49)
vector (p. 45)
vehicle transmission (p. 45)
vertical transmission (p. 45)
virulence (p. 36)
zoonotic disease (p. 48)

Review Questions

1. Describe the differences between
   A. infection and disease
   B. pathogenicity and virulence
   C. LD$_{50}$ and ID$_{50}$

2. Review the stages of an infectious disease.

3. Review the stages of an infection.

4. What is meant by direct versus indirect routes of infection?

5. What are the characteristics of a good reservoir for an infectious agent?

6. Name the various portals of entry for infectious agents, and name a disease associated with each.

7. What are some host factors that influence susceptibility to disease?

8. How have human activities affected the emergence of new pathogens?

Clinical Correlation Questions

1. Using the terms defined in this chapter, describe in a paragraph or two an illness you have had. Use words or phrases such as "portal," "transmission," "prodromal phase" (or "prodrome"), "acme," "viremia," "reservoir," "vector," or any other appropriate term found in Table 2.1 or 2.2. Then discuss how you might have prevented getting the disease.

2. How does knowing a pathogen's portal of exit help prevent disease?

Thought Questions: CHECK YOUR ANSWERS

Thought Question 2.1 Is a microbe with an LD_{50} of 5×10^4 more or less virulent than a microbe with an LD_{50} of 5×10^7? Why?

ANSWER: Because it takes fewer cells to cause disease, the microbe with the smaller LD_{50} (5×10^4) is the more virulent.

Thought Question 2.2 The genus of bacteria called *Salmonella* causes a wide range of illnesses in humans. *Salmonella enterica* serovar Typhimurium (*S.* Typhimurium) is usually transmitted by contaminated food and causes a diarrhea that typically resolves without medical intervention. The organism remains localized to the intestine of humans. *Salmonella enterica* serovar Typhi (*S.* Typhi) is also transmitted via contaminated food, but can penetrate the intestinal wall and enter the lymphatic system and bloodstream to cause high fever and in some cases death. Clearly, *S.* Typhi is more virulent than *S.* Typhimurium to humans. However, when tested for lethal dose (LD_{50}) in mice, *S.* Typhi fails to kill any mice, whereas *S.* Typhimurium has a very low lethal dose (it takes very few organisms to kill mice). How might you explain this apparent contradiction?

ANSWER: *S.* Typhi is host adapted to infect only humans. It will not infect other hosts because the bacterium either cannot attach to host cells in other animals or is unable to counter their immune mechanisms. *S.* Typhimurium has a broader host range but lacks some of the tools *S.* Typhi uses to cause severe disease in humans.

Thought Question 2.3 Why would quickly killing a host be a bad strategy for a pathogen?

ANSWER: The goal of any microbe is to maintain its species. If a microbe does not have an opportunity to easily spread to a new host, killing its host would be tantamount to suicide.

Thought Question 2.4 Chapter 1 explained Koch's postulates and how they are used to determine the etiology of a disease (see Figure 1.13). The postulates say that you must find the same organism in each case of a disease in order to say that the organism causes the disease. Say three individuals with septicemia have been admitted to the hospital. The clinical laboratory identifies *E. coli* in the blood of one, *Enterococcus* in the blood of another, and *Yersinia pestis* in the third. Do these findings defy Koch's postulates? After all, the same organism was not identified in each case.

ANSWER: The key in this case is that you know that blood should not harbor any microorganisms. So the question is not do you always find *E. coli* in someone with septicemia, but do you ever find a person without septicemia having bacteria in his or her blood? The answer is no. But the blood of anyone with clear signs of septicemia will contain some species of bacteria. The species can be different for different cases.

Thought Question 2.5 Yellow fever is a viral disease found primarily in South America and Africa. The symptoms include a high fever, headache, and muscle aches, and the disease can lead to kidney and liver failure. Liver damage produces a yellowing of skin (jaundice), giving the disease its name. The virus can be transmitted person-to-person by way of a mosquito whose bite can transmit the virus from an infected person to an uninfected person. However, HIV, another virus that is found in blood, cannot be transmitted by a mosquito bite. How might that be?

ANSWER: Mosquitoes that bite humans do not transfer blood (where the virus that spreads HIV would be) from a previously bitten person into the next person bitten. They transmit only saliva. HIV is quickly inactivated in mosquitoes because it does not replicate in mosquito cells, whereas yellow fever virus can replicate and survive for long periods. Another important factor is that once mosquitoes are fed, they do not immediately bite another person. HIV will be inactivated long before another meal is needed. However, because the yellow fever virus will replicate in the insect, the virus can be found in saliva and can be transmitted.

Thought Question 2.6 Unbeknownst to you, you have contracted the flu (influenza), a viral disease whose prodromal period is followed by fever, muscle aches, coughing, and lethargy. However, before you develop symptoms, you attend a party with 100 guests. During the evening, you greet, hug, and speak to each person for an equal amount of time. A week later, you find yourself quite ill from flu and are worried that you transmitted the disease to all the party guests. After another 10 days, you are feeling better and begin to ask about the health of the other guests. Much to your surprise, only 20 of them became ill. No one had been vaccinated with this year's flu shot. How could 80% of the guests have escaped disease?

ANSWER: Various factors can account for only 20% of your contacts getting ill. The main factors involve the underlying health and immune status of the person. Most of your contacts will have been "exposed," but some individuals' immune systems clear the pathogen more effectively than others. In addition, some of your contacts might use hand sanitizers liberally.

3

Observing
Microbes

[Pathogen in a Can]

SCENARIO Marianna and Julio, two children in Oklahoma aged 8 and 11, fell ill with food poisoning after consuming commercially canned hot dog chili sauce. The children complained of double vision and an inability to move their facial muscles.

SIGNS AND SYMPTOMS
Examination showed normal vital signs, symmetrical weakness (equal on right and left sides), and fixed pupils due to cranial nerve palsy (paralysis). Upper body paralysis gradually spread downward, and breathing became labored.

TESTING Botulinum toxin was identified in their blood by fluorescent monoclonal antibodies (antibodies produced by a genetically uniform cell culture).

TREATMENT The children were placed on mechanical ventilation (a machine that assists breathing) and treated with botulinum antitoxin. After several days, the children were removed from mechanical ventilation. They underwent physical rehabilitation for a full recovery.

FOLLOW-UP The chili sauce was traced to a canning facility where six swollen cans tested positive for botulinum toxin A. From the cans, inspectors cultured a bacterium that grew only without oxygen, as in a closed can. A microscope showed that the bacterium had a distinctive club-shaped appearance. The bacterium was *Clostridium botulinum*, which produces botulinum toxin, the cause of botulism. One end of the club-shaped cell contains an endospore, an inert form of the cell that can germinate and grow in a closed container of food. Growing cells produce botulinum toxin, leading to botulism, a life-threatening form of paralysis. Microscopy of the pathogen's unique form confirms the identification.

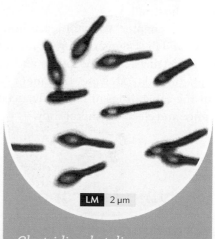

LM 2 µm

Clostridium botulinum
The distinctive club-shaped cells (purple) each show the endospore (pink bulge) near one end. Bright-field microscopy with Gram stain.

A can of food that looks harmless can hide deadly life-forms—too small for you to see. But a microscope reveals the food-borne contaminant *Clostridium botulinum*, a bacterium with a distinctive club-shaped cell. The swollen end of the "club" contains an endospore, an inert particle in which the cell's DNA is condensed with a few enzymes protected by a thick spore coat. Endospores of certain types of bacteria can survive many years until they encounter the right conditions for growth—for this species, access to food and moisture in the absence of oxygen. The club-shaped form of the sporulating cell is unique to *C. botulinum*. You can see *C. botulinum* by using bright-field microscopy with the simple technique of the Gram stain, which we will discuss later in the chapter.

Microscopy reveals the vast realm of bacteria, fungi, and protozoa invisible to our unaided eyes. The microbial world spans a wide range of sizes—over several orders of magnitude. For different size ranges, we use different instruments, from the simple bright-field microscope to the electron microscope. The microscope enables us to count the microbes in the human bloodstream or in natural environments such as the ocean. It detects emerging pathogens, from *Salmonella* bacteria in spinach to *Toxoplasma* parasites causing brain seizures. The microscope can actually follow pathogenic bacteria as they attach, invade, and move through a human host cell.

In this chapter, you will learn how to use the microscope to observe and distinguish different microbes. Microscopes are used in many settings, from hospitals and veterinary clinics to industrial plants and wastewater treatment facilities. You will also learn about the capabilities of more advanced techniques, such as fluorescence and electron microscopy.

3.1
Observing Microbes

CASE HISTORY 3.1
Brain Infection

Eleanor, a 28-year-old woman, presented in the emergency room with headache, fever, and neck stiffness. She said that light bothered her eyes, and she had difficulty bending her head forward. On examination, her temperature was 38°C (100.8°F) and her blood pressure was low, making her feel dizzy when she stood up. Eleanor underwent a lumbar puncture, or spinal tap, a procedure in which a spinal needle is inserted between the lumbar vertebrae in order to withdraw a sample of cerebrospinal fluid (CSF), the fluid that bathes the meningial lining of the brain. Normal CSF is clear and sterile, free of cells, but Eleanor's CSF sample was cloudy. The cloudy fluid revealed the presence of suspended cells, indicative of bacterial meningitis.

In Case History 3.1, why was the CSF sample cloudy? (Figure 3.1) Cerebrospinal fluid normally is clear and sterile, but Eleanor's sample was cloudy because it contained white blood cells responding

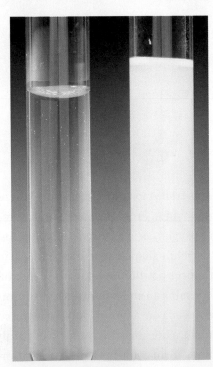

Figure 3.1   Cloudy Cerebrospinal Fluid

Cerebrospinal fluid was obtained through a lumbar puncture (spinal tap). The cloudy appearance is caused by suspended white blood cells infiltrating the cerebrospinal fluid in response to bacterial infection. Our eyes detect the presence of suspended cells but fail to resolve individual cells.

to a bacterial infection. Microscopy revealed infecting bacteria of a shape called diplococci (paired spherical cells), consistent with the species *Neisseria meningitidis*. But even before being viewed in a microscope, white blood cells and bacteria could be <u>detected</u> in a tube as suspended particles causing a cloudy appearance. Detection means to see that an object exists, even if you cannot see its details. For example, our eyes can detect a large group of microbes such as a spot of mold on a piece of bread (about a million cells).

Within the tube or the colony, the presence of cells may be detected, yet no individual cell can be <u>resolved</u>—that is, observed as a distinct entity separate from other cells. The cells cannot be resolved individually because they are too small.

In the preceding case history, microscopy was needed to resolve the size and shape of the diplococcus bacteria. The specific shape of the bacteria confirms the diagnosis of *N. meningitidis*. Most microbial cells are "microscopic," requiring the use of a **microscope** to be seen. In a microscope, **magnification** increases the apparent size of an image. "Useful magnification" resolves smaller separations between objects and thus increases the information obtained by our eyes. Why do we need magnification to resolve individual microbes? What makes a cell too small for unaided eyes to see? The answer is surprisingly complex. In fact, our definition of "microscopic" is based not on inherent properties of the organism under study but on the properties of the human eye. What is "microscopic" actually lies in the eye of the beholder.

The size at which objects become visible depends on the resolution of the observer's eye. **Resolution** is the smallest distance by which two objects can be separated and still be distinguished. In the eyes of humans and other animals, resolution is achieved by focusing an image on a retina packed with light-absorbing photoreceptor cells (rods and cones; **Figure 3.2**). A group of photoreceptors with

their linked neurons forms one unit of detection, comparable to a pixel on a computer screen. The distance between two retinal "pixels" limits resolution.

The resolution of the human retina (that is, the length of the smallest object most human eyes can see) is about 150 μm, or one-seventh of a millimeter. In the retinas of eagles, photoreceptors are more closely packed, so an eagle can resolve objects eight times as small or eight times as far away as a human can—hence the phrase "eagle-eyed," meaning sharp-sighted.

Note In this book, we use standard metric units for size:

> **1 millimeter (mm)** = one-thousandth of a meter
>   = 10^{-3} m
>
> **1 micrometer (μm)** = one-thousandth of a millimeter
>   = 10^{-6} m
>
> **1 nanometer (nm)** = one-thousandth of a micrometer
>   = 10^{-9} m
>
> **1 picometer (pm)** = one-thousandth of a nanometer
>   = 10^{-12} m

Some authors still use the traditional unit angstrom (Å), which equals a tenth of a nanometer, or 10^{-10} m.

Microbial Size and Shape

What is the actual size of a microbe? Different kinds of microbes differ in size, over a range of several orders of magnitude, or powers of 10 (**Figure 3.3**). Eukaryotic microbes are often large enough (10–100 μm) that we can resolve their compartmentalized structure under a light microscope. Prokaryotic microbes (bacteria and archaea) are generally smaller (0.4–10 μm); thus, their overall size can be seen

Figure 3.2   Defining the Microscopic

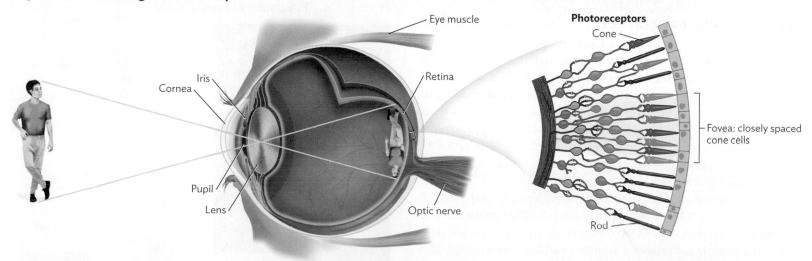

We define what is visible and what is microscopic in terms of the human eye. Within the human eye, the lens focuses an image on the retina. Light receptors are close enough to resolve the image of a human being, even at a distance where its apparent height shrinks to a millimeter.

Figure 3.3　Relative Sizes of Different Cells

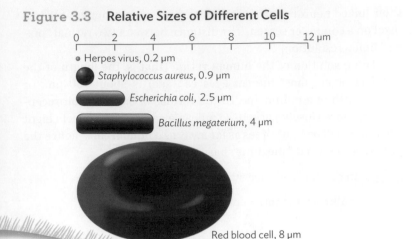

Microbial cells come in various sizes, most of which are below the threshold of resolution by the unaided human eye (150 µm). Most bacteria are much smaller than the human red blood cell, a small eukaryotic cell.

- **Light microscopy (LM) with stain.** The stained specimen absorbs light and appears dark against a bright background. The specimen is fixed (dead and attached to the slide), so no movement can be seen. Light microscopy with stain reveals the two nuclei and multiple flagella of *Giardia*.

- **Phase-contrast microscopy (PCM).** Different parts of the specimen with different refractive index appear darker or lighter. The light-dark pattern gives an illusion of three dimensions.

- **Fluorescence microscopy (FM).** The background and specimen appear dark, except for the fluorophores. Fluorophores are molecules that fluoresce (give off light at a specific wavelength). Only fluorescent-labeled structures are visible; in the example shown, the unlabeled nuclei cannot be seen.

- **Scanning electron microscopy (SEM).** Beams of electrons scatter from the stained surface of a specimen. The image shows much greater detail than with light microscopy, but only the specimen's external surface is enhanced. The appearance of three-dimensionality is approximately correct.

- **Transmission electron microscopy (TEM).** Electrons pass through a stained section. Very fine details may appear, such as individual ribosomes, but only a small fraction of the whole cell may be seen.

Note: Each micrograph in this book has a label for the type of microscopy (black) with a **scale bar** (yellow). The length of the scale bar corresponds to the actual size magnified, such as 5 µm. Like a ruler, the scale bar applies to length measurements throughout the micrograph.

under a light microscope, but their internal structures are too small to be resolved. Some eukaryotic microbes are as small as bacteria; for example, parasitic microsporidia are 1–40 µm. Thus, the actual size range of eukaryotic microbes overlaps that of prokaryotes.

EUKARYOTIC MICROBIAL CELLS Many eukaryotes, such as protozoa, can be resolved with **light microscopy (LM)** to reveal complex shapes and appendages and internal structures such as the nucleus. For example, the ameba in **Figure 3.4A**, from an aquatic ecosystem, shows a large nucleus and a pseudopod that can engulf prey. Pseudopods can be seen to move by the streaming of their cytoplasm. A very different eukaryotic microbe is the trypanosome *Trypanosoma brucei*, which causes sleeping sickness (**Figure 3.4B**). Again we see a prominent nucleus, but instead of pseudopods, the trypanosome has a whiplike flagellum.

The appearance of a microbe in an image depends on the type of microscopy. **Table 3.1** shows the protozoan *Giardia lamblia* imaged by common types of microscopy. *Giardia* is an intestinal parasite that often contaminates water supplies and spreads in child-care centers. The cell has two nuclei and multiple flagella that generate a whiplike movement. But different kinds of microscopy show different aspects of the cell.

Figure 3.4　Eukaryotic Microbial Cells

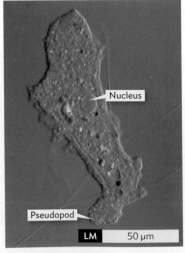

A. *Amoeba proteus.*

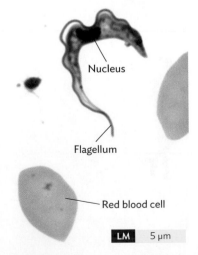

B. *Trypanosoma brucei* (cause of sleeping sickness) among blood cells.

Table 3.1

Giardia lamblia Visualized by Different Forms of Microscopy

Type of Microscopy	Characteristics of Image	Typical Image
Light microscopy (LM) with stain	Dark object (partly transparent) appears against bright background. Stain (methylene blue) darkens some parts more than others.	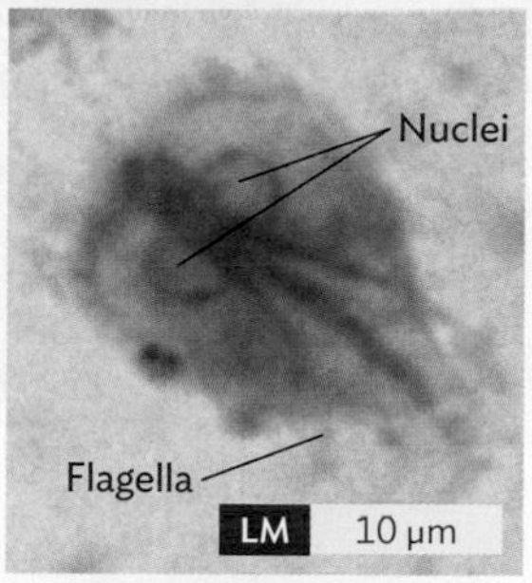
Phase-contrast microscopy (PCM)	Object looks transparent. Organelles appear as light-dark patterns caused by variation in refractive index. This type of imaging gives an illusion of three-dimensionality. Background may be dark or light.	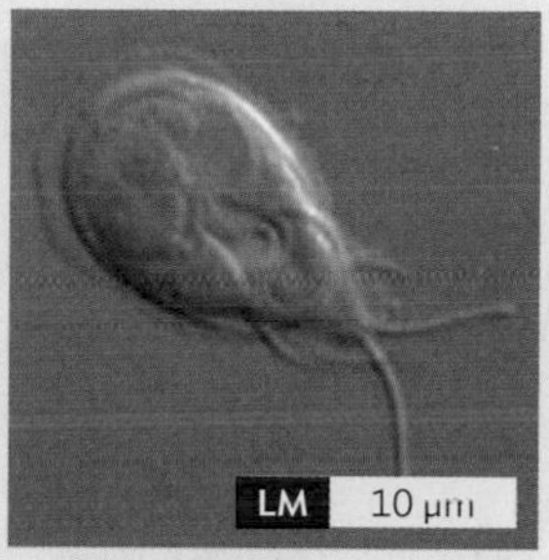
Fluorescence microscopy (FM)	Specific fluorescent molecules label parts of the cell—in this case, the cell surface and flagella. Fluorophore-labeled parts of the object show bright color against a dark background.	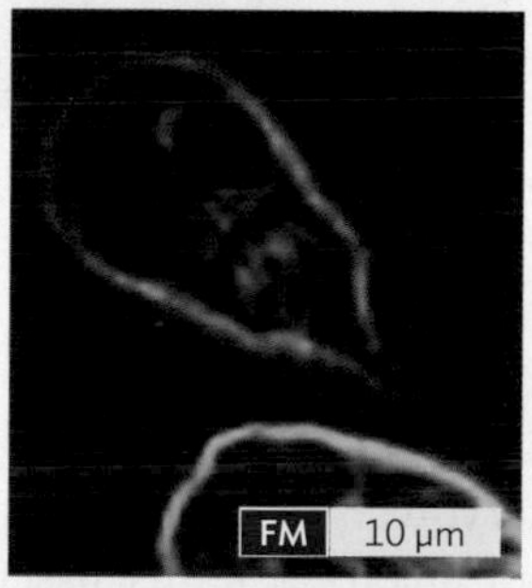
Scanning electron microsopy (SEM)	A high resolution of detail is present, much higher than with light microscopy. Shadowing of the object approximates actual three-dimensionality.	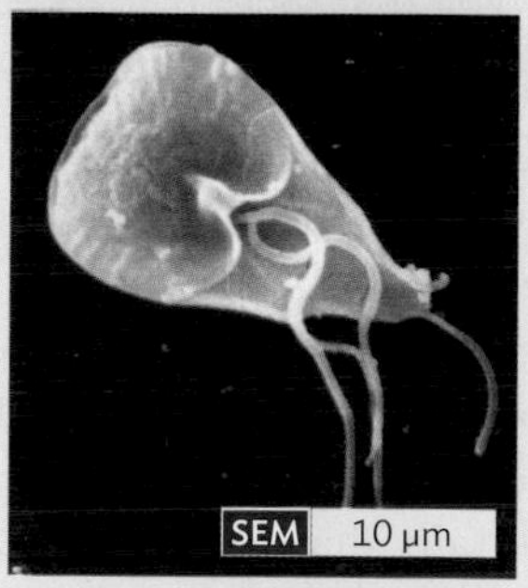
Transmission electron microscopy (TEM)	A thin section of the specimen appears. Only some of the cell's parts appear, those that happen to fall within the section. High resolution of detail reveals objects such as ribosomes and viruses (not shown here).	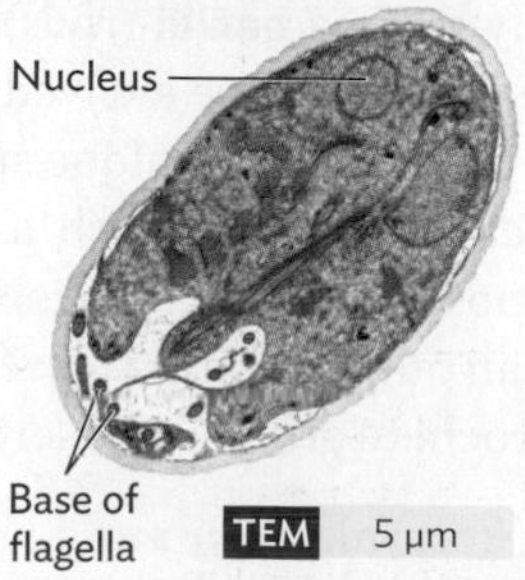

Figure 3.5   Common Shapes of Bacteria

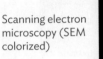
Light microscopy (LM) with stain

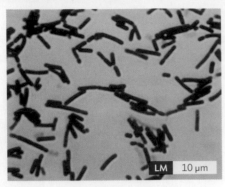

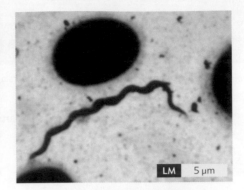

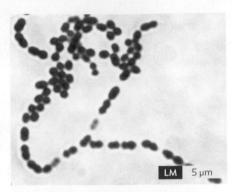

A. Filamentous rods (bacilli). *Lactobacillus lactis*, Gram-positive bacteria.

C. Spirochetes. *Borrelia burgdorferi*, cause of Lyme disease, among human blood cells.

E. Cocci in pairs (diplococci). *Streptococcus pneumoniae*, a cause of pneumonia. Methylene blue stain.

Scanning electron microscopy (SEM colorized)

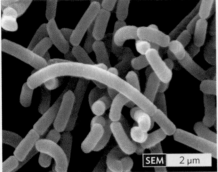

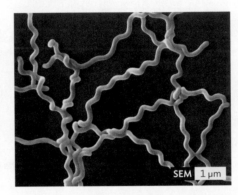

B. Rods (bacilli). *Lactobacillus acidophilus*, Gram-positive bacteria.

D. Spirochetes. *Leptospira interrogans*, cause of leptospirosis in animals and humans.

F. Cocci in chains. *Anabaena* spp., filaments of cyanobacteria. Producers for the marine food chain.

BACTERIAL CELLS   How do bacterial cells compare with cells of eukaryotes? Bacterial cell structures are generally simpler than those of eukaryotes (discussed further in Chapter 5). **Figure 3.5** shows representative members of several common cell types as visualized by either light microscopy or scanning electron microscopy (see Section 3.5). With light microscopy, the cell shape is barely discernible under the highest power. With scanning electron microscopy, the shapes appear clearer, but we still see no subcellular structures. Subcellular structures are best visualized with transmission electron microscopy and fluorescence microscopy (again, see Section 3.5).

Certain shapes of bacteria are common to other taxonomic groups (Figure 3.5). For example, both bacteria and archaea form similarly shaped **bacilli** (rods) and **cocci** (spheres). Thus, rods and spherical shapes have evolved independently within different taxa. On the other hand, a unique spiral shape is seen in the **spirochetes**, which cause diseases such as syphilis and Lyme disease. **Spirochetes** possess an elaborate spiral structure with internal flagella as well as an outer sheath. Spiral axial filaments are found only within the spirochete group of closely related species.

Link   You will learn more about the **spirochetes** in Section 10.4, including about some famous pathogens: *Treponema pallidum*, the

cause of syphilis, and *Borrelia burgdorferi*, the cause of Lyme disease. You'll also learn that many nonpathogenic spirochetes are digestive symbionts of hosts as diverse as termites and cattle, vital to their digestion of plant material.

Note   The genus name *Bacillus* refers to a specific taxonomic group of bacteria, but the term "bacillus" (plural, bacilli) refers to any rod-shaped bacterium or archaeon.

Microscopy for Different Size Scales

What kind of microscope do we need to observe microbes? The kind of microscope we need depends on the size of the object. **Figure 3.6** shows how different techniques resolve microbes and structures of various sizes. For example, a single paramecium can be resolved under a light microscope, but an individual ribosome requires electron microscopy.

- **Light microscopy (LM)** resolves images of individual bacteria based on their absorption of light. The specimen is commonly viewed as a dark object against a light-filled field or background; this is called **bright-field microscopy** (seen in Figure 3.6A and B). Advanced techniques, based on special properties of light, include dark-field, phase-contrast, and fluorescence microscopy.

Figure 3.6   Microscopy and X-ray Crystallography, Range of Resolution

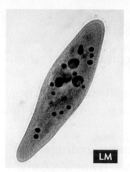 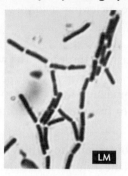 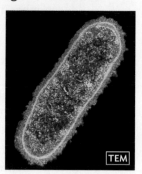 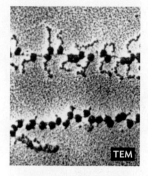 

A. Light microscopy reveals internal structures of a *Paramecium* (a eukaryote). Magnification 100×.

B. Rod-shaped *Pseudomonas* bacteria are barely resolved. Magnification 1,200×.

C. Internal structures of the bacterium *Escherichia coli* are revealed by transmission electron microscopy. Magnification 32,000×.

D. Individual ribosomes are attached to the messenger RNA molecules that are translated to make peptides. Magnification 150,000×.

E. A ribosome (diameter 21 nm) cannot be observed directly but can be modeled using X-ray crystallography.

10^{-3} m	10^{-4} m	10^{-5} m	10^{-6} m	10^{-7} m	10^{-8} m	10^{-9} m
1 mm	100 μm	10 μm	1 μm	0.1 μm = 100 nm	10 nm	1 nm = 10 Å

Human eye

Light microscopy

Scanning electron microscopy

Transmission electron microscopy

X-ray crystallography

- **Electron microscopy (EM)** uses beams of electrons to resolve details several orders of magnitude smaller than those seen with light microscopy. For example, EM can resolve the shape of ribosomes and viruses. In **scanning electron microscopy (SEM)**, the electron beam is scattered from the metal-coated surface of an object, generating an appearance of three-dimensional depth. In **transmission electron microscopy (TEM)** (Figure 3.6C and D), the electron beam travels through the object, where the electrons are absorbed by an electron-dense metal stain.

- **X-ray crystallography (X-ray)** detects the interference pattern of X-rays entering the array of molecules in a crystal. From the interference pattern, researchers build a computational model of the structure of the individual molecule, such as a protein or a nucleic acid, or even a molecular complex such as a ribosome (Figure 3.6E).

SECTION SUMMARY

- **Detection** is the ability to determine the presence of an object.
- **Resolution** is the smallest distance by which two objects can be separated and still be distinguished.
- **Magnification** means an increase in the apparent size of an image so as to resolve smaller separations between objects.

- **Eukaryotic microbes may be large enough to resolve subcellular structures** under a light microscope, although some eukaryotes are as small as bacteria.
- **Bacteria are usually too small for subcellular resolution.** Their shapes include characteristic forms, such as rods and spheres.
- **Different kinds of microscopy** are required to resolve cells and subcellular structures of different sizes.

3.2
Optics and Properties of Light

SECTION OBJECTIVES
- Explain how light interacts with an object.
- Describe how refraction enables magnification, and explain the limits to magnification and resolution.

When you use a microscope, how does it make some tiny object appear large? Light microscopy extends the lens system of your own eyes. Light is part of the spectrum of electromagnetic radiation (**Figure 3.7**), a form of energy that is propagated as waves associated with electrical and magnetic fields. Regions of the electromagnetic spectrum are defined by **wavelength**, the distance between one peak

Figure 3.7   Electromagnetic Energy

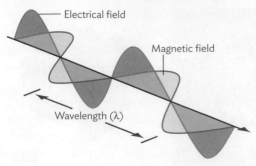

A. Electromagnetic radiation is composed of electrical and magnetic waves perpendicular to each other.

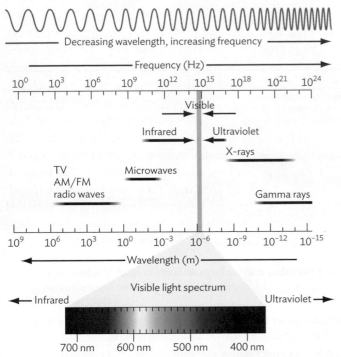

B. The electromagnetic spectrum includes the visible range, used in light microscopy.

of a wave and the next peak. The wavelength of visible light is about 400–750 nm. Radiation of longer wavelengths includes infrared and radio waves, whereas shorter wavelengths include ultraviolet rays and X-rays.

Light Carries Information

All forms of electromagnetic radiation carry information from the objects with which they interact. The information carried by radiation can be used to detect objects; for example, radar (using radio waves) detects a speeding car. All electromagnetic radiation travels through a vacuum at the same speed, about 3×10^8 meters per second (m/s), the speed of light. The speed of light (c) is equal to the wavelength (λ) of the radiation multiplied by its frequency (v), the number of wave cycles per second:

$$c = \lambda \, v$$

Because c is constant, the longer the wavelength λ, the lower the frequency v.

For electromagnetic radiation to resolve an object from neighboring objects or from its surrounding medium, certain conditions must exist:

- **Contrast between the object and its surroundings.** **Contrast** is the difference in light and dark. If an object and its surroundings absorb or reflect radiation equally, then the object will be undetectable. Thus, observing a cell of transparent cytoplasm floating in water is hard because the aqueous cytoplasm and the extracellular water tend to transmit light similarly, producing little contrast.

- **Wavelength smaller than the object.** For an object to be resolved, the wavelength of the radiation must be equal to or smaller than the size of the object. If the wavelength of the radiation is larger than the object, then most of the wave's energy will simply pass through it, like an ocean wave passing around a dock post. Thus, radar, with a wavelength of 1–100 centimeters (cm), cannot resolve microbes, though it easily resolves cars and people.

- **A detector with sufficient resolution for the given wavelength.** Your eye has a retina with photoreceptors that absorb radiation within a narrow range of wavelengths, 400–750 nm (0.40–0.75 μm), which we define as "visible light." But the distance between your retinal photoreceptors is 150 μm, about 500 times the wavelength of light. Thus, your eyes are unable to access all the information contained in the light that enters. The only way you can exploit the full information capacity of light is if the light rays from point sources of light are spread apart enough to be resolved by your retinal photoreceptors. The spreading of the light rays results in magnification of the image.

Light Interacts with an Object

How does light interact with an object? The physical behavior of light in some ways resembles a beam of particles and in other ways resembles a waveform. The particles of light are called photons. Each photon has an associated wavelength that determines how the photon will interact with a given object. The combined properties of particle and wave enable light to interact with an object in several different ways: absorption, reflection, refraction, and scattering.

- **Absorption** occurs when the photon's energy is acquired (absorbed) by the object (**Figure 3.8A**). The energy is converted to a different form, usually heat (infrared radiation), a form of electromagnetic radiation of longer wavelength than light. When a microbial specimen absorbs

Figure 3.8   **Interaction of Light with Matter**

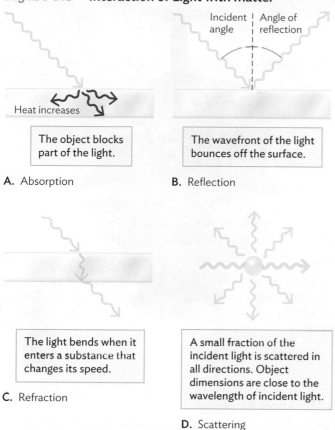

A. Absorption

B. Reflection

C. Refraction

D. Scattering

light, it can be observed as a dark spot against a bright field, as in bright-field microscopy. Some molecules that absorb light of a specific wavelength reemit energy not as heat but as light with a longer wavelength; this is called **fluorescence**. Fluorescence is used in microscopy to reveal specific parts of cells (see Section 3.5).

- **Reflection** occurs when the wavefront "bounces off" the surface of an object at an angle equal to its incident angle (**Figure 3.8B**). Reflection of light waves is analogous to the reflection of water waves. Reflection from a silvered mirror or a glass surface is used in the optics (light path) of microscopy.

- **Refraction** is the bending of light as it enters a substance that slows its speed (**Figure 3.8C**). Such a substance is said to be "refractive" and, by definition, has a higher **refractive index** than air. Refraction is the key property that enables a lens to magnify an image.

- **Scattering** occurs when a portion of the wavefront is converted to a spherical wave originating from the object (**Figure 3.8D**). If many particles simultaneously scatter light, a haze is observed—for example, the haze of bacteria suspended in a culture tube.

Refraction Enables Magnification

Magnification requires refraction of light through a medium of high refractive index, such as glass. As a wavefront of light enters a refractive material, the region of the wave that first reaches the material is slowed, while the rest of the wave continues at its original speed until it also passes into the refractive material (**Figure 3.9**). As the entire wavefront continues through the refractive material, its path continues at an angle from its original direction. At the opposite face of the refractive material, the wavefront resumes its original speed.

A refractive material with a curved surface can serve as a lens to focus light. Light rays entering a lens are each bent at an angle such that all rays meet at a common point. Parallel light rays entering the lens emerge at angles so as to intersect each other at the **focal point** (**Figure 3.10**). From the focal point, the light rays continue onward in an expanding wavefront. This expansion magnifies the image carried by the wave. The distance from the lens to the focal point (called the focal distance) is determined by the degree of parabolic curvature of the lens and by the refractive index of its material.

In Figure 3.10, the object under observation is placed just outside the **focal plane**, a plane containing the focal point of the lens. All light rays from the object are bent by the lens. Parallel rays converge at the opposite focal point. At the focal point, the light rays continue through until they converge with nonparallel light rays refracted by the lens. The plane of convergence generates a reversed image of the object. The image is expanded, or magnified, by the spreading out of refracted rays. When an image is magnified, the distance between parts of the image is enlarged, enabling our eyes to resolve finer details.

Figure 3.9   **Refraction of Light Waves**

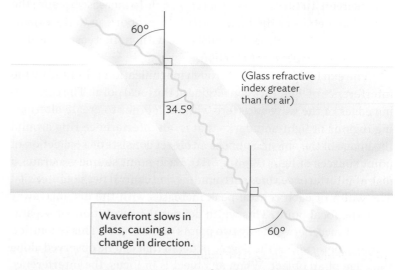

As one side of a wavefront enters the glass, which has a higher refractive index than air, the wave's speed decreases. But the rest of the wavefront continues at the original speed until it, too, enters the glass. The result is that the wavefronts of light shift direction as they enter the glass. As the light waves exit the glass at the same angle, they bend in reverse, traveling parallel to the original direction.

Figure 3.10   **Generating an Image with a Lens**

Curved shape of lens bends rays to converge at focal point.

Object

F ⸺ Focal point

F' ⸺ Focal point

Focal distance

Focal plane in front

Biconvex lens

Focal plane behind lens

Reversed and magnified image

The object is placed just outside the focal plane of the lens. All light rays from the object are bent by the lens. Parallel rays converge at the opposite focal point. The light rays continue through the focal point, generating an image of the object that is reversed and magnified.

Magnification and Resolution

How does magnification increase resolution? The resolution of detail in microscopy is limited by the degree to which details "expand" with magnification. The spreading of light rays does not necessarily increase resolution. For example, an image composed of dots does not gain detail when enlarged on a photocopier, nor does an image composed of pixels gain detail when enlarged on a computer screen. In these cases, resolution fails to increase because the individual details of the image expand in proportion to the expansion of the overall image. Magnification without increase in resolution is called **empty magnification**.

The expansion of detail through magnification is limited by the interference of light rays converging at the focal point. The converging edges of the wave interfere with each other to create alternating regions of light and dark—that is, an interference ring around the image of the object. Suppose an object consists of a collection of point sources of light (**Figure 3.11**). Each point source generates a disk of interference rings surrounding one central peak of intensity. The width of each central peak increases with the distance away from the focal point. This width defines the resolution, or separation distance, between any two points of the object. This resolution distance determines the degree of detail that can be observed along the edge of an object. When an object is in focus, the interference rings are narrow, and points of detail are well resolved. When an object is out of focus, the interference rings are broad, and details merge together, unresolved.

Figure 3.11   **Image Resolution**

In a magnified image, points of light generate interference rings. The wider the rings, the less resolved is the image.

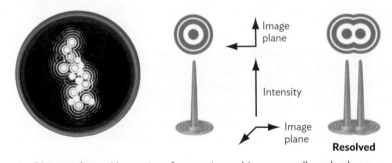

Image plane

Intensity

Image plane

Resolved

A. Object in focus. Narrow interference rings; objects are well resolved.

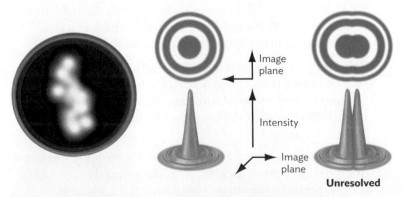

Image plane

Intensity

Image plane

Unresolved

B. Object out of focus. Wide interference rings; objects are unresolved.

SECTION SUMMARY

- **Electromagnetic radiation** interacts with an object and acquires information we can use to detect the object.
- **Contrast** between an object and its background makes it possible to detect the object and resolve its parts.
- **The wavelength of the radiation** must be equal to or smaller than the size of the object if we are to resolve the object's shape.
- **Absorption** means that the energy from light (or other electromagnetic radiation) is acquired by the object.
- **Reflection** means that the wavefront bounces off the surface of a particle at an angle equal to its incident angle.
- **Refraction** is the bending of light as it enters a substance that slows its speed.
- **Interference** limits the degree of resolution of details within an object.

Thought Question 3.1   Explain what happens to a refracted light wave as it emerges from a piece of glass of even thickness. How do its new speed and direction compare with its original (incident) speed and direction?

3.3
Bright-Field Microscopy

SECTION OBJECTIVES

- Explain the use of a bright-field microscope.
- Describe how to observe an object in focus by using a compound microscope.

What do bacteria look like under your microscope? In **bright-field microscopy**, an object such as a bacterial cell is perceived as a dark silhouette blocking the passage of light. Details of the dark object are defined by the points of light surrounding its edge. The optics of a bright-field microscope are designed to maximize detail under magnification by a lens.

To increase resolution, we must consider these factors:

- **Wavelength.** The increase in resolution with magnification is limited by the wavelength of light (to about $1{,}000\times$ magnification, for a typical student microscope). Any greater magnification expands only the width of the interference patterns; the image becomes larger but with no greater resolution (empty magnification).

- **Light and contrast.** For any given lens system, an optimal amount of light exists that yields the highest contrast between the dark specimen and the light background. High contrast is needed to achieve the maximum resolution at high magnification.

- **Lens quality.** All lenses possess inherent aberrations, shape defects that detract from perfect parabolic curvature. Instead

of trying to grind a "perfect" single lens, modern manufacturers construct microscopes with a series of lenses that multiply each other's magnification and correct for aberrations.

Achieving the highest magnification requires a lens of high curvature and high refractive index. **Figure 3.12** shows an **objective lens**, a lens situated directly above an object or specimen that we wish to observe at high resolution. An object at the focal point of the objective lens sits at the tip of an inverted light cone formed by rays of light through the lens converging at the object. The angle of the light cone is determined by the curvature and refractive index of the lens. The lens fills an aperture, or hole, for the passage of light; and for a given lens, the light cone is defined by an angle projecting from the midline. As the light cone angle increases, the horizontal width of the light cone increases, and a wider cone of light passes through the specimen. As the width of the light cone increases, the resolution improves. Thus, the greater the angle of the light cone from a given lens, the better the resolution.

Resolution also depends on the refractive index of the medium containing the light cone, which is usually air. The refractive index

Figure 3.12   Resolution and Numerical Aperture

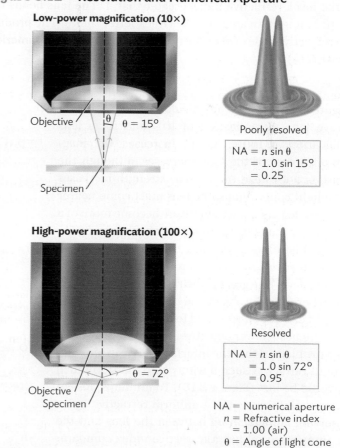

Low-power magnification (10×)

Objective   θ   θ = 15°

Specimen

Poorly resolved

$$NA = n \sin\theta$$
$$= 1.0 \sin 15°$$
$$= 0.25$$

High-power magnification (100×)

θ = 72°

Objective
Specimen

Resolved

$$NA = n \sin\theta$$
$$= 1.0 \sin 72°$$
$$= 0.95$$

NA = Numerical aperture
n = Refractive index
   = 1.00 (air)
θ = Angle of light cone

The resolution depends on numerical aperture (NA), which equals the refractive index (n) of the medium containing the light cone multiplied by the sine of the angle of the light cone (θ). Higher NA yields higher magnification.

Figure 3.13   Use of Immersion Oil in Microscopy

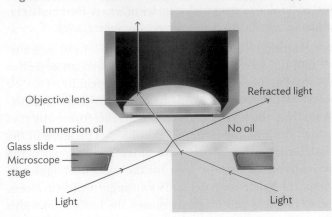

If the light rays exit the glass into air (lower refractive index), their direction bends away from the objective lens (red arrow), as explained in Figure 3.9. Immersion oil with a refractive index comparable to that of glass (n = 1.5) maintains a straight path of light rays (blue arrow). With immersion oil, more light is collected, NA increases, and resolution improves.

is the ratio of the speed of light in a vacuum to its speed in another medium. For air, the refractive index (*n*) is taken as 1, whereas lens material has a refractive index greater than 1. The lens bends the light spreading at an angle (θ, the Greek letter theta). The product of the refractive index (*n*) of the medium and sin θ is the **numerical aperture (NA)**:

$$NA = n \sin \theta$$

In Figure 3.12, we see the calculation of NA for an objective lens magnification of 10× and for a lens magnification of 100×. As NA increases, the magnification power of the lens increases, although the increase is not linear. As the lens strength increases and the light cone widens, the lens must come nearer the object. Defects in lens curvature become more of a problem, and focusing becomes more challenging. As the angle θ becomes very wide, too much of the light from the object is lost owing to refraction at the glass-to-air interface. The greater the refractive index of the medium between the object and the objective lens, the more light can be collected and focused.

The highest-power objective lens, generally 100×, poses a problem in that the interface between lens and air would refract at such a wide angle that too much light would be lost (**Figure 3.13**). To minimize loss of light, we maintain a zone of uniform refractive index by inserting **immersion oil** between the lens and the slide. Immersion oil has a refractive index comparable to that of glass (*n* = 1.5). Using immersion oil minimizes loss of light rays at the widest angles and makes it possible to reach 100× magnification with minimal distortion.

The Compound Microscope

The manufacture of higher-power lenses is difficult because of increased aberration. For this reason, we rarely observe through a single lens. Instead we use a **compound microscope**, a system of multiple lenses designed to correct or compensate for aberration. A typical arrangement of a compound microscope is shown in **Figure 3.14**. In this arrangement, the light source is placed at the bottom, shining upward through a series of lenses, including the condenser, objective, and ocular lenses.

The **condenser** lens focuses light rays from the light source onto a small area of the slide. Between the light source and the condenser sits a **diaphragm**, a device to vary the diameter of the light column. Lower-power lenses require operation at lower light levels because the excess light makes it impossible to observe the darkening effect of absorbance by the specimen (contrast). Higher-power lenses require more light and thus an open diaphragm.

The nosepiece of a compound microscope typically holds three or four objective lenses of different magnifying power, such as 4×, 10×, 40×, and 100× (requiring immersion oil). These lenses are arranged so as to rotate in turn into the optical column. A high-quality instrument will have the lenses set at different heights from the slide so as to be **parfocal**. In a parfocal system, when an object is focused using one lens, it remains in focus, or nearly so, when another lens is rotated to replace the first.

The image from the objective lens is amplified by a secondary focusing step through the **ocular lens**. The ocular lens sits directly beneath the observer's eye. In magnification, each light ray traces

Figure 3.14   The Anatomy of a Compound Microscope

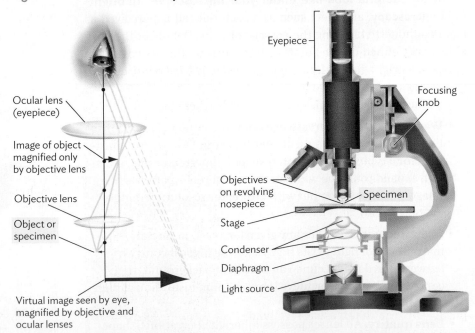

A. Light path through a compound microscope.

B. Cutaway view of a compound microscope.

a path toward a position opposite its point of origin, thus creating a mirror-reversed image (reversal of right and left). We need to keep this reversal in mind when exploring a field of cells.

The magnification factor of the ocular lens is multiplied by the magnification factor of the objective lens to generate the **total magnification** (power). For example, a 10× ocular times a 100× objective generates 1,000× total magnification.

Observing a specimen under a compound microscope requires several steps:

- **Position the specimen centrally in the optical column.** Only a small area of a slide can be visualized within the field of view of a given lens. The higher the magnification, the smaller the field of view will be seen.

- **Optimize the amount of light.** At lower power, too much light will wash out the light absorption of the specimen without contributing to magnification. At higher power, more light needs to be collected by the condenser. To optimize light, we need to set the condenser at the correct vertical position to focus on the specimen, and we need to adjust the diaphragm to transmit the amount of light that produces the best contrast.

- **Focus the objective lens.** The focusing knob permits adjustment of the focal distance between the objective lens and the specimen on the slide so as to bring the specimen into the focal plane. Typically, we focus first by using a low-power objective, which generates a greater **depth of field**—that is, a region along the optical column over which the object appears in reasonable focus. After focusing under low power, we can rotate a higher-power lens into view and then perform a smaller adjustment of focus.

Is the Object in Focus?

An object appears in focus (that is, is situated within the focal plane of the lens) when its edge appears sharp and distinct from the background. At higher power, however, recognition of the focal plane is a challenge because of interference. The shape of the dark object is defined by the points of light surrounding its edge. The partial resolution of these points of light generates extra rings of light surrounding an object whose dimensions are close to the resolution limit.

Figure 3.15 presents a microscopic image of *Rhodospirillum rubrum*, a species of photosynthetic purple bacteria that enriches wetland ecosystems. The bacteria are observed in a **wet mount** preparation, consisting of a drop of water on a slide with a

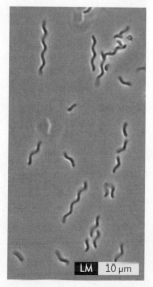

Wet mount of live bacteria swimming

coverslip. The advantage of the wet mount is that the organism is viewed in as natural a state as possible, without artifacts resulting from chemical treatment; and live behavior such as swimming can be observed. The disadvantage is that most living cells are transparent and therefore show little contrast with the external medium.

At highest magnification, the focal plane is so narrow that only part of each bacterium can be seen in focus. As *R. rubrum* cells swim in and out of the focal plane, their appearance changes through optical effects. When a bacterium swims out of the focal plane too close to the lens, resolution declines and the image blurs (Figure 3.15A). When the bacterium swims within the focal plane, its image appears sharp, with a bright line along its edge. Each helical cell shows well-focused segments alternating with hazier segments that are too close to the lens (Figure 3.15B). When the cell swims too far past the focal plane, the bright interference lines collapse into the object's silhouette, which now appears bright or "hollow" (Figure 3.15C). In fact, the bacterium is not hollow at all; only its image has changed.

Figure 3.15 *Rhodospirillum rubrum* **Observed at Different Levels of Focus: Wet Mount of Live Bacteria Swimming**

Too close to lens; image is blurred.

A. When a bacterium swims too close to the lens, its image blurs. LM 5 µm

In focus; image appears sharp.

B. When the bacterium lies within the focal plane, its image appears sharp. If the width of the cell crosses several focal planes, some parts appear sharp, whereas other parts appear blurred.

Too far beyond focal plane; image appears "empty," with interference rings.

C. When the bacterium lies too far from the lens, its image appears "empty" or "hollow," surrounded by rings of interference.

Object extends through focal planes.

D. When the spirillum extends through the focal plane, different parts show different focal effects (in focus, too near, or too far).

At high magnification, the depth of field narrows such that the bacterial cell extends across several focal planes. As a result, different parts of the cell appear out of focus (either too near or too far from the lens). When the end of a cell points toward the observer, light travels through the length of the cell before reaching the observer, so the cell absorbs more light and appears dark (Figure 3.15D). Observing motile bacteria swimming in and out of the focal plane presents a challenge even to experienced microscopists.

SECTION SUMMARY

- **In bright-field microscopy**, resolution depends on:
  - —The wavelength of light, which limits resolution to about 0.2 μm.
  - —The magnifying power of a lens, which depends on its numerical aperture ($n \sin \theta$).
  - —The position of the focal plane, the location where the specimen is "in focus" (that is, where the sharpest image is obtained).
- **A compound microscope** achieves magnification and resolution through the objective and ocular lenses.
- **A wet mount** specimen is the only way to observe living microbes.

Thought Question 3.2   In theory, which angle of the light cone would produce the highest resolution? What practical problem would you have in designing a lens to generate this light cone?

Thought Question 3.3   Under starvation conditions, bacteria such as *Bacillus thuringiensis*, the biological insecticide, repackage their cytoplasm into spores, leaving behind an empty cell wall. Suppose, under a microscope, you observe what appears to be a hollow cell. How can you tell whether the cell is indeed hollow or whether it is simply out of focus?

3.4
Stained Samples

SECTION OBJECTIVES

- Explain how a stain reveals additional information about a microscopic object.
- Describe how the Gram stain distinguishes two classes of bacteria.

CASE HISTORY 3.2

Blood Slide Reveals a Parasite

Purna, a 6-year-old boy in India's state of Assam, fell ill with fever. At the time, prolonged rainfall had left vast pools of water, breeding mosquitoes. First the boy felt intense chills, and his temperature rose to 40°C (104°F). He experienced nausea and intense headaches and vomited repeatedly. The fever subsided, only to return 3 days later. After several weeks of cycling fevers, Purna was at last brought to a clinic for examination. He showed extreme anemia, with enlarged spleen and liver. A blood film slide revealed *Plasmodium falciparum*, a parasite that causes malaria. In the Assam outbreak of malaria, 400,000 people were infected. Some travelers who returned to the United States experienced malarial symptoms months afterward.

The diagnosis of malaria offers a clinical application of microscopy using a blood sample that is fixed and treated with the Giemsa stain (Figure 3.16). Stained blood samples play a crucial role in malaria diagnosis, often a challenge to obtain in remote rural regions where malaria is prevalent. New technology such as the CellScope cell phone microscope aims to make such tests available in developing countries (see **inSight**).

The Giemsa stain contains eosin, a negatively charged dye that binds red blood cells, and methylene blue, a positively charged dye that binds the negatively charged phosphate groups of DNA. The methylene blue specifically stains the DNA-containing nucleus of the malarial parasite. Giemsa stain reveals both the "ring form" stage of the parasite, growing within a red blood cell, and the "gametocyte" stage that develops outside the host cell, ready for transmission to the next host.

Fixation and **staining** are procedures that enhance detection and resolution of microbial cells. These procedures kill the cells and stabilize the sample on a slide. In fixation, cells are made to adhere to a slide in a fixed position. Cells may be fixed with methanol or by heat treatment to denature cellular proteins, whose exposed side chains then adhere to the glass. A stain absorbs much of the incident

Figure 3.16   Blood Film Showing Red Blood Cells Infected with Malaria Parasite

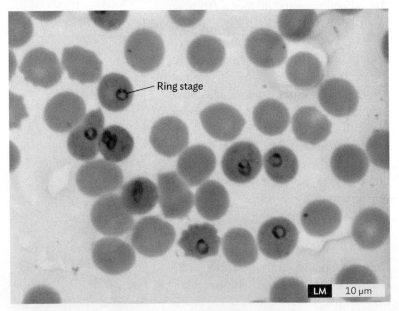

Plasmodium falciparum grows within red blood cells (ring stage). Emerging parasites develop into gametocytes. Bright-field, Giemsa stain.

Figure 3.17 Chemical Structure of Stains

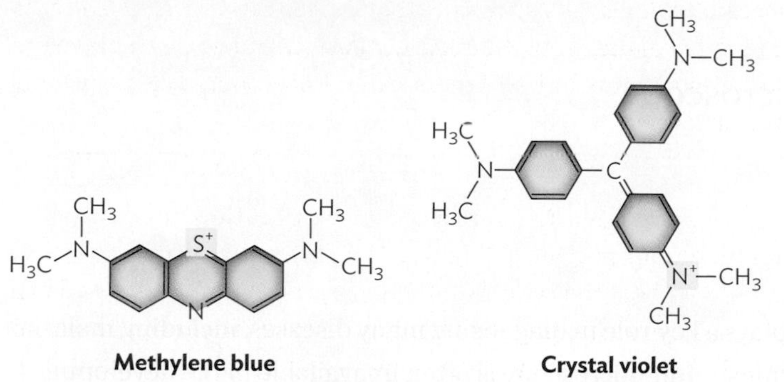

Methylene blue

Crystal violet

Methylene blue and crystal violet are cationic (positively charged) dyes. The positively charged groups react with the bacterial cell envelope, which carries mainly negative charge.

light, usually over a wavelength range that results in a distinctive color. The use of chemical stains was first developed in the nineteenth century, when German chemists used organic synthesis to invent new coloring agents for clothing. Clothing was made of natural fibers such as cotton or wool, so a substance that dyed clothing was likely to react with biological specimens.

How do stains work? Most stain molecules contain double bond systems that absorb visible light (**Figure 3.17**) and one or more positive charges that bind to negative charges on the bacterial cell envelope (to be discussed in Section 5.3). Different stains vary with respect to their strength of binding and the degree of binding to different parts of the cell.

Different Kinds of Stains

A simple stain adds dark color specifically to cells, but not to the external medium or surrounding tissue. The most commonly used simple stain is methylene blue, originally used by Robert Koch to stain anthrax bacteria. A typical procedure for fixation and staining is shown in **Figure 3.18**. First, a drop of culture is fixed on a slide by treatment with methanol or by heat on a slide warmer. Either treatment denatures cell proteins, exposing side chains that bind to the glass. The slide is then flooded with methylene blue solution. The positively charged molecule binds to the negatively charged cell envelope. After excess stain is washed off and the slide has been dried, it is observed under high-power magnification using immersion oil.

A **differential stain** stains two different kinds of cells differently. The most famous differential stain is the **Gram stain**, devised in 1884 by the Danish physician Hans Christian Gram (1853–1938). Gram first used the Gram stain to distinguish pneumococcus (*Streptococcus pneumoniae*) bacteria from human lung tissue. A similar use of the Gram stain is seen in **Figure 3.19A**, where *S. pneumoniae* bacteria appear dark purple against the pink background of human epithelial cells. Not all species of bacteria retain the Gram stain; for

Figure 3.18 Procedure for Simple Staining with Methylene Blue

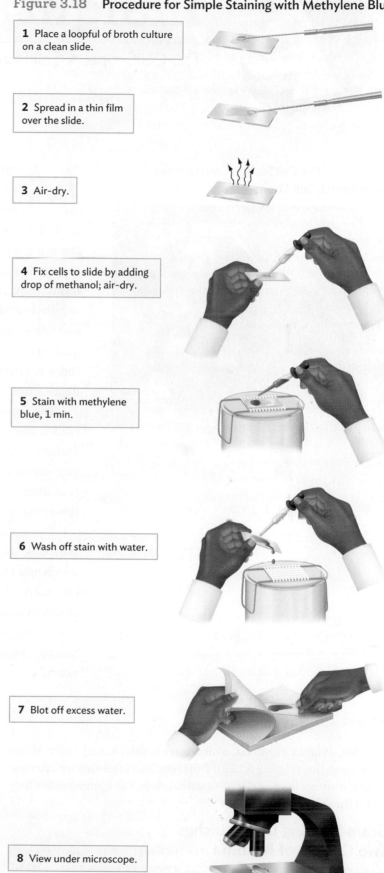

1 Place a loopful of broth culture on a clean slide.

2 Spread in a thin film over the slide.

3 Air-dry.

4 Fix cells to slide by adding drop of methanol; air-dry.

5 Stain with methylene blue, 1 min.

6 Wash off stain with water.

7 Blot off excess water.

8 View under microscope.

inSight

The CellScope: A Cell Phone Microscope

Figure 1   The CellScope: A Microscope Made from a Cell Phone

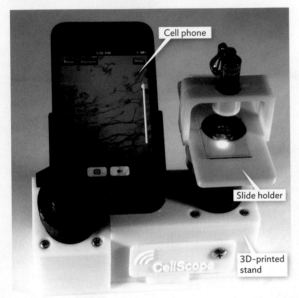

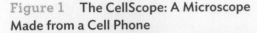

Microscopy plays a key role in diagnosing many diseases, including malaria and tuberculosis. But microscopy is often unavailable in the developing world, particularly in remote rural areas where few medical facilities exist. A solution to this problem was devised by Daniel Fletcher, professor of bioengineering at UC Berkeley. Fletcher and his students designed an inexpensive optical system that can be attached to a mobile cell phone camera (Figure 1). The phone is thus converted to a hand-held microscope, called the CellScope. The CellScope obtains remarkably high-quality images of blood films showing malaria infection and sickle-shaped cells. The optics can even include fluorescence filters allowing fluorescent antibody detection of tuberculosis bacteria in a sputum sample. Fluorescence requires illumination with light at one wavelength, followed by detection of light emitted by the sample at a different wavelength (explained in Section 3.5).

The optical attachment includes an eyepiece, an objective lens, and a sample holder. Illumination is provided by a low-energy light-emitting diode (LED). For fluorescence, the attachment includes special filters for detecting the excitation and emission wavelengths. Even with inexpensive lenses, the CellScope has achieved resolution of 1.2 µm, enabling detection of malaria parasites and *Mycobacterium tuberculosis*.

The cell phone's computer elements can manipulate images obtained through the CellScope. Users can transmit the images through the mobile wireless network, which now reaches many remote regions of the globe. Distant medical professionals can interpret the transmitted images. So far, workers have field-tested the device in several countries, including Kenya, Peru, and Benin, and clinical comparison trials are under way. In 2009, Rock Health funded a CellScope start-up company to promote mass-market production. In the future, this form of "telemicroscopy" may extend the reach of telemedicine to remote communities around the globe.

example, *Proteus mirabilis*, which infects the urinary tract, stains Gram-negative (Figure 3.19B). Different bacterial species are classified Gram-positive or Gram-negative, depending on whether they retain the purple stain.

Gram Staining Distinguishes Two Classes of Bacteria

In the Gram stain procedure (Figure 3.20), a dye such as crystal violet binds to the bacteria; it also binds to the surface of human cells,

but less strongly. After the excess stain is washed off, a mordant, or binding agent, is applied. The mordant used is iodine solution, which contains iodide ions (I^-). The iodide complexes with the positively charged crystal violet molecules trapped inside the cells. The crystal violet–iodide complex is now held more strongly within the cell wall. The thicker the cell wall, the more crystal violet–iodide molecules are held.

Next, a decolorizer, ethanol, is added for a precise interval (typically, 20 seconds). The decolorizer removes loosely bound crystal

Figure 3.19 Gram Staining of Bacteria (a Type of Differential Staining)

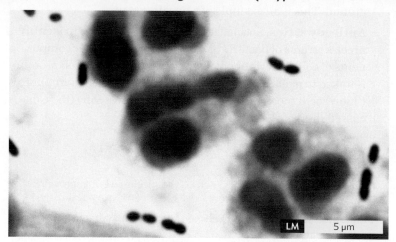

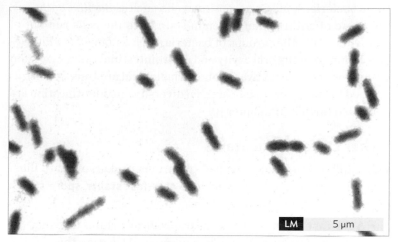

A. Gram stain of a sputum specimen from a patient with pneumonia, containing Gram-positive *Streptococcus pneumoniae* (purple diplococci) and Gram-negative epithelial cells of the patient (pink nuclei).

B. Gram-negative *Proteus mirabilis* (pink rods).

Figure 3.20 **The Gram Stain Procedure**

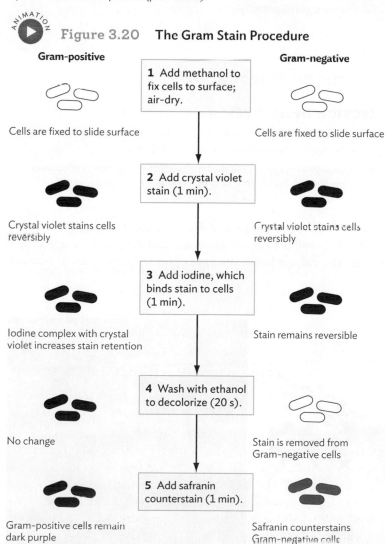

Gram-positive

Gram-negative

1. Add methanol to fix cells to surface; air-dry.

Cells are fixed to slide surface

Cells are fixed to slide surface

2. Add crystal violet stain (1 min).

Crystal violet stains cells reversibly

Crystal violet stains cells reversibly

3. Add iodine, which binds stain to cells (1 min).

Iodine complex with crystal violet increases stain retention

Stain remains reversible

4. Wash with ethanol to decolorize (20 s).

No change

Stain is removed from Gram-negative cells

5. Add safranin counterstain (1 min).

Gram-positive cells remain dark purple

Safranin counterstains Gram-negative cells

The Gram stain distinguishes between gram-positive cells, with thick cell walls, which retain the crystal violet stain, and gram-negative cells, with thinner cell walls, which lose the crystal violet stain but are counterstained by safranin.

violet–iodide molecules, but Gram-positive cells retain the stain tightly. The **Gram-positive** cells that retain the stain appear dark purple, whereas the **Gram-negative** cells are colorless. The decolorizer step is critical because if it lasts too long, the Gram-positive cells, too, will release their crystal violet stain. In the final step, a **counterstain**, safranin, is applied. This process allows the visualization of Gram-negative material, which the safranin stains pale pink.

The Gram stain procedure was originally devised to distinguish bacteria (Gram-positive) from human cells (Gram-negative). Microscopists soon discovered, however, that many important species, such as the intestinal bacterium *Escherichia coli* and the nitrogen-fixing symbiont *Rhizobium meliloti*, do not retain the Gram stain. It turns out that Gram-negative species of bacteria possess a thinner and more porous cell wall than Gram-positive species. A Gram-negative cell wall has only a single layer of peptidoglycan (sugar chains cross-linked by peptides), which more readily releases the crystal violet molecules. By contrast, a Gram-positive cell, such as that of *Staphylococcus aureus*, a cause of toxic shock syndrome and wound infections, has multiple layers of peptidoglycan. The multiple layers retain enough stain so that the cell appears purple.

Among bacteria, the Gram stain distinguishes many types of bacteria including two groups with distinctive cell wall structures: Proteobacteria (such as *E. coli* and *Proteus*), with a thin cell wall plus an outer membrane (Gram-negative), and Firmicutes (such as *Staphylococcus* and *Streptococcus*), with a multiple-layered cell wall but no outer membrane (Gram-positive). The outer membrane of Proteobacteria (discussed in Chapter 5) often possesses important pathogenic factors, such as the lipopolysaccharide endotoxins that cause toxic shock. During the Gram stain procedure, however, the decolorizing wash with ethanol disrupts the outer membrane, allowing most of the crystal violet–iodide complex to leak out.

Thus, the Gram stain emerged as a key tool for the biochemical identification of species, and it remains essential in the clinical laboratory. However, the Gram stain is reliable only for freshly cultured bacteria. Some Gram-positive cells, such as *Bacillus anthracis*, the cause of anthrax, fail to retain stain once the cells run out of nutrients. Still other groups of bacteria, such as *Lactobacillus* species common in the oral cavity and gastrointestinal tract, may stain either positive or negative (Gram-variable) and are thus not distinguished by the Gram stain. Bacterial diversity and identification are discussed further in Chapter 10.

Other Differential Stains

Other differential stains applied to various prokaryotes are illustrated in **Figure 3.21**. These include **acid-fast stains**, **spore stains**, **negative stains**, and **antibody stains**:

- **Acid-fast stain.** Carbolfuchsin specifically stains mycolic acids of *Mycobacterium tuberculosis* and *M. leprae*, the causative agents of tuberculosis and leprosy, respectively (Figure 3.21A).

- **Spore stain.** When samples are boiled with malachite green, the stain binds specifically to the endospore coat (Figure 3.21B). It detects spores of *Bacillus* species such as the insecticide *B. thuringiensis* and *B. anthracis*, the causative agent of anthrax, as well as spores of *Clostridium botulinum*, which produces botulinum toxin, and *C. tetani*, the causative agent of tetanus.

- **Negative stain.** Some bacteria synthesize a capsule of extracellular polysaccharide filaments, which protects the cell from predation or from engulfment by white blood cells. The capsule is transparent and invisible in suspended cells. However, adding a suspension of opaque particles such as India ink or Congo red can darken the surrounding medium. In the micrograph of *Flavobacterium capsulatum* shown in Figure 3.21C, the dye is excluded by the bacterium's thick polysaccharide capsule, which thus appears clear against the colored background. This is an example of a negative stain.

- **Antibody stains.** Specialized stains use antibodies to identify precise strains of bacteria or even specific molecular components of cells. The antibody (which binds a specific cell protein) is linked to a reactive enzyme for detection or to a fluorophore (fluorescent molecule) for fluorescence microscopy (to be discussed shortly, in Section 3.5).

SECTION SUMMARY

- **Fixation and staining** of a specimen kill the specimen but improve contrast and resolution.
- **Differential stains** distinguish between bacteria with different structural features.
- **The Gram stain** differentiates between two major bacterial taxa, Proteobacteria (Gram-negative) and Firmicutes (Gram-positive). Other bacteria and archaea vary in their Gram stain appearance. Eukaryotes stain negative.

3.5
Advanced Microscopy

SECTION OBJECTIVES

- Describe the different kinds of information obtained by dark-field, phase-contrast, and fluorescence microscopy.
- Describe how electron microscopy enables visualization of objects below the resolution limit of light, such as viruses and parts of cells.

Advanced optical techniques enable us to see microbes that are difficult or impossible to detect under a bright-field microscope, either because their size is below the limit of resolution of light or because

Figure 3.21   Differential Stains

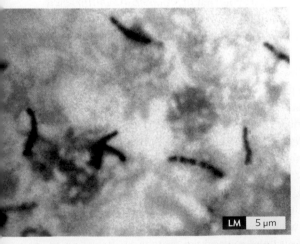

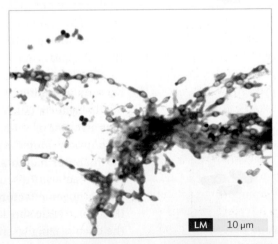

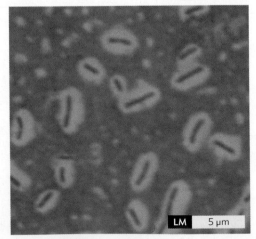

A. *Mycobacterium tuberculosis*, acid-fast stain (stained cells are red, 1–2 μm).

B. *Clostridium tetani*, endospore stain (stained endospores are blue-green, 2 μm long).

C. *Flavobacterium capsulatum*; negative stain with Congo red reveals translucent capsule surrounding each bacterium.

Figure 3.22 Dark-Field Observation of Bacteria

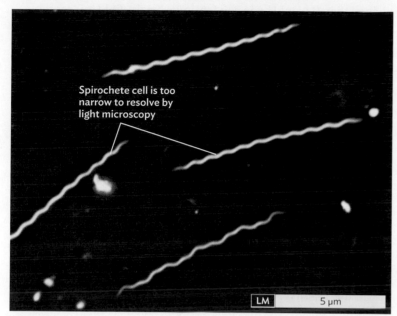

Spirochete cell is too narrow to resolve by light microscopy

LM   5 μm

Treponema pallidum specimen from a patient with syphilis. Note the detection of dust particles.

Figure 3.23 Phase-Contrast Microscopy

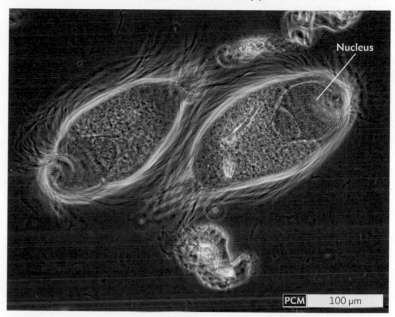

Nucleus

PCM   100 μm

Phase-contrast micrograph of *Trichonympha*, protozoa that live in the gut of a termite.

their cytoplasm is transparent. These techniques take advantage of special properties of light waves, including scattering and interference patterns.

Dark-Field Observation Detects Unresolved Objects

In **dark-field microscopy**, the condenser contains an opaque central disk that blocks transmitted light; thus, the field appears dark. The only light rays we detect are those scattered by the object, as shown earlier in Figure 3.8D. For dark-field optics, an intense light source allows a microbe smaller than the wavelength of light to scatter sufficient light for detection. Detection does not allow resolution; in effect, the microbes are visualized as halos of bright light against a dark field, just as stars are observed against the night sky.

One use of dark-field microscopy is to detect *Treponema pallidum*, the bacterium that causes syphilis. A *T. pallidum* cell is called a "spirochete," a spiral so narrow (0.1 μm) that its shape cannot be fully resolved by light microscopy. Nevertheless, the spiral form of *T. pallidum* can be detected by dark-field microscopy (Figure 3.22). In the dark-field image, the length of the cell is accurate, but the width of the cell body appears "overexposed," wider than the actual cell, because the scattered object lacks resolution.

A limitation of dark-field microscopy is that any tiny particle, including specks of dust, can scatter light and interfere with visualization of the specimen. Unless the medium is extremely clear, it can be difficult to distinguish microbes of interest from particulates. Other methods of contrast enhancement, such as phase contrast and fluorescence, avoid this difficulty.

Phase-Contrast Microscopy

Phase-contrast microscopy exploits small differences in refractive index between the cytoplasm and the surrounding medium or between different organelles. This technique is particularly useful for eukaryotic cells, such as parasites, which contain many intracellular compartments. For example, Figure 3.23 shows a phase-contrast image of *Trichonympha*, protozoa that live in the gut of a termite. These protozoa enable termites to digest wood that they consume. Within each cell, we can see the nucleus.

Phase-contrast microscopy makes use of the fact that living cells have relatively high contrast due to their high concentration of solutes. Given the size and refractive index of commonly observed cells, light is slowed by approximately one-quarter of a wavelength when it passes through the cell. In other words, after having passed through a cell, light exits the cell about one-quarter of a wavelength behind the phase of light transmitted directly through the medium. The optical system is designed to slow the refracted light by an additional one-quarter of a wavelength, so that the light refracted through the cell is slowed by a total of half a wavelength compared with the light transmitted through the medium.

When two waves are out of phase by half a wavelength, they interact so as to cancel each other's amplitude. Figure 3.24 illustrates the addition of wave amplitudes in phase and the subtraction of wave amplitudes out of phase. Phase-contrast optics generate alternating regions of light and darkness in the image. Even small differences in refractive index can produce dramatic differences in contrast between the offset phases of light.

Figure 3.24   Constructive (Additive) and Destructive (Subtractive) Interference of Light Waves

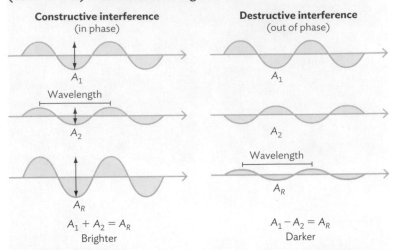

Constructive interference (in phase)

A_1

Wavelength

A_2

A_R

$A_1 + A_2 = A_R$
Brighter

Destructive interference (out of phase)

A_1

A_2

Wavelength

A_R

$A_1 - A_2 = A_R$
Darker

In constructive interference, the peaks of the two wave trains rise together; their amplitudes are additive, creating a wave of greater amplitude. In destructive interference, the peaks of the waves are opposite each other, so their amplitudes cancel, creating a wave of lesser amplitude.

Interference Microscopy

Other kinds of optical systems have been devised using light interference to enhance cytoplasmic contrast. Interference microscopy enhances contrast by superimposing the image of the specimen on a second beam of light that generates an interference pattern. The interference pattern produces an illusion of three-dimensional shadowing across the specimen. For example, Figure 3.25 shows the shape of the dust-borne bacterium *Bacillus subtilis* illuminated

Figure 3.25   Interference Micrograph of *Bacillus subtilis*

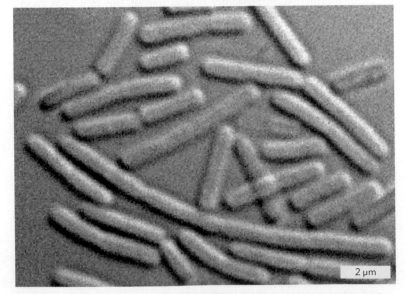

2 µm

The apparent three-dimensional effect is illusory. One bacterium appears as if crossed through another.

by interference contrast. The shape of the cells is more clearly defined than it would be in conventional bright-field microscopy.

Fluorescence Microscopy

CASE HISTORY 3.3

Infection from a Whirlpool Spa

Jared, a 48-year-old, previously healthy man was admitted to a hospital in Columbus, Ohio, after 6 days of increasing dyspnea (shortness of breath) and watery diarrhea. Jared also complained of a dry cough, myalgia (muscle pain), nausea, and vomiting. He appeared flushed and had dry mucous membranes. His temperature was 39°C (102.2°F), and his pulse rate 103 beats per minute. Chest radiographs showed pneumonia with consolidation (formation of a firm mass) of the left lower lobe. His C-reactive protein (CRP, a blood protein produced by the liver) was greater than 220 milligrams per liter (mg/l), a high level that indicates inflammation.

Before onset of symptoms, Jared recalled having cleaned the filter of an outdoor whirlpool spa. Culture from Jared's sputum and from the spa filter revealed Gram-negative bacilli consistent with *Legionella pneumophila*. The diagnosis was confirmed by fluorescence microscopy using direct fluorescent antibody (DFA) stain. The patient received intravenous amoxicillin and clarithromycin antibiotics for 8 days before his condition returned to normal.

The bacteria from Jared's sputum were shaped like small rods—just like many other species. How can we distinguish specific types of bacteria that look alike? **Fluorescence microscopy (FM)** is a tool that enables us to identify a single species, such as the Gram-negative rod *L. pneumophila* in Case History 3.3 (Figure 3.26). *L. pneumophila* is the cause of legionellosis, a form of pneumonia that can be rapidly fatal if not diagnosed in time. The bacteria grow intracellularly within amebas that contaminate water sources such as air-conditioning units or whirlpool spas (as in our case history). During human infection, *L. pneumophila* grows within lung macrophages.

In fluorescence microscopy, incident light is absorbed by the specimen and then emitted at a lower energy, thus longer wavelength. Fluorescence microscopy offers a powerful way to detect specific microbes and subcellular structures while avoiding signals from dust and other nonspecific materials. Bacteria such as *L. pneumophila* and *Mycobacterium tuberculosis* can be identified by fluorescence microscopy of samples stained with a fluorescent antibody specific to the bacterial species.

Fluorescence occurs when a molecule absorbs light of a specific wavelength (the excitation wavelength), exciting an electron to a higher energy level. The electron then falls to its original level by emitting a photon of lower energy and longer wavelength, the emission wavelength. The emitted photon has a longer wavelength (less energy) because part of the electron's energy of absorption was lost as heat.

Figure 3.26 *Legionella pneumophila* Visualized by Direct Fluorescent Antibody Stain

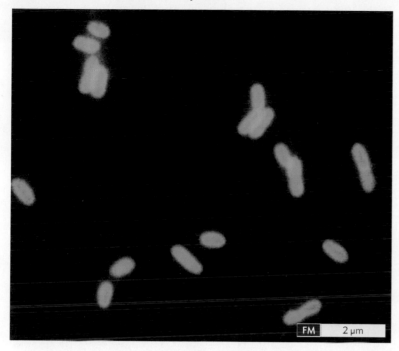

Figure 3.27 Phase-Contrast versus Fluorescence Microscopy

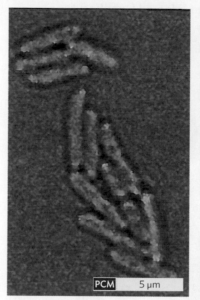 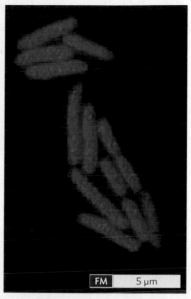

A. *E. coli* K-12 colony growing within a flow cell, visualized by phase contrast. The contours of individual cells appear distinct.

B. The same colony visualized by fluorescence of plasmid-expressed green fluorescent protein (GFP). Fluorescence appears in the cytoplasm; a darkened gap appears between two newly divided cells.

The wavelengths of excitation and emission are determined by the **fluorophore**, the fluorescent molecule used to stain the specimen. The optical system for fluorescence microscopy uses filters that limit incident light to the wavelength of excitation and limit emitted light to the wavelength of emission. The wavelengths of excitation and emission are determined by the fluorophore used. For detecting *L. pneumophila*, the fluorophore used is fluorescein, conjugated (attached) to an antibody specific for the cell surface of *L. pneumophila*. In Figure 3.26, the fluorescence is brightest around the rim of each cell, where the maximal "depth" of envelope fluoresces.

How do different kinds of microscopy offer different kinds of information? In **Figure 3.27**, compare the results of phase-contrast and fluorescence microscopy for the same sample of *E. coli* bacteria. In this case, the fluorophore is green fluorescent protein (GFP), a cytoplasmic protein expressed by the bacterial DNA. We can see that the phase-contrast image provides sharp outlines of cells, whereas the fluorescent image clearly marks the cytoplasm. As two daughter cells emerge, a dark space appears between them, where the newly formed envelopes separate the two.

Fluorescence microscopy has been used to develop advanced optical systems that reconstruct three-dimensional models of cells. An example is confocal laser fluorescence microscopy (CFM). Confocal microscopy is used to study the function of human cells and the ways that some pathogenic bacteria invade and propagate within human cells.

Electron Microscopy

In an electron microscope (**Figure 3.28**), beams of electrons generate images at resolution levels up to a thousandfold greater than that possible for light microscopy. The apparatus requires an evacuated sample chamber, an elaborate system of magnetic lenses, and computerized image reconstruction. Electron microscopy reveals microbes in remarkable detail. In research, the electron microscope can be used to observe virus infection and propagation and to study subcellular structures such as ribosomes and toxin-secreting organelles.

Figure 3.28 An Electron Microscope

Wah Chiu (standing) and Joanita Jakana, at Baylor College of Medicine, in Houston, Texas, using a JEOL 300-kiloelectron-volt (keV) electron microscope. Chiu and Jakana study virus structures by using cryo-electron tomography.

Figure 3.29   Transmission Electron Micrograph (TEM) of HIV Particles Emerging from an Infected Lymphocyte

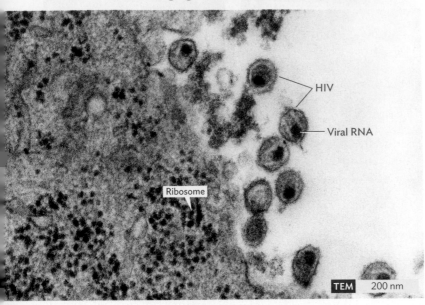

Helicobacter pylori, the major cause of gastritis. *H. pylori* inhabits the human stomach, a region long believed too acidic to harbor life. In the 1980s, however, Australian scientist Barry Marshall reported finding a new species of bacterium in the stomach. The bacterium, *Helicobacter pylori*, proved difficult to isolate and culture, and for over a decade, medical researchers refused to believe Marshall's report. Ultimately, electron microscopy (EM) confirmed the existence of *H. pylori* in the stomach and helped document its role in gastritis and stomach ulcers.

The scanning electron micrograph in **Figure 3.30** shows *H. pylori* colonizing the gastric crypt cells. *H. pylori* bacteria are helical rod cells. Their contours are well resolved by the scanning beam of electrons, an achievement far beyond that possible with a light microscope. In Figure 3.30, note also the use of **colorization**, or "false color," to emphasize particular aspects of the image. Colorization is a process by which the observer adds color to aid interpretation. The color added does not indicate any natural coloration, which cannot be observed through electron microscopy.

An important limitation of traditional electron microscopy, whether TEM or SEM, is that it usually can be applied only to fixed,

An important use of electron microscopy is observation of the life cycle of human immunodeficiency virus (HIV), the cause of acquired immunodeficiency syndrome (AIDS). **Figure 3.29** shows an example of **transmission electron microscopy (TEM)**, in which the electron beams penetrate a thin section of tissue stained with an electron-dense heavy-atom salt. The TEM shows HIV particles budding out of an infected lymphocyte. The dark spot within each virus particle is its RNA genome, compacted by special proteins. But only a slice through the virion is seen. The TEM resolution is high enough to reveal individual ribosomes (small dark spots) within the cytoplasm of the infected cell.

How does an electron microscope work? Electrons are ejected from a metal subjected to a voltage potential. The electrons travel in a straight line, like photons. And like photons, electrons interact with matter and exhibit the properties of waves. Beams of electrons are focused by means of a magnetic field. The shape of the magnet is designed to focus the beam of electrons in a manner analogous to the focusing of photons by a refractive lens. The transmission electron microscope closely parallels the design of a bright-field microscope (recall Figure 3.14) and includes a source of electrons (instead of light), a magnetic condenser lens, a specimen, and an objective lens. A projection lens projects the image onto a fluorescent screen, and the final images are obtained by a digital camera.

Another type of electron microscopy is **scanning electron microscopy (SEM)**. In SEM, the electron beams scan the specimen and are reflected by stain molecules to reveal the contours of its three-dimensional surface. SEM can be performed for a wide range of microbe sizes. SEM was used to define the shape and habitat of

Figure 3.30   *Helicobacter pylori* within the Crypt Cells of the Stomach Lining

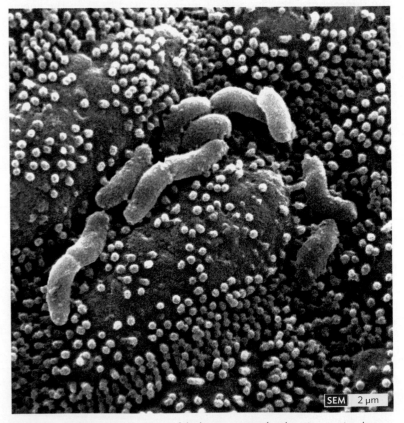

H. pylori bacteria grow on the lining of the human stomach, a location previously believed too acidic to permit microbial growth. Colorized to indicate bacteria (green).

stained specimens. The fixatives and heavy-atom stains can introduce **artifacts** (false structures not present in the original organism), especially at finer details of resolution. Different preparation procedures have sometimes led to substantially different interpretations of structure.

The development of exceptionally high-strength electron beams now permits low-temperature cryo-electron microscopy (cryo-EM). In cryo-EM, the specimen is flash-frozen—that is, suspended in water and frozen rapidly in a refrigerant of high heat capacity (ability to absorb heat). The rapid freezing avoids ice crystallization, leaving the water solvent in a glass-like phase. The specimen retains water content and thus closely resembles its viable form, although it is still ultimately destroyed by electron bombardment. The sample does not require staining because the high-intensity electron beams can detect smaller signals than in earlier instruments. Another innovation in cryo-EM is tomography, in which images are projected from different angles of a transparent specimen. Tomography avoids the need to physically slice the sample for thin-section TEM. The images from EM tomography are combined digitally to visualize the entire object.

SECTION SUMMARY

- **Dark-field microscopy** uses scattered light to detect objects too small to be resolved by light rays.
- **Phase-contrast microscopy (PCM)** superimposes refracted light and transmitted light shifted out of phase so as to show small differences in refractive index as distinct patterns of light and dark. This method is used to observe organelles that are transparent, with low contrast.
- **Interference microscopy** superimposes interference bands on an image, accentuating small differences in refractive index.
- **Fluorescence microscopy (FM)** detects specific cells or cell parts based on fluorescence by a fluorophore. Cell parts can be labeled by a fluorophore attached to an antibody stain.

- **Transmission electron microscopy (TEM)** is based on the focusing of electron beams that penetrate a thin sample, usually stained with a heavy-metal salt that blocks electrons. Much higher resolution is obtained than for light microscopy.
- **Scanning electron microscopy (SEM)** involves scanning a three-dimensional surface with an electron beam.

Thought Question 3.4　Compare fluorescence microscopy with dark-field microscopy. What similar advantage do they provide, and how do they differ?

Thought Question 3.5　You have discovered a new kind of microbe, never observed before. What kind of questions about this microbe might be answered by light microscopy? What questions would be better addressed by electron microscopy?

Perspective

You can use the tools of microscopy described in this chapter to discover microbes in our environment and to reveal pathogens in human samples. Identifying pathogens such as the malarial parasite or tuberculosis bacteria is a crucial aspect of diagnosis. In research, microscopy has shaped our understanding of microbial cells—how they grow and divide, organize their DNA and cytoplasm, and respond to antibiotics. Advanced microscopy reveals how pathogens infect their host cells: do they remain attached to the outside of the cell, or do they invade the cytoplasm and consume the host from within?

Next, in Chapter 4, you will learn the molecular building blocks of the microbes you observe by microscopy. You will see how small molecules such as phospholipids form flexible membranes for cell compartments and how large macromolecules such as DNA and proteins store information. These molecules all function together within a cell, as we explore further in Chapter 5.

LEARNING OUTCOMES AND ASSESSMENT

3.1 Observing Microbes

SECTION OBJECTIVES

- Explain how magnification improves resolution of a microscopic image.
- Explain what can be learned from different kinds of microscopy.

OBJECTIVES REVIEW

1. Which property of bacteria determines that they are "microscopic"?
   A. Bacteria can consist of isolated single cells.
   B. Bacteria can grow as chains or biofilms of cells.
   C. Bacteria are smaller than a centimeter in length.
   D. Bacteria are smaller than the resolution distance of the human retina.

3.2 Optics and Properties of Light

SECTION OBJECTIVES

- Explain how light interacts with an object.
- Describe how refraction enables magnification, and explain the limits to magnification and resolution.

OBJECTIVES REVIEW

3. Which condition is <u>not</u> required for an object to be seen under a microscope?
   A. The dimensions of the object must be greater than the wavelength of light.
   B. The object must be fixed to a slide.
   C. The object must be magnified to a size resolvable by the human retina.
   D. The object must show contrast with its surroundings.

3.3 Bright-Field Microscopy

SECTION OBJECTIVES

- Explain the use of a bright-field microscope.
- Describe how to observe an object in focus by using a compound microscope.

OBJECTIVES REVIEW

5. The part of the microscope that concentrates light onto the field of view in a slide is _____.
   A. the condenser
   B. the objective lens
   C. the ocular lens
   D. the optical column

3.4 Stained Samples

SECTION OBJECTIVES

- Explain how a stain reveals additional information about a microscopic object.
- Describe how the Gram stain distinguishes two classes of bacteria.

OBJECTIVES REVIEW

7. Giemsa stain can reveal _____.
   A. the Gram-positive cell envelope of *Bacillus anthracis*
   B. the ring form and gametocyte stage of the malaria parasite *Plasmodium falciparum*
   C. the spores of *Clostridium botulinum*
   D. the extracellular polysaccharide capsule of *Flavobacterium capsulatum*

3.5 Advanced Microscopy

SECTION OBJECTIVES

- Describe the different kinds of information obtained by dark-field, phase-contrast, and fluorescence microscopy.
- Describe how electron microscopy enables visualization of objects below the resolution limit of light, such as viruses and parts of cells.

OBJECTIVES REVIEW

9. In a patient's sample, the intracellular pathogen *Legionella pneumophila* can best be identified using _____.
   A. fluorescence microscopy, with a fluorophore attached to a cell-surface–specific antibody
   B. dark-field microscopy, using a dust-free medium
   C. phase-contrast microscopy
   D. electron microscopy

10. To visualize individual particles of human immunodeficiency virus (HIV) requires _____.
   A. fluorescence microscopy
   B. dark-field microscopy
   C. phase-contrast microscopy
   D. electron microscopy

2. Which kind of microscopy reveals microbes as dark objects against a field of light?
   A. Bright-field light microscopy
   B. Transmission electron microscopy
   C. Scanning electron microscopy
   D. Fluorescence microscopy

4. Which property of light allows magnification by a lens?
   A. Photons of light are absorbed by an opaque substance.
   B. When a light wave encounters a small particle, some of the light energy is scattered in the form of a spherical wave.
   C. Light bends, or "refracts," as it enters a substance that slows its speed.
   D. Light rays travel through a vacuum at a speed of about 3×10^8 m/s.

6. The part of the microscope that first magnifies the image of the object in a slide is _____.
   A. the condenser
   B. the objective lens
   C. the ocular lens
   D. the optical column

8. Which of these statements about the Gram stain is incorrect?
   A. The Gram stain procedure requires fixation of the sample to a slide.
   B. The Gram stain requires a dye such as crystal violet, followed by a mordant to strengthen the binding of the dye.
   C. To complete the Gram stain requires a counterstain such as safranin.
   D. The Gram procedure requires an antibody stain.

11. Which type of microscopy generated each of the images shown?

A. *Mycobacterium paratuberculosis,* associated with Johne's disease of cattle.

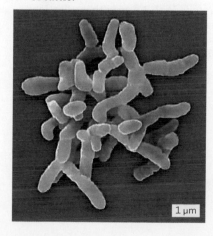

1 μm

B. Motile *E. coli* cells with flagella.

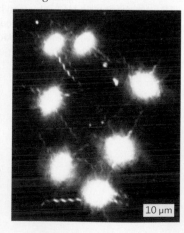

10 μm

C. *Streptococcus mutans,* bacteria found in dental plaque.

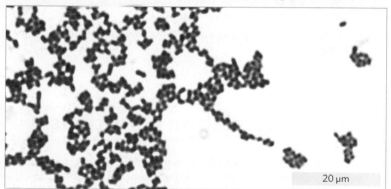

20 μm

D. *Bacillus anthracis,* cause of anthrax.

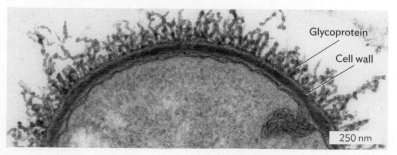

Glycoprotein

Cell wall

250 nm

Key Terms

absorption (p. 70)
acid-fast stain (p. 80)
antibody stain (p. 80)
artifact (p. 85)
bacillus (p. 68)
bright-field microscopy (p. 68)
coccus (p. 68)
colorization (p. 84)
compound microscope (p. 74)
condenser (p. 74)
contrast (p. 70)
counterstain (p. 79)
dark-field microscopy (p. 81)
depth of field (p. 75)

diaphragm (p. 74)
differential stain (p. 77)
electron microscopy (EM) (p. 69)
empty magnification (p. 72)
fixation (p. 76)
fluorescence (p. 71)
fluorescence microscopy (FM)
   (p. 82)
fluorophore (p. 83)
focal plane (p. 71)
focal point (p. 71)
Gram-negative (p. 79)
Gram-positive (p. 79)
Gram stain (p. 77)

immersion oil (p. 74)
light microscopy (LM) (pp. 66, 68)
magnification (p. 65)
microscope (p. 65)
negative stain (p. 80)
numerical aperture (NA) (p. 74)
objective lens (p. 73)
ocular lens (p. 74)
parfocal (p. 74)
phase-contrast microscopy (p. 81)
reflection (p. 71)
refraction (p. 71)
refractive index (p. 71)
resolution (p. 65)

scale bar (p. 66)
scanning electron microscopy (SEM)
   (pp. 69, 84)
scattering (p. 71)
spirochete (p. 68)
spore stain (p. 80)
staining (p. 76)
total magnification (p. 75)
transmission electron microscopy
   (TEM) (pp. 69, 84)
wavelength (p. 69)
wet mount (p. 75)
X-ray crystallography (p. 69)

Review Questions

1. What principle defines an object as "microscopic"?

2. Explain the difference between detection and resolution.

3. How do eukaryotic and prokaryotic cells differ in appearance under the light microscope?

4. Explain how electromagnetic radiation carries information and why different kinds of radiation can resolve different kinds of objects.

5. Define how light interacts with an object through absorption, reflection, refraction, and scattering.

6. Explain how refraction enables magnification of an image.

7. Explain how magnification increases resolution and why "empty magnification" fails to increase resolution.

8. Explain how angle of aperture and resolution change with increasing lens magnification.

9. Summarize the optical arrangement of a compound microscope.

10. Explain how to focus an object and how to tell when the object is in or out of focus.

11. Explain the relative advantages and limitations of wet mount and stained preparation for observing microbes.

12. Explain the significance (and limitations) of the Gram stain for bacterial taxonomy.

13. Explain the basis of dark-field, phase-contrast, and fluorescence microscopy. Give examples of applications of these advanced techniques.

14. Explain the difference between transmission and scanning electron microscopy and name the different applications of each.

Clinical Correlation Questions

1. After eating peanut butter, a large number of people get sick with vomiting and diarrhea. Samples of the peanut butter show a rod-shaped bacterium that stains Gram-negative. Unstained bacteria (live mount) swim with multiple flagella. Which pathogen do these results suggest?

2. A patient with HIV infection has failed to keep up with medications. The patient has a recurring cough with shortness of breath, and his sputum shows rod-shaped bacteria that stain acid-fast. The patient should be tested to confirm possible infection by which pathogen?

Thought Questions: CHECK YOUR ANSWERS

Thought Question 3.1   Explain what happens to a refracted light wave as it emerges from a piece of glass of even thickness. How do its new speed and direction compare with its original (incident) speed and direction?

ANSWER: The part of the wavefront that emerges first travels faster than the portion still in the glass, causing the wave front to bend toward the surface of the glass. Ultimately, the wave travels in the same direction and with the same speed as it did before entering the glass. The path of the emerging light ray is parallel to the path of the light ray entering the glass and is shifted over by an amount dependent on the thickness of the glass. This refraction will alter the path of the beam of light and decrease the amount of light reaching the lens of the microscope. Immersion oil has the same refractive index as glass and will limit the amount of light lost in this way.

Thought Question 3.2   In theory, which angle of the light cone would produce the highest resolution? What practical problem would you have in designing a lens to generate this light cone?

ANSWER: In theory, an angle of 90° for the light cone would produce the highest resolution. However, a 90° angle generates a cone of 180°—that is, a cone collapsed flat. Such a "cone" would require the object to sit in the same position as the objective lens—in other words, to have a focal distance of zero. In practice, the cone of light needs to be somewhat less than 180° to allow room for the object and to avoid substantial aberrations (light-distorting properties) in the lens material.

Thought Question 3.3   Under starvation conditions, bacteria such as *Bacillus thuringiensis*, the biological insecticide, repackage their cytoplasm into spores, leaving behind an empty cell wall. Suppose, under a microscope, you observe what appears to be a hollow cell. How can you tell whether the cell is indeed hollow or whether it is simply out of focus?

ANSWER: You can tell whether the cell is out of focus or actually hollow by rotating the fine-focus knob to move the objective up and down while observing the specimen carefully. If the hollow shape appears to be the sharpest image possible, it is probably a hollow cell. If the hollow shape turns momentarily into a sharp, dark cell, it was probably a full cell out of focus before.

Thought Question 3.4   Compare fluorescence microscopy with dark-field microscopy. What similar advantage do they provide, and how do they differ?

ANSWER: Both dark-field and fluorescence microscopy enable detection (but not resolution) of objects whose dimensions are below the wavelength of light. Dark-field technique is based on light scattering, which detects all small objects without discrimination. The technique is relatively simple and inexpensive. Fluorescence requires more complex optics involving wavelength filters. The fluorescence technique, however, provides a way to label specific parts of cells, such as cell membrane or DNA, or particular species of microbes by using fluorescent antibody tags.

Thought Question 3.5   You have discovered a new kind of microbe, never observed before. What kind of questions about this microbe might be answered by light microscopy? What questions would be better addressed by electron microscopy?

ANSWER: Light microscopy could answer questions such as: What is the overall shape of this cell? Does it form individual cells or chains? Is the organism motile? Only light microscopy can visualize an organism alive. Electron microscopy can answer questions about internal and external subcellular structures. For example, does a bacterial cell possess external structures such as flagella or pili? If the dimensions of the unknown microbe are smaller than the lower limits of a light microscope's resolution, EM may be the only way to observe the organism. Viruses are often characterized by shape, and this shape is observed by electron microscopy.

4

Living Chemistry: From Atoms to Cells

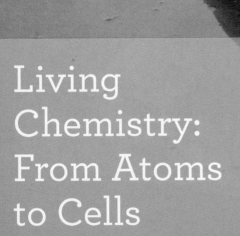

Lethal Chemistry from a Waterborne Pathogen

SCENARIO Water is the most abundant molecule of life—but too much water can be deadly. In the summer of 2010, heavy rains flooded Pakistan, submerging one-fifth of the country. Millions of people were displaced and exposed to water contaminated by microbial pathogens. Diarrheal disease threatened many, especially children.

PATIENT A 10-year-old boy named Akal developed profuse diarrhea. He showed extreme thirst, but clean drinking water was unavailable. His diarrhea became watery and cloudy with white flecks of intestinal mucus (known as rice-water stool). Akal became listless, his eyes sunken and his skin wrinkled from dehydration.

SIGNS AND SYMPTOMS A nurse observed a sample of Akal's diarrheal fluid under dark-field microscopy. Numerous bacteria swam rapidly, appearing like "twinkling stars." The bacteria, comma-shaped bacilli, stained Gram-negative.

DIAGNOSIS Later testing confirmed *Vibrio cholerae*, the cause of cholera. These bacteria produce cholera toxin, a protein complex that causes intestinal epithelial cells to secrete large amounts of chloride ion (Cl^-). The high chloride concentration draws out the positively charged sodium ions (Na^+). The increased extracellular salt (Na^+ and Cl^-, or $NaCl$) causes water to diffuse out of the epithelial cells, leading to dehydration and diarrhea. Without treatment, Akal's chance of death was 50%.

Oral Rehydration Therapy
Replaces the water and electrolytes lost during severe diarrhea.

TREATMENT Akal was treated with oral rehydration therapy (ORT), a method invented by Bangladeshi physicians. ORT involves gradual feeding of a solution containing glucose and $NaCl$ to replace ions and water lost through the diarrhea. Within hours, Akal's flesh had rehydrated, and within a week he made a full recovery.

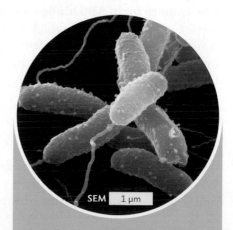

SEM   1 μm

Vibrio cholerae
A comma-shaped Gram-negative bacterium whose enterotoxin drives water out of the intestine.

Cholera threatens impoverished people living in crowded conditions such as refugee camps, where bacteria flushed out of an infected person by diarrhea spread rapidly to new hosts. The causative bacteria, *Vibrio cholerae* (shown on the preceding page), are tiny curved rods that stain negative in the classic Gram stain, as we learned in Chapter 3. Remarkably, the cause of mortality is a single type of bacterial toxin, cholera toxin (**Figure 4.1A**). Cholera toxin, a protein complex, kills the patient by extreme water loss (dehydration) and sodium loss (hyponatremia). The cellular loss of these simple chemicals leads to death.

How does one protein complex trick the body into losing its essential water and salt? The mechanism of cholera toxin demonstrates several key principles of cell biochemistry, such as protein interactions, ion flux, and membrane dynamics. **Figure 4.1B** shows how cholera toxin AB5 acts with a cell of the intestinal lining:

1. **The cholera toxin AB5 binds a ganglioside.** The AB5 toxin consists of one protein subunit A with five copies of subunit B. AB5 binds to a ganglioside, a special sugar-lipid molecule found on the host cell surface. (Sugars and lipids are discussed in Section 4.2.)

2. **AB5 toxin enters the cell by endocytosis.** In endocytosis, the membrane pinches in, forming a vesicle called an endosome (discussed in Sections 4.5 and 4.6).

3. **The endosome travels to the endoplasmic reticulum.** Within the endosome, the AB5 toxin releases part of its A subunit (called A1), which returns to the cytoplasm.

4. **A1 binds host G protein.** The A1-bound G protein overstimulates adenylate cyclase to convert adenosine triphosphate (ATP) to cyclic adenosine monophosphate (cAMP). (ATP reactions are presented in Section 4.4.)

5. **cAMP stimulates the chloride (Cl^-) transporter CFTR.** Chloride (Cl^-) buildup increases the negative charge outside the cell. The increased negative charge draws out positively charged ions such as potassium (K^+) and sodium (Na^+). The high concentration of all these ions outside causes osmotic imbalance, drawing water out of the cell. Water loss causes dehydration and death.

Under the dire circumstances of Pakistan's flood, how could a health worker save the lives of cholera patients suffering dehydration? By understanding ion transport and osmolarity (discussed in Section 4.5), we will see how doctors learned to reverse the life-threatening effects of cholera toxin by a surprisingly simple treatment of glucose and salt solution. The treatment is called oral rehydration therapy (discussed in this chapter's inSight).

This chapter introduces the biochemistry a health professional needs to know. Section 4.1 reviews the chemistry of small biological molecules, with an emphasis on the special properties of water. We then discuss macromolecules, including the large information-bearing molecules that define living organisms, such as proteins and nucleic acids. Finally, we explore important chemical reactions that keep cells alive and the membrane compartmentalization of these reactions inside a cell. These molecules and reactions are important both for microbes and for the human body.

Figure 4.1   Cholera Toxin Kills Cells

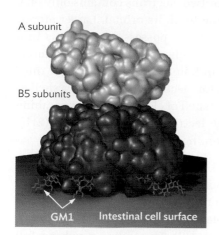

A. Cholera toxin is composed of protein A and five subunits of protein B. The AB5 complex binds a host surface sugar, ganglioside.

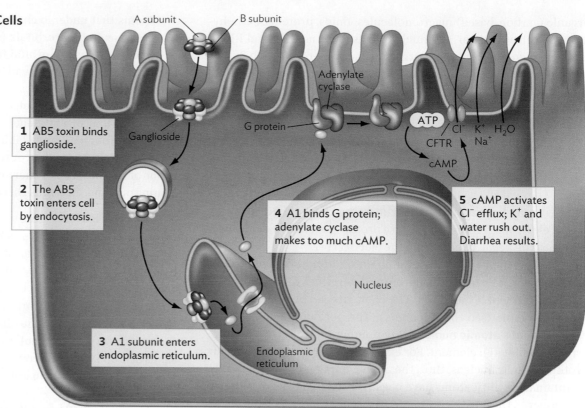

1 AB5 toxin binds ganglioside.

2 The AB5 toxin enters cell by endocytosis.

3 A1 subunit enters endoplasmic reticulum.

4 A1 binds G protein; adenylate cyclase makes too much cAMP.

5 cAMP activates Cl⁻ efflux; K⁺ and water rush out. Diarrhea results.

B. AB5 toxin binds gangliosides (step 1) and enters the cell by endocytosis (step 2). At the endoplasmic reticulum, an enzyme cleaves the A subunit, releasing A1 peptide (step 3). The A1 peptide enters the cytoplasm and binds G protein (step 4), thus activating adenylate cyclase. Adenylate cyclase converts ATP to cAMP, which stimulates the chloride efflux transporter CFTR (step 5). Chloride efflux increases negative charge outside the cell, drawing out positively charged ions such as potassium (K^+) and sodium (Na^+). High salt outside draws water out of the cell, causing diarrhea.

4.1
Elements, Bonding, and Water

SECTION OBJECTIVES

- Describe how atoms form the molecules of life.
- Describe the types of bonds atoms can form to make a molecule.
- Explain the special properties of the water molecule, and distinguish between hydrophilic and hydrophobic molecules.

What are living cells made of? Living cells are remarkably complex machines, able to integrate and respond to multiple stimuli, to catalyze (speed up) reactions, and to replicate themselves. Yet despite all the various tasks that cells perform, 98% of the mass of living organisms consists of just six elements: hydrogen (H), oxygen (O), nitrogen (N), phosphorus (P), sulfur (S), and carbon (C). For example, all the proteins and membranes involved in the cholera toxin cycle are composed of these elements.

The most abundant compound in cells is water (H_2O). The rest of the cell consists, for the most part, of just four different kinds of organic (carbon-based) macromolecules: lipids, proteins, carbohydrates (such as sugars), and nucleic acids (such as DNA and RNA). Cells are compartments that organize chemical reactions between these organic molecules. To understand life, we must understand the properties of water and organic molecules and of their building blocks, the chemical elements.

Cells consist mostly of water, inorganic ions (such as K^+ and Cl^-), and organic molecules. These cellular components are composed of atoms. An **atom** consists of a positively charged nucleus that contains protons and neutrons, surrounded by negatively charged electrons. The hydrogen nucleus consists of a single proton. Protons and neutrons have a mass approximately 2,000-fold greater than that of electrons. Yet the charge on a proton (+1) is of the same magnitude as the negative charge on the electron (−1).

Different types of atoms contain different numbers of protons and electrons. The number of protons determines the type of atom called an **element**. (Table 4.1). The elements can be organized into a periodic table (see the Appendix). The periodic table indicates each element's **atomic number** (the number of protons), the property that defines an element. For example, all carbon atoms have six protons in their nucleus; thus, the atomic number of carbon is 6. To maintain neutrality, atoms have negatively charged electrons equal in number to the positively charged protons.

The periodic table also shows each element's **mass number**. The mass number of an element is the mass of the nucleus, the sum of the number of protons and neutrons. The average mass number of the various isotopes of an atom (with different numbers of neutrons) gives the standard **atomic weight**.

Isotopes are forms of an element that differ in their number of neutrons. For example, the most abundant isotope of carbon is carbon-12 (with six neutrons), but naturally occurring isotopes of carbon-13 (seven neutrons) and carbon-14 (eight neutrons) also exist. Because carbon-13 and carbon-14 are rare, the average atomic mass is close to but not exactly 12. Although some isotopes are stable, others decay at a known rate. The decay of an isotope is known as radioactivity. Carbon-14 has a **half-life** (the amount of time it takes for half the sample to decay) of 5,700 years. The decay of carbon-14 is used in radiocarbon dating to determine the age of fossil bones and of ancient materials such as the Dead Sea Scrolls. Scientists use carbon-14 and shorter-lived isotopes such as tritium (hydrogen with a mass of 3—one proton and two neutrons) as tracers to follow specific atoms in metabolic pathways.

Chemical Bonds Form Molecules

How do atoms form molecules? The three-dimensional shape, or structure, of a molecule such as a protein determines how it functions in a cell. The structure depends on which atoms are present and how they are bonded together by interactions of their electrons. Pairs of electrons occupy regions of space called **orbitals**. The geometry of orbitals determines the shape of a molecule.

Atoms that undergo chemical reactions have unfilled "places" for electrons in their orbitals (see Table 4.1). Usually, the unfilled place occurs where an orbital for two electrons contains only one. Hydrogen has one place available to fill its orbital. Larger atoms, such as carbon, nitrogen, and oxygen, possess four orbitals in a shell that could hold eight electrons—the "octet rule." For example, carbon has four orbitals, each containing only one electron; thus, carbon has four places available for electrons to complete a shell of eight. To attain a complete outer shell of electrons, atoms combine in one of two ways: (1) by sharing electrons (covalent bond) or (2) by gaining or losing electrons (becoming ions).

COVALENT BONDS   Atoms that share electrons form a **molecule**. The sharing of electrons between two atoms is called a **covalent bond**. For example, an atom of carbon combines with four hydrogen atoms to form a molecule of methane, CH_4, the main component of natural gas (Figure 4.2). The carbon now has a full outer shell of eight electrons, and each hydrogen has a full shell of two electrons. Methane thus has four covalent bonds. Much larger molecules, such as the cholera toxin presented earlier, have thousands of atoms linked by covalent bonds. The bonds of a protein link together chains of carbon atoms as well as atoms of hydrogen, oxygen, and nitrogen.

The symbols CH_4 and H_2O are examples of **molecular formulas**. A molecular formula is a shorthand notation indicating the number and type of atoms present in a molecule. Atoms can also share more than one pair of electrons with another atom, forming double or triple bonds. In water, H_2O, each hydrogen shares two pairs of electrons with oxygen. The bonding in molecules can be represented by a **structural formula**, in which covalent bonds are shown as a line between two atoms (shown in Figure 4.2).

A structural formula represents the sequence of bonds in a molecule, but not its actual three-dimensional shape. The bonding orbitals in the outer shell of carbon are arranged so that they point to the vertices of a tetrahedron; methane, the simplest hydrocarbon, actually has the shape of a tetrahedron, with the hydrogen atoms

Figure 4.2   Covalent Bonding of Hydrogen, Carbon, and Oxygen

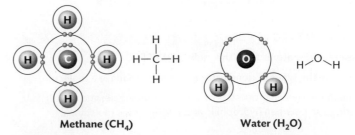

Methane (CH$_4$)   **Water (H$_2$O)**

Electron sharing in the four single bonds of methane, CH_4, and in the two single bonds of water (H_2O). In structural formulas, each line represents a covalent bond, and dots represent lone pairs of electrons in the outermost shell.

Table 4.1
Major Elements of Living Cells

Element	Symbol	Atomic Number	Mass Number[a]	Outer Electron Shell[b]	Number of Bonds Shared in Molecules	Ionic Charge	Example of Compound
Hydrogen	H	1	1	(H)	1	+1	H_2O; HCl (water; hydrochloric acid)
Carbon	C	6	12	(C)	4		CH_3-CH_2-OH (ethanol)
Nitrogen	N	7	14	(N)	3		NH_2-CH_2-COOH (glycine, an amino acid)
Oxygen	O	8	16	(O)	2		$C_6H_{12}O_6$ (six-carbon sugar such as glucose)
Sulfur	S	16	32	(S)	2		H_2S; SO_4^{2-} (hydrogen sulfide; sulfate ion)
Phosphorus	P	15	31	(P)	5		PO_4^{3+} (phosphate ion)
Sodium	Na	11	23	(Na)		+1	NaCl (Na^+ and Cl^-) (sodium chloride)
Chlorine	Cl	17	35	(Cl)		−1	$MgCl_2$ (Mg^{2+} and $2Cl^-$) (magnesium chloride)
Potassium	K	19	39	(K)		+1	K_3PO_4 ($3K^+$ and PO_4^{3+}) (potassium phosphate)
Calcium	Ca	20	40	(Ca)		+2	$CaCl_2$ (Ca^{2+} and $2Cl^-$) (calcium chloride)
Magnesium	Mg	12	24	(Mg)		+2	Mg-ATP (magnesium-ATP)
Iron[c]	Fe	26	56	(Fe)		+2 or +3	$Fe(OH)_3$ (Fe^{3+} and $3OH^-$) (iron III hydroxide)

[a]Mass number shown is that of the most common isotope of an element.

[b]Electrons in an outer, or "valence," shell of an atom can participate in forming chemical bonds with other atoms. For some atoms, an unpaired electron can share a covalent bond. For other atoms, the unpaired electron can ionize to form a cation.

[c]Iron can lose two electrons from its outer shell, plus a third electron from an inner shell (not shown), making a net charge of +3.

Figure 4.3   Molecular Models of Methane

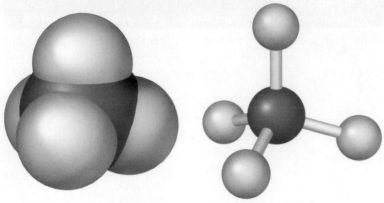

A. A space-filling model of methane.   **B.** A ball-and-stick model of methane.

at the vertices (**Figure 4.3A**). This **space-filling model** shows the volume filled by the outer shell of the atoms. **Figure 4.3B** shows a different kind of model used to depict the three-dimensional shape of molecules: the **ball-and-stick model**. Each stick shows the length and orientation of bonds between the nuclei of each pair of atoms.

IONIC BONDS   The molecules we have just described fill their orbitals by sharing electrons. A different way that atoms can obtain full outer shells is by gaining or losing electrons (**Figure 4.4A**). For example, Na loses its one outer-shell electron to become positively charged (Na^+) with its inner shell full. By contrast, Cl gains an electron to fill its outermost shell (Cl^-). Both Cl^- and Na^+ are charged atoms called **ions**. For an ion, the number of electrons and protons are unequal. **Anions** are negatively charged ions, and **cations** are positively charged ions. These ions are also known as **electrolytes** because their solutions conduct electricity. The appropriate concentrations of ions such as Cl^- and Na^+ are essential for microbial life and for the human body; for example, the excessive efflux of Cl^- causes the pathology of cholera. In the bloodstream, the appropriate Na^+ concentration is vital for blood pressure.

As water evaporates, anions and cations can form ionic crystals held together by **ionic bonds**, charge attractions between anions and cations. For example, sodium ion (Na^+) and chlorine ion, or chloride (Cl^-), interact to form crystals of table salt, NaCl (**Figure 4.4B**). Some kinds of crystal formation in the body can cause problems such as gout (uric acid crystals) and kidney stones (calcium oxalate or calcium phosphate).

As water content increases and the ion concentrations decline, ionic crystals are destabilized and begin to dissolve. To understand why, we need to understand the structure of water.

Water Is the Solvent of Life

Living organisms consist mostly of water; that is why the loss of water (dehydration) in cholera has fatal results. Water has many unique properties that make it especially suitable for sustaining life. For example, water molecules have uneven charge distribution, and their partial charges attract each other.

In some molecules, such as H_2 and O_2, no charge distribution exists because both atoms have equal electronegativity (tendency to hold electrons). Therefore, the electrons in the covalent bond are shared equally and form **nonpolar covalent bonds**. In water, however, the shared electrons of H_2O spend more of their time around the highly electronegative oxygen than near the less electronegative hydrogen (**Figure 4.5A**). A bond with unequal electron sharing is a **polar covalent bond**, so called because the molecule has partial positive and negative poles. In a water molecule, the polar covalent bonds point at an angle from each other; thus, their charge difference forms a partial charge separation within the molecule.

The charge separation within each water molecule causes water molecules to attract each other. This partial charge attraction, between the hydrogen of one water molecule and the oxygen of another molecule, is known as a **hydrogen bond** (**Figure 4.5B**). Hydrogen bonds occur not just in water but also in other molecules—in protein, for example, between a hydrogen bound to an oxygen or nitrogen and a second oxygen or nitrogen, either in the same or a

Figure 4.4   Formation of Ions and Ionic Crystals

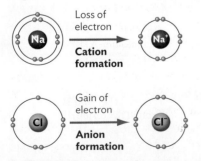

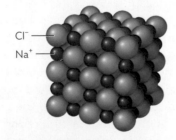

A. The loss of an electron from an atom of sodium to form the cation Na^+ and the gain of an electron by a chlorine atom to form the chloride anion Cl^-.

B. Oppositely charged anions and cations—in this case, Cl^- and Na^+—are attracted to one another and form crystals of table salt (sodium chloride).

Figure 4.5   Polar Covalent Bonds and Hydrogen Bonds in Water

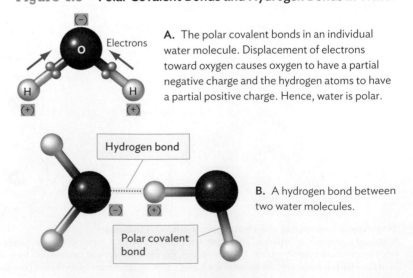

A. The polar covalent bonds in an individual water molecule. Displacement of electrons toward oxygen causes oxygen to have a partial negative charge and the hydrogen atoms to have a partial positive charge. Hence, water is polar.

B. A hydrogen bond between two water molecules.

different molecule. Hydrogen bonds are short-lived and constantly break and re-form in liquid water. Figure 4.5B shows the polar covalent bonds and hydrogen bonds present in water. Hydrogen bonds are weaker than covalent bonds.

The hydrogen bonds in water contribute to its unique properties that support life. Water is a liquid over a large temperature range because hydrogen bonds cause water molecules to attract each other, favoring the liquid state over the gas. The polar nature of water also defines its properties as a solvent. Compounds that are ionic or polar themselves tend to dissolve in water and are termed **hydrophilic** ("water loving"). For example, the ionic bonds in NaCl are very strong in the absence of water but are easily dissolved in the presence of water. This is because the polar water can surround and interact with the sodium and chloride ions and shield them from each other (**Figure 4.6A**). Water also dissolves polar compounds. In this case, water does not break the polar covalent bonds; rather, individual, intact polar molecules form hydrogen bonds with the surrounding water.

By contrast, compounds that are mostly nonpolar do not dissolve in water and are termed **hydrophobic** ("water fearing"). Examples of hydrophobic molecules include long-chain hydrocarbons such as vegetable oil and gasoline. Nonpolar molecules have no partial charges to attract water. Because water hydrogen-bonds with other water molecules, it tends to exclude nonpolar compounds,

Figure 4.6 Interactions between Water and Solutes

A. Water surrounds and interacts with ions or polar molecules, causing them to dissolve.

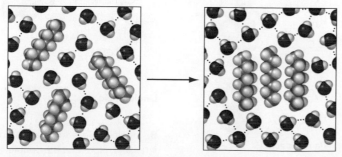

B. Nonpolar (hydrophobic) molecules do not dissolve in water. Instead, to minimize the disruption of hydrogen bonding among water molecules, nonpolar molecules in water aggregate.

forcing them together (**Figure 4.6B**). Nonpolar compounds are said to attract each other by "hydrophobic" forces. Hydrophobic attraction is weaker than hydrogen bonding, but it contributes to the stability of important cell features. For example, the components of cell membranes are held together by hydrophobic attraction (see Section 4.5).

SECTION SUMMARY

- **The major atoms of life are the elements carbon, hydrogen, oxygen, nitrogen, phosphorus, and sulfur.**
- **Atoms form bonds.** Bonds can be covalent or ionic. Weaker bonds include hydrogen bonds and hydrophobic interactions.
- **The molecular formula indicates atomic composition.** In a covalent molecule, the structural formula indicates order of covalent bonds.
- **Water is the solvent of life.** Water is the most prevalent molecule within most living organisms. Water molecules form hydrogen bonds with each other and with dissolved polar molecules or ions.
- **Molecules can be polar or nonpolar.** Polar molecules are hydrophilic (soluble in water), whereas nonpolar molecules tend to be hydrophobic—more soluble in a nonpolar medium such as a membrane.

Thought Question 4.1 Use the properties of molecules described in this section to explain why table salt (NaCl) dissolves readily in water but vegetable oil (a hydrocarbon) does not.

4.2
Lipids and Sugars

SECTION OBJECTIVES

- Describe organic molecules, and explain the role of functional groups and the formation of macromolecules.
- Describe the structure of lipids and their key functions within a cell.
- Describe the structure of sugars and polysaccharides.

What kinds of molecules form the structure of cells? Cells are largely made of **organic molecules**, defined as molecules that contain a carbon-carbon bond. Some organic molecules are small; for example, ethanol, CH_3–CH_2–OH, the active component of alcoholic beverages, has a single carbon-carbon bond. Other organic molecules, such as proteins and DNA, consist of thousands of atoms with many carbon-carbon bonds. Such large organic molecules are called **macromolecules**. Each type of macromolecule is a polymer composed of smaller units called monomers. Major macromolecules of cells are lipids, polysaccharides, nucleic acids, and proteins.

An important feature of any organic molecule is its functional groups. A **functional group** is a small group of atoms with characteristic bonding, shape, and reactivity (**Table 4.2**). Functional groups enable a molecule to undergo important chemical reactions. For example, in molecules of DNA, a chain-terminal hydroxyl group reacts with an additional nucleotide to add a "letter" onto the DNA

Table 4.2

Common Functional Groups of Organic Molecules[a]

Functional Group	General Structure	Example
Alkane	$R-\overset{\overset{H}{|}}{\underset{\underset{H}{|}}{C}}-H$	Ethane
Amino	$R-\overset{\overset{H}{\diagup}}{\underset{\underset{H}{\diagdown}}{N}}$	Ethylamine
Carboxyl	$R-\overset{\overset{O}{\|}}{C}-O-H$	Acetic acid
Ester	$R-\overset{\overset{O}{\|}}{C}-O-R$	Methyl formate
Hydroxyl	$R-O-H$	Ethanol
Carbonyl	$R-\overset{\overset{O}{\|}}{C}-R$	Dihydroxyacetone

[a]R stands for an organic group.

chain. Information about the functional groups present in a molecule is often indicated by a molecule's name; for example, amino acids (the building blocks of proteins) each contain an amino group (–NH₂) and a carboxylic acid group (–COOH).

Lipids

Lipids are organic molecules that serve as structural components of membranes and other parts of cells. Lipids store lots of energy (as in "calories") because their high number of C–H bonds release energy when metabolized with the oxygen we breathe. All lipids contain a substantial number of nonpolar C–H and C–C covalent bonds. Because lipids are nonpolar, they are hydrophobic and do not dissolve in water. Many toxic pollutants such as dioxins are also hydrophobic and therefore concentrate in the lipids of fat tissues.

Lipid building blocks include the fatty acids (Figure 4.7). Each fatty acid has a hydrophobic portion (the hydrocarbon tail) with a long chain of CH₂ units, like the hydrocarbon chains of gasoline. But the hydrocarbon tail of a fatty acid ends with a hydrophilic component, the carboxylic acid, which ionizes (releases a hydrogen ion, H⁺). The remaining ionized molecule, called a carboxylate, now bears a negative charge (R–COO⁻), where R stands for the rest of the molecule. The hydrocarbon tails avoid interacting with water, but the carboxylic acid end is highly water-soluble; such molecules interact with each other to form soap-like films or bubbles. Cells contain many different fatty acids, which differ in the number of carbons (usually an even number). Some fatty acids are "saturated"; that is, the carbon-carbon bonds are all single bonds, and the carbons are bonded to the maximum number of hydrogen atoms ("saturated" with hydrogen; see Figure 4.7A). Unsaturated fatty acids, on the other hand, contain one or more double bonds between adjacent carbons in the hydrocarbon tail. Monounsaturated fatty acids have one double bond; polyunsaturated fatty acids have multiple double bonds. In a membrane, saturated fatty acids can pack together in an orderly, stable arrangement. Thus, saturated fatty acids (including animal fats such as lard) tend to be solids at room temperature. In unsaturated fatty acids, the double bond causes a kink in the chain that prevents tight packing (see Figure 4.7B), so unsaturated fatty acids, such as vegetable oils, tend to remain fluid at room temperatures.

Lipids called "triglycerides" are composed of three fatty acids joined to glycerol by condensation reactions. In a **condensation reaction**, two molecules join to form a covalent bond while losing

Figure 4.7   Fatty Acids

The carboxyl group on the end of a fatty acid can act as an acid by releasing a proton (H⁺).

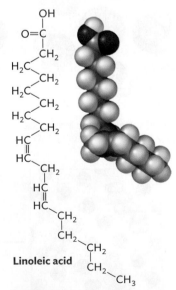

Palmitic acid

Linoleic acid

A. Palmitic acid is a saturated fatty acid; only single bonds connect the carbon atoms.

B. Linoleic acid is an unsaturated fatty acid. Note that the double bonds (highlighted in yellow) form kinks in the hydrocarbon tail of an unsaturated fatty acid.

Figure 4.8 Formation of a Triglyceride

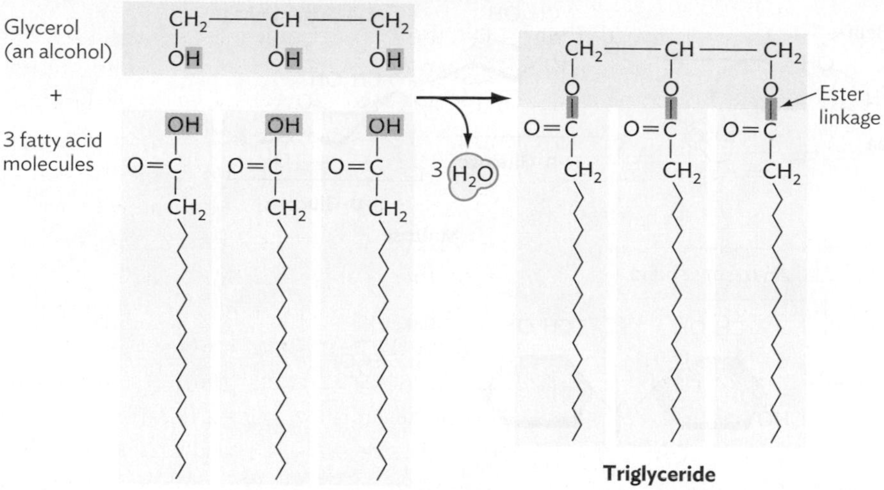

Condensation reactions form covalent ester linkages between the carboxyl groups of three fatty acids and the three hydroxyl groups of glycerol, with loss of three molecules of H_2O.

Figure 4.9 Structures of Two Hexoses: Glucose and Fructose

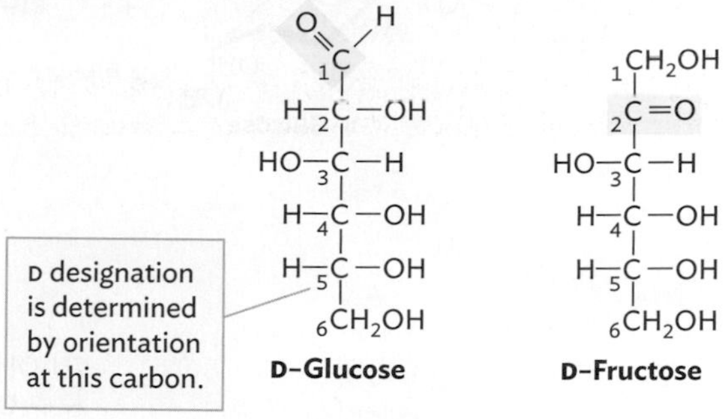

The numbering of the carbons starts at the carbon of the terminal carbonyl group (highlighted) or at the terminal carbon closest to the carbonyl (as in fructose). The D designation is determined from the orientation of the hydroxyl group on a chiral carbon (arrow) farthest from the carbonyl group (highlighted).

atoms that form H_2O. The three hydroxyl groups on glycerol can each undergo condensation with the carboxylic acid portion of fatty acids to form a triglyceride (Figure 4.8). Triglycerides are a compact energy source for cells; they are highly concentrated in fat tissue. Phospholipids, key components of membranes, contain fatty acids attached to two of the hydroxyl groups of glycerol and a phosphate attached to the third hydroxyl. Their structure determines how drug molecules get into cells. Phospholipid membranes are discussed further in Section 4.5.

Carbohydrates

Carbohydrates (sugars and polysaccharides) store energy, and they serve as part of cell surfaces. Sugar structures carry specific information; for example, the ganglioside is a sugar chain on the surface of intestinal epithelial cells that unfortunately binds cholera toxin (see Figure 4.1).

A **simple sugar**, or **monosaccharide**, has several linked carbons with multiple hydroxyl groups (R—OH) and typically one carbonyl group (R—C=O). Sugars differ from each other in the number of

carbons present, in the position of the carbonyl group, and in the orientation of the hydroxyl groups. In a sugar structure, the carbons are numbered starting with the carbon of the terminal carbonyl group or the terminal carbon closest to the carbonyl group. The six-carbon sugars (hexoses) are particularly important for energy storage and cell wall formation, whereas five-carbon sugars such as ribose form the backbone of nucleic acids (discussed in the next section). Hexose sugars include, for example, glucose and fructose (Figure 4.9). Hexoses all have the molecular formula $C_6H_{12}O_6$, but the atoms in different hexoses are linked differently. The difference between glucose and fructose lies in the position of the carbonyl, either at C-1 (glucose) or at C-2 (fructose).

Sugars with five or more carbons may also form a ring structure in which one of the hydroxyl groups reacts with the carbonyl group (Figure 4.10). Depending on how the bond between C-1 and C-2 was oriented during the formation of the ring, the cyclic form of glucose will be one of two types, alpha-glucose or beta-glucose. Alpha- and beta-linked sugars are found on cell surface proteins that bind to viruses. For example, deadly influenza viruses bind alpha-linked sugar proteins (glycoproteins) on respiratory cells.

Figure 4.10 Straight-chain and Cyclic Forms of Glucose

The straight-chain form of glucose can cyclize to form two different ring structures, alpha-glucose and beta-glucose. During cyclization, the carbonyl group (C=O) is rearranged to a carbon-hydroxyl (C—OH).

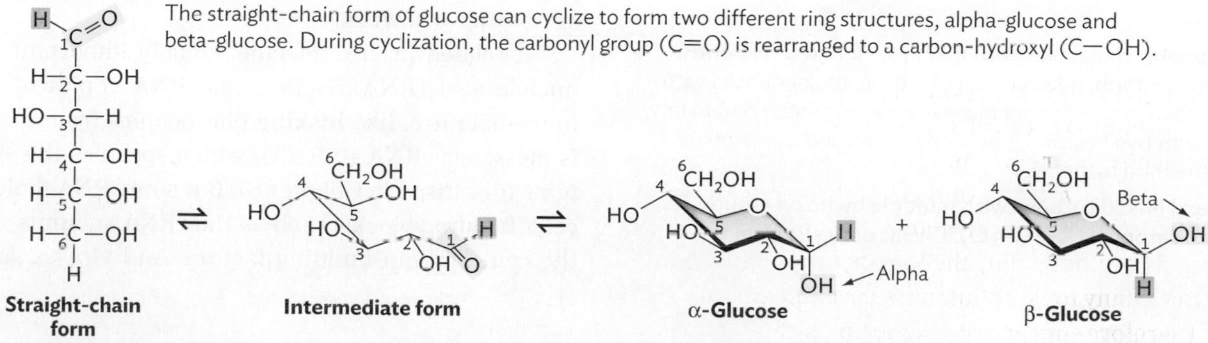

Figure 4.11   Formation of the Disaccharide Maltose

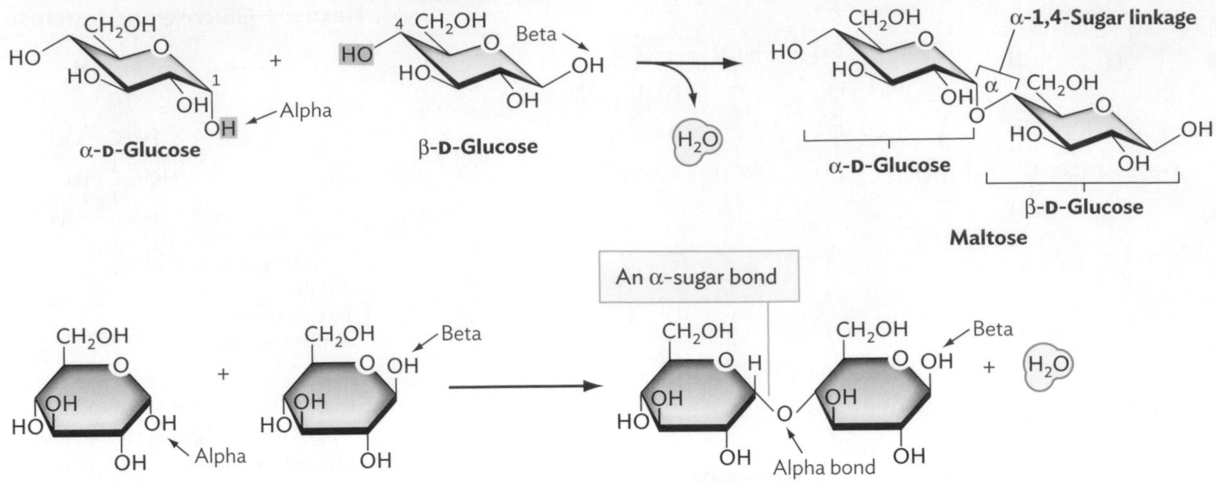

Maltose forms by a condensation reaction, in which two molecules of glucose join and H_2O is lost. Two molecules of glucose condense by linking the C-1 of one molecule of glucose via an oxygen atom to the C-4 of a second molecule of glucose. The hydroxyl groups of the second glucose can be either alpha or beta (the beta form is shown).

Two monosaccharides can link and form a **disaccharide** by a condensation reaction (**Figure 4.11**). In condensation, two molecules form a new covalent bond through loss of water (H_2O). The bond typically forms between the C-1 carbon of one monosaccharide and the C-4 carbon of another monosaccharide. Such a reaction requires an enzyme that recognizes the specific sugars (discussed shortly). The structure of the disaccharide varies, depending on which monosaccharides are involved and whether the C-1 carbon participating in the bond is in the alpha or beta form. Several sugars can condense to form short chains. Longer chains are called **polysaccharides**. Polysaccharides may be linear or branched. Sugars and polysaccharides are known as **carbohydrates**. Carbohydrates are an energy source for cells and also serve structural roles, as in cell walls.

SECTION SUMMARY

- **Organic compounds contain one or more carbon-carbon bonds.** Organic molecules make up much of a living organism.
- **Large biological organic molecules are called macromolecules.** Macromolecules contain functional groups that participate in various chemical reactions.
- **Lipids are hydrophobic molecules with mainly C–C and C–H bonds.** Triglycerides, a storage form of fat, consist of fatty acids condensed with glycerol. Phospholipids, a component of cellular membranes, contain fatty acids condensed with two hydroxyl groups of glycerol and a phosphate condensed with the third hydroxyl.
- **Sugars have several linked carbons with multiple hydroxyl groups (R—OH) and one carbonyl (R—C═O).** Chains of sugars form polysaccharides.

4.3
Nucleic Acids and Proteins

SECTION OBJECTIVES

- Describe the structure of DNA, and explain how DNA stores information in its sequence of base pairs.
- Describe the structure and diverse functions of RNA.
- Describe the fundamental structure of proteins.

How do living cells write information? Some macromolecules, particularly nucleic acids and proteins, contain complex information encoded in the pattern of their repeated units. The most famous kind of nucleic acid in cells is deoxyribonucleic acid (DNA). DNA is a cell's hereditary material, forming the chromosomes and containing the entire **genome** of a cell. The information content of DNA includes genes that encode proteins. In some cases, the DNA of a human genome can tell a physician which diseases we may get. And bacterial genomic DNA determines which antibiotics will work.

A related macromolecule, equally important for life, is ribonucleic acid (RNA). Cells make RNA copies of DNA genes for immediate use, like making photocopies from a book. Most RNA is messenger RNA (mRNA), which specifies the synthesis of proteins (discussed in Chapter 8). But some RNA molecules serve as a cell's building blocks, such as the rRNA subunits of the ribosome, the cell's protein-building factory. And viruses such as influenza

Figure 4.12 Nucleotide Structure

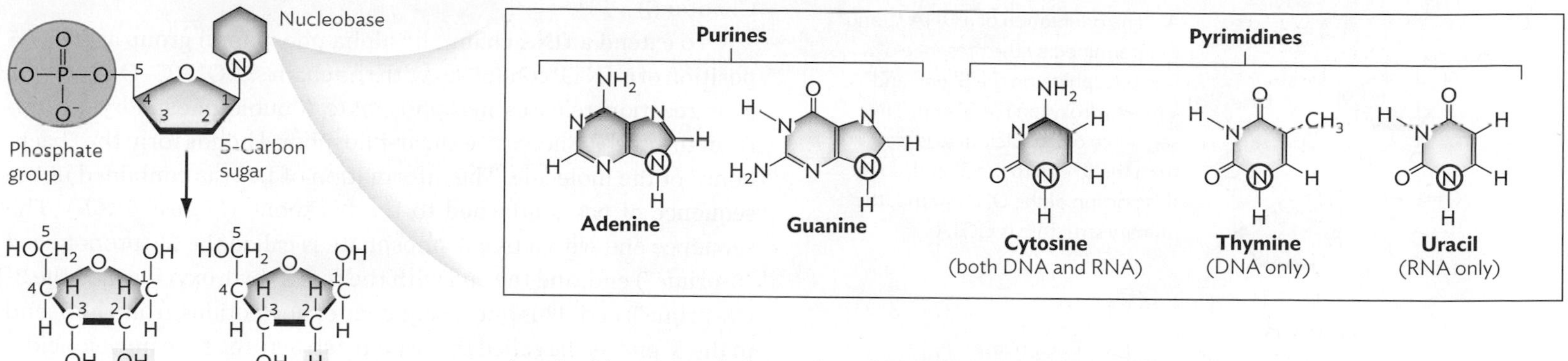

Nucleobase

Phosphate group

5-Carbon sugar

Ribose

Deoxyribose

Purines

Adenine

Guanine

Pyrimidines

Cytosine
(both DNA and RNA)

Thymine
(DNA only)

Uracil
(RNA only)

A nucleotide consists of a five-carbon sugar (ribose or deoxyribose), a phosphate group, and a nucleobase. The circled nitrogen (N) of the nucleobase bonds to the sugar.

and HIV actually have genomes of RNA. RNA viral genomes mutate faster than those of DNA, and these new strains can cause pandemic outbreaks (discussed in Chapter 12).

Nucleic Acids

Nucleic acids are long chains composed of nucleotides. A **nucleotide** has three components—a five-carbon sugar (pentose), a **nucleobase** (nitrogenous base, often just called a base), and a phosphoryl group (phosphate) (Figure 4.12). For RNA, the sugar is ribose, and for DNA, the sugar is 2-deoxyribose—that is, ribose with oxygen removed from the 2′ (two prime) carbon. The prime indicates that the numbering refers to the sugar portion of the nucleotide. In both DNA and RNA, the 1′ carbon of the sugar forms a covalent bond to a nitrogen of the base. Bases come in two structural classes: purines (adenine, A, and guanine, G) and pyrimidines (cytosine, C; thymine, T, found only in DNA; and uracil, U, found only in RNA). The 5′ (five prime) carbon usually bonds to a phosphoryl group, whereas the 3′ (three prime) carbon usually has a free hydroxyl group (OH).

A nucleotide has a phosphoryl group with a covalent ester bond (C–O–P–O–C) to the 5′ carbon of the sugar. Nucleotides can have either one, two, or three phosphoryl groups and are called monophosphate (MP), diphosphate (DP), or triphosphate (TP), respectively. Adenine nucleotides, for instance, are abbreviated AMP, ADP, or ATP (Figure 4.13). In addition to being the precursors for nucleic acid synthesis, the ribose nucleotides have cellular functions of their own. For example, ATP is an energy carrier in the cell (see Figure 4.13). Energy is released by the reaction of removing one or two phosphoryl groups from ATP. Other forms of nucleotides such as GTP and cyclic adenosine monophosphate (cAMP) act as signals. For example, cAMP activates a transporter that exports chloride ion from the intestine, and cholera toxin (discussed earlier) acts by inappropriately turning on an enzyme that forms cAMP.

Adenosine, shown in Figure 4.13, is a ribonucleotide capable of forming a unit of RNA. The formation of DNA requires deoxyribonucleotides, in which the 2′ carbon bonds to a hydrogen atom, not to OH. Deoxyribonucleotides are designated by a lowercase "d" in front of the nucleotide name. For example, cytosine diphosphate is CDP, and deoxycytosine triphosphate is dCTP. Nucleotide triphosphate monomers (collectively designated NTPs or dNTPs) form nucleic acids through reactions catalyzed by nucleic acid polymerase enzymes. The enzyme that assembles dNTPs into DNA is now used in the famous polymerase chain reaction (PCR)

Figure 4.13 ATP, an Important Energy-Carrying Molecule

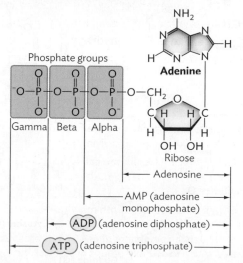

NH₂

Adenine

Phosphate groups

Gamma | Beta | Alpha

Ribose

Adenosine

AMP (adenosine monophosphate)

ADP (adenosine diphosphate)

ATP (adenosine triphosphate)

ATP is a nucleotide containing the nucleobase adenine. A nucleotide may be a monophosphate, diphosphate, or triphosphate. The phosphate attached to the sugar is designated alpha, the beta phosphate is in the middle, and the gamma phosphate is distal to sugar.

Figure 4.14   DNA Structure

A. The orientation of a DNA strand is determined by the presence of a free phosphate on the 5′ end and a free hydroxyl on the 3′ end. The sequence of a nucleic acid is read from the 5′ end to the 3′ end, so for this portion of the DNA strand, the primary structure is GCTA.

B. Hydrogen bonding (represented by red dots) between complementary base pairs. Two hydrogen bonds form between A and T, three hydrogen bonds between G and C. The DNA strands are antiparallel (point in opposite directions, 5′ to 3′).

to amplify DNA for forensic and paternity analysis (discussed in Chapter 8).

To extend a DNA chain, the alpha phosphoryl group at the C-5 position of a dNTP condenses with the chain's 3′ OH (C–O–P–O–C). This reaction releases pyrophosphate (double phosphoryl group) from the dNTP. Successive sugar-phosphoryl bonds form the "backbone" of the molecule. The information of DNA is contained in the sequence of bases attached to the backbone (Figure 4.14A). The sequence end with a free 5′ phosphate is called the 5′ (pronounced "5-prime") end, and the end with the free 3′ hydroxyl is called the 3′ ("3-prime") end. This linear sequence of nucleotides, from the 5′ end to the 3′ end, is the called the primary structure of the nucleic acid.

The DNA Double Helix

The chromosomes of all cells consist of DNA that is double-stranded, or duplex (Figure 4.14B). The two strands of DNA are held together by hydrogen bonds between a purine and a pyrimidine on opposite strands. Two hydrogen bonds form between each adenine and thymine, and three hydrogen bonds form between each cytosine and guanine. The strands twist around each other to form a helix, the famous double helix of DNA (Figure 4.15A). The double-helical form was first demonstrated in 1953 by Rosalind Franklin and Maurice Wilkins, whereas the nucleotide base pairing was proposed and reported by James Watson and Frances Crick. The sugar-phosphate backbone of DNA is on the outside of the helix (Figure 4.15B), its negative charges interacting with the partial positive charges on water molecules. Within the helix, the hydrophobic bases "avoid" water by stacking together like steps of a ladder.

Although the base pairs are mostly buried within the helix, portions of them remain accessible to proteins that regulate transcription (discussed in Chapter 8). The accessible portions take the form of two grooves in the helix—a wide major groove and a narrower

Figure 4.15   The DNA Double Helix

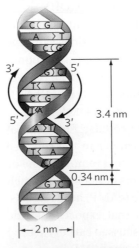

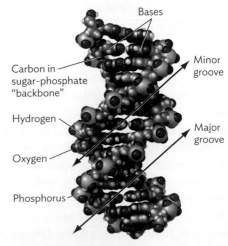

A. The DNA double helix, showing the purine-pyrimidine base pairs.

B. A space-filling model of DNA. Note that the base pairs stack on top of each other like the rungs of a ladder.

minor groove (see Figure 4.15B). Regulatory proteins play many roles in pathogenesis. For example, *Vibrio cholerae* proteins that detect environmental signals such as temperature and pH can tell when the pathogen has entered a host; these proteins then bind the pathogen's DNA to start the disease process.

RNA Takes Diverse Forms

RNA can form base pairs like those of DNA, except that in RNA, the nucleotide uracil replaces thymine. Some RNA molecules form a double helix like DNA; an example is the RNA of rotaviruses, the most common cause of diarrhea and "stomach flu" in young children. But in cells, RNA molecules are single-stranded. Single-stranded RNA can "double back" and form intramolecular base pairs, such as those in a tRNA molecule (Figure 4.16). These base-paired forms enable RNA to assume a globular protein-like form with enzymatic functions, such as protein synthesis within a ribosome. Base pairing between RNA and DNA is used to make mRNA from a DNA template.

Proteins

A cell makes thousands of different proteins and may contain more than 2 million protein molecules. Proteins perform many functions, such as catalyzing biochemical reactions, serving as receptors and transporters, providing structure, and aiding movement. Proteins fold into three-dimensional structures that may fit together in a

complex. The cholera toxin is an example of a protein complex composed of two types of protein chain (one A chain and five copies of B chain).

AMINO ACIDS All proteins, or peptide chains, are composed of the same building blocks: amino acids. All 20 of the amino acids used in proteins have the same general structure. Each **amino acid** contains a central carbon atom (the alpha carbon) covalently bound to four different parts—a hydrogen atom, an amino group, a carboxyl group, and a side chain (R, residue) that is unique for each amino acid (Figure 4.17A). The alpha carbon is an example of a **chiral carbon**, a carbon with four different groups attached to it. Chiral carbons exist in two different forms called **isomers** that are mirror images of each other, like your left and right hands, and the two forms cannot be superimposed (Figure 4.17B). Each amino acid isomer is designated L or D, based on the configuration at the alpha carbon. Only L-amino acids are used by ribosomes to make proteins. But D-amino acids, with the opposite configuration at the alpha carbon, are essential to build bacterial cell walls and peptide antibiotics. The amino acid D-alanine is a component of vancomycin, used to treat flesh-eating staphylococcus infections.

The side chains of amino acids can be grouped according to their hydrophobicity, charge, and/or presence of specific functional groups. Figure 4.18 shows the 20 amino acids used to make protein, along with their three-letter abbreviations and single-letter codes. The amino acid side chains vary greatly in their chemistry. They may be acidic (such as aspartic acid, whose carboxyl group deprotonates at neutral pH) or basic (such as lysine, whose amino group, $R-NH_2$, becomes protonated to $R-NH_3^+$). Other R groups are polar (serine) or hydrophobic (phenylalanine). Proline has a ring

Figure 4.16   tRNA Secondary Structure

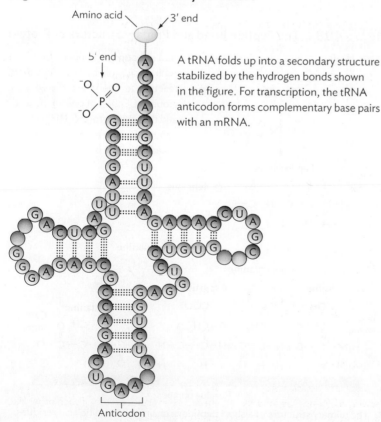

A tRNA folds up into a secondary structure stabilized by the hydrogen bonds shown in the figure. For transcription, the tRNA anticodon forms complementary base pairs with an mRNA.

Figure 4.17   Amino Acid Structure and Chiral Carbons

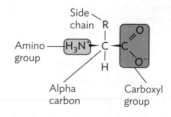

A. All amino acids contain a central carbon (the alpha carbon) bonded to an amino group, a carboxyl group, a hydrogen, and a variable group or side chain, designated R. At cellular pH, the amino and carboxyl groups are ionized. Because the alpha carbon is bound to four different molecules, it is a chiral carbon.

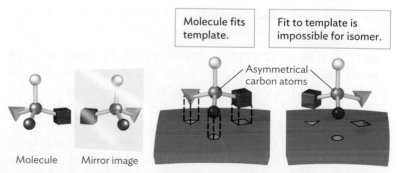

B. Chiral molecules exist in two forms that are mirror images of each other. The two forms cannot be superimposed, and only the correct mirror form can interact with a template (for example, an enzyme).

Figure 4.18   The 20 Amino Acids for Protein Synthesis

The categories of amino acids are based on their side chains, highlighted in yellow.

Charged amino acids

Basic alkaline amino acids

Lysine
(Lys or K)

Arginine
(Arg or R)

Histidine
(His or H)

Acidic amino acids

Aspartate
(Asp or D)

Glutamate
(Glu or E)

Polar amino acids

Asparagine
(Asn or N)

Glutamine
(Gln or Q)

Serine
(Ser or S)

Threonine
(Thr or T)

Hydrophobic amino acids

Alanine
(Ala or A)

Valine
(Val or V)

Isoleucine
(Ile or I)

Leucine
(Leu or L)

Methionine
(Met or M)

Phenylalanine
(Phe or F)

Tyrosine
(Tyr or Y)

Tryptophan
(Trp or W)

Special amino acids

Cysteine
(Cys or C)

Glycine
(Gly or G)

Proline
(Pro or P)

structure that interrupts the regular geometry of the polypeptide chain. The thiol (–SH) side chain of cysteine can form both inter-molecular and intramolecular disulfide (–S–S–) bonds. For example, in antibodies, disulfide bonds hold together the different combinations of heavy chains and light chains that generate millions of different antibodies (discussed in Chapter 16).

PROTEIN ORGANIZATION   The first level of organization, the **primary structure**, is simply the linear sequence of amino acids, specified by the DNA code. During protein synthesis, amino acids are connected by a condensation reaction to form a covalent peptide bond. The peptide bond forms between the carboxyl group of one amino acid and the amino group of a second amino acid (**Figure 4.19A**). The portion of the amino acid incorporated into the peptide after condensation is called an amino acid residue. The peptide chain has an amino terminus (N terminus) and a carboxyl terminus (C terminus); numbering of the amino acid residues starts at the

Figure 4.19   The Peptide Bond and Primary Structure of Protein

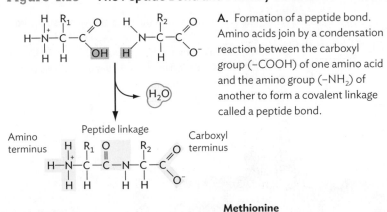

A. Formation of a peptide bond. Amino acids join by a condensation reaction between the carboxyl group (–COOH) of one amino acid and the amino group (–NH$_2$) of another to form a covalent linkage called a peptide bond.

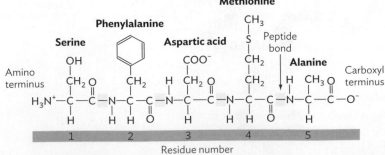

B. The primary structure of a short peptide, a chain of five amino acid residues.

Figure 4.20 **Secondary Structure of Proteins**
Both the alpha helix and beta sheet are held together by hydrogen bonds.

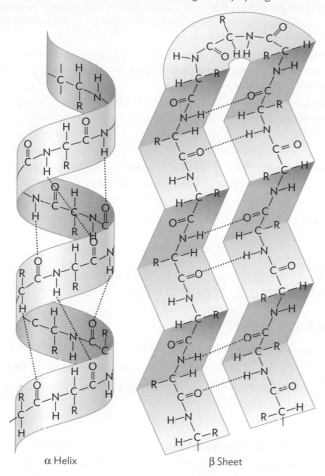

α Helix   β Sheet

Figure 4.21 **Prion Conformation Change**
A normal prion consists of mainly alpha helices (red coils). Under pathological conditions, the prion transitions to the abnormal, disease-causing conformation, consisting of beta sheets (cyan arrows).

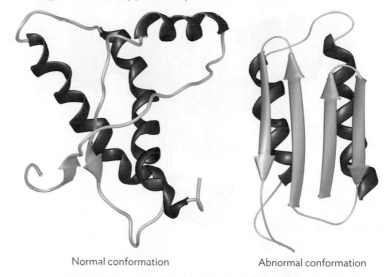

Normal conformation   Abnormal conformation

along the chain. The beta sheets form as a result of extensive hydrogen bonding between regions of the protein lying next to each other.

The **tertiary structure** of a protein is its unique three-dimensional shape (**Figure 4.22**). At the tertiary level of organization, regions distant in the primary structure may be brought close together. The nature and order of the side chains (the primary structure) determine how a protein folds into its final tertiary structure. The large number of amino acid side chains is responsible for the diversity of protein structures and hence functions. As proteins emerge from the ribosome, they are bound by chaperone proteins that help them fold into their functional shape, known as the **native conformation**. In soluble proteins, amino acids with hydrophobic

amino terminus (**Figure 4.19B**). A chain of five residues is shown in Figure 4.19B, but actual proteins may include hundreds of residues.

The **secondary structure** of a protein is the regular pattern of amino acid residues that repeat over short regions of the polypeptide chain. **Figure 4.20** depicts two common secondary structures, the **alpha helix** and the **beta sheet**. These secondary structures are extremely important for how antibodies recognize antigens, and for pathogens causing disease. An example is the prion, an infectious protein that causes a brain disease known as "mad cow disease" (bovine spongiform encephalopathy). The prion acts by causing an alpha-to-beta transition in the secondary structure of proteins within infected cells (**Figure 4.21**). The alpha helices are coiled structures formed by a series of hydrogen bonds between the oxygen in a carboxyl group and a hydrogen in an amino group four amino acids farther

Figure 4.22 **Tertiary Structure of a Protein**

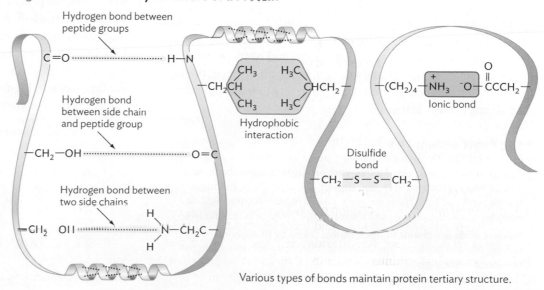

Hydrogen bond between peptide groups

Hydrogen bond between side chain and peptide group

Hydrogen bond between two side chains

Hydrophobic interaction

Ionic bond

Disulfide bond

Various types of bonds maintain protein tertiary structure.

Figure 4.23   Structure of an Antibody Complex: Two Light Chains and Two Heavy Chains

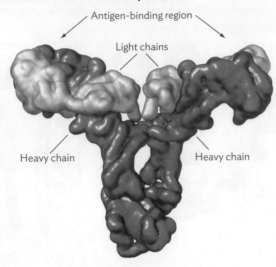

side chains tend to cluster inside the protein to minimize reactions with water, whereas polar amino acids are exposed on the surface. Tertiary structure is stabilized by hydrogen bonding between side chains and between side chains and the main chain. Ionic interactions between acidic (negatively charged) and basic (positively charged) side chains also contribute to tertiary structure. The one covalent interaction found stabilizing tertiary structure is disulfide bonds between two cysteine residues.

Some proteins form stable, functional complexes with other proteins, such as in the cholera toxin AB5 (see Figure 4.1). A multisubunit protein complex shows the fourth level of protein organization, quaternary structure. The forces that hold interacting polypeptide chains together are the same noncovalent interactions and disulfide bonds that maintain tertiary structure within a folded chain. An example of quaternary structure is shown in the space-filling model of an antibody, composed of two heavy chains and two light chains (Figure 4.23).

Proteins are dynamic structures with parts that move and bend. The weak bonds, such as hydrogen bonds and ionic bonds, contribute to protein structure but can come apart and re-form. Nevertheless, because these weak bonds do not break all at once, the large number of such bonds present act collectively to maintain surprisingly strong protein integrity.

SECTION SUMMARY

- **A nucleotide is made up of a five-carbon sugar, a nucleobase, and a phosphoryl group.** In RNA, the sugar is ribose; in DNA, the sugar is 2-deoxyribose (one fewer oxygen). The nitrogenous bases include adenine, guanine, cytosine, and thymine (for DNA) or uracil (for RNA).

- **Nucleic acids are chains of nucleotides** linked by phosphoryl groups. Nucleic acids such as DNA and RNA carry information in the sequence of their bases.

- **DNA in cells forms a double-stranded helix.** Adenine pairs with thymine, and cytosine pairs with guanine.

- **RNA is single-stranded in cells** but can form intramolecular base pairs. RNA molecules include mRNA (messenger RNA), tRNA (transfer RNA) and rRNA (ribosomal RNA).

- **Proteins are composed of amino acids.** In a protein, the 20 types of amino acids are condensed through peptide bonds, generating a unique order of amino acid residues.

- **The peptide chain (primary structure) is folded into secondary and tertiary structure.** A complex of multiple subunits has quaternary structure.

Thought Question 4.2   What kinds of amino acid residues would you find on the exterior surface of a membrane-embedded protein? What kinds of amino acid residues would you find in the middle of the protein? (Hint: The middle layer of the membrane is hydrophobic.)

4.4
Biochemical Reactions

SECTION OBJECTIVES
- Explain how energy and entropy determine the direction of reactions.
- Describe the oxidation-reduction changes that occur during reactions.
- Explain how enzymes catalyze reactions and how various factors influence the rate of reaction.

How do cells change shape and grow? A cell grows by breaking down molecules (catabolism) and building more complex ones (anabolism). Cellular catabolism and anabolism are described in Chapter 7.

Each change of a molecule occurs through chemical reactions. Chemical reactions are always catalyzed by the cell's enzymes, proteins that catalyze reactions; remember that, even when the enzyme is not shown. **Catalysis** means to greatly increase the rate of a reaction; the practical effect is to turn on a reaction that otherwise would be turned off. A cell's reactions all require enzymes to ensure that reactions occur only as needed by the cell.

An example of a biochemical reaction catalyzed by an enzyme is the formation of cAMP, the reaction that leads to excess Cl⁻ efflux in cholera (seen in Figure 4.1). cAMP forms by the conversion of ATP, catalyzed by adenylate cyclase:

The enzyme removes pyrophosphate, and the remaining adenosine monophosphate (AMP) is then cyclized at the 3′ OH, forming cAMP. cAMP mediates many important cell signals, such as that of the hormone adrenaline.

What determines whether a reaction goes forward? In the preceding reaction, the removed pyrophosphate is a high-energy compound that is rapidly hydrolyzed—that is, cleaved with input of H_2O. The splitting of ATP while consuming H_2O is called **hydrolysis**. In the hydrolysis shown, removal of pyrophosphate releases energy and thus drives the reaction forward. Overall, the reaction is said to have a favorable **free energy change (ΔG)**.

Energy Change Powers Life

Where do living cells get the fuel to drive their reactions? Like the engine of a car, reactions in organisms, such as the synthesis of proteins from amino acids by condensation, must spend energy. An endless cycle of reactions connects those that yield energy for cells (such as those that oxidize sugars or lipids) with reactions that spend energy (such as synthesis of DNA and proteins). When the energy chain fails, as when your heart stops pumping blood, the rest of your reactions fail, and your body undergoes rigor mortis—in other words, death.

All systems (including living systems, such as cells) must obey the laws of energy, known as the laws of thermodynamics (**Figure 4.24**). **Energy** is the ability to do work. The **first law of thermodynamics** states that *matter and energy are neither created nor destroyed*. In a closed system (a system that is not exchanging energy with its environment), the total amount of energy remains constant. Although the amount of energy remains constant, energy can be converted from one form to another. For example, the energy released by oxidizing food can be captured by enzymes building proteins and DNA for a cell to grow.

The **second law of thermodynamics** states that in energy transformations, some energy becomes unavailable to do work and is lost as disorder, or **entropy**. In other words, entropy tends to increase. For example, when you dilute a concentrated drug to the correct dosage for therapy, the diluted solution has increased entropy. Entropy is increased because the drug will never spontaneously become concentrated. The only way to increase the concentration again would be to add energy—for instance, by adding heat to evaporate the solvent.

Reactions that increase entropy may go forward. For example, the breakdown of food molecules into a larger number of smaller molecules increases entropy and yields energy. A living cell can couple energy-yielding processes to energy-spending processes such as building proteins. Thus, although life itself is a highly ordered process, it generates even greater disorder (entropy) outside the living organism.

For a reaction to go forward, there needs to be a change in free energy (G), designated ΔG. Reactions are favored when the free energy of the products is less than the free energy of the reactants; that is, when ΔG is negative. Reactions with a negative ΔG are termed "spontaneous," or **exergonic**, because they are energetically favorable and can occur without an input of energy. Reactions with a positive ΔG are **endergonic**. They are not energetically favorable, and energy needs to be added for them to proceed. Reactions with a ΔG of zero are at equilibrium, and neither products nor reactants are favored. For example, the following reaction, the oxidation, or "burning," of methane (CH_4) to release carbon dioxide and water, is spontaneous and yields free energy:

$$CH_4 + 2O_2 \rightarrow CO_2 + 2H_2O + \text{energy}$$

The bonds in methane and oxygen have more energy than do the bonds in CO_2 and H_2O. Because the products CO_2 and H_2O have less energy than the reactants, the ΔG is negative and the reaction goes forward. As it goes forward, the reaction releases heat.

A reaction does not need to release heat to go forward. Consider an ice cube placed at room temperature. The ice cube will spontaneously melt even though this reaction absorbs heat from the environment. Ice melting is spontaneous because of the increase in entropy (disorder) in going from a solid to a liquid. You can see a similar effect when a "cold pack" is activated to decrease swelling of an injured body part. Within the pack, a chemical reaction takes place in which entropy increases while heat is absorbed and the pack becomes cold. Thus, a large positive change in entropy may result in a net negative change (yield) of free energy, even though some heat is absorbed.

Free Energy Change and Chemical Equilibrium

The ΔG (net energy yield) of a reaction depends on the intrinsic energy and entropy changes of the molecules (just discussed) as well as the concentrations of reactants and products. Concentration refers to how much of something is present in a given volume.

Figure 4.24   Energy Can Be Changed into Different Forms but Can Never Be Created or Destroyed

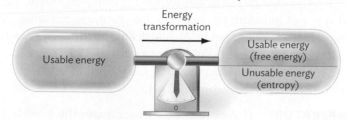

Energy transformation

Usable energy

Usable energy (free energy)

Unusable energy (entropy)

The first law states that in an energy transformation, the total amount of energy remains constant. Both sides have the same amount of energy (the balance reads zero). The second law states that in an energy transformation, the amount of usable energy, or free energy, decreases and the amount of unusable energy, or entropy, increases.

As reactant concentration increases, the reaction is more likely to proceed and yield energy. For example, a high concentration of sugar in grapes enables yeast to ferment the sugar, converting it to alcohol. The alcohol-producing reaction yields energy for the yeast to grow.

Concentration is reported in units of **molarity**. Molarity is defined as the number of moles of substance per liter of solution (usually water). One **mole** is Avogadro's number (6.022×10^{23}) of molecules. The weight of a mole of a substance (in grams) equals the average molecular mass in atomic mass units. For example, NaCl has a mass of 58.5 atomic mass units, so a mole of NaCl weighs 58.5 grams. To make a 1-molar solution of NaCl, we take 58.5 grams of NaCl and add water up to a liter.

A free energy change can be related to an equilibrium constant, K_{eq}, that indicates whether products or reactants will be favored at equilibrium. Reactions between biomolecules can be written as chemical equations. For example, in a given reaction, reactants A and B form products C and D. The value of K_{eq} at which the reaction is at equilibrium is given by

$$K_{eq} = [C][D]/[A][B]$$

The value of K_{eq} tells us how the concentrations of reactants and products affect the ability of a reaction to go forward. A K_{eq} of 1 means that neither products nor reactants are favored. A K_{eq} greater than 1 indicates that products are favored at equilibrium and corresponds to a negative free energy change that yields energy. A K_{eq} less than 1 means that reactants predominate and corresponds to a positive free energy change in which energy must be spent for the reaction to go forward.

A reaction can be made spontaneous and driven forward by keeping the concentration of reactants high or the concentration of products low. The cell uses this strategy in metabolism, where the product of one reaction is constantly removed by a subsequent reaction (discussed in Chapter 7). For example, the synthesis of cAMP is driven forward by the rapid hydrolysis of the product pyrophosphate to inorganic phosphate; thus, the concentration of the immediate product pyrophosphate is kept low, and cAMP formation is driven forward.

Activation Energy and the Rate of Reaction

Although a reaction with a large negative free energy change is spontaneous, it may be slow. The size of the energy change says nothing about the rate of a reaction. An everyday example of a spontaneous but slow reaction is the rusting of metal. A spontaneous reaction will be slow if it must pass through a high-energy transition state on the way to forming products. The **activation energy** (E_a) is the energy needed to reach this transition state (**Figure 4.25**). For reactions with low activation energies, random collisions between reactants may provide enough energy to boost them over the E_a. For reactions with high activation energies, collisions between molecules may rarely provide enough energy for them to reach the transition state.

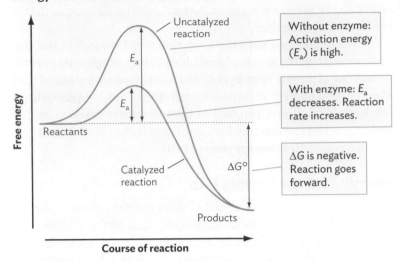

Figure 4.25   Enzymes Decrease Activation Energy and Increase Reaction Rate

The negative value of ΔG makes the reaction spontaneous, but the high activation energy makes the reaction extremely slow. The enzyme makes the reaction go fast enough to be useful.

Even though such a reaction goes forward, it is so slow that on the timescale of a cell, the reaction does not occur.

Enzymes

A living cell can increase the rate of a specific reaction by use of a specific enzyme. For example, the DNA polymerase enzyme of *E. coli* bacteria attaches nucleotides at the rate of 1,000 nucleotides per second. Most enzymes consist of protein, although some are RNA, such as the RNA subunits of the ribosome. An **enzyme** is a **catalyst**; that is, the enzyme participates in the reaction while increasing the reaction rate but is not itself consumed.

An enzyme catalyzes the conversion of substrates (reactants) to products (see Figure 4.25). The substrates fit into the "active site" of the enzyme, where functional groups of the enzyme stabilize the transition state. By stabilizing the transition state, holding the molecules in a specifically shaped binding pocket, the enzyme lowers the activation energy. The activation energy is the amount of energy needed to put in before the reaction goes forward, releasing more energy than the amount put in. Although enzymes increase reaction rates, they do not change the net ΔG of a reaction (see Figure 4.25).

The rate of reactions catalyzed by enzymes is affected by many factors, including temperature and pH.

TEMPERATURE   Higher temperature increases the rate of collision of reactants and thus the rate of reaction. Most bacteria grow faster as temperature increases; this is why warm milk spoils faster than milk in the refrigerator. At too high a temperature, however, the bacterium's enzymes lose their structure (become **denatured**), and growth slows or stops. Thus, the human body generates a fever to fight pathogenic bacteria.

pH (ACID OR BASE) Another factor affecting the rate of reaction is the hydrogen ion concentration, represented by pH, the negative logarithm of H^+ concentration. Most enzymes work only within a narrow range of pH. The pH of human blood is regulated very closely by our rate of breathing. When we hold our breath, carbon dioxide builds up in the blood, where it forms acid (decreasing pH). Decreased pH, or acidosis, may impair the brain and lead to coma. When we breathe too fast, or hyperventilate, the carbon dioxide is lost too fast; acid is lost, and blood pH rises (alkalosis). Alkalosis may cause fainting and seizures.

Hydrogen ions (H^+) are also referred to as protons because they have lost their single electron and consist of only a proton. Hydrogen ion concentration is reported using a pH (power of hydrogen) scale, where pH is the negative logarithm of the hydrogen ion concentration: $pH = -\log [H^+]$. The designation "log" refers to logarithm, or power of 10. In pure water, $[H^+]$ is 1×10^{-7} molar (M), a pH of 7, or neutral pH (Figure 4.26). A pH value below pH 7 (greater H^+ concentration) is acidic; an example of an acidic substance is lemon juice (pH 2). A pH value above pH 7 (lower H^+ concentration) is alkaline; an example of an alkaline substance is baking soda (pH 9).

Because the pH scale is logarithmic, every pH unit corresponds to a tenfold change in hydrogen ion concentration. Thus, a solution with a pH of 6 has a hydrogen ion concentration of 1×10^{-6} M, ten times the hydrogen ion concentration at pH 7 (1×10^{-7} M). Remember that 10^{-6} is a larger number than 10^{-7} whereas 10^{+7} is a higher number than 10^{+6}.

In living cells, where do hydrogen ions come from? The largest source of hydrogen ions is water (H_2O), which always exists in equilibrium with hydrogen ions (H^+) and hydroxide ions (OH^-). In water solution, the hydrogen ion concentration $[H^+]$ multiplied by the hydroxide ion concentration $[OH^-]$ always equals 1×10^{-14}. Hence, the concentrations of H^+ and OH^- are reciprocally related (if one goes up, the other goes down). If the pH is less than 7, a solution is acidic (more H^+, less OH^-). If the pH is greater than 7, the solution is basic, or alkaline (less H^+, more OH^-).

Acids (for example, carboxyl groups) donate protons, and bases (for example, amino groups) accept protons (Figure 4.27). At intracellular pH (near 7), the carboxyl group and the amino group of each amino acid are both ionized (charged), the carboxyl group carrying a negative charge and the amino group a positive charge (see Figure 4.18). The ionization of these groups can change if the pH changes. At lower pH (more acidic solution), protons move back onto the ionized carboxylic acid. For example, the side chain of an acidic amino acid such as glutamate (see Figure 4.18) is ionized (with a negative charge) at normal cell pH but may regain a proton and become glutamic acid at a lower pH. At higher pH values (lower proton concentrations), protonated amino groups (with a positive charge) lose their extra protons to become uncharged. Changes in the ionization state of carboxyl or amino groups on the side chains of amino acid residues can disrupt ionic bonds between these groups and lead to protein denaturation. This effect of pH on protein tertiary structure is one reason why cells can tolerate only a narrow range of intracellular pH and why acidosis or alkalosis in our own tissues is harmful.

Oxidation and Reduction

An important class of reactions catalyzed by enzymes are redox reactions (reactions involving reduction and oxidation). Redox reactions involve the transfer of electrons from one molecule to another or from one atom to another. Redox reactions yield the

Figure 4.26 pH values of Common Substances

Concentration of hydrogen ions compared with distilled water		Example of solutions at this pH
10,000,000	pH = 0	Battery acid (strong), hydrofluoric acid
1,000,000	pH = 1	Hydrochloric acid secreted by stomach lining
100,000	pH = 2	Lemon juice, gastric acid, vinegar
10,000	pH = 3	Grapefruit, orange juice, soda
1,000	pH = 4	Acid rain, tomato juice
100	pH = 5	Soft drinking water, black coffee
10	pH = 6	Urine, saliva
1	pH = 7	"Pure" water
1/10	pH = 8	Seawater
1/100	pH = 9	Baking soda solution
1/1,000	pH = 10	Great Salt Lake, milk of magnesia
1/10,000	pH = 11	Ammonia solution
1/100,000	pH = 12	Soapy water
1/1,000,000	pH = 13	Bleaches, oven cleaner
1/10,000,000	pH = 14	Liquid drain cleaner

The pH value equals the negative logarithm of the hydrogen ion concentration.

Figure 4.27 Organic Acids and Bases

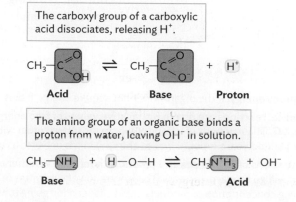

The carboxyl group of a carboxylic acid dissociates, releasing H^+.

Acid ⇌ **Base** + **Proton**

The amino group of an organic base binds a proton from water, leaving OH^- in solution.

$CH_3-NH_2 + H-O-H \rightleftharpoons CH_3N^+H_3 + OH^-$

Base **Acid**

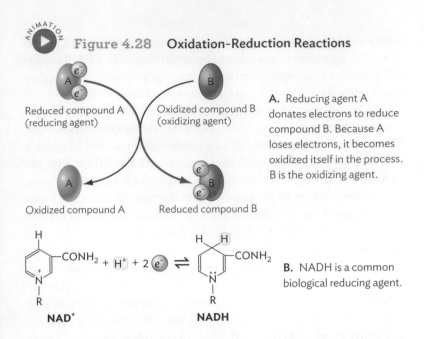

ANIMATION

Figure 4.28   **Oxidation-Reduction Reactions**

Reduced compound A
(reducing agent)

Oxidized compound B
(oxidizing agent)

A. Reducing agent A donates electrons to reduce compound B. Because A loses electrons, it becomes oxidized itself in the process. B is the oxidizing agent.

Oxidized compound A

Reduced compound B

$$-CONH_2 + H^+ + 2\,e^- \rightleftharpoons -CONH_2$$

NAD$^+$

NADH

B. NADH is a common biological reducing agent.

majority of the energy we obtain from oxidizing the food we eat (discussed in Chapter 7). Many cancer drugs and antiparasitic drugs target the cell's redox balance.

In a redox reaction, **reduction** refers to the gain of an electron, and **oxidation** refers to loss of an electron. The molecule that gains electrons becomes reduced, and the molecule that loses electrons becomes oxidized (**Figure 4.28A**). (A useful mnemonic device to remember this is "LEO the lion says GER." LEO: lose electrons, oxidation; GER: gain electrons, reduction.) An example from metabolism is the reduction of nicotinamide adenine dinucleotide (NAD$^+$) by a pair of hydrogen atoms removed from a food molecule such as glucose (**Figure 4.28B**). The reduction of NAD$^+$ (that is, addition of electrons) to form NADH enables NADH to reduce a protein of the electron transport chain, an important process by which cells acquire energy from food (discussed further in Chapter 7).

Oxidation and reduction reactions always occur together. That is, wherever one atom loses electrons, another atom must gain electrons. The oxygen gas that we breathe has a very strong tendency to gain electrons from the food molecules we eat. Oxygen usually gains electrons and becomes reduced in redox reactions. Molecules that become reduced are called **oxidizing agents** because they cause something else to become oxidized. In contrast, **reducing agents** can donate electrons, reduce other molecules, and become oxidized themselves in the process.

SECTION SUMMARY

- **Life requires energy.** If the organism cannot acquire energy, it dies.
- **Biochemical reactions are driven by negative values of free energy change, ΔG.** The reactions of a living organism must build order within the organism but decrease order (increase entropy) in the universe as a whole.
- **Concentrations of substrates and products determine whether a reaction will go forward.** Increasing concentration of substrates or decreasing concentration of products favors the reaction going forward.

- **Most biochemical reactions require catalysis by enzymes.** An enzyme catalyzes a specific reaction. Enzymes increase the rate of reactions by lowering the activation energy.
- **Reaction rates are affected by temperature and pH.**
- **Some reactions involve reduction and oxidation, the gain and loss of electrons.**

- -

Thought Question 4.3   Saline is a salt solution used for intravenous (IV) procedures; it contains approximately the same concentration of NaCl as in the blood. The concentration of NaCl in blood is approximately 0.154 mole per liter. How much salt would you weigh out to make 1 liter of saline? (*Hint*: See earlier definition of the molar weight of atoms.)

Thought Question 4.4   Your muscle cells contain ATP at concentrations of 1–10 mM (millimolar, thousandths of molar). The reaction of ATP hydrolysis (forming ADP and phosphate) has a large negative value of free energy change, ΔG. So why doesn't all the ATP hydrolyze?

- -

4.5
Membranes and Transport

SECTION OBJECTIVES

- Describe the structure of cellular membranes.
- Explain the concept of selective permeability of a membrane.
- Explain how membranes compartmentalize cells and control the movement of a cell's molecules.
- Describe the effect of osmosis on cells.

How do cells organize their molecules and their molecular reactions? How do they keep the substrates or reactants they need within the cell and exclude waste products or harmful toxins? All cells are enclosed by a **cell membrane** (also called the plasma membrane or cytoplasmic membrane) that creates an internal environment distinct from the external environment. The aqueous fluid inside the membrane is called **cytoplasm** (or cytosol). The cell membrane proteins regulate uptake of nutrient molecules, while pumping out toxins and waste products. Some membrane proteins receive signals from outside the cell while other membrane proteins conduct processes that yield energy. Some antibiotics act by disrupting bacterial membranes; an example is polymyxin B, found in the topical ointment Neosporin.

Membranes Consist of Lipids and Proteins

As shown in **Figure 4.29A**, membranes consist mainly of lipids and proteins. Many components have attached sugar chains; there are glycolipids (sugars joined to lipids) and glycoproteins (sugars joined to proteins). For example, we saw in the case how a membrane glycolipid ganglioside is used by cholera toxin to attach to the intestinal epithelial cell (see Figure 4.1). Membranes contain a greater number of lipid molecules than protein molecules, but the proteins are

Figure 4.29 The Cell Membrane

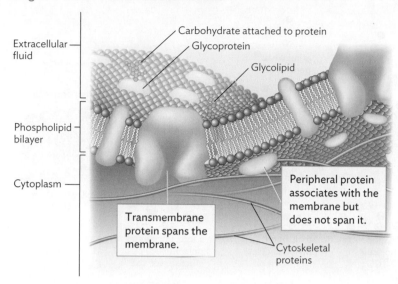

A. A cutaway view of the cell membrane.

B. Ribbon diagram from X-ray crystallographic data of a multidrug transporter protein. Red portions are alpha helices buried in the membrane, and yellow portions are loops within the watery cytoplasm or exterior.

larger and contribute about half the membrane mass. Proteins confer specific properties on membranes; for example, cell-surface proteins enable important communication with other cells. But viruses such as poliovirus and human immunodeficiency virus (HIV) have evolved to bind to these surface proteins to initiate infection.

In bacteria and eukaryotes, the predominant lipids in membranes are **phospholipids**. Phospholipids consist of a core of glycerol, to which two fatty acids and a modified phosphate group are attached via ester linkages (**Figure 4.30A**). Archaea, unlike eukaryotes and bacteria, have very different kinds of lipids that enable survival in extreme environments such as high temperature or acidity (discussed in Chapter 10). Phospholipids are **amphipathic**; the fatty acid hydrocarbon tails are hydrophobic, and the phosphate head group is hydrophilic. Amphipathic lipids are most stable in water when the hydrophilic portions interact with water and the hydrophobic portions cluster together away from water.

One way phospholipids can achieve stability is by forming a bilayer. Indeed, the cell membrane is a **phospholipid bilayer**, two

layers of phospholipids whose hydrocarbon fatty acid tails face the interior of the bilayer and whose charged phospholipid head groups face the aqueous cytoplasm and extracellular environment (**Figure 4.30B**). The phospholipid layer in contact with the cytoplasm is called the inner leaflet, and the layer in contact with the environment is called the outer leaflet. Cells contain several phospholipids that differ in how the phosphate head group is modified. The fatty acids composing phospholipids differ in the length of their fatty acid chains and the number of carbon-carbon double bonds (unsaturation).

Membranes also contain molecules that act as stiffening agents, providing mechanical stability to the phospholipid bilayer.

Figure 4.30 Phospholipids and the Lipid Bilayer

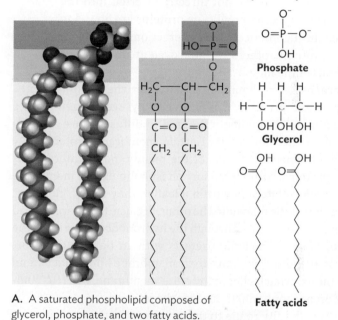

A. A saturated phospholipid composed of glycerol, phosphate, and two fatty acids.

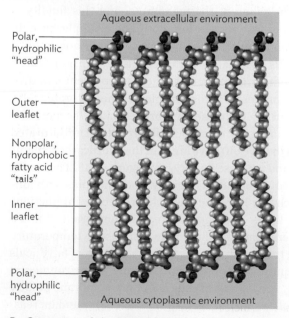

B. Orientation of phospholipids in the bilayer.

In eukaryotic membranes, the main stiffening agents are cholesterols; bacteria do not have cholesterol but have **hopanoids**, molecules of analogous structure that stabilize membranes.

Membrane proteins can be classified based on how they interact with membranes (Figure 4.29A). **Transmembrane proteins** (integral proteins) span the bilayer. Transmembrane proteins are amphipathic, meaning they have both hydrophobic and hydrophilic portions; the hydrophilic portions face the cytoplasm and the extracellular environment, and the hydrophobic components span the membrane. The transmembrane domains are often alpha helices containing 15–20 amino acid residues (Figure 4.29B). **Peripheral membrane proteins** are associated with the cell membrane but are not directly inserted into the bilayer. Some peripheral membrane proteins are bound to membranes through noncovalent interactions with transmembrane proteins, others by noncovalent interactions with the lipids of the membrane.

Membranes are dynamic structures; their components, such as lipids and proteins, can move rapidly, even rotating like a wheel. Membrane components are free to diffuse (move around) in the plane of the membrane. Certain membrane proteins may be restricted to specific regions of the membrane by interactions with other proteins. Although many phospholipids and membrane proteins can move laterally within a leaflet, they rarely "flip-flop" from one leaflet of the bilayer to the other, because the charged head groups would have to move through the hydrophobic interior of the membrane. Thus, pathogenic bacteria with an outer membrane, such as *Salmonella* species, have toxic membrane lipids that extend only from the external leaflet of the outer membrane, whereas completely different (nontoxic) lipids face the interior.

Membrane fluidity refers to the movement of membrane phospholipids within the plane of the membrane, and this fluidity is important for proper membrane function. For example, protein transport across the membrane is affected by membrane fluidity. Decreased fluidity is associated with decreased transport rates. Because a drop in temperature decreases fluidity, cold temperatures may slow transport processes across the membrane. Saturated fatty acids (such as in butter) decrease membrane fluidity because the linear hydrocarbon tails pack together well. In contrast, unsaturated fatty acids (such as in olive oil) have kinks in the hydrocarbon chains that limit packing and increase fluidity; see Figure 4.13B. The length of the fatty acid chains also affects fluidity. Phospholipids with longer hydrocarbon chains have increased hydrophobic interactions with neighboring lipids and thus decreased membrane fluidity. Organisms can alter membrane fluidity in response to temperature stress by changing the length and degree of saturation of fatty acids present in membrane phospholipids. For example, as environmental temperatures drop, some kinds of cells maintain membrane fluidity by replacing long-chain fatty acids with shorter chains and increasing the percentage of unsaturated fatty acids in their membranes.

Figure 4.31   Selective Permeability of Cell Membranes

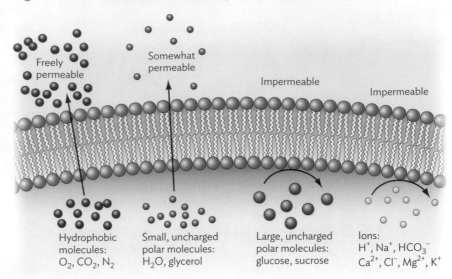

Hydrophobic molecules: O_2, CO_2, N_2

Small, uncharged polar molecules: H_2O, glycerol

Large, uncharged polar molecules: glucose, sucrose

Ions: H^+, Na^+, HCO_3^-, Ca^{2+}, Cl^-, Mg^{2+}, K^+

Membranes Are Selectively Permeable

How do cells allow some molecules to enter while excluding others? Membranes are selectively permeable; that is, they are permeable to some substances but not to others. In general, the cell membrane is permeable to hydrophobic molecules and impermeable to charged molecules (**Figure 4.31**). Diffusion across the membrane also depends on the size of the molecule. Small nonpolar molecules such as O_2 freely pass through the membrane, but larger nonpolar molecules diffuse across more slowly. Molecules that are polar but small (such as ethanol and water) can also diffuse across the membrane. The membrane is impermeable to large polar molecules such as glucose, and to charged molecules, regardless of their size. Thus, membranes help a cell keep its nutrients and enzymes while efflux proteins export wastes and toxins.

An interesting question in medicine is, how do drug molecules administered through the blood (a watery fluid) cross the hydrophobic membranes of cells to affect them? If a drug is soluble in water, how can it cross a hydrophobic membrane? Many kinds of drugs are **membrane-permeant weak acids** or **membrane-permeant weak bases** (**Figure 4.32**). Weak acids and weak bases exist in equilibrium between charged and uncharged forms:

Weak acid: $HA \rightleftharpoons H^+ + A^-$

Weak base: $B + H_2O \rightleftharpoons BH^+ + OH^-$

Weak acids and weak bases cross the membrane in their uncharged form: HA (weak acid) or B (weak base). On the other side, upon entering the aqueous cytoplasm, they dissociate (HA to A^- and H^+) or reassociate with H^+ (B to BH^+). Thus, the membrane-soluble HA conducts acid (H^+), or B conducts base (OH^-), across the membrane. The same effect occurs when fermenting bacteria release acids such as lactic acid. The lactic acid crosses the membrane in the hydrophobic protonated form (HA), then releases H^+ (strong acid) on the other side. Lactic acid acts as a natural food preservative; for

Figure 4.32 **Weak Acids and Weak Bases Cross the Cell Membrane**

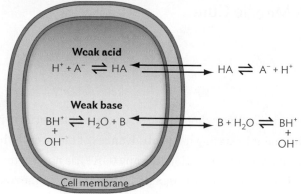

A weak acid crosses the membrane in the protonated form (HA). A weak base crosses the membrane in the deprotonated form (B).

Weak acid

$$H^+ + A^- \rightleftharpoons HA$$

$$HA \rightleftharpoons A^- + H^+$$

Weak base

$$BH^+ \rightleftharpoons H_2O + B$$
$$+$$
$$OH^-$$

$$B + H_2O \rightleftharpoons BH^+$$
$$+$$
$$OH^-$$

Cell membrane

example, during yogurt production, bacterial lactic acid builds up and limits further growth, preserving food value. Other membrane-permeant weak acids are commercial food preservatives such as benzoic acid (benzoate) and propionic acid (propionate).

Many key substances in cellular metabolism are membrane-permeant weak acids, such as acetic acid, and membrane-permeant weak bases, such as the nucleobases of DNA. Drugs that deprotonate within a cell (weak acids, acquiring a negative charge, A^-) include aspirin (acetylsalicylic acid) and penicillin. Drugs that protonate within a cell (weak bases, acquiring a positive charge, BH^+) include fluoxetine (Prozac) and tetracycline.

Transport across Cell Membranes

If large nutrient molecules such as glucose cannot diffuse directly across the cell membrane, how do cells acquire them? In both eukaryotes and prokaryotes, movement of molecules across the cell membrane occurs through specific transmembrane proteins such as channels and transporters. Many types of transporters exist, each differing in its energy requirements and in the types of molecules that it transfers across the membrane. For example, in the intestinal epithelium, a transporter protein cotransports glucose with sodium ions. Glucose-sodium transport is key to the success of oral rehydration therapy in treating cholera (see **inSight**). Transport proteins are discussed further in Chapter 5.

Link Section 5.2 explains how nutrients are transported into the cell and describes the **transport proteins** associated with passive, active, and facilitated transport.

DIFFUSION If a molecule (such as water) can penetrate the membrane or if a transport protein is available (such as for chloride ion), the rate of transport depends on the concentrations of the substance inside and outside the cell or cell compartment. **Diffusion** is the net movement of molecules from an area of high concentration to one of low concentration. Diffusion increases entropy and releases energy. Diffusion occurs by the random, thermal movement of molecules. Factors that influence the diffusion rate include:

- **Temperature.** Increased temperatures mean faster motion. The faster the molecules are moving, the faster they will arrive at the membrane and cross it.

- **Solubility of the molecules in the membrane.** To cross the membrane, the molecules must penetrate it. Hydrophobic molecules will dissolve in the membrane and cross it; charged molecules will not.

- **Size and shape of the molecule.** Friction between a molecule and its medium is a source of resistance that slows down motion. Larger molecules with more mass experience more resistance and cross the membrane more slowly.

- **Surface area of the membrane.** To cross the membrane, molecules must first encounter it. The chances of this happening are increased with greater membrane surface area.

- **Thickness of the membrane.** The thinner the membrane, the faster the molecules can get across.

- **Concentration gradient of the dissolved molecules.** A larger concentration difference between inside and outside increases the rate of diffusion. The more molecules there are in a given volume, the more will encounter the membrane and cross.

OSMOSIS **Osmosis** is the diffusion of water across a selectively permeable membrane from regions of high water concentration (low solute) to regions of low water concentration (high solute). Osmosis across the cell membrane must be tightly controlled to preserve the form of the cell. Thus, cells must maintain osmotic balance with the surrounding environment. The direction of water movement depends on the concentration of dissolved solutes in the cell in relation to the cell's environment (**Figure 4.33**). When a cell is in an **isotonic** or **isosmotic** environment (equal concentrations of dissolved solutes inside the cell and out), osmotic balance exists and water will enter and exit the cell at equal rates (that is, no net movement of water occurs). In a **hypertonic** or **hyperosmotic** environment (higher concentration of solutes outside the cell), osmosis causes net loss of water from a cell. The cell shrinks, and the concentration of cell solutes (inorganic ions as well as organic molecules) increases. This is what happens in cholera: the increased chloride concentration in the intestinal lumen draws water out of the epithelial cells and the blood circulation, ultimately from the body as a whole.

In a **hypotonic** or **hypoosmotic** environment (lower concentration of solutes outside the cell), osmosis causes net uptake of water. The cell swells with water, and the cell components are diluted (see Figure 4.33). If enough water enters, the cell is destroyed by **lysis**—a rupturing of the cell membrane and dispersal of cell contents. The sudden increase of osmotic pressure is called **osmotic shock**. That is why intravenous fluids must always include saline, not distilled water—to prevent blood cell lysis by osmotic shock.

Many free-living cells, such as microbes, live in environments that are hypotonic (lower concentration of solutes outside the cell than inside). To survive a dilute environment, most bacteria and many eukaryotes have a rigid cell wall external to the cell membrane

inSight

Oral Rehydration Therapy: A Miracle Cure

Figure 1   Dehydrated by Diarrheal Illness

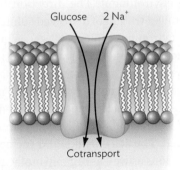

ADAM.

Pinched skin retains the shape.

Figure 2   Cotransport of Sodium Ions and Glucose

The sodium-glucose cotransporter protein carries glucose together with two sodium ions ($2\,Na^+$) across the plasma membrane. A gradient of glucose drives uptake of Na^+, and vice versa.

Glucose $2\,Na^+$

Cotransport

Figure 3   Oral Rehydration Therapy (ORT)

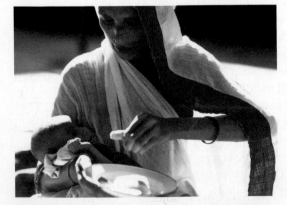

A mother provides ORT to an infant dehydrated by cholera, in Dhaka, Bangladesh.

A therapy so simple that schoolchildren are taught to perform it has become one of the major lifesaving medical advances of the twentieth century. This therapy, called oral rehydration therapy (ORT), is now saving hundreds of thousands of lives in countries such as Bangladesh that have pioneered widespread education about and access to ORT.

Diarrheal diseases cause 20% of all childhood deaths worldwide, particularly in communities lacking access to clean water (Figure 1). Causes of life-threatening diarrhea include cholera and pathogenic *E. coli*, viruses such as Norwalk and rotaviruses, and parasites such as *Giardia*. In the developing world, especially in remote communities lacking hospitals, dehydration from diarrhea is often a death sentence. Before 1950, it was thought that no oral treatment could reverse the dehydration and sodium loss (hyponatremia). Simply giving a saline solution (NaCl) by mouth has no benefit because the normal route of absorption of Na^+ is impaired in the presence of excess Cl^-, and if the Na^+ is not absorbed, then the water is not absorbed either. Thus, industrialized nations with advanced medical care developed intravenous fluids as the standard treatment for dehydration; this treatment is simply unavailable in parts of the world most vulnerable to diarrheal disease.

In the late 1950s, Dr. Hemendra Nath Chatterjee, in India, discovered a modified oral solution that promotes uptake of water. Chatterjee added glucose to the saline solution. The glucose molecules are absorbed by a membrane protein that cotransports glucose with $2\,Na^+$ (Figure 2). Normally, the transporter is used to take up glucose driven by a gradient of Na^+, but in diarrhea, the glucose drives sodium uptake. The increase in solute uptake then draws water back into the cells, reversing dehydration. Thus, a simple solution of salt and glucose can reverse the water loss of diarrhea while allowing the patient's immune system to eliminate the pathogenic bacteria.

In 1971, during the Bangladesh Liberation War, ORT was used to treat cholera patients. As a result, the death rate from cholera was less than 4%, compared with the 30% expected. Thereafter, Bangladesh and other countries with communities prone to waterborne disease adopted public education programs encouraging mothers and schoolchildren to learn to administer ORT. Figure 3 shows a mother in Bangladesh administering ORT to a severely dehydrated child. Within hours, the child recovers. In 2001, the Gates Award for Global Health was conferred on the Centre for Health and Population Research, in Dhaka, Bangladesh, for its role in development of ORT.

(discussed in Section 5.3). As water enters by osmosis and pushes against the cell wall (turgor pressure), the wall resists the tension and pushes back with an equal but opposite force known as wall pressure. The wall pressure is an inward pressure exerted by the cell wall against the cell membrane. When turgor pressure and wall pressure are equal in magnitude, the cell is at equilibrium with respect to water movement. Organisms that lack a cell wall employ other strategies to deal with hypotonic environments. For example, freshwater protists that lack a cell wall, such as amebas and paramecia, expel excess water through a contractile vacuole (an organelle shown in Figure 5.31).

What happens to marine microbes that live in high-salt conditions? Ocean-dwelling microbes such as *Vibrio cholerae* face the problem of water loss. These organisms accumulate solutes known as osmolytes, such as potassium ion, to ensure that they are isotonic to the external environment, thus preventing water loss.

Figure 4.33 **Osmosis and Water Balance**

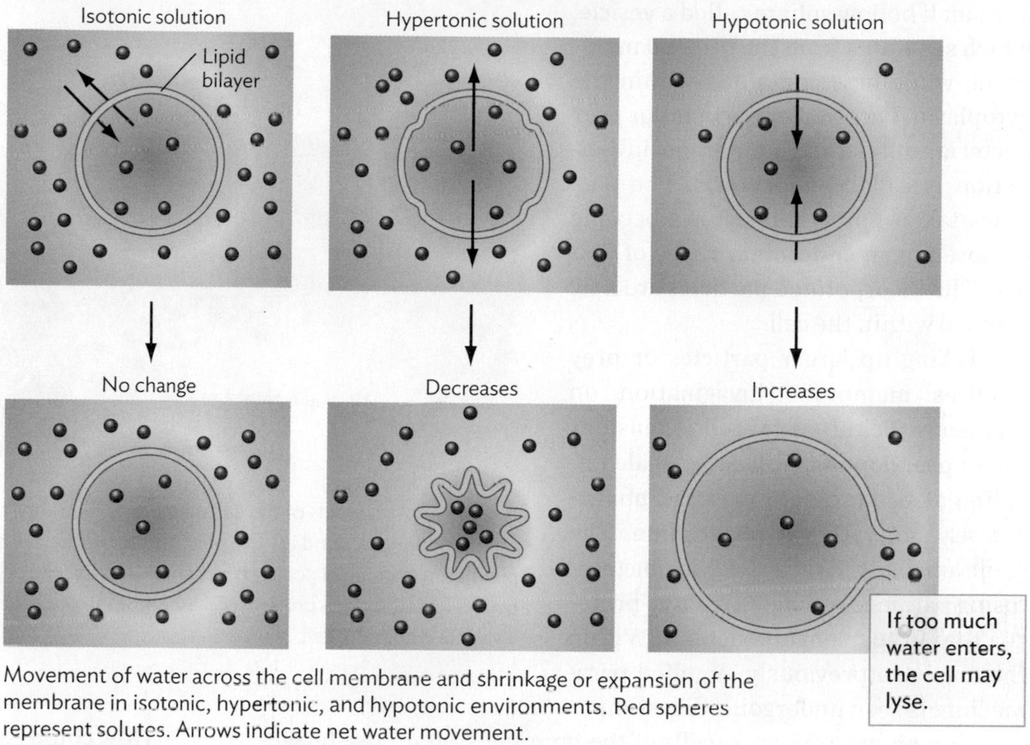

Movement of water across the cell membrane and shrinkage or expansion of the membrane in isotonic, hypertonic, and hypotonic environments. Red spheres represent solutes. Arrows indicate net water movement.

If too much water enters, the cell may lyse.

SECTION SUMMARY

- **Membranes consist of lipids, proteins, and small molecules.** The phospholipid bilayer includes inward- and outward-facing leaflets. The membrane is a fluid in which components can move in two directions but rarely flip sides.

- **Transmembrane proteins span the bilayer.** Peripheral membrane proteins are associated with one side of the bilayer.

- **Membranes are selectively permeable.** Small hydrophobic molecules can diffuse through a membrane, whereas polar or charged molecules cannot.

- **Membrane-permeant weak acids and membrane-permeant weak bases can cross the membrane in their hydrophobic form, but their ionized form is soluble only in the aqueous regions.**

- **Transport proteins can move molecules across the membrane.**

- **Osmosis is the diffusion of water across a membrane from a region of low solute concentration to a region of high solute concentration.** Cells that live in a hypotonic (hypoosmotic) environment, where water is diffusing in, need structures to exclude water.

4.6
Endocytosis and Phagocytosis

SECTION OBJECTIVES

- Describe how endocytosis and phagocytosis transport material into a eukaryotic cell.

- Explain how pathogens take advantage of endocytosis and phagocytosis

How many compartments may a cell have? The phospholipid bilayer of a cell membrane encloses a compartment of water solution (the cytoplasm) whose properties are different from those outside the membrane. Many bacteria have just the one compartment, the cytoplasm; some have a second compartment, the periplasm of Gram-negative bacteria (presented in Chapter 3). Photosynthetic bacteria such as cyanobacteria possess intracellular membrane compartments for photosynthesis.

Eukaryotic cells may hold volumes a thousandfold larger than that of typical bacteria. These larger eukaryotic cells face extra challenges of transport: how to get nutrients efficiently to all parts of the cell and how to export wastes and toxins. In a eukaryotic cell, extensive structures of intracellular membranes called **organelles** manage intracellular transport. Remarkably, these distinctive and intricate organelles are all composed of phospholipid membranes. While the function of the organelles is the subject of the next chapter, we consider here two eukaryotic cell processes by which cellular compartments form.

The cells of animals and eukaryotic microbes (such as amebas) need to take up objects that are too large for protein transporters, such as food particles or bacteria. The uptake mechanism requires a portion of the membrane phospholipids to rearrange in the bilayer, which bends and folds to form a pocket that contains the object (Figure 4.34). This pocket formation is called "invagination." Typically, the pocket contains a region of membrane in which cell-surface proteins have bound an object or particles useful for the cell. Ultimately, the pocket closes in as a small hollow sphere called a **vesicle**, which separates from the original membrane while now contained within the cytoplasm. For smaller particles or prey bacteria, uptake by membrane invagination is called **endocytosis**. The particles taken up by endocytosis become enclosed in an **endosome**, a type of vesicle. The incorporated particles are now trapped within the cell.

Taking up larger particles or prey requires membrane invagination on a larger scale, often by cell extensions called **pseudopods**. This large-scale engulfment of an object is called **phagocytosis**, generating a **phagosome**. The giant ameba in Figure 4.35 is phagocytosing paramecia, which are large protozoa that in turn engulf bacteria! Within the ameba, a previously engulfed paramecium is seen undergoing digestion.

Human macrophages (cells of the immune system) use phagocytosis or endocytosis to engulf pathogenic bacteria for digestion by **lysosomes** (organelles that break down pathogens). But some pathogens can subvert the uptake process—as we see in our next case.

 Figure 4.34   **Endocytosis in a Eukaryotic Cell**

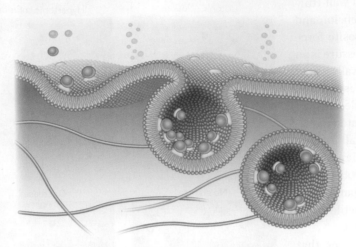

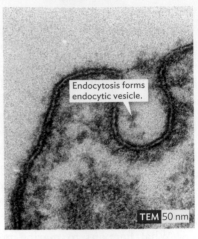

Endocytosis forms endocytic vesicle.

TEM 50 nm

A. The phospholipids of the cell membrane rearrange to allow the membrane to bend in. The bent portion of the cell membrane pinches in, forming a pocket that contains extracellular material bound to cell surface proteins. The membrane pocket closes off as a vesicle within the cell.

B. An endocytic vesicle forms within a rat capillary epithelial cell.

CASE HISTORY 4.1

Chicken Surprise

In October 2013, Kelsey was a 28-year-old computer programmer who worked for a prominent technology firm in Silicon Valley. In the middle of the night, Kelsey awoke with abdominal cramps and diarrhea. Her temperature spiked as high as 40°C (104°F). She had a bad headache and could not retain sufficient fluids. After 3 days, her fever subsided, but the diarrhea continued (about 20 times a day) and began showing blood. Kelsey was admitted to the hospital, where she was given intravenous fluids to control her dehydration. The physician asked her about her dietary history preceding the onset of symptoms. Kelsey recalled that the night before onset, she had eaten rotisserie chicken cooked at the supermarket. An antibody test confirmed the diagnosis of *Salmonella enterica* serovar Heidelberg. Because of the severity of her symptoms, Kelsey was treated with antibiotics, although sometimes antibiotics may prolong the persistence of *Salmonella*. When amoxicillin failed to control the infection because of bacterial resistance, ciprofloxacin was used. Ultimately, Kelsey's illness was one of more than 400 cases in 23 states traced to contaminated chicken from one California farm. Kelsey recovered, but 3 months later, *S.* Heidelberg could still be detected in her feces.

How could enteric bacteria make someone as healthy as Kelsey so sick, and persist in the body for months despite treatment with antibiotics? Unlike *Vibrio cholerae* (the culprit in our chapter-opening case history), *Salmonella* bacteria do more than alter the intestinal fluid balance. *Salmonella* actually invades the host cells of the intestinal lining, including enterocytes (absorptive epithelial cells) and macrophages (defensive white blood cells). Within the host cells, the bacteria grow and multiply in protected compartments called *Salmonella*-containing vacuoles.

Figure 4.35   Phagocytosis

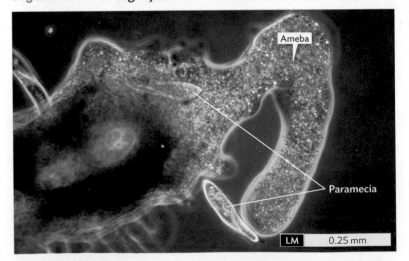

Ameba

Paramecia

LM   0.25 mm

Ameba engulfing paramecia.

Figure 4.36 *Salmonella*-Containing Vesicles

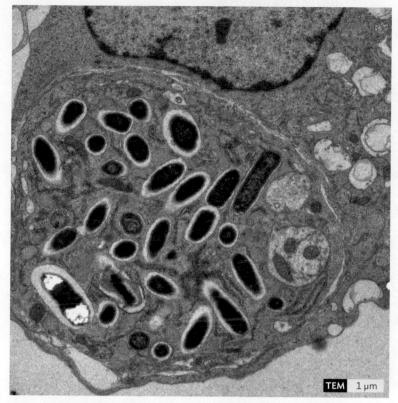

TEM 1 μm

Macrophage infected by *Salmonella* bacteria (dark rods) growing within individual vesicles.

To invade host cells, the bacteria take advantage of the host's own processes of uptake and transport mediated by membrane vesicles. These pathogens (called facultative intracellular bacteria) can stabilize the endosome to allow bacterial growth (Figure 4.36). The stabilized endosomes multiply as the bacteria grow, a protective vacuole surrounding each bacterium. Because the bacteria multiply within host cells, protected from antibiotics, they may persist for long periods. Other examples of pathogens that enter cells by phagocytosis include *Listeria* bacteria, a food-borne pathogen, and influenza virus.

Even when a pathogen remains outside the host cell, endocytosis can bring in toxins such as cholera toxin. As shown at the beginning of the chapter (see Figure 4.1), the pathogen *Vibrio cholerae* remains extracellular, but its toxin must gain entry to the intestinal cells. The cholera toxin first binds to the intestinal cell-surface molecule ganglioside; then it enters the cell by endocytosis, and the endocytic vesicle merges into the endoplasmic reticulum (transport membranes within the cell). In the ER membrane, a transport protein is co-opted to release a portion of the toxin (A1) into the cytoplasm, where it gains access to the protein complex that leads to Cl⁻ efflux.

How can a cell disable endocytosed pathogens and toxins? Normally, the endosome or phagosome must fuse with a lysosome.

Lysosomes help eukaryotic cells obtain nourishment from macromolecular nutrients, and they help destroy pathogenic bacteria and viruses. For digestion, lysosomes contain many hydrolytic enzymes (for example, proteases, nucleases, and lipases) and have an acidic pH of around 5. Ultimately, the digested contents of the lysosome are delivered to the endoplasmic reticulum, a series of membrane channels and organelles that make up the eukaryotic cell's **endomembrane system**.

Link Vesicle traffic, the **endomembrane system**, and intracellular organelles are discussed in Section 5.6.

SECTION SUMMARY

- **Endocytosis takes up small particles by membrane invagination.** Invaginating membranes form vesicles called endosomes that are enclosed within the cytoplasm.

- **Phagocytosis is the uptake of larger particles and bacteria.** Phagocytosis may include the formation and fusion of pseudopods.

- **Lysosomes fuse with endosomes and acidify their contents, adding factors that kill pathogenic bacteria.** Lysosomes deliver their contents to the endomembrane system.

Thought Question 4.5 Suppose that a pathogenic bacterium such as *Mycobacterium tuberculosis* gets taken up by a cell through phagocytosis. What do you think the bacterium has to do to avoid getting destroyed and instead reproduce within the host cell?

Perspective

The human body is made of molecules and chemical reactions. A simple change in salt concentration can dramatically impact health, causing the water loss and dehydration seen in cholera. A small change in blood pH, either too acidic or too alkaline, can lead to convulsions or fainting. Overall, we see how the various kinds of biological molecules, from water and inorganic ions to macromolecular complexes, interact with one another to compose a living cell. Small organic molecules such as sugars can be joined by enzymes to form polysaccharides, and various macromolecules come together to form cell walls and other components of the cell, including a wide range of proteins with various specialized functions. Phospholipids interact by hydrophobic forces to form membranes that compartmentalize the cell.

In Chapter 5, you will learn how the structures and processes discussed in Chapter 4 fit together within living cells. For example, you will see that sugars and amino acids (presented in Sections 4.2 and 4.3) are assembled to form the cage-like cell wall that encloses a bacterial cell. You will also learn how the membranes of a cell form its organelles and how organelles mediate the cell's interactions with pathogens.

LEARNING OUTCOMES AND ASSESSMENT

	SECTION OBJECTIVES	OBJECTIVES REVIEW

4.1 Elements, Bonding, and Water

- Describe how atoms form the molecules of life.
- Describe the types of bonds atoms can form to make a molecule.
- Explain the special properties of the water molecule, and distinguish between hydrophilic and hydrophobic molecules.

1. Which atom is <u>not</u> an essential part of most living cells?
   A. Carbon
   B. Hydrogen
   C. Oxygen
   D. Nitrogen
   E. Helium

4.2 Lipids and Sugars

- Describe organic molecules, and explain the role of functional groups and the formation of macromolecules.
- Describe the structure of lipids and their key functions within a cell.
- Describe the structure of sugars and polysaccharides.

4. Which part of a lipid is the most hydrophobic?
   A. Glycerol
   B. Hydrocarbon chain
   C. Carboxyl group
   D. Ester linkage of a triglyceride

4.3 Nucleic Acids and Proteins

- Describe the structure of DNA, and explain how DNA stores information in its sequence of base pairs.
- Describe the structure and diverse functions of RNA.
- Describe the fundamental structure of proteins.

7. A nucleotide contains all of these chemical structures, except for _____.
   A. a thiol (–SH)
   B. a phosphoryl group
   C. a nucleobase
   D. a pentose sugar

4.4 Biochemical Reactions

- Explain how energy and entropy determine the direction of reactions.
- Describe the oxidation-reduction changes that occur during reactions.
- Explain how enzymes catalyze reactions and how various factors influence the rate of reaction.

10. Which process does <u>not</u> yield energy for a cell?
    A. DNA synthesis
    B. Oxidation of sugar
    C. Oxidation of lipids
    D. Breakdown of carbohydrates into smaller molecules

4.5 Membranes and Transport

- Describe the structure of cellular membranes.
- Explain the concept of selective permeability of a membrane.
- Explain how membranes compartmentalize cells and control the movement of a cell's molecules.
- Describe the effect of osmosis on cells.

13. Which part of a phospholipid bilayer faces inward, toward the middle of the bilayer?
    A. Phosphoryl group
    B. Hydrocarbon chain
    C. Glycerol
    D. Carboxyl group

4.6 Endocytosis and Phagocytosis

- Describe how endocytosis and phagocytosis transport material into a eukaryotic cell.
- Explain how pathogens take advantage of endocytosis and phagocytosis.

16. Which pathogen does <u>not</u> enter a host cell by endocytosis?
    A. *Salmonella enterica*
    B. *Listeria monocytogenes*
    C. Influenza virus
    D. *Vibrio cholerae*

2. Which compound would dissolve poorly in water, compared with lipids?
    A. Butane ($CH_3CH_2CH_2CH_3$)
    B. NaCl
    C. Glycine
    D. CO_2

3. Which compound would dissolve in water by forming ions?
    A. NaCl
    B. O_2
    C. Glycine
    D. CO_2

5. A simple sugar contains all of these types of chemical structures, except for ____.
    A. a carbonyl group
    B. a hydroxyl group
    C. an amino group
    D. a carbon-carbon bond

6. Which molecule contains sugar components?
    A. Fatty acid
    B. Methane
    C. Glycine
    D. Glycoprotein

8. The alpha helix is a form of protein structure at which level?
    A. Primary structure
    B. Secondary structure
    C. Tertiary structure
    D. Quaternary structure

9. Which molecule is usually a component of RNA but not DNA?
    A. Uracil
    B. Cytosine
    C. Guanine
    D. Adenine

11. For an intestinal bacterium, which factor is most likely to increase the rate of a reaction catalyzed by an enzyme?
    A. Decreasing concentration of the substrate
    B. pH increase from pH 8 to pH 9
    C. Salt saturation
    D. Increase in temperature from 25°C to 35°C

12. Which reaction involves increasing entropy?
    A. Building a protein
    B. Replicating DNA
    C. Breaking down a protein
    D. Coiling DNA into a chromosome

14. Which molecule requires a transport protein in order to cross a membrane?
    A. H_2O
    B. O_2
    C. Alanine
    D. Ethanol

15. Which process involves osmosis?
    A. A transport protein carries a nutrient across a membrane.
    B. CO_2 crosses a membrane.
    C. A cell placed in distilled water takes up water and expands.
    D. A protein is synthesized in the cytoplasm.

17. An endocytic vesicle must be composed of which kind of molecules?
    A. Nucleic acids
    B. Phospholipids
    C. Sodium ions
    D. Sugars

18. Which cell can conduct phagocytosis?
    A. Muscle cell
    B. *Escherichia coli*
    C. Nerve cell
    D. Ameba

Key Terms

activation energy (p. 108)
alpha helix (p. 105)
amino acid (p. 103)
amphipathic (p. 111)
anion (p. 96)
atom (p. 94)
atomic number (p. 94)
atomic weight (p. 94)
ball-and-stick model (p. 96)
beta sheet (p. 105)
carbohydrate (p. 100)
catalysis (p. 106)
catalyst (p. 108)
cation (p. 96)
cell membrane (p. 110)
chiral carbon (p. 103)
complex (p. 103)
condensation reaction (p. 98)
covalent bond (p. 94)
cytoplasm (p. 110)
denatured (p. 108)
diffusion (p. 113)
disaccharide (p. 100)
electrolyte (p. 96)

element (p. 94)
endergonic (p. 107)
endocytosis (p. 116)
endosome (p. 116)
energy (p. 107)
entropy (p. 107)
enzyme (p. 108)
exergonic (p. 107)
first law of thermodynamics (p. 107)
free energy change (ΔG) (p. 107)
functional group (p. 97)
genome (p. 100)
half-life (p. 94)
hopanoid (p. 112)
hydrogen bond (p. 96)
hydrolysis (p. 107)
hydrophilic (p. 97)
hydrophobic (p. 97)
hyperosmotic (p. 113)
hypertonic (p. 113)
hypoosmotic (p. 113)
hypotonic (p. 113)
ion (p. 96)
ionic bond (p. 96)

isomer (p. 103)
isosmotic (p. 113)
isotonic (p. 113)
isotope (p. 94)
lipid (p. 98)
lysis (p. 113)
lysosome (pp. 116, 117)
macromolecule (p. 97)
mass number (p. 94)
membrane-permeant weak acid (p. 112)
membrane-permeant weak base (p. 112)
molarity (p. 108)
mole (p. 108)
molecular formula (p. 94)
molecule (p. 94)
monosaccharide (p. 99)
native conformation (p. 105)
nonpolar covalent bond (p. 96)
nucleobase (p. 101)
nucleotide (p. 101)
orbital (p. 94)
organelle (p. 115)

organic molecule (p. 97)
osmosis (p. 113)
osmotic shock (p. 113)
oxidation (p. 110)
oxidizing agent (p. 110)
peripheral membrane protein (p. 112)
phagocytosis (p. 116)
phagosome (p. 116)
phospholipid (p. 111)
phospholipid bilayer (p. 111)
polar covalent bond (p. 96)
polysaccharide (p. 100)
primary structure (p. 104)
pseudopod (p. 116)
reducing agent (p. 110)
reduction (p. 110)
secondary structure (p. 105)
second law of thermodynamics (p. 107)
simple sugar (p. 99)
space-filling model (p. 96)
structural formula (p. 94)
tertiary structure (p. 105)
transmembrane protein (p. 112)
vesicle (p. 116)

Review Questions

1. Which are the major elements of life? Which of life's molecules contain atoms of these elements?

2. Explain the different kinds of molecular bonding. Which types are stronger, and which are weaker?

3. Describe the composition of carbohydrates and lipids. What types of monomers must condense to form longer chains?

4. What is the composition of a nucleotide?

5. Explain how DNA and RNA are related and how they differ. Explain the difference between intramolecular and intermolecular base pairing.

6. For a protein, describe the primary, secondary, tertiary, and quaternary levels of structure.

7. Explain the role of energy and entropy in driving a chemical reaction. What is necessary for a reaction to power life?

8. Explain the role of enzymes in biochemical reactions. Why do all life's reactions need enzymes?

9. Describe the formation and composition of a membrane. How does a membrane participate in endocytosis?

Clinical Correlation Questions

1. A malnourished cholera patient is given clean water to drink but fails to improve. The nurse considers what additional components may be needed, such as sugar for energy, antibiotics to kill the bacteria, or vitamins to reverse malnutrition. Which additional component of oral therapy is needed, and why?

2. A patient is found in whom *Salmonella enterica* bacteria persist for several months in the digestive tract. By contrast, in cholera patients, *Vibrio cholerae* bacteria do not persist. Why not?

Thought Questions: CHECK YOUR ANSWERS

Thought Question 4.1 Use the properties of molecules described in this section to explain why table salt (NaCl) dissolves readily in water, but vegetable oil (a hydrocarbon) does not.

ANSWER: NaCl dissolves in the form of Na^+ and Cl^- ions. The charges on these ions readily interact with the partial charges on water molecules, which are polar. By contrast, vegetable oil consists of a hydrocarbon with no charge and no polarization. Hydrophobic forces cause hydrocarbons to associate with each other, but not water.

Thought Question 4.2 What kinds of amino acid residues would you find on the exterior surface of a membrane-embedded protein? What kinds of amino acid residues would you find in the middle of the protein?

ANSWER: The surface of a membrane-embedded protein has hydrophobic amino acid residues such as leucine and phenylalanine, which can interact favorably with the hydrophobic hydrocarbon chains of phospholipids. Within the middle of the protein, charged and polar side chains of amino acid residues will bind to each other, thus helping to hold the protein together.

Thought Question 4.3 Saline is a salt solution used for intravenous (IV) procedures; it contains approximately the same concentration of NaCl as in the blood. The concentration of NaCl in blood is approximately 0.154 mole per liter. How much salt would you weigh out to make 1 liter of saline?

ANSWER: According to the periodic table, the atomic weight of Na is 23 and that of Cl is 35.5. Thus, 1 mole of NaCl weighs 23 grams + 35.5 grams = 58.5 grams. The weight of 1 mole (58.5 grams) times 0.154 mole per liter gives 9 grams of NaCl to weigh out for 1 liter of saline.

Thought Question 4.4 Your muscle cells contain ATP at concentrations of 1–10 mM. The reaction of ATP hydrolysis (forming ADP and phosphate) has a large negative value of free energy change, ΔG. So why doesn't all the ATP hydrolyze?

ANSWER: The hydrolysis of ATP has a high energy of activation; that is, the reaction requires a large input of energy to reach the transition state between ATP and ADP. This activation energy can be decreased by an enzyme that binds ATP in its reaction site. Within a cell, such ATP-hydrolyzing enzymes are regulated to avoid spending all the ATP.

Thought Question 4.5 Suppose that a pathogenic bacterium such as *Mycobacterium tuberculosis* gets taken up by a cell through phagocytosis. What do you think the bacterium has to do to avoid getting destroyed and instead reproduce within the host cell?

ANSWER: When the bacterium *M. tuberculosis* is taken up by phagocytosis, the phagocytic vesicle fuses with a lysosome, which secretes enzymes of host defense. The lysosomal enzymes digest bacterial proteins and oxidize various cell components. *M. tuberculosis* uses several strategies to avoid destruction in the phagolysosome. One way is through an exceptionally thick, complex cell wall and envelope. The envelope contains molecules that the lysosomal enzymes cannot digest or oxidize.

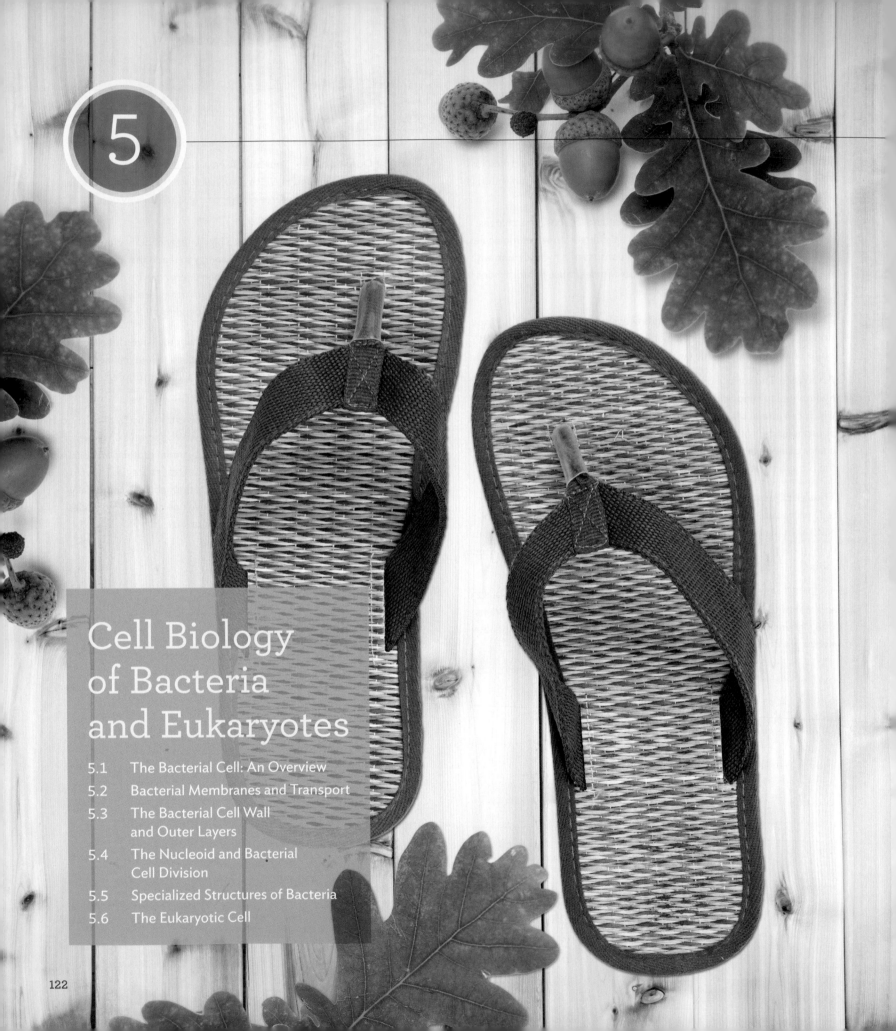

5

Cell Biology
of Bacteria
and Eukaryotes

[An Upscale Tick Bite]

SCENARIO Sharon, the CEO of a start-up company, lives in Westchester County, a wooded community north of New York City. She spends her summer weekends emailing her managers from the outdoor deck of her home, shaded by tall oak trees. The acorns attract mice and deer, and the leaf litter is full of ticks (*Ixodes scapularis*).

SIGNS AND SYMPTOMS One evening Sharon's husband noticed a red rash on her back, consisting of a ring shape several centimeters across, surrounding another red spot in the middle. Sharon recalled seeing this "bull's-eye" type of rash on the Internet. The rash was described as the hallmark of Lyme disease, or borreliosis, caused in the United States by the tick-borne bacterium *Borrelia burgdorferi* (in Europe, by the closely related species *Borrelia afzelii*). The distinctive rash, erythema migrans ("migrating redness"), begins at the site of a tick bite and expands concentrically as the bacteria migrate outward. Sharon recalled that a neighbor's child had suffered crippling arthritis caused by an undetected case of Lyme disease. Another neighbor who contracted Lyme disease had suffered from meningitis (inflammation of the brain lining) and neurological abnormalities, including facial paralysis, ultimately losing his job and his million-dollar home.

DIAGNOSIS AND TREATMENT The next day Sharon went to her doctor and was treated with the antibiotic doxycycline, a tetracycline derivative that targets the bacterial ribosome. Doxycycline is the antibiotic of choice for *B. burgdorferi*. Sharon was fortunate to make a full recovery before serious symptoms appeared.

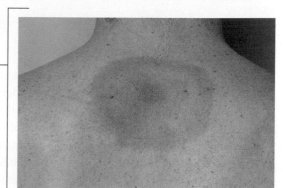

Erythema Migrans
The migrating red rash of Lyme disease, caused by the spiral-shaped bacterium *Borrelia burgdorferi*.

Ixodes scapularis
(adult females; actual size)

Ixodes scapularis
(nymphs; actual size)

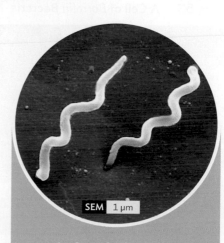

SEM 1 μm

Borrelia burgdorferi
The helical shape of *Borrelia burgdorferi* is typical of spirochetes.

CHAPTER OBJECTIVES

After reading this chapter, you will be able to:

- Describe the key parts of a bacterial cell.
- Explain the importance of a bacterial cell's parts for the function of antibiotics.
- Describe the key parts of a eukaryotic cell, and explain how pathogens take advantage of these to infect a eukaryotic host.

Lyme disease, or borreliosis, was first named for the town of Lyme, Connecticut, where an outbreak began in 1975; it is now the most common reported tick-borne disease in the Northern Hemisphere, with the heaviest concentration in New England. The causative agent, *Borrelia burgdorferi*, is transmitted to humans by the bite of an infected tick of the genus *Ixodes*. The *Ixodes* ticks are adapted to mice and deer, which maintain a permanent reservoir for human infection—especially in locales such as Westchester, where hunting is restricted and deer have no natural predators. In Sharon's case, the rash was observed early and the pathogen was eliminated. In other cases, where the rash is not observed, the clinical laboratory can reveal the bacteria by detecting serum antibodies (antibodies produced by the patient in response to bacteria) or by PCR amplification of bacterial DNA.

The causative bacterium, *Borrelia burgdorferi*, was first identified in 1982 by Swiss entomologist Willy Burgdorfer. *B. burgdorferi* show a flexible spiral shape, like a coiled audio cable. Bacterial flagella (rotary tails) propel the cell through body fluids in a corkscrew motion. The complex cell form, called a "spirochete," is unique to the Spirochete phylum of bacteria. As seen in **Figure 5.1**, the spiral cell body is surrounded by a thick outer sheath (colorized brown). Beneath the sheath extend helical flagella (violet) that are plugged into rotary motors embedded in the cell membrane. As the motors rotate, the helical flagella cause the entire cell body to flex in a corkscrew motion. The motion enables the spirochete bacterium to twist its way through host structures such as the capillary vesicles. The twisting motility enables the bacteria to migrate outward from the original site of the tick bite, causing the red rash called erythema migrans.

Bacterial cells have distinctive shapes as well as distinctive cell wall and cytoplasmic structures. These are some of the cell structures we will explore in this chapter. The structures are made of the proteins and membranes described in Chapter 4. Cell structures contribute to the disease potential of pathogens, and they offer targets for new antibiotics.

5.1
The Bacterial Cell: An Overview

SECTION OBJECTIVES

- Describe the structure and function of the bacterial cell wall, and explain its importance as a target for antibiotics.
- Explain the Gram-negative envelope structure and the role of LPS in pathogenesis.
- Describe the biochemical composition of a bacterial cell.

How does a microbial cell survive assault by host defenses or when stressed by extreme conditions in its natural environment? Bacterial cells build complex structures, such as a cell wall that acts like a molecular cage, protecting the cell membrane. The cell wall is a vital drug target; the penicillins and vancomycin block its synthesis. The bacterial cytoplasm also provides drug targets; for example, tetracyclines, aminoglycosides, and macrolides target the bacterial ribosome.

Two major classes of cell have fundamentally different forms: the **prokaryotic cells** (cells of bacteria and archaea, whose DNA lacks any nuclear membrane) and the **eukaryotic cells** (cells having a nucleus, such as those of plants and animals, including humans). Both kinds of cells show a wide range of form in different species; their diversity is explored in Chapters 10 and 11. In this chapter, however, we focus on the fundamental traits of bacteria and eukaryotes (**Table 5.1**).

Figure 5.1   A Cell of *Borrelia* Bacteria

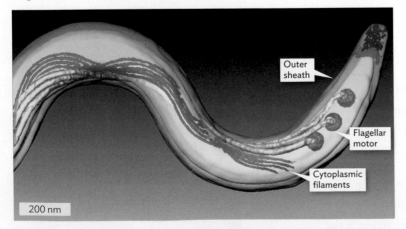

Outer sheath

Flagellar motor

Cytoplasmic filaments

200 nm

The spiral cell body is surrounded by a thick outer sheath (colorized brown). The cell membrane (colorized yellow), internal flagella (violet), cytoplasmic filaments (orange), and outer sheath (brown) are visualized by cryo-electron tomography.

Table 5.1
Bacteria and Eukaryotes: Cell Traits

Bacteria	Eukaryotes
Small cell size (usually 0.2–10 μm)	Wide range of cell size, from very small to very large (usually 0.2 μm–1 mm)
DNA organized in nucleoid throughout the cytoplasm	DNA contained in nucleus, enclosed by nuclear membrane
Small genome (0.5–15 million base pairs)	Wide range of genome size, including very large (0.5 million–20 billion base pairs)
Circular chromosome (usually), although may have multiple circular and linear chromosomes	Linear chromosomes (in nucleus); mitochondria (derived from bacteria) have one circular chromosome
Chromosomes replicate and segregate during cell growth	Chromosomes segregate by mitosis and meiosis, after replication during interphase
Few intracellular membranes (such as thylakoids of photosynthetic bacteria)	Many types of intracellular membranous organelles (such as endoplasmic reticulum, Golgi, and lysosomes)
No intracellular endosymbiotic organelles	Mitochondria and chloroplasts are organelles that evolved from endosymbiotic bacteria
Cell wall composed of peptidoglycan	Cell walls of plants and fungi composed of various carbohydrates (such as cellulose or chitin), but never peptidoglycan
Rotary flagella for motility, driven by proton motive force	Whiplike flagella for motility, with microtubule contraction driven by ATP

Most bacteria share the following traits:

- **Complex cell envelope.** The envelope (cell membrane, cell wall, and outer layers if present) protects the cell from environmental stress and predators. The cell membrane mediates exchange with the environment, such as toxin secretion and communication with fellow pathogens.

- **Small genome.** Prokaryotic genomes are compact, with relatively little noncoding DNA. Small genomes help a pathogen make as many offspring as possible from the limited resources of its host.

- **Tightly coordinated cell parts.** The cell's parts work together in a highly coordinated mechanism. Coordinated action helps a pathogen reproduce quickly, outracing the host's defensive response.

Link Membrane structure and function was introduced in Section 4.5.

In the early twentieth century, the cell was envisioned as a bag of "soup" full of floating ribosomes and enzymes. Today, microscopy shows that in fact the cell's parts fit together in a structure that is ordered, though flexible.

A Model of the Bacterial Cell

A model of the bacterial cell (**Figure 5.2**) shows how the major components of a bacterial cell fit together. This model represents *Escherichia coli*, but its general features apply to many kinds of bacteria. Remember that we cannot actually "see" the molecules within a cell, but microscopy and subcellular analysis generate a remarkably detailed model.

Within the bacterial cell, the cytoplasm consists of a gel-like network of proteins and other macromolecules. The cytoplasm is surrounded by a **cell membrane** (also called a **plasma membrane** or **cytoplasmic membrane**). As discussed in Chapter 4, the membrane is composed of phospholipids, hydrophobic proteins, and small organic molecules. The cell membrane prevents cytoplasmic proteins from leaking out, and it maintains concentration gradients of ions and nutrients. The cell membrane is also the site of oxidative phosphorylation (or respiration), the process in which electrons pulled from food molecules yield energy to pump hydrogen ions (H^+) across the membrane (discussed in Section 7.4). The resulting H^+ gradient then drives ATP synthesis by the ATP synthase complex (the knob-like structures shown in Figure 5.4).

The bacterial cell membrane is covered by a **cell wall**, a rigid, mesh-like structure called **peptidoglycan**. Peptidoglycan consists of sugar chains linked covalently by peptides. The cell wall is a single molecule that surrounds the cell like a cage. In Gram-negative bacteria such as *E. coli*, the entire cell wall is a single layer of peptidoglycan. (Gram-positive bacteria have multiple layers of peptidoglycan; discussed in Section 5.3.) Outside the cell wall, Gram-negative bacteria have an **outer membrane**. Between the outer membrane and cell membrane (or "inner membrane") lies a region of water solution called the **periplasm**, containing proteins such as sugar transporters.

The outer membrane consists of phospholipids and **lipopolysaccharides (LPS)**, a class of lipids attached to long polysaccharides (sugar chains). LPS and other polysaccharides can generate a thick **capsule** surrounding the cell. The capsule polysaccharides form a

Figure 5.2   Model of a Bacterial Cell (*Escherichia coli*)

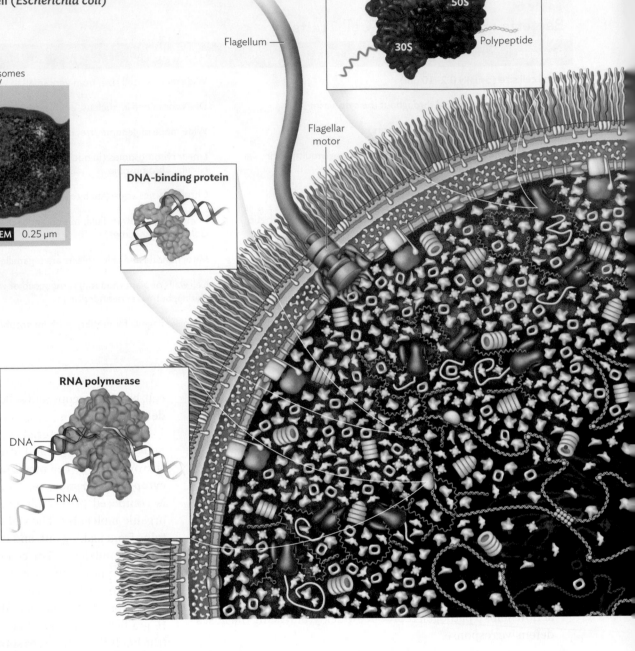

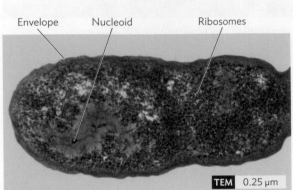

Envelope   Nucleoid   Ribosomes

TEM   0.25 μm

Envelope: The cell membrane contains embedded proteins for structure and transport. The cell membrane is supported by the cell wall. In this Gram-negative cell, the cell wall is coated by the outer membrane, whose sugar chain extensions protect the cell from attack by the immune system or by predators. The rotary motor of a flagellum is plugged into the membrane. *Cytoplasm:* Molecules of messenger RNA (mRNA) extend out of the nucleoid to the region of the cytoplasm rich in ribosomes. Ribosomes translate the mRNA to make proteins, which are folded into their active form by chaperones (proteins that assist folding). *Nucleoid:* The chromosomal DNA is wrapped around binding proteins. Replication by DNA polymerase and transcription by RNA polymerase occur at the same time within the nucleoid.

Ribosome

mRNA

50S

30S

Polypeptide

Flagellum

Flagellar motor

DNA-binding protein

RNA polymerase

DNA

RNA

slippery mucous layer that inhibits **phagocytosis** by white blood cells. Pathogens stripped of their capsule can become extremely vulnerable to our body's defenses.

Link **Phagocytosis** was introduced in Section 4.6 as a process of membrane invagination by which a cell may take up larger particles or prey. The process is further discussed in Chapter 15 in the context of immune system response to infection.

The cell membrane, cell wall, and outer membrane (for Gram-negative species) constitute the **cell envelope**. The envelope includes cell surface proteins that enable the bacteria to interact with specific host organisms. For example, some *E. coli* cell-surface proteins enable colonization of the human intestinal epithelium,

whereas *Sinorhizobium* cell-surface proteins enable colonization of legume plants for nitrogen fixation. Another common external structure is the **flagellum**, a helical protein filament whose rotary motor (seen in Figure 5.2) propels the cell in search of a more favorable environment. For *E. coli* and other Gram-negative bacteria, the flagella extend freely outside the cell. For spirochete bacteria such as *B. burgdorferi* (the cause of Lyme disease) or *Treponema pallidum* (the cause of syphilis), the flagella wrap around the cell body, enclosed by the outer sheath (see Figure 5.1).

Note In some texts, the term "envelope" refers only to structures outside the cell membrane. This book considers the cell membrane (or Gram-negative inner membrane) to be part of the bacterial envelope.

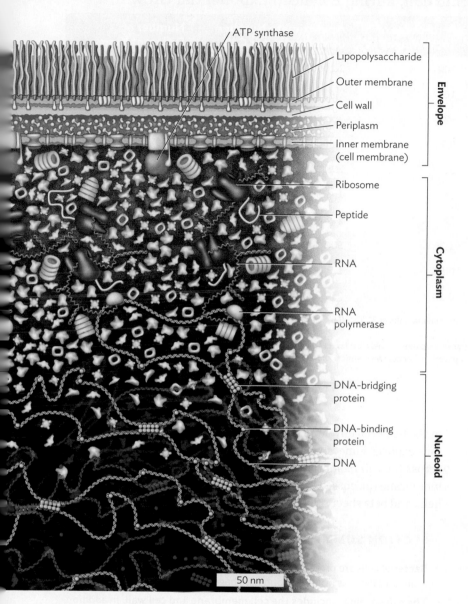

ATP synthase

Lipopolysaccharide

Outer membrane

Cell wall

Periplasm

Inner membrane
(cell membrane)

Envelope

Ribosome

Peptide

RNA

RNA
polymerase

Cytoplasm

DNA-bridging
protein

DNA-binding
protein

DNA

Nucleoid

50 nm

Biochemical Composition of Bacteria

Our model of the bacterial cell shows shape and size, but tells little about chemical composition. All cells share common chemical components introduced in Chapter 4:

- Water, the fundamental solvent of life

- Inorganic ions, such as potassium, magnesium, and chloride ions

- Small organic molecules, such as lipids and sugars that are incorporated into cell structures and that provide nutrition by chemical reactions

- Macromolecules, such as nucleic acids and proteins, which contain information, catalyze reactions, and mediate transport, among many other functions

Cell composition varies with species, growth phase, and environmental conditions. Table 5.2 lists the chemical components of a cell for the model bacterium *Escherichia coli* during exponential growth.

SMALL MOLECULES The *E. coli* cell consists of about 70% water. Inorganic ions such as potassium (K^+) store energy in the form of transmembrane gradients, and they serve essential roles in enzymes. For example, RNA polymerase requires a magnesium ion (Mg^{2+}) at its active site to help extend an RNA chain.

The cell also contains many kinds of small charged organic molecules, such as phospholipids and enzyme cofactors. Organic cations include **polyamines**, molecules with multiple amine groups that are positively charged. Polyamines balance the negative charges of the cell's DNA and stabilize ribosomes during translation. Some bacteria digest protein and release polyamines with names like "cadaverine" and "putrescine"; these polyamines cause the odors of decaying flesh or meat.

MACROMOLECULES A specific type of cell requires specific kinds of macromolecules (long-chain molecules), especially its proteins. Proteins vary among different species; and even a single species will make very different proteins when its environment changes. For example, a pathogen such as *Yersinia pestis* (the cause of bubonic plague) will make different proteins as it moves from the cold flea host to the warm human host and when it encounters an increase or decrease in nutrients such as iron. Individual proteins, encoded by specific genes, are found in very different amounts, from 10 per cell to 10,000 per cell. The total proteins encoded by a genome and capable of expression in the cell are known collectively as the **proteome**. The assemblage of proteins is unique to each bacterial strain.

Electron microscopy largely defines how we "see" the cell's interior as a whole (discussed in Chapter 3). A closer look at any of these components requires breaking the cell into parts, a process called **subcellular fractionation**. Subcellular fractionation provides protein complexes for study by tools such as X-ray crystallography. For example, subcellular fractionation and X-ray crystallography were used to reconstruct a model of the *E. coli* TolC protein complex embedded

Within the cell cytoplasm, the cell membrane and envelope provide an attachment point for one or more chromosomes. The chromosome is organized within the cytoplasm as a system of looped coils called the **nucleoid**. Unlike the round, compact nucleus of eukaryotic cells, the bacterial nucleoid is not enclosed by a membrane, and so the coils of DNA can extend throughout the cytoplasm. Loops of DNA from the nucleoid are transcribed by the enzyme RNA polymerase to form messenger RNA (mRNA). As the bacterial mRNA transcripts grow—before they are complete—they already bind ribosomes to start synthesizing polypeptide chains (proteins). This bacterial cell organization differs greatly from that of a eukaryotic cell, in which most mRNA molecules are completed and leave the nucleus before translation (see Section 5.6).

Table 5.2
Molecular Composition of a Bacterial Cell, *Escherichia coli*, during Balanced Exponential Growth[a]

Component		Percentage of Total Weight[b]	Approximate Number of Molecules per Cell	Number of Different Kinds
	Water	70%	20,000,000,000	1
	Proteins	16%	2,400,000	2,000[c]
RNA	rRNA, tRNA, and other small regulatory RNAs (sRNA)	6%	250,000	200
	mRNA	0.7%	4,000	2,000[c]
Lipids	Phospholipids (membrane)	3%	25,000,000	50
	Lipopolysaccharide (outer membrane)	1%	1,400,000	1
	DNA	1%	2[d]	1
	Metabolites and biosynthetic precursors	1.3%	50,000,000	1,000
	Peptidoglycan (murein sacculus)	0.8%	1	1
	Inorganic ions	0.1%	250,000,000	20

[a]Values shown are for a hypothetical "average" cell cultured with aeration in glucose medium with minimal salts at 37°C (98.6°F).
[b]The total weight of the cell (including water) is about 10^{-12} gram (g), or 1 picogram (pg).
[c]The number of kinds of mRNA and proteins is difficult to estimate because some genes are transcribed at extremely low levels and because some RNA and proteins are rapidly degraded.
[d]In rapidly growing cells, cell fission typically lags approximately one generation behind DNA replication, hence, two identical DNA copies per cell.

Figure 5.3   A Toxin Efflux Complex in *E. coli*

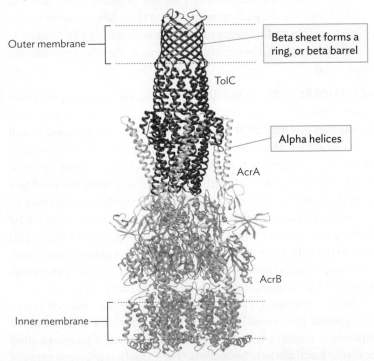

Outer membrane

Beta sheet forms a ring, or beta barrel

TolC

Alpha helices

AcrA

AcrB

Inner membrane

Model of the TolC efflux protein complex, based on X-ray crystallography of isolated components.

in the bacterial envelope (Figure 5.3). A TolC complex forms a tube that exports some toxins and antibiotics out of the cytoplasm and across the cell membrane (inner membrane), periplasm, and outer membrane (discussed in the next section). The proteins show alpha helix and beta sheet structures, as introduced in Chapter 4.

SECTION SUMMARY

- **Bacterial cells** are protected by a thick cell envelope. The envelope can protect pathogens from macrophage phagocytosis.

- **The cell envelope includes the cell membrane and cell wall.** In addition, the Gram-negative cell envelope includes an outer membrane outside the cell wall.

- **The nucleoid** has loops of DNA that extend throughout the bacterial cell. In the growing bacterial cell, DNA replication, RNA transcription, and protein synthesis occur together.

- **Proteins in the cell vary,** depending on the species and environmental conditions.

- **The biochemical composition of bacteria** includes nucleic acids, proteins, phospholipids, and other molecules. The chemistry of these molecules determines a pathogen's ability to cause disease.

Thought Question 5.1 Which molecules occur in the greatest number in a prokaryotic cell? The smallest number? Why does a prokaryotic cell contain 100 times as many lipid molecules as strands of RNA?

5.2
Bacterial Membranes and Transport

SECTION OBJECTIVES
- Describe the functions of a cell membrane.
- Explain how nutrients are transported and how energy is spent to drive transport.

What defines the existence of a cell? The defining structure is the cell membrane (**Figure 5.4**). As explained in Chapter 4, the membrane separates cytoplasm from the external medium and mediates transport between the two. The cell membrane is a phospholipid bilayer containing proteins. Membrane proteins transport molecules in and out of the cell. Other membrane proteins detect information from outside the cell, such as the level of iron in the blood of a host.

Bacterial Membrane Proteins

The cell membrane is a two-dimensional fluid within which float many proteins. The portion of the protein contained within the membrane must be hydrophobic (soluble in lipid), whereas portions extending into the cytoplasm are hydrophilic (water-soluble). Membrane proteins determine the capabilities of the cell, such as:

- **Structural support.** Some membrane proteins anchor together different layers of the cell envelope. Other proteins form the base of structures that extend out from the cell, such as pili and flagella, which enable adherence and motility.

- **Detecting environmental signals.** In *Vibrio cholerae*, the causative agent of cholera, ToxR is a transmembrane protein whose amino-terminal part reaches into the cytoplasm. When ToxR detects acidity and elevated temperature—signs of the host digestive tract—its amino-terminal domain binds to DNA at a sequence that activates expression of cholera toxin and other virulence factors (molecules involved in disease).

- **Secreting virulence factors and communication signals.** Some membrane proteins form secretion complexes to export toxins. Other membrane proteins enable cell signaling across the envelope. For example, symbiotic nitrogen-fixing bacteria use membrane proteins to signal their host plant roots to form nodules. The nodules incorporate the nitrogen-fixing bacteria into the plant cells, where they receive nutrients and provide the plant with fixed nitrogen.

- **Transport across the cell membrane.** Membrane proteins determine which substances move between the cytoplasm and the outside. They take up scarce nutrients and exclude harmful toxins.

- **Energy storage and transfer.** The membrane allows the cell to maintain different concentrations of molecules inside and outside the cell (a concentration gradient). The cell stores energy in these gradients, and membrane proteins harness that energy. For example, the TolC-AcrAB complex (see Figure 5.3) uses energy from a H^+ gradient to pump antibiotics out of the cell.

Transport of Nutrients

Whether a microbe is propelled by flagella toward a favorable habitat or, lacking motility, drifts through its environment, it must find nutrients and move them across the membrane into the cytoplasm. A few compounds, such as oxygen and carbon dioxide, can passively diffuse across the membrane, but most cannot. Nutrients may cross a membrane by binding a transporter protein, called a **permease**, that deposits the nutrient on the other side of the membrane. Alternatively, nutrients such as peptides and sugars may cross a membrane through a protein channel.

Microbes must also overcome the problem of low nutrient concentrations in the natural environment (for example, in lakes or streams). If the intracellular concentration of nutrients were no greater than the extracellular concentration, the cell would remain starved of most nutrients. To solve this dilemma, most organisms have evolved efficient transport systems that *concentrate* nutrients inside the cell relative to outside. However, moving molecules against a concentration gradient requires energy.

Some habitats have plenty of nutrients, but those nutrients may be locked in a form that cannot be transported into the cell. Starch is

Figure 5.4   Bacterial Cell Membrane and ATP Synthase

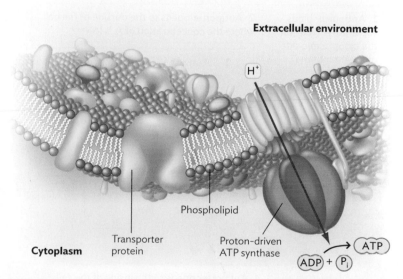

The cell membrane consists of a phospholipid bilayer, with hydrophobic fatty acid chains directed inward, away from water. Embedded proteins can transport nutrients. The proton motive force drives H^+ across the membrane through protein complexes such as ATP synthase, which makes ATP.

an example of a large, complex nutrient that cannot be transported across a membrane. To obtain complex nutrients, microbes secrete digestive enzymes that break them down into smaller compounds that are easier to transport.

PASSIVE AND ACTIVE TRANSPORT   A few nutrients, such as O_2, can be taken up by passive transport. **Passive transport** uses the concentration gradient of a compound (from higher to lower concentration) to move that compound across the membrane. The rate of passive transport can be increased by a membrane protein

ANIMATION

Figure 5.5   Coupled Transport

Symport

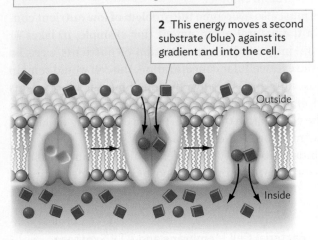

1 Energy is released as one substrate (red) moves down concentration gradient.

2 This energy moves a second substrate (blue) against its gradient and into the cell.

Outside

Inside

Antiport

1 Antiporter binds substrate A (red) on the cytoplasmic side of the membrane.

2 Antiporter opens to the outside of the cell, where the concentration of A is lower.

3 Substrate A leaves its binding site, and substrate B (blue) then binds to its site.

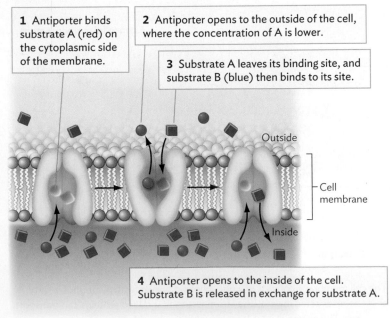

Outside

Cell membrane

Inside

4 Antiporter opens to the inside of the cell. Substrate B is released in exchange for substrate A.

The energy released by substrate A (red) traveling down its gradient drives uptake of substrate B (blue) against its gradient.

that carries the compound; this process is called **facilitated diffusion**. For most nutrients, however, the cell has to take up molecules present at low concentration outside the cell and concentrate them inside. Transport against a concentration gradient (that is, from lower to higher concentration) must spend energy. Transport that spends energy is called **active transport**. The energy spent can take the form of an energy carrier molecule, such as ATP, or an energy potential across the membrane, such as the **proton motive force** (the force driving H^+ across the membrane, discussed shortly). In either case, energy is spent to bring compounds into the cell.

The simplest way to use energy to move molecules across a membrane is to exchange the energy of one chemical gradient for that of another. The most common chemical gradients used are those of ions, particularly the positively charged ions Na^+ and K^+. These ions are kept at different concentrations on either side of the cell membrane. When an ion moves <u>down</u> its concentration gradient (from high to low), energy is released. Some transport proteins harness that released energy and use it to drive the transport of a second molecule <u>up</u>, or against, its concentration gradient. The use of energy from one gradient to drive transport up another gradient is called **coupled transport**.

Two types of coupled transport systems are **symport**, where the two molecules travel in the same direction, and **antiport**, in which the actively transported molecule moves in the direction opposite to the driving ion (**Figure 5.5**). An example of a symporter is the lactose permease LacY of *E. coli*, one of the first transport proteins whose function was elucidated. LacY brings lactose into the cell, coupled to H^+ moving inward down its gradient (symport).

TRANSPORT POWERED BY ATP   The largest family of energy-driven transport systems is the <u>A</u>TP-<u>b</u>inding <u>c</u>assette superfamily, also known as **ABC transporters**. Some ABC transporters take up nutrients; for example, *Mycobacterium tuberculosis* needs an ABC transporter to pick up iron during growth in the lungs. Others, called **efflux transporters**, expel hazardous wastes. The efflux ABC transporters are generally used as multidrug efflux pumps that help microbes expel antibiotics. *Lactococcus*, for example, can use one pump, LmrP, to export many types of antibiotics, including tetracyclines, quinolones, macrolides, and aminoglycosides.

An ABC transporter (**Figure 5.6**) consists of two proteins that form a membrane channel and two cytoplasmic proteins that contain a conserved amino acid motif that binds ATP (called an ATP-binding cassette). An additional membrane protein called a **substrate-binding protein** initially binds the substrate (also called a solute). The substrate-binding protein then transfers its substrate to the channel protein. **Hydrolysis** of ATP to ADP plus P_i yields energy to open the channel, allowing the solute to enter the cell.

Link   As discussed in Section 4.4, **hydrolysis** is a chemical process in which a molecule is split into two parts by the addition of a molecule of water. A great deal of energy is released in the hydrolysis of the high-energy molecule ATP.

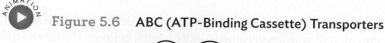

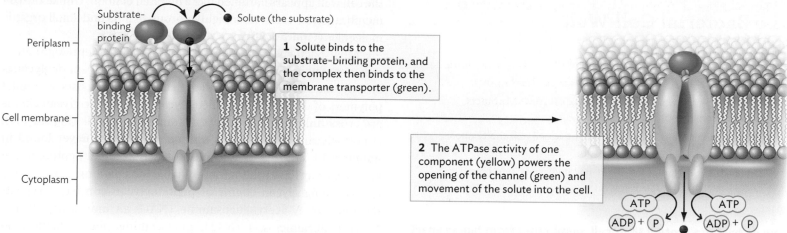

Figure 5.6   ABC (ATP-Binding Cassette) Transporters

Periplasm — Substrate-binding protein   Solute (the substrate)

1 Solute binds to the substrate-binding protein, and the complex then binds to the membrane transporter (green).

Cell membrane —

Cytoplasm —

2 The ATPase activity of one component (yellow) powers the opening of the channel (green) and movement of the solute into the cell.

ATP → ADP + P   ATP → ADP + P

SIDEROPHORES

SIDEROPHORES   Minerals needed by cells are often inaccessible in the environment outside. Iron, an essential nutrient of most cells, is mostly locked up as iron hydroxide, $Fe(OH)_3$, which is insoluble in water and therefore cannot be obtained by diffusion. Thus, in addition to membrane transporters, bacteria such as *M. tuberculosis* secrete iron-binding molecules called siderophores (Greek for "iron bearer"). The siderophores tightly bind whatever ferric iron is available in the environment. These iron scavenger molecules are produced and sent forth by cells when the intracellular iron concentration is low (**Figure 5.7**). In the example shown,

the siderophore binds iron in the environment, and the siderophore-iron complex then attaches to specific receptors in the bacterial outer membrane. The siderophore-Fe^{3+} complex may then be transported across the cell membrane by an ABC transporter.

PROTON MOTIVE FORCE   Energy from a cell's metabolism is stored in a form of potential energy across the membrane known as the proton motive force (PMF). PMF is generated when chemical energy is used to pump protons outside the cell, so that the proton concentration (H^+) is greater outside the cell than inside. Because protons are positively charged, proton movement across the cell membrane produces an electrical gradient, making the inside of the cell more negatively charged than the outside. The energy stored in the proton motive force (the H^+ concentration difference plus the charge difference) can be used to move nutrients into the cell via specific transport proteins, to drive motors that rotate flagella, and to drive synthesis of ATP (see Figure 5.4). The proton-powered ATP synthase is a target for antibiotics; for example, the diarylquinoline antibiotics target the ATP synthase of *Mycobacterium tuberculosis*. Proton motive force and energy use are discussed further in Chapter 7.

Figure 5.7   Iron Transport by a Siderophore and an ABC Transport Complex

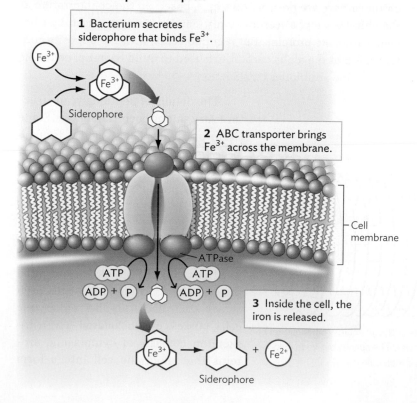

1 Bacterium secretes siderophore that binds Fe^{3+}.

Fe^{3+}

Fe^{3+}

Siderophore

2 ABC transporter brings Fe^{3+} across the membrane.

— Cell membrane

— ATPase

ATP → ADP + P   ATP → ADP + P

3 Inside the cell, the iron is released.

Fe^{3+} → + Fe^{2+}

Siderophore

SECTION SUMMARY

- **The bacterial cell membrane** contains proteins for functions such as transport and cell communication.

- **Active transport requires input of energy** from a chemical reaction or from an ion gradient across the membrane. Ion gradients generated by membrane pumps store energy for cell functions.

- **Antiporters and symporters are coupled transport systems** in which energy released by moving an ion from a region of high concentration to one of low concentration is used to move a desired solute against its concentration gradient.

- **ABC transporters use the energy from ATP hydrolysis** to move solutes "uphill" against their concentration gradients.

- **Siderophores are secreted to bind ferric iron and transport it into the cell.**

5.3
The Bacterial Cell Wall and Outer Layers

SECTION OBJECTIVES

- Describe the cell wall structure, and explain how it protects bacteria from osmotic shock.
- Explain the function of the Gram-positive cell wall and teichoic acids.
- Explain the function of the Gram-negative outer membrane, LPS, and periplasm.

How do bacteria protect their cell membranes from falling apart under pressure? Most bacteria have at least one structural supporting layer outside the cell membrane: the cell wall. A few classes of bacteria lack cell walls, most notably the mycoplasmas that cause pneumonia, and the chlamydias that cause the blinding eye disease trachoma and sexually transmitted infections. Mycoplasmas and chlamydias lost their cell walls through degenerative evolution (discussed shortly). Bacteria that do possess cell walls often have additional coverings outside, such as the outer membrane of Gram-negative bacteria (as seen in Figure 5.1).

The Cell Wall Is a Single Molecule

The cell wall makes a bacterial cell rigid and helps it withstand the intracellular turgor pressure that can build up as a result of osmotic pressure, or osmotic shock (discussed in Section 4.5). Historically, the cell wall was the target of the first antibiotic discovered, penicillin; the cell wall remains one of the most important antibiotic targets.

The bacterial cell wall consists of a single interlinked molecule that encloses the entire cell. The molecule has been isolated from *E. coli* and visualized by TEM (Figure 5.8A). In the image shown, the cell wall appears flattened like a deflated balloon. Unlike the cell membrane, the cell wall is highly porous to ions and small organic molecules (Figure 5.8B).

Most bacterial cell walls are composed of peptidoglycan, a polymer of peptide-linked chains of amino sugars. Peptidoglycan is synonymous with **murein** ("wall molecule") and consists of parallel polymers of disaccharides called **glycan** chains. The glycan chains are cross-linked with peptides of four amino acids (Figure 5.9). Peptidoglycan, or murein, is unique to bacteria, never found in human cells. For this reason, antibiotics that target peptidoglycan synthesis may have minimal side effects for a patient.

The long glycan chains of peptidoglycan consist of repeating units of *N*-acetylglucosamine (NAG, an amino sugar) and *N*-acetylmuramic acid (NAM, glucosamine plus a lactic acid group; see Figure 5.9). Parallel glycan strands are linked by short peptide cross-bridges containing four to six amino acid residues. The sequence of the cross-bridge-forming peptide is L-alanine, D-glutamic acid, *m*-diaminopimelic acid, and D-alanine (sometimes two of these). Note that this peptide contains two amino acids in the unusual D mirror form, D-glutamic acid and D-alanine. The L-alanine of a peptide cross-bridge is attached to NAG on one glycan strand, whereas the second amino group of *m*-diaminopimelic acid attaches to D-alanine from a neighboring peptide on a parallel glycan chain (attached to NAM). Formation of a cross-link removes the second D-alanine (fifth in the chain). The cross-linked peptides of neighboring glycan strands form the cage of the peptidoglycan.

Link As illustrated in Figure 4.17, a chiral molecule has a nonsuperimposable **mirror form**. Amino acid isomers, which contain a chiral carbon, are designated L or D, based on the configuration at the chiral (or alpha) carbon. Only L-amino acids are used by ribosomes to make proteins. But D-amino acids, with the opposite configuration at the alpha carbon, are used to build bacterial cell walls and peptide antibiotics (see Section 4.4).

The details of peptidoglycan structure vary among bacterial species. Some Gram-positive species, such as *Staphylococcus aureus* (a cause of toxic shock syndrome), have peptides linked by bridges of pentaglycine instead of the D-alanine link to diaminopimelic acid. In Gram-negative species, the diaminopimelic acid is linked to the outer membrane, as discussed shortly.

Does the cell wall alone determine the cell shape? In fact, recent research shows that bacteria, like eukaryotic cells, possess a **cytoskeleton**. The cytoskeleton consists of cytoplasmic proteins that mold the cell into a form

Figure 5.8　The Peptidoglycan Cell Wall

A. Entire cell wall isolated from *Escherichia coli*.

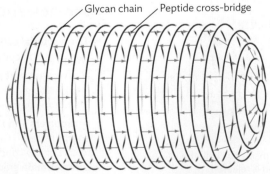

B. The cell wall consists of glycan chains (parallel rings) linked by peptides (arrows). The spaces between links are open, porous to large molecules.

Figure 5.9 Peptidoglycan Cross-Bridge Formation

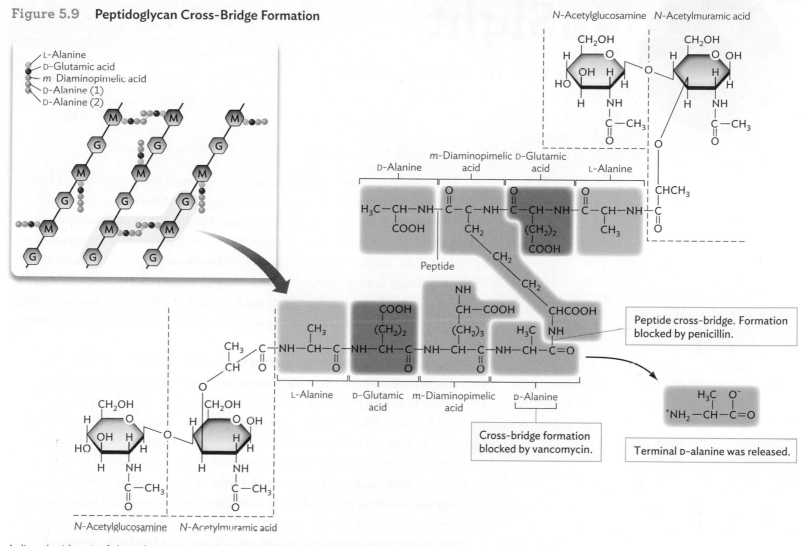

A disaccharide unit of glycan has an attached peptide of four to six amino acids. The peptide bonds with *N*-acetylmuramic acid. On the peptide, the extra amino group of *m*-diaminopimelic acid can cross-link to the carboxyl terminus of a neighboring peptide. The addition of D-alanine to the peptide is blocked by vancomycin, and the cross-bridge formation by transpeptidase is blocked by penicillin.

that is straight or curved, even helical. The bacterial proteins show intriguing homology (genetic similarity) to proteins of the better-known eukaryotic cytoskeleton (discussed in Section 5.6).

Peptidoglycan Synthesis as a Target for Antibiotics

Synthesis of peptidoglycan requires many genes encoding enzymes to make the special sugars, build the peptides, and seal the cross-bridges. These biosynthetic enzymes make great targets for antibiotics (discussed in Chapter 13). For example, the transpeptidase that cross-links the peptides is the target of penicillin (Figure 5.9). Vancomycin, a major defense against *Clostridium difficile* and drug-resistant staphylococci, prevents cross-bridge formation by binding the terminal D-Ala-D-Ala dipeptide, thus preventing release of the terminal D-alanine.

Unfortunately, the widespread use of such antibiotics selects for evolution of resistant strains. One of the most common agents of resistance is the enzyme beta-lactamase, which cleaves penicillin, preventing it from inhibiting transpeptidase. Strains resistant to vancomycin contain an altered enzyme that adds lactic acid to the end of the branch peptides in place of the terminal D-alanine. The altered enzyme is no longer blocked by vancomycin. Vancomycin is presently the "drug of last resort" for methicillin-resistant *Staphylococcus aureus* (MRSA).

Most of our antibiotics, including vancomycin, were actually discovered in harmless environmental bacteria, particularly actinomycete bacteria growing in soil (see **inSight**). As new forms of drug resistance emerge, researchers continue to seek new antibiotics that target cell wall formation.

inSight

Vancomycin: The Gift of Soil Bacteria

Vancomycin, which targets the bacterial cell wall, is the antibiotic of last resort for deadly infections such as pseudomembranous colitis, a common hospital-acquired infection caused by the Gram-positive bacterium *Clostridium difficile*, and abscesses caused by methicillin-resistant *Staphylococcus aureus* (MRSA). Deadly MRSA infections can arise from hospital complications or from a seemingly innocuous sports-related injury. Both *C. difficile* and MRSA are resistant to other drugs and therefore require treatment with front-line drugs such as vancomycin. Where did vancomycin originate—and where will other new drugs be found?

It may come as a surprise that most of the antibiotics we use against bacteria today were discovered out in the environment, in soil bacteria. A handful of soil is more than dirt: it contains one of the most complex natural environments (Figure 1A). Soil combines minerals from powdered bedrock with decaying leaves (detritus) and animal bodies, plus the waste products of many living organisms. These materials provide a habitat for countless species of microbes, including fungi, protists, and bacteria, as well as viruses that infect them. Any single sample of soil contains species of microbes unknown to science. And unknown microbes may produce unknown antibiotics.

Why do microbes produce antibiotic molecules? The limited nutrients of soil give rise to a fierce competition. And some bacteria use a particularly devious stratagem. The Actinomycetes (discussed in Chapter 10) are a class of bacteria that grow in the soil as long, fungus-like filaments of cells (Figure 1B). Actinomycetes

Figure 1
Soil Bacteria

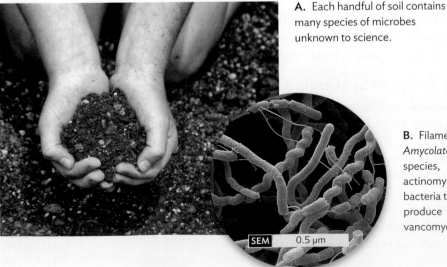

A. Each handful of soil contains many species of microbes unknown to science.

SEM 0.5 μm

B. Filaments of *Amycolatopsis* species, actinomycete bacteria that produce vancomycin.

have large genomes that encode many enzymes for digesting many kinds of organic molecules—including the complex constituents of fellow bacteria. As an actinomycete filament grows new cells at one end, the older cells at the other end age and die, lysing to release their contents. The lysed cell contents attract other soil bacteria to come and feed. But then the newer actinomycete cells release antibiotics that kill the feeding bacteria—so instead, the actinomycete feeds on them.

Of course, the other soil bacteria evolve resistance to the actinomycete antibiotics; so, actinomycete bacteria must continually evolve new ones. We can discover these antibiotics by isolating and culturing new species of actinomycetes. In 1953, a missionary in Borneo sent a sample of jungle soil to Eli Lilly and Company, a pharmaceutical firm. At Lilly, microbiologist Edmund Kornfeld isolated a species now called *Amycolatopsis orientalis* that produced a new antibiotic molecule (Figure 2). The molecule was called vancomycin because it vanquishes bacteria resistant to older antibiotics such as penicillins and metronidazole. Vancomycin binds to the terminal alanine dipeptide of a growing cell wall, preventing cross-bridge formation (see Figure 5.9). The structure of vancomycin is so complex that no organic chemist would have invented it; yet bacteria in nature continually evolve by natural selection to produce such new molecules.

But just as previous strains of *S. aureus* mutated and evolved resistance to methicillin, the previous wonder drug, today some MRSA strains have already grown resistant to vancomycin. So we must continue to discover new antibiotics.

Figure 2   **Vancomycin**

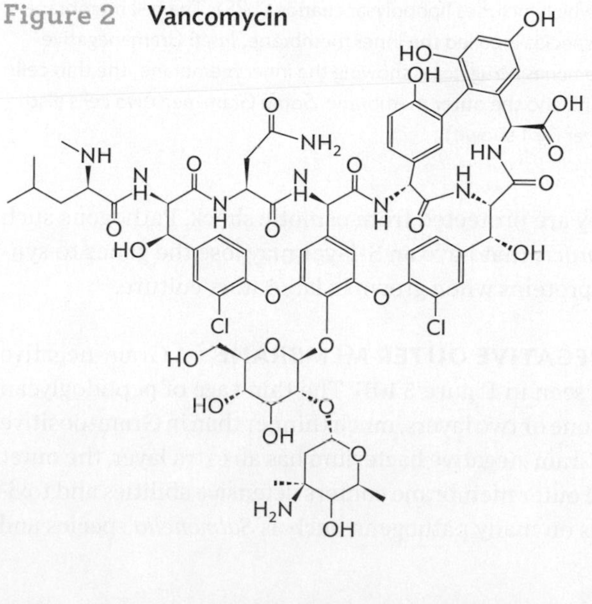

Gram-Positive and Gram-Negative Outer Layers

Outside the cell membrane, most bacteria have additional layers of cell envelope—materials that provide structural support and protection from predators and host defenses (Figure 5.10). Outer structures are attached to the cell wall and cell membrane, in some cases interpenetrating them. These structures define two major categories of bacteria distinguished by the Gram stain (discussed in Section 3.4):

- **Gram-positive bacteria** have a thick cell wall with multiple layers of peptidoglycan, threaded by teichoic acids. Gram-positive species include species in the phylum Firmicutes. The Firmicutes include many pathogens, such as *Bacillus anthracis*, the cause of anthrax, and *Streptococcus pyogenes*, the cause of "strep throat."

- **Gram-negative bacteria** have a thin cell wall (single layer of peptidoglycan) enclosed by an outer membrane. Gram-negative species include species in the phylum Proteobacteria, such as *Escherichia coli* and nitrogen-fixing rhizobia. Among the many Gram-negative pathogens is the species *Pseudomonas aeruginosa*, a cause of wound infections.

Besides these two groups, however, many kinds of bacteria do not fit the Gram stain models. For example, the mycobacteria have exceptionally complex cell walls containing waxy outer layers that exclude antibiotics (discussed shortly).

THE GRAM-POSITIVE CELL ENVELOPE A section of a Gram-positive cell envelope is shown in Figure 5.10A. The Gram-positive cell wall consists of multiple layers of peptidoglycan, up to 40 in some species. The peptidoglycan is reinforced by **teichoic acids** threaded through its multiple layers. Teichoic acids are chains of phosphodiester-linked glycerol or ribitol, with sugars or amino acids linked to the middle OH groups. The phosphodiester links are deprotonated, with negative charge. The negatively charged cross-threads of teichoic acids, as well as the overall thickness of the Gram-positive cell wall, help retain the Gram stain. Teichoic acids can help pathogens attach to host cells. They are "bacterial signatures" recognized by the immune system.

Outside the cell wall, Gram-positive cells may have a slippery capsule consisting of loosely bound polysaccharides (not shown in the diagram). The capsule can be visualized under the light microscope by using a negative stain consisting of suspended particles of India ink, which the capsule excludes.

An additional protective layer commonly found in Gram-positive cells is the surface layer, or **S-layer**. The S-layer is an important virulence factor for pathogens, such as *Bacillus anthracis*, the cause of anthrax. The S-layer is a crystalline layer of thick subunits consisting of protein or glycoprotein (proteins with attached sugars). Each subunit of the S-layer contains a pore large enough to admit a wide range of molecules. The subunits form a smooth layer on the cell wall or outer membrane (see Figure 5.10A). The proteins are arranged in a highly ordered array that can exclude predators

Figure 5.10   Gram-Positive and Gram-Negative Cell Envelopes

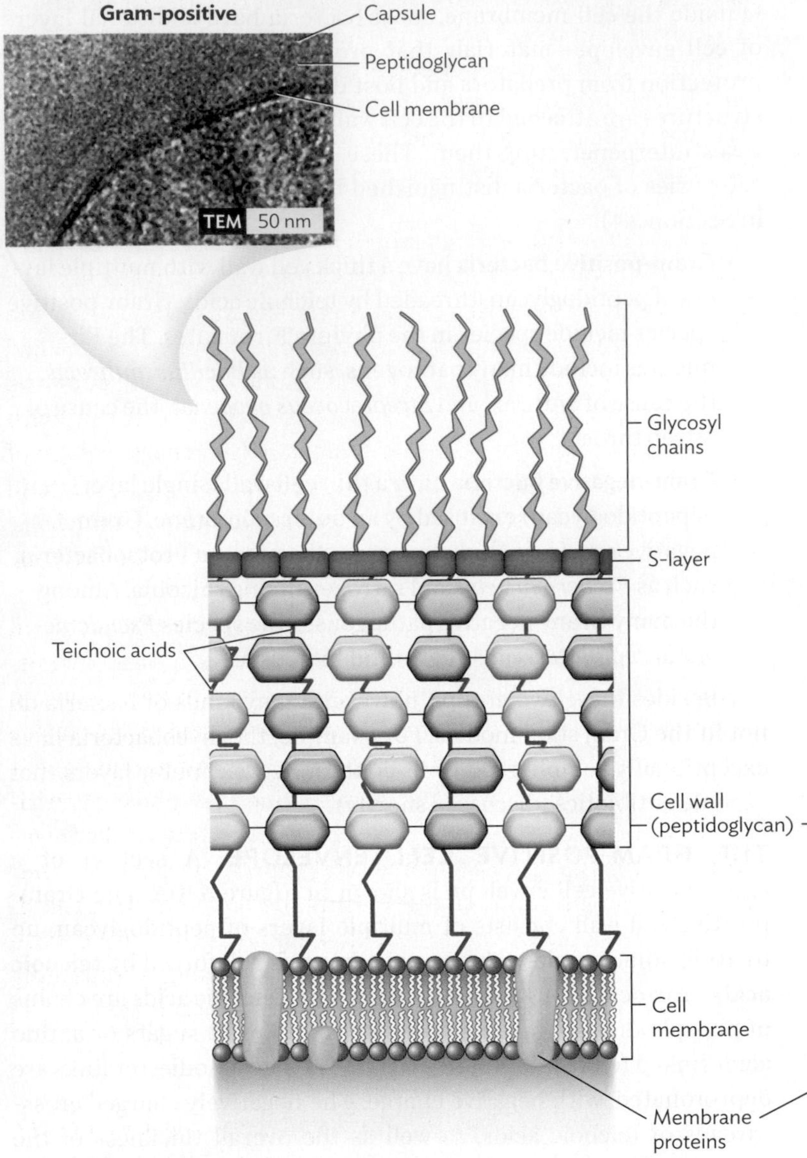

Gram-positive

Capsule
Peptidoglycan
Cell membrane

TEM 50 nm

Gram-negative

Outer membrane
Peptidoglycan
Inner membrane

TEM 50 nm

Glycosyl chains

S-layer

Teichoic acids

Cell wall (peptidoglycan)

Cell membrane

Membrane proteins

LPS

Outer membrane

Porin

Lipoproteins

Periplasm

Inner membrane

A. The Gram-positive cell has a thick cell wall with multiple layers of peptidoglycan, threaded by teichoic acids. The cell wall may be covered by an S-layer (surface array of proteins). In some Gram-positive species, carbohydrate filaments form a capsule. *Inset:* Gram-positive envelope of *Bacillus subtilis*, showing cell membrane, cell wall, and capsule.

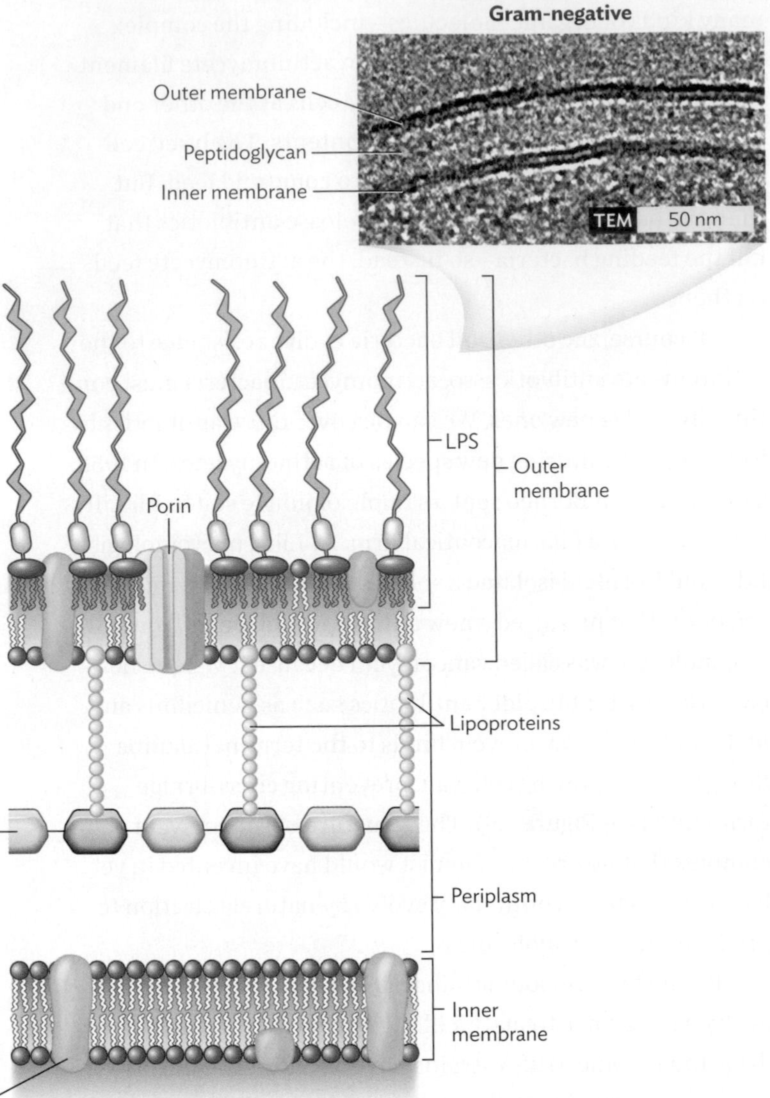

B. The Gram-negative cell has a single layer of peptidoglycan covered by an outer membrane, which includes lipopolysaccharide (LPS). The cell membrane of Gram-negative species is called the inner membrane. *Inset:* Gram-negative envelope of *Pseudomonas aeruginosa*, showing the inner membrane, the thin cell wall in the periplasm, and the outer membrane. Some Gram-negative cells also have an outer S-layer (not shown).

and parasites. The S-layer is permeable, allowing substances to pass through it in either direction.

Many bacterial species have lost parts of their ancestral outer layers through evolution. A trait can be lost in the absence of selective pressure for genes encoding the trait. This evolutionary loss is called **reductive evolution** or degenerative evolution (discussed in Chapter 10). For example, the mycoplasmas that cause pneumonia have permanently lost their cell walls. Mycoplasmas are close relatives of Gram-positive bacteria, but they have no need for cell walls because they are parasites living in host environments, such as the human

lung, where they are protected from osmotic shock. Pathogens such as *Bacillus anthracis* that have an S-layer may lose the genes to synthesize S-layer proteins when grown in laboratory culture.

THE GRAM-NEGATIVE OUTER MEMBRANE   A Gram-negative cell envelope is seen in Figure 5.10B. The thin cage of peptidoglycan consists of only one or two layers, much thinner than in Gram-positive bacteria. But a Gram-negative bacterium has an extra layer, the outer membrane. The outer membrane confers defensive abilities and toxigenic properties on many pathogens, such as *Salmonella* species and enterohemorrhagic *E. coli* (strains that cause hemorrhaging of the colon). Between the outer and inner (cell) membranes is the periplasm. The periplasm contains distinct proteins not found in the cytoplasm, such as proteins that protect the cell from dangerous oxidants produced by white blood cells.

The inward-facing leaflet of the outer membrane has a phospholipid composition similar to that of the cell membrane (in Gram-negative species, it is called the **inner membrane** or inner cell membrane). The outer membrane's inward-facing leaflet includes lipoproteins that connect the outer membrane to the peptide bridges of the cell wall. The major lipoprotein is called **murein lipoprotein** (Figure 5.11). Murein lipoprotein consists of a protein inserted in the inward-facing leaflet of the outer membrane that forms a peptide bond with the *m*-diaminopimelic acid of peptidoglycan (murein).

The outward-facing leaflet, however, consists of a special kind of membrane lipid called **lipopolysaccharide (LPS)**. LPS is of crucial

medical importance because its lipid A component acts as an **endotoxin**. An endotoxin is a cell component that is harmless so long as the pathogen remains intact, but when released by a lysed cell, the endotoxin (lipid A) overstimulates host defenses, inducing potentially lethal endotoxic shock (dangerously low blood pressure and high fever). Thus, antibiotic treatment of an LPS-containing pathogen can kill the cells but can also cause a reaction that kills the patient.

The LPS has three parts: the **lipid A** anchored in the outer membrane, a core polysaccharide, and the O antigen (O polysaccharide). The lipid A in LPS has shorter fatty acid chains than those of the inner cell membrane, and some are branched. Lipid A is attached to a core polysaccharide, a sugar chain that extends outside the cell. The core polysaccharide consists of about five sugars with side chains. It extends to the O antigen, a polysaccharide chain of as many as 200 sugars. The O antigen polysaccharide may extend longer than the cell itself. These polysaccharide chains form a layer that helps a pathogen resist phagocytosis by white blood cells. Some Gram-negative cells also have an outer S-layer.

The outer membrane also contains unique proteins not found in the inner membrane. Outer membranes contain a class of transporters called **porins** that permit entry of nutrients such as sugars and peptides. Cells express different outer membrane porins under different environmental conditions. In a dilute environment, cells express porins of large pore size, maximizing the uptake of nutrients. In a rich environment—for example, within a host—cells downregulate expression of large porins and express porins of smaller pore size, selecting only smaller nutrients and avoiding the uptake of toxins. For example, the porin regulation system of Gram-negative bacteria enables them to grow in the intestinal region containing bile salts—a hostile environment for Gram-positive bacteria.

The Mycobacterial Envelope

Mycobacteria such as *Mycobacterium leprae*, the cause of leprosy, and *Mycobacterium tuberculosis*, the cause of tuberculosis, are considered Gram-positive; but their envelopes are exceptionally thick and complex, including extra layers not found in other Gram-positive cells.

CASE HISTORY 5.1

A Rash Reveals an Ancient Disease

In the Indian state of Bihar, a school nurse noticed that a 13-year-old boy, Naranjan, had swellings on his nose and lips. When questioned, Naranjan complained of a pruritic (itchy) rash on his arms and legs. Upon examination, the nurse found pale patches of skin on his back. The pale patches lacked sensation, and there was partial loss of sensation in Naranjan's wrists and forearms. The patient was diagnosed with leprosy, caused by the bacterium *Mycobacterium leprae*. The bacteria stain acid-fast. The bacteria are spread primarily by

Figure 5.11 Lipoprotein and Lipopolysaccharide in the Gram-Negative Envelope

Murein lipoprotein consists of a protein embedded in the inward-facing leaflet of the outer membrane. The protein's C-terminal lysine bonds with the peptidoglycan (murein) cell wall.

nasal secretions. For unknown reasons, only 5% of people exposed to the bacteria are susceptible to infection. In a susceptible host, the bacteria can grow slowly for many years without symptoms. They infect the peripheral nerves, causing loss of sensation, and the body's immune reaction generates skin lesions. To halt the disease, Naranjan was started on a 2-year course of multidrug therapy combining dapsone, clofazimine, and rifampin. To monitor hepatotoxicity, liver function tests were ordered. Naranjan's family and neighbors were screened; one member, an uncle, showed skin lesions. The uncle was treated also. Leprosy is difficult to eradicate because of the long incubation period and because people hide their symptoms, fearing stigma as "lepers."

Figure 5.12A shows typical facial lesions of leprosy. The bacteria need to grow at a temperature lower than that controlled by the body; thus, lesions typically occur in superficial parts of the body with lower temperature, as shown by this infrared scan (Figure 5.12B). However, leprosy can develop in various forms in different parts of the body, often mistaken for other conditions. In most cases, lesions develop over many years because the bacteria grow so slowly; in laboratory culture, cell doubling may take as long as six weeks. Transmission is rare because only 5% of the human population is susceptible; thus, most industrial countries have eliminated leprosy simply by treating all known cases. But in societies where leprosy is well established, and where people hide their symptoms, the disease is hard to eradicate.

Why does *M. leprae* grow so slowly? The mycobacterial envelope is unusually thick, with several extra layers (Figure 5.13). The envelope allows identification by the acid-fast stain (discussed in Chapter 3). The outermost layer of capsule is a "waxy" coating (with benzene-like aromatic rings) that is extremely hydrophobic and

Figure 5.12   Leprosy Is Caused by *Mycobacterium leprae*

A. The characteristic skin lesions of leprosy take many years to develop.

B. Infrared image shows cooler areas of the skin (blue color), where leprosy lesions most commonly form.

Figure 5.13   Mycobacterial Envelope Structure

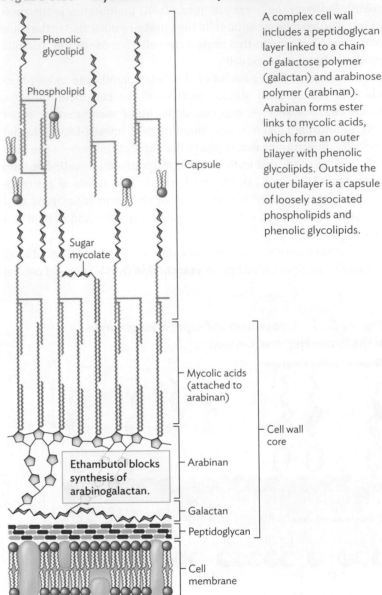

A complex cell wall includes a peptidoglycan layer linked to a chain of galactose polymer (galactan) and arabinose polymer (arabinan). Arabinan forms ester links to mycolic acids, which form an outer bilayer with phenolic glycolipids. Outside the outer bilayer is a capsule of loosely associated phospholipids and phenolic glycolipids.

retards phagocytosis by lymphocytes. The waxy coating excludes many antibiotics and offers exceptional protection from host defenses, enabling the pathogens of tuberculosis and leprosy to colonize their hosts over long periods. However, the thick, waxy layer also retards uptake of nutrients. As a result, *M. tuberculosis* and *M. leprae* grow extremely slowly. Similar waxy cell walls are found in actinomycetes, soil bacteria that produce many important antibiotics, such as streptomycin and vancomycin.

Mycobacterial cell envelopes include unusual lipids called **mycolic acids** (see Figure 5.13). Mycolic acids contain a backbone with two hydrocarbon chains—one comparable in length to typical membrane lipids (about 20 carbons), the other about threefold longer. Mycolic acids provide the basis for acid-fast staining, an important diagnostic test for mycobacteria and actinomycetes (described in Section 3.4). In *M. tuberculosis* and *M. leprae*, the mycolic acids form a kind of bilayer interleaved with "waxy" phenolic glycolipids. The mycolic acids and peptidoglycan are linked to chains of arabinogalactan, composed of galactose and the five-carbon sugar arabinose. Ethambutol, a major drug against tuberculosis, can inhibit arabinogalactan biosynthesis.

Other drugs effective against mycobacteria include dapsone (inhibits synthesis of nitrogenous bases for DNA), clofazimine (binds to guanine and blocks DNA synthesis), and rifampicin (inhibits synthesis of messenger RNA). Development of such drugs requires understanding the cell processes they target, as introduced in this chapter.

SECTION SUMMARY

- **The cell wall maintains turgor pressure.** The cell wall is porous, but its rigid network of covalent bonds protects the cell from osmotic shock.

- **The Gram-positive cell envelope** has multiple layers of peptidoglycan, interpenetrated by teichoic acids.

- **The S-layer**, composed of proteins, is highly porous but can prevent phagocytosis. In archaea, the S-layer serves the structural function of a cell wall.

- **The capsule**, composed of polysaccharide and glycoprotein filaments, protects cells from phagocytosis. Both Gram-positive and Gram-negative cells may possess a capsule.

- **The Gram-negative outer membrane** regulates nutrient uptake and excludes toxins. The envelope layers include protein pores and transporters with various degrees of selectivity.

- **Mycobacteria have an exceptionally thick and complex cell envelope, including mycolic acids.** The mycobacterial envelope excludes most antibiotics and prevents phagocytosis, but it also slows nutrient entry, leading to slow growth.

Thought Question 5.2 Why do antibiotics differ in their effectiveness against Gram-positive and Gram-negative bacteria?

5.4
The Nucleoid and Bacterial Cell Division

SECTION OBJECTIVES
- Describe how DNA is organized within the bacterial cell.
- Explain how DNA replication is coordinated with cell growth and division.

How does a bacterium protect its genomic DNA? An important function of the cell envelope is to contain and protect the cell's genome. **Figure 5.14** compares the organization of chromosomal material in enteropathogenic *E. coli* cells with that in a cultured human cell that they have colonized. In this thin-section TEM, each bacterium contains a filamentous nucleoid region that extends through the cytoplasm. In contrast, the nucleus of the eukaryotic

Figure 5.14   Bacterial Nucleoid and Eukaryotic Nucleus

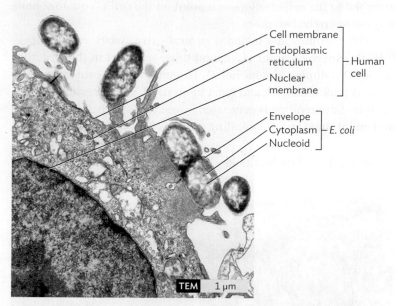

Enteropathogenic *Escherichia coli* bacteria attaching to the surface of a tissue-cultured human cell. The human cell has a well-defined nucleus surrounded by a nuclear membrane. By contrast, the *E. coli* cells lack a nuclear membrane; their DNA is more loosely arranged in the nucleoid region.

cell, only a fraction of which is visible in the figure, is many times larger than the entire bacterial cell, and the chromosomes it contains are separated from the cytoplasm by the nuclear membrane.

What is the physiological and reproductive state of a bacterial cell? A bacterial cell may be growing and reproducing new cells, or it may be nongrowing. A nongrowing cell may exist in a "stationary" state, metabolically active without expanding its cytoplasm; alternatively, its metabolism may be suspended completely, as in the case of a *Bacillus anthracis* spore that can persist for decades in the soil. But in bacteria, a growing cell is always a dividing cell. Unlike the cells of eukaryotes—which undergo mitotic division at a time separate from cell expansion—bacterial cells grow and divide as a continuous process. In a growing bacterial cell, DNA replicates in coordination with the expansion of the cell wall and ultimately the separation of the cell into two daughter cells.

Here we describe the organization of genetic material within a bacterial cell. The genetic processes involving DNA, RNA, and protein are discussed further in Chapter 8.

Bacterial DNA Is Organized in a Nucleoid

All living cells on Earth possess chromosomes consisting of DNA. In bacteria and archaea, the DNA genome usually consists of a single circular chromosome, but some bacterial species have a linear chromosome or multiple chromosomes. In this chapter, we focus on the simple case of a single circular chromosome.

Bacterial DNA is organized in a structure called the nucleoid (**Figure 5.15A**). The nucleoid contains loops of DNA held together by DNA-binding proteins (**Figure 5.15B**). All the loops connect back to a central point called the **origin of replication (*ori*)**, which is attached to the cell envelope at a point on the cell's "equator," halfway between the two poles.

The chromosome includes several extra twists, called supercoils or superhelical turns, beyond those inherent in the structure of the DNA duplex (double helix). The supercoiling causes portions of DNA to double back and twist upon itself, which results in compaction. Supercoiling is generated by enzymes such as DNA gyrase and maintained by DNA-binding proteins (shown in green in Figure

5.15B). The enzyme gyrase is a major target for antibiotics such as quinolones, which are used, for example, to treat bacterial pneumonia and urinary tract infections.

The information encoded in DNA is "read" by the processes of transcription and translation to yield gene products (discussed in Chapter 8). As bacteria grow, both transcription and translation occur at top speed while the DNA itself is being replicated. This remarkable coordination of replication, transcription, and translation is a hallmark of the bacterial cell; it explains why some bacteria can double in as little as 10 minutes.

Bacterial Cell Division

Most bacteria have a circular chromosome that begins to replicate at its origin, proceeding in both directions (bidirectionally) all around the circle. At the origin, the DNA double helix begins to unzip, forming two replication forks (discussed in Chapter 8). At each replication fork, DNA polymerase synthesizes DNA (**Figure 5.16**). Growing bacteria replicate their DNA continuously, with no resting phase. Note that the cell must expand its cytoplasm at the

Figure 5.15   The Nucleoid

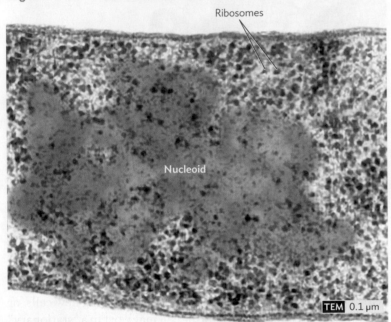

A. The *E. coli* nucleoid appears as clear regions that exclude the dark-staining ribosomes and contain DNA strands (colorized violet).

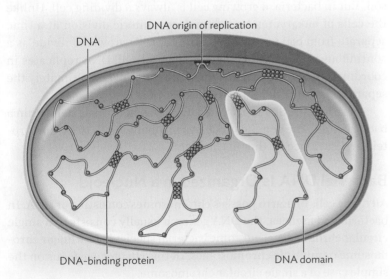

B. The nucleoid forms approximately 50 loops of chromosome called domains (shown shaded), which radiate from the center. Within each domain, the DNA is supercoiled and partly compacted by DNA-binding proteins (shown in green).

Figure 5.16   Replisome Movement within a Dividing Cell

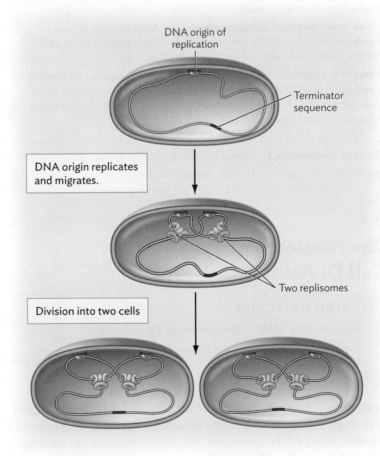

The DNA origin-of-replication sites (green) move apart in the expanding cell as the replisomes (DNA polymerase complexes) proceed in opposite directions around the circular chromosome until they meet at the terminator sequence. Replication of the termination sequence triggers septation.

same time that it synthesizes DNA. In a rod-shaped cell, the cell envelope and cell wall must elongate as well to maintain progeny of even girth and length.

Once DNA has been replicated completely to form two daughter genomes, the cell may divide to form two daughter cells. Replication of the DNA termination site triggers growth of the dividing partition of the envelope, called the **septum**. The septum grows inward from the sides of the cell, at last constricting and sealing off the two daughter cells in the process of **septation**. Envelope extension and septation require rapid biosynthesis of all envelope components, including membranes and cell wall. The pattern of septation differs among bacterial species, causing differences in the appearance of cells that provide clues to diagnosis.

CASE HISTORY 5.2

Deadly Infection Acquired from a Hospital

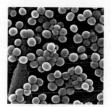

Jennifer, age 23, had just graduated from college with a Fulbright fellowship to study sociology in Africa. Before leaving the United States, she obtained a routine physical exam at a large city hospital. Upon returning home from the hospital, she developed a swelling in her leg (**Figure 5.17A**). When she returned to the hospital, she was told that the swelling represented an allergic reaction and was given anti-inflammatory agents. The swelling grew and within a day the skin ruptured with a bloody discharge. Upon return to the emergency room, Jennifer at last received the correct diagnosis of methicillin-resistant staphylococcus infection (MRSA; **Figure 5.17B**). A nurse commented that hospital visitors can acquire MRSA; the infectious agent is endemic at many hospitals in the United States and is very difficult to eradicate. She showed Jennifer a micrograph of the bacteria, *Staphylococcus aureus*, which septate in alternating division planes, thus forming clusters of cells. Jennifer required many weeks of treatment with the antibiotics doxycycline and clindamycin (inhibits protein synthesis) before the infection resolved, and she had to postpone her fellowship for a year.

The bacterial pattern that was observed in Jennifer's infection is characteristic of *Staphylococcus* species. Spherical cells (cocci), such as *Staphylococcus aureus*, do not elongate their cell walls as do rod-shaped bacilli. In cocci, the septation process generates the new cell envelope material to enclose the expanding cytoplasm (**Figure 5.18**). Furrows form in the cell envelope, in a ring all around the cell equator, as a new cell wall grows inward. The wall material must form two separable partitions. When the partitions are complete, the

Figure 5.17   Methicillin-Resistant *Staphylococcus aureus* (MRSA) Infection

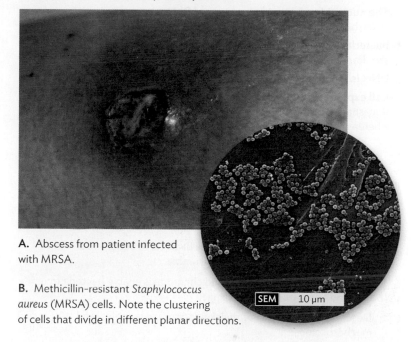

A. Abscess from patient infected with MRSA.

B. Methicillin-resistant *Staphylococcus aureus* (MRSA) cells. Note the clustering of cells that divide in different planar directions.

two progeny cells peel apart. The facing halves of each cell consist of entirely new cell wall.

The spatial orientation of septation determines the arrangement of cocci. Cells form chains when they septate in parallel planes, such as in *Streptococcus* species. If septation occurs in random orientations or if cells reassociate loosely after septation, they form compact hexagonal arrays similar to the grape clusters portrayed in classical paintings, hence the Greek derived term **staphylococci** (*staphyle* is Greek for "bunch of grapes") (see Figure 5.17B). Such grapelike clusters are found in colonies of *Staphylococcus aureus*. In other species, where septation occurs at precise right angles to the previous division, the cells form tetrads and even cubical groups of eight. *Micrococcus* bacteria, commonly found in dust and in the air we breathe, form tetrads.

 Figure 5.18   Septation in *Staphylococcus aureus*

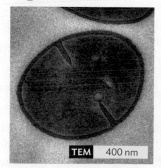 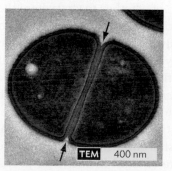 

A. Furrows appear in the cell envelope all around the cell equator as a new cell wall grows inward.

B. Two new envelope partitions are complete.

C. The two daughter cells peel apart. The facing halves of each cell contain entirely new cell wall.

SECTION SUMMARY

- **The nucleoid region contains loops of DNA,** supercoiled and bound to DNA-binding proteins.
- **Bacterial DNA is transcribed to RNA in the cytoplasm,** often simultaneously with DNA replication.
- **DNA is replicated bidirectionally by DNA polymerase.**
- **Cell expansion and septation are coordinated with DNA replication.** Rod-shaped cells (bacilli) elongate their cell walls before septation, whereas spherical cells (cocci) expand their cell walls during septation.

5.5
Specialized Structures of Bacteria

SECTION OBJECTIVES

- Describe how pili and stalks enable bacteria to adhere to a substrate where conditions are favorable.
- Explain how flagellar motility and chemotaxis enable bacteria to respond to environmental change.
- Describe the functions of thylakoids, storage granules, and magnetosomes.

Do bacteria have special structures for special tasks? The cell envelope, the nucleoid, and the gene expression complexes are needed for all bacterial cells. But other structures are needed to colonize different habitats, from the human colon to the hot springs of Yellowstone. In addition to the fundamental structures shared by all bacteria, different species have evolved specialized devices adapted to specific metabolic strategies and environments. Particularly important are structures for attachment and motility.

Pili and Stalks Enable Attachment

In a favorable habitat, such as a running stream full of fresh nutrients or the epithelial surface of a host, it is advantageous for a cell to adhere to a substrate such as a rock or an intestinal epithelial cell. **Adherence** is the ability to attach to a substrate, using specific structures such as pili (protein filaments). As bacteria grow and proliferate, however, they face the question of whether to stay where they are or leave their present habitat, with its dwindling nutrients and mounting waste products. In rapidly changing environments, cell survival requires **motility**, the ability to move and relocate. Motility requires structures such as rotary flagella. Some bacteria, such as the sewage bacterium *Caulobacter crescentus*, have it both ways by producing two kinds of progeny: one adherent, the other motile.

The most common structures that bacteria use to attach to a substrate are **pili** (singular, **pilus**); short attachment pili are called **fimbriae** (singular, **fimbria**). Pili are straight filaments of protein monomers called **pilin**. For example, pili attach the oral pathogen *Porphyromonas gingivalis* to gum epithelium, where they are associated with periodontal disease (**Figure 5.19**). A different kind of pili,

Figure 5.19   Pili: Protein Filaments for Attachment

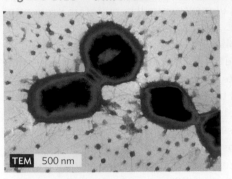

Porphyromonas gingivalis, a causative agent of gum disease, or gingivitis. The *P. gingivalis* cells show fimbriae (pili) along with vesicles budding from the cell's outer membrane.

TEM   500 nm

the **sex pili,** serve to attach a "male" donor cell to a "female" recipient cell before transferring DNA in a process called conjugation.

A different kind of attachment organelle is a membrane-bound extension of the cytoplasm called a **stalk**. The tip of the stalk secretes adhesion factors, called **holdfast**, which firmly attach the bacterium in an environment that has proved favorable. The mechanism of stalk and holdfast attachment has been extensively studied in iron-oxidizing bacteria such as *Gallionella ferruginea*, which clog the iron mines with massive biofilms of adherent cells.

Rotary Flagella Enable Motility and Chemotaxis

Bacteria and archaea that are motile generally swim using rotary **flagella** (singular, **flagellum**) (**Figure 5.20A**). Flagella are helical propellers that drive the cell forward like the motor of a boat. Different bacterial species have different numbers and arrangements of flagella. Flagella enhance virulence; for example, *Proteus* species use flagella to swim up the urethra and infect the bladder and kidneys.

Each flagellum is a spiral filament of protein monomers called flagellin. The filament is actually rotated by means of a rotary motor (**Figure 5.20B**). The flagellar motor is embedded in the layers of the cell envelope and is powered by the cell's proton motive force (PMF). Flagellar motility benefits the cell by dispersing progeny and thus decreasing competition. In addition, most flagellated cells have an elaborate sensory system known as **chemotaxis** that enables them to swim toward favorable environments (attractant signals, such as nutrients) and away from inferior environments (repellent signals, such as waste products).

 Chemotaxis: cellsalive.com

Chemotaxis requires a way for the cell to move toward attractants (positive chemotaxis) and away from repellents (negative chemotaxis). This is accomplished by flagellar rotation either clockwise or counterclockwise relative to the cell (**Figure 5.21**). When a cell is swimming toward an attractant chemical, the flagella rotate counterclockwise (CCW) and form a rotating helical bundle. The

Figure 5.20 Flagella on *Salmonella* Bacteria

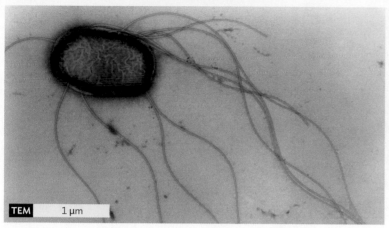

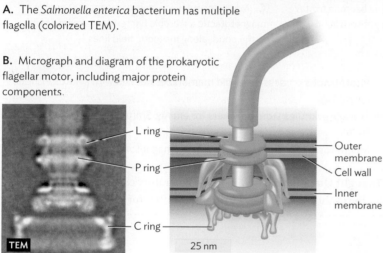

A. The *Salmonella enterica* bacterium has multiple flagella (colorized TEM).

B. Micrograph and diagram of the prokaryotic flagellar motor, including major protein components.

L ring

P ring

C ring

Outer membrane

Cell wall

Inner membrane

TEM   25 nm

Figure 5.21 Chemotaxis: Swimming toward Attractants

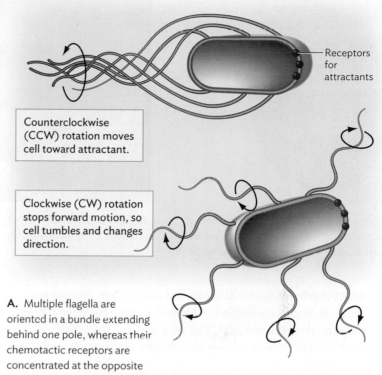

Receptors for attractants

Counterclockwise (CCW) rotation moves cell toward attractant.

Clockwise (CW) rotation stops forward motion, so cell tumbles and changes direction.

A. Multiple flagella are oriented in a bundle extending behind one pole, whereas their chemotactic receptors are concentrated at the opposite pole. When the cell veers away from the attractant, the receptors send a signal that allows one or more flagella to switch rotation from counterclockwise (CCW) to clockwise (CW). This switched rotation disrupts the bundle of flagella, causing the cell to tumble briefly before it swims off in a new direction.

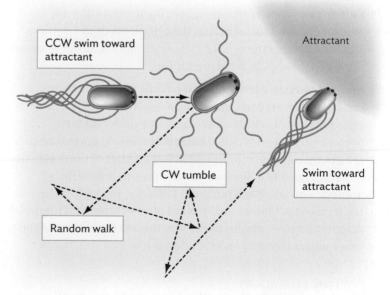

CCW swim toward attractant

Attractant

CW tumble

Swim toward attractant

Random walk

B. The pattern of movement resulting from alternating swimming and tumbling is a "biased random walk" in which the cell moves randomly but overall tends to migrate toward the attractant.

cell swims smoothly for a long stretch. When the cell veers away from the attractant, receptors send a signal that allows one or more flagella to switch rotation clockwise (CW), against the twist of the helix. This switch in the direction of rotation disrupts the bundle of flagella, causing the cell to tumble briefly, ending up pointed in a random direction. The cell then swims off in the new direction. The resulting pattern of movement generates a "biased random walk" in which the cell tends to migrate toward the attractant.

Note that bacterial flagella differ completely from the whip-like flagella and cilia of eukaryotes, to be discussed in Section 5.6.

Bacterial Structures for Different Habitats

Other structures that perform specialized functions for bacteria include photosynthetic membranes for light absorption and even magnetic structures for orientation of motility in a magnetic field (**Figure 5.22**). Here are some examples:

- **Thylakoids conduct photosynthesis.** Photosynthetic bacteria (phototrophs) need to collect as much light as possible to drive photosynthesis. To do this, they grow extensively folded intracellular membranes called thylakoids.

Figure 5.22   Structures for Different Habitats

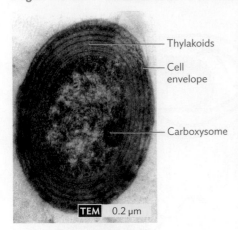

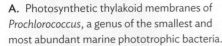

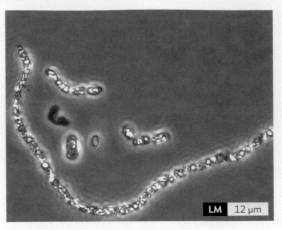

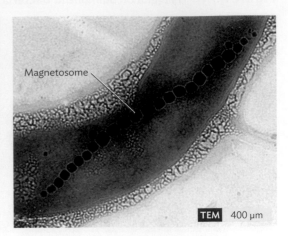

A. Photosynthetic thylakoid membranes of *Prochlorococcus*, a genus of the smallest and most abundant marine phototrophic bacteria.

B. Gas vesicles enable *Halochromatium* bacteria to stay at the surface of salt water, where they absorb light.

C. Magnetosomes (crystals of magnetite) enable a magnetotactic anaerobic bacterium to swim downward in the pond, along magnetic field lines.

- **Gas vesicles.** Aquatic phototrophs possess gas vesicles to increase buoyancy and keep themselves high in the water column, near the sunlight. The vesicles trap and collect gases such as hydrogen or carbon dioxide produced by the cell's metabolism.

- **Storage granules.** Some bacteria store energy in storage granules composed of glycogen or other polymers called polyhydroxyalkanoates (PHA). PHA polymers are of interest as a biodegradable plastic, which bacteria are engineered to produce industrially.

- **Sulfur granules.** Sulfur-metabolizing bacteria deposit granules of solid sulfur within the cytoplasm or as "globules" attached outside the cell. The presence of toxic sulfur granules may help these cells avoid predation.

- **Magnetosomes direct motility.** Some anaerobic pond-dwelling bacteria can swim along a magnetic field in a process called magnetotaxis, Magnetotactic species of bacteria possess tiny magnets called **magnetosomes**. Magnetosomes are microscopic membrane-enclosed crystals of the magnetic mineral magnetite. Because Earth's magnetic field lines point downward in the northern latitudes, magnetotactic bacteria swim "downward" toward magnetic north. This drives magnetosome-equipped anaerobes toward the bottom of the pond, where oxygen concentration is low.

SECTION SUMMARY

- **Adherence structures** such as pili, or fimbriae (protein filaments), and the stalk (a cell extension) enable prokaryotes to remain in a favorable environment.
- **Flagellar motility** of bacteria involves rotary motion of helical flagella.
- **Chemotaxis** involves a biased random walk up a gradient of an attractant substance or down a gradient of a repellent.

- **Phototrophs** possess thylakoid membranes packed with photosynthetic apparatus.
- **Storage granules** store polymers for energy. Sulfur granules store solid sulfur.
- **Magnetosomes** orient the swimming of magnetotactic anaerobic bacteria.

Thought Question 5.3   Why would laboratory culture conditions select for the evolution of cells lacking an S-layer? Or for cells lacking flagella?

5.6
The Eukaryotic Cell

SECTION OBJECTIVES
- Explain the structure and interconnection of membranous organelles in the endomembrane system.
- Describe the functions of the Golgi complex, endoplasmic reticulum, and nuclear membrane.
- Explain how the evolutionary process of endosymbiosis led to mitochondria and chloroplasts.

Eukaryotic cells include the cells of our own human body as well as the cells of all animals, plants, and fungi. They also include single-celled parasites that cause disease (such as malaria and giardiasis). How do eukaryotic cells compare with the cells of bacteria? Most well-known eukaryotic cells are much larger than bacteria; for instance, some amebas (alternative spelling, "amoebas") are visible to the unaided eye. The eukaryotic cells of animal bodies can contain a millionfold larger volume than that of a typical bacterium. A mammalian neuron (**Figure 5.23A**) can possess dendrites that extend for several centimeters and an axon that extends over meters.

Yet DNA-based surveys of natural environments, such as the ocean and soil, reveal other microbial eukaryotes as small as the smallest bacteria. An example is the marine alga *Ostreococcus tauri* (Figure 5.23B). *O. tauri* is barely longer than an *E. coli* cell, yet it manages to squeeze within its tiny volume all the major eukaryotic organelles: nucleus, mitochondria, endoplasmic reticulum (ER), and Golgi. Thus, eukaryotic cells have sizes over the range encompassed by bacteria and archaea, but also reach much larger sizes.

Organelles of the Eukaryotic Cell

Compared with prokaryotic cells, the larger eukaryotic cells have much more extensive systems of intracellular membranes. These membranes form various organelles consisting of vesicles and folded pockets of membrane (Figure 5.24). Membranous organelles provide many advantages to the eukaryotic cell. Membrane folds increase the membrane surface area without increasing the cell volume, thus expanding the area for membrane-bound complexes such as the electron transport system for respiration (discussed in Chapter 7). Membrane-enclosed organelles can allow different reactions to occur simultaneously under very different conditions. For example, proteins can be assembled by ribosomes in the cytoplasm at the same time that proteins are being broken down within the acidic lysosomes. Compartmentalization protects cytoplasmic components from harmful substances. For example, organelles called peroxisomes generate hydrogen peroxide (H_2O_2) to detoxify harmful substances. Localizing the reaction within the peroxisome keeps the toxic hydrogen peroxide away from other cell components such as proteins and DNA that are sensitive to oxidative stress.

DYNAMIC TRANSPORT BETWEEN MEMBRANOUS ORGANELLES As we saw in Section 4.6, a eukaryotic cell can take up larger objects such as prey bacteria by pinching in their cell membrane to enclose the object. The process

Figure 5.23 Eukaryotic Cells

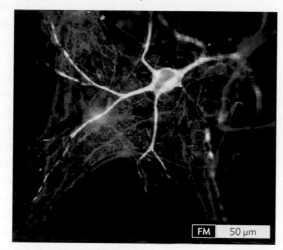

A. Rat spinal neuron, stained green with fluorescent anti-neurofilament antibody. Blue stain shows DNA; orange stain shows axons of surrounding neurons. Cell body is approximately 100 µm across; dendrites may extend several millimeters.

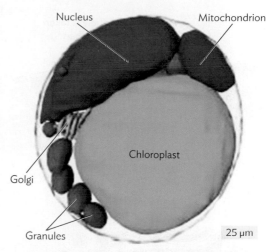

B. *Ostreococcus tauri*, a marine microbe barely bigger than *E. coli*, visualized by cryo-EM tomography.

Figure 5.24 Membranous Organelles of a Eukaryotic Cell

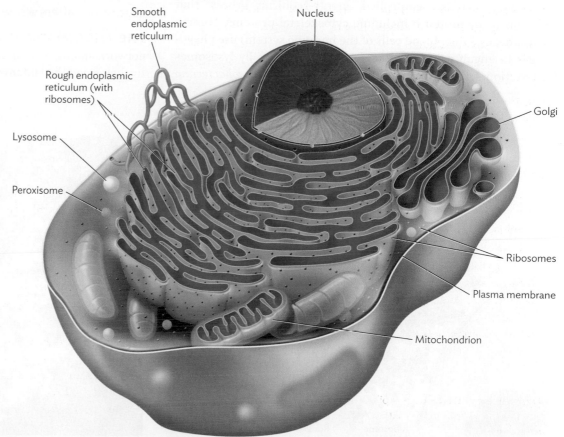

Organelles of the endomembrane system include the smooth endoplasmic reticulum (smooth ER), the rough ER (ER with attached ribosomes), lysosomes, and the Golgi complex. Other membranous organelles include the nucleus and the mitochondria.

Figure 5.25   Cell Trafficking

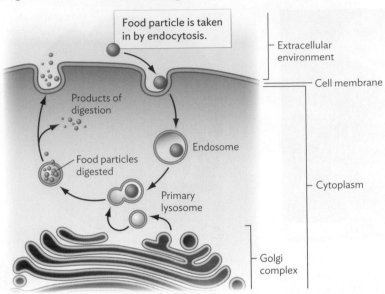

Lysosomes fuse with endosomes, providing enzymes to digest material brought into the cell by endocytosis.

of membrane invagination to form vesicles is called endocytosis (for small particles) or phagocytosis (for large particles or bacteria). Endocytosis is a controlled, energy-requiring process that relies on many proteins, including cytoskeletal proteins. Human macrophages (white blood cells of the immune system) use phagocytosis to engulf pathogenic bacteria for digestion by lysosomes (see Section 4.6). But pathogenic bacteria such as *Mycobacterium*

tuberculosis or viruses such as influenza can subvert phagocytosis or endocytosis to gain entry and multiply within a cell.

What happens to endocytic vesicles (endosomes) or phagosomes within the cell? Endosomes normally fuse with a **lysosome**, a membranous organelle containing enzymes that kill bacteria and process organic material for nutrition (**Figure 5.25**). Lysosomal vesicles containing nutrients can ultimately fuse with the **endoplasmic reticulum (ER)**. The ER is continuous with the outer nuclear membrane, and the **lumen** (interior) of the ER is continuous with the space between the two nuclear membranes (see **Figure 5.26**). This means that mixing can occur via vesicle fusion; thus, material in the ER does not need to cross a membrane to enter these other spaces. For example, fusion between the membranes of ER and the Golgi complex enables material contained within the ER to mix with the contents of the Golgi complex (Figure 5.24). The connected regions of ER lumen and Golgi are thus separated from the cytoplasm by endomembranes. The ER is used to sequester substances that must be held at low concentrations in the cytoplasm (for example, calcium ions for nerve signals).

Note that the reverse of endocytosis is **exocytosis**. In exocytosis, intracellular vesicles fuse with the cell membrane, and the contents of the vesicles are released to the extracellular environment. Exocytosis can expel waste materials from lysosome digestion. Also, exocytosis can mediate chemical communication, such as releasing neurotransmitters across a synapse to a receiving neuron.

THE ENDOMEMBRANE SYSTEM   The ER forms a transport network among several kinds of membrane compartments, collectively termed the **endomembrane system** (Figure 5.26). Organelles

Figure 5.26   The Endomembrane System

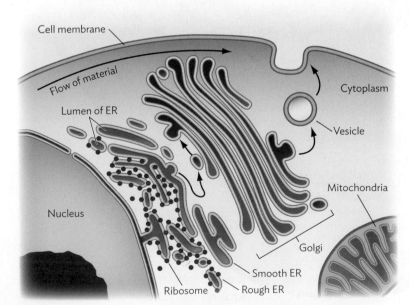

A. The relationship of the ER to other cellular membranes and the flow of material through vesicles from the rough ER to the cell membrane.

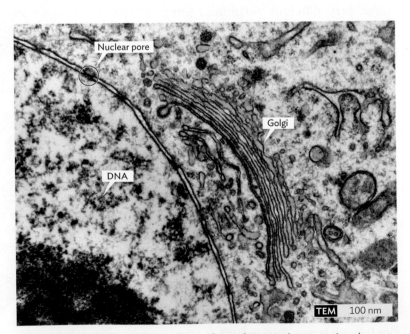

B. Cell section showing Golgi cisternae, the nuclear membrane, and nuclear pores.

of the endomembrane system include the endoplasmic reticulum (ER), the Golgi complex (also called the Golgi apparatus), lysosomes, and peroxisomes, as well as the outer membrane of the nucleus. Different organelles contain unique subsets of proteins that contribute to their function. Their membranes are distinct from one another and from the cell membrane, although vesicles may pinch off one organelle and dissolve into the membrane of another. The function of the ER and the Golgi complex is to direct to their proper cellular location proteins destined for lysosomes, for the cell membrane, or for secretion from the cell.

A special category of endoplasmic reticulum is the "rough ER," so called because of the "rough" appearance of the membrane surface studded with ribosomes. The rough ER is where ribosomes translate RNA sequence into membrane-embedded proteins and secreted proteins. Secreted proteins and proteins destined for the lumen of an organelle are threaded completely through the ER membrane and into the lumen of the ER. In contrast, transmembrane proteins are not threaded completely through, and part of the protein spans the membrane.

Proteins to be secreted are pinched off into vesicles which merge into the **Golgi**, or Golgi complex. The Golgi consists of separate membrane stacks (cisternae) that each contain enzymes to "tag" proteins with carbohydrate chains that direct them to organelles such as lysosomes. Proteins not targeted to lysosomes may be sent to the cell membrane. Vesicles leaving the Golgi fuse with the cell membrane, releasing their aqueous contents outside the cell. Membrane-embedded proteins from these vesicles then become part of the cell membrane.

The Nucleus Organizes DNA

Eukaryotes are so named because they possess a nucleus (the word is from the Greek for "true nucleus"). Indeed, the nucleus is often the most prominent feature of a eukaryotic cell viewed under a microscope (**Figure 5.27A**). The nucleus is an intracellular membrane-enclosed compartment containing chromatin, a complex of DNA and proteins. The nuclear membrane consists of two concentric phospholipid membranes. The outer nuclear membrane is continuous with the endomembrane ER, and the space between the two nuclear membranes is continuous with the lumen (inside) of the ER folds (**Figure 5.27B**). Nuclear function is important for infection by viruses, such as herpes and human immunodeficiency virus (HIV), which must manipulate nuclear processes to replicate their own genomes.

The nuclear membrane contains **nuclear pore complexes** that allow for transport of material into and out of the nucleus. Metabolites and small proteins can diffuse through the nuclear pores, but larger proteins and organelles require active transport. These selectively imported proteins contain a nuclear localization signal, a sequence of amino acids that acts like a ZIP code to direct them through a nuclear pore into the nucleus. Besides importing proteins, nuclear pores export mRNAs to the cytoplasm. But viruses

Figure 5.27   The Nucleus

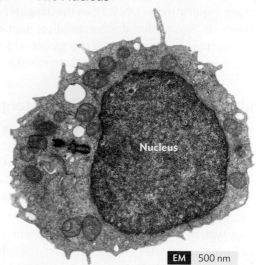

A. Electron micrograph of a eukaryotic yeast cell, showing the nucleus (colorized blue) and mitochondria (colorized red).

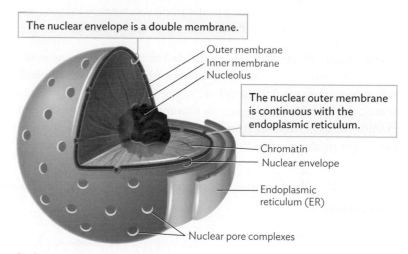

B. Diagram of a nucleus.

such as herpes can "hijack" a nuclear pore to inject the viral genome into the nucleus.

Within the nucleus, a structure called the **nucleolus** assembles ribosomes. At the nucleolus, multiple genes encoding ribosomal RNA (rRNA) are transcribed, and the resulting rRNA combines with ribosomal proteins imported into the nucleus from the cytoplasm to form the ribosomal subunits. Within the nucleus, the mRNA arising from transcription is translated by ribosomes, which check the transcript for errors and target faulty transcripts for destruction. Most of eukaryotic translation, however, occurs outside the nucleus, on mature mRNA using ribosomes exported to the cytoplasm.

In eukaryotes, unlike bacteria, cells divide at a different time from cell growth and gene expression. Eukaryotic transcription and translation occur only during interphase, an interval of the cell cycle during which the cell is not undergoing division. The cell grows during interphase, replicating its DNA. Then the daughter cells come apart through mitosis. Sexual reproduction requires meiosis, a modified form of mitosis that generates haploid (1n) gametes (Section 11.1). Gamete fertilization restores the 2n diploid chromosome number.

Mitochondria and Chloroplasts Yield Energy

Mitochondria and chloroplasts are organelles that obtain energy for the cell. Nearly all eukaryotes have mitochondria, which conduct oxidative phosphorylation and make ATP (discussed in Chapter 7). Chloroplasts are found only in photosynthetic eukaryotes, such as plants and algae. They perform photosynthesis (also making ATP).

Mitochondria and chloroplasts are not considered part of the endomembrane system; they undergo less membrane traffic with the endomembrane organelles. Mitochondria and chloroplasts evolved early in the history of life through a form of evolution called **endosymbiosis**. In endosymbiosis, one cell is engulfed by a larger cell, but not digested (**Figure 5.28A**). A mutualistic relationship evolves. In the evolution of mitochondria and chloroplasts, the larger eukaryotic cell provided protection to an intracellular bacterium, and the bacterium provided energy to the eukaryote.

MITOCHONDRIA   Mitochondria have two membranes: an outer membrane and an inner membrane (**Figure 5.28B**). The inner membrane has many infoldings called cristae that increase its surface area. The mitochondrial inner membrane is derived from the cell membrane of the ancestral bacterium; it contains the protein complexes for oxidative phosphorylation, including the ATP synthase.

The mitochondrial outer membrane is derived from the surrounding cell. Thus, the inner membrane has phospholipids similar to those of a bacterial membrane, whereas the outer membrane composition resembles that of the host cell.

Mitochondria contain two distinct compartments: the intermembrane space (between the two membranes) and the matrix enclosed by the inner membrane. Different stages of oxidative respiration occur in specific compartments. The arrangement of these processes is similar in bacteria and mitochondria. For example, in bacteria, the tricarboxylic acid cycle (TCA, or Krebs cycle) occurs in the cytoplasm; in mitochondria, the tricarboxylic acid cycle takes place inside the matrix, the metabolic equivalent of the prokaryotic cytoplasm. Mitochondria retain a small circular chromosome containing bacteria-like genes, as well as their own ribosomes. However, during evolution, most of the endosymbiont's original genes were transferred to the nuclear chromosomes of the host cell. Thus, inherited disorders of human mitochondria, such as Leigh's syndrome (failure of oxidative metabolism), may arise from mutations either in mitochondrial genes or in nuclear genes.

CHLOROPLASTS   Chloroplasts convert light energy from the sun to ATP and reduced NADPH in a process known as the "light reactions" or light-dependent reactions. In the subsequent light-independent reactions, the ATP and NADPH are used to reduce CO_2 to sugar. The endosymbiotic precursor of chloroplasts was a common ancestor of modern cyanobacteria. Like eukaryotic chloroplasts, cyanobacteria perform aerobic photosynthesis by using thylakoids, internal membrane pockets packed with chlorophyll, a light-absorbing pigment.

Figure 5.28   Endosymbiosis Led to Mitochondria and Chloroplasts

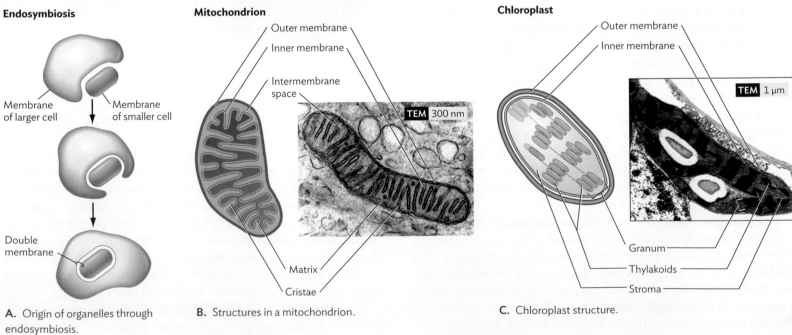

A. Origin of organelles through endosymbiosis.

B. Structures in a mitochondrion.

C. Chloroplast structure.

Chloroplasts have three membranes whose topology can be understood in light of endosymbiosis (Figure 5.28C). The outer membrane appears to be derived from the host eukaryotic cell, the inner membrane is equivalent to the bacterial cell membrane, and the thylakoid membrane is derived from the bacterial thylakoid membrane. The region inside the inner membrane is called the stroma and is equivalent to the bacterial cytoplasm. ATP and NADPH are produced in the stroma and used there in the light-independent reactions of CO_2 fixation. As a result of endosymbiosis, chloroplasts, like mitochondria, contain their own circular DNA and their own ribosomes.

The Cytoskeleton Maintains Shape

Eukaryotic cells contain a network of proteins that determine the shape of the cell, collectively termed the cytoskeleton. Cytoskeletal proteins are multifunctional, also involved in whole-cell movements and the movement of substances within the cell. Three major classes of cytoskeletal proteins are the microfilaments, intermediate filaments, and microtubules.

Microfilaments, also known as actin filaments, have a diameter of 7 nm (Figure 5.29A). They are formed when individual actin monomers (globular actin, or G-actin) polymerize, in a process powered by ATP hydrolysis, to form chains of filamentous actin (F-actin). Two F-actin chains twist around each other, forming a microfilament with a plus end and a minus end.

Microfilaments are dynamic structures, growing and shrinking in a controlled manner by polymerization and depolymerization of actin. For example, an ameba's pseudopod starts to move by polymerizing actin at the leading edge of growth. Actin polymerization can be manipulated by invading bacteria—for example, by *Listeria monocytogenes*, the cause of food poisoning (listeriosis) acquired from melons and other produce. The *Listeria* bacterium alters host actin polymerization in order to enter a host cell, and once inside, it induces host actin to polymerize behind one end of the bacterium. The actin tails then propel bacteria through the host and beyond, into neighboring host cells.

Intermediate filaments consist of various fibrous proteins that have a diameter of about 10 nm (Figure 5.29B). Intermediate filaments form a meshwork under the cell membrane, where they help maintain cell shape and strengthen the cell by resisting tension placed on the cell membrane.

Microtubules have a larger diameter (25 nm) than microfilaments and intermediate filaments (Figure 5.29C). The hollow microtubule structure consists of 13 dimers of tubulin: one alpha-tubulin protein plus one beta-tubulin protein form one tubulin dimer. Microtubules are dynamic structures that can polymerize and depolymerize. They segregate (pull apart) the duplicated chromosomes during mitosis. Other microtubules form parts of cilia and flagella (discussed shortly).

Microtubules help transport substances within the cell. Protein traffic through the endomembrane system relies on the controlled movement of vesicles from one cellular compartment to the next. Microtubules provide tracks that can move vesicles from one organelle to the next. Unfortunately, they also provide tracks for invading viruses. For example, particles of herpes virus (such as herpes simplex, the cause of cold sores and genital herpes) can enter the cytoplasm and travel down a track of microtubules to the nucleus. The virus then injects its DNA through a nuclear pore complex for replication within the nucleus.

Specialized Structures

As we saw for bacteria, different kinds of eukaryotic cells have different kinds of specialized structures, depending on their environment. Many eukaryotic microbes have motility organelles such as flagella or cilia, which help the microbe find and ingest food. Free-living

Figure 5.29   Cytoskeletal Proteins

Microfilament

Actin monomer

7 nm

A. Microfilaments consist of two strands of actin polymers twisted together.

Intermediate filament

8–12 nm

Fibrous subunit

B. Intermediate filaments are ropelike assemblages of various proteins.

Microtubule

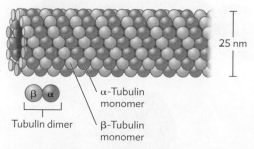

25 nm

α-Tubulin monomer

Tubulin dimer

β-Tubulin monomer

β   α

C. Microtubules are polymers of tubulin dimers.

Figure 5.30   Eukaryotic Flagella and Cilia

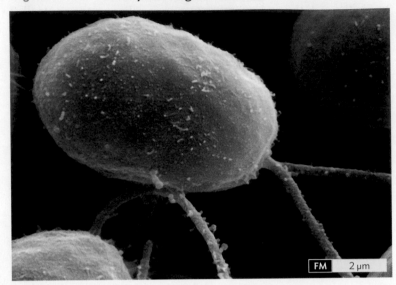

A. The protist *Chlamydomonas* has two long flagella.

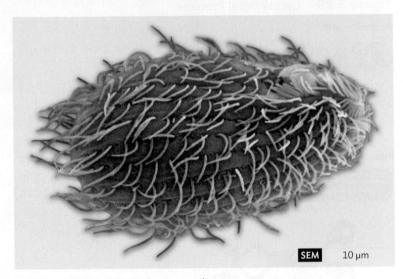

B. The protist *Tetrahymena* has many cilia.

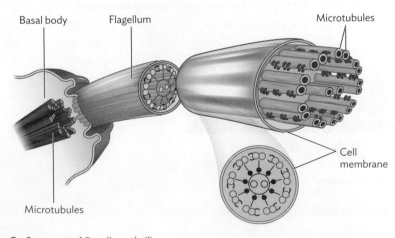

Basal body   Flagellum   Microtubules

Microtubules

Cell membrane

C. Structure of flagella and cilia.

eukaryotes that lack a cell wall require structures that counteract osmotic pressure; for example, contractile vacuoles pump out water. And parasites form invasive structures to penetrate host cells, such as the apical complex of the malaria-causing *Plasmodium*. Diverse eukaryotic parasites are discussed in Chapter 11.

EUKARYOTIC FLAGELLA AND CILIA   Flagella and cilia consist of bundles of microtubules within thin extensions of the cell membrane (Figure 5.30). Each flagellum, or each cilium, moves in a whiplike fashion, driven by interactions between microtubules and the motor protein dynein. Flagella of eukaryotes are relatively long, and a cell usually has only one or two of them; see, for example, the green alga *Chlamydomonas* (Figure 5.30A). Cilia (singular, cilium) are shorter and more numerous, as seen in the ciliate protist *Tetrahymena* (Figure 5.30B). Both flagella and cilia can move an entire cell. Cilia may also be used to capture food by sweeping extracellular fluid into the gullet of a protozoan such as a paramecium or a vorticella. Eukaryotic cilia and flagella use ATP hydrolysis by dynein to generate the whiplike motion along the length of the cilium or flagellum.

Note   Distinguish eukaryotic flagella from bacterial flagella, a completely different structure. Eukaryotic flagella are large, containing multiple microtubules enclosed by a membrane. They move with a whiplike motion, powered by ATP hydrolysis all along the flagellum. By contrast, bacterial flagella consist of a single coiled tube of protein, whose rotary motion is driven by the flagellar motor embedded in the cell envelope.

PELLICLE   Like bacteria, microbial eukaryotes typically possess a thick outer covering. Fungi possess cell walls of the polysaccharide chitin, whereas green algae have cell walls of cellulose. Protists such as amebas and ciliates lack a rigid cell wall, but they have a thick, flexible pellicle, consisting of membranous layers reinforced by protein microtubules. The pellicle allows great flexibility of shape; it enables uptake of large particles by endocytosis and of larger objects, even entire cells, by phagocytosis. Thus, protists have the nutritional option of engulfing prey, an option unavailable to prokaryotes and to eukaryotes with cell walls. The protist pellicle often contains additional structures such as extrusomes, which secrete toxins or digestive enzymes. Other kinds of eukaryotic cell surfaces are reinforced by inorganic materials, such as the silica shells of diatoms.

CONTRACTILE VACUOLE   Free-living microbial eukaryotes that lack a cell wall need special organelles to prevent osmotic shock (discussed in Section 4.5). An organelle that removes excess water from the cytoplasm is the **contractile vacuole** (Figure 5.31). As water enters the cell from a low-solute environment, the contractile vacuole takes up water through an elaborate network of intracellular channels extending throughout the cytoplasm. When the vacuole fills, it then contracts and expels the water through a pore, thus preventing osmotic shock and cell lysis.

Figure 5.31   Contractile Vacuole

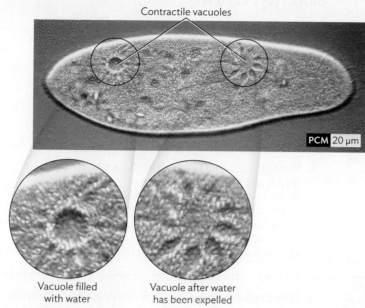

Contractile vacuoles

PCM 20 μm

Vacuole filled with water

Vacuole after water has been expelled

A. A paramecium, showing two contractile vacuoles. The inset images show radiating channels that collect water and expel it from the cytoplasm.

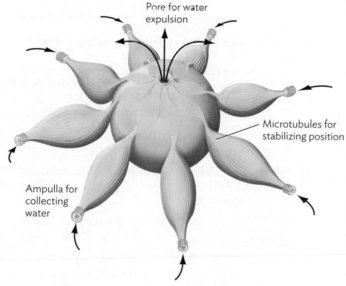

Pore for water expulsion

Microtubules for stabilizing position

Ampulla for collecting water

B. Diagram of a contractile vacuole.

SECTION SUMMARY

- **Membranous organelles** such as the endoplasmic reticulum, lysosomes, and the Golgi complex compartmentalize the functions of a eukaryotic cell.
- **Endocytosis and phagocytosis** mediate transport between the cell and the external environment and between organelles.
- **The nuclear membrane** encloses eukaryotic chromosomes. Nuclear pore complexes mediate transport in and out of the nucleus.
- **The nucleolus** assembles ribosomes, which are sent to the cytoplasm.
- **Mitochondria and chloroplasts** are organelles that evolved from endosymbiotic bacteria. Mitochondria conduct oxidative phosphorylation, and chloroplasts conduct photosynthesis (for plant cells).
- **The cytoskeleton of eukaryotic cells** is composed of microfilaments, intermediate filaments, and microtubules.
- **Flagella and cilia** of eukaryotes are whiplike organelles powered by ATP hydrolysis.

Thought Question 5.4   What are the advantages and disadvantages of a contractile vacuole, compared with a cell wall?

Thought Question 5.5   What kinds of human cells possess cilia or flagella, and what are their functions within the human body?

Perspective

Every new cell process that we discover offers new clues to the mystery of life—and new targets for lifesaving drugs. The structures of bacteria even reveal surprising insights into human cells, as our own mitochondria evolved from bacteria. Bacterial cells are adapted to grow in all kinds of environments, including human and animal hosts. But bacteria in external environments, such as the soil or ocean, provide new capabilities for our molecular therapies.

In the next chapter, we focus on growth and nutrition: How do bacteria obtain energy and materials from their environment, including host environments such as the human body? We also learn how bacteria use their energy to build their intricate cells. The chapters that follow reveal how genetics shapes microbial cells and populations and how evolution generates the bewildering diversity of microbial species.

LEARNING OUTCOMES AND ASSESSMENT

	SECTION OBJECTIVES	OBJECTIVES REVIEW

5.1
The Bacterial Cell: An Overview

- Describe the structure and function of the bacterial cell wall, and explain its importance as a target for antibiotics.
- Explain the Gram-negative envelope structure and the role of LPS in pathogenesis.
- Describe the biochemical composition of a bacterial cell.

1. Which cell component extends far outside a Gram-negative cell and helps prevent phagocytosis by blood cells?
   A. Cell wall
   B. Periplasmic proteins
   C. Polyamines
   D. Lipopolysaccharide (LPS)

5.2
Bacterial Membranes and Transport

- Describe the functions of a cell membrane.
- Explain how nutrients are transported and how energy is spent to drive transport.

4. Which is not a function of the cell membrane or membrane-embedded proteins?
   A. Protein synthesis in the cytoplasm
   B. Detecting changes in temperature and pH that tell a pathogen it is within a host
   C. Secreting toxins to counteract host defenses
   D. Transporting nutrients from the environment into a cell

5.3
The Bacterial Cell Wall and Outer Layers

- Describe the cell wall structure, and explain how it protects bacteria from osmotic shock.
- Explain the function of the Gram-positive cell wall and teichoic acids.
- Explain the function of the Gram-negative outer membrane, LPS, and periplasm.

7. Which cell component consists of a single macromolecule that surrounds the cell and protects it from osmotic shock?
   A. Cell wall
   B. Cell envelope
   C. Cell membrane or plasma membrane
   D. Lipopolysaccharide (LPS)

5.4
The Nucleoid and Bacterial Cell Division

- Describe how DNA is organized within the bacterial cell.
- Explain how DNA replication is coordinated with cell growth and division.

10. Which aspect of DNA organization does not occur in bacteria?
    A. DNA is double-stranded.
    B. DNA has an origin of replication attached to the cell envelope.
    C. DNA is contained within a nuclear membrane.
    D. Most species have a genome that is circular.

5.5
Specialized Structures of Bacteria

- Describe how pili and stalks enable bacteria to adhere to a substrate where conditions are favorable.
- Explain how flagellar motility and chemotaxis enable bacteria to respond to environmental change.
- Describe the functions of thylakoids, storage granules, and magnetosomes.

13. Which extracellular structure helps a bacterial cell attach to a substrate?
    A. Magnetosome
    B. Stalk
    C. Gas vesicle
    D. Thylakoid

5.6
The Eukaryotic Cell

- Explain the structure and interconnection of membranous organelles in the endomembrane system.
- Describe the functions of the Golgi complex, endoplasmic reticulum, and nuclear membrane.
- Explain how the evolutionary process of endosymbiosis led to mitochondria and chloroplasts.

16. Which organelle is not a part of the endomembrane system?
    A. Cell membrane
    B. Lysosome
    C. Golgi complex
    D. Mitochondrion

2. Which intracellular molecule is present in the greatest number?
   A. Proteins
   B. Messenger RNA
   C. Water molecules
   D. Potassium ions

3. Which part of the Gram-negative cell holds the ATP synthase?
   A. Cell wall
   B. Cell membrane (inner membrane)
   C. Periplasm
   D. Capsule

5. Which transport process would not require expenditure of energy?
   A. Uptake of sugar by an ABC transporter complex
   B. Facilitated diffusion of oxygen across a membrane
   C. Transport of an amino acid from a lower concentration to a higher concentration
   D. Symport of lactose coupled to H^+ moving inward down its gradient

6. Which structure can export antibiotics?
   A. TolC efflux complex
   B. Ribosome
   C. Amino acid transporter
   D. Lactose transporter

8. Which compound is found in the outer cell layers of *Mycobacterium tuberculosis*, but not in most other kinds of bacteria?
   A. Teichoic acids
   B. S-layer proteins
   C. Murein lipoprotein
   D. Mycolic acids

9. Which structure of a Gram-negative cell contains porins?
   A. Cell wall
   B. Inner membrane
   C. Periplasm
   D. Outer membrane

11. Which process can occur only after the bacterial DNA terminates replication?
    A. Completion of the septum to separate two daughter cells
    B. Transcription of genes to make messenger RNA
    C. Translation of RNA to make proteins
    D. Insertion of proteins into the cell membrane

12. How is DNA arranged inside a bacterial cell?
    A. Multiple pieces in the cytoplasm
    B. A circular molecule tangled throughout the cytoplasm
    C. A circular molecule with loops bound by DNA-binding proteins
    D. A circular molecule within a nuclear membrane

14. Which statement does not describe bacterial motility?
    A. Bacterial motility enables pathogens to colonize human organs.
    B. Helical flagella rotate, using a rotary motor embedded in the cell membrane.
    C. Flagella generate a whiplike motion powered by ATP.
    D. Cells alternate swimming and tumbling to migrate up the gradient of an attractant.

15. What is the process by which cells migrate toward nutrients?
    A. Chemotaxis
    B. Contraction of contractile vacuole
    C. Photosynthesis
    D. Active transport

17. Which structure does not contribute to the eukaryotic cytoskeleton?
    A. Microtubules
    B. Peptidoglycan cell wall
    C. Microfilaments
    D. Intermediate filaments

18. Which organelle evolved from a bacterium?
    A. Golgi
    B. Chloroplast
    C. Nucleus
    D. Endoplasmic reticulum (ER)

Key Terms

ABC transporter (p. 130)
active transport (p. 130)
adherence (p. 142)
antiport (p. 130)
capsule (p. 125)
cell envelope (p. 126)
cell membrane (p. 125)
cell wall (p. 125)
chemotaxis (p. 142)
contractile vacuole (p. 150)
coupled transport (p. 130)
cytoplasmic membrane (p. 125)
cytoskeleton (p. 132)
efflux transporter (p. 130)
endomembrane system (p. 146)
endoplasmic reticulum (ER) (p. 146)
endosymbiosis (p. 148)
endotoxin (p. 137)

eukaryotic cell (p. 124)
exocytosis (p. 146)
facilitated diffusion (p. 130)
fimbria (p. 142)
flagellum (pp. 126, 142)
glycan (p. 132)
Golgi (p. 147)
Gram-negative bacteria (p. 135)
Gram-positive bacteria (p. 135)
holdfast (p. 142)
inner membrane (p. 137)
intermediate filament (p. 149)
lipid A (p. 137)
lipopolysaccharide (LPS) (pp. 125, 137)
lumen (p. 146)
lysosome (p. 146)
magnetosome (p. 144)
microfilament (p. 149)

microtubule (p. 149)
motility (p. 142)
murein (p. 132)
murein lipoprotein (p. 137)
mycolic acid (p. 138)
nuclear pore complex (p. 147)
nucleoid (p. 127)
nucleolus (p. 147)
origin of replication (*ori*) (p. 140)
outer membrane (p. 125)
passive transport (p. 130)
peptidoglycan (p. 125)
periplasm (p. 125)
permease (p. 129)
pilin (p. 142)
pilus (p. 142)
plasma membrane (p. 125)
polyamine (p. 127)

porin (p. 137)
prokaryotic cell (p. 124)
proteome (p. 127)
proton motive force (p. 130)
reductive evolution (p. 136)
septation (p. 141)
septum (p. 141)
sex pilus (p. 142)
S-layer (p. 135)
stalk (p. 142)
staphylococcus (p. 141)
subcellular fractionation (p. 127)
substrate-binding protein (p. 130)
symport (p. 130)
teichoic acid (p. 135)

Review Questions

1. What are the major features of a bacterial cell, and how do they fit together for cell function as a whole?

2. What fundamental traits do most bacteria have in common with eukaryotic microbes? What traits are different?

3. Compare and contrast the structure of Gram-positive and Gram-negative cell envelopes. Explain the strengths and weaknesses of each kind of envelope.

4. Explain how DNA replication is coordinated with cell wall septation.

5. Compare and contrast bacterial flagella with eukaryotic flagella.

6. Describe the organelles of the endomembrane system. Explain the process of transport that transfers membranes between them.

7. Explain the function of mitochondria, chloroplasts, and the cytoskeleton.

Clinical Correlation Questions

1. A skin biopsy sample reveals bacteria with extremely narrow, flexible spiral cells. The cells have a thick sheath, beneath which rotary flagella embedded in the cell membrane cause the cells to twist in a corkscrew motion. The bacteria are associated with a tick bite. What disease may the patient have?

2. A skin biopsy sample reveals rod-shaped bacteria with a thick cell wall and capsule; the bacteria stain acid-fast. The thick outer layers of capsule prevent phagocytosis but may also decrease nutrient uptake; the bacteria can be cultured but require several weeks to grow colonies. The patient shows skin lesions with loss of sensation. What disease may the patient have?

Thought Questions: CHECK YOUR ANSWERS

Thought Question 5.1   Which molecules occur in the greatest number in a prokaryotic cell? The smallest number? Why does a prokaryotic cell contain 100 times as many lipid molecules as strands of RNA?

ANSWER: Inorganic ions occur in the greatest number in a prokaryotic cell (250 million per cell). They are also the smallest. DNA molecules are found in the lowest number (one large molecule, branched during replication). A prokaryotic cell contains 100 times as many lipid molecules as strands of RNA because lipids are small structural molecules, highly packed. RNA molecules are long macromolecules that are either packed into complexes (such as ribosomal RNA) or found as temporary information carriers (messenger RNA), present only as needed to make proteins.

Thought Question 5.2   Why do antibiotics differ in their effetiveness against Gram-positive and Gram-negative bacteria?

ANSWER: The cell envelopes differ greatly for Gram-positive and Gram-negative bacteria. Gram-positive bacteria have thick cell walls, which protect them from some antibiotics and toxins, but they lack the outer membrane of Gram-negative bacteria. The outer membrane contains porins that exclude small molecules, including antibiotics. Both Gram-positive and Gram-negative bacterial envelopes contain drug efflux pumps that exclude antibiotics, but they are different pumps that exclude different types of antibiotics. In addition, Gram-positive and Gram-negative bacteria have diverged deeply through evolution, so their fundamental structures differ in amino acid sequence even where the function is the same. For example, the ATP synthase of *Mycobacterium tuberculosis* is sufficiently distinct to be targeted by drugs that do not affect Gram-negative bacteria.

Thought Question 5.3   Why would laboratory culture conditions select for the evolution of cells lacking an S-layer? Or for cells lacking flagella?

ANSWER: Degeneration of protective traits is a common problem when conducting research on microbes that can produce 30 generations overnight. Their rapid reproductive rate gives ample opportunity for spontaneous mutations to accumulate over an experimental timescale. For the S-layer, in a laboratory test tube free of predators or viruses, mutant bacteria that fail to produce the thick protein layer would save energy compared with S-layer synthesizers and would therefore grow faster. Such mutants would quickly take over a rapidly growing population. For flagella, the motility apparatus requires 50 different genes generating different protein parts. Cells that acquire mutations eliminating expression of the motility apparatus gain an energy advantage over cells that continue to invest energy in motors. In a natural environment, the nonmotile cells lose out in competition for nutrients, despite their energetic advantage, but in the laboratory, cells are cultured in isotropic environments, such as a shaking test tube, where motility confers no advantage. These culture conditions lead to evolutionary degeneration of motility.

Thought Question 5.4   What are the advantages and disadvantages of a contractile vacuole, compared with a cell wall?

ANSWER: The disadvantage of a contractile vacuole is that it requires a continual input of energy to bail out the water. On the other hand, the contractile vacuole permits the existence of a cell that is flexible enough to engulf other cells as prey. But a cell wall does not have to spend energy (other than the initial energy to synthesize it). A cell wall is inflexible and does not allow another cell to be engulfed as prey.

Thought Question 5.5   What kinds of human cells possess cilia or flagella, and what are their functions within the human body?

ANSWER: The one type of human cell that has a flagellum is the spermatozoan (sperm cell). Sperm use the flagellum to swim up the cervix and reach the egg for fertilization. Many human tissue types have cilia, most notably the lining of the respiratory tract. The respiratory cilia contract in waves to push mucus and foreign particles up and out of the airways. In the auditory canal of the ear, very specialized cells called "hair cells" possess cilia that flex in response to sound waves.

6

Bacterial Growth, Nutrition, and Differentiation

[Metabolic Fingerprints of a Brain Pathogen]

SCENARIO A young mother who recently emigrated from Uganda brought her sick 18-month-old infant, Mirembe, to a New York City emergency room. The infant was crying uncontrollably, especially when the physician tried to move her head. Meningitis (an infection of the spinal fluid and membrane surrounding the brain) was suspected, and a spinal tap was performed to collect cerebrospinal fluid (CSF). The fluid, which appeared cloudy, was sent to the laboratory.

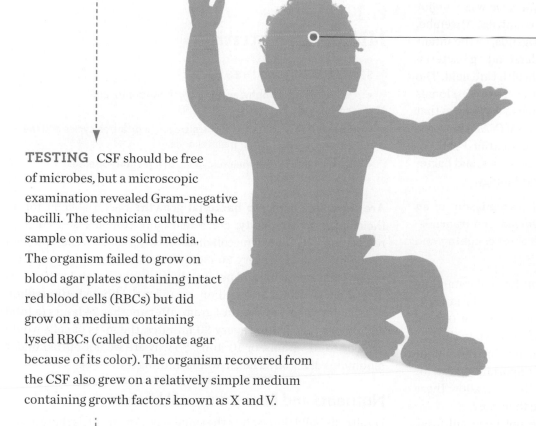

TESTING CSF should be free of microbes, but a microscopic examination revealed Gram-negative bacilli. The technician cultured the sample on various solid media. The organism failed to grow on blood agar plates containing intact red blood cells (RBCs) but did grow on a medium containing lysed RBCs (called chocolate agar because of its color). The organism recovered from the CSF also grew on a relatively simple medium containing growth factors known as X and V.

DIAGNOSIS The diagnosis was meningitis caused by the bacterium *Haemophilus influenzae*. The child received antibiotics and made a full recovery.

LM 6 μm

Haemophilus influenzae
Haemophilus influenzae (bottom) will not grow on blood agar (top) but will grow on chocolate agar (middle) in which red blood cells are lysed, releasing growth factors.

Haemophilus influenzae is a pathogenic bacterium that spreads from person to person by direct contact (**direct transmission**) or by **aerosolization** from someone with *H. influenzae* in his or her respiratory tract. Prior to 1988 and the advent of an effective capsular vaccine, *H. influenzae* was a major cause of childhood meningitis in developed countries. Mirembe, however, was not vaccinated prior to leaving Uganda, so the infant was susceptible. This case highlights why understanding bacterial physiology is important to those entering the health care field. The adage "know your enemy" applies to all health care professionals dealing with infectious diseases. Had the technician not known that *H. influenzae* requires two factors, X (hemin) and V (NAD), to grow, the cause of this infection might not have been identified and the child might have died. Clinical microbiology saves lives, and bacterial physiology is the foundation of clinical microbiology.

Link Recall from Section 2.3 that in **direct transmission** of an infectious agent, no inanimate objects are involved, and transmission of a lung pathogen by **aerosolization** involves coughing and sneezing.

In this chapter, we will build a foundation for understanding bacterial physiology by examining how and where bacteria grow. You might be surprised to learn, however, that only about 0.1% of the bacterial species that inhabit Earth have been grown in the laboratory. This fact extends to the native organisms inhabiting our intestines (called our gut microbiota). One important quest for microbiologists is to discover the conditions that will allow these microbes to grow on artificial medium, outside their host.

Although most of Earth's microbes have not been cultured, we have learned much from those that do grow in the laboratory, especially about the remarkably diverse ways they gain sustenance. Pathogens, for example, gain energy from consuming organic molecules such as glucose. In stark contrast, some nonpathogenic bacteria native to soil or water get their energy from light (photosynthesis), whereas others gain energy from oxidizing inorganic molecules such as sulfur. Indeed, there are so many microbes consuming so many different nutrients that you might wonder how Earth's biosystem avoids "eating" itself to death! As we explore these facets of microbiology, you will learn how microbes are experimentally grown and their numbers measured. We will also point out aspects of growth critical to finding, growing, and identifying pathogenic bacteria.

6.1
Microbial Nutrition

Are there any limits to bacterial growth? Bacterial cells, for all their apparent simplicity, are remarkably complex and efficient replication machines. One cell of *Escherichia coli*, for example, can divide into two cells every 20–30 minutes. At a rate of 30 minutes per division, one cell could potentially multiply to over 1×10^{14} cells in 24 hours—that is, 100 trillion organisms! Although 100 trillion cells would weigh only about 1 gram, after another 24 hours (a total of 2 days) of replicating every 30 minutes, the mass of cells would explode to 10^{14} grams, or 10^7 tons. So why are we not buried under mountains of *E. coli*? The answer is limiting nutrients.

Nutrients and Environmental Niches

Do all cells build themselves the same way? And if not, is there a core set of nutrients that all cells must use to grow? **Essential nutrients** are those compounds a microbe cannot make itself but must gather from its immediate environment if the cell is to grow and divide. Some nutrients, such as carbon (C), nitrogen (N), phosphorus (P), hydrogen (H), oxygen (O), and sulfur (S), are needed in large quantities (macronutrients) because they make up the carbohydrates, lipids, nucleic acids, and proteins of the cell. The enormous number of microbial species that exist today evolved to use different environmental sources of elements, so, no, all cells do not build themselves the same way. But all cells use the same essential nutrients.

Cations such as magnesium (Mg^{2+}), ferrous iron (Fe^{2+}), and potassium (K^+) are also needed in large amounts because they serve as enzyme cofactors (Mg^{2+}, Fe^{2+}) or help balance cell osmolarity (K^+). In addition, all cells require small amounts of cobalt, copper, manganese, molybdenum, nickel, and zinc—the so-called trace elements (micronutrients)—as essential components of some enzymes. Cobalt, for example, is part of the cofactor vitamin B_{12}, which aids the catalytic process of specific enzymes.

Some organisms, such as the gastrointestinal bacterium *E. coli*, make all their proteins, nucleic acids, and cell wall and membrane components from a very simple blend of chemical elements and compounds. Many microbes, however, cannot assemble all the compounds they need for growth on their own. Because of where they grow in nature, many bacteria evolved to require additional, specific **growth factors**, nutrients needed by some but not all microbial species. Table 6.1 presents the growth factors for some common microbes associated with disease. Consider *Haemophilus influenzae*, the causative agent of Mirembe's meningitis in our chapter-opening case history. This pathogen cannot synthesize NAD (V factor, a cofactor needed to harvest energy from nutrients) or heme (X factor, needed to make cytochromes for energy production). Thus, *H. influenzae* cannot grow on typical laboratory media unless those specific growth factors are added. Knowing what growth factors certain pathogenic bacteria require can help clinical laboratory technicians identify the agent causing an infection.

Another organism that requires extra growth factors is *Streptococcus pyogenes*, a bacterium that can cause a sore throat or a horrible skin infection. Why should *S. pyogenes* make glutamic acid or alanine using amino acid biosynthetic pathways if those amino acids are readily available in its normal environment (such as the human oral cavity)? Because the bacterium never needs to make these compounds, *S. pyogenes*, as a matter of efficiency, has lost the genes whose protein products synthesize glutamic acid and alanine. In the laboratory, we can grow many organisms in a **chemically defined minimal medium** that contains only those compounds needed for an organism to grow (Table 6.2). In the case of *S. pyogenes*, this would include glutamic acid and alanine in addition to the macro- and micronutrients mentioned earlier.

Link You will learn more in Chapter 7 about amino acid biosynthetic pathways and of the roles NAD and cytochromes play in energy production.

Although we can grow many bacteria in or on laboratory media, some bacteria have such complex nutrient requirements in their natural habitat that we still do not know how to grow them using artificial media. For instance, *Rickettsia* species that cause Rocky Mountain spotted fever and typhus grow only within the cytoplasm of eukaryotic cells (Figure 6.1). This obligate intracellular bacterium lost key pathways needed for independent growth because the host cell supplies them. We have yet to discover what those factors are.

Table 6.1
Growth Factors and Natural Habitats of Bacterial Pathogens

Organism[a]	Disease	Natural Habitats	Growth Factors
Shigella	Bloody diarrhea	Human intestine	Nicotinamide (NAD)[b]
Haemophilus influenzae	Meningitis	Humans and other animal species, upper respiratory tract	Hemin, NAD
Staphylococcus aureus	Boils, osteomyelitis	Nasal passages; widespread but short-lived in nature	Multiple requirements
Abiotrophia	Osteomyelitis	Humans and other animal species	Vitamin K, cysteine
Legionella pneumophila	Legionnaires' disease	Soil, refrigeration cooling towers	Cysteine
Bordetella pertussis	Whooping cough	Humans and other animal species	Glutamate, proline, cysteine
Francisella tularensis	Tularemia	Wild deer; rabbits	Multiple requirements, notably cysteine
Mycobacterium tuberculosis M. leprae	Tuberculosis Leprosy	Humans	Nicotinic acid (NAD),[b] alanine *M. leprae* is unculturable
Streptococcus pyogenes	Pharyngitis, rheumatic fever	Humans	Glutamate, alanine

[a]These are just examples. More complete coverage of the diseases caused by these organisms is found in Chapters 19–24.
[b]Both nicotinamide and nicotinic acid are derived from NAD, nicotinamide adenine dinucleotide.

Table 6.2
Composition of Some Commonly Used Media

Media	Ingredients per Liter	Organisms Cultured
Luria Bertani (complex)	Bacto tryptone[a] 10 g Bacto yeast extract 5 g NaCl 10 g pH 7	Many Gram-negative and Gram-positive organisms
M9 medium (defined)	Glucose 2.0 g Na_2HPO_4 6.0 g (42 mM) KH_2PO_4 3.0 g (22 mM) NH_4Cl 1.0 g (19 mM) NaCl 0.5 g (9 mM) $MgCl_2$ 2.0 mM $CaCl_2$ 0.1 mM pH 7	Gram-negative organisms such as *E. coli*
Azotobacter medium (defined)	Mannitol 2.0 g K_2HPO_4 0.5 g $MgSO_4$ • $7H_2O$ 0.2 g $FeSO_4$ • $7H_2O$ 0.1 g	*Azotobacter*
Sulfur oxidizers (defined)	NH_4Cl 0.52 g KH_2PO_4 0.28 g $MgSO_4$ • $7H_2O$ 0.25 g $CaCl_2$ 0.07 g Elemental sulfur 1.56 g CO_2 5% pH 3	*Acidithiobacillus thiooxidans*

[a]Bacto tryptone is a pancreatic digest of casein (bovine milk protein).

Obtaining Carbon: Autotrophy and Heterotrophy

How do microbes obtain carbon? Maintaining life on this planet is an amazing process and a delicate balancing act. For example, all Earth's life-forms are based on carbon. But if all life-forms used carbon in the same way, carbon would become locked in an unusable form. The carbon cycle involves two counterbalancing metabolic groups of organisms, heterotrophs and autotrophs, that use different **metabolic pathways**. Each group converts carbon to a form that the other group can use.

Heterotrophs (such as *E. coli*) rely on other organisms to make the organic compounds, such as glucose, that they use as carbon sources. All bacterial pathogens are heterotrophs. During heterotrophic metabolism, organic carbon sources are disassembled to generate energy and then reassembled to make cell constituents such as proteins and carbohydrates. This process converts a large amount of the organic carbon source to carbon dioxide (CO_2), which is then released to the atmosphere. Thus, if left alone, heterotrophs would deplete the world of organic carbon sources (converting them to unusable CO_2) and starve to death. For life to continue, CO_2 must be recycled.

Autotrophs use the CO_2 discarded by heterotrophs to make complex cell constituents made up of C, H, and O (for example, carbohydrates, which have the general formula $[CH_2O]_n$). The organic compounds synthesized by autotrophs can later be used as carbon sources by heterotrophs. The interdependence of autotrophs and heterotrophs is an integral part of the carbon cycle (**Figure 6.2**). It is important to realize that the metabolism of any one group of organisms depends on the metabolism of other groups of organisms. Even in the microbial world, "it takes a village."

Link Deeper discussion of **metabolic pathways**, such as the TCA cycle, used to harvest carbon and generate energy, will take place in Chapter 7. Here we provide an introductory look at how the growth of different bacterial life-forms interconnects.

Capturing Energy: Phototrophy and Chemotrophy

How do microbes obtain energy? The cellular synthesis of proteins, nucleic acids, and other structures from carbon requires an energy source. Depending on the organism, energy can be extracted from chemical reactions triggered by the absorption of light (**phototrophy**, or photosynthesis) or from **oxidation-reduction** reactions

Figure 6.1   *Rickettsia prowazekii* Growing within Eukaryotic Cells

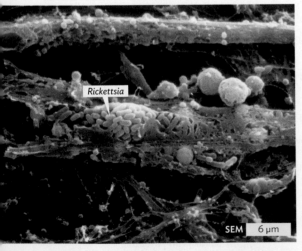

A. *R. prowazekii* growing within the cytoplasm of a chicken embryo fibroblast.

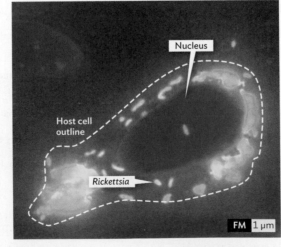

B. Fluorescent stain of *R. prowazekii* (green, 0.5 μm long) growing within a cultured human cell (outline marked by a dotted line). Host nucleus is blue; mitochondria are red.

Figure 6.2   The Carbon Cycle

The carbon cycle requires both autotrophs and heterotrophs.

C. *Nostoc* is an autotroph. These cyanobacteria live in a gelatinous sphere. The spheres shown are colonies (approx. 1-mm diameter) that are just beginning to divide. The individual cyanobacteria are the small cells forming the long strands *inside* the spheres.

A. Heterotrophs gain energy by degrading complex organic compounds, such as polysaccharides, to smaller compounds, such as glucose and pyruvate. The carbon from pyruvate moves through the tricarboxylic acid (TCA) cycle, or Krebs cycle, and CO_2 is released. In the absence of a TCA cycle, the carbon can end up as fermentation products, such as ethanol or acetic acid.

B. Autotrophs use light energy or energy derived from the oxidation of minerals to capture CO_2 and convert it to complex organic molecules.

that remove electrons from high-energy compounds to produce lower-energy compounds (**chemotrophy**). Chemotrophy comes in two forms. One form (**lithotrophy**) uses inorganic chemicals such as hydrogen (H_2), hydrogen sulfide (H_2S), ammonium (NH_4), nitrite (NO_2^-), and ferrous iron (Fe^{2+}) for energy. The other form (**organotrophy**) oxidizes organic compounds such as sugars to obtain energy. Most pathogenic bacteria use carbohydrates as energy *and* carbon sources, but some can also use inorganic compounds (for example, hydrogen gas) as sources of energy.

Link Recall from Section 4.4 that **reduction** refers to the gain of electrons, whereas **oxidation** refers to the loss of electrons. The molecule that gains electrons becomes reduced, and the molecule that loses electrons becomes oxidized.

In chemotrophy, a reduced compound, such as glucose, can donate one or more of its electrons to a less reduced (more oxidized) compound, such as nicotinamide adenine dinucleotide (NAD), releasing energy (in the form of donated electrons) and becoming oxidized in the process. NAD is a cell molecule critical to energy metabolism and is discussed, along with oxidation-reduction reactions, in Chapter 7.

Different species of Bacteria and Archaea can carry out these varied forms of carbon and energy acquisition. It is important to remember, however, that only Bacteria, not Archaea, cause disease, and among the Bacteria, only certain heterotrophic organotrophs are pathogenic.

Note The following prefixes for "-trophy" terms help distinguish different forms of energy-yielding metabolism.

Carbon Source for Biomass	
Auto-	CO_2 is fixed and assembled into organic molecules.
Hetero-	Preformed organic molecules are acquired from outside the cell, broken down for carbon, and the carbon reassembled to make biomass.

Energy Source	
Photo-	Light absorption captures energy.
Chemo-	Chemical electron donors are oxidized.

Electron Source	
Litho-	Inorganic molecules donate electrons.
Organo-	Organic molecules donate electrons.

Obtaining Nitrogen

How do different microbes obtain nitrogen, another essential nutrient? Nitrogen is a critical component of proteins, nucleic acids, and other cellular constituents and is required in large amounts by living organisms. Nitrogen gas (N_2) makes up nearly 79% of Earth's atmosphere, but the nitrogen in N_2 is unavailable for use by most

Figure 6.3   The Nitrogen Cycle

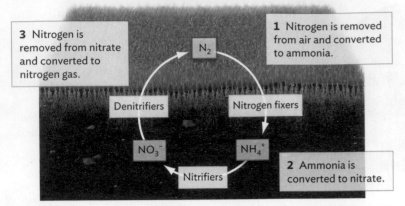

3 Nitrogen is removed from nitrate and converted to nitrogen gas.

1 Nitrogen is removed from air and converted to ammonia.

Denitrifiers

Nitrogen fixers

N_2

NO_3^-

NH_4^+

Nitrifiers

2 Ammonia is converted to nitrate.

Dinitrogen gas (N_2) is fixed in ammonium ions (NH_4^+) by species of bacteria (nitrogen fixers) that possess the enzyme nitrogenase. Other bacteria (nitrifiers) oxidize NH_4^+ to generate energy. Still others (denitrifiers) use oxidized forms of nitrogen, such as NO_3^-, as an alternative electron acceptor in place of O_2.

and archaea. Even bacterial pathogens can carry out nitrification, denitrification, or nitrogen fixation.

We have presented in this section the "big picture" of how bacterial cells of all stripes gather carbon, generate energy, and recycle nitrogen. This overview is preparation for the deeper discussions of bacterial metabolism to come in Chapter 7. For now, appreciate that all these life-forms coexist and collaborate to form vast nutrient cycles that sustain Earth. Without them, we humans would not exist.

SECTION SUMMARY

- **Essential nutrients fuel the growth** of all microorganisms.
- **Microbial genomes evolve in response to nutrient availability.** As a result, some disease-causing bacteria have developed requirements for growth factors supplied by their hosts.
- **Autotrophs use CO_2 as a carbon source.**
- **Microbes obtain energy through** phototrophy, lithotrophy, or organotrophy.
- **Heterotrophs consume organic compounds** made by autotrophs; the global carbon cycle is made possible by the interactions between heterotrophs and autotrophs.
- **Nitrogen fixers, nitrifiers, and denitrifiers** contribute to the nitrogen cycle.

Thought Question 6.1   Red blood cells contain NAD and hemin. Why, then, in our chapter-opening case history, did *H. influenzae* not grow on blood agar medium?

Thought Question 6.2   In a mixed ecosystem of autotrophs and heterotrophs, what happens when a heterotroph allows the autotroph to grow and begin to make excess organic carbon?

organisms because the triple bond between the two nitrogen atoms is highly stable and requires considerable energy to break. Nitrogen from N_2, to be used for growth, must first be "fixed," meaning it must be converted to ammonium ions (NH_4^+). As with the carbon cycle, various groups of organisms collaborate to convert nitrogen gas to ammonium, then to nitrate (NO_3^-), and back again to nitrogen gas in what is called the **nitrogen cycle** (Figure 6.3). One group "fixes" atmospheric nitrogen to make NH_4^+, while other groups of bacteria do the opposite, transforming ammonia to nitrate (**nitrification**) and then converting nitrate to N_2 (**denitrification**).

Nitrogen-fixing bacteria may be free-living in soil or water, or they may form symbiotic associations with plants or other organisms. A **symbiont** is an organism that lives in intimate association with a second organism. For example, *Rhizobium, Sinorhizobium,* and *Bradyrhizobium* species are nitrogen-fixing symbionts present in leguminous plants such as soybeans, chickpeas, and clover (**Figure 6.4**). These plant symbionts convert atmospheric nitrogen to the ammonia the plant needs to form proteins and other essential compounds. Although symbionts are the most widely known nitrogen-fixing bacteria, the majority of nitrogen in soil and marine environments is fixed by free-living bacteria

Figure 6.4   *Rhizobium* and a Legume

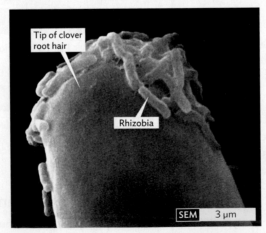

Tip of clover root hair

Rhizobia

SEM   3 µm

A. Symbiotic *Rhizobium* cells clustered on a clover root tip. The rhizobia shown here are clustered on the surface of the root. Soon they will start to invade the root and begin a symbiotic partnership that will benefit both organisms.

B. Root nodules (arrow). After the rhizobia invade the plant root, symbiosis between plant and microbe produces nodules.

6.2
Culturing and Counting Bacteria

SECTION OBJECTIVES

- Explain how pure cultures are obtained and why they are important in medicine.
- Distinguish synthetic, complex, selective, and differential media and their use in clinical microbiology.
- Describe the various ways bacterial growth is measured, and explain the advantages and disadvantages of each method.

How do we grow, culture, and separate microbes in the laboratory? And can all microbes be grown in "captivity"? Microbes in nature usually exist in complex, multispecies communities, but for detailed laboratory studies, they must be grown separately in pure culture. It is nothing short of amazing—and humbling—that after 130 years of trying to grow microbes in the laboratory, we have succeeded in culturing only 0.1% of the microorganisms around us. Since the time of Koch in the late nineteenth century, microbiologists have used the same fundamental techniques to culture bacteria in the laboratory. It is true that improvements have been made, but the vast majority of the microbial world has yet to be tamed.

Growing Bacteria in Culture Media

For those organisms that can be cultured, a variety of culturing techniques are used, each for a different purpose. Bacterial culture media may be either liquid or solid. A liquid, or broth, medium, in which organisms can move about freely, is useful when studying the growth characteristics of a single, genetically homogeneous strain of a single species (that is, a **pure culture**). Liquid media are also convenient for examining growth rates and microbial biochemistry at different phases of growth. Solid media, usually gelled with agar, are useful when trying to separate mixtures of different organisms as they are found in the natural environment or in clinical specimens.

Obtaining Pure Cultures

Solid media are basically liquid media to which a solidifying agent has been added. The most versatile and widely used solidifying agent is agar (the development of agar medium is covered in Section 1.5). Derived from seaweed, agar forms an unusual gel that liquefies at 100°C (212°F) but does not solidify until cooled to about 40°C (104°F). Liquefied agar medium poured into shallow, covered petri dishes cools and hardens to provide a large, flat surface on which a mixture of microorganisms can be streaked to separate individual cells. Each cell will divide and grow to form a distinct, visible colony of cells (Figure 6.5). As shown in Figure 6.6, a drop of liquid culture is collected using an inoculating loop and streaked across the agar plate surface in a pattern called **isolation streaking** (also known as **dilution streaking**). Organisms fall off the loop

as it moves along the agar surface. Toward the end of the streak, few bacteria remain on the loop, so at that point, individual cells will land and stick to different places on the agar surface. If the medium, whether artificial (for example, laboratory medium) or natural (dental plaque), contains the proper nutrients and growth factors, a single cell will multiply, producing many millions of offspring and forming a visible droplet called a **colony** (Figure 6.6B). A pure culture of the species can be obtained by touching a single colony with a sterile inoculating loop and inserting that loop into fresh liquid medium.

Another way to isolate pure colonies is the **spread plate** technique. Starting from a liquid culture of bacteria, a series of tenfold

Figure 6.5 Separation and Growth of Microbes on an Agar Surface

A. Colonies (1–5 mm diameter) of *Acidovorax aveane* separated on an agar plate. This organism is a plant pathogen that causes watermelon fruit blotch.

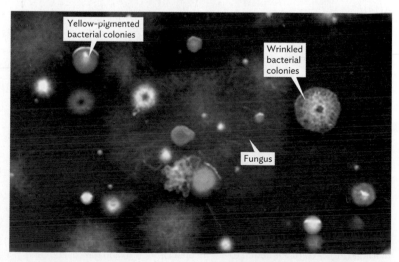

B. A mixed culture was separated by dilution on an agar plate. The culture yielded yellow-pigmented bacterial colonies, wrinkled bacterial colonies, and fungus.

Figure 6.6   Isolation (Dilution) Streaking Technique

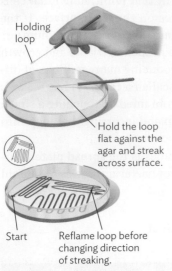

Holding
loop

Hold the loop
flat against the
agar and streak
across surface.

Start

Reflame loop before
changing direction
of streaking.

A. A liquid culture is sampled with an inoculating (holding) loop and streaked across the plate in three or four areas, with the loop flamed (sterilized) between areas. After the first section is streaked, the resterilized loop is passed three to five times through the original section and into an adjacent section, as demonstrated. Dragging the loop across the agar diminishes the number of organisms clinging to the loop until only single cells are deposited at a given location.

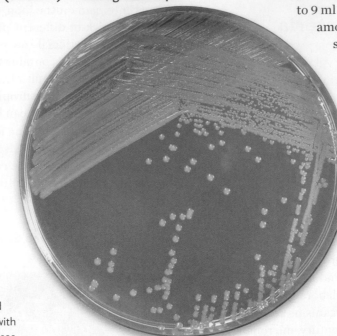

B. *Salmonella enterica* culture obtained by isolation streaking.

dilutions (transferring 1 milliliter of a broth with bacteria to 9 ml of sterile broth) is made, and a small, measured amount of each dilution is placed directly on the surface of separate agar plates (**Figure 6.7**). The sample is spread over the surface of the plate using a heat-sterilized, bent glass rod (cell spreader). The early dilutions, those containing the most bacteria, will produce **confluent** growth that covers the entire agar surface. The later dilutions, containing fewer and fewer organisms, yield individual colonies. Spread plates enable us to isolate pure cultures and can be used to enumerate the number of viable bacteria in the original growth tube. A **viable** organism is one that successfully replicates to form a colony. Thus, each colony on an agar plate represents one viable organism present in the original liquid culture.

It is important to note that the "one cell equals one colony" paradigm does not hold for all bacteria. Some organisms, such as *Streptococcus* or *Staphylococcus*, usually do not exist as single cells, but as chains or clusters of several cells. Thus, a cluster of ten *Staphylococcus* cells will form only one colony on an agar medium and is called a colony-forming unit (CFU).

Figure 6.7   Tenfold Dilutions, Plating, and Viable Counts

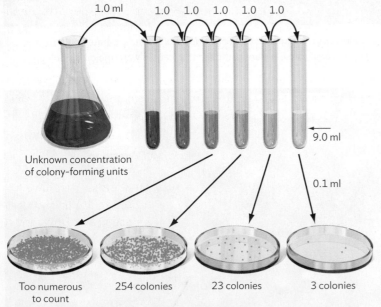

1.0 ml   1.0   1.0   1.0   1.0   1.0

Unknown concentration
of colony-forming units

9.0 ml

0.1 ml

Too numerous
to count

254 colonies

23 colonies

3 colonies

A. A culture containing an unknown concentration of cells is serially diluted. One milliliter (ml) of culture is added to 9.0 ml of diluent broth and mixed, and then 1 ml of this dilution is added to another 9.0 ml of diluent (10^{-2} dilution). These steps are repeated for further dilution, each of which lowers the cell number tenfold. After dilution, 0.1 ml of each dilution is spread onto an agar plate.

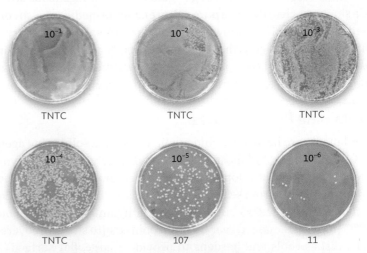

10^{-1} TNTC 10^{-2} TNTC 10^{-3} TNTC

10^{-4} TNTC 10^{-5} 107 10^{-6} 11

B. Plates prepared as in (A) are incubated at 37°C (98.6°F) to yield colonies. Generally, plates containing between 30 and 300 colonies are countable. Numbers outside that range are not statistically reliable. Multiplying the number of countable colonies (107 colonies per 0.1 ml on the 10^{-5} plate) by 10 gives the number of colonies in 1.0 ml of the 10^{-5} dilution. Multiplying that number by the reciprocal of the dilution factor ($1/10^{-5}$) gives the number of cells (colony-forming units, CFUs) per milliliter in the original broth tube ($107 \times 10 \times 10^5 = 1.07 \times 10^8$ CFUs per milliliter). TNTC = too numerous to count.

Growth in Complex Media and Synthetic Media

Bacteria can be grown in a precisely defined **synthetic medium**, or **minimal medium**, as mentioned earlier or in a nutrient-rich but poorly defined **complex medium**, also called **rich medium**. A synthetic medium starts with water; then various salts, carbon, nitrogen, and energy sources are added in precise amounts. For self-reliant organisms such as *E. coli* or *Bacillus subtilis*, those few chemicals are all that is needed. Other organisms, such as *Shigella* species or mutant strains of *E. coli* or *B. subtilis*, require additional ingredients to satisfy requirements imposed by the absence of specific metabolic pathways.

Recipes for complex media usually contain several poorly defined ingredients, such as yeast extract or beef extract, whose exact composition is not known (Table 6.2). These additives include a rich variety of amino acids, peptides, nucleosides, vitamins, and some sugars. Some organisms are particularly fastidious (such as *H. influenzae* or *Neisseria meningitidis*), requiring that components of blood (or blood itself) be added to a basic complex medium; the complex medium would then be called an **enriched medium**.

Complex, or rich, media provide many of the chemical building blocks that a cell would otherwise have to synthesize on its own. For example, instead of making proteins that synthesize tryptophan, all the cell needs is a membrane transport system to harvest prefabricated tryptophan from the medium. Likewise, fastidious organisms that require blood in their media may claim the heme released from red blood cells as their own, using it as an enzyme prosthetic group, a group critical to enzyme function (for example, the heme group in cytochromes). All of this saves the scavenging cell a tremendous amount of energy, and as a result, bacteria tend to grow fastest in complex media.

Selective and Differential Media

In a clinical diagnostic laboratory, it is important to quickly expose the presence of a pathogen in a patient's specimen. That is easy if the specimen should normally be sterile, such as cerebrospinal fluid, but challenging if it contains normal microbiota, as in fecal specimens.

Salmonella enterica is an important and widespread cause of diarrhea. But our intestines are crammed with many other bacteria that are considered normal. How, then, does the laboratory find *Salmonella* among all those other bacterial species in a sample of feces? Once again the answer lies in understanding bacterial physiology.

Microorganisms are remarkably diverse with respect to their metabolic capabilities and resistance to certain toxic agents. These differences are exploited by **clinical microbiology** through the use of **selective media**, which favor the growth of one organism over another, and **differential media**, which expose biochemical differences between two species that grow equally well.

Link You will learn more about enriched, selective, and differential media like blood, chocolate, CNA blood, and Hektoen agars and their uses in **clinical microbiology** in Chapter 25.

Several media used in clinical microbiology are both selective and differential. **MacConkey medium**, for example, selects for growth of Gram-negative bacteria because it contains bile salts and crystal violet, which prevent the growth of Gram-positives. Gram-negative bacteria will grow because they have outer membranes that keep the detergent away from the critical cytoplasmic membrane. The medium also includes lactose, neutral red, and peptones to differentiate lactose fermenters from nonfermenters. Lactose fermentors such as *E. coli* degrade lactose and secrete acidic products that lower pH around the colony. The acidic pH allows neutral red to enter cells and produce a red colony. In stark contrast, the pathogen *S. enterica* does not ferment lactose and grows only on the nonfermentable peptides. No acid is produced, neutral red does not enter cells, and the colonies are white. Thus, a culture grown on MacConkey medium will consist of Gram-negative organisms that can be identified as lactose fermenters (red colonies) or nonfermenters (white colonies; **Figure 6.8**). This medium is of particular benefit when diagnosing the etiology (cause) of diarrheal disease because normal intestinal bacteria that grow on MacConkey agar are lactose fermenters, whereas two important pathogens, *Salmonella* and *Shigella*, are lactose nonfermenters.

CASE HISTORY 6.1

Differential Media: Finding the Infectious Needle in the Haystack

A fecal sample from a patient with diarrhea was streaked onto MacConkey lactose agar medium, a medium that selectively prevents the growth of Gram-positive bacteria but will differentiate Gram-negative bacteria that can ferment lactose from those that cannot. The next day, several different types of bacterial colonies were visible on the agar plate. Some were red; others were white. The white colonies led the clinical microbiologist to suspect that the patient was infected with the Gram-negative bacillus *Salmonella enterica* serovar Typhimurium.

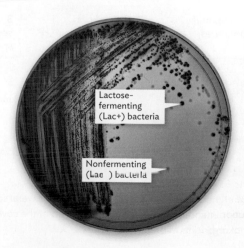

Figure 6.8

MacConkey Medium, a Culture Medium Both Selective and Differential

Isolation (dilution) streak of a mixture of lactose-fermenting (Lac+) and non-lactose-fermenting (Lac−) bacteria. Only Gram-negative bacteria grow on lactose MacConkey medium (selective). Only a species capable of fermenting lactose produces red colonies (differential).

Lactose-fermenting (Lac+) bacteria

Nonfermenting (Lac−) bacteria

CASE HISTORY 6.2

Urine Cultures (Where Numbers Count)

Melissa, a 25-year-old woman, visited her doctor's office complaining of a terrible burning sensation while urinating. The physician assistant (PA) suspected the patient was suffering from a urinary tract infection (UTI) in her bladder and sent a sample of urine to the laboratory. The laboratory technician took a very small but exact volume of urine and spread it onto agar medium. After 24 hours of incubation, the clinical microbiologist found large numbers of *Escherichia coli* colonies (a Gram-negative bacillus), equivalent to 100,000 bacteria per milliliter of the patient's urine. Because urine samples from healthy patients usually have fewer than 10,000 bacteria per milliliter, the PA determined that the woman did have a **urinary tract infection** and prescribed an antibiotic appropriate for the etiological (causative) agent, uropathogenic *E. coli*.

This case history illustrates how bacterial numbers can be determined in clinical samples as well as in laboratory culture. Counting or quantifying organisms that are invisible to the naked eye is surprisingly difficult because each of the available techniques measures a different physical or biochemical aspect of growth. Thus, a cell density value (given as cells per milliliter) derived from one technique will not necessarily agree with the value obtained by a different method.

Link You will learn how a clinician makes a presumptive diagnosis of **urinary tract infection** and how lab results help confirm that diagnosis in Chapter 23. Chapter 25 will describe how the clinical microbiologist determines whether a colony on an agar plate contains *E. coli* or some other bacterial species.

Direct Counting of Living and Dead Cells

Though not the technique used in Case History 6.2, microorganisms can be counted directly using a microscope. A dilution of a bacterial culture is placed on a special microscope slide called a hemocytometer (or, more specifically for bacteria, a Petroff-Hausser counting chamber; Figure 6.9). Etched on the surface of the slide is a grid of precise dimensions, and placing a coverslip over the grid creates a space of precise volume. The number of organisms counted within that volume is used to calculate the concentration of cells in the original culture.

However, "seeing" an organism under the microscope does not mean that the organism is alive. Living cells may be distinguished from dead cells by fluorescence microscopy using fluorescent chemical dyes, as discussed in Chapter 3. For example, propidium iodide, a red dye, inserts between DNA bases but cannot freely penetrate the energized membranes of living cells. Thus, only *dead* cells stain red under a fluorescence scope. Another dye, Syto-9, enters both living and dead cells, staining them both green. By combining Syto-9 with propidium iodide, microbiologists can distinguish living cells from dead cells: living cells will stain green, whereas dead cells appear orange or yellow because both dyes enter, and Syto-9 (green) plus propidium (red) appears yellow-orange (Figure 6.10).

Direct counting without microscopy can be accomplished by a sophisticated electronic instrument called a **fluorescence-activated cell sorter (FACS)**, which can count and separate bacterial cells with different properties. For instance, bacterial cells that synthesize a fluorescent protein or that have been tagged with a fluorescent antibody or chemical are streamed past a laser (Figure 6.11) that counts cell numbers and measures fluorescence intensity.

Viable Counts

Viable cells, as noted previously, are those that can replicate and form colonies on an agar plate. To obtain a viable cell count, dilutions of a liquid culture can be plated directly on an agar surface. After colonies form, they are counted, and the original cell number is calculated (see Figure 6.7). This procedure was used in diagnosing

Figure 6.9   The Petroff-Hausser Chamber for Direct Microscopic Counts

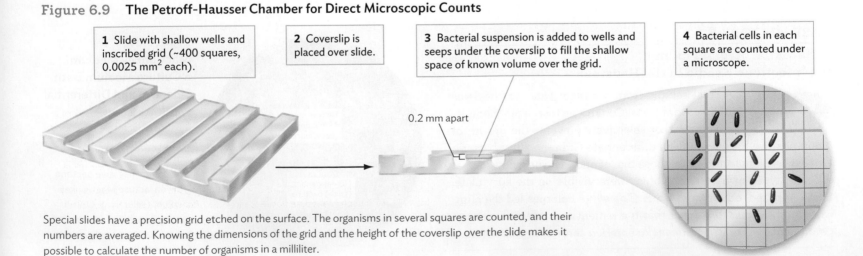

1 Slide with shallow wells and inscribed grid (~400 squares, 0.0025 mm² each).

2 Coverslip is placed over slide.

3 Bacterial suspension is added to wells and seeps under the coverslip to fill the shallow space of known volume over the grid.

4 Bacterial cells in each square are counted under a microscope.

0.2 mm apart

Special slides have a precision grid etched on the surface. The organisms in several squares are counted, and their numbers are averaged. Knowing the dimensions of the grid and the height of the coverslip over the slide makes it possible to calculate the number of organisms in a milliliter.

Figure 6.10 Live-Dead Stain

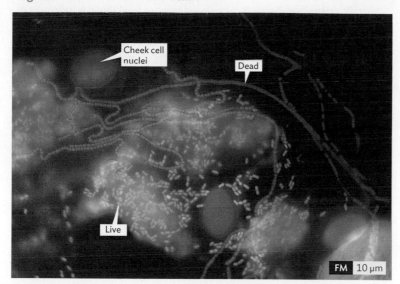

Live and dead bacteria visualized on freshly isolated human cheek epithelial cells using the LIVE/DEAD BacLight Bacterial Viability Kit. Dead bacterial cells fluoresce orange or yellow because propidium (red) can enter the cells and slip between the base pairs of DNA. Live cells fluoresce green because Syto 9 (green) enters the cell. The faint green smears are the outlines of cheek cells.

Figure 6.11 Fluorescence-Activated Cell Sorter

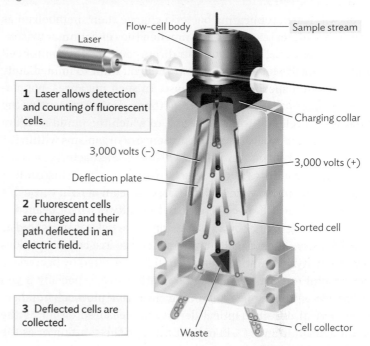

1 Laser allows detection and counting of fluorescent cells.

2 Fluorescent cells are charged and their path deflected in an electric field.

3 Deflected cells are collected.

A. Schematic diagram of a FACS apparatus (bidirectional sorting).

Melissa in Case History 6.2. For example, if 100 colonies are observed on an agar plate that was spread with 100 microliters (μl) of a 10^{-3} dilution of a culture, then the original culture broth contained 10^6 organisms per milliliter (100 colonies × 10 × 10^3). The actual formula is CFU/ml = (number of colonies per volume plated in μl) × (1,000 μl/volume plated) × (1/dilution factor).

Although viable counts are widely used in research, there are problems using this method to measure cell number. One issue is that viable counts usually underestimate the number of living cells in a culture. As noted earlier, metabolically active but damaged cells that do not form colonies on agar plates will typically not be counted as alive. Comparing a viable count with a direct count obtained from a live-dead stain can expose the presence of damaged cells. As mentioned earlier, organisms such as streptococci that grow in chains will cause actual cell numbers to be underestimated, which is why the results of this counting technique are reported as colony-forming units (CFUs) rather than cells.

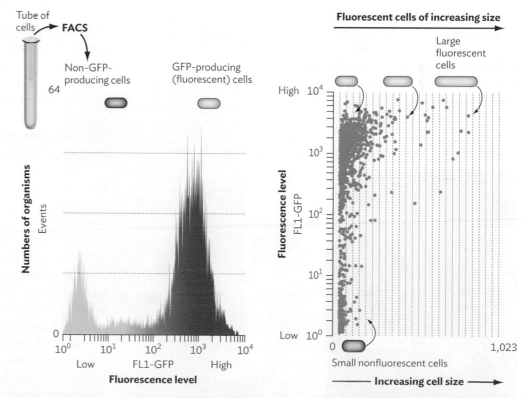

B. Separation of green fluorescent protein (GFP)–producing *E. coli* from non-GFP-producing *E. coli*. The low-level fluorescence in the cells on the left is baseline fluorescence (autofluorescence). The scatterplot displays the same FACS data, showing the size distribution of cells (*x*-axis) with respect to level of fluorescence (*y*-axis). The larger cells may be ones that are about to divide.

Optical Density

As you will learn, replicating bacteria change their metabolism as cell number increases (crowding) or when they encounter environmental stress. Using plate counts or direct counting to monitor cell number takes too long to carry out when you want to immediately study cells in a specific growth phase. In such instances, population size can be calculated by other methods, such as measuring the **turbidity** of a cell culture—the degree to which the liquid medium has become cloudy because of the presence of organisms within it.

The turbidity of liquid medium will increase as bacteria increase in number. Turbidity can be measured in real time by an instrument called a **spectrophotometer**, which passes a beam of light through a sample of the culture. The decrease in intensity of a light beam due to the scattering of light by a suspension of particles is measured as **optical density**. Thus, as cell numbers increase in a culture, so does optical density. The optical density of light scattered by bacteria is a very useful tool for estimating population size, especially if you have previously developed a standard curve that plots cell numbers versus optical density. Optical density is measured either as the percent transmittance (%T) of light through the solution or as the absorbance (A) of light by the solution. Each term can be converted to the other using the formula $A = 2 - \log \%T$.

SECTION SUMMARY

- **A pure culture** contains only a single bacterial species that is usually genetically homogeneous. Pure culture is needed for detailed molecular and biochemical studies of that species.
- **Bacteria can be cultured** on solid or liquid media.
- **Minimal defined media** contain only those nutrients essential for growth.

Figure 6.12   Symmetrical and Asymmetrical Cell Division

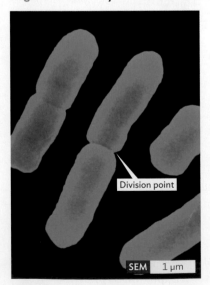

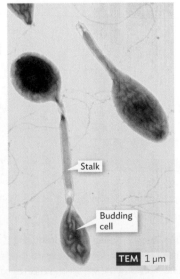

A. Symmetrical cell division, or binary fission, in *Lactobacillus* species.

B. Asymmetrical budding of the marine bacterium *Hyphomicrobium*.

- **Complex, or rich, media** contain many nutrients. Other media exploit specific differences between organisms and can be defined as selective or differential.
- **Microorganisms in culture may be counted directly** under a microscope, with or without staining, or by using a fluorescence-activated cell sorter.
- **Microorganisms can be counted indirectly**, as in viable counts and measurement of optical density.
- **A viable bacterial organism** is defined as a cell with the ability to replicate and form a colony on a solid-medium surface.

Thought Question 6.3   The addition of sheep blood to agar produces a very rich medium called blood agar. Do you think blood agar can be considered a selective medium? A differential medium? *Hint:* Some bacteria can lyse red blood cells.

6.3
The Growth Cycle

SECTION OBJECTIVES

- Discuss the phases of a typical bacterial growth curve.
- Explain how bacterial growth correlates to disease.
- Describe the purpose of continuous culture and how it correlates to the human digestive tract.

How do microbes grow? What determines their rate of growth? In nature, the answers to these questions are extremely complex. Most microbes in nature do not exist alone, but in complex communities of microbes or multicellular organisms that influence each other's growth. Nevertheless, all species at one time or another exhibit both rapid growth and nongrowth, as well as many phases in between. For clarity, we present here the principles of rapid growth, while bearing in mind the actual diversity of growth situations in nature.

The ultimate goal of any species is to make more of its own kind. This fundamental law ensures survival of the species. A typical bacterium (at least the typical bacterium that can be cultured in the laboratory) grows by increasing in length and mass, which facilitates expansion of its nucleoid as its DNA replicates (see Section 5.4). As DNA replication nears completion, the cell, in response to complex genetic signals, begins to synthesize an equatorial septum (cell wall) that ultimately separates the two daughter cells. In this overall process, called **binary fission**, one parent cell splits into two equal daughter cells (Figure 6.12A).

While the majority of culturable bacteria divide symmetrically into halves, some species divide asymmetrically. For example, the bacterium *Caulobacter* forms a stalked cell that remains fixed to a solid surface but reproduces by **budding** from one end to produce small, unstalked motile cells that then break off from the parent cell. The marine organism *Hyphomicrobium* also replicates asymmetrically by budding, releasing a smaller cell from a stalked parent (Figure 6.12B).

Eukaryotic microbes divide by a special form of cell fission involving **mitosis** (described in Chapter 11). Some eukaryotes also undergo more complex life cycles involving budding and diverse morphological forms. No matter how reproduction is accomplished, similar mathematical patterns of population growth are seen in all organismal populations, both prokaryotic and eukaryotic—including our own.

Link Recall that in a eukaryotic organism, DNA consists of paired chromosomes enclosed within a nucleus. Section 11.1 describes how pairs of newly replicated chromosomes are separated in the process of **mitosis** before the eukaryotic cell divides.

Exponential Growth

How does replication affect the growth of a population? Growth of any population obeys a simple law: as the cell number doubles, whatever parameter you use to measure growth also doubles (for example, optical density). Growth in which population size doubles at a *fixed* rate (say, every 20 minutes) is called **exponential growth** because the number of new cells produced after each cell division is larger each time the population divides. For example, 2 cells divide to become 4 cells (increase of 2 cells), 4 cells divide to become 8 cells (increase of 4 cells), 8 cells divide to become 16 cells (an increase of 8 cells). Plotting the increase in the number of cells over time generates an exponential curve, a curve whose slope increases continually. Exponential growth explains how ingesting only 10 cells of a pathogenic strain of *E. coli* can lead to bloody diarrhea within only 2 days.

How does binary fission of cells generate an exponential curve? As just illustrated, if each cell produces two cells per generation, then the population size at any given time is proportional to 2^n, where the exponent n represents the number of generations (replacement of parents by offspring) that have taken place between two time points. Thus, in both bacteria and eukaryotic microbes, binary fission will cause cell number to rise exponentially.

In an environment with unlimited resources, bacteria divide at a constant interval called the **generation time**. The generation time for any organism varies with respect to many parameters, including the bacterial species, type of medium, temperature, and pH. The generation time for cells in culture is also known as the **doubling time**, because the population of cells doubles over one generation. For example, one cell of *E. coli* placed into a complex medium will divide every 20 minutes. After 1 hour of growth (three generations), that one cell will have become eight (1 to 2, 2 to 4, 4 to 8). Because cell number (N) *doubles* with each division, the increase in cell number over time is exponential, not linear. (Imagine if your bank account started at $1 but doubled every day. In 20 days you would have over a million dollars.) A linear increase would occur if cell number rose by a *fixed* amount after every generation (for example, 1 to 2, 2 to 3, 3 to 4).

Starting with any number of organisms at time zero (N_0), the number of organisms after n generations will be $N_0 \times 2^n$. For example, a single cell after three generations ($n = 3$) will produce

$$1 \text{ cell} \times 2^3 = 8 \text{ cells}$$

The mathematics of exponential growth is relatively straightforward, but remember that microbes grow differently in pure culture (very rare in nature) than they do in mixed communities, where neighboring cells produce all kinds of substances that may feed or poison other microbes. In mixed communities, the microbes may grow individually floating in liquid, as in the open ocean, or as a biofilm on solid matter suspended in that ocean (biofilms are described in Section 6.6). In each instance, the mathematics of exponential growth applies at least until the community reaches a density at which different species begin to compete.

Phases of Growth

Exponential growth never lasts indefinitely because nutrient consumption and toxic by-products eventually slow the growth rate until it halts altogether. The simplest way to model the effects of changing conditions is to culture bacteria in liquid medium in a flask or test tube. This technique is called **batch culture**. In batch culture, no fresh medium is added during incubation; thus, nutrient concentrations decline and waste products accumulate during growth.

Batch cultures illustrate the remarkable ability of bacteria to adapt to their environment. As medium conditions deteriorate, alterations occur in membrane composition, cell size, and metabolic pathways, all of which impact generation time. Microbes possess intricate, genetic and metabolic mechanisms that slow growth before the cells lose viability. Plotting culture growth as the logarithm of the cell number versus incubation time allows us to see the effect of changing conditions on generation time and reveals several phases of growth (**Figure 6.13**).

Lag phase. Cells transferred from an old culture to fresh growth media need time to detect their environment, express specific genes, and synthesize components needed to institute rapid growth. As a result, bacteria inoculated into fresh media typically experience a lag period, or **lag phase**, during which cells do not divide. Various factors can contribute to the duration of lag phase. Cells taken from an aged culture may be damaged and require time for repair before they can replicate. Any change in carbon, nitrogen, or energy source must also be sensed by the microbe and appropriate enzyme systems synthesized before growth can resume. Transferring cells grown in a complex medium to a *fresh* complex medium results in a very short lag phase, whereas cells grown in a complex medium and then plunged into a minimal defined medium experience a protracted lag phase, during which the cells readjust to synthesize all the amino acids, nucleotides, and other metabolites originally supplied by the complex medium.

Log, or exponential, phase. Once cells have retooled their physiology to accommodate the new environment, they begin to grow exponentially and enter what is called **exponential phase**, or **logarithmic (log) phase**. Exponential growth is balanced growth, where all cell components are synthesized at constant rates in relation to each other. At this stage, cells grow

Figure 6.13   Bacterial Growth Curves

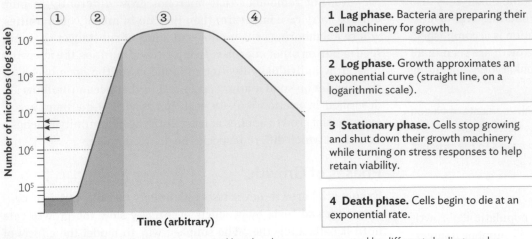

1 Lag phase. Bacteria are preparing their cell machinery for growth.

2 Log phase. Growth approximates an exponential curve (straight line, on a logarithmic scale).

3 Stationary phase. Cells stop growing and shut down their growth machinery while turning on stress responses to help retain viability.

4 Death phase. Cells begin to die at an exponential rate.

Phases of bacterial growth that occur in a typical batch culture are represented by different shadings under the curve. [Note that in this graph, the y-axis is a logarithmic scale that reflects order of magnitude differences between cell numbers, not simple increases in cell number. A one order of magnitude difference equals a tenfold increase in cell number. For example, growing from 2×10^6 cells (bottom arrow) to 4×10^6 cells (middle arrow) is a 100% increase in cell number. This increase is greater than the 50% change when growing from 4×10^6 cells to 6×10^6 cells (top arrow). In both cases, however, the increase in the number of cells is the same, 2×10^6. A logarithmic scale (the y-axis) presents these order of magnitude differences as distances between tick marks. In our example, the distance between the tick marks indicating 2×10^6 and 4×10^6 is greater than the distance between the tick marks indicating 4×10^6 and 6×10^6. However, when cell numbers double (from 2×10^6 cells to 4×10^6 cells or from 4×10^6 cells to 8×10^6 cells), the distances between those tick marks are equal. When we plot order of magnitude differences in this type of graph, exponential growth appears as a linear function.]

and divide at the maximum rate possible based on the medium and growth conditions provided (such as temperature, pH, and osmolarity). Cells are largest at this stage of growth.

As the cell density (number of cells per milliliter) rises during log phase, the rate of doubling eventually slows (called **late log phase**), and a new set of growth-phase–dependent genes are expressed. At this point, some bacteria can also begin to detect the presence of other members of their own species or even completely different species. Bacteria do this by sending and receiving chemical signals in a process known as quorum sensing (think of it as Twitter for microbes; discussed in Section 6.6 and Chapter 9).

Stationary phase. Eventually, viable cell numbers stop rising owing to lack of a key nutrient or buildup of waste products. This point occurs for bacteria grown in a complex medium when cell density rises above 10^9 cells per milliliter, but it can occur at lower cell densities if nutrients are limiting. At this point, the growth curve levels off and the culture enters what is called **stationary phase**. In stationary phase, the growth of individual cells slows and the numbers of cells dividing approximately equals the number of cells dying. As they enter stationary phase, many bacterial pathogens develop resistance to antibiotics and host defenses.

If they did not change their physiology, microbes entering stationary phase would be vulnerable to deteriorating medium conditions. Because cells in stationary phase are not as metabolically nimble as cells in exponential phase, damage from highly reactive by-products of oxygen metabolism (oxygen radicals such as superoxide; see Section 6.5) and other toxic metabolic by-products would readily kill them. As an avoidance strategy, some bacteria develop resistant spores (also called endospores) in response to nutrient depletion (see Section 6.6), whereas other bacteria undergo less dramatic but effective molecular reprogramming. The bacterium *E. coli*, for example, adjusts to stationary phase by decreasing its size. One benefit of these changes is that less nutrient is required to sustain the smaller cell. New stress resistance enzymes are also synthesized to detoxify toxic oxygen radicals, protect DNA and proteins, and increase cell wall strength through increased peptidoglycan cross-linking. As a result, *E. coli* cells in stationary phase become more resistant to heat, osmotic pressure, pH changes, and other stresses they might encounter while waiting for a new supply of nutrients.

Death phase. Without reprieve in the way of new nutrients, cells in stationary phase will eventually succumb to toxic chemicals present in the environment. Like the growth rate, the **death rate**, the rate at which cells die, is exponential. Although death curves are basically exponential, exact death rates are difficult to define because mutations arise that promote survival, and some cells grow by cannibalizing others. Consequently, the **death phase** is extremely prolonged. A portion of the cells will often survive for months, years, or even decades.

We have just described how bacteria live, grow, and die in batch culture. But how do these stages apply to growth in a human host? All growth stages do occur in a colonized or infected host, but their measurements are confounded by microbial competition, collaboration with resident microbiota, and the host's immune system. In vivo growth (growth in a live host) will be discussed further in Chapter 14.

Continuous Culture

In the classic growth curve that develops in closed systems (a test tube), the exponential phase spans only a few generations. In open systems, however, where fresh medium is continually added to a culture and an equal amount of culture is constantly siphoned off, bacterial populations can be kept in exponential phase at a constant

Figure 6.14   Chemostats and Continuous Culture

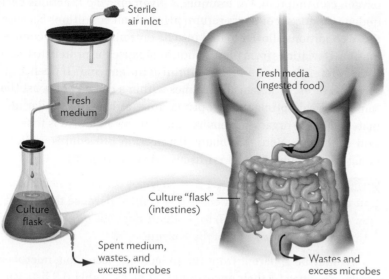

A. The basic chemostat ensures exponential growth by constantly adding and removing equal amounts of culture media.

B. The human gastrointestinal tract is engineered much like a chemostat in that new nutrients are always arriving from the throat while equal amounts of bacterial culture exit in fecal waste.

C. A modern chemostat.

cell mass for long periods of time. In this type of growth pattern, known as **continuous culture**, all cells in a population achieve a steady state, which permits detailed analysis of microbial physiology at different growth rates. The **chemostat** is a continuous culture system in which the diluting medium contains a growth-limiting amount of an essential nutrient (**Figure 6.14**). Increasing the flow rate increases the amount of nutrient available to the microbe. The more nutrient available, the faster a cell's mass will increase. The faster the cell's mass increases, the more quickly it will divide. (Cell division is triggered at a defined cell mass.) It follows, then, that the growth rate in a chemostat is directly related to the dilution rate, or flow rate (milliliters per hour divided by the vessel volume). The more nutrient a culture receives as a result of increasing flow rate, the faster those cells can replicate (that is, the shorter the generation time). Your intestine, which is normally populated by billions of bacteria, functions much like a chemostat.

Continuous cultures are used in the laboratory to study large numbers of cells at constant growth rate and cell mass for both research and industrial applications. Most bacteria in nature grow at very slow rates, a situation that can be mimicked in a chemostat. The physiology of these slow-growing cells is quite different from what is typically observed using batch culture.

SECTION SUMMARY

- **The growth cycle** of organisms grown in liquid batch culture consists of lag phase, log phase, stationary phase, and death phase.

- **The physiology of a bacterial population** changes with the growth phase.
- **Continuous culture** can be used to sustain a population of bacteria at a specified growth rate and cell density.

Thought Question 6.4   If bacteria divide every 20 minutes, why is the exponential growth curve (measured as viable count) smooth and not shaped like a series of steps?

Thought Question 6.5   Suppose 1,000 bacteria are inoculated into a tube of minimal salts medium, where they double once an hour, and 10 bacteria are inoculated into rich medium, where they double every 20 minutes. Which tube will have more bacteria after 2 hours? After 4 hours?

6.4
Environmental Limits on Microbial Growth

SECTION OBJECTIVES

- Describe how environmental changes can alter the shape of a growth curve.
- List different classes of microbes based on their preferred environmental niches (pH, temperature, salt).
- Identify the biological properties that allow different classes of microbes to grow in extreme environments.

What are the environmental limits beyond which microbes will not grow? The answer depends on the microbe. With our human frame of reference, we tend to think that "normal" growth conditions are those found at sea level with a temperature between 20°C (68°F) and 40°C (104°F), a near-neutral pH, a salt concentration of 0.9%, and ample nutrients. Any ecological niche outside this window is called *extreme* and the organisms inhabiting them **extremophiles**. Extremophiles are bacteria, archaea, and some eukaryotic microbes that can grow in extreme environments. For instance, one group of organisms can grow at temperatures above boiling, whereas another group requires a pH 2 acidic environment to grow. According to our definition of *normal* conditions, conditions on Earth when life began were certainly extreme. Consequently, the earliest microbes probably grew in these extreme environments. Organisms that grow under conditions that seem "normal" to humans probably evolved from an ancient extremophile that gradually adapted as the environment evolved to that of our present-day Earth.

Be aware that multiple extremes in an environment can be encountered simultaneously. For instance, in Yellowstone National Park, we can find an extreme acid pool next to an extreme alkali pool, both at extremely high temperatures. Thus, extremophiles typically evolve to survive multiple extreme environments. At Yellowstone, this means that some extremophiles can grow in highly alkaline pH at high temperature while others grow in highly acidic pH at high temperature.

Why do fundamental differences in physical conditions (temperature, pH, osmolarity) favor the growth of specific groups of organisms? The reason is that every protein and macromolecular structure within a cell is affected profoundly by changes in environ-

mental conditions and through reaction with the by-products of oxygen consumption. For example, a given enzyme functions best under a unique set of temperature, pH, and salt conditions because those conditions allow it to fold into its optimum shape, or conformation. Deviations from these conditions cause the protein to fold a little differently (denature) and diminish the enzyme's influence on **reaction rates**. Although all enzymes within a cell do not boast the same physical optima, these optima must at least be similar and matched to the organism's environment if the organism is to function effectively. Note that many disinfection procedures work by denaturing bacterial proteins (discussed in Chapter 13).

Link Recall from Section 4.4 that enzymes greatly increase **reaction rates**. The shape of an enzyme is critical because the conversion of substrates (reactants) to products depends on the substrate fitting into the "active site" of the enzyme.

As may be apparent from the preceding discussion, microbes can be classified by their environmental niche. Table 6.3 lists various classes of microorganisms and the environmental criteria that define them.

Temperature Extremes

Unlike humans (and mammals in general), microbes cannot control their temperature; thus, bacterial cell temperature matches that of the immediate environment. Because temperature affects the average rate of molecular motion, changes in temperature impact every aspect of microbial physiology, including membrane fluidity, nutrient transport, DNA stability, RNA stability, and enzyme structure and function. Every organism has an *optimum* temperature at which

---

Table 6.3

Basic Environmental Classification of Microorganisms

Environmental Parameter	Classification			
Temperature	Hyperthermophile[a] (growth above 80°C [176°F])	Thermophile[a] (growth between 50°C [122°F] and 80°C [176°F])	Mesophile (growth between 15°C [59°F] and 45°C [113°F])	Psychrophile[a] (growth below 15°C [59°F])
pH	Alkaliphile[a] (growth above pH 9)	Neutralophile (growth between pH 5 and pH 8)	Acidophile[a] (growth below pH 3)	
Osmolarity	Halophile[a] (growth in high salt, greater than 2 M NaCl)			
Oxygen	Strict aerobe (growth only in oxygen)	Facultative microbe (growth with or without oxygen)	Microaerophile (growth only in small amounts of oxygen)	Strict anaerobe (growth only without oxygen)
Pressure	Barophile[a] (growth at high pressure, greater than 380 atm)		Barotolerant (growth between 10 and 500 atm)	

[a]Considered extremophiles.

it grows most quickly, as well as minimum and maximum temperatures that define the limits of growth. These limits are imposed, in part, by the thousands of proteins in a cell, all of which must function within the same temperature range. The fastest growth rate for a species occurs at temperatures at which all the cell's proteins work most efficiently as a group to produce energy and synthesize cell components. Growth stops when rising temperatures cause critical enzymes or cell structures (such as the cell membrane) to fail. At cold temperatures, growth ceases because enzymatic processes become too sluggish and the cell membrane less fluid. The membrane needs to remain fluid so that it can expand as cells grow larger and so that proteins needed for solute transport can be inserted into the membrane.

Cell growth is limited to a narrow temperature range for a variety of reasons. For example, heat increases molecular movement within proteins. Too much or too little movement will interfere with enzymatic reactions. As a result, great biological diversity exists among microbes, as different groups have evolved to grow within different thermal niches. A species grows within a specific thermal range because its proteins have evolved to tolerate that range. Outside that range, proteins will denature or function too slowly for growth. The upper limit for protists is around 50°C (122°F), whereas some fungi can grow at temperatures as high as 60°C (140°F). Bacteria and archaea, however, have been found to grow at temperatures ranging from below 0°C (32°F) to above 100°C (212°F). Temperatures over 100°C are usually found around superheated thermal vents deep in the ocean, where water temperature can rise to 350°C (662°F) but the pressure is sufficient to keep water liquid.

Thermophiles, Psychrophiles, and Mesophiles

By their ranges of growth temperature, microorganisms can be classified as liking it hot (thermophiles), cold (psychrophiles), or in between (mesophiles; Figure 6.15).

Figure 6.15   Relationship between Temperature and Growth Rate

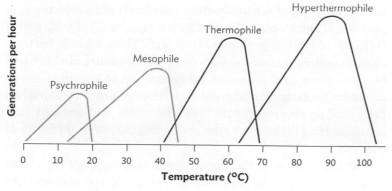

Each curve represents the temperature growth range for different classes of bacteria. Growth rate increases linearly with temperature until an optimum and then decreases. These curves are only an approximation. Growth rates and optimal ranges for individual species within a class will vary.

Figure 6.16   Thermophiles

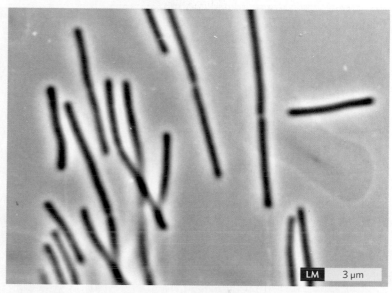

A. *Thermus aquaticus*, a hyperthermophile first isolated at Yellowstone National Park by Thomas Brock. Cell length varies from 3 to 10 μm.

B. Thermophile *Methanocaldococcus jannaschii* grown at 78°C (172.4°F) and 30 psi.

Thermophiles (Figure 6.16) have adapted to growth at high temperature, typically 55°C (131°F) and higher. Hyperthermophiles grow at temperatures as high as 121°C (250°F), which occur under extreme pressure (for example, at the ocean floor). These organisms flourish in hot environments such as composts or near thermal vents that penetrate Earth's crust on the ocean floor and on land (for example, hot springs; Figure 6.17). Most thermophiles have membranes and protein sequences specially adapted to life at high temperatures. Because enzymes in thermophiles do not unfold as easily as mesophilic enzymes, they more easily hold their shape at higher temperatures.

Figure 6.17   Thermophilic Environments

A. Yellowstone National Park hot spring.

B. A "smoker," a hydrothermal vent 2 miles deep in the Pacific Ocean. *Methanocaldococcus jannaschii* was isolated in 1983 near this vent.

One thermophile that has greatly benefited humans is *Thermus aquaticus*, discovered in a hot spring at Yellowstone National Park. A high-temperature thermophilic polymerase purified from this organism was critical to developing the polymerase chain reaction (PCR) used to amplify DNA in a test tube (discussed in Chapter 8). PCR has revolutionized molecular biology.

Figure 6.18   Psychrophilic Environment and Psychrophiles

A. An iceberg, in which psychrophilic organisms such as those shown in (B) can be found.

B. Bacteria from South Polar snow viewed on the surface of a membrane filter. These organisms are unclassified.

SEM   1 µm

Psychrophiles are microbes that grow at temperatures as low as –10°C (14°F), but their optimum growth temperature is usually around 15°C (59°F). Psychrophiles are prominent flora beneath icebergs in the Arctic and Antarctic (Figure 6.18). Psychrophilic microbes prefer cold because their proteins are more flexible than those of mesophiles and require less energy (heat) to function. The downside to increased flexibility is that it does not take much heat to denature psychrophilic proteins. As a result, psychrophiles grow poorly, if at all, when temperatures rise above 20°C (68°F). Another reason psychrophiles can grow in cold is that their membranes are more fluid than mesophilic membranes at low temperature. At higher temperatures, the membranes of psychropiles are *too* flexible and fail to maintain cell integrity.

Mesophiles include the typical "lab rat" microbes, such as *Escherichia coli* and *Bacillus subtilis.* Their growth optima range between 20°C (68°F) and 40°C (104°F), with a minimum of 15°C (59°F) and a maximum of 45°C (113°F). Because they are easy to grow and because human pathogens are mesophiles, much of what we know about protein, membrane, and DNA structure came from studying this group of organisms. Though they prefer growth at moderate temperatures, there are cold-resistant mesophiles of clinical importance. For instance, the modern practice of refrigeration has selected for cold-resistant mesophilic pathogens that can grow in refrigerated foods. One example is *Listeria monocytogenes* (one cause of food poisoning and septic abortions; Chapter 22).

Variations in Pressure

Creatures living at Earth's surface (sea level) are subjected to a pressure of 1 atmosphere (atm), which is equal to 0.101 megapascal (MPa) or 14 pounds per square inch (psi). These include bacteria that can destroy various foods. Because high pressure can kill these bacteria, many food processes are carried out at high pressure to minimize bacterial contamination. High-pressure processing has been used on cheeses, yogurt, luncheon meats, and oysters to kill contaminating bacteria without destroying the flavor or texture of the food.

At the bottom of the ocean, however, thousands of meters deep, hydrostatic pressure averages a crushing 400 atm and can go as high as 1,000 atm (101 MPa, or 14,600 psi) in ocean trenches (Figure 6.19). Microorganisms at locations deep within the ocean that have adapted to grow at oppressively high pressures are called barophiles or piezophiles. Barophiles actually *require* elevated pressure to grow, whereas barotolerant organisms grow well over the range of 10–500 atm (1–50 MPa), but their growth falls off thereafter. Many barophiles are also psychrophilic because the average temperature at the ocean's floor is 2°C (35.6°F).

Water Availability and Salt Concentration

Water is critical to life, but environments differ in terms of how much water is actually available to growing organisms; microbes can use only water that is not bound to ions or other solutes in solution. Water availability is measured as water activity (a_w), a quantity approximated by concentration. Because interactions with solutes (such as NaCl) lower water activity, the more solutes in a solution (for example, the higher the salt concentration), the less water is available for microbes to use for growth. **Osmolarity** is a measure of the number of solute molecules in a solution and is inversely related to a_w. The more particles in a solution, the greater the osmolarity and the lower the water activity.

Oceans, freshwater lakes, and estuarine environments (where ocean and freshwater mix) exhibit different concentrations of salt and water activity. Thus, the bacteria and archaea that inhabit these niches exhibit different salt tolerances. Many bacteria in these environments can be pathogenic. For example, *Vibrio vulnificus*, a Gram-negative species that causes fatal septicemia, and *Vibrio cholerae*, a cause of extreme diarrhea, are considered marine microorganisms that prefer growth at an elevated salt concentration (up to 1 M NaCl). Other diarrhea-causing organisms, such as *Escherichia coli* and *Salmonella*, prefer growth at much lower salt concentrations (approximately 0.2 M NaCl). These pathogens inhabit freshwater contaminated by fecal material from humans and other animals.

Link Recall from Section 2.4 that many diarrhea-causing organisms enter the body by ingestion but leave by defecation, where they can again be ingested when food, water, or hands become contaminated (oral-fecal route of transmission).

Some species of archaea have evolved to require high salt (NaCl) concentration; these are called **halophiles** (Figure 6.20). In striking contrast to most bacteria, which prefer salt concentrations in the range of 0.05–1 M (0.2%–5%) NaCl, the moderately halophilic microbes grow optimally at 0.85–3.4 M (5%–20%) NaCl. For comparison, seawater is about 3.5% NaCl. Extremely halophilic archaea require 3.4–5.1 M (20%–30%) NaCl to grow. Halotolerant organisms can grow in high or low salt (0%–30% NaCl).

All cells, even halophiles, prefer to keep a relatively low intracellular sodium ion (Na$^+$) concentration. To achieve a low internal

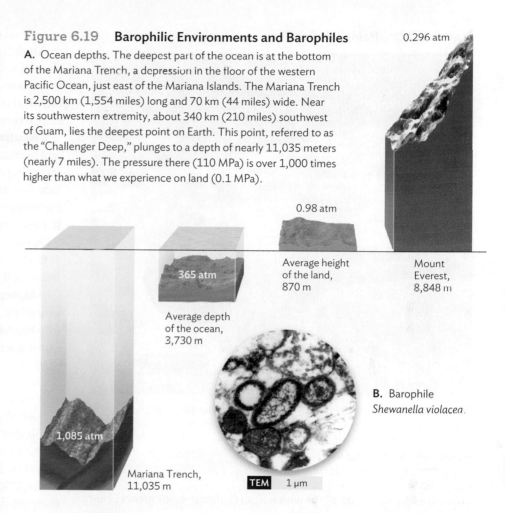

Figure 6.19 Barophilic Environments and Barophiles

A. Ocean depths. The deepest part of the ocean is at the bottom of the Mariana Trench, a depression in the floor of the western Pacific Ocean, just east of the Mariana Islands. The Mariana Trench is 2,500 km (1,554 miles) long and 70 km (44 miles) wide. Near its southwestern extremity, about 340 km (210 miles) southwest of Guam, lies the deepest point on Earth. This point, referred to as the "Challenger Deep," plunges to a depth of nearly 11,035 meters (nearly 7 miles). The pressure there (110 MPa) is over 1,000 times higher than what we experience on land (0.1 MPa).

0.296 atm

0.98 atm

Average height of the land, 870 m

Mount Everest, 8,848 m

365 atm

Average depth of the ocean, 3,730 m

1,085 atm

Mariana Trench, 11,035 m

B. Barophile *Shewanella violacea*.

TEM 1 μm

Na$^+$ concentration, halophilic microbes use special ion pumps to excrete sodium and replace it with other cations, such as potassium, which is a compatible solute (a solute that can accumulate to high levels intracellularly without affecting the functions of enzymes).

Microbial Responses to Changes in pH

As with salt and temperature, the concentration of hydrogen ions (H$^+$)—actually, hydronium (H$_3$O$^+$)—also has a direct effect on the cell's macromolecular structures. Extreme concentrations of either hydronium or hydroxide (OH$^-$) ions in a solution will limit growth. In other words, too much acid or base is harmful to cells. Despite this sensitivity to pH extremes, living cells can *tolerate* a greater range of H$^+$ concentration than of virtually any other chemical substance. *E. coli*, for example, tolerates an extracellular pH range of 2–9, a 10,000,000-fold difference!

How does pH influence growth? The intra-molecular bonds that shape proteins into active forms are highly sensitive to proton concentrations. Consequently, changing intracellular pH alters protein shape, which in turn changes protein activity. The result is that all enzyme activities exhibit optima, minima, and maxima with regard to pH, much as they do for temperature. As we saw with temperature, groups of microbes have evolved to inhabit diverse pH environments that can range from pH 0 to pH 11.5 (Figure 6.21).

Figure 6.20   Halophilic Salt Flats and Halophilic Bacteria

A. The halophilic salt flats along Highway 50, east of Fallon, Nevada, are colored pinkish red by astronomical numbers of halophilic bacteria.

B. *Halobacterium* species; cross-sectional cell width 0.5–0.8 μm.

However, the optimum pH at which a particular species grows is *not* dictated by the pH limits of critical cell proteins.

In general, most enzymes, regardless of the range of growth pH, tend to operate best between pH 5 and pH 8.5. Yet, many microbes grow in pH conditions that are even more acidic or alkaline than this range. How can that be? Growth above pH 8.5 or below pH 5 is possible because biological membranes are relatively impermeable to protons, a fact that allows the cell to maintain an internal pH compatible with protein function even when growing in extremely acidic or alkaline environments.

Even when extracellular pH is between 5 and 8, however, a cell can struggle to maintain an intracellular pH suitable for growth. The presence of membrane-permeant organic acids such as citric or propionic acids (also called weak acids) can accelerate the leakage of H^+ into cells, as discussed in Section 4.5. The food industry has taken advantage of this phenomenon by preemptively adding citric acid or sorbic acid to certain foods. This allows manufacturers to control microbial growth under pH conditions that do not destroy the flavor or quality of the food. (Food microbiology is discussed more fully in Chapter 27.)

Neutralophiles, Acidophiles, and Alkaliphiles

How do microbes grow at pH extremes, and how are they classified with regard to their growth pH? Cells have evolved to live under different pH conditions not by drastically changing the pH optima

of their enzymes but by using novel pH homeostasis strategies that maintain intracellular pH above pH 5 and below pH 8, even when the cell is immersed in pH environments well above or below that range (see Figure 6.21). Microbes are divided into three categories based on their success in maintaining intracellular pH under different extracellular pH extremes.

Neutralophiles are bacteria that generally grow between pH 5 and pH 8 and include most human pathogens. Many neutralophiles, including *E. coli* and *Salmonella enterica*, adjust their metabolism to maintain an internal pH slightly above neutrality, which is where their enzymes work best. Other species of neutralophiles allow their internal pH to fluctuate with external pH but usually maintain a pH difference (ΔpH) of about 0.5 pH unit across the membrane at the upper and lower limits of growth pH. The value of ΔpH is an important factor in energy production, as discussed in Chapter 7.

Acidophiles are bacteria and archaea that live in acidic environments. They are often **chemoautotrophs** (lithotrophs) that oxidize reduced metals and generate strong acids such as sulfuric acid. Consequently, they grow between pH 0 and pH 5. The ability to grow at this pH is due partly to altered membrane lipid profiles that decrease proton permeability and partly to ill-defined mechanisms that move protons out of the cell.

Alkaliphiles occupy the opposite end of the pH spectrum, growing best at values ranging from pH 9 to pH 11. They are commonly found in saline soda lakes, which have high salt concentrations and pH values as high as pH 11. Soda lakes, such as Lake Magadi, in Africa (**Figure 6.22A**), are steeped in carbonates, which explains their extraordinarily alkaline pH.

Figure 6.21   Classification of Organisms Grouped by Optimum Growth pH

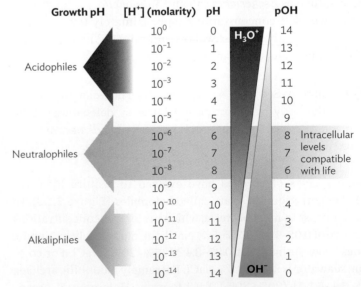

Note that pH and pOH are the reciprocal values of hydronium and hydroxide concentrations, respectively. Although the range of intracellular pH compatible with life is between pH 6 and pH 8, acidophiles grow well below that range and alkaliphiles grow well above it.

Figure 6.22 A Soda Lake Ecosystem

C. Pink flamingo colored by ingestion of *Spirulina*.

LM 10 µm

A. Lake Magadi, in Kenya. Its pink color is due to *Spirulina*.

B. *Spirulina* is a coiled, filamentous cyanobacteria that grows in the alkaline environment of Lake Magadi.

The cyanobacterium *Spirulina* is an alkaliphile that grows in soda lakes (**Figure 6.22B**). Its high concentration of carotene gives the organism a distinctive pink color (note the color of the lake in Figure 6.22A). *Spirulina* is also a major food for the famous pink flamingos indigenous to these African lakes and is, in fact, the reason pink flamingos are pink. After the birds ingest these organisms, digestive processes release the carotene pigment to the circulation, which then deposits it in the birds' feathers, turning them pink (**Figure 6.22C**).

Note Humans consume *Spirulina* as a health food supplement but do not turn pink because the cyanobacteria are only a small component of the diet.

The key to the survival of alkaliphiles is the cell membrane barrier that sequesters fragile cytoplasmic enzymes away from harsh extracellular pH. In contrast to proteins *within* the cytoplasm, enzymes *secreted* from alkaliphiles work in very alkaline environments. Similar base-resistant enzymes such as lipases and cellulases are included in laundry detergents to help get our "whites whiter and our brights brighter."

pH Homeostasis Mechanisms

When cells are placed in pH conditions below their optimum, protons can enter the cell and lower internal pH to lethal levels. Microbes can prevent the unwanted influx of protons in several ways. *E. coli*, for example, can reverse proton influx by importing a variety of cations such as K^+ or Na^+. At the other extreme, under extremely alkaline conditions, the cells can use Na^+/H^+ antiporters (see Section 5.2) to recruit protons into the cell in exchange for expelling Na^+. Some organisms can also change the pH of the medium by using various amino acid decarboxylases and deaminases (enzymes that remove a carboxyl or amine group from an organic molecule, producing alkaline or acidic products, respectively). For example, *Helicobacter pylori*, the causative agent of gastric ulcers, employs an exquisitely potent urease (an enzyme that catalyzes the hydrolysis of urea) to generate massive amounts of ammonia. The ammonia neutralizes acid pH and allows the organism to survive and grow along the stomach lining.

Many, if not all, microbes also have the ability to sense a change in environmental pH and alter cell physiology in response. The levels of many proteins increase, while the levels of others decrease. The result includes modifications in membrane lipid composition, enhanced pH homeostasis, and many other changes. Some pathogens, such as *Salmonella*, sense a change in external pH as part of the signal indicating the bacterium has entered a host cell environment (see **inSight**).

SECTION SUMMARY

- **Extremophiles** inhabit "fringe" environmental conditions that do not support human life.
- **The environmental niche** (such as high salt or acidic pH) inhabited by a particular species is defined by the tolerance of that organism's proteins and other macromolecular structures to the physical conditions within that niche.
- **Optimum growth values** of temperature, pH, and osmolarity vary among microbial species. Mesophiles, psychrophiles, and thermophiles are groups of organisms that grow at moderate, low, and high temperatures, respectively.
- **Barophiles (also known as piezophiles)** can grow at pressures up to 1,000 atm but fail to grow at low pressures.
- **Water activity (a_w)** is a measure of how much water in a solution is available for a microbe to use.
- **Osmolarity** is a measure of the number of solute molecules in a solution and is inversely related to a_w.
- **Halophilic organisms** grow best at high salt concentration.
- **Hydrogen ion concentration affects protein structure and function.** Thus, enzymes have pH optima, minima, and maxima.
- **Microbes use pH homeostasis mechanisms** to keep their internal pH near neutral when in acidic or alkaline media.
- **The addition of weak acids** to certain foods undermines bacterial pH homeostasis mechanisms, preventing food spoilage and killing potential pathogens.
- **Neutralophiles, acidophiles, and alkaliphiles** prefer growth under neutral, low, and high pH conditions, respectively.

Thought Question 6.6 How might the concept of water availability be used by the food industry to control spoilage?

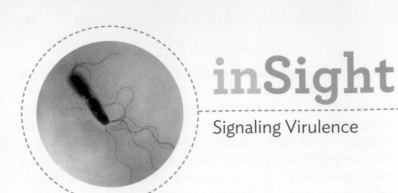

inSight

Signaling Virulence

Figure 1   *Salmonella* **within Human Epithelial Cell Phagosomes**

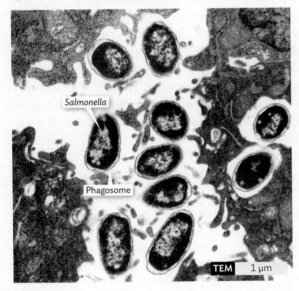

Salmonella

Phagosome

TEM   1 μm

One of the burning questions in the study of infectious disease is, How does a microbe know when it has entered a suitable host? Most, if not all, pathogens express certain genes only *after* infection, so they must sense that something in their environment has changed. A change in environment that causes increased virulence is called a virulence signal. One of these virulence signals is pH. On their way to causing disease, many pathogens travel through acidic host compartments. The most obvious are the stomach, which is extremely acidic (pH 1–3), and infected host cell phagosomes and phagolysosomes, which are mildly acidic (pH 4.5–6). *Salmonella enterica*, a major cause of diarrheal disease, provides an intriguing example of virulence that is triggered by low pH. After gaining entrance to the intestine via ingestion and subsequently penetrating intestinal epithelia, this pathogen tricks certain white blood cells, called macrophages (discussed in Chapter 15), into engulfing the bacterium, placing the pathogen within a phagosome (Figure 1). Recall from Section 3.6 that a phagosome is an intracellular membrane vesicle that holds material captured by phagocytosis, in this case *Salmonella*. *Salmonella* then waits for the phagosome (initially pH 6) to partially fuse with a lysosome and form a phagolysosome, whose purpose is to kill and digest the microbe. As the phagolysosome forms, the pH of the compartment drops to about pH 4.5. Many bacterial proteins are then expressed, many of which enable *Salmonella* to survive in the macrophage. However, drugs that prevent phagolysosome acidification (bafilomycin, for example) render *Salmonella* extremely susceptible to the various antimicrobial weapons wielded by the macrophage (described in Chapter 15). *Salmonella* "knows" it is in a macrophage phagolysosome in part because the pH drops. Many studies have mapped the complex bacterial regulatory circuits involved in sensing acid pH signals and translating those signals into defensive actions by the pathogen.

6.5
Living with Oxygen:
Aerobe versus Anaerobe

SECTION OBJECTIVES

- Differentiate anaerobes from aerobes and describe how they are cultured.
- Explain how both aerobes and anaerobes can cause disease.
- Discuss the basic differences between respiration and fermentation and how this impacts where an organism grows.

Earth's atmosphere contains considerable amounts of oxygen that we rely on to breathe, but do all bacteria grow in the presence of molecular oxygen (O_2)? Certainly many bacterial species do. Like us, some bacteria (aerobes) use oxygen as a terminal electron acceptor in the **electron transport chain**, a group of membrane proteins (cytochromes) that extract energy trapped in nutrients and help convert the energy to a biologically useful form. This use of O_2 as the terminal electron acceptor is called **aerobic respiration**. However, for many bacteria (anaerobes), including several important pathogens, oxygen is toxic.

Link You will learn more about the metabolism of aerobes and anaerobes in Chapter 7. This section serves as an introduction to these concepts.

CASE HISTORY 6.3

Attack of the Anaerobe

Robert, a 21-year-old software salesman from Dallas, Texas, presented to the Parkland Hospital emergency room with 3 days of cramping pain in the midgastric and lower abdominal areas. He had no nausea, vomiting, or diarrhea. Upon examination, the patient screamed in severe pain when the clinician pressed on his lower right abdomen. Blood samples were obtained and sent for bacteriological culture. A blood culture bottle incubated in air appeared sterile, whereas another blood culture bottle placed in a special chamber that removed oxygen grew thick with bacteria. A subsequent exploratory laparoscopy, in which a small incision is used to perform surgery and minimize scarring, revealed a grossly inflamed appendix that released a liter of pus when removed. The diagnosis was appendicitis caused by the anaerobic bacterium *Bacteroides fragilis*—a common intestinal microbe.

Not all bacteria can use or even tolerate oxygen. *Bacteroides fragilis*, for example, is eventually killed when exposed to air. The bacterium, a beneficial part of our normal intestinal flora that causes serious disease if it escapes to other tissues, grows in our intestine because the intestinal lumen lacks oxygen. In Case History 6.3, knowing that the cultured bacteria did not grow in the presence of oxygen enabled the laboratory technician to find the cause of Robert's disease.

So how do bacteria use oxygen, and why are some killed by it? Electrons pulled from various energy sources (for example, glucose) possess intrinsic, or potential, energy, an energy that electron transport proteins (the cytochromes mentioned earlier) can harness. Cytochromes, discussed more fully in Chapter 7, do this by extracting the energy in incremental stages and using it to move protons out of the cell, which produces proton motive force (introduced in Section 5.2). Once the cell has drained as much energy as possible from an electron, that electron must be passed to a diffusible, final electron acceptor molecule floating in the medium to clear the way for another electron to be passed down the electron transport chain. One such terminal electron acceptor is dissolved oxygen. The bad news is that certain by-products of oxygen metabolism, such as superoxide (O_2^-), hydroxyl radicals ($\cdot OH$), and hydrogen peroxide (H_2O_2), can be dangerously reactive, a serious problem for all cells. As a result, different microorganisms have evolved to either tolerate or avoid oxygen altogether. Table 6.4 gives examples of microbes that grow at different levels of oxygen.

Aerobes versus Anaerobes

We can illustrate the varied relationships between bacteria and oxygen by using a standing test tube filled with growth medium (Figure 6.23). The top of the tube, closest to air, is oxygenated; the bottom of the tube has almost no oxygen. Some microbes grow only at the top of the tube because they require oxygen to respire and make energy. Others "prefer" to grow at the bottom because they are killed by oxygen by-products.

A **strict aerobe** is an organism that not only exists in oxygen but also uses oxygen as a terminal electron acceptor. A strict aerobe grows *only* with oxygen present and consumes oxygen during metabolism (aerobic respiration). An aerobe will grow at the top of the tube shown in Figure 6.23. In contrast, a **strict anaerobe** dies in the least bit of oxygen (greater than 5 µM dissolved oxygen). The organism in Case History 6.3 was an anaerobe. Strict anaerobes do not use oxygen as an electron acceptor, but this is not why they die in air. Anaerobes die in air because they are vulnerable to the highly toxic, chemically reactive oxygen species produced by their own metabolism when exposed to oxygen. Reactive oxygen species (ROS) are essentially oxygen molecules or ions with one too few or one too many electrons. As a result, anaerobes will grow only at the bottom of the tube shown in Figure 6.23.

Aerobes, but not anaerobes, destroy ROS with the aid of enzymes such as superoxide dismutase (to remove superoxide) and peroxidase and catalase (to remove hydrogen peroxide). Aerobes also have resourceful enzyme systems that detect and repair macromolecules damaged by oxidation.

Table 6.4
Examples of Aerobic and Anaerobic Microbes

Aerobic Microbes	Anaerobic Microbes	Facultative Microbes	Microaerophilic Microbes
Neisseria spp. (causative agents of meningitis, gonorrhea)	*Azoarcus tolulyticus* (microbe that degrades toluene)	*Escherichia coli* (normal gastrointestinal flora; additional pathogenic strains)	*Helicobacter pylori* (causative agent of gastric ulcers)
Pseudomonas fluorescens (microbe found in soil and water; degrades pollutants such as TNT and aromatic hydrocarbons)	*Bacteroides* spp. (normal gastrointestinal flora)	*Saccharomyces cerevisiae* (yeast; used in baking)	*Lactobacillus* (microbe that ferments milk to form yogurt)
Mycobacterium leprae (causative agent of leprosy)	*Clostridium* spp. (soil microbes; causative agents of tetanus, botulism, and gas gangrene)	*Bacillus anthracis* (causative agent of anthrax)	*Campylobacter* spp. (causative agent of gastroenteritis)
Azotobacter (soil microbes that fix atmospheric nitrogen)	*Actinomyces israelii* (soil microbes that synthesize antibiotics)	*Vibrio cholerae* (causative agent of cholera)	*Treponema pallidum* (causative agent of syphilis)
Rhizobium spp. (soil microbes; plant symbionts)	*Desulfovibrio* spp. (microbes that reduce sulfate)	*Staphylococcus* spp. (found on skin; causative agent of boils, impetigo, toxic shock, others)	

Figure 6.23   Oxygen-Related Growth Zones in a Standing Test Tube

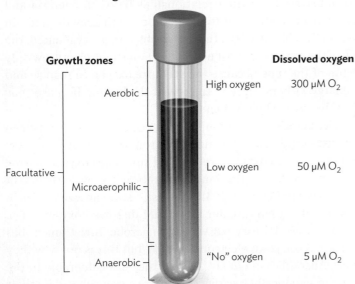

As oxygen diffuses and equilibrates with the air, oxygen levels in different areas of a simple nutrient broth tube will diminish the farther down the tube you look. Inoculating a series of tubes like this with bacteria having different oxygen requirements will result in anaerobes (not strict anaerobes) growing at the bottom of the tubes. Microaerophilic bacteria will grow in the middle and strict aerobes will grow only at the top. Facultative organisms can grow throughout.

Most microbes in the world are anaerobic, growing buried in the soil, within our anaerobic digestive tract, or within biofilms on our teeth. Anaerobic microbes fall into several categories. Some actually do respire using electron transport systems, but instead of using oxygen, they rely on terminal electron acceptors other than oxygen, such as nitrate, to conduct **anaerobic respiration** and produce energy. Anaerobes of another ilk do not possess electron transport chains and cannot respire and so must rely on carbohydrate **fermentation** for energy (that is, they conduct fermentative metabolism). In fermentation, ATP energy is produced through substrate-level phosphorylation in a process that does not involve cytochromes (as discussed in Section 7.3).

In addition to aerobes and anaerobes, there are several other categories of microbes distinguished by their oxygen tolerance. **Facultative** organisms can live with or without oxygen because they come equipped with enzymes that destroy toxic oxygen by-products (enzymes such as superoxide dismutase, catalase, and peroxidase). They will grow throughout the tube shown in Figure 6.23. Facultative organisms include the **facultative anaerobes** (such as *E. coli*) that in addition to having enzymes that destroy toxic oxygen radicals also possess the machinery for both fermentation and aerobic respiration. Whether a member of this group uses aerobic respiration, anaerobic respiration, or fermentation depends on the availability of oxygen and the amount of carbohydrate present. For example, the more carbohydrate present in a medium, the more likely a facultative anaerobe will ferment it, even in the presence of oxygen (see MacConkey medium in Figure 6.8). When carbohydrate becomes scarce, however, the facultative anaerobe prefers oxygen (aerobic respiration) over fermentation.

Another subclass of facultative organisms are the **aerotolerant anaerobes** that use only fermentation (not aerobic respiration) to provide energy but contain superoxide dismutase and catalase (or peroxidase) to protect them from ROS. Consequently, aerotolerant anaerobes will also grow throughout the tube in Figure 6.23. *Streptococcus pyogenes*, a common human pathogen, is an aerotolerant anaerobe. The final category of microbes based on oxygen tolerance is said to be **microaerophilic**, meaning they will grow only at low oxygen concentrations. Microaerophilic organisms possess *decreased* levels of superoxide dismutase and/or catalase and grow toward the middle of the tube in Figure 6.23.

Culturing Anaerobes

The discovery of anaerobic bacteria by pioneering microbiologists such as Louis Pasteur hinged on finding innovative ways to remove most if not all oxygen from culture environments. An ability to generate anaerobic conditions quickly led to the discovery that anaerobes are important human pathogens. Many anaerobic bacteria cause horrific human diseases, such as tetanus, botulism, and gangrene. Some of these organisms or their secreted toxins are even potential weapons of terror (for example, *Clostridium botulinum*).

Several techniques for culturing anaerobes are used today. One popular way to culture anaerobes, especially on agar plates, employs an anaerobe jar (**Figure 6.24A**). Agar plates streaked with the organism are placed into a sealed jar with a foil packet that releases H_2 and CO_2 gases. A palladium packet hanging from the jar lid catalyzes a reaction between the H_2 and O_2 in the jar to form H_2O, which effectively removes O_2 from the chamber. The CO_2 released is required by the anaerobe to produce key metabolic intermediates. An anaerobe jar was used in Case History 6.3 to grow *Bacteroides*. Some microaerophilic microbes, such as the pathogens *H. pylori* (the major cause of stomach ulcers) and *Campylobacter jejuni* (a major cause of diarrhea), require low levels of O_2 but elevated amounts of CO_2. These conditions are obtained by using similar gas-generating packets.

For a strict anaerobe exquisitely sensitive to oxygen, even more heroic efforts are required to establish an oxygen-free environment. A special anaerobic chamber with glove ports must be used in which the atmosphere is removed by vacuum and replaced with a precise mixture of N_2 and CO_2 gases (**Figure 6.24B**).

SECTION SUMMARY

- **Strict aerobes** require oxygen for growth, using it as a terminal electron acceptor to extract energy from nutrients. Aerobes cannot ferment or conduct anaerobic respiration but do have enzymes capable of destroying reactive oxygen molecules.

- **Oxygen is toxic to strict anaerobes**, cells that do not have enzymes capable of efficiently destroying reactive oxygen species.

- **Anaerobic metabolism can be either fermentative or respiratory.** Anaerobic respiration requires the organism to possess electron transport chains that can use compounds other than oxygen as terminal electron acceptors.

Figure 6.24   Anaerobic Growth Technology

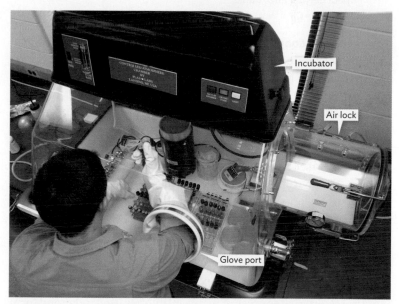

Catalyst in lid mediates reaction. $H_2 + \frac{1}{2}O_2 \rightarrow H_2O$

GasPak envelope generates H_2 and CO_2.

A. An anaerobe jar. Inoculated plates are placed in the jar. Water is added to the GasPak foil packet, and the top is sealed shut. Hydrogen and carbon dioxide emerge from the foil packet, and the hydrogen reacts with oxygen in the chamber to form water, effectively removing free oxygen from the chamber.

Incubator

Air lock

Glove port

B. An anaerobic chamber with glove ports. Materials such as plates and tubes are passed through the air lock and into the chamber, whose environment (fed from gas tanks) is mainly nitrogen with some hydrogen and carbon dioxide. Glove ports allow the investigator to manipulate the materials in the box without breaking a seal. An incubator is normally included in the box.

- **Facultative organisms** can grow in the presence or absence of oxygen because they possess enzymes that destroy reactive oxygen species.

- **Facultative anaerobes** grow with or without oxygen, have enzymes that destroy reactive oxygen molecules, and possess the ability for both fermentation and respiration (anaerobic and aerobic). They can use oxygen as a terminal electron acceptor.

- **Aerotolerant organisms** grow in either the presence or the absence of oxygen but use fermentation as their primary, if not only, means of gathering energy. These microbes also have enzymes that destroy reactive oxygen species, allowing them to grow in oxygen.

Thought Question 6.7 If anaerobes cannot live in oxygen, how do they incorporate oxygen into their cellular components?

Thought Question 6.8 How can anaerobes grow in the human mouth when so much oxygen is there?

6.6
Microbial Communities and Cell Differentiation

SECTION OBJECTIVES
- Discuss how biofilms develop.
- Explain the importance of biofilms to infection.
- Describe the process of sporulation, and explain how spores impact certain infections.

Can bacteria cooperate with each other? Bacteria were once thought of as unicellular organisms incapable of differentiation or communication; but we now know that many, if not most, bacteria form specialized, surface-attached communities called **biofilms**. Indeed, within aquatic environments, bacteria are found mainly associated with surfaces.

However, bacterial specialization is not limited to biofilms. Many bacteria undergo complex differentiation programs when faced with environmental stress. Some species, such as *E. coli*, undergo relatively simple changes in cell structure involving the formation of smaller cells or thicker cell surfaces that help them survive. Other species initiate elaborate cell differentiation processes. For example, endospore formers such as *Bacillus anthracis* (the cause of anthrax) generate heat-resistant spores that can remain dormant for thousands of years. Another group of bacteria, the actinomycetes, are rich sources of many antibiotics used today and form complex multicellular structures analogous to those of eukaryotic fungi. The production of these antibiotics is tied to the differentiation these organisms undergo as they adapt to growth on agar.

We begin this section with the landmark clinical case (Case History 6.4) that eventually led to the discovery that microbial biofilm communities can play critically important roles in the infectious disease process.

CASE HISTORY 6.4
Death by Biofilm

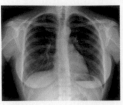

Cystic fibrosis (CF) is an inherited disease in which chloride ion (Cl⁻) transport is compromised in many organ systems. The lungs are especially affected by the production of thick, sticky mucous buildup. In 1951, two women with CF developed acute respiratory distress. From their chest X-rays, they were diagnosed as having an infectious bronchopneumonia. Treatment at the time consisted of intramuscular administration of the antibiotics streptomycin and penicillin. Repeated bronchial lavage (washing) was attempted in a futile bid to clear airway secretions. Tragically, the two women did not respond to therapy and died within hours. *Pseudomonas aeruginosa*, a Gram-negative bacillus common in soil and water, was subsequently isolated from both victims. This was the first report of *Pseudomonas* infection in CF patients.

We now know that lung infection with the Gram-negative organism *Pseudomonas aeruginosa* is a frequent cause of death in cystic fibrosis (CF) patients. Lung damage caused by CF provides a setting in which mucoid strains of *P. aeruginosa* can attach to injured tissue and form tenacious biofilm communities consisting of billions of bacteria. A *Pseudomonas* lung biofilm makes breathing more difficult for the CF patient and increases inflammation, which further damages the lung. Treating any biofilm infection is challenging because a biofilm, like a fortress, is nearly impenetrable to antibiotics, as you will see. Biofilms, in fact, are a major factor in many bacterial infections. All bacteria can make them. Consequently, understanding how biofilms form is a critical first step in devising treatments for many types of bacterial infections.

Biofilms: Multicellular Microbes?

A biofilm is a mass of bacteria that stick to and multiply on a solid surface, whether a stone in a lake or a lung in a cystic fibrosis patient. Biofilms can be constructed by a single species or by multiple, collaborating species and can form on a range of organic or inorganic surfaces. **Figure 6.25** shows a common mixed-species biofilm—called plaque—that forms on teeth. More serious examples include the single-species biofilm formed by *Pseudomonas aeruginosa* on the lungs of cystic fibrosis patients (as in Case History 6.4) or by staphylococci on medical implants (for instance, knee replacements).

Why would a microbe "want" to make a biofilm? The goal of biofilms in nature is to stay where food is plentiful. Why should a microbe travel off to hunt for food when it is already available? Once nutrients become scarce, however, individuals detach from the community to forage for new sources of nutrient.

Biofilms typically develop in the manner illustrated in **Figure 6.26**. When conditions are right, unattached (planktonic) single

Figure 6.25   Dental Biofilm

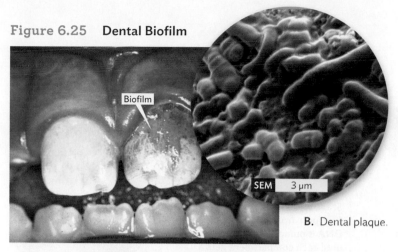

Biofilm

SEM   3 μm

B. Dental plaque.

A. The biofilm that forms on teeth is called plaque.

cells start to attach to nearby inanimate surfaces by means of flagella, pili, lipopolysaccharides, or other cell-surface appendages. These cells begin to coat that surface with an organic monolayer of polysaccharides or glycoproteins to which more planktonic cells can attach. At this point, some species can move along surfaces by using a twitching motility that involves the extension and retraction of a specific type of pilus. Ultimately, they stop moving and firmly attach to the surface. As more and more cells bind to the surface, they can begin to communicate with each other by sending and receiving chemical signals in a process called **quorum sensing**. Individual cells continually make these chemical signal molecules, but once the population reaches a certain number (analogous to an organizational "quorum"), the chemical signal reaches a concentration that the cells can sense. Reception of this signal triggers genetically regulated changes that cause bacteria to bind doggedly to the surface substrate and to each other.

Next, the cells form a thick extracellular matrix of polysaccharide polymers and entrapped organic and inorganic materials. These **exopolysaccharides (EPSs)**, such as alginate produced by *P. aeruginosa* and colanic acid produced by *E. coli*, increase antibiotic resistance by limiting antibiotic access to the center of the biofilm. The chemical signal molecules also stimulate production of some **antibiotic resistance** mechanisms. As the biofilm matures, the amalgam of adherent bacteria and matrix forms columns and streamers, creating channels through which nutrients flow.

A hallmark of all infectious diseases in which biofilms play a prominent role is their chronic nature. A biofilm infection may linger for months, years, or even a lifetime. It may not ever kill its victim, though the infection can compromise quality of life. Biofilm infections can smolder indefinitely, even in people with healthy immune systems.

Link A deeper discussion of biofilms and **antibiotic resistance** is provided in Chapter 13. You will learn that bacteria can become resistant to antibiotics by making enzymes that destroy the drug, by altering their cellular target, and by limiting their entry into the cell.

Endospores in Suspended Animation

Certain Gram-positive genera, including important pathogens such as *Clostridium tetani* (tetanus), *Clostridium botulinum* (botulism), and *Bacillus anthracis* (anthrax), have the remarkable ability to develop dormant spores that are heat and desiccation resistant. Resistance to heat and desiccation help make the spores of *B. anthracis* a potential bioweapon.

ANIMATION

Figure 6.26   Biofilm Development

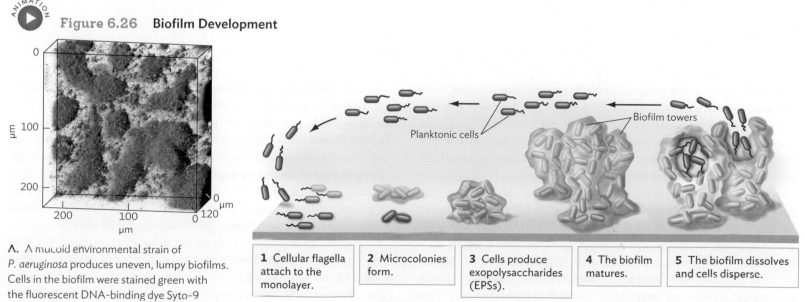

A. A mucoid environmental strain of *P. aeruginosa* produces uneven, lumpy biofilms. Cells in the biofilm were stained green with the fluorescent DNA-binding dye Syto-9 (3D confocal laser scanning microscopy).

Planktonic cells

Biofilm towers

| 1 Cellular flagella attach to the monolayer. | 2 Microcolonies form. | 3 Cells produce exopolysaccharides (EPSs). | 4 The biofilm matures. | 5 The biofilm dissolves and cells disperse. |

B. The stages of biofilm development in *Pseudomonas* generally apply to the formation of many kinds of biofilms.

ANIMATION

Figure 6.27   **The Seven Stages of Endospore Formation**

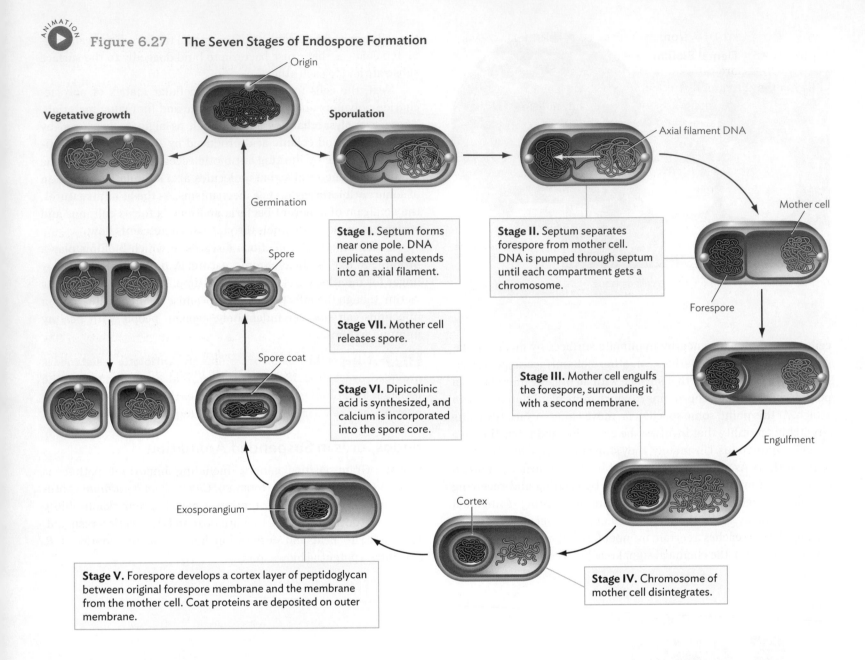

Origin

Vegetative growth

Sporulation

Germination

Axial filament DNA

Mother cell

Stage I. Septum forms near one pole. DNA replicates and extends into an axial filament.

Stage II. Septum separates forespore from mother cell. DNA is pumped through septum until each compartment gets a chromosome.

Forespore

Spore

Stage VII. Mother cell releases spore.

Spore coat

Stage VI. Dipicolinic acid is synthesized, and calcium is incorporated into the spore core.

Stage III. Mother cell engulfs the forespore, surrounding it with a second membrane.

Engulfment

Exosporangium

Cortex

Stage V. Forespore develops a cortex layer of peptidoglycan between original forespore membrane and the membrane from the mother cell. Coat proteins are deposited on outer membrane.

Stage IV. Chromosome of mother cell disintegrates.

Most of our knowledge of bacterial sporulation comes from the Gram-positive soil bacterium *B. subtilis*. When growing in rich media, this microbe undergoes normal vegetative growth and can replicate every 30–60 minutes. However, starvation initiates an elaborate 8-hour genetic program that directs an asymmetrical cell division process and ultimately yields a spore.

As shown in **Figure 6.27**, sporulation begins when the chromosome replicates and stretches into a long axial filament spanning the length of the cell (stage I). The cell makes a molecular "decision" to divide at one of the two cell poles (stage II) instead of in the middle, as normally occurs during growth. Septation divides the cell into two unequal compartments (stage III): the forespore, which will ultimately become the spore, and the larger mother cell, from which it is derived. Each compartment contains a chromosome.

The mother cell chromosome is eventually destroyed, and the mother cell membrane engulfs the forespore (stage IV). Next, a thick peptidoglycan layer (cortex) is placed between the two membranes surrounding the forespore (stage V). Layers of coat proteins are then deposited on the outer membrane. Stage VI completes the development of spore resistance to heat and chemical insults. This process includes the synthesis of dipicolinic acid and the uptake of calcium into the core of the spore. The mother cell, now called a sporangium, dies and releases the mature spore (stage VII).

Spores are resistant to many environmental stresses that kill vegetative cells. The nature of this resistance is due in part to desiccation of the spore (they have only 10%–30% of a vegetative cell's water content). But spores are also packed with small acid-soluble proteins (SASPs) that bind to and protect DNA. The SASP coat protects the spore's DNA from damage by ultraviolet light and various toxic chemicals.

A mature spore can exist in soil for at least 50–100 years, and some last thousands and even millions of years. Once proper nutrient conditions arise, another genetic program, called **germination**, is triggered that wakes the dormant cell, dissolves the spore coat, and releases a viable vegetative cell. How do we know spores can last millions of years? Because bacterial spores that were extracted from the stomachs of ancient bees trapped and preserved in amber millions of years ago have successfully germinated.

SECTION SUMMARY

- **Biofilms** are complex multicellular surface-attached microbial communities.

- **Chemical signals** enable bacteria to communicate (quorum sensing) and in some cases to form biofilms.

- **Biofilm development** involves adherence of cells to a substrate, formation of microcolonies, and ultimately formation of complex channeled communities that generate new planktonic cells.

- **Endospore development** in *Bacillus* and *Clostridium* involves production of dormant, stress-resistant endospores.

Thought Question 6.9 *Clostridium botulinum* is an anaerobic, Gram-positive, spore-forming bacillus, normally found in soil, that secretes a poisonous toxin. Why would this organism be a potential problem when Grandma decides to can vegetables such as green beans?

Perspective

Knowing how microbes grow and develop is an important part of understanding infectious diseases. Many facets of microbial physiology impact a pathogen's ability to sidestep a host's immune system, establish itself at certain body sites, survive at those sites, and collect nutrients—sometimes killing the host, sometimes not. As we proceed through future chapters, you will begin to understand how bacterial pathogens evolve and new ones suddenly emerge; why different organisms home in on specific body sites; and how bacteria "know" when to eject or inject specific proteins (toxins) that commandeer host cells. You will also appreciate that our ability to isolate and grow pathogens on (or in) laboratory media is a critical part of identifying a disease-causing pathogen and determining how to stop its rampage through a host.

LEARNING OUTCOMES AND ASSESSMENT

| | **SECTION OBJECTIVES** | **OBJECTIVES REVIEW** |

6.1 Microbial Nutrition

- Compare heterotrophy, autotrophy, phototrophy, and chemotrophy.
- Describe the importance of the nitrogen and carbon cycles and the role of microbes in their maintenance.
- Discuss biofilms and their relevance to infectious diseases.

1. A medium that contains only those compounds needed for an organism to grow is called a(n) _____.

2. A bacterial species that can make its own organic compounds from CO_2 is called a(n) _____.
   - A. organotroph
   - B. heterotroph
   - C. autotroph
   - D. phototroph
   - E. chemotroph

6.2 Culturing and Counting Bacteria

- Explain how pure cultures are obtained and why they are important in medicine.
- Distinguish synthetic, complex, selective, and differential media and their use in clinical microbiology.
- Describe the various ways bacterial growth is measured, and explain the advantages and disadvantages of each method.

4. From which of the following agar plate(s) can you directly obtain a pure culture?
   - A. Plate A
   - B. Plate B
   - C. Plate C
   - D. Plates A and B
   - E. Plates B and C

A.   B.   C.

6.3 The Growth Cycle

- Discuss the phases of a typical bacterial growth curve.
- Explain how bacterial growth correlates to disease.
- Describe the purpose of continuous culture and how it correlates to the human digestive tract.

8. With regard to the figure below, which phase exhibits balanced growth?
   - A. 1
   - B. 2
   - C. 3
   - D. 4

3. Label the following figure with these terms: nitrifiers, denitrifiers, nitrogen fixers, NH_4^+, NO_3^-, N_2.

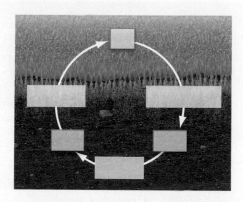

5. In question 4, image C best illustrates which of the following concepts?
   A. Selective medium
   C. Differential medium
   B. Viable count
   D. Enriched medium

6. A 21-year-old woman exhibits painful urination and a pain in her side. You are asked to determine how many bacteria are present in 1.0 ml of this patient's urine to help determine whether she has a urinary tract infection. You take 1 μl of her collected urine and streak it over the surface of a blood agar plate. You see 35 similar-looking colonies. How many bacteria are in 1.0 ml of this woman's urine?
   A. 3.5×10^2
   D. 3.5×10^5
   B. 3.5×10^3
   E. 3.5×10^6
   C. 3.5×10^4

7. Which of the following counting methods will yield the lowest number of bacteria per milliliter in a stationary-phase culture?
   A. Direct count
   B. Viable count
   C. Propidium iodide and Syto-9 staining
   D. Fluorescence-activated cell sorting
   E. They will all give approximately the same number.

9. The pathogen *Salmonella* Typhi has a generation time of 20 minutes in rich media. If you start with 10 bacteria per milliliter, *approximately* how many bacteria would be present in 1 ml after 12 hours (assuming unlimited nutrients)? (You can go online and use an anti-log calculator to get the answer.)
   A. 1×10^3
   D. 7×10^8
   B. 1×10^5
   E. 7×10^{10}
   C. 7×10^6
   F. 7×10^{11}

10. Decreasing the flow rate of media in a chemostat will have which of the following effects?
   A. Decrease growth rate
   B. Increase growth rate
   C. Shorten lag phase
   D. Increase exponential (log) phase
   E. Increase cell size

LEARNING OUTCOMES AND ASSESSMENT

	SECTION OBJECTIVES	OBJECTIVES REVIEW

6.4
Environmental Limits on Microbial Growth
	• Describe how environmental changes can alter the shape of a growth curve.
	• List different classes of microbes based on their preferred environmental niches (pH, temperature, salt).
	• Identify the biological properties that allow different classes of microbes to grow in extreme environments.
		11. Microorganisms that grow at very high pressures are called ____.
		12. With respect to their growth temperature requirements, pathogenic bacteria are considered ____.
		    A. extremophiles
		    B. psychrophiles
		    C. mesophiles
		    D. thermophiles
		    E. halophiles

6.5
Living with Oxygen: Aerobe versus Anaerobe
	• Differentiate anaerobes from aerobes and describe how they are cultured.
	• Explain how both aerobes and anaerobes can cause disease.
	• Discuss the basic differences between respiration and fermentation and how this impacts where an organism grows.
		15. Anaerobes fail to grow in the presence of oxygen for which of the following reasons?
		    A. Anaerobes cannot use oxygen to generate proton motive force.
		    B. Oxygen cannot be incorporated into an anaerobe's proteins.
		    C. Anaerobes lack enzymes that detoxify oxygen by-products.
		    D. Anaerobes lack enzymes that detoxify oxygen.

6.6 Microbial Communities and Cell Differentiation	• Discuss how biofilms develop. • Explain the importance of biofilms to infection. • Describe the process of sporulation, and explain how spores impact certain infections.	20. The transmission and sensing of chemical signal molecules between bacteria is called ____. 21. The secreted polymers that trap microbes within a biofilm are called ____.

13. Which of the following will take place when growth temperature for a bacterium is set below the organism's growth optimum?
    A. The culture will grow to a lower maximum cell density.
    B. The growth rate in exponential phase will be faster.
    C. Stationary phase will be shorter.
    D. The growth rate in exponential phase will be slower.
    E. Lag phase will be shorter.

14. Psychrophiles grow poorly at high temperatures for which of the following reasons?
    A. Ribosomes synthesize proteins too quickly for balanced growth.
    B. Proteins become less flexible.
    C. DNA replication becomes more error prone.
    D. Water activity becomes too low.
    E. Membranes become too flexible.

16. Anaerobes cannot conduct respiration. True or false?

17. In order to grow, anaerobes must carry out fermentation. True or false?

18. What site in the body has the most anaerobes that can cause infection?
    A. Skin   D. Intestine
    B. Lung   E. Oral cavity
    C. Vagina

19. A patient with appendicitis had his appendix removed. The surgeon removed almost 500 ml of pus from the infected appendix. Bacteriological examination of this material most likely found which of the following?
    A. Anaerobes only
    B. Strict aerobes only
    C. Facultative organisms only
    D. Anaerobes and strict aerobes
    E. Anaerobes and facultative organisms

22. Biofims are typically associated with which of the following types of infections?
    A. Acute   D. Latent
    B. Chronic   E. B and C only
    C. Septicemia

23. Spores are important to the infectious cycle of certain pathogens for which of the following reasons?
    A. They resist heat.
    B. They resist disinfectants.
    C. They remain viable for long periods.
    D. They inhibit the host's immune response.
    E. All of the above
    F. A, B, and C only

Key Terms

acidophile (p. 176)
aerobic respiration (p. 179)
aerotolerant anaerobe (p. 181)
alkaliphile (p. 176)
anaerobic respiration (p. 180)
autotroph (p. 160)
barophile (p. 175)
batch culture (p. 169)
binary fission (p. 168)
biofilm (p. 182)
budding (p. 168)
chemically defined minimal medium (p. 159)
chemoautotroph (p. 176)
chemostat (p. 171)
chemotrophy (p. 161)
colony (p. 163)
complex medium (p. 165)
confluent (p. 164)

continuous culture (p. 171)
death phase (p. 170)
death rate (p. 170)
denitrification (p. 162)
differential medium (p. 165)
dilution streaking (p. 163)
doubling time (p. 169)
electron transport chain (p. 179)
enriched medium (p. 165)
essential nutrient (p. 158)
exopolysaccharide (EPS) (p. 183)
exponential growth (p. 169)
exponential phase (p. 169)
extremophile (p. 172)
facultative (p. 180)
facultative anaerobe (p. 180)
fermentation (p. 180)
fluorescence-activated cell sorter (FACS) (p. 166)

generation time (p. 169)
germination (p. 185)
growth factor (p. 159)
halophile (p. 175)
heterotroph (p. 160)
isolation streaking (p. 163)
lag phase (p. 169)
late log phase (p. 170)
lithotrophy (p. 161)
logarithmic (log) phase (p. 169)
MacConkey medium (p. 165)
mesophile (p. 174)
microaerophilic (p. 181)
minimal medium (p. 165)
neutralophile (p. 176)
nitrification (p. 162)
nitrogen cycle (p. 162)
nitrogen-fixing bacterium (p. 162)
optical density (p. 168)

organotrophy (p. 161)
osmolarity (p. 175)
phototrophy (p. 160)
piezophile (p. 175)
psychrophile (p. 174)
pure culture (p. 163)
quorum sensing (p. 183)
rich medium (p. 165)
selective medium (p. 165)
spectrophotometer (p. 168)
spread plate (p. 163)
stationary phase (p. 170)
strict aerobe (p. 179)
strict anaerobe (p. 179)
symbiont (p. 162)
synthetic medium (p. 165)
thermophile (p. 173)
turbidity (p. 168)
viable (p. 164)

Review Questions

1. What nutrients do microbes need in order to grow?

2. Explain the differences between autotrophy, heterotrophy, phototrophy, and chemotrophy.

3. Briefly describe the carbon and nitrogen cycles.

4. Why is it important to grow bacteria in pure culture?

5. What factors define the growth phases of bacteria grown in batch culture?

6. Describe the important features of biofilms.

7. What are extremophiles, and why are these organisms important?

8. What parameters define any growth environment?

9. List and define the classifications used to describe microbes that grow in different physical growth conditions.

10. Why is water activity important to microbial growth? What changes water activity?

11. Why do changes in H^+ concentration affect cell growth?

12. How do acidophiles and alkaliphiles manage to grow at the extremes of pH?

13. If an organism can live in an oxygenated environment, does that mean that the organism uses oxygen to grow? Why or why not?

14. If an organism can live in an anaerobic environment, does that mean it cannot use oxygen as an electron acceptor? Why or why not?

15. What happens when a cell exhausts its available nutrients?

16. Why are biofilms important to infections?

17. Why are endospores important in medicine? Outline the stages of sporulation.

Clinical Correlation Questions

1. Brianna, a 4-year-old girl, is taken to the emergency room with a chief complaint of diarrhea. Bacteriological culture of her stool reveals the Gram-negative pathogen *Salmonella enterica*. The organism is known to contaminate poultry. When her mother is asked, she says that Brianna ate some homemade chicken salad at a party the day before but that her twin brother Brian had the same chicken salad 2 days earlier when it was first made, and he was fine. Can you explain why Brian might not have gotten ill but Brianna did after eating from the same preparation of chicken salad?

2. Two people happen to drink equal amounts of water while swimming in the Ganges River, in India. Unbeknownst to them, the water is contaminated with *Vibrio cholerae*, which causes the diarrheal disease cholera. Gagandeep, a 53-year-old man, has been taking aspirin to

prevent clots from forming in his blood; Padman, another 53-year-old man, has been taking a proton pump inhibitor for chronic indigestion. Three days later, Padman develops severe diarrhea but Gagandeep does not. Propose a plausible explanation for why one man became ill but the other did not.

3. *Helicobacter pylori*, a bacterial cause of stomach ulcers, can grow in media whose pH is above pH 5. It can survive but not grow in pH 2 media as long as the media contain urea. You inoculate a pH 2 broth with 1×10^7 *H. pylori* organisms and for 8 hours confirm that the organism does not grow. The next morning, however, you find that the medium has become quite turbid, suggesting that the organism has grown. Explain how this could happen.

Thought Questions: CHECK YOUR ANSWERS

Thought Question 6.1   Red blood cells contain NAD and hemin. Why, then, in our chapter-opening case history, did *H. influenzae* not grow on blood agar medium?

ANSWER: The bacterium cannot on its own break RBC membranes. Thus, X and V factors remain inaccessible. However, during the preparation of chocolate agar, heat treatment is used to lyse RBCs and release their contents. As a result *H. influenzae* can grow on chocolate agar but not blood agar.

Thought Question 6.2   In a mixed ecosystem of autotrophs and heterotrophs, what happens when a heterotroph allows the autotroph to grow and begin to make excess organic carbon?

ANSWER: At first the growth of the heterotroph might outpace the growth of the autotroph, using the carbon sources faster than the autotroph can make them. As the organic carbon sources diminish through consumption, growth of the heterotroph decreases, but the CO_2 formed by the heterotroph will allow the autotroph to grow and make more organic carbon. Ultimately, the ecosystem comes into balance.

Thought Question 6.3   The addition of sheep blood to agar produces a very rich medium called blood agar. Do you think blood agar can be considered a selective medium? A differential medium? *Hint:* Some bacteria can lyse red blood cells.

ANSWER: Blood agar can be considered differential because different species growing on blood have different abilities to lyse the red blood cells in the agar. Some do not lyse, others completely lyse red blood cells (secreted hemolysin produces complete clearing around a colony), and still others only partially lyse the blood (the secreted hemolysin produces a greening around the colony). The blood agar will therefore differentiate between hemolytic and nonhemolytic bacteria. The medium is very rich and does not prevent the growth of any organism and so is not considered selective.

Thought Question 6.4   If bacteria divide every 20 minutes, why is the exponential growth curve (measured as viable count) smooth and not shaped like a series of steps?

ANSWER: The answer is asynchronous growth. Not all cells replicate at the same instant. So, even a culture started with one cell will, over time, produce millions of offspring that begin division at slightly different times. The result is a smooth curve.

Thought Question 6.5   Suppose 1,000 bacteria are inoculated into a tube of minimal salts medium, where they double once an hour, and 10 bacteria are inoculated into rich medium, where they double every 20 minutes. Which tube will have more bacteria after 2 hours? After 4 hours?

ANSWER: After 2 hours, the minimal medium will contain $1,000 \times 2^2 = 4,000$ bacteria, whereas the rich medium will contain only $10 \times 2^{(2h)(3\ div/h)} = 640$ bacteria. After 4 hours, the minimal medium will contain 1.6×10^5 bacteria, whereas the rich medium will have surpassed this count, reaching nearly 4.1×10^5.

Thought Question 6.6   How might the concept of water availability be used by the food industry to control spoilage?

ANSWER: Food preservation traditionally includes water exclusion by salt, as seen in hams, back bacon, and salted fish, or by high concentration of sugar, as in canned fruit and jellies. The lower a_w prevents microbial growth. Dehydrating foods with an evaporating dehydrator will also prevent microbial growth.

Thought Question 6.7   If anaerobes cannot live in oxygen, how do they incorporate oxygen into their cellular components?

ANSWER: They incorporate oxygen from their carbon sources (for example, CO_2 and carbohydrates such as glucose), all of which contain oxygen. This form of oxygen will not damage the cells.

Thought Question 6.8   How can anaerobes grow in the human mouth when so much oxygen is there?

ANSWER: A synergistic relationship occurs between facultatives and anaerobes within a tooth biofilm. The facultatives consume oxygen within the biofilm microenvironment, which allows anaerobes to grow underneath them.

Thought Question 6.9   *Clostridium botulinum* is an anaerobic, Gram-positive, spore-forming bacillus, normally found in soil, that secretes a poisonous toxin. Why would this organism be a potential problem when Grandma decides to can vegetables such as green beans?

ANSWER: Home canning involves placing fresh fruit or vegetable preparations into a jar and heating the jar in a pressure cooker. The high pressure prevents water from boiling until it reaches 120°C (248°F), a temperature sufficient to kill spores. The heat also drives oxygen from the jar. Because *C. botulinum* is found in soil, so are its spores. If these heat-resistant spores contaminate the vegetables being canned and the canning process is incomplete, spores will survive. Once the preserves cool, the spores germinate in the anaerobic environment and the viable bacteria begin secreting toxin. If the food containing toxin is ingested, the unwitting victim will contract botulism.

7

Bacterial Metabolism

[A Spring Break Mishap]

SCENARIO Shane, a 21-year-old college student, was spending spring break at the beach in Cancún, where he frequently ordered drinks from the beachfront stands. Upon his return to classes, Shane fell ill with a fever of 38.9°C (102°F), severe abdominal cramps, and watery diarrhea.

SIGNS AND SYMPTOMS By the second day, Shane had blood in his stools, a condition known as dysentery. Upon admission to the hospital, Shane showed dehydration, and his rectal exam was very painful with bleeding.

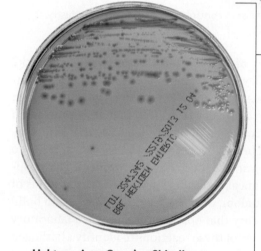

Hektoen Agar Growing *Shigella*
Shigella produces pale colonies (no sugar fermentation and no thiosulfate reduction) whereas *Salmonella* colonies would appear black (production of H₂S from thiosulfate reduction).

TESTING A fecal culture was performed on Hektoen agar, a selective medium for Gram-negative bacteria; bile salts exclude Gram-positives. Hektoen agar also differentiates among the enteric Gram-negative pathogens *Salmonella, Shigella,* and *Escherichia.* Most *Salmonella* species show black colonies due to H₂S formation from thiosulfate, whereas *Escherichia* species form colonies that are orange from lactose fermentation. Shane's culture produced translucent colonies, indicating *Shigella* species, which neither ferment lactose nor reduce thiosulfate to H₂S.

DIAGNOSIS AND TREATMENT Serotyping (determining which antibodies react with the pathogen) confirmed the species *Shigella flexneri,* a common cause of "traveler's diarrhea" from drinking contaminated water. Shane received intravenous rehydration, and was given the antibiotic quinolone, chosen based on laboratory tests of the pathogen's antibiotic sensitivity. Afterward, Shane made a full recovery.

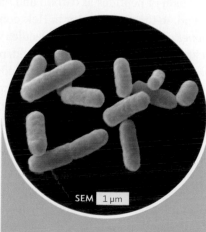

SEM 1 μm

Shigella flexneri
Shigella flexneri causes shigellosis, a form of dysentery.

CHAPTER OBJECTIVES

After reading this chapter, you will be able to:

- Describe the various kinds of biochemical reactions that provide energy for life.
- Explain ways that bacterial metabolism interacts with human health.
- Describe how microbial photosynthesis and nitrogen fixation support life and how these processes affect human health.

Shigellosis is a common illness in day-care centers and in developing countries where water supplies are contaminated. *Shigella* species are Gram-negative pathogens of the Enterobacteriaceae (see photo accompanying chapter-opening case history). The pathogen has an unusually low infective dose, requiring as few as ten organisms to cause infection. Because they are found in feces, among many normal biota, the culture and identification of *Shigella* requires selective and differential media (see Section 6.2). A selective medium for Gram-negative enteric organisms is Hektoen agar (see photo with case history). Hektoen agar includes indicators that use bacterial metabolism. Different bacteria conduct different kinds of metabolism to gain energy from food molecules. In the Hektoen formula, one such food molecule is lactose, a sugar fermented by *Escherichia coli* but not by most species of *Salmonella* and *Shigella*.

Fermentation is a process that bacteria use to obtain energy from sugars by breaking the molecules into smaller ones—a process that increases entropy (disorder) and therefore yields energy (discussed in Section 7.1). Often the products of fermentation include acids, such as lactic acid, which decrease the extracellular pH. The pH decrease causes a change in the indicator dye of Hektoen agar, hence the appearance of orange colonies of *E. coli*.

A different Gram-negative organism, *Salmonella*, fails to ferment lactose; instead, *Salmonella* species grow on peptides in the media. Metabolism of peptides as carbon and energy sources does not produce acid, so the pH indicator in Hektoen does not change. However, *Salmonella* can combine food molecules with the thiosulfate present in Hektoen; the food becomes oxidized, releasing H_2S. The oxidation of food by an inorganic molecule is called respiration, a different way to gain energy for growing cells (see Section 7.4). The H_2S produced by *Salmonella* respiration reacts with iron in Hektoen to produce a black FeS precipitate seen in the colonies. Thus, black colonies on Hektoen agar would indicate *Salmonella* as the pathogen tested.

Both fermentation and respiration are chemical reactions that yield energy to "do the work" of growing cells. The energy released by a reaction is transferred by enzymes to other reactions that build simple molecules into a complex cell. Chapter 7 presents the many kinds of reactions that enable bacteria to grow—and that reveal their identity.

7.1
Energy for Life

SECTION OBJECTIVES

- Define the major classes of energy-yielding metabolism, including photosynthesis, organotrophy, and lithotrophy.
- Explain the importance of energy and entropy for living cells.
- State the role of enzymes in controlling metabolic biochemistry.

Have you ever run out of money before your next paycheck? Energy is to life as money is to a household. If money stops coming in, the household goes bankrupt; if energy stops flowing through an organism, it dies. On a broader scale, if money falls short in an economy, the economy collapses. Similarly, if energy fails in an ecosystem, a community of organisms collapses. Thus, energy is key to survival of human beings and all life in our environment. This chapter presents the biochemical reactions by which different kinds of microbes obtain energy, along with the reactions they use to build biomass. In the case history that opened this chapter, a laboratory technician used knowledge of these reactions to identify a pathogen on the basis of its metabolic reactions.

Energy and Entropy for Living Cells

How do living organisms assemble themselves out of small, nonliving molecules? In everyday life, we know that simple objects do not spontaneously assemble themselves into something complex. As you learned in Chapter 4, the universe overall becomes more random, or less ordered; this disorder or randomness is called **entropy**. And yet, every living organism, from a soil microbe to a human, must assemble simple molecules such as H_2O and CO_2 into complex cells (**Figure 7.1**). To build something complex requires spending **energy**. In Chapter 4, we introduced energy as the property that enables a chemical reaction to go forward. For a living organism, energy enables the cell to perform all its functions, such as swimming or growing.

As a cell grows and builds complex molecules, entropy, or disorder, decreases. But the decrease in entropy is local to the cell. The cell releases heat and waste products to its surroundings, where disorder actually increases. Thus, the local, temporary spending of energy enables a cell to build order and grow, while increasing entropy (disorder) outside. Continued growth requires continual spending of energy and dissipation of heat.

Entropy increases beyond the cell's environment, throughout the entire biosphere. To overcome entropy, life on Earth captures energy from the Sun (**Figure 7.2**). Chapter 6 introduced the nutritional strategies by which microbes obtain the Sun's energy, either directly (as producers) or indirectly (by consuming the producers). Photosynthetic microbes and plants generate biomass, which is consumed by heterotrophs and decomposers. The consumers—including humans—store a small fraction of their energy in biomass. At each successive level of consumers, most of the energy is lost, radiated from Earth as heat. But a key fraction of energy is stored in the bodies of living organisms.

Figure 7.1   Living Organisms Assemble Simple Molecules into Complex Cells

To build complex cells, living organisms must spend energy.

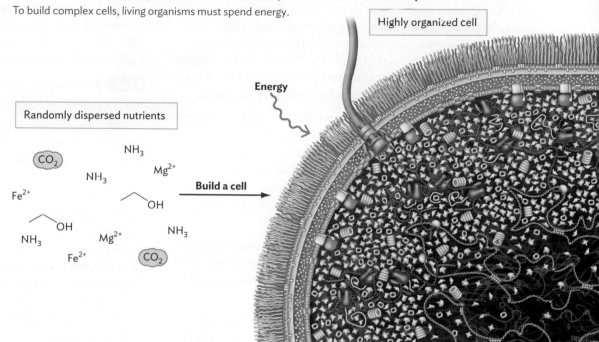

Figure 7.2   Energy Flows through the Biosphere

Solar radiation reaches Earth, where a small fraction is captured by photosynthetic microbes and plants. The microbial and plant biomass enters heterotrophs and decomposers, which convert a small fraction to biomass at each successive level while losing energy as heat.

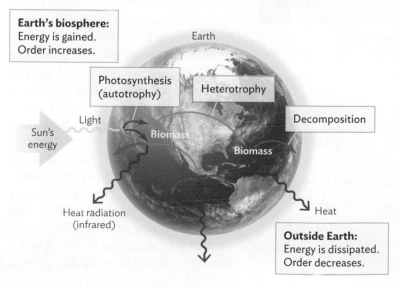

Metabolic Reactions That Yield Energy

Different kinds of metabolism are like different occupations: Each offers microbes a way to "make a living." The major categories of metabolism are summarized in **Table 7.1**. All these kinds of metabolism play important roles in the environment—and have consequences for human health.

- **Phototrophy** refers to biochemical reactions that capture energy from light. Light energy captured by chlorophyll is used to excite electrons from a stable molecule such as water (H_2O). In oxygenic photosynthesis, the excited electrons are used to reduce CO_2 to form sugar ($C_6H_{12}O_6$) while releasing oxygen gas (O_2). Oxygenic photosynthesis produces the food we eat and the oxygen we breathe. Most organisms conducting oxygenic photosynthesis are **autotrophs**, capable of forming all their own organic molecules from CO_2 plus minerals.

- **Chemotrophy** refers to reactions that yield energy stored in the chemical bonds of molecules, such as the biomass of a consumed prey organism. The transfer of an electron from one molecule to another (or to another part of the same molecule) releases a combination of products more stable than the original; this reaction yields energy. Classes of chemotrophy include organotrophy and lithotrophy, which are defined by the nature of their electron donors and acceptors.

- **Organotrophy** refers to metabolism in which the food molecules (electron donors) are organic. Most organisms that conduct organotrophy are **heterotrophs**, requiring preformed organic compounds.

— Aerobic organotrophy consists of reactions in which an organic molecule donates electrons to O_2, forming the stable molecule H_2O. Most forms of aerobic organotrophy are also classed as **aerobic respiration**, such as the oxidative breakdown of sugar releasing H_2O and CO_2. Human aerobic respiration requires us to breathe oxygen continually—and to release CO_2, which serves as the essential pH buffer for our blood.

— Anaerobic organotrophy involves reactions <u>without</u> oxygen. In anaerobic respiration, oxygen is replaced by an alternative electron acceptor such as nitrate. In fermentation, partial breakdown of food molecules releases small molecules such as ethanol or lactic acid. Microbial fermentation plays a key role in the production of foods such as cheese and alcoholic beverages.

- **Lithotrophy** refers to metabolism in which the food molecules (electron donors) are inorganic. Organisms that grow solely by lithotrophy are autotrophs, forming their own organic molecules from minerals.

— Aerobic lithotrophy involves reactions with oxygen. An example of aerobic lithotrophy is ammonia oxidation by O_2. Ammonia oxidation commonly occurs in the soil, especially where nitrogen-rich fertilizers are spread. A medical consequence is the formation of harmful levels of nitrites and nitrates that enter groundwater.

— Anaerobic lithotrophy is lithotrophic metabolism <u>without</u> O_2. For example, methanogenesis is a reaction by which certain archaea convert hydrogen and CO_2 to methane and water. Methanogens are ubiquitous in soil and in the digestive tract of animals.

To yield energy, a biochemical reaction must go forward from reactants to products. The direction of a reaction, whether it goes forward or in reverse, can be predicted by the **free energy change,** $\Delta G = \Delta H - T\Delta S$. If the value of free energy change (ΔG) is negative, a given metabolic reaction yields energy for the cell.

Link Reactions that yield energy require a negative value of **free energy change, ΔG** (see Section 4.4).

The free energy change, ΔG, includes two components: the change in chemical energy (ΔH) and the change in entropy multiplied by temperature ($-T\Delta S$). For example, the oxidation of glucose by O_2 (aerobic respiration) generates products with increased chemical stability (value of ΔH). Aerobic respiration also involves the breakdown of one molecule to a greater number of small molecules, thus increasing entropy (value of $-T\Delta S$). Other kinds of metabolism tend to rely on one component of free energy more than the other. For example, when baker's yeast in bread dough ferments sugar, the sugar molecule breaks down to two molecules of ethanol and two molecules of CO_2; the gas forms bubbles, causing the dough to rise. The bread-making reaction is driven more by the increase in entropy, and its efficiency is more temperature dependent than is respiration.

Enzymes Control Reactions

When a chemical reaction releases energy (such as by oxidizing a food molecule), the energy must be captured by another reaction that builds order (such as synthesizing protein). Enzymes manage this energy transfer from one reaction to another, as discussed in Chapter 4. An enzyme allows a specific reaction to occur only as needed, in the right amount at the right time. Enzymes act by lowering the activation energy, the input energy needed to generate the high-energy transition state on the way to products. Enzymes determine the precise rearrangement of chemical bonds and the rate at which biochemical reactions occur in the cell.

For an organism to grow, the chemical reactions of the food must be controlled so as to spend the energy in a biosynthetic reaction that builds the cell or that powers a cell function such as motility. What would happen if all the energy were spent at once? In uncontrolled oxidation, or combustion, all the energy of reaction is lost by the release of heat. Living cells instead oxidize food in a gradual process, subdivided into many steps, each of which transfers energy to a cellular process.

Table 7.1

Energy-Yielding Metabolism

Class of Metabolism	Source of Energy	Example (Simplified)
Phototrophy	Light-energy absorption makes high-energy molecule that donates electrons to acceptor	Photosynthesis Light, $H_2O + CO_2 \rightarrow$ sugar $+ O_2$
Chemotrophy	High-energy food molecule donates electrons to acceptor	
Organotrophy, aerobic	Organic molecule donates electrons to O_2	Aerobic respiration Sugar $+ O_2 \rightarrow H_2O + CO_2$
Organotrophy, <u>anaerobic</u>	Organic molecule donates electrons to itself or other molecule, <u>not</u> O_2	Fermentation Sugar $\rightarrow$ ethanol $+ CO_2$
Lithotrophy, aerobic	Inorganic molecule donates electrons to O_2	Ammonia oxidation $NH_3 + 2O_2 \rightarrow HNO_3 + H_2O$
Lithotrophy, <u>anaerobic</u>	Inorganic molecule donates electrons to other molecule, <u>not</u> O_2	Methanogenesis $4H_2 + CO_2 \rightarrow CH_4 + 2H_2O$

A cell's enzymes make sure the appropriate reaction occurs at the appropriate time and place for cell function.

Throughout the figures of Chapter 7, every reaction shown requires one or more enzymes. Be sure to remember this, even when no enzyme is shown.

SECTION SUMMARY

- **Different kinds of metabolism yield energy for different microbes.** Important classes of metabolism include phototrophy, organotrophy (aerobic and anaerobic), and lithotrophy (aerobic and anaerobic).

- **Energy-yielding reactions must show a negative value of free energy change (ΔG).** The free energy change includes changes in chemical energy and in entropy.

- **Enzymes catalyze all biochemical reactions.** Enzymes increase the rate of reaction and make sure the reaction occurs at the appropriate time and place for cell function.

7.2
Catabolism: The Microbial Buffet

SECTION OBJECTIVES

- Describe how catabolism converts many complex food molecules to a few kinds of catabolites.

- Explain how catabolism yields energy and how the energy is stored for use.

- Describe how the energy carriers ATP and NADH transfer energy between energy-yielding and energy-spending reactions.

What kind of metabolism yields energy for the human body— and for most of our intestinal bacteria? We and our bacteria break down complex organic food molecules, a process called **catabolism**. In principle, catabolism can break down any organic substance found in the environment. A given species, however, can catabolize only those substances for which its DNA encodes the right enzymes. Thus, humans express enzymes enabling catabolism of starches, lipids, and proteins, but not cellulose or lignin (a structural substance in woody plants). Many soil bacteria, however, have enzymes to catabolize cellulose and lignin. Other bacteria are pathogens that catabolize components of the human body. For example, the bacterium that causes acne, *Propionibacterium acnes*, catabolizes skin cell molecules such as sialic acids, matrix proteins, and lipids.

Certain kinds of food molecules are widely catabolized by many species. Carbohydrates, lipids, and proteins are common substrates for catabolism. In general, food molecules such as starch (a polymer of glucose units) are broken down by specific enzymes to smaller molecules called **catabolites** that feed into shared central pathways (**Figure 7.3**). For example, the catabolite glucose from starch breakdown enters **glycolysis**, a pathway of step-by-step catabolism in which energy is transferred to energy carriers ATP and NADH

(discussed shortly). Glucose also arises from the breakdown of cellulose, a glucose polymer found in plants. Cellulose is broken down by the enzyme cellulase; the glucose produced enters glycolysis, the same as the glucose from starch. Other materials, such as pectin, a sugar acid polymer in fruits, are broken down to sugar acids. Sugar acids are further catabolized by a modified pathway of sugar breakdown called the Entner-Doudoroff pathway.

Lipids and proteins are catabolized to products that enter the central pathways of glycolysis and the **tricarboxylic acid cycle (TCA cycle)**, also known as the **Krebs cycle** or the **citric acid cycle**. Lipids are broken down by hydrolysis (cleavage of a molecule while incorporating H_2O) to their major components glycerol and fatty acids (compounds introduced in Chapter 4). Glycerol is broken down to pyruvic acid (pyruvate). Fatty acids are broken down to two-carbon units of acetic acid, which is tagged with coenzyme A (acetyl-CoA) to enter the TCA cycle (discussed in Section 7.3). Proteins are hydrolyzed to peptides and ultimately to amino acids. The amino acids are broken down and converted to intermediates such as acetyl-CoA and TCA cycle intermediates such as 2-oxoglutarate (presented in Figure 7.10).

Note Many intermediates of metabolism are carboxylic acids (R–COOH) such as acetic acid, pyruvic acid, and succinic acid. Each acid exists in the cell as interchangeable forms that may be considered equivalent in our diagrams. The forms are shown here for acetic acid:

Acetic acid

Acetate
(deprotonated;
accompanied by H⁺)

Acetyl-CoA
(coenzyme A reversibly
replaces –OH)

Although many food molecules are digested by humans as well as by heterotrophic microbes, other substances, such as complex plant materials and petroleum compounds, are digested only by bacteria or fungi. Recent research shows that within our digestive tract, as much as 15% of our caloric intake is provided by bacteria such as *Bacteroides* species that catabolize complex plant fibers we have ingested. In the oceans, petroleum seeping from the ocean floor or from damaged wellheads is broken down by oil-eating bacteria. Most of the oil breakdown products eventually are converted to acetyl-CoA.

Energy Carriers: ATP and NADH

As food molecules break down and are oxidized, these reactions yield energy. The energy-yielding reactions are regulated by enzymes that transfer the energy into biosynthesis. Let's imagine a system in which each enzyme catalyzes an energy-yielding reaction coupled directly to a biosynthetic reaction, such as building an amino acid. In this manner, energy yield could be coupled directly

Figure 7.3   Many Carbon Sources Enter Central Pathways of Catabolism

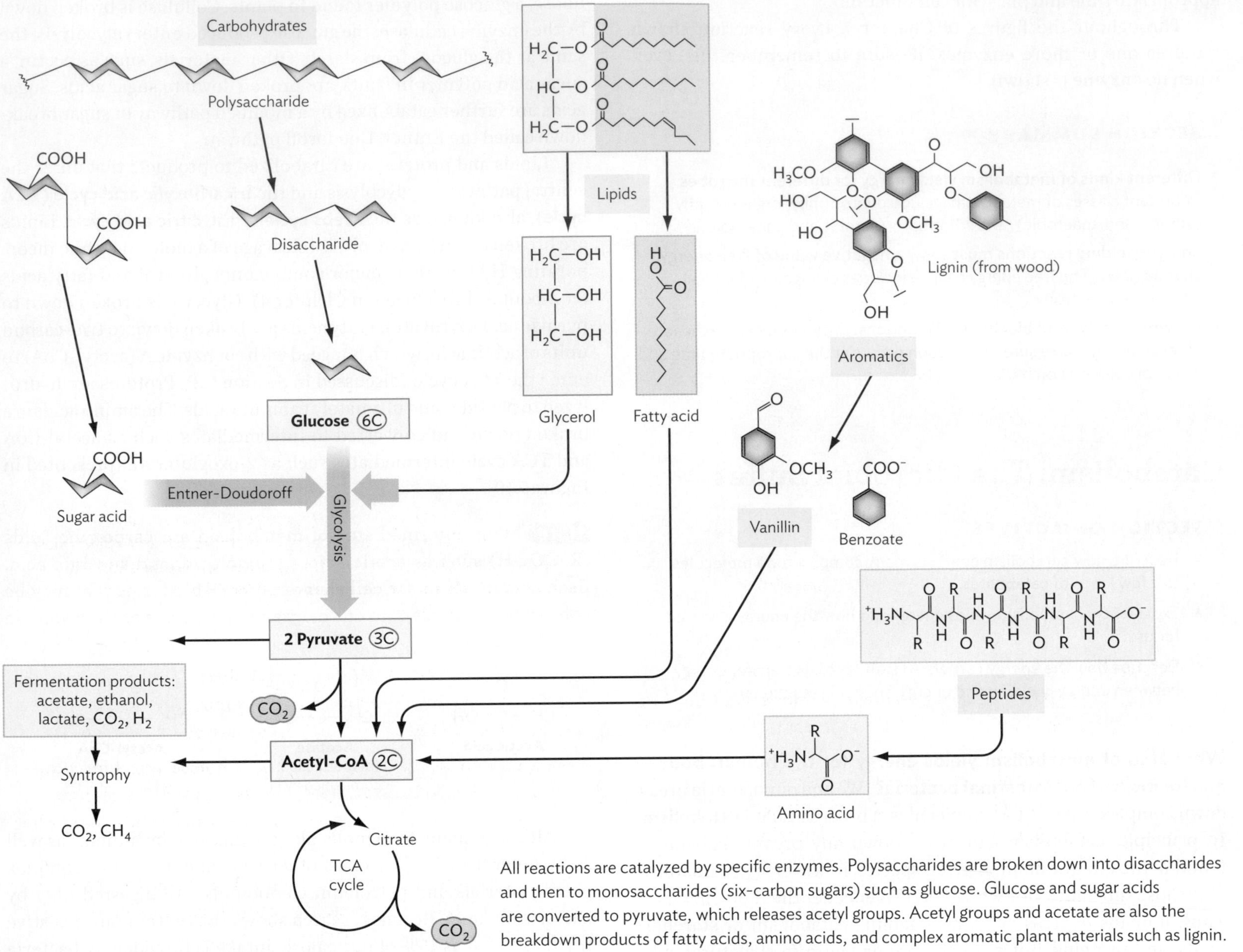

All reactions are catalyzed by specific enzymes. Polysaccharides are broken down into disaccharides and then to monosaccharides (six-carbon sugars) such as glucose. Glucose and sugar acids are converted to pyruvate, which releases acetyl groups. Acetyl groups and acetate are also the breakdown products of fatty acids, amino acids, and complex aromatic plant materials such as lignin.

to energy use, and thus little or no energy would be lost. How many enzymes would such a system require?

Suppose an organism could use one of 100 possible substrates to yield energy and needed to build 100 different compounds that make up its cell. For each substrate yielding energy, 100 different enzymes would be needed to couple that energy from any substrate to build each of the 100 different products. In all, the organism would need $100 \times 100 = 10,000$ different enzymes, each encoded by a specific gene. But a bacterium often has fewer than 5,000 genes in its entire genome. So how could all the needed enzymes be made?

Now suppose instead that all of the 100 substrates would be converted by enzymes to a common energy carrier such as ATP. Then this same type of energy carrier, holding the energy from any

of the 100 substrates, could now be used by an enzyme to build one of 100 products. Now this system requires only $100 + 100 = 200$ different enzymes. Thus, the use of common energy carriers enables a smaller genome to direct many kinds of metabolism.

Another important function of energy carriers is to gain and release energy in small amounts. If all the energy of food molecules were released at once (as in combustion), it would dissipate as heat without building biomass. In living cells, food molecules never break down in one step. Instead, the energy yield is divided among a large number of stepwise reactions with smaller energy changes. In this way, the cell can be thought of as "making change" by converting a large energy source to numerous smaller sources that can be "spent" conveniently for cell function and biosynthesis.

Figure 7.4   Energy Carriers: ATP and NADH

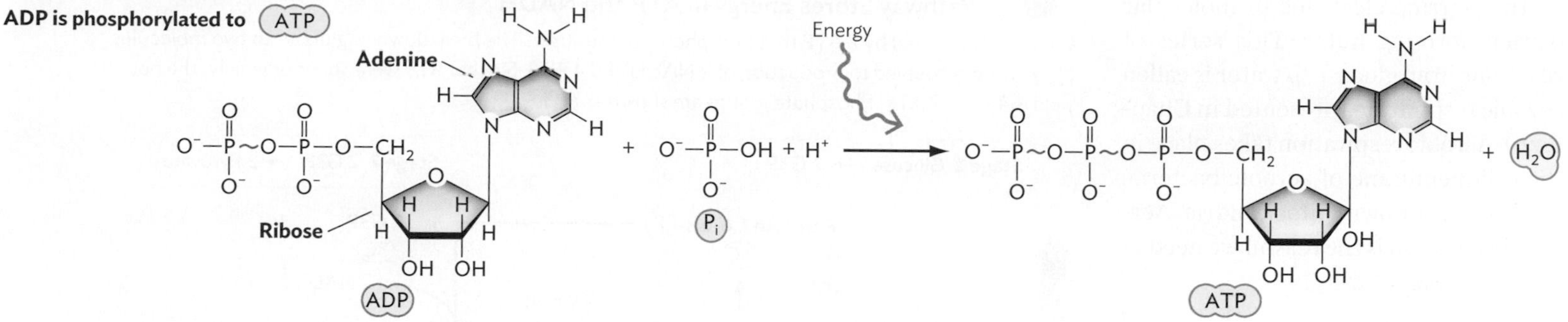

ADP is phosphorylated to ATP

A. The chemical reaction of ADP (adenosine diphosphate) with inorganic phosphate makes ATP (adenosine triphosphate). The reaction requires energy input because the negatively charged oxygens of the phosphates are forced near each other. Removal of each phosphate by hydrolysis (inserting H_2O) yields energy.

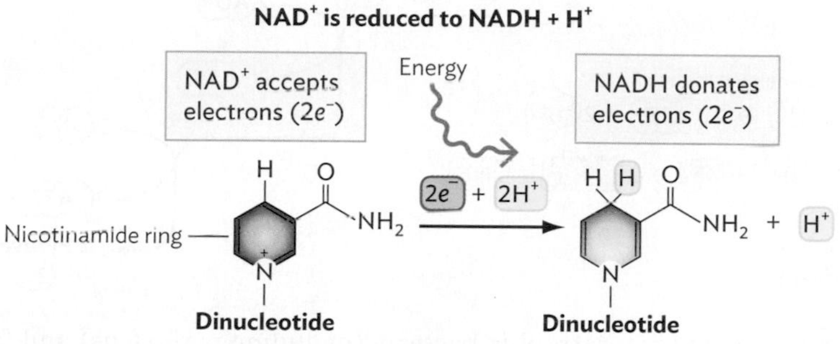

NAD^+ is reduced to $NADH + H^+$

B. In the reduction of NAD^+, the ring (shaded pink) loses a double bond as two electrons are gained from an electron donor; the resulting NADH carries energy. The reaction consumes two hydrogen atoms; one bonds to NAD^+, forming NADH, while the other ionizes in solution (H^+).

All organisms use common energy carriers such as **adenosine triphosphate (ATP)** and **nicotinamide adenine dinucleotide (NADH)**. Molecules of ATP and NADH are formed by many reactions of the catabolic pathways shown in Figure 7.3.

As described in Chapter 4, the energy carrier ATP is composed of a base (adenine), a sugar (ribose), and three phosphoryl groups. The adenine-ribose-phosphate molecule (adenosine nucleotide) is equivalent to a ribonucleotide, the monomer (unit) of RNA. The base adenine is a fundamental molecule of life, one that forms spontaneously from methane and ammonia in experiments simulating the origin of life on early Earth. ATP is an ancient component of cells, used by all living organisms; presumably, it was used by the ancestral cell from which all life, including humans, evolved.

During catabolism (see Figure 7.3), many of the enzymes couple an energy-yielding reaction to the formation of ATP (**Figure 7.4A**). ATP is formed by phosphorylation of adenosine diphosphate (ADP); the ADP molecule condenses with inorganic phosphate (P_i) to form ATP.

ATP formation requires spending energy (positive ΔG) because condensing ADP with phosphate brings negatively charged oxygen atoms together, where they repel each other. Because ATP formation requires spending energy, ATP breakdown yields energy. ATP breaks down by hydrolysis to form ADP and inorganic phosphate (P_i).

One molecule of NADH carries about three times as much energy as one molecule of ATP. Besides carrying energy, NADH carries two electrons ($2e^-$) from a food molecule, which participate in the oxidation-reduction reactions of metabolism. Oxidation-reduction reactions, or redox reactions, are those that involve a transfer of electrons between two molecules. As discussed in Section 4.4:

- **Oxidation is <u>loss</u> of an electron.** A molecule that loses an electron is said to be **oxidized** because oxygen has a strong ability to "take" electrons, but the electron can be lost to other atoms besides oxygen.

- **Reduction is <u>gain</u> of an electron.** A molecule that gains an electron is said to be **reduced** because the charge on the atom becomes slightly more negative.

In many catabolic reactions, a pair of electrons ($2e^-$) from a food molecule are transferred to the oxidized form of the carrier, NAD^+. Typically, the oxidized form NAD^+ receives two electrons ($2e^-$) plus a proton (H^+) from glucose, while the second H^+ from glucose enters water solution. Overall, reduction of NAD^+ consumes two hydrogen atoms to make NADH (**Figure 7.4B**):

$$NAD^+ + 2H^+ + 2e^- \rightarrow NADH + H^+ \qquad \Delta G = 62 \, kJ/mol$$

For this reaction, ΔG is positive; therefore, NADH formation requires an input of energy from the catabolized food molecule. The reduced energy carrier NADH can then donate two electrons ($2e^-$) to another molecule, restoring the oxidized form NAD^+. NADH is central to life because it provides a way of carrying electrons to the electron transport system (ETS), also known as the electron transport chain (introduced in Section 6.5 and further discussed in Section 7.4). The ETS then releases the energy from NADH by transferring electrons to molecular oxygen, forming water. This series of reactions from glucose to water is called aerobic respiration (presented in Chapter 6). Aerobic respiration takes place in the cell membrane of aerobic bacteria, and also in our own mitochondria. Aerobic respiration is the reason we need to breathe oxygen.

Note An atom of hydrogen removed from a C–H bond in a food molecule consists of a proton (H^+) plus an electron (e^-). In a metabolic reaction, the proton and electron may be transferred together to a molecule, or the proton may be released as a hydrogen ion (as in NADH + H^+). The electron transfer defines redox reactions; the loss of H^+ is reversible and does not affect the redox change.

 Figure 7.5   Embden-Meyerhof-Parnas (EMP) Pathway Stores Energy in ATP and NADH

Glucose is activated by 2 ATP that phosphorylate the sugar. The breakdown of glucose to two molecules of pyruvate is coupled to production of 2 NADH and 4 ATP. Since 2 ATP were spent originally, the net yield is 4 – 2 = 2 ATP. Phosphate groups are shown as P_i.

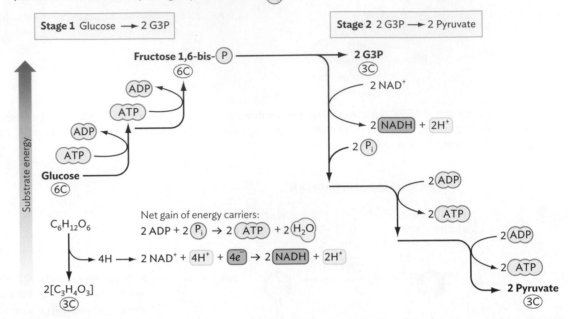

SECTION SUMMARY

- **Catabolic pathways organize the breakdown of large molecules** in a series of sequential steps coupled to reactions that store energy in small energy carriers such as ATP and NADH.
- **ATP stores energy in the bonds** formed by condensation of ADP with inorganic phosphate (P_i). Hydrolysis of these bonds yields energy.
- **NADH stores energy associated with a pair of electrons** that can reduce a substrate, such as a protein of the electron transport system (ETS). Oxidizing NADH to NAD^+ yields energy and transfers two electrons ($2e^-$) to a substrate molecule.

7.3
Glucose Catabolism, Fermentation, and the TCA Cycle

SECTION OBJECTIVES
- Describe how sugar is catabolized to pyruvate via glycolysis, and explain how these catabolic reactions generate ATP and NADH.
- Explain how pyruvate is further catabolized by fermentation or by the TCA cycle (when oxygen is available).
- Explain how bacterial catabolic reactions are used in clinical tests to identify a pathogen.

We now summarize the central pathways of catabolism that generate ATP and NADH from glucose. Glucose is converted to two molecules of pyruvic acid (glycolysis), which are decarboxylated to acetic acid (acetate). Acetate releases two molecules of CO_2 via the TCA cycle. These pathways of glycolysis and the TCA cycle are

essential for heterotrophic bacteria (including pathogens) and for human cells. Within human cells, the TCA cycle and respiration are conducted within mitochondria. Mitochondria are organelles that evolved from Gram-negative bacteria related to *Escherichia coli*, and their respiration mechanism is similar to that of bacteria. In external environments, bacteria obtain glucose for glycolysis by breaking down complex substrates (as shown in Figure 7.3). Bloodborne pathogens, however, such as the malaria parasite, take glucose directly from the plasma.

Glycolysis: From Glucose to Pyruvate

For many bacteria, the main pathway of glucose catabolism is glycolysis, also known as the Embden-Myerhof-Parnas (EMP) pathway. In glycolysis, one molecule of D-glucose (the "D" mirror form) breaks down into two molecules of pyruvic acid (Figure 7.5). Note that pyruvic acid exists in equilibrium with its anion, pyruvate, plus a hydrogen ion (H^+). For simplicity, pyruvic acid and other acids may be labeled as the anion ("-ate" suffix), although only some of the compound is ionized.

Glucose breaks down in two stages (see Figure 7.5). Stage 1 actually consumes energy (2 ATP) in order to prepare the glucose for a controlled breakdown that releases more energy than what was put in (4 ATP – 2 ATP = 2 ATP net). In stage 1, the six-carbon (6C) glucose molecule is primed for breakdown by two steps of sugar phosphorylation by 2 ATP. Each ATP phosphotransfer step requires spending energy. The sugar tagged by two phosphoryl groups (now called fructose 1,3-bisphosphate) then splits into two three-carbon (3C) molecules of glyceraldehyde 3-phosphate (G3P).

Figure 7.6   Glycolysis (EMP Pathway)

Glucose

2 ATP

2 ADP

Fructose
1,6-bisphosphate

Glyceraldehyde
3-phosphate

P_i

NAD^+

$NADH$ + H^+

1,3-bisphosphoglycerate

2 ADP

2 ATP

$CO_2^- - \overset{O}{\underset{}{C}} - CH_3$
Pyruvate

Glyceraldehyde
3-phosphate

P_i

NAD^+

$NADH$ + H^+

1,3-bisphosphoglycerate

2 ADP

2 ATP

$CO_2^- - \overset{O}{\underset{}{C}} - CH_3$
Pyruvate

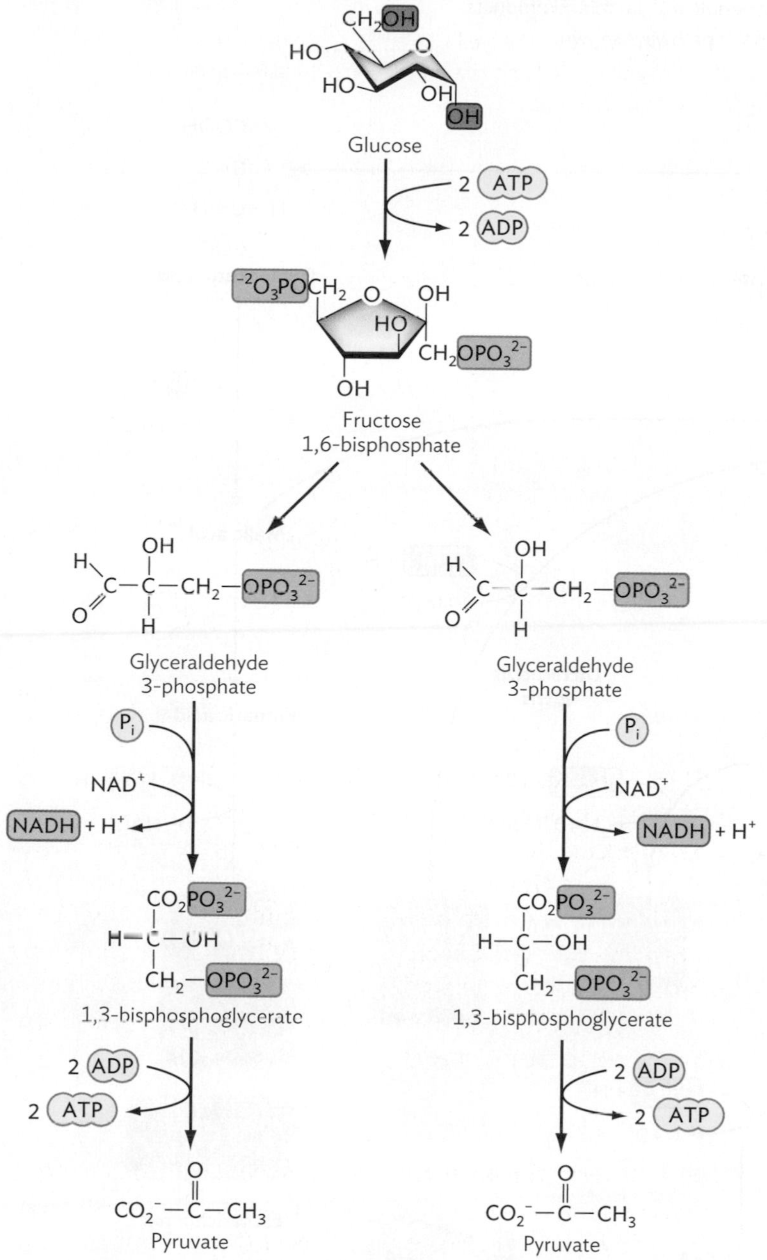

Glucose is phosphorylated by 2 ATP, which input energy to "prime" the pathway. The reactions include rearrangement to form fructose 1,6-bisphosphate. Fructose 1,6-bisphosphate is cleaved into two three-carbon molecules of glyceraldehyde 3-phosphate—a reaction that yields energy. Each molecule of glyceraldehyde 3-phosphate loses two hydrogen atoms by transferring $2e^-$ to NAD^+, forming NADH. Each three-carbon molecule now incorporates inorganic phosphate (P_i) to form 1,3-bisphosphoglycerate. Subsequent reactions transfer both phosphoryl groups to 2 ADP to form 2 ATP. (More details of glycolysis are presented in Figure A2.1.)

In stage 2, each of the two G3P molecules loses two electrons (from two hydrogen atoms), converting NAD^+ to NADH (in all, 2 NADH). Subsequent steps from each three-carbon molecule lead to pyruvate, yielding enough energy to form 2 ATP (in all, 4 ATP from the original glucose). Since 2 ATP were spent originally to phosphorylate glucose, the net gain of energy carriers is two molecules of ATP (2 ATP) plus two molecules of NADH (2 NADH).

Key intermediates of glucose metabolism are presented in Figure 7.6 (full details are presented in Figure A2.1). Most of the conversion steps have only a small change in energy—so small that the sign of ΔG depends on the concentrations of substrates or products; thus, some individual steps are reversible. In the cytoplasm, however, as intermediate products form, they are quickly consumed by the next step, and so the pathway flows in one direction.

Alternatives for Sugar Catabolism

Besides glycolysis (EMP pathway), microbes can use other pathways to break down sugar (Figure 7.7). Different species use different pathways. Also, different pathways are used for different environmental needs. For example, within our intestines, *E. coli* use the Entner-Doudoroff pathway to consume mucus secreted by the intestinal epithelium. Intestinal mucus contains sugar acids such as gluconic acid (deprotonated as gluconate). Gluconate can be broken down in a series of steps leading to glyceraldehyde 3-phosphate and ultimately pyruvate. The Entner-Doudoroff pathway enables *E. coli* and other bacteria to "farm" intestinal mucus.

Figure 7.7   From Glucose to Pyruvate

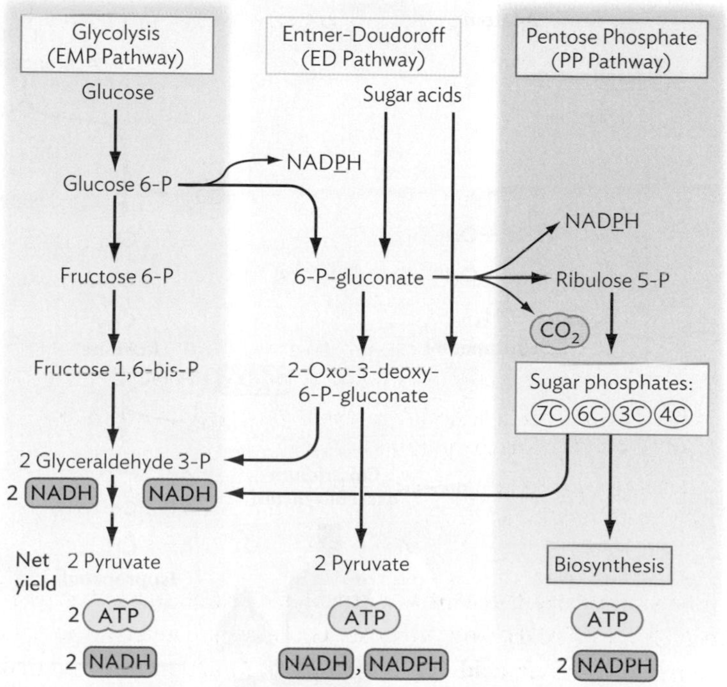

The Embden-Meyerhof-Parnas (EMP) pathway of glycolysis, the Entner-Doudoroff (ED) pathway, and the pentose phosphate (PP) pathway.

Figure 7.8   Fermentation Pathways

Alternative pathways from pyruvic acid (pyruvate) and the glycolysis intermediate phosphoenolpyruvate to end products, many of which we use for food or industry. Different species conduct different portions of the pathways shown.

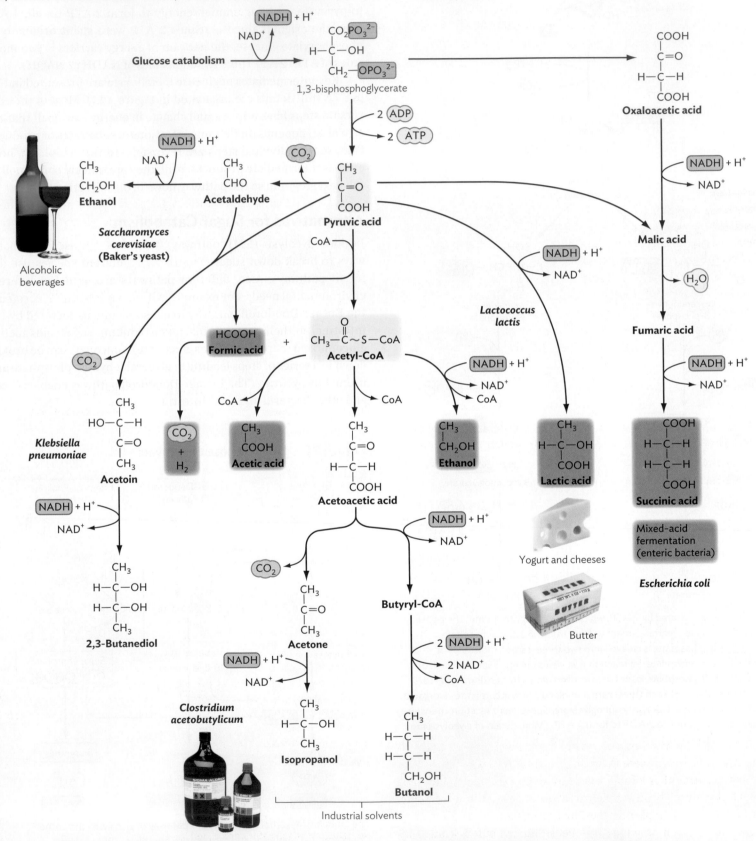

Another pathway, the **pentose phosphate pathway (PPP)**, allows bacteria to divert sugars into biosynthesis. In the pentose phosphate pathway, the early steps of glucose breakdown release one carbon as CO_2, forming a five-carbon sugar, ribulose 5-phosphate. The reaction also leads to formation of two molecules of NADPH, an NADH molecule with an extra phosphoryl group. The extra phosphate "tags" NADPH for use in biosynthesis reactions rather than for the electron transport system (Section 7.6). The NADPH can be used to hydrogenate substrates for building amino acids and cell walls, whereas the ribulose 5-phosphate can be used to construct nucleotide bases. Remaining steps of the pentose phosphate pathway lead back to G3P and pyruvate, but sugar intermediates of various sizes can be shunted off into biosynthesis (see Section 7.6).

Fermentation Completes Catabolism

So far, all our catabolic pathways have ended up with the product pyruvic acid (or its ionized form, pyruvate). Are these pathways complete? Actually, no, because along with pyruvate remain a large number of NADH molecules containing electrons removed from the sugars. For metabolism to be complete, the electrons, as well as the waste carbon from pyruvate, need to be released from the cell. As introduced in Chapter 6, catabolism can be completed in different ways, depending on the availability of oxygen (O_2):

- **Fermentation** occurs when the electrons of NADH are put back onto pyruvate, forming waste products such as lactic acid. No oxygen is consumed.

- **Respiration** occurs when the NADH donates electrons to the electron transport system (ETS), which ultimately transfers them to O_2 or to an alternative inorganic molecule serving as a terminal electron acceptor (discussed in Section 7.4).

Without O_2 or other inorganic electron acceptors, cells must recycle the electrons from NADH back onto the pyruvate in the process of fermentation. Fermentation forms partly oxidized products that have the same redox level (balance of O and H) as the original glucose (**Figure 7.8**). For example, glucose may be fermented to two molecules of lactic acid (lactate fermentation) or two molecules of ethanol plus two molecules of CO_2 (ethanolic fermentation) or to one molecule of lactic acid, one molecule of ethanol, and one molecule of CO_2 (heterolactic fermentation). These fermentation products are then excreted from the cell. This excretion of organic products seems wasteful, since products such as ethanol or lactic acid still contain a lot of food value—that is, potential for catabolism yielding energy. But without oxygen or other electron acceptors, the cell has no choice.

Fermentation has a long history in human civilization. Because fermentation is incomplete digestion, it offers us a way for limited microbial treatment of a food source to make a stable product that retains food value. The microbial "wastes," such as ethanol or acids, retard further growth of microbes, including pathogens; thus, they preserve foods such as wine and cheese for extended storage by

humans. Different species of microbes convert pyruvate to different combinations of products, yielding different properties and flavors; for example, *Propionibacterium* species generate propionic acid, the source of the flavor of Swiss cheese.

Because digestion is incomplete, fermenting microbes must catabolize a large amount of sugar to grow, producing a large amount of waste products. The waste products most commonly made by human-associated pathogens are acids such as lactic acid and acetic acid. The appearance of acids in diagnostic media readily changes the color of pH indicators such as the phenol red dye of phenol red broth (**Figure 7.9A**). Phenol red dye appears red above pH 7.4 but turns orange and then yellow in media acidified by fermentation acids (below pH 6.8). A culture of *Escherichia coli* turns phenol red broth yellow, showing fermentation to acid. In addition, the gas produced indicates production of CO_2 and H_2. A different enteric pathogen, *Alcaligenes faecalis*, ferments sugar poorly but converts peptides in the broth to alkaline amines; this culture turns deep red.

Other kinds of media show a color change indicating a strain's ability to metabolize a specific sugar (**Figure 7.9B**). For example, MacConkey agar (see Section 6.2) containing sorbitol allows *E. coli* K-12 to grow as red colonies (ferments sorbitol), whereas *E. coli* O157:H7 (the cause of hemorrhagic disease) forms white colonies (fails to ferment sorbitol).

Figure 7.9   Clinical Tests Based on Fermentation

Phenol red broth test

Gas

Sorbitol MacConkey agar

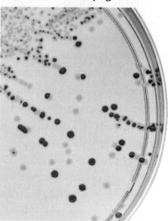

A. Organisms are cultured in a tube of broth containing phenol red. The dye turns yellow when protonated at low pH. Phenol red is orange-red at neutral pH, yellow at lower pH (acid), and red at higher pH (base). A small inverted tube (Durham tube) collects gaseous fermentation products (CO_2 and H_2). *Left to right: Escherichia coli* gives acidic fermentation products (yellow) and gas in a Durham tube; *Alcaligenes faecalis* does not ferment, and the tube turns red without gas; uninoculated control, red.

B. In the sorbitol fermentation test for pathogen *E. coli* O157:H7, white colonies (strain O157:H7) fail to ferment sorbitol, unlike red colonies (nonpathogenic *E. coli*).

Tricarboxylic Acid Cycle: Transferring All Electrons to NADH

The complete breakdown of glucose requires the breakdown of pyruvate to CO_2 and acetyl-CoA. The breakdown of pyruvate also removes two hydrogens, transferring $2e^-$ to form NADH. The NADH then transfers its $2e^-$ to the electron transport system (discussed in Section 7.4) for generating the proton motive force.

Link As is discussed in Section 5.2, energy from a cell's food reactions is stored in a form of potential energy across the membrane, known as the proton motive force.

Pyruvate breakdown is catalyzed by the pyruvate dehydrogenase complex (PDC). PDC is a key component of metabolism in bacteria and in mitochondria because it is the first enzyme to direct sugar catabolism into respiration instead of fermentation or biosynthetic pathways. In human mitochondria, PDC is especially important for organs with a high metabolic rate, such as the heart and brain. Defects in mitochondrial PDC cause heart failure and neurodegeneration.

Acetyl-CoA consists of acetic acid, in which the OH group is replaced (reversibly) with coenzyme A. Coenzyme A acts as a signal for enzymes. Like a "tag" for baggage at an airport, coenzyme A tells enzymes the correct pathway to send the acetyl group. In this case, enzymes must direct the acetyl group into the TCA cycle (**Figure 7.10**) instead of other reactions such as biosynthesis of fatty acids.

Aside from sugars, can bacteria use the TCA cycle to consume other foods that are broken down to acetyl-CoA? We saw in Figure 7.3 that acetyl-CoA can also arise from many nonsugar catabolites, such as fatty acids from lipids, and from complex aromatic derivatives such as lignin and petroleum (see inSight). Whatever its source, acetyl-CoA is directed by its "tag" into the TCA cycle.

In the TCA cycle (see Figure 7.10), the two carbons of acetyl-CoA are ultimately converted to $2CO_2$. One turn of the cycle yields enough energy to form ATP. The eight electrons from acetate's four hydrogen atoms are transferred to 3 NAD^+ (forming 3 NADH) and 1 FAD (an energy carrier similar to NAD, forming $FADH_2$). These carriers transfer the electrons to the electron transport system and ultimately to oxygen. The electron transfer reactions through the ETS produce a much larger number of ATP molecules than does fermentation (see Section 7.4). Thus, glucose respiration yields far greater energy than fermentation, which essentially "gives back" the potential energy of NADH and captures only the energy from 2 ATP.

The TCA cycle includes many intermediate molecules (see Figure 7.10). The acetyl group first gives up its CoA as it condenses with oxaloacetate, a four-carbon dicarboxylate (double acid) left over from a previous turn of the TCA cycle. The two-carbon acetyl plus four-carbon oxaloacetate form citrate, a six-carbon tricarboxylate (triple acid). Citrate, or citric acid, is a common energy storage molecule; it confers part of the flavor of citrus fruits such as oranges and lemons. Citrate loses CO_2 (originally the second

 Figure 7.10   Pyruvate Is Converted to Acetyl-CoA and Feeds into the TCA (Krebs) Cycle

The TCA cycle incorporates acetyl-CoA from the breakdown of pyruvate or other catabolites in the cell. The electrons from acetyl-CoA are used to make NADH and $FADH_2$, while the two carbons are oxidized to CO_2. (Full details of the TCA cycle are presented in Figure A2.5.)

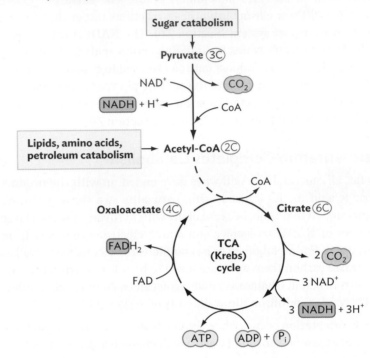

pyruvate carbon) and transfers $2H^+ + 2e^-$ to make NADH via reactions that form 2-oxoglutarate. The 2-oxoglutarate loses CO_2 (the third pyruvate carbon) and transfers $2e^-$ to form NADH. The remaining four-carbon molecule is succinate. Succinate transfers $2e^-$ to FAD, forming $FADH_2$. The subsequent reactions incorporate H_2O and transfer the final $2e^-$ to form NADH. Each $2e^-$ transferred to NAD^+ to form NADH is accompanied by the loss of two hydrogen atoms from pyruvate (and originally glucose). The final carbon skeleton is oxaloacetate, ready to condense with another acetyl group and start a new cycle.

Overall, in each round of the TCA cycle, oxaloacetate condenses with acetate (as acetyl-CoA) to form citrate; the citrate then loses two carbons as CO_2, regenerating oxaloacetate (**Figure 7.11**). The series of reactions store increments of energy in 3 NADH, $FADH_2$, and ATP. Each reaction step couples energy-yielding events (such as CO_2 release) to energy-storing events (such as NADH formation).

Figure 7.11 depicts key intermediates of the TCA cycle. Some of these intermediates serve multiple functions in the cell, such as substrates for biosynthesis. For instance, oxaloacetate can be converted to the amino acid aspartic acid (aspartate). The reaction requires only a single enzyme-catalyzed step, the incorporation of an amine group replacing the carboxylate (double-bonded oxygen atom). Similarly, one amine incorporation converts 2-oxoglutarate to glutamic acid (glutamate). Further details of the TCA cycle, including all intermediate molecules, are shown in Figure A2.5.

Although one full turn of the TCA cycle releases all the remaining carbons from the original pyruvate molecule, the metabolic pathway is still not complete. The NADH and $FADH_2$ produced from the TCA cycle must be recycled if more glucose or other carbohydrate is to be consumed. The energy carriers are recycled by donating their electrons onto a **terminal electron acceptor**. The final electron transfer, and energy capture for ATP, occurs through the ETS and the membrane-embedded ATPase (as described in Section 7.4).

Link As discussed in Section 6.5, once the cell has drained as much energy as possible from an electron, that electron must be passed to a diffusible, final electron acceptor molecule floating in the medium to clear the way for another electron to be passed down the electron transport chain. In aerobic respiration, the **terminal electron acceptor** is oxygen. Anaerobic microbes rely on alternative terminal electron acceptors, such as nitrate.

Suppose a cell were to obtain any of the TCA intermediates from a source other than glucose, instead of incorporating acetyl-CoA to form citrate. Could the molecules be processed by the TCA enzymes to yield energy? They could if the cell had transporters to take up such molecules from the environment. In fact, the ability of bacteria to catabolize TCA intermediates such as citrate is used in diagnostic media, along with sugar fermentations, to identify pathogens, as shown in Case History 7.1.

CASE HISTORY 7.1

Fermentation Products Identify a Meningeal Pathogen

In a Chicago hospital, a male infant, Mark, was born prematurely at 30 weeks (weight 1.7 kg, or 3.75 pounds) to a 15-year-old woman without prenatal care. The infant's physical scores at birth were normal, but within an hour he developed respiratory distress due to prematurity; he required intubation and mechanical ventilation (assisted breathing). A blood culture showed no bacterial growth, but ampicillin and gentamicin were begun in case of infection. By day 5, Mark showed metabolic acidosis (blood pH less

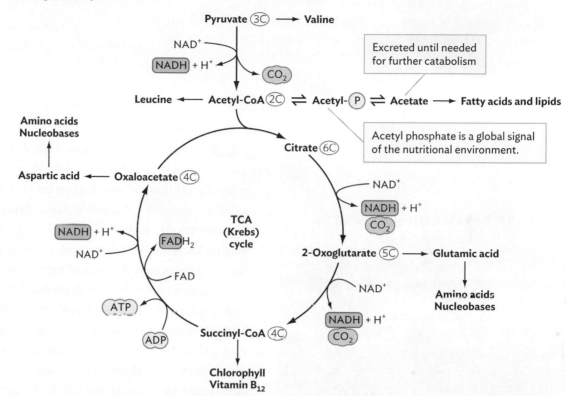
Figure 7.11   The TCA Cycle Intermediates Act as Substrates for Biosynthesis
TCA cycle intermediates are used to build amino acids, fatty acids, and nucleobases. (For other TCA intermediates, see Figure A2.5.)

than 7.35), thrombocytopenia (low platelet count), and cerebrospinal fluid (CSF) containing leukocytes and erythrocytes. The CSF was cultured and grew a Gram-negative enteric rod; the same organism was now isolated from blood. To identify the isolate, the clinician performed diagnostic tests. Simmons citrate agar tested positive (color change from raised pH, due to acid consumption). Phenol red tests for fermentation acid from dulcitol and melibiose were positive; sucrose was negative. Tests for arginine dihydrolase and ornithine decarboxylase were positive, but the organism did not possess lysine decarboxylase. These and other tests confirmed identification of *Citrobacter sedlakii*, a rare cause of septicemia and meningitis (inflammation of the brain lining). Antibiotic susceptibility was tested, and the ampicillin was replaced with cefotaxime. After 10 weeks of antibiotic therapy, Mark made a full recovery.

In Case History 7.1, an infant was born prematurely, a result of the young age of the mother and lack of prenatal care. Premature infants requiring tracheal intubation are susceptible to respiratory pathogens. In this case, the infectious agent was a rare opportunistic pathogen that normally inhabits the intestinal tract without causing problems (**Figure 7.12A**). Because the organism was so rare, its identification required multiple diagnostic tests. Simmons citrate agar tests the ability to respire by using a TCA intermediate (citric acid) as the sole carbon source. As the acidic carbon source is consumed, the pH of the agar rises, causing the color to change from

inSight

Bacteria Remediate Oil Spills

A dramatic case of bacterial catabolism on a massive scale occurred in 2010, when the Gulf of Mexico faced an unprecedented spill of oil from the Deepwater Horizon oil well blowout. The offshore oil platform exploded, releasing millions of barrels of oil into the gulf over 3 months. The leaked oil killed wildlife throughout the gulf, causing major environmental damage (Figure 1A). Workers tried to contain the spill, but ultimately most of the oil was consumed by marine bacteria. Only certain microbes possess the enzymes needed to degrade the complex organic mixture of petroleum.

Petroleum contains a high proportion of long-chain hydrocarbons, which are highly valued for their use as fuel. But unrefined petroleum, such as that released by the Deepwater rig, also contains significant amounts of aromatic molecules such as anthracene (Figure 1B). Such molecules are poisonous and carcinogenic; they contaminate shellfish and other organisms in the food chain. Aromatic molecules take a long time to degrade in the environment, but eventually bacteria and fungi catabolize most of them. Catabolism involves several enzyme pathways that generate the key intermediate catechol. Catechol is then catabolized to acetyl-CoA and succinyl-CoA, both of which can enter the TCA cycle (recall Figure 7.10). Thus, bacteria conducting aromatic catabolism can convert petroleum pollutants such as anthracene to CO_2, removing them from contaminated water. But aromatic catabolism is slower than sugar catabolism, and thus microbial remediation of spilled oil takes many years. And the release of CO_2, however necessary, adds to global warming.

A surprising discovery is that many of the marine bacteria involved in degrading spilled oil are related to pathogens. The Deepwater Horizon spill showed oil degradation by bacteria of the genus *Vibrio*, such as *Vibrio parahaemolyticus* and *Vibrio vulnificus*, which cause serious human infections. Thus, even a bacterium that has negative consequences for human health may have a positive role in our environment.

Figure 1   **Bacteria Catabolize Spilled Oil**

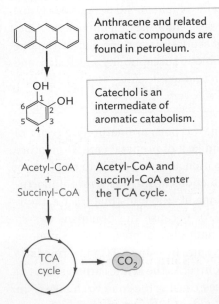

A. Marine bird contaminated by petroleum from an offshore wellhead.

Anthracene and related aromatic compounds are found in petroleum.

Catechol is an intermediate of aromatic catabolism.

Acetyl-CoA and succinyl-CoA enter the TCA cycle.

B. Petroleum contains aromatic components such as anthracene that can be digested by microbes through pathways generating the key intermediate catechol. Catechol is then catabolized to acetyl-CoA and succinyl-CoA, which enter the TCA cycle for conversion to CO_2.

green to blue (Figure 7.12B). The sugars dulcitol and melibiose showed positive reactions (yellow) in phenol red broth, as seen earlier in Figure 7.9A.

Another test, sucrose fermentation, addresses the ability of the organism to break down the disaccharide sucrose into glucose, then ferment to acids (lactic acid and/or acetic acid). In the case history, the result for sucrose fermentation was negative (red color, neutral pH). The decarboxylation of certain amino acids (arginine and ornithine) produced alkaline amines that raised pH for a color change. A similar test for lysine decarboxylase was negative. Combining the results of many such tests revealed a pattern highly consistent with a particular pathogen. Today, as genetic data are accumulated, species can also be identified by DNA tests, as will be discussed in Chapter 8.

Although we focus on sugar catabolism, recall that in nature microbes catabolize many different kinds of organic molecules. Within our intestines, bacteria such as *Bacteroides* species digest complex plant material into acetyl groups and other small molecules that our intestines can absorb. And in soil and water, bacteria consume toxic pollutants such as petroleum spilled by an oil well blowout (**inSight**).

Figure 7.12   Diagnostic Testing of a Respiratory Pathogen

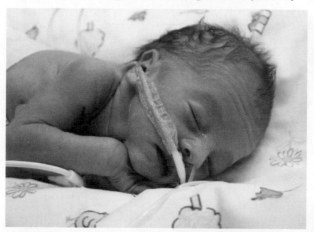

A. Premature infants requiring tracheal intubation are susceptible to respiratory pathogens.

Positive   Negative

B. Simmons citrate test of a Gram-negative pathogen. Growth on citrate (citric acid) as a sole carbon source consumes acid, thereby raising the pH, with a color change to blue.

SECTION SUMMARY

- **Glycolysis (EMP pathway) breaks down glucose to two molecules of pyruvic acid, storing energy in 2 ATP and 2 NADH.** In stage 1, glucose receives two phosphoryl groups from ATP and then is cleaved to two three-carbon sugars. In stage 2, both three-carbon sugars eventually are converted to pyruvate. The process generates 4 ATP; subtracting the 2 ATP spent priming glucose leaves a net yield of 2 ATP.

- **The pentose phosphate pathway generates 2 NADPH for biosynthesis plus a five-carbon sugar** that may be used to build nucleic acids. Further reactions yield other intermediate substrates for biosynthesis of amino acids and other cell components.

- **Fermentation completes catabolism** by donating electrons from NADH back to pyruvate to form waste products that are excreted from the cell.

- **Pyruvate decarboxylase and the TCA cycle complete sugar breakdown to CO_2 and H_2O.** Each acetyl group from pyruvate condenses with oxaloacetate to form citrate, which is successively broken down by enzymes until regenerating oxaloacetate. Energy is stored in the form of NADH, $FADH_2$, and ATP.

- **Sugar fermentation and organic acid respiration are used as tools in diagnostic media to identify pathogens.**

Thought Question 7.1   Why are glucose catabolism pathways ubiquitous, even in bacterial habitats where glucose is scarce? Give several reasons.

Thought Question 7.2   How is sugar catabolism used as a diagnostic indicator? Why are so many different sugars used as convenient indicators?

7.4
Respiration and Lithotrophy

SECTION OBJECTIVES

- Describe how NADH transfers electrons to electron transport proteins of the ETS and ultimately to the terminal electron acceptor, such as O_2.

- Explain how the transfer of electrons from one oxidoreductase to the next, yielding energy, is coupled to the pumping of H^+ across the membrane.

- Explain how lithotrophy yields energy by oxidizing mineral electron donors instead of organic molecules. State an example of how environmental lithotrophy affects human health.

Recall that the overall process of catabolism from substrate breakdown to reduction of a terminal electron acceptor (such as O_2) is called respiration (Figure 7.13). Respiration completes catabolism by donating electrons from NADH to the ETS in a process that stores energy by pumping protons (H^+) across the membrane to generate a gradient of hydrogen ions. The hydrogen ions return to the cell via ATP synthase, driving formation of ATP. The process of ATP formation by respiration is called **oxidative phosphorylation**. The "phosphorylation" part refers to phosphorylation of ADP to ATP by the ATP synthase complex driven by a proton gradient generated through the ETS.

The overall equation for the respiration of glucose with oxygen is

$$C_6H_{12}O_6 + 6H_2O + 6O_2 \rightarrow 12H_2O + 6CO_2$$

Figure 7.13   Respiration of Glucose

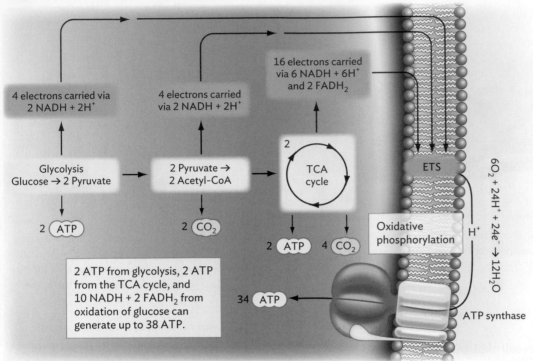

Bacterial glucose catabolism generates ATP through substrate-level phosphorylation and through the ETS pumping H^+ across the cell membrane to drive the ATP synthase. The complete oxidative breakdown of glucose to CO_2 and H_2O could theoretically generate up to 38 ATP. Under actual conditions, the number of ATP is smaller.

Glucose respiration can generate a relatively large number of ATP molecules per molecule of glucose, far more than in fermentation. In bacteria, however, the actual number of ATP molecules generated varies widely with the availability of a carbon source and oxygen. For example, as oxygen concentration decreases, the ability to oxidize NADH decreases, so the cell may make only 1 or 2 ATP per NADH.

Electron Transport System and Proton Motive Force

The redox energy carriers NADH and $FADH_2$ transfer their electrons to a membrane-embedded enzyme (electron transport protein) of the ETS. Each ETS enzyme is called an **oxidoreductase** because it receives electrons from a stronger electron <u>donor</u> (becoming reduced) and then transfers them to a stronger electron <u>acceptor</u> (becoming oxidized). These transfer reactions yield energy. The final oxidoreductase of the chain transfers $2e^-$ onto oxygen (the terminal electron acceptor). Two rounds of electron transfer ($4e^-$ in all) from 2 NADH allow incorporation of four hydrogen ions ($4H^+$) to form two molecules of H_2O:

$$O_2 + 4e^- + 4H^+ \rightarrow 2H_2O$$

The reduction of O_2 to water by all the electrons from glucose hydrogen atoms finally completes the aerobic catabolism of glucose.

Along the way, each redox step of an ETS yields energy. How is this energy stored? Surprisingly, at this point there is no other energy carrier molecule. Instead, some of the ETS oxidoreductases use the energy of electron transfer to pump hydrogen ions (protons) across the bacterial cell membrane. **Figure 7.14** shows an example of an ETS in the membrane of *E. coli*. In this ETS, NADH donates its pair of electrons ($2e^-$) to an oxidoreductase called NADH dehydrogenase (NDH-1). NDH-1 then transfers the $2e^-$ onto a quinone (Q), generating quinol (QH_2). The redox step from NADH through NDH-1 to quinol (QH_2) yields enough energy to pump an extra $4H^+$ across the membrane.

Quinones and quinols are small aromatic molecules soluble in the membrane. Quinols diffuse within the membrane, but do not cross the membrane between cytoplasm and exterior. Each quinol reduced by NDH-1 can transfer its $2e^-$ onto the next ETS complex, cytochrome *bo* oxidoreductase. Cytochrome *bo* oxidoreductase pumps $2H^+$ across the membrane and releases the $2H^+$ received from the quinol. Then, finally, this terminal oxidoreductase puts $2e^-$ onto molecular oxygen, incorporating one oxygen atom into water. Overall, for each NADH oxidized, up to $8H^+$ may be pumped across the membrane.

The ETS generates a gradient of hydrogen ions across the membrane—that is, a larger concentration of H^+ outside than inside. This hydrogen ion gradient, or **proton motive force (PMF)**, then drives the ATP synthase "machine" to phosphorylate ADP to ATP (**Figure 7.15**). As discussed in Section 5.2, the flow of protons (H^+) through the machine and back into the cell is driven both by the charge difference (more positive outside than inside) and by the H^+ concentration (greater outside). This proton potential (the PMF) functions as a current of protons, analogous to an electric current of electrons.

ATP synthase is a remarkable enzyme complex that actually rotates with proton flow, like a "water wheel" powered by a river (**Figure 7.16**). The "wheel" part is the F_o complex, a membrane-embedded circle of 12 c subunits. For every three protons ($3H^+$) that flow through the F_o, the F_1 rotates one-third of a turn and catalyzes ADP conversion to ATP. So, the flow of $9H^+$ generates 3 ATP, and the ATP synthase turns around to its starting position.

How does the mitochondrial ETS compare with that of bacteria? The mitochondrial inner membrane evolved from the bacterial cell membrane, and all its ETS complexes show high similarity with those of bacteria. The mitochondrial ETS includes

Figure 7.14 A Bacterial ETS for Aerobic NADH Oxidation

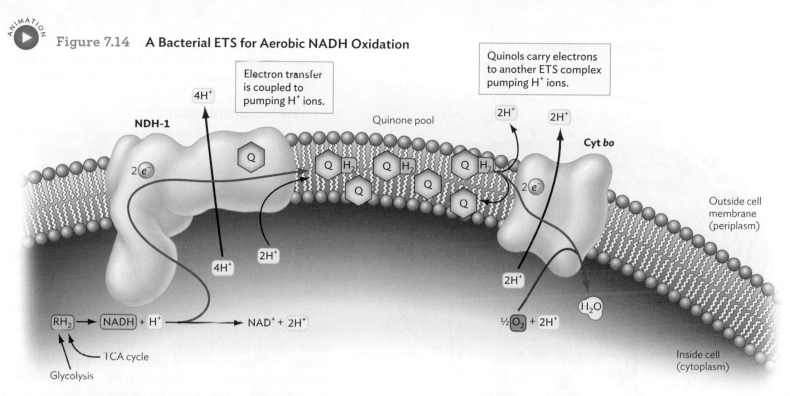

In *E. coli*, a pair of electrons ($2e^-$) from oxidoreductase NDH-1 are transferred onto a quinone (Q), generating quinol (QH_2). This redox step yields enough energy to pump $4H^+$ across the membrane. Each quinol can transfer its $2e^-$ onto the next oxidoreductase (cytochrome *bo*), which pumps $2H^+$ across the membrane in addition to $2H^+$ from the quinol. For each NADH oxidized, up to $8H^+$ may be pumped across the membrane.

Figure 7.15 The ETS Generates the Proton Motive Force (PMF)

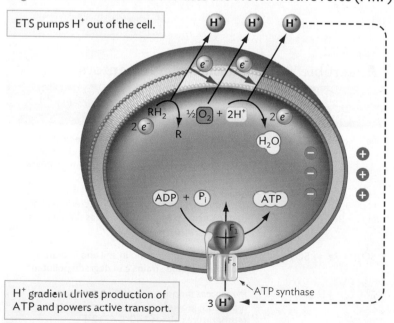

Some of the ETS oxidoreductases pump protons out of the cell. The resulting electrochemical gradient of protons (proton motive force) drives conversion of ADP to ATP through the ATP synthase.

Figure 7.16 Bacterial Membrane–Embedded ATP Synthase (ATP Synthase)

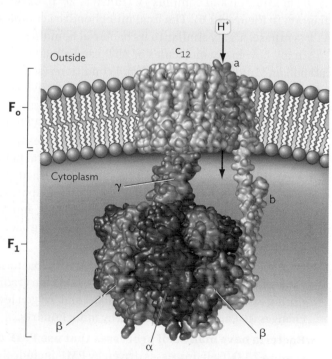

The F_1F_o complex is plugged into the *E. coli* plasma membrane. The three pairs of alpha (α) and beta (β) subunits rotate around the gamma (γ) axle, catalyzing formation and hydrolysis of ATP. The ring of 12 c subunits rotates while translocating three protons.

Figure 7.17   Processes Driven by the PMF

Besides ATP synthesis, the PMF directly drives flagellar rotation, uptake of nutrients, and efflux of toxic drugs.

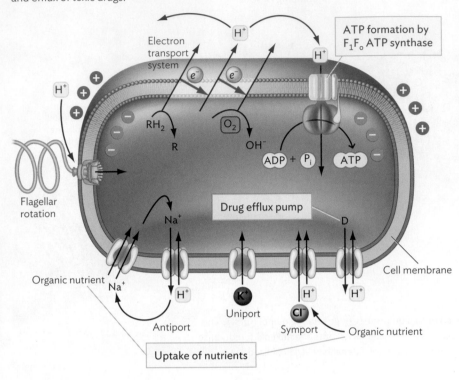

four oxidoreductases: complex I transfers $2e^-$ from NADH to a quinol; complex II receives $2e^-$ from $FADH_2$; complex III receives $2e^-$ from a quinol; and complex IV transfers $2e^-$ to oxygen (see the diagram in Figure A2.6). The final mitochondrial complex V is the ATP synthase, highly similar to its bacterial homolog.

In all, how many molecules of ATP can be made from glucose catabolism? In theory, under optimal conditions (see Figure 7.13), 1 NADH yields enough energy for the PMF to drive synthesis of 3 ATP; and 1 $FADH_2$ yields energy for 2 ATP. Overall, the PMF made from one molecule of glucose drives the membrane ATP synthase to synthesize as many as 34 ATP. Another 4 ATP comes from glucose breakdown via glycolysis and the TCA cycle (38 total per glucose). However, this theoretical total of 38 is never reached. Mitochondria need to spend several molecules of ATP to transport substrates from the cytoplasm, including pyruvate, phosphate, and ADP; their net ATP production is therefore about 30. Free-living bacteria make less ATP than mitochondria for several reasons:

- **Bacteria sacrifice efficiency for flexibility.** Bacteria spend some of their PMF energy to maintain stable ion gradients during extreme changes in external pH and oxygen levels, changes that are not experienced by mitochondria.

- **Bacteria have many cell processes that use PMF directly** (Figure 7.17). Cell processes driven by PMF include flagellar rotation, uptake of nutrients (such as H^+ uniport and symport), and efflux of toxic drugs. For example, bacteria resistant to tetracycline commonly have a membrane protein complex that

pumps tetracycline out of the cell, powered by an inwardly directed flow of H^+. The PMF spent on such processes is unavailable to power ATP synthase.

Anaerobic Respiration

So far, our discussion of electron transport assumes that bacteria "eat" the kind of food we do, consisting of organic carbon, and use the same electron acceptor that we breathe, gaseous O_2. But some bacteria—including pathogens—can use alternative electron acceptors instead of oxygen, such as nitrate (NO^{3-}) or iron (Fe^{3+}) (Table 7.2). For example, *Pseudomonas aeruginosa* is the most common pathogen infecting the lungs of cystic fibrosis patients. In cystic fibrosis, the lung sputum becomes anaerobic, thus requiring *P. aeruginosa* to donate electrons to nitrate. Soil bacteria use even more exotic electron acceptors. For example, *Geobacter metallireducens* transfers electrons to iron, uranium, and other metals (Figure 7.18A). *Geobacter* anaerobic respiration can remediate pollutant metals by converting them to soluble form. These bacteria can actually donate electrons to a metal electrode, generating electricity in a fuel cell.

Note that each electron acceptor (oxidant) must be reduced by a different terminal oxidoreductase. Oxidants Fe^{3+} and NO^{3-} are particularly useful electron acceptors in anaerobic soil and wetlands. In marine environments, especially near the ocean floor, sulfate (SO_4^{2-}) is commonly used as an electron acceptor because its concentration in seawater is high.

Table 7.2
Anaerobic Terminal Electron Acceptors

Terminal Electron Acceptor Reaction	Species and Environment
$Fe^{3+} + e^- \rightarrow Fe^{2+}$	*Geobacter metallireducens* transfers electrons from catabolism to iron, uranium, and other metals; can remediate pollutants; can generate electricity in a fuel cell
$NO^{3-} + 2e^- \rightarrow NO^{2-}$ $N_2O + 2e^- \rightarrow N_2$	*Pseudomonas aeruginosa*, a soil bacterium and opportunistic pathogen; colonizes the lungs of cystic fibrosis patients
$SO_4^{2-} + 2e^- \rightarrow SO_3^{2-}$	*Desulfovibrio* spp., found in soil and in marine sediments; some strains can degrade pollutants
$S^0 + 2e^- \rightarrow H_2S$	*Pyrococcus furiosus*, a hyperthermophilic archaeon; grows in marine thermal vents at temperatures above 100°C (212°F)
$S_4O_6^{2-} + 2e^- \rightarrow 2S_2O_3^{2-}$	*Salmonella enterica*, an enteric pathogen; induces white blood cells to produce tetrathionate ($S_4O_6^{2-}$)

A surprising medical example of anaerobic respiration is that of the enteric Gram-negative pathogen *Salmonella enterica* serotype Typhimurium. *Salmonella* bacteria can respire using tetrathionate ($S_4O_6^{2-}$) as an electron acceptor. Tetrathionate is not normally available in the host mucosa, and most bacteria cannot use it. But *Salmonella* bacteria induce an inflammatory response in white blood cells, which in turn produce reactive oxygen molecules, including tetrathionate. With tetrathionate available as an electron acceptor, *Salmonella* can grow in intestinal lumen, whereas other mucosal bacteria cannot; thus, *Salmonella* has a growth advantage during pathogenesis.

Lithotrophs Eat Rock

Suppose that bacteria have O_2 available but lack organic food such as sugars. Could they transfer electrons to their ETS from inorganic electron donors such as minerals or even hydrogen gas? Yes; this kind of "rock-eating" metabolism is called lithotrophy (see Table 7.1). In lithotrophy, an inorganic reduced molecule such as ferrous ion (Fe^{2+}), ammonium ion (NH_4^+), or hydrogen gas (H_2) donates electrons to the first oxidoreductase of the ETS (Table 7.3). Compared with catabolism of an organic compound, lithotrophy involves fewer intermediate reactions preceding the ETS. That is because an inorganic compound such as Fe^{2+} or H_2 can donate electrons directly to the ETS, without requiring catabolic breakdown of a complex molecule.

Lithotrophy involving metals such as iron is important for mining, where reduced metals from underground are uncovered. Exposed to oxygen, the reduced metals now feed lithotrophic bacteria and archaea that can break down the rock. This rock breakdown can lead to "leaching" of metals into solution, where they can be recovered by miners. Another form of lithotrophy involves oxidation of ammonia to nitrite and nitrate. An example of an ammonia-oxidizing bacterium in the soil is *Nitrosomonas* (Figure 7.18B). Ammonia oxidizers thrive in soil that is artificially fertilized with ammonia-rich material.

In our environment, bacterial lithotrophy and anaerobic respiration make many positive contributions. But certain kinds of bacterial metabolism can have unfortunate medical consequences, as shown in Case History 7.2.

CASE HISTORY 7.2
Nitrite Poisons an Infant

Brianna was a 2-month-old, formula-fed infant in a rural community in Nebraska. Brianna's mother noticed signs of blueness around the baby's mouth and fingers. One night after feeding, Brianna had trouble breathing, and her face had a peculiar lavender color. The infant became lethargic, salivated excessively, and had diarrhea and vomiting. An EMT came and took blood samples, which appeared brown and failed to turn pink when exposed to air. The EMT diagnosed methemoglobinemia, a condition in which toxic levels of nitrite oxidize hemoglobin and prevent it from receiving oxygen. The nitrite-oxidized hemoglobin is called methemoglobin; elevated levels can cause asphyxiation and death. A solution of methylene blue, a reducing agent, was administered by IV to reduce (add electrons to) the hemoglobin in Brianna's blood.

Nitrite forms in the digestive tract by bacterial reduction of nitrate during anaerobic respiration. The well water from which Brianna's formula was prepared was tested and found to contain high levels of nitrate, enough to cause methemoglobinemia. Brianna's parents were surprised because they had felt no ill effect from the water. But an infant's stomach pH is high, allowing bacteria to reduce nitrate to nitrite, which may then oxidize hemoglobin.

Figure 7.18 Anaerobic Respirers and Aerobic Lithotrophs

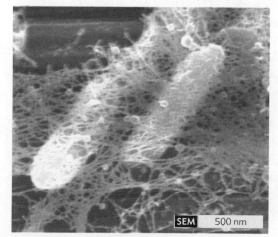

A. *Geobacter* conducts anaerobic respiration. A biofilm of *Geobacter metallireducens* transfers electrons to a graphite electrode.

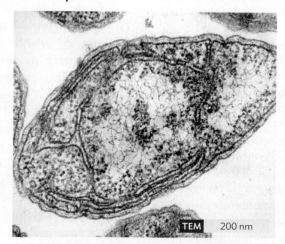

B. *Nitrosomonas* species oxidize ammonia to nitrite.

In Case History 7.2, the nitrite in the baby's blood was formed by her intestinal bacteria using nitrate (NO_3^-) as a terminal electron acceptor for anaerobic respiration, with nitrite (NO_2^-) as the product. But where did so much nitrate come from in the well water? Nebraska and other farm states have a problem with the runoff of fertilizer spread on fields, from which ammonia drains into the water supply. Much of the ammonia gets oxidized to nitrate by lithotrophic bacteria, which oxidize ammonia using O_2 (discussed earlier). The nitrate can permanently contaminate the drinking water, as in the case of the well water used for Brianna's formula.

Table 7.3
Lithotrophic Reactions

Lithotrophic Reaction	Species and Environment
$4Fe^{2+} + O_2 + 4H^+ \rightarrow 4Fe^{3+} + 2H_2O$	*Gallionella ferruginea*, a bacterium; oxidizes reduced iron in mine drainage and in water pipes; causes brown deposits in home plumbing
$FeS_2 + 14Fe^{3+} + 8H_2O \rightarrow 15Fe^{2+} + 2SO_4^{2-} + 16H^+$	*Ferroplasma acidarmanus*, an archaeon; oxidizes pyrite (iron-sulfur ore) in iron mines, generating strong acid
$O_2 + 2H_2 \rightarrow 2H_2O$	*Hydrogenomonas eutropha*, a soil bacterium; grows by oxidizing H_2 or an organic substrate
$2NH_3 + 3O_2 \rightarrow 2NO_2^- + 2H^+ + 2H_2O$	*Nitrosomonas europaea*, a bacterium; oxidizes ammonia to nitrite, causing nitrite pollution in lake water contaminated by ammonia-rich fertilizer
$CO_2 + 4H_2 \rightarrow 2H_2O + CH_4$	*Methanobrevibacter smithii*, an archaeon; grows in the human colon, in association with mixed-acid fermenting bacteria

A very different form of lithotrophy associated with normal human microbiota is **methanogenesis**, an anaerobic reaction of CO_2 and H_2 to form methane (see Table 7.3). Methanogenesis is performed solely by methanogenic archaea, or methanogens. Species of methanogens are highly diverse, ranging from deep-sea vent thermophiles to Antarctic psychrophiles. They are highly active in the anaerobic soil of wetlands and rice paddies. Surprisingly, some methanogens grow in the gut of humans and other animals, especially cattle. *Methanobrevibacter smithii* grows in the human colon by using the CO_2 and H_2 released by enteric bacteria through mixed-acid fermentation. The methanogens produce methane, which does not harm the host but does act as a greenhouse gas in global warming. Some methanogens also are isolated from dental plaque, but their contribution to disease is unknown.

SECTION SUMMARY

- **An electron transport system (ETS) transfers electrons from one membrane-embedded oxidoreductase to the next.** The final oxidoreductase transfers the electron to the terminal electron acceptor. Some of the oxidoreductases pump H^+ across the membrane, generating proton motive force.

- **Respiration is catabolism with transfer of electrons through an ETS,** whose terminal electron acceptor is O_2 or another inorganic oxidant. Anaerobic respiration involves terminal electron acceptors other than O_2.

- **Oxidative phosphorylation is the use of the proton motive force to drive the ATP synthase.** Three protons crossing through the ATP synthase power synthesis of one molecule of ATP.

- **Lithotrophy is the oxidation of an inorganic electron donor through an ETS.** Electron donors may be reduced metals, H_2, or NH_3.

Thought Question 7.3 The opportunistic pathogen *Pseudomonas aeruginosa* colonizes wounds and the lungs of cystic fibrosis patients. *P. aeruginosa* also grows in soil. It can respire aerobically or else anaerobically by using nitrate. In which habitat (the lung or the soil) would *P. aeruginosa* respire aerobically or with nitrate? What part of its ETS would need to change?

7.5
Photosynthesis and Carbon Fixation

SECTION OBJECTIVES

- Describe how photosynthesis harvests light to generate ATP and NADPH, while releasing O_2.
- Describe how a photosynthetic microbe fixes CO_2 into biomass.
- Explain the importance of photosynthesis for human health.

How long can you hold your breath? All the oxygen we breathe comes from **photosynthesis**—much of it by microbes. Photosynthesis fundamentally supports all the ecosystems of our biosphere. Every year, photosynthesis converts more than 10% of atmospheric carbon dioxide to biomass, most of which then feeds microbial and animal heterotrophs. As introduced in Chapter 6, most of Earth's carbon fixation, especially in the oceans, comes from phototrophic microbes such as cyanobacteria (**Figure 7.19**).

All organisms use electron transfer (redox reactions) to yield energy for growth and function. But the process must start with a high-energy molecule. A phototrophic organism obtains energy from light to excite (add energy to) electrons of molecules such as chlorophyll. The energy gathered from photons exciting chlorophyll is transferred to an ETS, remarkably similar to the ETS used to transfer electrons from organic food or from minerals. The process of light absorption by chlorophyll and energy transfer to build biomass is called photosynthesis. Some microbes, such as cyanobacteria, gain all their energy from photosynthesis; they are "photoautotrophs," existing entirely on water and inorganic ions. Other microbes only supplement their energy gained by catabolism with light absorption; they are "photoheterotrophs."

Some bacterial photosynthesis can use the energy of light absorption to split electrons from a sulfur compound such as hydrogen sulfide (H_2S). But the most advanced form of photosynthesis,

Figure 7.19 Cyanobacteria Conduct Photosynthesis

Green colonies of the cyanobacterium *Merismopedia*.

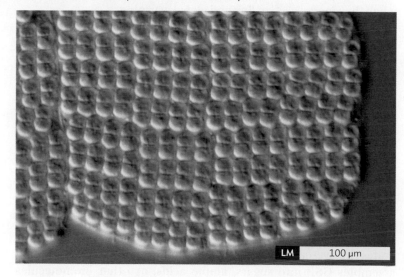

LM 100 µm

oxygenic photosynthesis, splits a water molecule (H_2O) (Figure 7.20A). The **photolysis** of water releases oxygen atoms, which form oxygen gas:

$$2H_2O \text{ (light absorbed)} \rightarrow O_2 + 4H^+ + 4e^-$$

Water-splitting photosynthesis is performed only by cyanobacteria and by plants and algae, whose chloroplasts evolved from an ancient cyanobacterium. The water-splitting reaction of oxygenic photosynthesis consumes large amounts of energy—a highly positive ΔG (energy change). But the energy consumed from light is then stored in biomass: four electrons are used to form NADPH, which then donates the electrons to fix CO_2. And besides fixing carbon into organic biomolecules, water-based photosynthesis produces oxygen gas, O_2.

Oxygen gas is a highly reactive substance that threatens to destroy photosynthetic cells unless it is released or detoxified by antioxidant proteins. The oxygen released by cyanobacteria and by plants is used by heterotrophic organisms for aerobic respiration. Virtually all the oxygen we breathe—all the O_2 available on our planet—was released as a "toxic waste" from photosynthetic bacteria and from algae and plant chloroplasts that evolved from bacteria.

Photolysis: Light Absorption and Electron Transfer

What is the mechanism of photolysis? Scientists eagerly seek to understand photolysis, looking for clues to build better light-harvesting devices, mechanisms that harvest light energy from splitting electrons from a molecule such as water (Figure 7.20B). Oxygenic photolysis involves chlorophyll photoexcitation in a reaction center (**photosystem II**), electron transfer through an ETS, a second photoexcitation and reaction center (**photosystem I**), and

generation of energy carriers ATP and NADPH. The overall reaction, absorption of 4 photons splitting $2H_2O$ to form O_2, drives formation of 3 ATP and 2 NADPH:

$$2H_2O + 2NADP^+ + 3[ADP + P_i] \rightarrow$$
$$O_2 + 2[NADPH + H^+] + 3ATP + 3H_2O$$

The stages of photolysis include:

- **Chlorophyll photoexcitation.** Light is absorbed by pigments called **chlorophylls**. Because there is no way to "reach out" to grab light rays, the cell needs to pack as many chlorophyll

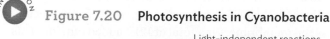

Figure 7.20 Photosynthesis in Cyanobacteria

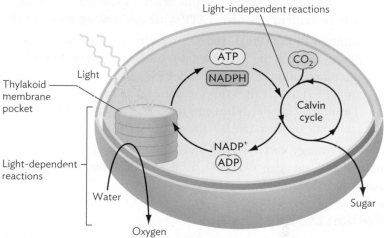

A. Light is absorbed by chlorophyll molecules in thylakoid membrane pockets. The absorbed energy splits water, releasing oxygen that forms O_2 gas. Light energy converts ADP to ATP, and the electrons from water reduce NAD^+ to NADPH. The ATP and NADPH provide energy and reducing power for the Calvin cycle to fix CO_2. The CO_2 is reduced and condensed to form sugars ($C_6H_{12}O_6$), amino acids, and other biomass for the cell.

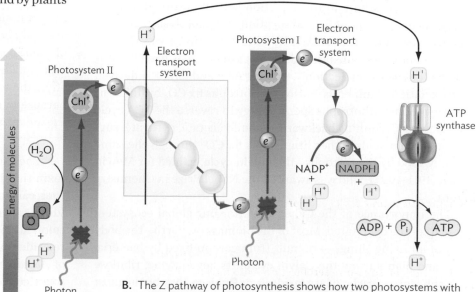

B. The Z pathway of photosynthesis shows how two photosystems with two ETS complexes obtain energy from light to make ATP and NADPH.

molecules as possible into its surface area in the hope of "capturing" photons. Most cyanobacteria have folded membranes called lamellae that multiply the surface area for light capture. The lamellae are packed with antenna complexes of proteins and chlorophyll that (1) absorb light and (2) transfer the absorption energy to the reaction center of the ETS.

- **Photosystem II (PS II).** In oxygenic photosynthesis, the photon energy absorbed by PS II is used to split $4H^+$ plus $4e^-$ from $2H_2O$. Each H^+ is released outside the cell, increasing the proton motive force (H^+ concentration and charge difference across the membrane). Each electron from H_2O reduces a quinol and enters an ETS in the bacterial cell membrane. Energy yield through the ETS is converted to pump $2H^+$ per electron across the membrane, for a total of $12H^+$ added outside. This proton motive force drives ATP synthase, making about 3 ATP.

- **Photosystem I (PS I).** The electrons from water that released energy through the ETS now get another energy boost from photoexcitation of chlorophyll in PS I. This transfer of electrons from PS II to PS I gives this mechanism the name "Z pathway." The newly energized electrons now enter another ETS, which transfers $4e^-$ onto 2 $NADP^+$ to form 2 NADPH + $2H^+$.

The energy carriers ATP and NADPH are now available to reduce and energize CO_2 (CO_2 fixation), building sugars, amino acids, and other biomass for the cell.

Carbon Dioxide Fixation

The incorporation of CO_2 into glucose and other metabolites such as amino acids is called **carbon fixation** or **carbon dioxide (CO_2) fixation**. The fixation of CO_2 completes oxygenic photosynthesis, as shown in Figure 7.20. The overall reaction to make glucose is

$$6CO_2 + 12\,[NADPH + H^+] + 18\,ATP + 18H_2O \rightarrow$$
$$C_6H_{12}O_6 + 6H_2O + 12\,NADP^+ + 18\,[ADP + P_i]$$

Note that nowhere in this equation is light absorbed. In fact, CO_2 can be fixed without light absorption and photolysis if another source of energy (ATP and NADPH) is available; that is why CO_2 fixation is sometimes called the "dark reaction." CO_2 is fixed by lithotrophs (described in Section 7.4) using ATP and NADPH derived from oxidation of minerals. Various lithotrophs fix CO_2 by different pathways. Some lithotrophs spend energy to reverse the TCA cycle and glycolysis, "running backward" to build acetate, pyruvate, sugars, and amino acids. Other lithotrophs fix CO_2 through the same pathway used by phototrophs—the **Calvin cycle**, named for Melvin Calvin (1911–1997), who was awarded the Nobel Prize in Chemistry in 1961.

The importance of the Calvin cycle for our global ecosystem can scarcely be overstated. Most of the biomass on Earth—the body parts of all living things—are built from carbon fixed by bacteria, algae, and plants using the Calvin cycle. Its key enzyme, ribulose 1,5-bisphosphate carboxylase/oxidase, or **Rubisco**, may be the most

abundant enzyme on Earth. The Calvin cycle plays the leading role in removing atmospheric CO_2; it is what climate scientists estimate when they model carbon flux and global warming.

Like the TCA cycle, the Calvin cycle starts with a key intermediate molecule to "recycle," in this case the five-carbon sugar ribulose 1,5-bisphosphate. The five-carbon sugar condenses with CO_2 and H_2O, catalyzed by the enzyme Rubisco. Because this sugar—like the six-carbon sugar in glycolysis—has two phosphoryl "tags," the newly formed six-carbon sugar can now split into two three-carbon molecules—each tagged with a phosphoryl group and each equivalent to the other in the rest of the cycle. The three-carbon molecules now become reduced to the sugar glyceraldehyde 3-phosphate (G3P), with energy and electrons from ATP and NADPH.

In six turns of the "cycle," six molecules each of CO_2 and H_2O are fixed and reduced by 6 NADPH (with energy from 6 ATP) to form 6 G3P. A complex series of sugar exchange reactions regenerates the five-carbon intermediate ribulose 1,5-bisphosphate, and assembles GSP into sugars, amino acids, or other biomolecules needed by the cell.

Photosynthetic microbes power our ecosystems, and they are rarely pathogens. Some marine algae, however, can generate toxins that cause human disease.

CASE HISTORY 7.3

Food Poisoned by Red Tide

Tonya, age 32, was spending a warm June weekend at a beach in Massachusetts. The water near shore looked clear, although farther offshore the water had a reddish tint. Tonya dug up a pailful of clams and cooked them thoroughly. After dinner, her lips grew numb and her fingers tingled. She felt nauseated and had difficulty walking across the room. Her speech became slurred, but she managed to dial 911. The ambulance got her to the hospital in time for Tonya to have her stomach pumped and receive assistance breathing. The physician explained to Tonya that she had contracted paralytic shellfish poisoning from consumption of clams contaminated by an algal toxin. The toxin, called saxitoxin, accumulates in the clams as they feed on the algae; cooking does not affect the toxin. At that time, Massachusetts had a ban on shellfish because of a high population density of saxitoxin-producing algae, *Alexandrium tamarense*, a cause of "red tide."

The cause of Tonya's poisoning was saxitoxin, a potent neurotoxin that inhibits the sodium channels of human and avian neurons, causing paralysis. Saxitoxin is produced by dinoflagellate algae (discussed in Chapter 11), which are photoheterotrophs whose red color causes the appearance of a "red tide" (**Figure 7.21A**). Not all red tides cause poisoning, but off the New England coast, blooms of *Alexandrium tamarense* (**Figure 7.21B**) produce the toxin. The toxin concentrates in durable cysts of the dinoflagellate, which

Figure 7.21   Red Tide Caused by Dinoflagellates

SEM | 5 μm

B. *Alexandrium* dinoflagellate that causes toxic red tide.

A. "Red tide," a bloom of dinoflagellate algae.

are consumed by filter feeders such as clams and other shellfish. The toxin concentrates in the clam's tissues but does not harm the clam—only the consumer of the clam's flesh.

Dinoflagellates are algae with an intriguing lifestyle that combines heterotrophy (flagellar motility and predation) with photosynthesis. They evolved from heterotrophic protozoa that engulfed green algae, whose own chloroplasts had evolved by a similar engulfing of a cyanobacterium. Such events of "endosymbiosis" contribute widely to the evolution of different eukaryotic microbes (discussed in Chapter 11). The oxygenic photosynthesis of these mixed metabolism microbes (called **mixotrophs**) contributes greatly to the O_2 production and CO_2 consumption of marine ecosystems.

SECTION SUMMARY

- **Photosynthesis is the absorption of light to excite an electron** and transfer the absorbed energy to an ETS. The energy is converted to make NADPH and ATP, which then reduce CO_2 to build biomass.
- **Photolysis is the use of energy absorbed from light to split a molecule** such as H_2O or H_2S.
- **CO_2 fixation is the reduction and incorporation of inorganic carbon** into the carbon skeleton of a biological molecule. Energy and reducing power come from ATP and NADPH, which are produced either by photolysis or by lithotrophy.

7.6
Biosynthesis and Nitrogen Fixation

SECTION OBJECTIVES
- Describe how intermediates of metabolism are reduced and assembled into biomolecules.
- Explain how bacteria and archaea fix nitrogen gas into ammonium ion and incorporate nitrogen into biomolecules.

How do microbes use their energy to grow? Cells connect their energy-yielding metabolism to reactions that use the energy. Building cell parts is known as **biosynthesis**. Biosynthesis generates new cells as well as secreted products such as toxins and antibiotics. Engineered microbes produce most of the antibiotics we use in medicine.

Biosynthesis is also called **anabolism**, the reverse of catabolism, which degrades complex molecules. For biosynthesis, some catabolic reactions, such as those of the TCA cycle and glycolysis, can run in reverse (with energy input) to build sugars and polysaccharides. Other anabolic reactions, such as those building amino acids, are distinct from those of catabolism.

What other kinds of biosynthesis do we find in pathogens? Pathogens, such as *Mycobacterium tuberculosis*, have many extra genes to make virulence factors, such as unusual envelope lipids that defeat host defenses. Biosynthetic pathways offer targets for antibiotics; for example, the tuberculosis antibiotic isoniazid interrupts the biosynthesis of mycolic acids in the envelope of *M. tuberculosis*. Other pathogens, such as intracellular pathogen *Chlamydia trachomatis*, the cause of sexually transmitted infections, have lost many genes for biosynthesis and thus depend on the host for many fundamental substrates.

Requirements for Biosynthesis

To build new parts, a cell must do the following:

- **Assemble carbon skeletons.** As discussed in Chapter 6, microbes obtain carbon either by fixing CO_2 (autotrophy) or by consuming organic molecules made by other organisms (heterotrophy). Autotrophs may reverse catabolic pathways such as the TCA cycle and glycolysis, whereas heterotrophs assemble products of catabolism such as acetyl-CoA and TCA cycle intermediates (Figure 7.22). Glycolysis forms glycerol and acetyl-CoA, which are assembled into fatty acids. Both glycolysis and the TCA cycle provide precursors for amino acids.

- **Fix nitrogen, sulfur, and phosphorus.** Besides carbohydrate, biosynthesis requires nitrogen, sulfur, phosphorus, and other essential elements. The availability of such elements often limits growth.

- **Spend energy.** Building complex structures requires energy. Biosynthetic enzymes spend energy by coupling

Figure 7.22   Biosynthesis: An Overview

Microbes obtain carbon skeletons through CO_2 fixation or through catabolism of compounds formed by other microbes. Nitrogen is obtained by N_2 fixation or by assimilation of organic amines. Biosynthesis spends energy from ATP and reduction energy from NADPH.

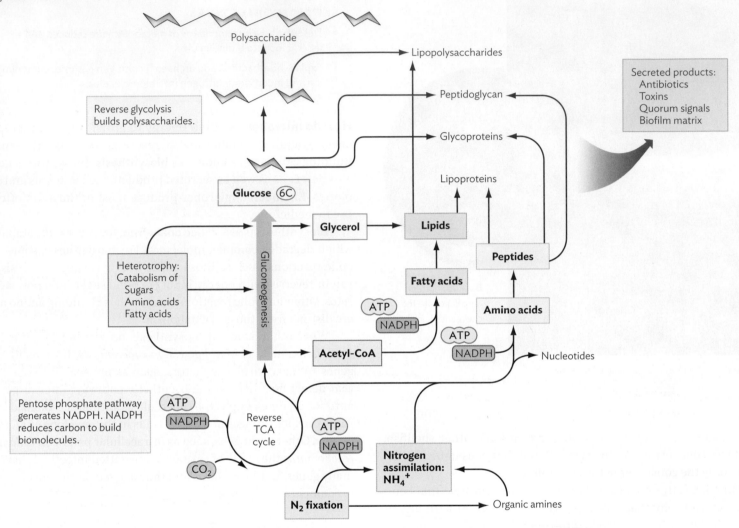

their reactions to the hydrolysis of ATP and the oxidation of NADPH. NADPH is produced from catabolism via the pentose phosphate pathway (Section 7.3).

- **Reduce carbon skeletons (add electrons).** Cell parts such as lipids and amino acids are more reduced than the substrates available, such as sugars. Thus, the biosynthesis of lipids and amino acids includes reduction of sugar-derived carbon skeletons by a reducing agent such as NADPH.

Nitrogen Fixation and Ammonia Incorporation

An important limit for biosynthesis is the incorporation of nitrogen. Many biological molecules, such as DNA and amino acids, contain nitrogen as well as carbon. This hunger for nitrogen has substantial impacts on human society and environmental quality. Human

agriculture is so intensive that most crops require the spreading of nitrogenous fertilizers. The production of such fertilizers incurs great costs, including the consumption of petroleum to reduce atmospheric nitrogen. And fertilizer runoff pollutes water supplies with ammonia and nitrate, as shown in Case History 7.2.

But a few crops, such as soybeans and other legumes, can grow without fertilizer because their nitrogen is fixed by symbiotic bacteria. These bacterial symbionts, called **rhizobia**, convert atmospheric nitrogen to ammonium ion that gets assimilated into plant proteins and nucleic acids. Exciting research focuses on the expansion of the rhizobial host range so that someday they may grow in symbiosis with other crop plants, such as wheat and corn.

Nitrogen gas (N_2) constitutes 78% of Earth's atmosphere; yet this form of nitrogen is surprisingly unavailable to organisms, except for certain species of bacteria and archaea. Thus, all life on

Earth depends on bacterial and archaeal nitrogen fixation. The process of nitrogen fixation requires a high energy input, because N_2 has a highly stable triple bond and because the process requires a large amount of reduction:

$$N_2 + 8H^+ + 8e^- + 16\,ATP \rightarrow 2NH_3 + H_2 + 16\,ADP + 16\,P_i$$

The reaction needs a total of eight electrons from 4 NADPH or from some other reducing factor, such as the protein ferredoxin. Of the eight electrons, six are needed to hydrogenate the nitrogens to ammonia, and two are needed to form H_2 in order to prime the reaction. The reaction occurs in four rounds of hydrogenation by the enzyme hydrogenase. Because each NADPH carries the energy equivalent of 3 ATP, the total amount of energy needed is $16 + (4 \times 3) = 28$ ATP equivalents.

A problem with nitrogen fixation is that its final product, ammonia, is toxic to cells. Thus, although cells need to incorporate ammonia (or the ionized form, ammonium ion, NH_4^+), they need to keep the concentration of free ammonia very low. Cells do so by combining ammonia immediately with one of the TCA cycle intermediates to form an amino acid. Most often, the TCA intermediate 2-oxoglutarate incorporates ammonium ion to form glutamic acid (Figure 7.23). In this reaction, NADPH donates two electrons to replace the ketone oxygen (R=O) with an amine group from ammonium ion. Once glutamate is formed, this amino acid can be maintained safely at a much higher concentration than ammonia. Furthermore, glutamate can transfer its amine group to other molecules to form other amino acids, a process known as "transamination." Thus, glutamate serves as a safe storage form for nitrogen.

Figure 7.23   Assimilation of NH_4^+ into Glutamate

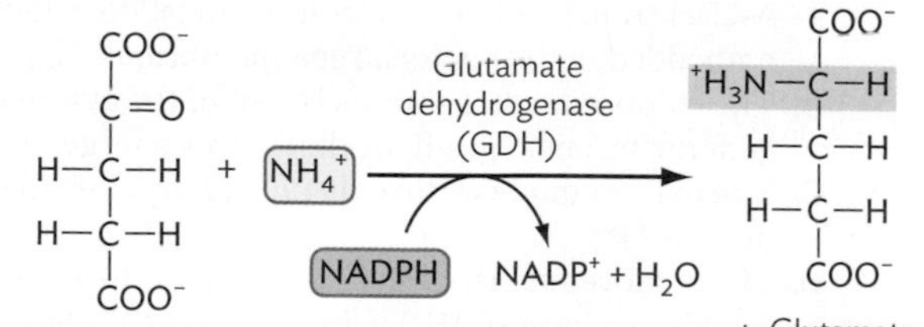

The key TCA cycle intermediate 2-oxoglutarate (alpha-ketoglutarate) incorporates one molecule of NH_4^+ at the ketone (C=O) to form glutamate.

SECTION SUMMARY

- **Biosynthesis, or anabolism, requires assembly of carbon skeletons** from intermediates of metabolism. All carbon skeletons derive from CO_2 fixed by autotrophs.
- **Biosynthesis requires energy from ATP and reduction by energy carriers such as NADPH.**
- **Nitrogen gas is fixed into ammonia,** which is assimilated into amino acids by condensation with a TCA intermediate.

Thought Question 7.4   Disease-causing bacteria vary widely in their ability to synthesize amino acids. What classes of pathogens are likely to make their own amino acids, and what kinds are not?

Thought Question 7.5   *Mycoplasma genitalium*, a cell wall–less bacterium that grows in human serum, lacks the ability to synthesize fatty acids. How do you think *M. genitalium* makes its cell membrane?

Perspective

In this chapter, we have seen how the energy-yielding reactions introduced in Chapter 4 enable living cells to build order out of disorder. Living organisms acquire energy by catalyzing reactions that form energy carriers and then spend the energy from these carriers to build cells. Within communities of organisms, some microbes absorb light energy for photosynthesis, whereas others consume biomass via catabolism. Much of human civilization is built on harnessing microbial catabolism for waste treatment and food production. In the diagnostic laboratory, clinicians use microbial metabolism for indicators to identify pathogens.

All forms of metabolism involve chemical exchange of electrons through oxidation and reduction. The transfer of electrons is fundamental to respiration, the oxidative catabolism of organic molecules to CO_2 and water. It also underlies the autotrophic means of building biomass: the gain of energy from photosynthesis and from lithotrophy. In living cells, the pathways of energy yield and of biosynthesis occur together, sharing common substrates and products. For example, acetyl-CoA produced by glycolysis is then used by pathways that synthesize fatty acids, amino acids, carbohydrates, and other biomolecules. Together, all of this energy yield and biosynthesis enables cells to grow and reproduce.

LEARNING OUTCOMES AND ASSESSMENT

7.1
Energy for Life

- Define the major classes of energy-yielding metabolism, including photosynthesis, organotrophy, and lithotrophy.
- Explain the importance of energy and entropy for living cells.
- State the role of enzymes in controlling metabolic biochemistry.

1. The class of metabolism most useful for food production because it yields important microbial "waste products" is _____.
   A. respiration
   B. fermentation
   C. lithotrophy
   D. photosynthesis

7.2
Catabolism: The Microbial Buffet

- Describe how catabolism converts many complex food molecules to a few kinds of catabolites.
- Explain how catabolism yields energy and how the energy is stored for use.
- Describe how the energy carriers ATP and NADH transfer energy between energy-yielding and energy-spending reactions.

4. Which of these food substrates must first be broken down to hexose sugars in order to enter glycolysis?
   A. Cellulose
   B. Lignin
   C. Lipids
   D. Proteins

7.3
Glucose Catabolism, Fermentation, and the TCA Cycle

- Describe how sugar is catabolized to pyruvate via glycolysis, and explain how these catabolic reactions generate ATP and NADH.
- Explain how pyruvate is further catabolized by fermentation or by the TCA cycle (when oxygen is available).
- Explain how bacterial catabolic reactions are used in clinical tests to identify a pathogen.

7. Which of the following statements does not describe glycolysis?
   A. Glycolysis spends 2 ATP to phosphorylate glucose during the breakdown to pyruvate, with reactions forming 4 ATP.
   B. The pathway of glucose breakdown to pyruvate is incomplete because it generates two NADH molecules that need to transfer electrons to pyruvate or to a terminal electron acceptor.

7.4
Respiration and Lithotrophy

- Describe how NADH transfers electrons to electron transport proteins of the ETS and ultimately to the terminal electron acceptor, such as O_2.
- Explain how the transfer of electrons from one oxidoreductase to the next, yielding energy, is coupled to the pumping of H^+ across the membrane.
- Explain how lithotrophy yields energy by oxidizing mineral electron donors instead of organic molecules. State an example of how environmental lithotrophy affects human health.

9. Which of the following statements does not describe an electron transport system (ETS)?
   A. NADH transfers two electrons to an oxidoreductase embedded in the bacterial cell membrane.
   B. The oxidoreductase pumps H^+ out of the cytoplasm, generating an H^+ gradient that stores energy.
   C. The oxidoreductase directly phosphorylates ADP to make ATP.
   D. The oxidoreductase transfers electrons to a quinol (Q), forming quinone (QH_2), which carries the electrons to a terminal oxidoreductase.

7.5
Photosynthesis and Carbon Fixation

- Describe how photosynthesis harvests light to generate ATP and NADPH, while releasing O_2.
- Describe how a photosynthetic microbe fixes CO_2 into biomass.
- Explain the importance of photosynthesis for human health.

12. Which major class of bacteria conducts oxygenic photosynthesis, releasing much of the oxygen we breathe?
   A. Cyanobacteria
   B. Spirochetes
   C. Proteobacteria
   D. Bacteroidales

7.6
Biosynthesis and Nitrogen Fixation

- Describe how intermediates of metabolism are reduced and assembled into biomolecules.
- Explain how bacteria and archaea fix nitrogen gas into ammonium ion and incorporate nitrogen into biomolecules.

15. Bacteria synthesize amino acids by obtaining carbon skeletons from _____.
   A. glycolysis
   B. the pentose phosphate pathway
   C. the TCA cycle
   D. all of the above

2. These compounds all require energy to build, except _____.
   A. DNA.
   B. peptidoglycan.
   C. fatty acid.
   D. carbon dioxide.

3. This chemical reaction does <u>not</u> require an enzyme:
   A. combustion.
   B. ATP formation.
   C. photosynthesis.
   D. sugar catabolism.

5. Which of these statements is true of NADH?
   A. NADH is a common food substrate for catabolism.
   B. NADH is an electron acceptor for catabolite electrons.
   C. NADH is an electron donor formed during catabolism.
   D. NADH is a waste product of fermentation.

6. Which catabolic pathway generates NADPH (but no NADH) for biosynthesis?
   A. Embden Meyerhof Parnas pathway (glycolysis)
   B. Entner-Doudoroff pathway
   C. Pentose phosphate pathway
   D. TCA cycle

   C. Glycolysis has a cycle of intermediates that incorporate a carbon skeleton and regenerate the original intermediate.
   D. Glycolysis occurs with an increase of entropy because one large molecule is broken down to two molecules of pyruvate.

8. Which molecule from glucose catabolism directly enters the TCA cycle?
   A. Acetyl-CoA
   B. Fructose 1,6-bisphosphate
   C. Glyceraldehyde 3-phosphate
   D. Glucose

10. Which of the following environmental reactions is <u>not</u> an example of lithotrophy?
    A. In a landfill, methanogens obtain energy by combining hydrogen gas plus CO_2 (products of bacterial fermentation) to form methane.
    B. *Geobacter* catabolizes small organic molecules and donates electrons to an electrode in a fuel cell.
    C. In soil spread with nitrogen-rich fertilizer, bacteria use O_2 to oxidize ammonia, forming nitrite (NO_2^-).
    D. In Yellowstone National Park, archaea use O_2 to oxidize hydrogen sulfide (H_2S), forming sulfuric acid.

11. Which metabolism can cause methemoglobinemia?
    A. Fermentation
    B. Respiration using oxygen
    C. Lithotrophy, oxidizing iron
    D. Anaerobic respiration, reducing nitrate to nitrite

13. Which portion of photosynthesis can occur in the dark and also occurs in many lithotrophic bacteria?
    A. Chlorophyll electron excitation
    B. ATP production from the H^+ gradient generated by the ETS
    C. NADPH production by electrons from the ETS
    D. CO_2 fixation via the Calvin cycle

14. Which kind of phototrophs cause food poisoning?
    A. Dinoflagellates in red tide
    B. Algal symbionts of coral
    C. Marine photosynthetic bacteria producing oxygen
    D. Purple sulfur-reducing bacteria

16. Bacteria incorporate ammonia into a TCA cycle intermediate to form the amino acid _____.
    A. cysteine
    B. glutamate
    D. tryptophan
    A. histidine

Key Terms

adenosine triphosphate (ATP) (p. 199)
aerobic respiration (p. 196)
anabolism (p. 215)
autotroph (p. 195)
biosynthesis (p. 215)
Calvin cycle (p. 214)
carbon dioxide (CO₂) fixation (p. 214)
carbon fixation (p. 214)
catabolism (p. 197)
catabolite (p. 197)
chemotrophy (p. 195)

chlorophyll (p. 213)
citric acid cycle (p. 197)
energy (p. 194)
entropy (p. 194)
fermentation (p. 203)
glycolysis (p. 197)
heterotroph (p. 195)
Krebs cycle (p. 197)
lithotrophy (p. 196)
methanogenesis (p. 212)
mixotroph (p. 215)

nicotinamide adenine dinucleotide
   (NADH) (p. 199)
organotrophy (p. 195)
oxidative phosphorylation (p. 207)
oxidized (p. 199)
oxidoreductase (p. 208)
oxygenic photosynthesis (p. 213)
pentose phosphate pathway (PPP)
   (p. 203)
photolysis (p. 213)
photosynthesis (p. 212)

photosystem I (p. 213)
photosystem II (p. 213)
phototrophy (p. 195)
proton motive force (PMF) (p. 208)
reduced (p. 199)
respiration (p. 203)
rhizobium (p. 216)
Rubisco (p. 214)
tricarboxylic acid cycle (TCA cycle)
   (p. 197)

Review Questions

1. Why must the biosphere continually take up energy from outside? Why can't all the energy be recycled among organisms, like the fundamental elements of matter?

2. How do organisms determine which of their catabolic pathways to use? How does catabolism depend on environmental factors?

3. Beer is produced by yeast fermentation of grain to ethanol. Why must the process of beer production be anaerobic? Why are such large quantities of ethanol produced, with relatively small production of yeast biomass?

4. Explain how glycolysis breaks down glucose to pyruvate. Why is it necessary to start by spending two molecules of ATP?

5. Explain the source of electrons and the sink for electrons (terminal electron acceptor) in respiration, lithotrophy, and photolysis.

6. What are the sources of substrates for biosynthesis? From what kinds of pathways do they arise?

7. How do different organisms obtain nitrogen? How is nitrogen assimilated into biomass?

Clinical Correlation Questions

1. A stool culture was obtained for a patient suffering from diarrhea and blood in his stool (dysentery). The cultured bacterium was used to inoculate Hektoen agar. The colonies on Hektoen agar were orange. Which bacterial pathogen was identified?

2. A family at the beach is having a seafood picnic. Two family members experience immediate numbness of the mouth and paralysis, followed by nausea and respiratory distress. What possible cause should be tested for immediately?

Thought Questions: CHECK YOUR ANSWERS

Thought Question 7.1   Why are glucose catabolism pathways ubiquitous, even in bacterial habitats where glucose is scarce? Give several reasons.

ANSWER: In habitats where free glucose monomers are scarce, there are still many glucose units linked in polymers such as starch and cellulose. If bacteria have enzymes such as amylase to break starch down into monomers, then the glucose can be catabolized by glycolysis or other pathways. Even if glucose is unavailable, enzymes may convert other sugar monomers, such as fructose or galactose, to glucose or other intermediates of glucose catabolism. Finally, it is always possible that glucose will suddenly become available to a microbe—for instance, the blood sugar of an animal host.

Thought Question 7.2   How is sugar catabolism used as a diagnostic indicator? Why are so many different sugars used as convenient indicators?

ANSWER: Many different kinds of sugars can be converted to glucose, which is then fermented to acids. The acids of fermentation provide a simple, inexpensive test by causing a color change in a titratable indicator dye. The different kinds of sugars are useful because different species of bacteria possess different combinations of enzymes to convert the sugars; many different combinations are possible and thus enable us to distinguish many species of bacteria.

Thought Question 7.3   The opportunistic pathogen *Pseudomonas aeruginosa* colonizes wounds and the lungs of cystic fibrosis patients. *P. aeruginosa* also grows in soil. It can respire aerobically or else anaerobically by using nitrate. In which habitat (the lung or the soil) would *P. aeruginosa* respire aerobically or with nitrate? What part of its ETS would need to change?

ANSWER: In the healthy lung alveoli, *P. aeruginosa* would have ample access to oxygen as a terminal electron acceptor; thus, the pathogen would conduct aerobic respiration. In the soil, the bacteria would quickly use up any oxygen in competition with neighboring microbes. With oxygen gone, *P. aeruginosa* would soon start expressing oxidoreductases that can transfer electrons to alternative electron acceptors—such as nitrate, if available. Similarly, in the lung of cystic fibrosis patients, infection by *P. aeruginosa* depletes oxygen; thus, the pathogen expresses its oxidoreductase to respire using nitrate.

Thought Question 7.4   Disease-causing bacteria vary widely in their ability to synthesize amino acids. What kinds of pathogens are likely to make their own amino acids, and what kinds are not?

ANSWER: Pathogens within the lumen of the digestive tract are most likely to synthesize their own amino acids because they compete intensively with other closely related species. As available amino acids get used up, the ability to make one's own amino acids might enable one species to outcompete its neighbors. By contrast, obligate intracellular pathogens are most likely to undergo degenerative evolution, losing their genes encoding enzymes of amino acid biosynthesis. Intracellular pathogens have an ample supply of amino acids from their host, and they have no competitors. However, there is selective pressure to lose unneeded genes because the supply of DNA monomers from the host is limited.

Thought Question 7.5   *Mycoplasma genitalium*, a cell wall–less bacterium that grows in human serum, lacks the ability to synthesize fatty acids. How do you think *M. genitalium* makes its cell membrane?

ANSWER: Without synthesizing its own fatty acids, *M. genitalium* must grow its membranes by using preformed fatty acids obtained from host cells within the blood. For mycoplasmas cultured in the laboratory, the composition of the membrane will vary, depending on which fatty acids are provided in the growth medium.

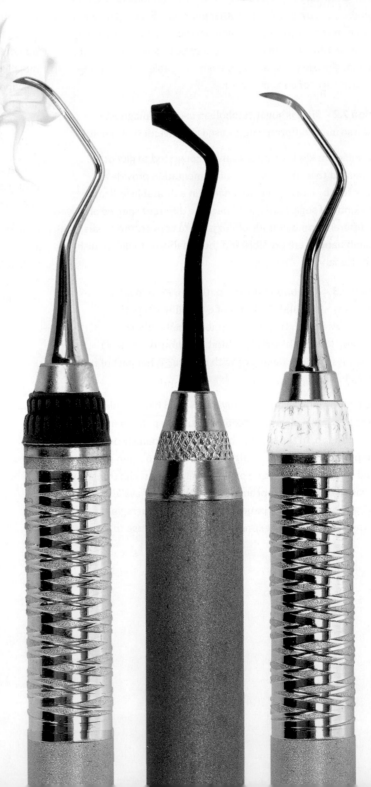

8

Bacterial Genetics and Biotechnology

Dental Biofilm Inflames the Gums

PATIENT HISTORY Tina, a 33-year-old store clerk from Peoria, Illinois, had her first dental exam in 5 years. She told the hygienist that her gums hurt and that stains appeared on her pillow where her jaw had rested. She also told the hygienist that she did not smoke or drink.

EXAMINATION The hygienist found that Tina's gums were swollen and bled upon probing. The gums had receded from Tina's teeth, forming pockets about 5 mm deep, and X-rays revealed some loss of bone.

DIAGNOSIS Tina then saw the dentist, who told her she had periodontitis, inflammatory disease of the gums and bone supporting the teeth. Periodontitis is caused by dental plaque, a biofilm of mixed bacterial species that grow on the teeth. Without regular oral hygiene, the biofilm grows beneath the gum and eventually causes loss of teeth.

Tina expressed surprise, as she thought that only elderly people suffered gum disease. The dentist asked her again whether she smoked, perhaps two packs a day? Tina denied smoking that much, but admitted to one pack a day. The dentist told her that smoking is a common factor in early gum disease, as are diabetes and genetic susceptibility. Under predisposing conditions, many different kinds of bacteria can cause gum disease.

TESTING To determine the bacterial species causing Tina's periodontitis, the dentist ordered a DNA test. The DNA test works by polymerase chain reaction (PCR), a technique in which a short piece of DNA is amplified (copied many times), making it possible to read the sequence of DNA base pairs. The DNA sequence reveals the bacterial species. Tina's DNA test revealed *Porphyromonas gingivalis* and *Aggregatibacter actinomycetemcomitans*, two Gram-negative anaerobes that are sensitive to metronidazole and amoxicillin, respectively.

TREATMENT The DNA test enabled the dentist to target the most effective antibiotics for Tina's condition. The dentist also enrolled Tina in a smoking cessation program.

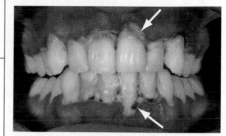

Gums Showing Periodontitis
Note the reddened gums and the receding gumline (arrows).

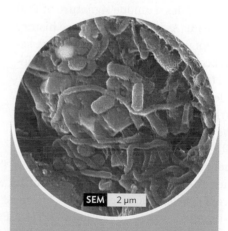

SEM 2 μm

Periodontal Biofilm
The diverse forms of bacteria can be identified based on their DNA sequence.

CHAPTER OBJECTIVES

After reading this chapter, you will be able to:

- Describe the structure and replication of bacterial genomes and plasmids.
- Describe techniques of biotechnology that aid medical diagnosis and treatment, including restriction enzymes, gel electrophoresis, recombinant DNA, and polymerase chain reaction (PCR).
- Explain how genes are transcribed to RNA and translated to protein.
- Explain how gene expression is regulated.

Some diseases, such as tuberculosis, are caused by a single pathogen. A single pathogen can be identified by a battery of tests based on colony appearance and metabolic tests, as shown in Chapter 7. But periodontitis involves a mixture of different species, which may be impossible to separate or culture. Each species has different virulence factors (factors associated with disease; discussed in Chapter 2). And each organism is sensitive to different antibiotics. For a given patient, how do we know which bacteria are present and which antibiotics to use?

A mixed-species sample can be sorted by its DNA. Each species of the mixture has a unique DNA sequence in its **genome**, the genetic material that defines the organism (**Figure 8.1**). The genomic DNA determines all the organism's inherited traits. The genome also provides a unique signature of that organism. As we saw in the case history that opened this chapter, a key gene sequence from the genome can be amplified (copied many times) by a technique called polymerase chain reaction (PCR) so that we can "read" the DNA sequence. The DNA sequence is then matched with a known database of bacterial sequences, and the species of the pathogen is identified. For the case of periodontitis, we can identify several different species in a mixed sample. The most prominent pathogens can be targeted by the right antibiotics. Thus, PCR helps identify multiple organisms in a condition with more than one infectious agent.

A living organism must copy its DNA accurately to retain traits that enable it to survive. Within cells, the DNA code must be "read" and copied to RNA, to make proteins specific to that organism.

Figure 8.1
Chemical Structure of the DNA Molecule

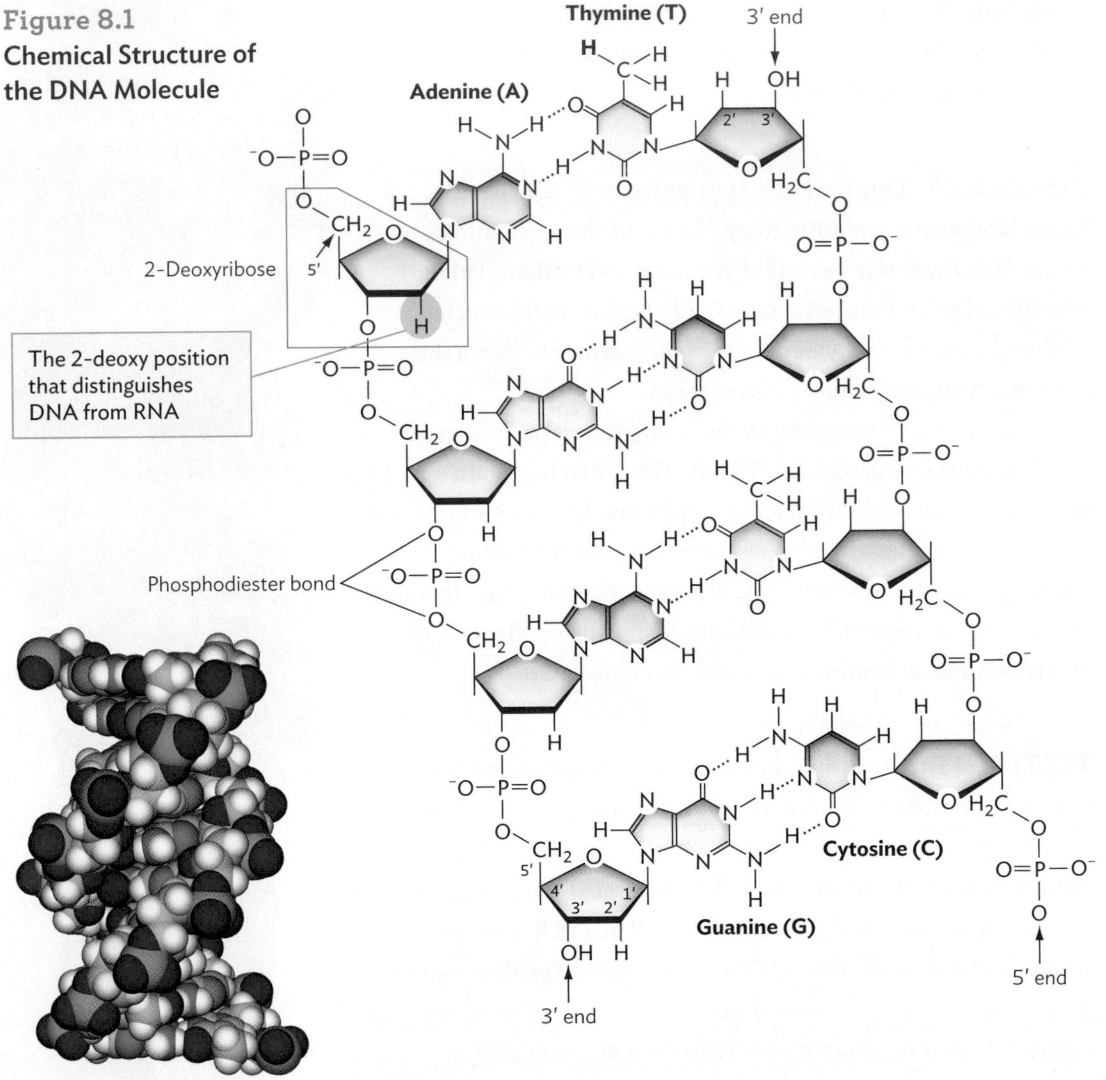

2-Deoxyribose

The 2-deoxy position that distinguishes DNA from RNA

Phosphodiester bond

Thymine (T)

Adenine (A)

Cytosine (C)

Guanine (G)

3' end

5' end

3' end

Yet over generations, the DNA sequence gradually changes. A few ancient sequences are shared by all life; but the sequence of the genome changes as organisms evolve into new forms with new traits, including new kinds of pathogens (discussed in Chapter 9).

Knowing how DNA works is an important part of understanding the interactions between microbes and humans. Pathogens entering a host, for instance, use their DNA information to make

Figure 8.2   DNA Spewing out of a Bacterial Cell Lysed by Osmotic Shock

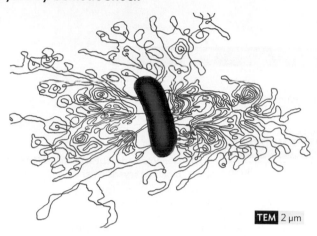

TEM 2 μm

Figure 8.3   Genes Organized within a Genome

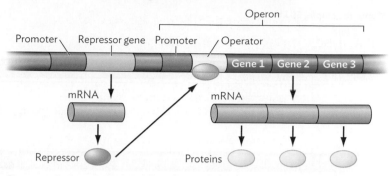

Diagram of a DNA sequence that specifies a single gene (yellow) with its own promoter and an operon of three collinear genes (green) that share a promoter.

virulence factors and secrete them into the host tissues. Outside a host, environmental bacteria use different genes to survive in different habitats.

In this chapter, we explain how bacteria replicate their DNA and express their genes, and we describe the consequences for human health and the environment. We also show how bacterial and viral components, such as enzymes and plasmids, provide tools for biotechnology—an exciting field that gives us powerful tools for diagnosis and therapy.

8.1
Bacterial Genomes

SECTION OBJECTIVES

- Describe the structure of a bacterial genome, and explain how it differs from a eukaryotic genome.
- Explain what a plasmid is, and describe the role of plasmids in the spread of antibiotic resistance genes.

How does DNA define a unique species of organism? As discussed in Chapter 5, all living cells maintain their "blueprint" genetic information in **chromosomes** that consist of double-stranded DNA. Each strand of DNA carries a sequence of bases of four types: adenine (A), thymine (T), cytosine (C), and guanine (G). The two strands form complementary base pairs, in which A pairs with T and G pairs with C (see Figure 8.1; also review Figure 4.12). The order of the base pairs determines the order of amino acids that form proteins.

Most bacteria have a genome in the form of a circular chromosome containing several million base pairs. The entire genome of the bacterium *Escherichia coli* consists of a circular chromosome of 4,600,000 base pairs. In a remarkable experiment, a cell of *E. coli* was lysed by osmotic shock, releasing its chromosome for electron microscopy (Figure 8.2). What spewed out of this single cell was a

chromosome of DNA 800 times longer than the bacterium itself. How can this enormous molecule fit into a single cell? And how can enzymes duplicate that much DNA without the whole molecule getting tangled up? These questions are addressed in Section 8.2.

Genome of a Species

Each species has a genome with a set of genes unique to the species. A **gene** is a unit of information encoded by DNA that can be expressed to form an RNA product (**Figure 8.3**). The RNA products of most genes are **messenger RNA (mRNA)**, an RNA molecule that directs the synthesis of a specific protein. The use of gene information to make RNA and protein is called **gene expression** (discussed in Sections 8.5 and 8.6). The information determining *when* a gene is expressed is contained in a DNA control sequence called a **promoter**. In the simplest case, a gene can stand alone, operating independently of other genes (yellow in Figure 8.3). The RNA produced from a stand-alone gene encodes a single protein.

In bacteria, genes often exist in a series, in a unit called an **operon**. All genes in an operon are lined up head to tail on the chromosome, and their expression (RNA synthesis) is controlled by a single promoter located in front of the first gene. The single RNA molecule produced from the operon contains all the information from all the genes in that operon.

Note For bacteria, each **structural gene** (gene encoding a product, either RNA or protein) is designated by an italic three-letter name. The name may have a fourth letter in upper case (*mdtA*), indicating the gene encodes a functionally related protein—in this case, a member of the Mdt complex. The corresponding protein name for each structural gene appears in roman type with the first letter also uppercase (MdtA).

Bacterial genomes are organized differently from those of eukaryotic organisms (such as humans). **Figure 8.4** compares the organization of a genome section of the bacterium *Escherichia coli* with a section of the human genome. The *E. coli* genome is packed with genes that encode proteins (green), with very little unused

Figure 8.4   Genome Structure in Bacteria and in Humans

Genome sample, 50 kb

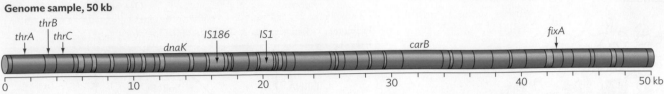

A. A genome section of the bacterium *Escherichia coli*. Sections include coding genes (green) and noncoding sequences (purple).

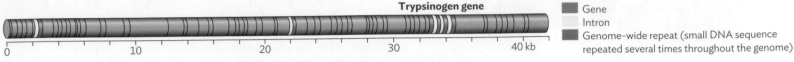

Trypsinogen gene

- Gene
- Intron
- Genome-wide repeat (small DNA sequence repeated several times throughout the genome)

B. A section of the human genome. Human coding genes are interrupted by introns (yellow).

sequence between them (purple) and only an occasional mobile element, such as an insertion sequence (IS). Coding genes often occur together in operons, such as the *thr* genes for threonine biosynthesis. By contrast, the human genome contains 95% noncoding sequences. The human coding genes are usually single, but they may be interrupted by noncoding introns (yellow). Eukaryotic genes are separated by large stretches of noncoding DNA.

Besides the main genomic DNA, bacteria may have additional smaller pieces of DNA called **plasmids**. A plasmid may enter or leave a bacterial cell without changing the species. Plasmids typically contain a small number of genes that are useful in a given environment or virulence genes that confer disease properties on the bacterium. For example, the plasmid shown in Figure 8.16 contains genes that confer resistance to two antibiotics, ampicillin and tetracycline.

Such drug resistance plasmids circulate in hospitals, carrying genes for resistance to five or more antibiotics at once. This is a problem because when a patient is treated with one antibiotic, the antibiotic enhances survival of pathogens that have acquired a plasmid conferring resistance to several other antibiotics.

Some plasmids may carry genes from one species to a completely unrelated species. When a hospital patient receives antibiotic therapy over the long term, the patient's normal biota (normally occurring harmless bacteria) will become antibiotic resistant by natural selection. A plasmid of a normally occurring organism may then acquire the antibiotic resistance genes and transfer them to disease-causing bacteria. The transfer of "mobile genes" from one species to another gives rise to dangerous pathogens, such as methicillin-resistant *Staphylococcus aureus* (MRSA)

Table 8.1
Genomes of Representative Bacteria

Species (strain)	Genomic Chromosome(s)[a] (kilobase pairs, kb)	Plasmid(s)[a] (kb)	Entire Genome (kb)
Agrobacterium tumefaciens Tumors in plants; genetic engineering vector	2,840 + 2,070	214 + 542	5,666
Borrelia burgdorferi Lyme disease	911	21 plasmids with sizes between 9 and 58	>1,250
Burkholderia cepacia	3,870 + 3,217 + 876	93	8,056
Escherichia coli K-12 Model strain for *E. coli* research	4,600		4,600
Mycobacterium tuberculosis Tuberculosis	4,400		4,400

[a]Purple ovals represent circular DNA; lines represent linear DNA.

and multidrug-resistant *Mycobacterium tuberculosis*. Transfer of a few genes from one organism to another is called **horizontal gene transfer**. Horizontal transfer of genes in evolution is discussed in Chapter 9.

Some naturally occurring bacteria have complex genomes that include multiple circular and linear pieces of DNA. Table 8.1 shows examples of bacterial species with different patterns of genome organization. In these species, the smaller DNA components are called plasmids, although the distinction between "plasmid" and "genome" is not always clear. We do not know, for example, why the genome of *Borrelia burgdorferi* is organized into 22 separate segments, nor how the replication of this multipart genome is coordinated.

By contrast, in eukaryotes, the genome always consists of linear chromosomes organized within the nucleus. This is true of microbial eukaryotic pathogens, such as the protist *Giardia* and the yeast *Candida*, as well as of their human hosts. Eukaryotic chromosome segregation during meiosis is discussed in Chapter 11.

Genomes Evolve and Change

How do organisms evolve into so many kinds, with unexpected new traits? A species is maintained by producing offspring that have the DNA sequence inherited from their parents. The DNA sequence is replicated by special enzymes (see Section 8.2). But DNA replication is never perfect. The error rate of DNA replication in enteric bacteria such as *E. coli* is estimated at 1 change per 10 million base pairs (bp). Thus, in a colony of bacteria whose genome is 5 million base pairs, about half the bacteria should contain a **mutation**—that is, a base pair different from that of the preexisting parent cell (discussed in Section 8.3). The "mutant" base pair now exists in the genome, just like other base pairs in the rest of the DNA sequence. Over many generations, such mutations add up, generating variant strains of the organism that are highly similar, but may differ in crucial ways such as their degree of virulence and their susceptibility to antibiotics.

SECTION SUMMARY

- **Genome** refers to the total DNA content of a cell.
- **Bacterial genomes are typically circular**, but can be linear or a combination of circular and linear DNA chromosomes.
- **Plasmids are small DNA circles** containing supplementary genes that may encode pathways that degrade materials from the environment or confer drug resistance. Plasmids may be transferable between cells.
- **Genomes evolve** by diverging from a common ancestor. Divergent strains of a pathogen can cause new forms of disease.

8.2
DNA Replication

SECTION OBJECTIVES

- Explain how bidirectional semiconservative DNA replication copies the circular chromosome of a bacterium during cell division.
- Explain the steps of replication by DNA polymerase, including initiation, elongation, and termination.

We've all heard about community meals, such as a holiday barbecue, after which everyone comes down sick with food poisoning. The culprit is typically *Salmonella* bacteria or Norwalk-type virus. How can a few bites of food make us sick within hours? The answer lies in DNA replication. Enteric pathogens have evolved to reproduce themselves and replicate their DNA as fast as possible to outcompete other gut biota. Replication efficiency explains why bacterial pathogens can cause disease so quickly after ingestion (usually within 24–48 hours). Viral replication can be faster yet, sacrificing accuracy for speed, as discussed in Chapter 12.

The process of bacterial replication involves an amazing number of proteins and genes working together. The molecular details are important because they provide targets for new antibiotics (Table 8.2) as well as tools for biotechnology, such as the

Table 8.2
Antibiotics That Target DNA Synthesis

Enzyme or Process of DNA Synthesis	Antimicrobial Molecule	Organism Affected	Source of Molecule
Helicase: Unwinds helix turns and superturns	Heliquinomycin	Human tumors	*Streptomyces* sp. MJ929-SF2
DNA polymerase, beta subunit: Adds nucleotides to growing DNA chain	Bleomycin	Human tumors	*Streptomyces verticillus*
DNA topoisomerase I	Topostins	Human cancer	*Flexibacter topostinus*
DNA stability	Metronidazole (redox reactions cause DNA breakage)	Anaerobic bacteria and protozoa	Artificial design
Gyrase: Adds superturns to newly completed DNA helix	Quinolones	Bacteria	Artificial design
Telomerase: Forms ends of eukaryotic linear chromosomes	Rubromycin	Human tumors	*Streptomyces* spp.

Figure 8.5   Pairing of a New Base: Correct versus Misaligned

The models shown are based on X-ray crystallography of exceptionally stable enzymes obtained from *Sulfolobus solfataricus*, an archaeon that grows at volcanic vents in near-boiling sulfuric acid.

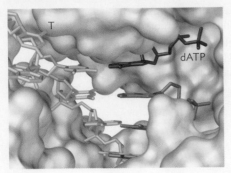

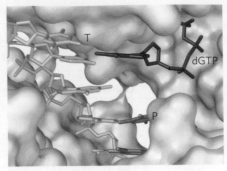

A. The DNA polymerase active site contains a template strand T (yellow) pairing with a new deoxy ATP (red) to be added to the growing complementary chain (brown).

B. Deoxy GTP can enter the binding site, but it is misaligned with the template. The DNA polymerase will reject the misaligned base.

polymerase chain reaction (PCR). In addition, the study of bacterial DNA replication and repair gives clues to DNA function in humans because the enzymes that replicate DNA are encoded by homologous genes, or homologs. **Homologs** are genes with shared ancestry and related function—in this case, shared from the common ancestor of bacteria and humans. The human homologs for bacterial DNA replication enzymes protect our own DNA from damage that causes cancer.

Semiconservative Replication

To replicate a molecule containing millions of base pairs poses formidable challenges. How does replication begin and end? How can two identical copies be generated? How is accuracy checked and maintained? In cells, DNA replication is semiconservative, meaning that each daughter cell receives one original strand (called the parental strand) and one newly synthesized strand. Synthesis of a new strand requires pairing each new nucleotide with a complementary base on the parental strand (see Figure 8.1B): adenine pairs with thymine (AT), and guanine pairs with cytosine (GC). The pairing of these bases is constrained by the active site of the

DNA polymerase, the enzyme that copies DNA (Figure 8.5A). Incorrect bases may enter the site but fail to pair properly (Figure 8.5B) and thus are rejected by the enzyme.

To replicate DNA, the double helix must unwind, exposing the base pairs to form new pairs with a growing strand (Figure 8.6). The original DNA strands are called **parental strands**; the growing strands, complementary to each parental strand, are called **daughter strands**. The unwinding requires initiation proteins and a "helicase" enzyme that spends ATP to unwind helical DNA. The partial unwinding of the helix generates a **replication fork** at each end of the replication "bubble." At each replication fork, the advancing DNA synthesis machine separates the parental strands while extending the new, growing strands. Each daughter duplex will be checked for accuracy against its parental strand—first by the DNA polymerase and then later by several types of proofreading and repair enzymes.

The process of DNA synthesis is divided into three phases:

- **Initiation**, which involves the unwinding of the helix, priming, and loading of the DNA polymerase enzyme complex

- **Elongation**, the sequential extension of DNA by adding DNA nucleotide triphosphates (dNTPs) with release of pyrophosphate, followed by proofreading

- **Termination**, in which the DNA helix is completely duplicated and replication stops

We now examine each phase of replication in turn. Note that all of the steps shown for DNA synthesis require catalysis by enzymes and DNA-binding proteins. Most of these proteins have homologs in the DNA synthesis of our own cells.

Initiation of DNA Synthesis

In most bacteria, replication begins at a single defined DNA sequence called the **origin (*oriC*)** (see Figure 8.6). Enzymes that synthesize DNA or RNA can connect nucleotides only in a 5'-to-3' direction. That is, every newly made strand begins with a 5' triphosphate and ends with a 3' hydroxyl group. A **DNA polymerase** (a chain-

Figure 8.6   Bidirectional Semiconservative Replication

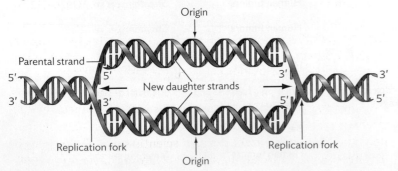

DNA replication with two replication forks. Parental strands are gray, and newly synthesized daughter strands are purple. Replication is called semiconservative because one parental strand is conserved and inherited by each daughter cell genome. It is called bidirectional because it begins at a fixed origin and progresses in opposite directions.

Figure 8.7   Nucleotide Condensation with 3′ OH End of DNA

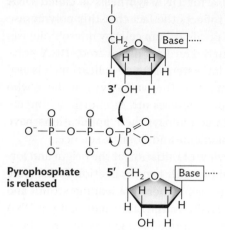

The nucleotide releases its terminal and middle phosphoryl groups as pyrophosphate. The phosphoryl group nearest the sugar condenses with the 3′ OH of the growing DNA chain.

Pyrophosphate is released

Nucleoside triphosphate

Figure 8.8   RNA Primers Provide the First 3′ OH for Each Strand

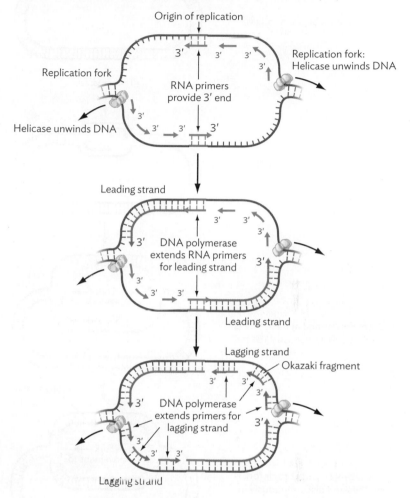

Each leading strand (pointing toward the replication fork) needs only one primer, but each lagging strand (pointing back from the fork) needs multiple primers.

lengthening enzyme complex) joins the 5′ phosphate of an incoming nucleoside triphosphate (NTP) (Figure 8.7). The new bond of the chain (C–O–P–O–C) is called a phosphodiester bond. To form this bond, the reaction releases a pair of phosphoryl groups (called pyrophosphate) from the NTP. Loss of pyrophosphate releases a lot of energy and thus drives the reaction forward.

If every new NTP must add onto a 3′ OH group, then how is the very first nucleotide added? A short **RNA primer** molecule with a 3′ OH end base-pairs with the DNA (Figure 8.8). The RNA primer is synthesized by a special enzyme called primase that does not need a 3′ OH end. The use of RNA for a primer instead of DNA may seem surprising, as eventually the RNA will have to be replaced by DNA nucleotides. The universal use of RNA primers by all cellular chromosomes suggests that the process is a vestige of replication in the ancient ancestral cells from which all life evolved. The chromosomes of those ancient cells consisted entirely of RNA.

Once replication has begun, the cell is committed to completing a full round of DNA synthesis. As a result, the decision of when to start is critical. If replication starts too soon, the cell accumulates unneeded copies of its DNA. If it starts too late, the dividing cell's septum cuts through the chromosome, killing both daughter cells.

Elongation of a DNA Strand

After initiation, a circular bacterial chromosome replicates bidirectionally. As illustrated in Figure 8.8, in bidirectional replication the helix unwinds and chain elongation proceeds in both directions away from the origin (Figure 8.9, step 1). The chromosome region of active DNA replication is called a "replication bubble." At each end of the replication bubble, the two replication forks (sites of helix unwinding and new-strand synthesis) move in opposite directions around the chromosome (Figure 8.9, step 2). Although not all enzymes are shown, remember that all the steps of DNA synthesis require catalysis by enzymes.

The 5′-to-3′ directed elongation of DNA leads to an interesting puzzle: if polymerases can synthesize DNA only in a 5′-to-3′ direction, and the two phosphodiester backbones of the double helix are antiparallel, how are both strands of a moving replication fork synthesized simultaneously?

At each fork, one strand is synthesized directly in a 5′-to-3′ direction toward the fork (Figure 8.9, step 3). But synthesizing the other new strand in a 5′-to-3′ direction requires that it move away from the fork. Thus, at each replication fork, one of the two daughter strands needs to elongate by "jogging back" toward the fork at regular intervals. The smoothly elongating strand is called the **leading strand**, whereas the jogging-back strand is called the **lagging strand**.

The fragments of DNA formed as the lagging strand is synthesized are called **Okazaki fragments** (Figure 8.9, step 4), named for Reiji and Tsuneko Okazaki, married scientists who discovered the fragments of the DNA lagging strand at Nagoya University in 1968.

ANIMATION ▶ Figure 8.9   **Replication of a Bacterial Genome**

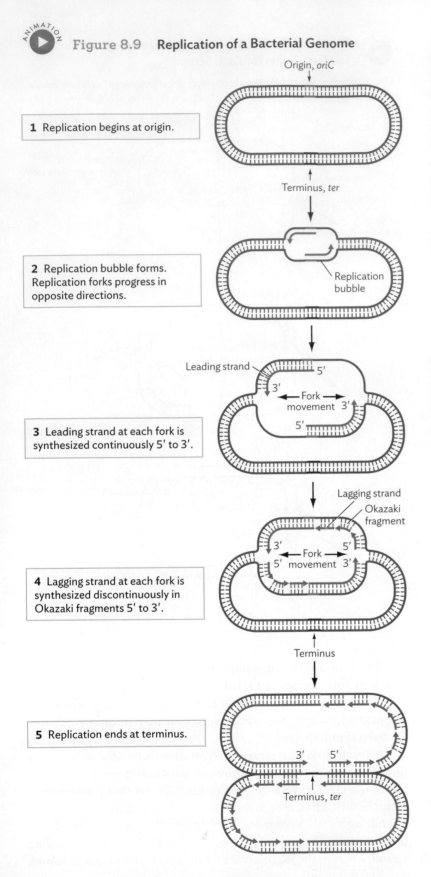

1 Replication begins at origin.

2 Replication bubble forms. Replication forks progress in opposite directions.

3 Leading strand at each fork is synthesized continuously 5′ to 3′.

4 Lagging strand at each fork is synthesized discontinuously in Okazaki fragments 5′ to 3′.

5 Replication ends at terminus.

Both leading and lagging strands are extended by a DNA polymerase. The main polymerase for DNA synthesis is called DNA polymerase III; the number reflects the fact that this polymerase happened to be the third one purified from cells by microbiologists early in the twentieth century. The first discovered, DNA polymerase I, actually catalyzes a later step of DNA replication; it is also known as the Kornberg enzyme for Arthur and Sylvy Kornberg, who discovered it in 1956. DNA polymerases are of enormous importance for all biology and biotechnology; their applications have transformed the practice of medicine and forensic science.

Figure 8.9 shows an overview of initiation at the origin and formation of the replication bubble with two replication forks. Note that the two replication forks each generate a leading strand and a lagging strand, oriented opposite to the other pair. Late in DNA replication, when most of the complementary strand has been elongated, the RNA primers need to be replaced with DNA. The RNA is removed by an enzyme that specifically recognizes RNA-DNA hybrid molecules and cleaves away only the RNA strand. After the removal of RNA, the enzyme DNA polymerase I "fills in" the complementary DNA nucleotides. Completion of DNA synthesis requires replication at the defined termination (*ter*) site located on the opposite side of the molecule from the origin.

ACCURACY OF DNA SYNTHESIS   An important problem for DNA synthesis is accuracy. The enzyme DNA polymerase III (abbreviated DNA pol III) places the correct complementary base only about 99.9% of the time. This may sound good; but one error in a thousand would lead to thousands of errors in the genome as a whole and many mutant genes. To avoid mutations, the newly made DNA must be proofread. Proofreading is done by several kinds of enzymes. For example, when DNA polymerase I runs into the 5′ backend of the next Okazaki fragment, it cleaves a base off while replacing it with a new base condensed with the 3′ OH of the DNA "behind." In this process of base replacement, any incorrectly paired bases get replaced by correct ones. Other repair enzymes conduct repair by recognizing a "bump" in the DNA where the incorrect base fails to make a smooth pair. The enzyme cleaves the backbone at either side of the incorrect base and removes a stretch of DNA that contains it, to be filled in correctly by polymerase I. Error-correcting enzymes are important for repairing mutations, as discussed in Section 8.3.

Because all DNA molecules look very similar, how do repair enzymes "know" which strand of an incorrect pair is the "new" DNA to be fixed? During DNA replication, each "old" DNA strand contains a methyl group on some of its cytosine bases. Repair enzymes recognize methylated DNA as "old" DNA, containing the original sequence. Thus, at the conclusion of replication, all the newly synthesized DNA needs to be methylated by methylase enzymes, a process akin to a "seal of approval" on the completed DNA.

Another requirement as DNA synthesis nears completion is to join the backbone of all the Okazaki fragments, to make a continuous, unbroken strand. The "nicks" (positions of unsealed backbone between two adjacent nucleotides) are joined by **DNA ligase**, an enzyme that condenses the 3′ hydroxyl "back end" of each fragment with the 5′ phosphoryl "front end" of the adjacent fragment.

Termination of DNA Synthesis

While the DNA molecule undergoes replication, an important step remains: methylating all the newly synthesized DNA. Once the repair enzymes have corrected all the errors they find, the new DNA becomes methylated so that it can be recognized as "old" DNA during the next round of replication.

In bacteria with circular chromosomes, the two replication forks eventually circle round to approach each other. The two forks meet at a DNA sequence called the terminus, or *ter* site (Figure 8.9, step 5). The terminus requires special proteins to complete the growing DNA strands while disengaging the DNA polymerases.

Another important process for completion of replication is that the DNA must regain its negative **supercoils**. Supercoils, or superturns, are extra turns in the DNA beyond those generated by the shape of the base pairs. DNA needs to be "unwound" slightly, that is, negatively supercoiled, for genes to be transcribed to RNA (see Section 8.5). Negative supercoils are generated by the enzyme **gyrase**—a particularly important drug target (**Figure 8.10**). Bacterial gyrase is the target of quinolones, a group of broad-spectrum antibiotics used to treat anthrax and pneumonia.

How does gyrase introduce a supercoil into a circular DNA molecule, lacking ends? The gyrase complex cleaves the backbones of both strands of DNA at the same position, causing a double-strand break. Holding onto both ends, the enzyme passes another double-stranded portion of the DNA through the break and then reseals the two backbones, a process that spends ATP. The net result is one turn "against" the twist of the helix—that is, a negative supercoil. Bacteria have several other kinds of enzymes that generate positive or negative supercoils for various functions; the general class of enzymes that change supercoiling is called topoisomerases.

So how long does it take for the bacteria in your contaminated chicken salad to double their DNA and undergo cell division? At optimum temperature and resources, the entire process of cell division in an enteric bacterium may take as few as 20 minutes (see Figure 5.16). As DNA replication terminates, the partition of the new cell wall and membranes (called the septum) forms at midcell. Once the septum is complete, the two daughter cells, with their new chromosomes, can separate.

SECTION SUMMARY

- **DNA is synthesized semiconservatively.** Each daughter helix contains one "old" parental strand (the template) and one newly formed complementary strand.
- **Initiation of DNA synthesis** begins at the replication origin and proceeds bidirectionally around the cell until both replication forks meet at the terminus.
- **Enzymes catalyze all steps of replication.** Helicase unwinds the helix; primase forms RNA primers with a 3′ OH end; DNA polymerase III

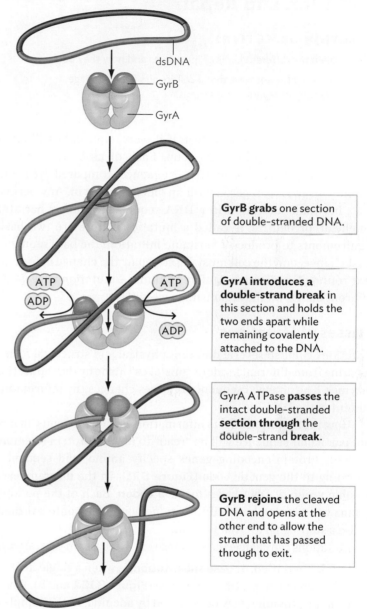

Figure 8.10   Gyrase Adds Negative Supercoils to DNA
The enzyme gyrase is the target of quinolone antibiotics.

dsDNA
GyrB
GyrA

GyrB grabs one section of double-stranded DNA.

ATP
ADP
ATP
ADP

GyrA introduces a double-strand break in this section and holds the two ends apart while remaining covalently attached to the DNA.

GyrA ATPase **passes** the intact double-stranded **section through** the double-strand **break**.

GyrB rejoins the cleaved DNA and opens at the other end to allow the strand that has passed through to exit.

extends the growing DNA strand; DNA polymerase I removes the primer RNA; proofreading enzymes correct mismatched bases; and DNA ligase seals the nicks between segments of each growing strand.

- **Leading and lagging strands are elongated at each growing fork.** The lagging strand requires backtracking at regular intervals in order to elongate from a 3′ OH end.
- **Completed DNA requires methylation and restoration of negative supercoils.** Gyrase, an important antibiotic target, forms the supercoils.

Thought Question 8.1   Why would a rapidly growing bacterial cell be killed by an antibiotic that blocks DNA replication? Suppose the bacteria form a biofilm on a catheter in a patient. Why might the bacteria have become resistant to the antibiotic?

8.3 Mutation and Repair

SECTION OBJECTIVES

- Explain the different kinds of mutations and how they occur.
- Describe different ways that a cell repairs DNA damage, preventing or reversing mutations.

Figure 8.11 The Genetic Code

1st Base	2nd Base: T Codon	T Amino acid	C Codon	C Amino acid	A Codon	A Amino acid	G Codon	G Amino acid
T	T T T	Phe	T C T	Ser	T A T	Tyr	T G T	Cys
T	T T C	Phe	T C C	Ser	T A C	Tyr	T G C	Cys
T	T T A	Leu	T C A	Ser	T A A	Stop	T G A	Stop
T	T T G	Leu	T C G	Ser	T A G	Stop	T G G	Trp
C	C T T	Leu	C C T	Pro	C A T	His	C G T	Arg
C	C T C	Leu	C C C	Pro	C A C	His	C G C	Arg
C	C T A	Leu	C C A	Pro	C A A	Gln	C G A	Arg
C	C T G	Leu	C C G	Pro	C A G	Gln	C G G	Arg
A	A T T	Ile	A C T	Thr	A A T	Asn	A G T	Ser
A	A T C	Ile	A C C	Thr	A A C	Asn	A G C	Ser
A	A T A	Ile	A C A	Thr	A A A	Lys	A G A	Arg
A	A T G	Met	A C G	Thr	A A G	Lys	A G G	Arg
G	G T T	Val	G C T	Ala	G A T	Asp	G G T	Gly
G	G T C	Val	G C C	Ala	G A C	Asp	G G C	Gly
G	G T A	Val	G C A	Ala	G A A	Glu	G G A	Gly
G	G T G	Val	G C G	Ala	G A G	Glu	G G G	Gly

3rd Base

Codons within a single box encode the same amino acid. Color-highlighted amino acids are encoded by codons in two boxes. Stop codons are highlighted pink. Note: In RNA, each thymine (T) is replaced by uracil (U).

DNA replication is never perfect; if it were, we would still all be single cells in the primordial soup. Each of us carries approximately 150 differences in DNA sequence compared with our parents—or even compared with an "identical" twin. Any permanent, heritable alteration in a DNA sequence, whether harmful, beneficial, or neutral, is called a mutation. There are two basic requirements to produce a heritable mutation: the base sequence must change, and the cell must fail to repair the change before the next round of replication. Different kinds of mutation have vastly different effects on the organism, as we will see.

Classes of Mutations

Mutations come in several different physical and structural forms, resulting from different kinds of "mistakes" made by the replication enzymes. The changed sequence of bases changes the information content of the DNA.

How the mutation affects information content depends on how that particular sequence will be "read" to make a protein (Sections 8.5–8.6). Protein-encoding genes specify amino acid sequences according to the genetic code (Figure 8.11). In the diagram, each possible group of DNA bases forms a **codon**. Each of the possible codons will ultimately specify an amino acid to add onto a protein or else a stop signal for that protein.

Mutations can change the sequence of DNA bases in several ways:

- A **point mutation**, or **base substitution**, is when a single nucleotide is changed in a DNA sequence (Figure 8.12A and B). For example, thymine may be replaced by adenine. In the complementary strand, the paired adenine is replaced by thymine.
- **Insertions** and **deletions** involve, respectively, the addition or subtraction of one or more nucleotides (Figure 8.12C and D), making the sequence either longer or shorter than it was originally.
- An **inversion** occurs when a fragment of DNA is flipped in orientation in relation to DNA on either side (Figure 8.12E).

Changes in DNA can affect the information content of the codons. For gene expression, DNA is copied to RNA that has the same bases as DNA, except that each thymine (T) is replaced by an equivalent uracil (U). But changes amongst the four DNA bases (A, C, G, T) are inherited as mutation:

- A **silent mutation** occurs when a base-pair substitution changes a codon but the new codon specifies the same amino acid as the original codon. For example, all the codons for leucine (Leu, in Figure 8.11) start with the same first and second bases. Thus, any point mutation changing the third letter of a codon for leucine will still encode leucine. The mutation is "silent" because no difference occurs in the protein.
- A **missense mutation** changes the codon and leads to the expression of an altered protein (see Figure 8.12A). For example, in the bacterium *Mycobacterium tuberculosis*, the *rpoD* gene for RNA polymerase can have a mutation replacing C with T, which changes a codon CAC (histidine) to TAC (tyrosine). The *rpoD* gene then expresses a protein containing tyrosine in place of histidine. The result is a rifampicin-resistant bacterium.
- A **nonsense mutation** replaces an amino acid–coding codon with a codon that specifies no amino acid and terminates translation (see Figure 8.12B). The result is a truncated protein, usually nonfunctional.
- A **frameshift mutation** occurs when insertion or deletion of one or more base pairs shifts the **reading frame** (see Figure 8.12C and D). The reading frame determines which base pairs are read in the first position of the codons. A shift of

Figure 8.12 Changes in a DNA Sequence That Result in Different Mutations

A. Missense point mutation: T-to-A is a point mutation that converts the phenylalanine codon TTT to the leucine codon TTA.

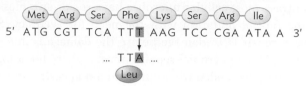

B. Nonsense point mutation: C-to-G converts the TCA codon, encoding serine, to the TGA stop codon.

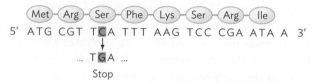

C. Insertion frameshift: The addition of AT into the middle of the TCC serine codon causes a shift in the reading frame that changes all the downstream amino acids.

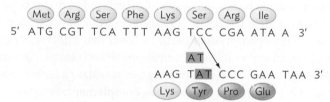

D. Deletion frameshift: Removing two nucleotides from the arginine codon shifts the frame down.

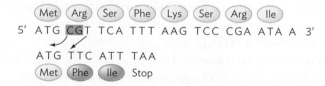

E. Inversion: Flipping a DNA sequence to face in the opposite direction changes the amino acids produced.

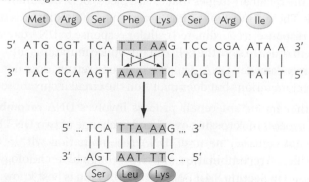

the reading frame leads to a sequence of completely different amino acids; the drastically altered protein is usually nonfunctional.

What causes mutation? Some mutations happen spontaneously, the result of a "mistake" by the DNA polymerase that fails to be repaired by a proofreading enzyme. Other mutations are caused by physical or chemical agents that damage DNA. Physical agents that damage DNA include cosmic rays, X-rays, and ultraviolet (UV) radiation. X-rays cause large mutations by breaking the DNA backbone and deleting a major part of a chromosome. Such major mutations can destroy the tissues. For example, in 2010, the U.S. Nuclear Regulatory Commission reported that a major hospital had accidentally irradiated cancer patients at levels so high that they suffered burns. That is why it is so important for an X-ray technician to correctly set the level of an X-ray machine so as to minimize the dose received by the patient.

UV radiation, such as that received from sunlight or from tanning salons, causes a different kind of DNA damage. UV absorption by an adjacent pair of pyrimidines (T or C) causes electrons to rearrange and form a covalent bond between the two bases. The fused pair of adjacent bases is called a **pyrimidine dimer** (Figure 8.13). The pyrimidine dimer causes a "bump" in the helix that fails to be read properly by the DNA polymerase, leading to a gap or an insertion of several incorrect bases in the complementary daughter strand. Such damage may appear subtle, but it can lead to deadly cancers such as melanoma. Melanoma resulting from tanning salons has been seen in people as young as age 18.

Figure 8.13 Pyrimidine Dimer Caused by UV Absorption

The energy from UV irradiation can be absorbed by pyrimidines (thymine and/or cytosine). Two adjacent pyrimidines react to form a covalent connection.

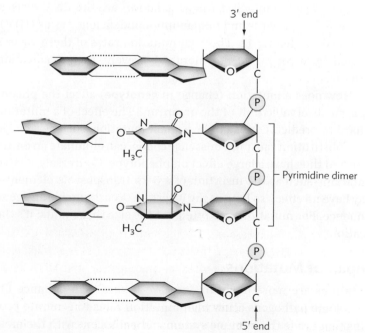

Figure 8.14   Colonies of Mutator Bacteria

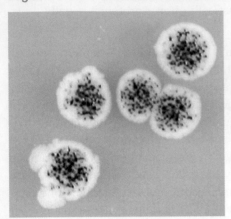

In an *E. coli* mutator strain, cells are lactose-negative mutants that do not ferment lactose and do not turn the indicator blue. But because the strain has a mutation in one of the mutator genes, a high rate of reversion to the lactose fermenter strain occurs, which shows up as blue papillae (tiny spots) on the white colony.

Chemical agents that can damage DNA include reactive oxygen molecules such as hydrogen peroxide (H_2O_2) and superoxide radicals ($\cdot O_2^-$) such as those produced by white blood cells to kill pathogens. Other agents, such as acridine orange, are "intercalating agents" that slip in between two base pairs. An intercalating agent can get "mistaken" for a base pair by the DNA polymerase. The polymerase may then insert an extra base in the daughter strand.

Biological processes can also increase the rate of mutation. Certain strains of a microbe are called **mutators** because their DNA replication process allows a high frequency of mutations. Mutators can be observed when they produce "sectored colonies" in which some cells have inherited a recent mutation causing a color change. Mutator strains of *Escherichia coli* are shown in **Figure 8.14**. A surprisingly high percentage of mutators arise in pathogens such as *Salmonella*. Mutators have the disadvantage of incurring lots of genetic defects; but within a host, they have the advantage of generating novel surface molecules that evade the host's immune system. The most highly mutable pathogens known are the RNA viruses, such as influenza virus and human immunodeficiency virus (HIV), discussed in Chapter 12. The high mutation rates of these viruses (caused by error-prone polymerases) prevent their eradication with vaccines.

How does a mutation (change in **genotype**) affect the **phenotype**, an observable trait of the organism? The effect of a mutation is hard to predict. For example, a small mutation, such as a single base substitution causing a missense codon, can eliminate an entire gene and thus have a large effect on phenotype. Conversely, a large mutation—such as the insertion of a 5-kb transposable element—may have no effect, so long as the insertion occurs outside any gene sequence. For mutations, it's often not the size that counts; it's the location.

Repair of Mutations

Mutations are more likely to decrease fitness than enhance fitness. Some pathogens allow high mutation rates to generate proteins that evade the immune system; but pathogens with the most extreme mutation rates produce surprisingly low proportions of infectious progeny. For example, RNA viruses such as influenza virus and HIV produce virions of which less than 1% may be able to infect the next cell.

By contrast, most cellular microbes have repair enzymes that balance the rate of mutation with the need for accuracy. Such enzymes can repair DNA damage before the damage is inherited as a mutation. Many microbial repair enzymes share homologs in the human body, descended from our common ancestral microbe. Inherited defects in the human DNA repair enzymes cause the disease xeroderma pigmentosum, in which the skin and eyes are completely sensitive to UV radiation, developing cancer at the slightest exposure.

The type of repair that bacteria use (and when it is used) depends on two things: the type of mutation needing repair and the extent of damage involved. Some repair enzymes excise a whole fragment of DNA that contains damaged bases, whereas others precisely excise just the damaged bases or directly reverse the damage. After extensive damage, special "emergency" DNA polymerases sacrifice accuracy to rescue the broken chromosome. Some examples of DNA repair include:

- **Base excision repair.** In **base excision repair**, an excision repair enzyme recognizes a specific damaged base and removes it from the DNA backbone (**Figure 8.15**). The base excision leaves a phosphoryl sugar (with no base) that a different repair enzyme, an endonuclease, can recognize and cleave. To complete the repair, DNA polymerase I synthesizes a replacement strand containing the proper base, complementary to the template DNA strand.

- **Methyl mismatch repair.** The process of **methyl mismatch repair** requires recognition of the methylation pattern in DNA bases. When DNA polymerase misincorporates a base during replication, a mismatch forms between the incorrect base in the newly synthesized but unmethylated strand and the correct base residing in the parental, methylated strand. Methyl-directed mismatch repair enzymes bind to the mismatch and excise the stretch of DNA <u>opposite</u> the methylated strand. DNA polymerase I then fills in the correctly matched nucleotides, and ligase seals the gaps.

- **SOS ("Save Our Ship") repair.** When DNA damage is extensive, the repair strategies that excise pieces of DNA cannot work. The cell must take more drastic measures known as the **SOS response**, a coordinated cellular response to DNA damage that, in order to save the cell, can introduce mutations into severely damaged DNA. This "error-prone repair" does not correct mutations but does maintain the circular chromosome.

Another important repair process involves DNA **recombination**, the process of "crossing over" and exchange of two DNA helices. (Do not confuse "natural" DNA recombination within cells with "artificial" recombination of DNA through biotechnology—to be discussed in Section 8.4.) DNA recombination is best known for

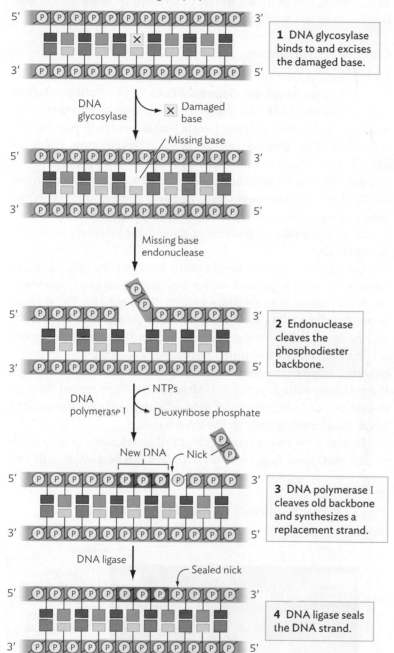

ANIMATION ▶ **Figure 8.15** **Base Excision Repair**
Specialized enzymes can recognize specific damaged bases and remove them from DNA without breaking the phosphodiester backbone. The result is an abasic site (a phosphoryl sugar with no base) that a specific endonuclease can recognize and cleave. After excision and removal, DNA polymerase I synthesizes a replacement strand containing the proper base.

1 DNA glycosylase binds to and excises the damaged base.

DNA glycosylase → ☒ Damaged base

Missing base

Missing base endonuclease

2 Endonuclease cleaves the phosphodiester backbone.

NTPs

DNA polymerase I

Deoxyribose phosphate

New DNA — Nick

3 DNA polymerase I cleaves old backbone and synthesizes a replacement strand.

DNA ligase

Sealed nick

4 DNA ligase seals the DNA strand.

its role in reassortment of eukaryotic chromosomes during meiosis (to be covered in Chapter 11). But the original function of recombination in our ancestral microbial cells was to enable replacement of a damaged DNA sequence by the "backup" of an intact homolog. Homologous sequences are available in bacteria because most growing cells maintain two or more copies of their genome before

the cell divides (see Section 8.2). The amazingly high radiation-resistant organism *Deinococcus,* isolated from irradiated food, retains eight copies of its DNA. The mechanism of DNA recombination is discussed in Chapter 9.

SECTION SUMMARY

- **Mistakes in DNA replication that fail to be corrected result in mutations.** Different kinds of mutation include base substitution, insertions and deletions, and inversions.

- **Pathogens may have high mutation rates, which help them evade the immune system.**

- **The effect of a mutation on the gene product depends on the genetic code.** A base change in a protein-coding gene may cause a silent mutation, a missense mutation, or a nonsense mutation. A base insertion may cause a frameshift.

- **A mutation's effect on phenotype depends on where the mutation occurs** and whether it affects the coding sequence or the expression of a gene.

- **Mutagens are chemical or physical agents that damage DNA and increase the rate of mutation.** Chemical agents include oxygen radicals that react with DNA. Physical agents include X-rays and UV radiation.

- **Repair systems include base excision repair and methyl mismatch repair.** Error-prone repair such as the SOS system allows base substitutions but preserves the DNA backbone to keep the chromosome intact.

Thought Question 8.2 When bacteria mutate and become resistant to an antibiotic, several different mechanisms of resistance are possible. How many different ways can you think of for a mutant protein to cause resistance?

Thought Question 8.3 Explain how it is possible for a very large mutation, involving a large part of a chromosome with many genes, to have no effect on the phenotype of the bacterium.

8.4
Biotechnology

SECTION OBJECTIVES

- Describe the use of plasmids, restriction enzymes, and gel electrophoresis to analyze DNA.

- Explain the formation of a recombinant DNA molecule, and describe applications in medical technology.

- Explain the use of DNA hybridization and DNA sequencing technology.

How can we use DNA to make discoveries in the laboratory? Isolated DNA can tell us many things, such as the identity of a participant in a crime scene or the presence of a pathogen in food. Analysis of DNA, together with construction of artificial DNA molecules for an applied purpose, is called **biotechnology**. The tools of biotechnology include physical and chemical methods, such as electrophoresis

Figure 8.16
Map of Plasmid pBR322

This plasmid (p) contains an origin of DNA replication and genes encoding resistance to ampicillin (*amp*) and tetracycline (*tet*).

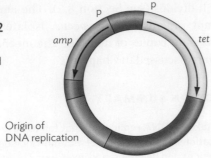

(discussed shortly), as well as enzymes isolated from bacteria, such as DNA polymerases. Here we describe key tools of DNA analysis: plasmid analysis, DNA hybridization, polymerase chain reaction (PCR) amplification, and DNA sequence analysis.

Plasmids in Biotechnology

Biotechnology makes use of plasmids, such as that shown in **Figure 8.16**. A plasmid has certain DNA sequences that can be cut by a specific enzyme called a restriction endonuclease. A **restriction endonuclease** is an enzyme that cuts DNA only at a specific DNA sequence called a **restriction enzyme site** (**Figure 8.17**). Usually, the restriction sequence is a "palindrome," in which the sequence of base pairs reads the same forward and back.

Figure 8.17   Bacterial DNA Restriction Endonucleases

Target sequences of sample restriction enzymes. The names of these enzymes reflect the genus and species of the source organism. For example, *Eco*RI comes from *Escherichia coli*.

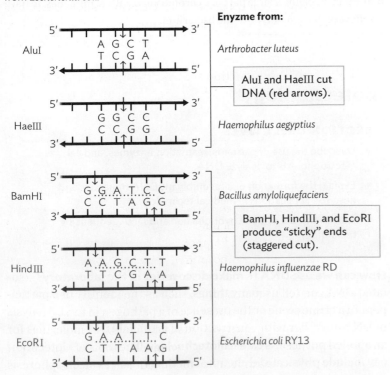

In nature, restriction enzymes are made by various kinds of bacteria to cleave foreign invading DNA that could be from a virus or an intracellular parasite. In biotechnology, we culture recombinant bacteria engineered to make the enzymes so we can use the enzymes to cut DNA at specific sequences. A cut piece of DNA, such as a plasmid, can be joined to DNA extracted from another organism, forming "recombinant DNA" (discussed shortly).

Gel Electrophoresis

Restriction enzymes generate DNA fragments that can be separated by a technique called **gel electrophoresis** to yield a reproducible pattern (**Figure 8.18**). The gel for DNA separation is formed from agarose, a long-chain polysaccharide that is soluble in water above 50°C (122°F) but forms a solid gel at room temperature. The agarose gel matrix has molecular holes through which a water solution can flow; but larger molecules, such as DNA, move through more slowly. The DNA fragments have negative charges on their phosphoryl groups; thus, the negatively charged fragments move toward a positive electrode, with a rate of movement inversely related to fragment size.

Gel electrophoresis begins with forming the gel. A warm agarose solution is poured into a tray within an electrophoresis chamber that leads to a voltage source (**Figure 8.19**). The gel tray includes a "comb," a piece of plastic with prongs that form wells within the gel. Once the gel has cooled and solidified, the comb is removed and DNA samples are added. The DNA sample buffer contains a dye that marks the movement of the sample through the gel under voltage potential. After the DNA has moved through the gel, driven by the electrical field, a fluorescent dye bound to the DNA allows photography of the DNA bands.

Figure 8.20 shows how a pattern of restriction cut sites leads to different-sized fragments of DNA. A DNA sequence with one

Figure 8.18   Restriction Fragment Length Analysis

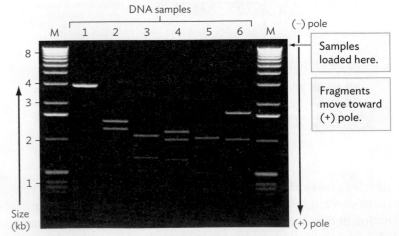

Agarose gel electrophoresis of EcoRI-restricted DNA fragments. M refers to marker fragments of known size. Smaller fragments move toward the positive charge faster than large fragments.

Figure 8.19   Agarose Gel Electrophoresis Separation of DNA Fragments

DNA samples are loaded into slots at the end of an agarose gel (step 1). When a voltage is applied (step 2), the DNA fragments (which have negative charge) are attracted toward the positive electrode (step 3). The fragments move according to size; the smaller fragments move fastest. When the fragments have run far enough to separate, they can be visualized by fluorescence when exposed to UV light (step 4).

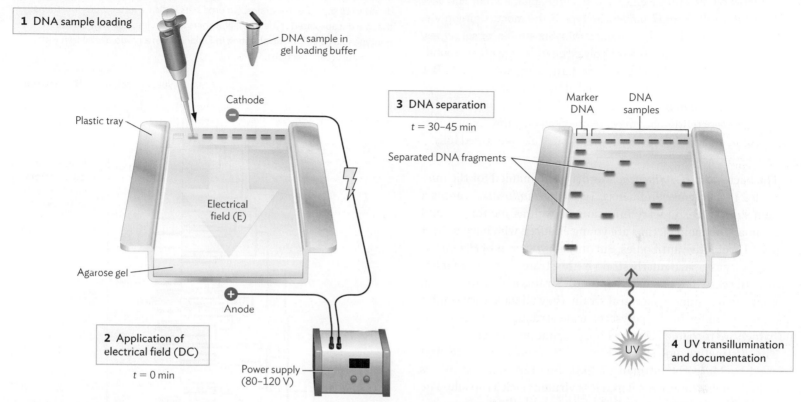

1 DNA sample loading

DNA sample in gel loading buffer

Plastic tray

Cathode

3 DNA separation

$t = 30–45$ min

Marker DNA   DNA samples

Separated DNA fragments

Electrical field (E)

Agarose gel

2 Application of electrical field (DC)

$t = 0$ min

Anode

Power supply (80–120 V)

UV

4 UV transillumination and documentation

Figure 8.20   EcoRI Cut Sites

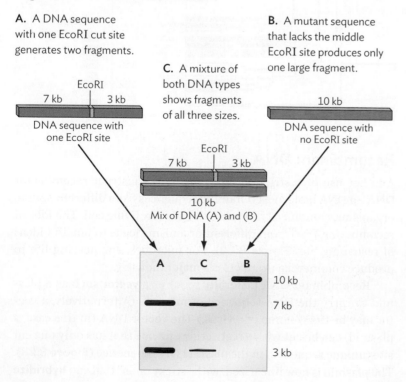

A. A DNA sequence with one EcoRI cut site generates two fragments.

B. A mutant sequence that lacks the middle EcoRI site produces only one large fragment.

C. A mixture of both DNA types shows fragments of all three sizes.

EcoRI

7 kb   3 kb

DNA sequence with one EcoRI site

10 kb

DNA sequence with no EcoRI site

EcoRI

7 kb   3 kb

10 kb

Mix of DNA (A) and (B)

A   C   B

10 kb

7 kb

3 kb

EcoRI cut site (A) generates two fragments separated in the electrophoretic gel. A mutant sequence that lacks the middle EcoRI site (B) produces only one fragment, equal in size to the sum of the two smaller fragments in (A). A mixture of both DNA types shows fragments of all three sizes (C).

How can we use restriction enzyme fragments of DNA for clinical diagnosis?

CASE HISTORY 8.1

Sepsis by a Rare Strain

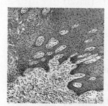

In San Francisco, a 3-year-old boy named Luke was admitted to the hospital after an abrupt onset of fever and a generalized seizure. The boy's cerebrospinal fluid was normal, and he had a normal chest radiograph, but his white blood cell count was 20,000/mm³ (normal is 5,000–10,000), with 85% polymorphonuclear leukocytes (PMNs). Luke's blood culture revealed bacteremia (bacteria in the blood). The bacteria were identified as "nontypeable" *Haemophilus influenzae* bacteria. The nontypeable *H. influenzae* strain was beta-lactamase-positive (expressed an enzyme to inactivate penicillin G, ampicillin, and amoxicillin) but proved sensitive to cefixime, a third-generation cephalosporin.

The diagnosis surprised the boy's two fathers. They showed the doctor Luke's full record of immunizations and booster shots, including the standard DTaP (diphtheria, tetanus, and pertussis), hepatitis B, pneumococcus, and *Haemophilus influenzae* b (Hib). Why was Luke not protected by the Hib vaccine? The doctor explained that Hib vaccine protected only from *H. influenzae* type b, the most common virulent strain—but not the only virulent strain. Six strains, or serotypes, a–f are "typed" based on the type of polysaccharide capsule surrounding the cell envelope. The Hib vaccine targets capsule type b. But some variant strains of *H. influenzae*, called "nontypeable," lack a capsule. These strains can be identified on the basis of other properties, such as enzyme activities, but they are unaffected by the vaccine. Fortunately, after 7 days of intravenous antibiotic, Luke recovered fully.

The bacterium *H. influenzae* was originally named for the mistaken idea that it causes influenza. In fact, *H. influenzae* causes a different kind of respiratory infection, as well as meningitis and bacteremia. The main victims are young children who have not yet developed protective antibodies. But only a few strains of *H. influenzae* cause disease, particularly strain b (the strain that caused meningitis in the case that opened Chapter 6). Strains a–f all arose from mutations in a common ancestral strain; they all have very similar genomes, but differ in the polysaccharide structures of their capsule, which are recognized by the human immune system.

The distinctive polysaccharides of *H. influenzae* were first observed by Margaret Pittman in 1931. But Pittman and others knew that other strains considered less virulent lack a capsule. The strains lacking a capsule are called "nontypeable," since they cannot be immunologically "typed" according to polysaccharide. Nontypeable, or unencapsulated, strains are unaffected by the body's immune response to the Hib vaccine, which targets capsule polysaccharide. Because most children in the United States now receive Hib vaccine, a growing proportion of the *H. influenzae* infections seen in hospitals are caused by nontypeable strains. Vaccines and immunization will be discussed more deeply in Section 17.6.

How can the "nontypeable" strains be identified? Metabolic tests can be used, but DNA-based tests are rapidly replacing them. The genetic relatedness of hundreds of strains can be measured by using restriction enzyme sites to compare the genomic DNA sequences of the *H. influenzae* strains. The method shown in Figure 8.21 is called "ribotyping," based on analysis of restriction enzyme sites in DNA sequence flanking the genes that specify ribosomal RNA (rRNA). The sites at which different individuals show different lengths of restriction-cut DNA are called **restriction fragment length polymorphisms (RFLPs)**. The number of differences in the DNA sequence was compared for all pairs of *H. influenzae* isolated and used to estimate the genetic distance between isolates. A genetic distance—that is, the length of each tree branch—is roughly a measure of the time since two given strains diverged. Although all these strains possess nearly the same genome, their tiny differences arising from mutation can cause huge differences in virulence—the difference between life or death of a child.

Figure 8.21 **Relatedness of *Haemophilus influenzae* Strains**
A phylogenetic tree was constructed by ribotyping 400 isolates of *H. influenzae*. The sequences were compared by computer analysis. The length of horizontal bars represents the sequence difference between each pair of strains, in base pairs per generation (cell division). The longer the bar, the more base differences separate the strains. Colored bars indicate letter types for strains that are encapsulated (typeable). Numbers at right indicate new isolates of nontypeable *H. influenzae* obtained in Finland from a group of children with otitis media (ear infection).

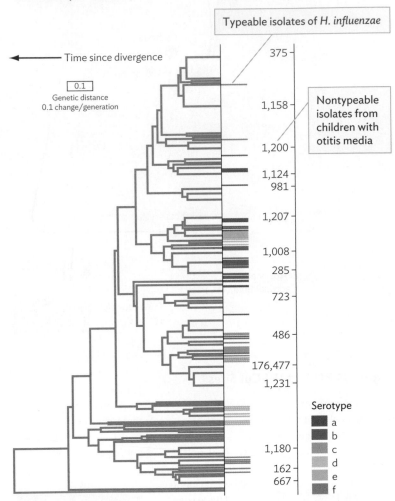

Recombinant DNA

Another use of restriction enzymes is to construct **recombinant DNA**—a DNA molecule containing sequences from different source organisms, combined "in a tube" outside any living cell. The idea of recombining DNA from different organisms leads to fanciful ideas of reshaping life (Figure 8.22). For microbes, engineering life to produce commercial products is a major industry.

Recombinant DNA generally requires a **vector** such as a plasmid to carry the DNA sequence of interest. (Alternatively, a vector may be the genome of a virus.) The vector DNA (in this case, a plasmid) can be cut with a restriction enzyme that has only one cut site (unique sequence) in the plasmid DNA sequence (Figure 8.23). The plasmid is now linearized, with "sticky ends" that can **hybridize**

Figure 8.22 Results of Recombinant DNA?

The imagined results of using recombinant DNA to modify animal genomes.

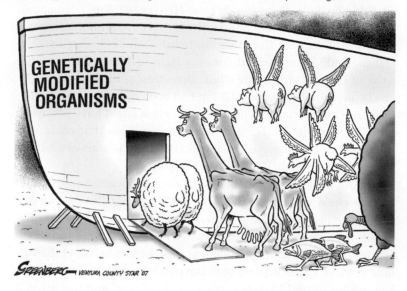

(bind to a complementary sequence of bases from a different helix). The ends hybridize to single-stranded ends of other DNA cut with the same enzyme.

A piece of foreign DNA—from any species of organism—can be cut with the same restriction enzyme; or alternatively, nucleotides can be added onto the ends of a DNA fragment to generate the same sticky ends. Now the piece of foreign DNA is combined in the same test tube with the cut vector plasmid. Some of the cut plasmids will hybridize to both ends of a piece of foreign DNA, re-forming a circle. The DNA circle is now completed by the enzyme DNA ligase, which seals the "nicks" in the backbone where the sticky ends hybridized. The recombinant plasmid can then be put back into a microbial cell by a process called **transformation**. Transformation generates a cell that can now express the protein product of a gene in the recombinant plasmid.

A somewhat different form of recombinant DNA is called **cDNA**, in which DNA is reverse-copied from a messenger RNA molecule and then inserted or cloned into a vector. The cDNA technology is presented in Section 8.5.

Recombinant DNA construction is one of many different technologies that use microbial DNA for important products of biotechnology. For example, the vaccine against human papillomavirus (HPV), now recommended for all boys and girls before puberty, consists of virus coat proteins expressed by viral genes in recombinant yeast (Gardasil, by Merck) or in recombinant baculovirus in insect cell culture (Cervarix, by GlaxoSmithKline). These vaccines are much safer than traditional vaccines using live or inactivated viruses, because no viruses are used for their manufacture—only coat proteins made from recombinant microbes that cannot infect humans.

DNA Hybridization

How can we use DNA to identify a pathogen? In the clinical lab, a quick test for certain pathogens can be performed by DNA hybridization with a known "probe" of a pathogen gene. For example, food can be tested for DNA from enteric pathogens such as *E. coli* O157:H7 or other strains having a gene that encodes Shiga toxin.

To analyze a DNA sample, researchers can hybridize the sample (called the "target") to a short sequence (the probe), which has an attached fluorophore (fluorescing molecule) (Figure 8.24). For example, the target DNA could be a sample from food, whereas the

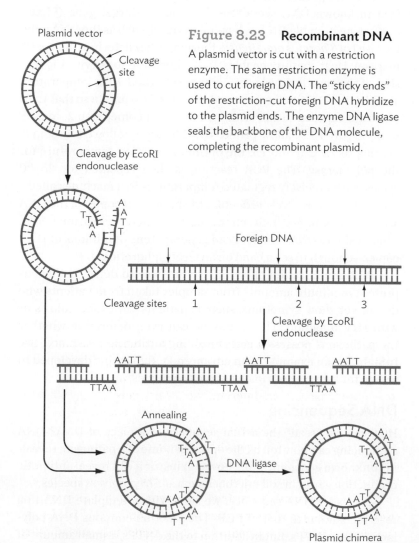

Figure 8.23 Recombinant DNA

A plasmid vector is cut with a restriction enzyme. The same restriction enzyme is used to cut foreign DNA. The "sticky ends" of the restriction-cut foreign DNA hybridize to the plasmid ends. The enzyme DNA ligase seals the backbone of the DNA molecule, completing the recombinant plasmid.

Figure 8.24 DNA Hybridization

Hybridization of a labeled DNA probe (short DNA sequence) to its complementary sequence in a target DNA.

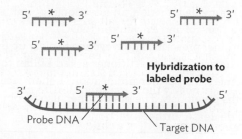

probe DNA could be a gene encoding Shiga toxin. The target DNA is first denatured (the helix melted) so that single strands come apart, exposing their bases. The DNA is then bathed in a solution containing the labeled probe DNA, which can base-pair specifically (hybridize) to a complementary sequence in the target DNA. The hybridization reaction is then detected by fluorescence. Alternative labels can be a radioisotope or an enzyme that reacts with a detectable substrate.

Hybridization techniques are rapid and inexpensive and can detect microbial pathogens in a large sample. But in clinical isolates, especially from infected patients, the pathogenic cells are few, with vanishingly small amounts of DNA. A more sensitive detection method is needed.

Polymerase Chain Reaction Amplification

Imagine how easily you could diagnose illness if you could detect one single copy of DNA in a patient's tissue sample. You might be able to do this by making a million copies in just 2 hours. A million copies of DNA is easy to detect by fluorescent dyes. This "amplification" of DNA offers a way to detect a pathogen.

CASE HISTORY 8.2

Spotting Surprise

In Wichita, Kansas, a 20-year-old college student named Lillian visited the college health practitioner for a routine gynecological exam. Lillian's periods had been regular, but she recently noted some spotting between periods. Her last menstrual period was 4 weeks before. The practitioner asked Lillian whether she was sexually active. Lillian reported two male partners within the past 6 months. She reported no vaginal discharge, dyspareunia (painful intercourse), genital lesions, or sores. Her breast, thyroid, and abdominal exam were within normal limits, as were her vital signs: blood pressure 118/68, pulse 74, respiration 18, temperature 37.1°C (98.6°F).

The genital exam revealed normal vulva and vagina. The practitioner found no cervical motion pain and no uterine or adnexal tenderness (areas that include and surround the fallopian tubes and ovaries). But Lillian's cervix appeared inflamed and bled easily, with a purulent discharge coming from the cervical os (opening).

The practitioner performed nucleic acid amplification tests (NAATs) for *Chlamydia trachomatis* and *Neisseria gonorrhoeae*. Within a day, the NAAT results were positive for *Chlamydia* and negative for *N. gonorrhoeae*. Wet mount microscopy revealed no pathogens but many WBCs (white blood cells). The diagnosis was confirmed as chlamydia infection, an infection reportable to the Centers for Disease Control and Prevention (CDC).

The practitioner prescribed doxycycline to clear up the infection. To curtail transmission, the practitioner asked Lillian to identify all her sexual partners for partner notification. Lillian recalled a third partner 7 months previously. The practitioner also recommended use of condoms to decrease further infections.

The capability of DNA amplification is now routine in thousands of laboratories around the world—such as the laboratory available for Lillian's testing in Case History 8.2. The DNA amplification in the NAAT did not require culture, or even microscopic detection of the pathogen, an obligate intracellular bacterium. The test enabled Lillian's practitioner to diagnose her condition within a day of Lillian's visit to the clinic. Detection of the correct pathogen is important because if Lillian had been treated with an inappropriate antibiotic, her chlamydia infection could have persisted, ultimately causing scarring of her fallopian tubes that could lead to infertility. Undiagnosed chlamydia infection is a major cause of infertility in the United States (see Chapter 23).

NAAT is based on a method of DNA amplification called **polymerase chain reaction (PCR)**. The DNA amplification technique makes use of heat-stable DNA polymerase isolated from thermophiles, bacteria, and archaea adapted to live in extreme heat. PCR was used in the case history opening this chapter to detect the various kinds of bacteria present in the patient's diseased gums. PCR has revolutionized biological research, medicine, and forensics; it commonly identifies or exonerates criminal suspects.

In PCR, specific DNA primers (usually 20–30 bases) are hybridized to known DNA sequences flanking the target gene (**Figure 8.25**). A heat-stable DNA polymerase is used, such as the DNA polymerase I of *Thermus aquaticus*, a thermophilic bacterium (Taq polymerase). Heat stability enables Taq polymerase to survive the first step of the reaction cycle: heating to 95°C (203°F), a temperature at which the DNA helix melts and the strands separate so that their bases are exposed to pair with primers. Next comes step 2, cooling to 55°C (131°F) to allow primer hybridization, followed by step 3, heating to 72°C (161.6°F), the optimum reaction temperature for the polymerase. The PCR reaction is then repeated for 25–30 cycles, with precisely regulated temperatures in a machine called a thermal cycler. Each cycle doubles the number of copies of the DNA sequence positioned between the forward and reverse primer ends. Thus, a single DNA template can generate tens of millions of DNA copies—enough to see a band on an electrophoretic gel.

PCR amplification offers an exciting way to detect pathogens present in minute amounts from samples taken from patients who do not yet show symptoms, such as patients with tuberculosis or with HIV infection. The technique can even determine whether the pathogens possess a gene encoding antibiotic resistance. See **inSight** for an example of an advanced PCR machine developed to screen for tuberculosis in developing countries.

DNA Sequencing

How can we "read" the actual sequence of a piece of DNA? DNA sequencing can now tell us the entire genome of a pathogen, revealing unknown genes encoding novel toxins; or it can reveal antibiotic production genes in soil microbes such as *Streptomyces* species.

To read a DNA sequence, we start with a template DNA in a reaction similar to that of PCR. The reaction contains DNA polymerase and dNTPs. But in addition to the dNTPs, a small amount of

Figure 8.25   Polymerase Chain Reaction (PCR)

The cyclic PCR reaction makes a large number of copies of a small piece of DNA.

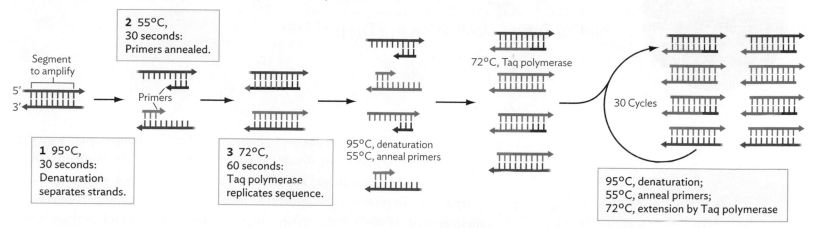

2 55°C, 30 seconds: Primers annealed.

Segment to amplify

1 95°C, 30 seconds: Denaturation separates strands.

Primers

3 72°C, 60 seconds: Taq polymerase replicates sequence.

95°C, denaturation 55°C, anneal primers

72°C, Taq polymerase

30 Cycles

95°C, denaturation; 55°C, anneal primers; 72°C, extension by Taq polymerase

Figure 8.26   DNA Sequencing

The DNA molecule is used as a template to synthesize complementary DNA in a reaction containing fluorescently tagged nucleotides of one of the four types (A, T, C or G). The tagged nucleotide terminates a DNA sequence as a short piece. The short pieces of DNA are separated by size by electrophoresis, appearing as bands in the gel. A laser detects the different fluorescent-tagged DNA pieces using their four different wavelengths of light emission.

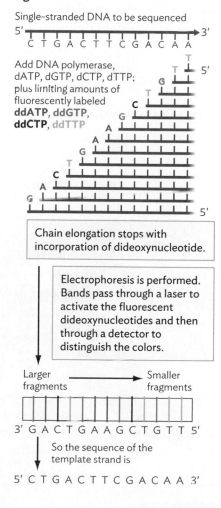

Single-stranded DNA to be sequenced

Add DNA polymerase, dATP, dGTP, dCTP, dTTP; plus limiting amounts of fluorescently labeled **ddATP, ddGTP, ddCTP, ddTTP**

Chain elongation stops with incorporation of dideoxynucleotide.

Electrophoresis is performed. Bands pass through a laser to activate the fluorescent dideoxynucleotides and then through a detector to distinguish the colors.

Larger fragments → Smaller fragments

3′ G A C T G A A G C T G T T 5′

So the sequence of the template strand is

5′ C T G A C T T C G A C A A 3′

"dideoxy" nucleotide is included for one of the four bases, such as dideoxy adenine. "Dideoxy" means that the 3′ OH is replaced by H. Thus, after the dideoxynucleotide is incorporated into DNA, there is no further 3′ OH available to add the next nucleotide. Four reactions are conducted in the same tube, with a different fluorescent label for each of the four dideoxy bases (A, T, C, or G).

The DNA polymerase then synthesizes DNA on the basis of the template, such as a PCR-amplified DNA sample from a bacterium. But each sequencing reaction contains just enough dideoxy ATP (or CTP, TTP, or GTP) to cause DNA polymerase occasionally to substitute the dead-end base for the natural one, at which point the chain stops growing. The result is a population of DNA strands of various size, each one cut short at a different adenine position. Now imagine using four different dideoxynucleotides corresponding to A, T, C, and G, with each dideoxy base tagged with a different-colored fluorescent dye (Figure 8.26). The result will be a series of different-sized strands tagged at their 3′ end with different colors, depending on the base incorporated. Electrophoresis for each of the four reactions will separate the various fragments according to size. A laser and detector positioned at the bottom of the electrophoretic gel read the individual fragments as they pass. The computer then prints a series of colored peaks whose order corresponds to the template DNA sequence.

The method described here to read DNA sequences is called Sanger sequencing. Today, even faster technologies, such as Illumina, allow rapid sequencing of whole genomes.

SECTION SUMMARY

- **A restriction endonuclease** is an enzyme that cleaves DNA at a specific sequence, which is usually a palindrome.
- **Gel electrophoresis** separates DNA fragments, enabling identification of patterns of restriction enzyme cut sites in a DNA sequence.
- **Recombinant DNA** is made by inserting a piece of DNA into a plasmid or viral DNA cut with the same restriction enzyme. The DNA backbone is sealed with DNA ligase. The inserted piece of DNA has a gene that can be expressed by a cell containing the recombinant plasmid or infected by a recombinant virus.
- **DNA hybridization** can identify a source of DNA by base-pairing between a target DNA and a labeled probe DNA. The probe DNA must be complementary to a sequence within the target.
- **Polymerase chain reaction (PCR)** amplifies (makes many copies of) a short DNA sequence between two primer sequences directed toward each other. A heat-resistant polymerase catalyzes the synthesis of DNA from each primer, and thermal cycling allows exponential amplification.

inSight

Finding Tuberculosis DNA in a Drop of Spit

Tuberculosis is a global plague; in India alone, 2 million people fall ill with the disease every year. But how do we detect this insidious disease, identify the victims, and treat them? The pathogen, *Mycobacterium tuberculosis*, grows slowly and can be present in extremely low numbers. The most common test, a sputum slide assay, misses half the cases of disease. Even when a case is identified, how do we provide the right antibiotics? It takes 2 weeks to culture *M. tuberculosis* to know which antibiotics will be effective. If the wrong antibiotic is prescribed, it will select only for the spread of a resistant strain.

Figure 1   Detecting Tuberculosis Bacteria with or without Rifampicin Resistance

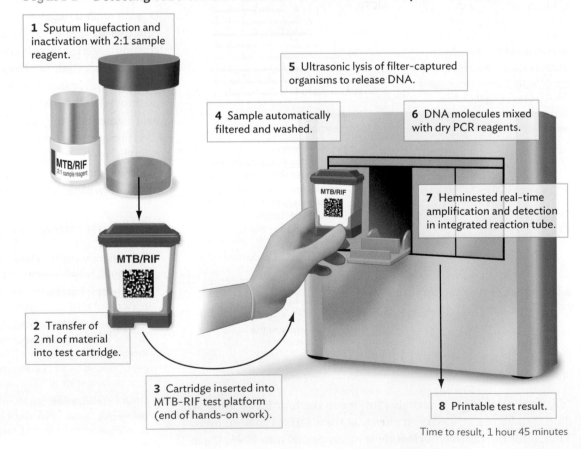

1 Sputum liquefaction and inactivation with 2:1 sample reagent.

5 Ultrasonic lysis of filter-captured organisms to release DNA.

4 Sample automatically filtered and washed.

6 DNA molecules mixed with dry PCR reagents.

7 Heminested real-time amplification and detection in integrated reaction tube.

2 Transfer of 2 ml of material into test cartridge.

3 Cartridge inserted into MTB-RIF test platform (end of hands-on work).

8 Printable test result.

Time to result, 1 hour 45 minutes

Figure 2 Real-Time PCR Detects Tuberculosis Bacteria

M. tuberculosis genomic DNA samples were amplified.

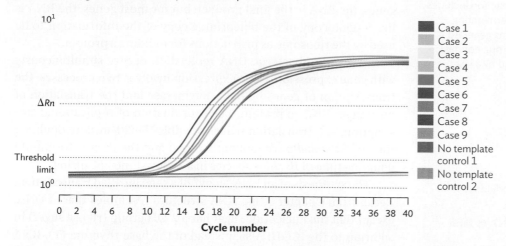

An international consortium of scientists, led by the Foundation for Innovative Diagnostics, in Geneva, Switzerland, has developed a new instrument to identify minute quantities of *M. tuberculosis* DNA. The instrument, called the Xpert MTB/RIF (Figure 1), uses advanced PCR technology to amplify tiny amounts of DNA in a drop of sputum (deep lung material that can be coughed up). As few as five bacterial genomes can provide enough DNA for amplification (Figure 2).

In the Xpert instrument, PCR primers are used that contain sequences flanking the gene *rpoD*, which encodes RNA polymerase. As discussed in Section 8.5, RNA polymerase is a central enzyme of gene expression, transcribing all bacterial genes to make RNA copies; messenger RNA copies are then decoded by ribosomes to make proteins. The unique *rpoD* sequence is amplified by a process called heminested real-time PCR. Heminested (half nested) means that a second round of amplification cycles replaces one of the primers with a different primer, while keeping the original primer. The introduction of one different primer helps to screen out extraneous amplifications. Real-time means that each round of amplification can be measured as it occurs, because the fluorescent label detection is extremely sensitive. Thus, we observe the DNA concentration as it rises.

Once the *rpoD* DNA is detected, its sequence is checked for a key mutation that confers resistance to the drug rifampicin. If the resistance sequence appears, then physicians know immediately to exclude rifampicin from the drug combination that is prescribed. The entire process of the Xpert instrument takes under 2 hours; all the steps are automated, requiring no advanced knowledge to run. In a study reported in the *New England Journal of Medicine*, the instrument detected tuberculosis in 98% of a patient group in Uganda and correctly identified all rifampicin-resistant cases. It can be run in remote villages by individuals with modest training. Could this process be made available widely? Providers such as the Bill and Melinda Gates Foundation are trying to do just that.

- **DNA sequencing** is a technique to read the sequence of base pairs in a DNA molecule. The sequence of a template DNA can be read by using reactions of DNA synthesis by a DNA polymerase. Each reaction includes dNTPs with a small amount of fluorescent-tagged chain-terminating dideoxy NTP for each of the four nucleotides. The reactions generate patterns of DNA fragments terminated at each of the four nucleotides; the mixtures are separated by electrophoresis and detected by a laser.

8.5
Transcription of DNA to RNA

SECTION OBJECTIVES

- Describe the different kinds and functions of RNA products that are made from DNA.
- Explain how RNA polymerase copies a sequence of DNA to form mRNA.
- For biotechnology, describe the use of reverse transcriptase to make a cDNA copy of an mRNA.

Figure 8.27   Operon for a Multidrug Resistance Protein Complex

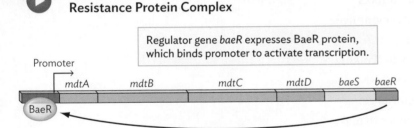

Regulator gene *baeR* expresses BaeR protein, which binds promoter to activate transcription.

A. Genes *mdtA*, *mdtB*, *mdtC*, and *mdtD* encode proteins of an efflux pump that exports antibiotics such as novobiocin. Gene *baeS* encodes an environmental sensor; *baeR* encodes a regulator that binds the promoter.

How does a gene form the product that it encodes? First, the enzyme **RNA polymerase** makes an RNA copy of the gene. For some genes, the RNA is the final product; but for most genes, the RNA is like a photocopy of the blueprint, a copy of the information to be used by the ribosome as instructions for building a protein.

In growing bacteria, DNA replication occurs simultaneously with gene expression. Gene expression involves two processes: the transcription of genes by RNA polymerase and the translation of messenger RNA to protein. The coordination of replication, transcription, and translation makes possible the 20-minute doubling time of *Salmonella* contaminating food and the record 10-minute doubling time of *Bacillus* species in a Yellowstone hot spring.

The synthesis of an RNA "copy" of a gene is called **transcription**. Recall from Chapter 3 that RNA structure resembles that of DNA, except that the nucleotide sugar has a 2′ OH group (ribose sugar) in addition to the 3′ OH. Also, instead of the base thymine (T), RNA uses uracil (U). In evolution, RNA evolved earlier in our ancestral cells (3.8 billion years ago), where it originally accomplished all the functions that DNA and protein serve in cells today. This model of early life, called the "RNA world," is discussed in Chapter 9.

Gene Expression: An Overview

How are genes organized on a chromosome, and what products are expressed? Many genes have as their final product an RNA molecule—for example, the ribosomal RNA molecules that form the main structural components of the small and large subunits of ribosomes (to be described in Section 8.6). Other "functional RNA" or "small noncoding RNA" molecules found in bacteria include transfer RNAs (tRNAs) and regulatory small RNAs (sRNAs). Eukaryotes have a much larger number of noncoding RNAs, including several thousand that serve as enzyme subunits.

The majority of genes, however, are transcribed to **messenger RNA (mRNA)**, which encodes one or more peptides (proteins) to

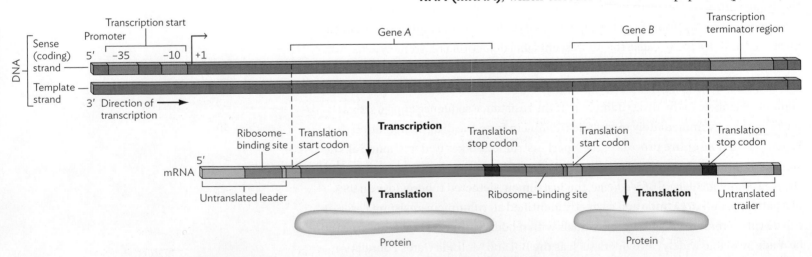

B. Alignment of a bacterial structural gene with its mRNA transcript. RNA polymerase binds to the –10 and –35 regions of DNA and, reading the template strand, begins RNA transcription at nucleotide +1. Transcription produces messenger RNA (mRNA). The mRNA molecule has an untranslated region of variable length and then a ribosome-binding site. A few nucleotides downstream, the ribosome begins translating at the translation start codon (usually AUG). The start codon specifies the amino terminus of the translated protein.

be translated by a ribosome. A series of genes in tandem, controlled by one promoter, makes an operon. **Figure 8.27A** shows an *E. coli* operon encoding proteins of a membrane complex that pumps out toxic molecules such as the antibiotic novobiocin. Within the operon, each structural gene (gene encoding a protein) is designated by an italic three-letter name, which may have a fourth letter in upper case (*mdtA*), as discussed in Section 8.1. All the genes of the operon shown are transcribed together, forming one mRNA. The genes are transcribed starting from the promoter, a short DNA sequence "upstream" of the gene *mdtA*.

How does the DNA sequence relate to the corresponding mRNA? The alignment between the DNA region comprising a structural gene and the resulting mRNA transcript is shown in **Figure 8.27B**. The operon shown here contains two genes (genes *A* and *B*). The double-helical DNA strands are shown separately in order to clarify which is the **sense strand** (containing the base sequence equivalent to that of the mRNA, substituting U's for T's) and its complement, the **template strand**. The template strand is the one that actually enters the RNA polymerase to base-pair with the growing chain of mRNA.

On the operon map in Figure 8.27, upstream (left) of the +1 start site are short DNA sequences that help bind the RNA polymerase. The +1 marks the DNA base where the mRNA transcript starts. The promoter sequences where RNA polymerase first binds are approximately –10 and –35 bases upstream of the +1 start. The sequence of the promoter can influence the amount of mRNA synthesized (discussed in Chapter 9).

RNA Synthesis

Like DNA synthesis, transcription to RNA includes stages of initiation, elongation, and termination. Each stage requires specific molecules to interact with the RNA and RNA polymerase. These steps are targets for antibiotics, such as rifamycin, which blocks RNA polymerase; rifamycin is an important drug for tuberculosis.

RNA polymerase is the enzyme complex that adds nucleotides onto the growing RNA chain. In bacteria, the enzyme complex is composed of four subunits, as seen in **Figure 8.28** (step 1). Before initiation, RNA polymerase complex includes a fifth subunit, called sigma (the Greek letter σ). The **sigma factor** can be one of several different versions that regulate genes responding to major environmental conditions such as heat shock (fever, within a host), oxidation stress (phagocytosis by a macrophage), or starvation (competition with other microbes in the colon). Genes whose products are required under most typical conditions are regulated by a "housekeeping" sigma factor.

When RNA polymerase complexes with a sigma factor, the composite RNA polymerase is known as a "holoenzyme," meaning that it includes all the subunits needed for initiation. The holoenzyme form of RNA polymerase binds loosely to any sequence of DNA. Thus, within the cell, the RNA polymerase binds and comes off DNA rapidly, as if "scanning" the sequence. When the RNA polymerase binds to a promoter sequence, however, the binding is tight

(Figure 8.28, step 2). To start transcribing a gene, the RNA polymerase–DNA complex must open through the unwinding of one helical turn of DNA, which causes DNA to become unpaired (Figure 8.28, step 3). Now the RNA polymerase can fit a ribonucleotide to base-pair with the DNA. To elongate the chain, the polymerase catalyzes the addition of new ribonucleotide triphosphates (rNTPs). Just as for DNA, each incoming rNTP condenses with a 3′ OH and releases pyrophosphate, the step that yields energy. Chain elongation is the step that can be blocked by rifamycin.

Only a few ribonucleotides can be added before the polymerase stops. Remember, the whole complex is still bound tightly to the promoter. So what happens? The sigma factor falls off the complex (Figure 8.28, step 4), allowing the core RNA polymerase to continue to move along the template, synthesizing RNA at approximately 45 bases per second. The unwinding of DNA ahead of the moving complex forms a 17-base-pair "transcription bubble" in which RNA is being synthesized, while behind it the two DNA strands "snap" back

Figure 8.28   Transcription Initiation and Elongation
The RNA polymerase holoenzyme scans the DNA sequence until it binds to a promoter. The promoter DNA helix unzips, exposing bases of the template to base-pair with incoming nucleotides. The polymerase catalyzes the addition of nucleotides. The sigma factor must fall off before elongation continues.

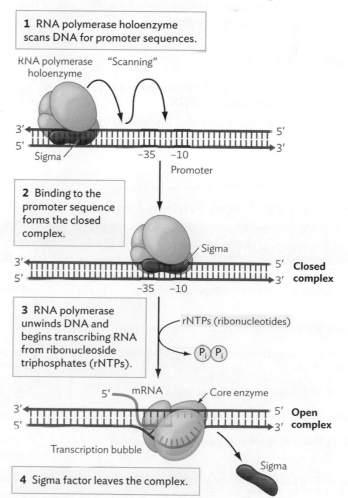

together. Because of the DNA unwinding, positive supercoils are formed ahead of the advancing bubble. These positive supercoils are removed by topoisomerase enzymes that return the DNA to its normal negative-supercoiled state. Finally, transcription must finish at the end of the mRNA, at a "pause site" in the DNA sequence that causes the RNA polymerase to fall off.

Biotechnology: Making DNA Copies of mRNA

For biotechnology (see Section 8.4), we often need to make a DNA copy of mRNA, called **cDNA**. The cDNA is useful because it provides a way to copy the protein-coding portion of a human gene (such as the gene for insulin) while omitting all the noncoding intron sequences that interrupt the gene. A cDNA can be cloned in a plasmid vector and then transformed into a bacterium such as *E. coli*, which then expresses large amounts of the therapeutic protein. Cloned cDNAs are used by the pharmaceutical industry to manufacture many important therapeutic agents.

How is cDNA made from mRNA? We "reverse-transcribe" the mRNA using an enzyme called **reverse transcriptase**. Reverse transcriptase is an enzyme derived from retroviruses (discussed in Chapter 12) such as human immunodeficiency virus (HIV). The HIV needs a reverse transcriptase in order to copy its RNA genome into DNA for integration into a host cell chromosome. Thus, we see an example of how even deadly retroviruses provide tools of biotechnology that can ultimately save lives.

SECTION SUMMARY

- Gene sequences contained in the DNA sense strand of an operon correspond to the sequence of an mRNA.
- The first gene of the operon is preceded (upstream) by a noncoding sequence called a promoter. RNA polymerase with a sigma factor

(holoenzyme) scans DNA (binds loosely and comes off) and then binds tightly to the promoter.

- **Promoter DNA unwinds, exposing bases to pair with incoming ribonucleotides.** The sequence of ribonucleotide triphosphates added is complementary to bases in the template DNA strand.
- **Sigma factor falls off, allowing RNA polymerase (core enzyme) to slide down the DNA and continue elongation.** Elongation terminates at a pause site, where the RNA polymerase falls off.
- **For biotechnology, DNA copies of mRNA are made using a viral reverse transcriptase.** The DNA copies are called cDNA.

8.6 Protein Synthesis

SECTION OBJECTIVES

- Explain the relationship between transcription and translation in bacteria.
- Describe the ribosomal steps of initiation, elongation, and termination. Explain how these steps may be targeted by antibacterial agents.
- Explain how proteins may require folding and secretion.

The translation of the mRNA message into an amino acid sequence by the ribosome is one of the most complex processes in molecular biology. This complexity helps explain why such a large number of antibiotics have been discovered to block the protein-making machinery of the ribosome (Table 8.3). Once formed, proteins determine most of the distinguishing traits of a cell. For instance, a thermophilic archaeon that grows at an ocean vent at temperatures above 100°C (212°F) makes proteins that can withstand extreme heat. On the other hand, an enteric pathogen that colonizes the intestinal lining

Table 8.3
Antibiotics That Target Protein Synthesis

Ribosome Subunit (30S or 50S): Process of Protein Synthesis	Antimicrobial Molecule	Organism Affected	Source of Molecule
30S subunit: Bind 16S rRNA at A site; interfere with incoming aminoacyl-tRNA. Also affects envelope stability.	Aminoglycosides (streptomycin)	Aerobic bacteria (Gram-positive and Gram-negative)	*Streptomyces griseus*
30S subunit: Block incoming aminoacyl-tRNA from binding A site	Tetracyclines	Bacteria	*Streptomyces* spp.
50S subunit: Bind peptide release tunnel and block translocation from A site to P site	Macrolides (erythromycin)	Bacteria	*Saccharopolyspora erythraea*
50S subunit: Inhibit peptidyltransferase activity	Lincosamides	Bacteria	*Streptomyces lincolnensis*
50S subunit: Inhibit complex formation	Everninomicins	Gram-positive bacteria	*Micromonospora carbonacea*
50S subunit: Bind 23S rRNA near P site; inhibit translation	Oxazolidinones (linezolid)	Gram-positive bacteria	Chemical synthesis
50S subunit: Bind peptidyltransferase domain and block elongation	Streptogramins	Bacteria	*Streptomyces* spp.

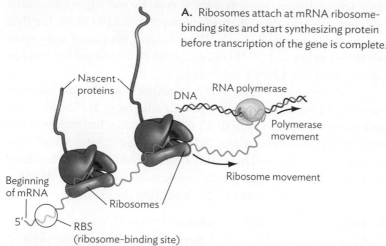

ANIMATION

Figure 8.29 **Coupled Transcription and Translation in Prokaryotes**

A. Ribosomes attach at mRNA ribosome-binding sites and start synthesizing protein before transcription of the gene is complete.

Nascent proteins

DNA

RNA polymerase

Polymerase movement

Ribosome movement

Beginning of mRNA

Ribosomes

5'

RBS (ribosome-binding site)

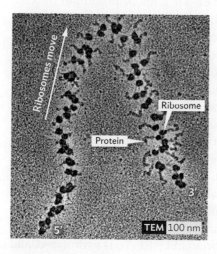

Ribosomes move

Ribosome

Protein

3'

5'

TEM 100 nm

B. Electron micrograph of a polysome (ribosome is 21 nm across). Several ribosomes may translate a single mRNA molecule at the same time. Each ribosome starts at the beginning (5' end) of the mRNA and travels toward the 3' end. Note that the synthesized protein molecule grows longer and longer the closer the ribosome gets to the 3' end of the mRNA. The protein molecule is seen most clearly at the end of the mRNA (right).

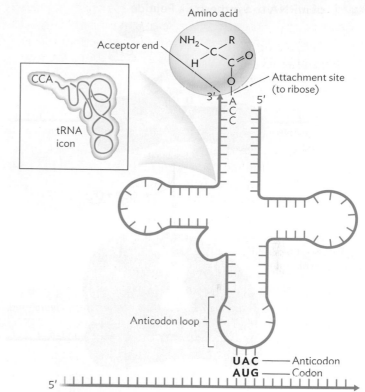

Figure 8.30 **Aminoacyl-tRNA Base-Pairs with mRNA**

Amino acid

Acceptor end

Attachment site (to ribose)

CCA

tRNA icon

Anticodon loop

UAC —— Anticodon
AUG —— Codon

5'

The tRNA anticodon consists of three nucleotides at the base of the anticodon loop. The anticodon hydrogen-bonds with the mRNA codon in an antiparallel direction. This tRNA is "charged" with an amino acid covalently attached to the 3' end.

secretes proteins that precisely signal host cells to allow bacterial adherence and colonization. Many deadly toxins, such as botulinum toxin and Shiga toxin, are proteins secreted by bacteria.

Structure of Messenger RNA (mRNA)

The mRNA transcript includes untranslated "leader" and "trailer" sequences that flank the protein-coding region. The leader and trailer sequences may help regulate the amount of gene product that is made. Just upstream of the coding region is the ribosome-binding site, an RNA sequence that aligns the ribosome on mRNA to properly initiate translation. The translation start site is a codon that specifies the amino acid methionine. The stop site is one of three codons not specifying an amino acid: UGA, UAA, or UAG. In an mRNA transcript containing multiple genes, each gene has its own translation start and stop codons.

Bacterial ribosomes start translating their mRNA even before transcription is complete (Figure 8.29A). As the 5' leader of the mRNA emerges from the RNA polymerase, ribosomes bind to the ribosome-binding site and start making the first peptide.

Meanwhile, the RNA polymerase continues along the DNA, elongating the mRNA. As the mRNA lengthens, additional ribosomes can hop on. In Figure 8.29B, the electron micrograph shows a "polysome," an mRNA with a series of ribosomes chugging along, synthesizing peptide. Notice how the ribosomes farthest from the mRNA upstream leader (those that have traveled longest) show the longest growing peptide chains.

Translation by the Ribosome

How are the three-letter codons of the genetic code "decoded" to add amino acids into a peptide? The decoder for the genetic code is transfer RNA (tRNA) (Figure 8.30). Each tRNA molecule has its own aminoacyl-tRNA synthetase, which attaches the correct amino acid to the correct tRNA. A tRNA attached to an amino acid is called a "charged" tRNA.

Each tRNA also possesses a loop, called the anticodon loop, that exposes three unpaired bases. The anticodon loop can base-pair with a codon in the mRNA—the codon that specifies the amino acid attached to the tRNA. The anticodon pairs with the codon at a special site deep within the ribosome. As each tRNA anticodon pairs with the codon of mRNA, an amino acid is added onto the growing peptide chain.

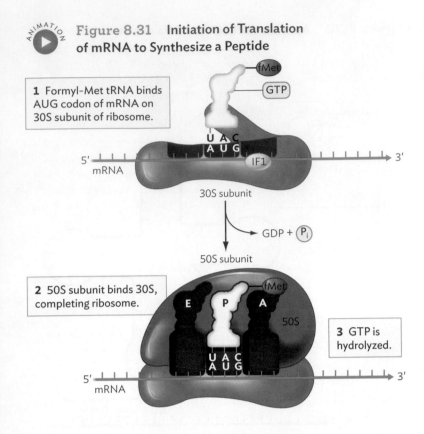

Figure 8.31   Initiation of Translation of mRNA to Synthesize a Peptide

1 Formyl-Met tRNA binds AUG codon of mRNA on 30S subunit of ribosome.

30S subunit

2 50S subunit binds 30S, completing ribosome.

3 GTP is hydrolyzed.

formyl (R–CHO) group. Peptides with formylmethionine are made only by bacteria and by our mitochondria, which evolved from bacteria. Interestingly, this difference is used by our white blood cells to detect the presence of invading bacteria and of mitochondria leaking from damaged tissue. White blood cells called leukocytes can detect as low as 10^{-12} M concentration of formylmethionyl peptides and migrate toward the source.

Next, the large subunit (50S subunit) of the ribosome joins the complex, fitting around it like a pot fits a lid (step 2). The 50S subunit encloses the initiator tRNA and completes formation of the three tRNA sites: the acceptor (A) site, also called "aminoacyl" site; the peptide (P) site; and the exit (E) site. To drive the reaction, GTP is hydrolyzed and initiation factors released (step 3). Now the ribosome is ready to add amino acids, elongating the chain.

ELONGATION   To add an amino acid to the chain, an aminoacyl-tRNA binds to the (A) site (**Figure 8.32**). The aminoacyl-tRNA is assisted by elongation factor EF-Tu, a major protein in the cell, often the most abundant protein appearing on electrophoretic gels of bacterial protein. The EF-Tu carries a GTP with energy to "spend" for elongation (Figure 8.32, step 1). EF-Tu can bring any tRNA to the A site, but the tRNA will bind only if its anticodon base-pairs correctly with the mRNA codon within the site; otherwise, the tRNA will slip out again. The anticodon allows some ambiguity at its third position; for instance, either U or C in the codon may be acceptable. Ambiguity at the third position means that fewer than 64 different tRNAs are actually needed to serve the entire code. Thus, the code has evolved to be more efficient.

The 50S subunit then catalyzes formation of a **peptide bond** (amide bond) between the amino end of the new amino acid and the carboxyl of the formylmethionine sitting in the P site. To drive peptide bond formation, the GTP brought by EF-Tu is hydrolyzed, and the resulting EF-Tu–GDP is expelled (Figure 8.32, step 2).

During peptide bond formation, the peptide is transferred from the tRNA in the P site to the tRNA in the A site. Finally, the mRNA must "move over" one position, so that each tRNA enters a new "pocket" in the ribosome (Figure 8.32, step 3). An elongation factor G (EF-G) with GTP inserts into the A site. The GTP is hydrolyzed to GDP, spending energy to shift the mRNA forward relative to the 50S subunit. Now the A site can receive the next amino acyl-tRNA. This move is a crucial step that can be blocked by antibiotics such as streptomycin. Now the peptidyl-tRNA again resides in the P site, ready to engage the next incoming aminoacyl-tRNA (Figure 8.32, step 4).

The ribosome is composed of two complex subunits, each of which includes rRNA and protein components. In bacteria, the subunits are named 30S and 50S for their "size" in Svedberg units (**Figure 8.31**). (A Svedberg unit reflects the rate at which a molecule sediments under the centrifugal force of a centrifuge.) Each subunit is an enormous complex consisting of an rRNA (16S rRNA for the 30S subunit; 23S and 5S rRNAs for the 50S subunit) plus a large number of proteins that stabilize the rRNA. Within the living cell, the two ribosomal subunits exist separately but come together on mRNA to complete the ribosome.

As we saw for DNA and RNA synthesis, protein synthesis occurs in three phases: initiation, elongation, and termination.

INITIATION   The key event of initiation is binding of the 30S subunit to the translation initiation sequence on the mRNA (see Figure 8.31). Completing the 30S initiation complex requires spending energy in the form of GTP. The initiation complex includes proteins called initiation factors (not shown). In bacteria, initiation also involves hybridization (base pairing) between the mRNA initiation sequence and the 3' terminal sequence of the 16S rRNA. In eukaryotes, the mechanisms of initiation are very different; that is one reason why so many antibiotics can disrupt function of the bacterial ribosome but not that of the human host (see Table 8.3).

When the 30S subunit binds the mRNA, the shape of the complex forms a "pocket" where the initiator tRNA (the start codon specifying methionine) can fit, while its anticodon base-pairs with the AUG codon of the mRNA (see Figure 8.31, step 1). The methionine attached to the tRNA is a special methionine with an extra

TERMINATION   The peptide chain terminates when a stop codon in the mRNA enters the A site. A stop codon is one of three codons (UGA, UAA, UAG) for which no tRNA anticodon exists. Instead, a protein called "release factor" enters the A site. The release factor triggers hydrolysis of the peptide-tRNA ester link, thus releasing the peptide. Finally, the 30S and 50S subunits fall off the mRNA and separate in the cytoplasm, available once again to form a new translation complex at some other mRNA.

ANIMATION ▶

Figure 8.32   **Translation: Elongation and Termination**

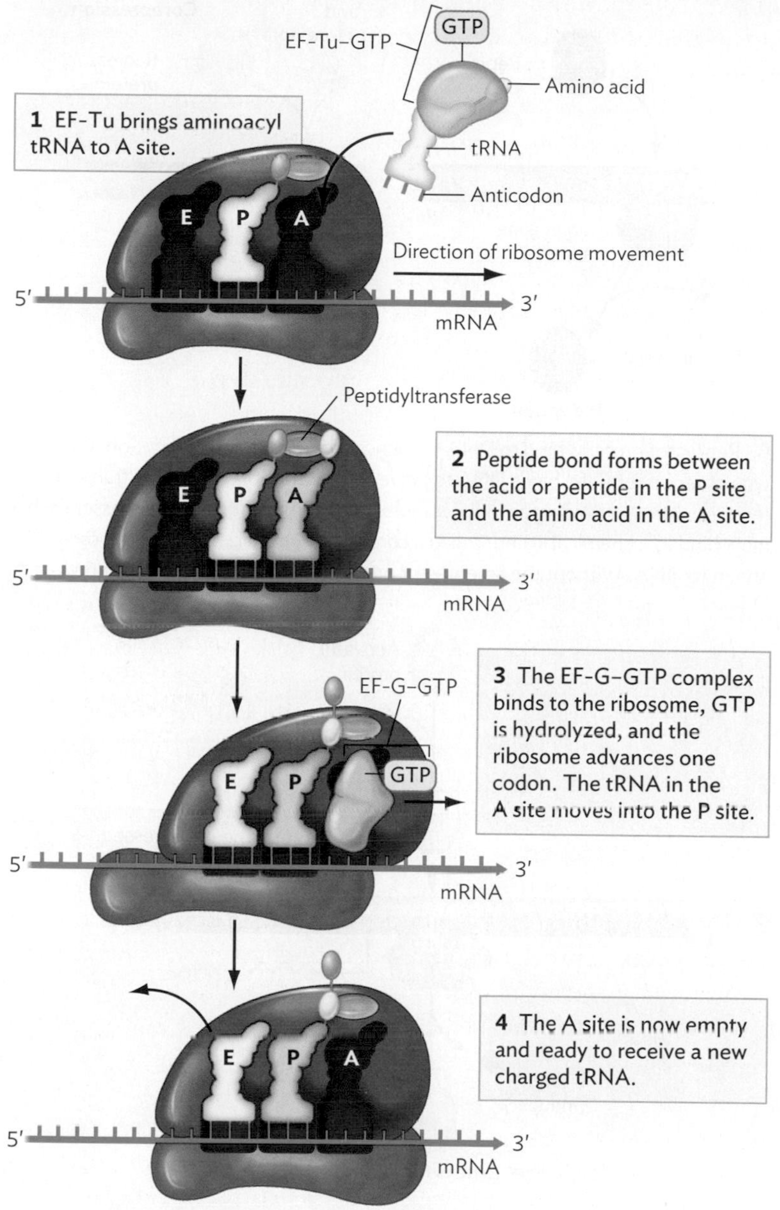

EF-Tu–GTP — GTP

Amino acid

1 EF-Tu brings aminoacyl tRNA to A site.

tRNA

Anticodon

E P A

Direction of ribosome movement

5′ ||||||||||||||||||| 3′
mRNA

Peptidyltransferase

E P A

2 Peptide bond forms between the acid or peptide in the P site and the amino acid in the A site.

5′ ||||||||||||||||||| 3′
mRNA

EF-G–GTP

3 The EF-G–GTP complex binds to the ribosome, GTP is hydrolyzed, and the ribosome advances one codon. The tRNA in the A site moves into the P site.

E P GTP

5′ ||||||||||||||||||| 3′
mRNA

4 The A site is now empty and ready to receive a new charged tRNA.

E P A

5′ ||||||||||||||||||| 3′
mRNA

As the peptide chain emerges from the ribosome, its various amino acid side chains interact very differently with the cytoplasmic water. Some peptides fold automatically, whereas others need the help of chaperones, proteins that assist the folding of newly formed peptides. Still other peptides cannot fold properly at all in the cytoplasm; they must be synthesized directly at the cell membrane, where they will reside as membrane-embedded proteins. Finally, an important group of peptides are **secreted**; that is, they are transported out of the cell in order to interact with the cell's environment. Secreted proteins include signaling molecules and toxins, which can help a pathogen cause disease. Secretion requires special protein complexes embedded in the membrane in order to transport the protein across the cell's outer layers and release it in the surrounding medium.

SECTION SUMMARY

- **Transfer RNA (tRNA) decodes the genetic code.** The tRNA is charged by an aminoacyl-tRNA transferase enzyme with a specific amino acid that corresponds to the anticodon of the tRNA.

- **Initiation of translation occurs when the 30S subunit binds mRNA.** The formylmethionyl-tRNA binds to an AUG codon of the first gene in the mRNA. Binding of the 50S site completes initiation.

- **Elongation comprises aminoacyl-tRNA entry and peptide bond formation.** Each new aminoacyl-tRNA binds the A site, with anticodon-codon pairing. If the pairing is correct, the 50S subunit catalyzes peptide bond formation. The mRNA translocates one position, so that now the "new" tRNA (holding onto the growing peptide) occupies the P site.

- **Termination occurs when a stop codon in the mRNA enters the A site.** The completed peptide is released, and the 30S and 50S ribosomal subunits fall off, to commence translation elsewhere.

- **The newly formed peptide must fold into its tertiary conformation.** Some peptides fold automatically, whereas others need help from chaperones.

- **Some proteins act in the membrane or are secreted.** Proteins destined for the membrane or for secretion require special export complexes.

Thought Question 8.4   In pathogenic bacteria, some genes contain multiple promoters that are used under different cell conditions, such as temperature or nutrient availability. Why might this be advantageous to the pathogen?

8.7
Regulation of Gene Expression

SECTION OBJECTIVES

- Describe how and why a cell regulates expression of a virulence gene.

- Outline the levels of gene regulation. Explain why it is useful to regulate gene expression at different levels.

A bacterial genome encodes thousands of different proteins needed to handle many different environments, including those of a host. How does the bacterium decide which proteins to make at what time? Some proteins, such as RNA polymerase, are always required for growth. Most proteins, however, are needed only under a limited set of conditions. For example, proteins that degrade the sugar lactose are useful only when lactose is present. Likewise, *Corynebacterium diphtheriae* needs to make diphtheria toxin only when it's growing inside the human body. To compete successfully with others, microbes cannot waste energy making unneeded proteins. Energy waste is minimized by elegant genetic controls that selectively increase or decrease transcription of a gene (that is, whether RNA is made) or control translation of mRNA to protein, mRNA degradation, or degradation of a regulatory protein. But how does a cell "know" when to alter a gene's expression? How does *C. diphtheriae* "know" whether it needs to make diphtheria toxin?

Figure 8.33   General Aspects of Transcriptional Regulation by Repressor and Activator Proteins

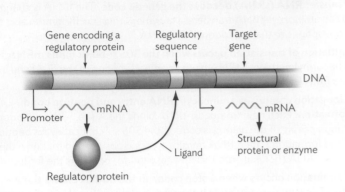

A. The regulatory gene encoding a regulatory protein may be located near or quite far away from the target gene on the chromosome. When it is located nearby, it is usually transcribed separately from the target gene. The regulatory protein will stick to DNA sequences near the promoter of the target gene and control whether transcription occurs.

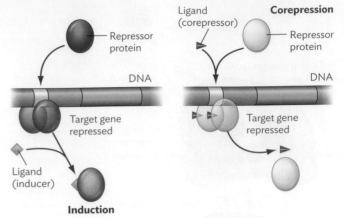

B. Repressor proteins bind to DNA sequences and prevent transcription. One type of repressor (left) binds DNA to prevent transcription. To induce transcription requires repressor binding an inducer molecule. A different type of repressor (right) must bind to a chemical ligand, called a corepressor, before the repressor can attach to DNA. Without the corepressor present, the gene will be transcribed.

Microbes use numerous mechanisms to sense their internal and external environments and then translate that information into action. The cell's surface, for example, contains an array of sensing proteins that monitor osmolarity, pH, temperature, and the chemical content of the surroundings or that detect the presence of a host or competitor. Sharing of chemical signal molecules makes it possible for members of microbial communities to communicate and cooperate in a process known as quorum sensing (more on this in Section 9.1). Even virulence genes have tapped into these sensing systems, using them to determine whether the microbe has entered a susceptible host.

Levels of Gene Regulation

Hundreds of intracellular and extracellular sensing mechanisms monitor the overall health of the cell and its environment. But how do these systems control gene expression? Expression of genes and operons and their mRNA products can be controlled at various levels. Different levels of control offer different advantages to the cell. Gene expression can be regulated by DNA sequence controls, transcriptional controls, translational controls, and posttranslational modifications. Here are some examples:

- **Changing the DNA sequence.** Some microbes can change the DNA sequence to activate or disable a particular gene. Phase variation is an example of such a mechanism. In phase variation, a DNA segment containing a promoter can reversibly flip so that the promoter is pointing toward or away from a downstream gene. Because the immune system targets bacterial surface proteins, turning off the genes encoding those proteins is one way the organism stays alive in a host. However, reversing a DNA change is a relatively rare event, so the rate of control is slow.

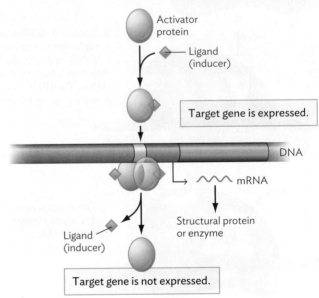

C. Induction of transcription may involve an activator protein. The activator protein must bind to a specific chemical ligand (an inducer) in the cytoplasm before the protein can bind to a DNA sequence near the target gene. Binding activates transcription of the target gene.

- **Control of transcription.** Many types of gene regulation occur at the level of transcription (discussed shortly). Transcription can be regulated by protein repressors, activators, and alternative sigma factors (the class of proteins that helps RNA polymerase recognize promoters).

- **Translational control.** Translation by ribosomes can be regulated by translation initiation sequences in the mRNA, which recognize specific translational repressor proteins.

- **Posttranslational control.** Once proteins are made, their activity can be controlled by modifying their structure—for example, by protein cleavage, phosphorylation (adding a phosphate), or methylation (adding a methyl group, –CH$_3$). These modifications can activate, deactivate, or even lead to destruction of the protein.

Transcriptional Control

Turning genes on (**induction**) or turning genes off (**repression**) basically means, to paraphrase Shakespeare, "to transcribe or not to transcribe." That is the question every cell must ask of every gene as it moves from one environment to the next. Some genes respond to changes inside the cell; others respond to outside influences. How do cells assign these tasks?

SENSING THE INTRACELLULAR ENVIRONMENT Sensing conditions within the cell is relatively simple. Different regulatory proteins bind to specific small-molecular-weight compounds (ligands) to determine the compound's concentration. **Figure 8.33A** shows a basic model of a regulatory protein that can sense an intracellular metabolite (not shown) binding to a regulatory (control) sequence and then determine whether the gene is transcribed. Depending on the system, the target gene could be induced (transcribed) or repressed (not transcribed).

Different regulatory proteins bind to different metabolite signal molecules (ligands). Take, for example, the expression of *dtx*, the diphtheria toxin gene. In diphtheria pathogenesis, expression of virulence genes is controlled by the repressor protein DtxR. DtxR senses the concentration of iron—a scarce metal ion that *Corynebacterium diphtheriae* needs to grow. DtxR binds iron, attaches to the regulatory region of the diphtheria toxin gene, and then represses its transcription (**Figure 8.34A**). When iron is plentiful, as can happen in the environment, no toxin is made. However, since much of the body's iron is tied up by the iron-binding proteins transferrin and lactoferrin, very little iron is left free for *C. diphtheriae* to use. Because iron is scarce, DtxR will not repress the *dtx* gene, and

diphtheria toxin will be made (**Figure 8.34B**). Host cells killed by the toxin release iron that the pathogen can now use to grow.

In contrast to repressor proteins that stop a gene's expression, other regulatory proteins called transcriptional **activators** can stimulate gene expression (**Figure 8.33C**). For example, a transcriptional activator protein might bind to a carbohydrate (for instance, arabinose) that has entered the cytoplasm. The activator protein alerts the cell that a new carbon source is available. The activator does this by binding to and activating the transcription of genes whose products metabolize that carbohydrate. Once the carbohydrate has been consumed, the activator cannot bind to the control region and the genes will not be expressed. There is no need to make those proteins if the carbohydrate is no longer available.

An inducer molecule can activate a given operon in two different ways. For some operons, the inducer causes an activator to bind the gene, as with the arabinose operon. Alternatively, gene expression for other operons is turned on when an inducer molecule relieves repression imposed by a repressor protein (see **Figure 8.33B**). A classic example is the lactose operon (**Figure 8.35**), whose gene products degrade lactose for use as a carbon and energy source. When lactose is provided in media, a form of lactose (called allolactose) binds to the Lac repressor protein (called LacI or LacR), causing the repressor to release from the *lac* operon. The *lac* operon, now free to be transcribed, will produce the enzymes needed to degrade lactose. The process is still called induction, even though a repressor protein is involved.

GLOBAL REGULATORS Some regulatory proteins control many different genes and operons throughout a genome. Proteins that affect the expression of many different genes are called **global regulators**. One example is the cyclic AMP receptor protein (CRP) of *Escherichia coli* and related species. Cyclic adenosine monophosphate (cAMP) is a chemical signal molecule synthesized from ATP. Its synthesis is, in part, a signal to the genome that the cell needs

Figure 8.34 **Transcriptional Regulation of the Diphtheria Toxin Gene (*dtx*) by the DtxR Repressor**

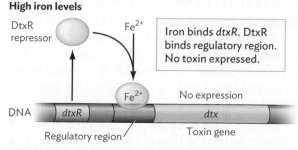

High iron levels

Iron binds *dtxR*. DtxR binds regulatory region. No toxin expressed.

A. Outside a human host, *Corynebacterium diphtheriae* will usually encounter sufficient iron for growth. It does not need to produce diphtheria toxin.

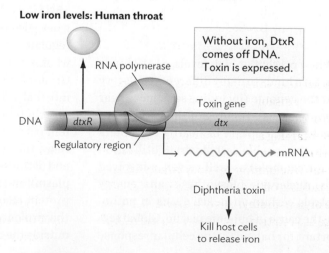

Low iron levels: Human throat

Without iron, DtxR comes off DNA. Toxin is expressed.

B. In the human throat, iron is sequestered in host cells. Diphtheria toxin will kill those cells and release iron that the bacteria can use.

Figure 8.35   Induction of the Lactose (*lac*) Operon

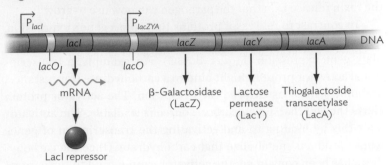

A. Organization of the *lac* operon. Bent arrows mark promoters. Green regions indicate *LacI* protein-binding sites on DNA. The products of the operon have different roles: LacY serves to transport lactose into the cell; LacZ works to degrade lactose to glucose and galactose; LacA's role in lactose metabolism is unknown.

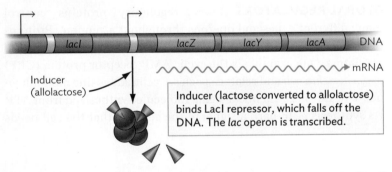

Repressor LacI binds to *lacO*. The bound protein overlaps the *lacZYA* promoter (P) and prevents transcription.

B. The LacI repressor binds to the operator *lacO* in front of the *lac* operon, hiding the promoter from RNA polymerase and thus preventing transcription.

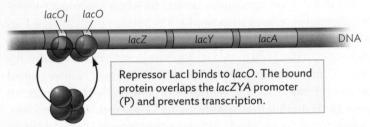

Inducer (lactose converted to allolactose) binds LacI repressor, which falls off the DNA. The *lac* operon is transcribed.

C. The inducer allolactose (an altered form of lactose made by low levels of LacZ (beta-galactosidase) removes the repressor LacI and allows expression of *lacZ*, *lacY*, and *lacA*.

more carbon and energy. When present at high levels, cAMP binds to the CRP protein, which in turn binds to many different regulatory DNA sequences throughout the genome (including sequences near promoters of the *lac* and *ara* operons). Binding of the CRP-cAMP complex to DNA will activate transcription of nearby promoters and genes, particularly those that enable carbohydrate utilization (of lactose and arabinose, for example) as well as genes involved in amino acid biosynthesis, nucleotide biosynthesis, and energy metabolism. CRP also controls certain virulence genes in pathogens such as *Yersinia pestis*, the cause of bubonic plague. Global regulators in general are important for balancing the cellular response

Figure 8.36   Two-Component Signal Transduction Systems Sense the External Environment

A transmembrane sensor kinase protein senses an environmental condition outside the cell. Binding of the environmental signal triggers phosphorylation of the sensor kinase. The phosphorylated sensor kinase will then transfer the phosphate to a response regulator protein in the cytoplasm. The response regulator then binds to target operator or activator DNA sequences and inhibits or stimulates gene expression depending on the gene. To downregulate the system, an enzyme removes phosphate from the regulator.

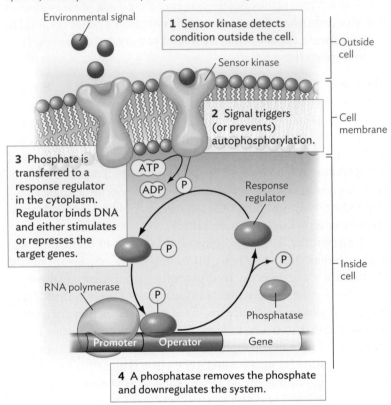

1 Sensor kinase detects condition outside the cell.

2 Signal triggers (or prevents) autophosphorylation.

3 Phosphate is transferred to a response regulator in the cytoplasm. Regulator binds DNA and either stimulates or represses the target genes.

4 A phosphatase removes the phosphate and downregulates the system.

to changes in the environment, such as the temperature increase when a bacterium enters the human body. Some global regulators sense the internal environment, whereas others report on what is happening outside the cell.

SENSING THE EXTRACELLULAR ENVIRONMENT Sensing what goes on outside the cell is challenging because intracellular regulatory proteins cannot reach through the membrane and touch what is outside. A common mechanism used by Gram-positive and Gram-negative organisms to sense the outside of the cell and transmit that information inside relies on a series of two-component, protein phosphorylation relay systems. Each two-component protein pair activates a different set of genes. The first protein in each relay, the sensor kinase, spans the membrane (Figure 8.36). One end of this sensor protein touches the outside environment (or periplasm), and the other end protrudes into the cytoplasm. Each sensor protein recognizes a different molecule or condition (for example, the virulence sensor PhoQ in *Salmonella* senses magnesium and pH outside the cell).

Stimulating the external sensory domain of a sensor kinase activates the cytoplasmic kinase domain. The kinase domain then transfers phosphate from ATP onto a specific histidine in the protein (autophosphorylation). Then, like a relay runner passing a baton to the next runner, the phosphorylated sensor kinase protein passes the phosphate to a cytoplasmic partner protein called a **response regulator** (transphosphorylation). The phosphorylated response regulator commonly binds to regulatory DNA sequences positioned in front of one or more specific genes and either activates or represses their expression. Regulatory relays of this type are called two-component signal transduction systems.

Note that in most two-component systems, only "information" is transmitted from outside the cell to inside; no molecule actually crosses the membrane. By contrast, inducer systems (Figure 8.33B and C) require uptake of an inducer molecule (such as lactose) from outside the cell.

SECTION SUMMARY

- **Microbes sense their environment and regulate gene expression** to optimize growth and survival.

- **Transcriptional regulation** is carried out by specialized DNA-binding proteins (called activators and repressors) that stimulate (activate) or inhibit (repress) the ease with which RNA polymerase initiates transcription at a given promoter.

- **Induction or repression** of a gene's transcription occurs when a signal molecule binds to a regulatory DNA-binding protein, which then, respectively, stimulates or inhibits transcription of that gene.

- **Two-component signal transduction** involves a **sensor kinase** protein that monitors one aspect of the external or internal cell environment and transmits a signal via transphosphorylation to a response regulator protein that then acts as an activator or repressor.

- **Additional mechanisms of control** include DNA rearrangements, regulating translation, posttranslational modifications, and degradation.

Thought Question 8.5   What would happen to the expression of the diphtheria toxin gene if the *dtxR* gene were deleted from the genome?

Perspective

In Chapter 8, we have learned how microbial DNA is maintained through reproduction—and how mutations occur and may be repaired. Mutation provides one of many mechanisms pathogens use to colonize a host and evade the immune system. Protein and RNA synthesis machines critical for bacterial growth are focused targets for many antibiotics, a subject discussed more in Chapter 13.

The genes and enzymes of bacterial and viral genetics provide tools for biotechnology. Restriction endonucleases enable us to define pieces of DNA, whereas PCR may generate a million copies of a gene. Genes that express valuable proteins can be cloned into plasmid vectors for production of the protein for therapy. In Chapter 9, we will see how bacteria transfer genes to other cells and how bacterial gene transfer mechanisms provide additional tools to manipulate DNA.

LEARNING OUTCOMES AND ASSESSMENT

	SECTION OBJECTIVES	**OBJECTIVES REVIEW**

8.1
Bacterial Genomes
	• Describe the structure of a bacterial genome, and explain how it differs from a eukaryotic genome.
	• Explain what a plasmid is, and describe the role of plasmids in the spread of antibiotic resistance genes.
		1. Which aspect of bacterial genome structure is the <u>same</u> in eukaryotes?
		  A. The genome is composed of double-helical DNA.
		  B. Genes are often arranged in tandem, within an operon.
		  C. The genome is compact, with genes interspersed by relatively short noncoding sequences.
		  D. Most genomes are circular.

8.2
DNA Replication
	• Explain how bidirectional semiconservative DNA replication copies the circular chromosome of a bacterium during cell division.
	• Explain the steps of replication by DNA polymerase, including initiation, elongation, and termination.
		4. Which step of DNA replication occurs <u>before</u> DNA polymerase extends the chain?
		  A. An RNA primer is synthesized.
		  B. A lagging strand is synthesized.
		  C. Proofreading enzymes check the DNA.
		  D. The RNA primers are replaced with DNA sequence.

8.3
Mutation and Repair
	• Explain the different kinds of mutations and how they occur.
	• Describe different ways that a cell repairs DNA damage, preventing or reversing mutations.
		7. Which kind of mutation has no effect on gene function?
		  A. A missense mutation
		  B. A silent mutation
		  C. A deletion mutation
		  D. A frameshift mutation

8.4
Biotechnology
	• Describe the use of plasmids, restriction enzymes, and gel electrophoresis to analyze DNA.
	• Explain the formation of a recombinant DNA molecule, and describe applications in medical technology.
	• Explain the use of DNA hybridization and DNA sequencing technology.
		10. Which technique is commonly used to reveal the size of a DNA fragment?
		  A. Gel electrophoresis
		  B. Recombinant DNA formation using a vector
		  C. DNA hybridization
		  D. DNA sequencing

8.5
Transcription of DNA to RNA
	• Describe the different kinds and functions of RNA products that are made from DNA.
	• Explain how RNA polymerase copies a sequence of DNA to form mRNA.
	• For biotechnology, describe the use of reverse transcriptase to make a cDNA copy of an mRNA.
		13. Which kind of RNA product of a gene will be translated by a ribosome to make protein?
		  A. Transfer RNA (tRNA)
		  B. Ribosomal RNA (rRNA)
		  C. Messenger RNA (mRNA)
		  D. Regulatory small RNA (sRNA)

8.6
Protein Synthesis
	• Explain the relationship between transcription and translation in bacteria.
	• Describe the ribosomal steps of initiation, elongation, and termination. Explain how these steps may be targeted by antibacterial agents.
	• Explain how proteins may require folding and secretion.
		16. Which step of translation is blocked by tetracycline?
		  A. The ribosome 30S subunit binds to mRNA.
		  B. The incoming aminoacyl-tRNA binds the A site of the ribosome.
		  C. The peptide bond forms, catalyzed by the ribosome.
		  D. The terminated peptide chain falls off the ribosome.

8.7
Regulation of Gene Expression
	• Describe how and why a cell regulates expression of a virulence gene.
	• Outline the levels of gene regulation. Explain why it is useful to regulate gene expression at different levels.
		19. Which kind of gene regulation does <u>not</u> change the rate of gene transcription to RNA?
		  A. A repressor binds a promoter sequence.
		  B. An activator binds a promoter sequence.
		  C. A two-component regulator binds a signal molecule.
		  D. A translation initiation sequence binds a translational repressor.

2. Which statement about plasmids is <u>incorrect</u>?
   A. Plasmids often carry genes encoding enzymes that confer resistance to an antibiotic.
   B. Plasmids are small circular or linear pieces of DNA.
   C. When a plasmid is transferred into a new cell, it usually changes the species of the host organism.
   D. Plasmids can incorporate mobile elements such as insertion sequences.

3. Which kind of sequences interrupt genes of eukaryotes but not prokaryotes?
   A. Plasmids
   B. Introns
   C. Primers
   D. Promoters

5. The function of gyrase is ____.
   A. priming DNA synthesis
   B. connecting Okazaki fragments
   C. generating negative superturns in DNA
   D. initiating RNA synthesis

6. The function of ligase is ____.
   A. priming DNA synthesis
   B. connecting Okazaki fragments
   C. generating negative superturns in DNA
   D. initiating RNA synthesis

8. Which kind of bacterial DNA repair leads to multiple mutations?
   A. Base excision repair
   B. SOS repair
   C. Methyl mismatch repair
   D. Proofreading of DNA during replication

9. A type of mutation that can cause a frameshift is ____.
   A. a deletion of a base pair
   B. a substitution of a base pair
   C. a silent mutation
   D. a missense mutation

11. Which technique allows same-day identification of a clinical isolate from a patient?
    A. Recombinant DNA formation
    B. Culturing the isolate
    C. Gel electrophoresis of a genome fragmented by restriction enzymes
    D. PCR amplification of a bacterial gene

12. For DNA sequencing, the nucleotides may be labeled with ____.
    A. Gram stain
    B. methylene blue stain
    C. electron-dense metal stain
    D. fluorophores (fluorescent molecules)

14. Which protein is needed only for initiation of transcription?
    A. The holoenzyme
    B. Ribosomal initiation factor
    C. DNA polymerase
    D. The sigma factor

15. The source of energy for DNA transcription to RNA is ____.
    A. opening of the DNA double helix
    B. RNA polymerase binding the promoter
    C. release of pyrophosphate by each nucleotide
    D. termination of RNA synthesis

17. What is the function of a chaperone protein?
    A. To synthesize RNA
    B. To initiate translation by the ribosome
    C. To fold a newly synthesized protein into its functional shape
    D. To secrete a protein from the cell

18. Which step of translation requires only the 30S subunit, not yet the entire ribosome?
    A. Binding the first f-Methionyl tRNA to the mRNA.
    B. Formation of the A site.
    C. Formation of a peptide bond.
    D. Movement of the peptide chain out of the P site.

20. Which kind of regulation has a long-term effect on a pathogen's genome and helps it evade the immune system?
    A. An activator binds a promoter.
    B. DNA phase variation occurs.
    C. A two-component regulator detects iron.
    D. A virulence protein is modified.

21. Which kind of gene regulation must involve transfer of protein phosphoryl groups?
    A. Corepressor binds a repressor to halt transcription.
    B. Inducer removes a repressor to enable transcription.
    C. Activator protein enhances transcription.
    D. Two-component regulator transmits information from an external signal to the DNA inside the cell.

Key Terms

activator (p. 251)
base excision repair (p. 234)
base substitution (p. 232)
biotechnology (p. 235)
cDNA (p. 239)
chromosome (p. 225)
codon (p. 232)
daughter strand (p. 228)
deletion (p. 232)
DNA ligase (p. 231)
DNA polymerase (p. 228)
elongation (p. 228)
frameshift mutation (p. 232)
gel electrophoresis (p. 236)
gene (p. 225)
gene expression (p. 225)
genome (p. 224)
genotype (p. 234)

global regulator (p. 251)
gyrase (p. 231)
homolog (p. 228)
horizontal gene transfer (p. 227)
hybridize (p. 238)
induction (p. 251)
initiation (p. 228)
insertion (p. 232)
inversion (p. 232)
lagging strand (p. 229)
leading strand (p. 229)
messenger RNA (mRNA) (p. 225)
methyl mismatch repair (p. 234)
missense mutation (p. 232)
mutation (p. 227)
mutator (p. 234)
nonsense mutation (p. 232)
Okazaki fragment (p. 229)

operon (p. 225)
origin (oriC) (p. 228)
parental strand (p. 228)
peptide bond (p. 248)
phenotype (p. 234)
plasmid (p. 226)
point mutation (p. 232)
polymerase chain reaction (PCR)
   (p. 240)
promoter (p. 225)
pyrimidine dimer (p. 233)
reading frame (p. 232)
recombinant DNA (p. 238)
recombination (p. 234)
replication fork (p. 228)
repression (p. 251)
response regulator (p. 253)
restriction endonuclease (p. 236)

restriction enzyme site (p. 236)
restriction fragment length
   polymorphism (RFLP) (p. 238)
reverse transcriptase (p. 246)
RNA polymerase (p. 244)
RNA primer (p. 229)
secreted (p. 249)
sense strand (p. 245)
sigma factor (p. 245)
silent mutation (p. 232)
SOS response (p. 234)
structural gene (p. 225)
supercoil (p. 231)
template strand (p. 245)
termination (p. 228)
transcription (p. 244)
transformation (p. 239)
vector (p. 238)

Review Questions

1. What is a genome? What is the structure of a genome in bacteria and in eukaryotes?

2. What are plasmids? How do they function in cells and in microbial communities? What kinds of traits are commonly encoded on plasmids?

3. What is a restriction endonuclease? Explain how restriction endonucleases can be used to define specific fragments of DNA.

4. Explain the polymerase chain reaction (PCR) and how it is used to identify a pathogen.

5. For DNA synthesis, explain the steps of initiation, elongation, and termination.

6. Explain how different kinds of mutation occur.

7. How do mutations change the information contained in DNA?

8. How are mutations repaired?

9. Explain how RNA is synthesized on the basis of the DNA gene. What is required to initiate, elongate, and terminate RNA synthesis?

10. How does the ribosome translate a message to make protein? Explain the steps of initiation, elongation, and termination.

11. How are proteins such as toxins secreted from a bacterial pathogen? What kind of protein complex is needed, and why?

Clinical Correlation Questions

1. A patient presents with bacteremia caused by a rare nontypeable strain of *Haemophilus influenzae*. Which test can identify the pathogen?

2. A patient reports spotting between periods, and she undergoes a genital exam. Her cervix is inflamed, with a purulent discharge. Which test will best identify the pathogen?

Thought Questions: CHECK YOUR ANSWERS

Thought Question 8.1 Why would a rapidly growing bacterial cell be killed by an antibiotic that blocks DNA replication? Suppose the bacteria form a biofilm on a catheter in a patient. Why might the bacteria have become resistant to the antibiotic?

ANSWER: Bacteria that are growing rapidly need to replicate their chromosomes in time for the cell to divide. If DNA replication is blocked, then the cells may try to divide while the two copies of the chromosome are not yet complete. The result is that the cells will lose their DNA as they divide and will die. On the other hand, when cells form a biofilm, most of them enter a state where they divide slowly or not at all. The thick layers of cells may restrict entry of the antibiotic. Furthermore, cells that are not dividing may be unharmed by an antibiotic that blocks DNA replication.

Thought Question 8.2 When bacteria mutate and become resistant to an antibiotic, several different mechanisms of resistance are possible. How many different ways can you think of for a mutant protein to cause resistance?

ANSWER: One way to become resistant to an antibiotic is for the genes encoding the target structure, such as the ribosome, to acquire a mutation resulting in an altered structure. The altered target structure now fails to bind the antibiotic. Another way to become resistant is for a cell to acquire a plasmid encoding an enzyme that cleaves the antibiotic. A third possibility is that a plasmid can encode a protein complex that exports the antibiotic out of the cell.

Thought Question 8.3 Explain how it is possible for a very large mutation, involving a large part of a chromosome with many genes, to have no effect on the phenotype of the bacterium.

ANSWER: One possibility is that a large segment of DNA could be inverted within a bacterial chromosome. Within the inverted region, most of the genes would be intact and could still be expressed. Although the map of the bacterial genome would be profoundly altered, the bacterium's phenotype (observed traits) might show little change.

Thought Question 8.4 In pathogenic bacteria, some genes contain multiple promoters that are used under different cell conditions, such as temperature or nutrient availability. Why might this be advantageous to the pathogen?

ANSWER: Multiple promoters enable a pathogen to respond to different environmental conditions. In particular, a pathogen needs to "know" when it finds itself within a host versus the outside environment. A vector-borne pathogen, such as the plague bacillus *Yersinia pestis*, needs to have promoters that receive signals of at least three environments: the human host, the flea host, and the outside (soil and air). Each environment requires the pathogen to express different proteins. For example, within the human host, signals such as elevated temperature and blood iron cause *Y. pestis* to secrete toxins that would be useless outside.

Thought Question 8.5 What would happen to the expression of the diphtheria toxin gene if the *dtxR* gene were deleted from the genome?

ANSWER: The *dtxR* gene specifies a transcriptional repressor. If the *dtxR* gene is deleted, the expression of diphtheria toxin may become constitutive (turned on continuously). Constitutive expression of the toxin may decrease fitness of the bacterium under conditions outside the host body.

9

Bacterial Genomes and Evolution

[Diphtheria, Toxins, and Natural Disasters]

SCENARIO When 15-year-old Oriel arrived at the clinic in Port-au-Prince, the doctors were not expecting diphtheria, an ancient disease almost never seen today. But this was Haiti. A major earthquake in 2010 had destroyed much of the country's infrastructure and many thousands of homes. The city lacked adequate sanitation, and tens of thousands of people, including Oriel, were living in tent cities—conditions ripe for spreading infectious diseases. Oriel had started feeling ill 6 days earlier with a sore throat and fever. Everyone thought it was a cold, but Oriel kept getting worse.

SIGNS AND SYMPTOMS Details are sketchy, but when he arrived at the clinic, Oriel had a fever of 38.9°C (102°F) and a "bull neck" from enlarged lymph nodes in his neck. He felt very weak and had a thick gray membrane (pseudomembrane) forming at the back of his throat, making breathing difficult.

DIAGNOSIS Diphtheria is caused by the Gram-positive rod *Corynebacterium diphtheriae*. This pathogen grows in the throat but secretes a powerful toxin that circulates and destroys the victim's cellular protein synthesis machinery. The gene encoding diphtheria toxin (*dtx*) is actually carried by a virus called beta phage, whose DNA inserts itself into the chromosome of *Corynebacterium*.

Diphtheria
A symptom of diphtheria is an inflammatory response in the throat that produces a grayish membrane. This pseudomembrane can obstruct airflow into the trachea.

TREATMENT As Oriel's condition worsened, a frantic search began for life-saving antitoxin. The search for antitoxin, through relief efforts led by actor Sean Penn, fell short. Oriel died May 7, 2010.

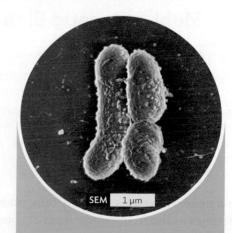

SEM | 1 μm

Corynebacterium diphtheriae
The causative agent of diphtheria, *Corynebacterium diphtheriae*, harbors DNA from a bacterial virus that carries the gene for diphtheria toxin.

CHAPTER OBJECTIVES

After reading this chapter, you will be able to:

- Describe how genes move between genomes by the processes of transformation, transduction, conjugation, and transposition.
- Explain the evidence for how microbial life originated and diversified.
- Describe how bioinformatics uses genomic information to predict traits of related organisms.
- Explain microbial taxonomy.

The unfortunate case of Oriel's diphtheria infection illustrates two major topics we will discuss in this chapter: the horizontal transfer of genes and traits between bacteria, and concepts of pathogen evolution. Recall from Chapter 8 that **horizontal gene transfer** involves the transfer of genes from one organism to another coexisting organism. Oriel succumbed to infection by *Corynebacterium*—a species that would be harmless without the "hitchhiking" beta phage DNA that had horizontally transferred the *dtx* toxin gene. The *dtx* gene is an example of a virulence gene, a gene that helps a pathogen cause disease. Horizontal transfer of virulence genes is a common way that pathogens evolve from otherwise harmless species.

This chapter will discuss how bacterial genes, such as *dtx*, are moved between bacteria by bacteriophages (bacterial viruses, also called "phages"). You will also learn how we identify genes that were transferred many thousands of years ago. We will consider as well how the horizontal transfer of genes is involved in the evolution of microbes, including modern-day pathogens and essential host microbiota.

9.1
Gene Transfer, Recombination, and Mobile Genetic Elements

SECTION OBJECTIVES

- Explain how genetic information moves between bacteria by transformation, phage-mediated transduction, and conjugation.
- Describe how DNA molecules recombine by homologous recombination and site-specific recombination.
- Describe how gene mobility and DNA recombination lead to the emergence of new pathogens.

How does nature play Frankenstein, making genomes change as microbes evolve? As we learned in Chapter 8, genomes change and evolve gradually through spontaneous mutations, such as errors in DNA replication or proofreading. Such mutations alter the DNA sequence that cells inherit from their parents. We discover genome change by comparing DNA sequences, a field of study called bioinformatics. Bioinformatic analysis of genomes shows that in addition

to small spontaneous mutations, microbes can gain and lose entire genes and undergo rearrangements and duplications. These rapid, large-scale changes are usually mediated by genetic elements such as viruses and transposons, which you will learn about in this chapter.

Large-scale gene traffic between species is known as horizontal gene transfer. Horizontal gene transfer, recombination, mutation, and DNA repair are all processes that accelerate natural selection. In this section, we will first describe the mechanisms used to move genes between microbial species and then explain how mobile genes contribute to microbial evolution.

Transformation

In 1928, a perceptive English medical officer, Fred Griffith (1879–1941), found that he could kill mice by injecting them with <u>dead</u> cells of a virulent pneumococcus (*Streptococcus pneumoniae*), a cause of pneumonia, but only if he also injected <u>live</u> cells of a nonvirulent mutant. Even more extraordinary was the discovery that the bacteria recovered from the dead mice were of the live, <u>virulent</u> type. Were the dead bacteria brought back to life? Unfortunately, Griffith was killed by a German bomb during an air raid on London in 1941 and never learned the answer.

In a landmark series of experiments published in 1944, Oswald Avery (1877–1955), Colin MacLeod (1909–1972), and Maclyn McCarty (1911–2005) proved that Griffith's experiment was not a case of reviving dead cells, but involved the transfer of DNA released from the virulent, dead strain into the harmless living strain of *S. pneumoniae*, an event that transformed the live strain into a killer. The process of importing free DNA into bacterial cells is now known as **transformation**. Transformation provided the first clue that gene transfer can occur in microorganisms.

Many bacteria can import DNA fragments and plasmids released from nearby dead cells via transformation. Natural transformation is a property inherent to many bacterial species and is carried out by specific protein complexes. Examples include Gram-positive bacteria such as *Streptococcus* and *Bacillus*, as well as Gram-negative species such as *Haemophilus* and *Neisseria* species. Transformation is a natural event during growth of some bacterial species. Such species of bacteria are said to be naturally **competent** for transformation.

Other bacteria, however, such as *E. coli* and *Salmonella*, do not possess the equipment needed to import DNA efficiently. Some bacteria require artificial manipulations to become competent; that is, to take up DNA into the cell. These laboratory techniques include perturbing the membrane by chemical and electrical methods. The chemical mechanism involves the use of calcium chloride ($CaCl_2$) and heat shock, which alters the cell membrane, allowing DNA to pass. **Electroporation**, on the other hand, uses a brief electrical pulse to "shoot" DNA across the membrane. These methods are used in the laboratory to move recombinant DNA plasmids (described in Section 8.1) between bacteria.

Why do some species, such as *Neisseria*, undergo natural transformation? Organisms that indiscriminately import DNA may consume the transformed DNA as food. More finicky species transform only compatible DNA sequences; that is, the two DNA molecules carry very similar (**homologous**) sequences. These microbes use DNA released from dead compatriots to repair their own damaged genomes. Yet other bacteria may use transformation to obtain new genes from different species. These new genes may help the transformed bacteria survive in a changing environment by providing a new function that confers a growth advantage.

Horizontal gene transfer is distinguished from **vertical gene transfer**, the generational passing of genes from parent to offspring, as in cell division. If horizontal acquisition of a gene or a group of genes improves the competitiveness of the cell, the new genes will be retained and passed on vertically to offspring. The descendants

will have gained new genes that may help them survive in new environments. Pathogens such as *Neisseria gonorrhoeae*, the cause of gonorrhea, have used transformation to acquire new genes whose protein products help the organism evade the host immune system.

Gram-positive organisms transform DNA by using a protein complex called a transformasome that spans the membrane (Figure 9.1). Assembly of the transformasome complex occurs only after a certain number of cells (called a quorum) have gathered. But how do they know? All cells synthesize a peptide called competence factor that progressively accumulates in the medium. Above a certain concentration threshold, competence factor binds to the sensory protein of a two-component signal transduction system (Section 8.7). The sensory protein transfers a signal (phosphate) to a transcription activator that then stimulates expression of the transformasome genes. The transformasome is made, and the cells are now competent, meaning they are ready to import DNA. Note that the more cells per volume there are in a culture, the more quickly competence factor will accumulate and trigger transformasome synthesis.

The accumulation of competence factor is yet another example of the phenomenon of quorum sensing (see Section 6.6), in which a small molecule made by the cell accumulates outside the cell before it affects transcription inside the cell. It is called quorum sensing because a certain number of cells (a quorum) must be reached before enough signal factor is made to evoke a change in gene expression.

Figure 9.1   Transformation in *Streptococcus*

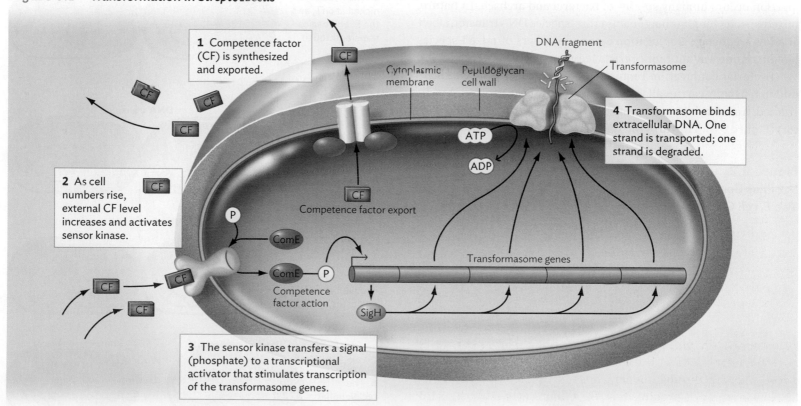

The process of transformation in this organism begins with the synthesis of a signal molecule (competence factor) and concludes with the import of a single-stranded DNA strand through a transformasome protein complex.

Quorum sensing is used by many pathogens to activate expression of virulence genes.

Once made, the transformasome captures extracellular DNA floating in the environment. A **nuclease** (DNA-cutting enzyme) degrades one strand of the incoming double-stranded DNA molecule while pulling the other strand intact into the cell. The energy for taking up DNA is supplied by hydrolysis of ATP. Once inside, the strand can be incorporated into the chromosome by recombination, a process we will discuss later.

Link As discussed in Section 8.4, a **nuclease** is an enzyme that functions to cut the strands of nucleic acids. An endonuclease cleaves the DNA or RNA backbone inside a stretch of sequence, whereas an exonuclease cleaves a base off the very end of a strand.

Some bacteria, such as Gram-negative species, transform without competence factors. Either they are always competent, like *Neisseria*, or they become competent when starved (for example, in *Haemophilus* species, a cause of pediatric sepsis and meningitis, presented in Case History 8.1).

Conjugation

Another way DNA can be transferred from one bacterium to another is by **conjugation**. DNA transfer by conjugation requires intimate cell-cell contact typically initiated by a special pilus (sometimes called a sex pilus) that protrudes from a donor cell (**Figure 9.2**). The pilus (plural, pili) is a filament of protein monomers (see Section 5.5). Conjugation occurs in many species of bacteria and archaea. In nature, mixed-species biofilms may enable cross-species DNA transfer, albeit at a low frequency. Conjugation among members of mixed-species biofilms is one way antibiotic resistance genes move between species.

Bacterial conjugation requires the presence of a special plasmid that usually carries all the genes needed for the process. A well-studied, transferable plasmid in *E. coli* is called **fertility factor (F factor)** (**Figure 9.3**). The tip of the specialized F-factor pilus attaches to

Figure 9.2
Sex Pilus Connecting Two *E. coli* Cells

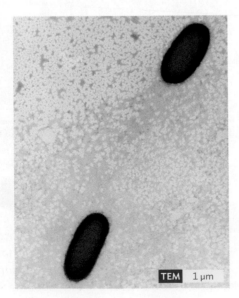

TEM | 1 μm

Figure 9.3   The Conjugation Process

Some plasmids, such as F factor in *E. coli*, can mediate cell-to-cell transfer of DNA. The plasmid is nicked at the nick site, *oriT* (step 3), and the 5′ end is transferred to the other cell. Purple arrowheads depict the 3′ end of replicating DNA. The 5′ end of the nick (black dot) will move through the pore and remain attached to the membrane while the rest of the single-stranded DNA passes into the recipient.

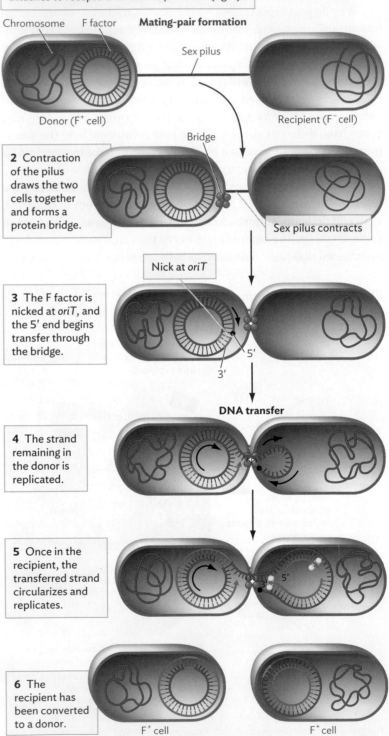

1 Sex pilus from the F⁺ plasmid donor (left) attaches to receptors on the recipient cell (right).

Chromosome   F factor   **Mating-pair formation**

Sex pilus

Donor (F⁺ cell)   Recipient (F⁻ cell)

2 Contraction of the pilus draws the two cells together and forms a protein bridge.

Bridge

Sex pilus contracts

3 The F factor is nicked at *oriT*, and the 5′ end begins transfer through the bridge.

Nick at *oriT*

5′

3′

DNA transfer

4 The strand remaining in the donor is replicated.

5 Once in the recipient, the transferred strand circularizes and replicates.

5′

3′

6 The recipient has been converted to a donor.

F⁺ cell   F⁺ cell

a receptor on the recipient cell and then contracts, drawing the two cells closer. The two cell envelopes fuse and generate a conjugation complex. The conjugation complex is composed of protein subunits, similar to the transformation complex (described earlier). Through the conjugation complex, single-stranded DNA passes from donor cell to recipient cell.

Conjugation begins when a donor cell (named the **F⁺ cell**) carrying the plasmid contacts a plasmidless recipient cell (named the **F⁻ cell**; Figure 9.3, step 1). The F⁺ cell forms a protein bridge to connect to the F⁻ cell (step 2). Bridge formation and membrane fusion triggers the nicking (cutting) of one strand of the plasmid DNA at the DNA transfer origin *oriT* (step 3). One end (the 5′ end) of the nicked strand begins to transfer into the recipient cell. DNA replication then replaces the transferred strand in the donor (step 4) and synthesizes a complementary strand for the transferred strand in the recipient (step 5). Once the F-factor DNA has been transferred and its complementary strand restored, the DNA circularizes to re-form a plasmid. The protein bridge joining donor and recipient cells spontaneously breaks, the membranes are repaired, and the two cells separate. The F⁻ recipient cell has now been converted to a new F⁺ donor cell (step 6).

Within the recipient cell, F factor can remain as a plasmid or integrate into the chromosome (**Figure 9.4**). When the chromosome contains an integrated F factor, the entire chromosome essentially becomes an F factor capable of transferring the host chromosome. Cells carrying integrated F factors are called **Hfr cells** because of the high likelihood (compared with F⁺ cells) that genes on the Hfr donor chromosome will be transferred and recombine into a recipient's chromosome (Hfr = <u>h</u>igh <u>f</u>requency of <u>r</u>ecombination of chromosomal genes). In other words, a donor gene will replace its counterpart gene in the recipient. The ability to move chromosomal genes by conjugation allowed us to understand gene linkage (how far genes are from each other), chromosome structure (linear or circular), and mutations.

An integrated F factor can be excised from the chromosome via host recombination mechanisms. Usually, the F factor is excised cleanly and restored to its original form. Occasionally, however, it is excised along with some flanking DNA from the host chromosome. A derivative F plasmid that contains host DNA is called an **F-prime (F′) plasmid** or F-prime (F′) factor (Figure 9.4). It, too, can be transferred.

Note Bacterial conjugation is completely different from the process called conjugation in eukaryotes. Microbial eukaryotes actually form a cytoplasmic bridge in which two genomes undergo <u>reciprocal</u> exchange (discussed in Chapter 11).

Transduction

Bacterial viruses (**bacteriophages**, or **phages**, to be discussed in Chapter 12) are composed of nucleic acid enclosed within a symmetrical protein structure called a capsid. A DNA phage injects its

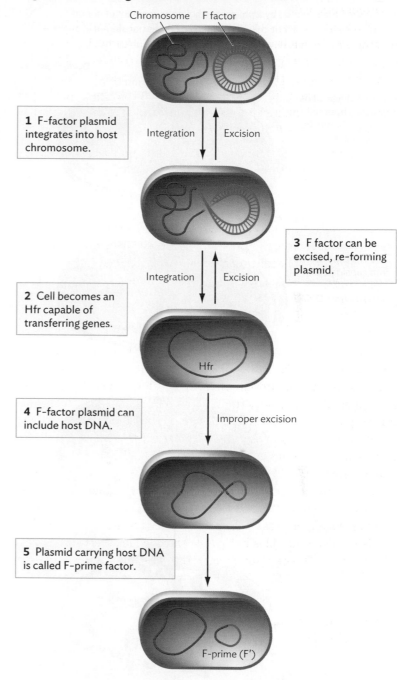

Figure 9.4   Integration and Excision of F Factor

Chromosome   F factor

1 F-factor plasmid integrates into host chromosome.

Integration   Excision

3 F factor can be excised, re-forming plasmid.

Integration   Excision

2 Cell becomes an Hfr capable of transferring genes.

Hfr

4 F-factor plasmid can include host DNA.

Improper excision

5 Plasmid carrying host DNA is called F-prime factor.

F-prime (F′)

DNA into a live bacterial cell. The phage DNA commandeers host protein synthesis machinery to make more capsid protein and uses bits of the host's DNA replication machinery to make more phage DNA. Newly made phage DNA is slipped into a capsid, which acts as a container. The completed phage particles are then released into the environment, often by lysing the host cell.

The process of forming a phage particle is not perfect. Some capsids may accidentally package a small piece of host bacterial DNA. The process by which a bacteriophage carries a payload of host DNA from one cell to another is known as **transduction**.

Figure 9.5   Generalized Transduction

Generalized transduction by a phage can move any segment of donor chromosome to a recipient cell. The number of genes transferred in any one phage capsid is limited, however, to what can fit in the phage head.

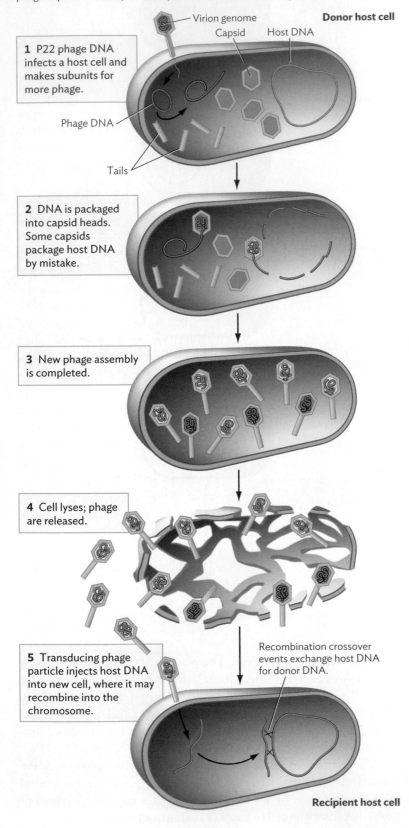

1 P22 phage DNA infects a host cell and makes subunits for more phage.

Virion genome
Capsid
Host DNA
Donor host cell
Phage DNA
Tails

2 DNA is packaged into capsid heads. Some capsids package host DNA by mistake.

3 New phage assembly is completed.

4 Cell lyses; phage are released.

5 Transducing phage particle injects host DNA into new cell, where it may recombine into the chromosome.

Recombination crossover events exchange host DNA for donor DNA.

Recipient host cell

In **generalized transduction**, the phage capsid mistakenly fills with a bacterial DNA fragment (**Figure 9.5**). The transducing phage particle thus contains only bacterial DNA, no phage DNA at all. When transducing phage particles are released, they can bind to another bacterial cell and inject the genes from the "donor" host. DNA recombination mechanisms can then replace the homologous DNA segment of the recipient host genome with the donor DNA. Any host genome segment can be packaged into a transducing phage particle and moved to a recipient cell.

In another process, called **specialized transduction**, the phage integrates its own genome into a specific DNA sequence in the host genome (**Figure 9.6**). The integrated phage genome is called a **prophage**. The phage DNA replicates along with the host DNA as the host cell multiplies. But at some point, the phage DNA may leave the host chromosome and start generating phage particles to be released by host cell lysis. In specialized transduction, the integrated phage DNA mistakenly carries a small piece of host DNA flanking the phage DNA insertion site. The flanking host DNA can thus be packaged with the phage genome and transferred with the phage to a new host cell.

DNA Restriction and Modification

As you might imagine, certain dangers await cells that indiscriminately accept transferred DNA. The most obvious risk involves bacteriophages whose replication will kill the target cells. A bacterium that can digest invading phage DNA while protecting its own chromosome would have a far better chance of surviving in nature than cells unable to make this distinction.

Phages are but one danger. The unrestricted incorporation of foreign DNA can be an energetic drain on the cell. Genes encoding competing or useless products would squander resources and lower the overall fitness of the cell. As a result, bacteria have developed a kind of "safe sex" approach to gene exchange. It is an imperfect approach, however, that still leaves room for beneficial genetic exchanges.

This protection system, called restriction and modification, involves the protective methylation (a chemical modification) of self DNA and the enzymatic cleavage (restriction) of foreign DNA (**Figure 9.7**). DNA restriction endonucleases (introduced in Section 8.4) recognize specific short DNA sequences (known as restriction sites) and cleave invading DNA at or near those sequences (**Figure 9.8**). Restriction enzymes purified from different bacteria can be used as DNA "scissors" in various applications of genetic engineering.

How do bacteria protect their own DNA from these endonucleases? During host DNA replication, DNA modification enzymes attach methyl groups to the host DNA sequences. Methylation makes the sequence invisible to the cognate (matched) restriction enzyme. Only one strand of the sequence needs to be methylated to protect the duplex from cleavage; thus, even a newly replicated DNA sequence in which one strand is temporarily unmethylated escapes restriction.

Figure 9.6   Specialized Transduction

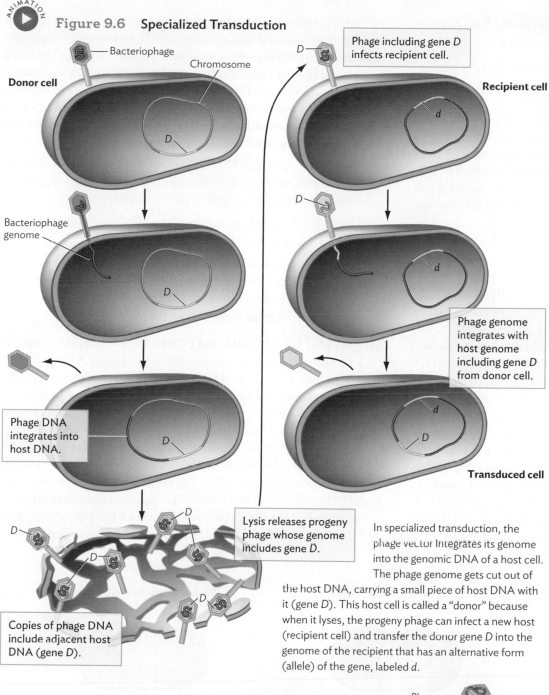

Donor cell

Bacteriophage

Chromosome

D

Bacteriophage genome

D

Phage DNA integrates into host DNA.

D

Lysis releases progeny phage whose genome includes gene *D*.

D *D* *D* *D*

Copies of phage DNA include adjacent host DNA (gene *D*).

Phage including gene *D* infects recipient cell.

Recipient cell

d

D

d

Phage genome integrates with host genome including gene *D* from donor cell.

d

D

Transduced cell

In specialized transduction, the phage vector integrates its genome into the genomic DNA of a host cell. The phage genome gets cut out of the host DNA, carrying a small piece of host DNA with it (gene *D*). This host cell is called a "donor" because when it lyses, the progeny phage can infect a new host (recipient cell) and transfer the donor gene *D* into the genome of the recipient that has an alternative form (allele) of the gene, labeled *d*.

Recombination between DNA Helices

Once a new piece of DNA has entered a cell, what happens to it? The answer depends on the nature of the acquired DNA. If the new DNA is in the form of a plasmid capable of autonomous replication, the plasmid will coexist in the cell separate from the host chromosome. If the foreign DNA is not capable of autonomous replication, it is either degraded by nucleases or incorporated into the chromosome through recombination.

Within a cell, two separate DNA molecules can recombine. **Recombination** is an enzyme-mediated process in which two DNA molecules exchange portions by breaking and re-forming their sugar-phosphate backbones. Mechanisms of recombination can be either "generalized" or "site-specific." **Generalized recombination**, also called **homologous recombination**, requires that the two recombining molecules have long stretches of DNA sequence homology. In contrast to generalized recombination, **site-specific recombination** requires very little sequence homology between the recombining DNA molecules, but does require a short (10–20 bp) sequence recognized by a special recombination enzyme.

Figure 9.8   EcoRI Restriction Site

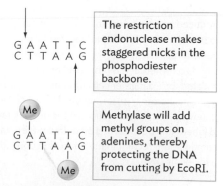

G A A T T C
C T T A A G

The restriction endonuclease makes staggered nicks in the phosphodiester backbone.

Me
G A A T T C
C T T A A G
Me

Methylase will add methyl groups on adenines, thereby protecting the DNA from cutting by EcoRI.

EcoRI is an example of a restriction-modification system. Shown here are cleavage (top) and methyl modification (bottom). The DNA sequence shown is specifically recognized by the endonuclease and methylase enzymes.

Figure 9.7   Restriction of Invading Phage DNA

Phage DNA is injected into a host where restriction endonucleases can digest it. Host DNA is protected because specific methylations of its own DNA prevent the enzymes from cutting it.

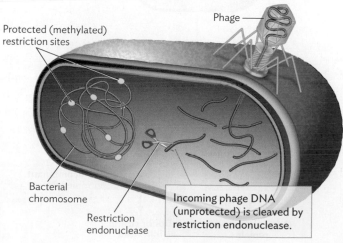

Protected (methylated) restriction sites

Phage

Bacterial chromosome

Restriction endonuclease

Incoming phage DNA (unprotected) is cleaved by restriction endonuclease.

Specialized recombination is what phages such as the beta phage of *C. diphtheriae* use to integrate into the host genome.

GENERALIZED RECOMBINATION   In generalized recombination, foreign DNA that enters a cell (by transformation, conjugation, or transduction) may recombine with the recipient genome if the two sequences are homologous (similar). The process uses the enzyme RecA to find and align homologous stretches of DNA sequence in two different DNA molecules. The enzyme then catalyzes an exchange of strands (similar to replacing a section of pipe in your home). Homologs of RecA are found in many bacteria and archaea. RecA coats a stretch of single-stranded DNA in the donor molecule, allowing "strand invasion" of the recipient duplex (**Figure 9.9**). RecA can recombine donor DNA that is single-stranded (entering by transformation) or double-stranded (entering by conjugation or transduction).

So, why should health care professionals care about all of this? Imagine your intestine contains millions of harmless *E. coli*—normal in a healthy intestine. Now what if your resident *E. coli* also contains a mutant DNA gyrase gene whose product (needed for DNA replication) is resistant to the antibiotic ciprofloxacin (an antibiotic used to treat serious bouts of diarrhea caused by pathogenic strains of *E. coli*). You are still healthy, but not if you eat a hamburger tainted with a pathogenic *E. coli* strain. In the mixed community of your intestinal microbiome, your *E. coli* Hfr strains conjugate with the pathogenic newcomers and transfer the ciprofloxacin resistance gene. The pathogen grows, you develop bloody diarrhea, and the clinician prescribes ciprofloxacin, but thanks to conjugation, the pathogenic strains are now resistant to that antibiotic. The disease persists and you end up hospitalized. Gene transfer and recombination mechanisms have human consequences.

When trying to understand recombination (and gene transfer overall), it is important to remember that the processes are generally random. The lucky bacterium that gets the right piece of DNA and recombines the right segment of that DNA will benefit. This is why a genetic experiment usually requires screening hundreds of millions of cells to find a few recombinants that grow into visible colonies.

Why is recombination advantageous to a microbe? There are three functions for generalized recombination in the microbial cell:

- Recombination probably first evolved as an internal method of DNA repair, useful to fix mutations or restart stalled replication forks. This role does not involve foreign DNA.

- Cells with damaged chromosomes use DNA donated by others of the same species to repair their damaged genes.

- Recombination is also part of a "self-improvement" program to obtain genes from other organisms that might enhance the competitive fitness of the cell. Thus, recombination makes major contributions to evolution.

SITE-SPECIFIC RECOMBINATION   In contrast to generalized recombination mechanisms that require RecA protein and can recombine any region of the chromosome, site-specific recombination does not use RecA and moves only a limited number of genes. This form of recombination involves short regions of homology between donor and target DNA molecules. Special recombinase enzymes recognize those sequences and catalyze a crossover between them to produce a cointegrate molecule. The integration of beta phage of *C. diphtheriae*, the organism in the opening case history, is one example of site-specific recombination. A short DNA sequence called the *att* site is present on both the beta phage DNA and the *Corynebacterium* chromosome. A phage-encoded protein

Figure 9.9   Generalized Recombination

In both single-strand and double-strand recombination, sections of the recipient strand are replaced by the donor DNA. The recipient DNA removed from the genome is degraded. A single-stranded molecule introduced by transformation will replace only one strand of the homologous region of the donor chromosome. The fate of the resulting hybrid molecule depends on whether it is replicated or repaired. If it is replicated, one daughter cell chromosome will look like the original before transformation. The other daughter cell, however, will have replicated the donor strand and inherit any traits carried on it. In double-strand recombination, recombination occurs in both strands of the recipient DNA.

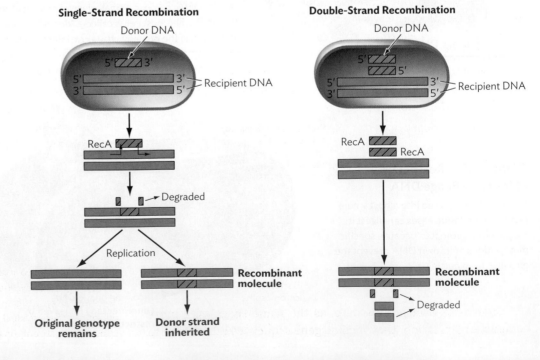

called integrase engages the two *att* sites and, through a strand breakage and rejoining mechanism, combines the two molecules. The process is similar to what happens when F-factor plasmid integrates (see Figure 9.4).

Mobile Genetic Elements

In 1948, Barbara McClintock (1902–1992) noticed that certain genetic traits of corn defied the laws of Mendelian inheritance. The genes responsible for these traits, sometimes called "jumping genes," seemed to hop from one chromosome to another. Although McClintock's theories were provocative at the time, we now know that these types of genes, referred to as **transposable elements**, or **transposons**, exist in virtually all life-forms and can move both within and between chromosomes by a process called **transposition**. Transposable elements can also transfer antibiotic resistance between bacteria.

Unlike plasmids, transposable elements cannot replicate outside a larger DNA molecule. They exist only as hitchhikers integrated into some other DNA molecule (**Figure 9.10**). All transposable elements include a transposase gene whose enzyme product (transposase) moves the element from one DNA molecule into another. The simplest transposable elements are short DNA sequences (700–1,500 bp) called **insertion sequences (ISs)**. An insertion sequence has a transposase gene flanked by short inverted DNA repeat sequences. The transposase, once made, binds to these inverted repeats and cuts the DNA. The transposase then creates staggered cuts in the target DNA to which the excised insertion sequence is ligated. The insertion sequence has now moved, or "hopped," from one segment of DNA to another. After transposition, the staggered ends flanking the insertion are replicated to produce short, duplicated DNA sequences that, like bookends, sit at both sides of the transposed element. Note that any transposable element, including insertion sequence elements, can hop to different sites within the same DNA molecule or between two different DNA molecules.

Note An inverted repeat is a DNA sequence identical to another downstream sequence in the opposite strand. The repeats have reversed sequences and are separated from each other by intervening sequences:

$$5'\text{-AATCGAT}\ldots\ldots\ldots\text{ATCGATT-}3'$$
$$3'\text{-TTAGCTA}\ldots\ldots\ldots\text{TAGCTAA-}5'$$

When no nucleotides intervene between the inverted sequences, it is called a palindrome. The inverted bottom strand reads the same as the top strand.

Composite transposons have, as the name suggests, a more complex organization; they include gene sequences in addition to

Figure 9.10 **Transposition and Origin of Target Site Duplication**

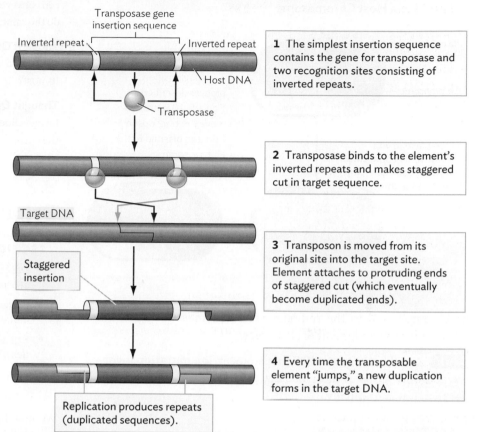

1 The simplest insertion sequence contains the gene for transposase and two recognition sites consisting of inverted repeats.

2 Transposase binds to the element's inverted repeats and makes staggered cut in target sequence.

3 Transposon is moved from its original site into the target site. Element attaches to protruding ends of staggered cut (which eventually become duplicated ends).

4 Every time the transposable element "jumps," a new duplication forms in the target DNA.

How a simple transposable element hops from one piece of DNA to another. Transposases (enzymes that catalyze transposition) generate duplications in the target site by ligating (attaching) the ends of the insertion element to the long ends of a staggered cut at the target DNA site.

insertion sequences. Tn*10*, for instance, consists of two insertion elements that flank a tetracycline resistance gene (**Figure 9.11**). This transposon is important because it mediates the spread of tetracycline resistance through populations of bacteria.

Figure 9.12 illustrates the transposition of a transposon from a plasmid into the chromosome. Note that the transposase randomly selects one of many possible target sequences where it will move the insertion element. If the transposon inserts within a structural gene, it produces an **insertion mutation** that can change the organism's phenotype. Today we know that these mobile DNA elements

Figure 9.11 **Example of a Composite Transposon**

Transposon Tn*10* is a composite transposon. Composite transposons contain genes flanked by two insertion elements. Tn*10* carries a gene for resistance to the antibiotic tetracycline. The yellow fragments represent inverted repeats. The internal repeats flanking the tetracycline resistance gene contain mutations preventing the individual IS elements from jumping off on their own. The transposase acts only on the external repeats.

Figure 9.12   A Transposon "Hops" from a Plasmid to a Host Chromosome

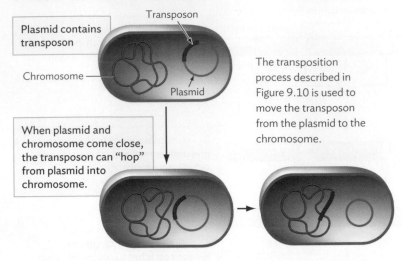

Plasmid contains transposon

Transposon

Chromosome

Plasmid

When plasmid and chromosome come close, the transposon can "hop" from plasmid into chromosome.

The transposition process described in Figure 9.10 is used to move the transposon from the plasmid to the chromosome.

have contributed to the remodeling of genomes throughout the evolutionary development of all species.

Link Recall from Section 8.3 that an **insertion mutation** involves the addition of one or more nucleotides, making the sequence longer than it was originally.

SECTION SUMMARY

- **DNA transfer mechanisms** enable microbes to obtain genes from their neighbors.
- **Horizontal gene transfers** between species occur by conjugation, transformation, and transduction.
- **Transformation** is the uptake by living cells of free-floating DNA from dead, lysed cells.
- **Conjugation** is a DNA transfer process mediated by a transferable plasmid that requires cell-cell contact and formation of a protein complex between mating cells.
- **Transduction** is the process whereby bacteriophages transfer fragments of bacterial DNA from one bacterium to another. In generalized transduction, a phage preparation can move any gene in a bacterial genome to another bacterium. In specialized transduction, a phage can move only a limited number of bacterial genes.
- **Restriction endonucleases** protect bacteria from invasion by foreign DNA.
- **Recombination** is the process by which DNA molecules are cleaved and reattached to other DNA. Recombination may be **general** (affects a long stretch of DNA homology) or **site-specific** (requires only short homology between ends of two DNA molecules).
- **Composite transposons and insertion sequences** ("jumping genes") move by site-specific recombination from one DNA molecule to another. Composite transposons can carry a variety of genes, including antibiotic resistance genes.

Thought Question 9.1 Can a *recA* mutant of *S. pneumoniae* experience horizontal gene transfer?

Thought Question 9.2 Transfer of an F factor from an F⁺ cell to an F⁻ cell converts the recipient to F⁺. Why does genome transfer by an Hfr not do the same?

Thought Question 9.3 How do you think phage DNA containing restriction sites evades degradation by the restriction screening system of its host?

Thought Question 9.4 How might a transposon containing a gene for tetracycline resistance make its way from the chromosome of one cell to the chromosome of a second cell?

9.2
Origins and Evolution

SECTION OBJECTIVES

- Describe how all life originated from microbes, how life-forms evolved over time, and how species continue to evolve.
- Explain how microbial genomes change via processes of random mutation and natural selection.
- Explain how a molecular clock works to measure phylogeny, and describe the key attributes of a useful molecular clock gene.
- Describe the three domains of life.

We now have the ability to sequence an organism's entire genome in a matter of days (DNA sequencing is discussed in Chapter 8). Scientists can use desktop computers and the Internet to quickly compare DNA sequences from thousands of sequenced genomes. This analysis led to startling discoveries about evolution and the interconnectedness of life. Archaea, for example, possess many traits in common with eukaryotes and bacteria. In fact, most proteins found in the archaeon *Methanocaldococcus jannaschii* are also found in Eukarya and/or Bacteria. As we just learned in Section 9.1, microbial genomes not only inherit parental genes but also may acquire genes from other species. The genome of *E. coli*, an organism under scrutiny for more than 100 years, turns out to be a mosaic of DNA segments acquired from other species. For example, *E. coli* O157:H7, a strain that causes bloody diarrhea, acquired several virulence operons from other species, including the genes for Shiga toxin, which causes kidney failure.

How do we know where genes come from? And how do we know that all organisms share a common ancestry? Our knowledge of the history of life is founded on a massive collection of evidence from geology, physiology, and DNA sequences.

Origins of Life

Earth itself formed approximately 4.5 billion years ago as the Sun and other planets condensed out of interstellar matter. The major elements of Earth—including the atoms of life, such as carbon, nitrogen, and oxygen—could not have originated in our Sun, which is too young to form elements heavier than helium. Heavier elements formed through nuclear reactions much earlier, from stars that were born and died billions of years before. Thus, when poets say "We are made of stardust," it's true.

How did inert atoms give rise to living organisms? We do not know; we can only point to evidence of life's processes in the geological record. Our earliest evidence for microbial life comes from carbon isotope ratios—that is, the ratio of carbon isotopes (atoms) of atomic mass 12 (^{12}C) versus 13 (^{13}C), which differ by one neutron. The reactions of carbon dioxide fixation through photosynthesis (discussed in Chapter 7) preferentially incorporate carbon-12 rather than carbon-13. Thus, a carbon deposit depleted of ^{13}C in relation to ^{12}C probably came from living cells that fixed CO_2 into biomass. The oldest rock on Earth showing depletion of carbon-13, the footprint of ancient life, is 3.7 billion years old (3.7 Gyr).

Some early rock formations called **microfossils** show organic patterns typical of living creatures (Figure 9.13). Ancient microfossils resemble microbial biofilms and stromatolites, towerlike layers of cyanobacteria that grow in shallow salty seas such as Shark Bay, in Australia. Well-preserved microfossils date back at least as far as 2 billion years in geological strata dated by radioisotope decay. More recent fossils, 500 million years old, reveal the multicellular animals such as trilobites that evolved from microbes. Ultimately, microbial ancestors gave rise to all multicellular plants and animals, including ourselves.

How did all the different species of life diverge from common ancestors? And how do we know that all modern life-forms possess a common microbial ancestor some billions of years ago? The answers to these questions come from a very different kind of "fossil": the remnants of millennia-old information that lie within modern sequences of DNA.

Microbial Divergence and Phylogeny

A unifying assumption of modern biology is that genetic relatedness arises from common ancestry. As organisms reproduce over many generations, their offspring acquire mutations (as discussed in Chapter 8). The mutant sequences become "fixed" in genomes, leading to **divergence**. Divergent organisms show less and less similarity the longer their lineage has diverged. Besides mutation, another source of divergence is gene mobility, discussed earlier. The gain or loss of a large number of genes can lead to a more sudden change, even to "instant speciation," in which a new strain of organism shows traits very different from the parent strain; for example, a bacterium might integrate DNA from a bacterial virus and, in the process, acquire a gene conferring virulence,

Figure 9.13 Geological Evidence for Early Life

Early in Earth's history, the geological record shows evidence of microbial life. Photos on the right include microfossils from the Proterozoic eon that show the presence of ancient microbes.

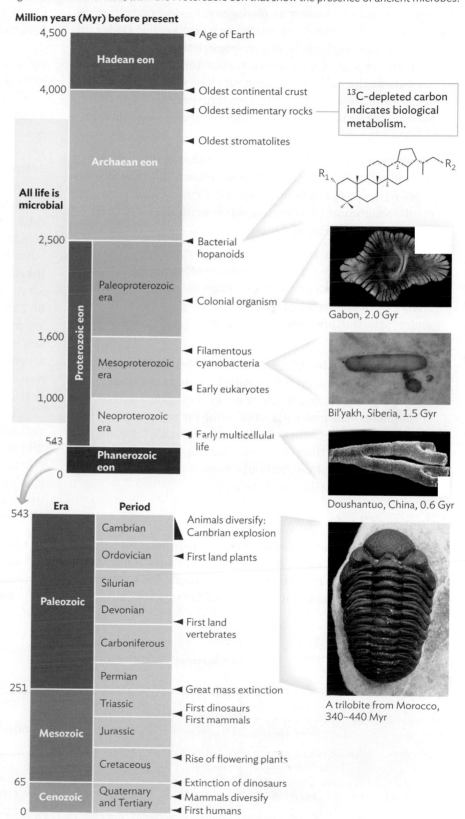

^{13}C-depleted carbon indicates biological metabolism.

Gabon, 2.0 Gyr

Bil'yakh, Siberia, 1.5 Gyr

Doushantuo, China, 0.6 Gyr

A trilobite from Morocco, 340–440 Myr

generating a pathogen. This is how *Corynebacterium diphtheriae* gained its toxin gene.

The gradual divergence of organisms generates a "family tree" of related organisms, known as **phylogeny**. Phylogeny comprises a series of branching groups of related organisms called **clades**. In classic phylogeny, each clade is a **monophyletic group**—that is, a group of species that share a common ancestor not shared by any species outside the clade. Large monophyletic groups branch into smaller monophyletic groups and ultimately species.

Several mechanisms of evolution determine how populations of organisms diverge:

- **Random mutation.** DNA sequence can change through rare replication mistakes (in bacteria and archaea, this typically happens to one of a million base pairs). Replication errors result in mutation. Most mutations are neutral; that is, they have no effect on gene function.

- **Natural selection.** In a given environment, natural selection favors the organism that produces more offspring in that environment (competitive advantage). Therefore, genes encoding traits that influence the survival of offspring are said to be under "selection pressure." A favorable mutation increases competitive advantage.

- **Reductive evolution (degenerative evolution).** A trait no longer needed for growth or survival is no longer under selection pressure. As a result, genes encoding the trait can accumulate mutations without affecting the organism's reproductive success. Because mutations that decrease function are more common than mutations that improve function, accumulating mutations without selection pressure leads to decline and ultimately loss of the trait.

Molecular Clocks

How do we tell time on an evolutionary scale (in the range of millions of years)? The answer, again, is found in DNA. Random mutations with neutral effects are not subject to selection and thus accumulate at a steady rate over generations. The result of random mutations

is **genetic drift**. In populations that do not interbreed, genetic drift causes genome sequences to diverge over time. If the mutation rate is <u>constant</u>, then genetic drift in a DNA sequence measures the time it took two species to diverge. A gene in which random mutations accumulate at a constant rate is called a **molecular clock**.

Molecular clocks have revolutionized our understanding of evolution, including that of human beings. The most important molecular clocks are genes that exist in all life-forms, such as ribosomal RNA (rRNA) genes. Ribosome structure and function are highly similar across all organisms. Carl Woese used the 16S rRNA gene to reveal the divergence of three domains of life (see Section 1.4). Other gene sequences are now used to measure divergence at different levels of classification.

A molecular clock measures time of divergence based on the acquisition of new random mutations in each round of DNA replication. For example, in Figure 9.14, each offspring in generation 2 acquires two new mutations; their sequences now differ from each other by 25%. In the next generation, each individual propagates the earlier mutations while acquiring two more random mutations. Strain 3A now differs by 50% from strains 3B and 3C, which differ by 25% from each other. (Actual mutation frequencies, of course, are much lower, about one base per million per generation.) The key assumption is that each generation of offspring acquires a consistent number of random mutations from the parents, and so the number of sequence differences between two species should be proportional to the time of divergence between them.

A gene used for a molecular clock must be one that is inherited vertically, from parent to offspring, and does not tend to be transmitted horizontally (by transformation or transduction, as discussed earlier). Genes horizontally transferred from an unknown species to another now known species would give inaccurate information about the divergence of the genome it entered.

In real genomes, the molecular clock works best if:

- **The molecular clock gene has the same function across all species being compared (no selection pressure).** That is, all versions of the gene serve the same function; they have avoided evolving differences due to different selection pressures.

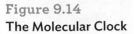
Figure 9.14
The Molecular Clock

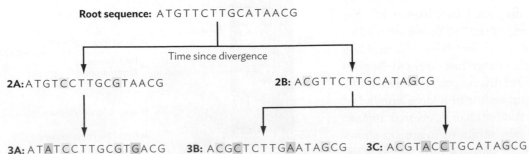

Root sequence: ATGTTCTTGCATAACG

Time since divergence

2A: ATGTCCTTGCGTAACG

2B: ACGTTCTTGCATAGCG

3A: ATATCCTTGCGTGACG   **3B:** ACGCTCTTGAATAGCG   **3C:** ACGTACCTGCATAGCG

As DNA replicates, mutations accumulate at random. The number of mutations accumulated is proportional to the number of generations and thus the time since divergence. In each sequence designation (for example, 2A), the numeral indicates the generation and the letter identifies a specific organism's DNA.

- **The generation time is the same for all species compared.** Mutations generally occur during replication. Consequently, species with shorter generation times (more frequent reproductive cycles) accumulate more mutations over a set length of time than a species that replicates less often (longer generation time). A large difference in generation time can lead to overestimates of how far back two species diverged because one species will have accumulated more mutations than the other.

- **The average mutation rate remains constant among species and across generations.** If different species mutate at different rates, species with more rapid rates of mutation will appear to have diverged over a longer time than is actually the case.

Using a molecular clock requires aligning similar genes (homologous sequences) from divergent species or strains. The number of differences between the sequences can be used to propose a tree of divergence.

Phylogenetic Trees

Once homologous sequences are aligned, the frequency of differences between them can be used to generate a **phylogenetic tree** that estimates the amount of evolutionary distance ("time") between the sequences (**Figure 9.15**). If mutation rate is the same for all sequences compared, then divergence data based on mutations can estimate how long it has been since two species diverged from a common ancestor.

A phylogenetic tree can compare any set of organisms. We can compare organisms from a single habitat, such as the human digestive tract or a Yellowstone hot spring. In **inSight**, you will see how phylogeny was used to discover surprising pathogens in soap scum from a shower curtain.

Alternatively, a set of related organisms may be compared from different habitats. **Figure 9.16** shows the divergence of selected gammaproteobacteria, a clade of Gram-negative bacteria (presented in Chapter 10). The tree is based on a set of **housekeeping genes**—that is, genes encoding functions essential for all cells and usually transmitted vertically (parent to offspring). *Escherichia, Shigella,* and *Salmonella* are closely related genera of intestinal bacteria. Their clade includes pathogenic species as well as commensal species normally present in the human body. The genus *Klebsiella* includes species that cause pneumonia. *Photorhabdus luminescens* is a bacterium that infects nematodes (roundworms) and helps them parasitize insects. *Yersinia pestis* causes bubonic plague. *Photobacterium profundum* is a marine barophile (high pressure) growing in a deep-sea trench. Diverse as they are, all these species diverged at some point from a common ancestor (which no longer exists). This common ancestor, in turn, diverged from the

Figure 9.15 **DNA Phylogenetic Tree**

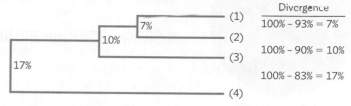

Divergence
100% – 93% = 7%
100% – 90% = 10%
100% – 83% = 17%

This phylogenetic tree was derived from divergence of four aligned DNA sequences. The length of each tree branch is proportional to the number of differences between two sequences. For instance, homologous sequences showing 10% difference diverged further back in time than homologous sequences showing a 7% difference.

"root" organism, which is an even more ancient common ancestor shared with the "outgroup" (most divergent sequence). The outgroup in Figure 9.16 is *Shewanella,* marine bacteria associated with fish spoilage. Thus, we see that closely related species, diverged from a recent common ancestor, can show surprisingly diverse traits and effects on humans.

The Three Domains of Life

The analysis of rRNA gene sequences shows that all life-forms on Earth, including humans and other animals, plants, and microbes, share a common ancestral cell (**Figure 9.17**). An amazing fact of life is that all living cells on Earth share profound similarities. From a molecular point of view, all cells consist of membrane-enclosed compartments that shelter a DNA-RNA protein synthesis machine. The fundamental components of this machine evolved before the three domains diverged from their last universal common ancestor. From a molecular standpoint, all cells on Earth appear more similar than they are different.

Figure 9.16 **Phylogeny of Selected Gammaproteobacteria**

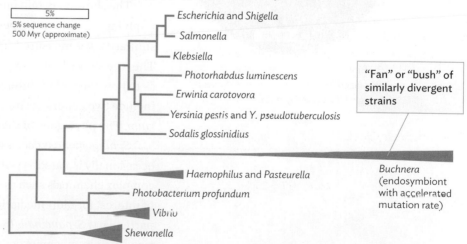

The phylogenetic tree was derived from sequences of highly conserved "housekeeping" proteins. The length of the scale bar corresponds to 5% amino acid sequence divergence, which takes 500 Myr to change. *Shewanella* is the beginning of an "outgroup" that originated from the same ancestral root organism as the *Vibrio* to *Escherichia* group.

inSight

A Shower Curtain Biofilm

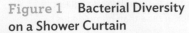

Figure 1   Bacterial Diversity on a Shower Curtain

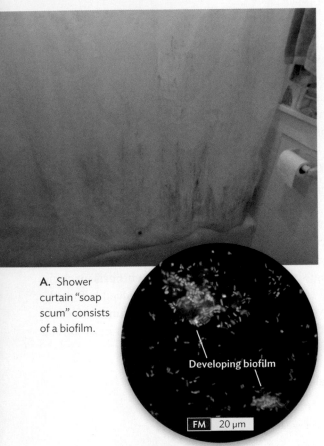

A. Shower curtain "soap scum" consists of a biofilm.

Developing biofilm

FM   20 µm

B. Biofilm from a shower curtain is visualized by epifluorescence microscopy using a DNA (DAPI) stain.

Suppose you finally get around to changing that shower curtain—and your allergy symptoms disappear. What was causing your allergic reaction? It just might have been a microbial inhabitant of the soap film—even a species never seen before. Novel microbes can be discovered surprisingly close to home—in the soil, on the roots of grass, in our own digestive tract. Even within our homes lurk microbial communities that include potentially dangerous pathogens.

Such organisms were discovered by Norman Pace and colleagues at the University of Colorado, where they usually study extremophile bacteria and archaea from the ocean floor or from thermal vents at Yellowstone National Park. This time, Pace explored a domestic aquatic habitat: the "soap scum" biofilm on a shower curtain (Figure 1). Domestic water systems such as cooling towers and whirlpool spas harbor pathogens such as *Legionella pneumophila*, the cause of legionellosis, a deadly form of pneumonia. Pathogens persist in biofilms growing within plumbing and air-conditioning units.

Lesser-known reservoirs for microbes are the biofilms that collect on household surfaces such as shower curtains, which are often neglected during cleaning. Viable cells of a shower curtain biofilm are revealed by fluorescence microscopy using a stain for DNA (see Figure 1B). Two of Pace's students, Scott Kelley and Ulrike Thiesen, extracted the microbial DNA, used PCR (see Section 8.4) to amplify the 16S rRNA genes, and then sequenced them. Comparison of observed gene sequences with those of known species identified the source organisms.

The shower curtain biofilm revealed 117 unique sequences for 16S rRNA, implying 117 different types of organisms. The sequences were aligned and their similarity was measured. Then the data were used to propose a phylogenetic tree. The tree shows how shower curtain samples relate to known bacteria, offering clues as to how they might interact with humans. The most abundant groups of species include two genera of Gram-negative bacteria: *Sphingomonas* and *Methylobacterium*. The phylogeny of the shower curtain sphingomonads is shown in Figure 2.

Sphingomonas species grow in a wide range of soil and water habitats and are occasionally isolated as yellow colonies from natural water supplies. They catabolize complex chemicals such as dibenzofuran and hexachlorocyclohexane and thus are of interest for bioremediation of environmental pollutants. Some species, however—particularly *S. paucimobilis* (highlighted in Figure 2)—cause opportunistic infections in immunocompromised individuals, including bacteremia, peritonitis, and abscesses.

Figure 2 *Sphingomonas* Phylogeny from a Shower Curtain Biofilm

A. Phylogeny of sphingomonad bacteria was based on rRNA gene sequence comparison.

B. Norman R. Pace characterized the first sequence phylogeny of thermophiles from high-temperature environments. His laboratory has since characterized the microbial diversity of other environments, such as a shower curtain biofilm.

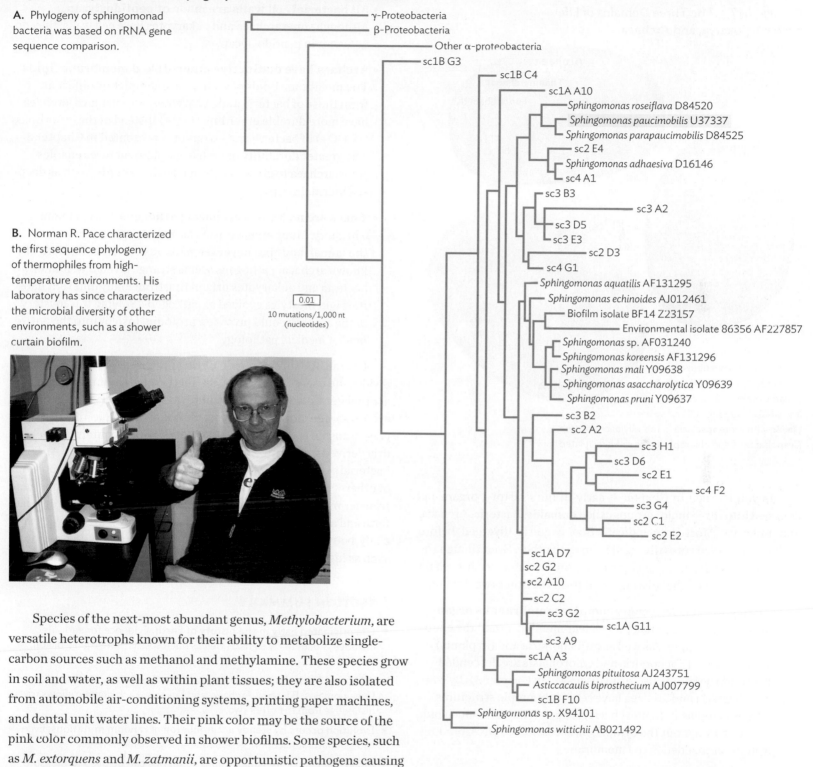

Species of the next-most abundant genus, *Methylobacterium*, are versatile heterotrophs known for their ability to metabolize single-carbon sources such as methanol and methylamine. These species grow in soil and water, as well as within plant tissues; they are also isolated from automobile air-conditioning systems, printing paper machines, and dental unit water lines. Their pink color may be the source of the pink color commonly observed in shower biofilms. Some species, such as *M. extorquens* and *M. zatmanii*, are opportunistic pathogens causing pneumonia, skin ulcers, and bacteremia. Growing numbers of immunocompromised individuals care for themselves at home, where they need to control their own micro-biological exposure. Thus, it is of concern to discover opportunistic pathogens in a home setting. Pace and colleagues conclude with a reminder that "exposure can be minimized by regular cleaning or by changing shower curtains."

Figure 9.17   The Three Domains of Life: Bacteria, Eukarya, and Archaea

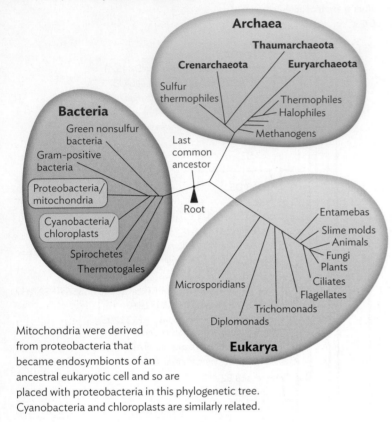

Mitochondria were derived from proteobacteria that became endosymbionts of an ancestral eukaryotic cell and so are placed with proteobacteria in this phylogenetic tree. Cyanobacteria and chloroplasts are similarly related.

As you learned in Chapter 1, early in life's history, organisms diverged into three major groups called **domains**: Bacteria, Archaea, and Eukarya. Most data indicate that bacteria diverged before archaea and eukaryotes diverged from each other. Even though the domains arose from an ancient common ancestor, each domain shows unique traits that distinguish it from the other two domains:

- **Eukaryotes possess many complex membranous organelles.** Organelles include the nuclear membrane and the endomembrane complex, as well as mitochondria and (in plants) chloroplasts. Mitochondria and chloroplasts are descended from endosymbiotic bacteria (discussed in Chapter 5). By contrast, bacteria and archaea have relatively simple structures in their cytoplasm. In most bacteria and archaea, the nucleoid extends throughout the cytoplasm, accessible to DNA-binding proteins embedded in the membrane.

- **Bacteria have a distinctive gene expression apparatus.** The DNA and RNA polymerases of bacteria are streamlined compared with those of eukaryotes and archaea, perhaps because the bacterial ancestor diverged first from the ancestor of the other two domains. Only bacteria use formylmethionine (instead of plain methionine) to fill the start codon in protein synthesis (discussed in Chapter 8).

- **All bacterial cell walls are made of peptidoglycan.** Although some archaea and eukaryotes have cell walls, they do not contain peptidoglycan.

- **Archaea have distinctive ether-linked membrane lipids.** The membrane lipids of archaea are completely different from those of bacteria and eukaryotes. For instance, archaea have more durable ether links (–O–) instead of the ester links (–COO–) of bacteria and eukaryotes presented in Chapter 3. The greater durability of archaeal lipid structures enables some archaea to grow in extreme environments, such as deep-sea thermal vents.

- **There are no known archaeal pathogens.** Why remains a mystery. Many archaea, such as methanogens, live within the human body but never seem to cause disease. The lack of known archaeal pathogens seems strange given that numerous bacteria and eukaryotes originating from many different habitats and lifestyles evolved as pathogens. Solving this mystery of the archaea could provide a profound insight into the entire field of medical pathology.

Overall, the three-domain phylogeny based on ribosomal genes divides life usefully into three distinctive clades. But when we look at other genes, we find remarkable evidence of horizontal gene flow even between domains. A most dramatic example of horizontal gene transfer is seen in the eukaryotes, which contain mitochondria derived from entire assimilated bacteria (see Section 1.4). Some bacterial genes persist within the mitochondria, whereas thousands of others "migrated" to nuclear chromosomes through mobile gene transfer mechanisms such as those described earlier. Between bacteria and archaea, many examples of gene transfer are known, especially between those organisms sharing a common environment, such as high-temperature habitats.

SECTION SUMMARY

- **Life originated early in the history of Earth.** Evidence for early life includes geochemical signs as well as microfossils more than 2 billion years old.

- **All life on Earth shares a common ancestral cell.** Early life diverged into three domains: Bacteria, Archaea, and Eukarya. Each domain shows traits distinctive from the other two.

- **Evolution occurs by steady accumulation of random mutations.** Even without natural selection, gene sequences gradually diverge.

- **Molecular clocks are genes whose sequence can be used to measure divergence time.** Molecular clock genes must have a constant mutation rate, and must share a common function in the organisms compared.

- **Phylogenetic trees show the relatedness between different kinds of organisms.** Phylogeny can be determined from DNA sequence comparison using molecular clocks.

- **Evolutionary history of organisms includes monophyletic descent for some genes and horizontal transfer of others.**

9.3
Natural Selection

How do new species arise? In Section 9.2, we described how the gradual accumulation of random mutations causes genomes to diverge and evolve. The rate of nonselected sequence changes enables us to measure evolutionary time. But for organisms, meaningful evolution occurs when a changed trait affects survival, which is the basis of **natural selection**. In natural selection, the variant strains that result from random mutation will differ in their chance of survival. Variants that survive and leave more offspring than other strains undergo "selection"; that is, their particular traits are overrepresented in the next generation. Over many generations, the descendants develop traits that are best suited for their environment and very different from those of their ancestors. At some point, traits can change enough for the descendant to be considered a new species.

Natural selection responds to environmental factors such as nutrient availability, temperature, and, in the medical realm, the presence of an antibiotic. Even over short periods, such as a few weeks, the natural selection of bacteria exposed to antibiotics can yield antibiotic-resistant descendants that will threaten a patient's life.

CASE HISTORY 9.1

The Bacteria Kept Coming Back

 Harrison was a 73-year-old man with kidney failure. He was admitted to the hospital with fever and hypotension, both of which are symptoms of bacteremia. When bacteria from Harrison's blood were grown on horse blood agar, it showed colonies typical of methicillin-resistant *Staphylococcus aureus* (MRSA) and confirmed the diagnosis of bacteremia (**Figure 9.18**). Further investigation uncovered his home dialysis line as the source of MRSA. He was treated with intravenous vancomycin (an antibiotic that targets cell wall synthesis), but the bacteremia persisted. The bacteria were tested and found to be partly resistant to vancomycin, though sensitive to rifampin (which targets RNA polymerase), ciprofloxacin (which targets DNA gyrase), and linezolid (which targets protein synthesis). The vancomycin was discontinued, and therapy was begun with rifampin, ciprofloxacin, and linezolid. Harrison's symptoms improved, but within 3 weeks, the bacteremia returned. Bacteria cultured from his blood now showed resistance to rifampin and ciprofloxacin. Linezolid was administered alone for 6 weeks, but 5 days after linezolid treatment ended, the bacteremia recurred. Linezolid was administered again, for another 6 weeks. Then a new blood culture revealed a form of MRSA with small colonies (see Figure 9.18). The small-colony variant was partly resistant to linezolid while fully resistant to all the other antibiotics. Harrison was now treated with trimethoprim-sulfamethoxazole. These final antibiotics had to be administered indefinitely, as they could not fully eliminate MRSA from Harrison's system.

In this case history, the lethal MRSA strain of *S. aureus* had already evolved resistance to drugs such as methicillin. The patient, Harrison, was an elderly man with a weakened immune system, unable to eliminate even small populations of an opportunistic pathogen such as MRSA. The original culture from the patient's blood showed bacteria sensitive to four antibiotics. But prolonged exposure resulted in natural selection for resistance. Ultimately, after 12 weeks' exposure to linezolid, the bacteria isolated showed resistance to three other antibiotics plus partial resistance to linezolid.

Fitness Depends on Environment

What determines the "fitness" of a trait? In Case History 9.1, the latest drug-resistant strain made smaller colonies in the laboratory (the small-colony variant). So how could "fitness" have increased? Natural selection yielded a population that was the "fittest" under a

Figure 9.18   Evolution of Antibiotic Resistance in MRSA

The initial MRSA isolate infecting the patient showed partial resistance to vancomycin but was sensitive to rifampin, ciprofloxacin, and linezolid. After exposure to these antibiotics, a new strain was isolated that grew more slowly (the small-colony variant) but was resistant to all the antibiotics.

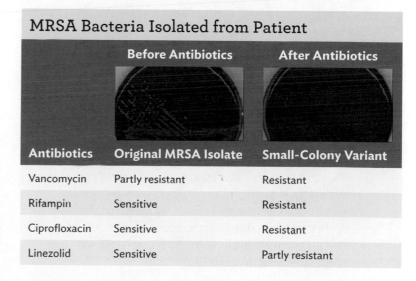

MRSA Bacteria Isolated from Patient

	Before Antibiotics	After Antibiotics
Antibiotics	**Original MRSA Isolate**	**Small-Colony Variant**
Vancomycin	Partly resistant	Resistant
Rifampin	Sensitive	Resistant
Ciprofloxacin	Sensitive	Resistant
Linezolid	Sensitive	Partly resistant

particular environmental condition—the presence of linezolid in an immunocompromised host where even a slow-growing strain could persist. In the absence of the drug, the original strain would have outcompeted the small-colony variant. Thus, natural selection for the "fittest" depends entirely on the environment in which selection occurs. Different environments lead to selection for different traits "fittest" for that particular environment.

How did the multidrug resistance come about? Researchers obtained DNA from Harrison's original MRSA as well as the later small-colony variant. The two genomes were sequenced and compared. Just three point mutations in the small-colony variant accounted for the drug resistance as well as the retarded growth. One of these mutations causes accumulation of the signal molecule ppGpp (guanosine tetraphosphate). The ppGpp molecule turns on a global regulator, similar to the CRP regulator discussed in the previous chapter (see Section 8.7). The ppGpp stress "regulon" includes many protective genes that enable a cell to survive in the presence of antibiotics. But the cost to the cell of constantly expressing this regulon is lowered growth rate; like a community continually under "terror alert," the cell's normal everyday processes are slowed by the demands of the stress response.

Evolution of Endosymbionts

We have seen how a pathogen can evolve multidrug resistance at the expense of growth rate in a short period of time. Over a longer period, pathogens evolve specific traits that enhance their ability to colonize a host, such as the production of toxins that disrupt the host's immune system. In addition, a pathogen can evolve by losing cellular functions that are supplied by the host cell. Ultimately, the pathogen may become completely dependent on its host. For

example, *Treponema pallidum*, the cause of syphilis, is an ancient pathogen of humans. The genome of *T. pallidum* shows loss of numerous genes found in related bacteria, including genes required for energy production and amino acid biosynthesis. The pathogen obtains carbon sources and amino acids from the bloodstream of its host; thus, there is no selection pressure for the pathogen to maintain these traits. Loss of traits in the absence of selection pressure is called **degenerative** or **reductive evolution**.

An even more remarkable kind of evolution occurs between parasites and their intracellular partner species, called **endosymbionts**. An intriguing medically important class of endosymbionts is the *Wolbachia* bacteria that live inside the cells of invertebrate hosts, such as the nematode worms that cause filariasis, also known as elephantiasis (see Figure 2.3). The filarial nematode *Brugia malayi* harbors *Wolbachia* endosymbionts. These *Wolbachia* strains may have entered the nematode originally as a pathogen of the parasite and then persisted because of its metabolic contributions to the host. A mutualistic relationship eventually evolved between the endosymbiotic *Wolbachia* and its host; presence of *Wolbachia* is required for the nematode's embryonic development. The bacteria grow within tissue layers beneath the nematode's skin and within the uterine tubes of females, where they enter the developing offspring (Figure 9.19).

Over time, the *Wolbachia* bacteria gained genetic traits that enabled growth within nematode hosts. But the *Wolbachia* genome also lost many genes through reductive evolution. With barely a million base pairs, the endosymbiont's genome retained genes for glycolysis and the TCA cycle but jettisoned the pathways needed to make amino acids and most vitamins. It nonetheless retains pathways to make purines, pyrimidines, and the coenzymes riboflavin and FAD—essential pathways not provided by its host nematode. Overall, *Wolbachia* appears to be evolving into an organelle of its host, like the ancestors of mitochondria and chloroplasts.

Filarial disease is very difficult to treat with antiparasitic agents. But because the nematodes require *Wolbachia* to live, human patients infected by the nematodes can be treated with antibiotics such as tetracycline. The *Wolbachia* disappear from worm tissues, and the worm burden gradually decreases. Antibacterial agents eliminate the worms sooner and more completely than does treatment with anti-nematode agents.

Figure 9.19   The Filarial Endosymbiont *Wolbachia*

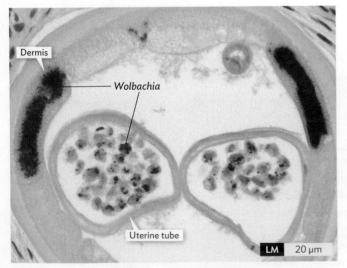

Cross section of the nematode *Brugia malayi*, showing endosymbiotic *Wolbachia* bacteria (stained pink) within the dermis and the uterine tubes.

SECTION SUMMARY

- **Natural selection favors traits that increase survival and reproduction.** Traits that increase the number of offspring inheriting the trait increase in the population.

- **Natural selection depends on the environment.** A trait favored under one set of conditions may be disfavored under other conditions.

- **Use of antibiotics leads to rapid selection of microbial strains resistant to the antibiotic.** Carefully limited use of antibiotics prolongs their period of effectiveness.

- Pathogens and parasites evolve specific traits enabling them to take advantage of a host and cause disease.

- Pathogens and parasites undergo degenerative (reductive) evolution. Degenerative evolution involves loss of traits that the parasite no longer needs because the functions are provided by its host.

Thought Question 9.5 Parasitic microbes, such as *Giardia* and *Plasmodium falciparum*, undergo degenerative evolution as well as selection for complex traits. Explain why both kinds of natural selection occur.

9.4
Bioinformatics

SECTION OBJECTIVES
- Explain how bioinformatic analysis helps reveal the pathogenesis and physiology of an emerging pathogen.
- Describe how gene duplication leads to different kinds of homologous genes, including orthologs and paralogs.

As revealed in Chapter 2, most bacteria that call humans "home" cannot be cultured in the laboratory. Nevertheless, through various DNA techniques, we know that they exist and can even sequence their genomes, or at least parts of them. Our knowledge about how microbes evolve can now be used to "interrogate" these DNA sequences and predict the traits of our hidden microbiota. The analysis of DNA involves comparing thousands of sequences through the use of computer algorithms, a field of study called **bioinformatics**. This section describes how bioinformatics can lead to startling discoveries about the nature of deadly pathogens and how they evolve.

CASE HISTORY 9.2
A Pathogen Evolves in a Single Patient

Nine-month-old Eva had 12 clinic visits over a 7-month period due to runny nose and/or ear infections. During each episode, nasopharyngeal swabs were obtained for bacterial culture and *Streptococcus pneumoniae* was obtained. Although all the infections were resolved with antibiotics, scientists gained insight into pathogen evolution by seeing how the isolates compared. Each isolate was first analyzed for seven genes that can indicate strain variability, a process called **multilocus sequence typing (MLST)**. The lab found two MLST types, indicating that the patient was infected with two divergent strains of *S. pneumoniae*. However, when the complete genomes of several of the isolates were sequenced and compared, four different genotypic strains were identified, all of which evolved as a result of 16 distinct recombination events between strains (**Figure 9.20**). Curiously, the technicians noticed that some recombined sequences did not seem to come from either of the two primary coinfecting strains. It appeared that this new DNA came from uncultured and undetected strains of *S. pneumoniae* that had colonized Eva's throat at one time or another. Horizontal gene transfers (transformation) between the undetected and detected isolates appear to have produced the new recombined strains of *S. pneumoniae*. The genome sequence data clearly support the idea that *S. pneumoniae* evolution is characterized by high rates of horizontal gene transfer and recombination. The scientists concluded that this and other related bacterial species use horizontal gene transfer to defeat the adaptive immunity of individual hosts, possibly explaining Eva's recurring infections. These types of bioinformatic comparisons suggest that horizontal gene transfers produce most of the genetic variation pathogens develop while persisting within a host.

Figure 9.20   Horizontal Gene Transfer and Evolution of *S. pneumoniae*

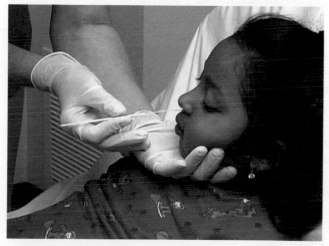

A. Nasopharyngeal sampling of a child.

Purple boxes indicate recombination events that led to pathogen strain ST13.

B. The patient's isolate *S. pneumoniae* ST13 genome (black) and the relative positions of recombination events that incorporated horizontally transferred DNA (purple boxes).

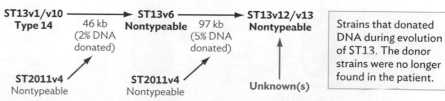

Strains that donated DNA during evolution of ST13. The donor strains were no longer found in the patient.

C. Possible recombination events that may have led to the creation of the strain ST13 isolated from the patient. Note the large size of the horizontally transferred DNA. The percentage of the whole genome sequence is shown in parentheses. The genome of *S. pneumoniae* is about 2.1 million bases (2.1 Mb).

Annotating the Genome Sequence

How do we compare genomes, such as those of the *S. pneumoniae* strains that colonized Eva's throat? Knowing the sequence is just a first step; meaningful information comes from **annotation** of the DNA sequence—that is, understanding what the sequence means. Annotation is analogous to identifying separate sentences and words in an unknown language. For most languages, you would look for punctuation marks like periods or exclamation points to identify sentences and then scan for blank spaces between letters to signify the beginning and end of a word. You may not initially know what the sentence says, but defining what makes a sentence is the first step.

Annotating a DNA sequence includes finding the start and stop sites of genes, as well as predicting the function of the gene

product. Computers use established rules to mark potential genes—called **open reading frames (ORFs)**. Then similarities are sought between the deduced amino acid sequences of those ORFs and the sequences of proteins with known functions. Similarities are used to infer the function of the unknown ORF. Genes for transfer RNA and ribosomal RNAs do not encode proteins but can be identified because their sequences are conserved across vast phylogenetic distances. The results of annotation can be represented in many different forms.

Lengthy DNA sequences are analyzed by computer programs such as ORF Finder. The program uses the universal genetic code to deduce all possible protein sequences that could be formed if all reading frames on RNA molecules were transcribed from either direction on the chromosome. An open reading frame (ORF), the equivalent of a sentence in our analogy, is defined as a DNA sequence that can potentially encode a string of amino acids of minimum length—say, 50 amino acids. An ORF must begin with a translation start codon (usually ATG or, more rarely, GTG or TTG) and end with a translation termination codon (in the DNA, these are TAA, TAG, and TGA).

Once an ORF is identified, computers use mathematical algorithms to determine whether the predicted protein resembles any other protein deposited in the worldwide databases or, even better, resemble proteins of known function. By "resemble" we mean that the unknown protein (often called the query protein) possesses amino acid sequences that are identical or functionally similar to those found in other proteins. As discussed in Section 9.1, sequences common to two different protein or DNA molecules are called homologous sequences.

Alignments can be done either with DNA sequences or with protein sequences deduced from the DNA sequences. Because of the degeneracy of the code, however, it is usually easier to pick up evolutionary relationships from distantly related species by comparing their protein products rather than the DNA sequence.

Homologs, Orthologs, and Paralogs

Homologies found between genes or proteins suggest an evolutionary relatedness. Genes or proteins that are homologous probably evolved from a common ancestral gene. Homologous genes or proteins can be classified as either **orthologs** or **paralogs**. Genes that are orthologs have similar functions but occur in two or more different species. For example, the gene for glutamine synthase (*gltB*) in *E. coli* is orthologous to *gltB* in *Vibrio cholerae*. Genes that are paralogs arise by duplication within the same species (or progenitor) but evolve to carry out different functions (**Figure 9.21**). For example, one type of protein secretion system (type III) that exports virulence proteins evolved from paralogous genes encoding flagella. The two sets of genes share a common ancestor but have since evolved completely different functions (in this case, secretion and motility).

Figure 9.21   Paralogous versus Orthologous Genes

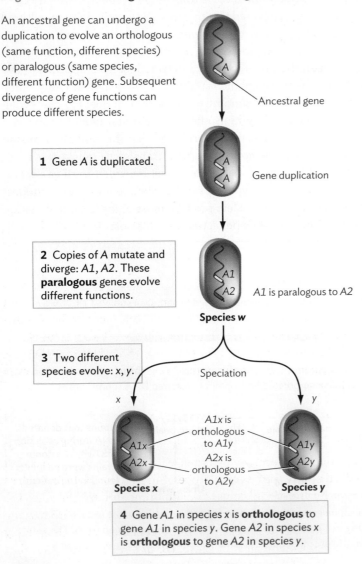

An ancestral gene can undergo a duplication to evolve an orthologous (same function, different species) or paralogous (same species, different function) gene. Subsequent divergence of gene functions can produce different species.

Ancestral gene

1 Gene A is duplicated.

Gene duplication

2 Copies of A mutate and diverge: *A1, A2*. These **paralogous** genes evolve different functions.

A1 is paralogous to A2

Species w

3 Two different species evolve: *x, y*.

Speciation

x   *y*

A1x is orthologous to A1y

A2x is orthologous to A2y

Species x   **Species y**

4 Gene *A1* in species *x* is **orthologous** to gene *A1* in species *y*. Gene *A2* in species *x* is **orthologous** to gene *A2* in species *y*.

Many computer programs and resources used to analyze DNA and protein sequences are freely available on the Internet. For example:

- **BLAST** (**B**asic **L**ocal **A**lignment **S**earch **T**ool) compares a sequence of interest with all other DNA or protein sequences deposited in sequence databases.

- **National Microbial Pathogen Data Resource** includes annotations of microbial pathogen genomes.

 National Microbial Pathogen Data Resource: nmpdr.org

A word of caution. It is enticing to make definitive proclamations about gene or protein function on the basis of the computer analysis of a DNA sequence. As good as a prediction may seem, it is only a prediction, a well-educated guess. Biochemical confirmation of function must be made, if possible. Nevertheless, the predictions we can make are powerful. As one example, a metabolic model was constructed by genome annotation for *Helicobacter pylori*, the causative agent of gastritis (ulcers). Scientists noted that genes for some amino acid biosynthesis pathways were missing from the *H. pylori* genome and were able to predict what amino acids or other compounds (for example, purines) the organism would require for growth (**Figure 9.22**). The metabolic requirements of *H. pylori* suggest possible target enzymes for new antibiotics. Predicting possible gene function(s) from bioinformatic data is an example of **functional genomics**. Functional genomics is an integrative process in which bioinformatic approaches allow scientists to make predictions about function for a set of genes and then test those predictions experimentally.

Bioinformatics provides new insight into pathogenesis, physiology, and evolution. For example, the DNA sequence of the bacterium *Rickettsia prowazekii* (the cause of epidemic typhus) revealed many similarities to genes within eukaryotic mitochondrial DNA. This finding led to the conclusion that eukaryotic mitochondria evolved from a rickettsial predecessor that became an endosymbiont (an intracellular symbiotic organism), rather than food, for a eukaryotic cell.

Evidence of horizontal transfer can be seen by comparing the proportion of GC base pairs along a chromosome (also known as GC content). A horizontally acquired gene often has a GC content very different from that of the rest of the chromosome. Within a chromosome, a **genomic island** is a section of DNA sequence that appears to have originated by horizontal transfer but no longer shows mobility. A genomic island generally contains a large number of genes serving a common function. For example, a **pathogenicity island** is a genomic island that encodes virulence factors (discussed in Section 18.1). They are found in bacteria pathogenic for plants and animals, but not in related nonpathogenic strains. *Salmonella enterica* serovar Typhimurium (previously called *Salmonella typhimurium*) contains three pathogenicity islands, each with a role specific to a different stage of disease.

SECTION SUMMARY

- **Annotation** requires computers that look for patterns in DNA sequence. Annotation predicts regulators and genes that specify protein and RNA products. Similarities in protein sequence (deduced from the DNA sequence) are used to predict protein structure and function.
- **An open reading frame (ORF)** is a sequence of DNA predicted by various sequence cues to encode an actual protein.
- **DNA alignments** of similar genes or proteins can reveal evolutionary relationships.
- **Paralogs and orthologs** arise from gene duplications. Paralogous genes coexist in the same genome but usually have different functions. Orthologous genes occur in the genomes of different species but produce proteins with similar functions.
- A DNA sequence with a **GC content** different from that of flanking chromosomal DNA is one sign of horizontal gene transfer.
- **Pathogenicity islands** are the result of horizontal gene transfers that improve the pathogenicity of a recipient.

Thought Question 9.6 An ORF 1,200 base pairs in length can encode a protein of what size and molecular weight? (*Hint:* Find the molecular weight of an average amino acid.)

9.5
Microbial Taxonomy

SECTION OBJECTIVES

- Describe how we define distinct species of bacteria, and explain why the definition of species is problematic.
- Explain how a microbe is identified, and describe the relative advantages and limitations of a dichotomous key, as compared with a probabilistic indicator.

What is a microbial species? Among eukaryotes, a species is defined by the principle that members of different species do not normally interbreed with each other. Bacteria and archaea, however, reproduce asexually, so interbreeding is not a basis for classification. And even without "interbreeding" in the traditional sense, microbes manage to transfer genes horizontally between distantly related clades.

Classifying Microbial Species

As genome sequence data became available, scientists hoped that quantitative measures of divergence could provide a consistent basis for defining species of microbes. But in some bacteria, such as *Helicobacter pylori*, the genomes of different strains that cause the same disease (gastritis) differ by as much as 7%. On the other hand, strains of *Bacillus* with nearly identical genomes cause completely different diseases, such as anthrax (*B. anthracis*) and caterpillar infection (the biological pesticide *B. thuringiensis*).

Figure 9.22 *Helicobacter pylori* Metabolism Predicted from DNA Sequence Information

Notice the vast amount of information you can learn about a cell just from sequencing the genome, a process that can now take just a few days. The large rectangle represents a cell. Many transport systems predicted from genome annotations are shown in the membrane. The main components of *H. pylori* central metabolism are presented within the rectangle. This scheme is based on using glucose as the sole carbohydrate source. A question mark is attached to pathways that could not be completely elucidated. Red arrows represent pathways or steps for which no enzymes were predicted from the sequence (for example, phenylalanine, methionine, and purine synthesis). Pathways for macromolecular biosynthesis (RNA, DNA, and fatty acids) were found but are not shown in this model.

- Amino acids, peptides, and amines
- Anions
- Carbohydrates, organic alcohols, and acids
- Cations
- Nucleosides, purines, and pyrimidines
- ATP-binding cassette (ABC) transporters

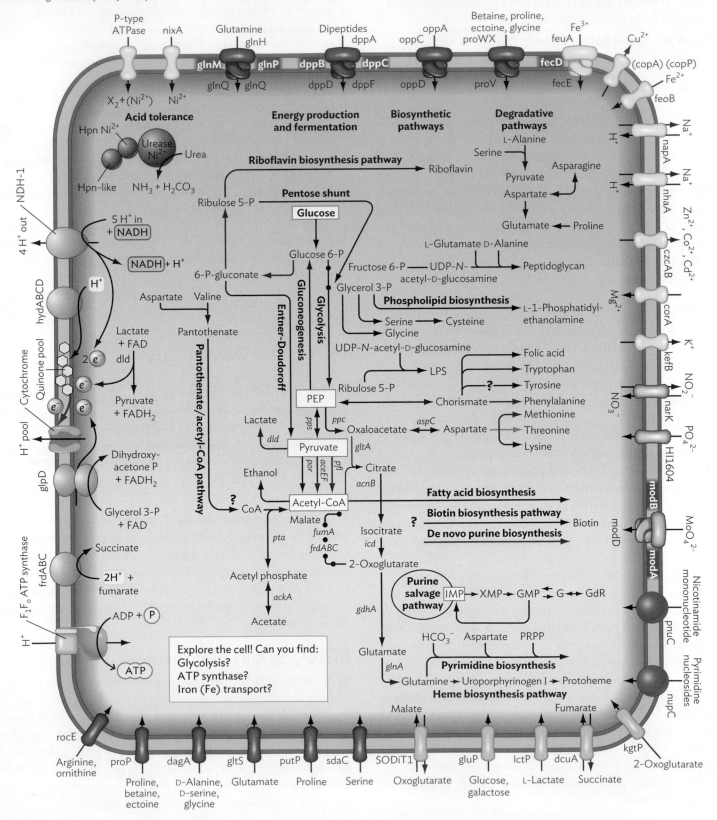

In defining species, microbiologists generally agree on the importance of two criteria: phylogeny (based on DNA relatedness) and ecology (based on shared traits and ecological niche). Phylogeny is used to define a species as a group of individuals whose sequences of key "housekeeping genes," such as ribosomal and transcriptional genes, are very similar, an indication that they were inherited vertically (from parent to offspring). In addition, a species should include individuals that share common traits and a common ecological niche, or "ecotype." Shared traits generally include cell shape and nutritional requirements, and a shared niche includes a common habitat and life history (for example, causing the same disease). By these criteria, highly divergent strains of *H. pylori* causing gastritis make up one species, whereas *B. anthracis* (anthrax) and *B. thuringiensis* (caterpillar infection) are different species despite their highly similar genomes.

Many microbiologists accept the following criteria for a working definition of a microbial species:

- **DNA hybridization ≥ 70%.** When DNA from two different genomes is denatured and mixed together, the strands reanneal to form a hybrid helix (helix of strands from two different DNA molecules). If the proportion of hybridization is 70% or greater, the two organisms are usually considered the same species.

- **Small-subunit (16S) rRNA gene sequence similarity ≥ 97%.** Two bacteria with 97% or greater similarity in 16S rRNA sequence are considered to share a species. The larger percentage reflects the relatively high conservation of rRNA sequence, compared with other genes. Hybridization and sequence similarity are related but not synonymous.

- **Average nucleotide identity of orthologs ≥ 95%.** If whole genomes are available and annotated, we can define all the orthologous genes (genes of the same function) shared by a pair of strains. If two strains share 95% or greater sequence identity for these genes, they are considered the same species.

Defining a species is part of the task of **taxonomy**, the description of distinct life-forms and their organization into different categories with shared traits. Taxonomy includes classification, the recognition of different classes of life; nomenclature, the naming of different classes; and identification, the recognition of the class of a given microbe isolated in pure culture. Taxonomy is critical for every microbiological pursuit, from diagnosing a patient's illness to understanding microbial ecosystems.

Classification generates a hierarchy of taxa (groups of related organisms; singular, taxon) based on successively narrow criteria. The fundamental basis of modern taxonomy is DNA sequence similarity, but historically, taxa were defined and named based on genetic and phenotypic traits. Levels of taxonomic hierarchy, or rank, were defined as phylum, class, order, and family (Table 9.1). Some levels of rank are designated by certain suffixes—for example, "-ales" (order) and "-aceae" (family). The ultimate designation of a

Table 9.1
Classification: Hierarchy of Taxa for an Antibiotic-Producing Bacterium

Taxon Rank	A Long-Studied Taxon
Domain	Bacteria
Division (phylum)	Actinobacteria Filamentous Gram-positive
Class	Actinobacteria Hi GC Gram-positive
Subclass	Actinobacteridae
Order	Actinomycetales Filamentous; acid-fast stain
Suborder	Streptomycineae
Family	Streptomycetaceae Filamentous; hyphae produce spores
Genus	*Streptomyces*
Species (date first described)	*Streptomyces coelicolor* (1908)

type of organism is that of species. A species name includes the capitalized genus name (group of closely related species) followed by the lowercase species name—for example, *Streptomyces coelicolor*, a soil bacterium that produces streptomycin and other antibiotics.

The nomenclature of microbes remains surprisingly fluid. As new traits are identified and genomes are sequenced, species are all too frequently renamed. For example, the species *Yersinia pestis*, the cause of bubonic plague, was called "*Bacillus pestis*" in 1896, "*Pasteurella pestis*" in 1923, and *Yersinia pestis* in 1944.

Note Taxonomic categories generally have two forms: formal and informal. The formal term is capitalized, with a latinized suffix: Actinomycetales, *Pseudomonas*, *Micrococcus*. The informal term is lowercase, in some cases with an anglicized ending; and informal references to genera are not italicized: actinomycetes, pseudomonads, micrococci.

Identifying a Microbe

Once a species has been described and classified, we need a way to identify it from natural and host environments. Identification poses special difficulties with microbes, which by definition are invisible to the unaided eye. Even under the electron microscope, thousands of different species may show similar shape and form.

The most conclusive way to identify an isolate is to sequence part or all of its genome. In clinical practice, DNA-based methods

Figure 9.23   Dichotomous Key for Wastewater Bacteria

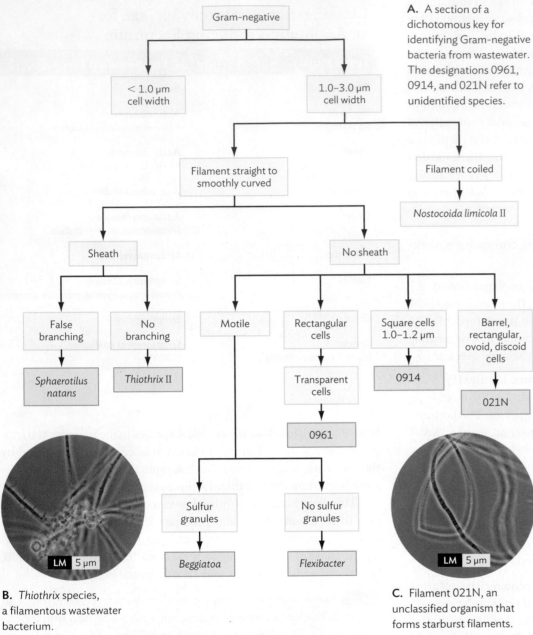

A. A section of a dichotomous key for identifying Gram-negative bacteria from wastewater. The designations 0961, 0914, and 021N refer to unidentified species.

B. *Thiothrix* species, a filamentous wastewater bacterium.

C. Filament 021N, an unclassified organism that forms starburst filaments.

during wastewater treatment. Filamentous bacteria are important in wastewater because their entangled filaments interfere with the settling of bacteria out of the treated water. In the key shown, most traits are phenotypic, such as cell size and motility. Most branch decisions have two choices, although one juncture (cell motility and shape) has four. The key identifies some organisms down to the species level (*Sphaerotilus natans*), others to the genus level (*Flexibacter*), and others only to numbered samples whose characterization remains incomplete (0914, 021N).

A disadvantage of the dichotomous key is that it requires a series of steps, each of which takes time. In the clinical setting, time is critical in identifying a potential pathogen and prescribing appropriate treatment. An alternative means of identification is the **probabilistic indicator**. A probabilistic indicator is a battery of biochemical tests performed simultaneously on an isolated strain. The indicator requires a predefined database of known bacteria from a well-studied habitat, such as Gram-negative bacteria from the human intestinal tract. The database can be used to identify a newly isolated specimen from a patient. When the test results are obtained, the probabilities for all results, negative or positive, for a given species are then multiplied to generate a probability score for the species. For each species in the database, a score is computed, and the scores are sorted. The top-scoring species is then taken as the most probable identification. **Figure 9.24** shows an example of a

are used more and more. Nevertheless, in the clinical lab—as well as in field and environmental microbiology—it is convenient to narrow down the "suspects" by using various easily determined traits, such as cell shape, staining properties, and metabolic reactions. Thus, practical identification is based on a combination of phylogeny (relatedness based on DNA sequence divergence) and phenotypic traits.

A common strategy of practical identification is the **dichotomous key**, in which a series of yes/no decisions successively narrows down the possible categories of species. **Figure 9.23** shows a dichotomous key used to identify filamentous bacteria found

BIOLOG test battery that identifies an isolate as *Stenotrophomonas maltophilia*, an antibiotic-resistant organism that frequently infects tracheostomy tubes and urinary catheters. By contrast, identifying this organism by a dichotomous key would have required several steps over several days. The rapid identification of such pathogens is critical in order to save the patient.

Another advantage of the probabilistic indicator is its inherent redundancy; if an isolate gives one odd result for its species, it may still show the highest probability score in the database. Note, however, that both the dichotomous key and the probabilistic indicator assume a known set of organisms. To explore and characterize a

Figure 9.24   Probabilistic Indicator: BIOLOG

71 carbon source plus 23 chemical sensitivity assays

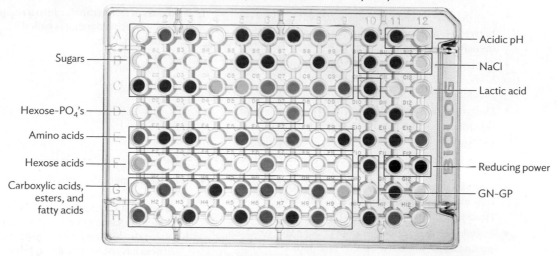

Sugars

Hexose-PO$_4$'s

Amino acids

Hexose acids

Carboxylic acids, esters, and fatty acids

Acidic pH

NaCl

Lactic acid

Reducing power

GN-GP

ID = *Stenotrophomonas maltophilia*

The BIOLOG GEN III plate performs multiple enzyme assays simultaneously. Each positive result involves a redox reaction that converts the dye to purple. The probabilities of each result are multiplied, and a computer program yields the most probable identification.

completely unknown organism—or an isolate with a new property, such as antibiotic resistance—requires sequencing of the organism's DNA.

SECTION SUMMARY

- Microbial species are defined based on sequence similarity of vertically transmitted genes.
- The species definition should be consistent with the ecological niche or pathogenicity.
- Taxonomy is the description and organization of life-forms into classes (taxa). Taxonomy includes classification, nomenclature, and identification.
- Classification is traditionally based on a hierarchy of ranks. Groups of organisms long studied tend to have many ranks, whereas recent isolates have few.
- Practical identification includes DNA-based tests and phenotypic traits. Methods of identification include the dichotomous key and the probabilistic test battery. Both methods assume a predefined set of organisms.
- Defining a new strain of a microbe requires sequencing its DNA.

Thought Question 9.7   According to Figure 9.23, how would you identify a straight, nonsheathed, Gram-negative bacterium that has sulfur granules, is motile, and is 1.0 μm wide? What would happen if you assigned the bacterium a width of 0.9 μm?

Perspective

Microbes regulate their gene expression in ways that maximize their growth in whatever habitat confronts them, from a geothermal spring to the human body. They even acquire new genes from other members of their habitat. Both vertical transmission of mutant sequences and horizontal acquisition of mobile genes contribute to the constant evolution of microbes. It is hard to say which is more astonishing: the overall commonalities of all living cells, evolved from their common ancestor, or the evolution of modern life-forms with vastly different adaptations to exploit every possible niche of our planet. In Chapter 10, we explore some of the amazingly diverse adaptations of different groups of bacteria, including the most important pathogens known to medicine.

LEARNING OUTCOMES AND ASSESSMENT

	SECTION OBJECTIVES	OBJECTIVES REVIEW

9.1
Gene Transfer, Recombination, and Mobile Genetic Elements

- Explain how genetic information moves between bacteria by transformation, phage-mediated transduction, and conjugation.
- Describe how DNA molecules recombine by homologous recombination and site-specific recombination.
- Describe how gene mobility and DNA recombination lead to the emergence of new pathogens.

1. A process of DNA transfer into a recipient cell that requires bacteriophage packaging is called _____.
   A. transformation
   B. transduction
   C. conjugation
   D. transposition

9.2
Origins and Evolution

- Describe how all life originated from microbes, how life-forms evolved over time, and how species continue to evolve.
- Explain how microbial genomes change via processes of random mutation and natural selection.
- Explain how a molecular clock works to measure phylogeny, and describe the key attributes of a useful molecular clock gene.
- Describe the three domains of life.

4. Which statement does not describe a process by which life-forms actually change?
   A. Mutations accumulate at random, even in populations under the same selection pressure.
   B. Natural selection favors the prevalence of traits with a fitness advantage in a given environment.
   C. When a pathogen associates with a host, the pathogen undergoes degenerative evolution, losing the genes that specify the functions provided by the host.
   D. Entirely new organisms emerge from environments, and they prevail over preexisting organisms.

9.3
Natural Selection

- Explain how fitness of a trait depends on the environment and how environment-dependent fitness leads to the evolution of antibiotic resistance in pathogens.
- Describe the evolution of endosymbiosis between a host cell and an internalized partner organism.

7. Suppose a pathogenic bacterium acquires a plasmid expressing an enzyme that modifies the target of an antibiotic. What may happen when the antibiotic disappears?
   A. Because the plasmid has a useful trait, the bacterium will increase copy number of the plasmid.
   B. The bacterium will outcompete other bacteria, even when the antibiotic disappears.
   C. The bacteria containing the resistance plasmid will decline, because replicating the plasmid costs energy.
   D. In the absence of antibiotic, the plasmid will mutate to provide resistance to a different antibiotic.

9.4
Bioinformatics

- Explain how bioinformatic analysis helps reveal the pathogenesis and physiology of an emerging pathogen.
- Describe how gene duplication leads to different kinds of homologous genes, including orthologs and paralogs.

10. Which statement describes an example of bioinformatic analysis to reveal the capabilities of a pathogen?
    A. *Streptococcus pneumoniae* was cultured on various agar media containing different combinations of nutrients in order to analyze the bacterium's metabolism.
    B. *S. pneumoniae* was cultured with bacteriophage in order to transfer genes between bacteria.

9.5
Microbial Taxonomy

- Describe how we define distinct species of bacteria, and explain why the definition of species is problematic.
- Explain how a microbe is identified, and describe the relative advantages and limitations of a dichotomous key, as compared with a probabilistic indicator.

12. Which statement describes evidence of highly divergent or distantly related organisms?
    A. The genomes of two organisms show greater than 70% DNA hybridization.
    B. Between two genomes, the gene sequences that are transcribed to form 16S rRNA show greater than 97% similarity.

2. All the following processes require a step of DNA recombination to integrate the incoming DNA into the host chromosome except ____.
   A. transfer of a multidrug resistance plasmid
   B. generalized transduction
   C. specialized transduction
   D. transposition of a mobile genetic element

3. The reason bacteria make restriction endonucleases is ____.
   A. to degrade excess genomic DNA
   B. to prevent replication of an invading bacteriophage
   C. to obtain nucleotides for DNA synthesis
   D. to synthesize DNA

5. Which property is required for a molecular clock gene?
   A. The gene must encode different functions in different organisms.
   B. The gene must acquire mutations at a steady rate over time in all organisms compared.
   C. The gene must encode functions under strong environmental selection, such as resistance to antibiotics.
   D. The gene must transfer readily between distantly related organisms.

6. Of the given genera, which genus is the most distantly related to *Escherichia coli*?
   A. *Salmonella*
   B. *Klebsiella*
   C. *Shigella*
   D. *Vibrio*

8. Which statement does <u>not</u> describe an example of degenerative (reductive) evolution?
   A. *Treponema pallidum*, the causative agent of syphilis, has lost many of the genes encoding enzymes of amino acid biosynthesis.
   B. *T. pallidum* has evolved virulence genes that enable it to take advantage of host cells.
   C. *Wolbachia* bacteria that live within host insect cells lack genes for biosynthesis of vitamins.
   D. *Chlamydia* bacteria are obligate intracellular pathogens that fail to synthesize a peptidoglycan cell wall.

9. When drug-resistant bacteria are isolated from a hospital patient, they may make smaller colonies than the original strain. The reason is ____.
   A. drug-resistant bacteria always grow more slowly
   B. smaller colony size provides an advantage in resisting antibiotics
   C. the random mutation that conferred resistance also conferred deleterious effects on growth
   D. the drug-resistant bacteria require different kinds of nutrients

   C. *S. pneumoniae* isolates were selected for genome analysis, which revealed which gene functions were shared and which genes were distinct to different species.
   D. Novel isolates of *S. pneumoniae* were constructed by genetic engineering.

11. Which statement does <u>not</u> describe a result of gene duplication within a genome?
    A. Two genes within an organism's genome show very similar DNA sequence, but they encode enzymes that act on different substrates.
    B. Two organisms each have a gene that encodes the same function.
    C. An organism has two gene copies that appear to have the same function but are expressed under different regulation.
    D. The presence of an antibiotic can select for gene amplification, the presence of several copies of an antibiotic resistance gene.

   C. Orthologous genes of two genomes share greater than 95% of the same sequence.
   D. Two genomes show very different proportions of GC versus AT content.

13. Pathogen identification in a clinical laboratory is most commonly based on ____.
    A. a combination of DNA-based tests and phenotypic traits
    B. sequencing the entire genome of an isolate directly from a patient
    C. microscopy and staining only, because these procedures are quick and inexpensive
    D. immunological tests only, because these tests are highly strain specific

Key Terms

annotation (p. 278)
bacteriophage (p. 263)
bioinformatics (p. 277)
clade (p. 270)
competent (p. 260)
conjugation (p. 262)
degenerative evolution (p. 276)
dichotomous key (p. 282)
divergence (p. 269)
domain (p. 274)
electroporation (p. 261)
endosymbiont (p. 276)
F⁺ cell (p. 263)
F⁻ cell (p. 263)

F-prime (F′) plasmid (p. 263)
fertility factor (F factor) (p. 262)
functional genomics (p. 279)
generalized recombination (p. 265)
generalized transduction (p. 264)
genetic drift (p. 270)
genomic island (p. 279)
Hfr cell (p. 263)
homologous (p. 261)
homologous recombination (p. 265)
horizontal gene transfer (p. 260)
housekeeping gene (p. 271)
insertion sequence (IS) (p. 267)
microfossil (p. 269)

molecular clock (p. 270)
monophyletic group (p. 270)
multilocus sequence typing (MLST)
   (p. 277)
natural selection (p. 275)
open reading frame (ORF) (p. 278)
ortholog (p. 278)
paralog (p. 278)
pathogenicity island (p. 279)
phage (p. 263)
phylogenetic tree (p. 271)
phylogeny (p. 270)
probabilistic indicator (p. 282)
prophage (p. 264)

recombination (p. 265)
reductive evolution (p. 276)
site-specific recombination (p. 265)
specialized transduction (p. 264)
taxonomy (p. 281)
transduction (p. 263)
transformation (p. 260)
transposable element (p. 267)
transposition (p. 267)
transposon (p. 267)
vertical gene transfer (p. 261)

Review Questions

1. What are the basic ways microorganisms exchange DNA?

2. Discuss horizontal versus vertical gene transfer.

3. Describe competence and how it comes about in a population.

4. What is an F factor, and how does it (and other factors like it) contribute to gene exchange?

5. Compare specialized versus generalized transduction.

6. Discuss how bacteria protect themselves from invading bacteriophages.

7. Explain the basic process of transposition. Why are insertion sequences always flanked by direct repeats of host DNA? How are transposons different from plasmids?

8. What is a pathogenicity island?

9. What is the common ancestor of all life, including humans?

10. What are the three domains of life? Which traits are unique to each domain, distinguishing it from the other two?

11. Explain how evolution generates phylogeny of organisms.

12. What is random mutation? How are the results of random mutation modified by natural selection?

13. Explain how a molecular clock works. What properties are required of a gene to be used for a molecular clock?

14. What is meant by homologous genes? Explain the difference between orthologs and paralogs.

15. What is endosymbiosis? How has endosymbiosis contributed to the evolution of eukaryotic organisms?

16. What is degenerative (reductive) evolution? Why does it occur?

Clinical Correlation Questions

1. Why might a pathogen strain that is defective in its ability to repair mutations (mutator phenotype; see Chapter 8) be a more successful pathogen than a related strain that is highly efficient in mutation repair?

2. *Salmonella enterica* is a Gram-negative species with more than 2,000 serovars (variant strains differing in surface proteins and carbohydrate antigens) that can cause gastrointestinal or bloodstream infections. Most serovars of *S. enterica* can infect human and nonhuman animals alike. However, one serovar, *S. enterica* serovar Typhi (the cause of a blood infection called typhoid fever), infects only humans. Propose an evolutionary model for how *S.* Typhi may have developed this narrow host range.

Thought Questions: CHECK YOUR ANSWERS

Thought Question 9.1   Can a *recA* mutant of *S. pneumoniae* experience horizontal gene transfer?

ANSWER: A *recA* mutant can be transformed and transduced, but whether the incoming DNA is maintained depends on the DNA. For example, transformation of plasmids can happen in a *recA* mutant because plasmid replication does not require plasmids to recombine with the chromosome. Transformation of chromosomal DNA cannot take place, however, because the donor DNA cannot be inherited without integrating into the recipient chromosome (requires RecA). The same is true for generalized transduction of chromosomal genes: no horizontal transfer is involved. Specialized transduction, however, can occur because the phage genome includes its own recombinase. Similarly, transposition can occur in a *recA* mutant because the transposon carries its own transposase gene. RecA is not needed.

Thought Question 9.2   Transfer of an F factor from an F⁺ cell to an F⁻ cell converts the recipient to F⁺. Why does genome transfer by an Hfr not do the same?

ANSWER: The last piece of an Hfr to transfer is the F factor and *oriT*. Only rarely will an entire chromosome transfer from one cell to another, so most Hfr transfers do not result in transfer of *oriT* and thus cannot initiate conjugation.

Thought Question 9.3   How do you think phage DNA containing restriction sites evades degradation by the restriction screening system of its host?

ANSWER: Phage DNA will survive because sometimes the modification enzyme gets to the foreign DNA before the restriction enzyme. Once methylated, the phage DNA will be shielded from restriction and the methylated molecule will replicate unchallenged. If, on the other hand, a foreign DNA fragment (not necessarily a phage) has been modified and the conditions are right, it might recombine into the host chromosome and convey a new character to the strain. Also, not all phage DNAs have the DNA sequence targeted by a host's particular set of restriction enzymes.

Thought Question 9.4   How might a transposon containing a gene for tetracycline resistance make its way from the chromosome of one cell to the chromosome of a second cell?

ANSWER: There are several possible routes. A bacteriophage can package host DNA containing the transposon. Once the DNA enters the recipient via transduction, the transposase can be expressed and mediate the "hop" from donor to recipient DNA. If the donor cell contains a conjugative plasmid, the transposon can "hop" into the plasmid, and the plasmid can move by conjugation into the recipient. The transposon can then hop into that cell's chromosome. Transformation is also possible. Finally, a few transposons are capable of conjugation themselves.

Thought Question 9.5   Parasitic microbes, such as *Giardia* and *Plasmodium falciparum*, undergo degenerative evolution as well as selection for complex traits. Explain why both kinds of natural selection occur.

ANSWER: Parasites undergo reductive evolution for traits whose functions are provided by the host, such as metabolism of complex food molecules and biosynthesis of amino acids available from the host. The loss of these traits, and the DNA that encodes them, saves energy and thus confers a competitive advantage compared with other members of the parasite species. At the same time, parasites evolve specialized traits that enhance their survival within the host. For example, the malaria parasite has evolved an apical complex that enables it to attach and penetrate a red blood cell. *Giardia* has evolved a strong flagellum that enables dissemination within the host intestine, as well as molecular defenses against the host immune system.

Thought Question 9.6   An ORF 1,200 base pairs in length can encode a protein of what size and molecular weight? (*Hint:* Find the molecular weight of an average amino acid.)

ANSWER: There are three bases per codon, so the ORF can encode 400 amino acids. The average molecular weight of an amino acid is 110. Therefore, the hypothetical protein will be approximately 44,000 daltons (44 kDa).

Thought Question 9.7   According to Figure 9.23, how would you identify a straight, nonsheathed, Gram-negative bacterium that has sulfur granules, is motile, and is 1.0 μm wide? What would happen if you assigned the bacterium a width of 0.9 μm?

ANSWER: The bacterium would key out as *Beggiatoa* species. If the cell width were measured as 0.9 μm, however, the identification would proceed down a completely different, wrong track at the first step of the key. This is one disadvantage of the dichotomous key.

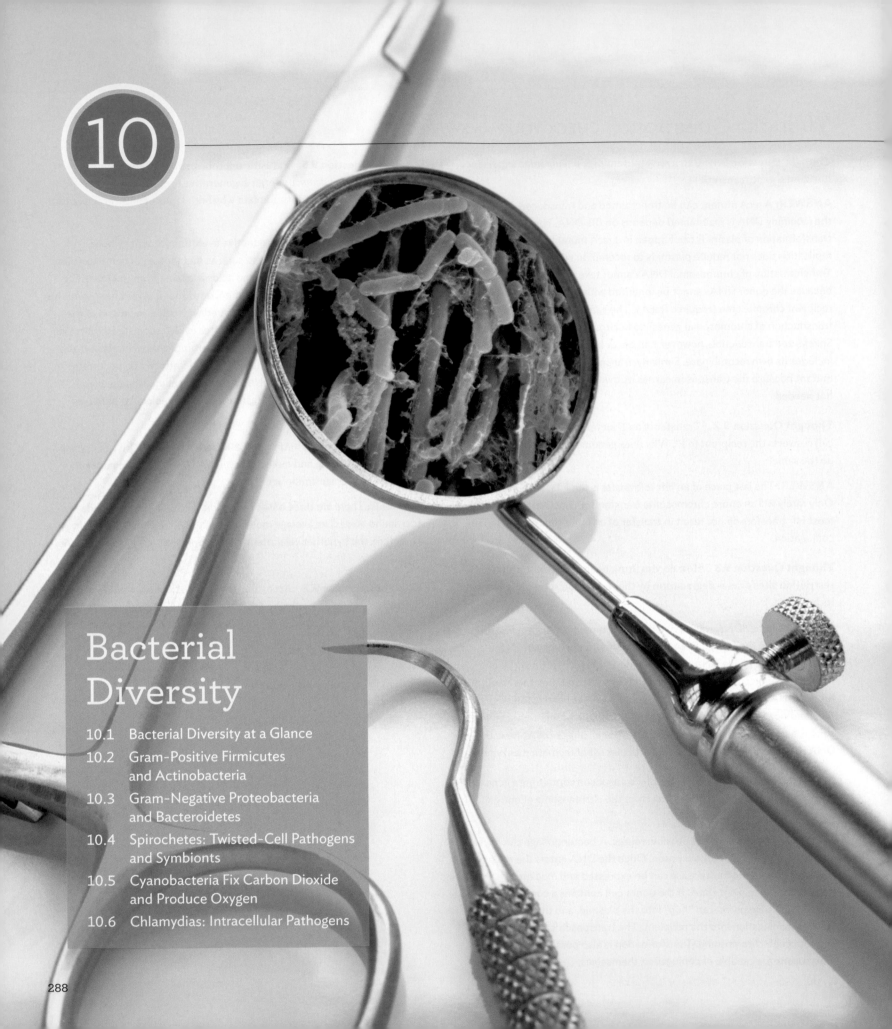

10

Bacterial Diversity

Poor Dental Care Comes Back to Bite

SCENARIO Denzel, a 52-year-old man without health insurance, worked on a farm in Texas. He visited a community clinic complaining that his jaw had been swollen for the past 8 months.

SIGNS AND SYMPTOMS Denzel reported no fever, chills, or weight loss, but now the facial swelling was causing pain when he opened his mouth. He had smoked for 40 years, a practice that leads to gum disease. The physician assistant found a region of facial swelling and hardened round shapes in the fascia (deep fibrous tissue) on the right side of his jaw. There was a slight reddening of the overlying skin and an open sore. Denzel's dentition (tooth development) was poor, with signs of untreated caries (tooth decay).

TESTING A computerized axial tomography (CAT) scan revealed soft-tissue swelling, dental caries, and an abscess on his right tonsil. Biopsy of the abscess revealed Gram-positive staphylococci and streptococci, Gram-negative rods, obligate anaerobes, and a radiating mass of Gram-positive filaments. Similar filaments were found within small yellow masses draining from the skin sore. The branching filaments contained rod-shaped bacteria called *Actinomyces israelii,* an occasional cause of wound infections and also a complication (sequela) of untreated dental caries.

DIAGNOSIS The condition is called actinomycosis, also known as "lumpy jaw." *Actinomyces israelii* is the main pathogen, but the long period without treatment enabled secondary infection by other kinds of bacteria.

TREATMENT The physician assistant treated Denzel with oral amoxicillin and doxycycline to clear the actinomycosis and referred him for dental work to eliminate caries and halt progression of polymicrobial gum disease. The physician assistant gave him information on low-cost health insurance available through the Health Insurance Exchange.

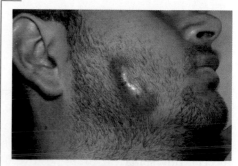

Facial swelling and sores characteristic of "lumpy jaw," or actinomycosis.

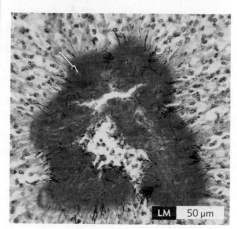

LM · 50 μm

Radial filamentous mass of *A. israelii.*

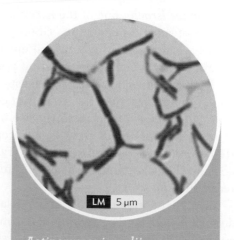

LM · 5 μm

Actinomyces israelii
Branching filaments of rod bacteria.

Denzel's condition is typical of people lacking regular dental care. Their infected teeth show **polymicrobial** infections— that is, infections involving many diverse kinds of bacteria. Denzel's polymicrobial infection included staphylococci and streptococci, tiny spherical Gram-positive bacteria of a phylum called Firmicutes. The Firmicutes are a vastly diverse group of organisms that include non-spore-forming genera such as *Staphylococcus* and *Streptococcus*, as well as spore-forming rods such as *Bacillus anthracis* and *B. thuringiensis* (discussed in Section 10.2). But besides the Gram-positives, Denzel's polymicrobial infection also included members of a very distantly related group, the Proteobacteria, which includes rod-shaped bacteria that stain Gram-negative (Section 10.3). The Proteobacteria are facultative anaerobes; they can grow with or without oxygen. Dental flora also contain obligate anaerobes such as *Bacteroides*, as well as *Treponema*. Oral *Treponema* species are tightly coiled spiral bacteria called spirochetes (see Section 10.4), a group that includes *Treponema pallidum*, the cause of syphilis.

Thus, a person's teeth may contain members of nearly all the major branches of the bacterial family tree. The most diversity is seen in those of us who fail to brush regularly and lack regular dental care.

Because Denzel's dental infection went untreated, a single pathogenic species was allowed to penetrate and spread through the deep tissues supporting his facial muscles. The pathogen was *Actinomyces israelii*, one of the Actinomycetes (see Section 10.2), a group of mostly filamentous Gram-positive rods. *Actinomyces* species are common in natural soil environments. But some can cause infections, usually by entering a wound. Untreated infection leads to connected abscesses that spread through surrounding tissues. In humans, the face is the most common site of infection, although *Actinomyces* species may also infect the abdominal cavity or the uterus, usually from a contraceptive intrauterine device (IUD). Antibiotics can lead to full recovery. Unfortunately, however, the condition is often misdiagnosed as tumors before appropriate treatment is applied.

The bewildering diversity of bacteria poses a challenge to the physician—and to the environmental microbiologist, who studies how microbes contribute to ecosystems. Deadly pathogens often turn out to be close relatives of bacteria that enhance our environment and improve human health. For example, *Bacillus anthracis*, the cause of anthrax, is very closely related to *Bacillus thuringiensis*, the inchworm insecticide (Bt) that doesn't infect humans. Similarly, the Shiga toxin–producing strains of *Escherichia coli* are closely related to *E. coli* strains that occur naturally in human intestines and help digest our food. In this chapter, we will explore the diversity of bacteria that have evolved by the processes outlined in Chapter 9.

10.1
Bacterial Diversity at a Glance

What are all the different kinds of bacteria? The genetic diversity of bacteria is overwhelming. Many differ more from each other than all the animals differ from all the plants! And yet another entire domain of microbes, the archaea, look superficially like bacteria but diverged from both eukaryotes and bacteria more than 3 billion years ago (**Figure 10.1**). The archaea include microbes in extreme habitats, such as hyperthermophiles in boiling hot springs and acidophiles in mine drainage acidic enough to dissolve a shovel. Some archaea, the methanogens, even live within our intestines, converting bacterial fermentation products into the methane that we release by passing gas. Archaea play interesting roles in our environment, as you'll learn in Chapter 27.

Even among the bacteria, some kinds diverged early, soon after bacteria diverged from archaea. These include "deeply branching" forms such as *Aquifex*, which grows in the hot springs of Yellowstone National Park. Another example is *Deinococcus*, bacteria that resist extremely high doses of ionizing radiation; they are being used to remediate (detoxify) waste from nuclear power plants.

Figure 10.1   Bacterial Phylogeny

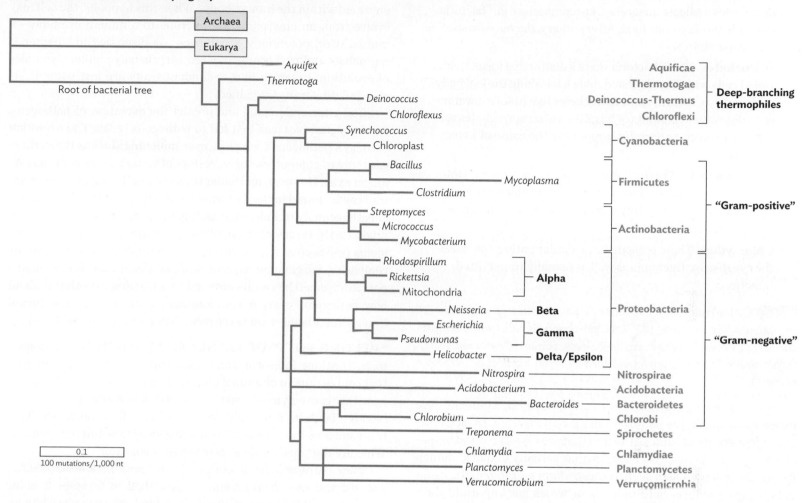

The Diversity of Life

How many different kinds of life are out there? We actually don't know how much diversity is out in the biosphere, because we can find only those organisms whose genes amplify by polymerase chain reaction (PCR) using the primer pairs that we design (discussed in Chapter 8). Primer pairs must be designed based on organisms whose genomes we already know. So by definition, any bacterial genomes we amplify from environmental samples must be related to organisms we've found before. And of those we find, less than 1% can be cultured to study their metabolism and physiology. Nevertheless, those 1% include a vast range of species, from psychrophiles growing in your refrigerator to thermophiles growing in your compost pile.

Here we outline some major categories of bacteria that you should know. Their evolutionary relationships are shown in Figure 10.1. Their relatedness is measured by comparing the sequence of their genes encoding 16S rRNA, the small-subunit ribosomal RNA. Because ribosomal RNA has the same function in all organisms, its sequence is highly conserved. The differences in gene sequence encoding rRNA approximately represent the time of divergence

since the common ancestor of two kinds of bacteria. The rRNA gene sequence can be determined from a microorganism that is "isolated" in pure culture from a clinical specimen or an environmental sample. The isolated organism is called an **isolate** or **strain**. Isolates or strains whose rRNA sequences are 97% similar usually share a common genus.

As we will see, nearly every major group of bacteria includes environmental organisms with positive contributions to ecosystems as well as pathogens that negatively impact humans and other hosts.

- **Gram-positive Firmicutes and Actinobacteria.** Most are heterotrophs with thick cell walls that resist drying. Many, such as *Clostridium*, form dormant endospores. Some Actinobacteria, called actinomycetes, grow as radial filaments (long chains of cells) and include *Streptomyces*, soil bacteria that produce antibiotics.

- **Gram-negative Proteobacteria and Bacteroidetes.** Proteobacteria encompass many species capable of a wide range of metabolism, from heterotrophs (such as the *E. coli* that inhabits animal digestive tracts) to soil and wetland lithotrophs and photosynthesizers. Many are used in biotechnology. *Bacteroides* species, obligate anaerobes, are commensal gut bacteria that help us digest our food. After surgery, *Bacteroides* species may cause abscesses.

- **Spirochetes.** These bacteria share a distinctive form: tightly coiled cells with flagella that double back along the cell body enclosed by a sheath. Some spirochetes live in soil or water or as digestive symbionts of termites. Other spirochetes are pathogens, such as *Borrelia burgdorferi*, the cause of Lyme disease.

- **Cyanobacteria.** These phototrophs fix carbon dioxide into biomass. They are the only bacteria that produce oxygen. Cyanobacteria include tiny marine bacteria as well as massive filamentous species.

- **Chlamydias.** These obligate intracellular pathogens cause the eye disease trachoma as well as sexually transmitted infections.

Note Formal names for taxa such as phyla, orders, and genera are capitalized (for example, phylum Cyanobacteria and genus *Streptococcus*). Informal names are lowercase roman ("cyanobacteria," "streptococci") as are adjectival forms ("cyanobacterial," "streptococcal").

In considering the different kinds of bacteria, we might have expected to find all the pathogens in a group together. In fact, however, most groups include pathogens that are closely related to species that have no negative effect or that actually enhance human health. Even a pathogen such as *Clostridium botulinum*, the cause of botulism, produces botulinum toxin, which has important therapeutic uses in medicine. The sections below include examples of how pathogens and commensals can emerge out of a mixed bacterial family.

Evolution and Emergence of New Species

If the genomes of today's life-forms reveal their past descent, what of their future? New species with new capabilities continue to emerge, driven by an organism's need to survive. Genomes change through mutations and horizontal gene transfer (gene transfer between cells). Through genetic change, a newly evolved cell may use resources more efficiently than its progenitor, thereby providing competitive advantage. Some genetic alterations even offer the chance of finding new sources of nutrient. For example, when an industry contaminates soil with pollutants such as chlorinated aromatics, at first no species of soil bacteria may be able to eat them. But suppose a mutation in one gene enables its product, an enzyme, to "fit" the chlorinated form and catabolize the pollutant. The mutant organism now has a competitive advantage because it can use a food source unavailable to others. Eventually, a population will evolve that can break down the pollutant completely. Such strains make valuable agents of bioremediation (discussed in Chapter 27).

What if the microbe's new food source happens to be nutrients held within the human body? When this happens, the cell may evolve from an environmental microbe to a human symbiont—a mutualist or, potentially, a pathogen. As discussed in Chapter 9, mutualists and pathogens undergo surprisingly similar dynamics of evolution, in which host-redundant traits are lost while host-manipulating traits are gained.

To better understand and predict the evolution of pathogens, we study past patterns that led to pathogens today. The evolution of today's pathogens took place over millennia. Just as typewriters gave way to computers, the ancestors of today's pathogens may no longer exist. However, molecular traces of a pathogen's evolutionary trail can be found in the pathogen itself. These traces take the form of pathogenicity islands, large multigene segments of a genome that originated by transfer from another organism via mobile genetic elements (see Section 9.1). Scientists discover pathogenicity islands by comparing the genome sequences of pathogens and closely related nonpathogens. The results have led to intriguing hypotheses about how pathogens evolved and continue to emerge in new forms. Emerging pathogens are a recurring theme throughout this book.

"MELTING POTS" OF PATHOGEN EVOLUTION Pathogens today continue to evolve, and occasionally a new version emerges that causes human or animal disease. Examples include the organisms that cause Lyme disease, cat scratch fever, and some forms of bloody diarrhea. You might wonder where these microbes come from and how they manage to sample genes from different sources. What are the "melting pots" of pathogen evolution?

New pathogens brew in three basic "pots": zoonoses, shelter species, and the environment. As described in Chapter 2, some organisms cause disease primarily in animals and can occasionally, given the opportunity, infect humans. These animal diseases are called **zoonoses**. Within the infected animal, the animal pathogen will interact with all kinds of bacterial species that make up natural animal microbiota. The genomes of these microbes can be sampled by the pathogen and, if they provide survival advantage in humans, could produce an organism better able to infect us.

SHELTER SPECIES Some hosts, called **shelter species**, are organisms that do not necessarily interact directly with humans but do provide shelter for bacteria. Shelter species include nematodes, earthworms, insects, protozoa, and even plants and larger animals. Mixed communities of bacteria and phages within a worm or other shelter species can lead to genetic exchange and evolution of the bacterium. Over time, some combination of genes could allow a bacterium to infect across a species barrier, to another animal, plant, or human.

Finally, the environment itself is an obvious genetic melting pot where genes are sampled from thousands of species. The environment is where some genes causing antibiotic resistance originated, where plasmids and transposons first evolved, and where phages became phages. These niches and opportunities all shape the evolution of pathogens and contribute to diversity.

SECTION SUMMARY

- **Bacteria and archaea** are two groups of prokaryotes that diverged very early in the history of Earth's biosphere.

- **Major phylogenetic groups of bacteria** include Gram-positive Firmicutes and Actinobacteria, Gram-negative Proteobacteria and Bacteroidetes, spirochetes, Cyanobacteria, and Chlamydiae.

- **New species and strains continue to emerge** through evolution. Organisms that acquire novel access to nutrients within animal bodies may evolve as mutualists or pathogens.

- **Pathogens evolve by sampling genes** spread by zoonoses, within shelter species, and from the environment.

Thought Question 10.1 What impact might global climate change have on bacterial diversity and pathogen evolution?

10.2
Gram-Positive Firmicutes and Actinobacteria

SECTION OBJECTIVES

- Describe the traits that enable Firmicutes to survive dryness and persist for long periods.

- Describe how diverse Actinobacteria have both positive and negative influences on human health.

Which bacteria are the toughest and can make spores that persist for thousands of years? A large group of related bacteria are called the "Gram-positive bacteria" because most members stain Gram-positive (Table 10.1). The Gram-positive bacteria include two major subgroups, the phyla Firmicutes and Actinobacteria. The name "Firmicutes" derives from Latin meaning "tough skin," and indeed, these bacteria have thick, tough cell walls, with multiple layers of peptidoglycan threaded by supporting molecules such as teichoic acids, a type of sugar-phosphate chain (see Figure 5.9). The thick, reinforced cell wall is what retains the Gram stain, especially in Firmicutes. Some Actinobacteria possess in addition a thick waxy coat that actually excludes the Gram stain, despite their genetic affiliation with Gram-positive bacteria.

Firmicutes and Actinobacteria differ genetically in their GC content—that is, the proportion of their genomes consisting of guanine-cytosine base pairs (as opposed to adenine-thymine base pairs). Firmicutes have a relatively low GC content (45%–60%), whereas many actinomycetes have a relatively high GC content (60%–70%). GC content differences commonly indicate evolutionary divergence.

Firmicutes

Firmicutes generally grow as well-defined rods or cocci, isolated or in short filaments consisting of cells that divide but remain attached end to end. Some species of Firmicutes form **endospores**, inert heat-resistant spores that can remain viable for thousands of years. The endospores also resist desiccation, high-energy radiation, freezing, and chemical disinfectants. Endospores are the most durable type of spore known to be formed by bacteria. By comparison, other bacteria, such as Actinomycetes, form spores that are starvation resistant but do not survive high temperature. Endospore-forming bacteria are common in soil and air.

ENDOSPORE FORMERS The spore-forming genus *Bacillus* includes species of major environmental and economic importance. For example, *Bacillus thuringiensis* (**Figure 10.2A**) kills caterpillar pests by infecting their digestive tract; the high pH of the insect gut interior activates the crystalline toxin released by the bacteria. *Bacillus* species grow best by aerobic respiration; they form large, cream-colored colonies on a nutrient agar plate exposed to air (**Figure 10.2B**). *Bacillus* species can be isolated from soil or food by suspending a sample in water and heating at 80°C (176°F) for half an hour. The elevated temperature kills vegetative cells (that is, cells undergoing binary fission). But endospores remain until the temperature cools, allowing them to germinate and grow.

If the plate of *B. subtilis* shown in Figure 10.2B sits for several days at room temperature, the cells will starve for nutrients; many of them will sporulate, developing endospores (discussed in Chapter 6). **Figure 10.3** shows some of the stages of sporulation of *B. subtilis*, revealed by fluorescence microscopy of cells stained with fluorophores that bind specific parts. To produce endospores, the cell divides near one of its poles, instead of at the cell equator. The polar compartment develops as the forespore, directed by unique regulatory proteins such as the one fluorescently labeled green. The larger compartment, called the motherspore, provides DNA and nutrients to the growing forespore; it will disintegrate after release of the endospore is complete. The released endospore may endure

Figure 10.2 *Bacillus* **Species**

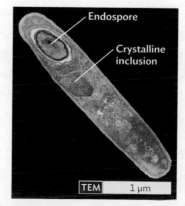

A. Sporulating cell of *Bacillus thuringiensis*, showing crystalline inclusion of insecticidal toxin (colorized TEM).

B. Colonies of *Bacillus subtilis* cultured on broth agar. Note flat, irregular form of each colony.

Table 10.1
Selected Gram-Positive Firmicutes and Actinobacteria

Firmicutes	Endospore Formers	Characteristics
	Bacillus anthracis	Rod-shaped soil bacteria; cause anthrax
	Bacillus subtilis	Rods; soil bacteria; major model system for research
	Bacillus thuringiensis	Rods; insecticide used against moth caterpillars
	Clostridium acetobutylicum	Rods; industrial use to produce butanol
	Clostridium botulinum	Rods with bulging spores; cause botulism; produce Botox
	Clostridium difficile	Rods with bulging spores; cause intestinal disease in patients whose normal biota are diminished by antibiotics
	Clostridium tetani	Rods with bulging spores; cause tetanus
	Non-Spore Formers	**Characteristics**
	Enterococcus	Cocci; commensal enteric biota
	Lactobacillus	Rods; dairy culture
	Lactococcus	Cocci; dairy culture
	Listeria	Rod-shaped; intracellular pathogens; grow at refrigerator temperature
	Ruminococcus	Cocci; cellulose-digesting symbionts of cattle
	Staphylococcus aureus	Cocci; infect skin, cause toxic shock syndrome and methicillin-resistant *S. aureus* (MRSA)
	Staphylococcus epidermidis	Cocci; commensal skin biota
	Mollicutes (Lack Cell Wall)	**Characteristics**
	Mycoplasma genitalium	Ameboid shape; commensal genital biota
	Mycoplasma pneumoniae	Ameboid shape; cause pneumonia
Actinobacteria	**Actinomycetes**	**Characteristics**
	Actinomyces israelii	Actinomycetes; form branched filaments; cause actinomycosis
	Streptomyces coelicolor	Actinomycetes; form mycelial filaments; produce many antibiotics
	Other Actinobacteria	**Characteristics**
	Corynebacterium diphtheriae	Irregular rods; cause diphtheria
	Micrococcus luteus	Cocci; common airborne bacteria; form yellow colonies
	Mycobacterium tuberculosis	Short rods; acid-fast; extremely thick cell wall with mycolic acids; cause tuberculosis
	Mycobacterium leprae	Short rods; acid-fast; cause leprosy
	Propionibacterium acnes	Rods; causes acne
	Propionibacterium freudenreichii	Rods; make Swiss cheese

Figure 10.3   Sporulation Visualized in *Bacillus subtilis*

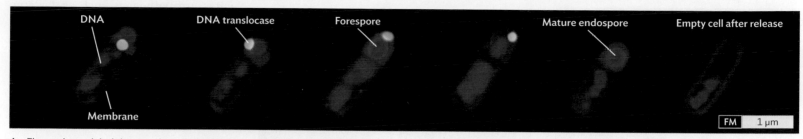

A. Fluorophores label the membranes (red) and the chromosomal DNA (blue). The small green patches indicate the position of a protein essential for forespore engulfment to form the endospore. Progression of the engulfment is shown from left to right. DNA translocase functions to move chromosomal DNA from mother cell into the forespore during polar septation.

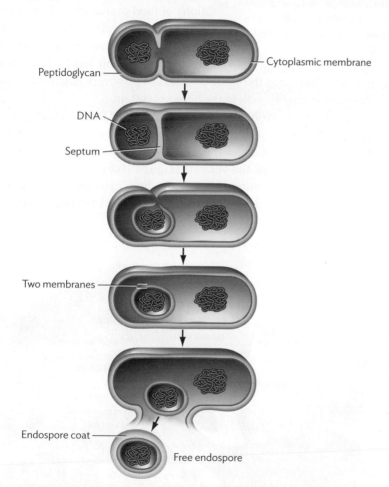

B. Unequal cell division leads to septation near one pole, forming the forespore, which develops into the endospore.

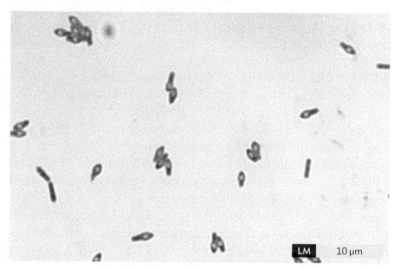

C. Malachite green stain reveals endospores (green).

for days, years, even centuries until it encounters favorable moisture and nutrients that allow it to germinate and restart vegetative growth.

Unlike *Bacillus*, most species of the genus *Clostridium* are obligate anaerobes. Their cells sporulate in a form distinct from that of *Bacillus*. The growing endospore swells the end of the cell, giving the cell a shape like a club (**Figure 10.4A**). *Clostridium* species include the causative agents of two well-known diseases, tetanus (*C.*

tetani) and botulism (*C. botulinum*). The botulism toxin, **botulinum**, or "Botox," is famous for its therapeutic use to relax muscle spasms and smooth wrinkles in the skin (**Figure 10.4B**). Other species, such as *C. acetobutylicum*, have economic importance as producers of industrial solvents such as butanol and acetone.

Clostridium spores are found in soil and water, ready to germinate and grow when the environment becomes anoxic. Many harmless species grow in the human colon, particularly in infants. But pathogenic *C. botulinum* can grow within the colon of very young infants, and **infant botulism** can cause floppy baby syndrome. Another species, *C. difficile*, is emerging in hospitals as a life-threatening intestinal pathogen, resistant to most antibiotics. Normal human gut biota outcompete *C. difficile*, but when patients are treated with antibiotics that eliminate normal biota, the pathogen can emerge and overwhelm the host (as in the disease **pseudomembranous enterocolitis**).

Link As discussed in Chapter 22, the disease **pseudomembranous enterocolitis** can arise after treatment with the antibiotic clindamycin, which kills most normal intestinal microbiota but not the naturally resistant Gram-positive anaerobe *C. difficile*.

Figure 10.4   *Clostridium* Species: Spore-Forming Anaerobes

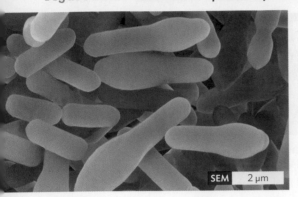

SEM   2 µm

A. As *Clostridium* cells sporulate, the endospore swells, forming a characteristic club-shaped appearance.

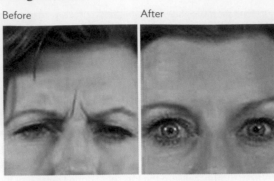

Before   After

B. The deadly botulinum toxin (Botox) from *C. botulinum* is used to relax muscle spasms. This woman was unable to open her eyelids fully (left frame)—a condition cured by injection of Botox (right frame).

NON-SPORE FORMERS   The Firmicutes include many species that do not form spores; they probably lost the spore-forming ability through degenerative evolution (see Section 9.3). For example, the lactic acid bacteria, especially *Lactococcus* and *Lactobacillus* species, ferment milk to make yogurt and cheese. Most lactic acid bacteria are obligate fermenters; that is, they generate ATP by glycolysis and substrate-level phosphorylation but cannot use oxygen to respire (see Section 7.4). They ferment by converting sugars to lactic acid. As the lactic acid builds up, the pH decreases until it halts bacterial growth. Thus, the carbon source retains much of its food value for human consumption.

Other non-spore-forming Firmicutes are pathogens such as *Listeria*. *Listeria* species are facultative anaerobic rods, named for the British surgeon Joseph Lister (1827–1912), who first promoted antisepsis during surgery. They include enteric pathogens such as *L. monocytogenes*, which may contaminate cheese and sauerkraut. Unlike most other food-associated organisms, *Listeria* grows at temperatures as low as 4°C (39.2°F). Under preindustrial conditions of food preparation, *L. monocytogenes* was generally outcompeted by other host-associated species. But the rise of refrigeration led to the emergence of *Listeria* as the cause of listeriosis, a severe gastrointestinal illness that can progress to the nervous system.

Listeria has evolved an intricate mode of intracellular pathogenesis that helps it evade the host immune system. Bacteria are taken up by macrophages into phagocytic vesicles, but they avoid digestion and escape the vesicles (**Figure 10.5**). The bacteria then multiply while traveling through the host cytoplasm, generating "tails" of actin. The actin tails eventually project the bacteria out of

Figure 10.5   *Listeria monocytogenes*: An Intracellular Pathogen That Travels on Tails of Actin

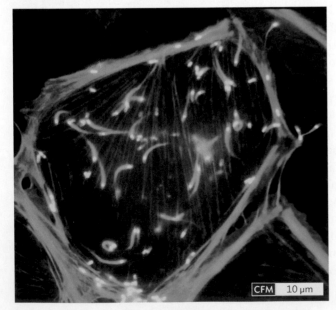

CFM   10 µm

A. Fluorescent antibodies mark the tails of polymerized actin (green) behind the Listeria cell bodies (red-yellow) traveling within an infected macrophage.

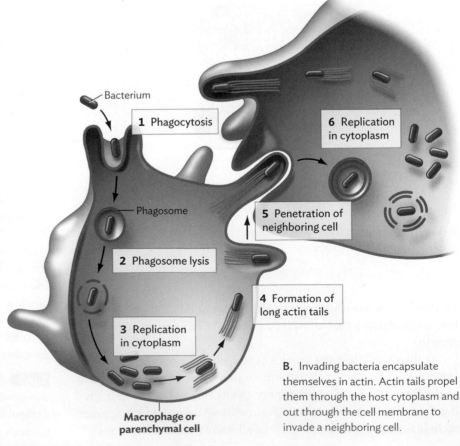

Bacterium

1 Phagocytosis

6 Replication in cytoplasm

Phagosome

5 Penetration of neighboring cell

2 Phagosome lysis

4 Formation of long actin tails

3 Replication in cytoplasm

Macrophage or parenchymal cell

B. Invading bacteria encapsulate themselves in actin. Actin tails propel them through the host cytoplasm and out through the cell membrane to invade a neighboring cell.

the original cell and enable it to penetrate a neighboring host cell. This intracellular projection strategy turns out to be surprisingly common among pathogens; for example, HIV (AIDS virus) disseminates by a similar mechanism (see Section 12.3).

Other non-spore-forming firmicutes form cocci, such as the paired cocci of *Enterococcus*. Normally found in the intestinal tract, *Enterococcus* species can become invasive in immune-compromised hospital patients, causing **urinary tract infections** and bacteremia. Enterococcal infections were commonly treated with the antibiotic vancomycin, but hospitals have now seen the rise of vancomycin-resistant enterococci (VRE).

Link Urinary tract infections are discussed in detail in Chapter 23. The most common etiologic agents of UTIs are the uropathogenic *E. coli* (75% of all UTIs), *Klebsiella, Proteus, Pseudomonas, Enterobacter, S. aureus, Enterococcus, Chlamydia*, fungi, and *Staphylococcus saprophyticus*.

Another group of non–spore formers, *Staphylococcus* species are facultative anaerobic cocci that grow in clusters, often hexagonal in arrangement (Figures 10.6A and B). They include common skin commensals such as *Staphylococcus epidermidis*. The staphylococci are generally salt tolerant, and their fermentation generates short-chain fatty acids that inhibit growth of skin pathogens. Diverse Gram-positive cocci favor different parts of the skin, as discussed in Chapter 19. For example, certain genera of Gram-positive anaerobes are prevalent in the folds of the umbilicus, or belly button (see the **inSight** box).

Certain Gram-positive cocci, however, are serious pathogens. *Staphylococcus aureus* causes impetigo and toxic shock syndrome, as well as pneumonia, mastitis (inflammation of the mammary gland), and osteomyelitis (bone infection). Disease involves several toxins, such as the toxic shock toxin (TSST-1). *S. aureus* is a major cause of nosocomial (hospital acquired) infections, especially of surgical wounds. The most dangerous strains, now resistant to most known antibiotics, are termed **MRSA** (methicillin-resistant *S. aureus*).

Link Bacterial infections, including those caused by *S. aureus*, are further discussed in Chapters 19–24. Staphylococcal skin infections, including **impetigo** and **MRSA** are, for example, covered in Section 19.3.

Streptococcus species generally form chains instead of clusters, as their cells divide in a single plane (Figures 10.6C and D). They are aerotolerant (grow in the presence of oxygen) but metabolize by **fermentation**. Many live on oral or dental surfaces, where they cause caries (tooth decay). Their fermentation of sugars produces such high concentrations of lactic acid that the pH at the tooth surface can fall to pH 4, which demineralizes teeth. *Streptococcus* species cause many serious diseases, including pneumonia (*S. pneumoniae*), strep throat, erysipelas, and scarlet fever (*S. pyogenes* or group A strep). Different species of *Streptococcus* can be identified by antibody reactions to their cell-surface glycoproteins. The antibody reactions with glycoproteins (bacterial antigens) are codified as the Lancefield grouping, named for the physician Rebecca Lancefield (1895–1981).

Link As is discussed in Section 6.5, many anaerobic microbes do not possess electron transport chains and thus cannot conduct aerobic respiration to produce energy. These anaerobic microbes instead perform carbohydrate **fermentation** to derive ATP.

Different strains of staphylococci and streptococci may be distinguished by their **hemolysis**, the ability to lyse red blood cells. Hemolysis can be tested on sheep blood agar (Figure 10.7). Colonies that partly hemolyze red blood cells, via hydrogen peroxide reaction, generate a greenish halo (alpha hemolysis). Colonies that fully hemolyze red blood cells, via a hemolysin enzyme, generate a clear halo (beta hemolysis).

Figure 10.6   Staphylococci and Streptococci

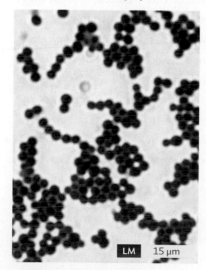

A. *Staphylococcus* species.

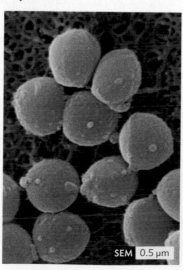

B. *Staphylococcus* species.

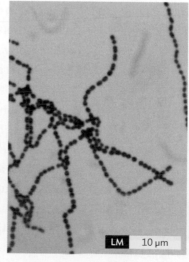

C. *Streptococcus* species (Gram stain).

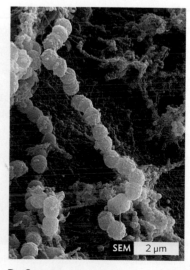

D. *Streptococcus* species.

inSight

Belly Button Biodiversity

Figure 1   The Umbilicus, or Belly Button: Host of Diverse Microbes

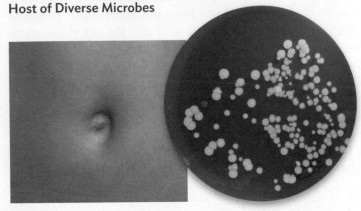

A. The indented scar of the umbilicus.

B. Plate culture of bacteria and fungi from an umbilical swab, sample 982 from the Belly Button Biodiversity project. The many colonies of different species vary in size, texture, and color.

Figure 2   *Anaerococcus* Species

This anaerobic Gram-positive coccus is often found within creased skin.

Everyone is aware of that mysterious part of our body known as the umbilicus or navel, or colloquially, the "belly button." The belly button is actually an indented scar left over from the removal of the umbilical cord at birth. In 90% of people, the belly button dips inward (Figure 1A), forming a moist, protective pocket inhabited by unique species of microbes—the umbilical microbiome. In very rare cases, the umbilicus can become infected; but most of the time, the microbial inhabitants simply maintain their own little ecosystem.

So what kinds of microbes live down there in the navel—and what might they be doing? Do different people have different umbilical microbiomes? To address these questions, scientists founded the Belly Button Biodiversity Project. For this project, a team of scientists led by Robert Dunn, at North Carolina State University, recruited more than 500 volunteers to provide samples of their umbilical microbiomes. Each volunteer sampled the umbilical indentation using a sterile cotton swab soaked in phosphate saline solution. The suspended microbes were collected and processed to isolate their genomic DNA. Another part of the sample was streaked on an agar plate in order to isolate colonies of growing strains (Figure 1B). The many different microbes formed colonies that varied in size, color, and texture—differences attributable to different DNA sequences in the microbial genomes.

Bacteria were identified based on the sequence of their genes encoding small-subunit ribosomal RNA (16S rRNA), a gene highly conserved among all life-forms. In all, more than 2,000 genetically unique strains were isolated, most of them appearing in only a few samples. But six isolates were found in more than 80% of human navels sampled. The most frequent isolates were typed to genera *Staphylococcus*, *Corynebacterium*, and *Bacillus*. Also common were actinobacteria such as *Micrococcus* and the clostridium-related bacteria *Anaerococcus*, *Finegoldia*, and *Peptidophilus*. The genus *Anaerococcus* (Figure 2) is an anaerobic Gram-positive coccus that commonly occurs as normal biota within creased regions of skin. The related anaerobic genus *Finegoldia* also includes normal commensal biota. But related species such as *Finegoldia magna* may cause infections of skin and bone. The presence of umbilical anaerobes suggests that the inner folds of the belly button consistently avoid exposure to air.

How do umbilical microbes interact with their host? The details remain unclear, but we hypothesize that as seen elsewhere in the skin, the dominant organisms maintain a stable community by helping exclude potential pathogens from the moist, protected habitat of the belly button.

Figure 10.7   Hemolysis Reactions

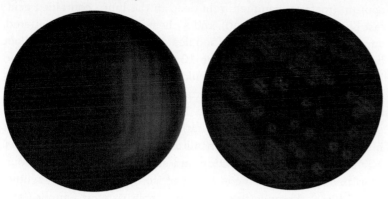

A. *Streptococcus pneumoniae* (alpha hemolysis).

B. *Staphylococcus aureus* (beta hemolysis).

Figure 10.8   Mycoplasmas: Bacteria without Cell Walls

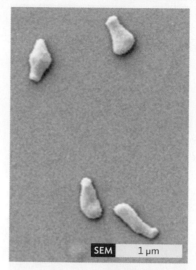

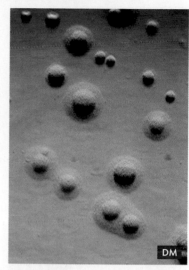

A. *Mycoplasma mobile*, a parasite of fish.

B. Mycoplasmas cultured on agar show a "fried egg" shape of colony.

BACTERIA LACKING A CELL WALL   A few Firmicutes, the mycoplasmas (class Mollicutes), entirely lack a cell wall. Mycoplasmas have completely lost their cell wall through reductive evolution, retaining only their cell membrane (**Figure 10.8**). Although they cannot stain "Gram-positive" like other Firmicutes, their genetic relatedness includes them in this category. In nature, mycoplasmas grow only within tissues of a host organism. The loss of the energy-expensive cell wall may have enhanced the cell's reproductive rate in a host environment protected from osmotic shock. On agar plates, some mycoplasmas form colonies with a characteristic "fried egg" appearance (Figure 10.8B). Every known class of multicellular organism is parasitized by mycoplasmas, including vertebrates, insects, and vascular plants; in humans, they cause pneumonia and meningitis

Actinobacteria

Actinobacteria, the "high-GC Gram-positives," include major pathogens and antibiotic producers, as well as essential decomposers in natural environments. Most actinobacteria can be identified under the microscope by the acid-fast stain, a procedure in which cells are penetrated with a dye that is retained in the cell even under treatment with acid alcohol (see Section 4.4). The acid-fast property is associated with unusual cell wall lipids, such as the mycolic acids of *Mycobacterium tuberculosis*.

A major group of Actinobacteria, the actinomycetes, undergo complex life cycles forming filamentous mycelia (branching filaments). **Figure 10.9A** shows colonies of *Streptomyces coelicolor*; the dark blue droplets on the colonies contain secreted antibiotics. *Streptomyces* bacteria grow as filaments (**Figure 10.9B**) that develop

specialized cells at their tips called arthrospores, which are released and disperse in the air to form new colonies. The mycelia of *Streptomyces* and other actinomycetes produce molecules we use as antibiotics, antitumor agents, and immunosuppressants. Pharmaceutical companies test soil samples all over the world for new actinomycete species that may produce the next new wonder drugs.

Other filamentous actinomycetes are pathogens. For example, *Nocardia* species are opportunists that cause pneumonia in hospital patients.

Actinobacteria that are *not* actinomycetes have thick acid-fast cell walls but lack the mycelial life cycle. They grow as isolated rods and cocci, such as the coccobacillus *Gardnerella vaginalis*, which causes bacterial vaginosis; or cells of variable shape, such as *Corynebacterium diphtheriae*, the cause of diphtheria. Another important actinobacterial genus is *Propionibacterium*, rod-shaped bacteria named for their fermentation pathway, which releases propionic acid—the component that gives Swiss cheese its distinctive taste.

Figure 10.9   *Streptomyces* Bacteria

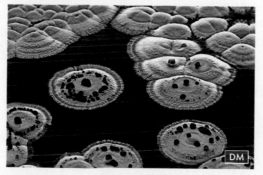

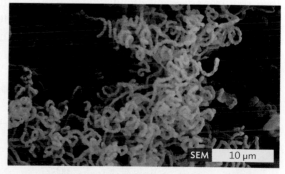

A. Colonies of *S. coelicolor* show sky-blue mycelia. The dark blue droplets on the colonies contain secreted antibiotics.

B. *Streptomyces* cells form coiled filaments.

Figure 10.10 *Mycobacterium tuberculosis*

A. Acid-fast stain of tissue sample containing *M. tuberculosis*.

B. Crinkled appearance of *M. tuberculosis* colonies.

had lost his appetite, and had been coughing for several weeks. His sputum (material coughed from deep in the lung) contained acid-fast bacilli (see Figure 10.10), and a chest X-ray revealed scattered, small (1–5 mm) granulomas (nodules of inflammation) throughout both lobes of the lung (**Figure 10.11**). The physician diagnosed Michael with secondary tuberculosis. Secondary tuberculosis is caused by bacteria reactivated from old lesions in which bacteria have persisted in a dormant state. The causative bacterium is *Mycobacterium tuberculosis*, a facultative intracellular pathogen that infects only humans and is transmitted though respiratory droplets. Michael was placed in isolation and started on a long-term, four-drug treatment regimen consisting of isoniazid (fatty acid synthesis inhibitor), rifampin (transcription inhibitor), pyrazinamide (fatty acid synthesis inhibitor), and streptomycin (protein synthesis inhibitor). The four-drug therapy continued for 2 months, followed by 4 months with isoniazid and rifampin only. Michael responded well to treatment. Luckily, he was not infected with one of the highly dangerous multidrug-resistant strains (MDR-TB).

Propionibacterium freudenreichii ferments milk during cheese production, whereas the close relative *Propionibacterium acnes* causes acne (skin infection). Thus, even when we have two closely related species of one genus, one can be a pathogen and the other a "friendly" life-form.

Actinobacteria include the causative agents of tuberculosis (*Mycobacterium tuberculosis*) and leprosy (*M. leprae*). Cells of *M. tuberculosis* can be detected by the acid-fast stain as tiny rods associated with sloughed cells in sputum (**Figure 10.10A**). The bacteria form crinkled colonies after 2 weeks of growth on agar-based media (**Figure 10.10B**). *M. tuberculosis* has exceptionally thick, complex membranes that include some of the longest-chain acids known, up to 90 carbons. The mycolic acids are linked to arabinogalactan, a polymer of the sugars arabinose and galactose built onto the peptidoglycan. This mycolyl-arabinogalactan-peptidoglycan complex forms a waxy coat that impedes the entry of nutrients through porins and thus slows the growth rate, but it also protects the bacterium from host defenses and antibiotics. For this reason, the cure for tuberculosis requires an exceptionally long course of antibiotic therapy.

Tuberculosis is one of the oldest documented diseases of humankind. How did *M. tuberculosis* evolve—and what clues may help us predict future emerging strains?

Evolution of *Mycobacterium tuberculosis*

CASE HISTORY 10.1

Tuberculosis: "I'll Be Back"

Michael, 45 years old, is a cook at a fast-food restaurant in Rome, Georgia. He went to see his physician because he was feeling fatigued, was unable to sleep,

Figure 10.11 **Lung X-Ray Showing Small Lesions Filled with *M. tuberculosis* Organisms**

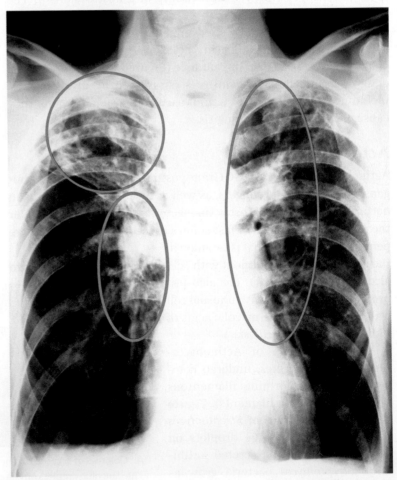

Tuberculosis is one of the most devastating and ancient of human diseases. The Greek physician Hippocrates (460–370 BCE) gives a description of a disease causing hard tubercules in the lungs and abscesses in the lumbar region of the spine. Modern molecular techniques such as PCR have found clear evidence of *M. tuberculosis* DNA in ancient mummies and in bones dating back to 3000 BCE. Because it has caused human disease for millennia, the evolutionary rise of *M. tuberculosis* is particularly interesting to examine.

How did *M. tuberculosis* evolve from a strictly environmental bacterium into a pathogen whose only reservoir is human? Evolution into a pathogen required several evolutionary steps. In natural environments, the ancestral mycobacteria were engulfed by predators such as amebas. By evolving to survive intracellularly, the bacterium essentially converted its predator to a host. The next step was to acquire the capacity to transfer between hosts. Once the intracellular niche was established, the host-pathogen relationship enabled *M. tuberculosis* to colonize new hosts and maintain a persistent chain of infection.

Genome comparisons between *M. tuberculosis* and nontuberculosis mycobacteria reveal a multistep process that led to *M. tuberculosis* (**Figure 10.12**). The evolutionary path began from an ancestral organism that produced actinomycete genera such as *Corynebacterium*, *Streptomyces*, and *Mycobacterium*. The ancestral mycobacterium diverged into species such as *Mycobacterium avium*, a pathogen of birds; *M. marinum*, a marine organism causing human skin ulcers; *M. kansasii*, a group of environmental organisms causing tuberculosis in mammals but rarely in humans; and *M. tuberculosis* and *M. bovis*, which cause tuberculosis in humans. Along the way, these strains acquired new genetic material via horizontal gene transfers from unknown sources (marked by arrows in Figure 10.12). Later, *M. tuberculosis* underwent deletion of genes that were no longer needed by the human-associated pathogen. Thus, the ancestral mycobacterial strains were environmental microbes, one of which adapted to inhabit an intracellular host environment and become *M. tuberculosis*.

It is interesting to note that *M. bovis*, an organism that produces tuberculosis-like disease in cattle, is not the progenitor of *M. tuberculosis*, as previously thought. It is, however, derived recently from a common, but unknown, predecessor. An attenuated strain of *M. bovis* called Bacillus Calmette-Guérin is used in some countries as a vaccine (the BCG vaccine) to protect against *M. tuberculosis*. In the United States, the Centers for Disease Control and Prevention (CDC) does not recommend BCG because the vaccine's protection rate is variable (from 0% to 60% in various trials) and because the vaccine stimulates antibody production that interferes with the tuberculin test.

M. tuberculosis continues evolving today. This is especially evident from the emergence of new forms of antibiotic resistance and from genomic studies comparing strains from around the world.

Note Certain bacterial species have names that are coincidentally similar. Distinguish between:

- *Streptococcus*, the Firmicute cocci, and *Streptomyces*, the actinomycetes that produce antibiotics.

- *Mycoplasma*, the Firmicutes that lack a cell wall, and *Mycobacterium*, the Actinobacteria with an extremely complex cell wall.

SECTION SUMMARY

- **Firmicutes** are Gram-positive bacteria that include endospore-forming genera such as *Bacillus* and *Clostridium*.

- **Nonsporulating firmicutes** include pathogenic rods such as *Enterococcus* and *Listeria*, as well as food-producing bacteria such as *Lactobacillus* and *Lactococcus*.

- ***Staphylococcus* and *Streptococcus*** are Gram-positive cocci that include normal human flora as well as serious pathogens causing toxic shock syndrome, pneumonia, and scarlet fever.

- **Mycoplasmas** belong phylogenetically to phylum Firmicutes, but they lack the cell wall and show ameboid motility. Species cause diseases such as meningitis and pneumonia.

- **Actinobacteria** include Actinomycetes, mycelial spore-forming soil bacteria such as *Streptomyces* and *Nocardia*, an opportunistic pathogen. Other actinobacteria have irregularly shaped cells; this group includes *Corynebacterium* species that causes diphtheria.

- **Mycobacteria** are acid-fast-staining actinomycete rods whose cell envelope contains a diverse assemblage of complex mycolic acids. Mycobacterial species cause tuberculosis and leprosy.

Figure 10.12 Evolution of "Slow-Growing Mycobacteria," Including *M. tuberculosis*

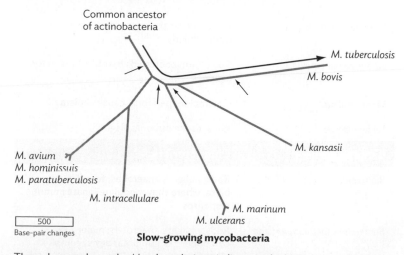

Common ancestor of actinobacteria

M. tuberculosis

M. bovis

M. kansasii

M. avium
M. hominissuis
M. paratuberculosis

M. intracellulare

500
Base-pair changes

M. marinum
M. ulcerans

Slow-growing mycobacteria

The red arrow shows the *M. tuberculosis* main lineage; the black arrows show where the lineage gained genes by horizontal transfer. Gene loss also occurred (not shown).

Thought Question 10.2 How do you think *Mycobacterium tuberculosis* manages to grow despite its thick envelope screening out most nutrients?

10.3
Gram-Negative Proteobacteria and Bacteroidetes

SECTION OBJECTIVES
- Describe the diverse environmental roles of Proteobacteria.
- Identify the diseases caused by diverse Gram-negative bacteria.

Which bacteria exhibit the widest range of shape and can "eat" nearly anything? The Proteobacteria show an amazing diversity of form and metabolism—from cocci to spiral cells, from heterotrophs and lithotrophs to photosynthesizers (Table 10.2). Yet all the Proteobacteria share a common form of cell envelope, which stains Gram-negative. As discussed in Chapter 5, the Gram-negative envelope includes an outer membrane, a cell wall of peptidoglycan permeated by the periplasm, and an inner membrane (comparable to the plasma membrane of other bacteria). The outer membrane is packed with receptor proteins and porins, making up two-thirds of the mass of the membrane. The outer membrane lipids contain long sugar polymer extensions (lipopolysaccharide, or LPS). In pathogens, LPS repels phagocytosis and has toxic effects when released by dying cells.

Enterobacteriaceae and Other Gram-Negative Rods

The best-studied family of Proteobacteria is the Enterobacteriaceae, of which *Escherichia coli* is the most famous. Some strains of *E. coli* grow normally in the human intestine, feeding on our mucous secretions and producing vitamins, such as vitamin K. But other strains, such as *E. coli* O157:H7, cause serious illness. A large proportion of the world's children die of *E. coli*–related intestinal illness before the age of 5. Other genera of Enterobacteriaceae that cause disease include *Salmonella* species, the cause of enteritis and typhoid fever; *Klebsiella*, a leading cause of hospital-acquired pneumonia; and *Yersinia pestis*, cause of bubonic plague. Many of the Enterobacteriaceae appear as rods, very similar under the microscope, but they can be distinguished by a variety of biochemical tests (Table 10.3). These tests, for example, may detect color changes arising from pH change during fermentation of particular food molecules, such as sugars and amino acids. Fermentation pathways distinguish between closely related species, such as *Escherichia coli*, which tests positive for methyl red, and the environmental organism *Enterobacter*, which tests positive for acetoin (Voges-Proskauer test).

All enterobacteria are facultative anaerobic rods that may grow in human or animal digestive tracts or in lakes and streams. They grow singly, in chains, or in biofilms. Many species are motile, with many flagella. Most strains grow well with or without oxygen, by either respiration (aerobic or anaerobic) or fermentation. They

Table 10.2
Selected Gram-Negative Proteobacteria and Bacteroidetes

Proteobacteria	Characteristics
Agrobacterium tumefaciens	Rods; induce plant tumors
Bdellovibrio bacteriovorus	Curved rods; invade periplasm and consume prey bacterium
Brucella melitensis	Rods; cause brucellosis disease of cattle and sheep
Coxiella burnetii	Rods; cause Q fever
Escherichia coli	Rods; main vehicle for biotechnology; strains cause intestinal and bladder infections
Geobacter metallireducens	Rods with electron-conducting nanowires; generate electricity in fuel cells
Helicobacter pylori	Wide spiral cells; grow in stomach acid, cause gastritis
Legionella pneumophila	Rods; grow within amebas or macrophages; cause legionellosis
Myxococcus xanthus	Rods; soil bacteria that aggregate to form fruiting bodies
Neisseria gonorrhoeae	Diplococci (pairs of cocci); cause gonorrhea
Neisseria meningitidis	Cocci; cause meningitis
Nitrobacter winogradskyi	Rods; nitrite-oxidizing lithotrophs active in soil
Rhodopseudomonas palustris	Rods; photoheterotrophs in pond water
Rickettsia rickettsii	Intracellular pathogens; cause Rocky Mountain spotted fever
Salmonella enterica	Rods; enteric pathogen, may grow intracellularly
Sinorhizobium meliloti	Rods; form bacteroids that fix nitrogen for legumes
Vibrio cholerae	Comma-shaped rods; cause cholera
Yersinia pestis	Rods; cause bubonic plague
Bacteroidetes	**Characteristics**
Bacteroides fragilis	Rods; obligate anaerobes; found in gut biota, where they modulate innate immune responses
Bacteroides thetaiotaomicron	Rods; obligate anaerobes; major component of gut biota; digest complex plant fibers
Porphyromonas gingivalis	Rods; obligate anaerobes; cause periodontal disease

Table 10.3

+Enteric Gram-Negative Bacteria: Examples of Biochemical Reactions

	Indole Production	Methyl Red	Voges-Proskauer	Simmons' Citrate	Hydrogen Sulfide	Urea Hydrolysis	Phenylalanine Deaminase	Lysine Decarboxylase	Arginine Dihydrolase	Ornithine Decarboxylase	Motility (36°C)	Gelatin Hydrolysis (22°C)	D-Glucose, Acid	D-Glucose, Gas	Lactose Fermentation	Sucrose Fermentation	D-Mannitol Fermentation	Dulcitol Fermentation	Adonitol Fermentation	D-Sorbitol Fermentation	L-Arabinose Fermentation
Citrobacter freundii	5	100	0	95	80	70	0	0	65	20	95	0	100	95	50	30	99	55	0	98	100
Enterobacter aerogenes	0	5	98	95	0	2	0	98	0	98	97	0	100	100	95	100	100	5	98	100	100
Escherichia coli	98	99	0	1	1	1	0	90	17	65	95	0	100	95	95	50	98	60	5	94	99
Klebsiella pneumoniae	0	10	98	98	0	95	0	98	0	0	0	0	100	97	98	99	99	30	90	99	99
Klebsiella oxytoca	99	20	95	95	0	90	1	99	0	0	0	0	100	97	100	100	99	55	99	99	98
Morganella morganii	98	97	0	0	5	98	95	0	0	98	95	0	100	90	1	0	0	0	0	0	0
Proteus mirabilis	2	97	50	65	98	98	98	0	0	99	95	90	100	96	2	15	0	0	0	0	0
Salmonella Choleraesuis	0	100	0	25	50	0	0	95	55	100	95	0	100	95	0	0	98	5	0	90	0
Salmonella Typhi	0	100	0	0	97	0	0	98	3	0	97	0	100	0	1	0	100	0	0	99	2
Salmonella, Most Serotypes	1	100	0	95	95	1	0	98	70	97	95	0	100	96	1	1	100	96	0	95	99
Serratia marcescens	1	20	98	98	0	15	0	99	0	99	97	90	100	55	2	99	99	0	40	99	0
Shigella sonnei	0	100	0	0	0	0	0	0	2	98	0	0	100	0	2	1	99	0	0	2	95
Shigella dysenteriae, Shigella flexneri, Shigella boydii	50	100	0	0	0	0	0	0	5	1	0	0	100	2	0	0	93	2	0	30	60

For each species, numbers indicate the percentage of clinical isolates showing a positive reaction for a given test.

ferment rapidly on carbohydrates, generating fermentation acids, ethanol, and gases (CO_2 and H_2) in various proportions, depending on the species. Their presence in the intestine supports the growth of organisms that use these gases, including methanogens (discussed in Chapter 19). Many strains form biofilms. Biofilm formation explains the persistence of drug-resistant infections, such as those associated with urinary catheters in long-term hospital patients.

Some Proteobacteria, such as *Geobacter* species, transfer electrons to iron or magnesium instead of oxygen. *Geobacter* species can be used to build a fuel cell in which the bacteria transfer electrons to an electrode, thus generating electricity. Such bacterial circuits can drive a clock or a research instrument.

Pathogenic Enterobacteriaceae

The Enterobacteriaceae include a wide variety of animal and plant pathogens. An interesting pathogen is *Proteus mirabilis*, which causes serous bladder and kidney infections in humans, particularly as a complication of surgical catheterization. *Proteus* species are heavily flagellated and display a remarkable swarming behavior (**Figure 10.13**). Upon an environmental signal, the flagellated rods grow into long-chain swarmer cells. The swarmers gather, forming "rafts" that swim together. Ultimately, they grow into a complex biofilm that resists antibiotics and can be hard to eradicate from a patient.

Closely related to Enterobacteriaceae are several genera of rod-shaped bacteria that are obligate respirers, such as *Pseudomonas*

Figure 10.13   The Enteric Rod *Proteus mirabilis*: Isolated Swimmer or Cooperative Swarmer

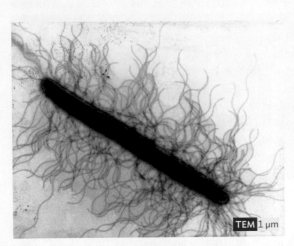

A. A thickly flagellated swarmer cell.

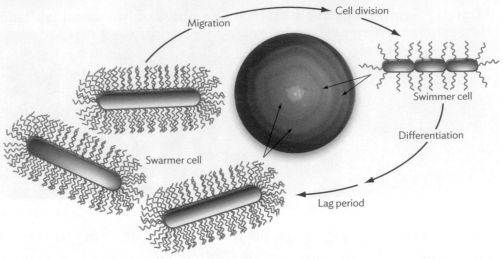

B. Waves of migration of *P. mirabilis* through blood agar.

(family Pseudomonadaceae). Many pseudomonads require oxygen for growth, although some can use alternative electron acceptors, such as nitrate. These genera metabolize an extraordinary range of natural compounds, including aromatic derivatives of lignin; thus, they have important roles in natural recycling and soil turnover. *Pseudomonas aeruginosa* commonly grows in soil as a decomposer. But in humans, *P. aeruginosa* can infect surgical wounds, and it forms biofilms on the pulmonary lining of cystic fibrosis patients.

An important intracellular pathogen related to the pseudomonads is *Legionella pneumophila*. Incapable of growth on sugars, *L. pneumophila* requires oxygen to respire on amino acids. The organism exhibits an unusual dual lifestyle, alternating between intracellular growth within human macrophages and intracellular growth within freshwater amebas (**Figure 10.14**). Growth within amebas facilitates transmission through aerosols into the human lung. *L. pneumophila* is an emerging pathogen that takes advantage of societal change—the prevalence of large-scale air-conditioning units with unfiltered water, which facilitate colonization by the bacteria.

Some Gram-negative rods are pathogens with complex life cycles involving more than one host. A famous example is *Yersinia pestis*, cause of the bubonic plague, which remains prevalent among prairie animals in the western United States. *Y. pestis* evolved from an enteric pathogen family, members of which persist today.

Figure 10.14   *Legionella pneumophila* Colonizing an Ameba

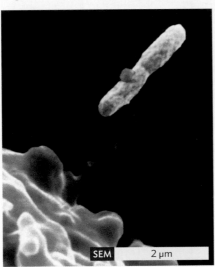

A. *L. pneumophila* cell is caught by an ameba's pseudopod.

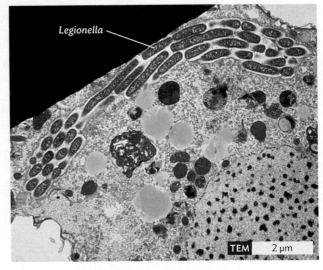

B. *L. pneumophila* cells have colonized the ameba.

Yersinia pestis and the Evolution of Flea-Borne Transmission

TALES OF TWO DISEASES I

CASE HISTORY 10.2

Yersinia and Pseudoappendicitis

Kimiko, a 6-year-old girl from Shodo Island, in Japan, had a 4-day history of fever, severe abdominal pain, and loose stools. She was diagnosed with appendicitis, but an exploratory laparotomy found a healthy appendix. She responded well to tetracycline antibiotic therapy. The doctors finally concluded she had enteritis (intestinal inflammation). Bacteriological examination of her stool eventually recovered *Yersinia pseudotuberculosis*, an occasional cause of enteritis. The organism, which is transmitted by ingesting contaminated food or water, was found in the family's well water and in the stool of animals found on their property.

CASE HISTORY 10.3

Yersinia and the Plague

 In July, Javier, a 28-year-old man from La Plata County, Colorado, went to a hospital emergency department with a 3-day history of fever, nausea, vomiting, and a lump in his groin area (an inguinal lymphadenopathy called a bubo). Javier was diagnosed with food poisoning (the lump was not considered important) and discharged without treatment. Three days later, Javier returned, gravely ill. He was hospitalized with a blood infection (sepsis), and both lobes of his lungs contained fluid and white blood cells (pneumonia). The presumed diagnosis was now plague. Once in the lung, the organism causing plague is easily transmitted to others by coughing, so the patient was placed in respiratory isolation. Javier was treated with the antibiotic gentamicin (protein synthesis inhibitor) and fortunately recovered. Cultures of blood and of the aspirate from the swollen lymph node in his groin grew *Yersinia pestis*, confirming the diagnosis. Medical investigators then found *Y. pestis* in fleas collected near the patient's home. Javier had contracted plague from the bite of a flea. Fleas serve as vectors that transmit the organism from infected wild rodents (such as prairie dogs) to pets and humans. It so happened that Javier lived in Colorado, where *Y. pestis*–infected wild rodents can be found.

Yersinia pseudotuberculosis and *Y. pestis* are both Gram-negative bacilli and members of the family Enterobacteriaceae. The first species causes gastrointestinal disease that can mimic appendicitis and is transmitted from animals to humans by contaminated food or water. The second species causes a deadly disease (plague) that can be transmitted from animals to humans by an infected flea or, when the disease progresses to the pneumonic (lung) stage, from one human to another by respiratory droplets produced from coughing. Despite large differences in pathogenesis, the two bacterial species are actually closely related.

Comparative genomic analyses indicate that *Y. pestis* is a variant of *Y. pseudotuberculosis* that diverged only within the last 1,500–20,000 years. Just a few discrete genetic changes separate the two species. But those changes were sufficient for the *Y. pseudotuberculosis* ancestor to evolve from food-borne and waterborne transmission, and the cause of mild infection, to flea-borne transmission and the agent (*Y. pestis*) of an aggressive, life-threatening infection (**Figure 10.15**). What small genetic changes could account for such a dramatically altered pathogenesis?

Horizontally transferred plasmids provide part of the answer. Plasmids are short circular pieces of DNA that can be acquired from the environment by **transformation** or from a donor cell by **conjugation**. Two plasmids acquired by *Y. pestis* encode important virulence factors:

- A phospholipase that digests membrane phospholipids

- A unique capsule made of peptide rather than the polysaccharides that compose most bacterial capsules

- A protease called plasminogen activator that dissolves clots in the human bloodstream

Figure 10.15 **Transmission of *Yersinia pestis* by a Flea Bite**

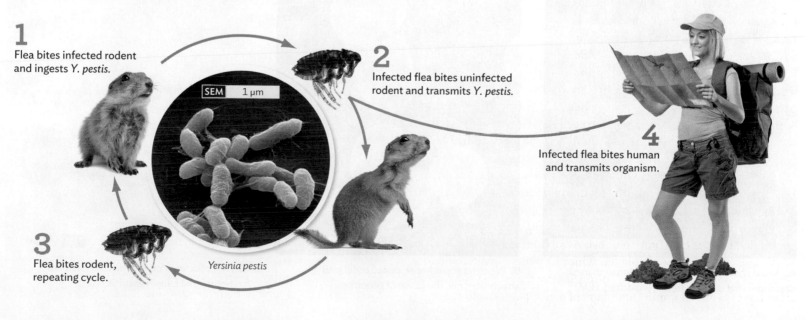

1 Flea bites infected rodent and ingests *Y. pestis*.

2 Infected flea bites uninfected rodent and transmits *Y. pestis*.

3 Flea bites rodent, repeating cycle.

4 Infected flea bites human and transmits organism.

SEM 1 μm

Yersinia pestis

Link As is discussed in Section 9.2, **transformation** and **conjugation** are mechanisms by which DNA can be transferred from one bacterium to another. Transformation refers to the assimilation of free DNA; DNA transfer by conjugation requires intimate cell-cell contact typically initiated by a special pilus that protrudes from a donor cell.

How does phospholipase enable flea-borne transmission? Once a flea bites an animal infected with *Y. pestis*, the phospholipase inactivates a toxic component of ingested blood and helps the microbe colonize the flea midgut, a critical step for transmission. Once established in the flea, the pathogen replicates to large numbers as a result of an unknown chromosomal gene and forms a large aggregate that blocks the flea's digestive tract. The flea constantly feels hungry, tries to take a blood meal on a new host, but regurgitates the bacteria into the bite. If the host is human, the human is infected.

Once transmitted to the host by the flea bite, *Y. pestis* becomes trapped at the site by blood clots. The host-expressed plasminogen activator protease helps dissolve clots and allows *Y. pestis* to disseminate in the bloodstream of the human host. Once within the bloodstream, the horizontally acquired peptide capsule of *Y. pestis* protects the bacterium from the human immune system.

From what we know of the two species of *Yersinia*, an evolutionary sequence has been proposed for the switch from food-borne/waterborne transmission to vector-borne transmission. We know that *Y. pseudotuberculosis* can cause a blood infection in rodents and will be taken up by fleas during a blood meal. But since the organism cannot colonize the flea, this is a dead-end transfer. Consequently, the first step in the switch must have been the horizontal transfer of the phospholipase gene enabling flea colonization. The second step in the process would have been acquisition of the unknown gene required for the flea-gut biofilm formation that blocks flea digestion. The third important step in the evolution of flea-borne transmission occurred when the progenitor cell captured the small plasmid containing the plasminogen activator gene. This new species, *Y. pestis*, can now dissolve clots, move from the flea bite to the bloodstream of its human host, and disseminate to the lung.

Diverse Gram-Negative Pathogens

Beyond Enterobacteriaceae, many other Gram-negative clades produce important pathogens. *Vibrio cholerae*, a comma-shaped organism with a flagellum (**Figure 10.16A**), causes the diarrheal disease cholera, a major problem for countries where sanitation is limited. For example, the epidemic of cholera in Haiti that followed the 2010 earthquake continues to hamper the island's recovery. A related organism, *Vibrio parahaemolyticus*, causes gastrointestinal illness through seafood contamination.

The Gram-negative rod *Bordetella pertussis* causes pertussis, or whooping cough, an illness prevented by immunization, but is a growing problem as our current vaccines lose effectiveness with age. *Neisseria gonorrhoeae*, which occurs as paired cocci (**Figure 10.16B**), causes the sexually transmitted infection gonorrhea; and *Neisseria meningitidis* causes meningitis, infection of the brain lining. *Haemophilus influenzae* and *Moraxella catarrhalis* cause pneumonia. *Francisella tularensis* causes the animal disease tularemia,

Figure 10.16   **Diverse Gram-Negative Pathogens**

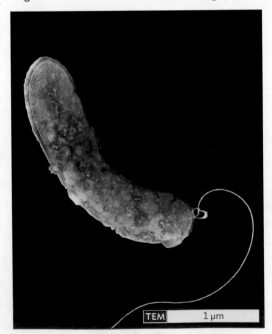 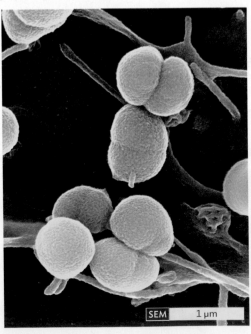 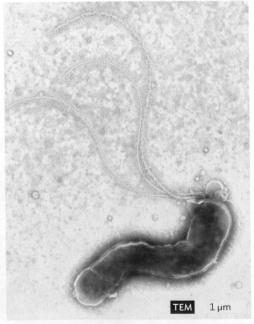

A. *Vibrio cholerae*, a flagellated, comma-shaped bacterium that causes cholera (colorized TEM).

B. *Neisseria gonorrhoeae*, paired cocci with attachment pili, the cause of gonorrhea.

C. *Helicobacter pylori*, a corkscrew-shaped bacterium that causes gastritis.

which can be transmitted to humans and is classified as a potential biowarfare agent.

The Epsilonproteobacteria (see Figure 10.1) include wide spiral-shaped organisms, some of which cause disease. *Helicobacter pylori* (Figure 10.16C) is one of the few bacteria that can colonize the gastric lining at extremely low pH. *H. pylori* causes chronic gastritis and is associated with stomach cancer. The related organism *Campylobacter jejuni* colonizes the intestinal tract and is a major cause of food poisoning.

Lithotrophs Eat Rock and Metals

Many kinds of Proteobacteria don't have to eat organic food; instead they oxidize minerals such as iron, sulfur, or nitrogen. Mineral oxidation for energy is called **lithotrophy**. Some lithotrophs, such as ammonia-oxidizing bacteria and methanogenic archaea, actually grow within human and animal digestive tracts. Lithotrophs in natural ecosystems have surprising uses for human technology. In wastewater treatment, for example, nitrogen lithotrophs (or nitrifiers) oxidize ammonia (NH_3) to nitrite (NO_2^-) or nitrite to nitrate (NO_3^-) (Figure 10.17A). Nitrifiers are of enormous economic and practical importance for wastewater treatment because they decrease the reduced nitrogen content of sewage. Special systems have been developed to retain nitrifier bacteria behind filters as one stage of water treatment. In the system shown in Figure 10.17B, nitrifying bacteria are encapsulated in pellets to retain them within the bioreactor while the treated water flows through a filter.

Acidithiobacillus is a genus of Proteobacteria that oxidize iron and sulfur. They produce extreme acids such as sulfuric acid. *Acidithiobacillus* metabolism can be used to solubilize metal ores during mining for iron or copper; this process is called "leaching" the metals. Metal leaching was performed by miners as early as the ancient Romans, although they did not understand the role of bacteria in their technology.

Intracellular Symbionts and Predators

Proteobacteria have evolved many intimate relationships with other organisms, both mutualistic and parasitic. Some of these relationships led to the bacteria evolving into organelles of eukaryotes. The chloroplasts in plant cells, for example, evolved from the lineage of cyanobacteria. The mitochondria of our own cells evolved from the lineage of the intracellular pathogen *Rickettsia*. Today, species of *Rickettsia* are carried by tick vectors, causing Rocky Mountain spotted fever. A similar tick-borne intracellular pathogen, *Ehrlichia chaffeensis*, causes ehrlichiosis. An obligate intracellular pathogen with a different transmission route (inhalation) is *Coxiella burnetii*, the cause of Q fever. Other pathogens may grow alternatively within a host cell or in extracellular fluids; these bacteria are called facultative intracellular pathogens. Examples of facultative intracellular pathogens include *Brucella* species, the cause of the livestock disease brucellosis, and *Salmonella enterica*, which may invade white blood cells in humans. Still other bacteria are capable of free-living existence but also engage in mutualism of vital importance to ecosystems and human agriculture. The most important such mutualism is nitrogen fixation. **Nitrogen fixation** is a process by which the nitrogen atoms in atmospheric nitrogen are made available for other uses.

Nitrogen is often a limiting nutrient for plants, which lack the enzymes needed to incorporate nitrogen gas into amino acids and proteins. Nitrogen-fixing endosymbionts of plants include bacterial genera such as *Rhizobium*, *Bradyrhizobium*, and *Sinorhizobium* (collectively known as rhizobia). Rhizobia can live freely in the soil, but they also colonize plants, usually legumes such as peas or alfalfa, by forming distinctive nodule structures (Figure 10.18A). In the nodules, the plant cells provide the bacteria with nutrients, as well as the pink colored "leghemoglobin," an oxygen-binding protein that maintains the anaerobic conditions needed to reduce (or "fix") N_2 to NH_4^+, which the plant can use. Each species of bacteria

Figure 10.17 Ammonia-Oxidizing Bacteria

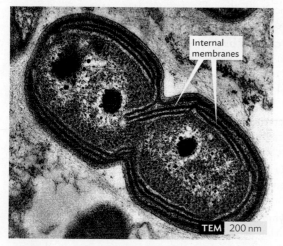

A. *Nitrosomonas europaea* oxidizes ammonia to nitrite. Internal membranes contain the electron transport complexes.

Internal membranes

TEM 200 nm

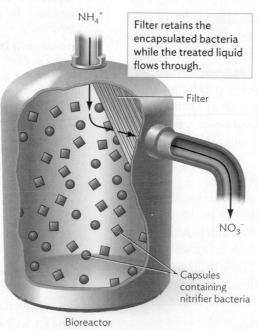

NH_4^+

Filter retains the encapsulated bacteria while the treated liquid flows through.

Filter

NO_3^-

Capsules containing nitrifier bacteria

Bioreactor

B. Wastewater treatment uses nitrifiers (ammonia-oxidizing bacteria) to remove ammonia.

Figure 10.18   Rhizobia: Legume Endosymbionts

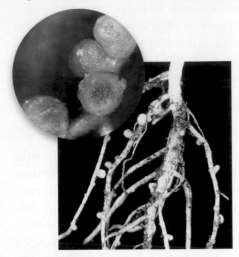

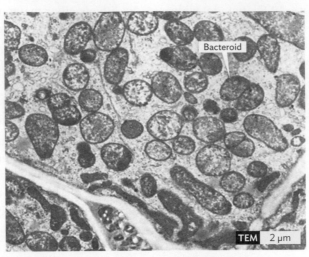

A. Legume nodules cut open to show pink regions where the plant cells produce leghemoglobin to maintain anaerobic conditions for bacteroid nitrogen fixation.

B. Within the host cells, *Sinorhizobium meliloti* cells form bacteroids that fix nitrogen while receiving nutrients from the plant.

colonizes a particular host range for infection. Within the host cell, the bacteria lose their own cell walls and become rounded "bacteroids," specialized for nitrogen fixation (**Figure 10.18B**).

Other bacteria have evolved as voracious predators that prey even on other bacteria. Bdellovibrios are bacteria that attack proteobacterial host cells such as *E. coli* (**Figure 10.19**). The structure of the "attack cell" is a small comma-shaped rod with a single flagellum. The attack cell attaches to the envelope of its host and then penetrates into the periplasm, where it uses host resources to grow. The growing bdellovibrio then produces enzymes that cross the inner membrane to degrade host macromolecules and make their components available to the bdellovibrio. Within the periplasm, the invading bdellovibrio elongates as a spiral filament while replicating its DNA. When most nutrients have been exhausted, the filament septates into multiple short cells. The cells develop flagella, and the periplasmic compartment bursts, releasing the newly formed attack cells. Bdellovibrios can be isolated from human sewage, a sign that vicious predatory behavior is going on in our intestines.

Bacteroidetes: Obligate Anaerobes

An important group of bacteria that stain Gram-negative but diverged earlier from the phylum Proteobacteria is the phylum Bacteroidetes. The most important genus of this group is *Bacteroides*. *Bacteroides* species, such as *B. fragilis* and *B. thetaiotaomicron*, are the major inhabitants of the human colon. These bacteria actually communicate with our immune system, telling our immune cells which responses to make while avoiding immune defenses that could harm them.

Bacteroides species are obligate anaerobes, incapable of growing in the presence of more than a tiny trace of oxygen. Their main

source of energy is fermentation of a wide range of sugar derivatives from plant material, compounds indigestible by humans and potentially toxic to us. *Bacteroides* species convert these substances to simple sugars and fermentation acids, some of which are absorbed by the intestinal epithelium. Thus, *Bacteroides* species serve two important functions for their host: they break down potential toxins in plant food and their fermentation products make up as much as 15% of the caloric value we obtain from food. Yet another benefit of *Bacteroides* species is their ability to remove side chains from bile acids, enabling return of bile acids to the hepatic circulation. In effect, our *Bacteroides* species and associated intestinal bacteria constitute a functional organ of the human body.

Bacteroides species cause trouble, however, when they reach parts of the body not designed to host them. During abdominal surgery, bacteria can escape the colon and invade the surrounding tissues. The displaced bacteria can form an abscess, a localized mass of bacteria and pus contained in a cavity of dead tissue. The interior of the abscess is anaerobic and often impenetrable to antibiotics.

SECTION SUMMARY

- **Proteobacteria** stain Gram-negative, with a thin cell wall and an LPS outer membrane. They show wide diversity of form and metabolism, including phototrophy, lithotrophy, and heterotrophy on diverse organic substrates.

- ***Yersinia pestis* (the cause of plague) evolved from *Y. pseudotuberculosis*** within the past 1,500–20,000 years. Flea-borne transmission of *Y. pestis* evolved following the horizontal transfer of virulence genes. Acquisition of these genes enabled *Y. pestis* to colonize the flea, block its intestine, and, once transmitted by flea bite, disseminate in the new host.

- **Gram-negative bacteria include many diverse pathogens,** such as *Vibrio cholerae*, the cause of cholera; *Neisseria gonorrhoeae*, the cause of gonorrhea; and *Helicobacter pylori*, the cause of gastritis.

- ***Bacteroides* bacteria are anaerobes that ferment complex plant materials** in the human colon. They may enter body tissues through wounds and cause abscesses.

- **Mutualistic endosymbionts** include nitrogen-fixing bacteria such as rhizobia, which form bacteroids within a legume root host cell.

Thought Question 10.3   The Gram-negative alphaproteobacterium *Rickettsia prowazekii* is an obligate intracellular pathogen that causes the disease typhus. It is transmitted by the body louse. What impact would an obligate intracellular lifestyle have on opportunities for horizontal gene transfer (transfer of genes between bacteria)?

Figure 10.19   Parasites of Bacteria: *Bdellovibrio*

Bdellovibrio bacteriovorus

A. *Escherichia coli* under attack by *Bdellovibrio bacteriovorus* (note predator cell within *E. coli* periplasm).

Bdellovibrio

E. coli periplasm

E. coli cytoplasm

AFM 1 μm

**B.
LIFE CYCLE
OF A
*Bdellovibrio***

1 *Bdellovibrio* finds host by chemotaxis.

2 *Bdellovibrio* binds to host receptors.

3 *Bdellovibrio* invades periplasm.

Spiral chain grows. **4**

5 Chain fragments into flagellated cells.

6 Host lysis releases *Bdellovibrio* cells.

10.4
Spirochetes: Twisted-Cell Pathogens and Symbionts

SECTION OBJECTIVES

- Describe the unique coiled cells of the spirochetes.
- Identify spirochetes that cause important human diseases.
- Identify spirochetes that grow as environmental symbionts.

Which spiral bacteria cause some of the world's most famous diseases? Spirochetes are a distinctive clade of bacteria with a cell structure consisting of a long, tight spiral that is flexible like a telephone cord (**Figure 10.20**). In many species, the spiral is so thin that its width cannot be resolved by bright-field microscopy. The best-known species are *Treponema pallidum*, the cause of syphilis; *Borrelia burgdorferi*, the cause of Lyme disease; and *Borrelia recurrentis*, which causes relapsing fever. Most spirochetes are slow-growing heterotrophs and difficult to culture. Besides the pathogens, however, many spirochetes are free-living organisms in soil or water. Many nonpathogenic spirochetes are digestive symbionts of hosts as diverse as termites and cattle, assisting their catabolism of plant material.

Figure 10.20   Spirochete Structure

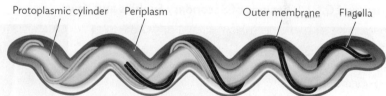

Protoplasmic cylinder   Periplasm   Outer membrane   Flagella

A. Spirochete cell structure, showing the arrangement of periplasmic flagella.

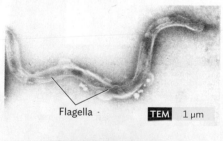

Flagella   TEM 1 μm

B. Spirochete with two periplasmic flagella, isolated from mouse blood.

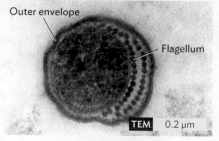

Outer envelope

Flagellum

TEM 0.2 μm

C. Cross section through a spirochete, showing outer envelope and flagella (axial fibrils). The unidentified spirochete was obtained from a human lesion of acute necrotizing ulcerative gingivitis.

The spirochete cell is surrounded by a thick outer sheath of lipopolysaccharides and proteins. The spirochete sheath is similar to a proteobacterial outer membrane, except that the periplasmic space completely separates the sheath from the plasma membrane. At each end of the cell, one or more flagella extend and double back around the cell body within the periplasmic space (Figure 10.20A). The periplasmic flagella rotate on proton-driven motors, as do extracellular flagella (see Section 5.5). But because the spirochete's flagella twine back around the cell body, their rotation forces the entire cell to twist around, corkscrewing through the medium. This corkscrew motion of the cell body turns out to have a physical advantage in highly viscous environments, such as human mucous secretions.

How have closely related species of the spirochete genus *Treponema* evolved to cause vastly different diseases? Consider the two diseases syphilis and yaws in Case Histories 10.4 and 10.5.

TALES OF TWO DISEASES II

CASE HISTORY 10.4

A Vaginal Ulcer That Seemed to Heal

Diane, a 20-year-old accountant from Cedar City, Utah, came to the Valley View Medical Center urgent-care clinic with a low-grade fever, malaise, and headache. She was sent home with a diagnosis of influenza but again sought treatment 7 days later with a macular rash (small, flat, red spots) on her trunk, arms, hands, and feet (**Figure 10.21A**). Further questioning revealed that 1 month previously, Diane had a painless ulcer on her vagina that healed spontaneously. Serology results indicated that Diane had contracted syphilis, caused by *Treponema pallidum* subspecies *pallidum*. Because this disease is sexually transmitted, Diane provided a list of her sexual partners over the past several months so the health department could contact them for testing and, if necessary, treatment.

CASE HISTORY 10.5

An Arm Wound That Seemed to Heal

Fabrice, an 11-year-old boy from the village of Ngbongbo Tokonzi, in the Democratic Republic of Congo (Africa), was seen at a nearby clinic. He had a cluster of raised, papular lesions on his arm (**Figure 10.21B**). When the doctor asked Fabrice whether he ever saw lesions like these before, Fabrice remembered having a single lesion about 2 months earlier, which spontaneously healed. Serology revealed that the boy had secondary yaws disease caused by the spirochete *Treponema pallidum* subspecies *pertenue*, which is a spirochete closely related to the strain that causes syphilis. However, *T. pallidum pertenue* is transmitted by direct contact with a cut or wound.

Figure 10.21   Lesions of Secondary Syphilis and Yaws

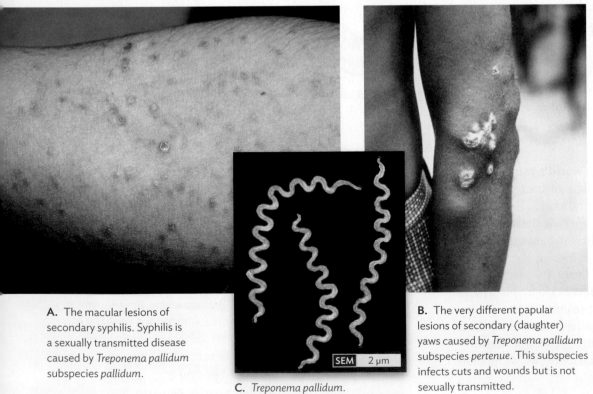

A. The macular lesions of secondary syphilis. Syphilis is a sexually transmitted disease caused by *Treponema pallidum* subspecies *pallidum*.

B. The very different papular lesions of secondary (daughter) yaws caused by *Treponema pallidum* subspecies *pertenue*. This subspecies infects cuts and wounds but is not sexually transmitted.

C. *Treponema pallidum*.

SEM   2 μm

As Diane found out, a strain of the spirochete *Treponema pallidum* causes the sexually transmitted disease syphilis (further discussed in Chapter 23); **Figure 10.21C**. Syphilis arises only from strains of the subspecies *pallidum* (*T. pallidum pallidum*). Strains of a different *T. pallidum* subspecies, subspecies *pertenue* (*T. pallidum pertenue*), produce a different disease called yaws, which is transmitted by direct skin contact. Yaws, like syphilis, is a multistage disease; thus, Fabrice thought his infection had gone away, but new lesions appeared 2 months later. Whereas syphilis is sexually transmitted and affects people worldwide, yaws is transmitted by direct skin contact predominantly in developing countries with a tropical climate. Besides their different modes of transmission and geographic distribution, these strains of *T. pallidum* differ dramatically in their abilities to cross the placenta and infect a developing fetus. Strains that cause syphilis can infect a fetus, but yaws strains cannot.

So what distinguishes the two subspecies? Are they perhaps a single species infecting hosts under different conditions? The *T. pallidum pallidum* and *T. pallidum pertenue* treponemes (spirochetes of the species *T. pallidum*) cannot be distinguished by morphology or physiology. Consequently, it had been proposed that syphilis and yaws were caused by the same organism and that climate or cultural differences resulted in the different diseases. However, a single strain causing both diseases seemed unlikely because syphilis and yaws have distinctive lesions and patterns of pathogenesis. Furthermore, an immune response to one will not fully cross-protect against the other in experimental animal infections.

What do we find in the subspecies genomes? The genomes of *T. pallidum pallidum* and *T. pallidum pertenue* are similar in both genome size and structure, but they differ in six regions, amounting to about 0.4% of the genome (about 4,500 bp out of 1,138,006 bp). The alterations involve small deletions and some repeated sequences. No plasmid differences or major gene differences are evident. It is likely that these minor differences cause the distinct infection patterns observed in syphilis and yaws. As of this writing, none of the *T. pallidum* subspecies have been cultured continuously in the laboratory; thus, we cannot yet mutate the genes to determine their potential roles in pathogenesis. It is remarkable that all of the pathogenic differences that distinguish *T. pallidum pallidum* from *T. pallidum pertenue* arise from a few very small and poorly understood differences in the genomic sequences of these subspecies.

In the course of evolution, which came first, *T. pallidum pertenue* or *T. pallidum pallidum*? Phylogenetic analysis examining the nature and frequency of differences in the two strains suggests that *T. pallidum pertenue* evolved first, followed by the syphilis subspecies. This suggests that syphilis evolved from the organism that causes yaws in tropical areas. This finding has strengthened the theory that, upon discovering the Americas, members of Columbus's crew contracted syphilis from sexual encounters with the native people and transported *T. pallidum pallidum* back to Europe more than 500 years ago.

Note Distinguish **spirochetes** (long, tightly coiled spirals with unique intracellular flagella) from **spirilla** (wide helical Gram-negative cells such as *Helicobacter* and *Rhodospirillum*).

SECTION SUMMARY

- **The spirochete cell** is a tight coil surrounded by a sheath and periplasmic space containing periplasmic flagella.
- **Spirochete motility occurs by a flexing motion** caused by rotation of the periplasmic flagella, propagated along the length of the coil.
- **Spirochetes include free-living fermenters** in water or soil. Others are endosymbionts of animal digestive tracts, such as the termite gut.
- **Spirochetes include important pathogens.** *Treponema pallidum* subspecies *pallidum* causes syphilis, and *Treponema pallidum* subspecies *pertenue* causes yaws.

Thought Question 10.4 For motility, what are the relative advantages of external flagella versus the flexible spiral cells of spirochetes?

10.5
Cyanobacteria Fix Carbon Dioxide and Produce Oxygen

SECTION OBJECTIVES

- Explain the key environmental contributions of cyanobacteria.
- Describe different forms of marine cyanobacteria.

Which bacteria throughout the biosphere make most of the oxygen we breathe? Cyanobacteria are the only bacteria that produce oxygen gas, a by-product of oxygenic photosynthesis. Cyanobacterial photosynthesis also fixes much of the carbon dioxide for consumers in ecosystems, particularly marine ecosystems supporting fish. These same cyanobacteria also fix much of the biosphere's nitrogen.

Cyanobacteria commonly appear green because of the predominant blue and red absorption by chlorophylls *a* and *b*. Essentially the same chlorophylls are found in plant chloroplasts, which evolved from internalized cyanobacteria that became endosymbionts. Oxygenic phototrophy is so successful that it drives all members of phylum Cyanobacteria, which account for about a quarter of all of Earth's biomass production. Despite their common metabolism of photosynthesis, cyanobacteria vary widely in cell form and behavior as well as in genetic diversity. Vast blooms of cyanobacteria occur in the ocean (Figure 10.22A). Cyanobacteria include species edible for humans and are components of spirulina salad and nostoc soup. No cyanobacterial pathogens are known, although some species produce substances that endanger aquatic consumers. For example, blooms of *Microcystis* produce microcystins, peptide toxins that damage the liver.

Cyanobacteria have evolved incredibly diverse forms. The simplest single-celled forms include *Synechococcus* and *Prochlorococcus*, the most abundant phototrophs in the oceans. Other genera, such as *Oscillatoria*, grow as long filaments of hundreds or even thousands of cells that can associate in thick biofilms. Oscillatoria cells are stacked like plates. Filamentous genera such as *Nostoc* develop special cells called heterocysts that fix nitrogen (Figure 10.22B). Other species, such as *Merismopedia*, form regular sheets of cells, while still others form globular aggregates.

Some cyanobacteria form a unique mutualism with fungi, called a **lichen** (Figure 10.23). In a lichen, the cyanobacterium fixes carbon dioxide and nitrogen into nutrients, whereas the fungus provides physical protection. Lichens form important portions of ground cover and essential food for animals, particularly in dry northern forests. Overall, cyanobacteria serve the ecosystem in several fundamental ways: by fixing carbon and nitrogen into biomass and by producing oxygen that heterotrophs can breathe.

Figure 10.22   Cyanobacteria

A. A ship plows through an ocean bloom of *Trichodesmium*.

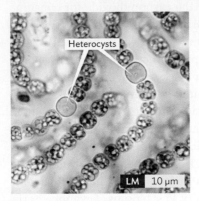

B. Filamentous *Nostoc* makes heterocysts (round cells).

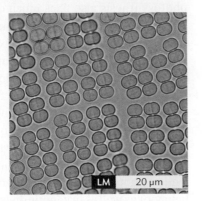

C. *Merismopedia* forms extended quartets, octets, and so on.

Figure 10.23   Lichens with Cyanobacteria

Lichens with symbiotic cyanobacteria provide a food source for forest animals.

SECTION SUMMARY

- Cyanobacteria produce much of the biosphere's oxygen gas.
- Cyanobacteria fix much of the biosphere's carbon and nitrogen.
- Lichens may include cyanobacteria in mutualism with fungi.

10.6
Chlamydias: Intracellular Pathogens

SECTION OBJECTIVES

- Describe the cell cycle of Chlamydias.
- Describe the diseases caused by Chlamydias.

Which bacteria lack a cell wall and are obligate intracellular pathogens? Chlamydias are a large group of bacteria with absent or diminished cell walls. These organisms evolved independently of the mycoplasmas (cell wall–less Firmicutes), and their alternative cell forms show very different environmental adaptations. Chlamydias are obligate parasites or pathogens. *Chlamydia trachomatis* (**Figure 10.24A**) is the causative agent of a major sexually transmitted disease in the United States. It is also a cause of trachoma, an eye disease dating back to ancient Egypt. The related species *Chlamydophila pneumoniae* causes pneumonia and has been implicated in cardiovascular disease. Chlamydias infect a wide range of host cell types, from respiratory epithelium to macrophages.

The Chlamydias show a complex developmental life cycle of parasitizing host cells (**Figure 10.24B**). Chlamydias alternate between two developmental stages with different functions: elementary bodies and reticulate bodies. The form of chlamydia transmitted outside host cells is called an elementary body. Elementary bodies resemble endospores in that they are metabolically inert, with a compacted chromosome. Though lacking a cell wall, they possess an outer membrane whose proteins are cross-linked by disulfide bonds, making a tough coat that provides osmotic stability. The elementary body adheres to a host cell surface and is endocytosed by the host cell.

To reproduce, the elementary body must transform itself into a reticulate body, named for the netlike appearance of its uncondensed DNA. The reticulate body has active metabolism and divides rapidly, but outside the cell it is incapable of infection and vulnerable to osmotic shock. To complete the infectious cycle, therefore, the reticulate bodies must develop into new elementary bodies before exiting the host. When the host cell lyses, the elementary bodies are released to infect new cells.

Without treatment, chlamydial infections may persist indefinitely. In the female reproductive tract, chlamydial infection may lead to inflammation of the fallopian tubes and ovaries (**Figure 10.24C**), a condition known as pelvic inflammatory disease (PID; see Section 23.5). PID can leave permanent adhesions (fibrous tissue across the organs), a common cause of infertility.

Figure 10.24 Chlamydia Life Cycle within a Cell of Urogenital Epithelium

Chlamydia trachomatis

A. *Chlamydia trachomatis* multiplying within a human cell (colorized TEM). Infected cell, equivalent to step 5 in part (B), contains reticulate bodies (yellow) growing and dividing, as well as newly formed elementary bodies (red).

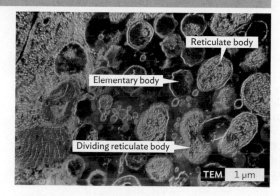

Reticulate body

Elementary body

Dividing reticulate body

TEM   1 μm

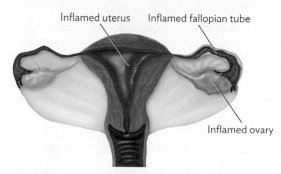

Inflamed uterus   Inflamed fallopian tube

Inflamed ovary

C. Pelvic inflammatory disease (PID). Chlamydial infection may cause permanent damage to the fallopian tubes and ovaries, resulting in infertility.

B. *Chlamydia* species persist outside the host as a spore-like elementary body that is metabolically inactive and contains compacted DNA. Upon endocytosis, the elementary body avoids lysosomal fusion and develops into a reticulate body in which the DNA, now uncondensed, has a reticular (netlike) appearance. The reticulate body replicates within the host cytoplasm and then develops into new elementary bodies that are released when the cell lyses.

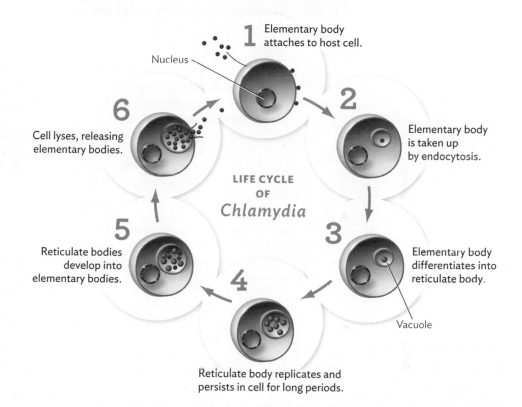

LIFE CYCLE OF *Chlamydia*

1 Elementary body attaches to host cell.

Nucleus

2 Elementary body is taken up by endocytosis.

3 Elementary body differentiates into reticulate body.

Vacuole

4 Reticulate body replicates and persists in cell for long periods.

5 Reticulate bodies develop into elementary bodies.

6 Cell lyses, releasing elementary bodies.

SECTION SUMMARY

- **Chlamydias are obligate intracellular parasites** that undergo a complex developmental progression, culminating in a spore-like form called an elementary body that can be transmitted outside the host cell. Chlamydias lack a cell wall.
- *Chlamydia trachomatis* causes a sexually transmitted infection as well as the eye disease trachoma.
- *Chlamydophila pneumoniae* causes pneumonia.

Perspective

Several themes emerge in the diversity of the domain Bacteria. Some phyla, such as Proteobacteria and Actinobacteria, have evolved highly diverse metabolism and cell form. Other phyla show remarkable uniformity in cell structure (Spirochetes) or metabolism (Cyanobacteria). From all these groups emerge pathogens, often closely related to organisms that enhance human health or contribute to environmental ecosystems. New species continually emerge through evolution, displaying new capabilities for environmental change or pathogenesis.

LEARNING OUTCOMES AND ASSESSMENT

	SECTION OBJECTIVES	OBJECTIVES REVIEW

10.1 Bacterial Diversity at a Glance

- Distinguish among five major groups of bacteria: Gram-positive bacteria, Gram-negative bacteria, spirochetes, Cyanobacteria, and Chlamydias.
- Describe the positive roles of bacteria in ecosystems.
- Describe how new pathogens emerge to cause new diseases.

1. Which group of organisms provides oxygen for us to breathe?
   A. Cyanobacteria
   B. Gram-negative bacteria
   C. Spirochetes
   D. Chlamydia

10.2 Gram-Positive Firmicutes and Actinobacteria

- Describe the traits that enable Firmicutes to survive dryness and persist for long periods.
- Describe how diverse Actinobacteria have both positive and negative influences on human health.

4. Which Gram-positive bacterium lacks a thick cell wall that excludes many antibiotics?
   A. *Clostridium botulinum*, the cause of botulism
   B. *Clostridium difficile*, an enteric organism that thrives when normal enteric biota have been eliminated by use of antibiotics
   C. *Mycobacterium tuberculosis*, the cause of tuberculosis
   D. *Mycoplasma pneumoniae*, a cause of pneumonia

10.3 Gram-Negative Proteobacteria and Bacteroidetes

- Describe the diverse environmental roles of Proteobacteria.
- Identify the diseases caused by diverse Gram-negative bacteria.

6. Which bacteria oxidize minerals instead of organic food?
   A. *Pseudomonas aeruginosa*, a common wound contaminant
   B. *Nitrobacter* species, growing in soil
   C. *Neisseria meningitidis*, a cause of meningitis
   D. *Bradyrhizobium*, a genus of nitrogen-fixing bacteria

10.4 Spirochetes: Twisted-Cell Pathogens and Symbionts

- Describe the unique coiled cells of the spirochetes.
- Identify spirochetes that cause important human diseases.
- Identify spirochetes that grow as environmental symbionts.

9. Which structural finding in a microscope would rule out the organism that causes Lyme disease?
   A. Round, coccoid cells
   B. Flexible, tightly coiled cells
   C. Under electron microscopy, intracellular flagella
   D. Cell motility

10.5 Cyanobacteria Fix Carbon Dioxide and Produce Oxygen

- Explain the key environmental contributions of cyanobacteria.
- Describe different forms of marine cyanobacteria.

12. Which trait do cyanobacteria lack?
    A. Producing oxygen through photosynthesis
    B. Nitrogen fixation
    C. Growth in globular clusters
    D. Intracellular flagella

10.6 Chlamydias: Intracellular Pathogens

- Describe the cell cycle of Chlamydias.
- Describe the diseases caused by Chlamydias.

14. Which bacterial structure do Chlamydias lack?
    A. Ribosomes
    B. Outer membrane
    C. Peptidoglycan cell wall
    D. Inner membrane

2. Which bacteria have tightly coiled cells with internal flagella?
   A. Cyanobacteria
   B. Gram-negative bacteria
   C. Spirochetes
   D. Chlamydias

3. View Figure 10.1. Which two genera are evolutionarily most closely related?
   A. *Escherichia* and *Pseudomonas*
   B. *Helicobacter* and *Nitrospira*
   C. *Helicobacter* and *Rickettsia*
   D. *Bacillus* and *Synechococcus*

5. Which Gram-positive bacteria endanger human health instead of providing benefits?
   A. *Listeria* bacteria, which grow in milk and cheese, even in the refrigerator
   B. *Lactobacillus* bacteria, which grow in milk during production of yogurt and cheese
   C. *Streptomyces* species, which produce antibiotics such as streptomycin
   D. *Staphylococcus epidermidis*, which is found in the human skin

7. Which human-associated Gram-negative bacteria grow only with extremely low oxygen?
   A. *Helicobacter pylori*, the cause of gastritis
   B. *Yersinia pestis*, the cause of plague
   C. *Salmonella enterica*, a cause of food poisoning
   D. *Bacteroides,* enteric bacteria that help digest our food but cause abscesses if they escape the intestine and enter surrounding tissues

8. Which of the following is unique to Gram-negative bacteria?
   A. Peptidoglycan
   B. Lipopolysaccharide
   C. Teichoic acid
   D. Flagella

10. Which organisms possess mutualistic spirochetes?
    A. Humans
    B. Lichens
    C. Termites
    D. Amebas

11. The agent of Lyme disease is transmitted by which vector?
    A. Ticks
    B. Termites
    C. Bats
    D. Amebas

13. Which organelle of eukaryotes descended from a common ancestor of cyanobacteria?
    A. Mitochondria
    B. Chloroplasts
    C. Golgi bodies
    D. Nucleus

15. Which life stage is absent from Chlamydias?
    A. Endospore
    B. Reticulate body
    C. Stage with netlike uncondensed DNA
    D. Elementary body

Key Terms

botulinum (p. 295)
endospore (p. 293)
hemolysis (p. 297)
infant botulism (p. 295)

isolate (p. 291)
lichen (p. 311)
lithotrophy (p. 307)
nitrogen fixation (p. 307)

polymicrobial (p. 290)
shelter species (p. 292)
spirillum (p. 311)
spirochete (p. 311)

strain (p. 291)
zoonosis (p. 292)

Review Questions

1. Which major phylum of bacteria shows a great variety of lifestyles, including disease, although all members have a similar, distinctive cell structure?

2. Which major phylum of bacteria all show the same kind of metabolism, yet vary widely in form and habitat?

Clinical Correlation Questions

1. If a patient has pneumonia, how might you discover whether the cause is a mycoplasma, a bacterium such as *Streptococcus*, or a virus?

2. A patient observes a round, red lesion that goes away after several weeks. If the lesion is caused by a spirochete, which of three diseases could it be? For each disease, what subsequent symptoms may occur?

Thought Questions: CHECK YOUR ANSWERS

Thought Question 10.1 What impact might global climate change have on bacterial diversity and pathogen evolution?

ANSWER: Climate change certainly influences bacterial diversity. As some environments warm and others cool, the microbes suited to those conditions will shift as well. As new opportunities arise for gene exchange, new competitions will dictate a need to alter biochemical pathways, and some species will disappear while others will rise. The exact impact of climate change on pathogen evolution is unknown, but the natural engineering and reengineering of genes in environmental species responding to climate change could produce the raw material needed to evolve new pathogens.

Thought Question 10.2 How do you think *Mycobacterium tuberculosis* manages to grow despite its thick envelope screening out most nutrients?

ANSWER: *M. tuberculosis* has porins in its envelope that enable nutrients to enter, albeit slowly. The bacterium grows slowly in a place where it has no competitors, namely, intracellularly within the macrophage.

Thought Question 10.3 The Gram-negative alphaproteobacterium *Rickettsia prowazekii* is an obligate intracellular pathogen that causes the disease typhus. It is transmitted by the body louse. What impact would an obligate intracellular lifestyle have on opportunities for horizontal gene transfer (transfer of genes between bacteria)?

ANSWER: Being limited to an intracellular lifestyle would severely limit opportunities for cross-species gene exchange. The likelihood of two bacterial species infecting the same host cell is small. However, the human pathogen *Legionella pneumophila* infects amebas that harbor several other symbiotic bacterial species that have donated DNA sequences to this organism. Whether a species such as *R. prowazekii* can incorporate DNA from bacteria engulfed and killed by its eukaryotic host is unknown.

Thought Question 10.4 For motility, what are the relative advantages of external flagella versus the flexible spiral cells of spirochetes?

ANSWER: External flagella, such as those of *E. coli*, propel the cell effectively in a dilute watery environment such as a pond. In water, the cell can freely swim and tumble to find a new direction. By contrast, in a viscous solution such as mucus, the cell cannot tumble. The twisting motion of the entire spirochete enables penetration of mucus and other viscous regions of a host organism.

11

Eukaryotic Microbes and Invertebrate Infectious Agents

[A Desert Sand Fly Parasite]

SCENARIO Mario, a pizza delivery worker in Indianapolis, age 29, had an open sore on his arm that would not heal. The sore developed a red depression in the middle, surrounded by a raised border. It was not painful, and Mario postponed seeking treatment because he was busy working two jobs and managing his family of three children.

SIGNS AND SYMPTOMS The sore began to heal, but after 2 months, additional sores formed on Mario's skin and within his mouth. He now experienced fever, swollen lymph nodes, and loss of weight.

TESTING AND DIAGNOSIS Mario finally went to a clinic and reported his symptoms to a physician assistant. The physician assistant asked whether he had traveled out of the country. Mario reported that 6 months before onset of symptoms, he had returned from a tour in Iraq as a sergeant in the U.S. Army. The physician assistant observed that the appearance of the sores was typical of cutaneous leishmaniasis. Microscopy and PCR confirmed the presence of the protozoan *Leishmania donovani,* a parasite commonly acquired by soldiers in Iraq, where it is transmitted by sand flies.

Sand Fly, the Vector That Transmits *Leishmania* Parasites

TREATMENT The physician assistant prescribed sodium stibogluconate (Pentostam), a mineral containing antimony, which kills the parasite. The sores healed, but Mario was told that he could never be sure the parasite was completely eradicated; symptoms might recur, even many years later. For this reason, all U.S. military personnel who have been deployed in Iraq and Afghanistan are restricted from future blood donations.

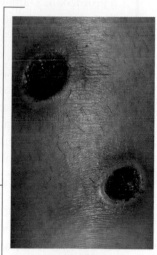

Skin Lesion Typical of Leishmaniasis

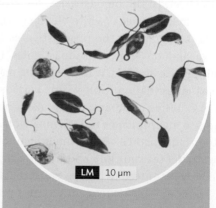

LM   10 μm

Leishmania, Single-Celled Parasites
Giemsa stain reveals the nucleus and whiplike flagellum.

CHAPTER OBJECTIVES

After reading this chapter, you will be able to:

- Define the major categories of eukaryotic microbes and invertebrate parasites.
- Describe the role of fungi in the environment and in human disease.
- Describe the role of protists in the environment and in human disease.
- Outline the infectious cycles of eukaryotic parasites.

The infectious agent Mario acquired, *Leishmania donovani*, is a protozoan, a single-celled eukaryotic parasite. *Leishmania* is commonly found in tropical and subtropical regions of South America and Asia. The protozoan (plural, protozoa) causes the disease leishmaniasis, which shows three different forms: cutaneous leishmaniasis, or skin infection; mucocutaneous leishmaniasis, in which infection invades the mucous membranes (mouth and nostrils) as well as the skin; and visceral leishmaniasis, or kala-azar, infection of internal organs, including the liver and spleen. The mucocutaneous and visceral forms are serious and can be fatal.

As discussed in Chapter 5, eukaryotic cells are very different from bacteria. The eukaryotic cell of *Leishmania* possesses a nucleus (see photo in chapter-opening case history) and a whiplike flagellum powered by ATP, unlike bacterial rotary flagella. Because the cell structure of eukaryotic microbes is so similar to that of human cells, there are far fewer antimicrobials against them than for bacteria, and the treatment agents, such as antimony-containing compounds, tend to have worse side effects on the host tissues.

Protozoan parasites such as *Leishmania* often have complex life cycles, sometimes involving multiple hosts (**Figure 11.1**). In *Leishmania*, the promastigote, or flagellated form of the protozoan, is carried by a sand fly (Figure 11.1, step 1), which injects it into the skin of the host. In the bloodstream, the promastigotes are phagocytosed by macrophages (white blood cells), which attempt to destroy them (step 2). But the promastigotes survive within the macrophage and develop into amastigotes (unflagellated forms) that multiply intracellularly (step 3). The macrophage bursts, releasing amastigotes that multiply within various host tissues, causing sores (step 4). When another sand fly takes a blood meal (step 5), it ingests infected macrophages full of amastigotes. The amastigotes develop into promastigotes within the midgut of the sand fly (step 6). The promastigotes multiply and migrate to the proboscis, where they can be transmitted to the next host that the sand fly bites.

Besides *Leishmania*, other protozoa cause diseases such as malaria and trypanosomiasis (sleeping sickness). Protozoan diseases cause tremendous mortality and morbidity among humans and animals. And another group of eukaryotes, the fungi, cause opportunistic infections among persons with weakened immune systems. At the same time, microbial eukaryotes, including protozoa, fungi, and algae, play key roles in all ecosystems. For example, protozoa and fungi participate in microbial catabolism during wastewater treatment, helping to break down organic wastes to carbon dioxide.

In this chapter, we will explore the diversity of microbial eukaryotes, including fungi, algae, and protozoa. We will describe the branching of multicellular animals and plants from the eukaryotic family tree, and we will explore the form and function of eukaryotes as pathogens and as partners in our ecosystem. In addition, we will introduce certain differentiated animals (helminths and arthropods) that act as infectious agents.

11.1
Eukaryotes: An Overview

SECTION OBJECTIVES

- Explain the processes of asexual (vegetative) reproduction and sexual reproduction.
- Distinguish the key traits of fungi, algae, amebas, alveolates, and trypanosomes.
- Distinguish between microbial and invertebrate parasites.

What is the structure of eukaryotic cells? As presented in Chapter 5, all eukaryotic cells, whether human or microbial, possess a nucleus and other key membranous organelles, and nearly all possess mitochondria. Membranous organelles enable some microbial eukaryotes to grow to a million times larger than a bacterial cell. Yet others, particularly marine algae, have been downsized by evolution to less than 2 μm, comparable in size to *Escherichia coli*. Even so, such tiny eukaryotes such as the alga *Ostreococcus tauri* (**Figure 11.2**) possess all the fundamental eukaryotic organelles, including nucleus, mitochondria, Golgi, and chloroplast.

Figure 11.1   Stages in the Life Cycle of *Leishmania*

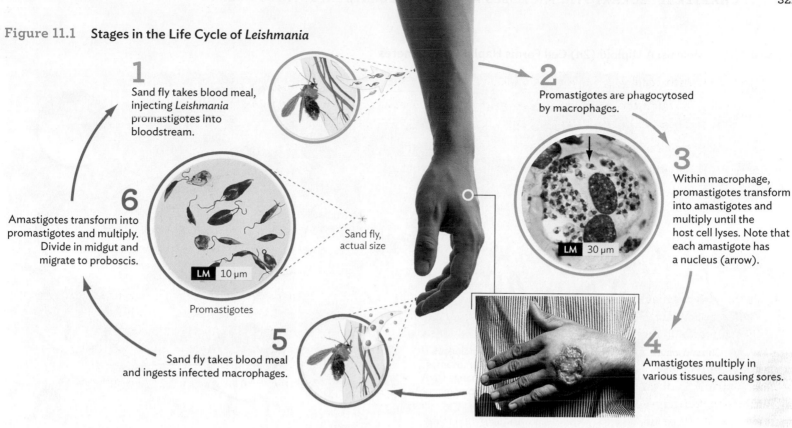

1 Sand fly takes blood meal, injecting *Leishmania* promastigotes into bloodstream.

2 Promastigotes are phagocytosed by macrophages.

3 Within macrophage, promastigotes transform into amastigotes and multiply until the host cell lyses. Note that each amastigote has a nucleus (arrow).

LM 30 μm

4 Amastigotes multiply in various tissues, causing sores.

5 Sand fly takes blood meal and ingests infected macrophages.

6 Amastigotes transform into promastigotes and multiply. Divide in midgut and migrate to proboscis.

LM 10 μm

Promastigotes

Sand fly, actual size

Reproduction of Eukaryotic Cells

All eukaryotes have linear chromosomes that must divide by mitosis. The "ends" of linear chromosomes require special means of replication not required for most bacteria with their circular chromosomes. DNA replication occurs during a portion of interphase, the active state of the cell. Once DNA replication is complete, the cell halts most of its enzymatic activities and prepares for mitosis. **Mitosis** ensures that each daughter cell receives a full set of daughter chromosomes. In most cases, mitosis involves four phases:

- **Prophase.** Paired chromosomes (postreplication) condense into short rods. In most species, the nuclear membrane dissolves.

- **Metaphase.** The chromosomes arrange along a plane across the equator of the cell.

- **Anaphase.** The pairs separate, pulled by spindle fibers composed of microtubules that contract. The separated chromosomes are pulled toward opposite poles of the cell.

- **Telophase.** The chromosomes decondense (become long and thin again), and a nuclear membrane forms around each set. The cell completes division of its cytoplasm, forming two daughter cells.

Many eukaryotic microbes, such as yeasts, can proliferate indefinitely by mitosis, a process called **asexual reproduction** or **vegetative reproduction**. But most eukaryotes, single or multicellular, also have the option of **sexual reproduction**. Sexual reproduction requires the reassortment of genetic material from different

Figure 11.2   Even a Tiny Eukaryote Has Organelles

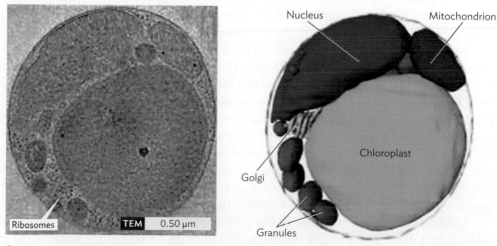

Nucleus
Mitochondrion
Chloroplast
Golgi
Granules
Ribosomes   TEM   0.50 μm

Ostreococcus tauri visualized by cryo-EM (**A**) and in 3D by tomography, colorized (**B**).

Figure 11.3   Meiosis: A Diploid (2n) Cell Forms Haploid (n) Gametes

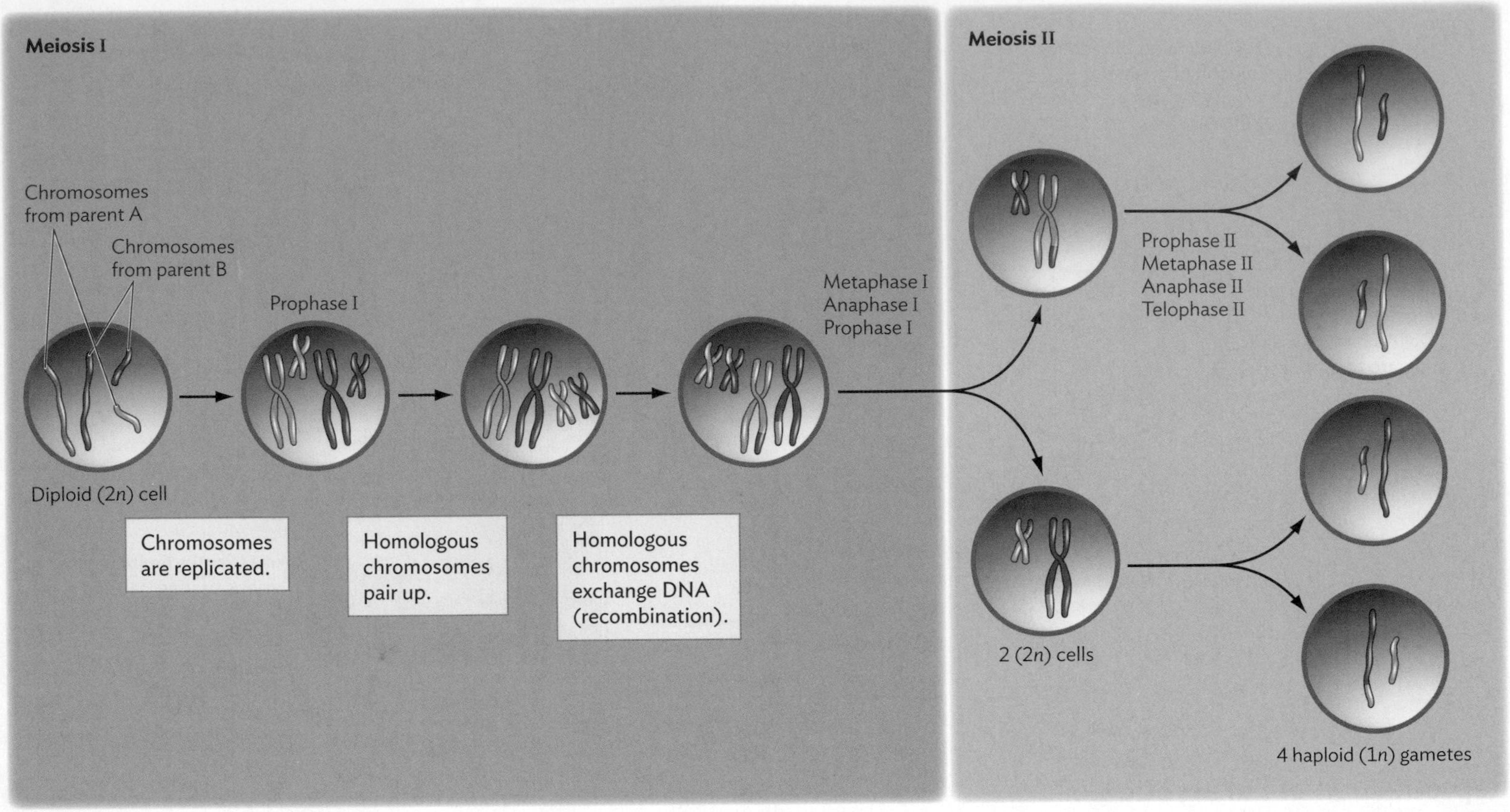

A. The chromosomes replicate, and then the homologous pairs line up and exchange DNA arms.

B. Two rounds of cell division yield four haploid (n) gamete cells.

chromosomes (**Figure 11.3**). A sexual life cycle alternates between cells that are **diploid** (2n, containing two copies of each chromosome) and sex cells that are **haploid** (n, containing a single copy of each chromosome). Two of the haploid sex cells (called gametes) can join each other by fertilization to regenerate a diploid cell (called a **zygote**). The diploid thus possesses two homologs of each chromosome—that is, two versions of the same chromosome from two different parents.

The process of gamete formation requires a special modification of mitotic cell division called **meiosis** (Figure 11.3). Meiosis includes two cell divisions, meiosis I and meiosis II. Like mitosis, meiosis I must be preceded by replication of all chromosomes (during interphase). Thus, a diploid (2n) cell becomes temporarily 4n. Unlike mitosis, prophase I requires that each replicated pair line up with its homolog—the homologous chromosome inherited from the other parent. Now the aligned homologs exchange portions of their DNA. This genetic exchange reassorts the gene versions from the two parents and increases genetic diversity in the next generation.

Meiosis I involves separation of the paired homologs during metaphase I and anaphase I. A short telophase occurs, in which each daughter cell now has 2n DNA helices, but as 1n pairs, each from one parent. The chromosome pairs now separate during meiosis II, which includes prophase II, metaphase II, anaphase II, and telophase II. The result is four haploid (n) cells, as seen in Figure 11.3B. For many microbes, the haploid forms may undergo asexual (vegetative) reproduction. At some point, the haploid forms may develop into specialized cells that reunite through fertilization, restoring a diploid form (the zygote).

What is the consequence of alternation between haploid (n) and diploid (2n) forms? Consider a protozoan parasite that grows within a human host. The haploid form of the parasite requires fewer resources and generates fewer varieties of coat proteins that might activate the host's immune system. But the diploid form generates novel combinations of genes that may provide an advantage when the environment changes or when the parasite enters a new host. Thus, most eukaryotic microbes maintain the option of a sexual cycle that alternates between haploid and diploid. In some cases, one form exists only briefly; for example, the malarial parasite proliferates mainly as a haploid but undergoes a brief cycle of fertilization and meiosis within the insect vector.

Diversity of Microbial Eukaryotes

Microbial eukaryotes are classified traditionally as fungi, protozoa, and algae. For most of human history, life was understood in terms of multicellular eukaryotes. There were animals (creatures that move to obtain food) and plants (rooted organisms that grow in sunlight). **Fungi**, which lack photosynthesis, were nonetheless considered a form of plant because they grow on the soil or other substrate. Thus, **mycology**, the study of fungi, was often included with botany, the study of plants. Later, microscopists came to recognize microscopic forms of fungi such as hyphae (filaments of cells) and yeasts (single cells); thus, fungi were studied also by microbiologists. Surprisingly, however, fungi and microsporidians show close genetic relatedness to multicellular animals.

Other microscopic life-forms, such as amebas and paramecia, are motile and appear more like microscopic animals. These animal-like organisms were called **protozoa**, meaning "first animals." Microscopic life-forms containing green chloroplasts were called **algae** (singular, alga) and were thought of as primitive plants. But some "protozoa" turned out to have chloroplasts; and some algae turned out to be motile with flagella. Today, the algae and protozoa are collectively termed **protists**, although the term "protozoa" remains in use for medical parasites.

The microscopic forms of fungi, protozoa, and algae are all considered "microbial eukaryotes." So how do these microbes relate to multicellular animals and plants? Plants show close genetic relationship with the "primary algae," green algae that evolved from a common chloroplast-bearing ancestor. Animals, however, are most closely related to fungi. The relatedness of animals to fungi came as a surprise because animals were thought more similar to protists, which seem to resemble animals in their motility. DNA sequence analysis shows that several different groups of protists are more different from each other—and from animals—than animals are from fungi.

The major kinds of microbial eukaryotes are summarized in Table 11.1. Although all these groups include microbial forms, they also include species that can be observed by the unaided eye (such as giant amebas) as well as multicellular forms (such as mushrooms and kelps). Thus, the term "microbe" serves as a working description but does not define a strict category of life.

Several major groups of microbial eukaryotes can be distinguished:

- **Fungi** grow in chains called hyphae (singular, hypha) or as single-celled yeasts. They have absorptive nutrition, digesting many complex plant polymers. Most are nonmotile.

- **Microsporidians** are spore-forming organisms closely related to fungi.

- **Algae** are protists containing chloroplasts that conduct photosynthesis, related to those of plants. Some algae are single-celled, whereas others grow as filaments or sheets.

- **Amebas** are single-celled protists of highly variable shape that form pseudopods, locomotor extensions of cytoplasm enclosed by the cell membrane.

- **Alveolates** are protists, mostly single-celled, with a complex multilayered outer covering called the cortex. Most alveolates have whiplike motility organelles called flagella (long, microtubular organelles, single or in doublets) or cilia (multiple short organelles).

- **Trypanosomes** and **metamonads** are flagellated protists with complex parasitic life cycles involving developmental stages within hosts.

These groups are considered "true" microbes, although they include large macroscopic forms, such as mushrooms (fungal fruiting bodies) and sheets of kelp (algae). Certain other multicellular parasites are not considered microbes, although they may be too small to see, such as mites and worms. These invertebrate animals have fully differentiated organ systems, sometimes comparable in complexity to those of vertebrates. The dynamics of transmission and infection of invertebrate parasites parallel those of microbial pathogens. Thus, these animal parasites are often covered by health professionals under the category of "eukaryotic microbiology." In particular:

- **Helminths** are multicellular worms, including the nematodes (roundworms), cestodes (tapeworms), and trematodes (flukes).

- **Arthropods** include insects such as fleas and lice, as well as noninsects such as mites.

Worms and arthropods are a major source of morbidity and mortality worldwide. A quarter of the world's population are parasitized by worms, including at least 10% of Americans. Their presence causes nutritional impairment and developmental delays. The horror of infection by worms and arthropods inspired the classic science fiction film *Alien*, which depicts an intelligent extraterrestrial parasite.

SECTION SUMMARY

- **Microbial eukaryotes** may undergo asexual (vegetative) reproduction. Alternatively, they may undergo a sexual cycle involving meiosis and fertilization.
- **Fungi** form hyphae with absorptive nutrition.
- **Algae** are protists that contain chloroplasts and conduct photosynthesis.
- **Amebas** are single-celled protists with pseudopod motility.
- **Alveolates** are single-celled protists with whiplike flagella or cilia.
- **Trypanosomes** are flagellated single-celled protists with complex parasitic cycles.
- **Invertebrate parasites** include helminths and arthropods.

Table 11.1
Eukaryotic Microbes and Multicellular Infectious Agents

Major Category	Example	Major Category	Example
Fungi and microsporidians Filaments or yeasts that usually lack motility; tough cell walls made of polysaccharides; heterotrophs by absorptive nutrition; include decomposers and parasites	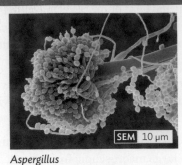 *Aspergillus*	**Algae** Chloroplast-containing protists that conduct photosynthesis	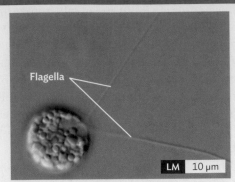 *Chlamydomonas*
Alveolates Protists with flagella (flagellates) or cilia (ciliates); free-living predators and host-dependent parasites; include ciliates, dinoflagellates, and apicomplexans	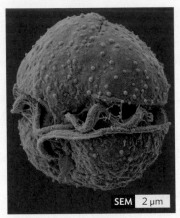 *Pfiesteria* (dinoflagellate)	**Helminths** Multicellular worms, including nematodes (roundworms), cestodes (tapeworms), and trematodes (flukes); include parasites as well as free-living members of soil ecosystems	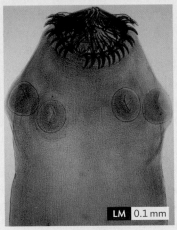 *Taenia solium* (scolex)
Amebas Heterotrophic protists with amorphous shape; motile with pseudopods	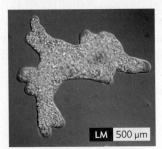 *Chaos carolinensis*	**Arthropods** Multicellular insects and related microscopic organisms; include parasites and free-living members of soil ecosystems	 Body louse
Trypanosomes and metamonads Parasites with complex life cycles	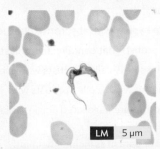 *Trypanosoma cruzi*		

11.2
Fungi and Microsporidians

SECTION OBJECTIVES
- Describe the structure of the fungal mycelium.
- Explain how fungi impact the environment.
- Describe how different types of fungi cause disease.

Fungi play a key role in all ecosystems. They recycle wood and leaves, including substances such as lignin that animals cannot digest. They decompose animal bodies—all of us someday will feed the fungi. They produce antibiotics such as penicillin and contribute to food products such as beer and cheese. Underground, fungal filaments called **mycorrhizae** extend the root systems of trees, forming a nutritional "Internet" that interconnects plant communities. Mycorrhizae could have inspired the fictional underground tree network shown in the 2009 film *Avatar*.

At the same time, some fungi are pathogens that threaten the lives of humans, animals, and plants. Inhalation of fungal spores can lead to aspergillus and **histoplasmosis**, conditions particularly dangerous for immunocompromised patients. Other fungi cause skin infections such as ringworm and athlete's foot (medical term, *tinea*). The fungi most commonly found in humans or in our local environment are presented in **Table 11.2**.

Link Section 20.4 describes several common fungal lung infections, including the endemic mycoses such as **histoplasmosis**, blastomycosis, and coccidioidomycosis.

CASE HISTORY 11.1

Fungus Growing in the Lung

Frank, a 45-year-old man, visited the hospital with respiratory distress and a high fever. Frank had diabetes mellitus and had smoked for 25 years. His white blood cell count was 24,000/µl, and his C-reactive protein was 24 mg/ml, both signs that his immune system was fighting an infection. His chest X-ray showed an abnormal shadow, and his chest computed tomography (CT scan) showed thickened bronchial walls (**Figure 11.4A**). He was coughing up black sputum. The physician prescribed an antibiotic, cefotiam hydrochloride, but Frank's condition worsened. By the 10th day, he was producing 100 ml of sputum a day, and the CT scan of his lung showed multiple lesions (clouded regions) (**Figure 11.4B**). Since antibacterial antibiotics had failed, Frank was tested for fungi. A culture of his sputum and PCR amplification of fungal DNA revealed the fungus *Aspergillus fumigatus* (**Figure 11.4C**). Frank was treated with the antifungal agents itraconazole, meropenem trihydrate, and amphotericin B, but he died on the 25th day.

Table 11.2
Fungi and Microsporidians

Phylum and Genus	Habitat or Disease
Ascomycetes	Form asci, pods of eight spores
Aspergillus	Plant saprophyte; produces aflatoxin; infects lungs of immunocompromised patients
Blastomyces	Blastomycosis; opportunistic infections of the skin
Candida	Yeast infections; thrush; opportunistic
Geomyces	White-nose syndrome in bats
Histoplasma	Histoplasmosis, lung disease
Magnaporthe	Rice blast, the most important disease of rice farming
Penicillium	Grows on bread and fruit; original source of penicillin
Saccharomyces	Ferments beer and bread dough
Stachybotrys	Black mold; contaminates dwellings
Trichophyton	Ringworm and athlete's foot skin infections
Basidiomycetes	Mushrooms and mycorrhizae
Lycoperdon	Edible mushrooms
Amanita	Extremely poisonous; produces alpha-amanitin toxin
Chytridiomycetes	Motile zoospores with single flagellum
Batrachochytrium	Frog epizootic in South America
Cryptococcus	Opportunistic infection of patients with AIDS
Neocallimastix	Digests plant material within bovine rumen
Zygomycetes	Saprophytes, insect pathogens, and mycorrhizae
Mucor	Soil mold; opportunistic infections
Rhizopus	Bread mold; food production (tempeh)
Microsporidia	Related to fungi
Encephalitozoon	Opportunistic infection of patients with AIDS

Aspergillus is a filamentous fungus of the ascomycete group (see Table 11.2). Filaments of *Aspergillus* species typically grow on soil or indoors as basement mold or on air conditioner coils. It forms spores that can reach significant concentrations in the air, where they are taken into the lungs and may cause allergic reactions. In patients with impaired immune function, *Aspergillus* spores may germinate and grow "fungus balls" that require surgical removal. The fungus is a common source of nosocomial (hospital acquired) infection. In Frank's case, however, the infection was community acquired. Frank's predisposing conditions (depressed immune function due to diabetes and smoking) had allowed the fungal infection to take hold.

Figure 11.4   Lungs Infected by *Aspergillus*

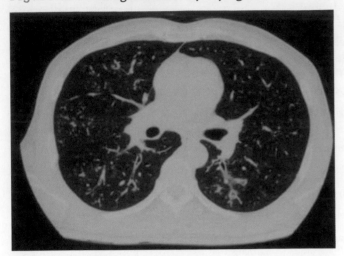

A. Chest computed tomogram, first day in hospital.

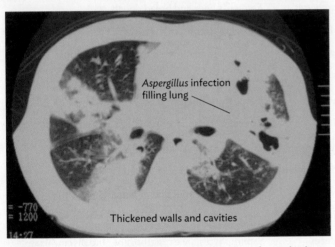

B. CT scan, tenth day in hospital. Large cavities have formed in lungs.

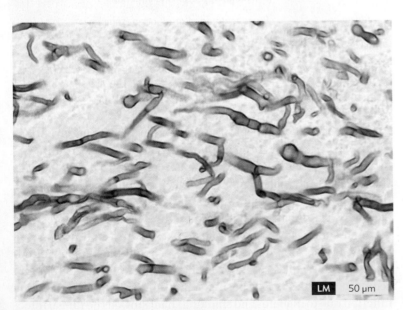

C. *Aspergillus* mycelium invading the lungs.

Filamentous Fungi

Fungi possess distinctive cell structures that enhance their growth. Filamentous fungi form branching tufts called a **mycelium** (Figure 11.5A). Their tough cell walls are composed of **chitin**, an acetylated amino-polysaccharide stronger than steel (Figure 11.5B). Its strength derives from multiple hydrogen bonds between fibers. Filamentous fungi grow by extending multinucleate cells called **hyphae** (singular, hypha) (Figure 11.5C). As a hypha extends, its nuclei divide mitotically without cell division. The mycelium may grow large enough to be seen as a fuzzy colony or puffball. As a fungal hypha expands, its chitinous cell wall enables it to penetrate softer materials, such as plant or animal cells. The biosynthesis of chitin provides a target for antifungal agents such as nikkomycin Z. Another drug target is the membrane lipid ergosterol, an analog of cholesterol not found in animals or plants. Ergosterol biosynthesis is inhibited by antifungal agents called triazoles.

Figure 11.5   Fungi Grow Hyphae with Cell Walls of Chitin

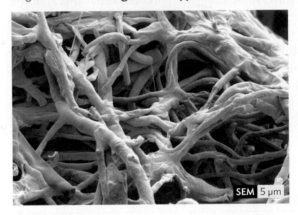

A. A mycelium: *Penicillium* fungal hyphae form the skin of Brie cheese.

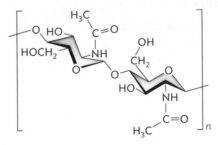

B. Chitin consists of polymers of *N*-acetylglucosamine.

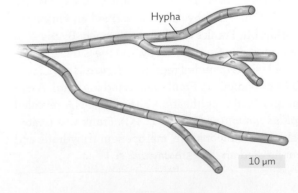

C. Fungal hyphae extend and form branches, generating a mycelium.

Most fungal forms are nonmotile (with the exception of chytridiomycete zoospores). Fungi cannot ingest particulate food, as do protists, because their cell walls cannot part and re-form like the flexible pellicle of amebas and ciliates (discussed next in Section 11.3). Instead, fungi secrete digestive enzymes and then absorb the broken-down molecules from their environment.

Filamentous fungi can expand at great length by asexual division (mitosis). Eventually, however, they run out of nutrients, and the cells undergo meiosis to form gametes. The gametes develop into spores for dispersal. An example of a spore-forming life cycle is that of ascomycete fungi such as *Aspergillus* (**Figure 11.6**). In *Aspergillus*, the tips of the diploid mycelium undergo meiosis and develop into haploid **ascospores** within a pod called an ascus. The ascospores disperse until they germinate in a region of fresh resources. Each ascus undergoes cell divisions to form a haploid mycelium, which is male or female. The male and female structures fuse, followed by migration of all the male nuclei into the female structure. The paired nuclei then undergo several rounds of mitotic division while migrating into the growing mycelium. In the mycelial tips, the paired nuclei finally fuse (becoming 2*n*) and re-form a diploid mycelium. Overall, the diploid-haploid life cycle enables a fungus to respond genetically to environmental change by reassorting its genes through meiosis and by recombining them through fertilization. Gene reassortment and recombination provide new genotypes, some of which may increase survival in the changed environment.

Some fungi pack their asci (or other spore-producing forms) into a larger structure called a **fruiting body**. Ascomycetes such as *Aspergillus* and *Penicillium* mold form relatively small fruiting bodies called conidiophores for airborne spore dispersal. Conidiospore-forming ascomycetes are the major type of mold associated with dampness in human dwellings; they caused massive damage to homes flooded in the wake of Hurricane Katrina (**inSight**).

The fruiting bodies of other ascomycete fungi can be impressively large and edible as prized foods. Morels (*Morchella hortensis*; see Figure 11.6) and truffles (*Tuber aestivum*) form large mushroom-like fruiting bodies whose ascospores are dispersed by animals attracted by their delicious flavor. The famous truffles grow underground, where human collectors traditionally use muzzled pigs to detect and unearth them.

Single-Celled Yeasts

Despite the advantages of multicellular hyphae, some fungi are single-celled, known as **yeasts**. Yeasts have the advantage of rapid growth and dispersal in aqueous environments. Many different single-celled yeasts evolved independently in different fungal taxa. Baker's yeast, *Saccharomyces cerevisiae*, is used to leaven bread and to brew wine and beer. Other yeasts, such as *Candida albicans* (**Figure 11.7**), are important members of human vaginal biota. But *C. albicans* is also an opportunistic pathogen, causing infections such as thrush in the mouth of patients with AIDS. Other yeasts that are opportunistic pathogens, commonly infecting AIDS patients, include *Pneumocystis carinii*, a yeast-form ascomycete, and *Cryptococcus neoformans*, a yeast-form basidiomycete.

Figure 11.6   The Life Cycle of an Ascomycete

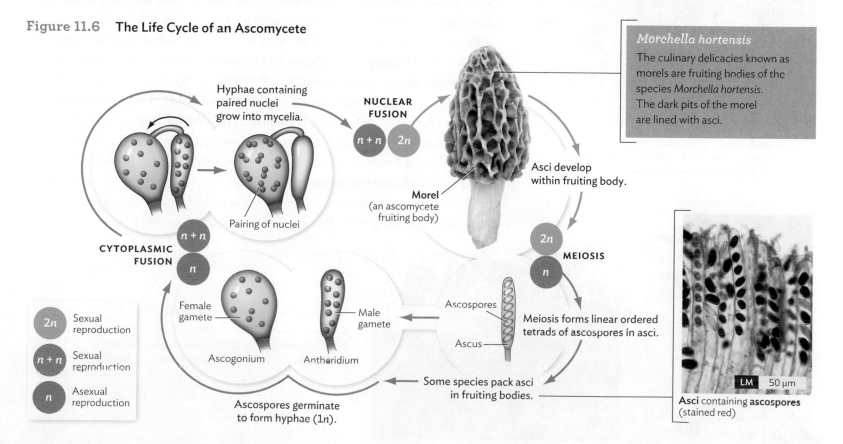

Morchella hortensis
The culinary delicacies known as morels are fruiting bodies of the species *Morchella hortensis*. The dark pits of the morel are lined with asci.

Hyphae containing paired nuclei grow into mycelia.

NUCLEAR FUSION

n + n   *2n*

Asci develop within fruiting body.

Pairing of nuclei

Morel (an ascomycete fruiting body)

CYTOPLASMIC FUSION

n + n

n

2n

n

MEIOSIS

Female gamete

Ascospores

Male gamete

Ascogonium   Antheridium

Meiosis forms linear ordered tetrads of ascospores in asci.

Ascus

2*n*   Sexual reproduction

n + n   Sexual reproduction

n   Asexual reproduction

Ascospores germinate to form hyphae (1*n*).

Some species pack asci in fruiting bodies.

LM   50 µm

Asci containing **ascospores** (stained red)

inSight

Mold after Hurricane Katrina

Figure 1   **Mold Growing above the Flood Line**

An undergraduate student volunteer indicates mold colonies growing above the flood line on a kitchen wall in a New Orleans home, 6 months after flooding.

In the wake of Hurricane Katrina in 2005, much of the city of New Orleans was flooded. Entire neighborhoods were submerged for days or weeks. The flooding of homes full of drywall and organic materials made ideal conditions for the growth of "mold," which generally consists of airborne ascomycete fungi. Mold grew not only on the materials submerged but also on the surfaces above, exposed to water-saturated air. **Figure 1** shows how mold grew up to 3 feet above the flood line (the highest level submerged) within the kitchen of a flooded home. Unfortunately, most homeowner insurance policies covered only damage "up to the flood line."

Mold growing on damp surfaces releases spores into the air. Inhaled mold spores cause respiratory problems, even at relatively low levels (such as 2,000 spores per cubic meter). In New Orleans, after the waters receded, the mold counts in the air inside the homes reached as high as 650,000 spores per cubic meter. Such spore levels cause asthma attacks and hypersensitivity pneumonitis, a condition in which fever rises and the lungs fail. All remediation workers were advised to wear an N95 respirator mask. The Natural Resources Defense Council (NRDC) found that even after flooded homes underwent full removal of contaminated items, including furniture, carpets, and drywall down to the studs, airborne spore counts still reached 70,000.

The most common types of mold found by the NRDC researchers were ascomycetes such as *Cladosporium* (**Figure 2**). *Cladosporium* species form black or dark green colonies of conidia that readily break off and become airborne. Other ascomycetes found at high levels included *Aspergillus* and *Penicillium* species. These organisms occur at low levels even in clean air. Their spores cause allergy attacks and infect immunocompromised individuals. Some ascomycete molds produce mycotoxins, fungal toxins that cause long-term health problems. In addition, some flooded homes showed significant levels of *Stachybotrys* species, known as "black mold," a possible cause of neurological problems and immune suppression. Once such molds establish growth, their eradication is a major challenge.

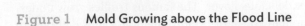

Figure 2   *Cladosporium* **Mold**

A. Conidiophores (branched hyphae) producing conidia (asexual spores) of *Cladosporium cladosporioides* (left).
B. *Cladosporium* mold growing on media in a petri dish (right).

Figure 11.7
***Candida albicans*: A Unicellular Fungus (Yeast)**

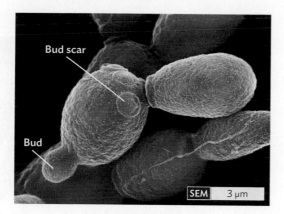

Bud scar

Bud

SEM   3 µm

Some yeasts are asexual, whereas others alternate between diploid (2*n*) and haploid (*n*) forms (**Figure 11.8**). Both haploid and diploid forms are single cells.

The haploid (*n*) cells undergo several generations of mitosis and budding. This asexual reproduction is referred to as vegetative growth. At some point, however, various stress conditions, such as nutrient limitation, may lead the haploid cells to develop into gametes. The gametes are of two mating types. Cells of the two mating types may fertilize each other by fusing together and combining their nuclei to make a diploid (2*n*) zygote. The diploid cell—like the haploid cell—may undergo vegetative budding and reproduction. Under stress, however, a 2*n* cell may undergo meiosis, regenerating

haploid cells. The haploid progeny possess chromosomes reassorted and recombined from those of their "parental" gametes.

Some fungi, such as *C. albicans*, can grow either as yeast or as mycelia, depending on the environment; these are known as "dimorphic" fungi. Another dimorphic fungal pathogen is *Blastomyces dermatitidis*, the cause of blastomycosis, a type of pneumonia. *B. dermatitidis* forms a mycelium in culture and in soil environments, but it grows as a yeast within the infected lung.

SECTION SUMMARY

- **Fungi have cell walls of chitin.** Chitin biosynthesis is a target for antifungal agents.

- **Fungi are heterotrophs that absorb nutrients.** Most fungi in nature absorb nutrients from decaying organisms, although some are pathogens that infect animal or plant hosts.

- **Filamentous fungi grow as multinucleate hyphae.** Hyphae are exceptionally strong filaments that can penetrate many materials.

- **Ascomycete fungi form asci that disperse spores.** Spores of fungi such as *Aspergillus* can spread disease or cause allergies.

- **Unicellular fungi are called yeasts.** The yeast *Saccharomyces cerevisiae* is used to make fermented drinks and to leaven bread. The yeast *Candida albicans* is a part of normal human biota, but certain strains can cause opportunistic infections.

Thought Question 11.1   For a fungus, what are the advantages of growing a mycelium? What are the advantages of growing as a single-celled yeast?

Figure 11.8
Yeast Alternation in Generations

Life cycle of *S. cerevisiae*. Haploid (1*n*) cells reproduce by budding (asexual). For sexual reproduction, a pair of opposite mating types fuse, forming a diploid (2*n*) cell. The 2*n* cell reproduces by budding (asexual) until it undergoes meiosis (sexual). Meiosis generates an ascus full of ascospores, which germinate as haploid cells.

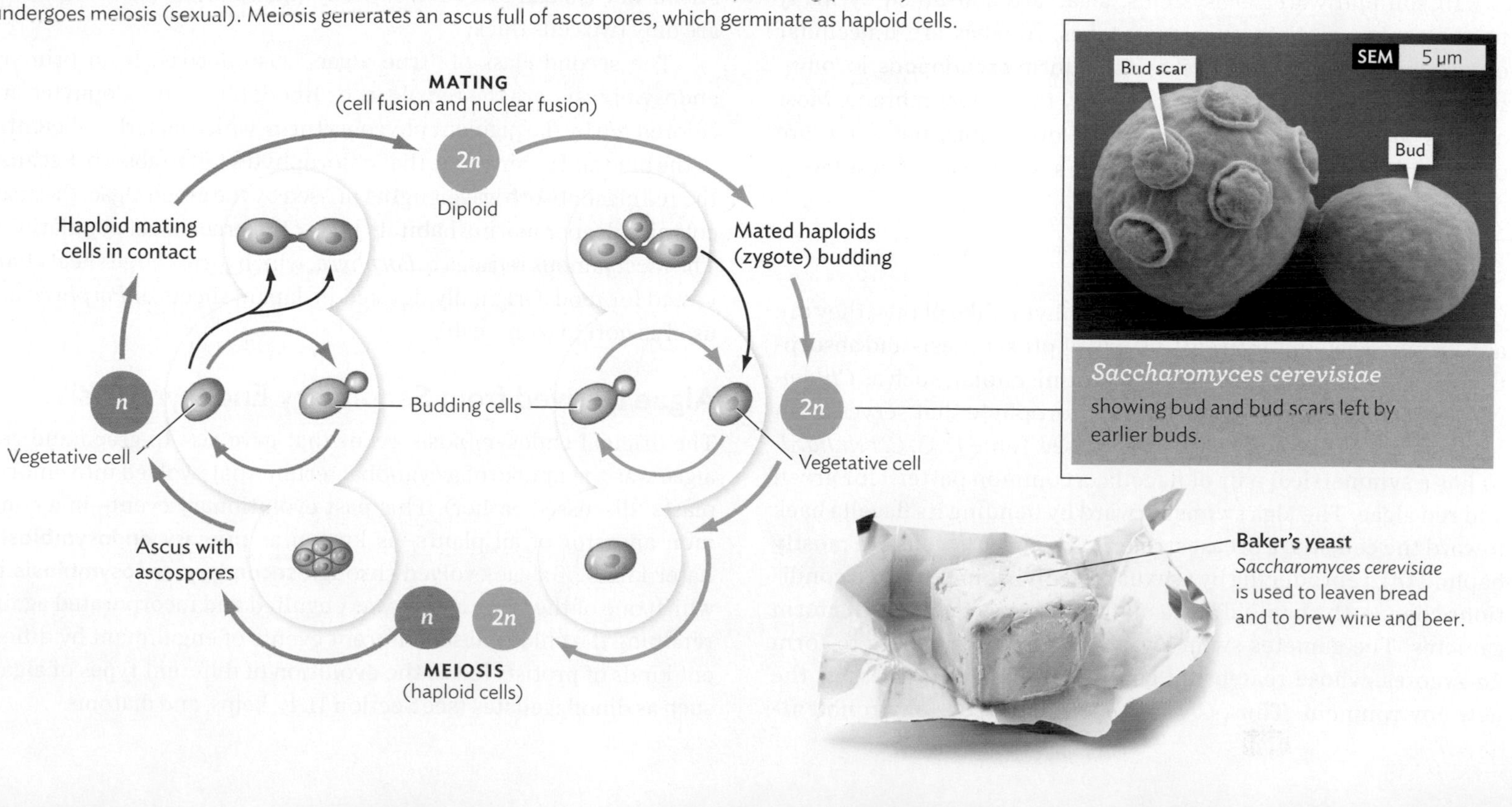

MATING
(cell fusion and nuclear fusion)

2*n*
Diploid

Haploid mating cells in contact

Mated haploids (zygote) budding

n

Budding cells

2*n*

Vegetative cell

Vegetative cell

Ascus with ascospores

n   2*n*

MEIOSIS
(haploid cells)

SEM   5 µm

Bud scar

Bud

Saccharomyces cerevisiae
showing bud and bud scars left by earlier buds.

Baker's yeast
Saccharomyces cerevisiae is used to leaven bread and to brew wine and beer.

11.3
Algae and Amebas Are Environmental Protists

SECTION OBJECTIVES

- Explain the environmental significance of algae and their relatedness to plants.
- Describe the form and motility of amebas.
- Describe two diseases caused by amebas.

How do algae affect our health and environment? Many eukaryotic microbes flourish in soil and water habitats, where they enrich the food web as producers and consumers. Major photosynthetic producers are the algae. In aquatic and marine ecology, the algae, together with photosynthetic bacteria, are known as phytoplankton (discussed in Chapter 27). They fix most of Earth's carbon and produce most of the oxygen we breathe.

Algae are protists that conduct photosynthesis by using chloroplasts. Chloroplasts evolved from ingested cyanobacteria through the process of endosymbiosis. An ancient proto-alga engulfed a photosynthetic cyanobacterium, perhaps with the aim of digesting it; but instead of being digested, the cyanobacterium provided photosynthetic carbon and received nutrients and protection in return from its host cell. Eventually, the cyanobacteria evolved into algal chloroplasts, organelles with genomes and structure greatly diminished compared with the original cyanobacteria. Some ancient algae then evolved into multicellular plants.

In soil and water ecosystems, algae are consumed by many heterotrophic protists, such as **amebas**. Amebas are unicellular organisms of highly variable shape that form **pseudopods**, locomotor extensions of cytoplasm enclosed by the cell membrane. Most amebas are free-living environmental organisms, but some are opportunistic pathogens that can cause severe cases of dystentery or meningitis.

Green and Red Algae

Green algae make up the group Chlorophyta. Like plants, they are autotrophs, growing entirely through photosynthesis and absorption of minerals. Many green algae are unicellular, such as *Chlamydomonas reinhardtii*, a unicellular chlorophyte that serves as an important model system for genetics (see Table 11.1). *C. reinhardtii* has a symmetrical pair of flagella, a common pattern for green and red algae. The alga swims forward by bending its flagella back toward the cell, like a breast stroke. *C. reinhardtii* cells are mostly haploid (n), reproducing by asexual cell division. But when conditions change, the haploid algae mate to undergo meiosis and form gametes. The gametes swim away and fuse with others to form $2n$ zygotes, whose reassorted genes may improve fitness in the new environment. Thus, *C. reinhardtii* undergoes environment-

Figure 11.9   Green (Primary Algae)

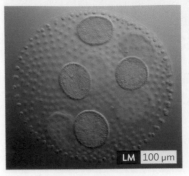

A. *Volvox* forms colonies of flagellated cells; the colony may grow to a size of 1–3 mm. Progeny colonies form within the larger sphere until the sphere bursts, releasing them.

B. *Cymopolia* forms calcified stalks of cells.

triggered alternation between diploid ($2n$) and haploid (n) forms. This alternation of diploid and haploid forms is similar to that of some fungal yeasts (see Section 11.2).

Other chlorophytes develop more complex multicellular forms. *Volvox* forms hollow spherical colonies, like a geodesic dome, composed of cells similar to those of *Chlamydomonas* (Figure 11.9A). *Spirogyra* is a filamentous pond dweller known for its spiral chloroplasts. In marine water, *Cymopolia* forms relatively large calcified stalks of cells (Figure 11.9B). Still other marine algae grow in undulating sheets, such as *Ulva*, the "sea lettuce" familiar to bathers. Sheets of *Ulva* can extend over many square meters, although they are only two cells thick.

The second class of "true algae," also derived from primary endosymbiosis, are the red algae, or Rhodophyta. Rhodophytes are colored red by the pigment phycoerythrin, which absorbs efficiently in the blue and green range that chlorophytes fail to absorb. Because the red algae absorb wavelengths missed by the green algae, they can colonize deeper marine habitats below the green algal populations. The most famous red alga is *Porphyra*, which forms large sheets harvested for food. Originally devised in Japan, sheets of *Porphyra* are used as nori to wrap sushi.

Algae Derived from Secondary Endosymbiosis

The original endosymbiosis event that gave rise to green and red algae was the uptake of a cyanobacterium that evolved into chloroplasts (discussed earlier). This past evolutionary event—in a common ancestor of all plants—is known as **primary endosymbiosis**. Later kinds of algae evolved through **secondary endosymbiosis** in which one of the "true algae" was engulfed and incorporated again, retaining the chloroplast. Different events of engulfment by different kinds of protists led to the evolution of different types of algae, such as dinoflagellates (see Section 11.4), kelps, and diatoms.

Figure 11.10   Kelp Forest

Sargassum natans forms the basis of the Sargasso Sea. The brown alga forms stalks with leaflike blades and round gas bladders to keep the alga afloat.

Algae derived from secondary endosymbiosis show "mixotrophic" nutrition, involving both photosynthesis and heterotrophy. For example, dinoflagellates are predators of smaller protists. Diatoms derived from a different secondary endosymbiotic event. Other kinds of secondary endosymbiont algae are structured more like plants. Among the most famous are the "brown algae," or kelps. Sargassum weed (Figure 11.10) forms the basis of the Sargasso Sea, a region of the Atlantic Ocean bounded by currents that maintain a relatively intact mass of algae. The sargassum weed forms stalks with leaflike blades and round gas bladders that keep the alga afloat, supporting a unique community of fish and invertebrates at the surface of the open ocean.

Amebas

Amebas are voracious predators, consuming other protists and bacteria. Some amebas can reach several millimeters in size—a single cell that is large enough to eat small invertebrates. There are two major groups of amebas: the lobed amebas, with large, bulky pseudopods (Figure 11.11A), and the filamentous amebas, with thin, needlelike pseudopods (Figure 11.11B). Some amebas with needlelike pseudopods, such as the **foraminiferans** (forams), form inorganic shells made of calcium carbonate (limestone). Fossil foraminiferan shells are common in rock formations derived from ancient seas. Foraminiferan shells formed the white cliffs of Dover in Britain and the stone used to build the Egyptian pyramids.

The lobed amebas (the more familiar kind) have pseudopods that extend lobes of cytoplasm through cytoplasmic streaming. Cytoplasmic streaming is the flow of nutrients and other components through the cytoplasm, mediated by actin proteins. Most lobed amebas are free-living in soil or water, where they prey on bacteria and small arthropods. They play an important role in the food web of soil and aquatic ecosystems. Some amebas are slime molds, in which individual amebas converge to form a fruiting body. Certain ameba species, however, cause diseases such as dysentery and meningitis.

Intestinal Amebiasis

CASE HISTORY 11.2

Spring Break Surprise

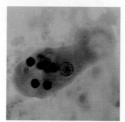

Brianna, a 22-year-old college student, had returned recently from spring break at the beach in Acapulco, Mexico. She presented at the college clinic complaining of severe abdominal cramps and fever. She also had nausea and bloody diarrhea with mucus. The nurse practitioner collected stool samples and sent them to the laboratory for analysis. Cultures were negative for bacterial pathogens. A trichrome stain was performed, revealing amebas of the trophozoite (active feeding) stage (Figure 11.12A). The stained amebas contained ingested red blood cells. The appearance of the amebas was characteristic of *Entamoeba histolytica*, the cause of amebic dysentery (bloody diarrhea), or intestinal amebiasis. Brianna was treated with paromomycin (an aminoglycoside that inhibits protein synthesis) for 7 days, and her symptoms resolved.

Figure 11.11   Amebas

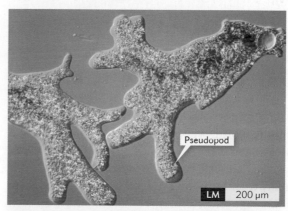

A. *Amoeba proteus* extends its lobed pseudopods (colorized LM).

Pseudopod

LM   200 µm

B. A foraminiferan extends needlelike pseudopods through its shell of calcium carbonate.

LM   500 µm

Figure 11.12
Entamoeba histolytica

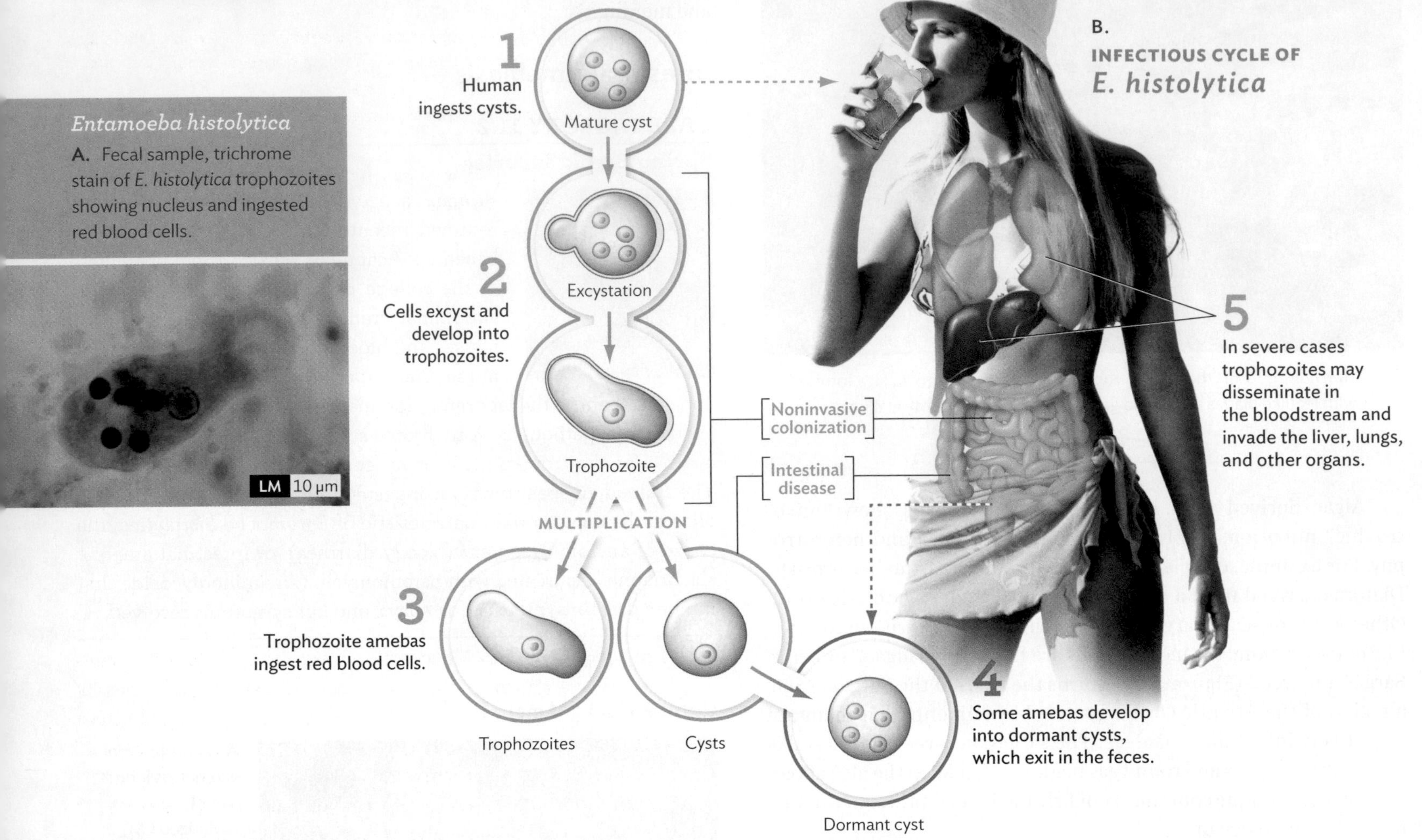

Entamoeba histolytica

A. Fecal sample, trichrome stain of *E. histolytica* trophozoites showing nucleus and ingested red blood cells.

1 Human ingests cysts.

Mature cyst

2 Cells excyst and develop into trophozoites.

Excystation

Trophozoite

MULTIPLICATION

3 Trophozoite amebas ingest red blood cells.

Trophozoites

Cysts

4 Some amebas develop into dormant cysts, which exit in the feces.

Dormant cyst

Noninvasive colonization

Intestinal disease

B.
INFECTIOUS CYCLE OF
E. histolytica

5 In severe cases trophozoites may disseminate in the bloodstream and invade the liver, lungs, and other organs.

Entamoeba histolytica is a serious cause of intestinal illness in much of the developing world, where it can lead to malnutrition and death. The amebas are usually ingested in the form of **cysts**, dormant cells encased in a tough coating (Figure 11.12). Cysts can remain viable for many weeks in contaminated water supplies, such as that which Brianna apparently drank. The ingested cysts reach the intestine, where they "excyst" and develop into trophozoites, amebas with pseudopods that phagocytose prey. In the intestinal lumen and in the bloodstream, the amebas ingest red blood cells. As they run short of food, the amebas develop into dormant cysts, which exit in the feces, a "fecal/oral" route of transmission. In severe cases, however, the trophozoites may invade the liver, lungs, and other organs.

The symptoms of dysentery can be caused by a number of different viruses and bacterial pathogens, which require different kinds of treatment; thus, the pathogen must be identified correctly. Common bacterial pathogens such as *Shigella* may be detected by immunological tests. To diagnose the presence of amebas and other intestinal parasites requires microscopy, as their cell size and shape

are distinctive. A procedure called Wheatley's trichrome stain reveals the parasites among human cells and fecal material (see Figure 11.12A). The Wheatley's trichrome stain involves iodine/alcohol treatment followed by a triple-stain solution. The cytoplasm of protozoa stains green or blue, whereas the nuclear membrane and ingested red blood cells stain bright red. Cyst forms of the ameba stain bluish purple. Nevertheless, it must be noted that the closely related species *Entamoeba dispar* is indistinguishable from *E. histolytica* under the microscope and is found ten times more often, so it is possible to miss the real pathogen.

Amebic dysentery is best avoided by careful choice of fluid consumption. When traveling in regions with limited sanitation, it is important to drink only filtered or bottled water. In most cases, an otherwise healthy host can eliminate amebic infection within 2 weeks. Repeated infections, however, may cause malnutrition. Sometimes the infection leads to chronic carrier status; and in a few cases, infection becomes systemic, involving the liver, lungs, and other organs. Systemic infection by amebas is difficult to treat and can be fatal. So, where good health care is available, cases of intestinal amebiasis are treated early with anti-amebic agents. The drug paromomycin specifically targets the small-subunit ribosomal RNA of microbial eukaryotes. An alternative drug, metronidazole, targets anaerobic protozoa and bacteria.

Other Diseases Involving Amebas

Several other diseases can involve different species of amebas. For example, primary amebic meningoencephalitis, an infection of the brain, can be caused by the amebas *Naegleria fowleri* and *Balamuthia mandrillaris*. The amebas may enter the body through the nose during swimming in a contaminated pool. Free-living soil amebas such as *Acanthamoeba* may contaminate contact lens cleaning solutions, causing keratitis, an infection of the cornea (Figure 11.13). Wearers of reusable contact lenses have a keratitis infection rate of two per thousand. Finally, free-living amebas may play host to bacterial pathogens such as *Legionella pneumophila*, the cause of legionellosis, an often fatal form of pneumonia (discussed in Section 20.3). Amebas carrying intracellular *L. pneumophila* often contaminate water supplies, air ducts, and even dental irrigation solutions. The host amebas enable the pathogen's persistence and transmission to human hosts.

SECTION SUMMARY

- **Algae are protists with chloroplasts that conduct photosynthesis.** They are important producers of biomass in all ecosystems. Green algae (Chlorophyta) and red algae (Rhodophyta) are autotrophic photosynthesizers, closely related to green plants.
- **Brown algae derived from secondary endosymbiosis.** Examples include diatoms and kelps.
- **Amebas move using pseudopods.** Lobed amebas have large, lobular pseudopods, whereas filamentous amebas have needle-shaped pseudopods.

Figure 11.13   Amebic Keratitis

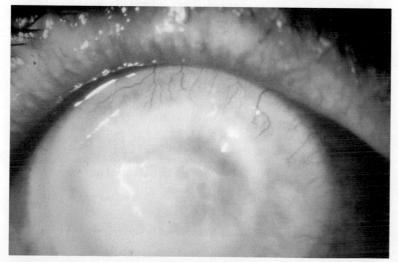

Acanthamoeba infects the keratin layers of the cornea, which becomes inflamed and clouded. Amebic keratitis commonly results from improperly cleaned contact lenses.

- **Free-living amebas are predators.** Predatory amebas play an important role in the food web of soil and aquatic communities. They may also harbor pathogenic bacteria.
- ***Entamoeba histolytica* causes intestinal amebiasis.** The amebas develop into cysts that are expelled in feces and remain dormant for long periods in water.
- ***Naegleria fowleri* causes meningoencephalitis.**

Thought Question 11.2   How do you think amebas evolved to carry symbiotic bacteria? How do these ameba-carried bacteria endanger humans?

11.4
Alveolates: Predators, Phototrophs, and Parasites

SECTION OBJECTIVES

- Describe the distinctive traits of ciliates, dinoflagellates, and apicomplexans.
- Outline the complex life cycle of *Plasmodium falciparum* and *P. vivax*, the parasites that cause malaria.

What other protists have interesting ways to catch prey? Alveolates include voracious ciliated predators that swallow prey nearly their own size, dinoflagellates that nourish coral reefs, and human parasites that cause malaria. In contrast to the amorphous shape of amebas, the alveolate cell form is highly structured. The term "alveolate" refers to the flattened vacuoles called alveoli within the outer cortex (Figure 11.14). Some alveoli contain plates of stiff material, such as protein, polysaccharide, or mineral. Besides alveoli, most alveolate protists possess other kinds of cortical organelles, such as extrusomes, organelles that extrude defensive enzymes or toxins; bands of microtubules for reinforcement; and whiplike cilia or flagella. Many alveolates engulf prey, but some species also possess chloroplasts for photosynthesis.

Alveolates comprise three major groups (Table 11.3):

- **Ciliates** possess large numbers of cilia, short projections containing microtubules, whose whiplike action is driven by ATP. They capture prey in the oral groove.
- **Dinoflagellates** possess two whiplike flagella, one of which is wrapped around the cell body. They both phagocytose prey and conduct photosynthesis.
- **Apicomplexans** lack flagella or cilia but have evolved specialized structures for parasitism. They are obligate parasites with complex life cycles.

Some parasitic protists exhibit life cycles involving multiple hosts. For example, the malaria parasite (or plasmodium) alternates between the human host and the Anopheles mosquito. Generally one type of animal constitutes the **definitive host**, the host

Figure 11.14   The Alveolate Cortex

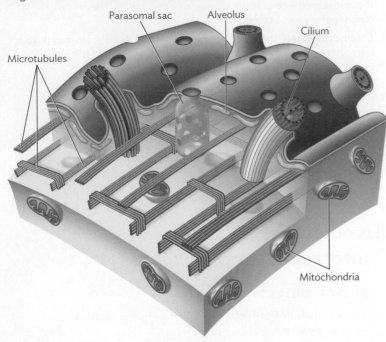

Beneath the pellicle (outermost cell layer) of a ciliated protist lie flattened sacs of fluid called alveoli. The cilia, composed of [9(2)+2] microtubules, are rooted in a network of lateral microtubules. Parasomal sacs take up nutrients and form endocytic vesicles.

endocytosis by specialized pores in their cortex, called parasomal sacs (see Figure 11.14). Paramecia in turn are consumed by larger ciliates, such as *Didinium* (**Figure 11.15B**).

Without a rigid cell wall, how do ciliates maintain osmotic balance? Ciliates generally grow in aqueous solutions with salt and organic concentrations that are lower than those of their cytoplasm. Thus, water concentration outside is higher, and water tends to run into the cell. The excess water is expelled by means of a contractile vacuole, a vacuole that withdraws water from the cytoplasm or contracts to expel it. Contractile vacuoles are widespread among protists and algae, but their mode of action has been studied mostly in paramecia. As water enters the cell from a low-solute environment, the contractile vacuole takes up water through an elaborate network of intracellular channels and then contracts to expel the water through a pore, thus preventing osmotic shock and lysis (see Section 5.6).

Ciliates, like other eukaryotes, have nuclei containing chromosomes that undergo division by mitosis and generate gametes by meiosis. Unique to ciliates, however, the cell maintains two types of nuclei: the micronucleus, whose genes maintain genetic fidelity for sexual reproduction; and the macronucleus, containing multiple gene copies that express gene products throughout the large cell. During asexual reproduction, both micronuclear and macronuclear chromosomes undergo replication.

where the parasite matures and may reproduce sexually; in the case of malaria, the definitive host is the mosquito. The alternate host (the human, in this case) is called the **intermediate host**. The intermediate host supports asexual proliferation of the parasite and is required for transmission to the next insect host.

Ciliates

Ciliates are heterotrophic protists that consume algae and smaller protists and are in turn consumed by amebas. A medically important ciliate is *Balantidium coli*, a cause of dysentery. A ciliate consists of a single cell covered with large numbers of stubble-like **cilia**. Cilia are short cytoplasmic projections containing microtubules. Powered by ATP, the cilia beat in coordinated waves that maximize the efficiency of motility.

Cilia serve two functions:

- **Cell propulsion.** Coordinated waves of beating cilia, usually covering the cell surface, propel the cell forward.

- **Food acquisition.** By generating water currents into the mouth of the cell, a ring of cilia around the mouth brings food into the cell.

One of the best-studied ciliates is the paramecium (**Figure 11.15A**). Paramecia feed on bacteria by trapping them in the oral groove and then consuming them through enzymes within digestive vacuoles. They can also take up smaller particles through

Table 11.3
Alveolate Protozoa

Phylum and Genus	Habitat or Disease
Ciliates	Ciliated protists
Paramecium	Pond predator of algae and bacteria
Stentor	Stalked ciliate predator; cilia draw prey into mouth; wastewater treatment community
Dinoflagellates	Flagellated protists; often grow in the ocean as marine plankton (floating organisms); secondary endosymbiotic algae; may also act as predators
Gymnodinium (Karenia)	Grows in algal blooms such as "red tide"; produces saxitoxins that contaminate shellfish
Noctiluca	Marine blooms generate bioluminescence
Symbiodinium	Coral endosymbiont that conducts photosynthesis; essential for reef formation
Apicomplexans	Parasites with apical complex; lack flagella or cilia; undergo complex life cycles
Babesia	Babesiosis, blood infection spread by ticks
Cryptosporidium	Intestinal infection, opportunistic; waterborne
Plasmodium	Malaria; carried by mosquito vector
Toxoplasma	Toxoplasmosis; transmitted to humans by cats

Figure 11.15 Ciliate Predators

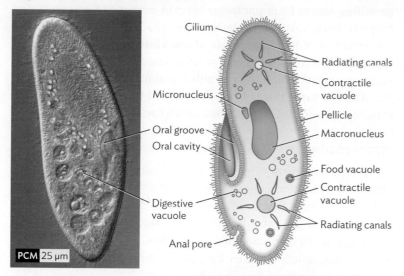

A. A paramecium has digestive vacuoles and an oral groove to ingest prey (colorized phase contrast).

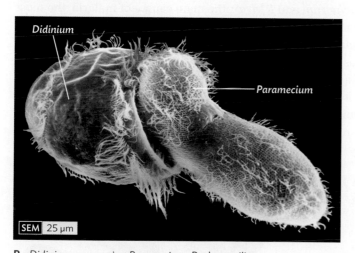

B. *Didinium* consuming *Paramecium*. Both are ciliates.

For sexual reproduction, a ciliate undergoes **conjugation** (Figure 11.16). During conjugation, two paramecia of opposite mating types form a cytoplasmic bridge. Within each conjugating cell, the 2n micronucleus undergoes meiosis to form four n gametic nuclei. Three out of four of the gametic nuclei disintegrate, as does the macronucleus. The haploid micronuclei undergo mitosis, forming two daughter micronuclei. Daughter micronuclei from each cell are exchanged across the cytoplasmic bridge and then fuse with their respective counterparts, restoring a 2n micronucleus to each cell. The bridged cells come apart, and each micronucleus generates a new macronucleus. The new macronucleus now expresses gene products.

Note Distinguish the conjugation of ciliates (with exchange of micronuclei) from the very different process of bacterial conjugation (one-way transfer of DNA from a donor bacterium to a recipient).

Figure 11.16 Conjugation

A. Two paramecia of opposite mating types form a cytoplasmic bridge. The 2n micronucleus of each cell undergoes meiosis. Each macronucleus, as well as three out of four meiotic products, disintegrates. The haploid micronuclei undergo mitosis, forming two daughter micronuclei. Daughter nuclei from each cell are exchanged across the cytoplasmic bridge and then fuse with their respective counterparts, restoring 2n micronuclei. The cells come apart, and each micronucleus generates a new macronucleus.

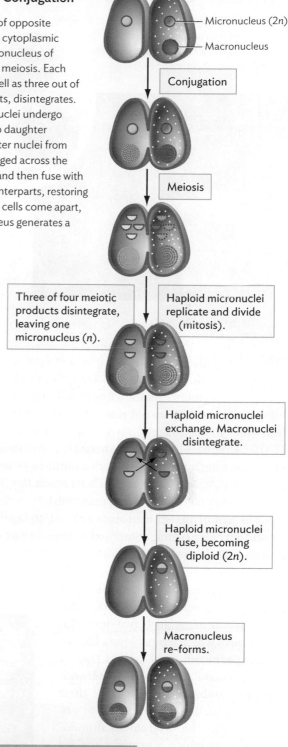

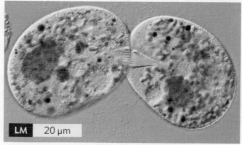

B. Pair of ciliates conjugating. Colored objects are food vacuoles and secretory granules.

Figure 11.17   Dinoflagellates

A. *Gymnodinium* dinoflagellates. Each cell has one flagellum wrapped around the cell, whereas the other flagellulm provides whiplike motion (colorized SEM).

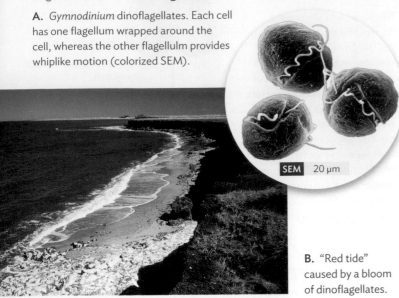

SEM   20 μm

B. "Red tide" caused by a bloom of dinoflagellates.

Dinoflagellates

The dinoflagellates are a major group of marine algae, essential to marine food chains (Figure 11.17A). Some dinoflagellates make carotenoid pigments that are bright red. Blooms of red dinoflagellates cause the famous red tide, which may have inspired the biblical story of the plague in which water turns to "blood" (Figure 11.17B). Dinoflagellates release toxins that can be absorbed by shellfish, poisoning human consumers months or years later. In aquatic and coastal regions, such as the Chesapeake Bay, *Pfiesteria* dinoflagellates cause algal blooms and poison fish.

The armor-plated appearance of a dinoflagellate results from its stiff alveolar plates, composed of proteins or calcified polysaccharides. The complex outer cortex includes various toxin-emitting extrusomes and endocytic pores, as well as a species-specific pattern of alveolar plates. Like ciliates, dinoflagellates are highly motile, but instead of numerous short cilia, dinoflagellates possess just two long flagella, one of which wraps along a crevice encircling the cell (see Figure 11.17A).

Dinoflagellates are called **secondary endosymbiont algae** because their evolution includes two events of endosymbiosis: the original algal endosymbiosis in which a protist engulfed a cyanobacterium, the precursor of chloroplasts, and a second endosymbiosis in which another protist engulfed the chloroplast-bearing alga. Dinoflagellates retain the predatory ability of the second engulfing protist as well as the chloroplast derived from the ancestral alga.

Certain dinoflagellates inhabit other organisms as mutualists, providing sugars from photosynthesis in exchange for a protected habitat. Their hosts include sponges, sea anemones, and, most important, reef-building corals. Coral endosymbionts, known as zooxanthellae, are vital to reef growth. The zooxanthellae are temperature sensitive, and their health is endangered by global warming. Rising temperatures in the ocean lead to coral bleaching (the expulsion of zooxanthellae), after which the coral dies.

Apicomplexans

Apicomplexans form a major group of parasites of humans and other animals. Their name derives from the "apical complex," a specialized structure that facilitates entry of the parasite into a host cell. Like the ciliates and dinoflagellates, apicomplexans possess an elaborate cortex composed of alveoli, pores, and microtubules. But as parasites, apicomplexans have undergone extensive reductive evolution, losing their flagella and cilia.

MALARIA   The best-known apicomplexan parasites are *Plasmodium falciparum* and *P. vivax*, the main causes of malaria, the most important parasitic disease of humans worldwide (Figure 11.18A). *P. falciparum* and *P. vivax* are carried by mosquitoes, which transmit the parasite to humans when the insect's proboscis penetrates the skin. The disease is endemic in areas inhabited by 40% of the world's population; it infects hundreds of millions of people and kills more than a million African children each year. As people travel, malaria increasingly shows up in the United States as well.

Figure 11.18
Malaria

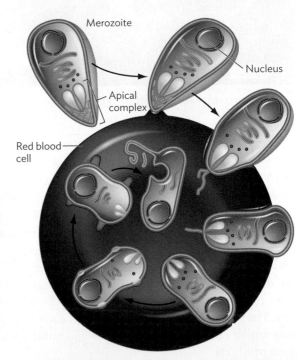

Merozoite

Nucleus

Apical complex

Red blood cell

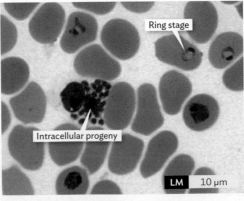

Ring stage

Intracellular progeny

LM   10 μm

A. Red blood cells infected with *Plasmodium falciparum*, an apicomplexan parasite that causes malaria (stained purple with a dye that interacts with DNA). One red blood cell can be seen bursting, unleashing parasites on surrounding cells.

B. Merozoite form of *P. falciparum* invades a red blood cell. The apical complex facilitates invasion and then dissolves as the merozoite transforms into an intracellular form.

The transmitted parasites invade the liver, where they develop into the "merozoite" form, which invades red blood cells (Figure 11.18B). The merozoite first contacts a red blood cell through interaction between its apical complex (the structure at the penetrating tip of the cell) and receptor proteins on the red blood cell. The apical complex contains a pair of secretory organelles that are capped by a ring of microtubules and that inject enzymes that aid entry of the parasite. The tip of the apical complex penetrates the host cell, enabling secretion of lipids and enzymes that facilitate invasion. Eventually, the entire merozoite enters the host cell, leaving no traces of the parasite on the host cell surface. Thus, the internalized parasite becomes invisible to the immune system until its progeny burst out. Some progeny are then picked up by mosquitoes, where they undergo a sexual life cycle and are transmitted to new hosts.

The malarial parasites acquire resistance rapidly and no longer respond to drugs such as quinine that nearly eliminated the disease half a century ago. The life cycle and molecular properties of *P. falciparum* have been studied extensively for clues to the development of new antibiotics and vaccines. The life cycle of *P. falciparium* and the etiology of malaria are discussed in Chapter 21.

TOXOPLASMOSIS An apicomplexan parasite common in the United States is *Toxoplasma gondii*, a parasite carried by cats and transmissible to humans. Approximately 30% of the world's population is infected with *T. gondii*. Fortunately, the infection is usually asymptomatic in humans because of our immune system, but the parasite can cross the placental barrier and harm a developing fetus. In rare cases, *T. gondii* can enter the brain. Some epidemiological evidence suggests a link between *T. gondii* infection and some cases of schizophrenia. The life cycle of *T. gondii* and the etiology of toxoplasmosis are discussed in Chapter 21.

BABESIOSIS An emerging human pathogen in the United States is the blood parasite *Babesia microti*. *B. microti* is spread by deer ticks, the same agents that spread Lyme disease—yet another unfortunate risk of a walk in the northeastern woods. *B. microti*, however, may also be spread by blood contact and transfusions. The parasite multiplies within red blood cells, similar to malaria. The symptoms of *B. microti* infection can mimic those of Lyme disease (caused by the tick-borne bacterium *Borrelia burgdorferi*) and are often misdiagnosed. Babesiosis is further discussed in Section 21.5.

WATERBORNE APICOMPLEXANS Several apicomplexans are emerging waterborne parasites. *Cryptosporidium parvum* infects the intestine of immunocompromised patients, causing watery diarrhea. An obligate intracellular pathogen, the organism generates a cyst form transmitted by the fecal-oral route. *Cryptosporidium* resists chlorination; it can cause major outbreaks, such as the one in 1993 when more than 400,000 people were sickened in Milwaukee, Wisconsin. Another cause of diarrhea is *Cyclospora*, also a danger for immunocompromised patients. *Cyclospora* causes occasional outbreaks from fecally contaminated produce.

SECTION SUMMARY

- **Alveolates have a complex outer cortex with alveoli.** Alveoli are flattened vacuoles that may contain plates of stiff material.
- **Ciliates are alveolates with cilia.** Cilia provide motility and generate currents that draw prey into the oral groove.
- **Dinoflagellates have a stiff plated cortex and a pair of flagella.** Dinoflagellates contain chloroplasts derived from secondary endosymbiosis, but they also conduct predation.
- **Apicomplexans are obligate parasites lacking cilia or flagella.** Important parasites cause malaria (*Plasmodium falciparum*) and toxoplasmosis (*Toxoplasma gondii*). Waterborne opportunistic pathogens include *Cryptosporidium parvum* and *Cyclospora*.

Thought Question 11.3 What are the relative advantages and limitations of flagellar motility, ciliate motility, and pseudopodial motility of amebas? Which kind of natural habitat or body environment might favor one kind of motility over another?

11.5
Trypanosomes and Metamonads

SECTION OBJECTIVES
- Describe the traits of trypanosomes.
- Describe the traits of giardias and trichomonads.

What other motile parasites cause human disease? A group of protists called the euglenids have whiplike flagella and chloroplasts arising from secondary endosymbiosis, similar to the dinoflagellates but more distantly related. Some of the euglenids, however, lost their chloroplasts and evolved into obligate parasites called **trypanosomes**. Trypanosomes consist of an elongated cell with a single flagellum. The trypanosome has a unique organelle called the "kinetoplast," consisting of a specialized mitochondrion that provides energy for the flagellum. Trypanosomes cause devastating diseases in humans and animals, such as sleeping sickness (*Trypanosoma brucei*), Chagas' disease (*Trypanosoma cruzi*), and leishmaniasis (*Leishmania donovani*; see chapter-opening case history).

Trypanosomes

Trypanosomiasis, also known as African sleeping sickness, is a major killer of humans and livestock; in some sub-Saharan countries, trypanosomiasis is the second biggest killer after AIDS. The parasite, *Trypanosoma brucei* (Figure 11.19A), is carried by the tsetse fly. *T. brucei* multiplies in the bloodstream of the host animal, causing repeated cycles of proliferation and fever that cause increasing drowsiness, ultimately leading to death if untreated. This trypanosome is known for its extraordinary degree of **antigenic variation**, the ability to generate numerous different versions of its coat protein and thereby evade antibodies of the host immune system. The

Figure 11.19
Trypanosomes Cause Sleeping Sickness

Trypanosoma brucei

A. Cells of the stumpy form and slender form.

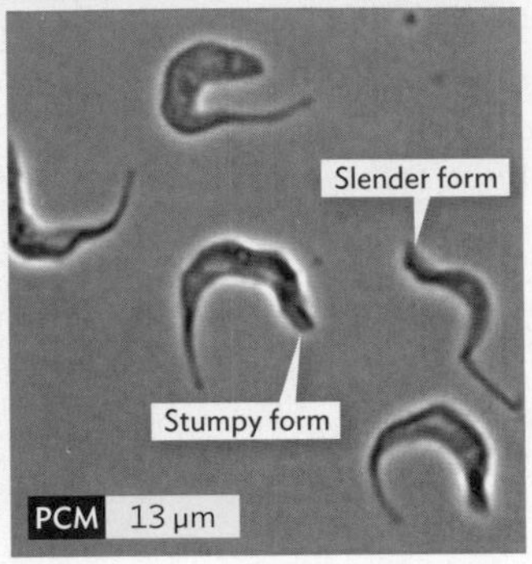

Slender form

Stumpy form

PCM   13 μm

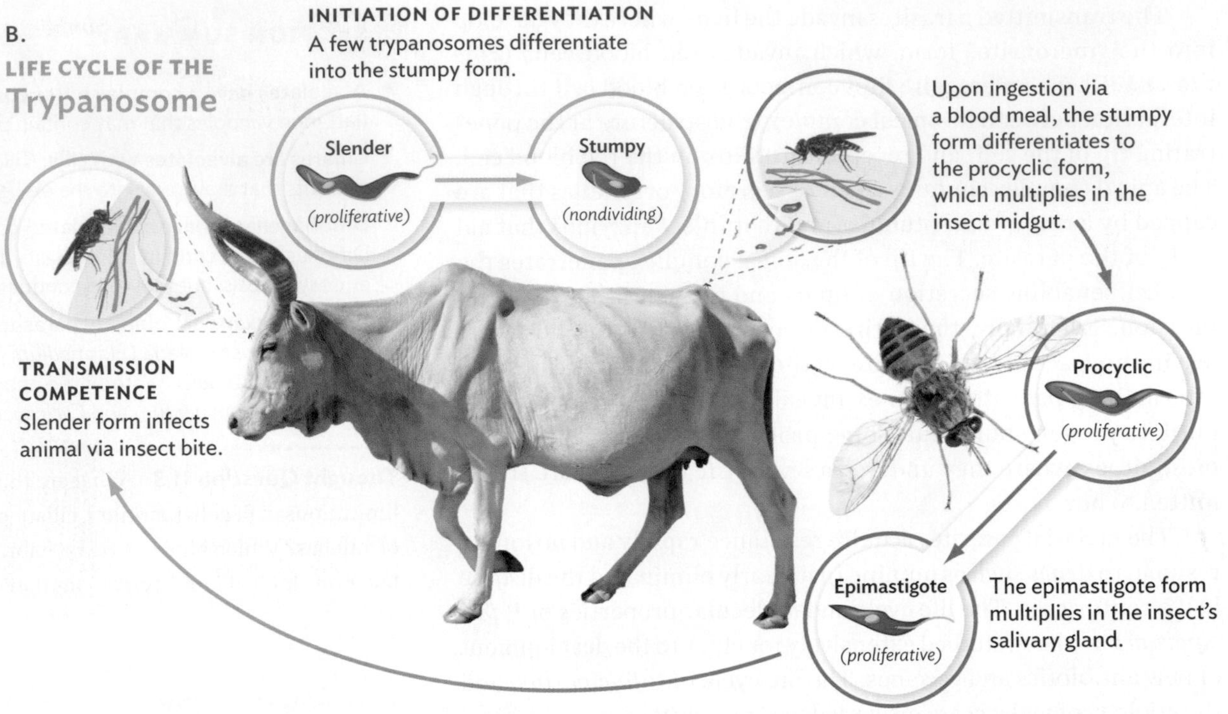

B.
LIFE CYCLE OF THE Trypanosome

INITIATION OF DIFFERENTIATION
A few trypanosomes differentiate into the stumpy form.

Slender
(proliferative)

Stumpy
(nondividing)

Upon ingestion via a blood meal, the stumpy form differentiates to the procyclic form, which multiplies in the insect midgut.

TRANSMISSION COMPETENCE
Slender form infects animal via insect bite.

Procyclic
(proliferative)

Epimastigote
(proliferative)

The epimastigote form multiplies in the insect's salivary gland.

trypanosome's genome includes 200 different active versions of its variant surface glycoprotein (VSG), the **antigen** inducing the immune response, as well as 1,600 different "silent" versions that can recombine with "active" VSG to make further variations. In effect, the trypanosome overwhelms the host immune system by continually generating new antigenic forms until the host repertoire of antibodies is exhausted.

Link Section 16.3 describes how a small segment of an **antigen** called an antigenic determinant is capable of eliciting an immune response. An immune response to a microbe is a composite of responses to different epitopes by thousands of individual B cells, the cells responsible for producing antibodies.

Diagnosis of trypanosomiasis requires microscopic detection of the parasite in centrifuged blood. The parasite invades many body systems, causing cardiac and kidney dysfunction. In a late stage, the trypanosomes cross the blood-brain barrier, invading the central nervous system. Neurological impairment leads to disruption of the sleep cycle—hence the term "sleeping sickness"—ultimately leading to death.

To infect its human host, the trypanosome needs to interconvert among several different forms of its life cycle. Several of these forms are shown in **Figure 11.19B**. For example, the "slender form" proliferates in the bloodstream of a human or other mammal (definitive host). But the "stumpy form" does not proliferate in the mammal; it needs to be ingested by the tsetse fly (intermediate host), where it lives as the asexual stage. Only the stumpy form can survive the tsetse midgut to develop into the forms that can grow there (procyclic and epimastigote). The salivary gland can transmit

the parasite to the next host, where the slender form develops again. Trypanosomiasis is further discussed in Section 21.5.

The molecular basis of conversion among developmental forms of a parasite offers targets for new antimalarial agents. Discovering new drugs against trypanosomes and malarial parasites is urgently needed to decrease the global impact of these devastating diseases.

Metamonads

The metamonads are a group of flagellated parasites distantly related to other flagellates. A common example is *Giardia lamblia*, the cause of intestinal giardiasis (**Figure 11.20**). In the United States, *Giardia* is a frequent nemesis of infant day-care centers, where a significant proportion of caregivers are carriers. *Giardia* is also endemic in wildlife and contaminates aquatic streams crossed by bears and other animals. The parasite can contaminate community water supplies and has become endemic in some Russian cities. *Giardia* and other metamonads are noted for their anaerobic metabolism and their complete absence of mitochondria, lost through degenerative evolution; their metabolism requires the anaerobic intestinal environment of their hosts.

Another common metamonad is *Trichomonas vaginalis*, the cause of trichomoniasis, one of the most common sexually transmitted infections. In North America, the asymptomatic infection rate among young adults is estimated at 50%. *T. vaginalis* grows in the urethra and in the vagina. Men rarely report symptoms, but women may experience pain on urination. Untreated infections lead to complications during pregnancy. The infection is treated with the antiparasitic drug metronidazole.

SECTION SUMMARY

SECTION SUMMARY

- **Trypanosomes are obligate parasites with a single flagellum.** Trypanosomes undergo complex life cycles. They cause sleeping sickness (*Trypanosoma brucei*), Chagas' disease (*Trypanosoma cruzi*), and leishmaniasis (*Leishmania donovani*).

- **Metamonads are parasites or endosymbionts with highly degenerate cells.** The intestinal parasite *Giardia lamblia* causes giardiasis. The urogenital parasite *Trichomonas vaginalis* causes trichomoniasis.

Thought Question 11.4 Do you think most eukaryotic parasites have genomes that are larger or smaller than those of free-living microbial eukaryotes? Why might parasite genomes be larger—and why might they be smaller?

11.6
Invertebrate Parasites: Helminths and Arthropods

SECTION OBJECTIVES

- Describe the parasitic cycles of worms, including roundworms, flukes, and tapeworms.
- Describe the parasitic cycles of arthropods, including ticks, mites, and bloodsucking insects.

What kind of parasites are actually multicellular animals? Infectious parasites include many invertebrate animals, primarily worms and arthropods. Although they are not, strictly speaking, microbes, some require a microscope to visualize. Their infectious cycles resemble those of microbial pathogens.

Invertebrate parasites, like the protozoan parasites discussed earlier, commonly cycle through more than one host, such as human and insect, or human and domestic animal. Typically there is a definitive host (where sexual reproduction occurs) and an intermediate host (asexual reproduction only). In some cases, the human or animal may be an **incidental host**—that is, a host in which the parasite may grow and cause serious morbidity but cannot be transmitted to other hosts.

Helminths

Parasitic worms are called **helminths**. The helminth category is not strictly monophyletic; that is, no one genetic branch contains only parasites, since most worms are nonparasitic. Nonetheless, worms that are parasitic cause a tremendous global burden of morbidity and mortality.

Worms are multicellular animals that possess fully differentiated organs. Although they are invertebrates, their body plans have surprising complexity, in some cases comparable to that of vertebrates. The free-living soil nematode *Caenorhabditis elegans* serves

Figure 11.20 ***Giardia lamblia*, a Metamonad Intestinal Parasite**

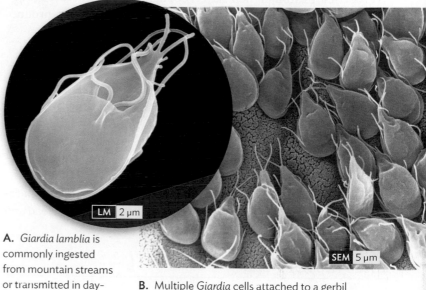

A. *Giardia lamblia* is commonly ingested from mountain streams or transmitted in day-care centers.

B. Multiple *Giardia* cells attached to a gerbil intestinal lining.

as a major model system for the study of human development (**Figure 11.21**). The entire genome of *C. elegans* has been mapped; and many of the worm's genes have homologs in the human genome involved in vital processes related to neural development, aging, and cancer. Helminths include three categories:

- **Nematodes (roundworms)** are cylindrical (hence the term "roundworm") with a digestive tube that ends in an anus. Their reproductive forms include hermaphrodites and males.

- **Trematodes (flukes)** are oval-shaped flatworms with a digestive tube that ends in a cecum (no outlet). They are hermaphrodites.

Figure 11.21 ***Caenorhabditis elegans***

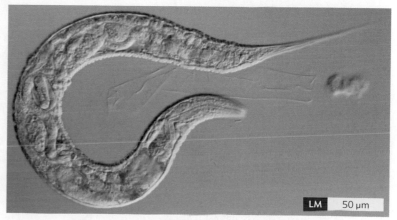

This free-living nematode is a famous model system for studying human development, cancer, and aging. Nematodes commonly grow in soil.

- **Cestodes (tapeworms)** are parasitic flatworms, absorbing nutrients through their skin. The sucker head structure grows a long "tail" of segments containing male and female reproductive structures.

Nematodes

Nematodes are diverse animals, with species adapted to all kinds of ecosystems; about half the known species are free-living. A new nematode species—dubbed "the worm from hell"—was identified in a gold mine more than 3 kilometers below Earth's surface, where the temperature reaches 48°C (118°F). These worms graze on biofilms of bacteria indigenous to the mine.

Unfortunately, many species of nematodes are human parasites. Nematode parasites include some of the most widespread and debilitating human infections worldwide.

CASE HISTORY 11.3 ⊚

Dotty's Itch

Dotty was the 2-year-old daughter of a Texas petroleum executive. The family included four children and three dogs. Dotty played outdoors with the other children and ranch animals. She developed a habit of scratching herself on her bottom. Her nanny tried to discourage the habit but did not check further. Within a month, Dotty's mother, an attorney for the petroleum firm, noticed that other members of the family had developed a similar habit, which they ascribed to discomfort with the humid summer. One evening when the nanny was out, Dotty's mother changed her daughter's diaper and noticed white threadlike worms a few millimeters long (**Figure 11.22**). A trip to the pediatrician revealed pinworms as well as pinworm eggs (see Figure 11.22). The eggs were observed on a "tape test," a piece of tape that was pressed to Dotty's anus and then placed on a slide for light microscopy. The physician prescribed mebendazole (a microtubule inhibitor) for the entire family, including the nanny and the maid who served dinner. Dotty's mother asked about the dogs; she was told that dogs do not carry pinworms, although they do carry other worms transmissible to humans. Since the drug mebendazole affects only mature worms, further treatment 2 weeks later was required to attack the eggs and immature worms that had grown in that time. Ultimately, it took 6 months of treatments to eliminate the infection from the household.

Pinworms are small, white nematodes of the species *Enterobius vermicularis*. They infect only humans and are the most prevalent helminth infection in the United States and Europe, possibly

Figure 11.22
Pinworms

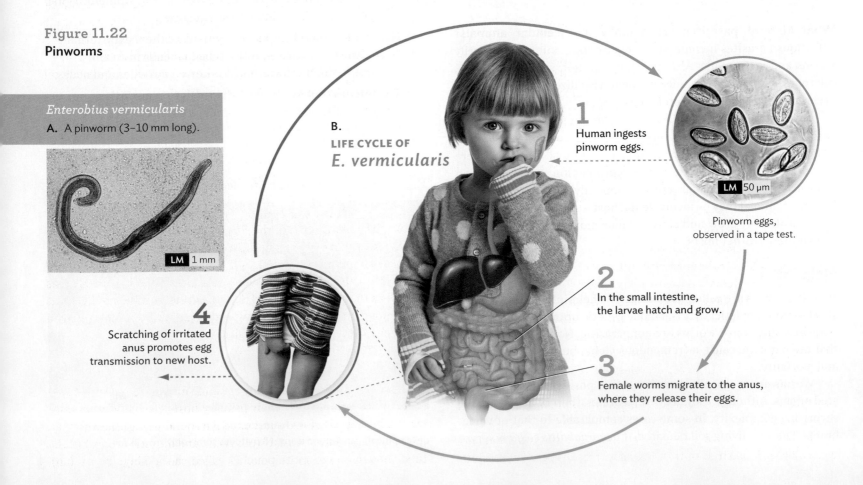

Enterobius vermicularis

A. A pinworm (3–10 mm long).

LM 1 mm

B.
LIFE CYCLE OF
E. vermicularis

1 Human ingests pinworm eggs.

LM 50 μm

Pinworm eggs, observed in a tape test.

2 In the small intestine, the larvae hatch and grow.

3 Female worms migrate to the anus, where they release their eggs.

4 Scratching of irritated anus promotes egg transmission to new host.

infecting 25%–50% of the population. Children commonly ingest pinworms by sucking dirty fingers. Eggs can also be ingested from airborne dust. In the small intestine, the larvae hatch and grow, absorbing nutrition from the host. Usually, larval growth is unnoticed unless there is an exceptionally large number of worms. The female worms migrate to the anus, where they release their eggs. The perianal movement of worms causes irritation, leading to scratching of the anus; in the process, the host's fingernails pick up eggs for transfer elsewhere.

When left untreated, pinworms can lead to complications, such as malabsorption of nutrients and anorexia related to the host's immune response. Also, secondary bacterial infection may occur at the perianal (around the anus) site of scratching or at the vulva or other places where the worms may migrate by mistake. The eggs persist for several weeks outside and are transmitted easily by contact with hands or with contaminated bedding. They are also transmitted by sexual contact.

Because transmission occurs readily within a household, all family members must be treated. The drug mebendazole, a selective inhibitor of worm microtubule biosynthesis, is effective against a broad spectrum of nematodes, including roundworm (*Ascaris*), whipworm, threadworm, and hookworm.

Pinworms are exclusive to humans, but other nematodes infect domestic animals and can be transmitted to humans. The roundworm *Toxocara* infects 14% of Americans and is contracted mainly from dogs. Toxocariasis is commonly without symptoms, but complications can occur, such as blindness. To avoid parasites, all dogs and cats should receive monthly medication.

Worms called **hookworms** (*Ancylostoma duodenale* and *Necator americanus*) are common in the southern United States and in tropical regions around the world. Their cycle differs from that of pinworms. Unlike pinworms, hookworm larvae grow in the soil, feeding on soil microbes until they penetrate the host skin, often through the soles of bare feet. The hookworms travel through the blood until they penetrate the alveoli in the lungs, then are coughed up to the throat and ingested. Within the small intestine, the hookworms latch onto the intestinal lining and suck blood. Their offspring then exit in the feces. Large numbers of hookworms can cause anemia and protein insufficiency.

Another parasitic nematode is *Trichinella spiralis,* the cause of trichinellosis. *T. spiralis* is named for the spiral-form cysts made by worm larvae embedded in muscle. Transmission to humans may occur from eating raw or undercooked pork; this is why pork traditionally is directed to be cooked "well done." Bear meat and horse meat can also carry *T. spiralis*. Once ingested, the larvae mature in the intestinal mucosa, mate, and produce progeny larvae. The new larvae now encyst in the host muscles, where they may cause pain and eventually death. In recent years, trichinellosis has become rare in industrial countries but remains a problem in developing countries.

In tropical regions, severely debilitating worm infections are caused by various kinds of tiny, **filarial** worms that grow within the

Figure 11.23
Roundworm (*Ascaris lumbricoides*) Infection of the Intestine
More than 400 roundworms were removed surgically from the intestine of a 5-year-old child in India.

blood, lymph, or subcutaneous fatty tissues. Most species are transmitted by biting flies, although some are transmitted by aquatic arthropods. Once introduced by an insect, the filaria reproduce and multiply to enormous numbers throughout the host's surface tissues. This large number of worms is necessary to ensure a high population within the tiny amount of blood meal taken by an insect, enabling transmission to the next host. For some species, such as *Brugia malayi*, the large numbers of worms in the lymphatic system may block the lymphatic ducts, leading to swelling of the body parts, a condition called elephantiasis. More will be said about elephantiasis in Chapter 21 (see Figure 21.28). Other filarial diseases include river blindness (*Onchocerca volvulus*), in which worm-induced corneal inflammation leads to blindness, and guinea worm disease (*Dracunculus medinensis*), in which worms form painful blisters on the skin until they are released outside into water.

Roundworms such as *Ascaris lumbricoides* can grow to relatively large sizes within the human digestive tract, and their effects may be lethal (**Figure 11.23**). *Ascaris* worms cause infection by ingestion of eggs in the soil. The worm eggs then develop into worms within the digestive tract, where they increase in size indefinitely and can obstruct the intestine. In some cases, worms may invade the lungs or other organs, causing life-threatening damage.

Within the intestine, the worms produce eggs that exit the anus. Defecation outdoors in soil may lead to transmission to other people who unknowingly pick up the eggs on hands or feet. Roundworms infect a billion people worldwide, primarily in tropical areas as well as the southern United States. Antihelminthic medications treat roundworms successfully when caught early, but the removal of larger worms may require surgical intervention.

Trematodes

Trematodes (flukes) infect many kinds of animals, though most species require a mollusk as their primary host. Trematodes are flatworms (phylum Platyhelminthes). Like roundworms, they have an internalized mouth, "throat" (pharynx), and digestive tube, but the tube ends in one or more pouches called caeca (singular, cecum).

Because the worm lacks an anus, it must expel its wastes back out the mouth. A fluke usually has two suckers, one near its mouth and one on the ventral side of the body. The mouth sucker ingests nutrients from the host.

Different fluke species preferentially attach to different internal organs, such as the liver, the lung, or the intestine. The phenomenon of a liver fluke was popularized on TV by an episode of *The X-Files* called "The Host." A real-life example of a liver fluke would be the Chinese liver fluke, *Clonorchis sinensis* (Figure 11.24). The fluke arises from eggs ingested by a snail, where the larvae develop into a free-swimming form. The swimming larvae then penetrate the skin and flesh of a fish, where they develop into a secondary form. If the fish is ingested raw or undercooked by a person, the parasite may attach to the person's bile duct, where it grows and matures, producing eggs that exit through the bowel and anus. This fluke can grow to 2.5 cm; others can grow as long as several meters. Prolonged growth of the fluke is associated with secondary infections and predisposition to cancer.

Link Section 22.6 covers parasitic infections of the digestive system and illustrates four general ways by which helminths infect humans (see Figure 22.29). The section describes several serious diseases caused by **flukes**, including opisthorchiasis—caused by the liver fluke *Opisthorchis*—and schistosomiasis—caused by species of *Schistosoma*, a genus of parasitic flatworms commonly known as blood flukes.

Cestodes

Tapeworms are transmitted through larvae embedded in uncooked meat. Different species are found in pork (*Taenia solium*), beef (*Taenia saginata*), and fish (*Diphyllobothrium*). Infection with fish tapeworms has historically been very common in regions of Asia and South America, where raw fish is consumed. In the United States, the growing popularity of raw fish dishes such as sushi and seviche has led to an increase in the incidence of fish tapeworm.

Tapeworms are flatworms (Platyhelminthes), and their form is very different from that of both flukes and roundworms. The tapeworm is composed of a head and successive segments, which form out of the head and then are gradually pushed down in a tail that may grow 2–15 meters in length (Figure 11.25). The head of the tapeworm has a special sucker device called the "scolex" (Figure 11.25B).

Each segment of the tapeworm contains both male and female reproductive organs, which may fertilize each other or cross-fertilize with another segment. Tapeworm segments packed with offspring pass out through the anus of the host, where the segments appear like moving bits of rice. The segments fall to the ground, where some other host may ingest them. A tapeworm may grow for many years inside the digestive tract of a host without notice until some of its offspring migrate to the muscle, brain, or liver. In these internal organs, the worms may form cysts, leading to muscle pain and neurological symptoms such as seizures.

From a global standpoint, parasitic worms cause substantial health problems, particularly for children, in whom malnutrition leads to stunted mental and physical development. At the same time, because worms are so prevalent, a hypothesis has been proposed that the human immune system evolved for optimal function in the presence of a modest number of helminths. This hypothesis is supported by the negative associations observed between the prevalence of helminths and prevalence of autoimmune disorders such as asthma, eczema, and Crohn's disease. Such diseases are common in urban areas with good hygiene but are rare in rural areas with poor hygiene and high worm prevalence. The hypothesis suggests that in the absence of helminths, components of the immune system that evolved to fight worms attack the body's own tissues instead. The concept has led to the highly controversial suggestion that modest exposure to helminths may help prevent autoimmune disorders (also discussed in Chapter 16). Although the hypothesis is unproved, it is being tested in clinical trials by exposing Crohn's patients to pig parasites that stimulate the human immune system but cannot complete an infectious cycle in humans.

Arthropods

Arthropods are invertebrate animals with an exoskeleton and jointed appendages; they include insects and arachnids (spiders and mites). Many arthropods are free-living, but others are parasites with complex life cycles. Some arthropods are **ectoparasites**, attaching to the surface of a human or other vertebrate. Others are parasites that burrow into the flesh. Some arthropod parasites suck blood, then fall off, whereas others inject their eggs to develop for many weeks or months inside the host. As we saw for worms, arthropod parasites of humans may be acquired from the environment, or they may be highly infectious, acquired from other infected humans.

ARACHNID PARASITES   Members of the arachnid group Acari are eight-legged **mites** and **ticks**. They include many diverse species, many of which are parasites of various animals and plants throughout terrestrial habitats.

Certain types of mites cause mange (in animals) and scabies (in animals or humans), a condition also known as acariasis (Figure 11.26). Scabies in humans is caused most often by the "human itch mite" species *Sarcoptes scabiei*. The mites attach to the skin by

Figure 11.24   Chinese Liver Fluke

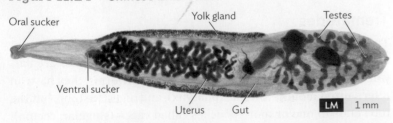

Oral sucker | Yolk gland | Testes
Ventral sucker
Uterus | Gut   LM   1 mm

Figure 11.25
Tapeworm

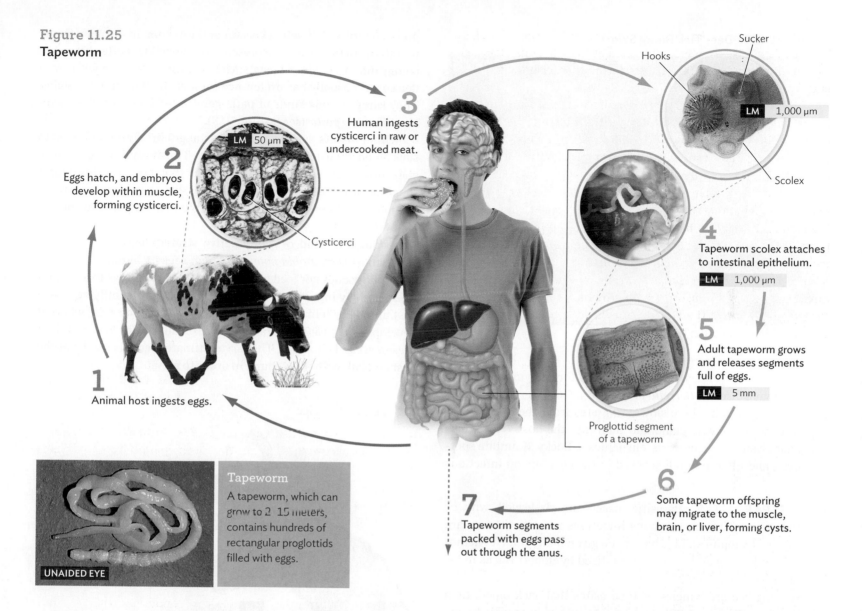

3 Human ingests cysticerci in raw or undercooked meat.

Hooks
Sucker
LM 1,000 µm
Scolex

2 Eggs hatch, and embryos develop within muscle, forming cysticerci.

LM 50 µm

Cysticerci

1 Animal host ingests eggs.

4 Tapeworm scolex attaches to intestinal epithelium.
LM 1,000 µm

5 Adult tapeworm grows and releases segments full of eggs.
LM 5 mm

Proglottid segment of a tapeworm

6 Some tapeworm offspring may migrate to the muscle, brain, or liver, forming cysts.

7 Tapeworm segments packed with eggs pass out through the anus.

Tapeworm
A tapeworm, which can grow to 2–15 meters, contains hundreds of rectangular proglottids filled with eggs.

UNAIDED EYE

suckers, then burrow under the skin by using special mouthparts and cutting surfaces on the forelegs. As the mites burrow into the skin, they lay eggs that hatch into larvae. Then the larvae come out of the skin to attach to a hair follicle, where they feed and molt until they reach the eight-legged adult stage; then they drop off to find new hosts. The movement of mites and larvae within the skin causes intense itching, as does the allergic reaction to the eggs. Itching and scratching lead to scabs as well as hair loss.

Mites are highly infectious, typically found at the wrists and other body parts that often come in contact with another person. Mites spread rapidly in groups of people in close quarters, such as a school dormitory or nursing home.

Ticks (order Ixodida) resemble mites in their eight-legged form, but they are larger and usually do not burrow completely in the skin. Ticks are ectoparasites that suck blood for nutrients to produce eggs, then fall off to disperse their progeny. Ticks often go unnoticed by their hosts, but are vectors for many serious disease agents, including bacteria, protozoa, and viruses. The deer tick,

Figure 11.26 Scabies

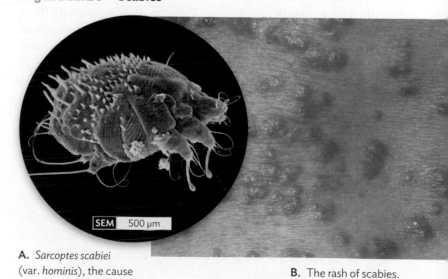

SEM 500 µm

A. *Sarcoptes scabiei* (var. *hominis*), the cause of scabies in humans.

B. The rash of scabies.

Figure 11.27   Deer Tick Biting Skin

Ixodes scapularis (**Figure 11.27**), is famous for carrying the spirochete of Lyme disease, *Borrelia burgdorferi*. Once the tick attaches to the host, the spirochete slowly migrates from the tick's digestive tract to its salivary gland and then into the host. Other ticks carry anaplasmosis, babesiosis, ehrlichiosis, Rocky Mountain spotted fever, and tularemia (all covered in the chapters on infectious disease).

INSECT PARASITES   Parasitic insects are six-legged, with or without wings. The most famous infectious parasitic insects are sucking lice (Anoplura). Lice have been given many colorful names throughout history and were immortalized by Robert Burns in his famous poem "To a Louse."

Sucking lice are wingless ectoparasites that suck blood, then produce eggs (**Figure 11.28**). Lice tend to be highly specific to one host species; and different species have preference for different body sites: head lice, body lice, and pubic lice (sexually transmitted). Head lice attach their eggs to a hair. Despite apparently good hygiene, head lice spread readily among people in close quarters, such as schools and army barracks; they are difficult to eliminate because the eggs are attached firmly and are not killed by normal shampoos, requiring special medication. Body lice lay their eggs in clothing and are more often associated with poor hygiene and economic status. Pubic lice, *Pthirus pubis*, commonly called "crabs," specifically infect the pubic hair (hair around the genitals). To a lesser extent, pubic lice infect the hair of eyebrows, eyelashes, and armpits. Their transmission is usually through sexual contact (although the dismayed sexual partner is sometimes assured that it "could have been the eyebrows").

Another human parasite common throughout history is the flea; as poet Ogden Nash wrote, "Adam had 'em." Fleas (*Siphonaptera*) are wingless insect parasites in which the adults jump instead of crawl; thus, they spend less time than lice directly on their host, although they can jump on and off. After sucking blood, fleas produce larvae that spin a cocoon and can remain dormant many months until they sense the presence of a host. Fleas often alternate among different animal hosts, such as cats or dogs, although a given flea species usually has preference for a particular animal species. They carry various kinds of pathogens, most famously the bacterium *Yersinia pestis* (see Section 10.3).

Other insects, such as mosquitoes, alight briefly on their host to suck blood, then depart to produce eggs. The mosquito eggs require stagnant water to develop into larvae that pupate. Mosquitoes themselves are technically not parasites, but they carry and transmit many kinds of pathogens, such as the malaria parasite and the yellow fever virus.

A different kind of reproductive strategy is that of the botfly (such as *Dermatobia hominis*, which seeks a human host). The adult botfly does not feed at all, but it attaches its eggs to a smaller fly, seemingly innocuous, which then deposits the botfly eggs on a host animal or human. Each egg burrows under the skin, where it develops as a larva. The parasitic larvae do not kill the host, but their presence can be painful. The larva ultimately crawls out from the host and falls off to the ground in order to pupate.

Figure 11.28   Lice

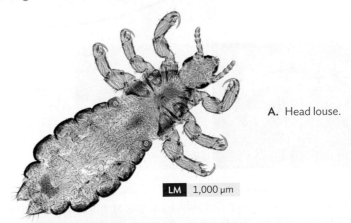

A. Head louse.

LM   1,000 µm

SEM 1 mm

B. Head lice laying eggs in hair.

Figure 11.29   Bedbugs

1 mm

A. Bedbugs were collected from an infested home.

B. Bedbug on skin.

A growing problem worldwide is the bedbug (Figure 11.29). Human bedbugs (*Cimex lectularius*) find their hosts at night while their human targets are asleep; they detect the host mainly from the exhaled carbon dioxide. Unnoticed, the bedbugs suck the host's blood and depart to lay eggs. Their bites lead to allergic reactions and psychological trauma. Bedbugs were nearly eradicated from the United States in the mid-twentieth century by the use of DDT and other pesticides. Today, bedbugs resist the older pesticides and can be killed only by newer chemicals that are restricted for their side effects. Elimination is possible but costly, and the persistent bugs spread especially fast through the homes of senior citizens who cannot afford home treatment. Hotels, movie theaters, and homes of all economic levels are now affected. Meanwhile, some bedbugs reportedly carry dangerous pathogens such as MRSA, although their ability to transmit disease remains unproved.

SECTION SUMMARY

- **Nematodes are roundworms with a complete digestive tube.** Many nematodes are free-living, but others are parasites of vertebrates, such as the roundworm *Ascaris*, pinworms, and hookworms.

- **Trematodes (flukes) are flatworms with suckers and a digestive tube but no anus.** Most flukes are parasites requiring multiple hosts, the primary host being a shellfish.

- **Cestodes (tapeworms) are segmented flatworms.** The scolex attaches within the intestine, growing a tail of segments, each of which contains reproductive organs. Larvae may encyst within muscle or brain.

- **Parasitic worms can be avoided by hygiene (washing hands), wearing shoes, and thorough cooking of meat or fish.**

- **Arthropod parasites include arachnids (eight-legged) and insects (six-legged).** Many are ectoparasites, attaching to the exterior of their host, although others burrow into their host.

- **Ticks suck blood, then fall off to lay eggs.** Ticks transmit numerous kinds of pathogens.

- **Human itch mites cause scabies.** Scabies mites burrow under the skin and lay eggs, which develop into larvae and feed on the skin.

- **Bloodsucking insects include lice, fleas, and bedbugs.**

Thought Question 11.5   The *X-Files* episode "The Host" depicts a human-sized worm as a "liver fluke" that is claimed to have a "scolex" and to regenerate from a cut piece. Which two kinds of worm are confused in this description?

Thought Question 11.6   Overall, why do eukaryotic parasites show such a wide range of size? What selective forces favor large size, and what favors small size?

Perspective

Although eukaryotic cells are generally larger and more complex than bacteria, there exist an extraordinary diversity of microbial eukaryotes. These organisms, including fungi, algae, and protists, serve many key roles in our ecosystems. They take part in food webs, especially in the ocean, that provide essential food for animals large and small. We will hear more about these ecosystems in Chapter 27.

At the same time, many fungi and protists parasitize humans, as do worms and arthropods, causing some of the most debilitating conditions worldwide. Diseases such as malaria and trypanosomiasis affect the habitability of major regions of the globe. And everywhere, people from all socioeconomic groups need to be aware of eukaryotic parasites as a medical concern.

LEARNING OUTCOMES AND ASSESSMENT

	SECTION OBJECTIVES	OBJECTIVES REVIEW
11.1 Eukaryotes: An Overview	• Explain the processes of asexual (vegetative) reproduction and sexual reproduction. • Distinguish the key traits of fungi, algae, amebas, alveolates, and trypanosomes. • Distinguish between microbial and invertebrate parasites.	1. Which eukaryotic microbes usually lack motility? A. Fungi B. Algae C. Amebas D. Trypanosomes
11.2 Fungi and Microsporidians	• Describe the structure of the fungal mycelium. • Explain how fungi impact the environment. • Describe how different types of fungi cause disease.	4. Most fungi are multicellular, but which of these fungi is single-celled (a yeast)? A. *Aspergillus* B. *Candida albicans* C. *Penicillium roquefortii* D. *Morchella hortensis*
11.3 Algae and Amebas Are Environmental Protists	• Explain the environmental significance of algae and their relatedness to plants. • Describe the form and motility of amebas. • Describe two diseases caused by amebas.	7. Which trait do many algae possess, but not plants? A. Chloroplasts B. Cell wall of cellulose or other polysaccharide C. Nucleus D. Flagella for motility
11.4 Alveolates: Predators, Phototrophs, and Parasites	• Describe the distinctive traits of ciliates, dinoflagellates, and apicomplexans. • Outline the complex life cycle of *Plasmodium falciparum* and *P. vivax*, the parasites that cause malaria.	10. Which organism swims by beating hairlike cilia? A. *Balantidium coli* B. *Didinium* species C. *Pfiesteria* species D. *Plasmodium falciparum*
11.5 Trypanosomes and Metamonads	• Describe the traits of trypanosomes. • Describe the traits of giardias and trichomonads.	13. Which is <u>not</u> a trait of the life cycle of *Trypanosoma brucei*? A. Motility via a single flagellum powered by the kinetoplast B. Antigenic variation, producing 200 different forms of its variant surface glycoprotein C. Development within the tsetse fly D. Transmission via the oral-fecal route
11.6 Invertebrate Parasites: Helminths and Arthropods	• Describe the parasitic cycles of worms, including roundworms, flukes, and tapeworms. • Describe the parasitic cycles of arthropods, including ticks, mites, and bloodsucking insects.	16. The highly prevalent infectious agent *Enterobius vermicularis* is what type of organism? A. Fluke B. Nematode C. Tapeworm D. Hookworm

2. Which infectious parasites are actually invertebrate animals, not true microbes?
   A. Fungi
   B. Trypanosomes
   C. Microsporidians
   D. Helminths

3. Which eukaryotic microbes have long, whip-like organelles?
   A. Dinoflagellates
   B. Ciliated protists such as paramecium
   C. Fungi
   D. Amebas

5. Which fungal process is inhibited by triazoles?
   A. Ergosterol biosynthesis
   B. Chitin biosynthesis
   C. Fruiting body formation
   D. Ascus development

6. Which pathogenic fungus commonly grows in a unicellular yeast form?
   A. Mycorrhyzae
   B. *Aspergillus*
   C. *Stachybotrys*
   D. *Candida*

8. Which kind of amebas do <u>not</u> commonly cause human disease?
   A. *Entamoeba histolytica*
   B. *Naegleria fowleri*
   C. Foraminiferans
   D. *Acanthamoeba* species

9. Which kind of amebas grow calcite shells that form limestone rocks?
   A. *Entamoeba histolytica*
   B. Foraminiferans
   C. *Naegleria*
   D. *Balamuthia*

11. Which apicomplexan pathogen is commonly harbored by cats?
    A. *Babesia microti*
    B. *Toxoplasma gondii*
    C. *Plasmodium falciparum*
    D. *Cryptosporidium parvum*

12. Which apicomplexan parasite invades blood cells?
    A. *Plasmodium vivax*
    B. *Trypanosoma cruzi*
    C. *Cryptosporidium*
    D. *Cyclospora*

14. *Giardia lamblia* is commonly transmitted via _____.
    A. infant day-care centers
    B. coughing and sneezing
    C. mosquitoes
    D. lactation

15. Which parasite is transmitted by sand flies?
    A. *Trypanosoma cruzi*
    B. *Giardia lamblia*
    C. *Leishmania donovani*
    D. *Toxoplasma*

17. Which insect ectoparasite is transmitted primarily through sexual contact?
    A. Ticks
    B. Fleas
    C. Bedbugs
    D. Pubic lice

18. Which parasite preferentially attaches to the liver?
    A. Tapeworms
    B. Nematodes
    C. Flukes
    D. Hookworms

Key Terms

alga (p. 323)
alveolate (p. 323)
ameba (pp. 323, 330)
antigenic variation (p. 337)
apicomplexan (p. 333)
arthropod (p. 323)
ascospore (p. 327)
asexual reproduction (p. 321)
cestode (p. 340)
chitin (p. 326)
ciliate (p. 333)
cilium (p. 334)
conjugation (p. 335)
cyst (p. 332)

definitive host (p. 333)
dinoflagellate (p. 333)
diploid (p. 322)
ectoparasite (p. 342)
filarial (p. 341)
fluke (p. 339)
foraminiferan (p. 331)
fruiting body (p. 327)
fungus (p. 323)
haploid (p. 322)
helminth (pp. 323, 339)
hookworm (p. 341)
hypha (p. 326)
incidental host (p. 339)

intermediate host (p. 334)
meiosis (p. 322)
metamonad (p. 323)
microsporidian (p. 323)
mite (p. 342)
mitosis (p. 321)
mycelium (p. 326)
mycology (p. 323)
mycorrhiza (p. 325)
nematode (p. 339)
pinworm (p. 340)
primary endosymbiosis (p. 330)
protist (p. 323)
protozoan (p. 323)

pseudopod (p. 330)
roundworm (p. 339)
secondary endosymbiont alga (p. 336)
secondary endosymbiosis (p. 330)
sexual reproduction (p. 321)
tapeworm (p. 340)
tick (p. 342)
trematode (p. 339)
trypanosome (pp. 323, 337)
vegetative reproduction (p. 321)
yeast (p. 327)
zygote (p. 322)

Review Questions

1. Summarize the key traits of fungi. What do fungi have in common with protists, and how do they differ?

2. Explain the alternation of diploid and haploid forms in yeasts and in algae. Why do eukaryotic microbes undergo a sexual cycle instead of just vegetative reproduction?

3. Explain the roles of fungi and protists in ecosystems. Why are they important in wastewater treatment? In animal digestion?

4. Suppose a patient has symptoms of dysentery, and a laboratory technician finds amebas in a microscope slide. The patient is treated with anti-amebic drugs, but the symptoms remain. What explanations are possible?

5. A physician diagnoses Lyme disease in a patient; then he advises testing as well for babesiosis. Why?

6. Outline the three major kinds of helminths. Explain what these parasites have in common and how they differ in terms of structure and life cycle.

7. Explain why some parasites need more than one host in their life cycle.

8. What is an ectoparasite? Which arthropods are considered ectoparasites, and which are not?

Clinical Correlation Questions

1. Two weeks after a beach vacation, a 25-year-old store clerk develops intestinal symptoms, including nausea and bloody diarrhea. Antibacterial antibiotics fail to resolve the symptoms, and stool studies rule out a bacterial diarrhea. A trichrome stain of the patient's blood reveals a parasite showing ingested red blood cells. What parasite is the likely cause of infection? What is its mode of transmission? Which antibiotic will most likely resolve the symptoms?

2. A 65-year-old woman has been hospitalized for 3 weeks for chronic obstructive pulmonary disease (COPD) when she develops lung pain and exacerbation of her shortness of breath. A CT scan shows a clouded region, indicating a mass in the left lung. Broad-spectrum antibacterial agents fail to resolve the infection. Which organism is a probable cause? What is the treatment?

Thought Questions: CHECK YOUR ANSWERS

Thought Question 11.1 For a fungus, what are the advantages of growing a mycelium? What are the advantages of growing as a single-celled yeast?

ANSWER: Mycelial growth allows coordinated expansion into a food resource, such as a host tissue. The large size of the mycelium enables it to avoid being engulfed by amebas or host macrophages. Single-celled yeasts, however, can disperse readily to new habitats, finding new resources to exploit.

Thought Question 11.2 How do you think amebas evolved to carry symbiotic bacteria? How do these ameba-carried bacteria endanger humans?

ANSWER: Amebas engulf bacteria as food. But some bacteria have molecular mechanisms to avoid digestion within the ameba. If the bacteria grow within the ameba, their respiration may provide nutrients such as vitamins that are useful to the ameba. At the same time, the ameba provides food molecules, and an endosymbiosis may evolve. However, some bacteria, such as *Legionella*, evolve a partnership with amebas but act as infectious agents within host cells such as macrophages. Inhaled water droplets may contain amebas with intracellular *Legionella* bacteria, which can cause legionellosis.

Thought Question 11.3 What are the relative advantages and limitations of flagellar motility, ciliate motility, and pseudopodial motility of amebas? Which kind of natural habitat or body environment might favor one kind of motility over another?

ANSWER: Flagellar motility is the most rapid, allowing rapid propulsion of a protist through a dilute watery habitat. However, if the flagellum gets caught or damaged, the cell cannot move. Cilia provide robust movement through a wide range of environments, including the presence of large particles to "dodge." Cilia also provide a way to generate currents that direct food into the cell's oral groove. However, cilia may be less effective in a highly viscous medium. Pseudopods enable an ameba to crawl through crevices and highly viscous material. Pseudopod motility is slow, however, and ineffective in open water.

Thought Question 11.4 Do you think most eukaryotic parasites have genomes that are larger or smaller than those of free-living microbial eukaryotes? Why might parasite genomes be larger—and why might they be smaller?

ANSWER: Natural selection tends to shrink the genome of a parasite, because the smaller the genome, the more progeny can be made from a given host, a limited resource. Also, the host can provide many complex materials that the parasite need not make, such as amino acids and vitamins. At the same time, however, the parasite needs special mechanisms to attach and manipulate the host and evade the host's defenses—mechanisms that require extra genes to encode. Thus, apicomplexans have evolved the apical complex, which penetrates the host cell, and trypanosomes have evolved a complex series of coat proteins that evade the host's immune system. Genes encoding these coat proteins expand the size of the parasite's genome.

Thought Question 11.5 The *X-Files* episode "The Host" depicts a human-sized worm as a "liver fluke" that is claimed to have a "scolex" and to regenerate from a cut piece. Which two kinds of worm are confused in this description?

ANSWER: The *X-Files* episode referred to the worm as a "fluke" that attached to the bile duct of a liver. But the episode depicted the parasite's head as a "scolex." The scolex is a specialized form with hooks and suckers found on a tapeworm, not a fluke. Also, the ability of a cut-off piece (a segment) to reproduce is characteristic of a tapeworm, not a fluke.

Thought Question 11.6 Overall, why do eukaryotic parasites show such a wide range of size? What selective forces favor large size, and what favors small size?

ANSWER: The advantage of large size, for a parasite, is that the larger the organism, the more resources it can mobilize to produce large numbers of progeny. On the other hand, the larger parasite is more likely to shorten the life span of the host. The smaller the parasite, the greater the number of individuals that can be made with finite resources. Thus, many kinds of parasites alternate between large and small developmental stages. An extreme example is tapeworms, which may grow to many meters in length, with a large number of independently reproductive segments. The progeny, however, are microscopic, so large numbers can be made.

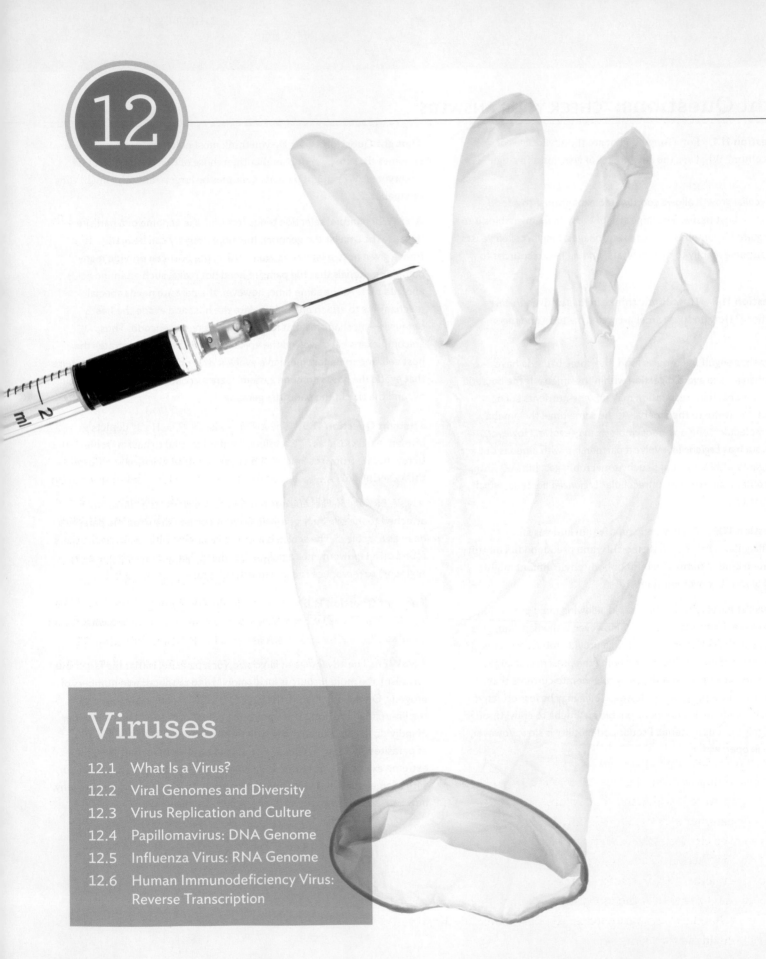

12

Viruses

SCENARIO Christine, age 25, was a critical care nurse in a hospital in Phoenix, Arizona. She sustained a needle stick injury while attempting to draw a blood sample from a patient chronically infected with hepatitis C virus.

Hepatitis C Virus (HCV)
Normal liver compared to liver with cirrhosis.

SIGNS AND SYMPTOMS Six weeks after the needle injury, Christine experienced fever and other flu-like symptoms. Because of the needle stick, hepatitis C infection of the liver was suspected.

TESTING Christine's liver function test (LFT) levels fluctuated, and the blood titer of virus rose and fell.

DIAGNOSIS AND TREATMENT
Christine was put on a course of ribavirin (an RNA synthesis inhibitor) and pegylated interferon-alpha (interferon protein with polyethylene glycol chains that stabilize interferon in the blood). She experienced a number of side effects, including hemolytic anemia (due to ribavirin buildup in the red blood cells) and fatigue, hair thinning, and depression (due to interferon injection). Her virus levels decreased to very low levels, as measured by viral RNA assay using PCR amplification. The therapy was discontinued because the side effects were greater than the effects of the virus.

RECURRENCE Ten years later, Christine experienced an unexplained fever. Given her prior history, she was tested for liver function. Liver enzymes aspartate aminotransferase and alanine aminotransferase were elevated (indicating liver damage). A PCR test of a blood sample revealed hepatitis C viral RNA. Christine was again treated with interferon and ribavirin, but the side effects were severe and the drugs were less effective than before. Over the next 20 years, her liver function declined, and the organ showed cirrhosis (liver deterioration with healthy tissue replaced by fibrosis). In 2013, she learned that two new antiviral drugs, sofosbuvir and simeprevir could stabilize her liver function until a donor might become available for liver transplant. But the drugs were enormously expensive, and Christine did not know how she could pay for them.

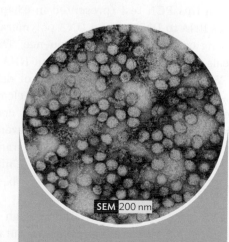

SEM 200 nm

Hepatitis C (HCV) Virions
(Colorized.)

CHAPTER OBJECTIVES

After reading this chapter, you will be able to:

- Define the nature of a virus and explain how a virus differs from a cell.
- Describe the major classes of viral genomes and cite an example of each.
- Explain how viruses are cultured.
- Describe the replication cycles of papillomavirus, influenza virus, and human immunodeficiency virus (HIV).

Hepatitis C virus (HCV), a cause of liver disease, is estimated to infect 200 million people worldwide, or more than 3% of the world's population. The prevalence of HCV in the United States is 2.5%; it is the country's most prevalent blood-borne disease. In some African countries, such as Egypt, prevalence is as high as 14%. Most people living with HCV have few or no symptoms; but the high prevalence of the virus means that the death rate due to HCV (in those patients with severe disease) will soon overtake the death rate of AIDS.

HCV is a virus. A **virus** is a noncellular particle with a genome contained by a **capsid** (protein coat). The virus takes over a cell to manufacture progeny. Viruses require a host cell to reproduce. The term "virus" refers to the particles, their life cycle, and the disease they cause, whereas the term **virion** refers specifically to the structure of the virus particle. Unlike cells, some viruses (such as HCV) have a genome of RNA. Other viruses (such as herpes) have DNA genomes. Because HCV genome consists of RNA, Christine's blood was tested for viral RNA using a PCR (polymerase chain reaction) test. In this PCR test (presented in Chapter 8), first an enzyme copies RNA to DNA; then a DNA polymerase makes exponentially growing copies of the DNA sequence. The amplified DNA then reveals sequences from HCV genomic RNA.

Without treatment, the course of disease with HCV varies greatly. About 15% of those infected clear the virus completely, whereas the rest may live for several decades before becoming aware of liver damage. However, 5% eventually experience liver cancer. Liver failure may necessitate a transplant; HCV is now the leading cause of liver transplant in the United States, but the waiting list is long, and the new liver always becomes infected with the virus still present in the blood.

RNA viruses such as HCV have high error rates during replication and therefore high rates of mutation. Within a patient, mutant strains rise and fall, depending on their virulence and on their resistance to the host immune system. No single drug can be used because the virions will quickly evolve resistance; instead, two or more antiviral agents must be used together. That is why Christine received dual treatment with ribavirin and interferon. Unfortunately, antiviral agents have severe side effects because they must retard the body's own cellular processes in order to attack viral activity; thus, normal cells suffer to some extent along with infected ones.

Ribavirin is an antiviral agent that is effective for many patients with HCV. Ribavirin is an RNA analog that inhibits replication of viral RNA (Section 13.7). Unfortunately, it can also inhibit function of erythrocytes (red blood cells). Interferon is another antiviral agent, often administered in tandem with ribavirin. Interferon is a host protein of the innate immune system (discussed in Chapter 15) whose gene is cloned in bacteria for pharmaceutical production. Cloned or recombinant interferon is a good example of a lifesaving **recombinant DNA** product of the biotechnology industry (discussed in Chapters 8 and 27). Interferon has many effects on viral proliferation as well as on normal cell growth. Although Christine was not helped by ribavirin and interferon, these drugs do control symptoms for many patients. We do not know why the drugs work better in some patients than in others. New drugs developed through research, such as sofosbuvir, control the disease in 90% of patients. But the expense of these drugs limits their use.

Link Section 8.4 introduces **recombinant DNA** technology, by which novel pieces of DNA can be created by splicing together DNA from different sources.

Hepatitis C is one of many important diseases caused by viruses. Viruses have enormous impact on human lives and populations and on life throughout all ecosystems. For instance, in the open ocean, viruses are the main source of predation that dissipates massive algal blooms. Every kind of cellular organism known, even the smallest of bacteria, is infected by viruses.

12.1
What Is a Virus?

Viruses are ubiquitous, infecting every taxonomic group of organisms, including bacteria, eukaryotes, and archaea. In marine ecosystems, viruses act as major predators and sequester significant amounts of nutrients in their structure. For humans, viruses cause many forms of illness, whose influence on our history and culture would be hard to overstate. More people died in the influenza pandemic of 1918 than in the battles of World War I. In the past 30 years, the AIDS pandemic caused by HIV has killed 25 million people worldwide, and the number continues to grow.

Viruses are part of our daily lives. The most frequent infections of college students are due to respiratory pathogens such as rhinovirus (the common cold) and Epstein-Barr virus (infectious mononucleosis), as well as sexually transmitted viruses such as herpes simplex virus (HSV) and papillomavirus (cutaneous warts, genital warts, and genital cancers). Viruses also impact human industry; for example, bacteriophages (literally, "bacteria eaters") infect industrial cultures of *Lactococcus* during the production of yogurt and cheese. Plant pathogens such as cauliflower mosaic virus and rice dwarf virus cause major losses in agriculture.

And yet, viruses have made astonishing contributions to medical research and biotechnology. Bacteriophages are used routinely as **cloning vectors**, small genomes into which foreign genes can be inserted and cloned for gene technology. Some of the most lethal human viruses, such as HIV, have been used to make harmless vectors for human gene therapy.

In natural environments, viruses fill important niches in all ecosystems. Viruses limit the population density of hosts, such as marine algae. Algal blooms grow so large that they show up in satellite photos from space; only viruses can dissipate such blooms. Because viruses dissipate dense populations, they also select for host diversity. Each viral species has a limited host range and requires a critical population density to sustain the chain of infection. The virus thus limits its host species to a population density far lower than that sustainable by the nutrient resources. The resources then support other species resistant to the given virus (but susceptible to others). Thus, overall, viruses prevent the dominance of any one host species. From bacteriophages in our colon to cold-tolerant viruses of Antarctica, viruses promote ecosystem diversity.

History of the Concept of a Virus

Our concept of a virus has developed over the past century. In the late eighteenth century, microbiologists puzzled over infectious agents that passed through a filter too small to permit bacterial cells. The Dutch microbiologist Martinus Beijerinck (1851–1931) proposed that a "virus" was some kind of infectious "fluid." But the "fluid" turned out to consist of small infectious particles that we now refer to as virions. Some types of virions are geometrically symmetrical particles that can actually be crystallized.

When the first viruses were discovered, they were defined as nonliving. The viruses known at that time lacked metabolism to use energy or conduct biosynthesis; and viruses such as TMV could be crystallized like inert chemicals. But some microbiologists have always questioned this position. Objections to the definition of viruses as nonliving entities include:

- Many bacteria, such as *Clostridium* and *Bacillus* species, form inert spores that can remain viable for thousands of years. Could a virus particle be compared to a spore?

- Obligate intracellular bacteria, such as *Rickettsia* and *Chlamydia* species, metabolize only during growth in a host cell. They develop inert spore-like forms that survive outside the host.

- Pathogens such as *Treponema pallidum* and *Mycobacterium leprae* show extensively degenerate genomes, relying on many host functions. Could viruses have evolved from a cellular ancestor that lost even more of its genes?

In 2003, Didier Raoult and colleagues described a new kind of virus that is larger than some bacteria and whose genome is larger than some bacterial genomes. The virus was named Mimivirus because it "mimics" a cell (Figure 12.1). Amebas phagocytose the virion as they do with bacteria that they eat. But once ingested, the virus takes over the ameba's cytoplasm, forming intracellular "virus factories" that resemble a contained living organism.

Figure 12.1 Mimivirus Infecting an Ameba

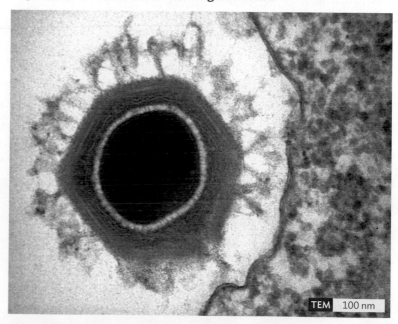

TEM 100 nm

About 300 nm across, the mimivirus is larger than some bacteria.

The mimivirus genome contains many genes typical of cellular genomes—a sign that the virus could have evolved from an ancestral cell. A similar cellular origin is proposed for the evolution of large DNA viruses such as herpes and smallpox. On the other hand, small RNA viruses such as feline leukemia virus (FeLV) have as few as three genes. FeLV and similar viruses might have evolved from a part of a cell such as telomerase, the enzyme that regenerates chromosome ends.

Another startling discovery about viral genomes—both large and small—is that many have been incorporated into the human genome, contributing a substantial part of our own DNA sequence. Viral evolution tells us much about human evolution.

Viruses Replicate in Host Cells

The effects of viral disease arise from their replication within host cells, which either destroys or debilitates the cells. Viruses also induce host responses that debilitate the host, such as overreaction by the immune system (discussed in Chapters 15 and 16). Some viruses alter host cell genomes to cause cancer, such as leukemia caused by human T-cell lymphotrophic virus (HTLV) and the skin and genital cancers caused by human papillomavirus (HPV; see Section 12.4).

The human gut teems with **bacteriophages** (phages), viruses that infect bacteria. An example is phage T2, which infects *Escherichia coli* (Figure 12.2A). Most bacteriophages insert their genome into the host cell, leaving the empty capsid outside. Within the bacterial cytoplasm, the phage genome directs the replication of progeny **virions** (virus particles). The virions are released when the host cell lyses.

In the laboratory, phages can be observed in a petri dish as a **plaque**, a clear spot against a lawn of bacterial cells (Figure 12.2B). The bacteria grow to completely cover the surface of the agar medium, rendering it opaque—except where bacteria are lysed by phages. Each plaque arises from a single virion or bacteriophage (virus that infects bacteria). The bacteriophage lyses a host cell and spreads progeny to infect adjacent cells. Eventually enough bacterial cells are lysed so that the plaque is visible to the naked eye. Plaques can be counted as representing individual infective virions from a phage suspension.

An example of a virus that infects humans is measles virus (Figure 12.2C). The measles virus has an envelope that is derived from the host cell plasma membrane during exit of the virus from the host cell. During entry of the virus into a new host cell, the envelope fuses with the host cell plasma membrane, releasing the viral contents into the cytoplasm of the cell. After replicating within the infected cell, newly formed measles virions become enveloped by host cell membrane as they bud out of the host cell. The spreading virus generates a rash of red spots on the skin of infected patients (Figure 12.2D). Untreated cases of measles can be fatal (1 in 500 cases).

Figure 12.2   Virus Infections

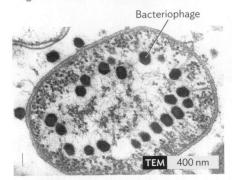
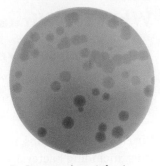

A. Bacteriophage T2 particles form a semicrystalline array within an *E. coli* cell.

B. Bacteriophage infection forms plaques of lysed cells on a lawn of bacteria.

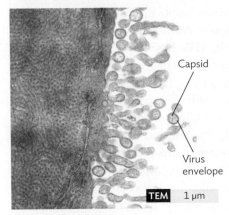
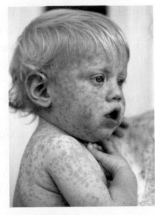

C. Measles virions bud out of human cells in tissue culture.

D. A child infected with measles shows a rash of red spots.

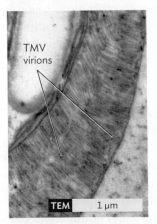

E. A tobacco leaf section is packed with tobacco mosaic virus particles.

F. A tomato leaf infected by tobacco mosaic virus shows mottled appearance.

Tobacco mosaic virus (TMV) infects a wide range of plant species, its virions accumulating to high numbers within plant cells (Figure 12.2E). The progeny virions then travel through interconnections to neighboring cells. Infection by TMV results in mottled leaves and stunted growth (Figure 12.2F). Plant viruses cause major economic losses in agriculture worldwide.

As new virions are produced, these virus particles need to find new cells to infect—and ultimately a new host. The process of reaching and infecting a new host is called **transmission**. Different viruses have different mechanisms and efficiencies of transmission. For example, the measles virus is transmitted by droplets of respiratory fluid. Measles transmission is highly efficient. By contrast, HIV (human immunodeficiency virus) is transmitted via blood or sexual contact. Transmission of HIV is relatively inefficient, especially via sexual contact; multiple contacts may occur before HIV infection is established in a new host.

Each species of virus infects a particular group of host species, known as the **host range**. Some viruses can infect only a single species; for example, HIV infects only humans. Close relatives of humans, such as the chimpanzee, are not infected, although they are susceptible to a virus that shares ancestry with HIV, simian immunodeficiency virus (SIV). In contrast, the West Nile virus, transmitted by mosquitoes, has a much broader host range, including many species of birds and mammals.

The term **tissue tropism** refers to the range of tissue types that a virus can infect. Some viruses have broad tropism; for example, Ebola virus infects many different tissues throughout the body. Other viruses have a narrow tropism; for example, rabies virus specifically infects nervous tissues, whereas influenza virus specifically infects cells of the respiratory epithelium. Both tropism and host range depend on various host factors, most importantly the surface receptor molecules, typically the specific proteins on a cell surface that the virus particle can bind to. For example, avian strains of influenza virus (the so-called avian flu) bind effectively to the host cell receptors in birds, but they bind less effectively to receptors in the upper respiratory tissues of humans.

In all cases, viruses use a relatively small number of virus-encoded proteins to commandeer the metabolism of their hosts. This situation has profound consequences for medical therapy:

- **Antiviral agents are hard to discover.** Because most viral pathogens have relatively few specific parts (compared with bacterial pathogens), there are relatively few targets for antibiotic design.

- **Antiviral agents have severe side effects.** Because viral replication involves so many host cell processes, in combination with the few viral components, agents that disrupt viral infection usually harm the host as well.

- **Viral genomes mutate fast, even faster than bacteria.** The small genomes of viruses enable them to mutate rapidly within the host, and thus no one antiviral agent will work for long. New agents with new targets must be found.

Fortunately, many viral diseases, such as colds and influenza, are soon eliminated by a healthy immune system—in most cases. But others, such as hepatitis C and human immunodeficiency virus (HIV), may lead to a lifetime infection. That is because these viruses are capable of hiding within host cells in a latent state, where they are inaccessible to the host immune system. Virology supports the old adage that prevention is better than cure. But so long as people are getting infected, we need new antiviral agents. And the only way to find them is by studying virus structure and function.

Note Virology has two uses of the word "latent." A virus may persist in the cell in a "latent state" ("latency")—that is, present but inactive until a molecular signal reactivates viral replication. But after infection, the latent period refers to the time during which a virus is multiplying within a host but cannot yet transmit infection to a new host.

Virus Structure

Viral genomes are more diverse in composition than the genomes of cells, which are always made up of double-stranded DNA. A viral genome is composed of either DNA or RNA, depending on the viral species; and in different species, the genome may be double-stranded or single-stranded (Table 12.1). As mentioned earlier, each viral genome (DNA or RNA) is enclosed within a protein coat, the capsid. The capsid keeps the viral genome intact outside the host. In some species, the capsid is further encased by an **envelope** formed out of host membranes with embedded viral envelope proteins. The capsid (and envelope, if present) must provide a mechanism of infection of the next host cell. The infection mechanism must be one of two types: (1) injection of viral genome into the host cell or (2) uptake into the host cell, followed by disassembly, or **uncoating**. In either case, the original virion loses its own structure and its own identity as such. Loss of individual identity is necessary to generate viral progeny.

Viral species typically show one of four structural types of capsid: icosahedral, filamentous, complex tailed, or amorphous (lacking symmetry).

ICOSAHEDRAL CAPSIDS Many viral capsids show radial symmetry, like a crystalline object. The advantage of symmetry is that it forms a package out of repeating protein units generated by a small number of genes. The smaller the viral genome, the more genome copies can be synthesized from the host cell's limited supply of nucleotides.

The form of a radially symmetrical virion is always based on an icosahedron, a polyhedron with 20 identical triangular faces. An example of an icosahedral capsid is that of the herpes simplex virus (Figure 12.3A). Each triangular face of the capsid is determined by the same genes encoding the same protein subunits. No matter what the pattern of subunits in the triangle is, the structure overall exhibits rotational symmetry characteristic of an icosahedron (Figure 12.3B). Within the icosahedral capsid, the herpes genome (composed of double-stranded DNA) is spooled tightly (Figure 12.3C), the DNA packed under high pressure. When the herpes capsid enters the cell, it is transported to a nuclear pore complex, where the pressure is released. The pressure release drives viral DNA into the host nucleus.

Table 12.1
Baltimore Groups of Viruses

Virus Example	Taxonomic Group with Examples	Virus Example	Taxonomic Group with Examples

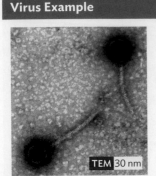

Phage lambda

± DNA

**Group I.
Double-stranded
DNA viruses**

Bacteriophage lambda infects *Escherichia coli*.
Herpes viruses cause chickenpox, genital infections, and birth defects.
Human papillomavirus strains cause warts and tumors.
Mimivirus, one of the largest known viruses, infects amebas.

Rabies virus

– RNA

**Group V.
(–) sense single-stranded
RNA viruses**

Orthomyxoviruses cause influenza.
Filoviruses such as Ebola virus cause severe hemorrhagic disease.
Paramyxoviruses cause measles and mumps.
Rhabdovirus causes rabies.

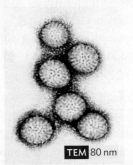

Geminivirus

+ DNA

**Group II.
Single-stranded
DNA viruses**

Bacteriophage M13 infects *E. coli*.
Parvoviruses cause disease in cats, dogs, and other animals.
Geminiviruses infect tomatoes and other plants.

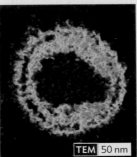

Human immunodeficiency
virus

+ RNA → DNA

**Group VI.
Retroviruses
(RNA reverse-
transcribing viruses)**

**Feline leukemia virus (FeLV), Rous sarcoma
virus (RSV),** and **avian leukosis virus (ALV)** cause cancer.
Lentiviruses include **human immuno-
deficiency virus (HIV),** the cause of AIDS,
and **simian immunodeficiency virus (SIV),** the cause of simian AIDS.

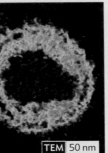

Rotavirus

± RNA

**Group III.
Double-stranded
RNA viruses**

Birnaviruses infect fish.
Reoviruses such as rotavirus cause severe
diarrhea in infants. Other reoviruses are in
clinical trials to fight tumors (oncolysis).
Varicosaviruses infect plants.

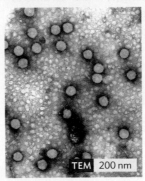

Caulimovirus

DNA ⇄ + RNA

**Group VII.
Pararetroviruses
(DNA reverse-
transcribing viruses)**

Caulimoviruses (such as cauliflower
mosaic virus, or CaMV) infect many kinds of
vegetables. CaMV provides the best vector
tools for plant biotechnology.
Hepadnaviruses such as hepatitis B virus
infect the human liver.

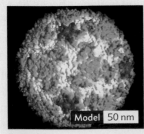

Poliovirus

+ RNA

**Group IV.
(+) sense single-
stranded RNA viruses**

Tobacco mosaic virus infects plants.
Coronaviruses such as SARS cause severe
respiratory disease.
Flaviviruses cause hepatitis C, West Nile
disease, yellow fever, and dengue fever.
Poliovirus infects human intestinal epithelium
and nerves.

Figure 12.3   Herpes: Icosahedral Capsid Symmetry

Depth into virion   0 10 30 50 65 nm

A. Icosahedral capsid of herpes simplex 1 (HSV-1), envelope removed. Imaging of the capsid structure is based on computational analysis of cryo–electron microscopy (cryo-TEM). Images of 146 virus particles were combined digitally to obtain this model of the capsid at 2-nm resolution.

Fivefold axis
Threefold axis
Twofold axis

Threefold   Fivefold   Twofold

B. Icosahedral symmetry includes fivefold, threefold, and twofold axes of rotation.

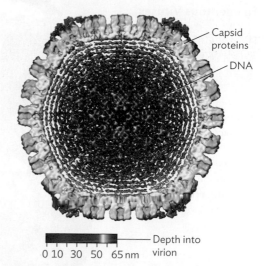

Capsid proteins
DNA

0 10 30 50 65 nm   Depth into virion

C. The icosahedral capsid contains spooled DNA.

Some icosahedral viruses, such as poliovirus and papillomavirus, have only a protein capsid. Herpes virions, however, possess in addition an envelope derived from the host nuclear or ER membrane (**Figure 12.4**). The envelope bristles with virus-encoded **spike proteins** that plug it onto the capsid. The spike proteins enable the virus to attach and infect the next host cell.

Figure 12.4   Envelope and Tegument Surround the Herpes Capsid

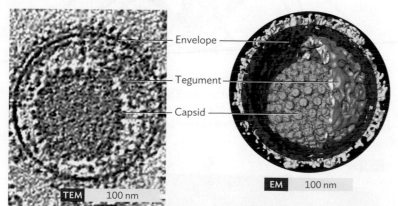

Envelope
Tegument
Capsid

TEM   100 nm

EM   100 nm

A. Section showing envelope and tegument proteins surrounding the capsid (cryo-TEM).

B. Cutaway reconstruction: spike envelope glycoproteins (yellow), envelope membrane (dark blue), tegument proteins (orange), and capsid (light blue) (cryo-EM tomography).

Between the envelope and the capsid may be additional proteins called **tegument** (see Figure 12.4). Tegument proteins are expressed during infection of a host cell and then packaged in the virion during envelope formation. Both viral and host proteins may be packaged. As the virus enters a new host cell, its envelope fuses with the host cell membrane in a way that opens the viral contents to the cytoplasm. The tegument proteins are released in the cytoplasm, where they help viral replication.

FILAMENTOUS VIRUSES   **Filamentous viruses** include tobacco mosaic virus (**Figure 12.5A**) as well as animal viruses such as Ebola virus, which causes a swiftly fatal disease of humans and related primates (**Figure 12.5B**). Many bacteriophages are filamentous; for example, filamentous phage CTXφ integrates its sequence into the genome of *Vibrio cholerae*, where it carries the genes for the deadly cholera toxin (discussed in Chapter 4).

Filamentous viruses show helical symmetry. The pattern of capsid monomers forms a helical tube around the genome, which is wound helically within the tube (see Figure 12.5A). The length of the helical capsid may extend up to 50 times its width, generating a flexible filament. Unlike the icosahedral capsid, which has a fixed size, the helical capsid can vary in length to accommodate different lengths of nucleic acid. This variable length is convenient for genetic engineering **vectors** (viruses with genomes artificially recombined to carry genes into cells).

Link Because viruses introduce their genetic material into the host cell as part of their replication cycle, they can be used as **vectors** to introduce genes into a cell. As discussed in Section 8.4, the process involves first inserting the desired gene sequence into the vector and then introducing the vector into the cell. Through this process, a cell is generated that can now express the protein product of the inserted gene sequence.

Figure 12.5
Filamentous Viruses

A. The helical filament of tobacco mosaic virus (TMV) contains a single-stranded RNA genome coiled inside. Image reconstruction is based on X-ray crystallography.

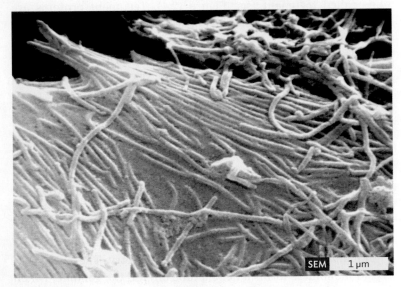

SEM   1 µm

B. Ebola virions.

TAILED BACTERIOPHAGES   Some bacteriophages have complex multipart structures. In a tailed phage, the icosahedral protein package, called the "head," is attached to an elaborate delivery device. For example, bacteriophage T4 (Figure 12.6) has an icosahedral head containing the pressure-packed DNA, attached to a helical "neck" that channels the nucleic acid into the host cell. The neck is surrounded by the tail sheath, with six jointed tail fibers that attach the bacteriophage to the host cell surface. Within the tail, an "injector" penetrates the host cell envelope, and the release of pressure within the head propels the DNA into the host cytoplasm.

AMORPHOUS VIRUSES   Some viruses have no symmetrical form (Figure 12.7). Poxviruses, such as vaccinia and smallpox, are large asymmetrical viruses with no symmetrical capsid. The size of vaccinia virus (360 nm) approaches that of some bacteria. Vaccinia, also known as "cowpox," is the virus used by Edward Jenner and others around 1800 to make the first vaccine against smallpox. Today, vaccinia is used as a genetic vector to develop vaccines against many animal diseases.

The DNA genome of a poxvirus is contained by a flexible "core wall" that also encloses a large number of enzymes and accessory proteins, similar to a cell's cytoplasm. The double-stranded DNA genome resembles that of a cell, except that it is stabilized by covalent connection of its two strands at each end. The core is enclosed loosely by a viral envelope studded with spike proteins. Large asymmetrical viruses are proposed to have evolved from degenerate cells.

Figure 12.6   Bacteriophage T4 Capsid

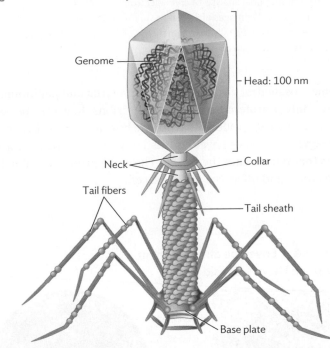

A. Phage T4 particle with protein capsid containing a packaged double-stranded DNA genome. The head is attached to a neck surrounded by a tail sheath, with tail fibers that attach to the surface of the host cell. After attachment, the sheath contracts and the core penetrates the cell surface, injecting the phage's genome.

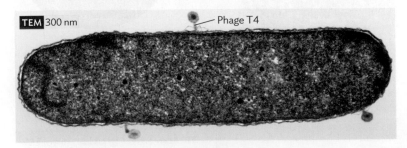

TEM 300 nm   Phage T4

B. *E. coli* infected by phage T4 (colorized blue).

- Viruses are noncellular particles (virions) containing a genome that replicates within a host cell. Viruses may have evolved from ancestral cells or from parts of cells.

- Viral genomes may consist of DNA or RNA.

- All kinds of organisms may be infected by viruses. A given virus has a host range that may be narrow or broad.

- The viral capsid is composed of repeated protein subunits—a structure that maximizes the structural capacity while minimizing the number of genes needed for construction. Capsids may be icosahedral, filamentous, tailed, or amorphous.

- Some types of viruses have an envelope derived from the host membrane. The envelope contains embedded spike proteins.

- Accessory proteins are contained within the capsid or as tegument between the capsid and envelope.

Thought Question 12.1 Which viruses have a narrow host range, and which have a broad host range?

Figure 12.7 Vaccinia Virus

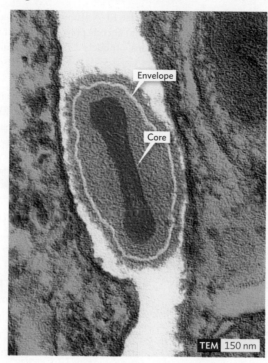

A. The vaccinia virus particle (colorized TEM).

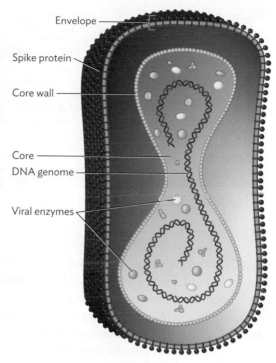

B. The pox virion consists of a core containing DNA and enzymes, surrounded by an envelope with embedded spike proteins.

12.2
Viral Genomes and Diversity

SECTION OBJECTIVES

- Describe the different classes of viral genomes, and give an example of each.
- Explain why different viruses infect different hosts.
- Describe the nature of viroids and prions.

How do viruses arrange their genetic information? The genomes of viruses, especially RNA viruses, can be surprisingly small; and the ability of such small entities to take over cells is remarkable. Viruses evolve rapidly, enabling them to evade host defenses and even to "jump" into new host species.

Viral Genomes

The genome of an RNA virus can be as small as three genes. **Figure 12.8A** depicts the genome of avian leukosis virus (ALV), a well-studied **retrovirus** (reverse-transcribing virus) that causes lymphoma in chickens. The ALV genome has three protein-encoding genes: *gag*, *pol*, and *env*. Each gene, however, can express two or three different products. The products from a given gene usually share a common functional category, such as core capsid, replicative enzymes, or envelope-embedded proteins. A gene's different products arise by different start sites of transcription and by cleavage of multiproduct proteins. Thus, the retroviral genome evolves for extreme economy of RNA content, maximizing the information potential of its tiny genome.

Other viral genomes can be large, approaching the size of cellular genomes. Double-stranded DNA viral genomes tend to be especially large, encoding numerous enzymes and regulatory proteins similar to those of cells. For example, the genome of herpes simplex virus 1, a cause of cold sores and genital herpes, spans 152 kilobases, encoding more than 70 gene products (**Figure 12.8B**). These herpes viral products include capsid and envelope proteins, DNA replication proteins, and accessory proteins that manage viral replication or interact with host immune cells. Other herpes genes encode latency proteins that can maintain the virus in a latent state within the host cell. Overall, herpes genomes have evolved to be highly regulated, with numerous extra products that have small but important effects on viral replication and host interaction.

Viroids and Prions

Some infectious agents are so limited in content that they cannot even be considered viruses. **Viroids** are virus-like infectious agents in which an RNA genome is itself the entire infectious particle;

Figure 12.8   Simple Viral Genomes

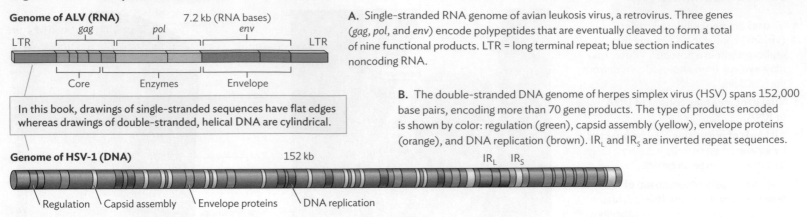

Genome of ALV (RNA)   7.2 kb (RNA bases)

A. Single-stranded RNA genome of avian leukosis virus, a retrovirus. Three genes (*gag*, *pol*, and *env*) encode polypeptides that are eventually cleaved to form a total of nine functional products. LTR = long terminal repeat; blue section indicates noncoding RNA.

In this book, drawings of single-stranded sequences have flat edges whereas drawings of double-stranded, helical DNA are cylindrical.

B. The double-stranded DNA genome of herpes simplex virus (HSV) spans 152,000 base pairs, encoding more than 70 gene products. The type of products encoded is shown by color: regulation (green), capsid assembly (yellow), envelope proteins (orange), and DNA replication (brown). IR$_L$ and IR$_S$ are inverted repeat sequences.

Genome of HSV-1 (DNA)   152 kb

there is no protective capsid. Most viroids infect plants, including many kinds of fruits and vegetables. For example, citrus viroids cause economic losses in the citrus industry. The viroid typically consists of a circular, single-stranded molecule of RNA that doubles back on itself to form base pairs interrupted by short unpaired loops. This unusual circularized form avoids breakdown by host RNase enzymes. The RNA folds up into a globular structure that interacts with host cell proteins. Some viroids have catalytic ability, comparable to enzymes made of protein. An RNA molecule capable of catalyzing a reaction is called a ribozyme. The small hepatitis D viral DNA acts like a viroid ribozyme (**Figure 12.9**). Hepatitis D coinfects human liver cells along with the hepatitis B virus, increasing the mortality rate of disease to 20%.

Figure 12.9   Hepatitis D Ribozyme

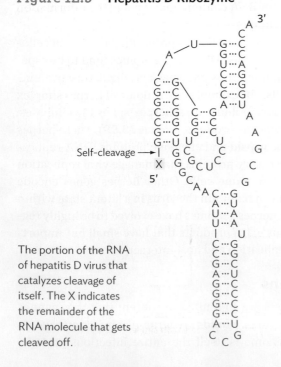

The portion of the RNA of hepatitis D virus that catalyzes cleavage of itself. The X indicates the remainder of the RNA molecule that gets cleaved off.

Yet another class of infectious agent has no nucleic acid at all. These agents, known as **prions**, are aberrant proteins arising out of a preexisting cell. Prions gained notoriety when they were implicated in transmissible Creutzfeldt-Jakob disease, a brain infection that became known as "mad cow" disease because it can be transmitted through consumption of beef from diseased cattle. Other diseases caused by prion transmission include scrapie, a brain disease of sheep; and kuru, a degenerative brain disease found in a small community of people who customarily consumed the brains of deceased relatives.

A prion is an abnormal form of a normally occurring brain cell protein called Prp^C (**Figure 12.10**). The prion form of the protein acts by binding to normally folded proteins of the same class and altering their conformation to that of the prion. The misfolded protein then alters the conformation of other normal subunits, forming harmful aggregates and ultimately killing the cell. In the brain, prion-induced cell death leads to tissue deterioration and dementia (discussed in Chapter 24).

Prion diseases can be initiated by infection with an aberrant protein. More rarely, the cascade of protein misfolding can start with the spontaneous misfolding of an endogenous host protein. The chance of spontaneous unfolding is greatly increased in individuals who inherit certain gene variants encoding the protein; thus, spontaneous prion diseases can be inherited genetically. Overall, prion diseases are unique in that they can be transmitted by an infective protein instead of by DNA or RNA; and they propagate conformational change of existing molecules without actually synthesizing new infective molecules.

Viral Diversity and Evolution

Like cellular organisms, viruses evolve through genome change and natural selection (discussed in Section 9.3). Their small genome size and small number of parts enable viruses to mutate tenfold or a hundredfold faster than their host cells. Rapid mutation and evolution of a virus leads to **antigenic drift**—a population of viruses whose mutant proteins are no longer recognized by host antibodies.

Figure 12.10 Prions Cause Disease

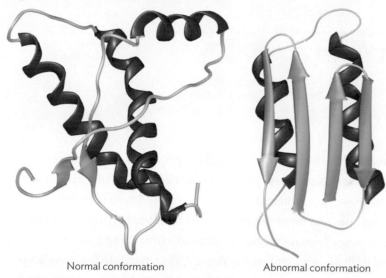

Normal conformation   Abnormal conformation

The conformation (folding pattern) of a normal, noninfectious, brain PrpᶜC protein (left) compared with that of an abnormal, infectious prion (right). The abnormal form "recruits" normally folded Prpᶜ proteins and changes their conformation into the abnormal form.

Antigenic drift generates new strains of virus that cause serious disease; for example, antigenic drift continually generates new influenza strains requiring repeated immunization.

Over extended time, viral evolution generates new kinds of viruses that cause different diseases in different hosts. **Figure 12.11** shows a phylogeny for herpes viruses, an ancient group of viruses infecting many animals. We see that over time, different diseases evolve infecting the same or different species. Members of the herpes family cause several important human diseases, including chickenpox (varicella-zoster), cold sores and genital lesions (herpes simplex 1 and 2), birth defects (cytomegalovirus), infectious mononucleosis (Epstein-Barr virus), and Kaposi's sarcoma (human herpes virus 8, not shown).

In general, viruses evolve at different levels, all of which are significant for medicine and epidemiology:

- **Within a host community.** Viruses evolve to preferentially infect different species—for example, equine herpes (in horses) versus murine herpes (in mice).

- **Within a viral species population.** Different viral strains arise that vary in infectivity and virulence. Closely related viruses may causes diseases that are similar (herpes simplex 1 and 2) or different (varicella-zoster versus cytomegalovirus).

- **Within an individual organism.** Viruses evolve variants that resist therapeutic agents. Extremely mutable viruses, such as hepatitis C and HIV, evolve a "quasispecies" of diverse strains that infect different tissues within a host.

Within natural environments, viruses evolve in response to host diversity. The most diverse communities tend to generate diverse viruses. A virus emerging from the environment into a human population may cause a serious epidemic among hosts whose immune systems have no prior exposure to the new pathogen. See **inSight** for a particularly compelling case of human viral pathogens emerging among African primate hosts.

Viral Classification: The Baltimore Model

How do we "classify" reproductive entities that don't even have cells and whose genomes differ radically in size and composition? The International Committee on Taxonomy of Viruses (ICTV) devised a classification scheme that includes factors such as capsid form (icosahedral or filamentous), envelope (present or absent), and host range. The most important classification factor, however, is the type of genome (see Table 12.1). Viruses of the same genome type (such as double-stranded DNA) are more likely to share ancestry with each other than with viruses of a different type of genome (such as RNA). The classification model based on genome type was devised by David Baltimore, who, with Renato Dulbecco and Howard Temin, was awarded the Nobel Prize in Physiology or Medicine in 1975 for discovering how tumor viruses cause cancer.

Figure 12.11 Phylogeny of Herpes Viral Genomes

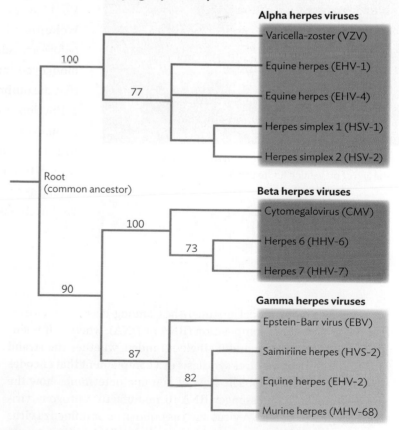

Phylogeny of human and animal herpes viruses, based on whole-genome sequence analysis comparing groups of genes with similar function. Numbers measure genetic divergence.

inSight

Monkey Business and Deadly Viruses

Figure 1 **Nathan Wolfe, Director of Global Viral Forecasting**

Wolfe hunts for emerging viruses in African monkeys, with the aim of preventing future pandemics.

In 2014, an outbreak of Ebola virus devastated the countries of Liberia, Sierra Leone, and Guinea; and some cases reached countries outside of Africa. Ebola virus, HIV, and monkeypox: where did they come from? These very different viruses all came from apes and monkeys, our nearest relatives in Africa, the birthplace of humans and other primate species. As primates (including humans) evolved, our viruses evolved with us—and continue to do so. A virus needs to find new hosts. It may jump from one species to a closely related one. Thus, a region such as Africa with rich diversity of our closest animal relatives is also a rich source of emerging viral pathogens. For example, Ebola virus causes outbreaks in gorillas and chimpanzees, killing large numbers of animals. The dead animals are then found by hunters, who bring them home to feed their families. This consumption of "bushmeat" can then lead to a human outbreak of Ebola.

Nathan Wolfe is a professor at Stanford University and director of Global Viral Forecasting, an organization that seeks to study emerging viruses (Figure 1). Wolfe's aim is to prevent future pandemics by actually predicting viral emergence before a pandemic occurs. For example, Wolfe's team reported an outbreak of monkeypox among humans in the Democratic Republic of the Congo. He showed that the outbreak correlated with the decline in smallpox vaccination following the extinction of smallpox; the vaccination had also cross-protected humans against monkeypox transmitted from monkeys. A slight mutation of monkeypox could lead to a global outbreak of smallpox-like disease. In fact, Wolfe argues, humanity has been extremely "lucky" with recent pandemics such as AIDS: if HIV could actually spread as easily as smallpox, think how many people would have died. In the next viral outbreak, we may not be so lucky.

The **Baltimore model** distinguishes among classes of viruses based on the genome composition (RNA or DNA); whether it is single or double-stranded; and if single-stranded, whether the strand encodes protein or requires synthesis of a complement that encodes proteins (Figure 12.12). The type of genome determines how the virus will make its messenger RNA to manufacture progeny virions. For example, an RNA virus such as poliovirus or influenza virus lacks DNA and therefore cannot use a cellular RNA polymerase to transcribe its genes; the virus must specify an RNA-dependent RNA polymerase. Assuming a common host, the different means of mRNA production generate distinct groups of viruses with shared ancestry.

The seven Baltimore groups of viruses, based on genome type and mRNA generation, are outlined in Figure 12.12. Examples of each group are presented in Table 12.1. Note, however, that within each genome type, very different viruses have evolved in hosts as different as human versus bacterium. It is likely that a given form of genome (such as single-stranded DNA) may have evolved independently in different viral lineages.

In his work, Wolfe travels among African villages, testing monkeys for new viruses. In one study, he found monkeys harboring new kinds of human T-lymphotrophic viruses (HTLVs), retroviruses related to HIV. He reported two strains, HTLV-3 and HTLV-4, never before seen, their effects on humans unknown. Many more such strains must exist in the wild. To avoid introducing such strains into humans, Wolfe urges the local hunters to avoid eating monkeys, especially dead monkeys they find in the forest—monkeys that may have died of viral disease. But the local people need these dead monkeys to feed their families; for them, the alternative may be starvation.

Although the circumstantial evidence is compelling, can we actually prove the host source of a given virus? At the University of Alabama at Birmingham, Beatrice Hahn used laboratory methods to do just that, tracing the precise origin of HIV. The cause of AIDS, HIV has one of the highest mutation rates of any virus, and it evolves quickly. Hahn performed statistical comparisons of various HIV strains with strains of simian immunodeficiency virus (SIV), viruses closely related to HIV that cause immunodeficiency in monkeys. She sequenced the genomes of 50 new SIV strains from chimpanzees in the Southern Cameroon. Some of the SIV strains she found were even more closely related to HIV than any SIV strain found before. Her work established the statistical likelihood that the major HIV strain affecting North America and Europe (group M) diverged specifically from SIV strains in chimpanzees of the Cameroon (Figure 2).

The emergence of HIV group M in humans occurred well before 1960. But in 2009, researchers reported a much more recent emergence of a novel HIV strain in a Cameroonian woman—a strain never before seen in humans. This strain showed closest relatedness to an SIV strain in gorillas. Thus, HIV was "caught in the act" of emerging from our primate relatives.

Although primates are a source of deadly viruses, they also provide our key source of information on how to combat these pathogens. For instance, SIV infects certain primate species without causing any symptoms. A monkey called the sooty mangabey shows no symptoms, despite high SIV titers, whereas the rhesus macaque develops simian AIDS. By studying the host response of monkeys that resist the virus, we may derive clues suggesting therapies for human AIDS.

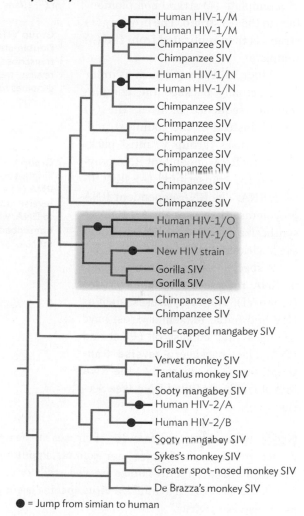

Figure 2 Beatrice Hahn's Phylogeny of HIV Strains

● = Jump from simian to human

The phylogeny of HIV strains shows that groups M and N diverged from SIV in chimpanzees, whereas a novel HIV strain arose from gorilla SIV (shaded orange). Strains of HIV-2 arose from SIV in sooty mangabeys.

The type of genome (DNA versus RNA, single- versus double-stranded) determines how the virus infects the cell and can influence the course of disease in the patient. Viruses such as the herpes and smallpox viruses (group I) make their own DNA polymerase or use that of the host for genome replication. Their genes can be transcribed directly by a host RNA polymerase or one encoded by a viral genome.

By contrast, single-stranded DNA viruses (group II), such as canine parvovirus, require the host DNA polymerase to generate the complementary DNA strand. The double-stranded DNA is then transcribed by host RNA polymerase to make mRNA.

Double-stranded RNA viruses (group III) require a viral **RNA-dependent RNA polymerase** to generate mRNA by transcribing directly from the RNA genome. Such viruses must package a viral RNA polymerase with their genome before exiting the host cell.

Single-stranded RNA viruses, such as hepatitis C and SARS coronavirus (group IV), have a positive-sense (+) strand (the coding strand) that can serve directly as mRNA to be translated to viral proteins. To replicate the RNA genome, however, requires synthesis of a template (–) strand complementary to the (+) strand. The progeny (+) strand is then replicated from the (–) template.

Other RNA viruses, such as influenza, actually package the complementary sequence of RNA, the (–) strand, instead of the coding strand. These viruses (group V) must package a viral RNA-dependent RNA polymerase for transcribing (–) RNA to (+) mRNA. The RNA-dependent RNA polymerase then uses the (+) RNA to synthesize genomic (–) RNA to package in virions.

A special case is the **retroviruses**, or **RNA reverse-transcribing viruses** (group VI). The retroviruses, such as HIV and feline leukemia virus, have genomes that consist of (+) strand RNA. They package a **reverse transcriptase**, which transcribes the RNA into a double-stranded DNA (see Section 12.6).

Figure 12.12   Baltimore Classification of Viral Genomes

There are seven categories of viral genome composition, reflecting RNA or DNA, single or double-stranded, and (+) or (–) strand. Each kind of genome requires a different route to mRNA and a different replication mechanism.

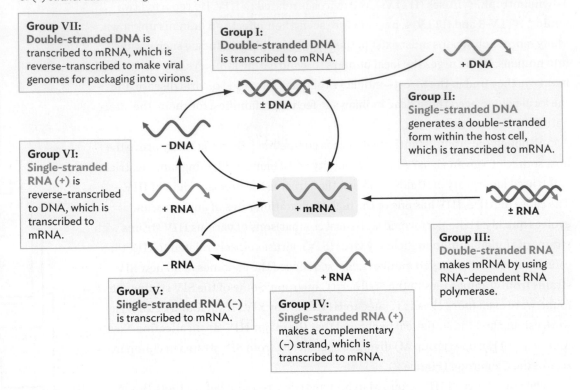

Group VII: Double-stranded DNA is transcribed to mRNA, which is reverse-transcribed to make viral genomes for packaging into virions.

Group I: Double-stranded DNA is transcribed to mRNA.

Group II: Single-stranded DNA generates a double-stranded form within the host cell, which is transcribed to mRNA.

Group VI: Single-stranded RNA (+) is reverse-transcribed to DNA, which is transcribed to mRNA.

Group III: Double-stranded RNA makes mRNA by using RNA-dependent RNA polymerase.

Group V: Single-stranded RNA (–) is transcribed to mRNA.

Group IV: Single-stranded RNA (+) makes a complementary (–) strand, which is transcribed to mRNA.

Note In nomenclature, families of viruses are designated by Latin names with the suffix "-iridae": for example, Papillomaviridae. Common forms of such family names are also used: for example, the papillomaviruses. Within a family, a virus species is simply capitalized, as in "Papillomavirus."

Now that we have introduced the major principles of virology, we are ready to examine the culture and infection cycle of a bacteriophage and the infection cycles of three different human-infecting viruses: papillomavirus (group I, dsDNA); influenza virus (group V, (–) sense RNA); and HIV (group VI, retrovirus).

SECTION SUMMARY

- **Viral genomes can be large or small.** RNA genomes usually have few genes, whereas double-stranded DNA viral genomes may approach the size of cellular genomes.
- **Viroids and prions are infectious particles that are not viruses.** Viroids consist of an RNA circle that self-hybridizes. Prions consist of aberrantly folded cellular proteins.
- **Viruses evolve into strains that cause diverse diseases in diverse hosts.**
- **Viruses evolve variant strains that resist therapeutic drugs.**
- **The Baltimore model classifies viruses according to their type of genome and means of generating mRNA.** Genomes may be RNA or DNA, single- or double-stranded.

Thought Question 12.2   What are the advantages to the virus of having a small genome? What are the advantages of a large genome?

12.3
Virus Replication and Culture

SECTION OBJECTIVES

- Explain how viral genomes enter cells.
- Describe the lytic and lysogenic cycles of bacteriophage infection.
- Explain how viruses are cultured using host cells.

How do viruses find their host cell, subvert its defenses, and take over its metabolism? Viruses show surprisingly diverse mechanisms for infection and replication. In this section, we examine bacteriophages as a model; subsequent sections present examples of human viruses with DNA and RNA genomes.

One way or another, all viral replication cycles must achieve the following tasks:

- **Host recognition and attachment.** Viruses must contact and adhere to a host cell that can support their particular reproductive mechanism.

- **Genome entry.** The viral genome (either isolated or within an endocytosed virion) must enter the host cell and gain access to the cell's machinery for gene expression.

- **Assembly of progeny virions.** Viral components must be expressed and assembled. Components usually "self-assemble"; that is, the joining of their parts is favored by the energy release of spontaneous assembly.

- **Exit and transmission.** Progeny virions must exit the host cell and then reach new host cells to infect. In the case of multicellular organisms, the virus must eventually reach other multicellular hosts (epidemiology is discussed in Chapter 26).

A virus needs to contact and attach to the surface of an appropriate host cell, one that its own genetics has evolved to take over. A virus binds specifically to a **surface receptor**, a protein on the host cell surface that is specific to the host species and that binds to a specific viral part. An example is the tail fibers of bacteriophage T4 (**Figure 12.13**).

Why would a host cell evolve to make a surface protein that attaches to a virus? Actually, the receptor for a virus is a protein with an important function for the host cell, but the virus has evolved to take advantage of the protein. For example, the CD4 surface protein of T cells (discussed in Chapter 16) is often called "HIV receptor," although it evolved as a signaling protein of the immune system.

Bacteriophage Replication

Bacteriophages interact with bacteria in all ecosystems, from soil and water to the human gut. Phages decrease population density, increase host diversity, and transfer genes among diverse bacteria. Phages are relatively easy to study in the laboratory, where their host bacteria can be grown in tube and plate cultures (methods presented in Chapter 6). Thus, phage-host systems provide important models for virus infectious cycles and for virus culture. We consider here the extensively studied bacteriophage lambda, which infects the common enteric bacterium *Escherichia coli*. Enteric bacteriophages are an important part of the ecology of the human intestine, where they transfer genes enabling our bacteria to digest different kinds of food materials that we consume.

How does bacteriophage lambda know which bacteria to infect within the mixed population of our colon? The phage virion binds specifically to the maltose porin in the outer membrane of *E. coli*. Although the maltose porin protein is often called "lambda receptor protein," it actually evolved in *E. coli* as a way to obtain the sugar maltose for catabolism. Thus, natural selection maintains the maltose porin in *E. coli*, despite the danger of phage infection.

So how does the inert phage particle penetrate the castle-like fortress of the bacterial cell? The phage injects its genome through the cell envelope into the cytoplasm, thus avoiding the need for the capsid to penetrate the tough cell wall. The sheath of the phage neck tube contracts, bringing the head near the cell surface to insert its DNA (see Figure 12.13). The pressure of the spooled

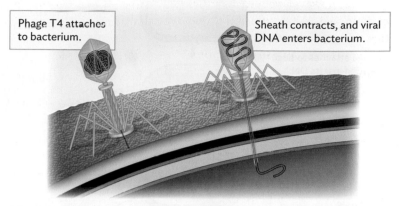

Figure 12.13 Bacteriophage Genome Insertion

Phage T4 attaches to bacterium.

Sheath contracts, and viral DNA enters bacterium.

A bacteriophage attaches to the cell surface by its tail fibers and then contracts to inject its DNA. The empty capsid remains outside as a "ghost."

DNA—as high as 50 atmospheres—is released, expelling the DNA into the cell. After inserting its genome, the phage capsid remains outside, attached to the cell surface. The empty capsid is termed a "ghost" because of its pale appearance in an electron micrograph. (The entry of animal viruses, in contrast, involves binding receptors that trigger endocytosis or fusion with the host cell membrane, to be discussed shortly.)

When a virulent phage injects its genome into a cell, it immediately begins a replicative cycle to form as many progeny phage particles as possible (**Figure 12.14**, left side). The process involves replicating the phage genome, expressing phage mRNA to make enzymes and capsid proteins, and phage assembly, followed by **lysis**, the rupture of the host cell, which releases progeny phage (**lytic infection**). Some phages, such as phage T4, digest the host DNA to increase the efficiency of phage production; others, such as phage lambda, retain the host DNA for potential **lysogeny**, a process in which the phage genome becomes integrated in the genome of the host cell (see Figure 12.14, right side). Lysogenic phages may carry toxin genes, such as the gene expressing Shiga toxin in *E. coli* O157:H7.

We now examine in more detail the alternative "fates" of lytic infection (lysis) and lysogeny. Although we focus on bacteriophages, similar alternative strategies for replication—for example, rapid virus replication versus hiding in a host cell—are exploited by human viruses such as HIV and herpes viruses.

LYTIC INFECTION After the phage inserts its DNA into the cell, the phage genes are expressed by the host cell RNA polymerase and ribosomes. "Early genes" are those expressed early in the lytic cycle—those needed for transcribing DNA and other early steps to produce progeny phage. Other phage-expressed proteins then work together with the host's cellular enzymes and ribosomes to replicate the phage genome and produce phage capsid proteins. The capsid proteins self-assemble into capsids and package the

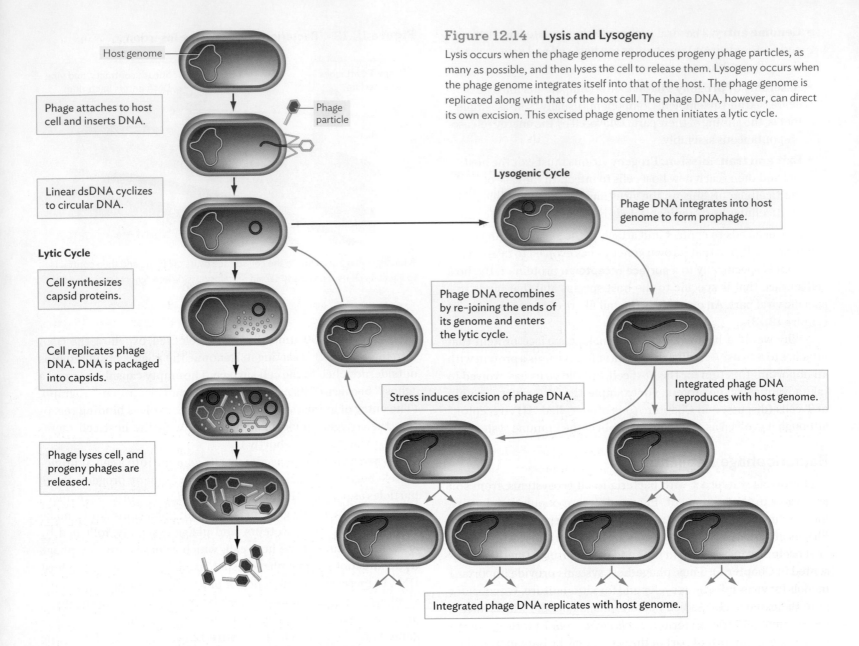

Figure 12.14   Lysis and Lysogeny
Lysis occurs when the phage genome reproduces progeny phage particles, as many as possible, and then lyses the cell to release them. Lysogeny occurs when the phage genome integrates itself into that of the host. The phage genome is replicated along with that of the host cell. The phage DNA, however, can direct its own excision. This excised phage genome then initiates a lytic cycle.

Host genome

Phage attaches to host cell and inserts DNA.

Phage particle

Linear dsDNA cyclizes to circular DNA.

Lysogenic Cycle

Phage DNA integrates into host genome to form prophage.

Lytic Cycle

Cell synthesizes capsid proteins.

Phage DNA recombines by re-joining the ends of its genome and enters the lytic cycle.

Cell replicates phage DNA. DNA is packaged into capsids.

Integrated phage DNA reproduces with host genome.

Phage lyses cell, and progeny phages are released.

Stress induces excision of phage DNA.

Integrated phage DNA replicates with host genome.

phage genomes—a process that takes place in defined stages, like a factory assembly line. At last, a "late gene" from the phage genome expresses an enzyme that lyses the host cell wall, releasing the mature virions. Lysis is also referred to as a burst, and the number of virus particles released is called the **burst size**.

Some phages (such as phage T4) reproduce entirely by the lytic cycle and thus are called **virulent phages**. Other phages (such as phage lambda) that can undergo lysogeny are called **temperate phages**. A temperate phage such as phage lambda can infect and lyse cells in the same manner as a virulent phage, but can also exploit the alternative pathway, lysogeny, which involves integrating its genome into that of the host cell.

LYSOGENY Phage lambda has a linear genome of double-stranded DNA, which circularizes upon entry into the cell. To

lysogenize, the circularized genome recombines into that of the host by site-specific recombination of DNA. In site-specific recombination, an enzyme aligns the phage genome with the host DNA and exchanges the sugar-phosphate backbone links with those of the host genome. The DNA exchange process integrates the phage genome into that of the host. The integrated phage genome is called a **prophage**. A similar integration event occurs in human viruses, such as when the DNA copy of an HIV genome integrates into the genome of a host T cell (discussed in Section 12.6).

Integration of the phage lambda genome as a prophage results in lysogeny, a condition in which the phage genome is replicated along with that of the host cell as the host reproduces. Implicit in the term "lysogeny," however, is the ability of such a strain to spontaneously generate a lytic burst of phage. For lysis to occur, the prophage (integrated phage genome) directs its own excision from the host genome. The two ends of the phage genome are cut, come apart from the host DNA molecule, and then are re-joined so that the phage DNA is once again circularized. The phage DNA then exits the host genome and initiates a lytic cycle, destroying the host cell and releasing phage particles.

For a temperate virus (a virus capable of lysogeny), the "decision" between lysogeny and lysis is determined by proteins that bind DNA and repress the transcription of genes for virus replication. Exit from lysogeny into lysis can occur at random, or it can be triggered by environmental stress such as UV light, which damages the cell's DNA. The regulatory switch of lysogeny responds to environmental cues indicating the likelihood that the host cell will survive and continue to propagate the phage genome. When a cell grows under optimal conditions, it is more likely that the phage DNA will remain inactive, whereas events that threaten host survival will trigger a lytic burst. An analogous phenomenon occurs in human viral infections such as herpes, in which environmental stress triggers reactivation of a virus that was dormant within cells (a latent infection). Reactivation of a latent herpes infection results in painful outbreaks of skin lesions.

During the exit from bacteriophage lysogeny or during the latent growth of animal viruses, the virus can acquire host genes and pass them on to other host cells. In bacteria, this process of viral transfer of host genes is known as **transduction** (discussed in Section 9.1). A transducing bacteriophage can pick up a bit of host genome and transfer it to a new host cell. In some forms of transduction, the entire phage genome is replaced by host DNA packaged in the phage capsid, resulting in a virus particle that transfers only host DNA. In other cases, only a small part of the host genome is transferred, attached to the viral genome.

The integration and excision of viral genomes make extraordinary contributions to the evolution of host genomes, including pathogens. For example, the CTXφ prophage encodes cholera toxin in *Vibrio cholerae*; and prophages encode toxins for *Corynebacterium diphtheriae* (diphtheria) and *Clostridium botulinum* (botulism). In the laboratory, the ability of phages to transfer genes is exploited in developing vectors (carrier genomes) for biotechnology.

Culturing Viruses

To study any living microorganism, we must culture it in the laboratory. A complication of virus culture is the need to grow the virus within a host cell. Therefore, any virus culture system must be a double culture of host cells plus viruses.

Bacteriophages may be produced in **batch culture**, a culture in an enclosed vessel of liquid medium. Batch culture enables growth of a large population of viruses for study. A phage sample is inoculated into a growing culture of bacteria, usually in a culture tube or a flask. The culture fluid is then sampled over time and assayed for phage particles. The growth pattern usually takes the form of a **one-step growth curve** (Figure 12.15). To observe one cycle of phage

Figure 12.15 One-Step Growth Curve for a Bacteriophage

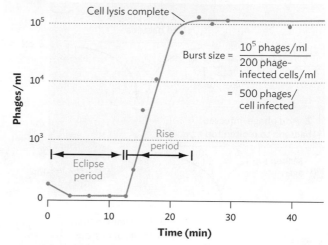

After initial infection of a liquid culture of host cells, the titer of virus drops near zero as all virions attach to the host. During the eclipse period, progeny phages are being assembled within the cell. As cells lyse (the rise period), phages are released until they reach the final plateau. The infectious cycle is typically complete within less than an hour.

reproduction, we must add phages to host cells at a multiplicity of infection (MOI, ratio of phage to cells) such that every host cell is infected. The phage particles immediately adsorb to surface receptors of host cells and inject their DNA. As a result, phages are virtually undetectable in the growth medium for a short period after infection called the **eclipse period**. The progeny phages within the bacterial cell before it lyses can be seen by transmission electron microcopy (see Figure 12.2A).

As cells begin to lyse and liberate progeny phages, the culture enters the rise period, in which phage particles begin appearing in the growth medium. The rise period ends when all the progeny phages have been liberated from their host cells. If the number of phages that go on to reinoculate other host cells is small, then the phage concentration at the end point divided by the original concentration of inoculated phage approximates the burst size—that is, the number of phages produced per infected host cell. We can estimate the burst size by dividing the concentration of progeny phages by the concentration of inoculated phages, assuming that all the original phage particles infect a cell.

An important culture technique is the growth of individual colonies on a solid substrate that prevents dispersal throughout the medium, as described in Chapters 1 and 4. Plate culture of colonies enables us to isolate a population of microbes descended from a common progenitor. But how can we isolate viruses as "colonies"? Although viruses can be obtained at incredibly high concentrations, they disperse in suspension. Even on a solid medium, viruses never form a solid visible mass comparable to the mass of cells that constitutes a cellular colony.

Figure 12.16   Pour-Plate Culture of Bacteriophages on Host Bacterial Lawn

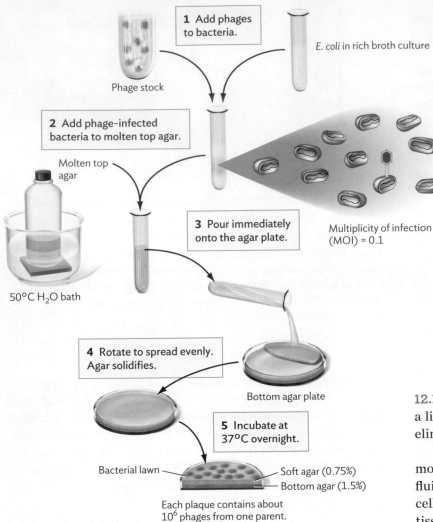

1 Add phages to bacteria.

E. coli in rich broth culture

Phage stock

2 Add phage-infected bacteria to molten top agar.

Molten top agar

3 Pour immediately onto the agar plate.

Multiplicity of infection (MOI) = 0.1

50°C H₂O bath

4 Rotate to spread evenly. Agar solidifies.

Bottom agar plate

5 Incubate at 37°C overnight.

Bacterial lawn — Soft agar (0.75%) — Bottom agar (1.5%)

Each plaque contains about 10⁶ phages from one parent.

A. A suspension of bacteria in rich broth culture is inoculated with a low proportion of phage particles (multiplicity of infection is approximately 0.1). This means that only a few of the bacteria become infected immediately, while the rest continue to grow. Each plaque arises from a single infected bacterium that bursts, its phage particles diffusing to infect neighboring cells.

Plaques

B. Phage lambda plaques on a lawn of *Escherichia coli* K-12.

12.17A). The host cells are usually an immortalized cell line, that is, a line of cells from a cancer in which genetic control of growth is eliminated, so the cells can double in culture indefinitely.

Tissue culture involves the growth of human or animal cells in a monolayer on the surface of a dish containing fluid medium. But the fluid medium would quickly disperse any viruses released by lysed cells. The tissue culture procedure is modified as follows. First, the tissue culture is inoculated with a virus suspension (step 1). Then, after sufficient time to allow for viral attachment to cells, the culture fluid is removed, including any unattached virions (step 2). The fluid is then replaced by a gelatin medium (step 3). The gel retards the dispersal of viruses from infected cells, and as the host cells die, plaques are observed (step 4). **Figure 12.17B** shows the appearance of plaques of human coronavirus (a cause of colds) cultured on colon carcinoma cells.

The remaining three sections of this chapter describe the replication and pathology of three viruses that infect humans: human papillomavirus (HPV), influenza virus, and human immunodeficiency virus (HIV). Each example reveals a different kind of replication cycle with very different implications for the patient.

SECTION SUMMARY

- **Virions attach to host cell surface receptors.** The receptors, proteins with important host function, determine which host cells the virus can infect.

- **Lytic replication lyses the host cell.** A virulent phage injects its DNA into a host cell, where it uses host gene expression machinery to produce progeny virions.

In viral plate culture (**Figure 12.16**), viruses from a single progenitor lyse their surrounding host cells, forming a clear area called a **plaque**. To perform a plaque assay of bacteriophages, we mix a diluted suspension of bacteriophages with bacterial cells in soft agar (agar at low concentration), then pour the mixture over an agar plate. Phage particles diffuse through the soft agar until they reach a host bacterium. Where no bacteriophages are present, the bacteria grow homogeneously as a "lawn," an opaque sheet over the surface (confluent growth). Where there is a bacteriophage, it infects a cell, replicates, and spreads progeny phages to adjacent cells, killing them as well. The loss of cells results in a round, clear area seemingly cut out of the bacterial lawn; the tiny clear areas are the plaques. Plaques can be counted and used to calculate the concentration of phage particles, or **plaque-forming units (PFUs)**, in a given suspension of liquid culture. The liquid culture can be analyzed by serial dilution in the same way we would analyze a suspension of bacteria.

CULTURING ANIMAL VIRUSES   For animal viruses, the plaque assay has to be modified for host cells in tissue culture (**Figure**

Figure 12.17 Culture of Human Coronavirus

A. Modified plaque assay using cultured host cells. The addition of gelatin medium (step 3) retards the dispersal of progeny virions from infected cells, restricting new infections to neighboring cells. The result is a visible clearing of cells (a plaque) in the monolayer.

1 Infect monolayer with virus.

2 Remove liquid medium.

3 Add gelatin medium.

4 Virus reproduces. Host cells lyse, forming plaques.

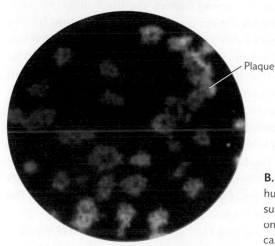

Plaque

B. Plaque assay in which human coronavirus suspension was plated on a monolayer of colon carcinoma cells.

- **Lysogeny maintains the phage genome within that of the host.** A phage that can undergo lysogeny is called a **temperate phage**. In lysogeny, the viral genome is integrated into that of the host cell, which then replicates the phage genome along with its own. Under stress, a temperate phage can initiate a lytic cycle.

- **Batch culture of viruses generates a step curve.** For a period of time, there are no new virions; then, as infected cells lyse, the viral titer rises rapidly.

- **Viruses requires culture within host cells.** Bacteriophages may be cultured either in batch culture or as isolated plaques on a bacterial lawn. Animal viruses are grown in tissue culture.

Thought Question 12.3 Given the mechanism by which viruses infect cells, how might an organism evolve to become resistant to a viral infection?

12.4
Papillomavirus: DNA Genome

SECTION OBJECTIVES
- Describe the genome of papillomavirus.
- Explain the infectious cycle of papillomavirus, and explain how papillomavirus causes cancer.

CASE HISTORY 12.1

Genital Warts from a Virus

Sean was a 19-year-old college sophomore at a large midwestern university. Sean visited the campus health center, where he told the nurse practitioner that the shaft of his penis had a raised spot. The spot caused no discomfort yet, but Sean knew it was abnormal. The practitioner told Sean that he probably had human papillomavirus (HPV), a cause of genital warts (**Figure 12.18A**). Although some HPV strains cause warts on external skin, genital HPV is transmitted only by sexual intercourse. The practitioner conducted a physical exam and discovered additional raised spots on Sean's anus. Sean told the practitioner that he had vaginal sex with women, and had also experimented sexually with his fraternity brothers. The practitioner told Sean that the warts would probably go away on their own if he maintained good sleep and health habits, keeping his immune system strong; but some HPV strains can persist for years without symptoms. The warts could be removed by freezing or chemical treatment, but the virus would persist. Meanwhile, the presence of HPV suggested that Sean was likely to contract related strains causing cancer of the penis or anus. The practitioner recommended use of condoms, although HPV may infect areas that condoms do not protect. She also suggested that Sean recommend the Gardasil HPV vaccine for his younger brother and sister before they became sexually active.

How prevalent is papillomavirus? In the United States, the genital strains of human papillomavirus (HPV), including those that cause cancers of the cervix, penis, and anus, infect 80% of adults. The discovery that cervical cancer is caused by HPV earned German researcher Harald zur Hausen the 2008 Nobel Prize in Physiology or Medicine (he shared the prize with Françoise Barré-Sinoussi and Luc Montagnier, the researchers who discovered that HIV causes AIDS; discussed in Section 12.6). The genital strains of HPV are highly infectious through sexual contact, including oral contact leading to throat cancers. Although condoms offer partial protection, greater protection is afforded by vaccines such as Gardasil, but the vaccine is effective only before exposure to the virus. Gardasil is now recommended by the U.S. Centers for Disease Control for all girls and boys by age 11–12. The vaccine protects against four of the most prevalent strains, including HPV-16, the leading cause of

Figure 12.18   Human Papillomavirus Infection

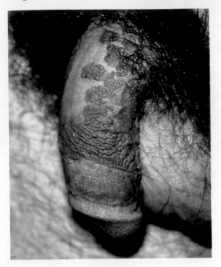

A. Certain strains of human papillomavirus (HPV) cause warts on the genitals or anus.

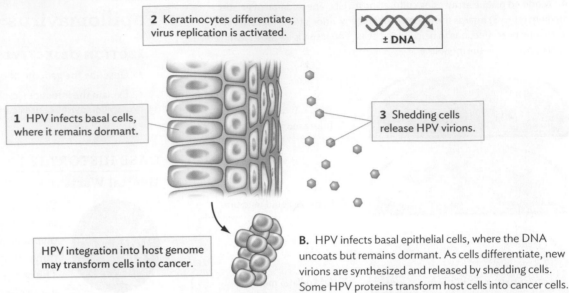

2 Keratinocytes differentiate; virus replication is activated.

± DNA

1 HPV infects basal cells, where it remains dormant.

3 Shedding cells release HPV virions.

HPV integration into host genome may transform cells into cancer.

B. HPV infects basal epithelial cells, where the DNA uncoats but remains dormant. As cells differentiate, new virions are synthesized and released by shedding cells. Some HPV proteins transform host cells into cancer cells.

genital cancer. The only sure way to avoid genital HPV, however, is to abstain from sexual contact or to have sex only in an exclusive relationship.

Like many viruses, HPV shows distinct tropism, the ability to infect specific organs or tissues. HPV has a narrow tropism: strains infect specifically the skin (cutaneous epithelium; **Figure 12.18B**) or the mucous membranes. Tropism is usually determined by specific cell-surface protein receptors (just as bacteriophage attachment requires specific host protein receptors). For HPV, the receptors for some strains are heparan sulfate proteoglycans (HSPGs), cell-surface proteins involved in wound healing. The cell needs these proteins for important functions, and HPV evolved to take advantage of them.

In the affected tissues, the virus must gain access to the actively dividing cells of the basal layer, usually through a tiny wound in the tissue. Virions are endocytosed by the basal cells. Their replication, however, is inhibited until the basal cells start to differentiate into keratinocytes (mature epithelial cells). As the epithelial layers reach the surface and slough off, progeny virions are shed. In other infected cells, however, the HPV virions become latent, persisting in the cell for months or years. These latent viral genomes may induce the host cells to form abnormal growths, such as warts or cancers. Different HPV strains cause skin warts, genital warts, and genital cancers; that is why, in Case History 12.1, Sean was told that the penile warts would not cause cancer, but similarly transmitted HPV strains could.

Structurally, the papillomaviruses are small icosahedral viruses, about twice the size of a ribosome. The virion has no envelope (**Figure 12.19A**). Its genome consists of a circular double-stranded DNA (see Figure 12.12, Baltimore Group I). The genome size of papillomavirus is relatively small, compared with those of herpes

or mimivirus; for example HPV-16 has fewer than 8,000 base pairs encoding only eight gene products (**Figure 12.19B**). The genes are expressed in overlapping **reading frames**, the three different positions to start defining triplet codons. Using three reading frames allows partial overlap of the genes and maximizes the efficiency of the information content in the shortest possible genome. The shorter the viral genome, the greater the number of virions that can be made with the DNA nucleotides available within a host cell.

Because HPV replication requires the developmental transition of basal cells into cells of the epidermis (outer skin layers) (see Figure 12.18B), HPV is difficult to grow in culture for study. HPV is often eliminated by the host immune system, but if infection persists, no effective drug therapies are known. To discover treatments for HPV, researchers must identify drug targets by studying the virus's replication cycle (**Figure 12.20**).

To infect a host cell and initiate replication, the HPV virion binds to cell-surface receptors (such as HSPGs) and becomes endocytosed by the host cell (see Figure 12.20, step 1). The virion then travels to the nucleus (step 2) by transport through the endomembrane system (described in Chapter 4). At a nuclear pore complex, the virus uncoats; that is, the capsid releases its genome (step 3). For HPV, the uncoated genome must enter the nucleus for all replication events. Nuclear replication is typical of many DNA viruses (such as herpes), but some DNA viruses, such as smallpox and mimiviruses, replicate entirely in the cytoplasm.

For HPV, host cell differentiation induces the viral DNA to replicate and undergo transcription by host polymerases (step 4). The viral mRNA transcripts then exit the nuclear pores, as do host mRNAs, for translation in the cytoplasm (step 5). The translated capsid proteins, however, return to the nucleus for assembly of the virions (step 6).

Figure 12.19 Human Papillomavirus HPV-16, a Cause of Genital and Anal Cancers

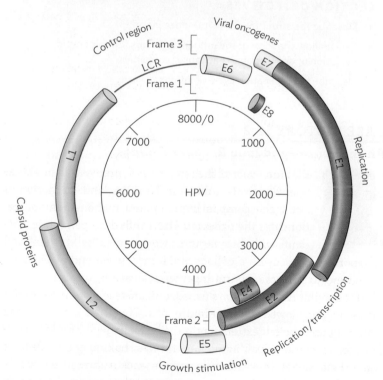

A. HPV virion, diameter 55 nm.

B. Genome of HPV-16. The DNA genome is circular; the "starting" base position (base 1) is collocated with base 7906. The genome encodes eight gene products, some of which overlap as shown.

Figure 12.20 Papillomavirus Replication Cycle

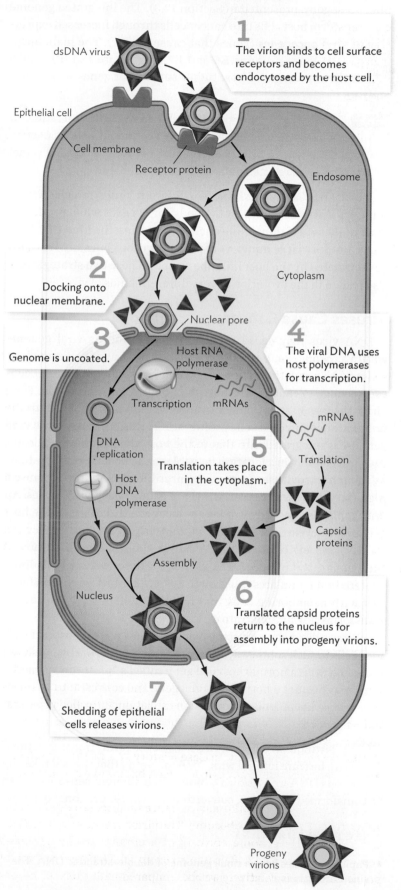

1 The virion binds to cell surface receptors and becomes endocytosed by the host cell.

2 Docking onto nuclear membrane.

3 Genome is uncoated.

4 The viral DNA uses host polymerases for transcription.

5 Translation takes place in the cytoplasm.

6 Translated capsid proteins return to the nucleus for assembly into progeny virions.

7 Shedding of epithelial cells releases virions.

As the keratinocytes complete differentiation, they start to come apart and are shed from the surface. The virions are released from the cell during this shedding process (step 7). In some cases, the virus eventually is eliminated by the body's immune system. In the basal cells, however, HPV may take an alternative pathway, integrating its genome into that of the host basal cells (analogous to phage lysogeny, presented in Section 12.3). The integrated genome can transform host cells into cancer cells through increased expression of viral **oncogenes** (genes that cause cancer). The main oncogenes of HPV are proteins E6 and E7 (see Figure 12.19B), which inhibit the expression of host tumor suppressor genes such as *p53* and *pRB*.

Link As described in Section 18.6, the protein **pRB** is a regulator that limits mitosis and cell division. HPV protein E7 helps destroy pRB, which removes a factor blocking cell division. The result is that cells replicate more often and can become cancerous.

Thus, HPV strains exhibit two different viral strategies. Oncogenic HPV strains persist in the body for decades while producing low numbers of virions. Non-oncogenic strains, the kind that form warts, produce large numbers of progeny virions but can eventually be eliminated by the host immune system. These two strategies are also shown in the evolution of very different kinds of viruses.

Viruses Cause Cancer

A DNA virus may cause cancer by integrating its viral genome within a chromosome of the host cell in such a way as to disrupt host cell growth regulation. The process of a virus inducing carcinogenesis (change to cancer) in a host cell is called **transformation**. For a virus, the advantage of cancer transformation is that it expands the population of infected cells proliferating virus particles. If the viral genome is integrated into that of the host, the host now replicates the viral genome in a location invisible to the host immune system.

Some retroviruses (reverse-transcribing RNA viruses) cause a high rate of virulent cancer; these are called oncogenic viruses. An oncogenic virus usually carries an oncogene that disrupts the host cell growth cycle. For example, human T-cell lymphotrophic virus (HTLV), closely related to HIV, causes abnormal proliferation of the T cells that it infects. HTLV carries the *tax* oncogene, encoding a protein that stimulates cell growth despite contact inhibition (contact with neighboring cells that halts growth). The *tax* oncogene transforms normal cells into cancer cells.

Where do oncogenes come from, and how do they interact with host cells in such subtle ways? In some cases, a viral oncogene shows homology with a normal host cell gene called a "proto-oncogene." A virus may acquire a host proto-oncogene and convert it to an oncogene whose uncontrolled expression causes uncontrolled proliferation of the cell.

Note Distinguish between two uses of the term "transformation." A cell can be transformed with DNA, meaning that the cell takes up exogenous DNA into its own genome. In a different sense, a cell can be transformed by a virus, converting to a cancer cell (oncogenesis).

SECTION SUMMARY

- **Papillomaviruses have a small genome of double-stranded DNA.** The virion is icosahedral with no envelope.

- **Human papillomavirus (HPV) shows tropism for the skin or for the genital mucosa.** Different strains cause warts or cancers of the skin or genitals. HPV-16 is the main cause of cervical cancer. Most sexually active people acquire HPV strains.
- **HPV virions bind specific cell-surface proteins of the basal epithelial layer.** The virion is then endocytosed, and the capsid is uncoated.
- **The HPV genome replicates within the nucleus of differentiating cells.** Progeny virions are transported to the cytoplasm, where they escape the host cell as it is sloughed off from the epithelium.
- **Some HPV genomes integrate into the host genome.** Integrated HPV genomes can persist for many years and transform the cell to cancer.
- **Viral oncogenes transform a host cell to cause cancer.**

Thought Question 12.4 What are the advantages and disadvantages to the virus of replication by the host polymerase, compared with using a polymerase encoded by its own genome?

12.5
Influenza Virus: RNA Genome

SECTION OBJECTIVES

- Describe the structure of the influenza virion.
- Explain how the form of the influenza genome enables rapid evolution of novel strains that cause pandemics.
- Describe the replication cycle of influenza virus.

CASE HISTORY 12.2

An Influenza Pandemic Reaches College

Eighteen-year-old Aisha was a first-year student at a small private college in New England. In the fall of 2009, the campus implemented management procedures for the expected H1N1 influenza pandemic. Students with flu symptoms were told to "self-isolate" and to report their status on a college webform. By the end of October, there were 50 students reporting per week.

One night, Aisha felt hot, flushed, and unsteady on her feet. Her thermometer showed a temperature of 39°C (102°F). She started coughing, and she felt extremely tired. She managed to find the webform on her computer, although her arm was shaking and made false starts. Finally, she typed her information and confirmed her status report. A college security officer arrived at her room wearing a face mask. The officer gave Aisha a face mask and told her to collect her essential personal items and books; Aisha did the best she could in her fevered state. The officer drove her to a separate residence reserved for suspected H1N1 cases. There Aisha lay in the bed and tried to sleep. In the morning a nurse's aide brought a tray of food and medication including oral oseltamivir (Tamiflu) and inhaled zanamivir (Relenza). The tray was left on the floor in the hallway, outside Aisha's isolation room. Aisha had to drag herself out of bed to open the door and get the tray. She was required to stay in the isolation room until 2 days after her temperature returned to normal. She and most other infected students made a full recovery, although one student was hospitalized with life-threatening complications.

Influenza A virus, an orthomyxovirus (Baltimore Group V; see Table 12.1), is one of the most common life-threatening viruses in the United States. The virus infects cells of the upper respiratory mucosa, causing fever, sore throat, headache, and other symptoms. Each year, influenza A infects approximately 10% of the U.S. population, causing about 36,000 deaths annually. The elderly are most susceptible, but in an epidemic year, mortality rises among young people.

Influenza strains show antigenic drift in that their envelope proteins continually mutate, evading the host immune system; thus, new strains emerge annually, the "seasonal flu." In addition, beyond the annual variant strains, at wider intervals extremely virulent strains of influenza emerge that cause pandemic mortality. For example, the famous influenza pandemic of 1918 infected 20% of the world's population and killed more people than World War I did. The 1918 strain arose as a mutant form of an influenza strain infecting birds. Such extremely virulent strains usually arise through **antigenic shift**, in which genes combine from two or more different influenza viruses. (Antigenic shift represents a larger change than antigenic drift, the smaller mutations observed in seasonal influenza strains.) Today, other "avian influenza" strains continue to emerge by viral recombination in animal hosts. As of this writing, however, the known avian strains have limited tropism for the human respiratory tract, and none have mutated to a form readily transmitted between humans (discussed further in Chapter 20).

Link As discussed in Section 20.2, **antigenic shifts and drifts** allow viruses to evade the host immune system. Influenza type A is most capable of undergoing genetic change, so it more commonly leads to seasonal disease. Type A influenza is classified into subtypes based on the kind of hemagglutinin (H) and neuraminidase (N) expressed on the viral surface, giving rise to some very well-known strains, such as H1N1 (swine flu) or H5N1 (avian flu).

In 2009, a highly transmissible strain related to swine influenzas ("swine flu") spread rapidly around the world. Public health officials feared that hospitals would overflow and that the severity of illness could resemble that of the 1918 pandemic. That is why colleges took measures for isolation, as experienced by Aisha in Case History 12.2. In fact, the 2009 flu did infect large numbers of people and strained our health systems, especially at institutions such as colleges. Fortunately, the symptoms turned out to be mild; the virus had limited virulence, compared with the strains in 1918 and 2006. A future strain, however, might emerge combining high transmission (as in 2009) with high human mortality (as for avian flu in 2006).

What causes the sudden appearance of new pandemic strains of influenza? The cause can be understood from the structure of the influenza virion and its mode of replication.

Influenza Virion Structure

The influenza virion is asymmetrical; it has no fixed capsid (Figure 12.21). Its genome consists of (–) sense RNA, in contrast to hepatitis C virus, which has (+) sense RNA.

A bizarre feature of the influenza genome is that its genomic RNA is divided among eight chromosome segments. Each segment contains a different essential gene. When influenza infects a cell, the eight RNA segments can **reassort** with segments from a coinfecting virion of a different strain. Human influenza RNA segments can reassort even with segments from a strain that normally infects a different host species and cannot complete its replication in the human cell (Figure 12.22). Such reassortment is particularly likely with influenza strains from birds and pigs, which are often raised in close proximity to humans. The reassortment may generate a new strain able to infect humans but possessing novel gene sequences unfamiliar to the host immune system. This interspecies reassortment is what leads to the most virulent pandemic strains of influenza.

Figure 12.21   Structure of Influenza

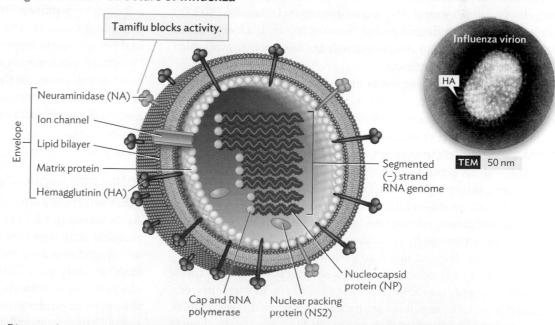

Diagram of virion structure showing the envelope (brown), envelope proteins (HA and NA), matrix protein (yellow), RNA segments with attached polymerase, and the nuclear packing protein NS2. The inset image shows an influenza A virion. The brush-like border coating the envelope consists of glycoproteins, hemagglutinin (HA), and neuraminidase (NA). Envelope proteins involved in receptor binding (HA and NA) are also called "spike" proteins.

ANIMATION

Figure 12.22   **Reassortment between Human and Avian Strains Generates Exceptionally Virulent Strains**

A. The reconstructed strain of the 1918 pandemic influenza virus is studied by Dr. Terrence Tumpey, microbiologist at the Centers for Disease Control and Prevention, Atlanta. The 1918 strain may hold clues to combating pandemic influenza today.

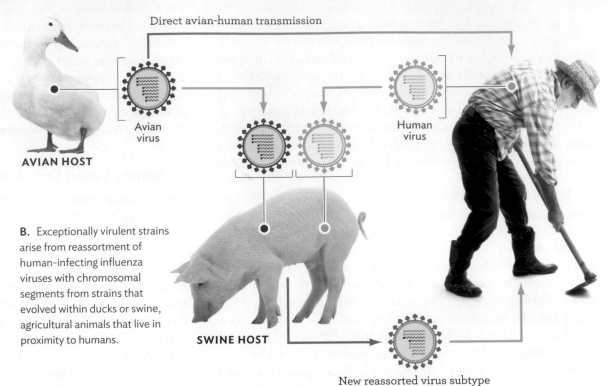

Direct avian-human transmission

Avian virus

AVIAN HOST

Human virus

B. Exceptionally virulent strains arise from reassortment of human-infecting influenza viruses with chromosomal segments from strains that evolved within ducks or swine, agricultural animals that live in proximity to humans.

SWINE HOST

New reassorted virus subtype

Within the influenza virion, the RNA segments are coated by nucleocapsid protein (NP). Each NP-coated RNA segment also possesses a bound RNA-dependent RNA polymerase complex, poised for synthesis. The prepackaged polymerase is necessary because, during infection, each (−) strand RNA segment must be transcribed to a (+) strand mRNA for translation—but the host has no RNA-dependent RNA polymerase. So the virus needs to bring its own, a copy of which is all set to go on each segment.

The virion's RNA segments are loosely contained by a shell of matrix proteins. The envelope derives from the phospholipid membrane of the host cell, which incorporates viral proteins such as hemagglutinin (HA) and neuraminidase (NA) that peg the membrane to the matrix, maintaining an intact but flexible structure. The HA and NA proteins differ slightly in sequence among different strains; these sequence differences give rise to the H and N numbers, such as strain H1N1, the so-called swine flu, and H5N1, an "avian flu" strain. Neuraminidase also acts as an enzyme whose activity can be blocked by the antiviral agent oseltamivir (Tamiflu), one of the main drugs available to treat influenza.

Given that viral infection requires all eight RNA segments, how does the assembly mechanism package exactly eight segments, one of each? In fact, the segments are packaged imperfectly, resulting in a majority of defective particles. As a result, less than 1% of influenza particles are capable of infection. This may sound inefficient, but a single infected cell can produce an astonishing 10,000 virions. Out of this number, there will be more than enough infective particles

to propagate the virus. Furthermore, the inexact packaging method allows great flexibility for reassortment among different virions infecting the same host cell.

It is interesting that influenza virus has responded to natural selection very differently than HPV. HPV has evolved the tiniest possible genome, generating as many virions as possible out of limited resources while persisting in the host. The influenza virus, on the other hand, has evolved a genome that maximizes gene reassortment in order to evade the immune response, while still generating large quantities of virions to infect the next host. Thus, evolution leads to different results, given different conditions of host cells and transmission.

Influenza Replication Cycle

What determines how influenza spreads—and how we can stop it? As discussed in Chapter 20, surprisingly small details of the host receptor structure may determine whether a strain of influenza will spread directly between humans. Thus, avian influenza strain H5N1 is rarely transmitted between humans; but a mutation in the HA gene could allow the envelope protein to bind human cells in the upper respiratory tract, which would increase transmission and possibly lead to an influenza pandemic. Much research addresses HA and other viral proteins as targets of possible drugs.

The influenza virus replication cycle is shown in **Figure 12.23**. An influenza virion attaches to a cell when its HA envelope protein

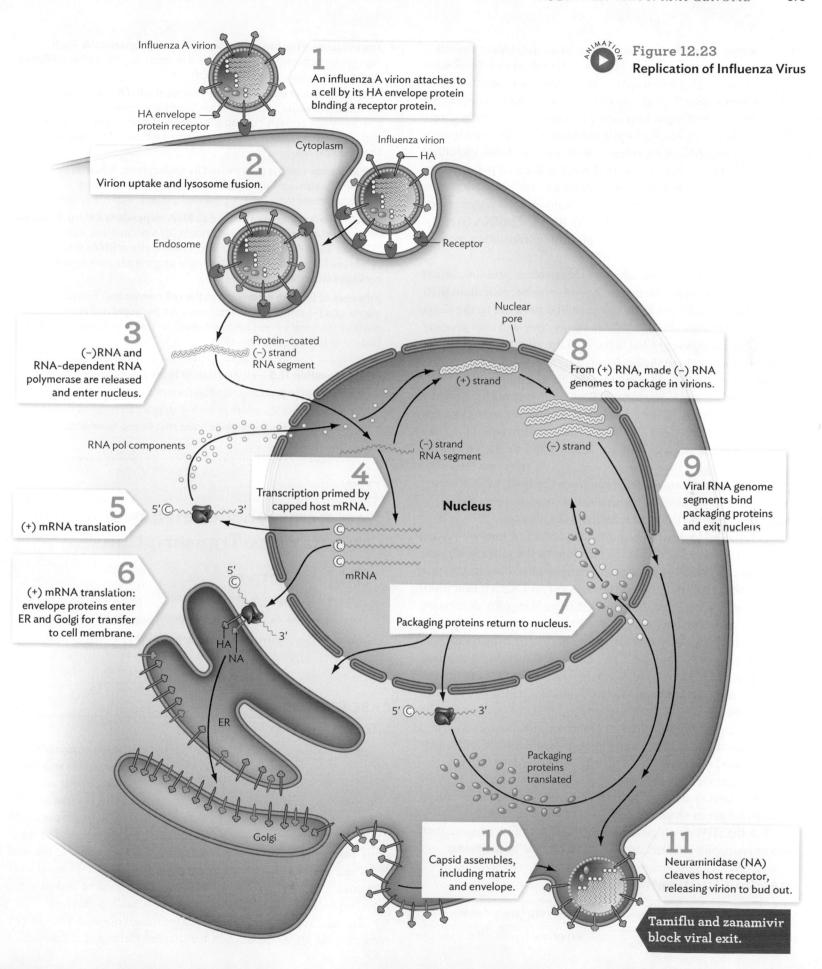

Figure 12.23
Replication of Influenza Virus

1 An influenza A virion attaches to a cell by its HA envelope protein binding a receptor protein.

Influenza A virion

HA envelope protein receptor

2 Virion uptake and lysosome fusion.

Cytoplasm

Influenza virion

HA

Endosome

Receptor

3 (−)RNA and RNA-dependent RNA polymerase are released and enter nucleus.

Protein-coated (−) strand RNA segment

Nuclear pore

(+) strand

8 From (+) RNA, made (−) RNA genomes to package in virions.

(−) strand RNA segment

(−) strand

RNA pol components

4 Transcription primed by capped host mRNA.

Nucleus

9 Viral RNA genome segments bind packaging proteins and exit nucleus

5 (+) mRNA translation

5′ C   3′

C

C

C

mRNA

6 (+) mRNA translation: envelope proteins enter ER and Golgi for transfer to cell membrane.

5′

C

HA

NA

3′

7 Packaging proteins return to nucleus.

ER

5′ C   3′

Packaging proteins translated

Golgi

10 Capsid assembles, including matrix and envelope.

11 Neuraminidase (NA) cleaves host receptor, releasing virion to bud out.

Tamiflu and zanamivir block viral exit.

binds to a host cell receptor, a glycoprotein (step 1). After the influenza virion undergoes endocytosis, the endocytic vesicle fuses with a lysosome (step 2). The low pH of the lysosome contents causes hydrogen ions to leak through the virion's ion channel, an important drug target. The hydrogen ions (low pH) cause the virion to disassemble and fuse with the endocytic membrane. As the membranes fuse, the virion contents are released into the cytoplasm, including all the viral (–) RNA segments with their prepackaged polymerases (step 3). The RNA segments pass through nuclear pore complexes into the nucleus, where the viral RNA-dependent RNA polymerases attached to each (–) RNA transcribe (+) RNA for mRNA (step 4). The mRNA then returns to the cytoplasm for translation by host cell ribosomes (step 5).

Some of the mRNA molecules encode envelope proteins, which are made by ribosomes attached to the endoplasmic reticulum (ER) (step 6). The ER will transport the envelope proteins to the Golgi and ultimately to the cell membrane, where they will coat progeny virions. Other proteins needed to package the viral RNA (packaging proteins) must return to the nucleus (step 7). In the nucleus, the (+) RNA strands now serve as templates for RNA-dependent RNA polymerase to make complementary (–) RNA segments for the progeny viral genomes (step 8). The (–) RNA viral genome segments become coated with proteins (step 9). At last, the protein-coated (–) RNA segments exit the nucleus and are transported to the cell membrane for packaging into capsids (step 10). As they assemble, the capsids acquire envelope membrane from the host, incorporating viral envelope proteins. The mature virions now bud out (step 11) in a massive release of virions that destroys the host cell.

The final budding step requires the action of neuraminidase (NA), an enzyme embedded in the envelope of the progeny virions; the enzyme cleaves host surface sugar molecules that bind exiting virions to the cell surface. NA is the target of oseltamivir (Tamiflu), and zanamivir, at present the most effective drugs for decreasing the symptoms of influenza.

Compared with the papillomavirus cycle, the infectious cycle of influenza is very rapid, causing rapid destruction of mucosal cells. But the host immune response is also rapid. If successful, the host immune response completely eliminates the flu virus within 1 or 2 weeks. Unlike HPV, there is no long-term persistence in the host cells. Thus, influenza virus is committed to rapid infection and highly efficient host-to-host transmission—as demonstrated during the 2009 H1N1 outbreak.

In the next section, we examine one more virus with another very different infection strategy: the retrovirus HIV, the cause of AIDS. HIV, like influenza virus, has an RNA genome with a high mutation rate. But the HIV virion generates a DNA copy of its genome that, like human papillomavirus, can integrate into the host genome.

SECTION SUMMARY

- **Influenza virus causes periodic pandemics of respiratory disease.** New virulent strains arise through reassortment of viral genome segments from human, avian, and swine strains.

- **The influenza virus consists of segmented (–) strand RNA.** Each segment is packaged with nucleocapsid proteins. Segments from different strains reassort by coinfection of a host cell.

- **Nucleocapsid and matrix proteins enclose the RNA segments of the influenza virus.** The matrix is enclosed by an envelope containing spike envelope proteins HA and NA. The HA proteins mediate virion attachment. NA is an enzyme that enhances release of progeny virions from the host cell.

- **The influenza virion is internalized by endocytosis, followed by lysosome fusion.** Low pH triggers viral envelope fusion with the endosome membrane, releasing the viral contents into the cytoplasm.

- **Viral (–) RNA segments attached to RNA-dependent RNA polymerase enter the nucleus.** The RNA-dependent RNA polymerase makes new (+) sense and (–) sense strands. The (+) sense viral mRNAs return to the cytoplasm for translation, while (–) sense progeny genome segments are packaged into virions.

- **Influenza virions are assembled at the cell membrane.** The capsid, matrix, and (–) strand RNA components are packaged and enclosed by host cell membrane containing viral envelope proteins. Completed virions bud out and destroy the host cell.

Thought Question 12.5   Researchers in Hong Kong are conducting a regular survey of the genomes of influenza strains emerging in a particular swine-processing facility, where swine are shipped from many different regions. What might the researchers learn that would be important for human public health?

12.6
Human Immunodeficiency Virus: Reverse Transcription

SECTION OBJECTIVES
- Describe the structure of the HIV virion.
- Explain the replicative cycle and epidemiology of HIV.
- Describe the positive roles of retroviruses in human health.

CASE HISTORY 12.3

Lifelong Infection by a Retrovirus

 At 24 years of age, Ralph had just been promoted at the car dealership where he worked in sales. He was ready to propose to his girlfriend, a bank clerk. It was several years since Ralph had engaged in risky behaviors, but he and his girlfriend both decided to get tested for STDs. That is how Ralph learned that he was HIV-positive. The diagnosis came as a surprise, since Ralph had no symptoms of disease. His T-cell count, however, had dipped to 500 cells/mm³ (normal level is about 1,200 cells/mm³). Now he faced the dilemma of how to avoid infecting his future wife and how to conceive healthy children.

The doctor informed Ralph of his treatment options. The current recommendation of the Centers for Disease Control and Prevention

(CDC) is to commence antiretroviral therapy as soon as an individual tests positive for the virus. A single daily pill was prescribed, containing tenofovir and emtricitabine (reverse transcriptase inhibitors, together marketed as Truvada). The advantage of early treatment with Truvada is the early halt in decline of T cells and the decrease in transmission risk. Some physicians, however, recommend starting antivirals only when T-cell counts fall below 500 cells/mm³, to delay the deleterious side effects of the drugs and to postpone selection for drug-resistant strains. Ralph was also told of options for conceiving children without infection, such as the "sperm washing" procedure that eliminates HIV from sperm for artificial insemination. He had a lot to discuss with his girlfriend.

What is human immunodeficiency virus (HIV), and what is its global impact today? HIV is the causative agent of acquired immunodeficiency syndrome (AIDS) and is a member of the lentivirus family. A **lentivirus** is a retrovirus that causes infections that progress slowly over many years. According to the United Nations, approximately 33 million people globally are estimated to be living with HIV, and 3 million people die of AIDS annually. In the United States, the first cases of AIDS were reported in 1981; three decades later, HIV infects more than a million Americans, and more than half a million have died of AIDS. Worldwide, HIV infects 1 in every 100 adults, equally among women and men. Countries in southern Africa have already experienced population decline; in South Africa, approximately 20% of adults test positive for HIV.

The experience of Ralph in Case History 12.3 is typical of HIV infection in developed countries. At some point, Ralph must have acquired HIV through high-risk behaviors such as sexual contact (genital or anal) or shared needles. The initial infection often goes unnoticed, as flu-like symptoms subside; or there may be no symptoms at all. During this early period (Figure 12.24), the number of CD4⁺T lymphocytes declines while the number of virions in the blood rises. T lymphocytes (or T cells) are white blood cells of the active immune system; they are needed to activate the B cells that generate antibodies (as will be explained in further detail in Chapter 16). During HIV infection, an immune response is mounted, leading to a rebound of T cells and fall of virion numbers. But for reasons not understood, the immune response is unable to prevent the next phase of HIV infection: a slow, steady loss of T cells (the period called "clinical latency"), until a level is reached at which the immune system fails (around 350 cells/mm³). At this critical level, without treatment the patient usually begins to experience the constitutional symptoms of AIDS, such as fever and swollen lymph nodes. Opportunistic infections soon follow, as the immune system can no longer respond (discussed in Chapter 21).

HIV Structure

HIV and its causative role in AIDS were discovered by French virologists Luc Montagnier and Françoise Barré-Sinoussi, at the Pasteur Institute, building on Robert Gallo's studies of retroviruses at

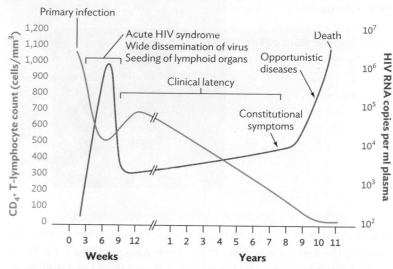

Figure 12.24 HIV Titer Increase and T-Cell Decline over Time

An initial rise in virus titer is fought off by the immune system, while the T-cell count declines and rebounds. Over years, the virus titer remains low while T-cell count steadily declines. At a T-cell level of about 350 cells/mm³, the constitutional symptoms of AIDS appear, followed by opportunistic infections and death.

NIH (Figure 12.25). The virion has an unusual conical-icosahedral capsid enclosed in an envelope with spike proteins (Figure 12.26). After three decades of research, there is no cure for infection and no vaccine. But study of the HIV life cycle has revealed a surprising array of effective therapeutic agents that slow the infection and prolong life for many years (Table 12.2). In the developed countries, treatment effectively prolongs the life of people with AIDS. Unfortunately, the cost leaves treatment out of reach for the majority of those infected worldwide.

The key polymerase of all retroviruses, including HIV, is **reverse transcriptase**. Reverse transcriptase is an enzyme that uses RNA template to synthesize DNA and then uses the new DNA as

Figure 12.25 HIV Discoverers

A. Françoise Barré-Sinoussi, at the Pasteur Institute, worked with Luc Montagnier to discover the virus that causes AIDS.

B. Luc Montagnier and Robert Gallo agree to collaborate on development of an AIDS vaccine, 2002.

Figure 12.26
HIV-1 Particles

HIV virions from infected human lymphoid tissue culture (thin section).

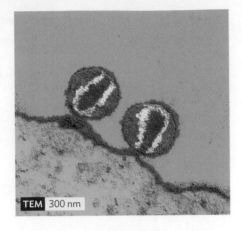

TEM 300 nm

occurs, and as people live for longer periods with HIV, new drugs are needed to replace the old ones.

The structure of the HIV virion is depicted in **Figure 12.27**. The conical core of capsid subunits encloses two copies of the RNA genome, together with reverse transcriptase and other enzymes. The presence of two genome copies is a unique feature of retroviruses. Each RNA genome copy is coated with nucleocapsid proteins similar in function to those of influenza virus. For priming DNA synthesis (reverse transcription), each RNA is complexed with a tRNA derived from the previously infected host cell. The capsid

template for the complementary DNA strand. Reverse transcriptase has the highest error rate of any known polymerase, possibly as high as one base in ten. For this reason, fewer than 1 in 1,000 progeny virions released in the blood are infectious. But the prodigious output of virions may reach 10 billion daily. So a large number will be infectious; and many of these will have mutations that evade the immune system and antiretroviral drugs.

To delay the loss of effectiveness due to viral mutation, HIV drugs are always taken in combinations of two or three, such as the Truvada combination of tenofovir and emtricitabine recommended for Ralph; thus, a mutant virion would need to have two or three lucky mutations to survive the drugs. But resistance inevitably

Figure 12.27   HIV-1 Structure and Genome

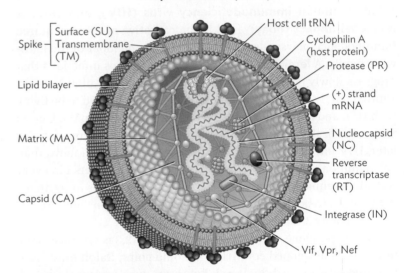

A. Internal structure of the HIV-1 virion, color-coded to match the genome. In the genome sequence, the staggered levels indicate three different reading frames. Each virion contains two copies of the RNA genome plus multiple copies of reverse transcriptase (RT) and protease enclosed within a conical capsid (capsid subunits plus subunits of a host protein, cyclophilin A). The capsid is surrounded by matrix subunits, which reinforces the host-derived phospholipid membrane, pegged by spike proteins (SU, TM).

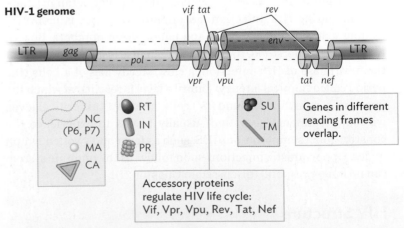

B. The genome encodes the virion proteins, as well as six accessory proteins that are expressed within the infected host cell and regulate the replicative cycle.

Table 12.2
Antiretroviral Drugs Active against HIV

Drug	Mode of Action
Azidothymidine (AZT)	Nucleoside reverse transcriptase inhibitor (base analog that halts elongation)
Emtricitabine (Truvada component)	Cytidine analog, reverse transcriptase inhibitor
Enfuvirtide	Fusion inhibitor (prevents viral envelope fusion with host cell membrane)
Maraviroc	Entry inhibitor (blocks spike protein binding to coreceptor CCR5)
Nevirapine	Nonnucleoside reverse transcriptase inhibitor (binds reverse transcriptase and inhibits activity)
Raltegravir	Integrase strand transfer inhibitor (blocks integration of DNA genome copy into host cell genome)
Ritonavir	Protease inhibitor (inhibits PR protease that cleaves initial protein to make final proteins)
Tenofovir (Truvada component)	Nucleoside reverse transcriptase inhibitor (base analog that halts elongation)

also contains about 50 copies of reverse transcriptase (RT) and protease, as well as a DNA integration factor—all of which are potential targets for drugs (see Table 12.2). The capsid is surrounded by matrix proteins, which reinforce the host-derived phospholipid membrane. The membrane is pegged by spike proteins composed of the envelope subunits TM and SU. As in influenza virus, the spike proteins play crucial roles in host attachment and entry.

The HIV genome includes three main genes that are found in all retroviruses: *gag*, *pol*, and *env*. The *gag* sequence encodes capsid, nucleocapsid, and matrix proteins; *pol* encodes reverse transcriptase (RT), integrase, and protease; and *env* encodes envelope proteins. The *gag* and *pol* sequences overlap, but (as in papillomavirus) they are translated in different reading frames.

HIV Replication Cycle

The first cells infected by HIV are T lymphocytes (T cells) that possess cell surface proteins called CD4 (the "receptor") and CCR5 (the "coreceptor"). As we have seen for other virus-binding host proteins, these receptors have important host functions. CD4 helps the T cell activate the immune response (discussed in Chapter 16). CCR5 is a receptor for a chemokine, an immune system signal (also discussed in Chapter 16). The CD4 function is essential for the host; but CCR5 is dispensable. Thus, people who carry a genetic defect for CCR5 have a functional immune system but resist infection by HIV-1, the main strain of HIV causing AIDS. The drug maraviroc was designed to prevent HIV infection by blocking CCR5.

As an HIV virion binds the CD4 and CCR5 receptors (Figure 12.28, step 1), its envelope fuses with the cell membrane of the T cell (step 2). This fusion step can be blocked by drugs such as enfuvirtide. Unlike influenza virus, the HIV core is released directly into the cytoplasm, without endocytosis. The two RNA genome copies are then reverse-transcribed by reverse transcriptase enzyme (step 3). Reverse transcription actually forms double-stranded DNA, a process that involves cleaving away the original RNA template and replacement with DNA. Reverse transcription can be inhibited by base analogs such as azidothymidine (AZT) and emtricitabine that cannot attach the next nucleotide.

The double-stranded DNA copy of the HIV genome enters the nucleus through a nuclear pore (step 4), where it integrates as a **provirus** (or episome) at a random position in a host chromosome (step 5). Integration is a critical step in that it ensures permanent infection of the host. The integrase enzyme can be blocked by inhibitors such as raltegravir. The entire length of the integrated HIV genome is transcribed to RNA by the host RNA polymerase II (step 6). Some of the RNA copies exit the nucleus (step 7) to serve as mRNA for translation to proteins. Proteins are synthesized in alternative versions, such as Gag-Pol. The formation of alternative versions involves cleavage by a viral protease. The viral protease breaks up the initial proteins translated from *gag* and *pol* into smaller, active proteins. Viral protease can be blocked by protease inhibitors, another major category of antiviral drugs (see Table 12.2).

Some full-length RNA transcripts exit the nucleus to be packaged as genomes for progeny virions (step 8). Meanwhile, Env (envelope) proteins are made within the endoplasmic reticulum (ER) (step 9). They pass through the Golgi for glycosylation (adding sugar chains) and packaging and are exported to the cell membrane (step 10). At the membrane, Env proteins plug into the core particle as it forms from the RNA dimers plus Gag-Pol peptides (step 11). The virions then bud out of the host cell, a process that requires cleavage of a host attachment protein called tetherin (step 12).

An alternative to viral budding is cell fusion, mediated by the binding of Env in the membrane to CD4 receptors on a neighboring cell. This occurs because, unlike influenza virus, which rapidly destroys the host cell, HIV can persist in a host cell for long periods while budding virions slowly and displaying Env proteins on the host cell membrane. Thus, an infected cell can fuse with an uninfected cell, and HIV core particles can enter the new cell through their fused cytoplasm. The fusion of many cells can form a giant multinucleate cell called a **syncytium**. Cell fusion with formation of syncytia enables HIV to infect neighboring cells without exposure to the immune system. Even more insidious, HIV-infected cells can form long "nanotubes" through which virions can transfer to uninfected T cells (Figure 12.29).

Retroviruses and the Human Genome

Suppose an integrated HIV genome mutated and lost the ability to produce progeny. What would happen to its genome? The integrated genome would be "trapped," replicating only as part of the host genome. Over many generations in its host, the viral sequence would inevitably accumulate more mutations. In fact, the human genome is riddled with remains of decaying retroviral genomes, collectively known as **retroelements**. Some retroelements contain all the genomic elements of a retrovirus, including *gag*, *env*, and *pol* genes, yet never generate virions. Presumably, the integrated retroviral genome lost this ability by mutation. Other elements, known as "retrotransposons," retain only partial retroviral elements but may maintain a reverse transcriptase to copy themselves into other genomic locations. Amazingly, about half the sequence of the human genome appears to have originated from viruses and retroelements.

Could a virus as deadly as HIV be used to improve human health? In fact, the exceptional ability of lentiviruses to deliver their genomes into a human genome makes them the most promising source of vectors for gene therapy. To be used for gene therapy, the integration sequences are maintained while the virulence genes are removed and replaced by human genes to be delivered by the vector. In 2009, a lentiviral vector was used to successfully treat two children with an inherited neurodegenerative disease, X-linked adrenoleukodystrophy (ALD). The vector transferred a corrected version of the gene into the children's bone marrow stem cells, which then migrated into the brain and halted disease progression. In 2013, the first child was cured of leukemia using an HIV-derived vector to carry genes into her T cells, causing her

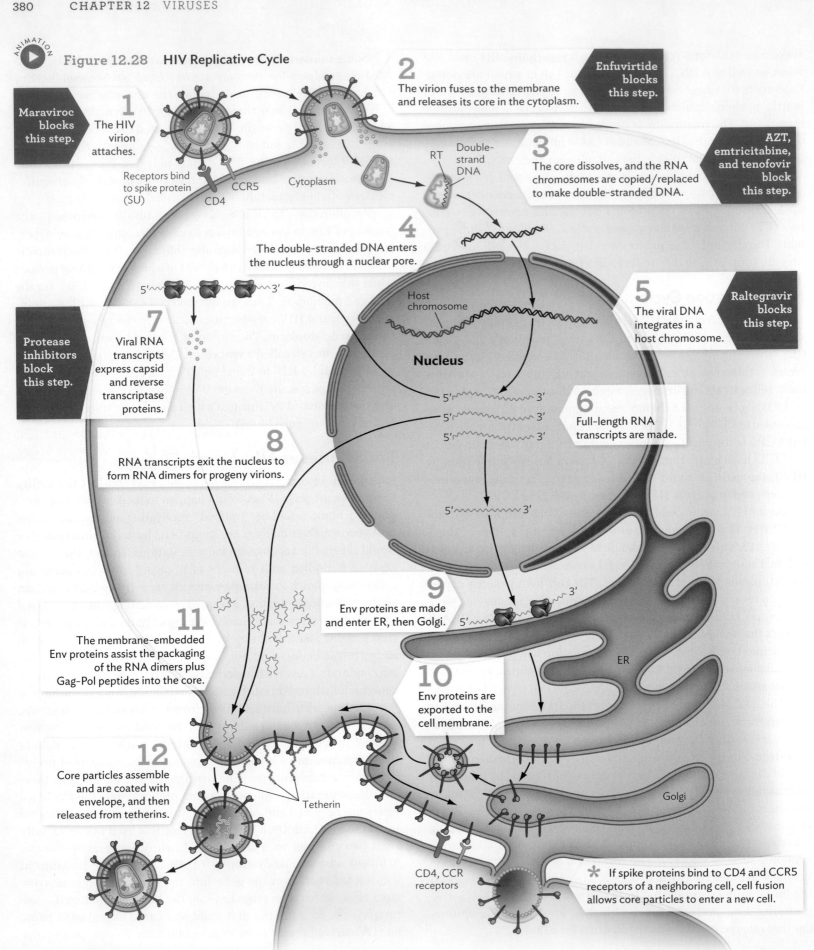

ANIMATION ▶

Figure 12.28   **HIV Replicative Cycle**

Maraviroc blocks this step.

1 The HIV virion attaches.

Receptors bind to spike protein (SU)
CD4   CCR5   Cytoplasm

2 The virion fuses to the membrane and releases its core in the cytoplasm.

Enfuvirtide blocks this step.

RT   Double-strand DNA

3 The core dissolves, and the RNA chromosomes are copied/replaced to make double-stranded DNA.

AZT, emtricitabine, and tenofovir block this step.

4 The double-stranded DNA enters the nucleus through a nuclear pore.

Host chromosome

Nucleus

5 The viral DNA integrates in a host chromosome.

Raltegravir blocks this step.

Protease inhibitors block this step.

7 Viral RNA transcripts express capsid and reverse transcriptase proteins.

5'   3'

6 Full-length RNA transcripts are made.

5'   3'
5'   3'
5'   3'

8 RNA transcripts exit the nucleus to form RNA dimers for progeny virions.

5'   3'

9 Env proteins are made and enter ER, then Golgi.

5'

ER

11 The membrane-embedded Env proteins assist the packaging of the RNA dimers plus Gag-Pol peptides into the core.

10 Env proteins are exported to the cell membrane.

Golgi

12 Core particles assemble and are coated with envelope, and then released from tetherins.

Tetherin

CD4, CCR receptors

✱ If spike proteins bind to CD4 and CCR5 receptors of a neighboring cell, cell fusion allows core particles to enter a new cell.

Figure 12.29   HIV Transfer through a Nanotube

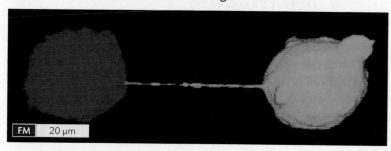

FM   20 µm

A T cell infected with HIV can transfer virions to an uninfected cell through a nanotubular connection. The two T cells are tagged by different fluorescent labels (red and green).

T cells to fight the cancer. Looking to the future, we can expect many health-transforming applications from these deadly viruses.

SECTION SUMMARY

- **Human immunodeficiency virus (HIV) causes an ongoing pandemic of acquired immunodeficiency syndrome (AIDS).** Although no cure exists, the molecular biology of HIV replication has led to drugs that extend life expectancy.

- **Reverse transcriptase uses the RNA genome of HIV as an initial template to synthesize double-stranded DNA, which integrates into the DNA of the host cell.** The HIV core particle contains two copies of its RNA genome, each bound to a primer (host tRNA) and reverse transcriptase (RT). The core is surrounded by an envelope containing spike proteins.

- **HIV envelope proteins bind the CD4 receptor** of T lymphocytes together with the chemokine receptor CCR5. Following virion-receptor binding and envelope membrane fusion, HIV virions are released into the cytoplasm.

- **Double-stranded DNA is synthesized from the HIV RNA by reverse transcriptase.** The retroviral DNA integrates into the host genome. This genome integration results in permanent infection of the host.

- **HIV mRNAs are expressed and exported to the cytoplasm for translation.** Envelope proteins are translated at the ER and exported to the cell membrane.

- **HIV virions are assembled at the cell membrane,** where virions are released slowly, without lysis. Alternatively, virions can infect other T cells through intercellular nanotubes or through cell fusion to form syncytia.

- **Ancient retroviral sequences persist within the human genome.** Retroviruses can be used to develop vectors for gene therapy, correcting the symptoms of inherited gene defects.

Thought Question 12.6   How do attachment and entry of HIV resemble attachment and entry of influenza virus? How do attachment and entry differ between these two viruses?

Perspective

Viral infections show important commonalities as well as intriguing diversity. Virus particles (and their genomes) range in size from a few components to assemblages approaching the complexity of cells. The advantage of simplicity is the minimal requirement for resources, whereas the advantage of complexity is the fine-tuning of mechanisms for evading host defenses. Viral genomes are remarkably diverse; they can be RNA or DNA, single- or double-stranded, linear or circular. Different viruses evolve to take advantage of different host cell proteins in order to recognize and gain entry into host cells and then replicate within the host cells. Finally, each virus needs a means of transmission to the next host.

Research in virology offers hope for new drugs and cures for humanity's worst plagues, as well as remedies for the devastating diseases of agricultural plants and animals. At the same time, it is sobering to note that despite the enormous volumes we now know about viruses such as HIV and influenza, the AIDS pandemic continues, and we face the likely emergence of new deadly flu strains. The pathology of viral diseases and epidemiology are discussed further in the chapters of Part V.

LEARNING OUTCOMES AND ASSESSMENT

	SECTION OBJECTIVES	**OBJECTIVES REVIEW**

12.1
What Is a Virus?

- Describe diverse forms of the virion, or virus particle.
- Explain the function of the viral envelope and accessory proteins.

1. Which of the following traits does <u>not</u> apply to viruses?
   A. Most virus particles pass through a filter.
   B. Isolated particles conduct their own metabolism.
   C. Viruses have a genome of nucleic acid contained by a protein capsid.
   D. Some classes of virus particles are enclosed by a phospholipid envelope.

12.2
Viral Genomes and Diversity

- Describe the different classes of viral genomes, and give an example of each.
- Explain why different viruses infect different hosts.
- Describe the nature of viroids and prions.

4. The genome of a virus may consist of any of these <u>except</u> _____.
   A. phospholipid
   B. single-stranded RNA
   C. single-stranded DNA
   D. double-stranded DNA

12.3
Virus Replication and Culture

- Explain how viral genomes enter cells.
- Describe the lytic and lysogenic cycles of bacteriophage infection.
- Explain how viruses are cultured using host cells.

7. The property of viruses that provides access to a host cell is _____.
   A. binding of a virion to a specific host receptor protein
   B. viral genome encoding of DNA polymerase
   C. self-assembly of virions within the cytoplasm
   D. viral use of an enzyme that lyses the host cell membrane

---

12.4
Papillomavirus: DNA Genome

- Describe the genome of papillomavirus.
- Explain the infectious cycle of papillomavirus, and explain how papillomavirus causes cancer.

10. Which of the following steps of a papillomavirus life cycle leads to cancer transformation?
   A. The HPV virion becomes endocytosed by the host cell.
   B. The virus uncoats, and its genome enters the nucleus for replication.
   C. The viral genome integrates into the host genome and causes increased expression of oncogenes.
   D. Progeny virions are released during host cell shedding.

12.5
Influenza Virus: RNA Genome

- Describe the structure of the influenza virion.
- Explain how the form of the influenza genome enables rapid evolution of novel strains that cause pandemics.
- Describe the replication cycle of influenza virus.

13. The form of the capsid of the influenza virus is _____.
   A. icosahedral
   B. filamentous
   C. complex tailed
   D. asymmetrical

12.6
Human Immunodeficiency Virus: Reverse Transcription

- Describe the structure of the HIV virion.
- Explain the replicative cycle and epidemiology of HIV.
- Describe the positive roles of retroviruses in human health.

16. Which structural feature of HIV determines the cell type that virions can infect?
   A. Virion envelope proteins specifically interact with the CD4 and CCR5 cell surface proteins of a T lymphocyte.
   B. The capsid dissolves within the T-cell cytoplasm, releasing packaged reverse transcriptase to generate a double-stranded DNA copy of the single-stranded RNA genome.
   C. The double-stranded DNA copy of the HIV genome enters the nucleus and integrates as a provirus.
   D. The integrated HIV genome is transcribed by host RNA transcriptase, expressing viral proteins.

2. Which of these traits of mimivirus differs from a cell?
   A. The mimivirus genome contains many cellular genes.
   B. The mimivirus particle is larger than some bacteria.
   C. Within a host ameba, mimivirus forms "viral factories" that resemble a contained living organism.
   D. Mimivirus has no ribosomes of its own; it must use the host ribosomes.

3. Tissue tropism is _____.
   A. the range of tissues a virus can infect
   B. the mechanism of viral transmission to a tissue
   C. the means of viral entry into cells
   D. the conversion of tissue to cancer

5. Each of these viruses has an icosahedral capsid except _____.
   A. poliovirus
   B. herpes virus
   C. Ebola virus
   D. papillomavirus

6. An example of an asymmetrical virus particle is _____.
   A. poliovirus
   B. papillomavirus
   C. smallpox virus
   D. herpes virus

8. Which of these steps is not part of a bacteriophage replication cycle?
   A. Virion binds to a host cell surface protein that provides an important host function.
   B. The virion injects its genome into the host cell.
   C. The capsid of the virion enters the host cytoplasm.
   D. The viral genome directs host ribosomes to synthesize viral components.

9. The function of bacteriophage tail fibers is _____.
   A. propel the virion toward a host cell
   B. attach the virion upon a host cell
   C. uncoat the virion within a cell
   D. help the progeny virion exit a host cell

11. How does HPV replicate its genome?
    A. Reverse transcriptase
    B. RNA-dependent RNA polymerase
    C. Host DNA polymerase
    D. Viral DNA polymerase

12. New HPV virions are released from the body by _____.
    A. exocytosis
    B. endocytosis
    C. becoming enveloped in cell membrane
    D. shredding of host cells

14. Which step of the influenza life cycle leads to evolution of virulent hybrid strains?
    A. The influenza virion is endocytosed by the cell and enters into the cytoplasm.
    B. The viral genome is composed of multiple segments that get transcribed by RNA-dependent RNA polymerase and reassort with segments from a coinfecting virus.
    C. Viral proteins are translated by ribosomes in the cytoplasm and are then transported by the ER to form progeny virions.
    D. The progeny virions bud out of the infected host cell with the help of neuraminidase (NA), which cleaves host surface molecules that capture the virions.

15. Which pathology is not caused by influenza?
    A. Cancer
    B. High fever
    C. Shortness of breath
    D. Coughing

17. Which drug is not used for HIV?
    A. Azidothymidine, which prevents reverse transcriptase from elongating DNA
    B. Raltegravir, which blocks integrase from integrating viral DNA copy into the host genome
    C. Oseltamivir, which binds neuramidase, preventing release of virions as they exit the host cell
    D. Protease inhibitors, which block the viral protease that cleaves Gag-Pol to form proteins with different functions

18. Following initial infection with HIV, the disease progresses as follows:
    A. HIV virions gradually increase in titer throughout the host lifetime.
    B. An initial burst of virions is followed by a long latent period during which T cells decline.
    C. The infected T cells decline, then increase and remain steady.
    D. T cells rapidly decline, and virions rapidly reach a high titer, which is sustained.

Key Terms

antigenic drift (p. 360)
antigenic shift (p. 373)
bacteriophage (p. 354)
Baltimore model (p. 362)
batch culture (p. 367)
burst size (p. 366)
capsid (p. 352)
cloning vector (p. 353)
eclipse period (p. 367)
envelope (p. 355)
filamentous virus (p. 357)
host range (p. 355)

lentivirus (p. 377)
lysis (p. 365)
lysogeny (p. 365)
lytic infection (p. 365)
oncogene (p. 372)
one-step growth curve (p. 367)
plaque (pp. 354, 368)
plaque-forming unit (PFU) (p. 368)
prion (p. 360)
prophage (p. 366)
provirus (p. 379)
reading frame (p. 370)

reassort (p. 373)
retroelement (p. 379)
retrovirus (pp. 359, 364)
reverse transcriptase (pp. 364, 377)
RNA-dependent RNA polymerase
   (p. 363)
RNA reverse-transcribing virus
   (p. 364)
spike protein (p. 357)
surface receptor (p. 365)
syncytium (p. 379)
tegument (p. 357)

temperate phage (p. 366)
tissue tropism (p. 355)
transduction (p. 367)
transformation (p. 372)
transmission (p. 355)
uncoating (p. 355)
virion (pp. 352, 354)
viroid (p. 359)
virulent phage (p. 366)
virus (p. 352)

Review Questions

1. Compare and contrast the form of icosahedral and filamentous (helical) viruses, citing specific examples.

2. How do viral genomes gain entry into cells in bacteria and animals?

3. Explain the structure and function of the seven Baltimore groups of viral genomes.

4. How do viral genomes interact with host genomes, and what are the consequences for host evolution?

5. Compare the lytic and lysogenic life cycles of bacteriophages. What are the strengths and limitations of each?

6. Compare the life cycles of RNA viruses and DNA viruses in human hosts. What are the strengths and limitations of each?

7. Explain the plate titer procedure for enumerating viable bacteriophages. How must this procedure be modified to titer animal viruses?

8. Explain how a pure isolate of a virus can be obtained. How do the procedures differ from those used for isolating bacteria?

9. Explain the generation of the step curve of virus proliferation. Why is virus proliferation generally observed as a single step, or generation, in contrast to the life cycles of cellular microbes, outlined in Chapter 6?

10. How do influenza virions gain access to the host cytoplasm?

11. How does influenza virus manage the replication and packaging of its segmented genome? What is the consequence of genome segmentation for virus evolution?

12. How does HIV provide ready-made components for replication of its RNA genome? How and why does reverse transcriptase make double-stranded DNA?

13. What is the role of protease in HIV replication? What is the significance of protease for AIDS therapy?

14. How did ancient retroviruses participate in the evolution of the human genome? What role do retroviruses play in gene therapy?

Clinical Correlation Questions

1. A 10-year-old child comes down with a fever and presents with large, raised sores typical of a pox disease. The child has recently acquired a pet rabbit from a store that also sells prairie dogs. The prairie dogs were obtained from an exotic pet dealer that also sells Gambian rats from Africa. What is the infectious agent?

2. During a routine physical checkup, a 50-year-old woman is advised to get tested for a virus with high prevalence in her age-group. To her surprise, the test comes back positive. The virus, which damages the liver, is known to be transmitted by sexual contact, needle stick, and tattoos conducted under unsanitary conditions. The patient can recall none of these behaviors, although it is hard to be certain of events from decades before. What is the pathogen?

Thought Questions: CHECK YOUR ANSWERS

Thought Question 12.1   Which viruses have a narrow host range, and which have a broad host range?

ANSWER: A virus with a narrow host range is human immunodeficiency virus, HIV. HIV causes AIDS only in humans, not even in chimpanzees, our nearest genetic relative. Poliovirus infects humans and chimpanzees but not nonprimate animals. A virus with a potentially broad host range is influenza virus. Influenza strains can infect different species of animals, but most strains cause disease in only one type of animal (humans, pigs, or birds). On the other hand, rabies virus has a broad host range, infecting many kinds of mammals.

Thought Question 12.2   What are the advantages to the virus of having a small genome? What are the advantages of a large genome?

ANSWER: For a virus, the advantage of a small genome is that the smaller the genome, the greater the number of progeny virions that can be formed out of the nucleotides available. The host cell contains a limited supply of nucleotides, which may determine the number of virions made. On the other hand, a large viral genome enables the virus to encode products with many extra functions that help control the host cell. For example, herpes viruses encode products that modulate the host immune system, preventing it from stopping the virus.

Thought Question 12.3   Given the mechanism by which viruses infect cells, how might an organism evolve to become resistant to a viral infection?

ANSWER: For viruses of animal cells, the crucial first step is for the virus to bind to a host cell surface. Surface binding requires a viral coat protein or spike protein to specifically recognize a receptor protein on the host cell membrane. But the host population may acquire a mutation in which the receptor protein is absent or has a changed peptide structure that no longer binds the viral spike protein. Mutation in the receptor gene can prevent infection by the virus.

Thought Question 12.4   What are the advantages and disadvantages to the virus of replication by the host polymerase, compared with using a polymerase encoded by its own genome?

ANSWER: The advantage of genome replication by the host polymerase is that the virus does not need to encode a polymerase in its own genome, and so its genome can be smaller. The advantage of the virus encoding its own polymerase is that the viral polymerase can be optimized for the needs of the virus; for example, the viral polymerase can be error prone, generating mutant progeny that escape the immune system. Also, RNA viruses must encode their own polymerase because the host cells lack any polymerase that synthesizes RNA from an RNA template.

Thought Question 12.5   Researchers in Hong Kong are conducting a regular survey of the genomes of influenza strains emerging in a particular swine-processing facility, where swine are shipped from many different regions. What might the researchers learn that would be important for human public health?

ANSWER: Swine are known to be an important source of reassorted strains of influenza virus that can cause human pandemics. For example, an avian strain and a human strain can coinfect swine and generate a strain that infects humans. The infectivity of influenza strains depends on the sequence and binding properties of their envelope proteins. Researchers can study the envelope proteins of influenza strains appearing in swine and predict whether these strains may "escape" into human populations. If a dangerous strain is found, it might be possible to devise a vaccine ahead of an outbreak.

Thought Question 12.6   How do attachment and entry of HIV resemble attachment and entry of influenza virus? How do attachment and entry differ between these two viruses?

ANSWER: Both HIV and influenza virus attach to their host cell by their envelope spike proteins binding to host receptors. For HIV-1, the Env protein binds receptors CD4 and CCR5; for influenza, the HA envelope protein binds to host glycoproteins. However, the two viruses enter their host cells by different routes. Influenza virus induces formation of an endocytic vesicle, whose acidification triggers membrane fusion and release of the core contents into the host cytoplasm. HIV virions, however, do not induce endocytosis and do not require acidification. The HIV envelope fuses with the cell membrane to release the core particle into the cytoplasm. In addition, HIV virions can travel between cells via thin tubes of cytoplasm (nanotubules) that connect an infected cell with an uninfected cell or through the merging of two cells to form a syncytium.

Sterilization, Disinfection, and Antibiotic Therapy

[A Needless Death]

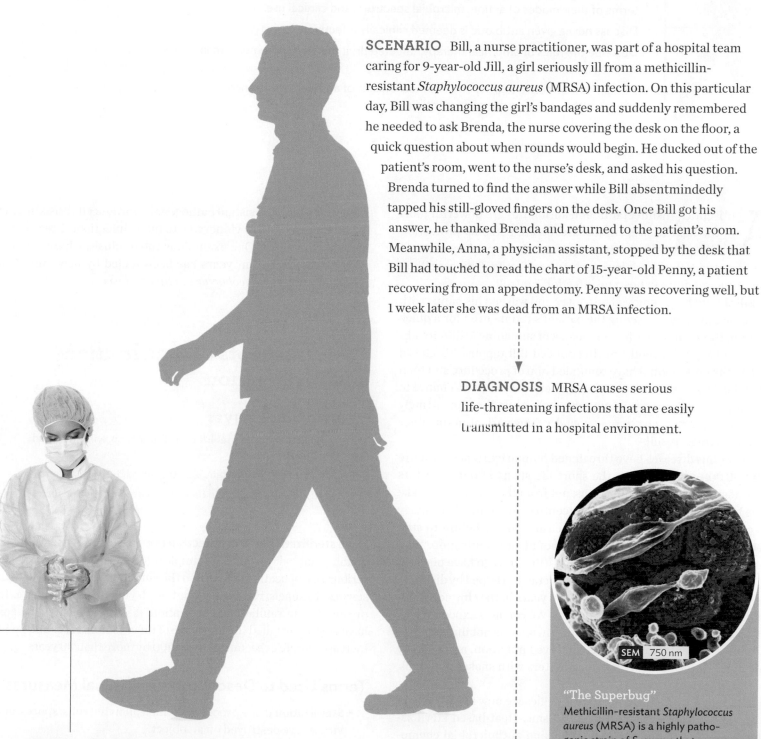

SCENARIO Bill, a nurse practitioner, was part of a hospital team caring for 9-year-old Jill, a girl seriously ill from a methicillin-resistant *Staphylococcus aureus* (MRSA) infection. On this particular day, Bill was changing the girl's bandages and suddenly remembered he needed to ask Brenda, the nurse covering the desk on the floor, a quick question about when rounds would begin. He ducked out of the patient's room, went to the nurse's desk, and asked his question. Brenda turned to find the answer while Bill absentmindedly tapped his still-gloved fingers on the desk. Once Bill got his answer, he thanked Brenda and returned to the patient's room. Meanwhile, Anna, a physician assistant, stopped by the desk that Bill had touched to read the chart of 15-year-old Penny, a patient recovering from an appendectomy. Penny was recovering well, but 1 week later she was dead from an MRSA infection.

DIAGNOSIS MRSA causes serious life-threatening infections that are easily transmitted in a hospital environment.

SEM 750 nm

"The Superbug"
Methicillin-resistant *Staphylococcus aureus* (MRSA) is a highly pathogenic strain of *S. aureus* that causes nosocomial and community-acquired infections. The photomicrograph depicts MRSA cells (shown in red) destroying a human neutrophil trying to kill the bacteria.

FOLLOW-UP Everyone who enters an MRSA patient's room must wash their hands with antiseptic soap, put on sterile gloves, and don a paper gown to cover their clothes. The procedure must be reversed when leaving the room.

Antiseptics, disinfectants, and antibiotics are powerful tools in the fight to prevent disease. So are personal protection devices, such as gloves and gowns. However, they all must be used conscientiously and properly to be effective. This case illustrates a major failure in hospital infection control.

Bill should have removed his gloves and washed his hands with antiseptic soap before leaving the patient's room—even for a quick question. By not doing so, he ran the risk of spreading MRSA to others in the hospital. Brenda, having noticed Bill tapping his gloved hand on the desk, should have reminded him of procedure and then wiped the desk down with disinfectant before Anna had a chance to pick up the MRSA contaminating the desk. Anna then unwittingly transmitted the microbe to the appendectomy patient in the other room, with tragic results.

Infectious diseases have threatened human existence for more than 200,000 years. In fact, the short life spans of our ancestors came mostly from deadly infections, not from the venom of a snake or the bite of a lion. Even in the eighteenth century, life expectancy was only 35–40 years, due in large measure to our failure to stop infections. Fortunately, the discoveries of disinfectants more than 200 years ago and of antibiotics almost 100 years ago have played a major role in increasing average life expectancy. Helped by disinfectants, life expectancy in 1900 was 45–50 years in the United States. Antibiotics made an even larger impact. We can now expect to live, on average, 75–80 years. Of course, the increase is not due solely to antibiotics. Other factors, such as improved nutrition, more widely available health care, and fewer encounters with snakes and lions, were important.

This chapter will describe the strategies we now use to control and (when necessary) kill microorganisms. Heat-based sterilization techniques, chemical disinfection, and antimicrobial chemotherapies are now part of our everyday routine. However, success in curbing infections is now threatened by the possibility that today's antibiotics will become useless. Decades-long abuse in how antibiotics were used has led bacteria to become antibiotic resistant. The surge of antibiotic-resistant pathogens has produced a crisis in which the human race is now vulnerable to many infectious diseases once thought conquered. One example is tuberculosis, whose dramatic resurgence over recent years has been fueled by new multidrug-resistant strains of *Mycobacterium tuberculosis*.

13.1
Basic Concepts of Sterilization and Disinfection

Have sterilization and disinfection improved our lives? Without a doubt. What today would be a simple infected cut in years past held a serious risk of death; and a trip to the surgeon, who worked with unsterilized scalpels, was tantamount to playing Russian roulette. Improvements in sanitation, sterilization, disinfection, and antisepsis have greatly curtailed the incidence of infectious diseases and helped increase our life expectancy since 1900 by more than 30 years.

Terms Used to Describe Antimicrobial Measures

- **Sterilization** is the process by which <u>all</u> living cells, spores, and viruses are destroyed on an object.

- **Disinfection** is the killing, or removal, of <u>disease-producing</u> organisms from <u>inanimate</u> surfaces; it does not necessarily result in sterilization. Pathogens are killed, but other microbes may survive.

- **Antisepsis** is similar to disinfection, but it applies to removing pathogens from the surface of <u>living</u> tissues, such as the skin. Antiseptic chemicals are usually not as toxic as disinfectants, which frequently damage living tissues.

- **Sanitation**, closely related to disinfection, consists of reducing the microbial population to safe levels and usually involves cleaning an object as well as disinfection.

Antimicrobial chemicals fall into two broad classes: those that kill microbes (cidal agents) and those that inhibit or control growth (static agents). Agents that kill microbes can be subcategorized as bactericidal, algicidal, fungicidal, or virucidal, depending on what type of microbe is killed. If the agent merely inhibits growth, then it would be called bacteriostatic, algistatic, fungistatic, or virustatic. For example, the antibiotic gentamicin is **bactericidal** in action, whereas chloramphenicol is **bacteriostatic**. You may have heard the term "germicidal." How does this term differ from what we just described? A **germ** is a microorganism, but especially one that causes disease. Consequently, **germicide** or **germicidal** describes any antiseptic or disinfectant chemical that kills germs (bacterial, viral, or fungal). Note that germicidal agents do not necessarily kill spores.

Although the emphasis is placed on antimicrobials acting on pathogens, it is important to note that these agents can also kill or prevent the growth of nonpathogens. In fact, many public health standards are based on <u>total</u> numbers of microorganisms found on an object, regardless of their pathogenic potential. For instance, to gain public health certification, the restaurants we frequent must demonstrate low numbers of total bacteria in their food preparation areas.

Sterilizing and Cidal Agents Kill Exponentially

How do we know when sterilization has been achieved? The challenge of sterilization is to be certain when every last microbe in a liquid or on a solid has been killed. Microbes in a culture exposed to a lethal chemical or condition (such as heat) do not all die instantaneously. Microbes die at an exponential rate, much the way they grow (considered in Chapter 6). Exponential death means that viable cell numbers decrease in equal fractions at constant intervals.

Why don't all cells die instantly when exposed to a lethal chemical or physical agent? Cells contain thousands of different proteins (and other types of targets) and thousands of molecules of each. Not all proteins in the cytoplasm, lipids in a membrane, or genes in a chromosome are damaged by an agent simultaneously. Damage accumulates over time. Only when enough molecules of an <u>essential</u> protein or gene are damaged will the cell die. Several factors influence the speed at which lethal damage accumulates:

- The initial population size (the larger the population, the longer it takes to decrease it to a specific number)

- Population composition (are spores involved?)

- Agent concentration, or dose for radiation

- Duration of exposure

- The presence of organic material (such as blood or feces) that can inhibit disinfectant action; also called organic load

Although the effect of agent concentration seems intuitively obvious, only over a narrow range will an increase in concentration produce an increase in death rate. Increases above a certain level might not accelerate killing at all. For example, a 70% ethyl alcohol solution is actually better than pure ethyl alcohol at killing organisms because some water is needed to help the ethanol penetrate cells, where the ethanol can dehydrate proteins.

The next question is, how do we compare the efficacy of different disinfecting or sterilizing agents? One way is to measure their decimal reduction times. **Decimal reduction time (D-value)** is the length of time it takes an agent (or condition) to kill 90% of the population, leaving 10% of the original population alive. For example, imagine that a culture of 100 million bacterial cells is heated to 100°C (**Figure 13.1**). After 1 minute, 90 million cells are killed and the number of viable cells remaining is now 10 million (10% of the original viable count). The D-value, therefore, is 1 minute. An agent (chemical or physical) with a D-value of 5 minutes takes longer to kill than an agent with a D-value of 1 minute.

You might then ask, "Why, if 90% of a population is killed in 1 minute, aren't the remaining 10% killed in the next minute?" It seems logical that all remaining microbes should have perished, but after the second minute, 1% of the original population still lives (the D-value remains equal to 1 minute). The agent still has the same D-value with respect to the surviving organisms as it did with

Figure 13.1   The Microbial Death Curve and the Determination of D-Values

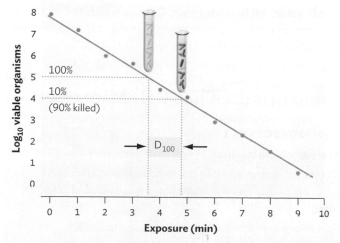

Bacteria were exposed to a temperature of 100°C, and survivors were measured by viable count. The D-value is the time required to kill 90% of cells (that is, for the viable cell count to drop by one $\log_{10}$ unit). In this example, the D-value at 100°C (D_{100}) is approximately 1 minute. The bacteria in the test tube are colored green if they are viable and red if they are nonviable (dead). Realize that the dead bacteria remain targets for the agent, which is why the live cells in the 90% killed tube do not die faster than those in the 100% living tube.

the original population because there is no change in the random chance (probability) that an agent will cause a lethal "hit" in a given live cell, even though more and more cells die. Remember, the concentration of the agent is the same. Moreover, the dead cells are still there in solution; their proteins, DNA, and so on, still absorb "hits" of the chemical agent. So, although there are fewer viable cells after 1 minute of treatment, each viable cell has the same random chance of suffering a lethal hit as when the treatment began. Thus, death rate is an exponential function.

In practice, a sterilization procedure is designed such that the length of treatment is several D-values longer than the time needed theoretically to decrease the population to one microbe. This overkill precaution ensures that few, if any, microbes survive. For example, if the sterilization time lasts three D-values beyond the predicted "last microbe," then the likelihood of a microbe remaining is approximately $(1/10)^3$, or 0.001.

SECTION SUMMARY

- **Sterilization** kills all living organisms, **disinfection** kills pathogens on inanimate objects, **antisepsis** is the removal of potential pathogens from the surfaces of living tissues, and **sanitation** combines disinfection with the cleaning of an object.

- **Antimicrobial compounds** can have a **static** (inhibitory) effect on growth or a **cidal** (killing) effect on viability. With respect to bacteria, agents can be bacteriostatic or bactericidal.

- **The D-value** is the time (or dose, in the case of irradiation) it takes an antimicrobial treatment to reduce the numbers of organisms to 10% of the original value.

- -

Thought Question 13.1   A hand-sanitizing solution inoculated with 2×10^6 organisms of *Staphylococcus aureus* exhibits a D-value of 5 minutes. What will the D-value be if ten times more *S. aureus* is added?

- -

13.2
Physical Agents That Kill Microbes

SECTION OBJECTIVES

- Compare autoclave sterilization with pasteurization.

- Explain the difference between refrigeration and lyophilization.

- Describe the uses of filtration and laminar flow biological safety cabinets.

- List conditions in which irradiation can be used to sterilize.

Can microbes be controlled without chemicals?   Physical agents are often used to kill microbes or control their growth. Commonly used physical control measures are temperature extremes, pressure (usually combined with temperature), filtration, and irradiation.

High Temperature and Pressure

Even though microbes were discovered less than 400 years ago, heating food products to render them safe has been practiced for more than 5,000 years. Moist heat is much more effective at killing than dry heat, thanks to the ability of water to penetrate cells. Many bacteria, for instance, easily withstand 100°C (212°F) dry heat but not 100°C in boiling water. We humans are not so different, finding it easier to endure a temperature of 32°C (90°F) in dry Arizona than in humid Louisiana.

Boiling water (100°C) can kill most vegetative (actively growing) organisms, but spores are built to withstand this abuse, and hyperthermophiles actually prefer it (see Table 6.3). Killing spores and thermophiles usually requires combining high pressure and temperature. At high pressure, the boiling point of water rises to a temperature rarely experienced by microbes living at sea level. Even endospores quickly die under these conditions. This combination of pressure and temperature is the principle behind sterilization using the steam autoclave (Figure 13.2). Standard conditions for steam sterilization are 121°C (250°F) at 15 psi (pounds per square inch) for 20 minutes (time can vary depending on the volume of material to be sterilized), a set of conditions that will kill all spores. These are also the conditions produced in pressure cookers used for home canning of vegetables.

Failure to adhere to the heat and pressure parameters of steam sterilization can have deadly consequences, even in your own home. Take, for instance, *Clostridium botulinum*, a spore-forming soil microbe that commonly contaminates fruits and vegetables used in home canning. The improper use of a pressure cooker while canning will allow clostridial spores to survive. Once the can or jar is cool, the spores will germinate and begin producing their deadly toxin. All of this happens while the can sits on a shelf waiting to be opened and consumed. Once ingested, the toxin makes its way to the nervous system and paralyzes the victim. Several incidents of this disease, called botulism, occur each year in the United States. (For more on botulism, see Chapters 18 and 24.)

The food industry uses several parameters to evaluate the efficiency of heat killing. The D-value has already been described. Another measure the industry uses is called 12D, the amount of time required to kill 10^{12} spores (or reduce a population 12 logs). Because finding 10^{12} spores in a food is highly unlikely, a 12D treatment will produce food that is highly safe. This measurement is extremely important to the canning industry, which must ensure that canned goods do not contain spores of *Clostridium botulinum*.

For example, the D-value for *C. botulinum* at 121°C (250°F) is 0.2 minute, so 12D is 2.4 minutes (12 × 0.2). If every can in a batch of 1 billion soup cans contained 1,000 spores, and all those cans were subjected to a 12D treatment at one time, a single can might still have a spore (1 billion cans × 1,000 spores = 10^{12} spores). Because much fewer than 1 billion cans are ever sterilized at one time, the likelihood of having any spores survive is nil.

Pasteurization

Originally devised by Louis Pasteur to save products of the French wine industry from devastating bacterial spoilage, **pasteurization** today involves heating a particular food (such as milk) to a moderately high temperature long enough to kill *Coxiella burnetii*, the causative agent of Q fever and the most heat-resistant, non-spore-forming pathogen known. The goal of pasteurization is not to sterilize milk (1 ml of pasteurized milk may still contain 20,000 bacteria); its aim is to kill pathogens without affecting the texture, color, or taste of the product. Too much heat can make milk or other foods inedible.

Many different time and temperature combinations can be used for pasteurization. The low-temperature long-time (LTLT) process involves bringing milk to a temperature of 63°C (146°F) for 30 minutes. In contrast, the high-temperature short-time (HTST) method, also called flash pasteurization, brings the milk to a temperature of 72°C (162°F) for only 15 seconds. Both processes accomplish the same thing: the destruction of *C. burnetii*, as well as *Mycobacterium tuberculosis*, *Salmonella*, and other potential pathogens.

A process known as ultra-high-temperature treatment (UHT; 134°C, or 273°F, for 1–2 seconds) can actually produce sterilized milk with an unrefrigerated shelf life of up to 6 months. This is important, especially in developing countries, where refrigeration is not always available.

Cold

Low temperatures have two basic purposes in microbiology: to temper growth and to preserve strains. Bacteria not only grow more slowly in cold but also die more slowly. Refrigeration temperatures (4–8°C, or 39–43°F) are used for food preservation because most pathogens are mesophilic: they grow best between 20°C (68°F) and 40°C (104°F) and grow poorly, if at all, at cold temperatures. One exception is the Gram-positive bacillus *Listeria monocytogenes*, which can grow reasonably well in the cold and causes disease when ingested (see Chapter 22).

Long-term storage of bacteria usually requires placing solutions in glycerol at very low temperatures (−70°C, or −94°F). Glycerol prevents the production of razor-sharp ice crystals that can pierce cells from without or within. This deep-freezing suspends growth altogether and keeps cells from dying. Another technique called **lyophilization** freeze-dries microbial cultures for long-term storage. In this technique, cultures are quickly frozen at very low temperatures (quick-freezing also limits ice crystal formation) and placed under vacuum, which causes frozen water to transition directly from solid to gas form (sublimation). The process removes all water from the medium and cells, leaving just the cells in the form of a powder. These freeze-dried organisms remain viable for years. Finally, viruses and mammalian cells must be kept at extremely low temperatures (−196°C, or −321°F), submerged in liquid nitrogen. Liquid nitrogen freezes cells so quickly that ice crystals do not have time to form.

Figure 13.2 The Steam Autoclave

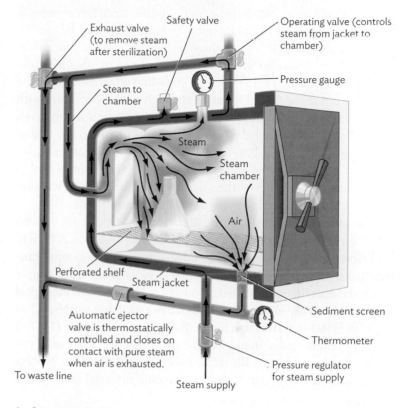

A. Steam supplied from an external boiler is stored in the autoclave's steam jacket. The opening of a valve allows steam to enter the steam chamber and expel air. When pure steam hits a valve in the air ejector line, the valve closes and chamber pressure and temperature build to the proper levels. After the appropriate period of time, another valve opens to allow steam to escape the chamber.

B. Two autoclaves in a microbiology laboratory.

Filtration

Before a liquid pharmaceutical drug is administered intravenously or orally to a patient in the hospital, the solution must be sterile. However, many drugs are sensitive to heat or chemical sterilization methods. These solutions can be sterilized by passing them through sterile filters with tiny pore sizes that effectively "sift" the microbes out of the fluid. Filtration through micropore filters with pore sizes of 0.2 μm can remove microbial cells, but not viruses, from solutions. To remove viruses, pore sizes of 20 nm are necessary (1 nm = 0.001 μm; 1 μm = 0.001 mm). Filter sterilization is also useful in standard microbiology laboratories. Samples from 1 ml to several liters can be drawn through a membrane filter by vacuum or can be forced through it using a syringe (Figure 13.3).

Air can also be sterilized by filtration. Infectious agents can become airborne in laboratory or hospital environments. A variety of personal protective devices depend on filtration to remove organisms from air before inhalation. These devices range from simple surgical masks to sophisticated air purifiers. **Laminar flow biological safety cabinets** are elaborate and effective ventilated workbenches in which air is forced through high-efficiency particulate air (HEPA) filters to remove more than 99.9% of airborne particulate material 0.3 μm in size or larger. Biosafety cabinets are critical to protect individuals working with highly pathogenic material (see Figure

Figure 13.3   Membrane Filtration

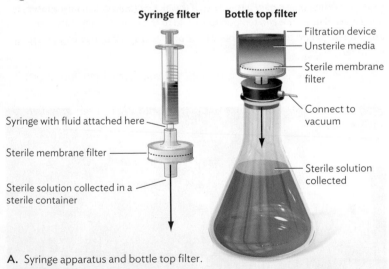

A. Syringe apparatus and bottle top filter.

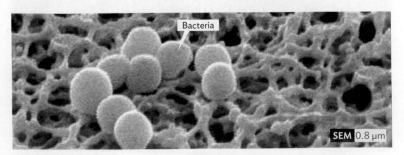

B. *Staphylococcus epidermidis* on Millipore filters (magnification 8,000×).

Figure 13.4   Immobilized Enzyme Filter

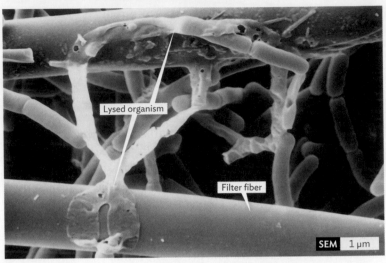

The primary function of this enzyme filter is to kill airborne microorganisms caught on the surface of the filter to protect against secondary contamination from microorganisms in air filtration systems. The photo shows lysed bacteria (*Bacillus subtilis*). The cell walls have been hydrolyzed by enzymatic action, and cell membranes are ruptured as a result of osmotic pressure pushing outward against the membrane.

2.16A and B). Newer technologies have been developed that embed antimicrobial agents or enzymes directly into the fibers of the filter (Figure 13.4). Organisms entangled in these fibers are not just trapped; they are attacked by the antimicrobials and lysed.

Irradiation

Public health authorities worldwide are increasingly concerned about food contaminated with pathogenic microorganisms such as *Salmonella* species, *Escherichia coli* O157:H7, *Listeria monocytogenes*, and *Yersinia enterocolitica*. Irradiation, in which food is bombarded with high-energy electromagnetic radiation, has long been a potent strategy for sterilizing food after harvesting. The food consumed by NASA astronauts, for example, has for some time been sterilized by irradiation as a safeguard against food-borne illness in space. Foods do not become radioactive when irradiated, and any reactive molecules produced when high-energy particles are absorbed by food dissipate almost immediately (these reactive molecules are also produced by cooking, by the way).

Ultraviolet (UV) light is one form of sterilizing radiation. However, owing to its poor penetrating ability, UV light is useful only for surface sterilization. Three more effective sources of irradiation are **gamma rays** (cobalt-60), electron beams, and X-rays. Radiation dosage of any of these treatments is usually measured in a unit called the gray (Gy), which is the amount of energy transferred to food, microbe, or other substance being irradiated. A single chest X-ray delivers roughly half a milligray (1 mGy = 0.001 Gy). To kill *Salmonella*, freshly slaughtered chicken can be irradiated at up to 4.5 kilograys (kGy), about 7 million times the energy of a single

Table 13.1

Examples of Foods Approved for Irradiation in the United States

Approval Year	Food	Dose	Purpose
1963	Wheat flour	0.2–0.5 kGy	Control mold
1964	White potatoes	0.05–0.15 kGy	Inhibit sprouting
1986	Pork	0.3–1.0 kGy	Kill trichina (*Trichinella*) parasites
1986	Fruit and vegetables	1.0 kGy	Control insects; increase shelf life
1986	Herbs and spices	30 kGy	Sterilize
1992	Poultry	1.5–3.0 kGy	Reduce numbers of bacterial pathogens
1997	Meat	4.5 kGy	Reduce numbers of bacterial pathogens
2006	Shellfish	5.5 kGy	Reduce numbers of *Vibrio* species and other pathogens
2010	Lettuce and spinach	4.0 kGy	Reduce numbers of bacterial pathogens

chest X-ray. Table 13.1 lists foods approved for irradiation in the United States.

Link As discussed in Section 8.3, **ultraviolet (UV) light** causes the dimerization of pyrimidine nucleotides in DNA, which can inhibit replication and transcription, whereas X-rays and **gamma rays** can break the DNA backbone.

When microbes in food are irradiated, water and other intracellular molecules absorb the energy and create short-lived reactive chemicals (oxygen radicals) that damage DNA. Unless the organism repairs this damage, it will die while trying to replicate. A microbe's sensitivity to irradiation depends on the size of its genome, the rate at which damaged DNA can be repaired, and other factors. It also matters if the irradiated food is frozen or fresh, as it takes a higher dose of radiation to kill microbes in frozen foods.

The size of the DNA "target" is a major factor in radiation efficacy. Parasites and insect pests, which have large amounts of DNA, are rapidly killed by extremely low doses of radiation, typically with D-values of less than 0.1 kGy (in this instance, the D-value is the dose of radiation needed to kill 90% of the organisms). It takes more radiation to kill bacteria (D-values in the range of 0.3–0.7 kGy) because they have less DNA per cell unit (less target per cell). It takes even more radiation to kill a bacterial spore (D-values on the order of 2.8 kGy) because they contain little water, the source of

most ionizing damage to DNA. Viral pathogens have the smallest amount of nucleic acid, making them resistant to irradiation doses approved for foods (viruses have D-values of 10 kGy or higher). **Prion particles** associated with bovine spongiform encephalopathy (BSE, also known as mad cow disease) do not contain nucleic acid and are inactivated by irradiation only at extremely high doses. Thus, irradiation of food is effective in eliminating parasites and bacteria but is woefully inadequate for eliminating viruses or prions.

Link **Prion particles** and the diseases they cause, such as mad cow disease, are discussed more fully in Section 24.6.

 CDC irradiation website: cdc.gov

Note Electromagnetic radiation emitted by microwave ovens does not directly kill bacteria. However, the heat generated when electromagnetic radiation excites water molecules in an organism will kill the organism if the temperature attained is high enough.

SECTION SUMMARY

- **The autoclave** uses high pressure to achieve temperatures that will sterilize objects or solutions.
- **Pasteurization** is a heating process designed to kill specific pathogens in milk and other food products.
- **Refrigeration** is used to prevent microbial growth in foods. Extreme cold (freezing) is used to preserve bacteria.
- **Lyophilization** is a method of freeze-drying that permits living cultures to remain viable for years.
- **Filtration** can remove cells from a solution and from the air, but it cannot remove the smallest viruses.
- **Irradiation** can kill pathogens in foods without damaging the food itself.

Thought Question 13.2   How would you test the killing efficacy of an autoclave?

13.3
Chemical Agents of Disinfection

SECTION OBJECTIVES

- List factors that can affect the efficacy of a disinfectant.
- Explain how disinfectants are compared.
- List the basic forms of commercial disinfectants.
- Describe the emergence of microbial resistance to disinfectants.

How can we kill microbes on items that would be damaged by physical agents? Physical agents such as heat are very effective, but numerous situations arise where their use is impractical (kitchen countertops, for instance) or plainly impossible (skin). In these instances, chemical agents are the best approach. There are

Table 13.2
CDC Levels of Disinfection Based on Range of Microbes Affected

CDC Disinfection Level[a]	Vegetative Bacterial Cells	Mycobacteria	Spores	Fungi	Viruses	Examples
High level	+	+	+	+	+	Ethylene oxide, glutaraldehyde, formaldehyde
Intermediate level	+	+	–	+	+	Phenolics, halogens
Low level	+	–	–	+	+/–	Alcohols, quaternary ammonium compounds

[a]High-level disinfection kills all organisms, except high levels of bacterial spores or prions. Intermediate-level disinfection kills mycobacteria, most viruses, and bacteria. Low-level disinfection kills some viruses and bacteria with a chemical germicide registered as a hospital disinfectant by the EPA.

a number of factors that influence the efficacy of a given chemical agent. These include:

- **The presence of organic matter.** A chemical placed on a dirty surface will bind to the inert organic material present, lowering the agent's effectiveness against microbes. It sometimes is not possible to clean a surface prior to disinfection (as in a blood spill), but the presence of organic material must be factored in when estimating how long to disinfect a surface or object.

- **The kinds of organisms present.** Ideally, the agent should be effective against a broad range of pathogens. Table 13.2 presents a classification of disinfectants based on their spectrum of activity.

- **Corrosiveness.** The disinfectant should not corrode the surface or, in the case of an antiseptic, damage skin.

- **Stability, odor, and surface tension.** The chemical should be stable upon storage, possess a neutral or pleasant odor, and have a low surface tension so it can penetrate cracks and crevices.

But how do you know whether you are using the most effective disinfectant? Two ways to compare the efficacy of disinfectants are the phenol coefficient and the use-dilution test. A third way, the disk diffusion test, will be described in Section 13.4 when we discuss antibiotics.

Measuring the Efficacy of Disinfectants

Phenol, first introduced by Joseph Lister in 1867 to reduce the incidence of surgical infections, is no longer used as a disinfectant, but its derivatives, such as cresols and *ortho*-phenylphenol, are still in use. The household product Lysol is a mixture of phenolics. Phenolics are useful disinfectants because they denature proteins, are effective in the presence of organic material, and remain active on surfaces long after application.

Although phenol is no longer used as a disinfectant, its potency is the benchmark against which other disinfectants are measured.

The **phenol coefficient test** involves inoculating a fixed number of bacteria—for example, *Salmonella enterica* serovar Typhi or *Staphylococcus aureus*—into dilutions of the test agent. At timed intervals, samples are withdrawn from each dilution and inoculated into fresh broth (which contains no disinfectant). The phenol coefficient is based on the highest dilution (lowest concentration) of a disinfectant that will kill all the bacteria in a test after 10 minutes of exposure, but leaves survivors after only 5 minutes of exposure. This concentration is known as the maximum effective dilution. Dividing the reciprocal of the maximum effective dilution for the test agent (for example, ethyl alcohol) by the reciprocal of the maximum effective dilution for phenol gives the phenol coefficient (Table 13.3). For example, if the maximum effective dilution for agent X is $1/_{900}$ and that of phenol is $1/_{90}$, then the phenol coefficient of X is the reciprocal of $1/_{900}$ divided by the reciprocal of $1/_{90}$, or $900 \div 90 = 10$. The higher the coefficient, the higher the efficacy of the disinfectant.

Table 13.3
Phenol Coefficients for Various Disinfectants

Chemical Agent	*Staphylococcus aureus*	*Salmonella Typhi*
Phenol	1.0	1.0
Chloramine	133.0	100.0
Cresols	2.3	2.3
Ethyl alcohol	6.3	6.3
Formalin	0.3	0.7
Hydrogen peroxide	—	0.001
Lysol	5.0	3.2
Mercury chloride	100.0	143.0
Tincture of iodine	6.3	5.8

The **use-dilution test** is especially useful for determining the ability of disinfectants to kill microorganisms dried onto a typical clinical surface (stainless steel). The test organism is air-dried onto a stainless steel surface and then exposed to different dilutions of a disinfectant for a fixed set of time (for example, 10 minutes). The entire surface is then placed in broth containing chemicals that neutralize the disinfectant. Successful disinfection results in no bacterial growth in the broth. Disinfectants that completely kill (or otherwise remove) microbes at the lowest dilutions of the disinfectant are considered the most effective.

Commercial Disinfectants

Many different types of disinfectants are available (Table 13.4). Ethanol, iodine, chlorine, and surfactants (such as detergents) are all used to reduce or eliminate microbial content from commercial products (see Appendix 3, Figure A3.1, for structures of various disinfectants). Ethanol, iodine, and chlorine are highly reactive compounds that damage proteins, lipids, and DNA. Iodine complexed with an organic carrier forms an iodophor, a compound that is water-soluble, stable, nonstaining, and capable of releasing iodine slowly to avoid skin irritation. Wescodyne and Betadine (trade names) are iodophors used, respectively, for the surgical preparation of skin and for wounds. Chlorine is a disinfectant recommended for general laboratory and hospital disinfection.

 CDC guidelines for disinfection and sterilization in health care facilities: cdc.gov

Detergents can also be antimicrobial agents. The hydrophobic (fat-loving) and hydrophilic ("water loving") ends of detergent molecules will emulsify fat into water. Cationic (positively charged) but not anionic (negatively charged) detergents are useful as disinfectants because the cationic detergents contain positive charges that can gain access to the negatively charged bacterial cell and disrupt membranes. Anionic detergents are not antimicrobial but do help in the mechanical removal of bacteria from surfaces.

Low-molecular-weight aldehydes (formula R-CHO) such as formaldehyde are highly reactive, combining with and inactivating proteins and nucleic acids. This makes them useful disinfectants, too.

GAS STERILIZATION Disposable plasticware such as petri dishes, syringes, sutures, and catheters are not amenable to heat sterilization or liquid disinfection. These materials are best sterilized using antimicrobial gases (or gamma irradiation). Ethylene oxide gas (EtO) is a very effective sterilizing agent; it destroys cell proteins by alkylation, is microbicidal and sporicidal, and rapidly penetrates packing materials, including plastic wraps. In an instrument resembling an autoclave called an ethylene oxide chamber (Figure 13.5), EtO at 700 mg/l will sterilize an object after 8 hours at 38°C (100.4°F) or 4 hours at 54°C (129.2°F). Unfortunately, EtO

Figure 13.5
Ethylene Oxide Chamber
Once the gas sterilization process is complete (usually 1–6 hours), the used EtO gas is vented from the chamber (before the door is opened) and destroyed via chemical "scrubbing" by passage through sulfuric acid solution or by passage over a catalytic oxidizer.

is explosive. A less hazardous gas sterilant is beta-propiolactone. It does not penetrate as well as EtO, but it decomposes after a few hours, which makes it easier to dispose of than EtO.

A new procedure known as gas discharge plasma sterilization may replace EtO because it is less harmful to operators. Gas discharge plasma is made by passing certain gases through a radio frequency electrical field to produce highly reactive chemical species that can damage membranes, DNA, and protein. It is not yet widely used.

ANTIMICROBIAL TOUCH SURFACES Despite widespread use of disinfectants and antibiotics by medical personnel, hospital-acquired infections remain a major concern. A promising, new antimicrobial technology that may help reduce infections in hospitals and elsewhere involves embedding antimicrobial compounds such as copper (Cu) in the surfaces people touch. Upon contact with bacteria, metallic copper releases toxic copper ions (Cu^+) that trigger the lysis of bacterial membranes within minutes, although the mechanism involved is unclear. It is thought that incorporating metallic copper into objects such as handrails, door releases, hospital bed rails, and the arms of visitor's chairs will kill microbes deposited by one person before those microbes can be transmitted to someone else.

Bacterial Resistance to Disinfectants

It is widely known that bacteria can develop resistance to antibiotics used to treat infections. This is a serious concern in the medical community. So, you might wonder whether bacteria can also develop resistance to chemical disinfectants used to prevent infections. The answer is yes—and no. It is difficult for a bacterium to develop resistance to chemical agents that have multiple targets and can easily diffuse into a cell. Iodine, for example, has both of these characteristics. However, disinfectants that have multiple targets at high concentrations may have only a single target at lower concentrations—a situation that can foster the development of resistance.

Table 13.4
Some Categories of Disinfectants

Disinfectant	Mode of Action	Uses	Toxicity	Comments	CDC Level of Activity[a]
Phenol	Denatures proteins	Standard for comparison	Can cause dermatitis; vapor affects lungs	Works in the presence of organics; rarely used except as a standard for comparison	Low to intermediate
Creosol *Ortho*-phenylphenol		Disinfection			
Bisphenols (two linked phenol groups) Lysol disinfectant (*o*-benzyl-*p*- chlorophenol)	Membrane and protein disruption	Disinfects solid surfaces	Very toxic if swallowed	Bacteria developing resistance	Intermediate
Triclosan (a chlorinated bisphenol)	Inhibits fatty acid synthesis at low concentrations	Soaps, cosmetics, medications			
Halogens Chlorine gas	Corrosive tissue damage; protein destruction	Disinfection of drinking water; sanitation; disinfection of pools, hospital equipment; wound antisepsis	Gas highly toxic	Sporicidal; inactivated by organics; corrosive to metals	Intermediate
Hypochlorite (bleach)	Toxic reaction products		Solution damages skin, eyes, mucous membranes		
Iodine tinctures (mixed with alcohol or water)	Membrane damage	Topical antiseptic	Can irritate tissue; toxic if ingested	Sporicidal; corrosive to metals	Intermediate
Iodophors (iodine mixed with solubilizing agent) Povidone-iodine		Antiseptic prep for surgery; hand scrubbing; disinfect equipment			
Chlorhexidine (complex of chlorine and two phenol rings)	Membrane disruption and protein denaturation	Hand scrubbing; prepare skin for surgical incision; dental hygiene; contact lens solutions; mouthwash; cleaning of skin	Low toxicity	Inactivated by anionic compounds; avoid food or drink for 1 hour after oral use	Intermediate
Alcohols 70% ethanol Isopropanol	Dehydration, protein denaturation; dissolve membrane lipids	Cleaning instruments; cleaning skin	Dries skin; toxic if ingested	Flammable; fast acting	Intermediate
Hydrogen peroxide	Hydroxyl radicals attack membrane lipids, DNA, protein	Wound cleaning; sterilize contact lenses	Will whiten skin	Sporicidal; works in presence of organic matter	High
Quaternary ammonium compounds (cationic agents)	Disrupts membrane	Sanitation of floors, walls, blood pressure cuffs	Toxic if ingested; irritates mucous membranes	Not sporicidal	Low
Glutaraldehyde	Alkylation damages protein, DNA, and RNA • Binds membranes, inhibits transport • Cross-links proteins and DNA	Disinfect medical equipment (e.g., endoscopes and respiratory therapy equipment)	Irritates skin, mucous membranes	Sporicidal; works in presence of organics	High
Formaldehyde	Reacts with protein, DNA, and RNA; cross-links proteins and DNA	Embalming agent; preserves anatomical specimens	Pungent odor, irritating fumes; carcinogen	Sporicidal; slow action	High
Ethylene oxide gas	Alkylation modifies proteins and DNA	Sterilize disposable plasticware	Carcinogen; eye and lung damage	Sporicidal; explosive	High

[a]High-level disinfection kills all organisms, except high levels of bacterial spores and prions. Intermediate-level disinfection kills mycobacteria, most viruses, and bacteria. Low-level disinfection kills some viruses and bacteria with a chemical germicide registered as a hospital disinfectant by the EPA.

For instance, triclosan (a halogenated bisphenol compound used in many soaps and deodorants) targets several cell constituents, making it bactericidal at high concentrations. However, at low concentrations triclosan only inhibits fatty acid synthesis and is merely bacteriostatic. Some organisms have evolved resistance to triclosan at low concentrations through mutations that altered the fatty acid synthesis protein normally targeted by triclosan. Consequently, in 2016 the FDA banned triclosan in consumer antiseptic washes.

Low-level resistance to disinfectants can be achieved through membrane-spanning, multidrug efflux pumps (see Section 13.6). For instance, an efflux system of *Pseudomonas aeruginosa*, a Gram-negative bacterium that causes infections in burn and cystic fibrosis patients, can pump several different biocides, detergents, and organic solvents out of the cell, thereby reducing their efficacy. This finding has led many clinicians to advocate caution in the widespread use of certain chemical disinfectants.

Finally, biofilms can impart resistance to disinfectants. Described in Section 6.6, biofilms are three-dimensional bacterial communities attached to solid surfaces. The extracellular matrix proteins and polysaccharides that bind these cells together can bind disinfectants, slowing their penetration to deeper recesses of the structure. Cells deep in the biofilm also have a stress-induced physiology better suited to resist disinfectant action. Finally, metabolic collaboration between multiple species in a mixed biofilm can be protective. For instance, enzymes from one species can shield a nonproducing species from a chemical insult, much like a big brother protecting a little brother.

SECTION SUMMARY

- **Factors that influence the efficacy of disinfectants** include the presence of organic matter; the variety of microbes present; and the agent's corrosiveness, stability, odor, and surface tension.
- **Phenolics, halogens (chlorine and iodine), alcohols, hydrogen peroxide, cationic agents, aldehydes, and gases** are general types of chemical disinfectants.
- **The phenol coefficient and use-dilution tests** are used to compare the efficacy of chemical disinfectants.
- **Ethylene oxide or beta-propiolactone** gases can be used to sterilize disposable plasticware.
- **Bacterial resistance** to some disinfectants is emerging.

13.4
Basic Concepts of Antimicrobial Therapy

SECTION OBJECTIVES

- Explain the concept of selective toxicity and its importance for clinically useful antibiotics.
- Explain what is meant by spectrum of activity, and discuss its clinical importance.

- Discuss how the effectiveness of an antibiotic is tested, and differentiate minimal inhibitory concentration (MIC) from minimal bacteriostatic concentration (MBC).
- Describe the chemotherapeutic index.

How do we eliminate pathogens from the body? We can't use disinfectants, which are often toxic to humans. Industrially synthesized, disinfectants lack specificity in their action and can affect animal as well as microbial cells. Instead we use antibiotics. **Antibiotics** (as made in nature) are chemical compounds synthesized by one microbe and able to selectively kill other microbes. When purified and administered to patients suffering from an infectious disease, antibiotics can produce seemingly miraculous recoveries (**Figure 13.6**). Although many clinically useful antibiotics are made biologically, others are synthesized chemically. The generic term for these natural and synthetic antibiotics is **chemotherapeutic agents**. The following case history illustrates several important principles of antimicrobial therapy.

CASE HISTORY 13.1
Molly's Meningitis

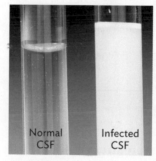

Normal CSF   Infected CSF

Three-year-old Molly was brought to the emergency department crying. She had a stiff neck and high fever (40°C, or 104°F). Gram stain of her cerebrospinal fluid revealed Gram-positive cocci, generally in pairs. The attending physician diagnosed Molly with meningitis and immediately prescribed intravenous ampicillin. Unfortunately, the child's condition worsened, so antibiotic treatment was changed to a third-generation cephalosporin (ceftriaxone). Molly began to improve within hours and was released after 2 days. A report from the clinical microbiology laboratory 2 days later identified the organism as *Streptococcus pneumoniae*. The report also included antibiotic susceptibility results, which revealed that this strain of *S. pneumoniae* was resistant to ampicillin but remained susceptible to cephalosporin.

Antibiotics Exhibit Selective Toxicity

The serendipitous discovery of penicillin in 1929 by Alexander Fleming marked the start of the golden age of antibiotic discovery (Section 1.3). But as early as 1904, the German physician Paul Ehrlich (1854–1915) realized that a successful antimicrobial compound would be a "magic bullet" that selectively kills or inhibits the pathogen but not the host. Ehrlich's "magic bullet" concept is now known as **selective toxicity**.

Selective toxicity is possible because key aspects of a microbe's physiology are strikingly different from those of eukaryotes. For example, bacterial antibiotic targets include peptidoglycan, which

Figure 13.6
The Power of Antibiotics

Pictures taken in 1942, shortly after the introduction of penicillin, show the improvement in a child suffering from an infection at different stages following treatment. Panels 5 and 6 show her fully recovered.

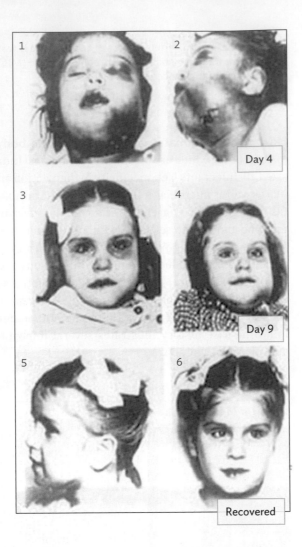

eukaryotic cells lack, and ribosomes, which are structurally distinct between these two domains of life. Thus, chemicals such as penicillin, which prevents peptidoglycan synthesis, and tetracycline, which binds to bacterial 30S ribosomal subunits, inhibit bacterial growth but are essentially invisible to host cells, since they do not interact with them at low doses. In Case History 13.1, Molly was given two different antibiotics, both of which selectively affect bacterial cell wall synthesis; only one, however, killed the bacteria.

Even with a high degree of selective toxicity, some antibiotics have unintended side effects that harm the patient. Antibiotic toxicities are unrelated to their antimicrobial modes of action. For example, chloramphenicol, a drug that targets bacterial 50S ribosomal subunits, can interfere with the development of blood cells in bone marrow, a phenomenon that may result in aplastic anemia (failure to produce red blood cells). The toxicity of an antibiotic can also depend on the age of the patient. Ciprofloxacin, for instance, can cause defects in human bone growth plates and should not be administered to pregnant women or children. Metabolism by the body can also be a problem. For instance, the liver metabolizes and detoxifies foreign chemicals, including antibiotics, and can become damaged by the toxic metabolic products.

Likewise, the kidney, which excretes drugs and their metabolites, can be damaged.

Problems can arise even if the drug does not directly impact mammalian physiology. For example, many people develop an extreme allergic sensitivity toward penicillin, a situation in which the treatment of an infection may end up being worse than the infection itself. Physicians must be aware of these allergies and use alternative antibiotics to avoid harming their patients.

As a student of microbiology, it is important that you distinguish between drug susceptibility and drug sensitivity. A microbe is susceptible to the drug's action, but a human can develop an allergic sensitivity to the drug.

Antimicrobials Have a Limited Spectrum of Activity

No single chemotherapeutic agent affects all microbes. As a result, antimicrobial drugs are classified based on the type of organisms they affect. Thus, we have antifungal, antibacterial, antiprotozoan, and antiviral agents. The term "antibiotic" is usually reserved for chemotherapeutic agents that affect bacteria. Even within a group, one agent may have a very narrow **spectrum of activity**, meaning it affects only a few species, whereas another antibiotic may inhibit many species. For instance, penicillin has a relatively narrow spectrum of activity, killing primarily Gram-positive bacteria. However, ampicillin, the drug first used in Case History 13.1, is penicillin with an added amino group that allows the drug to more easily penetrate the Gram-negative outer membrane. As a result of this chemically engineered modification, ampicillin kills Gram-positive and Gram-negative organisms and is described as having a broader spectrum of activity than penicillin. There are also antimicrobials that exhibit an extremely narrow spectrum of activity. One example is isoniazid, which is clinically useful only against *Mycobacterium tuberculosis*, the microbe that causes tuberculosis.

Antibiotics Are Classified as Bacteriostatic or Bactericidal

Patients typically believe that all antibiotics kill their intended targets. This is a misconception. Many drugs simply prevent the growth of the organism and let the body's immune system dispatch the intruding microbe. Thus, antimicrobials are also classified on the basis of whether or not they kill the microbe. As defined earlier, an antibiotic is bactericidal if it kills the target microbe; it is bacteriostatic if it merely prevents bacterial growth. Both antibiotics used in Molly's case are bactericidal when the organism is susceptible.

Minimal Inhibitory Concentration Reflects Antibiotic Efficacy

The in vitro effectiveness of an agent is determined by measuring how little of it is needed to stop growth. This is classically measured in terms of an antibiotic's **minimal inhibitory concentration (MIC)**, defined as the lowest concentration of the drug that will prevent the growth of an organism. But the MIC for any one drug will differ among different bacterial species. For example, the MIC of ampicillin needed to stop the growth of *S. aureus* will be different from that needed to inhibit *Shigella dysenteriae* or *S. pneumoniae*. The reasons that a drug may be more effective against one organism than another include the ease with which the drug penetrates the cell and the affinity of the drug for its molecular target.

So how do we measure MIC? As shown in **Figure 13.7**, an antibiotic is serially diluted along a row of test tubes containing nutrient broth. After dilution, the organism to be tested is inoculated at identical low population densities into each tube, and the tubes are usually incubated overnight. Growth of the organism is seen as turbidity. Note that in Figure 13.7, the tubes with the highest concentration of drug were clear, indicating no growth. The tube containing the MIC is the tube with the lowest concentration of drug that shows no growth. In this case, it was the 1.0-μg/ml tube. However, the MIC does not tell whether a drug is bacteriostatic or bactericidal.

MIC determinations are very useful for estimating a single drug's effectiveness against a single bacterial pathogen isolated from a patient, but until recently, they were not very practical

Figure 13.8 An MIC Strip Test

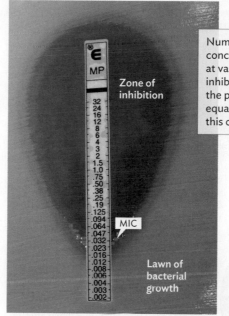

Numbers reflect the relative concentrations of antibiotic present at various points within the zone of inhibition. The concentrations along the periphery of the clear zone are equal and reflect the MIC, which in this case is 0.032 μg/ml.

The Etest (AB Biodisk) is a commercially prepared strip that produces a gradient of antibiotic concentration (μg/ml) when placed on an agar plate. The MIC corresponds to the point where bacterial growth crosses the numbered strip (white box).

Figure 13.7 Determining Minimal Inhibitory Concentration (MIC)

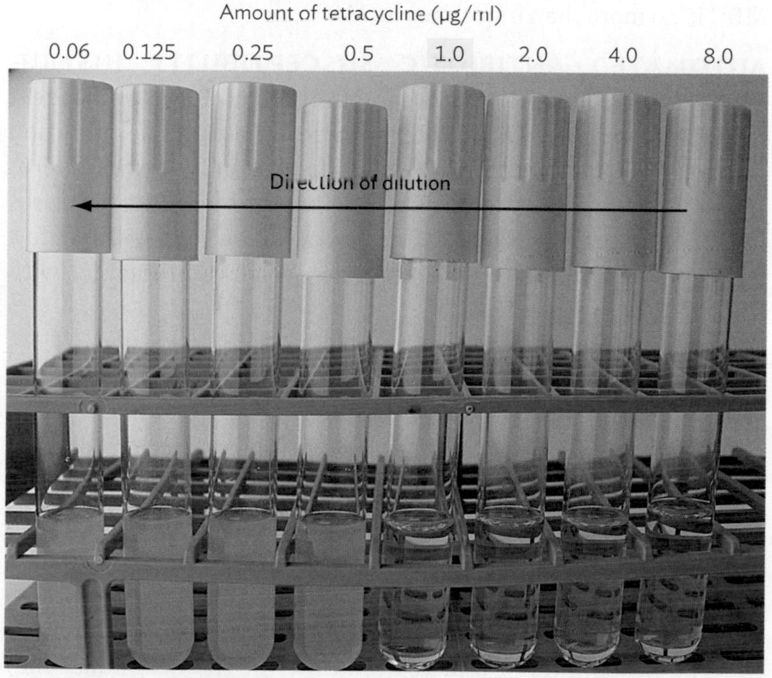

Amount of tetracycline (μg/ml)

| 0.06 | 0.125 | 0.25 | 0.5 | 1.0 | 2.0 | 4.0 | 8.0 |

Direction of dilution

In this series of tubes, tetracycline was twofold diluted serially starting at 8 μg/ml (tube at far right). To perform this serial twofold dilution, 1 ml of fluid from the 8-μg/ml drug-containing tube was mixed with 1 ml of media in the next tube (making 4 μg/ml); next, 1 ml of fluid from the 4-μg/ml tube was added to 1 ml in the next tube (making 2 μg/ml); and so forth. Each tube was then inoculated with an equal number of bacteria. Turbidity in a tube indicates that the antibiotic concentration was not sufficient to inhibit growth. The MIC in this example is 1.0 μg/ml.

when trying to screen 20 or more different drugs. Dilutions take time—time that the technician, not to mention the patient, may not have. The time required to evaluate antibiotic effectiveness can be reduced by using a strip test (like the Etest shown in **Figure 13.8**), which avoids the need for dilutions. The strip, containing a gradient of antibiotic, is placed over a fresh lawn of bacteria spread on an agar plate. While the bacteria are trying to grow, the drug diffuses out of the strip and into the medium. Drug emanating from the more concentrated areas of the strip will travel faster and farther through the agar than will drug from the less concentrated areas of the strip. Thus, the higher concentrations will kill or inhibit the growth of cells farther from the strip than the lower concentration areas. The result is a **zone of inhibition** where the antibiotic has stopped bacterial growth. The MIC is the point at which the elliptical zone of inhibition intersects with the strip.

KIRBY-BAUER DISK SUSCEPTIBILITY TEST   Clinical labs can process up to 100 or more bacterial isolates in one day, so individual MIC determinations as described earlier are not practical. Although the strip test eliminates the time and effort needed to make dilutions, it would take 20 or more plates to test an equal number of antibiotics for just one bacterial isolate. However, a simplified agar diffusion test that can test 12 antibiotics on one plate makes evaluating antibiotic susceptibility a manageable task.

Named for its inventors, the **Kirby-Bauer assay** uses a series of round filter paper disks impregnated with different antibiotics (**Figure 13.9**). A dispenser delivers up to 12 disks simultaneously to the surface of an agar plate covered by a bacterial lawn. The standardized medium used is called Mueller-Hinton agar. Each disk is marked to indicate the drug used. During incubation, the drugs diffuse away

Figure 13.9   The Kirby-Bauer Disk Susceptibility Test

A. Antibiotic disk dispenser used to deliver up to 12 disks to the surface of a Mueller-Hinton plate. The device is placed over the plate, and the plunger is depressed to deliver the disks.

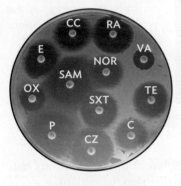

B. Disks impregnated with different antibiotics are placed on a freshly laid lawn of bacteria and incubated overnight. The clear zones around specific disks indicate growth inhibition. Shown are the results with normal methicillin-sensitive *Staphylococcus aureus* (MSSA).

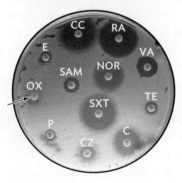

C. Methicillin-resistant *S. aureus* (MRSA). Note the lack of inhibition by the oxacillin disk (arrow). This strain is resistant to both methicillin and oxacillin.

C = chloramphenicol; CC = clindamycin; CZ = cefazolin; E = erythromycin; NOR = norfloxacin; OX = oxacillin; P = penicillin; RA = rifampin; SAM = sulbactam-ampicillin; SXT = sulfatrimethoprim; TE = tetracycline; VA = vancomycin.

from the disks into the surrounding agar and inhibit growth of the lawn to different distances. The zones of inhibition vary in width, depending on which antibiotic is used, the concentration of drug in the disk, and the susceptibility of the organism to the drug. The diameter of the zone around an antibiotic disk correlates to the MIC of that antibiotic against the organism tested. The clinical laboratory in Case History 13.1 used the Kirby-Bauer test to determine the susceptibility of Molly's pathogen to various antibiotics.

The Kirby-Bauer test has been standardized to ensure reproducibility so that results from a laboratory in California will match those in Alabama, Ohio, or any other state. The following parameters have been standardized:

- Size of the agar plate
- Depth of the medium
- Medium composition
- Number of organisms spread on the agar plate
- Size of the disks
- Concentrations of antibiotics in the disks
- Incubation temperature

Changing any one of these parameters will alter the size of the zone of inhibition and give an incorrect assessment of susceptibility.

Neither the Kirby-Bauer method nor the tube dilution techniques can tell you whether a drug is bactericidal because you don't know if the cells within the zone of inhibition (Kirby-Bauer) or in the tubes with no growth (MIC) are dead or if they remain viable but just stopped growing. To determine the **minimum bactericidal concentration (MBC)** of an antibiotic, a tube dilution test is performed and then the antibiotic is removed. Antibiotic can be removed by pelleting cells in the "no growth" tubes by centrifugation and replacing the antibiotic-containing medium with fresh medium. If cells grow, they were not killed. The lowest concentration tube that still does not show growth represents the MBC.

Almost any antibiotic can be both bactericidal at a high concentration (MBC) and bacteriostatic at a lower concentration (MIC). A drug, however, is deemed to have a bactericidal mode of action if its MBC is no more than four times its MIC.

AUTOMATED ANTIBIOTIC SUSCEPTIBILITY INSTRUMENTS Recent advances in computer technology have made it possible to automate MIC determinations using a microdilution format. Multiple antibiotics can be diluted in a 96-well microtiter plate and the plate optically scanned to monitor increases in cell numbers (**Figure 13.10**). Results can be seen within a few hours instead of overnight—precious time when dealing with life-threatening infections.

Determining Whether an Antibiotic Is Clinically Useful

Once obtained, the results of an MIC determination or a disk diffusion test must be evaluated to determine whether a given drug is clinically useful. A high concentration of drug may stop growth in the laboratory, but may not work in the patient. For an antibiotic to stop bacterial growth in the patient, the drug's concentration in tissue must remain higher than the MIC at all times.

The tissue level of a drug over time (or the drug's half-life) depends on how quickly the antibiotic is removed from the body via secretion by the kidney or destruction in the liver. Clinical usefulness of the drug also depends on whether side effects appear at tissue concentrations needed to affect the pathogen. As shown in **Figure 13.11**, as long as the concentration of the drug in the tissue or blood remains higher than the MIC (without side effects appearing), the drug will be effective. The concentration can be kept at

Figure 13.10 Automated MIC Determination

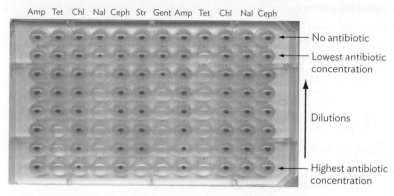

Amp Tet Chl Nal Ceph Str Gent Amp Tet Chl Nal Ceph

— No antibiotic

— Lowest antibiotic concentration

Dilutions

— Highest antibiotic concentration

Each column in the microtiter plate contains a concentration gradient of one antibiotic. An oxidation-reduction indicator in each well will turn red when bacteria in the well grow. The top row contains no antibiotic and serves as a growth control. The MIC of each antibiotic corresponds to the first well in each column (reading from top to bottom) that shows no growth (no red pellet). Concentration range for each drug from second well to last well: Amp = ampicillin (2–128 µg/ml); Tet = tetracycline (1–64 µg/ml); Chl = chloramphenicol (2–128 µg/ml); Nal = nalidixic acid (2–128 µg/ml); Ceph = cephalothin (2–128 µg/ml); Str = streptomycin (2–128 µg/ml); Gent = gentamicin (1–64 µg/ml). Note that in this example, the tetracycline and nalidixic acid MIC results varied between duplicates.

Figure 13.11 Correlation between MIC and Serum or Tissue Level of an Antibiotic

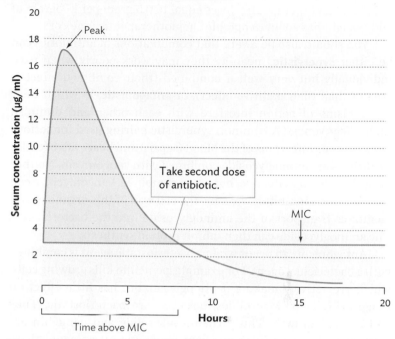

Peak

Take second dose of antibiotic.

MIC

Time above MIC

This graph illustrates the serum level of ampicillin over time. Notice how long the serum level of the antibiotic remains higher than the MIC (about 8 hours). Once the concentration falls below the MIC, owing to destruction in the liver or clearance through the kidneys and secretion, the infectious agent fails to be controlled by the drug. To maintain a serum level higher than the MIC, a second dose would be taken. The shaded area of the curve represents time above MIC. This explains why the doctor asks you to take three pills of ampicillin a day, one every 8 hours.

sufficient levels either by administering a higher dose, which runs the risk of side effects, or by giving a second dose before tissue levels from the first dose drop below the MIC. This is why patients are told to take doses of some antibiotics three times a day and other antibiotics only once a day.

Data from a disk diffusion test are evaluated using a table listing whether a zone size is wide enough (meaning the MIC is low enough) to be clinically useful. Table 13.5 shows susceptibility data for *S. aureus* against half a dozen antibiotics. The zone size chosen to indicate whether a pathogen is susceptible to the drug is based on knowing the average tissue level for each antibiotic that can be safely attained (average attainable tissue concentration).

THE CHEMOTHERAPEUTIC INDEX

One of the most important decisions a clinician must make when treating an infection is which antibiotic to prescribe. In addition to knowing whether the organism is susceptible to the antibiotic, the clinician must know whether the attainable tissue level of the antibiotic is higher than the MIC and must understand the relationship between the **therapeutic dose** and the **toxic dose** of the drug. Therapeutic dose is the <u>minimum</u> dose per kilogram of body weight that stops pathogen growth. Toxic dose is the <u>maximum</u> dose tolerated by the patient. The ratio of

toxic dose to therapeutic dose is called the **chemotherapeutic index**. Hopefully, the therapeutic dose is much lower than the toxic dose. Thus, the higher the chemotherapeutic index, the safer the drug. For example, drug A with a toxic dose of 1 gram per kilogram of

Table 13.5
Susceptibility Results for *Staphylococcus aureus*

Antibiotic	Quantity in Disk (µg)	Zone of Inhibition Diameter (mm)[a]		
		Resistant	Intermediate	Susceptible
Ampicillin	10	<12	12–13	>13
Chloramphenicol	30	<13	13–17	>17
Erythromycin	15	<14	14–17	>17
Gentamicin	10	≤12.5	—	>12.5
Streptomycin	10	<12	12–14	>14
Tetracycline	30	<15	15–18	>18

[a]The zone of inhibition is the diameter of no growth around an antibiotic disk on an agar plate.

body weight and a therapeutic dose of 0.1 gram per kilogram of body weight has a good chemotherapeutic index of 10. However, drug B with toxic and therapeutic doses equal to 0.5 gram per kilogram of body weight has an unacceptable chemotherapeutic index of 1.

You should also be aware that combinations of antibiotics can be either **synergistic**, meaning they may work poorly when used individually but very well if combined (their combined effect is greater than their additive effect), or **antagonistic**, in which their mechanisms of action interfere with each other and diminish their effectiveness. A common synergistic pairing used for serious enterococcus infections of the blood (sepsis) and heart valve (endocarditis) is an aminoglycoside combined with vancomycin. Neither drug alone is very effective, but when combined, vancomycin, a cell wall synthesis inhibitor, weakens the cell wall of this organism and facilitates transport of the aminoglycoside into the bacterial cell. The aminoglycoside can then inhibit protein synthesis.

Antagonism can occur when a bacteriostatic agent is combined with a bactericidal agent. For example, penicillin kills growing cells, which are actively making peptidoglycan, but has little effect on nongrowing cells. Macrolides, however, are bacteriostatic. They will prevent growth. Thus, when a macrolide and penicillin are combined, the macrolide's mechanism of action is antagonistic to the effect of penicillin.

SECTION SUMMARY

- **Antimicrobial agents** are produced naturally by microorganisms and synthetically in laboratories.
- **Selective toxicity** refers to the ability of an antibiotic to attack a unique component of microbial physiology that is missing or distinctly different from eukaryotic physiology.
- **Antibiotic side effects** on mammalian physiology can limit the clinical usefulness of an agent.
- **Antibiotic spectrum of activity** refers to the range of microbes that a given drug affects.
- **Bactericidal antibiotics** kill microbes; **bacteriostatic antibiotics** inhibit their growth.
- **Minimal inhibitory concentration (MIC)** of a drug is the minimum concentration that will inhibit growth. **Minimum bactericidal concentration (MBC)** is the minimum concentration that will kill the target cell. When correlated with average attainable tissue levels of the antibiotic, MIC can predict the effectiveness of an antibiotic in treating disease.
- **MIC is measured** using serial dilution techniques but can be approximated using the Kirby-Bauer disk diffusion technique.
- **Chemotherapeutic index** is the ratio of the toxic dose of an antibiotic to its therapeutic dose.

Thought Question 13.3   The concentration of the drug tobramycin in a tube of broth is 1,000 µg/ml. Serial twofold dilutions are made from this tube. Including the initial tube (tube 1), there are a total of ten tubes. Twenty-four hours after all the tubes are inoculated with *Listeria monocytogenes*, turbidity is observed in tubes 6–10. What is the MIC?

Thought Question 13.4   You are testing whether a new antibiotic will be a good treatment choice for a patient with a staph infection. The Kirby-Bauer test using the organism from the patient shows a zone of inhibition of 15 mm around the disk containing this drug. Clearly, the organism is susceptible. But you conclude from other studies that the drug will not be effective in the patient. What would make you draw this conclusion?

13.5
Antimicrobial Mechanisms of Action

SECTION OBJECTIVES

- Classify antibiotics by their molecular targets.
- Describe the process of cell wall synthesis, and indicate the targets for various cell wall antibiotics.
- Distinguish transcription inhibitors from translation inhibitors.

How do antibiotics work? As noted earlier, antibiotics exhibit selective toxicity because they disturb enzymes or structures unique to the bacterial target cell. Most antibiotics affect one of the following aspects of a microbe's physiology:

- Cell wall synthesis
- Cell membrane integrity
- DNA synthesis
- RNA synthesis
- Protein synthesis
- Metabolism

Chapters 5, 7, and 8 describe the cell structures and processes targeted by antibiotics. The following sections describe the salient features of these drugs and their modes of action. Summaries are provided in **Figure 13.12** and **Tables 13.6** and **13.7**.

Table 13.6
Targets of Antimicrobial Agents

Target	Antibiotic Examples
Cell wall synthesis	Penicillins, cephalosporins, bacitracin, vancomycin
Mycolic acid synthesis	Isoniazid
Protein synthesis	Chloramphenicol, tetracyclines, aminoglycosides, macrolides, lincosamides
Cell membrane	Polymyxin, amphotericin, imidazoles (antifungal)
Nucleic acid function	Nitroimidazoles, nitrofurans, quinolones, rifampin; some antiviral compounds, especially antimetabolites
Intermediary metabolism	Sulfonamides, trimethoprim

Figure 13.12 Bacterial Targets of Antimicrobial Drugs

The figure depicts the six broad categories of bacterial physiology that enable selective toxicity. It also identifies the general classes of antibiotics that affect these physiological processes.

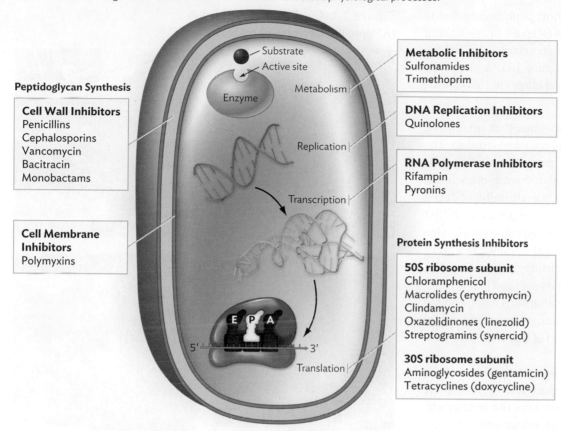

Peptidoglycan Synthesis

Cell Wall Inhibitors
Penicillins
Cephalosporins
Vancomycin
Bacitracin
Monobactams

Cell Membrane Inhibitors
Polymyxins

Substrate
Active site
Enzyme
Metabolism
Replication
Transcription
Translation
5' — 3'
E P A

Metabolic Inhibitors
Sulfonamides
Trimethoprim

DNA Replication Inhibitors
Quinolones

RNA Polymerase Inhibitors
Rifampin
Pyronins

Protein Synthesis Inhibitors

50S ribosome subunit
Chloramphenicol
Macrolides (erythromycin)
Clindamycin
Oxazolidinones (linezolid)
Streptogramins (synercid)

30S ribosome subunit
Aminoglycosides (gentamicin)
Tetracyclines (doxycycline)

Antibiotics That Target the Cell Wall

Bacterial cell walls are an obvious candidate for selective toxicity because peptidoglycan does not exist in mammalian cells; thus, antibiotics that target peptidoglycan synthesis should selectively kill bacteria. Both of the antibiotics used in Case History 13.1 kill bacteria by targeting cell wall (peptidoglycan) synthesis. An outline of the process is shown in **Figure 13.13A**. Basically, sugar molecules called *N*-acetylglucosamine (NAG) and *N*-acetylmuramic acid (NAM) are made by the cell and linked together into long peptidoglycan chains assembled at the cell wall (see Section 5.1). Disaccharide units of NAG and NAM are first made in the cytoplasm and shuttled across the membrane by a lipid carrier molecule (steps 1–4). The disaccharide units are then assembled outside the cytoplasmic membrane to form long strands of peptidoglycan (step 5). Adjacent strands are

Figure 13.13 Bacterial Cell Wall Synthesis Is a Target of Certain Antibiotics

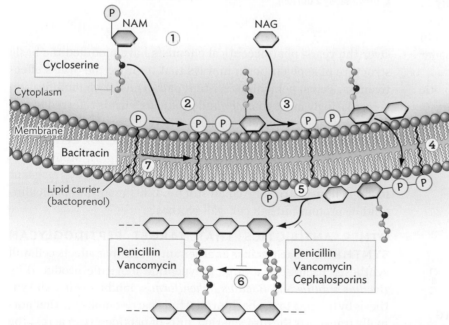

Cycloserine
Cytoplasm
Membrane
Bacitracin
Lipid carrier (bactoprenol)
P NAM
NAG
Penicillin Vancomycin
Penicillin Vancomycin Cephalosporins

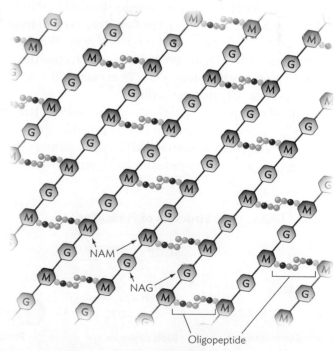

NAM
NAG
Oligopeptide

A. Cell wall (peptidoglycan) synthesis. NAG and NAM are made and joined as units in the cytoplasm (steps 1–3). Each unit is then transported across the cell membrane by a specialized lipid carrier molecule called bactoprenol (step 4). Enzymes then link the new unit to a preexisting chain of peptidoglycan (step 5) and catalyze a cross-link between peptide chains sticking out from NAM molecules on parallel chains (step 6). Penicillin, vancomycin, and cephalosporins inhibit the activity of these enzymes.

B. The cross-linking between strands shown in A provides the cage-like structure shown here. Antibiotics that inhibit cross-linking and chain elongation compromise the integrity of the cage.

then "snapped" together by cross-linking short peptide side chains that spring from individual NAM molecules (step 6). Cross-linking produces a rigid, cage-like macromolecular structure essential for maintaining cell shape (Figure 13.13B). Figure 13.13A also indicates where several antibiotics target various stages of this assembly process.

BETA-LACTAM ANTIBIOTICS The enzymes that attach the disaccharide units to preexisting peptidoglycan and produce the peptide cross-links are collectively called **penicillin-binding proteins (PBPs)** because penicillin, as well as related antibiotics, binds to them and inhibits their activity. Penicillin is a naturally produced antibiotic that contains a critical beta-lactam ring structure that chemically resembles a piece of the peptidoglycan peptide side chain (Figure 13.14A and B). Penicillin G itself is shown in Figure 13.14C. Because penicillin looks like the peptidoglycan side chain, the drug will bind to cell wall biosynthesis enzymes and halt synthesis of the chain. Note that different chemical groups can be added to the basic ring structure of penicillin to change the antimicrobial spectrum, stability, and intestinal absorption of the derivative penicillin (see Appendix 3, Figure A3.2).

The consequence of penicillin's action is a disaster for bacteria that are trying to grow larger and larger (Figure 13.15). A complete cell wall acts as a rigid "skeleton" that restrains the hydrostatic pressure from within the cell. Without an intact cell wall, the growing cell eventually bursts. Penicillin, then, is a bactericidal drug. Penicillin is most effective against Gram-positive organisms because the drug has difficulty passing through the Gram-negative outer membrane. Ampicillin, the first drug used to treat Molly in Case History 13.1, is a modified version of penicillin that more easily penetrates this membrane and is more effective than penicillin against Gram-negative microbes. Thus, ampicillin has a broader spectrum of activity than penicillin.

Cephalosporins are another type of beta-lactam antibiotic originally discovered in nature but modified in the laboratory (a type of **semisynthetic antimicrobial**) to fight microbes that are naturally resistant to penicillins (especially *Pseudomonas aeruginosa*).

Figure 13.14   The Structure of Penicillins

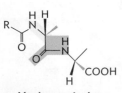
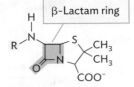
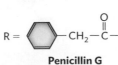

D-Alanine-D-alanine

A. The beta-lactam ring of all penicillin drugs resembles a part of the peptidoglycan structure (D-alanine-D-alanine of the side chain).

Basic structure

B. The beta-lactam core structure of all penicillins.

Penicillin G

C. Penicillin G. (Derivatives of penicillin vary in the R group shown in B; see Appendix 3, Figure A3.2.)

Figure 13.15   Effect of Ampicillin (a Penicillin Derivative) on *E. coli*

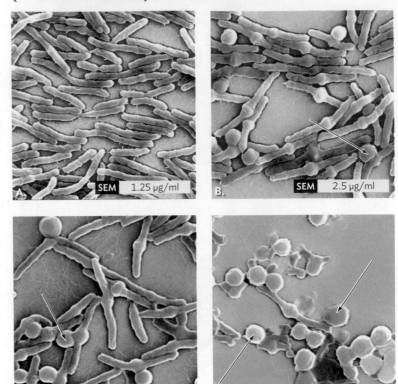

Cells were incubated for 1 hour at the antibiotic concentrations shown. Swollen areas of cells in panels B–D reflect weakening cell walls (arrows). Cells shown are approximately 2 μm long.

Over the years, pharmaceutical chemists have modified the basic structure of cephalosporin in ways that improve the drug's effectiveness against penicillin-resistant pathogens. Each modification is increasingly more complex and produces what is referred to as a new "generation" of cephalosporins. There are currently five generations of this semisynthetic antibiotic (Figure 13.16). Ceftriaxone, for example, is a third-generation cephalosporin. In Molly's case, even though the *S. pneumoniae* causing her meningitis was resistant to ampicillin, ceftriaxone could still bind the important penicillin-binding protein and halt cell wall synthesis.

OTHER ANTIBIOTICS THAT TARGET PEPTIDOGLYCAN SYNTHESIS **Bacitracin** is another antibiotic that affects cell wall synthesis. Bacitracin, a large polypeptide molecule produced by *Bacillus subtilis* and *Bacillus licheniformis*, inhibits cell wall synthesis by binding to the bactoprenol lipid carrier molecule that normally transports the disaccharide units of peptidoglycan across the cell membrane to the growing chain (see Figure 13.13). Resistance to bacitracin can develop if the organism possesses an efficient drug export system that pumps bacitracin out of the cell (discussed in Section 13.6). Normally, bacitracin is used only topically, usually in combination with polymyxin (for instance, in the over-the-counter

Figure 13.16   Cephalosporin Generations
Representative examples.

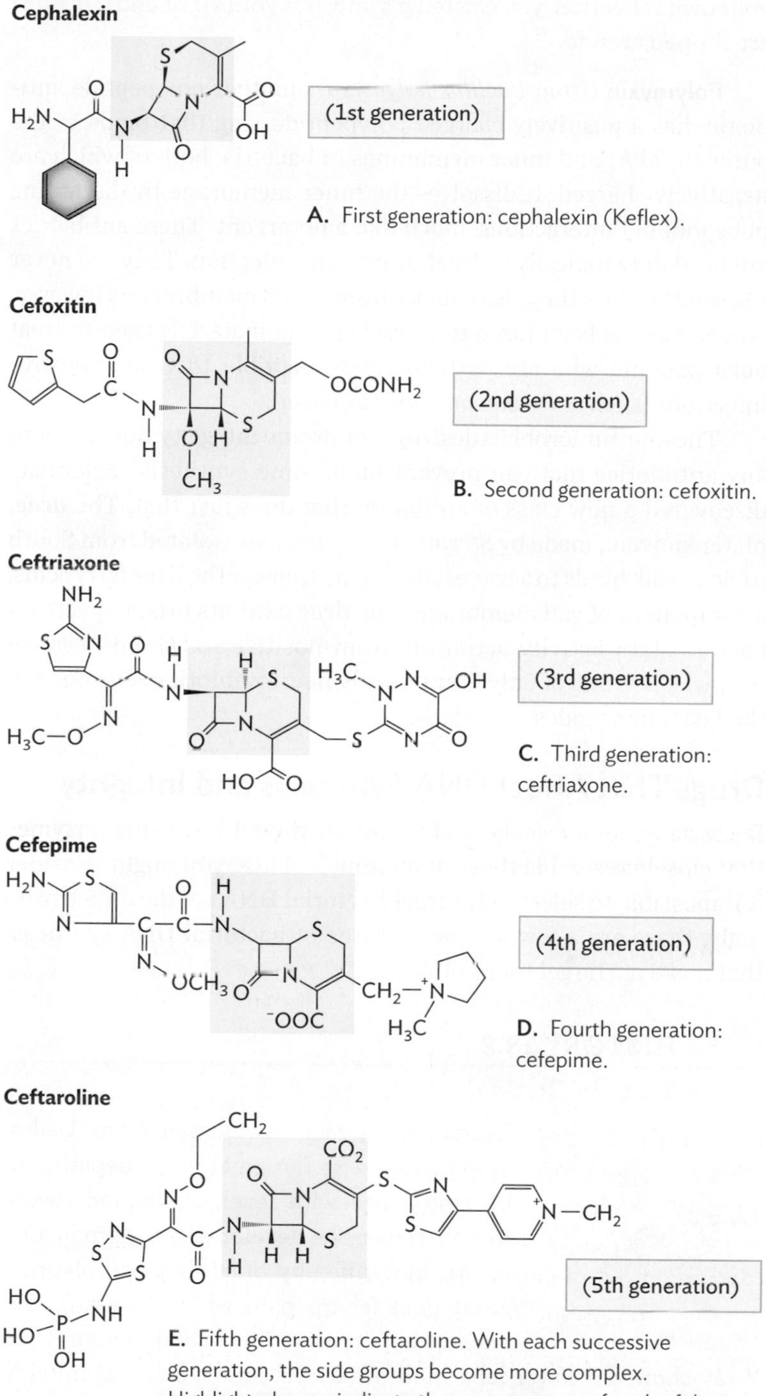

Cephalexin

(1st generation)

A. First generation: cephalexin (Keflex).

Cefoxitin

(2nd generation)

B. Second generation: cefoxitin.

Ceftriaxone

(3rd generation)

C. Third generation: ceftriaxone.

Cefepime

(4th generation)

D. Fourth generation: cefepime.

Ceftaroline

(5th generation)

E. Fifth generation: ceftaroline. With each successive generation, the side groups become more complex. Highlighted areas indicate the core structure of each of the cephalosporins, with beta-lactam rings.

cream Neosporin), because of serious side effects, such as kidney damage, that can occur if it is ingested.

Cycloserine (made by *Streptomyces garyphalus*) is one of several antimicrobials used to treat tuberculosis (see Appendix 3, Figure A3.3A). Relative to bacitracin, it acts at an even earlier step in peptidoglycan synthesis. Cycloserine inhibits two enzymes that

make part of the peptide side chain on NAM. As a result, the complete pentapeptide side chain cannot be made (see Figure 13.13). Incomplete side chains mean that cross-linking cannot occur and peptidoglycan integrity is compromised.

Vancomycin, a very large and complex glycopeptide produced by *Amycolatopsis orientalis* (see Appendix 3, Figure A3.3B), binds to part of the peptide end of a newly exported NAM-NAG disaccharide and blocks its addition to preexisting peptidoglycan (see Figure 13.13). The microbial mechanisms of resistance to vancomycin and penicillin are very different, which makes vancomycin particularly useful against penicillin-resistant bacteria. Potential side effects of vancomycin include kidney damage (nephrotoxicity) and hearing loss (ototoxicity), although these are very rare.

It is important to note that antibiotics targeting cell wall biosynthesis generally kill only growing cells. These drugs do not affect static, or stationary-phase, cells because in this state the cell has no need for new peptidoglycan.

MICROBIAL RESISTANCE TO CELL WALL–INHIBITING ANTIBIOTICS

The discovery of penicillin was considered by many to be the "magic bullet" Paul Ehrlich predicted in 1904. Unfortunately, we have learned repeatedly over the past 80 years that the microbial world does not sit idly by and politely die. Microbes constantly evolve and eventually become resistant to new antibiotics. Bacterial resistance to penicillin was first recognized in the 1940s as the cause of occasional treatment failures. Today, resistance to many different antibiotics is a growing concern throughout the world. Bacteria develop resistance to penicillin in two basic ways. The first is by inheriting a gene encoding one of the beta-lactamase enzymes, which cleave the critical ring structure of this class of antibiotics (Figure 13.17). The beta-lactamase enzyme is transported out of the cell and into the surrounding medium (for Gram-positives) or the periplasm (for Gram-negatives), where it will destroy penicillin before the drug even gets to the cell. Bacteria that produce beta-lactamase are still susceptible to certain modified penicillins and cephalosporins engineered to be poor substrates for the enzyme. Methicillin, for example, works well against

Figure 13.17   Penicillin Resistance
The enzyme beta-lactamase produced by bacteria carrying the beta-lactamase gene will destroy the critical beta-lactam ring and confer penicillin resistance. Some derivatives of penicillin (for example, oxacillin) block beta-lactamase action and will kill the bacteria.

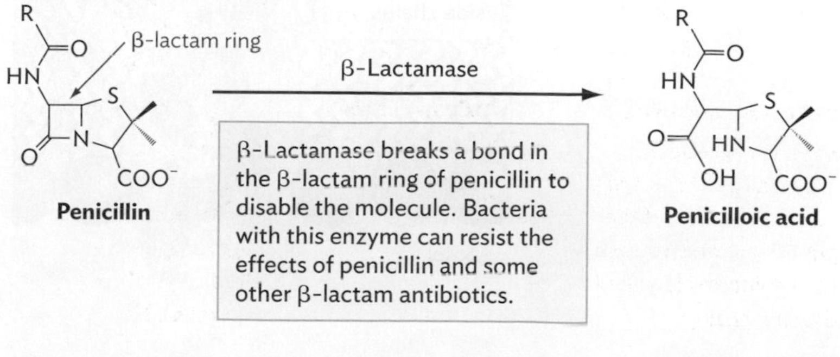

β-Lactamase breaks a bond in the β-lactam ring of penicillin to disable the molecule. Bacteria with this enzyme can resist the effects of penicillin and some other β-lactam antibiotics.

most beta-lactamase–producing microbes. However, a new enzyme called the New Delhi metallo-beta-lactamase, found predominantly in *Escherichia coli* and *Klebsiella pneumoniae*, destroys the beta-lactam ring of all of the penicillins, carbapenems, and cephalosporins.

Note Because of stability issues, methicillin is no longer used to treat infections and has been replaced by the more stable oxacillin. Organisms resistant to methicillin are also resistant to oxacillin. Nevertheless, the term "methicillin-resistant *S. aureus* (MRSA)" has been retained.

The second way a microbe can become resistant to penicillin is when a key penicillin-binding protein is altered by mutation. The resultant protein with its altered shape can still synthesize cell wall but no longer binds to the antibiotic. Most methicillin-resistant bacteria use this strategy. As described in the opening case history, hospitals take a special interest in methicillin-resistant *S. aureus* (MRSA) because very few drugs can kill it. One of the few remaining antibiotics effective against MRSA is vancomycin. Unfortunately, resistance to this drug is also developing.

In Molly's case, the penicillin (ampicillin)-resistant *S. pneumoniae* that infected her actually had an altered penicillin-binding protein, not a beta-lactamase. Inexplicably, no beta-lactamase–producing *S. pneumoniae* has ever been found.

Drugs That Target the Bacterial Membrane

Poking holes in a bacterial cytoplasmic membrane is another effective way to kill bacteria. A group of peptide antibiotics uses this strategy. **Gramicidin**, produced by *Brevibacillus brevis*, is a cyclic peptide composed of 15 alternating D- and L-amino acids. It inserts into the membrane as a dimer, forming a cation channel that disrupts membrane polarity (**Figure 13.18**).

Figure 13.18   Gramicidin Is a Peptide Antibiotic That Affects Membrane Integrity

Gramicidin forms a cation channel across cell membranes through which hydrogen, sodium, and potassium ions can freely pass.

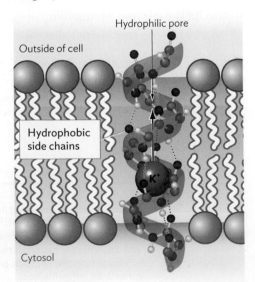

Outside of cell

Hydrophilic pore

Hydrophobic side chains

K+

Cytosol

Link As discussed in Section 4.4, D and L forms of any compound are essentially mirror images of each other. Much like looking at your own reflection, you can tell it's you, but your right and left sides are flipped around.

Polymyxin (from *Bacillus polymyxa*), another polypeptide antibiotic, has a positively charged polypeptide ring that binds to the outer (lipid A) and inner membranes of bacteria, both of which are negatively charged. It dissolves the inner membrane by disrupting phospholipid interactions, much like a detergent. These antibiotics are used only topically to treat or prevent infection. They are never ingested because they also damage human cell membranes. However, polymyxin has been fused to some bandage materials used to treat burn patients who are particularly susceptible to Gram-negative infections (such as *Pseudomonas aeruginosa*).

These antimicrobials destroy membrane integrity, but are there any antibiotics that can prevent membrane synthesis? Scientists discovered a new class of antibiotic that does just that. The drug, platensimycin, made by *Streptomyces platensis* isolated from South African soil, binds to a bacterial enzyme that synthesizes fatty acids, a component of cell membrane. The drug exhibits broad-spectrum bacteriostatic activity, acting on Gram-positive and Gram-negative bacteria. It is one of only four entirely new antibiotics developed in the last four decades.

Drugs That Affect DNA Synthesis and Integrity

Bacteria generally make and maintain their DNA using enzymes that closely resemble those of mammals. Thus, you might consider it impossible to selectively target bacterial DNA synthesis. Surprisingly, there are a few unique features of bacterial DNA synthesis that make it a target for antibiotics.

CASE HISTORY 13.2

When Time Is Critical

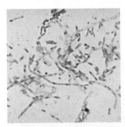

Lauren, a 23-year-old woman from Dallas, Texas, arrived at the emergency department by ambulance with fever, chills, and severe muscle aches. She developed a nonproductive cough, had difficulty breathing, had pleuritic chest pain (sharp pain when inhaling), and became hypotensive (low blood pressure). An X-ray showed a lower lobe infiltrate in the lungs. She was initially treated with ceftriaxone, but her condition worsened. Two days later, the clinical laboratory reported the presence of the Gram-negative anaerobe *Fusobacterium necrophorum* in blood cultures. The patient was diagnosed with pneumonia and treated with metronidazole, a DNA-damaging agent specific for anaerobes. She recovered fully.

This case illustrates a common approach to infectious diseases by clinicians. A critically ill patient clearly suffers from an infectious disease and needs immediate treatment. But because the identity of the organism is not yet known, the clinician will use an antimicrobial that has a broad spectrum of activity, hoping that it is broad enough to "cover" the usual pathogens that could cause the disease (in Lauren's case, pneumonia coupled with sepsis)—an approach called **empirical therapy**. However, in this instance the treatment did not "cover" this anaerobe well, so metronidazole was added to the treatment regimen. Metronidazole is one of several drugs that selectively affect the synthesis or integrity of DNA in microorganisms.

METRONIDAZOLE Also known by the trade name Flagyl, metronidazole is an example of a "prodrug"—a drug that is harmless until activated by reduction. Metronidazole is activated after being metabolized by microbial protein cofactors (ferridoxin) found in anaerobic or microaerophilic bacteria such as *Bacteroides* and *Fusobacterium* (Figure 13.19A). Aerobic microbes are resistant because they do not possess the electron transport proteins capable of reducing metronidazole. Once activated in anaerobes, the compound begins nicking DNA at random, thus killing the cell. Because the etiologic agent in Case History 13.2 was an anaerobe, metronidazole was an effective therapy. Metronidazole is also effective against protozoa such as *Giardia*, *Entamoeba*, and *Trichomonas*, where it is activated in special organelles (Figure 13.19B). Human cells also lack the anaerobiosis-associated cofactors and are thus unharmed by the drug.

SULFONAMIDES The sulfa drugs are known as antimetabolites because they interfere with the synthesis of metabolic intermediates. Ultimately, the sulfonamide (sulfa) drugs act to inhibit the synthesis of nucleic acids. Sulfa drugs such as sulfonilamide or sulfamethoxazole prevent the synthesis of folic acid, an important cofactor in the synthesis of nucleic acid precursors (Figure 13.20).

All organisms use folic acid to synthesize nucleic acids, so why are the sulfa drugs selectively toxic to bacteria? The selectivity occurs because mammals do not synthesize folic acid. Higher mammals generally rely on bacteria and green leafy vegetables as sources of folic acid. Bacteria make folic acid from the combination of PABA, glutamic acid, and pteridine. Sulfonilamide, a structural analog of PABA, competes for one of the enzymes in the bacterial folic acid pathway and inhibits folic acid production (see Figure 13.20C). Because humans lack that pathway, sulfa drugs are selectively toxic to bacteria. Sulfamethoxazole is typically used in combination with trimethoprim, a drug that inhibits a later step where folic acid (dihydrofolic acid) is converted to tetrahydrofolic acid. The combination, called bactrim, is very effective, especially in treating urinary tract infections.

Figure 13.19   Activation of Metronidazole

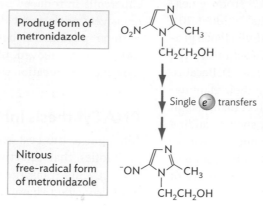

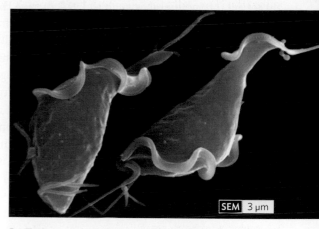

A. Metronidazole enters the anaerobe and is reduced (receives single-electron transfers) by anaerobic metabolism. The resulting free-radical form can attack DNA in the organism.

B. *Trichomonas vaginalis*, a protozoan that causes vaginal infections. Metronidazole is very effective at treating *Trichomonas* infections.

Figure 13.20   Mode of Action of Sulfonamides

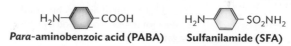

A. The structures of PABA and sulfanilamide (SFA) are very similar.

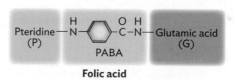

B. PABA, pteridine, and glutamic acid combine to make the vitamin folic acid.

Normal folic acid formation

Folic acid formation blocked by sulfanilamide (SFA)

C. Normal synthesis of folic acid requires that all three components engage the active site of the biosynthetic enzyme. The sulfa drugs replace PABA at the active site. The sulfur group, however, will not form a peptide bond with glutamic acid, and the size of sulfanilamide sterically hinders binding of pteridine, so folic acid cannot be made.

QUINOLONES This group of drugs inhibits DNA synthesis by targeting topoisomerases such as **DNA gyrase** that introduce negative supercoils into DNA ahead of replication forks (Figure 8.10). DNA gyrase bound to and inactivated by a quinolone will block progression of a DNA replication fork (the mechanism of action of DNA gyrase is discussed in Section 8.2 and shown in Figure 8.12). Because bacterial DNA gyrases are structurally distinct from their mammalian counterparts, quinolone antibiotics will not affect mammalian DNA replication. Various chemical modifications, such as adding fluorine and amine groups, have increased the antimicrobial spectrum and bloodstream half-life of these drugs. Ciprofloxacin and levofloxacin are quinolones commonly used to treat a variety of infections (Table 13.7).

Link As is discussed in Section 8.2, **DNA gyrase** relieves positive supercoils introduced by the DNA polymerase complex as it progressively peels apart double-stranded DNA at replication forks during replication. If these supercoils forming ahead of the moving fork are not removed, the buildup of torsional stress blocks fork movement, replication stops, and the cell dies.

RNA Synthesis Inhibitors

Rifampin is the best-known member of the rifamycin family of antibiotics that selectively binds to bacterial RNA polymerase and prevents transcription (see Appendix 3, Figure A3.4). Rifampin (also called rifampicin) is often used to treat tuberculosis or

Table 13.7
Clinical Uses, Possible Side Effects, and Modes of Action for Antimicrobials

Antimicrobial	Clinical Use	Possible Side Effects	Mode of Action
Aminoglycosides • Gentamicin • Kanamycin • Tobramycin • Amikacin • Streptomycin	Broad spectrum of activity but used mainly for infections caused by Gram-negative bacteria, such as *Escherichia coli* and *Klebsiella* and particularly *Pseudomonas aeruginosa*; effective against aerobic bacteria (not obligate/facultative anaerobes) and tularemia	Hearing loss, vertigo, kidney damage	Bactericidal; binds to the bacterial 30S ribosomal subunit, inhibits the translocation of peptidyl-tRNA, and causes misreading of mRNA
Carbapenems • Imipenem • Meropenem	Broad spectrum; bactericidal for both Gram-positive and Gram-negative organisms and therefore useful for empiric broad-spectrum antibacterial coverage (*Note:* MRSA resistance to this class)	Gastrointestinal upset, diarrhea, nausea, seizures, headache, rash and allergic reactions	Bactericidal; inhibits cell wall synthesis
Cephalosporins, 2nd generation • Cefoxitin	Extended spectrum; good Gram-positive coverage	Gastrointestinal upset, diarrhea, nausea (if alcohol taken concurrently), allergic reactions	Bactericidal; inhibits cell wall synthesis
Cephalosporins, 3rd generation • Cefixime • Ceftazidime • Ceftriaxone	Extended spectrum; improved Gram-negative coverage		
Cephalosporins, 4th generation • Cefepime	Extended spectrum; covers *Pseudomonas*		
Cephalosporins, 5th generation • Ceftobiprole	Extended spectrum; used for MRSA		
Chloramphenicol	Broad spectrum; *Staphylococcus, Neisseria meningitidis, Streptococcus pneumoniae, Haemophilus influenzae*	Aplastic anemia (decreases in red blood cells, white blood cells, and platelets)	Bacteriostatic; binds to ribosome and inhibits protein synthesis
Glycopeptides • Vancomycin • Teicoplanin	Narrow spectrum; Gram-positive coverage: *Staphylococcus* (MRSA), *Streptococcus, Enterococcus, Clostridium*	Kidney damage, tinnitus (ear ringing), diarrhea, nausea	Bactericidal; inhibits cell wall synthesis

(continued on next page)

Table 13.7

Clinical Uses, Possible Side Effects, and Modes of Action for Antimicrobials *(continued)*

Antimicrobial	Clinical Use	Possible Side Effects	Mode of Action
Lincosamides • Clindamycin • Lincomycin	Broad spectrum; used for serious staphylococcal, pneumococcal, and streptococcal infections in penicillin-allergic patients; also used for anaerobic infections; used topically for acne	Possible pseudomembranous enterocolitis	Bacteriostatic; inhibits protein synthesis
Lipopeptides • Daptomycin	Narrow spectrum; Gram-positive microbes	Muscle weakness, eosinophilic pneumonia, diarrhea	Bactericidal; binds to membranes, causes leaks and loss of membrane potential (energy drain)
Macrolides • Azithromycin • Clarithromycin • Erythromycin	Narrow spectrum; streptococcal infections, syphilis, respiratory tract infections, mycobacterial infections, Lyme disease, pertussis, chlamydia	Nausea, vomiting, diarrhea, jaundice	Bacteriostatic; binds to ribosome, inhibits protein synthesis
Monobactams • Aztreonam	Narrow spectrum; intravenous; Gram-negative bacteria, including *Pseudomonas*	Rash, diarrhea, vomiting	Bactericidal; inhibits cell wall synthesis
Oxazolidinones • Linezolid	Narrow spectrum; Gram-positive bacteria, including vancomycin-resistant enterococci, methicillin-resistant *Staphylococcus aureus*	Long-term use promotes bone marrow suppression and low platelet counts	Bacteriostatic; inhibits protein synthesis
Penicillins • Penicillin G • Amoxicillin • Cloxacillin • Oxacillin	Varies from narrow spectrum to broad; *Streptococcus*, *Treponema* (syphilis), *Borrelia* (Lyme disease)	Gastrointestinal upset, allergy	Bactericidal; inhibits cell wall synthesis
Polypeptides • Bacitracin • Polymyxin B • Gramicidin	Topical infections; bacitracin (Gram-positive), polymyxin B (Gram-negative), gramicidin (Gram-positive and Gram-negative)	Kidney damage, nerve damage	Bactericidal; Bacitracin (cell wall); polymyxin B and gramicidin (membrane integrity)
Quinolones • Ciprofloxacin • Levofloxacin	Broad spectrum; Gram-positive and Gram-negative coverage; used for urinary tract, sexually transmitted, gastrointestinal, respiratory tract, and bone infections	Inhibits bone growth in children, seizures (rare)	Bactericidal; inhibits topoisomerases (DNA gyrase and Topo IV)
Rifampin	Broad spectrum; Gram-positives, mycobacteria (tuberculosis)	Reddish-orange body fluids	Bactericidal; binds RNA polymerase, prevents transcription
Streptogramins • Quinupristin/dalfopristin	Narrow spectrum; vancomycin-resistant *S. aureus* (VRSA) and *Enterococcus* (VRE)	Nausea, diarrhea, headache	In combination, bactericidal; binds ribosome, inhibits protein synthesis
Sulfonamides • Trimethoprim-sulfamethoxazole	Narrow spectrum; mainly Gram-negatives; urinary tract infections	Nausea, allergy, kidney failure, sensitivity to light	Bacteriostatic; inhibits folic acid synthesis
Tetracyclines • Doxycycline • Minocycline	Broad spectrum; syphilis, chlamydia, Lyme disease, mycoplasmas, rickettsias	Gastrointestinal upset, sensitivity to light, depression of bone growth, staining of teeth	Bacteriostatic; binds ribosome, inhibits protein synthesis

Figure 13.21
Rifamycin in Urine

Rifampin is excreted in body fluids, turning them reddish orange. Shown is a urine bag from a hospital patient undergoing rifampin treatment for a cerebral abscess.

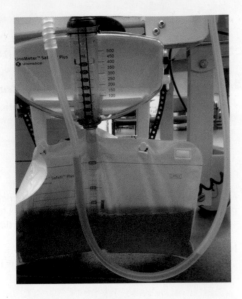

Figure 13.22   Modes of Action of Common, Clinically Useful Antibiotics That Inhibit Bacterial Protein Synthesis

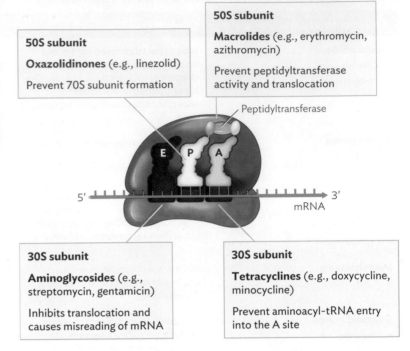

50S subunit

Oxazolidinones (e.g., linezolid)

Prevent 70S subunit formation

50S subunit

Macrolides (e.g., erythromycin, azithromycin)

Prevent peptidyltransferase activity and translocation

Peptidyltransferase

5′   mRNA   3′

30S subunit

Aminoglycosides (e.g., streptomycin, gentamicin)

Inhibits translocation and causes misreading of mRNA

30S subunit

Tetracyclines (e.g., doxycycline, minocycline)

Prevent aminoacyl-tRNA entry into the A site

meningococcal meningitis. Curiously, because rifampin is reddish orange, it turns bodily secretions, including breast milk, orange (Figure 13.21). The astute physician will warn the patient of this highly visible but harmless side effect to avoid unnecessary anxiety when the patient's urine changes color.

Protein Synthesis Inhibitors

Protein synthesis was initially described in Chapter 8. The molecular complexity of the process provides numerous sites that can be subverted by antibiotics. Fortunately, the structural differences between prokaryotic and eukaryotic ribosomes enable antibiotics to be selectively toxic, affecting only bacterial protein synthesis.

Inhibitors of protein synthesis can be classified into several groups according to their structures (see Appendix 3, Figure A3.5) and mechanisms of action (Figure 13.22). Note that most protein synthesis inhibitors are, by and large, bacteriostatic; these antibiotics generally work by binding to and interfering with the function of bacterial rRNA, which differs from eukaryotic rRNA. Features of these antibiotics are discussed in the next section.

CASE HISTORY 13.3

Treatment Meets Allergy

Sixteen-year-old Jamal arrived at the emergency department after 2 days of fever, malaise, chills, and neck stiffness. His most notable symptom was a painful, red, rapidly spreading rash covering the right side of his face. Throat cultures taken on admission revealed group A *Streptococcus pyogenes*, indicating that the rash was a case of erysipelas (described in Chapter 19). Although penicillin would be the drug of choice (no penicillin-resistant *S. pyogenes* has been seen clinically), Jamal is known to be allergic to this antibiotic. Instead, he was given azithromycin.

When a patient is known to be allergic to the usual drug of choice, a structurally distinct drug is best. Often, that drug will be one that inhibits protein synthesis—in this case, the macrolide azithromycin. Drugs that inhibit protein synthesis can be subdivided into different groups according to their structures and what part of the translation machine is targeted.

Drugs That Affect the 30S Ribosomal Subunit

The classification of antibiotics affecting protein synthesis is initially based on the bacterial ribosomal subunit targeted. Ribosomes have two major subunits (see Section 8.6). One class of antibiotics interferes with the 30S small subunit, and the other subverts the 50S large subunit.

AMINOGLYCOSIDES   Aminoglycosides include streptomycin, gentamicin (see Appendix 3, Figure A3.5A), and tobramycin. These broad-spectrum antibiotics are effective against Gram-positive and Gram-negative bacteria. The aminoglycosides bind to the 30S ribosomal subunit and cause misreading of the mRNA and inhibit translocation of peptidyl-tRNA from the A site to the P site. Unlike most protein synthesis inhibitors, aminoglycosides are bactericidal, but the reason for their killing effect is not clear (they may cause misreading of mRNA or generate reactive oxygen species). The potential toxic effects of these antibiotics include damage to the auditory nerve (hearing loss, ototoxicity) and damage to kidney function. Approximately 0.5%–3% of patients treated with gentamicin suffer from ototoxicity.

TETRACYCLINES This group includes doxycycline. Tetracyclines are broad-spectrum, bacteriostatic antibiotics that feature four fused cyclic rings, hence the name (see Appendix 3, Figure A3.5B). These antibiotics bind to the 30S subunit and prevent tRNAs carrying amino acids from entering the A site. Doxycycline is used to treat early stages of Lyme disease (*Borrelia burgdorferi*), acne (*Propionibacterium acnes*), and other infections. These antibiotics are not recommended for children under 8 years old because the drugs can be incorporated into developing teeth, giving them a yellowish-brown discoloration. Tetracyclines are also phototoxic, causing an increased tendency for sunburn.

GLYCYLCYCLINES Tigecycline is a commonly used member of this class. This broad-spectrum, bacteriostatic drug looks and acts like tetracycline. It prevents protein synthesis by binding to the 30S ribosomal subunit and inhibits the entry of aminoacyl-tRNA into the A site. Its major advantage over tetracycline is that it can function in tetracycline-resistant cells. Tetracycline resistance in pathogens is due mostly to the presence of specific efflux pumps in the bacterial membrane that remove the drug from the cell (see Section 13.6). Tigecycline is not readily pumped out of the cell by those pumps. A disadvantage of tigecyline is that it cannot be taken orally and must be slowly administered intravenously.

Drugs That Affect the 50S Ribosomal Subunit

CHLORAMPHENICOL This antibiotic prevents peptide bond formation by inhibiting peptidyltransferase in the 50S subunit (see Appendix 3, Figure A3.5E). It is bacteriostatic and broad spectrum but can have some toxic side effects. For instance, chloramphenicol can inhibit bone marrow development and lead to decreased production of red blood cells (called aplastic anemia). As a result, the use of this drug is reserved for situations in which suitable alternatives are unavailable.

MACROLIDES Macrolides include erythromycin, azithromycin, and clarithromycin. These bacteriostatic antibiotics are identifiable by their 16-member lactone rings (see Appendix 3, Figure A3.5C) and are effective mainly against Gram-positive bacteria (the drugs do not penetrate the outer membrane of Gram-negative bacteria). Macrolides bind to the 50S subunit and inhibit translocation of tRNA from the A site to the P site. Azithromycin was used to treat the *S. pyogenes* infection in Case History 13.3. Because they are structurally different from the beta-lactam antibiotics, such as penicillin, macrolides can be used safely in patients who are allergic to penicillin.

LINCOSAMIDES A commonly used member of this group is clindamycin (see Appendix 3, Figure A3.5D). These antibiotics are effective against Gram-positive bacteria and many anaerobes, both Gram-positive and Gram-negative. Their effect on protein synthesis is the result of binding to peptidyltransferase and preventing peptide bond formation.

OXAZOLIDINONES These wholly synthetic antibiotics (see Appendix 3, Figure A3.5F, linezolid) bind to the 50S subunit and prevent assembly of the 70S ribosome, an activity unique among protein synthesis inhibitors. The drugs are effective against Gram-positive organisms, but their use is restricted, serving as a last-resort antibiotic for MRSA infections. Gram-negative bacteria are resistant to this class of drugs because of multidrug efflux pumps and decreased permeability due to the outer membrane.

STREPTOGRAMINS This group includes quinupristin/dalfopristin (trade name Synercid). This antibiotic, with a Gram-positive spectrum of activity, is really a combination of two cyclic peptide drugs, quinupristin and dalfopristin. They bind to different sites on the 50S subunit and act synergistically. Dalfopristin blocks tRNA entry into the A site, while quinupristin blocks exit of a growing peptide from the ribosome. Individually, the drugs are bacteriostatic, but in combination they exhibit bactericidal activity. This drug is a last line of defense against life-threatening infections with vancomycin-resistant *Enterococcus faecium* and MRSA. To delay the development of resistance, quinupristin/dalfopristin is held in reserve for these infections.

SECTION SUMMARY

- **Selective toxicity** is achieved by targeting a process that occurs only in bacteria, not host cells.
- **Antibiotic targets** include cell wall synthesis, cell membrane integrity, membrane synthesis, DNA synthesis, RNA synthesis, protein synthesis, and metabolism.
- **Antibiotics targeting the cell wall** bind to the enzymes and lipid carrier proteins involved with peptidoglycan synthesis and cross-linking.
- **Antibiotics interfering with DNA** include the antimetabolite sulfa drugs that inhibit nucleotide synthesis; quinolones that inhibit DNA topoisomerases; and a drug, metronidazole, that when activated randomly nicks the phosphodiester backbone.
- **The RNA synthesis inhibitor** rifampin binds RNA polymerase and stops transcription.
- **Aminoglycosides, tetracyclines, and glycylcyclines** bind the 30S subunit of the prokaryotic ribosome.
- **A variety of antibiotics** bind the 50S ribosomal subunit and inhibit translocation (macrolides, lincosamides), peptidyltransferase activity (chloramphenicol, streptogramins), or formation of the 70S initiation complex (oxazolidinones).

Thought Question 13.5 When treating a patient for an infection, why would combining a drug such as erythromycin with a penicillin be counterproductive?

Thought Question 13.6 Why would a bacteriostatic antibiotic that only <u>inhibits</u> growth of a pathogenic bacterium be useful for treating an infection? (*Hint:* Is the human body a quiet bystander during an infection?)

13.6
The Challenges of Antibiotic Resistance

SECTION OBJECTIVES
- Describe the four basic mechanisms of antibiotic resistance.
- Explain how drug resistance develops.
- Outline strategies to counteract drug resistance.

Antibiotics can effectively kill or "stun" pathogens that infect us, but do these same pathogens have the means to defend themselves from attack? In this section, we discuss the various strategies microbes use to become resistant to antibiotics.

CASE HISTORY 13.4

Boy Meets Drug-Resistant Pathogen

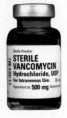

Brian, a 14-year-old boy with fever (39°C, or 102.2°F), chills, and left-sided pleuritic chest pain (pain when inhaling) was referred to a hospital emergency department by his general practitioner. A chest X-ray showed left lower lobe pneumonia. The boy reported that he was allergic to amoxicillin and cephalosporins (as a child he had developed a rash to these agents) and had been taking daily doxycycline (tetracycline) for the previous 3 months to treat mild acne. He was admitted to the hospital and treated with intravenous erythromycin because of his reported beta-lactam allergies, but he continued to feel sick. The day after admission, both sputum and blood cultures grew *Streptococcus pneumoniae*. After 48 hours, antibiotic susceptibility results indicated that the microbe was resistant to penicillin, erythromycin, and tetracycline. Armed with this information, the clinician immediately changed antibiotic treatment to vancomycin, a drug still effective in the face of these resistance mechanisms. The boy's fever resolved over the next 12 hours, and he made a slow but full recovery over the next week.

Unfortunately the scenario presented in this case is far too common and has become an extremely serious concern. The Centers for Disease Control and Prevention estimates that 23,000 people in the United States die every year from infections that cannot be cured because of antibiotic resistance. **Figure 13.23** illustrates the rapid rise of penicillin resistance of *S. pneumoniae* in the United States.

There are four basic ways a microbe can become resistant to an antibiotic (**Figure 13.24**). The resistant organism can:

- **Modify the target so that it no longer binds the antibiotic.** For example, a modified penicillin-binding protein enables methicillin resistance; an altered DNA gyrase produces resistance to quinolones; and an altered ribosomal protein can cause resistance to some aminoglycosides (discussed in Section 13.5).

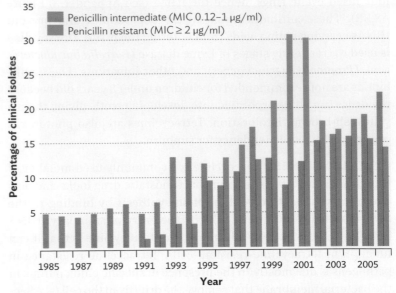

Figure 13.23   Penicillin-Resistant *Streptococcus pneumoniae* (United States, 1985–2006)

Penicillin was introduced in 1942. The first report of penicillin-resistant *S. pneumoniae* was in Spain in 1979. This graph tracks the increase in resistant strains in the United States. Strains resistant to intermediate levels of the drug have fewer altered penicillin-binding proteins than strains resistant to higher levels of the drug. Intermediate resistance means that the achievable tissue level of the drug may not be sufficient to inhibit growth of the strain in question. The numbers of penicillin-resistant isolates steadily increased until 2001, then dropped with the introduction of the pneumococcal vaccine for children.

- **Destroy the antibiotic before it gets into the cell.** One example is the enzyme beta-lactamase (or penicillinase), which is made exclusively to destroy penicillins (discussed in Section 13.5).

- **Add modifying groups that inactivate the antibiotic.** For instance, there are three different classes of enzymes that modify and inactivate aminoglycoside antibiotics.

- **Pump the antibiotic out of the cell using specific transport proteins (for example, tetracycline export) or nonspecific transport proteins.** The objective of this strategy is to pump drugs out of the cell faster than the drugs can get in. Protein pumps vary in structure. Single-component pumps reside in the cytoplasmic membrane of Gram-negative and Gram-positive bacteria. Multicomponent drug efflux pumps are present in Gram-negative bacteria only (discussed next). Tetracycline resistance, for instance, is typically due to an efflux pump.

A particularly dangerous type of drug resistance involves **multidrug resistance (MDR) efflux pumps** present in the cell membrane. A single pump in this class can export many different kinds of antibiotics with little regard for structure. MDR pumps evolved from pumps that normally expel all kinds of naturally occurring toxins, especially those that are membrane-soluble and thus enter the cell nonspecifically. Antibiotic MDR efflux pumps are now believed to contribute significantly to bacterial antibiotic resistance because of the variety of substrates they recognize and because of their expression in important pathogens. Strains of the pathogen *Mycobacterium tuberculosis*, for instance, have developed resistance to multiple drugs due in part to MDR pumps. Approximately 2 million people die from tuberculosis annually, mostly in developing nations. Chemists typically try to tweak the structure of an antibiotic to overcome a specific type of resistance mechanism, but the MDR pumps act on an exceptionally wide range of antibiotics, almost without regard to structure.

How Does Drug Resistance Develop?

Antibiotic resistance can arise spontaneously through mutation or through gene duplication followed by random mutations that "repurpose" the duplicated gene or genes. For instance, the MDR pumps evolved from genes encoding other transport mechanisms. Once an antibiotic resistance gene is established, gene transfer mechanisms such as conjugation, described in Chapter 9, can move these genes from one organism to another and from one species to another. In fact, several drug resistance genes now found in pathogenic bacteria actually had their start in the chromosomes of the original drug-producing organisms (as self-protection) and were passed on through **horizontal gene transfer**. Transfer of a drug resistance gene is particularly easy if the gene has been incorporated into a plasmid.

Link As is discussed in Section 9.1, **horizontal gene transfer** of a gene from one bacterium to another can involve the processes of transformation ("naked" DNA), conjugation (cell-cell contact), or transduction (phage transfer).

But why is antibiotic resistance so rampant? Consider the following case: A young mother brings her 4-year-old child to the physician. The child is screaming because he has an extremely painful sore throat. Simply looking at the throat is not diagnostic. The raw tissue could mean that the child is suffering from a bacterial infection, in which case antibiotics are needed. Alternatively, a virus could be the culprit, a situation where antibiotics do nothing but pacify the parent. More often than not, the clinician will prescribe an antibiotic without ever knowing the cause of disease. The problem is this: the more an antibiotic is used, the more opportunities there are to select for an antibiotic-resistant organism. The presence of the drug does not <u>cause</u> resistance but will kill off or inhibit the growth of competing bacteria that are sensitive, thereby allowing the resistant organism to grow to high numbers.

When should antimicrobials be administered? Certainly, in life-threatening situations where time is of the essence, antibiotics

Figure 13.24   Alternative Mechanisms of Antibiotic Resistance

Antibiotic resistance genes can be plasmid-borne or they can be part of the chromosome. A specific antibiotic resistance gene will carry out only one of the four mechanisms shown.

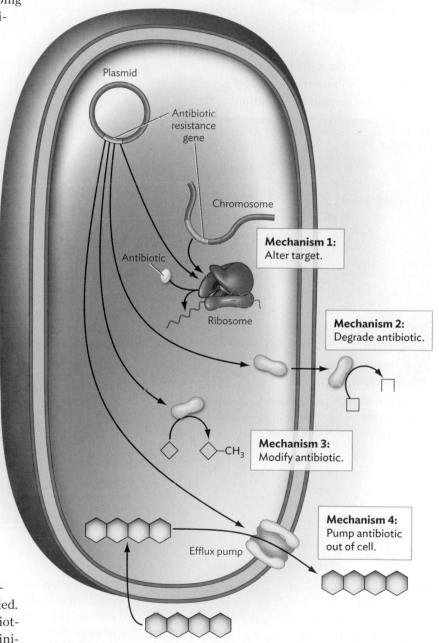

should be administered even before knowing the cause of the infection (empirical therapy). On the other hand, the most prudent course to take when a patient has a simple infection is to confirm a bacterial etiology before prescribing. An exception to this may be in an elderly or otherwise immunocompromised individual, who may be more susceptible to secondary bacterial infections that can occur subsequent to viral disease.

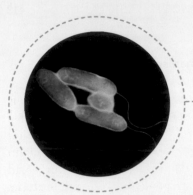

inSight

Antibiotic Treatment Failures: It's Not Only Resistance

Chronic infections present a formidable therapeutic challenge and a puzzling problem. They can be treated with antibiotics and appear to resolve; then, once the drug is stopped, the infection roars back. Curiously, though, the resurgent pathogen is not usually resistant to antibiotics in vitro. This type of infection becomes almost impossible to eradicate. The most vexing example is that of the incurable *Pseudomonas aeruginosa* infection of cystic fibrosis patients (Figure 1).

We now think that these resurging infections are explained by a long-known but unsolved puzzle of microbiology—antibiotic-tolerant **persister cells**—cells that neither grow nor die in the presence of bactericidal agents. Joseph Bigger, in 1944, noticed that penicillin would lyse a growing culture of *Staphylococcus aureus*, but a small number of persister cells always survived. These persisters were not mutants made permanently resistant through mutation. They acted as though dormant.

Figure 1   Chest X-Ray of Adult Cystic Fibrosis Patient with Chronic *Pseudomonas aeruginosa* Infection

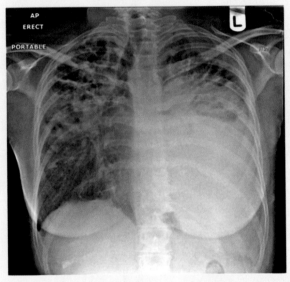

Note that the left lung (right side of image) is cloudy, indicating that it is filled with fluid. In medical terminology, this is called left-sided consolidation and pleural effusion.

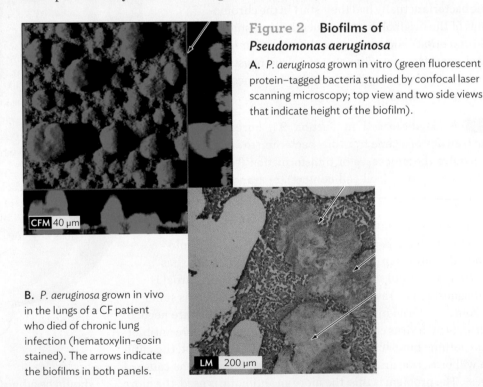

Figure 2   Biofilms of *Pseudomonas aeruginosa*

A. *P. aeruginosa* grown in vitro (green fluorescent protein–tagged bacteria studied by confocal laser scanning microscopy; top view and two side views that indicate height of the biofilm).

B. *P. aeruginosa* grown in vivo in the lungs of a CF patient who died of chronic lung infection (hematoxylin-eosin stained). The arrows indicate the biofilms in both panels.

Another proposed source of antibiotic resistance was the widespread practice of adding antibiotics to animal feed. No one quite knows why, but feeding animals antibiotics in their food makes for larger, and therefore more profitable, animals. Some estimates suggest that 70% of all antibiotics used in the United States are fed to healthy livestock. The consequence of this is that the animals may serve as incubators for the development of antibiotic resistance. Even if the resistance develops in nonpathogens, the antibiotic resistance genes produced can be transferred to pathogens by horizontal gene transfer. Finally, as of January 2017, the United States Department of Agriculture banned the use of medically important antibiotics in animal feed.

Figure 3 Persister Cells of *Pseudomonas aeruginosa*

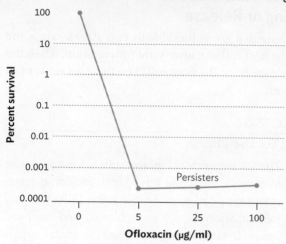

Dose-dependent killing of a clinical strain isolated from an 8-month-old patient. The antibiotic ofloxacin was added to stationary-phase cultures of this *P. aeruginosa* isolate. Concentrations were 0, 5, 25, or 100 μg/ml. After an 8-hour incubation, surviving persister cells were plated for colony count. The survivors of antibiotic treatment were not antibiotic resistant by standard tests.

In any biofilm or population of late-exponential-phase cells, persister cells that tolerate antibiotic treatment can be found. Figure 2 shows in vitro and in vivo biofilms of *P. aeruginosa*. It is now thought that persisters help explain antibiotic treatment failures and even latent bacterial infections, such as recrudescent typhus or latent tuberculosis. Figure 3 illustrates the presence of persister cells in a clinical isolate of *P. aeruginosa*.

How might dormancy explain antibiotic tolerance? If persisters are dormant and have little or no cell wall synthesis, translation, or topoisomerase activity, then even if antibiotics can bind to proteins or structures responsible for those activities, their functions cannot be corrupted. Therefore, antibiotics are tolerated at the price of not growing. An antibiotic can kill all susceptible bacteria in an infection, but the remaining persister cells serve as a source of regrowth and infection once the antibiotic is removed. You might say that persister cells "play dead" the way some animals do when threatened by a predator. The molecular mechanisms of persistence include ATP depletion and self-imposed slowing of translation.

Be aware that treatment failures for infectious diseases are not only due to genetically heritable antibiotic resistance mechanisms; there is a nongenetic cause called persistence in which infecting bacteria "play dead" when an antibiotic is present, only to grow again when the antibiotic is gone (see **inSight**).

Fighting Drug Resistance

Several strategies are being used to stay one step ahead of drug-resistant pathogens. In some instances, dummy target compounds that inactivate resistance enzymes have been developed. Clavulanic acid, for example, is a compound sometimes used in combination with penicillins such as amoxicillin. Clavulanic acid, a beta-lactam compound with no antimicrobial effect, competitively binds and ties up beta-lactamases secreted from penicillin-resistant bacteria. Amoxicillin, then, escapes destruction, enters the cell, and kills the bacterium. Sulbactam is another frequently used beta-lactamase inhibitor. A different strategy to combat drug resistance is to alter the structure of the antibiotic in a way that sterically hinders the access of modifying enzymes.

There are also a variety of actions that individuals can take to limit infections and thus the use of antibiotics. These include frequent hand washing, vaccinations, avoiding the use of antibiotics for viral infections, and refusing to use leftover antibiotics from your friends. Finally, when you do need to use an antibiotic, be sure to complete the full course. Stopping halfway through because you feel better will leave bacterial survivors that can regrow and develop resistance.

 Potential catastrophe of antibiotic resistance: youtube.com

SECTION SUMMARY

- **Antibiotic resistance** is a growing problem worldwide.
- **Mechanisms of antibiotic resistance** include modifying the antibiotic, destroying the antibiotic, altering the target to reduce affinity, and pumping the antibiotic out of the cell.
- **Multidrug resistance pumps** use promiscuous binding sites to bind antibiotics of diverse structure.
- **Antibiotic resistance can arise spontaneously** through mutation, can be inherited by gene exchange mechanisms, or can develop through gene duplication and mutational reengineering.
- **Indiscriminate use of antibiotics** has significantly contributed to the rise in antibiotic resistance.
- **Measures to counter antibiotic resistance** include chemically altering the antibiotic, using combination antibiotic therapy, and adding a chemical decoy.

Thought Question 13.7 How might bacteria such as *Streptomyces* and *Actinomyces* species avoid committing suicide when they make their antibiotics?

Thought Question 13.8 Jill goes up the hill with Jack, someone she has just met at a local bar, and engages in unprotected sex. The next morning, she becomes worried that Jack may have given her a sexually transmitted disease. Jill remembers that she still has some ciprofloxacin pills left over from an earlier respiratory tract infection (the original infection resolved after a few days and she forgot to take the rest of the treatment). Should she take the rest of the cipro prophylactically to prevent a possible STD? Explain why or why not.

13.7
Antiviral Agents

SECTION OBJECTIVES
- List the basic targets of antiviral agents.
- Discuss selective toxicity for antiviral drugs.
- Explain the spectrum of activity for antiviral drugs.

Why aren't there more antiviral agents? A father pleading with a physician to give his child antibiotics when the infant is suffering with a cold is an all too common dilemma faced by the general practitioner, but there is nothing of substance the physician can do. The common cold is caused by the rhinovirus, and no antibiotic designed for bacteria can touch it. So why are there so few antiviral agents in the clinician's arsenal? The reason is that selective toxicity is much harder to achieve for viruses than it is for bacteria. Viruses by their very nature use host cell functions to make copies of themselves. Thus, a drug that hurts the virus is likely to also harm the patient. Nevertheless, there are several useful antiviral agents in which selective viral targets have been found and exploited. Some of these agents are listed in Table 13.8. Select examples are discussed in this section.

Antiviral Agents That Prevent Virus Uncoating or Release

Membrane-coated viruses are vulnerable at two stages: when the virus is invading the host cell and after viral propagation, when the progeny viruses release from the host cell. The flu virus presents a good example of both.

CASE HISTORY 13.5
Infant Titus Fights the Flu

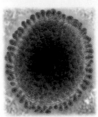

Titus, a 9-month-old infant, suffered from a series of chronic problems after birth, including respiratory syncytial virus bronchiolitis (infection and inflammation of the bronchioles) and neonatal group B streptococcal sepsis. (Neonatal sepsis happens soon after birth and is often caused by Lancefield group B *Streptococcus agalactiae*, passed to the newborn from an infected mother as described in Chapter 23.) As a result of his weakened state, Titus was fitted with a gastrostomy feeding tube so that his parents could feed him. Within a week of being sent home, however, Titus was rushed to the Johns Hopkins Hospital with an acute onset of fever, cough, regurgitation from his gastrostomy feeding tube, and dehydration. Physical exam confirmed that the boy had a severe cough resulting in respiratory

Table 13.8
Examples of Antiviral Agents

Virus	Agent	Mechanism of Action	Result
Influenza virus	Amantadine	Inhibits viral M2 protein	Prevents viral uncoating
	Zanamivir, oseltamivir (Tamiflu)	Neuraminidase inhibitor	Prevents viral release
Herpes simplex virus and varicella-zoster virus (shingles)	Acyclovir	Guanosine analog	Halts DNA synthesis
	Famciclovir, valacyclovir	Prodrugs of penciclovir and acyclovir, respectively; guanosine analogs	Halts DNA synthesis
Hepatitis B, chronic	Lamivudine	Cytosine nucleoside analog	Halts DNA synthesis
	Entecavir	Guanosine nucleoside analog	Halts DNA synthesis
Cytomegalovirus	Ganciclovir	Similar to acyclovir	Halts DNA synthesis
	Foscarnet	Analog of inorganic phosphate	Binds and inhibits virus-specific DNA polymerase
Respiratory syncytial virus and chronic hepatitis C	Ribavirin	RNA virus mutagen	Causes catastrophic replication errors
HIV	Zidovudine (AZT)	Nucleoside analog, resembles thymine	Inhibits reverse transcriptase
	Nevirapine	Nonnucleoside; binds to allosteric site	Inhibits reverse transcriptase
	Nelfinavir, indinavir, lopinavir, darunavir	Protease inhibitor	Prevents viral maturation
	Raltegravir	Integrase inhibitor	Prevents integration into host genome
	Maraviroc	CCR5 entry inhibitor	Prevents virus entry into host cells

distress, a rapid heart rate, and moderate dehydration. A nasopharyngeal aspirate (the nasopharynx is the nasal part of the pharynx) was positive for influenza A antigen. Titus was diagnosed with influenza and treated with the antiviral drug oseltamivir. He gradually improved and was discharged 4 days after admission.

Influenza can be a particularly dangerous disease. An unusually severe form of influenza A spread across the United States in 2003, killing higher than normal numbers of children and young adults. Vaccination helped limit disease transmission, but the availability of antiviral agents shortened the disease course. There are two selective antiviral targets. Recall that the membrane surrounding influenza virus contains neuraminidase, which cleaves sialic acid in host membranes to allow virus particles to escape from infected cells, and hemagglutinin, which binds the virus to host membrane sialic acid receptors during entry.

As described in Section 12.5, the membrane-enveloped influenza virus enters the cell in an endosome, which then acidifies. A virus protein called M2 then forms channels in the virus envelope, so the interior of the enveloped virus also acidifies. The drop in pH changes the structure of hemagglutinin on the viral membrane, allowing it to bind to membrane receptors on the endosome. As a result, virus and endosome membranes fuse and the virus is released into the cytoplasm. Amantadine (Figure 13.25A) is a specific inhibitor of the influenza M2 protein. Amantadine is useful only against influenza type A—not against type B strains. Unfortunately, many amantadine-resistant strains have developed, in part because of the widespread use of amantadine by Chinese poultry farmers. As a result, the drug is no longer recommended as a treatment for influenza.

The second target of the influenza virus is the envelope protein neuraminidase. The newer antiflu drugs, such as zanamivir (Relenza) or oseltamivir (Tamiflu), are **neuraminidase inhibitors** that act against types A and B influenza strains (Figure 13.25B).

Figure 13.25   Inhibitors of Influenza Proteins

Amantadine

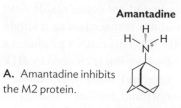

A. Amantadine inhibits the M2 protein.

Zanamivir

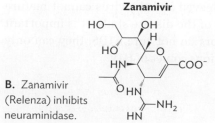

B. Zanamivir (Relenza) inhibits neuraminidase.

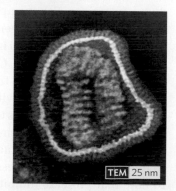

C. Influenza virus showing the viral envelope (brown) containing hemagglutinin and neuraminidase (proteins are too small to see).

Figure 13.26   Antiviral Agents That Prevent DNA Synthesis
Zidovudine (AZT) and acyclovir are analogs of thymine and guanine nucleotides, respectively. Because the analogs have no 3′ OH (yellow highlight) to which another nucleotide can add, DNA chain elongation ceases.

Zidovudine

Deoxyribonucleotide containing thymine

Acyclovir

Deoxyguanosine

Neuraminidase on the viral envelope allows virus particles to leave the cell in which they were made. Neuraminidase inhibitors prevent this release and cause the virus particles to aggregate at the cell surface, reducing the number of virus particles released.

The neuraminidase inhibitors, when used within 48 hours of disease onset, decrease viral shedding and reduce the duration of influenza symptoms by approximately 1 day. Given that flu symptoms generally last only 3–10 days, this treatment may not seem greatly beneficial, but for elderly or immunocompromised patients, as in Titus's case, shortening the course of the flu can minimize damage to the lungs, which in turn reduces the chance of developing life-threatening secondary bacterial infections such as pneumonia and bronchitis.

Antiviral DNA and RNA Synthesis Inhibitors

Most antiviral agents work by inhibiting viral DNA synthesis. These drugs chemically resemble normal DNA nucleosides but do not function as such. Many of these drugs contain structural analogs of deoxyribose attached to adenine, guanine, cytosine, or thymine. Viral enzymes add phosphate groups to these nucleoside analogs to form DNA nucleotide analogs. A DNA polymerase then inserts the DNA nucleotide analogs into the replicating viral DNA strand in place of a normal nucleotide. Once incorporated, however, the nucleotide analog cannot accept new nucleotides needed for chain elongation, and **DNA synthesis** stops (Figure 13.26). These DNA chain–terminating analogs are selectively toxic because viral polymerases more readily incorporate nucleotide analogs into their nucleic acid than do the more selective host cell polymerases.

Link Recall from Chapter 8 that for **DNA synthesis** to occur, the 5′ end phosphate of an incoming nucleotide triphosphate must attach to the 3′ OH end of the growing DNA chain. If the 3′ OH group is missing from the last base in the chain, as is the case with the nucleotide and nucleoside analogs, elongation and replication terminate.

In contrast to nucleoside analogs, acyclovir is a guanosine nucleobase analog that is active against the DNA viruses of chickenpox (varicella-zoster virus) and herpes (herpes simplex virus). Acyclovir is selectively converted to a mononucleotide by a viral enzyme (thymidine kinase) and then to the active trinucleotide form (acyclo-GTP) by host kinases. Although acyclo-GTP is incorporated into host and viral DNA during replication, it is selectively toxic to viral replication because host DNA polymerase has an effective proofreading activity (missing in the viral polymerase) that removes the analog from host DNA. Replication halts because acyclo-GTP lacks a 3′ hydroxyl end.

These antiviral DNA synthesis inhibitors work on DNA viruses such as herpes virus or on RNA retroviruses such as HIV, but not on RNA viruses such as influenza. Ribavirin, however, is a drug that when metabolized produces a compound resembling purine RNA nucleosides. Ribavirin will inhibit RNA metabolism by RNA viruses such as influenza and hepatitis C.

CASE HISTORY 13.6

HIV: The Honeymoon Is Over

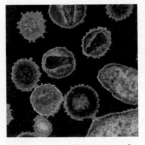

Bob and Carol, a recently married couple from Tacoma, Washington, came to the community clinic for prenatal care. Carol was 19 and reportedly 2 months pregnant with her first child. She denied intravenous (IV) drug use or a history of other sexual partners and had no history of sexually transmitted disease; however, a routine prenatal HIV antibody screen was reported as positive for HIV-1. Careful questioning revealed that Bob had a history of IV drug use. An HIV antibody screen for him was also positive. The laboratory results indicated that Carol did not yet require therapy (she had a low viral load—fewer than 1,000 copies per milliliter of blood—and a normal CD4 T-cell lymphocyte count), but since she was pregnant, a short course of antiretroviral therapy (the drug AZT) was recommended to help prevent transmission of the virus to her child. Bob had an HIV viral load of 10,000 copies per milliliter and was started on combination retroviral therapy, including a protease inhibitor.

Human immunodeficiency virus (HIV) is an RNA retrovirus that uses a **reverse transcriptase** to make DNA that then integrates into host nuclear DNA to form a provirus (discussed in Section 12.6). The virus targets specific white blood cells (T cells, discussed in Chapters 15 and 16) and slowly destroys parts of the immune system, leaving its victims helpless against secondary infections. HIV has killed hundreds of thousands of people in the United States since the HIV pandemic began, but being diagnosed with HIV is no longer a death sentence. Advances in antiretroviral therapy (ART) over the past 10 years have transformed HIV into a manageable chronic condition. At least half, if not most, of the HIV-infected people in the United States now live long enough to die from diseases of aging, such as heart attacks or strokes. A number of drugs are the heroes of this new paradigm.

Link As discussed in Section 8.5, **reverse transcriptase** is an enzyme that uses RNA as a template to make a complementary strand of DNA. HIV uses reverse transcriptase to copy its RNA genome into DNA and integrate the DNA copy into the host genome.

Nucleoside and Nonnucleoside Reverse Transcriptase Inhibitors

The antiretroviral nucleoside inhibitor zidovudine (abbreviated ZDV or AZT) used in Case History 13.6 is a nucleoside analog recognized by reverse transcriptase. Once incorporated into a replicating HIV DNA molecule, the DNA chain–terminating property of AZT prevents further DNA synthesis. Because HIV transmission from the mother to the neonate can occur at delivery or during breast-feeding, treatment of the mother and the child are important steps to prevent transmission. There are also nonnucleoside reverse transcriptase inhibitors. For example, the drug delavirdine binds directly to reverse transcriptase and inactivates the enzyme.

Protease Inhibitors

To make optimum use of its short provirus DNA sequence, HIV makes long, nonfunctional polypeptide chains that are cleaved into the functional proteins and enzymes needed to replicate and produce new virions. For example, the *gag* and *pol* genes reside next to each other in the HIV genome and are transcribed as a single mRNA molecule (discussed in Chapter 12). This mRNA produces a polyprotein called Gag-Pol. Once made, Gag-Pol is cleaved by HIV protease to make reverse transcriptase and integrase. Protease inhibitors, such as nelfinavir (Viracept) and lopinavir (Kaletra), are a powerful class of drugs that block the HIV protease. When the protease is inactivated, even though new virus particles are made, the polyprotein remains uncleaved and the virus cannot mature to infect other cells; progress of the disease stalls. It is important to note that protease inhibitors do not cure AIDS; they can only decrease the number of infectious copies of HIV.

Entry Inhibitors

Another way to stop HIV is to prevent the virus from infecting cells in the first place. Drugs called entry inhibitors do just that—they stop entry. There are two types of entry inhibitors. CCR5 inhibitors such as maraviroc block virus envelope protein gp120 (also known as SU) from binding to host surface protein CCR5, a coreceptor that, together with CD4, is needed for virus binding to certain T cells (see Section 12.6 and Figure 12.28 for details). As a result of CCR5 inhibition, the virus never attaches. Fusion inhibitors such as enfuvirtide, in contrast, do not prevent initial binding, but prevent HIV membranes from fusing with T-cell membranes and thereby halt viral entry. Imagine trying to enter a room through a door. CCR5 inhibitors are like removing the doorknob so there is nothing to grab onto. Fusion inhibitors, however, are like gluing the door shut. You can grab the knob but still cannot open the door.

HIV Treatment Regimens

Antiviral therapy with protease inhibitors is recommended for patients with symptoms of AIDS and for asymptomatic patients with viral loads above 30,000 copies per milliliter. Treatment should be considered even for patients with viral loads above 5,000, as for the husband in Case History 13.6.

Because HIV can mutate rapidly and become resistant to single-drug therapies, treatment today involves administering combinations of three or more antiretroviral drugs. This therapeutic strategy is called **highly active antiretroviral therapy (HAART)**. Most current HAART regimens include three drugs: usually two nucleoside reverse transcriptase inhibitors plus a protease inhibitor, a nonnucleoside reverse transcriptase inhibitor, or an integrase inhibitor. Integrase inhibitors, first approved in the United States in 2007, block the enzyme needed to insert viral DNA into the host genome. You can learn more about HIV and treatment regimens in Section 23.4.

SECTION SUMMARY

- **There are fewer antiviral agents** than antibacterial agents because it is harder to identify viral targets that provide selective toxicity. The spectrum of activity of an antiviral agent is linked to its molecular target: for instance, RNA virus inhibitors, DNA virus inhibitors, or retrovirus inhibitors.
- **Preventing viral attachment to, or release from, host cells** characterizes mechanisms of action for antiviral agents such as amantadine and zanamivir, used to treat influenza virus.
- **Inhibiting DNA synthesis** is the mode of action for most antiviral agents, although they work only for DNA viruses and retroviruses.
- **HIV treatments** include reverse transcriptase inhibitors, which prevent synthesis of DNA; protease inhibitors, which prevent the maturation of viral polyproteins into active forms; and entry inhibitors, which prevent HIV from infecting cells.

13.8
Antifungal Agents

SECTION OBJECTIVES
- Describe the major targets of antifungal agents.
- Discuss selective toxicity for antifungal agents.

We can kill bacteria and stop viral replication, but can we kill fungi that cause infection? Fungal infections are much more difficult to treat than bacterial infections—in part because fungal physiology is more similar to that of humans than is bacterial physiology. Fungi also have an efficient drug detoxification system that modifies and inactivates many antibiotics. Nevertheless, it is possible to treat fungal infections; but to have a fungistatic effect, repeated applications of antifungal agents are necessary to keep the level of unmodified drug above MIC levels.

CASE HISTORY 13.7

Blastomycosis

Enrique, a 37-year-old male, presented to the emergency department of a Florida hospital with persistent fever, malaise, and a painful mass in his right arm. He denied trauma to the arm. White blood cell count was elevated at 27,000/μl, and chest X-ray revealed a left lung infiltrate. Bronchoscopy revealed granulomatous (walled-off) inflammation containing a single yeastlike mass. Incision and drainage was performed on the arm mass, and cultures were obtained. Serum cryptococcal antigen tests were negative, as were tests for *Bartonella henselae* and *Toxoplasma*. Cultures from the right arm grew a fungal form similar to that identified from the bronchoscopy specimens. A tentative diagnosis of *Blastomyces dermatitidis* was confirmed using PCR. The patient was placed on intravenous amphotericin B, and his fever and leukocytosis (elevated white cell count) subsided within days. His medication was then changed to oral fluconazole for 6 months.

Fungal infections are described as being either superficial mycoses, such as athlete's foot, or systemic mycoses, such as blastomycosis (as in Case History 13.7). Because fungal and human biochemical pathways are very similar, finding selectively toxic antifungal agents has been difficult. Fortunately, there are some differences between a few fungal and human pathways that we can exploit. Currently, there are six categories of available antifungal agents:

- **Polyenes** (nystatin, amphotericin B), which disrupt membrane integrity

- **Azoles** (imidazoles and triazoles), which interfere with the synthesis of ergosterol, a cholesterol specific to fungi
- **Allylamines** (such as terbinafine, Lamisil), which also inhibit ergosterol synthesis
- **Echinocandins** (caspofungin), which block fungal cell wall synthesis
- **Griseofulvin**, which blocks cell division
- **Flucytosine**, which inhibits DNA synthesis

Superficial mycoses and systemic mycoses require very different treatments. Table 13.9 lists a number of clinically used antifungal agents and their use for different mycotic infections.

Superficial mycoses include infections of the skin, hair, and nails, as well as *Candida* infections of moist skin and mucous membranes (for example, vaginal yeast infections). Imidazole-containing drugs (clotrimazole and miconazole, with a fungistatic mode of action) are often used topically in creams for superficial mycoses (see Appendix 3, Figure A3.6A). These antifungal agents disrupt the fungal membrane by selectively inhibiting fungal enzymes of sterol synthesis but not inhibiting those of humans.

Triazoles such as itraconazole are administered orally, are fungistatic, and are less susceptible to metabolic degradation than are imidazoles. Lamisil (a terbinafine compound, fungicidal) is a different type of agent that selectively inhibits ergosterol synthesis in fungi. It is currently a popular antifungal agent used to treat superficial mycoses. More chronic dermatophytic infections typically require another antifungal agent called griseofulvin, produced by a *Penicillium* species (see Appendix 3, Figure A3.6B). Griseofulvin (a fungistatic) disrupts the mitotic spindle of fungi (but not humans) needed for cell division. This stops growth but does not kill the fungus. As the infected hair, skin, or nails grow and are replaced, the fungus is shed. Griseofulvin is typically taken orally.

Vaginal yeast infections caused by *Candida* are often treated with nystatin, a polyene fungicidal agent synthesized by *Streptomyces* that forms membrane pores (see Appendix 3, Figure A3.6C). The name "nystatin" came about because two of the people who discovered it worked for the New York State Public Health Department. Nystatin may be given orally or as a cream but is not available in injectable form.

The serious, sometimes fatal, consequences of systemic mycoses require more aggressive therapy. The drugs used in these instances include intravenous **amphotericin B** (produced by *Streptomyces*; see Appendix 3, Figure A3.6D) and oral fluconazole, a triazole compound. This was the regimen used in Case History 13.7. Amphotericin B binds to the ergosterol present in fungal but not mammalian membranes and destroys membrane integrity. Fluconazole, on the other hand, inhibits the synthesis of ergosterol, like the imidazoles. Thus, fungal cells grown in the presence of fluconazole make defective membranes.

Typically, curing systemic fungal infections requires long-term treatment to prevent disease relapse. Extended treatment is required because fungi grow slowly. However, even with long-term therapy, a fair number of patients do not respond. Fortunately, a new class of antifungal agents called echinocandins (caspofungin) has helped 70% of patients who do not tolerate other treatments.

Unfortunately, long-term treatment also increases the chances that the patient will exhibit toxic side effects of these drugs. For example, amphotericin B can damage kidneys, griseofulvin can trigger severe skin reactions, and caspofungin can damage the liver.

SECTION SUMMARY

- **Fungal infections** are difficult to treat because of similarities in human and fungal physiologies.
- **Azole-containing** antifungal agents (imidazoles and triazoles) and **allylamines** (Lamisil) inhibit ergosterol synthesis.
- **Griseofulvin** inhibits mitotic spindle formation.
- **Nystatin** and **amphotericin B** produce membrane pores and destroy membrane integrity.
- **Flucytosine** blocks DNA synthesis.

Perspective

Microbiology as a science was founded on the need to understand and control microbial growth in the hope of conquering human disease. As a result, the discoveries of disinfectants, antiseptics, and antibiotics have done much to improve the health and well-being of people around the globe. Infectious diseases dreaded for centuries (such as the plague and tuberculosis) were successfully defeated, or so we thought. Today, a crisis of antibiotic resistance looms because of the irresponsible use of antimicrobial agents. The fact that very few new, clinically useful antimicrobials have been discovered over the last quarter century should give us pause. There are several reasons for this failure. The first is that it is hard to discover new antibiotics capable of killing bacteria while remaining safe for humans. Second, once such a drug is found, bringing it to market is an arduous and expensive process involving many animal and human trials. The consequence is that the economics of developing a new antibiotic is not favorable. There are few incentives for pharmaceutical companies to develop new antibiotics. The impending crisis of antibiotic resistance, however, means we must redouble our efforts to find new drugs and impose measures that ensure the responsible use of existing antibiotics if we are to maintain our advantage over constantly evolving pathogens.

Table 13.9
Major Antifungal Agents and Their Common Uses

	Drug	Clinical Application								
		Systemic Mycoses				Opportunistic Mycoses				
		Coccidioido-mycosis	Histo-plasmosis	Blasto-mycosis	Paracoccidioido-mycosis	Aspergillosis	Candidiasis	Crypto-coccosis	Dermato-phytosis	Other
Polyenes (Fungicidal)	Amphotericin B (intravenous)	+[a]	+	+	+	+	+	+	–	
	Nystatin (oral; topical)	–	–	–	–	–	+ mc[b]	–	–	
	Natamycin (topical; eyedrops)	–	–	–	–	+ Superficial	+ Superficial	–	–	Mycotic keratitis
Azoles (Fungistatic), Imidazoles—Topical	Clotrimazole	–	–	–	–	–	+ mc	–	+	
	Miconazole (Monostat)	–	–	–	–	–	+ mc	–	+	
	Ketoconazole	–	–	+	–	–	+	–	+	Dandruff shampoo
Azoles (Fungistatic), Triazoles—Oral	Itraconazole	+	+	+	+	+	+ mc	+	+	Sporo-trichosis
	Fluconazole	+	+	+	?[c]	–	+	+	+	Sporo-trichosis
Allylamines—Topical or Oral (Fungicidal)	Terbinafine (Lamisil)	–	–	–	–	–	–	–	+	
Griseofulvin—Oral (Fungistatic)	Griseofulvin	–	–	–	–	–	–	–	+	
Echinocandins—Intravenous	Caspofungin	–	–	–	–	+	+	–	–	
Antimetabolites (Fungistatic or Fungicidal)	5-Fluoro-cytosine[d] (flucytosine)	–	–	–	–	+	+	+	–	Phaeohypho-mycosis

[a]+ indicates that the drug inhibits growth of the disease-causing agent; – indicates that the drug is not useful for the disease.
[b]mc, mucocutaneous but not systemic candidiasis.
[c]Insufficient data.
[d]Used only in combination with amphotericin B.

LEARNING OUTCOMES AND ASSESSMENT

SECTION OBJECTIVES

OBJECTIVES REVIEW

13.1
Basic Concepts of Sterilization and Disinfection

- Distinguish between sterilization, disinfection, antisepsis, and sanitation.
- Differentiate bacteriostatic agents from bactericidal agents.
- Discuss decimal reduction time and its use in sterilizing material.

1. _____ is the process by which a chemical kills disease-producing microbes on inanimate surfaces.
   - A. Sanitation
   - B. Disinfection
   - C. Sterilization
   - D. Antisepsis

13.2
Physical Agents That Kill Microbes

- Compare autoclave sterilization with pasteurization.
- Explain the difference between refrigeration and lyophilization.
- Describe the uses of filtration and laminar flow biological safety cabinets.
- List conditions in which irradiation can be used to sterilize.

4. Standard autoclave conditions are _____.
   - A. 101°C; 10 pounds of pressure, 30 minutes
   - B. 121°C, 15 pounds of pressure, 20 minutes
   - C. 72°C, atmospheric pressure, 15 sec
   - D. 121°C, atmospheric pressure, 15 sec
   - E. 121°C, 20 pounds of pressure, 15 minutes

5. An injectable drug solution filtered through a 0.2-μm filter should be considered sterile. True or false?

13.3
Chemical Agents of Disinfection

- List factors that can affect the efficacy of a disinfectant.
- Explain how disinfectants are compared.
- List the basic forms of commercial disinfectants.
- Discuss the emergence of microbial resistance to disinfectants.

8. Chemical agent A with a phenol coefficient of 10 is (more/less) effective at killing than chemical agent B with a phenol coefficient of 50.

9. Which of the following is an effective way to sterilize plasticware such as petri plates?
   - A. Steam sterilization
   - B. Chemical disinfection
   - C. Ethylene oxide gas
   - D. Irradiation
   - E. C and D only
   - F. B and C only

13.4
Basic Concepts of Antimicrobial Therapy

- Explain the concept of selective toxicity and its importance for clinically useful antibiotics.
- Explain what is meant by spectrum of activity, and discuss its clinical importance.
- Discuss how the effectiveness of an antibiotic is tested, and differentiate minimal inhibitory concentration (MIC) from minimal bacteriostatic concentration (MBC).
- Describe the chemotherapeutic index.

12. A series of culture tubes containing dilutions of an antibiotic are inoculated with a bacterial pathogen. Tube 1 has the highest concentration of drug and tube 6 has the lowest. Growth occurs in tubes 5 and 6; no growth is seen in tubes 1–4.
   - A. Which tube represents the MIC?
   - B. Which tube represents the MBC?

2. The effectiveness of a chemical agent is determined by _____.
   A. Kirby-Bauer test
   B. decimal reduction time
   C. complete death reduction time
   D. spore death time

3. Agent A acting on *Staphylococcus aureus* has a D-value of 5 minutes. If the population of microbes started at a cell density of 5×10^8 CFUs/ml, how many viable cells will remain after 15 minutes?
   A. 5×10^7 CFUs/ml
   B. 1×10^7 CFUs/ml
   C. 5×10^6 CFUs/ml
   D. 5×10^5 CFUs/ml
   E. 1×10^6 CFUs/ml

6. Pasteurization conditions are designed to kill which of the following microbes from milk?
   A. *Coxiella burnetti*
   B. *Staphylococcus aureus*
   C. *Mycobacterium tuberculosis*
   D. *Brucella abortus*
   E. *Clostridium botulinum*

7. You are developing a liquid food supplement for populations of developing nations following natural disasters. It must have a long shelf life without refrigeration. Which microbial control technique would best accomplish your goal?
   A. Filtration
   B. Pasteurization
   C. Gamma irradiation
   D. Steam sterilization
   E. Disinfection

10. Which of the following is useful as a disinfectant but not as an antiseptic?
    A. An iodophor
    B. Triclosan
    C. A cationic agent
    D. Glutaraldehyde
    E. Chlorhexidine

11. A patient just treated in the emergency department left a considerable amount of blood on the metal examination table. Which disinfectant is best for decontaminating the table?
    A. Glutaraldehyde
    B. Ethylene oxide
    C. Formaldehyde
    D. Hypochlorite bleach
    E. Chlorhexidine

13. The figure shows the effectiveness of two antibiotics on a bacterial pathogen isolated from spinal fluid. Which of the following could explain why a clinician might choose to treat the patient with drug A?
    A. Drug B may not penetrate into the brain.
    B. Drug B might be known to have a toxic effect on the patient (patient allergy).
    C. Drug A tissue levels might be higher and be maintained longer than drug B.
    D. All of the above
    E. A and B only

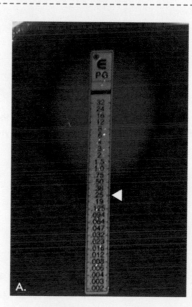

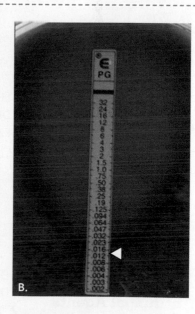

LEARNING OUTCOMES AND ASSESSMENT

	SECTION OBJECTIVES	OBJECTIVES REVIEW
13.5 **Antimicrobial Mechanisms of Action**	• Classify antibiotics by their molecular targets. • Describe the process of cell wall synthesis, and indicate the targets for various cell wall antibiotics. • Distinguish transcription inhibitors from translation inhibitors.	**14.** Which of the following antibiotic classes and host targets are matched correctly? A. Penicillin—membrane synthesis B. Aminoglycoside (e.g., gentamicin)—30S ribosome C. Quinolone (e.g., ciprofloxacin)—folic acid synthesis D. Sulfonamide—DNA gyrase E. Macrolide (azithromycin)—30S ribosome
13.6 **The Challenges of Antibiotic Resistance**	• Describe the four basic mechanisms of antibiotic resistance. • Explain how drug resistance develops. • Outline strategies to counteract drug resistance.	**17.** _____ is a single bacterial system that can provide resistance to multiple classes of antibiotics simultaneously. A. A multidrug efflux pump B. A beta-lactamase C. An enzyme that modifies antibiotic structure D. An enzyme that blocks a drug target from the drug
13.7 **Antiviral Agents**	• List the basic targets of antiviral agents. • Discuss selective toxicity for antiviral drugs. • Explain the spectrum of activity for antiviral drugs.	**24.** Which of the following classes of antivirals are used to treat influenza? A. DNA gyrase inhibitors B. Reverse transcriptase inhibitors C. DNA polymerase inhibitors D. Neuraminidase inhibitors E. Protease inhibitors
13.8 **Antifungal Agents**	• Describe the major targets of antifungal agents. • Discuss selective toxicity for antifungal agents.	**31.** Among the antifungal agents, _____ interferes with ergosterol synthesis, _____ inhibits DNA synthesis, and _____ disrupts the mitotic spindle. A. flucytosine B. a triazole or allylamine C. amphotericin B D. griseofulvin E. linezolid

15. A bacterial pathogen that produces the enzyme beta-lactamase will be resistant to which of the following antibiotics?
    A. Oxacillin
    B. Azithromycin
    C. Ceftriaxone
    D. Penicillin
    E. Oxazolidinone

16. Diagram where each of the following antibiotics affects cell wall synthesis: vancomycin, bacitracin, cephalosporins, cycloserine.

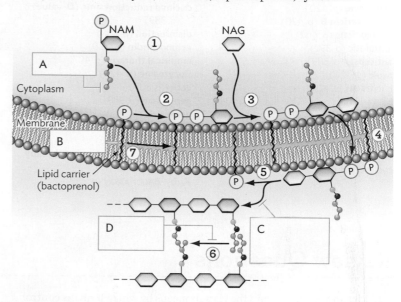

18. Clavulanic acid _____.
    A. inhibits protein synthesis
    B. acts as a decoy for beta-lactamase
    C. disrupts membrane integrity
    D. acts as decoy for enzymes that modify and inactivate aminoglycosides
    E. inhibits folic acid synthesis

Match which antibiotic resistance mechanisms are used by bacteria to resist the following antibiotics.

Antibiotic	Antibiotic Resistance Mechanism
19. Penicillin	A. Alter target
20. Methicillin	B. Degrade antibiotic
21. Quinolones	C. Modify antibiotic
22. Aminoglycosides	D. Pump antibiotic out of cell
23. Tetracyclines	

25. Which of the following best explains selective toxicity of antiviral nucleoside analogs?
    A. Viral polymerases are less selective than host polymerases.
    B. Viral polymerases catalyze different reactions than host polymerases.
    C. The nucleoside analogs do not enter the host nucleus.
    D. The antiviral nucleoside analogs function normally with host polymerases.
    E. There is a host enzyme that converts the analog incorporated in DNA to the proper nucleotide.

Match the antiviral drug to its target or mode of action.

Drug	Target or Modes of Action
26. Zanamivir	A. CCR5
27. Viracept	B. Fusion inhibitor
28. Enfuvirtide	C. Neuraminidase inhibitor
29. Ribavirin	D. Viral DNA replication
30. Acyclovir	E. Viral RNA synthesis
	F. Protease inhibitor

32. A male patient complains of an itchy irritation in his genital region. Your diagnosis is a superficial fungal infection. Which of the following drugs is the usual treatment for this disease?
    A. itraconazole
    B. miconazole
    C. griseofulvin
    D. amphotericin B
    E. nystatin

33. A female patient enters the hospital in severe respiratory distress. The laboratory report indicates that she has a severe lung infection caused by *Histoplasma*. Which of the following drugs would be most appropriate?
    A. Amoxicillin
    B. Nystatin
    C. Acyclovir
    D. Amphotericin B
    E. Terbinafine (Lamisil)

Key Terms

allylamine (p. 420)
amphotericin B (p. 420)
antagonistic (p. 402)
antibiotic (p. 397)
antisepsis (p. 389)
azole (p. 420)
bacitracin (p. 404)
bactericidal (p. 389)
bacteriostatic (p. 389)
cephalosporin (p. 404)
chemotherapeutic agent (p. 397)
chemotherapeutic index (p. 401)
cycloserine (p. 405)

decimal reduction time (D-value) (p. 389)
disinfection (p. 388)
echinocandin (p. 420)
empirical therapy (p. 407)
flucytosine (p. 420)
germ (p. 389)
germicidal (p. 389)
germicide (p. 389)
gramicidin (p. 406)
griseofulvin (p. 420)
highly active antiretroviral therapy (HAART) (p. 419)
Kirby-Bauer assay (p. 399)

laminar flow biological safety cabinet (p. 392)
lyophilization (p. 391)
minimal inhibitory concentration (MIC) (p. 398)
minimum bactericidal concentration (MBC) (p. 400)
multidrug resistance (MDR) efflux pump (p. 412)
neuraminidase inhibitor (p. 417)
pasteurization (p. 391)
penicillin-binding protein (PBP) (p. 404)
persister cell (p. 414)

phenol coefficient test (p. 394)
polyene (p. 419)
polymyxin (p. 406)
sanitation (p. 389)
selective toxicity (p. 397)
semisynthetic antimicrobial (p. 404)
spectrum of activity (p. 398)
sterilization (p. 388)
synergistic (p. 402)
therapeutic dose (p. 401)
toxic dose (p. 401)
use-dilution test (p. 395)
vancomycin (p. 405)
zone of inhibition (p. 399)

Review Questions

1. List and briefly explain the various means by which humans control microbial growth. What is a D-value? What is a phenol coefficient?

2. How are pasteurization and sterilization different?

3. What is the difference between a disinfectant and an antiseptic?

4. How is the effectiveness of a disinfectant measured?

5. How do microbes prevent the growth of other microbes?

6. What is selective toxicity? Provide examples.

7. Explain the difference between antibiotic susceptibility and antibiotic sensitivity.

8. What is meant by "spectrum of antibiotic activity"?

9. Provide examples of bacteriostatic and bactericidal antibiotics.

10. What is the Kirby-Bauer test? Does it tell whether a drug is bacteriostatic or bactericidal?

11. Give examples of drugs that target bacterial cell wall synthesis, RNA synthesis, protein synthesis, and DNA replication.

12. How do antibiotic-producing microorganisms prevent suicide?

13. Why is antibiotic resistance a growing problem?

14. What are the four basic mechanisms of antibiotic resistance?

15. What are some mechanisms used to combat the development of drug resistance?

16. Why are there few antiviral agents available to treat disease?

17. How does amantadine inhibit influenza?

18. Discuss the general modes of action of antifungal agents.

Clinical Correlation Questions

1. As part of your senior project, you have been isolating strains of *Escherichia coli* from your stool at 3-week intervals and have checked antibiotic resistance patterns on each isolate. The strains are sensitive to all antibiotics tested except ciprofloxacin (a quinolone antibiotic). Halfway through your project you develop a bloody diarrhea caused by *Shigella sonnei*, another Gram-negative rod. The physician prescribes amoxicillin for 7 days, which doesn't work, so he switches you to ciprofloxacin. Two weeks after stopping the treatment, you resume isolating *E. coli* from your stool. Now you find that several of the most recently isolated *E. coli* strains are resistant to ampicillin and ciprofloxacin. Explain how this could happen.

2. A patient is admitted with a systemic fungal infection. You want to give intravenous amphotericin B. Blood work shows elevated blood urea nitrogen (BUN) and creatinine levels. Explain why you should or should not administer amphotericin B. Use the Internet to search for the significance of elevated BUN and creatinine.

Thought Questions: CHECK YOUR ANSWERS

Thought Question 13.1 A hand-sanitizing solution inoculated with 2×10^6 organisms of *Staphylococcus aureus* exhibits a D-value of 5 minutes. What will the D-value be if ten times more *S. aureus* is added?

ANSWER: The D-value, the time required to decrease a microbial population to 10% of the original, will still be 5 minutes. D-value does not change with cell number. It will take 5 minutes longer to kill all the organisms, however.

Thought Question 13.2 How would you test the killing efficacy of an autoclave?

ANSWER: Construct a death curve by measuring the survival of a known quantity of spores (for example, of *Bacillus stearothermophilus*) after autoclaving for various lengths of time. Spores should be used because they are more resistant to heat than is any vegetative cell. Typically, autoclaves are regularly checked with spore strips that change color once the endospores are no longer viable.

Thought Question 13.3 The concentration of the drug tobramycin in a tube of broth is 1,000 µg/ml. Serial twofold dilutions are made from this tube. Including the initial tube (tube 1), there are a total of ten tubes. Twenty-four hours after all the tubes are inoculated with *Listeria monocytogenes*, turbidity is observed in tubes 6–10. What is the MIC?

ANSWER: 62.5 µg/ml, the concentration in tube 5, the last tube with no growth. (Tube 5 is a 16-fold dilution of tube 1: 1,000/500/250/125/62.5.)

Thought Question 13.4 You are testing whether a new antibiotic will be a good treatment choice for a patient with a staph infection. The Kirby-Bauer test using the organism from the patient shows a zone of inhibition of 15 mm around the disk containing this drug. Clearly, the organism is susceptible. But you conclude from other studies that the drug will not be effective in the patient. What would make you draw this conclusion?

ANSWER: You might determine that the drug's MIC is 100 µg/ml using a tube dilution test but find that the average attainable concentration of the drug in tissues or blood is only 10 µg/ml. Anything higher leads to toxic side effects in the patient. So, by the Kirby-Bauer test, the drug will inhibit growth of the bacterium, but only at a concentration beyond what the patient can tolerate. However, if the average attainable tissue level of the drug were greater than the MIC, the drug would be effective.

Thought Question 13.5 When treating a patient for an infection, why would combining a drug such as erythromycin with a penicillin be counterproductive?

ANSWER: Erythromycin, a bacteriostatic drug, will stop growth, which indirectly stops cell wall synthesis and renders the microbe insensitive to penicillin. This combination of drugs has antagonistic effects on in vivo activity.

Thought Question 13.6 Why would a bacteriostatic antibiotic that only inhibits growth of a pathogenic bacterium be useful for treating an infection? (*Hint:* Is the human body a quiet bystander during an infection?)

ANSWER: Inhibiting the growth of the pathogen buys time for the host's immune system to generate antibody and generate cytotoxic T cells that can clear the organism. Chapters 15 and 16 discuss the immune response.

Thought Question 13.7 How might bacteria such as *Streptomyces* and *Actinomyces* species avoid committing suicide when they make their antibiotics?

ANSWER: Along with possessing genes whose products synthesize the antibiotic, these organisms have genes whose products efflux the drug or protect their own sensitive cell targets. Some of these genes have been horizontally transferred to other bacteria and are the source of spreading antibiotic resistance.

Thought Question 13.8 Jill goes up the hill with Jack, someone she has just met at a local bar, and engages in unprotected sex. The next morning, she becomes worried that Jack may have given her a sexually transmitted disease. Jill remembers that she still has some ciprofloxacin pills left over from an earlier respiratory tract infection (the original infection resolved after a few days and she forgot to take the rest of the treatment). Should she take the rest of the cipro prophylactically to prevent a possible STD? Explain why or why not.

ANSWER: A physician can prescribe prophylactic antibiotics in certain situations. For instance, rifampin is prescribed for family members of a patient with meningitis. However, self-treatment using leftover or shared antibiotics does not ensure adequate treatment for an STD or any other infection. For example, different STDs may require different antibiotics to effect a cure. Self-treatment with inappropriate antibiotics can also lead to resistant STD microbes or medical complications requiring extended medical care.

Normal Human Microbiota: A Delicate Balance of Power

[Case of the Unsatisfactory Stool]

SCENARIO A few months ago, Jason, a 38-year-old software salesman, reported to his physician with complaints of "unsatisfactory stools." The frequency of his stools was about two to three times a day but fluctuated from week to week from being semisolid and covered with thick mucus to small, hard, elongated pellets. He also complained of having weekly episodes of severe cramping pain all over his abdomen that would come and go during the course of a day. He said he belched frequently and would sometimes experience an ineffectual urge to pass stools, occasionally passing only gas.

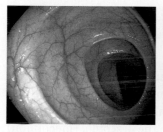 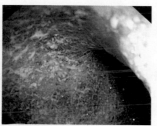

Colonoscopy
In this procedure, a fiber-optic endoscope is inserted from rectum to cecum. The endoscope can also take tissue samples. Although IBS is painful, the colons of patients, as viewed by colonoscopy, have a normal appearance. On the left is a normal colon, as viewed by colonoscopy. On the right is the colon of a person with inflammatory bowel disease, which is more serious than IBS. The mucosa is denuded, and active bleeding is seen.

Endoscope

DIAGNOSIS Jason was ultimately diagnosed with a nebulous disease called irritable bowel syndrome (IBS). IBS affects 10%–15% of the U.S. population. Although microbes are involved, IBS is not an infection. The cause of Jason's distress was most likely a shift in the microbial composition of his intestinal contents. Changing the mix of bacteria and archaea residing in the intestine, for example, can lead to inflammation in the bowel and the symptoms Jason experienced.

OUTCOME The clinician recommended that Jason ingest yogurt and other foods containing microorganisms that normally inhabit the intestine, such as *Bifidobacterium* and *Lactobacillus* species. The hope was that these bacteria could help restore the microbial composition of his intestine to a more healthy balance and thus reduce inflammation. In this case the treatment worked (it doesn't always), and within a week, Jason's gastrointestinal symptoms disappeared, much to his relief.

*****NOTE** Do not confuse IBS with inflammatory bowel disease (IBD). IBD is a more serious and complicated illness. Two common forms of IBD are ulcerative colitis, whose symptoms include small ulcers and abscesses in the colon and rectum, and Crohn's disease, in which any part of the gastrointestinal tract can become inflamed and ulcerated. Although less severe, IBS is nonetheless very disruptive and painful. IBS and IBD are both associated with altered microbiota, but IBD also appears to involve a more severe dysfunction of the immune system than does IBS.

SEM 5 μm

Intestinal Microbiome
Intestinal bacteria in the small intestine are essential to our survival. Among other things, they help train the immune system, digest carbohydrates that we cannot, and provide vitamins our bodies need.

After reading this chapter, you will be able to:

- Compare the composition of microbiomes from different areas of the body.

- List the benefits of our microbiota.

- Explain the intercommunications between human cells and microbiota.

- Discuss how the human body prevents infection by our microbiota.

- Describe the hygiene hypothesis and the potential role of microbiota in preventing allergies and other modern diseases.

L ike it or not, your body teems with microbial hitchhikers. You might even say that we are more microbe than human because the human body contains ten times more bacterial cells than human cells. All the bacteria, archaea, fungi, and protozoa that inhabit our bodies are called the human **microbiome** or **microbiota**. These microbes don't just sit there; they play critically important roles in our health and development.

Our microbiota is vast (100 trillion organisms), yet most species cannot be grown easily in the laboratory. Their growth in a host depends on unknown factors provided by the human microbiome itself.

You might also be surprised to learn that microbes and their human hosts chemically "talk" to each other. Chemical signals, including hormones, are sent back and forth, altering gene expression and physiology of all three domains that make a human: Eukarya, Archaea, and Bacteria. Today, considerable research is directed toward understanding the composition of a healthy microbiome and its relationship to human health.

This chapter will address several issues of the host-microbe relationship. For instance:

- What organisms make up our microbiota?

- Does the composition of an individual's microbiota change over time?

- Why do humans maintain normal microbiota if microbes also serve as sources of infection?

- How does our body keep members of our microbiota in their place?

- What cost-benefit analysis might explain the evolution of the host-microbe relationship?

14.1
Human Microbiota: Location and Shifting Composition

SECTION OBJECTIVES

- List body compartments that are sterile and discuss why they are sterile.

- List body compartments that are populated by microbiota, and name the key species found in each.

- Identify key features of populated body sites that influence microbiota composition.

How and when do we become colonized by microbes? We once thought that a fetus growing in the amniotic sac was sterile and that colonization by microbes began only after birth. That view is now challenged by this startling finding: the first stool (meconium) passed by a newborn <u>before</u> its first meal contains bacteria. To be present in meconium, these organisms must have colonized the fetal intestine in the womb. Another surprise is that the placenta, also once thought to be sterile, actually has a microbiome and may be one source of the fetal microbiome. It now appears that almost from the moment of our inception until the time of our death, we are constantly exposed to microbes.

Once the baby breaks out of the embryonic membrane, it is exposed to a whole new array of microbes residing in the birth canal and in the awaiting world. As a result, humans—and all other higher life-forms—are heavily populated with bacteria. Colonization typically occurs where our body interfaces with the external environment (for example, mouth, skin, and parts of the genitourinary tract). The more sequestered body sites are sterile. For example, most internal organs, as well as the blood and cerebrospinal fluid, should not harbor any bacteria. Microbes discovered at these sites usually have deleterious effects on the host. Their presence means an infection is under way.

 The assembly of an infant gut microbiome: youtube.com

Current research tells us that our microbiome is critically important to our health—even if the average person doesn't realize it's there. We now understand that the bacteria colonizing our bodies may collectively be as important to us as a kidney or liver. For instance, our body cells are constantly bathed in metabolites produced by gut bacteria. It is thought that these microbial metabolites, which circulate in our blood, significantly and favorably influence human health and development.

Figure 14.1   Examples of Normal Microbiota Present at Various Colonizing Sites

Highlighted species are notable for their predominance at the site or their potential for causing disease. *Staphylococcus* species are differentiated by whether they can produce coagulase, an enzyme that coagulates serum. *S. aureus* is generally coagulase-positive (there are some coagulase-negative strains), whereas other species of staphylococci normally found on skin are coagulase-negative. Some organisms, *S. pneumoniae*, for example, are pathogens but can be found as normal microbiota in some individuals. (*Note:* Viridans streptococci is not an actual species but represents a collection of streptococci that can turn blood agar green.)

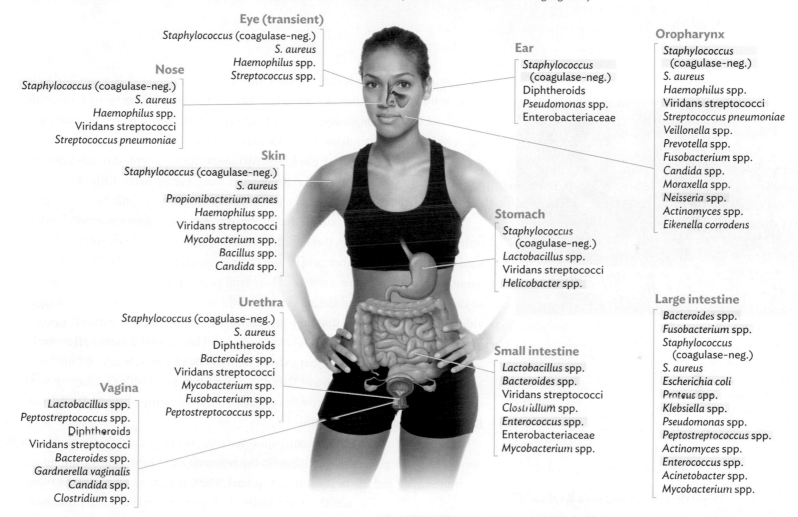

Eye (transient)
Staphylococcus (coagulase-neg.)
S. aureus
Haemophilus spp.
Streptococcus spp.

Nose
Staphylococcus (coagulase-neg.)
S. aureus
Haemophilus spp.
Viridans streptococci
Streptococcus pneumoniae

Skin
Staphylococcus (coagulase-neg.)
S. aureus
Propionibacterium acnes
Haemophilus spp.
Viridans streptococci
Mycobacterium spp.
Bacillus spp.
Candida spp.

Urethra
Staphylococcus (coagulase-neg.)
S. aureus
Diphtheroids
Bacteroides spp.
Viridans streptococci
Mycobacterium spp.
Fusobacterium spp.
Peptostreptococcus spp.

Vagina
Lactobacillus spp.
Peptostreptococcus spp.
Diphtheroids
Viridans streptococci
Bacteroides spp.
Gardnerella vaginalis
Candida spp.
Clostridium spp.

Ear
Staphylococcus (coagulase-neg.)
Diphtheroids
Pseudomonas spp.
Enterobacteriaceae

Oropharynx
Staphylococcus (coagulase-neg.)
S. aureus
Haemophilus spp.
Viridans streptococci
Streptococcus pneumoniae
Veillonella spp.
Prevotella spp.
Fusobacterium spp.
Candida spp.
Moraxella spp.
Neisseria spp.
Actinomyces spp.
Eikenella corrodens

Stomach
Staphylococcus (coagulase-neg.)
Lactobacillus spp.
Viridans streptococci
Helicobacter spp.

Small intestine
Lactobacillus spp.
Bacteroides spp.
Viridans streptococci
Clostridium spp.
Enterococcus spp.
Enterobacteriaceae
Mycobacterium spp.

Large intestine
Bacteroides spp.
Fusobacterium spp.
Staphylococcus (coagulase-neg.)
S. aureus
Escherichia coli
Proteus spp.
Klebsiella spp.
Pseudomonas spp.
Peptostreptococcus spp.
Actinomyces spp.
Enterococcus spp.
Acinetobacter spp.
Mycobacterium spp.

As mentioned earlier, the majority of species that make up our microbiota are unknown and have never been grown in the lab. New DNA sequencing technologies, however, allow scientists to sequence the genomes, or many pieces of them, from all the microbes at a particular body site. The process, called metagenomic DNA sequencing, has already provided valuable insights about our microbiome and host-microbe relationships. For instance, researchers who have compared microbiomes from the guts of lean and obese humans have found vast differences in gut microbiota. It is thought that different microbiota influence the efficiency of calorie harvesting from the diet and alter how the derived energy is stored. These findings are intriguing because they suggest that each person has a microbiome different from everyone else's; that is, microbiomes may be as unique as fingerprints. The composition of a person's microbiota is not set in stone, however. An individual's microbiota can evolve over time with a change in diet or after making new human (or animal) friends.

In this section, we describe some of the organisms that make up microbiota and discuss why they populate different sites. **Figure 14.1** illustrates various human body sites colonized by microbiota and provides examples of the organisms found there. **Table 14.1** lists the four prominent ecosystems of humans along with the number of microorganisms that typically inhabit each (called the **bioburden**). The table also includes the ratio of aerobes to anaerobes (there are anaerobic parts of the body, such as the intestine) and from where resident species originated. Note that the composition of our microbiota will vary throughout life as a result of continual reseeding from the environment. Viruses are also part of the human microbiome (called the **virome**), but because little is known of their identities or impact on human health, we will not dwell on them here.

 The human microbiome: youtube.com

inSight

Catching Crooks by Their Microbiomes

Figure 1

Imagine this: A man is shot in a dark alley (Figure 1). Police detectives find a gun nearby. It is the murder weapon, but there are no usable fingerprints. Forensic microbiologists, however, use new DNA sequencing technologies to identify the microbial community left on the weapon by the killer. They then match the microbial profile to a suspect, who later confesses. Although still in the realm of science fiction, a study led by microbiologists Rob Knight and Noah Fierer, of the University of Colorado, Boulder, suggests that such an investigative tool may be available one day.

Knight and Fierer reasoned that every person's microbiome is unique—as unique as a fingerprint. In their study, these researchers and their colleagues swabbed the hands of three people and their computer keyboards. They then generated DNA sequence profiles of bacterial populations from the hands and keyboards and compared them. As predicted, the microbial mix found on an individual's keyboard closely matched the bacteria residing on their hands (Figure 2). Furthermore, the bacterial DNA remained useful for at least 2 weeks after swabbing. You might ask why not just extract human DNA from a smudged fingerprint and identify the killer that way? There are very few human cells in a fingerprint (in contrast to the much larger number of bacteria), so obtaining human DNA from a fingerprint is an iffy proposition.

Of course, for bacterial "fingerprints" to be useful, they must distinguish a single person from the general population. So the research team took swabs from nine computer mice and the mouse owners' palms. The team then compared the bacterial DNA signatures of the mice and palms to a database of microbiomes taken from the hands of 270 other people as part of the Human Skin Microbiome project. In all nine cases, the bacterial profile on a computer mouse was more similar to that of the owner's palm than to any other hand in the database.

The question is whether the technique will be successful in the real world. At the moment, it isn't clear how unique a person's skin microbiome is, and then there is the problem of cross-contamination. In the published study, the researchers sampled objects handled by only one person, but if two or more people touch an object, they may leave behind a mix of bacteria that might resemble that of a third person. Despite these important questions, microbiome forensics is an intriguing prospect.

Figure 2   Depiction of "Germs on a Keyboard"

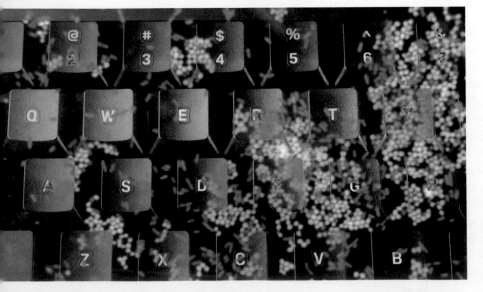

Table 14.1

The Prominent Bacterial Ecosystems of Humans

Microbial Reservoirs	Total Bioburden or Colony-Forming Units	Aerobe-to-Anaerobe Ratio[a]	How Acquired over a Lifetime
Skin	10^{4-6}/sweat gland	1:10	Initially birth canal; then oral and external environments
Mouth	10^{6-8}	1:10	Initially birth canal, caregiver; then food, water, fingers
Genitourinary tract	10^{8-9}/vagina/urethra	1:100	Surrounding external environment
Intestine	10^{11}/cm^3	1:1,000	Initially as a fetus; then baby formula, mother, ingestion of food and water

[a]Note that the more anaerobic the body site, the more anaerobes are present. Also note that even aerobic sites contain anaerobes.

Skin

The human adult, on average, is covered with 2 square meters (more than 21 square feet) of skin (**epidermis**) populated by 10^{12} microorganisms. Skin, as do all colonized body sites, harbors resident (normal) and transient microbiota. But even a resident microbe exhibits diversity in that different strains colonize at different times. Thus, our microbiota does not comprise a static population but represents a vibrantly changing mixture of strains and species, although some combinations of species may be unique and stable on a given individual (see **inSight**).

Several features of the epidermis make it difficult to colonize. Large expanses of skin are subject to drying, although some areas harbor enough moisture to support microbial growth; moist areas include the scalp, ear, armpit, and genital and anal regions. The skin also has an acidic pH (pH 4–6) owing to the secretion of organic acids by oil and sweat glands. As noted in Section 6.4, organic acids inhibit microbial growth by lowering bacterial cytoplasmic pH. Epidermal secretions are also high in salt and low in water activity, and they contain enzymes such as lysozyme that degrade bacterial peptidoglycan.

Despite these hurdles, many species of bacteria manage to colonize the epidermis. Most of these organisms are Gram-positive because they tend to be more resistant to salt and dryness. *Staphylococcus epidermidis*, various *Bacillus* species, and yeast such as *Candida* species are common examples of normal skin microbiota. Though normal, they are not always innocuous. One member of the skin's resident microbiota, the Gram-positive anaerobic rod *Propionibacterium acnes*, causes acne, a very visible plague of adolescence. Curiously, this is the same genus (but different species) that makes Swiss cheese (see Section 27.3). Increased hormonal activity during the teen years stimulates oil production by the sebaceous glands (Figure 14.2). *P. acnes* readily degrades the triglycerides in this oil, turning them into free fatty acids that then promote inflammation

Figure 14.2   Anatomy of the Skin and the Development of Acne

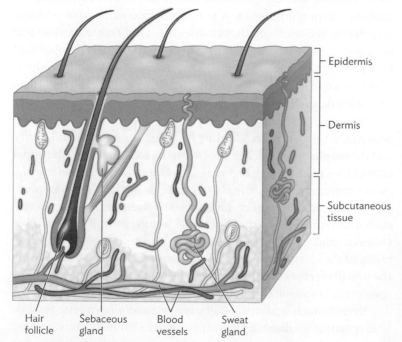

A. Location of sebaceous glands in the skin.

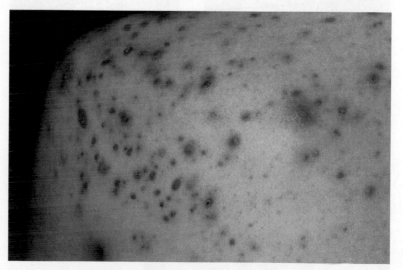

B. Acne on a patient's back.

of the gland. One consequence of the inflammatory response is the formation of a blackhead (comedo), a plug of fluid and keratin that forms in the gland duct (the oils turn black due to oxidation). The result is the typical skin eruptions of acne (see Chapter 19). Because of its bacterial basis, treatments for acne include tetracycline to kill the bacteria.

Figure 14.3   Examples of Anaerobic Oral Flora and Periodontal Disease

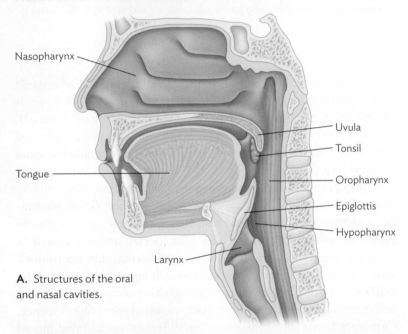

Nasopharynx

Uvula

Tonsil

Tongue

Oropharynx

Epiglottis

Hypopharynx

Larynx

A. Structures of the oral and nasal cavities.

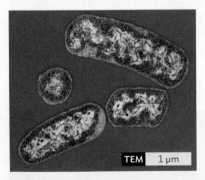

TEM   1 µm

B. *Prevotella* (colorized).

SEM   2 µm

C. *Fusobacterium*.

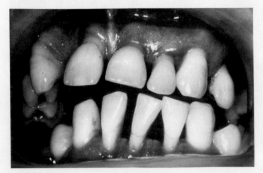

D. Symptoms of periodontal disease include red, swollen gums; bleeding gums; gum shrinkage; and teeth drifting apart.

The Eye

Because the eye is exposed to the outside environment, we might expect it to be heavily colonized. Actually, this is not the case; colonization is inhibited by the presence of antimicrobial factors such as lysozyme in the tears that continually rinse the eye surface (conjunctiva). Despite this, a few transient commensal bacteria can be found on the conjunctiva. Skin flora such as *Staphylococcus epidermidis* and diphtheroids (Gram-positive rods that look like clubs) as well as some Gram-negative rods, such as *Escherichia coli*, *Klebsiella*, and *Proteus*, manage to at least temporarily make the eye their home without causing damage. Occasionally, bacteria such as *Streptococcus pneumoniae* and *Haemophilus influenzae*, as well as some viruses, can cause an ocular disease known as pinkeye (Section 19.6), in which the inflamed conjunctiva becomes pink or red and emits a watery discharge.

Oral and Nasal Cavities

Within hours of birth, a human infant's mouth becomes colonized with nonpathogenic *Neisseria* species (Gram-negative cocci), *Streptococcus*, *Actinomyces*, *Lactobacillus* (all Gram-positive), and some yeasts. These organisms come from the environment surrounding the newborn, such as the mother's skin and garments. As teeth emerge in the newborn, the anaerobic nature of the space between teeth and gums supports the growth of anaerobes such as *Prevotella* and *Fusobacterium* (Figure 14.3). Whatever the organism, colonizers of the oral cavity must be able to adhere to surfaces such as teeth and gums to avoid mechanical removal and flushing into the acidic stomach. The teeth and gingival crevices are colonized by more than 500 species of bacteria.

Organisms such as *Streptococcus mutans*, which attaches to tooth enamel, and *Streptococcus salivarius*, which binds gingival surfaces, form a glycocalyx, a polysaccharide or peptide polymer secreted by the organism that enables them to firmly adhere to oral surfaces and to each other. They are but two of the microbes that lead to dental plaque formation. The acidic fermentation products of these organisms dissolve tooth enamel by demineralizing them and cause dental caries (tooth decay).

Important microbial habitats of the throat include the **nasopharynx**, which is the area leading from the nose to the oral cavity, and the **oropharynx**, which lies between the soft palate and the upper edge of the epiglottis (see Figure 14.3A). Organisms such as *Staphylococcus aureus* and *S. epidermidis* populate these sites. The nasopharynx and oropharynx can also harbor relatively harmless streptococci such as *Streptococcus salivarius* and *S. oralis*, as well as *S. mutans*. Other oropharyngeal organisms include a large number of diphtheroids and the small Gram-negative rod *Moraxella catarrhalis*. Within the tonsillar crypts (small pits along the tonsil surface) lie anaerobic species of *Prevotella*, *Porphyromonas*, and *Fusobacterium*.

Even though oral microbiota are normally harmless, they do hold potential for disease. Dental procedures, for instance, can cause these organisms to enter the bloodstream, transiently producing

Figure 14.4   Gross Pathology of Subacute Bacterial Endocarditis Involving the Mitral Valve

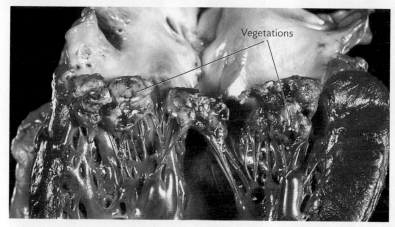

Vegetations

The left ventricle of the heart has been opened to show mitral valve fibrin vegetations due to infection. These growths are not present in a normal heart.

what is called **bacteremia** (presence of bacteria in the bloodstream, Section 21.3). Normal immune mechanisms typically clear these transient bacteremias quite easily, but in patients who have a heart mitral valve prolapse (heart murmur), microbes can occasionally become trapped in the defective valve and form bacterial vegetations. Vegetations are formed by a large number of bacterial cells encased within glycocalyx secreted by the organism (as a biofilm) and fibrin, produced by clotting blood. Because the onset of disease is often insidious (slow), it is called subacute bacterial endocarditis (Figure 14.4). Once buried deep within a vegetation, the microbes are extremely difficult to kill with antibiotics.

The Respiratory Tract

The respiratory tract has a surface area exposed to air of about 246 square feet. Originally thought to be sterile, the trachea and even deep areas of the lung are now known to contain low numbers of normal microbiota. Numbers range from approximately 10 to 100 bacteria per 1,000 human cells. The predominant species are from the Bacteroidetes and Firmicute families.

The upper respiratory tract, of course, contains numerous microbes, but many organisms entering the nasopharynx are trapped in the nose by cilia that beat toward the pharynx. The ultimate destination of these microbial interlopers is the acidic stomach and death. Microorganisms that make it into the trachea are trapped by mucus produced by ciliated epithelial cells that line the airways leading to the lungs, and the cilia themselves usher the microbes up and away from the lungs. The ciliated mucous lining of the trachea, bronchi, and bronchioles makes up the **mucociliary escalator** (Figure 14.5). The mucociliary escalator constantly sweeps foreign particles up and out of the lungs. This is extremely important for preventing respiratory infections. When the mucociliary escalator fails or is overwhelmed by inhaling too many infectious microbes, infections such as the common cold (for example,

rhinovirus) or pneumonia (for example, *Streptococcus pneumoniae*) can result.

As with the gut microbiome, the lung microbiome can contribute to the development of lung mucosal immunity. However, the lung microbiome can also add to the severity of chronic airway diseases such as COPD. Lungs damaged by disease (for example, by cystic fibrosis or emphysema) are populated by a number of different bacterial species that can exacerbate the disease state. Many of the organisms found are potentially pathogenic (*Streptococcus pneumoniae*, *Pseudomonas aeruginosa*, *Haemophilus influenzae*, and *Klebsiella pneumoniae*, for example), but others do not typically cause respiratory infections (for example, *Staphylococcus epidermidis*, some species of *Streptococcus*, *Corynebacterium* species, and some fungi, such as *Candida* species). Nevertheless, these organisms can form biofilm communities in the bronchioles of damaged lungs and interfere with lung function.

The Stomach

We have known for more than a century that the stomach contents are acidic and that gastric acidity can kill bacteria. Just how important that acidity is for protection against microbes is illustrated by the infection caused by *Vibrio cholerae*, the causative agent of cholera. Cholera is a severe diarrheal disease endemic to many of the poorer countries of the world. The toxin produced by the organism acts on the intestinal lining to cause voluminous diarrhea—as much as 10 liters a day (as described in the case history that opens Chapter 4). Death can occur rapidly as a result of dehydration and shock. Although cholera actually affects the intestines, not the stomach, the bacteria must survive passage through the stomach to reach the intestines.

Figure 14.5   Lung Structure and Mucociliary Escalator

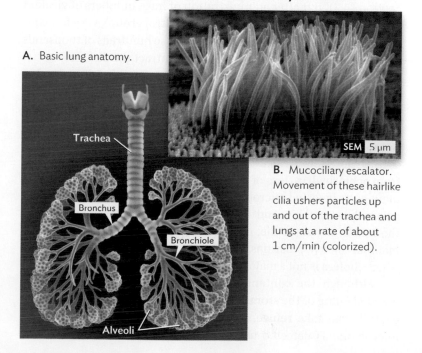

A. Basic lung anatomy.

Trachea

Bronchus

Bronchiole

Alveoli

SEM 5 μm

B. Mucociliary escalator. Movement of these hairlike cilia ushers particles up and out of the trachea and lungs at a rate of about 1 cm/min (colorized).

Figure 14.6   *Helicobacter pylori* **Defies Stomach Acidity**

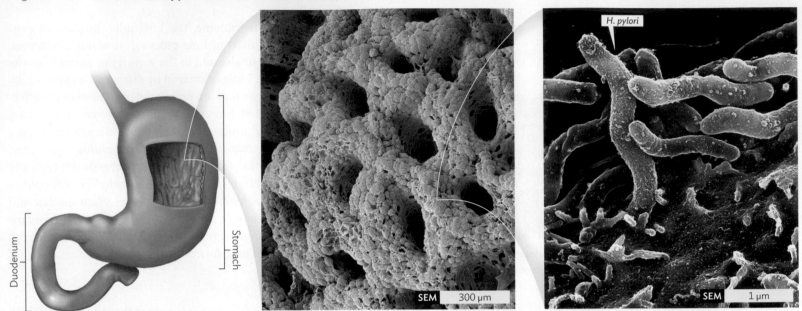

This figure shows the stomach at increasing levels of magnification to reveal *Helicobacter pylori* growing in the mucus lining the stomach epithelium. *Helicobacter pylori* bacteria attach to gastric epithelial cells and induce specific changes in cellular function, such as an increase in the expression of laminin receptor 1, a protein associated with the development of stomach cancer.

In the middle of the last century, the U.S. government was concerned that troops dispatched to endemic countries could develop cholera and become incapacitated in large numbers, so the U.S. Army decided to test potential cholera vaccines for their efficacy. The vaccines were living cultures of genetically weakened cholera bacteria that could not cause serious disease. Living strains were preferred because they were expected to grow in the gastrointestinal tract and stimulate natural mucosal immunity (discussed shortly and in Section 16.3). But after administration of huge numbers of virulent organisms to healthy volunteers (as a control group), very few contracted cholera. This was confusing because hundreds of thousands of malnourished people in India easily contract the disease.

Why did volunteers given live cholera generally fail to develop the disease? Variations in resistance to cholera were discovered to be related to differences in stomach acidity. The organism *V. cholerae* is extremely sensitive to low pH; even a pH of 4 will readily kill it. Because the pH in the stomachs of healthy volunteers was well below 4, it easily killed *V. cholerae*. Malnourished people, however, suffer from achlorhydria (loss of stomach acid). This permissive environment gives ingested microbes time to enter the less acidic intestine, where they can thrive and cause disease. Cholera epidemics in which thousands of people die, even today, typically occur among poor, malnourished populations. Because the U.S. populace is better nourished, cholera is not a major problem in the United States.

Although the contents of the stomach are very acidic, the mucous lining of the stomach is much less so. It is there that some bacteria can take refuge. In fact, the stomach harbors a diverse microbiota, as detected using cultural and molecular techniques

such as the polymerase chain reaction (PCR) to detect 16S rDNA (see Section 10.1). A classic stomach pathogen is *Helicobacter pylori* (see Section 10.3). Although this organism has a remarkable ability to resist acidic pH (it survives at pH 1 using the enzyme urease to generate ammonia), it will not grow at that pH. It can, however, grow and divide in the mucous lining of the stomach, where the pH is closer to 5 or 6 (**Figure 14.6**). The U.S. Centers for Disease Control and Prevention (CDC) estimates that *Helicobacter* colonizes the stomachs of half the world's population. Most of the time, *Helicobacter* does not cause any apparent problem, but on occasion, the organism can produce gastric ulcers and even cancer.

The Human Intestine

CASE HISTORY 14.1

Case of the Greasy Stool

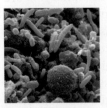

Renita, a 67-year-old woman, complained to her physician of having persistent diarrhea with weight loss, bloating, and excess flatulence (passing gas). Renita also said her stools looked greasy, which the physician knew as a sign that ingested fat was not being absorbed. A blood test revealed that the woman's vitamin B_{12} level was abnormally low. Cultures of fecal specimens failed to find any typical bacterial pathogens that could cause diarrhea. However, the decreased vitamin B_{12} level and excess fat in the stool suggested that the woman had a malabsorption disease. The physician, initially perplexed, suddenly realized that the cause could be small intestine bacterial overgrowth disease, a diagnosis confirmed by aspirating fluid from the jejunum (small intestine). The laboratory discovered abnormally high numbers of facultative and anaerobic bacteria in the fluid, including *Streptococcus*, *Escherichia coli*, *Lactobacillus*, and *Bacteroides*. Fermentation by these organisms produced the bloating and flatulence. Antibiotic treatment corrected the syndrome.

This case illustrates the remarkable balance that must be maintained between a host and its microbiota. Bacteria considered healthy members of one part of the intestine, the colon, can become irksome if they begin to populate another area, the jejunum. How does the healthy body avoid this?

The gastrointestinal tract beyond the stomach consists of an extremely long tube made of several sections, each of which provides a unique environment that supports the growth of different bacterial species. Pancreatic secretions, including bile (at pH 10), enter the intestine at a point just past the stomach and raise the intestinal pH to about pH 8, which immediately relieves the acid stress placed on microbes. The relatively high pH and bile content of the duodenum and jejunum (the section of small intestine just past the duodenum; Figure 14.7) allow colonization by only a few resident, mostly Gram-positive, bacterial genera (enterococci, lactobacilli, and diphtheroids). Gram-positive organisms are, in general, very sensitive to bile salts because they lack an outer membrane. However, the specific Gram-positive species that live in the

Figure 14.7   The Human Gastrointestinal Tract

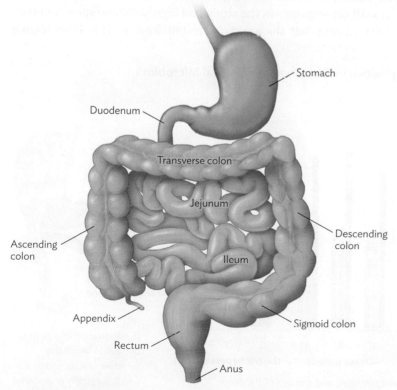

In addition to the structures shown here, the liver and gallbladder are considered parts of the gastrointestinal system.

Figure 14.8   Human Feces Showing Food Particles and Resident Bacteria (Colorized)

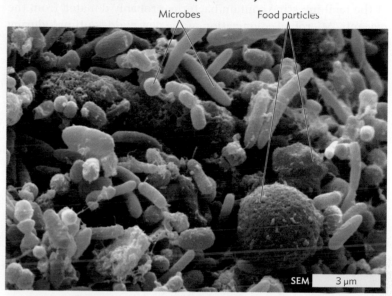

small intestine possess a bile salt hydrolase that destroys bile entering the cell.

There are only an estimated 10^7 bacteria per milliliter of jejunum aspirate. The problem in Case History 14.1 was that the bile content was decreased in the patient, allowing more bacteria to grow. Because there were so many bacteria, nutrients in the small intestine were consumed faster by bacteria than by the host, which caused the malabsorption symptoms. For instance, not enough vitamin B_{12} remained to be absorbed, hence the low blood level of B_{12}. Because bile also solubilizes ingested fat so it can be absorbed, the decreased bile content led to increased fat content in Renita's stools and the greasy appearance.

In contrast to the jejunum, the distal parts of the human intestine (ileum and colon) have a slightly acidic pH (pH 5–7) and less bile salts, conditions that support a more vibrant ecosystem. The intestine actually contains roughly 10^{11}–10^{13} bacteria per gram of feces (Figure 14.8). This generally anaerobic environment is populated by both anaerobic and facultative microbes in a ratio of 1,000 anaerobes to 1 facultative organism. Why is the intestine anaerobic? The small amount of oxygen that diffuses from the intestinal wall into the lumen is immediately consumed by the facultative microbes, rendering the environment anaerobic.

One anaerobe, *Bacteroides thetaiotaomicron*, provides us with a great benefit. Much of this organism's genome is dedicated to taking many of the complex carbohydrates we eat and breaking them down into products that can be absorbed by the body. We humans actually absorb 15%–20% of our daily caloric intake in this way. The gut microbiota perform a number of useful functions, such as fermenting unused energy substrates, training the immune system, preventing growth of pathogenic bacteria, regulating the development of the gut, producing vitamins for the host (such as biotin and vitamin K), and producing hormones to direct the host to store fats.

WHICH BACTERIA COLONIZE OUR INTESTINE?   The fetal intestine was thought to be sterile, but we now find that contents of the fetal intestine contain bacteria, probably donated from the placenta. Microbial colonization of the new child continues during birth and is influenced by the mode of delivery (birth canal versus cesarean) and type of feeding (formula versus breast milk). Formula-fed babies are initially colonized by a mixture of Enterobacteriaceae, streptococci, and staphylococci. The microbiota of breast-fed babies, however, consists primarily of *Bifidobacterium* species that easily digest human milk oligosaccharides (lactose with long side chains). One reason some people promote breast-feeding over bottle-feeding is that the metabolic products of *Bifidobacterium* have been shown to protect against infection by some intestinal pathogens. But there is much we do not know about colonization of the newborn human.

Only recently have we really started to learn what microbes inhabit our intestine (mainly because we cannot culture most of them). The use of culture-independent techniques such as PCR, which can identify nonculturable species through differences in their 16S RNA genes, reveals the identity, diversity, and variation among individual members of the human microbiome. For example, of 395 unique microbial "phylotypes" identified in humans, 60% were unknown at the species level, 80% have never been cultured, and huge differences in the populations of microbes were observed between individuals. (Phylotypes are organisms that can be assigned to different phyla; see Section 10.1).

All resident organisms normally present in the intestine are considered normal microbiota and are usually harmless if they stay where they belong. But if they accidentally escape into nearby tissues, they can cause serious disease (for instance, appendicitis). Intestinal bacteria have special talents that allow them to bind and colonize the intestinal wall. One reason why such a diverse mix of species is supported in the intestine is that different bacteria attach to different host cell receptors. The many different food sources within the intestine support diverse groups of microbes as well.

Organisms that inhabit the intestine include the family Enterobacteriaceae, of which *E. coli* is a member, and many anaerobes, such as *Bacteroides* (Gram-negative rods), *Peptostreptococcus* (Gram-positive cocci), and *Clostridium* (Gram-positive, spore-forming rods). In fact, the vast majority of our intestinal flora consists of the phyla Bacteroidetes and Firmicutes, of which *Bacteroides* species and *Clostridium* species, respectively, are members. Besides bacteria, other common inhabitants are the yeast *Candida albicans* and protozoa such as *Trichomonas hominis* and *Entamoeba hartmanni*.

The large intestine is both rich and diverse in terms of nutrient availability. The majority of intestinal bacteria require a fermentable carbohydrate for growth, so it has generally been assumed that carbohydrate metabolism is necessary for colonization by most species. In the large intestine, carbohydrates are derived from digested food and from host secretions, such as mucus. Mucus is a complex gel of sugar-coated proteins and lipids called glycoproteins and glycolipids. Contrary to what you might expect, *E. coli* in the intestine grows well on nutrients in mucus but very poorly in digested food contents. Thus, in the intestine, at least, *E. coli* doesn't grow on what we <u>eat</u> but on what we <u>secrete</u>.

Despite this buffet-like assortment of food, many species still end up vying for the same nutrient, so competition for food is fierce. Perturbing the balance of power with antibiotics, diet, or stress can lead to disease.

IS THERE A CONNECTION BETWEEN OUR MICROBIOTA AND OBESITY?   Obesity is a growing public health concern influenced by the increased consumption of high-energy foods and decreased physical activity. However, recent evidence suggests that gut microbiota may also play a role by increasing energy-harvesting efficiency. One study supporting this hypothesis used high-throughput DNA sequencing to compare the composition of microbiota in normal-weight, obese, and gastric bypass patients. Studies that use DNA sequencing to identify all the microbial species within a given environment are called **metagenomic studies**.

All three groups in the study had highly diverse species of bacteria present, but the archaea present were mainly H_2-oxidizing

Figure 14.9   Relationship between Obesity and Intestinal Microbiota

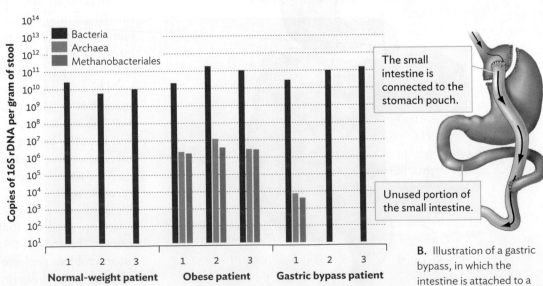

A. The relative numbers of bacteria, archaea, and archaea of the order Methanobacteriales in different normal-weight, obese, and gastric bypass patients measured as ribosomal RNA genes quantified by real-time PCR.

B. Illustration of a gastric bypass, in which the intestine is attached to a small, surgically engineered stomach pouch.

methanogens (order Methanobacteriales). The surprise was that obese individuals had much higher numbers of the H_2-utilizing **methanogenic archaea** than did normal-weight or gastric bypass individuals (**Figure 14.9**).

Link Recall from Section 7.1 that **methanogenic archaea** use the process of methanogenesis to convert hydrogen and CO_2 to methane and water.

The microbial collaboration leading to weight gain is thought to work as outlined in **Figure 14.10**. H_2 buildup resulting from bacterial fermentation normally feedback-inhibits fermentation, possibly by inhibiting NADH dehydrogenase. NADH dehydrogenase regenerates NAD+ needed for fermentation (discussed in Chapter 7). Oxidation of H_2 by methanogens would relieve that block and allow more efficient regeneration of NAD+. More efficient regeneration of NAD+ would allow bacteria to ferment more plant polysaccharides and produce more fermentation end products such as acetate and butyrate, which are readily utilized by human cells. The result is weight gain. Consequently, the weight loss following gastric bypass may reflect the double impact of altering the gut microbiota as well as decreasing food ingestion. Another intriguing study suggests a link between the intestinal microbiome, obesity, and diabetes. Might we one day treat these chronic diseases by altering our microbiota?

PROBIOTICS As already noted, the microbial composition of the intestine is complex but balanced. It is complex because there are many species residing there. It is balanced because these species manage to coexist without eliminating each other. The rival organisms help prevent infection by competing with pathogenic species for nutrients and can actually contribute to nutrition (for example, *E. coli* makes vitamin B_{12}, whereas other organisms help digest complex substrates). Emotional stress, a change in diet, or antibiotic therapy can throw off that balance and lead to poor digestion or disease, such as pseudomembranous enterocolitis (**Figure 14.11**). Some people favor restoring the natural microbial balance by orally ingesting living microbes such as the lactobacilli present in yogurt. Taking these supplements, called **probiotics** (from the Greek meaning "for life"), is thought to restore balance to the microbial community and return the host to good health. The most commonly used probiotic genera are *Lactobacillus* and *Bifidobacterium* (**Figure 14.12**).

The potential mechanisms by which resident microbiota improve intestinal health include competitive bacterial interactions (normal microbiota prevent growth of pathogenic bacteria), production of antimicrobial compounds, and immune modulation (an ability to induce, enhance, or suppress an immune response). The emerging use of probiotics in several gastrointestinal disorders (including inflammatory bowel disease and IBS) has led to increased interest in their use. Early results are promising, but much of the research remains controversial. An extremely exciting development is fecal transplants, in which the intestinal microbiome of a healthy person is transferred to a person with severe intestinal disease such as colitis. Fecal transplant is discussed in Section 14.5.

Figure 14.10 Model of Metabolic Collaboration
The presence of H_2-oxidizing methanogens in the human intestine removes H_2 produced by bacterial fermentation. The reduction of H_2 levels is thought to relieve H_2 end product inhibition of bacterial fermentation (possibly via NADH dehydrogenase). Enhanced fermentation produces excess short-chain fatty acids that are absorbed into the host's intestinal epithelial cells and metabolized as food.

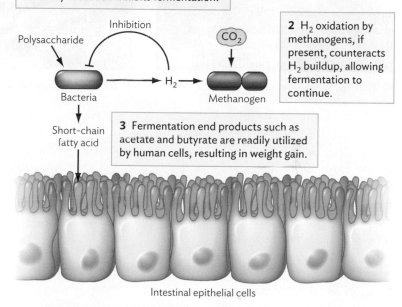

1 H_2 buildup from bacterial fermentation normally feedback-inhibits fermentation.

2 H_2 oxidation by methanogens, if present, counteracts H_2 buildup, allowing fermentation to continue.

Polysaccharide · Inhibition · CO_2 · H_2 · Bacteria · Methanogen

Short-chain fatty acid

3 Fermentation end products such as acetate and butyrate are readily utilized by human cells, resulting in weight gain.

Intestinal epithelial cells

Figure 14.11 Pseudomembranous Enterocolitis Caused by *Clostridioides difficile*
On the left is a normal colon as viewed by colonoscopy. On the right is the colon of a person with pseudomembranous enterocolitis. *C. difficile* is often a part of the normal intestinal microflora, but antibiotic treatment, especially following surgery, can kill off competing microbes, leaving *C. difficile* to grow unabated. A toxin produced by the organism growing at the epithelial surface damages and kills host cells, causing exudative plaques to form on the intestinal wall. The small plaques eventually coalesce to form a large pseudomembrane that can slough off into the intestinal contents.

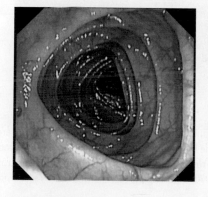
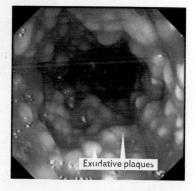

Exudative plaques

Figure 14.12   Commonly Used Probiotic Microorganisms

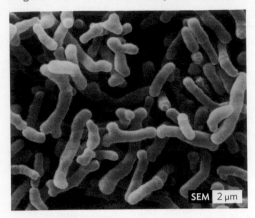

A. *Bifidobacterium* (colorized); note the Y shape of some cells.

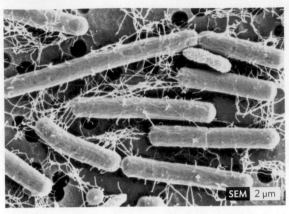

B. *Lactobacillus acidophilus* (colorized).

Genitourinary Tract

Much of the genitourinary tract is normally free from microbes (Figure 14.13). These areas include the kidneys, which remove waste products from the blood, and the ureters, which remove the urine from the kidneys. The urinary bladder, which holds urine until it is excreted, was thought to be sterile but is now known to harbor small numbers of microbes (mainly anaerobes). The distal urethra, because of its proximity to the outside world, normally contains *Staphylococcus epidermidis*, *Enterococcus* species, and some members of Enterobacteriaceae. Some of these organisms can cause urinary tract infections (UTIs) such as cystitis if they can make their way into the bladder (for example, via catheterization).

The large surface area and associated secretions of the female genital tract make it a rich environment for microbes. Composition of the vaginal microflora changes with the menstrual cycle, owing to changing nutrients and pH, and with the onset of menopause. The mildly acidic nature of vaginal secretions (approximately pH 4.5) discourages the growth of many microbes. As a result, the acid-tolerant *Lactobacillus crispatus* is among the most populous vaginal species. Healthy women appear to fall into two broad categories: 70% have lactobacilli as the primary type of vaginal flora, whereas 30% have mixed species with few lactobacilli. Women in the latter group appear to be more susceptible to sexually transmitted diseases.

The balance between competing species of the vaginal microbiota is crucial to preventing infection. Antibiotic therapy to treat an infection anywhere in the body can also affect the vaginal microbiota. The resulting imbalance can allow infection by *C. albicans* (yeast infection), which is not susceptible to antibiotics designed to kill bacteria.

 Jack Ravel; the vaginal microbiome: youtube.com

Figure 14.13   The Urogenital Tract in Women

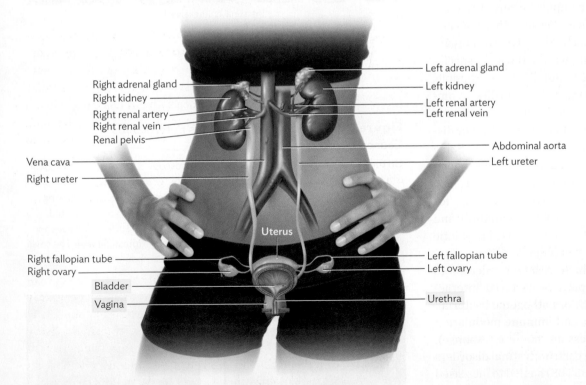

Blood in renal arteries carries waste products into the kidney. The waste products are removed by filtration and passed down the ureters as urine into the bladder. Urine is expelled through the urethra. The figure also shows the location of the adrenal glands, fallopian tubes, and uterus. Areas colonized by microbiota are the urinary bladder, vagina, and urethra (labels are shaded). All other organs represented are sterile.

SECTION SUMMARY

- The **normal microbiota** present on body surfaces constantly changes.

- **Normal microbiota can cause disease** if microbes breach the surfaces they colonize and gain access to deeper tissues or the circulation.

- **Colonization of skin** by microbes is difficult owing to surface dryness, an acidic pH, high salinity, and the presence of degradative enzymes.

- **Skin flora** consists primarily of Gram-positive microbes, including *Propionibacterium acnes*, which can cause acne.

- **Microbe-free areas of the body** are cerebrospinal fluid, the urinary tract excluding the bladder and urethra, and possibly deep areas of the lungs. The eyes are generally microbe-free, although some microbes can be present for short periods.

- **Oral and nasal surfaces** are colonized by aerobic and anaerobic microbes.

- **The large intestine** is populated by 10^9–10^{11} microbes per gram of feces in a ratio of approximately 1,000 anaerobes to 1 aerobe.

- **Collaboration between bacterial and archaeal species** in the intestine can form new, more efficient pathways that degrade food.

Thought Question 14.1 How can an anaerobic microorganism grow on skin or in the mouth, both of which are exposed to air?

Thought Question 14.2 What can an immunocompromised person with mitral valve prolapse do to prevent formation of subacute bacterial endocarditis when visiting the dentist?

Thought Question 14.3 Two men, both 55, ate raw oysters at a popular oyster bar. Two days later, one man is hospitalized with a life-threatening blood infection (septicemia) caused by the Gram-negative curved rod *Vibrio vulnificus*. The other man suffers no ill effects. The only difference between the two men is that the one infected was taking an over-the-counter proton pump inhibitor to reduce his stomach acidity and treat his persistent indigestion. What does this tell you about the infecting organism?

14.2
Benefits and Risks of a Microbiome

SECTION OBJECTIVES

- Describe ways that microbiota benefit the health of their host.
- Explain the risks of having a microbiome, especially to immunocompromised hosts.
- Discuss how host-microbiota relationships are studied in the laboratory.
- Describe how the compositional balance of our microbiome can be compromised and the potential effects on the individual.

Should we view colonizing bacteria as hostile armies encamped at our body's gates (also called portals of entry), just waiting for the opportunity to invade? The answer, of course, is no. Bacteria do not possess hostile intent, only the need to find food. Because the needs of our normal microbiota are met at the colonizing site, these microbes, for the most part, have not evolved tools to pass beyond their resident niche. In fact, members of our microbiota are of tremendous benefit to us, but, as with any relationship, there is risk.

Benefits of a Microbiome

Members of our microbiota are often beneficial (mutualistic). Normal microbiota help us digest food and synthesize compounds needed by the body (for example, vitamin B_{12}). They also interfere with the colonization of pathogens by competing for attachment receptors on host cells, competing for food sources, and synthesizing antimicrobial compounds such as fermentation end products. For example, the acidic fermentation products produced by lactobacilli in the female genitourinary tract help maintain a low pH. The low pH dissuades colonization by various pathogenic microbes.

Our microbiota also shape and enhance function of the immune system. The constant sampling of microbiota by mucosal cells of the immune system (as is done, for example, in Peyer's patches, discussed in Sections 14.3 and 15.2) keeps the immune system sharp. Bacterial proteins such as catalase can act as immunomodulins. **Immunomodulins** made by normal flora growing on mucosal surfaces modify the secretion of host proteins (such as cytokines and tumor necrosis factor) that influence the immune response. **Cytokines** are small secreted host proteins that bind to various cells of the immune system and regulate their function. (Cytokines will be discussed in Section 15.3 and Chapter 16.)

Some toxic bacterial products are "double-edged swords," having both positive and negative effects. **Enterotoxins**, for example, are proteins produced by some Gram-negative pathogens that damage the small intestine of the host and cause diarrhea. But they may also protect against colorectal cancer by activating membrane calcium channels in intestinal epithelial cells (cells lining the intestine). Increasing calcium transport through the membrane turns on an antiproliferative pathway that dampens host cell division, which provides resistance to colon cancer.

Note Do not confuse the terms "enterotoxin" and "exotoxin." An exotoxin refers to any protein with toxic activity that is secreted from the bacterium. An enterotoxin is an exotoxin that specifically affects the intestinal mucosa.

Emerging infectious diseases have spurred great interest in developing new vaccination strategies that can be developed quickly and deployed rapidly. A recently discovered benefit of commensal organisms has been their development as novel vaccine delivery systems. Select commensal species can be genetically engineered to produce proteins from pathogenic species and display them on their cell surfaces. Colonization of a host by the modified commensal strain will elicit an immune response to the cloned protein and provide immunity to the pathogen. Commensal microbes such as *Streptococcus salivarius* (part of the oral and nasopharyngeal microflora), *Lactococcus* species (residents of the intestinal tract),

Figure 14.14   Gnotobiotic Unit at North Carolina State University

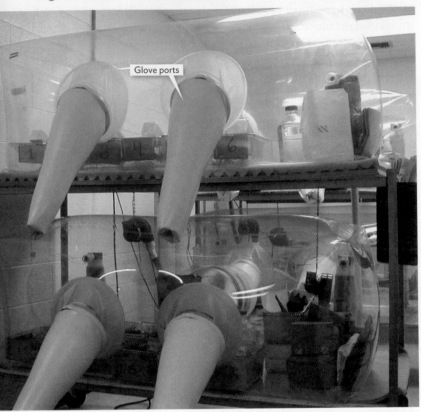

The room contains flexible film isolators for housing rats and mice in axenic, or gnotobiotic, conditions. Another room is used to sterilize supplies needed in the flexible film isolators.

specifically breach human defenses, accidental penetration beyond these sites can cause serious infections. A cancerous lesion in the colon, for example, can provide a passageway for microbes to enter deeper tissues, where they can begin to grow unabated. *Bacteroides fragilis*, a harmless anaerobe in the gut, can invade tissues through surgical wounds, causing abscesses that persist (and even lead to gangrene) after abdominal surgery (**Figure 14.15**).

By and large, the barriers of defense work well to prevent incursions by normal flora or, in the event of a breach, to kill the invader. These defenses sometimes break down in people whose immune systems are compromised by medical treatments (such as anti-cancer drugs) or by disease (such as a deficiency in **complement** factors). Such a person is described as an **immunocompromised host** and can be repeatedly infected by normal microbiota. Organisms causing disease in this situation are called **opportunistic pathogens**.

Link As described in Section 15.7, **complement** is a series of serum factors that work together to make holes in bacterial membranes. Some complement factors also attract phagocytic white cells to a site of infection.

Our intestine, with its diverse microbiota, can also be considered an incubator for antibiotic resistance. The nearly indiscriminate use of antibiotics, as when antibacterial antibiotics are used to treat viral diseases, selects for resident bacteria that spontaneously developed resistance. Antibiotic treatment can completely change the intestinal mix of microbes. Bacteria that develop resistance to an antibiotic can overgrow in the presence of antibiotic

and *Lactobacillus* species (present in the urogenital tract and rectal area) are being exploited in this way.

The benefit that microbiota offer to their hosts is especially apparent in gnotobiotic animals. A **gnotobiotic animal** is an animal that is germ-free or one in which all the microbial species present are known. Developing a gnotobiotic colony of animals involves delivering offspring by cesarean section under aseptic conditions in an isolator (**Figure 14.14**). The newborn is moved to a separate isolator where all entering air, water, and food are sterilized. Once gnotobiotic animals are established, the colony is maintained by normal mating between the members. Germ-free animals, however, often have poorly developed immune systems, lower cardiac output, and thin intestinal walls, and they are more susceptible to infection by pathogens. The lesson from these animals is that the presence of normal microbiota continually challenges the immune system and keeps it active.

Risks of a Microbiome

Despite the enormous benefits, maintaining resident microbiota is still a risky business. Even though microbes of our normal microbiota, unlike true pathogens, have not evolved mechanisms to

Figure 14.15   Anaerobic Gas Gangrene of the Abdominal Wall Caused by *Bacteroides*

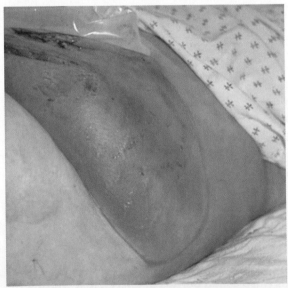

This infection was an unfortunate result of bowel surgery. Organisms escaped from the intestine and initiated infection in the abdominal wall.

and become a more dominant presence once the antibiotic treatment has stopped. Antibiotic resistance genes from these bacteria can then be passed to others, including pathogens, by transformation (see Figure 9.1), by conjugation (if the resistance genes are on a plasmid; see Figure 9.3), or by resident viruses via transduction (see Figure 9.5).

It is also clear that changes in the microbiota can produce some noninfectious diseases (for instance, irritable bowel syndrome and perhaps obesity). However, recent research suggests that what is important are not the members of the microbiota per se, but the presence of a core set of genes among the microbiome that are intrinsic to the host's well-being. An individual's microbiota is in a constant state of flux, like a hotel with guests constantly checking in and checking out. It is the hotel's staff with their individual core skill sets that keep the enterprise going. If the plumber quits the hotel, it is only a matter of time before the plumbing fails. Similarly, if a shift in the microbiome leads to the loss of some core bacterial genes, a human metabolic dysfunction can result. Thus, if our microbiota can be considered an "organ system," it is a potentially fickle one, apt to change with little provocation.

SECTION SUMMARY

- **Benefits of microbiota** include interfering with pathogen colonization, maintenance of our immune system, the stimulation of immunomodulatory protein production, and potential use as vaccine delivery vehicles.

- **Gnotobiotic animals** are germ-free or colonized with a known set of microbes.

- **Risks of normal microbiota** include accidental infection, especially in immunocompromised hosts, the potential for developing new antibiotic resistance mechanisms, and a susceptibility to some metabolic diseases when the balance of microbial species is disturbed.

Thought Question 14.4   People who take antibiotics for an upper respiratory tract infection often develop antibiotic-associated diarrhea. However, pathogenic microbes are not the usual cause of this diarrhea. Using what you have read here combined with Internet searches, formulate a general hypothesis that might explain this connection. (*Hint:* Look up osmotic diarrhea.)

14.3
Keeping Microbiota in Their Place

SECTION OBJECTIVES

- List and compare the various barriers that separate the microbiome from its host.
- Explain the function of tight junctions in preventing infection.
- Discuss the role of innate immunity in preventing infection by microbiota.

How are members of our microbiota kept from slipping past our own cells and infecting us? In the previous sections, we mentioned several ways in which normal microbiota benefit human health. However, members must be kept in their place, because those that escape can cause disease. Several physical barriers act to prevent escape.

Epithelial Barriers

The major barrier separating us from our microbiota is a layer of glycoprotein-covered epithelial cells lining the mucosa of our digestive, genitourinary, and respiratory tracts. The epithelium of most mucosal surfaces consists of interconnected, polarized epithelial cells whose basal membrane separates them from underlying connective tissue. The epithelial layers lining the mucosa are held together by so-called tight junctions formed between cells (Figure 14.16). The molecular "glue" of tight junctions is a series of molecules connected to the cytoskeleton of epithelial cells. Tight junctions act as a strictly regulated port of entry that opens and closes in response to various signals (for example, small secreted host peptides called cytokines) originating in the epithelium and lamina propria (a layer of connective tissue that lies beneath mucosal epithelium). Tight junctions are key elements controlling intestinal diffusion but also effectively prevent our microbiota from gaining access to deeper tissues.

Figure 14.16   Tight Junctions

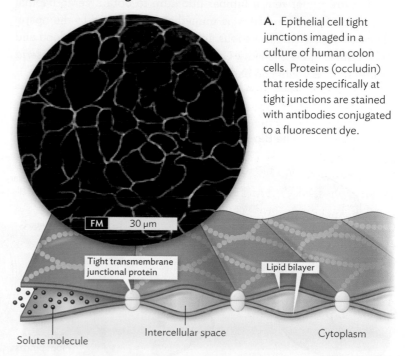

A. Epithelial cell tight junctions imaged in a culture of human colon cells. Proteins (occludin) that reside specifically at tight junctions are stained with antibodies conjugated to a fluorescent dye.

FM   30 μm

Tight transmembrane junctional protein

Lipid bilayer

Solute molecule   Intercellular space   Cytoplasm

B. Schematic representation of two adjoining epithelial cells fused to each other by tight junctions. Transmembrane junctional proteins in the membrane of two adjacent cells are clustered along the points of contact.

So how do bacteria get through the seemingly impervious epithelial barrier? Sometimes microtraumas such as small lesions that damage the epithelial layer will provide a temporary gateway to deeper tissues. Cancerous lesions, for example, can perforate this layer in the intestine; gunshot and knife wounds provide pathways for infection as well. But trauma and cancer are not the only means by which bacteria breach the epithelial barrier. The mucosa is also speckled with small patches of lymphoid tissue associated with the immune response. These areas, such as Peyer's patches in the intestine (see Section 15.2), sample normal bacteria to help maintain a low-level immune response. Normal microbiota may accidentally penetrate these areas. Many pathogens, on the other hand, actively invade these regions. We will discuss how the host copes with these breaches in Chapters 15 and 16.

CASE HISTORY 14.2

The Deadly "Hangover"

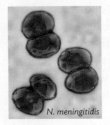
N. meningitidis

Hiroshi, a 19-year-old college student, arrived at the emergency department complaining of a fever and lethargy that started the previous evening. He originally suspected that the symptoms were due to a hangover from a fraternity party held 2 days earlier. However, Hiroshi became alarmed when turning his head became painful and when tiny red blotches suddenly appeared on his skin. The physician recognized this as a **petechial rash** (red spots on the skin that do not blanch, or turn white, when pressed). Blood cultures were obtained, and Hiroshi underwent a lumbar puncture to obtain cerebrospinal fluid (CSF) for culture. He was immediately admitted to the hospital and treated with intravenous antibiotics. The next day, blood and CSF cultures grew a deadly Gram-negative diplococcus—*Neisseria meningitidis*. Hiroshi had bacterial meningitis.

Neisseria meningitidis is a bacterium that can reside in the nasopharynx of some individuals, called carriers, without causing disease. Carriers are immune but can transmit the organism to a susceptible person. Transmission is more frequent among people housed in close quarters, such as on a college campus. *N. meningitidis* will initially colonize the nasopharynx of a newly infected person and cause a sore throat. From there the organism can gain access to the bloodstream. Usually this is where the infection stops as various immune mechanisms engage. However, sometimes the bacterium, while coursing through the bloodstream, can penetrate into the central nervous system through the nearly impervious blood-brain barrier, a brain lining that bars passage of most microbes and molecules. Once this barrier is breached, *N. meningitidis* can cause potentially deadly meningitis.

The Blood-Brain Barrier

"I think, therefore I am." This famous assertion by the fifteenth-century philosopher René Descartes underscores the importance of the human brain. Nature has taken note of the brain's importance by evolving extraordinary measures of protection. The skull, an obvious example, protects the brain from flying fists and falling branches. But there is an equally important, if less apparent, protective structure inside the brain called the **blood-brain barrier**. The blood-brain barrier is comprised of very tight junctions between the endothelial cells of capillaries feeding the brain. The section of the blood-brain barrier along the choroid plexus (called the blood-cerebrospinal fluid barrier) occurs between epithelial cells that surround capillaries (**Figure 14.17A**) and controls the passage of substances from the blood into the cerebrospinal fluid that bathes the brain and spinal cord. Cerebrospinal fluid, made primarily by the choroid plexus, has the composition of blood plasma but normally contains no blood cells—red or white.

The blood-brain barrier is a highly selective permeability barrier that lets essential metabolites, such as oxygen and glucose, pass from the blood to the brain but blocks most molecules with a

Figure 14.17 The Blood-Brain Barrier

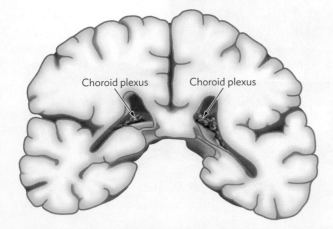

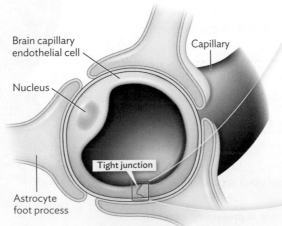

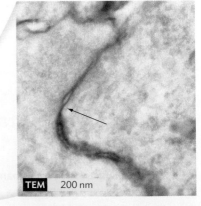

A. The choroid plexus in the brain is composed of a rich capillary bed and forms the blood-cerebrospinal fluid barrier.

B. Diagram of a brain capillary in cross section showing endothelial tight junctions.

C. Electron micrograph of boxed area in (B), showing the appearance of tight junctions between neighboring endothelial cells (arrows).

molecular mass greater than 500 daltons. This means that everything from hormones and neurotransmitters to viruses and bacteria are refused access to brain tissue (unless there is a specific transport system; see Table 14.2). It also means that many antibiotics, which would otherwise be capable of treating infections, cannot pass through the blood-brain barrier.

The blood-brain barrier is composed of thin, flat endothelial cells that form the walls of blood capillaries. In most parts of the body, endothelial cells overlap and form tight junctions. Though considered tight, cell-cell junctions in most of the body are leaky enough to let some materials slip between adjacent endothelial cells and thus through the wall of the blood vessel. Most endothelial cells also have active vesicle transport systems that can carry compounds from one side of the cell to the other. This is how hormones and nutrient molecules get to tissues. Unfortunately, potentially harmful agents, such as toxins, viruses, and the occasional bacterium, can slip through as well.

In brain capillaries, endothelial cells form extremely tight junctions with no breaks for leakage (Figure 14.17B and C). These junctions, which block the passage of most things except for very small molecules, are a crucial component of the blood-brain barrier. Brain endothelial cells also contain few intracellular vesicles, the lack of which adds to the barrier. The blood-brain barrier makes it very difficult for most pathogens to penetrate into the brain and keeps any microbiome organism that accidentally enters the bloodstream from gaining access to the central nervous system. However, some pathogenic bacteria (such as *N. meningitidis*) and viruses (such as eastern equine encephalitis virus) have developed ways to slip though the blood-brain barrier, usually at the choroid plexus, and cause disease, as we saw in Case History 14.2. The mechanisms used by pathogens to penetrate the blood-brain barrier will be revealed in Chapter 18.

The Maternofetal (Placental) Barrier

CASE HISTORY 14.3

Born Infected

L. monocytogenes

A tiny 1-kg (2.2 pound) female neonate (newborn) was born 2 months premature. The baby had extreme difficulty breathing and had to be intubated (a breathing tube inserted). The mother, at the time of admission, had complained of mild diarrhea and abnormal abdominal pain unrelated to her pregnancy. The infectious disease doctor who was called in to consult on the case immediately recognized the likely problem and ordered blood cultures be performed on the infant. The infant was also started on intravenous antibiotics. Two days later, the lab reported finding a Gram-positive bacillus—*Listeria monocytogenes*—in the infant's blood. This same organism was the cause of the mother's diarrhea. The mother had unwittingly ingested some unpasteurized cheese contaminated with this pathogen and developed listeriosis. The organism entered the mother's bloodstream and was transmitted across the placenta to the fetus prior to birth.

Table 14.2
Some Materials That Cross the Blood-Brain Barrier

Freely Diffusible	Requiring Carrier Systems in the Cells
Oxygen	Glucose
Carbon dioxide	Metabolic by-products
Alcohol	Tryptophan, tyrosine (precursors of neurohormones such as serotonin, which does not cross the barrier)
	L-DOPA (a precursor of dopamine, which does not cross barrier)

The human fetus is primarily germ-free (the fetal intestine may be an exception), even though areas of the pregnant mother teem with microorganisms. The placenta plays a major role in keeping a fetus pathogen-free. In mammals, the placenta separates the fetal and maternal circulations and keeps stray bacteria that find their way into the mother's bloodstream away from the fetus. Capillaries of the human placenta are the sites of maternofetal exchange (the area where substances are exchanged between the mother's blood and that of the fetus). The placental membrane (sometimes called the placental barrier) contains branches of fetal capillaries and is in direct contact with maternal blood (Figure 14.18). The placenta enables the transfer of oxygen, carbon dioxide, nutrients, and waste products between the mother and the fetus without maternal and fetal blood ever mixing. The endothelial cells that form fetal capillaries are, as in the case for other capillaries, connected by tight junctions. Very few species of bacteria that find their way into the mother's circulation can pass through the placental membrane and endothelial cells to infect the developing fetus. *Listeria monocytogenes*, however, is one of the few that can. How *Listeria* crosses the placenta will be discussed in Chapter 18.

Innate Immune Mechanisms

Mucosa, epithelial cells, and tight junctions are effective barriers that keep our microbiota in their proper place. However, breaches in these barriers do occur, and when they do, our immune system is poised to keep the bacterial threats at bay. As the next two chapters will discuss, there are two basic types of immunity. The first is nonspecific, innate immunity, in which host cells that form mucosal surfaces or circulate in the blood can recognize the presence of a microorganism and respond immediately. The second form is adaptive immunity, which takes longer to develop and includes the production of antibodies and specialized immune cells. Innate mechanisms actually sample the microbiome and then modulate the host's immune response to limit the numbers and diversity of microbes at the sites of colonization.

How do cells of the innate immune system recognize microbiota? These cells possess sensitive membrane and cytoplasmic

Figure 14.18   The Maternofetal Barrier

The placental barrier separates the maternal and fetal circulations.

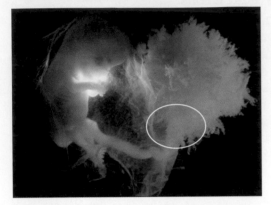

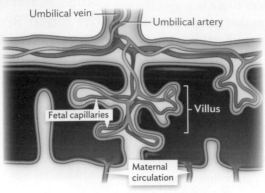

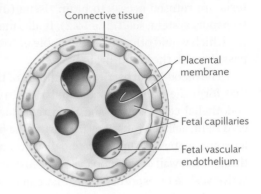

A. Fetus attached to the placenta. The circle indicates the location of the maternofetal barrier.

B. Overview of the maternal and fetal circulations at the placenta. Maternal blood enters the intervillous spaces that surround the villi containing the fetal capillaries. Fetal capillaries merge to form the umbilical veins and arteries that transport oxygen and nutrients to the fetus and carbon dioxide and wastes away from the fetus.

C. Cross section of the villus, showing the placental membrane and fetal capillaries. The cells in maternal and fetal blood never mix.

Figure 14.19   Toll-Like Receptors and Cytokines Help Keep Normal Microbiota in Balance

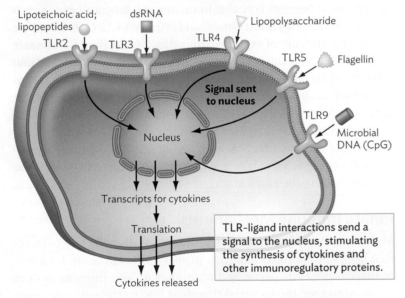

Different pieces of microbes, such as peptidoglycan, lipopeptides, lipopolysaccharides, flagellin, DNA (bacteria), and dsRNA (viruses), are called microbe-associated molecular patterns (MAMPs). MAMPs bind to different host cell pattern recognition receptors (PRRs), such as membrane-associated Toll-like receptors or intracytoplasmic NOD-like receptors. A PRR-MAMP interaction triggers a signal transduction pathway that stimulates the production of cytokines that either promote or inhibit inflammation (depending on the cytokine) or that stimulate antibody production. These processes control microbial proliferation at colonized body sites (the intestine, for instance) and help shape the diversity of our microbiota.

receptors, called pattern recognition receptors (PRRs), that bind various microbe-associated molecular patterns (MAMPs)—molecules such as peptidoglycan or lipopolysaccharide. PRRs include Toll-like receptors (TLRs) embedded in cell or vesicle membranes (**Figure 14.19**) and NOD-like receptors (NLRs) located in the cytoplasm. When bound to MAMPs, the PRRs direct host cells to synthesize small proteins called cytokines that attract phagocytic white cells to the area as well as stimulate the production of mucosa-specific antibodies. The cytokine-stimulated immune response then suppresses the growth of microbiota at mucosal surfaces. TLRs, NLRs, and other aspects of innate immunity will be described in Chapter 15.

SECTION SUMMARY

- **Tight junctions between epithelial cells** prevent bacteria from breaching the mucosal barriers that form the microbiome-human interface.
- **The blood-brain barrier** consists of extremely tight junctions between endothelial cells. Pathogens that cause meningitis possess special mechanisms enabling them to cross this barrier.
- **The placental membrane** forms a maternofetal barrier that helps protect the developing fetus from infection.
- **Innate immune cells possessing Toll-like or NOD-like receptors** help control the numbers and composition of our microbiota.

Thought Question 14.5   Why do many Gram-positive microbes that grow on the skin, such as *Staphylococcus epidermidis*, grow poorly or not at all in the gut? (*Hint:* Think about differences between skin and intestinal contents.)

Thought Question 14.6   What caused the petechial rash in Case History 14.2?

Thought Question 14.7 How might normal microbiota escape the intestine and cause disease at other body sites?

14.4
Microbiota-Host Communications

SECTION OBJECTIVES

- Describe mechanisms by which our microbiota communicate with our cells and tissues.
- Explain ways that our cells and tissues communicate with our microbiota.
- Provide examples of the benefits resulting from microbiota-host interactions.
- Argue the case for declaring our microbiome an "organ system" essential to our survival as a species.

Can we "talk" to our microbiota, and, if so, do they "talk back?" In sheer numbers, our microbiota outnumber our own cells 2 to 3 fold. With all the metabolic products produced by both sides of this partnership, surely we sense each other's existence. One might then hypothesize that after thousands of years of coevolution, these chemical exchanges would have become "conversations" in which each side attempts to influence the gene expression of the other to mutual benefit. Recent studies are starting to bear this out.

How the Microbiome Affects the Host

Studies that catalog small molecules present in cells or body fluids (the science of metabolomics) find that mammalians are awash in metabolic products produced by host microbiota. For example, a study with mice has shown that many end products of bacterial metabolism are present in blood (though the blood itself is sterile). Another study with humans discovered several microbial end products in urine. What are these microbial products doing there? Are they influencing gene expression in host cells? The answer appears to be yes.

A study of gnotobiotic mice colonized with only two members of the normal human microbiota, *Eubacteria rectale* (a firmicute bacterium) and *Bacteroides thetaiotaomicron*, resulted in a remarkable discovery. By monitoring the expression of all host cell genes via microarray analysis, researchers found that the presence of both bacteria changed the expression of 508 host genes. Of the 508 host genes affected by co-colonization, 112 were involved with host cell growth and proliferation and 130 were associated with cell death. Many of these genes are critically important in controlling self-renewal of the colonic epithelium. The data suggest that our microbial tenants actively participate in our colonic health and well-being.

However, this same study found that host gene expression did not change in animals colonized by only one of the species. How did

Figure 14.20 **Polysaccharide A (PSA) from *Bacteroides fragilis***

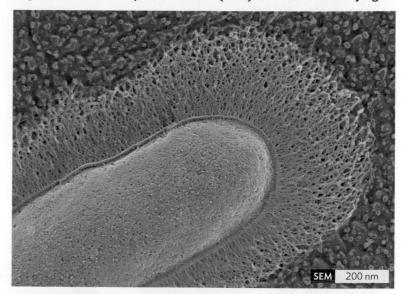

Bacteroides capsule composed of polysaccharide A.

the pair of bacteria alter gene expression in the host, whereas single-species colonization did not? The answer is not complete but is thought to be related to the production of butyrate by *E. rectale* and the increased expression of a host butyrate transporter that occurred only in the co-colonized host. Butyrate is known to profoundly alter the transcription of cultured human epithelial cells and is the preferred energy source for intestinal cells.

Other examples of bacteria influencing host expression are mounting. For example, one report suggests that the microbial product indole (produced when *Escherichia coli* degrades tryptophan) can "tell" host intestinal epithelial cells to tighten tight junctions and increase the production of anti-inflammatory products such as the cytokine interleukin 10 (IL-10). Cytokines are small, secreted host peptides that can be either pro-inflammatory or anti-inflammatory. Another report found that polysaccharide A made by *Bacteroides fragilis* also prevents intestinal inflammation by modulating the production of certain host cytokines (Figure 14.20).

How the Host Affects the Microbiome

Animal studies suggest that perturbations of behavior, such as stress, can change the composition of the microbiota; these changes may foster increased vulnerability to inflammatory stimuli in the gastrointestinal tract. Evidence also suggests that hormones such as epinephrine and norepinephrine, which are released following stress, can influence the composition of the intestinal microbiota. For example, these neurotransmitters have been shown to stimulate the growth of pathogenic and nonpathogenic *E. coli* in vitro and to influence their adherence to the mucosa. More than 50 different types of bacteria occupying major niches of the human body

Figure 14.21   The Spectrum of Known Neuroendocrine-Responsive Bacteria

Respiratory pathogens
Bordetella bronchiseptica
B. pertussis
Pseudomonas aeruginosa
Klebsiella pneumoniae

Cardiac pathogens
Enterococcus faecalis

Enteric microbiota and pathogens
Escherichia coli
Campylobacter jejuni
Proteus mirabilis
Salmonella enterica
Shigella sonnei
S. flexneri
Listeria monocytogenes
Vibrio parahaemolyticus
V. mimicus
V. vulnificus
Yersinia enterocolitica

Oral/peridontal pathogens
Actinomyces
Campylobacter
Capnocytophaga
Eikenella
Eubacterium
Fusobacterium
Leptotrichia
Neisseria
Peptostreptococcus
Prevotella
Streptococcus

Skin microbiota and pathogens
Staphylococcus epidermidis
S. capitis
S. saprophyticus
S. haemolyticus
S. hominis
S. aureus

have been shown to respond to neuroendocrine hormones (Figure 14.21). Figure 14.22 illustrates the dramatic effect norepinephrine has on biofilm formation by *Staphylococcus epidermidis*.

Microbial endocrinology, as the field is called, has focused mainly on how pathogens respond to target host's hormones. We are just beginning to question how hormones and other host products might influence local ecology of natural microbiota.

A final consideration in our discussion of interspecies interactions involves microbe-to-microbe communications that may take place among different members of the microbiome. Bacteria, in a process called quorum sensing (see Section 6.6), can communicate with each other using a variety of autoinducer molecules that they themselves secrete. These secreted molecules accumulate in the environment to the point where they can enter nearby bacteria of the same or different species. Sensing these chemicals alters the expression of specific genes in the receiving species (Section 9.1). Some studies have examined how quorum sensing shapes the composition of the microbiome and influences the generation of composite pathways in which gene products from different bacterial species form new metabolic systems.

Figure 14.23 summarizes the complex interspecies and interdomain communications that influence the composition and physiology of microbes inhabiting host niches. As tools become more sophisticated, we will gain a better understanding of what makes a healthy microbiome and what can be done to revitalize those that have tilted toward the unhealthy.

Microbiome as Organ System?

Our new understanding of the human microbiome and its importance to human health raises an interesting question: Should the microbiome be considered an organ system of the human body? An organ system can be defined as a relatively independent part of the body that performs one or more specific functions. Does your microbiome fit this description?

To start this line of reasoning, remember that your body has 10 times more microbial cells than human cells and 100 times more microbial genes than human genes. Thus, in terms of sheer numbers of cells and genes, your body is more microbial than animal. Microbiomes are also contained in defined areas and so are relatively independent parts of the body. As far as performing specific functions, microbes in the intestine, for example, contribute to our digestion, stimulate a healthy immune system, and promote human tissue development.

Figure 14.22   Effect of Norepinephrine on *Staphylococcus epidermidis* Biofilm Formation
Bacteria were added to slides and grown for 24 hours.

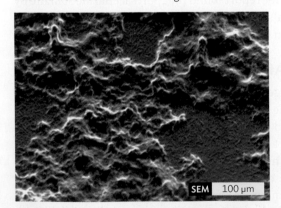

A. No norepinephrine; the control.

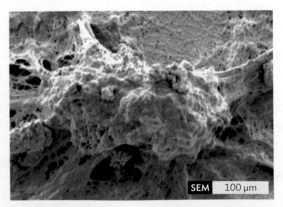

B. 0.1 mM norepinephrine. The slide prepared in the presence of hormone showed extensive biofilm formation and considerably more exopolysaccharide coat than the slide prepared without hormone.

Figure 14.23 Interspecies and Interdomain Influences on Gene Expression

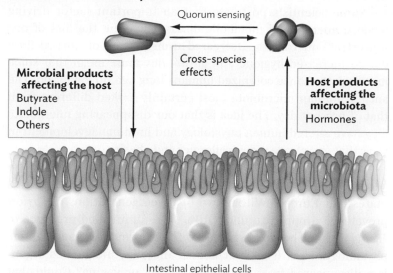

Intestinal epithelial cells

Communication at a colonized site can be viewed as a three-way "discussion." First, bacterial species communicate with each other (quorum sensing) to stimulate the production of chemicals or proteins that enhance survival. Second, some of those microbial products influence host gene expression and function. Finally, the host cells make compounds that affect the growth of microbiota. The result could be regarded as a three-way chemical conversation.

We can also measure the importance of an organ by noting the pathologies that develop when it is defective. Research shows that defects in microbiome composition have dire effects on human health; as already mentioned. For example, inflammatory bowel disease and obesity can result from microbiome imbalances.

Now let's consider inheritance of the microbiome. Genes for your traditional organs were vertically inherited from your parents. Your microbiome was inherited, at least in part, from your mother.

These arguments have led many scientists and health care professionals to now consider the microbiome as a "virtual organ system" that can no longer be ignored when considering the physiology of the human body. Treatments, such as fecal transplants, that target microbiota will become increasingly important tools in the medical toolbox.

 Microbiome as a human organ: onlinelibrary.wiley.com

SECTION SUMMARY

- **Products of the microbiome** can influence host gene expression.
- **Hormones** made by the host can alter the microbiome.
- **Autoinducers** made by different microbial species can influence cross-species gene expression and overall microbiome function via quorum-sensing mechanisms.
- **Many now consider the microbiome to be a virtual organ system.**

14.5 Natural Biological Control of Microbiota

SECTION OBJECTIVES

- Explain the role of probiotics in maintaining a beneficial microbiome.
- Describe the potential benefits of microbiome transplants, such as fecal transplants, in treating certain diseases.
- Discuss the prospects of phage therapy to control infectious diseases.
- Explain the disappearing microbiota hypothesis.

Are the members of our microbiota always one big, happy extended microbial family? Not exactly. As in any large community, conflicts will break out. In fact, pitting microbe against microbe is an effective way to prevent disease in humans and animals. One of the hallmarks of a healthy ecosystem is the presence of a diversity of organisms. This is true not only for tropical rain forests and coral reefs but also for the complex ecosystems of human skin and the intestinal tract. In these environments, the presence of harmless microbial species can retard the growth of undesired pathogens. Naturally occurring staphylococci on human skin, for example, produce short-chain fatty acids that inhibit the growth of pathogenic strains. *Lactobacillus acidophilus* colonizes the vaginal epithelium during childbearing years and establishes the low pH that inhibits the growth of pathogens. Another illustration is the human intestine, which is populated by as many as 500 microbial species. Most of these species are nonpathogenic organisms that exist in symbiosis with their human host. Vigorous competition between these organisms and the production of weak acids by fermentation help control the growth of numerous pathogens.

Applications of Probiotics

Microbial competition has been widely exploited to improve human health in a process known as probiotics. As mentioned in Section 14.1, a probiotic is a food or supplement that contains live microorganisms and improves intestinal microbial balance. Newborn baby chicks, for instance, are fed a microbial cocktail of normal flora designed to quickly colonize the intestinal tract and prevent colonization by *Salmonella*, a frequent contaminant of factory-farmed chicken. In another example, *Lactobacillus* and *Bifidobacterium* have been used to prevent and treat diarrhea in children.

As noted in the chapter-opening case history, yogurt is considered an excellent probiotic because it contains *Lactobacillus acidophilus* and a number of other lactobacilli that are part of a healthy gut microbiome. Yogurt is often useful in the treatment of irritable bowel syndrome and is recommended as a way to restore normal balance to gut microbes following antibiotic treatment.

Phage therapy is another biocontrol method, first described in 1907 by Félix d'Herelle at France's Pasteur Institute, long before antibiotics were discovered. Bacteriophages are viruses that prey on bacteria (see Section 12.1). Each bacterial species is susceptible to a limited number of specific phages. Because the culmination of a phage infection is often bacterial lysis, it was considered feasible to treat infectious diseases with a phage targeted to the pathogen. At one time, doctors used phages as medical treatment for illnesses including cholera and typhoid fever. In some cases, a liquid containing the phage was poured into an open wound. In other cases, phages were given orally, introduced via aerosol, or injected.

Sometimes the phage treatments worked; sometimes they did not. When antibiotics came into the mainstream, phage therapy largely faded. Now that more and more strains of bacteria are becoming resistant to known antibiotics, the idea of phage therapy has enjoyed renewed interest. But you might wonder whether bacteria can also develop resistance to phages. Won't the same resistance problem develop? Although bacteria can develop resistance to the phages that infect them, realize that both bacteria and phages coevolve. Because phages evolve and antibiotics do not, it is much easier to identify a new phage able to infect phage-resistant bacteria than it is to find a new antibiotic able to inhibit the growth of antibiotic-resistant bacteria. The potential of phage therapy has spurred several biotechnology companies to develop effective bacteriophage-based treatments. Several are in various stages of clinical trials.

An extreme form of probiotics is the **fecal transplant** (discussed in Chapter 22), where the intestinal microbiome of a healthy person is transferred to someone with a severe intestinal disease such as colitis. Restoring a "normal" microbiome in this way has successfully "cured" inflammatory bowel disease or ulcerative colitis caused by *Clostridioides difficile* (formerly *Clostridium difficile*) in patients who did not respond to antibiotics. Although the "fecal transplant" has been administered to patients rectally by colonoscopy or orally via endoscopy, experiments are under way to use ingestible capsules to selectively deliver a specific, purified cocktail of healthy microbiome microbes to the patient.

Effects of Disappearing Microbiota

As human health has improved in developing countries, unexplained new diseases have arisen. One hypothesis for the appearance of new diseases has been called the "disappearing microbiota hypothesis," otherwise known as the "hygiene hypothesis." Beginning in the nineteenth century and accelerating until today, dramatic changes in human behavior have influenced the makeup of our microbiota (Table 14.3). For example, the increased use of clean drinking water has interrupted the natural transmission of microorganisms to and between humans. Antibiotic use clears the body of much of its normal (as well as pathogenic) microbial species, leaving ecological voids that may be filled by species not as beneficial as those displaced.

Some scientists postulate that an important factor driving modern allergic and metabolic diseases may be the loss of our "ancestral" microbiota. Decreased transmission of normal flora due to increased hygiene and antibiotics must mean that some microbes that once colonized humans long ago no longer do. The ancient human microbiota most certainly looked different from that of present day. The idea is that our disappearing microbiota may have altered human physiology and immune development in ways that encourage today's allergies and obesity.

To give one example, the Gram-negative bacterium *Helicobacter pylori* at one time colonized 80% of humans in the United States and Europe. Within two or three generations, *H. pylori* prevalence dropped to under 10%. The diminished presence of *H. pylori* has altered gastric secretory, hormonal, and immune physiology. What if a lesser-known ancestral indigenous organism disappeared from the colon, mouth, or vagina? Could that change have contributed to some of the diseases (for example, diabetes, asthma, obesity) that are becoming more prevalent? The hope is that the Human Microbiome Project will begin to answer these and other questions surrounding the relationship between humans and their microbes.

Table 14.3
Changes in Human Behavior That Can Alter the Composition of Microbiota

Change	Consequence
Clean water	Reduced fecal transmission of bacteria
Cesarean deliveries	Reduced vaginal transmission
Use of preterm antibiotics	Reduced vaginal transmission
Reduced breast-feeding	Reduced cutaneous transmission and altered immunological environment
Smaller family size	Reduced early-life transmission
Widespread antibiotic use	Accelerates change of microbiome composition
Increased bathing/showering and use of antibacterial soaps	Accelerates change of skin microbiome composition
Use of mercury amalgam dental fillings	Accelerates change of oral cavity microbiome composition

SECTION SUMMARY

- **Biocontrol** is the use of one microbe to control the growth of another.
- **Probiotics** contain certain microbes that, when ingested, help restore balance to intestinal flora.
- **Phage therapy** offers a possible alternative to antibiotics in the face of rising antibiotic resistance.
- **The increase in diseases such as allergies** over the past century may be caused by changes in the human microbiome.

Thought Question 14.8 Your dorm mate, Teresa, has confided to you that she developed diarrhea after taking antibiotics for a few days to treat an upper respiratory tract infection. She still has a week to go on the antibiotics. To try to cure her diarrhea, she started eating active-culture yogurt, which in the past has helped her "cure" occasional episodes of diarrhea. Tell her why she has diarrhea and why the yogurt is not working this time.

Perspective

As we learn more and more about the human microbiome and its positive impact on human health and development, many have come to think of it as a new kind of organ system, albeit a constantly changing one. It has even been proposed that our adaptive immune system evolved, in part, to accommodate our microbiome. As a result, we reap the benefits of these microbes while keeping them from invading and killing us. One of the puzzles we will address in later chapters is how the intestine simultaneously accommodates the indigenous microbiome, maintains an intact mucosal barrier, and distinguishes symbiotic or mutualistic species from dangerous pathogens.

LEARNING OUTCOMES AND ASSESSMENT

14.1 Human Microbiota: Location and Shifting Composition

- List body compartments that are sterile and discuss why they are sterile.
- List body compartments that are populated by microbiota, and name the key species found in each.
- Identify key features of populated body sites that influence microbiota composition.

1. Which of the following body sites is considered sterile?
   - A. Urethra
   - B. Kidney
   - C. Small intestine
   - D. Stomach
   - E. Tonsils

14.2 Benefits and Risks of a Microbiome

- Describe ways that microbiota benefit the health of their host.
- Explain the risks of having a microbiome, especially to immunocompromised hosts.
- Discuss how host-microbiota relationships are studied in the laboratory.
- Describe how the compositional balance of our microbiome can be compromised and the potential effects on the individual.

5. The benefits of having a microbiome include which of the following?
   - A. The production of cytokines by intestinal microbes
   - B. Interfering with pathogen colonization
   - C. Stimulation of neuron development
   - D. All of the above
   - E. B and C only
   - F. None of the above

14.3 Keeping Microbiota in Their Place

- List and compare the various barriers that separate the microbiome from its host.
- Explain the function of tight junctions in preventing infection.
- Discuss the role of innate immunity in preventing infection by microbiota.

9. Endothelial cells are a major part of which barrier?
   - A. Blood-brain
   - B. Maternofetal
   - C. Peyer's patches
   - D. All of the above
   - E. A and B only
   - F. None of the above

14.4 Microbiota-Host Communications

- Describe mechanisms by which our microbiota communicate with our cells and tissues.
- Explain ways that our cells and tissues communicate with our microbiota.
- Provide examples of the benefits resulting from microbiota-host interactions.
- Argue the case for declaring our microbiome an "organ system" essential to our survival as a species.

12. Normal microbiota can influence host cell gene expression. True or false? _TRUE_

13. Hormones released from host cells can change bacterial gene expression. True or false?

TRUE

14.5 Natural Biological Control of Microbiota

- Explain the role of probiotics in maintaining a beneficial microbiome.
- Describe the potential benefits of microbiome transplants, such as fecal transplants, in treating certain diseases.
- Discuss the prospects of phage therapy to control infectious diseases.
- Explain the disappearing microbiota hypothesis.

16. Which of the following organisms is Y-shaped and considered a probiotic bacterial organism?
   - A. Salmonella enterica
   - B. Bifidobacterium
   - C. Staphylococcus aureus
   - D. Lactobacillus
   - E. Staphylococcus epidermidis

2. *Bifidobacterium* is a common inhabitant of the _____.
   A. skin
   B. oral cavity
   C. lung
   D. intestine
   E. urinary tract

3. The collection of microbes associated with a human body site is referred to as _____.

   microbiome

4. Which of the following factors will influence the types and relative numbers of microbes inhabiting the intestine?
   A. Antibiotic usage
   B. Types of ingested foods
   C. Housemates
   D. All of the above
   E. None of the above

6. An organism that is part of normal microbiota but can cause infection of an immunocompromised host is called a/an _____. *opportunist*

7. A germ-free animal is also called a _____ animal.

   axenic

8. A germ-free animal is characterized by _____.
   A. increased numbers of T cells
   B. diminished brain activity
   C. increased cardiac output
   D. thin intestinal walls
   E. neutrophils that exhibit enhanced phagocytic activity

10. Which of the following is a host structure that recognizes microbial structures?
    A. Tight junction
    B. Toll-like receptor
    C. Lipopolysaccharide
    D. Petechia
    E. Firmicute
    F. All of the above
    G. None of the above

11. Which of the following structures is designed to sample microbiota to maintain a low immune response to microbiota?
    A. Maternofetal barrier
    B. Blood-brain barrier
    C. Peyer's patch
    D. Petichiae
    E. Pattern recognition receptor

14. You perform an experiment using germ-free mice to study host-microbe interactions. Germ-free mice have an abnormally low level of T cells (a type of white blood cell) in their spleens. You find that transferring feces from a normal mouse to the intestines of germ-free mice restores normal T-cell levels over time. Which of the following is the most likely conclusion?
    A. Host cells secrete a factor that permits growth of the fecal transplant organisms.
    B. Donor T cells present in the transplant repopulated the recipient's spleen.
    C. Factors produced by the recipient's original microbiota inhibited production of T cells.
    D. Factors produced by donor microbiota stimulated production of T cells.

15. Our microbiome could be considered an organ system for which of the following reasons?
    A. Microbiota contribute to our caloric intake.
    B. The microbiome is located in specific parts of the body.
    C. Microbiota and human cells communicate with each other.
    D. A person's microbiome can be inherited.
    E. All of the above
    F. None of the above

17. Phage therapy is considered a viable alternative to antibiotics because _____.
    A. antibiotic-resistant bacteria are more susceptible to phage
    B. phages can evolve to infect phage-resistant bacteria
    C. antibiotics will not affect phage-resistant bacteria
    D. All of the above
    E. None of the above

18. Microbiome transplants are performed only between relatives. True or false? *false*

19. The improvement in hygiene practices over the past century has been associated with _____.
    A. increased cases of intestinal infections
    B. increased cases of allergies
    C. decreased life span
    D. an increase in antibiotic-resistant organisms
    E. increased microbiota

Key Terms

bacteremia (p. 435)
bioburden (p. 431)
blood-brain barrier (p. 444)
cytokine (p. 441)
enterotoxin (p. 441)

epidermis (p. 433)
fecal transplant (p. 450)
gnotobiotic animal (p. 442)
immunocompromised host (p. 442)
immunomodulin (p. 441)

metagenomic study (p. 438)
microbiome (p. 430)
microbiota (p. 430)
mucociliary escalator (p. 435)
nasopharynx (p. 434)

opportunistic pathogen (p. 442)
oropharynx (p. 434)
petechial rash (p. 444)
probiotic (p. 439)
virome (p. 431)

Review Questions

1. Name some sterile body sites.

2. What body sites are colonized by normal microbial flora?

3. Under what circumstances can commensal organisms cause disease?

4. Why are commensal organisms beneficial to the host?

5. What are probiotics? How do they help maintain health?

6. How does the lung avoid being colonized?

7. What is a gnotobiotic animal?

8. Describe the various barriers that separate the microbiome from host tissues.

Clinical Correlation Questions

1. A 50-year-old man, after returning to the United States from a trip to India, develops excessive watery diarrhea. The diagnosis is cholera caused by *Vibrio cholerae*, a very acid-sensitive, Gram-negative rod. Healthy individuals do not ordinarily develop cholera unless they ingest very large numbers of organisms. The patient's history reveals that he did not drink tap water or swim in lakes, bays, or seas during his visit. Previously he was quite healthy except for recurrent bouts of indigestion that he self-medicated with Prilosec (a medication that reduces production of stomach acid). Explain how this individual developed cholera.

2. You accidentally ingest food contaminated with a bacterial species harboring an antibiotic resistance plasmid. Fortunately, the organism does not survive in the intestine. Nevertheless, how might this organism influence your intestinal microbiota?

3. How would you explain the following set of experimental results? 1. Germ-free mice have an altered ratio of certain immune system white blood cells. 2. A fecal transplant from a normal mouse to a germ-free mouse restores the normal ratio of white cells. 3. Boiling the fecal transplant material for 10 minutes eliminates all culturable microbes, but transplanting this material still reverses the white cell defect. 4. Autoclaving the fecal material destroys its ability to restore the immune system.

Thought Questions: CHECK YOUR ANSWERS

Thought Question 14.1   How can an anaerobic microorganism grow on skin or in the mouth, both of which are exposed to air?

ANSWER: Facultative organisms living in proximity to the anaerobes will deplete the oxygen in the environment, especially around nooks and crannies (for example, between teeth and gums, gingival pockets) that would ordinarily prevent anaerobes from growing. These small spaces have limited access to oxygen.

Thought Question 14.2   What can an immunocompromised person with mitral valve prolapse do to prevent formation of subacute bacterial endocarditis when visiting the dentist?

ANSWER: Actually nothing needs to be done. Antibiotic premedication before simple dental procedures (tooth extraction or cleaning) has no effect on whether or not a patient with mitral valve prolapse develops endocarditis. Either way the risk is extremely low. However, antibiotic premedication (a high dose of amoxicillin or a cephalosporin) is highly recommended for someone about to undergo surgery at mucosal surfaces, like a tonsilectomy. In addition, immunocompromised individuals should premedicate even before simple dental procedures.

Thought Question 14.3   Two men, both 55, ate raw oysters at a popular oyster bar. Two days later, one man is hospitalized with a life-threatening blood infection (septicemia) caused by the Gram-negative curved rod *Vibrio vulnificus*. The other man suffers no ill effects. The only difference between the two men is that the one infected was taking an over-the-counter proton pump inhibitor to reduce his stomach acidity and treat his persistent indigestion. What does this tell you about the infecting organism?

ANSWER: *Vibrio vulnificus*, a normal inhabitant of brackish water where oysters are farmed, is a common contaminant of these bivalves. However, it is extremely acid sensitive. Normal stomach acidity readily kills this pathogen, even if large numbers of the bacteria are ingested. Taking drugs that decrease stomach acidity can allow *V. vulnificus* to survive long enough to reach the intestine, where it can penetrate the epithelium and gain entrance to the circulatory system.

Thought Question 14.4   People who take antibiotics for an upper respiratory tract infection often develop antibiotic-associated diarrhea. However, pathogenic microbes are not the usual cause of this diarrhea. Using what you have read here combined with Internet searches, formulate a general hypothesis that might explain this connection. (*Hint:* Look up osmotic diarrhea.)

ANSWER: Part of the answer is that antibiotic treatment lowers the numbers of intestinal microbes that digest nutrients such as complex carbohydrates. Left in the bowel, these carbohydrates set up an osmotic imbalance that draws water out of the intestinal cells. Excess water in the bowel produces loose stools and diarrhea. Bile acids that escape absorption in the small intestine are another factor. These bile acids are usually metabolized by bacteria in the large intestine. When bacterial composition and numbers are altered by antibiotic treatment, unmetabolized dihydroxy bile acids, which are potent secretory agents, contribute to the development of secretory diarrhea in the colon. Finally, though not proven, altering the composition of the microbiome may interfere with interdomain chemical signals that contribute to proper water absorption.

Thought Question 14.5   Why do many Gram-positive microbes that grow on the skin, such as *Staphylococcus epidermidis*, grow poorly or not at all in the gut? (*Hint:* Think about differences between skin and intestinal contents.)

ANSWER: Bile salts present in the intestine (not on the skin) easily gain access to and destroy cytoplasmic membranes of Gram-positive organisms (unless the organism possesses bile hydrolases). Gram-negative microbes have extra protection in the form of an outer membrane and so can survive better in the intestine.

Thought Question 14.6   What caused the petechial rash in Case History 14.2?

ANSWER: The rash was caused by lipopolysaccharide (called endotoxin) present in the outer membrane of the Gram-negative bacteria. The LPS is released from the organism and can trigger the coagulation system, which produces tiny clots that can block capillaries. When blockage occurs in skin capillaries, the tiny red rash is seen.

Thought Question 14.7   How might normal microbiota escape the intestine and cause disease at other body sites?

ANSWER: Normal microbiota can escape through intestinal perforations resulting from gunshot or knife wounds, surgery, or cancer.

Thought Question 14.8   Your dorm mate, Teresa, has confided to you that she developed diarrhea after taking antibiotics for a few days to treat an upper respiratory tract infection. She still has a week to go on the antibiotics. To try to cure her diarrhea, she started eating active-culture yogurt, which in the past has helped her "cure" occasional episodes of diarrhea. Tell her why she has diarrhea and why the yogurt is not working this time.

ANSWER: The antibiotic Teresa is taking probably altered her intestinal microbiota, which is now out of balance. This imbalance can lead to diarrhea, although why is not always clear (see Thought Question 14.4, osmotic diarrhea). The bacteria in yogurt can help restore balance and resolve diarrhea (it can also reverse constipation). Live and active yogurt products are made using *Lactobacillus* and *Streptococcus* species, organisms that can have a probiotic effect on the intestine. Unfortunately, because they are live bacteria, the antibiotics Teresa is taking may also kill these organisms, thereby negating any probiotic effect.

The Immune System: Inflammation and Innate Immunity

[A Tale of Two Tourists]

SCENARIO Two vacationing tourists, unknown to each other, visited the Alabama gulf coast over the same weekend in August and decided to take a swim in the Gulf of Mexico. One, a 12-year-old girl named Alexis, had a cut on her arm caused by a fall onto the sharp edge of a scooter. The other vacationer, Brad, a man 66 years of age, had recently cut his leg on a nail protruding from a wooden railing at his beach condo. Within moments of entering the Gulf of Mexico, several small Gram-negative microbes invaded both of their bodies through those cuts. Four days later, the young girl was heading back to Birmingham in the back of the family car, oblivious to the battle recently waged in her bloodstream. At the same time, Brad lay dead of an aggressive blood infection.

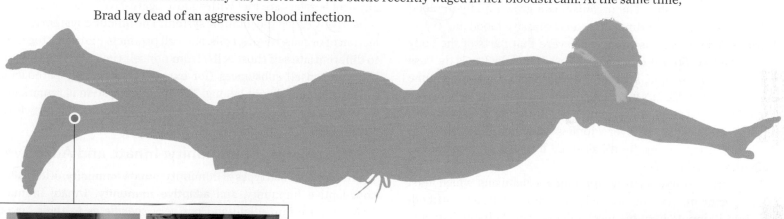

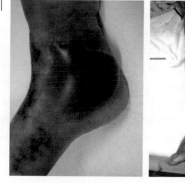 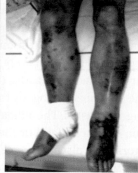

Cellulitis Caused by *Vibrio vulnificus*
V. vulnificus can rapidly produce hemorrhagic bullae (the fluid- and blood-filled vesicle on the left) and purpuric macules (large purple rashes on the right). The vesicle fluid and bloodstream will contain *V. vulnificus* bacteria.

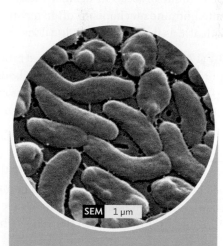

SEM 1 μm

Vibrio vulnificus
Vibrio vulnificus is a curved, Gram-negative rod that grows in saline environments. It can enter the body via wounds or ingestion of contaminated seafood such as oysters. The organism causes mild gastroenteritis in otherwise healthy people, but it can cause life-threatening blood and tissue infections in people with an immune system compromised by disease or drug therapy.

BRAD'S CLINICAL PROGRESSION Within a day after his swim, Brad's leg began to swell, was warm to the touch, and started to hurt. He put off going to the emergency department until day 2, when he realized that his leg was now extremely painful and had started to darken; and he felt feverish. At the hospital he had a fever of 39°C (102°F) and his blood pressure was abnormally low. Samples of wound material containing pus and a blood culture were collected and sent to the diagnostic laboratory to determine the identity of the pathogen. Results of the tests, however, would not be known for at least 48 hours. Meanwhile, he was admitted to the hospital and placed on empirical antibiotic therapy to try to stop the infection. By day 3, areas of his leg were turning black and becoming gangrenous, and the swelling had spread up and down his leg. His fever remained high and he became unresponsive. The antibiotics were not working. Brad died the next day. The lab report, now completed but too late to matter, indicated that both the wound and blood cultures contained *Vibrio vulnificus*.

Both swimmers in our opening case study were infected by the same pathogen, *Vibrio vulnificus*, a Gram-negative rod that lives in salt water. The organism can grow rapidly in blood (a 10-minute generation time) and cause massive blood and tissue infections. The infections can be so severe that parts of the body seem to dissolve. So why did Alexis live and Brad die? As is the case for most healthy people, a variety of nonspecific, innate immune factors present in Alexis's body killed the invading pathogen before it could multiply. Brad, on the other hand, was an alcoholic with liver disease. Because the liver supplies critically important non-specific defense proteins, Brad's diseased liver left him vulnerable to infection and disease.

In contrast to slow adaptive immune mechanisms, which must "see" a specific microbe before making immune cells and antibodies to fight it, innate immune mechanisms provide immediate and almost blind protection against many different microbes. These innate immune mechanisms often mean the difference between life and death, as it did for Alexis and does in our own daily lives.

Note Antibodies are small, Y-shaped proteins in blood that bind and tag invading pathogens. Different antibodies bind different bacteria and viruses. Innate immune mechanisms on their own can kill pathogens but are more aggressive if antibodies also coat the pathogen.

The previous chapter discussed innate physical and chemical barriers, such as skin and stomach acid, that separate us from our microbiota and help protect us from pathogens. This chapter provides an overview of the cellular makeup of our immune system, focusing on the nonadaptive, innate mechanisms that first engage pathogens after they breach the body's physical barriers.

15.1
Overview of the Immune System

What happens in our body after a bacterium, virus, fungus, or parasite enters? Microorganisms that slip through our body's physical barriers are met with a more aggressive series of defenses collectively called the immune system. The **immune system** is a collection of organs, tissues, cells, and cell products that work together to differentiate self (host cells) from nonself (invaders) and rid the body of nonself substances (for example, potentially pathogenic organisms) that do not belong. The immune system is remarkable in its capacity to respond to nearly any foreign molecular structure.

The Two Forms of Immunity: Innate and Adaptive

There are two broad types of immunity: **innate immunity**, often called nonadaptive immunity, and **adaptive immunity**. Innate immune responses are immediate and not very specific as to their target. Adaptive responses take time to develop but are highly specific to the given pathogen; and when the same pathogen reenters the body at a later date, the now primed adaptive response is swift. The protection these two systems afford the human body might be compared to the defenses used to protect medieval castles (Figure 15.1). For instance, the innate immune system is analogous to a moat stocked with alligators surrounding the castle. The alligators can indiscriminately attack any intruder, whether friend or foe (note, however, that it is a myth that alligators were placed in medieval moats). The adaptive immune system, on the other hand, is more like the king's archers. The archers, upon hearing a commotion outside the castle wall, climb to the rampart atop the castle, where they identify and selectively target hostile intruders with arrows (think antibodies).

The nonadaptive cellular responses (the alligators) of innate immunity are present even before birth. These innate mechanisms attack intruding microbes as soon as they enter the body (our first line of defense), and in a kind of molecular "call for backup," they send chemical signals to the more targeted adaptive immune system (the archers), alerting it that an infection is under way. Thus, innate immune mechanisms buy precious time for the more specific second line of defense to engage. Recall the stages of disease illustrated in Figure 2.8. The innate defense system operates continuously from the time of infection through the incubation period and until the infection ends. Adaptive immunity becomes evident only toward the end of the prodromal phase and is usually maintained for long periods after the infection clears.

Figure 15.1   The Immune System as Castle Defense

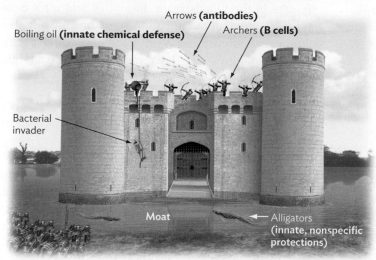

Arrows **(antibodies)**

Boiling oil **(innate chemical defense)**

Archers **(B cells)**

Bacterial invader

Moat

Alligators (innate, nonspecific protections)

In this analogy, the castle plays the role of our body under attack. The castle walls form a protective barrier, like skin. The barrels of boiling oil used to douse any who attempt to climb the castle walls can be compared to chemical defenses of innate immunity. The alligators swimming in the moat represent white blood cells, such as macrophages or neutrophils, that indiscriminately drown or attack anyone who tries to enter. The archers represent our adaptive immune system (B cells and plasma cells), which must first mount the towers, see the invaders, and then specifically target them with their arrows (antibodies). Swordsmen battling invaders inside the castle can be compared to cytotoxic T cells of the adaptive immune system (discussed in the next chapter), which specifically target infected host cells.

Link Recall from Section 2.2 that the incubation period of a disease marks the period of time between the initial infection and the early signs of disease. The period during which the first signs of disease are evident is known as the **prodromal phase**.

Although this chapter focuses on innate immunity, it is important to recognize that innate and adaptive immune mechanisms intertwine in ways that support and enhance each other's effectiveness. Adaptive immunity reacts to very specific structures called antigens. An **antigen** is any chemical, compound, or structure foreign to the body that will elicit an adaptive immune response (such as the production of antibodies). An adaptive immune response to a specific antigen does not occur until the body "sees" that antigen. Adaptive immune mechanisms can recognize billions of different antigenic structures and launch a directed attack against each one. Once such an attack is under way, the immune system stores a "memory" of the exposure in the form of specific memory cells. Any later encounter with the microbe will reactivate the appropriate memory cells, which then rapidly multiply, mobilize, engage, and destroy the intruder. Antibodies and adaptive immunity are discussed in more detail in Chapter 16.

The two types of immunity, innate and adaptive, are illustrated by the immune response to infection by the microorganism

Neisseria gonorrhoeae, which causes the sexually transmitted disease gonorrhea. A rapidly acting part of the innate immune response is **complement**, composed of numerous soluble protein factors constantly present in the blood. Within moments of an initial bloodstream infection, complement attacks and makes holes in the bacterial membrane, thereby killing the microbe (see Section 15.6). The second wave of the immune response, adaptive immunity, will then generate specific antibodies that bind any *N. gonorrhoeae* cells that escaped the innate mechanisms. These antibodies, however, are not made until well after the organism infects a person. Together, the forces of innate and adaptive immunity can ward off disease caused by *N. gonorrhoeae*. In fact, antibodies of the adaptive immune response when bound to *Neisseria* will actually stimulate killing of the pathogen by complement.

Although innate and adaptive immunity are often treated as separate entities, certain kinds of cells and organs participate in both. Our discussion of immunity begins by describing the various cells and organs that make up the whole of the immune system.

Cells of the Innate and Adaptive Immune Systems

Blood is composed of red blood cells (erythrocytes), white blood cells (leukocytes), and platelets (**Figure 15.2**). The different types of white blood cells develop from a precursor stem cell produced in bone marrow (**Figure 15.3**). The precursor stem cell can differentiate into myeloid stem cells and lymphoid stem cells. Myeloid stem cells further differentiate into the white blood cells (WBCs) of innate immunity, whereas the lymphoid stem cells generate the WBCs of adaptive immunity. The WBCs of innate immunity include:

- Polymorphonuclear leukocytes (PMNs)
- Monocytes
- Macrophages
- Dendritic cells
- Mast cells

Figure 15.2   Red and White Blood Cells

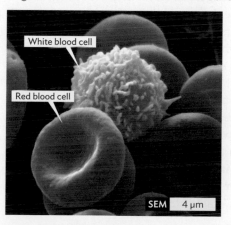

White blood cell

Red blood cell

SEM   4 µm

This micrograph illustrates the relative sizes and shapes of these cell types.

Figure 15.3   Development of White Blood Cells of the Immune System

Pluripotent stem cells in bone marrow divide to form two lineages. One lineage consists of the myeloid stem cells, which develop into polymorphonuclear leukocytes (PMNs) and monocytes, which function primarily as part of innate immunity. The lymphoid stem cells ultimately form natural killer cells, B cells, and T cells. B and T cells are involved in adaptive immunity. The B cells mature in bone marrow ("B" is for bone), and T cells mature in the thymus. Colors indicate groups of differentiated cells that arise from the same progenitor.

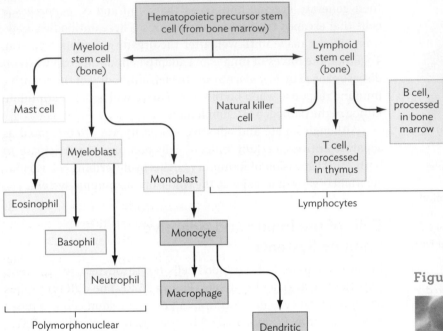

tissues and mucosa (mucus-secreting tissue) and do not circulate in the bloodstream. Basophils and mast cells have roles in inflammation and wound healing but also contain high-affinity receptors for a class of antibody (immunoglobulin E, or IgE) associated with allergic responses (to be described in Chapter 16).

Note Because neutrophils make up nearly all PMNs, the terms "neutrophil" and "PMN" are often used interchangeably.

MONOCYTES Monocytes (Figure 15.4C) are white blood cells with a single nucleus (not multilobed like a PMN); they engulf (phagocytize) foreign material and are the precursors of two other cell types: macrophages and dendritic cells (see Figure 15.3). Monocytes circulating in the blood migrate out of blood vessels into various tissues, where they then differentiate into macrophages and dendritic cells. **Macrophages** (Figure 15.6A) are phagocytic and form a major part of an amorphous **mononuclear phagocyte system (MPS)**, previously called

POLYMORPHONUCLEAR LEUKOCYTES **Polymorphonuclear leukocytes (PMNs)**, also called granulocytes, have multilobed nuclei, differentiate from an intermediate cell called the myeloblast, and contain enzyme-rich lysosome organelles (Figure 15.4A). PMNs are of several types named for their different staining characteristics. Each cell type has a different function. **Neutrophils**, making up nearly all white cells in the blood, can engulf microbes by a process called **phagocytosis**. Phagocytosis involves the extrusion of pseudopods that attach to and engulf the pathogen, which ends up in a **phagosome** vacuole (Figure 15.5). The phagocyte then kills microbes by fusing enzyme-engorged lysosomes with phagosomes. Enzymes spilling from the lysosome into the phagosome will destroy various components of the microbe and, ultimately, the microbe itself. **Basophils**, which stain with basic dyes, and **eosinophils** (Figure 15.4B), which stain with the acidic dye eosin, do not phagocytize (engulf) microbes but release products, such as major basic protein, that are toxic to the microbe. These white blood cells also release chemical mediators (called vasoactive agents) that affect the diameter and permeability of blood vessels, events important to inflammation (discussed in Section 15.3). **Mast cells** contain many granules rich in histamine and heparin. They are similar to basophils in structure but differentiate in a lineage separate from PMNs. Unlike PMNs, mast cells are residents of connective

Figure 15.4   Selected Types of White Blood Cells

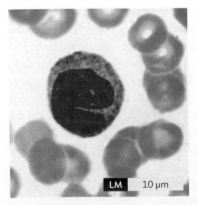

A. Neutrophil (PMN) (note the multilobed nucleus).

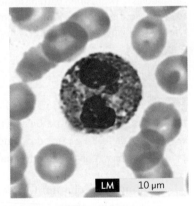

B. Eosinophils stain with eosin, an acidic dye.

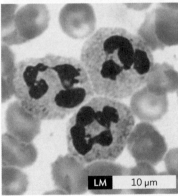

C. Monocyte with a single nucleus.

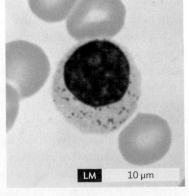

D. Lymphocyte (B cell or T cell) with its large single nucleus.

ANIMATION ▶ **Figure 15.5** **Phagocytosis**

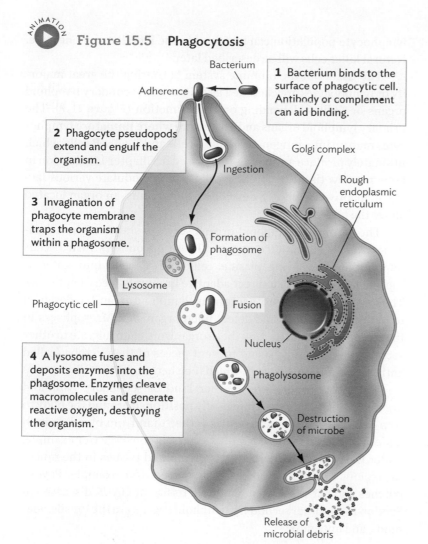

1 Bacterium binds to the surface of phagocytic cell. Antibody or complement can aid binding.

2 Phagocyte pseudopods extend and engulf the organism.

3 Invagination of phagocyte membrane traps the organism within a phagosome.

4 A lysosome fuses and deposits enzymes into the phagosome. Enzymes cleave macromolecules and generate reactive oxygen, destroying the organism.

the reticuloendothelial system, that is widely distributed throughout the body. The MPS is composed of monocytes that circulate in the blood as well as cells derived from monocytes, such as the macrophages that are dispersed in tissues. In addition to macrophages, the system is composed of specialized endothelial cells lining the sinusoids (special capillaries) of the liver (Kupffer cells), spleen, and bone marrow.

Macrophages are present in most tissues of the body and are the cells most likely to make first contact with invading pathogens. In many ways, they are like sentries with two functions. As part of innate immunity, they kill invaders directly. Protoplasmic protrusions from the macrophage surface extend and clasp nearby bacteria, pulling them into the cell (Figure 15.5, steps 1 and 2, and Figure 15.6B). Once ingested by the macrophage, the bacteria are destroyed by phagocytosis, as illustrated in Figure 15.5. Then, as the first step in adaptive immunity, the remnants are processed (degraded) into smaller peptide antigens and are presented on the macrophage cell surface. **T cells**, a type of lymphocyte produced by the adaptive immune system, have the ability to bind to antigens displayed on the macrophage and initiate an adaptive immune response. Different T cells are specific for different antigens. Cells such as macrophages that function to introduce or "present" bacterial antigens to cells of the adaptive immune system are called **antigen-presenting cells (APCs)**. (We will discuss how the specific immune response develops in Chapter 16.)

Macrophages are not the only antigen-presenting cells; **dendritic cells** (Figure 15.6C) also present antigens to T cells. Located in the spleen and lymph nodes, they, like macrophages, can take up, process, and present small antigens on their cell surface. Dendritic cells are different from macrophages in structure and because they take up small soluble antigens from the surroundings in addition to phagocytizing whole bacteria.

Figure 15.6 **Macrophages and Dendritic Cells**

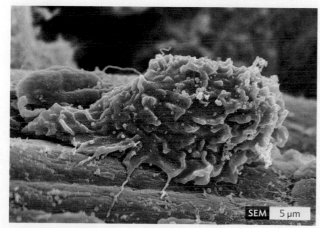

A. A macrophage (20 μm long) at the site of a skin wound (colorized).

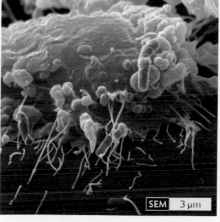

B. Membrane protrusions from a macrophage detecting and engulfing bacteria (pink: *E. coli*, 1.5 μm long; colorized).

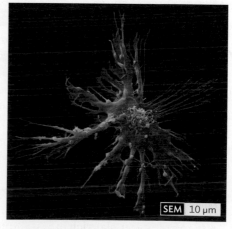

C. Dendritic cell (colorized).

WHITE BLOOD CELL DIFFERENTIALS IN NORMAL AND INFECTED PATIENTS   Today's clinical laboratory can take a blood sample and automatically count the total number of white blood cells and identify their various cell types—neutrophils, eosinophils, basophils, lymphocytes, and monocytes (Table 15.1), providing what is called a **white blood cell (WBC) differential**. The results can provide the clinician with important clues about the cause of a patient's illness. For example, elevated total WBC numbers indicates infection or allergy. Elevated neutrophils or their precursors, called band cells, suggest a bacterial infection, whereas elevated lymphocytes suggest a viral infection. An increase in eosinophils can indicate the presence of intestinal parasites or some blood parasites. Remember that there can be other explanations for a particular WBC profile (see lymphoid cancers in Chapter 17), so it is important to view the WBC differential results in the context of the patient's symptoms. For example, does the patient exhibit hallmark symptoms of viral or bacterial disease? If not, there could be a noninfectious cause for the abnormal profile.

Lymphoid Organs

The lymphoid stem cells noted in Figure 15.3 produce lymphocytes and natural killer cells. Lymphocytes (Figure 15.4D), which are the main participants in adaptive immunity, are present in blood at about 2,500 cells per microliter, accounting for more than one-third of all peripheral white blood cells. However, an individual lymphocyte spends most of its life within specialized solid tissues (lymphoid organs) and enters the bloodstream only periodically, where it migrates from one place to another, surveying tissues for possible infection or foreign antigens. No more than 1% of the total lymphocyte population can be found in the blood at any one time. Natural killer cells will be described later.

The tissues of the immune system harboring the great majority of lymphocytes are classified as primary or secondary lymphoid organs or tissues, depending on their function (Figure 15.7). The primary lymphoid organs and tissues are where immature lymphocytes mature into antigen-sensitive B cells and T cells. **B cells**, which ultimately produce antibodies (discussed in Chapter 16), develop in bone marrow tissue, whereas T cells, which modulate various facets of adaptive immunity, develop in the thymus, an organ located above the heart.

The secondary lymphoid organs serve as stations where lymphocytes can encounter antigens. These encounters turn B cells into antibody-secreting plasma cells and T cells into antigen-specific helper cells that direct adaptive immunity (see Chapter 16). The spleen is an example of a secondary lymphoid organ. It is designed to filter blood directly and detect microorganisms. Macrophages in the spleen engulf and destroy these organisms, then migrate to other secondary lymph organs and present pieces of the microbe (called antigens) to the B and T cells, which then become activated.

The **lymph nodes** are another kind of secondary lymphoid organ. Lymph nodes are strategically positioned around the body to trap organisms from local tissues rather than from blood. Lymph nodes are situated where lymphatic vessels converge (for example, under the armpits). There are also lymphoid tissues in the mucosal regions of the gut and respiratory tract (for example, Peyer's patches and gut-associated lymphoid tissue, or GALT, discussed in Section 15.2). Other secondary lymphoid organs are the tonsils, adenoids, and appendix.

Table 15.1
Guidelines for Interpreting White Blood Cell (WBC) Counts with Differential[a]

	Normal	Acute Bacterial Infection	Viral Infection	Parasitic Infection	Allergy
Total WBC Count	4,500–11,000/mm³	Elevated: 12,000–30,000	Elevated	Elevated	Elevated
Differential					
Neutrophils	54%–62%	Increased numbers and more immature forms (band cells)	Can be reduced		
Eosinophils	1%–3%			Increased	
Basophils	0%–0.75%				Increased
Lymphocytes	25%–33%		Increased		
Monocytes	3%–7%	Typically increase with chronic infections (e.g., tuberculosis, brucellosis, subacute bacterial endocarditis, *Rickettsia*, many protozoan infections			

[a]This table provides general guidelines. Individual organisms or certain noninfectious medical conditions or immunological defects can alter the findings. A blank space indicates little change in the parameter.

Figure 15.7 **Lymphoid Organs**

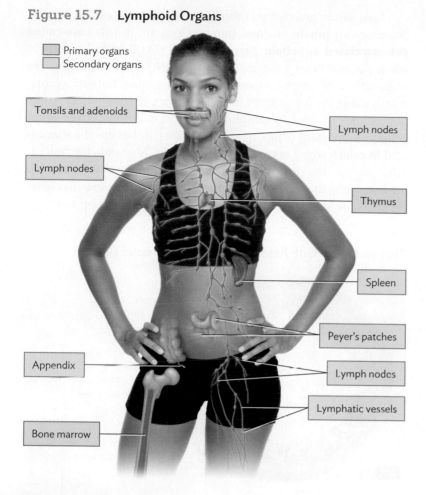

□ Primary organs
□ Secondary organs

Tonsils and adenoids

Lymph nodes

Lymph nodes

Thymus

Spleen

Peyer's patches

Appendix

Lymph nodes

Lymphatic vessels

Bone marrow

SECTION SUMMARY

- **The immune system** consists of both innate and adaptive mechanisms that recognize and contain or eliminate pathogens.

- **Innate immunity** includes physical barriers and some cellular responses to various microbial structures.

- **Adaptive immunity** is a cellular response to specific structures (antigens), which generates a memory of the first exposure. This memory allows a rapid, massive response upon later reexposure to the same antigen.

- **Myeloid bone marrow stem cells** differentiate to form cells of the innate immune system—phagocytic PMNs, monocytes, macrophages, antigen-presenting dendritic cells, and mast cells.

- **Lymphoid stem cells** differentiate into natural killer cells (part of the innate immune system) and lymphocytes (cells of the adaptive immune system). Lymphocytes are classified as B cells, which ultimately produce antibodies, and T cells, which regulate adaptive immunity.

- -

Thought Question 15.1 Invertebrates such as clams possess a strong innate immune system but do not possess adaptive immune mechanisms. Why do vertebrates, humans in particular, need an adaptive immune system in addition to the innate?

- -

15.2
Innate Host Defenses: Keeping the Microbial Hordes at Bay

SECTION OBJECTIVES

- Describe the physical barriers to infection.
- Explain the basic roles of SALT, GALT, and M cells in innate immunity.
- Describe chemical barriers employed by cells of the innate immune system.

We have broadly described the various cells and organs of the immune system, but what combination of cells, structures, and processes constitutes innate immunity? To continue our earlier analogy (see Figure 15.1), the first lines of castle defense include physical barriers (the watery moat and castle wall), chemical barriers (boiling oil tossed onto invaders trying to scale the wall), and hand-to-hand combat once the wall is breached. Similarly, the body's initial, innate defenses against infectious disease are composed of physical, chemical, and cellular barriers designed to prevent a pathogen's access to host tissues. Although generally described as nonspecific, you will see that some of our innate defense systems are more specific than others in recognizing invaders.

Physical Barriers to Infection

As described in Chapter 14, the first line of defense against any potential microbial invader (either commensal or pathogenic) is found wherever the body interfaces with the environment. These interfaces (skin, lung, gastrointestinal tract, genitourinary tract, and oral cavities) have similar defense strategies, although each has unique characteristics.

SKIN Skin is the protective shield separating you from the sea of microbes that inhabit your surroundings. Few microorganisms can penetrate skin because of the thick keratin armor produced by closely packed cells called keratinocytes. Keratin protein is a hard substance (hair and fingernails are made of this) that is not degraded by known microbial enzymes. An oily substance (sebum) produced by the sebaceous glands also covers and protects the skin. Its slightly acidic pH inhibits bacterial growth. Competition between species also limits colonization by pathogens, and microorganisms that manage to adhere are continually removed by the constant shedding of outer epithelial skin layers.

A consortium of specialized cells called **skin-associated lymphoid tissue (SALT)**, located just under the skin, recognize microbes that may slip through the physical barrier. **Langerhans cells**, specialized dendritic cells that can also phagocytize microbes, make up a significant portion of SALT. Once a Langerhans cell has ingested a microbe, the cell migrates by ameboid movement to nearby lymph

...d "presents" antigenic parts of the microbe to the immune ...n to activate adaptive immunity.

Note Do not confuse phagocytic Langerhans cells with the pancreatic "islets of Langerhans" that secrete insulin.

MUCOUS MEMBRANES Mucosal surfaces form the largest interface (200–300 square meters, roughly the area of a basketball court) between the human host and the environment. The intestine alone is 23–26 feet long. This large surface presents a huge containment problem. Mucous membranes must be selectively permeable to nutrients and waste components but form a staunch barrier against invading pathogens. Mucosal membranes are covered with special tight-knit epithelial layers that support this barrier function (see Figure 14.16). Mucus secreted from stratified squamous epithelial cells (box-shaped epithelium consisting of multiple cell layers) coats mucosal surfaces and traps microbes. Compounds within mucus can serve as a food source for some commensal organisms, but other secreted compounds—such as the enzymes **lysozyme**, which cleaves cell wall peptidoglycan, and lactoperoxidase, which produces toxic superoxide radicals—can kill an organism trapped in the mucus.

In addition to these nonspecific defense mechanisms, there are also semispecific innate immune mechanisms associated with mucosal surfaces. Host cells (even epithelial cells) in mucosa have evolved mechanisms to distinguish harmless compounds from dangerous microbes. Certain conserved structures on microbes, called **pathogen-associated molecular patterns (PAMPs)** or, more generally, **microbe-associated molecular patterns (MAMPs)**, are recognized by host cell surface receptors, such as various Toll-like receptors (discussed in Section 15.5) and the protein CD14. MAMPs include structures such as peptidoglycan and lipopolysaccharide that are unique to bacteria. Once a MAMP is recognized, the host cell sends out chemicals that can activate immune system cells involved with innate and adaptive immune mechanisms.

Link Recall from Section 5.1 that **peptidoglycan** is a cage-like macromolecule that surrounds the cell membranes of Gram-positive and Gram-negative bacteria. It is composed of cross-linked chains of *N*-acetylglucosamine and *N*-acetylmuramic acid. **Lipopolysaccharide** is present only in Gram-negative organisms and forms the major part of the outer membrane. Lipopolysaccharide is made of lipid linked to polysaccharide and is also called the endotoxin of Gram-negative bacteria because it elicits a dangerously strong immune response.

Mucosal surfaces throughout the body also contain a significant quantity of lymphoid tissue. **Mucosa-associated lymphoid tissue (MALT)** is scattered along mucosal linings. MALT is populated by lymphocytes such as T cells and B cells (part of the adaptive immune system) as well as plasma cells and macrophages, each of which is well situated to encounter antigens passing through the mucosal epithelium.

Like other mucous membranes, the gastrointestinal system possesses an innate mucosal immune system, in this case called **gut-associated lymphoid tissue (GALT)**. GALT includes tonsils, adenoids, and Peyer's patches (Figure 15.8A). **Peyer's patches** are aggregations of lymphoid tissues that dot the intestinal surface, primarily along the lower small intestine. These tissues help keep our microbiome in check by sampling intestinal microbes to "see" what is there. Important components of Peyer's patches are the specialized **M cells** wedged between epithelial cells. M stands for "microfold," which describes their appearance (Figure 15.8B). M cells are fixed cells that take up microbes from the intestine and release them, or pieces of them, into a pocket formed on the opposite, or

Figure 15.8   Gut-Associated Lymphoid Tissue (GALT)

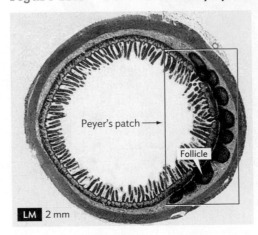

Peyer's patch →

Follicle

LM   2 mm

A. A Peyer's patch located on the intestine.

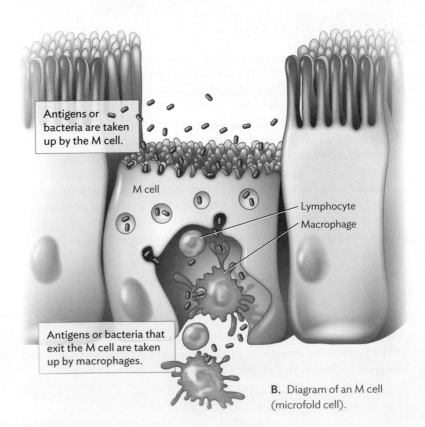

Antigens or bacteria are taken up by the M cell.

M cell

Lymphocyte

Macrophage

Antigens or bacteria that exit the M cell are taken up by macrophages.

B. Diagram of an M cell (microfold cell).

basolateral, side of the cell. Other cells of the innate immune system, such as macrophages, gather here and collect the organisms that emerge. As described earlier, macrophages engulf and try to kill the organism and place small, degraded components of the organism on the macrophage cell surface in a process called antigen presentation so that other immune system cells will recognize them.

Link As will be described in Chapter 16, **antigen presentation** by macrophages takes place as the macrophage moves to a regional lymph node. In the lymph node, the presented antigen will activate specific subsets of T lymphocytes. Some activated T lymphocytes stimulate B lymphocytes to become plasma cells and make antibodies against the microbe. Other T lymphocytes are activated to become cytotoxic T lymphocytes (CTLs) that can directly kill infected host cells.

M cells, acting as a phagocytic filter, are extremely important for the development of mucosal immunity to pathogens. However, M cells can also serve as a gateway for some pathogens, such as *Salmonella*, to gain entry to the body and cause disease.

LUNGS The lung also has a formidable defense. In addition to the respiratory mucociliary elevator discussed in Section 14.1, microorganisms larger than 100 μm become trapped by hairs and cilia lining the nasal cavity and trigger the forceful expulsion of air from the lungs (a sneeze). The sneeze is designed to clear the organism from the respiratory tract. Organisms that make it to the alveoli (see Figure 14.5A), however, are not easily expelled by a sneeze but are met by phagocytic cells called **alveolar macrophages**. These cells can ingest and kill most bacteria and send out chemical signals that attract other cells of the innate and adaptive immune systems to the site of infection.

Chemical Barriers to Infection

Chemical barriers to infection exist at numerous body sites. For instance, the acid pH of the stomach is lethal to most bacteria, and lysozyme in tears will degrade the cell walls of Gram-positive bacteria (see Chapter 14). Potentially pathogenic bacteria can also be killed by superoxide radicals generated by host enzymes such as lactoperoxidase, present in a variety of cell types. Another set of antimicrobial chemicals are small (29–47 amino acids) antimicrobial, cationic (positively charged) peptides called **defensins** that function as important components of innate immunity against microbial infections. Defensins are made by a variety of human cells. These antimicrobial peptides, which kill by destroying the microbial cytoplasmic membrane, are effective against Gram-positive and Gram-negative bacteria, fungi, and even some viruses (viruses, like HIV, that are wrapped with eukaryotic membranes as a result of budding from a host cell). Defensins are produced by many human cells, including cells of the skin, lung, genitourinary tract, and gastrointestinal tract.

SECTION SUMMARY

- **Skin defenses** against invading microbes include closely packed keratinocytes and a SALT lymphoid system made up largely of phagocytic Langerhans cells.

- **Mucous membrane defenses** involve secreted enzymes, cytokines, and GALT tissues, such as Peyer's patches, that contain phagocytic M cells.
- **M cells** in gut-associated lymphoid tissues sample bacterial cells at their surface and release pieces of them to immune system cells.
- **Phagocytic alveolar macrophages** inhabit lung tissues, contributing to nonspecific defense.
- **Chemical barriers against disease** include cationic defensins, acid pH in the stomach, and superoxide produced by certain cells.
- **Microbe-associated molecular patterns (MAMPs)** are recognized by Toll-like receptors (TLRs) found on many host cells such as macrophages. Binding triggers release of chemical signal molecules that activate innate and adaptive immune mechanisms.

Thought Question 15.2 Why do defensins have to be so small? Do defensins kill normal microbiota?

15.3
The Acute Inflammatory Response

SECTION OBJECTIVES
- List the five cardinal signs of inflammation.
- Outline the process of inflammation.
- Describe the roles of vasoactive factors and cytokines.
- Differentiate acute from chronic inflammation.

What is inflammation, and how does it develop? The boil shown in Figure 15.9 is an inflammatory response triggered by infection with the organism *Staphylococcus aureus*. Inflammation is a critical innate defense in the war between microbial invaders and their hosts. It provides a way for phagocytic cells (such as neutrophils) normally confined to the bloodstream to enter infected areas within tissues. Movement of these neutrophils out of blood vessels is called

Figure 15.9 Inflammation Caused by Infection

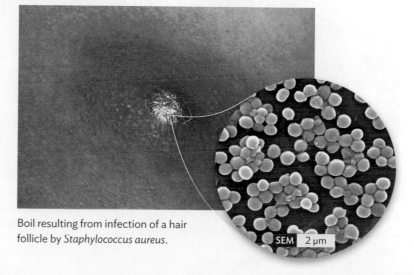

Boil resulting from infection of a hair follicle by *Staphylococcus aureus*.

SEM 2 μm

extravasation (discussed shortly). Once at the infection site, the neutrophils begin engulfing microbes. The white pus associated with an infection is teeming with these white blood cells. There are five cardinal signs of inflammation, first described more than 2,000 years ago: heat (warmth at the site), edema (swelling from fluid accumulation), redness, pain, and altered function or movement at the affected site (use the acronym HERPA to remember).

CASE HISTORY 15.1

A Case of Nonstick Neutrophils and Recurring Infections

Constance, a 7-month-old female, suffered from recurrent infections since the age of 2 weeks. She was frequently admitted to the hospital for *Staphylococcus aureus* sepsis (blood infection), hepatosplenomegaly (enlarged liver and spleen), oral thrush caused by *Candida albicans*, and poor weight gain. Repeated treatments with different antibiotics only temporarily resolved her infections. Two days ago, her alarmed mother and father once again brought Constance to the emergency department at the University of Chicago Medical Center. Constance was lethargic, didn't move much, and had a high fever. She also struggled for breath. Her blood work revealed high white blood cell counts (59,000/mm³; normal is 4,000–11,000) with a higher than normal level of neutrophils in her blood (neutrophilia). A chest X-ray showed inflammation in both lungs. (**Figure 15.10** shows a normal, clear lung compared with one showing an inflammatory infiltrate caused by infection.) It was no surprise, then, when the lab technician reported finding *Staphylococcus aureus* in Constance's blood. Her doctor ordered intravenous antibiotic therapy and was determined to find the cause of these repeated infections. He suspected a serious defect in Constance's immune system. Laboratory analyses found that her lymphocyte numbers, overall serum immunoglobulin (antibody) levels, as well as liver and kidney functions were all normal. No help there. Then flow cytometry analysis of surface proteins present on Constance's peripheral blood neutrophils revealed a complete absence of critically important membrane proteins. The doctor's heart sank. It was not good news.

Constance was repeatedly infected with *S. aureus* and other microbes because she had a rare but serious inherited defect in her neutrophils, one of the first responders to infection. The defect, called leukocyte adhesion deficiency (LAD), was not in the number of neutrophils she had; in fact, the excessive number of neutrophils in her blood was a clue to the problem. Her neutrophils lacked the "sticky" cell-surface **integrin** proteins needed to get these white cells out of the circulation and to a site of infection, a critical part of the inflammatory response. Unfortunately, patients with LAD usually do not survive beyond 2 years of age.

Note Flow cytometry was initially described in Chapter 6 as a way to count live versus dead cells. It can also be used to identify the presence of specific surface proteins if the cells are first treated with a fluorescent antibody that binds those proteins. Then, as the cells move past the laser in the flow cytometer, the instrument records how many of those proteins are present on the basis of fluorescence.

Signals Leading to Inflammation

Although many things can trigger inflammation, our focus is on how microbes cause the response. The process begins with the infection itself. Microorganisms introduced into the body—for example, on a wood splinter—will begin to grow and produce compounds that damage host cells (**Figure 15.11**). Resident macrophages (the sentries of the immune system) that wander into the infected area engulf these organisms and then release inflammatory mediators (chemoattractants) that "call" for more help. These mediators include **vasoactive factors** such as leukotrienes, platelet-activating factor, and prostaglandins, which act on blood vessels of the microcirculation, increasing blood volume and capillary permeability to help deliver white blood cells to the area. In addition, macrophages secrete small protein molecules called cytokines. Cytokines are made and secreted by cells as a way to communicate with other cells. These signaling peptides

Figure 15.10   X-Ray Images of Lung

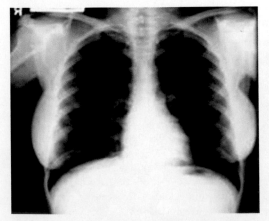

A. This image of a normal, clear lung is purportedly that of Marilyn Monroe, a famous actress circa 1950.

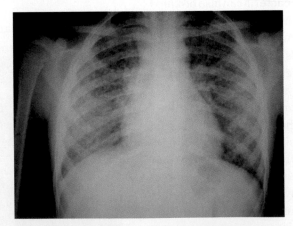

B. Bilateral inflammation (cloudiness) caused by *Mycobacterium tuberculosis*. The nodular cloudiness is due to the infiltration of white cells into affected areas.

Figure 15.11   Basic Inflammatory Response

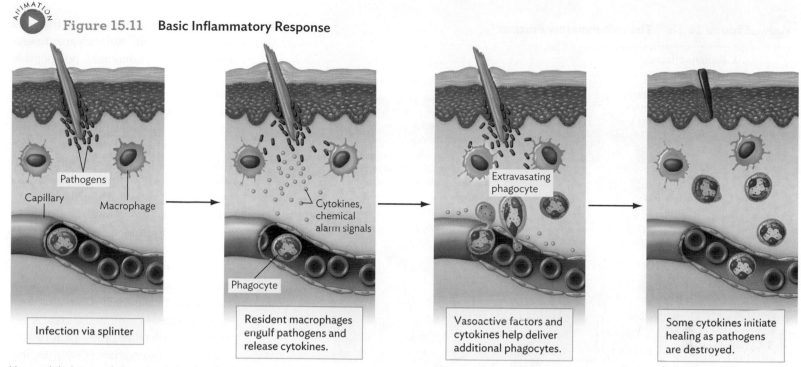

Neutrophils (a type of phagocyte) circulate freely through blood vessels and can squeeze between cells in the walls of a capillary (extravasation) to the site of infection. They then engulf and destroy any pathogens they encounter.

---

bind to specific membrane receptors on target cells and have powerful effects on the functions of those cells. The cytokines released from macrophages at the site of an infection diffuse to the vasculature and stimulate expression of specific receptors (called **selectins**) on the endothelial cells of capillaries and venules.

Vasoactive Factors, Cytokines, and the Extravasation Process

The movement of neutrophils through capillary vascular walls (**extravasation**; Figure 15.12) requires a relaxation of endothelial cell adhesion. The process is initiated by vasoactive factors released by macrophages; these vasoactive factors increase vascular permeability and stimulate **vasodilation** (widening of blood vessels). Vasodilation slows blood flow and, as a result, increases blood volume in the affected area. The more permeable vessel allows the escape of plasma into tissues. Both events cause localized swelling, redness, and heat. Vasoactive factors also stimulate local nerve endings, causing pain that alerts the patient to the infected area (remember HERPA).

Many kinds of cytokines exist, all of which are involved in regulating the immune response. Cytokines such as **interleukin 1 (IL-1)** and **tumor necrosis factor alpha (TNF-α)** are released by macrophages and stimulate the production of adhesion molecules (selectins) on the inner lining of the capillaries. The selectins snag neutrophils zooming by in the bloodstream, slow them down, and cause them to roll along the endothelium (see Figure 15.12 and the

Figure 15.12
Mechanism of Extravasation

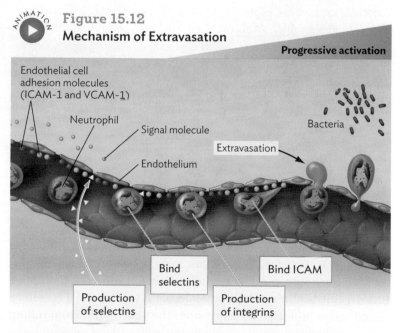

Extravasation is the process by which leukocytes move from the bloodstream into surrounding tissues. Signal molecules produced by damaged tissue cells induce the production of selectins (produced early in the process) on the surface of endothelial cells and integrins on the surface of the white blood cell (selectins and integrins not shown). Selectins capture leukocytes traveling through blood vessels. Leukocytes begin to roll along the vessel wall, and the integrins on the leukocyte surface lock onto the endothelial cell's adhesion molecules (ICAM-1 or VCAM-1). The leukocytes are progressively activated while rolling. The neutrophil ultimately squeezes through the wall between endothelial cells (extravasation).

Figure 15.13   The Inflammatory Process

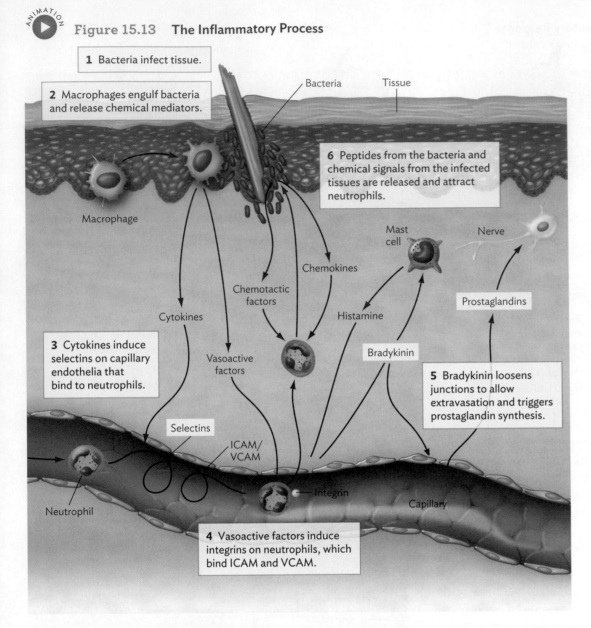

1 Bacteria infect tissue.

2 Macrophages engulf bacteria and release chemical mediators.

3 Cytokines induce selectins on capillary endothelia that bind to neutrophils.

4 Vasoactive factors induce integrins on neutrophils, which bind ICAM and VCAM.

5 Bradykinin loosens junctions to allow extravasation and triggers prostaglandin synthesis.

6 Peptides from the bacteria and chemical signals from the infected tissues are released and attract neutrophils.

Bacteria

Tissue

Macrophage

Mast cell

Nerve

Chemokines

Chemotactic factors

Prostaglandins

Cytokines

Histamine

Vasoactive factors

Bradykinin

Selectins

ICAM/VCAM

Integrin

Capillary

Neutrophil

Damaged tissue cells in the area of inflammation will release **brady-kinin**, a nine-amino-acid polypeptide that helps loosen the tight junctions between endothelial cells to promote extravasation (see Figure 15.13, step 5). Bradykinin molecules also bind to mast cells in the area, causing a calcium influx that triggers mast cell degranulation. The histamine released further loosens endothelial cell junctions, which allows more fluid and cells to move out (increased vascular permeability), adding to fluid accumulation (edema). Bradykinin will attach to capillary cells and induce synthesis of prostaglandins that in turn stimulate nerve endings, causing pain in the area. A key enzyme involved in prostaglandin synthesis is cyclooxygenase (COX). Aspirin as well as the popular prescribed anti-inflammatory agent celecoxib (Celebrex) are COX inhibitors that prevent the synthesis of prostaglandins and thus reduce inflammatory pain. (These drugs are also called nonsteroidal anti-inflammatory drugs, or NSAIDs.)

Once neutrophils have passed through the vascular wall, chemotactic factors are needed to lure them to the proper location. Some of these factors are made by the bacterium itself. The chemotactic factors bind to neutrophil surface receptors and stimulate pseudopod projections aimed toward the microbe. This causes the white blood cell to migrate in the direction of the infection. In addition, chemokine peptides IL-8 and CCL2 (formerly MCP-1) produced by damaged tissues can serve as chemoattractants for these white blood cells. **Chemokines** are cytokines that have chemotactic properties. Thus, much of what takes place at the site of an infection is not due directly to substances produced by the microbe but is the result of the body's reaction to the presence of the intruding organism. Once the cascade of inflammatory events delivers phagocytic cells to the site of infection, they begin devouring the microbes in an attempt to cure the infection.

Chronic Inflammation

Chronic inflammation results from the persistent presence of a foreign body. This type of inflammation inevitably causes permanent tissue damage, even though the body attempts repair. The causes of chronic inflammation are many. For example, infectious organisms such as *Mycobacterium tuberculosis*, *Actinomyces bovis*, and various

summary **Figure 15.13**, step 3). Rolling neutrophils that encounter inflammatory mediators are activated to produce and display integrin adhesion molecules on their surface. These were the molecules missing on Constance's neutrophils in Case History 15.1. The integrins on the neutrophils lock onto endothelial adhesion molecules ICAM-1 (intracellular adhesion molecule 1) and VCAM-1 (vascular cell adhesion molecule 1), which stops the neutrophils from rolling and initiates extravasation, where the white blood cells squeeze through the endothelial wall and into the tissues. There they can help macrophages attack the invading microbes.

Infections usually cause an increase in the number of neutrophils circulating in the bloodstream. However, the 59,000 neutrophils per cubic millimeter of blood in Case History 15.1 is exceptionally high. The patient's neutrophil count was high because, lacking integrin, her neutrophils could not leave the vessels to enter tissues. As a result, her neutrophils accumulated in the circulation, locked out from doing battle.

protozoan parasites can avoid or resist host defenses (Figure 15.14). As a result, they persist at the site and continually stimulate the basic inflammatory response. Nonliving, irritant material such as wood splinters, inhaled asbestos particles, and surgical implants can also lead to chronic inflammatory responses.

During a chronic infection, the body attempts to "wall off" the site of inflammation by forming a **granuloma**. A granuloma begins as an aggregation of mononuclear inflammatory cells surrounded by a rim of lymphocytes. The body then deposits fibroconnective tissue around the lesion, causing tissue hardening known as fibrosis.

SECTION SUMMARY

- **Inflammation** begins with a mechanism (extravasation) that moves neutrophils from the bloodstream into infected tissues.

- **Macrophages** in tissues engulf microorganisms then release vasoactive factors that increase vascular permeability and cytokines that stimulate production of selectin receptors on vascular endothelial cells.

- **Chemoattractant molecules** (for example, chemokines released from damaged cells) summon neutrophils.

- **Extravasation** is the process by which neutrophils pass between cells of the endothelial wall. Once out of the circulation, the neutrophils travel to the site of infection.

- **Bradykinin** is a nine-amino-acid polypeptide that helps loosen the tight junctions between endothelial cells to promote extravasation, and produces pain in the affected area.

- **Chronic inflammation** results from the persistent presence of a foreign object.

Thought Question 15.3   How does increasing the diameter of a blood vessel affect blood flow and blood pressure?

Thought Question 15.4   You have discovered a drug that can decrease the production of selectin molecules. What would this do to innate immunity, and for what noninfectious medical conditions might this be therapeutically useful?

15.4
Phagocytosis: A Closer Look

SECTION OBJECTIVES

- Describe how phagocytes recognize foreign cells.
- Explain the mechanisms phagocytes use to kill engulfed cells.
- List some ways pathogens avoid the consequences of phagocytosis.
- Discuss how our body ends the inflammatory process.

Although we have already discussed the basic process of phagocytosis and inflammation, several important questions remain. For example: How do phagocytes kill? Why don't phagocytes kill host cells in the absence of infection? Once phagocytes start killing, what makes them stop?

Figure 15.14   Chronic Inflammation

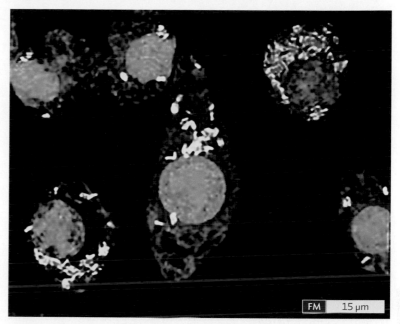

A. The fluorescent orange and yellow organisms shown here are *Mycobacterium tuberculosis* within macrophages in a tuberculosis abscess. A thick, waxy cell wall along with the ability to inhibit phagocyte killing mechanisms helps mycobacteria survive inside macrophages for prolonged periods. The continual stimulation of an inflammatory response leads to chronic inflammation.

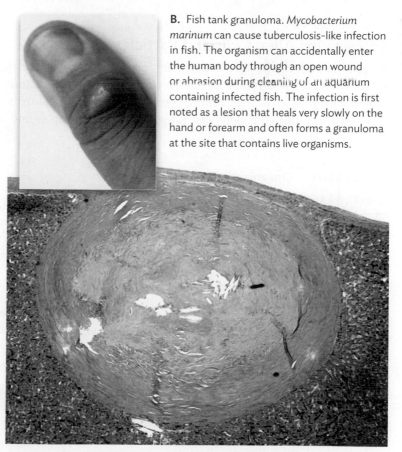

B. Fish tank granuloma. *Mycobacterium marinum* can cause tuberculosis-like infection in fish. The organism can accidentally enter the human body through an open wound or abrasion during cleaning of an aquarium containing infected fish. The infection is first noted as a lesion that heals very slowly on the hand or forearm and often forms a granuloma at the site that contains live organisms.

C. Cross section of liver showing a necrotizing granuloma caused by *M. tuberculosis* that spread through the bloodstream to this site after escaping the lung.

Figure 15.15 Phagocytosis

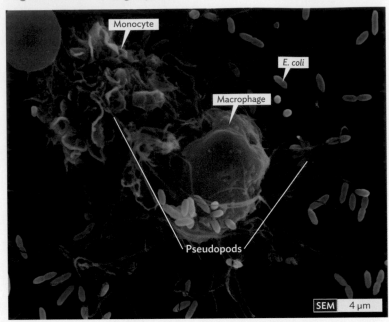

Lung pleural cavity macrophage and monocyte. The smooth cell with extended pseudopods is a macrophage engulfing *E. coli* (phagocytosis). The ruffled cell is a monocyte. Notice the pseudopods reaching toward the monocyte. Once contact is made between a macrophage pseudopod and the monocyte, further movement of the pseudopod stops to avoid attacking self.

Recognizing Alien Cells and Particles

Phagocytosis is an effective nonspecific immune response. However, it would be disastrous for white blood cells to indiscriminately phagocytize host as well as nonhost structures. Fail-safe controls built into our immune defenses prevent this from happening.

Without restraints, the mammalian immune system would constantly attack itself, causing more problems than it solves. When one of these fail-safe mechanisms fails, autoimmune diseases can arise.

For phagocytosis to proceed, macrophages and neutrophils must first "see" the surface of a particle as being foreign; note the macrophage pseudopods extending toward the monocyte (self) and *E. coli* (foreign) in Figure 15.15. When a phagocyte surface meets with the surface of another body cell, the phagocyte becomes temporarily paralyzed (inept at pseudopod formation). Paralysis allows the phagocyte to evaluate whether the other cell is friend or foe, self or nonself. Self-recognition occurs when glycoproteins located on the white blood cell membrane bind inhibitory glycoproteins that are present on all host cell membranes. The inhibitory glycoprotein on human cells is called CD47. Because invading bacteria lack the CD47 inhibitory surface molecules, they can readily be engulfed.

 Animation of macrophage cytokine release: youtube.com

Although many bacteria, such as *Mycobacterium* or *Listeria* species, are easily recognized and engulfed by phagocytosis, others, such as *S. pneumoniae*, a cause of pneumonia, possess polysaccharide capsules that are too slippery for pseudopods to grab (Figure 15.16A). This is where innate immunity and adaptive immunity join forces. Adaptive immunity produces antibodies that aid the innate immune mechanism of phagocytosis through a process known as **opsonization** (Figure 15.16B). The Y-shaped **antibodies** coat the surface of the bacterium, leaving the tail ends of the antibodies, called the Fc regions (see Section 16.3), pointing outward. Like two strips of Velcro, the Fc regions of these antibodies bind to specific

Figure 15.16 Opsonization

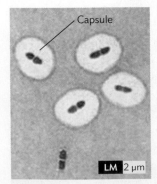

A. *Streptococcus pneumoniae* surrounded by capsule (India ink preparation). The slippery nature of the polysaccharide capsule makes phagocytosis more difficult.

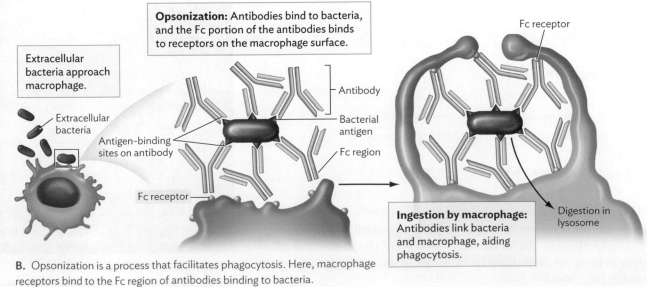

B. Opsonization is a process that facilitates phagocytosis. Here, macrophage receptors bind to the Fc region of antibodies binding to bacteria.

Fc receptors on phagocyte cell surfaces. As a result, the antibodies form a bridge that binds the bacteria to phagocyte Fc receptors, allowing the phagocyte to "grab" and engulf the invader.

Link You will learn in Section 16.3 that antibodies are Y-shaped molecules in which the two top arms of the Y bind to a specific antigen. Thus, one antibody has two identical antigen-binding sites.

Oxygen-Dependent and -Independent Killing Mechanisms

During phagocytosis, the cytoplasmic membrane of the phagocyte flows around, and then engulfs, the bacterium, producing an intracellular phagosome, as described in Figure 15.5. Subsequent phagosome-lysosome fusion (producing a phagolysosome) permits both oxygen-independent (anaerobic) and oxygen-dependent (aerobic) killing pathways. Oxygen-independent mechanisms include enzymes such as lysozyme that destroy the cell wall, compounds such as lactoferrin that bind iron and keep it away from the microbe, and defensins, small cationic antimicrobial peptides (described in Section 15.2).

Oxygen-dependent mechanisms kill through the production of various highly reactive oxygen radicals that bind and damage lipids, proteins, DNA, and RNA. Cellular enzymes in the phagosome membrane generate superoxide ion (O_2^-), hydrogen peroxide (H_2O_2), and hydroxyl ($\cdot OH$) radicals. Myeloperoxidase, present only in neutrophils, converts hydrogen peroxide and chloride ions to the highly reactive compound hypochlorous acid (HOCl).

Macrophages, mast cells, and neutrophils also generate reactive nitrogen intermediates that serve as potent cytotoxic agents. Nitric oxide (NO) is synthesized by NO synthetase. Further oxidation of NO by oxygen yields nitrite ions (NO_2^-) and nitrate ions (NO_3^-). These mechanisms are responsible for the large increase in oxygen consumption noted during phagocytosis, called the **oxidative burst**. The reactive chemical species formed during the oxidative burst do little to harm the phagocyte because the burst is limited to the phagolysosome and because the various reactive oxygen species, such as superoxide, are very short-lived.

Surviving Phagocytosis

Phagocytosis is very effective at killing microbes. However, some pathogens have evolved ingenious ways to avoid being killed by phagocytosis. Some bacterial intracellular pathogens, such as *Coxiella*, have evolved to live within the toxic phagolysosome. Others, such as *Shigella* and *Listeria*, escape from the phagosome before phagolysosome fusion can take place and live in the cytoplasm. Another group (*Salmonella*) remains in the phagosome but secretes proteins that prevent fusion with lysosomes. Still others, such as *Shigella*, enter the host cell (for instance, a macrophage) and kill it by triggering a host process called **apoptosis**. Death of the macrophage saves the bacterium from phagocytosis. Apoptosis is a normal host process in which a cell kills itself, but in such a way that

Figure 15.17 A Macrophage Ingesting an Apoptotic Neutrophil

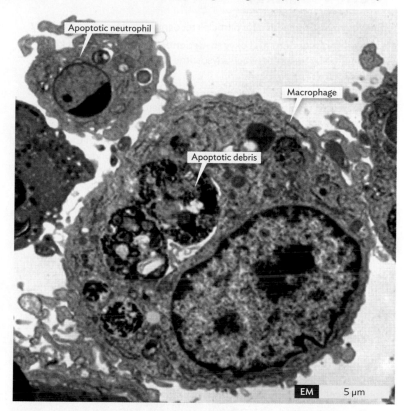

The electron micrograph was taken from a resolving lung inflammatory response and depicts an apoptotic neutrophil in contact with the macrophage surface and apoptotic debris within phagosomes.

limits damage to surrounding cells. Apoptosis is important during embryonic development. In a developing fetus, for example, cells that form webs between fingers and toes will apoptose before birth to release free digits. Apoptosis also plays a role in ending inflammation, which we discuss next.

Ending Inflammation

During an inflammatory response, many neutrophils converge on an infected area (commonly seen as pus). But what happens once the infection is cleared? Neutrophils are like tiny time bombs loaded with destructive enzymes and signal molecules. Because the life of a neutrophil is quite short (7 days) and there are no chemoattractants to draw them <u>away</u> from the former site of infection, cleanup crews of long-lived phagocytes, such as macrophages, are called on to quickly remove neutrophils from the area. Once a neutrophil has finished its job, the cell undergoes a programmed death process (apoptosis) that conserves membrane integrity (preventing the uncontrolled release of toxic compounds) and allows phagocytosis by macrophages (**Figure 15.17**). Once engulfed by the macrophage, the dead neutrophil can no longer release its toxic chemical ordnance into recovering tissues. Recovery from inflammation (drainage through the lymphatic system and wound healing) can begin.

15.5
Interferon, Natural Killer Cells, and Toll-Like Receptors

Can human cells under attack warn others of danger? When a residential community is threatened by a thief, a neighbor who has been robbed alerts others to take precautions. And if the thief is discovered, police are called to arrest him. Similarly, interferon peptides secreted by infected cells warn healthy host cells of a nearby infection, and natural killer cells seek out and then destroy cells already infected.

Interferons

Interferons are low-molecular-mass cytokines (15–20 kDa) produced by many eukaryotic cells in response to intracellular infection—that is, infection by viruses or by bacterial pathogens, such as *Listeria*, that can grow intracellularly. For instance, an uninfected host cell that encounters interferon will increase its antiviral defenses. The action of interferons is usually specific to the host species (that is, interferon from mice will not work on human cells) but virus nonspecific (human interferon will help protect against both poliovirus and influenza virus, for example).

There are two general types of interferons, which differ in the receptors they bind and the responses they generate (Table 15.2). Type I interferons have high antiviral potency; they consist of IFN-alpha (IFN-α), IFN-beta (IFN-β), and IFN-omega (IFN-ω). Type II interferon, IFN-gamma (IFN-γ), has more of an immuno-modulatory function that will be discussed in the next chapter.

Table 15.2
Classes of Interferons

Type	Examples	Function	Mechanism of Action
I	IFN-alpha IFN-beta IFN-omega	Antiviral	Induce dsRNA endonuclease, eIF2 kinase
II	IFN-gamma	Immuno-modulatory	Triggers signal cascades in macrophages, NK cells, and T cells; increases MHC II on cell surfaces

Type I interferons bind to specific receptors on uninfected host cells and induce the synthesis of two classes of proteins: double-stranded RNA-activated endoribonucleases, which can cleave viral RNA, and protein kinases (PKRs) that phosphorylate and inactivate eukaryotic initiation factor eIF2, which is required to translate viral RNA. These mechanisms affect both RNA and DNA viruses because protein synthesis is required for the propagation of all viruses.

Type II interferon functions by activating various white blood cells—for example, macrophages, natural killer cells, and T cells—to increase the number of **major histocompatibility complex (MHC)** antigens on their surfaces. MHC proteins are important for recognizing self and for presenting foreign antigens to the adaptive immune system. They will be discussed more fully in Chapter 16.

Natural Killer Cells

Body cells that are infected or that are cancerous can be a major problem for the host. Infected cells can hide a pathogenic microbe from the immune system, whereas cancer cells can take over and kill the host. A class of lymphoid cells called **natural killer (NK) cells** are able to identify and handle these situations. Instead of killing microbes, the mission of these cells is to destroy host cells that harbor microorganisms or that have been transformed into cancer cells (Figure 15.18). Natural killer cells recognize changes in cell-surface proteins of compromised cells and then degranulate to release chemicals that kill those cells.

Natural killer cells recognize their targets in two basic ways. One involves MHC class I molecules, and the other uses antibody Fc receptors. A normal host cell displays two classes of MHC molecules on the outside of the cell membrane. MHC I is an indicator of "self." (MHC II molecules will be discussed in Chapter 16.) NK cells have specific receptors that bind to self MHC I molecules on the surfaces of other cells in the body. An NK cell that "touches" a host cell with a compatible self MHC I protein will not attack that cell. However, if a host cell lacks MHC class I molecules, as can happen during infection or as a result of cancer, NK cells will perceive the host cell as foreign and a potential threat. The NK cell will insert a pore-forming protein called **perforin** into the

Figure 15.18   Natural Killer Cells

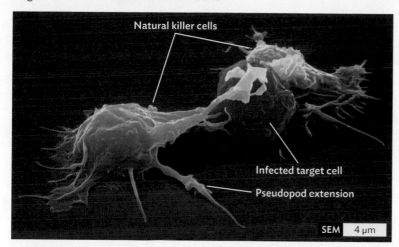

Natural killer cells

Infected target cell

Pseudopod extension

SEM   4 μm

Natural killer (NK) cells attack eukaryotic cells infected by microbes, not the microbes themselves. Perforin produced by the NK cell punctures the membrane of target cells, causing them to burst.

membrane of the target cell, through which cytotoxic enzymes are delivered (Figure 15.19).

Natural killer cells also contain antibody-binding Fc receptors on their cell surface. The second killing mechanism, called **antibody-dependent cell-mediated cytotoxicity (ADCC)**, occurs when the Fc receptor on the NK cell links to an antibody-coated host cell. Why would host cells be coated with antibodies? During their replication, many viruses place viral proteins in the membrane of the

infected cell. Because the viral proteins protrude from the host membrane, antibodies to those proteins will coat the compromised cell, tagging it for ADCC. Once the compromised cell is targeted, it is killed by the NK cell in the same way as described for cells lacking MHC, by insertion of a perforin molecule. This killing mechanism is another example of cooperation between innate immunity (NK cells) and adaptive immunity (antibody-producing lymphocytes).

Toll-Like Receptors, NOD Proteins, and Microbe-Associated Molecular Patterns (MAMPs)

CASE HISTORY 15.2

An Immune Defect That Takes a Toll

Bobby, an 11-year-old boy from Nebraska, suffered his entire life from recurring life-threatening infections caused mostly by the Gram-positive bacteria *Staphylococcus aureus* and *Streptococcus pneumoniae*. His infections included meningitis, osteomyelitis (bone infection), and arthritis caused by *S. pneumoniae*, as well as episodes of septicemia and osteomyelitis and recurrent boils caused by *S. aureus*. Tests to find a cause for these recurring infections were initially disappointing. The results of all standard immunological tests were normal, including T-cell lymphocyte responses, blood antibody levels, and antibody responses to injected proteins and polysaccharides. This means his adaptive immune system was functioning. The numbers of monocytes/macrophages in his blood were also normal. However, pro-inflammatory cytokine levels (indicators of innate immunity) measured during the latest infection were

Figure 15.19   Natural Killer Cell Killing a Target Cell

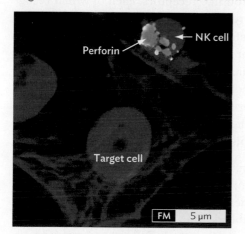

NK cell

Perforin

Target cell

FM   5 μm

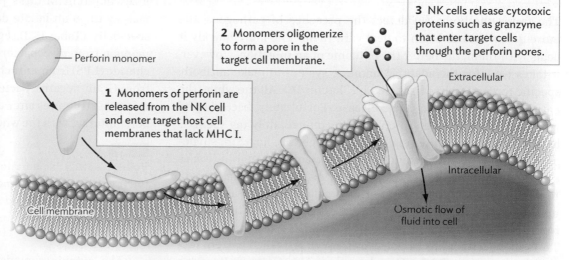

Perforin monomer

2 Monomers oligomerize to form a pore in the target cell membrane.

1 Monomers of perforin are released from the NK cell and enter target host cell membranes that lack MHC I.

3 NK cells release cytotoxic proteins such as granzyme that enter target cells through the perforin pores.

Extracellular

Intracellular

Osmotic flow of fluid into cell

Cell membrane

A. Natural killer cells recognize cells that have lost expression of MHC class I molecules. NK cells store cytotoxic molecules such as perforin in secretory lysosomes. Recognition of an infected cell triggers the release of these cytotoxic molecules (green fluorescence), which then kill the target cell. Red fluorescence indicates cytoskeleton.

B. Perforin forms a pore in the target cell membrane and destroys membrane integrity. Cytotoxic proteases, called granzymes, are also released from the NK cell, pass through the perforin pore, and enter the target cell. The granzymes trigger apoptosis of the target cell.

Table 15.3

Examples of Toll-Like Receptors and NOD-Like Receptors

Receptor	MAMPs Recognized	Source	Host Cells	Location
TLR1	Lipopeptides	Bacteria	Monocytes/macrophages, dendritic cells, B cells	Cell surface
TLR2	Glycolipids, lipoteichoic acids	Bacteria	Monocytes/macrophages, dendritic cells, mast cells	Cell surface
TLR3	Double-stranded RNA	Viruses	Dendritic cells, B cells	Cell compartment
TLR4	Lipopolysaccharide, heat-shock proteins	Bacteria	Monocytes/macrophages, dendritic cells, mast cells, intestinal epithelium	Cell surface
TLR5	Flagellin	Bacteria	Monocytes/macrophages, dendritic cells, intestinal epithelium	Cell surface
TLR6	Diacyl lipopeptides	Mycoplasma	Monocytes/macrophages, mast cells, B cells	Cell surface
TLR9	Unmethylated CpG residues in DNA	Bacteria	Monocytes/macrophages	Cell compartment
NOD1	Component of Gram-negative peptidoglycan	Gram-negative bacteria	Many cell types	Inflammasome
NOD2	Peptidoglycan component	Bacteria	Macrophages, dendritic cells, epithelia of lung and GI tract	Inflammasome
NLRP-3	Peptidoglycan	Bacteria	Many cell types	Inflammasome
NLRP-4	Flagellin, CpG, ATP, dsRNA	Bacteria	Many cell types	Inflammasome

considerably lower than expected. Further tests proved that Bobby inherited an innate immune defect in Toll-like receptor signaling. Currently, there is no cure for this immunodeficiency. To stem the tide of infections, Bobby was placed on long-term, preventive antibiotic treatment.

The faster the body can detect the presence of pathogens, the more quickly it can begin to deal with them. The more quickly it deals with them, the better the outcome of an infection. However, it takes time for the adaptive immune system to make antibody specific for a microbe (discussed in Chapter 16). All the while, the pathogen can grow and cause disease. Fortunately, bacteria and viruses possess unique structures that immediately tag them as being foreign. Structures such as peptidoglycan, flagellin, single- or double-stranded RNA, and lipoteichoic acids are not in tissues unless bacteria or viruses are present. As mentioned in Section 15.2, these structures have what are called microbe-associated molecular patterns (MAMPs) that can be recognized by **Toll-like receptors (TLRs)** present on various host cell types (Table 15.3). TLRs, along with NLRs (described shortly), are called **pattern recognition receptors (PRRs)**. Once bound to a MAMP, the TLR sends a signal to the interior of the cell to start making interferon and pro-inflammatory cytokines that will speed the inflammatory response (Figure 15.20).

First discovered in insects, Toll-like receptors are evolutionarily conserved glycoproteins present on the cell membranes of many eukaryotic genera. (Note, however, that some TLRs, such as TLR7, are located only in the membranes of intracellular vacuoles.) Humans have numerous Toll-like receptors, each of which recognizes different MAMPs present on pathogenic microorganisms, making them an innate defense mechanism with some degree of specificity (Table 15.3). For example, TLR2 binds to lipoarabinomannan from mycobacteria, zymosan from yeast, and lipopolysaccharide (LPS) from spirochetes. TLR4, on the other hand, binds LPS from Gram-negative bacteria. Note that these receptors bind fragments of structures after they are released from the microbe. They don't interact with the whole organism.

Note The term "Toll gene" came from Christiane Nüsslein-Volhard's 1985 exclamation, "That's crazy" (in German, "Das war ja toll") when shown the underdeveloped posterior of a mutant fruit fly. Thus, the gene was dubbed "Toll." Toll-like receptors in mammals display sequence similarities to the Toll genes involved with insect embryogenesis.

The cytokines made after a MAMP-TLR interaction float away, bind to receptors on various cells of the immune system, and direct them to engage the invading microbe (inflammation; see Figure 15.13). Bobby, in Case History 15.2, had a defect in one

Figure 15.20

Cell-Surface TLRs and Cytoplasmic NLRs Recognize MAMPs

Toll-like receptors (TLRs) bind extracellular microbe-associated molecular patterns (MAMPs) and transmit a signal through a cascade of other proteins to the nucleus. The signal cascade activates transcription of specific cytokine genes. (*Note:* TLR3, TLR7, and TLR9, not shown in this figure, are located in endosomal membranes.) Cytoplasmic NOD-like receptors (NLRs) bind MAMPs generated primarily by intracellular pathogens and trigger a signal cascade different from that used by TLRs. NLRs also trigger inflammasome assembly.

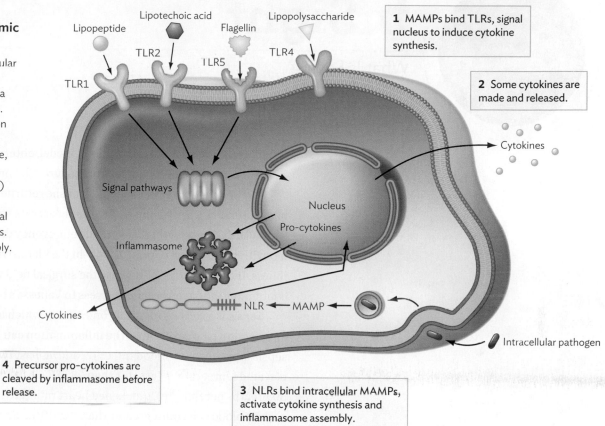

of the intracellular proteins that receives the signal from TLRs. Consequently, battle-ready levels of cytokines could not be made. Cytokines can influence the function of cells that are part of innate immunity, adaptive immunity, or both.

Although TLRs are restricted to membranes (some surface, some vacuolar), certain cytoplasmic proteins can also recognize MAMPs. **NOD-like receptor (NLR)** proteins, so named because of their similarity to a family of plant proteins (NOD factors) involved in pathogen resistance, are important intracellular sensors of MAMPs. When bound to a MAMP, NLRs trigger a signaling pathway different from that used by TLRs. The result, though, is the same—activation of cytokine genes in the nucleus. NLR activation also initiates assembly of inflammasomes that proteolytically cleave precursor pro-cytokines into active cytokines. Once made, the cytokines are secreted and bind to target immune system cells to initiate the inflammation and adaptive immune mechanisms needed to respond to intracellular pathogens.

Note The acronym "NOD" stands for "nucleotide-binding oligomerization domain."

The severity of Bobby's condition illustrates how important TLRs and innate immunity are for preventing serious disease—even when the adaptive immune system is functional. If Bobby lives long enough, however, his adaptive immunity (production of antibodies, for example) against these pathogens will build to a point where preventive antibiotic therapy can probably be stopped. TLRs

and NLRs also play important roles in the medical condition called sepsis (see **inSight**).

In addition to infection-driven inflammation triggered by MAMP interactions with TLRs and NLRs, there is also sterile inflammation in which human cells damaged by trauma or by decreased blood flow release damage-associated molecular pattern (DAMP) molecules, such as ATP, hyaluronan, and histones that also interact with certain PRRs. Like MAMPs, DAMPS trigger synthesis of pro-inflammatory cytokines that produce inflammation. The activation of innate immunity by DAMPs alerts the host to damage and leads to a repair response.

SECTION SUMMARY

- **Interferons are host-specific** molecules that can nonspecifically interfere with viral replication (type I) and modulate the immune system (type II).
- **Natural killer (NK) cells** are a class of white blood cell that destroy cancer cells or cells harboring microorganisms.
- **Natural killer cells target host cells** that have either lost MHC class I receptors as a result of infection or cancer or are coated with antibody (antibody-dependent cell-mediated cytotoxicity, ADCC).
- **Natural killer cells kill** by inserting perforin pores into the membranes of target cells.
- **Membrane-embedded Toll-like receptors (TLRs) and intracytoplasmic NOD-like receptors (NLRs)** recognize different microbe-associated molecular patterns (MAMPs) and activate synthesis and secretion of cytokines.

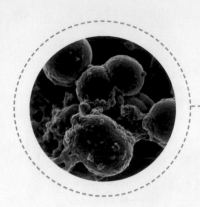

inSight

What Is Sepsis?

Vanessa, a 22-year-old fashion model, entered the hospital for what was to be a routine plastic surgery procedure. The procedure went well and she was released from the hospital, but she returned 3 days later extremely ill. Her blood pressure was alarmingly low, her heart rate was rapid, she was disoriented, and she had a dangerously high fever. The emergency department physician diagnosed her with sepsis, an infection, usually in the bloodstream, that causes life-threatening illness. It appears that in this case the surgical field was not thoroughly sterilized and contaminating bacteria gained access to Vanessa's bloodstream.

Sepsis is a serious medical condition in which a pro-inflammatory response to an infection is so severe that the inflammation can kill the patient. A patient with sepsis typically has a blood infection and at least two of the following symptoms: elevated temperature (greater than 38.5°C, or 101.3°F), fast respiration (more than 20 breaths per minute), quickened heart rate (greater than 90 beats per minute), or a white blood cell count greater than 12,000/μl. Severe sepsis can result in altered

Figure 1   The Triggering of Sepsis by LPS and Lipoteichoic Acid

Gram-negative bacilli — Lipopolysaccharide

Gram-positive cocci — Lipoteichoic acid

CD14 receptor   Toll-like receptor

Monocyte

TNF-α and IL-1 → Direct tissue injury

Tissue factor

Thrombin

Microvascular coagulopathy

Microclots form as blood proteins, such as thrombin, are exposed to tissue factor.

The MAMPs released by bacteria, such as lipoteichoic acid (Gram-positive bacteria) or LPS (Gram-negative bacteria), will bind TLRs on macrophages and other cell types. The activated cells release TNF-α and IL-1 pro-inflammatory cytokines. Damage to endothelial cells exposes tissue factor, which triggers the coagulation cascade. Fibrin and platelets form clots in the microvasculature and block the flow of oxygenated blood to various organs. TNF-α and IL-1 have other effects that can lead to direct tissue injury and elevated body temperature (fever).

mental status, decreased urine output, and decreased platelet counts. A patient who also exhibits extremely low blood pressure is said to have septic shock; 30% of these patients will die. Surprisingly, what causes these symptoms is the host response to the bacterial pathogen, not the pathogen itself.

The pathogens start the cascade of events leading to sepsis by interacting with TLRs and NLRs on host cells. For Gram-negative organisms, endotoxin, which is a component of lipopolysaccharide (LPS) in the outer membrane, stimulates innate immune cells via Toll-like receptor 4 (TLR4). For Gram-positive organisms (for example, *Staphylococcus aureus* and *Bacillus anthracis*), the triggers include cell wall components such as lipoteichoic acid (binds TLR2) and peptidoglycan (binds NLRs). Bacterial toxins called superantigens can produce septic shock as well (discussed in Section 18.3). Regardless of what starts the process, the result is that cells activated by MAMPs or by superantigens release a "storm" of pro-inflammatory cytokines such as TNF-α and IL-1 that affect the body's thermostat (discussed in Section 15.7) and damage endothelial cells lining blood vessels. Some of the trouble they cause was explained during our discussion of inflammation.

Figure 1 illustrates the basic mechanisms (or pathophysiology) of sepsis. Because blood vessels are damaged, proteins that are normally hidden from blood by endothelial cells (such as tissue factor, TF, the primary cellular initiator of blood coagulation) become exposed to circulating blood factors that mediate coagulation. Small blood clots (microclots) begin to form once exposed tissue factor binds to these blood proteins. The microclots block capillaries, causing diminished blood flow and poor oxygen delivery to affected organs. Coagulation occurs in many locations throughout the body, producing what is called disseminated intravascular coagulation (DIC). All of this coagulation causes a drop in blood platelet levels, which are important for clotting. The irony is that now, because platelet numbers are low, it takes longer for blood to coagulate. A consequence of damaged blood vessels and decreased clotting time is that blood begins seeping out of the damaged vessels and into tissues. The result is a rash that begins as a series of pinpoint red splotches on the skin called petechiae. The petechiae can enlarge and coalesce into large purple purpuric rashes (Figure 2). Massive leakage of blood into organs decreases the effective blood volume, which produces a severe drop in blood pressure. The patient goes into shock and often dies.

So how is sepsis treated? Intravenous fluids are initially used to try to raise blood pressure, but if that fails, which means the patient is in septic shock, vasopressor drugs are administered to constrict smooth muscles around blood vessels. The narrowing of vessels will increase blood pressure. Antibiotics are the only way to kill or inhibit growth of the organism. However, even dead organisms will release endotoxin or peptidoglycan, which can make the sepsis even worse before it gets better. The best hope for recovery is to recognize sepsis early before it becomes too severe and treatment becomes futile.

Figure 2 **Disseminated Intravascular Coagulation (DIC)**

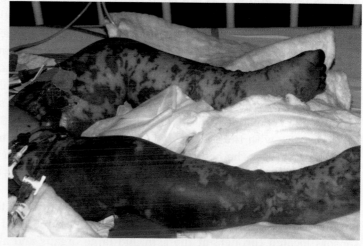

A 1-year-old boy has massive purpura as a result of DIC caused by *Neisseria meningitidis* septicemia.

Thought Question 15. 5   Both NK cells and neutrophils have Fc receptors on their cell membranes. NK cells successfully use these receptors to carry out ADCC on virus-infected cells that are coated with antibody bound to the viral proteins inserted in the host cell's membrane. Neutrophils can bind to these antibody-coated cells but will not phagocytize them. Why not?

Thought Question 15.6   *Shigella flexneri* is a Gram-negative rod that causes bacillary dysentery (bloody diarrhea). The organism is ingested and enters M cells in the intestine. After exiting the M cell, *S. flexneri* forcibly enters nearby macrophages and escapes from the phagocytic vacuole. *Shigella* triggers the activation of NLRs and cytokine release, but at a level much lower than you would expect. Propose a mechanism that might accomplish this.

Thought Question 15.7   Why haven't microbes evolved to avoid being recognized by Toll-like receptors?

Thought Question 15.8   Can you explain why the infections in Bobby's case (Case History 15.2) were caused by Gram-positive cocci and not Gram-negative bacteria (such as *Salmonella*) or viruses? You need to know that *S. aureus* and *S. pneumoniae* are extracellular pathogens, whereas viruses and *Salmonella* are intracellular pathogens.

15.6
Fever

SECTION OBJECTIVES

- Explain fever and its cause.
- Describe the advantages and disadvantages of fever.

What is fever, and why is it a good thing?   To understand fever, we first need to know how the body controls its temperature (thermoregulation). Heat in a human body is produced as a consequence of metabolic reactions. The liver and muscles are major generators of heat and will warm blood that passes through them. In a healthy person, body temperature is kept constant within a very small range, despite large differences in surrounding temperature and physical activity. The normal healthy adult oral temperature is 36°C–38°C (97°F–100°F), with the average being 37°C (98.6°F). Rectal temperature measures the body's core temperature and is about 1°C (1.8°F) higher than oral temperature. Fever is when a person's oral (or ear) temperature is above 38°C (100.4°F).

So, how do we normally maintain a 37°C (98.6°F) body temperature when the outside air temperature is 18.3°C (65°F)? Heat sensors located throughout the skin and large organs and along the spinal cord send information about the body's temperature to the thermoregulatory center in the hypothalamus, a small organ located in the brain near the brain stem. The hypothalamus acts as a thermostat by controlling blood flow through the skin and subcutaneous areas. Vasoconstriction (tightening of blood vessel diameter) prevents the release of body heat when we are cold, whereas vasodilation secures its quick release when we are hot. If skin temperature is too high, the hypothalamus directs vasodilation to accelerate heat release. If body temperature is too low, blood flow will decrease to conserve heat, and shivering begins as a way to generate more heat.

Fever (elevated body temperature) is a natural reaction to infection and is usually accompanied by general symptoms, such as sweating, chills, and the sensation of feeling cold. Substances that cause fever are known as **pyrogens**. External pyrogens (for example, certain bacterial toxins) originate outside the body, whereas internal pyrogens (such as interferon, tumor necrosis factor, and IL-6) are made by the body itself. External pyrogens generally cause fever by inducing the release of internal cytokine pyrogens. These pyrogenic cytokines bind to receptors on neurons near the thermoregulatory center in the hypothalamus. This stimulates production of phospholipase A2, an enzyme required to make prostaglandins, which then change the responsiveness of the thermosensitive neurons that make up the thermoregulatory center. In other words, pyrogens turn up the thermostat. The body thinks it is cold, so the hypothalamus sends signals via the autonomic nervous system (which acts below the level of consciousness) to constrict peripheral blood vessels (vasoconstriction). Heat is not released and builds up to cause fever. Involuntary muscle contractions (shivering and chills) also generate heat.

What are the advantages and disadvantages of fever? The obvious disadvantage of fever is the discomfort the patient feels and, if the fever rises too high (above 42°C, or 107.5°F), irreversible brain damage can occur. However, because the ideal growth temperature for many microbes is 37°C (98.6°F), elevated temperature can place the organism outside its "comfort zone" of growth. There is also evidence that fever reduces iron availability to bacteria. Slower growth of the pathogen allows the body's immune system time to subdue the infection before it is too late. Consequently, interventions that reduce a moderate fever caused by infection may be counterproductive to a speedy recovery.

SECTION SUMMARY

- **The hypothalamus** acts as the body's thermostat.
- **External and internal pyrogens** elevate body temperature by stimulating production of prostaglandins.
- **Prostaglandins** change the responsiveness of thermosensitive neurons in the hypothalamus.
- **Fever** raises body temperature above the optimal level for bacterial and viral growth and may reduce iron availability for bacterial pathogens.

Thought Question 15.9   If increased fever limits bacterial growth, why would bacteria make pyrogenic toxins that increase body temperature?

15.7
Complement

White blood cells engulf and kill pathogens, but can simple serum proteins also kill microbes? Yes, a series of 20 serum proteins called complement factors also attack bacterial invaders. Complement is extremely critical to preventing infections. But sometimes something goes wrong.

CASE HISTORY 15.3

A Little "Complement" Goes a Long Way

Derek, a 16-year-old high school chess prodigy, was rushed by his parents to the St. Michael's emergency department in Toronto, Canada. The staff found Derek very confused and disoriented. His parents said that Derek had complained of a sore throat a few days earlier but over the last 24 hours developed a very high fever and a severe headache. When examined, Derek had a stiff neck and many reddish spots, called petechiae, over his body. The attending physician performed a lumbar puncture, in which a needle is inserted into the spinal column between two lower back vertebrae to remove cerebrospinal fluid (CSF). Normally clear, Derek's CSF was cloudy, looking like dilute milk. The physician diagnosed meningitis. A few hours later, the hospital laboratory reported that the CSF contained Gram-negative diplococci and a rapid test indicated *Neisseria meningitidis*. The doctor was not surprised because this was Derek's third bout with meningitis caused by this organism. The other infections occurred when Derek was 9 and 13 years old. After this last episode, the physician suspected that Derek might have a genetic deficiency in one of his blood complement components. The laboratory confirmed that Derek's blood had a deficiency in complement factor C8.

Meningitis is a serious disease that can rapidly turn fatal. *Neisseria meningitidis*, only one of the possible causes of this disease, first infects the nasopharynx to cause a sore throat. Sometimes the bacteria penetrate into the bloodstream and are carried throughout the body. From the bloodstream, *N. meningitidis* can cross the blood-brain barrier and enter the brain and spinal column. The inflammation that results causes increased cranial pressure, which produces the headache, stiff neck, and disorientation. Over their lifetimes, most patients contend with only a single bout of meningitis caused by *N. meningitidis*. Recurring bouts of *Neisseria* meningitis are quite uncommon unless the patient has a defect in one or more of the complement cascade proteins.

Any localized bacterial infection can become deadly once an organism gains access to the bloodstream to cause septicemia. During septicemia, the organism can be spread (disseminated) to far-flung destinations throughout the body and cause secondary infections called metastatic infections. Even a trip to the dentist can result in oral bacteria gaining entrance to the blood. Why doesn't septicemia occur more often than it does? It turns out that a series of 20 proteins in the blood, collectively called complement, play a huge role in preventing blood infections. Complement was first discovered as a heat-labile component of blood that enhances (complements) the killing effect of antibodies on bacteria. Several complement components are proteases that sequentially cleave each other in a cascade. Triggering the complement cascade produces a number of outcomes. Pores called membrane attack complexes (MACs), composed of four complement components, are inserted into membranes and cause cytoplasmic leaks. In addition, pieces of the complement system can attract white blood cells and facilitate phagocytosis (opsonization).

Pathways of Complement Activation

There are three complement-activating pathways officially dubbed the classical pathway, the alternative pathway, and the lectin pathway. The classical complement pathway depends on antibody, so it links the innate immunity of complement with the adaptive immune system. However, because the adaptive immune system can take days to make antibody, this system does not immediately provide protection. The alternative pathway does not require antibody for activation and can therefore attack invading microbes long before a specific immune response can be launched. In terms of response time, the lectin pathway falls between the other two. The lectin pathway requires synthesis of mannose-binding lectin by the liver in response to macrophage cytokines released after a macrophage engages the pathogen. This lectin coats the surface of invading microbes and activates complement that bumps into it. A major goal of the complement cascade is to insert pores into target microbial membranes. The pores destroy membrane integrity, which kills the cell.

All pathways converge at the complement C3 protein (**Figure 15.21**). The steps are complex but simplified here for discussion.

1. **The alternative pathway.** The alternative pathway starts with the spontaneous cleavage of factor C3 into two parts, C3a and C3b. The C3b fragment combines with factor B to form a C3 convertase that quickens C3 cleavage. The C3b fragment also binds to carbohydrates on cell surfaces, such as LPS on Gram-negative outer membranes, and becomes part of a

Figure 15.21   Complement Cascades

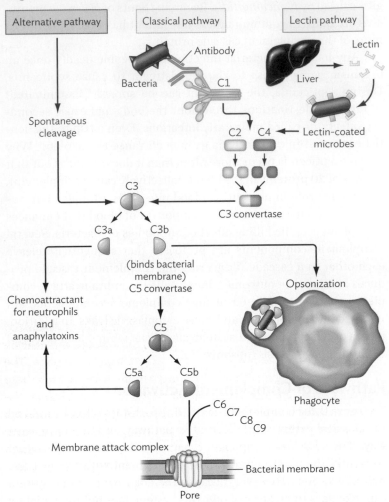

Three pathways can activate complement. The three pathways begin differently, but converge at C3. The alternative pathway starts with the spontaneous cleavage of C3. The C3b fragment (called C5 convertase) binds to bacterial surfaces (such as LPS) and cleaves C5. The C5b fragment enters the membrane and directs the assembly of the C6, C7, C8, and C9 pore, which kills the bacterial cell. The cascade from C3b to pore formation is common to all pathways. The classical pathway starts with antibody bound to a bacterial cell. The Fc tail of the antibody binds and activates the protease of C1. C1 mediates the cleavage of C4 and, indirectly, C2. Parts of C2 and C4 combine to form a C3 convertase, which speeds the cleavage of C3. In the lectin pathway, lectin, made by the liver, circulates in the blood and binds mannan (polymers of mannose) on bacterial surfaces. Lectin mediates cleavage of C4 and C2. As a result, C3 convertase forms and the cascade continues as before. C3a and C5a can act as chemoattractants and anaphylatoxins, whereas C3b can function as an opsonin as well as a C5 convertase.

complex (called C5 convertase) that cleaves a factor called C5 into C5a and C5b fragments. (You can think of C3b as a sticky "grenade" tossed at microbes in the bloodstream. Once it hits, an "explosion" of sequential reactions happens.) The C5a fragment made as a result of C5 convertase chemoattracts PMNs. The C5b fragment enters the bacterial membrane and directs the assembly of proteins C6, C7, C8, and C9 into a pore, the so-called membrane attack complex, which kills the bacterial cell. The cascade we have just described, from C5 convertase to pore formation, is <u>identical</u> in all complement activation pathways.

2. **The classical pathway.** The classical pathway starts with antibody bound to a bacterial cell. The Fc tail portion of the antibody binds and activates the protease of a large, spindly-looking complement factor called C1. C1 then cleaves two other factors, C4 and C2. Parts of C2 and C4 unite to form another C3 convertase (different from the one in the alternative pathway), which speeds the cleavage of C3 (again forming C3b), which propels the rest of the cascade.

3. **The lectin pathway.** In the lectin pathway, lectin, made by the liver, circulates in the blood and binds carbohydrate structures (mannose residues) on bacterial surfaces. Lectin mediates cleavage of C4—and, indirectly, C2—the same factors used in the classical pathway. As a result, C3 convertase once again forms, the C3b "grenade" is made, and the cascade continues as before.

When complement interacts with Gram-negative cells, lysozyme (present in serum) enters through outer membrane pores and cleaves peptidoglycan, making the cell membrane more susceptible to the membrane attack complex. Gram-positive bacteria are resistant to complement because they lack an outer membrane (no LPS to efficiently start the cascade) and have a thick peptidoglycan layer that hinders access of complement components. However, even in the absence of LPS, antibody-coated Gram-positive bacteria can activate complement via the classical pathway, and several complement factors can intensify inflammation. For instance, C3a and C5a fragments act as chemoattractants to call in neutrophils and work as anaphylatoxins, compounds that trigger the release of vasoactive factors such as histamine from endothelial cells, mast cells, or phagocytes. Also, C3b can function as an opsonin to facilitate phagocytosis. Because phagocytes contain C3b receptors on their surfaces, it is easier for them to grab and engulf cells coated with C3b.

Regulating Complement Activation

Membrane attack complexes are not selective about the membranes they attack. So how do normal body cells prevent self-destruction following complement activation? There are several places where regulatory factors can intervene. One such factor is the host cell surface protein CD59 (called protectin). This protein will bind any

C5b-C8 complex trying to form in the membrane and prevent C9 from polymerizing. Thus, no pore is formed and the host cell is spared. Another regulatory mechanism hinges on a normal serum protein called **factor H**. This protein prevents the inadvertent activation of complement in the absence of infection. Factor H normally binds to the surfaces of host cells, not pathogens, and protects the host cell from complement attack by inactivating C3b. Effective protection of host cells from complement requires CD59 and factor H. Some bacteria, such as certain strains of *Neisseria gonorrhoeae*, have evolved to protect themselves from complement by binding factor H—cloaking themselves in the host's own protective factor.

C-Reactive Protein and Complement

As noted earlier, inflammation is associated with the production of various cytokines by macrophages. Within hours of an infection some of these cytokines (IL-1, TNF-α, and IL-6) travel to the liver and stimulate synthesis of several so-called acute-phase reactant proteins, including **C-reactive protein**. This protein, named for its ability to activate complement, will bind to bacterial cell surfaces but not to host cell membranes. Once stuck to the bacterial cell surface, C-reactive protein binds C1 of the classical complement pathway and initiates the complement cascade. C-reactive protein essentially takes the place of antibody in the classical pathway.

Note The term "acute" in "acute-phase reactants" refers to the rapid, or acute, development of an infection. Despite their name, acute-phase reactants accompany both acute and chronic inflammatory states. They can also appear in association with trauma, infarction, inflammatory arthritis, and various neoplasms.

SECTION SUMMARY

- **Complement** is a series of 20 proteins naturally present in serum.
- **Activation of the complement cascade** results in a pore being introduced into target membranes.
- **The three pathways for activation** are the classical pathway, the alternative pathway, and the lectin pathway. They differ in the pathways leading to C3 cleavage to C3b.
- **The cascade of protein factors** C3 to C3b to C5 to C5b and the membrane assembly of C5b, C6, C7, C8, and C9 produce the membrane attack complex in target membranes.
- **C-reactive protein** in serum is activated when bound to microbial structures and will convert C3 to C3b, which can also start the complement cascade.

Thought Question 15.10 Why does complement have to be activated? Why is it not always active and ready to kill invading microbes?

Perspective

The various "hardwired" innate immune mechanisms described in this chapter keep our normal microbial flora at bay and provide an effective first line of defense against potential pathogens. The next chapter addresses what happens when microbes breach these innate defenses. Unlike the general protective mechanisms discussed in this chapter, the system of acquired immunity generates a molecular defense specifically tailored to a given pathogen. Keep in mind that the innate and adaptive immune systems "talk" to each other and collaborate to present the most effective response to a given pathogen, whether bacterial, viral, or parasitic.

LEARNING OUTCOMES AND ASSESSMENT

	SECTION OBJECTIVES	OBJECTIVES REVIEW
15.1 Overview of the Immune System	• Distinguish between innate immunity and adaptive immunity. • Describe the various cells and organs that constitute the immune system. • Discuss the white blood cell count differential and its role in diagnosing infectious disease.	1. Which of the following immune system components is considered part of the innate system? A. Antigen B. Complement C. Antibody D. B-cell lymphocytes E. T-cell lymphocytes
15.2 Innate Host Defenses: Keeping the Microbial Hordes at Bay	• Describe the physical barriers to infection. • Explain the basic roles of SALT, GALT, and M cells in innate immunity. • Describe chemical barriers employed by cells of the innate immune system.	4. The following cell type is part of SALT. A. B cell B. Macrophage C. Langerhans cell D. M cell E. Peyer's cell
15.3 The Acute Inflammatory Response	• List the five cardinal signs of inflammation. • Outline the process of inflammation. • Describe the roles of vasoactive factors and cytokines. • Differentiate acute from chronic inflammation.	Match the following: 7. Selectin A. Receptor on endothelial cells 8. Chemokines B. Attract macrophages 9. Bradykinin C. Edema 10. ICAM D. Fever
15.4 Phagocytosis: A Closer Look	• Describe how phagocytes recognize foreign cells. • Explain the mechanisms phagocytes use to kill engulfed cells. • List some ways pathogens avoid the consequences of phagocytosis. • Discuss how our body ends the inflammatory process.	14. The following process or protein prevents phagocytes from phagocytizing host cells. A. Apoptosis D. Fc receptors B. CD47 E. VCAM C. Integrin 15. Which of the following processes ends inflammation at a healed site of infection? A. Anti-neutrophil antibodies cause apoptosis. B. Dendritic cells engulf the dead neutrophil. C. Chemoattractants draw neutrophils away. D. Dead neutrophils release chemicals. E. Macrophages engulf dead neutrophils.

2. Which one of the following is considered a primary lymphoid organ?
   A. Peyer's patch
   B. Thymus
   C. Lymph node in the armpit
   D. Spleen
   E. Liver

3. A 35-year-old man complains of a fever and fatigue. His temperature is 38.3°C (101°F). You suspect a bacterial infection and order a CBC with differential. Which of the following is consistent with a serious bacterial infection?
   A. Leukocyte count 14,000/μl; 60% neutrophils; 10% eosinophils
   B. Leukocyte count 11,000/μl; 55% neutrophils; 30% lymphocytes
   C. Leukocyte count 15,000/μl; 60% neutrophils; 5% basophils
   D. Leukocyte count 19,000/μl; 80% neutrophils; 25% lymphocytes
   E. Leukocyte count 9,000/μl; 60% neutrophils; 25% lymphocytes

5. M cells _____.
   A. sample bacteria and other antigens from the intestine
   B. are a major producer of defensins
   C. contribute to the acidic intestinal environment
   D. are equivalent to alveolar macrophages
   E. are equivalent to Langerhans cells

6. Microbial structures recognized by host receptors are called _____.

11. A 1-year-old patient suffers from repeated infections caused by Gram-positive bacteria. Laboratory studies find 49,000 neutrophils per microliter of blood, but normal levels of antibodies. Which of the following most likely explains this result and the patient's symptoms?
    A. Excess integrin
    B. Lack of integrin
    C. Low levels of bradykinin
    D. Excess selectin
    E. Low monocyte level

12. List the five features of inflammation.

13. A granuloma is a sign of _____ infection.

16. A 5-year-old boy develops an infection with *Streptococcus pneumoniae*, an organism encased with a thick, slippery carbohydrate capsule. Which of the following helps phagocytes engulf these organisms?
    A. Defensin release
    B. NADH oxidase
    C. Fc receptors
    D. Integrin
    E. CD47

17. *Shigella dysenteriae*, a Gram-negative bacillus that causes bloody diarrhea, survives in a phagocyte by _____.
    A. preventing the oxidative burst
    B. growing in the phagolysosome
    C. preventing fusion between phagosomes and lysosomes
    D. triggering apoptosis of the phagocyte
    E. blocking integrin

LEARNING OUTCOMES AND ASSESSMENT

15.5
Interferon, Natural Killer Cells, and Toll-Like Receptors

- Describe the host-specific, antimicrobial function of interferons.
- Explain how natural killer cells first recognize and then kill infected host cells.
- Discuss Toll-like and NOD-like receptors and their contribution to innate immunity.

18. Interferon is a host-specific, but pathogen-nonspecific, defense mechanism. True or false?

19. Type I interferons can prevent viral infections by _____.
    A. activating natural killer cells
    B. triggering apoptosis of an infected host cell
    C. inducing production of antiviral antibodies
    D. stimulating phagocytosis
    E. inducing production of double-stranded RNA endonucleases

15.6
Fever

- Explain fever and its cause.
- Describe the advantages and disadvantages of fever.

22. Vasodilation will cause a decrease in body temperature. True or false?

15.7
Complement

- Define complement.
- Distinguish between the classical, alternative, and lectin pathways of complement activation.
- Discuss the role of complement in host defense.

24. The classical, alternative, and lectin pathways of complement share which one of the following?
    A. C1   D. C2
    B. Antibodies   E. C4
    C. C5

25. Some bacterial pathogens can prevent complement attack by coating themselves with which of these proteins?
    A. CD59   D. Perforin
    B. Factor H   E. C-reactive protein
    C. Factor C5

20. Which of the following mechanisms is shared by natural killer cells and phagocytes?
A. Fc receptors
B. Perforin
C. Phagocytosis
D. Selectins
E. Antibody-dependent cell cytotoxicity

21. Toll-like receptors bind to which of the following?
A. Interferon
B. Cytokines
C. Surface MHC I
D. Microbial structures
E. CD47

23. Which of the following organs or tissues serves as the body's thermostat?
A. Liver
B. Hypothalamus
C. Thymus
D. Bone marrow
E. Pituitary

26. Which of the following statements is true about complement?
A. Factor C3a acts as an opsonin.
B. Factor C3b directs assembly of the pore.
C. Factor C5b acts as an opsonin.
D. Factor C5b directs assembly of the pore.
E. Factor C1 is part of the alternative pathway.

27. An increase in blood C-reactive protein is an indication of inflammation, especially during an infection. Which of the following scenarios represents the most likely path to C-reactive protein increase during an infection?
A. MAMPs bind TLRs in macrophages and cytokines are released, triggering CRP production in the liver.
B. MAMPs bind to liver cells directly and trigger C-reactive protein production.
C. Bacterial infection triggers complement activation, and factor C3a travels to the liver and activates C-reactive protein production.

Key Terms

adaptive immunity (p. 458)
alveolar macrophage (p. 465)
antibody-dependent cell-mediated cytotoxicity (ADCC) (p. 473)
antigen (p. 459)
antigen-presenting cell (APC) (p. 461)
apoptosis (p. 471)
basophil (p. 460)
B cell (p. 462)
bradykinin (p. 468)
chemokine (p. 468)
complement (p. 459)
C-reactive protein (p. 481)
defensin (p. 465)
dendritic cell (p. 461)
eosinophil (p. 460)
extravasation (pp. 466, 467)
factor H (p. 481)

granuloma (p. 469)
gut-associated lymphoid tissue (GALT) (p. 464)
immune system (p. 458)
innate immunity (p. 458)
integrin (p. 466)
interferon (p. 472)
interleukin 1 (IL-1) (p. 467)
Langerhans cell (p. 463)
lymph node (p. 462)
lysozyme (p. 464)
macrophage (p. 460)
major histocompatibility complex (MHC) (p. 472)
mast cell (p. 460)
M cell (p. 464)
microbe-associated molecular pattern (MAMP) (p. 464)

monocyte (p. 460)
mononuclear phagocyte system (MPS) (p. 460)
mucosa-associated lymphoid tissue (MALT) (p. 464)
natural killer (NK) cell (p. 472)
neutrophil (p. 460)
NOD-like receptor (NLR) (p. 475)
opsonization (p. 470)
oxidative burst (p. 471)
pathogen-associated molecular pattern (PAMP) (p. 464)
pattern recognition receptor (PRR) (p. 474)
perforin (p. 472)
Peyer's patch (p. 464)
phagocytosis (p. 460)
phagosome (p. 460)

polymorphonuclear leukocyte (PMN) (p. 460)
pyrogen (p. 478)
selectin (p. 467)
sepsis (p. 476)
skin-associated lymphoid tissue (SALT) (p. 463)
T cell (p. 461)
Toll-like receptor (TLR) (p. 474)
tumor necrosis factor alpha (TNF-α) (p. 467)
vasoactive factor (p. 466)
vasodilation (p. 467)
white blood cell (WBC) differential (p. 462)

Review Questions

1. Name and describe various types of innate immunity.
2. What are Toll-like and NOD-like receptors, and how are they different?
3. How does the lung avoid being colonized?
4. Describe GALT and SALT.
5. Describe some chemical barriers to infection.
6. Discuss the different types of white blood cells.
7. What is a lymphoid organ?
8. Outline the process of inflammation.
9. How are neutrophils removed from a body site recovering from an infection?
10. Explain why phagocytes do not indiscriminately phagocytize body cells.
11. What is interferon?
12. Describe antibody-dependent cell cytotoxicity.
13. How is the complement cascade triggered?
14. How does complement kill bacteria?
15. Discuss the differences between the alternative and classical pathways of complement activation.
16. How does the host prevent membrane attack complexes from being formed in host cells?
17. Why might fever be helpful in fighting infection?

Clinical Correlation Questions

1. Danny, a 2-year-old boy with recurrent *Staphylococcus aureus* infections and oral thrush, was brought to the emergency department with a high fever and rapid breathing. A complete blood count showed 58,000 leukocytes/µl (normal is 4,000–11,000) and a neutrophilia. Bacterial cultures of his blood were positive for *Staphylococcus aureus*. Laboratory tests showed normal antibody and lymphocyte levels, but his white cells were deficient in integrin. Explain why, with such a high level of white cells in his blood, Danny could not fight off infections.

2. Patients with paroxysmal nocturnal hemaglobinuria develop dark urine at night and anemia caused by lysis of red blood cells. The defect in these patients is a lack of CD59 protein. Explain why this defect leads to lysis of red blood cells and what might be an effective treatment.

Thought Questions: CHECK YOUR ANSWERS

Thought Question 15.1 Invertebrates such as clams possess a strong innate immune system but do not possess adaptive immune mechanisms. Why do vertebrates, humans in particular, need an adaptive immune system in addition to the innate?

ANSWER: One hypothesis is that because most invertebrates produce many offspring and have short life spans, they have no need for a memory-based immunity. An alternative hypothesis is that vertebrates possess highly diverse microbiota, whereas invertebrates do not. Vertebrates derive great benefit from harboring thousands of resident microbes. Thus, vertebrate evolution has probably dampened the innate immune system compared with the lowly clam to allow for the existence of a microbiome. But that deficit has been balanced with an adaptive immune system that copes with members of their microbiome (as well as pathogens) that periodically breach the innate system.

Thought Question 15.2 Why do defensins have to be so small? Do defensins kill normal microbiota?

ANSWER: Defensins need to be small so they can get through the outer membrane of Gram-negative organisms and the thick peptidoglycan maze of Gram-positive organisms. Defensins do kill normal microbiota and in fact are part of what keeps levels of the normal intestinal microbiota in check.

Thought Question 15.3 How does increasing the diameter of a blood vessel affect blood flow and blood pressure?

ANSWER: The heart pumps at a constant force. Vessels can widen (dilate) and narrow (constrict) to change how much blood they can hold (capacity). When vessels constrict, their capacity to hold blood is reduced, allowing more blood to return faster to the heart, from which it is pumped. One consequence of this is that blood pressure increases. Conversely, when vessels dilate, their capacity to hold blood is increased, allowing less blood to return to the heart, and so the flow of blood through that vessel slows. As a result, blood pressure decreases.

Thought Question 15.4 You have discovered a drug that can decrease the production of selectin molecules. What would this do to innate immunity, and for what noninfectious medical conditions might this be therapeutically useful?

ANSWER: Selectin molecules on the surface of endothelial cells catch neutrophils passing in the bloodstream. Production of selectins increase in response to inflammatory cytokines secreted from a nearby infection site or during coronary artery disease (atherosclerotic vessels) or autoimmune diseases (for instance, lupus). Higher numbers of selectin molecules attract higher numbers of neutrophils to the site of inflammation, which can cause more damage. So decreasing selectin expression could help limit autoimmune or other inflammatory diseases. Selectin inhibitors are called cell migration inhibitors and have been approved for the treatment of asthma, psoriasis, and COPD (chronic obstructive pulmonary disease).

Thought Question 15.5 Both NK cells and neutrophils have Fc receptors on their cell membranes. NK cells successfully use these receptors to carry out ADCC on virus-infected cells that are coated with antibody bound to the viral proteins inserted in the host cell's membrane. Neutrophils can bind to these antibody-coated cells but will not phagocytize them. Why not?

ANSWER: One reason is that CD47 present on all host cell surfaces prevents neutrophils from phagocytizing host cells whether or not the host cell is coated by antibody. CD47 does not inhibit NK cell action. MHCI surface molecules will inhibit NK cell activity, but infected cells often have lower amounts of MHCI on their cell surface, so NK cells can carry out ADCC.

Thought Question 15.6 *Shigella flexneri* is a Gram-negative rod that causes bacillary dysentery (bloody diarrhea). The organism is ingested and enters M cells in the intestine. After exiting the M cell, *S. flexneri* forcibly enters nearby macrophages and escapes from the phagocytic vacuole. *Shigella* triggers the activation of NLRs and cytokine release, but at a level much lower than you would expect. Propose a mechanism that might accomplish this.

ANSWER: While in the macrophage, *Shigella* secretes a series of bacterial proteins that interfere with the regulatory cascades needed to activate cytokine gene expression in the host cell. The decrease in cytokine release dampens the inflammatory response.

Thought Question 15.7 Why haven't microbes evolved to avoid being recognized by Toll-like receptors?

ANSWER: The MAMPs recognized by Toll-like receptors are essential parts of the microbes. They cannot be altered sufficiently by mutation to remain functional and yet not be recognized by a TLR.

Thought Question 15.8 Can you explain why the infections in Bobby's case (Case History 15.2) were caused by Gram-positive cocci and not Gram-negative bacteria (such as *Salmonella*) or viruses? You need to know that *S. aureus* and *S. pneumoniae* are extracellular pathogens, whereas viruses and *Salmonella* are intracellular pathogens.

ANSWER: A possible reason is that the MAMPS of Gram-positive organisms like *S. aureus* and *S. pneumoniae* will bind mostly surface TLRs rather than intracellular NLRs. Viruses and *Salmonella* are intracellular pathogens whose MAMPs can interact with NLRs even if the TLR pathways are defective.

Thought Question 15.9 If increased fever limits bacterial growth, why would bacteria make pyrogenic toxins that increase body temperature?

ANSWER: The pyrogenic toxins have other effects that compromise and damage the host. The toxins can induce cytokines that damage local host cells or confuse the immune system. This can provide the pathogen with nutrients and help it hide from the immune system longer.

Thought Question 15.10 Why does complement have to be activated? Why is it not always active and ready to kill invading microbes?

ANSWER: If C3b were constantly made at significant levels, the complement cascade would target host cell membranes. The alternative pathway is constantly active but at very low levels, plus the functional life of C3b is very short (milliseconds) if it does not bind carbohydrates on membranes. The problem is that C3b can bind carbohydrates on host as well as bacterial membranes. So if C3a were constantly made at high levels, it would overwhelm the regulatory mechanisms that prevent low levels of C3b from starting a cascade on eukaryotic membranes.

16

The Immune System: Adaptive Immunity

[Born in a Bubble]

SEVERE COMBINED IMMUNODEFICIENCY DISEASE

David Philip Vetter was born in 1971 without an adaptive immune system. Better known as the "bubble boy," he was the only human known to live in a plastic, germ-free bubble for his entire 12 years of life (1971–1984). He suffered from a genetic disease known as severe combined immunodeficiency (SCID) in which the patient forms no T-cell lymphocytes and has dysfunctional (or sometimes no) B-cell lymphocytes. Severe SCID patients cannot launch a meaningful adaptive immune defense against any invading microbe—whether it be pathogen or normal microbiota.

PROTECTIVE ISOLATION

Because David's elder brother died of SCID (his sister was healthy), physicians knew he might have the disorder, too. Consequently, David was transferred to a sterile environment within seconds after birth to await a bone marrow transplant. Water, air, food, diapers, and clothes all were disinfected with special cleaning agents before entering his sterile plastic bubble. He was handled only through special plastic gloves attached to the wall. He lived for 12 years physically isolated in this plastic, sterile environment. His only ventures out of the bubble were in a NASA-designed sterile space suit.

DESPERATE TREATMENT

Fearing David's health could no longer endure, doctors desperately attempted a bone marrow transplant from his sister in 1984. What they didn't realize was that the transplanted tissue contained an undetectable amount of Epstein-Barr virus. Because David's defective immune system left him defenseless against the virus, he became extremely ill and developed multiple tumors. There was no choice but to remove him from his protective bubble for treatment. Finally, his mother was able to touch her son's skin, but it was to be the first and last time. He slipped into a coma and died—just before his 13th birthday.

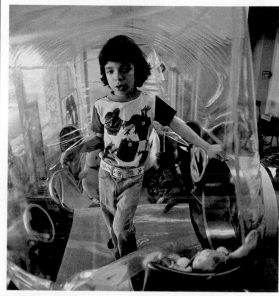

Life in a Bubble
For 12 years David lived inside this sterile plastic bubble, unable to touch the skin of another human being.

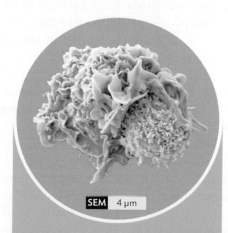

SEM   4 µm

T Cell Meets Antigen
A T cell (blue) interacts with antigens presented on an antigen-presenting cell (yellow). Lacking T cells, David's immune system was helpless against all manner of invading cells.

This case illustrates how utterly reliant we are on our adaptive immune system for survival. The innate immune system is certainly important, but alone it cannot protect us from the microorganisms that surround us. The key cells of adaptive immunity are the B-cell lymphocytes that make antibodies, and T-cell lymphocytes (first introduced in Chapter 15). T cells are divided into two general groups. One group, called helpers, orchestrate the overall adaptive immune response. The second group, called cytotoxic T cells, attack infected host cells. It is no wonder that SCID patients, who lack these lymphocytes, are at the mercy of microbes.

Remarkably, all it takes to subvert the entire adaptive immune process is a small defect in a single gene. The two most common forms of SCID result from genetic abnormalities linked to the X chromosome. In one form, patients lack an enzyme called adenosine deaminase and accumulate toxic amounts of deoxyguanosine triphosphate. In the second form, the patient cannot produce an important cytokine receptor that guides T-cell development and is used to communicate with B cells. (Cytokines were introduced in Section 15.3.) Fortunately, even though a small genetic change can destroy adaptive immunity, only one in every million people develops SCID.

This chapter begins by describing the two systems of adaptive immunity and the factors that influence how well foreign proteins, lipids, and microbes elicit adaptive immune responses. We then explore how B cells ultimately differentiate into plasma cells and make antibodies (defining what is called humoral, or circulating, immunity) and how certain T cells develop to directly kill infected host cells in what is called cell-mediated or cellular immunity. You will also learn about a type of T cell that controls the balance between humoral and cell-mediated responses to a given infection. Ultimately, you will appreciate why adaptive immunity is a major reason the human race still exists.

16.1
Adaptive Immunity: The Big Picture

Why do we call this form of immunity "adaptive"? As explained in Chapter 15, innate immune mechanisms are present from birth. Adaptive immunity, in contrast, develops as the need arises. For instance, adaptive immunity against malaria develops only after the individual has encountered the plasmodial parasite responsible for the disease; that is, the immune response adapts to the presence of the new organism. As a whole, the adaptive immune response is a complex, interconnected, and cross-regulated defense network. And it has a memory.

Link As seen in Figure 2.8, which illustrates the development of the adaptive immune response, an infecting microbe first increases in number before the adaptive system gears up. Then, as the adaptive system becomes increasingly more aggressive, the numbers of pathogenic microbes decrease and the infection resolves.

Note The terms "adaptive immune response" and "immune response" are often used interchangeably.

There are two types of adaptive immunity: humoral and cell mediated. In **humoral immunity**, **antibodies** are produced that

Figure 16.1 Antigens and Epitopes

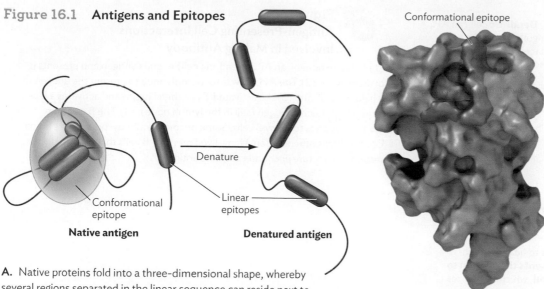

Native antigen

Conformational epitope

Denature

Linear epitopes

Denatured antigen

A. Native proteins fold into a three-dimensional shape, whereby several regions separated in the linear sequence can reside next to each other to form a conformational epitope. Denaturing the protein will unfold and destroy the conformational epitope by separating the various amino acid stretches that formed the epitope.

Conformational epitope

B. A three-dimensional protein structure showing, in red, four amino acids that form a conformational epitope.

directly target microbial invaders. The term "humoral" means "related to bodily fluids." Thus, antibodies are proteins that circulate in the bloodstream and recognize foreign structures called antigens. An **antigen** (also called an **immunogen**) is any molecule that will, when introduced into a person, elicit the synthesis of antibodies that specifically bind the antigen. From birth, the naive immune system is able to recognize billions of possible foreign antigens. Antigens stimulate B cells (B lymphocytes) to differentiate into antibody-producing **plasma cells**.

A **hapten** is a special form of antigen. Haptens are very small molecules, generally less than 1,000 daltons (molecular mass), that cannot elicit an immune response unless they are attached to a larger carrier protein. An example of a hapten is the antibiotic penicillin, a serious cause of allergic reactions in some individuals (see Chapter 17). The tiny molecule penicillin must bind to another protein before the immune system can recognize it. Thus, antigens include immunogens that elicit an immune response by themselves and haptens that must be attached to an immunogen in order to generate an immune response.

Cell-mediated immunity, the second type of adaptive immunity, employs teams of T cells (T lymphocytes) that can also recognize antigens. Some T cells help B cells become plasma cells; other T cells destroy host cells infected by the microbe possessing the antigen. In truth, the humoral and cellular immune responses are not separate, but intertwined, each relying on some facet of the other to work efficiently. T cells determine whether antibody- or cell-mediated mechanisms predominate in response to a specific antigen.

What triggers an immune response, and how long does it take to develop? Adaptive immunity develops over 3 or 4 days after you have been exposed to an invading microbe. Your immune system does not recognize the <u>whole</u> microbe—just innumerable tiny <u>pieces</u>

of it. Each small segment of an antigen that is capable of eliciting an immune response is called an **antigenic determinant**, or **epitope**. Thus, every protein will contain many different epitopes that the immune system "sees" when the larger protein antigen is broken into smaller segments after phagocytosis.

Even distinct three-dimensional shapes within a protein can be called antigenic determinants (or three-dimensional epitopes) if they produce a specific response. This happens when two or more stretches of amino acids that are far removed from one another in a protein's primary sequence become aligned side by side in three-dimensional space by protein folding (**Figure 16.1**). Such a three-dimensional structure may be recognized by the immune system as a single antigen. Besides proteins, other structures in the cell, such as complex polysaccharides, can have linear and three-dimensional epitopes (**Figure 16.2**). So the immune response to a microbe is really a composite of responses to different epitopes by thousands of individual B cells, the cells responsible for producing antibodies.

Figure 16.2 Breakdown of Antigens, Epitopes, and Immunogens

1 Bacterial cells are made of many antigens. Every surface or cytoplasmic structure is an antigen.

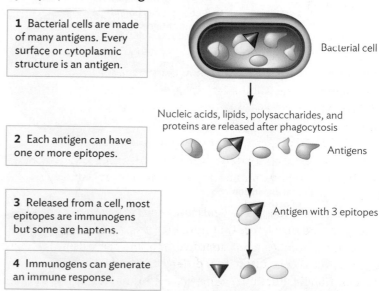

Bacterial cell

Nucleic acids, lipids, polysaccharides, and proteins are released after phagocytosis

Antigens

2 Each antigen can have one or more epitopes.

3 Released from a cell, most epitopes are immunogens but some are haptens.

Antigen with 3 epitopes

4 Immunogens can generate an immune response.

Every protein, nucleic acid, lipid, and carbohydrate in the cell can be recognized by the immune system, so they are all antigens. Antigens can have one or many different epitopes (a large protein, for example, will have many epitopes). Some epitopes are immunogenic; other epitopes act as haptens. So all immunogens are antigens, but not all antigens are immunogens (some are antigenic haptens). Each color and shape represents a different epitope.

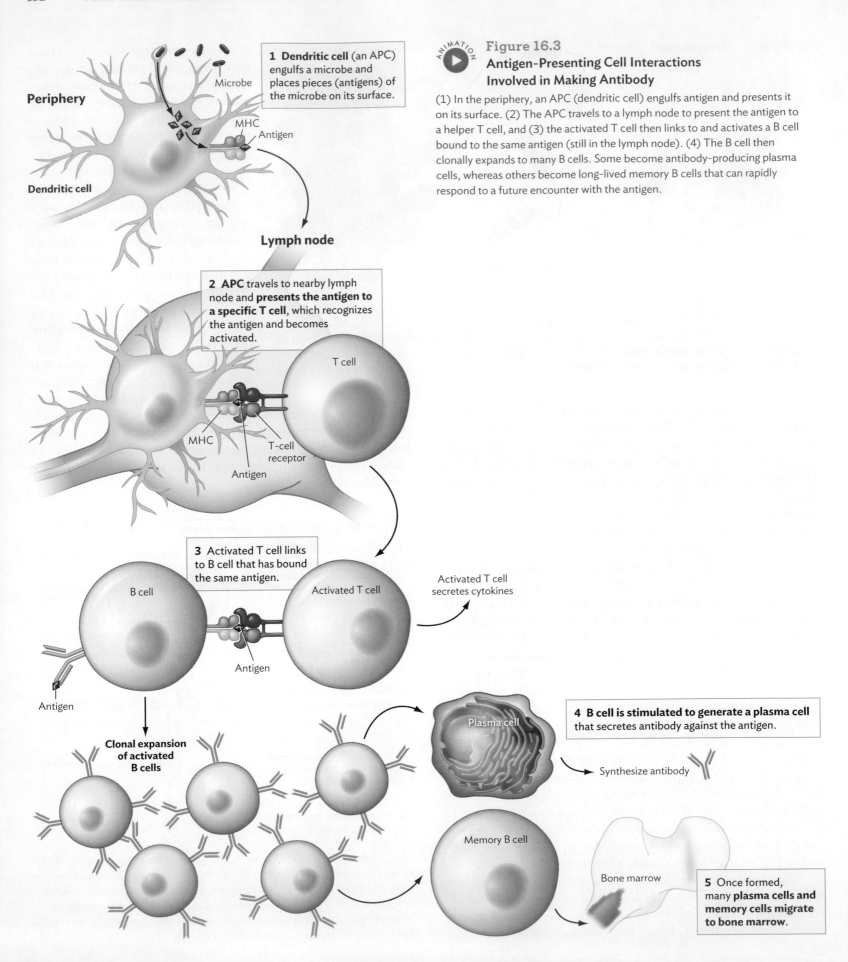

Periphery

1 **Dendritic cell** (an APC) engulfs a microbe and places pieces (antigens) of the microbe on its surface.

Microbe

MHC
Antigen

Dendritic cell

ANIMATION ▶

Figure 16.3
Antigen-Presenting Cell Interactions Involved in Making Antibody

(1) In the periphery, an APC (dendritic cell) engulfs antigen and presents it on its surface. (2) The APC travels to a lymph node to present the antigen to a helper T cell, and (3) the activated T cell then links to and activates a B cell bound to the same antigen (still in the lymph node). (4) The B cell then clonally expands to many B cells. Some become antibody-producing plasma cells, whereas others become long-lived memory B cells that can rapidly respond to a future encounter with the antigen.

Lymph node

2 **APC** travels to nearby lymph node and **presents the antigen to a specific T cell**, which recognizes the antigen and becomes activated.

T cell

MHC

T-cell receptor

Antigen

3 Activated T cell links to B cell that has bound the same antigen.

B cell

Activated T cell

Activated T cell secretes cytokines

Antigen

Antigen

Clonal expansion of activated B cells

Plasma cell

4 **B cell is stimulated to generate a plasma cell** that secretes antibody against the antigen.

Synthesize antibody

Memory B cell

Bone marrow

5 Once formed, many **plasma cells and memory cells migrate to bone marrow.**

Overview of Antibody-Dependent (Humoral) Immunity

The humoral immune response requires several cell types and cell-to-cell interactions. What are those interactions, and where do they take place? As illustrated in Figure 16.3, the humoral immune response begins with an infection somewhere in the body. Dendritic cells and macrophages patrolling the area gather up the foreign antigens and present them on their cell surface. Think of a stolen automobile entering a "chop shop," where the car is taken apart and individual parts are sold (presented) to someone waiting outside.

Phagocytic cells that degrade large antigens into smaller antigenic determinants and place those determinants on their cell surface are called **antigen-presenting cells (APCs)**. Many types of cells can be antigen-presenting. They include "professional" APCs, such as macrophages (monocytes), mast cells, dendritic cells, and B cells, and "nonprofessional" APCs (for example, endothelial cells or fibroblasts) under the right circumstances.

Presentation of antigens on APCs requires that the antigen be placed on a membrane surface protein structure called the **major histocompatibility complex (MHC)**, discussed later. The better an antigen can bind to these MHC surface proteins, the better it can trigger an immune response (or the more immunogenic it is). The stronger the binding, the easier it is for certain lymphocytes (T cells) to recognize the complex.

Once decorated with antigen, the professional antigen-presenting cell travels the lymphatic system to secondary lymphoid organs (lymph nodes), where B cells and T cells await (Figure 16.3, step 2). Specific T cells in the node then link to the antigen presented on the APC and become activated T cells. One type of activated T cell then finds, binds, and activates a lymph node B cell that encountered the same antigen. That interaction authorizes those B cells to replicate rapidly and develop into plasma cells able to pump out large amounts of specific antibody. (steps 3 and 4; note that plasma cells are not B cells.)

Each B cell is preprogrammed to make antibody for one and only one epitope. Thus, it takes ten different B cells to make antibodies to ten different epitopes. The antibody response to each individual epitope is **clonal**; that is, a single B cell that makes an antibody to a single epitope will, upon "seeing" that epitope, replicate to produce a large population (a clone) of B cells, all of which make antibodies to that epitope.

Once activated, B cells can also produce **memory B cells** that remember the exposure and stand ready to quickly generate plasma cells, should the antigen be encountered months or years later (Figure 16.3, step 5). Once formed, many memory B cells and most plasma cells leave the lymph node and migrate to the bone marrow. Other plasma cells remain in the lymph node. Wherever they reside, memory B cells patiently wait to be called to future sites of infection where they will generate a rapid antibody response.

 Antibody-mediated immune response: youtube.com

Overview of Cell-Mediated Immunity

The cell-mediated immune response shares some aspects of the humoral immune response. Another type of T cell, called **cytotoxic T cells (T_c cells)**, can also bind to microbial antigens presented on an APC and become activated. After leaving the lymph node, the cytotoxic T cell can seek out and directly kill any host cell infected with the microbe. The cytotoxic T cell recognizes an infected cell that places antigens from the invading microbe on its host cell surface. Besides directly killing infected cells, cytotoxic T cells also synthesize and secrete growth factors called cytokines (Section 15.2) that incite nearby macrophages to indiscriminately attack cells in the local area. Thus, cellular immunity, in general, is critical for dealing with intracellular pathogens such as viruses, whereas humoral immunity is most effective against extracellular bacterial pathogens such as *Streptococcus pneumoniae*, one cause of pneumonia.

As we proceed through the chapter, we will reveal how the immune system functions in layers, with each layer building on the previous one. As a result, we will periodically return to a particular aspect of the immune response—for example, B-cell differentiation into plasma cells—to integrate seemingly distinct parts of the immune system into a unified concept of immunity.

 Cell-mediated immune response: youtube.com

SECTION SUMMARY

- **An antigen** can elicit an antibody response. An antigen is usually made of many different epitopes (antigenic determinants), each of which binds to a different, specific antibody.
- **Humoral immunity** against infection is the result of antibody production originated by B cells.
- **Cellular immunity** involves cytotoxic T cells that directly kill infected host cells.
- **Antigen-presenting cells**, such as phagocytes, degrade microbial pathogens and present distinct pieces on their cell-surface MHC proteins.
- **T cells are activated** only by binding to APC-presented antigens.

Thought Question 16.1 Two different stretches of amino acids in a single protein form a three-dimensional antigenic determinant. Will the specific immune response to that three-dimensional antigen also respond to one of the two amino acid stretches alone?

16.2
Immunogenicity and Immune Specificity

SECTION OBJECTIVES

- Describe what makes one antigen more immunogenic than another.
- Discuss immunological specificity and its application to vaccine development.

Are some antigens better than others at generating an immune response? What about our own antigens? Do we react to them? Even though our immune system is capable of reacting to any molecular structure, there are limitations as to how strongly it reacts to different antigens, especially our own.

Antigenicity

Antigenicity, or **immunogenicity**, measures how well an antigen elicits an immune response. One antigen can be more immunogenic than another. For example, proteins are the strongest antigens, but carbohydrates can also elicit immune reactions. Nucleic acids and lipids are usually weaker antigens because they are made of relatively uniform repeating units and are very flexible. As such, they do not easily interact with antibodies. Proteins are more effective antigens for three reasons: they form a variety of shapes, they maintain their tertiary structure, and they are made of many different amino acids that can be assembled in many different combinations. These features provide stronger interactions with antibodies in the bloodstream and enable better recognition by lymphocytes, the cellular workhorses of the immune system.

Antigen presentation also plays a role in how immunogenic an antigen is. For example, the larger the antigen, the more likely it is that antigen-presenting phagocytic cells such as macrophages and dendritic cells will "see" and engulf it. The larger an antigen is, the more easily the APC can degrade it and present the epitopes on the APC's cell-surface MHC molecules, a necessary step before an immune response can occur. However, the tiny haptens mentioned earlier (such as penicillin) are too small to bind the MHC molecules. The inability of haptens to bind MHC proteins is why these molecules must be attached to larger antigens in order to generate an immune response.

Because antigens differ from one another, each specific antigen has a different **threshold dose** needed to generate an optimal response. A dose higher or lower than that threshold will not generate as strong an immune response. Lower doses activate only a few B cells. Exceedingly high doses of antigen can cause **B-cell tolerance** (anergy), a state in which B cells have been overstimulated to the point at which they do not respond to subsequent antigen exposures and make antibody. B-cell tolerance is part of the reason your immune system does not attack your own antigens.

As you might expect, the body must regulate the immune system carefully so that a response is not leveled against itself. In effect, the immune system must become "blind" to its own antigens; as a result, the host will often be blind to foreign antigens that resemble epitopes of its own cells. The more complex a foreign protein is, the more likely it will possess unique antigenic determinants that a lymphocyte can recognize as nonself. The further away an antigen structure is from "self," the greater its immunogenicity will be.

Immunological Specificity

Imagine a row of doors in a hotel corridor. Each door has a lock that is different from all the other locks. Each guest is given a key that unlocks only one door. You could say that each guest's key is highly specific. This analogy is similar to immunological specificity, in which an adaptive immune response to one antigen is generally not effective against a different antigen (that is, one immune response "key" fits only one antigen "door lock"). Stated in terms of infectious disease, an immune response to smallpox will not protect someone against the plague bacillus (*Yersinia pestis*), which is antigenically different from smallpox.

Although immunological specificity is important, it is not absolute. In the late eighteenth century, long before viruses were discovered, an English country physician named Edward Jenner (see Figure 1.15) learned to protect townsfolk from deadly smallpox disease by inoculating them with scrapings from lesions produced by a tamer disease, cowpox. By this process, Jenner unwittingly transferred the vaccinia virus (which we now know causes cowpox; see **Figure 16.4**) to villagers susceptible to smallpox (caused by the related but more dangerous variola virus). The resulting immune reaction to vaccinia produced an effective cross-protection against variola and thus prevented smallpox. This example illustrates that an immune reaction against one organism or virus may be sufficient to protect against an antigenically related, if not identical, organism. In honor of Jenner's discovery, the term **vaccination** describes the technique of exposing individuals to "tame" microbes to protect the person against the more dangerous, pathogenic strains. Vaccination has been used to protect humans against many microbial and viral pathogens (Table 16.1). Today, most **vaccines,**

Figure 16.4   Immunological Specificity Is the Basis of Vaccination

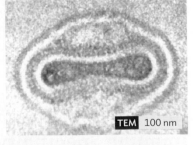

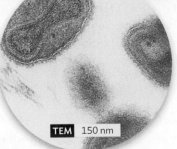

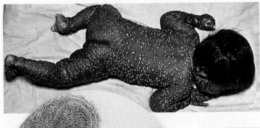

A. Photo of a smallpox patient, showing the white pox pustules.

B. The smallpox virus, variola major. The photo shows the dumbbell-shaped, membrane-enclosed nucleic acid core.

C. The vaccinia virus that causes cowpox. Edward Jenner recognized that disease symptoms were similar between the deadly smallpox disease and less severe cowpox disease and used cowpox scrapings to vaccinate humans against smallpox.

Table 16.1

Vaccines against Viral and Bacterial Pathogens[a]

Disease	Vaccine	Vaccination Recommended for:
Viral		
Chickenpox	Attenuated strain (will still replicate)	Children 12–18 months
Hepatitis A	Inactivated virus (will not replicate)	Children 12–18 months
Hepatitis B	Viral peptide antigen	Medical personnel; children 1–18 months
Influenza	Inactivated virus, attenuated virus, or antigen	Adults over 65 years
Measles, mumps, rubella	Attenuated viruses; MMR combined vaccine	Children 12–19 months
Polio	Attenuated (oral, Sabin); inactivated (injection, Salk)	Children 2–3 years
Rabies	Inactivated virus	Persons in contact with wild animals
Yellow fever	Attenuated virus	Military personnel
Bacterial		
Anthrax	*Bacillus anthracis*, components of toxin; unencapsulated strain	Agricultural and veterinary personnel; key health care workers
Cholera	Killed *Vibrio cholerae*, toxin components	Travelers to endemic areas
Diphtheria	Toxoid (inactivated toxin)	Children 2–3 months
Pertussis	Acellular *Bordetella pertussis*	Children 2–3 months
Tetanus	Toxoid	Children 2–3 months
Haemophilus influenzae type b (meningitis)	Bacterial capsular polysaccharide	Children under 5 years
Lyme disease	*Borrelia burgdorferi*, lipoproteins OspA and OspC surface antigens	Canines, human vaccine discontinued
Meningococcal disease	*Neisseria meningitidis*, bacterial capsular polysaccharides	Military personnel; high-risk individuals
Pneumococcal pneumonia	*Streptococcus pneumoniae*, bacterial capsular polysaccharides	Children and adults over 50 years
Tuberculosis (*Mycobacterium tuberculosis*)	Attenuated *Mycobacterium bovis* (BCG vaccine)	Exposed individuals
Typhoid fever	Killed or attenuated *Salmonella* Typhi	Individuals in endemic areas
Typhus fever	Killed *Rickettsia prowazekii*	Medical personnel in endemic areas; scientists

[a]There are three types of vaccines: killed whole pathogen, attenuated pathogen that still replicates in the host but does not harm the host, and component vaccines such as polysaccharide capsule or inactivated toxin (toxoid).

the material used to vaccinate someone, are killed pathogens, live but crippled (**attenuated**) strains of the pathogen, or inactivated microbial toxins (for example, diphtheria toxin).

Link You will learn more about **vaccines** and vaccination schedules in Chapter 17 and in the infectious disease chapters.

Cross-protection, as in the case of vaccinia and variola viruses, works only if key virulence proteins from the two different organisms share epitopes. No cross-protection occurs if these two proteins differ significantly. A good example in which cross-protection doesn't work is the common cold, caused by hundreds of closely related rhinovirus strains (rhinitis, a runny nose, is one of the symptoms of this viral disease). Rhinoviruses use a viral capsid protein to bind ICAM-1 on host cell membranes. Once bound, the membrane invaginates and pulls the virus into the cytoplasm. The cell is infected. Unfortunately, infection with one strain of rhinovirus will not immunize the victim against a second strain because the viral proteins that attach to the host's ICAM-1 surface

Figure 16.5   Neutralizing Antibodies Prevent Rhinovirus Attachment to Cell Receptors

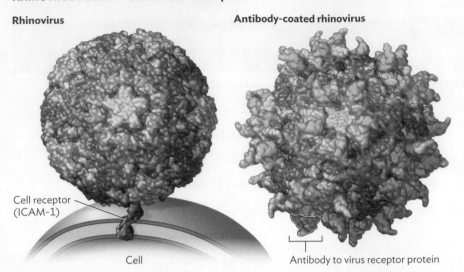

Rhinovirus

Antibody-coated rhinovirus

Cell receptor (ICAM-1)

Cell

Antibody to virus receptor protein

A. This figure (a real-world molecular model) illustrates the complexity of the rhinovirus capsid and shows the attachment of the virus to the cell-surface molecule ICAM-1 (intercellular adhesion molecule, shown in reddish brown).

B. This figure shows rhinovirus coated with protective (neutralizing) antibodies (green) that block the ICAM-1 receptors on the virus. As a result, the virus cannot attach to and infect the host cell.

protein differ dramatically in structure between different strains of rhinovirus (**Figure 16.5A**). As a result, so-called neutralizing antibodies, which bind to the attachment protein on one strain of rhinovirus, will prevent reinfection by the same virus strain (**Figure 16.5B**) but fail to bind an attachment protein from a different strain. In our analogy, a key that is specific to one lock will not work on a different lock. However, if scientists could devise an antibody with broad specificity—that is, one that acts like a master key in the hotel analogy—then a single vaccine might protect us from colds caused by numerous strains of rhinovirus.

SECTION SUMMARY

- **Proteins are better immunogens** than nucleic acids and lipids because proteins have more diverse chemical forms.

- **Antigen-presenting cells,** such as phagocytes, degrade microbial pathogens and present distinct pieces on their cell-surface MHC proteins.

- **Immunological specificity** means that antibody made to one epitope will not bind to different epitopes (although some weak cross-binding can occur).

Thought Question 16.2   How does a neutralizing antibody that recognizes a viral coat protein prevent infection by the associated virus?

16.3
Antibody Structure and Diversity

SECTION OBJECTIVES

- Distinguish immunoglobulin proteins from the immunoglobulin family of proteins.

- Diagram the structures and superstructures of antibodies and discuss their functions.

- Differentiate between the isotype, allotype, and idiotype of an antibody.

What are antibodies, and why are they so important?
Antibodies, also called **immunoglobulins**, are the keys to immunological specificity. They are glycoproteins made by the body in response to an antigen. Antibodies belong to the immunoglobulin superfamily of proteins (this superfamily includes the major histocompatibility proteins and the B-cell and T-cell receptors described later).

Like a miniature "smart bomb," an antibody circulates through blood and lymph, ignoring all antigens except for its specific target antigen. When an antibody finds its antigenic match, it binds to the antigen and initiates several events that destroy the target (see discussion of complement, Section 15.5). In addition to being free-floating, antibodies are also strategically situated on the surface of B cells, where they enable these lymphocytes to recognize specific antigens.

An antibody consists of four polypeptide chains. There are two long **heavy chains** and two short **light chains**, colored green and yellow, respectively, in **Figure 16.6**. The four polypeptides combine to form a Y-shaped, tetrameric (four subunit) structure held together by disulfide (—S—S—) bonds. Two disulfide bonds connect the two identical heavy chains to each other. One light chain is then attached near its carboxyl (COO⁻) end to the middle of each heavy chain by a single disulfide bond. The antigen-binding sites form at the amino-terminal (NH_3) ends of the light and heavy chains. One antibody molecule possesses two identical antigen-binding sites, one on each "arm" of the Y-shaped molecule. This bivalency enables antibodies to cross-link identical antigens, a feature critical to antibody function.

Antibodies Have Constant and Variable Regions

You might think that all antibodies are alike except for their antigen-binding sites. We actually produce five classes of antibodies, defined by five different types of heavy chains called alpha (α), mu (μ), gamma (γ), delta (δ), and epsilon (ε). Antibody classes are named according to their heavy chains. Thus, antibodies containing the alpha heavy chain are called IgA; those with mu, IgM; gamma, IgG; delta, IgD; and antibodies with an epsilon heavy chain are called IgE.

The heavy chains in each class differ one from another at a specific region in their amino acid sequence called the **constant region**, denoted C_H (Figure 16.7). For example, all alpha chain (IgA) antibodies have the same amino acid sequence in their constant region, but the constant-region sequence in all gamma chain (IgG) antibodies is different from that of IgA. The various heavy-chain constant regions impart distinct purposes to each class of antibody. One class (IgA) is secreted onto mucosal surfaces where pathogens try to sneak into our body. Other classes circulate only in blood. One type of antibody (IgM) is made quickly in response to an infection but binds to its target with only a moderately tight grip, only to be followed by another antibody type (IgG) that very tightly grips its target.

In contrast to the five heavy chains, there are only two classes of light chains: kappa (κ) and lambda (λ). An antibody of any heavy-chain class may contain two kappa chains or two lambda chains, but never one of each. Two-thirds of all antibody molecules carry kappa chains; the rest have lambda chains. Each light chain also has a constant region, designated C_L.

The antigen-binding part of an antibody, regardless of the heavy-chain class, is formed by highly variable amino acid sequences at the amino-terminal ends of the light and heavy chains. By "variable" we mean that the amino acid sequence of an antigen-binding site on an antibody that binds, say, peptidoglycan, is very different from the antigen-binding site on an antibody that binds a different antigen. The differences in these amino acid sequences in different antibodies alter the shape of the antigen-binding site and allow each antibody to bind a different antigen. The **variable regions** are referred to as the V_L regions and the V_H regions (Figure 16.7). Antibodies bind to antigens by using noncovalent interactions (for instance, ionic, or hydrophobic bonds), which allows for reversible antigen-antibody interactions.

Figure 16.6   Basic Antibody Structure

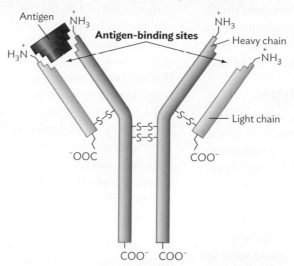
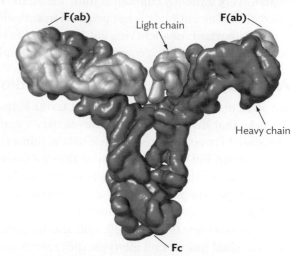

A. Each antibody contains two long heavy chains and two shorter light chains held together by disulfide bonds. The Y-shaped structure contains two antigen-binding sites [F(ab)], one at each arm of the molecule. The two antigen-binding sites are formed by the amino-terminal regions of the heavy- and light-chain pairs.

B. Three-dimensional structure of an antibody. The heavy chains are shown in green, the light chains in yellow. The F(ab) regions represent the antigen-binding sites. The Fc portion points downward and is used to attach the antibody to different cell-surface molecules.

Figure 16.7   Constant and Variable Regions in Antibody Structure

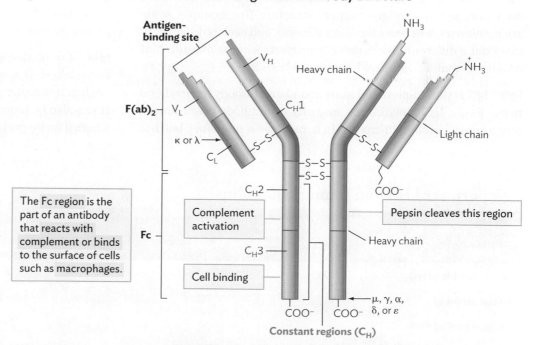

The variable amino acid sequences present at the amino termini of the heavy and light chains form the antigen-binding sites. The functional parts of the antibody (antigen binding versus Fc binding) can be separated from each other in the laboratory by pepsin, a protease that specifically cleaves inside the C_H constant region. This separates the Fc region from the antigen-binding sites. The two antigen-binding (ab) regions remain as one piece called F(ab')$_2$. Antibody isotype is defined by unchanging amino acid sequences in the heavy-chain constant regions. Allotype is defined by constant-region amino acid sequences that differ among individuals. Idiotype differences within a single person are defined by variable sequences in the antigen-binding site.

In addition to the two "arms" that bind antigen, called the **F(ab)₂ region**, every antibody contains a "tail," called the **Fc region**. The Fc tail region of an antibody is not involved in antigen recognition but is very important for anchoring antibodies to the surface of certain host cells and for binding components of the complement system.

Antibody Isotypes, Allotypes, and Idiotypes

We often discuss antibody diversity in terms of the amino acid differences that distinguish antibodies between mammalian species (**isotype**), between different individuals within a species (**allotype**), and between different antibodies within an individual (**idiotype**). Understanding the differences between antibody isotype, allotype, and idiotype is important because these terms reflect different levels of antibody diversity. Put most simply, a single person has five antibody isotypes: IgG, IgA, IgD, IgE, and IgM. But each isotype in an individual has several allotypic differences compared with the same isotype in another individual. Finally, a single antibody allotype in a person will contain millions of idiotypic differences that reflect antigen specificity. The synthesis and functional purpose of these different levels of antibody diversity will be described in the next several sections.

Antibody Isotype Functions and "Super" Structures

All antibody isotypes have the same basic structure. However, each isotype has a unique "super" structure (for example, some are monomers, whereas others are dimers), and each is designed to carry out a different task. Some key properties of the five different immunoglobulin classes are listed in Table 16.2.

IgG IgG is the simplest and most abundant antibody in blood and tissue fluids. IgG provides the majority of antibody-based immunity against invading microbes. It is made as a monomer but has four subclasses in humans. Each subclass varies in its amino acid composition and by the number of interchain cross-links. IgG molecules carry out several missions for the immune system. First, they bind and **opsonize** microbes; that is, they coat the microbe to make it more susceptible to phagocytes. Opsonizing IgG antibodies coat the microbe by attaching to microbe surface antigens. This binding "tags" the microbe and leaves the antibody Fc regions protruding outward. Phagocytes possess surface Fc receptors that can attach to the Fc region to gain a firmer "grip" on the microbe, facilitating phagocytosis (discussed in Section 15.3). IgG can also directly neutralize viruses by binding to virus attachment sites and is one of only two antibody types that can activate complement by the classical pathway (described in Section 15.7). IgG is also the only antibody capable of crossing the placenta to give passive immunity to the fetus.

IgA IgA, containing the alpha heavy chain, is called a secretory antibody because it is secreted across mucosal surfaces. It is most commonly found in secretions as a dimer whose monomeric parts are held together at their Fc regions by a protein called the J chain (Figure 16.8). Since each IgA monomer can bind two molecules of antigen, an IgA dimer can bind four molecules of antigen. Plasma cells within mucosal membranes synthesize the IgA dimer. Each dimer then passes through a mucosal cell, where a sixth protein, the secretory piece, is wrapped around the IgA dimer during the final secretion process. The secreted molecule, now called secretory IgA (sIgA), is found in tears, breast milk, and saliva and on other mucosal surfaces. The molecule sIgA is important for immunity against pathogens that use mucosal membranes as portals of entry.

IgM Circulating **IgM**, containing the mu heavy chain, is a huge, Ferris wheel–shaped molecule formed from five monomeric immunoglobulins tethered together by the J-chain protein (Figure 16.8). It can also be found in monomeric form on the surface of B cells, where it forms part of the B-cell antigen receptor (see Section 16.4).

Table 16.2
Properties of Human Immunoglobulins

Property	IgG				IgA	IgM	IgD	IgE
	IgG1	IgG2	IgG3	IgG4				
Serum half-life (days)	21	20	7	21	6	10	3	2
% Total serum Ig		70%			15%–20%	5%–10%	0.2%	0.002%
Antigen-binding sites		2			2–4	2–10	2	2
Produced by fetus		Poor, if at all			Poor, if at all	Yes	?	Poor, if at all
Transmitted across placenta		Yes			No	No	No	No
Binds complement		Yes			No	Yes	No	No
Opsonizing		Yes			No	No	No	No
Binds mast cells		No			No	No	No	Yes

Figure 16.8   Structures of IgA and IgM

The antibodies are made as multimers of two (IgA) or five (IgM) immunoglobulin molecules.

IgA is secreted as a dimer held together by the secretory piece and J chain and can bind four identical antigens.

IgM forms a pentamer held together by disulfide bonds and the J chain. IgM can bind ten antigens.

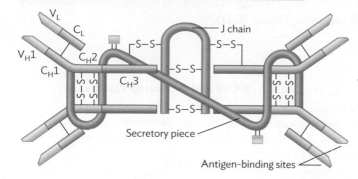

During the early stages of an immune response, IgM is the first antibody isotype detected in blood. Unlike the smaller IgG immunoglobulins, IgM is so large that it cannot cross the placenta (see Table 16.2).

IgD   IgD, containing the delta heavy chain, is a monomer present at very low levels in the blood but is abundant on the surface of B cells. Attached to the cell surface by their Fc regions, IgD, along with monomeric IgM, can bind antigen and signal B cells to differentiate and make antibody. IgD neither binds complement nor crosses the placenta.

IgE   IgE, defined by the epsilon heavy chain, is found mainly bound to the surface of mast cells and basophils. Mast cells and basophils contain granules loaded with inflammatory mediators. The

primary role of IgE is to amplify the body's response to invaders. Once secreted into the blood, IgE attaches to mast cells (Figure 16.9A and B), again by way of its Fc region, and, like a Venus flytrap, waits until its matched antigen binds to its antigen-binding site. When two surface IgE molecules on a mast cell are cross-linked by antigen, a signal is sent internally that triggers degranulation of the mast cell (see the discussion of hypersensitivity, Section 17.3). The

Figure 16.9   Mast Cells

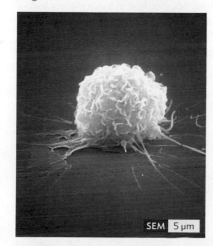

A. Scanning EM of a mast cell.

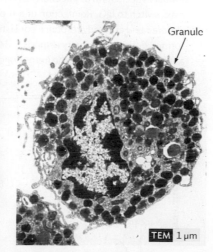

B. Transmission EM showing granules (arrow).

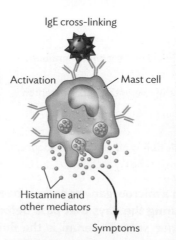

C. Hay fever is the result of degranulation of IgE-coated mast cells, which release histamine and other pharmacological mediators.

D. When released in the upper respiratory tract, these chemical mediators cause the symptoms of runny nose and sneezing.

subsequent release of histamine and other pharmacological media-tors from these granules helps orchestrate the acute inflammation that takes place during early host responses to microbial infection. The system is also responsible for severe allergic hypersensitivities, such as anaphylaxis, and milder forms, such as hay fever (Figure 16.9C and D).

SECTION SUMMARY

- **Antibodies, or immunoglobulins,** are members of the immunoglobulin superfamily.
- **Antibodies are Y-shaped** molecules that contain two heavy chains and two light chains.
- **There are five classes (isotypes) of heavy chains.** Each antibody isotype is defined by the sequence of the heavy chain.
- **Each antibody molecule contains two antigen-binding sites.** Each binding site is formed by the hypervariable ends of a heavy and light chain pair.
- **The Fc portion** of an antibody can bind to specific receptors on host cells. This binding is antigen independent.
- **Antibody isotype, allotype, and idiotype** refer to amino acid sequence differences found at the different levels of antibody diversity. **Isotypic differences** are found between species, **allotypic differences** are found within a single isotype (for instance, IgG) between two individuals of the same species, and **idiotypic differences** are found within a single antibody allotype in a single individual (usually at the antigen-binding site).

Thought Question 16.3 The mother of a newborn is diagnosed with rubella, a viral disease. Infection of the fetus can lead to serious consequences for the newborn. How can you use antibody production to determine whether the newborn was infected while in utero?

16.4
Humoral Immunity: Primary and Secondary Antibody Responses

SECTION OBJECTIVES

- Explain the difference between primary and secondary antibody responses.
- Discuss B-cell differentiation and the role of the B-cell receptor.
- Summarize how the immune system generates vast antibody diversity.
- Describe the role of memory B cells during an antibody response.

Once you have been infected with a microorganism or have been given a vaccine, what happens during the days that pass before antibodies begin to appear in your serum? Serum is the fluid that remains after your blood clots. During this lag period, a series of molecular and cellular events known as the **primary antibody response** take place. During the primary antibody response, an antigen binds to B cells that are only able to make antibodies against

that antigen. The activated B cells then proliferate and differentiate into high-output antibody-secreting plasma cells (Figure 16.10) and memory B cells. Recall that memory B cells do not secrete antibodies themselves but are primed to quickly become antibody-secreting plasma cells if the host is reinfected with the same pathogen at a later date, months or years after the initial infection. Thus, secondary exposure to the antigen triggers an almost instantaneous **secondary antibody response** (Figure 16.11).

Figure 16.10   B Cell and Activated Plasma Cell
Notice the increased size of the plasma cell compared with the B cell. The plasma cell cytoplasm is stuffed with endoplasmic reticulum needed to support the incredible increase in protein synthesis associated with antibody production.

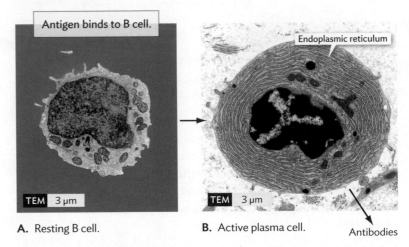

A. Resting B cell.   **B.** Active plasma cell.

Figure 16.11   Primary versus Secondary Antibody Response
Primary vaccination or infection leads to the early synthesis of IgM followed by a class switch to IgG. Reinfection or a second, booster dose of a vaccine results in a more rapid antibody response consisting mainly of IgG due to memory B cells formed during the primary response. Be aware that the time course and level of antibody made vary with the immunogen and the host.

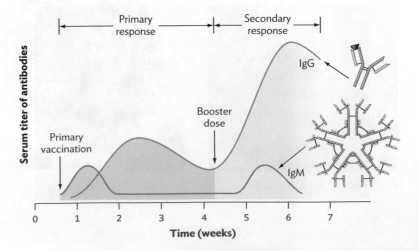

Figure 16.12 Clonal Selection Theory

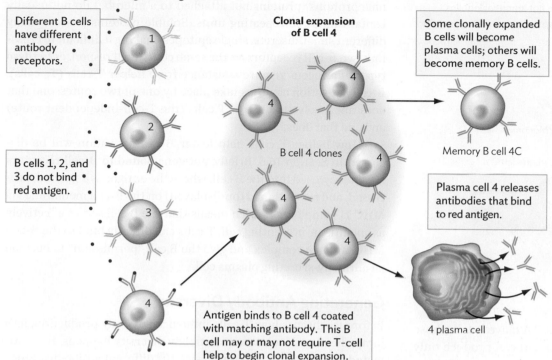

Antigen binding to the right B cell

Different B cells have different antibody receptors.

B cells 1, 2, and 3 do not bind red antigen.

Clonal expansion of B cell 4

B cell 4 clones

Antigen binds to B cell 4 coated with matching antibody. This B cell may or may not require T-cell help to begin clonal expansion.

Some clonally expanded B cells will become plasma cells; others will become memory B cells.

Memory B cell 4C

Plasma cell 4 releases antibodies that bind to red antigen.

4 plasma cell

The B-cell population is composed of individuals that have specificity for different antigens. When a B cell contacts its cognate antigen (the antigen to which the B cell antibody receptors bind), an intracellular signal is generated and the B cell is activated, leading to proliferation and differentiation of that clone (clonal expansion). Plasma cells and memory B cells are then generated from that clone.

The net result of the primary antibody response is the early synthesis and secretion of pentameric IgM molecules specifically directed against the antigen, or immunogen. During a later stage of the primary response, a process known as **isotype switching** (also called **class switching**) occurs in activated B cells, in which the predominant antibody type produced becomes IgG rather than IgM. Antibodies made during this primary response, though specific for the immunogen, are actually not of the highest affinity (the "tightness" by which the antibody binds to its antigen). Mechanisms to increase antibody affinity occur later in the immune response.

As the immunogen is cleared from the body, the levels of both IgG and IgM decline because the plasma cells that produced them die. Most plasma cells die within 3 months.

Memory B cells are maintained in the body because, unlike plasma cells, they continue to divide. If a person encounters the same antigen at a later time, memory B cells quickly proliferate and differentiate into plasma cells with no lag phase. Thus, memory B cells initiate the secondary antibody response (or "anamnestic response," from the Greek *anamnesis*, meaning "remembrance"). Memory B cells make up approximately 40% of the circulating B-cell population. During the secondary response, most memory B cells switch from IgM production to IgG production as they become plasma cells. Huge amounts of IgG antibody are secreted from these plasma cells. The new and improved IgG antibodies have a higher specificity for the antigen than the antibodies produced during the primary response (see "Making Memory B Cells").

The Secondary Response Is the Basis of Immunization

The speedy production of antibody during the secondary response is why immunization works to protect you against infectious diseases. During a natural infection, an aggressive pathogen can do considerable harm while the body is gearing up the primary response. Vaccination prior to infection can prevent this harm. In vaccination, an innocuous version of a pathogen (or a harmless piece of it) can be injected into a person to trigger the primary response without producing disease (or, at worst, producing only a mild form of the illness). Immunization primes the immune system to respond efficiently and without delay upon encountering the real pathogen. (Table 16.1 lists some viral and bacterial diseases for which immunization is available.)

Differentiation into Plasma Cells Occurs by Clonal Selection

As mentioned earlier, each B cell circulating throughout the body or harbored in a lymphoid organ is programmed to synthesize antibody that reacts with a single epitope. In a process called **clonal selection**, the antigen selects which B-cell clone proliferates and differentiates into antibody-producing plasma cells or memory B cells. The mechanism of clonal selection begins with an antigen binding to a B cell preprogrammed to bind to that antigen (**Figure 16.12**). Most of the resultant B-cell offspring then differentiate to become plasma cells that secrete large amounts of antibody specific for the antigen that triggered the process.

Figure 16.13   A B-Cell Receptor

The B-cell receptor is formed as a complex between a monomeric IgM and proteins Igα and Igβ in the membrane. Igα and Igβ are not immunoglobulins themselves, but associate with the antibody immunoglobulin part of the receptor.

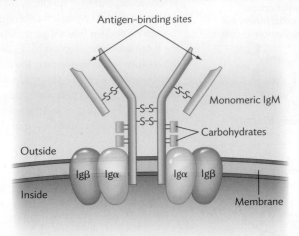

What starts the clonal selection process? A naive B cell (those that have not previously encountered antigen) can produce only IgM and IgD, both of which have <u>identical</u> antigen specificities. These two antibody classes are displayed like tiny satellite dishes on the B-cell surface, anchored by their Fc regions. These surface antibodies are the keys to stimulating proliferation. Upon binding to its corresponding antigen via these surface antibodies, the B cell is said to become **activated** and begins to multiply.

B-Cell Receptors

Each membrane-bound antibody on the B cell is associated with two other membrane proteins called Igα and Igβ (these are not immunoglobulins but are designated Ig because they <u>associate</u> with the surface antibody). The complex, shown in **Figure 16.13**, is called the **B-cell receptor (BCR)**. Each B cell may have upward of 50,000 B-cell receptors, all of which can bind identical antigens. Microbes generally have multiple copies of the same epitope on their surface (think about a virus capsid). Therefore, two adjacent B-cell receptors on one B cell can bind adjacent epitopes on the same microbe. Once bound, surface B-cell receptors begin to cluster in a process called capping. **Capping** initiates a signal cascade directed into the nucleus (**Figure 16.14**) that activates B-cell proliferation. Activated B cells multiply and differentiate into plasma cells that secrete IgM antibody as part of the primary immune response.

T-Cell-Dependent versus -Independent Antibody Production

As illustrated in Figure 16.14, the route to antibody production requires antigens with multiple repeating epitopes to directly cross-link B-cell receptors (the capping process) to trigger B-cell proliferation and differentiation into a plasma cell. However, soluble proteins (proteins not attached to a microbe) do not usually contain multiple repeating units. Soluble proteins possess many different small, discrete, single epitopes, making it difficult to cross-link two B-cell receptors on the same cell. B-cell responses to these types of antigens require assistance from **helper T cells (T$_H$ cells)**. B-cell activation can thus take place by one of two routes: one that does not involve the use of T cells (the T-cell-independent route) and one that does.

How helper T cells help foster B-cell activation will be discussed in Section 16.6. Briefly, antigens bound to B-cell receptors are endocytosed into the B cell, where the antigen is degraded, processed, and repositioned (or displayed) on the B-cell membrane via MHC II complexes—which means that the B cell serves effectively as an antigen-presenting cell. T cells can directly bind to the B-cell MHC-antigen complex and give the B cell "permission" to become an antibody-secreting plasma cell.

Generating Antibody Diversity

Before we discuss how T cells influence antibody production, let's jump ahead and look at how antibody diversity happens. It is estimated that each of us can synthesize 10^{11} different antibodies. Since each B cell displays antibodies to only one antigenic determinant,

Figure 16.14   Capping and Activation of the B cell

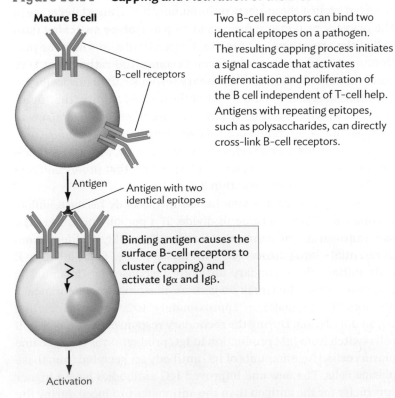

Two B-cell receptors can bind two identical epitopes on a pathogen. The resulting capping process initiates a signal cascade that activates differentiation and proliferation of the B cell independent of T-cell help. Antigens with repeating epitopes, such as polysaccharides, can directly cross-link B-cell receptors.

Binding antigen causes the surface B-cell receptors to cluster (capping) and activate Igα and Igβ.

it follows that the immune system can potentially produce 10^{11} different B cells displaying 10^{11} different B-cell antigen receptors, just not all at one time. Each person possesses only about 1,000 genes or gene segments involved in antibody formation. How are 10^{11} different antibodies made from only 10^3 genes? It turns out that antibody genes can rearrange within the genome of a differentiating B cell to make millions of possible combinations. Three steps are involved: rearrangement of antibody gene segments (or cassettes), the random introduction of mutations, and the generation of different codons during gene rearrangements. In humans, the process of generating antibody diversity takes place constantly over a lifetime.

For a deeper understanding of the actual process by which developing B cells generate antibody diversity, see **inSight**. Put simply, antibody diversity begins with progenitor B cells randomly rearranging a set of immunoglobulin genes to make an antigen-binding site (the V_L and V_H segments of the antibody mentioned earlier). Rearrangements happen at the 5′ end of the heavy-chain genes as well as lambda and kappa light-chain genes. The process is similar to randomly picking 3 blocks from a box of 200 differently shaped LEGO pieces and hooking them together. Millions of 3-block combinations can be made from the 200 pieces. You can think of one such combination as a single antigen-binding site. Once the DNA segment encoding the antigen-binding site is made, all of the offspring from that B cell will make antibodies with that same antigen-binding site.

How does the antigen-binding site then link to different heavy chains to make IgM, IgG, or IgA isotypes with the same antigen-binding specificity? Once made, the DNA encoding the antigen-binding site lies at the 5′ end of a cluster of five heavy-chain genes that sequentially encode the IgM, IgD, IgG, IgE, and IgA heavy-chain constant regions, in that order. Initially, only IgM and IgD antibodies are made. These antibodies are not secreted but are placed on the surface of the B cell to make B-cell receptors. After antigen binds to the B-cell receptor and the B cell interacts with a T cell, the B cell begins to proliferate, and the daughter cells eventually become antibody-secreting plasma cells or memory cells. During this time, a series of gene rearrangements will move a constant-region gene for IgG, IgE, or IgA next to the DNA encoding the antigen-binding site. Which constant region is moved depends on the cytokine cocktail present in the area. The activated B cell has undergone what is called an "isotype switch," or "class switch," so that either IgG, IgA, or IgE antibody is made, each with the same antigen-binding site. Only one isotype is made per B-cell clone.

Making Memory B Cells

Having encountered an antigen, the antigen-activated B cell will divide to make memory B cells as well as plasma cells (see Figure 16.12). Memory B cells are B cells that have already undergone class switching (that is, they are committed to making IgG, IgA, or IgE), and they are very long-lived. Their long lives provide immunological memory. But why do they survive longer than a regular B cell? B cells that do not encounter their matched antigen eventually die by programmed cell death (apoptosis). However, if the B-cell receptor binds to its matched (cognate) antigen, a signal is sent to the nucleus to make an anti-apoptosis protein (Mcl-1). The memory B-cell line can now propagate for many years, guarding us against microbial assault.

During its lifetime, a memory B cell, like other B cells, also hypermutates the DNA that encodes the antigen-binding region of its antibody genes (the VDJ and VJ regions described in inSight). These mutations can increase (or decrease) the affinity of antibody produced during the secondary response. Higher-affinity antibodies are selected because as an immune response clears an infection, antigen becomes scarce, so only B cells with the highest affinity B-cell receptor (surface antibody) can bind to the vanishing antigen and become activated. This process is known as **affinity maturation**, which means that with repeated exposures to the same antigen, a host will produce memory B cells and antibodies of successively greater affinities.

SECTION SUMMARY

- **The primary antibody response** to an antigen begins when B cells differentiate into antibody-producing plasma cells and memory B cells. IgM antibodies are generally the first class of antibodies secreted during the primary response.

- **Isotype switching** occurs late during the primary response when a subclass of B cells switches from making IgM to making other antibody isotypes (for instance, IgG).

- **The secondary antibody response** occurs during subsequent exposures to an antigen and arises because memory B cells are activated. IgG is the predominant antibody made.

- **Clonal selection** is the rapid proliferation of a subset of B cells during the primary or secondary antibody response.

- **A B-cell receptor (BCR)** consists of a membrane-bound antibody on a B cell in association with the Igα and Igβ proteins. Binding of antigen to the B-cell receptor triggers B-cell proliferation and differentiation.

- **T-cell-independent immunity,** the activation of B cells without the help of helper T cells, is possible when an antigen possesses multiple repeating epitopes that bind multiple B-cell receptors (capping).

- **Antibody diversity** is generated during B-cell development by recombining (mixing and splicing) about 1,000 DNA gene cassettes and introducing mutations within DNA segments corresponding to the antigen-binding site.

Thought Question 16.4 Why do immunizations lose their effectiveness over time?

Thought Question 16.5 IgM is the first antibody produced during a primary immune response. Of the five different types of antibodies produced, why would the body want IgM to be the first?

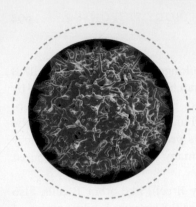

inSight

Building an Antibody: The Full Story

The first step in making a specific antibody occurs during the formation of a B cell from a progenitor stem cell in bone marrow. Obviously, this happens long before the B cell ever encounters antigen. Immunoglobulin genes in a bone marrow stem cell consist of many gene segments that can rearrange in many possible combinations. During differentiation of a stem cell into a mature B cell (which happens every day in a person's life), DNA segments are deleted in a process called **gene switching**, which <u>decreases</u> the number of gene segments in the mature B-cell DNA. The process starts at the 5′ end of an immunoglobulin gene cluster, which corresponds to the variable (V) end, or antigen-binding site, of the final peptide (Figure 1).

Making the antigen-binding site. In both the heavy- and light-chain genes, there are many tandem gene cassettes encoding potential variable regions separated by **recombination signal sequences (RSSs)**. These sequences allow recombination to bring together two widely separated gene cassettes and delete the intervening DNA. Approximately 170 V gene cassettes exist for the heavy and light chains. Heavy-chain genes start with the V regions followed by sets of D (diversity) genes, J (joining) region genes, and then the various heavy-chain constant regions. Light-chain genes are similar but lack the D regions. Light-chain V and J gene clusters are upstream of the light-chain constant-region genes (k and l).

A summary of the genetic processes leading to antibody formation is shown in Figure 1. Antibody formation begins in the heavy-chain gene with recombination events that take place at RSS sites in the D, J, and V segments (steps 1 and 2). The recombinational events delete all the intervening segments. The result is a joined VDJ DNA sequence and a mature (naive) B cell. Following transcription, the primary VDJ RNA transcript will then undergo RNA splicing to remove any J-segment RNA sequences that remain downstream of the VDJ RNA sequence. The result is a mature (naive) B cell. Remember, all DNA rearrangements to make the antigen-binding site happen before the B cell ever "sees" the antigen. A similar sequence of events occurs for the light chains, except that the product is VJ. A naive B cell that has not yet "seen" antigen has its assembled VDJ antigen-binding site DNA linked to the IgM gene, and the cell makes IgM B-cell receptors.

How does affinity of the antigen-binding site for antigen increase over time? The V region DNA is subject to high levels of somatic mutation that can alter the sequence of the antigen-binding site and increase its binding affinity. Hypermutation happens every time a memory B cell is exposed to the antigen and divides.

The isotype switch. Now that the antigen-binding site VDJ DNA sequence has been constructed, how does it become connected to <u>different</u> heavy chains? The next

Figure 1 Genetics of Antibody Formation

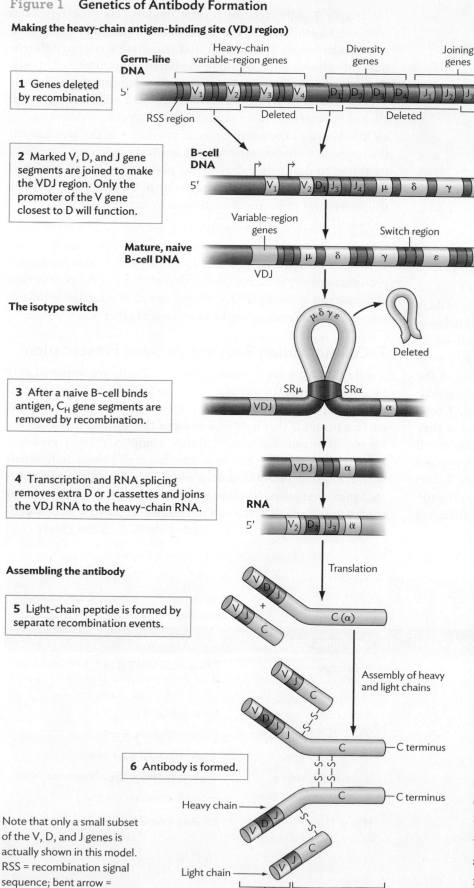

Making the heavy-chain antigen-binding site (VDJ region)

1 Genes deleted by recombination.

2 Marked V, D, and J gene segments are joined to make the VDJ region. Only the promoter of the V gene closest to D will function.

The isotype switch

3 After a naive B-cell binds antigen, C_H gene segments are removed by recombination.

4 Transcription and RNA splicing removes extra D or J cassettes and joins the VDJ RNA to the heavy-chain RNA.

Assembling the antibody

5 Light-chain peptide is formed by separate recombination events.

6 Antibody is formed.

Note that only a small subset of the V, D, and J genes is actually shown in this model. RSS = recombination signal sequence; bent arrow = promoter for transcription.

step in B-cell differentiation occurs during the actual primary immune response, when a mature (but naive) B-cell receptor finds its matched antigen.

In the early stages of the primary response, plasma cells produced from these B cells secrete only IgM. If this B cell also receives signals from a helper T cell, isotype switching will occur and the B cell may then make IgG, IgA, or IgE.

What "flips" the switch? Notice in Figure 1 (end of step 2) that the constant-region isotype gene segments are arranged in a row <u>after</u> a recombined VDJ region. The mechanism by which a B cell switches to make IgG, IgE, or IgA is similar to VDJ formation. Each constant segment, except delta, contains a sequence called a **switch region**. Recombination between these switch regions deletes the DNA sequences between the VDJ region and one of the constant regions (steps 3 and 4). Regardless of which C_H gene is selected, the antibody produced will have the same antigenic specificity (VDJ region) as the original IgM. The type of cytokines (see Section 15.3) secreted by helper T cells at the time of the switch will influence which C_H gene is selected.

To complete the process of antibody synthesis (before or after class switching), two assembled light-chain peptides combine with two heavy-chain peptides (steps 5 and 6). A single, mature B cell makes only one type of heavy chain and one type of light chain. In sum, a combination of gene rearrangements and random mutations generates the remarkable level of antibody binding site diversity we each possess.

16.5
T Cells: The Link between Humoral and Cell-Mediated Immunity

SECTION OBJECTIVES

- Explain the three major types of T cells.
- Discuss the role of major histocompatibility proteins in the adaptive immune response.
- Discuss the two routes of antigen processing and presentation.
- Describe processes that minimize autoimmunity.

Do the two branches of the immune system (antibody based and cell mediated) "know" of each other's existence? Different types of infections tilt the immune response toward one or the other, so the systems must communicate (the right hand does know what the left is doing). T cells hold the key to that balance by playing an integral role in antibody production and cell-mediated immunity. Although derived from the same progenitor stem cell as B cells, T cells develop in the thymus (rather than in the bone marrow, where B cells develop) and contain surface protein antigens different from those of B cells. In general, T cells are divided into two broad groups distinguished by whether they carry cellular differentiation proteins CD4 or CD8 on their cell surface (Table 16.3). Helper T cells display the surface antigen CD4, whereas the second major class of T cells, cytotoxic T cells (T_C cells), display CD8. CD4 T cells can stimulate either antibody- or cell-mediated immunity. CD8 T cells carry out cell-mediated immunity.

Helper T cells come in several models. The first type, T_H0 cells, can be stimulated by APCs to become the other types, such as follicular helper T cells (T_{FH}) that stimulate B cells to differentiate into plasma cells (antibody-mediated immunity), or T_H1 cells that activate cytotoxic T cells (cell-mediated immunity). Cytotoxic T (T_C) cells destroy the membranes of host cells infected with viruses or bacteria. The ratio of different T_H cells produced during an infection and the relative mix of cytokines in the area can tilt the immune response more toward antibody-mediated immunity (required to deal with extracellular pathogens) or cell-mediated immunity (used to handle intracellular pathogens). To be of any use, however, T cells must be activated by antigen.

Note T_H2 cells mainly stimulate B cells to class-switch and make IgE antibody useful against parasites such as helminths (worms). T_H17 mainly helps class-switch to IgG subtypes but also produces a pro-inflammatory cytokine (IL-17) that stimulates the recruitment of monocytes and macrophages to the site of an infection or inflammation. Treg cells suppress the responses of other immune cells.

T-Cell Activation Requires Antigen Presentation

T cells do not bind free-floating antigen. T cells are activated only by antigen bound to another cell's surface—that is, by antigen-presenting cells (APCs). As illustrated in Figure 16.15, the APC surface proteins that hold and present the antigen to T cells are known as major histocompatibility complex (MHC) proteins. MHC proteins differ between species and among individuals within a species. They help determine whether a given antigen is recognized as coming from the host (a self antigen) or from another source (a foreign antigen) in a phenomenon called histocompatibility (hence the name "major histocompatibility complex").

Table 16.3
Major Classes of T cells[a]

T-Cell Type	CD Coreceptor	MHC Restriction	Cytokines Produced	Major Function
T_H0 cell	CD4	Class II	Wide range	Differentiate into T_H1 or T_H2
Helper T cell				
T_{FH}	CD4	Class II	IL-12, IL-4	Activate B cells
T_H1	CD4	Class II	IFN-γ, IL-2, TNF-β	Activate cytotoxic T cells
T_H2	CD4	Class II	IL-4, IL-5, IL-6	B-cell helper; IgE class switch
T_H17	CD4	Class II	IL-17, IL-23, TNF-α	Pro-inflammatory; IgG subtype class switch
Treg	CD4	Class II	IL-10, TGF-β	Anti-inflammatory
Cytotoxic T cell (T_C)	CD8	Class I	IFN-γ, TNF	Kill virus-infected and cancer cells

[a]CD and MHC proteins are located on the surface of T cells.

Two classes of MHC molecules are found on cell surfaces. Both classes belong to the immunoglobulin family of proteins, but they are not immunoglobulins. **Class I MHC molecules** are found on all nucleated cells (not red blood cells). **Class II MHC molecules**, on the other hand, have a more limited distribution (see Figure 16.15). Class II MHC molecules are expressed only by antigen-presenting dendritic cells, macrophages, and B cells. MHC molecules are critical to the immune system because the T cell, to be activated, must first recognize a foreign antigen attached to an MHC molecule on an APC. Note that each MHC can bind a broad range of antigens and that each APC expresses multiple MHC alpha and beta chains. Multiple MHC molecules on an APC provide diversity of antigen recognition. Unlike antibodies, DNA rearrangements are not involved.

Figure 16.15   Major Histocompatibility Proteins

Major histocompatibility proteins

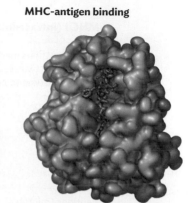

A. MHC class I molecules are composed of a large alpha chain and a small peptide called β_2-microglobulin. MHC class II molecules contain an alpha chain and a beta chain. The peptide-binding regions of both classes show variability in amino acid sequence that yields different shapes and grooves. Peptide antigens nestle in the grooves and are held there awaiting interaction with T-cell receptors. CD8 T cells recognize antigen peptides associated with class I molecules, whereas CD4 T cells recognize peptides bound to class II molecules.

B. Top view of an antigen (red) nestled in the MHC peptide-binding site.

Two Paths for Antigen Processing and Presentation

In the initial stages of an immune response, antigen-presenting cells internalize the pathogen, such as a virus or a bacterium, and degrade the pathogen into smaller pieces (epitopes). These epitopes are placed within MHC-binding clefts and transported back to the cell surface. Whether an antigen epitope binds to class I or class II MHC molecules generally depends on how the antigen initially entered the cell (Figure 16.16). Antigens synthesized by viruses and intracellular bacteria as they grow within the cytoplasm of an APC (called endogenous antigens) will attach to class I MHC molecules on the endoplasmic reticulum and are moved to the cell surface. Antigens produced outside of the APC (called exogenous antigens), as are most bacterial antigens, will enter the cell via phagocytosis and attach to class II MHC molecules affixed to the membranes of an acidic phagosome or lysosome (see Figure 16.16). The MHC class II–peptide complex is then carried to the cell surface.

Once the antigen is presented on the APC surface, T cells can interact with it via the T-cell receptors. Interactions between antigen-presenting cells and naive T cells occur within the lymph nodes, the spleen, or Peyer's patches in the gut, so APCs must make their way to those locations. How T cells then distinguish between MHC I and MHC II presentation is explained later.

Note Pathogens are not the only items processed by APCs. Large proteins, carbohydrates, and other molecules are engulfed, processed, and presented on surface MHC molecules.

You might be asking yourself, "Well, why don't my T cells see my own antigens? Are they tolerant somehow?" Actually, T cells that strongly bind self antigens are not really made tolerant; they are killed in a process called T-cell education.

T-Cell Education and Deletion

Greek mythology tells of Narcissus, a young man punished by the gods for scorning the women who fell in love with him. One day, Narcissus saw his beautiful reflection in a pool of water, fell in love, tumbled into the pool, and drowned. His inability to recognize himself is analogous to the danger posed by an immune system. Immune systems must be able to distinguish what is self (meaning antigens present in our own tissues) from what is not self. Innate immunity accomplishes this, in part, through the use of pattern recognition proteins such as the Toll-like and NOD-like receptors (see Section 15.5). How adaptive immune mechanisms avoid recognizing self is equally important. At the outset of development, the immune system is fully capable of reacting against self.

One mechanism the body uses to avoid attacking itself is to delete (kill) any T cells that react too strongly against self antigens (a process called self-tolerance). In humans, the T-cell deletion process occurs throughout life as new T cells are made. In this process, T cells undergo a two-stage selection, or "education" process, in the

Figure 16.16 **Processing and Presentation of Antigens by Antigen-Presenting Cells on Class I and Class II MHC Proteins**

Microbial proteins made in the host cytoplasm (upper left) are degraded, and peptides are placed on MHC class I molecules in the endoplasmic reticulum (ER). Microbial proteins made outside the cell (lower center) are endocytosed, degraded in the endosome, and placed on MHC class II molecules.

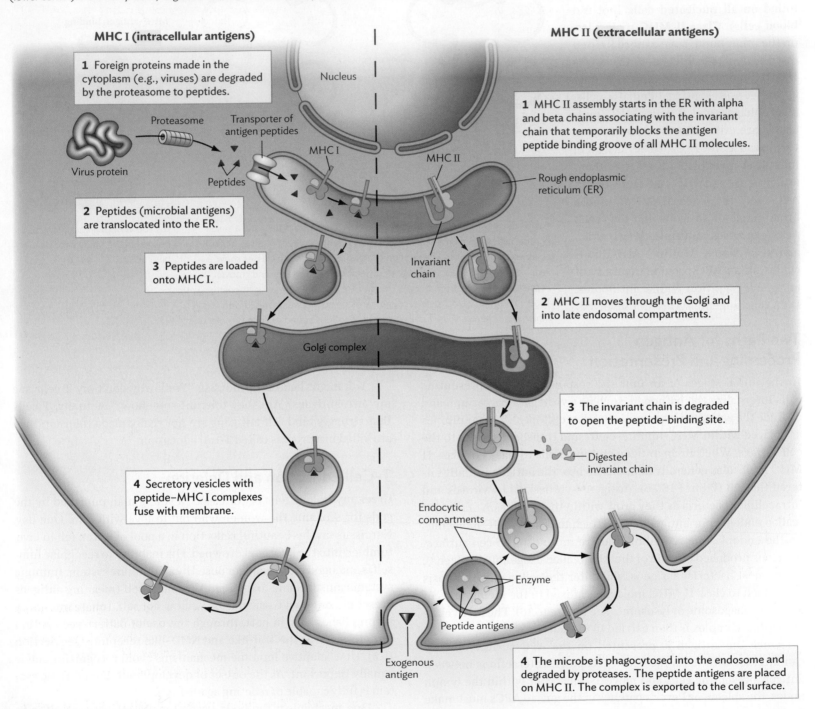

MHC I (intracellular antigens)

1 Foreign proteins made in the cytoplasm (e.g., viruses) are degraded by the proteasome to peptides.

Nucleus

Proteasome   Transporter of antigen peptides

Virus protein   MHC I   MHC II

Peptides

2 Peptides (microbial antigens) are translocated into the ER.

Rough endoplasmic reticulum (ER)

3 Peptides are loaded onto MHC I.

Invariant chain

Golgi complex

4 Secretory vesicles with peptide–MHC I complexes fuse with membrane.

Endocytic compartments

Enzyme

Peptide antigens

Exogenous antigen

MHC II (extracellular antigens)

1 MHC II assembly starts in the ER with alpha and beta chains associating with the invariant chain that temporarily blocks the antigen peptide binding groove of all MHC II molecules.

2 MHC II moves through the Golgi and into late endosomal compartments.

3 The invariant chain is degraded to open the peptide-binding site.

Digested invariant chain

4 The microbe is phagocytosed into the endosome and degraded by proteases. The peptide antigens are placed on MHC II. The complex is exported to the cell surface.

thymus (see Figure 15.7) to recognize self versus nonself. This education process involves the use of self antigens, not foreign antigens (Figure 16.17). T cells bearing T-cell receptors (TCRs, described later) that weakly recognize self MHC proteins displayed on thymus epithelial cells are allowed to live (**positive selection**). These T cells leave the thymus to seed secondary lymphoid organs such as the spleen and lymph nodes. T cells that recognize self MHC peptides too strongly, however, are killed in the thymus or deleted from the population (**negative selection**). Almost 95% of T cells entering the thymus die during the negative selection processes. This weeding out is important because if our T-cell repertoire included cells that bound self MHC too tightly, then our T cells would constantly react to our own MHC molecules, regardless of what antigen peptides were attached. Note, however, that some self-reactive T cells are allowed to survive and are converted to **regulatory T cells (Tregs)**. Regulatory T cells can block the activation of harmful, self-reactive (autoimmune) lymphocytes that escape deletion and enter the circulation.

The need for negative selection may seem obvious, but why is positive selection (promoting weak binding to MHC molecules) important? The positive selection process is needed because T cells must be able to recognize self MHC proteins before the T-cell receptor can bind the MHC-associated antigen.

In contrast to negative selection for B cells, which occurs in bone marrow, T-cell education is limited to the thymus. But if the thymus expresses only thymus antigens, how can T cells that respond to antigens expressed on other host cells (for example, heart cells) be removed? The answer is a special gene activator called AIRE in thymus cells that allows them to synthesize all human proteins in small amounts. This expression is necessary to complete T-cell education within the thymus.

You might also ask how someone who has had their thymus removed (a treatment for myasthenia gravis caused by a tumor in the thymus) can live if the organ is critical for T-cell maturation and for deleting self-reactive T cells. Actually, within a few years after birth, the thymus begins to lose its utility, such that very little function remains in adults. Fortunately, a large amount of T-cell education occurs during fetal development. Once a T cell is educated, reserve pools of these T cells are maintained throughout life outside of the thymus. Some T cells, apparently, can also mature in secondary lymphoid tissues. Thus, adults without a thymus can live relatively normal lives. Babies born without a thymus, however, have a severe, life-threatening T-cell deficit.

SECTION SUMMARY

- **T_H0 precursor cells,** which have surface CD4 molecules, can make T_H1, T_H2, and T_H17 cells that support B-cell differentiation into antibody-producing plasma cells. T_H1 cells also drive cell-mediated immunity mediated by CD8 T cells.

Figure 16.17 Recognizing Self

The process of T-cell education in the thymus. Lymphoid precursor cells enter the thymus to become T cells. The thymus can express most self antigens and present them on antigen-presenting cells (APCs). As part of its education process, an immature T cell whose T-cell receptors (TCRs) strongly recognize a self antigen is either killed through apoptosis or converted to a regulatory T cell that can inhibit the function of a self-reactive lymphocyte that may have escaped the thymus alive. A T cell whose TCRs recognize self only weakly is left alone (positively selected) to enter the bloodstream, where it can travel to secondary lymphoid organs.

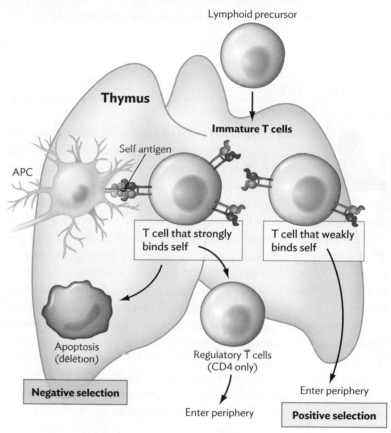

- **The major histocompatibility complex** consists of membrane proteins with variable regions that can bind antigens. Class I MHC molecules are on all nucleated cells, whereas antigen-presenting cells contain both class I and class II MHC molecules.

- **Antigen-presenting cells (APCs),** such as dendritic cells, present antigens synthesized during an intracellular infection on their surface class I MHC molecules, but place antigens from engulfed microbes or allergens on their class II MHC molecules. T cells recognize antigens bound to MHC.

- **T cells must be "educated"** to bind self MHC molecules weakly.

- -

Thought Question 16.6   How do viruses stimulate antibody production if viruses are intracellular pathogens and their intracellularly synthesized and processed epitopes are presented only onto MHC I surface receptors? Won't this trigger only T_c activation and cell-mediated immunity? (See Figure 16.16.)

- -

16.6
T-Cell Activation

CASE HISTORY 16.1

Bare Lymphocyte Syndrome

Fatima is the third child of Moroccan consanguineous healthy parents (cousins of first degree). From birth, she has suffered numerous infections, including varicella pneumonia, disseminated candidiasis, enterovirus encephalitis, and *E. coli* meningitis. When she was 3 years old, she underwent an extensive immunological workup to identify the cause of these recurring illnesses. Her white blood cell counts were in the normal range, as were her neutrophil and total lymphocyte counts. However, her total IgG, IgA, and IgM antibody levels were extremely low. This extreme Ig deficiency was the likely cause of her failure to fend off infections. However, the team of physicians treating Fatima still did not know the cause of the antibody deficiency. More tests were ordered to examine key protein surface markers on B cells (IgD), T_H0 cells (CD4), T_C cells (CD8), and APCs (MHC I and MHC II classes). Her B-cell numbers were normal, but her T cells were low, especially the CD4 T-cell set. The surprise came when the laboratory reported the absence of MHC class II proteins on her APC cells. The presence of MHC class I proteins, however, was quite normal. The physicians now understood the problem, but the cure would be risky. Fatima had a genetic disease called bare lymphocyte syndrome.

How do B cells lose MHC II receptors, and why does this defect change antibody levels? Fatima's lymphocytes (B cells) did not contain any MHC II receptors because of a defect in a transcriptional activator needed to express the MHC II genes. The lack of MHC II on these cells is why they are called bare lymphocytes, even though they have other surface proteins. Fatima's plight underscores the importance of MHC receptors to immunity. A defect in antigen presentation and thus T-cell activation was at the root of her condition. We will have more to say about her case later.

T-Cell Receptor Structure and Function

To be helpers or to gain cytotoxic activity, T cells must become activated. T-cell receptors are critically important for T-cell activation. T-cell receptors (TCRs) are antigen-binding molecules (not

Figure 16.18   The T-Cell Receptor (TCR) and CD3 Complex

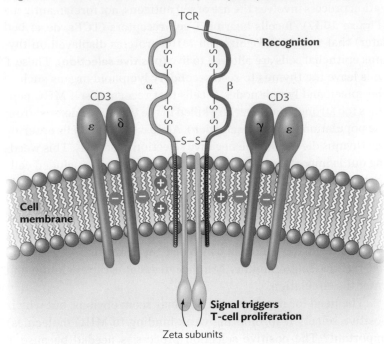

The T-cell receptor proteins are associated with CD3 proteins at the cell surface. Antigen binds to the alpha and beta subunits. The positive and negative charges holding the complex together come from amino acids in the peptide sequences. Once bound to antigen, the complex transduces a signal into the cell that triggers T-cell proliferation.

antibodies) present on the surface of T cells (**Figure 16.18**). These receptors bind peptides only (not lipids, polysaccharides, or DNA) and do not bind <u>soluble</u> antigen. A TCR will bind only to antigens attached to MHC surface proteins that are present on antigen-presenting cells.

The T-cell receptor is composed of several transmembrane proteins. The part used to recognize antigen is composed of two molecules, alpha and beta. Much like the immunoglobulins, the alpha and beta proteins of the TCR are formed from gene clusters that undergo gene rearrangements analogous to, but different from, the immunoglobulin genes.

The TCR alpha and beta proteins are found in a complex with four other peptides; together these form the CD3 complex. When stimulated, these ancillary CD3 complex proteins send a signal to the nucleus that triggers proliferation of the T cell. Thus, T cells, like B cells, are clonally selected to multiply.

Activation of T_H0 Helper Cells

Figure 16.19 summarizes the steps of antibody production starting from B-cell development in the bone marrow (steps 1–3, discussed in Section 16.4), T-cell activation in the lymph node (step 4), and finally, T-cell activation of B cells (step 5). We will refer to Figure 16.19 often during this discussion.

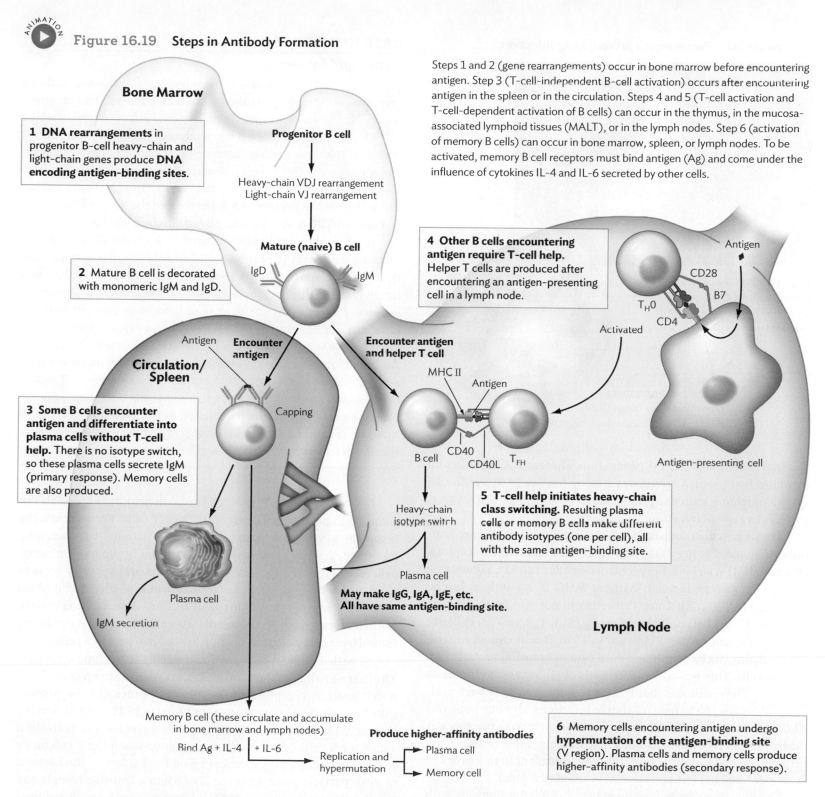

Figure 16.19   Steps in Antibody Formation

Bone Marrow

1 DNA rearrangements in progenitor B-cell heavy-chain and light-chain genes produce **DNA encoding antigen-binding sites**.

Progenitor B cell

Heavy-chain VDJ rearrangement
Light-chain VJ rearrangement

Mature (naive) B cell

2 Mature B cell is decorated with monomeric IgM and IgD.

IgD   IgM

Steps 1 and 2 (gene rearrangements) occur in bone marrow before encountering antigen. Step 3 (T-cell-independent B-cell activation) occurs after encountering antigen in the spleen or in the circulation. Steps 4 and 5 (T-cell activation and T-cell-dependent activation of B cells) can occur in the thymus, in the mucosa-associated lymphoid tissues (MALT), or in the lymph nodes. Step 6 (activation of memory B cells) can occur in bone marrow, spleen, or lymph nodes. To be activated, memory B cell receptors must bind antigen (Ag) and come under the influence of cytokines IL-4 and IL-6 secreted by other cells.

4 Other B cells encountering antigen require T-cell help. Helper T cells are produced after encountering an antigen-presenting cell in a lymph node.

Antigen

CD28
T_H0 B7
CD4
Activated

Circulation/ Spleen

Antigen   Encounter antigen

Encounter antigen and helper T cell

MHC II
Antigen

3 Some B cells encounter antigen and differentiate into plasma cells without T-cell help. There is no isotype switch, so these plasma cells secrete IgM (primary response). Memory cells are also produced.

Capping

B cell   CD40 T_{FH}
CD40L

Antigen-presenting cell

Heavy-chain isotype switch

5 T-cell help initiates heavy-chain class switching. Resulting plasma cells or memory B cells make different antibody isotypes (one per cell), all with the same antigen-binding site.

Plasma cell

Plasma cell

May make IgG, IgA, IgE, etc. All have same antigen-binding site.

IgM secretion

Lymph Node

Memory B cell (these circulate and accumulate in bone marrow and lymph nodes)

Bind Ag + IL-4 + IL-6

Produce higher-affinity antibodies

Replication and hypermutation → Plasma cell
                              → Memory cell

6 Memory cells encountering antigen undergo **hypermutation of the antigen-binding site** (V region). Plasma cells and memory cells produce higher-affinity antibodies (secondary response).

Two signals are required to activate T_H0 cells. The first is a meeting between a T_H0 cell and an APC, which happens when an antigen connects MHC II on the APC to the T-cell receptor (see Figure 16.19, step 4). Only a specific TCR that can bind the antigen will do. The CD4 molecule, whose presence is restricted to the surface of T_H0 cells, recognizes MHC II surface proteins (but not MHC I) and facilitates TCR binding. But linking a TCR to an antigen–MHC II is not enough to activate the T_H0 cell. The second

activation signal involves a CD28 molecule on the T-cell surface binding to a B7 protein on the APC cell surface (see Figure 16.19, step 4). Once that interaction happens, the T_H0 cell is activated.

As T_H0 cells become activated, different cytokines that are produced during an infection will influence T_H0 conversion to T_{FH}, T_H1, T_H2, or T_H17 cells. As mentioned earlier, T_{FH} cells are the primary T_H cells that stimulate B cells to become plasma cells and secrete antibody. T_H1 cells stimulate cytotoxic

Figure 16.20 *Pneumocystis jirovecii* **Lung Infection**

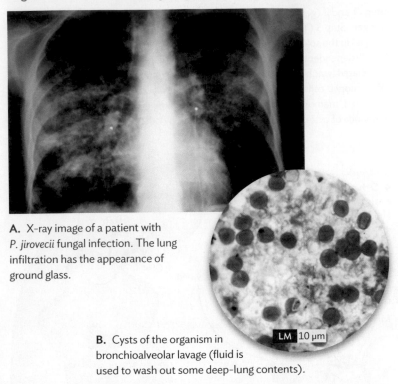

A. X-ray image of a patient with *P. jirovecii* fungal infection. The lung infiltration has the appearance of ground glass.

B. Cysts of the organism in bronchioalveolar lavage (fluid is used to wash out some deep-lung contents).

LM 10 μm

T cells. T_H2 cells promote IgE production, whereas T_H17 cells secrete cytokine IL-17, which stimulates nearby host cells to synthesize proinflammatory cytokines that, among other things, attract neutrophils to a site of infection.

At this juncture, let's revisit Fatima and the case of bare lymphocyte syndrome. Recall that her B cells had no MHC II receptors. When MHC II is lacking, the problem extends to all APCs (dendritic cells, macrophages, B cells). Without MHC II, no antigen can be presented to T_H0 cells. Thus, T_H0 cells will not proliferate (Fatima had a low CD4 T cell count), and no T_{FH}, T_H1, T_H2, or T_H17 cells are formed. Without helper T cells to "talk" to, B cells cannot initiate heavy-chain class switching and most cannot even differentiate into plasma cells. This was why blood levels of <u>all</u> her antibody classes were low. The result was that Fatima could not make enough antibody of any isotype to fend off simple infections. Her one hope was a **bone marrow transplant**. In this procedure, her bone marrow was destroyed and replaced with bone marrow containing healthy stem cells able to make MHC II proteins. The transplant in this case was successful, and her antibody levels returned to normal.

The question now is, how do helper T cells communicate with B cells?

Activated Helper T Cell Meets B Cell

The following case of a patient overproducing IgM provides a backdrop for understanding how helper T cells communicate with B cells—the next step in B-cell differentiation and antibody production.

Hyper IgM Syndrome

Nathan was a 10-month-old male who had suffered much in his short life. Since birth he's had recurrent serious extracellular bacterial infections, all of which were resolved by antibiotic treatment. On Christmas day, Nathan's mother discovered he had another high fever and a nonproductive cough (no mucus). He was also having trouble breathing. Fearing one more serious infection, Nathan's mother rushed him to the hospital. An X-ray showed the lung infiltrate seen in **Figure 16.20A**. A tube was placed through Nathan's nose and into his lung, and a solution was used to wash out deep-lung contents for microscopic examination (bronchoalveolar lavage). The fluid yielded the organisms seen in **Figure 16.20B**. Nathan was hospitalized with pneumonia caused by *Pneumocystis jirovecii*, a yeastlike fungus. Although most of us have been infected with this organism, our immune system almost always prevents disease. Finding the disease in Nathan suggested a serious immune dysfunction. An immunological workup revealed mild neutropenia (lower than normal numbers of neutrophils) but normal numbers of T and B cells, normal levels of complement, and normal complement activity. However, Nathan's blood exhibited exceptionally high levels of IgM but very low IgG and IgA levels. The physician suspected a specific defect in Nathan's T cells.

Disease caused by *P. jirovecii* is rare in healthy individuals but common in patients with defects in their T cells. For instance, the organism is a frequent cause of death for patients with acquired immunodeficiency syndrome (AIDS) caused by human immunodeficiency virus (HIV). HIV targets T cells and severely reduces their numbers in blood. Since Nathan had a normal T-cell blood count, HIV was not the culprit. This section will reveal the basis of Nathan's recurrent infections: a disease called hyper IgM syndrome caused by another communication breakdown between cells.

As with T cells, B cells need two signals to become activated. The first signal, described earlier, occurs when antigen cross-links B-cell antigen receptors on B-cell membranes (in the process called capping; see Figure 16.14 and Figure 16.19, step 3). During the <u>early</u> phase of the primary response, the second activation signal for B cells occurs in the lymph node when the B cell binds to complement factor C3 deposited on the surface of the bacteria or virus particle (see Section 15.7) or when a Toll-like receptor on the B cell interacts with a MAMP (see Section 15.5). This activation does not result in heavy-chain class switching; IgM only is secreted by the resulting plasma cell.

To become activated during the <u>later</u> primary immune response, B cells usually require helper T cells; usually T_{FH} cells. But the T_H cell must have a T-cell receptor that can bind to the same antigen that the B cell binds. How does a B cell gain specific T-cell help? Recall that B cells are also APCs.

Consequently, antigen bound by B-cell receptors is internalized, processed, and presented back on the B cell's surface MHC II receptors (see Figure 16.16 and Figure 16.19, step 4). Helper T cells with the proper TCR will then bind to the B-cell MHC II–antigen complex. This contact allows another very important cell-to-cell interaction. A surface protein called CD40 on the B cell binds to a protein called CD40 ligand (CD40L, or CD154) on the T cell. Interaction between CD40 and CD40 ligand triggers the second intracellular signal needed for B-cell activation. Once activated by the helper T cell, a B cell can undergo heavy-chain class switching to make IgG, IgA, or other antibodies and differentiate into plasma cells. (see Figure 16.19, step 5).

What do you suppose would happen in a person lacking CD40 ligand on their T_H cells because of a genetic mutation? The person's T-cell and B-cell numbers would remain normal, but the B cells would not undergo heavy-chain class switching. Thus, any plasma cells made will make only IgM. Serum levels of IgM will rise, but no other antibody type will be secreted; the result is hyper IgM syndrome. In Case History 16.2, Nathan had a mutation in the gene encoding CD40L. His B cells could not undergo a heavy-chain switch. Note that the IgM produced is not antigen specific because it does not result from clonal selection in response to antigen; thus, it is not effective in preventing disease.

Note How can high IgM levels be produced in hyper IgM syndrome if CD40-CD40L interactions are needed for plasma cell production? The short answer is that there are CD40-CD40L–independent mechanisms capable of triggering plasma cell formation, but they work only on certain IgM-producing B cells.

The Secondary Antibody Response

During the primary response, B cells generally need T-cell help to become plasma cells. But what about the secondary response? Do memory B cells also need T-cell help? During the secondary response, memory B cells still need help to become plasma cells but do not need direct contact with helper T cells. Memory B cells with antigen bound to their B-cell receptors can respond to the soluble IL-4 and IL-6 cytokines secreted by activated helper T cells without having direct contact with the T_H cell. IL-4 stimulates B-cell proliferation, while IL-6 directs differentiation into antibody-secreting plasma cells (see Figure 16.19, step 6).

SECTION SUMMARY

- **T-cell receptors (TCRs)** are antigen-specific T-cell surface proteins.
- **Activation of a T_H0 cell** requires two signals: T-cell receptor CD4 binding to an MHC II–antigen complex on an antigen-presenting cell, and CD28-B7 interaction.
- **T_H0 cell differentiation** to T_{FH}, T_H1, T_H2, or T_H17 cells is influenced by different cytokine "cocktails" secreted by macrophages and natural killer cells during an infection.

- **Activation of a B cell** into an antibody-producing plasma cell usually requires two signals. The first is antigen binding to a B-cell receptor. The second signal is binding to a T_H cell activated by the same antigen. This second signal involves the T cell receptor binding to an MHC II molecule bearing antigen on the B cell and the CD40 ligand on the T cell binding to CD40 on the B cell. Both interactions are required.

Thought Question 16.7   IL-10 is considered an anti-inflammatory cytokine because it down-regulates production of MHC II receptors. What consequences would down-regulating MHC II on macrophages have on the immune response?

Thought Question 16.8   Why does attaching a hapten to a carrier protein allow production of antihapten antibodies?

16.7
Cell-Mediated Immunity

SECTION OBJECTIVES

- Explain how cytotoxic T cells become activated.
- Discuss how activated T_C cells recognize their target host cells.
- Describe the role of cytokines in directing adaptive immunity toward humoral or cytotoxic responses.
- Distinguish antigens from superantigens.
- Diagram the basic ways that pathogens can avoid the immune system.

How does the development of cell-mediated immunity differ from antibody-dependent immunity? The other branch of the adaptive immune system, cell-mediated immunity, relies on another type of T cell, the cytotoxic T cell (T_C cell). Recall that cytotoxic T cells are marked by the surface protein CD8. Cell-mediated immunity is critically important for clearing infections caused by intracellular pathogens, such as viruses. However, engaging this branch of immunity can come at a price.

CASE HISTORY 16.3

The Run-Down Mechanic

Luke, a 24-year-old mechanic from Egypt Beach, Massachusetts, was feeling run-down. Over the past month he hadn't been eating much and was always tired, and he recently noticed pain just under his right rib cage. His urine had recently turned the color of tea, and his wife noticed that the whites of his eyes were taking on a yellowish tinge. Luke finally went to the doctor, who suspected Luke had cirrhosis, a liver disease. This preliminary diagnosis was based on Luke's symptoms. Luke's abdominal pain was located over the liver, and the color of his urine and the jaundice evident in his eyes all pointed toward liver disease. Luke did not drink alcohol and had never left the country, but when questioned, Luke admitted to taking intravenous drugs. The doctor suspected hepatitis and ordered a series of blood tests to check for the presence of liver enzymes, anti-hepatitis antibodies, and hepatitis DNA or RNA in his blood. The tests revealed a high level of hepatitis B viral DNA. Luke had hepatitis B disease.

There are several types of hepatitis viruses (types A–G). Hepatitis B virus is an enveloped DNA virus (discussed in Chapters 12 and 22) that can be transmitted by sexual intercourse, blood transfusion, or the use of contaminated needles. It seems likely that the virus was transmitted to Luke by a contaminated needle, given his intravenous drug use. The main target of hepatitis B virus is the liver, where the virus replicates in hepatocytes (liver cells).

Figure 16.21 Activation and Function of a Cytotoxic T Cell

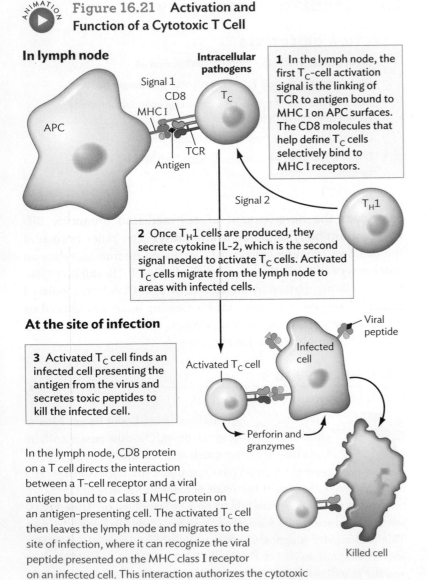

In lymph node

Signal 1
CD8
MHC I
APC
Antigen
TCR
T_C

Intracellular pathogens

1 In the lymph node, the first T_C-cell activation signal is the linking of TCR to antigen bound to MHC I on APC surfaces. The CD8 molecules that help define T_C cells selectively bind to MHC I receptors.

Signal 2
T_H1

2 Once T_H1 cells are produced, they secrete cytokine IL-2, which is the second signal needed to activate T_C cells. Activated T_C cells migrate from the lymph node to areas with infected cells.

At the site of infection

3 Activated T_C cell finds an infected cell presenting the antigen from the virus and secretes toxic peptides to kill the infected cell.

Activated T_C cell
Infected cell
Viral peptide
Perforin and granzymes
Killed cell

In the lymph node, CD8 protein on a T cell directs the interaction between a T-cell receptor and a viral antigen bound to a class I MHC protein on an antigen-presenting cell. The activated T_C cell then leaves the lymph node and migrates to the site of infection, where it can recognize the viral peptide presented on the MHC class I receptor on an infected cell. This interaction authorizes the cytotoxic T cell to kill the infected cell.

However, the damage to the liver is not caused by the virus per se. It is caused by cytotoxic T cells called in to fight the infection. This section will describe how cytotoxic T cells are activated and how they target infected cells.

T_H1 Cells Activate Cytotoxic T Cells

As with other lymphocytes, cytotoxic T cells circulate in the blood and lymph. Along the way T_C cells pause in lymph nodes to survey APCs for recognizable antigens. A T_C cell becomes activated when its T-cell receptor binds to an antigen presented on the APC's MHC I (**Figure 16.21**, step 1). The CD8 protein on T_C cells recognizes MHC I surface proteins, but not MHC II. To complete the activation process, T_C cells also need a little help in the form of a cytokine called IL-2, produced by the T_H1 class of helper T cells (step 2). Upon binding IL-2, the T_C cell gains cytotoxic activity, leaves the lymph node, travels to the site of infection, and kills any cell bearing the same peptide–MHC class I complex—for example, cells infected with the same virus that triggered T_C production (step 3).

Cytotoxic T cells kill infected target cells by releasing the contents of their granules containing the proteins granzyme and perforin. These proteins are delivered into the target cell. Perforin produces a pore in the target cell membrane through which granzymes can enter. In the cytoplasm, granzymes cleave and activate caspase proteins in the infected cell. The activated caspases then trigger apoptosis and cell death (**Figure 16.22**). Cytotoxic T cells also activate caspases when a T_C surface protein called Fas ligand (FasL) binds to Fas protein on infected cells.

Why are cytotoxic T cells more effective in clearing viral infections than B cells and antibodies? A major reason is that intracellular pathogens (for example, viruses) hide inside host cells, where they are protected from antibody. Consequently, these pathogens are best killed when the harboring host cell is also sacrificed (via cellular immunity). Because all nucleated cells have class I MHC molecules, any infected cell can be recognized by an appropriate T_C cell. The now-activated T_C cell will kill the infected host cell, sacrificing it for the good of the host.

Cellular immunity is a double-edged sword. For instance, during hepatitis B virus infection (see Case History 16.3), cytotoxic T cells play an important pathogenic role. They contribute to nearly all of the liver injury associated with the viral infection. Liver damage occurs because MHC I receptors on infected liver cells present hepatitis B virus epitopes on their surface, and these epitopes are recognized by the T_C cells. Liver cell death by cytotoxic T cells caused the symptoms noted in Luke's case history. Nevertheless, by killing infected cells and producing antiviral cytokines capable of purging hepatitis B virus from viable hepatocytes, cytotoxic T cells also eliminate the virus. However, hepatitis B virus is almost never completely eliminated, even with interferon-alpha therapy (which was done in Luke's case). Still, the therapy did reduce Luke's viral load, and his liver did recover some degree of health.

The Influence of Cytokines on the Immune Response

Throughout Chapters 15 and 16, we have mentioned that many different small peptides called cytokines are secreted by various immune, infected, or damaged cells. These cytokines have a profound influence over the direction and intensity of the immune response. Table 16.4 lists a fraction of the many different cytokines produced by various cell types and their influence over the immune system.

Some cytokines (called chemokines) are released from infected cells, and form a concentration gradient that attracts innate immune system cells to the site of infection. Other cytokines influence whether an immune response is more humoral or cell mediated by influencing the ratio of T_H1 cells to other T_H cells. Recall that cytokines secreted from T_H1 cells help activate cells involved with cell-mediated immunity (the CD8 T_C cells, macrophages, and natural killer cells) and that other cytokines trigger B cells to class-switch their immunoglobulin. Cytokines such as IL-3 can affect the innate immune system by stimulating the production of macrophages and neutrophils needed to clear an infection. The tumor necrosis family of cytokines can trigger fever (see Section 15.6) and various inflammatory processes.

A danger with cytokines is that the production of one cytokine can stimulate other cells to make more cytokines. This feed-forward loop can result in excessive cytokine production, called a **cytokine storm**, during some infections that can produce excessive tissue damage and contribute to the pathology of some diseases such as sepsis (see Chapter 15 inSight).

Superantigens Cross-Link MHC II and TCR

As described earlier, to activate our immune system, normal antigens must be processed and presented by antigen-presenting cells to T cells. Because the interaction is antigen specific, only a few T cells in our body will have the proper T-cell receptor able to recognize that antigen—in the range of 1–100 cells in a million. Proliferation of the T cells through antigen-dependent activation increases that number and increases the immune response to that antigen.

Some proteins, called **superantigens**, do not require antigen processing and presentation to stimulate the synthesis of massive amounts of cytokines (a cytokine storm). In fact, recognition as an antigen is not even involved. As illustrated in Figure 16.23, superantigens, such as staphylococcal toxic shock syndrome toxin (TSST), bind to the outside of T-cell receptors on T cells and to the MHC molecules on antigen-presenting cells (for example, macrophages). Because antigen specificity is not involved, superantigens can activate many more T cells than a typical, antigen-specific immune reaction (up to 20% of T cells in the body can be activated) and stimulate release of massive amounts of inflammatory cytokines from both T cells and APCs.

Figure 16.22 Cytotoxic T-Cell Killing

A cytotoxic T cell recognizes the target cell through TCR-antigen–MHC I interaction. The cytotoxic T cell introduces a perforin pore into the target cell membrane, which granzymes use as a portal. Granzymes start a cascade of caspase cleavages leading to mitochondrial damage that triggers apoptosis and DNA fragmentation. A FasL-Fas interaction also stimulates caspase cleavage and apoptosis.

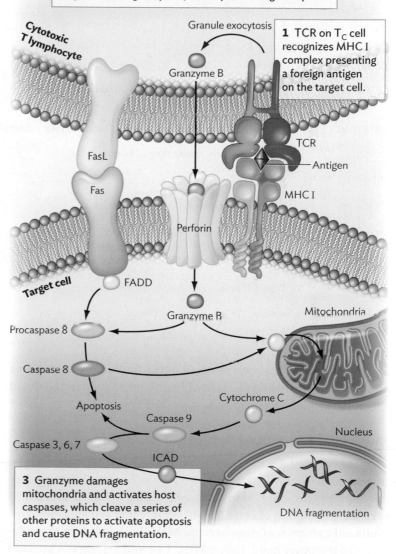

2 T_C cell secretes perforin, which makes a pore in the target cell, and granzymes, which pass through the pore.

1 TCR on T_C cell recognizes MHC I complex presenting a foreign antigen on the target cell.

Cytotoxic T lymphocyte

Granule exocytosis

Granzyme B

TCR

Antigen

FasL

Fas

MHC I

Perforin

Target cell

FADD

Granzyme B

Mitochondria

Procaspase 8

Caspase 8

Cytochrome C

Apoptosis

Caspase 9

Nucleus

Caspase 3, 6, 7

ICAD

DNA fragmentation

3 Granzyme damages mitochondria and activates host caspases, which cleave a series of other proteins to activate apoptosis and cause DNA fragmentation.

The effect of superantigens can be devastating because cytokines such as tumor necrosis factor can overwhelm the host immune system regulatory network and cause severe damage to tissues and organs. The result is disease and sometimes death. Jim Henson, the creator of Kermit the Frog and other Muppet characters, died in 1990 from complications of pneumonia caused by a potent superantigen produced by *Streptococcus pyogenes*. This example is but

Table 16.4
Select Cytokines That Modulate the Immune Response

Cytokine[a]	Sample Sources	General Functions
IL-1	Many cell types, including endothelial cells, fibroblasts, neuronal cells, epithelial cells, macrophages	Affects differentiation and activity of cells in inflammatory response; central nervous system; acts as endogenous pyrogen
IL-2	T cells	Stimulates T-cell and B-cell proliferation
IL-3	T cells	Stimulates production of macrophages, neutrophils, mast cells, others
IL-4	T_H2 cells, mast cells	Promotes differentiation of CD4 T cells into T_H2 helper T cells; promotes proliferation of B cells
IL-5	T_H2 cells	Acts as a chemoattractant for eosinophils; activates B cells and eosinophils
IL-6	T cells, macrophages, fibroblasts, endothelial cells, hepatocytes, neuronal cells	Stimulates T-cell and B-cell growth; stimulates production of acute-phase proteins
IL-8 (CXCL8)	Monocytes, endothelial cells, T cells, keratinocytes, neutrophils	Chemoattracts PMNs; promotes migration of PMNs through endothelium
IL-10	T_H2 cells, B cells, macrophages, keratinocytes	Inhibits production of IFN-γ, IL-1, TNF-α, IL-6 by macrophages
IL-17	T_H17 cells	Increases chemokine production to recruit monocytes and neutrophils
IFN-α/β	T cells, B cells, macrophages, fibroblasts	Promotes antiviral activity
IFN-γ	T_H1 cells, cytotoxic T cells, NK cells	Activates T cells, NK cells, macrophages
TNF-α	T cells, macrophages, NK cells	Exerts wide variety of immunomodulatory effects
TNF-β	T cells, B cells	Exerts wide variety of immunomodulatory effects

[a]IL = interleukin; IFN = interferon; TNF = tumor necrosis factor.

one of many ways that overreaction of the immune system causes morbidity (disease) and mortality (death), more so than the direct effects of the pathogen involved.

Microbial Evasion of Adaptive Immunity

Efficient as our immune system is, numerous viral and bacterial pathogens have developed effective means for evading the adaptive immune response. Many viruses produce proteins that downregulate production of class I MHC on infected cell surfaces. This limits antigen presentation because MHC I is needed for that process. On the other hand, losing MHC I should expose the infected host cell to natural killer cells because natural killer cells attack body cells that lack MHC I (discussed in Section 15.5). To surmount this obstacle, human cytomegalovirus places a decoy MHC I–like molecule on the surface of infected cells—an example of molecular mimicry. These decoys bind inhibitory receptors on natural killer surfaces that block natural killer cell cytotoxicity, thus sparing the infected cell.

Bacteria, like viruses, are masters of illusion when it comes to the immune system. A major cause of gastric ulcers, *Helicobacter pylori* expresses proteins from a cluster of pathogenicity genes that trigger apoptosis of T cells. Other bacteria manipulate the

expression of cytokines. Some, such as *Yersinia enterocolitica* (one cause of gastroenteritis), inhibit production of pro-inflammatory cytokines such as TNF, IL-1, and IL-8. Other bacterial pathogens, such as the mycobacteria that cause tuberculosis and leprosy, induce the production of anti-inflammatory cytokines, thereby dampening the immune response. Yet other ways bacterial and viral pathogens can evade the immune system will be described in later chapters. Fortunately, in most cases the human immune system catches on to these tricks and through redundant humoral and cellular mechanisms manages to resolve these infections.

SECTION SUMMARY

- **Activation of cytotoxic T cells** requires two signals: T-cell receptor–CD8 molecules that recognize MHC I–antigen complexes on APCs and cytokines such as IL-2 made from activated T_H1 cells. The activated cytotoxic T cells in turn destroy infected host cells.
- **T cells recognize antigen** presented only on "self" MHC–containing APCs.
- **Cytokines alter the balance** between cell-mediated (T_H1) and antibody-mediated (T_H2) immune responses.
- **Superantigens** stimulate T cells by directly linking the T-cell receptor on a T cell with MHC on an APC without undergoing APC processing and surface presentation.

Figure 16.23   Superantigens

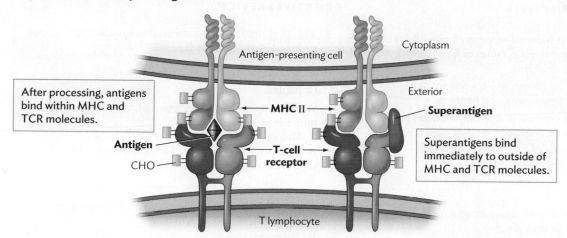

After processing, antigens bind within MHC and TCR molecules.

Antigen-presenting cell

Cytoplasm

MHC II

Antigen

CHO

T-cell receptor

T lymphocyte

Exterior

Superantigen

Superantigens bind immediately to outside of MHC and TCR molecules.

Antigen presentation of normal antigens by APCs requires the antigen to bind within the binding pockets of MHC and TCR molecules. Superantigens, however, do not require processing. They can bind directly to outer aspects of the TCR and MHC proteins, linking and activating the two cell types. Because antigen specificity is not involved, many T cells will be activated. CHO indicates carbohydrate additions to the protein.

• **Pathogens can evade immune responses** through various means (for instance, altering MHC production and manipulating pro- and anti-inflammatory cytokine production).

Thought Question 16.9   Transplant rejection is a major concern when transplanting most tissues because host T_C cells can recognize allotypic MHC on donor cells. So why are corneas easily transplanted from a donor to just about any other person? (*Hint:* Do corneas have blood vessels?)

Thought Question 16.10   Cytotoxic T cells (T_C cells) lyse the membranes of host cells carrying viruses or bacteria. Why doesn't this facilitate the spread of those organisms rather than help clear the infection?

Thought Question 16.11   Why does the immune system work so well against infection by some pathogens (for example, poliovirus) but not others (for example, *Staphylococcus aureus*)? Immunity raised by the poliovirus vaccine will prevent subsequent infection by poliovirus, but someone with an *S. aureus* infection (and the attendant immune response) can be reinfected many times by *S. aureus*.

Perspective

Immunity is a remarkable feat of nature. The realization that any one person's immune system can recognize and respond to virtually any molecular structure and yet remain selectively "blind" to his or her own antigens is hard to comprehend. But even more remarkable is that pathogenic microbes have found ways to outmaneuver the interconnected redundancies and safeguards of the immune system. In this chapter, we have outlined how a normal immune system functions. In the process, we provided a few examples of immune dysfunction, but there is more to discuss. The following chapter will tell more of what happens when the immune system fails (immunodeficiencies) or attacks itself to produce autoimmune diseases. In addition, Chapter 17 will explain some of the technological applications of immunology.

LEARNING OUTCOMES AND ASSESSMENT

16.1
Adaptive Immunity: The Big Picture

- Differentiate humoral immunity from cellular immunity and the basic roles of B cells and T cells in those systems.
- Explain the relationships between antigens, epitopes, immunogens, and haptens.
- Discuss the concept of a clonal immunological response to an epitope.
- Define antigen presentation and antigen-presenting cell.

1. A small segment of any antigen that can elicit an immune response is called a(n) _____.
   - A. Fc factor
   - B. epitope
   - C. hapten
   - D. cytokine
   - E. antibody

2. Which of the following secretes antibodies during an immune response?
   - A. Helper T cells
   - B. Macrophages
   - C. B cells
   - D. Plasma cells
   - E. Langerhans cells

16.2
Immunogenicity and Immune Specificity

- Describe what makes one antigen more immunogenic than another.
- Discuss immunological specificity and its application to vaccine development.

5. Which of the following are the most effective immunogens?
   - A. Lipids
   - B. Polysaccharides
   - C. Nucleic acids
   - D. Small chemicals such as penicillin
   - E. Proteins

16.3
Antibody Structure and Diversity

- Distinguish immunoglobulin proteins from the immunoglobulin family of proteins.
- Diagram the structures and superstructures of antibodies and discuss their functions.
- Differentiate between the isotype, allotype, and idiotype of an antibody.

7. Match the following to the numbered items in the accompanying structure:
   - A. Heavy chain
   - B. Light chain
   - C. Antigen-binding site
   - D. Fc region
   - E. Beta chain

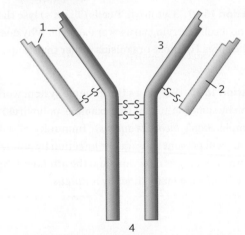

16.4
Humoral Immunity: Primary and Secondary Antibody Responses

- Explain the difference between primary and secondary antibody responses.
- Discuss B-cell differentiation and the role of the B-cell receptor.
- Summarize how the immune system generates vast antibody diversity.
- Describe the role of memory B cells during an antibody response.

12. Which of the following antibody isotypes is predominantly made following a secondary exposure to an antigen (a pathogen, for instance)?
   - A. IgA
   - B. IgD
   - C. IgE
   - D. IgG
   - E. IgM

13. B-cell receptors have which of the following as their primary component?
   - A. IgM or IgD
   - B. Major histocompatibility proteins
   - C. IgM only
   - D. IgG or IgD
   - E. IgE only

3. Which cell type is most directly responsible for coordinating humoral and cellular immune responses to an infection?
   A. Neutrophils
   B. T cells
   C. Macrophages
   D. B cells
   E. Plasma cells

4. All "professional" antigen-presenting cells _____.
   A. make antibody
   B. secrete perforin
   C. inject processed antigen into B cells
   D. present antigen on MHC surface receptors
   E. make complement

6. An antibody that binds only a specific, single structure is an example of _____.

8. Idiotypic sequences are found in which part of the structure shown in the previous question?
   A. Part 1
   B. Part 2
   C. Part 3
   D. Part 4

9. Two antibody class molecules, IgG and IgM, that recognize the same epitope and are from the same person differ from each other in their _____.
   A. light-chain constant regions
   B. heavy-chain constant regions
   C. antigen-binding sites
   D. variable regions

10. Which class of antibody has a form secreted across mucosal tissues of the intestine or lung?
    A. IgG
    B. IgM
    C. IgD
    D. IgA
    E. IgE

11. All immunoglobulin-family proteins are antibodies. True or false?

14. Clonal selection occurs when _____.
    A. B cells proliferate to produce new B cells with different B-cell receptor surface proteins
    B. B cells proliferate to produce new B cells with identical B-cell receptor surface proteins
    C. isotype switching occurs
    D. capping of B-cell receptor molecules takes place
    E. antigenic diversity is generated

15. Memory B cells undergo which of the following processes?
    A. Class switching of antibody isotype
    B. Affinity maturation of the antigen-binding site
    C. Antibody secretion
    D. The synthesis of IgG B-cell receptor surface proteins
    E. Switch from making kappa light chains to lambda light chains

LEARNING OUTCOMES AND ASSESSMENT

	SECTION OBJECTIVES	OBJECTIVES REVIEW

16.5
T Cells: The Link between Humoral and Cell-Mediated Immunity
	• Explain the three major types of T cells.
	• Discuss the role of major histocompatibility proteins in the adaptive immune response.
	• Discuss the two routes of antigen processing and presentation.
	• Describe processes that minimize autoimmunity.
		16. The following cells participate most directly in adaptive cell-mediated immunity.
		   A. Dendritic cells
		   B. Natural killer cells
		   C. Activated CD4 T_H2 cells
		   D. Neutrophils
		   E. Activated CD8 T cells

16.6 **T-Cell Activation**	• Describe a T-cell receptor (TCR), and explain its role in T-cell activation. • Explain how a TCR also participates in B-cell activation. • Discuss CD40 and CD40 ligand interaction. • Explain why memory B cells do not always need T-cell help for activation.	**20.** In addition to T-cell receptor complexes, T cells must express surface marker _____ in order to selectively recognize an MHC class II–antigen complex on antigen-presenting cells.

16.7
Cell-Mediated Immunity
	• Explain how cytotoxic T cells become activated.
	• Discuss how activated T_C cells recognize their target host cells.
	• Describe the role of cytokines in directing adaptive immunity toward humoral or cytotoxic responses.
	• Distinguish antigens from superantigens.
	• Diagram the basic ways that pathogens can avoid the immune system.
		27. Which of the following cell types are needed to activate CD8 T cells?
		   A. T_H2 cells
		   B. Dendritic cells
		   C. Plasma cells
		   D. Epithelial cells
		   E. Natural killer cells
		
		28. Cytotoxic T cells target infected cells that present antigen on _____.
		   A. MHC I receptors
		   B. MHC II receptors
		   C. MHC III receptors
		   D. Fas Ligand
		   E. CD8

17. A patient whose antigen-presenting cells lack the protein called the transporter of antigen peptides will be defective in the _____.
    A. generation of cytotoxic T cells
    B. generation of T_H1 cells
    C. generation of T_H2 cells
    D. stimulation of B cells
    E. deletion of self-reacting T cells

18. To function properly, all T cells must _____.
    A. bind strongly to some self antigen
    B. not bind to any self antigen
    C. express the AIRE protein
    D. bind weakly to some self antigen
    E. bind CD8 on antigen-presenting cells

19. MHC II complexes on antigen-presenting cells are loaded with _____ produced peptides.
    A. extracellularly
    B. intracellularly
    C. intracellularly or extracellularly

21. What happens in a person who lacks CD40 ligand on his or her T cells?
    A. Inability to make activated cytotoxic T cells
    B. Increased synthesis of IgG antibody
    C. Inability to synthesize antibody
    D. Inability to undergo isotype switching
    E. Hyperproliferation of activated B cells

For questions 22–26, match the immune response events with the location at which they occur:

22. Gene rearrangement to form the antigen-binding site (VDJ)

23. T-cell education

24. Isotype switching

25. B-cell activation by APCs

26. T-cell help for B cells

A. Bone marrow
B. Thymus
C. Bloodstream
D. Lymph node

29. To activate T cells, superantigens must _____.
    A. bind to the T-cell receptors and B-cell receptors in their antigen-binding sites
    B. bind to the antigen-binding region of MHC and outside the antigen-binding region of the T-cell receptors
    C. bind to T-cell receptors only
    D. bind MHC only
    E. bind T-cell receptors and MHC outside of their antigen-binding regions

30. Cytomegalovirus, which can cause a mononucleosis-type disease, prevents the immune system from destroying infected cells by _____.
    A. down-regulating MHC II on infected cells
    B. inhibiting production of anti-inflammatory cytokines
    C. placing a molecular mimic of MHC I on an infected cell surface
    D. placing a mimic of an Fc receptor on infected cells
    E. preventing phagolysosome fusion

Key Terms

activated (p. 502)
affinity maturation (p. 503)
allotype (p. 498)
antibody (p. 490)
antigen (p. 491)
antigenic determinant (p. 491)
antigenicity (p. 494)
antigen-presenting cell (APC) (p. 493)
attenuated (p. 495)
B-cell receptor (BCR) (p. 502)
B-cell tolerance (p. 494)
bone marrow transplant (p. 512)
capping (p. 502)
cell-mediated immunity (p. 491)
class I MHC molecule (p. 507)

class II MHC molecule (p. 507)
class switching (p. 501)
clonal (p. 493)
clonal selection (p. 501)
constant region (p. 497)
cytokine storm (p. 515)
cytotoxic T cell (T$_C$ cell) (p. 493)
epitope (p. 491)
F(ab)$_2$ region (p. 498)
Fc region (p. 498)
gene switching (p. 504)
hapten (p. 491)
heavy chain (p. 496)
helper T cell (T$_H$ cell) (p. 502)
humoral immunity (p. 490)

idiotype (p. 498)
IgA (p. 498)
IgD (p. 499)
IgE (p. 499)
IgG (p. 498)
IgM (p. 498)
immunogen (p. 491)
immunogenicity (p. 494)
immunoglobulin (p. 496)
isotype (p. 498)
isotype switching (p. 501)
light chain (p. 496)
major histocompatibility complex
    (MHC) (p. 493)
memory B cell (p. 493)

negative selection (p. 509)
opsonize (p. 498)
plasma cell (p. 491)
positive selection (p. 509)
primary antibody response (p. 500)
recombination signal sequence (RSS)
    (p. 504)
regulatory T cell (Treg) (p. 509)
secondary antibody response (p. 500)
serum (p. 500)
superantigen (p. 515)
switch region (p. 505)
threshold dose (p. 494)
vaccination (p. 494)
variable region (p. 497)

Review Questions

1. Define antigen, epitope, hapten, and antigenic determinant.
2. What is the basic difference between humoral immunity and cellular immunity?
3. Why are proteins better immunogens than nucleic acids?
4. What makes IgA antibody different from IgG?
5. Explain isotypic, allotypic, and idiotypic differences in antibodies.
6. What is antigen processing?
7. Discuss differences in the primary and secondary antibody responses.
8. Outline the basic steps that turn a B cell into a plasma cell.
9. What is isotype switching?
10. What signals are needed to activate helper T cells?
11. What signals activate cytotoxic T cells?
12. What influences T$_H$0 conversion to T$_H$1 or T$_H$2 cells?
13. How do superantigens activate T cells?

Clinical Correlation Questions

1. A 5-year-old patient presents to his pediatrician with another of what has been a series of infections. Flow cytometry of the boy's blood reveals that the patient is deficient in CD4 helper T cells. CD8 T cells and B cells are at normal levels. What do you expect to see if you measure his serum IgM, IgG, IgA, and IgE levels? How might the results explain the repeated infections?

2. A patient arrives from Africa with cycling fevers and chills. She is diagnosed with malaria. The malarial parasite grows inside red blood cells. Explain whether or not cytotoxic T cells will kill red blood cells infected with the malarial parasite.

Thought Questions: CHECK YOUR ANSWERS

Thought Question 16.1   Two different stretches of amino acids in a single protein form a three-dimensional antigenic determinant. Will the specific immune response to that three-dimensional antigen also respond to one of the two amino acid stretches alone?

ANSWER: Probably not. It is the three-dimensional shape formed by the two stretches that is recognized as an antigen. However, other specific immune responses involving different subsets of lymphocytes can recognize the separate amino acid stretches.

Thought Question 16.2   How does a neutralizing antibody that recognizes a viral coat protein prevent infection by the associated virus?

ANSWER: Neutralizing antibodies usually bind attachment proteins on the virus and sterically prevent them from binding to host cell receptors (see Figure 25.4). Some antibodies to enveloped viruses might trigger the complement cascade, destroying the membrane.

Thought Question 16.3   The mother of a newborn is diagnosed with rubella, a viral disease. Infection of the fetus can lead to serious consequences for the newborn. How can you use antibody production to determine whether the newborn was infected while in utero?

ANSWER: Since maternal IgM antibodies cannot cross the placenta, finding IgM antibodies to rubella antigens in the newborn's circulation indicates that the fetus was infected and initiated its own immune response.

Thought Question 16.4   Why do immunizations lose their effectiveness over time?

ANSWER: Because memory B cells eventually die. Without some exposure to antigen, those memory cells will not be replaced.

Thought Question 16.5   IgM is the first antibody produced during a primary immune response. Of the five different types of antibodies produced, why would the body want IgM to be the first?

ANSWER: IgM is a pentamer possessing ten antigen-binding sites. As a result, IgM can bind many antigens at once and rapidly clear those antigens from the bloodstream. The affinity of the antigen-binding sites, however, is not as strong as that made after a heavy-chain switch, which includes the generation and selection of mutations in the variable regions that more tenaciously bind antigen.

Thought Question 16.6   How do viruses stimulate antibody production if viruses are intracellular pathogens and their intracellularly synthesized and processed epitopes are presented only onto MHC I surface receptors? Won't this trigger only T_C activation and cell-mediated immunity? (See Figure 16.16.)

ANSWER: There are probably several ways this can happen. One way is if "dead" virus particles enter the cell by endocytosis rather than infection. The resulting processed epitopes can be placed on MHC II surface receptors, where they can promote activation of T_H0 cells. The resulting activated T_{FH} cells can stimulate B-cell differentiation into plasma cells that will produce antibodies.

Thought Question 16.7   IL-10 is considered an anti-inflammatory cytokine because it down-regulates production of MHC II receptors. What consequences would down-regulating MHC II on macrophages have on the immune response?

ANSWER: The macrophage, also an antigen-presenting cell, would be a poor presenter of epitopes to T_H0 cells. This would diminish the production of helper T cells and interfere with both humoral and cell-mediated responses.

Thought Question 16.8   Why does attaching a hapten to a carrier protein allow production of antihapten antibodies?

ANSWER: B cells with a B-cell receptor that can bind a hapten can take up hapten but cannot present the hapten to a helper T cell. The same B cell can also take up the hapten bound to a carrier molecule, and because the carrier molecule is larger than the hapten, the B cell will present the carrier epitope to the helper T cell. The helper T cell stimulates the B cell, which was already programmed to make antihapten antibody, to differentiate into plasma and memory B cells.

Thought Question 16.9   Transplant rejection is a major concern when transplanting most tissues because host T_C cells can recognize allotypic MHC on donor cells. So why are corneas easily transplanted from a donor to just about any other person? (*Hint:* Do corneas have blood vessels?)

ANSWER: The cornea is not normally vascularized (it lacks blood vessels). So even though corneal cells express MHC proteins, circulating host T cells do not have an opportunity to interact with them. The cornea will not be rejected. This is called an immune-privileged site.

Thought Question 16.10   Cytotoxic T cells (T_C cells) lyse the membranes of host cells carrying viruses or bacteria. Why doesn't this facilitate the spread of those organisms rather than help clear the infection?

ANSWER: For viruses, T_C-cell destruction of the target host cell destroys the machinery viruses use to replicate. So death of host cells prevents synthesis of more virus. Also, some bacterial pathogens and all viruses use host cells to hide from the immune system. Destroying the infected cell exposes pathogens to the rest of the immune system.

Thought Question 16.11   Why does the immune system work so well against infection by some pathogens (for example, poliovirus) but not others (for example, *Staphylococcus aureus*)? Immunity raised by the poliovirus vaccine will prevent subsequent infection by poliovirus, but someone with an *S. aureus* infection (and the attendant immune response) can be reinfected many times by *S. aureus*.

ANSWER: (1) The immune system might fail to prevent reinfection if a pathogen has many strains that are different in their surface antigen structure. Memory B or T cells from the last infection cannot recognize the new antigen. (2) A pathogen could also produce different antigenic types of toxin (e.g., botulism). (3) A pathogen may produce substances that interfere with a full-blown immune response so that effective or long-lasting immune memory is not established. (4) An infection (e.g., *S. aureus*) that infects skin and then walls itself off in an abscess can limit its interaction with the immune system. Once again, the body may not develop long-lasting immunity.

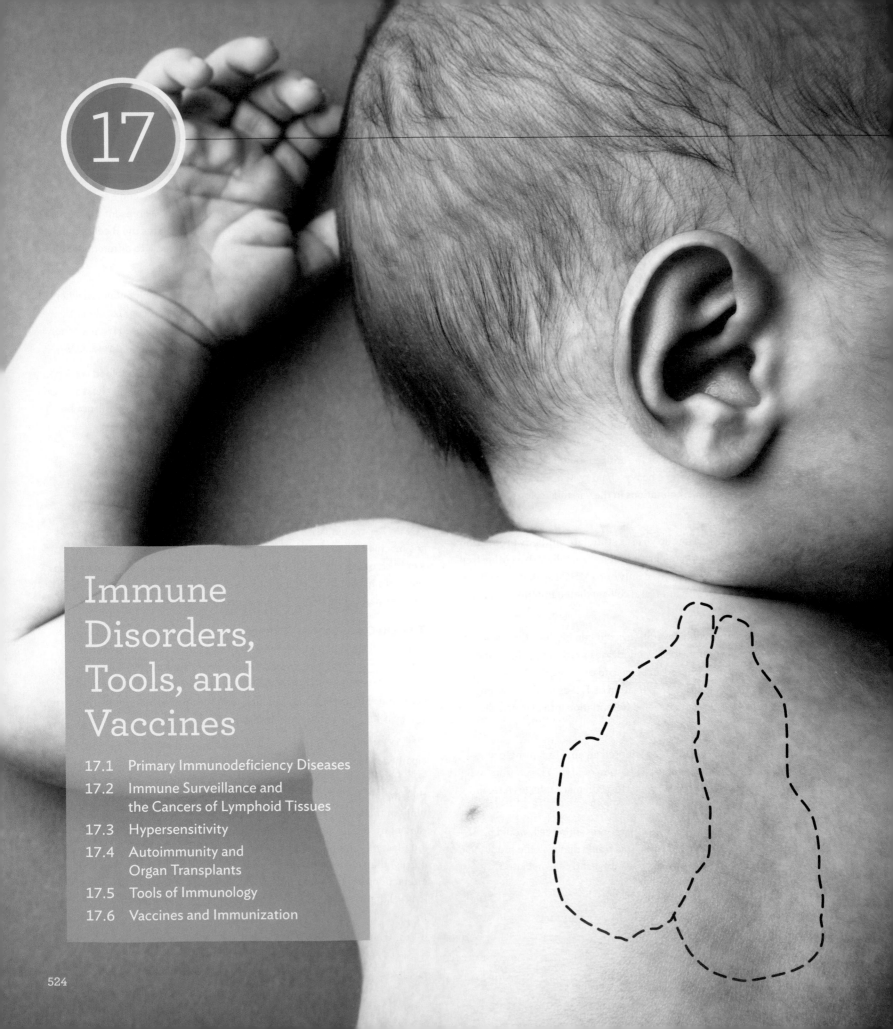

Immune Disorders, Tools, and Vaccines

[The Missing Thymus]

SCENARIO Erica's pregnancy went smoothly—as did the delivery of her child, Silas. Erica knew from a prenatal sonogram that her baby would be born with a cleft lip and palate, but those were easily correctable malformations.

TROUBLING LAB RESULTS However, after Silas was born, blood tests revealed he had a low white blood cell count. A more sophisticated test involving flow cytometry (to be discussed later in this chapter) discovered a lack of T cells in his blood, which led doctors to check his thymus. A computed tomography (CT) scan of his chest found no evidence of a thymus. Without a thymus, Silas could not make T cells.

DiGeorge Syndrome
One-year-old Silas was born with a rare congenital disease called DiGeorge syndrome that left him with a cleft palate and the inability to fend off infections. Left panel, before; right panel, after cleft palate surgery.

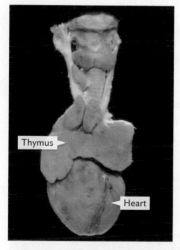

Thymus of an Infant, over the Heart
This organ is important for T-cell development. A patient with DiGeorge syndrome is left with little to no thymus.

DIAGNOSIS A subsequent DNA test, called fluorescence in situ hybridization (FISH), revealed a deletion in chromosome 22. The deletion, called 22q11.2, is known to cause **DiGeorge syndrome**, an immunodeficiency disease that can produce many congenital defects, including cleft palate and thymus aplasia (no thymus).

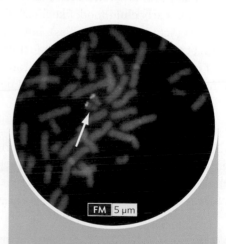

The FISH Assay
The fluorescence in situ hybridization (FISH) assay can detect deletions of the 22q11 region. Fluorescent DNA probes bind in the area of the deletion. If the deletion has occurred, the chromosome will show a red spot (arrow).

TRANSPLANT Unless Silas received a thymus transplant, he would probably die from an infection before he was 2 years old. Luckily, Erica's baby received a transplant from a matched donor and is alive today.

CHAPTER OBJECTIVES

After reading this chapter, you will be able to:

- Categorize immunodeficiencies of the innate and adaptive immune systems.
- Describe the various malignancies associated with the immune system.
- Explain the types of hypersensitivities, their treatments, and prevention.
- Discuss the positive and negative aspects of autoimmunity.
- Describe key concepts of vaccination.

Silas's case once again reveals an immune system that is both remarkably resilient yet frighteningly fragile. Its innate and adaptive layers, when fully functioning, provide a nearly impenetrable defense against infection. With one missing gene, however, the system breaks down, its owner now vulnerable to the microbial world. Immunodeficiency diseases (IDDs) are disorders in which part of a person's immune system is missing or diminished. Defects can occur in cell-mediated (T cell) and/or antibody-mediated (B cell) pathways. Either way, the afflicted person has trouble clearing infections (Table 17.1). For Silas, the lack of a thymus caused a severe depletion of his T cells. As summarized in Table 17.1, individuals with T-cell defects are generally susceptible to infections with viruses, fungi, and some protozoa. Patients with disorders in antibody formation (B cells) are chiefly prone to multiple infections with encapsulated bacterial pathogens such as streptococci, *Haemophilus*, and the meningococcus *Neisseria meningitidis*, and with the protozoan *Giardia*. Recurrent infections with encapsulated bacteria happen because the patient cannot make antibodies needed to coat the pathogen's capsule and aid phagocytosis.

Primary immunodeficiencies have a genetic basis and usually manifest in early childhood, like Silas's did. **Secondary immunodeficiencies** (acquired deficiencies), however, can be acquired at any age as a consequence of certain infections (HIV, for instance), the use of immunosuppressive drugs, or radiation therapy. In practice, an immunodeficiency disorder should be suspected in any person

experiencing recurrent, serious infections. Immunodeficiencies, however, are not the only form of immune dysfunction. This chapter will describe immune system–related diseases that include not only immunodeficiencies but also cancers, hypersensitivities, and autoimmune diseases that can increase susceptibility to infections. To conclude, we describe the tools used to examine a patient's immune status and discuss how vaccines stimulate immunity.

17.1
Primary Immunodeficiency Diseases

SECTION OBJECTIVES

- Distinguish primary and secondary immunodeficiencies.
- Describe the common forms of severe combined immunodeficiencies.
- Compare the underlying defects and outcomes of bare lymphocyte syndrome, X-linked agammaglobulinemia, hyper IgM syndrome, selective IgA deficiency, and complement deficiencies.

What are the major primary immunodeficiencies? More than 200 primary immunodeficiency diseases (also called disorders or dysfunctions) exist. B-cell defects causing antibody deficiencies are the most common, accounting for 50%–60% of primary

Table 17.1
Characteristic Infections of the Primary Immunodeficiencies

Immune Component	Primary Pathogen	Primary Site	Clinical Example
T cells	Intracellular bacteria, viruses, protozoa, fungi	Nonspecific	Severe combined immunodeficiency (SCID), DiGeorge syndrome
B cells	*Streptococcus pneumoniae, S. pyogenes, Haemophilus*	Lung, skin, central nervous system (CNS)	Deficiency in immunoglobulin G (IgG), IgM
	Enteric bacteria and viruses	Gastrointestinal tract, nasal, eye	Deficiency in IgA
Phagocytes	*Staphylococcus, Klebsiella, Pseudomonas*	Lung, skin, regional lymph node	Chronic granulomatous disease
Complement	*Neisseria, Haemophilus, S. pneumoniae, Streptococcus* spp.	CNS, lung, skin	C3, factors I and H, late C components

immunodeficiencies. T-cell disorders account for about 5%–10% of primary immunodeficiencies.

Chapter 16 described several immune system deficiencies. Examples included severe combined immunodeficiency (covered in the opening case history on SCID), bare lymphocyte syndrome (Case History 16.1), and hyper IgM syndrome (Case History 16.2). We will expand on those diseases and introduce some other common immunodeficiencies. Table 17.2 lists key features of these dysfunctions and indicates applicable case histories in the text. Figure 17.1 gives a schematic of the immune system and shows where various primary immunodeficiencies make their presence known.

T-Cell Disorders

Severe combined immunodeficiencies (SCIDs), introduced in the Chapter 16 opening case, are characterized by a lack of T cells. But B-cell and natural killer (NK) cell numbers can also be low, high, or even normal depending on the variety of SCID syndrome. Mutations in any one of 12 genes can lead to a SCID syndrome. The most common form involves a gene on the X chromosome (X-linked) producing a defective interleukin 2 (IL-2) receptor (the gamma chain) on T cells. Because other cytokine receptors share the gamma chain, this defect prevents many cytokines from signaling T cells to proliferate. The second most common form of SCID develops from

Table 17.2
Examples of Primary Immunodeficiencies[a]

Cell Involved	Disease	Cause	Gene Location	Case History	T-Cell Level	B-Cell Level	Immunoglobulin Levels
T cells	X-SCID	Defective gamma chain in receptors for interleukin 2 (IL-2)	X-linked	16.1	Absent	Normal but not functioning	All low
	ADA-SCID	Adenosine deaminase deficiency	Chromosome 20		Absent	Absent	All low
	DiGeorge syndrome	Defective development of thymus and parathyroid glands	Chromosome 22 (22q11) deletion (sporadic deletions during meiosis)	17.1	Absent	Variable	Variable
	Bare lymphocyte syndrome	Major histocompatibility complex (MHC) II deficiency; defective gamma chain IL-2 receptor	Chromosomes 1, 13, 16, and 19; regulators of MHC II genes	16.2	Low CD4 T cells	Normal	All low
B cells	Bruton's agammaglobulinemia	Defective tyrosine kinase	X-linked		Normal	Absent or low	All low
	Common variable immunodeficiency (CVID)	Defective B-cell development into plasma cells; cause unclear	Unknown		Normal	Normal or low	IgG and IgA low; IgM sometimes low
	Hyper IgM syndrome	T-cell CD40 ligand defect	X-linked or chromosome 12 (12p13)	16.3	Normal	Normal	Elevated IgM; others low
	Selective IgA deficiency	Unknown; failure to class switch; most common IDD	Unknown; rare in chromosomes 6, 14, 18, 19		Normal	Normal	IgA low
Phagocytes	Leukocyte adhesion deficiency	Lack of integrins; failure of extravasation	Chromosome 21 (21q22.3)	15.2	Normal	Normal	Normal
	IRAK deficiency	Toll-like receptor signaling, low cytokine secretion	Chromosome 12 (12q12)	15.3	Normal	Normal	Normal
	Chronic granulomatous disease	NADPH oxidase deficiency; no respiratory burst	X-linked (*CYBB* gene)		Normal	Normal	Normal
	Complement deficiency	Early components; late components	Chromosomes 1, 6, 11, 19	15.4	Normal	Normal	Normal

[a]Abnormal lab results are in color.

Figure 17.1
Correlation between Primary Immunodeficiencies and Defects in the Immune System
Follow text for details.

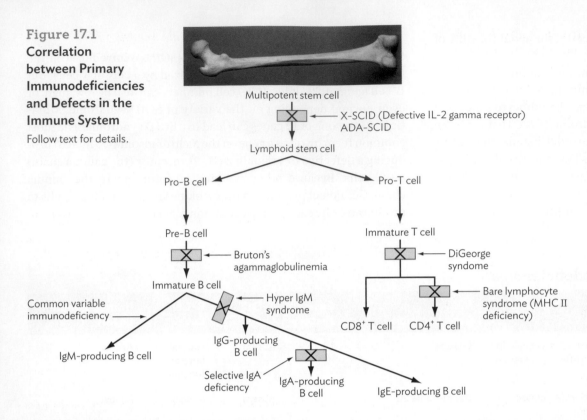

respectively. Recurrent infections begin soon after birth, but the degree of immunodeficiency varies considerably. Complete DiGeorge syndrome (no thymus) is fatal without transplantation of cultured thymus tissue.

MHC II deficiency (bare lymphocyte syndrome) is a rare disease (Case History 16.1) caused by a failure to express MHC II molecules on antigen-presenting cells. The mutations involved are *not* in the MHC II genes themselves but occur in genes encoding *regulators* of the MHC II genes. The deficiency produces severe defects in cellular and humoral immunity. Without MHC II, antigens cannot be presented to T_H0 cells (Section 16.6); T_H0 cells, then, will not form T_{FH}, T_H1, or other types of helper T cells. Without T_H cells to "talk" to, B cells cannot initiate heavy chain class switching or differentiate into plasma cells, explaining why no antibodies are made. Cell-mediated immunity is defective because without T_H1 helper cells, cytotoxic T cells cannot be activated. As you might guess, patients with this disease are extremely vulnerable to infections.

Link Recall from Section 16.5 that MHC (major histocompatibility complex) molecules come in two types. MHC type I molecules are on the surfaces of all cells except red blood cells. MHC type II molecules are specifically on antigen-presenting cells (APCs), such as macrophages, dendritic cells, and B cells. T_H cells, to become activated, must recognize antigen–MHC II complexes formed on an APC surface.

a deficiency in adenosine deaminase (ADA), an enzyme that breaks down and recycles adenosine. Diminished recycling of adenosine prevents DNA synthesis and causes apoptosis of T, B, and NK cell precursors. These patients lack all three subsets of lymphocytes.

By 6 months of age, most SCID infants develop infections caused by yeasts, viruses, or *Pneumocystis jirovecii*. Treatments can involve antibiotics, bone marrow stem cell transplantation to generate new T and B cells, ADA enzyme replacement by injection, or **IVIG** (**intravenous** immune globulin, or **immunoglobulin**). Immune globulin contains pooled immunoglobulin G (IgG) extracted from the plasma of more than 1,000 blood donors. The protective effect of IVIG against infection is only short-lived, lasting between 2 weeks and 3 months. Long-lasting ADA replacement is possible by introducing a functional ADA gene into the patient's genome (gene therapy).

Note Do not confuse blood plasma with serum. Plasma is the fluid portion of blood minus the cells. This material contains antibodies and fibrinogen, and it can clot. It is prepared by spinning a fresh tube of blood containing an anticoagulant. Serum is the clear fluid left after blood clots. It will contain antibodies.

DiGeorge syndrome results from mutations in one of several genes such as those on chromosome 22, chromosome 10p13, or other unknown genes. The mutations affect specific aspects of embryogenesis. Infants (such as Silas in the above case) can have low-set ears, midline facial clefts, a small receding mandible (jaw), and a congenital heart disorder (see chapter-opening photos). They also have incomplete development or absence of thymus and parathyroid glands, causing T-cell deficiency and hypoparathyroidism,

B-Cell Disorders

X-linked agammaglobulinemia (Bruton's disease) patients have mutations in the Bruton tyrosine kinase (*Btk*) gene on the X chromosome and produce very little immunoglobulin (agammaglobulinemia). The Btk enzyme is essential for B-cell development and maturation; without it, no B cells—and hence no antibodies—can form. As a result, male infants have small tonsils and do not develop lymph nodes. They suffer from recurrent pyogenic (pus producing) lung, sinus, and skin infections caused by encapsulated bacteria (for example, *Streptococcus pneumoniae*, *Haemophilus influenzae*). Treatment of infections requires IVIG and antibiotics.

Common variable immunodeficiency (CVID), or adult-onset hypogammaglobulinemia ("hypo" meaning low levels), is characterized by low immunoglobulin (Ig) levels. B cells look normal and can proliferate but do not develop into Ig-producing plasma cells. Although CVID is relatively common, the degree and type of serum immunoglobulin deficiency vary from patient to patient. Some patients' B cells will secrete IgM but will not class switch to IgG or IgA. B cells in other patients may produce almost no antibodies. In most patients the molecular defect is unknown. Infections tend to develop in adulthood and are most commonly those of the upper and lower respiratory tracts. Treatment requires IVIG.

Hyper IgM syndrome (Case History 16.2) patients have normal or elevated serum IgM levels but low levels (or complete absence) of other antibodies. The individual is susceptible to bacterial infections. Most cases are caused by mutations in a gene located on the X chromosome; this gene encodes a membrane protein (CD154, or CD40 ligand) decorating the surfaces of activated helper T cells (**Figure 17.2**). Normally, CD40 ligand on T_H cells interacts with B cells (in the presence of cytokines) and signals the switch from IgM to IgA, IgG, or IgE production. Without CD40L binding, B cells will not switch isotypes. For unknown reasons, patients also lack neutrophils (called neutropenia). The disease is characterized by recurrent pyogenic bacterial infections during the first 2 years of life (similar to X-linked agammaglobulinemia). Diagnosis comes from detecting normal or elevated serum IgM levels and low levels, or an absence, of other immune globulins. Treatment is monthly IVIG. Bone marrow transplantation can be curative for the X-linked form if an MHC-identical sibling donor is available.

Selective IgA deficiency is the most common primary immunodeficiency. Patients lack IgA, but the exact cause is largely unknown. The disease is called "selective" because the levels of other serum immunoglobulins, such as IgM and IgG, are normal. B cells can isotype class switch, just not to IgA. These patients have normal T-cell, neutrophil, and complement system function that can ward off many types of infections, but they still suffer recurrent **mucosal infections** of the ear, lung, or GI tract. This syndrome has no specific treatment other than antibiotics to cure infections.

Link Recall from Section 16.3 that IgA antibodies are secreted at mucosal surfaces. Patients with IgA deficiency can have frequent episodes of **mucosal infections**, such as bronchitis, pneumonia, and sinusitis, that mucosal immunity normally prevents.

The incidence of autoantibodies (antibodies to the body's own tissue) in IgA-deficient patients is high. Thus, autoimmune diseases are common and include rheumatoid arthritis and systemic lupus erythematosus (symptoms include sore and swollen joints of the hands or knees, a facial rash, and anemia).

Complement Deficiencies

Complement in serum is important for opsonizing and killing bacteria. **Complement deficiencies** are inherited as autosomal traits (not linked to the X chromosome) and can occur in the early complement components (C1–C4) or late complement components (C5–C9; Chapter 15 describes complement). Early complement components bind bacteria and facilitate phagocytosis (opsonization) of encapsulated bacteria, so patients with these defects are

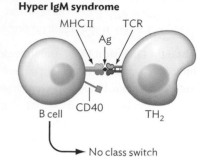

Figure 17.2   Defect Causing Hyper IgM Syndrome

A. T-cell receptors (TCR) on a helper T cell recognize which B cell to help by binding to antigen (Ag) presented on B-cell MHC II. Binding triggers differentiation into an antibody-secreting plasma cell. Interaction between T-cell CD40 ligand (CD40L) and CD40 on the B cell authorizes the B cell to class switch antibody isotype.

B. T cells that lack CD40L cannot authorize class switch. As a result, the plasma cells produce IgM only.

more susceptible to infections with *S. pneumoniae, S. pyogenes,* and *H. influenzae.* The late components form the bacteriolytic membrane attack complex (MAC). Defects in MAC formation will render a person more susceptible to Gram-negative infections, most notably *Neisseria meningitidis.* There is no specific cure for complement deficiencies. Infected patients are treated with antibiotics and can receive plasma that contains complement. Patients with complement deficiencies display normal humoral and T-cell responses and should receive normal immunizations.

SECTION SUMMARY

- **Immunodeficiency patients** usually present with recurrent infections.
- **T-cell deficiencies** account for 5%–10% of all immunodeficiencies. They include:
  —Severe combined immunodeficiencies (most commonly due to a defect in the IL-2 cytokine receptor or a defect in adenosine deaminase)
  —DiGeorge Syndrome (thymic aplasia)
  —Bare lymphocyte syndrome (MHC II deficiency on APCs)
- **B-cell deficiencies** are the most common immunodeficiencies. They include:
  —Bruton's agammaglobulinemia (lack of B-cell development)
  —Common variable immunodeficiency (defect in plasma cell development and failure to class switch antibody heavy chains)
  —Hyper IgM syndrome (failure of B cell to class switch, T-cell CD40 ligand defect)
  —Selective IgA deficiency (arrested IgA plasma cell development)
- **Immune globulin** treatment is useful for treating infections in immunodeficiency patients.
- **Complement deficiencies** can result in poor opsonization (C1–C4) or defective killing through MAC formation (C5–C9).

17.2

Immune Surveillance and the Cancers of Lymphoid Tissues

SECTION OBJECTIVES

- Discuss the relationship between our immune system and cancer.
- Distinguish lymphoma, leukemia, plasmocytoma, and multiple myeloma.
- Explain the defects leading to B-cell neoplasms.

Can defects in the immune system lead to cancer? **Cancer**, or a **neoplasm**, is defined as new growth of abnormal cells. **Tumors** are masses of abnormal cells that form within a tissue. A tumor can be **benign** (meaning a slow-growing, self-contained mass that does not infiltrate other tissues) or **malignant** (abnormal cells with rapid, uncontrolled growth that can spread to form tumors in other tissues). Spontaneous genetic damage can transform a normal cell into a malignant, cancerous one.

All people, including you, have probably had a cancerous cell develop in their body. Fortunately, an important function of our immune system is **immune surveillance**, a process that detects and eliminates cancer cells as they arise. This "search and destroy" function is possible because cancer cells are marked with surface antigens not present on normal cells. Our immune system can recognize and target those antigens.

Without immune surveillance, tumors would arise often. Consequently, dysregulation (that is, impaired regulation) of the immune system can result in a variety of cancers, particularly of lymphoid cells. Patients with primary immunodeficiencies, AIDS patients, and immunosuppressed transplant patients are particularly prone to cancer. The malignancies most commonly manifest as aggressive B-cell lymphomas and are often associated with Epstein-Barr virus (EBV) infection (Chapter 21). EBV, the cause of infectious mononucleosis, is a DNA virus that infects B cells and becomes latent by forming the equivalent of a circular plasmid. Expression of some EBV genes can cause a cell to become malignant.

This section discusses the cancers known as lymphomas, leukemias, plasmocytomas, and multiple myelomas. A **lymphoma** is a solid mass in a lymphoid organ (lymph node, spleen, thymus).

Leukemia refers to malignant lymphoid cells found primarily in the circulation or bone marrow. Plasma cells can also become malignant. Neoplasmic growth of a plasma cell at a single site is called a **plasmocytoma**. If cancerous plasma cells appear at multiple sites, mainly throughout bone, the condition is called **multiple myeloma**. The following section describes some lymphoid tissue cancers.

B-Cell Neoplasms

Cancers arising in B cells often come about when genes whose products normally control cell division accidentally move near the powerful promoter of an immunoglobulin gene. The division control gene will overexpress and either stimulate cell proliferation or prevent cell death (apoptosis). Either way, the affected cell continues to replicate. For instance, **Burkitt's lymphoma** (which can also manifest as a leukemia) develops when the Epstein-Barr virus infects B cells and causes the cell proliferation gene *c-myc* to move (translocate) next to highly expressed antibody heavy-chain or light-chain genes (Section 16.3). The result is uncontrolled cell division.

Follicular lymphoma is another example (Figure 17.3A). Under normal circumstances a B cell encountering a *poorly* matched antigen in a lymph node will undergo apoptosis and die. The *bcl-2* gene makes a protein that *prevents* apoptosis, but only under some circumstances. In follicular lymphomas, the *bcl-2* gene accidentally moves from its normal location to the immunoglobulin heavy-chain gene cluster and becomes highly expressed. Overexpression of *bcl-2* prevents cell death, which enables continued replication and tumor development. These cancers proliferate slowly but continuously, resulting in a long clinical course.

Chronic lymphocytic leukemia (Figure 17.3B) is the most common leukemia in North America and Western Europe. Most cases (65%) have a deletion in chromosome 13 at region q14.3. The

Figure 17.3   Lymphoid Cancers

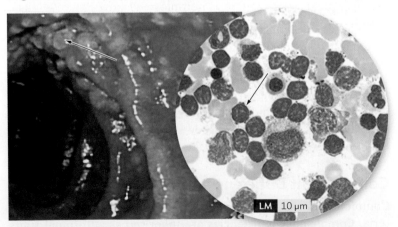

A. Follicular lymphoma of the duodenum. The follicles (arrow) consist of monoclonal IgA. A *bcl-2* gene translocation caused extensive expression of BCL-2 protein in the follicle.

B. Chronic lymphocytic leukemia (CLL). Blood smear showing abnormally high numbers of lymphocytes (arrow). These neoplastic lymphocytes show round to slightly irregular nuclei and scant cytoplasm. CLL is the most common leukemia in adults.

mutation eliminates a very short RNA molecule (microRNA) that normally *inhibits* expression of the anti-apoptosis *bcl-2* gene. Without the microRNA, *bcl-2* is overexpressed, cell death stops, and a B-cell malignancy results. To battle this malignancy, physicians inject the patient with antibodies against the B cell–specific surface antigen CD20. The antibody-CD20 complex initiates an immune response that kills B-cell lymphomas.

Hodgkin's lymphoma is a highly survivable cancer characterized by the spread of painless tumors from one lymph node group to another. Symptoms include fever (>38°C, or >100.4°F), night sweats, and weight loss. The tumors contain neoplastic, multinucleated Reed-Sternberg cells derived from B cells (**Figure 17.4**). Reed-Sternberg cells are so striking that any lymphomas lacking them are called **non-Hodgkin's lymphomas**. Non-Hodgkin's lymphomas range in severity from slow growing to aggressive. Patients with Hodgkin's lymphoma often have a history of infectious mononucleosis caused by Epstein-Barr virus (EBV). This fact is significant because EBV can stimulate synthesis of NF-κB, a regulatory factor that inhibits apoptosis and stimulates cell proliferation. The release of cytokines such as tumor necrosis factor (TNF) from Reed-Sternberg cells contributes to the fever and other features of chronic inflammation evident in these patients. Treatment of early-stage Hodgkin's involves a combination of chemotherapy and targeted radiation treatment. Survival rates approach 90%.

Plasma Cell Neoplasms (Multiple Myeloma)

Multiple myeloma accounts for about 1% of all cancers and typically occurs in older patients (>66 years). Because these tumors form in bone marrow, patients can experience excruciating bone pain. Neoplastic plasma cells multiply to large numbers and continue to make and secrete massive amounts of antibody (often incomplete). Secreted antibody light chains can form clumps, called amyloid, that deposit in, and damage, various organs, particularly the kidney. Free light chains secreted from multiple myeloma cells can even be secreted in urine (referred to as Bence-Jones proteins).

Figure 17.4   Reed-Sternberg Cell of Hodgkin's Lymphoma

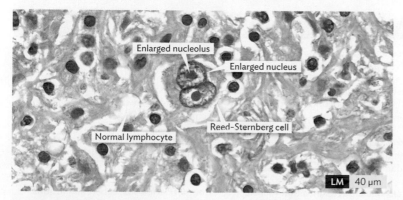

Classic neoplastic Reed-Sternberg cell is 20–50 μm with a large bilobed or multi-lobed nucleus. A large, round, eosinophilic central nucleolus is always present.

Figure 17.5   Serum Electrophoresis of Serum Immunoglobulins

A. Normal serum electrophoresis.   **B.** Multiple myeloma spike.

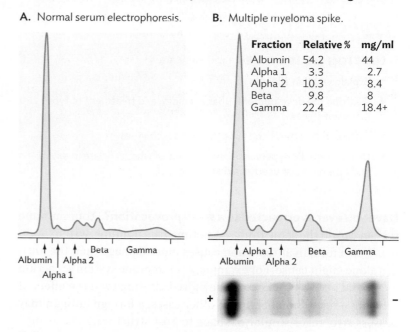

Fraction	Relative %	mg/ml
Albumin	54.2	44
Alpha 1	3.3	2.7
Alpha 2	10.3	8.4
Beta	9.8	8
Gamma	22.4	18.4+

Electrophoresis of serum (bottom panel) from a patient with multiple myeloma (an immunoglobulin-secreting tumor) shows an abnormally narrow gamma-globulin band (tracing in panel B), indicating the presence of a monoclonal (and therefore abnormal) immunoglobulin. Here the concentration of this protein is about 18 mg/ml, and the "+" on the printout indicates that this value is higher than normal for gamma globulin (normal range = 6–17 mg/ml). In other cases such a monoclonal protein may be present at much higher concentrations.

A single plasma cell that becomes malignant generates a clonal group of offspring, all of which make antibody reactive toward a single epitope. The dramatic overproduction of a monospecific antibody (also called a monoclonal antibody) can be identified when blood serum proteins are separated by electrophoresis (**Figure 17.5**). Excess immunoglobulin registers as a spike in the electrophoresis readout. The levels of all other antibodies are severely decreased, causing an immunosuppression that makes these patients more susceptible to infection.

SECTION SUMMARY

- **Immune surveillance** by the immune system normally eliminates cancer cells as they arise.
- **Immunodeficient patients** are more prone to cancers.
- **Lymphoid cancers** include lymphomas (tumors in a lymphoid organ), leukemia (cancerous lymphoid cells in circulation), plasmocytoma (cancerous plasma cell growth at a single site), and multiple myelomas (multiple plasmocytomas, usually in bone).
- **Lymphoid cancers** arise when certain natural cell cycle regulatory genes (*c-myc* or *bcl-2*) are overexpressed after moving to and coming under the control of a heavy- or light-chain gene. **Epstein-Barr virus** can cause these rearrangements.
- **Multiple myeloma cells** secrete high levels of their specific antibody.

17.3
Hypersensitivity

Have you ever overreacted to a small provocation? Your immune system can do the same thing. Sometimes the immune system overreacts to a foreign antigen and causes more damage than the antigen alone might cause. For example, your immune system's reaction to a certain type of pollen can be more irritating than the effect of the pollen itself (an allergy). In other cases, a foreign antigen may possess structures similar in shape to host structures. These antigens can trick the immune system into reacting against self. Both types of immune miscues are examples of allergic hypersensitivity reactions, and the antigen causing the reaction is called an **allergen**. There are four types of hypersensitivity reactions, as described in Table 17.3. The first three types are antibody-mediated; the fourth is cell-mediated.

CASE HISTORY 17.1

Sting of Fear: Type I Hypersensitivity

A bee stings a 9-year-old boy walking with his mother at the zoo. Within minutes, the boy begins sweating and itching. His chest then starts to tighten and he has tremendous difficulty breathing. Terrified, he looks to his equally frightened mother for help.

This is a classic example of severe **type I (immediate) hypersensitivity**, sometimes called **anaphylaxis**. Type I hypersensitivity reactions can also be mild and localized (as in hay fever, also called allergic rhinitis). Symptoms of type I hypersensitivities arise when an allergen (an antigen that elicits an allergic reaction) enters a susceptible person's body and triggers the release of pharmacologically active substances. For instance, inhaling pollen causes the release of these substances in the lungs of hay fever sufferers and produces symptoms that include rhinitis, watery and itchy eyes, and sneezing. By contrast, a bee sting causes systemic release of pharmacologically active substances as the venom circulates throughout the body. Thus, in these hypersensitive individuals, a bee sting can trigger anaphylaxis. Symptoms include swelling from fluid accumulation in tissues (edema), low blood pressure, cardiovascular collapse, and severe breathing difficulties, sometimes ending with suffocation. Anaphylaxis kills more than 1,500 people per year in the United States.

Stages of Type I Hypersensitivity

Type I hypersensitivity develops in two stages. The first stage is **sensitization**, which occurs when the patient first encounters the allergen. On *initial* exposure, an allergen elicits the production of IgE antibodies specific to the allergen (bee venom in the example above). The Fc portions of these antibodies bind to Fc receptors on the surfaces of mast cells. IgE-coated mast cells are then said to be *sensitized*. Recall that IgE primarily amplifies the body's response to invaders by causing mast cells to release inflammatory mediators (see Section 16.3). Allergens elicit higher levels of IgE in certain people; why some people are more affected than others is unclear, but a genetic predisposition appears to exist.

The second stage of type I hypersensitivity occurs within minutes of a *second* exposure to an allergen, when the allergen reacts with the IgE-coated mast cells. Most allergen molecules contain several identical antigenic sites. Thus, a single allergen particle during a second exposure can bind to adjacent IgE molecules

Table 17.3
Summary of Hypersensitivity Reactions

Type	Description	Time of Onset	Mechanism	Manifestations
I	IgE-mediated hypersensitivity	2–30 min	Ag induces cross-linking of IgE bound to mast cells with release of vasoactive mediators	Systemic anaphylaxis, local reactions, hay fever, asthma, eczema
II	Antibody-mediated cytotoxic hypersensitivity	5–8 h	Ab directed against cell surface antigens mediates cell destruction via ADCC or complement	Blood transfusion reactions, hemolytic disease of the newborn, autoimmune hemolytic anemia
III	Immune complex–mediated hypersensitivity	2–8 h	Ag-Ab complexes deposited at various sites induce mast cell degranulation via Fc receptor; PMN degranulation damages tissue (localized reaction)	Systemic reactions, disseminated rash, arthritis, glomerulonephritis
IV	Cell-mediated hypersensitivity	24–72 h	Memory T$_H$1 cells release cytokines that recruit and activate macrophages	Contact dermatitis, tubercular lesions

Ag = antigen; Ab = antibody; ADCC = antibody-dependent cell-mediated cytotoxicity; PMN = polymorphonuclear leukocyte or neutrophil.

ANIMATION ▶

Figure 17.6   Type I Hypersensitivity Reactions

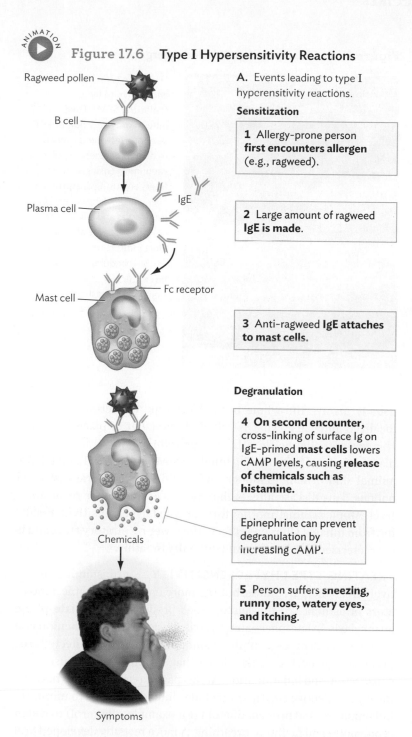

Ragweed pollen

B cell

Plasma cell

IgE

Mast cell

Fc receptor

Chemicals

Symptoms

A. Events leading to type I hypersensitivity reactions.

Sensitization

1 Allergy-prone person **first encounters allergen** (e.g., ragweed).

2 Large amount of ragweed **IgE is made.**

3 Anti-ragweed **IgE attaches to mast cells.**

Degranulation

4 On second encounter, cross-linking of surface Ig on IgE-primed **mast cells** lowers cAMP levels, causing **release of chemicals such as histamine.**

Epinephrine can prevent degranulation by increasing cAMP.

5 Person suffers **sneezing, runny nose, watery eyes, and itching.**

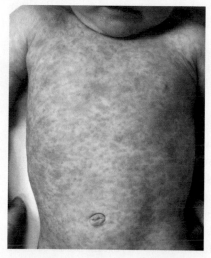

B. Peanut allergy. This case involved hives, but symptoms can include diarrhea, nausea, and life-threatening anaphylaxis.

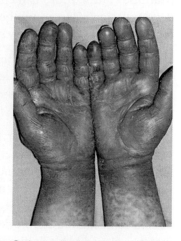

C. Latex allergy; a problem for some health care professionals. Symptoms range from itching and redness to breathing difficulties. Nonlatex, hypoallergenic gloves are available for these individuals.

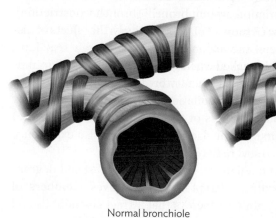

Normal bronchiole

Asthmatic bronchiole

D. Bronchoconstriction. During an asthma attack, smooth muscles in the bronchioles of the lung constrict and decrease airflow in the airways. Inflammation or excess mucus secretion can further decrease airflow.

protruding from the mast cell surface, creating a bridge between adjoining binding sites (**Figure 17.6A**). This cross-linked complex inhibits the mast cell enzyme adenylate cyclase, causing intracellular levels of cyclic adenosine monophosphate (cAMP) to fall. The lower levels of intracellular cAMP cause the mast cell granules to quickly migrate to the cell surface and release their contents in a process called **degranulation**.

Understanding this process is clinically important because anaphylaxis inhibitors such as epinephrine (also known as adrenaline) stimulate adenylate cyclase activity, increase cAMP levels, and stop the anaphylaxis cascade in its tracks. Epinephrine also relaxes

the smooth muscle surrounding bronchioles of the lung, easing breathing. Patients highly allergic to an allergen (such as bee venom or peanuts) will carry an EpiPen (an autoinjector) that can deliver an emergency dose of epinephrine.

PHARMACOLOGY OF TYPE I HYPERSENSITIVITY Mast cell degranulation releases chemicals with potent pharmacological activities. In humans, the most important are histamine, prostaglandins, leukotrienes, and chemotactic factors. The severity of an allergic reaction—whether it be life-threatening anaphylaxis as for the bee sting, or a less serious, more localized reaction such as contact dermatitis or hay fever—depends on several factors: the allergen, the route of allergen entry (injection, skin contact, inhalation), and the number of sensitized mast cells present. For example, an allergen injected (bee venom) or ingested (penicillin, peanuts; Figure 17.6B) can circulate via the blood and trigger the degranulation of larger numbers of more sensitized mast cells than an allergen that touches only skin (a latex glove; Figure 17.6C).

Note Although an allergy to latex can be caused by type I hypersensitivity reactions, latex allergies are more commonly the result of type IV hypersensitivity reactions (contact dermatitis) described later.

Histamine is a mediator synthesized and stored in mast cell granules before the encounter with allergen. Histamine causes the symptoms of allergy by binding to several different histamine receptors present on different body cells. Antihistamines, a common treatment for less severe forms of type I hypersensitivity, are structurally similar to histamine and work by blocking histamine receptors (for instance, Benadryl blocks the H1 receptor). Once released, histamine will produce itching by binding to receptors on certain nerve cells. It will loosen the endothelial junctions of blood vessels, causing vasodilation and increased vascular permeability. As a result, fluid leaves the blood and accumulates in tissues. Fluid that accumulates under skin will produce raised, itchy welts called hives. Histamine also causes smooth muscle constriction around bronchi (bronchoconstriction), which restricts airflow (Figure 17.6D). Collectively, the effects of histamine produce some of the major symptoms of allergic reactions: itching, raised welts (also called wheals); a reddening that surrounds the wheal (called a flare); and difficulty breathing (asthma). In systemic cases of anaphylaxis, vasodilation leads to hypotension (low blood pressure) because the loss of fluids from blood (into tissue) decreases blood volume. Anaphylactic patients will also experience bronchoconstriction so severe they may suffocate.

Other preformed mediators released during mast cell degranulation include a protease (tryptase) that cleaves members of complement, and chemotactic factors that attract eosinophils and neutrophils. (Recall that the terms neutrophil and PMN are often used interchangeably.)

Leukotrienes and **prostaglandins** are membrane-derived chemical mediators made after the mast cell binds to an allergen (see late-phase anaphylaxis later). Allergen binding to IgE on mast cells activates an enzyme (phospholipase A) that releases from the cell membrane a 20-carbon fatty acid called arachidonic acid. Arachidonic acid is then converted to leukotrienes by 5-lipooxygenase A and to prostaglandins by cyclooxygenase. Prostaglandins affect relaxation and contraction of smooth muscles. Prostaglandins and leukotrienes relax the smooth muscles of blood vessels to cause vasodilation. The increased blood flow produces reddening of the skin. However, if released near the lung, prostaglandins cause contraction of smooth muscle around bronchioles (bronchoconstriction), leading to restricted airflow and labored breathing.

 Toddler peanut allergy—how we found out: youtube.com

ATOPY Earlier we described a severe type I allergic reaction caused by a bee sting. Type I hypersensitivity reactions do not

Figure 17.7 Example of an Inhaled Allergen

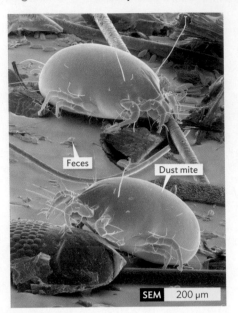

Dust mite (200–600 μm) and dust mite feces (colorized SEM). Dust mites roam every home, but not everyone is allergic to dust mite feces. Regular vacuuming (using a HEPA filter) and using dustproof or allergen-blocking covers on mattresses and pillows can control dust mite allergies. Dust mites can be killed by washing bedding weekly in hot water (54.5°C) and by freezing nonwashable items.

usually involve the whole body. Most type I reactions are more localized and cause what is called atopic ("out of place") disease. Hay fever, or allergic rhinitis, is a common manifestation of atopic disease that can be caused by inhaling dust mite feces (Figure 17.7), animal skin or hair (dander), and certain types of grass or weed pollens. This disease affects the eyes, nose, and upper respiratory tract. Atopic asthma, another form of type I hypersensitivity resulting from inhaled allergens, affects the lower respiratory tract and is characterized by wheezing and difficulty breathing.

TREATING TYPE I HYPERSENSITIVITY Antihistamines effectively treat allergic rhinitis, but the more important chemical mediators of asthma are the leukotrienes produced during **late-phase anaphylaxis**. In late-phase anaphylaxis, mast cells release chemotactic factors that call in eosinophils. Eosinophils entering the affected area produce high levels of leukotrienes that, like histamine, cause vasoconstriction and inflammation. At this point, antihistamines have little effect. Effective treatment includes inhaled steroids to minimize inflammation and bronchodilators (for example, albuterol) to widen bronchioles and facilitate breathing. A more recently developed oral medication, montelukast (Singulair), blocks leukotriene receptors in the lung. During severe asthma attacks, injected epinephrine is critical. Epinephrine will open airways to ease breathing through direct hormonal action.

Long-term treatment of an allergy requires identifying the cause, which is often not known. The offending allergen can be identified by individually injecting small amounts of various allergens into a patient's skin (Figure 17.8). The tested skin is then examined for reactions. Spots showing wheals indicate that the patient is sensitive to the corresponding allergen.

People sensitized to allergens are not necessarily doomed to suffer with the allergy their entire lives. A clinical treatment called desensitization can sometimes prevent anaphylaxis. Desensitization involves injecting small doses of allergen over a period of months. This process is thought to produce IgG molecules that circulate, bind, and neutralize allergens before they contact sensitized mast cells. Desensitization has been useful in cases of asthma and bee stings (but not food allergies), although results vary and are by no means guaranteed.

A relatively new treatment for allergic asthma, omalizumab (Xolair), can prevent sensitization. Omalizumab is a monoclonal antibody (administered by injection) that selectively binds IgE. It prevents IgE from binding to Fc receptors on mast cells, preventing sensitization.

Why don't all antigens generate type I hypersensitivity? Most antigens do not elicit high levels of IgE antibodies. What makes certain antigens (but not others) capable of eliciting high IgE levels, or what makes different individuals prone to different allergies, is unclear.

 Allergy tracker: weather.com

CASE HISTORY 17.2

Transfusion Confusion: Type II Hypersensitivity

 Richard, who is blood group B, was undergoing surgery after a car accident and accidentally received a transfusion with type A blood. Within hours, Richard experienced chills and lowered blood pressure. His urine turned red with blood. All these symptoms indicate an ABO blood group incompatibility.

Stages of Type II Hypersensitivity

Case History 17.2 describes a form of **type II hypersensitivity**. Type II hypersensitivity starts with antibody binding to cell surface antigens (donor red blood cells, in this case). The major characteristic that differentiates type II from type III hypersensitivity (considered next) is that type II hypersensitivity involves antibody binding to cell surface antigens, whereas type III hypersensitivity involves antibody binding to *soluble* antigens. The classic type II hypersensitivity reaction is illustrated by blood group incompatibilities arising from transfusions.

Human blood type falls into four major groups—A, B, AB, and O—in a classification system called the **ABO blood group system**. Blood group incompatibility occurs when specific antibodies in the serum of an individual of one blood type (A or B) bind to antigens

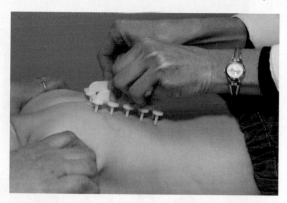
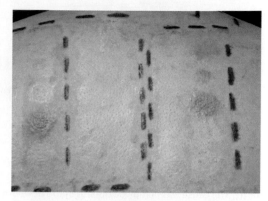

Figure 17.8 Allergy Skin Testing
Skin testing is generally the test of choice to evaluate allergies because it is a quick, accurate, nearly painless procedure. It can easily be performed in the clinic and provides results in less than 20 minutes.

A. In a skin test, drops of allergen extracts are placed on the forearms or the back and then pricked with a plastic device or a small needle.

B. Allergies reveal themselves as red, itchy, mosquito-like bumps called wheals. These itchy spots typically go away on their own in less than an hour.

on red blood cells (RBCs) of a different type (foreign RBCs). Type B individuals carry specific anti-A antibodies in their bloodstream, whereas type A individuals carry specific anti-B antibodies. These antibodies explain why a type A person cannot donate blood to a type B individual, and vice versa. RBCs that contain the A antigen are destroyed (lysed) by anti-A antibodies when transfused into a type B person. Lysis occurs as a result of classical complement activation (Section 15.7) or antibody-dependent, cell-mediated cytotoxicity (Section 15.5).

Type AB people carry neither antibody; and type O individuals, whose RBCs contain neither A nor B antigen, carry both anti-A and anti-B antibodies. Because RBCs from a type O person do not contain A or B antigens, type O RBCs can be donated to type A, B, AB, or O individuals. This is why type O people are sometimes called universal donors.

Note Researchers think that anti–group A and anti–group B antibodies arise early in life after encounters with environmental antigens whose structures mimic blood group antigens. For instance, anti–group A antibodies in a type B person are thought to arise after an encounter with influenza virus, which possesses epitopes similar to those of the A blood group. For a similar reason, anti–group B antibodies are believed to develop in type A individuals after encounters with Gram-negative bacteria in their intestine.

Another form of type II hypersensitivity is **Rh incompatibility disease** of the fetus and newborn, also known as erythroblastosis fetalis or hemolytic disease of the newborn (**Figure 17.9A**). Similar to ABO antigens, Rh antigens are chemical groups on red blood cells. A mother who is Rh-negative can carry a fetus that inherited the Rh$^+$ gene from the father. Incompatibility between mother (Rh$^-$) and fetus (Rh$^+$) can lead to a hypersensitivity reaction in the fetus, especially in a *second* pregnancy (**Figure 17.9B**).

Figure 17.9   **Type II Hypersensitivity**
Rh incompatibility disease of the newborn.

A. This infant presented with jaundice 8 weeks after birth. The cause was hemolytic disease of the newborn owing to Rh incompatibility. Severe cases can lead to intrauterine death.

When an Rh⁻ mother gives birth to an Rh⁺ firstborn child, fetal Rh⁺ RBCs enter the mother's bloodstream. The Rh⁻ mother's immune system will make anti-Rh⁺ IgG antibodies. The presence of anti-Rh⁺ antibodies (IgG) is a problem during a <u>subsequent</u> pregnancy because they can cross the placenta to attack an Rh⁺ fetus. Fortunately, giving the mother anti-Rh⁺ immunoglobulin (called RhoGAM) by intramuscular injection within a day after each birth of an Rh⁺ child can prevent Rh incompatibility disease. This antibody binds the fetal Rh⁺ antigen, preventing the mother's immune system from making Rh⁺ antibody and the attendant memory B cells. Passive immunity lasts only 4–6 weeks, long enough to clear the fetal antigens from the mother's bloodstream.

Why doesn't hemolytic disease of the newborn happen when a fetus having type AB blood is born to a type A mother? Doesn't the mother have anti-B antibodies? She does, but by and large they are IgM antibodies. IgM antibodies do not cross the placenta and cannot reach the bloodstream of the fetus. Anti-Rh⁺ antibodies are IgG and can cross the placenta.

CASE HISTORY 17.3

The Hives Have It:
Type III Hypersensitivity

A 16-year-old girl, Elaine, is treated for acne with minocycline (a tetracycline derivative). Two weeks later she develops a fever and an urticarial rash (**Figure 17.10**). Urticaria

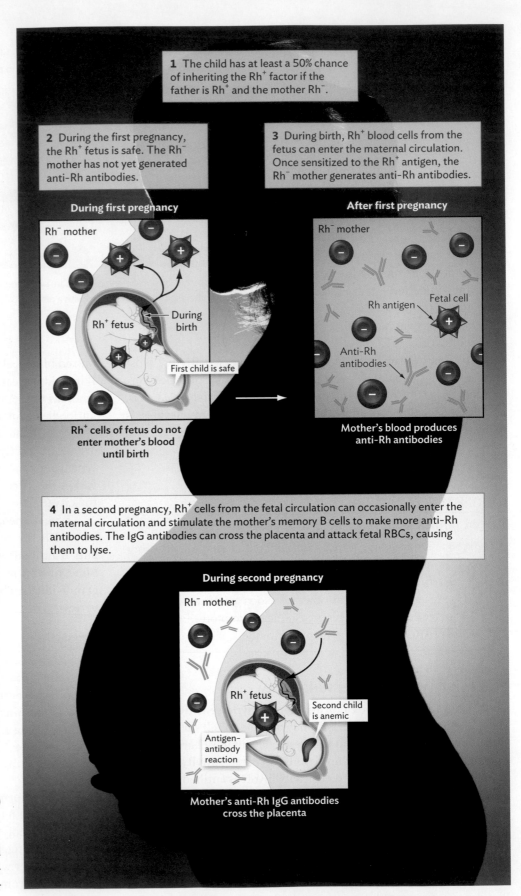

1 The child has at least a 50% chance of inheriting the Rh⁺ factor if the father is Rh⁺ and the mother Rh⁻.

2 During the first pregnancy, the Rh⁺ fetus is safe. The Rh⁻ mother has not yet generated anti-Rh antibodies.

3 During birth, Rh⁺ blood cells from the fetus can enter the maternal circulation. Once sensitized to the Rh⁺ antigen, the Rh⁻ mother generates anti-Rh antibodies.

During first pregnancy

Rh⁻ mother

Rh⁺ fetus

During birth

First child is safe

Rh⁺ cells of fetus do not enter mother's blood until birth

After first pregnancy

Rh⁻ mother

Rh antigen   Fetal cell

Anti-Rh antibodies

Mother's blood produces anti-Rh antibodies

4 In a second pregnancy, Rh⁺ cells from the fetal circulation can occasionally enter the maternal circulation and stimulate the mother's memory B cells to make more anti-Rh antibodies. The IgG antibodies can cross the placenta and attack fetal RBCs, causing them to lyse.

During second pregnancy

Rh⁻ mother

Rh⁺ fetus

Second child is anemic

Antigen–antibody reaction

Mother's anti-Rh IgG antibodies cross the placenta

B. Development of anti-Rh antibodies during pregnancy and Rh incompatibility during a second pregnancy.

Figure 17.10   **Type III Hypersensitivity**

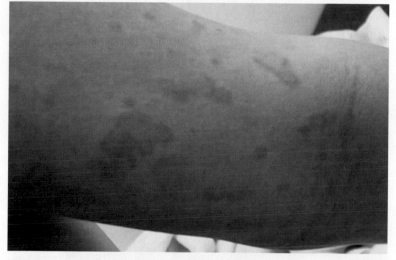

Urticarial rash caused by serum sickness, a form of type III hypersensitivity that can follow the administration of serum. The rash is similar to that formed after the administration of nonprotein drugs (for example, minocycline).

Figure 17.11   **Type III Hypersensitivity**

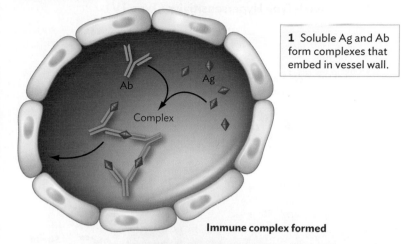

1 Soluble Ag and Ab form complexes that embed in vessel wall.

Immune complex formed

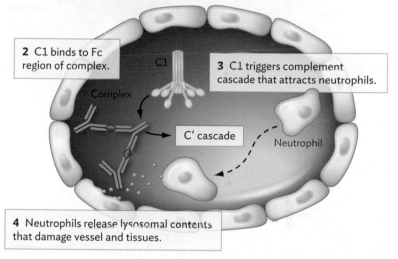

2 C1 binds to Fc region of complex.

3 C1 triggers complement cascade that attracts neutrophils.

4 Neutrophils release lysosomal contents that damage vessel and tissues.

Small complexes of antibody (Ab) bound to antigen (Ag) deposit on vascular walls. Complement factor C1 binds to the Fc regions of the antibodies and triggers a complement cascade. Products of the cascade attract neutrophils that deposit lysosomal components that damage the vessel wall endothelial cells.

are hives, an itchy rash caused by tiny amounts of fluid that leak from blood vessels just under the skin surface. Elaine also complains of muscle and joint pains. Her symptoms resolve after 5 days of anti-inflammatory steroid treatment.

Elaine's case is an example of **immune complex disease**, also called **type III hypersensitivity**. In contrast to type II hypersensitivity, type III hypersensitivity is initiated by IgG antibody binding to *soluble* antigen, which here is minocycline. Normally, macrophages readily engulf and clear antigen-antibody complexes, but excessive immune complexes may overwhelm the system and circulate in the blood (**Figure 17.11**). The complexes can embed in vessel walls (particularly in the kidneys and joints) and bind complement factor C1 of the classical pathway (Section 15.7). The binding of C1 triggers a cascade in which complement fragments recruit and activate polymorphonuclear leukocytes (PMNs). The complement-activated PMNs release lysosomal proteases and reactive oxygen species that damage host cells in the area. The resulting inflammation can cause skin eruptions (urticaria; Figure 17.10), damage the glomeruli of the kidneys (sites of blood filtration), and cause joint pain. **Serum sickness** is a type III hypersensitivity reaction caused by injecting a host with antiserum or other proteins derived from animals, such as antivenoms or hormones. Note that a common hypersensitivity to a different antibiotic, penicillin, is usually a type I anaphylaxis whose symptoms appear within minutes of antigen exposure. But penicillin can also cause a longer-term type III hypersensitivity reaction.

 Type III hypersensitivity immune complex animation HD

CASE HISTORY 17.4

TB or Not TB? Type IV Hypersensitivity

 The patient, Asuncion, is an 8-year-old girl from Argentina. She received the BCG vaccine for tuberculosis about 1 year before coming to the United States. Upon entering school in the United States, she must take a skin test for tuberculosis (TB). Asuncion tries to refuse but is told it is a requirement. She allows the nurse to apply the test to her arm. Three days later, the test site has a large red lesion and the skin is starting to slough (peel off). Does she have TB?

Type IV hypersensitivity, also known as **delayed-type hypersensitivity (DTH)**, is the only class of hypersensitivity triggered by

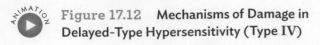

Figure 17.12   Mechanisms of Damage in Delayed-Type Hypersensitivity (Type IV)

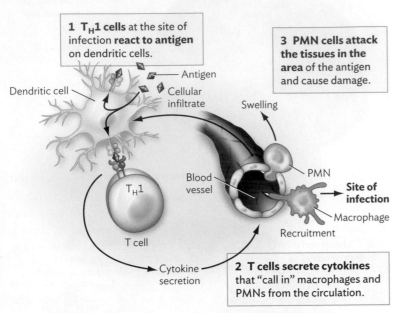

1 T$_H$1 cells at the site of infection **react to antigen** on dendritic cells.

3 PMN cells attack the tissues in the area of the antigen and cause damage.

Dendritic cell

Antigen

Cellular infiltrate

Swelling

T$_H$1

Blood vessel

PMN

Site of infection

Macrophage

T cell

Recruitment

Cytokine secretion

2 T cells secrete cytokines that "call in" macrophages and PMNs from the circulation.

A. T$_H$1 cells react to antigen on dendritic cells.

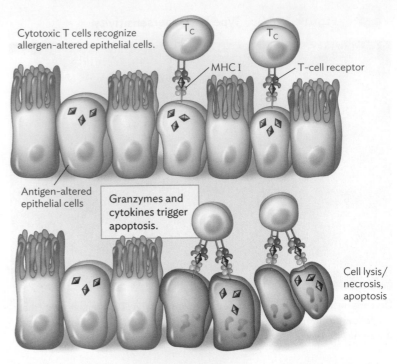

Cytotoxic T cells recognize allergen-altered epithelial cells.

T$_C$ T$_C$

MHC I   T-cell receptor

Antigen-altered epithelial cells

Granzymes and cytokines trigger apoptosis.

Cell lysis/ necrosis, apoptosis

B. Antigen-sensitized cytotoxic T cells recognize allergen haptens attached to proteins of other cells. This recognition triggers granzyme release and cytokine production that triggers apoptosis of the target cell.

antigen-specific T cells. Because T cells have to react and proliferate to cause a response, a reaction generally takes 24–48 hours to appear. This delay distinguishes type IV hypersensitivity from the more rapid antibody-mediated allergic reactions (types I–III). In the case study, Asuncion was initially sensitized by vaccination with BCG (an attenuated strain of *Mycobacterium bovis*, a close relative of *M. tuberculosis*). Because the organism grows intracellularly, the vaccination produced a cell-mediated immunity, complete with preactivated memory T cells (similar to memory B cells). When she was reinoculated by the skin test, the memory T cells activated and elicited a localized reaction at the site of injection.

Note Current evidence suggests that people vaccinated with BCG as children will not typically have a positive reaction as adults. The tuberculin test is used to screen adults for tuberculosis regardless of whether they were vaccinated with BCG. However, repeated tuberculin testing in BCG-vaccinated people could lead to this reaction.

Type IV sensitivity develops in two stages. The first stage, sensitization, involves processing and presentation of antigen on cutaneous dendritic cells (called Langerhans cells). These APCs travel to the lymph nodes, where T$_H$0 cells can react to them as described earlier (see Figure 16.19), generating activated T cells and a subset of memory T cells. On second exposure, two routes lead to DTH. *Memory T$_H$1 cells* can bind antigen presented on class II MHC receptors

on APCs at the site of infection. T$_H$1 cells release the inflammatory cytokines gamma interferon, TNF-β, and IL-2 (Figure 17.12A), which recruit macrophages and PMNs to the site and then activate macrophages and natural killer cells to release inflammatory mediators that damage innocent, uninfected bystander host cells. Alternatively, *memory cytotoxic T$_C$ cells* can bind antigen presented on class I MHC receptors of infected cells, become activated, and directly kill the host cell presenting the antigen (Figure 17.12B). A hallmark of type IV hypersensitivity is that transferring white blood cells, not serum, can transfer the sensitivity to a naive animal. This is because T cells, not serum antibody, are responsible for the reaction.

Forms of DTH reactions include allergic contact dermatitis (such as with poison ivy or latex) and allograft rejection after transplantation surgery. An allograft is a tissue graft from a donor of the same species (human) but who is not genetically identical to the recipient. The antigens involved with contact dermatitis are usually small haptens that must bind and modify normal host proteins to become antigenic (Figure 17.13). The hapten-modified protein is processed, and fragments containing the bound hapten are presented on MHC I complexes on the surfaces of antigen-presenting cells.

Allergies of all types have been on the increase for many years. Why that is, however, has remained an unsolved mystery. Recent research now suggests that the answer may be our obsession with cleanliness, described in **inSight**.

Figure 17.13 Contact Dermatitis

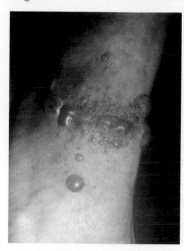

Pentadecylcatechols are chemicals present on the surface of poison ivy leaves. They are haptens that bind to proteins in dermal cells, where they can activate T cells. The result of the ensuing delayed-type hypersensitivity is contact dermatitis, whose hallmarks are insistent itching and fluid-filled blisters.

An unsaturated pentadecylcatechol

SECTION SUMMARY

- **Allergens** cause the host immune system to overrespond or react against self.
- **Type I hypersensitivity** involves IgE antibodies bound to mast cells by the antibody Fc region. Binding of antigen to the bound IgE causes mast cell degranulation. Symptoms occur within minutes of an exposure.
- **Type II hypersensitivity** (for example, Rh incompatibility) involves antibodies binding to cell surface antigens. This process can trigger cell-mediated cytotoxicity or activate the complement cascade. Symptoms occur within hours of an exposure.
- **Type III hypersensitivity** is an immune complex disease involving antibody complexes with small, soluble antigens. The complexes can activate complement or bind to mast cells. The result is the recruitment of PMNs that damage host cells in the area. Symptoms may take weeks to form.
- **Type IV hypersensitivity** (delayed-type hypersensitivity) involves antigen-specific T cells. First, T_H1 cells release cytokines that activate macrophages and NK cells. Macrophages and NK cells cause tissue damage. Second, activated T_C cells can directly kill any host cells that present the antigen. Reaction (for instance, skin) is seen within a few days of exposure.

Thought Question 17.3 Why do individuals with type A blood have anti-B and not anti-A antibodies?

Thought Question 17.4 Does a peanut allergy rely more on T_H1 or T_H2 cells?

Thought Question 17.5 Why would you consider blood group type O people "universal donors" but not universal acceptors of RBCs from type O, A, B, or AB individuals? Who might be considered a universal recipient?

Thought Question 17.6 Patients with primary immunodeficiencies can receive multiple IV immune globulin treatments with only minor side effects. Since IgG antibodies from different individuals have allotypic antigenic differences, why aren't serious hypersensitivity reactions to IVIG treatment more common?

17.4 Autoimmunity and Organ Transplants

SECTION OBJECTIVES

- Recall how the immune system differentiates self from nonself (foreign) antigens.
- Discuss how autoreactive antibodies can be formed.
- Describe the key features of the autoimmune diseases lupus, Graves' disease, diabetes, rheumatoid arthritis, and myasthenia gravis.
- Explain the basis of transplant rejection.

If my immune system is "blind" to my own antigens, how is it that I can develop autoimmune disease? The ability to distinguish between self antigens and foreign antigens is crucial to our survival; without this ability, our immune system would constantly attack us from within. Self-attack doesn't normally happen because the body develops tolerance to self; occasionally, however, an individual loses immune tolerance against some self antigens and mounts an abnormal immune response against his or her own tissues. The attack can involve antibodies or T cells and is called an **autoimmune response**.

Link As Section 16.5 describes, tolerance to self comes in large part from removing (deleting) T cells in the thymus that bind too strongly to self antigens. B-cell tolerance to self develops when a B cell is exposed to high doses of its matched self antigen in bone marrow. The exposure triggers apoptosis of that B cell.

Autoimmune responses may or may not produce disease. Almost 30% of the population, including those of you reading these words, will have an autoimmune antibody by age 65, but many will not exhibit disease.

You should realize that the mechanisms leading to autoimmune disease are essentially hypersensitivity reactions. An autoimmune response results in disease when an autoantibody or autoimmune T cell damages tissue components. Tissue damage can develop when antibody to a self antigen activates antibody-dependent cell cytotoxicity mechanisms (see Section 15.5). In this scenario, autoantibodies affixed to host cells can bind to natural killer (NK) cell Fc receptors and cause the NK cell to kill the tissue cell.

How might autoimmune antibodies form? One proposed mechanism starts with the occasional escape of self-reacting B cells from the negative selection process. The negative selection process normally kills B cells that react to host antigens. Self-reacting B cells that escape this process are usually not a problem because the specific helper T cells needed to activate them are successfully deleted from the T-cell population. Sometimes, however, the self-reacting B cell can be activated without *specific* T-cell assistance. The self-reactive B cell will, of course, take up and process a self antigen (or a foreign antigen that mimics a self antigen). If the antigen *also*

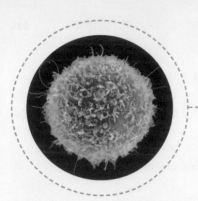

Let Them Eat Dirt! The Hygiene Hypothesis and Allergies

Figure 1   Child Undergoing Early Antigen Exposure

The incidence of asthma and other hypersensitivity diseases in children has skyrocketed since 1980. Today, an average of 1 of every 10 U.S. children has asthma. In 1980 it was closer to 1 in 30. How many of your own friends have asthma? What might account for this disturbing trend?

An early clue came after the reunification of Germany in the late 1990s when Erika von Mutius compared the allergy and asthma rates of East and West Germany. She expected that children growing up in the poorer, dirtier, less healthful cities of East Germany would have more allergies than youngsters in West Germany, with its cleaner and more modern environment. She found the opposite. Children in polluted areas of East Germany had less intense allergic reactions and fewer cases of asthma than children in West Germany. What does this mean?

Many studies now support what has been called the "hygiene hypothesis." The hygiene hypothesis states that children who play with many other children or animals early in life are exposed to more microbes (Figure 1), and their immune systems develop more tolerance for the irritants that cause asthma and other allergies. The hypothesis further suggests that the extremely clean household environments often found in the developed world derail the critical postnatal period of immune responses. In other words, the young child's environment can be "too clean" to effectively challenge a maturing immune system.

The potential problem with extremely clean environments is that they fail to expose children to microbes required to "educate" the immune system to properly launch its defenses against infectious organisms and other irritants. Without proper education, the immune system responses end up contributing to the development of asthma or other allergies.

The hygiene hypothesis is based in part on the observation that, before birth, the fetal immune system is suppressed to prevent it from rejecting maternal tissue. Such a low default setting is necessary before birth—when the mother is providing the fetus with her own antibodies. But immediately after birth, the child's own immune system must take over and learn how to fend for itself.

So why might more bacterial infections (asymptomatic) help protect a young child from allergies? Allergies are usually considered T_H2-driven diseases (recall that T_H2 cells stimulate IgE antibody production, a major player in allergies). Thus, altering the balance between T_H1 and T_H2 responses to a potential allergen may determine whether the person will develop an allergy (Figure 2). Cells with a lot of "say" in altering the balance between T_H1, T_H2, and regulatory T cell (also known as Tregs) development are antigen-presenting dendritic cells that, depending on whether bacteria or other antigens stimulate them, can alter whether a T_H0 cell becomes a T_H1, T_H2, or other helper T cell. Bacterial infections may *inhibit* allergies

Figure 2 How Microbes Might Prevent Allergies

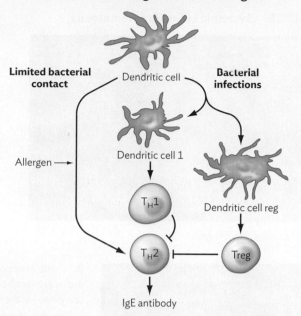

IgE antibody

Dendritic cells with limited bacterial contact may encourage T_H2 formation and responses that lead to allergies. Dendritic cells with more frequent bacterial contact can differentiate into cells that encourage T_H1 formation or stimulate Treg cells that suppress T_H2 responses, thereby limiting allergies.

by "educating" dendritic cells to promote T_H0 differentiation into T_H1 cells that trigger T_H1 responses (which, in turn, inhibit other T_H responses). The result is less chance of developing an allergy. Alternatively, the dendritic cell presenting bacterial antigens can stimulate the production of Tregs that produce anti-inflammatory cytokines such as IL-10 that can also inhibit T_H2 responses.

In addition to more infections, scientists now believe the normal microbiota of the gut play a major role in protecting individuals from developing some allergies. As one piece of evidence, killing neonatal gut flora with antibiotics has been associated with a greater risk of developing asthma. This result also suggests that changing the microbiome in one mucosal compartment (the intestine) can affect the immune system in another compartment (the lungs). These cross-organ effects show communication between different mucosal immune compartments. In another example, *intranasal* immunization with herpes simplex virus type 2 results in *vaginal* protection against genital infection with this virus. Thus, it appears that the mucosal immune system is a systemwide organ in which stimulation in one compartment (lung) can lead to changes in distal areas.

The take-home message from all of these studies is that, like a muscle, exercising your immune system by engaging microbes strengthens your immunity. No one is saying let babies eat dirt. However, it might not be as bad for them as you might think—and probably even helps.

contains a "nonself" epitope (possibly due to a mutation; Figure 17.14), this nonself epitope will also be presented on the B cell's MHC II molecule. When this happens, T cells reacting to the nonself epitope will recognize the nonself peptide and activate that B cell. Here, however, the B cell is programmed to make antibody to the *self* antigen, rather than *nonself* antigen—antibodies that set about attacking the host.

Some microbial antigens can also trigger autoimmune reactions. A good example is the M protein (a pilus-like, fuzzy coat) of *Streptococcus pyogenes*, a microbe that causes strep throat and the autoimmune disease rheumatic fever. M pili contain an epitope that resembles cardiac antigen, but the heartlike epitope is flanked

Figure 17.14 Formation of Autoimmune Antibodies

The model starts with a B cell expressing surface antibody (BCR) to self epitope A.

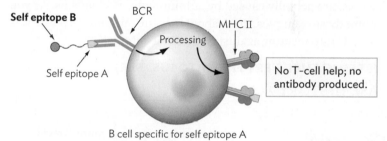

A. No T-cell help—no autoantibody production. Ordinarily, an antigen from self contains two epitopes: a normal self epitope A linked to a normal self epitope B. Helper T cells are not available to either self epitope, so no T cell can help to stimulate antibody production.

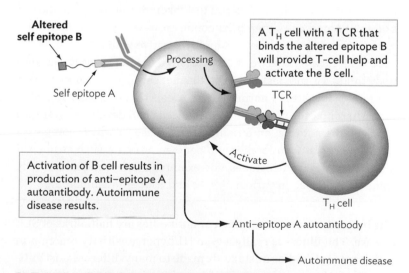

B. T_H2 cell available that can bind an altered self epitope. If the antigen from self contains a normal self epitope A and an abnormal (mutated) self epitope B, autoantibody production (that is, production of antibodies directed against self) may result. The B-cell receptor binds epitope A and processes the linked epitopes, presenting both on the B-cell MHC II receptors. If T_H cells with a TCR that can bind the altered epitope B are present, they can help activate the B cell, which differentiates into a plasma cell that makes autoantibodies to epitope A. Alternatively, the pentagon can be a foreign epitope (microbial?) that mimics a host self epitope, and the square another foreign epitope that does not mimic host.

by epitopes that are not related to the human host (streptococcal antigens). The heartlike epitope can bind surface antibody on a self-reacting B cell and be taken up. Because the nonself epitope and the heartlike epitope are contained within the same protein, the nonself (noncardiac) epitope will "piggyback" its way into the B cell and, as described above, can be used by helper T cells to activate the autoreactive B cells. Antibodies to the cardiac antigen are made, bind to cardiac tissue, and trigger autoimmune disease. With *S. pyogenes*, cardiac tissue is damaged and rheumatic fever results. The similarity between the epitope in M protein and cardiac tissue is an example of antigenic mimicry. The following section presents other examples of well-known autoimmune diseases.

Autoimmune Diseases

Many different human autoimmune diseases exist. You may be surprised to learn that some well-known diseases, such as type I diabetes, are actually caused by autoimmune mechanisms. As you explore these examples, think about how the immune response was "tricked" into reacting against its own self-interest.

SYSTEMIC LUPUS ERYTHEMATOSUS (SLE)

CASE HISTORY 17.5

Rash of a "Butterfly"

Tamil, a 26-year-old woman, recently developed joint pain in her wrists, fingers, and ankles. She also noticed a strange rash over her nose and cheeks that reminded her of a butterfly (Figure 17.15A). She told her physician that the redness grew worse after she was in the sun for only an hour. A blood test taken during her visit showed a normal blood cell differential count. However, her serum antinuclear antibody (ANA) test was highly positive with a 1:256 dilution; normal levels are negative at 1:8 dilution (Section 17.5 discusses antibody titers). Further tests revealed that her serum complement levels were much lower than normal. On the basis of the clinical and laboratory findings, the physician told Tamil she had developed systemic lupus erythematosus, an autoimmune disease in which complexes form between autoantibodies and her own proteins, such as nuclear material (DNA and histones, for example). The antigen-antibody complexes deposit in various tissues and trigger inflammation.

The butterfly rash and antinuclear antibodies are hallmarks of SLE disease. The illness is really a type III hypersensitivity reaction in which autoantibodies mistakenly made to many different host antigens bind those antigens and form circulating antigen-antibody complexes. The complexes deposit in the basement membranes of various organs, especially kidneys (Figure 17.15B), and activate the classical complement pathway (Chapter 15), causing inflammation and depleting serum complement levels. The inflammation of the kidney leads to leakage of protein and red blood cells into the urine.

What causes lupus remains a mystery. Viral infections leading to loss of regulatory T cells that help limit an inflammatory process as

Figure 17.15 Systemic Lupus Erythematosus

A. Classic butterfly rash of lupus.

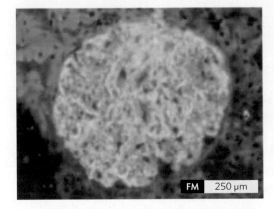

B. Granular deposits of antigen-antibody complexes in the glomerulus of the kidney. Viewed by immunofluorescence microscopy using fluorescently labeled antibody. The green areas would be absent in a normal kidney.

FM   250 μm

well as hormonal changes during pregnancy are suspected triggers. Although lupus has no cure, a variety of treatments can reduce symptoms, limit damage to vital organs, and reduce the risk of recurrence.

GRAVES' DISEASE, HASIMOTO'S DISEASE, AND MYASTHENIA GRAVIS   Other autoimmune diseases involve the production of autoantibodies that bind and block certain cell receptors. **Graves' disease** (hyperthyroidism), for example, occurs when an autoantibody is made that binds to the thyroid-stimulating hormone (TSH) receptor present on thyroid gland cells. Autoantibody binding mimics TSH binding and abnormally stimulates the production of the thyroid hormones triiodothyronine (T3) and thyroxine (T4), which increases metabolic rate and makes the patient hyperenergetic. The thyroid is not destroyed but instead enlarges to form a goiter (Figure 17.16). By contrast, **Hashimoto's disease** is a form of *hypo*thyroidism in which autoantibodies and T cells destroy thyroid follicle cells and thyroglobulin (the protein that produces T3 and T4) rather than stimulate thyroglobulin production. Consequently, Hashimoto patients are always tired or fatigued. The inflammation enlarges the thyroid to form a goiter that eventually atrophies.

Myasthenia gravis is a disease whose most striking symptom is profound muscle weakness that eventually hits all skeletal muscles. Figure 17.17 shows a patient with a drooping eyelid caused by this disease. The disease is caused by autoantibodies that bind

Figure 17.16 Graves' Disease

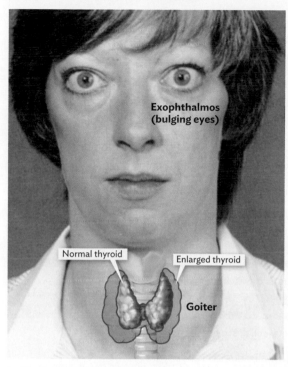

Exophthalmos (bulging eyes)

Normal thyroid

Enlarged thyroid

Goiter

In this Graves' disease patient, autoantibodies to thyroid-stimulating hormone receptors in the thyroid gland constantly stimulate the gland, causing its enlargement (goiter) and continual production of hormones T3 and T4. Hallmarks of the condition are an enlarged thyroid, bulging eyes (exophthalmos), heat intolerance, increased energy, difficulty sleeping, diarrhea, and anxiety.

to and block muscle receptors for acetylcholine. Thus, myasthenia gravis represents a type II hypersensitivity disease. Acetylcholine is a chemical released from nerves that is needed to transmit nerve impulses across a synaptic junction—here, to a muscle.

DIABETES Type 1 **diabetes** is an autoimmune disease that usually occurs during childhood when T cells attack and destroy the insulin-producing islet cells in the pancreas. The autoimmune reaction involves autoreactive CD4 T cells and cytotoxic T cells (CD8⁺).

Figure 17.17 Myasthenia Gravis

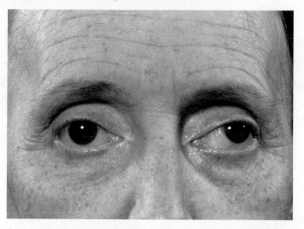

Autoantibodies against acetylcholine receptors on muscle cells inhibit the transmission of nerve impulses to muscle. The typical symptom is weakness of muscles that gets worse with activity and improves with rest. The muscles around the eyes are usually affected first. This patient's left eye does not move properly.

As such, it is a form of type IV hypersensitivity. (Unlike type 1 diabetics, type 2 diabetics continue to produce insulin, but their cells no longer respond to it.) The lack of insulin in type 1 diabetics leads to excessive levels of blood glucose, increased hunger, frequent urination, and a great thirst.

RHEUMATOID ARTHRITIS **Rheumatoid arthritis** is a chronic autoimmune disease characterized by fatigue and chronically inflamed synovium (soft tissue that lines the joints). One of every 100 people, mostly women, are afflicted. The inflammatory reaction erodes cartilage and bone around the joint, usually in the wrist, ankle, or cervical spine (Figure 17.18). The inflamed joint becomes painful, disfigured, and densely packed with lymphocytes, dendritic cells, NK cells, and plasma cells. Inflammation begins with autoantibodies (rheumatoid factors) and immune complex formation in joints that trigger type III hypersensitivity reactions. Eventually macrophages and autoreactive CD4 T cells produce proinflammatory cytokines, such as TNF-α, that worsen

Figure 17.18 Rheumatoid Arthritis

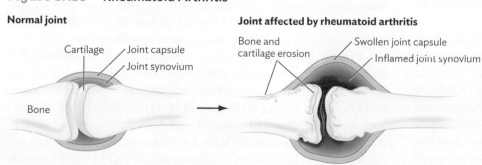

Normal joint

Cartilage

Joint capsule

Joint synovium

Bone

Joint affected by rheumatoid arthritis

Bone and cartilage erosion

Swollen joint capsule

Inflamed joint synovium

A. Inflammation of the synovium at a joint leads to erosion of bone and cartilage and swelling of the joint capsule.

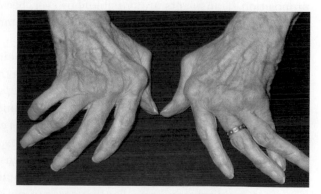

B. Deformation of hands caused by rheumatoid arthritis.

Table 17.4
Examples of Autoimmune Diseases

Autoimmune Disease	Autoantigen	Pathology
Type II Hypersensitivity—Mediated by Antibody to Cell Surface Antigens		
Acute rheumatic fever	Streptococcal M protein, cardiomyocytes	Myocarditis, scarring of heart valves
Autoimmune hemolytic anemia	Rh blood group	Destruction of red blood cells by complement, phagocytosis
Goodpasture's syndrome	Basement membrane collagen	Pulmonary hemorrhage, glomerulonephritis
Graves' disease	Thyroid-stimulating hormone receptor	Antibody stimulates T3, T4 production; hyperthyroidism
Myasthenia gravis	Acetylcholine receptor	Interrupts electrical transmission, progressive muscular weakness
Type III Hypersensitivity—Mediated by Antibody Complexes with Small Soluble Antigens (Immune Complex)		
Systemic lupus erythematosus	DNA, histones, ribosomes	Arthritis, vasculitis, glomerulonephritis
Type IV Hypersensitivity—Mediated by Antigen-Specific T Cells		
Type 1 diabetes	Pancreatic beta-cell antigen	Beta-cell destruction
Multiple sclerosis	Myelin protein	Demyelination of axons

the damage. Nonsteroidal anti-inflammatory drugs (NSAIDs) are typically used for early treatment, but more aggressive treatments for more severe disease involve TNF inhibitors methotrexate and etanercept (Enbrel). Table 17.4 lists other important autoimmune diseases.

 Pathogenesis of rheumatoid arthritis—hybrid medical animations: youtube.com

Organ Donation and Transplantation Rejection

One problem with having an immune system is that it makes transplanting an organ from one person to another a very dangerous proposition. Transplants are possible, however, because we have learned how to minimize the risk of the recipient's immune system rejecting the donated tissue.

Why are transplants dangerous? As a result of their "education" process, T cells that survive the selection process in the thymus by and large do not have T-cell receptors (TCRs) that "see" isogenic MHC proteins on other cells of the body (unless bound to a foreign antigen). However, about 5% of the surviving T cells can recognize and bind to allotypic MHC proteins on cells transplanted from other individuals (donors). This recognition happens because the allotypic MHC resembles a complex of self MHC bound to a foreign antigen. For this reason, transplanting organs from a donor with one type of MHC protein into a recipient with a different type of

MHC (called an **allograft**) typically ends badly, with the recipient rejecting the graft. The allotypic MHC proteins of the donor play a major role in this transplantation rejection process.

Note Do not confuse the terms "isogenic" and "isotype." Isogenic means genetically identical. MHC proteins on all cell types in your body are genetically identical; they are isogenic. The term "isotype" is generally reserved for the heavy chains of antibodies. For instance, although your gamma globulin heavy-chain isotype is very similar to gamma globulin from someone else, they are not isogenic.

To initiate transplant rejection, the recipient's T-cell receptors can *directly* bind either foreign MHC on donor cells (Figure 17.19A) or *pieces* of foreign MHC that were taken up, processed, and presented on recipient-cell MHC (Figure 17.19B). Once activated by one of these routes, cytotoxic T cells will destroy all donor cells containing the allotypic MHC proteins, rejecting the tissue.

Clearly, MHC type matching between donor and recipient is important for a successful transplant. The closer the match, the lower the likelihood of transplant rejection. However, blood transfusions do not require MHC matching. Type A blood from one person can be transfused to a second type A person regardless of MHC compatibility. The reason is that human red blood cells are not nucleated, and because only nucleated cells express MHC, human RBCs do not contain MHC proteins. Thus, rejection of RBCs via MHC incompatibility does not occur.

Figure 17.19 Transplant Rejection

A. In the direct route to tissue rejection, T cells from the recipient recognize the foreign MHC on a donor (graft) cell and become activated.

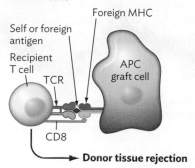

B. In the indirect route, pieces of foreign donor MHC (allo-MHC) are presented on the surface of recipient APCs. The recipient's cytotoxic T cells then respond and become activated. In either case, the activated cytotoxic T cells will subsequently kill any donor cells they encounter that express the foreign antigen.

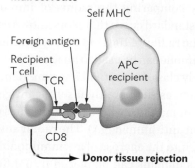

SECTION SUMMARY

- **Autoimmune disease** is caused by the presence of lymphocytes that react to self.

- **Autoreactive B cells** (B cells that make antibody directed against self epitopes) are activated when they present a coprocessed nonself epitope on their surface MHC. T cells that recognize the nonself epitope activate the B cell, which then secretes antibody against the self epitope.

- **Cytotoxic T cells can produce autoimmune disease** by killing host cells that make a self protein that closely resembles a foreign antigen.

- **Transplantation rejection** involves cytotoxic T cells that can recognize allotypic MHC proteins on transplanted tissues.

- **Recipient cytotoxic T cells can be activated** by directly binding allotypic MHC on graft cells or indirectly activated when proteins from the donor graft are presented on self APCs.

- **Red blood cells are not subject to transplant rejection** because they do not possess MHC proteins.

Thought Question 17.7 Do bone marrow transplants in a patient with severe combined immunodeficiency require immunosuppressive chemotherapy?

17.5
Tools of Immunology

SECTION OBJECTIVES

- Describe how immunoprecipitation assays are used to measure immunoglobulin levels.
- Explain agglutination and its use in blood typing and pathogen detection.
- Distinguish direct, indirect, and sandwich enzyme-linked immunosorbent assays (ELISAs).
- Discuss how monoclonal antibodies are made and their use in fluorescent antibody techniques (immunofluorescence microscopy, western blots, and flow cytometry).

How do we know when someone has an IgA deficiency . . . or a deficit of T cells? Many of the case histories presented in the last few chapters relied heavily on knowing the quantitative levels of certain antibodies, or the types of white blood cells, present in blood. Here we introduce some of the diagnostic tools used to measure those parameters. In addition to helping identify immune dysfunctions, these methods are crucial to diagnosing many infectious diseases.

Serology is the branch of immunology that reveals the contents of serum. It relies on the ability to visualize interactions between antibody and antigen, and it can be used to measure the concentrations of either. For any serological test, considering its **specificity** and **sensitivity** before trying to interpret the results is important. Specificity measures how selective an antibody is toward an antigen. An antibody that binds very dissimilar antigens has a very low specificity. Sensitivity, on the other hand, reflects how few antigen or antibody molecules a test can detect (described further in Section 25.6). A highly sensitive test detects very small amounts of antigen or antibody. Thus, the most effective serological test is both highly selective and highly sensitive.

Immunoprecipitation and Radial Immunodiffusion

Recall that a typical antibody molecule has two antigen-binding sites, each of which binds identical antigens. Because antibodies bind to more than one antigen molecule, antibodies can cross-link antigens in solution, ultimately forming complexes too large to remain soluble and thus fall out of solution (that is, precipitate; Figure 17.20). The phenomenon, called **immunoprecipitation**, is normally observed only in vitro, where the concentration of antigen and antibody can be manipulated experimentally. Immunoprecipitation occurs only with appropriate ratios of antigen and antibody molecules. Too many antigen molecules (antigen excess; Figure 17.20A) or too few antigen molecules (antibody excess; Figure 17.20B) results in complexes too small to immunoprecipitate. Large complexes are formed only at an appropriate antigen/antibody ratio called equivalence (Figure 17.20C). Equivalence is the point

Figure 17.20   Basis of Immunoprecipitation

Only when the numbers of epitopes and antigen-binding sites are roughly equivalent will a large complex form and fall out of solution.

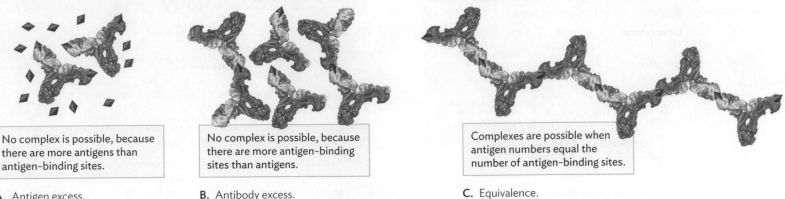

No complex is possible, because there are more antigens than antigen-binding sites.

A. Antigen excess.

No complex is possible, because there are more antigen-binding sites than antigens.

B. Antibody excess.

Complexes are possible when antigen numbers equal the number of antigen-binding sites.

C. Equivalence.

where the number of antigenic sites roughly equals the number of antigen-binding sites.

Radial immunodiffusion assay (RIA) is an application based on immunoprecipitation. RIA can be used to measure the concentration of an antigen. The antigen being measured could even be another antibody (**Figure 17.21**). In the RIA technique, a solution containing an antigen is placed into a well cut into an agarose gel that is impregnated with antibody to that antigen. Antigen in the well diffuses outward until reaching a zone of equivalence with the embedded antibody. At that point, antigen-antibody complexes precipitate to form a ring around the well. Remember that antigen concentration progressively *decreases* the farther the antigen diffuses

Figure 17.21   Radial Immunodiffusion

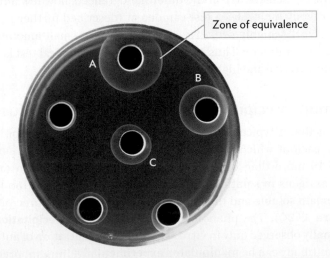

Zone of equivalence

The agarose plate shown is embedded with antibodies (anti-IgA antibodies) specific for a certain antigen—here, IgA. Different concentrations of the antigen (IgA) are placed in each well. The antigen diffuses into the agarose. A ring of precipitation occurs when the concentration of the diffusing antigen reaches a zone of equivalence with the antibody in the plate. The more antigen placed in the well, the larger the diameter of the precipitation ring. Well A was loaded with more antigen than well B or C.

away from the well. Consequently, the higher the concentration of antigen in the well, the farther that antigen will have to diffuse before the zone of equivalence is reached and a ring of precipitation forms. The concentration of antigen in an unknown solution is determined by comparing the radius of immunoprecipitation formed against a standard curve that plots known concentrations of antigen against the radius of the ring of precipitation.

The RIA technique can be used to measure the amount of IgA, or any other antibody class, in a patient's serum. Remember, antibodies are proteins and as such are themselves antigens. Thus, a rabbit injected with human IgA will make antibodies directed against human IgA (rabbit anti-human IgA). The rabbit anti-IgA antibodies can be embedded into the agarose of an immunodiffusion plate, and sera from various humans can be placed into different wells. The serum with the most IgA will produce the widest ring of precipitation. Serum from a patient with an IgA deficiency has very little IgA and will produce a very narrow ring of precipitation.

Agglutination

Agglutination and precipitation are based on the same basic cross-linking principle. But whereas precipitation measures small soluble antigens, **agglutination** measures insoluble antigens on whole cells (red blood cells or bacteria; **Figure 17.22A**). Because cells are large, agglutination reactions usually require pentameric IgM antibody for cross-linking. Agglutination reactions are used to determine the A, B, AB, and O blood groups of red blood cells. To perform the test, two drops of a patient's blood are placed on a slide. Antibody to the A antigen is mixed with one blood drop, and anti-B antibody is mixed with the second. Interaction between antigen and antibody causes the blood cells to clump, indicating a positive test (**Figure 17.22B**).

Agglutination can also be used to identify bacterial pathogens. **Figure 17.22C** shows colonies of group B beta-hemolytic streptococci responsible for serious neonatal infections. In nature, many strains of beta-hemolytic streptococci are not pathogenic; thus, it is of utmost importance to quickly determine whether a beta-hemolytic

Figure 17.22 Slide Agglutination

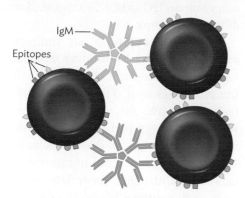

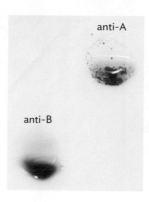

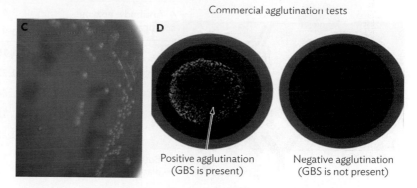

Commercial agglutination tests

Positive agglutination (GBS is present)

Negative agglutination (GBS is not present)

A. Illustration of red blood cells cross-linked with IgM against a surface antigen. Sizes of cells and antibodies are not proportional.

B. Blood typing in which a drop of blood is mixed with anti-B (left) or anti-A (right) antibodies. The blood type in this instance is type A.

C and D. Agglutination used to identify group B beta-hemolytic streptococci. Cells from suspected colonies (**C**) are mixed with antiserum against group B streptococci (**D**). Cells on the left (in panel D) are agglutination positive, meaning they are group B streptococci. Cells on the right (in panel D) are agglutination negative and not group B streptococci.

streptococcus isolated from an expectant mother is group B. Cells from one of those colonies will agglutinate when mixed with antibody directed against group B streptococci (**Figure 17.22D**).

Enzyme-Linked Immunosorbent Assay (ELISA)

Visualizing immunoprecipitation requires large amounts of antigen and antibody in just the right balance. More subtle antigen-antibody interactions can be detected by using an **enzyme-linked antibody** in which an enzyme that converts a colorless substrate to a colorful product is attached (conjugated) to a specific antibody. The enzymes most commonly used are horseradish peroxidase

and alkaline phosphatase. The technique that uses this approach is called an **enzyme-linked immunosorbent assay (ELISA)**. ELISA is sometimes referred to as an **enzyme immunoassay (EIA)**. EIA tests can detect antibody or antigen.

As shown in **Figure 17.23**, ELISAs come in three basic forms: direct, indirect, and sandwich capture. **Direct ELISA** starts by binding a known antigen to the well of a plastic microtiter dish. The specific antigen of interest (a virus protein, for instance) is detected by adding an enzyme-conjugated antibody specific to the antigen. The intensity of color change after the enzyme substrate is added reflects the amount of antigen that bound to the well and, thus, was present in the patient sample (usually tissue).

Figure 17.23 Basic Forms of Enzyme Immunoassays

Direct immunoassay is used to detect a known antigen (Ag). Indirect immunoassay is used to detect specific antibody (Ab) in the patient's serum. Capture assay sandwich assays can detect either Ag or Ab in a patient sample.

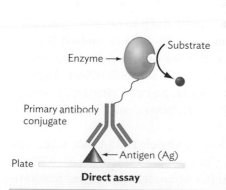

Direct assay

The known antigen is immobilized by direct adsorption to the assay plate and detected by adding enzyme-conjugated antibody.

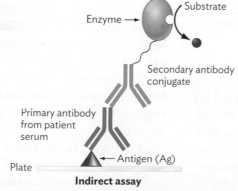

Indirect assay

A known antigen is directly adsorbed to the assay plate. Serum Ab to the antigen is captured by the plate antigen. Secondary enzyme-conjugated Ab is then used to detect the captured Ab.

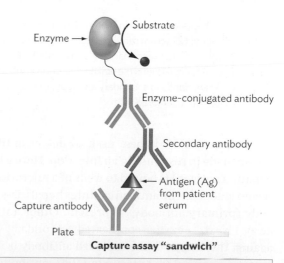

Capture assay "sandwich"

An antigen-specific Ab is directly adsorbed to the assay plate. Antigen from patient sample is captured by the plate antibody. Unlabeled secondary Ab to the Ag binds to the captured antigen. Enzyme-conjugated Ab is then used to detect the secondary Ab.

Figure 17.24   **Enzyme-Linked Immunosorbent Assays (ELISAs)**

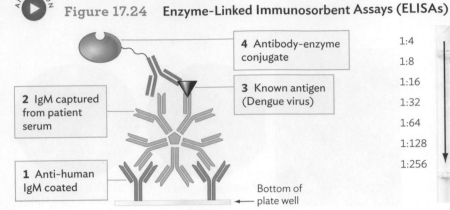

4 Antibody-enzyme conjugate

2 IgM captured from patient serum

3 Known antigen (Dengue virus)

1 Anti-human IgM coated

Bottom of plate well

A. In an IgM-capture ELISA, a plate is coated with antibody against human IgM constant region. Twofold dilutions of the patient's serum are added to the microtiter plate wells. All the IgM in serum is captured. Known antigen is then added (such as a dengue virus antigen), which will bind only to the dengue Ag–specific IgM. Anti-antigen antibody with bound enzyme is then added, followed by adding a chromogenic (pigment producing) substrate for the enzyme (not shown). A reaction with the substrate will cause a visible color change.

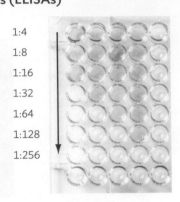

1:4
1:8
1:16
1:32
1:64
1:128
1:256

B. The antibody titer is determined by serially diluting (twofold) patient serum from top to bottom of dish (dilutions are marked). The titers from patient samples shown in columns 2–4 are higher than those in the first column.

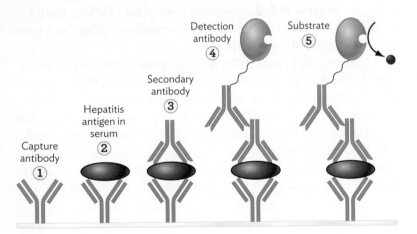

Detection antibody **④**

Substrate **⑤**

Secondary antibody **③**

Hepatitis antigen in serum **②**

Capture antibody **①**

C. Detecting hepatitis antigen in blood. (1) Plate is coated with an anti-hepatitis capture antibody; (2) serum sample is added. Any hepatitis antigen present binds to capture antibody; (3) secondary antibody is added (mouse IgG anti-hepatitis) and binds to antigen; (4) enzyme-linked detection antibody (rabbit anti-mouse IgG) is added and binds to secondary antibody; (5) substrate is added; enzyme converts substrate to colored form.

Indirect ELISA is best used to detect antibody in a patient's serum made in response to an infection. Here a known antigen (say, hepatitis antigen) is bound to wells of a microtiter plate and patient serum is added. The antigen captures hepatitis-specific patient antibody (primary antibody), if present. Other antibodies are washed away. Next, enzyme-conjugated secondary antibody directed against the *isotype* of the captured antibody is added (for instance, enzyme-conjugated anti-human IgG antibody). The amount of color change after adding the enzyme substrate is proportional to how much of the patient's IgG antibody was captured by the antigen. One advantage of this and the following indirect method is that one enzyme-conjugated detection antibody (anti-human IgG) can

be used to detect many different primary human IgG antibodies specific for different antigens.

Sandwich ELISA is an indirect assay that begins by binding what is called a capture antibody directly to the microtiter well. The capture antibody is specific for a given antigen, which could be a pathogen protein or a patient's antibody. Patient sample is added and the antigen of interest is captured. Unlabeled secondary antibody to the antigen is added, followed by enzyme-conjugated antibody to the secondary antibody.

How does the laboratory measure how much antibody is present in a patient's serum? An important measurement used to express the concentration of an antibody or antigen in a solution (serum) is called a **titer**. An antibody titer is a measurement of how much antibody in serum will recognize a particular epitope and how tightly the antibodies bind to that epitope. A titer is expressed as the greatest *dilution* of serum that still gives a positive result in an assay. To illustrate, imagine two bottles of water containing different intensities of blue dye. Bottle A is dark blue, but bottle B is much less blue and clearly contains less dye. You can estimate how much less by serially diluting the water from each bottle into a series of tubes containing clear water. With each dilution the water becomes less and less blue until, eventually, you do not see any blue. Bottle A, which started out the most blue, can be diluted more than bottle B before you no longer see color. If the dilution was carried out as serial twofold dilutions (1:2; 1:4; 1:8, etc.) and the last tube showing color for bottle A was a 1:128 dilution and 1:16 for bottle B, the titers would be the *reciprocal* values, 128 and 16. Bottle A, therefore, contained eight times more dye than bottle B (128/16).

The same basic procedure can be done to determine an antibody titer. The patient's serum is serially diluted twofold (1:2, 1:4; 1:8; 1:16, 1:32, 1:64; 1:128, etc.), and the dilutions are analyzed by ELISA. How much antibody is present in serum is reflected by how far you can dilute the serum and still see a color change. The reciprocal of the highest dilution that still shows a color change is called the antibody titer.

Figure 17.24 illustrates how sandwich ELISA can detect the quantity (titer) of an antibody (for example, IgM; panel A) in serum that binds to a specific antigen (for example, the dengue virus antigen). The procedure is performed in a plastic microtiter dish, a dish with many wells (Figure 17.24B). Capture antibodies against human IgM are first attached to the bottom of a well. The patient's serum is added and the serum IgM is allowed to bind to the anti-IgM stuck to the plastic. Everything not bound is washed off. Next, purified dengue virus antigen is added to the well. Dengue antigen will stick only to the captured anti-dengue IgM from the patient. An enzyme-linked, anti-dengue antibody is then added. It binds to the IgM-captured dengue antigen. After unbound enzyme-linked antibody is washed away, the enzyme substrate is added. A visible color is produced when the enzyme reacts with the substrate. The more anti-dengue IgM in the patient's serum, the more intense the color change.

Indirect ELISA can also detect an antigen in a patient's blood—for instance, hepatitis antigen (Figure 17.24C). Here the wells of a microtiter plate are coated with anti-hepatitis antibody (made from a rabbit, for example). The patient's serum is added, and antibodies in the well will capture any hepatitis antigen present. A second anti-hepatitis antibody (IgG made from a mouse, say) is added and binds to the captured hepatitis antigen (making a "sandwich"). To see how much hepatitis antigen was captured, enzyme-linked anti-mouse IgG is added, followed by the enzyme substrate. The more hepatitis antigen captured, the greater the color change.

One way to diagnose viral and some bacterial infections is to document a large increase in antibody levels (or titer) as a patient transitions from the acute phase of a disease to the convalescent phase. This increase reflects a specific antibody response to the causative agent. ELISA is a convenient way to determine antibody increases. A fourfold or greater increase in a specific titer measured from the acute to convalescent stage of a disease is diagnostic for a given microbial cause (for example, antibody titer against West Nile virus rises from 32 in acute stage to 128 in convalescent).

 ELISA.wmv: youtube.com

Monoclonal Antibodies

Many techniques in diagnostic microbiology require epitope-specific antibodies of a single isotype (for instance, IgG or IgM), known as monoclonal antibodies. As described in Section 17.2, monoclonal antibodies originate from a single antibody-producing B-cell clone. **Polyclonal antibody** is antiserum that contains all antibody isotypes with specificities directed against many different epitopes. A polyclonal mix of antibodies reduces the specificity and sensitivity needed for many immunological tests. So how do we generate the more specific monoclonal antibodies?

Figure 17.25 Production of Monoclonal Antibodies

Immunization of animals with a selected antigen stimulates antibody-forming B cells to produce a range of antibodies with various specificities. Collections of B cells are fused with tumor (myeloma) cells to produce immortalized hybridoma cells.

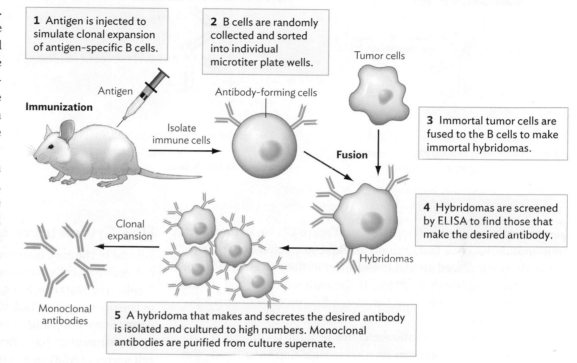

1 Antigen is injected to simulate clonal expansion of antigen-specific B cells.

2 B cells are randomly collected and sorted into individual microtiter plate wells.

3 Immortal tumor cells are fused to the B cells to make immortal hybridomas.

4 Hybridomas are screened by ELISA to find those that make the desired antibody.

5 A hybridoma that makes and secretes the desired antibody is isolated and cultured to high numbers. Monoclonal antibodies are purified from culture supernate.

The process is shown in Figure 17.25. Mice or rabbits are injected with a primary dose of antigen and then, some time later, a booster dose. B cells are then removed from the animal's spleen, sorted into separate wells of microtiter plates (one B cell per well), and stimulated to make antibody. Antibodies from each well can be screened for specificity against the desired epitope by using an ELISA. That is the easy part. The problem is keeping them alive and making monoclonal antibody over many years. Researchers solved the life expectancy problem by fusing the membranes of an antibody-producing B cell with those of myeloma cancer cells that divide indefinitely (these myeloma cells do not themselves make antibody). The resulting immortalized hybrid clone is called a **hybridoma**. Hybridoma cells divide indefinitely and make single-epitope, single-isotype, monoclonal antibodies.

Monoclonal antibodies are used in diagnostic tests to detect antibodies or antigens (see ELISA), treat viral diseases previously deemed untreatable, classify strains of a bacterial species (called typing), and identify cells or molecules in an organism.

Immunofluorescence Microscopy

Immunofluorescence microscopy was first described in Section 3.5 and was used in Figure 17.15B to see the deposition of a complement factor in the kidney of a lupus patient. The technique can also be used to find pathogenic microorganisms in tissue (direct immunofluorescence) or to reveal pathogen-specific antibodies in

Figure 17.26   Immunofluorescence Microscopy

Direct immunofluorescence test

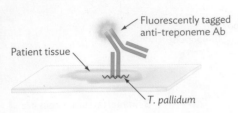

Patient tissue

Fluorescently tagged
anti-treponeme Ab

T. pallidum

Indirect immunofluorescence test

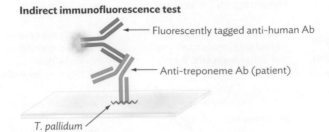

Fluorescently tagged anti-human Ab

Anti-treponeme Ab (patient)

T. pallidum

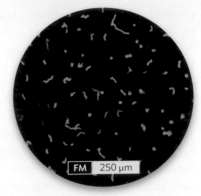

FM   250 μm

A. The direct test uses fluorescently labeled anti-pathogen antibodies (Ab) to directly detect the presence of a pathogen in host tissue.

B. The indirect test uses known pathogens placed on a slide to capture anti-pathogen antibodies in a patient's serum. Then fluorescently tagged anti-human antibodies are added to detect the captured antibodies, if any. (Inset is an indirect fluorescence assay detecting antibodies to *Helicobacter pylori*, a cause of stomach ulcers.)

serum (indirect immunofluorescence) (Figure 17.26). In the **direct immunofluorescence test**, a tissue or fluid suspected of harboring the pathogen is placed on a slide and specific, fluorescently labeled, anti-pathogen antibody is added. If the pathogen is in the tissue, it will fluoresce when viewed through a fluorescence microscope, indicating a positive test.

In the **indirect immunofluorescence test**, the technician places a sample of known pathogen on a glass slide and then floods the slide with patient serum. If the patient was infected and developed an immune response, the blood will contain pathogen-specific antibodies. Those antibodies will bind to the microbe stuck on the slide. Then, a fluorescently tagged, anti-human antibody is applied. If the patient serum had antibodies to the pathogen, the pathogen will now fluoresce. The etiologic agent for syphilis (*Treponema pallidum*) is often detected with the direct test, whereas the indirect test can identify patients with later stages of the disease, when the organism may be difficult to find but antibodies are still present.

Immunoblots (Western Blots)

Another important research technique involving antibodies is the **western blot**, which is used to detect the presence of a specific protein in cell extracts (Figure 17.27). Proteins from an extract are separated by using sodium dodecyl sulfate–polyacrylamide gel electrophoresis (SDS-PAGE). The proteins are transferred (that is, blotted) from the gel onto a nitrocellulose or other membrane. Next the membrane is flooded with a primary antibody that will bind to a specific protein. The membrane is then probed with a secondary, enzyme-linked antibody that specifically binds to the primary antibody (for example, anti-IgG antibody). Adding enzyme substrate sets off a luminescent reaction wherever an antibody sandwich has assembled. The light emitted can then be detected on X-ray film. Only protein bands to which the primary antibody has bound will be detected. The technique is used to determine whether, and at what levels, specific proteins are present in a cell extract.

Flow Cytometry versus FACS

Several of the case histories told you that a patient had low levels of B cells or T cells and sometimes even deficits in a specific class of T cells. For instance, Silas, the newborn with DiGeorge syndrome in the opening case, lacked T cells. How are these measurements done? Separating and counting cells in a patient's blood requires an instrument called a **flow cytometer** or **fluorescence-activated cell sorter (FACS)** in a process known as **flow cytometry** (Figure 17.28). The technique is possible because different types of cells are decorated with unique cell surface antigens. Thus, you can "tag" a specific cell type with a fluorescent monoclonal antibody that binds only to the unique antigen (for example, anti-CD8 for cytotoxic T cells).

The fluorescent antibody is added to blood, and after unbound antibody is removed, the blood cells are injected into the flow cytometer. The blood passes through a tiny orifice that delivers the cells single file past a laser (Figure 17.29A). The laser activates the fluorophore in the fluorescent antibody bound to the cell, and a detector measures the intensity of the emitted light. Multiple detectors can be used to measure fluorescence from different fluorophores. The results are graphically displayed in several ways. Figure 17.29B shows one approach and illustrates how the technique can detect aberrant T cells resulting from cancer. FACS can not only count the different cell types but also sort them into separate tubes for later analysis (Figure 17.29C). To separate cells, the standard flow cytometer will impart a charge on a passing cell, and deflecting plates move positively and negatively charged drops in opposite directions. A cell that fluoresces green (for example, a CD4 T cell tagged with anti-CD4 antibody) can be given a positive charge and a different cell type tagged with an antibody that emits red light can be given a negative charge. As the cells continue to move through the machine, the deflector plates pull the charged cells in opposite directions and the cells are routed to separate collection vessels. The sorted cells can then be used for other tests.

Figure 17.27 Western Blot Analysis

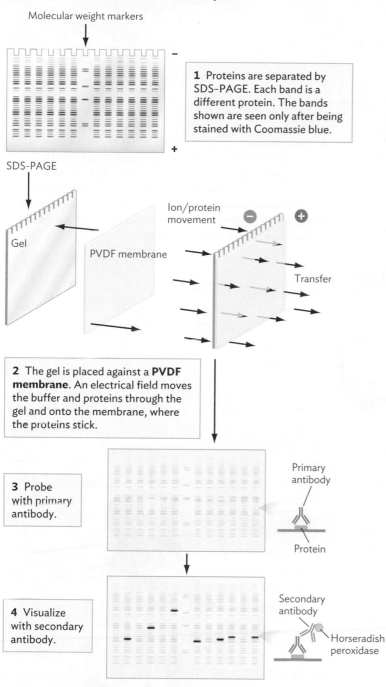

Molecular weight markers

1 Proteins are separated by SDS-PAGE. Each band is a different protein. The bands shown are seen only after being stained with Coomassie blue.

SDS-PAGE

Gel

PVDF membrane

Ion/protein movement

Transfer

2 The gel is placed against a **PVDF membrane**. An electrical field moves the buffer and proteins through the gel and onto the membrane, where the proteins stick.

3 Probe with primary antibody.

Primary antibody

Protein

4 Visualize with secondary antibody.

Secondary antibody

Horseradish peroxidase

A. Western blotting can identify specific proteins within a mixture of many proteins. Proteins in different sample extracts are applied to separate wells of a polyacrylamide gel (top of the gel). The proteins are separated by size after applying an electric field through the gel (smaller proteins travel farther down the gel). Separated proteins are transferred (blotted) onto a polyvinylidene difluoride (PVDF) membrane, which is then probed with a primary antibody against the protein of interest. An enzyme-linked secondary antibody against the primary antibody is used to find where on the gel the primary antibody bound. Adding enzyme substrate causes a color change in those bands.

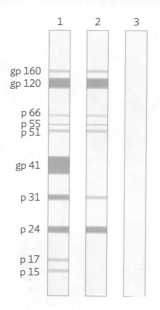

1 2 3

gp 160
gp 120

p 66
p 55
p 51

gp 41

p 31

p 24

p 17
p 15

B. Western blot to detect HIV proteins in patient sera. In this example, several anti-HIV antibodies are combined to detect multiple proteins. Lane 1 is a positive control showing many HIV proteins; lane 2 is an intermediate positive showing the presence of some HIV proteins; lane 3 is a negative control.

Figure 17.28 Fluorescence-Activated Cell Sorter

Dr. Robert Barrington (left) and Bryant Hanks at the University of South Alabama use a FACS machine to research systemic lupus erythematosus. Arrow shows where samples are injected into the machine.

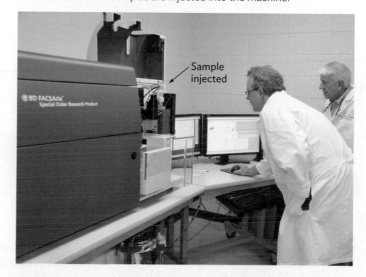

Sample injected

Figure 17.29   Using Flow Cytometry to Count and Sort CD4 and CD8 T Cells

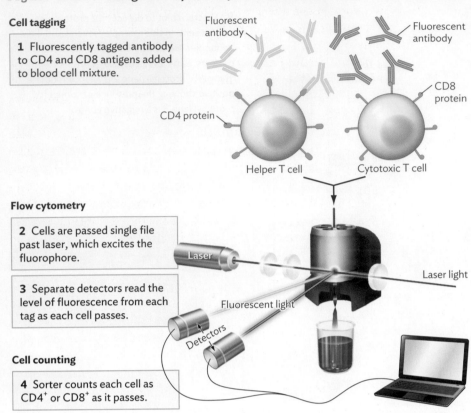

Cell tagging

1 Fluorescently tagged antibody to CD4 and CD8 antigens added to blood cell mixture.

Fluorescent antibody

Fluorescent antibody

CD8 protein

CD4 protein

Helper T cell   Cytotoxic T cell

A. Anti-CD4 and anti-CD8 antibodies tagged with different fluorescent dyes (PerCP and fluorescein isothiocyanate [FITC] in this example) are added to the mixture of cells. Each dye will fluoresce a different color. The mixture of cells is passed, single file, past the cell sorter's laser. Separate detectors read the level of fluorescence from the two dyes as each cell passes. In doing so, the machine counts the CD4 and CD8 T cells.

Flow cytometry

2 Cells are passed single file past laser, which excites the fluorophore.

3 Separate detectors read the level of fluorescence from each tag as each cell passes.

Laser

Laser light

Fluorescent light

Detectors

Cell counting

4 Sorter counts each cell as CD4⁺ or CD8⁺ as it passes.

A negative sign (−) indicates the absence of CD protein; a positive sign (+) indicates the presence of a CD protein.

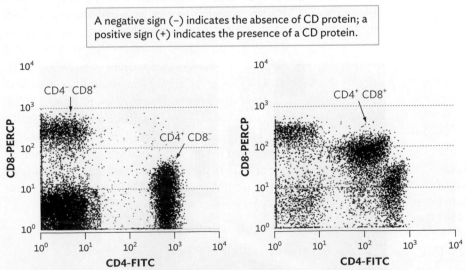

B. Results of CD4 (x-axis)/CD8 (y-axis) counts from two patients. The left-hand plot shows a normal situation. Lower right quadrant cells express CD4 but not CD8 (the helper T-cell set), whereas upper left quadrant cells express CD8 but not CD4 (cytotoxic T cells). Lower left quadrant cells express neither antigen. In the right-hand plot you can see the presence of cells that express both the CD4 and the CD8 antigen (arrow, upper right quadrant), which is highly irregular. Cells with abnormal antigen expression may indicate a T-cell malignancy.

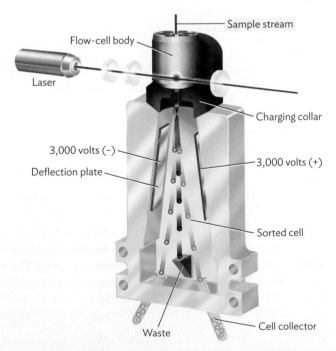

C. Schematic of a FACS apparatus (bidirectional sorting). A tiny positive or negative charge is added to each cell as it passes the detector. Deflector panels will then move (sort) the charged cells into separate collection vessels for further study.

SECTION SUMMARY

- **Immunological tools** can detect antigens or antibodies (which are themselves antigens) in a patient's serum or tissues.

- **Radial immunodiffusion** measures the concentration of an antibody or soluble antigen in a serum sample. Molecules from a serum sample radially diffuse outward from a well cut into agarose embedded, respectively, with either antigen or antibody. Molecule concentration in a well is proportional to the diameter of the immunoprecipitation ring that forms around the well.

- **Agglutination** is an antigen-antibody reaction that detects insoluble antigens on whole cells. Antibody binding will link cells, causing them to aggregate.

- **Enzyme-linked immunosorbent assays (ELISAs)** use enzyme-linked antibodies to detect specific antibodies or antigens in a patient's serum. The serum target molecule is first captured by matched antigen or antibody attached to wells of a microtiter dish. Next, enzyme-conjugated secondary antibody is added to the well followed by substrate for the enzyme.

- **Monoclonal antibodies** are secreted by immortalized hybridomas made by fusing a B cell with a myeloma cell. Hybridomas produce single-isotype antibodies with single-epitope specificity. Monoclonal antibodies are used for ELISA, immunofluorescence microscopy, or immunoblotting.

- **Immunofluorescence microscopy** uses fluorescently tagged monoclonal antibodies to identify antigens in blood or tissue samples.

- **Immunoblotting** uses monoclonal or polyclonal antibodies to detect specific proteins in a sample after polyacrylamide gel electrophoresis.

- **Flow cytometry** uses fluorescently tagged monoclonal antibodies to count specific cell types in a mixed sample. Fluorescence-activated cell sorting can then separate and collect the different cell types.

17.6
Vaccines and Immunization

SECTION OBJECTIVES

- Distinguish active from passive immunization.
- List key features of an effective vaccine.
- Explain the different types of vaccines.
- Discuss the concept of herd immunity.

Have you ever wondered what vaccines are and how they work? Their origin can be traced back more than 1,000 years ago to China and India (Section 1.3). The early attempts were crude and fraught with peril. Healthy people were intentionally infected with material from a diseased person in an attempt to protect them from later disease—for example, material from a smallpox pustule containing pathogenic variola virus. Unfortunately, about 1 of every 100 deliberately infected patients died from smallpox. Things improved 700 years later, when the less virulent but closely related vaccinia virus (cowpox) was used. The process was thereafter called vaccination or, now, immunization. The material used to vaccinate someone is called a **vaccine**.

The type of immunization just described is called **active immunization**. Injecting an antigen such as a weakened or killed pathogen stimulates the immune system to actively make antibody and antigen-specific T cells. The resulting cellular and humoral responses then retain a memory of the antigen. Another type of immunization, termed **passive immunization**, is really a form of immunotherapy. Examples of passive immunization include the injection of immune globulin or hyperimmune globulin to treat the infections of immunodeficient patients (Section 17.1). **Antitoxins** (antibodies that inactivate microbial toxins) are another form of passive immunization. Antitoxin sera are usually obtained from animals, such as horses. Preparations of horse sera effective against diphtheria, tetanus, and botulism toxins are available. However, injecting horse sera into humans can cause serious allergic reactions such as serum sickness or anaphylaxis (Section 17.3). Breast-feeding is another form of passive immunization because mother's milk contains IgA antibodies that can protect her child from ingested pathogens.

Link You will learn in Section 18.3 that microbial toxins are proteins (exotoxins) secreted from pathogenic bacteria and some fungi. After secretion, most exotoxins enter eukaryotic target cells and interfere with the function of important host cell molecular pathways. The result is disease and sometimes death.

Requirements of an Effective Vaccine

What makes a good vaccine? The best vaccine in terms of long-term protection is a natural infection—if you survive, that is. In a natural infection the microbe engages all aspects of the immune system. However, an effective commercial vaccine should not harm the person being vaccinated. It should stimulate B-cell (antibody) and T-cell (cell mediated) responses and produce long-term memory (that is, it should have a lasting effect). Vaccination should not require many boosters to achieve protection, and it must protect against the natural pathogen. (The reason for booster doses is that secondary exposure to an antigen provides a more robust and long-lasting immunity.) The most useful vaccines should be inexpensive, be easy to administer, and have a long shelf life. These last points are especially true for vaccines used in developing countries, where storage and access to the population are serious challenges.

What form should a vaccine take? This depends on the pathogen and which of its antigens generate a protective immune response. Even though every antigen of a pathogen can produce an immune response, most do not produce a *protective* immune response that will prevent disease. Vaccines come in four basic forms:

1. Killed whole cells or inactivated viruses

2. Live, attenuated bacteria or viruses

3. Antigen molecules purified from the pathogen (acellular or toxoid)

4. DNA vaccines (not yet commercially available)

Table 17.5 lists many vaccines currently used. It notes their form, route of delivery (injection, oral, etc.), and when they are used.

Table 17.5
Approved Vaccines

Vaccine Type/Disease	Route of Delivery	Routine or Targeted Use	Comments on Use
Whole-Cell, Killed Bacteria			
Cholera	Subcutaneous (SubQ)	Travel to endemic areas	Short-term protection
Plague	SubQ	No longer available	
Live, Attenuated Bacteria			
Tuberculosis (BCG)	Intradermal	Only high-risk groups in U.S., routine in many other countries	
Typhoid fever	Oral	Travel to endemic areas	Low effectiveness
Acellular Vaccines, Bacteria			
Anthrax (protective antigen; PA)	Intramuscular (IM)	Only high-risk groups	
Meningococcal meningitis (capsule)	SubQ	Routine, children	
Haemophilus influenzae type b meningitis (capsule)	IM	Routine in children	
Pneumococcal pneumonia (capsule)	IM or SubQ	Routine in young and elderly	
Pertussis (aP; inactive pertussis toxin, filamentous hemagglutinin, pertactin)	IM	Routine, children	Given in a mixture with tetanus and diphtheria toxoids (DTaP vaccine)[a]
Toxoids			
Diphtheria	IM	Routine, children	All three given in combination with *B. pertussis* pertactin and filamentous hemagglutinin
Tetanus	IM	Routine, children	
Pertussis	IM	Routine, children	
Botulism toxin	IM	Only highly exposed individuals	
Viruses, Inactivated			
Hepatitis A	IM	Routine, children	
Influenza	IM	Routine, seasonal	Recombinant vaccines are under development
Japanese encephalitis	SubQ	Travel to endemic areas	
Polio (Salk)	IM	Routine, children	
Rabies	IM	Victim of animal bite	Killed virus and recombinant vaccine are available

(continued on next page)

Table 17.6 shows the schedule, recommended by the Centers for Disease Control and Prevention, for administering vaccinations, including boosters to children.

Killed whole-cell or inactivated-virus vaccines (cholera and rabies, for example) use microbes inactivated by formalin or phenol. For these vaccines to work, the organism must be dead/inactivated, but the antigenicity of the protective antigens must remain intact.

As effective as killed vaccines are, vaccines made from living microbes are usually better. Live, "crippled" (or attenuated) versions of a pathogen can arise when the organism is repeatedly grown on artificial growth media for long periods. The attenuated strains arise because mutations develop in pathogenicity genes required for growth in the host only (the "use it or lose it" principle). Attenuated strains make better vaccines than killed microbes because

Table 17.5
Approved Vaccines (*continued*)

Vaccine Type/Disease	Route of Delivery	Routine or Targeted Use	Comments on Use
Viruses, Attenuated			
Adenovirus	Oral	Military	
Chickenpox/shingles (varicella/herpes zoster)	SubQ	Routine, children/adults	In children, given at the same time as the measles, mumps, rubella vaccine (MMRV); adults receive different dose
Influenza	Intranasal	Routine, children	
Measles (rubeola)	SubQ	Routine, children	Given as part of MMRV[b]
Mumps	SubQ	Routine, children	Given as part of MMRV
Poliovirus (Sabin)	Oral		Not currently recommended in U.S.
Rotavirus	Oral	Routine, children	
Rubella	SubQ	Routine, children	Given as part of MMRV
Smallpox (vaccinia)	Punctures	Voluntary for high risk; some military	
Yellow fever	Inhaled	Travel to endemic area	
Virus Subunit			
Hepatitis B (HepB surface antigen; HBsAg)	IM	Routine, at birth	
Influenza			Under development
Human papillomavirus (Gardasil)	IM	Indicated for males and females aged 9–26	Capsid proteins from types 6, 11, 16, 18

[a]DTaP = Diphtheria, tetanus, and acellular pertussis multivalent vaccine.
[b]MMRV = Measles, mumps, and rubella multivalent vaccine.

the act of replicating in a host, even a little, leads to a more robust immune response. The response is better because replication generates increased antigen load and stimulates immune responses more targeted to the natural pathogen (for example, mucosal immunity is different from circulating immunity). A word of caution: live, attenuated vaccines are very dangerous for immunocompromised patients (see earlier discussion of immunodeficiencies).

Toxoid vaccines are made by inactivating the toxic activity of a microbial toxin without compromising its antigenicity. These vaccines do not directly eliminate a pathogen; they inactivate a toxin made by the pathogen. Of course, without the damage caused by the toxin, host defenses may well destroy the pathogen. As with most vaccines, toxoid vaccines usually require several booster shots to achieve a protective level of immunity, which eventually wanes. For example, the tetanus and diphtheria vaccines given during childhood should be readministered every 10 years to maintain effective immunity. Because many older adults have not received boosters, they are prone to these diseases.

Subunit vaccines use only fragments of a microorganism or toxin. Subunit vaccines can be made by purifying the antigen (for instance, a toxin subunit) directly from the pathogen, or the genes encoding these subunits can be introduced into another microbe that then produces large amounts of the desired antigen (**recombinant subunit vaccine**). For example, the viral capsid antigens that make up Gardasil, one of the vaccines given to prevent cervical cancer by human papillomavirus, are made by a genetically modified yeast.

Conjugated vaccines physically link a highly immunogenic protein (such as diphtheria toxoid) with a poorly immunogenic capsule polysaccharide (as from *Haemophilus influenzae* type b) to boost the immune reaction to the polysaccharide. Why is this necessary? Polysaccharides can bypass T cells and directly activate B cells; therefore, they are called T-cell-independent antigens. But activation is not complete without T-cell help. Conjugated vaccines are made to solicit T-cell help. Similar to what Figure 17.14 showed, the B cell takes up the T-cell-dependent diphtheria part of a conjugated vaccine along

Table 17.6
Recommended Immunization Schedule for Children and Adolescents[a]

Vaccine Age →	Birth	1 Month	2 Months	4 Months	6 Months	9 Months	12 Months	15–18 Months	24 Months	4–6 Years	11–12 Years
Hepatitis B[b]	HepB	HepB			HepB					HepB series	
Rotavirus			Rota	Rota	Rota						
Diphtheria, tetanus, acellular pertussis[c]			DTaP	DTaP	DTaP			DTaP		DTaP	Tdap
Haemophilus influenzae type b[d]			Hib	Hib	Hib[c]		Hib				
Inactivated poliovirus			IPV	IPV	IPV					IPV	
Measles, mumps, rubella							MMR			MMR	MMR
Varicella							Varicella			Varicella	
Meningococcal[e]							Administer to high-risk children				MCV4
Pneumococcal[f]			PCV13	PCV13	PCV13		PCV13		PPSV23		
Influenza[g]					Influenza (yearly)						
Hepatitis A[h]							HepA series			Hep A	
Human papillomavirus[i]											HPV

[a]This schedule indicates the recommended ages for routine administration of currently licensed childhood vaccines, as of January 1, 2013.

▬ Range of recommended ages   ▬ Catch-up immunization   ▬ Assessment at age 11–12 years

[b]Hepatitis B vaccine (HepB). *At birth:* All newborns should receive monovalent HepB, administered soon after birth and before hospital discharge.

[c]Diphtheria and tetanus toxoids and acellular pertussis vaccine (DTaP). Tdap is a modified vaccine with lower doses of diphtheria and pertussis toxoids.

[d]*Haemophilus influenzae* type b conjugate vaccine (Hib).

[e]Meningococcal conjugate vaccine (MCV4). MCV4 should be administered to all children at age 11–12 years, as well as to unvaccinated adolescents at high school entry (age 15 years). Vaccine contains four types of capsules.

[f]Pneumococcal vaccine. The 13-valent pneumococcal conjugate vaccine (PCV) is recommended for all children aged 2–23 months and for certain children aged 24–59 months. The final dose in the series should be administered at age ≥12 months. Pneumococcal polysaccharide vaccine (PPSV) is a 23-valent vaccine recommended in addition to PCV for certain high-risk groups. PPSV is also recommended for people over 65.

[g]Inactivated influenza vaccine should be administered annually starting at age 6 months. Live, attenuated influenza vaccine should not be given until 2 years, and not to immunocompromised individuals.

[h]Hepatitis A vaccine (HepA). HepA is recommended for all children at age 1 year (12–23 months).

[i]Human papillomavirus (HPV) vaccine. (HPV4 [Gardasil] and HPV2 [Cervarix]). Administer a 3-dose series of HPV vaccine on a schedule of 0, 1–2, and 6 months to all adolescents aged 11–12 years. Either HPV4 or HPV2 may be used for females, and only HPV4 may be used for males.

with the polysaccharide part of the vaccine when the latter binds to the B-cell receptor. Diphtheria protein epitopes are then processed and presented on a B-cell MHC receptor. Helper T cells that recognize the diphtheria epitope can then offer help and fully activate the B cell, even though the B cell is programmed to make anti-capsule antibody. This type of vaccine is particularly important for producing effective immunity to capsular antigens in children.

DNA vaccines represent a new, emerging technology. Instead of injecting an antigen into a person, we will eventually inject DNA encoding that antigen directly into muscle (**Figure 17.30**). In this

Figure 17.30
Bio-Rad's Gene Gun

A blast of helium fires tiny gold pellets carrying harmless snippets of a pathogen's DNA into the skin. The DNA infiltrates the nucleus of the skin and muscle cells. The DNA forces the cell to produce pathogen proteins, triggering the immune system. Although a promising technology, no commercial DNA vaccines are yet available.

technique, currently limited to animal testing, the muscle cell makes the microbial protein from the foreign DNA, and a healthy immune response ensues. Why is this approach better than simply injecting the protein directly? Although the direct approach works for many vaccines, active synthesis of the protein within a host cell may be the best way to generate a potent immune response. The reason is that a small amount of any foreign protein synthesized in a host cell will be degraded and the pieces presented on the surface of that cell or are picked up by APCs and presented to helper T cells. This process is akin to what occurs during natural infection (discussed in Chapter 16). By contrast, proteins simply injected into a person are not processed the same way and may not generate a response as complete or long-lasting. We should stress that DNA vaccines are still experimental, although their future looks promising.

Administering Vaccines

Vaccines can be administered by intramuscular injection, subcutaneous injection, oral ingestion, inhalation, or gene gun. Choosing the proper route of administration is important for getting the appropriate response for the pathogen in question. For instance, do you need primarily a circulating antibody response or a mucosal response? Oral vaccines will generate a greater gut mucosa immune response than a circulatory antibody response, for instance.

Multiple vaccines can be given simultaneously and safely. Most are administered during childhood, when the diseases can be the most devastating (Table 17.6). Almost all these vaccines are given in multiple doses, called booster doses, to provide longer-lasting immunity. The exception is the influenza vaccine, which is given in a single dose but changes every year. Most vaccines are not administered until at least 2 months of age, because maternal antibody crossing through the placenta to the fetus (or to a newborn through breast milk) will persist for a short time in the newborn, temporarily protecting the baby from disease and possibly dampening the response to a vaccine antigen administered during that time.

Herd Immunity

Must everyone in a community be vaccinated to be protected? Not really. In practice it is impossible to immunize all humans against any given disease, even in developed countries such as the United States. But generally the infection rate for a given infection after an aggressive vaccination program is much lower than you would expect from the number of people vaccinated. For instance, early in the last century, measles (caused by the rubeola virus) was a common and potentially deadly disease. In 1954, before measles vaccination, about 4,000 cases of measles occurred for every 1 million unprotected people. Today, with 90% of the U.S. population being vaccinated, we still have 31 million *unvaccinated*, and thus vulnerable, people. So you might still expect to see 4,000 cases of measles for every 1 million *unprotected* individuals. Surprisingly, fewer than one case occurs per million unvaccinated people, and most of those

originated with someone returning from an endemic country. So, how are the millions of unvaccinated people protected?

The answer is, in large part, due to herd immunity. **Herd immunity** occurs when a large portion of a population is immunized (**Figure 17.31**). Because the vaccinated individuals in the "herd," or community, will not become infected, they cannot spread the disease to others. At the very least, this slows progression of the disease

Figure 17.31   Herd Immunity

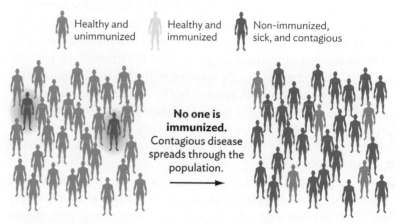

A. Two infected and contagious people can rapidly spread disease throughout an uninfected population.

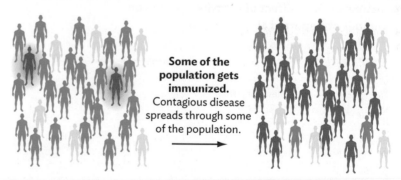

B. Immunizing only a few healthy people in a population does not slow the spread of disease among the unvaccinated (greater than 90% in the example still get ill), but doing so will protect those who were vaccinated.

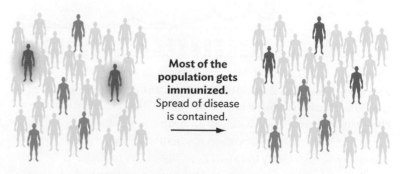

C. Vaccinating a large fraction of the population, however, will prevent most of those unvaccinated (75% in the example) from encountering sick and contagious individuals.

throughout the populace. It is estimated that immunizing three-fourths or more of a population can greatly reduce the incidence of the disease by cutting off transmission. Be aware, though, that the level of coverage needed for a given pathogen varies with the disease involved. Herd immunity works only for diseases that are contagious. Herd immunity is not possible for infections, like tetanus, that are not transmitted directly between humans.

Opposition to Vaccines

The data are incontrovertible. Vaccines have literally saved millions of people from certain death. Study after study shows that the incidence of illness and death by a given pathogen dramatically declines once an effective vaccine program is under way. Take measles, in the example above. Figure 17.32 illustrates how precipitously the numbers of people infected with measles dropped after the advent of the measles vaccine in the early 1960s. Widespread use of measles vaccine has led to a greater than 99.99% reduction in measles cases in the United States compared with the prevaccine era, and in 2012, only 55 cases of measles were reported in the United States.

Nearly all people who receive vaccines suffer no, or only mild, reactions, such as fever or soreness at the injection site. Very rarely do more serious side effects, such as allergic reactions, occur.

Vaccines are *extremely* safe. However, unsettling, erroneous information is circulating on the Internet purporting a link between vaccinations and other diseases, such as diabetes or autism. No well-controlled scientific study supports these claims. Clearly, the risks associated with contracting preventable infectious diseases far outweigh the minimal risk associated with being vaccinated against them. As proof of this point, the undervaccination of children in the United States starting in the 1980s and 1990s has led to an increase in cases of whooping cough caused by *Bordetella pertussis* (5,000 in 1990; 25,616 in 2005; 48,277 in 2012). Whooping cough will be described in Chapter 20. This example shows that failing to vaccinate produces a human population highly susceptible to these preventable infections and an inevitable resurgence of serious diseases with often deadly outcomes.

Note Live, attenuated vaccines, such as the one for measles, should not be given to immunocompromised patients, or to a person who lives with an immunocompromised patient. A competent immune system is needed to stop growth of the live, albeit crippled, microbe. Even a crippled virus can cause disease in patients having an immune deficiency. Even so, many immunocompromised patients can still receive and respond to subunit vaccines, depending on the nature of the immune deficiency.

 List of available vaccines: fda.gov; cdc.gov
CDC vaccine schedule: cdc.gov

Figure 17.32   Effect of Measles Vaccine on Number of Reported Measles Cases

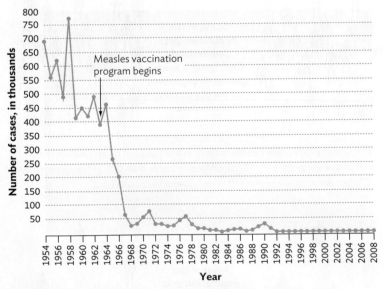

Widespread measles vaccination began in 1963. The result was a greater than 99% reduction in measles cases in the United States compared with the prevaccine era.

SECTION SUMMARY

- **Active vaccination** is the process of introducing an antigen into a person to activate an immune response. If the antigen is a pathogen, or part of a pathogen, the individual will be protected from future infection.

- **Passive vaccination or immunotherapy** involves injecting protective antibodies into an individual. **Antitoxins** are a form of passive immunization.

- **Vaccine types include** killed or inactivated pathogens, live attenuated pathogens, inactivated toxins (toxoids), fragments of a pathogen (subunit vaccines), conjugated vaccines, and DNA vaccines.

- **Vaccines can be administered** orally, intranasally, by injection, or by DNA gun.

- **Herd immunity** is when enough people in a population become vaccinated to a communicable disease, so the transmission of that disease to an unvaccinated individual becomes less likely.

Thought Question 17.8   Explain why the immune response to a ribosomal protein of *Streptococcus pneumoniae* is not protective, but an immune response to its capsule is protective. That is, why isn't *S. pneumoniae* ribosomal protein a good vaccine?

Perspective

As complex and as highly regulated as our immune system is, it should be no surprise that things can go wrong. Hypersensitivities, immunodeficiencies, cancer, autoimmune attacks—all these dysfunctions arise as a consequence of our immune system having so many moving parts. And yet, despite all its complexity and unintended consequences, our very existence depends on the immune system remaining vigilant 24 hours a day, 7 days a week, for 80–100 years. When it fails, we die. What other machine can stand that level of pressure?

This chapter concludes our unit on the immune system, but it does not end our discussion of immunity. Upcoming chapters describe the many infectious agents that challenge our immune system and, often, turn it against us to inflict damage (immunopathology). As we describe the conflicts between host and pathogen, continue to think about which parts of the immune system are engaged and when. Recall, too, the immunological tools that can measure which of us, human or microbe, is winning.

LEARNING OUTCOMES AND ASSESSMENT

17.1
Primary Immunodeficiency Diseases

- Distinguish primary and secondary immunodeficiencies.
- Describe the common forms of severe combined immunodeficiencies.
- Compare the underlying defects and outcomes of bare lymphocyte syndrome, X-linked agammaglobulinemia, hyper IgM syndrome, selective IgA deficiency, and complement deficiencies.

1. Human immunodeficiency virus causes _____.
   A. a primary immunodeficiency
   B. a secondary immunodeficiency
   C. a tertiary immunodeficiency
   D. agammaglobulinemia
   E. IgA deficiency

2. The most common form of severe combined immunodeficiency is due to _____.
   A. adenosine deaminase deficiency
   B. diminished production of IL-2
   C. decreased surface CD154 (CD40 ligand)
   D. lack of IL-2 receptor
   E. tyrosine kinase deficiency

17.2
Immune Surveillance and the Cancers of Lymphoid Tissues

- Discuss the relationship between our immune system and cancer.
- Distinguish lymphoma, leukemia, plasmocytoma, and multiple myeloma.
- Explain the defects leading to B-cell neoplasms.

5. A solid mass in a lymphoid organ is called a _____.
   A. multiple myeloma
   B. plasmocytoma
   C. leukemia
   D. lymphoma
   E. All of the above
   F. None of the above

17.3
Hypersensitivity

- Explain what an allergen is.
- Describe the development, pharmacology, and treatment of type I hypersensitivity.
- Differentiate type II and type III hypersensitivities.
- Discuss type IV hypersensitivity in terms of what cells are involved and how it can be used as a diagnostic tool.

9. Which of the following hypersensitivity reactions is antibody-independent?
   A. Type I
   B. Type II
   C. Type III
   D. Type IV

10. Epinephrine stops anaphylaxis by which of the following means?
    A. Increasing cAMP production to stop mast cell degranulation.
    B. Decreasing cAMP production to stop mast cell degranulation.
    C. Relaxing smooth muscle around bronchioles.
    D. Contracting smooth muscle around bronchioles.
    E. A and C only
    F. B and C only
    G. C and D only

3. A 1-month-old child with a low white blood cell count, low-set ears, cleft palate, and an underdeveloped thymus will most likely have which one of the following?
   A. IgA deficiency
   B. X-linked agammaglobulinemia
   C. DiGeorge syndrome
   D. Bare lymphocyte syndrome
   E. Hyper IgM syndrome

4. Treatment of patients with X-linked agammaglobulinemia includes which one of the following?
   A. IV administration of immune globulin
   B. Thymus transplant
   C. T-cell transfusion
   D. B-cell transfusion
   E. Complement transfusion

6. The cell shown to the right is diagnostic for _____.
   A. follicular lymphoma
   B. Burkitt's lymphoma
   C. chronic lymphocytic leukemia
   D. plasmocytoma
   E. Hodgkin's lymphoma

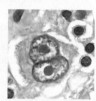

7. Infection with Epstein-Barr virus is linked to _____.
   A. follicular lymphoma
   B. non-Hodgkin's lymphoma
   C. chronic lymphocytic leukemia
   D. plasmocytoma
   E. Burkitt's lymphoma

8. The process by which lymphocytes scan for cancerous cell growth is called _____.

This figure illustrates the initiation of an allergic reaction to a bee sting. Label the items marked.

11.

12.

13.

14.

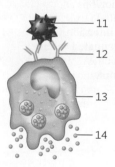

Matching: Match the following lettered responses to the numbered questions below. Each question has two correct answers.

A. Type I hypersensitivity	E. T_H1 cells
B. Type II hypersensitivity	F. Immune complexes
C. Type III hypersensitivity	G. IgE
D. Type IV sensitivity	H. Cell lysis

15. Patient injected with antiserum made from horse blood; 12 days later, patient develops fever, urticarial rash, and joint pains.

16. Mouse A has an allergy to a chemical placed on its skin; mouse B does not. However, when mouse B is transfused with white blood cells from mouse A, mouse B now becomes allergic to the same chemical.

17. A 5-year-old boy with type A blood was in a serious car accident and required a transfusion. Mistakenly he received type AB blood. Within hours he was experiencing chills, his blood pressure dropped, and his urine was tinged red.

LEARNING OUTCOMES AND ASSESSMENT

	SECTION OBJECTIVES	OBJECTIVES REVIEW

17.4 Autoimmunity and Organ Transplantation	• Recall how the immune system differentiates self from nonself (foreign) antigens.
	• Discuss how autoreactive antibodies can be formed.
	• Describe the key features of the autoimmune diseases lupus, Graves' disease, diabetes, rheumatoid arthritis, and myasthenia gravis.
	• Explain the basis of transplant rejection.
		18. Autoantibodies are formed when _____.
		  A. molecular mimicry enables a mismatched T_H cell to activate a B cell that escaped deletion
		  B. T and B cells that are autoreactive to the same antigen escape negative selection
		  C. a B cell undergoes isotype switching
		  D. a superantigen activates a T cell and an APC

17.5 Tools of Immunology	• Describe how immunoprecipitation assays are used to measure immunoglobulin levels.
	• Explain agglutination and its use in blood typing and pathogen detection.
	• Distinguish direct, indirect, and sandwich enzyme-linked immunosorbent assays (ELISAs).
	• Discuss how monoclonal antibodies are made and their use in fluorescent antibody techniques (immunofluorescence microscopy, western blots, and flow cytometry).
		The radial immunoprecipitation gels shown here are impregnated with different anti-immunoglobulin antibodies as indicated. Sera from a control healthy individual and three patients were placed in the wells. Match the immunoprecipitation pattern for each patient with the following immunodeficiency diseases.
		
		Choices: complement deficiency; hyper IgM deficiency; X-linked agammaglobulinemia; selective IgA deficiency
		
		  21. Patient A _____
		  22. Patient B _____
		  23. Patient C _____

17.6 Vaccines and Immunization	• Distinguish active from passive immunization.
	• List key features of an effective vaccine.
	• Explain the different types of vaccines.
	• Discuss the concept of herd immunity.
		26. Which of the following is an example of passive immunization?
		  A. Inactivated virus
		  B. Natural infection
		  C. Live, attenuated vaccine
		  D. DNA vaccine
		  E. Breast milk

19. The pathophysiology of Graves' disease involves _____.
    A. antibody that targets acetylcholine receptor
    B. cytotoxic T cells that destroy pancreas beta islet cells
    C. cytotoxic T cells that destroy thyroid follicle cells
    D. antibody that targets myelin protein
    E. antibody that targets thyroid-stimulating hormone receptor

20. Why can type A red blood cells be transplanted (transfused) between allogeneic individuals?
    A. Absence of MHC type II complexes on red blood cells.
    B. Absence of MHC type I complexes on red blood cells.
    C. Absence of a nucleus in red blood cells.
    D. Type A antibodies in the recipient do not recognize the donor's red cells.
    E. The recipient's T cells recognize the donor's blood cells as self.

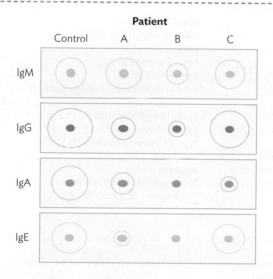

24. An ELISA was performed on the sera of three patients to detect antibody against hepatitis B virus. The titer results are shown below. Which patient's serum had the most anti–hepatitis B antibody?
    A. Patient 1   1:30
    B. Patient 2   1:1,350
    C. Patient 3   1:640
    D. Not enough information

25. Which of the following is the best method for counting the numbers of different populations of T cells?
    A. Western blot
    B. Radial immunoprecipitation
    C. Immunofluorescent microscopy
    D. Flow cytometry
    E. ELISA

27. The best protection against a second infection by a pathogen is which of the following?
    A. Inactivated-virus vaccine
    B. IV immunoglobulin
    C. Subunit vaccine
    D. Live, attenuated vaccine
    E. Natural infection

28. The concept whereby unvaccinated persons are protected from infection by those vaccinated in the community is called _____.

Key Terms

Review Questions

1. Name B-cell and T-cell immunodeficiencies and list key features of each.
2. What are the differences between a lymphoma, leukemia, plasmocytoma, and multiple myeloma?
3. How are the genes *c-myc* and *bcl-2* involved in lymphocytic cancers?
4. What are Reed-Sternberg cells?
5. What is immune surveillance?
6. Describe the differences between hypersensitivity types I, II, III, and IV.
7. How does a B cell programmed to make an antibody against self become activated in the absence of *specific* T-cell help?
8. How does atopy differ from anaphylaxis?
9. How is IgE involved in allergic hypersensitivity?
10. Describe autoimmune diseases that are based on type II, type III, and type IV hypersensitivities.
11. Discuss the differences between transplant rejections of nucleated cells and rejection of blood cells.
12. What is immunoprecipitation and radial immunodiffusion?
13. How is an antibody titer determined?
14. How can immunofluorescence help diagnose an infectious disease?
15. What are monoclonal antibodies?
16. How can the number of CD4 and CD8 T cells be counted in blood?
17. What are the features of an effective antibody?
18. Describe the different types of vaccines.

Clinical Correlation Questions

1. An Rh-negative mother is pregnant with an Rh-positive fetus. This is her second Rh-positive child. The fetus, however, is showing signs of distress because its red blood cells are lysing. You want to transfuse the fetus through its umbilical cord with new red blood cells to replace those lost. What cells should you use (Rh-positive or Rh-negative) and why?

2. A 6-month-old girl has suffered with recurrent bacterial infections almost since birth. Her white blood cell count when she was healthy was 85,000/µl (normal: 5,000–10,800/µl) with 70% neutrophils. Flow cytometric analysis was performed on her peripheral blood leukocytes. Cell markers Fcγ (Fc receptor, a neutrophil surface marker) and CD18 (an integrin) were measured. Results from a healthy control and the patient are shown at the right:

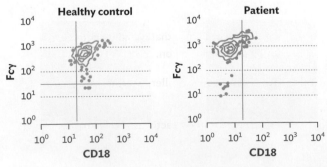

Which immunodeficiency disease does the patient have: hyper IgM, syndrome, bare lymphocyte syndrome, leukocyte adhesion deficiency, or chronic granulomatous disease?

Thought Questions: CHECK YOUR ANSWERS

Thought Question 17.1 Which cell type of the immune system would be missing in a patient lacking surface MHC I molecules? *Hint:* What lymphocyte requires antigen presentation by MHC I?

ANSWER: CD8 cytotoxic T cells are low in number when MHC I is missing because they cannot be activated to proliferate. CD8 T-cell activation requires interaction with MHC I on an antigen-presenting cell. These patients have normal antibody levels and overall white blood cell counts, however. In addition to CD8 cells, NK cells are also deficient and cannot effectively kill host cells missing MHC I molecules (see Section 15.5). Though they lack surface MHC I, the APCs actually do contain *intracellular* MHC I. The mutation for this form of bare lymphocyte syndrome is actually in the *TAP* gene, whose product helps transport antigens into the endoplasmic reticulum for loading on MHC I. Without loading, the MHC I molecule is not transported to the cell surface. Patients with this disorder typically develop chronic lung infections in late childhood.

Thought Question 17.2 Immune globulin used to treat or prevent infections in immunodeficient patients is IgG pooled from 1,000–2,000 individuals. The donors were not chosen because of documented prior exposure to the infectious agent being treated. So why would this IgG protect the patient?

ANSWER: The use of many donors helps to ensure that the preparation contains a broad spectrum of protective antibodies and is relatively uniform. Most people are naturally infected by the same pathogens that commonly infect the immunodeficient patients. However, the donor immune systems effectively fought these infections. Therefore, the donated serum will contain the appropriate antibodies. The content of specific antibodies in pooled immune globulin is routinely performed only for antibodies to measles, poliovirus, and hepatitis B surface antigen.

In contrast to immune globulin, hyperimmune globulin (HIG) comes from individuals whose sera contain high titers (levels) of antibodies to specific pathogens. HIG preparations are available for hepatitis B (HBIG), cytomegalovirus (CMVIG), varicella zoster (VariZIG), rabies, vaccinia, tetanus toxin, and botulinum toxin.

Thought Question 17.3 Why do individuals with type A blood have anti-B and not anti-A antibodies?

ANSWER: Because the B-cell population that would react to type A antigen was deleted during B-cell maturation.

Thought Question 17.4 Does a peanut allergy rely more on T_H1 or T_H2 cells?

ANSWER: T_H2 cells, which are needed to activate B cells to make the IgE antibodies.

Thought Question 17.5 Why would you consider blood group type O people "universal donors" but not universal acceptors of RBCs from type O, A, B, or AB individuals? Who might be considered a universal recipient?

ANSWER: Blood group O red blood cells do not contain A or B antigens. They can be given to group A or B patients because the cells will not be attacked by anti-B or anti-A antibodies, respectively, present in those patients. Blood group O blood can be universally given to A, B, AB, or O patients. (Some minor blood groups could cause problems.) Blood group O people cannot receive blood from any other group because their serum contains antibodies against A and B antigens. Those patients would experience a type II reaction. Universal recipients are people with blood group AB. Because they have neither anti-A nor anti-B antibodies, they can receive red blood cells from A, B, or O type donors.

Thought Question 17.6 Patients with primary immunodeficiencies can receive multiple IV immune globulin treatments with only minor side effects. Since IgG antibodies from different individuals have allotypic antigenic differences, why aren't serious hypersensitivity reactions to IVIG treatment more common?

ANSWER: First, severely immunodeficient patients will not develop an antibody response to foreign proteins. Second, each lot of immune globulin will come from different individuals, so even if the patient developed an immune response to antigens in one lot, those antigens will be different in a second lot.

Thought Question 17.7 Do bone marrow transplants in a patient with severe combined immunodeficiency require immunosuppressive chemotherapy?

ANSWER: No. Since the patient has no T cells to recognize foreign antigens, the transplant is not rejected.

Thought Question 17.8 Explain why the immune response to a ribosomal protein of *Streptococcus pneumoniae* is not protective, but an immune response to its capsule is protective. That is, why isn't *S. pneumoniae* ribosomal protein a good vaccine?

ANSWER: APCs can present pieces of the bacterial ribosomal proteins on their surface. B cells and T cells can recognize those antigens to produce humoral and cell-mediated responses. However, the resulting antibodies and cytotoxic T cells will not "see" that ribosomal antigen in a living *S. pneumoniae* because the antigen is internal. Antibodies will be able to attack capsular antigen because it is a surface structure. (Cytotoxic T cells are not as important because *S. pneumoniae* is not an intracellular pathogen.)

18

Microbial Pathogenesis

SCENARIO Robert Stevens, a 63-year-old photo editor for a Florida tabloid, was looking forward to a nature trip to North Carolina later in the week. Before he left, someone asked him to examine a strange love letter to Jennifer Lopez that the paper's offices received. Coworkers later said the letter was sprinkled with an unknown white powder. The letter was ultimately deemed a hoax and discarded. Eight days later, while in North Carolina, Robert started feeling ill with muscle aches, nausea, and fever. The symptoms waxed and waned over the 3-day trip. Then on October 2, the day after he arrived home, he awoke from sleep feverish, vomiting, and mentally confused. His wife rushed him to the emergency department.

2001 Bioterror Attack
Robert Stevens, a photo editor from Florida, was the first victim of the anthrax attack of 2001.

SIGNS AND SYMPTOMS His temperature there was 39°C (102.2°F). Cerebrospinal fluid (CSF) and blood cultures were collected for microbiological examination, but the clinician who performed the spinal tap immediately knew something was terribly wrong. The CSF in the tube was cloudy when it should have been clear.

TEST RESULTS The lab later reported the CSF had a protein level of 666 mg/dl (too high), a glucose level of 57 mg/dl (within normal limits), and a white blood cell count of 4,700/microliter, mostly polymorphonuclear leukocytes (there shouldn't be any). An X-ray of Robert's lungs also showed an infiltrate. Both the CSF and blood cultures grew Gram-positive rods.

OUTCOME Within hours after admission, Robert had a generalized grand mal seizure and was intubated to help him breathe. On October 5, his third day in the hospital, Robert developed unresponsive hypotension (low blood pressure), underwent cardiac arrest, and died despite desperate attempts at resuscitation. The 63-year-old photo editor had become the first victim of the anthrax bioterrorism attack of 2001.

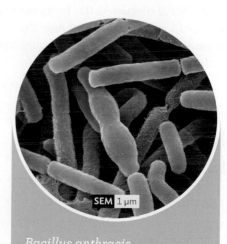

SEM 1 μm

Bacillus anthracis
Vegetative and spore-forming (pink) *Bacillus anthracis* cells.

CHAPTER OBJECTIVES

After reading this chapter, you will be able to:

• Correlate the genetics of pathogenicity with pathogen evolution.

• Describe the major molecular "tools" microbes use to cause disease.

• Discuss the strategies different pathogens use to survive in a host.

• Compare the pathogenic mechanisms of viral, bacterial, fungal, and protozoan pathogens.

The 2001 anthrax attack, which came on the heels of the 9/11 airliner attacks on New York and Washington, DC, led to widespread panic across the country. Someone had spread weaponized spores of *Bacillus anthracis* via the U.S. mail system (weaponized spores can float in air longer than natural spores). Who was responsible? To date many people have been investigated but no one has been brought to trial.

Robert's inhalational anthrax developed after endospores of *B. anthracis*, presumably packed with the letter, deposited deep into the alveolar spaces of his lung. (Someone must inhale many spores—more than 10,000—to get sick.) In the lung, macrophages engulf dormant endospores and transport them to regional lymph nodes. There the spores germinate, becoming a growing form of the bacteria that causes toxemia (toxin release in the bloodstream) and massive septicemia.

The genome (chromosome) of *B. anthracis* is very similar to that of another common soil organism, *Bacillus cereus*. So, two Gram-positive rods, identical in appearance, are both found in soil. Why is one a killer and the other not? For *B. anthracis* and *B. cereus* the differences between friend and foe primarily involve two plasmids present in *B. anthracis*, pXO1 and pXO2, encoding the anthrax toxin and capsule biosynthesis genes, respectively. We will describe how the toxin works in Section 18.3.

Chapter 2 discussed the basic concepts of infectious disease; Chapters 15 and 16 explained how the immune system deals with infections. Now we describe the many tools pathogens use to grow and ultimately cause disease in a host.

How do pathogens cause disease and what distinguishes a pathogen from a nonpathogen? Recall from Chapter 2 that the process pathogens use to cause disease is called **pathogenesis**, which includes the following steps:

1. Pathogen entry
2. Tissue attachment and colonization
3. Immune avoidance
4. Host damage
5. Pathogen exit

Pathogens can be distinguished from nonpathogens by the presence of **virulence factors**, which help establish the pathogen in the host and disturb host functions. Many virulence genes reside in **pathogenicity islands**, whereas others reside in plasmids (for example, the genes for the diarrhea-producing labile toxin of certain *Escherichia coli* strains) or in phage genomes (such as the diphtheria toxin gene of *Corynebacterium diphtheriae*). All these factors contribute to disease, but keep in mind that the degree of harm an infection causes depends not only on the pathogen's virulence mechanisms but also on immunopathology—how aggressively the immune response tries to destroy the pathogen.

Link As is discussed in Section 9.4, **pathogenicity islands** are segments of a pathogen's chromosome acquired by horizontal gene transfers such as transformation, transduction, and conjugation. They are found in both plant and animal pathogens and encode genes responsible for causing disease.

What do pathogenicity, or virulence, genes do? Some encode molecular "grappling hooks," such as pili that attach to host cells. Once attached, microbes can secrete toxins that injure the host cell. Other virulence genes encode enzymes (such as coagulase) that encase the pathogen in fibrin nets to prevent damage by host inflammatory responses. Virulence genes in some bacterial pathogens can even "reprogram" host cells. These organisms inject proteins *directly* into the host cell to disrupt normal signaling pathways. This reprogramming causes the target cell to engulf the bacterium, undergo apoptosis (commit suicide), or facilitate an even more intimate host-pathogen attachment. All these factors enhance the disease-producing capability, or pathogenicity, of the pathogen.

18.1
The Tools and Toolkits of Bacterial Pathogens

SECTION OBJECTIVES

• Differentiate between pathogenicity genes and pathogenicity islands.

• Correlate horizontal gene transfer with pathogen evolution.

Pathogenicity Islands

Most pathogenicity island genes in well-studied pathogens were acquired by horizontal transmission from other organisms. New, emerging pathogens continue to evolve this way. Horizontal gene transfers move whole blocks of DNA (more than 10 kb, or kilobases) from one organism to another, placing the blocks directly in the chromosome in what is called a **genomic island**. If the island increases the "fitness" of a microorganism (pathogen) during interaction with a host, it is called a **pathogenicity island**. Genomic islands in general have several signature features:

- The ratio of nucleobases (GC/AT) of a genomic island is different from that of the chromosome. In Figure 18.1, for example, the GC content of the host chromosome averages 50%. But somewhere along the genome, a 50-kb region sticks out on the graph, showing a content of 40%. This area probably reflects the GC content of the microbe that donated the island—a donor microbe that probably no longer exists.

- Genomic islands are typically flanked by genes with homology to phage or plasmid genes. These genes were probably once associated with the transfer vector that moved the island from one organism to another.

- Genomic islands encode clusters of genes that contribute to the fitness of the organism while in a host, such as protein export systems that secrete toxins (for example, type III secretion systems that inject toxic proteins directly into target host cells, which Section 18.4 covers).

 Virulence factors of pathogenic bacteria: mgc.ac.cn

Pathogen Evolution through Horizontal Gene Transmission

The Gram-positive organism *Streptococcus agalactiae*, also called group B streptococcus for its capsular antigen type, is the leading pathogen affecting neonates in the developed world. It is part of the normal vaginal microbiota of some women and can be transferred to a baby passing through the birth canal. The organism is a major cause of newborn septicemia, which can lead to long-term illness and even death.

At least five clusters of *S. agalactiae* strains are recognized, which vary in the severity of disease they can cause. Scientists at the Pasteur Institute discovered how these different strain clusters may have evolved. The scientists showed that different *S. agalactiae* strains exchange large portions (20–330 kb) of their chromosome via conjugative elements (discussed in Chapter 9) located at different positions around the chromosome. The distribution of these genes among strains of *S. agalactiae* has produced strains with varied degrees of virulence. This work demonstrated that large conjugal exchanges can contribute significantly to genome evolution and pathogenicity.

The *S. agalactiae* example illustrates that the infection process is like a chess match, with each side, human and microbe, trying to outmaneuver the other. The result is that host and pathogen coevolve.

SECTION SUMMARY

- **Genomic islands** are DNA sequences within a species that are acquired by horizontal gene transfer (conjugation, transduction, transformation) from a different species. Genomic islands contain features, such as altered GC content and the remnant genes of phages or plasmids, that mark them as different from the rest of the genome.

- **Pathogenicity islands** are genomic islands that encode virulence factors.

- **Virulence genes** encode proteins that enhance the disease-causing ability of the organism. Many virulence genes are within pathogenicity islands, but some are located outside an obvious genomic island or reside in plasmids.

Thought Question 18.1   What is the significance of the CSF results in the case that opens this chapter?

Figure 18.1 Model Pathogenicity Island

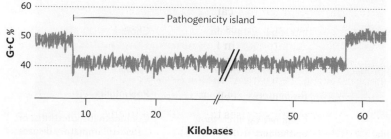

A. The guanine + cytosine (GC) content of the island (about 40%) is different from that of the core genome (50%).

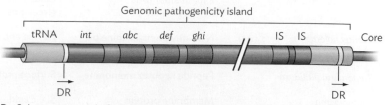

B. Schematic model of a pathogenicity island. The DNA block is linked to a tRNA gene and flanked by direct repeats (DR) that may be "footprints" of a transposon or a virus-mediated transfer. The genome also typically includes the integrase gene (*int*) and insertion sequences (IS), which may also be remnants of transposition.

18.2
Microbial Attachment

Why is it so difficult for an infected body to expel pathogens?
Regardless of the disease, pathogens must reach a tissue colonization site, whether by swimming there, accidentally bumping into it, or by hitchhiking with a vector. Once at the site, invading pathogens need attachment mechanisms to stay put.

The human body has many ways to exclude pathogens. The lungs use a mucociliary escalator (see Figure 14.5) to rid themselves of foreign bodies, the intestine uses peristaltic action to ensure that its contents are constantly flowing, and the bladder uses contraction to propel urine through the urethra with tremendous force. How do bacteria ever manage to stay around long enough to cause problems? Like a person grasping a telephone pole during a hurricane, successful pathogens moving through the body manage to grab host cells and tenaciously hold on. Thus, the first step toward infection is attachment, also called adhesion. An **adhesin** is the general term for any microbial factor that promotes attachment.

We described in Chapter 12 how viruses use capsid or envelope proteins to attach to specific host cell receptors. Bacteria have similar strategies to bind a host. As Section 5.5 describes, some bacteria can sprout hairlike appendages called **pili** (singular, pilus; also called **fimbriae**), whose tips contain receptors for mammalian cell surface structures. Or they can use various adherence proteins or other molecules on the bacterial surface. Sometimes they use both pili and nonpilus adhesins. Table 18.1 summarizes bacterial attachment strategies. We consider here two basic types of pili: type I and type IV.

Table 18.1
Examples of Bacterial Adhesins

Bacterium	Adhesin	Host Receptor	Attachment Site	Disease
Streptococcus pyogenes	Protein F	Amino terminus of fibronectin	Pharyngeal epithelium	Sore throat
Streptococcus mutans	Glucan	Salivary glycoprotein	Pellicle of tooth	Dental caries
Streptococcus salivarius	Lipoteichoic acid	Unknown	Buccal epithelium of tongue	None
Streptococcus pneumoniae	Cell-bound protein	N-acetylhexosamine galactose disaccharide	Mucosal epithelium	Pneumonia
Staphylococcus aureus	Cell-bound protein	Amino terminus of fibronectin	Mucosal epithelium	Various
Neisseria gonorrhoeae	N-methylphenylalanine pili	Glucosamine galactose carbohydrate	Urethral/cervical epithelium	Gonorrhea, pelvic inflammatory disease
Enterotoxigenic *E. coli*	Type I fimbriae (pili)	Species-specific carbohydrate(s)	Intestinal epithelium	Diarrhea
Uropathogenic *E. coli*	Type I fimbriae (pili)	Complex carbohydrate	Urethral epithelium	Urethritis
Uropathogenic *E. coli*	P pili (pyelonephritis-associated pili)	P blood group	Upper urinary tract	Pyelonephritis
Bordetella pertussis	Pili ("filamentous hemagglutinin")	Galactose on sulfated glycolipids	Respiratory epithelium	Whooping cough
Vibrio cholerae	N-methylphenylalanine pili	Fucose and mannose carbohydrate	Intestinal epithelium	Cholera
Treponema pallidum	Peptide in outer membrane	Surface protein (fibronectin)	Mucosal epithelium	Syphilis
Mycoplasma	Membrane protein	Sialic acid	Respiratory epithelium	Pneumonia
Chlamydia	Lipooligosaccharide, OmcB surface protein	Sulfonated glycosaminoglycans	Conjunctival or urethral epithelium	Conjunctivitis, urethritis, or pelvic inflammatory disease
Corynebacterium diphtheriae	Pili	Unknown	Pharyngeal epithelium	Diphtheria

Figure 18.2 Attachment Pili

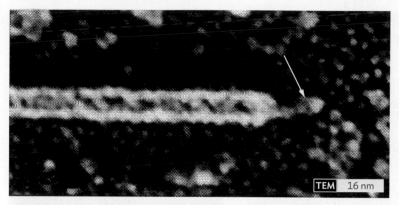

High-resolution micrograph showing a type I pilus. The FimH adhesin at the tip (arrow) is the protein that binds to the cell receptor.

Pilus Assembly

How bacteria assemble pili on their cell surfaces is an engineering marvel. Regardless of pilus type, the process is similar. The shafts of pili are cylindrical structures composed of identical pilin protein subunits. Adorning the tip of the pilus is the adhesin protein that binds to host receptors (Figure 18.2). In addition to these structural components, many other proteins work together as a machine to assemble the structure. As shown in Figure 18.3, assembly of a pilus starts with the tip protein, which will ultimately bind to carbohydrates on host membranes after the pilus is complete. Once the tip is set at the bacterial surface, identical pilin subunits are strung together to form the shaft forcing the tip protein farther from the surface.

Figure 18.3 Assembly of Type I Pili

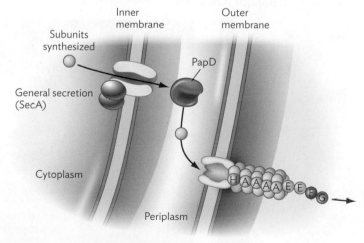

The figure illustrates pyelonephritis-associated pilus (Pap) assembly in *E. coli* but also represents other type I pili such as Fim in Figure 18.2. The Sec system secretes proteins to the periplasm, where PapD chaperones them to the site of assembly. Assembly starts with the tip protein, PapG, marked at the far right, which ultimately binds to carbohydrates on host membranes. The subunits fit together like puzzle pieces at the base of a growing pilus. The arrow at the head of the elongating pilus indicates the direction of pilus growth.

Pili are more than simple microbial "stick pins" affixing a bacterium to a host cell membrane. Pathogens and normal microbiota are often subjected to strong shear forces from fast-moving bodily fluids such as urine in the urethra, mucus during a sneeze or cough, or intestinal contents moved by peristalsis. To counter these shear forces, some pili tighten their "grip" on target cells (acting more like Velcro than a pin), making it extremely difficult to pry the organism from its target host cell (Figure 18.4). In this way the organism resists being flushed away.

Figure 18.4 Bacterial Attachment by Pili

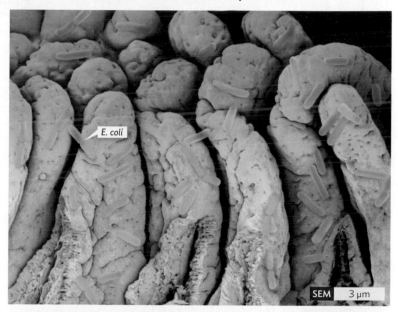

A. *E. coli* attached to the surface of the intestine.

B. Elementary student hangs from a Velcro wall. Much as the stress of gravity strengthens the individual Velcro connections, sheer stresses strengthen the connection between pilus tip adhesion proteins and their host receptor molecules.

Figure 18.5   Type IV Pili

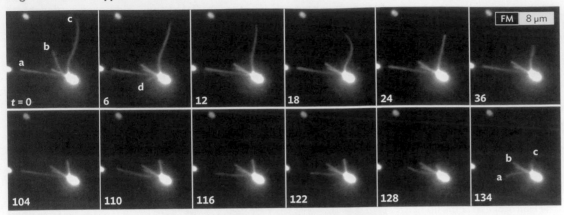

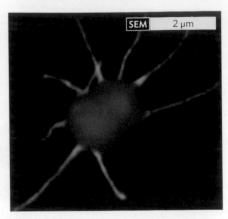

A. Photographic evidence of type IV pilus extension and retraction in cells of *Pseudomonas aeruginosa*. Filament b retracts; then filament d extends at 6 seconds and retracts. Filament c attaches briefly at its distal tip (note straightening at 24 seconds) and then begins to retract. Time (*t*) in seconds.

B. Type IV pili (green) are essential for *Neisseria meningitidis* (red) to interact with brain endothelial cells. Diplococcal cells are approximately 1.6 μm in diameter.

Types of Pili

The classification of pili is now based largely on protein sequence information (deduced from DNA sequence). **Type I pili** are static, hairlike appendages with specialized tip proteins that can bind to specific substrates. For example, some type I pili on *E. coli* (members of our normal microbiota) bind to mannose residues present on many host cells. Other type I pili, such as the pyelonephritis adhesion pili (Pap) of uropathogenic *E. coli*, bind to a digalactoside sugar group present on urinary tract host surfaces (called the P blood group antigen).

Whereas type I pili are static structures "patiently" waiting to bind their receptor, **type IV pili** are more dynamic. Type IV pili are thin and flexible, and repeatedly extend and retract, giving its bacterial owner a spastic movement called twitching motility. Type IV pili have an important role in pathogenesis and are found in a broad spectrum of Gram-negative bacteria. Pathogens with type IV pili include *Vibrio cholerae*, *Pseudomonas aeruginosa* (Figure 18.5A), certain pathogenic strains of *E. coli*, *Neisseria meningitidis*, and *N. gonorrhoeae*. All type IV pili use similar secretion and assembly machinery involving at least a dozen proteins.

"Twitching motility" occurs when the pilus elongates, attaches to a surface, and then disassembles from the base, which shortens the pilus and pulls the cell forward. This mechanism is akin to using a grappling hook to scale a building. The type IV pili of *Neisseria meningitidis*, shown in Figure 18.5B, are essential for crossing the blood-brain barrier and causing bacterial meningitis.

Nonpilus Adhesins

Bacteria also carry nonpilus adhesin proteins that stick to host tissues by way of host proteins such as integrin, which normally promotes interactions between host cells, and fibronectin, a host protein that mediates cell-cell adhesion in tissues (Figure 18.6). Examples of nonpilus adhesins include *Bordetella* surface pertactin

(which binds to host integrin), *Streptococcus* protein F (binds to fibronectin), *Streptococcus* M protein (binds to fibronectin and complement regulatory factor H), and the enteropathogenic *E. coli* surface protein intimin (binds to Tir, another *E. coli* protein that is injected into the host cell membrane).

Fimbriae (pili) often initiate binding between bacterium and host (think Velcro on a stick), after which a nonpilus attachment protein creates a more intimate attachment. For *Neisseria gonorrhoeae*, once the type IV pilus has attached to the mucosal epithelial cell, the filamentous pilus retracts, pulling the bacterium down onto

Figure 18.6   Nonpilus Adhesins

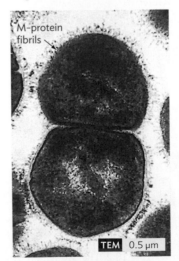

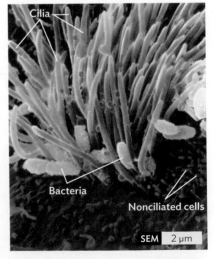

A. M-protein surface fibrils on *Streptococcus pyogenes*.

B. *Bordetella pertussis* colonizing tracheal epithelial cells. This organism uses a surface protein called pertactin, as well as a pilus called filamentous hemagglutinin to bind bronchial cells.

the host cell membrane. Tight secondary binding is then mediated by the neisserial Opa membrane proteins, another example of a non-pilus adhesin (Opa is so named because of the *opa*city it adds to colony appearance). Type IV pili also help *N. meningitidis* cells stick to each other in the throat, making a biofilm. After a while, some pili are modified and detach from surfaces. Detachment gives individual diplococci the freedom to disseminate through the bloodstream and cause sepsis or meningitis (**inSight**). Inhibiting adhesion can make a pathogenic organism non-virulent. Therefore, making anti-adhesin antibodies after an infection (or vaccination) is an important part of protective immunity.

Biofilm Development

As discussed in Section 6.6, bacteria in most environments form biofilms: organized, high-density communities of cells that embed themselves in self-made exopolymer matrices. Biofilms allow microorganisms to adhere to any surface, living or nonliving, and survive in hostile environments. Within a single biofilm you can find localized differences in the expression of surface molecules, antibiotic resistance, nutrient use, and virulence factors. Bacteria in biofilms also coordinate their behavior through cell-cell communication using secreted chemical signals (quorum sensing).

Biofilms are important features in chronic infections of oral, lung, and urogenital (bladder) tissues. For example, *Pseudomonas aeruginosa* causes a life-threatening, chronic lung infection in individuals with cystic fibrosis (CF). This microbe grows as biofilm aggregates within mucus from CF patients and causes inflammation. Insufficient mucociliary clearance of the CF lung contributes to *P. aeruginosa* biofilm formation. Biofilms are also important in periodontitis (gum disease), indwelling catheter infections, infections of artificial heart valves, chronic urinary tract infections,

Figure 18.7 A Bacterial Biofilm Infection

A pressure ulcer showing bacteria of different morphotypes (rods and cocci) colonizing the wound.

Figure 18.8
Biofilm Forming inside a Needleless Connector
Staphylococci forming a biofilm secrete a polysaccharide matrix that holds cells together. Needleless connectors connect syringes or bags of medication to an indwelling catheter tube inserted into a patient's vein.

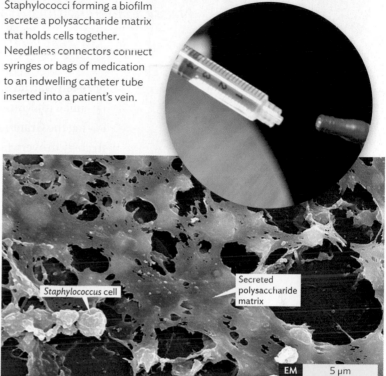

recurrent tonsillitis, rhinosinusitis (inflammation of nasal sinuses), chronic otitis media (middle ear infection), chronic wound infections, and osteomyelitis (bone infection).

Figure 18.7 shows a chronic infection of skin (a pressure ulcer). The chronic presence of these organisms in biofilms will continually stimulate innate immune mechanisms through interactions with **Toll-like receptors**, causing chronic inflammation. A biofilm infection may linger for months, years, or even a lifetime. If associated with an implanted medical device such as an artificial knee or heart valve, the device may have to be replaced. Today, catheter-related bloodstream infections (caused by biofilms forming on indwelling catheters) are the most serious and costly health care–associated infections (**Figure 18.8**).

Link As Section 15.5 discusses, **Toll-like receptors (TLRs)** are part of our innate defense system. TLRs present on various host cell types can recognize and bind various microbe-associated molecular patterns, such as lipopolysaccharide or flagellar proteins, and initiate the synthesis of cytokines that stimulate the immune system.

Biofilm infections are important clinically. Bacteria in biofilms can tolerate antimicrobial compounds and persist despite sustained host defenses. White blood cells, for instance, cannot easily penetrate biofilms, so biofilm infections are hard to cure. Tolerance to antibiotics may be caused by poor nutrient penetration into the deeper regions of the biofilm, leading to a stationary phase–like

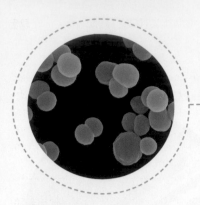

inSight

Neisseria meningitidis: Training Single-Cell Infiltrators

In any war, an invading army establishes a base of operations (a beachhead), from which small infiltration teams are sent deeper into enemy locations. Many pathogens, including the Gram-negative bacterium *Neisseria meningitidis*, carry out a strikingly similar strategy to overwhelm their host. *Neisseria meningitidis* is a leading worldwide cause of fatal sepsis and meningitis. At any one time, 10%–20% of the human population asymptomatically carries the organism in their nasopharynges, the pathogen's major reservoir. To establish asymptomatic infection, *N. meningitidis* uses type IV pili to adhere to throat epithelial cells (Figure 1A), where the bacterium multiplies and forms biofilm aggregates, also mediated by type IV contacts, this time between bacteria (Figures 1B and C). These asymptomatic carriers develop an antibody response that kills the occasional organism entering the bloodstream. However, if the occasional meningococcal cell enters the bloodstream before an effective immune response can be launched, a rapidly progressive invasive disease can develop. But why do individual cells only occasionally leave the aggregate, pass through the throat into blood, and from the blood into the brain—essentially sending scouts to establish serious infection elsewhere in the body? Guillaume Duménil and colleagues at the Université Paris Descartes think they have part of the answer—and it has to do with those pili.

N. meningitidis has an enzyme (PptB) that adds phosphoglycerol to certain residues in pilin. This modification prevents type IV pili from mediating bacterial aggregation. As a consequence, some bacteria detach from the aggregate and disseminate. These lone cell scouts can better migrate across the epithelium and enter the bloodstream. More intriguing is that the modification is a regulated process that does not happen all the time. Transcription of the *pptB* gene increases only *after* bacteria attach to and grow on epithelial cells. Consequently, the pathogen's base of operations is established in the throat before the infiltration teams are released. More on the pathogenesis of *Neisseria meningitidis* is found in Section 24.4.

Figure 1   Building a Biofilm on Host Cells

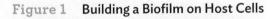

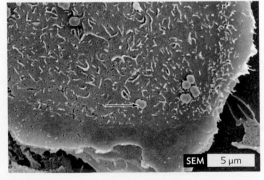

A. Initial adhesion occurs with individual or small groups of bacteria (arrow points to a diplococcus).

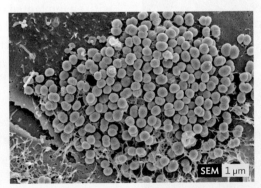

B. After initial adhesion, bacteria proliferate and form large bacterial aggregates known as microcolonies.

C. A cluster of *Neisseria* (pseudocolored in orange) attached to a human epithelial cell. A network of type IV pili cover the bacteria. When the fibers retract, they pull with tremendous force on the surface of a human cell, changing the cell surface. Type IV fiber retraction stimulates the formation of microvilli beneath the bacteria.

dormancy. Bacterial factors important to biofilm formation include type IV pili, quorum-sensing structural genes and regulators, and extracellular matrix synthesis. Interfering with cell-cell signaling can prevent or limit biofilm formation and may offer a target for new antimicrobial therapies.

SECTION SUMMARY

- **Bacteria use pili and nonpilus adhesins** to attach to host cells.
- **Type I pili** produce a static attachment to the host cell, whereas **type IV pili** continually assemble and disassemble.
- **Nonpilus adhesins** are bacterial surface proteins or other molecules that can tighten interactions between bacteria and target cells.
- **Biofilms** play an important role in chronic infections by enabling persistent adherence to host cells and resistance to bacterial host defenses and antimicrobial agents.

18.3
Toxins: A Means to Seduce, Hijack, or Kill Host Cells

SECTION OBJECTIVES

- Describe the nine basic cellular targets for bacterial toxins.
- Explain the modes of action for staphylococcal alpha toxin, cholera toxin, diphtheria toxin, Shiga toxin, and anthrax toxin.
- Differentiate endotoxin from exotoxin.

CASE HISTORY 18.1

The Telltale Cough

Will, a formerly rambunctious 5-year-old from San Francisco, was taken to the emergency department after having a severe cough for 2 weeks. His illness started as a runny nose, dry cough, and low-grade fever. But within days the coughing would come in violent fits lasting up to 1 minute (called a paroxysm). The nurse practitioner attending Will observed one of these fits. Unable to breathe between coughs, the boy became cyanotic (turned blue). When the staccato of coughs finally ended, Will gasped, desperate for breath. The air rushing to fill his lungs made an ear-piercing whooping sound. This telltale "whoop" led the nurse practitioner to suspect whooping cough, a disease highly contagious in its early stages. In the later stages (>4 weeks) the patient is no longer contagious because the bacterial agent, *Bordetella pertussis*, has succumbed to the immune response. However, the telltale cough will persist for weeks while the lungs repair the damage caused by the pertussis toxins. The nurse practitioner, now wearing a mask, took a nasopharyngeal swab sample (**Figure 18.9**) for bacteriological culture and prescribed the antibiotic azithromycin (a member of the macrolide class of antibiotics that inhibit protein synthesis; Section 13.5). Because the organism takes so long to grow in the laboratory (5–12 days) and because the

organism is hard to find late in the disease, a blood sample was also drawn to test for anti–*B. pertussis* immunoglobulin A antibody. By the next day the serum test was positive for *B. pertussis* antibodies. The pathogen eventually also grew in the culture. Despite extensive efforts at vaccination, more than 41,000 cases of whooping cough were reported in the United States in 2012 (33,000 in 2014).

Can pathogens damage host tissues directly, or is all damage from immunopathology? Immunopathology is important, of course. But after attaching to the host, many pathogens, including *B. pertussis*, secrete protein toxins (called **exotoxins**) that alter host cell function, disrupt the immune system, or outright kill the host cell to unlock their nutrients (because dead host cells ultimately lyse). Bacterial pathogens have developed an impressive array of toxins that take advantage of different key host proteins or structures. Gram-negative bacteria also possess a nonprotein yet toxic compound called **endotoxin** that can hyperactivate host immune systems to harmful levels. Endotoxin is part of the outer membrane and will be described later.

Exotoxin Modes of Action

Microbial exotoxins are categorized on the basis of their cellular targets and mechanisms of action, from lysing cell membranes to disrupting neurotransmitter secretion. The basic modes of action are summarized in **Figure 18.10** and **Table 18.2**. Notice how exotoxins can target many different host mechanisms.

- **Plasma membrane disruption.** Toxins in this group, exemplified by the alpha (α) toxin of *Staphylococcus aureus*, form pores in host cell membranes that cause cell contents to leak.

Figure 18.9
Nasopharyngeal Swab

Figure 18.10   Actions of Bacterial Exotoxins
The figure summarizes the modes of action for many bacterial exotoxins.

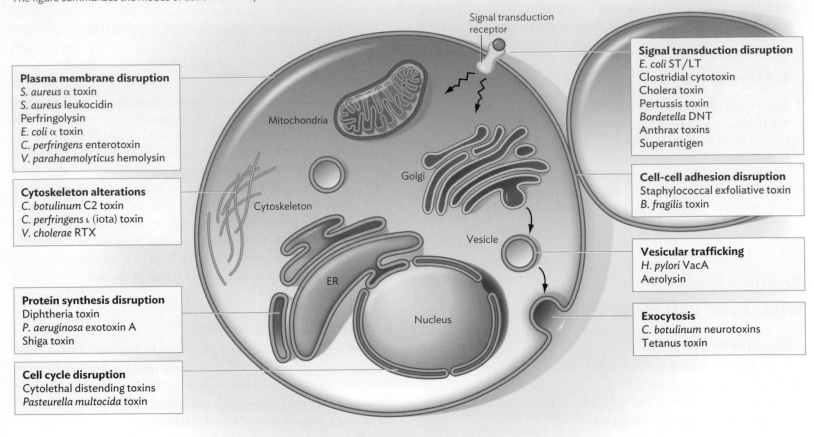

Plasma membrane disruption
S. aureus α toxin
S. aureus leukocidin
Perfringolysin
E. coli α toxin
C. perfringens enterotoxin
V. parahaemolyticus hemolysin

Cytoskeleton alterations
C. botulinum C2 toxin
C. perfringens ι (iota) toxin
V. cholerae RTX

Protein synthesis disruption
Diphtheria toxin
P. aeruginosa exotoxin A
Shiga toxin

Cell cycle disruption
Cytolethal distending toxins
Pasteurella multocida toxin

Signal transduction receptor

Signal transduction disruption
E. coli ST/LT
Clostridial cytotoxin
Cholera toxin
Pertussis toxin
Bordetella DNT
Anthrax toxins
Superantigen

Cell-cell adhesion disruption
Staphylococcal exfoliative toxin
B. fragilis toxin

Vesicular trafficking
H. pylori VacA
Aerolysin

Exocytosis
C. botulinum neurotoxins
Tetanus toxin

Mitochondria

Golgi

Cytoskeleton

Vesicle

ER

Nucleus

- **Cytoskeleton alterations.** Toxins in this group can cause host cell actin to polymerize or depolymerize. Actin is a protein in the cytoplasmic microfilaments that maintain cell shape and act like "train tracks" along which vesicles filled with protein cargo move. *Vibrio cholerae* RTX toxin, for example, depolymerizes actin and alters cell shape. Other cytoskeleton-altering exotoxins enable pathogens to break into the host cell. These toxins force host cell membranes to wrap around the bacterium, pulling the invader into the cytoplasm.

- **Protein synthesis disruption.** Toxins in this group, exemplified by diphtheria and Shiga toxins, target eukaryotic ribosomes and prevent protein synthesis.

- **Cell cycle disruption.** These toxins either stop (*E. coli* cytolethal distending toxin) or stimulate (*Pasteurella multocida* toxin) host cell division.

- **Signal transduction disruption.** This type of toxin subverts host cell secondary messenger pathways by increasing or decreasing the synthesis of critical signal molecules such as cyclic AMP (cholera and *Bordetella pertussis* toxins increase cAMP synthesis).

- **Cell-cell adhesion.** Some protease exotoxins cleave proteins that bind host cells together. One such toxin, exfoliative toxin of *Staphylococcus aureus*, breaks the connection between dermis and epidermis, giving victims a gruesome scalded skin appearance.

- **Vesicle traffic.** The major toxin in this class (VacA of *Helicobacter pylori*) has several modes of action depending on the host cell. The most visually striking effect is its ability to cause vacuolization, the fusion of many intracellular vesicles.

- **Exocytosis.** This group includes two protease exotoxins that interrupt the movement (or exocytosis) of host cytoplasmic vesicles to membranes. One example is tetanus toxin, which prevents nerve cells from releasing the inhibitory neurotransmitter GABA (gamma-aminobutyric acid). The frightening result is paralyzing muscle spasms.

- **Excessive activation of the immune response (superantigens).** These exotoxins activate the immune system without being processed by antigen-presenting cells (APCs) and are potent triggers of fever. The pyrogenic (fever producing) toxins of *Staphylococcus aureus* (such as toxic shock syndrome toxin) and *Streptococcus pyogenes* are examples of superantigen exotoxins.

Table 18.2
Characteristics of Bacterial Exotoxins[a]

Toxin	Organism	Mode of Action	Host Target/ Target Tissue	Disease	Toxin Implicated in Disease[b]
Damage Plasma Membranes					
Aerolysin	*Aeromonas hydrophila*	Pore former	RBCs, leukocytes	Diarrhea	(Yes)
Perfringolysin O	*Clostridium perfringens*	Pore former	RBCs, leukocytes	Gas gangrene[c]	Unknown
Hemolysin[d]	*Escherichia coli*	Pore former	RBCs, leukocytes	UTIs	(Yes)
Listeriolysin O	*Listeria monocytogenes*	Pore former	RBCs, leukocytes	Food-borne systemic illness, meningitis	Yes
Alpha toxin	*Staphylococcus aureus*	Pore former	RBCs, leukocytes	Abscesses[c]	(Yes)
Panton-Valentine leukocidin	*Staphylococcus aureus*	Pore former	RBCs, leukocytes	Abscesses, necrotizing pneumonia	(Yes)
Pneumolysin	*Streptococcus pneumoniae*	Pore former	RBCs, leukocytes	Pneumonia[c]	(Yes)
Streptolysin O	*Streptococcus pyogenes*	Pore former	RBCs, leukocytes	Strep throat, scarlet fever	Unknown
Disrupt Cytoskeleton					
Vibrio cholerae RTX	*Vibrio cholerae*	Actin depolymerization	Cross-links actin fibers	Cholera	Unknown
C2 toxin	*Clostridium botulinum*	ADP-ribosyltransferase	Monomeric G-actin	Botulism	Unknown
Iota toxin	*Clostridium perfringens*	ADP-ribosyltransferase	Actin	Gas gangrene[c]	(Yes)
Inhibit Protein Synthesis					
Diphtheria toxin	*Corynebacterium diphtheriae*	ADP-ribosyltransferase	Elongation factor 2	Diphtheria	Yes
Shiga toxins	*E. coli/Shigella dysenteriae*	N-glycosidase	28S rRNA	HC and HUS	Yes
Exotoxin A	*Pseudomonas aeruginosa*	ADP-ribosyltransferase	Elongation factor 2	Pneumonia[c]	(Yes)
Disrupt Cell Cycle					
CLDT	*E. coli, Campylobacter, Haemophilus ducreyi,* others	DNase	DNA damage, triggers G_2 cell cycle arrest	Diarrhea chancroid, others	Unknown
Pasteurella multocida toxin	*Pasteurella multocida*	Mitogen (also activates Rho GTPases)	Nucleus (encourages cell division)	Wound infection	(Yes)
Activate Second Messenger Pathways					
CNF	*E. coli*	Deamidase	Rho G proteins	UTIs	Unknown
Cholera toxin (CT) and labile toxin (LT)	*Vibrio cholerae* (CT) and *E. coli* (LT)	ADP-ribosyltransferase	G proteins	Diarrhea	Yes
ST[d]	*E. coli*	Stimulates guanylate cyclase	Guanylate cyclase receptor	Diarrhea	Yes
EAST	*E. coli*	ST-like?	Unknown	Diarrhea	Unknown

(continued on next page)

Table 18.2
Characteristics of Bacterial Exotoxins[a] (continued)

Toxin	Organism	Mode of Action	Host Target/ Target Tissue	Disease	Toxin Implicated in Disease[b]
Activate Second Messenger Pathways					
Edema factor	*Bacillus anthracis*	Adenylate cyclase	ATP	Anthrax	Yes
Dermonecrotic toxin	*Bordetella pertussis*	Deamidase	Rho G proteins	Rhinitis	(Yes)
Pertussis toxin	*B. pertussis*	ADP-ribosyltransferase	G protein(s)	Pertussis (whooping cough)	Yes
C3 toxin	*Clostridium botulinum*	ADP-ribosyltransferase	Rho G protein	Botulism	Unknown
Toxin A	*Clostridium difficile*	Glucosyltransferase	Rho G protein(s)	Diarrhea/PC	(Yes)
Toxin B	*C. difficile*	Glucosyltransferase	Rho G protein(s)	Diarrhea/PC	Unknown
Cholera toxin	*V. cholerae*	ADP-ribosyltransferase	G protein(s)	Cholera	Yes
Lethal factor	*B. anthracis*	Metalloprotease	MAPKK1/MAPKK2	Anthrax	Yes
Disrupt Cell-Cell Adhesion					
Exfoliative toxins	*S. aureus*	Serine protease, superantigen	Desmoglein; TCR and MHC II	Scalded skin syndrome[c]	Yes
Bacteroides fragilis toxin	Enterotoxigenic *B. fragilis*	Metalloprotease	E-cadherin	Diarrhea, inflammatory bowel disease	Yes
Redirect Vesicle Traffic					
VacA	*Helicobacter pylori*	Large vacuole formation, apoptosis	Receptor-like protein tyrosine phosphatase, sphingomyelin	Gastric ulcers, gastric cancer	(Yes)
Interrupt Exocytosis					
Neurotoxins A–G	*C. botulinum*	Zinc metalloprotease	VAMP/synaptobrevin, SNAP-25 syntaxin	Botulism	Yes
Tetanus toxin	*Clostridium tetani*	Zinc metalloprotease	VAMP/synaptobrevin	Tetanus	Yes
Activate Immune Response (Superantigens)					
Enterotoxins	*Staphylococcus aureus*	Superantigen	TCR and MHC II; medullary emetic center (vomit center)	Food poisoning[c]	Yes
Exfoliative toxins	*S. aureus*	Serine protease, superantigen	Desmoglein; TCR and MHC II	Scalded skin syndrome[c]	Yes
Toxic shock toxin	*S. aureus*	Superantigen	TCR and MHC II	Toxic shock syndrome[c]	Yes
Pyrogenic exotoxins	*Streptococcus pyogenes*	Superantigens	TCR and MHC II	Toxic shock syndrome, scarlet fever	Yes Yes
Lethal factor	*B. anthracis*	Metalloprotease	MAPKK1/MAPKK2	Anthrax	Yes

[a]Abbreviations: CLDT, cytolethal distending toxin; CNF, cytotoxic necrotizing factor; EAST, enteroaggregative *E. coli* heat-stable toxin; HC, hemorrhagic colitis; HUS, hemolytic uremic syndrome; LT, heat-labile toxin; MAPKK, mitogen-activated protein kinase kinase; MHC II, major histocompatibility complex class II; PC, antibiotic-associated pseudomembranous colitis; SNAP-25, synaptosomal-associated protein; ST, heat-stable toxin; TCR, T-cell receptor; UTI, urinary tract infection; VacA, vacuolating toxin; VAMP, vesicle-associated membrane protein.
[b]Yes, strong causal relationship between toxin and disease; (Yes), animal model or cell culture shows role in pathogenesis.
[c]Other diseases are also associated with the organism.
[d]Toxin is also produced by other genera of bacteria.

Link As first discussed in Section 15.1, **antigen-presenting cells** such as macrophages introduce, or "present," bacterial antigens to cells of the adaptive immune system.

GENERAL CONCEPT: TWO-SUBUNIT AB TOXINS Many, but not all, bacterial toxins have two subunits, usually called A and B. These two-subunit complexes are termed **AB toxins**. The A subunit is toxic; the B subunit binds host cell receptors. Thus, the B subunit for each toxin delivers the A subunit to the host cell. Many AB toxins have five B subunits arranged as a ring, in the center of which is nestled a single A subunit (**Figure 18.11A**).

Several AB subunit toxins have an **ADP-ribosyltransferase** enzymatic activity. These enzyme toxins transfer the ADP-ribose group from an NAD molecule to a target protein (**Figure 18.11B**). The now modified protein has an altered function. Sometimes the function is destroyed (for example, protein synthesis is disrupted when diphtheria toxin ADP-ribosylates a eukaryotic ribosomal protein called EF2); other times, the target protein is locked into an active form that is insensitive to regulatory feedback control (for example, cAMP synthesis continues unchecked in the presence of cholera toxin).

This section describes mechanisms of some toxins. Later chapters will cover the diseases they influence.

STAPHYLOCOCCUS ALPHA TOXIN *Staphylococcus aureus*, an organism that causes boils and blood infections, produces a hemolysin called **alpha toxin**. Alpha toxin forms a transmembrane, seven-member pore in target cell membranes. The resulting leakage of cell constituents and influx of fluid cause the target cell to burst. **Figure 18.12A** and **B** show a completed pore and a cutaway view exposing the channel, respectively. Diagnostic microbiology laboratories visualize hemolysins (proteins that lyse red blood cells) such as alpha toxin by inoculating bacteria onto agar plates containing sheep red blood cells (**Figure 18.12C**). The clear, yellow zones of lysed RBCs around the *S. aureus* colonies indicate that the microbe secretes a hemolysin.

Figure 18.11 AB Toxins

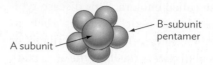

A. A typical AB toxin consists of an A subunit and a pentameric B subunit joined noncovalently.

B. Many AB toxins are ADP-ribosyltransferase enzymes that modify protein structure and function.

Figure 18.12 Alpha Hemolysin of *Staphylococcus aureus*

Alpha hemolysin

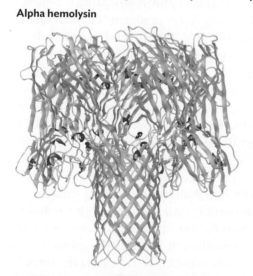

A. Three-dimensional figure of the pore complex comprising seven monomeric proteins.

Cross section of alpha hemolysin

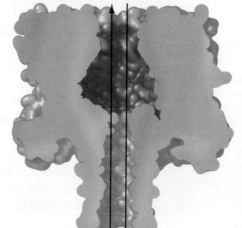

B. Cross section showing the channel.

Beta hemolysis by *S. aureus*

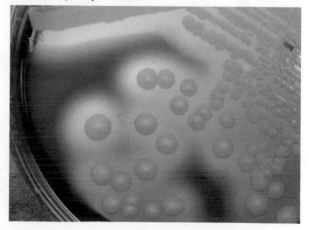

C. A blood agar plate inoculated with *S. aureus*. The organism secretes the alpha toxin, which diffuses away from the producing colony. It forms pores in the red blood cells embedded in the agar, causing them to lyse, which is visible as a clear area surrounding each colony. The clear form of hemolysis is called beta, or complete, hemolysis even though the toxin is called alpha toxin. See Section 25.3 for more information.

CHOLERA AND *E. COLI* LABILE TOXINS *Vibrio cholerae* produces a severe diarrheal disease called cholera. This disease typically afflicts malnourished people of poor countries such as Bangladesh (as the Chapter 4 opening case history describes), or countries in which war or natural disasters (such as the aftermath of the 2010 earthquake in Haiti) disrupt access to clean water. *V. cholerae* produces an enterotoxin nearly identical to one produced by some strains of *E. coli* associated with "traveler's diarrhea" (these strains are called enterotoxigenic *E. coli*, or ETEC). Enterotoxins specifically affect the intestine. The *E. coli* enterotoxin is called **labile toxin (LT)** because heat easily destroys it. Cholera toxin (CT) and LT are both AB toxins with identical modes of action, which is to increase the level of cAMP made inside the host cell. The mode of action for these toxins was first discussed in Chapter 4 and illustrated in Figure 4.1.

Recall that normal intestinal transport mechanisms absorb NaCl, other ions (electrolytes), and water from material flowing through the intestine (Figure 18.13). This absorption produces well-formed feces with little water and salt content. CT and LT, however, reverse this process.

Figure 18.13   Intestinal Absorption and Secretion of Electrolytes and Water

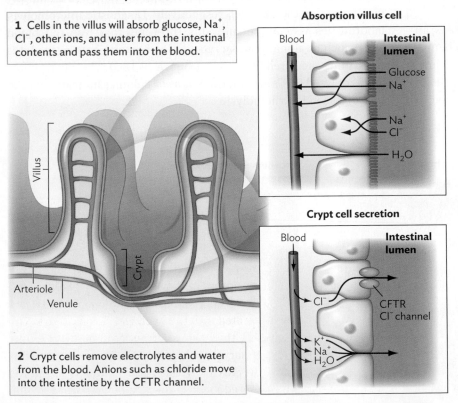

1 Cells in the villus will absorb glucose, Na$^+$, Cl$^-$, other ions, and water from the intestinal contents and pass them into the blood.

Absorption villus cell

Blood — **Intestinal lumen**
— Glucose
— Na$^+$
— Na$^+$
— Cl$^-$
— H$_2$O

Crypt cell secretion

Blood — **Intestinal lumen**
— Cl$^-$
— CFTR Cl$^-$ channel
— K$^+$
— Na$^+$
— H$_2$O

2 Crypt cells remove electrolytes and water from the blood. Anions such as chloride move into the intestine by the CFTR channel.

Intestinal villi and crypt cells, respectively, absorb and secrete electrolytes and water from or into the intestinal lumen. This figure shows some of the mechanisms used. CT and LT increase cAMP levels, which then stimulate the CFTR chloride channel in crypt cells and inhibit the NaCl-absorptive mechanism in absorptive cells. The resulting secretion and accumulation of chloride ions causes a charge imbalance across the membrane that brings other ions out. The resultant salt imbalance across the membrane draws water into the intestinal lumen, producing diarrhea.

Figure 18.14   Pathogenesis of Enterotoxigenic *E. coli*

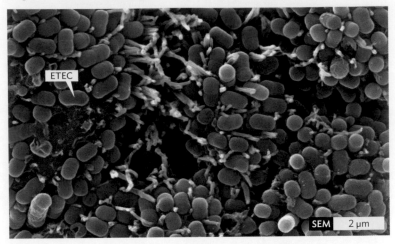

Adherence of enterotoxigenic *Escherichia coli* to the intestinal epithelium. ETEC binds to the fingerlike villi on the apical surface. Like *Vibrio cholerae*, ETEC does not invade the host cell.

After *V. cholerae* or ETEC attaches to the cells lining the intestinal villi (Figure 18.14), the organisms secrete their AB toxins (Figure 4.1B). The B subunits stick to a carbohydrate surface structure called ganglioside GM1 molecules on eukaryotic cell membranes and deliver the A subunit to the target cell. The A subunit possesses the toxic part of the molecule, an ADP-ribosyltransferase.

As Figure 4.1B outlines, the binding of CT or LT to GM1 triggers endocytosis. After the endocytic vacuole travels to the endoplasmic reticulum, the toxic ADP-ribosyltransferase part of the A subunit is exported into the cytoplasm. There it modifies (ADP-ribosylates) a membrane-bound regulatory protein (called a G protein) that controls the cyclic AMP–producing enzyme adenylyl cyclase. Once modified, the G factor overstimulates adenylyl cyclase to make excessive amounts of cAMP.

Where does the diarrhea come from? Excessive cAMP *activates* secretion of Cl$^-$ via the cystic fibrosis transmembrane conductance regulator (CFTR) present in crypt cells. The Cl$^-$ transporter is called CFTR because a defect in this protein causes the lung disease cystic fibrosis. CFTR activation causes chloride (negative charge), sodium (positive charge), and other ions to *leave* the cell and move into the intestinal lumen. This movement causes an osmotic imbalance between the inside and outside of the cell. In an attempt to balance osmolarity, water also leaves the cell. Because the affected cells line the intestine, the escaping water enters the intestinal lumen, producing watery stools (diarrhea).

How does diarrhea benefit the pathogen? Diarrhea helps propagate the species. The more diarrhea produced and expelled, the more organisms are made and disseminated throughout the environment. This cycle increases the chance that another host will ingest the organism, ensuring survival of the species. Diarrhea can also benefit a pathogen by decreasing competition with other organisms as they are swept away. In fact, nearly all organisms in the diarrheal fluids of cholera patients are *V. cholerae* bacteria.

Different pathogens have discovered other ways to alter host cAMP levels. The Gram-negative pathogen *Bordetella pertussis* discussed in Case History 18.1 causes a childhood respiratory infection called whooping cough, so named for the whooping sound a child makes when trying to take a breath after a fit of coughing. This microbe also secretes an ADP-ribosylating toxin that activates adenylyl cyclase activity in lung cells. Runaway cAMP synthesis in this system causes fluid accumulation, or edema, in the lung.

In addition to pertussis toxin, *B. pertussis* secretes a "stealth" adenylyl cyclase that remains inactive until it enters host cells, where it becomes active by binding the calcium-binding protein calmodulin. The resulting increase in cAMP levels inappropriately triggers certain host cell signaling pathways and damages lung cells.

DIPHTHERIA TOXIN   When an invader tries to conquer a new land, an early goal is to destroy the industrial base of the foe. If the target cannot make materials and weapons needed to survive, it will fall. In much the same way, pathogens deploy some bacterial exotoxins to destroy a host's synthetic base, which is its ability to synthesize proteins. *Corynebacterium diphtheriae*, for example, produces a potent toxin that inhibits protein synthesis in human cells and causes the disease diphtheria (described in the case history opening Chapter 9 and in Section 20.3) (Figure 18.15). The organism remains in the pharynx while the toxin spreads via the circulatory system throughout the body causing many symptoms, including heart damage. It is the toxin, not the organism's growth, that is responsible for these symptoms. The gene encoding the toxin actually resides in the genome of a bacteriophage that lysogenizes *C. diphtheriae*. The phage, called beta phage, remains functional and

is a prime example of how a phage genome may, over many generations, degenerate to become a pathogenicity island, losing its former identity as a phage genome.

Diphtheria toxin is also an ADP-ribosyltransferase and another example of an AB toxin. Diphtheria toxin is synthesized as a single polypeptide that is cleaved into a B subunit that binds to the target cell receptor on the host membrane and an A subunit containing the ADP-ribosylase activity. The two peptides cling to each other by a disulfide bond.

The toxin (A + B) enters the cell by endocytosis. When the inside of the endosome vesicle acidifies (via H^+-pumping ATPases), the disulfide bonds holding the AB subunits together are reduced (broken), allowing subunit A to pass through the vesicle membrane. Once inside the cytoplasm, the A subunit modifies an important protein synthesis factor called elongation factor 2 (EF2). This act halts protein synthesis and kills the cell. Today, thanks to intense vaccination efforts using inactivated toxin (called a toxoid), diphtheria is rare in the United States.

Why is diphtheria toxin made? Iron availability appears to be a key factor. The body holds its iron very tightly in proteins such as lactoferrin and ferritin. So to an invading microbe, the body seems an iron-poor environment. This apparent lack of iron makes it hard for the pathogen to grow. Diphtheria toxin, whose production is iron-regulated, offers the organism a way to rob the host's iron stores by killing the cell. The pathogen then uses transport mechanisms to "vacuum" up the released iron.

Link   As described in Section 8.7, diphtheria toxin is iron-regulated: When plentiful, iron binds to the repressor protein DtxR, which binds and represses the dtx toxin gene because the toxin isn't needed. Given an iron deficit in the body, DtxR can no longer repress dtx and diphtheria toxin is made. Many pathogens use the quest for iron to control the expression of pathogenicity genes.

SHIGA TOXIN: *SHIGELLA* AND *E. COLI* O157:H7   Shiga toxin also destroys host protein synthesis, but it does so by a mechanism different from that of diphtheria toxin. *S. flexneri* and *E. coli* O157:H7 (also known as enterohemorrhagic *E. coli*) cause foodborne diseases whose symptoms include bloody diarrhea. These organisms produce an important AB-type toxin called **Shiga toxin** (or Shiga-like toxin). The gene (*stx*) encoding Shiga toxin is part of a phage genome integrated into the bacterial chromosome. The A subunit, upon entering the host, destroys protein synthesis by cleaving 28S rRNA in eukaryotic ribosomes (ricin toxin has a similar mode of action). This rRNA is needed to identify the beginning of an mRNA that has entered a ribosome. Strains that produce high levels of Shiga toxin are associated with acute kidney failure known as hemolytic uremic syndrome. Like diphtheria toxin, iron availability is also a key factor for inducing the expression of *stx*.

ANTHRAX   A century ago, anthrax (caused by *Bacillus anthracis*; pictured in the chapter opener) was mainly a disease of cattle and sheep. Humans acquired the disease only by accident. Today we

Figure 18.15
Corynebacterium diphtheriae

Methylene blue stain of *Corynebacterium diphtheriae*. Note the club-like appearance of the organisms.

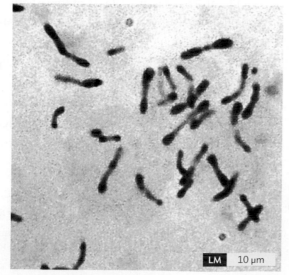

LM   10 µm

fear the deliberate shipment of *B. anthracis* through the mail or its dispersion from the air ducts of heavily populated buildings (see the opening case of this chapter). What makes this Gram-positive, spore-forming microbe so dangerous is the secretion of a three-component toxin (encoded in a plasmid). Anthrax toxin is composed of a membrane-binding protein (B subunit) and two enzyme components (A subunits). The B subunit of the toxin is called **protective antigen (PA)** because antibodies to this protein *protect* hosts from disease. The pathogen secretes PA, which binds to host cell surface receptors and autoassembles in the membrane as a seven-membered pore (**Figure 18.16**, steps 1–3). The toxic components (A subunits) of anthrax toxin, called **edema factor (EF)** and **lethal factor (LF)**, bind to the PA pores, and endocytosis carries them into the cell (Figure 18.16, steps 4 and 5). The two proteins then pass through the vacuole's pore into the host cytoplasm.

Figure 18.16 *Bacillus anthracis* **Anthrax Toxin Mode of Entry**
Anthrax toxin from *B. anthracis* enters the host cell (EF = edema factor; LF = lethal factor).

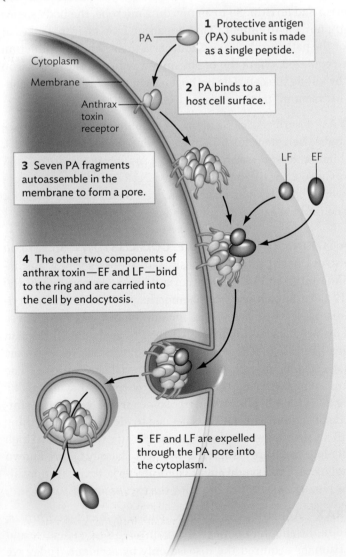

1 Protective antigen (PA) subunit is made as a single peptide.

PA

Cytoplasm

Membrane

Anthrax toxin receptor

2 PA binds to a host cell surface.

3 Seven PA fragments autoassemble in the membrane to form a pore.

LF   EF

4 The other two components of anthrax toxin—EF and LF—bind to the ring and are carried into the cell by endocytosis.

5 EF and LF are expelled through the PA pore into the cytoplasm.

EF and LF are enzymes that attack the signaling functions of the host cell. Edema factor is an adenylate cyclase that remains inactive until entering the cytoplasm, where it binds host calmodulin (much like the exotoxin from *Bordetella* pertussis discussed above). This binding inactivates calmodulin but activates adenylyl cyclase, resulting in the overproduction of cAMP. The result is the secretion of electrolytes and water that accumulate outside cells in affected tissues (edema).

The third component of anthrax toxin, LF, is a protease that cleaves components of critical regulatory cascades that affect cell growth and proliferation. One consequence of subverting these regulatory cascades is a failure to produce chemokines that call immune cells to the affected area to fight the infection. Lacking chemokines, the affected cell is essentially "gagged" and cannot call for help.

We have examined only a few of the many toxins that pathogens employ. The chapters on infectious disease will describe some of the others, including tetanus and botulism toxins. What our brief sampling reveals, however, is the evolutionary "ingenuity" pathogens use to tame the human host.

 Anthrax video: youtube.com

Only Gram-Negative Bacteria Make Endotoxin (LPS)

Endotoxin is an important virulence factor common to all Gram-negative bacteria. Not to be confused with secreted *exotoxins*, endotoxin is part of the **lipopolysaccharide (LPS)** that forms the outer leaflet of the Gram-negative outer membrane (Section 5.3). LPS is composed of modules of lipid A (the endotoxic part of LPS), core glycolipid, and a repeating polysaccharide chain (**Figure 18.17**). When Gram-negative bacteria die, endotoxin is released. Endotoxin can then bind macrophages or B cells to trigger the release of cytokines such as TNF-α, interferon, and interleukin 1 (IL-1) (Sections 15.3 and 16.7). The massive release of these active agents, known as a cytokine storm, causes a variety of symptoms, such as:

- Fever
- Activation of clotting factors leading to disseminated intravascular coagulation
- Activation of the alternate complement pathway
- Vasodilation, leading to hypotension (low blood pressure)
- Shock due to hypotension
- Death when other symptoms are severe

Note Lipopolysaccharides are also the outer-membrane structures called O antigens used to classify different strains of *E. coli* as well as other Gram-negative organisms (for example, *E. coli* O157 versus *E. coli* O111).

Figure 18.17   Endotoxin

LPS membrane

Water   Membrane   LPS   Water

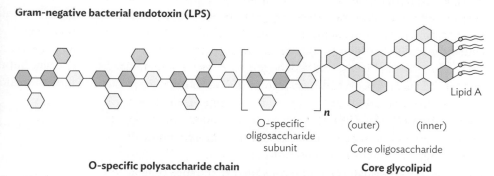

Gram-negative bacterial endotoxin (LPS)

Lipid A

O-specific oligosaccharide subunit

(outer)   (inner)

Core oligosaccharide

O-specific polysaccharide chain

Core glycolipid

A. Model of a lipopolysaccharide (LPS) membrane of *Pseudomonas aeruginosa*, consisting of 16 LPS molecules (red) and 48 ethylamine phospholipid molecules (white).

B. Basic structure of endotoxin, showing the repeating O-antigen side chain (farthest away from the microbe) and the membrane-proximal core glycolipid and lipid A (contains endotoxic activity).

You can see the role of endotoxins in disease in infections with the Gram-negative diplococcus *Neisseria meningitidis* (**Figure 18.18A**), a major cause of bacterial meningitis. *N. meningitidis* has, as part of its pathogenesis, a septicemic phase where the organism can replicate to large numbers in the bloodstream. The large amount of endotoxin present massively depletes clotting factors. This depletion leads to internal bleeding, most prominently displayed to a physician as small pinpoint hemorrhages called **petechiae** on the patient's hands and feet (**Figure 18.18B**; see also Chapter 15 inSight). Capillary bleeding near the surface of the skin causes the petechiae. Gram-negative sepsis is always treated with antibiotics; however, this strategy comes with a calculated risk. During treatment, the enormous release of endotoxin from dead bacteria could kill the patient. Untreated Gram-negative sepsis is, however, almost always fatal, so treating it, albeit risky, is imperative.

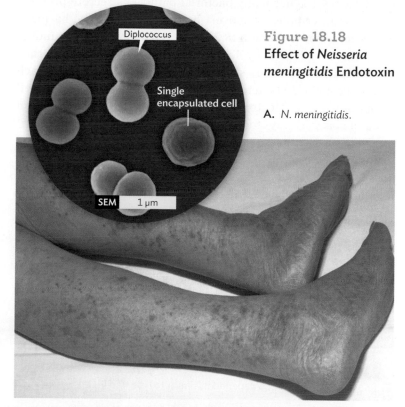

Figure 18.18
Effect of *Neisseria meningitidis* Endotoxin

Diplococcus

Single encapsulated cell

SEM   1 μm

A. *N. meningitidis*.

B. Petechial rash caused by *N. meningitidis*.

SECTION SUMMARY

- **Nine categories of protein exotoxins exist,** based on their host cell target. These include toxins that disrupt membranes, inhibit protein synthesis, act as superantigens, restructure the cytoskeleton, cleave cell-cell adhesion proteins, alter host cell signaling, redirect vacuolar traffic, prevent exocytosis, and disrupt the cell cycle.

- *S. aureus* **alpha toxin** forms pores in host cell membranes.

- **AB subunit toxins** are common among bacterial toxins. The B subunit always promotes penetration through host cell membranes, and the A subunit has toxic activity. Some AB toxins add ADP-ribose to specific host proteins. These include **cholera toxin**, *E. coli* **labile toxin**, and **pertussis toxins**, which alter host cAMP production, and **diphtheria toxin**, which inactivates eukaryotic protein synthesis.

- **Shiga toxin** is an AB toxin that cleaves host cell 28S rRNA in host cell ribosomes.

- **Anthrax toxin** is a three-part AB toxin with one B subunit (protective antigen) and two different A subunits that affect cAMP levels (edema factor) and cleave host signal proteins (lethal factor).

- **Lipopolysaccharide**, also known as endotoxin, is an integral component of Gram-negative outer membranes and an important virulence factor that triggers massive release of cytokines from host cells. The indiscriminate release of cytokines can cause fever, shock, and death.

18.4
Secretion of Virulence Proteins

SECTION OBJECTIVES

- Differentiate type II, type III, and type IV secretion systems used to export exotoxins/effector proteins.
- Discuss the evolutionary relationships between secretion systems and other microbial structures.

How do exotoxins get out of the bacterial cell? The previous section stressed that many bacterial pathogens secrete proteins that destroy, cripple, or subvert host target cells. The bacterial toxins described in Section 18.3 are secreted into the surrounding environment, where they float until chance intervenes and they hit a membrane-binding site. However, other pathogens attach to a tissue and *inject* bacterial proteins directly into the host cell cytoplasm. These effector proteins may not kill the cell but can redirect host pathways to benefit the microbe.

Here we describe some secretion systems crucial to delivering toxins or effector proteins to the external environment or to target host cells. A particularly interesting aspect of these secretory systems is that they evolved from, and bear structural resemblance to, other, more innocuous systems. Secretion systems that are evolutionarily related to other molecular machines include:

- Type II protein secretion, paralogous to type IV pili
- Type III protein secretion paralogous to flagellar synthesis machinery
- Type IV protein secretion paralogous to conjugation machinery

Link Recall from Section 9.5 that **paralogous** genes are a class of homologous genes (genes whose sequences are similar). Paralogous genes arise from a gene duplication that occurs in a single species, but the duplicated genes diverge via mutation to have different functions.

Table 18.3 gives examples of virulence proteins associated with these export systems. All these systems are associated with Gram-negative bacterial pathogens.

Before describing specific virulence protein export systems, we should briefly discuss the **general secretion pathway**. Present in both Gram-negative and Gram-positive bacteria, the general secretion pathway (called SecA-dependent) moves many proteins, some involved with virulence, from the cytoplasm across the cytoplasmic membrane and into the external environment for Gram-positive bacteria, or into the periplasm of Gram-negative bacteria (Figure 18.3). From the periplasm, more specific transport systems can export some of these proteins across the outer membrane. Some virulence protein secretion systems capture and secrete periplasmic virulence proteins across the outer membrane. Other secretion systems snatch virulence proteins directly from the cytoplasm, move them across the two bacterial membranes (bypassing the general secretion system), and expel them from the cell.

Type II Secretion Systems Are Pistons

Type II secretion offers a clear example of how nature has modified the blueprints of one system to do a different task. DNA sequence analysis shows that the genes used to make a type IV pilus (see

Table 18.3
Secretion Systems for Bacterial Toxins

Secretion Type	Features	Examples
I	SecA dependent,[a] one effector per system	*E. coli* alpha hemolysin, *Bordetella pertussis* adenylate cyclase
II	SecA dependent, similar to type IV pili	*Pseudomonas aeruginosa* exotoxin A, elastase, cholera toxin
III	SecA independent, syringe, related to flagella, secrete multiple effectors from the bacterial cytoplasm directly into the target host cell cytoplasm	*Yersinia* Yop proteins, *Salmonella* Sip proteins, enteropathogenic *E. coli* (EPEC) EspA proteins, TirA
IV	Related to conjugational DNA transfers, secrete multiple effectors	*B. pertussis* toxin, *Helicobacter* CagA
V	Autotransporter, SecA dependent to periplasm, self-transport through outer membrane, one effector per system	Gonococcal and *Haemophilus influenzae* IgA proteases
VI	Related to phage tails, single effector	*Burkholderia* and *Vibrio cholerae* VgrG

[a]The general secretory system, which requires SecA protein, transports some virulence proteins to the periplasm of Gram-negative bacteria. These toxins are then transported across the outer membrane by a separate transporter.

Section 18.2) were duplicated at some point during evolution and redesigned to serve as a protein secretion mechanism. Recall that type IV pili have the unusual ability to extend and retract from the outer membrane, producing the twitching motility of *Neisseria* and *Pseudomonas*. Type II protein secretion mechanisms mirror this property. Virulence proteins to be secreted via type II systems first make their way to the periplasm via the general secretion pathway. Type II secretion systems then use a pilus-like structure as a piston to ram the folded proteins through an outer membrane pore structure and into the surrounding void (**Figure 18.19**). Piston action occurs via cyclic assembly and disassembly of pilus-like proteins. Cholera toxin is a well-known example of a protein that a type II secretion mechanism expels.

Type III Secretion Is a Syringe

The etiological agents of Black Death and various forms of diarrhea (caused by species of *Yersinia*, *Salmonella*, and *Shigella*) can take bacterial virulence proteins made in the cytoplasm and *inject* them directly into the eukaryotic cell cytoplasm without the protein ever getting into the extracellular environment. Direct delivery is a good idea because it eliminates the dilution that happens when a toxin is secreted into media. This strategy also avoids the need to tailor the toxin to fit an existing host receptor. The same export system can deliver several different proteins.

How does direct delivery happen? Some microbes use tiny molecular syringes, called **type III secretion** systems, embedded in their membranes to inject proteins directly into the host cytoplasm. **Figure 18.20** shows type III secretion needles and a model of the complex. Cell-to-cell contact between host and bacterium triggers secretion through type III systems. Type III systems are evolutionarily related to flagellar genes, whose products export flagellin proteins through the center of a growing flagellum.

Virulence proteins secreted by type III systems hijack normal host cell signaling pathways. Some secreted proteins can cause dramatic, actin-driven rearrangements of host membranes that ultimately engulf the microbe (**Figure 18.21**). Engulfment is not, however, the only virulence strategy associated with type III systems. Some pathogenic strains (also called pathovars) of *E. coli* use type III secretion to *avoid* entering the host cell. These systems deliver effector proteins that alter host actin polymerization and cause the host cell to erect an elevated membrane pedestal on which the *E. coli* perch (**Figure 18.22**).

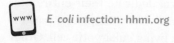
E. coli infection: hhmi.org

Figure 18.19   Type II Secretion

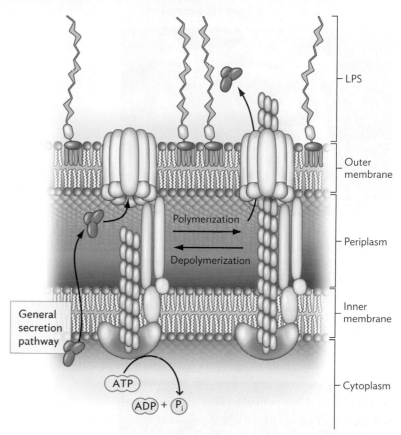

A general secretion system, called Sec-dependent, transports a type II secretion protein from the cytoplasm to the periplasm of a Gram-negative bacterium. The protein is then loaded into the type II secretion mechanism, where polymerization of the pilus-like ramrod propels the protein across the cytoplasm and into the surrounding medium. After secretion, the ramrod depolymerizes from the base.

Figure 18.20   Needle Complex of the *Salmonella* Type III Secretion System

Unlike other secretion systems, the type III mechanism injects proteins directly from the bacterial cytoplasm into the host cytoplasm. The proteins in these systems are related to flagellar assembly proteins.

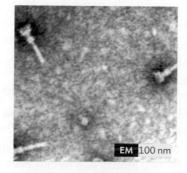

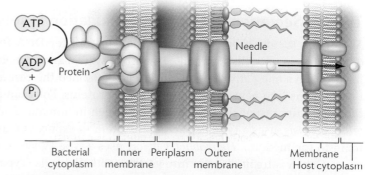

A. Purified type III needle complexes from *Salmonella*.

B. Schematic of the *S.* Typhimurium needle complex penetrating a host cell membrane.

Figure 18.21 *Shigella* **Invades a Host Cell Ruffle Produced as a Result of Type III Secretion**

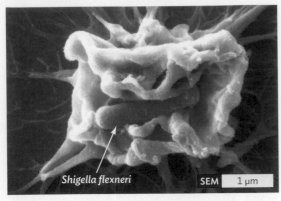

Shigella flexneri   SEM   1 µm

Shigella flexneri entering a human cell. The bacterium interacts with the host cell surface and injects (via its type III secretion apparatus) its invasin proteins, which choreograph a local actin-rich membrane ruffle at the host cell. The ruffle engulfs the bacterium and eventually disassembles, internalizing the bacterium within an endosome vacuole.

Figure 18.22   *E. coli* **Engineers Membrane Pedestals**

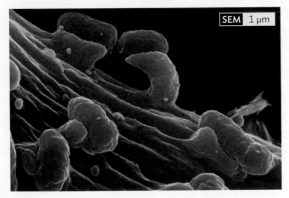

SEM   1 µm

Some pathovars of *E. coli* use type III secretion systems to inject effector proteins into host cells that redirect host actin polymerization to build a membrane pedestal. The pedestal helps prevent engulfment of this extracellular pathogen.

Type IV Secretion Resembles Conjugation Systems

As Section 9.2 describes, many bacteria can transfer DNA from donor to recipient cells via a cell-cell contact system known as conjugation. In some pathogens, evolution has modified the conjugation systems into new systems that export proteins. For example, the bacterium responsible for whooping cough in humans, *Bordetella pertussis*, uses a **type IV secretion** system to export pertussis toxin to the medium. *Helicobacter pylori* uses a type IV system to inject CagA exotoxin directly into gastric epithelial cells. Type IV systems move proteins directly out of the bacterial cytoplasm and across the cytoplasmic and outer membranes, but not necessarily across the host cell membrane.

SECTION SUMMARY

- **Many pathogens use specific protein secretion pathways** to deliver exotoxins. Types II, III, and IV secretion systems are restricted to Gram-negative rods.
- **Type II secretion** systems use a pilus-like extension/retraction mechanism to push proteins from the periplasm out of the cell. These systems evolved from type IV pilus genes.
- **Type III secretion** uses a molecular syringe to inject proteins from the bacterial cytoplasm into the host cytoplasm. These systems evolved from flagellar genes.
- **Type IV secretion** systems use conjugation-like machinery to export proteins from the cytoplasm. These systems evolved from conjugation genes.

18.5
Surviving within the Host

SECTION OBJECTIVES

- Distinguish extracellular pathogens from facultative and obligate intracellular pathogens.
- Discuss the three main ways intracellular pathogens avoid intracellular destruction.
- Describe various strategies pathogens use to avoid the immune system.
- Explain why and how intracellular pathogens distinguish between intracellular and extracellular existence.

If the immune system is so good at killing microbes, how do we ever get sick? Can pathogens somehow hide from the immune system? Once inside a human host, a successful pathogen must avoid detection and destruction for as long as possible. Many products of the virulence arsenal help the microbe escape or resist innate immune mechanisms. Others are dedicated to stealth, that is, hiding from the immune system. But how do pathogens know they are in a host?

The microbe uses the same regulatory mechanisms that sense environmental conditions in a pond to learn its whereabouts in a host. Using two-component signal transduction systems, discussed in Section 9.1, the microbe monitors conditions (iron, magnesium, and pH) in a host cell vacuole. These conditions help inform the pathogen of its whereabouts and tell it whether virulence genes should be engaged.

Intracellular Pathogens

Intracellular pathogens seek refuge from innate and humoral immune mechanisms by invading and living inside host cells as unwanted guests. Hiding intracellularly offers the pathogen temporary safe harbor because antibodies and phagocytic cells will not penetrate live host cells. Some bacteria dedicate their entire lifestyle to intracellular parasitism. *Rickettsia*, for example, for reasons still unknown, will not grow outside a living eukaryotic cell. Other microbes, such as *Salmonella* and *Shigella*, are considered **facultative intracellular pathogens** because they can live either inside host cells or free from them. We have discussed how intracellular pathogens get into cells, but how do they withstand intracellular attempts to kill them?

CASE HISTORY 18.2

A Threatened Fetus

Shakia was in her fourth month of pregnancy and excited to become a mother. A 33-year-old African-American from Indianapolis, Shakia had been in good health until about a week ago, when she started to experience abdominal pain, fever, muscle aches, and diarrhea. She visited her obstetrician to make sure it was nothing serious. When the obstetrician asked whether she had eaten anything unusual in the past few weeks, Shakia recalled eating some soft Mexican cheese about 3 weeks earlier during a visit to Oregon with her husband. The physician sent a sample of Shakia's blood and stool to the local diagnostic laboratory and prescribed the antibiotic ampicillin as treatment. The lab reported finding a Gram-positive, non-spore-forming bacillus called *Listeria monocytogenes* in both the fecal and blood samples. Shakia had mild gastroenteritis and septicemia. The septicemia placed her unborn child at serious risk of infection because this bacterium can cross the placental barrier. If not treated quickly, an infected fetus has a 70% chance of dying from meningitis and multiple organ system failure. The doctor hoped she treated Shakia in time.

L. monocytogenes is a facultative intracellular bacterial pathogen that causes a mild gastrointestinal illness in healthy adults but can infect a fetus in the womb or a newborn during parturition (birth). The organism, which grows at both body and refrigerator temperatures, is usually ingested in contaminated soft cheeses or luncheon meat and avoids the immune system by growing inside intestinal epithelial cells and macrophages. Using similar tactics, the pathogen can cross the placental barrier and infect a fetus.

Any bacterial pathogen that seeks refuge in a host cell initially enters by endocytosis, ending up in a phagosome. Once inside the phagosome, intracellular pathogens have three options to avoid being killed by a phagolysosome (Figure 18.23). They can survive in a phagolysosome (Fate 1), prevent phagosome-lysosome fusion (Fate 2), or escape the phagosome (Fate 3).

ESCAPE FROM THE PHAGOSOME The Gram-negative bacillus *Shigella dysenteriae* and the Gram-positive bacillus *Listeria monocytogenes*, both of which cause food-borne gastrointestinal disease (Section 22.5), use hemolysins to break out of the phagosome vacuole before it fuses with a lysosome that contains lethal enzymes. Once free in the cytoplasm, these facultative intracellular bacilli are thought to enjoy unrestricted growth, having found ways to redirect host cell function to their own ends.

A fascinating aspect of escaping the phagosome involves motility. In vitro, *Shigella* and *Listeria* are both nonmotile (no flagella) at 37°C but manage to move around inside the host cell. How do they move? These species are equipped with a special device at one end of the cell that mediates host cell actin polymerization behind the bacterium. The result is called an actin tail that propels the organism forward through the cell until it reaches a membrane (Figure 18.24). The membrane is then pushed into an adjacent cell, where the organism once again ends up in a vacuole. This strategy allows the microbe to spread from cell to cell without ever encountering the extracellular environment, where it would be vulnerable to attack.

Another bacterial pathogen that breaks out of the phagosome is the obligate intracellular, Gram-negative rod *Rickettsia prowazekii*, the cause of epidemic typhus. But unlike *Shigella* or *Listeria*, *R. prowazekii* appears to be an "energy parasite" that can transport ATP from the host cytoplasm and exchange it for spent ADP in the bacterium's cytoplasm. However, this ability does not explain its obligate intracellular status, since giving *Rickettsia* ATP outside a host does not allow the bacterium to grow. Other factors that prevent *Rickettsia* from growing outside a eukaryotic cell remain to be discovered.

INHIBITING PHAGOSOME-LYSOSOME FUSION Some intracellular pathogens avoid the hazard of lysosomal enzymes by preventing the lysosome from fusing with the phagosome. *Salmonella*, *Mycobacterium*, *Legionella*, and *Chlamydia* are good examples. For example, *Legionella pneumophila* grows inside alveolar macrophage phagosomes and produces the potentially fatal Legionnaires' disease, so named for the veterans group that suffered the first recognized outbreak in 1976 (Section 20.3). The organism, once inside a phagosome, secretes proteins through the vesicle membrane and into the cytoplasm. These bacterial proteins interfere with the cell-signaling pathways that cause phagosome-lysosome fusion. The result is that *L. pneumophila* can grow in a friendlier, less toxic vesicle. The organism survives in the macrophage, which then shuttles (disseminates) the organism to the lymph nodes and circulation (Figure 18.23; Fate 2).

THRIVING UNDER STRESS In what could be called the "grin and bear it" strategy, some intracellular pathogens are adapted to grow in the harsh environment of the phagolysosome. *Coxiella burnetii*, for example, grows well in the acidic phagolysosome environment (Figures 18.23, Fate 1, and 18.24B). This obligate intracellular organism causes a flu-like illness called Q fever (query fever), with symptoms including sore throat, muscle aches, headache, and high fever. It has a mortality rate of about 1%, so most people recover. The organism allows phagosome-lysosome fusion because the acidic environment that results enables *Coxiella* to survive and grow.

Extracellular Immune Avoidance

Can extracellular pathogens also avoid the immune system? Chapters 14 and 15, on host defense, first discussed this topic. Many bacteria, such as *Streptococcus pneumoniae* and *Neisseria meningitidis*, produce a thick polysaccharide capsule that envelops the

Figure 18.23   Alternative Fates of Intracellular Pathogens

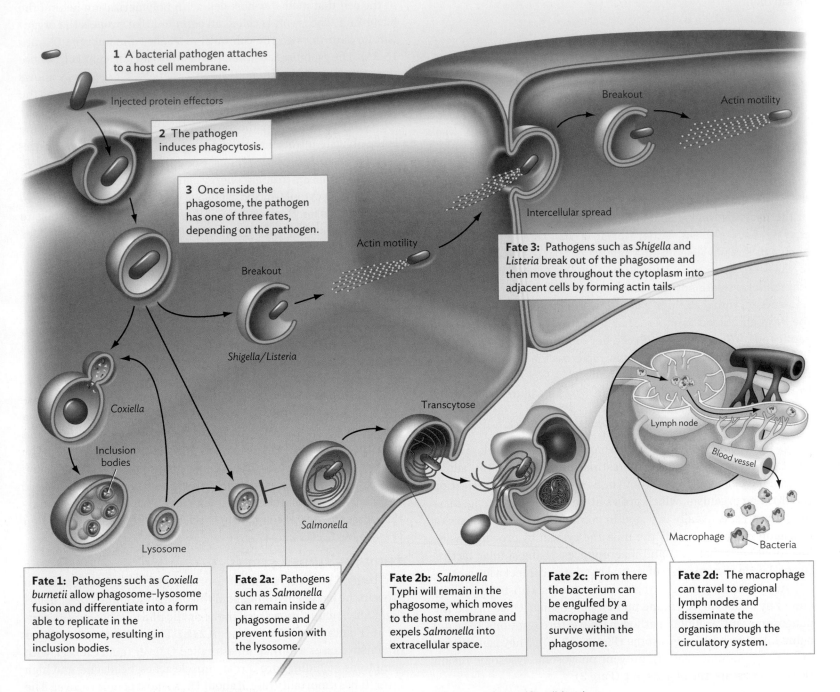

1 A bacterial pathogen attaches to a host cell membrane.

Injected protein effectors

2 The pathogen induces phagocytosis.

3 Once inside the phagosome, the pathogen has one of three fates, depending on the pathogen.

Breakout

Actin motility

Breakout

Intercellular spread

Actin motility

Fate 3: Pathogens such as *Shigella* and *Listeria* break out of the phagosome and then move throughout the cytoplasm into adjacent cells by forming actin tails.

Shigella/Listeria

Coxiella

Inclusion bodies

Transcytose

Lymph node

Blood vessel

Lysosome

Salmonella

Macrophage   Bacteria

Fate 1: Pathogens such as *Coxiella burnetii* allow phagosome-lysosome fusion and differentiate into a form able to replicate in the phagolysosome, resulting in inclusion bodies.

Fate 2a: Pathogens such as *Salmonella* can remain inside a phagosome and prevent fusion with the lysosome.

Fate 2b: *Salmonella* Typhi will remain in the phagosome, which moves to the host membrane and expels *Salmonella* into extracellular space.

Fate 2c: From there the bacterium can be engulfed by a macrophage and survive within the phagosome.

Fate 2d: The macrophage can travel to regional lymph nodes and disseminate the organism through the circulatory system.

Different pathogens have different strategies to survive in a host cell. Some tolerate phagolysosome fusion (*Coxiella*), others prevent phagolysosome fusion (*Salmonella*, *Legionella*), and still others escape the phagosome to replicate in the cytoplasm (*Shigella* and *Listeria*). *Salmonella enterica* serovar Typhi, the cause of typhoid fever, invades intestinal cells, prevents phagolysosome fusion, exits the intestinal cell, and invades awaiting macrophages. The bacterium can survive inside the macrophage, which then shuttles the pathogen to lymph nodes and to the circulation.

cell. Capsules coat bacterial cell wall components that phagocytes normally use for attachment (see Section 15.4). The slippery nature of capsules makes "grabbing" the bacterial cell difficult for phagocytes. Fortunately, immune defense mechanisms can circumvent this avoidance strategy by producing opsonizing antibodies (immunoglobulin G) against the capsule itself (see Section 15.4). The **Fc regions** of antibodies that bind to the capsule stick out away from the bacterium, where the Fc receptors on phagocyte membranes can recognize them. Binding of the Fc region to the phagocyte's Fc receptor can then trigger phagocytosis.

Link Recall from Section 16.3 that an antibody is shaped like a Y. At the top of the Y are two antigen-binding sites (called Fab sites). The tail of the Y, called the **Fc region**, can bind to Fc receptors on phagocyte membranes.

Certain cell-surface proteins enable some pathogens to avoid phagocytosis. *Staphylococcus aureus* has a cell wall protein called **protein A** that binds to the Fc region of antibodies, hiding the bacteria from phagocytes. This works in two ways. Protein A can bind to the Fc region of an antibody and cause the antigen-binding sites to point *away* from the microbe. However, even if the antigen-binding site binds to one bacterium, protein A from a second bacterium can bind to the Fc region and block phagocyte recognition.

Other microbes can trigger apoptosis (programmed cell death) in target host cells. The pathogen makes proteins that enter the host cell and trigger this "suicide" program. If the macrophage self-destructs, it cannot destroy the microbe.

Many pathogens can alter their antigenic structure to avoid detection by the immune system. Some bacterial pathogens have multiple genes encoding alternative proteins for surface structures that our immune system easily recognizes, such as flagella and pili. Sequential expression of alternative proteins that are antigenically different from each other will confuse the immune system; an immune system that has been "tuned" to flagellin A will not recognize flagellin B. So, *Salmonella* that switched from flagellin A to flagellin B will escape the initial immune response, at least for a time.

Many bacteria and viruses also use imitation (mimicry) to confuse the immune system. Some microbes make proteins that look like cytokines. These factors manipulate the balance of helper T cells and send immunity down the wrong path for combating the microbe. All these strategies buy the microbe more time to overwhelm the host.

Cell-cell communication via quorum sensing can also help pathogens evade the immune system. Recall that in quorum sensing the concentration of a secreted chemical **autoinducer** molecule rises as the number of bacteria in an infected space increases. Once the pathogen detects these autoinducer molecules, the pathogen increases virulence gene expression. Why would a pathogen employ quorum sensing to regulate virulence factors? One reason may be to prevent alerting the host that it is under attack before enough microbes can accumulate. Tripping the host's alarms too early would

Figure 18.24   Intracellular Pathogens

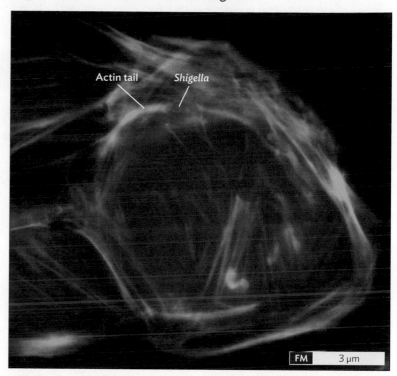

A. Actin tails (fluorescently stained green) propel intracellular *Shigella flexneri* (stained red) through the host cytoplasm. *Shigella* is a facultative intracellular pathogen.

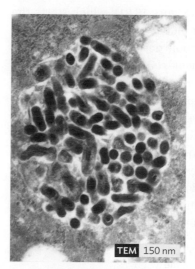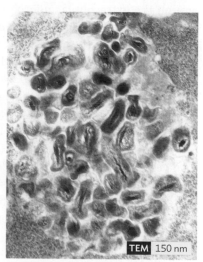

B. A typical vacuole in mouse macrophage cells infected with *Coxiella burnetii* at 2 hours (left) and 6 hours (right) after infection. The organism, which prefers to live in the acidified vacuole, undergoes a form of differentiation that changes its shape and alters its interactions with the host cell. *Coxiella* is an obligate intracellular pathogen.

make eliminating infection easy. However, releasing toxins and proteases once many bacteria have amassed could overwhelm the host.

Salmonella: A Model of Bacterial Pathogenesis

The enteric pathogen *Salmonella enterica* uses type III secretion systems to invade the eukaryotic host cell and become an intracellular parasite. After someone ingests contaminated food or water, this bacterium attaches to and invades M cells interspersed along the intestinal wall within Peyer's patches. M cells are specialized intestinal epithelial cells (see Figure 15.8B) that sample the intestinal microbiota for the immune system. However, many intestinal pathogens use M cells as a portal to cross the epithelial barrier. *Salmonella*, after invading M cells, causes an inflammatory response that leads to diarrhea.

To break into the M cell, *Salmonella* uses a type III protein secretion system (encoded by pathogenicity island 1 [SPI-1]) to deliver a cocktail of at least 13 different protein toxins (called effector proteins) directly into the cytosol of the M cell (Figure 18.25). These effector proteins interfere with signal transduction cascades and modulate the host response. One mission of these effectors is to rearrange the cytoskeleton, ruffling the eukaryotic membrane around the extracellular microbe (Figure 18.21), which starts the process of engulfment. *Salmonella* induces this response to avoid the normal endocytic process that could destroy it.

Once *Salmonella* enters an epithelial cell or a macrophage, it finds itself in a vacuole (phagosome). Normally, an enzyme-packed lysosome would then fuse with the phagosome and release its contents in an effort to kill the invader. *Salmonella*, however, possesses a second pathogenicity island called SPI-2 that subverts this host response. SPI-2 uses another type III secretion system to inject proteins that alter vesicle trafficking (Fate 3, as described in Figure 18.23), thereby reducing phagosome-lysosome fusion and sparing the intracellular bacteria. *S.* Typhi, but not *S.* Typhimurium, survives well inside macrophages and can use them to disseminate throughout the body.

SECTION SUMMARY

- **Intracellular bacterial pathogens** try to avoid the immune system by growing inside host cells. Pathogens use different mechanisms to avoid intracellular death.
- **Certain pathogens use hemolysins** to escape from the phagosome and grow in the host cytoplasm.
- **Some microbes use actin tails** to move within and between host cells.
- **Inhibiting phagosome-lysosome fusion** is one way pathogens can survive in phagosomes.
- **Survival in a phagolysosome** requires that the intracellular pathogen have a specialized physiology that can counter the chemical attacks levied against it.

Figure 18.25   Schematic Overview of *Salmonella* Pathogenesis
Salmonella injects effector proteins into a host M cell to affect the activity of host proteins that trigger cell death (apoptosis), influence gene expression (via host proteins Cdc42 and JNK), influence electrolyte movements, and rearrange actin. Actin rearrangement ruffles the host membrane to engulf the organism.

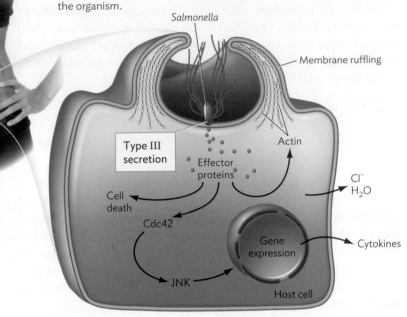

- **Pathogens can evade the immune system** by hiding in capsules, changing their surface proteins, triggering apoptosis, or using molecular mimicry.
- **Two-component signal transduction systems** can regulate virulence gene expression in response to the environment.
- **Quorum sensing** will prevent pathogens from synthesizing virulence proteins too early during infection.

Thought Question 18.4   How can one determine whether a bacterium is an intracellular parasite?

Thought Question 18.5   Why might killing a host be a bad strategy for a pathogen?

18.6
Viral Pathogenesis

SECTION OBJECTIVES

- Discuss antigenic variation in rhinovirus and influenza viruses.
- Explain how human immunodeficiency virus (HIV) targets T cells and describe the roles of Nef and Tat proteins in HIV pathogenesis.
- Outline how human papillomavirus (HPV) causes unregulated cell division.
- Explain latency of herpes virus and the role of microRNAs in helping the virus reactivate.

Are virulence mechanisms of viruses different from those of bacteria? The pathogenic strategies of viruses are similar to and yet distinctly different from those of bacteria. On their own outside a host cell, viruses are inert objects, lacking motility and unable to replicate. Like tiny molecular water mines, they float in the environment until bumping into a suitable target cell. Once in a host cell, viruses have no need to transport food, expel waste, or generate energy, because the host does all that. But each virus must still attach to a host cell and subvert its biochemistry, directing it to make more virus particles. Chapter 12 discussed basic virus replication strategies. However, pathogenesis involves more than replication. The progression of disease depends on where in the body the virus replicates, how it interacts with the immune system, and how the virus subverts host cell functions. This section will describe some of these concepts. Chapters 19–24 will then build on those concepts to describe the infectious diseases of various organ systems.

Rhinovirus versus Influenza Virus

You have probably wondered why you repeatedly get colds if the immune system is so effective. Shouldn't you be protected from recurring infections after you have one cold? The problem is that the common cold is caused by more than 100 known serotypes of rhinovirus, a small (30 nm), nonenveloped, single-stranded RNA virus. Serotypes of a virus (or bacterial species) differ from one another in the antigenic structure of a specific protein.

The host receptor for these viruses is ICAM-1 (see Figure 16.4), which is also important for the extravasation (movement) of leukocytes through blood vessels (see Section 15.3). Many differences among rhinovirus strains can be traced to the structure of one of the rhinovirus capsid proteins, VP1, which binds to host ICAM-1. Because of strain differences in VP1, antibodies that neutralize one strain of rhinovirus will not neutralize a different strain. Thus, we repeatedly contract a cold because we are infected with a different strain.

Unlike many viruses, rhinoviruses replicate most efficiently at 33°C. They grow in the cooler regions of the respiratory tract, such as nasal passages. These viruses cause a runny nose because virus-infected cells release bradykinin and histamine, both of which cause fluid loss from local blood vessels (discussed in Section 15.3).

Certain strains of coronavirus also cause common cold symptoms. However, recent variants of coronavirus are responsible for the newer, deadlier diseases known as severe acute respiratory syndrome (SARS-CoV) and Middle East respiratory syndrome coronavirus (MERS-CoV), which Chapter 26 will discuss.

Influenza infection proceeds differently from rhinovirus infections (Chapter 20), even though the public often confuse the two. Influenza virus (an RNA virus introduced in Section 12.5) first establishes a local upper respiratory tract infection, where it targets and kills mucus-secreting, ciliated epithelial cells, destroying this primary defense. If the virus spreads down to the lower respiratory tract, it can cause shedding of bronchial and alveolar epithelium down to the basement membrane (a thin sheet of fibers lying below epithelial cells that line mucosal surfaces of organs, such as

in the lung). This process compromises oxygen and CO_2 exchange, and breathing becomes difficult. Systemic symptoms, such as muscle aches, are due to the release of interferon and lymphokines in response to viral interactions with Toll-like and NOD-like receptors. Because their natural defenses are compromised, persons infected with flu are susceptible to superinfection with bacterial pathogens such as *Haemophilus influenzae* or *Streptococcus pneumoniae*. What makes flu so dangerous is that every 10–15 years the virus undergoes a major genetic alteration, called an **antigenic shift**, that results in a new potential pandemic strain to which almost no one on Earth is immune (*actual* pandemics typically occur in 30-year cycles).

Antigenic shift occurs when two strains of influenza that usually infect different animals infect the same cell, usually in animals such as pigs or chickens. Recall that influenza has a segmented RNA genome (Section 12.5). Thus, as two influenza viruses disgorge their eight nucleic acid segments in the same cell, reassortment can occur between them, generating a chimeric virus that may have new pairings between the hemagglutinin and neuraminidase genes (as well as other genes), which reside on different genome segments. The new influenza virus may then "jump" to humans. A new pandemic begins if hemagglutinin or neuraminidase in the new virus has not previously been in a human influenza strain. Antibodies that recognize the old hemagglutinin or neuraminidase will not recognize and neutralize the new one.

Link As Section 12.5 discusses, hemagglutinin and neuraminidase are influenza virus proteins located in the virion's envelope. Hemagglutinin is so named because it binds to and agglutinates red blood cells. The virus uses it to attach to host cell receptors. Neuraminidase is an enzyme that cleaves host surface sugar molecules (sialic acid) that prevent virus release from an infected cell. Hemagglutinin is required for virus entry and neuraminidase is required for virus release.

Random mutations caused by the error-prone viral RNA polymerase can cause minor changes in hemagglutinin antigen structure as well, in a process called **antigenic drift**. Antigenic shifts and drifts are why a new vaccine (the flu shot) is required each year. These shots are highly recommended for the elderly, who are susceptible to the disease, and are a good idea for everyone else.

An estimated 21 million people worldwide died of the flu in the pandemic of 1918 (Figure 18.26A). Many flu epidemics have taken place since then, but none have been so deadly. Though different from influenza, what was so frightening about the SARS outbreak that began in 2003 is that it had a fatality rate of greater than 10%, making it deadlier than the 1918 influenza strain, which had a fatality rate of 2%–4%.

The 1918 influenza virus was thought lost, and with it the opportunity to learn why it was so deadly. However, scientists sequenced influenza genes recovered from the preserved lung tissue of a U.S. soldier who died from influenza in 1918 and from an Alaskan influenza victim who was buried in the permafrost that same year. In

Figure 18.26   Flu Epidemic of 1918

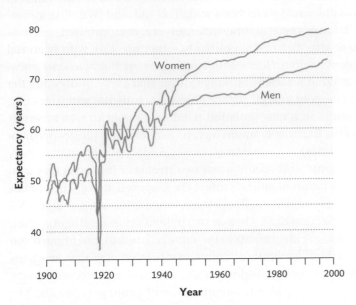

A. Life expectancy in the United States in the twentieth century. Notice the dramatic dip in 1918, the result of deaths from influenza.

B. Pathogenesis of the 1918 influenza virus. The virus is shown with its eight genome segments and the viral protein (letter designations) each encodes. Proteins contributing to the extreme virulence of the 1918 virus are shown with their effects. Inset is of a letter carrier on his mail route in 1918 wearing a cloth face mask designed to prevent spread of the disease. The gauze mask would have worked had the agent been a bacterium, which was suspected at the time, but it was ineffective at stopping transmission of the tiny influenza virus.

2005, scientists regenerated an active version of the 1918 pandemic virus, which is kept under tight security and containment.

Why was the 1918 flu strain so virulent? To find out, scientists spliced genes from the resurrected strain into a common-variety seasonal flu strain and infected mice with the recombinant viruses. Mice infected with a virus containing the hemagglutinin (HA), neuraminidase (NA), or polymerase (PB1) gene from the 1918 virus had much more severe symptoms than those of the seasonal flu (Figure 18.26B). At least two other genes from the 1918 virus proved important as well. One gene triggered infiltration and apoptosis (cell suicide) of macrophages, increasing inflammation; another (called NS1) is a potent inhibitor of type I **interferon**, which allowed the virus to escape an important part of antiviral innate immunity.

Link Recall from Section 15.5 that type I **interferons** are cytokines that virus-infected cells secrete. Once interferon enters a nearby uninfected host cell, it renders that host cell resistant to viral infection by inhibiting viral replication. Type I interferons also stimulate production of surface MHC I molecules that can present viral antigens to cytotoxic T cells.

Human Immunodeficiency Virus

Acquired immunodeficiency syndrome (AIDS) became an important disease in the late twentieth century and continues to grip us today. Some 30 million–40 million cases exist worldwide. Infection by the retrovirus HIV (human immunodeficiency virus) causes the patient to become severely immunocompromised and susceptible to a wide variety of infectious diseases. Death in patients with AIDS is usually the result of these secondary infections. Section 12.6 discusses the molecular biology of HIV replication, and Section 21.4 will describe the disease; here we will deal with mechanisms of HIV pathogenesis.

In contrast to rhinovirus, which exclusively binds to one cell receptor, ICAM-1, HIV binds multiple cell receptors via its spike protein, glycoprotein 120 (gp120) (Section 12.6; see Figure 12.28). HIV spike protein binds host cell CD4 protein (meaning T cells are the usual targets), as well as the chemokine receptor (CCR5; see Figure 12.28). By binding to the chemokine receptor, HIV can block chemokine binding and disconnect communication between the target cell and the immune system.

The virus also makes a protein (negative factor, Nef) that has several functions, one of which is to prevent T-cell apoptosis, which would destroy the virus. Apoptosis is a process triggered in distressed human cells, as would occur during an infection. Apoptosis is a protective measure; by killing the host cell, it also destroys viruses attempting to replicate. You could say the apoptotic cell "takes one for the team." HIV, however, has evolved to keep its host cell alive, at least for a while. Nef protein inhibits a regulatory cascade that leads to apoptosis. As a result, the T cells live longer so virus can replicate and bud from them (Figure 18.27A).

Figure 18.27 HIV and AIDS

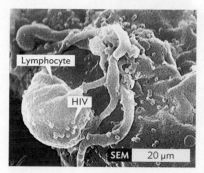

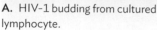

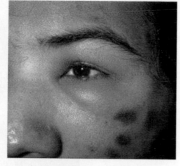

A. HIV-1 budding from cultured lymphocyte.

B. Kaposi's sarcoma (oval spots) and periorbital cellulitis infection.

Host lymphocytes eventually do die from attack by **CD8 cytotoxic T cells** and **NK cells** despite the presence of Nef. Because CD4 T cells are the primary target of HIV, the levels of these T cells decline dramatically in an infected patient and cause an immunodeficiency. Monitoring T-cell numbers, therefore, is a crucial diagnostic tool.

Link Recall from Section 16.7 that activated, viral antigen-specific **CD8 cytotoxic T cells** will recognize and kill any host cells presenting that viral antigen on their MHC I receptors. HIV-infected T cells present HIV antigens on their MHC I receptors and become targets for antigen-specific cytotoxic T cells. **NK cells** kill HIV-infected cells primarily via antibody-dependent cell cytotoxicity, described in Section 15.5.

Another important cause of AIDS symptoms is the HIV Tat protein (trans-activator of the HIV promoter). A trans-activator is a protein that can diffuse through the cytoplasm to bind and regulate a target gene sequence. The virus needs the Tat protein in order to replicate, but infected monocytes also release Tat. The protein then enters other cells, where it alters host regulatory cascades and gene expression. Tat decreases MHC I synthesis, upregulates and downregulates various cytokines, and stimulates apoptosis, all of which help to disassemble the immune system. A weakened immune system renders the victim more susceptible to infections by other pathogens such as human herpes virus 8 (HHV8) that can cause Kaposi's sarcoma (Figure 18.27B). Kaposi's sarcoma is a malignancy of endothelial or lymphatic cells and can arise anywhere—gastrointestinal tract, mouth, lungs, skin, or brain.

Because Tat protein plays a central role in viral replication and subduing the immune system, some believe that a vaccine using inactivated Tat will help prevent AIDS. Table 18.4 presents other HIV proteins important to the pathogenesis of HIV and AIDS.

Simply having detectable HIV does not mean someone is overtly ill. Only when T-cell numbers fall to around 300 per microliter of blood (the normal range is 500–1,500) will the signs of early symptomatic HIV infection appear, such as recurrent infections (see Section 23.4).

Human Papillomavirus

CASE HISTORY 18.3

Sex, Warts, and Cancer

Jeneen is a 45-year-old migrant crop worker from Belize. Over the past 2 months she has noticed abnormal vaginal bleeding accompanied by a foul-smelling yellowish vaginal discharge. She was also experiencing increasing pain in her pelvis. She visited a free clinic near the California farm where she worked, hoping a simple medicine would help. The clinic doctor performed a pelvic exam and colposcopy to view her cervix. He saw an advanced stage of cervical cancer (Figure 18.28). He took a sample for DNA testing and a Pap smear to confirm the neoplasm. Jeneen admitted having six sexual partners over the years. She also recalled seeing a wart-like growth in her genital area about 20 years ago. At the time she was too embarrassed to go to the doctor—but the warts disappeared after a few months, so she thought nothing of it. A week later the Pap smear results confirmed his diagnosis: late-stage cervical cancer. The DNA test revealed the presence of human papillomavirus DNA. Tragically, Jeneen would never return to Belize.

Table 18.4
HIV Proteins That Affect Pathogenesis

Protein	Role in Pathogenesis
Tat (transcriptional trans-activator)	Accelerates HIV transcription by host polymerase; is also secreted from host cells and, upon entering an uninfected cell, can trigger apoptosis (programmed cell death)
Vpr (viral protein R)	Regulates the import of HIV into the nucleus before the viral cDNA integrates into the host chromosome
Nef (negative factor)	Downregulates CD4 and MHC class I proteins, protecting the infected cell from killing by CD8 cytotoxic T cells; slows apoptosis in infected T cells; secreted Nef activates apoptosis in uninfected lymphocytes
Vif (virion infectivity factor)	Counteracts a host cellular protein that naturally inhibits HIV replication
Vpu (viral protein U)	Counteracts a host cellular protein that inhibits virus assembly and export

Figure 18.28   HPV-Associated Cervical Cancer

Normal   Cancerous

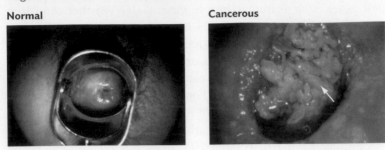

Colposcopic view of a normal and cancerous cervix. A colposcope is an instrument used to examine the cervix and vagina.

Figure 18.29
Human Papillomavirus

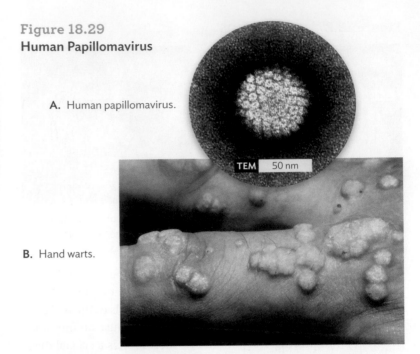

A. Human papillomavirus.

TEM   50 nm

B. Hand warts.

Human papillomavirus (HPV) is a common DNA virus that infects skin cells, causing them to divide excessively (Figure 18.29A). The result is warts (Figure 18.29B). HPV can produce warts on the feet, hands, vocal cords, mouth, and genital organs. More than 60 types of HPV have been identified so far. Each type infects different parts of the body, with specific strains of HPV associated with cervical and penile cancer. Genital HPV is, in fact, one of the most common sexually transmitted diseases among college students (Chapters 19 and 23 further discuss the diseases HPV causes).

Why does HPV cause excessive cell division? The answer is self-preservation. Because HPV needs host replication machinery to make progeny virus, the more often HPV makes host cells divide, the more often virus particles can be made. But how does HPV stimulate host cell division? Several mechanisms are involved. In one, HPV synthesizes a viral protein called E7 that inactivates the host protein pRb, a protein that normally *limits* host cell division. The host pRb protein limits cell division by preventing the transcription of genes needed for mitosis and for cell cycle progression toward mitosis (c-Myc and cyclin E [CcnE], respectively). The properly regulated cell will stop dividing (Figure 18.29C).

HPV protein E7 disrupts this control by sending pRb to the cell's **proteasome**, which degrades damaged or unneeded cellular proteins. These unneeded proteins are selectively degraded because a protein modification called **ubiquitination** "tags" them. The HPV E7 protein stimulates ubiquitination of pRb and sends it to its doom. With pRb removed, cell division control is reduced. Virus and host cell division is at least partially unrestrained. Skin warts are the result, or, in the worst-case scenario, cervical or penile cancer develops (usually 15 years or more after the initial lesion). Even after the initial infection resolves (90% resolve within 2 years), the virus may remain latent in tissues and can reactivate years later if the immune system becomes impaired. Fortunately, researchers have developed an effective HPV vaccine that will protect against most cervical cancer if given before exposure (Section 23.4 will discuss this vaccine).

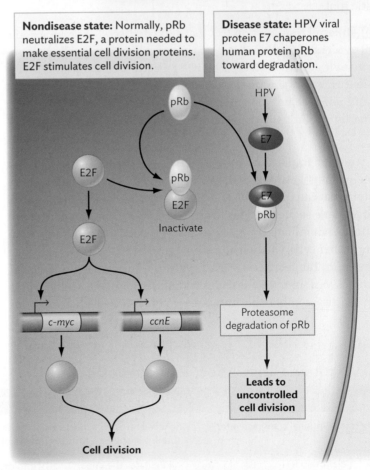

Nondisease state: Normally, pRb neutralizes E2F, a protein needed to make essential cell division proteins. E2F stimulates cell division.

Disease state: HPV viral protein E7 chaperones human protein pRb toward degradation.

C. Part of the strategy HPV uses to stimulate cell division in infected cells. Protein pRb normally helps limit cell division by inactivating E2F. Virus E7 protein, however, inactivates pRb. EF2, therefore, keeps stimulating cell division so the cell divides uncontrollably and becomes cancerous.

Latent Herpes Virus

Herpes viruses are double-stranded DNA viruses that cause a variety of diseases, including cold sores (herpes simplex type 1, HSV-1), genital herpes (herpes simplex type 2), mononucleosis, and some cancers (Burkitt's lymphoma caused by Epstein-Barr virus). Later chapters will discuss many of these further. A key feature of herpes virus infections is the development of a latent state in neural ganglia or leukocytes, depending on the virus. Like a bear waking from hibernation, active virus can emerge from latency years after the primary infection. Reactivation of latent herpes virus is most commonly noticed as recurrent cold sores around the mouth or as genital herpes lesions.

HSV DNA circularizes after entering the cell nucleus and becomes latent by shutting down viral replication genes. The proviral DNA then floats freely in the nucleus (the HSV genome rarely integrates into the host DNA). The latent virus DNA can reactivate by unknown mechanisms after exposure to UV light, emotional stress, fever, or immune suppression. The results are cold sores or recurrent genital lesions. However, viral replication can trigger apoptosis in which the infected cell kills itself to destroy the virus and spare nearby host cells. Herpes virus, however, avoids triggering host cell apoptosis (and its own demise) during virus reactivation. How?

Latent herpes virus DNA produces several small RNA molecules called microRNAs (miRNA, 20–25 nt long) that interfere with the host cell apoptosis program. This process is especially important when a latent virus reactivates. The miRNAs inhibit the two known pathways that trigger apoptosis, although the inhibitory mechanism remains unclear. In a clever turn, the latent virus DNA makes only miRNA and no viral proteins. Because no foreign proteins are present to trigger an immune response, the host cannot detect the infected cell as long as the virus remains latent. And by virtue of inhibiting apoptotic pathways, once the virus reactivates, the infected cell will live long enough to make more virus.

SECTION SUMMARY

- **Multiple reinfection** of the same person by a virus can occur because different strains of the virus have minor sequence differences in attachment proteins (for example, capsid or envelope). Because of these differences, neutralizing antibodies made during a previous infection are ineffective against the new strain.
- **Viral infection** can diminish the immune response by lowering MHC I levels on host cells or by altering cytokine production (HIV, for example). The result can increase a patient's susceptibility to other, less virulent microbes.
- **Antigenic shifts** in a viral antigen (as in the hemagglutinin of influenza) can lead to a new pandemic of the disease.
- **Animals coinfected with two viruses** (either similar strains or even different viruses) can serve as incubators for antigenic shifts (influenza) or the evolution of new viruses.
- **Some pathogenic viruses can become latent** after infection. Herpes virus, for example, becomes latent, does not express viral proteins, and escapes immune surveillance mechanisms. Reactivated herpes viruses produce microRNA molecules that prevent the host cell from undergoing apoptosis.

Thought Question 18.6 Why do rhinovirus infections fail to progress beyond the nasopharynx?

Thought Question 18.7 How do you think an infection with one type of pathogen could influence coinfection with a second pathogen?

18.7
Protozoan Pathogenesis

SECTION OBJECTIVES
- Describe different ways protozoans invade host cells.
- Correlate the immune response to protozoan-induced damage to host tissues.
- Explain several mechanisms protozoans use to avoid the immune system.

We have seen how bacteria and viruses cause human disease, but what about eukaryotic microbes? Is there, in fact, eukaryote-on-eukaryote virulence? Protozoans cause many different human diseases, including malaria (*Plasmodium*), sleeping sickness (*Trypanosoma*; Figure 18.30A), amebic dysentery (*Entamoeba*), and trichomoniasis, a sexually transmitted vaginitis. Protozoan infections produce tissue damage, but these pathogens, in general, do not produce toxins as potent as those of bacteria. With few exceptions (for example, *Entamoeba*), damage from protozoan infections is due primarily to persistence of the organism in tissues and the resulting immunopathology. The immune reaction causing the pathology can be triggered by protozoan antigens or by host antigens released as a result of pathogen growth.

Although they lack potent toxins, these parasites employ several other clever strategies to avoid or escape the immune system, a subterfuge that leads to persistent growth of the protozoan. Their immune-avoidance strategies include antigenic masking, antigenic variation, intracellular location, and immunosuppression.

Antigenic Masking

Some parasites escape immune detection by covering themselves with host antigens. Trypanosomes, for instance, appear to bind the Fc region of host antibodies. Binding antibody Fc regions enables the parasite to hide from the Fc receptors on phagocytes, thereby hindering phagocytosis.

Antigenic Variation

Some protozoan parasites tinker with their surface antigens, changing them during an infection. Parasites covered with the new antigens will escape the immune response made against the original antigens. Trypanosomes, for example, do not have an intracellular stage, which makes them an easy target for the immune system. Indeed, macrophages clear many trypanosomes from blood. Trypanosomes that

Figure 18.30

African Sleeping Sickness

A. Patients with African sleeping sickness. The inset shows the protozoan cause of the disease, *Trypanosoma brucei*, in blood. A painful nodule (lower right) on the leg of one girl shows the initial site of infection.

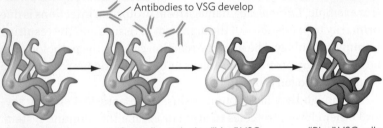

Trypanosome coated with VSG

One cell switched to "blue" VSG. The "green" VSG protozoa are killed.

"Blue" VSG cells repopulate blood.

B. Antigenic variation as an immune-avoidance mechanism. The figure illustrates *T. brucei* coated with one type of VSG antigen (green). Eventually antibodies build up that can attack the "green" form of VSG and kill the cells. However, a few protozoa will begin expressing a different VSG (blue) that the antibody does not recognize. These variants survive and repopulate the blood. This cycle continues because the *T. brucei* genome contains hundreds of silent *vsg* genes that can become activated—one at a time.

survive, however, do so because they have swapped to a different antigenic form of variant surface glycoprotein (VSG; **Figure 18.30B**). Each trypanosome contains a large repertoire of VSG genes with different sequences, but only one VSG gene is expressed at a time. Switching to a new VSG allows the pathogen to avoid the antibody response and produce successive surges of parasitemia (parasites in the blood). Antigenic variation and recurrent bouts of parasitemia

cause progressively serious symptoms in the patient, making the development of a preventive vaccine extremely hard.

Intracellular Location

The intracellular lifestyle of some protozoan parasites helps delay their recognition by the immune system. Even once they are discovered, being inside a host cell can still protect the parasite from the direct effects of the host's immune response. How do protozoans enter host cells? Different pathogens use different mechanisms. For example, *Leishmania*, a protozoan transmitted by tiny sand flies, engages normal phagocytic receptors on macrophages to engineer its entry. By contrast, *Toxoplasma* species, which probably infect 10%–20% of the U.S. population, use an actin-myosin motor of their own to forcibly drive themselves into a host cell (**Figure 18.31**). Although *Toxoplasma* infects at least an estimated 50% of the world's population, the disease toxoplasmosis is rare unless the patient is immunocompromised (for instance, an AIDS patient or the fetus of a pregnant mother).

Fortunately, the intracellular lifestyle offers protozoans only temporary refuge from the immune system. Some *Plasmodium* parasites that cause malaria, for instance, first grow in hepatocytes and then inside red blood cells, temporarily hiding the protozoan from the immune system. Eventually, however, the pathogen must place special proteins on the host cell surface that are needed for the parasite's survival or for causing the infected red blood cell to stick to blood vessel walls. The cell-mediated immune system will now detect and mount a response to these surface antigens.

WWW *Toxoplasma* invasion 5X speed

Immunosuppression

Another way protozoans survive in a host is to suppress the host's immune system, which parasitic protozoan infections generally do to some degree. This suppression delays the detection of antigenic variants and reduces the ability of the immune system to kill or hinder growth of the parasites. One good example is *Leishmania donovani*, which, as stated above, infects macrophages as part of its life cycle. As a rule, cell-mediated immunity (CMI; Chapter 16) best handles intracellular pathogens. Normal CMI develops when macrophages secrete cytokine IL-12 to stimulate helper T cells to differentiate into T_H1 cells. T_H1 cells then secrete the cytokine interferon-gamma needed to activate cytotoxic CD8 T cells. Cytotoxic T cells will then kill host cells harboring an intracellular pathogen. *L. donovani* avoids this fate by keeping infected macrophages from producing IL-12, undermining the CMI mechanism.

Plasmodium, on the other hand, makes a protein that mimics human macrophage migration inhibitory factor. This malarial protein alters the cytokine profile produced by antigen-presenting macrophages. The inappropriate levels and mix of cytokines inhibit the

Figure 18.31
Toxoplasma Entry into Host Cells

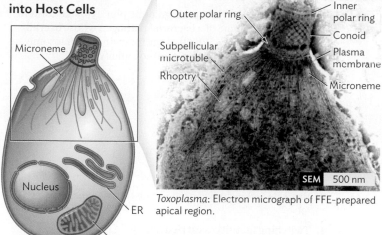

Toxoplasma: Electron micrograph of FFE-prepared apical region.

A. View of *Toxoplasma*. The apical end of *Toxoplasma* (see inset) sports a myosin motor attached to the parasite's actin filaments. This motor is used to carry out gliding motility. Microneme organelles secrete a number of microneme proteins (MIC) that facilitate attachment.

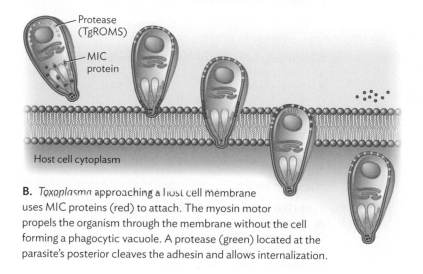

B. *Toxoplasma* approaching a host cell membrane uses MIC proteins (red) to attach. The myosin motor propels the organism through the membrane without the cell forming a phagocytic vacuole. A protease (green) located at the parasite's posterior cleaves the adhesin and allows internalization.

production of memory CD4+ T cells, so the recovering patient does not develop protective immunity and becomes susceptible to subsequent plasmodium infections.

SECTION SUMMARY

- **Protozoan parasites cause tissue damage** that is primarily the result of immunopathology.
- **Antigenic masking** by coating the parasite with host proteins can hide the organism from the immune system.
- **Growing inside host cells** offers parasites temporary sanctuary from immune recognition.
- **Periodically changing cell surface antigens** allows a parasite to evade the immune response to the previous antigen.
- **Altering the pattern of host cytokine production** is one way parasites can actively suppress the immune response to their presence.

Perspective

By now, it should be clear that attachment, immune avoidance, and subversion of host signaling pathways are common goals of most successful pathogens, be they bacterial, viral, or parasitic. But even with all we know, many aspects of infectious disease remain a mystery. Chapters 19–24 describe a variety of infectious diseases. Understanding the basic strategies of pathogenesis and the types of molecular weapons pathogens can wield will help you appreciate how infectious diseases are diagnosed, treated, and—sometimes—prevented.

You will learn, too, in the coming chapters, that new pathogens are constantly emerging. Over the last few decades, we have seen the development of HIV, SARS, avian flu, hantavirus, West Nile virus, and *E. coli* O157:H7, as well as the reemergence of "flesh-eating" streptococci, to name but a few. Can we ever stop pathogens from emerging? Probably not. For every countermeasure we develop, nature designs an effective counter-countermeasure. Our hope is that continuing research into the molecular basis of pathogenesis and antimicrobial pharmacology will keep us one step ahead.

LEARNING OUTCOMES AND ASSESSMENT

	SECTION OBJECTIVES	OBJECTIVES REVIEW
18.1 The Tools and Toolkits of Bacterial Pathogens	• Differentiate between pathogenicity genes and pathogenicity islands. • Correlate horizontal gene transfer with pathogen evolution.	1. The graph to the right plots guanine plus cytosine content across the genome of a bacterial pathogen whose overall GC content is 55%. Which of the following chromosome segments could represent a pathogenicity island? A. Segment A B. Segment B C. Segment C D. Segment D E. Segments A and C F. Segments B and D
18.2 Microbial Attachment	• Explain why pathogens need to attach to host cells. • Describe various microbial attachment techniques. • Differentiate the structure and function of type I and type IV pili. • Discuss the role of biofilms in pathogenicity.	3. Bacteria lacking pili cannot attach to host cells. True or false? 4. Twitching motility is carried out by _____. 5. Chronic infections are usually associated with _____. A. antibiotic-resistant bacteria B. biofilm-producing bacteria C. capsule-forming bacteria D. Gram-negative bacteria E. exotoxin-producing bacteria
18.3 Toxins: A Means to Seduce, Hijack, or Kill Host Cells	• Describe the nine basic cellular targets for bacterial toxins. • Explain the modes of action for staphylococcal alpha toxin, cholera toxin, diphtheria toxin, Shiga toxin, and anthrax toxin. • Differentiate endotoxin from exotoxin.	9. Which of the following toxins does calmodulin activate? A. Diphtheria toxin D. Shiga toxin B. Cholera toxin E. Anthrax EF toxin C. Staphylococcal alpha toxin 10. Which of the following toxins ADP-ribosylates a G-factor? A. Diphtheria toxin D. Shiga toxin B. *Escherichia coli* LT E. Anthrax EF toxin C. Staphylococcal alpha toxin
18.4 Secretion of Virulence Proteins	• Differentiate type II, type III, and type IV secretion systems used to export exotoxins/effector proteins. • Discuss the evolutionary relationships between secretion systems and other microbial structures.	13. Type III secretion systems are paralogous to _____. A. chromosome replication systems B. conjugation systems C. transduction systems D. flagellum synthesis systems E. type IV pili

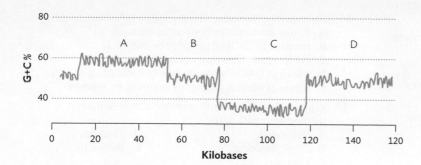

6. _____ pili continually assemble and disassemble.

7. *Bordetella* pertactin is an example of a(n) _____.
   A. pilus
   B. capsule
   C. nonpilus adhesion
   D. secreted effector protein
   E. endotoxin

2. Which of the following mechanisms contributes directly to pathogen evolution?
   A. Conjugation
   B. Type III secretion
   C. Exocytosis
   D. ADP-ribosyltransferase
   E. All of the above

8. Pilus contact between a pathogen and a host cell _____.
   A. delivers exotoxins into cells via the pilus
   B. enables the pathogen to avoid the immune system
   C. extracts nutrients from the host cell
   D. prevents evacuation of the pathogen
   E. All of the above
   F. None of the above

11. The following toxin stimulates cytokine production through a superantigen mechanism:
    A. Diphtheria toxin
    B. Shiga toxin
    C. Toxic shock syndrome toxin
    D. Tetanus toxin
    E. Staphylococcal alpha toxin

12. Endotoxin is associated with _____.
    A. *Escherichia coli*
    B. *Staphylococcus aureus*
    C. *Bacillus anthracis*
    D. *Streptococcus pyogenes*
    E. influenza virus

14. Type II secretion systems _____.
    A. export DNA and protein
    B. inject proteins directly into the host cytoplasm
    C. secrete proteins into the surrounding environment
    D. import secreted host proteins into the bacterium
    E. are homologous to conjugation systems

LEARNING OUTCOMES AND ASSESSMENT

18.5
Surviving within the Host

- Distinguish extracellular pathogens from facultative and obligate intracellular pathogens.
- Discuss the three main ways intracellular pathogens avoid intracellular destruction.
- Describe various strategies pathogens use to avoid the immune system.
- Explain why and how intracellular pathogens distinguish between intracellular and extracellular existence.

15. *Salmonella enterica*, while causing an infection, _____.
    A. moves intracellularly via actin tails
    B. remains in the host cell phagosome
    C. prefers growth within a phagolysosome
    D. uses a polysaccharide capsule to avoid phagocytosis
    E. makes protein A that binds antibody Fc regions

16. Which one of the following bacteria is an obligate intracellular pathogen?
    A. *Salmonella*   D. *Shigella*
    B. *Staphylococcus*   E. *Rickettsia*
    C. *Listeria*

18.6
Viral Pathogenesis

- Discuss antigenic variation in rhinovirus and influenza viruses.
- Explain how human immunodeficiency virus (HIV) targets T cells and describe the roles of Nef and Tat proteins in HIV pathogenesis.
- Outline how human papillomavirus (HPV) causes unregulated cell division.
- Explain latency of herpes virus and the role of microRNAs in helping the virus reactivate.

19. Which of the following best explains why one person can develop multiple colds over a lifetime?
    A. Ineffective B-cell memory for the rhinovirus
    B. Multiple rhinovirus strains that possess different VP1 structures
    C. A constantly changing ICAM-1 structure in hosts
    D. Ineffective antibody response to the rhinovirus
    E. Different strains of rhinovirus attach to different host cell receptors

18.7
Protozoan Pathogenesis

- Describe different ways protozoans invade host cells.
- Correlate the immune response to protozoan-induced damage to host tissues.
- Explain several mechanisms protozoans use to avoid the immune system.

21. Match the following organisms to the method of entry into host cells:

    Toxoplasma   A. Antibody-mediated entry
    Leishmania   B. Actin-myosin motor
                     C. Phagocytic cell receptors

17. *Listeria monocytogenes* _____.
    A. moves intracellularly via actin tails
    B. remains in the host cell phagosome
    C. prefers growth within a phagolysosome
    D. uses a polysaccharide capsule to avoid phagocytosis
    E. makes protein A that binds antibody Fc regions

18. Which of the following strategies helps a pathogen avoid the immune system?
    A. Delay pilus formation until physical contact with a target cell
    B. Immediately synthesize virulence factors upon entering a host
    C. Delay synthesis of virulence factors until a certain bacterial cell number is reached
    D. Turn off the secretion of all bacterial proteins
    E. Maintain an extracellular presence in the bloodstream

---

20. Matching

    HIV glycoprotein 120   A. Accelerates HIV transcription
    HIV Tat protein   B. Prevent(s) MHC I processing
    HPV E7 protein   C. CD4 binding
    Herpes microRNAs   D. Prevent(s) apoptosis
                                 E. Stimulate(s) host cell division

---

22. Half the world's population suffers from the disease toxoplasmosis. True or false?

23. Trypanosomes avoid the immune system by varying their surface antigens. True or false?

24. By which of the following mechanisms do *Leishmania* protozoans avoid the immune system?
    A. Upregulate TNF-α in infected macrophages
    B. Vary their surface antigens
    C. Inhibit plasma cell formation
    D. Invade red blood cells
    E. Downregulate IL-12 production in infected macrophages
    F. None of the above

Key Terms

AB toxin (p. 579)
adhesin (p. 570)
ADP-ribosyltransferase (p. 579)
alpha toxin (p. 579)
antigenic drift (p. 591)
antigenic shift (p. 591)
autoinducer (p. 589)
edema factor (EF) (p. 582)
endotoxin (pp. 575, 582)

exotoxin (p. 575)
facultative intracellular pathogen
   (p. 587)
fimbria (p. 570)
general secretion pathway (p. 584)
genomic island (p. 569)
intracellular pathogen (p. 586)
labile toxin (LT) (p. 580)
lethal factor (LF) (p. 582)

lipopolysaccharide (p. 582)
pathogenicity island (p. 569)
pathogenesis (p. 568)
petechia (p. 583)
pilus (p. 570)
proteasome (p. 594)
protective antigen (PA) (p. 582)
protein A (p. 589)
Shiga toxin (p. 581)

type I pilus (p. 572)
type II secretion (p. 584)
type III secretion (p. 585)
type IV pilus (p. 572)
type IV secretion (p. 586)
ubiquitination (p. 594)
virulence factor (p. 568)

Review Questions

1. Describe the basic features of a pathogenicity island.

2. Explain various ways bacteria can attach to host cell surfaces.

3. Describe the basic steps by which pili are assembled on the bacterial cell surface. How do type I and type IV pili differ?

4. Explain the broad categories of toxin mode of action.

5. What is ADP-ribosylation and how does it contribute to pathogenesis?

6. Explain the differences between exotoxins and endotoxins.

7. Explain the mechanisms of secretion that type II and type III protein secretion systems carry out. What are the paralogous origins of these systems?

8. Describe the key features of *Salmonella* pathogenesis.

9. What different mechanisms do intracellular pathogens use to survive within the infected host cell?

10. Describe different molecular strategies that microbes use to avoid the immune system.

11. How do bacteria determine whether they are in a host environment?

12. What is antigenic shift?

13. How is attachment of influenza virus different from HIV attachment? Why does HIV produce an immune deficiency, whereas influenza does not?

14. Explain in basic terms how human papillomavirus can trigger cancer.

15. Describe two basic mechanisms protozoan pathogens use to enter host cells.

16. What mechanisms do protozoans use to avoid the immune system?

17. Which phyla of protozoans include human pathogens? (You may need to consult Chapter 11.)

Clinical Correlation Questions

1. You have just learned that your cousin's newborn boy, only 2 days old, has listeriosis caused by *Listeria monocytogenes*. What can you tell her family about how the child most likely acquired the disease?

2. In the above case, which of the following test results would best indicate the child contracted the disease in utero? (You should review Table 16.1.) Positive indicates an elevated antibody level of the indicated antibody class.

	Anti-*Listeria* Antibodies			
	Child		Mother	
Scenario	IgM	IgG	IgM	IgG
1	Positive	Negative	Negative	Positive
2	Negative	Positive	Positive	Positive
3	Negative	Negative	Negative	Positive
4	Positive	Positive	Negative	Positive

3. In 2011 a new form of pathogenic *E. coli* emerged in Europe that caused bloody diarrhea and a kidney disease called hemolytic uremic syndrome in many victims. The organism (O104:H4) was unusual in that it carried a gene for a distinct form of enteroaggregative pilus, produced high levels of Shiga toxin, and carried a distinct set of other virulence genes never seen assembled in the same strain. What is the most likely explanation for the appearance of this new organism?

4. Cytomegalovirus can cause a life-threatening disease of newborns. During infection the virus tags MHC I molecules with ubiquitin in infected cells. Ubiquitination causes MHC I molecules to degrade before they can be transported to the cell surface. What is the most likely immunological consequence of this action?

Thought Questions: CHECK YOUR ANSWERS

Thought Question 18.1   What is the significance of the CSF results in the case that opens this chapter?

ANSWER: The results clearly indicated bacterial meningitis. Cerebrospinal fluid in a healthy person is sterile. It contains few, if any, white blood cells and almost no protein (<40 mg/dl), but it will contain a fair amount of glucose (40–70 mg/dl). In bacterial meningitis, white blood cells (mostly neutrophils) enter the CSF to attack the bacteria, raising WBC levels in CSF. The PMNs and the many bacteria present all contain protein, which dramatically increases the CSF protein level. Bacteria and PMNs will consume glucose; if the blood-brain barrier is disrupted, glucose transport is faulty. CSF glucose levels will thus decrease. Although an obvious decrease in CSF glucose did not happen in this case, the other indications were clear.

Thought Question 18.2   Antibodies to which subunit of cholera toxin will best protect a person from the toxin's effects?

ANSWER: Antibodies to the B subunit will be more protective. Inactivating the B subunit will prevent the toxin from binding to cell membranes. The A-subunit active site is typically sequestered in these toxins and inaccessible to antibody. Furthermore, once the A subunit has entered a host cell, antibodies cannot enter and neutralize it.

Thought Question 18.3   Would patients with iron overload (excess free iron in the blood) be more susceptible to infection?

ANSWER: With a few exceptions, withholding iron from potential pathogens is a host defense strategy because when iron is plentiful, the microbe does not have to expend energy to get it and so can readily grow. On the other hand, low iron levels can also be a signal to express various virulence genes, so for some organisms, a high iron level might hinder infection.

Thought Question 18.4   How can one determine whether a bacterium is an intracellular parasite?

ANSWER: Microscopic examination to see whether bacteria are found within cultured mammalian cells is usually not satisfactory. The difficulty lies in determining whether the organism is inside the host cell or just bound to its surface—or, if it is inside, whether it is a live or dead bacterium. One common approach is to add to infected cell monolayers an antibiotic that can kill the microbe but will not penetrate the mammalian cells—such as the protein synthesis inhibitor gentamicin (we did not discuss this previously). A bacterium that invades a host cell will gain sanctuary from gentamicin and grow intracellularly. Extracellular bacteria and bacteria attached to the host cell are killed. Intracellular bacteria are then released by lysing the mammalian cell membranes with gentle detergent and plating the lysate on agar media. Each colony that forms represents a single intracellular bacterium. Of course, this approach will work only if the microorganism is not an obligate intracellular parasite.

Thought Question 18.5   Why might killing a host be a bad strategy for a pathogen?

ANSWER: The goal of any microbe is to maintain its species. If a microbe does not have an opportunity to easily spread to a new host, killing its host would cause its own death.

Thought Question 18.6   Why do rhinovirus infections fail to progress beyond the nasopharynx?

ANSWER: For one thing, rhinoviruses are susceptible to acid pH (pH 3.0), so they cannot replicate in the gastrointestinal tract. They also grow best at 33°C, which may help explain their predilection for the cooler environs of the nasal mucosa.

Thought Question 18.7   How do you think an infection with one type of pathogen could influence coinfection with a second pathogen?

ANSWER: It will depend on the combination of pathogens. A pathogen that shares an important antigenic feature with a second pathogen could elicit an immune response that will dispatch the second pathogen. However, a pathogen that suppresses the immune response could make the patient more susceptible to secondary infection by other pathogens. For example, HIV depletion of T cells will make the patient more susceptible to bacterial, fungal, protozoan, and viral pathogens.

19

Infections of the Skin and Eye

[Rash Ruins Vacation]

SCENARIO Nate, a 69-year-old type 2 diabetic, awoke in his Naples, Italy, hotel room with a tingling sensation that extended from his spine to the middle of his rib cage, like a thin invisible ribbon.

SIGNS AND SYMPTOMS The discomfort increased until the next morning, when it became severe pain. Small blisters revealed themselves along an irritated red swath of skin. The hotel concierge referred Nate to the local doctor.

DIAGNOSIS The Italian doctor diagnosed Nate with St. Anthony's fire, a disease, the doctor said, caused by the same virus that causes chickenpox. Nate was surprised to learn that the chickenpox virus he had as a small boy had long ago nestled at the root of one of his spinal nerves. It stayed there until this trip, when it began replicating, causing the painful skin eruptions along the route of that nerve (dermatome).

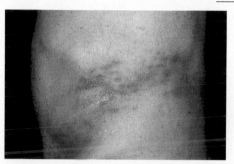

Shingles
Latent VZV emerges from spinal nerve to cause skin lesions along a dermatome.

TREATMENT The doctor prescribed antiviral and pain relief medications and told Nate to get plenty of rest and make sure that his blood sugar levels were well controlled. He advised Nate to follow up with his regular physician when back in the States.

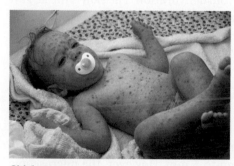

Chickenpox
Varicella-zoster virus emerges from skin cells to cause chickenpox vesicles. These are very infectious.

FOLLOW-UP Nate cut short his trip to Italy and returned home to the United States, where his doctor confirmed the diagnosis and the treatment plan. His doctor told Nate that in Italy, St. Anthony's fire is another name for shingles, a disease caused by the same virus that causes chickenpox.

RESOLUTION Nate eventually recovered from shingles with the help of his antiviral medication and rest, and he was finally able to return to Italy for a trip that turned out very nicely this time.

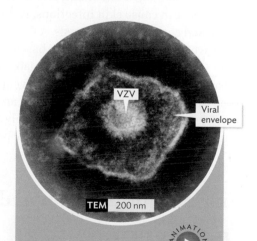

TEM 200 nm

Varicella-Zoster Virus
VZV is a DNA virus that gains a membrane envelope when it buds from an infected cell. The viral envelope fuses with the membrane of a new target cell.

CHAPTER OBJECTIVES

After reading this chapter, you will be able to:

- Correlate the anatomy and physiology of skin and eyes to the infectious processes affecting them.
- Describe symptoms that can help differentiate skin and eye infections caused by viruses from those caused by bacteria and parasites.
- Relate pathogenic mechanisms to disease prevention strategies for microbes that infect skin and eyes.
- Correlate the physiology of each pathogen with its antimicrobial treatment and preventive modalities.

Taken for granted as just the covering for other body tissues, human skin is not only a barrier deflecting continuous assaults on our body but also a mirror that can reflect the status of our health. Many infectious diseases manifest themselves as skin rashes or eruptions. Varicella-zoster virus (VZV), the virus that interrupted Nate's Italian vacation, causes the itchy, crusty, vesicular lesions of childhood chickenpox and the painful rash of shingles (also called herpes zoster) experienced by older adults. Although the virus erupts from the skin, the most common portal of entry is the respiratory tract. VZV, however, does not typically damage other tissues of the body but disseminates through blood to the dermis, where the virus actively replicates.

Varicella-zoster is not, of course, the only microbe that can cause skin infections. Measles, smallpox, warts, and cold sores, among other skin diseases, are all viral in origin. Various bacteria can cause boils, burn infections, and so-called flesh-eating disease. Fungi, too, can cause skin infections, including athlete's foot, ringworm, and cutaneous candidiasis. A few of these diseases and their causative agents can be diagnosed based solely on signs and symptoms. Other skin infections require laboratory testing. Some skin diseases, such as shingles, can be prevented through vaccination, some can be treated by antimicrobials, and some must simply run their course. This chapter focuses not only on skin but also on infections of the eye because of the eye's close proximity to the skin.

19.1
Anatomy of the Skin and Eye

SECTION OBJECTIVES

- List the different cellular components of the skin and their functions.
- Describe the structural similarities and differences between dermis and epidermis.

- Describe the role of different components of the skin in disease prevention.
- Distinguish between the internal and external parts of the eye.

Why is the skin considered an organ? Our skin is made of a combination of cells that work together to provide a physical barrier that protects us from pathogenic microbes and helps us regulate our body temperature and avoid dehydration. The **skin** is the largest human organ, covering, on average, 16–22 square feet of surface. It serves as an effective barrier for blocking microbial access to deeper tissues (see Section 15.2) even though the skin itself serves as host to 1 trillion microbes.

Structure of the Skin

The skin, with its superficial layer (**epidermis**) and deeper layer (**dermis**), covers the human body in varying thicknesses ranging from approximately 0.05 mm to 5 mm. However, the microscopic structure of the skin is the same regardless of its thickness.

The epidermis consists of five layers of epithelial cells that produce and secrete substances that act as waterproofing agents to reduce water loss from the body surface (**Figure 19.1**). The outermost layer, stratum corneum (or the horn layer), consists of dead keratinocytes that are shed roughly every 2 weeks. Keratinocytes are the cells that make the tough, fibrous, structural protein keratin. The deepest layer of epidermis, stratum basale (also called stratum germinativum), consists mostly of stem cells called basal cells as well as melanocytes, cells that produce melanin, the pigment that gives skin its color. Basal cells continuously divide and produce keratinocytes and more basal cells. The newly made keratinocytes push the older keratinocytes upward through the other epidermal layers, where the keratinocytes mature and produce increasing amounts of keratin. The mature keratinocytes eventually reach the surface, where they maintain the superficial layer of the skin by replacing

Figure 19.1   Structure of the Skin

The skin is composed of two main layers, epidermis and dermis. The epidermis contains five layers featuring progressively maturing keratinocytes.

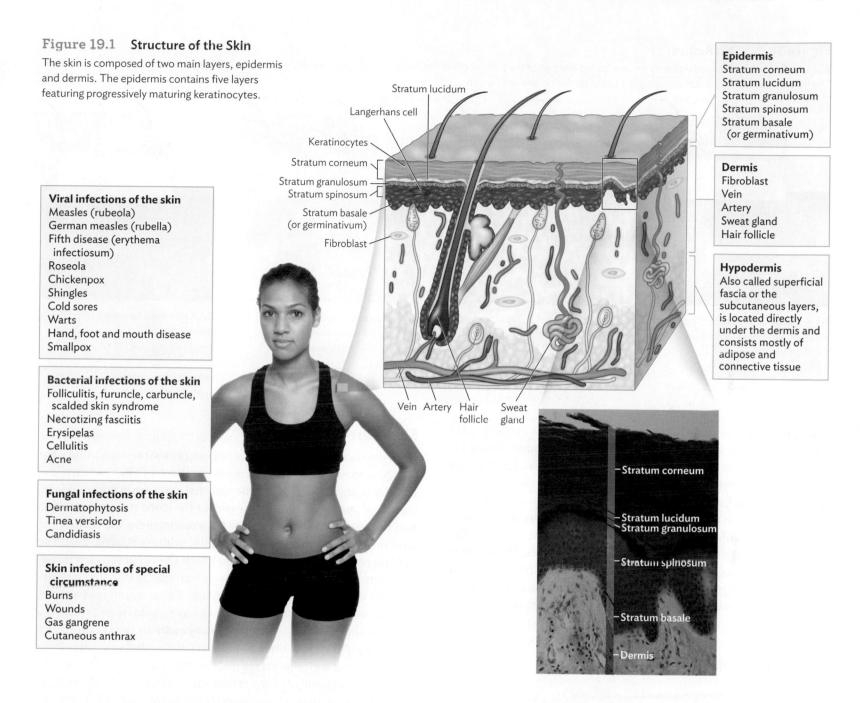

Epidermis
Stratum corneum
Stratum lucidum
Stratum granulosum
Stratum spinosum
Stratum basale
  (or germinativum)

Dermis
Fibroblast
Vein
Artery
Sweat gland
Hair follicle

Hypodermis
Also called superficial fascia or the subcutaneous layers, is located directly under the dermis and consists mostly of adipose and connective tissue

Stratum lucidum
Langerhans cell
Keratinocytes
Stratum corneum
Stratum granulosum
Stratum spinosum
Stratum basale (or germinativum)
Fibroblast

Vein   Artery   Hair follicle   Sweat gland

Viral infections of the skin
Measles (rubeola)
German measles (rubella)
Fifth disease (erythema infectiosum)
Roseola
Chickenpox
Shingles
Cold sores
Warts
Hand, foot and mouth disease
Smallpox

Bacterial infections of the skin
Folliculitis, furuncle, carbuncle, scalded skin syndrome
Necrotizing fasciitis
Erysipelas
Cellulitis
Acne

Fungal infections of the skin
Dermatophytosis
Tinea versicolor
Candidiasis

Skin infections of special circumstance
Burns
Wounds
Gas gangrene
Cutaneous anthrax

Stratum corneum
Stratum lucidum
Stratum granulosum
Stratum spinosum
Stratum basale
Dermis

dead keratinocytes that are shed. Langerhans cells are star-shaped, macrophage-like dendritic cells that reside in the epidermis above the basal layer. Langerhans cells survey the epidermis and remove foreign antigens (for example, from an infection), carrying them to a nearby lymph node for **presentation** to T lymphocytes.

Link As discussed in Section 16.2, antigen-presenting cells, such as phagocytes, degrade microbial pathogens and present distinct pieces on their cell-surface MHC proteins. Antigen **presentation** is an essential part of adaptive immunity.

The dermis is the layer under the epidermis and consists mostly of connective tissue. Here you can find the types of cells found in connective tissue, such as white blood cells and fibroblasts as well as

mast cells and macrophages. The dermis hosts blood vessels, nerves, hair follicles, and sweat glands. The hypodermis, also called superficial fascia or the subcutaneous layer, is located directly under the dermis and consists mostly of adipose and connective tissues.

What Is a Rash?

A change in color and texture of the skin is usually referred to as a rash. **Exanthem** is the name given to any widespread skin rash accompanied by systemic symptoms such as fever, malaise, and headache. **Enanthems** are rashes on mucous membranes. Exanthems and enanthems are usually caused by an infectious agent, such as a virus, and represent a reaction to a toxin produced by the organism, damage to the skin by the organism, or an immune response.

Figure 19.2   Skin Rashes

A skin rash consists of lesions that may be localized to one part of the body or cover the entire surface (generalized). The lesions that make up a rash are classified based on their shape and contents.

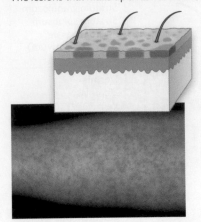  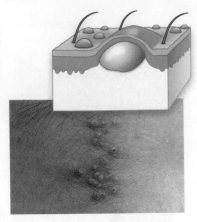 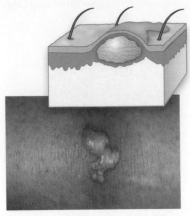

A. A macule is a flat lesion that cannot be palpated, like a freckle.

B. A vesicle forms from accumulation of fluid under the epidermis.

C. A papule is a raised lesion resulting from accumulation of material, infectious or otherwise, in the dermis.

D. A pus-filled raised lesion on the skin is called a pustule. It is usually the result of a buildup of the cellular debris of inflammatory cells, with or without microorganisms, under the epidermis.

Aside from infections, a variety of other situations can affect the skin and produce a rash. Allergies to drugs can produce noninfectious rashes (see Section 17.3). A number of factors, such as the form and type of the lesion, its location, whether it is spreading, and the pattern of spread, as well as other associated symptoms have to be considered in order to identify the cause as infectious or otherwise.

There are different types of rashes. A macular rash is a flat, often red segment of skin, called a **macule**, that is less than 1 cm in diameter, whereas a papular rash is a small, circumscribed, solid, elevated lesion, a **papule**. When a papule fills with pus, the lesion is called a **pustule**. A **maculopapular rash** is essentially a papule that is reddened. Sometimes, as in herpes infections of the skin, the rash consists of **vesicles**, small blisters, and is called a vesicular rash. Figure 19.2 illustrates various types of lesions.

 Database of rashes: hardinmd.lib.uiowa.edu

Mucous Membranes

Mucous membranes (called **mucosae**), which are epithelial in origin, line the inside of the body and are continuous with the skin in several places. The inside of your eyelids, gastrointestinal tract, and urogenital tract, for example, are covered by mucous membranes. Not all mucous membranes produce mucus (for example, those in the eye or mouth), but they all serve as a barrier to protect the structures they cover. We can think of mucous membranes as "the skin for inside the body." In this chapter, we focus on the infections of the skin and the external parts of the eye. Infections of the mucosa will be discussed in Chapters 22 (gastrointestinal diseases) and 23 (urogenital diseases).

Structure of the Eye

Our eyes are our organs of sight; they help us see the world, and what we see shapes our actions and reactions. Each eye has external and internal parts (Figure 19.3). The external parts, from the outside in, consist of our eyelids and outer part of the globe (the eyeball). The part of the eyeball that is outside the skull includes the cornea, the lens, the iris, the pupil, and the sclera (the white part of the eye). The outside parts of the eye are covered with the conjunctiva, a moist, transparent mucous membrane that is continuous with the conjunctiva that covers the inside of the eyelids. The conjunctiva becomes red and feels uncomfortable when we have a cold or an attack of a seasonal allergy, a condition called **conjunctivitis** (see Figure 19.3). The inside of the eye is lined by neuronal tissue that originates from the optic nerve and is called retina. The retina is orange in color. The macula is in the center of the retina and is where the best vision, or central vision, occurs. A viscous and transparent fluid called vitreous humor fills the inside of the globe, mostly to maintain the shape of the eye. Eye infections are always considered very serious because they have the potential to cause blindness.

SECTION SUMMARY

- **Skin** is the largest organ of the body.
- **Epidermis** consists of five layers of epithelial cells. Keratinocytes made in the deepest layer move upward through the other layers and replace dead cells shed from the outermost layer.
- **Langerhans cells** in the epidermis sample antigens for the immune system.
- **Dermis** is made of connective tissue and contains blood vessels, hair follicles, nerves, mast cells, and macrophages.

Figure 19.3   Structure of the Eye

The eye is a "ball"-shaped structure called a globe situated inside the eye socket of the skull. Muscles connected to the exterior of the globe help move it in different directions as we desire.

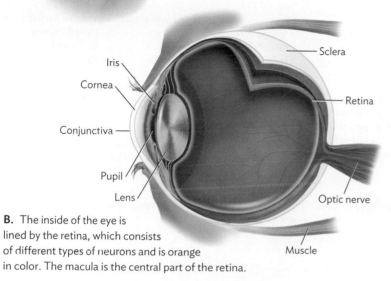

A. The external eye consists of the eyelids, the sclera, the iris, and the corneal opening called the pupil.

B. The inside of the eye is lined by the retina, which consists of different types of neurons and is orange in color. The macula is the central part of the retina.

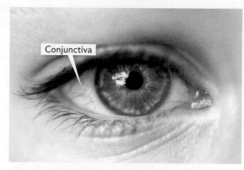

C. Conjunctiva.

Viral infections of the eye
Conjunctivitis
Keratitis
Trachoma

Bacterial infections of the eye
Conjunctivitis
Trachoma
Ophthalmia neonatorum

Fungal and parasitic infections of the eye
Fungal keratitis
Parasitic keratitis

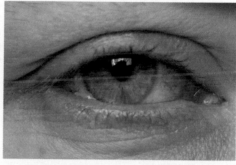

D. Inflamation of the conjunctiva, or conjunctivitis.

- **Enanthems and exanthems** are rashes associated with systemic symptoms. The forms of rashes include macules (flat red), papules (solid, elevated), vesicles (fluid filled), and pustules (papule filled with pus).

- **Eye anatomy** includes external structures (the conjunctiva and the cornea) and internal structures: the **retina**, orange-colored neuronal tissue lining the inside of the eye, and **macula**, the center of the retina, where the best vision occurs.

19.2
Viral Infections of the Skin

SECTION OBJECTIVES

- Classify viruses that cause skin rashes on the basis of their genomic structure.
- Explain the pathogenesis of viral skin infection sequelae.
- Outline how to diagnose the causative agent of a viral skin infection according to the patient's signs and symptoms.
- Recommend treatment and prevention options for different viral skin infections.

Why do we get viral skin infections? Many viruses prefer replicating in our skin cells, and we see the evidence of their visit as a rash. Some viruses, such as varicella, come for repeat visits, each time producing a different kind of rash (for example, chickenpox

or shingles). Some produce small infected blisters, called vesicles; others produce growths such as warts or tumors. The majority of viruses that cause a rash access our body first through our respiratory tract, examples are varicella-zoster virus (VZV) and rubella (German measles). But others, such as papillomavirus (warts), directly attack the epithelial cells of the skin. Table 19.1 lists several examples of viruses that affect the skin and eye.

Viruses That Produce Macular Rashes

CASE HISTORY 19.1

Moyo's Macular Rash

Assaggi, a young mother from Mudzengerere village, Zimbabwe, carried her young son, Moyo, 2 miles to the Mosomo clinic. When Moyo arrived, he appeared confused and nonresponsive. He had a high fever and even seized once during the examination. The clinic doctor diagnosed probable viral encephalitis. The physician, noticing that Moyo had a macular rash over much of his body, asked Assaggi whether Moyo had ever been vaccinated for measles. She was evasive at first but finally admitted that her son had not been vaccinated. The clinic doctor had been seeing many cases of measles recently, and a significant number of the affected children had died. He heard that the death toll from the measles epidemic in Zimbabwe stood at 570, 5% of all

Table 19.1

Viral Diseases of the Skin and Eye

Disease	Symptoms	Complications	Virus	Transmission	Diagnosis	Treatment	Prevention[a]
Measles (rubeola)	High fever, conjunctivitis, Koplik's spots, maculopapular rash	Myocarditis, pericarditis, ADEM, SSPE	Measles virus	Respiratory droplets, direct contact	Clinical presentation; serology	Treat symptoms; antiviral ribavirin is used for the immuno-compromised	Vaccine available (MMR routine childhood vaccination)
German measles (rubella)	Low-grade fever, maculopapular rash	Congenital rubella syndrome	Rubella virus	Respiratory droplets, direct contact	Clinical presentation; viral culture and RT-PCR	Treat symptoms	Vaccine available (MMR routine childhood vaccination)
Erythema infectiosum (fifth disease)	Slapped-cheek rash	Sickle cell disease patient: severe anemia; pregnant women: miscarriage, spontaneous abortion, or fetal defect	Parvovirus B19	Respiratory droplets; vertical and transfusion transmission also possible but not common	Clinical presentation; PCR	Treat symptoms; immunoglobulin treatment for immuno-compromised	No vaccine available
Roseola infantum (exanthem subitum, sixth disease)	High fever, macular or maculopapular red rash on body (not face), erythematous tympanic membrane	Possible neurological symptoms; rarely, encephalitis	Herpes simplex virus-6 (HSV-6) and HSV-7	Respiratory droplets	Clinical presentation	Treat symptoms	No vaccine available
Chickenpox	Generalized pruritic vesicular rash	Shingles	Varicella-zoster virus	Respiratory droplets	Clinical presentation; FAMA, ELISA, and PCR	Treat symptoms, avoid ibuprofen	Vaccine available (routine childhood vaccination)
Shingles	Painful vesicular rash with a dermatomal distribution	Keratitis	Varicella-zoster virus	The virus can be transmitted from a person who has shingles to a person who has not had chickenpox and has not been vaccinated; such transmission can cause chickenpox (not shingles) in the recipient	Clinical presentation; Tzanck smear; viral culture; PCR; direct fluorescent antibody staining	Antiviral treatment: valacyclovir, famciclovir, acyclovir	Vaccine available
Cold sores	Blisters usually around mouth (or genitals)	Herpetic whitlow, herpes gladiatorum, herpetic keratitis, encephalitis	HSV-1 (sometimes HSV-2)	Contact	Clinical presentation; Tzanck smear; viral culture; serological assays; PCR	HSV-1, acyclovir and famciclovir, valacyclovir; HSV-2, valacyclovir	No vaccine available
Common warts (verruca vulgaris)	Small, hard growths on the skin; some subtypes cause genital warts	Cosmetic complications	Human papillomavirus (HPV)	Contact	Clinical presentation	Removal by freezing, burning, or surgical resection	Vaccine is now available for some strains of HPV that cause genital warts (see Section 23.4)

(continued on next page)

Table 19.1
Viral Diseases of the Skin and Eye (continued)

Disease	Symptoms	Complications	Virus	Transmission	Diagnosis	Treatment	Prevention
Hand, foot and mouth disease	Fever, painful oral lesions, rash on palms of hands and soles of feet	Myocarditis, meningitis, and pneumonia	Coxsackievirus A16, enterovirus 17	Contact with respiratory or oral fluids, blisters, or feces of infected persons	Clinical presentation; RT-PCR of throat or stool samples	Supportive therapy	No vaccine available in U.S.
Smallpox	Enanthem on oral mucosa followed by a pustular, red, and crusted, exanthem on skin	Generalized sepsis, pneumonia, encephalitis, blindness, death from multiorgan failure, severe scarring in survivors	Variola major	Respiratory droplets	Eradicated and considered a biological weapon; suspected cases must be reported immediately, and diagnosis and treatment is per CDC protocols		Vaccine available

*Childhood vaccination schedule is found in Table 17.6.

suspected cases. He started Moyo on the antiviral medication ribavarin, but he was not optimistic. A week later Moyo was dead.

Mr. Fibion Rupiya, the village elder, sought assistance from the police and health officials after realizing that several children who recently died had symptoms of measles. The children belonged to a religious sect that disdained immunization. As a result of Mr. Rupiya's alert, more than 100 children in Mashonaland Central Province were forcibly vaccinated against measles. Some children were locked up in bedrooms by their parents, and others were hidden in the bush, but health officials managed to find the children and vaccinate them.

 Zimbabwe measles epidemic: promedmail.org, search archive 20100928.3515

MEASLES (RUBEOLA)   **Measles** is one of the six exanthem diseases of childhood and is sometimes called first disease (Table 19.2). Although we generally think of it as non-life-threatening, measles caused an estimated 530,000 fatalities among children younger than 15 years in 2003, even though an effective vaccine has been available since 1963. The measles virus is a member of the Paramyxoviridae family (negative-sense, single-stranded RNA viruses) that replicates inside the host cytoplasm (see Section 12.1) and causes a very contagious disease. The portal of entry for measles virus is through either the respiratory system or the conjunctiva. The virus replicates in the lungs, moves to the regional lymph nodes, and produces a viremia (presence of virus in the blood) that spreads the virus throughout the body. An 8- to 10-day incubation period culminates in the **prodromal phase**, with symptoms resembling those of the common cold or flu. The patient runs a fever as high as 40°C (104°F), has no appetite, and feels tired and run down (malaise). Conjunctivitis is a common problem and can be severe in some cases.

Link   Recall from Section 2.2 that an infectious disease proceeds through the following stages: incubation period, **prodromal phase**, invasive phase, and decline phase, followed by either convalescence or death.

The maculopapular rash of measles is mediated by the immune response of the patient. Cytotoxic T cells of the immune system target the infected endothelial cells of small blood vessels in the dermis. In an effort to kill infected cells, the cytotoxic T cells damage the vessels, resulting in the rash. This immune reaction, however, is critical to controlling the virus. T-cell-deficient individuals do not have the rash but do have uncontrolled disease and usually die.

Table 19.2
The Exanthems of Childhood

Disease	Old Name	Section
Measles	First disease	19.2
Scarlet fever	Second disease	20.3
Rubella	Third disease	19.2
Filatow-Dukes' disease, believed to be the scalded skin syndrome	Fourth disease	19.3
Erythema infectiosum	Fifth disease	19.2
Roseola infantum	Sixth disease	19.2

Measles usually starts with an enanthem, called Koplik's spots, which are white spots on the buccal mucosa (inner lining of the cheeks) (**Figure 19.4A**). Koplik's spots are **pathognomonic**—a sign characteristic of a particular condition or disease—for measles. The enanthem is then followed by an exanthem that appears first at the hairline and then spreads to the head, neck, body, arms, legs, and sometimes, although rarely, even the palms of the hands and soles of the feet (**Figure 19.4B**). The fever peaks over the next 2–3 days, and the patient experiences respiratory symptoms (including sore throat and cough) and conjunctivitis. The rash starts to fade about 4 days after it appears and is gone by day 7. The fever also subsides 4 days after the exanthem appears, but the cough can persist for another 2 weeks.

Immunocompromised patients infected with measles virus can develop extremely serious complications such as viral encephalitis (brain inflammation) that can cause death, blindness due to ulceration of the cornea, myocarditis (inflammation of heart muscle) or pericarditis (inflammation of the pericardial membrane covering the heart), and a variety of gastrointestinal maladies.

Serious complications of measles in immune-competent patients include **acute disseminated encephalomyelitis (ADEM)**, a disease that strips off the myelin covering of affected nerves, and a delayed complication called **subacute sclerosing panencephalitis (SSPE)**. ADEM is an autoimmune disease that presents during the recovery phase. SSPE is a progressive degenerative disease of the central nervous system that occurs 7–10 years after natural measles infection. Its cause is not clear, but it may be due to a variant strain of measles virus that produces a persistent infection. ADEM and SSPE are not caused by attenuated vaccine strain.

Diagnosing measles is based first on clinical presentation and confirmed by serological tests that measure the presence of anti–measles virus antibodies in the patient's blood. Diagnosis is more challenging in immunocompromised patients because they present atypically (possibly no rash) and have a less than optimal immune response (low antibody level). In this situation, viral cultures and the histological study of affected tissues are required.

Treatment of measles is generally supportive. Supportive therapy (or comfort care) increases patient comfort rather than producing a cure. The virus itself is eliminated by the patient's immune system. Fortunately, the disease can easily be prevented through the use of the multivalent MMR vaccine containing attenuated measles, mumps, and rubella viruses (see Table 16.1). Vaccination nearly eliminated measles from the United States. Only 50 cases a year were seen, mostly in people returning from overseas. However, the irrational fear some parents have about vaccinating their children has caused a resurgence in measles cases in the United States. More than 600 cases were seen in 2014.

GERMAN MEASLES (RUBELLA)   **German measles** (called third exanthem disease of childhood, third disease, or three-day measles) was initially publicized in German literature, earning rubella its common name—German measles. German measles produces a rash similar to measles (rubeola) and was thought to be a type of measles when first described in the 1750s. However, the disease is actually caused by a completely different virus—rubella, a togavirus, which is a single-stranded RNA (positive sense) enveloped virus. Rubella was pronounced eliminated from the United States in 2004 but is still endemic in many parts of the world.

Little is known about the pathogenesis of the rubella virus. We know that infection is acquired by inhalation of aerosolized respiratory particles. Initial replication takes place in the cytoplasm of cells lining the nasopharynx and in the neighboring lymph nodes. A viremia ensues during the first week of the 12- to 23-day incubation period. The pinpoint maculopapular pink rash (smaller than in rubeola) erupts suddenly on the head and spreads to the body and extremities, much like measles. This rash, like the rash of measles, is thought to be immune related, does not darken and scab over,

Figure 19.4   The Rash of Measles

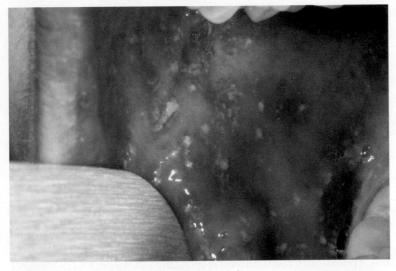

A. Koplik's spots, a white enanthem in the mouth.

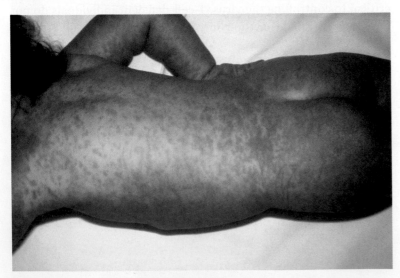

B. The exanthem of measles, a maculopapular rash, on the back.

spreads much faster than the rash of measles, and usually has a short duration of 1–3 days (leading to the colloquial name "three-day measles"). A low-grade fever and enlargement of head and neck lymph nodes usually accompany the appearance of the rash.

German measles is very mild, usually subclinical (without clinical symptoms), and self-limiting (resolves on its own) in children but has a more difficult course in adults, often involving joint pain, particularly in the knees, wrists, and fingers. Bacterial superinfections and other complications are extremely rare. Rubella virus can, however, pass through the placenta and cause a variety of defects in a developing fetus, a complication known as congenital rubella syndrome. Infants born with this syndrome usually suffer from hearing loss, but mental retardation, eye, or cardiovascular defects can also occur. The risk of fetal infection, along with severity of congenital problems and number of fetal organs involved, is highest when exposed during the first trimester of pregnancy.

Because of the mild nature of the disease, diagnostic tests are not indicated unless there is a suspicion of congenital rubella syndrome. Enzyme-linked immunosorbent assays (ELISAs) or enzyme immunocapture techniques (see Section 17.5) can detect rubella-specific antibodies in blood. Reverse transcription polymerase chain reaction (RT-PCR; see Section 25.4) can detect the virus RNA in cord blood or placenta samples to confirm congenital rubella infections. Positive viral cultures obtained from maternal nasopharyngeal secretions are also diagnostic. Supportive and symptomatic treatments are the only available options for patients suffering from rubella infections. The disease and its potential complications are easily prevented through vaccination with the MMR multivalent vaccine.

FIFTH DISEASE (ERYTHEMA INFECTIOSUM) **Erythema infectiosum**, also known as fifth disease, is caused by human parvovirus B19. The disease was first reported by Anton Tschamer in 1889, but its cause was discovered, accidentally, in 1975 in sample 19 of panel B of a routine screening of blood for hepatitis B. Parvovirus B19 is a small, nonenveloped, single-stranded DNA virus of the Parvoviridae family. Exposure to the virus via the respiratory tract is followed by viral

attachment to and replication in erythrocyte progenitor cells (cells that grow and change into red blood cells). The infection results in a viremia within 7–10 days. The virus binds the P antigen, globoside, expressed abundantly on the membrane of red blood cells and their progenitors, but another unknown receptor is probably required to infect respiratory cells. The body responds to the viremia by producing antibodies to the virus and an inflammatory response with production of tumor necrosis factor alpha (TNF-α), interferon gamma (IFN-γ), and interleukins 2 and 6 (IL-2 and IL-6).

Prodromal symptoms of mild fever, flu-like symptoms, and sometimes arthralgia (joint pain) coincide with viremia. The initial rash is pathognomonic for erythema infectiosum and is often described as a "slapped cheek" rash because it is a malar (on the cheek), erythematous (reddened), warm, and nonpruritic (nonitching) rash that appears as the viremia is resolving (Figure 19.5A). The rash is immune mediated. The slapped-cheek rash is followed by the appearance of a red (or gray) papular enanthem on the palate and throat. The third stage is a maculopapular rash that forms on the body and the limbs, with areas of clearing that gives it the appearance of lace (reticular rash; Figure 19.5B). The rash may also present in young adults as a papular purpuric rash (flat, dark purple spots) in a "glove" and "sock" distribution on the hands and feet (Figure 19.5C). The disease is benign in most individuals but can become life-threatening in some patients—for example, in people with chronic anemias or those who are immunosuppressed.

The slapped-cheek rash is pathognomonic, but detecting IgM to B19 or using quantitative PCR to identify viral DNA is definitive. Treatment is supportive except in immunocompromised patients, in which case intravenous (inside the vein, IV) immune globulin is administered. There is no vaccine.

Note Immune globulin is produced by pooling donated blood from at least 1,000 people who have been tested to be free of bloodborne diseases such as HIV or hepatitis. The antibody proteins are then separated out of the whole blood. Immune globulin can be used to treat a number of diseases but runs the risk of hypersensitivity reactions.

Figure 19.5 Erythema Infectiosum (Fifth Disease)

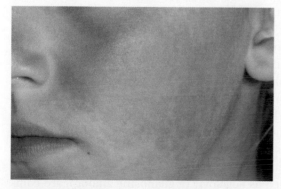

A. The initial rash appears on the cheek and has a slapped-cheek appearance.

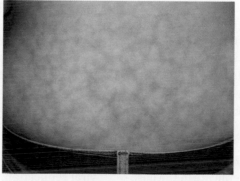

B. Reticular rash (lacelike appearance).

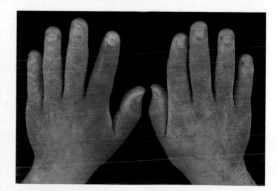

C. The rash of fifth disease on the hand of a young adult infected with parvovirus B19.

ROSEOLA INFANTUM **Roseola infantum**, also called roseola, exanthem subitum (meaning sudden rash), pseudorubella, sixth disease, and the three-day fever, is a viral disease of very young children, usually under 3 years of age. It is transmitted by respiratory secretions or saliva. The disease is described as 3–5 days of very high fever (over 40°C, or 104°F), prompting many an emergency department visit. Resolution of the fever is followed by a sudden macular or maculopapular red rash all over the body that blanches (turns white) when touched. The rash is probably caused by antigen-antibody complexes. Human herpes virus 6 (HHV-6) and, less frequently, human herpes virus 7 (HHV-7), both double-stranded DNA viruses related to the herpes virus that causes cold sores, are usually responsible for the illness. Herpes viruses are described in Section 12.1. Roseola cases are sporadic. The disease is mild and is contagious only during the febrile stage ("with fever" stage). Once the fever has dissipated, the child does not need to be isolated from the day-care population. The virus is actually present in most people, remaining **latent** (hidden and not manifesting disease) within myeloid and bone marrow progenitor cells.

Diagnostic tests are not done for roseola unless there is an unusual presentation of the disease (when, for example, the rash and fever happen simultaneously). Complications are rare. There is no vaccine.

Viruses That Produce Papular and Pustular Rashes

Some viral infections make a patient's skin erupt into a rash that resembles small pimples filled with clear liquid or pus. Some of these rashes leave a permanent round scar on the skin called a "pockmark," whereas others resolve without scarring their victims. Some viral diseases can start with a simple raised rash but end up producing life-threatening systemic symptoms affecting more than one organ system. One of these, the dreaded smallpox, has fortunately been eradicated as a disease, yet the virus is still preserved in secure American and Russian freezers. Other viral diseases have not been eradicated but can be prevented by vaccination (chickenpox) or treated more effectively than in the past (cold sores).

CHICKENPOX **Chickenpox** is caused by a very contagious member of the Herpesviridae family, varicella-zoster virus (VZV; see case history opening this chapter). The VZV capsid contains a linear double-stranded DNA genome and is surrounded by a lipid envelope acquired when the virus buds from an infected cell. Initial exposure to the virus causes chickenpox, usually a mild childhood disease that resolves by itself. The disease resolves, but the virus does not. Instead, viral DNA takes refuge in the dorsal root ganglia and in 20% of patients can emerge later in life, causing the painful disease known as shingles (herpes zoster).

The virus is contracted by inhaling infected particles from skin lesions. Once inhaled, the virus is presumed to replicate in the nasopharynx and then infects the regional lymph nodes, leading to viremia. A second round of viral replication takes place in the liver and the spleen, followed by a secondary viremia 14–16 days postinfection. VZV can then invade capillary endothelial cells and the deepest layer of epidermis. VZV infection of epidermal cells produces fluid accumulation that results in the characteristic vesicles (see Figure 19.2B). Children do not usually have prodromal symptoms. The disease first announces itself as an itchy rash on the child's face, back, chest, and belly and includes maculopapules, vesicles, pustules, and scabs. The clear liquid inside the vesicles contains the viral particles that can be transferred either by contact or inhalation when aerosolized.

The course of the disease is usually mild in children but can become life-threatening in immunocompromised victims. Because the immune system in these patients cannot subdue the virus, a persistent viremia develops that can result in pneumonia (lung infection), encephalopathy (brain disorder), and dissemination of the virus, a condition called "progressive varicella." However, even immunocompetent patients can develop a complication as they age. During the acute phase of the childhood disease, viruses in the vesicles infect the nerve endings of the skin and travel along these nerves to the ganglia, where they hide in a dormant state. Dormancy, or latency, is established when the viral DNA integrates into host DNA, and it can last for decades. Under certain ill-defined situations (including stress), the neurotropic (nerve loving) virus reactivates and begins replicating.

After reactivation, the replicating virus particles travel along the sensory nerves of the skin to produce a localized, painful, dermatomal rash known as shingles (illustrated in the chapter opening). A **dermatome** is the area of the skin served by a specific nerve. The rash usually runs in a strip from the back around to the chest (the term *shingles* is derived from the Latin word for "girdle"). Whereas chickenpox is usually a mild disease of childhood, the agony of shingles is typically reserved for older adults. Varicella-zoster-specific cell-mediated immunity decreases as the person gets older, which may explain why there is a progressive increase in the incidence of shingles with advancing age.

A diagnosis of chickenpox or shingles is usually made clinically, but a variety of other techniques may be used, including culturing the virus from the vesicular fluid, using fluorescent antibody to membrane antigen (FAMA) to detect virus proteins on host membranes, using enzyme-linked immunosorbent assay (ELISA) to quantitate antiviral antibodies (see Section 17.5), and using polymerase chain reaction (PCR) tests to detect viral DNA (see Section 8.4). These tests are used to investigate a difficult case or to establish previous exposure or immunity.

Chickenpox is ordinarily treated using antihistamines, oatmeal baths, and calamine lotion to reduce and control the intense itching and acetaminophen to reduce fever and pain. Aspirin is not used to treat viral infections because it is associated with Reye's syndrome, involving sudden encephalopathy and liver damage. Inhibitors of viral replication, such as acyclovir (a guanosine analog), are commonly used to treat shingles but are used only for severe or complicated cases of chickenpox.

Vaccination against varicella is part of the routine childhood immunization schedule. The vaccine is a live, attenuated form of VZV. Protection from this vaccine (or infection) wanes, so persons over 60 years of age should be vaccinated by a zoster vaccine to prevent shingles regardless of whether they remember having had chickenpox in the past.

COLD SORES AND GENITAL HERPES **Cold sores** and **genital herpes** infections are caused by herpes simplex viruses 1 and 2 (HSV-1 and HSV-2), which are distant relatives of the varicella-zoster virus. HSV-1 and HSV-2 are enveloped, linear, double-stranded DNA viruses. The envelope in this case derives from the host nuclear membrane as virus nucleocapsids bud from the nucleus. Best known for infecting the skin and mucous membranes, they are also able to infect the central nervous system and occasionally the visceral organs (the internal organs). These viruses are transmitted by direct contact. They replicate in mucosal surfaces or epidermis and eventually enter the local sensory or autonomic nerve endings that serve as a path to the virus's final destination: the neuronal cell in the ganglia, where the virus becomes latent. The primary infections may be subclinical or symptomatic, but reactivation of the latent virus always results in symptoms as the virus moves along the nerve toward the skin or mucous membrane. Severity and recurrence of infection with either virus is dictated by the immune status of the person who has acquired it. HSV infections are serious and difficult to manage in immunocompromised transplant patients or in people infected with human immunodeficiency virus (HIV).

Cold sores (also called fever blisters or herpes labialis) are primarily caused by HSV-1 (**Figure 19.6A**). HSV-1 is very contagious and is present in either active or latent form in 60%–90% of older adults. Currently, there is no vaccine against HSV-1, and the virus is easily spread from person to person by contact. The virus remains dormant in the trigeminal nerve ganglion that links the central nervous system and the face. The major symptom of initial or reactivated infections is the appearance of several painful vesicles on **erythematous** (red and inflamed) skin.

Note A canker sore is a painful, flat, round or oval, open sore in the mouth. Canker sores do not contain virus, are not the same as cold sores, and are not contagious. The etiology of canker sores is unknown but they can be caused by emotional stress, hormonal changes, vitamin deficiencies, or as a consequence of viral infections.

Primary infections with HSV-1 are usually associated with fever, sore throat, muscle aches, and cervical lymphadenopathy (inflammation of lymph nodes of the neck), along with the vesicular lesions, lasting up to 2 weeks. Reactivation of the infection is not associated with systemic symptoms. Some patients sense the reactivation of the disease because they have prodromal symptoms, including pain, tingling, or burning sensation before the characteristic vesicles appear. The vesicles most commonly form at the outer border of the lips, but the virus present in these lesions can also infect other parts of the body where the skin is disrupted. In children and some health care professionals (dentists in particular), abrasions of the skin of the fingers result in an HSV-1 infection referred to as **herpetic whitlow** (**Figure 19.6B**). Rugby players and wrestlers have suffered outbreaks of HSV-1 infections on their neck, face, or arm that have been aptly named **herpes gladiatorum**. HSV-1

Figure 19.6   Herpes Virus Lesions

A. Cold sores on the upper lip.

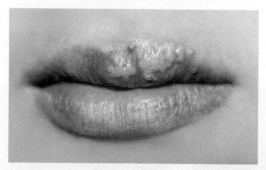

C. Erythema multiforme target lesions in a patient infected with HSV-1.

B. A herpetic whitlow. Herpes simplex infection around the fingernails is caused in children by thumb-sucking when they have a cold sore and in health care workers who fail to use latex gloves.

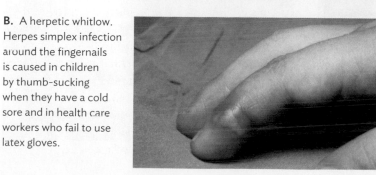

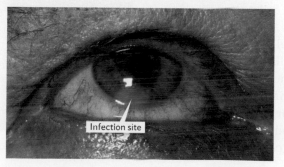

Infection site

D. Herpes corneal infection stained with a fluorescent dye. The blood vessels penetrate the infected cornea to deliver neutrophils to the area, which is normally devoid of blood vessels.

Figure 19.7   Tzanck Smear

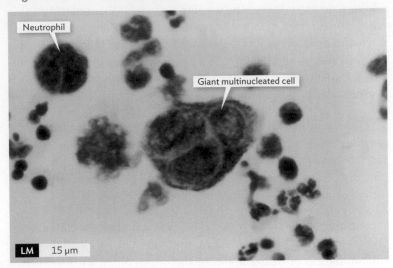

A Giemsa stain of cells scraped from the base of a herpes simplex vesicle shows a giant multinucleated cell in the center of the picture. Varicella-zoster virus can also give rise to giant multinucleated cells.

DNA has also been detected in lesions of erythema multiforme, an immune-mediated, target-shaped lesion (Figure 19.6C), suggesting an association between the two. In immunocompetent persons, cold sores heal by crusting over within a few days to a week.

HSV-1 can set up a chronic infection that may involve the deeper parts of the skin in immunocompromised patients. The virus is also the leading infectious cause of blindness in the United States. In **herpetic keratitis**, virus reactivated in the trigeminal ganglion travels the ophthalmic branch of the trigeminal nerve and infects the cornea. HSV-1 replication then triggers a severe inflammatory response that can destroy the cornea (Figure 19.6D).

Definitive diagnosis of HSV-1 disease is typically made when the virus is cultured from a vesicle, but now real-time PCR assays provide sensitive and rapid identification. A variety of serological tests, immunofluorescent staining (see Section 17.5), and the Tzanck smear test can also be used to verify the presence of the virus. HSV-1 and VZV alter the shape of the cells they infect. In the Tzanck smear test, the base of a lesion is scraped, and the cells are smeared on a microscope slide, stained, and examined microscopically for telltale giant multinucleated cells (Figure 19.7).

Antiviral medications, such as acyclovir, valacyclovir, and famciclovir, are used during primary infections to reduce the pain and duration of lesions, shorten the fever period, and decrease

viral shedding. Reactivation infections can be treated with topical antiviral drugs.

The other medically important herpes virus, HSV-2, causes primarily genital herpes, a sexually transmitted disease that will be discussed in Chapter 23. It is important to note that the tissue specificity of HSV-1 and HSV-2 is not absolute. It is possible for HSV-1 to infect the genital skin and mucosa and for HSV-2 to infect the oral tissue.

WARTS Warts (or **papillomas**) have long appeared on children's fingers (Figure 19.8). How to remove them has been the subject of some colorful discussions. Mark Twain gives an account of Tom Sawyer and Huck Finn discussing the validity of touching a dead cat or plunging their hand in spunk water (rain that pools in a dead tree stump) to get rid of warts! Myths about the cause of warts rival those about their cure; frogs have often been designated as one of the supposed culprits. We now know that warts are caused by some subtypes of human papillomavirus (HPV). Papillomaviridae is a several-hundred-subtype family of small, nonenveloped viruses with a double-stranded circular DNA genome that infect mammals, birds, and even tortoises and turtles.

HPV is transmitted to skin by contact, but there are host- and tissue-specific strains, so the subtype that causes warts on the soles of the feet is different from the one responsible for warts on the palms of the hands. HPV probably enters the cell via an endosome by a receptor-mediated mechanism. Once inside, HPV DNA leaves the endosome and enters the nucleus. Virus-synthesized proteins then interfere with cell proliferation controls (discussed in Sections 12.4 and 18.6), and the infected epithelial cell replicates uncontrollably, producing those unsightly (but noncancerous) warts. Treatments for cutaneous warts include freezing, burning (electrical or with chemicals), and surgical removal—all more effective than "spunk water" but none as imaginative!

Other subtypes of the virus, particularly subtypes 6, 11, 16, and 18, are sexually transmitted (see Section 23.5 and Case History

Figure 19.8   Warts

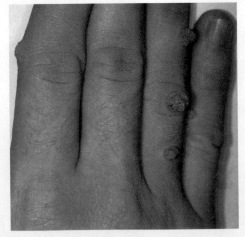

A. A common presentation of hand warts.

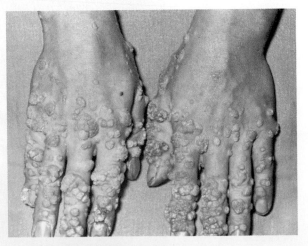

B. Extreme hand warts.

18.3). HPV-6 and HPV-11 infect the mucous membranes of the ano-genital region and cause 90% of genital warts (**condyloma acuminata**). Human papillomavirus subtypes 16 and 18 have more serious consequences, as they are linked to 70% of human cervical cancers. Vaccines prepared from hollow virus capsids are now available. The vaccine, however, must be administered before the recipient (male or female) is sexually active to ensure effectiveness.

HAND, FOOT AND MOUTH DISEASE   **Hand, foot and mouth disease (HFMD)** is a contagious disease caused by positive-sense, single-stranded RNA enteroviruses of the Picornaviridae family. The majority of HFMD cases are caused by coxsackievirus A16 and enterovirus 71 (EV-71) although other coxsackieviruses may rarely cause the disease as well. HFMD is often a mild disease that usually affects very young children (under 5 years of age). The infected child has a fever and vague symptoms (such as not feeling well and poor appetite). A painful rash that starts as red spots begins appearing in the back of the mouth, on the palms of the hands, and on the soles of the feet 2 days after the fever starts. The red spots become blister-like and painful, and the child does not eat enough food or drink enough water. The major complication of the disease, therefore, is dehydration, which in some cases is severe enough to require giving the child intravenous fluids. Life-threatening complications such as severe neurological or heart disease may rarely occur. Neurological complications are usually related to EV-71 infection. Heart-related complications are seen when the disease is caused by coxsackievirus B (see Section 21.2).

Diagnosis of HFMD is usually made based on patient history and the presenting clinical signs and symptoms, but isolation of the virus from throat or stool samples may be required in some cases. Treatment is limited to reducing fever and pain with over-the-counter medications except aspirin, which is associated with Reye's syndrome. Hand washing, keeping surfaces clean, and avoiding contact with an infected person are the only means of prevention because there are no vaccines available for this disease.

Note HFMD is different from the foot-and-mouth disease of livestock (also called hoof-and-mouth disease) that is caused by the genus *Aphthovirus*.

SMALLPOX   **Smallpox** is the dreaded disease caused by the large, enveloped, DNA-containing, "dumbbell"-shaped variola virus (Figure 19.9). Its brutal history includes killing almost 10% of the world's population at one time. Humans are the only known reservoir of variola, which was eradicated from the human population in 1979 thanks to an effective vaccine and aggressive vaccination campaigns.

Variola is a member of the Poxviridae family. The virus is divided into two variants, variola major and variola minor, based on the severity of the symptoms that they cause. The two variants are genetically very similar to one another and to the vaccinia virus. Vaccinia virus causes cowpox and is the virus used to make a vaccine against smallpox (see Figure 16.4C). Smallpox produces a

Figure 19.9   Smallpox Virus

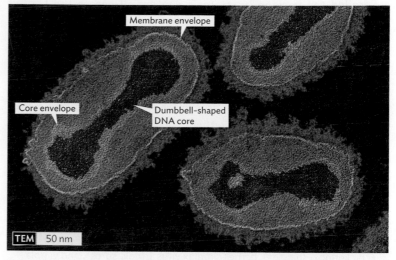

The image shows variola major, the etiologic agent of smallpox, in color. The virus is unusual among DNA viruses because it replicates in the cytoplasm of host cells rather than in the nucleus.

characteristic vesicular rash on the skin (**Figure 19.10**), from which dried particles can aerosolize and transmit the disease.

The smallpox virus was transmitted from person to person by direct or indirect contact. The infective particles would drop onto the mucosal membranes of the upper airways or alveoli. This was the direct, and most common, mode of disease transmission. Indirect transmission of the disease was also possible, but less common, when a person inhaled small aerosolized particles or handled fomites containing the virus. The virus would replicate in the lymph nodes and lymphoid organs and through two sequential viremias infect internal organs and bone marrow before showing up as small spots on the oral mucosa (enanthem) and finally the pox pustule on the skin (exanthem). The virus could be isolated from the kidneys, liver, and bone marrow as well as from the vesicular rash that finally appeared on the skin (see Figure 19.10).

There are no FDA-approved treatments for smallpox, and the virus is considered a possible bioterrorism threat. We no longer

Figure 19.10   The Rash of Smallpox

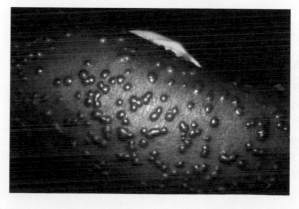

The rash of smallpox starts as vesicular and progresses to a pustular rash.

vaccinate against smallpox. Government officials, however, have stockpiled the vaccine in strategic centers all across the United States. Definitive diagnosis of smallpox can be made based on viral cultures and serology. The Centers for Disease Control and Prevention (CDC) has developed specific algorithms for diagnosis of variola and necessary responses if a suspected case presents. Closely related monkeypox remains endemic in Africa, infecting animals and occasionally people. The disease produces smallpox-like symptoms in humans but is not as lethal.

SECTION SUMMARY

- **Viruses causing macular rashes** include measles (rubeola), rubella (German measles), parvovirus B19 (erythema infectiosum), and human herpes virus 6 (roseola).

- **Viruses causing vesicular/pustular** rashes include varicella-zoster (chickenpox, shingles), herpes simplex virus 1 (cold sores), and variola (smallpox).

- **The respiratory portal of entry** is used by VZV, measles, rubella, and variola. The viruses are transmitted by inhalation of respiratory droplets or aerosols; they replicate in lymph nodes and lymphoid tissue and pass via lymphatics and blood to the skin.

- **Skin is the portal of entry** for HSV-1 and papillomavirus, which are transmitted by direct skin contact.

- **Diagnosis of viral skin diseases** is usually based on clinical presentation, but laboratory tests can be used in certain instances, especially in cases involving severe disease in immunocompromised patients.

- **Treatment of viral skin diseases** usually involves supportive therapy, although antivirals can be used for treating shingles and complications of most of the other viruses. Vaccination is recommended for measles, rubella, and VZV.

Thought Question 19.1 A child brought to a U.S. clinic exhibits the symptoms of German measles, yet the disease was declared eliminated from within the United States in 2004. How might the child have contracted the disease?

Thought Question 19.2 Explain why cytotoxic T cells do not attack nerve cells containing latent viruses such as VZV or herpes viruses.

19.3
Bacterial Infections of the Skin

SECTION OBJECTIVES

- Name the bacterial pathogens that cause common skin infections.
- Categorize skin infections by the physiology of the bacteria causing them.
- List clinical and laboratory means of diagnosing bacterial skin infections.
- Develop a prevention and treatment plan for bacterial skin infections.

How do bacteria infect our skin if the skin is supposed to be a protective barrier? Minor insults to the skin, such as an abrasion (superficial skin damage) or even a paper cut, can give the bacteria a way in. Healthy individuals develop infections of the skin only rarely. People with underlying immunosuppressive diseases such as diabetes, however, are at much higher risk. Skin infections caused by bacteria range from simple boils to severe, complicated "flesh-eating" diseases (see Table 19.3).

A number of different bacterial species can cause infections of the skin and mucous membranes, but Gram-positive *Staphylococcus* and *Streptococcus* species are the major contributors. Staphylococci resemble microscopic clusters of grapes, whereas streptococci appear as strings of pearls (Figure 19.11). The pathogens from each are capable of invading and causing disease in different tissues. *Bacillus anthracis* is another Gram-positive organism that can infect our skin to cause cutaneous anthrax, a zoonotic disease (transmitted from animals to humans). Ingestion of anthrax spores may cause gastrointestinal disease in rare cases. Dissemination of *B. anthracis* to the lungs causes a deadly disease, pulmonary anthrax (see Section 20.3).

Staphylococcal Skin Infections

Staphylococci associated with skin infections include *Staphylococcus epidermidis*, a normal member of skin microbiota (see Chapter 14), and *S. aureus*, which can cause serious skin infections as well as other infections. *S. aureus*, often a normal inhabitant of the nares (nostrils), can infect a cut or gain access to the dermis via a hair follicle. Once *S. aureus* infects the skin, its growth triggers an inflammatory response that attracts macrophages and neutrophils. Pus forming at infected sites is composed of bacteria and white cells, most of which are dead. *S. aureus* possesses a number of enzymes that contribute to disease, including coagulase, which helps coat the bacterium with fibrin and walls off the infection from the immune system and antibiotics. The result is an **abscess**. *S. aureus* produces coagulase and *S. epidermidis* does not, making the former a pathogen and the latter an **opportunistic pathogen**. As a result of coagulase, "staph" infections generally require surgical drainage as well as antibiotic therapy. Note that coagulase production correlates with the organism's ability to produce exotoxins that cause host tissue damage and weaken host defenses.

Link As discussed in Section 2.1, **opportunistic pathogens** are microbes that cause disease only in an immunocompromised host.

Staphylococcus aureus infection of the hair follicles can be superficial or deep. A superficial infection is referred to as **folliculitis** (Figure 19.12A). A **boil**, or **furuncle**, is a painful abscess of a hair follicle (Figure 19.12B). **Carbuncles** are lesions that result when several boils join together (Figure 19.12C). Diagnosis of these infections is usually made clinically and is based on their appearance. Folliculitis and most furuncles normally do not require antibiotic treatment and can resolve on their own. Carbuncles and some furuncles have to be drained and treated with antibiotics to resolve the infection.

Other staphylococcal diseases are caused by toxin-producing strains in which the organism remains localized, perhaps in an abscess, but the toxin disseminates via the bloodstream. Some strains of *S. aureus* produce toxic shock syndrome toxin (TSST), a **superantigen** that can lead to serious disease (Staphylococcal toxins are discussed in Section 18.2). Some other strains of *S. aureus* produce a toxin called exfoliative toxin that causes a blistering disease in children called staphylococcal **scalded skin syndrome** (Figure 19.12D). Exfoliative toxin, like TSST, is a superantigen, but it also cleaves a skin cell adhesion molecule. Without adhesion, the epidermis will separate from the underlying skin

Figure 19.11 *Staphylococcus* and *Streptococcus*

A. Staphylococci grow in clusters.

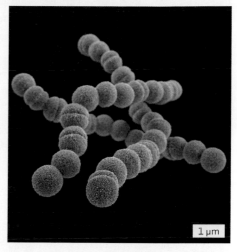

B. Streptococci grow in chains, as shown in this illustration.

layer to produce blisters in which the skin appears to be scalded by hot water. Remember, the exfoliated skin is not infected. The toxin disseminates from a staphylococcal infection located elsewhere in the body.

Link As discussed in Sections 16.6 and 16.7, antigens normally require processing by antigen-presenting cells to activate cells of our immune system. Some proteins, called **superantigens**, are able to indiscriminately stimulate large number of T cells because antigen recognition is not involved.

A particularly dangerous strain of *S. aureus* called **methicillin-resistant *S. aureus* (MRSA)** has emerged over the past decade or so. *S. aureus* infections are commonly treated with penicillin-like drugs such as methicillin that prevent cell wall synthesis. (Virtually all *S. aureus* strains are resistant to penicillin.) A majority of *S. aureus* strains (about 60% of clinical isolates) have now evolved to resist methicillin. MRSA has developed resistance to methicillin and many other penicillin-like drugs through a mutation that alters one of the proteins (called a penicillin-binding protein; PBP) involved in cell wall synthesis. Because methicillin is normally used as a first line of defense against staphylococcal infections, treatment

failure of MRSA can be life-threatening, and alternative drugs, such as vancomycin (a cell wall synthesis inhibitor unrelated to penicillin), must be used.

MRSA first appeared as nosocomial infections—infections that occur after a patient enters a health care setting (Section 26.2). These infections are also known as hospital-acquired infections or health care–associated infections. Today, these antibiotic-resistant organisms are no longer confined to hospitals. Individuals who have not been in a hospital are being infected with MRSA in what are called community-acquired infections. This is occurring at an epidemic rate in the United States (an incidence of approximately 20 per 100,000 persons). Seventy percent of staphylococcal skin infections are now caused by MRSA. A health care provider can no longer assume that a patient walking into the office with a staphylococcal infection will respond to methicillin (oxacillin). Vancomycin and linezolid are the antibiotics typically used to treat MRSA infections.

Note Methicillin is no longer used in the United States. Oxacillin, which is from the same class of antibiotics, is used instead. The strains tested for resistance, therefore, are tested against oxacillin, but the name MRSA has not been changed. Strains that are resistant to oxacillin are also resistant to methicillin.

Figure 19.12 **Staphylococcal Infections of the Skin**

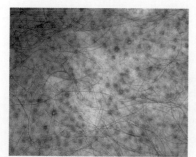

A. Folliculitis.

B. A boil, or furuncle.

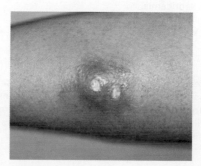

C. A carbuncle.

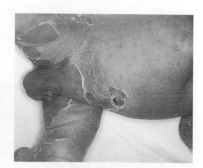

D. Scalded skin syndrome.

Table 19.3

Bacterial Diseases of the Skin and Eye

Disease	Symptoms	Complications	Bacteria	Transmission	Diagnosis	Treatment	Prevention
Folliculitis	Infected hair follicles	Skin abscess	*Staphylococcus aureus*	Disease not transmitted from person to person[a]	Clinical presentation; Gram stain (usually not necessary); specimen culture (usually not necessary)	Warm compress for folliculitis and uncomplicated furuncles; topical antibiotics for uncomplicated nonbullous impetigo; oral or systemic antibiotics as warranted, usually inhibitors of cell wall synthesis such as oxacillin, or vancomycin for MRSA; hospitalization and IV antibiotics for complicated cases	No vaccine available
Furuncle (boil)	Red, swollen painful skin mass	Scarring, sepsis, necrotizing fasciitis	Usually polymicrobial, *Staphylococcus aureus*	Disease not transmitted from person to person[a]	Clinical presentation; Gram stain (usually not necessary); specimen culture (usually not necessary)	Warm, moist compress; incision and drainage; systemic antibiotic sometimes indicated	No vaccine available
Carbuncle	Red, painful, draining mass (coalescence of boils), fever, malaise	Scarring, sepsis, necrotizing fasciitis	Usually polymicrobial, *Staphylococcus aureus*	Disease not transmitted from person to person[a]	Clinical presentation; Gram stain (usually not necessary); isolation of bacteria from specimen in culture (usually not necessary)	Warm, moist compress; incision and drainage; systemic antibiotic sometimes indicated	No vaccine available
Impetigo	Small papular lesions on an erythematous base that often weep (ooze) and crust	Complications are rare; include scarring, cellulitis, acute post–streptococcal glomerulo-nephritis	*Staphylococcus aureus* and *Streptococcus pyogenes*, Gram-positive cocci	Disease not transmitted from person to person[a]	Clinical presentation; Gram stain if confirmation needed; culture (rarely done)	Topical or systemic antibiotics, depending on patient; oral antibiotic in complicated cases: dicloxacillin or amoxicillin-clavulanate; if high prevalence of community MRSA (>10%), then clindamycin, trimethoprim-sulfamethoxazole, and others are considered	No vaccine available
Bullous impetigo	Large blister-like lesions full of a clear yellowish liquid	Cellulitis, staphylococcal scalded skin syndrome	*Staphylococcus aureus*	Disease not transmitted from person to person[a]	Clinical presentation; Gram stain for confirmation; culture (rarely done)		
Staphylo-coccal scalded skin syndrome (SSSS)	Generalized peeling of skin, usually in infants (generalized form of bullous impetigo), fever, chills, and weakness	Dehydration, secondary bacterial infections of skin, septicemia, cellulitis, and shock	Exfoliative toxin–producing strains of *Staphylococcus aureus*	Disease not transmitted from person to person[a]	Clinical presentation; histology showing skin damage; Gram stain; culture	IV antibiotics with penicillinase-resistant penicillins such as cloxacillin	No vaccine available
Erysipelas	Fiery red, raised, well-demarcated rash with an uneven surface (orange peel)	Systemic symptoms, necrotizing fasciitis, meningitis	*Streptococcus pyogenes*	Disease not transmitted from person to person[a]	Clinical presentation; culture of needle aspirates of skin lesions; blood culture	Oral amoxicillin for very mild cases; IV or IM (intramuscular) penicillin V or G	No vaccine available

(continued on next page)

Table 19.3

Bacterial Diseases of the Skin and Eye (continued)

Disease	Symptoms	Complications	Bacteria	Transmission	Diagnosis	Treatment	Prevention
Cellulitis	Red, hot swollen area of the skin that is tender to palpation	Skin abscess, fever, inflammation of lymphatic vessels (lymphangitis), sepsis	*Staphylococcus aureus* and *Streptococcus pyogenes* most common	Autoinfection; disease not transmitted from person to person[a]	Clinical presentation; blood culture if systemic symptoms (such as fever) are present	Oral antibiotics; dicloxacillin or cephalexin usually used	No vaccine available
Necrotizing fasciitis	Rapidly progressing cellulitis; lesion spreads along the fascia (deep to the skin); systemic symptoms	Sepsis, scarring and disfigurement, loss of limb, death	*Streptococcus pyogenes*, majority of cases; others: *Staphylococcus aureus*, *Clostridium perfringens*, and *Vibrio vulnificus*	Transmission differs, depending on the organism	Clinical presentation; skin biopsy and culture; blood tests; CT scan	Surgical debridement and leaving the wound open to heal from inside out; antibiotics as per culture and sensitivity	No vaccine available
Burn infections	Pus, possibly with green hue with *Pseudomonas*	Delayed, incomplete, or lack of wound healing; sepsis; death	*Pseudomonas aeruginosa*, Gram-negative; various Gram-positive cocci and Gram-negative bacilli	Transmitted from environmental sources or autoinfection	Clinical presentation; culture sensitivity	IV antibiotics and hospitalization; surgical debridement as indicated	No vaccine available
Gas gangrene	Necrotic tissue including muscle, pus, foul-smelling odor, much gas and pain	Disfigurement, amputation, liver damage, kidney failure, sepsis, shock, coma	*Clostridium perfringens*	Environmental sources; autoinfection	Clinical presentation; culture and sensitivity	Surgical debridement and IV antibiotics, clindamycin	No vaccine available
Cutaneous anthrax	Ulceration with blackened center	Sepsis, pneumonia, death	*Bacillus anthracis*	Environmental sources; direct contact with infected animals or persons	Isolation of bacteria from clinical specimens per CDC; serology	Doxycycline, penicillin, and ciprofloxacin	Avoid contact with contaminated animals; no vaccine available
Pink eye, bacterial conjunctivitis	Red, watery, irritated eye with purulent discharge, sticky eyelids	Conjunctival scarring and obstruction of tear ducts	*Haemophilus influenzae*, *Streptococcus pneumoniae*	Person-to-person contact	Clinical presentation; Gram stain of the purulent discharge (rarely done)	Stop contact lens use; topical ophthalmic antibiotics: erythromycin, fluoroquinolones, azithromycin, tobramycin, or gentamicin; exclusion from school	No vaccine available[b]
Trachoma	Eye pain and redness, itching eye, turned-in eyelids	Corneal opacification and blindness	*Chlamydia trachomatis*	Person-to-person contact	Clinical presentation using WHO simple grading system; tissue culture of *C. trachomatis*; PCR	Oral azithromycin; topical tetracycline or topical plus oral sulfonamides are used as alternatives if azithromycin cannot be used	No vaccine available
Ophthalmia neonatorum	Red eye with ocular discharge	Corneal ulceration and blindness	*Chlamydia trachomatis*, *Neisseria gonorrhoeae*, other bacteria possible	Transmitted from infected mother during birth	Culture of ocular specimen; Gram stain of smear of an ocular sample; direct fluorescence antibody test; PCR; rapid enzyme immunoassay	Oral erythromycin for *C. trachomatis*; IV or IM ceftriaxone for *N. gonorrhoeae*; other antibiotics as indicated for other bacteria	No vaccine available

[a]Organisms can be transferred from person to person, but the disease is generally not. Once an organism is transferred it can colonize the new host.
[b]Vaccines are available to protect against *H. influenzae* and *S. pneumoniae* respiratory tract infections.

CASE HISTORY 19.2

Leili's Crusted Lip

 Leili, a 3-year-old girl, was brought to the pediatrician by her mother, who was worried about the rash on Leili's lip. It started 4 days earlier as a little bump above her lip that spread to the corner of her mouth (**Figure 19.13A**). The base of the rash was red and covered by pustules. The little girl said that her rash hurt, and the clinician noticed a honey-colored crust on the ruptured pustules. Leili's mother said that her daughter had not been trying to scratch the rash nor was the area around it hot to the touch. Leili's mother was told that her little girl had impetigo, a skin infection often caused by bacteria called staphylococci, although streptococci can also cause it. The doctor gently swabbed a sample from one of Leili's sores and sent it to the lab, where the causative agent would be identified and its antibiotic susceptibility determined. She explained that impetigo is a very contagious disease, and she gave Leili's mother a topical antibiotic to use. Treating topically rather than orally would accomplish two goals. Topical application would reduce the chance that the bacteria would develop drug resistance, and because Leili had a very mild case of impetigo, the antibiotic in the gel should easily penetrate the pustules and kill the organism. Leili and her mother left but returned in a week for a follow-up exam. The "rash" was gone, replaced by Leili's smile. The pediatrician explained that the cause was in fact methicillin-sensitive *Staphylococcus aureus* (MSSA) not MRSA. Leili's mother in turn reported that another child at Leili's day care had impetigo before Leili, and that is where Leili probably acquired the infection.

Impetigo is a contagious skin infection that affects mostly young children. The majority of cases are caused by *Staphylococcus aureus* and occasionally by group A beta-hemolytic streptococci (discussed later).

There are two main types of impetigo: **nonbullous impetigo** (see Figure 19.13A), the most common form (the kind that Leili had), seen as honey-crusted lesions, and **bullous impetigo** (Figure 19.13B), fluid-filled and blister-like. The nonbullous form starts as superficial "bumps" that become papules and vesicles before they form the characteristic pustules on erythematous skin. The fluid from the pustules leaks out as they break open ("weep") and forms a honey-colored crust. In bullous impetigo, the vesicles grow larger and become **bullae** (blisters) that are full of clear yellow fluid. Strains of *S. aureus* that produce exfoliative toxin A can cause bullous impetigo. Scalded skin syndrome is a generalized form of bullous impetigo seen mainly in infants (see Section 18.3).

A minority of cases of impetigo result from streptococci, mostly *Streptococcus pyogenes*, a microbe normally associated with sore throat infections (pharyngitis). Among the many strains of *S. pyogenes* that cause infections, some will put the patient at higher risk of complications after the infection is resolved. These complications are referred to as post-streptococcal sequelae (conditions resulting from another disease) and include post-streptococcal acute glomerulonephritis (inflammation of the kidney caused by antigen-antibody complexes) and rheumatic fever caused by self-reactive antibodies (discussed in Chapter 20). The risk for acute glomerulonephitis appears to be higher for patients infected with impetigo-producing strains of *S. pyogenes* than for patients infected with strains associated with pharyngitis (see Section 20.3). It is unclear whether there is strain specificity for rheumatic fever.

Diagnosis of impetigo is made by clinical signs and symptoms and can be confirmed by isolating the bacteria from the lesions. It is important to treat impetigo to avoid complications such as scalded skin syndrome. In Case History 19.2, Leili was treated with a topical antibiotic, since her case was mild. Oral antibiotics are better suited for more severe cases. The antibiotic used should be active against Gram-positive cocci.

Figure 19.13 Impetigo

Impetigo is usually caused by *Streptococcus aureus*, although *S. pyogenes* can also be an agent.

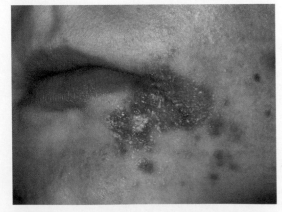

A. Nonbullous impetigo presents with honey-colored crusted lesions on a violaceous (purplish red) base without any fluid content.

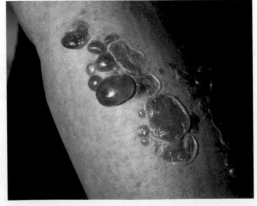

B. Bullous impetigo is uncommon and has fluid-filled, blister-like lesions.

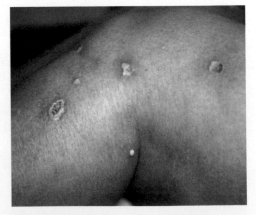

C. Ecthyma is a rare form of impetigo that is invasive.

Streptococcal Skin Infections

CASE HISTORY 19.3

The Voracious Bacteria

One weekend in June, Cassi was camping with her three children. She suffered a minor cut on her finger, which she bandaged properly. She also injured the left side of her body while playing sports with her kids. Not thinking much of either of her minor injuries, she went to bed. Two days later, Cassi was extremely ill. Her symptoms included vomiting, diarrhea, and a fever. She was also in severe pain where she had injured her side, and the area had begun to bruise (the skin did not look broken). By the next day, she could barely get out of bed, and by the end of the night, she was breathing with difficulty and could not see. Her side began to leak fluid and blood. Cassi was admitted to the hospital in septic shock, with no detectable blood pressure. She was given vasopressors (medications that constrict the blood vessels) to raise her blood pressure to the normal level. An infectious disease specialist diagnosed the problem as necrotizing fasciitis, and Cassi was rushed into surgery. In an effort to save her life, about 7% of her body surface was removed. Because the large wound infection in her side would need to resolve before a skin graft could be performed to repair it, the hole in Cassi's body was left wide open (**Figure 19.14A**). Meanwhile, the continued use of vasopressors caused gangrene in her fingers and lower extremities (**Figure 19.14B**). After nearly 3 months and several operations, including amputation, Cassi recovered.

What kind of organism can cause this horrific disease? **Necrotizing fasciitis**, also known as flesh-eating disease, is rare and is often caused by *Streptococcus pyogenes*, which may be the reason it used to be called streptococcal gangrene. The disease is caused when microbes penetrate the skin through a microscopic cut or abrasion. The bacteria spread in the subcutaneous tissue, causing death and destruction of fat and fascia without harming the skin itself, so patients often do not get help until it is too late. Fascia is the sheath of thin, fibrous tissue that covers muscles and organs. There are two different types of the disease, types 1 and 2. Generally speaking, type 1 is polymicrobial, whereas type 2 is due to one microorganism, usually *Streptococcus pyogenes* and sometimes *Staphylococcus aureus*. Other organisms that can cause necrotizing fasciitis include *Clostridium perfringens* (a Gram-positive, anaerobic rod found in soil) and, in immunocompromised

hosts (for example, alcoholics or diabetics), *Vibrio vulnificus* (a Gram-negative marine bacillus).

Treatment of necrotizing fasciitis caused by any bacterial species requires rapid and aggressive surgical removal of the affected tissue (debridement) coupled with aggressive IV antibiotic therapy. Antibiotic treatment alone is almost always ineffective because of insufficient blood supply to the affected tissues. Therapy includes antibiotics such as clindamycin and metronidazole, which act against anaerobes and Gram-positive cocci, and gentamicin, a drug particularly effective against Gram-negative microbes.

The incidence of necrotizing fasciitis may have risen recently owing to the increased use of nonsteroidal anti-inflammatory drugs, which accentuate a person's susceptibility to infection by *S. pyogenes*. In Case History 19.3, Cassi probably had *S. pyogenes* on her skin when the injury to her side occurred. The injured area probably suffered an invisible microabrasion, providing a good growth environment for the organism and enabling the cocci to secrete potent toxins that trigger inflammation and mediate death of surrounding tissues. Why some patients infected with *S. pyogenes* develop necrotizing fasciitis while others do not may be related to the immunological status of the patient.

Different strains of *S. pyogenes* have different virulence factors, giving these microbes much versatility. Virulence factors include:

- A capsule that when thick will help the organism avoid phagocytosis by macrophages

- The pilus-like M protein (more than 80 different serotypes), which binds a complement regulatory protein (factor H) and protects the microbe from opsonization by complement (see Section 15.7)

- A cell wall containing lipoteichoic acid, thought to facilitate adherence to pharyngeal epithelial cells

Figure 19.14 **Necrotizing Fasciitis by *Streptococcus pyogenes***

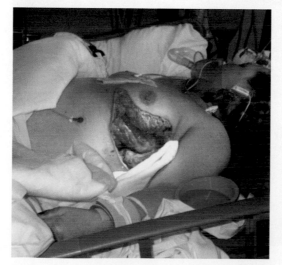

A. Flesh removed from a patient in an effort to stop the spread of necrotizing fasciitis.

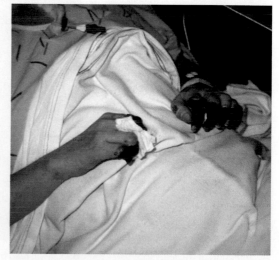

B. Gangrene of the fingers caused as a result of vasopressors given to maintain the patient's blood pressure.

- Enzymes that lyse blood cells (streptolysins)

- Enzymes that degrade DNA (dexoyribonucleases), fibrin (streptokinase), and connective tissue (hyaluronidase), making pus less viscous and allowing the organism to spread rapidly

- Peptidoglycan, capable of activating the alternative complement pathway and a microbe-associated molecular pattern (MAMP) that binds NOD-like receptors, causing inflammation (see Section 15.5)

S. pyogenes is also capable of producing several exotoxins called **streptococcal pyogenic exotoxins (SPEs)**. SPEs are superantigens. The massive amounts of cytokines released in response to SPEs can produce impressive levels of inflammation and lead to shock (cardiovascular collapse, extremely low blood pressure). Formerly called erythrogenic toxins, SPEs are associated with scarlet fever (see Section 20.3), streptococcal toxic shock syndrome, and necrotizing fasciitis.

The natural reservoir for *S. pyogenes* is the human nasopharynx and parts of the skin. The innate and adaptive immune systems of carriers normally protect the host from disease. Of course, not everyone harbors the pathogenic strains. The virulent strains, however, can be spread between carriers and non-carriers.

Why are these streptococci classified as group A beta-hemolytic? A large number of streptococci produce and secrete a hemolysin (an enzyme capable of lysing red blood cells). When plated onto agar containing red blood cells, hemolysin-secreting strains lyse red blood cells around the colony to produce a clear zone called beta hemolysis. These streptococci are subclassified in a clinical laboratory

into groups A–O according to their cell wall antigens (see Section 25.3). *S. pyogenes* is the main pathogen among the group A beta-hemolytic streptococci (also called GAS) and is often used synonymously with the term group A strep, although this is not completely accurate. It is true that all *S. pyogenes* bacteria are group A streptococci, but not all group A streptococci are *S. pyogenes*.

Our final point about GAS involves the role of the pilus-like **M protein** in post-streptococcal sequelae. As noted earlier, rheumatic fever can develop after the resolution of a primary GAS infection. This sequela is thought to be the result of immunological cross-reactivity between specific GAS M protein antigens and host antigens. In a case of mistaken identity, the autoreactive B cells that make antibodies against cardiac antigen appear to be activated by the bacterial M-protein antigen. Once made, the autoreactive antibodies begin to attack the self antigen of the heart. The ensuing antigen-antibody complexes trigger an inflammatory reaction that damages those tissues and produces rheumatic fever (see Section 17.4).

Link As discussed in Section 17.4, *S. pyogenes* uses the pilus-like **M protein** to attach to host cells. Some strains of *S. pyogenes* possess M proteins with epitopes similar to those of some human cells, especially those of the heart; infections by these strains can result in production of autoreactive antibodies.

ERYSIPELAS Erysipelas is another disease instigated by *S. pyogenes*. The disease involves the upper layer of dermis and then spreads to the superficial lymphatics. The "rash" of erysipelas may appear on the face but is most commonly seen on the lower extremities (Figure 19.15A). Erysipelas is an acute infection that begins typically as a small erythematous patch and progresses to a fiery-red, shiny plaque. Erysipelas lesions on the face cover the cheeks and the bridge of the nose, sometimes giving them a "butterfly" appearance (Figure 19.15B). The patient has swollen lymph nodes (lymphadenopathy) and often has fever and other systemic symptoms. Streptococci are usually the cause, but the bacteria cannot be isolated from the surface of the affected area because the infection is in the deeper dermis. Patients must be treated with antibiotics, since septicemia can occur quickly.

Note In England and the United States, St. Anthony's fire refers to erysipelas, whereas in Italy and Malta, the colorful name is used for herpes zoster (shingles).

CELLULITIS Cellulitis is a term commonly used to indicate an uncomplicated non-necrotizing inflammation of the dermis related to acute infection.

Figure 19.15   Erysipelas

Streptococcus pyogenes, and rarely other bacteria, can spread to the deeper layers of skin, causing a dangerous infection.

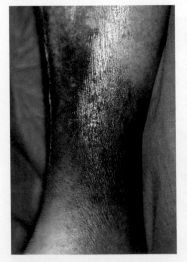

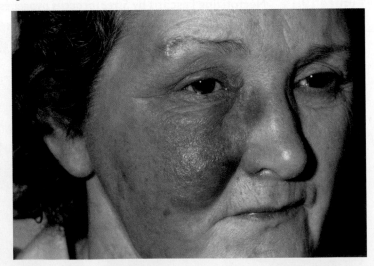

A. Infection of the leg. Note the clear demarcation between infected and noninfected areas.

B. Classic rash on the face.

Figure 19.16   Cellulitis

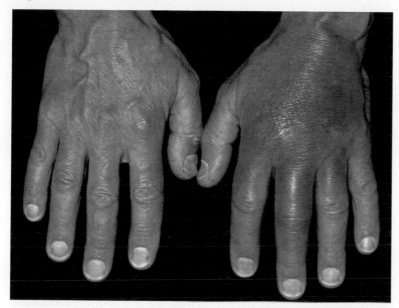

Note the swelling of the affected hand and lack of epidermal involvement.

Cellulitis does not involve the fascia or muscles, but is characterized by localized pain, swelling, tenderness, erythema, and warmth (Figure 19.16). Cellulitis is a relatively common infection. It develops more slowly than erysipelas and involves the deeper levels of dermis, including the subcutaneous fat and connective tissue. The infection is often a complication of a wound infection. Although *S. pyogenes* is the most frequent cause of cellulitis in immunocompetent adults, a number of bacterial species, including *S. aureus*, Gram-negative bacilli, and anaerobes, can cause this skin infection. Diagnosis is by evaluation of clinical symptoms and history and bacterial cultures of the original infection. Cellulitis is treated by aggressive antibiotic therapy administered by mouth, although intravenous administration of the antibiotic may be required, depending on the severity of the case.

ACNE   Acne vulgaris is America's most common disease, affecting 60%–70% of Americans at some time during their lives. Twenty percent will have severe acne, resulting in permanent physical scarring. Acne usually affects the face, but any area of skin with a dense population of follicles, such as the upper part of the chest or the back, can also be affected. Not all acne is the same. Some acne cases result from blocked hair follicles or pores called **comedones**. Comedones can be open (blackheads) or closed (whiteheads). Inflammatory acne arises from inflamed macules, papules, pustules, and nodules (aggregations of cells resembling papules but larger than 0.5 cm). The most severe type of acne is cystic or nodular acne, consisting of painful fluid-filled cysts or nodules, respectively, under the skin surface (Figure 19.17).

Several factors contribute to acne development. The first is a genetic predisposition in which epithelial cells in a follicle replicate more often than normal (hyperproliferation), which contributes to plugging of the follicle. A child has a much greater chance of developing acne if both parents had acne than if neither parent had the condition. High levels of certain hormones can cause hyperproliferation and excessive sebum production (nodular acne). Sebum is a waxy, oily substance secreted by sebaceous glands that plugs the follicle (see Figure 19.1). Finally, the Gram-positive, anaerobic bacterium *Propionibacterium acnes* plays an important role. A normal member of skin microbiota, *P. acnes* uses the triglycerides in sebum as a nutrient. The growing numbers of this organism promote inflammation by binding to Toll-like receptor 2 on macrophages and neutrophils, which then release proinflammatory cytokines such as TNF-α (Section 15.5 and Table 15.3).

Topical retinoids, which help unplug clogged skin pores, and topical antibiotics are the first line of treatment for acne. Oral retinoids and antibiotics are sometimes used for severe acne.

A cautionary word about bacterial skin infections: Skin lesions are a rich source of bacteria. Without medical intervention, the bacteria in these lesions can enter the bloodstream and cause bacteremia. Once in the blood, the bacteria can disseminate to other body sites and cause focal or metastatic lesions or produce life-threatening sepsis (see Chapter 15 inSight).

Note Proper cleaning of wounds and cuts will reduce the number of microorganisms in the site, which will in turn enhance healing.

SECTION SUMMARY

- *Streptococcus pyogenes* **(also known as group A streptococci; GAS) and** *Staphylococcus aureus* are common bacterial causes of skin infections.
- **Folliculitis, scalded skin syndrome, and impetigo** are usually caused by *S. aureus*.
- *S. aureus* **virulence factors involved with skin infections** include coagulase, toxic shock syndrome toxin, and exfoliative toxin.
- **Methicillin-resistant** *S. aureus* **(MRSA)** has become an important cause of community-acquired staphylococcal infections.

Figure 19.17   Cystic Acne

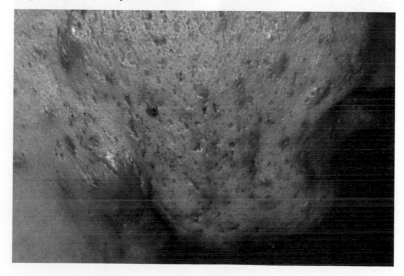

- **Necrotizing fasciitis** is usually caused by *S. pyogenes* but can be caused by other pathogens as well.
- ***S. pyogenes* virulence factors involved with skin infections** include capsule, M protein, and streptococcal pyogenic exotoxins (SPEs).
- **Cellulitis** is a skin infection caused by a variety of bacterial species.
- **Erysipelas** is a skin infection caused by *S. pyogenes*.
- **Acne** is influenced by genetics, hormone levels, and the bacterium *Propionibacterium acnes*.
- **Infections of the skin** can disseminate via the bloodstream to other sites in the body.

Thought Question 19.3 Why do viral skin infections often lead to secondary bacterial infections, whereas bacterial skin infections do not lead to secondary viral infections?

Figure 19.18   Various Forms of Ringworm

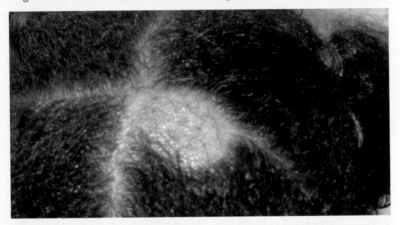

A. Gray patch tinea capitis.

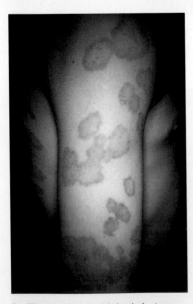

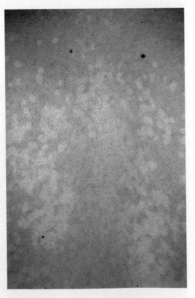

B. Tinea corporis. Multiple lesions can come together to form a "flower petal" shape.

C. The hypopigmented lesions of tinea versicolor caused by the yeast *Malassezia*.

19.4
Fungal Infections of the Skin

SECTION OBJECTIVES

- Outline common fungal skin infections according to their etiology.
- Compare dermatophytes with other fungi that commonly infect the skin.
- Explain how common fungal infections are diagnosed clinically.

Fungi, which include molds and yeasts, are eukaryotic microbes, either filamentous or single-celled (as in yeasts), as presented in Section 11.2. **Dermatophytes**, from the Greek for "skin plants," are fungi that "love" human skin. Dermatophytes find cool, moist, keratinized tissues (skin, hair follicles, nails) a perfect growth environment. Dermatophytic fungi invade the epidermis and elicit an inflammatory response. *Epidermophyton, Trichophyton,* and *Microsporum* cause the majority of dermatophytic infections (Table 19.4). Diseases caused by these fungi are named after the location of the infection and not the fungus itself. Medieval doctors, ignorant of microscopic fungi but well versed in Latin, called the infections "tinea" (Latin for "gnawing worm"). So if the lesion sits on the scalp, it is called **tinea capitis**; on the body, **tinea corporis**; in the fold of the groin, **tinea cruris** (jock itch); and on the foot, **tinea pedis** (athlete's foot). Collectively, these infections are also called **ringworm**. Fungal infection of the nails is called **tinea unguium** or onychomycosis.

Tinea capitis is a fungal infection of the scalp (primarily in small children) that has two different varieties: black dot tinea capitis (BDTC) and gray patch tinea capitis (GPTC). The black dot form of tinea capitis, caused by *Trichophyton tonsurans*, is the most common form in the United States. The name comes from the fact that infected hair may break off at the scalp, leaving black dots. Gray patch tinea capitis, caused by *Microsporum canis*, is transmitted to humans primarily from cats and dogs. Again, the hair breaks off above the scalp but frosts over (Figure 19.18A).

Tinea corporis refers to ringworm on the body surface, excluding the groin (Figure 19.18B). The lesion is round or ring-shaped (annular) and itches (is pruritic). The ring expands as the fungus grows outward from the initial site of infection. Jock itch, or tinea cruris, is a form of tinea corporis that affects the crural fold (the groin). Jock itch is more prevalent in men. The main cause of jock itch in the United States is *Trichophyton rubrum*. Athlete's foot (or tinea pedis) is caused by either *Trichophyton mentagrophytes* (acute athlete's foot) or *Trichophyton rubrum* (see Figure 2.2). Spores of these fungi are shed from the skin of an infected person and stick to the skin of uninfected individuals that may be using the same gym.

Tinea versicolor is a long-term (chronic) infection of the skin, prevalent mostly in warm, moist climates. Its cause is a genus of yeast (*Malassezia*), not a dermatophyte. This yeast is actually part of the normal skin biota and is not contagious. *Malassezia* is a dimorphic yeast, meaning it can have both yeast and mycelial (hyphal)

Table 19.4

Fungal and Parasitic Diseases of the Skin and Eye

Disease	Symptoms	Complications	Fungus or Parasite	Transmission	Diagnosis	Treatment	Prevention
Dermatomycosis	Dry, scaly lesions	Scarring, secondary bacterial infections, and cellulitis	Dermatophytes: *Epidermophyton*, *Trichophyton*, and *Microsporum*	Environmental sources; person-to-person contact	KOH preparation	Topical azole compounds (such as clotrimazole); systemic azole compounds if indicated (fluconazole)	No vaccine available
Tinea versicolor	Small, flat, discolored lesions of skin, variable colors, does not itch	Cosmetic complications	*Malassezia* genus of yeast	Environmental sources	Clinical presentation; KOH preparation	Oral or topical azole antifungals	No vaccine available
Candidal intertrigo	Patchy, smooth-looking lesions	Can become invasive in the immuno-compromised	*Candida albicans* (major cause), other *Candida* spp.	Autoinfection in the immuno-compromised	Clinical presentation; microscopic evaluation of clinical specimen; culture	Treatment depends on patient age, immune status, and disease presentation; clotrimazole and nystatin as indicated; maintain oral hygiene	No vaccine available
Fungal keratitis	Eye pain and redness, sensitivity to light, blurred vision, eye discharge	Corneal scarring, blindness	*Fusarium*, *Aspergillus*, and *Candida* spp.	Environmental sources	Clinical presentation; isolation of fungus in culture from corneal scrapings for definitive diagnosis; microscopic examination of corneal sample; PCR	Antifungal medications: topical natamycin for superficial infections (*Aspergillus* and *Fusarium*); systemic antifungals for deep infections: amphotericin B, fluconazole, or voriconazole	No vaccine available
Parasitic keratitis	Corneal lesions	Encephalitis	*Acanthamoeba* spp.	Environmental sources	Direct visualization of corneal ulcer under slit lamp; microscopy; isolating *Acanthamoeba* in culture; PCR	6–12 months of antiseptic treatment of corneal lesions; pain management	No vaccine available

forms. Pathogenesis begins when the round yeast form converts to hyphal form and invades the stratum corneum. The lesions of tinea versicolor are usually small with a sharp border and hypopigmented (Figure 19.18C). Hypopigmentation is thought to occur when azelaic acid produced by the infecting yeast damages the melanocytes of the skin.

Candida species are dimorphic yeasts that form part of the normal biota of the gastrointestinal tract, vaginal tract, oral cavity, and even skin. Best known for causing vaginal infections, *Candida* (usually *C. albicans*) can also infect the skin (in its mycelial form) as well as mucous membranes and body organs. In healthy individuals,

Candida can overgrow and cause disease in the oral cavity or vagina as a result of changes in the local microbiota. The fungus can cause serious problems in patients who are immunocompromised. On the skin, *Candida* infects areas where the skin touches and rubs together, such as between the fingers, under the arm, or the groin. The most common of these skin infections is referred to as **candidal intertrigo** (Figure 19.19). Normally kept at bay by macrophages, neutrophils, and T-cell immunity, defects in these cell types (as in neutropenia, abnormally low numbers of neutrophils) provide *C. albicans* an opportunity to penetrate skin microabrasions and gain a "foothold."

Figure 19.19   Candidal Intertrigo

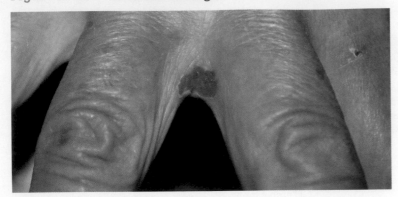

Candida infection between the fingers.

Diagnosis of fungal skin diseases is made from their clinical appearance and by microscopic examination of potassium hydroxide (KOH) preparations of skin flakes or hair. The KOH destroys the skin cells but not the more resilient walls of mycelia or spores, which can be seen under a light microscope. These fungi can also be cultured on a special selective medium called Sabouraud agar (an acidic, pH 5.6 medium containing antibiotics to thwart the growth of bacterial contaminants and culture fungi). A variety of antifungal medications are used to treat these diseases. The most common, the imidazole compounds (such as clotrimazole), can be bought over the counter. The antimicrobials either directly interrupt fungal membrane integrity or interfere with the synthesis of sterols required for fungal membrane synthesis. (See antifungal agents in Section 13.8.)

Link As discussed in Section 25.2, **potassium hydroxide (KOH) preparations** are made by placing a few drops of 10%–30% KOH on a hair follicle or nail or skin sample on a slide. The slide is heated gently to dissolve skin or nail cells, which makes fungal components more visible.

SECTION SUMMARY

- **Dermatophytes** are fungi that cause skin infections.
- **Fungal skin diseases are named by location,** not by fungal species; tinea capitis (scalp), tinea corporis (body), tinea cruris (groin), tinea pedis (feet).
- **Ringworm** refers to the circular appearance of lesions as the fungus extends in all directions.
- *Malassezia* and *Candida* **are dimorphic yeasts** that can cause skin infections.
- **KOH preparations** are used to diagnose fungal diseases of the skin.

Thought Question 19.4. Why do some people develop fungal infections but others do not?

19.5
Skin Infections of Special Circumstance

SECTION OBJECTIVES

- Discuss why *Pseudomonas aeruginosa* is a common cause of wound infections.
- Categorize organisms that are commonly isolated from diabetic foot wounds on the basis of their Gram stain status.
- Explain the roles of anaerobic and aerobic bacteria in the pathogenesis of gas gangrene.

Why does physical injury, even when it is planned, like surgery, increase the risk of skin infections? Some microorganisms infect human skin only when its integrity is compromised. This section discusses skin infections that involve unusual precipitating factors (such as burns or traumatic injuries) or a chronic disease (such as diabetes; see **inSight**).

Burn Wound Infections

Each year in the United States, about 40,000 people are hospitalized with burn infections; about 4,000 die. Roughly 500 of those deaths are due to infection. The skin is a primary barrier against infection, so when this barrier is damaged, as in a burn, pathogens have a direct route to infiltrate the body and cause disease. Victims with burns covering more than 10% of their body are most likely to develop serious infections (Figure 19.20).

Burns obviously alter the innate immune features of the skin but also subvert other arms of the immune system. T-cell activity

Figure 19.20   Burn Wound Infected with *Aspergillus*

Burn wound of a 7-year-old boy who developed an infection with the fungus *Aspergillus* 3 weeks after the burn. Sadly, the boy died despite aggressive medical treatment.

declines as a result of an increase in the number of regulatory T cells and a decrease in the number of helper T cells. The levels of inflammatory cytokines and complement also decrease. In addition, burns decrease neutrophil chemotaxis and phagocytosis, diminishing their bactericidal activity. All of these factors make the exposed tissues of the victim more susceptible to infection, while destruction of local blood and lymphatic vessels prevents immune cells and systemically administered antibiotics from being delivered to the site of infection.

Immediately after a severe burn, the surface of the wound is essentially heat sterilized. So why is it that covering a burn immediately with sterile bandages often fails to prevent infection? For one thing, deep structures that survive the initial burn injury, such as sweat glands and hair follicles, often contain staphylococci, which will quickly colonize the wound surface. Then 5–7 days after the burn, other Gram-negative and Gram-positive bacteria colonize the wound. These microbes typically come from the patient's gastrointestinal tract, upper respiratory tract, or the hospital environment and are inadvertently transferred to the wound by health care workers. Serious burn wound infections can produce cellulitis, necrotizing fasciitis, and even sepsis. Methicillin-resistant *Staphylococcus aureus* (MRSA) is perhaps the most common cause of burn wound infections. Gram-negative bacteria and fungi such as *Candida* and *Aspergillus* species can also cause dangerous infections.

Pseudomonas aeruginosa (a Gram-negative, aerobic rod) is a major cause of serious wound infections owing to its wealth of virulence factors and its high level of antibiotic resistance. *P. aeruginosa* is nearly ubiquitous in the environment. It is not a primary pathogen but is an opportunistic pathogen able to infect tissues when one of the body's defenses is down. Most clinical specimens produce a green pigment called pyocyanin, which gives its colonies, and the infected pus, a greenish hue. The organism's toxin repertoire includes **exotoxin** A and exotoxin S, which are **ADP-ribosyltransferases** that, respectively, inactivate host protein synthesis (the target is EF2) and interfere with intracellular host signaling pathways. Both can cause cell death. The organism also secretes phospholipases and proteases that contribute to tissue invasion and disease. Many of the genes encoding the virulence factors are controlled by quorum-sensing mechanisms discussed in Chapter 6 (see Section 6.6). *P. aeruginosa* is resistant to a great many antibiotics, making treatment of burn wounds challenging.

Link As discussed in Section 18.3, some **exotoxins** have **ADP-ribosyltransferase** enzymatic activity, meaning that they can transfer an ADP-ribose group from an NAD molecule to a target protein. Addition of the ADP-ribose will modify the target protein and alter or inactivate its function. When the target protein is EF2 (elongation factor 2), protein synthesis is halted and the cell is killed.

Gas Gangrene

Gangrene is localized necrosis (tissue death). There are two types of gangrene: dry gangrene, which is caused when the blood supply to a tissue is cut off, and wet gangrene, which is caused by an infection. Dry gangrene usually affects fingers and toes and is seen in cases of severe frostbite and sometimes in individuals who have diabetes or certain autoimmune diseases. Wet gangrene is a necrosis due to an infection and can affect an internal or external part of the body. Anaerobic infections affecting the genital area (Fournier's gangrene) and infections caused by *Clostridium* species that produce large amounts of gas (gas gangrene) are also classified as wet gangrenes.

CASE HISTORY 19.4

Shot in the Black Hills of South Dakota

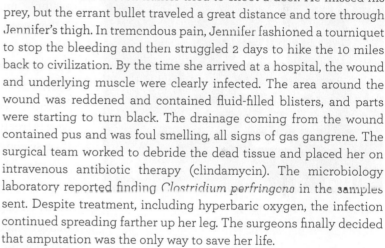

Jennifer remembered the day she was shot, but not the day they had to remove her leg. She was camping in the middle of the Black Hills of South Dakota when an unseen hunter tried to shoot a deer. He missed his prey, but the errant bullet traveled a great distance and tore through Jennifer's thigh. In tremendous pain, Jennifer fashioned a tourniquet to stop the bleeding and then struggled 2 days to hike the 10 miles back to civilization. By the time she arrived at a hospital, the wound and underlying muscle were clearly infected. The area around the wound was reddened and contained fluid-filled blisters, and parts were starting to turn black. The drainage coming from the wound contained pus and was foul smelling, all signs of gas gangrene. The surgical team worked to debride the dead tissue and placed her on intravenous antibiotic therapy (clindamycin). The microbiology laboratory reported finding *Clostridium perfringens* in the samples sent. Despite treatment, including hyperbaric oxygen, the infection continued spreading farther up her leg. The surgeons finally decided that amputation was the only way to save her life.

Gas gangrene, also called clostridial myonecrosis because muscle tissue is affected, is an extremely serious, life-threatening disease caused mainly by *Clostridium perfringens*, a Gram-positive, spore-forming, anaerobic bacillus (**Figure 19.21**). *C. perfringens* is a true saprophyte, growing only on dead tissue. The organism is widely distributed throughout the environment and is also found in the intestine. Being ubiquitous in soil, *C. perfringens* commonly infects traumatic wounds (knife or gun wounds, for example), but infections can occur spontaneously with no obvious precipitating event. Infections usually start out mixed with aerobic organisms that consume oxygen and provide an anaerobic environment for the clostridia. *C. perfringens* is highly fermentative and produces considerable amounts of foul-smelling gases that can be trapped by skin and produce bullae. Pockets of gas in muscle can choke off blood supply (ischemia) and contribute to necrosis, which produces more food for the pathogen.

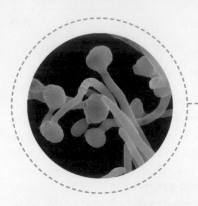

inSight

Of Skin and Wounds and Diabetes

Figure 1 *Candida albicans*

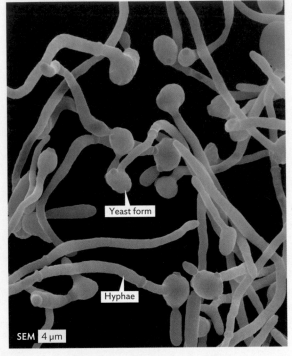

Yeast form

Hyphae

SEM 4 µm

Colorized SEM of yeast and hyphal forms of *Candida albicans*.

Figure 2 **Diabetic Foot Ulcer**

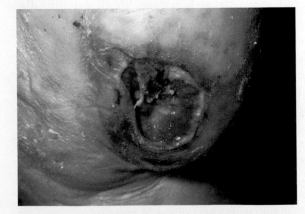

Diabetic foot ulcers often result from minor trauma that goes unnoticed. Patients suffering from a neuropathic ulcer often do not feel the pain and can use the affected foot without any difficulty. Note the pus in the wound and the redness in the periphery, signifying cellulitis.

Most of us have suffered an occasional blister from a tight-fitting shoe or a minute cut when we accidentally stepped on a small object while walking barefoot out in the yard. We simply protect the small injury, and soon it is healed and forgotten. Remarkably, the same blister or cut may cost a diabetic patient the amputation of a foot. Wounds take longer to heal in diabetics and are more prone to infection than in nondiabetics. But why is this so?

An increase in plasma glucose above normal physiological levels (such as after a meal) is normally met with the production of additional insulin by beta cells of the pancreas. The insulin binds to its receptors on the surfaces of other body cells (for example, muscle cells) and helps those cells take up the additional glucose, which is then used as fuel. Type 1 diabetes is a disease in which the production of insulin is stopped as a result of damage to the pancreatic beta cells. In type 2 diabetes, insulin is made but the insulin receptors on target cells become less responsive to the hormone. The result in either case is widely fluctuating levels of blood glucose (because the glucose is not transported and burned efficiently), even when the disease is treated and well controlled. Elevated blood glucose levels are responsible for diverse complications that affect all body systems, including the skin.

Diabetic patients are more susceptible to fungal skin infections caused by *Candida* (**Figure 1**) and *Rhizopus* species. Because tissues in diabetic patients cannot easily transport glucose, serum glucose levels remain high after a meal. Despite being bathed in glucose, cells are actually starving and must resort to fatty acid degradation for energy production. Fatty acid degradation leads to ketone production. *Rhizopus* species possess a ketone reductase that allows them to grow in high levels of ketone and cause infections. Why diabetics experience higher rates of *Candida albicans* infection is not known, but may be a result of the suppressed immune status of these patients.

Neuropathy (nerve damage) is a common complication of diabetes and can affect a variety of nerves. Neuropathy may result when elevated glucose levels cause microvascular damage to vessels that supply nerves. The most commonly affected are the sensory nerves of the feet. A patient with sensory neuropathy in the foot may lose the ability to feel pain in that foot. As a result, a minor foot trauma can go unnoticed and quickly become a serious infected foot wound (**Figure 2**). In fact, the majority of diabetic foot wounds result from simple traumas (such as a tight-fitting shoe) that may become seriously infected, making diabetes the number-one cause of non-traumatic lower-extremity amputation in the United States.

Diabetic foot wounds are also more susceptible to infection because the patient is immunocompromised. Although we do not know exactly how the immune system is damaged by diabetes, we know that essential neutrophil functions, such as chemotaxis, adhesion, and phagocytosis, are affected. In addition, C3-mediated opsonization of bacteria is defective (see Section 15.7).

Infection also inhibits wound healing. Some diabetic foot ulcers become chronic and heal very slowly. Chronic wounds can be infected repeatedly, making prolonged antibiotic therapy a necessity. A chronic wound can extend to the bone if untreated (or if the treatment is unsuccessful) and cause a bone infection (osteomyelitis). Gram-positive cocci (*Staphylococcus*, *Streptococcus*, *Enterococcus*) as well as Gram-negative *Pseudomonas* and anaerobic species, including *Bacteroides* and *Clostridium* species, are commonly isolated from these infected wounds.

Treatment of diabetic foot wounds includes debriding dead tissue, providing antibiotic therapy, and reducing the pressure on the wound (called off-loading) using specially designed shoes and shoe inserts. Health care providers are attempting to reduce the incidence of diabetes-related amputations by aggressively promoting the prevention and treatment of diabetic foot wounds. Prevention is promoted by screening patients for neuropathy and by patient education. Among the treatment techniques are custom-made off-loading devices, artificial skin, and even sterile maggot therapy (Figure 3).

Infection is one reason diabetes is such a serious health concern. It is predicted that one-third of all Americans could suffer from diabetes by 2050. A majority of these will be type 2 cases that could be prevented through diet and exercise.

Figure 3   Maggot Treatment of a Diabetic Foot Infection

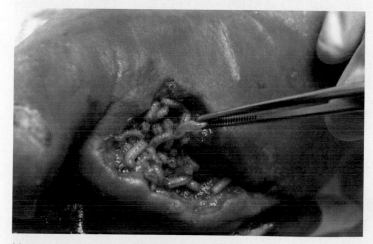

Maggots can rid a diabetic patient of an infection caused by methicillin-resistant *Staphylococcus aureus* (MRSA). Note that this person has already had several toes amputated.

Figure 19.21   Gas Gangrene Caused by *Clostridium perfringens*

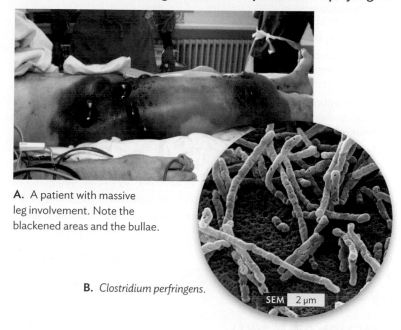

A. A patient with massive leg involvement. Note the blackened areas and the bullae.

B. *Clostridium perfringens*.

SEM   2 μm

C. perfringens produces a number of toxins (up to 17) that promote spread of the organism through the host's tissues. The secreted toxins basically kill host tissue ahead of the advancing pathogen. The many strains of *C. perfringens* are classified into types A–E based not on their antigenic properties but on the combination of toxins they produce. The toxins include lecithinases (alpha toxin), collagenase, hyaluronidase, fibrinolysin, and leukocidins that destroy neutrophils. Some strains also produce *C. perfringens* enterotoxin (CPE), which can lead to a gastrointestinal disease (see Section 22.5).

Treatment requires debridement and excision of surrounding tissue. Amputation is necessary in many cases. Antibiotics alone are not sufficient because they do not penetrate ischemic muscles enough to be effective. Penicillin or clindamycin is given along with surgical treatment to prevent spreading of the microbe. In addition to surgery and antibiotics, hyperbaric oxygen therapy, in which the affected limb is placed in an oxygen chamber, is used to inhibit the growth of and kill the anaerobic *C. perfringens*.

C. perfringens infections are often so severe that by the time a diagnosis is made, curing the infection with antibiotics is impossible. Furthermore, the extent of injury caused by the infection may leave muscle tissues so damaged that the body cannot replace the lost structures. It should be noted that gas gangrene is not associated only with limbs. Any body part with necrotic muscle can be affected (for example, the abdominal wall after surgery).

SECTION SUMMARY

- **Burns affect several aspects of the immune system.** The physical barrier of skin is destroyed, and T-cell and neutrophil responses to infection are undermined.

- **Microbes that cause burn infections** include staphylococci and streptococci residing in hair follicles and Gram-negative bacteria originating from the gastrointestinal tract (for example, *E. coli*, *Klebsiella*) or from the environment (*Pseudomonas aeruginosa*).

- ***Pseudomonas aeruginosa* is an opportunistic pathogen** that expresses many virulence factors, including exotoxins A and S, which kill host cells, and enzymes that promote tissue invasion.

- **Wound injuries** can host life-threatening infections by organisms such as *Clostridium perfringens*, a Gram-positive, anaerobic, spore-forming rod commonly found in soil and the gastrointestinal tract.

- **Gas gangrene** (clostridial myonecrosis) is a disease associated with necrotic tissue, including muscle. *C. perfringens* is the usual cause.

Thought Question 19.5   Why is it that wounds tend to be polymicrobial, meaning you usually can find more than one species of bacteria in a wound?

19.6
Eye Infections

SECTION OBJECTIVES

- Identify viral, bacterial, and fungal causes of eye infections.
- Describe the symptoms and etiologies of "pink eye."
- Explain how the pathogenesis of infectious keratitis differs from that of endophthalmitis.
- Outline the relationship between corneal epithelium and fungal keratitis.

Are all eye infections serious?   Actually, some infectious diseases of the eye are self-limiting, but others can cause blindness.

Infections of the eye are quite common. Infections of the external parts of the eye include conjunctivitis and keratitis. Pink eye is a general, nonspecific term for conjunctivitis, inflammation of the conjunctiva, which can be due to an infection, trauma, or an allergic reaction (**Figure 19.22**). **Keratitis** is inflammation of the cornea, which is a sight-threatening process. A hallmark of bacterial keratitis is its rapid progression; corneal destruction may be complete in 24–48 hours with some of the more virulent bacteria. Infections of the inner structures of the eye (**endophthalmitis**) are extremely uncommon and almost always result either from direct spread of a superficial eye infection or from seeding of the internal eye structures with bacteria carried by blood from another infected site (hematogenous extension).

Viral Eye Infections

Viral conjunctivitis is seen as a diffuse pinkness of the conjunctiva, often called "pink eye." Conjunctivitis is usually associated with a respiratory tract infection (typically an adenoviral cold or sore throat). The eyelid may be swollen, and the eye itself is red and produces a lot of tears and sometimes a mucoid (mucus-like) discharge. Contact with a person with viral conjunctivitis can easily spread the virus, but how readily a subsequent infection forms depends on the virus. Viral conjunctivitis is a self-limiting condition that has no treatment. Patients are advised to frequently wash hands and change towels, not share towels, avoid touching their eyes with their hands, and to stay home as long their eye (or eyes) have a discharge so that they do not spread the infection to others or reinfect themselves.

Viral keratitis can lead to blindness and is a medical emergency. Herpes viruses are a major cause of keratitis and keratoconjunctivitis in the United States, with herpes simplex 1 responsible for the majority of cases. The primary lesion may not cause much damage, but recurrence of the lesion due to reactivation of the virus can cause corneal scarring and blindness (**Figure 19.23**). The antiviral medications acyclovir and trifluridine are used to treat herpes

Figure 19.22   Pink Eye, the Common Name for Conjunctivitis

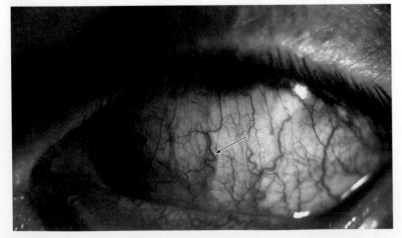

An inflamed conjunctiva is red, giving rise to the term "pink eye." The arrow shows one of the blood vessels that have become prominent. When the eye shows such prominent vessels, it is said to be "injected." Pink eye is commonly caused by adenoviruses or allergies.

Figure 19.23   Scarring Caused by Viral Keratitis

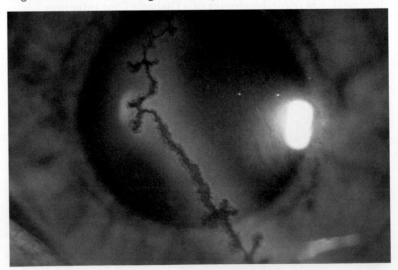

Yellow fluorescein dye added to the affected eye reveals a branching (dendritic) scarring ulcer on the cornea.

Figure 19.24   Herpes Zoster Ophthalmicus

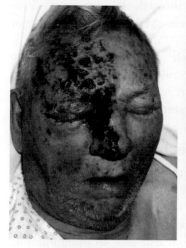
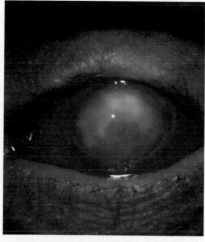

A. An outbreak of shingles along the ophthalmic division of the trigeminal nerve results in eruption of vesicular lesions on the forehead, eyelids, and nose.

B. Corneal inflammation and scarring in herpes zoster ophthalmicus.

keratitis. A 14-day treatment course is reported to improve healing of the ulcers, and their long-term use decreases the rate of recurrence of the lesions.

The trigeminal nerve is a cranial nerve (a nerve originating from the brain) with several divisions. One of the divisions, the ophthalmic, extends into the eye and to the tip of the nose to gather sensory information and send it back to the brain. An outbreak of shingles along the ophthalmic division of the trigeminal nerve (Figure 19.24) will result in the eruption of vesicular lesions on the forehead, eyelids, and nose and may even spread to the eye itself, causing corneal inflammation and eye pain and sensitivity, a condition called **herpes zoster ophthalmicus**. The virus can spread deep into the inner structures of the eye and cause necrosis (death) of the retina (acute retinal necrosis) in some immunocompromised patients. VZV infection of the eye threatens a patient's sight and is considered a medical emergency. Oral antiviral medications such as acyclovir, valacyclovir, and famciclovir are used to treat herpes zoster infection.

Bacterial Eye Infections

Bacterial conjunctivitis usually affects one eye and is acute, painful, and purulent. *Haemophilus influenzae* eye infections, however, initially look like viral or allergic conjunctivitis. Conjunctivitis due to the common pyogenic (pus producing) bacteria, such as staph and strep, causes marked irritation and a stringy, opaque, grayish or yellowish mucopurulent discharge that may cause the lids to stick together, especially after sleep (**Figure 19.25**). *Chlamydia trachomatis* and *Neisseria gonorrhoeae*, two species that cause serious infections in the reproductive system (see Section 23.4), also cause

the majority of acute bacterial conjunctivitis. How this happens will be discussed shortly. Potential complications of bacterial conjunctivitis include spreading of infection to the cornea and blindness. With proper antibiotic treatment, however, these infections usually resolve before they spread to the cornea.

Chlamydia trachomatis, an obligate intracellular bacterial pathogen, is an important cause of conjunctivitis. Inclusion conjunctivitis can be contracted by newborns passing through an infected vagina (these particular strains cause an important sexually transmitted disease). Fortunately, inclusion conjunctivitis is usually self-limiting. The disease is prevented by prophylactically treating all newborns with antibiotic eyedrops (erythromycin) as soon as they are born.

A second, more serious disease caused by some strains of *C. trachomatis* is **trachoma**. Trachoma is the most frequent cause of infectious blindness in the world. The organism is spread by direct contact with eye, nose, and throat secretions. The patient initially notices a mild conjunctivitis, but soon the cornea becomes vascularized, and leukocytes infiltrate the affected area. The conjunctiva of the inner eyelid becomes pebbled and causes a turning in of the eyelashes (trichiasis). As the patient blinks, these eyelashes rake across the cornea. The cornea can become ulcerated and scarred, causing blindness. If diagnosed early, trachoma can be cured by oral (azithromycin) and topical (tetracycline) antibiotic treatment.

Neisseria gonorrhoeae can cause a serious eye disease of infants called **ophthalmia neonatorum** (neonatal conjunctivitis). As with

Figure 19.25   Purulent Bacterial Conjunctivitis

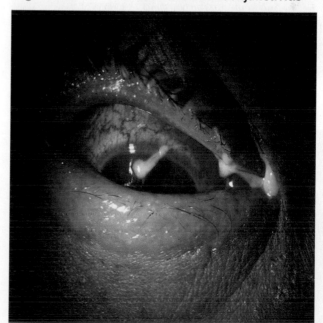

Bacterial infection of the eye can result in severe conjunctivitis. The purulent material on the cornea is extending to the corner of the eye. The eyelashes are covered in pus as well and appear wet and sticky.

Figure 19.26   Fungi Causing Ocular Infections

Fungi can infect an injured cornea when they are accidentally introduced into the eye. *Fusarium* (**A**) and *Aspergillus* (**B**) are molds naturally found on plants, in water, and in soil. *Candida* (**C**) is part of human microbiota.

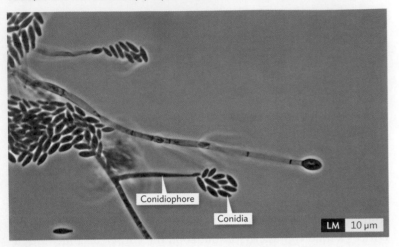

A.

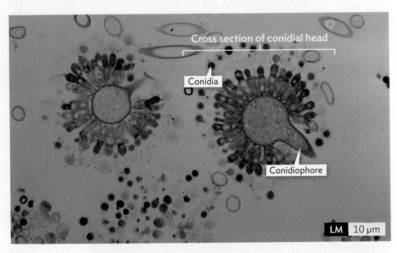

B.

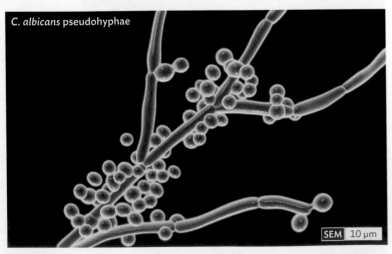

C.

inclusion conjunctivitis, the organism is passed from an infected mother to newborn during passage through the birth canal. Prophylactic treatment with erythromycin eyedrops will also prevent this disease.

This year, many millions of people worldwide will develop eye infections. Trachoma alone affects 6 million people. Almost all of these infections are treatable. Unfortunately, because of the cost, those who are poor will fail to get treated, and many will go blind.

Fungal and Parasitic Eye Infections

Fungi and parasites usually have a very hard time infecting our eyes because they cannot penetrate the intact corneal epithelium. But even a very small injury to the eye, such as a minor abrasion, can provide these organisms with an opportunity to cause ocular infections. The resulting corneal infection can lead to inflammation, necrosis, and even blindness. Wearing a contact lens or getting a foreign body (an eyelash, a grain of sand, a small piece of dirt) in your eye may cause an abrasion sufficient for fungi to gain entry. A large portion of ocular fungal and parasitic infections are seen among those practicing poor contact lens hygiene.

Fungal keratitis is most often caused by species of *Fusarium, Aspergillus* (both found in the environment), and *Candida* (part of the human microbiota) (**Figure 19.26**). The incidence of fungal keratitis is extremely variable in different regions, but it seems to be higher in warm, moist climates.

Definitive diagnosis of fungal keratitis is made by culturing the fungi from the corneal scraping specimen. PCR amplification and confocal microscopy (see Section 3.5), if available, are also useful. How fungal eye infections are treated depends on whether the infection is superficial or invasive. Invasive lesions require systemic therapy (oral or intravenous delivery of antifungal agents) and often surgery. Natamycin 5% solution (for filamentous fungi) and amphotericin B (for yeast) are currently drugs of choice for superficial lesions (see polyene antifungals, Section 13.8).

Parasitic keratitis is due mostly to *Acanthamoeba* species (see Section 11.3), protozoa that are found ubiquitously in nature (air, water, and soil); in heating, ventilation, and air-conditioning equipment; and even in sewage systems. The parasite can exist as a cyst for years when conditions are unfavorable. Ocular *Acanthamoeba* infection makes the eye red, painful, and teary. The patient will avoid light (photophobia) and suffer visual loss. If untreated, the infection may cause permanent visual impairment (**Figure 19.27A**) or spread to the brain (**Figure 19.27B**).

Acanthamoeba is increasingly recognized as an important cause of keratitis in non–contact lens wearers, but wearing contact lenses is currently the main risk factor for this disease. Contaminated lens solution and lack of proper contact lens hygiene are the main contributors to the risk in patients who wear contact lenses.

Diagnosis of *Acanthamoeba* ocular infections can be difficult, mostly because of a low degree of suspicion (it is not the most likely cause of keratitis), but chronic inflammation of the cornea

Figure 19.27   *Acanthamoeba* keratitis

Acanthamoeba species are free-living protozoa that are found in air, water, and soil and that can cause corneal infection and inflammation.

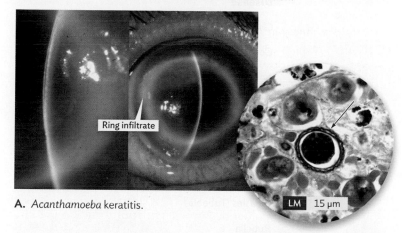

A. *Acanthamoeba* keratitis.

B. *Acanthamoeba* cyst (arrow) in brain tissue.

and resistance to other antimicrobial treatments is an indication. Definitive diagnosis requires culturing the organisms from a corneal specimen. PCR and confocal microscopy, in conjunction with clinical evaluation, are also very useful. Prolonged treatment, 6–12 months, with antiseptic agents and ocular and systemic pain medications are used to help eradicate the parasite from the cornea, but surgery may be required to restore vision.

SECTION SUMMARY

- **Eye infections** usually start as a conjunctivitis and, if unresolved, develop into keratitis.
- **Pink eye** is usually caused by an adenovirus or *Haemophilus influenzae*.

- **Conjunctivitis caused by staphylococci and streptococci** includes a mucopurulent discharge that can produce crusting of the affected eye.
- **The leading cause of bacterial infectious blindness is *Chlamydia trachomatis*,** an obligate intracellular pathogen. Blindness results from trichiasis of the cornea.
- ***C. trachomatis* and *Neisseria gonorrhoeae* can cause blindness in newborns.** The organisms are transmitted during passage through an infected birth canal, but disease is prevented by applying erythromycin eyedrops at birth.
- ***Acanthamoeba*** species, protozoa found ubiquitously in nature, can cause parasitic keratitis.

Thought Question 19.6   Why is eye infection with *Acanthamoeba* often associated with the use of contact lenses?

Perspective

Skin truly is a mirror that reflects the status of our health. The color, tone, and texture of skin can change in response to nutrition, health, hormonal status, and the state of our immune system. Skin appearance also changes as a consequence of many microbial diseases. The infections discussed in this chapter manifest themselves as blatant skin rashes or lesions. However, keep in mind that infections elsewhere in the body, such as streptococcal sore throat caused by GAS or typhoid fever caused by *Salmonella typhi*, also produce striking skin manifestations (scarlet fever and rose spots), even though the skin itself is not infected. These diseases are described in other chapters. Other infectious diseases, such as anthrax, tularemia, and plague, include infected skin lesions but were not included here because the etiologic agent is more typically associated with disseminated disease. These, too, will be covered later. Clearly, skin rashes and lesions, unlike beauty, often reflect something more than skin deep.

Figure 19.28   Summary: Microbes That Infect the Skin and Eyes

Viruses	Genome	Disease
Varicella-zoster	DNA, ds	Chickenpox
HSV-1 (sometimes HSV-2)	DNA, ds	Cold sores/genital lesions
HHV-6 and HHV-7	DNA, ds	Roseola infantum
Papillomavirus	DNA, ds	Warts
Variola	DNA, ds	Smallpox
Adenoviruses	DNA, ds	Pink eye
Parvovirus B19	DNA, ss	Erythema infectiosum
Measles virus	RNA, ss, negative sense	Measles (rubeola)
Rubella virus	RNA, ss, positive sense	Rubella

Bacteria	Morphology	Disease
Bacillus anthracis	Rods, Gram-positive, facultative	Cutaneous anthrax
Enterococcus spp.	Cocci, Gram-positive, anaerobic	Wound infections
Staphylococcus aureus	Cocci, Gram-positive, facultative	Impetigo, folliculitis, furuncles, carbuncles, scalded skin syndrome, cellulitis
Streptococcus pyogenes	Cocci, Gram-positive, facultative	Erysipelas, necrotizing fasciitis, cellulitis, ecthyma
Haemophilus influenzae	Coccibacillus, Gram-negative, aerobic	Pink eye
Chlamydia trachomatis	Pleomorphic,[a] obligate intracellular parasite	Eye infections
Pseudomonas aeruginosa	Rods, Gram-negative, aerobic	Wound infections
Bacteroides	Rods, Gram-negative, anaerobic	Wound infections
Vibrio vulnificus	Rods, Gram-positive, aerobic	Necrotizing fasciitis
Clostridium perfringens	Rods, Gram-positive, anaerobic	Gas gangrene, necrotizing fasciitis
Propionibacterium acnes	Rods, Gram-positive, anaerobic	Acne vulgaris

Fungi	Type	Disease
Epidermophyton spp.	Dermatophytes	Tinea capitis, tinea corporis including tinea cruris (jock itch), tinea pedis (athlete's foot)
Trichophyton spp.	Dermatophytes	
Microsporum spp.	Dermatophytes	
Rhizopus spp.	Mold	Diabetes-related wound infections
Malassezia spp.	Yeast	Tinea versicolor
Candida spp.	Yeast, dimorphic, can form true hyphae and pseudohyphae	Diabetes-related wound infections

[a]Chlamydia have a structure resembling the Gram-negative cell wall that is devoid of peptidoglycan despite having the genes for peptidoglycan synthesis. These bacteria do not stain well, and if they do stain, they stain Gram-negative-like.

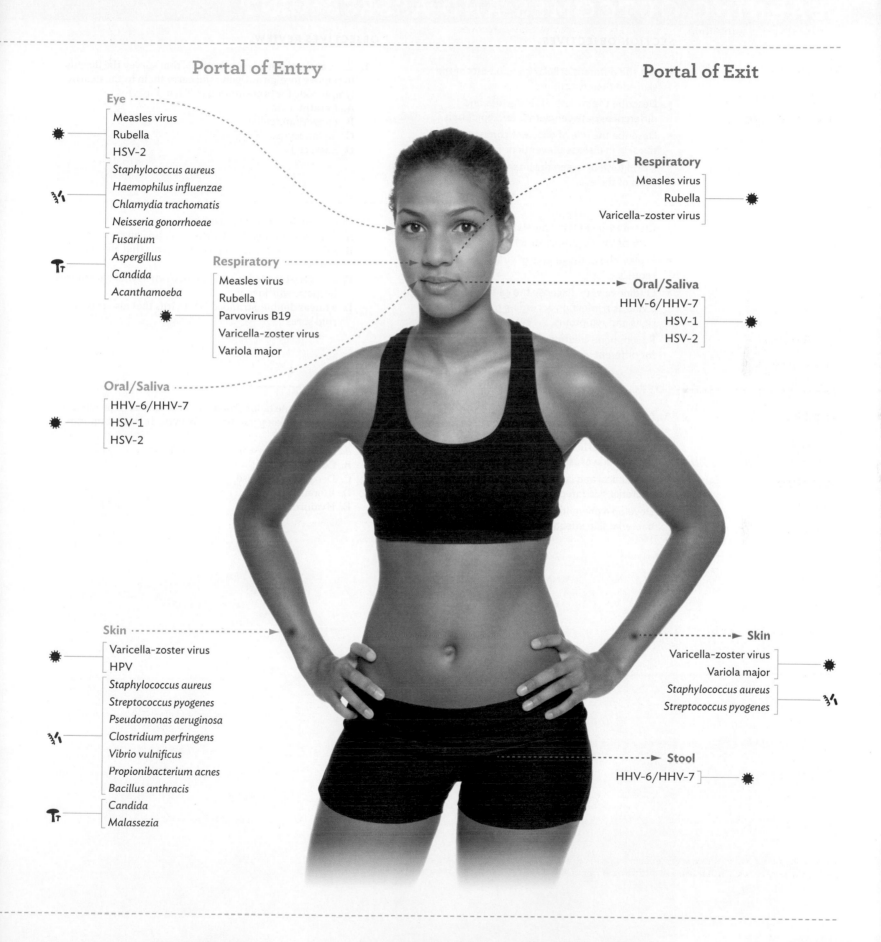

Portal of Entry

Eye

- Measles virus
- Rubella
- HSV-2

- *Staphylococcus aureus*
- *Haemophilus influenzae*
- *Chlamydia trachomatis*
- *Neisseria gonorrhoeae*

- *Fusarium*
- *Aspergillus*
- *Candida*
- *Acanthamoeba*

Respiratory

- Measles virus
- Rubella
- Parvovirus B19
- Varicella-zoster virus
- Variola major

Oral/Saliva

- HHV-6/HHV-7
- HSV-1
- HSV-2

Skin

- Varicella-zoster virus
- HPV

- *Staphylococcus aureus*
- *Streptococcus pyogenes*
- *Pseudomonas aeruginosa*
- *Clostridium perfringens*
- *Vibrio vulnificus*
- *Propionibacterium acnes*
- *Bacillus anthracis*

- *Candida*
- *Malassezia*

Portal of Exit

Respiratory

- Measles virus
- Rubella
- Varicella-zoster virus

Oral/Saliva

- HHV-6/HHV-7
- HSV-1
- HSV-2

Skin

- Varicella-zoster virus
- Variola major

- *Staphylococcus aureus*
- *Streptococcus pyogenes*

Stool

- HHV-6/HHV-7

LEARNING OUTCOMES AND ASSESSMENT

SECTION OBJECTIVES

OBJECTIVES REVIEW

19.1
Anatomy of the Skin and Eye

- List the different cellular components of the skin and their functions.
- Describe the structural similarities and differences between dermis and epidermis.
- Describe the role of different components of the skin in disease prevention.
- Distinguish between the internal and external parts of the eye.

1. _____ are antigen-presenting cells that survey the dermis to remove foreign antigens and carry them to the nearby lymph node for presentation to T lymphocytes.
   A. Keratinocytes
   B. Langerhans cells
   C. Melanocytes
   D. Basal cells

19.2
Viral Infections of the Skin

- Classify viruses that cause skin rashes on the basis of their genomic structure.
- Explain the pathogenesis of viral skin infection sequelae.
- Outline how to diagnose the causative agent of a viral skin infection according to the patient's signs and symptoms.
- Recommend treatment and prevention options for different viral skin infections.

5. Herpes gladiatorum is caused by _____.
   A. a single-stranded RNA virus that also causes warts
   B. a double-stranded circular DNA virus that also causes warts
   C. a single-stranded segmental RNA virus that also causes herpetic whitlow
   D. a linear double-stranded DNA virus that also causes cold sores

19.3
Bacterial Infections of the Skin

- Name the bacterial pathogens that cause common skin infections.
- Categorize skin infections by the physiology of the bacteria causing them.
- List clinical and laboratory means of diagnosing bacterial skin infections.
- Develop a prevention and treatment plan for bacterial skin infections.

8. Which enzyme helps *Staphylococcus aureus* to produce an abscess to keep itself away from the human immune system's reach?
   A. Gyrase
   B. Coagulase
   C. Deoxyribonuclease
   D. Ribonuclease
   E. Hyaluronidase

2. Keratinocytes are continuously produced from _____ cells.

3. Vesicular lesions are formed in the _____ layer of skin.

4. Tissue originating from the optic nerve lines the outside structures of the eye. True or false?

6. _____ is a feared bioterrorism DNA virus and the etiologic agent of a disease that was eradicated in 1979.
   A. Roseola
   B. Variola
   C. Rubella
   D. HSV-1
   E. HPV

7. A male infant born to a mother who suffered rubella infection during the first trimester of her pregnancy suffers from severe hearing loss. The infant's hearing loss is probably due to _____.
   A. horizontal transmission of the latent rubella virus to the newborn during birth
   B. vertical transmission of the latent rubella virus to the newborn during birth
   C. horizontal transmission of rubella virus to the fetus during the mother's active infection
   D. vertical transmission of rubella virus to the fetus during the mother's active infection

9. Which of the following is the most common type of bacteria causing skin infections?
   A. Gram-negative rods
   B. Gram-positive cocci
   C. Facultative anaerobes
   D. Facultative aerobes
   E. Acid-fast bacteria

10. A patient is brought to the hospital suffering with severe necrotizing fasciitis. Which of the following is the best treatment option?
   A. Oral antibiotics
   B. IV antibiotics
   C. Surgical removal of affected tissue
   D. Surgical removal of affected tissue and oral antibiotics
   E. Surgical removal of affected tissue and IV antibiotics

LEARNING OUTCOMES AND ASSESSMENT

	SECTION OBJECTIVES	OBJECTIVES REVIEW
19.4 Fungal infections of the Skin	• Outline common fungal skin infections according to their etiology. • Compare dermatophytes with other fungi that commonly infect the skin. • Explain how common fungal infections are diagnosed clinically.	11. Production of azelaic acid by _____ damages skin melanocytes and causes hypopigmentation. A. *Malassezia* B. *Candida* C. *Epidermophyton* D. *Trichophyton* E. *Microsporum*
19.5 Skin Infections of Special Circumstance	• Discuss why *Pseudomonas aeruginosa* is a common cause of wound infections. • Categorize organisms that are commonly isolated from diabetic foot wounds on the basis of their Gram stain status. • Explain the roles of anaerobic and aerobic bacteria in the pathogenesis of gas gangrene.	14. Which of the following contributes to an increased risk of wound infection in burn patients? A. Decreased humoral response B. Increased inflammatory response C. Decreased cell-mediated immunity D. Increased superficial vascularity
19.6 Eye Infections	• Identify viral, bacterial, and fungal causes of eye infections. • Describe the symptoms and etiologies of "pink eye." • Explain how the pathogenesis of infectious keratitis differs from that of endophthalmitis. • Outline the relationship between corneal epithelium and fungal keratitis.	18. A 7-year-old child is suffering from red, watery eyes and a runny nose. He feels very well otherwise and has no other symptoms. His mother reports that "he had no symptoms last night and woke up this morning with both eyes looking like this." Which of the following best describes the condition of his eyes? A. Acute allergy attack causing conjunctivitis B. Acute bacterial eye infection C. Acute viral eye infection D. Acute viral keratitis E. Acute fungal keratitis

12. The most common laboratory method used for initial diagnosis of a fungal infection is _____.
    A. Tzanck preparation
    B. culture and sensitivity
    C. KOH preparation
    D. Gram stain
    E. fluorescence microscopy

13. A patient complains of "a patch of hair loss." Your examination shows a small area of hair loss that has numerous black dots. You diagnose black dot tinea capitis and suspect that this patient has a fungal infection he might have acquired _____.
    A. wearing an old pair of athletic shoes
    B. playing with a household dog or cat
    C. wearing tight underwear and pants
    D. swimming in a gym's crowded pool

15. *Pseudomonas aeruginosa* _____.
    A. is a pathogen found only in hospitals
    B. is a ubiquitous opportunistic pathogen
    C. infects burn wounds only
    D. is a Gram-positive anaerobe

16. What types of microorganisms are commonly isolated from diabetic foot wounds?
    A. Fungi
    B. Protozoa
    C. Bacteria
    D. Viruses

17. Which of the following is true regarding gas gangrene?
    A. It may be caused by aerobes.
    B. It results from infection of superficial wounds with any anaerobe.
    C. It occurs only in trauma (knife or gunshot) wounds.
    D. It often is due to polymicrobial wound infection with aerobes and *C. perfringens*.

19. Inflammation of _____ can have a variety of etiologies, including infections, and is often referred to as "pink eye."

20. A patient suffering from inflammation of the inner structure of the eye after a traumatic eye injury (penetrating foreign body or surgery) is said to have _____.
    A. conjuctivitis
    B. scleritis
    C. endophthalmitis
    D. keratitis
    E. keratoconjunctivitis

Key Terms

abscess (p. 618)
acne vulgaris (p. 625)
acute disseminated encephalomyelitis
  (ADEM) (p. 612)
boil (p. 618)
bulla (p. 622)
bullous impetigo (p. 622)
candidal intertrigo (p. 627)
carbuncle (p. 618)
cellulitis (p. 624)
chickenpox (p. 614)
cold sore (p. 615)
comedone (p. 625)
condyloma acuminata (p. 617)
conjunctivitis (p. 608)
dermatome (p. 614)
dermatophyte (p. 626)
dermis (p. 606)

enanthem (p. 607)
endophthalmitis (p. 632)
epidermis (p. 606)
erysipelas (p. 624)
erythema infectiosum (p. 613)
erythematous (p. 615)
exanthem (p. 607)
folliculitis (p. 618)
furuncle (p. 618)
gangrene (p. 629)
gas gangrene (p. 629)
genital herpes (p. 615)
German measles (p. 612)
hand, foot and mouth disease (HFMD)
  (p. 617)
herpes gladiatorum (p. 615)
herpes zoster ophthalmicus (p. 633)
herpetic keratitis (p. 616)

herpetic whitlow (p. 615)
impetigo (p. 622)
keratitis (p. 632)
latent (p. 614)
macule (p. 608)
maculopapular rash (p. 608)
measles (p. 611)
methicillin-resistant *S. aureus* (MRSA)
  (p. 619)
mucosa (p. 608)
necrotizing fasciitis (p. 623)
nonbullous impetigo (p. 622)
ophthalmia neonatorum (p. 633)
papilloma (p. 616)
papule (p. 608)
pathognomonic (p. 612)
pustule (p. 608)
ringworm (p. 626)

roseola infantum (p. 614)
scalded skin syndrome (p. 619)
skin (p. 606)
smallpox (p. 617)
streptococcal pyogenic exotoxin (SPE)
  (p. 624)
subacute sclerosing panencephalitis
  (SSPE) (p. 612)
tinea capitis (p. 626)
tinea corporis (p. 626)
tinea cruris (p. 626)
tinea pedis (p. 626)
tinea unguium (p. 626)
tinea versicolor (p. 626)
trachoma (p. 633)
vesicle (p. 608)
wart (p. 616)

Review Questions

1. What are keratinocytes and Langerhans cells?

2. Describe the basic types of skin rashes and lesions. Give an example of each.

3. What is the pathophysiology of most skin rashes?

4. Discuss the pathogenesis of congenital rubella syndrome.

5. Discuss complications of varicella and rubeola diseases.

6. What is fifth disease?

7. What causes warts? Can warts be prevented?

8. What microbe can cause cold sores and an ocular disease? How does it cause both diseases?

9. What organism causes furuncles? Explain the pathogenesis of the disease.

10. Name the causes of impetigo. What virulence factors do these organisms produce?

11. Why do some strains of *S. pyogenes* cause necrotizing fasciitis, whereas others do not?

12. What is the difference between erysipelas and impetigo?

13. What is gas gangrene?

14. What is the cause of tinea versicolor? Describe the appearance of the disease.

15. List important etiologic agents of burn infections. What is their source?

16. Why is *Pseudomonas aeruginosa* such an important pathogen?

17. What is the difference between conjunctivitis and keratitis?

Clinical Correlation Questions

1. A 7-year-old child who had a nonpruritic malar rash on her skin now has white papular enanthem on a red base on her palate and throat. What is the most likely causative agent, and how do we prevent disease with this pathogen?

2. A 5-year-old girl is brought to the clinic by her mother because of a rash on her face. Examination reveals a few small pustules on an erythematous base. You send a sample for culture and sensitivity. Microscopic and biochemical examination of the bacteria in the resulting culture show catalase-negative, Gram-positive cocci that look like a string of pearls under the microscope. Is *S. aureus*, *S. epidermidis*, *S. pyogenes*, *C. perfringens*, or *P. aeruginosa* responsible for the small pustules on the face, and why?

3. A patient presents to the ophthalmologist complaining of eye pain. Examination shows a dendritic lesion on the cornea of the affected eye. The ophthalmologist suspects that a virus has caused the lesion and obtains corneal scrapings for microscopic examination. What is the ophthalmologist likely to see?

Thought Questions: CHECK YOUR ANSWERS

Thought Question 19.1   A child brought to a U.S. clinic exhibits the symptoms of German measles, yet the disease was declared eliminated from within the United States in 2004. How might the child have contracted the disease?

ANSWER: A patient history would reveal whether the child was born in an endemic area outside the United States, had recently traveled to an endemic area, or had contact with someone else recently returned from an endemic area. Individuals born in the United States who have not been vaccinated against measles will be at high risk of contracting measles if they come in contact with an infected person or travel to endemic areas.

Thought Question 19.2   Explain why cytotoxic T cells do not attack nerve cells containing latent viruses such as VZV or herpes viruses.

ANSWER: The integrated viral DNA does not express any proteins to present on the host membrane. Thus, cytotoxic T cells are blind to the presence of these latently infected cells.

Thought Question 19.3   Why do viral skin infections often lead to secondary bacterial infections, whereas bacterial skin infections do not lead to secondary viral infections?

ANSWER: The integrity of the skin is a major deterrent to bacterial infections. Viral skin lesions produce physical breaks in that defense. The breaks allow potentially pathogenic bacteria that may be colonizing the outermost layer of epidermis to gain access to living skin tissues. The reverse doesn't generally happen because viruses capable of causing skin infections are not normal biota of the skin

Thought Question 19.4   Why do some people develop fungal infections but others do not?

ANSWER: As with any infection, there are several factors involved. Differences in various aspects of innate and adaptive immunity are important, as are the number of breaks in the skin through which the fungus can enter (opportunity). Immunocompromised patients generally have a higher risk of infection, fungal or otherwise.

Thought Question 19.5   Why is it that wounds tend to be polymicrobial, meaning you usually can find more than one species of bacteria in a wound?

ANSWER: Wounds are surrounded by healthy tissue colonized by normal microbiota. The injured tissue may then be infected by multiple species of bacteria originating from the surrounding microbiota.

Thought Question 19.6   Why is eye infection with *Acanthamoeba* often associated with the use of contact lenses?

ANSWER: The main source of *Acanthamoeba*, a free-living protozoan, is contaminated water. No one knows why the risk of infection with this organism is higher in contact lens wearers, but using contaminated water for hand washing prior to putting in contact lenses or contaminating the lens solution may be the reason.

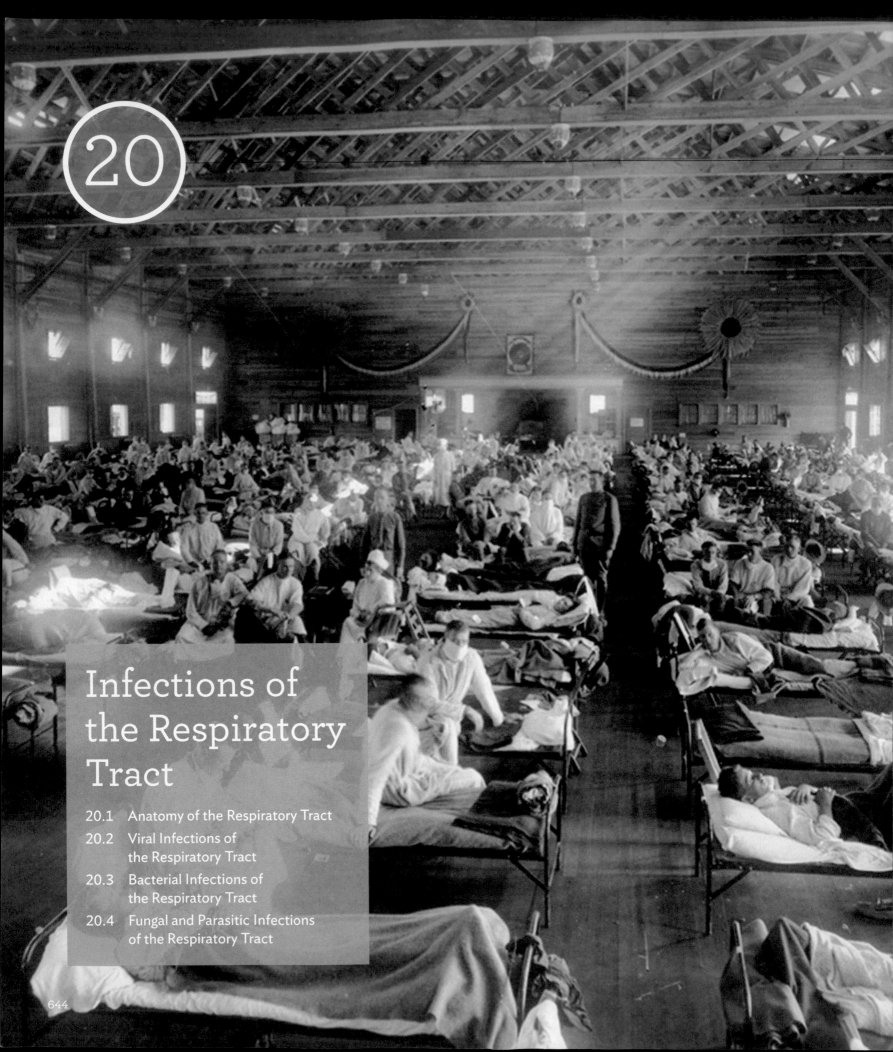

20

Infections of the Respiratory Tract

SCENARIO It was the summer of 1918, and 20-year-old Floyd, a U.S. army private stationed in Kansas, began to feel ill. He visited the camp doctor complaining of a cough, a headache, and some chills. Within a day, Floyd's fever reached 40°C (104°F). He became incoherent and lost consciousness. When Floyd awoke, he found himself in a large barracks lying on a cot among a long line of cots filled with sick soldiers.

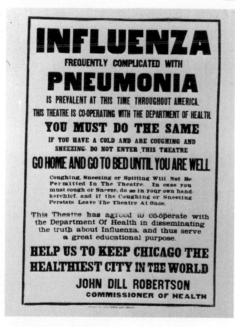

1918 Warning Poster
A public health poster in 1918 warning that flu can develop into what we now know as secondary bacterial pneumonia. No effective treatment was known.

SIGNS AND SYMPTOMS His cough was severe. Floyd saw dark, purplish brown spots appear on the face of the GI lying on the cot next to his. The soldier began to cough blood and eventually his face turned blue from lack of oxygen—a sign that death was imminent. Doctors could do little to help.

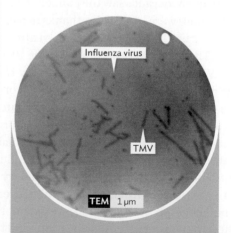

First Electron Micrograph of Influenza Virus (1941)
This sample was from someone who died of flu. The tiny gray dots are the dreaded virus. The tube-like tobacco mosaic virus (TMV, Figure 12.5) was added to the sample for comparison.

RESOLUTION Fortunately, after about 2 weeks, Floyd started to improve. Many others were not so lucky and died. No one knew what caused the disease raging through the camp or how to stop it.

What started with a sudden cough and chills for nearly 500 million people—one-third of the world's population at the time—ended with pneumonia and death for more than 50 million worldwide. They all suffered from the viral influenza pandemic that came to be known as the "Spanish flu." The pandemic of 1918 is thought to have begun with 18 cases in early March of 1918 in Haskell County, Kansas, and spread to European civilians and soldiers who were fighting in the Great War (World War I) by May of 1918. The outbreak in Spain, which led to the pandemic's name, actually happened later. At the time, the United States and other allied countries were reluctant to report the illness among the soldiers in an effort to maintain morale. Consequently, the first news of the epidemic wasn't reported until newspapers in Spain broke the story, and the epidemic of 1918 came to be known as the Spanish flu. Subsequent decades saw the pandemic flus of 1957 and of 1968. Later pandemics were less deadly, thanks in part to a better understanding of the virus, the infection, and its treatment. But no one knows what would happen if a strain similar to the 1918 flu were to strike again.

Viral infections of the respiratory tract continue to challenge the medical community owing to a lack of definitive treatments and the occurrence of secondary bacterial infections. Today, preventive measures such as vaccination, much improved treatment offered in intensive care units (ICUs), and the availability of antiviral medications have lessened the mortality of influenza virus. But although antibiotics to treat dangerous secondary bacterial infections give flu patients a fighting chance, their overuse has encouraged the emergence of new antibiotic-resistant bacteria, threatening the effectiveness of current medications and technology.

Lung and upper respiratory tract infections are among the most common diseases of humans. Many different viruses, bacteria, fungi, and parasites can cause pneumonia. Successful lung pathogens come equipped with attachment mechanisms and countermeasures against various lung defenses (such as alveolar macrophages). The most prevalent respiratory diseases are of viral origin, and most viral infections of the respiratory tract (such as the common cold) do not spread beyond the lung. Fortunately, most viral diseases of the respiratory tract are self-limiting and typically resolve within 2 weeks. Lung damage caused by a primary viral infection, however, can lead to secondary infections by bacteria, endangering the weakened patient.

20.1
Anatomy of the Respiratory Tract

Are all respiratory tract infections the same? No. Respiratory tract infections are often classified by whether they affect the upper or lower respiratory tract. The **upper respiratory tract** includes the upper airways, consisting of the nasal passages, oral cavity, and pharynx (**Figure 20.1A**). Other structures in the head that have direct or indirect communication with the upper airways include the passages to the sinuses (small air-filled cavities in the skull), lacrimal (tear) ducts, and the middle ear. The lowest structure in the upper respiratory tract is the larynx (the voice box). Infections in the upper airways can easily spread to these associated structures, and vice versa.

The **lower respiratory tract** (**Figure 20.1B**) consists of a series of diverging tubes that eventually terminate in alveoli, small sacs where oxygen from the inhaled air is exchanged for carbon dioxide in the blood in a process called respiration. The tubes leading to the

Figure 20.1   Anatomy of the Human Respiratory Tract

A. The upper respiratory tract. Oral and nasal cavities open into the pharynx, forming the upper part of the human respiratory tract. The pharynx is the hollow space shared by the nasal and oral cavities. Both the esophagus, the tube that carries food from the mouth to the stomach, and the trachea, or windpipe (part of lower respiratory tract), are continuous with the pharynx. Different portions of the pharynx are named for the structure closest to them. The nasopharynx, for example, is the part of the pharynx attached to and adjacent to the nose. Inflammation of the pharynx is a common respiratory problem called pharyngitis, or sore throat.

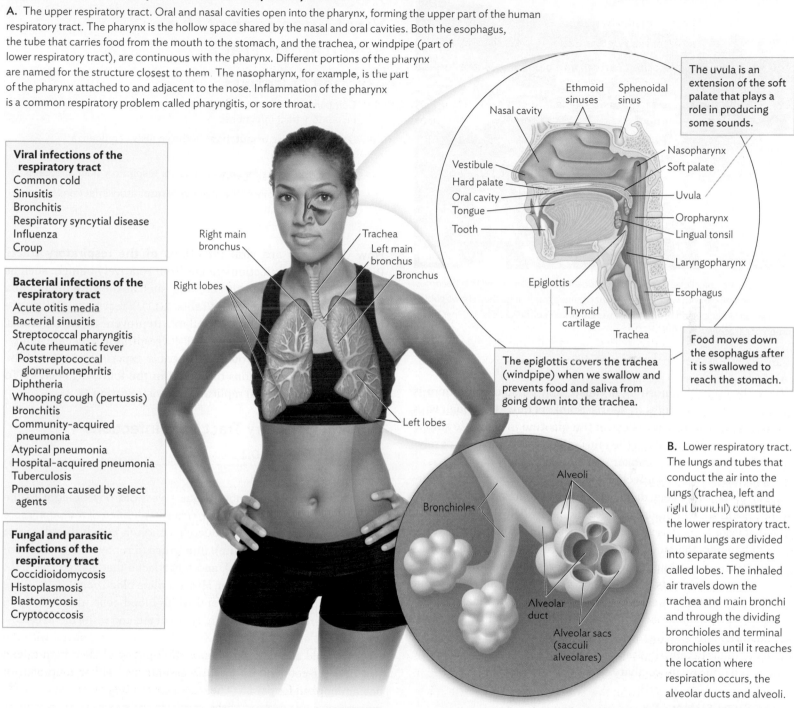

Viral infections of the respiratory tract
Common cold
Sinusitis
Bronchitis
Respiratory syncytial disease
Influenza
Croup

Bacterial infections of the respiratory tract
Acute otitis media
Bacterial sinusitis
Streptococcal pharyngitis
  Acute rheumatic fever
  Poststreptococcal glomerulonephritis
Diphtheria
Whooping cough (pertussis)
Bronchitis
Community-acquired pneumonia
Atypical pneumonia
Hospital-acquired pneumonia
Tuberculosis
Pneumonia caused by select agents

Fungal and parasitic infections of the respiratory tract
Coccidioidomycosis
Histoplasmosis
Blastomycosis
Cryptococcosis

The uvula is an extension of the soft palate that plays a role in producing some sounds.

Ethmoid sinuses
Sphenoidal sinus
Nasal cavity
Nasopharynx
Vestibule
Soft palate
Hard palate
Oral cavity
Uvula
Tongue
Oropharynx
Tooth
Lingual tonsil
Laryngopharynx
Epiglottis
Esophagus
Thyroid cartilage
Trachea

The epiglottis covers the trachea (windpipe) when we swallow and prevents food and saliva from going down into the trachea.

Food moves down the esophagus after it is swallowed to reach the stomach.

Right main bronchus
Trachea
Left main bronchus
Bronchus
Right lobes
Left lobes

B. Lower respiratory tract. The lungs and tubes that conduct the air into the lungs (trachea, left and right bronchi) constitute the lower respiratory tract. Human lungs are divided into separate segments called lobes. The inhaled air travels down the trachea and main bronchi and through the dividing bronchioles and terminal bronchioles until it reaches the location where respiration occurs, the alveolar ducts and alveoli.

Bronchioles
Alveoli
Alveolar duct
Alveolar sacs (sacculi alveolares)

alveoli include the trachea, right and left main bronchi, bronchioles, terminal bronchioles, and alveolar ducts. Note that some people assign trachea as being part of the upper respiratory tract. All of the respiratory tubes are warm and moisturize and remove impurities from inhaled air as it travels to the alveoli. The cells that line the interior of the respiratory tubes (the lumen) are epithelial cells sporting appendages called cilia that are extensions of the cellular membrane. Cilia remove foreign particles, including microorganisms that are trapped in the sticky mucus. Mucus is formed when mucin, which is secreted by goblet cells interspersed among the epithelium, combines with water (**Figure 20.2**). The cilia "beat" like a whip upward, moving the mucus and its trapped particles up toward the oral and nasal cavities for clearance and expulsion. The ciliated epithelium, referred to as the **mucociliary escalator** (or mucociliary elevator), is responsible for keeping the lungs devoid of microorganisms. Any impairment of the mucociliary escalator results in the accumulation of inhaled particles in the airways and a reflexive cough. Heavy smokers, for example, have a tendency to cough when

Figure 20.2   Tracheal Lining

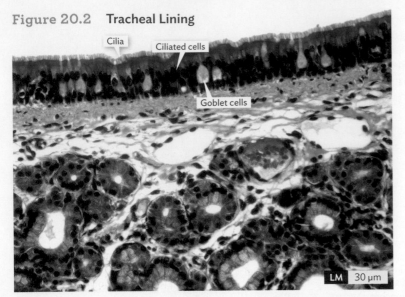

Goblet cells are unicellular glands that produce mucin, a compound that becomes mucus when mixed with water. This colorized micrograph of trachea shows goblet cells (green) interspersed between ciliated columnar epithelial cells that line the lumen of a trachea.

they wake up in the morning because their ciliated epithelium is trying to clear their lungs of debris. Smokers quiet the cough with the first smoke of the day because the nicotine in cigarette smoke paralyzes the cilia and stops the mucociliary escalator. This paralysis of the cilia leaves the smoker susceptible to many diseases.

Secondary bacterial infections of the lung often arise in the airways after they have been damaged by a viral infection. Patients who do not drink enough fluids will become mildly dehydrated, a condition that leads to increased mucus viscosity. Thick mucus or bacterial growth hampers the movement of the mucociliary escalator. *Bordetella pertussis*, for example, the bacterium responsible for whooping cough, directly interacts with the mucociliary escalator and slows it down (Figure 18.6B). A sluggish mucociliary escalator makes it difficult to expel the microbe, so the patient's susceptibility to secondary bacterial infection increases. Cold sufferers are advised to drink plenty of fluids because hydration helps decrease the viscosity of the mucus and improve mucociliary escalator function.

SECTION SUMMARY

- **Respiratory infections** affect either the upper or lower respiratory tract.
- **The upper respiratory tract** consists of nasal and oral cavities, the pharynx, and associated structures (ear, sinuses, larynx).
- **The lower respiratory tract** consists of the trachea, bronchi, bronchioles, terminal bronchioles, alveolar ducts, and alveoli.
- **Alveolar ducts** lead to the alveoli, where the process of respiration takes place.
- **Microorganisms in the lung** are trapped by mucus produced by goblet cells, swept out by the cilia of the mucociliary escalator, or killed by alveolar macrophages. These mechanisms are part of the innate immune system.

20.2
Viral Infections of the Respiratory Tract

SECTION OBJECTIVES

- Compare and contrast the signs and symptoms of upper and lower respiratory tract infections.
- Describe the role of mutations in the genesis of influenza pandemics.
- List the risk factors for upper and lower respiratory tract infections.
- Identify the etiology of the major viral respiratory infections based on their presentation.

How dangerous are viral infections of the respiratory tract? Respiratory tract infections range from relatively benign, such as the common cold, to life-threatening, such as the viral pneumonia caused by the influenza virus (Table 20.1). Treatment options are limited, and few vaccines are available to prevent these infections. The result is that viral infections of the respiratory tract are annoyingly common. In fact, you have probably experienced most of the viral respiratory infections discussed in the following section. To review the basics of viral replication, revisit Section 12.2.

Upper Respiratory Tract Viral Infections

CASE HISTORY 20.1

Jacob's Runny Nose

Jacob, a rambunctious 5-year-old boy brought to the clinic by his mother, was complaining of a scratchy sore throat, a nonproductive cough (a cough that does not produce sputum), rhinorrhea (a runny nose), nasal congestion (a stopped-up nose), and a headache that started the night before. He did not have a fever. His complete blood count (CBC)—the number of platelets and red and white blood cells in blood, along with its differential (the different types of red and white blood cells)—was normal, as was his chest X-ray. Jacob was diagnosed with the common cold and advised to rest, drink plenty of clear liquids, and use over-the-counter cough suppressants and either ibuprofen or acetaminophen for pain.

Why do we keep catching colds? The **common cold** is the most frequent viral infection of the upper respiratory tract. We remain susceptible to "catching" a cold because the symptoms are caused by more than 200 viral subtypes. The average person suffers a few colds a year—more episodes for those of us who are younger than for those of us who are older. The annual economic impact of simple colds is enormous. Children can catch as many as eight and adults as many as five colds every year. It is very difficult to accurately calculate the cost because no one is actually keeping records, but the estimates, based on small-population studies of several thousand

Table 20.1
Viral Diseases of the Respiratory System

Disease	Symptoms	Complications	Virus	Transmission	Diagnosis	Treatment	Prevention
Common cold	Runny, congested nose, sore throat	Rare; infection of sinuses, ear, throat, or lung	Rhinoviruses	Person-to-person contact	Clinical symptoms	Supportive treatment	Hand washing; no vaccine available
Sinusitis	Sinus congestion, facial pain, headache	Bacterial sinusitis that has purulent, sometimes malodorous, nasal discharge and fever	Viruses that cause common cold	Typically not spread from one person to another	Clinical presentation	Pain relievers; nasal corticosteroids; antibiotics when progresses to bacterial infection	No vaccine available
Respiratory syncytial disease	Fever, chills, cough, chest pain, respiratory wheezing	In very young infants and children, heart failure, low blood pressure, and seizures	Respiratory syncytial virus (RSV)	Respiratory droplets	Clinical presentation; rapid viral testing (rapid immunoassay)	*For infants and children:* hydration; oxygen; bronchodilators (to relieve contraction and constriction of bronchi); palivizumab (monoclonal antibody to F surface glycoprotein of RSV) for premature infants, per criteria *For adults:* supportive therapy; ribavirin as indicated	No vaccine available
Croup	Rare in adults; seal-like bark, hoarseness, cough, fever, respiratory distress	Bacterial respiratory infections	Parainfluenza (80%), other viruses (20%)	Contact	Clinical presentation; chest X-ray	Corticosteroids; bronchodilators; humidified air; oxygen	Hand washing; no vaccine available
Influenza	Fever, chills, cough, chest pain, sore throat, muscle pain	Secondary bacterial pneumonia	Influenza A and B viruses	Respiratory droplets or contact	Clinical presentation; viral culture; rapid antigen testing; serology; RT-PCR	Zanamivir (Relenza), oseltamivir (Tamiflu); less commonly: amantadine, rimantadine	Seasonal vaccines available
Severe acute respiratory syndrome	High fever, chills, cough, chest pain, stiffness, confusion	Kidney failure, heart failure, blood clots in lungs, respiratory distress	SARS virus	Close contact with infected person	According to criteria set by CDC, including: positive viral culture; positive serum antibody by test validated by CDC	Supportive therapy; ribavirin in some cases	Strict personal hygiene; no vaccine available

U.S. households, are in the billions of dollars. Our economy spends approximately $17 billion in direct medical costs (buying over-the-counter medications, for example) and $22.5 billion in indirect costs (workdays lost) every year!

Rhinoviruses are the most common cause of colds, with more than 100 subtypes, but other prominent players include coronaviruses, influenza, parainfluenza, and respiratory syncytial viruses. Presentation of colds in adults is milder and of shorter duration than in children. A scratchy or sore throat accompanied by a runny, congested nose are usually the only symptoms in adults and clear up within a week. In children, the course of a cold can take as long as 2 weeks. The good news is that the common cold is almost always self-limiting and does not cause any serious illness. Complications of a cold, which are uncommon except among the very young, the immunocompromised, or patients with lung disease, include inflammation or infection of the sinuses, middle ear, throat, or lungs.

Note Contrary to common belief, the color of nasal discharge is not an indicator of the presence or absence of bacteria. Yellow or white nasal discharge may reflect an increase in the number of polymorphonuclear cells (PMNs). A green discharge is thought to result from an enzymatic activity of these cells.

The viruses that cause a cold can invade through upper respiratory tract openings or through ocular mucosa, which means that cold viruses can easily spread from person to person by contact or by inhaling large or small aerosols. Given these modes of transmission, the only effective preventive measures against spreading a cold are frequent hand washing and using the lower arm (rather than the hand) to cover the nose or mouth when sneezing or coughing.

There are time-honored remedies aimed at alleviating the symptoms of a cold, such as drinking plenty of liquids, but there are

no FDA-approved medications for treating the common cold and no vaccine. A variety of over-the-counter medications that promise prevention or treatment are on the market, but their efficacy, contraindications, and side effects are unknown.

Symptoms of the common cold are mimicked by a number of other viral pathogens at the onset of their infective cycle. These include respiratory syncytial virus (RSV) and influenza (discussed shortly), as well as early stages of rubella and rubeola viral infections (see Section 19.2).

Another common viral infection of the upper respiratory tract is **viral sinusitis**. Viral infections of the upper airways often spread to the sinuses and cause inflammation and congestion of the sinuses (sinusitis). The sinus cavities are lined with mucous membranes, which become inflamed when infected. The inflammation and congestion of the sinuses, however, can narrow passages that connect them to the nasal cavity, impeding the flow of mucus. Sinus congestion commonly causes the headache that we feel when our nose is "stopped up" from a bad cold. Viral sinusitis usually resolves on its own, as does a cold.

Lower Respiratory Tract Viral Infections

Infections of the lower respiratory tract, regardless of whether their etiology is viral, bacterial, or fungal, are called **bronchitis** when the bronchial tubes are involved and **pneumonia** when the lungs are affected. Viral bronchitis is a self-limiting disease. The most common viral causes of bronchitis are influenza A and B, parainfluenza, coronavirus (types 1–3), rhinovirus, respiratory syncytial virus, and human metapneumovirus. A cough (dry or sputum producing) that lasts more than 5 days is the main symptom of the disease, which is diagnosed clinically and treated with cough suppressants and occasionally an inhaled medication to reduce inflammation of the bronchial tubes.

Clinical presentation of pneumonia is somewhat different, depending on the causative agent, but all include shortness of breath and cough. Pneumonia can result from numerous viruses, but respiratory syncytial virus (RSV) and influenza viruses are responsible for the majority of cases.

RESPIRATORY SYNCYTIAL VIRUS (RSV) DISEASE Respiratory syncytial disease is an important viral lung infection caused by **respiratory syncytial virus (RSV)**, an enveloped, negative-sense, single-stranded RNA virus of the Paramyxoviridae family (**Figure 20.3**). RSV derives its name from the fusion of adjacent infected cells into a **syncytium**, a giant cell containing many nuclei. Cell fusion is caused by viral proteins that travel to the infected cell's membrane, causing fusion of membranes between neighboring cells.

RSV is transmitted from person to person. Infections occur when the virus comes in contact with mucous membranes of the eyes, mouth, or nose or through the inhalation of droplets generated by a sneeze or cough. The virus causes disease mostly in the very young. By age 2, all children are thought to have been infected by it. The course of infection in healthy children is relatively benign,

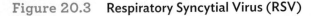

Figure 20.3   Respiratory Syncytial Virus (RSV)

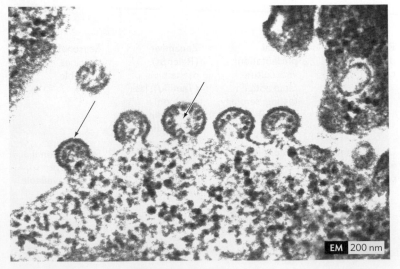

EM 200 nm

Five virions (red arrow) can be seen budding from an infected cell. The black arrow points to the viral nucleic acid cores.

although recent data suggest that frequent RSV infections are linked to childhood asthma and allergic wheezing.

RSV illness usually begins with fever, runny nose, cough, and sometimes wheezing. The virus moves down the respiratory tract to cause more severe disease in some children. RSV is the most common cause of **bronchiolitis** (inflammation of the bronchioles) and pneumonia among infants and children under 1 year of age. Infected infants who are 6 months of age or younger often require hospitalization. There is some evidence suggesting that the severity of RSV infection in infants may be due to coinfection with metapneumovirus (another member of the Paramyxoviridae family) or with the bacterium *Bordetella pertussis* (cause of whooping cough).

RSV attachment to respiratory epithelial cells triggers an immune response that includes production of chemokines and recruitment of inflammatory cells but fails to confer immunity against reinfection. The extent of the inflammatory response impacts the severity of presenting symptoms and depends on the patient's age, overall health, and genetic background. Generally speaking, healthy adults will have only cold-like symptoms, mainly rhinitis; but some adults experience dangerous lower respiratory tract disease. Risk of a more severe illness is greater among the elderly or people with compromised cardiac, pulmonary, or immune systems. As yet, there is no vaccine.

Link As discussed in Section 15.3, **chemokines** are small glycoproteins produced by damaged tissues that serve as chemoattractants for white blood cells.

The exact mechanism of the immune response to RSV is not well delineated. Failed clinical trials of an RSV vaccine in the 1960s seem to agree with more recent experimental evidence that the virus induces an inadequate immune memory, resulting in more severe disease with the second exposure. Vaccine production will have to wait until we know how RSV misdirects our adaptive immune system.

Presumptive diagnosis of RSV is usually made based on clinical presentation. A number of other viruses, including adenovirus, influenza, and parainfluenza, can cause signs and symptoms similar to RSV. A list of conditions with similar signs and symptoms is called a differential diagnosis. A definitive diagnosis confirms the exact etiology of a disease (in this case RSV) among those listed in a differential diagnosis. Definitive diagnosis of RSV requires isolating the virus from the patient's respiratory secretions, which can be collected from a variety of sources including the nose, throat, or even the bronchi and lungs. Rapid immunoassays (see Section 17.5), PCR (see Section 8.3), and serology tests (see Section 17.5) are also available. Serological testing is not a good choice in infants, since the presence of maternal antibodies is likely to result in a false positive.

Treatment options for RSV infections are currently limited to relieving the symptoms and using the one medication available. A nucleoside analog (ribovarin, Section 13.7) that inhibits RSV replication has been approved by the FDA to treat severe cases of RSV, but it is contraindicated in pregnant women.

INFLUENZA **Influenza** is caused by single-stranded, negative-sense enveloped RNA viruses of the Orthomyxoviridae family. Although some influenza strains attain pandemic status (such as the one that caused the Spanish flu described at the beginning of this chapter), most strains cause seasonal outbreaks in limited regions. Seasonal influenza can infect people of all ages. Influenza viruses are divided into types A, B, and C, on the basis of their antigenic determinants. Types A and B influenza viruses can both result in "the flu," but type A is responsible for the majority of cases. Type A viruses infect a variety of animals as well as people. This virus is constantly changing and is the cause of large flu epidemics and occasional pandemics. Type B viruses are found primarily in humans and usually result in a milder form of illness than type A. Type B viruses do not cause pandemics. Type C causes only a mild respiratory disease and no epidemics or pandemics, so it is not classified based on subtypes and there is no vaccine to protect against it.

Link As discussed in Section 16.1, your immune system does not recognize a microbe as a whole but instead recognizes tiny pieces of it. Each small segment of a microbe that is capable of eliciting an immune response is called an **antigenic determinant**, or epitope.

The pathogenicity of influenza viruses is largely dependent on three of their proteins: hemagglutinin (HA), neuraminidase (NA), and M2 (Sections 12.5 and 18.6). Hemagglutinin is a transmembrane protein that forms the spikes on the viral surface (**Figure 20.4A**). The HA spikes bind the *N*-acetylneuraminic (sialic) acid residues present on the outer surface of the respiratory epithelial cells to gain entry (**Figure 20.4B**). The host cell tries to defend itself by engulfing the bound virus via endocytosis. Captivity in the vacuole, however, is short-lived.

The low pH of the endoplasmic vacuole helps the influenza virus remove proteins covering its genome (uncoating), making its RNA accessible and ready for replication.

Figure 20.4   Influenza

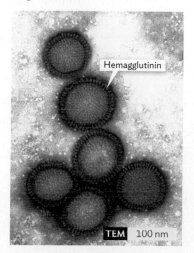

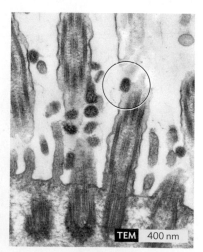

A. Hemagglutinin (HA) spikes on the viral surface.

B. Influenza virion (circled) attached to an epithelial cell's cilia.

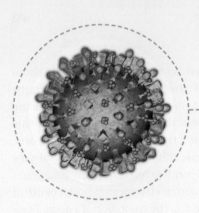

inSight

The Search for a Universal Flu Shot

Wouldn't it be nice to have a flu vaccine that lasts for as long as we live—or at least much longer than a year? A universal flu shot would free us from fear of the flu and reduce the overall cost of flu vaccine production and administration, not to mention decreasing flu-associated deaths and health care costs. It would also make the scientist who discovers it the next Edward Jenner.

The problem is the flu virus and its ever-changing outer protein, hemagglutinin, which looks like a series of lollipops anchored to the viral coat. Current vaccines are directed against the most prominently displayed, but highly mutable, part of the virus—the head of hemagglutinin. The stalk of the HA protein is a lot less mutable than the tip and can generate an immune response. This genetic stability should mean that the stalk is a much better vaccine candidate than the seasonally changing HA tip (Figure 1). But the stalk is mostly hidden from our immune system by the ball-shaped head of the protein, which presents difficulties in using it as a vaccine target. Despite this limitation, Wayne Marasco, of Harvard University, and colleagues have discovered antibodies that bind to the stalk of HA and protect test animals from many different types of flu viruses. So a vaccine made from the HA stalk could possibly serve as a universal flu vaccine. It is also feasible that antibodies to the HA stalk made in the laboratory could be an effective way to treat flu victims.

Proteins inside the viral envelope are in principle much better candidates for a universal vaccine because the genes producing them mutate much less frequently than do the genes for surface proteins when the virus replicates. The viral envelope, however, sequesters these proteins (such as the HA stalk) from the reach of antibodies made against them by our immune system. But flu vaccine designers may have identified a vulnerability of the influenza virus: one of the internal viral proteins, influenza nucleoprotein, is stripped off of viral nucleic acid and displayed on infected cell surfaces. Antibodies made to target this protein should recognize the surface nucleoprotein and mark the infected cell for destruction by cytotoxic

T cells. Elimination of infected cells would also make virus replication within them impossible and therefore reduce the severity of the disease.

Another vaccine candidate is the protein M2. M2 is mostly hidden inside the virion envelope, but a portion of M2 is exposed outside the virus. This exposed part is immunogenic and genetically stable and thus might be considered a candidate for a universal vaccine. However, the M2 of animal influenza viruses is slightly different from the M2 of human strains. Consequently, a hybrid virus, such as swine flu (H1N1), may escape antibodies raised against the human M2.

The answer to the dilemma of how to make a universal flu vaccine may be to design a multivalent version that stimulates production of antibodies to the HA stalk of hemagglutinin, the viral nucleoprotein, and the different M2s. Research on this is under way and promising, but for now roll up your sleeves once a year if you wish to be protected.

Figure 1 Potential Targets for a Universal Vaccine against Influenza Virus

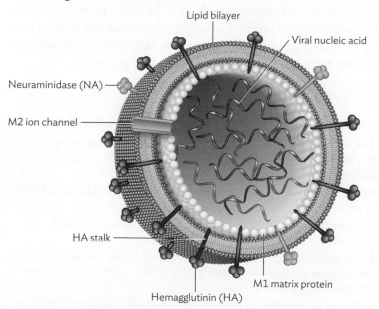

The genes for the hemagglutinin stalk and M2 proteins of the influenza virus mutate much less frequently than the gene for hemagglutinin, making them good targets for generating a universal flu vaccine.

Acidification of the vacuole also changes the conformation of the viral protein M2, leading to formation of a pore in the membrane of the vacuole through which the viral genetic material can exit the endosomal compartment and enter the host cell cytoplasm. When inside the cytoplasm, the viral RNA is replicated, producing progeny that will be transported to the cell surface, once again bound to sialic acid residues of the host cell.

As the influenza virion buds out of the host cell membrane, viral neuraminidase cleaves the sialic acid residues that anchor the budding virus particles to the infected cell surface while also preventing the viruses from clumping together. The progeny are now free to find and bind an uninfected cell of their own.

Influenza viruses are designated as being avian flu strains or swine flu strains. The difference in targeting an animal host is related to the ability of different flu viruses to bind to different sialic acid residues in the lungs of these animals. Two forms of sialic acid residues are associated with cell surface glycoproteins. The two forms differ based on their linkages to galactose. One form is called alpha 2,3, the second form is called alpha 2,6. The human respiratory tract contains both receptors. Avian flu strains are generally those that bind to alpha 2,3 sialic acid residues. Birds, such as ducks, principally have the alpha 2,3 sialic acid receptors in their guts and lack the alpha 2,6 receptors. Pigs, on the other hand, have both types of sialic acid receptors. Flu viruses that bind to the alpha 2,6 form are generally called swine flu strains, although both types of viruses can infect pigs. Human lungs also contain alpha 2,3 and alpha 2,6 receptors, so both avian and swine flu strains can infect humans.

Why do some influenza strains cause more severe disease in humans than other strains? The reason is not entirely clear but recent studies indicate several features of the virus affect virulence, such as the type of hemagglutinin used (Section 18.6). An earlier hypothesis correlating virulence of the virus strain with the distribution of the different host sialic acid receptors in the human respiratory tract is likely incorrect.

Because influenza viruses can infect humans, pigs, and birds, different viruses can sometimes infect the same cell (for example, in a duck landing on a pig farm in Thailand). When this happens, the eight RNA segments of each **segmented influenza genome** will uncoat, mix in the cytoplasm, and re-sort to form a new influenza virus containing a new form of the hemagglutinin gene. This process of RNA segment mixing produces what is called an **antigenic shift**. Antigenic shifts typically happen every 10–15 years. In between these large antigenic shifts, smaller so-called **antigenic drifts** in HA structure can occur because the influenza RNA polymerase is very error prone, introducing numerous mutations that can ultimately alter the structure of the HA protein.

Link Section 12.5 describes how each of the eight **segments of the influenza virus** genome has its own prepackaged polymerase. The segments pass through a nuclear pore complex and enter the host nucleus. Within the nucleus, each segment generates both positive-sense and negative-sense RNAs. The positive-sense viral RNA strands are transported back to the host cytoplasm for translation, whereas the negative strands are packaged into viral progeny.

Antigenic shifts and drifts allow viruses to evade the host immune system. Influenza type A is most capable of undergoing genetic change, so it more commonly leads to seasonal disease. Type A influenza is further classified into subtypes based on the kind of hemagglutinin (H) and neuraminidase (N) expressed on the viral surface, giving rise to some very well-known strains, such as H1N1 (swine flu) or H5N1 (avian flu).

Every year, the Centers for Disease Control and Prevention (CDC) and the World Health Organization (WHO) identify influenza strains that cause the most cases of flu during that year. Flu outbreaks are monitored all over the world, including areas where living conditions foster close contact between humans and the animals that serve as incubators for new influenza strains. Close contact (for example, a farmer that shares living quarters with his ducks) can facilitate the virus "jumping" from the animal to humans and the start of a possible new pandemic. The identified strains are then used to generate a multivalent vaccine for the coming year. Research is currently under way to develop a single universal vaccine effective against seasonal flu from year to year (see **inSight**).

Influenza viruses used for inactivated (killed-virus) vaccines are usually grown in fertilized chicken eggs (**Figure 20.5**). Two different preparations of the flu vaccine are made: an inactivated vaccine that can be injected intramuscularly, and an intranasal live-attenuated vaccine that is administered as a nasal spray. The attenuated, intranasal vaccine offers more effective protection because the attenuated virus actually replicates in the host. The intranasal vaccine is used for healthy people or those with egg allergies. The injectible vaccine is used for the immunocompromised and those who care for them. Flu vaccines must be administered annually in advance of (usually October) and during the flu season. The effectiveness of a flu vaccine varies from 50% to 80%, depending on how well the vaccine matches the infecting strain. The length of the flu season varies from year to year but commonly starts in November and ends in May, with February claiming the majority of cases.

Antiviral agents able to effectively treat the flu include those that inhibit the M2 surface protein (amantadine and ramantadine) and those that inhibit neuraminidase activity. Amantadine and ramantadine are rarely used anymore because most flu viruses have become resistant to them and because of their central nervous system side effects. Antiviral agents now used preferentially are the neuraminidase inhibitors zanamivir (Relenza) and oseltamivir (Tamiflu, Section 13.7).

CROUP   Some viruses can infect the larynx and then spread to the trachea and even bronchi, a condition called **laryngotracheobronchitis (LTB)**, or **croup**. Young children (6 months to 5 years of age) are at highest risk for contracting the disease. Typically, the croup patient has had a fever and a runny nose for a few days before coming to the clinic with a "barking" cough. The barking cough (or the seal-like bark) is caused by swelling of the larynx. The most common causes of croup are the parainfluenza viruses, especially types 1 and 2. Note that viruses that cause bronchitis in adults can also cause croup in children. Treatment of croup depends on the severity of the major signs and symptoms of the disease. Children who have only mild disease (no barking cough at rest) are treated at home with humidified air and often steroids to reduce airway inflammation. Those with severe croup are usually hospitalized and treated with oxygen. The majority of croup cases are mild and resolve within a week, but severe cases may lead to respiratory failure.

SECTION SUMMARY

- **Viruses cause the majority of upper respiratory tract infections.**
- **The common cold is the most common viral infection** of the upper respiratory tract. A majority of the 200 viral subtypes that cause the common cold are rhinoviruses.
- **Influenza viruses and respiratory syncytial virus (RSV)** can cause both upper and lower respiratory infections.
- **Pneumonia** is a lung infection. Influenza virus is a major cause of viral pneumonia.
- **Bronchiolitis,** inflammation of bronchioles, is the most frequent presentation of a lower respiratory RSV infection in infants and children younger than 1 year of age.
- **RSV-infected cells fuse to form syncytia,** giant cells containing many nuclei.
- **Influenza viruses affect people of all ages,** and their pathogenicity is dependent on three of their proteins: hemagglutinin (HA), neuraminidase (NA), and the M2 protein.
- **Influenza virus subtypes** are based on their HA and NA proteins.
- **Antigenic shifts and drifts** in the HA gene are caused by frequent mutations and result in different influenza strains. Shifting and drifting HA antigens help the virus evade the host's immune system.
- **Effectiveness of the annual multivalent flu vaccines** depends on how well vaccine antigens match those of the infecting strain.

Thought Question 20.1   For which populations would the intranasal flu vaccine not be recommended, and why?

Figure 20.5   Growing Viruses to Make the Flu Vaccine

Factory workers in China examine the eggs used to grow the influenza virus for production of the inactivated flu vaccine. The live-attenuated vaccine is prepared in mammalian cells.

20.3
Bacterial Infections of the Respiratory Tract

SECTION OBJECTIVES

- List the bacterial etiologies of respiratory tract diseases and the disease's anatomical location.
- Describe the relationships between host and bacteria that contribute to the pathogenesis of respiratory tract infections.
- Compare and contrast clinical presentations of bacterial infections of the respiratory tract.
- Develop a prevention plan for different bacterial respiratory pathogens.

Is there a relationship between viral and bacterial infections of the respiratory tract? Viral infections often lead to congestion, dehydration, and weakening of the immune system. The damaged tissue is more susceptible to bacterial colonization and disease. A variety of bacteria cause respiratory infections (Table 20.2), with impressive economic and personal consequences. For example, it is projected that U.S. children 0–4 years of age have 13.7 million cases of acute otitis media (middle ear infection) per year with an estimated economic burden of $3.8 billion. Lung infections cause approximately 1 million hospitalizations in the United States annually. Development of vaccines against some bacterial pathogens, especially *Streptococcus pneumoniae* and *Haemophilus influenzae*, has reduced the U.S. incidence of lung infections and epiglottitis (inflammation of the epiglottis, the cartilage covering the opening of the windpipe when we swallow). Those who come down with these infections still face high mortality rates, currently about 10% for streptococcal pneumonia, despite antibiotic therapy.

Upper Respiratory Tract Bacterial Infections

CASE HISTORY 20.2

A Big Pain in a Little Ear

Chloe is a 2-year-old generally healthy child who was brought to the clinic by her father. The child had a fever since the previous night, was irritable, and kept pulling on her left ear. Chloe also had a runny nose for the past 2 weeks. Her temperature was 38.5°C (101.3°F) at the clinic, and the physical exam showed a red and swollen ear, an erythematous (reddened) ear canal, and a bulging tympanic membrane. Chloe was diagnosed with acute otitis media and prescribed an antibiotic that she had to take for 7 days.

What is the biggest risk factor for getting a bacterial upper respiratory infection (URI)? Viruses that infect and damage the upper respiratory tract create conditions suitable for a secondary bacterial infection. Some URIs occur more often in certain age

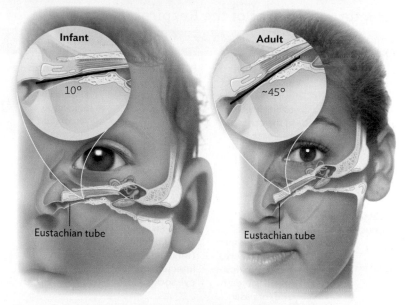

Figure 20.6 **The Changing Angle of the Eustachian Tube**

The Eustachian tube drains from the middle ear into the nasal cavity. In infants and young toddlers, the tube is shorter and almost horizontal, which makes the drainage less optimal. As a result, bacteria can get trapped in the middle ear and cause an infection. As we get older, the Eustachian tube lengthens and its path becomes angled, making drainage of the middle ear much easier and ear infections much less likely.

groups; for example, infections of the middle ear are most common in infants and toddlers like Chloe, thanks to the angle of the Eustachian tube, a tube that connects the middle ear to the nasal cavity (Figure 20.6). We discuss bacterial URIs, including the risk factors for acquiring them, in this section.

OTITIS MEDIA The most common diagnosis made during a sick-child visit to the pediatrician's office is middle ear infection, or **acute otitis media (AOM)** (Figure 20.7). Most children (80%–90%) will have at least one episode of AOM by the time they are 3 years old. The number of episodes is higher for those who attend a day-care center or who live with a sibling who has a history of frequent AOM.

Signs and symptoms of AOM are wide-ranging, and diagnosis is generally made based on clinical presentation and physical exam. Definitive diagnosis can be made by culturing middle ear fluid obtained through tympanocentesis (removal of fluid behind the eardrum), but the procedure is rarely performed. This is because of the difficulty in interpreting the results, given the natural colonization of the upper respiratory tract with microbes that cause AOM and the close proximity of the middle ear to the nasopharynx.

Infections of the middle ear are caused by viruses as well as bacteria, both Gram-positive and Gram-negative. Only 20% of AOM cases are purely viral, but viral upper respiratory tract infections are the most common antecedent event to bacterial AOM. Bacteria migrate from the nasopharynx to the middle ear usually following a common cold to cause what is generally a self-limiting disease.

Table 20.2
Bacterial Diseases of the Respiratory System

Disease	Symptoms	Complications	Bacteria	Transmission	Diagnosis	Treatment	Prevention[a]
Acute otitis media (AOM)	Earache, ear discharge, fever, irritability, sense of blockage in the ear, decreased hearing	Rare; hearing loss, meningitis	*Streptococcus pneumoniae, Haemophilus influenzae, Moraxella catarrhalis*	Spread of bacteria from throat to middle ear	Clinical presentation; definitive diagnosis by culture	Pain relievers; oral amoxicillin or amoxicillin-clavulanate, intramuscular ceftriaxone; antibiotic eardrops; surgical placement of tube in the ear to enhance drainage	No vaccine available[b]
Bacterial sinusitis	Sinus congestion, feeling pressure behind the eye, facial pain, foul-smelling breath and/or nasal discharge	Meningitis, brain abscess, orbital cellulitis (inflammation and infection of the eye socket)	Same as acute otitis media	Spread of bacteria from adjacent infected structures to sinuses	Clinical presentation	Pain relievers; antipyretics (medications to bring down fever); amoxicillin or amoxicillin-clavulanate	No vaccine available[a]
Pharyngitis	Sore throat, fever, pain swallowing, tender lymph nodes, headache, nausea and vomiting	AOM in children, abscess behind tonsils, acute rheumatic fever (some strains), glomerulonephritis (some strains)	*Streptococcus pyogenes*	Person-to-person contact	Clinical diagnosis based on Centor score[c]; rapid antigen detection test; definitive diagnosis by culture	Penicillin V or amoxicillin preferred; cephalosporins, clindamycin, azithromycin when penicillins not an option (e.g., penicillin allergies); pain relievers and antipyretics	No vaccine available
Diphtheria	Sore throat, low-grade fever, formation of a membrane over tonsils and soft palate, toxin-related symptoms within 6–10 days: irregular heartbeat, stupor, coma, death	Airway obstruction, septicemia, endocarditis (nontoxigenic strains)	*Corynebacterium diphtheriae*	Respiratory droplets, person-to-person contact	Culture; PCR; amplification of *dtx* gene from clinical specimen	Equine antitoxin (available through CDC); erythromycin or penicillin G	Vaccine available
Whooping cough (pertussis)	Prolonged (>2 weeks) paroxysmal (sudden and rapidly repeated) cough; "whoop" noise when breathing in after the cough paroxysm, followed by vomiting	Pneumonia, seizure, severe respiratory failure	*Bordetella pertussis*	Respiratory droplets	Clinical presentation; PCR	Azithromycin, clarithromycin; trimethoprim-sulfamethoxazole used as alternative but contraindicated for pregnant or nursing women and infants <2 months old	Vaccine available

(continued on next page)

Table 20.2
Bacterial Diseases of the Respiratory System (continued)

Disease	Symptoms	Complications	Bacteria	Transmission	Diagnosis	Treatment	Prevention[a]
Bacterial bronchitis	Cough, shortness of breath on exertion, wheezing, may have fever or sore throat	Acute disease associated with asthma; long-term respiratory decline if chronic	*Mycoplasma pneumoniae, Chlamydophila pneumoniae, Bordetella pertussis, Haemophilus influenzae, Streptococcus pneumoniae, Moraxella catarrhalis*	Respiratory droplets	Clinical presentation	Inhaled medications to dilate and reduce inflammation of bronchial wall; antibiotics, depending on patient and microbial agent	No vaccine available[b]
Lobar pneumonia (pneumococcal pneumonia)	Cough, shortness of breath, fever, chills, chest pain	Meningitis, kidney failure, heart failure	*Streptococcus pneumoniae*	Person-to-person contact, respiratory droplets	Clinical presentation; chest X-ray	Antibiotics; macrolides, fluoroquinolones; beta-lactam, cephalosporin as alternative antibiotics	Vaccine available
Walking (atypical) pneumonia	Fever, sore throat, nonproductive (dry) cough, sweating, chest pain	Ear infections, skin rashes, anemia	*Mycoplasma pneumoniae*	Person-to-person contact	Clinical presentation	Macrolides; fluoroquinolones, tetracyclines	No vaccine available
Legionellosis (Legionnaires' disease)	Fever, chest pain, coughing up blood-tinged sputum, muscle pain, abdominal pain, vomiting and diarrhea	Respiratory failure, shock (sudden drop in blood pressure), kidney failure	*Legionella pneumophila*	Inhalation of aerosols, mists, sprays, or droplets of contaminated water sources	Culture; detection of *L. pneumophila* antigen in urine	Macrolides	No vaccine available
Pseudomonas pneumonia	Cough, fever, shortness of breath, chest pain	Declining lung function, death	*Pseudomonas aeruginosa*	Contaminated water droplets and instruments	Culture; clinical history and presentation	Quinolones, aminoglycosides	No vaccine available
Tuberculosis	Productive cough progressing to coughing up blood; fever, loss of appetite, night sweats, weight loss, fatigue	Lung abscess, spread of TB to other organs	*Mycobacterium tuberculosis*	Inhalation of droplets	Definitive diagnosis by culture; tuberculin (Mantoux) skin test; interferon gamma release assay	Numerous treatment protocols, depending on patient specifics; initial simultaneous four-drug therapy: rifampin, isoniazid, ethambutol, pyrazinamide	Vaccine available
Inhalation anthrax (woolsorter's disease)	Common cold symptoms progressing to fever, muscle aches, weakness, shortness of breath	Shock, death	*Bacillus anthracis*	Inhalation of spores	Culture; serology	Ciprofloxacin, doxycycline	Vaccine available

[a]The childhood vaccine schedule is shown in Table 17.6.

[b]Pneumococcal vaccines protect against a minority of the more than 90 strains of *Streptococcus pneumoniae* (23 in PPSV for adults and 13 in PCV for children). Vaccine against *H. influenzae* type b (Hib) is also available, but it does not offer protection against the nontypeable *H. influenzae*.

[c]W. J. McIsaac, D. White, D. Tannenbaum D, and D. E. Low. 1998. A clinical score to reduce unnecessary antibiotic use in patients with sore throat. *Canadian Association Medical Journal* **158**(1):79.

Figure 20.7 Acute Otitis Media

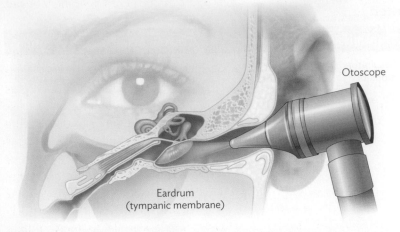

Otoscope

Eardrum
(tympanic membrane)

A. Viewing the tympanic membrane with an otoscope.

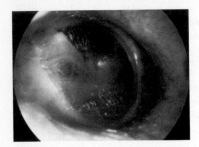

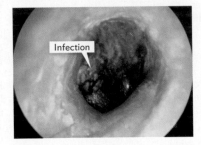

Infection

B. Otoscopic view of a healthy tympanic membrane.

C. View of an infected tympanic membrane.

Immunocompromised patients, however, can have a complicated course leading to the more invasive disease process of **epiglottitis** (inflammation of the epiglottis), pneumonia, or meningitis.

Three bacterial species, *Streptococcus pneumoniae*, *Haemophilus influenzae*, and *Moraxella catarrhalis*, are the most frequent causes of AOM in otherwise healthy children. All three species are capable of colonizing a child's nasopharynx early in life, but *S. pneumoniae* is the number one culprit. *S. pneumoniae* is a facultative anaerobe, Gram-positive alpha-hemolytic coccus possessing a thick polysaccharide capsule (**Figure 20.8**). The makeup of the polysaccharide capsule differs from strain to strain and is thought to be responsible for determining the extent of the bacterium's invasiveness (the propensity to cause bacteremia and more severe disease). More than 90 different serotypes of *S. pneumoniae* have been identified based on capsule structure, all of which are capable of causing infections. The invasiveness of the different serotypes is variable. *S. pneumoniae* strains causing AOM are generally of the noninvasive serotypes but seem to be associated with an increased risk of childhood asthma. A multivalent polysaccharide vaccine that protects against more invasive serotypes of *S. pneumoniae* appears responsible for reducing annual cases of AOM in recent years. The important role of *S. pneumoniae* as a cause of pneumonia will be described later.

The second most common cause of AOM is *Haemophilus influenzae*, a Gram-negative bacillus once thought to be the cause of influenza. The organism is fastidious but can be grown on chocolate agar, which contains growth factors X (hemin) and V (NAD), in a CO_2 incubator (see the case history at the beginning of Chapter 6). Some strains have an outer capsule and some do not. Six serotypes of encapsulated strains (serotypes a–f) can be detected by antisera that react with the polysaccharide of the capsule. *H. influenzae* type b, Hib, is the most invasive and can cause severe diseases such as meningitis, bacteremia, pneumonia, and epiglottiitis in infants and very young children. These infections are now rare thanks to an Hib polysaccharide vaccine included routinely in childhood vaccinations.

The nonencapsulated varieties of *H. influenzae* are said to be nontypeable because they do not react to any capsule antisera. The genomes of nontypeable *H. influenzae* strains differ greatly from each other. Some strains, for example, have the genes for the capsule but do not express them. Others do not have the capsule genes at all. Traditionally, the nontypeable strains have been classified, or "biotyped," based on their ability to produce indole, urease, and ornithine decarboxylase. Eight biotypes have been identified, some of which are linked to clinical syndromes such as Brazilian purpuric fever (biotype 3) and neonatal sepsis (biotype 4). The majority of *H. influenzae* infections of the upper and lower respiratory tracts are now caused by the nontypeable strains; no particular biotype has been identified as the main culprit.

Because the Hib vaccine does not confer any immunity toward nonencapsulated strains, nontypeable *H. influenzae* has become the second most common cause of AOM and conjunctivitis (inflammation of the conjunctiva) in children. These bacteria are capable of invading and surviving inside respiratory tract epithelial cells. Nontypeable strains can also form biofilms in the middle ear. As described in Chapter 6, bacteria growing in biofilms are more resistant to antibiotics than the same bacteria growing individually, which may explain why some cases of AOM are refractory to antibiotic treatment.

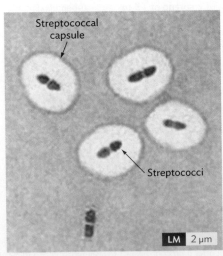

Streptococcal capsule

Streptococci

LM 2 µm

Figure 20.8
Streptococcus pneumoniae
Photomicrograph of *Streptococcus pneumoniae*, one of the most common causes of acute middle ear infections, pneumonia, and sinusitis.

Transmission of *H. influenzae* from person to person occurs through contact with infected respiratory secretions or by inhaling aerosolized droplets. A number of different classes of antibiotics are effective against *H. influenzae*, but the beta-lactam antibiotics, particularly amoxicillin, remain the treatment of choice. Some strains of *H. influenzae*, however, are resistant to beta-lactam antibiotics because they produce an enzyme (beta-lactamase) that inactivates the antibiotic.

Link As discussed in Section 13.5, beta-lactam is a major component of the penicillin antibiotics, and **beta-lactamase** enzymes cleave this critical ring structure. The beta-lactamase enzyme is transported out of the cell and into the surrounding environment, where it destroys the penicillin before the drug even gets to the cell.

The third most common cause of AOM in children, *Moraxella catarrhalis* is a Gram-negative diplococcus, but it causes far less disease than *S. pneumoniae* and *H. influenzae*. Other bacteria, such as *Chlamydia trachomatis*, have also been implicated in AOM. Treatment of AOM usually involves systemic or local antibiotics and depends on the severity of infection, type of bacteria, and patient's age and health status.

BACTERIAL SINUSITIS The same pathogens that cause bacterial middle ear infections and pneumonia (discussed shortly) can infect the sinuses. This is why **bacterial sinusitis** often accompanies middle ear infection in children or pneumonia in adults. Inflammation and congestion of the sinuses obstruct the sinuses and the mucus flow. Accumulation of mucus in a warm and relatively airy cavity supplies bacterial pathogens with a nice breeding ground. The pressures behind the eyes, pain in the face, and foul-smelling nasal discharge or breath are common symptoms of bacterial sinusitis. Diagnosis is usually made based on clinical presentation, and the patient is treated with antibiotics.

PHARYNGITIS

CASE HISTORY 20.3

The Contagious Football Player with GAS

Dallin, a 16-year-old high school athlete, came to the clinic complaining of sore throat. He did not have a cough or any cold or flu symptoms. His cervical nodes (lymph nodes in the neck) were enlarged and tender to palpation, there was an exudate visible on his tonsils and pharynx (pharyngotonsilar region), and his temperature was 39.8°C (103.6°F). The physician assistant (PA) told Dallin that he probably had a strep throat but she could not perform a rapid strep test to confirm this because they had run out of the strips. The PA then swabbed the back of Dallin's throat and sent the sample to the lab for culture and antibiotic sensitivity testing. Dallin was given an antibiotic empirically (based on clinical experience and in the absence of definitive diagnosis). He was told that he was contagious and should rest and avoid contact with others until better. He followed the recommendations, recovered, and was back playing football in a few days. The lab isolated *Streptococcus pyogenes*, also called group A streptococcus (GAS), from Dallin's pharyngeal swab, confirming the diagnosis.

A patient complaining of a sore throat is usually suffering from **pharyngitis** (inflammation of the pharynx), although the cause may instead be **tonsillitis** (inflammation of the tonsils), **laryngitis** (inflammation of the larynx), or **peritonsillar abscess** (an abscess in the pharynx). Differential diagnosis for a patient complaining of sore throat includes a variety of viral and bacterial etiologies (Table 20.3), including infectious mononucleosis (see Section 21.2) and diphtheria (discussed shortly). A sore throat may also be due to a noninfectious cause, such as gastroesophageal reflux disease (GERD, frequent heartburn) or thyroid dysfunction, but the associated signs and symptoms of those diseases easily distinguish them from pharyngitis caused by infectious agents.

Pharyngitis is caused primarily by viruses, particularly those of the common cold and flu. The disease usually resolves without intervention regardless of viral or bacterial origin. Bacteria such as the *S. pyogenes* infecting Dallin cause less than 20% of the pharyngitis cases, but those cases are the most severe and have the highest

Table 20.3
Etiology of Infectious Pharyngitis

Viruses	
Respiratory viruses	• Rhinoviruses • Coronavirus • Respiratory syncytial virus (RSV) • Parainfluenza virus • Influenza A and B viruses • Adenovirus
Other viruses	• Herpes simplex virus types 1 and 2 (HSV-1 and HSV-2) • Enteroviruses (coxsackie A) • Epstein-Barr virus (infectious mononucleosis) • Cytomegalovirus (CMV) • Human immunodeficiency virus (HIV)
Bacteria	
Gram-positives	• *Streptococcus*, group A (most common bacterial cause) • *Streptococcus*, groups C and G • *Corynebacterium diphtheriae* • *Arcanobacterium haemolyticum*
Gram-negatives	• *Neisseria gonorrhoeae* • *Neisseria meningitidis* • *Francisella tularensis* (rare)
Other	• *Mycoplasma pneumoniae*
Fungi	
	• *Cocccidiodes immitis* (retropharyngeal abscess) • *Candida albicans* (thrush)

Figure 20.9 "Strep" Throat

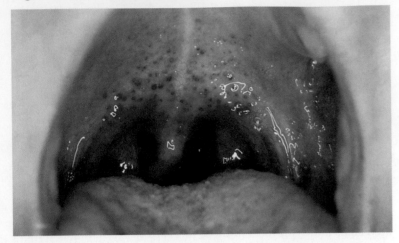

Note the reddened area on the pharynx and the enlarged tonsils.

potential for detrimental outcomes if untreated. Bacterial infections of the throat can be caused by *S. pyogenes, Neisseria gonorrhoeae, Chlamydophila pneumoniae,* and *Chlamydophila psittaci.* Streptococcal species other than *S. pneumoniae* are divided into groups A–U based on the antigenic composition of their carbohydrate peptidoglycans (discussed further in Section 25.3). Streptococcal groups A, B, and C are the most common streptococcal causes of pharyngitis. Group A streptococci (GAS)—*Streptococcus pyogenes* in particular—are the bacteria often equated to sore throat in the public's mind. The notoriety is well deserved because group A streptococci cause the more severe cases of the disease.

Streptococcal pharyngitis, or strep throat, is contagious and is spread through person-to-person contact or by indirect contact with items contaminated by secretions containing the bacteria. The hallmarks of streptococcal pharyngitis (Figure 20.9) are sudden onset of high fever and sore throat, tender enlarged cervical lymph nodes, exudate on the tonsils, and absence of a cough. These symptoms will usually resolve within 5 days even when untreated. Treatment with antibiotics, however, will quickly stop replication of the pathogen. A set of criteria based on hallmark signs and symptoms (Centor criteria) may be used to identify the likelihood of a bacterial infection in patients complaining of a sore throat.

Some strains of *S. pyogenes* produce exotoxins called streptococcal pyogenic exotoxins (SPEs) that can cause fever and a red rash called scarlet fever (described in Section 19.3). The rash usually starts on the head and neck and eventually spreads to the trunk of the body and then to the arms and legs (Figure 20.10). Some patients with scarlet fever may also have a very distinctive red "bumpy" tongue referred to as a "strawberry tongue" (Figure 20.11).

Definitive diagnosis of streptococcal pharyngitis is made from culturing the organism from a patient's pharyngeal exudates (see Section 25.2). A rapid strep test using immunoassay technology is available and widely used (Figure 20.12). The tests are highly specific for *S. pyogenes*. Patients with *S. pyogenes* pharyngitis are

considered noncontagious 24 hours after initiation of antibiotic treatment.

Antibiotics that inhibit cell wall synthesis, the beta-lactams, are commonly the first line of therapy. Macrolide antibiotics (for example, erythromycin) are used for patients with a penicillin allergy. The relative ease by which streptococcal infections can be cured has reduced complications of streptococcal pharyngitis to a remarkable degree.

STREPTOCOCCAL SEQUELAE Sequelae are pathological conditions that can result after a primary disease has run its course. Sequelae from streptococcal infections are caused by the immune response to the bacteria. *S. pyogenes* uses pilus-like surface structures called M protein to attach to host cells (see Section 18.2). Some strains of *S. pyogenes* possess M proteins with epitopes similar to those of some human cells, especially cardiac or kidney cells. When our body produces antibodies against these M proteins, those antibodies can bind (cross-react) with our own cells

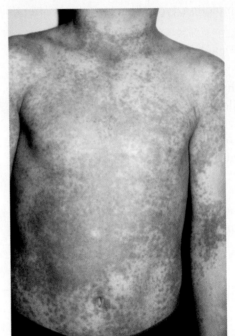

Figure 20.10
The Rash of Scarlet Fever
The rash starts on the trunk and spreads to the extremities but spares the palms of the hands and soles of the feet.

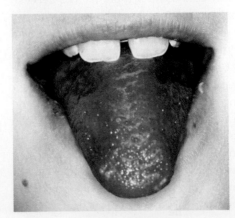

Figure 20.11
The Strawberry Tongue of Scarlet Fever

Figure 20.12   A Rapid Strep Test

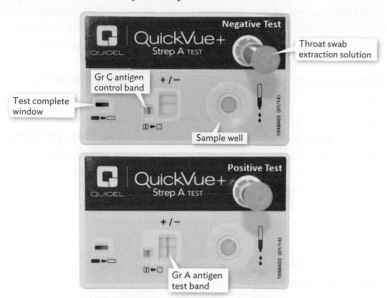

The rapid strep test is an in-office immunochromatography test done by a clinician to determine the presence of group A beta-hemolytic streptococci (*S. pyogenes*). The test begins when streptococcal cell wall antigens extracted from a throat swab are added to a sample well containing red-tagged anti-strep-A antibody. The Gr A antigen binds to the antibody and the complex moves along a membrane to an area embedded with a different anti-strep-A antibody (large, rectangular area). Any red-tagged antigen-antibody complexes present will be captured by the membrane-fixed antibody and generate a vertical red line (a positive test). The horizontal blue control line indicates that the reagents were added properly. Another control is the small red line (small, rectangular area), indicating the presence of test kit group C antigen also added to the sample well. Note that a positive group A strep test is not an absolute indication of a strep infection, as some people are strep carriers, in which case the strep bacteria are growing in their throat without causing any symptoms.

or tissue components in an **autoimmune reaction** (our antibodies acting against us), resulting in serious sequelae. Post-streptococcal sequelae include rheumatic fever (cardiac disease) and glomerulonephritis (kidney disease), which develop long after the causative organism has been cleared from the body. Whether an individual develops rheumatic fever or glomerulonephritis depends on the antigenic type of M protein made by the infecting GAS strain.

Link Recall from Section 17.4 that the M protein (pili) of *Streptococcus pyogenes* contains a cardiac-like epitope that is flanked by epitopes (streptococcal antigens) unrelated to the human host. The cardiac-like epitope can bind surface antibody on a self-reacting B cell, be taken up, and processed. The

nonself epitope "piggybacks" its way into the B cell and is presented on the B cell MHC II protein. Specific helper T cells can recognize the MHC II-nonhuman epitope and activate the autoreactive B cell, resulting in an **autoimmune reaction**.

Patients with **acute rheumatic fever (ARF)** are generally young children, 4–9 years of age, with a recent history of sore throat approximately 3 weeks prior to presentation. The illness starts with a high fever and one or more symptoms resulting from damage to the patient's heart, joints, skin, or nervous system. The initial presenting symptom of rheumatic fever is pain in a large joint (arthritis), which typically starts in the knees and spreads to the ankles, elbows, and wrists successively. The joints are not affected all at the same time. Instead, improvement of symptoms in the affected joints coincides with the onset of symptoms in other joints (for example, the ankles), a phenomenon called migrating, or migratory, arthritis. The X-ray of the affected joints may show some effusion (fluid accumulation), but the inflammation is normally minimal compared with the pain, and the synovial fluid is aseptic.

Heart tissue is affected through complex interactions between T cells, cross-reacting anti-heart antibodies (triggered by streptococcal M protein) and complement. The result is inflammation of the different layers of tissue covering the heart, including the innermost layer, which also forms the heart valves (see Section 21.1). Pancarditis, simultaneous inflammation of different tissues of the heart, is almost always accompanied by valvulitis (inflammation of a valve) and onset of a heart murmur (an abnormal sound) or changes in an existing one (Figure 20.13). Patients suffering from rheumatic heart disease will complain of shortness of breath and fatigue as well as varying degrees of chest discomfort or pain.

The skin manifestations associated with ARF include a rash that usually appears early in the course of disease and is referred to as the **erythema marginatum** (*marginatum*, "having a distinct margin") or erythema annulare (*annulare*, "ring-shaped"). The circular edge of the lesion is erythematous (red), and the center is paler by

Figure 20.13   Heart Valve Damaged by Rheumatic Fever

Inflammation caused by heart cross-reacting anti-M-protein antibodies damages the heart valve, making it unable to close as needed.

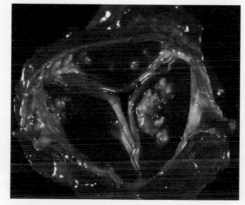

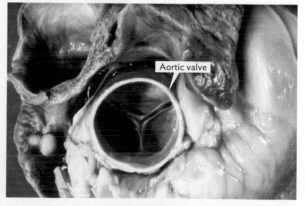

A. A damaged heart valve.

B. A healthy heart valve.

Figure 20.14   Erythema Marginatum

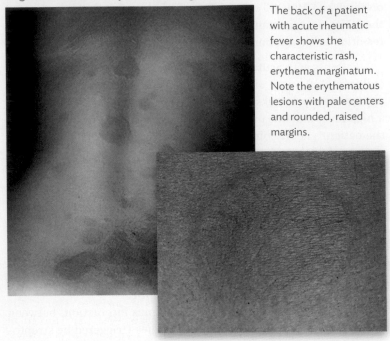

The back of a patient with acute rheumatic fever shows the characteristic rash, erythema marginatum. Note the erythematous lesions with pale centers and rounded, raised margins.

comparison (Figure 20.14). Skin manifestations may also include subcutaneous nodules (Figure 20.15). The nodules are firm and painless and usually resolve within a month. The skin overlying the the nodules does not have a rash or ulceration.

Streptococcal pharyngitis may also result in an acute glomerulonephritis (inflammation of renal tissue), which can be asymptomatic with microscopic hematuria (blood in urine). Alternatively, the disease may manifest as acute renal failure, hypertension, edema, and proteinuria (excessive protein in the urine) accompanied by reddish brown urine (Coca-Cola colored), collectively referred to as the nephritic syndrome. Strains causing symptoms of **post-streptococcal glomerulonephritis** are said to be nephritogenic and are thought to cause immune complex–mediated glomerulonephritis. The exact mechanism of damage to the kidneys is unknown. Structural similarities between streptococcal antigens and components of human renal tissue supports the occurrence of molecular mimicry, a phenomenon in which the immune system attacks self antigens that resemble components of a pathogen.

Reemerging Upper Respiratory Tract Infections

Many diseases over the last century were close to being eliminated but are now experiencing an unwelcome resurgence, even in developed countries such as the United States. The anti-vaccination movement has significantly contributed to the resurgence of such diseases as whooping cough. Fear of vaccination is fueled, in part, by the unfounded notion that vaccination can lead to autism. Unfortunately, as more and more families deny their children vaccines, we will continue to see once deadly diseases reemerge.

CASE HISTORY 20.4

The "Bull Neck"

Ramona, a 23-year-old Hispanic mother, brought her 3-year-old daughter into the emergency department. The child was lethargic, had a fever of 40°C (104°F), and was having difficulty breathing. The mother explained that the family arrived in the United States from El Salvador the previous week. The attending physician noted an extreme swelling of the child's cervical lymph nodes, giving the girl a thick, "bull neck" appearance (Figure 20.16). She also noticed the beginnings of a membranous growth at the back of the child's throat that was beginning to obstruct the trachea. It was grayish in color and bled when scraped. The distraught mother admitted that the child had not received any vaccinations before arriving in New Mexico. Suspecting the nature of the child's illness, the physician immediately admitted the child to the hospital and ordered administration of penicillin and a specific antitoxin. The culture results of a throat swab sent to the microbiology lab confirmed the physician's suspicion. The root of the child's disease was *Corynebacterium diphtheriae*, which causes diphtheria.

Figure 20.15
Rheumatic Subcutaneous Nodules

These nodules are usually located near tendons or over a bony surface or prominence. The overlying skin is not inflamed and typically can be moved over the nodules. The diameter varies from a few millimeters to 1–2 cm.

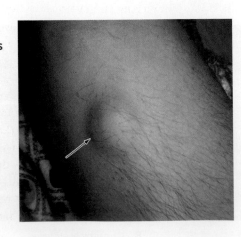

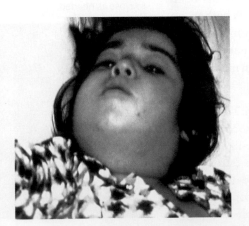

Figure 20.16
Bull Neck

Swollen lymph nodes give the characteristic "bull neck" appearance in a patient suffering from diphtheria.

DIPHTHERIA   Among the many bacterial causes of sore throat, *Corynebacterium diphtheriae* is one that was almost eliminated from the developing world, thanks to routine immunization practices; but it remains a threat from global travel to regions lacking effective immunization. In the United States, the disease this microbe causes has launched a comeback in recent years, as some parents have decided against vaccinating their children. **Diphtheria** is a reportable disease caused by *C. diphtheriae*, a Gram-positive rod that exists in two major forms, lysogenized and nonlysogenized. The toxigenic form is lysogenized and produces a toxin (diphtheria toxin) whose gene, *dtx*, is encoded by the beta phage genome embedded in the *C. diphtheriae* chromosome (see Section 18.3). The nontoxigenic form of the bacterium is not lysogenized with the phage. Both forms of this species can infect the upper respiratory tract and the skin. Toxigenic strains infecting the throat produce toxin that can cause cardiac and nervous system dysfunction. The bacteria remain localized in the pharynx, but the toxin is free to disseminate via the circulation. Cutaneous infections with toxigenic diphtheria, however, rarely result in systemic manifestations.

Upper respiratory tract infection with *C. diphtheriae* results in a sore throat, cervical lymphadenopathy, a low-grade fever, and a "growing" thick, grayish membrane (made of dead cells, debris, and fibrin) at the back of the throat. This **pseudomembrane** gives the disease its name, diphtheria, which is derived from the Greek word for "leather" (**Figure 20.17A**). Infections of the skin are seen as lesions that do not heal (**Figure 20.17B**) and are sometimes covered by a gray membrane. Cardiac manifestations often start as the local symptoms begin to improve and can vary from shortness of breath and irregular heartbeat to heart blocks (problems with the electrical system of the heart) and cardiac dilation (enlargement of heart chambers due to muscle thinning and weakness). The severity of neurological symptoms depends on how quickly the disease is treated with antitoxin. Symptoms can include weakness, paralysis of the local tissue (particularly the soft palate), paralysis of cranial nerve III (the oculomotor nerve, causing double vision or an inability of the pupils to react to light), or total paralysis.

Transmission of *Corynebacterium diphtheriae* is person-to-person, with asymptomatic carriers serving as the only reservoir. Presumptive diagnosis is made when an erythematous throat develops exudates that coalesce to form the pseudomembrane. The pseudomembrane adheres avidly to the underlying tissue and bleeds when scraped to remove a sample. Definitive diagnosis is made by culturing the bacteria from the throat or skin lesions and verifying toxin production. Toxin production can be verified by PCR amplification (see Section 8.2) of the bacterial *dtx* gene or by laying a filter paper strip impregnated with diphtheria antitoxin over a streak of the suspect bacterium. If toxin is made, the antibody against the toxin (the antitoxin) will react with its antigen (the toxin) and form a precipitate called an immunoprecipitate. In the latter procedure, if toxin is made, a line of precipitation will form (see Section 17.5).

Antibiotics and antitoxin are simultaneously used to treat the disease and should be administered early for best outcomes. The antitoxin is a hyperimmune serum produced in horses and carries a 10% risk of hypersensitivity and serum sickness. Of course, the best medicine is prevention by vaccination. The vaccine is inactivated diphtheria toxin (called a toxoid), whose toxic activity has been destroyed while leaving its antigenic properties intact. Diphtheria toxoid is part of the **DTaP series** of vaccinations, which also include tetanus toxoid and an acellular pertussis vaccine (Table 17.6).

Link As discussed in Section 17.3, **serum sickness** is a type III hypersensitivity reaction mediated by immune complex deposition in tissues with subsequent complement activation. The classic syndrome can develop when a person is injected with antiserum derived from animals (heterologous serum).

WHOOPING COUGH   **Pertussis**, or **whooping cough**, is another disease reemerging in developed countries. It is a highly contagious illness caused by the Gram-negative bacillus *Bordetella pertussis*. The infection is acquired when a person inhales aerosolized bacteria-containing droplets coughed by someone with the illness. After an incubation period of 7–21 days, the disease progresses through catarrhal, paroxysmal, and convalescent phases. Symptoms of the catarrhal phase are indistinguishable from a viral URT infection. The paroxysmal phase is marked by a violent paroxysmal (rapidly repeated) cough. The cough is sometimes followed by vomiting but always by a struggling deep breath that makes a whooping noise, giving the disease its name. The paroxysms of coughing gradually disappear during the next 2–3 weeks as the patient recovers in the convalescent phase.

Figure 20.17   Diphtheria

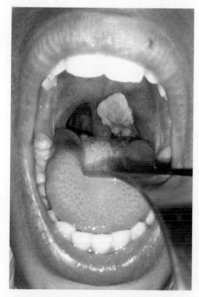

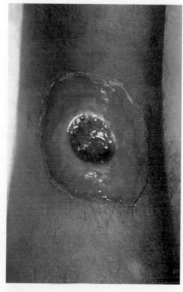

A. Pseudomembrane produced by *Corynebacterium diphtheriae* is thick and leather-like.

B. Cutaneous lesion caused by *C. diphtheriae*.

Figure 20.18   *Bordetella* Binding to Ciliated Epithelial Cells

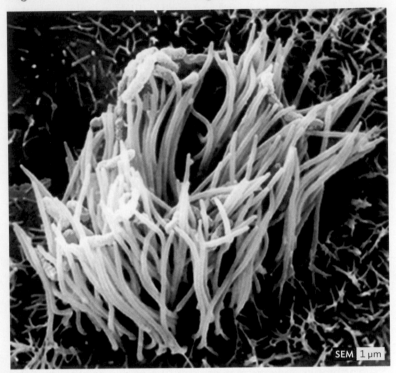

SEM | 1 µm

Rabbit tracheal epithelial cells infected with virulent wild-type *Bordetella bronchiseptica*, a close relative of *B. pertussis*. Wild-type *B. bronchiseptica* preferentially adheres to cilia and is rarely seen bound to aciliated cells or aciliated portions of ciliated cells. This adherence impedes the ciliary escalator that normally removes bacteria from the lung. *B. bronchiseptica* is a common respiratory pathogen of cats and dogs but can cause whooping cough in immunocompromised humans. Bacteria have been pseudocolored pink.

among adults and better diagnostic (particularly PCR) and surveillance practices may account for some of the increase. Macrolides are the antibiotics of choice for treatment, but trimethoprim-sulfamethoxazole can be used under special conditions.

Lower Respiratory Tract Bacterial Infections

Bacteria account for the majority of lower respiratory tract infections. Aerobic bacteria find the human lung particularly desirable to assault. The bacteria that attack our lungs scare us for a very good reason: they cause severe diseases with high mortality rates.

BRONCHITIS   Inflammation of the bronchi (bronchitis) is usually an acute and self-limiting disease with multiple viral and bacterial etiologies. A productive cough is generally the only presenting symptom. At the beginning, an upper respiratory tract infection is indistinguishable from acute bronchitis. Acute bronchitis is usually suspected when a patient's cough persists for more than 5 days. Viruses are the usual cause of acute bronchitis, but some bacteria that cause pneumonia, such as *Mycoplasma*, *Chlamydophila*, and *Bordetella* species can produce acute bronchitis as well.

Chronic bronchitis affects more than 40% of smokers and is associated with increased production of mucus and a decline in pulmonary function. Smoking is the major but not the only risk factor for chronic bronchitis. Long-term exposure to cigarette smoke, environmental exposure to air pollutants, and repeated lower respiratory tract infections during childhood can also increase the risk of chronic bronchitis. The viscous mucus that is not removed can then serve as a breeding ground for bacterial pathogens. Bacteria

The pathogenesis of *B. pertussis* is still the subject of much research. Its entire genome has been sequenced, and a number of biologically active molecules and virulence factors have been identified. Among them are adhesion factors such as filamentous hemagglutinin (FHA) and tracheal colonization factor, two surface proteins that latch the bacterium onto cilia in the respiratory tract (**Figure 20.18**). The pathogen also produces a number of toxins, including an adenylyl cyclase activated by contact with calmodulin in host cells, a tracheal cytotoxin that is a component of *Bordetella*'s peptidoglycan, and the two-component pertussis toxin that activates host adenylyl cyclase (toxins are described in Section 18.3). The functions of some of these toxins are known, but some are unclear. Increased cAMP production, for example, causes the accumulation of fluid in the lungs in much the same way that cholera toxin causes diarrhea.

The incidence of whooping cough had been declining for decades, thanks to effective vaccines. The original vaccine was made from killed cells but has been replaced by an acellular vaccine composed only of purified pertussis toxin and FHA. Unfortunately, the incidence of disease has been increasing (**Figure 20.19**), due mainly to parents' refusal to vaccinate their children. Waning immunity

Figure 20.19   Incidence of Whooping Cough in the United States

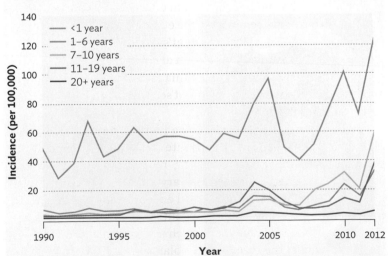

In the early 1990s, parents started to refuse to vaccinate their children because of some studies that seemed to link vaccination to autism. The link has been scientifically rejected. This graph shows the reported pertussis incidence (per 100,000 persons) by age group in the United States from 1990 to 2012.

associated with chronic bronchitis are the same as those causing other upper and lower respiratory tract infections. Patients who suffer from chronic bronchitis are at an increased risk for chronic obstructive pulmonary disease (COPD) and death. Around 70%–80% of COPD patients develop respiratory tract infections, of which nearly half are due to bacteria.

COMMUNITY-ACQUIRED PNEUMONIA The term "pneumonia" refers to an infection that causes inflammation in the lung, but does not infer a specific infectious agent. Most cases of pneumonia are attributed to a handful of bacteria (see Table 20.2). And most cases of pneumonia, regardless of etiology, occur in the winter months because the infectious agents are more easily transmitted from person to person when people are confined indoors. Cases of pneumonia that are acquired outside a hospital or health care setting are said to be community acquired.

Pneumonias and their bacterial causes are often classified as "typical" or "atypical." The term *atypical* was coined by Hobart Reimann in 1938 to distinguish the first diagnosed human cases of pneumonia caused by *Mycoplasma pneumoniae* from cases caused by *Streptococcus pneumoniae*. Other bacteria, such as *Legionella* and *Chlamydophila*, produce atypical pneumonias as well. What makes an atypical pneumonia atypical? Generally, atypical pneumonias affect multiple organ systems, usually produce a normal white blood cell (WBC) count, and produce symptoms that appear gradually, often mimicking upper respiratory infections at the onset. The terms *typical* and *atypical*, however, do not have a solid clinical basis, despite their common use. Organisms that produce typical pneumonias can also present atypically.

Lung infections can be contracted during a hospital stay (hospital-acquired, or nosocomial, pneumonias) or outside a hospital setting (community-acquired pneumonias). Community-acquired pneumonias result from a somewhat different set of bacteria than hospital-acquired pneumonias. How a particular pneumonia presents depends on the etiologic agent, the length and severity of infection, and the patient's overall health status. However acquired, lung infections almost always present with fever and cough. Some patients with pneumonia are able to continue normal activities for quite some time without seeking help, giving rise to the term *walking pneumonia*.

Access of microorganisms to the lungs is well restricted. Several defense mechanisms contribute to lung innate immunity, including the mucociliary escalator, respiratory epithelium, and alveolar macrophages. The mucociliary escalator continually traps and removes offending agents. Upon encountering a pathogen, the respiratory epithelium produces a number of cytokines, chemokines, and antimicrobial peptides, such as lysozymes and defensins, to defend against the invaders. Alveolar macrophages provide further help deeper in the lungs, assisted by efficient recruitment of neutrophils (PMNs) when needed. Organisms that have developed the tools to survive these defenses are able to gain access to the lungs and set up infection (see Sections 18.1 and 18.5).

CASE HISTORY 20.5

Night Sweats in a Nursing Home

 In March, James, an 80-year-old resident of a New Jersey nursing home, had a fever accompanied by a productive cough with brown sputum (mucous secretions of the lung that can be coughed up). James reported to the attending physician that he was short of breath, had pain on the right side of his chest, and suffered from night sweats. Blood tests revealed that his white blood cell (WBC) count was 14,000/μl (normal value: 4,500–10,000), composed of 77% segmented forms (polymorphonuclear leukocytes, PMNs; normal value: 40%–60%) and 20% bands (immature PMNs; normal value: 0%–3%). The chest radiograph revealed a right upper lobe infiltrate (**Figure 20.20A**). From this information, the clinician made a diagnosis of pneumonia. Microscopic examination of the patient's sputum revealed Gram-positive cocci surrounded by a capsule in pairs and short chains (**Figure 20.20B**). Bacteriological culture of his sputum and blood yielded *Streptococcus pneumoniae*.

Streptococcus pneumoniae (also called pneumococcus) accounts for about 25% of community-acquired cases of pneumonia. **Pneumococcal pneumonia** occurs mostly among the elderly (like James in our case history) as well as in smokers and immunocompromised individuals, including diabetics and alcoholics. A breakdown of pneumonia cases by causative organism is shown in **Figure 20.20C**. *S. pneumoniae* is also able to cause upper respiratory tract infections such as otitis media, sinusitis, and meningitis, discussed earlier.

Our upper airways are frequently colonized by one of the 91 serotypes of *S. pneumoniae*. Colonization is usually a benign process that leads to a carrier state. The noses and throats of 30%–70% of a given population can contain *S. pneumoniae*. The pneumococcus is highly host adapted, making humans its only natural reservoir. Initial colonization is facilitated by a slippery polysaccharide capsule that allows the organism to avoid engulfment by phagocytic cells and removal by the mucociliary escalator, which helps *S. pneumoniae* gain access to the respiratory epithelium. Once these bacteria reach the epithelium, the thick capsule becomes a hindrance to attachment and changes to a thin, transparent one. A large number of virulence factors will then contribute to the invasion of the host and onset of the disease.

The microbe reaches the lung, usually after being aspirated, and will grow in the nutrient-rich edema fluid of the alveolar spaces. Neutrophils and alveolar macrophages then arrive to try to stop the infection. They are called into the area from the circulation by chemoattractant chemokines released by damaged alveolar cells. The polysaccharide capsule of the pneumococcus, however, makes phagocytosis very difficult.

In an otherwise healthy adult, pneumococcal pneumonia usually involves one lobe of the lungs; thus, it is sometimes called lobar pneumonia. The infiltration of PMNs and fluid leads to the

Figure 20.20   Pneumonia Caused by *Streptococcus pneumoniae*

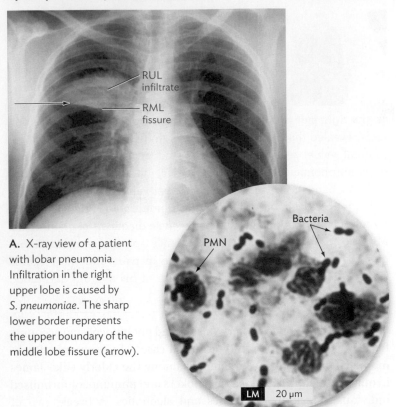

A. X-ray view of a patient with lobar pneumonia. Infiltration in the right upper lobe is caused by *S. pneumoniae*. The sharp lower border represents the upper boundary of the middle lobe fissure (arrow).

B. Micrograph of *S. pneumoniae*. Sputum sample showing numerous PMNs and extracellular diplococci in pairs and short chains.

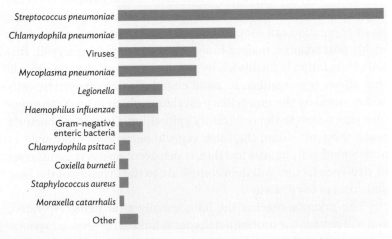

C. Relative incidence of pneumonia caused by various microorganisms.

characteristic diffuse cloudy areas seen on X-rays (see Figure 20.20A). In contrast, infants, young children, and the elderly more commonly develop an infection in other parts of the lungs, such as around the air vessels (bronchi), causing bronchopneumonia.

The white cell count in James's case history is telling. The patient had an elevated WBC count (normal is 4,500–10,000/μl) and elevated band cells (normal is 0%–3%)—increases indicative of a bacterial, rather than viral, infection. The number of neutrophils (PMNs), the front-line combatants against bacterial infection, rises in response to the presence of microbes. The sudden need for a larger quantity of neutrophils causes an increase in release of immature neutrophils from the bone marrow. The immature neutrophils have band-shaped nuclei and are called "bands." Historically, the "bands" were first depicted on the left side of a linear diagrammatic classification of white blood cells. Ever since, the increase in percentage of bands in the peripheral blood has been noted as a "left shift." Elevated white blood cells and a "left shift" are indicative of a bacterial infection.

In recent years, several outbreaks of pneumococcal pneumonia have occurred in nursing homes, affecting numerous residents. Such outbreaks can be avoided when the elderly receive the pneumococcal polysaccharide vaccine (PPSV) as a hedge against infection. There are more than 90 antigenic types of pneumococcal capsular polysaccharide; the vaccine contains the 23 polysaccharide types that most often cause disease and is recommended for individuals over 65 as well as for those who are immunocompromised. James in Case History 20.5 failed to receive the vaccine.

Young children are also vaccinated against pneumococcus, but not with PPSV because a child's immune system is not developed enough to recognize and respond to the free polysaccharide antigens in PPSV. The polysaccharide antigens must be joined (conjugated) to a protein so that they can be recognized by the child's immune system (see Section 17.6). So far, polysaccharides from 13 of the more dangerous 23 *S. pneumoniae* strains have been conjugated to proteins. The result is a 13-valent vaccine called polyvalent conjugate vaccine, or PCV. PCV is used to immunize children 2–59 months of age or children 5–18 years of age if they have certain medical conditions (such as sickle cell disease, a compromised immune system, or HIV infection) and have not received the vaccine before.

In addition to causing pneumonia, *S. pneumoniae* can invade the bloodstream and the membrane covering of the brain (meningitis). Individuals with health problems such as liver disease, AIDS, and organ transplants are even more likely to die from these diseases because of their compromised immune systems.

An emerging infectious disease problem throughout the United States and the world is the increasing resistance of *S. pneumoniae* to antibiotics. At least 30% of the strains isolated are already resistant to penicillin, the former drug of choice for treating the disease. Why antibiotic resistance is on the rise for *S. pneumoniae* and other microbes is discussed in Chapter 13.

ATYPICAL PNEUMONIA *Mycoplasma pneumoniae* is perhaps the best-known cause of "atypical" bacterial pneumonia, but other bacteria, such as *Legionella* and *Chlamydophila*, also produce atypical pneumonia. Mycoplasma lack a rigid cell wall (no peptidoglycan), usually exist in a filamentous form, and are the smallest known species of bacteria capable of independent growth. *M. pneumoniae* can be grown on media supplemented with serum in the presence or absence of oxygen, but the organism is fastidious and takes 3 weeks to grow in culture. Transmission is person-to-person, involving the inhalation of aerosolized infected droplets. *M. pneumoniae* uses a specialized tip containing adherence proteins to attach itself to the epithelium of the host cell (**Figure 20.21**). From this perch it can acquire nutrients from the host cell.

Symptoms of *M. pneumoniae* infection are insidious, starting with headache and malaise and progressing to a cough that is usually nonproductive. A low-grade fever may be present, and chills are common. Pneumonia resulting from an infection with *Mycoplasma* is more common in school-age children and young adults attending college and relatively uncommon in infants and young children, in sharp contrast to the disease caused by *S. pneumoniae* that predominantly affects the very young and very old. Mycoplasma infections activate the immune system, but the response does not protect against reinfection. The exact mechanism of pathogenesis is not known, but the host's immune reaction to the organism is thought to be the cause of pathology, rather than damage directly inflicted by the bacteria.

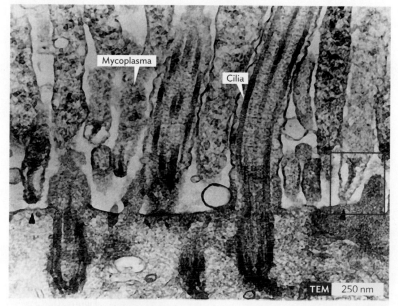

Figure 20.21 *Mycoplasma pneumoniae*

Hamster trachea ring infected with *M. pneumoniae*. Note the orientation of the mycoplasmas through their specialized tip-like organelle, which permits close association with the respiratory epithelium. The area within the red rectangle shows a *M. pneumoniae* structure, called the terminal organelle, that helps the bacterium attach to the host cell.

Patients infected with *Mycoplasma pneumoniae* may have subclinical anemia with a positive Coombs test or develop a positive cold agglutinin test 2 weeks postinfection. A Coombs test detects IgG antibodies in the patient's blood that can react with antigens on the same patient's red blood cell membrane and destroy it. Some diseases (including *M. pneumoniae* infections) can result in production of such antibodies. **Cold agglutinins** are antibodies that can agglutinate red blood cells at 4°C (39.2°F), but not at 37°C (98.6°F) and are produced only in response to certain diseases. White blood cell count is typically normal but can be slightly elevated. Diagnosis usually involves ELISA immunoassay to detect a >4-fold increase in anti-mycoplasma IgM and IgG titers when comparing acute (active disease) and convalescent (patient recovering) sera. Real-time PCR testing to detect presence of the organism is also available, but the sensitivity varies from 40% to 90%, depending on the timing of the sample collection. Treatment, usually macrolide antibiotics, is based on clinical presentation and serology.

Although we think of respiratory infections as being typically spread from person to person through aerosolization of respiratory secretions, that is not always the case. The natural environment, for example, can serve as a reservoir for some lung pathogens. Knowing where disease agents may reside is important both in prevention and in diagnosis.

LEGIONELLOSIS (LEGIONNAIRES' DISEASE) *Legionella pneumophila* is an aerobic Gram-negative bacterium that causes an atypical pneumonia. The first recognized outbreak developed after a meeting of the American Legion in Philadelphia, Pennsylvania, in 1976, which is how the disease came to be known as Legionnaires' disease, or legionellosis. The infected legionnaires presented with fever, various systemic symptoms, shortness of breath, and an initial productive cough that often progressed to production of a blood-tinged sputum. Tests subsequently showed that a less virulent form of the organism caused an earlier outbreak in Pontiac, Michigan, and was dubbed **Pontiac fever**. Today we know that legionellosis is a community-acquired illness that is not spread person-to-person.

L. pneumophila contaminates various water sources, ranging from lakes to the hot-water or air-conditioning distribution systems of hospitals, hotels, and other large buildings (the sources of the infection in the Philadelphia and Michigan outbreaks). Transmission is through the inhalation of contaminated water droplets. The organism is an intracellular pathogen capable of growing inside alveolar macrophages and free-living amebas. After being phagocytized, the bacterium prevents fusion of the phagosome with the lysosome (see Section 18.5).

Pathogenesis is facilitated by a number of virulence factors, including flagella, exotoxins, endotoxin, and proteins that help the organism prevent fusion of the phagosome and lysosome—particularly the gene products of *dot* (defective organelle trafficking) and *icm* (intracellular multiplication). Diagnosis of legionellosis is difficult because these bacteria are difficult to grow in the laboratory and not all laboratories have the capability to culture *L. pneumophila*. As a result, the number of cases reported is thought to be an underestimate. Diagnosis requires isolating the organism on buffered charcoal yeast extract (BCYE) agar supplemented with antibiotics to prevent growth of normal microbiota or detecting a bacterial antigen shed in urine via enzyme immunoassay (EIA). Legionellosis can be treated with a number of antibiotics, including macrolides or quinolones.

HOSPITAL-ACQUIRED PNEUMONIA   The CDC defines **health care–associated infections** (**HAI**, formerly known as nosocomial infections) as "infections that patients acquire during the course of receiving healthcare treatment for other conditions." A variety of bacteria, both pathogenic and opportunistic, cause HAI in almost every body system. *Pseudomonas aeruginosa*, described in Chapter 19 as an important cause of burn infections, is also the most common cause of hospital-acquired pneumonia.

P. aeruginosa is an opportunistic Gram-negative pathogen with a large arsenal of virulence factors that can be used once the bacterium breaches primary host defenses. The bacterium is a rare agent of community-acquired pneumonia but a major contributor to pneumonia and mortality in patients with cystic fibrosis (CF), an inherited disease that affects secretory glands producing mucus and sweat, and in those who are hospitalized. Hospital-acquired pneumonia caused by *P. aeruginosa* is an acute disease with a high mortality rate, and yet the organism is able to colonize the lungs of CF patients for long periods of time, causing a chronic infection. *P. aeruginosa* achieves this long-term colonization by forming a biofilm in the lungs of CF patients. Biofilm formation is facilitated by decreased sialylation (addition of sialic acid to cell-surface structures) of CF lung cells, which exposes more bacterial receptors, and by the bacteria's shortening their "O" lipopolysaccharide side chain (see Sections 10.3 and 18.3) and down-regulating the production of their toxins. *Pseudomonas* strains associated with CF also produce a thick, gelatinous slime layer made of alginate that severely retards cilial movement. The alginate layer also gives colonies of these strains a very mucoid appearance. Colonization as a biofilm results in chronic inflammation, reduction of lung function, and finally death due to respiratory failure.

Other Gram-negative bacilli that cause hospital-acquired pneumonias include *Escherichia coli*, *Klebsiella pneumoniae*, *Enterobacter* species, and *Acinetobacter* species. Gram-positive cocci associated with hospital-acquired pneumonias are *Staphylococcus aureus*, including methicillin-resistant *S. aureus* (MRSA), and *Streptococcus* species. Treatment of hospital-acquired lung infections is difficult because most of the organisms exhibit resistance to multiple antibiotics.

TUBERCULOSIS (TB)   Until recently, **tuberculosis**, caused by the acid-fast bacillus *Mycobacterium tuberculosis*, was considered of passing historical significance to physicians practicing in the developed world. In 1985, however, owing primarily to the newly recognized HIV pandemic and a growing indigent population, TB resurfaced in developed countries, especially in inner-city hospitals. In 1991, highly virulent **multidrug-resistant (MDR)** tuberculosis strains were reported. These strains not only produce fulminant (rapid-onset) and fatal disease among patients infected with HIV (the time between TB exposure and death is 2–7 months) but also have proved highly infectious to healthy people without HIV. The **tuberculin skin test**, discussed shortly, indicates exposure to tuberculosis bacilli when positive. Conversion rates (changing from a negative to a positive test) of up to 50% are reported in health care workers exposed to MDR strains. An immune system exposed to *M. tuberculosis* generates a T-cell-mediated response to *M. tuberculosis*, providing the basis for the skin test. This is true whether the exposed individual subsequently develops disease or not.

M. tuberculosis causes primarily a respiratory infection but can disseminate through the bloodstream to produce abscesses in many different organ systems. Disseminated disease is called "miliary" TB because the size of the infected nodules, called **tubercles**, approximates the size of millet seeds in bird feed. *M. tuberculosis* bacteria are spread from person to person (no animal reservoir) through aerosolization of respiratory secretions. *Mycobacterium bovis*, which normally causes TB disease in cattle, can also infect and cause TB in humans, especially children, but this accounts for a minority of TB cases. Once either species enters the lung, the bacteria are phagocytized by macrophages and survive sheltered within modified phagolysosomes. A delayed-type hypersensitivity response results, and small, hard tubercles (granulomas) form around the site of infection. Over time, the tubercles develop into caseous lesions that have a cheese-like consistency and can calcify into the hardened Ghon complexes seen on typical X-rays (**Figure 20.22**). The trapped bacilli are usually killed by the immune system. If not, they are contained by the immune system, often for extended periods of time, a condition called **latent tuberculosis infection (LTBI)**. A person with LTBI is not infectious.

In some LTBI cases, however, *M. tuberculosis* overcomes the imposed confinement of the immune system, multiplies, and after

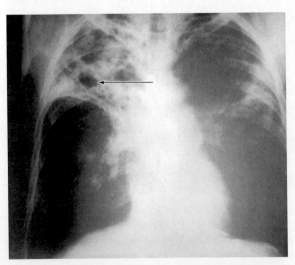

Figure 20.22
Calcified Ghon Complex of Tuberculosis
The arrow points to a Ghon complex in a patient's right upper lobe. Note the difference in appearance compared with Figure 20.20A.

a period of 4–12 weeks causes primary disease. People with a weakened immune system and the very young are often among those who manifest primary disease, exhibiting a productive cough that generates sputum and experiencing fever, night sweats, and weight loss. These patients are said to have **active tuberculosis disease**, and they are contagious. They also become tuberculin test positive owing to delayed-type hypersensitivity.

Latent TB can be reactivated months or years after initial infection, a condition called **secondary tuberculosis** or **reactivation tuberculosis**. Secondary TB commonly occurs in people who become immunocompromised. The tubercles described earlier form during secondary TB (reactivation TB). The symptoms of secondary TB are more serious than those of primary TB, mainly due to uninhibited replication of the bacteria. They include severe coughing, greenish or bloody sputum, low-grade fever, night sweats, and weight loss. The gradual wasting of the body led to the older name for tuberculosis—consumption. Among patients with secondary TB, 40% show cavity formation on a chest X-ray (**Figure 20.23**). Some new and extremely powerful anti-inflammatory medications used to treat other conditions may weaken the patient's immune system just enough to cause reactivation of TB.

TB bacilli usually cause pulmonary disease, but they can disseminate and gain access to other parts of the body by invading the lymph nodes of the lungs. From there they spread into the blood. The bacilli are then carried throughout the body, giving them a chance to infect other organs, such as the adrenal glands or bones. When an organ other than the lungs is infected with *M. tuberculosis*, the disease is said to be extrapulmonary. Patients who have only extrapulmonary TB are usually not contagious. TB bacilli can cause a disseminated infection, often in the immunocompromised.

About 10% of people who have a healthy immune system and have been exposed to TB may develop TB. The infection risk depends on the frequency and duration of exposure to people with active TB disease. Health care workers, immunocompromised patients, people who live or work in high-risk congregate settings (such as group homes or intermediate-care facilities), and children and adolescents who live with a patient with active disease are at higher risk of contracting the infection. The elderly and patients who take immunosuppressive drugs, on the other hand, are at higher risk of developing secondary (reactivation) TB.

Screening for LTBI or TB disease involves chest X-ray, the Mantoux tuberculin skin test (TST), interferon gamma release assays (IGRAs), nucleic acid amplification testing (NAAT), and examination of sputum by microscopy and culture. The CDC has developed extensive guidelines for screening and interpretation of test results based on exposure risk, general health, and immune status of the individual tested. A chest X-ray is helpful in a patient with signs and symptoms of disease but very difficult to interpret as a screening tool. A positive **Mantoux tuberculin skin test (TST)** indicates a delayed-type hypersensitivity reaction to *M. tuberculosis* proteins (called purified protein derivative, PPD) injected under the skin.

Figure 20.23 Cavity Formation in Tuberculosis

A. A cavity is seen inside the white rectangle on a chest X-ray.

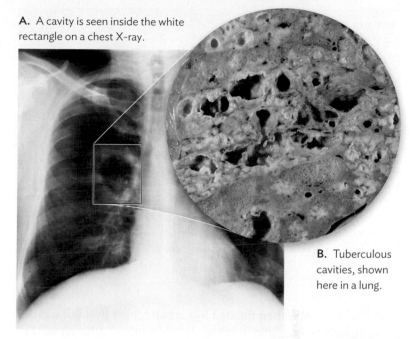

B. Tuberculous cavities, shown here in a lung.

The result of the test is read 48–72 hours after placement of the antigen. An induration (area of hard, raised skin) larger than 5–15 mm at the injection site is interpreted as a positive test, depending on the health and risk factors of the tested patient.

Note A positive tuberculin reaction does not mean that the patient has tuberculosis but does mean that the patient was at one time exposed to tubercle bacilli. In most cases, the immune response will kill the organism when it infects the lung. But memory T cells generated during the response will persist and can respond years later to the injection of PPD, producing a positive tuberculin test reaction.

Interferon gamma release assays (IGRAs) are blood tests that measure white blood cells' release of interferon gamma when exposed to mycobacteria. IGRAs are the preferred method of TB screening for persons who may not return for TST reading.

The tuberculin skin test and IGRAs do not distinguish between latent infection (LTBI) and active disease. Amplification of the mycobacterial DNA from patient blood and/or visualization of *M. tuberculosis* bacilli in patient sputum under the microscope is used to detect disease. Confirmed diagnosis of the disease is made by culturing the TB bacilli from a patient's sputum or tissue. Culturing is also used to monitor treatment progress. Negative cultures are a required criterion to declare the patient free of TB disease.

Mycobacterium species have a Gram-positive cell wall but cannot be Gram stained because about 60% of their cell wall is lipid (this is called a waxy coat, Section 5.3). The Ziehl-Neelsen stain (also called acid-fast stain) is used instead of the Gram stain when a sample is suspected to contain mycobacteria. The bacteria are fixed

Figure 20.24   Ziehl-Neelsen Staining of Mycobacteria

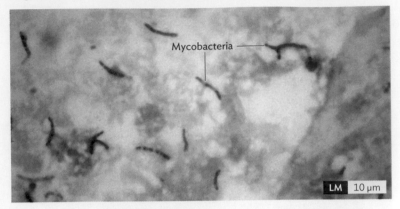

Mycobacteria are able to resist decoloration when washed with acidified ethanol and under the microscope appear as red bacilli in tissue specimens or samples stained from a culture. Ziehl-Neelsen staining is also referred to as acid-fast staining.

to a glass slide and then flooded by carbolfuchsin dye and washed subsequently with acidified ethanol. The acidified ethanol washes away the dye from other bacteria but not from mycobacteria, which retain the dye and appear as red bacilli under the microscope (Figure 20.24), which is why mycobacterial species are called **acid-fast bacilli (AFB)**. Isolation of AFB in culture is required to confirm diagnosis. Other laboratory techniques used to grow or identify *Mycobacterium* are discussed in Sections 25.3 and 25.4.

Ten drugs are currently approved by the Food and Drug Administration (FDA) for treatment of tuberculosis. Initial treatment of active disease is aggressive and involves a four-drug regimen called first-line drugs (or drugs of choice): isoniazid, rifampin, pyrazinamide, and ethambutol, given over a course of several months. MDR strains are defined as being resistant to two or more first-line drugs. MDR strains are treated with a nine-drug regimen. **Extreme drug-resistant tuberculosis (XDR-TB)** strains are resistant to three or more second-line drugs (drugs used when drugs of choice fail) as well as two or more first-line drugs. These strains are almost untreatable.

Pneumonia Caused by Select Agents

Bioterrorism is defined as the intentional release of biological agents to bring about disease or death to humans, livestock, or crops. The fear that some infectious agents may be used for such nefarious purposes has led the CDC to develop a list of select agents (Table 2.4) whose use and distribution among researchers are highly restricted. One such agent is *Bacillus anthracis*.

B. anthracis is a spore-forming, Gram-positive rod that primarily causes infections of animals (for example, cattle). The spores can survive for decades in soil and under proper conditions can even germinate and multiply there. Infection of humans usually results from accidental exposure to infected animals or animal products contaminated with spores (for instance, animal hides). *B. anthracis* can infect a number of tissues, among them skin (wound

infections), lungs, and gastrointestinal tract (rare). Spores can be inhaled if aerosolized and produce inhalation anthrax. Weaponized spores are specially treated to increase the time they can remain airborne. Despite the fact that pulmonary anthrax is contracted by inhalation, transmission does not usually occur from person to person (too few organisms or spores are aerosolized to support person-to-person transmission). The lethality of the bacterium is due to its three-subunit toxin (see Section 18.3). The core subunit, called protective antigen (PA), helps the trivalent toxin enter the host cell. The core subunit is called protective antigen because antibodies to this protein will protect hosts from disease. The other two subunits, the edema factor (EF) and lethal factor (LF), attack internal host metabolic components to disrupt host cell function and growth. The severity of disease produced as well as the ability to store and aerosolize spores makes *B. anthracis* a possible agent for bioterrorism (as described in the case that opens Chapter 18).

Inhalation anthrax has a prodromal phase that lasts 5 days, during which the patient suffers from mostly nonspecific symptoms resembling those of influenza. Some may have additional symptoms of hemoptysis (coughing up blood), dyspnea (shortness of breath), chest pain, and nausea. Once there is a clinical suspicion of inhalation anthrax, rapid diagnosis and antibiotic treatment is critical, as death can occur within 48 hours. Local and state health authorities must be informed and follow prescribed containment protocols to limit spread of the disease. Treatment is most effective during this time, but the rapid disease course makes diagnosis before death very difficult.

Many microbes designated as select agents cause deadly pneumonias and can be transmitted by aerosolization (though not always by person-to-person contact), including *Coxiella burnetii* (the most infectious organism known), *Yersinia pestis* (the cause of plague), *Francisella tularensis*, *Coccidioides immitis* (fungal infection), Lassa fever virus, and Ebola virus.

SECTION SUMMARY

- **Three bacterial species,** *Streptococcus pneumoniae, Haemophilus influenzae,* and *Moraxella catarrhalis*, are the most frequent causes of acute otitis media (AOM) in otherwise healthy children.

- ***S. pneumoniae,* the most frequent etiology of AOM,** is identified based on the structure of its antiphagocytic polysaccharide capsule. All identified serotypes can cause an infection, but the relative invasiveness of *S. pneumoniae* strains depends on capsule structure.

- ***H. influenzae* strains may be encapsulated (typeable) or lack a capsule (nontypeable).** Encapsulated *H. influenzae* subtype b (Hib) is the most invasive and can cause epiglottitis and meningitis in addition to AOM.

- **Unencapsulated, nontypeable *H. influenzae*** causes the majority of *H. influenzae*-related infections. Hib vaccine does not protect against infections with nontypeable strains.

- ***Streptococcus pyogenes*** is the most common bacterial cause of pharyngitis.

- **Rheumatic fever and post-streptococcal glomerulonephritis** are sequelae of infections by some serotypes of *S. pyogenes*.

- **Scarlet fever is a consequence of pharyngitis** caused by certain serotypes of *S. pyogenes* that produce streptococcal pyogenic exotoxin (SPE).

- ***Streptococcus pneumoniae* is the most common cause of bacterial pneumonia.** The pneumococcus organism normally infects one of the lobes of the lungs, which has given rise to the term *lobar pneumonia*.

- ***Mycoplasma pneumoniae*** lacks a cell wall, usually exists in a filamentous form, and results in a pneumonia with an atypical presentation.

- ***Corynebacterium diphtheriae*** is a Gram-positive rod that can be lysogenized by the phage carrying the *dtx* gene that encodes the diphtheria toxin. ***C. diphtheriae* can cause upper respiratory tract and skin infections.** Circulating toxin can cause systemic disease.

- **Whooping cough** is caused by the Gram-negative bacillus *Bordetella pertussis*, which produces several toxins including the pertussis toxin.

- ***Pseudomonas aeruginosa*** is an opportunistic Gram-negative bacillus that is the most common cause of infection in burn units and the leading cause of hospital-acquired infections.

- ***Legionella pneumophila*** is an aerobic Gram-negative bacillus that contaminates various environmental and industrial water sources.

- ***Mycobacterium tuberculosis*** is an acid-fast bacillus that causes TB, a chronic form of pneumonia that is reemerging as a result of an increase in the number of immunosuppressed patients.

- **Highly virulent multidrug-resistant (MDR) strains of *M. tuberculosis*** can produce a fulminant and fatal disease among HIV-infected individuals and others who are severely immunosuppressed.

- ***Bacillus anthracis*** is a spore-forming Gram-positive bacterium that normally causes disease in animals and can occasionally cause skin lesions in humans who come in contact with an infected animal.

- **Inhalation anthrax** is a highly uncommon, often lethal form of pneumonia caused when spores of *B. anthracis* are inhaled.

Thought Question 20.2 Why are smokers at a higher risk for bacterial bronchitis?

Thought Question 20.3 The TB skin test is usually negative in AIDS patients. Why?

Thought Question 20.4 Which have the higher risk of a bad outcome, hospital- or community-acquired pneumonias, and why?

20.4
Fungal and Parasitic Infections of the Respiratory Tract

SECTION OBJECTIVES

- Identify the most common etiologies of fungal lower respiratory tract infections.

- Describe the differences between fungal and bacterial lower respiratory tract infections.

- Name the risk factors associated with fungal respiratory tract infections.

CASE HISTORY 20.6
A Boxer's Fight to Survive

 Tyrrell, a 35-year-old male boxer who installs home insulation for a living, was recently admitted to a Maryland hospital when he presented at the emergency department complaining of difficulty walking, fever, chills, night sweats, and a recent 10 kg (22-lb) weight loss. He denied having prior pneumonia, sinus infection, arthritis, hematuria (blood in the urine), numbness, or muscle weakness. Tyrrell also denied any history of intravenous drug use and reported to be in a monogamous relationship for the past 4 years. There was no history of travel outside the area for the past year. A review of Tyrrell's past medical history showed that he had visited the emergency department 6 months earlier with flu-like symptoms, a chronic cough that produced blood-tinged white sputum, shortness of breath, loss of appetite, and weight loss. One month prior to current admission, he developed some painless subcutaneous nodules and became so short of breath that he could no longer continue boxing. At that time, an X-ray taken in the emergency department showed right upper lobe infiltrate, indicating pneumonia (**Figure 20.25A**). A

Figure 20.25 Pneumonia and Metastatic Disease Caused by *Blastomyces dermatitidis*

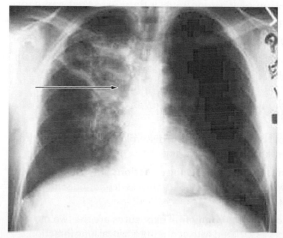

A. Diffuse infiltrate in the right lung (arrow).

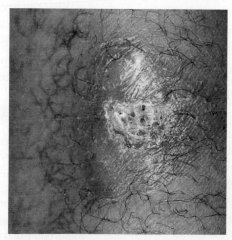

B. Metastatic leg lesion at the tibia.

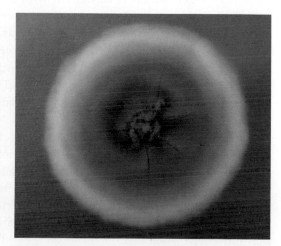

C. Fungal colony of *Blastomyces dermatitidis*.

tuberculosis skin test was negative. He was given a prescription for the antibiotic azithromycin (a commonly used antibiotic to treat bacterial infections of the respiratory tract) and discharged. The cough and weight loss continued. Physical exam at the time of current admission revealed several painful subcutaneous nodules, one filled with pus, and a tender tibia that prevented him from walking (Figure 20.25B). A CBC obtained at the time of current admission showed normal counts and differential.

What are the main symptomatic differences between respiratory infections caused by fungi and those caused by viruses or bacteria? Respiratory infections caused by viruses and bacteria have clearly defined clinical presentations. In contrast, fungal infections of the respiratory tract present with a constellation of vague and sometimes seemingly unrelated symptoms. In Tyrrell's case, the expression of frankly purulent material (pus) from the left leg nodule suggests an infectious process. His recent history of chronic cough suggests that the infection in this patient probably started in the lung, after which the organism disseminated throughout the body via the bloodstream. Fungus is a probable cause, given the chronic nature of the patient's symptoms. Tuberculosis, caused by the bacterium *Mycobacterium tuberculosis*, also causes a chronic lung infection and might have been suspected except for the negative TB skin test.

The most likely fungal causes of infection in this case history are the endemic mycoses, such as histoplasmosis, blastomycosis, and coccidioidomycosis. **Coccidioidomycosis**, caused by a variety of *Coccidioides* species, including *Coccidioides immitis*, is endemic in the western United States. Tyrrell had never traveled to the western United States, ruling out exposure to *Coccidioides*. **Histoplasmosis** is caused by *Histoplasma capsulatum* and most commonly presents as a flu-like pulmonary illness, with erythema nodosum (tender red bumps on the skin) and arthritis (swollen joints) or arthralgia (joint pain), none of which the patient had. **Blastomycosis** can disseminate to the lung, skin, bone, and genitourinary tract, which is consistent with the pattern of organ involvement seen in this patient. *Cryptococcus neoformans*, an encapsulated yeast that causes **cryptococcosis**, could also be considered. The most prevalent clinical form of cryptococcosis, however, is meningoencephalitis in AIDS patients, although disease can also involve the skin, lungs, prostate gland, urinary tract, eyes, myocardium, bones, and joints. Cryptococcosis, therefore, is considered but rejected.

Amphotericin B, an antifungal agent (Section 13.8), was ultimately given to Tyrrell. His fever lowered almost immediately, and the skin nodules diminished. After 2 weeks, a fungus was found in the cultures of the nodule biopsy, bronchoalveolar lavage (washes), and urine (Figure 20.25C). This fungus was identified by a DNA probe as *Blastomyces dermatitidis*, confirming the diagnosis of blastomycosis.

Blastomyces dermatitidis is a dimorphic fungus, meaning it can exist in two different forms (yeast at 25°C or hyphal at 37°C). The fungus resides in the soil of the Ohio and Mississippi River valleys and the southeastern United States. The portal of entry is the respiratory tract, and infection is usually associated with occupational and recreational activities in wooded areas along waterways, where there is moist soil with a high content of organic matter and spores. The incubation period ranges from 21 to 106 days. This patient most likely inhaled conidia (fungal spores) from the soil while crawling underneath houses installing insulation. Fungal respiratory tract infections are acquired from the environment and not transmitted person-to-person by respiratory aerosols. The physician learned that Tyrrell had used only a T-shirt to cover his mouth and nose, an ineffective method of keeping spores from entering the respiratory tract. He should have worn a respirator.

Several critical features of this case help differentiate it from the preceding case of pneumococcal pneumonia. First, the initial macrolide antibiotic, azithromycin, should have killed most bacterial sources of infection. Second, the X-ray finding of diffuse infiltrate is more indicative of fungal lung infection than pneumococcal infection. The blood count was also a clue. Bacterial infections usually cause an increase in WBCs, especially in the percentage of band cells (immature neutrophils), but Tyrrell's blood count and differential (percentage of each type of white blood cell in the sample) were within normal limits. Finally, the metastatic lesions (infectious lesions that develop at a secondary site away from the initial site of infection) on the leg were in no way consistent with *S. pneumoniae*. Many infectious diseases start out as a localized infection but end up disseminating throughout the body to cause metastatic lesions.

In addition to viral, bacterial, and fungal pneumonias, parasites such as *Strongyloides stercoralis* and *Toxoplasma gondii* can (but rarely do) cause pneumonia. Parasitic pneumonias are highly uncommon and present almost exclusively in HIV patients or result from dissemination from another infected site of the body. *Dirofilaria*, *Paragonimus*, and *Entamoeba histolytica* infections of the lung are also seen in patients with compromised immune systems but are even rarer.

With the exception of *Paragonimus*, a flatworm that can cause a lung infection if ingested in undercooked crayfish or crab, worms do not usually infect the lungs. Hookworms and *Ascaris* (see Section 22.6) can, however, trigger infiltration of eosinophils into the lungs, causing an eosinophilic pneumonia.

SECTION SUMMARY

- **Fungal infections of the lungs** result from dissemination of the fungi from the primary site of infection to the lung.
- **The most common causes of fungal lung infections** are *Histoplasma capsulatum*, *Cryptococcus neoformans*, *Blastomyces dermatitidis*, and *Coccidioides immitis*.
- **Immunosuppression and environmental exposures are** the two most important risk factors associated with acquiring a fungal lung infection.

Thought Question 20.5   What is an indication that a chronic cough may be due to a fungal infection?

Perspective

It is hard to overstate the significance of respiratory tract infections today and, indeed, throughout history. Respiratory diseases have resulted in major changes in our behavior and major shifts in population dynamics. There is cause to fear the next big pandemic. The possibility that an infectious agent could spread quickly and globally was demonstrated by the SARS virus, first reported as the cause of a severe respiratory illness in Asia in February 2003. The virus spread rapidly to Europe and North and South America, causing 8,098 cases of disease and 774 deaths worldwide according to the World Health Organization (WHO). Although global interventions by health officials prevented a pandemic, the outbreak dramatically illustrated the potential for air travel and globalization to spread deadly infections. A coordinated global response to disease outbreaks is needed not only to prevent the resurgence of diseases once considered eliminated from the developed world, but also to contain the costs imposed by infectious epidemics. Even minor respiratory tract diseases have a major effect on our productivity and our economy. The estimated total economic impact of cold-related work loss, for example, exceeds $20 billion per year in the United States.

Figure 20.26   Summary: Microbes That Infect the Respiratory Tract

Viruses	Classification	Associated Diseases
Adenoviruses	DNA	Common cold
Cytomegalovirus	DNA, ds	CMV disease
Respiratory syncytial virus	RNA, ss, negative sense	RSV disease, sometimes common cold
Metapneumovirus	RNA, ss, negative sense	Common cold
Influenza virus	RNA, ss, negative sense	Influenza, sometimes common cold
Parainfluenza viruses	RNA, ss, negative sense	Croup, common cold
Rhinoviruses	RNA, ss, positive sense	Common cold
Enteroviruses	RNA, ss, positive sense	Common cold
SARS virus	RNA, ss, positive sense	Severe acute respiratory syndrome
Coronaviruses	RNA, ss, positive sense	Common cold

Bacteria	Classification	Associated Diseases
Mycobacterium tuberculosis	Acid-fast bacilli	Tuberculosis
Streptococcus pneumoniae	Gram-positive cocci	Otitis media, sinusitis, pneumonia, meningitis
Bacillus anthracis	Gram-positive rod	Inhalation anthrax
Corynebacterium diphtheriae	Gram-positive rod	Diphtheria
Bordetella pertussis	Gram-negative coccobacilli	Whooping cough
Pseudomonas aeruginosa	Gram-negative rod	Pneumonia
Legionella pneumophila	Gram-negative rod	Legionellosis (Legionnaires' disease)
Chlamydophila pneumoniae	N/A	Pneumonia
Chlamydophila psittaci	N/A	Pneumonia
Mycoplasma pneumoniae	N/A	Pneumonia

Fungi	Classification	Associated Diseases
Histoplasma capsulatum	Dimorphic yeast	Pneumonia
Coccidioides immitis	Dimorphic yeast	Pneumonia
Cryptococcus spp.	Dimorphic yeast	Pneumonia
Blastomyces dermatitidis	Dimorphic yeast	Pneumonia

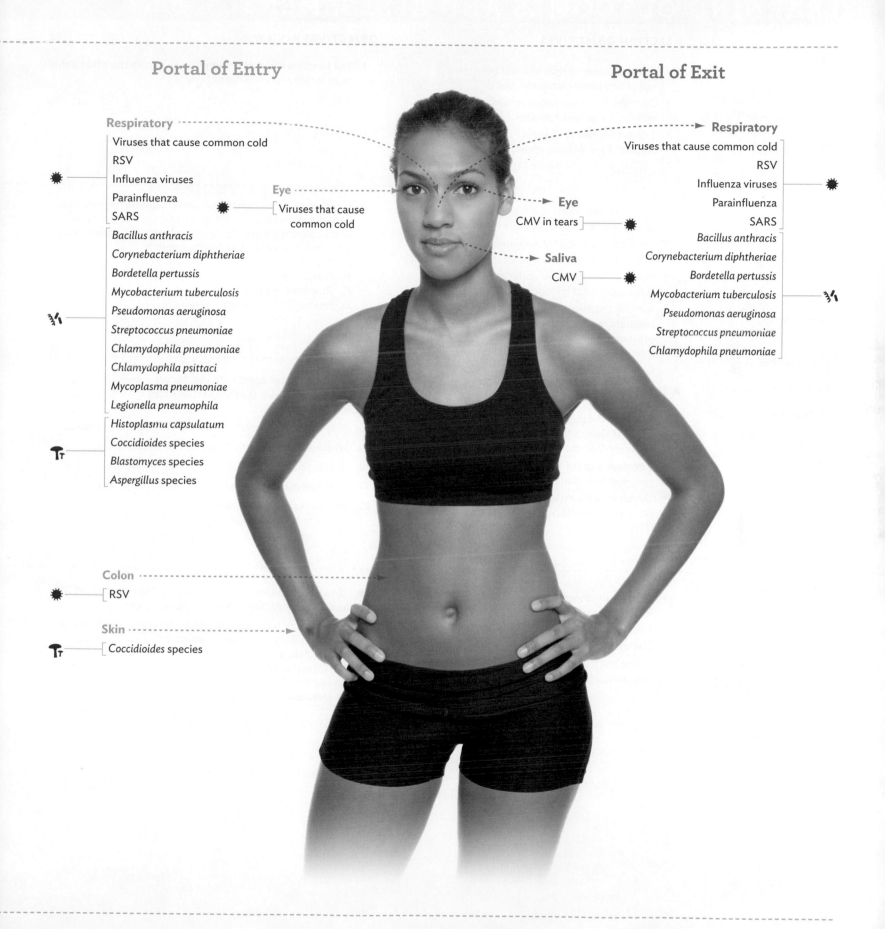

Portal of Entry

Respiratory
Viruses that cause common cold
RSV
Influenza viruses
Parainfluenza
SARS

Bacillus anthracis
Corynebacterium diphtheriae
Bordetella pertussis
Mycobacterium tuberculosis
Pseudomonas aeruginosa
Streptococcus pneumoniae
Chlamydophila pneumoniae
Chlamydophila psittaci
Mycoplasma pneumoniae
Legionella pneumophila

Histoplasma capsulatum
Coccidioides species
Blastomyces species
Aspergillus species

Eye
Viruses that cause
common cold

Colon
RSV

Skin
Coccidioides species

Portal of Exit

Respiratory
Viruses that cause common cold
RSV
Influenza viruses
Parainfluenza
SARS

Bacillus anthracis
Corynebacterium diphtheriae
Bordetella pertussis
Mycobacterium tuberculosis
Pseudomonas aeruginosa
Streptococcus pneumoniae
Chlamydophila pneumoniae

Eye
CMV in tears

Saliva
CMV

LEARNING OUTCOMES AND ASSESSMENT

	SECTION OBJECTIVES	OBJECTIVES REVIEW
20.1 Anatomy of the Respiratory Tract	• List the anatomical differences between the upper and lower respiratory tracts. • Describe the relationship of eyes, ears, and nose with the upper airways. • Explain the role of the mucociliary escalator in respiratory infection processes.	1. The larynx is a part of the upper respiratory tract that helps us generate sounds. True or false?
20.2 Viral Infections of the Respiratory Tract	• Compare and contrast the signs and symptoms of upper and lower respiratory tract infections. • Describe the role of mutations in the genesis of influenza pandemics. • List the risk factors for upper and lower respiratory tract infections. • Identify the etiology of the major viral respiratory infections based on their presentation.	4. Nasal congestion, sore throat, and cough can be present in both upper and lower respiratory tract infections. Which of the following will differentiate a lower respiratory tract infection in an otherwise healthy adult? A. Itchy eyes B. Runny nose C. Shortness of breath D. Production of sputum
20.3 Bacterial Infections of the Respiratory Tract	• List the bacterial etiologies of respiratory tract diseases and the disease's anatomical location. • Describe the relationships between host and bacteria that contribute to the pathogenesis of respiratory tract infections. • Compare and contrast clinical presentations of bacterial infections of the respiratory tract. • Develop a prevention plan for different bacterial respiratory pathogens.	Match the infectious agent with the disease it causes, choosing from the following list. (Some questions may have more than one answer.) 8. *Streptococcus pyogenes* A. Acute otitis media 9. *Mycobacterium tuberculosis* B. Pharyngitis 10. *Haemophilus influenzae* C. Bronchitis 11. *Pseudomonas aeruginosa* D. Pneumonia 12. *Streptococcus pneumoniae*
20.4 Fungal and Parasitic Infections of the Respiratory Tract	• Identify the most common etiologies of fungal lower respiratory tract infections. • Describe the differences between fungal and bacterial lower respiratory tract infections. • Name the risk factors associated with fungal respiratory tract infections.	18. Blood-tinged sputum can be a presenting symptom of TB as well as fungal pneumonias. True or false? 19. Which of the following laboratory test results indicates a fungal etiology in a patient suffering from pneumonia? A. A left shift B. Normal values C. Increased neutrophils D. Decreased neutrophils

2. A child has swallowed a large coin that is now lodged in his pharynx. He was barely able to breathe. The paramedics punctured a hole in the child's neck to insert a tube and help him breathe. Which structure was punctured by the paramedics?
   A. Bronchial tube
   B. Trachea
   C. Alveolar duct
   D. Bronchiole

3. The mucociliary escalator is responsible for _____.
   A. enhancing movement of air in and out of the bronchi
   B. preventing foreign materials from reaching the lungs
   C. moving CO_2 out of the lungs so it can be exhaled
   D. producing mucus only when we have a respiratory tract infection

5. Influenza pandemics are attributed to antigenic drift in the HA gene. True or false?

6. The error-prone replication of genomic material of influenza virus A necessitates production of a new flu vaccine annually. True or false?

7. _____ are at highest risk for developing RSV pneumonia.
   A. The elderly
   B. Adolescents
   C. Adults
   D. Infants

13. _____ is the most effective prevention for whooping cough.
    A. Childhood vaccination
    B. Respiratory isolation
    C. Antibiotic prophylaxis
    D. Wearing a respirator

14. Scarlet fever is _____.
    A. a sequela of pharyngitis caused by certain serotypes of *S. pyogenes*
    B. a consequence of pharyngitis caused by toxin-producing *S. pyogenes*
    C. a skin rash accompanying most streptococcal upper respiratory tract infections
    D. a common manifestation of streptococcal infections in the elderly

15. A small bacterium that has no cell wall has been isolated from the sputum of a patient suffering from atypical pneumonia. This patient's lung is infected with _____.
    A. *Bacillus anthracis*
    B. *Streptococcus pneumoniae*
    C. *Legionella pneumophila*
    D. *Mycoplasma pneumoniae*
    E. *Mycobacterium tuberculosis*

16. Patients who have TB are treated with a combination of four different first-line anti-tuberculosis drugs. True or false?

17. *Mycobacterium bovis* infects cattle and cannot cause tuberculosis in humans. True or false?

20. A patient with a _____ lung infection is probably infected with HIV.
    A. *Histoplasma*
    B. *Blastomyces*
    C. *Cryptococcus*
    D. *Coccidioides*

21. Coccidioidomycosis is endemic in the _____.
    A. northern United States
    B. southern United States
    C. eastern United States
    D. western United States

Key Terms

acid-fast bacillus (AFB) (p. 670)
active tuberculosis disease (p. 669)
acute otitis media (AOM) (p. 655)
acute rheumatic fever (ARF) (p. 661)
antigenic drift (p. 653)
antigenic shift (p. 653)
bacterial sinusitis (p. 659)
blastomycosis (p. 672)
bronchiolitis (p. 651)
bronchitis (p. 650)
coccidioidomycosis (p. 672)
cold agglutinin (p. 667)
common cold (p. 648)
croup (p. 654)
cryptococcosis (p. 672)

diphtheria (p. 663)
DTaP series (p. 663)
epiglottitis (p. 658)
erythema marginatum (p. 661)
extreme drug-resistant tuberculosis
  (XDR-TB) (p. 670)
health care–associated infection
  (HAI) (p. 668)
histoplasmosis (p. 672)
influenza (p. 651)
interferon gamma release assay
  (IGRA) (p. 669)
laryngitis (p. 659)
laryngotracheobronchitis (LTB)
  (p. 654)

latent tuberculosis infection (LTBI)
  (p. 668)
lower respiratory tract (p. 646)
Mantoux tuberculin skin test (TST)
  (p. 669)
mucociliary escalator (p. 647)
multidrug-resistant (MDR) (p. 668)
peritonsillar abscess (p. 659)
pertussis (p. 663)
pharyngitis (p. 659)
pneumococcal pneumonia (p. 665)
pneumonia (p. 650)
Pontiac fever (p. 667)
post-streptococcal glomerulonephritis
  (p. 662)

pseudomembrane (p. 663)
reactivation tuberculosis (p. 669)
respiratory syncytial virus (RSV)
  (p. 650)
secondary tuberculosis (p. 669)
streptococcal pharyngitis (p. 660)
syncytium (p. 650)
tonsillitis (p. 659)
tubercle (p. 668)
tuberculin skin test (p. 668)
tuberculosis (p. 668)
upper respiratory tract (p. 646)
viral sinusitis (p. 650)
whooping cough (p. 663)

Review Questions

1. What anatomical structures form the respiratory tract?

2. Describe the mucociliary escalator and its function.

3. What are the differences between primary and secondary bacterial infections of the respiratory tract?

4. List the symptoms and etiologic agents of the common cold.

5. What is the relationship of RSV to bronchiolitis in infants?

6. How are appropriate strains of influenza selected each year for production of flu vaccine?

7. Why does croup cause a "barking cough"?

8. Name the top three causes of acute otitis media.

9. What distinguishes bacterial sinusitis from viral sinusitis?

10. What respiratory infections are caused by *Streptococcus pyogenes*?

11. Describe the sequelae related to infections with some *S. pyogenes* serotypes.

12. How are streptococcal species (other than *S. pneumoniae*) divided into different groups?

13. What types of vaccines are available against *S. pneumoniae*? Who is vaccinated with which vaccine?

14. What is atypical pneumonia, and what organisms cause it?

15. What are cold agglutinins, and what is their relationship to walking pneumonia?

16. Who is susceptible to pneumonia resulting from *Pseudomonas aeruginosa*, and how does the organism colonize some patients' lungs?

17. Describe the differences between latent, active, primary, secondary, multidrug-resistant, and extremely drug-resistant TB.

18. Describe the Mantoux skin test and IGRA tests.

19. What is an acid-fast bacillus?

20. What organism causes inhalation anthrax?

Clinical Correlation Questions

1. A 65-year-old female is in your office complaining of a chronic cough. She says she started coughing a few months ago. The cough was dry and did not bother her much at first, but now she is coughing a lot and brings up blood-tinged sputum. She takes only one medication, the new arthritis pills you prescribed a few months back. She denies any shortness of breath or any travel in the past year and says she has been mostly home taking care of her grandchildren. She also denies any other complaints. She is overweight but has no history of any other medical problems. Her temperature, heart rate, and respiratory rate are within normal limits, and there are no lesions on her skin. What organism do you suspect is causing the problem, and why?

2. The chest X-ray of a patient suffering from cough, high fever, and shortness of breath shows infiltrates in only the middle lobe of her right lung. Complete blood count with differential shows markedly elevated white blood cells and a decided left shift. What is the most likely diagnosis: bacterial, viral, fungal, or parasitic pneumonia?

3. Charles is an 8-year-old boy who is brought to the clinic complaining of a persistent dry cough for the past 2 months. His mother reports that Charles had what appeared to be a cold with a sore throat and a low-grade fever 2 months ago. Charles says he recovered from his cold and sore throat after about a week or so but has not stopped coughing. Charles has one of his paroxysmal cough episodes while in the office. You witness him coughing hard and hear a loud sound when he finally is able to stop coughing and inhales. Charles has received no childhood immunization—his parents refused. Charles's temperature, respiration, and heart rate are within normal limits; his chest X-ray shows clear, healthy lungs; and he does not appear to be in any distress. What organism is most likely the cause of Charles's symptoms?

[Too Tired for Soccer]

SCENARIO Mira, a 16-year-old soccer player, complained of a sore throat and woke up in a cold sweat two days in a row. She felt very tired but wanted to go to soccer practice anyway. Her mother was reluctant to give permission unless she went to her pediatrician first.

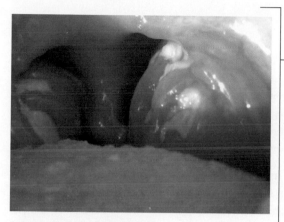

Infectious Mononucleosis
Enlarged tonsils with overlying exudates are one of the signs of infectious mononucleosis, caused by Epstein-Barr virus. In young children the tonsils can become large enough to touch each other and obstruct air flow.

LABS Mira had a low fever when they got to the clinic. The nurse practitioner (NP) examined her and found that Mira had a mild fever, pharyngitis, swollen tonsils, and some swollen neck lymph nodes. She swabbed Mira's throat to check for infectious mononucleosis (mono) and strep throat.

LAB RESULTS The mono test was positive. It was at this point that Mira recalled her boyfriend complaining of a sore throat a couple of weeks back.

TREATMENT Mira was instructed to rest, take ibuprofen for pain, and call the office if any of her symptoms got worse. She was told not to take aspirin because it could cause a severe reaction. The NP also insisted that Mira not play soccer, or have any other strenuous physical activity, because her spleen could be enlarged as a result of infection and an enlarged spleen could burst (rupture) if she was hit, or fell, on the left side of her body. The NP further explained that a ruptured spleen causes internal bleeding, a life-threatening condition.

FOLLOW-UP By week 4 Mira called the clinic to see whether she could start playing soccer again. She was asked to go to the clinic for a follow-up first. The NP examined Mira and told her she could go back to play soccer since her spleen was not enlarged. Mira was happy to go back to her soccer practice that afternoon and resume her normal life.

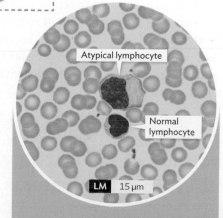

Atypical Lymphocyte
EBV infection of B cells produces the large atypical lymphocytes found in blood smears of patients with infectious mononucleosis.

CHAPTER OBJECTIVES

After reading this chapter, you will be able to:

- Correlate the anatomy and physiology of the circulatory system with the infectious processes affecting them.

- Describe symptoms that can help differentiate circulatory system infections caused by viruses from those caused by bacteria and parasites.

- Relate prevention strategies based on transmission routes for microbes that cause systemic infections.

- Choose appropriate treatment and prevention modalities for microbes that cause systemic infections.

Mira's infectious mononucleosis (IM or mono for short) is a good example of a **systemic infection**, an infection that starts in one part of the body but eventually affects many other sites. Microbes that cause systemic infections travel from an initial infection site throughout the body, using blood or lymph as a highway. The signs and symptoms of systemic disease can arise from dissemination of an infection or from the generalized response of the immune system to a localized infection. The virus that causes mono, Epstein-Barr virus, originally infected the epithelium of Mira's oropharynx, causing a sore throat. Then the virus spread through her body and caused her systemic problems: enlarged lymph nodes, mild fever, fatigue, and night sweats.

Many systemic infections are bacterial diseases such as plague (*Yersinia pestis*), tularemia (*Francisella tularensis*), and septicemia (caused by many species of bacteria). But viruses can cause systemic infections affecting many organ systems as well. Systemic viral diseases include mononucleosis, as in the case just described, but also more frightening diseases, such as the hemorrhagic fevers (Ebola, Marburg), and even some childhood diseases discussed in other chapters (chickenpox, measles). Recall that in this book, diseases are introduced in the context of the organ system that displays the disease's most prominent sign or symptom (in the context of the skin for chickenpox and measles, for example). This chapter focuses on diseases caused by organisms that primarily infect cells of the circulation or that ride the system to infect multiple organs simultaneously.

21.1
Anatomy of the Cardiovascular and Lymphatic Systems

SECTION OBJECTIVES

- List the structural components of the cardiovascular and lymphatic systems and their relationship to one another.

What is the most important role of the cardiovascular system? The cardiovascular system provides nutrients and oxygen to all organs of the body, but its most important role is to deliver oxygen to the brain. A quote from Rainer Maria Rilke reads, "All the soarings of my mind begin in my blood." In fact, no one's brain can function, let alone soar, without an ample supply of oxygenated blood. Without a functioning brain, all other organs will fail to function properly, or at all. Fortunately, our body is equipped with a highly efficient circulatory system of blood dispersal, retrieval, and reoxygenation. Fluids and white cells that leave the circulation to nourish tissues or fight infections are collected by another network called the **lymphatic system**. To understand how pathogens disseminate via the circulatory highways and byways, we must first review the basic anatomy of these systems.

Anatomy of the Heart

The organ that delivers oxygenated blood is the heart, a pump with four chambers attached to pipes (blood vessels) that form two closed circuits: one between the heart and the lungs, called **pulmonary circulation**, and one between the heart and all of the body (including the lungs), called **systemic circulation** (Figure 21.1). Pulmonary circulation is responsible for taking oxygen-depleted blood coming into the heart from all locations of the body and moving that blood to the lungs. In the lungs, blood is replenished with oxygen and then returned to the heart. Systemic circulation then delivers oxygen-repleted blood to the entire body and collects oxygen-depleted blood and carries it back to the heart.

The four chambers of the human heart are the right and left atria, which receive the blood coming into the heart, and the right and left ventricles, which pump the blood out. Each atrium is separated from its ventricle by a one-way valve that allows the blood to flow only

Figure 21.1   The Cardiovascular System

Systemic viral infections
Infectious mononucleosis
Burkitt's lymphoma
Cytomegalovirus infections
Dengue fever

Viral infections of the heart
Cardiomyopathy
Pericarditis
Myocarditis

Systemic bacterial infections
Systemic inflammatory
 response syndrome (SIRS)
Bacteremia
Sepsis
Septicemia
Septic shock
Plague
Lyme disease
Typhoid fever
Rocky Mountain spotted fever
Tularemia
Cat scratch fever

**Bacterial infections of
 the heart**
Subacute bacterial
 endocarditis
Infectious endocarditis
Acute bacterial endocarditis
Prosthetic valve endocarditis

Systemic parasitic infections
Malaria
Babesiosis
Chagas' disease
Toxoplasmosis
Leishmaniasis
Filariasis

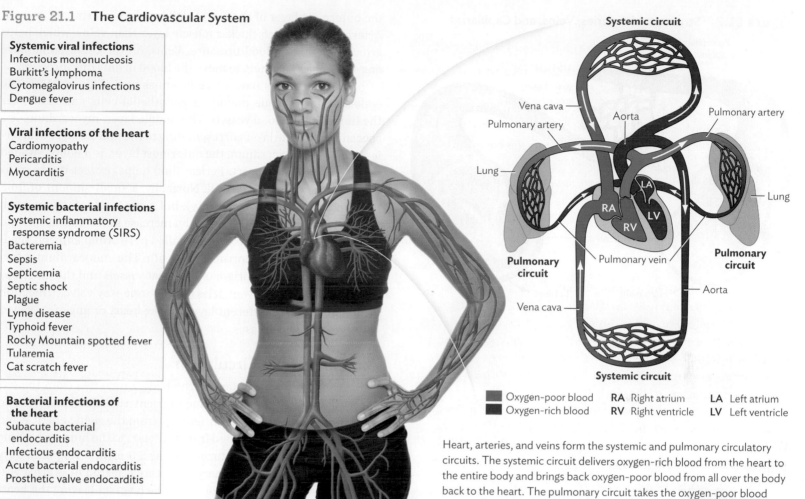

Heart, arteries, and veins form the systemic and pulmonary circulatory circuits. The systemic circuit delivers oxygen-rich blood from the heart to the entire body and brings back oxygen-poor blood from all over the body back to the heart. The pulmonary circuit takes the oxygen-poor blood from the heart to the lungs and returns oxygen-rich blood from the lungs to the heart.

from atrium to ventricle. Large vessels bring the blood into or out of the heart. Vessels that move the blood away from the heart are called arteries, and those that bring the blood back to the heart are called veins. Blood flowing in the systemic arteries is oxygenated, but the blood in the systemic veins is oxygen depleted and has a high concentration of CO_2.

In the pulmonary circulation, the relationship between veins, arteries, and oxygenated blood is the exact opposite of the relationship seen in the systemic circulation. The pulmonary artery takes oxygen-depleted blood away from the heart to the lungs, and the pulmonary vein brings oxygenated blood back from the lungs to the heart. The heart's left ventricle then pumps the blood into the systemic circulation.

Although arteries and veins have different functions, they have the same basic structure (Figure 21.2). Each vessel is a hollow tube. The lumen, the hollow inside part, is lined with a specialized epithelium, called endothelium, and separated from the dense connective tissue layer that forms

Figure 21.2   Structure of Arteries, Veins, and Capillaries

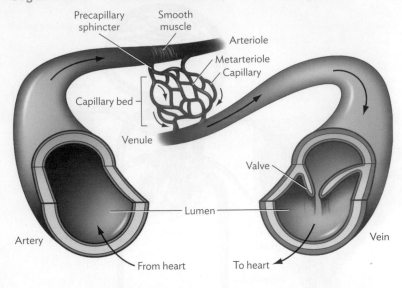

Figure 21.3   The Heart Wall

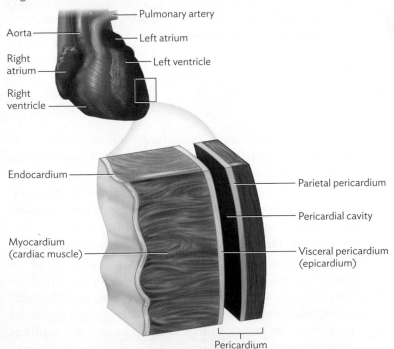

Three tissue layers form the wall of the heart chambers: pericardium, myocardium, and endocardium. Pericardium is the outermost layer, with a small space between the part that is next to the heart (visceral) and the outer part (parietal) called pericardial cavity. Infections affecting the heart can cause inflammation and accumulation of fluids inside the pericardial cavity, which puts pressure on the heart muscle and impedes proper contraction.

the outermost layer of the vessel wall by a layer of smooth muscle. Arteries have a much thicker muscle layer than veins, which helps arteries maintain blood pressure. Veins, on the other hand, have one-way valves that help to move the blood in only one direction.

Heart chambers have three layers as well (Figure 21.3). The endocardium is made mainly of endothelial cells and resembles the lining of the blood vessels. The middle layer, myocardium, is a specialized muscle layer corresponding to the muscle layer of arteries and veins. Pericardium, the outermost layer, is a double-walled sac that encases the heart. Pericardium helps protect the heart as it beats against the chest wall. Normally, a small amount of fluid (about 25 ml) fills the sac (cavity), but some viral infections, such as cytomegalovirus (CMV), cause an increase in the volume of fluid in the pericardial sac, a condition called pericardial effusion, leading to chest pain and shortness of breath. The endocardium is further specialized at the origin of the great vessels and the junctions between the atria and ventricles to make one-way valves. Infection or inflammation of different layers of the heart or any of its valves results in serious diseases.

Anatomy of the Circulation

The aorta, the largest artery of the body, originates from the left ventricle and delivers oxygen- and nutrient-rich blood to the entire body. The pulmonary artery originates from the right ventricle and takes the deoxygenated blood from the heart to the lungs to become replete with oxygen and returned to the left atrium via pulmonary veins. The aorta and pulmonary artery are referred to as great vessels. A one-way valve at the origin of each vessel leaving the heart prevents blood from coming back into the ventricles after it has been pumped out. Deoxygenated blood collected from all body tissues returns to the right atrium by two large veins, the superior and inferior vena cava.

As vessels branch out, they become successively smaller until they become tissue capillaries consisting of just one layer of endothelial cells. The pressure in the arterial part of capillaries is still high enough to move plasma out of the vessel and into the tissue so that the tissue can get food and oxygen. The oxygen- and nutrient-depleted plasma and some waste products generated by tissues then flow back into the venous capillaries. The arterial and venous sides of capillaries are attached to each other and make up what is called the capillary bed.

Note Plasma and serum are both blood derivatives. Plasma is the fluid that remains after cells and platelets have been removed. Serum is the fluid that remains after blood coagulates. So, serum lacks cells, platelets, and factors needed for coagulation.

Not all the fluid that flows out of the arterial side of capillaries is picked up by the venous side. The remaining fluid, along with large waste molecules and white blood cells, flows into the adjacent small lymphatic vessels that originate next to the capillary beds (Figure 21.4). Small lymphatic vessels join each other to form larger ones. The lymph travels through lymph nodes and becomes

Figure 21.4   The Lymphatic System

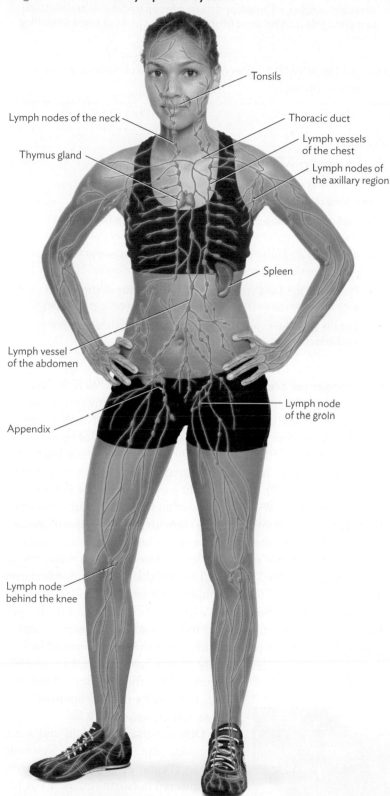

Tonsils

Lymph nodes of the neck

Thymus gland

Thoracic duct

Lymph vessels of the chest

Lymph nodes of the axillary region

Spleen

Lymph vessel of the abdomen

Lymph node of the groin

Appendix

Lymph node behind the knee

A. The lymphatic system is a network of organs, lymph nodes, lymph ducts, and lymph vessels that collect and transport lymph from tissues to the bloodstream.

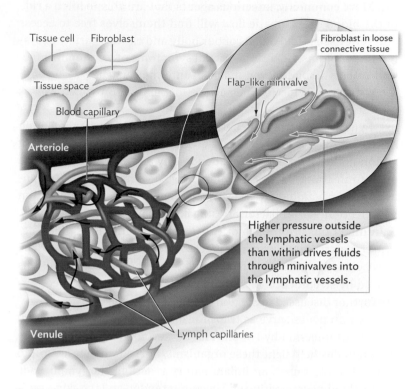

Tissue cell   Fibroblast

Tissue space

Blood capillary

Arteriole

Fibroblast in loose connective tissue

Flap-like minivalve

Higher pressure outside the lymphatic vessels than within drives fluids through minivalves into the lymphatic vessels.

Venule   Lymph capillaries

B. The lymphatic capillary beds are adjacent to the capillary beds. Pressure in the interstitial space drives fluid through minivalves into the lymphatic vessels.

richer in **lymphocytes**. Eventually, the lymph reaches the larger lymphatic vessels that pour their contents back into the venous circulation just before it reaches the right atrium.

Link   As discussed in Section 15.1, **lymphocytes**, the main participants in adaptive immunity, spend most of their lives within lymphoid organs and enter the bloodstream only periodically, migrating from one place to another and surveying tissues for possible infection or foreign antigens.

The constant flow and high pressure of the blood makes it exceedingly difficult for microbes to attach themselves to the structures of the heart, but cardiac infections do occur. Viral and bacterial infections of the cardiac muscle often inflict irreparable damage. The infected muscle can lose its ability to contract properly (a condition called **cardiomyopathy**) and even fail.

General Aspects of Systemic Infections

How common are direct infections of cardiovascular and lymphatic tissues? Cardiovascular tissue infections are uncommon; when they occur, they usually develop as an existing localized infection spreads. Microorganisms or their toxins can affect any of the three layers of the heart, cause inflammation, and inflict damage. Inflammation of endocardium, **endocarditis**, myocardium, **myocarditis**, or pericardium, **pericarditis**, are often seen as complications of infectious diseases. On rare occasions, all three tissue layers of the heart are affected at the same time, a condition called **pancarditis**.

More commonly, infectious agents that are able to hitch a ride in the blood or lymphatic flow will find themselves free to access different tissues and organs of our body and cause a variety of systemic symptoms. General signs and symptoms of a systemic infection usually include fever, malaise, and muscle aches, along with more organ-specific manifestations, depending on which organs are affected.

Lymph flows from the periphery to (and through) the lymph nodes before it is poured into the venous circulation. Swelling of lymph nodes is termed **lymphadenopathy** and can have infectious or noninfectious causes. Inflammation of lymph nodes (**lymphadenitis**) or lymphatic vessels (**lymphangitis**) usually results from an infection—viral, bacterial, fungal, or parasitic (Figure 21.5). Skin infections are the major cause of lymphangitis as the infection travels from the primary infected site to the lymph nodes. The most common causes of lymphangitis are acute skin infections caused by streptococci or staphylococci (bacterial skin infections are further discussed in Section 19.3).

Lymph nodes serve as traps for organisms carried away from the site of infection by lymph. The numerous lymphocytes in the lymph nodes help fight these organisms. Spread of an infection to a lymph node causes an inflammatory reaction: the lymph node swells (lymphadenopathy) and becomes tender, and the skin over it may be red (lymphadenitis). Cellulitis, abscess formation, and blood infections are serious complications of infectious lymphadenitis.

SECTION SUMMARY

- **The circulatory system** consists of the heart and blood vessels and is divided into pulmonary and systemic circuits.
- **Endocarditis, myocarditis, and pericarditis** are terms describing inflammation of various layers of the heart.
- **Capillaries** are where nutrients and oxygen are delivered from blood to the tissues and waste products generated by tissues are returned to the blood.
- **Lymphatic vessels collect waste products, tissue fluids, and white blood cells,** channeling them through lymph nodes and ultimately back into the venous system prior to reaching the heart.

Figure 21.5   Infections of the Lymphatic System

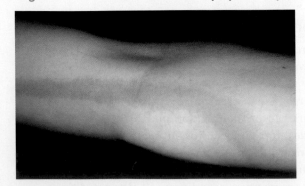

A. Lymphangitis.

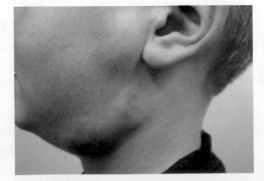

B. Lymphadenitis.

- **Lymphadenopathy** describes swollen lymph nodes. **Lymphadenitis** describes swollen, inflamed, and painful lymph nodes that are infected. **Lymphangitis** is an inflamed lymph vessel seen as a red streak extending from a localized skin infection.

Thought Question 21.1   Why do the red streaks of lymphangitis always move toward the heart?

21.2
Systemic Viral Infections

SECTION OBJECTIVES

- Explain the pathogenesis of each systemic viral infection discussed.
- Relate the diagnoses of systemic viral infections to the pathogenesis of the specific virus causing the disease.
- Relate the signs and symptoms of each systemic viral infection with the effect of the respective virus on different organ systems.
- List the treatment and prevention options for different systemic viral infections.

How dangerous are systemic viral infections? Well, it depends on the health of your immune system. Systemic viral infections are uncommon in the general healthy population in North America, but when they occur, they are caused mostly by one of three rather famous members of the herpes family of viruses: Epstein-Barr virus (EBV), the cause of infectious mononucleosis (mono); herpes simplex virus (HSV), the cause of cold sores and genital herpes; and cytomegalovirus (CMV), the cause of an infection similar to mono. All three viruses establish latency (remain present without producing any symptoms) after infection regardless of a person's immune status, and all three cause severe disease and complications in immunocompromised patients. Immunocompetent individuals, on the other hand, tend to be asymptomatic or have a milder presentation of these viral diseases. We limit our discussion of the herpes family of viruses in this chapter to EBV and CMV. Herpes simplex viruses 1 and 2 cause primary infections that are discussed in Sections 19.2 and 23.5, respectively. Systemic dissemination of HSV occurs in immunocompromised individuals and results in viral meningitis, as discussed in Section 24.3.

Other viruses, some old and some emerging, can also cause systemic disease (Table 21.1). Most are rarely encountered in North America, but the ease of travel and the wide-reaching routes of international commerce are open doors that may allow any of these illnesses to reach our shores. Knowing about them is the single most effective preventive tool.

Table 21.1

Systemic Viral Diseases

Disease	Symptoms	Complications	Virus	Transmission	Diagnosis	Treatment	Prevention
Infectious mononucleosis	Pharyngitis, fever, lymphadenopathy, fatigue, splenomegaly	Bacterial throat infections, spleen rupture, anemia, nervous system problems, death	Epstein-Barr virus	Saliva	Heterophile antibodies; serology	Treatment of symptoms; activity restriction	No vaccine available
CMV mononucleosis-like syndrome	Similar to those of infectious mononucleosis	Transplacental transmission causing cytomegalic inclusion disease (CID); retinal disease and a variety other complications possible in newborns and immunocompromised	Cytomegalovirus	Saliva	"Owl's eye"; inclusion bodies; serology; viral culture; DNA studies		No vaccine available
Cytomegalic inclusion disease (CID)	Premature birth; small head size; seizures; lung, liver, or spleen problems at birth	Deafness, mental retardation, seizures, jaundice, purpura	Cytomegalovirus	Transmitted vertically from mother to fetus			
Dengue fever, dengue hemorrhagic fever (DHF)	Fever, nausea, and vomiting; rash; aches and pains Additional symptoms in DHF: severe hemorrhage, respiratory distress, organ failure, and shock	Progression to severe dengue with intense severe abdominal pain, liver inflammation, persistent vomiting, hemorrhage, abnormally low blood pressure, death	Dengue virus	Mosquito; vertical transmission possible but rare; exposure to infected blood or transplantation	Virus isolation; PCR; serology; positive tourniquet test[a]	Hospitalization; supportive care, including hydration, pain management, and blood transfusion in DHF cases	Vaccine candidates currently in development
Yellow fever, jungle yellow fever	High fever, chills, joint and muscle pain, nausea, vomiting	Progression to toxic form of the disease; vomiting blood; jaundice; petechiae or purpura; bleeding from mouth, nose, eyes; multiorgan failure and death	Yellow fever virus	*Aedes aegypti* mosquito	Virus isolation; serology	Hospitalization to control bleeding and prevent shock and provide hydration and supportive care	Live, attenuated vaccine for travelers, residents, and laboratory staff
Lassa fever	Fever, retrosternal and generalized pain, vomiting, diarrhea, encephalitis, neurological symptoms, mucosal bleeding	Various degrees of hearing loss that becomes permanent in ⅓ of cases; spontaneous abortion	Old World arenaviruses	Contact with *Mastomys* rodent urine or droppings or with the blood or body fluids of an infected person	Virus isolation, detection of anti-Lassa virus antibodies, Lassa antigen	Hospitalization, supportive care, antiviral therapy with ribavirin	No vaccine available
Chikungunya fever (CHIK fever), also known as knuckle fever	High fever, rigors, severe joint pain, rash, nausea, vomiting	Persistent joint pain for several months or years in some patients; occasional neurological, heart, or GI problems	Chikungunya virus	*Aedes aegypti* or Asian tiger (*Aedes albopictus*) mosquito	Virus isolation; serology	Treatment of symptoms	Experimental vaccine in clinical trials

(continued on next page)

Table 21.1
Systemic Viral Diseases (continued)

Disease	Symptoms	Complications	Virus	Transmission	Diagnosis	Treatment	Prevention
Crimean Congo hemorrhagic fever (CCHF)	High fever; back, joint, and stomach pain; vomiting, sore throat, petechiae on the palate, sore eyes; bleeding and jaundice in severe cases	Persistent joint pain for months to years	Nairovirus, also known as Asian Ebola virus and Congo virus	Ixodid (hard) ticks; contact with blood, body fluids, and excretion of infected animals or humans	ELISA, RT-PCR; isolation of virus in culture	Supportive treatment; ribavirin is considered reasonable therapy by CDC; no direct evidence available	Mouse brain–derived vaccine developed and used in Eastern Europe only; no worldwide vaccine available
Rift Valley fever	*Mild form:* flu-like symptoms; some mild cases with retinal lesions *Severe form:* meningo-encephalitis, heart failure, jaundice, purpura, hemorrhage	Inflammation of the retina, leading to visual loss in 1%–10% of cases	Rift Valley fever virus	Direct or indirect contact with blood or organs of infected animals; *Aedes* mosquito; consuming unpasteurized or uncooked animal product from infected animal	Virus isolation; enzyme-linked immunoassay (ELISA) for presence of IgM antibodies specific to virus; RT-PCR	Supportive therapy	Experimental vaccine has been used for exposed laboratory workers and veterinarians working in sub-Saharan Africa
Marburg hemorrhagic fever	Acute high fever with severe headache, severe malaise, severe hemorrhage from several sites, shock	Prolonged recovery; inflammation of the testis, spinal cord, eye, parotid gland; prolonged hepatitis, death	Marburg virus	Contact with unknown reservoir in nature or close contact with people with Marburg virus infection or with their body fluids	Antigen detection by ELISA; RT-PCR; serology in patients who are recovering; skin biopsy for postmortem diagnosis	Supportive care	No vaccine available
Ebola hemorrhagic fever	Acute high fever with severe headache, severe malaise, nausea, vomiting; lymphadenopathy; altered mental status; severe hemorrhage from several sites; shock	Death; prolonged recovery for the rare survivors, with possible sensory problems, headaches, hepatitis, and testicular inflammation	Ebola virus	Contact with unknown reservoir in nature or close contact with people with Ebola infection or with their body fluids	Clinical presentation; ELISA; antigen detection; serology; isolation of virus in culture	Supportive care, passive immunization (see Section 17.6); an experimental drug made of three neutralizing antibodies, ZMapp	Vaccine in clinical trials

[a]Blood pressure cuff is left inflated at halfway between systolic and diastolic pressure for 5 minutes. Appearance of 20 or more petechiae on a 2.5-cm² patch of skin on forearm is interpreted as a positive test.

Infectious Mononucleosis: The "Kissing Disease"

What is the "kissing disease"? The virus causing **infectious mononucleosis**, the Epstein-Barr virus (EBV), is shed in the saliva of the person harboring it. Exchange of saliva, as often happens with kissing, passes the virus from one person to another, earning this infection the title "kissing disease." Once Mira, the patient in the opening case history, realized she had mononucleosis, it did not take her long to recall a certain kiss with a certain boy who was not feeling well. The disease was first described as an infectious disease by Nil Filatov in 1887 (and independently again by Emil Pfeiffer in 1899). Pfeiffer coined the name "glandular fever" because of its hallmark symptoms of fever, sore throat, and enlarged lymph nodes (Figure 21.6A). In 1920, Thomas Sprunt and Frank Evans described the mononuclear leukocytosis (really a lymphocytosis) that accompanies the glandular fever, and the disease became known as infectious mononucleosis in United States. EBV was identified as the etiologic agent of infectious mononucleosis in 1967 when a technician working on a cell line harboring EBV developed infectious mononucleosis and subsequently developed antibodies to the virus.

Figure 21.6   Infectious Mononucleosis and Epstein-Barr Virus

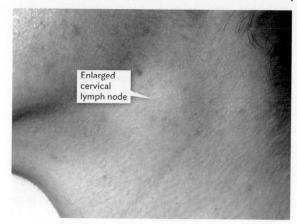

Enlarged cervical lymph node

A. Symptoms of mononucleosis include swollen cervical lymph nodes (often bilateral), sore throat, fever, and fatigue.

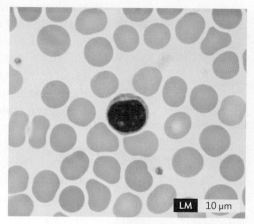

LM   10 μm

B. A normal lymphocyte in peripheral blood.

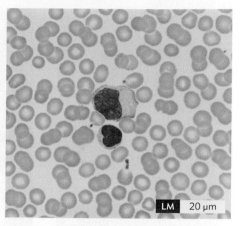

LM   20 μm

C. Large, irregular, atypical lymphocytes (resembling monocytes) seen in the peripheral blood of this patient with infectious mononucleosis.

Epstein-Barr virus is a member of the herpes family of viruses, also known as human herpesvirus 4 (HHV-4). More than 95% of adults worldwide are infected with EBV. In fact, nearly all children in developing countries are exposed to EBV by age 4. The majority of those infected (more than 70%) will have no symptoms thanks to an effective cellular immune response, but all who do become infected with EBV will shed the virus in their saliva for many weeks until the infection becomes latent. The latent infection will last the rest of the victims' lives, making them a reservoir for the virus. Humans are the only known reservoir.

EBV initially infects and replicates in the oropharyngeal epithelium, tonsils, and salivary glands. B cells become infected when they come in contact with the epithelium of the oropharynx or crypts of the tonsils. The virus is not found in the blood, but the infected B cells spread the infection throughout the body. Infected B cells in blood smears have an atypical appearance that resembles that of monocytes. This mistake led to the disease name, mononucleosis (Figure 21.6B and C). Normally, the immune system controls EBV infection using mainly cytotoxic T cells. The proliferation of B and T cells results in the enlargement of lymph nodes (lymphadenitis) and of organs such as the liver and spleen that remove old or abnormal cells from the peripheral circulation.

What kind of laboratory test might distinguish mononucleosis from other lymphadenopathies? A rapid blood test called the mononucleosis "spot test" can usually determine if someone has mono. EBV non-specifically stimulates all infected B cells to proliferate and differentiate into plasma cells and long-lived memory B cells. Latently infected memory B cells are the reason EBV produces a lifelong infection. The plasma cells produced from EBV-infected B cells will secrete whatever antibody their B-cell progenitors were programmed to make. Since a variety of B cells are infected, a variety of different antibodies will be produced. These include EBV-specific as well as a variety of nonspecific antibodies. Among the nonspecific antibodies are those that react against horse and sheep red blood cells, collectively known as **heterophile antibodies**. The presence of heterophile antibodies is the basis of the monospot test used to screen for mono, as was the case for Mira. The monospot test is a latex agglutination reaction in which patient serum is added to latex beads coated with horse red blood cell antigens. Unlike many serological tests that remain positive long after an infection has resolved, a positive monospot test will revert to negative as the symptoms wane. There are also EBV-specific tests for EBV-specific IgM and IgG antibodies.

Link As is discussed in Section 16.4, having encountered an antigen, the antigen-activated B cell will divide to make memory B cells and plasma cells. Memory B cells are very long-lived, serving to provide immunological defense against the antigen if encountered in the future.

Link As discussed in Section 17.5, agglutination is an in vitro process by which antibodies in a patient's serum cross-link antigens that are naturally present on whole cells (red blood cells or bacteria) or that were artificially coated onto latex beads (the monospot test). Antibody cross-linking of the cells or beads causes them to clump (agglutinate) into visible precipitates. A precipitate indicates that the test is positive (the antibody is present).

In the United States, the typical EBV infection occurs in young adults and produces a sore throat, fever, and generalized lymphadenopathy that often includes the lymph nodes behind the ear, on the neck, under the arm, or in the groin. Disease symptoms appear about 5 weeks after infection. Most patients with infectious mononucleosis also suffer from extreme fatigue and difficulty concentrating. Examination of the throat usually shows enlarged tonsils with exudates (see the chapter-opening case history), which can lead to a mistaken diagnosis of strep throat.

Figure 21.7   Burkitt's Lymphoma

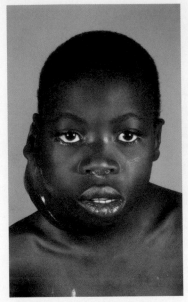

A. Front view of the face of a child showing a large tumor in the jaw due to Burkitt's lymphoma.

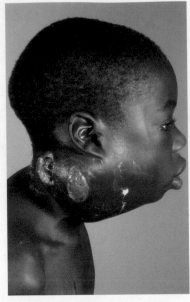

B. The surface of the skin has begun to ulcerate over the growth. It is thought that malaria in childhood alters the body's response to the virus, causing cells to become cancerous.

Figure 21.8   Making the Connection between a Cancer and a Virus

A. Denis Parsons Burkitt first described the lymphoma that bears his name.

B. Michael Epstein (shown), along with Yvonne Barr and Bert Achong, discovered the viral cause of Burkitt's lymphoma. The link to mononucleosis was sheer chance, when a technician handling a Burkitt's lymphoma cell line developed mono.

Most cases of infectious mononucleosis are mild and resolve by themselves after 2–4 weeks. The fatigue and difficulty concentrating associated with most cases of the disease, however, can linger months after recovery from the acute illness.

Dissemination of EBV can cause a variety of other complications, including enlargement of the liver (**hepatomegaly**) and jaundice (especially in patients over 40 years of age) or enlargement of the spleen (**splenomegaly**), which is seen in 50%–60% of patients. The possibility of splenic rupture explains why the nurse practitioner warned Mira about playing sports. In addition to affecting the liver and spleen, EBV can cause neurological problems, such as meningitis, encephalitis, or Guillain-Barré syndrome (see Chapters 22 and 24), or even kidney failure. About 5% of patients may develop a rash, usually on their arm and body.

Symptoms of mono are very similar to bacterial pharyngitis (sore throat). A patient with pharyngitis may be empirically treated with ampicillin. Patients with mono who are mistakenly treated with ampicillin almost always develop a rash, but this does not mean that they will have a reaction to penicillin in the future. Once the antibiotic is stopped, the rash subsides. There are some indications that protein impurities in the preparation of ampicillin may contribute to the ampicillin-induced nonallergic rash of mono, but the exact mechanism is still unknown. Mononucleosis is typically a self-limiting disease requiring only supportive care, such as activity restriction and pain management. Death due to infectious mononucleosis is rare but can happen as a result of splenic rupture or bacterial superinfections.

Cancers Associated with Epstein-Barr Virus

Epstein-Barr virus is associated with a number of cancers of lymphoid origin, including Hodgkin's disease, nasopharyngeal carcinoma, and perhaps tonsillar carcinoma, but the strongest association is with endemic **Burkitt's lymphoma** (Figure 21.7). A lymphoma is a cancer of lymphocytes, and Burkitt's lymphoma is a cancer of B lymphocytes in which tumors form primarily in the upper and lower jaws and orbit of the eye. EBV, as a result of its integration into human DNA, generates DNA rearrangements that can move a gene that activates cell proliferation next to a highly expressed antibody promoter (see Section 17.1).

Burkitt's lymphoma was first described by Denis Parsons Burkitt (Figure 21.8A), a surgeon who in 1961 presented the results of his research in Uganda in a talk titled "The Commonest Children's Cancer in Tropical Africa—A Hitherto Unrecognized Syndrome." The "endemic pediatric" cancer he described eventually became known as Burkitt's lymphoma. Among attendees of the lecture was a pathologist and electron microscopist, Michael Anthony Epstein. In 1964, Michael Epstein (Figure 21.8B), Bert Achong, and Yvonne Barr reported finding a new strain of virus in cultured lymphoblasts from Burkitt's lymphoma. This new virus is what we now know as the Epstein-Barr virus.

There are three different categories of Burkitt's lymphoma based on disease presentation: endemic (found exclusively in Africa), sporadic (found worldwide), and immunodeficiency associated. The EBV virus is found in 95% of the endemic variety, in 10%–20% of the sporadic cases, and in 40%–50% of the immunodeficiency-related cases (HIV-infected individuals). The endemic variety of Burkitt's lymphoma (eBL) is the most common type of childhood cancer in equatorial Africa, where malaria is also prevalent. There are a number of observations and hypotheses regarding the relationship between malaria and Burkitt's lymphoma. We know that malaria increases the risk of endemic Burkitt's lymphoma, but we do not know why. The immunodeficiency-related variety is seen mostly in HIV-positive patients, even those with high CD4$^+$ counts (see Section 23.5), suggesting a different mechanism for development of these tumors than immunosuppression alone.

Burkitt's lymphoma grows rapidly and should be aggressively treated with a variety of anticancer chemotherapies. Treatment for EBV itself is not necessary because once the cancer is detected, the chromosomal rearrangement has already occurred and the virus plays no further role.

Cytomegalovirus Infections

How are cytomegalovirus infections different from other herpes virus infections? **Cytomegalovirus (CMV)**, a double-stranded DNA virus also called human herpesvirus 5, initially replicates in epithelial cells (typically in the salivary glands) but eventually infects circulating lymphocytes and monocytes, the kidneys, and different glands. As with most herpes viruses, CMV can integrate into host chromosomes and remain latent for life. The difference between CMV and many other herpes virus infections lies in the severity of disease. CMV infections are usually asymptomatic or mild, except when they affect fetuses or someone who is immunocompromised.

CMV infections are incredibly common worldwide. In the United States, 50%–80% of adults are infected with the virus by the time they are 40 years old. We know this because the virus leaves "footprints" in the form of antibodies. CMV infection usually happens in childhood. People who are infected with CMV are usually asymptomatic but will shed the virus in different body fluids for a long time (months to years in urine and saliva). The viral shedding period is generally longer for children than it is for adults. The main mode of transmission is contact with contaminated saliva, particularly in day-care centers and among family members, but transmission by sexual contact or from transplanted tissue is also possible. Primary CMV infection is mostly asymptomatic in immunocompetent people, but a minority of healthy people may develop an illness resembling mono with similar nonspecific symptoms.

THE IMMUNE RESPONSE TO CMV INFECTION Cell-mediated immunity is the main defense against primary CMV infection following initial contact, and it stops secondary infections when latent CMV virus is reactivated. Consequently, suppressing cellular immunity can lead to serious primary as well as secondary infections. Reactivation of latent CMV viral DNA can happen anytime in life. Secondary CMV infection can also develop when a person harboring a latent CMV becomes infected with a *new* strain of virus, a process called **reinfection**.

Anything that suppresses the immune system is a major risk factor for CMV reactivation. So patients on immunosuppressive therapy, such as those on certain medications after an organ transplant or with medical conditions such as AIDS, are at higher risk of reactivation and symptomatic infection. Symptomatic CMV disease in immunocompromised individuals can affect almost every organ of the body, resulting in a fever of unknown origin, pneumonia, hepatitis, encephalitis, myelitis, colitis, uveitis (inflammation of the eye's interior), retinitis, and neuropathy.

CONGENITAL CYTOMEGALIC DISEASE Unlike most viral pathogens, CMV can transplacentally pass from infected mother to fetus. The reported vertical transmission rate is nearly 75%, but the great majority of infected newborn infants (80%) are asymptomatic. Symptomatic infected newborns are said to have **congenital cytomegalovirus infection**, or **cytomegalic inclusion disease (CID)**. Generalized infection can occur in the infant, causing complications such as low birth weight, microcephaly (abnormally small head), seizures, a petechial rash similar to the rash of congenital rubella syndrome, and moderate hepatosplenomegaly (enlarged liver and spleen) with jaundice. Though severe cases can be fatal, with supportive treatment most infants with CMV disease will survive. However, from 80% to 90% of severe cases will have complications within the first few years of life. The most common abnormality associated with CID is hearing loss, but it also remains one of the leading causes of mental retardation in United States.

Note CMV is a member of the so-called TORCH microbes that can cross the placenta and infect a fetus to cause severe congenital defects. The **TORCH complex** includes toxoplasma, rubella, cytomegalovirus, and herpes simplex 2. The O can also stand for others, such as coxsackievirus, syphilis, varicella-zoster virus, HIV, parvovirus B19, and *Listeria*.

CID was initially named in 1905 when researchers noticed that cells obtained from infants with the disease looked distinctly different from other cells under the microscope. The infected cells were much larger than normal (sometimes called **giant cells**) and had inclusion bodies that gave them the appearance of an owl's eyes (Figure 21.9). The owl's eye appearance remains an indication of

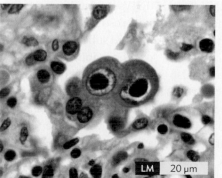

Figure 21.9
Cytomegalic Inclusion Disease

A hematoxylin- and eosin-stained tissue sample from a lung infected with CMV shows the resulting enlargement of an infected cell and two large inclusion bodies that have the appearance of an owl's eyes.

LM   20 μm

Figure 21.10   Myocarditis

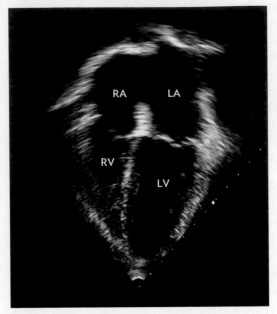

A. An echocardiogram showing normal-sized heart chambers (LV = left ventricle, RV = right ventricle, LA = left atrium, RA = right atrium).

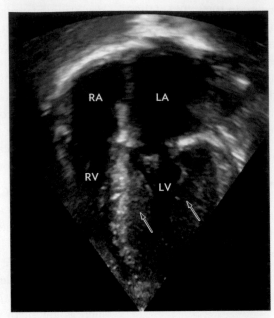

B. An echocardiogram showing a small left ventricle with thickened walls. The inflammatory process caused by a viral infection can result in thickening of the left ventricular wall. Thickening of myocardium reduces the size of the left ventricle, which means that the heart is not able to pump a sufficient volume of blood into the circulation or even to its own muscle tissue. Eventually, the diminished blood supply severely damages the heart muscle.

Reemergence of certain systemic viral diseases is especially troubling. The "chik" (Chikungunya) virus of the Togaviridae family, for example, is reemerging. CHIKV causes millions of cases of debilitating polyarthralgias (pain in several joints) and fever in and around the Indian Ocean and has now spread to Europe. Ebola virus, a lipid-enveloped, twisted, threadlike RNA virus (Figure 2.5A), has a frightening reputation for the hemorrhagic fever it causes. Endemic to Africa, disease outbreaks of Ebola are characterized by acute (rapid) onset of fever, severe muscle pains, and horrible bleeding from multiple orifices (nose, mouth, anus, and vagina; Figure 2.18) and often result in death (50%–60% mortality). The virus is easily transmitted person-to-person by contact with contaminated body fluids. As of February 2015, the most recent outbreak of Ebola in West Africa included 24,000 victims, about half of whom had died. The disease could have easily spread worldwide when a few foreign volunteers who treated the victims unknowingly became infected with the virus, brought it back to their home countries, including the United States, and developed the disease. Fortunately, major containment efforts prevented spread of the disease in these countries.

A large number of systemic disease–producing viruses are transmitted by either mosquitoes or rodents and often cause high fevers and hemorrhage (see Table 21.1). Simple, practical measures, such as mosquito and rodent control, have tremendous potential to improve the lives of millions around the world, save billions of dollars, and help keep us healthy.

CMV infection even today. Congenital CMV infection results in permanent problems in nearly 20% of those affected. Epstein-Barr virus and the cytomegalovirus infections can be distinguished from each other using specific diagnostic approaches. A major distinction between the two is the absence of heterophile antibodies in CMV infections and their presence in EBV infection. CMV infections are treated with a variety of antiviral medications. Table 21.1 summarizes the diagnostic and treatment options for CMV infections.

Other Systemic Viral Diseases

While uncommon in the United States, a number of serious systemic viral diseases are seen in other parts of the world. Some of the viral agents—such as the four closely related viruses that cause dengue fever (DENV-1, DENV-2, DENV-3, and DENV-4) from the Flaviviridae family—impose a tremendous health and resource burden. Dengue fever is endemic throughout the tropics and subtropics. Victims of dengue often have contortions due to intense joint and muscle pain. For this reason, dengue is also known as "breakbone fever." Dengue viruses are called arboviruses because they are transmitted by an arthropod (an insect vector, a mosquito). There were 2.2 million cases of dengue fever in 2010, 1.2 million of which occurred in the Americas (mostly in South America and the Caribbean). Among reported cases of dengue fever, 49,000 involved the most severe variant of the disease: **dengue hemorrhagic fever (DHF).**

Viral Infections of the Heart

As noted earlier, pericarditis, myocarditis, and endocarditis can result when the various layers of the heart are infected by microbes that have entered the circulation. Viral pericarditis is more common than bacterial, which will be discussed later. Viral infection of the pericardium leads to an inflammation and a collection of fluid in the pericardial cavity that increases intrapericardial pressure, resulting in chest pain and low fever. Among the many viral causes of pericarditis and myocarditis are coxsackieviruses (particularly coxsackievirus B; see Section 19.2), echovirus, Epstein-Barr, and HIV (HIV is covered further in Section 23.4). Infections of the myocardium can make heart muscle lose its ability to contract properly (Figure 21.10). Treatment of viral heart infections is usually supportive and may require draining the pericardial sac in the case of pericarditis. Antiviral therapy has not proved effective and is not typically recommended. Most patients, however, recover.

SECTION SUMMARY

- **Infectious mononucleosis** is caused by the Epstein-Barr virus. Symptoms are extreme tiredness and swollen cervical lymph nodes. Transmission is usually via saliva.

- **Epstein-Barr virus also causes Burkitt's lymphoma,** a B-cell cancer. EBV can integrate into the DNA of a host cell and cause a rearrangement that dramatically stimulates proliferation.

- **Cytomegalovirus** is a herpes virus that infects epithelial cells, leukocytes, monocytes and lymphocytes, as well as the kidneys and various glands. CMV disease is typically mild unless the patient is immunocompromised.

- **CMV can be transmitted vertically to a fetus,** causing a generalized infection.

- **Dengue fever virus, Ebola virus, and Chikungunya virus** also cause systemic disease.

- **Cardiomyopathies** can be caused by echovirus, Epstein-Barr, and HIV.

Thought Question 21.2   What is the significance of a patient having a positive EBV IgM test but a negative EBV IgG test?

21.3
Systemic Bacterial Infections

SECTION OBJECTIVES

- Compare the systemic signs and symptoms of different bacterial circulatory infections.

- Generate a list of organ malfunctions according to the pathogenesis of each bacterial pathogen discussed.

- Relate the symptoms of the systemic bacterial infections discussed with the progression of each disease.

- Summarize the treatment and prevention plans for the various systemic bacterial infections discussed.

How do bacteria gain access to our circulatory systems? Bacterial infections of the circulation fall into two broad categories: those that primarily grow in the bloodstream to cause **sepsis** and those that disseminate through the blood to infect a variety of other organs. The latter group often produces lymphadenitis along the way, although not all organisms that produce lymphadenitis produce systemic disease.

Sepsis and Septic Shock

The majority of systemic bacterial infections start innocuously enough when a few microorganisms enter the bloodstream. This happens more often than you think. For example, when a dental hygienist cleans your teeth, bacteria can enter the bloodstream. The result of this breach is called **bacteremia**. Fortunately, your immune

Figure 21.11   Overlap of Infection, Sepsis, and Systemic Inflammatory Response Syndrome (SIRS)

Not all infections result in sepsis, and not all SIRS patients have sepsis.

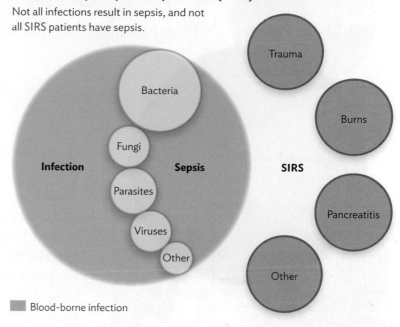

Infection — Bacteria, Fungi, Parasites, Viruses, Other

Sepsis

SIRS — Trauma, Burns, Pancreatitis, Other

Blood-borne infection

system will quickly dispatch these invaders before they can begin to grow and cause disease.

Septicemia is said to occur when a pathogen that has entered the bloodstream overcomes our innate defensive mechanisms, such as complement, and begins to replicate to high numbers. Pathogens that cause septicemia can enter the bloodstream in a number of ways. They can escape from a local infection (an abscess, for example) situated anywhere in the body, gain access via a bleeding wound, or enter by transfusion (see **inSight**) with contaminated blood or blood products. Gram-positive, Gram-negative, aerobic, and anaerobic bacteria can all produce septicemia under the right conditions. Any septicemia will lead to serious medical emergencies.

One such emergency is **systemic inflammatory response syndrome (SIRS)**, a set of symptoms produced by a variety of infectious and noninfectious causes (Figure 21.11). A patient with SIRS has either too high or too low a body temperature, rapid heart and breathing rates, and an abnormal white blood cell count. SIRS caused by an infection is referred to as **sepsis** (commonly known as blood poisoning; see Chapter 15 inSight). When sepsis progresses to involve the kidneys or the heart, these organs begin to fail, and the patient is said to be in severe sepsis. Severe sepsis can progress to septic shock.

"Shock" is a general term used to describe cardiovascular collapse, meaning the heart fails and there is not enough blood pressure to perfuse the tissues (a condition called **hypotension**, or low blood pressure). **Septic shock**, the catastrophic fall of blood pressure due to severe sepsis, is a complex process in which host macrophages or T cells interact with circulating microbial exotoxins called **superantigens** or **pathogen-associated molecular patterns (PAMPs)**, such as LPS in Gram-negative organisms or teichoic acid

inSight

Transfusion-Associated Infections

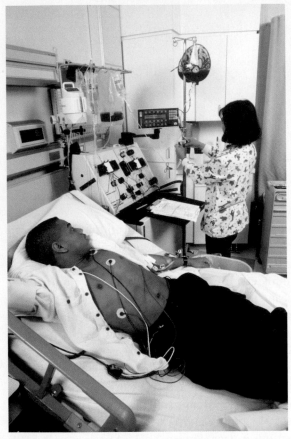

A boy with sickle-cell disease is being transfused with blood.

One of the greatest gifts you can give is blood. Donated blood has saved countless lives worldwide. If you lose excessive blood from a car accident or during surgery, blood is life. But as you can see from this chapter and from others to follow, infections by many microorganisms arise from dissemination in blood. How can we be sure that the blood we receive in an emergency is safe? Consider the case of Travis.

Travis, a 12-year-old African-American male with sickle-cell disease, was brought to the clinic by his older sister for a checkup. He appeared healthy, but lab results of his blood revealed that the boy had a hemoglobin value of 6 (normal is 13–17) and a hematocrit of 18% (normal is 41%–53%). On the basis of these findings, the clinic doctor decided a transfusion was required. Travis had had transfusions before and knew what to expect (**Figure 1**). He laid back and started texting his friend about weekend plans. Thirty minutes after starting the transfusion, Travis started feeling ill. He was hot and progressively light-headed and had chills. The nurse took his temperature and found he had a fever of 39.9°C (103.8°F). When the doctor examined Travis, he noted that Travis was diaphoretic (profusely sweating), his blood pressure was 60/30, and he was pale, agitated, and tachycardic (heart beat at a faster than normal rate). What was happening?

Travis was not having a transfusion reaction, which is what you might expect. Travis was suffering from transfusion-related sepsis by a Gram-negative rod, *Yersinia enterocolitica*. How could this happen? Don't hospitals test donated blood for infectious diseases? Well, they do test for some diseases, but not all. In this case, testing the blood immediately after donation would probably have missed the culprit pathogen. *Y. enterocolitica* is a food-borne pathogen that can penetrate the intestinal wall, cause lymphadenitis, and disseminate. Most healthy people to whom this happens are asymptomatic because the number of bacteria in the blood is typically very low and the immune system quickly kills the organism. However, if this person donates blood before the bacteremia is cleared, the donated blood could contain low numbers of *Y. enterocolitica*. But a low number of bacteria in blood will not trigger sepsis within 30 minutes. So what happened?

Y. enterocolitica is one of a few pathogens that can actually grow at refrigeration temperature (*Listeria* is another). As the bag of donated blood hung in the refrigerator awaiting use, bacteria grew to high levels, and the amount of LPS present in the blood transfused into Travis's arm was enough to quickly trigger septic shock. Travis was treated with antimicrobials and recovered.

Laboratory testing of donated blood is intended to ensure that transfusion recipients receive the safest possible blood products. As of 2011, such testing consists of determining the ABO blood group and Rh blood type of the donated unit and testing for red cell antibodies to avoid transfusion reactions. In addition, screening for the infectious disease agents shown in Table 1 is also performed. Nucleic acid testing (PCR) and enzyme immunoassay (EIA) are the methods used to screen for these agents. Notice that not all pathogens are screened, and some are screened only under select circumstances. For example, blood is routinely screened for the malarial parasite when donated in an endemic country, but not in nonendemic countries such as the United States. This is due, in part, to cost-benefit calculations. Transmission of many diseases by transfusion is rare, and even when present in blood, the low numbers of some agents preclude detection. In a case like Travis's, testing is done after a problem arises. The clinic staff immediately disconnected Travis's transfusion bag and sent it to the laboratory for bacteriological examination, which revealed the presence of high numbers of *Y. enterocolitica*.

In this age of rapid transportation and mobility, a U.S. citizen might travel to a country endemic for a disease such as malaria, be infected, and return to donate blood before symptoms arise. But if you have ever given blood, you know that donors must fill out a fairly lengthy questionnaire packed with queries about a donor's travel, sexual habits, drug use, and so on—all kept extremely confidential. A person who just returned from the Amazon rain forest where malaria is endemic will declare his travel and then learn that a donation cannot be made for at least 6 months. When disease is acquired through blood transfusion, as in Travis's case, the blood bank will check the donor's records, and all units of that donor's blood will be destroyed.

The CDC is one of the federal agencies responsible for ensuring the safety of the U.S. blood supply through investigations and surveillance. The U.S. Food and Drug Administration (FDA) is responsible for regulating how blood donations are collected and how blood is transfused.

Table 1
Infectious Agents Screened in Donated Blood

Routinely Screened		Discussed in Chapter
HIV-1 and HIV-2	Human immunodeficiency viruses	Chapter 23
HTLV-I and HTLV-II	Human T-lymphotropic viruses[a]	
HBV	Hepatitis B virus	Chapter 22
HCV	Hepatitis C virus	Chapter 22
WNV	West Nile Virus	Chapter 24
Treponema pallidum	Agent of syphilis	Chapter 23
Screened in Endemic Countries		
Trypanosoma cruzi	Chagas' disease	Chapter 21
Plasmodium	Malaria	Chapter 21
Screened in Blood Intended for Immunosuppressed, Neonates, and Pregnant Women[b]		
CMV	Cytomegalovirus	Chapter 21

[a]HTLV are retroviruses known to cause T-cell leukemia and lymphoma.
[b]CMV screening of blood is not required for immunocompetent donors, but donated blood intended for immunocompromised recipients should be screened prior to release for clinical use.

Figure 21.12   Petechial and Purpuric Rashes

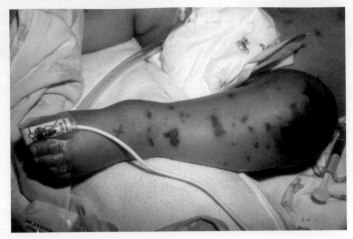

A. A petechial rash on a patient's lower leg.

B. A larger, more extensive purpuric rash on the leg of a 9-month-old patient.

in Gram-positives. The activated host cells then release massive amounts of cytokines that promote inflammation, cause vasodilation, and trigger coagulation.

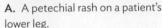 As discussed in Section 15.2, host cell surface receptors can identify conserved microbial structures called **pathogen-associated molecular patterns (PAMPs)**. PAMPs can be recognized by a variety of host cells. **Superantigens**, however, can bypass the normal route of antigen processing and simultaneously bind to the outside of T-cell receptors and the MHC molecules of antigen-presenting cells (see Section 16.7). In this way, superantigens nonspecifically activate a large number of T cells. Host cell recognition of either PAMPs or superantigens stimulates an intense, nonspecific immune response.

Vasodilation during sepsis allows fluid to leave the blood vessels, contributing to hypotension. Excessive coagulation causes clotting in the small blood vessels throughout the body (**disseminated intravascular coagulation**, or **DIC**), a process that damages organs and can lead to death. The excessive clotting ultimately uses up a large portion of platelets. As a result, observing a low platelet count (a normal count is 130,000–400,000 per microliter of blood) is an indicator, along with fever and low blood pressure, of septic shock and DIC. Ironically, the low platelet count results in slower clotting times for these patients. Blood, unable to clot, can leak from vessels below the skin to produce a petechial rash (**Figure 21.12**).

Organisms commonly associated with septicemia and septic shock include Gram-negative bacteria (*E. coli*, *Klebsiella*, *Yersinia enterocolitica*, *Salmonella* Typhi, *Franciscella tularensis*, *Neisseria meningitidis*) and Gram-positive bacteria (*Staphylococcus aureus*, *Streptococcus pneumoniae*, enterococci, and *Bacillus anthracis*). Even anaerobes that are normal inhabitants of the intestine (for example, *Bacteroides fragilis*) can be lethal if they escape the intestine and enter the blood, as might happen following surgery. This is one reason surgical patients are given massive doses of antibiotics immediately before and after surgery.

How do we diagnose and treat septicemia? The obvious way to identify the cause of septicemia is to identify the presence of a bacterium in the blood. Blood cultures involve taking samples of a patient's blood from two different locations (such as two different arms; see Section 25.2). Growth of the same organism in cultures taken from two body sites rules out inadvertent contamination with skin flora, which would probably yield growth in only one culture. The blood is added to liquid culture medium and incubated at 37°C (98.6°F). Incubation should be done aerobically and anaerobically. Treatment of septicemia depends on the organism identified, although a patient with suspected septicemia is empirically treated, usually with a broad-spectrum cephalosporin antibiotic.

Systemic Infections Involving Multiple Organs

Several organisms, some of which can also produce sepsis, use the bloodstream as a railway to infect other organs. We describe only selected multiple-organ diseases here; others are listed in Table 21.2.

CASE HISTORY 21.1

Prairie Dogs Are Not Good Companions

 A 25-year-old New Mexico rancher was admitted to an El Paso hospital because of a 2-day history of headache, chills, and fever (40°C, 104°F). The day before admission, he began vomiting. The day of admission, an orange-sized, painful swelling in the right groin area was noted (**Figure 21.13A**). A lymph node aspirate and a smear of peripheral blood were reported to contain Gram-negative rods that exhibited bipolar staining (the ends of the bacilli stained more densly than the middle; **Figure 21.13B**). The patient's white blood cell count was 24,700/μl (normal is 4,300–10,800/μl), and platelet count was 72,000/μl (normal is 130,000–400,000/μl). In the 2 weeks prior to becoming ill, the patient had trapped, killed, and skinned two prairie dogs, four coyotes, and one bobcat. The patient had cut his left hand shortly before skinning a prairie dog. PCR and typical biochemical testing of a Gram-negative rod isolated from blood cultures identified the organism as *Yersinia pestis*, the organism that causes plague. The patient received an antibiotic cocktail of gentamicin and tetracycline. He eventually recovered after 6 weeks in intensive care.

Table 21.2
Major Systemic Bacterial Diseases

Disease	Symptoms	Complications	Bacteria	Transmission	Diagnosis	Treatment	Prevention
Lyme disease	*Stage 1*: rash *Stage 2*: chills, headache, malaise, systemic involvement *Stage 3*: neurological changes	*Untreated patients*: neurological problems, arthritis, heart rhythm abnormalities *Treated patients*: post-treatment Lyme disease syndrome[a]	*Borrelia burgdorferi*	Deer tick	*In endemic areas*: erythema migrans ("bull's-eye" rash) *In nonendemic areas*: erythema migrans and positive antibodies against *B. burgdorferi*; positive antibodies against *B. burgdorferi* plus at least one organ involvement	Penicillin; tetracycline	No vaccine available
Brucellosis (undulant fever, Mediterranean fever, Malta fever)	Fever, weakness, sweats	Splenomegaly, osteomyelitis, endocarditis, others	*Brucella abortus*	Contact with animal products or consuming unpasteurized milk	Positive culture; serology; PCR	Tetracycline plus rifampin is treatment of choice	No vaccine available
Leptospirosis	Fever, photophobia, headache, abdominal pain, skin rash	Liver involvement and jaundice	*Leptospira interrogans*	Contact with urine of infected animals	IgM antibodies against *Leptospira* detected by enzyme immunoassay; microscopic agglutination test (MAT); dark-field microscopy	Erythromycin, penicillin	Vaccine for animals available
Epidemic typhus	Chills, fever, cough, headache, muscle pain, maculopapular rash	Splenomegaly, coma	*Rickettsia prowazekii*	Human louse, flying squirrel, flea	Serology; culture	Tetracycline, chloramphenicol	No vaccine available
Tularemia (rabbit fever, deer fly fever)	Fever, chills, headache, muscle pain, abdominal pain, pneumonia, death	No long-term complications	*Francisella tularensis*	Rabbits, rodents, deer flies, ticks, mosquitoes	Serology; PCR; histology of tissue samples	Gentamicin, streptomycin	Vaccine available, but not currently in the U.S.
Typhoid fever	Septicemia, chills, fever, hypotension, rash (rose spots)	Intestinal holes (perforation) and/or intestinal bleeding in 3rd week of illness	*Salmonella enterica* serovar Typhi	Fecal-oral route	Blood culture; serology (Widal test)	Fluoro-quinolones, ceftriaxone	Vaccines available
Salmonellosis (typhoid-like fever)	Septicemia, chills, fever, hypotension	Infection, aneurysm	*Salmonella* Cholerasuis	Fecal-oral route	Blood culture	Fluoro-quinolones, ceftriaxone	No vaccine available
Vibriosis	Serious with immunocompromised patients; fever, chills, multiorgan damage, death	No long-term complications	*Vibrio vulnificus*	Contact with seawater, raw oysters	Blood culture	Tetracycline plus aminoglycoside	No vaccine available
Bubonic plague	Buboes (swollen lymph glands), high fever, chills, headache, chest pain, shortness of breath, cough	Pneumonia, septicemia	*Yersinia pestis*	Rodents, rodent fleas, human respiratory aerosol; potential bioterrorism agent	Culture	Levofloxacin; streptomycin or tetracycline	Vaccine available, but not currently in the U.S.

[a]Fatigue and muscle and joint pain that can last for more than 6 months; sometimes referred to as chronic Lyme disease.

Figure 21.13 Plague

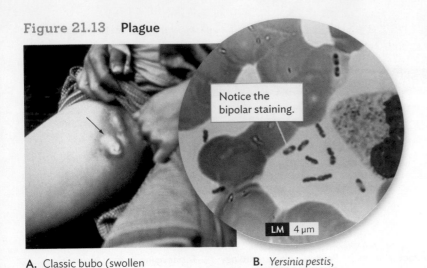

A. Classic bubo (swollen lymph node) of bubonic plague.

Notice the bipolar staining.

LM 4 μm

B. *Yersinia pestis*, bipolar staining.

PLAGUE **Plague** is caused by the bacterium *Yersinia pestis*, an organism that can infect both humans and animals. During the Middle Ages, the disease, known as the Black Death, killed more than a third of the population of Europe. Such was the horror it evoked that invading armies would actually catapult dead plague victims into embattled fortresses—perhaps the first use of biowarfare.

Yersinia pestis is present in the United States and is endemic in 17 western states. It is normally transmitted from animal to animal, typically rodents such as rats and prairie dogs, by the bite of infected fleas. Figure 21.14 illustrates the infective cycles of the plague bacillus, the sylvatic cycle (transmission in the wild), and the urban cycle (transmission in urban settings). Humans are not usually part of the natural infectious cycle. However, in the absence of an animal host, the flea can take a blood meal from humans and thereby transmit the disease to them (example of a zoonotic disease). During the Middle Ages, urban rats venturing back and forth to the countryside became infected by the fleas of wild rodents that served as a reservoir. Upon returning to the city, the rat fleas passed the organism on to other rats, which then died in droves. The rat fleas, deprived of their normal meal, were forced to feed on city dwellers, passing the disease on to them.

Figure 21.14 The Cycles of Plague

The sylvatic cycle occurs in the wild, where fleas transmit the organism between rodents. An accidental interaction with urban rats can trigger a similar urban cycle. Humans can be infected through contact with infected fleas coming from either cycle. Flea bite transmission initiates bubonic plague symptoms that can progress to pneumonic plague. Pneumonic plague is highly infectious, which can cause epidemic spread of the disease.

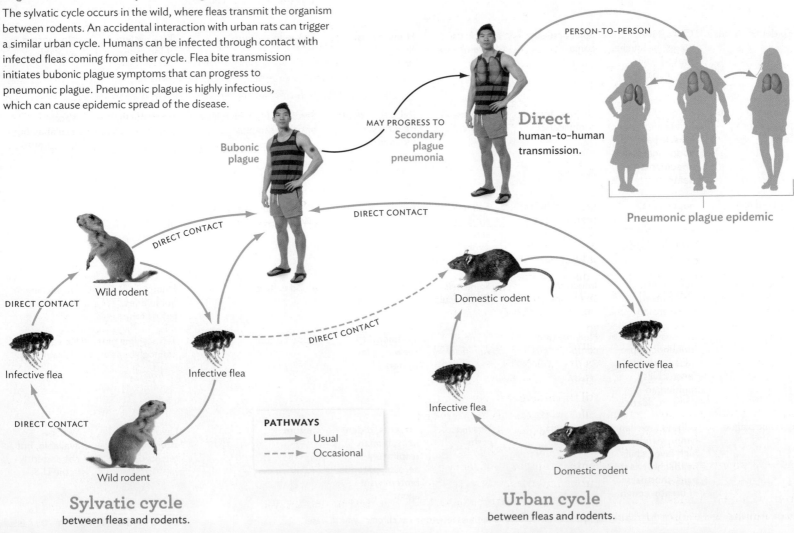

PERSON-TO-PERSON

MAY PROGRESS TO Secondary plague pneumonia

Direct human-to-human transmission.

Bubonic plague

Pneumonic plague epidemic

DIRECT CONTACT

DIRECT CONTACT

Wild rodent

DIRECT CONTACT

DIRECT CONTACT

Infective flea

Infective flea

DIRECT CONTACT

Domestic rodent

Infective flea

Infective flea

DIRECT CONTACT

Wild rodent

PATHWAYS
→ Usual
--→ Occasional

Domestic rodent

Sylvatic cycle
between fleas and rodents.

Urban cycle
between fleas and rodents.

Individuals bitten by an infected flea or infected through a cut while skinning an infected animal first exhibit the symptoms of **bubonic plague**. Bubonic plague emerges as the organism moves from the site of infection to the lymph nodes, producing characteristically enlarged nodes called buboes (Figure 21.13A). From the lymph nodes, the pathogen can enter the bloodstream, causing **septicemic plague**. In this phase, the patient can go into shock from the massive amount of endotoxin in the bloodstream. Neither bubonic nor septicemic plague is passed from person to person. As the organism courses through the bloodstream, however, it will invade the lungs and produce **pneumonic plague**, which can be easily transmitted from person to person through aerosol droplets generated by coughing. Pneumonic plague is the most dangerous form of the disease because it can kill quickly and spread rapidly through a population. Pneumonic plague is so virulent that an untreated patient can die within 24–48 hours. The organism is usually identified postmortem.

The organism has numerous virulence factors, including so-called V and W cell-surface lipoprotein antigens that inhibit phagocytosis. Another factor, the F1 protein capsule surface antigen, is partly responsible for blocking phagocytosis in mammalian hosts. Certain biofilms formed by *Y. pestis* are also important. An extracellular matrix synthesized by *Y. pestis* produces an adherent biofilm in the flea midgut that contributes to flea-to-mammal transmission. The biofilm blocks flea digestion, making the flea feel "starved" even after a blood meal. Therefore, the flea jumps from host to host in a futile effort to feel full. As the flea tries to take a blood meal, the blockage causes the insect to regurgitate bacteria into the wound. This curious effect of *Y. pestis* on the insect vector is another unique reason for how plague spreads so quickly.

Link As discussed in Section 6.6, bacteria can form specialized, surface-attached communities called **biofilms**. Cells in a biofilm are differentiated into distinct types that complement each other's functions, as in multicellular organisms.

Y. pestis uses type III secretion systems (see Section 18.4) to inject virulence proteins (YopB and YopD) into host cell membranes. Injection of the Yop proteins disrupts the actin cytoskeleton and helps the organism evade phagocytosis. By evading phagocytosis, the organism avoids triggering an inflammatory response and produces massive tissue colonization.

Plague has disappeared from Europe; the last major outbreak occurred in 1772. The reason for its disappearance is not known but was probably the result of multiple factors, including human intervention. Although it wasn't until the nineteenth century that doctors understood how germs could cause disease, Europeans recognized by the sixteenth century that plague was contagious and could be carried from one area to another. Beginning in the late seventeenth century, governments created a medical boundary, or *cordon sanitaire*, between Europe and the areas to the east from which epidemics came. Ships traveling west from the Ottoman Empire were forced to wait in quarantine before passengers and cargo could be unloaded. Those who attempted to evade medical quarantine were shot.

LYME DISEASE

CASE HISTORY 21.2

Bull's-Eye Rash

 Brad, a 9-year-old from Connecticut, developed a fever and a large (8 cm) reddish rash with a clear center (**erythema migrans**) on his arm (**Figure 21.15A**). He also had some left facial nerve palsy. Brad had returned a week previously from a Boy Scout camping trip to the local woods, where he did a lot of hiking. When asked by his physician, Brad admitted finding a tick on his stomach while in the woods but thought little of it. The doctor ordered serological tests for *Borrelia burgdorferi* (the organism that causes Lyme disease), *Rickettsia rickettsii* (which produces Rocky Mountain spotted fever), and *Anaplasma phagocytophilum* (which causes ehrlichiosis). The ELISA for *B. burgdorferi* came back positive, confirming a diagnosis of Lyme disease. The boy was given a 3-week regimen of doxycycline (a tetracycline derivative), which led to resolution of the rash and palsy.

Lyme arthritis was first reported in Lyme, Connecticut, in the 1970s, but the causative organism, *Borrelia burgdorferi*, was not identified until 1982. Since then, **Lyme disease** (also known as **Lyme borreliosis**) has become the most common vector-borne illness in the United States and is considered an emerging infectious disease. The main endemic areas are the northeastern coastal area from Massachusetts to Maryland, Wisconsin and Minnesota, and northern California and Oregon. Lyme disease is also common to parts of Europe and can be found in Sweden, Germany, Austria, Switzerland, and Russia.

Figure 21.15 Lyme Disease

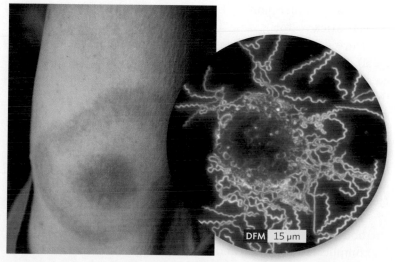

A. Erythema migrans rash.

B. *Borrelia burgdorferi*, the agent of Lyme disease.

Figure 21.16
Ixodes scapularis **Vector and Its Host Associations**

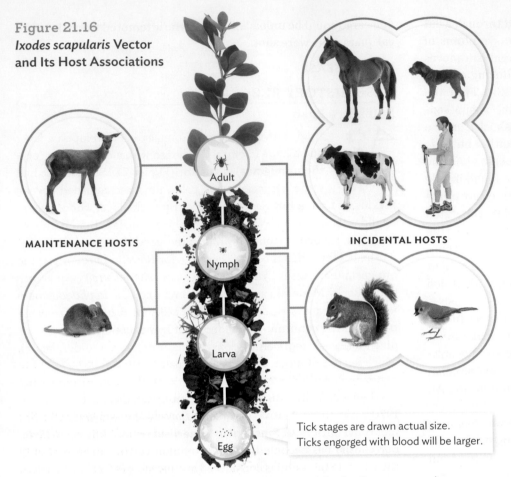

MAINTENANCE HOSTS   INCIDENTAL HOSTS

Adult

Nymph

Larva

Egg

Tick stages are drawn actual size.
Ticks engorged with blood will be larger.

A. Life cycle of *Ixodes scapularis*. The life cycle from eggs to adult takes 2 years to complete. As the ticks develop, they are attracted to the barberry bush (shown). Removing the bush from wooded areas associated with Lyme disease can dramatically lower the incidence of disease.

SEM   1 mm

B. *Ixodes scapularis* (deer tick, also known as blacklegged tick).

B. burgdorferi is a spirochete (**Figure 21.15B**) transmitted to humans by ixodid ticks (deer ticks or hard ticks; **Figure 21.16**). In the northeastern and central United States, where most cases occur, the deer tick *Ixodes scapularis* transmits the spirochete, usually during the summer months. In the western United States, *I. pacificus* is the tick vector.

During its nymphal stage (the stage after taking its first blood meal), *I. scapularis* is the size of a poppy seed. Its bite is painless, so it is easily overlooked. Infection takes place when the tick feeds because the spirochete is regurgitated into the host. However, the organism grows in the tick's digestive tract and takes about 2 days to make its way to the tick's salivary gland, so if the tick is removed before that time, the patient will not be infected. Once it is transferred to the human, the microbe can travel rapidly via the bloodstream to any area in the body, but it prefers to grow in skin, nerve tissue, synovium (joint lining), and the conduction system of the heart.

How *Borrelia* infects the human host and spreads through the body so quickly is not yet well understood, largely because of its complicated life cycle and its unusual genome. We know that the organism uses a variety of surface proteins, adhesion proteins, and other factors to establish infection and to neutralize the complement system of the host. *Borrelia* has enzymes that may contribute to the clinical symptoms of Lyme disease, including an enzyme that cleaves the proteoglycans found in the host connective tissues. Human connective tissues include tendons, ligaments, and cartilage (like that found in joints).

The *B. burgdorferi* genome, which consists of a large linear chromosome and 21 linear and circular plasmids, has been sequenced. The organism's genomic sequence has helped identify two plasmids important in *Borrelia* infections. The complete array of *Borrelia* virulence factors and pathogenic mechanisms, however, has yet to be determined.

Lyme disease has three general stages. The first two stages are part of early Lyme disease. Stage 1 is localized dissemination of *B. burgdorferi*, which occurs 3–30 days after the initial exposure. Approximately 75% of patients experience a "bull's eye" or "target" rash called erythema migrans, usually at the site of the tick bite, that varies in appearance. The lesions contain *B. burgdorferi* and are infiltrated with lymphocytes. This stage is often associated with constitutional symptoms, such as fever, myalgias (muscle pain), arthralgias (joint pain), and headache.

Stage 2 occurs weeks to months after the initial infection, after *B. burgdorferi* spreads from the blood to other organs (hematogenous spread). In this stage, the patient can be quite ill with malaise, myalgias, arthritis (joint inflammation), and arthralgias, as well as neurological or cardiac involvement. The inflammation causing the joint pain of Lyme disease is thought to be an autoimmune reaction (our antibodies reacting against us). Lyme arthritis affects mostly large joints, especially the knees. Interestingly, only one knee is affected at a time, but the pain can alternate between knees. Common neurological manifestations include **Bell's palsy** (facial paralysis), inflammation of spinal nerve roots, and chronic meningitis. The most common cardiac manifestation is an irregular heart rhythm.

Late Lyme disease, stage 3, occurs months to years later and can involve the synovium, nervous system, and skin, though skin involvement is more common in Europe than in North America. Arthritis occurs in the majority of patients with borreliosis and lasts from weeks to months in any given joint. Joint fluid analysis typically shows a white blood cell count of 10,000–30,000/μl. Late neurological involvement may include peripheral neuropathy and encephalopathy, manifested as memory, mood, and sleep disturbances.

Treatment with antibiotics is recommended for all stages of Lyme disease but is most effective in the early stages. Treatment for early Lyme disease (stage 1) is a course of doxycycline for 14–28 days. Lyme arthritis is usually slow to respond to antibiotic therapy. Despite antimicrobial drug treatment, patients with persistent active arthritis and persistently positive PCR tests may have incomplete microbial eradication and may be the most likely to benefit from repeated treatment with injected antibiotics.

Curiously, 25% of people infected with *B. burgdorferi* never develop erythema migrans, and many infected individuals are also unsure of tick bites. Patients with Lyme disease may present with arthritis as their first complaint. Therefore, knowing that the patient lives or recently traveled to an endemic area can provide a critical clue. Recent studies also show that ticks are commonly coinfected with other bacterial pathogens such as *Ehrlichia* or *Anaplasma* species. Patients with Lyme disease that are coinfected with these organisms often present with symptoms that can complicate diagnosis and may contribute to chronic Lyme disease.

Systemic Infections Caused by Intracellular Pathogens

Plague and borreliosis are diseases caused by *extracellular* pathogens, bacteria that grow outside of host cells. Some bacteria producing systemic infections are *intracellular* pathogens, with various modes of transmission and disease course.

Salmonella enterica serovar Typhi, for example, is a Gram-negative rod acquired through ingestion. As described in Section 18.5, *S.* Typhi (and other, nontyphoidal salmonellae) invade intestinal M cells (growing inside phagosomes) and exit into the lamina propria, where they invade tissue macrophages. The macrophages head for the circulation and taxi *S.* Typhi through the bloodstream. Along the way, the organism can infect a variety of organ systems (lymph nodes, liver, spleen, kidney, bone marrow). The result is **typhoid fever**, a potentially fatal disease marked by lymphadenopathy, a fever of 39.4°C–40°C (103°F–104°F), confusion, and a distended abdomen, as well as a short-lived but characteristic maculopapular rash on the skin (rose spots) that contains bacteria (Figure 21.17). From the bloodstream, the organisms can sometimes infect the gallbladder, converting a person who survived infection into a chronic carrier (Typhoid Mary, Section 22.5). The gallbladder of a chronic carrier will constantly secrete bile containing *S.* Typhi into the intestines, which is how the disease is transmitted. Food preparers who chronically shed *S.* Typhi in their feces

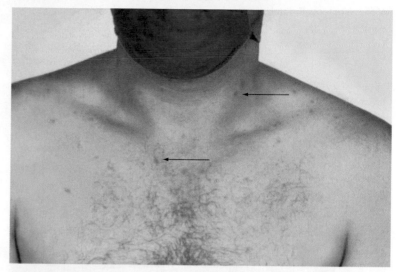

Figure 21.17   **Rose Spots of Typhoid Fever**

Note that the rash is limited to the trunk of the body.

and fail to wash their hands after bathroom use can introduce the organism into food and infect their patrons. Humans are the only reservoirs for this pathogenic variety of *Salmonella*.

The bacteria that cause **Rocky Mountain spotted fever** (RMSF; *Rickettsia rickettsii*; Figure 21.18A) or epidemic typhus (*Rickettsia prowazekii*) are *obligate* intracellular bacteria, replicating only inside of a host cell's cytoplasm. RMSF and epidemic typhus are two of several diseases collectively called **rickettsioses**. *R. rickettsii* and *R. prowazekii* are transmitted from reservoir animals to humans by the bite of arthropod vectors; ticks are the vector for RMSF, and human body lice are the vector for epidemic typhus (see Table 21.2). While biting, these ticks and lice also defecate, depositing the rickettsia at the bite site. Subsequent scratching by the host forces the organism into the bite, where the rickettsia can invade endothelial cells of the blood vessels. After phagocytosis, the organisms escape the phagosome and replicate to large numbers in the cell's cytoplasm. Eventually, the infected cells burst and the organism disseminates to infect additional endothelial cells and organs. Symptoms for both diseases include chills, confusion, fever, headache, muscle aches, and sometimes photophobia. A major sign differentiating the two diseases is the rash that develops. The rash for RMSF usually begins on the wrists and ankles, then spreads toward the center of the body (Figure 21.18B). In contrast, epidemic typhus rashes begin on the trunk. Both diseases are treated with antibiotics, usually doxycycline.

Other facultative intracellular pathogens that can cause systemic disease include the Gram-negative rods *Francisella tularensis* (**tularemia** or **rabbit fever**), transmitted by ingestion, wound infection, or ocular contamination from infected small wild animals; three *Brucella* species transmitted by ingesting milk or meat from infected cattle (*Brucella abortus*), pigs (*Brucella suis*), or goats (*Brucella melitensis*); and *Bartonella henselae*, a pathogen of cats that can

Figure 21.18   Rocky Mountain Spotted Fever
RMSF is a potentially fatal infection transmitted to humans by ticks.

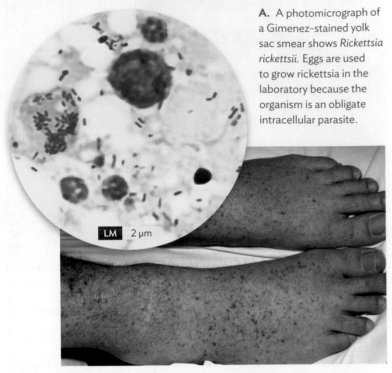

A. A photomicrograph of a Gimenez-stained yolk sac smear shows *Rickettsia rickettsii*. Eggs are used to grow rickettsia in the laboratory because the organism is an obligate intracellular parasite.

B. The classic rash of RMSF often begins on the wrists and ankles and spreads rapidly toward the center of the body. The rash may also be present on the palms and soles.

be transmitted to humans by cat scratch or flea bite (it's not clear which), causing **cat scratch disease**. *Ehrlichia* and *Anaplasma* are two obligate intracellular pathogens that produce flu-like illnesses that are transmitted from deer to humans by tick bite.

SECTION SUMMARY

- **Septicemia** is caused by many Gram-positive and Gram-negative bacterial pathogens and can lead to disseminated, systemic disease. **Septic shock** is sepsis in which the patient becomes hypotensive.
- **Blood cultures** are useful in diagnosing septicemia and endocarditis.
- **Plague** has sylvatic and urban infection cycles involving transmission between fleas and rats. **Y. pestis–infected flea** bites lead to bubonic plague. Aerosolized respiratory secretions will directly spread *Y. pestis* pneumonic plague from person to person (no insect vector).
- **Lyme disease** is caused by the spirochete *Borrelia burgdorferi*, which is transmitted from animal reservoirs to humans by the bite of *Ixodes* ticks. The initial presentation of the disease is a "bull's-eye" rash, also called erythema migrans.
- **Typhoid fever,** caused by *Salmonella* Typhi and transmitted by ingestion, causes high fever and is marked by rose spots forming on the patient's trunk. The organism is a facultative intracellular pathogen.
- **Rocky Mountain spotted fever** (*Rickettsia rickettsii*) and **epidemic typhus** (*Rickettsia prowazekii*) are caused by obligate intracellular pathogens that grow in the cytoplasm of endothelial cells. The rashes produced by these

organisms either start on the wrists and ankles and spread to the trunk (RMSF) or start on the trunk and move outward (epidemic typhus).

Thought Question 21.3   A patient diagnosed with SIRS has blood samples taken from different body sites, including percutaneous (through the skin) and two different IV sites. The laboratory reports the isolation of *Staphylococcus epidermidis* from blood samples drawn from one of the IV sites. The remaining cultures, including the second IV site, show no growth. The patient is noted to have improved since being admitted 2 days ago. The attending physician does not change the diagnosis of the patient from SIRS to sepsis but orders continued daily blood cultures from percutaneous and IV sites. What may be the reason for this decision?

Thought Question 21.4   LPS from Gram-negative bacteria is a potent cause of septic shock. Consequently, is it a good idea to treat septic patients with a bacteriocidal antibiotic, which could cause the release of more LPS from killed bacteria?

Thought Question 21.5   From what you know about the transmission of RMSF and epidemic typhus, which one is associated with war-torn countries?

21.4
Bacterial Infections of the Heart

SECTION OBJECTIVES
- Name and define the different manifestations of bacterial infections of the heart.
- Compare and contrast different types of endocarditis.
- Discuss the possible links between bacterial infections and atherosclerosis.

Are heart infections considered systemic infections? No, but they result from the circulatory spread of microbes originating from other body sites. In addition, organisms infecting the heart can break free to circulate and infect other organs. Bacterial infections of the heart include pericarditis (infection of the sac surrounding the heart), myocarditis (infection of heart muscle), and endocarditis (inflammation of the inner layer of the heart) (see Figure 21.3). The inflammation of the heart known as rheumatic fever (discussed in Chapter 19) is not an infection per se but an autoimmune reaction that follows streptococcal pharyngitis caused by certain strains of *S. pyogenes* (discussed in Chapters 17 and 19).

Pericarditis, when it occurs, usually develops after a respiratory tract infection. The advent of antibiotics has made pericarditis rare. Bacterial myocarditis is also rare in immunocompetent patients and is usually seen as a consequence of overwhelming sepsis. Endocarditis is by far the most common form of bacterial cardiac infection and is the focus of the following section. Table 21.3 lists bacteria that typically cause heart infections.

Bacterial Endocarditis (Infectious Endocarditis)

CASE HISTORY 21.3

Dental Procedure Leads to "Heartache"

Elizabeth was 58 years old and had a history of mitral valve prolapse (a common congenital condition in which a heart valve does not close properly). She was on immunosuppressive therapy following a kidney transplant. She had recently been admitted to the hospital complaining of fatigue, intermittent fevers for 5 weeks, and headaches for 3 weeks—symptoms the physician recognized as possible indications of endocarditis. Elizabeth reported having a dental procedure a few weeks prior to the onset of symptoms but forgot to take antibiotic beforehand. A sample of her blood placed in a liquid bacteriological medium grew Gram-positive cocci, which turned out to be *Streptococcus mutans*, a bacterial species associated with dental caries. With the finding of bacteria in the bloodstream, the diagnosis of bacterial endocarditis was confirmed. Elizabeth began a 1-month course of intravenous penicillin G and gentamicin therapy and eventually recovered to normal health.

Infectious endocarditis can be viral, fungal, or bacterial in origin. The viral and fungal forms of the disease are very rare and mostly limited to patients with transplanted heart valves. Endocarditis caused by bacteria, on the other hand, can be a consequence of many bacterial diseases, such as brucellosis, gonorrhea, psittacosis, staphylococcal and streptococcal infections, and Q fever caused by *Coxiella burnetti*. The most common causes are found in Table 21.3.

Endocarditis is traditionally classified as acute or subacute, based on the pathogenic organism involved and the speed of clinical presentation. **Subacute bacterial endocarditis (SBE)** has a slow onset with vague symptoms. It is usually caused by a bacterial infection of a heart valve. Subacute bacterial endocarditis infections are usually (but not always) caused by a viridans streptococcus from the oral microbiota (for example, *Streptococcus mutans*, a common cause of dental caries). **Viridans streptococci** is a general term for normal microbiota streptococci whose colonies produce green alpha hemolysis on blood agar (viridans, from the Greek *viridis*, "to be green"). Most patients who develop infective endocarditis have mitral valve prolapse (90%), although this is frequently not the case when patients are intravenous drug abusers or have hospital-acquired (nosocomial) infections.

As in Case History 21.3, bacterial endocarditis can begin at the dentist's office, although very rarely. Following a dental procedure (such as tooth restoration) or in someone with gingivitis, oral microbiota can transiently enter the bloodstream and circulate. *S. mutans*, which is not normally a serious health problem, can become lodged onto damaged heart valves, grow as a biofilm, and secrete a thick layer of polysaccharides (called glycocalyx) that encases the microbes and forms a vegetation on the valve, damaging

it further (**Figure 21.19A** and **B**). If untreated, the condition can be fatal within 6 weeks to a year. Other bacteria that are important include intestinal bacteria such as *Streptococcus gallolyticus* or *Enterococcus* species. These organisms can escape the intestine of patients with colorectal cancer and enter the bloodstream.

When virulent organisms, such as *Staphylococcus aureus*, gain access to cardiac tissue, a rapidly progressive and highly destructive infection ensues: **acute bacterial endocarditis (ABE)**. Symptoms of acute endocarditis include fever, pronounced valvular regurgitation (backflow of blood through the valve), and abscess formation.

Most patients with subacute bacterial endocarditis present with a fever that lasts several weeks. They also complain of nonspecific symptoms, such as cough, shortness of breath, joint pain, diarrhea, and abdominal or flank pain. Endocarditis is suspected in any patient who has a heart murmur and an unexplained fever for at least 1 week. Sharing needles or using a contaminated needle can introduce bacteria directly into the bloodstream, so subacute bacterial endocarditis should be considered in any intravenous drug user who has a fever, even in the absence of a heart murmur. Definitive diagnosis requires blood cultures that grow bacteria, even in the presence of strong clinical suspicion for bacterial endocarditis.

Curing endocarditis is difficult because the microbes are usually sealed away in a nearly impenetrable glycocalyx. Consequently, eradicating microorganisms from the vegetation almost always requires hospitalization, where high doses of intravenous antibiotic therapy can be administered and monitored. Antibiotic therapy usually continues for at least a month, and in extreme cases, surgery may be necessary to repair or replace the damaged heart valve.

Table 21.3
Organisms That Cause Infections of the Heart

Pericarditis (rare)	Myocarditis (rare)	Endocarditis
Haemophilus influenzae	Streptococcus pyogenes	Streptococcus mutans (oral source)
Neisseria meningitidis	Staphylococcus aureus	Streptococcus gallolyticus (formerly S. bovis; intestinal source)
Streptococcus pneumoniae	Corynebacterium diphtheriae (disseminated toxin)	Enterococcus (intestinal source)
Streptococcus pyogenes	Borrelia burgdorferi	Staphylococcus spp.
Staphylococcus aureus	Leptospira interrogans	**HACEK** organisms[a]
	Rickettsia	

[a]HACEK organisms include Haemophilus (Haemophilus parainfluenzae, Haemophilus aphrophilus, Haemophilus paraphrophilus); Aggregatibacter actinomycetemcomitans (Actinobacillus actinomycetemcomitans); Cardiobacterium hominis; Eikenella corrodens; and Kingella kingae.

Figure 21.19   View of Bacterial Endocarditis

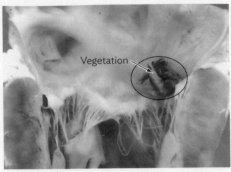

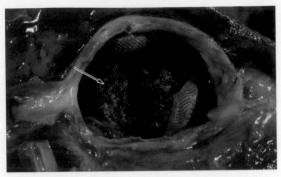

A. An open normal mitral valve shows the cords that tether it to the heart wall.

B. Close-up of native mitral valve endocarditis, showing vegetation, a microbial biofilm encased in a thick glycocalyx coating (arrow).

C. An infected prosthetic heart valve showing bacterial endocarditis (granulated tissue at center, arrow). When infection occurs early (within 2 months after surgery), it is likely that organisms gained entry during the operative period.

As a preventive measure, patients with certain heart conditions (an artificial heart valve, for instance) or on immunosuppressive therapy (such as Elizabeth in Case History 21.3) are prophylactically treated with penicillin or azithromycin 1 hour before a dental procedure. The idea is that the antibiotic will kill any oral bacteria that enter the bloodstream during the procedure and prevent the development of endocarditis. Although prophylactic therapy was previously recommended for patients with heart valve prolapse (a common condition), recent data suggest that there is no significant preventive benefit.

Prosthetic Valve Endocarditis

For patients with a severely defective heart valve, replacing it with a prosthetic valve is critical. However, the surgical process can result in the artificial valve becoming infected, a condition called **prosthetic valve endocarditis (PVE)**. *Staphylococcus epidermidis* and *S. aureus* are the most common organisms responsible for causing prosthetic valve endocarditis (Figure 21.19C). Many of these strains are methicillin resistant and extremely hard to treat. As with native valve endocarditis, the symptoms of PVE are vague. The most common symptoms are fever and chills accompanied by fatigue, anorexia, back pain, and weight loss. Prolonged intravenous antibiotics coupled with possible surgical removal and replacement of the valve are important treatment options. As noted earlier, patients with prosthetic heart valves should be prophylactically treated with penicillin or azithromycin 1 hour before certain dental procedures.

Atherosclerosis and Coronary Artery Disease

Microbes may also be involved in the process of **atherosclerosis** (hardening of the arteries). Atherosclerosis is an inflammatory reaction in response to the deposition of "bad" cholesterol (low-density lipoprotein, or LDL) inside the arterial wall. Atherosclerosis leads to narrowing or blockage of arteries and can be life-threatening. Coronary artery disease, for instance, is the result of atherosclerotic deposits in the arteries of the heart, which can produce a heart attack if blood flow is occluded.

The causes of atherosclerosis have been studied and debated for many years. It now appears that microbes may have an important role in the development of this disease. *Chlamydophila pneumoniae* (see Section 20.3), for example, is suspected of involvement. Infection of alveolar macrophages by *C. pneumoniae* may lead to circulating "activated" macrophages that generate inflammatory cytokines and factors that result in lowered high-density cholesterol (HDL, the "good" cholesterol) and increased clotting—both of which can damage endothelial cells and form plaques. Some studies found evidence of periodontal disease pathogens such as *Porphyromonas gingivalis* within atherosclerotic plaques.

SECTION SUMMARY

- **Pericarditis** is an infection of the outer covering of the heart that sometimes follows a respiratory infection.
- **Myocarditis** is an infection of the heart muscle.
- **Endocarditis,** an infection of the inner heart lining, typically involves heart valves.
- **Infectious endocarditis** can have acute or subacute onsets.
- **Subacute bacterial endocarditis** is usually an endogenous infection caused by *S. mutans*.
- *Chlamydophila pneumoniae* **infection** may contribute to atherosclerosis.

Thought Question 21.6   A blood culture was taken from a 63-year-old previously healthy patient with suspected subacute bacterial endocarditis. The laboratory reports *Streptococcus gallolyticus* present in three out of three bottles. Seeing this result, the physician suspects that the patient has another underlying condition. What is it? (*Hint:* Where is the organism normally found in the body?)

Thought Question 21.7   Why is stroke a complication of bacterial endocarditis, especially endocarditis of the mitral valve?

21.5
Systemic Parasitic Infections

SECTION OBJECTIVES

- Construct a list of systemic parasitic infections according to the type and mode of transmission of each parasite.
- Relate the different stages of systemic parasitic diseases to the growth cycle of the infecting agent.
- List the organ systems affected by each systemic parasitic infection discussed.
- Devise a plan for prevention of each systemic parasitic infection discussed.

What are the major risk factors for acquiring a systemic parasitic infection? Poverty, a compromised immune system, and travel to endemic areas are among the most common risk factors. Although systemic infections caused by protozoa and helminths (see Section 11.1) are uncommon in developed countries, these diseases are among the most prevalent and devastating worldwide. Be mindful that within a matter of hours, someone infected with one of these diseases can fly into your city's airport. Table 21.4 summarizes important features and treatments of systemic parasitic infections.

Malaria and Mosquitoes

Why is malaria the most devastating infectious disease known? Each year, 300–500 million people develop malaria worldwide, and 1–3 million of these people, mostly children, die. Fortunately, the disease is relatively rare in the United States; only around 1,000 cases occur annually, and almost all are acquired as a result of international travel to endemic areas (Figure 21.20). Once again, ease of travel illustrates the diagnostic importance of knowing a patient's history.

Malaria is caused by four species of *Plasmodium*: *P. falciparum* (the most deadly), *P. malariae*, *P. vivax*, and *P. ovale*. The life cycle of *Plasmodium* is complex and involves two cycles—an asexual erythrocytic cycle in the human (intermediate host) and a sexual cycle in the mosquito, the definitive host. Figure 21.21 summarizes the key stages in the life cycle and transmission of *Plasmodium*.

In the erythrocytic cycle, the organisms enter the bloodstream through the bite of an infected female *Anopheles* mosquito (the mosquito injects a small amount of saliva containing *Plasmodium*). The haploid sporozoites travel immediately to the liver, where they undergo asexual fission to produce merozoites. Released from the

Figure 21.20 Malaria, a Major Disease Worldwide

Endemic areas of the world where malaria is prevalent.

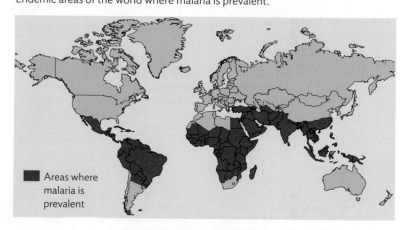

■ Areas where malaria is prevalent

Table 21.4
Selected Systemic Diseases Caused by Protozoa and Helminths

Disease	Symptoms	Complications	Parasite	Transmission	Diagnosis	Treatment	Prevention
Protozoan							
Malaria	Fever (may be every 48 or 72 hours but usually is random), chills, headache, vomiting	Inflammation of the brain, difficulty breathing, low blood glucose, anemia, death	Plasmodia; *Plasmodium falciparum* most common cause	*Anopheles* mosquitoes; via blood transfusion; can be transmitted from mother to child	Microscopy, thin and thick blood smears of patient; rapid antigen diagnostic testing (RDT) to detect malaria antigens	Therapy depends on type of infection; artemisinin-based combination therapy (ACT) recommended by WHO for uncomplicated cases; chloroquine, mefloquine, quinine, and quinidine are also used	No vaccine available; chemoprophylax offered to travelers to endemic areas

(continued on next page)

Table 21.4

Selected Systemic Diseases Caused by Protozoa and Helminths (*continued*)

Disease	Symptoms	Complications	Parasite	Transmission	Diagnosis	Treatment	Prevention
Protozoan							
Babesiosis	Flu-like symptoms, then fever, shaking chills, sweats, muscle and joint pain, shortness of breath, hemolytic anemia, hemoglobinuria	Low and unstable blood glucose, low platelet count, disseminated intravascular coagulation leading to blood clots and hemorrhage, multi-organ failure, death	*Babesia* spp.: *Babesia microti* in northeastern U.S., *B. equi* in California, unidentified species in Washington, *B. divergens* in Europe	Hard-bodied ticks: *Ixodes scapularis* (*I. dammini*), *I. ricinus* in Europe	Microscopy, thin and thick blood smears of patient	Antimicrobial choice depends on severity of disease and age of patient; atovaquone, azithromycin, clindamycin, and quinine are used	No vaccine available
Chagas' disease (American trypanosomiasis)	Fever, malaise, nausea, vomiting, and diarrhea; if progresses to cardiac form, then palpitations, chest pain, shortness of breath	Heart failure, enlargement of esophagus, enlargement of colon	*Trypanosoma cruzi*	Reduviid bug, blood transfusion, vertical transmission, organ transplant, food-borne or laboratory exposure (rare)	Microscopy; serology; PCR for acute phase; culture for acute phase	Benznidazole and nifurtimox are effective but are available only from CDC	No vaccine available
Leishmaniasis	Mucocutaneous leishmaniasis: skin ulcer can disseminate and cause ulceration on other mucosal surfaces	Disfigurement, hemorrhage, infections, death, preterm delivery or stillbirths	*Leishmania mexicana*, *L. braziliensis* (New World), *Leishmania tropica* (Old World)	Sand fly bite (cannot penetrate clothing)	Biopsy and culture of fresh tissue from the margin of the ulcer	IV pentavalent antimonies (sodium stibogluconate or meglumine antimoniate), amphotericin B	No vaccine available
Toxoplasmosis	Headache, seizures, abnormal gait, jaundice, lymphadenitis, fever in adults	Myocarditis, neurological problems, transmission of organism to fetus; gestational stage of development at time of fetal infection determines extent of complications and may include mental retardation, neurological abnormalities, microcephaly, hydrocephalus, anemia in newborns	*Toxoplasma gondii*	Inhaling airborne spores, contact with contaminated cat feces, or ingestion of contaminated food; transfusion, organ transplant, vertical transmission to fetus	CT with contrast to detect characteristic lesions; serology (serology is positive in 60% of cases)	Different complex regimens; drug of choice: pyrimethamine plus sulfadiazine 3–4 weeks	No vaccine available
Helminth							
Schistosomiasis, sometimes called bilharziasis	Abdominal pain, fever, chills, diarrhea, hepatitis, dysentery, pneumonitis	Dermatitis, cellulitis, lymphangitis	*Schistosoma haematobium*, *S. mansoni* and *S. japonicum*	Contact with contaminated water containing the free-living infectious organisms	Biopsy showing eggs; eggs in feces	Diethyl-carbamazine, albendazole	No vaccine available
Lymphatic filariasis	Fever, inguinal or axillary lymph-adenopathy (painful), limb or genital swelling		*Wuchereria bancrofti*	Mosquito-borne (*Aedes*; *Anopheles*)	Microfilariae in peripheral blood smear, biopsy	Albendazole and ivermectin	No vaccine available

707

Figure 21.21   **Malaria: Life Cycle of *Plasmodium falciparum* between Mosquito and Human Host**

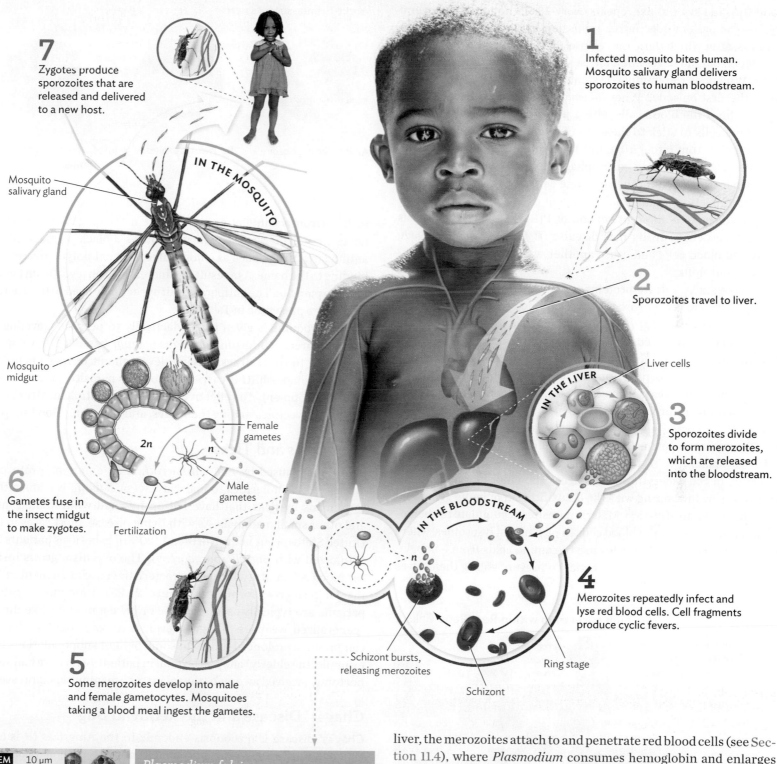

7   Zygotes produce sporozoites that are released and delivered to a new host.

1   Infected mosquito bites human. Mosquito salivary gland delivers sporozoites to human bloodstream.

IN THE MOSQUITO

Mosquito salivary gland

Mosquito midgut

2n

n

Female gametes

Male gametes

Fertilization

6   Gametes fuse in the insect midgut to make zygotes.

2   Sporozoites travel to liver.

Liver cells

IN THE LIVER

3   Sporozoites divide to form merozoites, which are released into the bloodstream.

IN THE BLOODSTREAM

n

Schizont bursts, releasing merozoites

Schizont

Ring stage

4   Merozoites repeatedly infect and lyse red blood cells. Cell fragments produce cyclic fevers.

5   Some merozoites develop into male and female gametocytes. Mosquitoes taking a blood meal ingest the gametes.

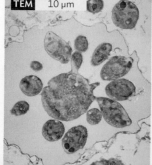

TEM   10 μm

Plasmodium falciparum

Schizont after completion of division. A residual body of the organism (yellow-green) is left over after division. The erythrocyte has lysed and only a ghost cell remains; no cytoplasm is seen surrounding the merozoites just being released. Free merozoites are seen outside the membrane.

liver, the merozoites attach to and penetrate red blood cells (see Section 11.4), where *Plasmodium* consumes hemoglobin and enlarges into a trophozoite. The protist nucleus divides, so that the cell, now called a schizont, contains up to 20 or so nuclei. The schizont then divides to make the smaller, haploid merozoites (see Figure 21.21, step 4). The glutted red blood cell eventually lyses, releasing merozoites that can infect new red blood cells.

Sudden, synchronized release of the merozoites and red cell debris triggers the telltale symptoms of malaria—violent, shaking chills followed by high fever and sweating. The erythrocytic cycle (and thus the symptoms) repeats every 48–72 hours. After several cycles, the patient goes into a remission lasting several weeks to months, after which there is a relapse.

Much of today's research focuses on why malarial relapse happens. Why does the immune system fail to eliminate the parasite after the first episode? When *Plasmodium* invades the red blood cells, it lines the blood cells with a protein, PfEMP1, that causes the blood cells to stick to the sides of blood vessels. This removes the parasite from circulation, but the protein cannot protect the parasite from patrolling macrophages, which eventually detect the invader and recruit other immune cells to fight it. So, during a malarial infection, a small percentage of each generation of parasites switches to a different version of PfEMP1 that the body has never seen before. In its new disguise, *P. falciparum* can invade more red blood cells and cause another wave of fever, headaches, nausea, and chills.

These sticky surface proteins are the antigens that the body's immune system recognizes and attacks. Once the immune system fights off one version of *P. falciparum* malaria, the parasite alters which gene is expressed. The resulting antigenic variation blinds the immune system, allowing a new wave of illness. The body now has to repeat the recognition and attack responses all over again.

Diagnosis of malaria involves microscopic demonstration of the protist within erythrocytes (the Wright stain; Figure 21.22 and see Figure 11.18A) or through serology to identify antimalarial antibodies. Treatment regimens include chloroquine or mefloquine, which kill *Plasmodium* in erythrocytic asexual stages, and primaquine, effective in the exoerythrocytic stages. The chloroquine family of drugs acts by interfering with the detoxification of heme generated from hemoglobin digestion. Malaria parasites accumulate the hemoglobin released from red blood cells in plasmodial lysosomes, where digestion occurs. The parasites use the amino acids from hemoglobin to grow but find free heme toxic. To prevent eating themselves

Figure 21.23   *Babesia*-Infected Erythrocytes

Wright-Giemsa–stained smears.

 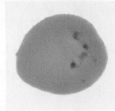 

A. Ringlike trophozoite.   **B.** Two merozoites.   **C.** Maltese cross (tetrad).

to death from accumulating too much heme, the organism detoxifies heme via polymerization, which produces a black pigment. Many antimalarial drugs, such as chloroquine, prevent polymerization by binding to the heme. As a result, the increased iron level (from heme) kills the parasite. The antiparasitic drugs actually used to treat active disease are presented in Table 21.4.

Chloroquine is given prophylactically to persons traveling to endemic areas. Unfortunately, *Plasmodium* has been developing resistance to these drugs, forcing development of new ones. The antigenic shape-shifting carried out by this parasite has so far stymied development of an effective vaccine. Prevention involves eliminating the vector, avoiding their bites, and screening blood supply.

Babesiosis and Ticks

Babesiosis, caused by *Babesia microti*, is an emerging protozoan disease with similarities to malaria, except *Babesia* is transmitted to humans by ticks that have fed on infected white-footed mice and other small mammals. As with Lyme disease, most cases in the United States occur in the Northeast (20% of babesiosis patients are coinfected with *Borrelia burgdorferi*). The organism grows inside red blood cells and produces a characteristic cell arrangement called the Maltese cross formation (Figure 21.23). Immunocompetent patients are typically asymptomatic or have a mild flu-like illness (generalized weakness, fatigue, fever, anorexia). But the disease can produce prolonged illness in some patient subpopulations (for example, the elderly) and can be fatal in patients who have had splenectomy, received an organ transplant, or are immunosuppressed.

Chagas' Disease and the Reduviid Bug

Chagas' disease is a zoonosis endemic to the Americas (it is also called American trypanosomiasis). Almost 11 million people in South America, Central America, and Mexico have Chagas' disease. The illness is caused by the protozoan hemoflagellate *Trypanosoma cruzi* (Figure 21.24A), which was discovered in 1910 by Carlos Chagas, a Brazilian microbiologist. *T. cruzi* is found in both wild and domestic animals at scattered locales from southern Argentina to the southern United States.

Figure 21.22   *Plasmodium falciparum* within Red Blood Cells

Note the multiple ring forms (arrows) of the protozoan within red blood cells. The dark dots are nuclei, and the "empty" space is the food vacuole (Wright stain).

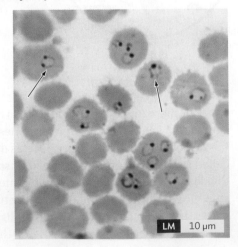

LM   10 µm

Figure 21.24 **Chagas' Disease**

A. *Trypanosoma cruzi*, the cause of Chagas' disease.

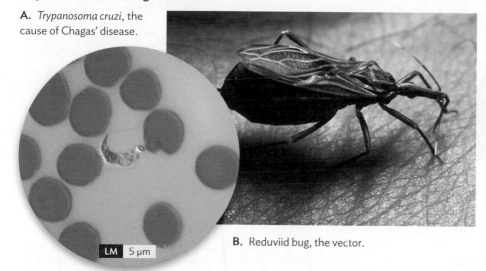

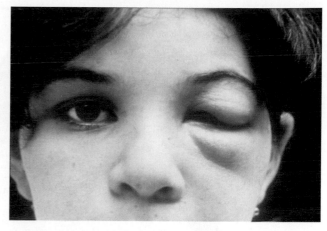

B. Reduviid bug, the vector.

C. Romaña sign.

T. cruzi lives and grows in the gut of a bloodsucking insect (hematophagous triatomine), the reduviid bug (genus *Reduvius*), that likes to make its home near a source of blood—for example, in the walls of mud or stone huts (**Figure 21.24B**). The insect's meal is blood, and its nasty habit of defecating while eating means that it leaves its feces at the site of the bite (usually on the face). The person or animal, in response to the sting of the insect, inadvertently rubs the feces along with its trypanosome content into the broken skin of the bite site. Because the reduviid often bites people around the mouth, it has earned the nickname "the kissing bug." the sclera of the eye can also serve as a portal of entry when it is rubbed with contaminated fingers. This leads to a painless edema of the tissue around the eye, called the Romaña sign (**Figure 21.24C**), and may be followed by edema of the face.

Within a week after transmission, most patients develop a mild acute form of Chagas' with symptoms of fever and lymphadenopathy. A minority of infected people will have a more violent form of acute disease involving systemic symptoms that include enlargement of liver and spleen, neurological symptoms, and even myocarditis. These symptoms resolve spontaneously in almost all cases of acute disease, but the parasite remains in most victims for an indefinite period of time. The acute phase of trypanosomiasis (Chagas' disease) is treated with nifurtimox or benznidazole, whose modes of action remain unclear.

After 4–6 weeks, patients who have not been cured develop the chronic form of the disease. Chronic disease presents itself in some 20%–30% of patients. The organism attacks different organs, including the heart, the gastrointestinal tract, and the nervous system, long after the initial infection (sometimes more than 20 years later). Death from Chagas' disease is usually due to heart failure caused when myocarditis destroys the heart muscle. The management of chronic Chagas' disease is primarily supportive, although some studies suggest that benznidazole may be beneficial.

Toxoplasmosis and Cats

Do you own a cat? If so, pay close attention to this section. **Toxoplasmosis** is a disease caused by *Toxoplamsa gondii* (**Figure 21.25** and **Figure 18.31**), a coccidian protozoan with a complicated life cycle (**Figure 21.26**). The parasite starts its life cycle in its definitive host, a cat (the feline stage), and completes it in an intermediate host, a bird or a mammal (the nonfeline stage). Cats do not become sick when they ingest spores of *T. gondii*, but their intestinal tract serves as a "breeding ground" where spores undergo the changes necessary to produce gametes that fuse together to form zygotes. The zygotes then form a rigid wall around themselves.

Infected cats can release up to 100 million of these walled-off gametes each day for 7–21 days when they defecate. Ambient temperature and exposure to air induce sporulation and maturation of the spores. Cats, other mammals, and birds ingest the spores, called sporozoites, when eating contaminated food or drinking contaminated water. Stomach acidity of the intermediate host dissolves the

Figure 21.25 *Toxoplasma gondii* **Cysts**

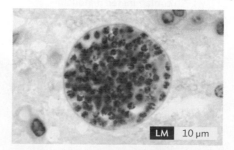

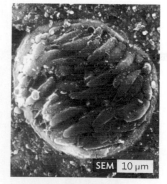

A. Microscopic cysts containing *Toxoplasma gondii* develop in the tissues of many vertebrates. Here, in mouse brain tissue, thousands of resting parasites (stained red) are enveloped by a thin parasite cyst wall.

B. A tissue cyst with its many bradyzoites.

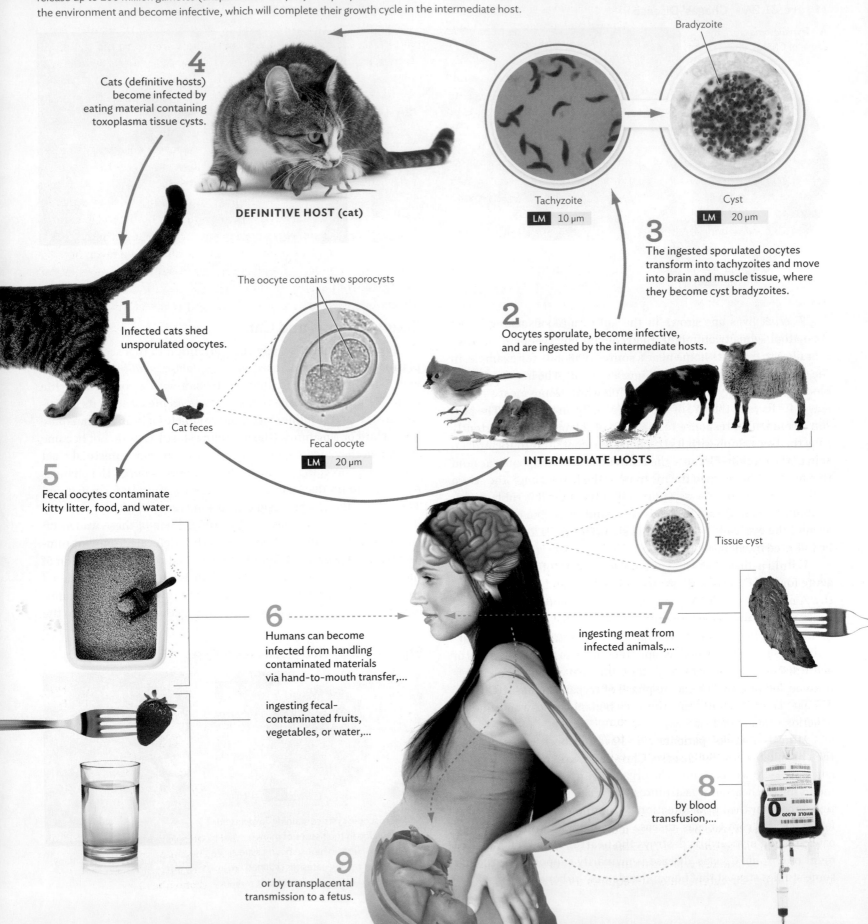

Figure 21.26 Life Cycle of *Toxoplasma gondii*

The only known definitive hosts for *Toxoplasma gondii* are domestic cats and their relatives. Infected cats can release up to 100 million gametes (unsporulated oocytes) every day for 7–21 days. The oocytes will sporulate in the environment and become infective, which will complete their growth cycle in the intermediate host.

Bradyzoite

4 Cats (definitive hosts) become infected by eating material containing toxoplasma tissue cysts.

DEFINITIVE HOST (cat)

Tachyzoite
LM 10 μm

Cyst
LM 20 μm

3 The ingested sporulated oocytes transform into tachyzoites and move into brain and muscle tissue, where they become cyst bradyzoites.

The oocyte contains two sporocysts

1 Infected cats shed unsporulated oocytes.

2 Oocytes sporulate, become infective, and are ingested by the intermediate hosts.

Cat feces

Fecal oocyte
LM 20 μm

INTERMEDIATE HOSTS

5 Fecal oocytes contaminate kitty litter, food, and water.

Tissue cyst

6 Humans can become infected from handling contaminated materials via hand-to-mouth transfer,...

ingesting fecal-contaminated fruits, vegetables, or water,...

7 ingesting meat from infected animals,...

8 by blood transfusion,...

9 or by transplacental transmission to a fetus.

wall of the cyst containing the spores and releases them into the small intestine. The spores enter the epithelial cells of the small intestine and rapidly divide until the numbers of parasites are so great that the cell ruptures and releases them. The rapidly growing protozoan cells are called **tachyzoites** because of their rapid ("tachy") division. Tachyzoites can invade all *nucleated* cells, so they spread rapidly through blood (without infecting red blood cells) and all organs. The tachyzoites change into a slow-dividing version of themselves (**bradyzoites**) and become encased in a cyst as the host immune system fights back and eliminates the tachyzoites. The major sites of cyst formation are the host's muscles and central nervous system (brain and spinal cord). The cysts represent a chronic stage of the infection.

If the intermediate host is an immunocompetent human, humoral and cell-mediated immune responses will eliminate most of the tachyzoites (but not the cysts). Infected individuals suffer no or very mild symptoms. However, suppression of a host's immune system, as in patients with AIDS, can cause spontaneous release of parasites from these cysts and their transformation to the invasive tachyzoite form and more serious disease.

Congenital disease is another serious consequence of toxoplasmosis. *T. gondii* can cross the placenta from a parasitemic (with parasites in blood) mother to her developing fetus. The severity of fetal disease depends on the gestational age at transmission. Generally, earlier infections are more severe and result in a classic triad of chorioretinitis (inflammation of the retina and vascular layer of the eye), hydrocephalus (abnormal accumulation of cerebrospinal fluid in the brain), and intracranial calcifications. A large percentage of infected newborns are asymptomatic but can develop symptoms months after birth.

Congenital toxoplasmosis is a preventable disease. Expectant mothers should be screened for antibodies to *T. gondii* and treated if the test is positive. The currently recommended drugs for *T. gondii* infection act primarily against the tachyzoite form but do not eradicate the encysted form (bradyzoite). Pyrimethamine is the most effective agent and is included in most drug regimens.

Do *Toxoplasma* cysts in the brain have any effect on a host's behavior? Oxford University zoologists showed that the parasite *Toxoplasma gondii* appears to alter the brain chemistry of rats, making them more likely to seek out cats. Infection thus makes a rat more likely to be killed and the parasite more likely to end up in a cat—the only host in which it can complete the reproductive step of its life cycle. In humans, exposure to *Toxoplasma* has been linked to schizophrenia, although no connection has yet been proven.

Leishmaniasis and the Sand Fly

Another group of protozoan diseases, collectively called **leishmaniasis**, come in two basic forms: a cutaneous form produced by *Leishmania tropica* and *L. braziliensis* and a visceral form caused by *L. donovani*. The visceral form can grow at core body temperatures, whereas the cutaneous form requires the cooler temperature of skin. The protozoa causing both of these clinical syndromes are transmitted by the female sand fly (**Figure 21.27A**). Smaller than a mosquito, the sand fly is found in tropical areas, around the Mediterranean, and in the Middle East. Its tiny size allows this vector to penetrate the mesh of standard mosquito netting.

Leishmania spends part of its life cycle in the gut of the sand fly, but its life cycle is completed in a vertebrate host. The sand fly carries the motile promastigote form of the protozoan (**Figure 21.27B**). Multiplication of the pathogen blocks the insect's esophagus. When taking a blood meal, the infected sand fly regurgitates the organism with saliva into the bite. After being introduced into the skin by a bite, the promastigote loses its flagellum and becomes an amastigote that multiplies inside tissue macrophages. In visceral leishmaniasis, macrophages carry the organism throughout the body.

The symptoms of visceral leishmaniasis include fever, hepatosplenomegaly (enlarged spleen and liver) resulting from accelerated production of phagocytic blood cells (**Figure 21.27C**), wasting and weakness, lymphadenopathy (often), and a characteristic darkening

Figure 21.27   Visceral Leishmaniasis

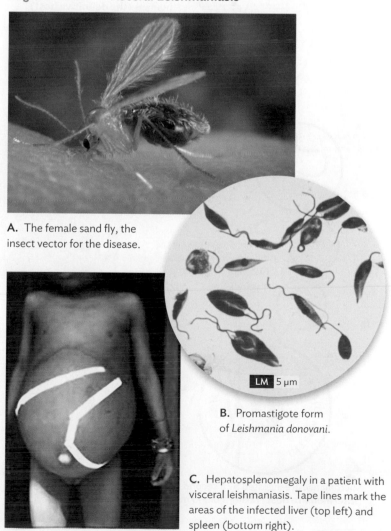

A. The female sand fly, the insect vector for the disease.

LM 5 μm

B. Promastigote form of *Leishmania donovani*.

C. Hepatosplenomegaly in a patient with visceral leishmaniasis. Tape lines mark the areas of the infected liver (top left) and spleen (bottom right).

of the skin (from which the Indian name for the disease, kala-azar, or black fever, originated).

In visceral leishmaniasis, patients may die within a matter of months from hemorrhage, severe anemia, secondary bacterial infections of mucous membranes, bacterial pneumonia, septicemia, tuberculosis, dysentery, or measles. Antiparasitic drugs containing antimony are commonly used for treatment, although an oral drug, miltefosine (an inhibitor of protein kinase B and C), is used in some countries.

Lymphatic Filariasis and Filarial Worms

Filariasis describes a group of diseases caused by nematodes in the family Filariidae. More than 170 million people worldwide are infected by filarial worms, primarily in Africa, Asia, the eastern Pacific, and Indonesia. The disease comes in three forms involving different filarial species: a cutaneous group (*Loa loa*, *Onchocerca volvulus*, and *Mansonella streptocerca*), a lymphatic group (*Wuchereria bancrofti* and *Brugia malayi*), and a body cavity group (*Mansonella perstans* and *Mansonella ozzardi*).

Our focus here is on lymphatic filariasis because of its systemic nature. Understanding the life cycles of filarial nematodes is important for understanding transmission (**Figure 21.28**). An infection starts when larvae are transmitted from mosquitoes or flies to humans (the definitive host) and develop into adult worms that can reproduce. For the lymphatic group of filarial worms, this occurs in the lymphatic system. The adults mate in the lymphatics and produce thousands of microfilariae per day that enter the bloodstream. Adult worms can live and produce microfilariae for at

Figure 21.28   Lymphatic Filariasis

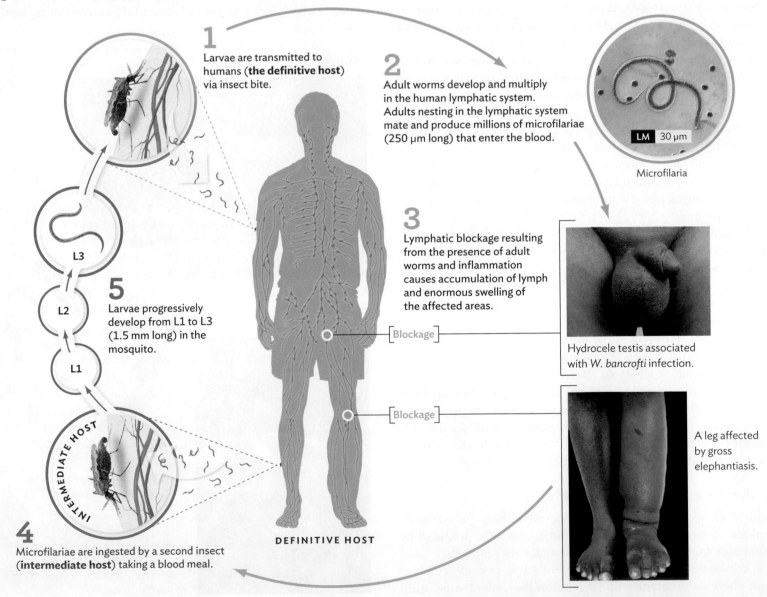

1 Larvae are transmitted to humans (**the definitive host**) via insect bite.

2 Adult worms develop and multiply in the human lymphatic system. Adults nesting in the lymphatic system mate and produce millions of microfilariae (250 µm long) that enter the blood.

LM   30 µm

Microfilaria

3 Lymphatic blockage resulting from the presence of adult worms and inflammation causes accumulation of lymph and enormous swelling of the affected areas.

Blockage

Blockage

Hydrocele testis associated with *W. bancrofti* infection.

A leg affected by gross elephantiasis.

5 Larvae progressively develop from L1 to L3 (1.5 mm long) in the mosquito.

L3

L2

L1

INTERMEDIATE HOST

4 Microfilariae are ingested by a second insect (**intermediate host**) taking a blood meal.

DEFINITIVE HOST

least 5 years. As they course through the bloodstream, microfilariae can be ingested by a second mosquito (intermediate host) taking a blood meal. Microfilariae in the mosquito take 10–14 days to develop into the stage 3 larvae that can be transmitted to the human host by insect bite.

The presence of adult worms in the lymphatic system stimulates an impressive inflammatory response that can produce lymphatic obstruction. Because the lymphatic circulation moves lymph (white cells and recovered tissue fluid) from the extremities toward the heart, a lymphatic obstruction means that lymph will accumulate behind the blockage. The result can be a grotesque swelling of a limb (**elephantiasis**) or, in males, the scrotum (called **hydrocele**); see Figure 21.28. Lymphatic filariasis is first contracted in childhood.

Diagnosis of lymphatic filariasis is based on finding microfilariae in blood smears or detecting filarial antigen in blood with or without seeing microfilariae. Antihelminthic treatments may include ivermectin (paralyzes the parasite), diethylcarbamazide (action unclear), and albendazole (decreases ATP production), among others. Surgical intervention can correct hydrocele but has been less successful with limb elephantiasis.

Many other pathogenic helminths infect multiple organ systems after migrating in the blood. We have chosen to discuss these worms in whatever system they primarily infect or that serves as the portal of entry and exit (for example, schistosomiasis is covered with gastrointestinal infections).

SECTION SUMMARY

- **Malaria** is caused by *Plasmodium* species. These are transmitted from mosquitoes (the definitive host) to humans (intermediate host), where the organism invades and replicates within red blood cells.
- **Babesiosis,** caused by *Babesia microti*, is transmitted by ticks to humans. The infection is potentially lethal in those who are immunodeficient.
- **Chagas' disease** develops following infection by the protozoan *Trypanosoma cruzi*. The tiny reduviid bug transmits the parasite from animal to humans, usually during a bite to the face.

- **Toxoplasmosis** is caused by *Toxoplasma gondii*. *T. gondii* is usually transmitted from cats (definitive host) to humans (intermediate host) who accidentally ingest sporozoites present in infected cat feces.
- **Congenital toxoplasmosis** results from transplacental transmission of *T. gondii* from a parasitemic mother to a developing fetus.
- **Leishmaniasis** is caused by several species of *Leishmania* protozoa transmitted to humans via the sand fly.
- **Lymphatic filariasis** is a disease caused by a variety of nematode species including *Wuchereria bancrofti* and is potentially disfiguring. Disease starts when a mosquito (the intermediate host) transmits larvae of the worm to the human (the definitive host). Larvae enter the lymphatic system and develop into adults that block lymphatic circulation.

Thought Question 21.8 Sickle-cell disease occurs in people who have *two* copies of a defective gene for hemoglobin production, *HbS*. This defective hemoglobin will polymerize under low-oxygen conditions, causing red blood cells to morph from round cells to cells shaped like sickles. Sickle-cell trait occurs in people who have one normal and one mutant copy of the hemoglobin gene. The red blood cells of people with sickle-cell trait do not change shape under low oxygen because 50% of their hemoglobin is normal and the rest abnormal. People with sickle-cell trait are resistant to malaria. What could explain this phenomenon? (*Hint: Plasmodium* parasites consume a lot of oxygen during growth.)

Perspective

Many of the bacteria, viruses, fungi, protozoa, and worms we discuss with specific organ systems also disseminate systemically when given the chance. In fact, many microbial pathogens could be considered "explorers" in which "scouts" from an initial infection are sent through the circulatory system to seek new sources of food (organ systems) and possible shelter from an immune system bent on their destruction. For example, *Neisseria meningitidis* causes meningitis but requires sepsis to do so. Keep dissemination in mind as you progress further through these infectious disease chapters. Take note of which organisms can move between organ systems.

Figure 21.29   Summary: Microbes That Infect the Circulatory System

🦠 **Viruses**	**Genome**	**Disease**
Epstein-Barr virus	DNA, ds	Infectious mononucleosis
Cytomegalovirus	DNA, ds	Cytomegalic inclusion disease, CMV mononucleosis-like syndrome
Lassa virus	RNA, ss, ambisense[a]	Lassa fever
Nairovirus, also known as Asian Ebola virus and Congo virus	RNA, ss, negative sense	Crimean-Congo hemorrhagic fever (CCHF)
Marburg virus	RNA, ss, negative sense	Marburg hemorrhagic fever
Ebola virus	RNA, ss, negative sense	Ebola hemorrhagic fever
Rift Valley fever virus	RNA, ss, negative sense (large and medium segments), ambisense (small segment)	Rift Valley fever
Dengue virus	RNA, ss, positive sense	Dengue fever/dengue hemorrhagic fever (DHF)
Yellow fever virus	RNA, ss, positive sense	Yellow fever, jungle yellow fever
Chikungunya virus	RNA, ss, positive sense	Chikungunya fever (or CHIK fever), also known as knuckle fever

🦠 **Bacteria**	**Morphology**	**Disease**
Borrelia burgdorferi	Spirochete	Lyme disease
Leptospira interrogans	Spirochete	Leptospirosis
Brucella abortus	Gram-negative rod	Brucellosis (undulant fever, Mediterranean fever, Malta fever)
Rickettsia prowazekii	Gram-negative rod	Epidemic typhus
Francisella tularensis	Gram-negative rod	Tularemia (rabbit fever, deer fly fever)
Salmonella Typhi	Gram-negative rod	Typhoid fever
Salmonella enterica serotype Choleraesuis	Gram-negative rod	Typhoid fever
Yersinia pestis	Gram-negative rod	Plague (bubonic, pneumonic)
Vibrio vulnificus	Gram-negative curved rod	Vibriosis

🪱 **Parasites**	**Type**	**Disease**
Plasmodia, *Plasmodium falciparum* most common cause	Protozoan	Malaria
Toxoplasma gondii	Protozoan	Toxoplasmosis
Babesia spp.; *Babesia microti* in northeastern U.S.; *B. equi* in California; unidentified species in Washington; *Babesia divergens* in Europe	Protozoan	Babesiosis
Trypanosoma cruzi	Protozoan	Chagas' disease
Leishmania mexicana, *L. braziliensis* (New World), *Leishmania tropica* (Old World)	Protozoan	Leishmaniasis
🪱 *S. haematobium*, *S. mansoni*, *S. japonicum*	Helminth	Schistosomiasis, sometimes called bilharziasis
🪱 *Wuchereria bancrofti*	Helminth	Lymphatic filariasis

[a]Ambisense = an RNA strand that has partially positive polarity (positive sense) and partially negative polarity (negative sense).

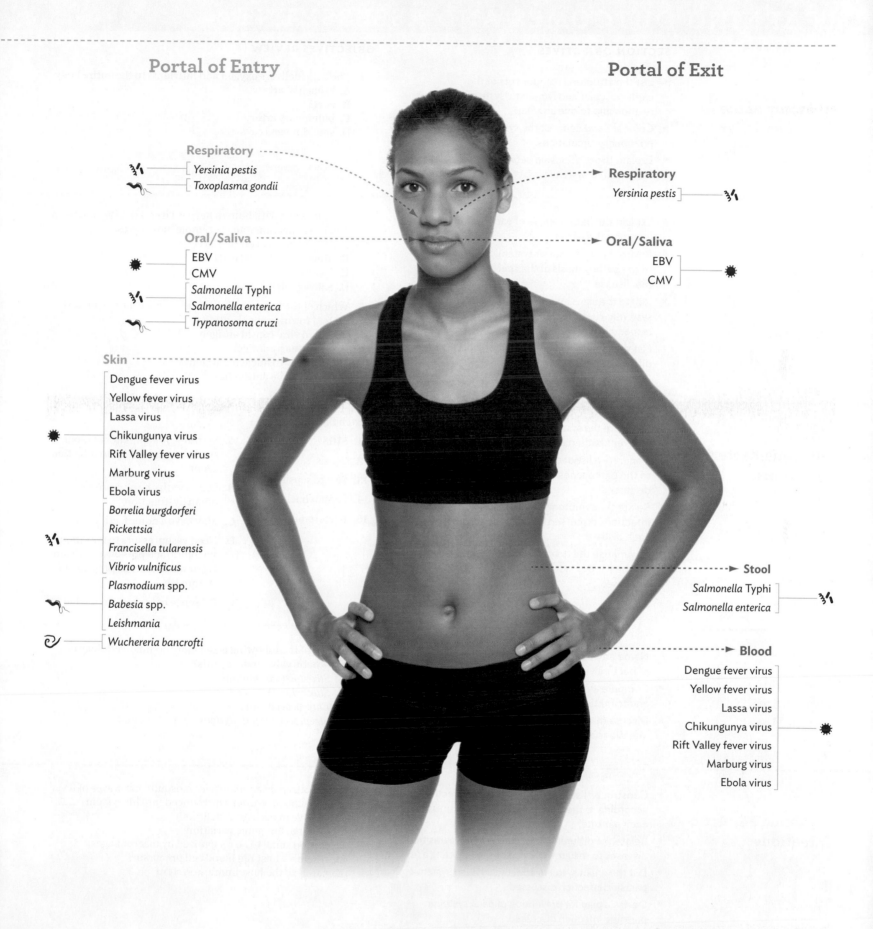

Portal of Entry

Respiratory
- Yersinia pestis
- Toxoplasma gondii

Oral/Saliva
- EBV
- CMV
- Salmonella Typhi
- Salmonella enterica
- Trypanosoma cruzi

Skin
- Dengue fever virus
- Yellow fever virus
- Lassa virus
- Chikungunya virus
- Rift Valley fever virus
- Marburg virus
- Ebola virus
- Borrelia burgdorferi
- Rickettsia
- Francisella tularensis
- Vibrio vulnificus
- Plasmodium spp.
- Babesia spp.
- Leishmania
- Wuchereria bancrofti

Portal of Exit

Respiratory
- Yersinia pestis

Oral/Saliva
- EBV
- CMV

Stool
- Salmonella Typhi
- Salmonella enterica

Blood
- Dengue fever virus
- Yellow fever virus
- Lassa virus
- Chikungunya virus
- Rift Valley fever virus
- Marburg virus
- Ebola virus

LEARNING OUTCOMES AND ASSESSMENT

	SECTION OBJECTIVES	OBJECTIVES REVIEW

21.1
Anatomy of the Cardiovascular and Lymphatic Systems
	• List the structural components of the cardiovascular and lymphatic systems and their relationship to one another.
	• Compare and contrast the systemic and pulmonary circulations.
	• Explain the relationship between the circulatory systems and systemic infections.
		1. The _____ delivers oxygen and nutrients to the entire body.
		  A. bronchial artery
		  B. aorta
		  C. pulmonary artery
		  D. superior vena cava

21.2
Systemic Viral Infections
	• Explain the pathogenesis of each systemic viral infection discussed.
	• Relate the diagnoses of systemic viral infections to the pathogenesis of the specific virus causing the disease.
	• Relate the signs and symptoms of each systemic viral infection with the effect of the respective virus on different organ systems.
	• List the treatment and prevention options for different systemic viral infections.
		4. _____ infected with human herpes virus 4 (HHV-4) have an atypical appearance and resemble monocytes.
		  A. Oropharyngeal epithelial cells
		  B. Tonsilar epithelial cells
		  C. B cells
		  D. Salivary gland cells
		
		5. Which of the following is the basis for the monospot test?
		  A. IgM to viral capsid antigen
		  B. IgG to viral capsid antigen
		  C. IgM to early antigen
		  D. Antibody to nuclear antigen
		  E. Heterophile antibodies

21.3 **Systemic Bacterial Infections**	• Compare the systemic signs and symptoms of different bacterial circulatory infections. • Generate a list of organ malfunctions according to the pathogenesis of each bacterial pathogen discussed. • Relate the symptoms of the systemic bacterial infections discussed with the progression of each disease. • Summarize the treatment and prevention plans for the various systemic bacterial infections discussed.	Match the following: 11. SIRS 12. Sepsis 13. Severe sepsis 14. Septic shock 15. Bacteremia A. Systemic inflammatory response syndrome caused by an infectious agent B. Sepsis with one or more signs of organ dysfunction C. May have a noninfectious etiology D. Need vasopressors to maintain systolic blood pressure of 90 mm Hg or mean arterial pressure of 70 mm Hg E. Presence of bacteria in the blood

21.4
Bacterial Infections of the Heart
	• Name and define the different manifestations of bacterial infections of the heart.
	• Compare and contrast different types of endocarditis.
	• Discuss the possible links between bacterial infections and atherosclerosis.
		18. Which of the following is the more common etiology of prosthetic valve endocarditis?
		  A. *Streptococcus mutans*
		  B. *Staphylococcus aureus*
		  C. *Enterococcus* spp.
		  D. *Streptococcus gallolyticus*

21.5
Systemic Parasitic Infections
	• Construct a list of systemic parasitic infections according to the type and mode of transmission of each parasite.
	• Relate the different stages of systemic parasitic diseases to the growth cycle of the infecting agent.
	• List the organ systems affected by each systemic parasitic infection discussed.
	• Devise a plan for prevention of each systemic parasitic infection discussed.
		21. A man bitten by an *Anopheles* mosquito has waves of fever because the protozoans introduced into his system _____.
		  A. survive in his liver indefinitely
		  B. undergo antigenic variation
		  C. are too small to be recognized by macrophages
		  D. resemble host red blood cell precursors
		  E. paralyze the host immune system

2. The main function of the pulmonary circulation is to provide lungs with oxygenated blood. True or false?

3. A patient is in your office with an inflamed lymph node under her arm. The swollen node is warm and painful to the touch. Which of the following best explains this condition?
   A. This is an example of simple noninfectious lymphadenopathy.
   B. This is an example of lymphangitis.
   C. This is an example of lymphadenitis.
   D. This is an example of a systemic infection.
   E. This is an example of pulmonary circulatory insufficiency.

6. What is the most dangerous form of transmission of CMV from a mother to a child?
   A. Breast-feeding
   B. Transplacental
   C. Contact
   D. None of the above

Match the following:

7. Dengue fever virus   A. Owl's eyes

8. Kissing disease   B. Splenomegaly

9. Congenital CMV   C. Burkitt's lymphoma

10. EBV and malaria   D. Hemorrhagic fever

16. Bell's palsy is more likely to be a manifestation of _____ than of any other stage of Lyme disease.
    A. stage 1
    B. stage 2
    C. stage 3
    D. stage 4

17. _____ is a contagious disease caused by Gram-negative intracellular bacteria that can colonize the gallbladder of an infected person, making that person a carrier.
    A. Rocky Mountain spotted fever
    B. Tularemia
    C. Lyme disease
    D. Infective endocarditis
    E. Typhoid fever

19. Inflammation of the three structural layers of the heart is referred to as _____.
    A. pancarditis
    B. endocarditis
    C. myocarditis
    D. epicarditis

20. Activation of alveolar macrophages by *Chlamydophila pneumoniae* is linked to decreased plasma HDL levels. True or false?

22. Which of the following is a symptom of a disease caused by a nematode?
    A. Darkening of the skin (kala-azar, or black fever)
    B. Chorioretinitis (inflammation of the membrane covering the retina)
    C. Hydrocele (grotesque swelling of the scrotum)
    D. Romaña sign (painless edema of the tissue around the eye)
    E. Waves of fever and chills repeating every 48–72 hours

Key Terms

acute bacterial endocarditis (ABE) (p. 703)
atherosclerosis (p. 704)
babesiosis (p. 708)
bacteremia (p. 693)
Bell's palsy (p. 700)
bradyzoite (p. 711)
bubonic plague (p. 699)
Burkitt's lymphoma (p. 690)
cardiomyopathy (p. 685)
cat scratch disease (p. 701)
Chagas' disease (p. 708)
congenital cytomegalovirus infection (p. 691)
cytomegalic inclusion disease (CID) (p. 691)
cytomegalovirus (CMV) (p. 691)

dengue hemorrhagic fever (DHF) (p. 692)
disseminated intravascular coagulation (DIC) (p. 696)
elephantiasis (p. 713)
endocarditis (p. 685)
erythema migrans (p. 699)
filariasis (p. 712)
giant cell (p. 691)
HACEK (p. 703)
hepatomegaly (p. 690)
heterophile antibody (p. 689)
hydrocele (p. 713)
hypotension (p. 693)
infectious mononucleosis (p. 688)
leishmaniasis (p. 711)
Lyme borreliosis (p. 699)
Lyme disease (p. 699)

lymphadenitis (p. 686)
lymphadenopathy (p. 686)
lymphangitis (p. 686)
lymphatic system (p. 682)
malaria (p. 705)
myocarditis (p. 685)
pancarditis (p. 685)
pericarditis (p. 685)
plague (p. 698)
pneumonic plague (p. 699)
prosthetic valve endocarditis (PVE) (p. 704)
pulmonary circulation (p. 682)
rabbit fever (p. 701)
reinfection (p. 691)
rickettsioses (p. 701)
Rocky Mountain spotted fever (p. 701)

sepsis (p. 693)
septicemia (p. 693)
septicemic plague (p. 699)
septic shock (p. 693)
splenomegaly (p. 690)
subacute bacterial endocarditis (SBE) (p. 703)
systemic circulation (p. 682)
systemic infection (p. 682)
systemic inflammatory response syndrome (SIRS) (p. 693)
tachyzoite (p. 711)
TORCH complex (p. 691)
toxoplasmosis (p. 709)
tularemia (p. 701)
typhoid fever (p. 701)
viridans streptococcus (p. 703)

Review Questions

1. What is systemic infection?
2. What is the function of the lymphatic system?
3. Compare and contrast systemic and pulmonary circulation.
4. Relate the structure of the heart and the layers of the heart wall to organisms that can affect them.
5. What are the most common causes of systemic viral infections?
6. What organs are affected in infectious mononucleosis?
7. What are the signs and symptoms of infectious mononucleosis?
8. Describe the relationships between EBV, Burkitt's lymphoma, and malaria.
9. What is TORCH, and why is it important?
10. Describe cytomegalic inclusion disease (CID).
11. What is a hemorrhagic fever, and what viruses can cause it?
12. Differentiate between bacteremia, SIRS, sepsis, and septic shock.
13. What is the mechanism of disseminated intravascular coagulation (DIC)?
14. Describe how *Y. pestis* uses its type III secretion system differently than *Salmonella*.
15. Which is more contagious, bubonic plague or pneumonic plague? Why?
16. Describe the different stages of Lyme disease.
17. Name intracellular pathogens and parasites that cause systemic infections and the diseases they cause.
18. Describe why symptoms of malaria follow a cycle of 48–72 hours.
19. Name the mode of transmission for each of the parasites discussed in this chapter.
20. Which of the systemic diseases described in this chapter can be prevented by vaccination?

Clinical Correlation Questions

1. A patient presents with a solitary enlarged lymph node in his groin. The bacteria isolated from lymph node aspirates of this patient are reported as Gram-negative bacilli with bipolar staining. Is this patient infected with *Yersinia pestis*, *Borrelia burgdorferi*, *Salmonella* Typhi, or *Rickettsia rickettsii*? Explain.
2. A patient is diagnosed with sepsis. You have ordered blood cultures, but you have to give antibiotics until the blood culture results come back. You order IV antibiotics. The antibiotics you order should be effective against what class or classes of microorganisms? Explain.
3. A 22-year-old otherwise healthy marathon runner presents to the clinic complaining of palpitations along with pain and swelling in his right ankle for the past 3 weeks. He reports that he has had similar episodes on and off for the past 6 months. His ECG shows an abnormal heart rhythm. Examination of his feet reveals neuropathy in the right foot. Synovial fluid aspirate shows 25,000 white blood cells, but the culture is negative. His medical history is benign and his vaccinations are up to date. He regularly travels to Wisconsin on business and recalls a circular rash on his left arm following a morning run when he was there about a year ago. What is the most likely bacterial cause of this young man's symptoms, and how is it transmitted?

Thought Questions: CHECK YOUR ANSWERS

Thought Question 21.1 Why do the red streaks of lymphangitis always move toward the heart?

ANSWER: Because the lymph in lymph vessels moves toward the heart so it can reenter the venous system near the heart.

Thought Question 21.2 What is the significance of a patient having a positive EBV IgM test but a negative EBV IgG test?

ANSWER: Because IgM is the first antibody that appears after an infection, the patient probably has an acute infection with EBV. Had the pattern been reversed, the patient would probably be recovering from the disease or had already recovered.

Thought Question 21.3 A patient diagnosed with SIRS has blood samples taken from different body sites, including percutaneous (through the skin) and two different IV sites. The laboratory reports the isolation of *Staphylococcus epidermidis* from blood samples drawn from one of the IV sites. The remaining cultures, including the second IV site, show no growth. The patient is noted to have improved since being admitted 2 days ago. The attending physician does not change the diagnosis of the patient from SIRS to sepsis but orders continued daily blood cultures from percutaneous and IV sites. What may be the reason for this decision?

ANSWER: Diagnosis of sepsis requires positive blood cultures. In this case, only one blood sample exhibits growth, which indicates contamination, not sepsis. The continuation of daily blood cultures is used to monitor the patient closely, making sure SIRS does not change to sepsis.

Thought Question 21.4 LPS from Gram-negative bacteria is a potent cause of septic shock. Consequently, is it a good idea to treat septic patients with a bacteriocidal antibiotic, which could cause the release of more LPS from killed bacteria?

ANSWER: Treatment with an antibiotic is not a good choice, but it is usually the only choice.

Thought Question 21.5 From what you know about the transmission of RMSF and epidemic typhus, which one is associated with war-torn countries?

ANSWER: Epidemic typhus. Body lice are associated with decreased sanitary conditions and require close contact between victims in order to migrate from one host to another. Both of these unsettling requirements are features of war.

Thought Question 21.6 A blood culture was taken from a 63-year-old previously healthy patient with suspected subacute bacterial endocarditis. The laboratory reports *Streptococcus gallolyticus* present in three out of three bottles. Seeing this result, the physician suspects that the patient has another underlying condition. What is it? (*Hint:* Where is the organism normally found in the body?)

ANSWER: *Streptococcus gallolyticus* and enterococci are typical members of the intestinal microbiome. The physician would suspect intestinal cancer or some other breach in the integrity of the intestine that would allow *S. gallolyticus* access to the bloodstream.

Thought Question 21.7 Why is stroke a complication of bacterial endocarditis, especially endocarditis of the mitral valve?

ANSWER: Bacteria infecting the heart valves (especially the mitral valve) form a mass, called vegetation. A piece of the vegetation can break off and be transported by the bloodstream to the brain's blood vessels and occlude them. The occlusion prevents blood perfusion at the site and causes an ischemic (inadequate blood supply) stroke.

Thought Question 21.8 Sickle-cell disease occurs in people who have *two* copies of a defective gene for hemoglobin production, *HbS*. This defective hemoglobin will polymerize under low-oxygen conditions, causing red blood cells to morph from round cells to cells shaped like sickles. Sickle-cell trait occurs in people who have one normal and one mutant copy of the hemoglobin gene. The red blood cells of people with sickle-cell trait do not change shape under low oxygen because 50% of their hemoglobin is normal and the rest abnormal. People with sickle-cell trait are resistant to malaria. What could explain this phenomenon? (*Hint: Plasmodium* parasites consume a lot of oxygen during growth.)

ANSWER: The answer is not entirely clear. One hypothesis is that plasmodia infecting RBCs in heterozygous individuals consume so much oxygen that these cells sickle and are marked for destruction by phagocytes, which will also destroy the parasite. Another hypothesis is that sickle-cell trait RBCs produce higher levels of oxygen radicals that can directly kill the parasite.

Infections of the Digestive System

[Dehydration: A Toddler's Plight]

SCENARIO Dao Ming is a 2-year-old who woke up lethargic with watery, non-bloody diarrhea. His mother kept him home from the day-care center he normally attends in downtown Portland. Ming's diarrhea became more pronounced as the day wore on, he seemed to be urinating less, and he started to develop a fever. By the next morning, Ming's eyes started to appear sunken and his diarrhea persisted. His parents rushed him to the emergency department (ED) of a nearby hospital. When asked, Ming's parents said their son had not received any vaccinations.

SIGNS AND SYMPTOMS Ming had a fever of 39.5°C, an elevated heart rate, and rapid breathing. The ED physician determined that Ming was dehydrated. The doctor ordered a complete blood count (CBC).

LABS The laboratory reported Ming's white cell counts were normal, so the doctor began to suspect Ming's diarrhea had a viral origin. An enzyme immunoassay of the boy's stool sample identified the presence of rotavirus, a virus that can be prevented by vaccination.

0.9% Sodium Chloride
Injection USP
500 mL

Intravenous (IV) Rehydration
Excessive diarrhea can produce dangerous levels of dehydration, especially in children. Rehydration by IV is often used to replace the fluids and electrolytes lost.

RESOLUTION The boy was admitted to the hospital, where he received intravenous fluids to help rehydrate him and oxygen to help his breathing. With the supportive therapy, Ming's diarrhea disappeared within a day and by week's end he was back playing in day care.

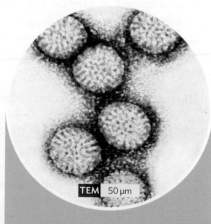

TEM   50 µm

Electron Micrograph of Rotaviruses
Rotavirus is the most common cause of diarrheal disease in children around the globe.

Raise your hand if you have <u>never</u> had diarrhea! Nearly everyone has experienced diarrhea—a condition characterized by frequent loose bowel movements accompanied by abdominal cramps. Hundreds of millions of cases occur each year in the United States. The loose stools usually result from inflammation of the gastrointestinal tract (a condition called gastroenteritis) due to viral or bacterial growth or toxin production. These infections or intoxications cause large amounts of water and electrolytes to leave the intestinal cells and enter the intestinal lumen. Consequently, the patient not only suffers diarrhea, but can become dangerously dehydrated; this is particularly true of the very young, like Ming, and the very old. Most diarrheal disease is viral in origin, with rotavirus and norovirus being the primary culprits. Among bacteria, *Salmonella*, a Gram-negative bacillus, and *Campylobacter*, a Gram-negative helical bacillus, vie for being the most frequent cause. Immunoassay revealed that rotavirus caused Ming's symptoms.

Link As discussed in Section 17.5, enzyme immunoassay (EIA) involves taking an enzyme that converts a colorless substrate to a colorful product and attaching that enzyme to a specific antibody. These antibodies are then used to detect specific antigen or antibody (see Figure 17.23).

Rotavirus is the most common cause of severe diarrhea among children worldwide, resulting in the hospitalization of approximately 55,000 children each year in the United States and 250,000 emergency department visits annually among children under 5 years of age. Commonly spread in day-care centers, it is estimated that every child will have been infected with rotavirus by the age of 5. Fortunately, most children have a mild case, and in countries with easily accessible health care, few children die of rotavirus infections. In developing countries, however, rotavirus infections cause approximately half a million deaths among children younger than 5 years of age every year. The availability of vaccines is expected to reduce the number of rotavirus cases in the United States.

Diarrhea is the ailment that comes to mind when we think of digestive tract diseases, but not all diseases of the digestive system involve diarrhea. For example, mumps is an infection of the parotid gland (one of the salivary glands), and hepatitis arises from infections of the liver, which is considered part of the digestive system. We will start our discussion by reviewing the anatomy of the digestive system, including the oral cavity and gastrointestinal tract, and by describing general features of relevant infectious syndromes.

22.1
Anatomy of the Digestive System

What structures of the human digestive system do microbial pathogens encounter? The human digestive system includes the oral cavity, the gastrointestinal tract (the GI tract, also called the alimentary canal), and its associated structures: salivary glands, liver, gallbladder, and pancreas (Figure 22.1). The gastrointestinal tract is a long tube extending from mouth to anus. It is divided into structural and functional sections to deny ingested bacteria access to the rest of the body, to process food and absorb nutrients, and, ultimately, to expel the unwanted digested material as feces, also called stool. Realize that wherever food goes, microbes go. Pathogens that gain access to the GI tract in food can colonize anywhere from the mouth to the anus, and some can penetrate certain tissues to gain access to the circulation.

Figure 22.1   The Gastrointestinal Tract

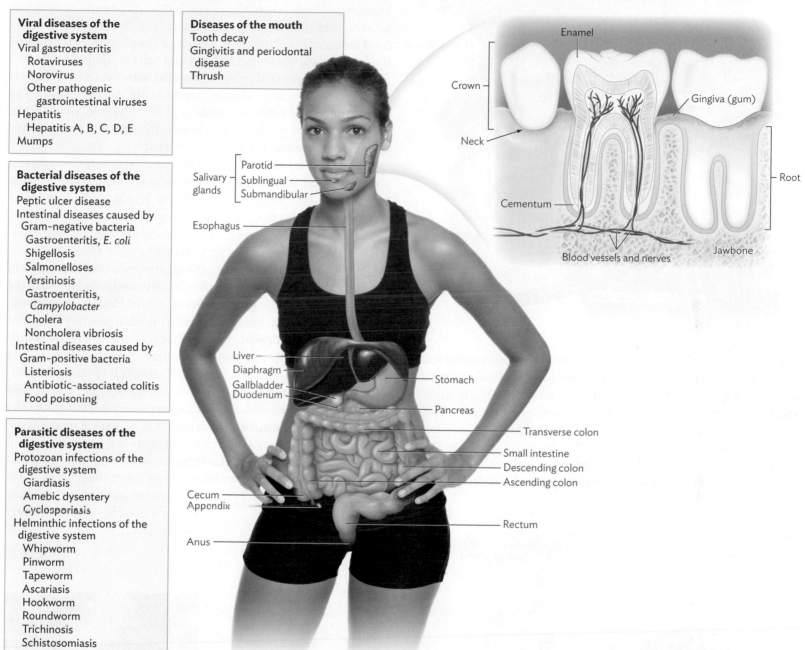

Viral diseases of the digestive system
Viral gastroenteritis
  Rotaviruses
  Norovirus
  Other pathogenic gastrointestinal viruses
Hepatitis
  Hepatitis A, B, C, D, E
Mumps

Bacterial diseases of the digestive system
Peptic ulcer disease
Intestinal diseases caused by Gram-negative bacteria
  Gastroenteritis, *E. coli*
  Shigellosis
  Salmonelloses
  Yersiniosis
  Gastroenteritis, *Campylobacter*
  Cholera
  Noncholera vibriosis
Intestinal diseases caused by Gram-positive bacteria
  Listeriosis
  Antibiotic-associated colitis
  Food poisoning

Parasitic diseases of the digestive system
Protozoan infections of the digestive system
  Giardiasis
  Amebic dysentery
  Cyclosporiasis
Helminthic infections of the digestive system
  Whipworm
  Pinworm
  Tapeworm
  Ascariasis
  Hookworm
  Roundworm
  Trichinosis
  Schistosomiasis

Diseases of the mouth
Tooth decay
Gingivitis and periodontal disease
Thrush

Salivary glands — Parotid, Sublingual, Submandibular
Esophagus
Liver
Diaphragm
Gallbladder
Duodenum
Cecum
Appendix
Anus
Stomach
Pancreas
Transverse colon
Small intestine
Descending colon
Ascending colon
Rectum

Enamel
Crown
Neck
Cementum
Blood vessels and nerves
Gingiva (gum)
Root
Jawbone

The mouth opens to the oral cavity, whose basic anatomy was introduced in Section 20.1 (see Figure 20.1). Figure 22.1 shows the anatomy of teeth, an important site of infection. Teeth protrude through a mucosal tissue called gingiva, or gums, which are vulnerable to infection. The part of a tooth protruding outward through the gum is called the crown; the part below the gum is called the root. The crown is covered in enamel (made of hydroxyapatite, a calcium phosphate polymer), whereas the root is encased in cementum (also a calcium substance). The hydroxyapatite of enamel is arranged in parallel crystalline rods (enamel rods) that add to enamel's incredible strength, but it can be dissolved by cariogenic (tooth decay–producing) bacteria. Underneath the hard enamel and cementum shells lies a porous dentin layer, and under that, pulp. Nerves and blood supplies penetrate the pulp through a root canal.

Once food enters the mouth (the oral cavity), it is mixed with saliva produced by the salivary glands and chopped into small pieces via the process of mastication (chewing). Saliva helps lubricate food to protect teeth and soft tissue while salivary amylase and

Figure 22.2   Salivary Glands

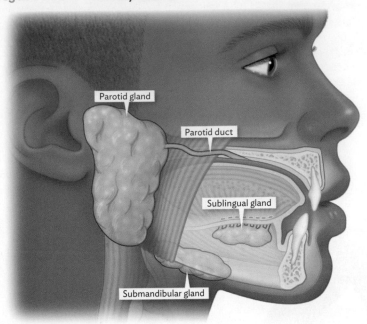

that extends from the pharynx through an opening in the diaphragm and connects to the stomach. The diaphragm is a muscular structure that separates the contents of the abdominal cavity from the contents of the chest. The job of the esophagus is to propel the food toward the stomach. The esophagus is equipped with two sphincters that prevent reflux of digested material from unexpectedly passing up from the stomach and possibly into the airways. The upper esophageal sphincter, made of skeletal muscle, sits near the oropharynx. Farther down, just before the stomach, is the lower esophageal sphincter made of smooth muscle.

After passing through the esophagus, food enters the stomach, a muscular sac shaped like a J. The serious business of digestion begins here with digestive enzymes and highly acidic (pH 1.5–2) stomach secretions (gastric juice) that prevent growth of most (but not all) microbial pathogens. *Vibrio cholerae* and *Shigella*, for example, have developed mechanisms that allow them to survive stomach acidity and cause severe gastrointestinal disease (discussed in Section 22.5). The interior stomach lining is made of epithelial cells and is often called the **gastric mucosa** (Figure 22.3A). The gastric mucosa is not smooth but is dotted with innumerable tiny holes that are the openings of **gastric pits**. Gastric pits are formed by invaginations of the gastric mucosa and are lined with several specialized cell types shown in Figure 22.3B. Different cell types secrete different digestive enzymes: chief cells secrete pepsin, parietal cells secrete hydrochloric acid, and mucous neck cells secrete mucus and bicarbonate. The mucus and bicarbonate form a thick mucus layer, called the gastric mucosal barrier, that coats the epithelial cells of the gastric mucosa and protects them from the harsh acidity of gastric juice. *Helicobacter pylori* bacteria can infect the mucosa of the human stomach and are responsible for a majority of gastric ulcers (Figure 22.3C).

lipase break down starches and some fats. There are several salivary glands, of which the major ones are parotid, submandibular, and sublingual glands (Figure 22.2). The parotids are the largest of the salivary glands and sit just forward (anterior) of the ears. Their microbiological significance comes from being the infection site of mumps virus.

Once leaving the mouth, masticated food passes through the pharynx to enter the esophagus. The esophagus is a muscular tube

Figure 22.3
The Gastric Mucosa and Gastric Pits

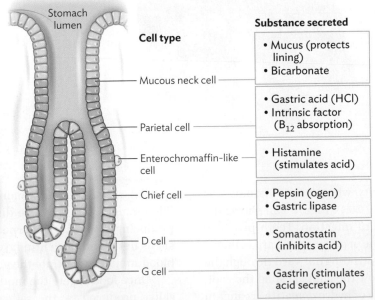

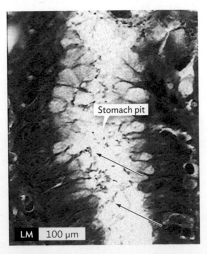

A. The epithelial cells that form the gastric mucosa. The holes are the openings of gastric pits.

B. Diagram of a gastric pit and the various cell types that are involved in secreting mucus and bicarbonate (mucous neck cells), acid (parietal cells), and digestive enzymes (chief cells) and that control acid secretion.

C. Arrows show *Helicobacter pylori* bacteria in a stomach pit.

Food in the stomach is broken down physically by stomach muscle contractions and chemically by gastric juice to form "chyme." Chyme eventually moves through another ring-shaped orifice named the pyloric sphincter and enters the duodenum, the first part of the human small intestine. The 22 feet of small intestine is coiled serpent-like through the abdomen. The small intestine is divided into three sections: the duodenum, closest to the stomach and shortest in length; the jejunum, the middle section; and the ileum, the longest section, which joins to the large intestine, or colon. Viral and bacterial infections of the intestines usually lead to diarrheal diseases, some of which are a nuisance, whereas others are life-threatening. *Salmonella*, for example, is an important gastrointestinal pathogen that attaches to specialized cells (M cells) in the Peyer's patches to gain access to the deeper tissues of the gut. M cells in Peyer's patches normally sample and transport antigens (including bacteria) from the gut lumen and deliver them to the immune cells.

Link As discussed in Section 15.2, Peyer's patches are aggregates of lymphoid tissue found in the lower small intestine as part of the gut-associated lymphoid tissue (GALT).

The liver, gallbladder, and pancreas connect to the duodenum, where they deposit bile salts and acids and pancreatic digestive enzymes that further break down chyme, especially the fatty components. A wavelike motion produced by intestinal muscles (peristalsis) propels the chyme through the intestine. Most of the digestion and absorption of nutrients takes place in the small intestine, so when the chyme enters the large intestine, it is mostly devoid of nutrients.

The colon is also divided into three sections based on structural properties: the ascending colon, transverse colon, and descending colon, which ends in the rectum and anus. The appendix is located at the beginning of the colon, where it is connected to the small intestine. The job of the large intestine is to absorb whatever water and nutrients remain in chyme, a process by which chyme and its bacterial contents are concentrated into a semisolid waste called feces. Feces are finally expelled into the rectum and out of the body through the anal sphincter. Expulsion of feces provides a key transmission route for enteric pathogens. For this reason, sanitary disposal of feces is a major public health concern.

SECTION SUMMARY

- **Teeth consist of a crown covered in enamel above the gum and a root below the gum,** encased in cementum. The tooth's interior contains dentin and, at the center, pulp. Nerves enter through a root canal.
- **Food entering the mouth is masticated** by teeth and mixed with saliva made by salivary glands.
- **The esophagus propels food to the stomach,** where acidic secretions released from the gastric mucosa begin breaking food down to chyme.
- **Chyme moves through the pyloric sphincter** into the duodenum.

- **The gallbladder, liver, and pancreas deliver bile and various enzymes** to chyme in the duodenum, which helps break down the chyme to more digestible particles as it moves through the jejunum and ileum of the small intestine.
- **The large intestine (colon) removes any remaining nutrients and water,** concentrating the chyme into feces, which exit through the anal sphincter.

22.2
Infections of the Oral Cavity

SECTION OBJECTIVES

- List microbes comprising normal microbiota of the oral cavity.
- Analyze how biofilms contribute to infections of the oral cavity.
- Compare the signs and symptoms of common infections of the oral cavity.
- Explain how oral cavity infections impact diseases elsewhere in the body.

Does our oral cavity have any natural defense against infections? Yes. Human saliva contains proteins, such as lysozyme, that help inactivate the microscopic invaders. The mucosa of the digestive system, including that of the oral cavity, produces immunoglobulin A (IgA), which is thought to also have some antimicrobial activity. The major defense against microbial infections, however, is left to the extremes of pH in the stomach and small intestine. In this section, we will discuss common infectious diseases of the oral cavity caused by bacteria and fungi. The major viral infection of the oral cavity, cold sores, caused by herpes simplex virus, was already discussed in Section 19.3).

Tooth Decay

How do microbes infect the gum and teeth? Tooth and gum disease is associated with the formation of "plaque," the stuff dental hygienists scrape off your teeth (Figure 22.4A). Plaque starts as a biofilm formed by microbiota of the oral cavity at the margin of the teeth and gum. The normal microbiota in plaque include *Streptococcus*, *Lactobacillus*, *Peptostreptococcus*, *Veillonella*, and diphtheroid species (nonpathogenic corynebacteria). Bacteria in the biofilm produce a substance that covers and protects the biofilm from antibiotics and other chemicals. As the biofilm thickens, it is referred to as **dental plaque**. If not removed, plaque can build up, and the bacteria living there will ferment any sugars left on the teeth and make acidic end products. Acid is the one thing that can dissolve enamel. We develop **dental caries**, or tooth decay (also called a cavity), when acid eats away at the enamel, creating crevices for more bacteria to grow and make more acid that dissolves more enamel (Figure 22.4B). Eventually, the decay penetrates to the nerve and we feel pain. Two major types of bacteria, *Streptococcus mutans* and *Lactobacillus* species, cause most dental caries. Lactobacillus is also an important probiotic. So, please, before giving up yogurt reread the Chapter 14 opening case.

Dental caries are treated by removing decay (drilling) and filling the cavity with resins. Eating less sugar, brushing, and flossing regularly all help prevent the development of caries. A dentist can also seal your teeth with a resin coating to prevent the formation of surface pits produced by grinding teeth. Surface pits serve as a niche where plaque can form. Finally, drinking fluoride-treated water (provided by most municipalities) helps harden tooth enamel.

Gingivitis and Periodontal Disease

Plaque left on the teeth eventually absorbs minerals and thickens into dental caries. As tartar forms below the gumline, it causes inflammation of the gums, called **gingivitis** (Figure 22.4C), which can lead to more serious **periodontal disease**, or periodontitis (Figure 22.5). In periodontitis, the gum bleeds easily and recedes from the tooth. As the disease progresses, the density of bone holding the tooth in place also decreases. Periodontitis is most often caused by *Aggregatibacter* (formerly *Actinobacillus*) *actinomycetemcomitans*

Figure 22.4   Dental Caries and Gingivitis

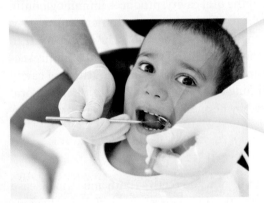

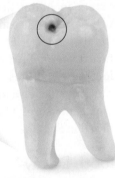

B. Bacteria in plaque produce acids that dissolve enamel to form a cavity (dental caries).

A. Dental hygienists use sharp tools to remove plaque from your teeth.

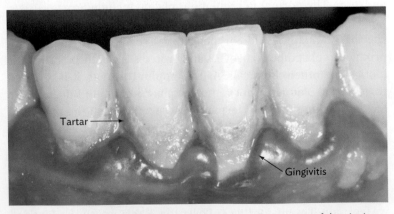

C. Heavy plaque buildup (called tartar) can trigger inflammation of the gingiva around your teeth, the beginning of gingivitis.

Figure 22.5   Periodontal Disease

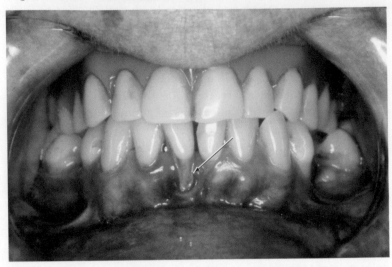

Gingivitis can lead to periodontitis (periodontal disease), where the gums withdraw from the teeth (arrow).

and *Porphyromonas gingivalis*, but other species, particularly *Bacteroides forsythus*, are also involved.

Dental caries and periodontal disease can cause tooth loss or abscess formation when the infection spreads to the pulp. Treatment of these infections depends on the type of infection and almost always includes systemic antibiotic therapy. As you may recall from the case history that opened Chapter 8, the symptoms of periodontal disease include painful, swollen gums that easily bleed when touched.

In its most serious form, gingivitis can become a necrotizing infection (necrotizing ulcerative gingivitis) commonly called **trench mouth** or **Vincent's angina**. Trench mouth is easy to recognize. Symptoms include painful, bleeding, bright red gums covered with a gray film. Craterlike canker sores form between the teeth and gums, and the patient has bad breath, a foul taste in the mouth, and fever. Swollen lymph nodes can appear around the head, neck, or jaw. Eating and swallowing will also be painful. Treating trench mouth includes removing dead gum tissue followed by hydrogen peroxide mouthwash and antibiotics.

Thrush

Thrush, a disease common to neonates and infants, is first noticed as a white coating in the child's mouth resembling cottage cheese. The disease is caused by a species of yeast, *Candida albicans*. Thrush lesions often start as tiny areas that enlarge to white patches on oral mucosa (Figure 22.6). *C. albicans* can colonize the adult mouth, vagina, and other areas, some of which can be sources of infection during birth or after. Topical treatment (oral spray or a swab) with a nystatin solution is curative. Adults with severely compromised immune systems are also likely to develop thrush.

Figure 22.6 Thrush

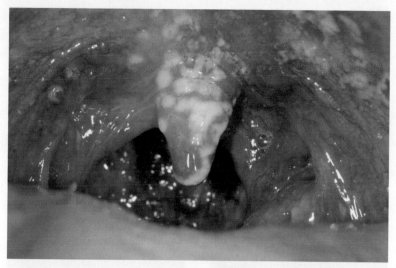

Lesions caused by *Candida albicans* often start as tiny areas that enlarge to white patches on oral mucosa. In adults, thrush is a sign of a compromised immune system.

SECTION SUMMARY

- **Plaque forms as a biofilm** at the border of teeth and gums.
- **Cavities (caries) form as bacteria in plaque ferment carbohydrates** and make acid that dissolves enamel. *Streptococcus mutans* and *Lactobacillus* species cause most acid production.
- **Gingivitis forms when plaque builds to tartar,** which starts an inflammation at the gumline.
- **Periodontal disease eventually develops if gingivitis is left untreated.** Gums recede and bone density declines.
- **Trench mouth is a serious necrotizing infection** that develops from neglected gingivitis.
- **Thrush is an infection of oral mucosa** caused by the yeast *Candida albicans*.

22.3
Gastrointestinal Syndromes

SECTION OBJECTIVES

- List the most common signs and symptoms associated with gastrointestinal infections.
- Correlate microbial pathogenic mechanisms with the presenting signs and symptoms of gastrointestinal infections.
- Outline the general mechanisms of diarrhea.

Why does the gastrointestinal tract get so many infections? The oral cavity and the gastrointestinal tract are portals between the inner us and the outer world; thus, numerous microbes pass through them. Some of these microbes overstay their welcome to erode teeth and destroy gums, as just described, or they infect gastrointestinal tissues and impair their function. Each gastrointestinal pathogen has what you might call its own tissue proclivities or specificities (tropisms), so different microbes infect different parts of the digestive tract. M cells, for example, are used by *Shigella* or *Salmonella* as a conduit to cause infection. Before we delve into the variety of viruses, bacteria, and parasites involved, a general description of the attendant diseases will set the stage.

Link As discussed in Section 15.2, specialized M cells (or microfold cells) are wedged between epithelial cells of Peyer's patches of the intestinal surface. These cells take up microbes from the intestinal lumen and release them, or pieces of them, into a pocket formed on the opposite side, where they are collected by other cells, such as macrophages.

Types of Diarrhea

What is diarrhea? **Diarrhea** is defined by the World Health Organization as having three or more loose (not formed) stools per day. Normally, intestinal mucosal cells absorb water (8–10 liters per day) *from* the intestinal contents, essentially drying out stool. Diarrhea occurs when water is not absorbed or when it actually *leaves* intestinal cells and *enters* the intestinal lumen. The excess water loosens stool and *voilà*—diarrhea. Several situations can cause water to be drawn from the intestinal lining:

- **Osmotic diarrhea** follows the intake of nonabsorbable substrates, such as lactulose, a synthetic sugar, which is used to treat constipation. Since lactulose is not absorbed, it remains within the intestinal lumen and increases osmolarity. When the osmolarity of intestinal contents is greater than internal osmolarity of mucosal cells, water leaves the cells and enters the lumen in an attempt to equalize osmotic pressure. The result is osmotic diarrhea. Infectious organisms (such as rotavirus) that prevent nutrient absorption can cause osmotic diarrhea. Osmosis is discussed in Section 4.5.

- **Secretory diarrhea** develops when microbes cause mucosal cells to increase ion secretion, as seen with cholera toxin. The secretion (exit) of electrolytes produces an electrolyte imbalance across the cell membrane (more electrolytes outside than inside the cell). Water then leaves the cells to equilibrate the imbalance.

- **Inflammatory diarrhea** occurs when the mucosal lining of the intestine is inflamed (red and swollen), as through an infection. In response to infection, the release of inflammatory cytokines attracts PMNs that inadvertently damage the epithelial layers. Damage to the intestinal wall diminishes the intestine's ability to absorb nutrients and water. If the damage is severe enough, red and white blood cells can enter the stool (called bloody diarrhea, or dysentery). *Shigella* and *Salmonella* species, among others, can cause infections leading to inflammatory diarrhea.

- **Motility-related diarrhea** occurs when food moves so quickly through the intestine (hypermotility) that there isn't sufficient time to absorb water or nutrients. Among various causes of intestinal hypermotility are enterotoxins (toxins that are specific for intestinal mucosa) produced by some microbes such as rotavirus (described later).

Stress, food allergies, and a number of microbial etiologies cause inflammation of different parts of the gastrointestinal tract, which often manifests as diarrhea. The following general terms are used to describe the inflammation of different parts of the GI tract:

- **Gastritis** refers to inflammation of the stomach lining (for instance, ulcers).
- **Gastroenteritis** has evolved into a nonspecific term for any inflammation along the gastrointestinal tract.
- **Enteritis** is defined as inflammation mainly of the small intestine.
- **Enterocolitis** is inflammation of the colon and small intestine.
- **Colitis** refers to inflammation of the large intestine (colon).

You might wonder why some microbes prefer the chaos of diarrhea to the relative stability of a nice commensal relationship. After all, isn't finding a niche and sticking to it the goal of every microbe? The simple answer is that diarrhea enables the dissemination of microbes that might otherwise kill their host or be killed themselves after too aggressively provoking the host's immune system. Dissemination allows the microbe to reach new hosts and proliferate. A diarrhea-causing microbe is sort of like a serial bank robber fleeing from city to city to avoid capture and find more banks to rob.

Signs and Symptoms of Hepatitis

Hepatitis is an inflammation of the liver (Figure 22.7) in response to infection, overconsumption of alcohol, exposure to toxic chemicals (such as the dry-cleaning solvent carbon tetrachloride), some medications (including over-the-counter medications such as Tylenol), or autoimmune reactions. Several viruses from different virus families can infect liver cells, as can some amebas and worms. The inflammatory response to these infections damages the liver and is ultimately the cause of hepatitis. Symptoms (see Figure 22.7) include an enlarged, painful liver; jaundice (yellowing of eyes or skin); dark urine; clay-colored stools; low-grade fever; nausea and vomiting; and weight loss. Because the liver is damaged, bilirubin (an orange-yellow breakdown product of heme that is normally cleared by the liver) spills over into the bloodstream (hyperbilirubinemia) and is secreted in urine. Hyperbilirubinemia manifests in the patient as jaundice and dark urine. The liver also uses bilirubin to make bile, which is deposited into the small intestine and produces the brown color of feces. A damaged liver fails to deliver adequate amounts of bile, resulting in clay-colored stools. A key laboratory indicator of liver damage is the elevation of liver enzymes, such as aspartate aminotransferase (AST) and alanine aminotransferase (ALT), in serum. The various causes of viral hepatitis will be described in the next section.

Figure 22.7   Hepatitis

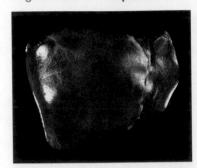

A. Normal liver.

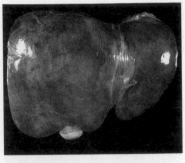

B. Diseased liver resulting from chronic hepatitis.

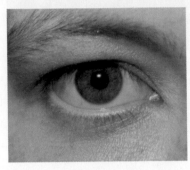

C. Yellowing of the conjunctiva of the eye in jaundice.

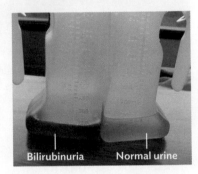

D. Bilirubinemia is a symptom of hepatitis in which urine turns brown (bilirubinuria) from excess bilirubin secreted from a damaged liver.

SECTION SUMMARY

- **Infectious diarrhea can be classified** as osmotic, secretory, inflammatory, and motility related.
- **Inflammation at different intestinal locations** produces gastritis (stomach), enteritis (small intestine), colitis (colon), enterocolitis (distal ileum and proximal colon), and gastroenteritis (stomach and small intestine; also a general term for gastrointestinal symptoms).
- **Hepatitis is inflammation of the liver.** Symptoms include jaundice, dark urine, and clay-colored stools. The elevation of serum liver enzymes AST and ALT are indicators of liver damage.

22.4
Viral Infections of the Gastrointestinal Tract

SECTION OBJECTIVES

- Match gastrointestinal symptoms to the viral infections that cause them.
- List routes of transmission of specific viral diseases of the gastrointestinal tract.
- Discuss viral pathogenic mechanisms responsible for specific disease signs and symptoms in the gastrointestinal tract.

Table 22.1
Viral Diseases of the Digestive System

Disease	Symptoms	Complications	Virus	Transmission	Diagnosis	Treatment	Prevention[b]
Mumps (parotitis)	Prodromal chills, fever, headache, swollen parotid glands, pain with chewing and swallowing	Meningitis, encephalitis, orchitis	Mumps virus	Aerosolization, fomites	EIA,[a] PCR, serum IgM or rise in IgG titer	Supportive treatment	Vaccine available (childhood immunizations)
Acute gasteroenteritis (mostly children under 3)	Watery, dark green diarrhea	Dehydration	Rotavirus	Fecal-oral route	EIAs, PCR for RNA	Rehydration, treatment of symptoms	Vaccines available
Acute gastroenteritis, also called stomach flu, (older children and adults)	Watery diarrhea, vomiting, headache, fever, chills, myalgia	Dehydration	Norovirus	Fecal-oral route	EIA, PCR for RNA	Rehydration, treatment of symptoms	No vaccine available
Acute gastroenteritis (mostly adults)	Watery diarrhea, vomiting, headache, fever, chills, myalgia	Dehydration	Sapoviruses	Fecal-oral route	Electron microscopy, RT-PCR	Rehydration, treatment of symptoms	No vaccine available
Acute gastroenteritis	Diarrhea, vomiting, abdominal pain, fever	Dehydration	Adenovirus, types 40 and 41	Fecal-oral route	EIA, PCR for RNA	Self-limited disease; rehydration	No vaccine available
Hepatitis	Variable degree of nausea, vomiting, abdominal pain, jaundice, dark urine, clay-colored stools	Vary, depending on causative viral agent; see Table 22.2	Hepatitis A–E viruses (from different families)	Varies, depending on the virus; see Table 22.2	Varies, depending on the causative viral agent; see Table 22.2	Varies, depending on the causative viral agent	Vaccines available against hepatitis A and B viruses; experimental vaccine with contradictory results against hepatitis E

[a]EIA = enzyme immunoassay to detect viral protein.
[b]The childhood vaccine schedule is shown in Table 17.6.

How do viruses affect your gastrointestinal tract? A wide range of viruses can affect your digestive system. Their effects range from swollen salivary glands to diarrhea and hepatitis. Some viral infections of the gastrointestinal tract can even spread to the brain. In fact, viral infections cause the majority of diarrheal disease worldwide as well as deaths from diarrheal disease. Some viruses that affect the digestive system are spread by aerosols, others rely on fecal-oral transmission, and some hepatitis viruses are sexually transferred. Table 22.1 presents features of common viruses that affect the digestive system.

Link As discussed in Section 2.4, **fecal-oral transmission** is used by those microbes that enter the body when ingested and leave when defecated.

Viral Gastroenteritis

Many people wrongly think that most cases of diarrhea are caused by a bacterial agent. Actually, viruses cause more intestinal disease than any bacterial species. Rotavirus has been the main cause of infectious diarrhea, but its incidence is declining now that we have an effective vaccine. Lesser-known viruses that cause gastroenteritis include members of the positive-sense, single-stranded RNA viruses of the Caliciviridae family (similar to norovirus); adenovirus strains 40 and 41 (double-stranded DNA), which can cause diarrhea that lasts 10–14 days; astroviruses (single-stranded RNA); and toroviruses (positive-sense, single-stranded RNA), which affect primarily infants under 1 year of age.

ROTAVIRUS GASTROENTERITIS **Rotavirus** is a nonenveloped, segmented, double-stranded RNA virus spread by the fecal-oral route. The virus is highly infectious, is endemic around the globe, and affects all age groups, although children between 6 and 24 months are most severely affected. The CDC estimates that by age 5, all children have had a rotavirus infection.

The rotavirus incubation period is approximately 2 days, after which the patient suffers frequent episodes of watery, dark green, explosive diarrhea. The patient can also feel nauseated, vomit, and have abdominal cramping. Diarrhea is caused in part by the virus destroying the intestinal epithelial cells, thus disrupting the absorptive function that these cells normally perform (osmotic diarrhea). In addition, a virus-encoded **enterotoxin** (a toxin that affects the intestines) stimulates intestinal nerves to overactivate intestinal motility to produce hyperperistalsis (increased intestinal motility). This enterotoxin also stimulates the vagus nerve, a long cranial nerve (cranial nerve 10) connecting the brain to many organs, including the stomach. Overstimulation of the vagus nerve makes stomach muscles involuntarily contract, making the patient vomit.

Severe dehydration and electrolyte loss due to the diarrhea will cause death unless supportive measures, such as fluid and electrolyte replacement, are undertaken. There is no curative treatment, but most patients recover if rehydrated properly, as did Dao Ming in the opening case history of this chapter. Few deaths from rotavirus occur in the United States, but each year more than 600,000 children worldwide die from this viral diarrhea. The mortality and incidence of this disease has declined since 2006 in the United States because of the availability of rotavirus vaccines. The vaccine is given in three doses at 2, 4, and 6 months of age as part of the routine childhood immunization schedule (Table 17.6). Protective antibodies conferred are primarily IgA (Chapter 16) and can be transferred in mother's milk.

NOROVIRUS GASTROENTERITIS With the advent of the rotavirus vaccine, **norovirus** (formerly Norwalk virus) is becoming the most common nonbacterial cause of gastroenteritis worldwide. Norovirus is a nonenveloped, positive-sense, single-stranded RNA virus transmitted by the fecal-oral route (**Figure 22.8A**). The classic scenario for norovirus outbreaks is the cruise ship in which tens of people get violently ill simultaneously as a result of contaminated food and direct person-to-person contact (**Figure 22.8B**). Norovirus outbreaks are actually more common wherever there are many people in a small area. The virus can spread quickly in nursing homes, dormitories, hotels, etc. About 24 hours after infection, patients experience a sudden onset of vomiting, abdominal cramps, and watery diarrhea. As with rotavirus infections, diarrhea results when norovirus destroys the absorptive capacity of the cells that line the intestinal lumen. Norovirus infections are sometimes called "stomach flu" because in addition to diarrhea, patients suffer from headache, fever, chills, and muscle aches (myalgia)—symptoms similar to influenza. The symptoms mercifully last only

Figure 22.8   Norovirus

A. The virus particles display icosahedral symmetry.

TEM 100 nm

B. An outbreak of norovirus diarrhea hit the cruise ship *Lady Anne* on the Rhine River in Germany in November 2008. The photo shows an elderly woman leaving the cruise ship with two Red Cross paramedics in full protective gear.

about 12–24 hours. All age groups are affected, although the principal victims are adults. Treatment is supportive, and no vaccine is available.

Hepatitis

CASE HISTORY 22.1

"Hurling" in Hawaii

Three weeks ago, Katherine, a 17-year-old from Waikiki, Hawaii, developed nausea, fever, and aches. She initially thought it was the flu. She felt extremely tired, dozed through school auditorium events, and was unable to participate in golf tournaments or the school debating team. About 2 days ago, Katherine could not keep food down and was taken to the emergency department (ED), where she was given an antinausea shot. The shot didn't work; she continued to vomit and made a second visit to the ED. She noted that her urine seemed dark in color, and she told the physician that she had not been hospitalized prior to this event and had not traveled outside of the United States. She denied using drugs or having any sexual partners. The practitioner noticed a yellow tinge to the white part of her eyes. A blood test showed that she was positive for anti-hepatitis A virus (HAV) antibodies.

Table 22.2
Characteristics of Hepatitis Viruses

Feature	Hepatitis A	Hepatitis B	Hepatitis C	Hepatitis D	Hepatitis E
Common Name	Infectious hepatitis	Serum hepatitis	Non-A, non-B posttransfusion hepatitis	Delta agent	Enteric non-A, non-B hepatitis
Virus Family and Structure	Picornavirus; nonenveloped; positive-sense, ssRNA	Hepadnavirus; enveloped; circular partially dsDNA	Flavivirus; enveloped; positive-sense, ssRNA	Viroid-like; enveloped; circular ssRNA	Calicivirus-like capsid; nonenveloped; positive-sense, ssRNA
Transmission	Fecal-oral	Parenteral, sexual (blood/body fluid transmission)	Parenteral, sexual (blood/body fluid transmission)	Parenteral, sexual (blood/body fluid transmission)	Fecal-oral
Onset	Abrupt	Insidious	Insidious	Abrupt	Abrupt
Incubation Period (Days)	15–50	45–160	14–180	15–64	15–60
Severity	Mild	Occasionally severe	Usually subclinical; 70%–80% chronicity (become chronic)	Usually mild when it coinfects with HBV; often severe when it superinfects an HBV carrier	Nonpregnant patients, mild; pregnant women, severe
Diagnosis	Clinical presentation and history; anti-hepatitis A virus immunoglobulin M (IgM) test	Clinical presentation; serum levels of HBsAg, HBeAg, and anti-HBc IgM	Serology; molecular assay for HCV RNA	Detection of HDV RNA by reverse transcription-polymerase chain reaction (RT-PCR); detection of serum immunoglobulin G (IgG) and IgM antibodies against hepatitis D	Clinical presentation and history; IgM anti-HEV in the serum by ELISA
Treatment	Supportive care	Interferon-alpha; nucleotide or nucleoside inhibitors; vaccinate against hepatitis A to prevent secondary HAV infection	Treatment plans are individualized and may include interferon-alpha and/or ribavirin; vaccinate against hepatitis A and B	Interferon-alpha; antiviral lamivudine has been tried, but efficacy is questionable; liver transplant	Usually self-limiting; nucleoside reverse transcriptase inhibitors in severe cases
Mortality	<0.5%	1%–2%	34%	High to very high	Nonpregnant patients, 1%–4%; pregnant women, 10%–40%
Chronic/Carrier State	No (acute)	Yes (acute or chronic)	Yes (acute or chronic)	Yes (acute or chronic)	No (acute or chronic)
Associated with	No other diseases	Primary hepatocellular carcinoma, cirrhosis	Primary hepatocellular carcinoma, cirrhosis	Cirrhosis, fulminant hepatitis	No other diseases

Katherine's symptoms are typical of acute viral hepatitis caused by one of the hepatitis viruses: A, B, C, or E. (There is also a hepatitis D virus, but HDV is defective and produces disease only if it coinfects a patient with hepatitis B virus.) The initial disease symptoms of fever, nausea, vomiting, dark urine, clay-colored stools, and often jaundice can be seen in all patients suffering from hepatitis, regardless of the viral cause. Despite producing similar symptoms, the different viruses that cause hepatitis do not even belong to the same virus family (Table 22.2). They have different morphologies, different genomic structures, and ultimately produce different clinical pictures. The good news about hepatitis is the ongoing decline in disease incidence. The not-so-good news is that hepatitis diseases are still dangerous and often incurable.

HEPATITIS A   Hepatitis A virus (HAV) is a nonenveloped, positive-sense, single-stranded RNA virus transmitted by the fecal-oral route, so overcrowding and lack of personal hygiene can increase the risk of getting this highly contagious disease. Eating undercooked seafood collected from contaminated waters and traveling to developing countries are also risk factors for HAV infection. Once inside the body, the virus replicates in the liver during an incubation period of up to 4 weeks. By the time jaundice appears, the number of virus particles in the blood and feces has started to drop, making the virus difficult to find. Fortunately, our immune system eventually comes to our rescue, and hepatitis A infections almost always go away without medical intervention (are self-limiting) in healthy people. There is no asymptomatic carrier state or links to future liver disease.

Health care professionals use serological tests to determine whether a person is *currently* infected with hepatitis A virus or if the patient has *previously* been infected. Acutely infected individuals have anti-HAV antibodies of the IgM type (as in Katherine's case), whereas those who were previously infected with the virus (or vaccinated) will have anti-HAV antibodies of the IgG type (and probably little or no IgM). People infected with this virus acquire long-term immunity. Children are now routinely immunized with HAV-inactivated viral vaccines.

Link As discussed in Section 16.3, the net result of the primary antibody response is the early synthesis and secretion of IgM molecules specifically directed against the antigen or immunogen. During a later stage of the primary response, a process known as **isotype switching**, or **class switching**, occurs, in which the predominant antibody type produced becomes IgG rather than IgM.

HEPATITIS B   Hepatitis B virus (HBV), an enveloped DNA virus, is transmitted through the exchange of blood (for instance, sharing needles between drug users) or body fluids (sexual intercourse). HBV does not have a fecal-oral transmission. It is the only hepatitis virus that has a DNA genome, which is small but codes for four viral proteins. Health care professionals can identify the extent and severity of HBV disease by immunologically detecting either the proteins (for example, HBsAg) or the antibody response to them. HBsAg is an HBV surface antigen made in such high amounts during acute infection that it accumulates as tubular structures in the blood (Figure 22.9).

Once infected, there is a relatively long incubation period of 8–12 weeks before typical hepatitis symptoms slowly (insidiously) appear. Only 30%–50% of infected immunocompetent individuals older than 5 years of age become symptomatic with this acute disease, and only 1% of them develop acute liver failure. Most people who are infected develop long-term chronic infections lasting years. Chronically infected but otherwise asymptomatic individuals essentially serve as the reservoir for HBV. Immunosuppressed patients and children younger than 5 years of age rarely develop acute disease because they lack an adequate inflammatory response. They are asymptomatic but remain chronically infected.

HBsAg as well as IgM to HBsAg can be found in the blood as soon as symptoms appear. In acute infections, both decrease as symptoms decrease. If HBsAg persists in blood samples for longer than 6 months, the patient has chronic HBV. IgM antibody will decrease even in chronically infected patients, but IgG antibodies to hepatitis will remain high.

During chronic infection, HBV DNA and antigens are found in organs and tissues other than the liver and may be the reservoirs for chronic infection. Two-thirds of people with chronic HBV show no symptoms, but the other third have a chronic *active* infection in which the liver is slowly and continuously damaged over time. The accumulated damage leads to cirrhosis (scarring of the liver) and liver failure. Chronic active infection also increases the risk of liver cancer. Note that patients chronically infected with HBV can develop a rapidly fatal disease if infected with hepatitis D virus.

The acute infection is treated only with supportive measures, but treatment of chronic infections includes long-term antiviral medications (currently 1 year) and interferon-alpha (currently 4 months) to suppress viral replication. The best weapon against getting HBV infection is prevention. Vaccination against this virus is now part of routine childhood immunizations. The vaccine consists of purified HBsAg prepared from recombinant yeast that were engineered to produce this protein in quantity.

Hepatitis D virus (HDV) is a positive-sense, single-stranded RNA virus that has an HDV core and an HBsAg coat derived from a coinfecting (simultaneously infecting) hepatitis B virus. HDV cannot replicate unless it coinfects with HBV or infects a chronic carrier of HBV. The result is a rapidly progressive disease that destroys the liver. Fortunately, infections with this virus are uncommon in the United States. There are no vaccines available for hepatitis D.

Figure 22.9   Hepatitis B

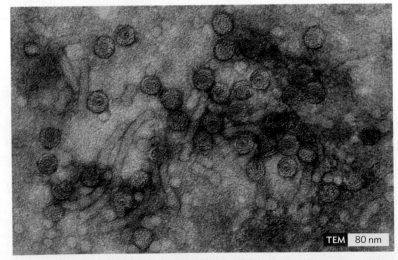

Colored electron micrograph of hepatitis B virions shows the dark brown circular infectious virions as well as the long tubular structures made from Hep B surface antigen (HBsAg).

HEPATITIS C   **Hepatitis C virus (HCV)** is an enveloped, positive-sense, single-stranded RNA virus transmitted by blood and body fluids (as described in the case history that opens Chapter 12). HCV causes an acute infection within 2–12 weeks of exposure. The acute infection is mild and self-limited, so the patient often does not show any symptoms. Liver function (and serum enzyme levels) in some patients can even return to normal after the initial acute disease, but the majority of those infected with hepatitis C virus (70%–85%) develop a chronic infection, even with normal liver function.

Most people who have chronic hepatitis C infection (60%–70%) have active liver disease, including fatty liver (accumulation of fat in the liver) leading to liver failure, and are at risk of developing liver cancer (hepatocellular carcinoma). The less intense inflammation process associated with HCV means that most chronic infections take a long time to cause liver failure. The stronger the immune response, the more rapidly liver disease develops.

Diagnosis for both acute and chronic disease is made by serology to detect antibody to hepatitis C and molecular assays that detect or quantify HCV viral DNA. The CDC recommends that persons at risk be screened for this infection and that a specialist be consulted about the treatment of those who have the disease. HCV treatment is based on the guidelines set by the Infectious Diseases Society of America (IDSA) and the American Association for the Study of Liver Diseases (AASLD). Acute HCV infections are effectively treated with interferon (IFN) and the antiviral drug ribavirin, which reduces synthesis of viral DNA. Three recently approved FDA drugs are used to treat chronic hepatitis C: sofosbuvir (Sovaldi), a nucleotide analog inhibitor of the HCV polymerase enzyme; simeprevir (Olysio), a protease inhibitor; and a combination of sofosbuvir and ledipasivir (Harvoni). Ledipasivir has a similar mode of action to sofosbuvir.

HEPATITIS E   **Hepatitis E virus (HEV)** is a positive-sense, single-stranded RNA virus that is rare in the United States but common in many other parts of the world. The virus is transmitted through the fecal-oral route, usually when a person drinks contaminated water. After an incubation period of 2–8 weeks, a usually mild and self-limiting disease develops. The patient experiences muscle aches, fatigue, mild fever, nausea, loss of appetite, and mild icteric (jaundice) symptoms. Fulminant disease—one that develops rapidly and is serious—is more likely in those who are pregnant, have preexisting liver disease, or are malnourished. There is some evidence of vertical transmission (from mother to fetus) when the mother is infected during the third trimester of pregnancy. Chronic HEV infection is increasingly seen in solid organ transplant recipients and other immunosuppressed patients in developed countries. Hepatitis E infection can be diagnosed by serology. Detection of the virus in bile, stool, or liver cells by RT-PCR or immune electron microscopy is also possible, but there are currently no commercial tests available, so samples can be processed only by specialized labs. A vaccine has been developed but is not yet available. Treatment is supportive.

Mumps

Mumps is a highly infectious, self-limiting, viral infection of the parotid gland caused by a **paramyxovirus**, a single-stranded RNA virus of the Paramyxoviridae family (a different paramyxovirus causes measles). The virus is shed in saliva and can be spread through sneezing, coughing, and other direct contact with another person's infected saliva. The virus initially replicates in the upper respiratory tract, then hitchhikes through the bloodstream to the parotid gland. Prodromal symptoms become noticeable 2 or 3 weeks after infection and include chills, headache, loss of appetite, and a lack of energy. Swelling of the salivary glands in the face (**parotitis**) generally occurs within 12–24 hours of the prodromal symptoms and usually disappears by the seventh day (Figure 22.10). The patient will complain of pain when chewing or swallowing and can have a fever as high as 40°C (104°F). Fortunately, once a person has contracted mumps, he or she becomes immune to the disease, regardless of how mild or severe symptoms may have been.

Routine childhood vaccination has made mumps an uncommon occurrence. Unvaccinated children and adults, however, are still at risk. Although most cases of mumps are uncomplicated, complications can occur. Surprisingly, adults who contract mumps are more prone to disease sequelae. In one complication, seen in 15% of adult cases, the covering of the brain and spinal cord becomes

Figure 22.10   Mumps

A. Child with a swollen parotid gland due to the mumps virus.

GREG CAN'T PLAY
MUMPS

B. Before the advent of the mumps vaccine in 1967, children with mumps were generally quarantined at home as a way to contain an epidemic. The head wrap was used to limit jaw movement that caused pain.

inflamed, a condition known as meningitis. Symptoms of viral meningitis include a stiff neck, headache, vomiting, pain with bending or flexing the head, and a lack of energy. Mumps meningitis usually resolves within 7 days with no lingering effects. However, the mumps virus can spread into the brain itself, causing inflammation (encephalitis). Symptoms of mumps encephalitis include seizures, an inability to feel pain, and high fever. The patient typically recovers completely within 14 days, although some are left with seizure disorders. About 1 in 100 patients with mumps encephalitis die from the complication.

Swelling of the testicles (**orchitis**) is a painful complication in 25% of postpubescent males who contract mumps. Orchitis can develop about 7 days after the parotitis stage. Painful swelling can occur in one or both testicles, accompanied by fever, nausea, and headache. Symptoms usually subside after 5–7 days, although the testicles can remain tender for weeks. Mumps orchitis in some cases causes long-term complications of testicular atrophy and infertility. Girls occasionally suffer an inflammation of the ovaries, or **oophoritis**, as a complication of mumps, but this condition is less painful than orchitis in boys.

The first vaccine against mumps was licensed in the United States in 1967 and is now a component of the MMR (measles, mumps, and rubella) trivalent vaccine. The vaccine is given in two doses at 1 and 4 years of age. Because of widespread vaccination, the disease rate in the United States has fallen 99%.

SECTION SUMMARY

- **Rotavirus is the most common cause of diarrhea,** especially among children. Transmitted by the fecal-oral route, the virus causes osmotic (anti-absorptive) diarrhea and enterotoxin-induced hyperperistalsis. A vaccine is available and recommended for children.

- **Norovirus is a major cause of diarrhea among adults** (anti-absorptive mechanism). Fecal-oral and person-to-person transmission can cause outbreaks among people in close quarters. No vaccine is available, but disease resolves quickly.

- **Hepatitis is caused by five different viruses.** Hepatitis A virus (HAV) is transmitted by the oral-fecal route (often via contaminated seafood). A vaccine is available and recommended for children.

- **Hepatitis B virus (HBV) can cause acute or chronic hepatitis.** It is transmitted by blood or other body fluids. A vaccine containing HBsAg is recommended for older children. Antiviral treatments are used for active disease. **Hepatitis D virus** is defective and infects only with HBV.

- **Hepatitis C virus (HCV) can cause a mild, acute infection that resolves quickly but often becomes chronic** and will slowly, over years, produce liver damage. There is no vaccine, but ribavarin, pegylated interferon, sofosbuvir (a viral polymerase inhibitor), and simeprevir (a protease inhibitor) are used as treatments.

- **Hepatitis E virus (HEV) is usually a mild disease** transmitted through the fecal-oral route. A vaccine has been developed but is not yet available.

- **Mumps is a viral infection of the parotid gland (parotitis)** that spreads through aerosolization and disseminates via blood to the parotid gland. Mumps vaccine is recommended for children.

Thought Question 22.1 If hepatitis viruses all replicate in the liver and are released into the intestine in bile, why is it that only HAV and HEV can be transmitted by the fecal-oral route?

Thought Question 22.2 Why are persons immunized against HBV also protected from HDV?

22.5
Bacterial Infections of the Gastrointestinal Tract

SECTION OBJECTIVES
- Categorize GI infections by bacterial pathogenic mechanisms.
- Correlate bacterial virulence mechanisms to GI disease symptoms.
- Relate the treatment of GI infections to virulence of GI pathogens.
- Evaluate treatment and prevention methods used for bacterial infections of the gastrointestinal tract.

How do bacterial gastrointestinal infections compare with infections caused by viruses? Bacterial infections of the gastrointestinal tract tend to be more severe than viral infections. A large part of this difference is due to the toxins bacteria produce and the inflammatory response they provoke. On the other hand, bacterial infections can usually be treated with antibiotics, whereas viral infections cannot. **Table 22.3** summarizes the bacterial diseases described in this section.

CASE HISTORY 22.2

It's Not What You Eat

Gary, a 34-year-old accountant who immigrated to Nebraska from Poland 7 years ago, has since his teenage years been bothered periodically by episodes of epigastric pain (pain around the stomach), nausea, and heartburn, especially after he ate certain foods. Gary was told to avoid those foods and take antacids as needed. Antacids usually alleviated the symptoms. Over the years, he received several courses of treatment with cimetidine (Tagamet) or famotidine (Pepcid) to reduce acid secretion and provide relief. Recently, the physician ordered an upper-GI endoscopy, in which a long, thin tube tipped with a camera and light source was inserted into Gary's mouth and into his stomach (**Figure 22.11**). The view through the endoscope showed some reddened areas in the antrum (bottom part) of the stomach. The endoscope was also equipped with a small clawlike structure that obtained a small tissue sample (biopsy) from the lining of Gary's stomach. A urease test performed on the antral biopsy turned positive in 20 minutes.

Table 22.3

Bacterial Diseases of the Digestive System

Disease	Symptoms	Complications	Bacteria	Transmission	Diagnosis	Treatment	Prevention
Gram-Negative Bacteria							
Peptic ulcer disease	Pain associated with eating	GERD, increased risk of esophageal and gastric cancer	*Helicobacter pylori*	Fecal-oral route	Urea breath test; detection of *H. pylori* antibody in blood; detection of *H. pylori* antigen in stool; detection of urease in biopsy samples	Depends on patient and type of ulcer; reduce stomach acid production (proton pump inhibitors); two or three antibiotics	No vaccine available
Gastroenteritis, *E. coli* (traveler's diarrhea)	Watery diarrhea, no fever or low-grade fever, abdominal cramps, nausea	Dehydration	Enterotoxigenic *E. coli*	Fecal-oral route	Clinical presentation; isolation from stool	Treat symptoms; fluid and electrolyte replacement; antibiotic treatment 1–3 days with fluoroquinolones or azithromycin	No vaccine available
Colitis, *E. coli*[a] (bacillary dysentery)	Bloody diarrhea, fever	Dehydration, chronic diarrhea, septicemia, pneumonia	Enteroinvasive *E. coli*	Fecal-oral route	Isolation from stool	Possibly fluoroquinolones; hydration	No vaccine available
Enterocolitis, *E. coli*	Acute (often bloody) diarrhea, abdominal cramps, painful defecation	HUS, TTP	Entero-hemorrhagic *E. coli* (*E. coli* O157:H7)	Fecal-oral route	Isolation from stool	Hydration; treat symptoms; antidiarrheal medications may increase risk of HUS in children; narcotics and NSAIDs not recommended	No vaccine available
Shigellosis (bacillary dysentery)	Bloody diarrhea, abdominal pain and cramps, tenesmus,[b] fever	May produce HUS, TTP	*Shigella* spp., especially *S. dysenteriae*	Fecal-oral route	Isolation from stool	Oral rehydration; antibiotics may be indicated in some cases	Vaccine being developed but not available
Salmonellosis (*Salmonella* ileocolitis)	Diarrhea with some blood and white cells, fever, abdominal pain and cramps, nausea, sometimes vomiting	Sepsis, osteomyelitis	*Salmonella* spp. (nontyphoidal)	Fecal-oral route	Isolation from stool	Treat symptoms; antibiotics not indicated as they increase risk of carrier state	No vaccine available
Yersiniosis	Severe right lower quadrant abdominal pain, fever, diarrhea, possible blood in stool	Liver and spleen abscesses, serious GI problems such as perforation	*Yersinia enterocolitica*	Fecal-oral route	Isolation from stool; isolation of *Y. enterocolitica* DNA from stool by PCR	Hydration; antibiotics may be required in cases of very high fever	No vaccine available
Enterocolitis, *Campylobacter*	Prodrome of fever and malaise leading to watery and bloody diarrhea, periumbilical[c] abdominal pain	Guillain-Barré syndrome (*C. jejuni*)	*Campylobacter* spp.	Fecal-oral route	Isolation from stool; dark-field microscopy; Gram stain with carbol fuchsin counter stain	Hydration; antibiotics may be required in severe cases and cases with high fever	No vaccine available

(continued on next page)

Table 22.3

Bacterial Diseases of the Digestive System (continued)

Disease	Symptoms	Complications	Bacteria	Transmission	Diagnosis	Treatment	Prevention
Cholera	Acute large-volume (10–14 liters) watery diarrhea ("rice water" stool); thirst; altered mental status	Electrolyte imbalance leading to cardiac and renal failure and death	*Vibrio cholerae*	Fecal-oral route	Isolation from stool	Replace fluids and electrolytes; antibiotics in cases of severe dehydration; antibiotic choice and dosage depend on age (adult vs. children of different ages): doxycycline, azithromycin, ciprofloxacin	Vaccine available but not in the United States
Vibriosis	Acute watery diarrhea, sometimes bloody, abdominal cramps and pain, sometimes fever	Soft tissue infections, organ failure, shock	*Vibrio parahaemolyticus*	Fecal-oral route (particularly contaminated seafood, especially raw oysters)	Isolation from stool	Usually self-limiting; replace fluids; tetracycline and ciprofloxacin	No vaccine available
Gram-Positive Bacteria							
Listeriosis	Watery diarrhea	Meningitis in immunocompromised; systemic disease in neonates	*Listeria monocytogenes*	Fecal-oral route (refrigerated pâté and meat spreads, soft cheeses, hot dogs and lunch meats, unpasteurized milk)	Isolation from stool, blood, or CNS	Oral rehydration; IV antibiotics for septicemia or meningitis	No vaccine available
Antibiotic-associated diarrhea and pseudo-membranous colitis	Watery or mucoid diarrhea, fever, abdominal pain, and/or cramps	Bowel perforation, SIRS, sepsis, others	*Clostridium difficile*	Fecal-oral route	Clinical presentation; PCR to identify *tox* gene in stool specimen; EIA to detect toxins in stool	Depends on severity of symptoms; remove offending antibiotic in all cases; supportive treatment may be sufficient for mild cases; vancomycin and metronidazole as indicated for severe cases	No vaccine available
Gastroenteritis, *C. perfringens* (food poisoning)	Antibiotic-associated diarrhea, watery	Segmental necrosis of jejunum (enteritis necroticans or pigbel), which is a life-threatening disease	*Clostridium perfringens*	Fecal-oral route	Detection of toxin in stool samples	Remove offending antibiotic; oral hydration	No vaccine available
Gastroenteritis, *S. aureus* (food poisoning)	Acute vomiting and diarrhea, abdominal pain and cramps, sometimes fever	Dehydration	*Staphylococcus aureus*	Ingestion of food contaminated with toxin	Clinical presentation; examination of food for presence of toxin	Self-limiting; hydration	No vaccine available

[a]Disease presentation is indistinguishable from shigellosis.
[b]Feeling like you are going to have a bowel movement without being able to have one.
[c]Periumbilical = around the umbilicus (belly button).

Histological examination of the biopsy confirmed moderate chronic active gastritis (inflammation of the stomach lining) and revealed the presence of numerous spiral-shaped organisms. Cultures of the antral biopsy were positive for the bacterium *Helicobacter pylori*.

Peptic Ulcers

Painful and sometimes life-threatening gastric and duodenal ulcers, collectively known as **peptic ulcers** (**Figure 22.12**), were for many years blamed on spicy foods and stress. These factors were believed to cause increased acid production that ate away at the stomach lining, even though the gastric mucosa was known to be well protected from stomach acid, which can fall as low as pH 1.5. Then, in the 1980s, there was a seismic shift in our understanding of ulcers. Australians Robin Warren and Barry Marshall discovered unusual Gram-negative, helical bacteria present in the biopsies of gastric ulcers and proposed that bacteria, not pepperoni, caused ulcers. Their hypothesis was viewed with skepticism and declared as heresy by the established medical community. Faced with disbelief bordering on ridicule, Marshall drank a vial of the helical organisms and waited. A week later, he began vomiting and suffered other painful symptoms of gastritis. The young intern could not have been happier. He had proved his point. We now know that this curly-shaped microbe, *Helicobacter pylori*, causes the vast majority of stomach and duodenal ulcers (**Figure 22.13**). In fact, *H. pylori* is the only known cause of infectious gastritis.

The discovery of *Helicobacter pylori*, a Gram-negative member of Helicobacteriaceae, and its association with gastric ulcer disease

Figure 22.11   **Endoscopy**

Endoscopy is used to view the stomach lining.

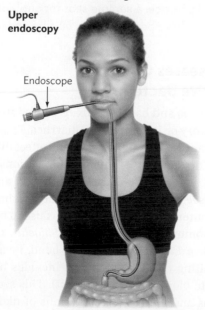

Upper endoscopy

Endoscope

Figure 22.12   **Peptic Ulcers**

Erosion of the lining of the stomach or duodenum is called an ulcer. Continued erosion of the ulcer can lead to bleeding. Patients with peptic ulcers often experience pain on the left upper side of their abdomen.

A. The white arrow indicates a gastric ulcer.

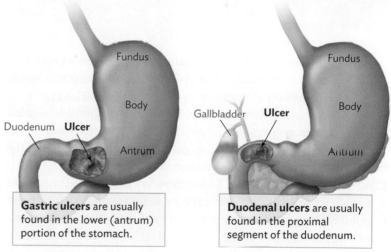

| **Gastric ulcers** are usually found in the lower (antrum) portion of the stomach. | **Duodenal ulcers** are usually found in the proximal segment of the duodenum. |

B. Location of gastric ulcers. Pain usually occurs when the stomach is empty and may temporarily be relieved after eating, which dilutes stomach acidity.

C. Location of duodenal ulcers. Pain typically starts about 2 hours after eating as the acidic contents of the stomach and bile acids empty into the duodenum.

led to an upheaval in gastroenterology. Prior to this discovery, treatment focused on suppressing acid production, which did not provide long-term relief. Sadly, within 1 year after acid suppression therapy, up to 80% of patients suffered a relapse of their ulcer. Therapy now includes antimicrobial treatment to kill the bacteria and acid suppression therapy to prevent further inflammation while the ulcer heals. Warren and Marshall (who recovered from his gastritis) received the 2005 Nobel Prize in Physiology or Medicine for their groundbreaking work.

Figure 22.13   A Bacterial Cause of Gastric Ulcers

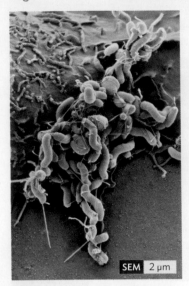

A. View of *Helicobacter pylori* in stomach crypts (colorized SEM).

B. Australian physician Barry Marshall was so sure he was right about the cause of stomach ulcers that he swallowed bacteria to prove his point.

We now know that *H. pylori* causes one of the most common chronic infections in adults. The risk of infection is 10% in young adults and increases to more than 50% for those who are 60 or older. *Helicobacter pylori* can inhabit the human stomach because of a bacterial enzyme called urease that converts urea to ammonia. The alkaline ammonia neutralizes gastric acidity in and around the organism and allows *H. pylori* to survive (but not grow) in the pH 1.5–2 stomach environment. This temporary protection from stomach acidity gives *Helicobacter* time to swim (using a tuft of polar flagella) to the relative safety of the mucus layer covering the stomach lining. In contrast to the stomach itself, the mucus layer has a hospitable pH of about 6. The bacteria then colonize the stomach by attaching to receptors on gastric epithelial cells. The presence of urease, lipopolysaccharides, and some other antigenic macromolecules produced by *H. pylori* initiates an immune response, which in turn leads to inflammation (gastritis), which in some people can progress to ulcers. The signs and symptoms of gastritis vary from person to person but often include pain in the upper abdomen and bloating. Victims also complain of nausea and/or vomiting and a lack of appetite. Of those infected, 15%–20% develop gastric or duodenal ulcers. Individuals with gastric ulcer disease can suffer with dyspepsia (indigestion, upset stomach) and a hunger-like pain in the upper abdomen that temporarily abates after eating or using antacids. Patients with duodenal ulcer disease develop upper abdominal pain and fullness, bloating, belching, nausea, and sometimes even vomiting <u>after</u> they eat. The pain results from ulcerated tissue coming in contact with the bile acids injected into the duodenum by the gallbladder.

Ulcers can become severe enough to "eat" through the lining of the stomach and cause bleeding and may completely perforate the wall of the stomach to form a hole. Perforated ulcers are serious because the contents of the stomach and blood will leak into the abdominal cavity. The patient with a perforated ulcer will have a sudden sharp pain in the epigastric region of the abdomen that radiates to the back. The patient's condition worsens rapidly within 12 hours of onset and without surgical intervention will end in a catastrophic loss of blood pressure and collapse. Patients who have a bleeding ulcer may also have dark (tar-colored) stools and in some cases have low red blood cell counts. In addition to ulcers, *H. pylori* infection is a risk factor for stomach cancer and may be involved in the development of gastroesophageal reflux disease (GERD, repeated partial regurgitation of gastric acid that damages the esophagus). Long-term *H. pylori* infection causes chronic inflammation of the stomach lining, leading to precancerous changes that may increase the risk of gastric malignancy.

Diagnosis of *H. pylori* infection can be achieved in several ways. In one, a small sample of stomach tissue is removed by endoscopy (as in Gary's case) and analyzed for the presence of bacteria, urease, or abnormal changes in the stomach tissue. The retrieved sample can then be sent for culture and antibiotic susceptibility testing. Other common procedures include testing the stool for the presence of *H. pylori* antigen and using assays (enzyme-linked immunosorbent assay, ELISA) to detect serum antibodies against *Helicobacter* CagA protein. ELISA methodology is described in Section 17.5.

Patients with *H. pylori* infection are treated with proton pump inhibitors to reduce stomach acidity and a combination of high doses of antibiotics (for example, metronidazole, amoxicillin, and clarithromycin) for 14 days. The most serious concern about *H. pylori* infection is the strong link between the infection and an increased risk of developing stomach cancer or MALT lymphomas.

> **Note** In clinical practice, a sample sent for antibiotic susceptibility testing is said to have been sent for "culture and sensitivity."

Intestinal Diseases Caused by Gram-Negative Bacteria

A number of bacteria and bacterial toxins cause intestinal diseases marked by secretory or inflammatory diarrhea. Figure 22.14 illustrates how different pathogens colonize and affect different parts of the intestines. The majority of intestinal infections are caused by Gram-negative bacteria. Enterotoxigenic *E. coli* and *Vibrio cholerae* produce a secretory diarrhea with large volumes of watery stools, whereas enterohemorrhagic *E. coli* causes a bloody diarrhea despite being noninvasive. *Salmonella*, *Shigella*, *Campylobacter*, and enteroinvasive *E. coli*, on the other hand, invade the intestinal lumen, resulting in a diarrhea that includes blood and fecal leukocytes (white blood cells in the stool). This section describes the pathogenesis and clinical presentations of diarrheal diseases caused by Gram-negative bacteria.

Figure 22.14 Intestinal Locations of Bacterial Infections
Different bacteria show a preference for different parts of the human small and large intestines to cause severe morbidity and mortality.

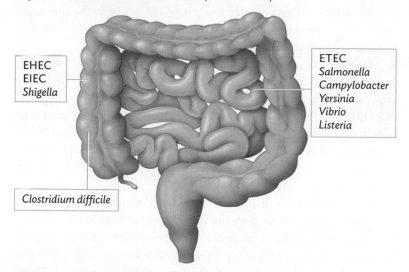

EHEC
EIEC
Shigella

ETEC
Salmonella
Campylobacter
Yersinia
Vibrio
Listeria

Clostridium difficile

CASE HISTORY 22.3

Field Trip Diarrhea

Tammy, a 6-year-old girl from Montgomery County, Pennsylvania, arrived at a hospital's emergency department with bloody diarrhea, a temperature of 39°C (102.2°F), abdominal cramping, and vomiting. She was admitted to the hospital 5 days after a kindergarten field trip to the local dairy farm. The child's health history was otherwise unremarkable. At the time of hospital admission, her parents were asked about Tammy's activities during the trip. They confirmed that Tammy bought a snack while at the farm. The laboratory reported that her fecal smear was positive for leukocytes and contained Gram-negative rods that produced Shiga toxins 1 and 2. Subsequent testing of the isolate's serotype revealed *E. coli* O157:H7. By this time, Tammy had developed further problems. Her face and hands had become puffy, she had decreased urine output despite being given IV fluids (suggesting kidney damage), and she was beginning to develop some neurological abnormalities. Laboratory analyses of blood samples revealed thrombocytopenia (reduced blood platelet count) and confirmed hemolytic uremic syndrome (HUS, renal failure). The child was given intravenous fluid and electrolyte replacement. Antibiotics were not administered, and Tammy eventually recovered.

Pathogenic bacteria that cause diarrhea (called **enteric pathogens**) are often more aggressive than viral pathogens, so they can cause more extensive epithelial damage. Some bacteria can invade and destroy the intestinal epithelial cells, whereas others remain stuck to the surface of the host cell and secrete toxins that damage the underlying cells. In either case, cytokines produced by the damaged cells can call in neutrophils and trigger inflammation. If the damage and inflammation are severe enough, as in Tammy's

case, white blood cells or blood mixed with stool can be seen by the patient or discovered in the laboratory. White blood cells in stool are rarely seen when the case is viral.

You might wonder why antibiotics were not administered to Tammy. Although antibiotic treatment of infectious gastroenteritis seems intuitive, it is rarely used (there are exceptions). Most gastrointestinal infections are viral, so antibiotics are ineffective. Gastroenteritis caused by bacteria usually resolves spontaneously, so antibiotics are unnecessary. Of course, severe systemic disease stemming from gastroenteritis (for example, *Salmonella* infection; see Section 21.3) can develop, often in the young or elderly. In addition, diseases such as typhoid fever (*Salmonella* Typhi) or bacillary dysentery (*Shigella sonnei*) will respond well to antibiotics. The most important treatment for diarrheal disease is the replacement of fluid and electrolytes either intravenously or through oral rehydration. We begin with *Escherichia coli*, the cause of Tammy's distress, and will describe what caused Tammy's puffy face, renal failure, and neurological signs.

ESCHERICHIA COLI GASTROENTERITIS Although numerous varieties of *E. coli* are part of our normal microbiota, there are at least six different classes (**pathovars**) of diarrheagenic (diarrhea-causing) *E. coli* strains. A pathovar is a group of serovars within a species (for instance, having different O and H Ag) that possesses the same combination of virulence genes whose products produce a similar pathology and similar symptoms. A single pathogenic species can have one or many pathovars.

E. coli pathovars can be differentiated by telltale **O and H cell-surface antigens**, which may be identified using agglutination tests. Thus, O157:H7 denotes the specific version of LPS (O157) and flagellar protein (H7) found on *E. coli* O157:H7. Three notable categories of diarrheagenic *E. coli* are described here.

Link As discussed in Section 18.3, the "O antigen" is part of the bacterium's LPS, and H antigen is a flagellar protein. Both **H and O antigens** can contribute to bacterial pathogenicity.

- **Enterotoxigenic *E. coli* (ETEC)** strains, as the name implies, secrete enterotoxins. One such toxin, called **labile toxin** (heat-labile toxin, LT), has a mode of action identical to that of cholera toxin, which was initially discussed in Sections 4.1 (see Figure 4.1) and 18.3 (see Figure 18.14). ETEC causes a secretory, watery diarrhea that can quickly dehydrate a patient. The organism has a fecal-oral route of transmission and no animal reservoirs.

- **Enteroinvasive *E. coli* (EIEC)** is the only pathovar of *E. coli* that invades epithelial host cells. EIEC is genetically very similar to *Shigella* species and even possesses the same plasmid required for invasion. As with ETEC strains, enteroinvasive *E. coli* does not have an animal reservoir and is transmitted by the oral-fecal route. The invasive nature of EIEC means that the patient will present with a bloody diarrhea, much like that caused by *Shigella*, which we will discuss later.

- **Enterohemorrhagic *E. coli* (EHEC)** O157:H7 and other strains are recently recognized pathogens that cause serious disease in humans. An important difference from other diarrheagenic *E. coli* is that EHEC strains have animal reservoirs; that is, the bacteria reside in the intestines of cattle and other animals. The animal reservoirs increase the chances of human contact. Transmission is fecal-oral, as you will see.

Case History 22.3 did not involve Tammy only. Numerous other children visiting the farm that day also developed diarrhea with similar symptoms. Using epidemiological investigative tools described in Chapter 26, the Centers for Disease Control and Prevention (CDC) identified the source of the disease outbreak. Fifty-one infected patients and 92 controls (children who visited the farm but did not become ill) were interviewed. Infected patients were more likely than controls to have had physical contact with cattle. All 216 cattle on the farm were sampled by rectal swab, and 13% yielded *E. coli* O157:H7 with a DNA restriction pattern indistinguishable from that isolated from the patients. This finding indicated that the cattle were the source of infection. Activities that promoted hand-mouth contact, such as nail biting and purchasing food from an outdoor concession, were more common among the children who contracted disease (fecal-oral route of infection). Furthermore, separate areas were not established for eating and interactions with farm animals. Visitors could touch cattle, calves, sheep, goats, llamas, chickens, and a pig while eating and drinking. Hand-washing facilities were unsupervised and lacked soap, and disposable hand towels were out of the children's reach. All of these situations provided ample opportunity for infection.

E. coli O157:H7 rarely affects the health of the reservoir animal. But when an infected steer is slaughtered, fecal matter can contaminate the carcass despite the manufacturer's considerable efforts

Figure 22.15 Attaching and Effacing Lesion Produced by *E. coli*

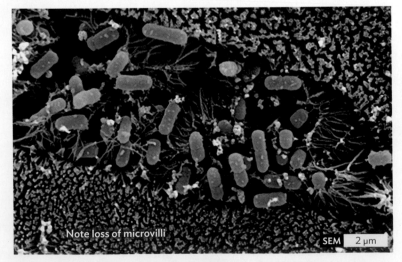

Note loss of microvilli

SEM | 2 µm

The image shows a lesion produced by an EPEC strain. Identical effacing lesions are produced by EHEC.

to prevent it. Grinding the tainted meat into hamburger distributes the microbe throughout. As a result, cooking burgers to 71.1°C (160°F) is essential to kill any existing EHEC. Also be aware that cross-contamination between foods is possible. Using the same cutting board to prepare meat and salad is highly likely to contaminate the salad, which will not be cooked.

Despite its common association with hamburger, vegetarians are not safe from this organism. During heavy rains, waste from a cattle farm can easily wash into nearby vegetable fields. If the cattle waste contains *E. coli* O157:H7, the crops are contaminated. One such outbreak occurred in 2006, when spinach from certain areas of California became contaminated with this pathogen, prompting a nationwide recall of bagged spinach and a month without spinach salad.

E. coli O157:H7 and, indeed, most EHEC strains have a remarkably low oral infectious dose because these pathogens are particularly capable of living in the acidic maelstrom of the stomach. *E. coli* survives stomach acidity using an impressive level of acid resistance that rivals that of the gastric pathogen *Helicobacter pylori*. These acid resistance mechanisms permit survival of *E. coli* in the acidic stomach and enable a mere 10–100 individual organisms to cause disease.

Once entering the intestine, the attachment of all pathovars of *E. coli* begins with specialized pili acting as grappling hooks that tether the bacteria to the host cell. Some pathovars, such as EHEC and enteropathogenic *E. coli* (EPEC), go further. These strains inject bacterial proteins directly into the host cytoplasm using syringe-like type III secretion systems (discussed in Section 18.4), establishing one-way communication. Some of these proteins instruct the host to form a membrane **pedestal** that hoists the organism up and away from the cell while maintaining intimate contact with it (see Figure 18.22). The pedestal prevents engulfment of the organism, keeping it away from intracellular defenses.

The host, for its part, emits a "cry" for help by way of cytokine secretion that initiates an inflammatory response. The resulting loss of microvilli produces what is called an **attaching and effacing lesion** (Figure 22.15). An anti-absorptive inflammatory diarrhea results, complete with blood and white cells entering the stool.

HEMOLYTIC UREMIC SYNDROME (HUS) AND THROMBOTIC THROMBOCYTOPENIC PURPURA (TTP) In Case History 22.3, Tammy's bloody diarrhea was accompanied by decreased urine output, a sign of kidney damage. The kidney damage was caused by two toxins (Shiga toxins 1 and 2, described in Section 18.3) that are encoded by bacteriophage genes embedded in the bacterial chromosome. These toxins are identical to the Shiga toxins produced by *Shigella dysenteriae* (discussed shortly). The toxins cleave host ribosomal RNA, which halts protein synthesis and kills the host cell. Endothelial cells in the kidney and brain are the most vulnerable. Endothelial damage triggers the formation of tiny platelet-fibrin clots that block blood vessels, leading to two major syndromes—**hemolytic uremic syndrome (HUS)** and

thrombotic thrombocytopenic purpura (TTP). HUS occurs when the microclots are limited to the kidney. The microclots clog the tiny blood vessels in this organ and cause decreased urine output, which can lead to kidney failure and death. Decreased urine output means that fluid is being retained in the body. Fluid retention explains Tammy's puffy face and hands.

In TTP, the clots occur throughout the circulation, causing reddish skin hemorrhages called petechiae and purpura (see Figure 21.12). Petechiae (small pinpoint purplish red spots) and purpura (purple patches) result from hemorrhages under the skin. Neurological symptoms (including confusion, severe headaches, and possibly coma) then arise from microhemorrhages in the brain. The hemorrhaging occurs because platelets needed for normal clotting have been removed from circulation as they form the microclots. The decreased number of platelets is called **thrombocytopenia**.

The development of HUS, as in Tammy's case, is a common consequence of *E. coli* O157:H7 infection, especially in children. Unfortunately, HUS can be treated only with supportive care, such as blood transfusions and dialysis throughout the critical period until kidney function resumes. It is important to note that antibiotic treatment can actually increase the release of Shiga toxins from these organisms and trigger HUS. Thus, antimicrobial therapy is not recommended.

Note that many Shiga toxin–producing strains of *E. coli* are not EHEC strains; that is, they fail to cause hemorrhagic disease. Thus, the term **Shiga toxin *E. coli* (STEC)** is a blanket term used for any *E. coli* that carries a Shiga toxin gene.

We have focused discussion in this section on three diarrheagenic pathovars of *E. coli*, but there are others, including **enteroaggregative *E. coli* (EAEC)**, which possesses unique bundle-forming pili, and **enteropathogenic *E. coli* (EPEC)**, similar to EHEC but without Shiga toxin. Each of these pathovars causes diarrhea, but they are isolated less frequently than EHEC.

SHIGELLOSIS Pathogenic species of *Shigella* (a Gram-negative bacillus, member of the family Enterobacteriaceae) cause **shigellosis**, a bloody diarrhea also called **bacillary dysentery**. The organisms are genetically very similar to EIEC (taxonomists, in fact, argue that *Shigella* is really a strain of *Escherichia*). *Shigella* lacks flagella (so no H antigen) and therefore is nonmotile. The structures of LPS O antigens divide *Shigella* into 45 antigenic serotypes, separated into four groups, or species (A–D). Two species are responsible for the majority of shigellosis cases. *Shigella sonnei* (group D) is the most common pathovar in the United States, whereas *Shigella dysenteriae* (group A) is predominant elsewhere in the world.

The symptoms of shigellosis begin after a short incubation period of 1–4 days with severe abdominal cramping, high-grade fever, vomiting, loss of appetite (anorexia), and large-volume watery diarrhea. Subsequently, the patient experiences an uncontrollable urge to defecate (tenesmus) and fecal incontinence and defecates a small volume of mucoid diarrhea with visible red blood present. *Shigella dysenteriae*, but none of the other *Shigella* species, produces

Shiga toxins that can lead to HUS and TTP. Treatment includes supportive oral rehydration therapy, and in contrast to EHEC, antibiotic treatment (usually a quinolone) is recommended to prevent a transient carrier state. Without antibiotic treatment, *Shigella* can continue to shed in stools for weeks after symptoms disappear, and the recovered patient will serve as a source of infection.

The pathogenesis of *Shigella* is similar to that of EIEC and is outlined in **Figure 22.16**. The organisms are very acid resistant, which means they easily survive stomach acidity (infectious dose is 10–100 ingested organisms). Once in the intestine, the bacteria are carried into the colon and attach to microfold (M) cells in Peyer's patches (see Figure 15.8). The microbe then engineers its entry (invasion) into the M cell (Figure 22.16, step 1). M cells are a common conduit for bacterial pathogens that invade the intestinal mucosa. *Shigella* (as well as EIEC) uses a virulence plasmid–encoded type III secretion system to inject bacterial effector proteins into the host cell. These effector proteins alter host actin polymerization and cause a membrane ruffle to form around the organism, bringing *Shigella* into the host cell (shown in Figure 18.21). The bacillus, now within a phagosome, uses an enzyme to lyse the phagosome membrane and escape to the cytoplasm (step 2).

A remarkable thing then happens in the cytoplasm. These bacteria, lacking flagella, nevertheless begin to move. As described in Section 18.5, proteins at one end of the bacillus begin to polymerize actin tails that move the organism through the cytoplasm. This helps propel the organism directly from one host cell to another. Because *Shigella* species can grow outside or inside a eukaryotic host cell, they are considered facultative intracellular pathogens.

The organism can exit the basolateral side of the M cell, where it is met by a macrophage (step 3). *Shigella* continues to grow in the macrophage and triggers apoptosis. The dying macrophage sends out the pro-inflammatory cytokine IL-1 beta, and the bacteria are released into the lamina propria. *Shigella* then enters the basolateral side of epithelial cells, escapes the vacuole, and uses actin motility (actin tails) to move into adjacent cells, where the phagosome escape process repeats itself (step 4). Stressed by the invader, host epithelia release another pro-inflammatory cytokine, IL-8.

The infected epithelial cells begin to die, destroying the integrity of the mucosal surface (step 5). At the same time, PMNs arrive, having been called to the scene by the cytokines. The PMNs engulf and destroy the bacteria, thereby preventing dissemination (the infection remains localized). These white cells add to the intestinal damage and pass through compromised epithelial cell-cell junctions into the intestinal lumen (step 6). The rampant destruction also allows blood to enter the stool, resulting in the bloody diarrhea.

Shigella is transmitted by the fecal-oral route from person to person or via contaminated food or water. However, *Shigella* does not have an animal reservoir (other than apes and monkeys). Because of its ability to resist gastric acid, it takes few organisms to cause disease. This low infectious dose means that *Shigella* is easily transferred in day-care facilities between children who, by most accounts, have limited hygiene skills.

Figure 22.16 Pathogenesis of *Shigella* Infection

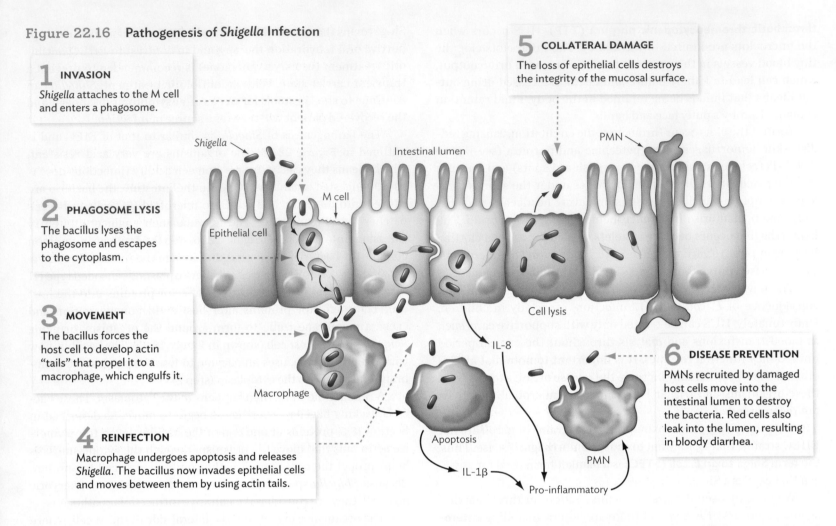

1 INVASION
Shigella attaches to the M cell and enters a phagosome.

2 PHAGOSOME LYSIS
The bacillus lyses the phagosome and escapes to the cytoplasm.

3 MOVEMENT
The bacillus forces the host cell to develop actin "tails" that propel it to a macrophage, which engulfs it.

4 REINFECTION
Macrophage undergoes apoptosis and releases *Shigella*. The bacillus now invades epithelial cells and moves between them by using actin tails.

5 COLLATERAL DAMAGE
The loss of epithelial cells destroys the integrity of the mucosal surface.

6 DISEASE PREVENTION
PMNs recruited by damaged host cells move into the intestinal lumen to destroy the bacteria. Red cells also leak into the lumen, resulting in bloody diarrhea.

Labels in figure: Shigella, Intestinal lumen, PMN, M cell, Epithelial cell, Cell lysis, Macrophage, IL-8, Apoptosis, IL-1β, Pro-inflammatory, PMN

SALMONELLOSIS   The genus *Salmonella* (Gram-negative bacillus, member of the family Enterobacteriaceae) consists of only two species: *S. bongori* (found in cold-blooded animals, or ectotherms) and the more clinically significant *S. enterica*. The problem is that *S. enterica* includes more than two <u>thousand</u> infectious serovars (equivalent to serotype, based on O and H antigens). Fortunately, all these serovars cause only two basic clinical diseases, **typhoid fever** (or **enteric fever**) and enterocolitis. Collectively, *Salmonella* infections are called **salmonelloses**. The nomenclature for these organisms has undergone extensive revision over the years. Serovars were previously given species status, such as *typhimurium, heidelberg, dublin*; note that these strains were often named after the city in which they were first isolated. Today the strains are referred to as follows: *S. enterica* serovar Typhimurium (no italics), or simply *S.* Typhimurium for convenience.

Note Typhoid fever is an ancient disease whose name comes from the Greek *typhos*, meaning an ethereal smoke or cloud that was believed to cause disease and madness. In the advanced stages of typhoid fever, the patient's mental state is "clouded."

Enteric fevers caused by *S.* Typhi, *S.* Paratyphi, and *S.* Choleraesuis are actually systemic infections that sometimes include abdominal symptoms (Section 21.3). Regardless of GI involvement, the disease starts with the ingestion of bacteria in contaminated food or water (fecal-oral transmission), which is why it is included in this chapter. *S.* Typhi and *S.* Paratyphi do not have animal reservoirs. Transmission, therefore, relies on symptomatic as well as asymptomatic human carriers contaminating food or water with fecal material containing these organisms.

The symptoms of enteric fever occur between 1 and 3 weeks postingestion. The classic presentation begins with a stepwise increase in fever to over 40°C (104°F) (**Figure 22.17**) and bacteremia during the first week followed by abdominal pain, continued fever, and chills. In some individuals, at fever's peak, a salmon-colored rash (called rose spots) containing *Salmonella* can appear on the patient's trunk and abdomen (see Figure 21.17). Diarrhea can also develop, especially in children, although some patients report constipation. The spleen and liver can become enlarged (**hepatosplenomegaly**), and the patient may develop septic shock (see Chapter 15 inSight), although the disease usually resolves after about 3–4 weeks. Treatment includes antibiotic therapy: fluoroquinolones for susceptible strains or ceftriaxone for multidrug-resistant strains. Disease caused by *S.* Typhi can be prevented with vaccines—a live, oral attenuated vaccine or a purified capsule polysaccharide vaccine (Vi antigen, discussed shortly). However, because there are only 200–300 cases of typhoid fever reported each year in the United

Figure 22.17 Typhoid Fever

Stepwise increase in fever over the first week. A patient's temperature was monitored twice daily (closed circles) and graphed. The graph shows daily rises and falls of the patient's fever. The fever crescendos to its highest by day 7 and then gradually recedes to normal range by day 21.

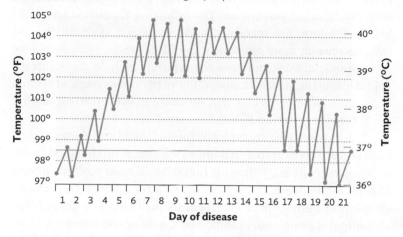

105°
104°
103°
102°
101°
100°
99°
98°
97°
Temperature (°F)
40°
39°
38°
37°
36°
Temperature (°C)
1 2 3 4 5 6 7 8 9 10 11 12 13 14 15 16 17 18 19 20 21
Day of disease

States, the vaccines are reserved for high-risk populations, such as military personnel and travelers to endemic areas.

Gastrointestinal disease caused by other serovars of *Salmonella enterica*, such as Typhimurium, Heidelberg, or Dublin (so-called nontyphoidal strains), does not typically develop into septicemia or septic shock unless the host is immunocompromised. (Case History 6.1 presented a patient with salmonellosis.) These organisms do have prominent animal reservoirs and currently represent the most common bacterial cause of diarrhea in the United States. Poultry, including chicken, chicken eggs, and turkey, are the most common sources of infection; the presence of these pathogens in reptiles such as pet store turtles results in some cases of direct transmission to pet owners.

Symptoms of enteritis are indistinguishable from those caused by other enteric pathogens. Nausea, vomiting, and diarrhea occur 6–48 hours after ingestion. Stools are normally loose and non-bloody, although dysentery may develop in some. Fever usually resolves within 3 days and diarrhea within a week. Uncomplicated enteritis caused by *Salmonella* should not be treated with antibiotics because the infection is usually short-lived and self-limiting. However, septicemia and invasive disease caused by some *Salmonella* serovars must be treated with antibiotics (often fluoroquinolones). *S.* Choleraesuis, for example, causes little if any diarrhea but rapidly penetrates the intestinal mucosa and invades the bloodstream (septicemia) to cause disseminated disease.

Salmonella is not as acid resistant as *E. coli* or *Shigella* and as a result has a higher infectious dose (10,000–100,000 organisms ingested). Once in the small intestine, *Salmonella* binds to M cells present in Peyer's patches in the ileum (**Figure 22.18**, step 1). Section 17.5 described *Salmonella*'s use of type III secretion systems to engineer its entry into M cells and limit phagolysosomal fusion.

Figure 22.18
***Salmonella* Pathogenesis**

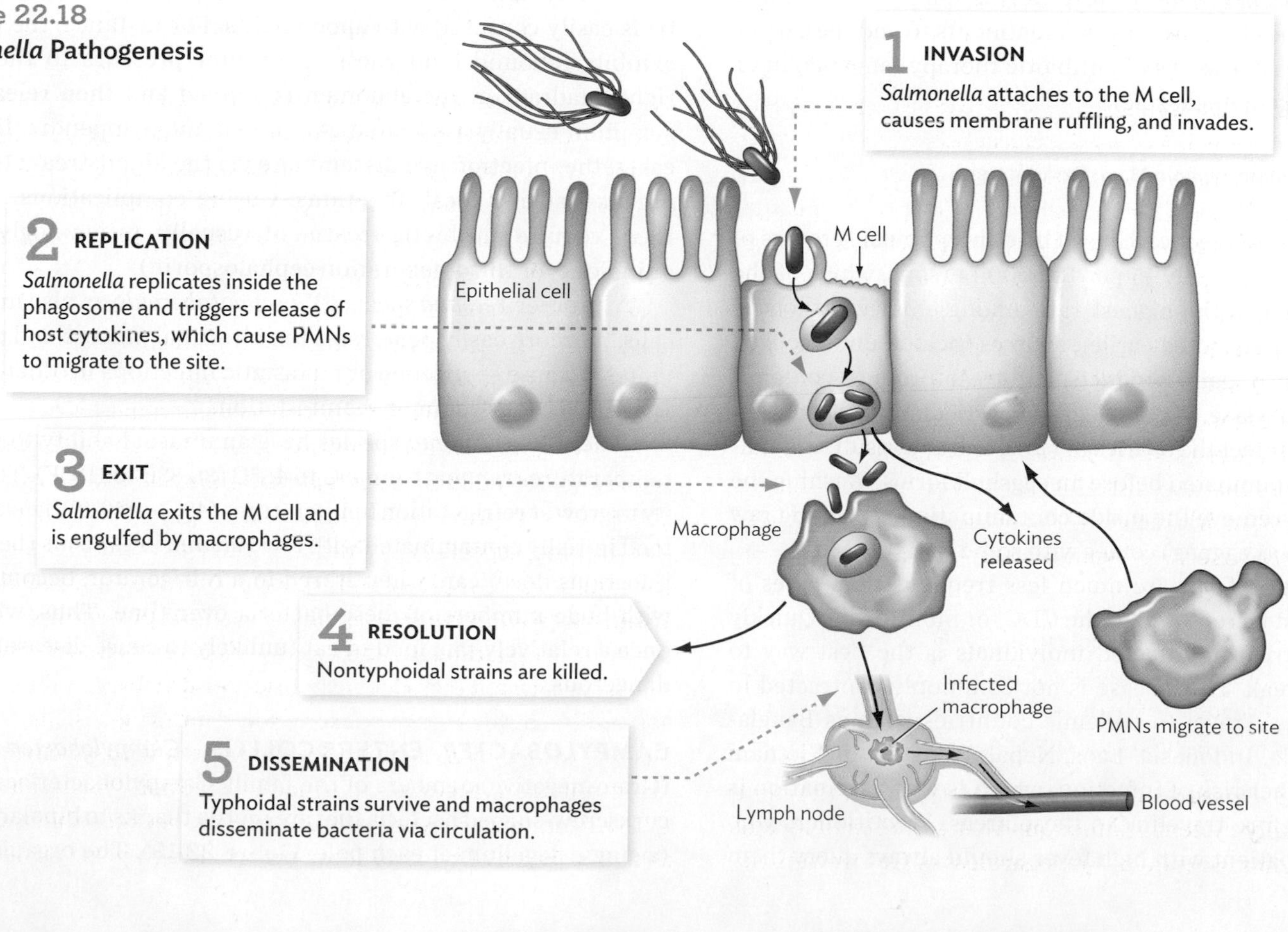

1 INVASION
Salmonella attaches to the M cell, causes membrane ruffling, and invades.
2 REPLICATION
Salmonella replicates inside the phagosome and triggers release of host cytokines, which cause PMNs to migrate to the site.
3 EXIT
Salmonella exits the M cell and is engulfed by macrophages.
4 RESOLUTION
Nontyphoidal strains are killed.
5 DISSEMINATION
Typhoidal strains survive and macrophages disseminate bacteria via circulation.
M cell
Epithelial cell
Macrophage
Cytokines released
Infected macrophage
PMNs migrate to site
Lymph node
Blood vessel

Salmonella, unlike *Shigella*, remains in the phagosome and replicates there. As the organism replicates, the phagosome migrates to the basolateral side of the M cell (step 2). Cytokines released by the infected cells start the inflammatory process (described in Section 15.3), which includes a call to attract PMNs. *Salmonella* bacteria are ultimately released from the basolateral side of the M cell (step 3), where they are met by macrophages and PMNs.

The nontyphoidal strains of *S. enterica* are killed once they are engulfed by these phagocytic cells (step 4) and do not survive the inflammatory process. But inflammation severely compromises the integrity of the intestinal mucosa and causes the loss of fluid (diarrhea) as well as some blood and white cells (inflammatory diarrhea), although the stools usually are not overtly bloody. In contrast, the typhoidal strains of *Salmonella* can survive within macrophages, which then serve as cellular "Trojan horses" carrying the organism to the regional lymph nodes and eventually the bloodstream (step 5). At this stage, the patient's fever begins to rise and sepsis becomes a risk.

During its travels, *S.* Typhi is protected from the immune system by a capsule called **Vi Ag (virulence antigen)**, unique to *S.* Typhi. As a result, the organism can ride the bloodstream unassailed and disseminate around the body. The pathogen eventually takes refuge in the gallbladder and periodically reenters the colon to be shed in feces. The colonized patient, after recovery, becomes an asymptomatic carrier who can transmit the organism to others through contaminated food. Up to 10% of untreated typhoid patients can shed *Salmonella* in feces for as long 3 months, and 1%–4% can become chronic carriers. Treatments to end the carrier state involve at least 4 weeks of antibiotic therapy (amoxicillin or trimethoprim-sulfamethoxazole).

 True Story Behind Typhoid Mary; Dark Matters

The incidence of infection caused by nontyphoidal serovars of *Salmonella* is approximately 15 per 100,000 persons, which at the time of this writing is the highest rate among enteric pathogens under surveillance. As noted earlier, sources include undercooked poultry, ground beef, dairy products, and fresh produce contaminated with animal waste. Chicken eggs in particular are a problem. *S.* Enteritidis can infect the ovaries of hens, and as a result, egg contents become contaminated before an eggshell forms. Washing the eggshell does not remove the inside contamination, so adding raw eggs to food (such as eggnog) comes with some risk.

Cases of enteric fever are much less frequent than cases of enteritis and must be reported to the CDC for monitoring. Quickly identifying and treating infected individuals is the best way to prevent an outbreak. The disease is not commonly contracted in the United States. Travel to endemic countries such as Bangladesh, China, India, Indonesia, Laos, Nepal, Pakistan, or Vietnam comes with a higher risk of infection, which is why vaccination is recommended before traveling to these areas. Practitioners who are evaluating a patient with high fever should always query them about international travel because it can be an invaluable clue to the cause.

YERSINIOSES Diseases caused by *Yersinia* species (Gram-negative bacilli, members of the family Enterobacteriaceae) are collectively called **yersinioses**. The infamous *Y. pestis* is transmitted by flea bite, not ingestion, and produces a systemic disease called **plague** or, more ominously, the Black Death. Because plague is a systemic disease and does not typically include gastrointestinal involvement, it was discussed in the previous chapter (Section 21.3). Two *Yersinia* species that do cause gastrointestinal disease are *Y. enterocolitica* and *Y. pseudotuberculosis*.

These organisms are typically found in wild and domestic animals and are transmitted to humans via contaminated food (milk, meat, or vegetables). Following ingestion, *Y. enterocolitica* infects M cells in the ileum and colon, where, after a few days, it causes symptoms of enteritis or enterocolitis involving low-grade fever, abdominal cramps, and, in some cases, nausea and vomiting. Diarrhea, which may contain blood and white cells, can linger for up to 2 weeks and then end. As with most uncomplicated diarrheas, antibiotic therapy is not recommended.

Y. enterocolitica infections can cause problems outside the intestine if they make it to the mesenteric lymph nodes. These nodes are embedded in the membrane (mesentery) that suspends the jejunum and ileum from the posterior wall of the abdomen. The result is mesenteric adenitis (**adenitis** is inflammation of a gland or lymph node). Pain associated with mesenteric adenitis is easily confused with appendicitis. For instance, the patient exhibits rebound tenderness—pain after pressure to the lower right quadrant of the abdomen is applied and then released—a symptom usually associated with an inflamed appendix. In some cases, the infection can disseminate via the bloodstream to other organs, causing focal infections. Curing complications such as these require antibiotic treatment (usually an aminoglycoside, quinolone, or third-generation cephalosporin).

The other *Yersinia* species, *Y. pseudotuberculosis*, also infects M cells but more easily penetrates the intestinal mucosa and disseminates to cause septicemia or metastatic infections in other organs. Diarrhea is not a common manifestation.

These two *Yersinia* species have an unusual ability to grow at temperatures ranging from 1°C to 45°C (38.8°F to 113°F). This ability to grow at refrigeration temperatures (4°C, or 39.2°F) means that food initially contaminated with few *Yersinia* organisms (below the infectious dose) can, when stored in a refrigerator, become laden with huge numbers of these bacteria over time. Thus, what was once a relatively safe food to eat (unlikely to cause disease) is now dangerous.

CAMPYLOBACTER ENTEROCOLITIS *Campylobacter* species (Gram-negative members of the family Campylobacteriaceae) are corkscrew-shaped bacteria that are motile thanks to bipolar flagella (a single flagellum at each pole; Figure 22.19). The organisms are

Figure 22.19   *Campylobacter*

Scanning EM of *Campylobacter* shows the helical shape of the bacterium. Inset shows the organism's bipolar flagella.

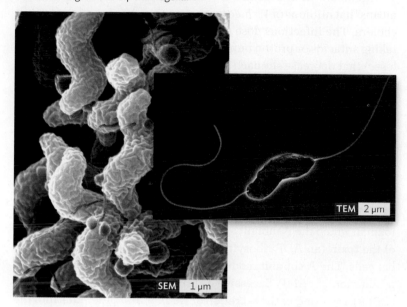

microaerophilic (they prefer low oxygen conditions, 5%, for growth) and are sensitive to low pH, yet they somehow survive a variety of inhospitable environments to cause infection. The infectious dose is around 500 bacteria.

Campylobacter species (mostly *Campylobacter jejuni* and *Campylobacter coli*) represent the number one cause of diarrhea worldwide. These bacteria are part of the normal intestinal microbiota of poultry and cattle and are transmitted to humans through consumption of contaminated and undercooked meat, contaminated water, or unpasteurized milk. *Campylobacter* harbors a large plasmid, called pVir, and is capable of causing bloody diarrhea. The organism uses flagella and adhesins to colonize the epithelium in the distal small intestine, then invades the mucosa through M cells to produce invasive disease.

The incubation period following ingestion is about 7 days, followed by a brief (24-hour) prodromal phase consisting of high fever, generalized pain, and sometimes dizziness. Severe abdominal pain, sometimes localized to the area around the belly button (periumbilical), and diarrhea with up to ten bowel movements a day begin suddenly. Nausea is common, and vomiting may also be present. These symptoms, as with enterocolitis caused by *Yersinia enterocolitica*, mimic appendicitis and can be mistaken for it. Blood and mucus can appear in the stool of infected patients. The signs and symptoms of infection in children vary somewhat, depending on the age of the child. Severe complications such as meningitis and seizures or convulsion can be seen but are not common. Neonates can become infected during birth if their mother is a carrier or may acquire the organism as a nosocomial infection while in the hospital nursery.

As with most gastrointestinal pathogens, our immune system is quite effective at confining *Campylobacter* to the GI tract, and

the disease usually ends within a week. Oral rehydration therapy is thus usually treatment enough, and antibiotics are unnecessary. However, antibiotic therapy can be used for patients suffering with prolonged symptoms (diarrhea more than a week) and severe abdominal pain with bloody stools.

Despite normally being self-limiting, *Campylobacter*-associated disease in immunocompromised patients can take a more complicated course. These patients not only experience a more severe gastrointestinal disease but also can develop disseminated infections resulting in **pancreatitis** (inflammation of the pancreas), **cholecystitis** (gallbladder infection), or **peritonitis** (infection of the peritoneum) as well as **myocarditis** (infection of heart muscle) or **pericarditis** (infection of the heart sac, or pericardium). These infections *must* be treated with antibiotics.

Campylobacter species are naturally resistant to a number of antibiotics, including penicillins, trimethoprim, and most cephalosporins. The drug of choice for treating severe gasteroenteritis is erythromycin (a macrolide), but the organisms are also sensitive to fluoroquinolones, clindamycin, tetracyclines, aminoglycosides, and chloramphenicol. Unfortunately, pathogenic strains resistant to macrolides and fluoroquinolone have already emerged and are prevalent in some parts of the world.

Following *Campylobacter* infection, a small portion of patients develop an ascending muscle weakness known as **Guillain-Barré syndrome (GBS)**, an autoimmune disease that appears to be triggered by cross-reactivity between *Camplylobacter* lipopolysaccharide structures and gangliosides (lipid-carbohydrate molecules) on nerve fibers. Antibodies to the *Campylobacter* LPS bind to nerve gangliosides and begin an immune complex disease process in which cytotoxic lymphocytes and macrophages infiltrate the nerves and remove the protective myelin sheath (Figure 22.20). Most GBS patients make a good recovery but may require hospitalization. About 10% die. Although other infectious agents, mostly

Figure 22.20   Guillain-Barré Syndrome

Infiltration of cytotoxic lymphocytes (T cells) and macrophages demyelinates the nerve, causing nerve impulses to dissipate before they can reach a muscle.

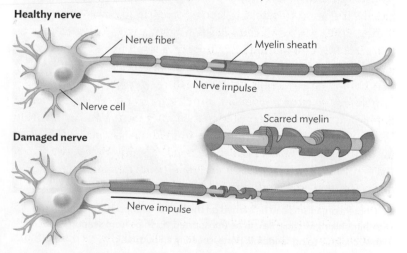

viral, have been associated with Guillain-Barré syndrome, *Campylobacter* is linked to 40% of cases and is the number one antecedent cause of this disease.

CHOLERA   The case history that opened Chapter 4 described the predicament of Akal, a 10-year-old Pakistani with watery diarrhea caused by *Vibrio cholerae*. Members of *Vibrio* (a Gram-negative curved rod belonging to the family Vibrionaceae) are primarily aquatic organisms that grow well in brackish waters found in estuaries (where ocean and fresh water mix, such as bays). These halophilic ("salt loving") bacteria are motile via a single polar flagellum (**Figure 22.21A**). *Vibrio* species that cause diseases (called **vibrioses**), are divided into cholera and noncholera vibrios. **Cholera**, caused by *V. cholerae*, is a severe, watery diarrhea in which the patient can lose 10–15 liters of fluid a day. Serotyped by their LPS O antigens, strains of *V. cholerae* that cause epidemic cholera are designated serotype O1 (classical and El Tor biotypes) and O139. Non-O1/O139 strains are occasionally associated with outbreaks of cholera but more commonly cause wound infections or sepsis in immunocompromised hosts. Following the 2010 earthquake in Haiti, an outbreak of cholera infected more than 640,000 people and left 8,000 dead. The disease is thought to have been brought to the island by Nepalese soldiers who joined the United Nations peacekeeping force deployed to Haiti following the quake. Although cholera is typically found in developing countries with poor hygiene facilities, occasional outbreaks caused by the O1 serotype have appeared along the Gulf Coast of the United States.

Members of the genus are extremely acid sensitive, which means that millions of *V. cholerae* bacteria must be ingested to cause cholera. The infectious dose can be dramatically lower in patients taking antacids or proton pump inhibitors such as omeprazole (Prilosec) that decrease stomach acidity. Survivors that make their way to the small intestine attach to epithelia there and secrete a toxin called **cholera toxin**, which causes a secretory diarrhea. Cholera toxin stimulates secretion of electrolytes from intestinal epithelial cells, followed by water. The result is a watery diarrhea that has the appearance of rice water (the water drained from boiled rice). Because *V. cholerae* is not invasive, there is no fever, and fecal blood and white cells are absent.

Link   As discussed in Section 18.3, cholera toxin (CT) is an AB toxin in which the B subunit structure helps introduce the A subunit of the toxin (an ADP ribosyltransferase) into the intestinal epithelial cells. The A subunit modifies adenylyl cyclase, leading to overproduction of cAMP. Excessive cAMP activates the Cl^- transporter (CFTR), resulting in the secretion of Na^+, Cl^-, and other ions into the feces. The intestine secretes water in an effort to maintain intracellular osmolarity. The excessive amount of water secreted into the colon results in the "rice water" stool appearance of cholera.

Treatment requires rehydration via an oral or intravenous route. Because large volumes of water and electrolytes are lost, blood can become extremely viscous and organs will not function properly. Without rehydration therapy, death comes to 50% of patients. **Figure 22.22** shows the approach used during large cholera outbreaks in endemic areas. Patients lie in "cholera cots" containing a hole under which a bucket is placed. Because the victim is too weak to move, diarrheal fluid is collected in the bucket, its volume measured, and an equal volume of replacement fluid given intravenously.

When treated with fluid replacement only, cholera is generally self-limiting, and mortality drops to less than 1%. Nevertheless, antibiotic treatment (doxycycline) will shorten the duration and volume of fluid loss and hasten clearance of the organism from stool. Two different oral vaccines are available outside the United States. The CDC, however, does not recommend cholera vaccination for U.S. travelers because the risk of getting cholera is low in U.S. travelers and the vaccines offer only partial protection (up to 50%). Prevention strategies in poor, endemic areas can be rather crude but effective. For example, *V. cholerae* tends to cling to the legs of copepods that live in brackish water (**Figure 22.21B**). The simple act of filtering water collected from these areas through the finely woven sari cloth worn by Indian women removes the copepods and with them the majority of the vibrios.

Several species of **noncholera Vibrio** cause noncholera diseases, the most important of which are *V. parahaemolyticus* and *V. vulnificus*. Others, described earlier, are classified as non-O1 *V. cholerae*. *V. parahaemolyticus* is widespread in aquatic environments. Outbreaks are associated with eating undercooked seafood, especially

Figure 22.21   *Vibrio cholerae*

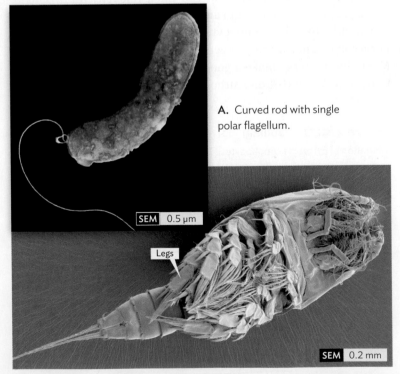

A. Curved rod with single polar flagellum.

SEM   0.5 µm

Legs

SEM   0.2 mm

B. Cholera organisms can be carried on the legs of copepods. Copepods and their hitchhiking *V. cholerae* can be transported via ships from seaports endemic with *V. cholerae* to nonendemic ports, sparking an outbreak.

raw oysters. Its pathogenic mechanism is unclear, but the organism can cause a watery or, less commonly, bloody diarrhea.

Vibrio vulnificus infections include sepsis (which occurs in patients with liver disease, such as cirrhosis) and wound disease that can affect people even with no underlying disorder (as described in the case history that opens Chapter 15). As with all waterborne, pathogenic vibrios, infections with *V. vulnificus* typically occur in coastal areas in the warm summer months, when the organism grows best. Wounds can become infected from swimming or fishing in endemic coastal areas. Alternatively, individuals with decreased stomach acid can contract disease through ingestion (raw oysters, for instance). The species grows extremely fast in blood (10-minute generation time); as a result, full-blown, life-threatening sepsis can develop quickly. Wound infections can begin to swell within hours of infection and become an extremely painful cellulitis that can develop into necrotizing disease (**Figure 22.23**). Left untreated, the mortality rate is an incredible 50%. Treatment must include doxycycline.

Intestinal Diseases Caused by Gram-Positive Bacteria

There are only a few species of Gram-positive bacteria worthy of mention that cause gastrointestinal diseases. These include *Listeria monocytogenes*, *Clostridium difficile*, *Clostridium perfringens*,

Figure 22.22   Mass Treatment of Cholera Victims

A. Cholera cots used to collect the copious watery diarrhea from incapacitated patients.

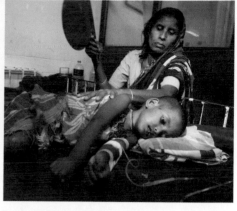

B. A child being treated on a cholera cot at a clinic in Bangladesh. Note the IV lines used to replace fluids and electrolytes.

and *Staphylococcus aureus*. We first discussed *C. perfringens* in Section 19.5 because it is the etiologic agent of gas gangrene. Some species of *C. perfringens*, however, produce an enterotoxin called *C. perfringens* enterotoxin (CPE) that causes abdominal cramps and diarrhea that usually resolve in 24 hours. A major symptom caused by intestinal pathogens is diarrhea, and most diarrheal diseases caused by Gram-positive pathogens are related to the toxins they produce. Therefore, definitive diagnosis of these pathogens is made by culturing the offending agent or, in some cases, detecting the toxin produced.

LISTERIOSIS

CASE HISTORY 22.4

Premature Arrival

 A 2.1-kg (4.7-pound) female neonate was born at 32 weeks. A normal neonate averages 3.17 kg, or 7 pounds, and is born at 40 weeks. The baby was clearly in distress and was intubated at birth because of poor respiratory effort. A chest X-ray showed an infiltrate in the baby's right lung. Her mother, Linda, presented at the time of delivery with lower abdominal pain, a temperature of 39°C (102°F), and a white cell count of 25,000/μl (normal is 4,000–10,000). Blood cultures from Linda and her child grew Gram-positive coccobacilli (**Figure 22.24**). Mother and baby fully recovered after antibiotic treatment. The physician later learned that Linda had recently visited Mexico, where she ate unpasteurized soft cheese called *queso blanco*.

Linda initially experienced a mild gastrointestinal distress caused by *Listeria monocytogenes* (a Gram-positive, non-spore-forming rod, member of the family Listeriaceae), an organism found in the gastrointestinal tracts of many animals and in natural environments. At some point, the organism entered her bloodstream

Figure 22.23   Cellulitis Caused by *Vibrio vulnificus* Infection

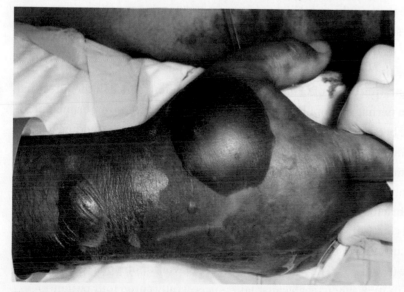

The bubbles under the skin, called bullae, are actually large blisters filled with fluid. The fluid is a result of the inflammatory response to the infection and contains blood and *V. vulnificus*.

inSight

Fecal Transplant (Bacteriotherapy)

Figure 1 **Dr. Alexander Khoruts**

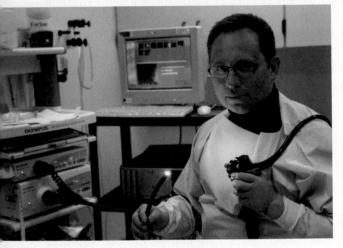

Dr. Khoruts, a gastroenterologist at the University of Minnesota, is shown here holding a colonoscope. Dr. Khoruts used the gut bacteria of a patient's husband to cure her chronic and debilitating *C. difficile* infection.

So, you could not believe your eyes when you read the title of this inSight, and you read it again. Indeed, fecal transplant does sound strange, to say the least, but it is a real approach to help cure some very dangerous gastrointestinal ailments, such as recurrent *Clostridium difficile* infections. And it makes sense, given the close relationship between nutrition, normal gut microbiota, and gastrointestinal health. We are born with minimal microbiota in our GI tract, but the GI tract rapidly becomes colonized by desirable microbes that help us use food and stave off some diseases. As discussed in Chapter 14, the number of bacteria residing in the human intestine is smallest in the duodenum and grows larger until it reaches a remarkable 10^9–10^{12} colony-forming units per milliliter (CFU/ml) in the distal colon. This microbial menagerie includes a diverse array of known and unknown microbes. Changes in the population of bacteria colonizing our intestines can lead to disease—for example, pseudomembranous colitis caused by *Clostridium difficile* after normal gut microbiota have been killed by antibiotics. The fecal matter of patients suffering from pseudomembranous colitis has significantly fewer firmicutes and *Bacteroides* species, for example, than samples from healthy individuals.

Patients are at risk of developing hospital-acquired, or nosocomial, infections during a hospital stay. Among the more dangerous hospital-acquired infections are those involving antibiotic-resistant strains of *C. difficile*. Patients who develop a *C. difficile* infection in the hospital are at higher risk for serious complications and frequent relapse than nonhospitalized patients. In some instances, the only way to save the patient's life is to remove the patient's colon (colectomy), a surgical procedure that carries its own risks, such as injury to other nearby organs (including the bladder and small intestine) and blood clot formation.

If the precipitating cause of *C. difficile* infection is an imbalance in the intestinal microbiota, then perhaps repopulating the intestines of patients with normal microbiota may be the better answer. Normal microbiota from a healthy individual can be "transplanted" into the patient through a variety of methods, including a nasogastric tube (inserted into the stomach through the nose) or a colonoscope, depending on the patient's condition (**Figure 1**). Fecal transplantation is not a new procedure. Hieronymus Fabricius ab Aquapendente (1537–1619), a surgeon in one of Europe's most prestigious institutes of learning, the University of Padova, Italy, performed the first transplantation of enteric bacteria, although he didn't know what bacteria were at the time (**Figure 2**). The procedure fell out of favor for obvious reasons until 300 years later, in 1958, when fecal transplants successfully cured four patients suffering from pseudomembranous colitis. *C. difficile* had yet to be identified as the cause of this disease in 1958.

Figure 2　Hieronymus Fabricius ab Aquapendente

Figure 22.24　Listeriosis and *Listeria monocytogenes*

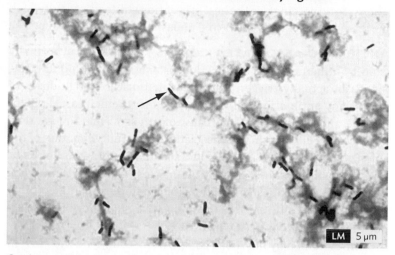

Cerebrospinal fluid in a patient with listeriosis shows characteristic Gram-positive rods of *Listeria*. Listeriosis in adults is much more common among patients infected with HIV than among the general population.

There were other sporadic uses of the technique until recent years, when the emergence of drug-resistant and virulent strains of C. *difficile* began to give rise to more and more treatment failure of pseudomembranous colitis and popularized bacteriotherapy. How effective is this method? According to a January 2013 report in the *New England Journal of Medicine* online, fecal transplant cured 81% of patients suffering from pseudomembranous colitis, compared with 31% of those who received antibiotics.

 Duodenal infusion of donor feces for recurrent *Clostridium difficile*: nejm.org

A number of other serious GI diseases, including colon cancer and inflammatory bowel diseases, are also linked to changes in nutrition and the composition of microbes that colonize the intestine. Repletion of normal gut microbiota has shown impressive success rates in patients who suffer from Crohn's and ulcerative colitis, two chronic and very painful inflammatory bowel diseases. This is very good news. The better news is the mounting evidence that proper nutrition, which supports a healthy microbiome, protects against the onset of some of these gastrointestinal diseases, and even against colon cancer. A healthy gut means a healthy you. So if you eat your fruits and vegetables to keep the microbes in your bowel happy and active, they will protect you.

(as indicated by her fever and elevated WBC count) and crossed the placental barrier into her fetus (vertical transmission). Her premature baby was born with septicemia and pneumonia and could have developed meningitis.

In healthy adults, *L. monocytogenes* usually causes mild gastrointestinal illness called **listeriosis**. In newborns, as well as in the elderly or immunocompromised people such as HIV patients, however, the pathogen can cause life-threatening septicemia, meningitis, and/or pneumonia. Approximately 2,500 cases occur each year in the United States, leading to about 500 deaths. Children and adults become infected from eating contaminated foods. Person-to-person passage is rare (no fecal-oral transmission). However, infected mothers can vertically transmit the pathogen to their fetus (as in Linda's case). *L. monocytogenes* grows not only at body temperature but also in the cold. As a result, the organism is found in processed and unprocessed foods, especially soft cheeses, hot dogs, lunch meats, milk, and salads stored in the refrigerator. These foods can be contaminated if made with unpasteurized raw milk or when processed foods come in contact with the organism in processing plants. Because *Listeria* multiplies in the cold, the risk of infection increases with prolonged refrigeration of these foods.

Listeria engineers its entry into epithelial cells of the small intestine, breaks out of the phagosome, and zooms around the host cell using actin-based motility, just as *Shigella* does. Actin motility allows the microbe to spread from cell to cell without having to encounter the immune system. Actin-based motility also explains *Listeria*'s ability to cross the blood-brain barrier to cause meningitis as well as the maternal-fetal barrier to cause neonatal septicemia. The organism can survive in macrophages, which can, like taxicabs, shuttle the pathogen through the bloodstream. The immune system effectively halts most infections at this stage, but dissemination will produce serious disease in immunocompromised patients.

Ampicillin is the treatment of choice for *Listeria* infections. There are no vaccines, so the best prevention strategy is to fully cook meats, wash vegetables, and avoid unpasteurized dairy products.

ANTIBIOTIC-ASSOCIATED COLITIS Although we normally expect antibiotics to cure infections, sometimes antibiotic treatment can actually trigger gastrointestinal disease. High doses of almost any antibiotic (including quinolones, penicillins, cephalosporins, and clindamycin) may cause diarrhea or sometimes a serious disease called antibiotic-associated colitis (inflammation of the colon). This condition is also called pseudomembranous colitis or pseudomembranous enterocolitis if the small intestine is involved. Taking oral antibiotics will kill most normal intestinal microbiota except the naturally resistant Gram-positive, spore-forming anaerobe *Clostridium difficile* (see **inSight**).

Unrestrained by microbial competition, *C. difficile* will grow in the intestine and produce its specific toxins that can damage intestinal cells. The organism's growth can lead to inflammation of the intestinal mucosa and the formation of plaques along the intestinal wall (see Figure 14.9). The plaques are yellowish white and made of sloughed-off intestinal epithelial cells, inflammatory cell debris, fibrin, and mucin. The plaques coalesce to form a pseudomembrane. Because they block the intestinal mucosa, pseudomembranes cause malabsorption of nutrients and water, which results in diarrhea. As the pseudomembrane enlarges, it begins to slough off and pass into the stool.

C. difficile produces three different toxins: A, B, and CDT. CDT is related to *C. perfringens,* and its role in the pathogenesis of *C. difficile* is not clear. Toxins A and B, which are closely related to each other (63% amino acid homology), can glucosylate (add a glucose molecule to) host cell proteins that make up the cytoskeleton. Altering the cytoskeleton changes the shape of the cell and weakens tight junctions, leading to leakage of fluids and hemorrhage. The toxins also cause an increase in production of tumor necrosis factor alpha (TNF-α), which in turn causes a large inflammatory response. The genes for *C. difficile* toxins and their regulators are all located on a pathogenicity island.

Why some people suffer from diarrhea while others have the life-threatening pseudomembranous colitis is not clear. Diagnosis of this disease involves PCR identification of the organism or immunological identification of the toxin in fecal samples. Metronidazole or vancomycin are appropriate treatments for this organism, although nonpharmacological therapies such as probiotics are also used (see Section 14.1).

Link As discussed in Section 9.4, large, horizontally transferred DNA sequences on a chromosome that contain a large number of genes encoding virulence factors are a class of genomic island called a pathogenicity island.

STAPHYLOCOCCUS AUREUS FOOD POISONING Like a ventriloquist who can make her voice sound like it is coming from another room, *Staphylococcus aureus* can cause intestinal disease without really being in the intestine. Everyone has heard of the local church or school picnic where scores of people become violently ill within hours of eating unrefrigerated potato salad. *S. aureus* is the usual culprit in these disasters. Not all strains of *S. aureus* cause food poisoning, but those that secrete enterotoxins into tainted foods, such as pies, turkey dressing, or potato salad, can cause serious GI distress. *S. aureus* heat-stable enterotoxins are a cause of food poisoning. **Food poisoning** is defined as an illness caused by eating food contaminated with viruses, bacteria, and bacterial and other toxins.

After ingestion, the *S. aureus* enterotoxin travels to the intestine, where it enters the bloodstream and stimulates nerves leading to the vomit center in the brain. Because the toxin is preformed, symptoms occur quickly after ingestion. Within 2–6 hours, the poisoned patient begins vomiting and experiences severe stomach cramps and possibly diarrhea. The disease, though violent, is not life-threatening and usually resolves spontaneously within 24–48 hours. In contrast, diarrheas caused by infectious agents, such as *Salmonella enterica,* that must first grow in the victim, do not occur until 12–24 hours after ingestion, sometimes longer. A clinician noting quick onset of symptoms in a patient will immediately suspect staphylococcal food poisoning. Antibiotic treatment is not needed for staph food poisoning.

SECTION SUMMARY

- *Helicobacter pylori* **causes gastric and duodenal ulcers.** The organism uses urease to survive gastric acidity before attaching to the gastric mucosa, where it is protected by the gastric mucus layer.

- **Key pathovars of** *Escherichia coli* **include enterotoxigenic** *E. coli,* **enteroinvasive** *E. coli,* **and enterohemorrhagic** *E. coli.*

- **Invasive** *Shigella* **species produce colitis.** *Shigella dysenteriae* produces Shiga toxins, moves in and between cells by actin-mediated motility, and does not disseminate beyond the intestine.

- *Salmonella enterica* **includes thousands of serovars** that invade M cells in the small intestine and remain in vacuoles, limiting phagolysosome fusion. *S.* Typhi enters macrophages and disseminates to produce typhoid fever with little GI involvement. Other serovars elicit inflammation, cause enteritis with bloody diarrhea, and usually do not disseminate.

- *Yersinia enterocolitica* **and** *Y. pseudotuberculosis* **invade M cells,** produce enteritis, sometimes disseminate to produce adenitis, and grow at refrigeration temperatures.

- *Campylobacter* **species invade M cells,** cause inflammatory diarrhea, and can trigger autoimmune **Guillain-Barré syndrome.**

- *Vibrio* **species are marine organisms** that include *V. cholerae* (cause of cholera), *V. parahaemolyticus* (found in undercooked seafood), and *V. vulnificus* (may cause necrotizing wound infections and sepsis in immunodeficient patients).

- *Listeria monocytogenes* **causes mild gastroenteritis and more serious diseases** by invading M cells in the small intestine and moving from cell to cell, crossing placental and blood-brain barriers; the organism can cause meningitis in the elderly and immunocompromised.

- *Clostridium difficile* causes antibiotic-associated diarrhea and pseudomembranous colitis. The disease follows antibiotic treatment and causes anti-absorptive diarrhea. Antibiotic treatment is generally not recommended for uncomplicated bacterial diarrhea; oral rehydration therapy is critical for treatment of bacterial diarrhea.

- **Staphylococcal food poisoning is a toxigenic disease** caused by a staphylococcal enterotoxin; vomiting and diarrhea begin 2–6 hours after ingestion.

Thought Question 22.3 You are treating a patient for severe, sharp abdominal pain that lessens after eating. How would you design an ELISA to detect serum antibodies to CagA?

Thought Question 22.4 How might the ability of *Y. enterocolitica* to grow in a cold environment be used to more effectively isolate the organism from a patient's stool?

Thought Question 22.5 A college friend tells you she has been suffering with watery diarrhea for 2 days and wonders what could be causing it. You find out she has also experienced abdominal pain. What viruses or bacteria would you place in your list of differential diagnoses (list of possible causes)? What if she tells you there is blood in her stool?

Thought Question 22.6 A stool culture of a patient afflicted with *S. aureus* food poisoning reveals no *S. aureus*. Where could you find the *S. aureus* that caused the food poisoning?

22.6
Parasitic Infections of the Gastrointestinal Tract

SECTION OBJECTIVES

- Distinguish GI infections caused by protozoa from those caused by helminths.
- Describe the role of parasitic life cycles in the transmission and pathogenesis of parasitic GI diseases.
- Generate a list of the diagnostic, prevention, and treatment options available for the parasitic diseases discussed.

What infections are caused by eukaryotic pathogens? As we learned in Chapter 11, many microbial pathogens are eukaryotes. In the developed nations of the world, we tend to associate gastrointestinal disease with viruses and bacteria; but in fact, waterborne parasites such as *Giardia*, *Cryptosporidium*, and roundworms cause numerous infections in the United States. The developing world contends with millions upon millions of cases of parasitic diseases caused by protozoan and helminths (worms). In this section, we will discuss several diseases caused by eukaryotic microbes (summarized in Table 22.4).

Lectures on parasites: virology.ws

Protozoan Infections of the Digestive System

Most students of biology are familiar with protozoa such as paramecia and amebas, but many are surprised to learn that some species of protists cause serious human disease. For instance, *Entamoeba histolytica* and *Cryptosporidium parvum* cause the diarrheal diseases amebic dysentery and cryptosporidiosis, respectively. Some diseases have great importance to public health. "Notifiable diseases" are those that must be reported to the local health department or to the CDC when diagnosed by a health care provider or a laboratory (further discussed in Section 26.3). Cryptosporidiosis, for example, is a notifiable disease. The CDC reported an increase in the number of confirmed and probable cases of cryptosporidiosis in the United States from 7,656 in 2009 to 8,951 in 2010.

GIARDIASIS The flagellated protozoan *Giardia intestinalis* (formerly *Giardia lamblia*) is a major cause of diarrhea throughout the world. In the United States alone, *G. lamblia* was responsible for more than 11,000 reported cases of **giardiasis** (diarrhea caused by *Giardia*) in 2005 and probably caused thousands more that were not reported. *G. intestinalis* is very infectious and can be found in various rodents, deer, cattle, and even household pets in addition to humans. *Giardia intestinalis* enters a human or other host as a cyst present in drinking water contaminated by feces (Figure 22.25A). Ingestion of as few as 25 cysts can lead to disease. Following ingestion, the hard, outer coating of the cyst is dissolved by the action of digestive juices to produce a **trophozoite** (Figure 22.25B), which attaches itself to the wall of the small intestines and reproduces. Offspring quickly encyst and are excreted from the host's body.

Asymptomatic carriers of *G. intestinalis* are common. It has been estimated that anywhere from 1% to 30% of children in U.S. day-care centers are carriers. Disease usually manifests as greasy stools alternating between a greasy or watery diarrhea, loose stools, and constipation. However, some patients experience explosive diarrhea. Diagnosis is usually made by observing the cysts or trophozoite forms of the protozoan in feces. Metronidazole is a drug often used to cure the disease. As for prevention, proper treatment of community water supplies is essential.

Figure 22.25 *Giardia intestinalis*
This protist is a major cause of diarrhea in the world.

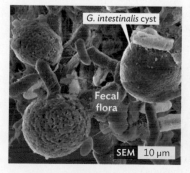
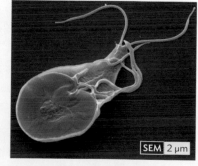

A. Cysts present in fecal matter.   B. Trophozoite form

Table 22.4

Parasitic Diseases of the Digestive System

Disease	Symptoms	Complications	Parasite	Transmission	Diagnosis	Treatment	Prevention
Protozoa							
Amebic dysentery (amebiasis)	Bloody diarrhea, fever, abdominal pain	Bowel obstruction, intestinal perforation, liver abscess	*Entamoeba histolytica*	Fecal-oral route	Microscopic detection of trophozoites or cysts; isolation in culture; isolation of *E. histolytica* antigen and DNA in stool (PCR, RT-PCR); clinical presentation with travel history	Amebicides (paromomycin, diloxanide, iodoquinol) given intraluminally (into the bowel lumen)	No vaccine available
Giardiasis (beaver fever, traveler's diarrhea)	Diarrhea, flatulence, bloating, weight loss, abdominal pain, greasy stools	Malabsorption of nutrients	*Giardia intestinalis*	Fecal-oral route (contaminated water)	Visualization of cysts by microscopy; antigen detection in stool or intestinal samples by EIA; *Giardia*-specific DNA (RT-PCR)	Asymptomatic cases may not be treated; antiparasitic medications in all symptomatic cases; metronidazole, tinidazole, others	No vaccine available
Helminths							
Ascariasis (roundworm)	Eosinophilic pneumonia, lung irritation, cough; small-bowel obstruction, abdominal pain, vomiting, diarrhea or constipation, peritonitis	Malabsorption of nutrients, bowel obstruction, hepatic and biliary complications	*Ascaris lumbricoides*	Fecal-oral route	Visualization of eggs in stool (may also be found in vomitus or sputum)	Albendazole, mebendazole, ivermectin	No vaccine available
Pinworm	Itching anus, sometimes diarrhea	Secondary bacterial infections, sleep disturbance from itching	*Enterobius vermicularis*	Fecal-oral route	Eggs on rectal swab; worms on cellophane (Scotch) tape applied to anus in morning before bathing	Pyrantel pamoate, mebendazole, albendazole	No vaccine available
Whipworm (trichuriasis)	Nausea, fatigue, abdominal pain, loss of appetite	Anemia	*Trichuris trichiura*	Fecal-oral route	Visualization of eggs in stool	Albendazole, mebendazole, pyrantel pamoate	No vaccine available
Hookworm	Recurrent epigastric pain, nausea, flatulence, sometimes itching at site of entry	Impaired growth in children, increased maternal mortality in pregnant women, pneumonitis (inflammation of the lung)	*Necator americanus*	Penetration of skin; eggs released in stool	Visualization of eggs in stool	Albendazole, mebendazole, pyrantel pamoate	No vaccine available

(continued on next page)

Table 22.4

Parasitic Diseases of the Digestive System (continued)

Disease	Symptoms	Complications	Parasite	Transmission	Diagnosis	Treatment	Prevention
Strongyloidiasis (threadworm)	Asymptomatic or burning pain in abdomen, nausea and vomiting, alternating watery diarrhea and constipation, itching skin rash	Sepsis and pneumonitis in immuno-compromised, with high mortality (60%–85%)	*Strongyloides stercoralis*	Penetration of skin; eggs released in stool	Serial stool examination for larvae (gold standard); serology; marked eosinophilia (increased eosinophils)	Ivermectin, albendazole, thiabendazole	No vaccine available
Trichinosis	Nausea, vomiting, diarrhea, abdominal pain, fever, muscle weakness	Arthritis, CNS and cardiac involvement, death due to heart or respiratory paralysis	*Trichinella* spp.	Ingestion of contaminated raw pork	Serology; muscle biopsy; elevated muscle enzymes and eosinophilia	Steroids; mebendazole, albendazole	No vaccine available
Tapeworm (pork)	Diarrhea, nausea, vomiting, visual disturbances	Meningitis, increased risk of epilepsy	*Taenia solium*	Fecal-oral route	Antigen detection in stool sample by ELISA; visualization of eggs in stool	Antiparasitic medications with steroids, depending on type of infection	No vaccine available
Tapeworm (fish)	Abdominal pain or discomfort, diarrhea, enlarged liver, weakness, weight loss	Anemia, intestinal blockage, vitamin B_{12} deficiency	*Diphyllobothrium latum*	Fecal-oral route	Visualization of eggs or parasite in stool	Antiparasitic medication: praziquantel; niclosamide can also be used	No vaccine available
Liver fluke (Oriental or Chinese liver fluke)	Mostly asymptomatic; enlarged liver, upper abdominal pain and anorexia	Liver inflammation, cancer of bile ducts	*Opisthorchis* and *Clonorchis*	Fecal-oral route	Visualization of eggs in stool; liver biopsy	Antiparasitic medications: praziquantel, albendazole	No vaccine available
Liver fluke (fascioliasis, the common liver fluke)	Can be asymptomatic; fever, abdominal pain, enlarged liver, anemia, malaise	Inflammation and obstruction of bile duct, jaundice	*Fasciola hepatica*	Fecal-oral route	Visualization of eggs in stool, more than one specimen needed; EIA with excretory-secretory (ES) antigens combined with confirmation of positives by immunoblot	Antiparasitic medications; triclabendazole is drug of choice	No vaccine available
Blood fluke (schisto-somiasis)	Dysentery, abdominal pain, fever, chills, GI bleeding, pneumonitis, enlarged liver and spleen	Damage to small intestine, causing death; encephalitis	*Schistosoma mansoni*	Penetration of skin; eggs eliminated in feces and urine	Visualization of eggs in biopsy specimen; antibody detection by ELISA followed by immunoblot for confirmation	Antiparasitic medications: praziquantel	No vaccine available

AMEBIC DYSENTERY (AMEBIASIS)   Although you rarely hear about this organism, 10% of the world's population are infected with *Entamoeba histolytica*. *E. histolytica* is an ameba that alternates between a motile trophozoite form and an infectious cyst form that survives for weeks in the environment. These forms can be differentiated in the microscope by the number of nuclei they contain. The cysts contain four nuclei, whereas trophozoites contain only one (see Figure 11.12).

Following ingestion of cysts, trophozoites are released in the small intestine and in most individuals remain in the large intestine as harmless commensals. The trophozoites replicate and some become infectious cysts that are shed in stool. Trophozoites are sensitive to acid and are not infectious. In some unfortunate individuals, 2–6 weeks after ingesting the cysts and the trophozoites emerge, the protozoan penetrates the intestinal mucosa, causing colitis (**amebic dysentery**, or **amebiasis**), or invades the bloodstream to produce abscesses in the liver, lungs, or brain. Amebic dysentery is marked by diarrhea, bloody stools, fever, abdominal pain, and weight loss. Liver abscesses will result in hepatitis symptoms. Note that amebic infections can also be transmitted by anal-oral sexual contact.

Diagnosis of amebic dysentery is tricky because visually similar nonpathogenic *Entamoeba* species are among the normal intestinal microbiota. The laboratory must identify trophozoites that have ingested red blood cells in stool or colonic abscesses to make the diagnosis. Prevention of disease relies on adequate water sanitation. Treatment of active disease includes iodoquinol, which will kill intestinal cysts and trophozoites by an unknown mechanism but will not penetrate the intestinal mucosa. Thus, extraintestinal infections are treated with metronidazole, which will penetrate to deeper tissues and kill disseminated organisms.

CRYPTOSPORIDIOSIS   *Cryptosporidium* species (*Cryptosporidium hominis* and *C. parvum*) are widely distributed in the world and cause self-limiting diarrheal disease (**cryptosporidiosis**) in immunocompetent hosts. Following ingestion, oocysts dissolve (excyst) and release sporozoites that then enter intestinal epithelial cells. There the parasites undergo asexual and then sexual cycles to produce infective oocysts. These oocysts are excreted in stool (Figure 22.26). The oocysts are quite hardy and resist killing even in chlorinated water. Thus, both drinking water and recreational water, which can be contaminated by animal waste containing *Cryptosporidium*, are sources of infection.

About a week after ingestion, the patient develops a watery diarrhea that lasts 1–2 weeks. The mechanism underlying this secretory diarrhea is unknown. Symptoms can include abdominal pain, nausea, anorexia, fever, and weight loss. In contrast to this self-limiting disease, immunocompromised AIDS patients with CD4 counts less than 100/μl will develop a chronic, remarkably profuse, life-threatening diarrhea (1–25 liters per day) and suffer severe fluid and electrolyte depletion.

Figure 22.26   *Cryptosporidium* Oocysts

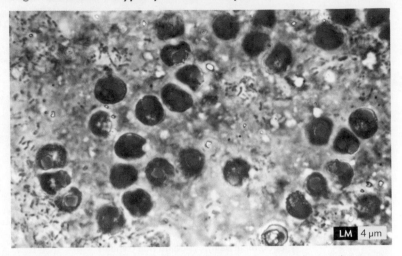

Rounded oocysts stained with safranin stain (pink, 4–6 μm in diameter) are seen in feces and contain the infective sporozoite stage.

Because oocytes of *Cryptosporidium* are tiny (4–6 μm in diameter), their discovery in stool relies on a variety of specific tests (the lab should know you are looking for this parasite). A modified acid-fast stain, direct immunofluorescence, or immunoassay can be used. Treatment of immunocompetent patients includes oral rehydration and the drug nitazoxanide, which kills the protozoan by interfering with its anaerobic metabolism. Unfortunately, the drug is ineffective in AIDS patients. Retroviral treatments leading to increased CD4 counts can resolve the infection in these patients.

CYCLOSPORIASIS   *Cyclospora cayetanensis* causes illness (**cyclosporiasis**) ranging from asymptomatic to diarrhea as well as flu-like symptoms, flatulence, belching, and fatigue. The disease is generally self-limiting but can be treated with sulfamethoxazole-trimethoprim if needed. The organism makes oocysts in the intestine (Figure 22.27), but the oocyst is not infective, which means that direct fecal-oral transmission cannot occur (this differentiates *Cyclospora* from *Cryptosporidium*, whose oocytes show fecal-oral transmission). The *Cyclospora* oocytes need a few days or weeks in the environment to sporulate and produce sporocysts so that they can become infectious and cause disease. Each sporulated oocyst contains two sporocysts, each harboring an elongated sporozoite. Fresh produce and water can then serve as vehicles for transmission, and the sporulated oocysts are ingested. The oocyst shell dissolves in the gastrointestinal tract, freeing the sporozoites to invade the epithelial cells of the small intestine. Once inside the intestinal epithelial cells, the sporozoites undergo asexual multiplication followed by sexual development to mature into spherical oocysts that can be detected in the stools of infected patients.

Figure 22.27 **Stages in the Life Cycle of *Cyclospora cayetanensis***

Oocysts are excreted in feces unsporulated (noninfective). After 1–2 weeks outside the host, they sporulate and become infective.

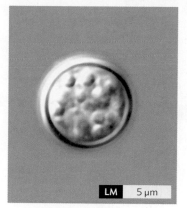

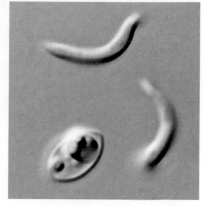

A. An undifferentiated oocyst.

B. A sporulating oocyst containing two maturing sporocysts. An ingested sporulated oocyst can lead to disease.

C. A ruptured oocyst releasing one of its two immature sporocysts.

D. One free sporocyst is shown with two free sporozoites.

Helminthic Infections of the Digestive System

CASE HISTORY 22.5

Worming Through

Ileana, a high-ranking U.S. government official, recently traveled to the Near East on a diplomatic mission. During a formal reception, she ate some delicious steak tartare (raw beef), a traditional dish in that region. However, 3 months later, she noticed thin white rectangular segments in her stool (approximately 1 × 2 × 0.2 cm). She experienced nausea, in part from the parasite and in part from the sight of worms in her stool. Laboratory studies revealed that the segments were proglottids (body segments) of *Taenia saginata*, a tapeworm (**Figure 22.28A**). Her stool also contained eggs of this worm (**Figure 22.28B**). She was reassured by her physician, who told her that this infection is unlikely to have clinical consequences in a healthy person. The physician prescribed niclosamide, which eliminated the rest of the tapeworm.

Helminths (worms) were first described in Section 11.6. The parasitic helminths, ranging in size from incredibly tiny (0.3 mm) to ridiculously huge (25 m, or 82 feet), fall into three groups: **roundworms (nematodes)**, **flukes (trematodes)**, and **tapeworms (cestodes)**. Trematodes and cestodes are classified together as **flatworms (platyhelminths)**. Symptoms of helminthic infections are caused primarily by the immune response provoked by these worms rather than by pathogenicity factors such as toxins. A telltale sign that practitioners look for when diagnosing worm infections is an increase in the numbers of eosinophils present in peripheral blood counts (**eosinophilia**). The role of eosinophils in clearing helminth infections remains controversial, but these white blood cells are clearly drawn to sites of worm infections and are responsible for a considerable amount of the inflammatory pathology.

HELMINTH LIFE CYCLES AND TRANSMISSION TO HUMANS
So how does one manage to contract worm infections? Based on the individual worm's life cycle, there are four general ways by which helminths infect humans, in this book classified as life cycles A–D. Life cycle A involves the fecal-oral route of transmission; life cycle B is a fecal-environment-skin route; life cycle C involves the fecal–intermediate host–oral route; and life cycle D is a combination of

Figure 22.28 **Tapeworm**

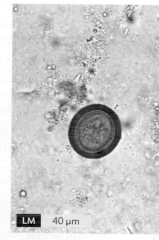

A. *Taenia saginata* can grow to several meters in length.

B. *T. saginata* egg found in stool.

Figure 22.29

Life Cycles and Transmission of Helminthic Parasites

Panels depict the different paths worms take to develop from eggs to larvae to adults back to eggs.

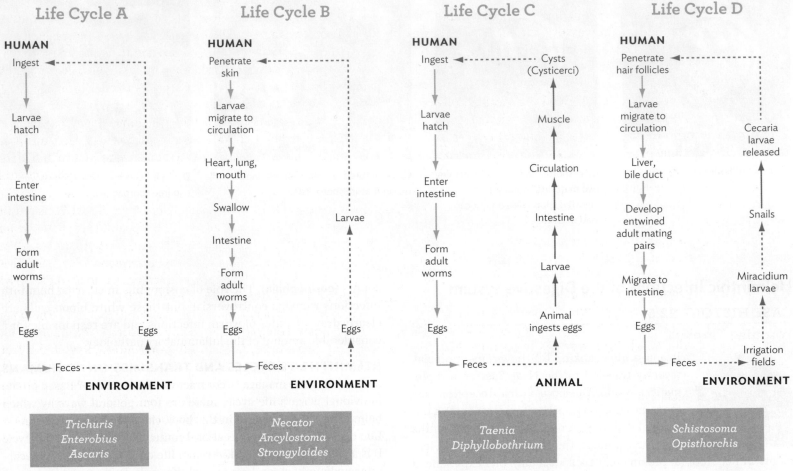

life cycles A and B. Be aware that some helminths require an **intermediate host** to complete their life cycle. A **definitive host** is one in which the adult worms develop and lay eggs. All worms have one or more definitive hosts, but only a few worms have intermediate hosts. **Figure 22.29** provides a summary of the four primary ways in which helminths infect humans.

Some helminths use life cycle A. Eggs from certain roundworm species (such as *Trichuris*, *Enterobius*, and *Ascaris*) can contaminate food or water and be ingested. The larvae hatch in the stomach, attach to the intestine, and enter intestinal tissue to become male and female worms. Worm sex then produces more eggs, usually in huge quantities. The eggs are then excreted in feces and await digestion by a new host. The *Ascaris* life cycle is a little more complicated. Before *Ascaris* larvae can develop into adults, they first migrate through the circulation, enter the lungs, break into the alveoli, and ascend to the oral cavity, where they are again swallowed. Only then can adults develop and lay eggs.

Life cycle B is preferred by other species. After being excreted in feces, eggs from hookworms such as *Necator americanus* and *Ancylostoma duodenale* hatch in the environment and develop into larvae. Infection results when the larvae actively penetrate human skin (now you know one of the reasons your mother told you not to play outside in bare feet). Once in the lymphatic and blood circulations, larvae travel to the heart and lungs and ultimately the throat. Larvae in the oral cavity are then swallowed with saliva (as with *Ascaris*) and travel to the small intestine, where they anchor (attach) and feed on blood. The larvae develop into adults that produce eggs, which are then excreted in feces to begin the next cycle of transmission. Eggs will not hatch in the intestine.

A third group of helminths prefers to go through an intermediate host and use life cycle C. Tapeworms such as *Taenia*, *Diphyllobothrium*, and *Echinococcus* mature in the human intestine (definitive host), releasing eggs in feces. Grazing animals (intermediate hosts) ingest the eggs, and stomach acid releases the larvae. Liberated larvae travel to the animal's GI system, enter the circulatory system, invade muscle, and become encased in cysts (called **cysticerci**). Humans eating meat containing cysticerci become infected. Cooking potentially contaminated foods such as pork and fish will break this cycle.

Finally, there are those helminths that "cannot make up their minds" and use an either-or approach, life cycle D. The flatworms *Schistosoma* and *Opisthorchis* have fork-tailed larvae called **cercariae**, which are present in contaminated water (often irrigated fields or ponds) and can directly burrow through the hair follicles of a wading human or be ingested. The larvae migrate through the circulation to the liver and bile ducts, where they develop into male and female flatworms that become permanently entwined as mating pairs. The writhing pairs migrate to blood vessels in the small intestine, where they lodge, feed on blood, and lay eggs. The eggs are then released into irrigation fields through fertilization, defecation, or urination. Once in the water, the eggs release ciliated larvae called **miracidia** that seek out nearby snails (intermediate hosts) and burrow into them. In the snail, the larvae morph into the large, fork-tailed cercaria that started this cycle. Cercaria are released from snails by the thousands into the surrounding water, where they await new, unsuspecting human hosts.

Clinical manifestations of helminthic diseases can be divided into four basic categories: intestinal distress alone, intestinal distress associated with migratory symptoms (caused by migrating larvae), muscle involvement, and liver disease.

HELMINTHS THAT CAUSE INTESTINAL DISTRESS The **whipworm**, *Trichuris trichiura*, is so named because it resembles a whip (**Figure 22.30A**). Infecting humans via life cycle A, adult worms form in the intestine. When these worms burrow into the

intestinal wall, the wall hemorrhages, making gateways for secondary bacterial infection (**Figure 22.30B**). A heavily infected individual can experience dysentery, loss of muscle tone, and even a prolapsed (protruding) rectum.

Pinworm, *Enterobius vermicularis*, is the most common worm disease of children; it was described in Case History 11.3, "Dotty's Itch." The worms are very small and use life cycle A for transmission (**Figure 22.31A**). Females tend to migrate from the anus at night and lay their eggs in the perianal region. A simple piece of Scotch tape applied to the anus for a few seconds at night will catch eggs that the lab can microscopically identify (**Figure 22.31B**). Symptoms of pinworm infection include pronounced anal itching and sometimes diarrhea. Itching can interrupt the child's sleep and promote self-reinfection. As the child scratches, eggs stick to the child's fingers, and when those fingers are placed back in the child's mouth, the eggs are reingested.

Tapeworms include *Taenia saginata* from beef (discussed in Case History 22.5), *Taenia solium* from pig, and *Diphyllobothrium latum* from fish. We become infected with these parasites when we eat contaminated raw or undercooked beef, pork, or fish. Tapeworms are visually impressive segmented worms, whose head (**scolex**) contains a central ring of hooklets flanked at four corners by suckers that attach to the intestine (**Figure 22.32**). The body segments, called **proglottids**, form eggs and break off from the main worm (quite disconcerting to the person confronted with 2–3 feet of worm parts exiting the anus). Adult worms can release up to 1 million eggs per day this way. For such a large parasite, tapeworms cause relatively few symptoms, which include nausea, vague complaints of abdominal pain, and the occasional discovery of proglottids in feces (as Ileana discovered in our case history). The nutrient demands of the large worm can also cause the host to lose weight.

Figure 22.30 Whipworm, *Trichuris trichiura*

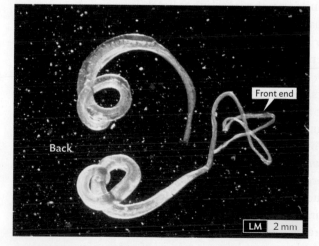

A. The narrow end of the worm attaches to the intestine, while the fatter end hangs in the intestinal lumen to sample nutrients.

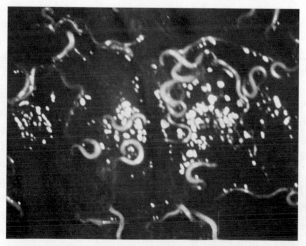

B. Colon heavily infested with whipworms.

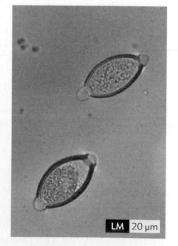

C. Whipworm eggs, easily identified by their plugged ends. Whipworm eggs measure 50–55 μm in length.

Figure 22.31   Pinworm, *Enterobius vermicularis*

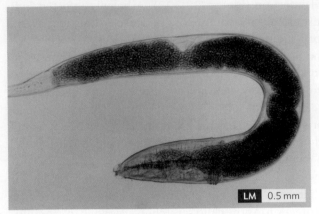

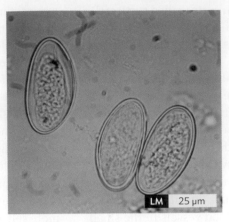

A. *Enterobius* up close.

B. Pinworm egg capsule.

Figure 22.32   Tapeworm

Notice the head with hooklets and suckers and the segmented proglottids.

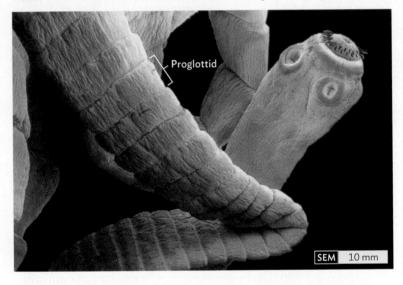

HELMINTHS THAT CAUSE INTESTINAL DISTRESS WITH LARVAL MIGRATION SYMPTOMS Ascariasis is caused by the largest of the roundworms—*Ascaris lumbricoides*, 40 cm long (Figure 22.33A)—and is the most common nematode infection in the world. Symptoms of larval migration to the lung include an irritating nonproductive cough and a burning sensation. Intestinal infection is usually asymptomatic unless a large clump of entangled worms causes a small-bowel obstruction, in which case symptoms can include abdominal pain, vomiting, constipation or diarrhea, appendicitis, and peritonitis (Figure 22.33B). It is estimated that 1.2–1.5 billion people in tropical and subtropical regions have *Ascaris* in their GI tract. The number of cases in the United States is estimated at 4 million. Diagnosis involves finding appropriately shaped eggs in stool (Figure 22.33C).

Hookworms, such as the *Necator americanus* (Figure 22.34A) and *Ancylostoma duodenale*, are nematodes that get their common

name both from the vicious-looking oral cutting plates they use to anchor themselves to the intestinal wall and from the hook on their tails (Figure 22.34B). Following the course of life cycle B, larvae from these species hatch in the environment and penetrate the skin, usually the feet, then follow a visible serpentine path to the lungs and mouth (Figure 22.34C). Most infections are asymptomatic, but the larvae can cause itchy dermatitis (ground itch) at the site of penetration and a mild, transient pneumonia as they migrate into the lungs. Once the larvae are in the intestine (after being swallowed), there may be gastrointestinal pain and inflammatory diarrhea. The main symptom is iron deficiency.

Strongyloides stercoralis (Figure 22.35) is a nematode (roundworm) that has a dramatically modified B-type life cycle. The added wrinkle is that adult worms can form in the human intestine or in the soil and lay eggs in both environments. To illustrate, let's start a cycle with larvae shed in feces from an infected human. The larvae in soil can do one of two things: stay as they are and develop into adults that mate and lay eggs (they maintain a soil cycle of replication) or change into a different larval form (a **filariform**) that is able to penetrate human skin to start a new infection. Eggs produced in soil can also hatch filariform larvae capable of penetrating human skin. Larvae that end up penetrating the skin, again usually the feet, migrate through the bloodstream to the lungs and mouth, are swallowed, and make their way to the intestine. There they develop into females that can reproduce without males (**parthenogenesis**). These adult worms release eggs that hatch in the intestinal mucosa, and the larvae are again shed in feces. This contrasts with hookworms, whose eggs must leave the intestine to hatch.

Disease caused by *S. stercoralis* is called **strongyloidiasis**. It starts with an intensely itchy, red serpentine skin rash at the initial infection site made by larvae as they migrate. The threadlike eruption can move at an alarming pace (to the patient), 10 cm per hour. Most patients, however, have few other symptoms. Sometimes as the worms burrow into the duodenum or jejunum, the patient experiences gastric pain. A heavy infection can lead to small-bowel obstruction, but host immunity usually keeps the disease contained, even with autoreinfection. However, treating a patient with immunosuppressive agents (corticosteroids) allows hyperinfection and the production of enormous numbers of larvae. The larvae can disseminate and cause symptoms of pneumonia as well as liver, kidney, and central nervous system involvement.

HELMINTH DISEASE WITH MUSCLE INVOLVEMENT Trichinosis, caused by *Trichinella spiralis*, is transmitted by eating pork or other meat in which cysts of this parasite are embedded. The

Figure 22.33 *Ascaris lumbricoides*

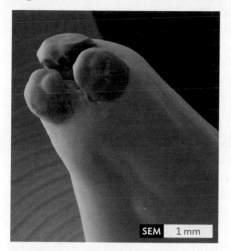

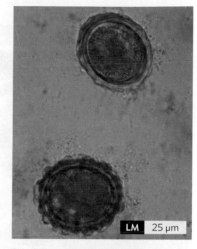

A. The head of *Ascaris lumbricoides*.

B. Cluster of *Ascaris* worms in a resected bowel (part of the bowel that has been removed).

C. *Ascaris* eggs. Note the thick, bumpy coat.

Figure 22.34 **Hookworm**

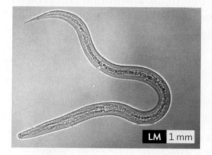

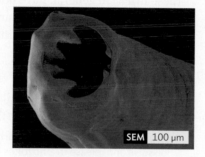

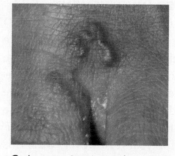

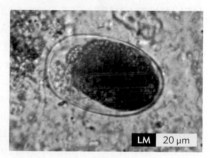

A. *Necator americanus*. Note the hook on the tail.

B. Hookworm sucker.

C. Larvae migrating under the skin.

D. Hookworm egg.

Trichinella life cycle is different from the others we've described in that it all takes place in a mammalian host (no need for an intermediate host). After ingestion, the envelopes of ingested cysts dissolve in the stomach, freeing the larvae, which then burrow into intestinal mucosa. There they develop into adults, mate, and produce eggs that hatch, and the resulting larvae enter the circulatory systems (blood and lymph). All tissues are subject to larval invasion, but final development happens when coiled larvae enter muscle and become encased as cysts (Figure 22.36). Transmission takes place when a new host eats the infected muscle of a dead host. Because humans generally do not engage in cannibalism, we contract disease after eating contaminated meat from another animal, such as pig or wild boar.

Disease severity ranges from unnoticeable to life-threatening, depending on how many larvae were ingested. Symptoms, when present, start with nausea, abdominal pain, diarrhea, fever, and sweating. Once larvae migrate into muscles, the patient experiences intense muscle pain. Heart and brain involvement can be lethal. In most cases, symptoms slowly subside, but once the larvae have encysted, there is no cure. The best prevention is to thoroughly cook pork and wild meats.

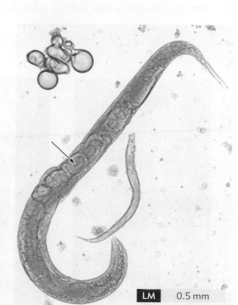

Figure 22.35
Strongyloidiasis Caused by *Strongyloides stercoralis*

An adult free-living female *S. stercoralis* (about 3 mm) alongside a smaller rhabditoid larva (noninfectious, produced in patient). Notice the developing eggs in the adult female (arrow).

Figure 22.36   Encysted Larvae of *Trichinella spiralis* in Muscle
Cysts are about 1 mm in length.

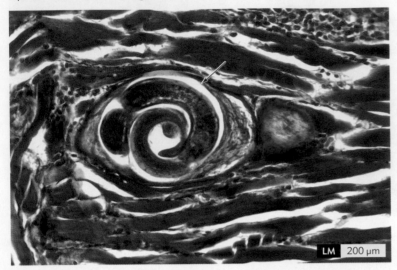

HELMINTH DISEASES WITH LIVER INVOLVEMENT Schistosomes (species of the genus *Schistosoma*; Figure 22.37A–C) and the liver flukes *Opisthorchis* (Figure 22.37D) and *Fasciola* are trematodes that replicate via life cycle D, as noted earlier, and use snails as their intermediate hosts. The patient with **schistosomiasis** will first notice itchiness at the site where the worm entered the body. This is followed by fever, chills, diarrhea, and a cough. Because schistosomes coat their bodies with host blood proteins, the parasite can avoid the immune system for years. However, with this chronic infection comes liver enlargement.

Opisthorchis differs from *Schistosoma* in that *Opisthorchis* does not enter a human host by penetrating the skin but enters when the person ingests cercarial larvae present in poorly cooked or raw fish (intermediate host). In **opisthorchiasis**, larvae enter the intestine and migrate to the bile duct, where they mature and shed eggs into the intestine. Unless heavily infected, most victims are asymptomatic. Moderately to heavily infected patients complain of vague right upper quadrant abdominal pain. Symptoms come from toxic irritation of the bile duct and granulomas in the liver.

DIAGNOSING AND TREATING HELMINTH DISEASES Most of the worm infections just discussed are relatively rare in countries such as the United States that have extensive sanitation systems and effective hygiene practices. These diseases are, however, common in developing countries still building the infrastructure and health standards needed for their prevention. Easy travel between countries means that travel history in a patient with diarrheal disease can suggest possible worm infestations. In all cases, diagnosis includes a finding of eosinophilia in peripheral blood and discovering appropriately shaped eggs in stool samples. Treatment varies with the organism. Antihelminthic drugs are available; however, these drugs, in general, have poor systemic absorption and tend to either paralyze the worm or starve it (see Table 22.4). Thus, it is difficult to ensure complete elimination of worms.

SECTION SUMMARY

- **Protozoa causing diarrhea** include *Giardia intestinalis*, *Entamoeba histolytica*, *Cryptosporidium*, and *Cyclospora*.
- **Protozoan disease typically starts with the ingestion of oocysts,** which dissolve and release trophozoites that attach and invade the intestinal wall. From this locale more cysts are made that pass in feces.
- **Helminth diseases are caused by three families of worms:** nematodes, trematodes, and cestodes.

Figure 22.37   Liver Flukes

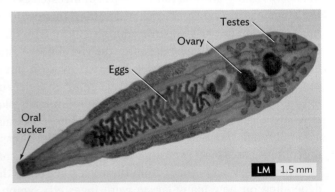

D. *Opisthorchis sinensis*. Shown is the hermaphroditic larva seen in a mammalian host. The tail of *Opisthorchis* is not forked.

Testes

Ovary

Eggs

Oral sucker

A. A mated pair of *Schistosoma mansoni*, the human blood fluke. The female fits inside a groove on the larger male's body.

B. Egg of *S. mansoni*. Note the lateral spine, a hallmark of this organism recognized by clinical parasitologists.

C. Fork-tailed cercaria of *S. mansoni*.

- **A helminth life cycle has three major stages:** eggs, larvae, and adults. Transmission to humans can be fecal-oral, transdermal, vector-borne, and through a life cycle using both an intermediate and a definitive host.

- **Worms that cause intestinal distress** include *Trichuris*, *Enterobius*, and *Taenia*.

- **Worms that cause intestinal distress with migratory symptoms** include *Ascaris*, *Necator*, *Ancylostoma*, and *Strongyloides*.

- *Trichinella* **causes disease with muscle involvement.**

- **Worms that cause liver disease** include *Schistosoma*, *Opisthorchis*, and *Fasciola*.

- **Eosinophilia** is a key marker for helminthic disease.

- **Stool exams** looking for eggs and parasites are important for diagnosing protozoan and worm diseases.

Perspective

Because of its open access to the outside world, its rich nutrients, and its abundant nooks and crannies, the gastrointestinal tract is like a four-star hotel to microbes. Their lengths of stay vary, depending on what we eat, what attachment tools are available, and which attachment sites are open (any room at the inn?). How are these and other factors that influence the composition of our normal GI microbiota (discussed in Chapter 14) affected by the presence of pathogens? Although competition with abundant normal microbiota and effective mucosal immunity pose daunting obstacles that pathogens must overcome, the GI tract plays host to more infectious microbes than any other organ system. One reason infections are so frequent in the GI tract is the open-access aspect of the system. Another reason involves the weapons pathogens wield to "soften the battlefield." Toxins that increase intestinal motility flush less tenacious organisms from the intestine. And the ability of some pathogens to invade host cells removes competition altogether.

Be aware, too, that a pathogen's ability to perturb normal microbiota and intestinal function has far-reaching consequences for the host. For example, cholera toxin's ability to pump electrolytes out of the host will affect muscle contraction (heart), nerve impulses, and other aspects of host physiology you wouldn't intuitively connect to diarrhea. Finally, you should realize that because of inadequate sanitation, diarrhea is one of the greatest causes of death, especially of children, throughout the world.

Figure 22.38 Summary: Microbes That Infect the Digestive System

Viruses	Genome	Disease
Hepatitis B	DNA, ds (partially), circular	Hepatitis
Adenoviruses; types 40 and 41	DNA, ds, linear	Acute gastroenteritis primarily in children
Rotavirus	RNA, ds, linear, segmented	Acute diarrhea in children under 3 years of age
Mumps virus	RNA, ss, negative sense	Mumps
Hepatitis D[a]	RNA, ss, negative sense	Hepatitis
Norovirus	RNA, ss, positive sense	Acute gastroenteritis, also called stomach flu
Sapoviruses	RNA, ss, positive sense	Acute gastroenteritis in adults
Hepatitis A	RNA, ss, positive sense	Hepatitis
Hepatitis C	RNA, ss, positive sense	Hepatitis
Hepatitis E	RNA, ss, positive sense	Hepatitis

Bacteria	Morphology	Disease
Helicobacter pylori	Gram-negative	Peptic ulcer
E. coli ETEC	Gram-negative	Traveler's diarrhea
E. coli EIEC	Gram-negative	Bacillary dysentery
E. coli EHEC (e.g., E. coli O157:H7)	Gram-negative	Diarrhea, may progress to bloody
Shigella	Gram-negative	Shigellosis (bacillary dysentery)
Salmonella	Gram-negative	Salmonellosis (enterocolitis)
Yersinia enterocolitica	Gram-negative	Yersiniosis
Campylobacter	Gram-negative	Enterocolitis
Vibrio cholerae	Gram-negative	Cholera
Vibrio parahaemolyticus	Gram-negative	Vibriosis
Listeria monocytogenes	Gram-positive	Listeriosis
Clostridium difficile	Gram-positive	Antibiotic-associated diarrhea (pseudomembranous colitis)
Clostridium perfringens	Gram-positive	Gastroenteritis (due to intoxication)
Staphylococcus aureus	Gram-positive	Gastroenteritis (food poisoning due to intoxication)

Parasites		Disease
Protozoa		
Entamoeba histolytica	No flagella/NA[b]	Amebic dysentery
Giardia intestinalis	Flagellated/NA	Giardiasis
Helminths	**Type/Life Cycle (A–D, as Defined in Figure 22.29)**	
Ascaris lumbricoides	Nematode/A	Ascariasis
Enterobius vermicularis	Nematode/A	Pinworm
Trichuris trichiura	Nematode/A	Whipworm (trichuriasis)
Necator americanus	Nematode/B	Hookworm
Strongyloides stercoralis	Nematode/modified B	Strongyloidiasis
Trichinella spp.	Nematode/wholly within mammalian host	Trichinosis
Taenia solium	Cestode/C	Tapeworm (pork)
Diphyllobothrium latum	Cestode/C	Tapeworm (fish)
Opisthorchis and Clonorchis	Trematode/D	Liver fluke
Fasciola hepatica	Trematode/D	Liver fluke
Schistosoma mansoni	Trematode/D	Blood fluke, schistosomiasis

[a]Capable of replication and causing disease only when coinfecting with hepatitis B virus.
[b]NA = not applicable.

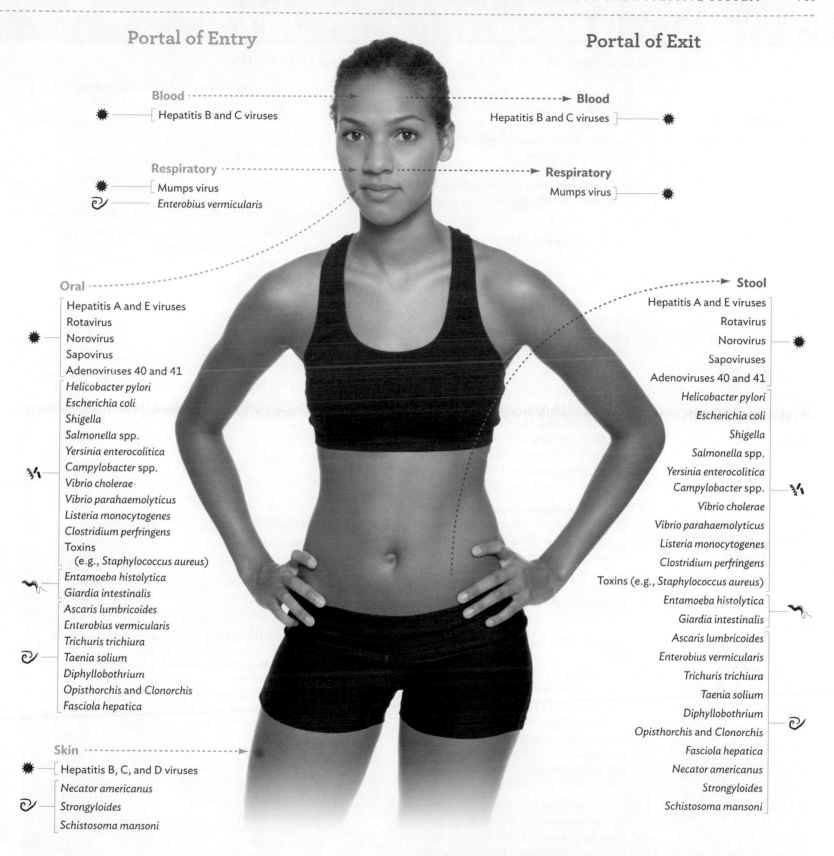

Portal of Entry

Portal of Exit

Blood
Hepatitis B and C viruses

Blood
Hepatitis B and C viruses

Respiratory
Mumps virus
Enterobius vermicularis

Respiratory
Mumps virus

Oral
Hepatitis A and E viruses
Rotavirus
Norovirus
Sapovirus
Adenoviruses 40 and 41
Helicobacter pylori
Escherichia coli
Shigella
Salmonella spp.
Yersinia enterocolitica
Campylobacter spp.
Vibrio cholerae
Vibrio parahaemolyticus
Listeria monocytogenes
Clostridium perfringens
Toxins
   (e.g., *Staphylococcus aureus*)
Entamoeba histolytica
Giardia intestinalis
Ascaris lumbricoides
Enterobius vermicularis
Trichuris trichiura
Taenia solium
Diphyllobothrium
Opisthorchis and *Clonorchis*
Fasciola hepatica

Stool
Hepatitis A and E viruses
Rotavirus
Norovirus
Sapoviruses
Adenoviruses 40 and 41
Helicobacter pylori
Escherichia coli
Shigella
Salmonella spp.
Yersinia enterocolitica
Campylobacter spp.
Vibrio cholerae
Vibrio parahaemolyticus
Listeria monocytogenes
Clostridium perfringens
Toxins (e.g., *Staphylococcus aureus*)
Entamoeba histolytica
Giardia intestinalis
Ascaris lumbricoides
Enterobius vermicularis
Trichuris trichiura
Taenia solium
Diphyllobothrium
Opisthorchis and *Clonorchis*
Fasciola hepatica
Necator americanus
Strongyloides
Schistosoma mansoni

Skin
Hepatitis B, C, and D viruses
Necator americanus
Strongyloides
Schistosoma mansoni

LEARNING OUTCOMES AND ASSESSMENT

22.1 Anatomy of the Digestive System

- List the structural components of the digestive system and describe the relationship between them.
- Name the functions of different structures of the digestive system.
- Explain the roles of specialized cells of the intestinal tract (chief cells, M cells, parietal cells, and mucous neck cells).

1. The _____ is physically attached to the stomach.
   A. esophagus
   B. duodenum
   C. ileum
   D. jejunum
   E. A and B only

22.2 Infections of the Oral Cavity

- List microbes comprising normal microbiota of the oral cavity.
- Analyze how biofilms contribute to infections of the oral cavity.
- Compare the signs and symptoms of common infections of the oral cavity.
- Explain how oral cavity infections impact diseases elsewhere in the body.

5. Which of the following bacterial genera is part of the normal microbiota of the oral cavity?
   A. *Streptococcus*
   B. *Lactobacillus*
   C. *Peptostreptococcus*
   D. All of the above
   E. None of the above

22.3 Gastrointestinal Syndromes

- List the most common signs and symptoms associated with gastrointestinal infections.
- Correlate microbial pathogenic mechanisms with the presenting signs and symptoms of gastrointestinal infections.
- Outline the general mechanisms of diarrhea.

11. _____ refers to inflammation of the small intestine.
    A. Gastritis
    B. Enteritis
    C. Colitis
    D. Hepatitis
    E. Esophagitis

22.4 Viral Infections of the Gastro-intestinal Tract

- Match gastrointestinal symptoms to the viral infections that cause them.
- List routes of transmission of specific viral diseases of the gastrointestinal tract.
- Discuss viral pathogenic mechanisms responsible for specific disease signs and symptoms in the gastrointestinal tract.

Match the following:

14. Hepatitis B	A. Diarrhea
15. Hepatitis C virus	B. RNA genome
16. Mumps virus	C. Orchitis
17. Norovirus	D. DNA genome

22.5 Bacterial Infections of the Gastrointestinal Tract

- Categorize GI infections by bacterial pathogenic mechanisms.
- Correlate bacterial virulence mechanisms to GI disease symptoms.
- Relate the treatment of GI infections to virulence of GI pathogens.
- Evaluate treatment and prevention methods used for bacterial infections of the gastrointestinal tract.

20. An enterotoxin secreted by _____ growing in food is an important cause of food poisoning.
    A. *Salmonella enterica* serovar Typhimurium
    B. *Vibrio cholerae*
    C. *Staphylococcus aureus*
    D. *Streptococcus pyogenes*
    E. Enteroinvasive *E. coli*

22.6 Parasitic Infections of the Gastrointestinal Tract

- Distinguish GI infections caused by protozoa from those caused by helminths.
- Describe the role of parasitic life cycles in the transmission and pathogenesis of parasitic GI diseases.
- Generate a list of the diagnostic, prevention, and treatment options available for the parasitic diseases discussed.

25. _____ is caused by protozoa.
    A. Cryptosporidiosis
    B. Strongyloidiasis
    C. Trichinosis
    D. Schistosomiasis
    E. Opisthorchiasis

2. A patient is diagnosed with appendicitis. Which of the following parts of his digestive system is affected?
   A. Esophagus
   B. Stomach
   C. Small intestine
   D. Large intestine

3. The lining of the human stomach is protected by a layer of mucus and bicarbonate produced by the parietal cells of stomach. True or false?

4. Removing a large section in the middle of the small intestine will result in _____.
   A. inadequate absorption of nutrients
   B. increased absorption of water and electrolytes
   C. loss of bile and pancreatic juices
   D. increased gastroesophageal reflux

6. Which of the following statements is true?
   A. Biofilm buildup results in the formation of dental plaque, which if not removed can result in tooth decay.
   B. Bacteria in the biofilm produce a substance that covers and protects the biofilm from antibiotics and other chemicals.
   C. Both A and B

Match the following:

7. Bleeding, red gums   A. *Candida albicans*

8. Tooth decay   B. Trench mouth

9. Thrush   C. Plaque formation

10. Which of the following diseases is linked to gingivitis?
    A. Pneumonitis   C. Gastroenteritis
    B. Vincent's angina   D. Colitis

12. A patient with cholera will have _____.
    A. an inflammatory diarrhea
    B. increased absorption of water and electrolytes
    C. secretory diarrhea
    D. intestinal hypermotility

13. Bacterial infections of the gastrointestinal tract are the only cause of inflammation in the digestive system. True or false?

18. Inflammation of the _____ may result in hyperbilirubinemia and clay-colored stools.
    A. liver
    B. intestinal mucosa
    C. salivary glands
    D. gastric mucosa

19. Hepatitis D can replicate only inside a cell that is coinfected with _____.
    A. hepatitis E
    B. hepatitis B
    C. hepatitis C
    D. hepatitis G

21. A patient who has bloody diarrhea is diagnosed with shigellosis. *Shigella* avoided this patient's immune system by producing _____.
    A. attaching and effacing lesions   D. pedestals
    B. Shiga toxin 1   E. actin tails
    C. Shiga toxin 2

22. Patients suffering from _____–induced diarrhea are treated with antibiotics to reduce the possibility of a carrier state.
    A. *Vibrio cholerae*   D. *Salmonella* Typhi
    B. *Salmonella* Typhimurium   E. *Escherichia coli*
    C. *Vibrio vulnificus*

23. The most effective preventive measure against cholera is _____.
    A. good sanitation
    B. vaccination
    C. prophylactic antibiotics
    D. adequate intake of fluids

24. *Salmonella* Typhi has an animal reservoir. True or false?

26. A patient complains of a nonproductive cough accompanied by a burning sensation and is suffering from abdominal pain and vomiting. She is diagnosed with bowel obstruction. Examination of her stool reveals roundworms and worm eggs. The reason for the patient's cough is _____.
    A. migration of worms to the lungs
    B. migration of larvae to the lungs
    C. a systemic reaction to the infection
    D. an associated pneumonia

27. *Strongyloides stercoralis* can cause infection following _____.
    A. ingestion of larvae   D. A and C only
    B. larvae penetration of the skin   E. A and B only
    C. ingestion of the small roundworms

28. A 25-year-old day-care teacher is diagnosed with parasitic diarrhea after stool examination shows flagellated protozoa and trophozoites. What is the treatment for this infection?
    A. Proper hydration   D. Vancomycin
    B. Electrolyte replacement   E. Penicillin
    C. Metronidazole

Key Terms

adenitis (p. 744)
amebiasis (p. 754)
amebic dysentery (p. 754)
ascariasis (p. 758)
attaching and effacing lesion (p. 740)
bacillary dysentery (p. 741)
cercaria (p. 757)
cestode (p. 755)
cholecystitis (p. 745)
cholera (p. 746)
cholera toxin (p. 746)
class switching (p. 732)
colitis (p. 728)
cryptosporidiosis (p. 754)
cyclosporiasis (p. 754)
cysticercus (p. 757)
definitive host (p. 756)
dental caries (p. 725)
dental plaque (p. 725)
diarrhea (p. 727)
enteric fever (p. 742)
enteric pathogen (p. 739)
enteritis (p. 728)
enteroaggregative E. coli (EAEC) (p. 741)
enterocolitis (p. 728)
enterohemorrhagic E. coli (EHEC) (p. 740)

enteroinvasive E. coli (EIEC) (p. 739)
enteropathogenic E. coli (EPEC) (p. 741)
enterotoxigenic E. coli (ETEC) (p. 739)
enterotoxin (p. 730)
eosinophilia (p. 755)
feces (p. 725)
filariform (p. 758)
flatworm (p. 755)
fluke (p. 755)
food poisoning (p. 750)
gastric mucosa (p. 724)
gastric pit (p. 724)
gastritis (p. 728)
gastroenteritis (p. 728)
giardiasis (p. 751)
gingivitis (p. 726)
Guillain-Barré syndrome (GBS) (p. 745)
hemolytic uremic syndrome (HUS) (p. 740)
hepatitis (p. 728)
hepatitis A virus (HAV) (p. 732)
hepatitis B virus (HBV) (p. 732)
hepatitis C virus (HCV) (p. 733)
hepatitis D virus (HDV) (p. 732)
hepatitis E virus (HEV) (p. 733)
hepatosplenomegaly (p. 742)

hookworm (p. 758)
inflammatory diarrhea (p. 727)
intermediate host (p. 756)
isotope switching (p. 732)
labile toxin (p. 739)
listeriosis (p. 749)
miracidium (p. 757)
motility-related diarrhea (p. 728)
mumps (p. 733)
myocarditis (p. 745)
nematode (p. 755)
noncholera Vibrio (p. 746)
norovirus (p. 730)
oophoritis (p. 734)
opisthorchiasis (p. 760)
orchitis (p. 734)
osmotic diarrhea (p. 727)
pancreatitis (p. 745)
paramyxovirus (p. 733)
parotitis (p. 733)
parthenogenesis (p. 758)
pathovar (p. 739)
pedestal (p. 740)
peptic ulcer (p. 737)
pericarditis (p. 745)
periodontal disease (p. 726)
peristalsis (p. 725)
peritonitis (p. 745)

pinworm (p. 757)
plague (p. 744)
platyhelminth (p. 755)
proglottid (p. 757)
rotavirus (p. 730)
roundworm (p. 755)
salmonellosis (p. 742)
schistosomiasis (p. 760)
scolex (p. 757)
secretory diarrhea (p. 727)
Shiga toxin E. coli (STEC) (p. 741)
shigellosis (p. 741)
strongyloidiasis (p. 758)
tapeworm (pp. 755, 757)
thrombocytopenia (p. 741)
thrombotic thrombocytopenic purpura (TTP) (p. 741)
thrush (p. 726)
trematode (p. 755)
trench mouth (p. 726)
trichinosis (p. 758)
trophozoite (p. 751)
typhoid fever (p. 742)
Vi Ag (virulence antigen) (p. 744)
vibriosis (p. 746)
Vincent's angina (p. 726)
whipworm (p. 757)
yersiniosis (p. 744)

Review Questions

1. What are the different parts of the digestive system?

2. What is the importance of the acidic pH of the stomach?

3. What is the relationship between biofilm formation in the mouth and dental plaques?

4. What are differences between gingivitis, periodontitis, dental caries, and trench mouth?

5. Which bacteria often cause periodontitis?

6. Describe thrush and its causative agent.

7. How do gastritis, gasteroenteritis, colitis, and enterocolitis differ from each other?

8. What are the types of diarrhea?

9. List the signs and symptoms of hepatitis.

10. How do the viruses that cause hepatitis differ from each other?

11. Describe how rotavirus increases intestinal motility.

12. What part of the digestive system is infected with mumps virus?

13. Name viral infections of the digestive system that can be prevented by vaccination.

14. Why is hepatitis D virus found only in coinfection with hepatitis B virus?

15. How does Helicobacter pylori cause an infection in the stomach?

16. How is H. pylori infection diagnosed?

17. What is hemolytic uremic syndrome (HUS)?

18. Which enteric pathogens can cause a bloody diarrhea?

19. List enteric pathogens that produce an enterotoxin.

20. How do different bacterial toxins produce GI disease?

21. How are various pathovars of E. coli different from each other?

22. Describe how Shigella causes shigellosis.

23. How do various salmonelloses differ from each other, and how are they similar?

24. What are the complications of C. jejuni infections?

25. Name risk factors associated with diseases caused by members of different Vibrio species.

26. How are treatments for V. cholerae and V. vulnificus different from each other, and why?

27. What is listeriosis, and how is it diagnosed and treated?

28. What are the complications of listeriosis?

29. How is Listeria's ability to grow in the cold related to its pathogenesis?

30. How does S. aureus cause food poisoning?

31. What is giardiasis, and who is at risk of acquiring it?

32. What organism causes amebic dysentery, and how?

33. Define cryptosporidiosis and describe its diagnosis and treatment.

34. List the gastrointestinal helminth infections, their modes of transmission, and how they are treated.

35. Describe the four life cycles used by helminths as related to their transmission.

36. List the signs and symptoms for different helminth diseases.

37. Express the best preventive methods used against viral, bacterial, and parasitic infections of the digestive system.

Clinical Correlation Questions

1. A patient who has just returned from Mexico presents to the emergency department complaining of voluminous watery diarrhea. Vitals are normal and stool sample does not contain any blood or leukocytes. The PA tells the patient that he most likely has "traveler's diarrhea." What is the causative organism, and how does it cause diarrhea?

2. A 24-year-old male presents to the ER with diarrhea, high fever (40°C, or 104°F) and chills, malaise, and small red spots on his trunk. The patient reports that he has had a fever for days and that the fever was initially low grade but has been getting higher and higher. Blood culture reveals a facultative intracellular organism. What is the most likely diagnosis? Explain.

3. A patient with a 2-week history of watery diarrhea presents to the ER complaining of right lower quadrant pain and a low-grade fever. He is diagnosed with appendicitis and taken to surgery. The patient's appendix is found to be normal during the surgery, but he is found to have mesenteric lymphadenitis and terminal ilcitis. What bacterial pathogen is most likely to cause this patient's symptoms? Explain.

Thought Questions: CHECK YOUR ANSWERS

Thought Question 22.1   If hepatitis viruses all replicate in the liver and are released into the intestine in bile, why is it that only HAV and HEV can be transmitted by the fecal-oral route?

ANSWER: HAV and HEV are nonenveloped viruses. Nonenveloped viruses are more resistant than enveloped viruses to the effects of intestinal bile and stomach acid.

Thought Question 22.2   Why are persons immunized against HBV also protected from HDV?

ANSWER: Because HDV requires HBV to replicate. Someone immunized against HBV can be infected with HDV, but no disease can develop.

Thought Question 22.3   You are treating a patient for severe, sharp abdominal pain that lessens after eating. How would you design an ELISA to detect serum antibodies to CagA?

ANSWER: ELISA can be performed by coating wells in a plastic microtiter dish with *Helicobacter* CagA antigen. Serum from the patient is then added to the well. If antibodies to CagA are present, they will bind to the CagA antigen. Unbound antibody is removed by washing, and a secondary antibody that binds human IgG is added to the well. These anti-antibodies have an enzyme linked to them. A sandwich is formed as follows: [plastic dish]–[CagA protein]–[anti-CagA antibody]–[anti-IgG antibody]–[enzyme]. When the appropriate enzyme substrate is added to the well, the enzyme acts on it to produce light or a chromogenic product that can be detected. The more anti-CagA antibody present in the serum, the more light or product is produced by the linked enzyme.

Thought Question 22.4   How might the ability of *Y. enterocolitica* to grow in a cold environment be used to more effectively isolate the organism from a patient's stool?

ANSWER: The large number of normal microbiota present in infected stool could obscure the presence of low numbers of *Yersinia* causing disease.

Placing the stool in a refrigerator for several days (called cold enrichment) can increase the numbers of *Yersinia* present (because they grow) while inhibiting growth of normal microbiota. Cold enrichment combined with various selective media make it easier for the laboratory to identify the presence of *Yersinia* in stool samples.

Thought Question 22.5   A college friend tells you she has been suffering with watery diarrhea for 2 days and wonders what could be causing it. You find out she has also experienced abdominal pain. What viruses or bacteria would you place in your list of differential diagnoses (list of possible causes)? What if she tells you there is blood in her stool?

ANSWER: For watery diarrhea, norovirus is a prime candidate (because of the patient's age), but rotovirus, ETEC, *V. cholerae* (if she visited an endemic area), and *Salmonella* (it may be hard to see blood in stool without a laboratory test) should be considered as well. Absence of vomiting and the duration of diarrhea place staphylococcal enterotoxin low on the differential list; absence of blood in stool reduces the likelihood of *Shigella*. For bloody diarrhea, the following organisms should be considered: *Shigella, Salmonella* (though not always), EHEC, *Yersinia enterocolitica, Campylobacter* (though not always), *Entamoeba*, and some of the helminths (although unlikely if she has not traveled outside the country).

Thought Question 22.6   A stool culture of a patient afflicted with *S. aureus* food poisoning reveals no *S. aureus*. Where could you find the *S. aureus* that caused the food poisoning?

ANSWER: *Staphylococcus aureus* food poisoning is caused by the ingestion of preformed toxin present in food contaminated with *S. aureus*. You could find the organism in the offending food (if any remains). Remember that the enterotoxin is heat stable. If the food was stored and then heated, the toxin will remain active but the organism may be killed, so you may not find the organism in the food.

23

Infections of the Urinary and Reproductive Tracts

[Frequent Urination]

SCENARIO Lois woke up with a headache one morning, a bit disoriented and not feeling well. She is 87 years old and lives by herself. She had not slept well the night before because she had awakened several times to urinate. Lois is diabetic, but her blood sugar the night before had been 120 mg/dl (normal is 100 mg/dl) and it measured 180 mg/dl in the morning. She went about her day, taking her insulin shot and eating her breakfast. The breakfast did not stay with her long, and after several bouts of vomiting, she called her son.

Urine Specimen Cup
Patients can use these sterile cups to collect their own urine. Lois, however was poorly responsive and needed to have a catheter inserted into her bladder.

LABORATORY RESULTS Gram stain of the urine sample showed Gram-negative rods. Cultures of urine and blood showed bacteria and confirmed everyone's suspicions: she had a severe urinary tract infection, and the bacteria had spread to her bloodstream, causing sepsis (blood infection; see Chapter 21). Lois was diagnosed with urosepsis, a term often used to describe a urinary tract infection complicated by sepsis. The organism in this case was found to be a uropathogenic strain of *Escherichia coli*.

TREATMENT Lois was transferred to the intensive care unit (ICU), where she was treated with IV antibiotics, fluids, and her routine medications. Lois fully recovered and was sent home with extensive patient education given to her, her son, and her daughter-in-law.

SIGNS AND SYMPTOMS
Lois was almost nonresponsive by the time her son and daughter-in-law arrived. She had no fever—not unusual for the elderly. Her family members were familiar with this; it had happened before, but never so severely. They took Lois to the ER, where her temperature measured below normal; her heart rate was faster than normal and her blood pressure dangerously low. Lois was immediately given fluids, and blood and urine samples were taken for laboratory analysis.

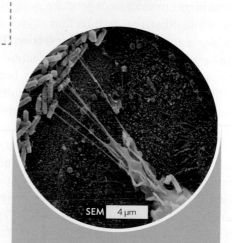

SEM 4 μm

Bacterial Biofilm in the Urinary Bladder
Uropathogenic *Escherichia coli* (green) cover the surface epithelium of a mouse bladder (purple) while a leukocyte (blue) attacks the organism. Uropathogenic *E. coli* can form a biofilm in the bladder, making it very difficult to rid the bladder of infection.

CHAPTER OBJECTIVES

After reading this chapter, you will be able to:

• Correlate the anatomy and physiology of the urinary and reproductive tracts with the infectious diseases affecting them.

• Distinguish genitourinary infections from one another according to the presenting signs and symptoms.

• Describe similarities and differences between sexually and non-sexually transmitted infections of the genitourinary system.

• Explain effective prevention and treatment methods for infectious diseases affecting the urinary and reproductive tracts.

L ois's problem started with a bladder infection with no clear symptoms. The bladder infection progressed rapidly, and by the time she got to the ER, she had **urosepsis** and was in danger of losing her life. How could this happen?

Urine made in the kidneys is normally sterile, so the presence of bacteria in urine (called **bacteriuria**) is considered abnormal. Bacteriuria might or might not produce symptoms like a burning sensation while urinating. An infection anywhere from the kidney to the bladder is called a **urinary tract infection (UTI)**. UTIs are the second most common type of bacterial infection in humans, ranking just behind respiratory infections such as bronchitis and pneumonia. UTIs have a large economic impact, resulting in innumerable office visits, more than 100,000 hospital admissions, and $1.6 billion in medical expenses each year.

Long-time diabetics, like Lois, often do not have the burning pain usually associated with a UTI because of the nerve damage (neuropathy) caused by elevated blood glucose levels. The process of aging itself also seems to reduce the severity of UTI symptoms. For example, up to 50% of women over 65 years of age report having no symptoms even when they have confirmed bacteriuria.

The main job of the urinary tract is to rid the body of metabolic waste products, whereas the primary role of the reproductive tract is to propagate the human species. Infections in either system can undermine vital functions and have serious consequences. The formation and excretion of urine, for instance, appears so mundane, yet its interruption can cause extreme pain and even threaten life.

23.1
Anatomy of the Urinary Tract

SECTION OBJECTIVES

• List the function of each structural component of the male and female urinary tracts.

• Explain the relationship between the upper and lower urinary tract anatomy and physiology.

• Discuss the significance of structural differences between the male and female urinary tract as they relate to infections.

What is the job of our urinary tract? Normal metabolism generates a lot of harmful waste. The urinary tract, which includes the kidneys, ureters, urinary bladder, and urethra, removes this waste and any excess water and electrolytes from the bloodstream and then expels the product: urine (**Figure 23.1**). Large waste molecules (for example, unneeded proteins or lipids) are eliminated mostly by the liver or gastrointestinal tract. Kidneys act as a filter to eliminate unwanted small molecules from the blood while maintaining plasma electrolytes and blood pH in the normal physiological range.

The urinary system is divided into upper and lower tracts. Kidneys and ureters are considered the upper tract; the urinary bladder and urethra form the lower tract.

Figure 23.1   The Urinary Tract

Renal arteries bring unfiltered blood from the heart to each kidney via the aorta and renal veins return filtered blood from each kidney to the heart via the inferior vena cava. A ureter originates from each kidney and ends in the urinary bladder. Urine collects in the bladder until it is full; the urine is expelled through the urethra.

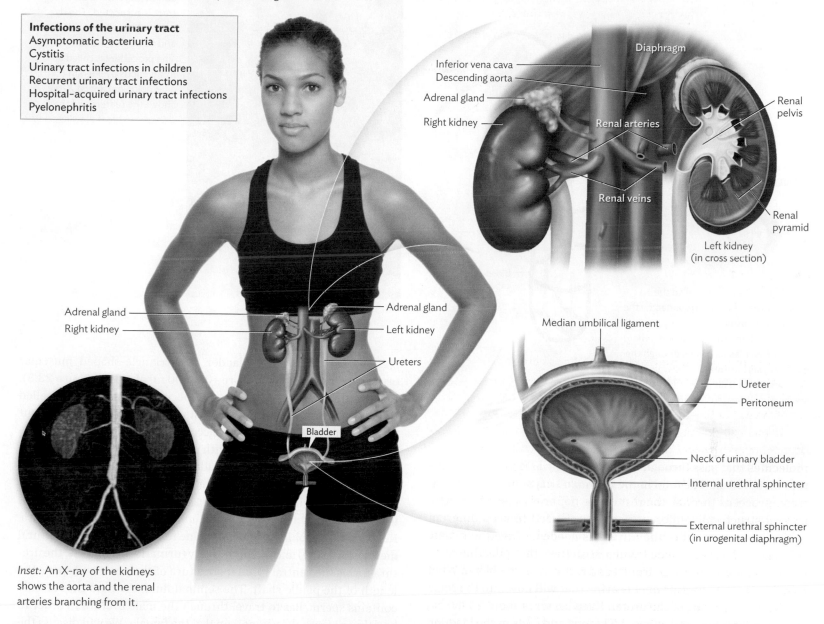

Infections of the urinary tract
Asymptomatic bacteriuria
Cystitis
Urinary tract infections in children
Recurrent urinary tract infections
Hospital-acquired urinary tract infections
Pyelonephritis

Diaphragm

Inferior vena cava
Descending aorta
Adrenal gland
Right kidney
Renal arteries
Renal veins
Renal pelvis
Renal pyramid
Left kidney (in cross section)

Adrenal gland
Right kidney
Adrenal gland
Left kidney
Ureters
Bladder

Median umbilical ligament
Ureter
Peritoneum
Neck of urinary bladder
Internal urethral sphincter
External urethral sphincter (in urogenital diaphragm)

Inset: An X-ray of the kidneys shows the aorta and the renal arteries branching from it.

The kidneys are two bean-shaped organs that sit in the abdominal cavity on either side of the descending aorta with an adrenal gland perched on top of each. The renal artery and vein enter at the concave part of each kidney, called the renal hilum. The renal pelvis, the beginning part of the ureter, exits at the renal hilum (see inset in Figure 23.1).

Each kidney is made of many tubular structures called nephrons, roughly a million of them, parts of which make the renal pyramids. Each nephron has a horseshoe-shaped structure called a Bowman's capsule that surrounds a ball of blood capillaries known as a glomerulus (**Figure 23.2**). It is the glomerulus (plural, glomeruli) that becomes damaged in acute **post-streptococcal glomerulonephritis**, an autoimmune sequela of some streptococcal infections. Small molecules such as water, electrolytes, glucose, CO_2, and protons move from the blood into the Bowman's capsule, leaving large molecules and blood cells behind. Among the small

Figure 23.2 Nephron Structure

A. Nephrons are the filtering units of the kidneys. Each nephron starts at the Bowman's capsule and continues as a long tube that ends in the collecting duct.

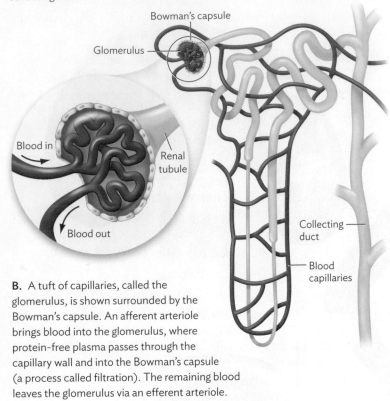

Bowman's capsule

Glomerulus

Blood in

Renal tubule

Blood out

Collecting duct

Blood capillaries

B. A tuft of capillaries, called the glomerulus, is shown surrounded by the Bowman's capsule. An afferent arteriole brings blood into the glomerulus, where protein-free plasma passes through the capillary wall and into the Bowman's capsule (a process called filtration). The remaining blood leaves the glomerulus via an efferent arteriole.

Figure 23.3 Ureters, Urethra, and the Urinary Bladder

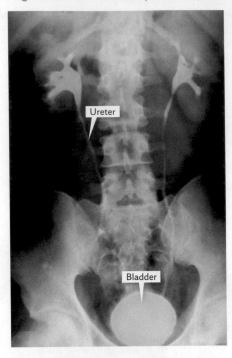

Ureter

Bladder

Contrast dye is given intravenously when structural abnormality in the urinary tract is suspected. The dye concentrates in the kidneys and flows into the bladder. The procedure is called an intravenous pyelogram (IVP). This color-enhanced X-ray shows the ureters and the urinary bladder following an IVP.

molecules that pass through is urea, a toxic waste product of protein metabolism. The fluid in the Bowman's capsule, called the filtrate, proceeds through the rest of the nephron, a long tube with several structurally different parts specialized to keep different electrolytes in or out of the filtrate as needed. When the filtrate reaches the last part of the tubular structure, the collecting duct, the extra water is removed from it and returned to the blood. What remains in the collecting duct is urine that will flow into the renal pelvis, the beginning of the ureter. Each ureter is about 25–30 cm (10–12 inches) long and starts at a kidney and ends in the bladder. Ureters have muscular walls that continuously contract and relax to propel urine down to the urinary bladder.

Link As discussed in Section 20.3, some strains of *Streptococcus pyogenes* (the nephritogenic strains) have antigens with structural similarities to components of human renal tissue. The immune system of patients suffering from sore throat caused by the nephritogenic *S. pyogenes* attacks self antigens, causing immune complex–mediated inflammation and damage to the renal tissue, a condition called acute **post-streptococcal glomerulonephritis**.

The human urinary bladder is a triangle-shaped muscular organ that can collect and store 400–600 ml of urine (**Figure 23.3**). Inside, the bladder is lined with specialized epithelium (called transitional epithelium) that consists of several layers of cells put together in a way that can easily accommodate the changes in bladder volume as the bladder fills with urine and empties. In some people, uropathogenic *E. coli* can invade and form a biofilm inside these cells and serve as a source for recurrent bladder infections.

Two small circular muscles, called the internal and external urethral sphincters, mark the beginning and end of the urethra. Relaxing these sphincters allows urine to flow through the urethral meatus (opening) and out of the body (urination). In males, the urethra forms the central tubular structure of the penis and spans the length of the penile shaft. The seminal fluid, the fluid that usually contains sperm, has to travel through the male urethra in order to be delivered into the vaginal canal of the female. We will discuss the anatomy of the reproductive tract later in this chapter.

SECTION SUMMARY

- **The urinary tract** includes the kidneys, ureters, urinary bladder, and urethra.
- The **kidneys filter waste products** from blood to make urine, the **ureters deliver urine** to the bladder, the **bladder stores** the urine, and the **urethra passes urine** outside the body.
- **Nephrons** include the **Bowman's capsule**, which surrounds capillaries of the **glomerulus**, where filtration occurs.

- **Urine leaves the nephron** through the collecting duct and passes through the renal pelvis and the ureters to reach the bladder.
- **Urethral sphincters** control urine flow from the bladder.
- The opening of the urethra to the outside of the body is **called the urethral meatus.**

23.2
Urinary Tract Infections

SECTION OBJECTIVES

- Discuss the role of physical and hormonal factors in the development of urinary tract infections in women.
- Explain the relationships between lower and upper urinary tract infections.
- Compare and contrast the signs and symptoms of common urinary tract infections.
- List treatments for the most common etiologies of urinary tract infections.

How big a problem are urinary tract infections? A huge one! Everyone is at risk of getting a UTI, but women suffer most cases. In fact, women have a greater than 50% chance of getting a urinary tract infection sometime during their lifetime, and 20%–40% of those infected will have recurrent infections. UTI is the most common nosocomial infection and the 15th most common diagnosis made by family physicians during office visits. There are two types of urinary tract infections: bladder infections, or **cystitis**, and kidney infections, or **pyelonephritis**.

Why do women get urinary tract infections more often than men? The perineum (the distance between anus and vagina or the base of the penis) of the male is roughly twice as long as the female's. The short female perineum places the female urethra adjacent to the vaginal and rectal openings, which are colonized by some potential pathogens. The combination of a short urethra and proximity to pathogens is thought to be responsible for the increased risk of UTIs for women. The longer male urethra makes it more difficult for bacteria to reach the urinary bladder.

The risk of a bladder infection in young women is also related to sexual activity and is proportional to the frequency of intercourse. In postmenopausal women, bladder incontinence, partially due to lower estrogen (female hormone) levels, increases the risk for UTI. Hormonal changes of pregnancy and menopause also reduce the normal muscle tone in the urethral wall, causing it to become wider and shorter and further increasing the chances of getting a lower urinary tract infection in these women. For men, instrumentation, such as the insertion of a catheter through the urethra to retrieve urine from the bladder, is the main risk factor for getting a UTI. Diabetes increases the risk of UTI for everyone.

Pathogenesis of Urinary Tract Infections

How do we get urinary tract infections? The newly discovered, but mostly unculturable, microbiome of the bladder is not responsible for cystitis (Section 14.1). Pathogenic microorganisms from other sources, however, can "find" their way into the urinary tract and establish an infection in one of three basic ways:

- **Infection from the urethra to the bladder.** Bacteria residing along the superficial urogenital membranes of the urethra can ascend to the bladder. This is more common in women than in men.

- **Ascending infection to the kidney.** Organisms from an established infection in the bladder ascend along the ureter to infect the kidney.

- **Descending infection from the kidney.** Descending infection occurs when an infected kidney sheds bacteria that descend via the ureters into the bladder. Kidney infections may arise when microorganisms are deposited in the kidneys from the bloodstream.

Several organisms can cause a UTI. In Lois's case, the laboratory found the organism to be a Gram-negative bacillus that ferments lactose, suggesting *E. coli* as the likely culprit. Given that the normal habitat of *E. coli* is the gastrointestinal tract, the initial UTI suffered by the patient was probably the result of an inadvertent introduction of the microbe into the urethra (which then made its way up the urethra and into the bladder). Patients may suffer recurring UTIs as a result of organisms entering the bladder by a descending route or organisms emerging from a pod-like biofilm inside the bladder (see **inSight**).

Urine is bacteriostatic to most of the commensal Gram-positive organisms inhabiting the perineum and the vagina, such as lactobacilli, *Corynebacterium*, diphtheroids, and *Staphylococcus epidermidis*. In contrast, Gram-negative rods not only thrive in urine but also may be adapted to cause urinary tract infections through specialized pili. As a result, most urinary tract infections are caused by facultative Gram-negative rods from the GI tract. Uropathogenic strains of *Escherichia coli*, for example, typically have P-type pili, with a terminal receptor for the P antigen (a rather appropriate name for a bladder-specific virulence factor). The P antigen is a blood group marker found on cells lining the perineum and urinary tract; it is expressed by approximately 75% of the population. Individuals expressing P antigen are particularly susceptible to UTIs. The most common etiologic agents of UTI are the uropathogenic *E. coli* (75% of all UTIs), *Klebsiella, Proteus, Pseudomonas, Enterobacter, S. aureus, Enterococcus, Chlamydia*, fungi, and *Staphylococcus saprophyticus*.

Although some UTIs resolve spontaneously, others can progress to destroy the kidney or, via Gram-negative septicemia, the

inSight

Recurrent UTIs: The "Hide and Seek" Strategy of Uropathogens

Figure 1 **Bladder Pods of Uropathogenic *E. coli***

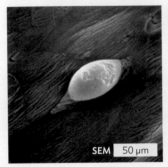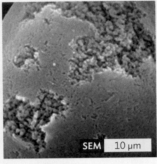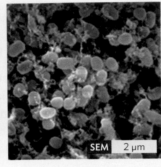

A. Increasing magnifications of large intracellular communities of uropathogenic *E. coli* (UTI89) inside pods on the surface of a mouse bladder infected for 24 hours.

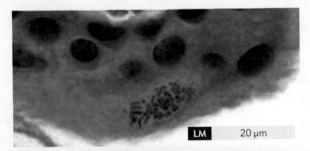

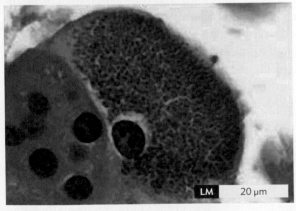

B. Hematoxylin- and eosin-stained sections of *E. coli*–infected mouse bladders show a bacterial factory 6 hours after inoculation (top panel) and a pod 24 hours after inoculation (bottom panel). Bacteria in the pod were densely packed and shorter and filled the host cell.

Doctors used to think that women suffering from recurrent bladder infections were repeatedly infected as a result of sexual activity or poor hygiene—a presumption many women with chronic infections found frustrating and offensive. Evidence now indicates that bacteria may hide inside bladder cells, forming a reservoir for reinfection.

Specialized colonies of *E. coli* have been found living in the surface layer of cells lining the bladders of infected mice (Figure 1A). Though the mouse served as a model system, the bacteria used were originally isolated from humans. These strains of *E. coli* cause about 80% of all urinary tract infections. The bacteria use pili to latch onto proteins coating the bladder cells. Those proteins form a protective substance, known as uroplakin, that strengthens the epithelial cells and shields them from toxins that may build up in the urine. The bacteria can slip under the protective barrier and penetrate superficial cells lining the bladder. Once inside the cell, the microbes begin to multiply, and the epithelial cell fills with bacteria also coated with uroplakin, taking on the appearance of a pod (Figure 1B).

The pod-like *E. coli* biofilms we describe here are notoriously resistant to antibiotic treatment and attacks by the immune system. Living inside host cells and surrounded by uroplakin, the *E. coli* in the bladder pods have devised a clever mechanism for surviving. Bacteria living on the edges of the biofilm may break free, leading to subsequent infection. Understanding the cycle of infection and pod formation may help researchers design drugs to prevent new urinary tract infections or stop established ones.

Although this work was done with mice, intracellular bacterial communities of *E. coli* have also been identified in some women with recurrent cystitis. Elsewhere in the body, biofilms hiding inside cells may also act as reservoirs for other hard-to-treat chronic or recurring infections, such as ear infections.

C. Scott Hultgren (at Washington University in St. Louis) discovered the intracellular bacteria-filled pods in the bladder lumen.

host. As a result, antibiotic therapy is recommended. Infections of the lower urinary tract are usually afebrile (without fever) except in young children. In older patients, like the elderly Lois, UTIs frequently show atypical symptoms, including delirium, which disappears when the UTI is treated.

Identification of culturable microorganisms in the urine makes the definitive diagnosis of a UTI. A urine sample must be collected and sent for urinalysis and culture. Urinalysis has three parts: determining the color and clarity of the sample, microscopic examination, and biochemical studies performed with a dipstick (Figure 23.4). A urine dipstick is used to check for the presence of leukocyte esterase and nitrites. Leukocyte esterase is an enzyme of white blood cells and is present in the urine only when white blood cells are present, even if not intact. Presence of nitrite in the urine points to the presence of bacteria, which can form nitrite from the nitrate normally present in the urine. A noncontaminated urine sample from a healthy individual should not contain any white blood cells. Clinicians often make a presumptive diagnosis of UTI on the basis of the patient's signs and symptoms, starting empiric therapy to reduce the risk of serious complications.

In most patient populations, the treatment of asymptomatic bacteriuria is not necessary, and consequently, screening is not recommended. An important exception is pregnant women, for whom asymptomatic infection carries significant risks and treatment provides obvious benefits.

Symptomatic bacterial UTIs are treated with antibiotics such as trimethoprim-sulfamethoxazole (Bactrim). Choosing the appropriate antibiotic depends on the infecting bacteria and their drug resistance profile as well as the age, sex, and health status of the patient. Most clinicians agree that increased intake of liquids to "flush" the system is helpful, and cranberry juice is a favorite among some because it contains proanthocyanidins (tannins) that have antimicrobial qualities. The enthusiasm for cranberry juice as a cure is not universal, but it is often recommended nevertheless.

Bacterial Infections of the Urinary Tract

As mentioned, two types of urinary tract infections (UTIs) exist: cystitis (bladder infection) and pyelonephritis (kidney infection). Several organisms can cause either type of infection with similar signs and symptoms (Table 23.1). Inflammation of the urethra, or **urethritis**, is signified by pain or burning upon urination and urethral discharge. For two reasons, our main discussion of urethritis will take place in the section on reproductive diseases (Section 23.4). First, women and men with urethritis (gonococcal and nongonococcal) typically contract the disease as a result of sexual intercourse. Second, in men, the urethra is shared by the urinary and reproductive tracts.

ASYMPTOMATIC BACTERIURIA Interestingly, the presence of bacteria in the urine (bacteriuria) often does not produce any

Figure 23.4   Urinalysis

Urine is collected in a sterile cup by using the clean-catch technique. A strip is then inserted into the patient's urine, and diagnoses are made based on the color changes. Normal results are on the far left. Progressively abnormal results are shown toward the right.

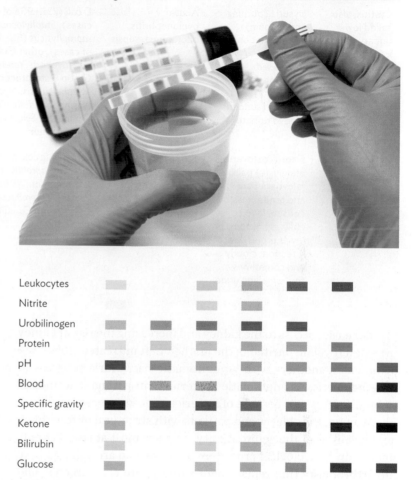

signs or symptoms. When this happens, the person is said to have asymptomatic bacteriuria. How can patients with bacteriuria be asymptomatic? That's a good question with no definitive answer. The microbiology of asymptomatic bacteriuria is the same as for other uncomplicated UTIs. In some cases, the lack of symptoms may be due to changes in the pathogens, making them less virulent, or the immune system may be less responsive, or a combination of the two.

Although asymptomatic bacteriuria is typically not considered problematic, it can be a serious concern for pregnant women and the elderly. Asymptomatic bacteriuria in pregnant women is associated with low birth weight of the newborn, preterm labor, and perinatal death (fetal death just prior to birth). Production of massive amounts of female hormones during pregnancy loosens the walls of the ureters (tonic relaxation of ureters), allowing bacteria in the

Table 23.1
Infections of the Urinary Tract

Disease	Symptoms	Complications	Organism	Transmission	Diagnosis	Treatment	Prevention
Cystitis, also called lower urinary tract infection	Dysuria (burning and/or pain) upon urination, frequency (increased urination frequency), and urgency (sudden urge to urinate)	Acute pyelonephritis, bacteremia, sepsis	*E. coli* (75%–90% of cases), *Staphylococcus saprophyticus* (5%–15% of cases), other Gram-negative rods, rarely: Group B streptococci, *M. tuberculosis*, *Mycoplasma, Candida*, viruses, *Chlamydia, Schistosoma haematobium*	Self-inoculation or hospital acquired	Clinical presentation; urinalysis; urine dipstick for nitrite and leukocyte esterase; urine culture	Oral antibiotics (trimethoprim-sulfamethoxazole, fluoroquinolones); special regimens recommended for men and pregnant women	No vaccine available
Pyelonephritis	Flank (costovertebral angle) pain, mid-back pain and tenderness, nausea, fever, chills, symptoms of lower urinary tract infection, cloudy and sometimes foul-smelling urine	Kidney necrosis or abscess, sepsis, preterm labor in pregnant women	*Escherichia coli, Staphylococcus saprophyticus*; less common: *Klebsiella, Proteus, Enterobacter, Clostridium, Candida* in diabetics	Ascending infection from lower urinary tract or hematogenous spread from a distant focus of infection	Urine culture; clinical symptoms; imaging (rarely)	Oral or IV antibiotic therapy, depending on disease presentation; fluoroquinolones are drugs of choice except in pregnant women and children	No vaccine available

bladder easier access to the kidney and increasing the risk of kidney infection (pyelonephritis) in the mother. Left untreated, 20%–30% of pregnant mothers with asymptomatic bacteriuria progress to pyelonephritis. Asymptomatic bacteriuria in pregnant women is treated aggressively because of its serious consequences.

Women, the elderly, and patients with structural or functional abnormalities of the genitourinary tract are most at risk. Patients undergoing a urological procedure or those who are going to have hip arthroplasty (hip replacement surgery) are also screened and treated in order to reduce specific postsurgical infections. The prevalence of asymptomatic bacteriuria increases with age and depends on the patient population.

CYSTITIS

CASE HISTORY 23.1

When It Hurts to Pee

Lashandra is 24 years old and was experiencing increased frequency of urination and dysuria (pain or burning with urination) over the past 3 days. This was the first time Lashandra had ever suffered from persisting dysuria. She consulted her general practitioner, who requested a midstream specimen of urine. Upon microscopic examination, the urine was found to contain more than 50 leukocytes/µl (normal is fewer than 5) and 35 red blood cells/µl (normal is 3–20). Epithelial cells from the urogenital area contaminate a midstream specimen if the specimen is not collected properly. No squamous epithelial cells (skin cells) were seen, indicating a well-collected midstream catch. The urine culture plated on agar medium

yielded more than 10^5 colonies/ml (meaning more than 10^5 organisms/ml in the urine) of a facultative, anaerobic, Gram-negative bacillus that can ferment lactose.

The first question to ask in this case is whether the patient had a significant UTI. The answer is yes. The purpose of the midstream urine collection is to provide laboratory data to make this determination. Urine becomes contaminated with normal microbiota that adhere to the urethral wall. In a midstream collection, the patient urinates briefly, stops to position a sterile collection jar, then resumes urinating in the jar (see Section 25.2). The initial brief period of urination minimizes the number of contaminating microbes in the sample by washing away organisms clinging to the urethral opening before the sample is collected. Nevertheless, the collected sample will still contain low numbers of organisms representing normal flora. A diagnosis of UTI is made when the number of bacteria in the sample becomes greater than 10^5/ml. (However, that number can be as low as 1,000/ml in some symptomatic women.) The patient in the case history was symptomatic and had sufficient numbers of bacteria in her urine to indicate a UTI.

Do you know what a bladder infection feels like? Chances are you answered yes, because by young adulthood most of us have either known someone who has suffered a urinary tract infection or have experienced it ourselves. Uncomplicated (no fever) acute cystitis starts with a sudden onset of **dysuria** (painful or difficult urination), increased frequency of urination, urgency, burning upon urination, and nocturia (waking up several times at night to urinate). Some patients may have fever and lower abdominal pain, but these

Table 23.2
Signs and Symptoms of Urinary Tract Infections in Different Populations

Patient Population	Dysuria	Frequency	Urgency	Nocturia	Fever	Other
Newborn	+	N/A	–	–	+	Failure to thrive, irritability, sepsis, jaundice, vomiting
Infants–preschoolers	+ (preschoolers)	–	+	–	+ (mostly infants)	Failure to thrive, diarrhea, vomiting, abdominal or flank pain, urgency and new-onset urinary incontinence (preschoolers)
School-aged children	+	+	+	–	Possible	New-onset incontinence, vomiting, abdominal or flank pain
Adults (cystitis)	+	+	+	+	–	Hematuria and suprapubic pain; systemic symptoms may be present in complicated cases
Adults (pyelonephritis)	–	–	–	–	+	Costovertebral tenderness, flank pain, chills, oliguria, hematuria, nausea, vomiting
Elderly	Only in 12%	Only in 10%	Only in 2%	+ (may be chronic due to other causes)	–	Nonspecific symptoms in majority, new or worsening confusion, new or worsening incontinence, functional decline, falls

are uncommon symptoms. Most elderly patients (77%), like Lois, whom you met in the chapter-opening case, usually have nonspecific symptoms. They feel confused, their incontinence gets worse, they may suffer a fall, or they may be unable to function at their usual levels. Children's symptoms are different, depending on their age, and often include a fever when they are very young (Table 23.2).

URINARY TRACT INFECTIONS IN CHILDREN The main risk factor for developing a urinary tract infection in children is incomplete emptying of the bladder. Some children do not empty their bladder properly because of congenital abnormalities of the urinary tract. Others (the majority) may postpone emptying their bladder ("hold it") for a variety of reasons: they do not want to stop playing or they do not want to use a public toilet, for example. In male infants less than 6 months of age, the lack of circumcision increases the risk of urinary tract infections. Colonization of the foreskin during the early stages of life may increase the chance of bacteria accessing the urethra. The risk profile, however, seems to equalize between circumcised and uncircumcised infant populations at about 6 months of age.

Childhood urinary tract infections can lead to serious damage and scarring of the kidney if not treated promptly, which can cause kidney-related disease in adulthood. Fortunately, urinary tract infections are uncommon in children except in the very young (less than 6 months old). Prevalence (number of cases in a population) of UTI in children decreases as children get older; for example, the overall prevalence of febrile (with fever) UTI in female children

decreases from 7.5% in infants 0–3 months old to 2.1% in girls older than 12 years of age.

RECURRENT URINARY TRACT INFECTIONS Some women suffer from recurrent urinary tract infections. In fact, recurrent UTIs (defined as either two or more UTIs in 6 months or three or more UTIs in 12 months) are common among young and postmenopausal women. Incomplete emptying of the bladder is thought to be the culprit after menopause, but sexual activity and use of spermicides are risk factors for recurrent UTIs in young women. Prophylactic (preventive) continuous antibiotic therapy is used in both groups to prevent the recurring infections. Postcoital (after intercourse) antibiotic prophylaxis is an alternative to continuous therapy for treating recurring infections in young women. Other factors are also thought to be involved (see inSight).

HOSPITAL-ACQUIRED URINARY TRACT INFECTIONS Urinary tract infections account for 30%–40% of hospital-acquired (nosocomial) infections, making them the most common infection developed during a hospital stay. Most of these infections are associated with urinary catheterization, which is performed on nearly 25% of hospitalized patients (Figure 23.5A and B). The organisms isolated from catheter-associated infections (CAIs) are more likely to have some antibiotic resistance. *Escherichia coli* causes most of the hospital-acquired UTIs, but several other bacteria are also implicated, including *Proteus mirabilis*, *Pseudomonas aeruginosa*, *Enterococcus faecalis*, and *Staphylococcus epidermidis*. Figure 23.5C

Figure 23.5   Urinary Catheters

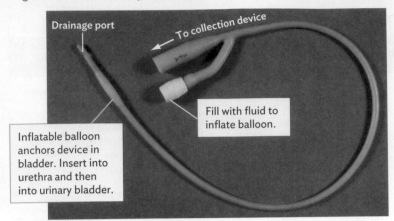

A. A urinary catheter.

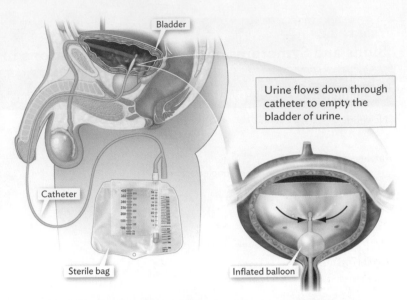

B. Urinary catheters are inserted through the urethra into the bladder. The balloon at the end of the catheter is inflated to keep it anchored inside the bladder. The urine flows through the catheter and is collected inside a sterile bag.

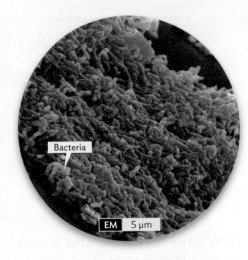

C. Electron micrograph shows biofilm formed by *P. aeruginosa* on the surface of a urinary catheter.

shows a biofilm of bacteria forming on the inner surface of a recently used urinary catheter. The biofilm protects the bacteria from the mechanical flow of urine, host defenses, and antibiotics, making bacterial elimination difficult. While in the patient, the catheter can shed organisms into the bladder and produce an infection.

PYELONEPHRITIS   Pyelonephritis (kidney infection) is not a common disease, but like cystitis, it is more prevalent in women than in men. Patients suffering from pyelonephritis typically have a high fever (38°C, or 104°F, or higher), chills, **oliguria** (producing a less-than-normal volume of urine), flank pain, tenderness at the point between the last rib and the lumbar vertebrae (**costovertebral tenderness**; Figure 23.6), and nausea.

The most common way bacteria reach the kidneys is to ascend from the lower urinary tract along the ureters. Bacterial flagella and various pili, such as type IV pili responsible for **twitching motility** in Gram-negative bacteria, allow bacteria to swim or climb up the ureters. Septicemic spread from infections at other body sites is also possible, but rare.

Link   As discussed in Section 18.2, type IV pili can actually make cells move because the assembly process involves repeated elongation and retraction of the pili. The process, called "**twitching motility**," occurs when the pilus elongates, attaches to a surface and then depolymerizes from the base, which shortens the pilus and pulls the cell forward.

The presentation of pyelonephritis varies from mild to severe. A mild disease can be treated with an antibiotic on an outpatient basis. Severe pyelonephritis, on the other hand, can cause collapse of the patient's cardiovascular system (shock), threatening the patient's life. Hospitalization in an intensive care unit (ICU) and

Figure 23.6   Costovertebral Angle and Costovertebral Tenderness
The point between the last rib and the vertebrae is called the costovertebral angle.

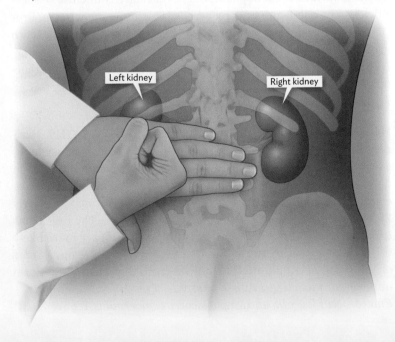

intravenous administration of antibiotics are required to resolve the infection.

SECTION SUMMARY

- **Urinary tract infections** are extremely common, especially in women. **Diagnosis of a UTI** includes finding a significant number of bacteria in urine (typically greater than 10^5 organisms/ml of urine). Common causes include uropathogenic *Escherichia coli*, *Klebsiella*, *Staphylococcus aureus*, *Enterococcus*, and *Staphylococcus saprophyticus*.

- **Bladder infections** can arise from a urethra contaminated with perianal skin microbes or from bacteria descending from an infected kidney. **Cystitis** is a simple bladder infection whose symptoms include dysuria, increased frequency of urination, urgency, and burning upon urination.

- **UTIs in children** are more prevalent in children less than 6 months old and uncommon in children over 12 years of age.

- **Catheter-associated UTIs** are the most common hospital-acquired infection.

- **Asymptomatic UTIs** occur when significant numbers of organisms appear in the urine (greater than 10^5 organisms/ml of urine) but symptoms are not evident.

- **Kidney infections (pyelonephritis)** can arise from bladder infection by microbes that ascend the ureter or occasionally from septicemic blood. Patients with pyelonephritis usually have high fever, oliguria, costovertebral tenderness, and nausea.

Thought Question 23.1 Why are urinary tract infections among the most commonly acquired nosocomial (hospital-acquired) infections?

Thought Question 23.2 Why would someone with pyelonephritis have a lower urine output?

23.3
Anatomy of the Reproductive Tract

SECTION OBJECTIVES

- List the structural components of the male and female reproductive tracts.
- Explain the function of each component of the male and female reproductive tracts.
- Discuss the significance of structural differences between the male and female reproductive tracts as they relate to infections.

Both male and female reproductive systems include gender-specific reproductive organs as well as endocrine and exocrine glands (Figure 23.7). Endocrine glands secrete hormones into the bloodstream and affect multiple organ systems. Exocrine glands secrete enzymes

Figure 23.7   The Male and Female Reproductive Systems

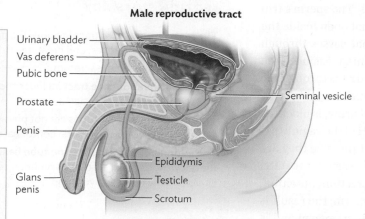

Viral infections of the reproductive tract
Genital warts
Genital herpes
Neonatal herpes
HIV
Hepatitis B and C infections

Bacterial and protozoan infections of the reproductive tract
Syphilis
Gonorrhea
Chlamydial infection
Lymphogranuloma venereum
Chancroid
Trichomoniasis
Pelvic inflammatory disease (PID)

Non-sexually transmitted infections of the reproductive tract
Bacterial vaginosis
Fournier's gangrene
Vulvovaginal candidiasis

Male reproductive tract

Urinary bladder, Vas deferens, Pubic bone, Prostate, Penis, Glans penis, Seminal vesicle, Epididymis, Testicle, Scrotum

Female reproductive tract

Ovary, Fallopian tube, Uterus, Urinary bladder, Pubic bone, Crus of clitoris, Labium minus, Cervix, Vagina

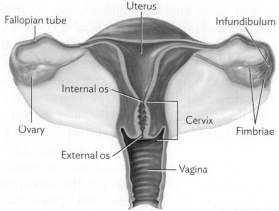

Uterus, Fallopian tube, Infundibulum, Internal os, Ovary, Cervix, Fimbriae, External os, Vagina

A. Ovaries are located at either side of the uterus but not attached to it.

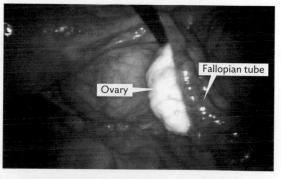

Fallopian tube, Ovary

B. A normal uterine (fallopian) tube is shown draping over a normal ovary.

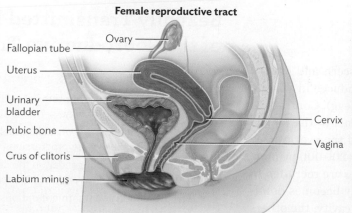

and some chemical hormones into the immediate external vicinity, where their effects are localized. The different morphologies of the male and female reproductive tracts notwithstanding, the ultimate physiological function of the two systems is the same: to produce a human offspring, which occurs through the most intimate of human contacts. Unfortunately, the warm and moist environment of the genital mucosa, so necessary for procreation, also provides an ideal environment for viruses, bacteria, fungi, and parasites to grow, wreaking havoc with the reproductive system. Once an organism establishes an infection in one person's reproductive tract, it can usually be transferred to others via sexual contact. Infections of the human reproductive system are all communicable, and most have to be reported to public health officials upon diagnosis.

Male Reproductive System

The main job of the male reproductive system is to produce sperm and deliver it to the female reproductive system. The male reproductive tract has both external and internal parts; the external parts include the penis, scrotum, testes (singular, testis) housed in the scrotum, and epididymides (singular, epididymis).

Spermatozoa (sperm) are produced in the testes and mature as they wend their way through the epididymides, where they are stored until delivered to the female reproductive tract via the penis or are degraded and absorbed. The internal parts of the male reproductive tract include a long tube, the vas deferens, that transfers mature sperm from the epididymides to the penis. The sperm's trip through the "vas" ends in the ejaculatory ducts that open inside the prostatic urethra, the section of male urethra that passes through the prostate gland. The ejaculatory duct forms from the fusion of the vas deferens with the end of the seminal vesicle and runs through the prostate gland to empty into the male urethra. During a male erection, the urinary bladder, and therefore the flow of urine, is blocked.

Sperm must be protected against the acidic pH of the vagina (pH 3–4) and be provided with nutrition and lubrication once they are sent on their quest to find the ovum. The seminal vesicles, prostate, and bulbourethral glands provide the needed protection, nutrition, and lubrication by adding different fluids to the sperm; the result is the ejaculate that travels through the urethra upon orgasm and is delivered to its destination.

Female Reproductive System

The female reproductive system is built to discern and choose the best sperm candidate from the millions introduced in a single ejaculate to fertilize the one existing ovum (the egg). Contrary to its male counterpart, the entire system is internal to the female body and consists of the vagina, the uterus, and the ovaries (see Figure 23.7). Each ovary contains approximately 200,000 primordial follicles. Every month, some of these follicles are recruited to mature. One of the recruited follicles eventually becomes dominant and releases an oocyte into the peritoneal cavity; the others are degraded and absorbed.

Many oocytes are lost in the peritoneal cavity of the pelvis because there is no direct contact between the uterine tubes (also called the fallopian tubes) and the ovaries (see Figure 23.7). As soon as the oocyte is released, the funnel-shaped end of the uterine tube (the infundibulum) bends over the ovary, and the fingerlike projections on its edge (called fimbriae) perform a sweeping movement to capture the released ovum and move it into the tube. The fallopian tube is the site of fertilization of the ovum.

The fertilized egg will eventually move through the long fallopian tube and enter the uterus, a muscular organ whose chamber is lined with a specialized layer of tissue called endometrium. The uterus ends at the cervix (neck of the womb), which is also highly muscular and attached to the top of the vagina, or birth canal. The vagina is a muscular and distensible tubular structure, 8–10 cm (3–4 inches) long and lined with a mucous membrane. Several glands located around the vaginal entrance, the vaginal introitus, secrete lubricating material necessary to facilitate vaginal intercourse and delivery of a baby.

Note The older term "sexually transmitted disease (STD)" has been replaced by "sexually transmitted infection (STI)" because not all infections transmitted by sexual contact result in disease. A particularly important example is human immunodeficiency virus (HIV). A person infected with HIV usually does not develop the disease called acquired immunodeficiency syndrome (AIDS) for many years.

SECTION SUMMARY

- **The male reproductive tract** includes the penis, testes, epididymides, vas deferens, seminal vesicles, and prostate.
- **Testes** are housed in the scrotum.
- **The female reproductive tract** includes the ovaries, fallopian (uterine) tubes, uterus, cervix, and vagina.
- **The uterine (fallopian) tubes** are not physically connected to the ovaries, and many oocytes are lost in the peritoneal cavity of the pelvis.
- **The infundibulum of the uterine tube** bends over the ovary as soon as the oocyte is released, so that its fimbriae perform a sweeping movement to capture the released ovum and move it into the tube.

23.4
Sexually Transmitted Infections of the Reproductive Tract

SECTION OBJECTIVES

- Relate the signs and symptoms of sexually transmitted infections to their microbial etiology.
- Separate complications of each sexually transmitted infection according to the disease progression.
- Explain the impact of gender on the development of sexually transmitted infections and their complications.
- List prevention and treatment options available for sexually transmitted infections.

Is sexual intercourse the only way to get a sexually transmitted infection? No. **Sexually transmitted infections (STIs)** are defined as infections transmitted primarily through sexual activity, which may include genital, oral-genital, or anal-genital contact. Some infections of the reproductive tract, however, are not exclusively sexually transmitted and will be discussed separately in Section 23.5. Some STIs cause symptoms, and some are asymptomatic. The asymptomatic nature of certain STIs (for instance, HIV) may be transient, the microbe harbored for long periods before disease appears; or the infection may completely resolve before ever becoming symptomatic.

Organisms or viruses that cause STIs are generally very susceptible to drying and require direct physical contact with mucous membranes for transmission. Because sex can take many forms in addition to intercourse, these microbes can initiate disease in the urogenital tract, rectum, or oral cavities. Microbes that infect the human reproductive tract include viruses, bacteria, fungi, and protozoa. Generally speaking, risk-taking behaviors, such as unprotected sex, having multiple sexual partners, and drug abuse (including alcohol), are major contributors to acquiring these infections.

The best way to prevent sexually transmitted infections is to abstain from sexual activity—not a popular option for most. Staying in a long-term, mutually monogamous relationship with someone who is uninfected is the next best thing. Correct and consistent use of latex male condoms is very effective in reducing the risk of transmitting STIs transferred by genital fluids, such as HIV, gonorrhea, chlamydia, and trichomoniasis. Condom use, however, reduces the risk of ulcer-producing STIs (syphilis, herpes, genital warts) only if the condom covers the infected site or the ulcer.

A patient who has symptoms of one STI is very likely to have other STIs as well. For example, patients who have gonorrhea are highly likely to be infected with another type of bacteria, *Chlamydia*.

Viral Infections of the Reproductive System

Most viral sexually transmitted infections are caused by three viruses, as shown in Table 23.3: human papillomavirus (HPV), herpes simplex virus (HSV), and human immunodeficiency virus (HIV). The initial infection with these viruses varies from asymptomatic (often the case) to severe. The infected person, even when asymptomatic, can pass the virus to a healthy person through sexual contact.

GENITAL WARTS

CASE HISTORY 23.2

Freeze What?

Jamie, a 20-year-old college student from New Mexico, was in the clinic complaining of a "growth" around her "private area." She said she had no other symptoms and denied any itching or discomfort in the affected area. Her medical history was benign. She said she likes to drink a lot and admitted to having unprotected sex with multiple partners. Physical examination showed warts near the vaginal introitus (vaginal opening) but no edema (swelling) or erythema (reddening). The doctor diagnosed genital warts and discussed different treatment options, including cryotherapy, which freezes the warts. Jamie was shocked but picked cryotherapy anyway, even though the procedure had to be repeated every 1–2 weeks. Her warts disappeared after the third session, but she was told to come back to the clinic every 3 months for 1 year to check for recurrence. She followed up as directed and was lucky enough to be free of any recurrence at 1 year and later.

Genital warts, or **condylomata acuminata** (singular, **condyloma acuminatum**), is the most common sexually transmitted viral infection. Specific subtypes of the human papillomavirus produce cauliflower-shaped, soft, flesh-colored lesions on infected patients' genitals or rectum (Figure 23.8). Genital warts resemble warts found elsewhere on the skin (see Section 19.2). Most HPV infections occur in women, but the risk of infection is also significant for men who have sex with other men (MSM; homosexual or bisexual men). For some patients, like Jamie, there are no uncomfortable symptoms. Others, particularly those with many lesions, can experience itching, pain, or tenderness.

Note Epidemiologists in the 1990s invented the term "men who have sex with other men (MSM)" because some males choose not to accept a homosexual or bisexual identity.

HPV subtypes causing genital warts are part of a large virus family whose members are responsible for warts on the upper and lower extremities (see Section 19.2). More than 40 subtypes of HPV exclusively infect the anogenital area. Subtypes 6 and 11 are the most

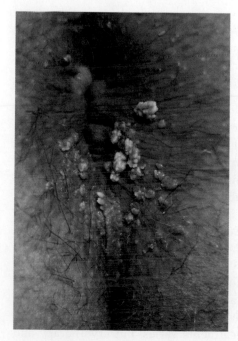

Figure 23.8
Anal Condylomata Acuminata (Genital Warts) Caused by Human Papillomavirus (HPV)

Table 23.3
Viral Sexually Transmitted Infections of the Reproductive Tract

Disease	Symptoms	Complications	Virus	Transmission	Diagnosis	Treatment	Prevention
Genital warts	Soft tissue growths (wartlike), recurrent	Laryngeal or respiratory infections in infants born vaginally to infected mothers	Human papillomavirus (HPV)	Sexual contact	Clinical presentation of the lesion; biopsy	Removal of the warts by surgery, cryotherapy, medications (podofilox), other methods	Recombinant quadrivalent human papillomavirus (HPV) vaccine (Gardasil)
Genital herpes	Clusters of erythematous papules and vesicles, burning, pain, itching, and sometimes fever, headache, and muscle pain	Genital scarring, infection in infants born vaginally to infected mothers with active lesions or those shedding the virus (neonatal herpes)	Herpes simplex virus (HSV-1 and HSV-2)	Sexual contact	Classic clinical presentation; viral culture, PCR, serology	Antiviral medications: acyclovir, famciclovir, valacyclovir; different treatment regimens for first episode, recurrent episodes, and suppression; sexual partners should be screened and treated	No vaccine available
HIV/AIDS	Primary infection: majority will have no symptoms after acquiring the virus; some will have acute illness with fever, muscle and joint pain, lymphadenopathy, and rash on the trunk of the body; symptoms vary, depending on progression of disease and the patient	Opportunistic infections, malignancies, other morbidities defined by CDC as AIDS-defining conditions	Human immunodeficiency virus (HIV)	Blood or body fluids, transplacental, sexual contact	Extensive guidelines for diagnosis recommended by CDC, WHO, and Infectious Disease Society of America; criteria different for patients <18 months of age; ELISA, immunofluorescence assay, nucleic acid (RNA and DNA) tests, HIV p24 antigen test, viral cultures	Antiretroviral therapy based primarily on CD4$^+$ T cell count and monitoring viral load; prophylactic treatment to prevent opportunistic infections; other treatments as needed	No vaccine available

common and responsible for nearly 90% of the cases. Infection with some subtypes of HPV is also linked to cervical, penile, anal, and throat cancers. These subtypes are divided into a "low risk" and a "high risk" group, depending on the strength of their association with cervical malignancy. Subtypes 16 and 18 of the "high risk" group are the two most commonly isolated from cervical cancer tissue.

Diagnosis of HPV is usually made by inspecting the affected area, as was the case for Jamie, but biopsy may be required in rare complicated cases. Several treatment options are available for the infected person, including topical medications (for example, interferon-alpha), cryotherapy, and surgical methods. Treatments are designed to remove the warts but do not cure the viral infection. The success rate of the treatments varies, but cryotherapy and

electrosurgery (burning the tissue by delivering high-frequency electric current to it) are the most effective methods.

The recombinant HPV vaccine, Gardasil, offers protection against serotypes 6, 11, 16, and 18; vaccination is recommended for girls and boys at age 11 or 12 years. A 2013 report shows a remarkable 50% decrease in the rate of cervical cancer in vaccinated women.

GENITAL HERPES Infection of mucosal surfaces and skin of the anogenital area by herpes simplex viruses 1 and 2 is called **genital herpes** or herpes genitalis (**Figure 23.9**). The infection is a worldwide problem, with some 50 million people thought to have it. HIV-infected persons have a higher prevalence of seropositivity for HSV-2 than those who are not HIV infected. Clinical presentation

ANIMATION

Figure 23.9   Genital Herpes

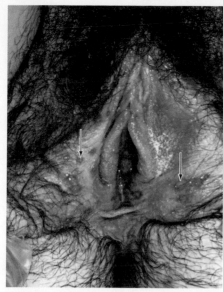

HSV-2 lesions on the genitals of an infected woman.

and course of HSV-2 infection is also different in patients who are HIV-positive.

Link As discussed in Section 19.2, **HSV-1** causes primarily cold sores, whereas **HSV-2** is responsible for most of the genital herpes cases. The tissue specificity of HSV-1 and HSV-2 is not absolute. It is possible for HSV-1 to infect the genital skin and mucosa and for HSV-2 to infect the oral tissue.

When someone without antibodies to HSV-1 or HSV-2 (meaning no prior HSV-1 or HSV-2 infection) has a positive PCR test for herpes viral DNA, he or she is said to have a **primary infection**. The hallmark of primary infection is that the lesions, usually painful, start as erythematous papules (red, slightly raised area of skin) that progress to vesicles and finally pustules (see Figure 23.9). Nearly half of all primary genital herpes cases are HSV-1, but recurrent infections are primarily HSV-2.

A person with a primary infection may be asymptomatic (also called subclinical), severely ill, or exhibit every gradation between those two extremes. The signs and symptoms of a primary infection, in addition to the characteristic lesions, can include burning, pain, itching, dysuria, and even systemic symptoms such as headache or muscle ache and fever. A person with a severe primary genital herpes infection suffers from almost all of these signs and symptoms.

After the initial outbreak, the lesion resolves, and the virus moves to a bundle of nerves at the base of the spine. There the virus remains inactive for a time called the **latent period**, during which there are no symptoms (see Sections 18.6 and 19.2). The latent period sets the stage for recurrent episodes of the disease.

A recurrent episode of genital herpes occurs when the virus becomes activated (sometimes triggered by stress) and travels through nerves to the skin's surface, causing an outbreak of ulcers. These recurrent episodes tend to be milder than the initial outbreak

but are still infectious. Recurrent disease can also involve the brain, which is known as herpes encephalitis (see Section 24.3).

Diagnosis of HSV is made clinically when the characteristic lesions are observed. Samples from the vesicles or pustules are then collected and tested for presence of virus to confirm the clinical **diagnosis of herpes**. Confirmation of diagnosis is particularly important in HIV-positive patients because the disease presents differently in these patients. There is no cure for genital herpes and no vaccine to help prevent the disease. Antiviral medications (for instance, acyclovir) are used to help reduce both the severity of the symptoms and the frequency of recurring episodes.

Link As discussed in Section 19.2, a **diagnosis of herpes** viruses is definitively made when virus is isolated from the clinical sample in culture. Serological tests, real-time PCR, and observation of giant multinucleated cells in skin samples from the base of vesicles (Tzanck smear test) are also used to verify the presence of herpes viruses.

NEONATAL HERPES   Pregnant mothers with genital herpes can transmit the virus to their baby before (transplacental), during or after delivery; a baby so infected is said to have **neonatal herpes**. Some mothers may not even be aware that their disease is active if they are shedding the virus but have no lesions. The incidence of neonatal disease is approximately 1 of every 3,200 births. Neonatal herpes may appear only as a skin infection with small, fluid-filled vesicles that can rupture, crust over, and finally heal (**Figure 23.10**). However, the virus can also spread (disseminate) through the infant and affect many parts of the infant's body, including the brain (herpes encephalitis), liver, lungs, and kidneys. These neonates are extremely ill. They can appear blue (cyanosis), have rapid breathing (tachypnea), appear jaundiced, have a low body temperature (hypothermia), and bleed easily.

Figure 23.10   Neonatal Herpes

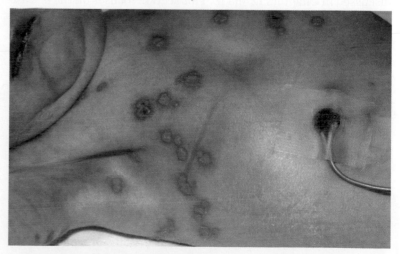

Pregnant mothers who have genital herpes may pass the virus to their baby as the newborn is passing through the birth canal.

Guidelines for preventing vertical HSV transmission include the following:

- In the absence of active lesions or prodromal symptoms, allow vaginal delivery.

- At the time of delivery, consider obtaining a herpes culture from the mother's cervix and from the site of recurrence, as well as from the newborn baby.

- If there is an active herpetic lesion, perform cesarean delivery, preferably within 4–6 hours of membrane rupture.

- Treat the mother with suppressive antiviral agents such as acyclovir or famciclovir prior to delivery.

Untreated newborns diagnosed with disseminated HSV have a mortality rate of 85%. Treatment with antiviral agents reduces mortality to 35%.

HUMAN IMMUNODEFICENCY VIRUS (HIV) INFECTION

The first human case of HIV is now reported to have occurred in the Belgian Congo (today's Democratic Republic of the Congo) in 1959. Recent evidence suggests that HIV first entered the United States in 1970–71 from Haiti (not in 1980 by the Canadian male flight attendant known as "patient zero"). The virus is thought to have originated when the simian immunodeficiency virus jumped (via contaminated blood) from chimpanzees to humans that were hunting the animals and subsequently mutated to HIV sometime in the 1930s.

HIV caused the greatest pandemic of the late twentieth century and remains a serious problem today, especially in Africa. The World Health Organization (WHO) estimates that 22.4 million people in sub-Saharan Africa live with HIV, nearly 68% of all global cases. The Centers for Disease Control and Prevention (CDC) estimates that 1.1 million Americans are infected with the virus. HIV has claimed the lives of more than 22 million people worldwide, over half a million in the United States alone. The molecular biology and virulence of HIV, as well as its targeting of CD4$^+$ T cells, are discussed in Chapters 12 and 18. This section focuses on the disease that HIV causes—**acquired immunodeficiency syndrome (AIDS)**.

HIV is a lentivirus with an RNA genome, a member of the Retroviridae family, and is a prominent example of viruses that can be transmitted either sexually (vaginally, orally, anally) or through direct contact with body fluids, such as occurs with blood transfusion or the sharing of hypodermic needles by intravenous drug users. HIV is not transmitted by kissing, tears, or mosquito bite. It can, however, be transferred from mother to fetus through the placenta (transplacental transfer). We discuss HIV infection here because sexual contact remains the major route of transmission. The current available data from the CDC for the United States show that gay, bisexual, and other men who have sex with men (MSM) constitute the largest group of infected patients (52%), followed by heterosexuals (27%) and IV drug abusers (16%). Women, overall, account for 20% of cases. The global incidence of HIV infection, however, has been trending downward by and large, in part due to multinational funding for "test and treat" policies to provide education, screening, and treatment.

There have been two reports of "functional cure," or remission, of HIV infection (no detectable HIV in peripheral blood). The first report was about a Mississippi toddler who had been born HIV-positive. The infected newborn had been treated with high doses of antiviral medications within 30 hours of birth and remained virus free at 2 years of age in March 2013. Unfortunately, the child has since become HIV-positive, proving the "functional cure" transient in nature. The second report, by French researchers, announced "HIV remission" in 14 patients who had been diagnosed in the late 1990s to early 2000. The patients had been given standard antiviral treatment within 10 weeks of infection. The 14 French patients remain off antiviral therapy and have not had a major viral rebound. Whether this phenomenon of "functional cure," or "remission," will translate into actual cure remains to be seen.

Having entered the bloodstream, HIV infects **CD4$^+$ T cells** and replicates very rapidly, producing a billion particles a day. The concentration of virus in the blood, expressed as the number of viral RNA copies per milliliter of blood, is called the **viral load**. Although the body normally replaces the entire CD4$^+$ population every 2 weeks, the rate of replenishment cannot keep up with the pace of destruction caused by rapid viral replication. The number of CD4$^+$ cells will eventually start to decline. The CD4$^+$ count is now recognized as an important prognostic factor and is used by the CDC to classify the progression of HIV infection for clinical treatment and epidemiological purposes. The CDC classification is similar but not identical to that of the World Health Organization.

Link As discussed in Section 18.6, HIV glycoprotein 120 (gp120, also called the spike protein) infects **CD4$^+$ T cells** by binding to two surface proteins, CD4 and CCR5, the chemokine receptor. This essentially selects the helper T cells as the primary target of the virus while blocking chemokines from binding their receptor. The event effectively disconnects communication between the target cell and the immune system. The infected CD4$^+$ cells are destroyed by cytotoxic T cells and natural killer (NK) cells.

Clinical progression of HIV infection is divided into four stages: a primary stage beginning with seroconversion (finding antibodies to the virus); clinical latency (slow, steady loss of T cells without any symptoms); early symptomatic disease; and AIDS. Most patients are asymptomatic during the initial stage of HIV infection. Nearly half of infected individuals have nonspecific symptoms defined as "mononucleosis-like," including headache, low-grade fever, lymphadenopathy, and a maculopapular rash (see Chapter 19) on the trunk. This is called **primary (acute) HIV infection**. The acute symptoms usually last 1–2 weeks, after which the disease enters the **latent stage**. Most patients seroconvert to HIV-positive within 4–10 weeks of exposure, and by 6 months after exposure, more than 95% of exposed individuals have seroconverted. The period between seroconversion and disease manifestation varies.

Figure 23.11   AIDS Indicator Opportunistic Infections

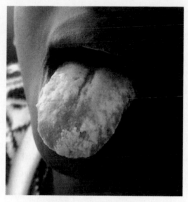

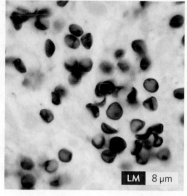

A. Oral candidiasis (thrush) is normally not seen in adults. The white patches are caused by secondary infection by the yeast *Candida albicans*.

B. *Pneumocystis jirovecii* infection of the lung will garner an AIDS diagnosis. Note the cuplike appearance of the fungus, almost like crushed Ping-Pong balls. Organisms are 2–6 μm in diameter.

Table 23.4
AIDS-Defining Conditions in Adults and Adolescents

- Bacterial infections, multiple or recurrent[a]
- Candidiasis of bronchi, trachea, or lungs
- Candidiasis of esophagus[b]
- Cervical cancer, invasive[c]
- Coccidioidomycosis, disseminated or extrapulmonary
- Cryptococcosis, extrapulmonary
- Cryptosporidiosis, chronic intestinal (>1 month's duration)
- Cytomegalovirus disease (other than liver, spleen, or nodes), onset at age >1 month
- Cytomegalovirus retinitis (with loss of vision)[b]
- Encephalopathy, HIV related
- Herpes simplex: chronic ulcers (>1 month's duration) or bronchitis, pneumonitis, or esophagitis (onset at age >1 month)
- Histoplasmosis, disseminated or extrapulmonary
- Isosporiasis, chronic intestinal (>1 month's duration)
- Kaposi's sarcoma[b]
- Lymphoid interstitial pneumonia or pulmonary lymphoid hyperplasia complex[a,b]
- Lymphoma, Burkitt's (or equivalent term)
- Lymphoma, immunoblastic (or equivalent term)
- Lymphoma, primary, of brain
- *Mycobacterium avium* complex or *Mycobacterium kansasii,* disseminated or extrapulmonary[b]
- *Mycobacterium tuberculosis* of any site, pulmonary,[b,c] disseminated,[b] or extrapulmonary[b]
- *Mycobacterium*, other species or unidentified species, disseminated[b] or extrapulmonary[b]
- *Pneumocystis jirovecii* pneumonia[b]
- Pneumonia, recurrent[b,c]
- Progressive multifocal leukoencephalopathy
- *Salmonella* septicemia, recurrent
- Toxoplasmosis of brain, onset at age >1 month[b]
- Wasting syndrome attributed to HIV

[a]Only among children less than 13 years of age.
[b]Condition that might be diagnosed presumptively.
[c]Only among adults and adolescents at least 13 years of age.

Although most patients exhibit no symptoms during the latent phase, some patients do have swollen lymph nodes (lymphadenopathy). Patients with two or more swollen lymph nodes outside the groin region for three months or more during the latent period are said to have "persistent generalized lymphadenopathy," or PGL. Specialized cells in the lymph node "trap" the virus, resulting in the enlargement of the nodes; the number of virus particles in the peripheral blood (the viral load) is low at this point, and there is a high concentration of extracellular HIV in these nodes. The enlarged lymph nodes inevitably are destroyed by the ever-increasing concentration of virus, and they release their viral content into the peripheral blood.

The subsequent decrease in CD4⁺ T cells eventually leads to the **early symptomatic HIV infection stage**, formerly called AIDS-related complex. This stage is defined by the occurrence of specific clinical conditions, which include unusual infections, high fever, and diarrhea that persist for over a month, some cancers, and peripheral neuropathy. The debilitated immune system leaves the victim susceptible to secondary infections such as thrush (**Figure 23.11A**), persistent vaginal candidiasis, listeriosis, or shingles in more than one dermatome (an area of skin that is mainly supplied by a single spinal nerve).

After several years, the disease can progress to AIDS. The CDC guidelines define AIDS as a CD4⁺ cell count of less than 200 cells/μl, regardless of the presence or absence of any other symptoms; CD4⁺ T-lymphocyte percentage of total lymphocytes less than 14%; or the presence of an AIDS indicator condition (**Table 23.4**). Once the CD4⁺ T-cell population falls below 500 cells/μl, the indicator conditions may present as an opportunistic infection such as *Pneumocystis jirovecii* (**Figure 23.11B**) or candidiasis of the esophagus, trachea, bronchi, or lungs. Various unusual cancers are also on the list of AIDS indicator conditions. For instance, **Kaposi's sarcoma**, which is a common cancer seen in AIDS patients (see Figure 18.27B) and is caused by the human herpes virus type 8 (HHV8), is a devastating condition associated with AIDS.

Diagnosis of an HIV infection requires detecting anti-HIV antibodies, determining CD4⁺ T-cell count, and quantifying the viral load of a patient. Quantitative assay to determine viral load is done via quantitative PCR to detect HIV-specific genes such as *gag, pol,* or *nef,* a gene that encodes a protein that interacts with host cell signal transduction components. (Because HIV is an RNA virus, quantitative PCR starts with a reverse transcriptase to convert HIV RNA to DNA, followed by a thermostable DNA polymerase amplification; see Section 8.4.) Diagnosis of AIDS is also based on CD4⁺ count and/or presence of AIDS indicator conditions as discussed earlier.

Link As discussed in Section 12.6, HIV has the main open reading frame containing the *gag, pol,* and *env* genes found in other retroviruses. The *gag* gene encodes capsid and nucleocapsid proteins; the *pol* gene encodes the reverse transcriptase; and the *env* gene encodes the viral envelope protein.

Treatment of HIV infections currently consists of giving the patient a combination of antiviral medications, a strategy named **highly active antiretroviral therapy**, or **HAART** (see Section 13.7). Knowing the viral load of a patient is the most important guiding factor in starting antiviral therapy. The viral load is also used to monitor the effectiveness of antiviral therapy or to determine whether the combination of drugs being used need be changed.

Treatment regimens follow complicated guidelines that take individual patient history, viral load, and CD4+ counts into account. Several classes of pharmaceuticals are now available, including nucleoside (and nucleotide) reverse transcriptase inhibitors (NRTIs) and non-nucleoside reverse transcriptase inhibitors (NNRTIs); drugs that inhibit processing of the viral proteins, protease inhibitors (PIs); drugs that inhibit integration of viral cDNA into the host chromosome, integrase strand transfer inhibitors (INSTIs); and CCR5 antagonists that are designed to inhibit the entry of virus into the cell. More information on some of these drugs can be found in Section 13.7.

HIV-infected individuals are routinely screened and treated for other infections per CDC guidelines. Prophylactic antibiotic treatment, for example, is routinely used to prevent some recurrent infections in affected patients.

Some HIV-infected individuals can control the initial viremia of the infection. These patients remain asymptomatic for years without any antiviral therapy and are called **long-term nonprogressors**. An estimated 4%–7% of HIV-infected individuals qualify for long-term nonprogressor status. A low viral load of 1,000–10,000 viral copies/ml in these patients is thought to contribute to their lack of progression to more severe disease. This may be due to an effective cytotoxic T-cell response. A very small percentage of HIV-seropositive patients maintain a high CD4+ cell count and show negligible or no viremia (fewer than 50–75 viral copies/ml) and are called "**elite controllers**." The credit for this also goes to the infected person's highly effective cytotoxic T cells.

Remember, a person who is HIV-positive does not necessarily have AIDS; it may take years to develop. A vaccine is not yet available to prevent AIDS, in part because the envelope proteins of the virus (see Figure 12.27) typically change their antigenic shape.

 CDC information on HIV and AIDS: cdc.gov

HEPATITIS B AND C INFECTIONS   Hepatitis viruses and the diseases they cause were described in connection with gastrointestinal tract infections because the primary target of the infection is the liver. However, two of the viruses, HBV and HCV, can be sexually transmitted. Chapter 22 presents details of these viruses and their treatments.

Bacterial and Protozoan Infections of the Reproductive Tract

The latest available estimates of the global prevalence of STIs indicate that every year 340 million individuals acquire a curable STI. Most of these STI cases (about 90%) occur in developing countries, which house 75% of the world's population. Reduced access to health care, a direct result of the extreme poverty in these regions; lack of reproductive health services; war; and population growth are all to blame. In the United States, at least half of all new cases of STIs occur in young men and women, according to CDC estimates. Undoubtedly, having unprotected sex plays a prominent role here as well in the developing world.

SYPHILIS

CASE HISTORY 23.3

The Great Imitator

 Odonna, a pregnant 18-year-old woman, came to the Nassau County, New York, urgent-care clinic with a low-grade fever, malaise, and headache. She was sent home with a diagnosis of influenza. She again sought treatment 7 days later after she discovered a macular rash (flat, red) developing on her trunk, arms, palms of her hands, and soles of her feet. She thought maybe she was allergic to the new soap she was using, but the rash was not itchy or painful. Further questioning of the patient revealed that 1 year ago, she had a painless ulcer on her vagina that healed spontaneously. She was diagnosed with secondary syphilis, a stage of the disease that can mimic many other conditions. She was given a single intramuscular injection of penicillin and told that her sexual partners had to be treated as well.

The vaginal ulcer, the long latent period, and secondary development of rash on the hands and feet described in the case history are classic symptoms of syphilis (Table 23.5). **Syphilis** was recognized as a disease as early as the sixteenth century, but the organism responsible, the spirochete *Treponema pallidum*, was not discovered until 1905 (Figure 23.12A). The illness has several stages. The disease has often been called the "great imitator" because its symptoms in the second stage, as exhibited in the case history, can mimic those of many other diseases. The incubation stage can last from 2 to 6 weeks after transmission, during which time the organism multiplies and spreads throughout the body. **Primary syphilis** is an inflammatory reaction at the site of infection called a **chancre** (Figure 23.12B). About a centimeter in diameter, the chancre is painless and hard, and it contains spirochetes. Patients are usually too embarrassed to seek medical attention, and because it is painless, they hope that it will "go away."

The initial lesion does go away after several weeks, and without scarring. The disease has now entered the latent stage.

Figure 23.12   **Syphilis**

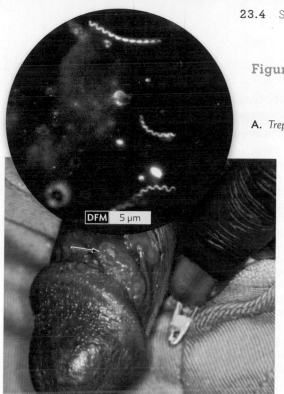

A. *Treponema pallidum.*

B. Chancre of primary syphilis.

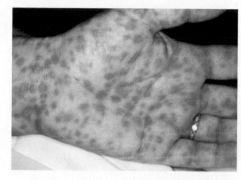

C. Macular rash of secondary syphilis.

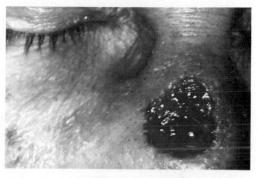

D. A gumma on the face of a patient with tertiary syphilis.

Table 23.5

Bacterial and Protozoan Sexually Transmitted Infections of the Reproductive Tract

Disease	Symptoms	Complications	Organism	Transmission	Diagnosis	Treatment	Prevention
Syphilis	*Primary:* chancre (painless ulcer) *Secondary:* generalized rash, including palms of hands and soles of feet, generalized nontender lymphadenopathy, fatigue, fever *Tertiary:* Loss of coordination of movement (tabes dorsalis), progressive dementia, stroke and other CNS abnormalities; abnormal pupillary reaction to light; gummas, thoracic aortic aneurysm	*Complications of primary syphilis:* secondary disease *Complications of secondary syphilis:* latent disease leading to tertiary stage *Complications of tertiary syphilis:* disability and death	*Treponema pallidum*	Sexual contact	Definitive diagnosis based on dark-field microscopy and detection of *T. pallidum* in lesion or tissue samples; presumptive diagnosis based on positive results of two types of serology tests; clinical presentation	Alternative treatments for pregnant women and children; sexual partners should be screened and treated; penicillin G benzathine is the drug of choice for nonpregnant women and adult males *Primary and secondary syphilis:* 2.4 million units single dose IM (intramuscular), 2 g of azithromycin orally is used as well *Tertiary syphilis:* 2.4 million units IM once a week for 3 weeks if no neurosyphilis; complicated regimens for tertiary stage with neurosyphilis	No vaccine available

(continued on next page)

Table 23.5

Bacterial and Protozoan Sexually Transmitted Infections of the Reproductive Tract (continued)

Disease	Symptoms	Complications	Organism	Transmission	Diagnosis	Treatment	Prevention
Gonorrhea	*Gonococcal cervicitis in women:* usually no symptoms; if symptoms are present, they include vaginal discharge, dyspareunia (pain when having vaginal intercourse), vaginal bleeding after sex *Gonococcal urethritis in males:* mucopurulent or purulent discharge, dysuria, urethral itching	*Complications of gonococcal cervicitis in women:* pelvic inflammatory disease (PID) *Complications of gonococcal urethritis in males:* epididymitis, disseminated infection	*Neisseria gonorrhoeae* (also called gonococcus, has >100 serotypes)	Sexual contact	>5 WBC in Gram stain of urethral discharge, WBC containing Gram-negative diplococci; clinical presentation; positive leukocyte esterase test on first-void urine; culture; nucleic acid amplification test from urethral swab or urine	Ceftriaxone 250 mg IM single dose plus either azithromycin 1 g single dose orally or doxycycline 100 mg orally twice a day for 7 days; alternative treatments available if needed; sexual partners should be screened and treated	No vaccine available
Nongonococcal urethritis	Dysuria, urethral discharge and itching	Prostatitis and epididymitis in men; PID in women	*Chlamydia trachomatis* is the most common cause (20%); *Mycoplasma genitalium* (9%); adenovirus (4%); herpes simplex virus (2%)	Sexual contact	>5 WBC in Gram stain of urethral discharge; absence of WBC containing Gram-negative diplococci; nucleic acid amplification test from urethral swab or urine; positive leukocyte esterase test on first-void urine; clinical presentation	Azithromycin 1 g orally in single dose or doxycycline 100 mg orally twice daily for 7 days; alternative regimens if needed; sexual partners should be screened and treated	No vaccines available
Lymphogranuloma venereum (LGV)	Unilateral tender inflammation of groin lymph nodes, anal pain and discharge if anal exposure, constipation	Chronic disease, lymphatic scarring leading to genital (scrotal) elephantiasis, penile ulcers, neuropathy	*Chlamydia trachomatis*	Sexual contact	Culture of swab of the lesion or aspirate of lymph node; direct immunofluorescence test; nucleic acid tests	Doxycycline 100 mg orally twice daily for 21 days (recommended treatment); other regimens; prolonged treatment may be needed for HIV-positive patients	No vaccine available

(continued on next page)

Over the next 5 years, symptoms may be absent, but at any time, as described in the case history, the infected person can develop the rash typical of **secondary syphilis** (Figure 23.12C). The rash can be similar to that produced by many diseases. The patient remains contagious in this stage. The symptoms eventually resolve, and the patient reenters a latent stage of syphilis. Some patients eventually progress over years to **tertiary syphilis** and develop many cardiovascular or nervous system–associated symptoms. Soft growths resembling tumors, called **gummas**, are often found during later stages of tertiary syphilis. A gumma has necrotic (dead) tissue in the center

Table 23.5
Bacterial and Protozoan Sexually Transmitted Infections of the Reproductive Tract (continued)

Disease	Symptoms	Complications	Organism	Transmission	Diagnosis	Treatment	Prevention
Chancroid (also called soft chancre)	Acute and extremely painful multiple genital ulcers that start as papules or pustules	Secondary infections and fistulas (hollow passages between two epithelium-lined body organs); high risk of having HIV or other STI	*Haemophilus ducreyi*	Sexual contact (highly infectious)	Isolation of *H. ducreyi* in culture from genital ulcer samples; clinical presentation	Four antibiotic regimens recommended: azithromycin 1 g orally single dose, ceftriaxone 250 mg single dose IM, ciprofloxacin orally multiple doses (contra-indicated during pregnancy and lactation), erythromycin multiple doses; sexual partners should be examined and treated within 10 days	No vaccine available
Trichomoniasis	*Women:* copious malodorous vaginal discharge, vulvar itching and burning *Men:* either asymptomatic or only urethritis symptoms	Adverse pregnancy outcome, including premature labor and low birth weight in newborns	*Trichomonas vaginalis*	Sexual contact	Culture for *T. vaginalis* from vaginal discharge or from urethral swab, urine, or semen; motile *T. vaginalis* on wet mount; PCR	Metronidazole 2 g orally in single dose or tinidazole 2 g orally in single dose; alternative treatments available	No vaccine available

surrounded by inflammatory and fibrotic cells (Figure 23.12D). Gummas are usually found in the liver, but they can form in the mouth, bone, brain, heart, and skin as well. The patient may develop dementia and eventually dies from the disease.

Neurological symptoms resulting from syphilis at any stage of the disease are referred to as **neurosyphilis**. Syphilis patients whose cerebrospinal fluid (CSF) shows an elevated white blood cell count (more than 20 cells/µl; normal is 0–5 cells/µl) or who have a reactive CSF **Venereal Disease Research Laboratory (VDRL) test** have developed neurosyphilis. The VDRL test and the **rapid plasma reagin (RPR) test**, a modification of the VDRL test, are nonspecific tests that detect antibodies against a nontreponemal heart antigen called cardiolipin. A positive VDRL or RPR is followed by a specific test, such as direct or indirect fluorescent antibody staining (fluorescent treponemal antibody test, FTA-ABS; see Section 17.5, Figure 17.26). The symptoms of neurosyphilis vary widely. Early neurosyphilis develops during the primary and secondary phases of the disease and may be asymptomatic. Alternatively, the meninges (membranes enveloping the brain) and structures of the eyes or ears may

be affected. The results can be meningitis, ocular complications, or tinnitus and hearing loss. Late neurosyphilis involves damage to the central nervous system, particularly the brain. Neurological manifestations during tertiary syphilis are mostly **general paresis** or **tabes dorsalis**. General paresis of late neurosyphilis refers to problems of mental function associated with late syphilis. The neurological problems that arise include defects in personality, affect, reflexes, eye, sensorium (illusions, delusions, or hallucinations), intellect, and speech. The first letters of these traits spell the word "paresis." *Tabes dorsalis* refers to demyelination of axons in certain parts of the spinal cord called the dorsal root and dorsal root ganglia, which causes abnormal gait and degeneration of the weight-bearing joints (Charcot's joints).

According to the CDC, a presumptive diagnosis of syphilis can be made during any stage of disease based on positive results from two serology tests (described later). A positive PCR (looking for bacterial genes) or dark-field examination of the material from the lesion (looking for highly motile spirochetes) will make the definitive diagnosis during the early and second phases of the disease.

A diagnosis of tertiary syphilis is based solely on clinical presentation and serology. The diagnosis of latent syphilis, however, begins with serological testing but has several additional criteria. One criterion, for example, is detecting the organism in tissues using fluorescent antibody.

Syphilis is particularly dangerous in pregnant women. The treponeme can cross the placental barrier and infect the fetus, causing **congenital syphilis**. At birth, infected newborns may have notched teeth visible on X-ray (**Figure 23.13**), perforated palates, and other congenital defects. Women should be screened for syphilis as part of their prenatal testing to prevent these congenital infections.

Common serological tests for syphilis include nontreponemal tests used for screening and treponemal-specific tests used for definitive diagnosis. An example of a nontreponemal test is the rapid plasma reagin (RPR) test (**Figure 23.14A**). Reagin is an antibody to the lecithin-cholesterol antigen called cardiolipin, which is not a treponemal antigen. Infection with *T. pallidum* elicits reagin antibodies that react against cardiolipin. The RPR test involves adding a mixture of charcoal particles coated with cardiolipin to a patient's serum. If anti-cardiolipin antibodies (reagin) are present, the charcoal beads clump, representing a positive test (reactive circles in Figure 23.14A). The RPR test is very sensitive, but false positives are common, which is why a more specific serological test should be performed if an RPR test is positive.

Treponemal-specific tests more definitively determine whether a patient is infected with *T. pallidum*. Some are direct immunofluorescence tests to detect treponemal antigen in patient tissue; others are indirect immunofluorescence (Figure 17.26B) or immunochromatography (**Figure 23.14B**) tests that examine patient serum for the presence of anti-treponemal antibodies (see Section 17.5).

Penicillin G benzathine is the antibiotic of choice for syphilis, given at 2.4 million units in a single intramuscular injection for primary and secondary disease and once weekly for 3 weeks (7.2 million units) in tertiary disease without evidence of neurosyphilis. Other antibiotics are used in patients with penicillin allergies. Treatment regimens for patients with neurosyphilis always include high doses of antibiotics, usually penicillin or ceftriaxone given intravenously or by daily injections. Patients who are allergic to penicillin can be

Figure 23.13   Notched Teeth in an Adult Caused by Congenital Syphilis

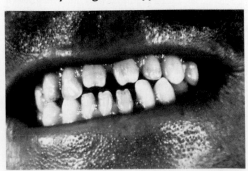

The treponeme can cross the placenta and infect the fetus. Notched teeth (also called Hutchinson's teeth) indicate congenital syphilis.

Figure 23.14   Examples of Serological Tests for Syphilis

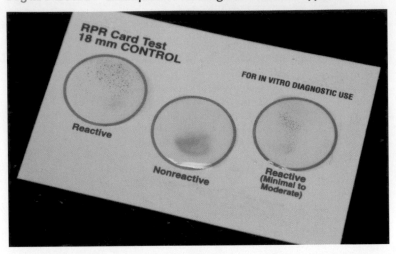

A. The RPR (rapid plasma reagin) test screens for nonspecific antibodies that may indicate infection with *Treponema pallidum*. A positive test shows agglutination of cardiolipin-coated charcoal particles

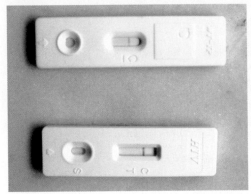

B. A positive enzyme immunoassay for a specific *T. pallidum* antigen is compared with one done for HIV. The red line indicates a positive reaction.

desensitized to penicillin first (Section 17.3) or given other antibiotics. Antibiotics are useful for eradicating the organism, but there is no vaccine, and cure does not confer immunity.

A serious consequence of antibiotic treatment is the **Jarisch-Herxheimer reaction**. After treatment begins, the dying treponemes release inflammatory molecules that trigger a cytokine cascade leading to the symptoms, which include myalgias, fever, headache, and tachycardia. Sometimes the Jarisch-Herxheimer reaction exacerbates whatever current syphilitic lesions are manifested (for example, rash or chancre). The reaction is common, develops within several hours after beginning antibiotic treatment, and usually clears within 24 hours after onset. Its exact etiology is unclear, although it may be due to an immunological reaction to the rupture of spirochetes.

Columbus and the New World theory of syphilis. An outbreak of syphilis that spread throughout Europe soon after Columbus and his crew returned from the Americas (1493) led to the theory that Columbus brought the treponeme to Europe from the New World. Whether Columbus's crew carried syphilis to or from the

New World has been debated for 500 years. A recent genetic study may have resolved the issue. By sequencing 26 geographically distributed strains of pathogenic syphilis, scientists discovered that the sexually transmitted syphilis strains most closely resemble the non-sexually transmitted yaws-causing strains of *T. pallidum* from South America. Thus, the crew of the Columbus voyages may have brought smallpox and measles to America but returned to Europe with syphilis.

The Tuskegee experiment. Unfortunately, much of what we know of syphilis is the result of the infamous Tuskegee experiment conducted in the 1930s in Alabama. The study was entitled "Untreated Syphilis in the Negro Male." Through dubious means and deception, a group of African-American males were enlisted in a study that promised treatment but whose real purpose was to observe how the disease progressed without treatment. Today, such experiments are barred thanks to strict institutional review board (IRB) oversights in which human subjects must sign informed-consent forms. An interesting treatise on the Tuskegee experiment can be found on the Internet at the National Center for Case Study Teaching in America (search term: Bad Blood).

GONORRHEA

CASE HISTORY 23.4

Symptom: Urethral Discharge

Rick, a 22-year-old mechanic from Honolulu, Hawaii, saw his family doctor for treatment of painful urination and urethral discharge. His medical history was unremarkable. Rick was sexually active, with three regular and several "one time–good time" partners. Physical examination was benign except for a prevalent urethral discharge. The discharge was Gram stained and sent for culture. The Gram stain revealed many pus cells, some of which contained many phagocytosed Gram-negative diplococci (**Figure 23.15A**), a finding consistent with *Neisseria gonorrhoeae* infection. Blood was drawn for syphilis serology, which proved negative. Rick was given an intramuscular injection of ceftriaxone (250 mg) to treat the gonococcus, and oral doxycycline (100 mg, two times a day) was prescribed for 7 days to empirically treat any possible coinfection with chlamydia. The bacteriology lab recovered the bacteria seen in the Gram-stained smear of the urethral discharge. The organism produced characteristic colonies on chocolate agar and was identified as *Neisseria gonorrhoeae*. Chocolate agar plates contain heat-lysed red blood cells that turn the medium chocolate brown (**Figure 23.15B**). The case was subsequently reported to the state public health department. Upon his return visit, Rick's symptoms had resolved, and a repeat culture was negative.

The disease here is classic **gonorrhea** (also known as "the clap") caused by *Neisseria gonorrhoeae*, a Gram-negative diplococcus. A characteristic that distinguishes *Neisseria* infections from *Chlamydia* infections (described later) is that bacterial cells are seen in gonorrheal discharges but not in chlamydial discharges. Gonorrhea has been a problem for centuries and remains epidemic in the United States today. Symptoms generally occur 2–7 days after infection but can take as long as 30 days to develop. Most infected men (85%–90%) exhibit symptoms such as painful urination, yellowish white discharge from the penis, and sometimes swelling of the testicles and penis. The Greek physician Galen (129–ca. 199 CE) originally mistook the discharge for semen. This led to the name gonorrhea, which means "flow of seed."

In contrast to men, most infected women (80%) do not exhibit symptoms and constitute the major reservoir of the organism

Figure 23.15 *Neisseria gonorrhoeae*

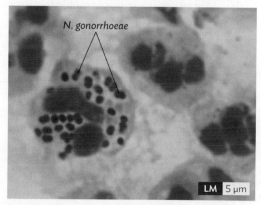

A. Within pus-filled exudates, the Gram-negative diplococci are often found intracellularly inside polymorphonuclear cells (PMNs). The intracellular bacteria in this case are no longer viable, having been killed by the antimicrobial mechanisms of the white cell.

B. Colonies of *N. gonorrhoeae* growing on chocolate agar (agar plates containing heat-lysed red blood cells that turn the medium chocolate brown).

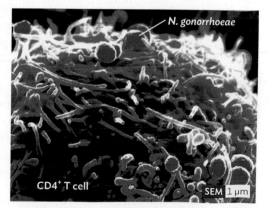

C. *N. gonorrhoeae* binding to CD4⁺ T cells (colorized SEM), inhibiting T-cell activation and proliferation, which may explain the ease of reinfection.

Figure 23.16   Gonococcal Infection

Gonococcal infection in women can be asymptomatic even when cervicitis is present.

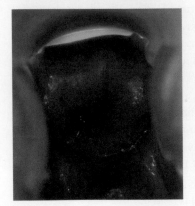

A. Normal cervix.

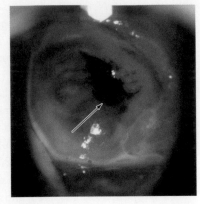

B. Gonococcal cervicitis. Notice the reddening and the purulent exudates (arrow).

(Figure 23.16). If they are asymptomatic, they have no reason to seek treatment and thus can spread the disease. When symptoms are present, they are usually mild. A symptomatic woman will experience a painful burning sensation when urinating and will notice vaginal discharge that is yellow or occasionally bloody. She may also complain of cramps or pain in her lower abdomen, sometimes with fever or nausea. As the infection spreads throughout the reproductive organs (uterus and fallopian tubes), pelvic inflammatory disease (PID) occurs.

N. gonorrhoeae is generally serum sensitive (does not survive the serum) owing to its sensitivity to **complement**. However, certain serum-resistant strains can make their way to the bloodstream and cause disseminated infections throughout the body. As a result, both sexes can develop purulent arthritis (joint fluid containing pus), endocarditis, or even meningitis. An infected mother can also infect her newborn during parturition (birth), leading to a serious eye infection called **ophthalmia neonatorum**. Because of this risk and because most infected women are asymptomatic, all newborns receive antimicrobial eyedrops at birth.

 Recall from Section 15.6 that serum contains 20 host proteins collectively referred to as **complement**. Pores (called membrane attack complexes), consisting of four complement components, are inserted into membranes of Gram-negative bacteria and cause cytoplasmic leaks. In addition, pieces of the complement system can attract white blood cells and facilitate phagocytosis (opsonization).

Because adults engage in a variety of sexual practices, *N. gonorrhoeae* can also infect the anus or the pharynx, where it can develop into a mild sore throat. These infections generally remain unrecognized until a sexual partner presents with a more typical form of [gen]itourinary gonorrhea. Because no lasting immunity is built up, [infecti]on with *N. gonorrhoeae* is possible. This is due in part to

periodic changes in the shapes (immunogenicity) of various surface antigens and because the organism can apparently bind to CD4⁺ T cells, inhibiting their activation and proliferation to become memory T cells (Figure 23.15C).

Treatment for gonorrhea is complicated because of widespread resistance to penicillin, tetracycline, and now quinolones. *N. gonorrhoeae* infections are now typically treated with a single intramuscular (IM) injection of ceftriaxone plus either oral azithromycin or doxycycline to empirically treat possible chlamydia infection. No serological test or vaccine for gonorrhea exists because the organism frequently changes the structure of its surface antigens.

Gonorrhea is epidemic in the United States and is a notifiable disease (a disease that must be reported to federal, state, or local officials when diagnosed). Syphilis and chlamydial infections are also notifiable diseases. Patients with STIs are questioned about their sexual contacts, which enables public health officials to track, and slow, the spread of STIs by treating potentially infected people.

CHLAMYDIAL INFECTION   *Chlamydia trachomatis* is a Gram-negative-like obligate intracellular pathogen that has several serotypes infecting the reproductive tract and the eye (trachoma; Section 19.6). According to the CDC, **chlamydia** is the most frequently reported sexually transmitted infectious disease in the United States, but many people are completely unaware that they are infected. Three-fourths of infected women, for instance, have no symptoms. *Chlamydia* is the major cause of nongonococcal urethritis in the United States.

Chlamydia are unusual Gram-negative-like organisms with a unique developmental cycle. They are obligate intracellular pathogens that start as small, nonreplicating infectious elementary bodies that enter target eukaryotic cells. Once inside vacuoles, they begin to enlarge into replicating reticulate bodies (Figure 23.17). As the vacuole fills, now called an inclusion, the reticulate bodies divide to become new nonreplicating elementary bodies. *Chlamydia trachomatis* can cause STIs, pneumonia, and trachoma.

People most at risk of developing genitourinary tract infections with *Chlamydia* are young, sexually active men and women; anybody who has recently changed sexual partners; and anybody who has recently had another sexually transmitted infection. The spread of *Chlamydia* has also been linked to its ability to bind to sperm in a process called "hitchhiking." It is thought that this interaction helps the organism spread to females. Left untreated, *Chlamydia*, like *N. gonorrhoeae*, can cause pelvic inflammatory disease. Men left untreated can suffer urethral and testicular infections and a serious form of arthritis.

Nongonococcal urethritis is diagnosed when the Gram stain of a patient's urethral mucopurulent or purulent discharge (≥5 white blood cells per oil immersion field) reveals no organisms (*Chlamydia* does not stain). Differential diagnosis of nongonococcal urethritis includes several etiologies, both infectious and noninfectious. When a chlamydial infection is suspected, samples of the patient's urethral discharge are sent for nucleic acid amplification tests such

Figure 23.17
Replication Cycle of *Chlamydia*

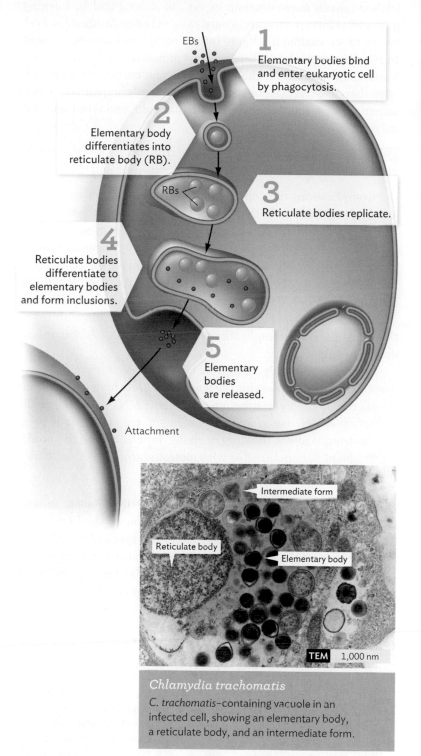

EBs

1 Elementary bodies bind and enter eukaryotic cell by phagocytosis.

2 Elementary body differentiates into reticulate body (RB).

RBs

3 Reticulate bodies replicate.

4 Reticulate bodies differentiate to elementary bodies and form inclusions.

5 Elementary bodies are released.

Attachment

Intermediate form

Reticulate body

Elementary body

TEM 1,000 nm

Chlamydia trachomatis

C. trachomatis–containing vacuole in an infected cell, showing an elementary body, a reticulate body, and an intermediate form.

Figure 23.18 Lymphogranuloma Venereum

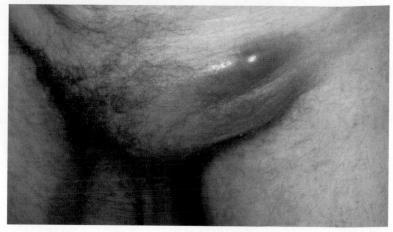

Painful lymphadenopathy caused by *C. trachomatis*.

as PCR to confirm a diagnosis. *Chlamydia* infections can be treated with either azithromycin (single 1-gram dose orally) or doxycycline (100 mg twice a day orally for 7 days). Doxycycline cannot be used during pregnancy because tetracycline derivatives interfere with bone development, so for pregnant women the choice is either the single-dose azithromycin or amoxicillin (500 mg three times a day orally for 7 days).

LYMPHOGRANULOMA VENEREUM **Lymphogranuloma venereum (LGV)** is a chlamydial disease involving only the L1, L2, and L3 serotypes of *C. trachomatis*. These serotypes are more virulent and invasive than other chlamydial serotypes. The organism gains entrance through breaks in mucosal surfaces and travels via the lymphatics to multiply within mononuclear phagocytes (macrophages and monocytes) in regional lymph nodes. LGV is initially characterized by self-limited genital papules or ulcers followed in several weeks by painful inguinal and/or femoral lymphadenopathy, which may be the only clinical manifestation at presentation (**Figure 23.18**). Patients with LGV may also present with rectal lesions, especially among patients participating in receptive anal intercourse. In these cases, rectal pain, discharge, and bleeding may be confused with other gastrointestinal conditions, such as colitis.

CHANCROID *Haemophilus ducreyi* is a Gram-negative, fastidious (with complex nutritional requirements for growth) coccobacillus that causes painful genital ulcers and inflammation of the regional lymph nodes (**Figure 23.19**). The painful lesions are usually on the external genitals and are called **chancroid**, or soft chancre. Chancroid is common in the tropics and developing countries but uncommon in the United States. Only 15 cases were reported to the CDC in 2012. Infection with *H. ducreyi* is strongly associated with illicit drug use. Patients with chancroid are likely to have syphilis simultaneously, and there is an increased risk of this infection in the HIV-1-infected population.

Figure 23.19   Chancroid

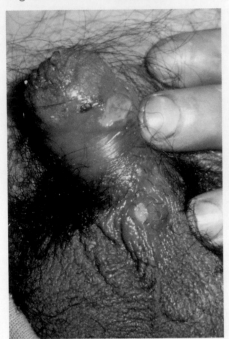

Infection with *Haemophilus ducreyi* causes a painful lesion called chancroid, in contrast to the painless lesion of primary syphilis, which is called a chancre.

TRICHOMONIASIS *Trichomonas vaginalis* is a flagellated protozoan (**Figure 23.20**) that causes an unpleasant sexually transmitted vaginal disease called **trichomoniasis**. Approximately 2 million–3 million infections occur each year in the United States. *Trichomonas* can infect both men and women, but men are usually asymptomatic. Even among infected women, 25%–50% are considered asymptomatic carriers.

There is no cyst in the life cycle of *T. vaginalis*, and transmission is via the **trophozoite** stage only. Growth of this protozoan can

Figure 23.20
Trichomonas vaginalis

T. vaginalis is a flagellated protozoan that causes a common sexually transmitted disease.

SEM 1 μm

cause vaginal pH to rise above pH 4.5 (normal vaginal pH varies from 4.2 to 5.0 depending on ethnicity). The female patient with trichomoniasis may complain of vaginal itching and/or burning, pain during intercourse, and a musty vaginal odor. An abnormal yellowish green vaginal discharge (often frothy) may also be present. In 2% of patients, physical exam shows punctate (spotty or mottled) hemorrhagic lesions on the vaginal wall and cervix, referred to as "strawberry" cervix (**Figure 23.21**). Males may complain of painful urination (dysuria), urethral or testicular pain, and lower abdominal pain.

Link As discussed in Section 11.3, the **trophozoite** is a stage of the protozoan life cycle correlating with active feeding and multiplying.

Complications of trichomoniasis include an increased risk of infection with HIV in both men and women. *T. vaginalis* infections may lead to prostatitis, epididymitis, and infertility in men and pelvic inflammatory disease and a type of cervical cancer in women. Pregnant women who are infected have an increased risk of premature rupture of the amniotic sac, preterm labor, and low-birth-weight infants. Pregnant women who are coinfected with HIV have a higher risk of transmitting the virus to their infant in addition to the other possible complications associated with trichomoniasis during pregnancy.

Diagnosis is usually made from clinical presentation; definitive diagnosis requires observing the flagellated protozoan in secretions by microscopy. Polymorphonuclear cells (PMNs), which are the primary host defense against the organism, are also usually present. Metronidazole is the drug of choice for treating trichomoniasis. Pregnant patients treated with metronidazole, however, have a higher risk of preterm labor.

PELVIC INFLAMMATORY DISEASE (PID) Pelvic inflammatory disease (PID) is a damaging infection of the uterus and fallopian tubes caused by several microbial species, including *N. gonorrhoeae* and *Chlamydia* when these organisms directly spread from the vagina to these other structures. PID is often subclinical, but symptomatic patients can present with severe low abdominal

Figure 23.21   Strawberry Cervix

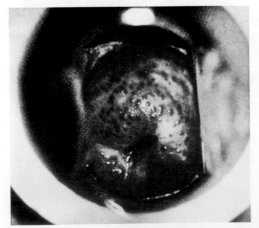

T. vaginalis infection causes a purulent discharge and lesions on the cervix referred to as "strawberry cervix." The discharge seen is usually thin but can appear white, yellow, or green. Compare to Figure 23.16A.

pain (constant, dull, aching pain or cramping pain), fever, nausea, and vomiting. Pelvic exam of these patients shows an inflamed cervix that is extremely tender and painful when moved (see Figure 23.16). A vaginal discharge containing white blood cells is common. The damage produced by PID can lead to tubal pregnancies, infertility, and chronic pelvic pain.

SECTION SUMMARY

- **Sexually transmitted infections (STIs)** are transmitted through genital, oral-genital, or anal-genital contact and must be reported to public health officials.

- **Microbes that cause STIs are susceptible to drying** and require direct contact with mucous membranes for transmission.

- **Patients with one STI** are likely to be infected with multiple sexually transmitted microbes.

- **Genital warts** (condylomata accuminata), the most common sexually transmitted viral infection, are caused by specific strains of HPV (usually subtypes 6, 11, 16, and 18). A multivalent vaccine is available for those subtypes.

- **Genital herpes** is marked by anogenital ulcerations caused by herpes virus types 1 and 2. From an initial infection, the virus can become latent in spinal nerves and reemerge to cause recurrent episodes.

- **Babies can acquire neonatal herpes** by vertical transmission from pregnant mothers infected with HSV before, during, or after birth.

- **Human immunodeficiency virus (HIV) infection** proceeds through four stages: primary (coincident with seroconversion), latent, early symptomatic, and then AIDS. The number of circulating CD4$^+$ T cells in a patient as well as viral load are important prognostic indicators of the disease.

- **A CD4$^+$ T-cell count less than 200 cells/μl is defined as AIDS.** HIV depletion of CD4$^+$ T cells results in lethal secondary infections and cancers.

- **Syphilis, gonorrhea, and chlamydial diseases** are the most common sexually transmitted bacterial infections.

- **Syphilis has three stages:** primary (localized, painless chancres), secondary (widespread lesions over the body that can mimic other skin rashes), and tertiary (cardiovascular and nervous system lesions). Penicillin is the treatment of choice.

- **Gonorrhea is marked** by a purulent urethral discharge in men and a vaginal discharge in women, although women are often asymptomatic. Cephalosporins are the antibiotic treatments of choice.

- *Chlamydia* **is the major cause of nongonococcal urethritis.** *Chlamydia* species are Gram-negative, obligate intracellular bacteria. Azithromycin or doxycycline is used for treatment.

- *Trichomonas vaginalis* is a flagellated protozoan that causes trichomoniasis, a sexually transmitted vaginal disease. The reservoir for this organism is the male urethra and the female vagina.

Thought Question 23.3   Why would early symptomatic HIV disease lead to shingles in more than one dermatome?

Thought Question 23.4   What aspect of the immune system is affected by diminished numbers of CD4$^+$ T cells?

Thought Question 23.5   *Neisseria meningitidis* readily causes meningitis, yet *N. gonorrhoeae* does not. What is the explanation for this?

Thought Question 23.6   How would you treat pelvic inflammatory disease?

Thought Question 23.7   A patient in your office complains of a high fever and diarrhea of 3 weeks' duration. Upon physical exam, you notice he has swollen lymph nodes. You also discover that he is an IV drug user. Blood work that you order reveals a negative RPR test and a CD4$^+$ count of 500 cells/ml (normal is 1,000 cells/ml). What might you conclude from this, and what other test should be run?

23.5
Non-Sexually Transmitted Infections of the Reproductive Tract

SECTION OBJECTIVES
- Relate signs and symptoms of non-sexually transmitted infections to their microbial etiology.
- Explain the role of the vaginal microbiome in prevention and/or development of non-sexually transmitted infections of the reproductive tract.
- List treatment options available for non-sexually transmitted infections of the reproductive tract.

How are infections transmitted to the reproductive tract if not sexually? Some genital tract infections in women can develop when sexual activity introduces microbes present on the perineum into the vagina or by overgrowth of certain organisms present in low numbers among the vaginal microbiome. The resulting vaginal infection (**vaginosis**) and inflammation (**vaginitis**) are commonly seen in women of childbearing years. Hormonal changes that take place over the menstrual cycle or during pregnancy as well as sexual activity can change the composition of resident microbiota. Antibiotic treatment can also alter the balanced microbial population of the vagina, leading to a takeover by resident yeasts or anaerobes. A common example is a *Candida* yeast infection suffered by women when their vaginal microbiota is inadvertently killed by treating another infection with a short course of antibiotics. *Candida* then finds itself unopposed for growth (Table 23.6).

In males, infection of the epididymis (**epididymitis**), testes (**orchitis**), or prostate gland (**prostatitis**) can develop in several ways. Surgery or instrumentation (for example, the insertion of a urinary catheter) can obstruct urine flow normally used to flush organisms from the system. Systemic diseases or immune suppression can provide an opportunity for organisms to spread from nearby areas. The main cause of bacterial prostatitis (inflammation and infection of the prostate) is *E. coli* (87.5%), which either invades the gland

Table 23.6
Bacterial and Fungal Non-Sexually Transmitted Infections of the Reproductive Tract

Disease	Symptoms	Complications	Organism	Transmission	Diagnosis	Treatment	Prevention
Bacterial vaginosis (BV)	Foul-smelling discharge	Premature delivery, postpartum endometritis (infection and inflammation of the endometrium)	*Gardnerella vaginalis* (major cause); several other minor etiologies, including *Mycoplasma hominis* and *Peptostreptococcus* spp.	Self-inoculation or overgrowth of existing *G. vaginalis*	Clinical diagnosis (Amsel criteria) requires three of the following: • homogeneous thin white or gray discharge coating vaginal walls • >20% clue cells present on wet mount microscopy of vaginal saline preparation • vaginal pH >4.5 • fishy odor of vaginal discharge before or after adding 10% potassium hydroxide, KOH, a positive "whiff" test Gram stain also used in diagnosis	Topical or oral metronidazole or clindamycin; treatment regimen is different for pregnant women	No vaccine available
Fournier's gangrene	Prodromal fever 2–7 days prior to other symptoms; severe pain; tenderness in the genital area can be accompanied by itching and edema of the overlying skin, progressing to wounds with drainage and gangrene	Variable systemic effects, including septic shock	Polymicrobial necrotizing fasciitis of genitals (particularly penis and scrotum), including anaerobic organisms, Enterobacteriaceae, streptococcal and staphylococcal species, and fungi	Surgical or other trauma to the penis, scrotum, perineum, or anal areas	Culture of the wound; blood and urine cultures; clinical presentation	Surgery to remove necrotic tissue; antibiotic therapy; tetanus prophylaxis	No vaccine available
Vulvovaginitis	Acute vaginal discharge and vulvar itching and pain that can result in dysuria	Chronic candidal vulvovaginitis	*Candida albicans* (85%–95%) and other *Candida* species; overgrowth of personal flora secondary to immune suppression, hormonal changes, trauma, or antibiotic therapy	Self-inoculation or overgrowth of existing *C. albicans*	Clinical presentation; wet prep or Gram stain showing yeasts, hyphae, or pseudohyphae; culture	Topical antifungals (butoconazole, clotrimazole, miconazole, nystatin, tioconazole, terconazole) or oral antifungals (fluconazole); treatment depends on the type of infection: uncomplicated, complicated, or recurrent	No vaccine available

from the adjacent rectum or extends from a urinary tract infection. Chronic prostatitis is often nonbacterial, but *E. coli* is the most common cause when the condition is secondary to an infection.

Bacterial Vaginosis (BV)

The large surface area and associated secretions of the female reproductive tract make it a rich environment for microbes. Composition of the vaginal microflora changes with the menstrual cycle, owing to changing nutrients and pH. The mildly acidic nature of vaginal secretions (approximately pH 4.5) discourages the growth of many microbes. As a result, the acid-tolerant *Lactobacillus acidophilus* is among the most populous vaginal species. As with all microbial ecosystems associated with the human body, the balance between competing species is crucial to preventing infection. Antibiotic therapy to treat an infection can also cause the loss of normal flora. In the vagina, this allows infection by *C. albicans* (yeast infection), which is not susceptible to antibiotics designed to kill bacteria.

A decrease in the numbers of lactobacilli among the vaginal microbiota is thought to be the major culprit in the pathogenesis of bacterial vaginosis, a common infection in women of childbearing years. *Lactobacillus* species produce lactic acid (a strong microbiocide that lowers the pH of the vagina to an acidic 3.4–4.5) and antimicrobial compounds such as hydrogen peroxide and bacteriocins. An acidic pH makes the vaginal canal an inhospitable environment for many pathogens. Lactic acid is an effective microbiocide against both HIV and *Neisseria gonorrhoeae*. Bacteriocins secreted by lactobacilli are narrow-spectrum bactericidal proteins that increase the permeability of the target microbe's cell membrane.

Not all women have lactobacilli as the major species of their vaginal microbiota. Recent research using culture-independent methods shows that several microbial species other than lactobacilli can be the dominant members of normal vaginal microbiota in some women. The percentage of women colonized by non-lactobacilli-dominated vaginal microbiota appears to vary with ethnic background.

Bacterial vaginosis (BV) is not sexually transmitted, but sexual activity is a risk factor. Women who have multiple sexual partners or a new sexual partner, who douche (rinse out their vagina with water or a cleansing agent), or who lack lactobacilli are at higher risk for BV. Clinicians should use their own judgment in determining whether patients at high risk of BV should be screened. Low-risk patients, however, should not be screened.

Bacterial vaginosis can be polymicrobial or result from overgrowth of several indigenous anaerobic bacteria. The vaginal tissue of the patient affected with BV does not show any sign of inflammation (this is why it is called vaginosis and not vaginitis), and in most cases, the patient has no symptom other than a thin discharge. In some, particularly those cases caused by overgrowth of *Gardnerella vaginalis*, the discharge is grayish and foul smelling (musty or fishy) and the patient complains of dysuria and pruritis (itchiness). Pregnant women with BV may develop serious complications, such as

Figure 23.22   Clue Cells

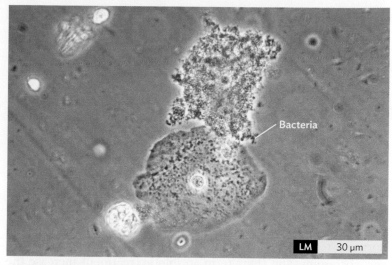

Samples obtained from the grayish discharge of bacterial vaginosis show "clue cells": vaginal epithelial cells with bacteria, usually *Gardnerella vaginalis*, stuck to them, similar to the one shown here.

premature delivery, spontaneous abortion or postpartum (after delivery) endometritis (inflammation of the endometrium).

Diagnosis of BV is based on clinical presentation, a vaginal pH greater than 4.5, presence of a film of thin grayish white discharge coating the vaginal wall, and more than 20% clue cells. Clue cells in a vaginal sample are vaginal epithelial cells covered with bacteria, which gives them a stippled look (Figure 23.22). Oral and topical antibiotics (usually metronidazole) are used to treat BV in both pregnant and nonpregnant women. Antibiotic therapy of pregnant women with BV reduces the risk of preterm delivery and spontaneous abortion. Oral or vaginal probiotics containing *Lactobacillus* species are also used to treat bacterial vaginosis.

Fournier's Gangrene

Fournier's gangrene is a form of necrotizing fasciitis infection that most often involves the soft tissues of the male genitals, although women can also contract the disease. The infection involves a mixed population of aerobic and anaerobic microorganisms—typically streptococci, clostridia, staphylococci, *Klebsiella*, and other enteric bacteria. The most common sources of these agents are skin and urogenital areas. Patients with Fournier's gangrene usually have an underlying immune dysfunction, such as the patient in Figure 23.23. Risk factors for Fournier's gangrene include diabetes, IV drug use, trauma, recent surgery, immune suppression (for example, cirrhosis or malignancy), peripheral vascular disease, and morbid obesity. Treatment of this disease includes broad-spectrum antibiotics (imipenem or meropenem) and surgical debridement. The mortality rate for Fournier's gangrene, a urological emergency, varies from 3% to 45%, depending on how far the disease progressed before diagnosis.

Figure 23.23   Fournier's Gangrene

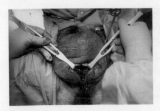

This patient presented to the emergency department with exquisite (intense) scrotal pain. A foul-smelling, purulent fluid emerged when the scrotum was surgically opened, exposing necrotic tissue inside.

Vulvovaginal Candidiasis

Candida is part of the normal vaginal microbiota of most women. **Vulvovaginal candidiasis** (vulvovaginitis), usually called a yeast infection, is neither an opportunistic nor a sexually transmitted infection; rather, it is a *Candida* overgrowth problem. Diabetes, hormone replacement therapy (or taking birth control pills), taking corticosteroids, and short-term antibiotic treatment increase the risk of vulvovaginitis caused by *Candida* because these conditions alter the balance of the normal vaginal microbiota.

Yeast infections are very common during childbearing years. Most premenopausal women (75%) have suffered at least one episode of yeast infection and recall the intense vulvar itching and irritation that causes external dysuria (painful urination without urethritis). Different members of the *Candida* species can cause the infection, but *Candida albicans* is responsible for most (80%–92%) cases. The infection may or may not be accompanied by a curd (or cottage cheese)-like discharge, but there is no foul odor (Figure 23.24).

About half of all patients who get a yeast infection will have a recurrence. If a woman gets four or more *Candida* infections, she is said to have recurrent yeast infections; these are thought to be caused by species other than *Candida albicans* (for instance, *Candida glabrata*). HIV-infected women have a higher incidence of vulvovaginal candidiasis and suffer a longer duration of symptoms, but the severity of symptoms is not any different from that in non-HIV-infected women. Vaginal pH remains unchanged (about 4.5), in contrast to what is seen in *Trichomonas* infection.

Diagnosis is made by adding 10% potassium hydroxide to a wet mount of the discharge (or material scraped from the infected area) and examining it under the microscope. Budding yeast will be seen in positive cases.

Azole drugs, described in Chapter 13, are believed to block the biosynthesis of ergosterol present in yeast membranes. Over-the-counter topical antimycotic drugs (vaginal suppositories) can achieve cure rates in excess of 80%. The only oral azole agent approved by the U.S. Food and Drug Administration (FDA) for treating vulvovaginitis is fluconazole, which also achieves a high rate of cure. Therapeutic concentrations of this drug are maintained in vaginal secretions for at least 72 hours after the ingestion of a single 150-mg tablet.

SECTION SUMMARY

- **Vaginosis** is a bacterial infection of the vagina (often caused by the anaerobe *Gardnerella vaginalis*) produced following an imbalance in the vaginal microbiota.

Figure 23.24   Vulvovaginal Candidiasis

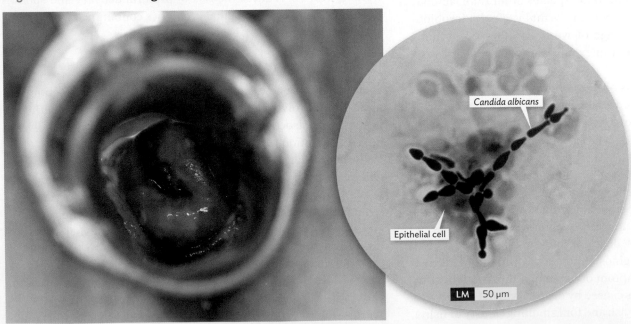

A. Vulvovaginal candidiasis as seen through a speculum. Compare to Figure 23.16A.

B. Budding *Candida albicans* cells are seen in a Gram stain of vaginal secretions.

- **Epididymitis, orchitis, and prostatitis** are caused by local bacteria when urine flow is obstructed or when the immune system is suppressed.

- **Fournier's gangrene** is a form of necrotizing fasciitis that most often involves the soft tissues of the male genitals, although women can also contract the disease. A mixed population of aerobic and anaerobic microorganisms, typically enteric bacteria, are commonly responsible for causing this infection.

- **Candidiasis,** commonly called a yeast infection, is neither an opportunistic nor a sexually transmitted infection; it results from overgrowth of *Candida* species.

Perspective

Cystitis, pyelonephritis, and other infections of the urinary tract are common in medical practice. Clinicians realize that UTI presentations can vary, depending on the age of the patient and their comorbidities. Remember, too, that infections starting in the urinary tract can spread to other organ systems (urosepsis), and infections focused elsewhere in the body can also end up causing UTIs. The toll on the lives of patients, especially women, and on the nation's economy makes UTIs an important, though somewhat embarrassing, burden on society.

STIs, however, have an added social dimension. When you have sex with someone, it's the same as having sex with everyone that person had sex with before you. Even a single encounter can bring exposure to multiple STIs. There are scary cases, too, in which people with HIV have willfully engaged in unprotected sex with others, knowing full well that they would probably transmit HIV to their partners.

Of course, the impact of STIs on society is not new. Humans have been afflicted with these diseases for thousands of years. So why is it that we have been so ineffective in preventing these infections? Our failure is due in part to the ingenuity of microbes in evading our immune system and their ability to mutate and become resistant to the medications we have. We have also helped them to survive and flourish by our desire to have sex and our propensity to ignore effective preventive methods, simple as they are, such as the use of a condom.

Figure 23.25   Summary: Microbes That Infect the Genitourinary System

Viruses	Genome	Disease
Herpes simplex 2	DNA, ds	Genital herpes
Herpes simplex 1	DNA, ds	Genital herpes
Human papillomavirus	DNA, ds	Genital warts
Human immunodeficiency virus	RNA, ss, positive sense (two copies of the genome in a core capsid)	HIV/AIDS

Bacteria	Morphology	Disease
Neisseria gonorrhoeae	Gram-negative, cocci	Gonorrhea, also called gonococcal urethritis (STI)
Chlamydia trachomatis	Gram-negative-like,[a] coccoid or rod-shaped	Chlamydial diseases; nongonococcal urethritis and lymphogranuloma venereum (STI)
Haemophilus ducreyi	Gram-negative, coccobacilli	Chancroid
Escherichia coli	Gram-negative, bacilli	Epididymitis, cystitis, prostatitis, pyelonephritis
Proteus mirabilis	Gram-negative, bacilli	Cystitis, pyelonephritis
Klebsiella	Gram-negative, bacilli	Cystitis, pyelonephritis
Pseudomonas	Gram-negative, bacilli	Cystitis, pyelonephritis
Enterobacter	Gram-negative, bacilli	Cystitis, pyelonephritis
Enterococcus	Gram-positive, cocci[b]	Cystitis, pyelonephritis
Treponema pallidum	Spirochete	Syphilis (STI)
Staphylococcus aureus	Gram-positive, cocci	Cystitis, pyelonephritis
Staphylococcus saprophyticus	Gram-positive, cocci	Cystitis, pyelonephritis
Gardnerella vaginalis	Gram-variable, coccobacilli	Bacterial vaginosis

Fungi	Type	Disease
Candida albicans	Yeast	Vulvovaginitis, cystitis in diabetic patients

Parasites	Type	Disease
Trichomonas vaginalis	Protozoa	Trichomoniasis

[a]Chlamydia trachomatis is an obligate intracellular aerobic organism, very difficult to stain but closely related to Gram-negative bacteria.
[b]Enterococci usually occur in pairs and sometimes in short chains.

Portal of Entry

Mouth, Oral Cavity
HSV-1 and HSV-2

Blood
Hepatitis B and hepatitis C
HIV

Anus, Perineum, Vagina, Penis, or Self-Inoculation
HSV-1 and HSV-2
HBV
HCV
HPV
HIV
Neisseria gonorrhoeae
Chlamydia trachomatis
Treponema pallidum
Haemophilus ducreyi
Escherichia coli
Proteus mirabilis
Klebsiella
Pseudomonas
Enterococcus
Enterobacter
Staphylococcus aureus
Staphylococcus saprophyticus
Gardnerella vaginalis
Candida albicans
Trichomonas vaginalis

Portal of Exit

Mouth, Oral Cavity
HSV-1 and HSV-2

Blood
Hepatitis B and hepatitis C
HIV

Anus, Perineum, Vagina, or Penis
HSV-1 and HSV-2
HBV
HCV
HPV
HIV
Neisseria gonorrhoeae
Chlamydia trachomatis
Treponema pallidum
Haemophilus ducreyi
Escherichia coli
Proteus mirabilis
Klebsiella
Pseudomonas
Enterococcus
Enterobacter
Staphylococcus aureus
Staphylococcus saprophyticus
Gardnerella vaginalis
Candida albicans
Trichomonas vaginalis

LEARNING OUTCOMES AND ASSESSMENT

23.1
Anatomy of the Urinary Tract

- List the function of each structural component of the male and female urinary tracts.
- Explain the relationship between the upper and lower urinary tract anatomy and physiology.
- Discuss the significance of structural differences between the male and female urinary tract as they relate to infections.

1. Which structure of the urinary tract acts as a filter to remove unwanted material from blood?
   - A. Kidneys
   - B. Ureters
   - C. Urinary bladder
   - D. Urethra
   - E. Urethral meatus

23.2
Urinary Tract Infections

- Discuss the role of physical and hormonal factors in the development of urinary tract infections in women.
- Examine the relationships between lower and upper urinary tract infections.
- Compare and contrast the signs and symptoms of common urinary tract infections.
- List treatments for the most common etiologies of urinary tract infections.

4. Which of the following is a risk factor for developing bladder infections in young nonpregnant women?
   - A. Sexual activity
   - B. Low plasma estrogen concentration
   - C. Bladder incontinence
   - D. Elevated plasma estrogen concentration

5. A male hospitalized patient who had a urinary catheter inserted to enable urination is now running a high fever and feels a lot of pain when tapped at the costovertebral angle. The patient's blood pressure is within normal limits and his heart rate is 110 (normal is 60–100). Urine culture

23.3
Anatomy of the Reproductive Tract

- List the structural components of the male and female reproductive tracts.
- Explain the function of each component of the male and female reproductive tracts.
- Discuss the significance of structural differences between the male and female reproductive tracts as they relate to infections.

9. The _____ is (are) external to the human body.
   - A. ovaries
   - B. vagina
   - C. testes
   - D. seminal vesicles

23.4
Sexually Transmitted Infections of the Reproductive Tract

- Relate the signs and symptoms of sexually transmitted infections to their microbial etiology.
- Separate complications of each sexually transmitted infection according to the disease progression.
- Explain the impact of gender on the development of sexually transmitted infections and their complications.
- List prevention and treatment options available for sexually transmitted infections.

12. Infection with certain serotypes of _____ increases the risk of cervical cancer.
   - A. human papillomavirus
   - B. human immunodeficiency virus
   - C. herpes simplex 1
   - D. herpes simplex 2

23.5
Non-Sexually Transmitted Infections of the Reproductive Tract

- Relate signs and symptoms of non-sexually transmitted infections to their microbial etiology.
- Explain the role of the vaginal microbiome in the prevention and/or development of non-sexually transmitted infections of the reproductive tract.
- List treatment options available for non-sexually transmitted infections of the reproductive tract.

20. A young woman visiting her clinician complains of intense vulvar itching and burning upon urination. Physical exam reveals a whitish, odorless, thick discharge and inflamed and erythematous external genital tissue. Her cervix is normal and not tender. The most likely diagnosis is _____.
   - A. candidal vulvovaginitis
   - B. bacterial vaginosis
   - C. gonococcal urethritis
   - D. nongonococcal urethritis
   - E. trichomoniasis

2. The urine is transported to the urinary bladder via _____.
A. nephrons
B. urethra
C. ureters
D. kidneys

3. Which of the following is a major difference between the male and female urinary tracts?
A. The number of ureters
B. The size of the perineum
C. The structure of the urethral meatus
D. The length of the urethra
E. The bladder lining

---

shows *Proteus mirabilis*; blood cultures are negative. Which of the following statements is probably true?
A. This patient has pyelonephritis resulting from an ascending bladder infection.
B. This patient has a bladder infection resulting from a descending kidney infection.
C. This patient has a systemic infection that has spread to his kidneys.
D. Continuous flow of urine from kidneys to the bladder prevents the infection from spreading between these organs.
E. Urinary tract infections never cause a fever.

6. The majority of urinary tract infections are caused by Gram-negative bacteria because urine is bacteriostatic to Gram-positive bacteria. True or false?

7. The majority of cystitis cases in children are due to _____.
A. congenital structural abnormalities that prevent complete bladder emptying
B. children postponing emptying their bladder
C. the shorter length of urethra in younger children
D. continuous changes in the normal microbiota of children's perineum

8. Which of the following urinary tract infections is a risk factor for developing cardiovascular shock?
A. Cystitis
B. Urethritis
C. Pyelonephritis
D. Acute glomerulonephritis

---

10. Which of the following statements about the female reproductive tract is correct?
A. The female reproductive tract has internal and external components.
B. Every month only one follicle is recruited to mature into an ovum.
C. There is no direct contact between the ovaries and the fallopian tubes.
D. The cervix forms the end of the birth canal.

11. Sexually transmitted infections are passed from one person to another only through intercourse. True or false?

Match choices A–E to items 13–17.

13. Vesicular lesions

14. Acute HIV infection

15. HPV vaccine

16. Kaposi's sarcoma

17. Host protein that HIV binds to

A. "Mononucleosis-like" symptoms

B. CD4$^+$

C. AIDS

D. Gardasil

E. HSV

18. Women serve as the natural reservoir for _____ because they are usually asymptomatic when infected with this microbe.
A. *Chlamydia trachomatis*   C. *Haemophilus ducreyi*
B. *Treponema pallidum*   D. *Neisseria gonorrhoeae*
                                     E. *Trichomonas vaginalis*

19. A grayish yellow discharge with a fishy odor along with punctate hemorrhagic lesions on the cervix (strawberry cervix) is indicative of _____.
A. *Chlamydia trachomatis*   C. *Haemophilus ducreyi*
B. *Treponema pallidum*   D. *Neisseria gonorrhoeae*
                                     E. *Trichomonas vaginalis*

---

21. A 45-year-old truck driver with a 10-year history of diabetes presents to the clinic complaining of a wound in his groin that extends into his scrotum. He denies using ilicit drugs or having any trauma in the area. You send a sample from the wound to the lab for identification and diagnose him with Fournier's gangrene. Which of the following is the lab most likely to report?
A. Gram-negative rods normally found in the GI tract
B. Gram-positive rods normally found on the skin
C. A mixed population of aerobic and anaerobic organisms
D. An obligate intracellular parasite
E. A flagellated anaerobic protozoan

22. Which of the following is used to treat bacterial vaginosis?
A. Metronidazole
B. Probiotics with lactobacillus
C. Topical azole compounds
D. A and B only
E. A and C only

Key Terms

acquired immunodeficiency syndrome (AIDS) (p. 784)
bacteriuria (p. 770)
chancre (p. 786)
chancroid (p. 793)
chlamydia (p. 792)
condyloma acuminatum (p. 781)
congenital syphilis (p. 790)
costovertebral tenderness (p. 778)
cystitis (p. 773)
dysuria (p. 776)
early symptomatic HIV infection stage (p. 785)
elite controller (p. 786)
epididymitis (p. 795)

Fournier's gangrene (p. 797)
general paresis (p. 789)
genital herpes (p. 782)
gonorrhea (p. 791)
gumma (p. 788)
highly active antiretroviral therapy (HAART) (p. 786)
Jarisch-Herxheimer reaction (p. 790)
Kaposi's sarcoma (p. 785)
latent period (p. 783)
latent stage (p. 784)
long-term nonprogressor (p. 786)
lymphogranuloma venereum (LGV) (p. 793)
neonatal herpes (p. 783)

neurosyphilis (p. 789)
oliguria (p. 778)
ophthalmia neonatorum (p. 792)
orchitis (p. 795)
pelvic inflammatory disease (PID) (p. 794)
primary (acute) HIV infection (p. 784)
primary infection (p. 783)
primary syphilis (p. 786)
prostatitis (p. 795)
pyelonephritis (p. 773)
rapid plasma reagin (RPR) test (p. 789)
secondary syphilis (p. 788)
sexually transmitted infection (STI) (p. 781)

syphilis (p. 786)
tabes dorsalis (p. 789)
tertiary syphilis (p. 788)
trichomoniasis (p. 794)
urethritis (p. 775)
urinary tract infection (UTI) (p. 770)
urosepsis (p. 770)
vaginitis (p. 795)
vaginosis (p. 795)
Venereal Disease Research Laboratory (VDRL) test (p. 789)
viral load (p. 784)
vulvovaginal candidiasis (p. 798)

Review Questions

1. Describe areas of the urinary and reproductive tracts associated with infectious disease. Explain how microbes gain access to these structures.

2. Describe the difference between cystitis, pyelonephritis, vaginosis, and vulvovaginitis. What organisms are the most common causes of these diseases?

3. When can asymptomatic bacteriuria be a dangerous condition?

4. Discuss how the presentation of urinary tract infections can differ between children, adults, and the elderly.

5. Define recurrent UTIs. What are some possible causes of recurrent UTIs?

6. Explain how UTIs and pyelonephritis can be differentiated in a patient.

7. What are the most common viral diseases of the reproductive tract? Describe key features of these viruses (structure and target cells).

8. Explain how you would differentiate the diseases you cited in question 7 based on patient presentation.

9. Which sexually transmitted infections have a latent period sandwiched between symptomatic periods?

10. Discuss neonatal disease involving sexually transmitted agents. How can you distinguish a neonate infected with one or the other of these agents?

11. Describe the stages of HIV infection and the parameters used to characterize those stages. What is the significance of a CD4$^+$ count less than 400 cells/ml? Less than 200 cells/ml?

12. How do the lesions of syphilis, chancroid, and LGV differ?

13. Discuss how the diagnosis and treatment of gonorrhea differ from the diagnosis and treatment of syphilis.

14. What is pelvic inflammatory disease, and what unique dangers does it pose?

15. What is the most common sexually transmitted infectious disease? What is unique about the growth of this pathogen?

16. What is a characteristic feature of lymphogranuloma venereum?

17. For which STIs are vaccines available?

18. Describe features of Fournier's gangrene. Who is at risk?

19. Explain how changes in the vaginal microbiome can be a risk factor for bacterial vaginosis.

Clinical Correlation Questions

1. A young woman is complaining of a fishy-smelling vaginal discharge. Physical exam does not reveal any inflammation. The cervix is normal and nontender. Microscopic examination of the discharge reveals clue cells. What organism is the most likely etiology of this woman's infection? How did this patient acquire this infection?

2. A 25-year-old pregnant female is in the clinic complaining of a maculopapular rash that extends into the palms of her hands and soles of her feet. Physical exam shows an abnormal gait. You explain to the patient that when her infant is born, it may have several diseases and abnormalities. What physical abnormalities is the newborn likely to have?

CHAPTER REVIEW   805

Thought Questions: CHECK YOUR ANSWERS

Thought Question 23.1 Why are urinary tract infections among the most commonly acquired nosocomial (hospital-acquired) infections?

ANSWER: There are several contributory factors. The organisms can piggyback on a catheter and into the bladder at the time of insertion. Presence of a catheter in the urethra keeps the urethral meatus open, allowing microbes better access to the bladder. Patients who are in the hospital and catheterized, such as postsurgical patients or patients in the intensive care unit, have a compromised immune system and are more susceptible to infections.

Thought Question 23.2 Why would someone with pyelonephritis have a lower urine output?

ANSWER: Infection of the kidney can damage the nephrons, which will decrease the filtration through the kidney. Less water from blood enters the nephrons, so less urine is produced.

Thought Question 23.3 Why would early symptomatic HIV disease lead to shingles in more than one dermatome?

ANSWER: Shingles, described in Chapter 19, involves reemergence of HSV from a spinal nerve. Each spinal nerve serves a different section of skin (dermatome). In a typical patient, the immune system keeps outbreaks in check, but an outbreak of shingles limited to a single dermatome or adjacent dermatomes is possible. Immunosuppression in early AIDS allows activation of latent HSV that may be present in multiple spinal nerves.

Thought Question 23.4 What aspect of the immune system is affected by diminished numbers of CD4$^+$ T cells?

ANSWER: CD4$^+$ T cells are helper T cells associated with cell-mediated and antibody responses, depending on whether they are T_H1 or T_H2 cells. See Chapter 16.

Thought Question 23.5 *Neisseria meningitidis* readily causes meningitis, yet *N. gonorrhoeae* does not. What is the explanation for this?

ANSWER: *N. meningitidis* possesses a polysaccharide capsule that protects the organism from complement and phagocytosis in serum. The organism is "serum resistant" and can disseminate through the blood to the brain. *N. gonorrhoeae* does not possess this capsule and is usually sensitive to serum complement.

Thought Question 23.6 How would you treat pelvic inflammatory disease?

ANSWER: Broad-spectrum antibiotics should be used for all potential causes, but always include antibiotics that treat *Neisseria* and *Chlamydia*. Surgical intervention may be necessary.

Thought Question 23.7 A patient in your office complains of a high fever and diarrhea of 3 weeks' duration. Upon physical exam, you notice he has swollen lymph nodes. You also discover that he is an IV drug user. Blood work that you order reveals a negative RPR test and a CD4$^+$ count of 500 cells/ml (normal is 1,000 cells/ml). What might you conclude from this, and what other test should be run?

ANSWER: The patient is likely to have an HIV infection, and the gastrointestinal symptoms and diarrhea may be due to a secondary infection. This patient should be tested for HIV to determine the appropriate course of treatment.

24

Infections of the Central Nervous System

[Diagnosis Comes Too Late]

SCENARIO Miguel, a 19-year-old young man from Michoacán, Mexico, worked at a sugarcane plantation in Louisiana. One day Miguel felt extremely tired, his left shoulder hurt, and his left hand was numb. Over the next 3 days his condition worsened and he was taken to a New Orleans hospital.

DISEASE PROGRESSION AND ADDITIONAL TESTS Miguel developed a fever of 38.4°C (101.1°F) and had trouble breathing, which prompted intubation (insertion of a tube into the trachea). During the next several days, Miguel became unresponsive to external stimuli, developed fixed and dilated pupils, and had episodes of bradycardia (slow heart rate) and hypothermia (low core body temperature). His CSF now had 87 WBCs/ml with 97% lymphocytes and an elevated protein level of 233 mg/dl (normal: 15–45 mg/dl). An electroencephalogram (EEG) was consistent with encephalitis. Bacterial, viral, and fungal cultures of blood and CSF were negative, as were serological tests for HIV, syphilis, herpes simplex virus, arboviruses, Lyme disease, and autoimmune neuropathies.

PHYSICAL EXAMINATION AND LAB WORK Physical examination at the hospital revealed hyperesthesia (numbness, tingling, pinprick sensation) of his left shoulder, weakness of his left hand, generalized areflexia (absence of reflexes), and drooping of his left upper eyelid. Cerebrospinal fluid (CSF) obtained from a lumbar puncture (see Figure 24.4) had a mildly elevated white blood cell count of 8 cells/ml (normal: 0–5), a normal glucose level, and no organisms on staining. Miguel was admitted to the intensive-care unit with suspected viral encephalitis or meningitis.

DIAGNOSIS AND CONCLUSION
A diagnosis of rabies was suspected based on the clinical history, even though animal exposure was unknown. On day 18, rabies was confirmed by finding rabies virus–specific IgG and IgM in Miguel's CSF and serum. After the doctor discussed Miguel's prognosis with the family and a subsequent EEG showed severe cortical impairment, Miguel was extubated and died shortly thereafter. Rabies virus antigen was detected in postmortem brain tissues, and antigenic typing determined it was a vampire bat rabies virus variant, which was confirmed by nucleic acid amplification and sequencing.

Adapted from: CDC. 2011. *Morbidity and Mortality Weekly Report* (MMWR) 60(31):1050–1052.

TEM 50 nm

Rabies Virus
Rabies virus is an enveloped, bullet-shaped, negative-sense RNA virus (Rhabdoviridae family) transmitted to humans by the bite of a rabid animal. The virus infects the peripheral nervous system, moves to the central nervous system, and causes encephalitis (brain inflammation), which ultimately results in death.

CHAPTER OBJECTIVES

After reading this chapter, you will be able to:

- Correlate the anatomy and physiology of the nervous system with the infectious processes affecting them.
- Describe symptoms that can help differentiate nervous system infections caused by viruses from those caused by bacteria and parasites.
- Relate pathogenic mechanisms to disease prevention strategies for microbes that infect the nervous system.
- Correlate the physiology of each pathogen with its antimicrobial treatment.

Rabies, caused by the rabies virus, is one of the deadliest of all infectious diseases known to humans; fortunately, it is rarely seen in the United States. Rabies virus is endemic in different mammalian animal populations, including foxes, dogs, cats, and bats. The infected animals usually transfer the virus to us by depositing saliva with rabies virus in our flesh when they bite us or following casual contact with our injured skin. Once within the human body, rabies virus attacks the central nervous system and almost always causes death. Anyone who has had a dog or a cat knows that pets must be vaccinated against rabies. It is because of pet vaccination programs that the number of rabies cases has drastically declined in the United States. Other microorganisms can also attack our nervous system, but thanks to a vigilant immune system and our knowledge of infectious disease, we can fight these other microbes more effectively than we can rabies. That is not to say that other infections of the nervous system are harmless; on the contrary, most have the potential to cause immense damage and even death.

24.1
Anatomy of the Central Nervous System

SECTION OBJECTIVES

- Describe the functional organization of the central nervous system.
- Explain how meninges and cerebrospinal fluid protect the central nervous system.
- Describe the differences between two major types of neurons.
- Correlate the anatomy of the central nervous system to the symptoms of CNS infections.

How is the nervous system organized? The central nervous system is the control center of the body. It houses our cognition, makes sense of our environment, formulates and helps dictate responses to all stimuli, enacts reflexes, and regulates every voluntary and involuntary action, from walking and talking to body temperature and heart rate, and it does it all seamlessly. Our nervous system has two main parts (**Figure 24.1**): the **central nervous system (CNS)** and the **peripheral nervous system (PNS)**. The CNS controls and coordinates everything, while the PNS collects data from the periphery (parts of the body outside the brain) and from the outside environment and communicates that information to the CNS. You might envision the CNS as a computer's hard drive with different kinds of memories and operating systems to classify, store, and analyze data received from the PNS. The PNS provides the needed circuitry and instruments to collect and input data from the body and the environment and then conveys orders from the CNS to the periphery. Damage to either system can have dire consequences.

Peripheral nerve damage (neuropathy) is a complication of systemic diseases such as diabetes and several infectious diseases, such as leprosy (see Case History 5.1), syphilis, and Lyme disease. In this chapter, however, we will focus on diseases of the central nervous system.

The nervous system has, of course, defensive measures in place. The CNS is protected from most harm by several mechanisms. The hard bony structures that house the CNS, the skull and vertebrae, are physical barriers to possible trauma from the outside world, such as a blow to the head. Three thin membranes (the **meninges**; singular **meninx**) separate the brain and spinal cord from the bony structures surrounding them. The meninges include an inner layer next to the brain called pia mater, a middle layer named arachnoid, and an outer layer next to the skull called dura mater (**Figure 24.2**). Specialized projections of pia mater, called the choroid plexus, secrete a clear fluid derived from blood called **cerebrospinal fluid (CSF)** into the subarachnoid space.

We produce approximately 500–700 ml of CSF every day. The optimal CSF volume in an average adult is around 140–270 ml. The extra CSF produced by a person's choroid plexus is taken up by specialized structures called arachnoid villi, which will discard it into the venous circulation. The pulsation of local arteries helps circulate the CSF around the CNS. This means that CSF has a normal

Figure 24.1 The Nervous System

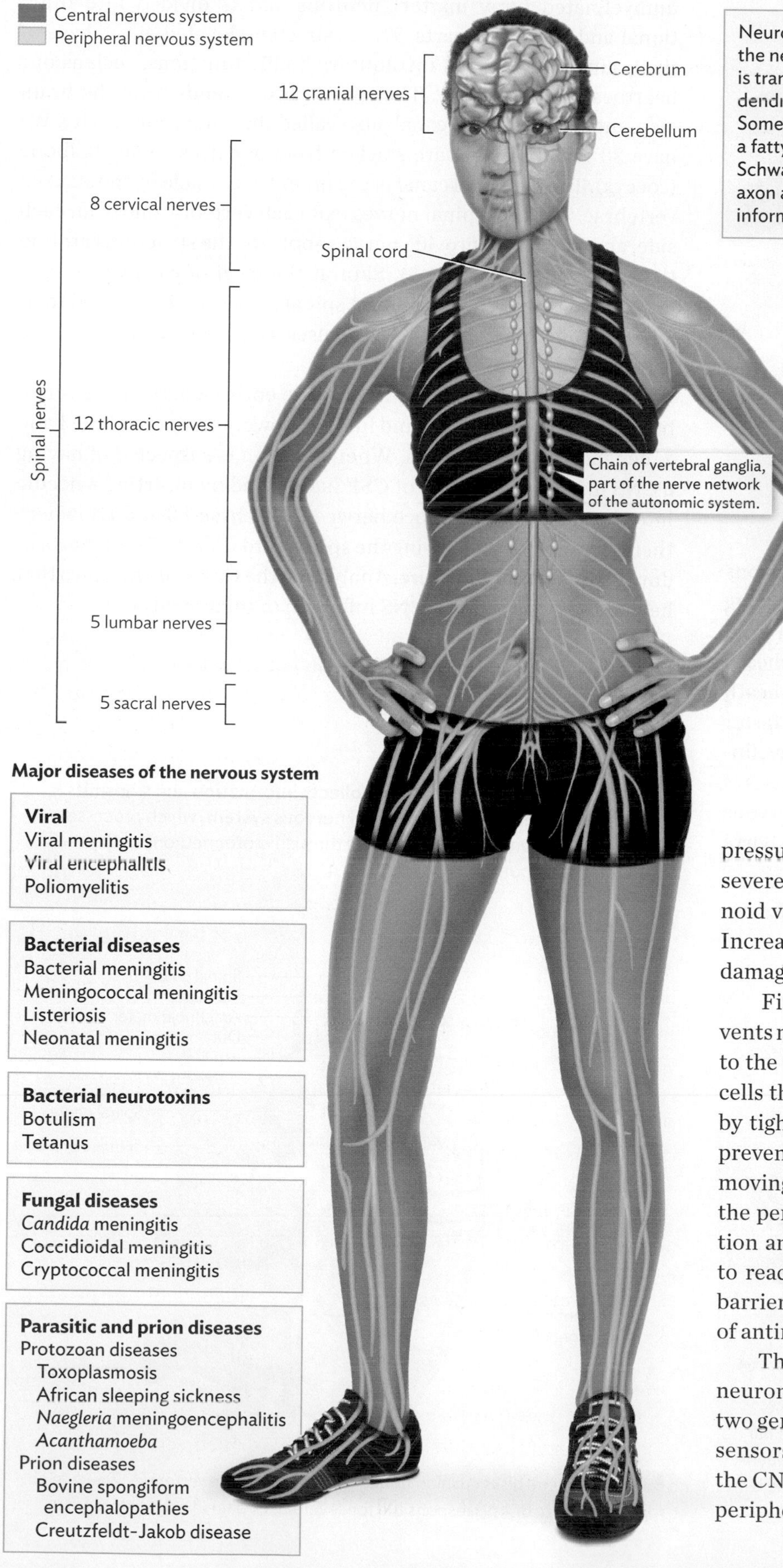

■ Central nervous system
■ Peripheral nervous system

12 cranial nerves

Cerebrum

Cerebellum

8 cervical nerves

Spinal cord

Spinal nerves

12 thoracic nerves

Chain of vertebral ganglia, part of the nerve network of the autonomic system.

5 lumbar nerves

5 sacral nerves

Neurons are specialized cells of the nervous system. Information is transferred between axons and dendrites through the cell body. Some axons are covered in myelin, a fatty molecule produced by Schwann cells that protects the axon and accelerates information transfer.

Axon terminal

Node of ranvier

Myelin sheath

Schwann cell

Axon

Cell body

Nucleus

Dendrite

FM 100 µm

Major diseases of the nervous system

Viral
Viral meningitis
Viral encephalitis
Poliomyelitis

Bacterial diseases
Bacterial meningitis
Meningococcal meningitis
Listeriosis
Neonatal meningitis

Bacterial neurotoxins
Botulism
Tetanus

Fungal diseases
Candida meningitis
Coccidioidal meningitis
Cryptococcal meningitis

Parasitic and prion diseases
Protozoan diseases
  Toxoplasmosis
  African sleeping sickness
  Naegleria meningoencephalitis
  Acanthamoeba
Prion diseases
  Bovine spongiform
    encephalopathies
  Creutzfeldt-Jakob disease

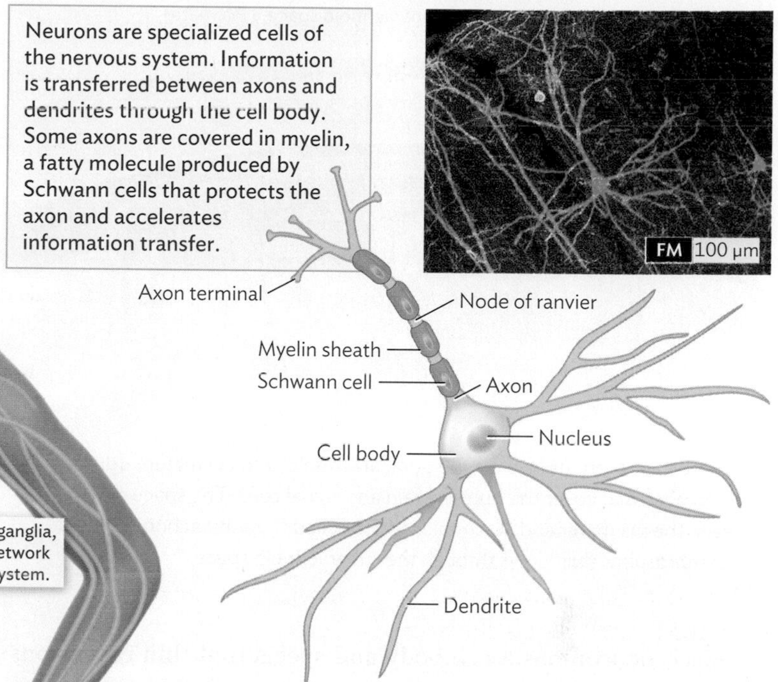

pressure associated with it. Infections in the CNS can result in severe inflammation of the meninges and occlusion of the arachnoid villi, which in turn can result in an increase in CSF pressure. Increased CSF pressure is a very serious condition that can cause damage to the brain or spinal cord.

Finally, a firewall of sorts, called the **blood-brain barrier**, prevents most microorganisms and large molecules from gaining access to the CNS. The blood-brain barrier exists because the endothelial cells that make up the CNS capillaries are attached to one another by tight junctions (see Section 14.3). These specialized structures prevent bacteria, large molecules, and hydrophilic molecules from moving out of the circulating blood and into the CSF. Unfortunately, the permeability of the barrier can be compromised by inflammation and other impairments, allowing bacteria, fungi, or parasites to reach the CNS and cause infection. Moreover, the blood-brain barrier has the unintended consequence of hindering the delivery of antimicrobials and medicines that are not lipid-soluble.

The CNS and PNS consist mainly of specialized cells called neurons. Neurons come in different shapes and sizes but fall into two general categories: **sensory neurons** send information from the sensors in the periphery (skin, gastrointestinal tract, and so on) to the CNS, and **motor neurons** send information from the CNS to the periphery.

Figure 24.2   The Meninges

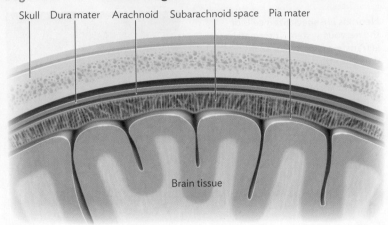

The three membranous layers—the dura, arachnoid, and pia mater—form the meninges that cover the human brain and spinal cord. The space deep between the pia mater and arachnoid layer is called the subarachnoid space. The cerebrospinal fluid flows through the subarachnoid space.

Each neuron has a cell body and specialized thin extensions called axons and dendrites (see Figure 24.1 inset). Dendrites bring information to the cell body, whereas axons send information from the cell body. Some axons are covered by a whitish-colored sheath composed of a fatty molecule called myelin. The myelin sheath insulates and protects the axons it covers. Information travels faster along myelinated axons than it does along those that are unmyelinated. When myelin is destroyed, transmission of nerve impulses is impaired. Loss of the myelin sheath contributes to certain nervous system disorders, including multiple sclerosis and Guillain-Barré syndrome (see Figure 22.20).

The human brain consists of myelinated (white matter) and unmyelinated (gray matter) neurons and is divided into functional and structural parts. The brain stem, located at the base of the brain, controls vital involuntary bodily functions, such as our heartbeat and breathing. The spinal cord extends from the brain stem into the vertebral canal, also called the spinal cord cavity. We have 33 vertebrae that are stacked from our neck to our tailbone (coccyx). The vertebral canal is the inner space made by the stacked vertebrae. A pair of spinal nerves exit each vertebra, one from each side, and innervate (provide nerve supply to) the structures around that vertebra (Figure 24.3). Skin at the level of each vertebra is innervated by the corresponding spinal nerve and is referred to as a dermatome (see Section 19.3, Viruses That Produce Papular and Pustular Rashes).

The spinal cord extends almost the entire length of the vertebral canal until it stops around lumbar (lower torso) vertebrae 1 and 2, depending on the person. When a person is suspected of having a CNS infection, a sample of CSF is removed by inserting a needle into the subarachnoid space between vertebrae L3 and L4, where there is no chance of injuring the spinal cord (Figure 24.4), a procedure called **lumbar puncture**. Analysis of the CSF will give clues that help identify the cause of CNS infection or inflammation.

 Lumbar puncture (Brown University): youtube.com

SECTION SUMMARY

- **The peripheral nervous system collects information and transmits it via sensory neurons to the central nervous system,** which processes that information and returns signals through motor neurons of the PNS, directing a response.

Figure 24.3   Vertebrae

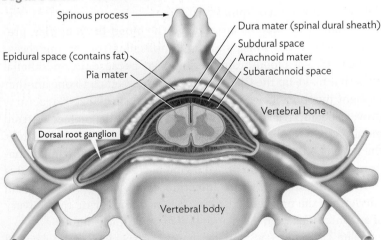

A. Each vertebra has a body, a spinous process, and a central opening. The spinal cord is separated from the bony structure of the vertebrae by the spinal meninges, which are continuous with the brain meninges. The space between the outermost layer of the meninges, the dura mater, and the spinous bone is called the subdural space.

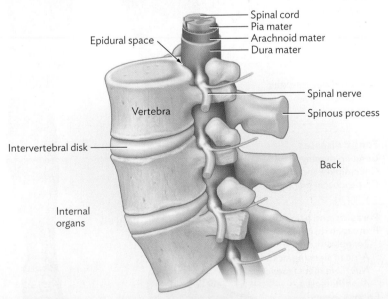

B. Intervertebral disks cushion the contact between the vertebrae. Damage or rupture of these disks is responsible for severe back pain in many people.

Figure 24.4 Lumbar Puncture

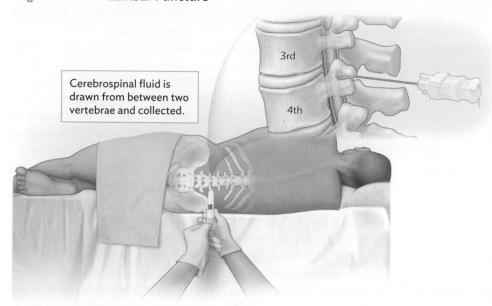

Cerebrospinal fluid is drawn from between two vertebrae and collected.

3rd

4th

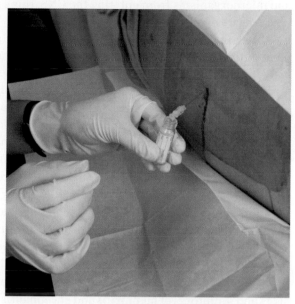

A. A patient lies on his or her side (knees up, head down). A spinal tap needle is inserted between the third and fourth lumbar vertebrae and into the subarachnoid space using aseptic technique. The procedure is normally performed under local anesthesia.

B. CSF is allowed to drip from the needle into a collection vial.

- **Neurons** are made of dendrites that bring information to the cell body and axons that send information from the cell body. Some axons are myelinated.
- **The CNS** includes the brain, encased in the skull, and the spinal cord, encased in vertebrae.
- **The meninges** are three membranes (dura mater, arachnoid, and pia mater) that cover the brain and spinal cord. Cerebrospinal fluid (CSF) fills the space between the arachnoid and pia mater.
- **The choroid plexus** are projections of pia mater that secrete CSF.

24.2
Overview of the Infectious Diseases of the Central Nervous System

SECTION OBJECTIVES
- Categorize central nervous system infections by the anatomical structures affected.
- Use key laboratory information to distinguish types of CNS infections.
- Compare and contrast the roles of meninges and cerebrospinal fluid in the pathogenesis of CNS infections.

How serious are brain infections? A brain infection must be diagnosed quickly and properly to avoid severe disease and possible death. A timely diagnosis can make the difference between life and death—or at least better quality of life versus worse. Yet, CNS infections are easily misdiagnosed because of vague initial presenting signs and symptoms. Most CNS infections start with flu-like symptoms and nonspecific neurological manifestations such as a headache, dizziness, or fatigue. In this section, we will discuss the general concepts of CNS infections and lay the groundwork for the rest of the chapter.

How do microorganisms reach the CNS? The path to the CNS varies, depending on the invading agent. Most microbes first infect distal sites in the body and then travel to the CNS. Others reach the CNS via an infection in an adjacent bone or one of the structures closely associated with the CNS, such as eyes, sinuses, pharynx, or ears. Toxins made by organisms embedded elsewhere in the body can travel along and up the peripheral nerves to reach the CNS. A few organisms target the CNS directly; for example, some mosquito- and tick-borne viruses enter the circulation as a result of an insect bite and travel directly to the CNS.

Damage to the CNS as a result of infection can range from mild to severe. In most cases, as with our patient, Miguel, in the opening case history, symptoms become more serious. The rabies patient initially presented with relatively mild and vague symptoms (generalized fatigue, left shoulder pain, and left-hand numbness), but as the infection progressed, his symptoms worsened and he lost some normal reflexes, such as the reactive narrowing of his pupils to light. Eventually, he lost the ability to breathe on his own.

The signs and symptoms of infectious diseases of the CNS do not usually point to a specific offending agent (bacteria versus virus, for example). Health care providers must identify compositional changes in blood or cerebrospinal fluid to make a correct diagnosis. Ironically, the major obstacle in treating CNS infections is the natural blood-brain barrier that prevents harmful substances and microorganisms from entering the CNS; this same barrier also blocks entry of much-needed antimicrobials.

Three major types of diseases are associated with CNS infections: meningitis, encephalitis (plural, encephalitides), and myelitis. **Meningitis**, or inflammation of meninges, develops when the meninges are infected by a microbe or irritated by a chemical or biochemical. Symptoms include severe headache (due to increased intracranial pressure), fever, and confusion. **Encephalitis**, or inflammation of the brain, can also result from an infection or a reaction to a chemical or biochemical agent. Encephalitis can occur as a primary or secondary disease. **Primary encephalitis** is the result of direct attack on the brain, as for rabies. **Secondary encephalitis** ensues when the infection or inflammation spreads from the meninges to the brain. Bacterial and fungal encephalitides are relatively rare and usually occur when an infection from another site, usually the meninges, extends from that location into the brain itself.

Clinically, the distinguishing difference between the presenting signs and symptoms of meningitis and encephalitis is that encephalitis will cause disruption and deficiencies in the patient's cognitive, motor, and sensory functions. The nature of the deficiency depends on the area of the brain that has been damaged. Imaging studies of a patient's brain are usually performed to exclude other causes of brain dysfunction that may have similar symptoms (for example, a brain abscess or tumor). Figure 24.5 illustrates the difference between a normal brain and one damaged by encephalitis.

Viruses can also cause **myelitis** (inflammation of the spinal cord) and **radiculitis** (inflammation of the spinal nerve root, where the spinal nerve exits the vertebra). A combination of terms is used when the disease affects more than one site. For example **meningoencephalitis** denotes simultaneous inflammation of the meninges and brain, whether by infection or chemical insult.

CNS infections are generally rare. When they do occur, however, viruses are the most common cause, especially of the primary encephalitides. Most of these viral infections, however, are mild. We become infected with these viruses when we cross paths with their reservoirs, such as ticks, birds, or other humans. Transmission of viral encephalitides is often through an insect bite; but depending on the virus, it can also happen after inhaling contaminated respiratory droplets or eating or drinking contaminated food or liquids as well as through direct contact. Typically, an incubation period of various length is followed by symptomatic disease, which may be be mild or severe. Even when presentation is mild, encephalitis is a potentially serious disease with neurological manifestations that can persist after the infection has resolved.

Inflammation of the brain may lead to cerebral edema (swelling and excessive fluid retention in the brain) that produces an increase in intracranial pressure. The patient will experience fever, headache, fatigue, nausea, vomiting, and neurological changes. Neurological changes can be as benign as **paresthesia** (numbness, tingling, pinprick sensation), **hyperesthesia** (extreme sensitivity), and **areflexia** (absence of reflexes), as suffered by our rabies victim at the beginning of this chapter. Changes in mental status, memory, and temper are also common, as are confusion, disorientation, and irritability. The symptoms of encephalitis can progress to severe deficits that may include muscle weakness, an unsteady gait, paralysis, and an inability to breathe.

Diagnosis of CNS infections often requires examination of CSF for chemical and biochemical composition and the presence of infective agents, blood, or other cells (Table 24.1). CSF from a healthy person is usually clear, containing no cells. **Pleocytosis** (increased number of white blood cells in CSF) is a common feature of almost any meningitis. The type of white blood cells found may offer a clue as to cause. Finding polymorphonuclear leukocytes (PMNs) in the CSF suggests a bacterial infection, whereas the presence of lymphocytes typically reflects a viral infection. CSF is best obtained <u>before</u> initiating antibacterial treatment so as to avoid killing bacterial pathogens before the laboratory can grow them. As noted earlier (see Figure 24.4), a spinal tap (lumbar puncture) involves inserting a needle between the third and fourth lumbar vertebrae into the subarachnoid space and collecting samples of the cerebrospinal fluid. Usually four sequential 1-ml sterile samples are obtained and sent to the laboratory to determine glucose and protein concentration, white and red blood cell count with differential (identifying different types of cells), Gram stain, and cultures as indicated. (We discuss specimen collection and processing in Section 25.2.)

Treatment of CNS viral infections is largely supportive and aimed at reducing the inflammation of the brain and intracranial pressure. Some antiviral agents, such as ribavirin, are used to treat selected viral encephalitides on an investigative basis, but no medications are currently approved for treatment of viral cases.

Figure 24.5   A Computed Tomography (CT) Scan of a Brain Displaying Encephalitis

In computed tomography, computers are used to generate a three-dimensional image of an object from a large series of two-dimensional X-rays taken around a single axis of rotation.

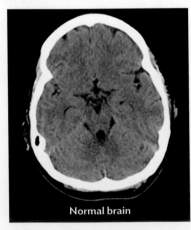

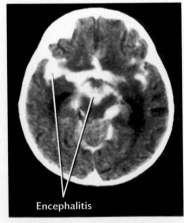

A. CT scan of a normal brain.

B. CT scan showing an accumulation of contrast dye around areas damaged by encephalitis.

Table 24.1
Typical CSF Specimen Values

CSF Specimen	Number of WBC/µl	Glucose, mg/dl	Protein, mg/dl
Normal[a]	0–5	40–70	15–50
Viral infection	<100, mostly lymphocytes	Usually normal	Usually normal or mildly elevated
Bacterial infection	100–>1,000, mostly PMNs	Usually decreased	Mildly or markedly elevated
Fungal infection	Varies	Low	Usually mildly elevated

[a]Normal CSF is colorless and clear, with ≤5 white blood cells, primarily lymphocytes, some monocytes, and no neutrophils.

In contrast, bacterial, fungal, and parasitic infections of the central nervous system are aggressively treated with antimicrobials in combination with supportive measures. Permanent damage or death remains a significant threat with these latter organisms, even with quick diagnosis and immediate treatment. As always, an ounce of prevention is worth a pound of cure.

Vaccination is the best means of prevention. Unfortunately, vaccines are available for only a minority of viruses and bacteria that affect the CNS; and no vaccines exist for parasitic or fungal culprits (as you will see in Tables 24.2 and 24.3). So, preventive measures largely aim to avoid contact between us and potential infectious agents. For example, reducing insect populations by fumigation, venturing outdoors only at certain times to minimize insect or arthropod exposures, and reducing contact with infected persons and animals help reduce incidence of infection. State and local health departments, in cooperation with the CDC, help monitor and publicize outbreaks, helping enormously both in reducing rates of infection and in ensuring that the public knows when to seek medical treatment.

SECTION SUMMARY

- **CNS infections** generally start with flu-like symptoms, including headache, dizziness, and fatigue.
- **Microbial routes to the CNS** can be direct (for example, following insect vector bites) or indirect (via dissemination from a distal infected site).
- **Analysis of the composition of CSF** can aid in diagnosing whether a disease is of bacterial, viral, or other origin. Pleocytosis (increased number of white blood cells in CSF) is a sign of meningitis.
- **The three types of CNS infection** are meningitis (inflammation of meninges), encephalitis (inflammation of the brain), and myelitis (inflammation of the spinal cord).
- **Viruses cause most CNS infections,** especially encephalitides, but generally produce mild disease.
- **Treatment of viral CSF infections** is largely supportive. Bacterial CSF infections can be treated with antibiotics.

- **Vaccines** are available for only a limited number of organisms that cause CNS infections.

Thought Question 24.1   What will cause a CSF sample to contain a large number of red blood cells?

24.3
Viral Infections of the Central Nervous System

SECTION OBJECTIVES
- Classify viral CNS infections by their mode of transmission.
- Explain the role of reservoirs in pathogenesis and prevention of viral diseases of the central nervous system.
- Outline clinical diagnoses of viral CNS infections according to the patient's history, signs, and symptoms.
- Recommend treatment and prevention options for different viral CNS infections.

What are the different kinds of viral CNS infections? Meningitis and encephalitis are the two major categories of viral CNS infections. CNS infections caused by viruses can be extremely dangerous. Fortunately, viral infections of the CNS are uncommon and typically resolve favorably in patients with a healthy immune system. Most viruses that attack the CNS tend to infect the meninges and cause what is called **aseptic meningitis** because no bacteria are present. Meningitis is, in fact, one of the presenting symptoms in approximately 30% of initial herpes simplex virus type 2 infections. But some viruses that attack CNS prefer the brain tissue instead of meninges and produce symptoms of encephalitis. Herpes simplex virus type 1 (HSV-1), for example, is responsible for most fatal cases of viral encephalitis. Herpes viruses are also associated with a strange neurological manifestation called Bell's palsy (see **inSight**).

Although viral infections of the meninges and brain tissue are more common, viruses can also attack the myelin sheath of myelinated nerves. Damage to the myelin sheath is irreversible. Table 24.2 summarizes key features of many of the viruses that affect the CNS; some of the more common viral CNS diseases are discussed here. You have already heard about the most famous one: rabies.

Viral Encephalitis

Viral encephalitis is categorized as either primary or secondary. Among the causes of secondary encephalitis are viral meningitis caused by HSV-1, reactivation of latent human herpesviruses (except HHV8, which causes Kaposi's sarcoma), postviral infection autoimmune reactions, and childhood diseases such as measles and rubella (which are now rare, thanks to vaccination). The deadliest of all infectious diseases, rabies, is a primary viral encephalitis.

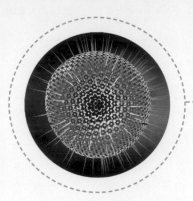

inSight

Why Does Only Half of My Forehead Wrinkle?

Figure 1   Bell's Palsy

The muscles on the left side of this patient's face do not receive the commands sent from his brain—a result of Bell's palsy.

A. The patient cannot raise his left eyebrow.

B. The patient cannot raise the left corner of his mouth.

C. The patient cannot close his left eye, and the left corner of his mouth is drooping.

How strange must it feel to all of a sudden start losing control of half of your face? You try to smile, but only half of your face will show it. You are now frowning, because you are mad at your face not taking orders from your brain, but you notice that only half of your forehead will wrinkle. Is this a bad dream? No, worse: it is reality and it is called Bell's palsy, a condition that affects nearly 40,000 people every year in this country.

What is Bell's palsy? Bell's palsy is weakness or paralysis of the muscles of one side of the face (Figure 1). Some viral infections can result in inflammation of cranial nerve VII (CN VII), the facial nerve. The inflammation damages the nerve so that it no longer can relay messages from the brain to the muscles of the face. So when the affected person wants to smile, only one side of the face (the side with a functioning CN VII) will smile; the other side remains expressionless.

The cause of Bell's palsy is often not known, and the disorder does not have a favorite race or sex. Viral infections are not the only cause of facial nerve palsy. Other causes include autoimmune diseases, trauma, and congenital abnormalities. Among viruses, the one most strongly associated with Bell's palsy is herpes simplex virus type 1 (HSV-1), the same virus that causes the cold sore (see Figure 12.3 and Section 19.2). Other members of the herpes family of viruses, such as herpes zoster virus, cytomegalovirus, and Epstein-Barr virus, have also been implicated.

Patients feel like their face is pulled to one side and may drool. Those affected may be sensitive to sound, lose their sense of taste, have difficulty drinking or swallowing, have food fall out of their mouth, and have difficulty closing one of their eyes. The onset of symptoms is sudden and usually takes no more than 2–3 days to manifest. The recovery is usually rapid but can take as long as weeks or months.

No specific diagnostic tests exist for Bell's palsy, although tests are often run to exclude other causes of facial paralysis (for example, tumors, multiple sclerosis, other brain lesions, or Lyme disease). No specific treatments for this disorder are available, but eye patches and/or eye lubricants (eyedrops or eye ointments) are used to prevent eye dryness, corneal lesions, and corneal scarring.

So what should you do if half of your face suddenly starts feeling numb? You should immediately seek medical care to make sure it is not something a lot more serious, and a lot less exotic, than Bell's palsy.

Table 24.2

Viral Infections of the Central Nervous System

Virus	Incubation Period	Meningitis[a]	Encephalitis[a]	Transmission	Diagnosis	Treatment	Prevention
Rabies virus	2 weeks–12 months	Rare	+	Mammalian bite or contact	Isolation of virus from clinical samples; serology	Therapy using postexposure prophylactic rabies vaccine	Avoiding infected mammals; the available vaccine is used for postexposure treatment only
Western equine encephalitis virus (WEEV)	5–10 days	+	+	Mosquitoes	IgM-based serology including ELISA; demonstration of virus or viral antigen	Supportive care; management of symptoms	Avoiding mosquitoes; no vaccine for human use
Eastern equine encephalitis virus (EEEV)	5–10 days	Rare	+	Mosquitoes	Clinical presentation; serology; viral culture; detection of viral antigen in brain postmortem	Supportive care; management of symptoms	Avoiding mosquitoes; no vaccine available
Venezuelan equine encephalitis virus	1–5 days		+	Mosquitoes	Clinical presentation; isolation of virus, from blood or throat swab, in culture	Supportive care; management of symptoms	Avoiding mosquitoes; human and horse vaccines available but human vaccine not available in the U.S.
West Nile virus (WNV)	1–5 days	Uncommon	+	Mosquitoes	Clinical presentation; serology; nucleic acid amplifications; tests; viral isolation from blood or CSF	Supportive care; interferon-alpha 2b and immune globulin IV	Avoiding mosquitoes; no vaccine available
Saint Louis encephalitis virus	5–15 days	+	+	Mosquitoes	Clinical presentation: difficulty urinating (dysuria), tremors in the elderly; leukocytosis with left shift; serology	Supportive care	Avoiding mosquitoes; no vaccine available
Japanese encephalitis virus	5–15 days	Rare	+	Mosquitoes	Detection of viral-specific antibody in serum or CSF by ELISA; viral isolation	Supportive care	Avoiding mosquitoes; vaccine available for travelers to Asia
Murray Valley encephalitis virus	1–4 weeks		+	Mosquitoes	Isolation of virus or viral RNA in clinical specimen serology	Supportive care	Avoiding mosquitoes; no vaccine available
Dengue virus	2–7 days		+	Mosquitoes	Serology; RT-PCR; viral isolation during acute phase of hemorrhagic fever	Supportive care; blood transfusion as needed	Avoiding mosquitoes; vaccine available in some countries (limited effectiveness)
Powassan virus	7–30 days		+	Ticks	Clinical presentation; serology CSF analyses	Supportive care	Avoiding ticks; no vaccine available

(continued on next page)

Table 24.2

Viral Infections of the Central Nervous System (*continued*)

Virus	Incubation Period	Meningitis[a]	Encephalitis[a]	Transmission	Diagnosis	Treatment	Prevention
LaCrosse agent of California encephalitis: California group of viruses	5–15 days		+	Mosquitoes	IgM-capture assays of serum and CSF; CBC with differential shows WBC count of ≥20,000 with a left shift; CSF analyses	Supportive care	Avoiding mosquitoes; no vaccine available
Colorado tick fever virus	7–10 days		+	Ticks	Serology; isolation of virus from blood or CSF	Supportive care	Avoiding ticks; no vaccine available
Coxsackievirus, echovirus	3–10 days	+	Rare	Person to person	Clinical presentation: conjunctivitis and a rash; viral isolation from blood, CSF, stool, or respiratory secretions; PCR; serology	Supportive care	No vaccine available
Human immuno-deficiency virus (HIV)	5–70 days for those with acute HIV infection	+	Meningo-encephalitis	Person to person	Extensive guidelines for diagnosis recommended by CDC, WHO, and Infectious Disease Society of America; testing criteria different for patients <18 months of age; ELISA, immunofluorescence assay, nucleic acid (RNA and DNA) tests, HIV p24 antigen test, viral cultures	Antiretroviral therapy based primarily on CD4+ cell count and monitoring viral load; prophylactic treatment to prevent opportunistic infections; other treatments as needed	No vaccine available
Herpes simplex virus type 1 (HSV-1)	Variable, subsequent to primary lesion	Rare	+	Not applicable	HSV lesions on skin or mucous membranes; detection of HSV-1 DNA by PCR	Supportive care in ICU (intensive-care unit); acyclovir IV	No vaccine available
Herpes simplex virus type 2 (HSV-2)	2–4 weeks after delivery	+		From mother to fetus or from mother to newborn during birth	Clinical presentation; PCR of CSF sample; PCR of blood or skin lesions as supportive evidence	Supportive care in ICU; acyclovir IV	No vaccine available
Lymphocytic choriomeningitis virus (LCMV)	1–3 weeks	+		Exposure to rodent secretions or excretions	Viral cultures; serology; ELISA available through CDC; PCR amplification of viral RNA (not commercially available)	Supportive care; anti-inflammatory medications	Rodent control; no vaccine available
Poliovirus	3–35 days	+	+	Fecal-oral; direct person-to-person contact; contact with infected phlegm or mucus	Isolation of virus from throat washings, spinal fluid, or cultures; PCR; serology	Supportive care	Vaccine available

[a]A + sign designates that the virus causes the specified disease.

RABIES Rabies is a zoonotic disease causing some 30,000–70,000 human fatalities annually worldwide. The agent is an enveloped, single-stranded, negative-sense RNA virus and a member of the genus *Lyssavirus* in the Rhabdoviridae family (see Table 12.1). Viruses in this family have an elongated bullet shape (Figure 24.6) and include two famous members, rabies virus and vesicular stomatitis virus (VSV). VSV is a major pathogen of cattle that can be transferred to humans. VSV infection in humans produces flu-like symptoms and sometimes oral lesions that resemble cold sores, but the infection usually resolves by itself—a stark contrast to rabies.

Rabies virus can infect all mammals, but the endemic reservoir species varies with geographic locale. The highest incidence of human rabies infections is found in rural areas of Asia and Africa, mostly among children. Dogs are the main reservoir on these continents, with dog bites the main route of transmission. In contrast, canine infections in the United States are almost nonexistent because of vaccination and animal control programs. Despite those efforts, 35 confirmed cases of rabies have occurred in the United States between 2000 and 2014. Twenty-five percent of those were linked to travel abroad. Because rabies is endemic in certain animal populations outside the United States, rabies should be included in the differential diagnosis of any patient who presents with symptoms of encephalitis and who has recently traveled abroad.

 Tracking outbreaks of infectious disease, including rabies: healthmap.org

Rabies virus is a **neurotropic virus**, meaning it prefers to replicate in nerves. The virus spreads through the body of an infected animal or person and then into the salivary glands in the later stages of the disease. When the infected animal bites someone, the virus is transferred with saliva into the bite wound. The virus replicates in the wound and infects the local nerve and eventually travels up that nerve to the CNS. Although usually transmitted by animal bite, humans can also acquire the infection through small cuts or abrasions or mucous membranes that come in contact with an infected animal or their saliva. Contact or even bites from infected animals can be so minor that they don't concern the person. The wound heals and the incident is often forgotten, as happened with the migrant worker in the case history opening this chapter. Appearance of symptoms finally drives the patient to seek medical attention.

Initial symptoms of the disease during the prodromal stage are very nonspecific. The area near the wound or contact site is usually

Figure 24.6 Rabies

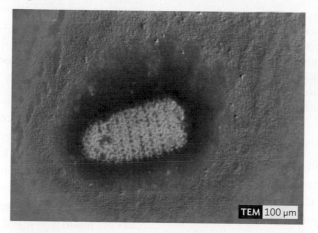

A. Rabies virus has an elongated shape frequently referred to as "bullet shaped."

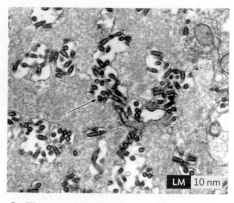

B. The virus (arrow), shown here in great numbers in hamster brain, initially replicates in the bite wound and subsequently moves along the nerve to the brain. Accumulations of large numbers of the virus in intracellular inclusions are called Negri bodies.

(50%–80% of the time) pruritic (itchy), painful, and hypersensitive to any stimulus (hyperesthesia). The patient will feel tired and have a fever and a headache—all typical signs and symptoms of a viral infection. Some may have nausea and vomiting, and some may feel and act anxious, even agitated. The wound will heal. After an incubation period that varies, depending on the virus variant (generally 2–3 months), neurological symptoms will ensue and force the patient to seek medical care.

Rabies appears in three basic forms: encephalitic, paralytic, and an atypical form usually associated with bat bites. Most rabies cases (80%) are encephalitic, or "furious," the form typically portrayed in motion pictures, in which the affected animal or person fears water (hydrophobia), is aggressive, and may bite. In paralytic rabies, the patient exhibits a weakness in all four limbs (quadriparesis) that can be misdiagnosed as Guillain-Barré syndrome (see Section 22.5).

A few rabies victims present atypically with neuropathic pain (pain due to nerve damage, like the shoulder pain in the chapter-opening case), sensory or motor nerve manifestations (our patient had numbness in his left hand), abnormal movements of the bitten limb, and eventual signs of brain stem damage, such as difficulty breathing. Regardless of initial presentation, rabies will progress to coma and finally death.

The rabies virus has several variants that can be identified (as in our opening case history) by using antigenic typing or nucleic acid techniques. Identifying the variant is particularly important, since that knowledge allows public health workers to identify the infected animal population. With this strategy, most U.S. cases have been linked to contact with silver-haired bats (Figure 24.7). Bites from these bats are also a problem in Mexico and Canada. Foxes were once a major reservoir in Canada, but the distribution of rabies vaccine in bait significantly reduced transmission in that wild population.

Figure 24.7   Silver-Haired Bats

The silver color of the fur on their backs has earned *Lasionycteris noctivagans* the alias "silver-haired bats." Silver-haired bats are small mammals and the main transmitters of rabies virus to humans in the United States.

Rabies is often misdiagnosed, probably owing to the rarity of cases. In 2004, misdiagnosed rabies among organ donors in the United States led to three fatal rabies infections in organ recipients. The most common transplant-related rabies infections have been traced to the corneas of donors who died from encephalitis. More stringent screening procedures are now in place to ensure that a donated organ is free of rabies virus. Unfortunately, no matter what the virus variant or length of the incubation, the appearance of symptoms is ominous because the disease is almost always fatal.

Treatment of rabies includes administering rabies immune globulin and rabies vaccine. Unfortunately, rabies treatment is successful only if given soon after the exposure and before appearance of the symptoms. Inquiring about animal bites or contacts is therefore very important in patients who complain about sudden onset of neurological symptoms indicating CNS infection. Health care providers use an algorithm to determine whether a bite victim is at risk of rabies and needs treatment.

MOSQUITO-BORNE ENCEPHALITIS   Many of the viruses that cause encephalitis are transmitted to humans by arthropod vectors—that is, mosquitoes and ticks. We discuss the two best recognized here and summarize the others in Table 24.2.

CASE HISTORY 24.1

Death on Golden Pond

In August, Mr. C brought his 21-year-old son, Rich, to a New Jersey emergency department. Rich appeared dazed and had trouble responding to simple commands. When questioned about his son's activities over the past few months, Mr. C told the physician that Rich planned to enter veterinary school in the fall and had spent the month of July relaxing and sunning himself on New Jersey beaches. Aside from the shore, his favorite locale was a pond in a wooded area near a horse farm. On the afternoon before admission, Rich became lethargic and tired. He returned home and went to bed. That evening, his father woke him for supper, but Rich was confused and had no appetite. By 11 p.m., Rich had a fever of 40.7°C (105.3°F) and could not respond to questions. A few hours later, when his father had trouble rousing him, he brought Rich to the ER. Over the next week, Rich's condition worsened to the point where his limbs were paralyzed. Two weeks later, he died.

Serum samples taken upon entering the hospital and a few days before he died showed a sixfold rise in antibody titer to eastern equine encephalitis (EEE) virus. Brain autopsy showed many small foci of necrosis in both the gray and white matter. The serology and autopsy results confirmed the diagnosis of eastern equine encephalitis virus infection as the cause of death.

Eastern equine encephalitis (EEE) virus, a member of the family Togaviridae (positive-sense, single-stranded RNA genome), is transmitted from bird to bird by mosquitoes. Horses and humans contract the disease the same way (**Figure 24.8**). Viruses transmitted by insect bites are generally referred to as **arboviruses**. Rich contracted it inadvertently while visiting the pond. The disease, an encephalitis, is often fatal (35% mortality) but fortunately rare. One reason human EEE disease is rare is that the species of mosquito that usually transmits the virus between marsh birds does not prey on humans. But sometimes a human-specific mosquito bites an infected bird and then transmits EEE virus to humans. Another reason human disease is rare is that the virus generally does not fare well in the human body. Many persons infected with EEE virus have no apparent illness because the immune system thwarts viral replication. However, as already noted, mortality is high in those individuals who do develop disease.

An interesting point in Case History 24.1 is that diagnosis relied on detecting an increase in antibody titer to the virus. Typically, the body has not had time to generate large amounts of specific antibodies when disease symptoms first appear. After a week or so, often when the patient is nearing recovery (convalescence), antibody titers have risen manyfold. The general rule is that a greater than fourfold rise in antibody titer between acute disease and convalescence (or in this case death) indicates that the patient has had the disease. Although this knowledge could not help the patient in Case History 24.1, it was valuable for public health and prevention strategies.

Link As discussed in Section 17.5, the amount (or **titer**) of a specific antibody in serum is determined by how far serum can be diluted and still produce a visible antigen-antibody reaction. Serological tests, such as ELISA, discussed in Section 25.5, are used to determine titer.

Eastern equine encephalitis virus is one of several viruses transmitted by mosquitoes. Other members of Togaviridae and viruses from two other families of the arboviruses (flaviviruses and bunyaviruses) also cause mosquito-borne primary encephalitis (see Table 24.2). West Nile virus holds an equal place of prominence and causes as much concern as EEE.

WEST NILE ENCEPHALITIS   Five major diseases are caused by flaviviruses (positive-sense, single-stranded RNA), all transmitted by mosquitoes. **West Nile virus (WNV)** and three other flaviviruses cause encephalitis or meningitis. The fifth virus, the dengue fever virus (DFV), which has four serotypes (DENV 1–4), causes dengue

Figure 24.8
Eastern Equine Encephalitis Virus

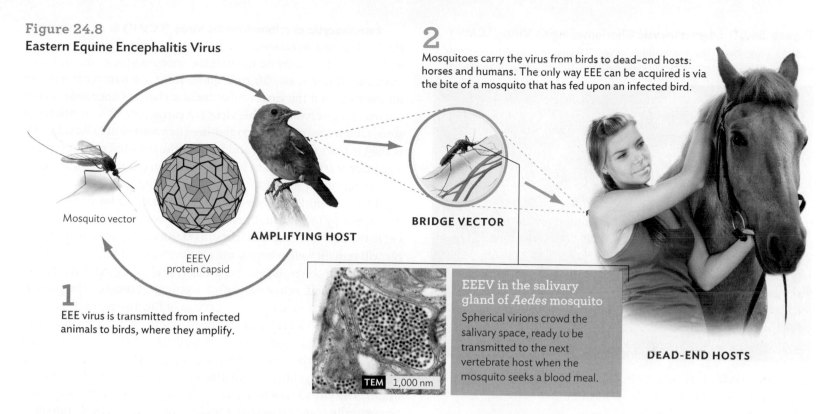

1 EEE virus is transmitted from infected animals to birds, where they amplify.

Mosquito vector

EEEV protein capsid

AMPLIFYING HOST

2 Mosquitoes carry the virus from birds to dead-end hosts: horses and humans. The only way EEE can be acquired is via the bite of a mosquito that has fed upon an infected bird.

BRIDGE VECTOR

EEEV in the salivary gland of *Aedes* mosquito
Spherical virions crowd the salivary space, ready to be transmitted to the next vertebrate host when the mosquito seeks a blood meal.

TEM 1,000 nm

DEAD-END HOSTS

fever (DF). Any one of the DENV serotypes can lead to dengue hemorrhagic fever (DHF) or dengue shock syndrome (DSS). Exposure to any one serotype fails to confer immunity to the others (see Section 21.2).

West Nile virus (**Figure 24.9**), also endemic in bird populations and transmitted by mosquito, was first discovered in the West Nile district of Uganda in 1937 and remained mainly in Eastern Europe, Asia, West Africa, and the Middle East. A European outbreak of West Nile encephalitis in 1962 was followed nearly 40 years later by its arrival and outbreak in New York. The virus subsequently spread through the country. Now only a few states can boast being West Nile–free. The strain of West Nile that caused the New York epidemic is almost identical to one found in a goose farm in Israel in 1968, making travel a likely mode of introduction of the virus into the United States.

Most people infected with the virus do not develop disease, thanks to their immune system. Most of those who do develop disease have a flu-like illness, but some develop encephalitis. Rigorous surveillance and insect control programs have reduced cases of WNV to fewer than 2,085 in 2014.

A virus related to WNV, Zika virus, is transmitted from monkeys (the Zika reservoir) to humans by an *Anopheles* mosquito. Zika usually produces mild disease in humans unless it is transmitted from a pregnant mother to her fetus. Zika infection of the fetus causes abnormal development of the brain (microencephaly or small head).

Viral Meningitis

Viral meningitis is normally not as severe as bacterial meningitis and often tends to be a mild, self-limiting disease in adults

with healthy immune systems. The patient with viral meningitis has a headache and a fever, which may be accompanied by nausea and vomiting, photophobia, muscle ache, and malaise. The symptoms of viral meningitis are largely the same as those of meningitis caused by other etiologies and generally persist for 7–10 days before they resolve. Most cases of viral meningitis (80%–85%) are caused by enteroviruses (positive-sense, ssRNA viruses, members of the Picornaviridae family). Poliovirus is a prominent enterovirus that infects the CNS and causes paralysis but not meningitis. Non-polio enteroviruses, particularly echoviruses and coxsackieviruses, are the major causes of viral meningitis in both children and adults.

Figure 24.9   Molecular Model of WNV

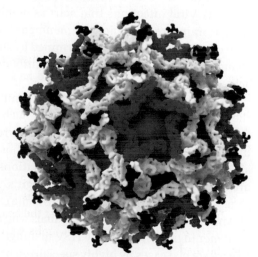

The genetic material of WNV (RNA) is surrounded by a nucleocapsid (smooth red area), which is enclosed in an outer protein envelope containing attachment proteins (multicolored).

Figure 24.10   Lymphocytic Choriomeningitis Virus (LCMV)
LCMV is a member of the Arenaviridae family. The glycoprotein protuberances on the enveloped surface of the virus are used to attach to and enter target cells.

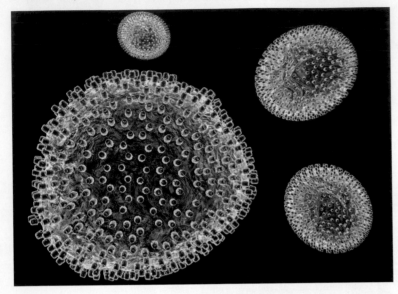

Arboviruses (insect-borne viruses) that cause encephalitis, such as EEE and West Nile, commonly cause meningitis as well. Many other viruses, such as influenza, HSV-2 (see Table 12.1), and the mumps and measles viruses, can also cause meningitis (see Table 24.2). Childhood vaccinations have all but eliminated meningitis cases caused by viruses responsible for mumps and measles.

A combination of patient history, epidemiology, and diagnostic information may point to a viral cause as opposed to a bacterial (or other) cause. However, because bacterial meningitis is a severe disease with a bad prognosis, any suspected case of meningitis is initially treated as bacterial and antimicrobials are administered empirically.

Meningitis is classified as aseptic if CSF and blood cultures are negative for bacteria. As with encephalitis, most cases of meningitis are viral. Viral infections that start as meningitis and then spread to the brain are classified as meningoencephalitis. The specific viral etiologic agent can be determined by subjecting CSF, blood, or sometimes tissues to viral culture, antibody detection, or nucleic acid–based methods.

Treatment for viral meningitis is limited mostly to providing the patient with symptomatic relief, such as reducing pain and fever, and offering supportive care, such as rest in a quiet room with dim lighting and intravenous fluids for those with prolonged emesis (vomiting). Several antiviral medications are available, but they are not indicated to treat viral meningitis. Acyclovir is the exception. Acyclovir is used empirically in some cases involving neonates and immunocompromised individuals who are suspected of having viral meningitis, particularly HSV meningitis, but its use is limited to very special cases by highly specialized physicians.

Lymphocytic choriomeningitis virus (LCMV) is an enveloped, single-stranded, negative-sense RNA virus of the Arenaviridae family that can cause aseptic meningitis, encephalitis, or meningoencephalitis (Figure 24.10). Household mice are natural reservoirs for the virus, but the virus causes them no harm. Other rodents are not natural reservoirs for this virus but can acquire it from mice. Pet store rodents have also been identified as a source of LCMV. LCMV is shed in the saliva, urine, and feces of the infected rodent for life. The virus is transmitted to humans through inhalation of aerosolized secretions or through direct contact with them.

LCMV meningitis is distinguished from other forms of aseptic meningitis by the unusually elevated number of lymphocytes present in the CSF of the infected person (>1,000/μl). Some infected people will remain well, whereas others develop a flu-like disease with fever, muscle aches, headache, nausea, and vomiting; the disease resolves by itself. A few days after seeming to recover, the patient will suffer a second, more severe phase of the disease that includes fever, a bad headache, and a stiff neck. Acute encephalitis may also be part of the secondary phase of the disease. LCMV patients with acute encephalitis present with sensory or motor deficits, including paralysis, myelitis, and hydrocephalus (increased fluid on the brain). The virus is not transmitted from person to person but can be vertically transmitted from mother to fetus. Vertical transmission in humans, an increased incidence in urban areas, and the severe second-phase symptoms have gained this virus prominence in recent years. Patients affected by LCMV are given supportive therapy, mainly analgesics, because currently no effective antiviral treatments are available.

Poliomyelitis

Poliovirus is a very contagious enterovirus of the Picornaviridae family that easily spreads either by the fecal-oral route or directly from person to person via saliva exchange. Once inside a new host, the virus multiplies in the throat or gastrointestinal tract and soon is shed in the saliva or stool of the infected person. Most of those infected remain asymptomatic but still shed the virus and transmit it to others. In a few cases, some of the virions produced in the intestine get into the circulation, eventually reach the CNS, and cause one of three types of **poliomyelitis** (polio), whose symptoms range in severity. The patient can exhibit mild disease with nonspecific symptoms such as headache and malaise; a nonparalytic form of the disease; or the more serious paralytic polio. Less than 1% of clinical cases of polio are paralytic. The incubation period for any of these diseases can be as short as 3 days or as long as 35 days, but symptoms usually appear after 1–3 weeks.

Patients with mild polio syndrome will have the symptoms typical of most viral infections—fever, malaise, headache, and muscle ache. Nonparalytic disease starts with a prodromal phase with symptoms resembling those of the mild form of disease, which then progresses to aseptic meningitis. Nonparalytic disease lasts up to 10 days before it resolves.

Patients with the paralytic form of polio suffer from unilateral muscle weakness and flaccid paralysis that reaches its full presentation in about 3 days. Paralytic polio can affect almost any part of the CNS, including the cranial nerves, and cause severe disability or even death. Among the critical nerves that can be affected is the phrenic nerve, which controls the movements of the diaphragm, the muscle that separates the chest from the abdomen and plays a major role in ventilation. Damage to the phrenic nerve will interfere with breathing. During the peak years of polio infections (1940s and 1950s), a negative-pressure ventilator was used to help patients breathe. The machine was dubbed the "iron lung." Twenty-five iron lungs are still in use in the United States, helping victims of polio (Figure 24.11).

Thirty to forty years after recovery, victims of paralytic polio may experience new weakness or sudden worsening of existing weakness, along with increased muscle pain. These symptoms define a condition called post-polio paralytic disease. Post-polio syndrome is not an infection process, and people experiencing the syndrome do not shed poliovirus. The weakness can occasionally spread to muscle groups that were not originally affected. The appearance of the post-polio syndrome (PPS) is insidious, its progression is slow, and the prognosis depends on severity of the original disease. Patients who originally had a mild case of polio will probably have a mild case of PPS. Patients who suffered from a severe case of polio are at risk of a severe case of PPS.

Diagnosis of polio includes isolation of poliovirus from the throat, stool, or CSF. Finding the virus in a symptomatic patient does not mean the patient has polio. So, other causes of muscle weakness and acute paralysis, such as non-polio enteroviruses, Japanese encephalitis virus, and coxsackievirus A24, have to be ruled out. Observing virus-specific changes to infected cell cultures (cytopathic effect) is also very helpful.

Treatment for polio patients and those with post-polio syndrome includes rehabilitation and exercises to help prevent further atrophy of affected muscles. Pain is relieved using different pain relievers.

Fortunately, polio has been eradicated from the Western Hemisphere, thanks to vaccination and the efforts of the Pan American Health Organization. Two types of polio vaccines exist. Jonas Edward Salk developed the first polio vaccine, which contains inactivated forms of all three serotypes of poliovirus. The inactivated vaccine does not contain active virus. The **Salk vaccine** must be injected for immunity to develop. The second vaccine, developed by Albert Bruce Sabin, is a live, **attenuated** (diminished virulence) vaccine that contains mutated forms of all three serotypes of the poliovirus. The **Sabin vaccine** is given orally, which affords these mutated viruses a chance to briefly replicate in intestinal cells and generate a more natural immunity. Unfortunately, attenuated viruses can, at times, revert (change back through a mutation) to their active, virulent form. For the Sabin vaccine, the result is **vaccine-associated paralytic poliomyelitis (VAPP)**. Because of VAPP, the Advisory

Figure 24.11 The Iron Lung

Martha Mason, of Lattimore, North Carolina, became paralyzed after contracting polio at age 11. She passed away in 2009 after living in an iron lung for 60 years. Her friend's grandson is shown reading aloud to her during a visit.

Committee on Immunization Practices, in the year 2000, recommended exclusive use of the inactivated polio (Salk) vaccine in the United States. The last case of VAPP in the United States was seen in 1999.

Is polio still a significant disease? Visit the HealthMap.org site and do an advanced search for cases of polio worldwide over the past year. You will see that many cases still occur in Africa and the Indian subcontinent. The new cases shown for South America are vaccine associated. Vaccination, however, has dramatically reduced the annual numbers of diagnosed cases around the world from hundreds of thousands to around a thousand. This feat offers hope that one day we will eradicate this devastating disease.

SECTION SUMMARY

- **Viral encephalitis** can occur as a primary infection (rabies) or as a secondary consequence to an infection elsewhere (HSV-1).

- **Rabies virus** has a variety of animal reservoirs. Its pathogenesis starts with an animal bite, after which the virus infects nerves in the area and then moves along those nerves to the spinal cord and to the brain. In later stages, the virus travels to the salivary glands, where it can be transmitted with saliva by a bite.

- **Symptoms of rabies** are initially vague (headache, fever, nausea). Then 2–3 months later, neurological symptoms begin (hydrophobia, the fear of water; aerophobia, fear of fresh air; aggression; confusion) and may include quadriparesis. Disease progresses to coma and death. Disease can be prevented if rabies immune globulin and rabies vaccine are administered before symptoms begin.

- **Eastern equine encephalitis virus** is an arbovirus endemic to birds. It is transmitted between birds by mosquito vectors, which can also transmit the virus to humans. Disease in humans (encephalitis) is rare but serious.
- **West Nile virus** is also endemic to birds and transmitted by mosquito bite. The disease is similar to EEE but typically less severe.
- **Viral meningitis** (also called aseptic meningitis) is less severe than bacterial meningitis. Major causes are enteroviruses, arboviruses, HSV-2, lymphocytic choriomeningitis virus, as well as mumps and measles viruses. Treatment is supportive.
- **Poliomyelitis** is a very contagious disease caused by poliovirus, an enterovirus of the Picornaviridae family. The disease can manifest in either nonparalytic or paralytic forms; paralytic polio can affect almost any part of the CNS.

Thought Question 24.2 What steps could you take to limit your exposure to arboviruses?

24.4
Bacterial Infections of the Central Nervous System

SECTION OBJECTIVES
- Identify risk factors for each of the bacterial CNS diseases.
- Describe how different bacteria overcome host defenses to cause CNS disease.
- List patient history information and clinical signs and symptoms specific to each bacterial pathogen of the central nervous system.
- Relate pathogenesis of bacterial CNS diseases to prevention and treatment modalities used.

How do bacteria infect the central nervous system? CNS infections involving bacteria affect the meninges primarily, but from the meninges bacteria can spread to the brain to cause meningoencephalitis and brain abscesses (**Figure 24.12**). The initial symptoms of bacterial meningitis, headache and fever, are nonspecific, but the disease progresses quickly into a dangerous, life-threatening situation. Infection of the meninges and the resulting inflammation, for example, will interfere with the natural flow of CSF, leading to increased intracranial pressure and brain dysfunction (**encephalopathy**), despite the lack of a brain infection. The very young and the very old are particularly at risk of contracting bacterial meningitis. Effective treatments are available in most cases, but the diagnosis must be made rapidly. Unfortunately, the mortality rate remains significant even with timely diagnosis and treatment. **Table 24.3** lists various bacteria that can cause meningitis.

Note Encephalopathy (disease of the brain tissue) is not the same as encephalitis (inflammation of the brain).

Figure 24.12 **Magnetic Resonance Imaging of a Brain Abscess**

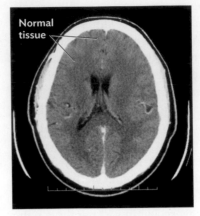
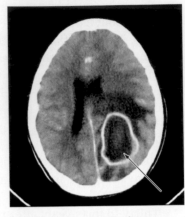

A. MRI of normal brain tissue.

B. MRI showing a brain abscess (arrow).

Bacterial Meningitis

Meningitis is considered a medical emergency. Hallmark symptoms include sudden onset of severe headache accompanied by neck pain, fever, and **nuchal rigidity** (stiffness of the neck resulting in an inability to touch chin to chest). Meningitis can be caused by bacteria, viruses, fungi, and parasites, but symptoms alone cannot determine the cause. Patient history can provide clues (for instance, was the patient exposed to a certain agent?), but laboratory results are needed for a definitive diagnosis.

Because bacteria cause the most serious forms of meningitis, any person presenting with the aforementioned symptoms is presumed to have bacterial meningitis until proven otherwise and is immediately treated with broad-spectrum antibiotic therapy (**empirical therapy**). If it is determined that the offending agent is not bacterial, antibiotic treatment can be stopped. The severity of bacterial meningitis is due in part to the fact that bacteria can cause other systemic, life-threatening problems in addition to inflammation of the meninges. *Neisseria meningitidis*, for example, can cause sepsis and disseminated intravascular coagulation. Complications are a major reason antimicrobial treatment of meningitis too often fails.

Three bacteria—*Streptococcus pneumoniae*, *Neisseria meningitidis*, and *Haemophilus influenzae* type b—have always received top billing as major etiologies of bacterial meningitis. *S. pneumoniae* is the number-one cause of meningitis in infants, children, and the elderly, whereas *N. meningitidis* holds the number-one spot for causing the disease in adolescents and young adults. *H. influenzae* type b may cause meningitis in young children. Aggressive vaccination programs have, however, significantly reduced the number of meningitis cases caused by these organisms (see Section 20.3). *Listeria monocytogenes*, which has no vaccine, is a relatively new arrival to the scene that continues to gain notoriety, thanks mostly to increasing numbers of immunocompromised patients who are generally more susceptible to infections by this and other organisms.

Table 24.3

Bacterial Infections of the Central Nervous System

Bacterial Species	Group at Risk	Transmission	Diagnosis	Treatment[a]	Prevention
Neisseria meningitidis	All ages[b]	From nasopharynx via bacteremia	Clinical presentation; isolation of *N. meningitidis* from CSF and/or blood; Gram stain of CSF; PCR of clinical samples is possible	Penicillin G; ampicillin when penicillin sensitive; ceftriaxone or cefotaxime when penicillin resistant	Vaccine available
Streptococcus pneumoniae	All ages	Extension from other foci in respiratory tract	Clinical presentation; CSF analyses; bacterial culture, Gram stain; PCR; CBC with differential	Penicillin G; ceftriaxone or cefotaxime when penicillin sensitive; ceftriaxone, cefotaxime, or vancomycin when penicillin resistant; vancomycin or levaquin when resistant to cephalosporins	Vaccine available
Streptococcus agalactiae (group B streptococci)	Neonates (infants <3 months of age)	Vertical transmission from mother; infection of mucous membranes causing sepsis and spreading to CNS	Clinical presentation; CSF analyses; blood culture	Multiple CDC treatment guidelines, including penicillin G, ampicillin, vancomycin, meropenem, cephalosporins; screening and antimicrobial treatment of pregnant women colonized with group B streptococci	No vaccine available
Listeria monocytogenes	Neonates and elderly	Vertical transmission through placenta; fecal-oral route	Clinical presentation; CSF analyses; CSF, blood, or amniotic fluid culture; Gram stain	Ampicillin or penicillin G; alternatively: trimethoprim-sulfamethoxazole; may add an aminoglycoside	No vaccine available
Escherichia coli (neonatal meningitis *E. coli*, NMEC)	Neonates and elderly	Various	Clinical presentation; CSF analyses; CSF and/or blood culture; Gram stain	Third-generation cephalosporins; alternatively: aztreonam, fluoroquinolone, meropenem, trimethoprim-sulfamethoxazole, ampicillin	No vaccine available
Pseudomonas aeruginosa	Hospitalized patients; the immunocompromised	Bacteremia	Clinical presentation; CSF analyses; CSF and/or blood culture; Gram stain	Cefepime or ceftazidime, aztreonam, ciprofloxacin, meropenem	No vaccine available
Coagulase-negative *Staphylococcus*	All ages	Bacteremia originating from a foreign body or wound	Clinical presentation; CSF analyses; CSF and/or blood culture; Gram stain	Methicillin susceptible: nafcilllin or oxacillin; methicillin resistant: vancomycin	No vaccine available
Staphylococcus aureus	All age groups	Direct and indirect contact and hematogenous spread	Clinical presentation; CSF and/or blood culture; Gram stain	Third-generation cephalosporins, daptomycin, linezolid; methicillin resistant: vancomycin	No vaccine available
Haemophilus influenzae	Unvaccinated adults and children	Extension from upper respiratory tract foci	Clinical presentation; CSF analyses; CSF and/or blood culture; Gram stain	Ampicillin, ceftriaxone, cefotaxime; third-generation cephalosporins if beta-lactamase-positive	Vaccine available
Mycobacterium tuberculosis	HIV infected	Disseminated TB	Clinical presentation; CSF analyses; CSF and/or blood culture; chest X-ray	Variable, per CDC guidelines	No vaccine available

(continued on next page)

Table 24.3

Bacterial Infections of the Central Nervous System (*continued*)

Bacterial Species	Group at Risk	Transmission	Diagnosis	Treatment[a]	Prevention
Treponema pallidum (syphilitic aseptic meningitis)	Those infected with syphilis	Vertical transmission causing congenital syphilis; complication of syphilis	Clinical presentation; CSF analyses; cerebral angiography[c]; VDRL or RPR blood test; FTA-ABS test to confirm syphilis	Variable, per CDC guidelines	No vaccine available
Borrellia burgdorferi[d]	All ages	Complication of Lyme disease	Clinical presentation; CSF analyses	IV antibiotic, choice and dosage dependent on patient's age; ceftriaxone, cefotaxime and penicillin G usually used	No vaccine available
Bacteroides fragilis	The immuno-compromised	Bacteremia	Clinical presentation; CSF analyses	Metronidazole	No vaccine available
Fusobacterium spp.	The immuno-compromised	Bacteremia	Clinical presentation; CSF analyses	Metronidazole	No vaccine available

[a]Suspected cases of meningitis are treated empirically until etiology is determined. Empirical antibiotic therapy usually includes a combination of a ceftriaxone and cefotaxime + vancomycin + doxycycline (for bacterial tick-borne etiology during tick season) and acyclovir.
[b]Exposure to infected individuals requires prophylactic antibiotic treatment.
[c]Angiography is a procedure in which a dye is injected into the blood to study blood and/or lymph vessels by X-ray.
[d]Autoimmune disease; not an actual infection of the CNS.

As noted earlier, patient history—particularly travel and exposure history—as well as findings on the physical exam can help suggest causes of bacterial meningitis, but a definitive diagnosis requires laboratory analysis of CSF samples. CSF from cases of bacterial meningitis will contain increased numbers of white cells (PMNs), an increase in protein content, and a decrease in glucose concentration (all of which argue against viral causes). Definitive diagnosis is based on identifying the bacterial agent from CSF and/or blood. Because CSF is drawn from a sterile body site, a quick, direct Gram stain of CSF material can be extremely informative. For instance, finding a Gram-positive diplococcus in CSF strongly suggests infection by *S. pneumoniae*. Gram-negative cocci or Gram-negative rods would suggest *N. meningitidis* or *H. influenzae*, respectively. Even though the results are quick and strongly indicate cause, Gram staining must be followed by culture and identification (see Section 25.2).

CASE HISTORY 24.2

The Purple Rash

In April 2001, Abdul, a 4-month-old infant from Saudi Arabia, was hospitalized with fever, tender neck, and purplish, nonblanching spots (petechiae and purpuric spots) on his trunk (**Figure 24.13A**). Suspecting meningitis, the clinician took a CSF sample and examined it by Gram stain. The smear revealed Gram-negative diplococci inside PMNs. The CSF was turbid with 900 leukocytes/μl, and *Neisseria meningitidis* was confirmed by culture. The child was treated with cefotaxime and made a full recovery. His father, the person who brought him in, was clinically well. However,

meningococcus was isolated from his oropharynx, as well as from the throat of the patient's 2-year-old brother. Isolates from the patient, his father, and his brother were positive by agglutination with meningococcus A, C, Y, W135 polyvalent reagent. The father's vaccination certificate confirmed that he had received a quadrivalent meningococcal vaccine. All three isolates were sent to the WHO (World Health Organization) Collaborating Center, which confirmed meningococcus serogroup W135. DNA analysis of the three isolates found them to be indistinguishable, meaning that the father and his children were infected with the same strain of *N. meningitidis*. Why did Abdul's brother not have meningitis? And why was Abdul's vaccinated father colonized?

MENINGOCOCCAL MENINGITIS　The meningococcus, *Neisseria meningitidis* (**Figure 24.13B**), is a Gram-negative diplococcus that can colonize the human oropharynx (throat), where it causes mild, if any, disease. At any given time, 10%–20% of the healthy population can be colonized and asymptomatic, like Abdul's 2-year-old brother. The organism spreads directly by person-to-person contact or indirectly via fomites or droplet nuclei from sneezing. In the throat, *N. meningitidis* uses type IV pili to adhere to and then invade epithelial cells. The organism moves from one side of the host cell to the other (**transcytosis**), exits the epithelial cell, and enters the bloodstream, where the real problem arises. The purple spots (**petechiae** and purpura) on Abdul's skin were the result of dissemination of *N. meningitidis*, whose **endotoxin** causes blood to leak out of the capillaries. This type of rash does not lose color when pressed (it is nonblanching).

Link As discussed in Section 18.3, **endotoxin**, a part of the lipopolysaccharide (LPS) of Gram-negative bacteria, triggers the release of cytokines that increase capillary permeability. LPS also causes depletion of clotting factors. The combination causes leakage of blood from capillaries, which if under the skin produces pinpoint spots called **petechiae**. **Petechiae** may enlarge into purple patches called **purpura**.

Unlike the gonococcus (*Neisseria gonorrhoeae*; see Section 23.4), *N. meningitidis* is very resistant to complement owing to its production of a polysaccharide capsule. Complement, part of our innate immunity, puts a hole in the bacterial membrane so that it is destroyed (see Section 15.7). The capsule prevents the complement from accessing the bacterial membrane, which allows the microbe to produce a transient blood infection (bacteremia) and reach the blood-brain barrier.

Sinus and ear infections with *N. meningitidis* can extend directly to the meninges, whereas septicemic spread requires passage through the blood-brain barrier. Once confronted with the blood-brain barrier, *N. meningitidis* again uses type IV pili to adhere to endothelial cells but does not enter the host cell. Instead, the interaction loosens the tight junctions between cells and disrupts the blood-brain barrier. The organism then squeezes through these spaces (like a child squeezing through the hole in a fence) and encounters the meninges. Once in the CSF, microbes can multiply almost at will. **Figures 24.13C** and **D** show the remarkable damage that *N. meningitidis* and other microbes, such as *S. pneumoniae*, cause in the brain.

Several antigenic types of capsules, called type-specific capsules, are produced by different strains of pathogenic *N. meningitidis*: types A, B, C, W135, and Y. Types A, C, Y, and W135 are usually associated with epidemic infections seen among people kept in close quarters, such as college students or military personnel. Type B meningococcus is typically involved in sporadic infections. However, large outbreaks of type B in the United States occurred at Princeton University and the University of California, Santa Barbara, in December 2013. Antibodies to capsular antigens are used to type the organisms causing an outbreak. Knowing the capsular type of organism involved in each case helps determine whether the disease cases are related and where the infection may have started. In Abdul's case, a polyvalent reagent containing antibodies to all four capsular types was used to confirm *N. meningitidis*. The test could not distinguish between the four serotypes tested. Further testing was necessary to see whether the strains from the father and his children were the same.

Meningococcal meningitis is highly communicable by the respiratory route. Transmission is most common among people in closed communities, such as families, college dormitories, and military barracks. As a result, close contacts, such as the parents or siblings of any patient with meningococcal disease, should receive antimicrobial prophylaxis within 24 hours of diagnosis; a single dose of ciprofloxacin (a quinolone antibiotic, discussed in Section 13.5) can be given to adults, and 2 days of rifampicin can be given to children.

Figure 24.13 Bacterial Meningitis

Meningococcal disease is dreaded by parents and medical practitioners alike for its rapid onset and the difficulty of obtaining a timely and accurate diagnosis.

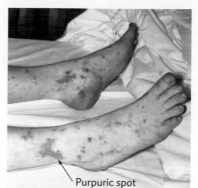

A. Purpuric spots produced by local intravascular coagulation due to *Neisseria meningitidis* endotoxin. The rash in meningitis typically has petechial (small) and purpuric (large) components.

Purpuric spot

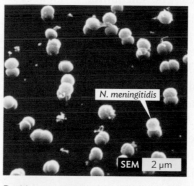

N. meningitidis

SEM 2 μm

B. *N. meningitidis*.

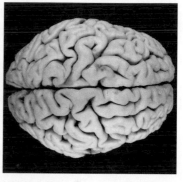

C. Normal brain.

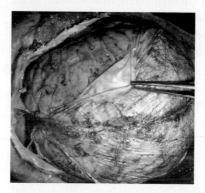

D. Autopsy specimen of meningitis due to *Streptococcus pneumoniae*. Note the greening of the brain, compared with the pink normal brain.

A tetravalent vaccine containing the A, C, Y, and W135 capsular structures is available to prevent disease by these strains. Because of growing concern about the transmission of meningitis among teenagers, the tetravalent vaccine is now recommended for children between ages 11 and 12 and is required by many colleges. Vaccination does not prevent colonization (as with Abdul's father in Case History 24.2), but it does prevent dissemination and disease.

Vaccine to group B *capsule* is not available because the structure of this polysaccharide is similar to that of host tissue surface polymers and so is seen by the immune system as an autoantigen. To curb the 2013 type B outbreak, the CDC approved a unique type B *N. meningitidis* vaccine that was already available in Europe. This vaccine contains outer-membrane vesicles and proteins from *N. meningitidis* type B but not its capsular antigen. The vaccine can now be given in the United States to teens 16 years old or older.

Although the capsule can protect the meningococcus against serum complement, once antibody against the capsule is made, the organism is easily killed via the classical complement cascade (see Section 15.7). This is important because patients with complement deficiencies are susceptible to repeated *N. meningitidis* (or *S. pneumoniae*) infections. As a result, children as young as 2 years old with diagnosed complement deficiencies are given the meningitis vaccine to boost other protective immune mechanisms.

LISTERIOSIS *Listeria monocytogenes* seems like just another food-borne pathogen that causes a few days of diarrhea and fever. That is true if you have a competent immune system. The newborn, elderly, and immunocompromised who encounter *L. monocytogenes* are, however, at increased risk of a much more invasive disease course that includes meningitis, rhombencephalitis (encephalitis of the brain stem), and brain abscess.

Listeria monocytogenes is a Gram-positive, facultative anaerobic rod and a facultative intracellular pathogen that is a major cause of zoonotic disease. Disease caused by *L. monocytogenes* is called **listeriosis** (see Sections 18.5 and 22.5 and Case Histories 18.2 and 22.4. Eating contaminated foods, especially soft foods such as some unpasteurized soft cheeses, can introduce the organism into our gastrointestinal tract. In the intestine, the organism invades epithelial cells, breaks out of the engulfing phagosome, and moves in and between cells by using host cell actin tail formation, much like *Shigella* (see Figure 18.24A). *Listeria* invades macrophages and disseminates in the bloodstream, where it can encounter the blood-brain barrier. *L. monocytogenes* causes up to 10% of all bacterial community-acquired meningitis and has a high mortality rate (up to 26%).

Listeriosis during pregnancy is associated with a very high percentage of fetal infection (up to 90%) and fetal death (stillbirth or spontaneous abortion). The death rate of infected, live-born infants is much lower than that of their fetal counterparts but still is about 20%. Pregnant women who have listeriosis will have fever and often a backache and do not feel well. *L. monocytogenes* infection must be investigated by taking a detailed dietary history from any pregnant woman who has a fever and flu-like symptoms. Diagnosis of meningitis or neonatal infection by *Listeria* can be made by isolating the organism from blood or CSF cultures. No vaccine is available, but antibiotic treatment (ampicillin) is effective.

NEONATAL MENINGITIS Pathogens are usually transferred from mother to newborn through contact with contaminated vaginal secretions during birth and sometimes by directly crossing the maternofetal barrier before birth (*Listeria*, for instance, can infect the newborn by either of these routes). Neonatal meningitis (meningitis in a newborn) can also be nosocomial or community acquired after birth.

Meningitis in patients less than a week old is often attributable to transmission of bacteria during birth. *N. meningitidis* is rarely responsible because the microbe is not found in vaginal secretions, nor does it cross the maternofetal barrier. Rather, neonatal meningitis commonly involves infection by *Escherichia coli* (NMEC) and group B streptococcus (*Streptococcus agalactiae*). Other Gram-negative organisms (such as *Klebsiella*) and some Gram-positive cocci (such as *Enterococcus*, *S. aureus*, and *L. monocytogenes*) can cause meningitis in the newborn as well.

The infected infant will have neurological problems such as lethargy, irritability, poor muscle tone, and (occasionally) febrile seizures due to an abnormally high body temperature. Infants with neonatal meningitis feed poorly and may have nausea and vomiting, diarrhea, and even difficulty breathing. Definitive diagnosis is made based on CSF examination, but infants suspected of having meningitis are empirically treated with ampicillin and an aminoglycoside antibiotic until the offending agent is identified. Lumbar puncture is repeated at 24–48 hours after initiation of treatment to make sure that sterility of CSF is restored.

The incidence of neonatal meningitis has remained relatively unchanged for the past 40 years, but the mortality rate has dropped significantly. The prognosis seems to depend mainly on the type of organism involved, whether the infant is preterm or full term, and the age at diagnosis. Neurological sequelae such as seizures and developmental delays continue to characterize a high percentage of long-term outcomes.

Bacterial Neurotoxins That Cause Paralysis

Imagine a disease that causes complete loss of muscle function. Using secreted exotoxins, two microbes cause lethal paralytic diseases. In one instance, the victim suffers a flaccid paralysis in which the muscles go limp, causing paralysis and respiratory difficulty. Voluntary muscles fail to respond to the mind's will because the botulinum toxin interferes with neural transmission. The disease, called **botulism**, is typically food-borne and is caused by an obligate anaerobe, the Gram-positive, spore-forming bacillus named *Clostridium botulinum*.

CASE HISTORY 24.3

Botulism—The Vicious Green Beans

In June, Mitch, a 47-year-old resident of Oklahoma, was admitted to the hospital with rapid onset of progressive dizziness, blurred vision, slurred speech, difficulty swallowing, and nausea. Findings on examination included absence of fever, drooping eyelids, facial paralysis, and impaired gag reflex. He developed breathing difficulties and required mechanical ventilation. The patient reported that during the 24 hours before onset of symptoms, he had eaten home-canned green beans and a stew containing roast beef and potatoes. Analysis of the patient's stool detected botulinum type A toxin, but no *Clostridium botulinum* organisms were found. The patient was hospitalized for 49 days, including 42 days on mechanical ventilation, before being discharged.

Figure 24.14   The Basic Structure of Tetanus and Botulinum Toxins

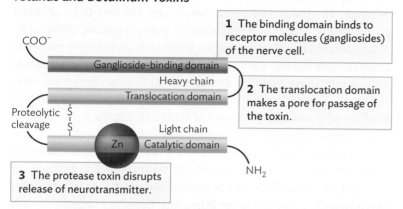

1 The binding domain binds to receptor molecules (gangliosides) of the nerve cell.

2 The translocation domain makes a pore for passage of the toxin.

3 The protease toxin disrupts release of neurotransmitter.

Schematic diagram of tetanus and botulinum toxins.

Figure 24.15   Mechanism of Action of Botulinum Toxin

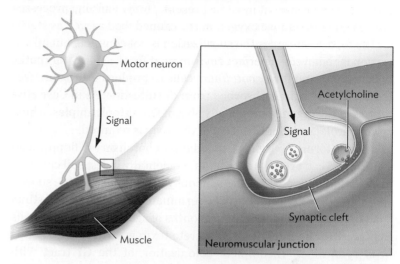

A. The neuromuscular junction. Inset shows vesicles filled with neurotransmitters.

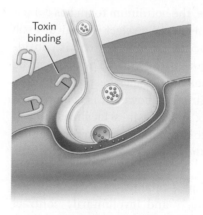

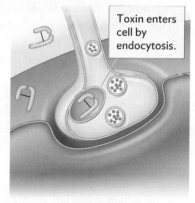

Toxin binding

Toxin enters cell by endocytosis.

B. Toxin binds via the heavy chain.

C. Toxin is endocytosed into the nerve terminal.

D. The small subunit of the toxin enters the cytoplasm, cleaves its target, and makes the nerve terminal unable to release acetylcholine.

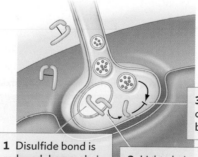

1 Disulfide bond is reduced; heavy chain binds to membrane.

2 Light chain enters cytoplasm.

3 Light chain cleaves proteins, blocks exocytosis.

Paralysis is among the most feared symptoms of CNS damage. Two bacteria of the same genus produce two very potent neurotoxins, botulinum and tetanus toxins, that cause paralysis and often lead to death. The toxins are produced by *Clostridium botulinum* and *Clostridium tetani*, respectively. The two toxins share 30%–40% identity and have similar structures and nearly identical modes of action. Each toxin is initially made as a single peptide that is cleaved after secretion to form two fragments, heavy and light chains (analogous to the AB toxins; Section 18.3) that remain tethered to each other by a disulfide bond (Figure 24.14). Each heavy chain has a binding domain that binds to a receptor molecule (ganglioside) on the nerve cell membrane and a translocation domain that makes a pore in the nerve cell. The toxin (the light chain) then passes through the pore and into the cell. The light chains are proteases that disrupt the movement of exocytic vesicles containing neurotransmitters needed for contraction or relaxation of muscles.

Both botulinum and tetanus toxin enter at peripheral nerve ends where nerve meets muscle (the neuromuscular junction). The toxin binds to target membranes and is brought into the cell by **endocytosis**. The low pH that forms in the endosome cleaves the disulfide bonds holding the two halves of the toxin together, a channel is assembled through the membrane, and the light chain (toxin) is released into the cytoplasm. The toxic subunit then cleaves key host proteins involved in the **exocytosis** of vesicles containing neurotransmitters (Figure 24.15), disrupting release of the neurotransmitters and causing disease.

With such a high degree of mechanistic similarity, why do tetanus and botulinum toxins have such drastically different effects? The answer is based on where each toxin acts in the nervous system. Botulism is an intoxication resulting from the ingestion of preformed toxin in poorly prepared, contaminated foods (whether the organism is also ingested is immaterial), as in Case History 24.3. Once botulinum toxin enters a peripheral nerve, it will immediately cleave peptides that help with exocytic release of the neurotransmitter acetylcholine (see Figure 24.15D). Without acetylcholine to activate

nerve transmission, muscles will not contract and the patient is paralyzed. Botulism is a disease of the peripheral nervous system.

Occasionally, germination of *C. botulinum* in an infected wound or in the anaerobic environment of the gastrointestinal tract will allow these organisms to produce exotoxin and cause disease. In food, the spores of *C. botulinum* germinate, and the cells that emerge produce toxin. Because the organism is an anaerobe, an anaerobic environment must be present. Home-canning processes are designed to remove oxygen in the canned food as well as sterilize the food. Sometimes the sterilization is not complete but all the oxygen is removed—a perfect environment for the surviving spores to germinate and for *C. botulinum* cells to produce exotoxin. After ingestion, the toxin is absorbed from the intestine. As in Case History 24.3, the organism is usually absent from stool samples, which is why serological identification of the toxin is required.

A rare form of botulism, called infant botulism or "floppy head syndrome," can arise when infants accidentally ingest soil containing *C. botulinum* spores or are fed honey that may harbor the spores. Ingested spores can potentially germinate in the gastrointestinal tract of infants younger than 12 months of age. The vegetative cells that grow in the intestine and secrete toxin will cause the "floppy head syndrome." Incomplete colonization of the GI tract with microbiota and a less than optimal immune system may be why only very young infants are susceptible to this condition.

Although botulism is now rare, the botulinum toxin has gained renewed interest because it is considered a select biological toxic agent of potential use to bioterrorists. Botulinum toxin is considered to be one of the deadliest poisons in the world. The toxin can be absorbed through the eyes, mucous membranes, respiratory tract, or cuts in the skin. Commercial interest, however, is even higher. Botulinum toxin can safely relax muscles in a localized area if injected in small doses. Botulism toxin administered in this manner is known commercially as Botox and is used cosmetically by plastic surgeons to reduce facial wrinkles.

In striking contrast to botulism is **tetanus**, a very painful disease in which muscles continuously and involuntarily contract (called tetany or spastic paralysis). Tetanus is caused by **tetanospasmin**, a potent exotoxin made by another species of *Clostridium—Clostridium tetani* (**Figure 24.16A**). Tetanospasmin interferes with

neural transmission, but in contrast to botulinum toxin, it causes *excessive* nerve signaling to muscles, forcing the victim's muscles to contract. The jaw muscles (masseters) are the first to be affected, giving the disease tetanus its common name, "lockjaw." Eventually, other muscles are affected; the back will arch grotesquely while the arms flex and the legs extend. The patient remains locked this way until death. Spasms can be strong enough to fracture the patient's vertebrae. **Figure 24.16B** shows the result of injecting a mouse's hind leg with just a tiny amount of tetanus toxin. In both botulism and tetanus, death can result from asphyxiation.

Tetanus, unlike botulism, is not a food-borne disease. *C. tetani* spores are introduced into the body by trauma (such as by stepping on a dirty rusty nail). Necrotic tissue then provides the anaerobic environment required for germination. Growing, vegetative cells release tetanospasmin, which enters the peripheral nerve cells at the site of injury. But instead of cleaving targets here, the toxin travels up axons toward the spinal column, where it becomes fixed on presynaptic inhibitory motor neurons (presynaptic neurons communicate with other neurons and not with the muscle directly). The function of vesicles in these neurons is to release *inhibitory* neurotransmitters that dampen nerve impulses. By cleaving vesicle proteins, tetanus toxin blocks release of the inhibitory neurotransmitters GABA and glycine into the synaptic cleft, leaving excitatory nerve impulses unchecked. As a result, excitatory impulses come too frequently and produce the generalized muscle spasms characteristic of tetanus (**Figure 24.17**).

Botulism and tetanus are preventable diseases. No vaccine exists for botulism; therefore, proper cooking and canning of food is essential to prevent botulism. Tetanus vaccine (toxoid) is part of routine childhood vaccinations. The vaccine does not confer lifelong immunity, and a booster is needed every 10 years. So, people who may have been exposed to the bacterium—for example, by cutting themselves on a rusty sharp object—have their wounds debrided and are given a booster shot.

Antibiotics are of no use in treating either disease because bacteria themselves are not the cause. Injection of antitoxins and supportive care to the patient consitute the mainstay treatment for both diseases.

Figure 24.16 Tetanus and Tetany

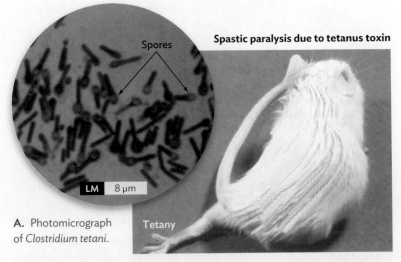

Spores

LM 8 µm

Spastic paralysis due to tetanus toxin

Tetany

A. Photomicrograph of *Clostridium tetani*.

B. Mouse injected with tetanus toxin in left hind leg.

SECTION SUMMARY

- **Hallmarks of meningitis** include severe headache accompanied by neck pain, nuchal rigidity, and fever followed by lethargy and confusion. Empirical antibiotic treatment is imperative when meningitis is suspected.

- **The neurological deficits of encephalitis,** such as uneven gait, do not occur unless the infection has spread to the brain (meningoencephalitis).

- **Common agents of bacterial meningitis** include *Neisseria meningitidis*, *Streptococcus pneumoniae*, *Haemophilus influenzae*, and *Listeria monocytogenes*.

- **CSF results indicating bacterial meningitis** include pleocytosis (usually neutrophils), increased protein concentration, decreased glucose concentrations, and a Gram stain showing bacteria. Definitive diagnosis, however, requires isolating the organism from CSF.

- *Neisseria meningitidis,* a Gram-negative diplococcus, spreads from person to person by aerosolization, colonizes the oropharynx, and occasionally penetrates the epithelium to gain entrance to the bloodstream. Vaccination with *N. meningitidis* A, C, Y, and W135 capsular antigens is recommended for children between ages 11 and 12.

- *Listeria monocytogenes,* a Gram-positive rod, is a food-borne, facultative intracellular pathogen that penetrates the intestinal mucosa, invades macrophages, and disseminates through the bloodstream. It typically causes mild diarrhea but can cross the blood-brain barrier and infect meninges in very young, very old, and immunocompromised patients.

- **Common agents of neonatal meningitis** are *Streptococcus agalactiae* (group B streptococcus, transmitted during birth), *Escherichia coli* (NMEC, transmitted after birth), and *Listeria monocytogenes* (transmitted through the placenta before birth, by vaginal secretions during birth, or other means after birth).

- **Botulism and tetanus** are toxigenic diseases caused by the Gram-positive, spore-forming bacteria *Clostridium botulinum* and *Clostridium tetani*, respectively. Toxins produced by these bacteria prevent the release of neurotransmitters from nerves. In botulism, flaccid paralysis arises from lack of excitatory neurotransmitters (acetylcholine); in tetanus, spastic paralysis arises from lack of inhibitory neurotransmitters (GABA).

Thought Question 24.3 How do the actions of tetanus toxin and botulinum toxin help the bacteria colonize or obtain nutrients?

Thought Question 24.4 If *Clostridium botulinum* is an anaerobe, how might botulinum toxin get into foods?

Figure 24.17 Retrograde Movement of Tetanus Toxin to an Inhibitory Neuron

Tetanus toxin enters the nervous system at the neuromuscular junction and travels up the axon to an inhibitory neuron of the central nervous system. There the toxin cleaves proteins associated with the exocytosis of vesicles containing inhibitory neurotransmitters. In the absence of inhibitory signals (gamma-aminobutyric acid, GABA), excitatory signals (acetycholine, ACh) that stimulate muscle contraction prevail.

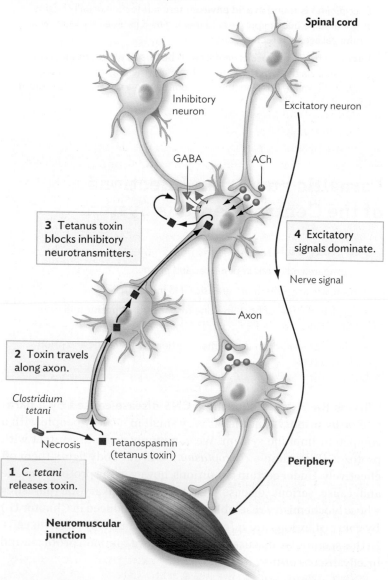

24.5 Fungal Infections of the Central Nervous System

SECTION OBJECTIVES

- Name fungi that cause CNS disease.
- Describe the role of host immunity in fungal infections of the CNS.
- Define fungal CNS infections in the context of environmental exposure risks.
- Identify diagnostic methods and treatment options for different fungal CNS infections.

Fungal infections (**mycoses**) of the CNS are rare and caused mostly by dissemination of a primary infection elsewhere in the body. The patients most commonly affected have a compromised immune system (as in AIDS or in the elderly) or have undergone a neurosurgical procedure to implant a shunt, a device used to reduce intracranial pressure. The shunt can serve as a conduit into the brain for bacterial and fungal pathogens.

The most common fungus to invade the CNS is *Candida*, particularly *Candida albicans*. CNS infections caused by *Candida* include meningitis and brain abscesses. ***Candida* meningitis** is often seen in premature neonates, whereas brain abscesses occur predominantly in immunocompromised patients of any age. Diagnosis is made by isolating *Candida* from the CSF. Standard treatment is combination therapy with amphotericin B and flucytosine until all signs and symptoms of the disease have resolved.

Coccidioides species *C. immitis* and *C. posadasii* are fungi that grow in dry desert soil. The dry climate makes their mycelia easily breakable. Slight air movements break the dry mycelia into single-cell spores called **arthroconidia** that can float in air for extended periods. Inhaling the spores causes a primary fungal infection, pneumonia, 1–3 weeks after exposure (see Section 20.4). Fatigue

Figure 24.18   *Cryptococcus neoformans*

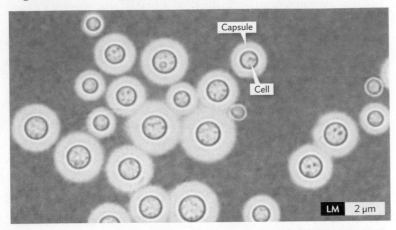

An India ink stain of cerebrospinal fluid shows the thick capsule of *C. neoformans*.

and joint pain lasting for extended periods (months) may be the only symptoms of the infection, giving it the nickname "desert rheumatism." The fungus can disseminate and cause meningitis after a delay lasting weeks or months.

The symptoms of **coccidioidal meningitis** are very nonspecific, with a persistent headache being the most prominent. *Coccidioides* should be in the differential diagnosis when a patient with severe, progressing headache, blurred vision, and altered mental status has a history of travel to arid areas. The travel history does not need to be extensive. For instance, a short stopover (hours to days) in Phoenix, Arizona, may qualify. Coccidioidal meningitis is extremely lethal unless it is diagnosed and treated quickly; about 95% of patients who are not treated die within 2 years. The definitive diagnosis is made by isolating the fungus from the CSF. Patients with confirmed coccidioidal CNS infection are treated with fluconazole or itraconazole for life.

Cryptococcus neoformans var. *neoformans* is a fungus often present in pigeon and chicken droppings and finds its way into the soil. *C. neoformans*, notable for its thick, protective capsule, can spread to the CNS through the bloodstream after initially infecting the respiratory tract when it is inhaled. Extension of the disseminated fungus through the bloodstream often causes meningoencephalitis, especially in immunocompromised patients. More than 80% of cryptococcus meningitis cases are associated with AIDS, making **cryptococcal meningoencephalitis** an AIDS-defining infection. The incidence of the disease has decreased in HIV patients who have access to antiviral medication (because their immune systems recover), but worldwide nearly a million cases are diagnosed annually, with a 60% mortality rate.

Another variety of cryptococcus, *Cryptococcus neoformans* var. *gattii*, is an emerging cause of meningitis in the northwestern United States, particularly Oregon. This fungus also infects birds, is spread by bird droppings, and causes pulmonary disease and meningitis when inhaled, as does *C. neoformans*. But in striking contrast to *C. neoformans*, *C. gattii* readily causes disease in immunocompetent individuals. The organism infects a variety of trees around the world, including the River Red gum trees (such as eucalyptus) in Australia where most cases of *C. gattii* infection occur.

Identification of *Cryptococcus* with its prominent capsule in the CSF sample makes the definitive diagnosis (Figure 24.18). Patients suffering from cryptococcal meningoencephalitis are currently treated with amphotericin B, fluconazole, and flucytosine, on the basis of a complicated triphasic regimen that ends with maintenance therapy.

SECTION SUMMARY

- **Fungal infections of the CNS generally result from dissemination of the agent from a primary infection elsewhere,** typically in immunocompromised patients.
- ***Candida albicans* is the most common fungal cause of meningitis.**
- ***Coccidioides* is found in arid environments.** Infection usually begins as pneumonia but can progress to meningitis. Lifetime treatment with antifungal agents is recommended.
- **Infection by *Cryptococcus neoformans* begins as a pneumonia** that can spread via the bloodstream to the brain (meningoencephalitis). Diagnosis is made upon seeing yeast containing a prominent capsule in CSF preparations.

24.6
Parasitic and Prion Infections of the Central Nervous System

SECTION OBJECTIVES

- Compare and contrast parasites and prions.
- Evaluate acquisition of parasitic CNS infections.
- Explain the role of host immune responses in parasitic and prion CNS disease.
- Name prevention and treatment options for parasitic and prion CNS disease.

What is the risk of acquiring a CNS disease caused by a parasite or by prions? The risk is very small in healthy people with a competent immune system. We commonly come in contact with protozoa (for example, *Toxoplasma*), and our body fights them off effectively. Under certain conditions, however, the protozoa invade and cause serious disease. (Protozoa are eukaryotic parasites whose biochemistry resembles our own; introduced in Chapter 11.) By contrast, prions are infectious proteins made by our own cells. In this section, we discuss some of the more common protozoa and briefly discuss prions.

Protozoan Diseases

Toxoplasma gondii is a parasitic protozoan that can infect most mammals, including humans. It is one of the world's most common parasites. Its primary host is the cat, which can be infected by eating contaminated meat or the feces of infected animals or by vertical transmission from mother to offspring. Contact with cats, especially with cat feces, is the primary mode of transmission to humans. Over half of the world's population is thought to be infected with this parasite. Most infections are asymptomatic, but the disease, **toxoplasmosis**, typically begins as a respiratory infection that resolves (see Section 21.5, Figure 21.26). The organism can, however, form cysts in muscles and nervous tissues, including the brain, and remain latent for years unbeknownst to the host. If the patient becomes immunocompromised (for example, develops AIDS), the organism can begin to grow and damage the brain. Symptoms include confusion, fever, headache, retinal inflammation that causes blurred vision, and seizures. Antiparasitic treatment is available, as is a vaccine for cats.

Pregnant women are usually asymptomatic when they contract toxoplasmosis. *Toxoplasma,* however, can be vertically transmitted from an infected mother to her fetus. Transmission during the first trimester (first 3 months) can cause stillbirth or serious sequelae in the newborn. The infected fetus may not have any symptoms at birth (in 70%–90% of cases) but can later develop complications such as mental disability and blindness.

Deep in rural sub-Saharan Africa, the dreaded tsetse fly roams and carries with it the protozoa that cause **African sleeping sickness**, *Trypanosoma brucei rhodesiense* and *T. brucei gambiense* (Figure 24.19). Reservoirs for *T. brucei rhodesiense* include wild animals (lions, pigs, hyenas), domestic animals (cows and goats), and humans. Humans are the main reservoir for *T. brucei gambiense*. The tsetse fly becomes infected after taking a blood meal from an infected person. The trypanosome replicates in the fly's gut and migrates to the salivary glands, where the microbe develops into its infective stage. When the fly bites a new host, trypanosome-laden saliva is deposited in the wound, where the microbe then multiplies and produces a painful lesion. Eventually, the parasite migrates to the lymphatic and blood circulations, triggering fever and lymphadenopathy (swollen lymph nodes). The later stage of trypanosomiasis involves CNS invasion, at which point the patient develops daytime sleepiness but nighttime insomnia, a listless gaze, tremors, halting speech, and a shuffling gait. In the final phase, the progressive neurological damage ends in coma and death.

Trypanosomes such as *T. brucei* and *T. cruzi*, the cause of Chagas' disease (see Section 21.5), are successful parasites that manage to escape the host's immune response by a complex mechanism of **antigen switching**. The knowledge of antigen switching has led us to the first steps in developing an anti-trypanosome vaccine.

Link As discussed in Section 18.7, **antigen switching** (or antigenic variation) in trypanosomes involves many variant surface glycoprotein (VSG) genes, only one of which is expressed at any one time. Each glycoprotein has a different antigenic structure. Switching from one VSG gene to another changes the antigenic characteristic of the organism.

Eflornithine is a highly effective anti-trypanosome drug, although others are available. African sleeping sickness is not seen elsewhere in the world (unless an infected tourist has returned from Africa). But 10,000 cases are reported each year in Africa, although more are suspected.

Naegleria fowleri is an ameba that produces an acute, and usually lethal, central nervous system infection (Figure 24.20). The organism is found in freshwater but also in geothermal pools, heated pools, aquaria, and sewage. The *Naegleria* developmental cycle includes cysts, trophozoites (the amebic form that infects humans), and flagellated forms. Trophozoites infect humans or animals by entering the olfactory nerve, following the path of the nerve directly to the brain. The organism then literally eats through the blood-brain barrier and causes **Naegleria meningoencephalitis**.

 Naegleria infection: Brain-eating ameba, CNN

Symptoms of an infection with *Naegleria fowleri* start with loss of taste and smell, after which headaches, nausea, and stiff neck develop. The patient experiences hallucinations and death, usually within 14 days of exposure. Diagnosis, made by finding *N. fowleri* trophozoites in CSF, is usually a death sentence, in part because the disease is rare and its progress rapid. If diagnosis is made early, amphotericin B is the drug of choice for treatment, however, a new drug, miltefosine, is now available from the CDC. Fortunately, contact with *Naegleria fowleri* does not always result in infection. Our immune system fends off *Naegleria* infections quite well. Only 30 or so cases have been reported since 2000 despite common exposure to this protozoan. *Naegleria* was once a pathogen of warmer climates, but global warming appears

Figure 24.19 African Sleeping Sickness

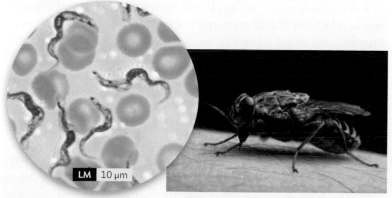

LM 10 μm

A. *Trypanosoma brucei gambiense* in a peripheral smear of blood.

B. The tsetse fly transmits the trypanosomes from infected people or animals to healthy people.

Figure 24.20   *Naegleria fowleri*

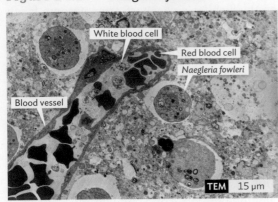

A. *Naegleria fowleri* in infected brain tissue.

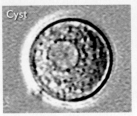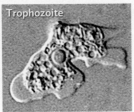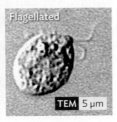

B. *Naegleria fowleri* in various stages.

to be causing this pathogen to migrate to lakes farther and farther north. Minnesota saw its first two cases in 2010 and 2012.

Acanthamoeba (Figure 24.21) is another ameba that can infect the brain. It is found worldwide in water and soil and can be transmitted into the eye by contaminated contact lenses or can enter the body through inhalation or breaks in the skin. Most everyone is exposed to this organism at some time in life, but few people contract disease. The most common symptom is keratitis of the eye, but in immunocompromised individuals, the organism can move up the optic nerve into the brain to cause encephalitis. Although few, most cases of **Acanthamoeba encephalitis** are fatal.

Onset of encephalitis is slow (weeks to months) and the symptoms are similar to those of *Naegleria*. Ketoconazole and trimethoprim-sulfamethoxazole can be effective against this organism, but diagnosis is usually made too late.

Figure 24.21
Acanthamoeba

This unusual-looking ameba can infect a corneal epithelium damaged with microscopic abrasions and travel along the optic nerve to cause CNS disease.

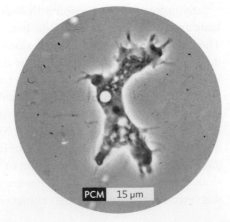

Prion Diseases

Try to imagine slowly losing your mind and knowing there is nothing that you or your loved ones can do about it. An unusual infectious agent called the prion has been implicated as the cause of a series of relatively rare but invariably fatal brain diseases (Table 24.4). Prions are infectious agents that do not have a nucleic acid genome (see Section 12.2). The idea that a protein alone could not mediate an infection used to be universally accepted. The **prion** is now recognized as an infectious, misfolded protein that resists inactivation by procedures that destroy protein. The discovery that proteins alone can transmit an infectious disease has come as a surprise to the scientific community. Diseases caused by prions are especially worrying, since prions resist destruction by many chemical agents and remain active after heating at extremely high temperatures. In some documented cases, sterilized surgical instruments, originally used on a prion-infected person, still held infectious agent and transmitted the disease to a later surgery patient.

How can a nonliving entity without nucleic acid be an infectious agent? Prions associated with human brain disease are thought to be aberrantly folded forms of a normal brain protein (see Section 12.2). The thought is that when a prion is introduced into the body and manages to enter the brain, it interacts with normally folded forms of the protein, causing them to refold incorrectly. The improperly folded proteins fit together like Lego blocks to produce damaging aggregated structures (amyloid) within the brain (Figure 24.22).

Some investigators suggest that living infectious agents, such as *Spiroplasma* (a spiral-shaped genus of bacteria whose members lack cell walls), can somehow change the shape of normal brain protein, converting it to an infectious prion protein. This proposal is not, however, universally accepted and has yet to be proved.

Prion diseases are often called **transmissible spongiform encephalopathies (TSEs)** because the postmortem appearance of the brain includes large, spongy vacuoles in the cortex and cerebellum. These plaques are visible in a brain sample from a victim of one of these diseases, **Creutzfeldt-Jakob disease (CJD)** (Figure 24.22C). Typically, onset of CJD symptoms occurs about age 60, and about 90% of individuals die within 1 year. In the early stages of disease, people may have failing memory, behavioral changes, lack of coordination, and visual disturbances. As the illness progresses, mental deterioration becomes pronounced, and involuntary movements, blindness, weakness of extremities, and coma may occur.

Most mammalian species appear to develop prion diseases. Since 1996, mounting evidence points to a causal relationship between outbreaks in Europe of a disease in cattle called **bovine spongiform encephalopathy (BSE)**, or "mad cow disease," and a disease in humans called **variant Creutzfeldt-Jakob disease (vCJD)**. Both disorders are invariably fatal brain disorders. They have unusually long incubation periods (measured in years) and are caused by an unconventional transmissible agent.

As of this writing, only three cases of BSE have been detected in the United States. Thanks to aggressive surveillance efforts in the

Table 24.4
Prion Diseases

Disease	Susceptible Animal	Incubation Period	Disease Characteristics
Creutzfeldt-Jakob (sporadic, familial, new variant [vCJD])	Human	Months (vCJD) to years	Spongiform encephalopathy (degenerative brain disease)
Kuru	Human	Months to years	Spongiform encephalopathy
Gerstmann-Straussler-Scheinker syndrome	Human	Months to years	Genetic neurodegenerative disease
Fatal familial insomnia	Human	Months to years	Genetic neurodegenerative disease with untreatable insomnia
Mad cow disease	Cattle	5 years	Spongiform encephalopathy
Scrapie	Sheep	2–5 years	Spongiform encephalopathy
Wasting disease	Deer	Months to years	Spongiform encephalopathy

United States and Canada (where only 19 cases have been found), it is unlikely, but not impossible, that BSE will be a major food-borne hazard to humans in this country. The CDC monitors the trends and current incidence of typical and variant CJD in the United States.

SECTION SUMMARY

- ***Toxoplasma gondii*** is a protozoan parasite typically transmitted from cats to humans. Infection is common, but disease is rare and usually confined to immunocompromised patients. Latent brain cysts of infected individuals can excyst, and the organisms can grow, destroying brain tissue.

- ***Trypanosoma brucei gambiense*** is the protozoan that causes African sleeping sickness. The organism is transmitted by the bite of the tsetse fly in sub-Saharan Africa.

- ***Naegleria fowleri*** is a common environmental ameba that causes occasional cases of meningoencephalitis.

- ***Acanthamoeba*** is a parasitic ameba that can cause keratitis, but in an immunocompromised patient, it can move into the brain to cause encephalitis.

- **Transmissible spongiform encephalopathies,** such as Creutzfeldt-Jakob disease, are believed to be caused by nonliving, proteinaceous agents called prions.

Perspective

Any infectious disease comes with some degree of angst. The common cold, for instance, is usually suffered with only a mild sense of annoyance, whereas a bout of bloody diarrhea causes anxiety. Gonorrhea, too, produces anxiety, but also embarrassment and anger. However, few infections are met with the level of fear brought by a CNS infection. The severity of bacterial meningitis and the slow mental and physical decline caused by Creutzfeldt-Jakob disease help drive this fear. What most people do not realize is that the CNS is the most heavily fortified and protected system in the body. Our immune system and the blood-brain barrier make the CNS almost impenetrable to microbes. As a result, CNS infections are infrequent and serious disease is extremely rare. For example, fewer than 500 cases of meningococcal meningitis occurred in the United States in 2014 and only 8 cases of neuroinvasive eastern equine encephalitis that same year—out of 319 million people. Nevertheless, adhering to simple precautions such as hand washing, wearing insect repellent outdoors, and vaccinating pets can lower the risk even more.

Figure 24.22   Transmissible Spongiform Encephalopathies
Improperly folded proteins stack to form aggregates called plaques. Plaques appear as vacuoles in the brain tissue.

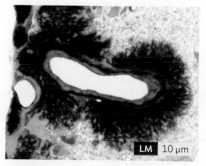

A. Prions form an amyloid plaque around a blood vessel in the brain.

B. Normal brain section.

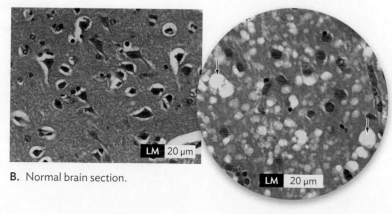

C. Section of brain taken from a victim of Creutzfeldt-Jakob disease. Note the "Swiss cheese" appearance, indicating brain damage. The arrows point to two of the larger vacuoles.

Figure 24.23   Summary: Microbes That Infect the Nervous System

Viruses	Genome	Disease
Herpes simplex virus 1	DNA, ds	Encephalitis
Herpes simplex virus 2	DNA, ds	Meningitis
Colorado tick fever virus	RNA, ds, segmented	Encephalitis
LaCrosse agent of California encephalitis	RNA, segmented, negative sense	Encephalitis
Lymphocytic choriomeningitis virus	RNA, segmented	Meningitis
Rabies	RNA, ss, negative sense	Rabies
Western equine encephalitis virus	RNA, ss, positive sense	Encephalitis
Eastern equine encephalitis virus	RNA, ss, positive sense	Encephalitis/meningitis (rare)
Venezuelan equine encephalitis virus	RNA, ss, positive sense	Encephalitis
West Nile fever virus	RNA, ss, positive sense	Encephalitis/meningitis (uncommon)
Saint Louis encephalitis virus	RNA, ss, positive sense	Encephalitis/meningitis
Japanese B encephalitis virus	RNA, ss, positive sense	Encephalitis
Murray Valley encephalitis virus	RNA, ss, positive sense	Encephalitis
Dengue viruses	RNA, ss, positive sense	Encephalitis
Powassan virus	RNA, ss, positive sense	Encephalitis
Coxsackievirus, echovirus	RNA, ss, positive sense	Encephalitis/meningitis (rare)
Human immunodeficiency virus	RNA, ss, positive sense	Meningitis

Bacteria	Morphology	Disease
Mycobacterium tuberculosis	Acid-fast, bacilli	Meningitis
Escherichia coli	Gram-negative, bacilli	Meningitis
Haemophilus influenzae	Gram-negative, bacilli	Meningitis
Bacteroides fragilis	Gram-negative, bacilli	Meningitis
Pseudomonas aeruginosa	Gram-negative, bacilli	Meningitis
Listeria monocytogenes	Gram-negative, bacilli	Meningitis
Neisseria meningitidis	Gram-negative, bacilli	Meningitis
Streptococcus pneumoniae	Gram-positive, cocci	Meningitis
Streptococcus agalactiae, group B streptococci	Gram-positive, cocci	Meningitis
Coagulase-negative *Staphylococcus*	Gram-positive, cocci	Meningitis
Staphylococcus aureus	Gram-positive, cocci	Meningitis
Treponema pallidum	Spirochete	Meningitis
Borrelia burgdorferi	Spirochete	Meningitis

Parasites	Type	Disease
Toxoplasma gondii	Protozoan	Encephalopathy
Trypanosoma brucei gambiense	Protozoan	African sleeping sickness
Naegleria fowleri	Ameba	Meningoencephalitis
Acanthamoeba	Ameba	Keratitis, encephalitis

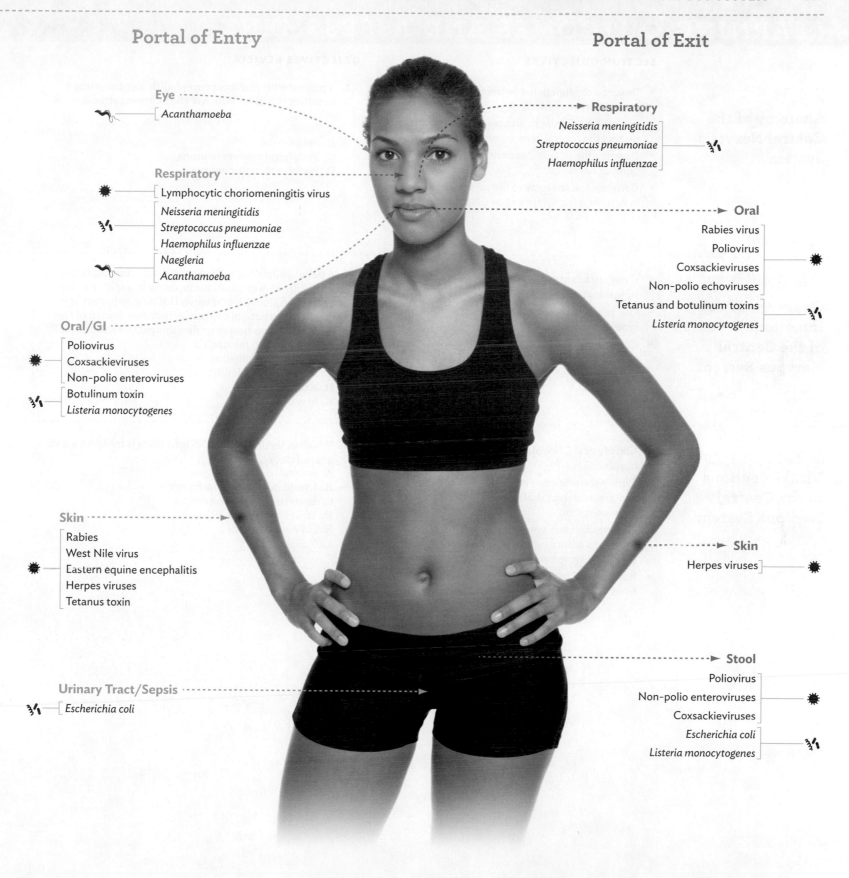

Portal of Entry

Eye
[*Acanthamoeba*

Respiratory
[Lymphocytic choriomeningitis virus
[*Neisseria meningitidis*
Streptococcus pneumoniae
Haemophilus influenzae
[*Naegleria*
Acanthamoeba

Oral/GI
Poliovirus
Coxsackieviruses
Non-polio enteroviruses
[Botulinum toxin
Listeria monocytogenes

Skin
Rabies
West Nile virus
Eastern equine encephalitis
Herpes viruses
Tetanus toxin

Urinary Tract/Sepsis
[*Escherichia coli*

Portal of Exit

Respiratory
Neisseria meningitidis
Streptococcus pneumoniae
Haemophilus influenzae

Oral
Rabies virus
Poliovirus
Coxsackieviruses
Non-polio echoviruses
Tetanus and botulinum toxins
Listeria monocytogenes

Skin
Herpes viruses

Stool
Poliovirus
Non-polio enteroviruses
Coxsackieviruses
Escherichia coli
Listeria monocytogenes

LEARNING OUTCOMES AND ASSESSMENT

| SECTION OBJECTIVES | OBJECTIVES REVIEW |

24.1
Anatomy of the Central Nervous System

- Describe the functional organization of the central nervous system.
- Explain how meninges and cerebrospinal fluid protect the central nervous system.
- Describe the differences between two major types of neurons.
- Correlate the anatomy of the central nervous system with the symptoms of CNS infections.

1. A patient suffering from encephalitis has diminished cognitive function. Which of the following structures is affected?
   A. Brain
   B. Spinal cord
   C. Peripheral sensory neurons
   D. Peripheral motor neurons

24.2
Overview of the Infectious Diseases of the Central Nervous System

- Categorize central nervous system infections by the anatomical structures affected.
- Use key laboratory information to distinguish types of CNS infections.
- Compare and contrast the roles of meninges and cerebrospinal fluid in the pathogenesis of CNS infections.

5. A patient admitted to the hospital with meningitis becomes unable to answer questions properly in addition to his initial symptoms. You identify this new symptom as a deficit in the patient's cognitive function. Which of the following terms best describes this patient's condition?
   A. Complicated meningitis
   B. Primary encephalitis
   C. Meningoencephalitis
   D. Radiculitis
   E. Myelitis

24.3
Viral Infections of the Central Nervous System

- Classify viral CNS infections by their mode of transmission.
- Explain the role of reservoirs in pathogenesis and prevention of viral diseases of the central nervous system.
- Outline clinical diagnoses of viral CNS infections according to the patient's history, signs, and symptoms.
- Recommend treatment and prevention options for different viral CNS infections.

8. Which of the following CNS infections is transmitted via aerosolization?
   A. Rabies
   B. Lymphocytic choriomeningitis
   C. Eastern equine encephalitis
   D. West Nile encephalitis
   E. HSV-1 meningitis

2. CNS infections are relatively rare, mostly because of _____.
   A. meninges
   B. cerebrospinal fluid
   C. the blood-brain barrier
   D. the skull and bony structures

3. To obtain CSF for analysis, you have to enter the space between _____.
   A. brain and spinal cord
   B. the subdural space
   C. arachnoid and dura mater
   D. arachnoid and pia mater
   E. arachnoid and subarachnoid mater

4. Severe meningitis can cause damage to the brain because of increased CSF pressure resulting from _____.
   A. damage to the choroid plexus
   B. damage to arachnoid villi
   C. damage to the blood-brain barrier
   D. damage to bony structures

6. A CSF sample from a patient suffering from a viral meningitis will most likely show _____.
   A. an increased number of red blood cells
   B. an increased number of polymorphonuclear cells
   C. an increased number of lymphocytes
   D. an increased number of macrophages
   E. an increased number of endothelial cells

7. A 20-year-old is in the hospital with a high fever, severe headache, neck pain, nausea, vomiting, and confusion. Examination of a sample of his CSF shows many PMNs. What is the most likely cause of his headache?
   A. Inflammation of his meninges
   B. Disruption of his blood-brain barrier
   C. Inflammation of his spinal cord
   D. Demyelination of his sensory neurons
   E. Demyelination of his motor neurons

9. Poliomyelitis is transmitted from person to person through the fecal-oral route. True or false?

10. Why is eastern equine encephalitis (EEE) a relatively rare disease in humans?
   A. Because close human and horse contacts are not common
   B. Because mosquitoes that transfer the virus from horses to humans are rare
   C. Because ticks that transfer the virus do not usually prey on humans
   D. Because mosquitoes that transfer the virus between birds do not usually prey on humans

LEARNING OUTCOMES AND ASSESSMENT

	SECTION OBJECTIVES	OBJECTIVES REVIEW

24.4
Bacterial Infections of the Central Nervous System

- Identify risk factors for each of the bacterial CNS diseases.
- Describe how different bacteria overcome host defenses to cause CNS disease.
- List patient history information and clinical signs and symptoms specific to each bacterial pathogen of the central nervous system.
- Relate pathogenesis of bacterial CNS diseases to prevention and treatment modalities used.

11. *Neisseria meningitidis* can attach to the endothelial cells of the blood-brain barrier. How does the organism penetrate the barrier later?
   A. Attachment of the organism loosens the tight junctions between endothelial cells.
   B. Attachment of the organism leads to its endocytosis by the endothelial cells.
   C. Attachment of the organism results in its phagocytosis.
   D. Attachment of the organism reduces the local complement activity.

24.5
Fungal Infections of the Central Nervous System

- Name fungi that cause CNS disease.
- Describe the role of host immunity in fungal infections of the CNS.
- Define fungal CNS infections in the context of environmental exposure risks.
- Identify diagnostic methods and treatment options for different fungal CNS infections.

14. A major risk factor for acquiring fungal CNS infections is _____.
   A. environmental exposure
   B. being immunocompromised
   C. proximity to arthropod vectors
   D. A and B only
   E. A and C only

24.6
Parasitic and Prion Infections of the Central Nervous System

- Compare and contrast parasites and prions.
- Evaluate acquisition of parasitic CNS infections.
- Explain the role of host immune responses in parasitic and prion CNS disease.
- Name prevention and treatment options for parasitic and prion CNS disease.

16. A patient diagnosed with a parasitic keratitis is now suffering from meningitis secondary to spread of the parasite from his eye to his brain via the optic nerve. Which of the following is the most likely parasite?
   A. *Toxoplasma gondii*
   B. *Trypanosoma brucei*
   C. *Acanthamoeba* species
   D. *Cryptococcus neoformans*
   E. *Coccidioides posadasii*

12. An 8-month-old infant is brought to the emergency department by her parents, who report that she started having constipation and poor feeding, which progressed to lethargy and now difficulty breathing. The parents report that symptoms progressed over the past 24 hours. History shows that the child's last meal included a few bites of a natural peanut butter and honey. Which of the following organisms' spores might have contributed to this disease process?
   A. *Neisseria meningitidis*
   B. *Streptococcus pneumoniae*
   C. *Clostridium botulinum*
   D. *Escherichia coli*
   E. *Clostridium tetani*

13. A woman who is 4 months' pregnant is in the clinic complaining of a fever, flu-like symptoms, and not feeling well. Physical exam of the patient and laboratory analysis of her urine rule out any respiratory or urinary tract infections. She admits that her diet has consisted of only cheeses and crackers; everything else makes her nauseated. You are concerned about an infection by which of the following organisms?
   A. *Clostridium botulinum*
   B. *Escherichia coli*
   C. *Staphylococcus aureus*
   D. *Listeria monocytogenes*
   E. *Streptococcus pneumoniae*

15. A patient develops a fungal pneumonia a week after his vacation in Arizona to take pictures of the desert areas. He is now suffering from severe headache, blurred vision, and altered mental status. This patient is likely to be infected with _____.
   A. *Coccidioides immitis*
   B. *Cryptococcus neoformans*
   C. *Candida albicans*
   D. *Toxoplasma gondii*
   E. *Naegleria fowleri*

17. _____ is the drug of choice to treat *Naegleria* meningoencephalitis.

18. The etiologic agent of transmissible spongiform encephalitis (TSE) is _____.
   A. a protozoan
   B. a protein
   C. a small ribonucleic acid
   D. an ameba
   E. a nucleic acid

Key Terms

Acanthamoeba encephalitis (p. 832)
African sleeping sickness (p. 831)
arbovirus (p. 818)
areflexia (p. 812)
arthroconidium (p. 829)
aseptic meningitis (p. 813)
attenuated (p. 821)
blood-brain barrier (p. 809)
botulism (p. 826)
bovine spongiform encephalopathy (BSE) (p. 832)
Candida meningitis (p. 829)
central nervous system (CNS) (p. 808)
cerebrospinal fluid (CSF) (p. 808)
coccidioidal meningitis (p. 830)
Creutzfeldt-Jakob disease (CJD) (p. 832)

cryptococcal meningoencephalitis (p. 830)
eastern equine encephalitis (EEE) virus (p. 818)
empirical therapy (p. 822)
encephalitis (p. 812)
encephalopathy (p. 822)
endocytosis (p. 827)
exocytosis (p. 827)
hyperesthesia (p. 812)
listeriosis (p. 826)
lumbar puncture (p. 810)
lymphocytic choriomeningitis virus (LCMV) (p. 820)
meningitis (p. 812)
meningoencephalitis (p. 812)
meninx (p. 808)

motor neuron (p. 809)
mycosis (p. 829)
myelitis (p. 812)
Naegleria meningoencephalitis (p. 831)
neurotropic virus (p. 817)
nuchal rigidity (p. 822)
paresthesia (p. 812)
peripheral nervous system (PNS) (p. 808)
petechia (p. 825)
pleocytosis (p. 812)
poliomyelitis (p. 820)
primary encephalitis (p. 812)
prion (p. 832)
purpura (p. 825)
rabies (p. 808)
radiculitis (p. 812)

Sabin vaccine (p. 821)
Salk vaccine (p. 821)
secondary encephalitis (p. 812)
sensory neuron (p. 809)
tetanospasmin (p. 828)
tetanus (p. 828)
toxoplasmosis (p. 831)
transcytosis (p. 824)
transmissible spongiform encephalopathy (TSE) (p. 832)
vaccine-associated paralytic poliomyelitis (VAPP) (p. 821)
variant Creutzfeldt-Jakob disease (vCJD) (p. 832)
West Nile virus (WNV) (p. 818)

Review Questions

1. What is the basic anatomy of the CNS, and how does it relate to the PNS?

2. What is the source of cerebrospinal fluid?

3. What is the blood-brain barrier?

4. What is a lumbar puncture? How is it done?

5. What are three paths microbes can take to gain access to the CNS?

6. What are the three basic types of CNS infections?

7. How do the symptoms of encephalitis and meningitis differ?

8. What findings in an analysis of CSF would suggest viral versus bacterial meningitis?

9. What is the difference between primary and secondary encephalitis?

10. Discuss the pathogenesis, symptoms, treatment, and prevention of rabies, eastern equine encephalitis, and West Nile virus encephalitis.

11. Compare the presenting symptoms of viral meningitis with those of bacterial meningitis.

12. List the major causes of viral meningitis.

13. For which patient populations are viral CNS infections most dangerous?

14. List the major causes of bacterial meningitis and neonatal meningitis. When should antibiotics be administered?

15. Travel history can yield important clues as to the identity of which organisms that cause CNS infections?

16. For *Neisseria meningitidis* and *Listeria monocytogenes*, discuss the pathogenesis, symptoms, treatment, and prevention of disease.

17. Discuss the pathogenesis of botulism and tetanus. How can these diseases be prevented?

18. Discuss key features of major fungal infections of the CNS.

19. Discuss the nature of prions and how they are thought to cause disease.

Clinical Correlation Questions

1. A 1-week-old infant presents with a temperature of 38.9°C (102°F). The child is listless but cries when attempts are made to move her head. Her cerebrospinal fluid contains 180 white blood cells/mm³ (normal is ≤5), with 94% neutrophils. From this information, what do you suspect is the causative agent? What treatment would you recommend?

2. A 63-year-old female with a history of occasional back and knee pain presents to the emergency department of a local hospital complaining of having a crooked smile since receiving an injection at her doctor's office earlier in the morning. When asked to smile, the patient complies, but only half of her mouth opens to a proper smile, giving her smile a crooked appearance. Physical exam shows droopy eyelids and cockeyed eyebrows. The patient denies having any similar experience in the past or any trauma to her face. Which medical specialist did she see earlier, and which injection did she receive?

Thought Questions: CHECK YOUR ANSWERS

Thought Question 24.1   What will cause a CSF sample to contain a large number of red blood cells?

ANSWER: A traumatic lumbar puncture can cause overt bleeding at the site and contaminate the sample. Presence of many red blood cells in a CSF sample obtained without any trauma often indicates malignancy.

Thought Question 24.2   What steps could you take to limit your exposure to arboviruses?

ANSWER: Avoid the outdoors if possible; but when outdoors, use insect repellent and wear long-sleeved shirts and long pants when hiking in endemic areas.

Thought Question 24.3   How do the actions of tetanus toxin and botulinum toxin help the bacteria colonize or obtain nutrients?

ANSWER: This is a difficult question to answer. Few scientists have speculated. Recall that the toxins are encoded by genes in resident bacteriophages that became part of the clostridial genome through horizontal transfer from some other source. Because the organisms (vegetative as well as spores) normally reside in soil, the function of these toxins may have something to do with survival in that habitat. The toxin's effect on humans may simply be an unfortunate accident. However, with tetanus toxin, there may be benefit to the organism in that muscle spasms could limit oxygen delivery to infected tissues, enabling a more anaerobic environment for growth. Cell death may also release iron or other nutrients useful to *Clostridium tetani*.

Thought Question 24.4   If *Clostridium botulinum* is an anaerobe, how might botulinum toxin get into foods?

ANSWER: The most common way botulinum toxin gets into food today is via home canning. The canning process involves heating food in jars to very high temperatures. The heat destroys microorganisms and drives out oxygen; both processes help to preserve foods. If the jars are not heated sufficiently to ensure sterilization, spores of *C. botulinum* will survive. When the jars are cooled for storage at room temperature, the spores germinate; the organism then grows in the anaerobic medium and releases the toxin. When the food is eaten, the toxin is eaten, too.

25

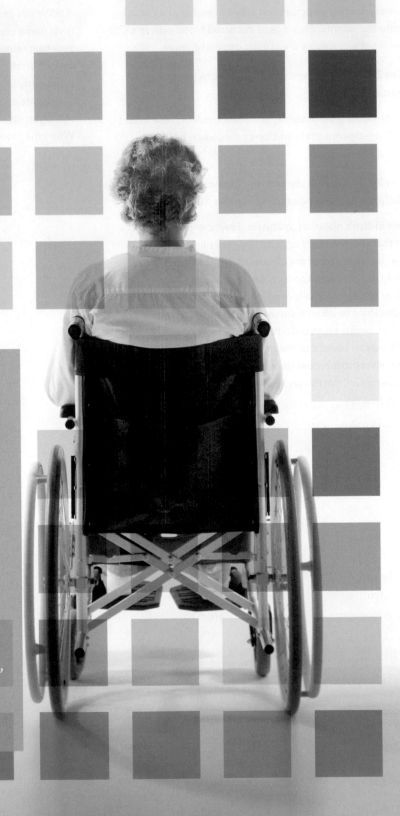

Diagnostic Clinical Microbiology

[Dazed and Confused in New Jersey]

PRESENTATION Jean, a 63-year-old woman from Plainfield, New Jersey, was brought by ambulance to the Muhlenberg Regional Medical Center, accompanied by her husband. On arrival she was disoriented and nonresponsive. Her husband said that Jean is normally quite active, but over the past 24 hours she had become increasingly tired and eventually could not answer simple questions. A stroke was suspected.

PHYSICAL EXAM AND PRELIMINARY LAB RESULTS There were no obvious signs of headache or unilateral muscle weakness (weakness on one side), which argued against a stroke. Blood was drawn and, because she could not walk to the bathroom, she was catheterized to collect urine. The laboratory reported that her white blood cell count was normal, but the nurse attending to Jean noticed that the urine in her collection bag was cloudy. A fresh urine specimen was obtained directly from her catheter and subjected to a rapid dipstick test, which in minutes revealed high levels of nitrites, indicating bacterial growth, and the enzyme leukocyte esterase, signifying the presence of white blood cells. Jean had a urinary tract infection (UTI).

CONFIRMATORY LABORATORY DIAGNOSIS
The urine specimen contained more than 100,000 bacteria per milliliter. Pink colonies growing on MacConkey agar indicated the organism was a Gram-negative rod that ferments lactose and was probably *E. coli,* a common cause of UTIs. On day 3, tests showed that the biochemical capabilities of the isolate matched those of *E. coli,* confirming the presumptive diagnosis. However, Jean was discharged long before these results were known.

TREATMENT AND OUTCOME The physician, knowing Jean had a UTI, started her on an intravenous quinolone antibiotic a mere 4 hours after admission. Within 24 hours Jean had regained full mental ability and was able to walk out of the hospital.

Urine Collection Bag
A nurse empties a urine collection bag. Because it may have been collecting for hours, this urine is unsuitable for bacteriological culture. Suitable urine is taken straight from the catheter or by suprapubic puncture.

Urinalysis
A urine dipstick used for rapid analysis. The plastic strip is dipped into fresh urine. Patches that monitor pH, nitrate, leukocyte esterase, protein, blood, glucose, bilirubin, and ketones change color. The presence of nitrite, leukocyte esterase, or both in the urine indicates urinary tract infection.

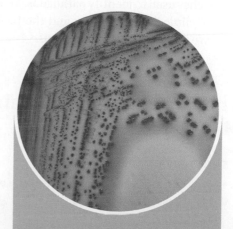

Escherichia coli
Escherichia coli growing on a MacConkey lactose plate. The colonies turn red from acid produced by lactose fermentation.

The symptoms in this case may appear strange for a urinary tract infection, but they are quite common among older patients (Chapter 23). Older patients with UTIs often do not notice the usual symptoms of flank pain or burning when urinating. Their altered mental status is due to the toxicity of the infection, which may spread into the bloodstream (sometimes called urosepsis; see the opening case history for Chapter 23). Without quick action, the patient could have died. The nurse's keen awareness and the rapid urine dipstick result certainly avoided a tragic outcome.

Many of the case histories sprinkled throughout this textbook name a pathogenic culprit without describing how it was identified. In this chapter we focus on identifying pathogens, a major objective of clinical microbiology. For instance, you may wonder how clinical microbiologists know which organisms to suspect in a given case. And you might be curious about what tests unequivocally identify the right pathogen. Another puzzle is how the *real* pathogen is found among the thousands of normal microbiota. The task would seem similar to finding the proverbial needle in a haystack.

Our coverage of clinical microbiology is, by necessity, selective. Not all techniques, nor all diseases, can be described. Our goal is simply to demonstrate the general principles and problem-solving approaches used to identify pathogens. You will see that the modern tools of clinical microbiology and the technologists that brandish them are indispensable to clinicians.

25.1
The Importance of Clinical Microbiology

What is the first step in investigating any infectious disease? As in any good detective mystery, the first step is to identify the most likely suspects. This can be accomplished, in part, from observing the disease symptoms in a patient and knowing what organisms typically produce those symptoms. It also helps if the physician is aware of a similar disease outbreak under way in the community. Beyond these clues, the etiological agent must be identified by using biochemical, molecular, serological, or antigen detection strategies.

Why take the time to identify an infectious agent? This is the first question many students ask when contemplating the effort and expense required to identify the genus and species of an organism causing an infection. Why not simply treat the patient with an antibiotic and be done with it? Although this approach sounds appealing, identifying an infectious agent is important for several compelling reasons:

- **Many bacteria are resistant to certain antibiotics.** As discussed in Chapter 13, antibiotic resistance is an increasingly serious global problem. Characterizing a microbe's antibiotic resistance profile must be part of identifying any microbe. Understanding which antibiotics can treat an infectious agent helps the physician avoid prescribing an inappropriate drug. In addition, life-threatening emergencies such as meningitis force physicians to make their best guess as to which antibiotic will target (or "cover") the likely pathogens *before* knowing the identity of the pathogen. This approach is called **empirical therapy**. But sometimes the physician's selection is wrong. Thus, the clinical laboratory must identify the causative organism and its antibiotic resistance pattern before the clinician knows whether the correct treatment was prescribed.

- **Antibiotic-resistant pathogens are spreading across the world.** A major task of national and international health organizations such as the U.S. Centers for Disease Control and Prevention (CDC) is to track the global spread of antibiotic-resistant pathogens. Consequently, knowing the identities of pathogens that cause disease and to which antibiotics they are resistant is crucial to this task. We must be aware, for instance, of which antibiotics work best on a pathogen in order to minimize treatment failures. A case in point is *Neisseria gonorrhoeae*. Before 1970 most clinical isolates were susceptible to penicillin, a drug widely used to treat gonorrhea. Today, most clinical isolates are penicillin-resistant.

Figure 25.1   Outbreak of Diarrhea in Babies

A. Infants attending the same day-care center can spread shigellosis between them via the fecal-oral route. However, parents will take each child to a different doctor.

B. Diagnostic laboratories will identify *Shigella* in each case and report this to local public health officials. Here the organism is growing on Hektoen agar, described later.

- **Specific pathogens are associated with secondary disease complications.** Clinicians often seek to identify a pathogen because many diseases have serious secondary complications linked to a given organism. For example, children whose sore throats are caused by certain strains of *Streptococcus pyogenes* can develop serious complications such as rheumatic fever long after the infection has resolved. These complications (disease sequelae) are the immunological consequence of bacterial and host antigen cross-reactivity (Section 17.4) Knowing early on that *S. pyogenes* has caused a child's sore throat allows the physician to prescribe antibiotics that will quickly eradicate the infection and prevent dangerous sequelae from developing.

- **Tracking the spread of a disease can lead to its source.** Consider a situation in which ten infants scattered throughout a city develop bloody diarrhea (**Figure 25.1A**). The clinical laboratory identifies the same strain of *Shigella sonnei*, a Gram-negative bacillus, as the cause in each case (**Figure 25.1B**).

Finding the same strain in all cases suggests that the pathogens probably originated from the same source. Shigellosis is transferred person to person by the fecal-oral route (Sections 2.4 and 22.5). Uninfected persons can become infected after eating contaminated food or by placing their contaminated fingers in their mouth. As soon as public health officials determine that the same strain of *Shigella* infected several patients, officials will question parents and might learn that all the children attend the same day-care center. By testing the other children and workers in that center, officials can stop the infection from spreading. Chapter 26 covers this investigative process, called **epidemiology**, more completely.

Link You will learn in Chapter 26 that **epidemiology** is the study of the transmission and incidence of disease.

 Overview of a medical microbiology lab: youtube.com

SECTION SUMMARY

- **Identifying a pathogen** enables us to:
  —**Use appropriate antibiotics,** if necessary.
  —**Anticipate possible sequelae.**
  —**Track and limit the spread** of the disease.

25.2
Specimen Collection

SECTION OBJECTIVES

- Identify sterile and nonsterile body sites.
- List procedures and precautions used to collect specimens from sterile body sites.
- List procedures used to collect specimens from body sites containing normal microbiota.

How do microbiologists "catch" pathogens hiding in clinical samples? To identify disease-causing microbes, clinicians and nurses must collect, and the laboratory must process, a wide variety of clinical specimens. Specimens can include simple cotton swabs of sore throats, in which the swab is placed into a liquid transport medium before being sent to the laboratory, as well as urine and fecal samples that are transported directly. Table 25.1 lists some common pathogens and diseases from earlier chapters, also presenting the general procedures labs use to detect these pathogens and the clinical samples involved. Each procedure requires a hypothesis as to the possible agent, and the savvy technologist has to know how the likely organisms grow and survive, as discussed in Chapter 6. We will explain these procedures later in this chapter. Here we discuss how the clinical specimens are collected and processed.

Table 25.1
Identification Procedures for Selected Diseases [Reference Table]

Agent	Disease	Examples of Diagnostic Tests
Bacteria		
Corynebacterium diphtheriae	Diphtheria	Material from nose and throat cultured on a special medium; in vivo or in vitro tests for toxin
Bordetella pertussis	Pertussis (whooping cough)	Smears of nasopharyngeal secretions stained with fluorescent antibody; ELISA for toxin in respiratory secretions; culture on special media; PCR
Legionella pneumophila	Legionellosis; Legionnaires' disease; Pontiac fever	Culture on special medium; antigen detection in urine by enzyme immunoassay; PCR and nucleic acid probes can identify *Legionella* nucleic acid
Campylobacter spp.	Campylobacteriosis	Isolate bacteria from stool on selective media (CAMPY blood agar plates) incubated at 42°C; PCR of stool samples
Leptospira interrogans	Leptospirosis	Serological tests early in the illness and after 2–3 weeks to detect rise in antibody titer; PCR
Listeria monocytogenes	Listeriosis	Culture of blood and spinal fluid; selective and enrichment cultures performed on food samples to grow potential pathogens; DNA probe to rapidly identify colonies
Chlamydia trachomatis	Chlamydial genital infections	Using monoclonal antibody to identify *C. trachomatis* antigen in urine or pus; nucleic acid probes
Treponema pallidum	Syphilis	Direct fluorescent antibody staining from tissue samples; serological tests
Francisella tularensis	Tularemia	Cultures from blood that use cysteine-containing media; fluorescent antibody stain of pus; detection of rise in antibody titer (specialized lab)
Yersinia pestis	Plague	Using fluorescent antibody or ELISA to identify capsular antigen; PCR; IgM serology (specialized lab)
Viruses		
Rhinovirus	Common cold	Using specific antibodies to identify strain from nasal secretions (specialized lab)
Influenza	Flu	Nasal secretions used for rapid antigen detection or viral culture; serology to compare influenza antibody levels in blood samples taken during acute and convalescent stages of illness; RT-PCR
Hantavirus	Hantavirus pulmonary syndrome	Using electron microscopy or monoclonal antibody to detect antigen in tissues; ELISA and Western blot tests for IgG and IgM antibodies in victim's blood (specialized lab)
Herpes simplex	Different strains cause cold sores, ocular lesions, and genital lesions	Viral culture; using fluorescent antibody or DNA probes to identify the viral antigen in clinical material (specialized lab)
Mumps	Mumps	Rise in serum antibody titer, or presence of IgM antibody to mumps virus in victim's blood; viral culture (specialized lab)
Rotavirus	Diarrhea	ELISA of diarrheal stool for virus; RT-PCR
HIV	AIDS	Detection of antibody to HIV-1 in patient's blood; Western blot; PCR
Fungi		
Coccidioides immitis	Coccidioidomycosis; infection in lung can disseminate to almost any tissue	Observing large, thick-walled, round spherules from clinical specimens; PCR identification
Histoplasma capsulatum	Histoplasmosis; infection of lung; sometimes disseminates	Stained material from pus, sputum, tissue, etc., examined for intracellular *H. capsulatum* yeast phase; blood tests for antibody to the organism; antigen detection in urine or blood

CASE HISTORY 25.1

A Literal Pain in the . . . Rectum

 A 4-year-old boy, Evan, was admitted to the hospital to evaluate and treat persistent pain in the rectal area. His problem began about 1 week earlier with ill-defined pain in the same region. He had a white blood cell count of 24,900 with 87% granulocytes (neutrophils). An abdominal computed tomography scan revealed an abscess (walled-off region of infectious bacteria) next to his rectum. A needle aspiration of the abscess drained 20 ml of yellowish, foul-smelling fluid. The specimen was plated on blood agar under aerobic conditions, but the culture was negative despite evidence of bacteria in the abscess drainage. (Blood agar is a rich medium containing 5% sheep red blood cells.) Why didn't the infectious agent grow?

This problem is related to specimen collection and processing. Internal abscesses located near the gastrointestinal tract are often anaerobic infections, in this case caused by the Gram-negative rod *Bacteroides fragilis*, a strict anaerobe (**Figure 25.2A**). Intestinal microbes, most of which are anaerobic, can sometimes escape the intestine if the organ is damaged. If the body's immune system cannot eliminate the infection, the tissues form a fibrous layer that walls off the infected region (an abscess). The specimen here should have been collected under anaerobic conditions by aspiration into a nitrogen-filled tube before transport to the clinical laboratory. Alternatively, a swab of the abscess material can be inserted into a special transport tube that has a built-in system to eliminate oxygen (**Figure 25.2B**). However, because the specimen was collected, transported, and handled in air, the oxygen probably killed many anaerobic microbes.

B. fragilis has a stress response system that permits the organism to survive for one or two days in oxygen, so some bacteria may have survived transport. The laboratory still had a chance to find the organism. This raises the second problem in the case study. Once the lab received the specimen, they cultured it only under aerobic conditions. The laboratory should have also incubated a series of plates anaerobically (see Figure 6.24). This contrived case shows the importance of both proper specimen collection and proper processing.

Two broad categories of specimens exist: specimens taken from body sites that are normally sterile and specimens taken from sites teeming with normal microbiota. Collecting specimens from each type of site comes with its own special considerations.

Samples from Sterile Body Sites

Some body sites should be sterile (contain no microorganisms) when collected from a healthy individual. These sites include blood, cerebrospinal fluid, pleural fluid from the space that lines the outside of the lung, synovial fluid from joints, peritoneal fluid from the abdominal cavity, and any tissues from internal organs. Because these sites are sterile, specimens can be plated onto nonselective agar media (blood or chocolate agar, for example) as well as selective media. Nonselective media can be used because any organism found in these specimens is considered significant.

BLOOD CULTURES Blood, of course, should be sterile. Blood cultures are used to evaluate possible blood infections (septicemia) or sepsis (Chapter 15 inSight and Section 21.3). Sepsis must be considered in any patient who presents with a very high fever, is experiencing chills, and appears disoriented. How, then, is a blood sample taken and processed for blood culture?

Figure 25.2 Anaerobic Infection

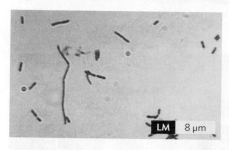

A. Gram stain of *Bacteroides fragilis* (1.5–4 µm long).

LM   8 µm

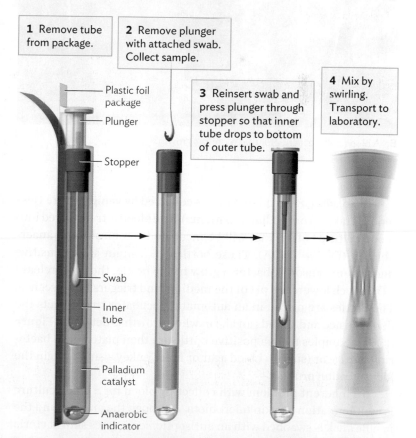

1 Remove tube from package.

2 Remove plunger with attached swab. Collect sample.

3 Reinsert swab and press plunger through stopper so that inner tube drops to bottom of outer tube.

4 Mix by swirling. Transport to laboratory.

- Plastic foil package
- Plunger
- Stopper
- Swab
- Inner tube
- Palladium catalyst
- Anaerobic indicator

B. Vacutainer anaerobic specimen collector. Plunging the inner tube to the bottom will activate a built-in oxygen elimination system. The anaerobic indicator changes color when anaerobiosis has been achieved.

Figure 25.3   Blood Cultures

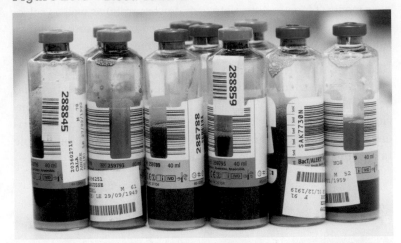

A. Typical blood culture bottles.

B. A blood culture incubator that monitors growth.

In an adult, 8–10 ml of blood is collected by **venipuncture** (less blood is taken from pediatric patients). The blood is transferred into blood culture bottles that will be incubated aerobically and anaerobically (Figure 25.3A). These bottles all contain a pH-sensitive fluorescent sensor. If bacteria grow in the bottle, they will release CO_2, which lowers the pH of the medium and triggers fluorescence. The bottles are placed in an automated incubator that detects the fluorescence and sounds an alarm when growth is detected (Figure 25.3B). Samples from a positive bottle are then plated onto bacteriological agar (such as blood agar or MacConkey agar) to begin the identification process.

An inherent problem with collecting blood for a blood culture is contamination by skin microbiota. Although the area around the needle stick is swabbed with an antiseptic, sometimes skin bacteria survive and contaminate the needle as it is inserted. The organism is then transferred with blood into the culture bottle. Thus *Staphylococcus epidermidis*, a prominent member of the skin microbiome, commonly contaminates blood cultures. One way to screen for

contaminants is to take blood from two body sites (left arm and right arm, for instance). If the same organism grows in both cultures, it is most likely the agent. However, contamination is the likely conclusion if an organism is found in one culture but the other culture is sterile, or if each culture yields a different organism.

 Blood culture procedure, Dalhousie: youtube.com

CEREBROSPINAL FLUID CULTURES   Meningitis is a rapidly progressive disease that can kill quickly (Chapter 24). Thus, identifying the microbial cause is crucial and time-sensitive. The lumbar puncture, commonly called a spinal tap (Figure 24.4), is the most common method for collecting cerebrospinal fluid (CSF). As Section 24.1 first described, the patient lies on his or her side, with knees pulled up toward the chest and chin tucked down (Figure 24.4A). After disinfecting and anaesthetizing the skin, a health care worker inserts a spinal tap needle, usually between the third and fourth lumbar vertebrae (in the lower spine). Once the needle is properly positioned in the subarachnoid space (the space between the spinal cord and its covering, the meninges), CSF pressures can be measured and fluid collected for testing.

A portion of CSF is spun in a centrifuge at low speed to concentrate any cells that may be present, and the sediment is Gram stained. No cells should be present unless an infection is under way (Figure 25.4). Any bacteria or white blood cells seen should be reported immediately. Other important tests performed on CSF include determining the amount of protein (normally low) and glucose (normally high). Higher than normal levels of protein and lower than normal glucose is a sign of bacterial infection. The CSF is then cultured onto blood agar and chocolate agar media, and/or rapid molecular tests such as PCR are performed to identify the agent.

Figure 25.4   *Neisseria meningitidis* in a CSF Sample

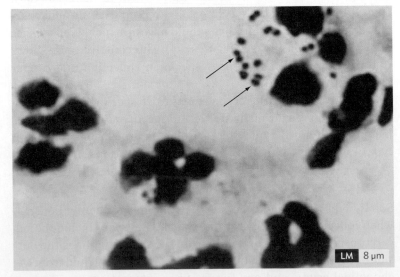

Gram stain of CSF from a meningitis patient infected with *N. meningitidis*. Arrows indicate bacteria found within infiltrating polymorphonuclear leukocytes.

PLEURAL, SYNOVIAL, AND PERITONEAL FLUID CULTURES

Infection or inflammation of lungs, joints, or the peritoneal cavity can cause excessive fluid accumulation at these body sites. If a clinician suspects an infection at these sites, samples of the fluids are aspirated using a sterile syringe and then cultured for the presence of bacteria or fungi. Aspirated fluids can be viewed microscopically for inflammatory cells but are typically cultured on agar plates and incubated aerobically and anaerobically.

URINE COLLECTION
Urine in the bladder of a healthy individual was once considered sterile. Studies now show that the bladder does, indeed, have normal microbiota, although their role in human health is unclear. Why didn't we see them before? Members of the urinary bladder microbiota are primarily anaerobic and not culturable under the typical aerobic conditions used to process urine. Thus, for all practical purposes clinical microbiologists consider urine in the bladder to be sterile, or nearly so. Symptoms of urinary tract infections appear only when significant numbers of aerobically culturable bacteria appear in urine.

When collected from a catheterized patient, urine should contain few, if any, aerobically culturable bacteria unless an infection is present. Catheterization involves passing a thin, sterile tube through the urethra and directly into the bladder (Figure 25.5A and Section 23.2). Catheterization is used primarily to assist urination by immobilized patients, but it also offers a convenient way to collect urine for bacteriological examination. Unfortunately, inserting the catheter through the nonsterile urethra can introduce culturable organisms into the bladder and cause an infection. Also, once a patient is catheterized, the urine should be collected only from the catheter, *never* from the collection bag. Urine may sit for hours in the collection bag, so organisms initially present at small, insignificant numbers have time to replicate to large numbers even if the patient does not have a UTI.

To determine how many bacteria are present in the urine, the clinical lab spreads a small sample (1–10 μl) over the entire surface of a blood agar plate, which is allowed to incubate aerobically for 24 hours (Figure 25.5B and C).

Each colony observed came from a single cell in the urine. If 10 μl was spread on the plate and 10 colonies grew, then 10 bacterial cells were present in the 10 μl of urine. Results, however, are reported as "cells per milliliter of urine," which here would be 1,000 colony-forming units per ml of urine (10 cells/0.01 ml = 1,000 cells/ml). Finding 100 or more cells of a single species per ml of *catheterized* urine indicates a UTI if the patient is symptomatic.

When a catheter is not in place, urine is commonly collected through a **midstream clean-catch technique**, which the patient performs. In this procedure, the patient cleans the external genitalia with a sterile wipe containing an antiseptic. The patient then partially urinates to wash as many organisms as possible out of the urethra and then collects 5–15 ml of the midstream urine in a sterile cup. This urine sample will probably not be truly sterile because of urethral contamination, but the number of culturable bacteria will be low.

Figure 25.5   Collecting a Urine Specimen

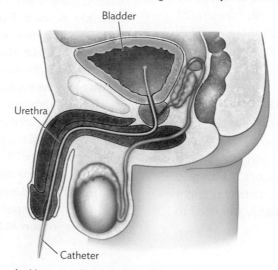

A. Urinary catheter showing placement in the urethra. A balloon attached to the end of the catheter (not shown) is inflated in the bladder to keep the catheter in place.

B. Calibrated loops used to streak urine samples onto agar media.

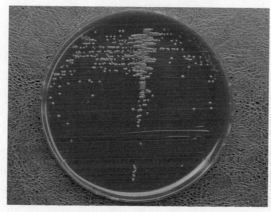

C. Result of plated urine sample. Each colony arose from a single organism present in the urine at collection.

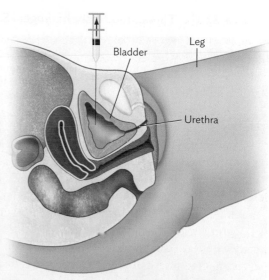

D. Suprapubic puncture. A sterile needle is inserted through the skin above the pubic arch and into the bladder.

The clinical laboratory determines how many organisms per milliliter are present in the midstream catch and informs the physician whether an infection is present. Finding more than 1,000 organisms of a single colony type per milliliter of urine from a midstream clean catch indicates an infection in a patient with symptoms of a UTI (burning pain upon urination, increased frequency of urination). These patients should be treated. However, 10,000 or fewer bacteria in a clean-catch specimen is usually considered normal in an *asymptomatic* patient. In fact, asymptomatic patients with 100,000 or more culturable bacteria per milliliter of urine (called asymptomatic bacteriuria) are usually not treated unless they are pregnant or will be undergoing an invasive urinary tract procedure.

Urine collection in babies is a special problem. They cannot perform a clean catch and are unlikely to leave a typical Foley balloon catheter in place. Short-term catheterization without inflating a balloon is usually tried first, but in rare instances the situation may call for a **suprapubic puncture**, in which a sterile needle is inserted through the abdomen into the bladder and urine is removed by aspiration (Figure 25.5D). Any bacterial number greater than 1,000 per ml (of a single species) suggests a UTI. Because of very high false-positive results, affixing a sterile bag to the baby's genitals with a sticky strip is no longer recommended.

Collections from Sites with Normal Microbiota

Identifying pathogens at sites that contain normal microbiota is harder. A stool (fecal) sample, for instance, is normally teeming with microbial species. These specimens are typically plated onto selective media (for example, MacConkey, Hektoen, or CNA agar) to eliminate or decrease the number of normal microbiota that might contaminate the specimen. The following techniques are used to collect specimens from nonsterile body sites:

- Swabs are used to collect samples from draining abscesses and from inflamed throat tissue (Figure 25.6A). Swabs should be placed in specialized liquid nutrient medium for transport to the laboratory. Upon arrival, the sample will be used to inoculate blood and chocolate agars.

- Nasopharyngeal swab samples collect secretions from the nasopharynx (Figure 25.6B). Some viruses and some bacterial pathogens, including *Bordetella pertussis*, infect this area. A special cotton-tipped flexible wire is passed through the nose and into the nasopharynx. Again, the swab should be placed in specialized liquid media for transport to the laboratory.

 Nasopharyngeal swab procedure: nejm.org

- Sputum comes from deep lung secretions and is the result of inflammation. It is not saliva. A lung infection results in a considerable amount of sputum, which can be expectorated for oral collection. Thus, sputum is an appropriate fluid to sample from patients suspected of having tuberculosis or pneumonia. Collection is usually performed in the morning with the patient sometimes sequestered in a sputum collection booth to prevent aerosolization (Figure 25.7A). Generally, a 2-ml volume is adequate. The sample should contain recently discharged material from the bronchial tree with minimal saliva content. If the patient has difficulty producing a sputum specimen (Figure 25.7B), sputum production may be induced by inhaling a warm aerosol of sterile 5%–10% sodium chloride in water produced by a nebulizer. Figure 25.7C shows an acid-fast stain of sputum containing *M. tuberculosis* (described further in Section 25.3).

Figure 25.6   Throat and Nasopharyngeal Samples

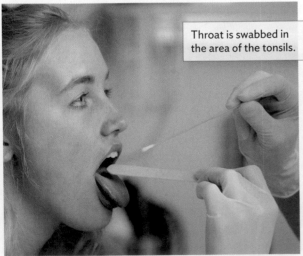
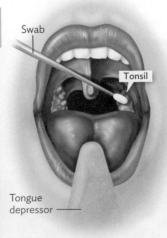
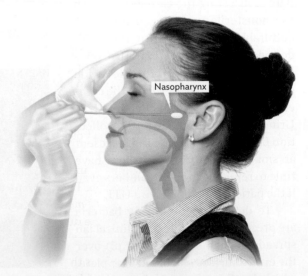

A. A sterile cotton-tipped applicator is used to swab the back of the throat around the tonsils and uvula. A sterile tongue depressor allows unobstructed access to the throat.

B. A nasopharyngeal swab. Sterile cotton on a flexible wire is inserted through the nose and back to the nasopharynx.

- Fecal stool samples are collected from patients complaining of diarrhea. These samples can be either cup samples (a teaspoon amount of feces) or a rectal swab. The laboratory will plate stool samples on a variety of selective and differential media (MacConkey, Hektoen; Section 6.2) or on more specialized media designed for *Campylobacter* or *Vibrio* species. The stool may also be examined for parasite eggs (ova) and the parasites themselves.

- Needle aspiration of deep wound abscesses is appropriate when an anaerobic infection is suspected (Case History 25.1). The needle of a syringe is inserted into the abscess and a portion of pus is withdrawn (aspirated) into the syringe. The aspirate should be transferred quickly to a transport vial containing no oxygen.

- Skin samples are taken when a mycotic or viral infection (papillomavirus, for example) is suspected. Skin scrapings are taken from the active border areas of lesions with the help of a sterile scalpel. When a fungal infection is suspected, several scrapings are placed in a drop of 10% KOH (potassium hydroxide) to dissolve tissue. The preparation is then examined under a bright-field microscope to detect fungal elements (hyphae, conidia, ascospores; Section 11.2). Samples from suspected viral lesions are transferred to 5 ml of phosphate-buffered glycerin virus transport medium at pH 7.6 and analyzed for viral antigens by a variety of immunological tests.

Handling Biological Specimens

CASE HISTORY 25.2

Microbe Hunter Infects Self

On July 15, an Alabama microbiologist was taken to the emergency department with acute onset of generalized malaise, fever, and diffuse myalgias. She received a prescription for oral antibiotics and was released. On July 16, she became tachycardic and hypotensive and returned to the hospital. She died 3 hours later. Blood cultures were positive for *N. meningitidis* serogroup C. Three days before symptoms appeared, the patient had prepared a Gram stain from the blood culture of a patient later shown to have meningococcal disease; the microbiologist also handled agar plates containing CSF cultures from the same patient. Coworkers reported that in the laboratory, aspirating fluids from blood culture bottles was typically performed at the open laboratory bench—without using biosafety cabinets, eye protection, or masks. Testing at CDC indicated that the isolates from both patients were indistinguishable. The laboratory at the hospital infrequently processed isolates of *N. meningitidis* and had not processed another meningococcal isolate during the previous 4 years.

Section 2.4 described the precautions that research and clinical laboratory personnel must take to avoid infecting themselves with the pathogens they handle. Recall that pathogens are grouped by their

Figure 25.7 Collecting Sputum

A. A patient with tuberculosis has coughed up sputum and is spitting it into a sterile container. (The patient is sitting in a special sputum collection booth that prevents the spread of tubercle bacilli. The booth is decontaminated between uses.)

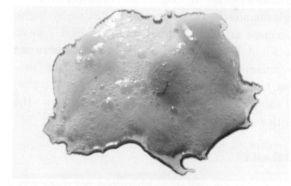

B. Sputum sample.

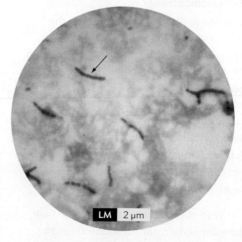

C. Acid-fast stain of *M. tuberculosis* in sputum. Note the reddish acid-fast bacilli (arrow) among the blue normal microbiota and white blood cells in the sputum, which are not acid-fast.

level of risk from low to high (risk levels 1–4). Risk level, in turn, dictates the biological safety level needed to protect the scientist and to prevent the pathogen from escaping the lab (BSL 1–4; Table 2.4). However, clinical laboratory workers do not know which, if any, pathogen lurks in the samples they process. Consequently, they must expect the worst. Then again, typical community hospitals rarely, if ever, encounter risk level 3 or 4 pathogens such as *Brucella* species or Ebola virus, respectively. So, routinely requiring level 4 containment procedures including airlocks and positive-pressure suits to process urine, blood, and stool would be overreacting.

In general, most clinical specimens are handled using level 2 biosafety procedures (including the use of a biosafety cabinet) if aerosolization is unlikely (streaking an agar plate, for example). When risk level 3 pathogens are suspected, such as *Mycobacterium tuberculosis,* culture manipulations must be performed under level 3 biosafety conditions, including negative-pressure airflow into the room. Biosafety level 4 conditions are used only when the severity of a patient's symptoms suggest infection by a risk level 4 agent such as hantavirus or Ebola virus.

What about biosafety procedures for nurses or technologists collecting patient samples? The nurse or phlebotomist (the person who collects blood) *must* thoroughly wash the hands and wear latex gloves when collecting or handling a specimen. This measure not only protects the hospital employee from contracting a disease but also will prevent transmitting potential pathogens from the provider to the patient. Chapter 26 talks more about hospital-acquired infections.

In addition to gloves, *sterile* collection devices such as syringes, catheters, collection cups, and cotton swabs must be used. A swab or collection cup accidentally contaminated by touch or sneeze can introduce microorganisms into the sample and complicate laboratory identification. Conversely, contaminated syringes or catheters used to collect samples can introduce new pathogens into the patient, causing septicemia or urinary tract disease.

SECTION SUMMARY

- **Special collection precautions** are necessary when collecting specimens from abscesses of suspected anaerobic etiology.
- **Common specimens** include sputum, throat, stool, urine from the bladder, and skin, as well as sterile ones such as blood and CSF.
- **Specimens** from sites containing normal microbiota must be processed on selective media to find the potential bacterial pathogen.
- **Appropriate biosafety protective measures** are necessary when collecting or handling potentially infectious biological materials.

Thought Question 25.1 Two blood cultures, one from each arm, were taken from a patient with high fever. One culture grew *Staphylococcus epidermidis,* but the other blood culture was negative (no organisms grew). Is the patient suffering from septicemia caused by *S. epidermidis*?

Thought Question 25.2 A 30-year-old woman with abdominal pain went to her physician. After the examination, the physician asked the patient to collect a midstream clean-catch urine sample that the office staff would send to the lab across town for analysis. The woman complied and handed the collection cup to the nurse. The nurse placed the cup on a table at the nurse's station. Three hours later, the courier service picked up the specimen and transported it to the laboratory. The next day, the report came back "greater than 200,000 CFUs/ml; multiple colony types; sample unsuitable for analysis." Why was this determination made?

25.3
Pathogen Identification Using Biochemical Profiles

SECTION OBJECTIVES

- Explain how biochemical profiles can be used to identify microbial pathogens.
- Discuss the major biochemical features that broadly differentiate Gram-negative and Gram-positive pathogens.
- Describe the Lancefield classification and its importance to clinical microbiology.
- Explain the difference between selective and differential media and their use in clinical microbiology.
- Discuss differences between the acid-fast stain and Gram stain.

Since thousands of bacterial species and viruses can cause disease, how can a laboratory quickly, sometimes within hours, identify which one is causing an infection? The solution, in part, comes from knowing the biochemical and enzymatic features of these bacteria. Because no two species have the same biochemical "signature," the clinical lab can look for reactions, or combinations of reactions, unique to a given species. This section discusses how to identify bacterial pathogens from their biochemistry.

CASE HISTORY 25.3

Medical Detective Work

Cindy, a 38-year-old woman with no significant previous medical history, came to the emergency department complaining of a mild sore throat persisting for 3 days. Her symptoms included arthralgia (joint pain), myalgia (muscle pain), and low-grade fever. The day before, she had a severe headache with neck stiffness, nausea, and vomiting. She was not taking medications, had no known drug allergies, and did not smoke. Cindy lived with her husband and two children, all of whom were well. Cerebrospinal fluid (CSF) was collected from a spinal tap and sent to the diagnostic laboratory. The CSF appeared cloudy (it should be clear) and contained 871 white blood cells (mainly neutrophils) per microliter (normal is $0–10/\mu l$); the glucose level was 1 mg/dl (normal is 50–80 mg/dl); and the total protein level was 417 mg/dl (normal is under 45 mg/dl). Gram stain of a CSF smear revealed Gram-negative rods. The sample was streaked onto chocolate, blood, and Hektoen agars for microbial identification.

As discussed in Sections 24.4 and 25.2, low glucose and elevated protein levels in CSF indicate *bacterial* infection, not viral. The increase in white blood cells revealed that the woman's immune system was trying to fight the disease. The presence of Gram-negative rods in the CSF smear confirmed a diagnosis of bacterial meningitis, since CSF should be sterile. Now it was up to the clinical laboratory to determine the etiological agent.

Problem-Solving Algorithms to Identify Bacteria

How does a clinical microbiology laboratory handle incoming specimens such as the one considered in Case History 25.3? What are the first steps toward identifying the etiological agent of a disease? Over the years, clinical microbiologists have developed algorithms (step-by-step problem-solving procedures) that expose the most likely cause of a given infectious disease. For instance, only a limited number of microbes cause meningitis. The microbiologist poses a series of binary yes-or-no questions about the clinical specimen in the form of biochemical or serological tests. Typical questions in this case might include the following:

- Is an organism seen in the CSF of a patient with symptoms of meningitis?
- Is the organism Gram-positive or Gram-negative?
- Does it stain acid-fast?

Answers to a first round of questions will dictate the next series of tests to be used.

Because speed is essential in deciding how to treat the patient, a series of tests is not always carried out sequentially. To save time, a slew of tests are carried out simultaneously, but the results are *interpreted* sequentially based on the algorithm. In our case history, for instance, consider the most common causes of bacterial meningitis (discussed in Chapter 24): *Neisseria meningitidis*, *Streptococcus pneumoniae*, *Haemophilus influenzae*, and *Escherichia coli*. A microbiologist Gram-stained the CSF sample and simultaneously plated it onto three media—chocolate agar, blood agar, and Hektoen agar. Chocolate agar is an extremely nutrient-rich medium that looks brown owing to the presence of heat-lysed red blood cells (**Figure 25.8A** and **B**). Because it is so nutrient-rich, all four organisms will grow on chocolate agar. However, nutritionally fastidious organisms such as *N. meningitidis* and *H. influenzae* will not grow well, if at all, on ordinary blood agar because these bacteria cannot lyse red blood cells and release required nutrients. Less fastidious organisms, such as *S. pneumoniae* and *E. coli*, will grow on blood agar, but only *E. coli* can grow on Hektoen agar (**Figure 25.8C–E**), which is a

Figure 25.8   Chocolate Agar and Hektoen Agar: Two Widely Used Clinical Media

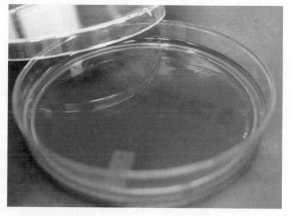

A. Uninoculated chocolate agar. Its color is due to gently lysed red blood cells that provide a rich source of nutrients for fastidious bacteria.

B. Chocolate agar inoculated with *Neisseria gonorrhoeae*. This organism will not grow well on typical blood agar because important nutrients remain locked within intact red blood cells, which *N. gonorrhoeae* cannot lyse.

C. Uninoculated Hektoen agar, which contains lactose, peptone, bile salts, thiosulfate, an iron salt, and the pH indicators bromothymol blue and acid fuchsin—the bile salts prevent growth of Gram-positive microbes.

E. Hektoen agar inoculated with *Salmonella enterica*. This organism does not ferment lactose but instead grows on the peptone amino acids. The resulting amines are alkaline and produce a more intense blue color with bromothymol blue. *Salmonella* species also produce hydrogen sulfide gas from the thiosulfate. Hydrogen sulfide reacts with the medium's iron salt to produce an insoluble, black iron sulfide precipitate visible in the center of the colonies.

D. Hektoen agar inoculated with *Escherichia coli*. This organism ferments lactose to produce acidic fermentation products that give the medium an orange color owing to the pH indicators acid fuchsin and bromophenol blue.

Figure 25.9   API 20E Strip Technology to Biochemically Identify Enterobacteriaceae

Uninoculated strip

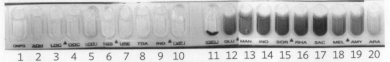

1 2 3 4 5 6 7 8 9 10   11 12 13 14 15 16 17 18 19 20

A. Uninoculated API strip. Each well contains a different medium that tests for a specific biochemical capability. The numbers correspond to those in Table 25.2. The color of the media after 24-hour incubation indicates a positive or negative reaction (see Table 25.2).

E. coli results after 24 hours

B. API results for _E. coli_. Plus (+) and minus (−) indicate positive and negative reactions, respectively.

P. mirabilis results after 24 hours

C. API results for _Proteus mirabilis_.

selective and differential medium for enteric Gram-negative rods. Section 6.2 describes **differential and selective media**.

Link ⟩ As discussed in Section 6.2, a **selective medium** favors the growth of one organism over another, whereas a **differential medium** exposes biochemical differences between two species that grow equally well. Hektoen agar is both selective and differential. Bile salts in the medium prevent the growth of Gram-positive organisms (selective). If the organism ferments the sucrose or lactose in the medium and makes acid, the pH indicators that give the plate its green color will turn the colony orange (differential).

In Case History 25.3, the Gram stain of the CSF revealed Gram-negative rods, which ruled out _Neisseria_ (a Gram-negative diplococcus) and _Streptococcus pneumoniae_ (a Gram-positive diplococcus). The organism in CSF did grow on blood agar, which eliminated _Haemophilus influenzae_ (a Gram-negative, nonenteric rod) as a candidate. It also grew on Hektoen, where it produced orange, lactose-fermenting colonies. Thus, the organism was a Gram-negative, enteric rod—probably _E. coli_. Additional biochemical tests confirming the identity of the organism had to be carried out, but this simple example shows how simultaneous tests can be interpreted.

IDENTIFYING GRAM-NEGATIVE BACTERIA   The Gram-negative bacterium afflicting Cindy in Case History 25.3 was subjected to a battery of 36 biochemical tests in an automated microbial

identification instrument described below. A simplified example of the identification process is shown in **Figure 25.9**. The analytical profile index (API 20E) strip can test 20 metabolic processes. Such a test battery is called a **probabilistic indicator**. The results of all the individual tests are compared with those from a database of pathogen tests, which lists probabilities of a positive or negative result for each test and for each organism. The strip requires overnight incubation and tests whether the organism can ferment a series of carbon sources. It also provides evidence of specific products of fermentation. The results appear as colored reactions in each chamber and are scored as positive or negative, depending on the color (**Table 25.2**). For example, in the indole chamber (ninth well from the left in Figure 25.9B), a red reaction is positive and indicates that the organism can produce indole from tryptophan. A colorless chamber would be a negative result.

The results of the API chambers can be interpreted as a dichotomous key, with lab personnel making a stepwise interpretation starting with a key reaction, such as lactose fermentation (tagged ONPG in Table 25.2). The reaction is read as positive or negative on the basis of the color. Then, following a printed flowchart, the technologist would go to the next key reaction—say, indole production, and read it as positive or negative. If the organism is a lactose fermenter and the indole test is positive, the choices have been narrowed to _E. coli_ or _Klebsiella_ species. Another reaction is read to distinguish between the remaining choices. The process continues until a single species is identified.

A simplified example is shown in **Figure 25.10** using a limited number of enteric Gram-negative species. In reality, it takes many more reactions than the 11 shown to make a definitive identification because of species differences with respect to a single reaction. For example, Figure 25.10 shows that _K. pneumoniae_ and _K. oxytoca_ exhibit opposite indole reactions, even though they are of the same genus. Even within a given species, only a certain percentage of strains might be positive for a given reaction. The inherent danger in using a dichotomous key is that one anomalous result can lead to an incorrect identification.

Most clinical laboratories in the United States and Europe now use more sophisticated automated identification systems. These include BioMerieux's Vitek; Siemens' WalkAway Plus; and Becton, Dickinson, and Company's Phoenix systems (**Figure 25.11A**). All use cards or plates with 30 or more biochemical or enzymatic reactions (**Figure 25.11B** and **C**). As with the API chambers, the results appear as colored reactions readable by computer. The BD Phoenix system, for example, will convert results into a ten-digit code used to identify genus and species.

IDENTIFYING NONENTERIC GRAM-NEGATIVE BACTERIA
The procedures we have outlined will accurately identify members of Enterobacteriaceae, but pathogenic Gram-negative bacilli can also be found in other phylogenetic families. One possibility in Cindy's case history is that the Gram-negative bacillus seen in CSF smears would _not_ grow on blood agar or on the other selective

Table 25.2
Reading the API 20E [Reference Table]

Test	Substrate	Reaction Tested	Negative Results	Positive Results
1 ONPG	ONPG	Beta-galactosidase	Colorless	Yellow
2 ADH	Arginine	Arginine dihydrolase	Yellow	Red/orange
3 LDC	Lysine	Lysine decarboxylase	Yellow	Red/orange
4 ODC	Ornithine	Ornithine decarboxylase	Yellow	Red/orange
5 CIT	Citrate	Citrate utilization	Pale green/yellow	Blue-green/blue
6 H$_2$S	Na thiosulfate	H$_2$S production	Colorless/gray	Black deposit
7 URE	Urea	Urea hydrolysis	Yellow	Red/orange
8 TDA	Tryptophan	Deaminase	Yellow	Brown-red
9 IND	Tryptophan	Indole production	Yellow	Red (2 min)
10 VP	Na pyruvate	Acetoin production	Colorless	Pink/red (10 min)
11 GEL	Charcoal gelatin	Gelatinase	No diffusion of black	Black diffuse
12 GLU	Glucose	Fermentation/oxidation	Blue/blue-green	Yellow
13 MAN	Mannitol	Fermentation/oxidation	Blue/blue-green	Yellow
14 INO	Inositol	Fermentation/oxidation	Blue/blue-green	Yellow
15 SOR	Sorbitol	Fermentation/oxidation	Blue/blue-green	Yellow
16 RHA	Rhamnose	Fermentation/oxidation	Blue/blue-green	Yellow
17 SAC	Sucrose	Fermentation/oxidation	Blue/blue-green	Yellow
18 MEL	Melibiose	Fermentation/oxidation	Blue/blue-green	Yellow
19 AMY	Amygdalin	Fermentation/oxidation	Blue/blue-green	Yellow
20 ARA	Arabinose	Fermentation/oxidation	Blue/blue-green	Yellow

media, but it would grow as small, glistening colonies on chocolate agar. This result would implicate the Gram-negative rod *Haemophilus influenzae*.

Growth of *H. influenzae* requires hemin (X factor) and NAD (V factor), so confirming the identity as *H. influenzae* involves growing the organism on agar medium containing hemin and NAD. This can be done by placing small filter paper disks containing these compounds on a nutrient agar surface (not blood agar) covered with the organism. *H. influenzae* will grow only around a strip containing both X and V factors. Alternatively, X and V factors can be incorporated into Mueller-Hinton agar, as shown in **Figure 25.12A**. The organism grows on chocolate medium because the lysed red blood cells release these factors. Although XV growth phenotype is still used for identification, fluorescent antibody staining (discussed shortly) is more specific.

Link As discussed in Section 6.1, hemin is a heterocyclic compound found in hemoglobin and cytochromes that bind iron. Section 4.4 explains that NAD (nicotinamide adenine dinucleotide) is a compound that removes hydrogen and electrons from compounds undergoing catabolism.

The laboratory would have used a completely different identification scheme in the preceding meningitis case had workers found a Gram-negative diplococcus (rather than a Gram-negative rod). A Gram-negative diplococcus would suggest *Neisseria meningitidis*. *N. gonorrhoeae* is also possible, although less likely because the gonococcus lacks the protective capsule that meningococci use to survive in the bloodstream. Without this capsule, *N. gonorrhoeae* is not as resistant to serum complement and cannot disseminate to the meninges.

Figure 25.10

Simplified Biochemical Algorithm to Identify Gram-Negative Rods

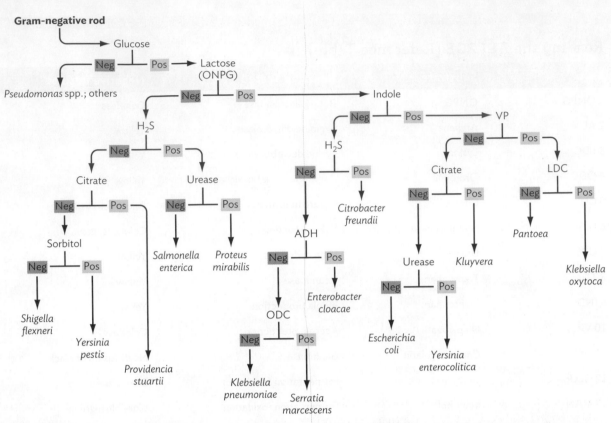

The diagram presents a dichotomous key using a limited number of biochemical reactions and selected organisms to illustrate how biochemistry is used to identify species. For example, if our specimen was positive for glucose and lactose and indole and VP and LDC, it is identified as *Klebsiella oxytoca*. Abbreviations and reactions: ADH (arginine dihydrolase): stepwise degradation of arginine to citrulline and ornithine; Citrate: use of citrate as a carbon source; H_2S: production of hydrogen sulfide gas; Indole: indole production from tryptophan; Lactose (ONPG): lactose fermentation to produce acid (orthonitrophenyl beta-D-galactoside, cleaved by beta-galactosidase); LDC (lysine decarboxylase): cleavage of lysine to produce CO_2 and cadaverine; ODC (ornithine decarboxylase): cleavage of ornithine to make CO_2 and putrescine; Sorbitol: sorbitol fermentation to produce acid; Urease: production of CO_2 and ammonia from urea; VP (Voges-Proskauer test): production of acetoin or 2,3-butanediol. (*Note:* "Neg" for *Pseudomonas* in terms of glucose indicates an inability to ferment glucose. *Pseudomonas* can still use glucose as a carbon source.)

Figure 25.11   Automated Microbiology Identification System

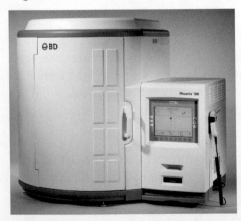

A. The BD Phoenix system uses plates with many reaction wells and a computerized plate reader to automatically identify pathogenic bacteria. Such an instrument automatically generates and evaluates the combination of reactions that develop as the bacteria grow.

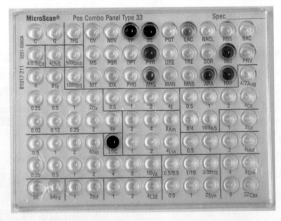

B. Microbial identification plate.

C. Loading a multiwell ID plate with a multichannel pipettor. All wells are simultaneously loaded with the same volume and number of bacteria.

The first test to determine whether the organism is a species of *Neisseria* is the **cytochrome oxidase test** (Figure 25.12B). In this test, a few drops of the colorless reagent *N,N,N′,N′*-tetramethyl-*p*-phenylenediamine dihydrochloride are applied to the suspect colonies. The reaction, which takes place only if the organism possesses both cytochrome oxidase and cytochrome *c*, turns the *p*-phenylenediamine reagent (and the colony) a deep purple/black. Although many bacteria possess cytochrome oxidase, *Neisseria* is one of only a few genera whose members also contain cytochrome *c* in their membranes. Other oxidase-positive bacteria include *Pseudomonas*, *Haemophilus*, *Bordetella*, *Brucella*, and *Campylobacter* species—all Gram-negative rods. All Enterobacteriaceae, however, are oxidase-negative because they lack cytochrome *c*.

Finding an oxidase-positive, Gram-negative diplococcus in Cindy's case would strongly indicate that the pathogen is a member of *Neisseria*. Differentiating between species of *Neisseria* then involves testing their ability to grow on certain carbohydrates. Once identified as *N. meningitidis,* its capsule type can be determined using immunofluorescent antibody staining.

IDENTIFYING GRAM-POSITIVE PYOGENIC COCCI Recall the woman with necrotizing fasciitis (see Case History 19.3). How did the laboratory determine that the etiological agent was *Streptococcus pyogenes*? A sample algorithm, or flowchart (Figure 25.13), shows how. The physician sends a cotton swab containing a sample from a lesion to the clinical laboratory. The laboratory technologist streaks the material onto several media: (1) blood agar (on which both Gram-positive and Gram-negative organisms will grow), (2) blood agar containing the inhibitors colistin and nalidixic acid (called a **CNA plate**; *only* Gram-positives will grow on this agar), and (3) MacConkey agar (on which only Gram-negative bacilli will grow) (see Section 6.2). The suspect organism in this case grows on the CNA and blood plates. Because they grow in the presence of the compounds in CNA that inhibit growth of Gram-negative organisms, one would immediately suspect the organism to be Gram-positive, an assumption borne out by the actual Gram stain.

Note Though skin is normally populated by many different microorganisms that could potentially contaminate a wound (as in Case History 19.3), clinical samples from an infected lesion are populated primarily by the etiological agent. This occurs because the pathogenic microbe outgrows normal microbiota. Using selective media to isolate the infectious agent will further simplify diagnosis by reducing growth of any normal microbiome members still present.

Figure 25.12 *Haemophilus influenzae* **Growth Factors and** *Neisseria meningitidis* **Oxidase Reaction**

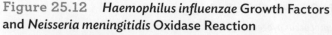

A. *H. influenzae* will grow on an agar plate (here, Mueller-Hinton agar) only when the medium has been fortified with both X factor (hemin) and V factor (NAD), but not either one alone.

B. Oxidase-positive reaction for *Neisseria meningitidis*. Oxidase reagent (which is colorless) was dropped onto colonies of *N. meningitidis* grown on chocolate agar. The test is called the cytochrome oxidase test, but it really tests for cytochrome *c*.

The algorithm tells the laboratory technologist that since the organism is a Gram-positive coccus, the next step is to test for **catalase** production. Catalase, which converts hydrogen peroxide (H_2O_2) to O_2 and H_2O, clearly distinguishes staphylococci from streptococci (remember, the morphology, or shape, of Gram-stained cells alone is not reliable enough to make that distinction). The **catalase test** is performed by mixing a colony with a drop of H_2O_2 on a glass slide. Effusive bubbling due to the release of oxygen indicates catalase activity (see Figure 25.13). Staphylococci are catalase-positive, whereas streptococci are catalase-negative. Many other organisms possess catalase activity, including the Gram-negative rod *E. coli*. However, on the basis of the algorithm, *E. coli* would not be considered because it does not grow on CNA agar and is not Gram-positive.

Note When performing a catalase test from colonies grown on blood agar, be sure not to transfer any of the agar, since red blood cells also contain catalase.

Having established that the organism is catalase-negative, the technologist examines the blood plate for evidence of **hemolysis**, the microbe's ability to directly lyse red blood cells. Three types of colonies are possible: nonhemolytic, alpha-hemolytic, and beta-hemolytic. **Nonhemolytic** streptococci do not produce a lytic zone around their colonies. **Alpha-hemolytic** strains produce large amounts of hydrogen peroxide that oxidize the heme iron within intact red blood cells to produce a green product. As a result, alpha-hemolytic streptococci produce a green zone around their colonies called **alpha hemolysis**—even though the red blood cells remain intact. (For example, *Streptococcus mutans*, a cause of dental caries and subacute bacterial endocarditis, is alpha-hemolytic.)

Figure 25.13   Algorithm to Identify Gram-Positive Pathogenic Cocci

The red arrows follow the identification of *Streptococcus pyogenes*. The bacitracin and optochin results are designated "positive" if the organism is susceptible (note the zone of no growth around the disk) and "negative" if the organism is resistant to the agent.

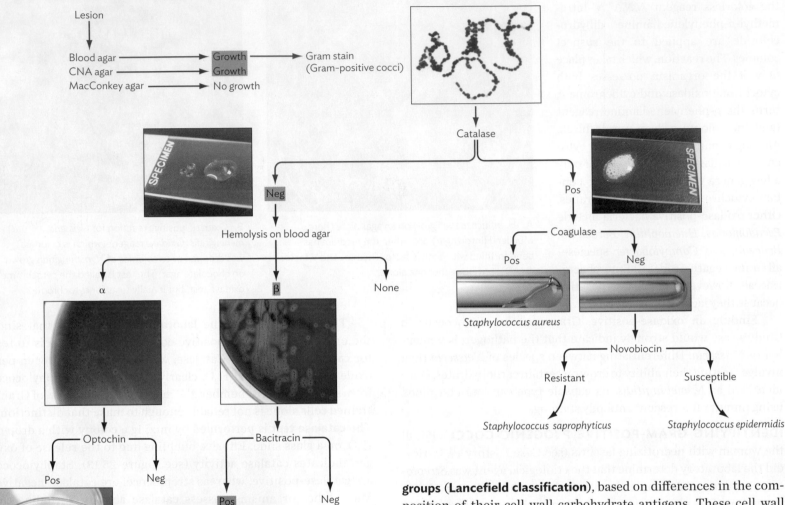

Streptococcus pneumoniae

Streptococcus pyogenes

Staphylococcus aureus

Staphylococcus saprophyticus

Staphylococcus epidermidis

groups (**Lancefield classification**), based on differences in the composition of their cell wall carbohydrate antigens. These cell wall differences are distinguished from each other immunologically and divide the streptococci into Lancefield groups A–U. Rebecca Lancefield, for whom the classification scheme is named, was the first to use immunoprecipitation to group the streptococci. Group A beta-hemolytic streptococci (also called GAS), defined as the species *Streptococcus pyogenes*, cause most streptococcal diseases.

Unfortunately, the Lancefield classification procedure is somewhat time-consuming and not readily amenable as a rapid identification method. However, the group A beta-hemolytic streptococci are uniformly susceptible to the antibiotic bacitracin. Thus, a simple bacitracin disk susceptibility test can be used to indicate the group A beta-hemolytic streptococci (that is, *S. pyogenes*; Figure 25.13). But beware: many bacteria are bacitracin sensitive. So like the catalase test, the bacitracin test must be used as part of an algorithm to be useful for identification. The technologist must follow the appropriate algorithm before assigning importance to this or any other test result. It is irrelevant, for instance, if an alpha-hemolytic organism is bacitracin susceptible. Some such organisms may

Still other streptococci produce a completely clear zone of true hemolysis surrounding their colonies, called **beta hemolysis**. Complete hemolysis of red blood cells occurs owing to the export of bacterial enzymes, called hemolysins, that lyse red cell membranes. The flowchart indicates that the organism from Case History 19.3 was beta-hemolytic.

The final relevant test in this flowchart involves susceptibility to the antibiotic bacitracin, which identifies the most pathogenic group of beta-hemolytic streptococci. The beta-hemolytic streptococci are subdivided into many groups, known as the **Lancefield**

exist, but they are not associated with disease. The organism in our case of necrotizing fasciitis, however, was beta-hemolytic, so bacitracin susceptibility indicated that the organism was *S. pyogenes*.

The other tests named in Figure 25.13 are equally important for identifying Gram-positive infectious agents. For example, *Streptococcus pneumoniae* is an important cause of pneumonia. Like *S. pyogenes*, *S. pneumoniae* is a catalase-negative, Gram-positive coccus; unlike *S. pyogenes*, it is alpha-hemolytic. Optochin susceptibility is a property closely associated with *S. pneumoniae*, whereas other alpha-hemolytic strains of streptococci are resistant to this compound. Thus, an **optochin susceptibility disk test** is a useful tool to identify *S. pneumoniae*.

Coagulase is a key reaction used to distinguish the pathogen *Staphylococcus aureus*, which causes boils and bone infections, from other staphylococci, such as the normal skin species *S. epidermidis*. To conduct a coagulase test, a tube of plasma is inoculated with the suspect organism. If the organism is *S. aureus*, it will secrete the enzyme coagulase. Coagulase will convert fibrinogen to fibrin and produce a clotted, or coagulated, tube of plasma. (More recently, diagnostic labs are using an immunochromatographic enzyme immune assay method to more quickly detect coagulase [Section 25.6].) Coagulase-negative staphylococci can still be medically important, however. *S. saprophyticus*, for instance, is an important cause of urinary tract infections. It can be distinguished from *S. epidermidis* on the basis of resistance to the antibiotic novobiocin.

IDENTIFYING ACID-FAST BACTERIA A few pathogenic bacteria cannot be Gram stained, most notably *Mycobacterium tuberculosis*, the legendary agent of tuberculosis (Section 20.3). Mycobacteria do not Gram stain well because the organisms sport a waxy outer coat composed of **mycolic acid** that resists penetration by most dyes. However, a harsher procedure called the **acid-fast stain**, a technique that German scientists Franz Ziehl and Friedrich Neelsen described in 1882, can stain the organism. The original Ziehl-Neelsen acid-fast stain used phenol and heat to drive carbol fuchsin (a red dye) into mycobacterial cells on glass slides. Destaining with an acid alcohol solution removes the stain from all bacterial pathogens *except* mycobacteria. After rinsing, the slide is counterstained with methylene blue. Mycobacteria will be seen as curved, red rods (called acid-fast bacilli), whereas everything else looks blue (Figure 25.7C).

A more modern version of the acid-fast stain uses fluorochrome auramine O to stain the mycolic acid. This fluorescent dye also resists removal by an acid alcohol wash, and when observed under a fluorescence microscope, the dye causes mycobacteria to fluoresce bright yellow (Figure 25.14A). A presumptive diagnosis of tuberculosis is made after finding acid-fast bacilli in a patient's sputum.

Because *M. tuberculosis* grows slowly, taking 4 weeks to form colonies on agar media, other fast-growing organisms present in sputum can easily overgrow tubercle bacilli on ordinary media. Löwenstein-Jensen (LJ) medium is a selective medium designed to grow *M. tuberculosis*. LJ medium contains antibiotics to inhibit the

Figure 25.14   Fluorescent Acid-Fast Stain and Growth of *Mycobacterium tuberculosis*

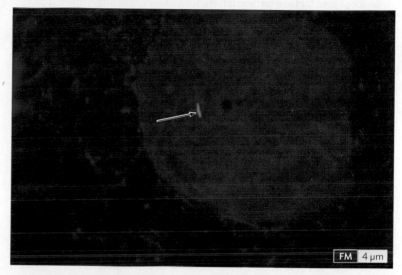

A. Auramine O acid-fast stain of *M. tuberculosis*. Figure 25.7C shows a standard acid-fast stain.

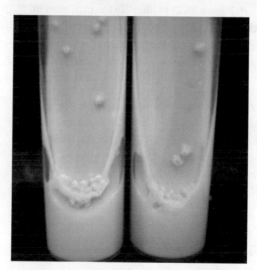

B. Löwenstein-Jensen medium enables growth of mycobacterial species, some of which grow extremely slowly. The colonies look like bread crumbs.

growth of other bacteria and uses coagulated eggs and potato starch as nutrients. The medium is typically used as a slanted agar in a test tube rather than a petri plate (Figure 25.14B). Biochemical tests can identify *M. tuberculosis* from among nontuberculosis mycobacteria, but newer nucleic acid–based methods can do this much faster than traditional culture-based techniques (described later).

SECTION SUMMARY

- **Selective media** inhibit growth of one group of organisms while permitting the growth of others (such as Gram-positive bacteria versus Gram-negative bacteria). This technique is often used to prevent growth of normal flora while permitting the growth of pathogens.

- **Differential media** exploit the unique biochemical properties of a pathogen to distinguish it from similar-looking nonpathogens.

- **Identification through biochemical tests** uses a dichotomous key in which each test eliminates possible pathogens until only one remains. **Glucose and lactose fermentation,** for instance, are often the first tests used to separate species of Gram-negative bacilli.

- **Automated systems** can simultaneously evaluate 30 or more biochemical tests and determine the probability that the biochemical profile of a pathogen recovered from a patient matches that of a given species.

- **The catalase test** distinguishes the **Gram-positive cocci** recovered from a patient into *Staphylococcus* (catalase positive) or *Streptococcus* (catalase negative) genera. The **coagulase test** differentiates *Staphylococcus aureus* (coagulase positive) from *Staphylococcus epidermidis* (coagulase negative).

- **Hemolysis** differentiates catalase-negative, Gram-positive cocci (*Streptococcus*) into nonhemolytic, alpha-hemolytic, and beta-hemolytic groups. **Bacitracin susceptibility** of a beta-hemolytic streptococcus indicates the strain is *S. pyogenes*.

- **The Lancefield classification** differentiates beta-hemolytic streptococci into serological groups. Group A streptococci (*S. pyogenes*) and group B streptococci (*S. agalactiae*) are the most common pathogens.

- **The acid-fast stain** differentiates *Mycobacterium* and some *Actinomyces* species from all other bacteria.

Thought Question 25.3 Use Figure 25.13 to identify the organism in the following case. A sample was taken from a boil located on the arm of a 5-year-old boy. Bacteriological examination revealed the presence of Gram-positive cocci that were also catalase-positive, coagulase-positive, and novobiocin-resistant.

Thought Question 25.4 Staphylococci from blood cultures were isolated from two body sites, but the laboratory says the organisms are probably not the cause of the patient's disease. Why might the laboratory technologist reach that conclusion?

25.4
Pathogen Identification by Genetic Fingerprinting

SECTION OBJECTIVES

- Discuss how the polymerase chain reaction can be used to identify pathogens in a clinical sample and the role of real-time PCR in clinical diagnosis.
- List the advantages and disadvantages of PCR, restriction fragment length polymorphisms, and biochemical strategies of pathogen identification.

What is the quickest way to identify a pathogen from a clinical specimen? From the time the clinical lab receives a specimen, determining the biochemical profile of the likely agent can take days, or sometimes weeks. This can be a problem if the clinician has chosen the wrong empiric antimicrobial therapy to treat the patient.

Because we now know the genome sequences of many pathogens, clinical labs can use rapid, nucleic acid–dependent techniques to identify a particular pathogen. Detecting DNA (for example, PCR; Section 8.3) takes only a few hours. DNA/RNA detection methods are especially useful for viruses, which otherwise require tissue culture techniques to grow the virus or serology to detect an increase of antiviral antibodies. Recall that antibodies may not appear in serum until the patient is recovering. Also, molecular detection techniques may provide the *only* means to identify newly recognized, or emerging, pathogens.

PCR

The polymerase chain reaction (PCR) is the most widely used molecular method in the clinical laboratory's diagnostic arsenal. DNA primers that bind to unique genes in a pathogen's genome can be used to specifically amplify DNA or RNA present in a clinical specimen. Successful amplification can be seen as an appropriately sized fragment in agarose gels after electrophoresis or, more typically, by real-time techniques, which we will discuss later.

Why do we need PCR to detect these nucleic acids? Without the PCR amplification steps, clinical samples usually provide too little nucleic acid from infecting microorganisms to be detected. For example, *Mycobacterium tuberculosis*, the cause of tuberculosis, can take weeks to grow on standard bacteriological media. Detecting its DNA in the specimen, however, yields a rapid diagnosis. Unfortunately, clinical samples have only minute amounts of pathogen DNA to detect. PCR, however, amplifies *M. tuberculosis* nucleic acid, turning one copy of DNA into billions of copies.

Preparing a clinical specimen for PCR by extracting the DNA or RNA usually takes less than an hour. PCR itself is completed in 2–3 hours. What might take 2–3 days (or sometimes weeks) using biochemical algorithms may take less than a day with molecular strategies. Most clinical laboratories are now equipped with **thermocyclers** (the instrument that precisely and rapidly cycles the temperature during PCR) and can carry out this type of identification protocol for organisms for which specific DNA primers are available.

Table 25.3 lists instances in which DNA detection tests are useful. An example in Figure 25.15 shows using PCR to type different strains of the anaerobic pathogen *Clostridium botulinum*, the cause of food-borne botulism (Section 24.4). Unlike *S. pneumoniae*, these organisms are not typed serologically. Rather, *C. botulinum* is divided into types based on what neurotoxin genes they possess. In this example, **multiplex PCR** was used to simultaneously search for these toxin genes. Multiplex PCR uses multiple sets of primers, one pair for each gene, combined in one tube with a specimen. One must ensure that the primers chosen yield different-sized products, do not interfere with each other, and do not produce artifactual PCR products that can confound interpretation. Multiplex PCR can help identify sets of specific genes present in a single species or can screen for multiple pathogens in a clinical sample. In the latter case, primer sets are designed to amplify genes unique to each pathogen.

Table 25.3
Some Nucleic Acid–Based Detection Tests

Test	Intended Use	Specimen for PCR	Transport Conditions (Temperature and Maximum Time for Arrival at Lab)
Detect HIV proviral DNA	Diagnose HIV infection in newborns; to resolve indeterminate serological results	>2 ml whole blood in EDTA tube; EDTA prevents coagulation by chelating (binding) ions	Room temperature, within 24 hours after collection
Detect *Bordetella pertussis*	Diagnose whooping cough (pertussis)	Nasopharyngeal swab	Room temperature, within 24 hours after collection
Detect *Borrelia burgdorferi*	Diagnose/monitor Lyme disease	Tick; Skin biopsy; >1 ml spinal fluid; or urine	Wrap tick in moist tissue in small plastic bag, 4°C (cold packs), within 24 hours after collection
Detect *Rickettsia rickettsii*	Diagnose/monitor Rocky Mountain spotted fever	Tick; >2 ml whole blood in EDTA tube; Skin biopsy	Wrap tick in moist tissue in small plastic bag. Blood: room temperature. Skin lesion biopsy within 24 hours after collection; 4°C
Detect *Ehrlichia chaffeensis*	Diagnose/monitor human monocytic ehrlichiosis	Tick; >2 ml whole blood in EDTA tube	Wrap tick in moist tissue in small plastic bag, room temperature, within 24 hours after collection
Detect norovirus and other viruses	Diagnose viral gastroenteritis, investigation of food-borne and waterborne outbreaks	>1 ml diarrheal stool in sterile container	4°C (cold packs), within 72 hours after collection
Typing of *Mycobacterium tuberculosis*	Determine relatedness of *M. tuberculosis* isolates; investigation of suspected outbreaks	Sputum, urine	Room temperature
Typing of various Gram-negative and Gram-positive bacteria	Determine relatedness of isolates; investigation of suspected nosocomial or food-borne outbreaks	Food, feces	4°C (cold packs), within 24 hours after collection (if *Campylobacter* suspected, transport at room temperature)

Figure 25.15   Using Multiplex PCR to Identify *Clostridium botulinum*

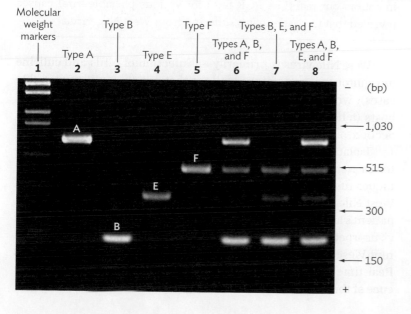

C. botulinum cells are typed by which toxin genes a strain possesses, not on the basis of surface antigens. Isolates can be typed in a single round of PCR that includes primer pairs for each of the four major toxin genes: types A, B, E, and F. Because multiple products are sought in a single reaction, this is called multiplex PCR. Each lane in the agarose gel was loaded with multiplex products from different isolates of *C. botulinum* and subjected to electrophoresis. Recall from Section 8.3 that DNA, which is negatively charged because of its phosphate backbone, is loaded into the top of the gel and moves toward a positive electrode at the bottom. The slower-moving fragments (toward top of gel) are larger than those that have moved farther down the gel toward the positive pole. Lane 1, DNA size markers; lanes 2–5, individual strains of *C. botulinum*; lanes 6–8, mixtures of strains as indicated. bp = base pairs.

CASE HISTORY 25.4

Tracking a Slow-Growing Pathogen

From August through December, 45 patients in the pediatric surgery unit of a New Delhi hospital developed postoperative wound infections. Of these, 42 were outpatients, whereas 3 were inpatients who had undergone major surgery. The diseases ranged in severity from chronic ear infections to bacteremia from using hemodialysis equipment. Thirty-two clinical samples of pus and wound exudates were stained for acid-fast bacilli (Figure 25.7C). The same smear samples were cultured on Löwenstein-Jensen slants (Figure 25.14B) and examined over several weeks for growth. Biochemical tests indicated that the organism was a mycobacterium in each case. PCR identified the slow-growing organism *Mycobacterium abscessus*. From the DNA fingerprint, infection control workers traced the source of the outbreak to the tap water in the operating room and to a defective autoclaving process (the result of a leaky vacuum pump and faulty pressure gauge in the autoclave).

As noted in Section 25.3, one can use the acid-fast stain to make a presumptive diagnosis of *M. tuberculosis*. Although the acid-fast stain is very useful, sometimes organisms cannot be found in a sputum sample—and even if they are found, a definitive diagnosis requires confirmatory tests. Case History 25.4 points out an underlying problem with some biochemical identifications. Growing enough organisms to perform these biochemical tests took weeks. An alternative PCR test that probed for mycobacterium-specific genes encoding 16S rRNA could have been performed directly on the clinical specimen and would have confirmed the diagnosis immediately, much faster than biochemical methods. Instead, in the case presented, the PCR test was not performed until *after* the cells were grown in the laboratory.

RFLP Analysis

The DNA-based method used in this case amplified a small DNA sequence located at the end of a gene highly conserved among all mycobacterial species. The sequence can be amplified by PCR from all species. However, DNA fragments amplified from different species of mycobacteria will have somewhat different DNA sequences. The alterations in DNA sequence were exposed by a process called **restriction fragment length polymorphism** (RFLP) analysis. Sequences from different species will produce uniquely sized DNA fragments following restriction enzyme digestion. The clinical laboratory compared the restriction patterns produced from mycobacteria isolated from different areas throughout the New Delhi hospital in the case history. The results allowed the infection control staff to trace the source of all the infections to the tap water in the operating room.

RFLP "DNA fingerprinting" can also be used to compare entire genomes of related strains of a single species. This technique relies on many small differences in DNA sequence that occur between strains. Because of these small differences in sequence, the number and location of restriction enzyme sites in the chromosome will differ. In **Figure 25.16A**, DNA from a strain of *Mycobacterium tuberculosis* is cut with **restriction enzymes**, and the digested fragments are separated by **gel electrophoresis** (**Figure 25.16B**). The gel is then probed with a radioactively labeled fragment of insertion sequence IS6110 (a fluorescent probe can also be used). IS6110 is a repetitive DNA sequence present at multiple sites in the chromosome. The radioactive probe can hybridize only to DNA fragments that contain IS6110 sequences. Strains of *M. tuberculosis* obtained from different outbreaks would probably display different hybridization patterns, similar to a bar code. Strains isolated from the same outbreak would look similar, if not identical.

Link Recall from Section 8.4 that **restriction enzymes** cleave DNA at sequence-specific sites to generate DNA fragments. The fragments, whose sizes reflect the positions of the sites in the DNA, can be separated by **gel electrophoresis** to yield a reproducible pattern. Negative charges on the DNA fragments induce them to move toward a positive electrode at the far end of the gel. The smallest fragments move fastest through the pores of the gel.

Real-Time PCR

CASE HISTORY 25.5

A Stiff Neck and a Stealthy Mosquito

A 55-year-old man complaining of headache, high fever, and neck stiffness was admitted to a local hospital. The man appeared confused and disoriented. He also complained of muscle weakness. History indicated that he had received several mosquito bites approximately 2 weeks before. CSF and blood specimens were sent to the laboratory. The report the next day indicated no bacterial infection, but real-time PCR tests for various possible viral causes revealed that the patient was suffering from West Nile virus.

West Nile virus is primarily an infection of birds and culicine mosquitoes (a group of mosquitoes that can transmit human diseases), with humans and horses serving as incidental, dead-end hosts (a host that is not part of the organism's normal life cycle). Section 24.3 describes other features. Here we focus on diagnosis.

Isolating a disease-causing virus is challenging. Many laboratories are not equipped to carry out the special tissue culture techniques required to grow viruses. Viral infections, including human West Nile virus infections, are usually diagnosed by measuring the patient's antibody response. For instance, the presence of West Nile virus–specific IgM in cerebrospinal fluid is a good indicator of current West Nile virus infection, but it is indirect and not conclusive. Real-time PCR is a molecular test that can quickly reveal the presence of the virus itself.

Figure 25.16   DNA Fingerprinting of *Mycobacterium tuberculosis*

A. Genetic fingerprinting of *M. tuberculosis* isolates. All DNA fragment sizes produced by restriction enzyme cleavage are shown. Fragments containing IS6110 are marked yellow.

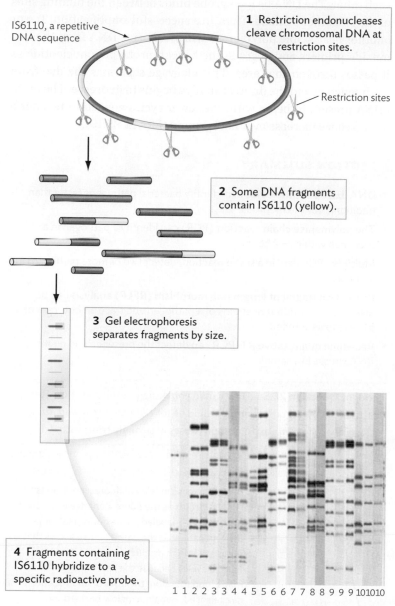

IS6110, a repetitive DNA sequence

1 Restriction endonucleases cleave chromosomal DNA at restriction sites.

Restriction sites

2 Some DNA fragments contain IS6110 (yellow).

3 Gel electrophoresis separates fragments by size.

4 Fragments containing IS6110 hybridize to a specific radioactive probe.

1 1 2 2 3 3 4 4 5 5 6 6 7 7 8 8 9 9 9 101010

B. Each lane in the gel contains DNA extracted from different isolates. Fragments containing IS6110 hybridize to a specific radioactive probe and become visible. A characteristic banding pattern (fingerprint) appears for each isolate. Isolates with similar banding patterns are assigned the same number in this example.

Reverse transcription quantitative RT-PCR (qPCR) is used routinely for the high-throughput diagnosis of viral pathogens such as West Nile virus. Because the genome of West Nile virus is made of single-stranded RNA, its genome must first be converted to DNA by using **reverse transcriptase** before PCR can be attempted. The quantitative advantage of qPCR is that you can estimate the number of virus particles present in the sample from the number of viral RNA molecules there. Another advantage is that qPCR does not require gel electrophoresis.

Link Recall from Sections 8.5 and 12.6 that the enzyme **reverse transcriptase** uses RNA as a template to make double-stranded DNA, also called complementary DNA (cDNA).

The basic technique is as follows: Viral RNA is extracted from the sample and converted to cDNA by using a single, virus-specific primer and the enzyme reverse transcriptase (**Figure 25.17**). The more virus particles in the sample, the more cDNA product is made. The cDNA is then amplified by PCR using two specific primers and a heat-stable DNA polymerase such as Taq polymerase (Figure 8.25 describe how PCR exponentially increases copies of DNA). The trick with qPCR is in the method used to detect and quantify amplification.

Figure 25.17   Reverse transcriptase synthesis of DNA

Conversion of mRNA to cDNA by reverse transcription

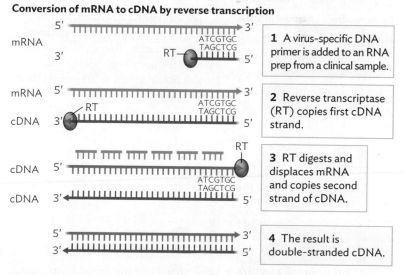

1 A virus-specific DNA primer is added to an RNA prep from a clinical sample.

2 Reverse transcriptase (RT) copies first cDNA strand.

3 RT digests and displaces mRNA and copies second strand of cDNA.

4 The result is double-stranded cDNA.

Reverse transcriptase (RT) uses mRNA as a template to make DNA. The DNA can then be amplified by using standard PCR methods. A specific oligonucleotide primer added to the reaction tube will anneal to the 3′ end of the viral RNA and RT will synthesize the first strand of a complementary DNA (cDNA). Subsequent amplification by PCR requires adding a thermostable DNA polymerase (Taq) that can withstand the denaturing and DNA synthesis temperatures required to amplify the cDNA.

Figure 25.18　Real-Time PCR

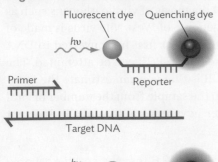

1 The reporter probe contains a fluorescent dye and a quenching dye, so no fluorescence is emitted.

2 Target DNA is denatured at 95°C (only ssDNA shown).

3 Temperature is lowered to 55°C. The reporter probe anneals downstream of a DNA primer. Still no fluorescence.

4 Temperature is raised to 72°C. Taq polymerase extends upstream DNA and degrades reporter. The release of the fluorescent dye from the vicinity of the quencher allows fluorescence, which is measured.

Taq DNA polymerase extends an upstream primer and reaches the downstream reporter probe where a 5′-to-3′ exonuclease activity degrades the probe, releasing the fluorescent dye from the vicinity of the quencher so it can now fluoresce.

Figure 25.19　Results of Real-Time PCR

Cycle number	Amount of DNA
0	1
1	2
2	4
3	8
4	16
5	32
6	64
7	128
8	256
9	512
10	1,024
11	2,048
12	4,096
13	8,192
14	16,384
15	32,768
16	65,536
17	131,072
18	262,144
19	524,288
20	1,048,576
21	2,097,152
22	4,194,304
23	8,388,608
24	16,777,216
25	33,554,432
26	67,108,864
27	134,217,728
28	268,435,456
29	536,870,912
30	1,073,741,824
31	1,400,000,000
32	1,500,000,000
33	1,550,000,000
34	1,580,000,000

A. The exponential increase in PCR products after each cycle of hybridization and polymerization. The switch to yellow indicates the point where the increase in product plateaus because the primers have been exhausted.

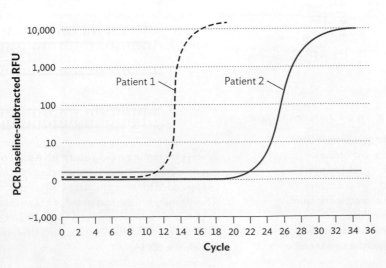

B. An increase in relative fluorescence units (RFUs) occurs as the fluorescent dye is released from the dual-labeled probe during real-time PCR. In patient 1, the dashed curve shows that fluorescent PCR product remained flat for the first 10 cycles because the amount of DNA made, and therefore the level of fluorescent dye released, remained below background level (blue line). After cycle 10, fluorescence increased logarithmically. For patient 2, the solid red curve representing PCR product remained below detection for 22 cycles and then increased. The more DNA present at the start, the sooner RFU values will increase over background (that is, the fewer heating cycles will be needed to see the increase over background). The slope eventually decreases because the fluorescent probe has become limiting.

In one method, a third, fluorescent oligonucleotide (called a **molecular beacon probe**) is added to the PCR tube (Figure 25.18). The DNA probe contains a fluorescent dye at its 5′ end and a chemical-quenching dye at the 3′ end that absorbs the energy emitted from the fluorescent dye. Like a pillow covering a flashlight, no light is emitted so long as the intact probe keeps the two chemicals close to each other. The DNA beacon probe binds between the binding sites of the two other primers. Thus, in a successful amplification, Taq or another heat-stable polymerase will synthesize DNA from the two outside primers and degrade the beacon probe oligonucleotide as it passes through that area. This cleavage separates the dye from the quencher, and the fluorescent dye begins to fluoresce. The more cDNA present to begin with, the fewer cycles needed to register a fluorescence increase over background (Figure 25.19).

SECTION SUMMARY

- **DNA-based technologies** can identify bacterial pathogens faster than traditional culture methods.
- **The polymerase chain reaction (PCR)** can identify a pathogen in a specimen within 1–2 hours.
- **Multiplex PCR** can, in a single reaction, detect two or more pathogens in a sample.
- **Restriction fragment length polymorphism (RFLP) analysis** can be used to identify different strains of a pathogen and trace an outbreak of a disease to its source.
- **Real-time quantitative RT-PCR** can detect the presence and quantity of RNA viruses in a sample.

Thought Question 25.5　How would you design a multiplex PCR test for real-time qPCR? A good example would be for *Chlamydia trachomatis* and *Neisseria gonorrhoeae*, two common causes of sexually transmitted disease.

Figure 25.20 **Antibody-Capture Enzyme-Linked Immunosorbent Assay (ELISA)**

Enzyme-linked immunosorbent assay (ELISA)

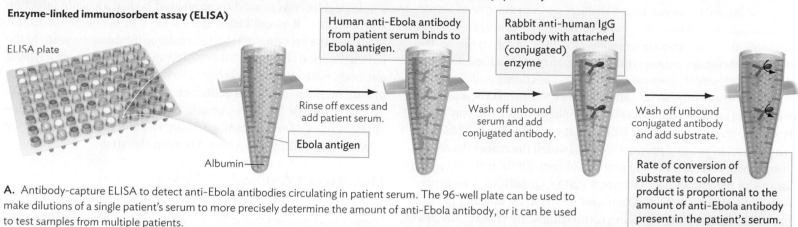

A. Antibody-capture ELISA to detect anti-Ebola antibodies circulating in patient serum. The 96-well plate can be used to make dilutions of a single patient's serum to more precisely determine the amount of anti-Ebola antibody, or it can be used to test samples from multiple patients.

25.5
Pathogen Identification by Serology, Antigenic Footprints, and Other Immunological Methods

SECTION OBJECTIVES

- Describe how enzyme-linked immunosorbent assays can identify microbial antigens present in clinical samples or antimicrobial antibodies present in serum.
- Discuss the fluorescent antibody staining technique.

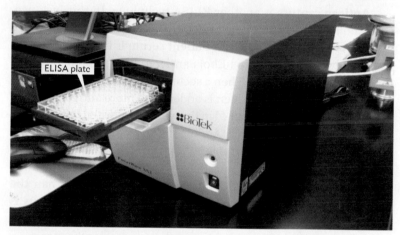

B. An ELISA plate reader.

Beside biochemical and nucleic acid–based tests, immunological strategies are quite useful in detecting a pathogen or identifying the "footprints" of a pathogen's past or recent presence. When direct detection of a pathogen is not possible, detection of specific IgM antibodies or microbial antigens may also help diagnose an acute infection.

ELISA

CASE HISTORY 25.6
Bloody Death in West Africa

During March 2014, about 60 people in the southeastern forest regions of Guinea, Africa, became ill with high fever, diarrhea, headache, vomiting, and gastrointestinal bleeding from the rectum. At least 23 people died. The presumptive diagnosis was Ebola hemorrhagic fever. Laboratory confirmation included viral antigen and antibody detection by enzyme-linked immunosorbent assay (ELISA). Laboratory-confirmed Ebola patients were defined as patients who were either positive for Ebola virus antigen or Ebola IgG antibody. Once identified, rigorous quarantine mechanisms were implemented in the hope of preventing outbreaks in other villages and countries. Despite these efforts, by August 2014 the Ebola epidemic had spread into the neighboring countries of Sierra Leone, Liberia, and (by air travel) Nigeria. By then the outbreak involved approximately 1,300 presumed or confirmed cases, more than half of whom died. As of January 2015, a mere 9 months after it started, the Ebola crisis in West Africa had claimed 7,000 lives and showed little sign of stopping.

Ebola is a horrific disease with a death rate of more than 50% and no vaccine or antiviral agent able to cure the infection (Sections 12.1, 12.2, and 21.2). Nevertheless, early laboratory diagnosis using ELISA can lead to quick implementation of supportive measures (fluid replacement, blood pressure maintenance) that can improve chances of survival. As first described in Section 17.5, an ELISA (enzyme-linked immunosorbent assay) detects antibodies indirectly or antigens directly. **Antibody-capture** ELISA detects serum antibodies. It is carried out in a 96-well microtiter plate, which allows testing multiple patient serum samples simultaneously. An antigen from the virus (here, Ebola) is attached (or adsorbed) to the plastic of the wells (**Figure 25.20A**). Patient serum is then added. If Ebola-specific antibodies are present in the serum, they will bind to the antigen attached to the microtiter plate. The antigen-antibody complex is then reacted with a detection antibody, rabbit anti–human IgG, to which an enzyme (for example, horseradish peroxidase) has been attached (or conjugated). This antigen-antibody-antibody complex attaches the enzyme to the well. The chromogenic substrate for the enzyme is added next (for example, tetramethylbenzidine). If enzyme-conjugated antibody has bound to any human IgG, the enzyme will convert the substrate to a colored product (blue for tetramethylbenzidine). Enzyme activity can be measured with an ELISA plate reader (Figure 25.20B). The amount of colored product formed, which the reader detects as absorbance, correlates to the amount of anti-Ebola antibody present in the patient sample (Figure 25.20A). Quantitative measurements of patient antibodies can also be made by diluting patient sera in the ELISA plate wells and comparing acute versus convalescent serum (Section 17.5).

Antigen capture is another ELISA technique. In this instance, anti-Ebola antibody, not viral antigen, is adsorbed to the wells of a microtiter plate (Figure 25.21). Patient serum is then added to the wells. If the serum contains Ebola antigen, the antibody tethered to the well will capture it. Then a second, enzyme-conjugated antibody against the Ebola antigen is added. The more antigen present in the serum, and thus captured in the well, the more enzyme-linked antibody will bind to the well. Adding the appropriate chromogenic substrate will yield a measurable colored product.

Antibody against Ebola may be easier to detect than viral antigen because antibodies will be present at higher levels than the virus itself. But because a delay occurs between the time virus is first present in serum and when the body manages to make antibody, directly detecting viral antigen offers a speedier diagnosis.

A previous case history also shows the utility of ELISAs in differentiating between acute and chronic infections. During a discussion of cytotoxic T cells, Case History 16.3 introduced Luke, a 24-year-old man who was jaundiced, had pain in his upper right abdomen (liver), and was extremely tired. The case mentioned that serological tests were used to prove that he had an acute infection with hepatitis B virus. Those tests were ELISAs. A direct ELISA identified the presence of HBsAg (a soluble viral coat protein) in the patient's serum, and an indirect ELISA discovered IgM anti-HBsAg antibody, both of which develop soon after infection. The presence of both markers is a sign of acute hepatitis B disease (Section 22.4). When the disease becomes chronic, however, the IgM antibody disappears, but IgG anti-HBsAg antibody and the HBsAg protein (the latter indicating active infection) remain elevated.

Fluorescent Antibody Staining

In Case History 20.5, an 80-year-old nursing home resident contracted pneumonia caused by *Streptococcus pneumoniae*. The laboratory diagnosis was probably made by using the biochemical algorithm described earlier. However, more than 80 serological types of *S. pneumoniae* exist, each containing a different capsular antigen. How can the lab identify which antigenic type has caused the infection? One way is to stain the organism with antibodies.

Figure 25.22A shows the result of staining a smear of the isolated streptococcus with fluorescently tagged antibodies directed against a specific antigenic type of capsule. This approach is called **fluorescent antibody staining**. Viewed under a fluorescence microscope, the organism is "painted" green when the right antibody binds to the capsule (**direct immunofluorescence microscopy**). For pneumonia, this knowledge probably will not help in treating the individual patient, but its broader value is in determining whether a *single type* of organism is responsible for an outbreak of pneumonia, which in turn is of epidemiological value for identifying the source of the bacterium.

On the other hand, fluorescent antibody staining techniques are crucial for rapidly identifying organisms that are difficult to grow. Infected tissues can be subjected to direct fluorescent antibody staining. Figure 25.22B, for example, shows a direct fluorescent antibody stain of pleural fluid from a patient with Legionnaires' disease.

Link Recall from Section 17.5 that **direct immunofluorescence microscopy** involves applying fluorescently tagged antimicrobial antibodies directly to a patient sample to reveal the presence of the microbe or microbe antigen. Indirect immunofluorescence microscopy identifies antimicrobial antibodies present in a patient's serum.

Other Procedures for Identifying Pathogens

Space does not permit a complete listing of the various methods used to identify the etiologic agents of all infectious diseases, but Table 25.1 will show you some additional examples. In general, you will notice that easily cultured bacteria are grown in the laboratory, after which biochemical tests are performed. Difficult-to-grow bacterial, viral, and fungal species are typically identified through immunological or nucleic acid–based techniques. Immunological techniques can directly identify microbes in infected tissues or indirectly measure

Figure 25.21   Antigen-Capture ELISA

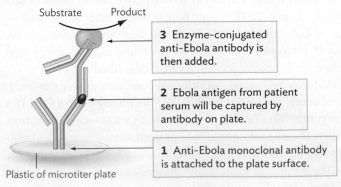

4 If Ebola antigen is present, the conjugated antibody will be captured by the complex. Adding substrate will yield a colored product.

Substrate   Product

3 Enzyme-conjugated anti-Ebola antibody is then added.

2 Ebola antigen from patient serum will be captured by antibody on plate.

1 Anti-Ebola monoclonal antibody is attached to the plate surface.

Plastic of microtiter plate

Antigen-capture, or direct, ELISA captures Ebola antigens circulating in patient serum. The steps of this procedure are similar to those shown in Figure 25.20.

a rise in antibody titer (Chapter 17). DNA-based detection methods include PCR, qPCR, and RFLP analysis. Eukaryotic microbial parasites such as *Plasmodium* species (the cause of malaria), *Giardia intestinalis* (which causes giardiasis, a diarrheal disease), and *Entamoeba histolytica* (the cause of amebic dysentery) can be identified through their telltale morphologies under the microscope, making biochemical tests unnecessary.

A new technology being used for pathogen detection, **mass spectrometry (MS)**, is gaining wide acceptance among clinical microbiologists. Mass spectrometry (the molecular technique used is called matrix-assisted laser desorption ionization–time of flight mass spectrometry [**MALDI-TOF MS**]) can provide results in minutes compared to the many hours required by biochemical profile methods. In brief, a MALDI-TOF instrument bombards microbes with photons, which cause release of bacterial proteins and break each protein into fragments. The fragments move through a positively charged electrostatic field toward an ion detector. The time it takes a fragment to hit the detector (called *time of flight*) is proportional to the molecular weight of the fragment. The molecular weights of a pathogen's ribosome fragments, for instance, can be compared with those cataloged in databases to identify the genus and species of the pathogen. Many large and small laboratories are rapidly moving toward use of this new technology.

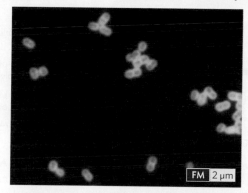

Figure 25.22   Fluorescent Antibody Stain

A. *Streptococcus pneumoniae* capsule. The capsule is the green halo. The center of the halo is the cell.

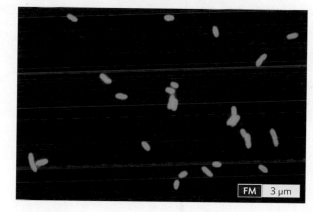

B. *Legionella pneumophila* (approx. 1 μm) from a respiratory tract specimen.

SECTION SUMMARY

- **The presence of serum antibodies made in response to a specific pathogen** can be revealed and quantified by using indirect ELISA techniques.
- **The presence of virus antigens in serum** can be detected by using direct ELISA or real-time quantitative PCR.
- **Fluorescent antibody staining** can rapidly identify organisms or antigens present in tissues.

Thought Question 25.6   Why does finding IgM to West Nile virus indicate current infection? Why wouldn't finding IgG do the same?

Thought Question 25.7   Specific antibodies against an infectious agent can persist for years in the bloodstream. So how can antibody titers be used to diagnose diseases such as infectious mononucleosis? Couldn't the antibody be from an old infection?

25.6
Point-of-Care Rapid Diagnostics

SECTION OBJECTIVES
- Differentiate between assay specificity and sensitivity.
- Explain the basic immunochromatography assay used in point-of-care diagnostic kits.
- Discuss the advantages and disadvantages of point-of-care diagnostics.

Is there a quick, inexpensive way to screen sick patients for infectious agents? Conventional diagnosis of an infection often requires sending a clinical specimen to a far-away laboratory, followed by a considerable delay in obtaining results. Some patients inevitably lose patience and fail to attend follow-up appointments. **Point-of-care (POC) laboratory tests** address these problems because they are designed to be used directly at the site of patient care, such as physicians' offices, outpatient clinics, intensive-care units, emergency departments, hospital laboratories, and even patients' homes. Most patients are happy to wait 40–50 minutes for a rapid POC test result in order to receive immediate treatment or reassurance. But how do we know a POC test is reliable?

Sensitivity and Specificity of Diagnostic Assays

Before we describe types of POC assays, you must first understand the difference between an assay's **sensitivity** and **specificity**. These measurements address two questions you must ask when analyzing a new diagnostic test:

1. If a patient has a disease, how often will the test be positive (a measure of *sensitivity*)?
2. If a patient does *not* have a disease, how often will the test be negative (a measure of *specificity*)?

An assay with high sensitivity and high specificity is a valuable diagnostic tool, whereas an assay with low sensitivity or low specificity is useless.

Figure 25.23   Principle of Immunochromatographic Rapid Diagnostic Tests

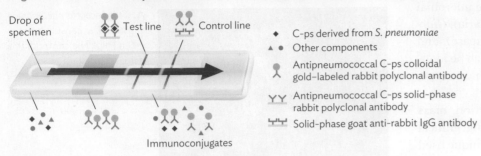

- ◆ C-ps derived from *S. pneumoniae*
- ▲ ● Other components
- ⚥ Antipneumococcal C-ps colloidal gold–labeled rabbit polyclonal antibody
- ⋎⋎ Antipneumococcal C-ps solid-phase rabbit polyclonal antibody
- ⊔⊔ Solid-phase goat anti-rabbit IgG antibody

A. The example is a test for the presence of *Streptococcus pneumoniae* in sputum. An extract of sputum is placed onto one end of the strip where *S. pneumoniae* capsular polysaccharide antigen (C-ps), if present, will bind antipneumococcal C-ps polyclonal antibodies tagged with colloidal gold. The resulting immunoconjugates move by capillary action to the upper membrane, where antipneumococcal C-ps solid-phase polyclonal antibodies capture them, thereby forming sandwich conjugates in the sample. The presence of both a test and a control line indicates a positive result, whereas the appearance of only a single control line indicates a negative result.

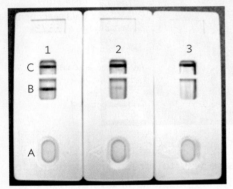

B. Immunochromatography test for anti–*Treponema pallidum* antibodies. Serum samples from three patients. Serum from patient 1 contained anti–*T. pallidum* antibodies. C = control; B = test; A = site where sample is loaded.

A test's *sensitivity* reflects how *small* a concentration of a microbe, antibody, or antigen a test can detect. A highly sensitive assay can detect the presence of minute quantities of hepatitis B antigen, for example. A less sensitive test may require larger amounts of that antigen before the test turns positive. Any sample containing less than that amount of antigen would yield a **false negative reaction**, meaning the antigen was present but the test could not detect it. A person with a false negative reaction actually has the disease, but the rapid test suggests the person doesn't. Sensitivity is measured by using the new assay on a group of people who you know *have* the disease. A test sensitivity of 0.95 indicates that of 100 people with the disease, the test will turn positive for 95 of them but will be negative for 5 patients with the disease (false negatives).

In contrast to sensitivity, *specificity* measures how well an assay can distinguish between two closely related targets (think antigens). A test highly specific for a *Mycobacterium tuberculosis* antigen, for example, is positive *only* for the *M. tuberculosis* antigen but negative for a similar antigen from a different *Mycobacterium* species. A less

specific test might turn positive when *either* antigen is tested. The antigen from the other species could produce a so-called **false positive reaction**. Someone with a false positive reaction does *not* have the disease, but the new assay suggests the person does. Specificity is measured by testing the new assay on a group of people who you know *do not have* the disease. A test whose specificity is 0.95 means that the test will be *negative* for 95 of 100 patients without disease, but 5 disease-free patients will falsely test positive. The higher the sensitivity and specificity values, the better the test.

Commercial POC Tests

Commercial POC tests are widely available to diagnose bacterial and viral infections and parasitic diseases, including malaria (Table 25.4). However, as convenient as these tests are, sensitivity may be compromised in the quest for a speedy result. Infectious-disease specialists and clinical microbiologists should be aware of the indications and limitations of each rapid test, so that they can use them appropriately and correctly interpret their results. Some tests exhibit insufficient sensitivity and should therefore be coupled with confirmatory tests when the results are negative (false negatives, such as the *Streptococcus pyogenes* rapid antigen detection test), whereas the results of other POC tests need to be confirmed when positive (false positives, such as malaria).

The typical POC test involves an **immunochromatographic assay**. The test for *Streptococcus pneumoniae* capsular antigen shown in Figure 25.23A is one example. This particular immunochromatographic test, called a "red colloidal gold" test, involves extracting the relevant antigen (for example, capsule antigen) from a clinical specimen and placing a few drops of the extract on a test strip containing rabbit antibodies to the antigen. The antibodies have red colloidal gold particles attached to them. The antigen-antibody complexes that form move by capillary action to the upper level of the strip, where a line of more anti-antigen antibodies embedded in the strip capture them, forming a sandwich. The colloidal gold particles accumulate and eventually produce a red test line, indicating the presence of the antigen. By contrast, rabbit antibodies not bound to the antigen pass through the test line, but goat anti–rabbit IgG antibodies on a control line capture them, once again forming a red control line showing that the strip components are working. Figure 25.23B shows results of a test for *T. pallidum* antibodies.

 QuickVue rapid strep A test.mov: youtube.com

Table 25.4
Some Point-of-Care Rapid Diagnosis Test Kits for Infectious Diseases [Reference Table]

Pathogen/ Disease/Antigen	Test Type	Sample	Indication	Performance		Notes
				Sensitivity (%)	Specificity (%)	
Bacterial						
Chlamydia	ICT	Vaginal swab, urine	Screening, suspicion of pelvic inflammatory disease	83	99	Detects antigen
N. gonorrhoeae	PCR	Urine	Urethral exudate	95	99	Detects antigen
Syphilis	ICT	Blood	Screening	90–95	90–95	Detects antibody
Group A streptococci	EIA	Pharyngeal swab	Sore throat	53–99	62–100	Confirms negative swabs
Legionella antigen	ICT	Urine	Severe pneumonia/risk factors for legionellosis	76	99	Only serotype 1 reliably detected
Pneumococcal capsule antigen	ICT	Urine (pleural fluid, CSF)	Severe pneumonia (empyema, meningitis)	66–70	90–100	Detects antigen
Clostridium difficile toxin	ICT	Stool	Antibiotic-associated diarrhea	49–80	95–96	Notably less sensitive than cultures or PCR
S. agalactiae	PCR	Vaginal swab	Peripartum detection of colonization	92	96	Detects antigen
Protozoan						
Malaria	ICT	Blood	Fever in returning traveler	87–100	52–100	Sensitivity better for *Plasmodium falciparum*
Trichomonas	ICT	Vaginal swab	Symptoms of vaginitis	83	98	
Viral						
Influenza	ICT	Nasopharyngeal swab	Flu-like symptoms	20–55	99	Low sensitivity; probably not helpful during outbreaks; lower in adults
RSV	ICT	Nasopharyngeal swab	Viral symptoms, especially during the winter season	59–97	75–100	Detects antigen
HIV	ICT	Blood (oral fluid)	Screening, preventing vertical transmission	99–100	99–100	Detects anti-HIV antibodies
Dengue fever IgG-IgM	ICT	Blood	Screening in endemic regions	90	100	Detects antibody
Epstein-Barr virus/ mononucleosis	ICT	Blood	Symptoms (swollen lymph nodes, sore throat)	100	90	Detects IgM heterophile antibodies
Rotavirus	ICT	Stool	Diarrhea in child	88	99	Detects antigen
Hepatitis B (HBsAg)	ICT	Blood	Jaundice, pain over liver	95	100	Detects antigen
Rubella IgG	ICT	Blood	Pregnancy	99	99	Detects antibody

CSF, cerebrospinal fluid; EIA, enzyme immunoassay; HIV, human immunodeficiency virus; ICT, immunochromatographic test; MRSA, methicillin-resistant *Staphylococcus aureus*; PID, pelvic inflammatory disease; RSV, respiratory syncytial virus.

inSight

Using Lab-on-a-Chip to Detect Pathogens

Figure 1 Model of a PCR-Based Lab-on-a-Chip Used to Detect Pathogens

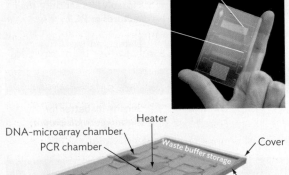

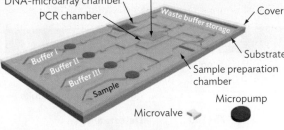

DNA-microarray chamber
PCR chamber
Heater
Waste buffer storage
Cover
Buffer I
Buffer II
Buffer III
Sample
Substrate
Sample preparation chamber
Microvalve
Micropump

New technologies offer hope that the power of an entire laboratory can be miniaturized to a series of chambers on a small chip.

Imagine a day when all the agar plates, test tubes, culture flasks, and incubators used in the clinical microbiology laboratory are shrunk to the size of a single credit card. Sounds like science fiction, but that day is not too far off. In fact, in some respects it's already here.

Conventional laboratory diagnostic procedures can take several days to identify a pathogen. Meanwhile, the pathogen can multiply in the patient and spread. The rapid molecular techniques we have discussed can be costly and require experienced technologists to handle the specimen, mix the reagents, and evaluate the results. This sophisticated specimen handling poses a challenge for rural health care clinics around the world.

The solution to this situation, and the hottest topic among clinical microbiologists, involves microfluidics and a **lab-on-a-chip (LOC)**. The idea is to develop a small, simple device that can identify pathogens in minutes rather than days and that can deliver results locally with minimal expense (Figure 1).

The lab-on-a-chip concept is based on the integrated circuit, in which wires and circuits are integrated into a semiconductor chip with electricity flowing through it. In a lab-on-a-chip, instead of electricity, liquid flows through tiny channels etched into a small silicon chip. Strategically placed electrodes on the chip move the sample along a predefined route. In one LOC used to detect pathogens, part of the slide is filled with nanoparticles (10–1,000 nanometers) that adhere to pathogens applied in a sample drop of water. The fluid moves to a chamber containing antibodies that bind the pathogen and cause the particles to agglutinate (see Chapter 17), a result that an optical sensor detects as scattered light. Similar LOC technologies attach fluorescent dyes to the antibodies.

Another chip under development uses PCR. In this case, capture DNA probes specific for certain viruses or bacteria are immobilized (solid phase) in microarray chambers in a glass slide. The ideal chip will have three components: an extraction chamber that extracts DNA from samples, a PCR chamber that uses fluorescent probes to amplify a specific pathogen gene from the extracted DNA, and a detection microarray chamber where the amplicons (the amplified fragments) anneal to the solid-phase capture DNA. An external detector would detect the fluorescent light emissions. Figure 1 shows a prototype of a pathogen detection chip. Some LOC chips have already been designed to detect up to nine pathogens simultaneously.

Microbiologist Phil Tarr, lamenting the same old agar-based technology still being used in clinical labs, once said that if Louis Pasteur were to rise from the grave and walk into a clinical micro lab, he would feel right at home. With these new LOC technologies, that is no longer the case.

POC rapid tests have several advantages and some disadvantages. Advantages of POC tests include:

- Culturing is not required.
- The clinician can immediately initiate specific antibiotic therapy.
- They help avoid unnecessary antibiotic use in the case of a viral infection.
- They enable the clinician to quickly recognize **chain of infection** among patients with similar symptoms.
- They enable rapid notification of patients who are difficult to reach.

Link Recall from Chapter 2 (and described further in Section 26.1) that the **chain of infection** has four major links: the etiologic agent (its virulence and invasiveness), its reservoir (animal, human, or environment), its mode of transmission, and its host (portal of entry, immune competence).

Disadvantages of POC tests include:

- No data about pathogen antibiotic sensitivity.
- An increased risk of the technologist becoming infected.
- Double or multiple infections are more likely to be overlooked than in culture.

The newer molecular tests described earlier exhibit better sensitivity and specificity than most immunochromatographic assays, but they require more expensive instrumentation and training. In the coming years, further evolution of POC tests may lead to new diagnostic approaches, such as panel testing that targets all possible pathogens suspected in a specific clinical setting. The development of next-generation multiplexed tests based on serology or molecular techniques will certainly facilitate quicker diagnosis and improved patient care (see **inSight**).

SECTION SUMMARY

- **Point-of-care diagnostic tests** can rapidly identify the cause of an infectious disease.
- **Sensitivity** measures the likelihood that an assay will be positive for a patient who has the disease. Specificity reflects the likelihood that an assay will be negative for a patient free from disease.
- **Immunochromatography** is the primary platform for POC testing.
- **An advantage of point-of-care tests is that specific therapy** can be initiated quickly after a positive result.
- **A drawback to point-of-care tests** is that they may miss simultaneous, multiple infections.

Perspective

In 1880, Robert Koch used the surface of a potato slice to culture and isolate the agent of anthrax. Ever since, clinical microbiologists have tried to design faster and more accurate ways to collect, detect, and identify pathogenic microorganisms. Their efforts shaped the strategies currently used to diagnose infectious diseases. The clinical microbiology laboratory today is undergoing another new and exciting technological revolution in which culture-based methods are slowly giving way to molecular techniques such as PCR, DNA sequencing, and mass spectrometry. Someday we may even be able to use a single chip to identify every possible pathogen in any clinical specimen.

As this chapter ends, reflect on earlier case histories and try to envision which techniques were used to identify the pathogen. Then think ahead about how these methods might help epidemiologists predict local outbreaks and the spread of pandemics (Chapter 26). Finally, know that whatever your eventual role in health care—as a clinician, technician, nurse, parent, or patient—understanding the ways of specimen collection, sample processing, and the diagnostic strategies used to reveal a pathogen's identity will benefit you.

LEARNING OUTCOMES AND ASSESSMENT

SECTION OBJECTIVES

OBJECTIVES REVIEW

25.1
The Importance of Clinical Microbiology

- List the reasons for identifying the cause of an infectious disease.
- Explain the basic epidemiologic purpose of identifying the cause of an infectious disease.

1. Of the following, which is the best clinical reason to identify the etiologic agent causing an infection?
   A. Accurately predicts disease severity
   B. Alerts the clinician to possible disease sequelae
   C. Pinpoints the source of the infection
   D. Indicates whether gloves should be worn when examining the patient

25.2
Specimen Collection

- Identify sterile and nonsterile body sites.
- List procedures and precautions used to collect specimens from sterile body sites.
- List procedures used to collect specimens from body sites containing normal microbiota.

3. Which of the following specimens when collected properly will contain normal microbiota in a healthy patient?
   A. Cerebrospinal fluid
   B. Pleural fluid
   C. Throat swab
   D. Blood culture
   E. Tissue biopsy sample

4. _____ is the procedure used to collect cerebrospinal fluid.

25.3
Pathogen Identification Using Biochemical Profiles

- Explain how biochemical profiles can be used to identify microbial pathogens.
- Discuss the major biochemical features that broadly differentiate Gram-negative and Gram-positive pathogens.
- Describe the Lancefield classification and its importance to clinical microbiology.
- Explain the difference between selective and differential media and their use in clinical microbiology.
- Discuss differences between the acid-fast stain and Gram stain.

7. Which of the following organisms will grow on chocolate agar but not blood agar?
   A. *Neisseria meningitidis*
   B. *Escherichia coli*
   C. *Salmonella* Typhimurium
   D. *Staphylococcus aureus*
   E. *Treponema pallidum*

8. A urine specimen from a patient experiencing a burning pain upon urination contained catalase-positive, Gram-positive cocci that were coagulase positive. Which of the following bacterial species has these characteristics?
   A. *Staphylococcus epidermidis*
   B. *Escherichia coli*
   C. *Streptococcus pyogenes*
   D. *Staphylococcus aureus*
   E. *Streptococcus pneumoniae*

2. Twenty-five students from two schools were stricken with bacterial diarrhea. All the patients went to a countywide fair 3 days ago that 400 students attended. Which of the following would best confirm that the county picnic was the source of infection?
A. Different bacterial species were isolated from the sick students but all had the same antibiotic resistance pattern.
B. The same bacterial species was found in each of the 400 students and all specimens had the same antibiotic resistance pattern.

C. The same bacterial species with the same antibiotic resistance pattern was found in all the sick students and in a vendor at the picnic.
D. The same bacterial species but with different antibiotic resistance patterns was found in all the sick students and in a vendor at the picnic.
E. C and D only

5. A 13-year-old boy has a high fever and is disoriented, but he has a normal blood pressure. Blood cultures are obtained. After overnight incubation, growth is observed in all blood culture bottles. Which of the following is the most defensible conclusion?
A. The sample was contaminated because an aerobe and an anaerobe are unlikely to both be present in the patient's blood.
B. The patient has septicemia caused by an organism that is facultative with respect to oxygen.
C. The patient has septicemia and the organism is aerobic.
D. The patient has septicemia caused by a Gram-negative rod.
E. The antibiotic resistance patterns of the organisms isolated from all blood culture bottles will be identical.

6. Which of the following specimens should undergo centrifugation and the sediment subjected to Gram stain?
A. Blood
B. Fecal
C. Urine
D. CSF
E. Throat

9. A fecal sample from a patient with diarrhea contained a bacterium that produced white colonies on MacConkey medium, produced colonies with black centers on Hektoen agar, and is catalase positive but did not grow on CNA blood agar. Which of the following bacterial species has this biochemical profile?
A. Streptococcus gallolyticus
B. Shigella flexneri
C. Escherichia coli
D. Staphylococcus aureus
E. Salmonella enterica

10. An organism isolated from the bloodstream of a person with necrotizing fasciitis grows on CNA blood agar, is beta-hemolytic on blood agar, does not grow on MacConkey agar, is catalase negative, and is serologically classified as group A. The most likely organism is _____ and the serologic test used is called _____.

11. _____ prevents the Gram staining of Mycobacterium tuberculosis.

Diagnostic Microbiology Matching
Use the choices that follow to answer questions 12–17.
12. Oxidase-positive coccus
13. Lactose-fermenting bacillus
14. Hydrogen sulfide–producing bacillus
15. Coccus with serologically typed capsule
16. Catalase-positive, coagulase-positive coccus
17. Acid-fast bacillus

A. Mycobacterium tuberculosis
B. Streptococcus pneumoniae
C. Staphylococcus aureus
D. Neisseria gonorrhoeae
E. Escherichia coli
F. Salmonella enterica
G. Yersinia pestis

For each of the six diseases below, choose which pathogen from the choices for questions 12–17 most likely caused each disease.
18. Abscess
19. Diarrhea
20. Vaginitis
21. Cystitis
22. Pneumonia
23. Reactivation pneumonia

LEARNING OUTCOMES AND ASSESSMENT

25.4
Pathogen Identification by Genetic Fingerprinting

- Discuss how the polymerase chain reaction can be used to identify pathogens in a clinical sample and the role of real-time PCR in clinical diagnosis.
- List the advantages and disadvantages of PCR, restriction fragment length polymorphisms, and biochemical strategies of pathogen identification.

24. Bacterial pathogens can be typed in three ways: _____, _____, and/or _____.

25.5
Pathogen Identification by Serology, Antigenic Footprints, and Other Immunological Methods

- Describe how enzyme-linked immunosorbent assays can identify microbial antigens present in clinical samples or antimicrobial antibodies present in serum.
- Discuss the fluorescent antibody staining technique.

26. Draw the scheme for an indirect fluorescent antibody staining technique.

25.6
Point-of-Care Rapid Diagnostics

- Differentiate between assay specificity and sensitivity.
- Explain the basic immunochromatography assay used in point-of-care diagnostic kits.
- Discuss the advantages and disadvantages of point-of-care diagnostics.

28. Which of the following statements can you make once you know that a detection assay has a specificity of 0.95?
    A. A patient *with* disease has a 95% chance of testing positive with the assay.
    B. A patient *with* disease has a 5% chance of testing negative with the assay.
    C. A patient *without* disease has a 95% chance of testing negative with the assay.
    D. A patient *without* disease has a 95% chance of testing positive with the assay.

25. Three patients in the same hospital developed pneumonia caused by *Streptococcus pneumoniae* during their stay. The infection control specialist suspected that the same strain of *S. pneumoniae* infected all three patients and that a common source existed. PCR assays used to examine the isolates from each patient produced DNA fragment products of identical size, but after further study, the infection control expert concluded that the three patients were not infected with the same strain. Which of the following would best support that conclusion?

A. No common source of *S. pneumoniae* was identified.
B. The DNA sequences of the PCR fragments were different.
C. The isolates formed different-sized colonies on blood agar.
D. The patients were present in the hospital on different days.

27. The following ELISA was used to measure hepatitis A antibody levels in five patients (viewed in a microtiter plate). From the results shown, which patient(s) probably have the disease? Each patient's serum underwent serial twofold dilutions. (In the diagram at right, a = serum collected during the acute phase of the disease; c = serum collected during the convalescent phase of the disease.)

A. Patient 3 only
B. Patients 3 and 4 only
C. Patients 3, 4, and 5 only
D. Patients 1, 3, 4, and 5 only
E. Patient 2 only

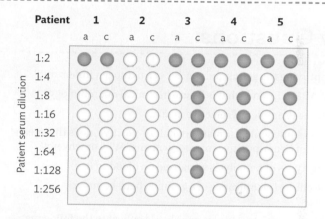

29. A point-of-care test will provide useful information about the antibiotic susceptibility of a pathogen. True or false?

30. Evaluate the immunochromatography results below. Which figure represents a proper positive result for the test antigen?

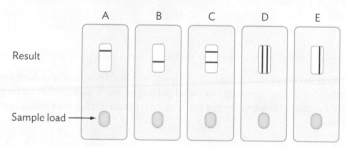

31. You are given two tubes containing different amounts of the same antigen and a third control tube containing no antigen. You have designed two different ELISAs to test for the presence of that antigen. You find the following:

- Both tests are negative when the control tube is used.
- Assay A turns positive when both antigen-containing tubes are analyzed.
- Assay B turns positive only with sample from antigen-containing tube 1.

Assay B is _____ than Assay A.
A. more sensitive
B. more specific
C. less sensitive
D. less specific

Key Terms

acid-fast stain (p. 859)
alpha hemolysis (p. 857)
alpha-hemolytic (p. 857)
antibody capture (p. 865)
antigen capture (p. 866)
beta hemolysis (p. 858)
catalase (p. 857)
catalase test (p. 857)
CNA plate (p. 857)
cytochrome oxidase test (p. 857)
empirical therapy (p. 844)

false negative reaction (p. 868)
false positive reaction (p. 868)
fluorescent antibody staining (p. 866)
hemolysis (p. 857)
immunochromatographic assay (p. 868)
lab-on-a-chip (LOC) (p. 870)
Lancefield classification (p. 858)
Lancefield groups (p. 858)
MALDI-TOF MS (p. 867)
mass spectrometry (MS) (p. 867)

midstream clean-catch technique (p. 849)
molecular beacon probe (p. 864)
multiplex PCR (p. 860)
mycolic acid (p. 859)
nonhemolytic (p. 857)
optochin susceptibility disk test (p. 859)
point-of-care (POC) laboratory test (p. 867)
probabilistic indicator (p. 854)

restriction fragment length polymorphism (p. 862)
reverse transcription quantitative RT-PCR (qPCR) (p. 863)
sensitivity (p. 867)
specificity (p. 867)
suprapubic puncture (p. 850)
thermocycler (p. 860)
venipuncture (p. 848)

Review Questions

1. Why is identifying the genus and species of a pathogen important?

2. What is an API strip, and what is its use in clinical microbiology?

3. Describe three examples of selective media.

4. If a colony on a nutrient agar plate is catalase-positive, does this mean it is made up of Gram-positive microorganisms? Why or why not?

5. Describe the types of hemolysis visualized on blood agar.

6. What is the clinical significance of a group A, beta-hemolytic streptococcus?

7. Describe the appropriate samples one should collect in the following cases:
   a. Severe headache, stiff neck, high fever
   b. High fever, altered mental status, but flexible neck
   c. Abdominal pain in lower-right quadrant, nausea

8. How does one distinguish *S. aureus* from *S. epidermidis*?

9. Why are PCR identification tests preferable to biochemical approaches?

10. How is qPCR performed?

11. Describe an ELISA.

12. Name some sterile and nonsterile body sites.

13. List seven common types of clinical specimens collected for bacteriological examination.

14. How can genomics help identify nonculturable pathogens?

15. What is a point-of-care test? What are the limitations of point-of-care tests?

Clinical Correlation Questions

1. In the case history that opens Chapter 2, Brandon had a small, round, painless lesion on his penis. The lesion exuded a clear fluid. The physician sent a sample of the fluid to the clinical laboratory, which found highly motile, corkscrew-shaped bacteria. How would the laboratory definitively diagnose *Treponema pallidum*?

2. Construct a short case history in which the causative agent is respiratory syncytial virus. Describe the patient, the signs and symptoms, the clinical specimens you would collect, what tests you would run, and the results needed to prove the patient had RSV. (This exercise may be done in groups or individually.)

3. Describe how you can use DNA-based methods to quickly determine whether a slow-growing bacterial pathogen, such as *Mycobacterium tuberculosis* isolated from a patient, carries an antibiotic resistance gene.

Thought Questions: CHECK YOUR ANSWERS

Thought Question 25.1 Two blood cultures, one from each arm, were taken from a patient with high fever. One culture grew *Staphylococcus epidermidis*, but the other blood culture was negative (no organisms grew). Is the patient suffering from septicemia caused by *S. epidermidis*?

ANSWER: Probably not. *S. epidermidis* is a common inhabitant of the skin and could easily have contaminated the needle when blood was taken from the patient. The fact that only one of the two cultures grew this organism supports this conclusion. If the patient had really been infected with *S. epidermidis*, both blood cultures would have grown this organism.

Thought Question 25.2 A 30-year-old woman with abdominal pain went to her physician. After the examination, the physician asked the patient to collect a midstream clean-catch urine sample that the office staff would send to the lab across town for analysis. The woman complied and handed the collection cup to the nurse. The nurse placed the cup on a table at the nurse's station. Three hours later, the courier service picked up the specimen and transported it to the laboratory. The next day, the report came back "greater than 200,000 CFUs/ml; multiple colony types; sample unsuitable for analysis." Why was this determination made?

ANSWER: Although the CFU number is high enough to consider relevant, UTIs are typically caused by a single organism. The fact that the lab found many different colony types suggests a problem with specimen collection. In this case, not refrigerating the sample allowed the small number of urethral contaminants to overgrow the specimen.

Thought Question 25.3 Use Figure 25.13 to identify the organism in the following case. A sample was taken from a boil located on the arm of a 5-year-old boy. Bacteriological examination revealed the presence of Gram-positive cocci that were also catalase-positive, coagulase-positive, and novobiocin-resistant.

ANSWER: *Staphylococcus aureus*. The novobiocin test is irrelevant in this situation.

Thought Question 25.4 Staphylococci from blood cultures were isolated from two body sites, but the laboratory says the organisms are probably not the cause of the patient's disease. Why might the laboratory technologist reach that conclusion?

ANSWER: To be a valid cause of septicemia, two strains isolated from blood samples taken from two sites should be identical. For the lab to shed suspicion of these strains of staphylococci, the two strains might have had different antibiotic susceptibility patterns or polymorphisms in DNA restriction patterns, suggesting that they are not identical.

Thought Question 25.5 How would you design a multiplex PCR test for real-time qPCR? A good example would be for *Chlamydia trachomatis* and *Neisseria gonorrhoeae*, two common causes of sexually transmitted disease.

ANSWER: The sample would be processed to extract DNA. Two primer sets would be designed to amplify specific genes from *C. trachomatis* and *N. gonorrhoeae*. Two beacon reporter probes that bind between each primer set are made with different fluorescent dyes attached to one end, and a quencher is attached to the other. Real-time PCR is performed on the sample, using the two primer sets and two reporter probes. Different color emissions will result depending on which organism is present.

Thought Question 25.6 Why does finding IgM to West Nile virus indicate current infection? Why wouldn't finding IgG do the same?

ANSWER: Upon infection with any organism, IgM antibodies are the first to rise (discussed in Chapter 15). After a short time the levels of IgM decline as IgG levels rise. IgG, however, can remain in serum for years (produced by memory B cells), making it a poor prognosticator of current infection.

Thought Question 25.7 Specific antibodies against an infectious agent can persist for years in the bloodstream. So how can antibody titers be used to diagnose diseases such as infectious mononucleosis? Couldn't the antibody be from an old infection?

ANSWER: During the course of a disease, the body's immune system increases the amount of antibody made specifically against the infectious agent. Thus, one compares the antibody titer in a blood sample taken from a patient in the active, or acute, phase of disease with the antibody titer several weeks later, when the patient is in the recovery, or convalescent, stage. Seeing a greater-than-fourfold rise in a specific antibody titer (for example, in mononucleosis) indicates that the patient's immune system was responding to the specific agent. Remember, simply finding IgG against an organism or virus in serum indicates only that the patient was exposed to that microbe at some time in the past. One could look for a specific increase in a specific IgM (as in Thought Question 25.6), but those reagents are not available for all infectious diseases.

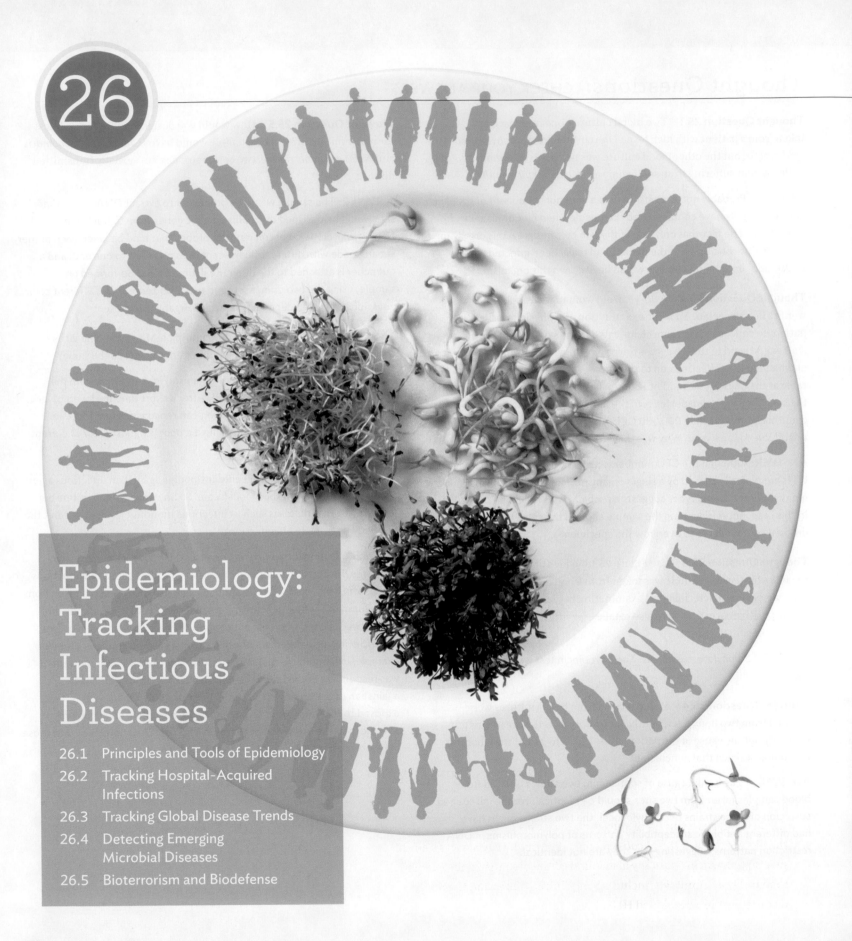

26

Epidemiology: Tracking Infectious Diseases

[Case of the "Killer" Sprouts]

OUTBREAK In May 2011, people in Germany started to die from hemolytic uremic syndrome. By June, 3,228 cases and 35 deaths had been reported. By then a massive epidemiological hunt was under way for the cause of the disease and its source.

LABORATORY FINDINGS
The organism isolated in each case was a form of *E. coli* rarely seen before. Its lipopolysaccharide and flagellar serotype was O104:H4, an enteroaggregative *E. coli* (EAEC). However, unlike EAEC this organism secreted massive amounts of type 2 Shiga toxin (contributing to hemolytic uremic syndrome [HUS]), making it a Shiga toxin–producing *E. coli* (STEC), like O157:H7 but without the attaching and effacing proteins of O157:H7. The new pathovar was called EAEC-STEC O104:H4.

EPIDEMIOLOGY In early June the finger-pointing began. Germany initially blamed Spanish cucumbers, which led to a huge economic loss in Spain as other countries stopped importing these vegetables. Later evidence showed that Spain was not the source of this organism. Finally, the agricultural minister of Lower Saxony (Germany) announced that an organic farm near Uelzen, which produces a variety of sprouted foods, was the likely source of the *E. coli* outbreak. Before the source was found, several tourists in Germany also ingested the organism and returned to their home countries, including the United States and Canada, where they then developed HUS.

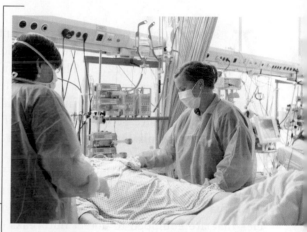

Outbreak of HUS
Hospital staff caring for a patient with hemolytic uremic syndrome at University Hospital Schleswig-Holstein in Lübeck.

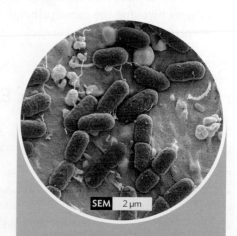

SEM | 2 µm

Emergent Strain of *E. coli*
The outbreak was linked to a new pathovar of *E. coli* called EAEC-STEC O104:H4 that contaminated fenugreek bean sprouts grown at an organic farm in Germany.

CHAPTER OBJECTIVES

After reading this chapter, you will be able to:

- Explain the epidemiologic basis for outbreaks, endemics, epidemics, and pandemics.
- Describe how epidemiologists track nosocomial and community outbreaks of a disease.
- Discuss the basic methods used to trace epidemics and pandemics to a source.
- Explain the impact of technology and climate on the epidemiology of infectious disease.
- Outline factors that influence the emergence of new pathogens.

As of 2016, the case that opens this chapter remains the deadliest outbreak of enterohemorrhagic *E. coli* (EHEC; discussed in Section 22.5). Thousands of stories like this one, involving many diseases, underscore the integral roles of clinical microbiology and epidemiology in containing disease outbreaks. The organism in this case was quickly identified and, despite some early unfortunate missteps, the source of the infection was eventually pinpointed.

Throughout history, infectious diseases have killed more people than all our wars combined. Our present success in controlling the spread of disease is due largely to worldwide surveillance agencies that are equipped to detect outbreaks quickly, before major epidemics develop. These agencies rely on smaller clinical microbiology laboratories, scattered throughout the world, that help clinicians diagnose infectious diseases. As in the opening case, finding the *source* of an outbreak is also crucial. This chapter will describe the basic concepts of epidemiology and discuss how epidemiologists track infectious diseases, identify "patient zero," and recognize and contain emerging diseases. Finally, we'll consider bioterrorism. Clinical scientists and epidemiologists are trained to detect bioterrorist attacks by using the same principles they employ to detect naturally occurring infectious diseases.

26.1
Principles and Tools of Epidemiology

SECTION OBJECTIVES

- Explain the four major links in a chain of infection.
- Differentiate between descriptive and analytical epidemiology.
- Define "endemic," "epidemic," and "pandemic," and explain the difference between disease prevalence and incidence.
- Explain the three steps of surveillance and describe how molecular approaches can contribute to surveillance.

What is epidemiology, and why is it important? The word "epidemiology" is derived from Greek, loosely translated as "the study of that which befalls man" (epi = on or upon; demos = people; logos = the study of). Epidemiology examines the distribution and determinants of disease frequency in human populations. Put more simply: epidemiologists determine the source of a disease outbreak and identify the factors that influence how many individuals will contract the disease. Epidemiological principles are also used to determine the effectiveness of treatments and to identify emerging disease syndromes, such as SARS (severe acute respiratory syndrome), MERS (Middle East respiratory syndrome), Lyme disease, and now Zika virus. We already covered some concepts of epidemiology in Chapter 2 when we discussed the basics of infectious disease, including routes of transmission. Now we will explore how those principles are used to track disease.

To an epidemiologist, data are clues: clues that include locations of outbreaks, where infected persons traveled, what they ate, and whom they met. The infectious disease epidemiologist scrutinizes these clues, looking for trends that will pinpoint the source of an infectious agent and suggest where it will spread. With this knowledge comes the power to stop or limit a disease's march through a community, a country, or even the world. Remember from Chapter 2 that the **chain of infection** (Figure 26.1) has four major links: the etiologic agent (its virulence and invasiveness), the agent's reservoir (animal, human, or environment), its mode of transmission (via person to person, air, vectors, or inanimate objects, including food), and its host (portal of entry, immune competence). Influencing any one link can break the chain of infection.

John Snow (1813–1858), Father of Epidemiology

The first known case in which the source of a disease outbreak was methodically investigated took place in the mid-nineteenth century. During a serious outbreak of cholera in London in 1854, John Snow (Figure 26.2A) used a map to plot the locations of all the diarrheal cases he learned about. The source of the infection was unknown ("experts" believed it was transmitted through a contagious power in the mist, or miasma), but Snow thought that if the cases clustered geographically, he might gain a clue as to the actual source. Water in that part of London was pumped from separate wells located in various neighborhoods. Snow's map revealed a close association between the density of cholera

Figure 26.1   Links in the Chain of Infection

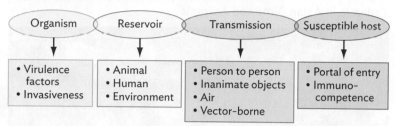

cases and a single well located on Broad Street (**Figure 26.2B**). Simply removing the pump handle of the Broad Street well ended the epidemic, proving that the well water was the source of infection (removing the pump handle broke the transmission link in the chain of infection). This approach succeeded brilliantly, even though the infectious agent that causes cholera, *Vibrio cholerae*, was not recognized until 1905, more than 50 years later. Today, epidemiologists still identify clusters of patients afflicted with a given disease to trace potential sources of infectious disease outbreaks. The science of medical statistics, founded by **Florence Nightingale**, is also critical to this effort.

Link Recall from Section 1.3 that **Florence Nightingale** (1820–1910), a British nurse and statistician, developed a valuable set of data collection techniques upon which modern epidemiology was built. Her techniques proved that soldiers fighting in the Crimean war died primarily from infectious disease and not from bullets or poor nutrition. Her heroic exploits led to improved hygiene at many army hospitals and saved the lives of untold numbers of soldiers.

Types of Epidemiology

Epidemiologists are detectives, but they are reporters, too. And as any reporter will tell you, a good story describes the what, who, where, when, and why (or how) of an event. For the epidemiologist, the 5 Ws are case definition (what), person (who), place (where), time (when), and causes/risk factors/modes of transmission (why and how).

DESCRIPTIVE EPIDEMIOLOGY

Every infectious disease has patterns. Sometimes the disease is most frequent during certain times of the year (influenza in winter; West Nile virus in summer), is focused in certain geographic locations (Lyme disease in the northeastern United States; Dengue fever in Central and South America), or occurs among certain types of people (respiratory syncytial virus in children; Creutzfeldt-Jakob disease in older adults; pneumocystis among the immunocompromised). By collecting data relating to person, place, and time (called **descriptive epidemiology**), an investigator can identify geographical areas or population groups that have high rates, or elevated risks, for a particular disease.

ANALYTICAL EPIDEMIOLOGY
An epidemiologist will take the person, place, and time data of descriptive epidemiology and formulate hypotheses about the cause of the disease and the possible risk factors involved. **Analytical epidemiology** tests those hypotheses. Testing can involve either experimental or observational studies. In an **experimental study**, for instance, you might compare a new vaccine against a **placebo** (an inactive substance used as a control) in two groups of patients. In this type of experimental study you would ask whether the vaccinated group contracted disease less often than the placebo group.

An **observational study** can also test hypotheses. Here an epidemiologist will observe the exposure and disease status of each study participant. John Snow's study of cholera in London was observational. Two main types of observational studies exist. In a **cohort study**, an investigator observes whether each participant is exposed to an agent and then notes whether each develops disease. The exposed group is compared with the unexposed group. In a **case-control study**, investigators enroll a group of people with a disease and a control group of people without disease. They then compare

Figure 26.2   Early Epidemiology

A. John Snow.

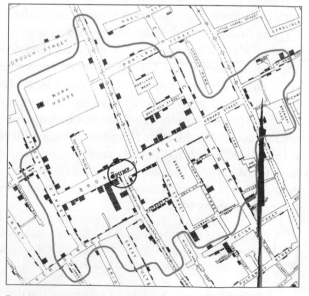

B. Map of London that Snow used to pinpoint the source of the cholera outbreak in 1854. The Broad Street well is found within the red circle. Each black bar represents a death from cholera.

Figure 26.3 Difference between Endemic and Epidemic Disease

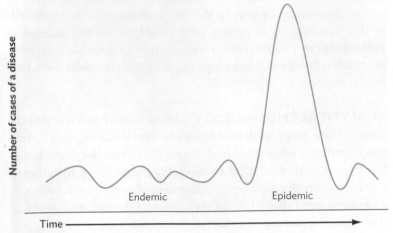

An endemic disease is continually present at a low frequency in a population of constant size and composition. A sudden rise in disease frequency constitutes an outbreak or epidemic (if the upswing is not due to many infected patients moving into the area).

previous exposures between the groups. For instance, during an outbreak of hepatitis A, you might ask whether the case patients had exposure to a type of food more often than did the control group. If so, that food could be the source of the infection.

Endemic, Epidemic, or Pandemic?

The terms "endemic," "epidemic," and "pandemic" reflect how many cases of a disease are observed over time in a given geographical locale (county, state, country, or world). As noted in Chapter 2, certain diseases are considered **endemic**, or native to certain areas (Figure 26.3). An endemic disease exhibits a steady (usually low) number of cases in a location. For example, *Campylobacter jejuni* is endemic in much of the United States, as are chickenpox and *Shigella*. Other diseases, such as those caused by *Borrelia burgdorferi* (Lyme disease) or West Nile virus, are endemic to smaller areas (such as the northeastern United States). A disease endemic to an area usually has a reservoir there (for instance, birds are the reservoir for West Nile virus). Recall that a reservoir is an animal, bird, or insect that harbors the infectious agent and is indigenous to a geographical area. Humans become infected when they come in contact with the reservoir.

 Planning to travel? Worried about disease incidence where you're going? Visit *The Yellow Book*, Chapter 3, for advice on endemic disease in various locales: cdc.gov

A nonendemic area is free of a disease (the disease has no reservoir). However, even a locale considered nonendemic may still develop sporadic cases of the disease. This can happen when a person living in a nonendemic area receives a contaminated package

Figure 26.4 The Avian Flu Virus H5N1 Is Endemic in Asian Birds and Animals

Live poultry for sale in Vietnam.

from an endemic area, or when people travel to an endemic area where they become infected, only to return home and develop disease. For instance, even though measles is not considered endemic in the United States, sporadic cases of measles (50–100 per year) can occur when someone lacking immunity returns from a country where measles is endemic, such as the United Kingdom. Outbreaks there are thought to be caused by decreased MMR vaccination of children, a consequence of baseless speculation that vaccines are linked to autism and allergies. In 2015, a similar situation developed in the United States after someone with measles visited Disneyland. The virus infected several tourists, who returned home and spread the disease among 176 mostly unvaccinated children in 17 states.

Endemic disease is important, but when the number of cases of a disease rises above the endemic level we have an **outbreak**. Local outbreaks can develop into more serious **epidemics** (Figure 26.3) when the case numbers rise rapidly and the affected geographical area widens. Epidemics develop, in part, because of rapid and direct human-to-human transmission. An endemic disease can become epidemic if the population of the reservoir increases, which allows for more frequent human contact, or if the infectious agent evolves to spread directly from person to person, bypassing the need for a reservoir. The latter is the concern with the H5N1 avian flu virus, which is endemic in animals and birds in Asia (Figure 26.4). Although H5N1 does not easily spread between humans now, it could evolve to do so.

Out-of-control epidemics that spread across continents are called **pandemics** (AIDS and influenza, for example). Pandemics may be long-lived—such as the bubonic plague pandemic in the fourteenth century and the AIDS pandemic in the late twentieth and early twenty-first centuries—or they may be short-lived, as with the 1918 flu pandemic. Table 2.3 lists some pandemics that struck the world. The fear of a pandemic can lead to extreme public health measures. When the cholera pandemic hit Chicago in the 1860s, for instance, it led to quarantine of the entire city and spawned new laws governing drinking water and sewage treatment.

Note "Outbreak" is also a generic term applied to everything from small local increases in case numbers all the way to large case increases across the globe.

Disease Prevalence versus Incidence

When discussing a disease, epidemiologists distinguish between prevalence and incidence. **Prevalence** describes the total number of active cases of a disease in a given location regardless of when the case first developed (think of it as the disease's burden on the society). **Incidence**, however, refers to the number of *new* cases of a disease in that location over a specified time (the risk of acquiring a disease). Take a city of 100,000 people. For chronic diseases, such as tuberculosis, incidence rates can provide better insight into whether efforts to limit disease are succeeding. In our imaginary city, prevalence might remain at 0.1% (100 cases per 100,000) over each of 2 years even though new cases develop. The steady prevalence rate happens because some patients who contracted the disease in year 1 might have been cured (or died) before year 2 but were replaced by newly developed cases.

However, if the *incidence* of *new* cases per year rises from 10 cases/50,000 previously uninfected persons in year 1 to 30 new cases/50,000 previously uninfected persons in year 2, then efforts to stem the disease are failing. By contrast, if the incidence drops from 10 new cases to 3 per 50,000 previously uninfected, then control measures are working. An endemic disease will maintain a relatively constant prevalence and incidence rate. Outbreaks are marked by an increase in both incidence and prevalence. This is what happened during the 2014 Ebola epidemic in West Africa.

Incidence Statistics

Incidence statistics can be broken down by sex, race, age, occupation, and many other categories. The data can then be presented to reflect geographical or yearly trends. For example, **Figure 26.5** presents the U.S. geographical distribution of Lyme disease and syphilis for 2011. Note the geographical clusters of disease. Lyme disease occurs mostly in the Northeast (Figure 26.5A), corresponding to the geographical location of the *Ixodes* tick vector. In contrast, syphilis is more prominent in the South (Figure 26.5B).

Plotting case incidence over time can yield other insights. **Figure 26.6A**, for instance, tracks U.S. measles cases from 1977 to 2012. Measles vaccine was introduced in 1963, after which the annual number of new cases per year fell from 26 per 100,000 to only 2–3. An unexpected spike of cases in 1990 (resulting from a decrease in vaccinations [75% of victims were unvaccinated]), prompted a change in U.S. vaccination practices. Measles consequently is no longer considered endemic to the United States, although traveler-imported cases from other countries do occur.

Figure 26.6B tracks the incidence of mumps, which decreased after the introduction of mumps vaccine in 1967. The institution of a second dose of mumps vaccine in 1990 led to historically low **morbidity** until 2006. The 2006 outbreak of more than 6,000 cases affected primarily college students aged 18–24 years in the Midwest. The outbreak was surprising, given the high level of vaccine coverage. Health officials concluded that the immune response for mumps faded more rapidly than was thought and that some people did not receive the second dose of vaccine. (Immunity wanes when memory B and T cells are not periodically stimulated by exposure to antigen.) This situation left many immunized persons at risk. As a result, the Advisory Committee on Immunization Practices updated its vaccination recommendations in 2006 to ensure that children received

Figure 26.5   Geographical Representation of Disease Incidence

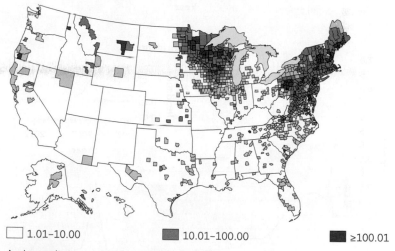

| 1.01–10.00 | 10.01–100.00 | ≥100.01 |

A. Lyme disease in 2012. Incidence is highest in the northeastern United States.

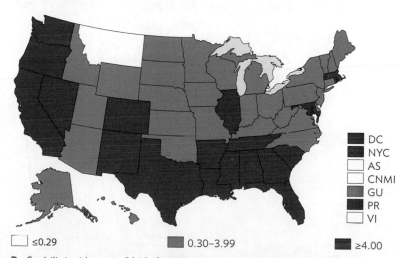

DC
NYC
AS
CNMI
GU
PR
VI

| ≤0.29 | 0.30–3.99 | ≥4.00 |

B. Syphilis incidence in 2012. Small squares represent District of Columbia (DC), New York City (NYC), American Samoa (AS), Commonwealth of Northern Mariana Islands (CNMI), Guam (GU), Puerto Rico (PR), and U.S. Virgin Islands (VI). Incidence rate for both charts = numbers of cases per 100,000 population.

Figure 26.6   Disease Incidence of Measles, Mumps, and Pertussis in the United States

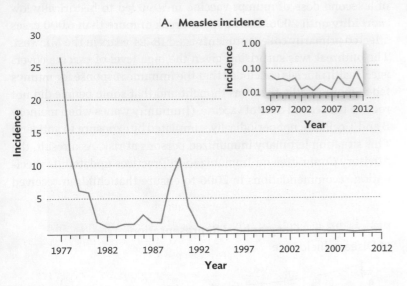

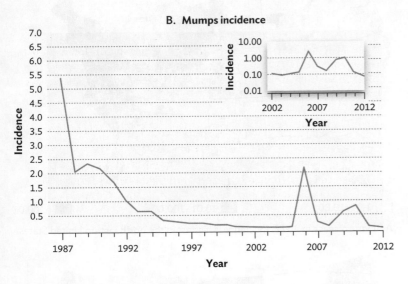

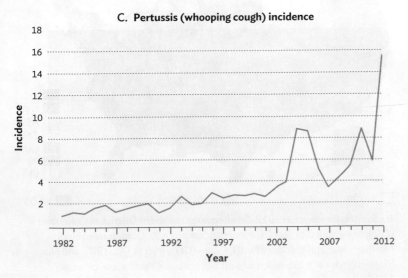

A. Measles incidence (per 100,000 population) from 1977 to 2012. Inset magnifies the 1997–2012 region of larger graph. Measles vaccine was introduced in 1963. Measles is no longer considered endemic in the United States.

B. Mumps incidence (per 100,000 population) from 1987 to 2012. Inset magnifies the 2002–2012 region of larger graph. Mumps vaccine was introduced in 1967. The institution of a second dose of mumps vaccine in 1990 led to historically low morbidity until 2006. A 2006 outbreak of more than 6,000 cases, affecting primarily college students aged 18–24 years in the Midwest, caused the Advisory Committee on Immunization Practices to update its vaccination recommendations.

C. Pertussis incidence (per 100,000 population) from 1982 to 2012. The recent spikes in cases appear related to diminished coverage by vaccination and the waning of protection among older adults. The acellular pertussis vaccine was introduced in 1996 as part of the DTaP series, replacing the older, killed, whole-cell vaccine used since 1948.

two doses of mumps vaccine and that health care workers were immune or revaccinated.

Link Recall from Section 2.2 that **morbidity** refers to the incidence of a disease state and that mortality measures how many patients died from the disease.

The increasing incidence of pertussis, or whooping cough, shown in **Figure 26.6C**, is not entirely clear. Factors thought to be involved include decreased rates of vaccination, a possibly less effective vaccine, and improved detection. It is also recommended that older people, whose immunity to pertussis is waning, receive a booster vaccination with Tdap. Tdap is a multivalent vaccine containing full-strength tetanus toxoid combined with diphtheria toxoid and acellular pertussis antigens in doses lower than standard DTaP. These examples illustrate how epidemiologists monitoring incidence and prevalence trends can anticipate epidemics and contain them.

Thought Question 26.1   In 2011, British Columbia, Canada, reported that 77 people contracted mumps. Seventy percent of those cases were aged 18–35 years. How would you have linked these cases, and what do you think led to the outbreak?

Tools of Epidemiology

The principles just outlined combined with knowledge of disease transmission, the presence of effective diagnostic laboratories, and vigilant surveillance programs are all vital for tracking diseases on local and global scales. An interactive network linking health care professionals to local, state, and national agencies forms the early-warning system we rely upon to keep us one step ahead of pathogenic microbes. This section describes some of the tools they employ.

CLINICAL MICROBIOLOGY LABORATORIES   Infectious disease epidemiology depends on dedicated clinical microbiology laboratories to identify infectious agents. These laboratories are located in or are associated with thousands of hospitals around the world. In addition, most (if not all) hospitals send samples of difficult cases or unusual organisms to regional reference laboratories equipped to run technically challenging tests and handle extremely dangerous or hard-to-grow pathogens (for instance, *Francisella*). Though important for individual patient care, this information also yields a wealth of data useful for epidemiological studies. Processing this information is part of surveilling an infectious disease.

SURVEILLANCE   Surveillance is a three-stage process. The first stage is to see whether a patient's disease fits the Centers for Disease Control and Prevention (CDC) case definition of a specific infection. Validating an infection requires an infection control practitioner (ICP) to review a patient's chart and see whether the patient had the signs and symptom of an infection. Surveillance definitions cover infections of the bloodstream, urinary tract, lung (pneumonia), and surgical sites. The second stage of surveillance is to identify the case's geographic location and the patient's demographic profile (such as age, sex, and race). The third stage is to use this information to identify trends.

Surveillance traditionally required laborious manual data entry and assessment. Now, integrated software solutions assess incoming data from microbiology laboratories and other online sources. By reducing the need for data entry, this software significantly reduces the data workload of ICPs, freeing them to concentrate on clinical surveillance.

Epidemiological early-warning systems require an extensive organization that coordinates information from many sources. In the United States, that duty falls to the CDC. On the world stage, it is the World Health Organization (WHO). Any disease considered highly dangerous or infectious is first reported to local public health centers, usually within 48 hours of diagnosis. The local centers forward that information to their state agencies, which then report to the CDC in Atlanta. This is how authorities in 2013 recognized that an unusual *Neisseria meningitidis* group B outbreak occurred at the University of California, Santa Barbara.

 Public health at local, state, national, and global levels: ashp.org

MOLECULAR SURVEILLANCE   Molecular approaches to infectious disease surveillance can also prevent or limit the international spread of disease. A pandemic of pulmonary tuberculosis currently affects more than 2 billion people. Many *Mycobacterium tuberculosis* infections are caused by multidrug-resistant (MDR) strains that are difficult, if not impossible, to kill with existing antibiotics (see Section 20.3). MDR tuberculosis is especially serious among refugee populations attempting to flee war-torn countries. Measures to ensure that refugees are screened for tuberculosis as they enter neighboring countries are vital to contain the pandemic. Although chest X-rays are often mandatory, a positive image will be obtained only if the disease is at a relatively advanced stage. X-rays will not identify actively infected individuals who have not developed the characteristic lung tubercles (Section 20.3).

Unfortunately, the acid-fast staining of sputum samples (Section 25.2) also fails to detect individuals at an early stage of infection. However, PCR amplification techniques are much more sensitive for identifying these individuals. With a much lower threshold of detection, PCR amplification can detect far fewer mycobacterial cells than an acid-fast stain can. As time progresses, PCR surveillance strategies will be used more often to track the worldwide ebb and flow of microbial diseases. Epidemiologists already use PCR and other DNA strategies for epidemiological purposes to type (that is, determine relatedness of) different microbial isolates by generating a complex DNA profile specific for a particular strain.

SECTION SUMMARY

- **Epidemiology** is a field that examines factors that determine the distribution and source of disease.

- **Descriptive epidemiology** identifies geographical areas or groups of people that have a high incidence of a disease.

- **Analytical epidemiology** uses experimental or observational studies (cohort or case–control) to test hypotheses about the cause of a disease or the risk factors associated with that disease.

- **Endemic, epidemic, and pandemic** refer, respectively, to a disease that is always present in a locale, an increase of disease cases (outbreak) in a geographic area, and the worldwide outbreak of a disease.

- **Surveillance techniques** help epidemiologists detect new outbreaks of a disease and predict when and where a disease may spread. International, national, and local health organizations and their laboratories work together to develop case descriptions and collect incidence data needed to track an infectious disease.

Thought Question 26.2   A recent outbreak of *E. coli* O157:H7 disease has been detected. How long will it take from the time the first victim eats contaminated food before a laboratory confirmation of this organism is established and a public health lab determines that strains from several cases are identical? As you work through this timeline, consider every event that must occur and how long each takes.

26.2
Tracking Hospital-Acquired Infections

SECTION OBJECTIVES

- Discuss the nature of nosocomial infections and list the two most common forms.

- List key questions that can help identify the source of a hospital-acquired infection.

- Discuss procedures used to prevent hospital-acquired infections.

CASE HISTORY 26.1

Death by Negligence

 Stephanie was a bright 7-year-old girl who came to her rural Manitoba hospital for a tonsillectomy. She was a little scared, but her mother was with her and she liked the surgeon, Dr. Ryan, and his funny Winnie-the-Pooh ties. Sometimes she would even grab his tie while Dr. Ryan was examining her (Figure 26.7). However, she didn't much like the IV catheter Dr. Ryan had to insert in her arm. It hurt a little, but she was brave and didn't cry much. The surgery went well and Stephanie was recovering in her hospital room a few hours later. However, that night the nurse noted her temperature had risen to 102°F. By the next day, Stephanie's heart was racing and her breathing was labored. She was rushed to the pediatric intensive care unit and empiric antibiotic therapy was initiated. Despite this treatment, her blood pressure began to fall. Stephanie was diagnosed with postsurgical septic shock. To everyone's horror, Stephanie died two days later of MRSA sepsis. The organism recovered from her blood was a MRSA strain identical to one causing a wound infection in a 57-year-old man two floors down.

How could this happen? Infections contracted in a hospital setting are called **nosocomial infections**, hospital-acquired infections, or **health care-associated infections (HAIs)**. The classic example is a patient initially admitted for medical treatment, here a tonsillectomy, who then develops a serious infection. The source of the microbe is somewhere in the hospital. But where? You will later learn the unfortunate circumstance that led to Stephanie's deadly MRSA infection.

Health care-associated infections are a major problem for hospitals, both nationally and internationally. The gravity of the situation is evident in the following statistic: 5%–10% of hospital patients develop nosocomial infections. What is worse is that approximately one-third of health care-associated infections are preventable, with appropriate surveillance and preventive action by hospital staff. Table 26.1 lists major types of health care-associated infections and risk factors. Of course, a single patient who contracts an HAI is tragic, but even more troubling is when several patients become infected with the same organism.

Figure 26.7   Opportunity for a Health Care-Associated Infection

As the doctor examines this child, he could accidentally transmit an infectious agent if his lab coat, tie, or stethoscope is contaminated with pathogens from another patient.

Ignaz Semmelweis, "Savior of Mothers"

The best-known and perhaps earliest case of hospital epidemiology was initiated by Ignaz Semmelweis, a Hungarian physician practicing in a Vienna hospital in 1847 (Figure 26.8). All hospital obstetric wards of the day struggled with a deadly disease called puerperal sepsis, or childbed fever, a lethal blood infection that many women contracted after giving birth in the hospital (10%–35% mortality). Semmelweis noticed that the incidence of the disease, now known to be caused by *Streptococcus pyogenes*, increased during certain times of the year. He then realized that these times coincided with an infusion of medical students into the clinic. Tracking the movements of these students, he found they came straight from the autopsy room to the maternity ward to assist in deliveries. Semmelweis made the connection that medical student and physician hands were probably

Table 26.1
Typical Nosocomial Infections

Infection Type (% Nosocomial)	Risk Factors	Common Etiologic Agents
Urinary tract (40–50)	Indwelling catheters, other instrumentation	*Escherichia coli, Candida* spp.
Pneumonia (15–20)	Aspiration of endogenous microbiota, ventilators	*Streptococcus pneumoniae, Haemophilus, Pseudomonas, Staphylococcus aureus, Klebsiella pneumoniae, Acinetobacter*
Surgical wounds (20–30)	Surgeon's technical skill, patient's underlying diseases (diabetes, obesity), age, presence of surgical drains	*Staphylococcus aureus*, coagulase-negative staphylococci, enteric bacteria, anaerobes
Septicemia (5–10)	Intravascular devices (catheters)	Derived from microbiota at the site, *Staphylococcus aureus*, enterococci, Gram-negative bacilli, *Candida*
Diarrhea (15–20)	Antibiotic treatment	*Clostridium difficile*

transmitting "cadaverous particles" (germs) from the dead bodies in the autopsy room to the women in the maternity ward.

The value of hand washing was not recognized in the mid-1800s. Germ theory was still in its infancy and highly controversial. Physicians scoffed at hand washing unless they were covered in blood and guts. Nevertheless, Semmelweis insisted that all people entering the clinic wash their hands with an antiseptic (a chlorinated lime solution) before touching any patient. The audacity of this physician, a Hungarian nonetheless, caused an uproar in the Viennese medical community.

Although Semmelweis's edicts led to a sharp decline in the incidence of puerperal sepsis, his success went unrewarded. It is not clear whether rejection of his work contributed to his slow mental decline, but this "savior of mothers"—as he was called by those he saved—eventually developed dementia and ended up confined to a darkened cell in a mental institution, where he died in 1865 from, ironically, a blood infection. But in the long run, his advocacy of antisepsis became a key precept of hospital care.

Figure 26.8
Ignaz Semmelweis
(1818–1865)

Identifying the Source of Hospital-Acquired Infections

So how do infection control personnel identify and contain sources of infection in hospitals? Infectious disease epidemiologists are essentially detectives who must ask probing questions in the midst of an outbreak:

- **Is each patient infected with an identical isolate of the pathogenic organism?** If the isolates of MRSA are different, for example, there may be multiple sources. Immunological or DNA approaches are typically used to determine relatedness. In Case History 26.1, both MRSA isolates were identical by DNA analyses.

- **Has any member of the hospital staff made contact with all the patients?** If a hospital employee harbors the organism as part of the person's normal microbiota, he or she could be the common source. In the above case, someone attending to the 57-year-old MRSA patient may have become contaminated with MRSA and accidentally transferred it to Stephanie.

- **Did all patients come in contact with the same medical instrument, the same lot of medical supplies (such as IV fluid, indwelling catheters, intubation tubes), or the same food or water?** Any of these might be contaminated.

- **Were the affected patients in contact with each other at any time, enabling person-to-person transmission?**

- **If the infectious agent is airborne, could air ducts linking the rooms have problems?**

- **Have hospital staff, sinks, surfaces, and carts been tested for presence of the organism?** Water is the most common source of nosocomial infections.

- **Have standard infection control measures been maintained?**

Answers to these questions will often, but not always, reveal the source of the infections. Once identified, the problem can be rectified and recurrence prevented using effective infection control measures.

Infection Control Measures

So how do hospitals prevent nosocomial infections? Hand sanitation is one important factor. Most hospitals have posted dispensers of foam hand sanitizer outside and inside each patient's room (Figure 26.9A). Everyone must use these dispensers when entering or exiting.

In addition to hand washing, disposable paper gowns and latex gloves (Figure 26.9B) are worn when entering the rooms of

Figure 26.9   Important Infection Control Measures

A. Foam hand sanitizer outside a patient's room.

B. Disposable gown and gloves.

Figure 26.10   Peripherally Inserted Central Catheter (PICC Line)

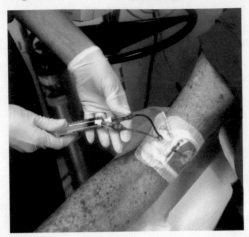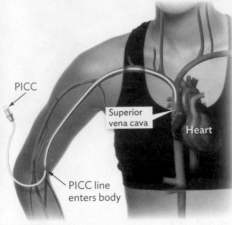

PICC

Superior
vena cava

Heart

PICC line
enters body

This is a form of long-term, intravenous access used for several purposes, including nutrition or extended antibiotic therapy. A PICC is inserted into a peripheral vein (for instance, arm) and then advanced to the heart, where it rests in the superior vena cava.

immunocompromised patients or of patients with extremely dangerous infectious diseases (such as MRSA). These measures protect the susceptible patient and, in the case of "gowning," protect the visitor (health care worker or relative). Disposing of gowns and gloves before leaving the room will also prevent spread of the organism to others in the hospital.

Aseptic technique is also crucial to good patient care. Asepsis must be maintained whenever a procedure such as inserting a catheter into the bladder or a vein is performed. The site must be prepared with antiseptic, and sterile surgical drapes and gloves should be used when inserting intravascular lines such as a peripherally inserted central catheter (Figure 26.10). Equipment used for procedures should be sterilized by heat if possible or come from undamaged sterile packaging. Surfaces of tables, doorknobs, and the like should be disinfected and linens cleaned to prevent accidental transfer of pathogens to other persons.

So how did Stephanie end up with sepsis in Case History 26.1, and how could it have been prevented? Before Stephanie's surgery, Dr. Ryan was examining the 57-year-old MRSA patient's wound when his tie accidentally grazed the wound area (he did not don a protective gown before entering the room). *Staphylococcus aureus* was transferred to the tie. Later, while Stephanie was playing with Dr. Ryan's tie, the organisms moved from the tie to Stephanie's arm. Inserting the catheter in her arm inadvertently introduced the pathogen into her bloodstream. Two days later Stephanie was dead. Hospital surveillance procedures later identified five more of Dr. Ryan's patients who were contaminated with the same strain of MRSA. This tragedy could have been prevented if Dr. Ryan had simply not worn a tie or kept his tie tucked behind a lab coat or under a surgical gown while examining patients. This and other scenarios raised awareness that clothing is an important potential vector for transmitting disease in hospital settings.

"Daisy Chain" Nosocomial Infections

Patients, especially elderly, long-term-care patients, are often moved between hospitals and nursing home facilities. Transferred patients take with them any bacteria they harbor, which may initiate **daisy-chain nosocomial infections**: an insidious series of linked infections. For example, in Florida, a severely wounded trauma victim developed a lung infection with a multidrug-resistant strain of *Salmonella enterica* serovar Senftenberg. As diagrammed in Figure 26.11, 19 other cases of *S. enterica* Senftenberg developed in that hospital. One of those patients was transferred to a nursing home facility, after which 14 new cases developed, some of whom were transferred to additional facilities, leading to yet more cases. Repeated breakdown of standard infection control measures by hospital personnel apparently caused the daisy-chain transmission of the organism to patients in different hospitals. The case also illustrates how organisms, such as MRSA, previously restricted to hospital environments can escape to cause community-acquired infections.

Where do we stand today with hospital-acquired infections? Have things improved? Available statistics through 2012 for U.S. hospitals indicate that incidence of such infections has fallen an average of about 20% since 2008. The only infection category that increased in incidence was catheter-associated urinary tract infections. So although the situation is improving, much work remains.

SECTION SUMMARY

- **Nosocomial, or hospital-acquired, infections** are a major concern for hospital staff.
- **The source of single or multiple nosocomial infections can be identified** by answering questions such as:
  —Are all strains of the organism identical?
  —Did any hospital staff have access to all the patients?
  —Did all the patients use a particular medical device?
  —Were standard infection control measures enforced?
- **Infection control measures** can prevent most nosocomial infections. These measures include frequent hand washing, using gowns and sterile gloves to interact with infected or immunocompromised patients, and using aseptic technique when inserting intravascular lines or urinary catheters.
- **Transferring an infected patient** between health care facilities can spread nosocomial infections.

Thought Question 26.3   A nosocomial outbreak of MRSA occurred in a community hospital. The infection control team identified a nurse harboring the same strain of MRSA as part of his normal microbiota. The nurse, however, refused treatment because he wasn't sick. Should the hospital administrators fire the person? Should all hospital personnel be screened for MRSA as a condition of employment?

Figure 26.11   "Daisy Chain" Movement of an Antibiotic-Resistant Strain of *Salmonella enterica* Serovar Senftenberg between Hospitals

The diagram portrays how a single infected patient transferred between two health care facilities can initiate a string of infections in multiple facilities as infected patients are transferred to other facilities. Each arrow indicates a patient transfer. A breakdown in aseptic procedures most likely led to the secondary infections.

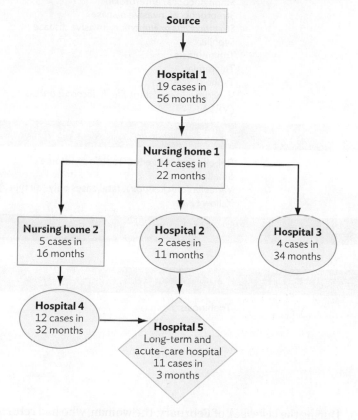

Thought Question 26.4   Imagine a large metropolitan hospital in which eight serious nosocomial infections with MRSA have occurred, and you must determine the source of infection so it can be removed. Using common bacteriological and molecular techniques, how would you accomplish this task?

26.3
Tracking Global Disease Trends

SECTION OBJECTIVES
- Discuss how global disease trends are recognized.
- Explain the importance of finding the index case of an outbreak.
- Employ online sources to track worldwide disease trends.

How do we track global outbreaks of infectious diseases and predict when one may be coming our way? Procedures to track global outbreaks of disease are similar to what is used at local and national levels but are more complex. For instance, the chapter opening case history described an intercontinental outbreak of EHEC disease. Stopping the spread of that disease began with data collection at the local level, but communication and cooperation between the health agencies of several nations was crucial.

Local Physician Notification of Health Organizations

How do epidemiologists first recognize that an epidemic is under way and then identify the agent and its source? Certain diseases, because of their severity and transmissibility, are called reportable, or notifiable, diseases (Table 26.2). Physicians must report instances of these diseases to a central health organization, such as individual state health agencies, as well as the CDC in the United States, or the WHO in Geneva, Switzerland, so that incidences of certain diseases within a population can be tracked and upsurges noted. Diseases of *unknown* etiology are also reported to health authorities because a new emerging disease not on the list of notifiable diseases can be detected as a cluster of patients with unusual symptoms or combinations of symptoms. This approach is known as **symptomatic** or **syndromic surveillance**. A new disease could manifest with common symptoms (for example, the cough and fever of SARS) but cannot be linked to a known disease agent by clinical tests. An upsurge in cases of either a reportable or an emerging disease will set off institutional "alarms" that initiate efforts to determine the source and cause of the outbreak.

Finding Patient Zero

When trying to contain the spread of an epidemic, it is vital to track down the first case of the disease (known as the **index case** or **patient zero**) and then identify all people who had contact with the individual so that the person can be treated or separated from the general population (that is, **quarantined**). When a new disease arises, the epidemiological search for the index case starts only after several patients have been diagnosed and a new disease syndrome declared. This is what happened with AIDS in the 1970s.

Identifying an index case within a specific community is easier if the disease syndrome is already recognized, as was the case with the 2003 severe acute respiratory syndrome (SARS) outbreak in Singapore. According to the World Health Organization, a suspected case of SARS is defined as an individual who has a fever greater than 38°C, exhibits lower respiratory tract symptoms, and has traveled to an area of documented disease or has had contact with a person afflicted with SARS. The index case in Singapore was a 23-year-old woman who had stayed on the ninth floor of a hotel in Hong Kong while on vacation. A physician visiting from southern China who stayed on the same floor of the Hong Kong hotel during this period is believed to have been the source of her infection, as well as that of the index patients who later caused outbreaks in Vietnam and Canada.

Table 26.2

Notifiable Infectious Diseases (Centers for Disease Control and Prevention)

Bacterial

Anthrax	Hansen's disease (leprosy)	Salmonellosis (nontyphoid fever types)
Botulism	Legionellosis	Shigellosis
Brucellosis	Leptospirosis	Staphylococcal enterotoxin
Campylobacter infection	Listeriosis	Streptococcal invasive disease
Chlamydia infection	Lyme disease	*Streptococcus pneumoniae*, invasive disease
Cholera	Meningitis, infectious	Syphilis
Diphtheria	Pertussis	Tuberculosis
Ehrlichiosis	Plague	Tularemia
E. coli O157:H7 infection	Psittacosis	Typhoid fever
Gonorrhea	Q fever	Vancomycin-resistant *Staphylococcus aureus*
Haemophilus influenzae, invasive disease	Rocky Mountain spotted fever	(VRSA)

Viral

Dengue fever	Measles	Rubella and congenital rubella syndrome
Hantavirus infection	Mumps	Smallpox
Hepatitis, viral	Poliomyelitis	Varicella (chickenpox), fatal cases only (all types)
HIV infection	Rabies	Yellow fever

Fungal

Coccidioidomycosis

Parasitic

Amebiasis	Giardiasis	Trichinosis
Cryptosporidiosis	Malaria	
Cyclosporiasis	Microsporidiosis	

Figure 26.12   Severe Acute Respiratory Syndrome (SARS)

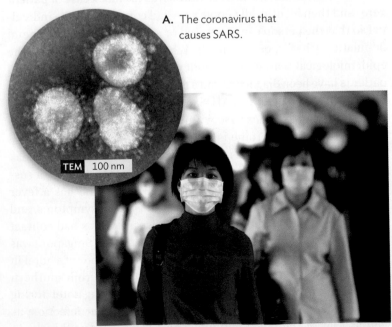

A. The coronavirus that causes SARS.

TEM 100 nm

B. Citizens of China, including the military, donned surgical masks in 2003 to slow the spread of SARS.

During the last week of February, the woman, who had returned to Singapore, developed fever, headache, and a dry cough. She was admitted to Tan Tock Seng Hospital, Singapore, on March 1 with a low white blood cell count and patchy consolidation in the lobes of the right lung. Tests for the usual microbial suspects (*Legionella*, *Chlamydia*, *Mycoplasma*) were negative. Electron microscopy of nasopharyngeal aspirates showed virus particles with widely spaced clublike projections (Figure 26.12A). At the time of her admission to the hospital, the clinical features and highly infectious nature of SARS were not known. Thus, for the first 6 days of hospitalization, the patient was in a general ward, without barrier infection control measures. During this period, the index patient infected at least 20 other individuals, including hospital staff, nearby patients, and visitors. Figure 26.13 presents a network diagram showing how the disease spread between patients.

Within weeks, the WHO named the disease in China severe acute respiratory syndrome (SARS) and issued travel alerts (discussed further in Section 26.5). These alerts allowed Singapore health officials to rapidly identify the index patient and her contacts, limiting spread of the illness. Ultimately, SARS killed fewer than 1,000 victims worldwide, although thousands more became ill and recovered. Implementation of infection control procedures and

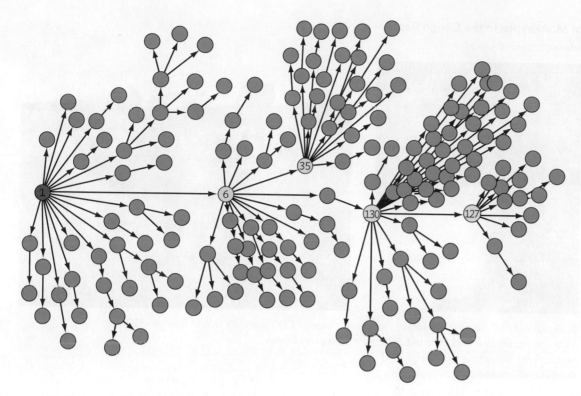

Figure 26.13
Network Diagram of SARS in Singapore
The diagram maps the connections between SARS patients during the initial outbreak. Patient 1 is "patient zero," or the index case, in this instance (red circle). The numbered cases were considered "superspreaders" because they infected more than 10 contacts (yellow circles). For example, patient 6 was a nurse who attended to "patient zero." Patient 35, a diabetic with heart disease, initially stayed in the same room as patient 6.

the voluntary wearing of surgical masks by the public helped ward off a potential pandemic (**Figure 26.12B**). Today, SARS remains a threat, but one of lesser concern. Methicillin-resistant *Staphylococcus aureus* (MRSA) and H5N1 avian flu are considered more pressing dangers.

Obstacles to Global Surveillance

In a perfect world, every province in every country on every continent would have modern clinical laboratories, efficient case reporting, and effective political systems needed to quickly detect and limit outbreaks of infectious disease. Unfortunately, the resources of developing countries are too limited to reach public health surveillance goals. CDC and WHO often mobilize teams of scientists and medical personnel, sending them to remote parts of the world to help, especially when global health is threatened. Teams often encounter desperate conditions in which local medical personnel struggle against overwhelming odds (**Figure 26.14**).

Do-It-Yourself, Online Epidemiology Research

You can do your own crude epidemiological research online. After logging in to your Google account, go to Google Trends and type "influenza." You will see a graph depicting how many searches were made for influenza over several years (adjust the time range to start at 2004). A significant search peak is seen around 2009, corresponding to the last influenza pandemic. Try the same thing for Ebola or Zika. The idea behind this approach is that as people become ill with a certain disease, they will search for information

online. An outbreak can be detected almost as quickly as through epidemiological approaches. Of course, media reports can also drive a spike in searches. Consequently, the Google Trends strategy will not replace rigorous epidemiological practices. The link below will take you to Healthmap, a website that constantly updates outbreaks of infectious diseases throughout the world. See what's happening where you live and where you might be going.

 Healthmap. World locations of current infectious disease outbreaks: healthmap.org

SECTION SUMMARY

- **Tracking a global outbreak** begins with physicians and clinical laboratories notifying local health agencies, the CDC, and the WHO of reportable diseases and diseases of unknown cause.
- **Finding patient zero** is important to contain the spread of disease.
- **Online search engines** can help identify an increased incidence of an infectious disease by monitoring worldwide searches and blogs that mention the disease.

Thought Question 26.5 What are some reasons why some diseases spread quickly through a population, whereas others take a long time?

Thought Question 26.6 On Google Trends, when you look at the search frequency for *Streptococcus pneumoniae*, you will notice a periodicity to the number of searches, especially if you focus on U.S. searches. What biological phenomenon might account for that trend?

Figure 26.14   CDC Mission to Monitor Monkeypox in the Congo Basin
Monkeypox is a close relative of smallpox and is endemic to central Africa.

A. A local market in Boende, at the equator in the Democratic Republic of Congo, sells typical "bush meat." Electric power in this area is inconsistent, so no refrigeration exists.

B. A regional clinic/hospital pharmacy that supports the local area of up to 5,000 people. They have about four syringes and needles that are used over and over again.

C. Team of CDC scientists sent to Boende in 2011 to survey monkeypox.

26.4
Detecting Emerging Microbial Diseases

SECTION OBJECTIVES

- Describe the relationship between animal health and human disease.
- Explain how molecular techniques can identify previously unrecognized and unculturable pathogens.
- Discuss how technology contributes to the evolution of emerging pathogens and the spread of infectious disease.
- Correlate the rise in reemerging diseases to the AIDS pandemic.
- Explain how ecology and climate change influence the spread of infectious diseases.

How do scientists predict and stop the emergence of a new infectious disease? A group of factors have simultaneously converged over recent decades to create the perfect incubator for new emerging and ancient reemerging infectious diseases. The most important of these factors include genetic adaptability of microbes, increased global travel and transportation, altered host susceptibility, climate change, aggressive land use, and human demographics and behavior. Also, both public and animal health infrastructures have broken down, poverty has increased, and social inequality persists. Humans are responsible for most of these factors. As a result, the struggle between humans and microbes has intensified. Microbes now have more opportunities to inhabit new niches, cross species boundaries, and travel worldwide to establish infections in new populations of people and animals. Public health initiatives and epidemiology are crucial to keep the struggle balanced in our favor.

SARS: An Epidemiological Success Story

CASE HISTORY 26.2

The Stowaway

In 2003, a 48-year-old man was hospitalized in Dutchess County, New York, with a 101°F (38.4°C) fever, headache, and body aches. He also had difficulty breathing. He had just returned from a business trip to China. Health care workers were stumped for a diagnosis. Serological tests, PCR analysis, and fluorescent antibody stains failed to reveal a cause. Was this something new?

The man in the above case history had contracted severe acute respiratory syndrome (SARS), a disease unknown at the time. Within 7 weeks in early 2003, epidemiologists armed with global technologies and rapid DNA-sequencing techniques tracked, named, identified, completely sequenced, and contained a newly emerging disease with a scary death rate. This feat was remarkable, especially when we consider that after the first cases of AIDS appeared in 1981, it took more than 3 years just to identify the virus, and it took 7 years to track down the Lyme disease spirochete, *Borrelia burgdorferi*, after Lyme disease was first recognized in 1975.

We now know that SARS first developed in Asia and then began to spread by jet to other countries, such as Canada. Although the death rate might seem modest at about 5%, it was higher than the

4% reached during the 1918 flu pandemic, which killed 25 million–40 million people worldwide. SARS clearly posed a formidable threat. Unprecedented cooperation between WHO, CDC, and many other health organizations around the world played a major role in tracking and containing the disease. Patient information, case histories, and possible treatment regimens flooded into a secure WHO website. The most exciting posting was the 30,000-bp sequence of the SARS genome, taken from one-millionth of a gram of genetic material isolated from a Toronto patient. The following 2003 timeline shows how quickly this all this took place:

- February 14—China first reports 305 cases of atypical pneumonia to WHO.
- March 15—WHO issues emergency travel alert, names illness SARS.
- March 17—Leading epidemiologists join WHO in conference call, agree to unprecedented cooperation.
- April 4—U.S. President George W. Bush authorizes quarantine of SARS patients.
- April 12—Canadian scientists in Vancouver post the completed genome on the Internet.
- April 16—WHO announces that SARS is caused by a pathogen never before seen in humans, a new coronavirus.

Genomic Strategies to Identify Nonculturable Pathogens

Microbiologists have developed many strategies to identify the causes of infectious diseases. Robert Koch in the late 19th century devised a set of postulates that, when followed, can identify the agent of a new disease (Section 1.3). One important feature of Koch's postulates calls for growing the suspected organism in pure culture. However, for some bacterial diseases the agent cannot be cultured. How might they be identified?

CASE HISTORY 26.3

Curious Case of the 6-Year Diarrhea

In April, Joe, a 50-year-old man from Topeka, Kansas, had an abrupt onset of watery diarrhea with a stool frequency of up to ten times every 24 hours. This was the latest episode in a 6-year history of illness beginning with recurring fevers, flu-like symptoms, profuse night sweats, and painful joint swelling. The current bout of diarrhea was associated with gripping lower-abdominal pain, especially after meals. No blood or mucus was found in the stools. Blood and stool cultures tested negative for known infectious agents. Serology was also negative for syphilis, brucellosis, toxoplasmosis, and leptospirosis. His weight fell rapidly from 182 to 160 pounds within 4 weeks of onset. A flexible sigmoidoscopy showed only diffuse mild erythema in the bowel. However, the appearance of the small bowel was consistent with malabsorption, a condition in which the intestine does not properly absorb nutrients.

A duodenal biopsy to look for the organisms of disease showed large macrophages in the lamina propria (a layer of loose connective tissue beneath the epithelium of an organ). Bacterial rods characteristic of Whipple's disease were seen with electron microscopy. PCR analysis of tissue samples confirmed the diagnosis.

George Whipple first diagnosed this disease in 1907. Symptoms include malabsorption, weight loss, arthralgia (joint pain), fevers, and abdominal pain. Any organ system can be affected, including the heart, lungs, skin, joints, and central nervous system. The mysterious cause of Whipple's disease went undiscovered for 85 years but was suspected to be of bacterial etiology even though an organism was never cultured. The agent was finally identified in 1992—not by culturing, but by blindly amplifying bacterial genes that encode 16S ribosomal RNA from biopsy tissues.

All bacterial 16S rRNA genes have some sequence regions that are highly homologous across species and other sequences that are unique to a species. Tissues from many patients diagnosed with Whipple's disease were subjected to PCR analysis using the common 16S rDNA primers. If a bacterial agent was present, it was predicted that PCR would amplify a DNA fragment corresponding to the agent's 16S rRNA gene. All tissues produced such a fragment, indicating that bacteria were in the tissues. DNA sequence analysis of these fragments indicated that the organism was similar to actinomycetes but unlike any of the known species.

The organism is a Gram-positive, soil-dwelling actinomycete named *Tropheryma whipplei* in honor of the physician who first recognized the disease. Because of its bacterial etiology, Whipple's disease can be treated with antibiotics, usually trimethoprim-sulfamethoxazole (trade names, Bactrim, Septra, Cotrim). Figure 26.15 shows in situ antibody staining for *T. whipplei* RNA in a tissue biopsy sample.

This story reveals that Koch's postulates (see Figure 1.12) must sometimes be modified when identifying the cause of a new disease. Here, the organism could not be cultured in pure form in the laboratory but was found, via molecular techniques, in all instances of the disease. However, more recent studies have cultured *T. whipplei* in vitro. The medium used was formulated based on nutritional requirements deduced from knowing the DNA sequence of the *T. whipplei* genome.

Technology's Influence over the Emergence and Spread of Infectious Agents

In the 1970s, medical science was claiming victory over infectious disease. Yet today, despite all we know of microbes—despite the many ways we have to combat microbial diseases—our species, for all its cleverness, still lives at the mercy of the microbe. Lyme disease, MRSA, SARS, Ebola, *E. coli* O157:H7, HIV, "flesh eating" streptococci, hantavirus—all these and many other new diseases have emerged over the last 30 years. Worse yet, forgotten scourges, such as tuberculosis, have reappeared. What happened?

Figure 26.15   Whipple's Disease

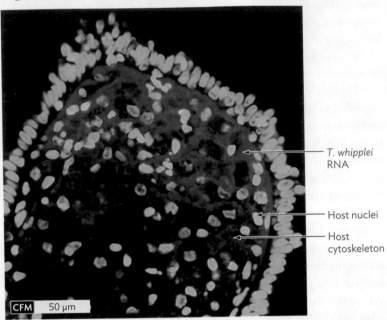

Fluorescent in situ hybridization of a small intestinal biopsy in a case of Whipple's disease. In this test, a fluorescently tagged DNA probe that hybridizes specifically to *Tropheryma whipplei* RNA is added to the tissue. Other fluorescent probes are used to visualize host nuclei and cytoskeleton. Blue = *T. whipplei* rRNA; green = nuclei of human cells; red = intracellular cytoskeletal protein vimentin. Magnification approx. 200×.

Part of the answer is progress itself. Travel by jet, the use of blood banks, and suburban sprawl have all opened new avenues of infection (**Figure 26.16**). People unwittingly infected by a new disease in Asia or Africa can, traveling by jet, bring the pathogen to any other country in the world within hours. The person may not even show symptoms until days or weeks after the trip. This means that diseases can spread faster and farther than ever. Newly emerged blood-borne pathogens can also spread by blood transfusion—a major problem with HIV before an accurate blood test was developed to screen all donated blood.

Although human encroachments into the tropical rain forests have often been blamed for the emergence of new pathogens, one need go no farther than the Connecticut woodlands to find such developments. *Borrelia burgdorferi*, the spirochete that causes Lyme disease, lives on deer and white-footed mice and is passed between these hosts by the deer tick (Figure 21.16). Humans have crossed paths with these animals for years. Why, then, have we suddenly become susceptible to disease? The answer appears to be suburban development. In the wild, foxes and bobcats hunt the mice that carry the Lyme agent. As developers clear land and build roads and houses, numbers of predators shrink, leaving the infected mice and ticks to proliferate. Humans in these developed areas are more likely to be bitten by an infected tick and contract the disease than in prior decades.

Many other examples show the unintended consequences of technology and development in breeding disease. Here are just a few:

- **Mad cow disease.** Modern farming practices (North America and Europe) of feeding livestock the remains of other animals helped spread transmissible spongiform encephalopathies similar to Creutzfeldt-Jakob disease associated with prions (Section 24.6). Because the prion is infectious, the brain matter from one case of mad cow disease could end up infecting hundreds of other cattle, which in turn increases the chance that the disease could spread to humans.
- **Babesiosis.** Often mistaken for Lyme disease, the protist *Babesia microti* is also carried by the deer tick, which transmits the agent from mice.
- **Hepatitis C.** Transfusions and transplants spread this blood-borne viral disease.
- **Influenza.** Live poultry markets in Asia serve as breeding grounds for avian flu viruses, which can jump to humans.
- **Enterohemorrhagic *E. coli* (for example, *E. coli* O157:H7).** Modern meat-processing plants can accidentally grind trace amounts of these acid-resistant, fecal organisms into beef while making hamburgers.

Natural environmental events can also trigger upsurges in the incidence of unusual diseases. A good example involves the hantavirus (also known as Sin Nombre virus). An unprecedented outbreak of hantavirus pulmonary disease occurred in the Four Corners area of Arizona, New Mexico, Colorado, and Utah in 1993, and then again

Figure 26.16   Visualizing Human Mobility Networks

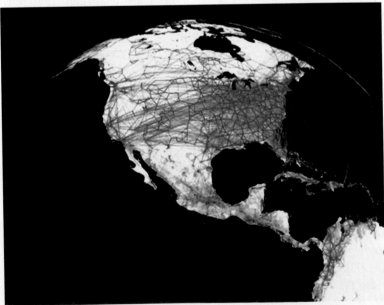

Lines show short-range transportation and long-range airline connections capable of spreading disease. Commuting patterns strongly correlate with how a pandemic evolves.

in Yosemite National Park in 2012. Heavy rains in each case led to greater than normal growth of plants that serve as food for deer mice. The resulting increases in deer mice, which can carry the virus, made it more likely that humans and infected mice would come in contact.

You might ask whether a zoonotic pathogen that leaps from animals to humans can be transmitted between humans. Lucky for us, many pathogens such as rabies virus, West Nile virus, and *Borrelia burgdorferi* find humans to be dead-end hosts, incapable of spreading the disease to other humans. Zoonotic diseases that cannot transmit from one person to another fail for one or more of the following reasons:

- No opportunity to transmit

- No convenient portal of exit

- An inability to reach a high enough concentration in human blood or secretions to reach an infectious dose

For example, for *Borrelia burgdorferi* to transfer from one person to another, a deer tick would have to bite an infected victim and then hop to a new human host. This scenario is highly unlikely because tick transfer between two people is rare. Likewise, people with rabies seldom have an opportunity to bite another person—and even if they did, the concentration of the virus in human saliva is

thought to be too low. West Nile Virus is not transmitted between people because the virus does not reach a concentration in a victim's blood high enough for a mosquito bite to transfer an infectious dose. However, epidemiologists do worry about whether a zoonotic pathogen might evolve to transfer from person to person.

Reemerging Pathogens

Figure 26.17 is a world map showing the general locations of emerging and reemerging diseases. A reemerging disease is one thought to be under control but whose incidence has risen. For example, the incidence rate of tuberculosis in the United States dropped sharply in the 1950s, but the number of cases over the past 30 years has increased just as dramatically. The trigger for its reemergence was the AIDS pandemic beginning in 1980. Immunocompromised AIDS patients are highly susceptible to infection by many organisms, including *Mycobacterium tuberculosis*.

M. tuberculosis is also reemerging among non-AIDS patients owing to the development of drug-resistant strains. Drug resistance in *M. tuberculosis* developed largely because patients failed to complete their full courses of antibiotic treatment. Treatment usually involves three or more antibiotics to reduce the risk that resistance to any one drug will develop. Many patients failed to take all three

Figure 26.17   Locations of Some Emerging and Reemerging Infectious Diseases

Examples shown represent extreme increases in reported cases. Many of these diseases, such as HIV and cholera, are widespread but show alarming increases in the areas indicated.

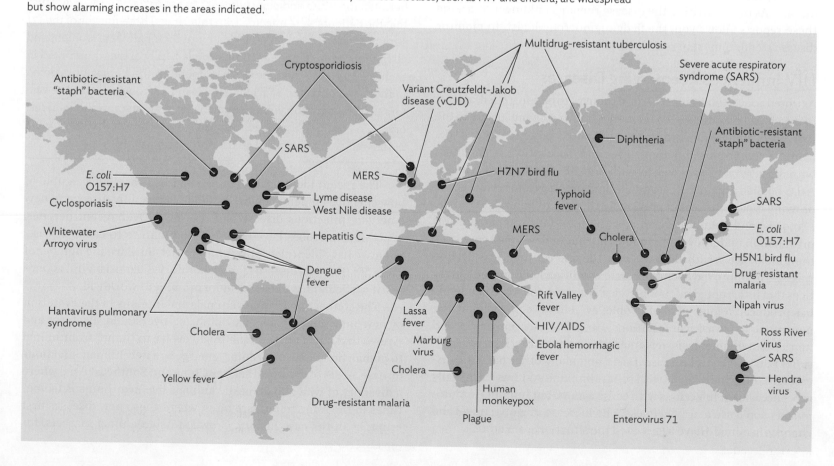

drugs simultaneously, allowing the organism to develop resistance to one drug at a time until it became resistant to all of them. These highly drug-resistant strains are almost impossible to kill. The link between noncompliance and the development of drug resistance is the primary reason why many tuberculosis patients must now take their drugs in the presence of the physician or another member of the medical staff.

As described in Section 26.3, highly infectious diseases are identified by using aggressive epidemiological surveillance. Communications between local, national, and world health organizations are crucial and helped expose the rise in tuberculosis. To see how emerging diseases are monitored, search the Internet for a website called ProMED-mail. There you will find daily reports from around the world that describe new outbreaks of infectious diseases. For example, on April 16, 2014, the site reported on an Ebola outbreak, in Guinea, Africa, that involved 197 people and 122 deaths. Twenty-four health care workers fell ill, 13 of whom died. The source of the virus was not known, but it is a new strain (we now know that this developed into a major Ebola epidemic in Western Africa). On April 8, 2014, another ProMed post reported on a mysterious tuberculosis-like disease threatening to wipe out the Ayoreo-Totobiegosode tribe from Paraguay's Chaco region. The Ayoreo is the last uncontacted tribe living outside the Amazon. Cattle ranchers have been burning the Indians' land to force them out, which forces contact between the displaced Indians and outside people. Soon after contact the Indians are stricken by this disease. As of this writing the disease remains unknown. Tracking these reports allows you, the fledgling microbiologist, to monitor disease trends almost in real time.

HIV Influences Reemerging Diseases

As noted above, tuberculosis—once thought to be controlled, at least in developed countries—is a reemergent threat owing in part to the increased number of immunocompromised patients with AIDS. Immunocompromised patients are more susceptible to infection with *M. tuberculosis*, as well as with non-TB mycobacteria. In fact, AIDS has had a major impact on the epidemiology and reemergence of many types of diseases previously thought to be under control. Protozoal and helminthic parasitic diseases, for example, are still prevalent in underdeveloped countries, but in developed countries their incidence greatly declined over the past century. This decline was due to improved socioeconomic conditions and hygiene practices that interrupted infection cycles. However, the AIDS pandemic has produced a huge pool of people, even in developed countries, now prone to parasitic infestations. Hospitals are seeing more and more cases of amebiasis, leishmaniasis, pneumocystis, cryptosporidiosis, trypanosomiasis, ascariasis, and many other parasitic diseases. Healthy individuals with robust immune systems can usually repel occasional infections with these agents, but immunocompromised people are nearly helpless. The increase in global travel and changes in climate have aggravated the situation even more.

Ecology and Climate Influence Pathogens' Epidemiology and Evolution

Over the last three decades, approximately 75% of emerging human infectious diseases have been zoonotic diseases (infections of animals transmitted to humans). At the same time, environmental insults continue to create favorable environments for new infectious diseases to develop.

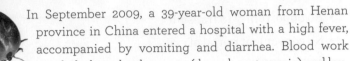

CASE HISTORY 26.4

It Started in Henan

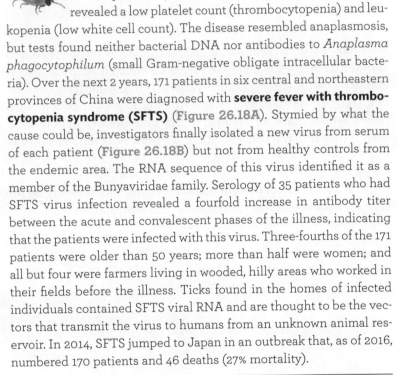

In September 2009, a 39-year-old woman from Henan province in China entered a hospital with a high fever, accompanied by vomiting and diarrhea. Blood work revealed a low platelet count (thrombocytopenia) and leukopenia (low white cell count). The disease resembled anaplasmosis, but tests found neither bacterial DNA nor antibodies to *Anaplasma phagocytophilum* (small Gram-negative obligate intracellular bacteria). Over the next 2 years, 171 patients in six central and northeastern provinces of China were diagnosed with **severe fever with thrombocytopenia syndrome (SFTS)** (Figure 26.18A). Stymied by what the cause could be, investigators finally isolated a new virus from serum of each patient (Figure 26.18B) but not from healthy controls from the endemic area. The RNA sequence of this virus identified it as a member of the Bunyaviridae family. Serology of 35 patients who had SFTS virus infection revealed a fourfold increase in antibody titer between the acute and convalescent phases of the illness, indicating that the patients were infected with this virus. Three-fourths of the 171 patients were older than 50 years; more than half were women; and all but four were farmers living in wooded, hilly areas who worked in their fields before the illness. Ticks found in the homes of infected individuals contained SFTS viral RNA and are thought to be the vectors that transmit the virus to humans from an unknown animal reservoir. In 2014, SFTS jumped to Japan in an outbreak that, as of 2016, numbered 170 patients and 46 deaths (27% mortality).

The recent identification of SFTS virus is a prime example of how effective epidemiological methods can rapidly discover a new emerging infectious disease and its cause. Over recent decades, many new microbes have emerged from nature to cause disease. Some we have identified, others not. In fact, an alarming number of cases of infectious disease lack a laboratory diagnosis. Viral zoonoses in particular have become more prominent worldwide.

The connection between animals and humans in the evolution of new infectious organisms cannot be overstated. Bacteria and viruses that infect only animals might evolve to "jump" from animal to human hosts and, in so doing, emerge as a new human infectious disease. Areas of the world such as China and Southeast Asia, where a diversity of wild and domestic animals live near humans (sometimes in the same room), are often where these new diseases first erupt, as in the case history. Crowded Asian animal markets, for

Figure 26.18 Emerging Viral Disease Caused by SFTS Bunyavirus

A. Areas in China shown in green indicate where SFTS bunyavirus was isolated from patients.

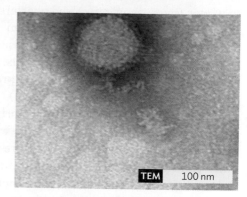

B. SFTS virus grown in Vero cells (epithelial cell line obtained from the kidney of an African green monkey).

instance, are opportune places for the transmission of pathogens from animals to humans (Figure 26.19).

Many examples show how the natural environment can harbor pathogens and foster the emergence of new ones. Knowledge that pathogen reservoirs exist in animals, arthropods, and plants has given rise to a new collaborative effort among clinicians, scientists, veterinarians, and ecologists called the **One Health Initiative**. The goal of the One Health Initiative is to control human health through animal health and vice versa. For example, using a vaccine-loaded bait to vaccinate wild rodents can decrease Lyme disease in humans. The idea has recently been extended to plant pathology because some bacteria are pathogens of both plants and humans (for example, *Pantoea agglomerans*, formerly called *Enterobacter agglomerans*, can cause leaf blight on rice and other plants or sepsis in immunocompromised humans). Also, some enteric bacteria can live within the vascular system of some edible plants (such as *E. coli*, *Klebsiella*, and *Salmonella* in lettuce).

Weather and climate change also have a considerable influence over when and where diseases happen. For example, many diseases

Figure 26.19 A Situation That "Invites" Bird Pathogen Transmission to Humans

are seasonal in appearance. Influenza and pneumonia are prevalent in winter (more people in closed quarters), whereas Lyme disease from *B. burgdorferi* and diarrhea from *Vibrio parahaemolyticus* occur more often in summer (Lyme disease because of more frequent contact between humans, animals, and vectors, and *V. parahaemolyticus* because warm gulf water promotes bacterial growth, which increases seafood contamination).

Climate change appears to be bringing more severe and extreme weather to many parts of the world. The new normal is more rain, heavier snowstorms, and more and greater floods, among other changes. Decreased sanitation in the wake of flooding will propagate gastrointestinal infections. Increasing temperatures and flooding will also influence animal migrations—and the migration of the microbes they carry. Thus, temperate regions of the world are warming, and diseases previously restricted to tropical climates have begun appearing in these once-temperate regions.

Only by integrating our knowledge of the environment and ecology with epidemiology can we completely understand how disease trends will change. Anticipating these changes will be crucial to devise and implement appropriate disease prevention strategies. Outbreaks such as the 2006 *E. coli* O157:H7–contaminated spinach episode underscore the fact that human health, animal health, and global health are inextricably linked. The spinach outbreak started when wild hogs traipsed through cow dung contaminated with O157:H7 and then raced through spinach fields, contaminating the leafy vegetable with the pathogen. Eating salads made with this spinach caused bloody diarrhea in several unfortunate people. Clearly, we need a holistic approach to understand, protect, and promote the health of all.

SECTION SUMMARY

- **Cooperation between governmental health agencies** can lead to rapid recognition of an emerging disease.
- **Molecular methods, such as PCR, can help identify new pathogens** that cannot be cultured.
- **Emerging diseases** can spread quickly around the world as a result of air travel.
- **Modern technology and urban growth** have provided opportunities for new diseases to emerge.
- **The increasing number of HIV-associated immunocompromised patients** has played an important role in reemerging infections.
- **Multidisciplinary collaboration** among ecologists, veterinarians, clinicians, and other scientists is necessary to devise appropriate strategies aimed at disease intervention and epidemic prevention.

Thought Question 26.7   In Case History 26.4, describing the new disease SFTS, which one of Koch's postulates remained unfulfilled?

Thought Question 26.8   What would make the SFTS virus a global concern?

26.5
Bioterrorism and Biodefense

SECTION OBJECTIVES
- Define the term "select agent."
- List some potential bioweapons.
- Discuss the properties of an effective bioweapon.

"A wide-scale bioterrorism attack would create mass panic and overwhelm most existing state and local systems within a few days," said Michael T. Osterholm, director of the Center for Infectious Disease Research and Policy at the University of Minnesota. "We know this from simulation exercises." These words were spoken on October 9, 2001.

Less than a month after the September 11, 2001, attack on the World Trade Center, some unknown person(s) sent weapons-grade anthrax spores through the U.S. mail (see the Chapter 18 opening case). Thankfully, only five people died and a mere 25 became ill. Despite the inefficiency of the attack, the impact was enormous. More than 10,000 people took a 2-month course of antibiotics after possible exposure, and mail deliveries throughout the country were affected. The simple act of opening an envelope suddenly became risky business.

After this attack and other events, the CDC and the National Institutes of Health (NIH) assembled a list of select agents (marked in blue in Table 2.4) that could potentially be used as bioweapons. A **select agent** is an infectious agent (or toxin) that has a high virulence or mortality rate, is highly contagious, lacks a protective vaccine, or cannot be effectively treated with antimicrobials or other pharmaceuticals. Microorganisms considered bioweapons can be used to conduct **biowarfare**, which aims to inflict massive casualties, or bioterrorism, which may have few casualties but causes widespread psychological trauma.

Although the list of select agents is recent, biowarfare is not new. In the Middle Ages, victims of the Black Death (plague caused by *Yersinia pestis*) were flung over castle walls with catapults; during the French and Indian War, in the eighteenth century, a British officer, Colonel Henry Bouquet, appears to have distributed two smallpox-infested blankets and a similarly infested handkerchief to Native Americans; and during World War II, the Japanese Imperial Army experimented with infectious disease weapons, using Chinese prisoners as guinea pigs. Even the United States has participated through developing weapons-grade anthrax spores. The strain of *B. anthracis* used in the 2001 anthrax attack purportedly had the same genetic signature as a strain originally researched at Fort Detrick, a U.S. Army base where research into possible bioweapons was conducted in the 1950s and 1960s. It now hosts elements of the U.S. biological defense program.

The first documented act of bioterrorism in the United States occurred in 1984, when followers of the cult leader Bhagwan Shree Rajneesh tried to control a local election in the city of The Dalles, Oregon, by infecting restaurant salad bars with *Salmonella* (Figure 26.20). The goal was to keep non-Rajneesh voters away from the voting booths. More than 700 people became ill. Eventually, vials of *Salmonella* were found in a clinic laboratory on the cult's compound. Dr. Michael Skeels, director of the Oregon State Public Health Laboratory, discovered the vials during a surprise raid by local, state, and federal agencies. The CDC in Atlanta confirmed that the strain found matched the strain that sickened the restaurant patrons. Rajneesh was never charged with the crime (although his subordinates were), but he pleaded guilty to immigration violations and received a 10-year suspended sentence and a fine of US$400,000. Deported and barred from reentering the United States for 5 years, he died in India in 1990.

 Homegrown bioterror attack (1984 Rajneeshee bioterror attack): en.wikipedia.org

How effective are bioweapons? The method by which a biological agent is dispersed plays a large role in its effectiveness as a weapon. Only a few people became ill during the 2001 anthrax attack (Figure 26.21A), in part because of effective epidemiological surveillance but mainly because anthrax is difficult to disperse and cannot easily be transmitted from one person to another. Someone must inhale thousands of spores to contract disease, which means that effective dispersal of the spores is critical for the use of anthrax as a weapon. Once spores hit the ground, the threat of infection is limited. Weapons-grade spores are very finely ground so that they stay airborne longer. But as we saw, the letter-borne dispersal system did not generate many victims. Nevertheless, the potential threat of weapons-grade anthrax on the battlefield led the U.S. military in 2006 to resume vaccinating all soldiers serving in Iraq, Afghanistan, and South Korea.

An effective bioweapon would be one that capitalizes on person-to-person transmission. In an easily transmitted disease, one infected person could disseminate disease to scores of others within 1 or 2 days. So for generating massive numbers of deaths, anthrax was a poor choice. But the goal of most terrorists is not to kill many people but rather to terrorize them. In that regard, the anthrax attack succeeded.

The most effective bioweapon in terms of inflicting death (biowarfare) would have a low infectious dose, be easily transmitted

Figure 26.20 Bioterror Attack in the United States, 1984

A. Four of the eleven restaurants in The Dalles, Oregon, targeted in the 1984 Rajneeshee bioterror attack.

B. Mug shot of Rajneesh in 1985 taken by the Oregon Department of Corrections, Multnomah County, Oregon.

Figure 26.21 Dealing with Bioterrorism

A. Members of a hazardous materials team near Capitol Hill during the anthrax attacks in 2001.

B. An Illinois man suffering from smallpox, 1912.

between people, and be one to which a large percentage of the population is susceptible. Smallpox (variola major virus) fits these criteria and would be the bioweapon of choice (Figure 26.21B). Fortunately, however, smallpox has been eradicated (almost) from the face of the earth and is not easily obtained. Only two WHO-sanctioned laboratories still harbor the virus—one in the United States and one in Russia. It is believed that the virus has been destroyed in all other laboratories (although in 2014 some long-forgotten, unsecured, sealed vials of smallpox virus were found stored in a closet at an FDA lab in Washington, DC). Since smallpox is a serious biowarfare agent, the last two smallpox repositories must remain secure.

Although the good news is that smallpox disease has been eradicated, the bad news is that no civilian born after 1970 has been vaccinated (except some military and laboratory workers). As a result, most anyone under 45 years of age is susceptible to smallpox. Even

inSight

What's Blowing in the Wind? Quick Pathogen
Detection Systems Guard against Bioterrorism

**Figure 1   The BioFlash-E Biological
Identifier Detection System**

In April 1979, workers at a secret Soviet biological weapons facility neglected to replace a filter in a laboratory ventilation system, and a cloud of highly weaponized *Bacillus anthracis* spores quickly spewed into the air outside. The deadly plume drifted downwind, infecting humans and cattle across a wide area. The resulting outbreak ultimately killed more than 64 people. Had a rapid pathogen detection system been available and deployed at the facility, all those lives could have been saved. Unfortunately, 35 years ago such technology was only the stuff of science fiction. Today, that technology exists in small, portable forms.

Current surveillance systems include the Biowatch program run by the CDC and the Biohazard Detection System (BDS) employed by the U.S. Postal Service. The Biowatch system places air samplers in undisclosed cities to collect particulates. The filters are then processed in designated Biowatch laboratories that use PCR technology to test for several biological pathogens. The BDS system collects air samples from mail-canceling equipment and uses an automated system to screen for *Bacillus anthracis*. What is needed in addition to these safeguards are small, portable devices that can identify the presence of multiple pathogens within minutes.

Figure 2   Principle of the Bioelectronic Sensor Used in the BioFlash-E Biological Identifier System

1 B cells are exposed to bioagents in test sample.

Antibodies

2 B-cell antibodies bind to the agent.

Antigen cross-linked antibodies

B cell with added aequorin gene

Tyrosine kinase

3 Cross-linking triggers a phosphorylation cascade.

Phospholipase C

PIP_3

IP_3 + DAG

4 Calcium influx activates aequorin.

Ca^{2+}

Open Ca^{2+} channels

Aequorin

Light emission

Ca^{2+}

5 Photons released hit the photon detector.

Endoplasmic reticulum

DAG = diacylglycerol; IP3 = 1,4,5-inositol trisphosphate; PIP3 = phosphatidylinositol 3,4,5-trisphosphate.

Indeed, several multipathogen molecular detection platforms have been developed. Some involve machines that detect antigen-antibody interactions; others use automated PCR to amplify specific pathogen genes and detect the products through hybridization to immobilized oligonucleotides. As described in the Chapter 25 inSight, the wet chemistries for all these detection systems are carried out in a roughly 2-inch square called lab-on-a-chip (LOC).

An interesting addition to the field of quick pathogen detection is the BioFlash-E Biological Identifier (previously known as PANTHER, for PAthogen Notification for THreatening Environmental Releases), a device developed in 2008 that can detect and identify a pathogen in as little as 3 minutes—significantly faster than traditional methods requiring isolation and growth of the pathogen (which can take 48 hours or longer). The BioFlash-E sensor (Figure 1) uses a cell-based technology that can detect as few as a dozen particles of a pathogen per liter of air. Currently, it can detect 24 pathogens, including the potential bioterrorism agents of anthrax, plague, smallpox, and tularemia.

The device uses an array of B cells, each displaying antibodies to a particular bacterium or virus. The cells are engineered to emit photons (particles of light) when they detect their target pathogen. How do they do this? The B cells have been bioengineered to express the gene that encodes aequorin, a calcium-sensitive bioluminescent protein. When a pathogen surface antigen cross-links the appropriate B-cell surface antibodies, a signal transduction cascade produces a rapid influx of calcium (Figure 2). Aequorin in the B cell will luminesce within seconds of calcium influx, and a photon detector detects the light emitted. The device then lists any pathogens found.

Quick pathogen detection technologies are constantly being improved in the effort to defend against bioterrorism. Small, quick pathogen detection devices could be used in buildings, subways, and other public areas. Eventually, such technology will supplant the classical clinical microbiology practices that must grow the microbe to identify it. There is hope, for example, that such devices can be used on farms or in food-processing plants to test for contamination by *E. coli*, *Salmonella*, or other food-borne pathogens. Another potential application is in medical diagnostics, where the technology could test patient samples, giving rapid results without having to send samples to a laboratory. Wherever it is used, successful quick pathogen detection should enable local and global public health centers to quickly detect potential bioterrorism threats and rapidly mobilize response teams.

people who received the smallpox vaccination more than 45 years ago are at risk, since their protective antibody titers have diminished. A terrorist attack with smallpox would cause terrible numbers of deaths. The vaccine (varicella virus) does, however, exist. Were a smallpox attack to be launched, the vaccine would be rapidly administered to limit the spread of disease. Nevertheless, the economic and psychological impact of a smallpox epidemic would be devastating.

Research with organisms considered select agents is tightly regulated. Because *Yersinia pestis*, for instance, is a select agent, laboratory personnel working with it must now possess security clearance with the Department of Justice. The laboratory must also register with CDC to legally possess this pathogen. Furthermore, access to the lab and the organism must be tightly controlled.

Recall from Section 2.4 that biosafety containment has four levels. Agents designated level 1 require the lowest level of containment, whereas level 4 agents are highly virulent and require the strictest containment procedures.

Much has improved since Michael Osterholm offered his dire assessment of a wide-scale bioterrorism attack. Education and surveillance procedures have been bolstered, and new detection technologies are being developed (**inSight**). We will probably never be fully protected from attack, biological or otherwise, but recent efforts, including the Strategic National Stockpile of medicines and medical supplies, have improved the situation.

 Bioterrorism Preparedness Act: fda.gov

SECTION SUMMARY

- **Select agents** are bacterial and viral pathogens that CDC considers potential bioweapons.
- **Bioweapons,** when they have been used, typically kill few people but create great fear.
- **Effective bioweapons** would be transmitted person to person, have a low infectious dose, be easily transmissible, and have a large fraction of the target population susceptible.
- **Research with select agents** is tightly regulated.

Perspective

This chapter has examined the basic epidemiological principles used to detect outbreaks of infectious disease, whether they were naturally acquired, hospital-acquired, or acquired as a result of malicious intent. As you can see, controlling the spread of disease is a daunting and ever-changing task. This challenge is upon us for several reasons, not the least of which is because pathogens evolve. Centuries-old pathogens evolve to elude the immune system and to avoid eradication by antibiotics. Meanwhile, new pathogens keep emerging from the shrinking jungles and impoverished areas of the world. To stop the parade of new pathogens that threaten us, the world must take a more considered approach toward land use, confront the issue of climate change, and move more aggressively to improve sanitation around the globe.

LEARNING OUTCOMES AND ASSESSMENT

SECTION OBJECTIVES

OBJECTIVES REVIEW

26.1 Principles and Tools of Epidemiology

- Explain the four major links in a chain of infection.
- Differentiate between descriptive and analytical epidemiology.
- Define "endemic," "epidemic," and "pandemic," and explain the difference between disease prevalence and incidence.
- Explain the three steps of surveillance and describe how molecular approaches can contribute to surveillance.

1. The four links in the chain of infection are _____, _____, _____, and _____.

Which link in the chain of infection will the following items interrupt?

2. Hand washing: _____

3. Antibiotics: _____

4. Insect eradication: _____

5. Vaccine: _____

26.2 Tracking Hospital-Acquired Infections

- Discuss the nature of nosocomial infections and list the two most common forms.
- List key questions that can help identify the source of a hospital-acquired infection.
- Discuss procedures used to prevent hospital-acquired infections.

9. Which of the following is the most common nosocomial infection in the United States?
   A. Sepsis
   B. Pneumonia
   C. Meningitis
   D. Urinary tract infections
   E. Diarrhea

26.3 Tracking Global Disease Trends

- Discuss how global disease trends are recognized.
- Explain the importance of finding the index case of an outbreak.
- Employ online sources to track worldwide disease trends.

12. What is the usual first step in identifying a pandemic?
   A. Local hospitals and physicians notify city or state health agencies about patients with a specific disease.
   B. State agencies report cases to the national health agencies (such as CDC).
   C. National health agencies report cases to the World Health Organization.
   D. Epidemiologists use data sent to WHO to identify the global locations and the numbers of cases of a disease.

26.4 Detecting Emerging Microbial Diseases

- Describe the relationship between animal health and human disease.
- Explain how molecular techniques can identify previously unrecognized and unculturable pathogens.
- Discuss how technology contributes to the evolution of emerging pathogens and the spread of infectious disease.
- Correlate the rise in reemerging diseases to the AIDS pandemic.
- Explain how ecology and climate change influence the spread of infectious diseases.

14. Climate change can widen the spread of an infectious disease. True or false?

15. A man who contracted Dengue fever in Brazil and traveled by jet to Canada would be the index case for an outbreak of Dengue fever in Canada. True or false?

26.5 Bioterrorism and Biodefense

- Define the term "select agent."
- List some potential bioweapons.
- Discuss the properties of an effective bioweapon.

23. NIH designates an agent that might be used for biowarfare as a(n) _____.

6. An increase in disease incidence above the endemic level is called a/an _____.
   A. epidemic
   B. outbreak
   C. pandemic
   D. nosocomial

7. Patients with Lyme disease were asked how often they walk through woods, use insect repellent, and wear shorts or pants. A second set of patients without Lyme disease from the same practice were asked the same questions, and the responses from the two groups were compared. This is an example of _____.
   A. experimental epidemiology
   B. a case cohort study
   C. a case–control study
   D. descriptive epidemiology

8. The number of new cases of a disease arising in a population over a given period is called _____.

10. Twelve cases of MRSA occurred over the course of a year in the hospital. What data will most definitively tell you whether a common source exists?
   A. One doctor and three nurses attended to all the affected patients.
   B. All affected patients ate food from the hospital cafeteria.
   C. The genetic analysis found that all the MRSA strains were identical.
   D. All the affected patients were treated by the same nurse.
   E. All the affected patients had contact with each other.

11. You are about to enter the hospital room of a patient who had an appendectomy. If you planned to walk to the bed but not touch the patient, which of the following should you do?
   A. Because you will not touch the patient, no special precautions are needed.
   B. Don a protective gown.
   C. Put on sterile gloves.
   D. Sanitize hands before entering the room.
   E. C and D
   F. B, C, and D

13. In any epidemic or pandemic, why is identifying the index case important?
   A. To treat the patient
   B. To identify where the index case traveled and whom that person contacted
   C. To quarantine that person
   D. To identify the food that person has eaten
   E. A and B
   F. B and C

True or false? The following are recognized contributors to emerging or reemerging infectious diseases:

16. Antibiotic resistance

17. Climate change

18. Acquired immunodeficiency syndrome

19. Air travel

20. Deforestation

21. Vaccination

22. Fifteen people developed symptoms of muscle aches, high fever, and vomiting. You suspect a blood-borne pathogen. But evidence of typical agents known to cause these symptoms was not found, and cultures of blood for bacterial, viral, and fungal agents failed to grow anything. Which of the following strategies would most likely identify a new agent?
   A. Screen each patient for elevated antibody titers.
   B. Screen blood by PCR for rRNA genes and sequence the products.
   C. Determine whether the patients crossed paths.
   D. Test whether a blood transfusion will transfer the disease to another human.
   E. All of the above.
   F. None of the above.

24. *Salmonella enterica* has been used as a bioweapon but is not considered a select agent because _____.
   A. infections can be treated with antibiotics
   B. a vaccine exists
   C. diagnostic laboratories can easily identify it
   D. it has no animal reservoirs
   E. it is difficult to grow

25. Which of the following characteristics contributed to the decision to add *Yersinia pestis* to the list of select agents?
   A. Infections cannot be treated with antibiotics.
   B. It is easily transmitted from person to person.
   C. Diagnostic laboratories can easily identify it.
   D. It has no animal reservoirs.
   E. It is difficult to grow.

Key Terms

analytical epidemiology (p. 881)
biowarfare (p. 898)
case–control study (p. 881)
chain of infection (p. 880)
cohort study (p. 881)
daisy-chain nosocomial infection (p. 888)
descriptive epidemiology (p. 881)

endemic (p. 882)
epidemic (p. 882)
experimental study (p. 881)
health care–associated infection (HAI) (p. 886)
incidence (p. 883)
index case (p. 889)

nosocomial infection (p. 886)
observational study (p. 881)
One Health Initiative (p. 897)
outbreak (p. 882)
pandemic (p. 882)
patient zero (p. 889)
placebo (p. 881)

prevalence (p. 883)
quarantined (p. 889)
select agent (p. 898)
severe fever with thrombocytopenia syndrome (SFTS) (p. 896)
symptomatic surveillance (p. 889)
syndromic surveillance (p. 889)

Review Questions

1. Who is John Snow?

2. How is a pandemic different from an epidemic?

3. Explain the difference between the prevalence and incidence of disease.

4. How can an endemic disease become an epidemic?

5. Discuss three ways to graphically represent epidemiological data and what information can be gleaned from each.

6. What are the three stages of disease surveillance?

7. What is a nosocomial infection? How can it be contracted?

8. Discuss five questions that infectious disease personnel should ask in response to a nosocomial infection.

9. Describe some common infection control measures.

10. How are global outbreaks of disease identified?

11. Briefly describe four emerging diseases.

12. How can genomics help identify nonculturable pathogens?

13. What factors contribute to the development and spread of emerging infectious diseases?

14. Why are diseases reemerging that were long thought to be on the decline?

15. What is a select agent? Name four select agents and the diseases they cause.

16. Discuss whether a bioweapon needs to be efficient to be effective.

Clinical Correlation Questions

1. A local hospital received nine patients with symptoms of diarrhea, cramps, and low-grade fever. *Listeria monocytogenes* was isolated from the feces of all victims. All victims had attended an outdoor community block party on April 19. Five people who attended the party but were not ill were also surveyed. From the table below, list trends you find in the data and indicate what is the likely source of the infection.

2. Graphically represent an epidemic with respect to time and cases.

3. On the ProMED-mail website (promedmail.org), click on Links to view outbreaks recorded by WHO. What outbreaks happened throughout the world during the current year?

| Patient | Age | Sex | Cohabitation | Illness | Onset Date Time | Foods Ingested | | | | |
						Baked Ham	Potato Salad	Soft Cheese	Milk	Ice Cream
1	55	M		N		Y	N	Y	N	Y
2	49	M		Y	4.21 9 AM	Y	N	Y	Y	Y
3	85	F		Y	4.22 12 PM	N	Y	Y	N	N
4	62	M	Lives with 5	Y	4.24 11 AM	Y	Y	N	Y	Y
5	61	F	Lives with 4	Y	4.21 6 AM	N	N	Y	N	Y
6	56	M		Y	4.22 12 AM	N	Y	Y	N	N
7	69	M		Y	4.22 1 PM	Y	N	Y	N	N
8	35	F	Lives with 9	N		N	N	N	Y	Y
9	45	F	Lives with 8	Y	4.24 3 PM	N	Y	Y	Y	Y
10	39	F		Y	4.21 4 AM	Y	N	Y	Y	Y
11	25	M		N		N	Y	N	Y	N
12	48	M		Y	4.20 11 PM	Y	N	Y	N	Y
13	24	F		N		N	Y	N	Y	Y
14	55	M		N		Y	Y	N	N	N

Thought Questions: CHECK YOUR ANSWERS

Thought Question 26.1   In 2011, British Columbia, Canada, reported that 77 people contracted mumps. Seventy percent of those cases were aged 18–35 years. How would you have linked these cases, and what do you think led to the outbreak?

ANSWER: Each case must match an official case description of mumps (fever, headache, muscle aches, tiredness, and loss of appetite, followed by swelling of salivary glands); each patient would be tested for IgM antibody against mumps virus (why IgM? See Section 16.4); epidemiologists would then look for whether the patients had been in contact with each other. The outbreak resulted from a Fraser Valley religious sect that opposes vaccination.

Thought Question 26.2   A recent outbreak of *E. coli* O157:H7 disease has been detected. How long will it take from the time the first victim eats contaminated food before a laboratory confirmation of this organism is established and a public health lab determines that strains from several cases are identical? As you work through this timeline, consider every event that must occur and how long each takes.

ANSWER: Once a patient eats contaminated food, it can take 2–3 days to become ill, and 1–5 days before he or she seeks treatment. At that point a stool sample will be collected and sent to the hospital laboratory. It will take the organism overnight to grow on agar media and lab personnel another day to perform the biochemical tests needed to identify the organism as *E. coli* O157:H7. The organism is then shipped to a public health laboratory (0–4 days in shipping) and then 1–4 days for the public health lab to perform DNA "fingerprinting" to determine the relatedness between strains received from different cases. So it takes at *least* 1 week from the time of ingestion for public health agencies to declare an outbreak with a common source.

Thought Question 26.3   A nosocomial outbreak of MRSA occurred in a community hospital. The infection control team identified a nurse harboring the same strain of MRSA as part of his normal microbiota. The nurse, however, refused treatment because he wasn't sick. Should the hospital administrators fire the person? Should all hospital personnel be screened for MRSA as a condition of employment?

ANSWER: These are clinical dilemma questions. The hospital cannot allow the nurse to continue treating patients without confirming that he was rid of the organism. The hospital might offer the nurse a position without patient contact, but he could still transmit the organism to patient-contact staff. No one has addressed the idea of screening hospital staff for MRSA, although some states screen all ICU patients for the organism.

Thought Question 26.4   Imagine a large metropolitan hospital in which eight serious nosocomial infections with MRSA have occurred, and you must determine the source of infection so it can be removed. Using common bacteriological and molecular techniques, how would you accomplish this task?

ANSWER: Samples from affected patients and hospital staff would be screened for MRSA. DNA from each strain would be analyzed by multilocus sequence typing (MLST) or DNA sequencing of conserved genes such as *coa* (coagulase). Strains from patients will have identical patterns or sequences if they came from the same source. The source would harbor MRSA with the same pattern. The source may be a staff member or inanimate objects, such as surgical equipment or a ventilation apparatus. A connection between patients and the colonized staff member or instruments must also be established.

Thought Question 26.5   What are some reasons why some diseases spread quickly through a population, whereas others take a long time?

ANSWER: One factor is mode of transmission; airborne diseases can spread more quickly than food-borne diseases, for instance. Sexually transmitted diseases spread more slowly still. Herd immunity is another factor. Herd immunity is based on the number of individuals within a population that are resistant to a disease. Someone immune to the disease cannot pass it on. The more immune people in a population (or herd), the slower the spread of the epidemic to susceptible people.

Thought Question 26.6   On Google Trends, when you look at the search frequency for *Streptococcus pneumoniae*, you will notice a periodicity to the number of searches, especially if you focus on U.S. searches. What biological phenomenon might account for that trend?

ANSWER: The trend: searches are more frequently made in the winter months, probably because frequency of *S. pneumoniae* disease is highest in the winter and lowest in the summer months. Search for Lyme disease, whose incidence is highest in the summer, and you will see the opposite trend.

Thought Question 26.7   In Case History 26.4, describing the new disease SFTS, which one of Koch's postulates remained unfulfilled?

ANSWER: The most difficult postulate to fulfill is to reproduce the disease in a suitable animal model. Scientists also have to confirm that the tick *Haemaphysalis longicornis* is a vector for the virus and identify potential animal reservoirs for the virus.

Thought Question 26.8   What would make the SFTS virus a global concern?

ANSWER: Person-to-person transmission, which has not been observed.

Environmental and Food Microbiology

SCENARIO In Huntington, West Virginia, the U.S. Army Corps of Engineers had a problem. For decades, the army had produced the compound 2,4,6-trinitrotoluene (TNT), an explosive used in the manufacture of munitions. Inevitably, some of the TNT entered the soil.

120-mm Mortar Bomb
Contains explosive TNT.

Windrow Composting
A machine turns the soil at regular intervals to balance anaerobic and aerobic microbial digestion.

CONTAMINATION In 1988, after the ordnance works was shut down, the Environmental Protection Agency (EPA) determined that the manufacturing plant had leaked unsafe levels of TNT into the surrounding environment. TNT is toxic to humans and wildlife, causing anemia and liver disease. For many years, Huntington's water and soil had to be contained without effective treatment.

WEST VIRGINIA

Huntington

TREATMENT In 2003, the army chose to conduct bioremediation, a process in which living organisms decompose and remove contaminating material. The bioremediation process used is called windrow composting. Windrows are elongated hills of soil that are turned by a machine to regulate the availability of oxygen for compost. In composting, the soil is combined with carbohydrate-rich material such as vegetable waste or wood chips. Environmental bacteria catabolize these materials, along with the TNT. The TNT is catabolized to carbon dioxide.

OUTCOMES At the Huntington base, windrow composting removed more than 99% of the contaminating TNT. The bacterial process helped restore the local environment and saved the army more than half a million dollars on alternative containment measures that would not have removed the contamination.

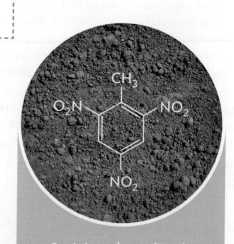

2,4,6-trinitrotoluene (TNT)
Explosive from bomb manufacture contaminates soil.

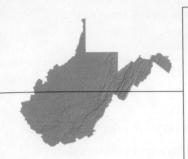

CHAPTER OBJECTIVES

After reading this chapter, you will be able to:

- Describe the essential roles of microbes in ecosystems that support all life, including humans.

- Describe how we use microbes to produce food and beverages and how we protect food from microbial spoilage.

- Explain how microbes cycle essential materials, including human wastes, throughout the biosphere.

In the case history that opens this chapter, the "patient" is not an individual person, but a sickened habitat—a polluted environment in Huntington, West Virginia. A Cornell University study estimates that, worldwide, environmental pollution causes 40% of human deaths. Our health depends on the quality of our environment, which provides the water we drink and the soil in which we grow our food. The contributions of an environment to human health are called **ecosystem services**. When an environment is polluted, its ecosystem services are lost or degraded.

 Water, air, and soil pollution causes 40 percent of deaths worldwide, Cornell research survey finds: news.cornell.edu

How can a contaminated environment be restored? Restoration can involve the metabolism of living microbes, a technology called **bioremediation**. Bioremediation is a waste management technique that involves the use of organisms to remove or neutralize pollutants from a contaminated site. During bioremediation, surprisingly, microbes can decompose even highly toxic compounds such as TNT. In fact, the very property that makes TNT useful as an explosive—the huge release of energy as it breaks down—enables bacteria to use it as food. First, under anaerobic conditions (lack of oxygen within the compost pile), bacteria use the TNT molecule to accept electrons from the carbohydrate-rich compost. Next, the compost material is aerated by the rotation machine (see the photo in the chapter-opening case history), allowing aerobic respiration to occur—a key cell metabolism you learned about in Chapter 7. Aerobic respiration breaks down the composted TNT via the **TCA cycle**, releasing CO_2 and H_2O and yielding energy for microbes to build biomass. Thus, TNT composting joins the environment's endless cycle of food webs in which organisms form living bodies and decompose them. And microbes are the key players.

Link Section 7.3 describes how the **TCA cycle** completes catabolism of organic food molecules.

Microbial communities are the foundation of Earth's biosphere, shaping the environments inhabited by plants and animals. Of all organisms, microbes grow in the widest range of habitats, from Antarctic lakes to the acid drainage of iron mines, as well as the air, water, and soil that surround us. In the ocean, vast quantities

of microbes produce the biomass that ultimately feeds fish and humans. In forests and fields, microbes decompose plant material, generating fertile soil. And in all environments, microbes associate with animals and plants. Our skin microbes protect us from pathogens, while our intestinal microbes digest complex plant fibers (discussed in Chapter 14). Food production involves the creation of artificial microbial ecosystems. On a global scale, microbes cycle the elements, consuming CO_2 and fixing nitrogen for all living things. This chapter explores how microbes interact with each other and with their many diverse habitats on Earth in ways that affect our lives and health.

27.1
Microbes in Ecosystems

SECTION OBJECTIVES

- Describe the various roles of microbes in natural ecosystems and the kinds of ecosystem services that microbes provide.

- Explain how physical factors determine which microbes grow in a given habitat.

- Outline how microbes participate in the food chains of ocean, soil, and wetland.

- Explain the phenomenon of biochemical oxygen demand and the cause of dead zones.

Can any organism exist by itself? "No man is an island," nor is any microbe. All organisms exist within an **ecosystem** consisting of populations of species plus their habitat or environment. A **population** is a group of individuals of one species living in a common location. Within an ecosystem, each population of organisms fills a specific **niche**. The niche is a set of traits that defines the conditions for an organism to live and the organism's relations with other members of the ecosystem. For example, the niche of *Staphylococcus epidermidis* is that of a human symbiont, a heterotroph that catabolizes hydrocarbons from human perspiration. The habitat of *S. epidermidis* is the human epidermis, and the bacterium's metabolism releases organic acids that suppress the growth of pathogens.

Microbes Fill Unique Niches in Ecosystems

In every ecosystem, certain key roles are filled only by microbes. Only bacteria and archaea fix nitrogen gas (N_2) into ammonia, which plants incorporate into protein and nucleic acids. Without bacterial nitrogen fixation, life on Earth would grind to a halt, starved of nitrogen. Legume plants such as soybeans actually harbor symbiotic bacteria called rhizobia that fix nitrogen within the plant cells (Figure 27.1). In a different example, only bacteria and fungi degrade lignin (wood biomass) and other complex plant polymers, releasing their carbon for uptake into food chains (Figure 27.2). Without decomposition, plant biomass would pile up (think autumn leaves), smothering new life.

In addition, microbial metabolism evolves so fast and in so many diverse ways that microbial genomes can respond to new challenges in our environment—even the alarming rise of waste plastics in our oceans. If an energy-yielding reaction exists, some microbe must have evolved—or will evolve—to use it. For this reason, we can search contaminated environments (such as Huntington's TNT-contaminated soil) for species of microbes that we could use for bioremediation.

What determines the environmental contribution of a microbe? What a microbe gives to and takes from its ecosystem depends on two things:

- **The microbe's genome.** Each genome encodes a set of enzymes that can interconvert molecules (discussed in Chapters 7 and 8). For example, the genomes of cyanobacteria encode the enzymes and electron transport proteins for photosynthesis. Microbial genomes share content through mobile genes (acquired by transfer mechanisms, discussed in Chapter 9).

- **The microbe's environment.** The environment's physical and chemical factors, such as temperature, nutrient supplies, oxygen availability, and pH, determine the range of metabolic options. In a wetland, for example, if hydrogen gas is available from decomposing biomass, *Rhodopseudomonas palustris* will oxidize hydrogen to water. On the other hand, if light is available without oxygen, as in a sewage-polluted lake, *R. palustris* will

Figure 27.2   White Rot Fungi Decompose a Tree Stump

use anaerobic photosynthesis to split hydrogen sulfide. Either way, *R. palustris* forms biomass for consumers in the food web.

All organisms depend, directly or indirectly, on the presence of other organisms. Some populations cooperate by reciprocally supplying each other's needs. Cooperation by organisms may be incidental, as when hydrogen-oxidizing bacteria use H_2 from fermenting bacteria. Or cooperation may involve mutualism, a highly developed partnership in which two species have coevolved to support each other (discussed in Section 27.2).

Carbon Assimilation and Dissimilation: The Food Web

The interactions between microbes and their ecosystems include two common roles of metabolic input and output, often referred to as assimilation and dissimilation, respectively. The processes of metabolism were discussed in Chapter 7, but ecology sees them from a different point of view.

Figure 27.1   Soybeans Carry Bacterial Partners That Fix Nitrogen

B. Cross section through a soy root cell that contains nitrogen-fixing bacteroids of *Bradyrhizobium japonicum*.

TEM   0.5 µm

A. Soybean farm.

Figure 27.3   Food Web: All Levels Include Microbes

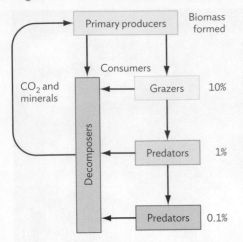

Primary producers fix CO_2 into biomass; grazers (primary consumers) feed on primary producers, and predators consume the grazers and smaller predators. Decomposers consume dead bodies of organisms from every level, recirculating their minerals to the producers. Percentages indicate fraction of original CO_2 converted to biomass at each trophic level.

Assimilation refers to processes by which organisms acquire an element, such as carbon from CO_2, to build into body parts. Organisms that produce biomass from inorganic carbon (usually CO_2 or bicarbonate ions) are called **primary producers**. Producers are a key determinant of productivity for other members of the ecosystem, particularly the **consumers** that feed on producers.

Dissimilation is the process of breaking down organic nutrients to inorganic minerals such as CO_2 and NO_2^-, usually through oxidation. An example of a nutritive process causing dissimilation is aerobic respiration. Microbial dissimilation releases minerals for uptake by plants, and it provides the basis of wastewater treatment (discussed in Section 27.5). Waste removal is key for every life-form, and wastewater treatment is fundamental to modern human societies.

The major interactions among organisms in the biosphere are dominated by the production and transformation of **biomass**, the bodies of living organisms. To obtain energy and materials for biomass, all organisms participate in a **food web** (Figure 27.3). A food web describes the ways in which various organisms consume each other as well as each other's products. Levels of consumption are called **trophic levels**. Each trophic level represents the consumption of biomass of organisms from the previous level, lower in the food web. At each trophic level, the fraction of biomass retained by the consumer is small; most is dissipated as CO_2 to provide energy.

Every food web depends on primary producers for two things:

- **Absorbing energy from outside the ecosystem.** In most cases, the ultimate source of energy is the sun, and the producers are phototrophs.

- **Assimilating minerals into biomass.** The biomass of producers is then passed on to subsequent trophic levels.

The majority of carbon in Earth's biosphere is assimilated by oxygen-producing phototrophs such as cyanobacteria, algae, and plants. Certain important ecosystems are founded on carbon fixation by lithotrophs (microbes that oxidize minerals). For example, in hydrothermal vent communities, bacteria oxidize hydrogen

sulfide to fix CO_2, capturing both gases as they well up from Earth's crust. The vent communities, however, require oxygen generated by phototrophs from above.

In addition to producers, all ecosystems include consumers that acquire nutrients from producers and ultimately dissimilate biomass by respiration, returning carbon to the atmosphere (see Figure 27.3). Consumers constitute several trophic levels based on their distance from the primary producers. The first-level consumers, generally called **grazers**, directly feed on producers. Grazers (including protists, insects, and vertebrate herbivores) convert 90% of the carbon back to CO_2 through respiratory metabolism, yielding energy. Consumers occupying the next level, often called **predators**, feed on the grazers, again converting 90% back to atmospheric CO_2. In microbial ecosystems, the trophic relationships are often highly complex, as a given species may act as both producer and consumer.

At each trophic level, some of the organisms die, and their bodies are consumed by **decomposers**, returning carbon and minerals to the environment for use by producers. Decomposers are invariably microbes (fungi or bacteria). Decomposers have particularly versatile digestive enzymes capable of breaking down complex molecules such as lignin. Without decomposers, carbon and minerals needed by phototrophs would be locked away by ever-increasing mounds of dead biomass. Instead, all biomass is recycled, and all the energy acquired by ecosystems is eventually converted to heat.

Environmental Limiting Factors

In all environments, key physical factors such as oxygen and temperature limit the ability of microbes to grow. Physical factors have different effects on different species; thus, physical factors determine which microbial species inhabit a given environment.

Link Sections 6.4 and 6.5 describe how factors such as oxygen availability and temperature limit the growth of microbes.

OXYGEN The availability of oxygen determines whether microbes use molecular oxygen as an electron acceptor during cellular respiration on organic compounds produced by other organisms (discussed in Chapter 7). Aerobic respiration is favored by well-tilled soil, where maximal surface area of soil particles is exposed to oxygen. Aerobic respiration enables microbes to maximize the breakdown of food substrates to CO_2.

In anaerobic environments, such as waterlogged soil and deep sediment, microbes use minerals such as iron (Fe^{3+}) and nitrite (NO_2^-) to oxidize organic compounds supplied by other organisms (anaerobic respiration). But these anaerobic electron acceptors are soon used up, so breakdown to CO_2 takes much longer without oxygen. Anaerobic environments allow much slower rates of assimilation and dissimilation than occurs in the presence of oxygen. The total volume of anaerobic microbial communities, however, includes vast regions underground that exceed the volume of our aboveground oxygenated biosphere. Underground communities of

Figure 27.4   Microbes in a Drop of Pond Water

LM   100 µm

Figure 27.5   Algal Bloom

Fertilizer runoff causes an algal bloom in Toledo, Ohio, in 2014.

anaerobic bacteria and archaea filter water for our wells—an ecosystem service crucial for human health.

SALINITY   High salinity (salt concentration) limits the growth of microbes adapted to freshwater conditions. Salt-ridden land is also excluded from agriculture, influencing human health. On the other hand, many microbial species have adapted to high salinity (they are called halophiles). The haloarchaea, for instance, bloom in population as a body of water shrinks and becomes hypersaline. Haloarchaea grow in salt ponds that process salt for human consumption.

pH   Acidity (low pH) is important geologically because a high concentration of hydrogen ions accelerates the release of reduced minerals from exposed rock. Extreme acidity is often produced by lithotrophic acidophiles, such as those at iron mines. The increasing acidity enhances further release of minerals to be oxidized, while excluding acid-sensitive competitors. Acid mine drainage leaks into streams and lakes, killing fish and endangering human health.

Microbes in Freshwater

Freshwater habitats include lakes, rivers, and wetlands. Freshwater is usually less than 0.1% salt. The water is close to the sediment, so lakes tend to contain organic nutrients that support a moderate density of microbial life (**Figure 27.4**). That microbial growth is why lake water more often appears green or brown rather than clear.

In the upper layers of water, the concentration of heterotrophic microorganisms determines the **biochemical oxygen demand (BOD)**, the amount of oxygen removed from the water by aerobic respiration. The upper layers of a lake may have such a low concentration of organisms that the BOD is low and thus the dissolved oxygen content is high. The BOD rises, however, when excess sewage is introduced. Human and industrial sewage contains organic compounds rich in nitrogen as well as carbon. Microbes readily oxidize

these organic compounds by aerobic respiration. Sewage microbes also release ammonia, which nitrifying microbes oxidize to nitrates. Oxygen use by microbial respiration deprives fish of oxygen, leading to a "zone of hypoxia" where fish cannot grow.

A lake that receives excessive nutrients, such as runoff from agricultural fertilizer or septic systems, shows increased BOD and may become **eutrophic**. In a eutrophic lake, the nutrients support growth of algae to high densities, causing an algal bloom (**Figure 27.5**). The algal bloom is consumed by heterotrophic bacteria, whose respiration removes all the oxygen. But all fish and invertebrates need to breathe oxygen, so in a eutrophic lake, fish die off owing to lack of oxygen.

Algal blooms increasingly threaten our water supplies. For example, in 2014, fertilizer runoff in Lake Erie caused a bloom of algae that released a toxin called microcystin. If swallowed, microcystin causes intestinal illness, skin rashes, and liver damage. The toxin contaminated the water supply of Toledo, a major city in Ohio. For several days, drinking water was banned for half a million people. Algal blooms associated with climate change increasingly threaten water systems worldwide.

Microbes of the Oceans

The oceans cover more than two-thirds of Earth's surface, reaching depths of several kilometers and forming an immense habitat. The salt concentration is much higher than in freshwater, typically 3.5%. This high salt level prevents growth of many aquatic bacteria, although some salt-tolerant pathogens, such as *Vibrio cholerae*, grow well.

In the oceans, most CO_2 fixation and biomass production are performed by the smallest inhabitants, phototrophic bacteria (**Figure 27.6**). Marine photosynthetic bacteria and algae fix a large portion of the world's carbon dioxide into biomass, and they

Figure 27.6   Microbes within Food Webs

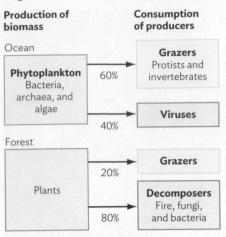

Production of biomass

Ocean

Phytoplankton Bacteria, archaea, and algae

Consumption of producers

Grazers Protists and invertebrates

60%

Viruses

40%

Forest

Plants

Grazers

20%

Decomposers Fire, fungi, and bacteria

80%

In marine ecosystems, the primary producers are bacteria, archaea, and algae. Viruses break down both producer and consumer microbes. In a forest, the major producers are trees, much of whose biomass is poorly digested by grazers; 80% of biomass is decomposed by fungi and bacteria.

produce much of the oxygen we breathe (**Figure 27.7A**). Many are free-floating, single-celled microbes, whereas others grow as tiny biofilms on bits of shell and detritus called "marine snow" (**Figure 27.7B**). Still other "microbes" may form long sheets such as kelps (eukaryotic algae, discussed in Chapter 11). Bacteria and algae support vast marine ecosystems with many trophic levels, ultimately including whales, the largest animals on Earth. Fish depend on a vast invisible food web of microbes, and thus their numbers remain limited despite the seemingly huge volume of ocean. Thus, marine fish populations are especially vulnerable to overharvesting and to any environmental change that disrupts the food chain.

An important determinant of marine habitat is the **thermocline**, a depth at which temperature decreases steeply and water density increases. At the thermocline, a population of heterotrophs will peak, feeding on organic matter that settles from above. Coastal regions show the highest concentration of nutrients and living organisms and the least light penetration. The open ocean is largely oligotrophic (extremely low concentration of nutrients and organisms).

The ocean floor experiences extreme pressure beneath several kilometers of water. Nevertheless, the deep ocean supports vast, unimagined realms of life supported by unique forms of bacteria. Organisms that live there are barophiles, adapted to high pressure (discussed in Chapter 6). The reduced minerals from thermal vents support huge ecosystems of lithotrophic bacteria and archaea. Some are mutualists of giant clams and tube worms. The clams and tube worms evolved in symbiosis with bacteria that oxidize hydrogen sulfide emanating from the vent. That means that these animals have blood full of hydrogen sulfide—a poison that would kill most animals. The ocean floor provides so many reduced inorganic minerals that bacterial redox reactions generate electricity. As a result, the entire interface between floor and water acts as a charged battery.

Dead Zones

In the upper layers of the ocean, as we saw for freshwater lakes, the concentration of heterotrophic microorganisms determines the BOD. When sewage is introduced, oxygen-consuming bacteria multiply, and BOD rises. Until recently, BOD was considered a local issue, affecting the health of individual lakes and rivers. Today, however, large regions of ocean have become **dead zones**, or zones of hypoxia, devoid of most fish and invertebrates.

A well-known dead zone is a region in the Gulf of Mexico off the coast of Louisiana where the Mississippi River releases about 40% of the U.S. drainage to the sea (**Figure 27.8A**). Over its long, meandering course, which includes inputs from the Ohio and Missouri Rivers, the Mississippi builds up high levels of organic wastes, as well as nitrates from agricultural fertilizer. When these nitrogen-rich substances flow rapidly out to the gulf in the spring, they lift the nitrogen limitation on algal growth and feed massive algal blooms. The algal population then crashes, and their sedimenting cells are consumed by heterotrophic bacteria. The heterotrophs use up the available oxygen, raising the BOD. The resulting hypoxia (lack of oxygen) kills off the fish, shellfish, and crustaceans over a region equivalent in

Figure 27.7   Marine Microbes

A. Bioluminescent dinoflagellates at the ocean shore, Vaadhoo Island, Maldives.

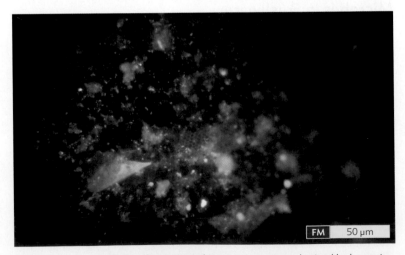

FM   50 µm

B. Marine snow, particles of organic and inorganic matter colonized by bacteria.

Figure 27.8   The Dead Zone in the Gulf of Mexico

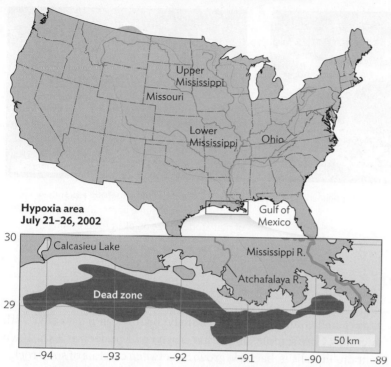

A. Map of the Mississippi River drainage, which empties into the Gulf of Mexico. Every summer, drainage high in organic carbon and nitrogen causes algal blooms, leading to hypoxia and death of fish.

B. Aerial view of the oil leaked from the Deepwater Horizon oil wellhead in the Gulf of Mexico, 2010.

size to the state of New Jersey. A dead zone can be expanded by a petroleum spill such as that caused by the Deepwater Horizon oil well blowout in 2010 (**Figure 27.8B**).

Another effect of dead zones is the anaerobic metabolism of nitrogen. As mentioned earlier, nitrogen-rich effluents from fertilizer runoff are oxidized by lithotrophs to nitrate. Then, in the dead zone, where oxygen is low, some bacteria use nitrate as an alternative electron acceptor for anaerobic respiration. These anaerobic respiring bacteria now release some of the reduced nitrogen in the form of a gas, nitrous oxide (N_2O). Nitrous oxide is a highly potent greenhouse gas; it also reacts catalytically with ozone, depleting the ozone layer.

What can be done about dead zones and about the polluted drainages that lead to them? In the long run, prevention is far better than cleanup. To prevent dead zones, the oceanic release of industrial pollutants such as petroleum must be avoided. In addition, all the communities throughout the river drainage area must treat their wastes to eliminate nitrogenous wastes before disposal.

Microbes in Soil

In contrast to the ocean, where the base of producers is almost entirely microbial, on land the major producers and fixers of CO_2 are macroscopic plants. Plants generate detritus, discarded biomass such as leaves and stems, that requires decomposition by fungi and bacteria. Detritus decomposition helps to form soil, one of the

densest sources of microbes on Earth. Soil is a complex mixture of decaying organic and mineral matter that feeds vast communities of microbes, many of which grow as biofilms on soil particles (**Figure 27.9**). And soil-based agriculture is the major source of food

Figure 27.9   A Soil Particle Colonized by Microbes

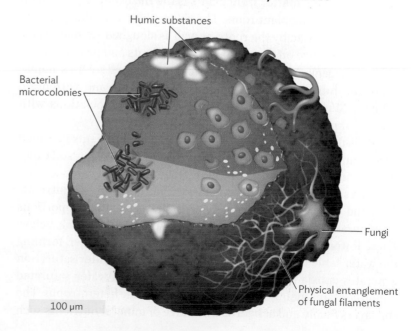

Humic substances

Bacterial microcolonies

Fungi

Physical entanglement of fungal filaments

100 µm

Figure 27.10   Soil Microbes

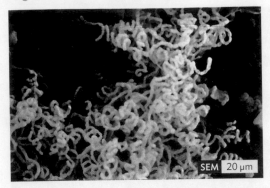

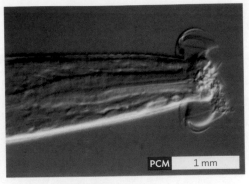

A. Hyphae of actinomycete bacteria, *Streptomyces griseus*. *Streptomyces* bacteria give the soil its characteristic odor.

B. Bacteria grow on the surfaces of fungi.

C. A bacteria-feeding nematode has special mouthparts to consume bacteria. Other species of nematodes graze on plant roots or fungi.

for our planet's human inhabitants. The qualities of a given soil—oxygenated or water saturated, acidic or alkaline, salty or fresh, nutrient-rich or nutrient-poor—define what food can be grown and whether the human community will eat or starve.

The surface layer of soil consists of dark, organic detritus, such as shreds of leaves fallen from plants. The detritus is in the earliest stages of decomposition by microbes (**Figure 27.10**), primarily fungi such as *Mycena* species, and by actinomycete bacteria (filamentous bacteria that produce antibiotics; discussed in Chapter 10) such as *Streptomyces*. *Streptomyces* generates chemicals whose odors give soil its characteristic smell.

The top layers of soil feature a food web of extraordinary complexity (**Figure 27.11**). The major producers are green plants, whose root systems feed predators, scavengers, and mutualists. Another source of organic matter from plants is the **rhizosphere**, the region of soil surrounding plant roots. The rhizosphere contains proteins and sugars released by the roots, as well as sloughed-off plant cells. Some bacteria live off the rhizosphere nutrients and protect plants from pathogens. Microbes feed diverse predators such as nematodes and fungi, which in turn feed invertebrate and vertebrate animals. Many soil microbes have mutualistic associations with animals or plants (discussed in Section 27.2).

Well-drained soil is full of oxygen, as well as nutrients liberated by decomposing microbes and used by plants. The deeper soil experiences periods of water saturation from rain. Rainwater leaches (dissolves and removes) some of the organic and mineral nutrients from the upper layers. Further below lie increasing proportions of minerals and rock fragments broken off from bedrock below. These lower layers are permanently saturated with water, forming the water table that we tap with wells. Prolonged water saturation generates anoxia (oxygen depletion). This anoxic, water-saturated region contains mainly lithotrophs and anaerobic heterotrophs. The soil layers finally end at bedrock, a source of mineral nutrients such

as carbonates and iron. Interestingly, bedrock is permeated with microbes. Core samples show that crustal rock as deep as 3 km down contains endoliths, bacteria growing between crystals of solid rock.

Microbes in Wetlands

A **wetland** is defined as a region of land that undergoes seasonal fluctuations in water level, so that sometimes the land is dry and oxygenated, at other times water saturated and anaerobic (**Figure 27.12**). Wetlands give us many ecosystem services; for example, the Everglades filter much of the water supply for Florida communities. Because wetlands have access to rich minerals, oxygen, and sunlight all at once, they produce the highest rate of biomass of all ecosystems.

CASE HISTORY 27.1

Desert Kidneys: A Constructed Wetland

In 1994, the city of Phoenix, Arizona, faced challenges to their water supply. In a desert, water is scarce, and quality is hard to maintain. The city's water department needed to decrease their wastewater's organic nitrogen content, which threatened the river downstream. A study proposed the construction of a wetland environment to receive the treated effluent from the Phoenix sewage plant. Over the next 10 years, the city constructed a series of experimental wetlands, leading to the large-scale Tres Rios Constructed Wetlands Project (**Figure 27.13A**). By 2010, the constructed wetland encompassed 800 acres and filtered 170 million gallons per day. Besides controlling nitrogen and organic content, the wetland restores groundwater, offers a carbon offset (net CO_2 removed from the atmosphere), and generates a wildlife-rich recreational area. The Tres Rios wetland now attracts beavers, bald eagles, and bobcats, as well as many human visitors.

Figure 27.11   The Soil Food Web

Plants are the major producers, although some production also occurs from lithotrophs such as ammonia oxidizers. Detritus from plants is decomposed by fungi and bacteria, which feed protists and small invertebrates such as nematodes. Protists and small invertebrates are consumed by larger invertebrates and vertebrate animals.

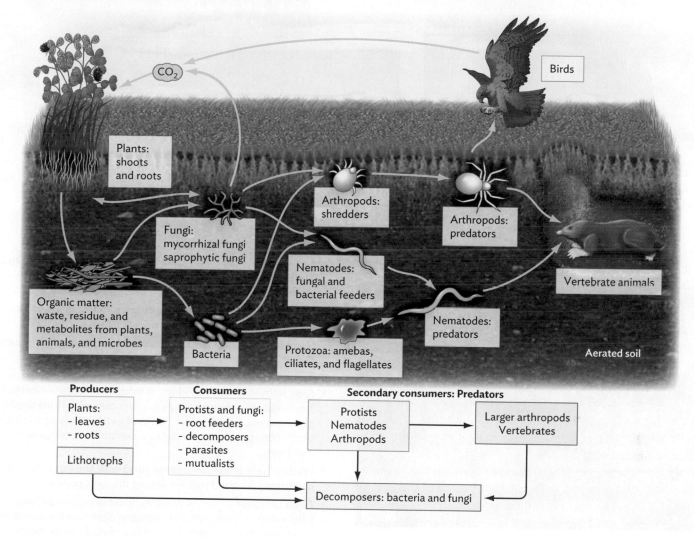

Wetlands are called the "kidneys" of our environment, because they filter organic wastes and recycle nutrients. In the case of Tres Rios, a constructed wetland filters excess nitrogenous compounds from the Phoenix effluent that could otherwise harm the downstream river by causing excess BOD and loss of fish. The wetland was constructed by excavating shallow basins and installing pipes that allow wastewater to disperse gradually, giving environmental microbes time to metabolize (**Figure 27.13B**). The organic carbon feeds anaerobic respirers, and the nitrogen feeds denitrifying bacteria (bacteria that use nitrate as an electron acceptor). Soil bacteria and archaea release nutrients that encourage growth of wetland plants.

How does a wetland manage change in water flow? Wetland soil undergoes periods of anoxic water saturation, leading to what is known as hydric soil. Hydric soil is characterized by "mottles," patterns of color and paleness (**Figure 27.14**). The reddish brown portions (for example, surrounding a plant root) result from oxidized

Figure 27.12   Wetland Soil

A true wetland experiences periods of water saturation alternating with dry soil. Soil that is water saturated becomes anaerobic because O_2 diffuses slowly through water.

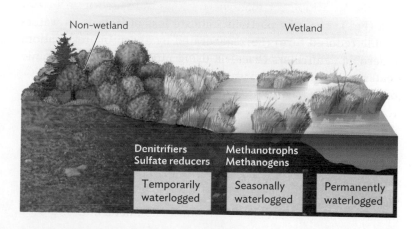

Figure 27.13   Constructed Wetlands Purify Wastewater

A. Tres Rios, Arizona, constructed wetland purifies municipal wastewater and enriches the river ecosystem.

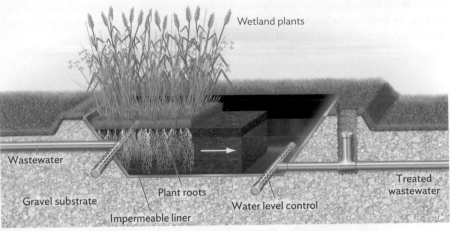

B. Wetland plants and associated microbes purify wastewater as it flows through.

Figure 27.14   Wetland Mottled Soil

1 cm

Alternating periods of water saturation and dryness give the soil a mottled color. The reddish brown portions around root holes result from oxidized iron (Fe^{3+}), whereas the gray portions indicate Fe^{2+} generated by anaerobic respiration and leached out by water.

iron (Fe^{3+}). The gray portions indicate loss of iron in its water-soluble reduced form (Fe^{2+}) generated by anaerobic respiration. Anaerobic denitrifiers (bacteria using nitrate to oxidize organic food) remove nitrate from water before it enters the water table—one of the many ways that wetlands protect our water supply.

SECTION SUMMARY

- **Microbial populations fill unique niches in ecosystems.** Every environment on Earth, including those below ground, contains microbes. In principle, every chemical reaction that yields energy might be used by some kind of microbe.

- **Assimilation occurs when microbes fix essential elements into biomass,** which recycles within ecosystems, including human agriculture. Some key elements, such as nitrogen, are fixed solely by bacteria and archaea.

- **Dissimilation occurs when consumers catabolize the bodies of producers,** generating CO_2 and releasing heat energy.

- **Microbial activity depends on oxygen level, temperature, salinity, and pH.** The largest volume of our biosphere contains anaerobic (living without oxygen) bacteria and archaea.

- **Food webs involve many kinds of microbes.** In marine and freshwater food webs, microbes are the main producers, with many trophic levels of consumers, including fish. Soil food webs include a complex array of microbial producers, consumers, predators, decomposers, and mutualists.

- **Wetland soils alternate aerated (dry) with anoxic (water-saturated) conditions.** Wetland soils are among the most productive ecosystems.

- **Dead zones** are caused by massive runoff of organic effluents, rich in carbon and nitrogen, into the rivers and ultimately the ocean. High BOD accelerates heterotrophic respiration and depletes oxygen needed by fish.

27.2
Microbial Symbiosis with Animals

SECTION OBJECTIVES

- Explain the different kinds of microbe-animal symbioses, including mutualism, synergism, commensalism, amensalism, and parasitism.
- Describe the digestive communities of animals.
- Explain how microbial communities contribute to human health.

Why do some microbes live within animals, including our own body? Outside our body, environmental microbes provide benefits at a distance. For example, aquatic cyanobacteria produce oxygen that is breathed by organisms around the globe. But other kinds of interactions require intimate association between microbes or between a microbial population and animals or plants. Intimate association between organisms of different species is called **symbiosis**. Symbiotic associations include a full range of positive through negative relationships. Multicellular animals and plants provide unique habitats for symbiotic microbes. All animals coevolve with their specific associated microbes, some of which can grow nowhere else. Many are beneficial to the human body. Others, of course, can be highly adapted pathogens, as presented in Chapters 18–24.

Classes of Symbiosis

The most highly developed forms of symbiosis involve partner species that require each other for survival, a phenomenon termed **mutualism**. A famous example of mutualism is that of termites, which harbor bacteria and protists that digest the cellulose from wood. Termites absolutely require their digestive microbes, and the microbes cannot live outside the termite. The termite–gut microbe mutualism is an obligate association for both partners.

Less intimate cooperation is called **synergism**, in which both species benefit but can grow independently and show fewer specific cellular interactions. For example, on human skin, *Staphylococcus epidermidis* and related species produce short-chain fatty acids whose uptake inhibits growth of yeasts and pathogens. The staphylococci receive nutrients from human skin secretions, and they outcompete inhibited microbes for scarce water and nutrients. Nevertheless, staphylococci and human beings can live without each other.

An interaction that benefits one partner only is called **commensalism**. Commensalism is difficult to define in practice, since "commensal" microbes often provide a hidden benefit to their host. For example, gut bacteria such as *Bacteroides* species were considered commensals until it was discovered that their metabolism aids human digestion. *Bacteroides* species break down complex plant polysaccharides to sugars that humans can digest—much as termite microbes digest wood and bovine rumen microbes digest grass (discussed shortly).

An interaction that harms one partner nonspecifically, without harming the other, is called **amensalism**. An example of amensalism is actinomycete production of antimicrobial peptides that kill surrounding bacteria. The dead bacterial components are then catabolized by the actinomycete. Finally, **parasitism** is an intimate relationship in which one member (the parasite) benefits while harming a specific host. Many microbes have evolved specialized relationships as parasites, including intracellular parasitic bacteria such as the rickettsias, which cause diseases such as Rocky Mountain spotted fever.

The distinction between mutualist and parasite is often subtle. Lichens consist of a mutualistic association between fungus and algae; but environmental changes can convert the fungus to a parasite. On the other hand, parasitic microbes may coevolve with a host to the point that each depends on the other for optimal health. For example, the incidence of certain allergies and immune disorders appears to correlate with lack of exposure to pathogens and parasites.

Animals Host Microbial Digestive Communities

As discussed in Chapter 14, many kinds of microbes inhabit parts of the human body, as well as other animal bodies. The human skin, digestive tract, and mucous membranes harbor complex communities that have food webs with competitive and cooperative interactions, much like the free-living communities in soil or water.

Our understanding of human-associated microbial communities leads to surprising new treatments for human disease. An example is fecal transplant, or bacteriotherapy (discussed in the Chapter 22 inSight). Fecal bacteriotherapy produces "miracle cures" for many patients with recurrent *Clostridium difficile* infection and may help others with irritable bowel syndrome (see the case history that opens Chapter 14). In these cases, we don't even know the identity of the bacteria involved, but a balanced community of microbes from a healthy person can colonize a diseased colon and return it to normal function. The finding is particularly remarkable given that individuals differ substantially in their microbial populations; there must be many possible combinations compatible with health. Yet a given population from a given healthy person can restore health to another person—even with a colon severely distressed by inflammation and bleeding. Physicians now argue that the colonic microbial community constitutes an "organ" of the human body, as essential as our other human organs.

Human microbiota also interact positively with our immune system and generate defenses against pathogens. For example, *Bifidobacterium* down-regulates human secretion of cytokines, molecules that cause inflammation in ulcerative colitis. Other colonic bacteria produce bacteriocins, molecules that prevent pathogens from adhering to the gut lining.

Many vertebrate and invertebrate animals host digestive communities of microbes, including thousands of species of bacteria, protists, and archaea. The human colon contains many fermenters and methanogens, some of which feed on intestinal mucus, whereas others digest complex plant fibers, thus providing up to 15% of our caloric intake. Our gut microbiota enable us to acquire nutrition from complex plant fibers that we could not otherwise digest. The most extensively studied digestive communities are those of ruminants, such as cattle and sheep. Throughout most of human civilization, ruminants have provided protein-rich food, textile fibers, and mechanical work. A historical reference is the biblical injunction to consume an animal that "is cleft-footed and chews the cud"—that is, "ruminates," or redigests its food in the fermentation chamber known as the rumen (**Figure 27.15**).

The bovine gut system has four chambers (Figure 27.15A). The rumen initially digests the feed and then passes it to the reticulum. The reticulum breaks the feed into smaller pieces and traps indigestible objects, such as stones or nails. After initial digestion, feed is regurgitated for rechewing and then returned to the rumen, by far the largest of the chambers. In the rumen, feed is broken down to small particles and fermented slowly by thousands of species of microbes. The partially digested feed passes to the omasum, which

Figure 27.15   The Bovine Rumen

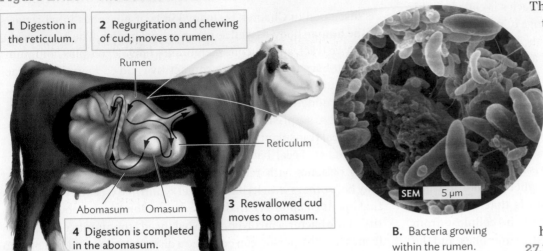

1 Digestion in the reticulum.

2 Regurgitation and chewing of cud; moves to rumen.

Rumen

Reticulum

Abomasum   Omasum

3 Reswallowed cud moves to omasum.

4 Digestion is completed in the abomasum.

SEM   5 μm

A. The rumen is the largest of four chambers in the bovine stomach.

B. Bacteria growing within the rumen.

absorbs water and short-chain acids produced by fermentation. The abomasum then decreases pH and secretes enzymes to digest proteins before sending its contents to the colon for further nutrient absorption and waste excretion.

Different microbes fill different niches in ruminal metabolism. Cattle grown on relatively poor forage (that is, forage high in complex plant content) show a high proportion of ruminal fungi. By contrast, cattle fed a high-cellulose diet, such as hay, grow faster and show cellulose-fermenting bacteria such as *Ruminococcus* and *Fibrobacter*. Today, commercial agriculture has largely shifted from hay to grain, whose starch content leads to even more rapid digestion and faster growth of the animal. Unfortunately, rapid starch digestion favors sugar fermenters such as *Prevotella* and *E. coli*, which generate higher acid levels and gases, leading to "starch bloat." Furthermore, rumen acidity selects for acid-resistant pathogens such as the *E. coli* strain O157:H7.

One problem with ruminal fermentation is the bacterial production of H_2 and CO_2 through mixed-acid fermentation. The H_2 and CO_2 from fermentation are converted by methanogens to methane. Cows make so much methane that a cannula inserted into the rumen liberates enough of the gas to light a flame. Methane wastes valuable carbon from feed and adds a large amount of greenhouse gas to global warming.

Diseases of Animals and Humans

Animals, like humans, are subject to many microbial pathogens. Animal pathogens affect agriculture; for example, in 2007 in the United Kingdom, an outbreak of hoof-and-mouth disease (caused by aphthovirus) occurred in cattle. In hoof-and-mouth disease, the virus causes blisters on the mucous membranes and hooves,

impairing the animal's ability to feed and stand. The outbreak required farmers to slaughter thousands of dairy cattle to contain the spread of disease.

Hoof-and-mouth disease does not infect humans; but other pathogens, such as *E. coli* O157:H7, may be carried by animals and then transferred to humans, where they cause disease. An animal population that carries pathogens transmissible to humans is called a reservoir of infection (discussed in Chapter 2). Many kinds of animal pathogens can be picked up by humans, even if humans are not the preferred host (Table 27.1). An important example is *Toxoplasma*, a cat parasite that can infect humans and cause congenital defects in a developing fetus (the replication cycle is presented in Figure 21.26). Emerging diseases can arise from "exotic" pets whose microbiota are unfamiliar to physicians. For example, in 2003, the United States experienced an outbreak of monkeypox transmitted by pet prairie dogs. The prairie dogs contracted the virus from a Gambian giant rat imported by the same pet distributor. Monkeypox virus is closely related to smallpox and could someday mutate into a similar virulent strain.

A disease acquired by humans from animals is called a **zoonosis**. Typically, a zoonotic strain, such as "avian flu" (avian-adapted strains of influenza), fails to transmit effectively between humans. But the danger is that once established in humans, an animal pathogen may mutate into a strain exceptionally deadly to humans—and exceptionally virulent—because the humans lack immunity. The H1N1 influenza strain of 2009 is an example of a newly emerged, virulent strain. Fortunately, H1N1 turned out to have a relatively low mortality rate; but had it mutated to a deadly form, the result could have been a pandemic similar to the flu pandemic of 1918. Human diseases can also be transmitted to animals. For example, diseases such as polio, measles, and respiratory viruses have been transmitted to endangered apes under study by researchers. Some researchers are attempting to vaccinate endangered apes against human viruses as well as endemic ape viruses such as Ebola.

The incidence of vector-borne and zoonotic diseases is challenging to predict. However, the Centers for Disease Control and Prevention (CDC) warns that environmental change can increase the range of animals and pathogen vectors such as ticks and mosquitoes. Increasing global temperatures are already increasing the spread of human-animal pathogens. An important example of such a pathogen is West Nile virus, the cause of West Nile encephalitis, previously uncommon in the United States. West Nile virus is now spreading among birds and horses, transmitted by a tropical mosquito whose range is spreading northward thanks to global warming. These mosquitoes increasingly transmit the virus to humans in the Northern Hemisphere.

Table 27.1
Animal Diseases Transmitted to Humans

Disease	Pathogen	Transmission to Humans
Anthrax	*Bacillus anthracis*	Cattle or sheep transmit through contact with hide or wool
Ascariasis	Roundworm (*Ascaris*)	Ingestion of contaminated soil, fruits, or vegetables; transmitted via feces, often of dogs or cats
Brucellosis	*Brucella*	Cattle, goats, pigs, and dogs can transmit the pathogen through infected meat or milk
Bubonic plague	*Yersinia pestis*	Fleas from prairie dogs or rats transmit the bacteria to humans
Cat scratch fever	*Bartonella*	Cats transmit through bite or scratch, usually to children
Monkeypox	*Orthopoxvirus*	African monkeys and rodents, including giant rats sold in the United States as pets, transmit a milder smallpox-like disease to humans
Rabies	*Lyssavirus*	Dogs, raccoons, bats, and other mammals transmit through bite to humans
Salmonellosis	*Salmonella enterica*	Pet turtles and other reptiles spread pathogen through contact, usually to children; also spread by contact with raw chicken and eggs
Toxoplasmosis	*Toxoplasma gondii*	Cat feces and undercooked contaminated pork transmit to humans
Tularemia	*Francisella tularensis*	Rodents transmit pathogen to humans via fly or tick bite, more rarely transmitted through consumption of contaminated food or water
West Nile encephalitis	*Arbovirus*	Horses and birds spread pathogen via mosquitoes to humans

SECTION SUMMARY

- **Symbiosis is an intimate association** between organisms of different species. The association may cause benefits or harm.
- **Mutualism is a form of symbiosis in which each partner species benefits from the other.** The relationship is usually obligatory for growth of one or both partners.
- **Synergism is a relationship in which both species benefit.** The partners are easily separated, and either partner can grow independently of the other.
- **Commensalism is an association in which one partner benefits and the other is unharmed.** It may, however, be difficult to prove that the second partner receives zero benefit.
- **Parasitism is a form of symbiosis in which one species grows at the expense of another host organism.**
- **The rumen of ruminant animals is a complex microbial digestive chamber.** Rumen microbes digest complex plant materials. Gut fermenters contribute to digestion in many vertebrate animals, including humans.
- **Environmental conditions facilitate disease transmission between humans and animals.** An animal population that can transmit disease to humans is called a reservoir of infection.

Thought Question 27.1 Explain what mutualism and parasitism have in common and how they differ. Describe an example of a relationship that combines aspects of both.

Thought Question 27.2 Why do the microbes of digestive communities leave food value for the animal host? How can the animal obtain nourishment from waste products that the microbes could not use?

27.3
Fermented Foods and Beverages

SECTION OBJECTIVES
- Describe the roles of bacteria and fungi in the production of cheese and fermented soy foods.
- Outline the process of ethanolic fermentation in the production of beer and wine.
- Explain why fermented foods and beverages retain high food value.

If "you are what you eat," are you a microbe? You consume microbes every day as part of the food you put in your mouth. Some foods, such as mushrooms and seaweed (sushi covering), are considered microbial products. Others—including plant foods, from salad to fruit—contain microbes growing within the plant's vasculature. Still other foods are produced by the intentional culturing of microbes upon a food substrate. These foods are called **fermented foods**, referring to anaerobic fermentation, the predominant class of metabolism in microbial food production. They include some of our most famous food staples, such as bread and cheese, as well as beverages such as wine and beer. One of the most complex food fermentations is that of cocoa bean fermentation for chocolate (see **inSight**). In effect, fermented foods are the product of controlled ecosystems that we establish for our own nutrition.

inSight

Chocolate: The Mystery Fermentation

Figure 1   Cocoa Beans

A. Cocoa fruit, showing beans encased in mucilage.

B. Heap of beans covered by plantain leaves. Fruit pulp ferments, liquefies, and drains away, while the beans acidify and turn brown.

Chocolate, the product of the cocoa bean *Theobroma cacao*, or "food of the gods," requires one of the most complex fermentations of any food. For all the sales of chocolate, totaling 2.5 billion kilograms per year worldwide, no "starter culture" has yet been standardized to ferment the cocoa bean. Instead, cocoa beans harvested in Africa or South America are heaped in mounds upon plantain leaves for fermentation by indigenous microorganisms, essentially the same way cocoa has been processed for thousands of years (Figure 1). The beans must ferment immediately where they are harvested; they cannot be exported and fermented later.

The fermentation occurs outside the cocoa bean, within the pulp that clings to the beans after they are removed from the cocoa fruit. The pulp contains 1% pectin (a complex polysaccharide found in fruits) plus amino acids and minerals. These nutrients support growth of many kinds of microbes. Brazilian microbiologist Rosane Schwan, of the University of Lavras, Brazil, analyzed the microbial community (Figure 2). Schwan defined three stages of succession:

1. **Yeast fermentation (anaerobic).** The pulp initially is full of citric acid (pH 3.6), which favors growth of yeasts such as *Candida* and *Saccharomyces*. The yeasts consume citric acid, increasing pH. They also degrade pectin into glucose and fructose, allowing the pulp to liquefy. As the liquefied pulp drains, its remaining sugars are fermented to ethanol, CO_2, and acetate. These products eventually rise to levels that inhibit yeast growth. The ethanol and acetate penetrate the bean embryo, preventing germination and disrupting cell membranes. The disrupted cells release enzymes that generate key molecules of chocolate. For example, theobromine is a stimulant similar to caffeine, and 2-phenylethylamine is a neurotransmitter associated with the pleasure response.

2. **Lactic acid bacteria (anaerobic).** The consumption of citric acid by yeast increases the pH to 4.2, enabling growth of lactic acid bacteria. *Lactobacillus* species ferment sugars and citric acid to acetate, CO_2, and lactic acid. When fermentable substrates are used up, *Lactobacillus* stops growing.

3. **Acetic acid bacteria (aerobic).** The beans are turned over and mixed periodically to permit access to oxygen. With oxygen, aerobic bacteria such as *Acetobacter* respire, converting the ethanol and acids to CO_2. The consumption of acids neutralizes undesirable acidity. The bean is also penetrated by oxygen,

Figure 2 Microbial Succession during Cocoa Pulp Fermentation

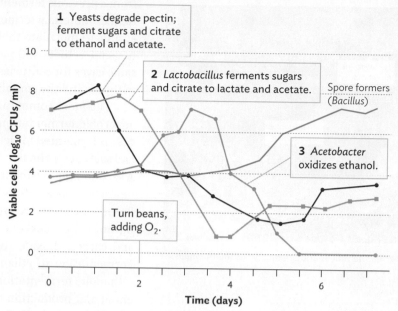

1 Yeasts degrade pectin; ferment sugars and citrate to ethanol and acetate.

2 *Lactobacillus* ferments sugars and citrate to lactate and acetate.

Spore formers (*Bacillus*)

3 *Acetobacter* oxidizes ethanol.

Turn beans, adding O_2.

A. Rosane Schwan harvests cocoa beans for her research on cocoa fermentation.

B. Yeasts generate alcohol, which lactic acid bacteria convert to lactate and acetic acid bacteria oxidize to CO_2. CFUs = colony-forming units.

which oxidizes key cocoa components, such as polyphenols. Polyphenol oxidation generates the brown color of cocoa and contributes flavor.

Oxidative respiration within the mound of beans generates heat faster than it can dissipate, raising the temperature as high as 50°C. The high temperature halts bacterial fermentation. After the pulp drains, the beans are dried and roasted. The roasting process completes the transformation of cocoa. Without the preceding fermentation process, no flavor would develop. Cocoa liquor and cocoa butter are extracted from the fermented beans and then recombined with sugar and other components to make "cocoa mass" (Figure 3). The cocoa mass is stirred for several days to achieve a smooth texture; then it is molded into the decorative forms known as chocolate.

The quality of chocolate rests largely on the original process of pulp fermentation, whose details remain poorly understood. Schwan is working to develop a defined starter culture, a community of microbes to produce a predictable high quality of flavor. At present, however, one of the world's most refined and highly prized commercial food products still depends on the indigenous microbial fermentation of cocoa pulp.

Figure 3 Chocolate Manufacture

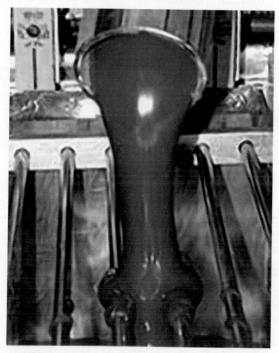

Cocoa mass contains cocoa butter and liquor extracted from the cocoa beans, mixed with sugar and other ingredients.

Figure 27.16   Bacterial Community within Emmentaler Cheese

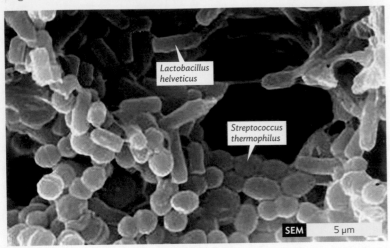

Bacterial species include *Lactobacillus helveticus* and *Streptococcus thermophilus*.

Aims of Fermentation

Virtually all human cultures have developed fermented foods, food products that are modified biochemically by microbial growth. Fermented foods that are produced commercially include dairy products such as cheese and yogurt, soy products such as miso (from Japan) and tempeh (from Indonesia), vegetable products such as sauerkraut and kimchi, and various forms of cured meats and sausages. Alcoholic beverages are made from grapes and other fruits (wine), grains (beer and liquor), and agave (tequila).

Why do we eat foods that have been altered by microbial growth? Food fermentation bestows certain benefits:

- **To preserve food.** Certain bacteria, particularly *Lactobacillus* and *Lactococcus* species, metabolize only a narrow range of nutrients (such as sugars) before their waste products (acids) build up and inhibit further growth. Thus, acid buildup keeps the product stable for much longer than the original food substrate. That is why yogurt and cheese last much longer than fresh milk. Other fermentation products that preserve food include ethanol (alcoholic beverages) and ammonia (fermented soybean products).

- **To improve digestibility.** Microbial action breaks down fibrous macromolecules and tenderizes the product, making it easier for humans to digest. Meat and vegetable products are tenderized by fermentation.

- **To add nutrients and flavors.** Microbial metabolism generates vitamins, particularly vitamin B_{12}, as well as flavor molecules, such as esters and sulfur compounds.

The nature of fermented foods depends on the quality of the fermented substrate, as well as on the microbial species and the type of biochemistry they perform. Traditional fermented foods depend on indigenous microbiota—that is, microbes found naturally in

association with the food substrate—or on starter cultures derived from a previous fermentation, as in yogurt or sourdough fermentation. Commercial fermented foods use highly engineered microbial strains to inoculate their cultures, although in some cases indigenous microbiota still participate. For example, cheeses aged in the same caves for centuries include fermenting organisms that persist in the air and the containers used.

The most common metabolism in food fermentation involves anaerobic fermentation of glucose (discussed in Chapter 7). Glucose is fermented to lactic acid (lactic acid fermentation) in cheeses and sausages, primarily by lactic acid bacteria such as *Lactobacillus*. A second-stage fermentation of lactic acid to propionic acid (propionic acid fermentation) by *Propionibacterium* generates the special flavor of Swiss and related cheeses. Some kinds of vegetable fermentation, as in sauerkraut, involve production of lactic acid, ethanol, and carbon dioxide (heterolactic fermentation) by *Leuconostoc*. Fermentation to ethanol plus carbon dioxide without lactic acid (ethanolic fermentation) is conducted by yeast during bread leavening and production of alcoholic beverages. In some food products, particularly those fermented by *Bacillus* species, proteolysis and amino acid catabolism generate ammonia in amounts that raise the pH (alkaline fermentation). For example, alkaline fermentation forms the soybean product natto.

Other products require the growth of mold, such as the mold-spiked Roquefort cheese. Mold growth requires some oxygen for aerobic respiration. Respiration must be limited, however, to avoid excessive decomposition of the food substrate and loss of food value.

Fermented Dairy Products

Microbial fermentation of milk converts a liquid suspension of proteins and carbohydrates to a semisolid or solid product, such as yogurt or cheese. The practice of milk fermentation probably arose among ancient herders who collected the milk of their pack animals but had no way to prevent the rapid growth of bacteria. The milk was stored in a container such as the stomach of a slaughtered animal. After hours of travel, the combined action of stomach enzymes (rennet) and lactic acid–producing bacteria caused the milk proteins to denature (misfold) and coagulate into **curd**. The curd naturally separates from the liquid portion, called **whey**. Both curds and whey can be eaten, as in the nursery rhyme "Little Miss Muffet." The curds, however, are particularly valuable for their concentrated protein content.

A cheese is any milk product from a mammal (usually cow, sheep, or goat) in which the milk protein coagulates to form a semisolid curd. How does milk coagulate? The major organic components of cow's milk are milk fat (about 4% unless skimmed), protein (3.3%), and the sugar lactose (4.7%). Milk starts out at about pH 6.6, very slightly acidic. At this pH, the milk proteins are completely soluble in water; otherwise, they would clog the animal's udder as the milk came out. Fermentation generally begins with bacteria such as *Lactobacillus* and *Streptococcus* (**Figure 27.16**). As bacteria ferment

Figure 27.17   Cheese Varieties

A. Cottage cheese, an unripened perishable cheese.

B. Emmentaler Swiss cheese, with eyes produced by carbon dioxide fermentation.

C. Feta cheese, a soft cheese from goat's milk, preserved in brine.

D. Roquefort, a medium hard cheese ripened by spiking with *Penicillium roqueforti.*

milk sugars to lactic acid, the pH starts to decline. Because no aerobic respiration occurs, only lactic acid is made in large quantities that acidify the milk until the bacteria can no longer grow. Halting bacterial growth minimizes the oxidation of amino acids, thereby maintaining food quality.

Milk contains suspended droplets of hydrophobic proteins called **caseins**. As the pH of milk declines below pH 5, the casein molecules misfold (or "denature"). The tangling of denatured caseins generates a gel-like network throughout the milk, trapping other substances, such as droplets of milk fat. This protein network generates the semisolid texture of **yogurt**, a simple product of milk acidified by lactic acid bacteria.

In soft cheeses such as cottage cheese (**Figure 27.17A**), bacterial lactic acid acidification is the only cause of coagulation. Other kinds of cheese, such as Swiss cheese (**Figure 27.17B**), include an additional step of casein coagulation by proteases such as rennet. Rennet derives from the fourth stomach of a calf. The water-soluble portion, about one-third of the total casein, enters the whey and is lost from the curd. To retain the whey protein, the cheese is treated at high temperature, which denatures even the whey protein so it is retained in the curd. Aging, that is, incubation under controlled humidity for several months, further develops flavor. Some cheeses are brined (treated with concentrated salt), as in feta cheese (**Figure 27.17C**), to limit bacterial growth and develop flavor. Other cheeses are "ripened" by inoculation with mold, either to form a surface crust, as in Brie and Camembert, or to spike deep into the cheese, as in blue cheese or Roquefort (**Figure 27.17D**).

Soybean Fermentation

Soybeans offer one of the best sources of vegetable protein and are indispensable for the diet of millions of people, particularly in Southeast Asia. But soybeans also contain substances that decrease their nutritive value, such as phytate, a sugar phosphate that inhibits iron absorption by the intestine. Soybeans also contain lectins,

proteins that bind to cell-surface glycoproteins within the human body. At high concentration, lectins may upset digestion and induce autoimmune diseases. Protease inhibitors interfere with chymotrypsin and trypsin, thus decreasing the amount of protein that can be obtained from soy-based food. But all these biochemical drawbacks of soybeans are diminished by microbial fermentation, while the protein content remains comparable to that of the unfermented bean (40%).

A major fermented soy product is tempeh, a staple food of Indonesia and other countries in Southeast Asia (**Figure 27.18**). Tempeh consists of soybeans fermented by *Rhizopus oligosporus*, a common bread mold. The mold grows as a white mycelium that permeates the beans, joining them into a solid cake. The final product has a mushroom-like taste and is served fried or grilled like a hamburger. Besides decreasing the negative factors of soy, the mold growth breaks down proteins into more digestible peptides and amino acids. During World War II, tempeh was fed to American prisoners of war held by the Japanese. The tempeh was later credited with saving the lives of prisoners whose dysentery and malnutrition had impaired their ability to absorb intact proteins.

Figure 27.18   Tempeh, a Mold-Fermented Soy Product

A. Fried tempeh.

B. *Rhizopus oligosporus* mold, used to make tempeh.

Figure 27.19   Beer Production: Ancient and Modern

A. Making beer in ancient Egypt, circa 3000 BCE. The mash is stirred in earthen jars.

B. Fermenters in a modern brewery.

Alcoholic Beverages

Ethanolic fermentation of grain or fruit was important to early civilizations because it yielded a source of water free of pathogens. Traditional forms of beer also provided essential vitamins supplied by the unfiltered yeasts. Ethanol provides a significant source of caloric intake, but it is also a toxin that impairs mental function. A modest level of ethanol enters the human circulation naturally from intestinal biota, equivalent to a fraction of a drink per day. For protection, the human liver produces the enzyme alcohol dehydrogenase, which detoxifies ethanol. In a healthy liver, alcohol dehydrogenase can metabolize small amounts of alcohol without harm. However, excess alcohol consumption can overload the liver's capacity for detoxification and permanently damage the liver and brain.

BEER: ALCOHOLIC FERMENTATION OF GRAIN   Beer production is one of the most ancient fermentation practices and is depicted in the statuary of ancient Egyptian tombs dated to 5,000 years ago (Figure 27.19A). The earliest Sumerian beers were made from bread soaked in water and fermented. Today, most beer is produced commercially by using giant vats to ferment barley (Figure 27.19B). Production of high-quality beer involves complex processing, with many steps. The barley grains must germinate (sprout as a seedling) in order for enzymes from the grain to break down starch (long-chain carbohydrates) into maltose (two molecules of glucose). The fermenting grains are mashed in water and cooked down. Flowers of the hops plant are added for flavor and for their antibacterial properties that increase the beer's shelf life. The glucose then is fermented by the yeast, producing ethanol and CO_2. Finally, the beer is filtered to remove yeast and other solids.

The yeast in the beer mash ferment most of the maltose to ethanol (Figure 27.20). In addition, side products in lower concentration (such as long-chain alcohols and esters) contribute flavors. But they also make unpleasant off-flavors if present in too great an amount. The off-flavors result from the presence of oxygen, which is needed because yeast requires a slight amount of aerobic metabolism to synthesize some of its cell components. The presence of oxygen diverts some pyruvate into acetaldehyde and diacetyl (shaded green), two compounds that cause off-flavors.

WINE: ALCOHOLIC FERMENTATION OF FRUIT   The fermentation of fruit gives rise to wine, another class of alcoholic products of enormous historical and cultural significance. Grapes (*Vitis vinifera*) produce the best-known wines, but wines

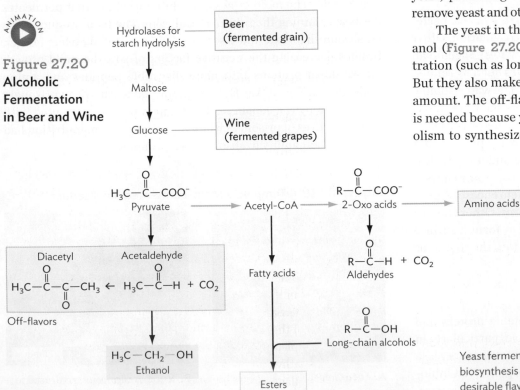

Figure 27.20
Alcoholic Fermentation in Beer and Wine

Yeast fermentation generates ethanol in substantial quantities. The biosynthesis of amino acids generates by-products that contribute both desirable flavors (long-chain alcohols) and off-flavors (acetaldehyde and diacetyl).

and distilled liquors are also made from apples, plums, and other fruits. The key difference between fermentation of fruits and fermentation of grains is the exceptionally high monosaccharide content in fruits. Grape juice, for example, can contain concentrations of glucose and fructose as high as 15%. The availability of simple sugars allows yeast to begin fermenting immediately, without preliminary breakdown of long-chain carbohydrates, as in the fermentation of grain for beer.

SECTION SUMMARY

- **Fermented foods preserve food value** because the anaerobic fermentation products (acids or alcohol) limit the growth of the microbes and leave most of the protein intact.

- **Milk curd** forms by lactic acid fermentation and rennet proteolysis. The cleaved casein peptides coagulate to form a semisolid curd. The main fermentative organisms are lactic acid bacteria.

- **Cheese varieties** include unripened cheese, as well as ripened cheeses involving subsequent steps such as brining or mold growth. Cheese flavors are generated by minor side products of fermentation.

- **Soy fermentation** to tempeh and other products improves digestibility and decreases undesirable soy components.

- **Beer derives from alcoholic fermentation of grain.** Barley grains are germinated, allowing enzymes to break down the starch to maltose for yeast fermentation. Secondary products of grain fermentation, such as long-chain alcohols and esters, generate the special flavors of beer.

- **Wine derives from alcoholic fermentation of fruit,** most commonly grapes. The grape sugar (glucose) is fermented by yeast to alcohol.

Thought Question 27.3 In food production, why does oxygen allow excessive breakdown of a food substrate, compared with anaerobic fermentation?

Thought Question 27.4 Compare and contrast the role of fermenting organisms in the production of cheese and alcoholic beverages.

27.4
Food Spoilage and Preservation

SECTION OBJECTIVES

- Explain the difference between food spoilage and food contamination.

- Outline the different processes of spoilage of dairy products, meat, seafood, and plant foods. Explain why different types of food spoil differently.

- Describe the various physical and chemical approaches to food preservation.

If microbes help us make fine foods, then why does food spoil?

Excess microbial growth can decrease nutrients and form toxins. We humans have always competed with microbes for our nutrients. When early humans killed an animal, microbes commenced immediately to consume its flesh. Because meat perished so fast, it made economic sense to share the kill immediately and consume all of it as soon as possible. Vegetables might last longer, but eventually they succumbed to mold and rot. Yet how does one define "rot" versus "food production"? To a surprising degree, the definition of spoilage depends on cultural practice. What is sour milk to one person may be buttermilk to another; and meat that one society considers spoiled, another considers merely aged.

From our earliest history, societies developed practices that control and minimize microbial degradation of food. Practices such as drying, smoking, and canning enabled humans to survive winters and dry seasons on stored food. Such practices are called **food preservation**. Modern food preservation includes antimicrobial agents, chemical substances that either kill microbes or slow their growth, and physical treatment with heat and pressure.

Food Spoilage and Food Contamination

After food is harvested, several kinds of chemical changes occur. Some begin instantly, whereas others take several days to develop. Some changes, such as meat tenderizing, may be considered desirable; others, such as putrefaction, render food unfit for consumption. The major classes of food change include:

- **Enzymatic processes.** After an animal dies, its flesh undergoes proteolysis by its own enzymes. Limited proteolysis tenderizes meat. Plants after harvest undergo other changes; in harvested corn, for example, the sugar rapidly converts to starch. That is why vegetables taste sweetest immediately after harvest, before the sugar is lost.

- **Chemical reactions with the environment.** The most common abiotic chemical reactions involve oxidation by air—for example, lipid autooxidation, which generates rancid odors.

- **Microbiological processes.** Microbes from the surface of the food begin to consume it—some immediately, others later in succession—generating a wide range of chemical products. In meat, internal organs of the digestive tract are an important source of microbial decay.

Microbial activity can aid food production, but it can also have various undesirable effects. Two classes of microbial effects are distinguished: food spoilage and food contamination with pathogens.

Food spoilage refers to microbial changes that render a product obviously unfit or unpalatable for consumption. For example, rancid milk or putrefied meats are unpalatable and contain metabolic products that may be harmful to human health, such as oxidized fatty acids or organic amines. Food that is capable of spoilage is referred to as perishable food. Signs of food spoilage include a change in appearance from that of the normal form, such as a color change, a change in texture, an unpleasant odor, or an undesirable taste.

Different pathways of microbial metabolism lead to different kinds of spoilage. Sour flavors result from acidic fermentation products, as in sour milk. Alkaline products generate bitter flavor.

Oxidation, particularly of fats, causes rancidity, whereas general decomposition of proteins and amino acids leads to putrefaction. The particularly noxious odors of putrefaction derive from amino acid breakdown products that often have apt names, such as the amines cadaverine and putrescine and the aromatic product skatole.

Food contamination, or **food poisoning**, refers to the presence of microbial pathogens that cause human disease—for example, rotaviruses that cause gastrointestinal illness. Pathogens usually go unnoticed as food is consumed because their numbers are very low, and they may not even grow in the food. Even the freshest-appearing food may cause serious illness if it has been contaminated with a small number of pathogens.

How Food Spoils

Different foods spoil in different ways, depending on their nutrient content, the microbial species, and environmental factors such as temperature.

DAIRY PRODUCTS Milk and other dairy products contain carbon sources, such as lactose, protein, and fat. In fresh milk, the nutrient most available for microbial catabolism is lactose, which commonly supports anaerobic fermentation to sour milk. Fermentation by the right mix of microbes, however, leads to yogurt and cheese production, as previously described. Under certain conditions, bitter off-flavors may be produced by bacterial degradation of proteins. The release of amines causes a rise in pH. Protein degradation is most commonly caused by psychrophiles, species that grow well at cold temperatures, such as those of refrigeration.

Cheeses are less susceptible than milk to general spoilage because of their solid structure and lowered water activity. However, cheeses can grow mold on their surface. Historically, the surface growth of *Penicillium* strains led to the invention of new kinds of cheeses. But other kinds of mold, such as *Aspergillus*, produce toxins and undesirable flavors.

MEAT AND POULTRY Meat in the slaughterhouse is easily contaminated with bacteria from hide, hooves, and intestinal contents. Muscle tissue offers high water content, which supports microbial growth, as well as rich nutrients, including glycogen, peptides, and amino acids. The breakdown of peptides and amino acids produces the undesirable odorants that define spoilage, such as cadaverine and putrescine.

Meat also contains fat, or adipose tissue, but the lipids are largely unavailable to microbial action because they consist of insoluble fat (triacylglycerides). Instead, meat lipids commonly spoil abiotically by autooxidation (reaction with oxygen) of unsaturated fatty acids, independent of microbial activity. Thus, when meats are exposed to air during storage, they turn rancid—particularly meats such as pork, which contains highly unsaturated lipids. Autooxidation can be prevented by anaerobic storage, such as vacuum packing, which also prevents growth of aerobic microorganisms.

In industrialized societies, the most significant determinant of microbial populations in meat spoilage is the practice of refrigeration. Refrigeration prolongs the shelf life of meat because contaminating microbes from a thermoregulated animal are inhibited at low temperature (Chapter 6). But some bacteria are **psychrotrophs**, which grow at lower temperature as well as body temperature. Typical psychrotrophs are *Pseudomonas* and *Listeria* species. *Listeria monocytogenes* causes serious illness, such as seen in the multistate listeriosis outbreak from ice cream in 2015.

SEAFOOD Fish and other seafood contain substantial amounts of protein and lipids, as well as amines such as trimethylamine oxide. Fish spoils more rapidly than meat and poultry for several reasons. First, fish do not thermoregulate, and they inhabit relatively low-temperature environments. Because fish grow in low-temperature environments, their surface microorganisms tend to be psychrotrophic and thus grow during refrigeration. In addition, marine fish contain high levels of the osmoprotectant trimethylamine oxide (TMAO), which bacteria reduce to trimethylamine as an alternative to oxygen—a form of anaerobic respiration. Trimethylamine is a volatile amine that gives seafood its "fishy" smell. Finally, the rapid microbial breakdown of proteins and amino acids leads to foul-smelling amines and sulfur compounds.

PLANT FOODS Fruits, vegetables, and grains spoil differently from animal foods because of their high carbohydrate content and their relatively low water content. The low water content of plant foods usually translates into considerably longer shelf life than for animal-based foods. Carbohydrates favor microbial fermentation to acids or alcohols that limit further decomposition, and this microbial action can be managed to produce fermented foods, as described in Section 27.3.

Plant pathogens rarely infect humans but may destroy the plant before harvest. Most plant pathogens are fungi or fungus-like pathogens, such as the oomycete that caused the Irish potato famine. Plant pathogens continue to devastate local economies and cause shortages worldwide; for example, the witches'-broom fungus *Moniliophthora perniciosa* causes a fungal disease of cocoa trees that has drastically cut Latin American cocoa production.

After harvest, various molds and bacteria can soften and wilt plant foods by producing enzymes that degrade the pectins and celluloses that give plants their structure. In general, the more processed the food, the greater the opportunities for spoilage. For example, citrus fruits generally last for several weeks, but peeled oranges are susceptible to spoilage by Gram-negative bacteria.

Pathogens Contaminate Food

Intestinal pathogens spread readily because many pathogens can be transmitted through food without any outward sign that the food is spoiled. The CDC estimates that there are 76 million cases of gastrointestinal illness a year in this country, usually spread through water or food. Thus, one in four Americans experiences gastrointestinal illness in a given year. Table 27.2 summarizes the top pathogens that contaminate our food.

Table 27.2
Food-Borne Pathogens in the United States[a]

Pathogen	Incidence and Transmission	Course of Illness
Norovirus (Norwalk and Norwalk-like viruses)	Most common cause of diarrhea; also called "stomach flu" (no connection with influenza). An estimated 180,000 cases occur per year. Transmitted mainly by virus-contaminated food and water. Infection rates are highest under conditions of crowding in close quarters, such as inside a ship or a nursing home.	Disease lasts 1 or 2 days. Includes vomiting, diarrhea, and abdominal pain; headache and low-grade fever may occur.
Salmonella	Most common food-borne cause of death; more than 1 million cases per year; estimated 600 deaths per year. Transmission nearly always through food—raw, undercooked, or recontaminated after cooking, especially eggs, poultry, and meat; also contaminates dairy products, seafood, fruits, and vegetables.	Causes gastrointestinal disease that includes diarrhea, fever, and abdominal cramps lasting 4–7 days. Fatal cases are most common in immunocompromised patients.
Campylobacter	More than 1 million cases of campylobacteriosis per year; estimated 100 deaths per year. Grows in poultry without causing symptoms. Transmission is mainly through raw and undercooked poultry; contaminates half of poultry sold. Occurs less often in dairy products or in foods contaminated after cooking.	In humans, usually causes severe bloody diarrhea, fever, and abdominal cramps lasting 7 days. Fatal cases are most common in immunocompromised patients.
Escherichia coli O157:H7	An emerging pathogen, first recognized in a hamburger outbreak in 1982; now known to infect 73,000 people yearly, including 60 deaths per year. Grows in cattle without causing symptoms. Transmitted through ground beef; also through unpasteurized cider and from produce, where it grows as an endophyte.	In humans, usually causes severe bloody diarrhea and abdominal cramps lasting 5–10 days. About 5% of patients, especially children and elderly, develop hemolytic uremic syndrome, in which the red blood cells are destroyed and the kidneys fail.
Clostridium botulinum	Causes about 100 cases per year of botulism, with a 50% fatality rate if untreated. C. botulinum grows in improperly home-canned foods, more rarely in commercially canned low-acid foods and improperly stored leftovers such as baked potatoes. Spores occur in honey, endangering infants under 2 years of age.	Botulinum toxin from growing bacteria causes progressive paralysis, with blurred vision, drooping eyelids, slurred speech, difficulty swallowing, and muscle weakness. Infant botulism causes lethargy and impaired muscle tone, leading to paralysis.
Listeria monocytogenes	Listeria bacteria grow in animals without causing symptoms. Animal feces may contaminate water, which is then used to wash vegetables. In the United States, there are 1,600 cases of L. monocytogenes infection including 260 deaths per year. Transmission occurs mainly through vegetables washed in contaminated water and through soft cheeses. Listeria is psychrotrophic, growing at refrigeration temperatures.	Listeriosis involves fever, muscle aches, and sometimes gastrointestinal symptoms. In pregnant women, symptoms may be mild but lead to serious complications for the unborn child.
Shigella	Shigella infects about 18,000 people a year in the United States; in developing countries, Shigella infections are endemic in most communities. Transmission occurs through fecal-oral contact or from foods washed in contaminated water.	Shigellosis involves gastrointestinal symptoms such as diarrhea, fever, and stomach cramps, usually lasting 7–10 days. Complications are rare.
Staphylococcus aureus	S. aureus is best known as the cause of skin infections transmitted through open wounds. However, S. aureus can also be transmitted through high-protein foods such as ham, dairy products, and cream pastries.	S. aureus causes toxic shock syndrome. Can also cause food poisoning via preformed toxins.
Toxoplasma gondii	T. gondii is a parasite believed to infect 60,000 people annually, most with no symptoms. In a few cases, serious disease results. T. gondii is transmitted through contact with feces of infected animals, particularly cats, or through contaminated foods such as pork.	Toxoplasmosis causes mild flu-like symptoms; but in pregnant women, its transmission to the unborn child can lead to severe neurological defects, including death. Neurological complications also occur in immunocompromised patients.
Vibrio vulnificus	V. vulnificus is a free-living marine organism that contaminates seafood or open wounds. About 40 cases per year are reported.	V. vulnificus can infect the bloodstream, causing septic shock. Threatens mainly people with preexisting conditions such as liver disease.

[a]Ten major food-borne pathogens highlighted by the U.S. Public Health Service.

Figure 27.21 *Salmonella enterica* **Outbreak from Contaminated Peanut Butter**

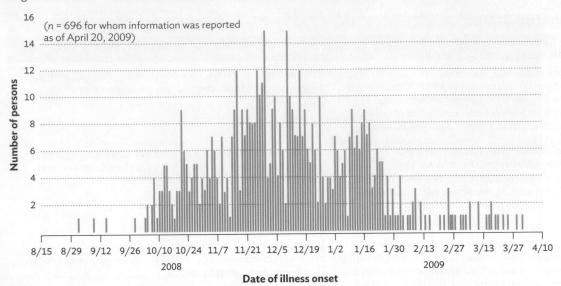

(n = 696 for whom information was reported as of April 20, 2009)

A. Infected individuals reported to the CDC from September 1, 2008, through April 20, 2009.

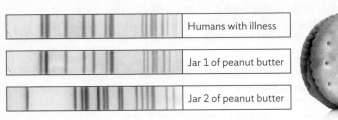

Humans with illness

Jar 1 of peanut butter

Jar 2 of peanut butter

B. Electrophoretic separation of restriction-digest DNA fragments from bacterial strains isolated from humans with illness. The peanut butter in jar 1 contained the same strain of *Salmonella enterica*; the peanut butter in jar 2 contained a different strain.

CASE HISTORY 27.2

Peanut Cracker Surprise

In the fall of 2008, several individuals, many of them children, were sickened from the common food pathogen *Salmonella enterica*. Most infected individuals developed diarrhea, fever, and abdominal cramps 12–72 hours after infection, and symptoms lasted 4–7 days. Over the next 6 months, cases were reported from nearly all U.S. states, ultimately totaling 700 people (**Figure 27.21A**). The curve of the outbreak (cases rising and then falling) followed the profile of a single-source epidemic, in which all infections are ultimately traced to one source. All cases of illness showed a common strain of the pathogen *S. enterica* serovar Typhimurium (**Figure 27.21B**).

CDC researchers investigated the outbreak by comparing the food intake histories of ill persons with that of matched controls. They found a statistical association between illness and intake of peanut butter, eventually narrowed to a specific brand of peanut butter sold to institutions. As the epidemic grew, cases emerged in which the contaminated food product was crackers filled with peanut butter cream. Ultimately, the peanut butter and cream were traced back to peanuts from a single factory in Georgia.

This case of *Salmonella* contamination is a typical example of a **point-source epidemic** in which food is the common source of a pathogen. But this case illustrates how modern food industries pose new challenges to finding the source. The peanuts from the Georgia factory ended up in many secondary products, from peanut butter to crackers filled with peanut butter cream. The cream contained peanuts from a secondary supplier to the company that had made the crackers.

At the original food plant, the source of *Salmonella* contamination could not be identified precisely, because too much time had passed since the peanuts had shipped—but the plant records showed that product samples had tested positive for *Salmonella*. Instead of discarding the product, the plant had retested the samples until they "tested negative." Many health violations were cited, including gaps in the walls and dirt buildup throughout the plant.

An important question is, How could the CDC know which *Salmonella* cases reported from around the country were part of this one outbreak? DNA analysis was used to show that all cases shared a common strain of the pathogen *S. enterica* serovar Typhimurium. (A serovar is a strain whose surface proteins elicit a distinctive immune response.) The strain was identified by analyzing its genomic DNA cleaved by **restriction endonucleases**.

Link Section 8.4 explains how **restriction endonucleases** cleave DNA at sequence-specific positions. DNA cleaved by restriction endonucleases generates fragments of defined length that are separated on an electrophoretic gel.

The *Salmonella* DNA samples from infected patients were cleaved with restriction endonucleases, and the fragments were separated by pulsed-field electrophoresis (see Figure 27.21B). The pulsed field optimizes separation of the largest fragments. The distance each fragment moves is visualized as a band in the gel. The band pattern, or "fingerprint," of *Salmonella* DNA from infected patients showed the same fragment lengths as *Salmonella* DNA from the peanut butter sample (labeled 1 in Figure 27.21B). The band pattern from peanut butter sample 1 was different from that of another contaminated peanut butter sample (labeled 2); the bacterium in sample 2 proved to be unrelated to the *Salmonella* outbreak.

This case illustrates several troubling features of food contamination in modern society. It shows the consequence of a food production plant's failure to follow regulations and the failure of health inspectors to enforce them. The contaminated product shipped out to a diverse array of institutions, such as schools, and to secondary producers, such as cookie manufacturers, who incorporated the peanut butter ingredient. The bacteria then remained viable in contaminated food products for many months, sickening people long after the contamination event had occurred.

Food contamination events, such as the peanut *Salmonella* contamination, commonly lead to point-source epidemics, in which each infection is traced directly to a common source. By contrast, a **person-to-person epidemic**, or **chain of infection**, occurs when one infected individual goes on to infect one or more others. An example of a chain of infection would be a norovirus outbreak, in which contaminated food can sicken a few people who then spread the infection to others.

Food Preservation

Cultural practices and cuisines have long evolved so as to limit food spoilage. Such practices include cooking (heat treatment), addition of spices (chemical preservation), and fermentation (partial microbial digestion). In modern commercial food production, spoilage and contamination are prevented by many methods based on fundamental principles of physics and biochemistry that limit microbial growth (discussed in Chapter 6).

PHYSICAL MEANS OF PRESERVATION Specific processes that preserve food based on temperature, pressure, or other physical factors include:

- **Dehydration and freeze-drying.** Removal of water prevents microbial growth. Water is removed either by application of heat or by freezing under vacuum (known as freeze-drying or lyophilization). Drying is especially effective for vegetables and pasta. The disadvantage of drying is that some nutrients are broken down.

- **Refrigeration and freezing.** Refrigeration temperature (typically 4°C–16°C, or 39°F–61°F) slows microbial growth, as shown in an experiment comparing bacterial growth in ground beef at different temperatures (**Figure 27.22**). Nevertheless, refrigeration also selects for psychrotrophs, such as *Listeria*. Freezing halts the growth of most microbes, but preexisting contaminant strains often survive to grow again when the food is thawed. This is why deep-frozen turkeys, for example, can still cause *Salmonella* poisoning, especially if the interior is not fully thawed before roasting.

- **Controlled or modified atmosphere.** Food can be packed under vacuum or stored under atmospheres with decreased oxygen or increased CO_2. Controlled atmospheres limit abiotic oxidation as well as microbial growth. For example, CO_2 storage is particularly effective for extending the shelf life of apples.

- **Pasteurization.** Invented by Louis Pasteur, pasteurization is a short-term heat treatment designed to decrease microbial contamination with minimal effect on food value and texture.

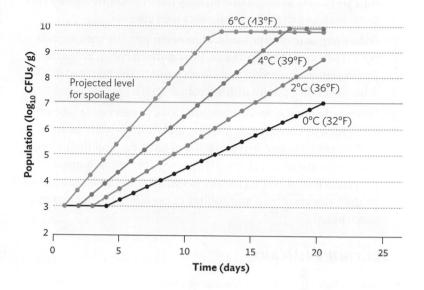

Figure 27.22 Bacterial Growth in Ground Beef
The growth rate of total aerobic bacteria in ground beef declines at lower storage temperatures. CFUs = colony-forming units.

For example, milk is commonly pasteurized at 72°C (161°F) for 15 seconds followed by quick cooling to 4°C (39°F). Pasteurization is most effective for extending the shelf life of liquid foods with consistent, well-understood microbial flora, such as milk and fruit juices.

- **Canning.** In canning, the most widespread and effective means of long-term food storage, food is cooked under pressure to attain a temperature high enough to destroy endospores (typically 121°C, or 250°F). Commercial canning effectively eliminates microbial contaminants, except in very rare cases. The main drawback of canning is that it incurs some loss of food value, particularly that of labile biochemicals such as vitamins, as well as loss of desirable food texture and taste.

- **Ionizing radiation.** Exposure to ionizing radiation, known as food irradiation, effectively sterilizes many kinds of food for long-term storage. The main concerns about food irradiation are its potential for unknown effects on food chemistry and the hazards of the irradiation process itself for personnel involved in food processing. Nevertheless, irradiation is highly effective at eliminating pathogens that would otherwise cause serious illness—for example, pathogenic *E. coli* strains in ground beef.

CHEMICAL MEANS OF PRESERVATION Many kinds of chemicals are used to preserve foods. Major classes of chemical preservatives include:

- **Acids.** While microbial fermentation can preserve foods by acidification, an alternative approach is to add acids directly. Organic acids commonly used to preserve food include benzoic acid, sorbic acid, and propionic acid. The acids are generally added as salts: sodium benzoate, potassium sorbate, sodium propionate. These acids act by crossing the cell membrane in the protonated form (at low external pH) and then releasing their protons at the higher intracellular pH. For this reason, they work best in foods that already have moderate acidity (pH 5–6), such as dried fruits and processed cheeses.

- **Other organic compounds.** Numerous organic compounds, both traditional and synthetic, have antimicrobial properties. For example, cinnamon and cloves contain the benzene derivative eugenol, a potent antimicrobial agent.

- **Inorganic compounds.** Inorganic food preservatives include salts, such as phosphates, nitrites, and sulfites. Nitrites and sulfites inhibit aerobic respiration of bacteria, and their effectiveness is enhanced at low pH. These substances, however, may have harmful effects on humans; nitrites can be converted to toxic nitrosamines, and sulfites cause allergic reactions in some people.

SECTION SUMMARY

- **Food spoilage** refers to chemical changes that render food unfit for consumption. Food spoils through degradation by enzymes within the food, through spontaneous chemical reactions, and through microbial metabolism.

- **Food contamination, or food poisoning,** refers to the presence of microbial pathogens that cause human disease or the presence of toxins produced by microbial growth. Food harvesting, processing, and shared consumption are all activities that spread pathogens.

- **Different foods spoil in different ways.** Dairy products can be soured by excessive fermentation or made bitter by bacterial proteolysis. Meat and poultry are putrefied by decarboxylation of amino acids. Fish and other seafood harbor psychrotrophic bacteria that grow under refrigeration. Vegetables spoil by excess growth of bacteria and molds.

- **Food preservation** includes physical treatments such as freezing and canning, as well as the addition of chemical preservatives such as benzoates and nitrites.

Thought Question 27.5   If you were a food safety regulator, which pathogen in Table 27.2 would you consider your top priority? Defend your answer based on factors such as numerical incidence of infection, severity of disease, and economic losses due to illness.

27.5
Wastewater Treatment

SECTION OBJECTIVES

- Describe the processes of the water cycle, and explain how human and industrial wastewater enter it.
- Explain the stages of standard wastewater treatment, including the key participation of microbial ecosystems.

What is the fate of all the food we eat—and where do our wastes go? Human society generates tremendous amounts of organic wastes, from human sewage as well as industrial effluents. These wastes are rich in organic compounds containing reduced carbon and nitrogen, such as short-chain fatty acids and urea. Introduction of these compounds into an ecosystem disrupts the ecological balance, leading to eutrophication and dead zones, as discussed in Section 27.1.

The Water Cycle

Untreated human and industrial wastes eventually enter the water cycle (Figure 27.23). In the water cycle, atmospheric water condenses in clouds and then precipitates as rain or snow. The precipitated water flows down to lakes and the oceans. In water, bacteria quickly consume the organic molecules as food, combining them with oxygen. But oxygen diffuses slowly through water. When the waste is highly concentrated, the bacteria oxidize it faster than the oxygen is replaced. Microbial respiration then competes with that of fish, invertebrates, and amphibians for the limited supply of oxygen dissolved in water, raising the BOD. High BOD can cause algal blooms leading to die-off of fish and other aquatic animals and forming dead zones, as discussed in Section 27.1. To prevent such disruption, human and industrial wastes must undergo treatment before being released into the ecosystem.

Water Treatment

Two common approaches to community wastewater treatment are **wastewater treatment plants** and **wetland filtration**. Both approaches depend on microbes to remove organic carbon and nitrogen from water before it returns to aquatic systems and ultimately the oceans. However, more specialized polluters require additional, specialized treatments. For example, hospital wastes include large amounts of pathogens and dangerous objects such as needles. Such wastes may require more drastic methods, such as incineration.

In industrialized nations, all municipalities use some form of wastewater treatment plant that includes microbial metabolism (Figure 27.24). The purpose of wastewater treatment is to decrease the BOD and the level of human pathogens before water is returned to local rivers. The wastewater treatment plant is the final destination for all household and industrial liquid wastes passing through the municipal sewage system. It may come as a surprise to learn that a modern wastewater plant can convert sewage to water that exceeds all government standards for humans to drink—and in many cities is in fact returned to the source of drinking water.

A typical plant includes the following stages of treatment:

- **Preliminary treatment** consists of screens that remove solid debris, such as sticks, dead animals, and hygiene items.

- **Primary treatment** includes fine screens and sedimentation tanks that remove insoluble particles. The particles eventually

Figure 27.23   **The Hydrologic Cycle Interacts with the Carbon Cycle**

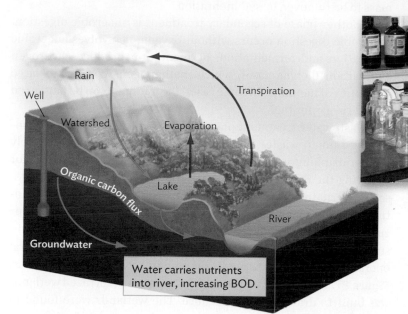

B. Bottled water samples are measured for dissolved oxygen over time; the rate of decrease of dissolved oxygen indicates biochemical oxygen demand (BOD).

A. The hydrologic cycle carries bacteria and organic carbon into groundwater and aquatic systems.

anaerobic digestion. In aerobic digestion, the microbes form particulate flocs (aggregate particles) of biofilm.

- **Tertiary (advanced) treatment** includes chemical applications such as chlorination or ozone or UV treatment to eliminate pathogens.

The microbial ecosystems of aerobic digestion require continual aeration to maximize the microbes' aerobic respiration, which breaks down pollutants to carbon dioxide and nitrates. The microbes typically include bacilli such as *Pseudomonas*, as well as filamentous species such as *Nocardia* (Figure 27.25A). Optimal treatment depends on the ratio of filamentous to single-celled bacteria: enough filaments to hold together the flocs for sedimentation, but not so many as to trap air and cause flocs to float and foam, preventing sedimentation. The flocs are sedimented as sludge, also known as **activated sludge**, referring to their microbial activity.

are recombined with the solid products of wastewater treatment (known as sludge). The sludge ultimately is used for fertilizer or landfill or is incinerated.

- **Secondary treatment** consists primarily of microbial ecosystems that decompose the soluble organic content of wastewater. Two phases of secondary treatment are aerobic and

Figure 27.24   **Wastewater Treatment Plant**

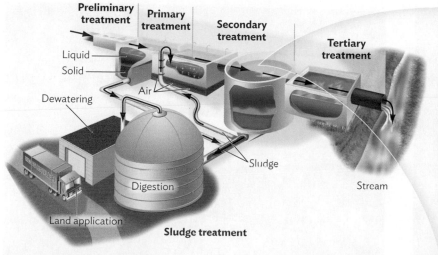

A. In a typical municipal treatment plant, wastewater undergoes primary treatment (filtering and settling), secondary treatment (microbial decomposition), and tertiary treatment (chlorination or other chemical treatments).

B. Aeration basin for secondary treatment.

Figure 27.25   Microbes in Wastewater Treatment

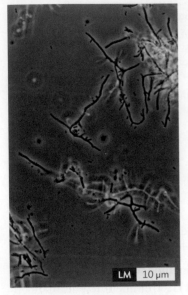

A. *Nocardia* species, filamentous bacteria from flocs formed during secondary treatment.

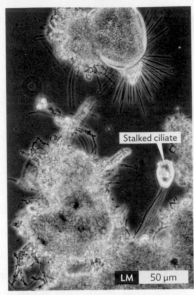

B. Flocs with a stalked ciliate, which preys on bacteria.

Besides bacteria, the ecosystem of activated sludge includes filamentous methanogens, which metabolize short-chain molecules such as acetate within the anaerobic interior of flocs. Thus, wastewater treatment generates methane, often in quantities that can be recovered as fuel. In addition, the bacteria are preyed upon by protists such as stalked ciliates (**Figure 27.25B**), swimming ciliates, and amebas, as well as invertebrates such as rotifers and nematodes. These predators serve a valuable function of limiting the free-floating population of bacteria, enabling the bulk of the biomass to be removed by sedimentation.

Another phase of secondary treatment is anaerobic digestion. Anaerobic digestion occurs without oxygen; it involves anaerobic respiration and fermentation. Fermentation produces hydrogen and carbon dioxide gases, which methanogens convert to large quantities of methane. In some cases, the methane can be harvested as natural gas for fuel.

Wastewater treatment plants are remarkably effective at converting human wastes to ecologically safe water and ultimately human drinking water. The plants are, however, impractical for purifying the runoff from large agricultural operations. For large-scale alternatives to treatment plants, communities and agricultural operations are looking to **wetland restoration**. Much of our current water supply is filtered and purified by natural wetlands, such as the Florida Everglades (**Figure 27.26A**). And agricultural operations replace treatment plants with constructed wetlands. **Figure 27.26B** shows a hog farm where a series of terraced wetlands was built to drain liquefied manure. The wetlands were found to produce fewer odors and to remove organics more efficiently and at a lower cost than do traditional filtration plants.

SECTION SUMMARY

- **In the water cycle,** water precipitates from the atmosphere into rivers, lakes, and oceans, where it enters living organisms. Organisms expel water containing wastes. The water ultimately evaporates, thus returning to the atmosphere.

- **Wastewater treatment** involves primary treatment (filtration and sedimentation), secondary treatment (microbial decomposition of organic matter), and tertiary treatment (chemical killing of microbes).

Figure 27.26   Water Filtration by Wetlands

A. The marshes of the Everglades act as natural filters for the aquifers of southern Florida.

B. A filtering system installed by Steve Kerns on his hog farm in Taylor County, Iowa. A series of hillside terraces form wetlands containing bacteria that purify hog manure and wastewater.

- **Secondary treatment includes aerobic and anerobic digestion.** Aerobic digestion requires aeration and converts organic matter to carbon dioxide. Anaerobic digestion occurs without aeration and generates methane as well as carbon dioxide.

- **Wastewater treatment cuts down BOD.** Microbial communities decompose the soluble organic content that causes algal blooms and dead zones.

- **Wetlands filter water naturally.** Wetland filtration helps purify groundwater entering aquifers.

Thought Question 27.6 What would happen if wastewater treatment lacked microbial predators? Why would the result be harmful?

27.6
Biogeochemical Cycles

SECTION OBJECTIVES

- Describe how microbes participate in the global cycling of carbon by converting carbon between its reduced and oxidized forms.

- Explain the implications of global carbon cycling for human health.

- Outline the unique roles of microbes in cycling different forms of nitrogen.

- Explain how human technology perturbs the global nitrogen cycle.

How do microbes help cycle elements of life throughout Earth's biosphere? Even microbes have limits: no microbe, nor any other living creature, can convert one element to another. Living organisms conduct chemical reactions, not nuclear reactions. Because organisms cannot create their own elements, they need to get them from their environment. So organisms acquire their elements either from nonliving components of their environment, such as by fixing atmospheric CO_2, or from other organisms, by grazing, predation, or decomposition. Furthermore, all organisms recycle their components to the biosphere. The partners in this recycling include abiotic entities such as air, water, and minerals, as well as biotic entities such as predators and decomposers. Collectively, the metabolic interactions of microbial communities with the biotic and abiotic components of their ecosystems are known as biogeochemistry or geomicrobiology.

Cycles of Elements for Life

Which elements need to be recycled and made available for life? In most organisms, six elements predominate: carbon, oxygen, nitrogen, hydrogen, phosphorus, and sulfur (discussed in Chapter 6). All these elements flow in biogeochemical cycles of nutrients throughout the biotic and abiotic components of the biosphere. The environmental levels of key elements can limit biological productivity. For example, the concentration of iron limits the marine populations of phytoplankton, which depend on iron supplied by wind currents.

Biogeochemical cycles include both biological components (such as phototrophs that consume CO_2 and heterotrophs that release CO_2) and geological components (such as volcanoes that release CO_2 and oceans that absorb CO_2). The major part of the biosphere containing significant amounts of an element needed for life is called a **reservoir** of that element. Each reservoir acts both as a source of that element for living organisms and as a sink to which the element returns. For example, the ocean is an important reservoir for carbon (CO_2 in equilibrium with HCO_3^-, bicarbonate ion).

How do we study microbial cycling on a global scale? How do we figure out whether ecosystems are net sources or sinks of CO_2? **Figure 27.27A** shows a tower used for sampling atmospheric CO_2 as

Figure 27.27

Global Atmospheric CO_2 Is Rising

A. At Austin Cary Memorial Forest, a FLUXNET tower is used for atmospheric CO_2 sampling as part of a global effort to monitor carbon flux.

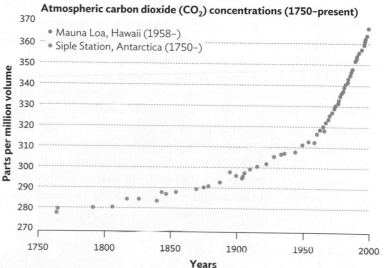

B. Atmospheric CO_2 levels since 1750 reveal the increase of this greenhouse gas accompanying the rise in industrial consumption of fossil fuels. Microbial CO_2 fixation helps limit CO_2 increase in the atmosphere.

Figure 27.28　The Carbon Cycle: Aerobic and Anaerobic

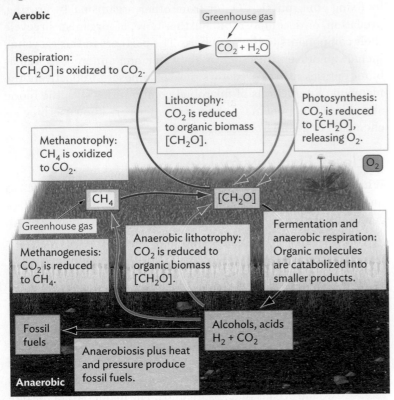

Aerobic

Greenhouse gas
$CO_2 + H_2O$

Respiration:
$[CH_2O]$ is oxidized to CO_2.

Lithotrophy:
CO_2 is reduced to organic biomass $[CH_2O]$.

Photosynthesis:
CO_2 is reduced to $[CH_2O]$, releasing O_2.

Methanotrophy:
CH_4 is oxidized to CO_2.

O_2

CH_4　$[CH_2O]$

Greenhouse gas

Methanogenesis:
CO_2 is reduced to CH_4.

Anaerobic lithotrophy:
CO_2 is reduced to organic biomass $[CH_2O]$.

Fermentation and anaerobic respiration: Organic molecules are catabolized into smaller products.

Fossil fuels

Alcohols, acids $H_2 + CO_2$

Anaerobiosis plus heat and pressure produce fossil fuels.

Anaerobic

Aerobic and anaerobic conversions of carbon. In an aerobic environment, photosynthesis generates molecular oxygen (O_2), which enables the most efficient metabolism by heterotrophs and lithotrophs. In an anaerobic environment, photosynthesis generates oxidized minerals, but no oxygen gas. Fermentation generates organic carbon products as well as CO_2 and H_2. In the absence of oxygen, methanogens convert carbon dioxide (CO_2) and molecular hydrogen (H_2) to methane (CH_4), one of the most potent greenhouse gases. Blue = reduction of carbon; red = oxidation of carbon; black = fermentation; $[CH_2O]$ = organic biomass.

part of a global effort to monitor carbon flux. The CO_2 is measured by infrared absorption spectroscopy, a method in which the pattern of infrared radiation absorbed by the gas molecules reveals which gases are present. The results of such readings at stations around the globe have led to a consensus among scientists that CO_2 levels are rising rapidly, faster than ever before on Earth (**Figure 27.27B**). The CO_2 traps solar radiation as heat—a process known as the greenhouse effect. The greenhouse effect of CO_2 and other gases is raising Earth's temperature at an unprecedented rate. This observation has enormous political and economic implications.

Will the human-induced global warming cause mass extinctions of Earth's species? Or can we use our knowledge of microbial ecology to channel microbial activities into recovering the balance—for example, by increasing microbial CO_2 fixation? Throughout most of this book, we focus on the growth of individual organisms. Now we examine how the collective metabolic activities of microbial populations contribute to global cycles of elements throughout Earth's biosphere. And we consider the ways that we humans enlist microbes to manage our environment.

The Carbon Cycle

All food webs involve influx and efflux of carbon. From a global standpoint, where is carbon found on Earth? The major source of ...on is carbonate rock in Earth's crust. But Earth's crust is the

source least accessible to the biosphere above ground. Crustal rock provides carbon only to organisms at the surface and to subsurface microbes that grow extremely slowly. Thus, subsurface carbon turnover is very slow.

The carbon reservoir that cycles most rapidly through living bodies is that of the atmosphere, a source of CO_2 for photosynthesis and lithotrophy (**Figure 27.28**). The atmosphere also acts as a **sink** (a place that accepts an element) for CO_2 produced by heterotrophy and by geological outgassing from volcanoes. But the atmospheric reservoir is much smaller than other sources, such as crustal rock and the oceans. The ocean acts as both a reservoir and a sink; it can absorb a lot of extra CO_2, and convert the gas to carbonate ions. In addition, marine algae such as diatoms trap a substantial amount of carbon in biomass. A portion of their biomass sinks to the ocean floor through the weight of their silicate or carbonate exoskeletons. Nevertheless, despite these important buffering effects of the ocean, atmospheric CO_2 continues to increase at a rate of about 1% per year.

However, the fact that CO_2 is a greenhouse gas does not make it inherently "bad" for the environment. In fact, if heterotrophic production of CO_2 were to cease altogether, phototrophs would run out of CO_2 in roughly 300 years, despite the vast quantities of carbon present in the ocean and crust. Thus, both CO_2 fixers and heterotrophs need each other.

Oxygen Interacts with the Carbon Cycle

The global cycle of carbon in the biosphere is closely linked to the cycles of oxygen and hydrogen, elements to which most carbon is bonded. Overall, carbon cycles between CO_2 and various reduced forms of carbon, including biomass (living material). But the results of carbon cycling depend on the presence of molecular oxygen (O_2). Oxygen-rich ecosystems include the photic (light-receiving) zone of oceans and the surface of terrestrial habitats, such as soil. In an aerated (or oxic) habitat, such as the open ocean, the ecosystem absorbs enough light for the rate of photosynthesis to exceed the rate of heterotrophy. Microbial and plant photosynthesis fixes CO_2 into biomass (designated by the shorthand $[CH_2O]$). Phototrophs include bacteria and protists as well as plants. Aerobic CO_2 fixation by bacteria and plants releases O_2. The O_2 is then used by heterotrophs (such as bacteria, protists, and animals) to convert $[CH_2O]$ back to CO_2. In the presence of light, a net excess of O_2 is released.

Anoxic environments support lower rates of biomass production than do oxygen-rich environments. Anaerobic conversion of CO_2 to biomass is done mainly by bacteria and archaea. Vast, permanently anaerobic habitats extend several kilometers below Earth's surface, encompassing greater volume than the rest of the biosphere put together. In these habitats, endolithic bacteria inhabit the interstices of rock crystals. In soil and water, anaerobic metabolism includes fermentation of organic carbon sources, as well as respiration and lithotrophy with alternative electron acceptors such as nitrate, ferric iron (Fe^{3+}), and sulfate. Anaerobic decomposition by microbes is one stage in the formation of fossil fuels such as oil and natural gas (primarily methane).

Anoxic environments near the surface also favor production of methane from the H_2, CO_2, and other fermentation products of anaerobes. A major source of concern for global greenhouse gases is the methane hydrates accumulating in deep marine sediments, generated by huge deep-ocean communities of methanogens. Warming the methane hydrate releases methane gas, which quickly rises to the atmosphere. Geological evidence suggests that rapid methane release accompanied the retreat of the glaciers during ice age transitions. A rapid methane release today could accelerate global warming.

The Nitrogen Cycle

Another major element that cycles largely by microbial conversion is nitrogen (Figure 27.29). Nitrogen is found in more different oxidation states than any other major biological element. Conversion between these oxidation states requires several metabolic processes that occur only in bacteria and archaea. These processes include nitrogen fixation, ammonia conversion to nitrate, and the reduction of gases such as nitrogen dioxide (NO_2), nitrogen monoxide (NO), and nitrous oxide (N_2O). Without bacteria and archaea, the global nitrogen cycle would not exist.

Where is nitrogen found on Earth? Much nitrogen is found in Earth's crust in the form of ammonium salts in rock, but this form is largely inaccessible to microbes. The major accessible source of nitrogen is the atmosphere (79% nitrogen gas, N_2). However, N_2 is a highly stable molecule that requires enormous input of reducing energy before assimilation is possible. Thus, for many natural ecosystems and most forms of agriculture, nitrogen is the limiting nutrient for primary productivity (production of biomass by photosynthesis).

Until recently in Earth's history, N_2 entered ecosystems only via nitrogen-fixing bacteria and archaea. In the twentieth century, however, the Haber process was invented for artificial nitrogen fixation to generate fertilizers for agriculture. In the Haber process, N_2 is hydrogenated to ammonia (NH_3) by methane under extreme heat and pressure. Today, the Haber process for producing fertilizers accounts for approximately 30%–50% of all nitrogen fixed

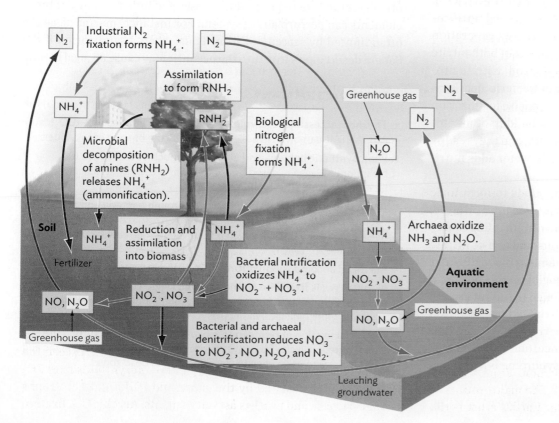

Figure 27.29
The Global Nitrogen Cycle
Prokaryotic conversions of nitrogen occur throughout the biosphere.
Blue = reduction; red = oxidation; black = redox-neutral; yellow highlight = greenhouse gas.

on Earth. Other human activities, such as fuel burning and use of nitrogenous fertilizers, contribute to oxidized nitrogen pollutants such as nitrous oxide (N_2O), a potent greenhouse gas. The nitrogen cycle is now the most perturbed of the major biogeochemical cycles.

As discussed in Chapter 7, bacterial nitrogen fixation occurs by the overall reduction of N_2 into two molecules of ammonium ion (NH_4^+). Plants and bacteria can combine ammonium ion with carbon skeletons (such as TCA cycle intermediates) to form amino acids. But given the energy expense and the need to exclude oxygen when fixing N_2, many species of bacteria, as well as all plants and animals, have lost the nitrogen fixation pathway by degenerative evolution (the loss of a trait in the absence of need). The need to fix nitrogen and make ammonia disappeared in these life-forms because other bacteria and archaea present in every ecosystem still fix nitrogen. Nitrogen-fixing bacteria in the soil include anaerobes such as *Clostridium* species and facultative Gram-negative enteric species of *Klebsiella* and *Salmonella*, as well as obligate respirers such as *Pseudomonas*. In ocean and freshwater, the major nitrogen fixers are cyanobacteria. Within an ecosystem, nitrogen fixers ultimately make the reduced nitrogen available for assimilation by nonfixing microbes and plants, either directly through symbiotic association (such as that between rhizobia and legumes) or indirectly through predation (marine cyanobacteria) and decomposition (soil bacteria). Symbiotic nitrogen fixation by rhizobial bacteria is important for agriculture, as it increases the yield of crops such as soybeans.

The fixed form of nitrogen (ammonium ion and ammonia) in soil or water is quickly oxidized for energy by **nitrifiers**, bacteria that can oxidize ammonia to nitrite (NO_2^-), or nitrite to nitrate (NO_3^-). The oxidation of ammonia to nitrite and nitrate is called **nitrification**. Nitrification is a form of lithotrophy, an energy generation pathway involving oxidation of minerals. Production of both nitrite and nitrate generates acid, which can acidify the soil. Nitrate produced in the soil is assimilated by plants and bacteria nearly as quickly as ammonium ion. But in agriculture, intensive fertilization generates a large excess of ammonia and thus a buildup of nitrites and nitrates. These ions are highly soluble in water, and they readily diffuse into rivers and lakes. Nitrogen buildup in water adds nutrients that contribute to algal blooms.

Aquatic nitrate reacts with organic compounds to form toxic nitrosamines. Nitrate influx also relieves the nitrogen limit on algae, causing algal blooms and raising BOD. Chronic nitrate influx causes eutrophication and die-off of fish. Human consumption of nitrate in drinking water can lead to **methemoglobinemia**, a blood disorder in which hemoglobin is inactivated. Methemoglobinemia occurs in infants whose stomachs are not yet acidic enough to inhibit growth of bacteria that convert nitrate to nitrite. Nitrite oxidizes the iron in hemoglobin, eliminating its capacity to carry oxygen. The failure to carry oxygen leads to a bluish appearance, one cause of "blue baby syndrome." Nitrite-induced blue-baby syndrome is a problem

Figure 27.30 Nitrate Runoff Contaminates Groundwater

Nitrate in drinking water is especially prevalent in agricultural regions of the United States. Bacterial ammonification of fertilizer, followed by nitrification, generates nitrate and nitrite. Nitrate and nitrite runoff from oxidized nitrogenous fertilizers pollutes streams and groundwater.

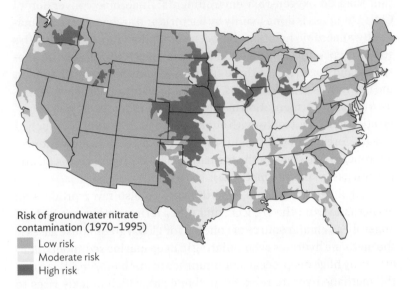

Risk of groundwater nitrate contamination (1970–1995)
- Low risk
- Moderate risk
- High risk

in intensively cultivated agricultural regions, such as Kansas and Nebraska (**Figure 27.30**).

The global cycles we have described for carbon and nitrogen have counterparts in the cycling of sulfur, phosphorus, iron, and other essential elements. Sudden losses and influxes of any of these elements can perturb an ecosystem. For instance, marine algae are often limited by scarcity of iron. Thus, when a sandstorm off the Sahara desert blows iron into the ocean, it may cause an algal bloom.

SECTION SUMMARY

- **The most accessible reservoir for carbon is the atmosphere (CO_2).** Atmospheric carbon is severely perturbed by the burning of fossil fuels.
- **Photosynthetic bacteria and algae cycle much of the CO_2 in the biosphere.** Oxygen released by phototrophs is used by aerobic heterotrophs and lithotrophs.
- **Anaerobic environments cycle carbon through bacteria and archaea.** Bacteria conduct fermentation and anaerobic respiration. Methanogens release methane, much of which is removed by methane-oxidizing bacteria and archaea.
- **Microbial decomposition returns CO_2 to the atmosphere.**
- **Nitrogen gas is fixed into ammonium ion (NH_4^+) only by bacteria and archaea.** Other organisms can incorporate ammonium ion into amino acids and other biological molecules.
- **Industrial fixation of nitrogen to make fertilizers has perturbed the global nitrogen cycle.** Excessive nitrogen fixation destabilizes aquatic and marine ecosystems, leading to algal blooms and dead zones.

Perspective

The function of our entire biosphere depends on our microbial partners to cycle key elements and acquire energy to drive the food web. Microbes colonize a vast array of habitats, from the deserts and oceans to the roots of plants and the digestive tracts of animals. Wherever found, microbes both respond to and modify the environment that surrounds them. They provide key ecosystem services, such as biomass production and waste filtration. More directly, we use bacteria and fungi to make fermented foods, from cheese to champagne. On a global scale, we depend on microbial participation in biogeochemical cycles. Yet today, for the first time in history, human technology now rivals the ability of microbes to perturb those cycles (**Figure 27.31**). In effect, our planet is a patient who needs better health management. To manage and moderate our alterations—for our own survival and that of the biosphere—our fate still depends on the microbes.

Figure 27.31 Planet Earth Isn't Well

LEARNING OUTCOMES AND ASSESSMENT

27.1 Microbes in Ecosystems

SECTION OBJECTIVES

- Describe the various roles of microbes in natural ecosystems and the kinds of ecosystem services that microbes provide.
- Explain how physical factors determine which microbes grow in a given habitat.
- Outline how microbes participate in the food chains of ocean, soil, and wetland.
- Explain the phenomenon of biochemical oxygen demand and the cause of dead zones.

OBJECTIVES REVIEW

1. A habitat in which the major phototrophs consist of the smallest unicellular bacteria is _____.
   A. forest
   B. ocean
   C. desert
   D. wetland

27.2 Microbial Symbiosis with Animals

SECTION OBJECTIVES

- Explain the different kinds of microbe-animal symbioses, including mutualism, synergism, commensalism, amensalism, and parasitism.
- Describe the digestive communities of animals.
- Explain how microbial communities contribute to human health.

OBJECTIVES REVIEW

4. A relationship in which one partner benefits while the other partner receives neither benefit nor harm is called _____.
   A. mutualism
   B. synergism
   C. commensalism
   D. parasitism

27.3 Fermented Foods and Beverages

SECTION OBJECTIVES

- Describe the roles of bacteria and fungi in the production of cheese and fermented soy foods.
- Outline the process of ethanolic fermentation in the production of beer and wine.
- Explain why fermented foods and beverages retain high food value.

OBJECTIVES REVIEW

7. A soft cheese in which milk proteins are coagulated solely by bacterial lactic acid fermentation is _____.
   A. cottage cheese
   B. Gouda cheese
   C. Swiss cheese
   D. Brie cheese

27.4 Food Spoilage and Preservation

SECTION OBJECTIVES

- Explain the difference between food spoilage and food contamination.
- Outline the different processes of spoilage of dairy products, meat, seafood, and plant foods. Explain why different types of food spoil differently.
- Describe the various physical and chemical approaches to food preservation.

OBJECTIVES REVIEW

10. Which process is especially common in seafood spoilage?
    A. Carbohydrates are fermented to acids or alcohols by indigenous microbes.
    B. Lipids spoil by oxidation of unsaturated fatty acids, causing rancidity.
    C. Sugars rapidly ferment to lactic acid, turning the food sour.
    D. Anaerobic respiration by bacteria converts trimethylamine oxide to trimethylamine, causing a distinctive unpleasant odor.

27.5 Wastewater Treatment

SECTION OBJECTIVES

- Describe the processes of the water cycle, and explain how human and industrial wastewater enter it.
- Explain the stages of standard wastewater treatment, including the key participation of microbial ecosystems.

OBJECTIVES REVIEW

13. Which process of the water cycle leads to increased BOD in a freshwater lake?
    A. Water condenses in clouds.
    B. Water precipitates as rain or snow.
    C. Agricultural runoff enters streams.
    D. From the lake, water evaporates in the atmosphere.

27.6 Biogeochemical Cycles

SECTION OBJECTIVES

- Describe how microbes participate in the global cycling of carbon by converting carbon between its reduced and oxidized forms.
- Explain the implications of global carbon cycling for human health.
- Outline the unique roles of microbes in cycling different forms of nitrogen.
- Explain how human technology perturbs the global nitrogen cycle.

OBJECTIVES REVIEW

16. Which part of the global carbon cycle is performed <u>only</u> by microbes?
    A. Respiration of organic compounds, releasing CO_2
    B. CO_2 fixation into biomass
    C. Fermentation, releasing organic acids
    D. Methanogenesis, reducing CO_2 to methane

2. Which of the following is <u>not</u> a function of wetland microbes?
   A. Decomposition and recycling of dead animal and plant bodies
   B. Filtering nitrogenous organic wastes from human wastewater
   C. Providing mineral nutrients for vascular plants
   D. Serving as the base of an open-water food chain for the largest marine mammals

3. Biochemical oxygen demand is increased by ____.
   A. excessive bacterial nitrogen fixation
   B. excessive organic substrates for microbial aerobic respiration
   C. sulfur pollution involving oxidation
   D. aquatic nitrates and nitrites

5. Which statement is <u>not</u> true of the bovine rumen?
   A. The rumen supports many species of bacteria, archaea, and protozoa.
   B. The rumen is well supplied with oxygen from the breathing animal.
   C. Rumen bacteria catabolize plant polysaccharides that the bovine enzymes cannot otherwise digest.
   D. Methanogens convert CO_2 and H_2 (products of bacterial fermentation) to methane.

6. Which bacteria do <u>not</u> contribute positively to human health?
   A. *Shigella flexneri*
   B. *Bacteroides* spp.
   C. *Staphylococcus epidermidis*
   D. *Bifidobacterium* spp.

8. The following is <u>not</u> a step of beer production.
   A. Germination of barley grains
   B. Mashing the grains in water
   C. Removal of toxic ethanol
   D. Filtration to remove the yeast

9. Cheese provides exceptional food value because ____.
   A. it is fermented by nitrogen-fixing bacteria
   B. bacterial fermentation breaks down all the organic molecules
   C. bacterial fermentation breaks down mainly sugars, leaving most protein intact
   D. fermentation adds valuable minerals to the food product

11. Which of these food preservation methods requires subsequent cold storage?
    A. Ionizing radiation
    B. Canning
    C. Pasteurization
    D. Packaging under CO_2 atmosphere

12. Which microbial pathogen can grow in food under refrigeration?
    A. *Salmonella enterica*
    B. Norovirus
    C. *Campylobacter jejuni*
    D. *Listeria monocytogenes*

14. Which stage of water treatment does <u>not</u> involve microbes?
    A. Primary treatment by screens and sedimentation
    B. Secondary treatment by aerobic respiration
    C. Secondary treatment by <u>anaerobic</u> respiration
    D. Floc sedimentation as sludge

15. Which microbial process decreases the BOD of wastewater?
    A. Nitrogen fixation
    B. Aerobic or anaerobic respiration
    C. Lithotrophy
    D. Photosynthesis

17. Which part of the global nitrogen cycle is performed by many organisms <u>besides</u> microbes?
    A. Nitrogen gas fixation into ammonium ion
    B. Ammonium ion incorporation into amino acids
    C. Oxidation of ammonia to form nitrite and nitrate (nitrification)
    D. Reduction of nitrate to nitrous oxide by anaerobic respiration (denitrification)

18. Which form of nitrogen is a potent greenhouse gas?
    A. Nitrogen gas (N_2)
    B. Nitrous oxide (N_2O)
    C. Nitrate (NO_3^-)
    D. Nitrite (NO_2^-)

Key Terms

activated sludge (p. 931)
amensalism (p. 917)
assimilation (p. 910)
biochemical oxygen demand (BOD) (p. 911)
biomass (p. 910)
bioremediation (p. 908)
casein (p. 923)
chain of infection (p. 929)
commensalism (p. 917)
consumer (p. 910)
curd (p. 922)
dead zone (p. 912)
decomposer (p. 910)

dissimilation (p. 910)
ecosystem (p. 908)
ecosystem service (p. 908)
eutrophic (p. 911)
fermented food (p. 919)
food contamination (p. 926)
food poisoning (p. 926)
food preservation (p. 925)
food spoilage (p. 925)
food web (p. 910)
grazer (p. 910)
methemoglobinemia (p. 936)
mutualism (p. 917)
niche (p. 908)

nitrification (p. 936)
nitrifier (p. 936)
parasitism (p. 917)
person-to-person epidemic (p. 929)
point-source epidemic (p. 928)
population (p. 908)
predator (p. 910)
preliminary treatment (p. 930)
primary producer (p. 910)
primary treatment (p. 930)
psychrotroph (p. 926)
reservoir (p. 933)
rhizosphere (p. 914)
secondary treatment (p. 931)

sink (p. 934)
symbiosis (p. 917)
synergism (p. 917)
tertiary (advanced) treatment (p. 931)
thermocline (p. 912)
trophic level (p. 910)
wastewater treatment plant (p. 930)
wetland (p. 914)
wetland filtration (p. 930)
wetland restoration (p. 932)
whey (p. 922)
yogurt (p. 923)
zoonosis (p. 918)

Review Questions

1. What unique functions do microbes perform in ecosystems?

2. What kinds of microbial metabolism are favored in aerated environments? In anoxic environments?

3. Can a microbe be both a producer and a consumer? Explain.

4. Explain the microbial relationships in various forms of symbiosis, including mutualism, commensalism, and parasitism. Outline an example of each, detailing the contributions of each partner.

5. Compare and contrast the marine food web with the soil food web. What kinds of organisms are the producers and consumers? How many trophic levels are typically found?

6. What are the advantages of fermentation for a food product? How does acidic fermentation contribute to the formation of different cheeses? What is the role of different kinds of metabolism performed by different microbial species?

7. Compare and contrast acidic and ethanolic fermentation processes. What different kinds of foods are produced by each type of fermentation?

8. Explain the difference between food spoilage and food poisoning.

9. What are the most important food-borne pathogens based on infection rates? Based on mortality rates?

10. What are the major means of preserving food? Compare and contrast their strengths and limitations.

11. Explain the role of biochemical oxygen demand (BOD) in water quality and how it may be perturbed by human pollution.

12. Outline the function of a wastewater treatment plant. Include the phases of primary, secondary, and tertiary treatment. Explain the roles of microbes in these phases of water treatment.

13. Explain how the carbon cycle differs in oxygenated and anoxic environments.

14. Describe the various roles of microbes in the nitrogen cycle. Explain which roles are unique to microbes.

Clinical Correlation Questions

1. A community is cited by the EPA for discharging excess nitrogen into the local river. The community decides to construct a wetland system to filter their wastewater. Over a period of years, the wetland grows a dense array of plants and attracts wildlife. The water that flows out of the wetland into the stream shows low organic carbon and nitrogen content, well below the acceptable EPA limits. Explain why this system enhances the health of the human community.

2. Shortly after takeoff, six passengers in a plane fall ill with diarrhea and vomiting. These passengers are all members of a tour group that has shared dining experiences. The passengers vomit into air sickness bags and have to travel through the aisle to the restroom with their air sickness bags and diarrhea. The source of their illness is confirmed to be norovirus. Subsequently, over the next 12 days after the flight, seven passengers who were not members of the tour group report illness that points to norovirus. Does this case represent a point-source epidemic, a chain of infection, or both? Explain.

Thought Questions: CHECK YOUR ANSWERS

Thought Question 27.1   Explain what mutualism and parasitism have in common and how they differ. Describe an example of a relationship that combines aspects of both.

ANSWER: In mutualism, both partner species benefit from an intimate relationship. In parasitism, one species grows at the expense of a host. A relationship that combines both would be that of a lichen. Lichens consist of a mutualism between algae and fungi, in which the algae provide organic carbon for the fungus, whose structure protects the algae. But depending on environmental conditions, the fungus may actually consume the algae faster than they grow. Another example is the role of *Bacteroides* bacteria in the human digestive tract. Some species of *Bacteroides* enhance human digestion, but when they escape the colon and invade surrounding tissues, they may cause abscesses.

Thought Question 27.2   Why do the microbes of digestive communities leave food value for the animal host? How can the animal obtain nourishment from waste products that the microbes could not use?

ANSWER: The digestive community is usually anaerobic. Anaerobic catabolism lacks sufficient electron acceptors (such as O_2 and nitrate) to accept all the electrons from organic food. Therefore, organic molecules such as acetate and sugars may remain to be absorbed by the intestinal lining and enter the bloodstream. The blood contains oxygen, which enables complete respiration to carbon dioxide.

Thought Question 27.3   In food production, why does oxygen allow excessive breakdown of a food substrate, compared with anaerobic fermentation?

ANSWER: Oxygen oxidizes food molecules such as sugars and amino acids. Oxidation decreases the energy yield available for the consumer of the food and generates toxic products such as nitrosamines that cause cancer. The oxidation of amino acids decreases the usable protein content of food. By contrast, bacterial anaerobic fermentation breaks down sugars but leaves most of the protein intact.

Thought Question 27.4   Compare and contrast the role of fermenting organisms in the production of cheese and alcoholic beverages.

ANSWER: In cheese production, the fermenting bacteria generate lactic acid, which lowers the pH of the food product. The acidity limits microbial growth and preserves the cheese. In the production of alcoholic beverages, the fermentation products of yeast are ethanol and carbon dioxide. The ethanol is retained as a major part of the product. In some beverages, the carbon dioxide gas remains dissolved under pressure but forms bubbles once the bottle is opened.

Thought Question 27.5   If you were a food safety regulator, which pathogen in Table 27.2 would you consider your top priority? Defend your answer based on factors such as numerical incidence of infection, severity of disease, and economic losses due to illness.

ANSWER: This question has no one correct answer. A top priority might be *Campylobacter* because it infects a million people annually, causing substantial morbidity and economic loss. On the other hand, *Clostridium botulinum* contamination is rare, but botulism has a particularly high death rate. Thus, the regulated standards for canning foods are very high. Still another case to consider is norovirus. While norovirus infections are rarely fatal, the person-to-person transmission rate is high, leading to large outbreaks associated with institutions such as cruise lines or nursing homes. Thus, the social and economic consequences of norovirus are severe.

Thought Question 27.6   What would happen if wastewater treatment lacked microbial predators? Why would the result be harmful?

ANSWER: Within the ecosystem of an aerobic digester in secondary treatment, bacteria are consumed by microbial predators such as ciliated protists. The protists convert most of the bacterial biomass to CO_2, which dissipates in the atmosphere. Without microbial predators, the bacterial cellular biomass could build up to unacceptable levels, trapping carbon and clogging the treatment system.

Figure A1.1   The Periodic Table of the Elements

Main-group elements

Atomic number (number of protons)

Chemical symbol

Atomic mass (average of all isotopes)

Transitional elements

Main-group elements

Metals
Nonmetals
Metalloids
Noble gases

1 1A	2 2A	3 3B	4 4B	5 5B	6 6B	7 7B	8	9 8B	10	11 1B	12 2B	13 3A	14 4A	15 5A	16 6A	17 7A	18 8A
1 **H** 1.00794																	2 **He** 4.00260
3 **Li** 6.941	4 **Be** 9.01218											5 **B** 10.811	6 **C** 12.011	7 **N** 14.0067	8 **O** 15.9994	9 **F** 18.9984	10 **Ne** 20.1797
11 **Na** 22.9898	12 **Mg** 24.3050											13 **Al** 26.9815	14 **Si** 28.0855	15 **P** 30.9738	16 **S** 32.066	17 **Cl** 35.4527	18 **Ar** 39.948
19 **K** 39.0983	20 **Ca** 40.078	21 **Sc** 44.9559	22 **Ti** 47.88	23 **V** 50.9415	24 **Cr** 51.9961	25 **Mn** 54.9381	26 **Fe** 55.847	27 **Co** 58.9332	28 **Ni** 58.693	29 **Cu** 63.546	30 **Zn** 65.39	31 **Ga** 69.723	32 **Ge** 72.61	33 **As** 74.9216	34 **Se** 78.96	35 **Br** 79.904	36 **Kr** 83.80
37 **Rb** 85.4678	38 **Sr** 87.62	39 **Y** 88.9059	40 **Zr** 91.224	41 **Nb** 92.9064	42 **Mo** 95.94	43 **Tc** (98)	44 **Ru** 101.07	45 **Rh** 102.906	46 **Pd** 106.42	47 **Ag** 107.868	48 **Cd** 112.411	49 **In** 114.818	50 **Sn** 118.710	51 **Sb** 121.76	52 **Te** 127.60	53 **I** 126.904	54 **Xe** 131.29
55 **Cs** 132.905	56 **Ba** 137.327	57 **La** 138.906	72 **Hf** 178.49	73 **Ta** 180.948	74 **W** 183.84	75 **Re** 186.207	76 **Os** 190.23	77 **Ir** 192.22	78 **Pt** 195.08	79 **Au** 196.967	80 **Hg** 200.59	81 **Tl** 204.383	82 **Pb** 207.2	83 **Bi** 208.980	84 **Po** (209)	85 **At** (210)	86 **Rn** (222)
87 **Fr** (223)	88 **Ra** 226.025	89 **Ac** 227.028	104 **Rf** (261)	105 **Db** (262)	106 **Sg** (263)	107 **Bh** (262)	108 **Hs** (265)	109 **Mt** (266)	110 **Ds** (281)	111 **Rg** (280)	112 **Cn** (285)	113 **Uut** (284)	114 **Fl** (289)	115 **Uup** (288)	116 **Lv** (293)	117 **Uus** (294)	118 **Uuo** (294)

Lanthanide series

58 **Ce** 140.115	59 **Pr** 140.908	60 **Nd** 144.24	61 **Pm** (145)	62 **Sm** 150.36	63 **Eu** 151.965	64 **Gd** 157.25	65 **Tb** 158.925	66 **Dy** 162.50	67 **Ho** 164.930	68 **Er** 167.26	69 **Tm** 168.934	70 **Yb** 173.04	71 **Lu** 174.967

Actinide series

90 **Th** 232.038	91 **Pa** 231.036	92 **U** 238.029	93 **Np** 237.048	94 **Pu** (244)	95 **Am** (243)	96 **Cm** (247)	97 **Bk** (247)	98 **Cf** (251)	99 **Es** (252)	100 **Fm** (257)	101 **Md** (258)	102 **No** (259)	103 **Lr** (260)

Symbol	Name	Symbol	Name	Symbol	Name	Symbol	Name	Symbol	Name
Ac	Actinium	Cu	Copper	Fe	Iron	P	Phosphorus	S	Sulfur
Al	Aluminum	Cm	Curium	Kr	Krypton	Pt	Platinum	Ta	Tantalum
Am	Americium	Ds	Darmstadtium	La	Lanthanum	Pu	Plutonium	Tc	Technetium
Sb	Antimony	Db	Dubnium	Lr	Lawrencium	Po	Polonium	Te	Tellurium
Ar	Argon	Dy	Dysprosium	Pb	Lead	K	Potassium	Tb	Terbium
As	Arsenic	Es	Einsteinium	Li	Lithium	Pr	Praseodymium	Tl	Thallium
At	Astatine	Er	Erbium	Lv	Livermorium	Pm	Promethium	Th	Thorium
Ba	Barium	Eu	Europium	Lu	Lutetium	Pa	Protactinium	Tm	Thulium
Bk	Berkelium	Fm	Fermium	Mg	Magnesium	Ra	Radium	Sn	Tin
Be	Beryllium	Fl	Flerovium	Mn	Manganese	Rn	Radon	Ti	Titanium
Bi	Bismuth	F	Fluorine	Mt	Meitnerium	Re	Rhenium	W	Tungsten
Bh	Bohrium	Fr	Francium	Md	Mendelevium	Rh	Rhodium	Uuo	Ununoctium
B	Boron	Gd	Gadolinium	Hg	Mercury	Rg	Roentgernium	Uup	Ununpentium
Br	Bromine	Ga	Gallium	Mo	Molybdenum	Rb	Rubidium	Uus	Ununseptium
Cd	Cadmium	Ge	Germanium	Nd	Neodymium	Ru	Ruthenium	Uut	Ununtrium
Ca	Calcium	Au	Gold	Ne	Neon	Rf	Rutherfordium	U	Uranium
Cf	Californium	Hf	Hafnium	Np	Neptunium	Sm	Samarium	V	Vanadium
C	Carbon	Hs	Hassium	Ni	Nickel	Sc	Scandium	Xe	Xenon
Ce	Cerium	He	Helium	Nb	Niobium	Sg	Seaborgium	Yb	Ytterbium
Cs	Cesium	Ho	Holmium	N	Nitrogen	Se	Selenium	Y	Yttrium
Cl	Chlorine	H	Hydrogen	No	Nobelium	Si	Silicon	Zn	Zinc
Cr	Chromium	In	Indium	Os	Osmium	Ag	Silver	Zr	Zirconium
Co	Cobalt	I	Iodine	O	Oxygen	Na	Sodium		
Cn	Copernicium	Ir	Iridium	Pd	Palladium	Sr	Strontium		

Figure A2.1   Embden-Meyerhof-Parnas Pathway (Glycolysis)

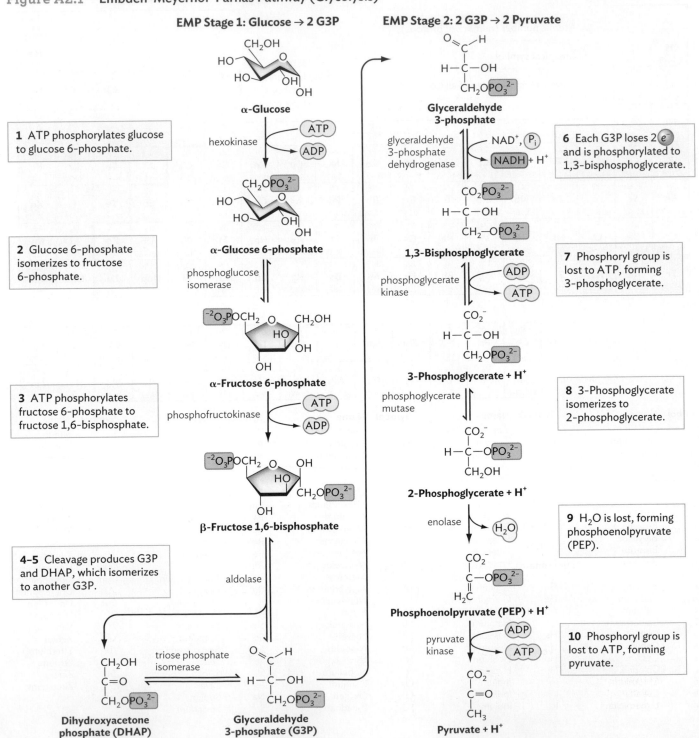

EMP Stage 1: Glucose → 2 G3P

α-Glucose

1 ATP phosphorylates glucose to glucose 6-phosphate.

hexokinase

α-Glucose 6-phosphate

2 Glucose 6-phosphate isomerizes to fructose 6-phosphate.

phosphoglucose isomerase

α-Fructose 6-phosphate

3 ATP phosphorylates fructose 6-phosphate to fructose 1,6-bisphosphate.

phosphofructokinase

β-Fructose 1,6-bisphosphate

4–5 Cleavage produces G3P and DHAP, which isomerizes to another G3P.

aldolase

triose phosphate isomerase

Dihydroxyacetone phosphate (DHAP)

Glyceraldehyde 3-phosphate (G3P)

EMP Stage 2: 2 G3P → 2 Pyruvate

Glyceraldehyde 3-phosphate

glyceraldehyde 3-phosphate dehydrogenase

6 Each G3P loses 2 e^- and is phosphorylated to 1,3-bisphosphoglycerate.

1,3-Bisphosphoglycerate

phosphoglycerate kinase

7 Phosphoryl group is lost to ATP, forming 3-phosphoglycerate.

3-Phosphoglycerate + H⁺

phosphoglycerate mutase

8 3-Phosphoglycerate isomerizes to 2-phosphoglycerate.

2-Phosphoglycerate + H⁺

enolase

9 H_2O is lost, forming phosphoenolpyruvate (PEP).

Phosphoenolpyruvate (PEP) + H⁺

pyruvate kinase

10 Phosphoryl group is lost to ATP, forming pyruvate.

Pyruvate + H⁺

Figure A2.2   Entner-Doudoroff Pathway of Glucose Catabolism

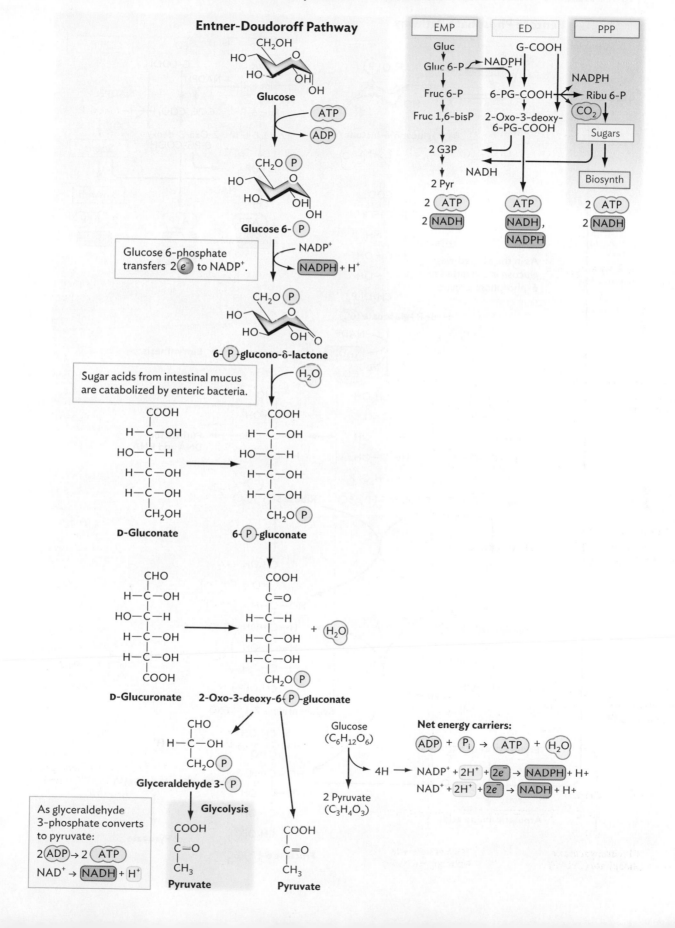

Figure A2.3   Pentose Phosphate Pathway of Glucose Catabolism

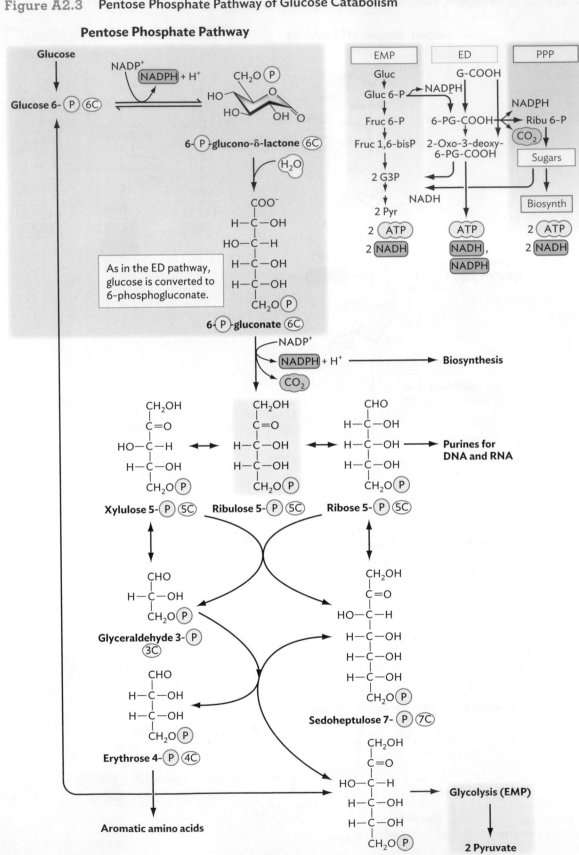

Figure A2.4   The Tricarboxylic Acid (TCA) Cycle

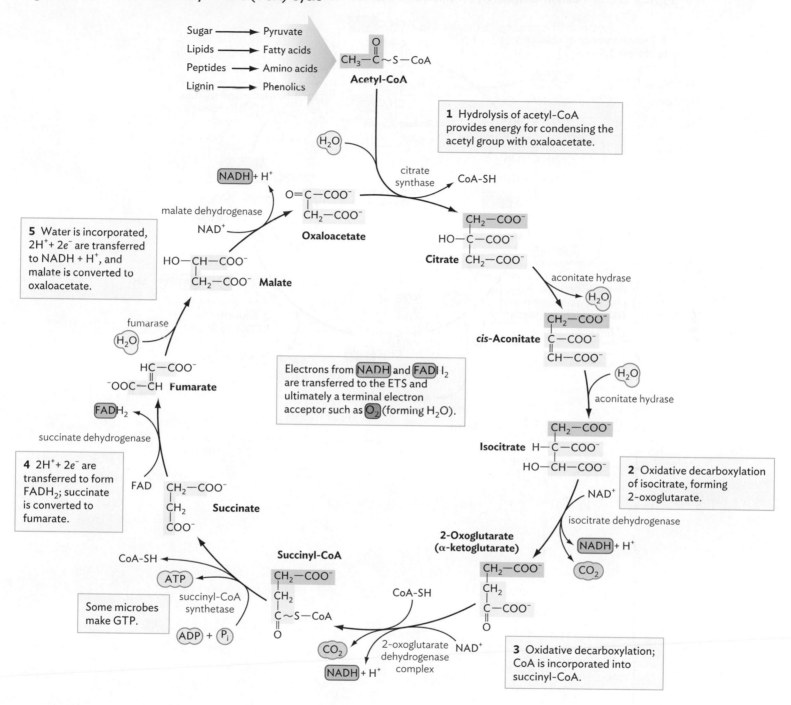

Figure A2.5   **Calvin-Benson Cycle of Carbon Dioxide Fixation**

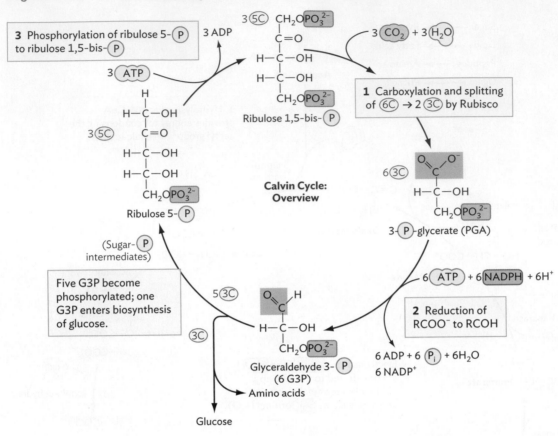

Figure A2.6   The Mitochondrial Electron Transport System (ETS)

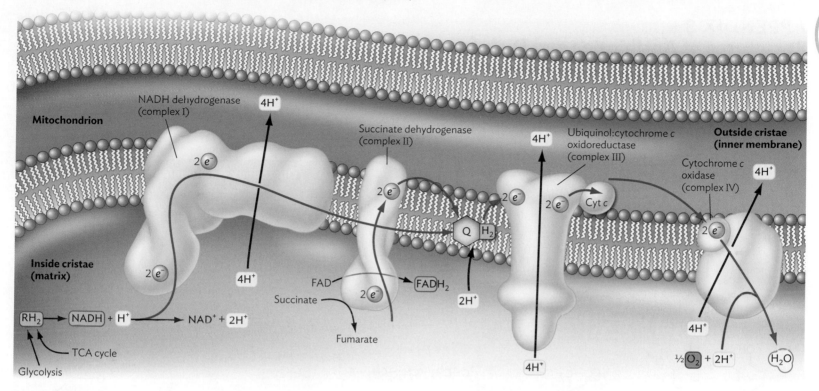

APPENDIX 3
Structures of Antibiotics and Disinfectants

Figure A3.1 **Structures of Some Common Disinfectants, Antiseptics, and Surfactants**
Refer to Section 13.3.

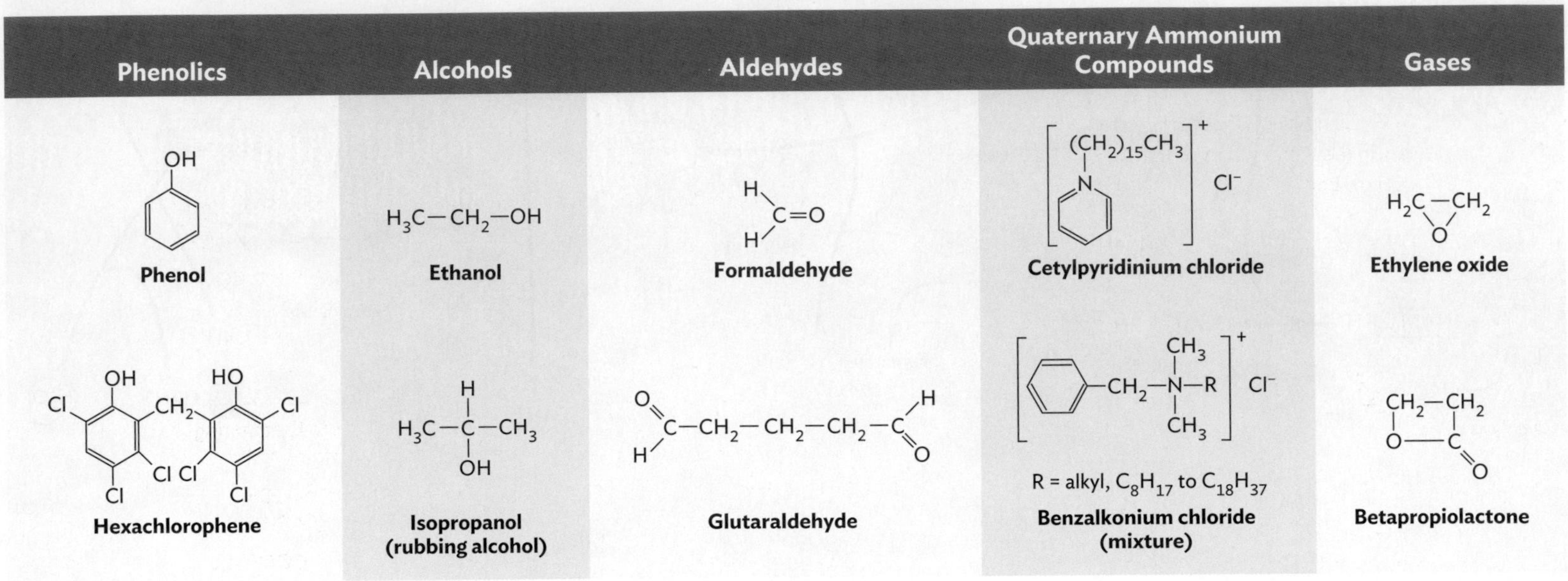
Phenolics
Alcohols
Aldehydes
Quaternary Ammonium Compounds
Gases
Phenol
Ethanol
Formaldehyde
Cetylpyridinium chloride
Ethylene oxide
Hexachlorophene
Isopropanol (rubbing alcohol)
Glutaraldehyde
R = alkyl, C8H17 to C18H37
Benzalkonium chloride (mixture)
Betapropiolactone

Figure A3.2 The Structures of Penicillins

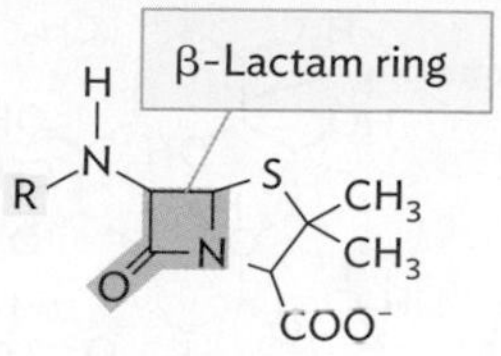

Basic structure

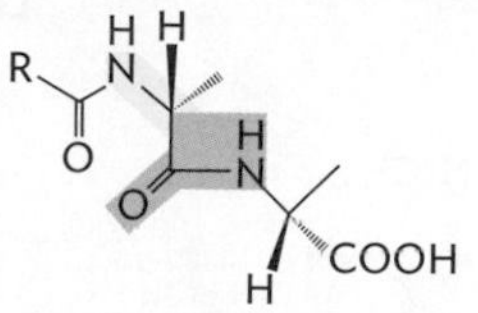

D-Alanine-D-alanine

6-Aminopenicillanic acid

A. The beta-lactam ring of all penicillin drugs resembles a part of the peptidoglycan structure (D-alanine-D-alanine of the side chain).

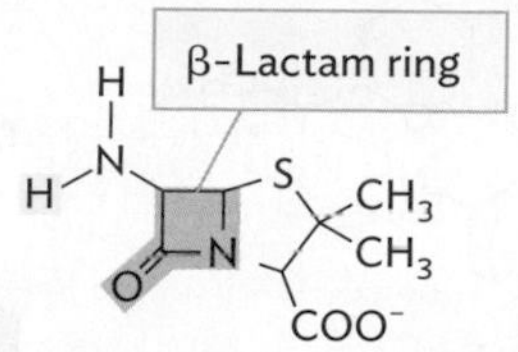

B. 6-Aminopenicillanic acid is the core penicillin molecule from which derivatives are made.

C–G. Derivatives of penicillin that vary in the R group as highlighted by a yellow box.

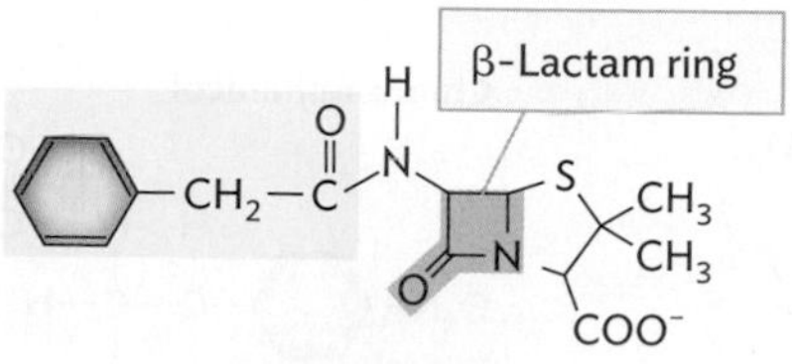

Penicillin G

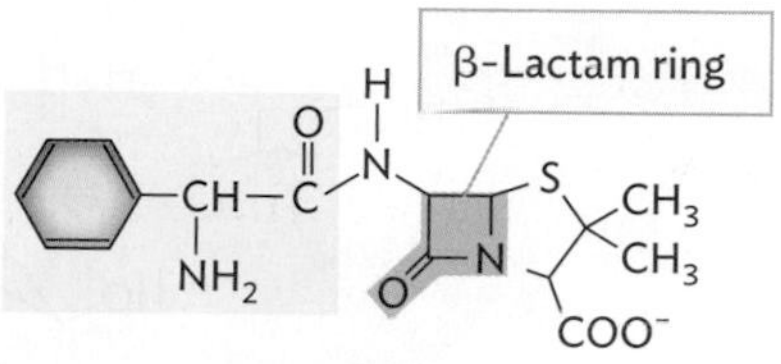

Ampicillin

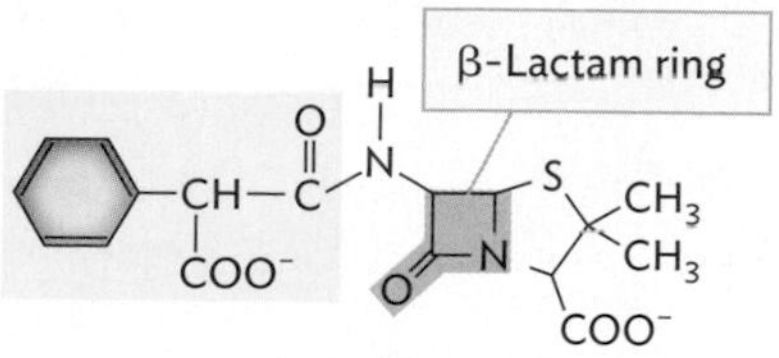

Carbenicillin

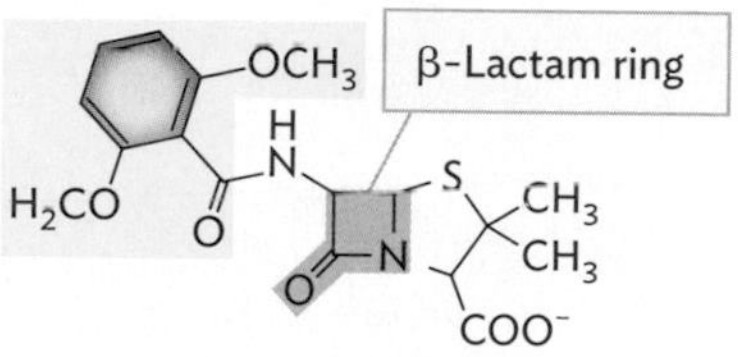

Methicillin

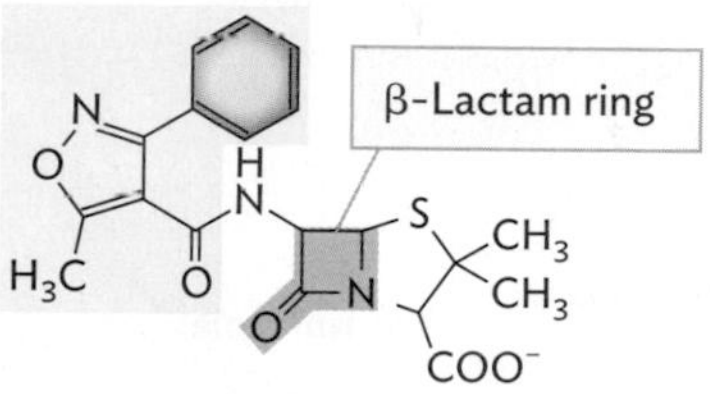

Oxacillin

Figure A3.3 Structure of Cycloserine and Vancomycin

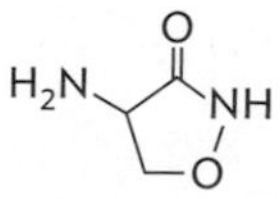

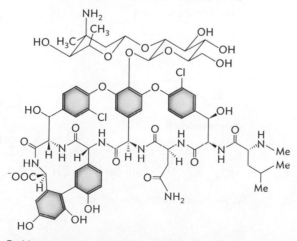

A. Cycloserine, an analog of D-alanine, is one of several drugs used to treat tuberculosis.

B. Vancomycin is a cyclic glycopeptide made by *Amycolatopsis orientalis*. This antibiotic is synthesized by an exceedingly complex biochemical pathway. Me = methyl.

Figure A3.4 Rifamycin

Rifampin (a member of the rifamycin group) binds to the RNA polymerase beta subunit to halt transcription.

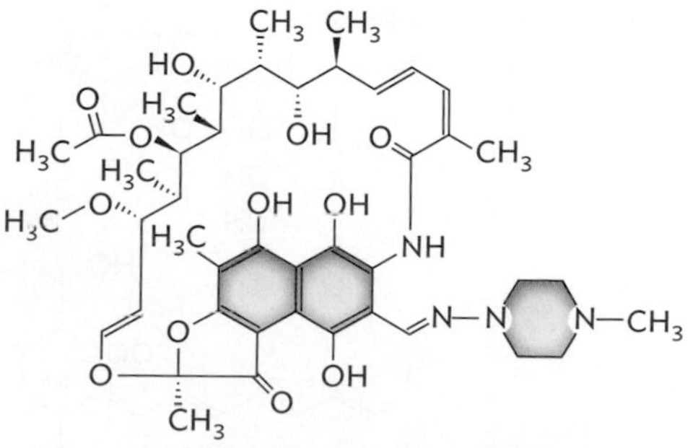

Figure A3.5
Protein Synthesis Inhibitors

Gentamicin

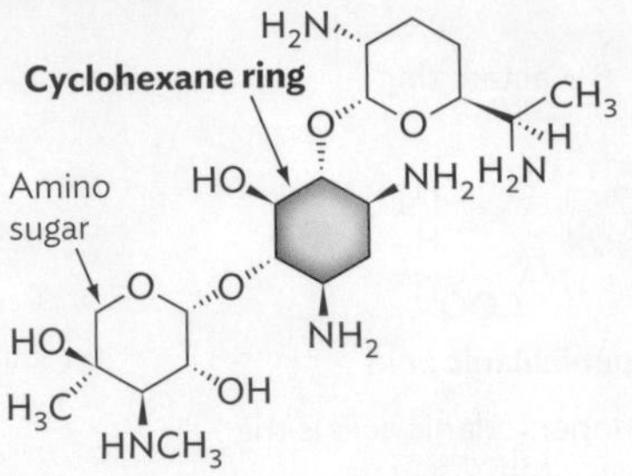

A. The aminoglycoside gentamicin.

Doxycycline

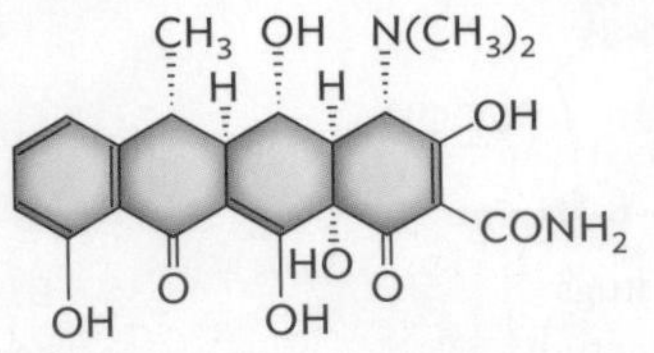

B. The tetracycline doxycycline.

Erythromycin D

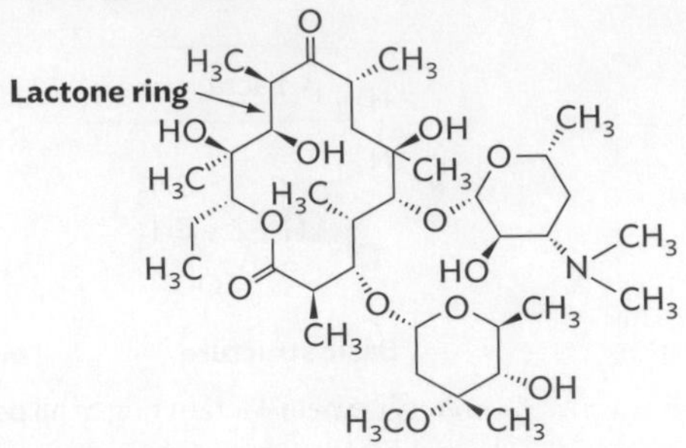

C. The macrolide antibiotic erythromycin D.

Clindamycin

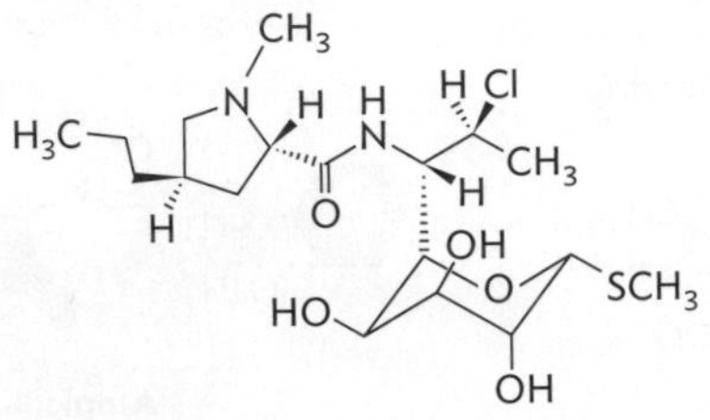

D. The lincosamide antibiotic clindamycin.

Chloramphenicol

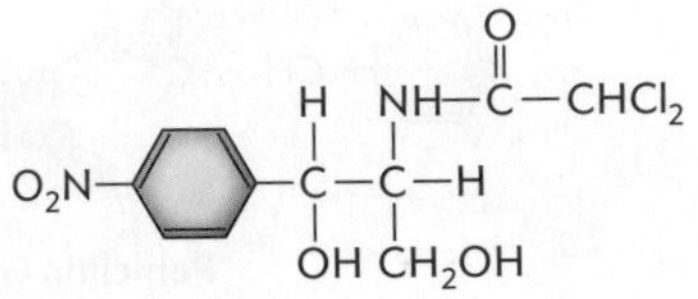

E. Chloramphenicol.

Linezolid

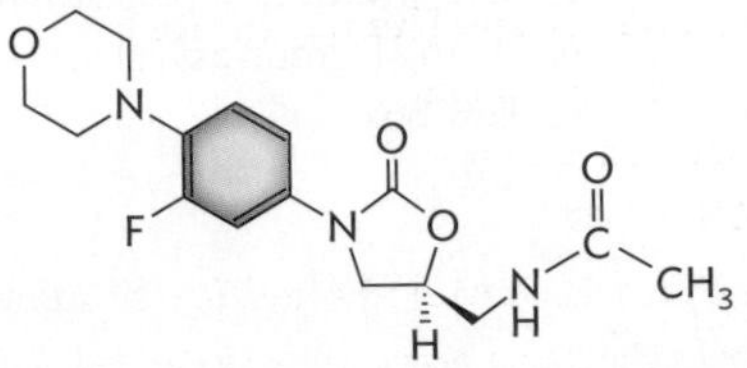

F. The oxazolidinone linezolid.

Figure A3.6
Examples of Antifungal Agents

Clotrimazole

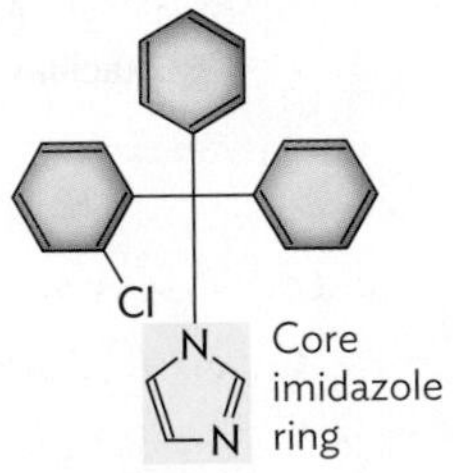

Core imidazole ring

A. Clotrimazole belongs to the group of imidazole antifungals, so named because they all contain an imidazole ring.

Griseofulvin

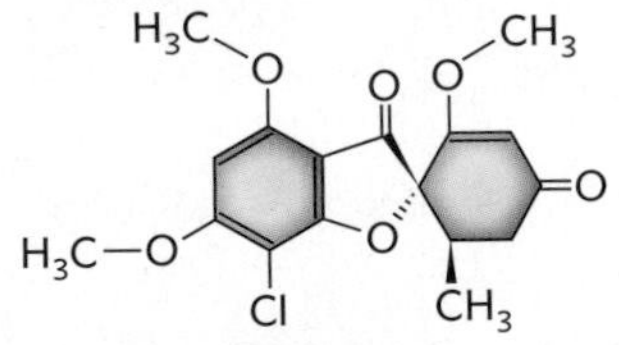

B. Griseofulvin is produced by *Penicillium griseofulvum*.

Nystatin

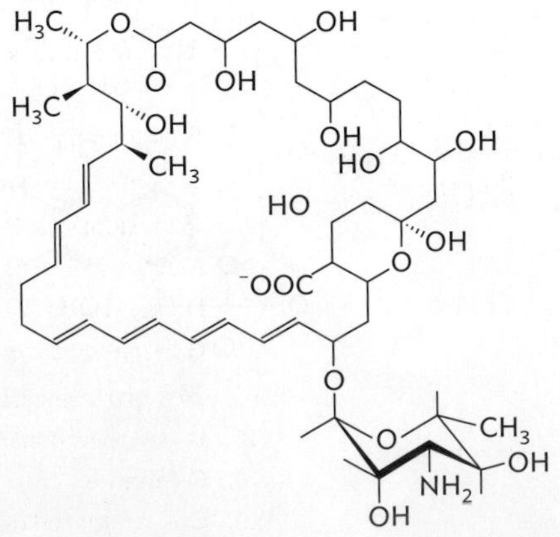

C. Nystatin is a polyene macrolide produced by *Streptomyces noursei*.

Amphotericin B

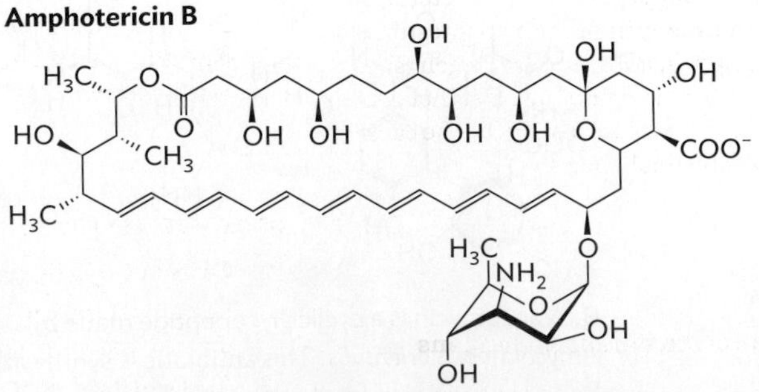

D. Amphotericin B is a polyene produced by *S. nodosus*.

[Answers to Objectives Review and Clinical Correlation Questions]

Chapter 1

OBJECTIVES REVIEW

1.1 From Germ to Genome: What Is a Microbe?

1. A. A nucleus with a nuclear membrane
2. D. These microbes have never been shown to cause disease.
3. A. A genome that undergoes replication

1.2 Microbes Shape Human History

4. C. Catherine of Siena
5. D. Louis Pasteur
6. C. AIDS

1.3 Medical Microbiology and Immunology

7. B. Both charts represent the rise and fall of an infectious disease.
8. C. When the microbe is introduced into a healthy, susceptible host, the same disease occurs.
9. A. Antibiotics reproduce themselves to form more antibiotics.

1.4 Microbes in Our Environment

10. B. Lithotrophs
11. C. Cyanobacteria
12. B. Nitrogen fixation

1.5 The DNA Revolution

13. D. Rosalind Franklin
14. C. Claire Fraser-Liggett
15. C. Polymerase chain reaction (PCR) amplification of DNA

CLINICAL CORRELATION QUESTIONS

1. The tests reveal no sign of bacteria, and the symptoms are typical of viral pneumonia. Antibiotics do not affect viruses, only bacterial pathogens. Therefore, antibiotics would not be prescribed; instead, the patient is offered supportive therapy that minimizes the damage caused by the virus and helps the body recover.

2. Koch's first postulate states that if a microbe is the causative agent of disease, it is not found in healthy hosts. Many types of *E. coli* bacteria, however, grow normally in the human intestine. If the particular strain is a known pathogen (such as *E. coli* O157:H7) or an emerging strain not commonly found, then it may turn out to be the cause of the patient's intestinal illness.

Chapter 2

OBJECTIVES REVIEW

2.1 Normal Microbiota versus Pathogens

1. C. Virulence
2. B. Immune avoidance
3. B. The infectious dose 50% is 1,000 organisms.

2.2 Basic Concepts of Disease

4. nosocomial infection
5. D. sign and symptom
6. A secondary infection is an infection that follows a primary infection. Secondary infections are typically caused by an organism different from the one that caused the primary infection.
7. Mortality
8. C. the illness phase

2.3 Infection Cycles and Disease Transmission

9. C. The HIV virus does not replicate in mosquito cells.
10. transplacental transmission or vertical transmission
11. B. Immune system
12. endemic

2.4 Portals of Entry and Exit

13. E. Parenteral
14. D. Level 4
15. D. Positive-pressure room ventilation

2.5 Host Factors in Disease

16. E. All of the above
17. A. Respiratory
18. immunopathogenesis
19. All are true

2.6 Global Change and Emerging Infectious Diseases

20. D. Leptospirosis
21. E. A and C only

CLINICAL CORRELATION QUESTIONS

1. Answers will vary.
2. Knowing the portal of *entry* for a particular microbe enables individuals to protect themselves against contracting disease from someone else. Knowing the portal of *exit* enables a patient with disease to take measures to prevent transmitting the disease to a healthy person. For instance, knowing that the portal of exit for influenza virus is the respiratory tract could lead the infected patient to wear a sterile mask when venturing out.

Chapter 3

OBJECTIVES REVIEW

3.1 Observing Microbes

1. D. Bacteria are smaller than the resolution distance of the human retina.
2. A. Bright-field light microscopy

3.2 Optics and Properties of Light

3. B. The object must be fixed to a slide.
4. C. Light bends, or "refracts," as it enters a substance that slows its speed.

3.3 Bright-Field Microscopy

5. A. the condenser
6. B. the objective lens

3.4 Stained Samples

7. B. the ring form and gametocyte stage of the malaria parasite *Plasmodium falciparum*
8. D. The Gram procedure requires an antibody stain.

3.5 Advanced Microscopy

9. A. fluorescence microscopy, with a fluorophore attached to a cell-surface–specific antibody
10. D. electron microscopy
11. A. SEM. B. Dark-field microscopy. The cell bodies appear "overexposed" by scattered light, but the technique reveals flagella, whose width would be too narrow for resolution. C. Gram stain, bright-field microscopy. D. TEM of thin section.

CLINICAL CORRELATION QUESTIONS

1. *Salmonella enterica*
2. *Mycobacterium tuberculosis*

Chapter 4

OBJECTIVES REVIEW

4.1 Elements, Bonding, and Water

1. E. Helium
2. A. Butane ($CH_3CH_2CH_2CH_3$)
3. A. NaCl

4.2 Lipids and Sugars

4. B. Hydrocarbon chain
5. C. an amino group
6. D. Glycoprotein

4.3 Nucleic Acids and Proteins

7. A. a thiol (–SH)
8. B. Secondary structure
9. A. Uracil

4.4 Biochemical Reactions

10. A. DNA synthesis
11. D. Increase in temperature from 25°C to 35°C
12. C. Breaking down a protein

4.5 Membranes and Transport

13. B. Hydrocarbon chain
14. C. Alanine
15. C. A cell placed in distilled water takes up water and expands.

Endocytosis and Phagocytosis

16. D. *Vibrio cholerae*
17. B. Phospholipids
18. D. Ameba

CLINICAL CORRELATION QUESTIONS

1. A patient with cholera often has multiple bio-chemical needs because of the effects of cholera or of preexisting malnutrition. Vitamins and sugar may well be needed, but will not address the urgent need to restore electrolyte balance. Antibiotics could kill the bacteria, but in this instance the patient's physiological recovery is a higher priority. The top priority is to provide NaCl to replace the ions lost via the transport proteins that were inappropriately activated by the cholera toxin.

2. When *S. enterica* bacteria infect a patient, the bacteria are endocytosed by the cells of the intestinal lining. Within the cell, the bacteria form intracellular membrane vesicles that protect the bacteria from host defenses.

Chapter 5

OBJECTIVES REVIEW

5.1 The Bacterial Cell: An Overview

1. D. Lipopolysaccharide (LPS)
2. C. Water molecules
3. B. Cell membrane (inner membrane)

5.2 Bacterial Membranes and Transport

4. A. Protein synthesis in the cytoplasm
5. B. Facilitated diffusion of oxygen across a membrane
6. A. TolC efflux complex

5.3 The Bacterial Cell Wall and Outer Layers

7. A. Cell wall
8. D. Mycolic acids
9. D. Outer membrane

5.4 The Nucleoid and Bacterial Cell Division

10. C. DNA is contained within a nuclear membrane.
11. A. Completion of the septum to separate two daughter cells
12. C. A circular molecule with loops bound by DNA-binding proteins

5.5 Specialized Structures of Bacteria

13. B. Stalk
14. C. Flagella generate a whiplike motion powered by ATP.
15. A. Chemotaxis

5.6 The Eukaryotic Cell

16. D. Mitochondrion
17. B. Peptidoglycan cell wall
18. B. Chloroplast

CLINICAL CORRELATION QUESTIONS

1. A skin biopsy could reveal *Staphylococcus* species as a cause of infection; but the cells would be small, round cocci, without flagella. *Proteus mirabilis* is unlikely to be found in skin. *Mycobacterium leprae* might be found in skin, but the bacteria

consist of small nonmotile rods. *Borrelia burgdorferi*, however, consist of narrow spiral-shaped cells called spirochetes. Their internally embedded flagella cause the cell to twist, generating motility. The patient may have borreliosis.

2. The skin biopsy reveals bacteria that stain acid-fast. Acid-fast rods are usually mycobacteria, either *M. tuberculosis* or *M. leprae*. The skin lesions with loss of sensation suggest the disease leprosy. *M. leprae* bacteria are one of the slowest-growing pathogens known; they can take several weeks to grow colonies in culture.

Chapter 6

OBJECTIVES REVIEW

6.1 Microbial Nutrition

1. defined minimal medium
2. C. autotroph
3. Clockwise from top: Nitrogen, Nitrogen fixers, NH_4^+, Nitrifiers, NO_3^-, Denitrifiers

6.2 Culturing and Counting Bacteria

4. E. Plates B and C
5. C. Differential medium
6. C. 3.5×10^4
7. B. Viable count

6.3 The Growth Cycle

8. B. 2
9. F. 7×10^{11}
10. A. Decrease growth rate

6.4 Environmental Limits on Microbial Growth

11. barophiles or piezophiles
12. C. mesophiles
13. D. The growth rate in exponential phase will be slower.
14. E. Membranes become too flexible.

6.5 Living with Oxygen: Aerobe versus Anaerobe

15. C. Anaerobes lack enzymes that detoxify oxygen by-products.
16. False
17. False
18. D. Intestine
19. E. Anaerobes and facultative organisms

6.6 Microbial Communities and Cell Differentiation

20. quorum sensing
21. exopolysaccharides
22. B. Chronic
23. F. A, B, and C only

CLINICAL CORRELATION QUESTIONS

1. Brian ate the salad when it was fresh and there were few live *Salmonella* organisms present. If the salad was not refrigerated properly or if it was left out for hours at room temperature, the pathogen could have grown to numbers high enough to cause disease in Brianna. (Recall infectious dose from Chapter 2.)

2. The proton pump inhibitor that Padman takes decreases his stomach acidity. *V. cholerae*, which is very sensitive to acid, is better able to survive in his stomach than in Gagandeep's. The number of organisms that reached Padman's intestine surpassed the infectious dose, so he became ill.

3. The urease produced by *H. pylori* converted enough urea in the medium to ammonia to change the pH of the medium from pH 2 to over pH 5, a pH in which this organism can grow.

Chapter 7

OBJECTIVES REVIEW

7.1 Energy for Life

1. B. fermentation
2. D. carbon dioxide.
3. A. combustion.

7.2 Catabolism: The Microbial Buffet

4. A. Cellulose
5. C. NADH is an electron donor formed during catabolism.
6. C. Pentose phosphate pathway

7.3 Glucose Catabolism, Fermentation, and the TCA Cycle

7. C. Glycolysis has a cycle of intermediates that incorporate a carbon skeleton and regenerate the original intermediate.
8. A. Acetyl-CoA

7.4 Respiration and Lithotrophy

9. C. The oxidoreductase directly phosphorylates ADP to make ATP.
10. B. *Geobacter* catabolizes small organic molecules and donates electrons to an electrode in a fuel cell.
11. D. Anaerobic respiration, reducing nitrate to nitrite

7.5 Photosynthesis and Carbon Fixation

12. A. Cyanobacteria
13. D. CO_2 fixation via the Calvin cycle
14. A. Dinoflagellates in red tide

7.6 Biosynthesis and Nitrogen Fixation

15. D. all of the above
16. B. glutamate

CLINICAL CORRELATION QUESTIONS

1. The enteric pathogen formed orange colonies on Hektoen agar. It was thus identified as *Escherichia coli*.

2. The physician should test immediately for the presence of saxitoxin, a toxic substance produced by red tide algae.

Chapter 8

OBJECTIVES REVIEW

8.1 Bacterial Genomes

1. A. The genome is composed of double-helical DNA.
2. C. When a plasmid is transferred into a new cell, it usually changes the species of the host organism.
3. B. Introns

8.2 DNA Replication

4. A. An RNA primer is synthesized.
5. C. generating negative superturns in DNA
6. B. connecting Okazaki fragments

8.3 Mutation and Repair

7. B. A silent mutation
8. B. SOS repair
9. A. a deletion of a base pair

8.4 Biotechnology

10. A. Gel electrophoresis
11. D. PCR amplification of a bacterial gene
12. D. fluorophores (fluorescent molecules)

8.5 Transcription of DNA to RNA

13. C. Messenger RNA (mRNA)
14. D. The sigma factor
15. C. release of pyrophosphate by each nucleotide

8.6 Protein Synthesis

16. B. The incoming aminoacyl-tRNA binds the A site of the ribosome.
17. C. To fold a newly synthesized protein into its functional shape
18. A. Binding the first f-Methionyl tRNA to the mRNA.

8.7 Regulation of Gene Expression

19. D. A translation initiation sequence binds a translational repressor.
20. B. DNA phase variation occurs.
21. D. Two-component regulator transmits information from an external signal to the DNA inside the cell.

CLINICAL CORRELATION QUESTIONS

1. No capsule stain would be possible, because the nontypeable strains lack a capsule. A chest radiograph could show the presence of infection, but not identify the infectious agent. The only way to identify the strain of *H. influenzae* would be through DNA analysis, such as ribotyping analysis of RFLPs.

2. Tests such as palpation of the uterus and Gram stain of the discharge could suggest the presence of a sexually transmitted infection. To identify the causative organism, such as *Chlamidia* or *Neisseria gonorrhoeae*, the clinician could perform a standard DNA test called the nucleic acid amplification test (NAAT).

Chapter 9

OBJECTIVES REVIEW

9.1 Gene Transfer, Recombination, and Mobile Genetic Elements

1. B. transduction
2. A. transfer of a multidrug resistance plasmid
3. B. to prevent replication of an invading bacteriophage

9.2 Origins and Evolution

4. D. Entirely new organisms emerge from environments, and they prevail over preexisting organisms.

5. B. The gene must acquire mutations at a steady rate over time in all organisms compared.
6. D. *Vibrio*

9.3 Natural Selection

7. C. The bacteria containing the resistance plasmid will decline, because replicating the plasmid costs energy.
8. B. *T. pallidum* has evolved virulence genes that enable it to take advantage of host cells.
9. C. the random mutation that conferred resistance also conferred deleterious effects on growth

9.4 Bioinformatics

10. C. *S. pneumoniae* isolates were selected for genome analysis, which revealed which gene functions were shared and which genes were distinct to different species.
11. B. Two organisms each have a gene that encodes the same function.

9.5 Microbial Taxonomy

12. D. Two genomes show very different proportions of GC versus AT content.
13. A. a combination of DNA-based tests and phenotypic traits

CLINICAL CORRELATION QUESTIONS

1. Pathogens are exposed to changing stressful environments owing to host defenses and antibiotic treatments—conditions expected to favor mutators. Therefore, mutator phenotypes might accelerate the evolution of pathogenic strains by increasing variation of surface antigens to avoid the immune system, as well as by facilitating acquisition of pathogenic determinants and antibiotic resistance.

2. The short answer is that *S*. Typhi underwent reductive evolution to limit its host range. You could generally hypothesize that an ancestor of *S*. Typhi lost one or more genes encoding proteins needed to attach to, or replicate in, nonhuman cells. For instance, some research indicates that the pili of *S*. Typhi include mutations that reduce binding to nonhuman cell receptors. A more recent study also found that a protein injected by many serovars of *Salmonella* into nonhuman macrophage cells is missing from *S*. Typhi. Macrophages are white blood cells that wander through tissues and kill bacteria. To cause disease, pathogenic *Salmonella* must survive and grow within macrophages by injecting proteins that alter macrophage function. *S*. Typhi, however, lacks a protein needed to grow in nonhuman macrophages but is fully able to survive in human macrophages because that protein is not needed.

Chapter 10

OBJECTIVES REVIEW

10.1 Bacterial Diversity at a Glance

1. A. Cyanobacteria
2. C. Spirochetes
3. A. *Escherichia* and *Pseudomonas*

10.2 Gram-Positive Firmicutes and Actinobacteria

4. D. *Mycoplasma pneumoniae*, a cause of pneumonia
5. A. *Listeria* bacteria, which grow in milk and cheese, even in the refrigerator

10.3 Gram-Negative Proteobacteria and Bacteroidetes

6. B. *Nitrobacter* species, growing in soil
7. D. *Bacteroides*, enteric bacteria that help digest our food but cause abscesses if they escape the intestine and enter surrounding tissues
8. B. Lipopolysaccharide

10.4 Spirochetes: Twisted-Cell Pathogens and Symbionts

9. A. Round, coccoid cells
10. C. Termites
11. A. Ticks

10.5 Cyanobacteria Fix Carbon Dioxide and Produce Oxygen

12. D. Intracellular flagella
13. B. Chloroplasts

10.6 Chlamydias: Intracellular Pathogens

14. C. Peptidoglycan cell wall
15. A. Endospore

CLINICAL CORRELATION QUESTIONS

1. Bacteria such as *Streptococcus pneumoniae* can be observed from the patient's sputum using a Gram stain for the Gram-positive cell wall. The mycoplasma, however, has no cell wall, and the virus is not a cell. Mycoplasma or virus would have to be detected by a test for serum antibodies or by a PCR test for the pathogen's DNA.

2. The disease could be syphilis (caused by *Treponema pallidum* subsp. *pallidum*), Lyme disease (caused by *Borrelia burgdorferi*), or yaws (caused by *Treponema pallidum* subsp. *pertenue*). The lesion of syphilis (acquired by sexual contact) disappears, but later a macular rash of multiple spots appears on other parts of the body; the pathogen can also cross the placenta and infect a fetus. Yaws is caused by a different subspecies of *T. pallidum*; it is transmitted by casual contact, but not across the placenta. The original lesion can be followed by multiple lesions distributed over the body. If the disease is Lyme disease, the pathogenic spirochete *Borrelia burgdorferi* can later spread to the joints and the nervous system.

Chapter 11

OBJECTIVES REVIEW

11.1 Eukaryotes: An Overview

1. A. Fungi
2. D. Helminths
3. A. Dinoflagellates

11.2 Fungi and Microsporidians

4. B. *Candida albicans*
5. A. Ergosterol biosynthesis
6. D. *Candida*

11.3 Algae and Amebas Are Environmental Protists

7. D. Flagella for motility
8. C. Foraminiferans
9. B. Foraminiferans

11.4 Alveolates: Predators, Phototrophs, and Parasites

10. A. *Balantidium coli*
11. B. *Toxoplasma gondii*
12. A. *Plasmodium vivax*

11.5 Trypanosomes and Metamonads

13. D. Transmission via the oral-fecal route
14. A. infant day-care centers
15. C. *Leishmania donovani*

11.6 Invertebrate Parasites: Helminths and Arthropods

16. B. Nematode
17. D. Pubic lice
18. C. Flukes

CLINICAL CORRELATION QUESTIONS

1. The parasite is an ameba, *Entamoeba histolytica*, which engulfs red blood cells by phagocytosis. The amebas are transmitted through contaminated water. The antiparasitic drug paromomycin will eliminate the infection.
2. *Aspergillus* species is the most probable cause. *Aspergillus* may grow as a fungal ball localized in the lung, which may become invasive. Antifungal agents such as itraconazole may be attempted, but often surgical removal is required.

Chapter 12

OBJECTIVES REVIEW

12.1 What Is a Virus?

1. B. Isolated particles conduct their own metabolism.
2. D. Mimivirus has no ribosomes of its own; it must use the host ribosomes.
3. A. the range of tissues a virus can infect

12.2 Viral Genomes and Diversity

4. A. phospholipid
5. C. Ebola virus
6. C. smallpox virus

12.3 Virus Replication and Culture

7. A. binding of a virion to a specific host receptor protein
8. C. The capsid of the virion enters the host cytoplasm.
9. B. attach the virion upon a host cell

12.4 Papillomavirus: DNA Genome

10. C. The viral genome integrates into the host genome and causes increased expression of oncogenes.
11. C. Host DNA polymerase
12. D. shedding of host cells

12.5 Influenza Virus: RNA Genome

13. D. asymmetrical
14. B. The viral genome is composed of multiple segments that get transcribed by RNA-dependent RNA polymerase and reassort with segments from a coinfecting virus.
15. A. Cancer

12.6 Human Immunodeficiency Virus: Reverse Transcription

16. A. Virion envelope proteins specifically interact with the CD4 and CCR5 cell surface proteins of a T lymphocyte.
17. C. Oseltamivir, which binds neuramidase, preventing release of virions as they exit the host cell
18. B. An initial burst of virions is followed by a long latent period during which T-cells decline.

CLINICAL CORRELATION QUESTIONS

1. The infectious agent is monkeypox, a virus endemic in Africa among monkeys and rodents, also frequently transmitted to humans. The infection is prevented by smallpox vaccination, but because smallpox is eliminated in the wild, most people no longer receive vaccination and therefore remain vulnerable to other highly contagious poxviruses.
2. Hepatitis C virus shows elevated prevalence in individuals born between 1945 and 1964. Most infected individuals show no symptoms, but hidden damage may occur to the liver, ultimately requiring liver transplant.

Chapter 13

OBJECTIVES REVIEW

13.1 Basic Concepts of Sterilization and Disinfection

1. B. Disinfection
2. B. decimal reduction time
3. D. 5×10^5 CFUs/ml

13.2 Physical Agents That Kill Microbes

4. B. 121°C, 15 pounds of pressure, 20 minutes
5. False
6. A. *Coxiella burnetti*
7. C. Gamma irradiation

13.3 Chemical Agents of Disinfection

8. less
9. E. C and D only
10. D. Glutaraldehyde
11. A. Glutaraldehyde

13.4 Basic Concepts of Antimicrobial Therapy

12A. Tube 4
12B. Can't tell. Not enough information.
13. D. All of the above

13.5 Antimicrobial Mechanisms of Action

14. B. Aminoglycoside (e.g., gentamicin)—30S ribosome
15. D. Penicillin
16. vancomycin (C and D), bacitracin (B), cephalosporins (C), cycloserine (A)

13.6 The Challenges of Antibiotic Resistance

17. A. A multidrug efflux pump
18. B. acts as a decoy for beta-lactamase
19. B. Degrade antibiotic
20. A. Alter target
21. A. Alter target
22. C. Modify antibiotic
23. D. Pump antibiotic out of cell

13.7 Antiviral Agents

24. D. Neuraminidase inhibitors
25. A. Viral polymerases are less selective than host polymerases.
26. C. Neuraminidase inhibitor
27. F. Protease inhibitor
28. B. Fusion inhibitor
29. E. Viral RNA synthesis
30. D. Viral DNA replication

13.8 Antifungal Agents

31. B. a triazole or allylamine; A. flucytosine; D. griseofulvin
32. B. miconazole
33. D. Amphotericin B

CLINICAL CORRELATION QUESTIONS

1. The most likely scenario is that a plasmid expressing an ampicillin-resistance gene was transferred from *Shigella* to some of your resident *E. coli*. Those strains of *E. coli*, already expressing ciprofloxacin resistance, survived when you were being treated with ampicillin.
2. An increase in BUN and creatinine levels in blood are signs of kidney damage. A damaged kidney does not efficiently filter or excrete urea and creatinine, which causes these compounds to accumulate in blood. Amphotericin B is known to exhibit nephrotoxicity, so administering this drug to someone already having a damaged kidney could lead to even greater damage.

Chapter 14

OBJECTIVES REVIEW

14.1 Human Microbiota: Location and Shifting Composition

1. B. Kidney
2. D. intestine
3. microbiota or microbiome
4. D. All of the above

14.2 Benefits and Risks of a Microbiome

5. B. Interfering with pathogen colonization
6. opportunistic pathogen
7. gnotobiotic
8. D. thin intestinal walls

14.3 Keeping Microbiota in Their Place

9. E. A and B only
10. B. Toll-like receptor
11. C. Peyer's patch

14.4 Microbiota-Host Communications
12. True
13. True
14. D. Factors produced by donor microbiota stimulated production of T cells.
15. E. All of the above

14.5 Natural Biological Control of Microbiota
16. B. *Bifidobacterium*
17. B. phages can evolve to infect phage-resistant bacteria
18. False
19. B. increased cases of allergies

CLINICAL CORRELATION QUESTIONS

1. The man avoided drinking or swimming in water that could potentially be contaminated with large numbers of *V. cholerae*. Hence, he most likely ingested only small numbers of this pathogen from other sources. Small numbers of ingested vibrios are easily killed by the acidic stomach. However, the man was treating his indigestion with Prilosec, a proton pump inhibitor. The medication slows stomach acidification, which means that the stomach will have a less acidic pH that *V. cholerae* can survive. As a result, he was at increased risk of developing cholera, even if only small numbers of this pathogen were ingested.

2. The antibiotic resistance plasmid could be transferred by transformation or transduction to a bacterial member of your normal microbiota—one capable of maintaining that plasmid. If you are later treated with the corresponding antibiotic, the drug-resistant bacterial species will survive and grow while other members are killed. When the antibiotic is removed, the antibiotic-resistant strain, now present at high numbers, could limit the growth of returning microbiota and change the balance of competing species.

3. There is a spore-forming bacterium present in mouse feces that cannot be grown in the laboratory. It will survive boiling but not autoclave temperatures. As part of the mouse microbiome, this organism (or set of organisms) stimulates the innate immune system, which balances the ratio of white blood cells. This is a true scenario. Google "segmented filamentous bacteria and the immune system."

Chapter 15

OBJECTIVES REVIEW

15.1 Overview of the Immune System
1. B. Complement
2. B. Thymus
3. D. Leukocyte count 19,000/µl; 80% neutrophils; 25% lymphocytes

15.2 Innate Host Defenses: Keeping the Hordes at Bay
4. C. Langerhans cell
5. A. sample bacteria and other antigens from the intestine
6. microbe-associated molecular patterns (MAMPs)

15.3 The Acute Inflammatory Response
7. A
8. B
9. C
10. A
11. B. Lack of integrin
12. Heat, edema, redness, pain, altered function
13. chronic

15.4 Phagocytosis: A Closer Look
14. B. CD47
15. E. Macrophages engulf dead neutrophils.
16. C. Fc receptors
17. D. triggering apoptosis of the phagocyte

15.5 Interferon, Natural Killer Cells, and Toll-Like Receptors
18. True
19. E. inducing production of double-stranded RNA endonucleases
20. A. Fc receptors
21. D. Microbial structures

15.6 Fever
22. True
23. B. Hypothalamus

15.7 Complement
24. C. C5
25. B. Factor H
26. D. Factor C5b directs assembly of the pore.
27. A. MAMPs bind TLRs in macrophages and cytokines are released, triggering CRP production in the liver.

CLINICAL CORRELATION QUESTIONS

1. The high level of white cells in his blood was the result of their increased production in his bone marrow in response to cytokines produced at the site of infection and because, once in the circulation, these white cells, lacking integrin, could not extravasate out of capillaries to the site of infection. Unable to get to the site of infection, these cells could not fight the infection.

2. CD59 is a cell-surface protein that binds to complement factor C5a and prevents formation of the MAC pore complex on host cells. Red blood cells left unprotected by CD59 will lyse as a result of MAC formation. The current best way to treat this deficiency is by administering antibody to C5a. Removing C5a in this way will prevent MAC formation in red blood cells. Eculizumab is a commercially available anti-C5a monoclonal antibody used to treat these patients. Monoclonal antibodies are described in Chapter 17.

Chapter 16

OBJECTIVES REVIEW

16.1 Adaptive Immunity: The Big Picture
1. B. epitope
2. D. Plasma cells
3. B. T cells
4. D. present antigen on MHC surface receptors

16.2 Immunogenicity and Immune Specificity
5. E. Proteins
6. immunological specificity

16.3 Antibody Structure and Diversity
7. 1 C; 2 B; 3 A; 4 D
8. A. Part 1
9. B. heavy-chain constant regions
10. D. IgA
11. False

16.4 Humoral Immunity: Primary and Secondary Antibody Responses
12. D. IgG
13. A. IgM or IgD
14. B. B cells proliferate to produce new B cells with identical B-cell receptor surface proteins
15. B. Affinity maturation of the antigen-binding site

16.5 T Cells: Link between Humoral and Cell-Mediated Immunity
16. E. Activated CD8 T cells
17. A. generation of cytotoxic T cells
18. D. bind weakly to some self antigen
19. A. extracellularly

16.6 T-Cell Activation
20. CD4
21. D. Inability to undergo isotype switching
22. A. Bone marrow
23. B. Thymus
24. D. Lymph node
25. D. Lymph node
26. D. Lymph node

16.7 Cell-Mediated Immunity
27. B. Dendritic cells
28. A. MHC I receptors
29. E. bind T-cell receptors and MHC outside of their antigen-binding regions
30. C. placing a molecular mimic of MHC I on an infected cell surface

CLINICAL CORRELATION QUESTIONS

1. His blood would contain normal or elevated levels of IgM but very low levels of the other classes of antibodies. IgM is present because some B cells can become plasma cells and make some IgM antibody even without T-cell help. T-cell help, however, is required for isotype switching to other antibody types. Because of this, IgM is made but a switch to IgG or other antibody types is rare. Repeated infections happen because IgM is not a high-affinity antibody (T-cell help is needed to trigger affinity maturation of the antigen-binding sites). Also, IgM does not contain Fc regions that can bind Fc receptors on phagocytes. The lack of IgG, which has these features, means that bacterial pathogens can overwhelm the minimal protection afforded by IgM. In addition, without CD4 T_H1 cells, CD8 cytotoxic T cells are not properly activated.

2. Cytotoxic T cells will not kill infected red blood cells. Red blood cells, which are not nucleated, will not have surface MHC I receptors that can present antigen to CD8 cytotoxic T cells.

Chapter 17

OBJECTIVES REVIEW

17.1 Primary Immunodeficiency Diseases

1. B. a secondary immunodeficiency
2. D. lack of IL-2 receptor
3. C. DiGeorge syndrome
4. A. IV administration of immune globulin

17.2 Immune Surveillance and the Cancers of Lymphoid Tissues

5. D. lymphoma
6. E. Hodgkin's lymphoma
7. E. Burkitt's lymphoma
8. immune surveillance

17.3 Hypersensitivity

9. D. Type IV
10. E. A and C only
11. Allergen
12. IgE
13. Mast cells or basophil
14. Histamine, prostaglandins, and leukotrienes
15. C, F
16. D, E
17. B, H

17.4 Autoimmunity and Organ Transplants

18. A. molecular mimicry enables a mismatched T_H cell to activate a B cell that escaped deletion
19. E. antibody that targets thyroid-stimulating hormone receptor
20. B. Absence of MHC type I complexes on red blood cells.

17.5 Tools of Immunology

21. Hyper IgM syndrome
22. X-linked agammaglobulinemia
23. Selective IgA deficiency
24. B. Patient 2   1:1,350
25. D. Flow cytometry

17.6 Vaccines and Immunization

26. E. Breast milk
27. E. Natural infection
28. herd immunity

CLINICAL CORRELATION QUESTIONS

1. Rh-negative cells because the mother's anti-Rh$^+$ antibodies have crossed the placenta and will continue to attack any newly transfused Rh$^+$ red blood cells in the fetus. An Rh-positive fetus does not make antibodies to Rh-negative blood cells (no foreign antigen is present on Rh-negative cells).
2. Leukocyte adhesion deficiency. The patient was deficient in CD18 integrin but had plenty of Fc receptor. Presence of the Fc receptor on leukocytes suggests that the leukocytes are plentiful. However, the loss of CD18 means those leukocytes (PMNs and lymphocytes) cannot extravasate from the bloodstream into tissues when called. Therefore, many leukocytes accumulate in blood. So although the elevated WBC count indicates that the patient is responding to an infection, leukocytes cannot get to the site of infection. The patient, therefore, is more susceptible to recurrent infections.

Chapter 18

OBJECTIVES REVIEW

18.1 The Tools and Toolkits of Bacterial Pathogens

1. E. Segments A and C
2. A. Conjugation

18.2 Microbial Attachment

3. False
4. type IV pili
5. B. biofilm-producing bacteria
6. Type IV
7. C. nonpilus adhesion
8. D. prevents evacuation of the pathogen

18.3 Toxins: A Means to Seduce, Hijack, or Kill Host Cells

9. E. Anthrax EF toxin
10. B. *Escherichia coli* LT
11. C. Toxic shock syndrome toxin
12. A. *Escherichia coli*

18.4 Secretion of Virulence Proteins

13. D. flagellum synthesis systems
14. C. secrete proteins into the surrounding environment

18.5 Survival within the Host

15. B. remains in the host cell phagosome
16. E. *Rickettsia*
17. A. moves intracellularly via actin tails
18. C. Delay synthesis of virulence factors until a certain bacterial cell number is reached

18.6 Viral Pathogenesis

19. B. Multiple rhinovirus strains that possess different VP1 structures
20. C, A, E, D

18.7 Protozoan Pathogenesis

21. *Toxoplasma*, B; *Leishmania*, C
22. False
23. True
24. E. Downregulate IL-12 production in infected macrophages

CLINICAL CORRELATION QUESTIONS

1. The pathogen infected the pregnant mother and passed through the placenta to the fetus.
2. Scenario 4.

3. Horizontal transfer of genes from other strains of *E. coli*.
4. Decreased killing of infected cells by cytotoxic T cells.

Chapter 19

OBJECTIVES REVIEW

19.1 Anatomy of the Skin and Eye

1. B. Langerhans cells
2. basal
3. epidermis
4. False

19.2 Viral Infections of the Skin

5. D. a linear double-stranded DNA virus that also causes cold sores
6. B. Variola
7. D. vertical transmission of rubella virus to the fetus during the mother's active infection

19.3 Bacterial Infections of the Skin

8. B. Coagulase
9. B. Gram-positive cocci
10. E. Surgical removal of affected tissue and IV antibiotics

19.4 Fungal Infections of the Skin

11. A. *Malassezia*
12. C. KOH preparation
13. B. playing with a household dog or cat

19.5 Skin Infections of Special Circumstance

14. C. Decreased cell-mediated immunity
15. B. is a ubiquitous opportunistic pathogen
16. C. Bacteria
17. D. It often is due to polymicrobial wound infection with aerobes and *C. perfringens*.

19.6 Eye Infections

18. A. Acute allergy attack causing conjunctivitis
19. the conjunctiva
20. C. endophthalmitis

CLINICAL CORRELATION QUESTIONS

1. The most likely cause of infection is the measles virus. The white papular enanthem that appears on the palate and throat is referred to as Koplik spots. Vaccination with attenuated virus is the means by which this disease is prevented (see Table 17.5).
2. *S. pyogenes* is responsible for the pustules. *C. perfringens* (a Gram-positive anaerobe) and *P. aeruginosa* (a Gram-positive aerobe) both are rod-shaped bacteria (bacilli). *S. aureus* and *S. epidermidis* are catalase-positive. *S. pyogenes* is catalase-negative. Staphylococci appear as clusters, much like a cluster of grapes under the microscope, but *Streptococcus pyogenes* looks like pearls on a string. The only bacterial species consistent with the lab results is *S. pyogenes*.
3. The ophthalmologist is likely to see multinucleated giant cells (Tzank cells). Herpes simplex

viruses can infect the cornea and cause a dendritic lesion. Scrapings from the lesion spread on a glass slide and viewed under the microscope (Tzank smear) reveal multinucleated giant cells called Tzank cells.

Chapter 20

OBJECTIVES REVIEW

20.1 Anatomy of the Respiratory Tract
1. True
2. B. Trachea
3. B. preventing foreign materials from reaching the lungs

20.2 Viral Infections of the Respiratory Tract
4. C. Shortness of breath
5. False
6. True
7. D. Infants

20.3 Bacterial Infections of the Respiratory Tract
8. B. Pharyngitis
9. D. Pneumonia
10. A. Acute otitis media, C. Bronchitis, D. Pneumonia
11. D. Pneumonia
12. A. Acute otitis media, C. Bronchitis, D. Pneumonia
13. A. Childhood vaccination
14. B. a consequence of pharyngitis caused by toxin-producing *S. pyogenes*
15. D. *Mycoplasma pneumoniae*
16. True
17. False

20.4 Fungal and Parasitic Infections of the Respiratory Tract
18. True
19. B. Normal values
20. C. *Cryptococcus*
21. D. western United States

CLINICAL CORRELATION QUESTIONS
1. The most likely organism causing the clinical picture given is *Mycobacterium tuberculosis*. There is no history of travel or immune suppression, so fungal causes are not the prime suspect. *Mycoplasma pneumoniae* and *Bordetella pertussis* can also cause a chronic cough but will not progress to coughing up a blood-tinged sputum.
2. The decided left shift (increased number of neutrophils) indicates a bacterial pneumonia. *Streptococcus pneumoniae* is the most likely cause based on the clinical presentation and the typical involvement of one lobe of the lung (lobar pneumonia).
3. *Bordetella pertussis*. The distinct sound of inspiration made after a paroxysm of cough has given the disease its name: whooping cough. Because the paroxysmal cough of whooping cough continues for quite some time after the bacteria are gone, the disease was called the 100-day cough.

Chapter 21

OBJECTIVES REVIEW

21.1 Anatomy of the Cardiovascular and Lymphatic Systems
1. B. aorta
2. False
3. C. This is an example of lymphadenitis.

21.2 Systemic Viral Infections
4. C. B cells
5. E. Heterophile antibodies
6. B. Transplacental
7. D. Hemorrhagic fever
8. B. Splenomegaly
9. A. Owl's eyes
10. C. Burkitt's lymphoma

21.3 Systemic Bacterial Infections
11. C. May have a noninfectious etiology
12. A. systemic inflammatory response syndrome caused by an infectious agent
13. B. Sepsis with one or more signs of organ dysfunction
14. D. Need vasopressors to maintain systolic blood pressure of 90 mm Hg or mean arterial pressure of 70 mm Hg
15. E. Presence of bacteria in the blood
16. B. stage 2
17. E. Typhoid fever

21.4 Bacterial Infections of the Heart
18. B. *Staphylococcus aureus*
19. A. pancarditis
20. True

21.5 Systemic Parasitic Infections
21. B. undergo antigenic variation
22. C. Hydrocele (grotesque swelling of the scrotum)

CLINICAL CORRELATION QUESTIONS
1. *Yersinia pestis*. *Borrelia burgdorferi* is a spirochete, not a bacillus. *Salmonella* Typhi and *Rickettsia rickettsii* cause generalized lymph adenopathy and do not exhibit bipolar staining.
2. Antibiotics effective against Gram-positive, Gram-negative, and anaerobic bacteria. Viruses and parasites do not cause septicemia.
3. *Borrelia burgdorferi*, which is transmitted by tick bite.

Chapter 22

OBJECTIVES REVIEW

22.1 Anatomy of the Digestive System
1. E. A and B only
2. D. Large intestine
3. False
4. A. inadequate absorption of nutrients

22.2 Infections of the Oral Cavity
5. D. All of the above
6. C. Both A and B
7. B. Trench mouth
8. C. Plaque formation
9. A. *Candida albicans*
10. B. Vincent's angina

22.3 Gastrointestinal Syndromes
11. B. Enteritis
12. C. secretory diarrhea
13. False

22.4 Viral Infections of the Gastrointestinal Tract
14. D. DNA genome
15. B. RNA genome
16. C. Orchitis
17. A. Diarrhea
18. A. liver
19. B. hepatitis B

22.5 Bacterial Infections of the Gastrointestinal Tract
20. C. *Staphylococcus aureus*
21. E. actin tails
22. D. *Salmonella* Typhi
23. A. good sanitation
24. False

22.6 Parasitic Infections of the Gastrointestinal Tract
25. A. Cryptosporidiosis
26. B. migration of larvae to the lungs
27. B. larvae penetration of the skin
28. C. Metronidazole

CLINICAL CORRELATION QUESTIONS
1. The causative agent is enteroxigenic *E. coli*. The organism produces a heat labile toxin that increases cAMP production in the host, which results in secretion of electrolytes followed by water into the lumen of the intestine.
2. Typhoid fever. Typhoid fever is caused by *Salmonella enterica* serovar Typhi, a facultative intracellular organism. A fever that originally is low grade and keeps going up and up every day (crescendos) and the red spots on the trunk (likely to be rose spots) indicate typhoid fever.
3. *Y. enterocolitica*. The differential diagnosis of this patient's symptoms includes a number of different bacteria. *Y. enterocolitica*, however, is the only etiology that would cause mesenteric lymphadenitis and terminal ileitis and symptomatically mimic appendicitis.

Chapter 23

OBJECTIVES REVIEW

23.1 Anatomy of the Urinary Tract
1. A. Kidneys
2. C. ureters
3. D. The length of the urethra

23.2 Urinary Tract Infections

4. A. Sexual activity
5. A. This patient has pyelonephritis resulting from an ascending bladder infection.
6. True
7. B. children postponing emptying their bladder
8. C. Pyelonephritis

23.3 Anatomy of the Reproductive Tract

9. C. Testes
10. C. There is no direct contact between the ovaries and the fallopian tubes.
11. False

23.4 Sexually Transmitted Infections of the Reproductive Tract

12. A. human papillomavirus
13. E. HSV
14. A. "Mononucleosis-like" symptoms
15. D. Gardasil
16. C. AIDS
17. B. CD4$^+$
18. D. *Neisseria gonorrhoeae*
19. E. *Trichomonas vaginalis*

23.5 Non-Sexually Transmitted Infections of the Reproductive Tract

20. A. candidal vulvovaginitis
21. C. A mixed population of aerobic and anaerobic organisms
22. D. A and B only

CLINICAL CORRELATION QUESTIONS

1. *Gardnerella vaginalis* is the most likely cause. A decrease in the number of lactobacilli of the normal vaginal microbiota leads to an increase in the vaginal pH. *G. vaginalis*, another member of the normal vaginal microbiota, grows better in slightly alkaline pH. Overgrowth of *G. vaginalis* causes bacterial vaginosis and the symptoms described in the case.
2. Notched teeth, a perforated palate that will need surgery to repair, and other defects

Chapter 24

OBJECTIVES REVIEW

24.1 Anatomy of the Central Nervous System

1. A. Brain
2. C. the blood-brain barrier
3. D. arachnoid and pia mater
4. B. damage to arachnoid villi

24.2 Overview of the Infectious Diseases of the Central Nervous System

5. C. Meningoencephalitis
6. C. an increased number of lymphocytes
7. A. Inflammation of his meninges

24.3 Viral Infections of the Central Nervous System

8. B. Lymphocytic choriomeningitis
9. True

10. D. Because mosquitoes that transfer the virus between birds do not usually prey on humans

24.4 Bacterial Infections of the Central Nervous System

11. A. Attachment of the organism loosens the tight junctions between endothelial cells.
12. C. *Clostridium botulinum*
13. D. *Listeria monocytogenes*

24.5 Fungal Infections of the Central Nervous System

14. B. being immunocompromised
15. A. *Coccidioides immitis*

24.6 Parasitic and Prion Infections of the Central Nervous System

16. C. *Acanthamoeba* species
17. Amphotericin B
18. B. a protein

CLINICAL CORRELATION QUESTIONS

1. This is neonatal meningitis. Gram-positive bacteria, especially *Streptococcus agalactiae*, and Gram-negative bacteria, often *Escherichia coli*, are the most likely etiologies because of the infant's age. The infant should be treated with antibiotic(s) to cover both Gram-positive and Gram-negative bacteria until the results of culture and sensitivities are obtained from the clinical laboratory.
2. A dermatologist who gave the patient a botox (botulinum toxin) injection in the face. The botulinum toxin relaxes the facial muscles, which in turn reduces the appearance of wrinkles. A number of factors, including administration of a larger-than-needed dose or injury to a nearby nerve, could have contributed to the facial paralysis of this patient.

Chapter 25

OBJECTIVES REVIEW

25.1 The Importance of Clinical Microbiology

1. B. Alerts the clinician to possible disease sequelae
2. C. The same bacterial species with the same antibiotic resistance pattern was found in all the sick students and in a vendor at the picnic.

25.2 Specimen Collection

3. C. Throat swab
4. Lumbar puncture or spinal tap
5. B. The patient has septicemia caused by an organism that is facultative with respect to oxygen.
6. D. CSF

25.3 Pathogen Identification Using Biochemical Profiles

7. A. *Neisseria meningitidis*
8. D. *Staphylococcus aureus*
9. E. *Salmonella enterica*
10. *Streptococcus pyogenes*; Lancefield classification
11. A waxy coat or mycolic acid
12. D

13. E
14. F
15. B
16. C
17. A
18. C
19. F
20. D
21. E
22. B
23. A

25.4 Pathogen Identification by Genetic Fingerprinting

24. biochemically, antigenically (or serologically), genetically
25. B. The DNA sequences of the PCR fragments were different.

25.5 Pathogen Identification by Serology, Antigenic Footprints, and Other Immunological Methods

26.

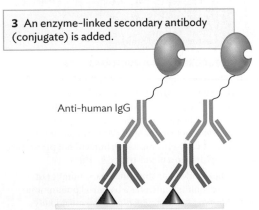

1 Microbial antigen is dried on a glass slide and treated with a chemical fixative.

2 Dilutions of patient serum are incubated with the antigen on the slide and then rinsed. Specific antibodies in the serum bind to the microbial antigen.

3 An enzyme-linked secondary antibody (conjugate) is added.

Anti-human IgG

27. C. Patients 3, 4, and 5 only

25.6 Point-of-Care Rapid Diagnostics

28. C. A patient *without* disease has a 95% chance of testing negative with the assay.
29. False
30. C
31. C. less sensitive

CLINICAL CORRELATION QUESTIONS

1. The laboratory could use darkfield microscopy to see live, motile, corkscrew-shaped, spirochete bacteria, but this would have to be done soon after collection because these anaerobic organisms die quickly in air. Serological tests would be used to detect anti-treponemal antibody in the patient's blood (RPR, VDRL; Section 23.4). The laboratory could also use fluorescent anti-treponemal antibody and immunofluorescence microscopy to "stain" the organisms present in tissue scraped from the chancre (Section 17.4). This test would provide the most definitive diagnosis. Culture is not possible because, as mentioned in Chapter 10, *T. pallidum* cannot grow in laboratory medium.

2. Answers will vary.

3. Use PCR to amplify the suspected antibiotic resistance gene and monitor its production by using real-time PCR. If the gene, for example, is *rpoD* encoding RNA polymerase, you can sequence the PCR product and look for mutations known to produce a rifampin-resistant protein.

Chapter 26

OBJECTIVES REVIEW

26.1 Principles and Tools of Epidemiology

1. the etiologic agent, the reservoir, the mode of transmission, and the host

2. Mode of transmission

3. Etiologic agent

4. Reservoir

5. Host

6. B. outbreak

7. C. a case–control study

8. incidence

26.2 Tracking Hospital-Acquired Infections

9. D. Urinary tract infections

10. C. The genetic analysis found that all the MRSA strains were identical.

11. D. Sanitize hands before entering the room.

26.3 Tracking Global Disease Trends

12. A. Local hospitals and physicians notify city or state health agencies about patients with a specific disease.

13. E. A and B

26.4 Detecting Emerging Microbial Diseases

14. True

15. True

16. True

17. True

18. True

19. False

20. True

21. False

22. B. Screen blood by PCR for rRNA genes and sequence the products.

26.5 Bioterrorism and Biodefense

23. select agent

24. A. infections can be treated with antibiotics

25. B. It is easily transmitted from person to person.

CLINICAL CORRELATION QUESTIONS

1. Everyone who was ill either ate soft cheese or lived with someone who ate soft cheese at the block party. Older people were more likely to get the disease.

2. Answers will vary.

3. Answers will vary.

Chapter 27

OBJECTIVES REVIEW

27.1 Microbes in Ecosystems

1. B. ocean

2. D. Serving as the base of an open-water food chain for the largest marine mammals

3. B. excessive organic substrates for microbial aerobic respiration

27.2 Microbial Symbiosis with Animals

4. C. commensalism

5. B. The rumen is well supplied with oxygen from the breathing animal.

6. A. *Shigella flexneri*

27.3 Fermented Foods and Beverages

7. A. cottage cheese

8. C. Removal of toxic ethanol

9. C. bacterial fermentation breaks down mainly sugars, leaving most protein intact

27.4 Food Spoilage and Preservation

10. D. Anaerobic respiration by bacteria converts trimethylamine oxide to trimethylamine, causing a distinctive unpleasant odor.

11. C. Pasteurization

12. D. *Listeria monocytogenes*

27.5 Wastewater Treatment

13. C. Agricultural runoff enters streams.

14. A. Primary treatment by screens and sedimentation

15. B. Aerobic or anaerobic respiration

27.6 Biogeochemical Cycles

16. D. Methanogenesis, reducing CO_2 to methane

17. B. Ammonium ion incorporation into amino acids

18. B. Nitrous oxide (N_2O)

CLINICAL CORRELATION QUESTIONS

1. When organic nitrogen-rich molecules enter a river, they can lead to blooms of algae that produce toxins harmful to people who drink the water. Also, the organic nitrogen can be released as ammonia, which nitrifiers convert to nitrites and nitrates; these ions are toxic, especially to infants. Instead, in the case reported here, wetland bacteria break down organic molecules from wastewater and facilitate incorporation of carbon and nitrogen into the biomass of living organisms in the wetland ecosystem. The water that comes out is purer, enhancing the health of the river and contributing to healthy water supplies.

2. The initial six passengers probably all became sick following ingestion of a shared food source. But the subsequent illness of other passengers over the next two weeks is the result of a chain of infection transmitted by the first ill passengers. Norovirus is often acquired from contaminated food, but also transmits readily from person to person.

ABC transporter   An ATP-powered transport system that contains an ATP-binding cassette and a transmembrane channel. (5)

aberration   An imperfection in a lens. (3)

ABO blood group system   A system that classifies blood on the basis of the presence of antigenic carbohydrates A and B on the surface of red blood cells. (17)

abscess   A localized region of pus surrounded by inflamed tissue. (19)

absorption   In optics, the capacity of a material to absorb light. (3)

AB toxin   An exotoxin composed of two subunits. The toxic A subunit is delivered to the host cell by the B subunit. (18)

Acanthamoeba encephalitis   A rare but often fatal inflammation of the brain caused by the ameba *Acanthamoeba*. (24)

acid-fast bacillus (AFB)   A descriptive name for mycobacteria, because with their mycolic acid–rich cell wall they stain positive with the Ziehl-Neelsen stain (or acid-fast stain). (20)

acid-fast stain   A diagnostic stain for mycobacteria and actinomycetes, which retain the dye carbolfuchsin because of mycolic acids in the cell wall. (3, 25)

acidophile   An organism that grows fastest in acidic environments (generally defined as below pH 5). (6)

acme   The point during the illness phase of an infectious disease when symptoms are most severe. (2)

acne vulgaris   A common chronic disease of the skin resulting from inflammation and/or blockage of hair follicles and their associated glands; involves mainly the face, chest, and back and is characterized by comedones, inflamed papular and pustular lesions, and inflamed cysts. (19)

acquired immunodeficiency syndrome (AIDS)   A disease caused by HIV that leads to the destruction of T cells and the inability to fight off opportunistic infections. (23)

activated   With respect to B cells, the process that follows binding of antigen to surface B-cell receptors, leading to B-cell clonal expansion. (16)

activated sludge   Sedimented flocs consisting of wastewater and a combination of filamentous and single-celled microbes and invertebrates designed to reduce the biochemical oxygen demand (BOD) and pathogen load in wastewater. (27)

activation energy   The energy needed for reactants to reach the transition state between reactants and products. (4)

activator   A regulatory protein that can bind to a specific DNA sequence and increase transcription of genes. (8)

active tuberculosis disease   In tuberculosis patients, the stage when they are contagious and exhibit primary disease symptoms, including coughing sputum, fever, night sweats, and weight loss. (20)

active immunization   Injection of a weakened or killed pathogen or of an antigenically active part of a pathogen with the intent of stimulating the adaptive immune system to produce specific antibodies and antigen-specific T cells. (17)

active transport   An energy-requiring process that moves molecules across a membrane against their electrochemical gradient. (5)

acute bacterial endocarditis (ABE)   An acute bacterial infection of the heart by a virulent microbe that causes severe inflammation of the heart lining (endocardium) and valves; compare with *subacute bacterial endocarditis*. (21)

acute disseminated encephalomyelitis (ADEM)   An autoimmune disease that follows an acute viral infection (such as measles), causing inflammation of the spinal cord and brain. (19)

acute otitis media (AOM)   Bacterial infection of the middle ear; typically found in children following an upper respiratory viral infection; usually caused by *Streptococcus pneumoniae*, *Haemophilus influenzae*, or *Moraxella catarrhalis*. (20)

acute rheumatic fever (ARF)   An autoimmune sequela of streptococcal infection; results from cross-reactivity of antibodies produced against certain streptococcal M protein epitopes with epitopes of the heart, joints, and nervous system. (20)

adaptive immunity   Immune responses activated by a specific antigen and mediated by B cells and T cells. (15)

adenitis   Inflammation of a lymph node or gland. (22)

adenosine triphosphate (ATP)   A ribonucleotide with three phosphates and the base adenine. It has many functions in the cell, including as a precursor for RNA synthesis and as an energy carrier. (7)

adherence   The ability of an organism to attach to a substrate. (5)

adhesin   Any cell-surface factor that promotes attachment of an organism to a substrate. (2, 18)

ADP-ribosyltransferase   An activity of some bacterial toxins that enzymatically transfers the ADP-ribose group from NAD^+ to target proteins, altering the target protein's structure and function. (18)

aerobic lithotrophy   The metabolic oxidation of inorganic compounds by the transfer of electrons to O_2 in reactions that yield energy. (7)

aerobic organotrophy   The metabolic oxidation of organic compounds to yield energy by the transfer of electrons to O_2. Most forms are also a type of aerobic respiration. (7)

aerobic respiration   The use of oxygen as the terminal electron acceptor in an electron transport system. A proton gradient is generated and used to drive ATP synthesis. (6, 7)

aerosol   A suspension of particles in air. This concept includes bacterial and viral pathogens that are transmitted via sneezing. (2)

aerotolerant anaerobe   An organism that does not use oxygen for metabolism but can grow in the presence of oxygen. (6)

affinity maturation   The process by which the antigen-binding site of an antibody population gains increased affinity for its target antigen or epitope during an adaptive immune response. (16)

African sleeping sickness   A protozoan disease caused by *Trypanosoma brucei* transmitted via the bite of the tsetse fly. The disease begins with a localized lesion, followed by lymphadenopathy, fever, and damage to the central nervous system leading to coma and death. (24)

agar   A polymer of galactose that is used as a gelling agent. (1)

agglutination   A technique for measuring insoluble antigens on whole cells. For example, ABO blood groups are typed using antibodies against the A and B antigens and looking for clumping of cells. (17)

AIDS   See **acquired immunodeficiency syndrome**.

alga *pl.* **algae**   A microbial eukaryote that contains chloroplasts and conducts photosynthesis. (1, 11)

alkaliphile   An organism with optimal growth in an environment that is above pH 9. (6)

allergen   An antigen that causes an allergic hypersensitivity reaction. (17)

allograft   Transplanted tissue from a donor with different MHC genes from those of the recipient. Unless the recipient is immunosuppressed, the transplant is rejected. (17)

allotype   An amino acid difference in the antibody constant region that distinguishes individuals within a species. (16)

allylamine   Member of a class of antifungal agents, such as Lamisil (terbinafine), that inhibits ergosterol synthesis. (13)

alpha helix   A protein secondary structure in which a portion of the polypeptide chain forms a helix held in place by hydrogen bonding between an amino hydrogen and the carbonyl of the fourth next aminoacyl group along the peptide backbone. (4)

alpha hemolysis   A diagnostic test used to identify bacteria by their ability to incompletely lyse red blood cells in a blood agar plate, generating green color. (25)

alpha-hemolytic   Refers to a bacterial species that carries out alpha hemolysis, such as *Streptococcus mutans* or *S. pneumoniae*. (25)

alpha toxin   A toxin produced by *Staphylococcus aureus* that lyses red blood cell; a hemolysin. (18)

alveolar macrophage   A type of macrophage, located in the lung alveoli, that phagocytoses foreign material. (15)

alveolate   A member of the eukaryotic group Alveolata, single-celled organisms with complex cortical structure that includes the ciliates, dinoflagellates, and apicomplexans. (11)

amastigote   A stage in the life cycle of certain protozoan parasites, such as *Leishmania*. It is an unflagellated form of the parasite that grows in macrophages and associated tissues of the host. Ingestion of infected macrophages by a sand fly continues the parasite life cycle. (11)

ameba   A protist that moves via pseudopods; also spelled amoeba. (11)

amebiasis   See **amebic dysentery**. (22)

amebic dysentery   Also called *amebiasis*. A gastrointestinal disease caused by *Entamoeba histolytica* invasion of the intestinal mucosa or into the bloodstream. Symptoms include colitis, diarrhea, fever, bloody stools, abdominal pain, and sometimes hepatitis. (22)

amensalism   An interaction between species that harms one partner but not the other. (27)

amino acid   The monomer unit of proteins. Each amino acid contains a central carbon covalently bonded to a hydrogen, an amino group, a carboxyl group, and a side chain. An exception is proline, in which the side chain is cyclized with the central carbon. (4)

amphipathic   Refers to a molecule having both hydrophilic and hydrophobic portions. (4)

amphotericin B   An antifungal drug that binds the fungus-specific sterol ergosterol and destroys membrane integrity. (13)

amyloid   1. Protein aggregates deposited in tissue, generally associated with pathology. In the brain, beta-amyloid deposits are associated with transmissible spongiform encephalopathy (prions) and Alzheimer's disease. (24) 2. In the kidney and other organs, amyloid accumulation refers to clumps of secreted antibody light chains, generated by neoplastic plasma cells. (17)

anabolism   Also called *biosynthesis*. The synthesis of complex biomolecules from smaller precursors. (9)

anaerobic lithotrophy   The metabolic oxidation of inorganic compounds in energy-yielding reactions that occur with an electron acceptor that is not O_2. An example is methanogenesis, in which H_2 donates electrons to CO_2 to form CH_4 and H_2O. (7)

anaerobic organotrophy   The metabolic oxidation of organic compounds to yield energy via the transfer of electrons to an acceptor other than O_2. When an electron transport chain is involved, the terminal electron acceptor is an inorganic molecule other than O_2, such as nitrate (anaerobic respiration). In fermentation, an organic molecule is the terminal electron acceptor. (7)

anaerobic respiration   The use of a molecule other than oxygen as the final electron acceptor of an electron transport chain. (6)

analytical epidemiology   The testing of hypotheses using experimental or observational methods to establish the cause of disease and associated risk factors based on descriptive epidemiological data. (26)

anaphylaxis   A type I hypersensitivity reaction due to introduction of an allergen into the bloodstream, causing sensitized mast cells to release large amounts of histamine and other immune mediators. Responses can include rapid drop in blood pressure, constricted airways, and swelling of the epiglottis. (17)

angstrom (Å)   A tenth of a nanometer, or 10^{-10} m. (3)

anion   A negatively charged ion. (4)

annotation   Interpreting the information in genome sequences, including identifying genes and predicting gene function. (9)

antagonistic   Referring to antibiotics used in combinations whose mechanisms of action interfere with one another and hence diminish their effectiveness; compare with *synergistic*. (13)

antibiotic   A molecule that can kill or inhibit the growth of selected microorganisms. (1, 13)

antibody   A host defense protein (a type of immunoglobulin) produced by differentiated B cells that binds to a specific antigen. (16)

antibody capture   An immunoassay in which antigen coats the surface of the ELISA microtiter wells, and patient serum is added to look for antibodies in the serum that will bind to the antigen. (25)

antibody-dependent cell-mediated cytotoxicity (ADCC)   The process by which natural killer cells destroy viral protein–expressing, antibody-coated host cells. (15)

antibody stain   An imaging procedure that uses an antibody to visualize cell components recognized by the antibody with high specificity. The stain is an antibody either linked to an enzyme that converts a colorless reactant to a colored product or linked to a fluorescent molecule that can be seen using fluorescence microscopy. (3)

anticodon loop   Region of a tRNA molecule containing three unpaired nucleotides that bind to a mRNA codon via complementary base pairing during translation. (8)

antigen   A compound, recognized as foreign by the cell, that elicits an adaptive immune response. See also *immunogen*. (16)

antigen capture   A type of immunoassay in which a specific antibody coats the wells of the ELISA plate, and patient serum is added to look for antigens in the serum that will bind to the antibody. (25)

antigenic determinant   See **epitope**. (16)

antigenic drift   Random mutations in a viral genome that cause minor changes in the structure of viral surface antigens, such as influenza hemagglutinin and neuraminidase. The consequence is a new viral strain that might better evade the host immune system. (12, 18, 20)

antigenicity   See **immunogenicity**. (16)

antigenic shift   A major change in a viral antigen that occurs when two (or more) strains of a virus with segmented genomes infect the same host cell and re-sort their genome segments. The resulting chimeric virus can express a new combination of surface antigens (such as influenza hemagglutinin and neuraminidase) that can alter virulence or enable the infection of new hosts. (12, 18, 20)

antigenic variation   The ability of some protozoan parasites to express multiple variants of their surface protein antigens, which overwhelms the host's ability to produce antibodies against the parasite, enabling it to evade the host immune response. (11)

antigen-presenting cell (APC)   An immune cell that can process antigens into epitopes and display those epitopes on the cell surface for recognition by naive T cells. (15, 16)

antiport   Coupled transport in which the molecules being transported move in opposite directions across the membrane. (5)

antisepsis   The removal of pathogens from living tissues. (13)

antiseptic   A chemical that kills microbes. (1)

antitoxin   Antibodies that bind to and inactivate microbial toxins. (17)

apicomplexan   A member of the eukaryotic phylum Apicomplexa, parasitic alveolates that possess an apical complex used for entry into a host cell. (11)

apoptosis   Programmed cell death of eukaryotic cells. (2, 15)

arabinogalactan   A polymer of galactose and arabinose; an important structural component of the mycobacterial cell wall. (5)

arbovirus   Any of a group of viruses transmitted by arthropod vectors, especially insects. (24)

archaeon *pl.* **archaea**   A prokaryotic organism that is a member of the domain Archaea, distinct from bacteria and eukaryotes. (1)

areflexia   The absence of reflexes. (24)

arthroconidium *pl.* **arthroconidia**   A fungal spore produced from the fragmentation of specialized hyphae. It is the infectious agent of *Coccidioides immitis* and *C. posodasii*, which can cause coccidioidomycosis. (24)

arthropod   A member of the phylum Arthropoda, characterized by an exoskeleton, segmented body, and appendages with joints. Many are parasites. (11)

artifact   A structure that is incorrectly interpreted when viewed through a microscope. (3)

ascariasis   Disease caused by *Ascaris lumbricoides*, a parasitic roundworm. After ingestion, the larvae migrate to the lungs first and then to the intestines. (22)

ascospore   The spore produced by an ascomycete fungus. (11)

aseptic   Free of microbes. (1)

aseptic meningitis   Inflammation of meninges due to nonbacterial causes, including viruses. (24)

asexual reproduction   Also called *vegetative reproduction*. Reproduction of a cell by fission or mitosis to form identical daughter cells. (11)

assimilation   An organism's acquisition of an element, such as carbon from CO_2, to build into body parts. (27)

atherosclerosis   Hardening of the arteries caused by an inflammatory response to the accumulation of LDL inside the arterial walls, leading to narrowing or blockage of arterial lumen. (21)

atom   The fundamental particle of an element. It is made of a positively charged nucleus, composed of protons and neutrons, and of negatively charged electrons that encircle the nucleus. (4)

atomic number   The number of protons in an atom; it is unique for each element. (4)

atomic weight   The average mass number of the isotopes of an atom. (4)

attaching and effacing lesion   Loss of intestinal microvilli as a result of an inflammatory response to the presence of enteropathogenic or enterohemorrhagic *E. coli*. (22)

attenuated   Referring to vaccines, live but crippled virus or bacteria injected into a host to stimulate an adaptive immune response. (16, 24)

autoclave   A device that uses pressurized steam to sterilize materials by raising the temperature above the boiling point of water at standard pressure. (1)

autoimmune response   A pathology caused by lymphocytes that can react to self antigens. (17)

autoinducer   A secreted molecule that induces quorum-sensing behavior in bacteria. (18)

autotroph   An organism that can reduce CO_2 to produce organic carbon for biosynthesis. (6, 7)

azole   An antifungal agent (e.g., imidazole) that interferes with the synthesis of ergosterol. (13)

babesiosis   An emerging disease caused by *Babesia microti*, an apicomplexan parasite that is transmitted by the bite of an infected tick. (21)

bacillary dysentery   See **shigellosis**. (22)

bacillus *pl.* **bacilli**   A bacterium with a linear shape. (3)

bacitracin   A topical antibiotic that affects cell wall synthesis. (13)

bacteremia   A bacterial infection of the blood. (14, 21)

bacterial sinusitis   Inflammation and congestion of the paranasal sinuses caused by bacterial infection. (20)

bactericidal   Having the ability to kill bacterial cells. (13)

bacteriophage   Also called *phage*. A virus that infects bacteria. (9, 12)

bacteriostatic   Having the ability to inhibit the growth of bacterial cells. (13)

bacterium *pl.* **bacteria**   A prokaryotic organism that is a member of the domain Bacteria, distinct from archaea and eukaryotes. (1)

bacteriuria   The presence of bacteria in the urine. (23)

ball-and-stick model   A molecular model in which stick lengths represent the distances between bonded pairs of atomic nuclei. (4)

Baltimore model   A classification scheme that organizes viruses by genome type and method of replication; was devised by David Baltimore. (12)

barophile   Also called *piezophile*. An organism that requires high pressure to grow. (6)

base excision repair   A DNA repair mechanism that cleaves damaged bases off the sugar-phosphate backbone. After endonuclease cleaves the phosphodiester backbone at the damaged site, a new, correct DNA strand is synthesized complementary to the undamaged strand. (8)

base substitution   See **point mutation**. (8)

basophil   A white blood cell that stains with basic dyes and secretes compounds that aid innate immunity. (15)

batch culture   The growth of bacteria in a closed system without additional input of nutrients; oxygen may be provided. (6, 12)

B cell   An adaptive immune lymphocyte that gives rise to antibody-producing plasma cells. Along with T cells, they are antigen-specific cells responsible for the adaptive immune response. (15)

B-cell receptor (BCR)   A B-cell membrane protein complex containing an antibody in association with the proteins Igα and Igβ. (16)

B-cell tolerance   The exposure of B cells to a high antigen dose, preventing future antibody production against that antigen; particularly important in preventing the immune system from attacking self antigens. (16)

Bell's palsy   Paralysis on one side of the face. (21)

benign   Referring to tumors that are slow growing and self-contained; they do not spread to infiltrate other tissues. (17)

beta hemolysis   A diagnostic test used to identify species and strains of bacteria on the basis of their ability to completely lyse red blood cells in a blood agar plate. (25)

beta sheet   A protein secondary structure that arises from hydrogen bonding between amino and carbonyl groups along neighboring stretches of the polypeptide backbone to form a sheet. (4)

binary fission   The process of replication in which one cell divides to form two genetically equivalent daughter cells of equal size. (6)

bioburden   The number of microorganisms found in a pharmaceutical product or in a body part. (14)

biochemical oxygen demand (BOD)   The amount of oxygen removed from an environment (usually water or wastewater) by aerobic respiration; also called biological oxygen demand. (27)

biofilm   A community of microbes growing on a solid surface. (6)

bioinformatics   A discipline at the intersection of biology and computing that analyzes gene and protein sequence data. (9)

biomass   The dry weight of organic matter found in the bodies of living organisms. (27)

bioremediation   The use of living organisms to detoxify environmental contaminants. (27)

biosynthesis   See **anabolism**. (7)

biotechnology   An applied field of biology in which DNA and genomes are analyzed and artificial molecules of DNA are constructed for various purposes, such as forensics, agriculture, food safety, and medical research. (8)

biowarfare   The utilization of pathogenic microbes to inflict mass casualties or cause widespread panic and fear. (26)

blastomycosis   Primary fungal lung infection caused by inhalation of *Blastomyces dermatitidis* spores, which can be followed by disseminated disease affecting the skin, bone, and genitourinary organs. (20)

blood-brain barrier   A selectively permeable membrane made up of tightly packed capillaries that supply blood to the brain and spinal cord. Large molecules and most pathogens cannot permeate the narrow spaces. Fat-soluble (lipophilic) molecules and oxygen can dissolve through the capillary cell membranes and are absorbed into the brain. (14, 24)

boil   See **furuncle**. (19)

bone marrow transplant   Medical procedure in which a person's bone marrow is destroyed and then replaced with bone marrow that contains healthy stem cells for the proper production of lymphocytes and other blood cells. (16)

botulinum   The protein toxin produced by the bacterium *Clostridium botulinum*; causes botulism and is used to relax muscle spasms and reduce wrinkles. (10)

botulism   A food-borne disease involving flaccid muscle paralysis; caused by a *Clostridium botulinum* toxin. (24)

bovine spongiform encephalopathy (BSE)   A prion disease that infects cattle, characterized by a slow deterioration of brain functions. It is invariably fatal. (24)

bradykinin   A cell signaling molecule that promotes extravasation, activates mast cells, and stimulates pain perception. (15)

bradyzoite   The slowly dividing cyst-enclosed form of sporozoan protozoans such as *Toxoplasma gondii*. (21)

bright-field microscopy   A type of light microscopy in which the specimen absorbs light and appears dark against a light background. (3)

bronchiolitis   Inflammation of the bronchioles. (20)

bronchitis   Inflammation of bronchi caused by a viral or bacterial infection. (20)

bubonic plague   A disease caused by the bacterium *Yersinia pestis*; characterized by swollen lymph nodes that often turn black. (21)

budding   A form of reproduction in which mitosis of the mother cell generates daughter cells of unequal size. (6)

bulla *pl.* **bullae**   A fluid-filled sac, or a blister, such as those associated with bullous impetigo. (19)

bullous impetigo   A contagious skin infection caused by *Staphylococcus aureus* strains that produce exfoliative toxin A; characterized by fluid-filled blisters. (19)

Burkitt's lymphoma   A B-cell cancer associated with Epstein-Barr virus infection. (17, 21)

burst size   The number of virus particles released from a lysed host cell. (12)

Calvin cycle   The metabolic pathway of carbon dioxide fixation in which the CO_2-condensing step is catalyzed by Rubisco. Found in chloroplasts and many bacteria. (7)

cancer   Also called *neoplasm*. Uncontrolled division of abnormal cells. (17)

candidal intertrigo   Skin infection caused by *Candida* species where skin rubs together, such as between toes or under the arms. (19)

Candida **meningitis**   A *Candida* infection of the meninges, often seen in premature neonates. (24)

capping   The clustering of B-cell receptor molecules at one pole of a B cell after receptors have been cross-linked by antigens or epitopes. (16)

capsid   The protein shell that surrounds a virion's nucleic acid. Within an enveloped virus, such as HIV, the capsid may be called a core particle. (12)

capsule   A slippery outer layer composed of polysaccharides that surrounds the cell envelope of some bacteria. (5)

carbohydrate   A class of biomolecules made of the atoms carbon, hydrogen, and oxygen; composed of simple sugars and sugars covalently bound together to form larger structures, including disaccharides, oligosaccharides, and polysaccharides. (4)

carbon dioxide (CO_2) fixation   Also called *carbon fixation*. The enzymatic covalent incorporation of inorganic carbon dioxide (CO_2) into an organic compound. (7)

carbon fixation   See **carbon dioxide (CO_2) fixation**. (7)

carbuncle   A painful lesion resulting from the joining of several furuncles. (19)

cardiomyopathy   A decrease in the ability of the heart muscle to contract. (21)

carrier   A person who harbors a disease agent but does not have symptoms of the disease. (2)

case–control study   An epidemiological study comparing a disease (or affected) group with a control group that is disease-free (or unaffected). Each group is assessed for previous exposure to a possible causal agent. (26)

casein   A family of hydrophobic milk proteins found as suspended droplets in milk. (27)

catabolism   The cellular breakdown of large molecules into smaller molecules, releasing energy. (7)

catabolite   A molecule, formed by the enzymatic digestion of food molecules, that is fed into metabolic energy pathways. (7)

catalase   An enzyme that catalyzes the conversion of hydrogen peroxide to molecular oxygen and water. (25)

catalase test   A rapid diagnostic test that looks for bubbles due to the release of O_2 when hydrogen peroxide is added to a bacterial colony. (25)

catalysis   The process of accelerating the rate of a chemical reaction; in cells, it is always accomplished with the assistance of enzymes. (4)

catalyst   A molecule or substance that increases the rate of a chemical reaction without being consumed in the reaction. (4)

cation   A positively charged ion. (4)

cat scratch disease   An infection caused by *Bartonella henselae* following a bite or scratch from an infected cat. (21)

cDNA   A synthetic copy of DNA made using messenger RNA as the template for reverse transcriptase. (8)

cell envelope   The cell membrane, cell wall, and (for Gram-negative species) outer membrane of a bacterium. (5)

cell fusion   A method of viral spreading in which an infected host cell fuses with an uninfected cell, allowing the virus particles to enter the uninfected cell through their common cytoplasm. (12)

cell-mediated immunity   A type of adaptive immunity employing mainly T-cell lymphocytes. (16)

cell membrane   Also called *plasma membrane* or *cytoplasmic membrane*. The phospholipid bilayer that encloses the cytoplasm. (4, 5)

cellulitis   A spreading superficial infection (inflammation) of the skin characterized by localized pain, swelling, tenderness, erythema, and warmth . (19)

cell wall   A rigid structure outside the cell membrane. The molecular composition depends on the organism; it is composed of peptidoglycan in bacteria. (5)

central nervous system (CNS)   The brain and spinal cord. (24)

cephalosporin   A class of semisynthetic beta-lactam antibiotics used to fight penicillin-resistant microbes. (13)

cercaria   The fork-tailed larva of the parasitic flatworms *Schistosoma* and *Opisthorchis*. (22)

cerebrospinal fluid (CSF)   A clear fluid, derived from blood, that bathes the spinal cord and brain. (24)

cestode   See **tapeworm**. (11, 22)

Chagas' disease   A zoonotic illness caused by the protozoan *Trypanosoma cruzi* that is transmitted by the bite of the bloodsucking reduviid bug. (21)

chain of infection   Factors that affect the spread of disease, which include the organism, reservoir, mode of transmission, and host. Also used to imply *person-to-person transmission*. (1, 26, 27)

chancre   A painless, hard lesion due to an inflammatory reaction at the site of infection with *Treponema pallidum*, which causes syphilis. (23)

chancroid   A sexually transmitted infection of *Haemophilus ducreyi* that causes painful ulcers on the genitals. (23)

chaperone   A protein that helps other proteins fold into their correct tertiary structure. (8)

chemically defined minimal medium   See **minimal medium**. (6)

chemoautotroph   An organism that oxidizes inorganic compounds to yield energy and reduce carbon dioxide; also called chemolithotroph, lithotroph, or chemolithoautotroph. (6)

chemokine   A protein produced by damaged or infected tissues that stimulates the migration of white blood cells toward the site of infection. (15)

chemostat   A continuous culture system in which the introduced medium contains a limiting nutrient. (6)

chemotaxis   The ability of an organism to move toward or away from specific chemicals. (5)

chemotherapeutic agent   Antibiotic made by microbes or modified or synthesized in the laboratory. (13)

chemotherapeutic index   A measure of an antibiotic's safety; it is the ratio of the toxic dose to the therapeutic dose. (13)

chemotrophy   Metabolism that yields energy from oxidation-reduction reactions without using light energy. (6, 7)

chickenpox   A childhood disease, caused by the varicella-zoster virus, that presents as an itchy rash over much of the upper body. (19)

chiral carbon   A carbon bonded to four types of functional groups; it can thus take two forms, exhibiting mirror symmetry. (4)

chitin   A polymer of *N*-acetylglucosamine that is a major structural component of fungal cell walls. (11)

chlamydia   1. An obligate, intracellular parasitic bacterium of the phylum Chlamydiae. 2. The most frequently reported sexually transmitted disease in the United States. Symptoms range from none, to a burning sensation upon urination, to sterility. (23)

chlorophyll   A magnesium-containing porphyrin pigment that captures light energy at the start of photosynthesis. (7)

cholecystitis   Inflammation of the gallbladder. (22)

cholera   A serious gastrointestinal illness caused by *Vibrio cholerae* and characterized by excessive watery diarrhea. (22)

cholera toxin   The *Vibrio cholerae* protein that enters host intestinal cells, alters cAMP production, and is responsible for the symptoms of cholera. The protein is an example of an AB toxin. (22)

chromosome   Part or all of an organism's genome; it is packaged into a highly compacted form. (8)

chronic lymphocytic leukemia   The most common type of leukemia, it is a B-cell cancer often associated with an abnormality in specific microRNA production. (17)

cidal agent   An antimicrobial chemical that kills microbes. (13)

cilia *sing.* **cilium**   Short, hairlike structures of eukaryotes that beat in waves to propel the cell; structurally similar to flagella. (11)

ciliate   Motile protozoan of the Alveolates, typically identified by having many cilia on the cell surface and two different-sized nuclei. (11)

citric acid cycle   See **tricarboxylic acid (TCA) cycle**. (7)

clade   Also called *monophyletic group*. A group of organisms that includes an ancestral species and all its descendants. (9)

class I MHC molecule   Membrane surface protein found on all nucleated cells of the human body (absent from red blood cells and platelets); presents cytoplasmically synthesized foreign antigen epitopes (e.g., virus epitopes) to cytotoxic T cells. (16)

class II MHC molecule   Membrane surface protein found on antigen-presenting cells (dendritic cells, macrophages, and B cells); presents phagocytized foreign antigen epitopes to helper T cells. (16)

class switching   See **isotype switching**. (16, 22)

clonal   Giving rise to a population of genetically identical cells, all descendants of a single cell. (16)

clonal selection   The rapid proliferation of a subset of B cells during the primary or secondary antibody response. (16)

cloning vector   A small genome, often a plasmid, into which foreign DNA can be inserted for cloning. (12)

CNA plate   Blood agar medium containing the inhibitors colistin and nalidixic acid. It is a selective medium that permits the growth of Gram-positive organisms. (25)

coccidioidal meningitis   An infection of the meninges caused by the fungus *Coccidioides immitis* or *C. posadasii*. (24)

coccidioidomycosis   A fungal respiratory infection caused by inhalation of the spores of a variety of *Coccidioides* species, including *Coccidioides immitis*. (20)

coccus *pl.* **cocci**   A spherically shaped bacterial or archaeal cell. (3)

codon   A set of three nucleotides within mRNA that encodes a particular amino acid. (8)

cohort study   An epidemiological study comparing an exposed group with a control group that is not exposed to a particular disease-causing agent. Each individual in the study is assessed for whether he or she contracts the disease. (26)

cold agglutinin   An antibody that can agglutinate red blood cells at 4°C but not at 37°C. (20)

cold sore   A collection of small blisters occurring on the lips that is caused by the herpes simplex virus 1. (19)

colitis   Inflammation of the large intestine. (22)

colonization   The ability of a microorganism to affix itself to a host surface or an environmental habitat, where it can then replicate. (2)

colony   A visible cluster of microbes on a plate, all derived from a single founding microbe. (1, 6)

colorization   An imaging technique in which color is added to a microscopic image to provide contrast and/or highlight components of the image. (3)

comedone   A type of acne resulting from blocked hair follicles or pores. (19)

commensalism   An interaction between two different species that benefits only one partner. (27)

common cold   The most frequent viral infection of the upper respiratory tract, it is a self-limiting infection, usually caused by a rhinovirus. (20)

common variable immunodeficiency (CVID)   A disorder characterized by various types of serum immunoglobulin deficiencies, including low levels of immunoglobulin isotypes (IgG, IgA, and IgE), and an impaired ability to produce specific antibodies after antigen exposure. (17)

competence factor   A species-specific secreted bacterial protein that induces competence for transformation. (9)

competent   Able to take up DNA from the environment. (9)

complement   Innate immunity proteins in the blood that form holes in bacterial membranes, killing the bacteria. Some components attract phagocytes; others can coat bacteria and promote phagocytosis. (15)

complement deficiency   Autosomally inherited defects in the complement pathway, resulting in increased susceptibility to infections with *Neisseria* and other pyogenic bacteria. (17)

complex   A macromolecule made of multiple protein subunits that work in a coordinated manner. (4)

complex medium   Also called *rich medium*. A nutrient-rich growth solution including undefined chemical components such as beef broth. (6)

composite transposon   A transposon containing genes in addition to those for transposition, such as genes for antibiotic resistance or catabolic functions. (9)

compound microscope   A microscope with multiple lenses to compensate for lens aberration and increase magnification. (3)

condensation reaction   In biochemistry, the formation of a covalent bond between two molecules that in the process releases a water molecule. (4)

condenser   In a microscope, a lens that focuses parallel light rays from the light source onto a small area of the specimen to improve the resolution of the objective lens. (3)

condyloma acuminatum *pl.* **condyloma acuminata**   Genital warts, caused by the human papillomavirus. (19, 23)

confluent   Describing a mode of growth that results in a lawn of organisms covering a surface. (6)

congenital cytomegalovirus infection   Also called *cytomegalic inclusion disease (CID)*. A cytomegalovirus infection contracted in utero. (21)

congenital syphilis   Syphilis contracted in utero. (23)

conjugated vaccine   A vaccine that combines a highly immunogenic protein with a poorly immunogenic bacterial capsule polysaccharide to boost the immune reaction to the polysaccharide. (17)

conjugation   Horizontal gene transmission involving cell-to-cell contact. In bacteria, pili draw together the donor and recipient cell envelopes, and a protein complex transmits DNA across. In ciliated eukaryotes, a conjugation bridge forms between two cells, connecting their cytoplasm and allowing the exchange of micronuclei. (9, 11)

conjunctivitis   Inflammation of the conjunctiva of the eye. (19)

constant region   The region of an antibody that defines the class of a heavy chain (isotype) or light chain. (16)

consumer   An organism that acquires nutrients from producers, either directly or indirectly. (27)

continuous culture   A culture system in which new medium is continuously added to replace old medium. (6)

contractile vacuole   An organelle in eukaryotic microbes that pumps water out of the cell. (5)

contrast   Differential absorption or reflection of electromagnetic radiation between an object and its background that allows the object to be distinguished from the background. (3)

convalescence   The stage of an infectious disease after symptoms have disappeared and the patient is recovering back to health. (2)

core polysaccharide   A sugar chain that attaches to the glucosamine of lipopolysaccharides and extends outside the cell. (5)

costovertebral tenderness   Tenderness between the last rib and the lumbar vertebrae; indicator of pyelonephritis. (23)

counterstain   A secondary stain used to visualize cells that do not retain the first stain. (3)

coupled transport   The movement of a substance against its electrochemical gradient (from lower to higher concentration or from opposite charge to like charge) using the energy provided by the simultaneous movement of a different chemical down its electrochemical gradient. (5)

covalent bond   A bond between two atoms that share a pair of electrons. (4)

C-reactive protein   A protein that stimulates the complement cascade; it is induced in the liver by cytokines. Elevated levels in the blood are associated with inflammation of any kind, including infection. (15)

Creutzfeldt-Jakob disease (CJD)   A prion disease that infects humans, characterized by a slow deterioration of brain functions. It is invariably fatal. (24)

croup   Also called *laryngotracheobronchitis* (*LTB*). Viral infection of the larynx, trachea, and bronchi resulting in a "barking" cough; usually caused by parainfluenza viruses. (20)

cryptococcal meningoencephalitis   A fungal infection of the CNS caused by *Cryptococcus neoformans* var. *neoformans*. It is a serious concern among immunocompromised patients. (24)

cryptococcosis   A serious, potentially fatal fungal disease of the lung or meninges caused by *Cryptococcus neoformans*. Cryptococcal meningitis is usually a hallmark infection for AIDS. (20)

cryptosporidiosis   An intestinal disease caused by the parasite *Cryptosporidium*. (22)

curd   Coagulated milk proteins produced by the combined action of lactic acid–producing bacteria and stomach enzymes of certain mammals, such as cattle. (27)

cycloserine   An antibiotic that inhibits peptidoglycan synthesis. (13)

cyclosporiasis   A gastrointestinal disease caused by infection with *Cyclospora cayetanensis* sporocysts; presentation varies from asymptomatic to diarrhea as well as flu-like symptoms. (22)

cyst   1. A tough, protective capsule in tissue that encloses the larva of a parasitic worm within an infected animal; 2. an infectious form of many protozoan parasites, usually passed in feces. (11)

cysticercus *pl.* **cysticerci**   The cystic phase of certain tapeworms, such as *Taenia*. The cysts are found in the muscle of infected cattle, pigs, and goats. (22)

cystitis   Inflammation of the urinary bladder such as with infection. (23)

cytochrome oxidase test   A diagnostic test for the presence of *Neisseria*. The colorimetric test is positive only if the bacteria contain both cytochrome oxidase and cytochrome *c*. (25)

cytokine   Any of a group of small signaling proteins secreted by human cells that bind to receptors on endothelial and immune system cells, altering the behavior of those cells. (14)

cytokine storm   Massive production of cytokines, leading to excessive tissue damage, inflammation, and shock. (16)

cytomegalic inclusion disease (CID)   See **congenital cytomegalovirus infection**. (21)

cytomegalovirus (CMV)   Human herpes virus 5, a double-stranded DNA virus; cause of a mononucleosis-like disease in adults and cytomegalic inclusion body disease of neonates. (21)

cytoplasm   Also called *cytosol*. The aqueous solution contained by the cell membrane in all cells and outside the nucleus (in eukaryotes). (4)

cytoplasmic membrane   See **cell membrane**. (5)

cytoskeleton   A collection of filamentous proteins that impart structure to cells, aid movement of cells, and facilitate transport of components within cells. In a eukaryote, components of the cytoskeleton include microfilaments, intermediate filaments, and microtubules. (5)

cytosol   See **cytoplasm**. (4)

cytotoxic T cell (T$_C$ cell)   A T cell that expresses CD8 on its cell surface, binds MHC I–presented epitopes, and can secrete toxic proteins such as perforin and granzymes. (16)

daisy-chain nosocomial infection   An infection that begins within one hospital or medical facility that is then spread to other medical centers by infected patients who are moved from place to place. (26)

daughter strand   One of the two growing strands formed during DNA replication; it is complementary to the parental strand, which serves as its template. (8)

dead zone   An anoxic region of an ocean, devoid of most fish and invertebrates; also called zone of hypoxia. (27)

death phase   A period of cell culture, following stationary phase, during which bacteria die faster than they replicate. (6)

death rate   The rate of halving of population density under a deleterious condition. (6)

decimal reduction time (D-value)   The length of time it takes for a treatment to kill 90% of a microbial population, and hence a measure of the efficacy of the treatment. (13)

decline phase   In the course of an infectious disease, the stage in which symptoms begin to subside. (2)

decomposer   An organism that consumes dead biomass. (27)

defensin   A type of small, positively charged peptide, produced by animal tissues, that destroys the cell membranes of invading microbes. (15)

definitive host   The host organism in which a parasitic worm develops and undergoes sexual reproduction. (11, 22)

degenerative evolution   See **reductive evolution**. (9)

degranulation   A process in which intracellular vesicles (granules) within mast cells, neutrophils, natural killer cells, and cytotoxic T cells release their toxic molecules to the external environment. (17)

delayed-type hypersensitivity (DTH)   See **type IV hypersensitivity**. (17)

deletion   The loss of one or more nucleotides from a DNA sequence. (8)

denatured   Loss of protein structure (and hence function) due to extreme temperature or a chemical agent. (4)

dendritic cell   An antigen-presenting white blood cell that primarily takes up small soluble antigens from its surroundings. (15)

dengue hemorrhagic fever (DHF)   The most severe form of dengue fever, with symptoms that include hemorrhaging, respiratory distress, extreme bone pain, and organ failure. (21)

denitrification   Energy-yielding metabolism in which nitrate (NO_3^-) is reduced to nitrite (NO_2^-), diatomic nitrogen (N_2), and in some cases ammonia (NH_3). (6)

dental caries   Tooth decay, which is caused by acidic products of bacterial fermentation that destroy tooth enamel. (22)

dental plaque   A thickening biofilm of microorganisms that forms at the margin of the teeth and gums. (22)

depth of field   In a microscope, a region of the optical column over which a specimen appears in reasonable focus. (3)

dermatome   An area of skin served by a specific spinal nerve. (19)

dermatophyte   A fungal pathogen that colonizes human skin. (19)

dermis   The deeper layer of skin located below the epidermis and consisting mostly of connective tissue. (19)

descriptive epidemiology   The collection of data concerning the people, places, and times of events related to a disease. (26)

desensitization   A clinical treatment to decrease allergic reactions by exposing patients to small doses of an allergen. (17)

diabetes   Chronic systemic disease caused by abnormal plasma glucose levels due to faulty glucose metabolism. (12)

diaphragm   A device in a microscope to vary the diameter of the light column, changing the amount of light admitted. (3)

diarrhea   A malady of the gastrointestinal tract characterized by having three or more loose stools a day. (22)

dichotomous key   A tool for identifying organisms, in which a series of yes/no decisions successively narrows down the possible categories of species. (9)

differential   A list of possible causes of an infection or disease. (2)

differential medium   A growth medium that can distinguish between various bacteria on the basis of metabolic differences. (6)

differential stain   A stain that differentiates among objects by staining only particular types of cells or specific subcellular structures. (3)

diffusion   The energy-independent net movement of a substance from a region of higher concentration to a region of lower concentration. (4)

DiGeorge syndrome   A recessive genetic immunodeficiency disorder causing underdevelopment or complete lack of the thymus, underdevelopment of some facial features, and twitching of extremities. (17)

dilution streaking   See **isolation streaking**. (6)

dinoflagellate   A member of the eukaryotic group Dinoflagellata—tertiary endosymbiont algae, alveolates with two flagella, one of which is wrapped distinctively around the cell equator. (11)

diphtheria   A systemic disease caused by the dissemination of a toxin produced *Corynebacterium diphtheriae*. (20)

diploid   An organism (or cell) that contains two copies of each chromosome in its cells. (11)

direct contact transmission   Spreading of a pathogen either through intimate contact or by aerosolization. (2)

direct ELISA   An immunoassay that detects a specific antigen bound to a microtiter plate with enzyme-conjugated antibodies against that specific antigen and a colorless substrate that the enzyme converts to a colored product. (17)

direct immunofluorescence test   A diagnostic method for detecting microbial pathogens in a patient sample by adding fluorescently labeled anti-pathogen antibodies to a sample of tissue or fluid (blood, saliva, CSF) and then observing under fluorescence light microscopy to see the now-fluorescent pathogen in the sample. (17)

disaccharide   A carbohydrate composed of two sugar molecules joined by a covalent bond. (4)

disease   The disruption of the normal structure or function of any body part, organ, or system that is recognized by a characteristic set of symptoms. (2)

disinfection   The removal of pathogenic organisms from inanimate surfaces. (13)

disseminated intravascular coagulation (DIC)   A consequence of septic shock in which excessive coagulation occurs, causing clots to form in organs throughout the body. (21)

dissimilation   An organism's catabolism or oxidation of nutrients to inorganic minerals that are released into the environment. (27)

divergence   The accumulation of genetic differences between two lineages after evolutionary separation; the differences arise from mutations, gene mobility, and natural selection. It can lead to the formation of new species. (9)

DNA ligase   An enzyme that cells use to form a covalent bond at a nick in the phosphodiester backbone; also used in molecular biology laboratories to join pieces of DNA. (8, 12)

DNA polymerase The group of enzymes responsible for replicating DNA. (8)

DNA polymerase III The enzyme in bacteria responsible for synthesizing the daughter strands during DNA replication. (8)

DNA sequencing A technique to determine the order of bases in a DNA sample. (1)

DNA vaccine An experimental vaccine that contains the DNA for the antigen rather than the fully formed antigen. (17)

domain In taxonomy, one of three major subdivisions of life: Archaea, Bacteria, and Eukarya. In protein structure, a domain is a portion of a protein that possesses a defined function, such as binding DNA. (9)

doubling time The generation time of bacteria in culture; the amount of time it takes for the population to double. (6)

DTaP series A vaccine against diphtheria toxin, tetanus toxin, and pertussis antigens. (20)

dysuria Painful or difficult urination. (23)

early symptomatic HIV infection stage A phase of HIV infection characterized by decreased CD4⁺ T-cell numbers, unusual secondary infections, as well as high fever and prolonged diarrhea. (23)

eastern equine encephalitis (EEE) virus A positive-sense, single-stranded RNA virus transmitted to birds, humans, and horses via a mosquito vector. (24)

echinocandin Antifungal agent that blocks fungal cell wall synthesis. (13)

eclipse period The time after viral genome injection into a host cell but before complete virions are formed. (12)

ecosystem A community of species plus their environment (habitat). (27)

ecosystem service The benefits for human society conferred by a natural ecosystem. (27)

ectoparasite A harmful organism that colonizes the surface of a host. (2, 11)

edema factor (EF) A component of anthrax toxin with adenylate cyclase activity. (18)

efflux transporter A type of ATP-binding cassette transporter that expels waste products from cells. (5)

electrolyte Another name for ion; a charged atom. (4)

electron microscopy (EM) A form of microscopy in which a beam of electrons accelerated through a voltage potential is focused by magnetic lenses onto a specimen. (3)

electron transport chain A series of membrane-embedded proteins that converts the energy of redox reactions into a proton potential; also referred to as the electron transport system (ETS). (6)

electroporation A laboratory technique that temporarily makes the cell membrane more leaky to allow the uptake of DNA. (9)

element A pure chemical substance composed of one type of atom; it is given a unique identifier, its atomic number, which equals the number of protons in its nucleus. (4)

elephantiasis A symptom of several diseases of the lymphatic system, including infection by filarial worms, characterized by blockage of lymphatic vessels leading to grotesque swelling of the limbs. (21)

elite controller A rare HIV-positive individual who maintains a high CD4⁺ T-cell number and shows little or no viremia. (23)

elongation The middle phase of DNA synthesis, in which DNA polymerase adds dNTPs to the growing daughter strands. (8)

empirical therapy The administration of antibiotics according to a patient's symptoms and one's understanding of infectious disease before the pathogen has been identified; used in life-threatening situations, such as meningitis. (13, 24, 25)

empty magnification Magnification without an increase in resolution. (3)

enanthem A rash on a mucous membrane. (19)

encephalitis Inflammation of the brain. (24)

encephalopathy Diseased or damaged brain such as that resulting from an infection. (24)

endemic See **endemic disease**. (26)

endemic disease A disease that is always present in a population, although the prevalence of infection may be low. (2)

endergonic Describing a reaction that requires an input of energy to proceed. (4)

endocarditis Inflammation of the endocardium (heart's inner lining), usually involving the heart valves. (21)

endocytosis The invagination of the cell membrane to form a vesicle that contains extracellular material. (4, 24)

endomembrane system A series of membranous organelles that organize uptake, transport, digestion, and expulsion of particles through a eukaryotic cell. Includes endosomes, lysosomes, the endoplasmic reticulum, the nuclear envelope, and the Golgi complex. (5)

endoparasite A harmful organism that lives inside a host. (2)

endophthalmitis Infection of the inner eye. (19)

endoplasmic reticulum (ER) A complex membranous organelle of eukaryotic cells where protein and lipid synthesis occur. (5)

endosome A membrane-bound compartment inside a eukaryotic cell. (4)

endospore A durable, inert, heat-resistant bacterial spore that can remain viable for thousands of years. (10)

endosymbiont An organism that lives as a symbiont inside another organism. (1, 9)

endosymbiosis An intimate association between different species in which one partner grows within the body or cell of another organism. (5)

endotoxin A lipopolysaccharide in the outer membrane of Gram-negative bacteria that becomes toxic to the host after the bacterial cell has lysed. (5, 18)

energy The ability to do work. (4, 7)

enriched medium A growth solution for fastidious bacteria, consisting of complex medium plus additional components. (6)

enrichment culture The use of selective growth media and specific incubation conditions to allow only certain microbes to grow. (1)

enteric fever See **typhoid fever**. (22)

enteric pathogen Pathogenic bacteria that cause diarrhea. (22)

enteritis Inflammation of the small intestines. (22)

enteroaggregative E. coli (EAEC) Diarrheagenic *E. coli* strains that have unique bundle-forming pili to adhere to and colonize the intestinal mucosa. (22)

enterocolitis Inflammation of the colon and small intestines. (22)

enterohemorrhagic E. coli (EHEC) Diarrheagenic *E. coli* strains that secrete Shiga toxin and induce intestinal cells on which they attach to form pedestals; they are found in animal reservoirs and cause hemorrhagic disease (bloody stools). (22)

enteroinvasive E. coli (EIEC) Diarrheagenic *E. coli* strains that invade epithelial cells. (22)

enteropathogenic E. coli (EPEC) Diarrheagenic *E. coli* strains that form pedestals similar to those that occur in EHEC but typically lack toxins. (22)

enterotoxigenic E. coli (ETEC) Diarrheagenic *E. coli* strains that secrete enterotoxins. (22)

enterotoxin Toxic protein made by microbes that damages the intestine of the host and causes diarrhea and/or vomiting; produced by some Gram-negative and Gram-positive pathogens. (14, 22)

entropy A measure of the disorder in a system. (4, 7)

envelope For a virus, a membrane enclosing the capsid, or core particle. (12)

enzyme A biological catalyst; a protein or RNA that can speed up the rate of a reaction without itself being changed. (4)

enzyme immunoassay (EIA) See **enzyme-linked immunosorbent assay (ELISA)**. (17)

enzyme-linked antibody An antibody conjugated to an enzyme that converts a colorless substrate to a colored product. (17)

enzyme-linked immunosorbent assay (ELISA) A set of diagnostic techniques that use enzyme-linked antibodies to detect specific antigens, pathogens, or antibodies. (17)

eosinophil A white blood cell that stains with the acidic dye eosin and secretes compounds that facilitate innate immunity. (15)

eosinophila An abnormally large number of eosinophils in the blood. (22)

epidemic A disease outbreak in which many individuals in a population become infected over a short time. (2, 26)

epidermis The outer protective cell layer in most multicellular animals. (14, 19)

epididymitis Inflammation of the epididymis, the coiled tube that carries semen from the testes. (23)

epiglottitis Inflammation of the epiglottis. (20)

epitope Also called *antigenic determinant*. A small segment of an antigen that can elicit an immune response. An antigen can have many epitopes. (16)

equivalence The antigen/antibody ratio that leads to immunoprecipitation of large, insoluble complexes. It is the state in which the number of antigenic sites roughly equals the number of antigen-binding sites. (17)

erysipelas An acute infection of the upper layer of dermis that spreads to the superficial lymphatics and is caused by streptococcus bacteria, typically *Streptococcus pyogenes*, and that is characterized by a fiery-red rash, lymphadenopathy, and fever. (19)

erythema infectiosum Fifth disease, which is caused by parvovirus B19 and causes a nonpruritic malar rash, "slapped-cheek" rash, along with mild fever and flu-like symptoms. (19)

erythema marginatum A skin rash usually on the trunk or extremities that occurs in some children who have contracted acute rheumatic fever. (20)

erythema migrans A bull's-eye rash characteristic of borreliosis (Lyme disease). (21)

erythematous Reddened and inflamed (describes skin). (19)

essential nutrient A compound that an organism cannot synthesize and must acquire from the environment in order to survive. (6)

etiologic agent The organism, virus, or toxin that causes a disease. (2)

eukaryote An organism whose cells contain a nucleus. All eukaryotes are members of the domain Eukarya. (1, 5)

eukaryotic cell A cell that possesses a nucleus, or that differentiated from a nucleated stem cell. (5)

eukaryotic microbe A microorganism that is a member of the domain Eukarya. (1)

eutrophic Referring to an aquatic or marine environment in which oversupply of electron donors (usually organic matter) leads to oxygen depletion and hence loss of animal life. (27)

exanthem Any widespread rash on the skin that is accompanied by fever, malaise, or headache. (19)

exergonic Describing a spontaneous reaction that releases free energy. (4)

exocytosis Fusion of vesicles with the cell membrane to release vesicle contents into the extracellular milieu. (5, 24)

exopolysaccharide (EPS) A thick extracellular matrix of polysaccharides and entrapped materials that forms around microbes in a biofilm. (6)

exotoxin A protein toxin, secreted by bacteria, that kills or damages host cells. (18)

experimental study A controlled analysis designed to test a question or hypothesis. (26)

exponential growth Growth in which the population size doubles at a fixed rate. (6)

exponential phase Also called *logarithmic (log) phase*. A period of cell culture during which bacteria grow exponentially at their maximum possible rate based on the growth conditions. (6)

extravasation The movement of white blood cells out of blood vessels and into surrounding infected tissue. (15)

extreme drug-resistant tuberculosis (XDR-TB) Strains of *Mycobacterium tuberculosis* that are resistant to at least two first-line tuberculosis drugs and at least three second-line drugs. (20)

extremophile An organism that grows only in an extreme environment—that is, an environment including one or more conditions that are "extreme" in relation to the conditions for human life, such as very high or low pH. (6)

F(ab)₂ region The two linked antigen-binding arms of an antibody molecule. (16)

facilitated diffusion A process of passive transport across a membrane that is facilitated by transport proteins. (5)

factor H A normal serum protein that prevents the inadvertent activation of complement in the absence of infection. (15)

facultative Having two alternative growth states or conditions, such as aerobic or anaerobic. (6)

facultative anaerobe Capable of growth under either anaerobic or aerobic conditions. (6)

facultative intracellular pathogen A pathogen that can live either inside host cells or outside host cells. (18)

false negative reaction Negative result from a diagnostic test of a patient who has the disease or disease-causing pathogen, but the test is not sensitive enough to detect its presence. See also *sensitivity*. (25)

false positive reaction Positive result from a diagnostic test of a patient who is free of the disease or disease-causing pathogen, but the test is not specific enough to detect only the relevant parameter. See also *specificity*. (25)

F⁺ cell The DNA donor cell that transmits the fertility factor F to an F⁻ cell during conjugation. (9)

F⁻ cell The DNA recipient cell during conjugation. (9)

Fc region The region of an antibody that binds to specific receptors on host cells in an antigen-independent manner. It is found in the carboxy-terminal "tail" region of the antibody. (16)

fecal-oral route A route of transmission of pathogenic microbes in which organisms are ingested in contaminated food or water and then defecated, so they again contaminate the food or water supply. (2)

fecal transplant Transfer of the intestinal microbiome of a healthy donor to a patient suffering from a severe intestinal disease; an example of using probiotics. (14)

feces A mixture of chyme and intestinal bacteria concentrated into a semisolid waste material in the large intestines. (22)

fermentation Also called *fermentative metabolism*. The production of ATP via substrate-level phosphorylation, using organic compounds as both electron donors and electron acceptors. Industrial fermentation is the production of microbial products that are made by microbes grown in fermentation vessels; it may include respiratory metabolism to maximize microbial growth. (1, 6, 7)

fermentative metabolism See **fermentation**. (6)

fermented food A food product that is biochemically modified by microbial growth. (27)

fertility factor (F factor) A specific plasmid (transferred by an F⁺ donor cell) that contains the genes needed for pilus formation and DNA export. (9)

filamentous virus A virus consisting of a helical capsid surrounding a single-stranded nucleic acid. (12)

filarial Referring to a parasitic roundworm. (11)

filariasis A variety of diseases caused by nematodes (roundworms) in the family Filariidae. (21)

filariform Larval form of hookworms and parasitic nematodes that can penetrate human skin, migrate to the intestines, and establish a new infection. (22)

fimbria *pl.* **fimbriae** See **pilus**. (5, 18)

first law of thermodynamics Energy can be transferred from place to place, and it can be transformed, but energy cannot be created or destroyed. It is a statement about the conservation of energy. (4)

fixation The adherence of cells to a slide by a chemical or heat treatment. (3)

flagellum *pl.* **flagella** A filamentous structure for motility. In prokaryotes, a helical protein filament attached to a rotary motor; in eukaryotes, an undulating cell membrane–enclosed complex of microtubules and ATP-driven motor proteins. (5)

flatworm See **platyhelminth**. (22)

flow cytometer See **fluorescence-activated cell sorter (FACS)**. (17)

flow cytometry A diagnostic and analytical method for counting, sorting, and detecting cells by suspending them in a stream of liquid that forces cells to move in single file between a laser and a detector. (17)

flucytosine Antifungal agent that inhibits DNA synthesis. (13)

fluke Also called *trematode*. Member of a family of parasitic flatworms that cause infections of the blood, lungs, liver, and intestines. (11, 22)

fluorescence The emission of light from a molecule that absorbs light of a specific wavelength and emits light of a longer wavelength. (3)

fluorescence-activated cell sorter (FACS) Also called *flow cytometer*. A device that can count cells and sort them according to differences in fluorescence. (6, 17)

fluorescence microscopy (FM) Microscopy in which a fluorescent specimen is illuminated with light in the excitation range of wavelength, and observed in the wavelength range of emission. (3)

fluorescent antibody staining A diagnostic and analytical method of identifying microorganisms by visualizing them under a fluorescence microscope after they have been exposed to antibodies conjugated to a fluorophore. (25)

fluorophore A fluorescent molecule used to stain specimens for fluorescence microscopy. (3)

focal plane A plane that contains the focal point for a given lens. (3)

focal point The position at which light rays that pass through a lens intersect. (3)

follicular lymphoma A non-Hodgkin's lymphoma in which the *bcl-2* gene translocates near an immunoglobulin heavy-chain promoter, causing overexpression of *bcl-2*, which prevents cell death. (17)

folliculitis   Inflammation of hair follicles, often caused by a *Staphylococcus aureus* infection. (19)

fomite   An inanimate object on which pathogens can be transmitted from one host to another. (2)

food contamination   See **food poisoning**. (27)

food poisoning   Also called *food contamination*. A disease caused by ingesting food contaminated with microbial pathogens or toxins. (22, 27)

food preservation   Methods for keeping food from spoiling owing to microbial contamination; includes drying, smoking, canning, acidification (pickling), addition of antimicrobial agents, and killing of microbes with heat or pressure. (27)

food spoilage   Microbial changes that render a food unfit or unpalatable for consumption. (27)

food web   A network of interactions in which organisms obtain nutrients from or provide nutrients for each other—for example, by predation or by mutualism. (27)

foram   See **foraminiferan**. (11)

foraminiferan   Also called *foram*. An ameba with a calcium carbonate shell and a helical arrangement of chambers. (11)

Fournier's gangrene   A necrotizing fasciitis infection of the perineum and genitals. (23)

F-prime (F′) factor   See **F prime (F′) plasmid**. (9)

F-prime (F′) plasmid   Also called *F-prime (F′) factor*. A fertility factor plasmid that contains some chromosomal (host) DNA. (9)

frameshift mutation   A gene mutation involving the insertion or deletion of one or more nucleotides that causes a shift in the codon reading frame. (8)

free energy change   In a chemical reaction, a measure of how much energy that is available to do work is released (negative free energy change) or required (positive free energy change) as the reaction proceeds. (4)

fruiting body   A multicellular fungal or bacterial reproductive structure. (11)

functional genomics   A field of molecular biology that uses data from genomic research to describe the relationship between genes, proteins, and protein functions. (9)

functional group   A cluster of covalently bonded atoms that behaves with specific properties and functions as a unit. (4)

fungus *pl.* **fungi**   A heterotrophic opisthokont eukaryote with chitinous cell walls. Fungi include Eumycota but traditionally may refer to fungus-like protists such as the oomycetes. (1, 11)

furuncle   Also called a *boil*. An abscess of a hair follicle. (19)

gangrene   A condition in which a substantial portion of tissue (often in the extremities) becomes necrotic. (19)

gas gangrene   Form of gangrene caused by bacterial infection, usually by *Clostridium perfringens*; also called clostridial myonecrosis. (19)

gastric mucosa   The layer of epithelial cells that line the stomach. (22)

gastric pit   One of many invaginations of the gastric mucosa lined with secretory cells involved in digestion. (22)

gastritis   Inflammation of the stomach lining. (22)

gastroenteritis   Inflammation of the gastrointestinal tract. (22)

gel electrophoresis   An analytical method for separating a mixture of nucleic acids or proteins on the basis of their size and charge, by subjecting them to an electric field as the molecules migrate through a sieving gel medium. (8)

gene   A distinct series of nucleotides within DNA that has a distinct function (regulatory) or whose product has a distinct function (protein). The functional unit of heredity. (8)

gene expression   Transcription and translation of one or more specific genes. (8)

generalized recombination   Also called *homologous recombination*. Recombination between two DNA molecules that share long regions of DNA homology. (9)

generalized transduction   A phage-mediated gene transfer process in which any bacterial donor gene can be transferred to a recipient cell. (9)

general paresis   A condition of progressive dementia and partial paralysis in the later stages of syphilis. (23)

general secretion pathway   Mechanism in bacteria for moving proteins from the cytoplasm across the cell membrane (or inner membrane of Gram-negative bacteria). (18)

generation time   The species-specific period for doubling of a population (e.g., by bacterial cell division) in a given environment, assuming no depletion of resources. (6)

gene switching   Switching between two (out of five) classes of immunoglobulin genes (e.g., from IgM to IgG) during B-cell development. (16)

genetic drift   The process by which two species diverge as a result of random mutations. (9)

genital herpes   An infection of the external genitals caused by herpes simplex viruses 1 and 2. (19, 23)

genome   The complete genetic content of an organism; the sequence of all the nucleotides in a haploid set of chromosomes. (1, 4, 8)

genomic island   A region of DNA sequence whose properties indicate that it has been transferred from another genome. Usually comprises a set of genes with shared function, such as pathogenicity or symbiosis support. (9, 18)

genotype   All or part of the genetic constitution of an individual organism or cell. (8)

geochemical cycling   The global interconversion of various inorganic and organic forms of elements. (1)

germ   A microorganism that causes disease. (13)

German measles   A childhood disease, caused by the rubella virus, that produces a rash, fever, and swollen lymph nodes. (19)

germicidal   Able to kill cells but not spores. (13)

germicide   A chemical that kills germs. (13)

germination   The activation of a dormant spore to generate a vegetative cell. (6)

germ theory of disease   The theory that specific diseases are caused by specific microbes. (1)

giant cell   An abnormally large cell formed by fusion of several cells, as seen in cytomegalovirus infection. (21)

giardiasis   Diarrheal disease caused by the parasitic protozoan *Giardia intestinalis* (formerly *G. lamblia*). (22)

gingivitis   Inflammation of the gums. (22)

global regulator   Protein that controls the expression of multiple genes. (8)

glycan   Polysaccharide chain composed of oxygen-linked (O-linked) monosaccharides. (5)

glycolysis   The catabolic pathway of glucose oxidation to pyruvate, in which glucose 6-phosphate isomerizes to fructose 6-phosphate, ultimately yielding 2 pyruvate, 2 ATP, and 2 NADH; also called the Embden-Meyerhof-Parnas (EMP) pathway. (7)

gnotobiotic animal   An animal that is germ-free or colonized by a known set of microbes. (14)

Golgi   See **Golgi complex**. (5)

Golgi complex   Also called *Golgi* (or Golgi apparatus). Membranous organelle that modifies proteins and helps sort them to the correct eukaryotic cell compartment. (5)

gonorrhea   Sexually transmitted disease caused by *Neisseria gonorrhoeae*. (23)

gramicidin   A peptide antibiotic that disrupts membrane integrity. (13)

Gram-negative   Describing cells that do not retain the Gram stain. (3)

Gram-negative bacteria   Bacterial species that due to their thin cell wall of peptidoglycan lose the crystal violet stain during the decolorizing step; possess inner (cytoplasmic) and outer membranes. (5)

Gram-positive   Describing cells that retain the crystal violet Gram stain and appear dark purple after staining. (3)

Gram-positive bacteria   Firmicutes; bacterial species that due to their thick cell wall of peptidoglycan retain the crystal violet stain despite the decolorizing step; do not have an outer membrane. (5)

Gram stain   A differential stain that distinguishes cells that possess a thick cell wall and retain a positively charged stain (Gram-positive) from cells that have a thin cell wall and outer membrane and fail to retain the stain (Gram-negative). (3)

granuloma   A thick lesion formed around a site of infection. (15)

Graves' disease   Autoimmune disorder in which an excess of thyroid hormone is produced, causing hyperthyroidism. (17)

grazer   A first-level consumer, feeding directly on producers. (27)

griseofulvin   An antifungal antibiotic that inhibits cell division. (13)

growth factor   A compound needed for the growth of only certain cells. (6)

Guillain-Barré syndrome (GBS)   Autoimmune disease with ascending muscle weakness due to nerve damage, often resulting from a prior *Campylobacter* infection. (22)

gumma   A soft, tumor-like growth resulting from the effects of tertiary syphilis. (23)

gut-associated lymphoid tissue (GALT)   Lymphatic tissues such as tonsils and adenoids that are found in conjunction with the gastrointestinal tract and contain immune cells. (15)

gyrase   A topoisomerase that introduces negative supercoils in double-stranded DNA; enables DNA to untwist during DNA replication. (8)

HACEK   A collection of fastidious Gram-negative species that can cause endocarditis. (21)

half-life   The amount of time it takes for one-half of a radioactive sample to decay. (4)

halophile   An organism that requires a high extracellular salt (NaCl) concentration for optimal growth. (6)

hand, foot and mouth disease (HFMD)   A relatively mild childhood disease usually caused by coxsackievirus A16 or enterovirus 71, characterized by fever and rash of the hands, feet, and mouth. (19)

haploid   An organism (or cell) that contains one copy of each chromosome in its cells. (11)

hapten   A small compound that must be conjugated to a larger carrier antigen in order to elicit an antibody response. (16)

Hashimoto's disease   A type of hypothyroid disease caused by an autoimmune disorder in which antibodies and T cells destroy self thyroid cells. (17)

health care–associated infection (HAI)   An infection contracted by a patient receiving treatment for other conditions within a health care setting. (20, 26)

heavy chain   The larger of the two protein types that make up an antibody. Each antibody contains two heavy chains and two light chains. (16)

helminth   Multicellular worm that is a parasite. (11)

helper T cell (T$_H$ cell)   A T cell that expresses CD4 on its cell surface and secretes cytokines that modulate B-cell isotype, or class, switching. (16)

hemolysis   The ability to lyse red blood cells; used diagnostically to identify species of streptococci and staphylococci. (10, 25)

hemolytic uremic syndrome (HUS)   A life-threatening sequela of infections with Shiga toxin–producing pathogens such as *E. coli* O157:H7 in which microclots in the renal blood supply decrease kidney function and reduce urine output. (22)

hepatitis   An inflammation of the liver, caused by infection or by exposure to a toxic substance. (22)

hepatitis A virus (HAV)   A picornavirus transmitted by the fecal-oral route that causes a self-limiting infectious disease. (22)

hepatitis B virus (HBV)   An orthohepadnavirus transmitted by blood and other bodily fluids that causes acute and chronic infections of the liver. (22)

hepatitis C virus (HCV)   A flavivirus transmitted primarily by blood, but also by sexual contact, that causes chronic liver infections; it can cross the placenta. (22)

hepatitis D virus (HDV)   A deltavirus that exacerbates HBV infections; HDV is an incomplete virus requiring HBV to replicate. (22)

hepatitis E virus (HEV)   A hepevirus that is transmitted by the fecal-oral route, most often through contaminated water, causing a self-limiting infection. (22)

hepatomegaly   Enlargement of the liver. (21)

hepatosplenomegaly   Enlargement of the liver and spleen. (22)

herd immunity   Concept that an unvaccinated individual is protected from person-to-person transmission of a pathogen because most members of the community were vaccinated against that pathogen, lessening the likelihood that the unvaccinated person will come in contact with an infected person. (17)

herpes gladiatorum   A skin infection of the neck, face, or arm caused by herpes simplex virus 1. (19)

herpes zoster ophthalmicus   A disease caused by reactivation of dormant herpes zoster virus specifically along the first division of the trigeminal nerve, resulting in eruption of vesicles over the forehead, eyelids, nose, and possibly eye. (19)

herpetic keratitis   Inflammation of the cornea caused by herpes simplex infection of the eye. (19)

herpetic whitlow   Herpes simplex virus 1 infection of the skin of the fingers. (19)

heterophile antibody   A random collection of antibodies produced in response to Epstein-Barr virus infection of B cells; react against diverse antigens from different species; anti horse red blood cell antibodies are the basis of the monospot test for infectious mononucleosis. (21)

heterotroph   An organism that uses external sources of organic carbon compounds for biosynthesis. (6, 7)

Hfr cell   A bacterial cell capable of high-frequency recombination caused by the presence of a chromosomally integrated F factor. (9)

highly active antiretroviral therapy (HAART)   A three-drug antiretroviral cocktail highly effective at inhibiting the replication of HIV in patients. (13, 23)

histamine   A cell-mediator molecule released by mast cells that dilates blood vessels and causes some of the symptoms associated with type I hypersensitivity. (17)

histoplasmosis   A fungal infection of the lungs caused by *Histoplasma capsulatum*. (20)

Hodgkin's lymphoma   A cancer of the lymphocytes sometimes associated with prior infectious mononucleosis caused by Epstein-Barr virus and characterized by the presence of Reed-Sternberg cells. (17)

holdfast   Adhesion factors secreted by the tip of a stalk to firmly attach an organism to a substrate. (5)

homolog   One of a group of genes or proteins with shared ancestry and whose DNA sequences and amino acid sequences are similar. (8)

homologous   Similar, referring to DNA sequences derived from a common ancestral gene. (9)

homologous recombination   See **generalized recombination**. (9)

hookworm   A soil-dwelling parasitic nematode that penetrates the host skin and migrates to the intestines, where it multiplies. (11, 22)

hopanoid   A five-ringed hydrocarbon lipid found in bacterial cell membranes; also called a hopane. (4)

horizontal gene transfer   The passage of genes from one genome into another, nonprogeny genome. (8, 9)

horizontal transmission   In disease, the transfer of a pathogen from one organism into another, nonprogeny organism. (2)

host range   The species that can be infected by a given pathogen. (2, 12)

housekeeping gene   A gene that encodes an essential function found in all cells; e.g., a gene for ribosomal RNA. (9)

humoral immunity   A type of adaptive immunity mediated by antibodies.(16)

hybridize   For a nucleic acid strand, to form base pairs with a complementary strand from a different source. (8)

hybridoma   An immortal cell line created by the fusion of a monoclonal antibody–producing B cell and a myeloma cell. (17)

hydric soil   Soil that undergoes periods of anoxic water saturation. (27)

hydrocele   A buildup of fluid in the scrotum that causes it to swell. (21)

hydrogen bond   An electrostatic attraction between a hydrogen bonded to an oxygen or nitrogen and a second, nearby oxygen or nitrogen. (4)

hydrolysis   The cleaving of a bond by the addition of a water molecule. (4)

hydrophilic   Soluble in water; either ionic or polar. (4)

hydrophobic   Insoluble in water; nonpolar. (4)

hyperesthesia   Extreme sensitivity to stimuli of one or more senses, especially skin. (24)

hyper IgM syndrome   A group of genetic diseases characterized by an overproduction of IgM antibodies and a lack of other antibody classes due to defects in class-switching proteins. (17)

hyperosmotic   See **hypertonic**. (4)

hyperthermophile   An organism adapted for optimal growth at extremely high temperatures, generally above 80°C (176°F) and as high as 121°C (250°F). (6)

hypertonic   Also called *hyperosmotic*. Having more solutes than another environment separated by a semipermeable membrane. Water will tend to flow toward the hypertonic solution. (4)

hypha *pl.* **hyphae**   Threadlike filament forming the mycelium of a fungus. (11)

hypoosmotic   See **hypotonic**. (4)

hypotension   Low blood pressure. (21)

hypotonic   Also called *hypoosmotic*. Having fewer solutes than another environment separated by a semipermeable membrane. Water will tend to flow away from the hypotonic solution. (4)

idiotype   A difference in amino acid sequence, usually in the antigen-binding site of an antibody, that distinguishes different antibodies within any isotype class of antibodies in an individual. (16)

IgA   An antibody isotype that contains the alpha heavy chain. It can be secreted and found in tears, saliva, breast milk, and so on. (16)

IgD   An antibody isotype that contains the delta heavy chain. It is found on B-cell membranes. (16)

IgE   An antibody isotype that contains the epsilon heavy chain. It is involved in degranulation of mast cells. (16)

IgG   An antibody isotype that contains the gamma heavy chain. It is found in serum. (16)

IgM   The first antibody isotype detected during the early stages of an immune response. It contains the mu heavy chain and is found as a pentamer in serum. (16)

illness phase   Stage of an infectious disease when signs and symptoms are evident and the immune system is combating the infection. (2)

immersion oil   An oil with a refractive index similar to that of glass and that minimizes light ray loss at wide angles, thereby minimizing wavefront interference and maximizing resolution. (3)

immune complex disease   See **type III hypersensitivity**. (17)

immune surveillance   Identification and elimination of tumor cells by the immune system before they become clinically detectable. (17)

immune system   An organism's cellular defense system against pathogens. (1, 15)

immunity   Resistance to a specific disease. (1)

immunization   The stimulation of an immune response by deliberate inoculation with a weakened or killed pathogen or isolated antigen from that pathogen, in hopes of providing immunity to the disease caused by the pathogen. (1)

immunochromatographic assay   A simple rapid assay to detect specific antigens or antibodies in a patient sample (blood, urine, or CSF). (25)

immunocompromised host   An animal with a weakened immune system. (14)

immunogen   An antigen that, by itself, can elicit antibody production. (16)

immunogenicity   Also called *antigenicity*. A measure of the effectiveness of an antigen in eliciting an immune response. (16)

immunoglobulin   A member of a family of proteins that includes antibodies and B-cell receptors. (16)

immunomodulin   A protein made by normal microbiota that influences the host immune response by modifying the secretion of host proteins, such as a cytokine. (14)

immunopathology   Damage caused to tissues as the result of an immune response. (2)

immunoprecipitation   The antibody-mediated cross-linking of antigens to form large, insoluble complexes. Immunoprecipitation is used in research labs and is normally seen only in vitro. (17)

impetigo   Contagious skin infection of children caused by *Staphylococcus aureus* and by group A beta-hemolytic streptococci; characterized by red sores around the nose, mouth, and neck that break open and form a honey-colored crust; also called nonbullous impetigo. (19)

incidence   In referring to a disease, the number of new cases in a given sample or location over a specified period; compare with *prevalence*. (26)

incidental host   A host in which a pathogen or parasite can grow and cause disease but not be transmitted to other hosts. (11)

incubation period   The time between initial infection of a host by a microbe and the manifestation of signs and symptoms of disease. (2)

index case   Also called *patient zero*. The first case of an infectious disease in a location and an important piece of data for helping contain the spread of disease. (26)

indirect ELISA   An immunoassay in which an antigen fixed to a plastic surface is used to capture specific antibody from a patient's serum; a secondary antibody able to bind the captured antibody is added next; the secondary antibody is conjugated either to a fluorescent molecule or to an enzyme that can catalyze a reaction that turns a colorless substrate to a colored product. (17)

indirect immunofluorescence test   A diagnostic method for detecting specific anti-pathogen antibodies in a patient by attaching the pathogen to a slide, adding patient serum followed by fluorescently labeled anti-human antibodies, and looking under fluorescence light microscopy for fluorescence, which indicates that the serum contains antibodies against the pathogen. (17)

indirect transmission   Spreading of a pathogen from host to host through an intermediary such as an inanimate object, animal vector, or contaminated food and water. (2)

induction   Increased transcription of target genes due to an inducer binding a repressor, which prevents repressor-operator binding. (8)

infant botulism   Infant paralysis by botulism toxin following the ingestion and germination of *Clostridium botulinum* spores. (10)

infection   The growth of a pathogen or parasite in or on a host. (2)

infectious disease   A disease that can be transferred from host to host, caused by pathogenic bacteria, viruses, or fungi or parasitic helminths and protists. (2)

infectious dose 50% (ID$_{50}$)   The number of bacteria or virions required to cause disease symptoms in 50% of an experimental group of hosts. (2)

infectious mononucleosis   Highly contagious disease with flu-like symptoms caused by the very common Epstein-Barr virus. (21)

inflammatory diarrhea   Diarrhea caused by inflammation and inflammation-induced damage of the intestinal mucosa that diminishes the intestines' ability to absorb nutrients and water. (22)

influenza   A respiratory tract infection caused by strains of the influenza virus, a member of the orthomyxoviruses. (20)

initiation   The first phase of DNA synthesis, in which the DNA helix unwinds, RNA primers attach, and DNA polymerase and accessory proteins assemble on the promoter; also a term to describe the initial events in transcription and in translation. (8)

innate immunity   Also called *nonadaptive immunity*. The system of nonspecific mechanisms the body uses for protecting against pathogens. (15)

inner leaflet   The layer of the cell membrane phospholipid bilayer that faces the cytoplasm. (4)

inner membrane   In Gram-negative bacteria, the membrane in contact with the cytoplasm, equivalent to the cell membrane. (5)

insertion   The addition of nucleotides to the middle of a DNA sequence. (8)

insertion sequence (IS)   A simple transposable element consisting of a transposase gene flanked by short, inverted-repeat sequences that are the target of transposase. (9)

integrin   Member of a family of host cell membrane proteins involved in adhesion of cells to one another and to the extracellular matrix. (15)

interference microscopy   An optical imaging method that uses interference between two light beams to generate an image with enhanced contrast. (3)

interferon   A family of cytokines that inhibit viral replication. (15)

interferon gamma release assay (IGRA)   A blood test that measures release of interferon gamma by leukocytes upon exposure to mycobacteria; a tuberculosis screening method. (20)

interleukin 1 (IL-1)   A cytokine released by macrophages. (15)

intermediate filament   A eukaryotic cytoskeletal protein that is composed of various proteins, depending on the cell type. (5)

intermediate host   A transient host in which a parasite undergoes asexual reproduction before being transmitted to a definitive host. (11, 22)

intracellular pathogen   A pathogen that lives within a host cell. (18)

intravenous immunoglobulin (IVIG)   Pooled IgG extracted from the plasma of many blood donors. The antibodies are delivered intravenously to patients who cannot make their own antibodies. (17)

invasion   The ability of a pathogen to enter and survive inside host cells. (2)

invasiveness   The ability of a pathogen to spread rapidly through host tissues. (2)

inversion   A reversal of the orientation of a DNA fragment within a chromosome. It may allow or repress the transcription of a particular gene. (8)

ion   An atom or molecule containing a charge—that is, a number of electrons greater or smaller than the number of protons. (4)

ionic bond   A bond between ions of positive and negative charge. (4)

isolate   A microbe that has been obtained from a specific location and grown in pure culture. (10)

isolation streaking   Also called *dilution streaking*. A method for creating individual colonies of bacteria on an agar plate that can be used to generate pure cultures. (6)

isomer   In chemistry, two or more compounds or ions that contain the same number of atoms but differ in structural arrangement and properties. (4)

isosmotic   See **isotonic**. (4)

isotonic   Also called *isosmotic*. Being in osmotic balance; having equal concentrations of solutes on both sides of a semipermeable membrane. A cell in an isotonic environment will neither gain nor lose water. (4)

isotype   A species-specific antibody class, defined by the structure of the constant region of the heavy chain. IgG, IgA, IgM, IgD, and IgE are the five isotypes. (16)

isotype switching   Also called *class switching*. A change in the predominant antibody isotype produced by a B cell. (16, 22)

Jarisch-Herxheimer reaction   An acute reaction that occurs within hours after initiating treatment for syphilis; an inflammatory response triggered by dying treponemes, resulting in fever, headaches, tachycardia, and myalgias. (23)

Kaposi's sarcoma   A cancer caused by human herpes virus 8. It is a hallmark of AIDS. (23)

keratitis   Inflammation of the cornea. (19)

Kirby-Bauer assay   A method for determining antibiotic susceptibility. Antibiotic-impregnated disks are placed on an agar plate whose surface has been confluently inoculated with a test organism. The antibiotic diffuses away from the disk and inhibits growth of susceptible bacteria. The width of the inhibitory zone is proportional to the susceptibility of the organism. (13)

Koch's postulates   Developed by Robert Koch; four criteria that should be met for a microbe to be designated the causative agent of an infectious disease. (1)

Krebs cycle   See **tricarboxylic acid (TCA) cycle**. (7)

labile toxin (LT)   An *E. coli* enterotoxin, destroyed by heat, that increases host cellular cAMP concentrations. (18, 22)

lab-on-a-chip (LOC)   A diagnostic or analytical instrument that integrates multiple laboratory functions into a tiny silicon chip with incorporated microfluidics. (25)

lagging strand   During DNA replication, the daughter strand of DNA that is synthesized as a series of short fragments that are then ligated. (8)

lag phase   A period of cell culture, occurring right after bacteria are inoculated into new media, during which there is slow growth or no growth. (6)

laminar flow biological safety cabinet   An air filtration appliance that removes pathogenic microbes from within the cabinet. (13)

Lancefield classification   See **Lancefield groups**. (25)

Lancefield groups   Also called *Lancefield classification*. Classification scheme for beta-hemolytic streptococci that is based on differences in their cell wall carbohydrate antigens. (25)

Langerhans cell   A specialized, phagocytic dendritic cell that is the predominant cell type in skin-associated lymphatic tissue. (15)

laryngitis   Inflammation of the larynx. (20)

laryngotracheobronchitis (LTB)   See **croup**. (20)

late log phase   Period of cell culture growth where exponential growth rate slows. (6)

latent   Referring to a virus or infection that is hidden and not manifesting symptoms of disease. (19)

latent period   The time in the viral life cycle when progeny virions have formed but are still within the host cell. (23)

latent stage   Referring to an HIV infection, it is the period following infection by the virus but before symptoms characteristic of HIV/AIDS appear. (23)

latent state   A period of the infection process during which a pathogenic agent is dormant in the host and cannot be cultured. (2)

latent tuberculosis infection (LTBI)   A condition in which a person is infected with *Mycobacterium tuberculosis* but who is asymptomatic and is not infectious to others. (20)

late-phase anaphylaxis   A stage of anaphylaxis in which mast cells in the affected area release leukotrienes that cause inflammation and vasoconstriction. (17)

leading strand   During DNA replication, the daughter strand of DNA that is formed by continuous elongation in the 5′-to-3′ direction. (8)

leishmaniasis   Disease caused by infection with *Leishmania* species. (21)

lentivirus   A member of a family of retroviruses with a long incubation period. An example is HIV. (12)

lethal dose 50% (LD$_{50}$)   A measure of virulence; the number of bacteria or virions required to kill 50% of an experimental group of hosts. (2)

lethal factor (LF)   A component of anthrax toxin that cleaves host protein kinases. (18)

leukemia   Disease in which malignant lymphoid cells circulate in the blood or bone marrow. (17)

leukotriene   Lipid mediator of inflammation derived from arachidonic acid that is secreted by mast cells, macrophages, and other cells; causes smooth muscle contraction, mucus secretion, and activation of leukocytes leading to inflammation and allergic reaction symptoms. (17)

lichen   A symbiotic organism composed of a fungal partner and an algal or cyanobacterial partner species. (10)

light chain   The smaller of the two protein subunits that make up an antibody. Each antibody contains two heavy chains and two light chains. (16)

light microscopy (LM)   Observation of a microscopic object based on light absorption and transmission. (3)

lipid   Member of a large class of organic molecules characterized by being partially or completely hydrophobic and often containing long hydrocarbon chains. (4)

lipid A   Portion of lipopolysaccharide that is anchored in the outer membrane of the Gram-negative bacterial envelope; also functions as an endotoxin. (5)

lipopolysaccharide (LPS)   Any of a class of structurally unique phospholipids found in the outer leaflet of the outer membrane in Gram-negative bacteria. Many are endotoxins. (5, 18)

listeriosis   Food-borne disease caused by *Listeria monocytogenes*. (22, 24)

lithotroph   See **chemoautotroph**. (1)

lithotrophy   The metabolic oxidation of inorganic compounds to yield energy used to reduce carbon dioxide; also called chemoautotrophy. (6, 7, 10)

logarithmic (log) phase   See **exponential phase**. (6)

long-term nonprogressor   An HIV-positive person who has remained asymptomatic for years without antiviral therapy. (23)

lower respiratory tract   Lower airways of the body, which include the trachea, bronchi, and lungs. (20)

lumbar puncture   A spinal tap; medical procedure used to withdraw cerebrospinal fluid from a patient. (24)

lumen   1. The interior of an intracellular membrane-enclosed compartment (phagosome). (5) 2. The cavity of a tubular organ (intestine) or part (blood vessel).

Lyme borreliosis   Also called *Lyme disease*. A tick-borne disease caused by *Borrelia burgdorferi* that may involve skin lesions and arthritis. (21)

Lyme disease   See **Lyme borreliosis**. (21)

lymphadenitis   Inflammation of lymph nodes. (21)

lymphadenopathy   Swelling of lymph nodes. (21)

lymphangitis   Inflammation of a lymphatic vessel, often seen as a red streak extending from a localized bacterial skin infection. (21)

lymphatic system   Network of lymph-carrying vessels and peripheral lymphoid tissues through which extracellular fluid passes before it is collected and returned to the venous circulation. (21)

lymph node   A secondary lymphatic organ, formed by the convergence of lymphatic vessels, where antigen-presenting cells and lymphocytes interact. (15)

lymphocytic choriomeningitis virus (LCMV)   An arenavirus transmitted from rodents to humans that causes meningitis, encephalitis, and meningio-encephalitis. (24)

lymphogranuloma venereum (LGV)   Sexually transmitted disease caused by especially virulent serotypes of *Chlamydia trachomatis*. (23)

lymphoma   Solid tumor found in a lymphoid organ. (17)

lyophilization   The removal of water from food, by freezing under vacuum, to limit microbial growth; also called freeze-drying. (13)

lysis   The rupture of the cell by a break in the cell wall and membrane. (4, 12)

lysogeny   A viral life cycle in which the viral genome integrates into and replicates with the host genome but retains the ability to initiate host cell lysis. (12)

lysosome   An acidic eukaryotic organelle that aids digestion of molecules. Not found in plant cells. (4, 5)

lysozyme   A hydrolytic enzyme secreted by eukaryotic cells that degrades bacterial cell wall peptidoglycan. (15)

lytic infection   A viral life cycle in which progeny virions are released from the host cell by virus-induced rupturing of the cell membrane. (12)

MacConkey medium　A differential, selective medium that selects for Gram-negative bacteria and can differentiate between lactose fermenters and nonfermenters. (6)

macromolecule　A large organic molecule, such as a protein or nucleic acid, typically comprising thousands of carbon-carbon bonds. (4)

macrophage　A mononuclear, phagocytic, antigen-presenting cell of the immune system found in tissues. (15)

macule　A flat, red skin rash less than 1 cm across. (19)

maculopapular rash　A reddened papule. (19)

magnetosome　An organelle containing the mineral magnetite that allows microbes to sense a magnetic field. (5)

magnification　An increase in the apparent size of a viewed object as an optical image. (3)

major histocompatibility complex (MHC)　Transmembrane cell proteins important for recognizing self and for presenting foreign antigens to the adaptive immune system. (15, 16)

malaise　Sensations of fatigue and mild discomfort in various locations of the body. (2)

malaria　A disease caused by the apicomplexan *Plasmodium falciparum*, transmitted by mosquitoes. (21)

MALDI-TOF MS　A technique of mass spectrometry in which macromolecules such as proteins are fragmented and ionized, and their time of flight (TOF) provides information on the molecular mass. (25)

malignant　Referring to a tumor whose cells are undergoing rapid, unchecked growth. (17)

Mantoux tuberculin skin test (TST)　See *tuberculin skin test*. (20)

mass number　The mass of an atom's nucleus, which is equal to the sum of the number of protons and neutrons. (4)

mass spectrometry (MS)　Spectrometry in which defined fragments of a macromolecule are analyzed by particle mass. (25)

mast cell　A white blood cell that secretes proteins that aid innate immunity. Mast cells reside in connective tissues and mucosa and do not circulate in the bloodstream. (15)

M cell　Specialized epithelial cell (microfold cell) within Peyer's patches lining the intestine; transports antigens and microbes from the intestines across the epithelial barrier to macrophages. (15)

MDR efflux pump　See **multidrug resistance efflux pump**. (13)

measles　Childhood disease caused by a paramyxovirus; characterized by flu-like symptoms and a maculopapular rash caused by cytotoxic T-cell damage to small blood vessels; also known as rubeola. (19)

meiosis　A form of cell division by which a diploid eukaryotic cell generates haploid sex cells that contain recombinant chromosomes. (11)

membrane-permeant weak acid　An acid that exists in charged and uncharged forms, such as acetic acid. The uncharged form can penetrate the cell membrane. Also called membrane-permeant organic acid. (4)

membrane-permeant weak base　A base that exists in charged and uncharged forms, such as methylamine. The uncharged form can penetrate the cell membrane. (4)

memory B cell　A long-lived type of lymphocyte pre-programmed to produce a specific antibody. After encountering their activating antigen, memory B cells differentiate into antibody-producing plasma cells. (16)

meningitis　Inflammation of the meninges. (24)

meningoencephalitis　Inflammation of the meninges and the brain. (24)

meninx *pl.* **meninges**　Set of three membranes that envelop the brain and spinal cord; the membranes are (from the inside going out) the pia mater, arachnoid mater, and dura mater. (24)

mesophile　An organism with optimal growth between 20°C (68°F) and 40°C (104°F). (6)

messenger RNA (mRNA)　An RNA molecule that encodes a protein. (8)

metagenomic study　An analytical method that uses DNA sequencing to identify all the organisms directly from an environmental sample. (14)

metamonad　A member of a phylum of flagellated protozoa that have no mitochondria. Several of them, such as *Giardia intestinalis*, parasitize humans. (11)

methanogen　An organism that carries out methanogenesis, using hydrogen to reduce CO_2 and other single-carbon compounds to methane, yielding energy. (1)

methanogenesis　An energy-yielding metabolic process that produces methane, commonly from hydrogen gas and oxidized one- or two-carbon compounds. It is unique to archaea. (7)

methemoglobinemia　A disorder in which red blood cells contain excess methemoglobin, a form of hemoglobin that transports oxygen to tissues less efficiently, causing tissues to starve (i.e., become hypoxic). (27)

methicillin-resistant *S. aureus* (MRSA)　Strains of *Staphylococcus aureus* that are resistant to first-line antibiotics methicillin and oxacillin. (19)

methyl mismatch repair　A DNA repair system that fixes misincorporation of a nucleotide after DNA synthesis. The unmethylated daughter strand is corrected to complement the methylated parental strand. (8)

MHC II deficiency (bare lymphocyte syndrome)　Rare genetic disease in which MHC II molecules are absent on antigen-presenting cells, resulting in greatly increased susceptibility to many kinds of infection. (17)

microaerophilic　Requiring oxygen at a concentration lower than that of the atmosphere, but unable to grow in high-oxygen environments. (6)

microbe　An organism or virus too small to be seen with the unaided human eye. (1)

microbe-associated molecular pattern (MAMP)　Formerly called *pathogen-associated molecular pattern (PAMP)*. Molecules associated with groups of microbes, both pathogenic and nonpathogenic, that are recognized by cells of the innate immune system. (15)

microbiome　The total community of microbes found within a specified environment. (14)

microbiota　The normally occurring microbes of a body part or environmental habitat. (2, 14)

microfilament　A eukaryotic cytoskeletal protein filament composed of polymerized actin; also called an actin filament. (5)

microfossil　A microscopic fossil in which calcium carbonate deposits have filled in the form of ancient microbial cells. (9)

microscope　A tool that increases the magnification of specimens to enable viewing at higher resolution. (3)

microsporidian　Member of a large phylum of obligate intracellular fungal parasites that produce spores. (11)

microtubule　A eukaryotic cytoskeletal protein filament composed of polymerized tubulin. (5)

midstream clean-catch technique　A sample collection method that yields urine with a reduced number of contaminating bacteria from the urethra. (25)

minimal inhibitory concentration (MIC)　The lowest concentration of a drug that will prevent the growth of an organism. (13)

minimal medium　Also called *chemically defined minimal medium*. A bacterial growth solution that contains a defined set of known components limited to only what a particular organism (or group of organisms) needs to grow. (6)

minimum bactericidal concentration (MBC)　The lowest concentration of an antibiotic that kills the test cells. (13)

miracidium　A ciliated larva of pathogenic flatworms. (22)

missense mutation　A point mutation that alters the sequence of a single codon, leading to a single amino acid substitution in a protein. (8)

mite　A small arachnid, including parasites that burrow into the skin and lay eggs. When inhaled, some may cause allergic reactions. (11)

mitosis　The orderly replication and segregation of eukaryotic chromosomes, usually prior to cell division. (11)

mixotroph　An organism capable of both photosynthetic and heterotrophic metabolism. (7)

molarity　A unit of concentration measured as the number of moles of solute per liter of solution. (4)

mole　The amount of a chemical substance that contains 6.02×10^{23} molecules of that substance; Avogadro's number equals 6.02×10^{23}. (4)

molecular beacon probe　An oligonucleotide with a fluorescent dye on one end and a quencher on the other that fluoresces only when the dye and quencher molecules are separated by degradation; a molecular biology reagent used to quantify the results of qPCR. (25)

molecular clock　The use of DNA or RNA sequence information to measure the time of divergence among different species. (9)

molecular formula　A notation indicating the number and type of atoms in a molecule. For example, H_2O is the molecular formula for water. (4)

molecule   A substance composed of two or more atoms that are sharing electrons covalently. (4)

monocyte   A white blood cell with a single nucleus that can differentiate into a macrophage or a dendritic cell. (15)

mononuclear phagocyte system (MPS)   A collection of cells that can phagocytose and sequester extracellular material. (15)

monophyletic group   See **clade**. (9)

monosaccharide   Also called *simple sugar*. The monomer unit of sugars. Monosaccharides have the molecular formula $(CH_2O)_n$. (4)

morbidity   The state of being diseased; the relative incidence of disease in a population. (2)

mordant   A chemical that causes specimens to retain stains better by combining with the stain to form an insoluble compound. (3)

mortality   The number of people who have died in a population in a given area or over a given period, often measured with respect to a specific disease. (2)

motility   The ability to generate self-directed movement. (5)

motility-related diarrhea   Diarrhea caused by hypermotility of the intestinal tract, causing food to move faster than normal in the intestines, so there isn't enough time to absorb water and nutrients. (22)

motor neuron   A neuron that sends signals from the central nervous system to muscles and glands to produce movement. (24)

mucociliary escalator   The ciliated mucous lining of the trachea, bronchi, and bronchioles that sweeps foreign particles up and away from the lungs; also known as the mucociliary elevator. (14, 20)

mucosa   A mucous membrane. (19)

mucosa-associated lymphoid tissue (MALT)   System of concentrated sites of lymphoid tissue closely associated with mucosal surfaces, whose function includes surveying the antigens and pathogens that pass through the mucosae. (15)

multidrug resistance (MDR) efflux pump   A transmembrane protein pump that can export many kinds of antibiotics largely independent of structure. (13)

multidrug-resistant (MDR)   Referring to a strain of pathogen that is resistant to at least two first-line drugs. (20)

multilocus sequence typing (MLST)   A diagnostic DNA sequencing method that types multiple genetic loci from patient-isolated bacterial cultures in order to identify strain variability. (9)

multiple myeloma   Cancerous growth of plasma cells at multiple sites in the body, especially in the bone marrow. (17)

multiplex PCR   A polymerase chain reaction that uses multiple pairs of oligonucleotide primers to amplify several different DNA sequences simultaneously. (25)

mumps   Childhood disease caused by a paramyxovirus; it infects the major salivary glands, causing them to swell as a result of inflammation. (22)

murein   See **peptidoglycan**. (5)

murein lipoprotein   The major lipoprotein that connects the outer membrane of Gram-negative bacteria to the peptidoglycan cell wall. (5)

mutation   A heritable change in a DNA sequence. (8)

mutator   A strain of a microorganism that has an abnormally high frequency of mutations. (8)

mutualism   A symbiotic relationship in which both partners benefit. (2, 27)

myasthenia gravis   Autoimmune disease characterized by severe muscle weakness; caused by autoantibodies against acetylcholine receptors at neuromuscular junctions. (17)

mycelium *pl.* **mycelia**   A fungal hypha that projects into the air (aerial mycelium) or into the growth substrate (substrate mycelium). (11)

mycolic acid   One of a diverse class of sugar-linked fatty acids found in the cell envelopes of mycobacteria such as *Mycobacterium tuberculosis*. (5, 25)

mycology   The study of fungi. (11)

mycorrhiza *pl.* **mycorrhizae**   Fungus involved in an intimate mutualism with plant roots, in which nutrients are exchanged. (11)

mycosis   A fungal infection. (24)

myelitis   Inflammation of the spinal cord. (24)

myocarditis   Inflammation of the heart myocardium. (21, 22)

Naegleria **meningoencephalitis**   A life-threatening infection of the central nervous system caused by the pathogenic ameba *Naegleria fowleri*. (24)

nasopharynx   The passage leading from the nose to the oral cavity. (14)

native conformation   The fully folded, functional form of a protein. (4)

natural killer (NK) cell   A lymphocyte that kills some tumor cells and cells infected with a virus or bacteria; an important component of innate immunity. (15)

natural selection   The mechanism by which a change occurs in the frequency of genes in a population under environmental conditions that favor some genes over others. (9)

necrotizing fasciitis   Also known as flesh-eating disease, a severe skin infection usually caused by the Gram-positive coccus *Streptococcus pyogenes*. (19)

negative selection   In immunology, the destruction of T cells bearing T-cell receptors (TCRs) that bind strongly to self MHC proteins displayed on thymus epithelial cells. (16)

negative stain   A stain that colors the background and leaves the specimen unstained. (13)

nematode   See **roundworm**. (11, 22)

neonatal herpes   Herpes infection of a newborn who receives the virus from its mother, who has genital herpes. (23)

neoplasm   See **cancer**. (17)

neuraminidase inhibitor   Any of a class of anti-influenza drugs that target neuraminidase on the viral envelope and decrease the number of virus particles produced. (13)

neurosyphilis   Neurological symptoms resulting from syphilis at any stage of the disease, caused by infection of the central nervous system by *Treponema pallidum*. (23)

neurotropic virus   A virus that preferentially infects neurons. (24)

neutralophile   An organism with an optimal growth range in environments between pH 5 and pH 8. (6)

neutrophil   A white blood cell of the innate immune system that can phagocytose and kill microbes. (15)

niche   An organism's environmental requirements for existence and its relations with other members of the ecosystem. (27)

nicotinamide adenine dinucleotide (NADH)   An energy carrier in the cell that can donate (NADH) or accept (NAD^+) electrons. (7)

nitrification   The oxidation of reduced nitrogen compounds to nitrite or nitrate. (6, 27)

nitrifier   An organism that converts reduced nitrogen compounds to nitrite or nitrate. (27)

nitrogen cycle   The biogeochemical transformations of nitrogen compounds in the biosphere. (6)

nitrogen fixation   The ability of some prokaryotes to reduce inorganic diatomic nitrogen gas (N_2) to two molecules of ammonium ion (NH_4^+). (1, 10)

nitrogen-fixing bacterium   A bacterium that can reduce diatomic nitrogen gas (N_2) to two molecules of ammonium ion (NH_4^+). (6)

nitrosamine   A member of a group of toxic chemical compounds that are found in several household items, tobacco, and food products. (27)

NOD-like receptor (NLR)   A eukaryotic cytoplasmic protein that recognizes particular MAMPs present on microorganisms. (15)

nonadaptive immunity   See **innate immunity**.

nonbullous impetigo   See **impetigo**. (19)

noncholera *Vibrio*   Any pathogenic *Vibrio* species other than *V. cholerae*. (22)

nonhemolytic   Inability of a bacterial species to lyse red blood cells in a blood agar plate. (25)

non-Hodgkin's lymphoma   Any lymphoma that does not contain Reed-Sternberg cells and hence is not a Hodgkin's lymphoma. (17)

nonpolar covalent bond   A covalent bond in which the electrons in the bond are shared equally by the two atoms. (4)

nonsense mutation   A mutation that changes an amino acid codon into a premature stop codon. (8)

norovirus   Formerly known as Norwalk virus; fecal-orally transmitted virus that is a major cause of gastroenteritis. (22)

nosocomial infection   Hospital-acquired infection. (26)

nuchal rigidity   An inability to bend the neck forward touching chin to chest; the symptom is a sign that the meninges are irritated and is a hallmark of meningitis. (24)

nuclear pore complex   Large complex of proteins that traverse the nuclear membranes, providing selective movement of molecules between the cytoplasm and the nucleus within a cell. (5)

nucleobase   A planar, heteroaromatic nitrogenous base that forms a nucleotide of nucleic acids; determines the information content of DNA and RNA. (4)

nucleoid   The region within a prokaryotic cell that contains the compacted loops of a bacterial chromosome. (5)

nucleolus *pl.* **nucleoli**   A region inside the nucleus where ribosome assembly begins. (5)

nucleotide   The monomer unit of nucleic acids, consisting of a five-carbon sugar, a phosphate, and a nitrogenous base. (4)

numerical aperture   The product of the refractive index of the medium and sin θ. As numerical aperture increases, the magnification increases. (3)

O antigen   See **O polysaccharide**. (5)

objective lens   In a compound microscope, the lens that is closest to the specimen and generates magnification. (3)

observational study   Epidemiological approach in which the exposure and disease status of each study participant is monitored and recorded. (26)

ocular lens   In a compound microscope, the lens situated closest to the observer's eye; also called eyepiece. (3)

Okazaki fragment   A short fragment of DNA that is synthesized on the lagging strand during DNA synthesis. (8)

oliguria   The production of abnormally low amounts of urine. (23)

oncogene   A gene that, through mutation or inappropriate expression, can lead to cancer. (12)

oncogenic virus   A virus that causes cancer. (12)

One Health Initiative   A collaboration dedicated to improving the lives of all species by integrating human medicine, veterinary medicine, and environmental science. (26)

one-step growth curve   Graphical representation of a lytic virus life cycle—from attachment through infection, formation of progeny, and cell lysis—in a system with a fixed number of host cells. (12)

oophoritis   Inflammation of the ovaries. (22)

open reading frame (ORF)   A DNA sequence predicted to encode a protein. (9)

operon   A collection of genes that are in tandem on a chromosome and are transcribed into a single RNA. (8)

ophthalmia neonatorum   Serious eye infection of a newborn caused by *Neisseria gonorrhoeae* or *Chlamydia trachomatis* transferred from the mother to the newborn during parturition; also called neonatal conjunctivitis. (19, 23)

opisthorchiasis   Disease caused by liver flukes, *Opisthorchis* species, following ingestion of larvae; infected hosts are often asymptomatic but can develop diarrhea, abdominal pain, and mild fever. (22)

O polysaccharide   Also called *O antigen*. A sugar chain that connects to the core polysaccharide of lipopolysaccharides. (5)

opportunistic pathogen   A microbe that normally is not pathogenic but can cause infection or disease in an immunocompromised host organism. (2, 14)

opsonization/opsonize   The process by which phagocytosis is aided by coating pathogens with IgG antibodies or complement. (15, 16)

optical density   A measure of how many particles are suspended in a solution, based on light scattering by the suspended particles. (6)

optochin susceptibility disk test   A diagnostic method for identifying *Streptococcus pneumoniae* by its susceptibility to the antibiotic optochin. (25)

oral-fecal route   Ingestion of pathogens through the mouth. (2)

orbital   A region of space occupied by electrons around an atomic nucleus. (4)

orchitis   Inflammation of the testicles. (22, 23)

organelle   A membrane-enclosed compartment within eukaryotic cells that serves a specific function. (4)

organic molecule   A molecule that contains a carbon-carbon bond. (4)

organotrophy   The metabolic oxidation of organic compounds to yield energy without absorption of light. (6, 7)

orlgln (*orlC*)   The region of a bacterial chromosome where DNA replication initiates. (8)

origin of replication (*ori*)   A DNA sequence at which DNA replication initiates. In a bacterial chromosome, this site is also attached to the cell envelope. (5)

oropharynx   The area between the soft palate and the upper edge of the epiglottis. (14)

ortholog   A gene (or protein) present in more than one species that derived from a common ancestral gene and encodes the same function in each of the species. (9)

osmolarity   A measure of the number of solute molecules in a solution. (6)

osmosis   The diffusion of water from regions of high water concentration (low solute) to regions of low water concentration (high solute) across a semipermeable membrane. (4)

osmotic diarrhea   Diarrhea caused by the inability of the intestine to absorb nutrients from chyme, resulting in a hypertonic intestinal lumen compared with the mucosal cells of the intestines. (22)

osmotic shock   A sudden increase in osmotic pressure within a cell that is moved rapidly into a hypotonic environment. (4)

outbreak   A situation in which the number of cases of a disease exceeds the endemic level of that disease. (26)

outer leaflet   The layer of the cell membrane phospholipid bilayer that faces away from the cytoplasm. (4)

outer membrane   In Gram-negative bacteria, a membrane outside the cell wall. (5)

oxidation   Refers to the loss of an electron. (4)

oxidative burst   A large increase in the oxygen consumption of immune cells during phagocytosis of pathogens as the immune cells produce oxygen radicals to kill the pathogen. (15)

oxidative phosphorylation   An electron transport chain that uses diatomic oxygen as a final electron acceptor and generates a proton gradient across a membrane for the production of ATP via ATP synthase. (7)

oxidized   The state of a molecule that has lost an electron to an oxidizing agent during a redox reaction. (7)

oxidizing agent   A molecule that accepts electrons during a redox reaction. (4)

oxidoreductase   An electron transport system protein that accepts electrons from one molecule (oxidizing that molecule) and donates electrons to a second molecule (reducing the second molecule). (7)

oxygenic photosynthesis   A form of phototrophy in which the excited electrons reduce CO_2 to form sugars and release O_2. (7)

pancarditis   Inflammation of all three layers of the heart. (21)

pancreatitis   Inflammation of the pancreas. (22)

pandemic   An epidemic that occurs over a wide geographic area. (2, 26)

papilloma   See **wart**. (19)

papule   A small, solid, elevated skin lesion. (19)

paralog   A gene that arises by gene duplication within a species and evolves to carry out a different function from that of the original gene; also the protein expressed by that gene. (9)

paramyxovirus   A member of the single-stranded RNA virus family Paramyxoviridae, which includes the mumps and measles viruses. (22)

parasite   Any bacterium, virus, fungus, protozoan (protist), or helminth that colonizes and harms its host; the term commonly refers to protozoa and to invertebrates. (2)

parasitism   A symbiotic relationship in which one member benefits and the other is harmed. (27)

parental strand   One of the two strands of a DNA double helix used as a template for the synthesis of a daughter strand during DNA replication. (8)

parenteral route   The introduction of materials into the body other than through the gastrointestinal system; typically via intravenous or intramuscular injection. (2)

paresthesia   A sensation of tingling, prickling, burning, or numbness on the skin due to pressure on or damage to peripheral nerves. (24)

parfocal   In a microscope with multiple objective lenses, having the objective lenses set at different heights so that focus is maintained when switching between lenses. (3)

parotitis   Inflammation of the major (parotid) salivary glands. (22)

parthenogenesis   Reproduction due to development of unfertilized (usually female) gametes. (22)

passive immunization   Injection of antibodies or antiserum into a naive patient to provide immediate, but temporary, protection against a toxin or pathogen or to help a patient coping with certain autoimmune diseases. (17)

passive transport   Net movement of molecules across a membrane without energy expenditure by the cell. (5)

pasteurization   The heating of food at a temperature and time combination that will kill pathogens such as *Mycobacterium tuberculosis*. (13)

pathogen   A bacterial, viral, fungal, protozoan, or helminthic agent of disease; among health professionals, it typically is limited to bacteria, viruses, and fungi. (1, 2)

pathogen-associated molecular pattern (PAMP)   See **microbe-associated molecular pattern**. (15)

pathogenesis   The process microbes use to cause disease in a host. (18)

pathogenicity   The ability of a microorganism to cause disease. (2)

pathogenicity island   A type of genomic island, a stretch of DNA that contains virulence factors and may have been transferred from another genome. (9, 18)

pathognomonic   Referring to a sign or symptom characteristic of a particular disease. (19)

pathovar   A group of strains within a bacterial species that have the same virulence factors, induce a similar pathology, but differ in traits such as antigen specificity. (22)

patient history   Written summary of a conversation between a health care professional and a patient that aids the health care professional in developing a diagnosis; also called a medical history. (2)

patient zero   See **index case**. (26)

pattern recognition receptor (PRR)   Receptor on a cell of the innate immune system that recognizes common molecular patterns on microbial surfaces. (15)

pedestal   An actin-based, raised surface structure produced by a host epithelial cell in response to binding certain bacterial pathogens. (22)

pellicle   A protective cover around some protists; it is made of flexible, membranous layers reinforced by microtubules. (5)

pelvic inflammatory disease (PID)   Inflammation of the uterus and fallopian tubes, usually due to infection by either *Neisseria gonorrhoeae* or *Chlamydia trachomatis*. (23)

penicillin-binding protein (PBP)   A member of a group of bacterial proteins, involved in cell wall synthesis, that is the target of penicillin and other beta-lactam antibiotics. (13)

pentose phosphate shunt (PPS)   An alternative glycolytic pathway in which glucose 6-phosphate is first oxidized and then decarboxylated to ribulose 5-phosphate, ultimately generating 1 ATP and 2 NADPH. (7)

peptic ulcer   A break in the lining of the stomach or duodenum resulting from inflammation associated with *Helicobacter pylori* infection of the gastric mucosa. (22)

peptide bond   The covalent bond that links two amino acid monomers. (8)

peptidoglycan   Also called *murein*. A polymer of peptide-linked chains of amino sugars; a major component of the bacterial cell wall. (5)

perforin   A cytotoxic protein, secreted by cytotoxic T cells and natural killer cells, that forms pores in target cell membranes. (15)

pericarditis   Inflammation of the heart pericardium. (21, 22)

periodontal disease   A progressive bacterial disease of the gums; can include inflammation and bleeding of the gums, gum recession around the roots of the teeth, and erosion of the teeth; also called periodontitis. (22)

peripheral membrane protein   A protein that is associated with a membrane but does not span the phospholipid bilayer. (4)

peripheral nervous system (PNS)   Nervous system outside the brain and spinal cord that delivers sensory information to the central nervous system (CNS) and returns motor commands to muscles and other parts of the body. (24)

periplasm   Region between the outer and inner membranes of Gram-negative bacteria; it contains water, solutes, and a variety of transport proteins. (5)

peristalsis   Waves of contraction passing along the walls of the intestines that move chyme through the intestines. (22)

peritonitis   Inflammation of the peritoneum; a life-threatening condition if not treated promptly. (22)

peritonsillar abscess   An abscess in the pharynx. (20)

permease   A substrate-specific carrier protein in the membrane. (5)

persister cell   A pathogenic cell within a host that neither grows nor dies when exposed to appropriate antimicrobials; this tolerance is due not to genetic mutation but to the cell's existence in a transient, dormant state. (13)

person-to-person epidemic   See **chain of infection**. (27)

pertussis   See **whooping cough**. (20)

petechia   *pl.* petechiae   A small red spot on the skin due to blood leaking from capillaries under the skin, resulting from depletion of clotting factors. (18, 24)

petechial rash   Tiny red spots on the skin that do not blanch when pressed. (14)

petri dish   A round dish with vertical walls covered by an inverted dish of slightly larger diameter. The smaller dish can be filled with a substrate for growing microbes. Also called petri plate. (1)

Peyer's patch   Aggregates of lymphoid tissue found in the lower small intestine. (15)

phage   See **bacteriophage**. (9)

phagocytosis   A form of endocytosis in which a large extracellular particle is brought into the cell. (4, 15)

phagosome   A large intracellular vesicle that forms as a result of phagocytosis. (4, 15)

pharyngitis   Inflammation of the pharynx; also called a sore throat. (20)

phase-contrast microscopy   Microscopy in which contrast is generated by variable values of refractive index within the specimen, leading to additive and subtractive interference patterns between the light that penetrates the specimen and the light transmitted outside the specimen. (3)

phenol coefficient test   A test of the ability of a disinfectant to kill bacteria; the higher the coefficient, the more effective the disinfectant. (13)

phenotype   The outward expression of a genotype. (8)

phospholipid   The major component of membranes. A typical phospholipid is composed of a core of glycerol to which two fatty acids and a modified phosphate group are attached. (4)

phospholipid bilayer   Two layers of phospholipids; the hydrocarbon fatty acid tails face the interior of the bilayer, and the charged phosphate groups face the cytoplasm and extracellular environment. The cell membrane is a phospholipid bilayer. (4)

photolysis   The first energy-yielding phase of photosynthesis; the light-driven separation of an electron from a molecule coupled to an electron transport system. (7)

photosynthesis   The metabolic ability to absorb solar energy and convert it to chemical energy for biosynthesis. Autotrophic photosynthesis, or photoautotrophy, includes CO_2 fixation. (7)

photosystem I   A protein complex that harvests light from a chlorophyll, donates an electron to an electron transport system and receives an electron from a small molecule such as H_2S or H_2O, and stores energy in the form of NADPH. (7)

photosystem II   A protein complex that harvests light from a bacteriochlorophyll, donates an electron to an electron transport system, and stores energy in the form of a proton potential. (7)

phototrophy   The use of chemical reactions powered by the absorption of light to yield energy. (6, 7)

phylogenetic tree   A diagram depicting estimates of the relative amounts of evolutionary divergence among species. (9)

phylogeny   The classification of organisms on the basis of their genetic relatedness; a measurement of genetic relatedness. (9)

piezophile   See **barophile**. (6)

pilin   The protein monomer that polymerizes to form a pilus. (5)

pilus *pl.* **pili**   Also called *fimbria* (pl. *fimbriae*). A straight protein filament composed of a tube of pilin protein monomers that extend from the bacterial cell envelope. (5, 18)

pinworm   Small, white nematode of the species *Enterobius vermicularis* that infects humans. (11, 22)

placebo   An inactive substance used as a control in an experiment to determine the effectiveness of a medicine or drug. (26)

plague   A life-threatening disease caused by *Yersinia pestis*. (21, 22)

plaque   A cell-free zone on a lawn of bacterial cells caused by viral lysis. (12)

plaque-forming unit (PFU)   A measure of the concentration of phage particles in liquid culture. (12)

plasma cell   A fully differentiated antibody-producing cell. (16)

plasmocytoma   A plasma cell tumor found at a single site in the body, typically in the bone or soft tissue; can develop into multiple myeloma. (17)

plasma membrane   See **cell membrane**. (5)

plasmid   An extrachromosomal genetic element that may be present in some cells. (8)

platyhelminth   Also called *flatworm*. Member of the phylum Platyhelminthes, comprising morphologically simple invertebrate animals, including the parasitic trematodes and cestodes. (22)

pleocytosis   Elevated number of white blood cells in cerebrospinal fluid. (24)

pneumococcal pneumonia   Pneumonia caused by *Streptococcus pneumoniae*. (20)

pneumonia   Infection of the lungs that can be caused by a variety of bacteria, viruses, fungi, and parasites. (20)

pneumonic plague   A highly virulent and contagious *Yersinia pestis* lung infection. (21)

point mutation   Also called *base substitution*. A change in a single nucleotide within a nucleic acid sequence. (8)

point-of-care (POC) laboratory test   A rapid diagnostic test that is run at the site of patient care. (25)

point-source epidemic   An epidemic in which multiple cases of a disease appear over a short period due to exposure to a common source of contaminated food or water. (27)

polar covalent bond   A covalent bond in which the electrons are distributed unequally between two atoms. (4)

poliomyelitis   A group of neurological diseases caused by migration of poliovirus into the cerebrospinal fluid. (24)

polyamine   A molecule containing multiple amine groups that is positively charged at neutral pH. (5)

polyclonal antibodies   A complex mixture of epitope-specific antibodies directed against the many epitopes of an antigen; produced after injection of the antigen into a laboratory animal (e.g. rabbit). Each epitope-specific antibody is made by a different B cell/plasma cell. (17)

polyene   An antifungal agent that disrupts the integrity of the cell membrane. (13)

polymerase chain reaction (PCR)   A method to amplify DNA in vitro by using many cycles of DNA denaturation, primer annealing, and DNA polymerization with a heat-stable polymerase. (8)

polymicrobial   Referring to an infection involving several or many types of pathogens. (10)

polymorphonuclear leukocyte (PMN)   A granulocyte; a white blood cell with a multilobed nucleus and cytoplasmic granules. PMNs are classified as neutrophils, basophils, and eosinophils. (15)

polymyxin   A topical antibiotic that binds to and dissolves the inner membrane of bacteria. (13)

polysaccharide   A polymer of sugars. See also *glycan*. (4)

Pontiac fever   Acute upper respiratory tract infection caused by certain strains of *Legionella* bacteria. (20)

population   Members of one species that live in a common location. (27)

porin   A transmembrane protein complex that allows movement of specific molecules across the cell membrane, and in Gram-negative bacteria, across the outer membrane. (5)

positive selection   In immunology, the survival of T cells bearing T-cell receptors (TCRs) that do not recognize self MHC proteins displayed on thymus epithelial cells. (16)

post-streptococcal glomerulonephritis   Sequela of a *Streptococcus pyogenes* infection that can cause symptoms from inflammation of renal tissue to kidney failure. (20)

predator   A consumer that feeds on grazers or on other lower-level predators. (27)

preliminary treatment   The first stage of wastewater treatment, which uses screens and traps to remove solid debris from the sewage. (27)

prevalence   In referring to a disease, the total number of active cases in a given sample or location; compare with *incidence*. (26)

primary (acute) HIV infection   Stage in which a patient first contains detectable levels of antibodies against HIV antigens. (23)

primary antibody response   The production of antibodies upon first exposure to a particular antigen. B cells become activated and differentiate into plasma cells and memory B cells. (16)

primary encephalitis   Inflammation of the brain due to a virus or other pathogen that directly attacks the brain. (24)

primary endosymbiosis   An ancient endosymbiotic event between a cyanobacterium and a eukaryotic cell, which led eventually to the evolution of green and red algae after the cyanobacterium evolved into a chloroplast. (11)

primary immunodeficiency   A disease or disorder of the immune system that is the result of a genetic or developmental error. (17)

primary infection   Refers to a patient who has become infected by a pathogen for the first time. (23)

primary pathogen   A disease-causing microbe that can breach the defenses of a healthy host. (2)

primary producer   An organism that produces biomass (reduced carbon) from inorganic carbon sources such as CO_2. (27)

primary structure   The first level of organization of proteins, consisting of the linear sequence of amino acid residues. (4)

primary syphilis   The initial inflammatory reaction (chancre) at the site of infection with *Treponema pallidum*. (23)

primary treatment   An early stage of wastewater treatment designed to mechanically remove insoluble particles from sewage. (27)

prion   An infectious agent that causes propagation of misfolded host proteins; consists of a defective version of the host protein. (12, 24)

probabilistic indicator   A means of quickly identifying microbes in the clinical setting, based on a battery of biochemical tests performed simultaneously on an isolated strain. (9, 25)

probiotic   A food or nutritional supplement that contains live microorganisms and aims to improve health by promoting beneficial bacteria. (14)

prodromal phase   Stage of a disease characterized by vague, generalized symptoms. (2)

prodrome   Generalized symptoms, such as headache, fever, and malaise, that precede disease-specific symptoms. (2)

proglottid   Segment of a tapeworm containing both male and female reproductive organs. (22)

prokaryote   An organism whose cell or cells lack a nucleus. Prokaryotes include both bacteria and archaea. (1, 5)

prokaryotic cell   Cell that does not possess a nucleus. (5)

promastigote   Stage of a trypanosomatid protozoan parasite, such as *Leishmania*, found in both the insect and vertebrate hosts. (11)

promoter   A noncoding DNA regulatory region immediately upstream of a structural gene; needed for transcription initiation. (8)

prophage   A phage genome that has integrated into a bacterial host genome. (9, 12)

prostaglandin   Lipid product of arachidonic acid metabolism that is secreted by mast cells activated by an allergen; causes smooth muscle contraction, mucus secretion, and activation of leukocytes leading to inflammation and allergic reaction symptoms. (17)

prostatitis   Inflammation of the prostate gland. (23)

prosthetic valve endocarditis (PVE)   A life-threatening bacterial infection of a prosthetic heart valve following heart valve surgery. (21)

proteasome   Intracellular protein complex where unneeded and damaged proteins are degraded. (18)

protective antigen (PA)   The core subunit of anthrax toxin, so called because immunity to this protein protects against disease. (18)

protein A   *Staphylococcus aureus* cell wall protein that binds to the Fc region of antibodies, hiding the *S. aureus* cells from phagocytes. (18)

proteome   All the proteins expressed in a cell at a given time. The "complete proteome" includes all the proteins that the cell can express under any condition. The "expressed proteome" represents the set of proteins made under a given set of conditions. (5)

protist   A single-celled eukaryotic microbe, usually motile; not a fungus. (1, 11)

proton motive force   The potential energy of the concentration gradient of protons (hydrogen ions, H^+) plus the charge difference across a membrane. (5, 6, 7)

protozoan *pl.* **protozoa**   A heterotrophic eukaryotic microbe, usually motile, that is not a fungus. (1, 11)

provirus   A viral genome that is integrated into the host cell genome. (12)

pseudomembrane   A membrane-like material such as the one made of dead cells, bacteria, and fibrin that forms at the back of the throat in a patient with diphtheria. (20)

pseudopod   A locomotor extension of cytoplasm bounded by the cell membrane. (4, 11)

psychrophile   An organism with optimal growth at temperatures below 20°C (68°F). (6)

psychrotroph   An organism such as *Listeria* that grows well below 20°C but grows equally well, or faster, at higher temperatures. (27)

public health   Profession in which the health and disease of large populations are monitored and assessed. (1)

pulmonary circulation   The system of vessels that moves blood between the heart and the lungs. (21)

pure culture   1. A culture containing only a single strain or species of microorganism. (6) 2. A large number of microorganisms that all descended from a single cell. (1)

purpura   Patches of purplish discoloration on the skin due to hemorrhaging of capillaries. (24)

pustule   A papule that has filled with pus. (19)

pyelonephritis   Infection and inflammation of the kidney. (23)

pyrimidine dimer   A form of DNA damage caused by exposure to ultraviolet light, which results in the covalent bonding of two adjacent pyrimidines on a DNA strand. If not repaired, the dimer will be read incorrectly during DNA replication, resulting in a mutation to the DNA. (8)

pyrogen   Any substance that induces fever. (15)

quarantined   Separated from the general population to limit the spread of infection. (26)

quaternary structure   The highest level of organization of proteins, in which multiple polypeptide chains interact and function together. (4)

quorum sensing   The ability of bacteria to sense the presence of other bacteria via secreted chemical signals called autoinducers. (6)

rabbit fever   See **tularemia**. (21)

rabies   A deadly infectious disease caused by the rabies virus, which invades the nervous system, including the brain. (24)

radial immunodiffusion assay   A technique in which a ring of precipitation is visualized in an agarose gel impregnated with antibody. Antigen placed within a well diffuses outward until reaching a zone of equivalence where antigen-antibody complexes precipitate and form a ring. (17)

radiculitis   Inflammation of a spinal nerve root. (24)

rapid plasma reagin (RPR) test   A nonspecific test for syphilis that looks for the presence of anti-cardiolipin antibodies, which form as a result of infection by the syphilis spirochete. (23)

reactivation tuberculosis   See **secondary tuberculosis**. (20)

reading frame   A sequence of nucleotides in DNA or RNA whose nonoverlapping triplets are potentially translatable into a polypeptide. (8, 12)

reassort   The process by which genome segments come together from different influenza strains infecting the same host cell. (12)

receptor   Molecule on the extracellular surface of a host cell that can be bound by a microbial adhesin protein. (2)

recombinant DNA   A DNA molecule, prepared in a laboratory, that contains sequences from two or more source organisms (often different species). (8)

recombinant subunit vaccine   A vaccine containing synthetic antigen made in the laboratory by using recombinant DNA technology. (17)

recombination   The enzyme-mediated process by which two DNA molecules exchange sections of DNA by cutting and splicing their helix backbones. (8, 9)

recombination signal sequence (RSS)   A DNA sequence downstream of antibody heavy- and light-chain genes that allows recombination between widely separated antibody gene segments. (16)

reduced   The state of a molecule that has gained an electron from a reducing agent during a redox reaction. (7)

reducing agent   A molecule that donates electrons during a redox reaction. (4)

reduction   Gain of an electron. (4)

reductive evolution   Also called *degenerative evolution*. The loss or mutation of DNA encoding traits that are not under any selection pressure. (5, 9)

reflection   Deflection of an incident light ray by an object at an angle equal to the incident angle. (3)

refraction   The bending and slowing of light as it passes through a substance. (3)

refractive index   The ratio of the speed of light in a vacuum divided by the speed of light through a medium, such as water or glass. (3)

regulatory T cell (Treg)   A T cell that regulates the activity of another T cell, usually by suppressing its activity. (16)

reinfection   A subsequent infection by the same microorganism, such as infection with a new strain of cytomegalovirus (CMV) in a person who is CMV-seropositive. (21)

rennet   A mixture of stomach enzymes from ruminants that is used to curdle milk in the production of yogurt and cheese. (27)

replication fork   During DNA synthesis, the region of the chromosome that is being unwound. (8)

repression   The down-regulation of gene transcription. (8)

reservoir   1. The major part of the biosphere that contains a significant amount of an element needed for life. (27) 2. An organism that maintains a virus or bacterial pathogen in an area by serving as a host. (2)

resolution   The smallest distance between two objects at which they can still be distinguished as separate objects. (3)

respiration   The oxidation of reduced organic electron donors through a series of membrane-embedded electron carriers to a final electron acceptor. The energy derived from the redox reactions is stored as an electrochemical gradient across the membrane and may be harnessed to produce ATP. (7)

respiratory route   Inhalation of airborne pathogens that then enter the respiratory system of the host. (2)

respiratory syncytial virus (RSV)   A virus that causes bronchitis, bronchiolitis, and pneumonia. (20)

response regulator   A cytoplasmic protein that is phosphorylated by a sensor kinase, and once phosphorylated, modulates gene transcription by binding to DNA regulatory sequences. (8)

restriction endonuclease   A bacterial enzyme that cleaves double-stranded DNA within a specific short sequence, usually a palindrome. (8)

restriction enzyme site   A DNA sequence recognized and cleaved by a restriction endonuclease. (8)

restriction fragment length polymorphism   A difference in DNA sequence between strains or individuals that is identified as a difference in the length of DNA fragments cut by restriction endonucleases. (8, 25)

retroelement   A mobile genetic element in the genome of an organism. (12)

retrovirus   Also called *RNA reverse-transcribing virus*. A single-stranded RNA virus that uses reverse transcriptase to generate a double-stranded DNA copy of its genome. (12)

reverse transcriptase   An enzyme that produces a double-stranded DNA molecule from a single-stranded RNA template. (8, 12)

reverse transcription quantitative RT-PCR (qPCR)   A PCR technique that uses an oligonucleotide probe, such as a molecular beacon, which is cleaved during each PCR cycle, providing a signal (such as fluorescence) that can be measured and correlated with the amount of product generated after each PCR cycle. (25)

rheumatoid arthritis   Chronic autoimmune disease causing inflammation and pain in many joints, especially those of hands and feet. (17)

Rh incompatibility disease   A form of type II hypersensitivity caused when an Rh⁻ mother produces antibodies that destroy her fetus's Rh⁺ red blood cells; also known as erythroblastosis fetalis. (17)

rhizobium *pl.* **rhizobia**   A bacterial species of the order Rhizobiales that forms highly specific mutualistic associations with plants in which the bacteria form intracellular bacteroids that fix nitrogen for the plant. (7)

rhizosphere   The soil environment surrounding plant roots. (27)

rich medium   See **complex medium**. (6)

rickettsioses   A collection of diseases presenting with similar symptoms that are each caused by a species of *Rickettsia*. (21)

ringworm   Various forms of dermatophytosis. (19)

rise period   During viral culture, the time when cells lyse and viral progeny enter the medium. (12)

RNA-dependent RNA polymerase   An enzyme that produces an RNA complementary to a template RNA strand. (12)

RNA polymerase   Enzyme responsible for synthesizing an RNA copy from a DNA template during transcription. (8)

RNA primer   A short segment of RNA, complementary to a specific region of a chromosome, that is required for DNA polymerase to carry out DNA replication. (8)

RNA reverse-transcribing virus   See **retrovirus**. (12)

Rocky Mountain spotted fever   Tick-borne disease caused by *Rickettsia rickettsii*. If not treated promptly, it can be fatal. (21)

roseola infantum   A disease of young children in which a high fever is followed by a rash, caused by human herpes virus 6 or, less frequently, human herpes virus 7. (19)

rotavirus   One of a group of nonenveloped dsRNA viruses that cause severe diarrhea in children. (22)

roundworm   Also called *nematode*. A member of a large family of tubular worms that includes parasitic pinworms, hookworms, and whipworms. (11, 22)

Rubisco   Ribulose 1,5-bisphosphate carboxylase oxygenase, the enzyme that catalyzes carbon fixation in the Calvin cycle. (7)

Sabin vaccine   Oral polio vaccine made of live, attenuated virus. (24)

Salk vaccine   Injected polio vaccine made of inactivated virus. (24)

salmonellosis   Any infection caused by *Salmonella*. (22)

sandwich ELISA   An ELISA designed to capture the antigen of interest between an immobilized anti-target capture antibody and the same or another anti-target antibody in solution. (17)

sanitation   Killing or removing of disease-producing organisms through the safe disposal of waste material and by disinfection and proper cleaning. (13)

scalded skin syndrome   The destructive consequence of *Staphylococcus aureus* exfoliative exotoxin, which destroys adhesion molecules that hold cells in the host epidermis to the underlying dermis, leaving the skin severely blistered. (19)

scale bar   The length of which corresponds to the actual size magnified, such as 5 μm. Like a ruler, a scale bar applies to length measurements throughout a micrograph. (3)

scanning electron microscopy (SEM)   Electron microscopy in which the electron beams scan across the specimen's surface to reveal the three-dimensional topology of the specimen. (3)

scattering   Interaction of light with an object resulting in propagation of spherical light waves at relatively low intensity. (3)

schistosomiasis   A serious helminthic disease caused by *Schistosoma* species of trematodes (flukes); it can cause damage to a wide variety of organs, including intestines, bladder, liver, lungs, and spleen. (22)

scolex   The head of a tapeworm. (22)

secondary antibody response   A memory-B-cell–mediated rapid increase in the production of antibodies in response to a repeat exposure to a particular antigen. (16)

secondary encephalitis   Inflammation of the brain caused by a spreading infection or inflammation of the meninges. (24)

secondary endosymbiont alga   An algal species that evolved from a secondary endosymbiotic event. (11)

secondary endosymbiosis   An ancient endosymbiotic event between a descendant of primary endosymbiosis and another eukaryotic cell. (11)

secondary immunodeficiency   Immunodeficiency acquired as a result of infection, radiation treatment, or use of immunosuppressive drugs. (17)

secondary structure   The second level of organization of proteins, consisting of repeating patterns, especially the alpha helix and the beta sheet. (4)

secondary syphilis   The second stage of syphilis, signified by a rash that may appear at some point after the primary latent stage of syphilis. (23)

secondary treatment   An intermediate stage of wastewater treatment that uses microbiological processes to decompose soluble organic matter in the wastewater and thus reduce its biochemical oxygen demand (BOD). (27)

secondary tuberculosis   Also called *reactivation tuberculosis*. A new round of serious disease that is caused by *Mycobacterium tuberculosis* in immunocompromised patients with latent tuberculosis. (20)

second law of thermodynamics   In every transfer or transformation of energy, some energy is lost, and as a result, the entropy of the system will increase. (4)

secreted   Referring to molecules that are transported out of a cell for a specific purpose, such as to signal another cell. (8)

secretory diarrhea   Diarrhea caused by microbial action (usually a toxin) that induces mucosal cells to increase secretion of ions, making the intestinal lumen hypertonic. (22)

select agent   A pathogen or toxin from a pathogen that could serve as a bioweapon because it meets some or all of the following: it is highly virulent, it causes a high rate of mortality, it is very contagious, no vaccine exists against it, and treatment with antimicrobials or other drugs is largely ineffective. (26)

selectin   One of a family of cell adhesion molecules. (15)

selective IgA deficiency   Immunodeficiency disease in which the patient fails to produce IgA but produces relatively normal levels of IgG and IgM. (17)

selective medium   A medium that allows the growth of certain species or strains of organisms but not others. (6)

selective toxicity   The ability of a drug, at a given dose, to harm the pathogen and not the host. (13)

semiconservative   Describing the mode of DNA replication whereby each new double helix contains one old, parental strand and one newly synthesized daughter strand. (8)

semisynthetic antimicrobial   A drug made by chemically modifying a precursor antimicrobial that was isolated from a natural source. (13)

sense strand   The strand of DNA containing the sequence of nucleotides that is equivalent to the mRNA sequence synthesized from the template strand of DNA. (8)

sensitivity   For a diagnostic test, a parameter that describes how small a concentration of a microbe, antibody, or antigen a test can detect if it is present in the sample; compare with *specificity*. (17, 25)

sensitization   First stage of type I hypersensitivity, in which a naive patient encounters the allergen, which elicits production of anti-allergen IgE that binds to mast cells, thus sensitizing the mast cells. (17)

sensor kinase   A transmembrane protein that phosphorylates itself in response to an extracellular signal and then transfers the phosphoryl group to a response regulator protein. (8)

sensory neuron   Neuron that transmits sensory information about the environment and state of the organism to the central nervous system. (24)

sepsis   A systemic inflammatory response, triggered by an infection, that is so extreme it can kill the patient. (15, 21)

septation   The formation of a septum, a new section of cell wall and envelope that separates two prokaryotic daughter cells. (5)

septicemia   An infection of the bloodstream. (21)

septicemic plague   Infection of the bloodstream by *Yersinia pestis*. (21)

septic shock   Catastrophic fall in blood pressure owing to severe sepsis. (21)

septum *pl.* **septa**   A plate of cell wall and envelope that forms to separate two daughter cells. (5)

sequela *pl.* **sequelae**   A serious, harmful immunological consequence of disease that occurs after the infection itself is over. An example is rheumatic fever. (2)

serology   The branch of immunology that analyzes the contents of serum. (17)

serum   The noncell, liquid component of blood that remains after blood coagulation. (16)

serum sickness   Type III hypersensitivity reaction following injection of animal-derived protein, such as antiserum or antivenom. (17)

severe combined immunodeficiency   An immune deficiency in which T-cell responses are absent. There are multiple genetic causes, but it is invariably fatal unless T-cell immunity can be restored by transplantation or gene therapy. (17)

severe fever with thrombocytopenia syndrome (SFTS)   An emerging viral disease caused by a bunyavirus that appears to be transmitted by a tick. (26)

sex pilus   A pilus specialized to attach a DNA-donating bacterium to a recipient bacterium, enabling DNA transfer through a conjugation apparatus. (5)

sexually transmitted infection (STI)   An infection transmitted primarily through sexual activity. (23)

sexual reproduction   Reproduction involving the joining of gametes generated by meiosis. (11)

shelter species   An organism, often an invertebrate or protist, that harbors microbes that exchange genetic information and undergo evolution within the host, occasionally leading to changes that allow a microbe to cross species and infect a new host, such as a human. (10)

Shiga toxin   An exotoxin produced by *Shigella dysenteriae* and *Escherichia coli* O157:H7 (and some other *E. coli* pathovars) that disables host cell protein synthesis. (18)

Shiga toxin *E. coli* (STEC)   Any *Escherichia coli* pathovar that can produce Shiga toxin. (22)

shigellosis   Also called *bacillary dysentery*. Bloody diarrhea caused by pathogenic *Shigella* species. (22)

sigma factor   A protein needed to bind RNA polymerase to a DNA promoter for the initiation of transcription in bacteria. (8)

sign   A consequence of a disease in a patient that can be observed by another, such as a runny nose or fever; compare with *symptom*. (2)

silent mutation   A mutation that does not change the amino acid sequence encoded by an open reading frame. The changed codon encodes the same amino acid as the original codon. (8)

simple stain   A stain that makes an object more opaque, increasing its contrast with the external medium or surrounding tissue. (3)

simple sugar   See **monosaccharide**. (4)

sink   A part of the biosphere that can receive or assimilate significant quantities of an element; may be biotic (as in plants fixing carbon) or abiotic (as in the ocean absorbing carbon dioxide). (27)

site-specific recombination   Recombination between DNA molecules that do not share long regions of homology but do contain short regions of homology specifically recognized by the recombination enzyme. (9)

skin   The largest organ in the human body, providing a physical barrier to infection, protecting underlying organs, aiding in thermoregulation and water loss, synthesizing vitamin D, and sensing various aspects of the outside world, such as heat, cold, and pressure. (19)

skin-associated lymphoid tissue (SALT)   Immune cells, such as dendritic cells, located under the skin that help eliminate bacteria that have breached the skin's surface. (15)

S-layer   A crystalline protein surface layer replacing or outside the cell wall in many species of archaea and bacteria. (5)

smallpox   A deadly human disease caused by the variola virus. (19)

SOS response   A coordinated cellular response to extensive DNA damage; includes error-prone repair. (8)

space-filling model   A molecular model that represents the volume of the electron orbitals of the atoms, usually to the limit of the van der Waals radii. (4)

specialized transduction   Transduction in which the phage can transfer only a specific, limited number of bacterial donor genes to the recipient cell. (9)

species   A single, specific type of organism, designated by a genus and species name. (1)

specificity   For a diagnostic test, a parameter that describes how well it can distinguish among closely related targets; compare with *sensitivity*. (17, 25)

spectrophotometer   An instrument for measuring the amount of light absorbed by a solution when different wavelengths of light are passed through that solution; used to calculate the concentration of substances in solution. (6)

spectrum of activity   The group of pathogens for which an antimicrobial agent is effective. (13)

spike protein   A viral glycoprotein that connects the membrane to the capsid or the matrix and may be involved in viral binding to host cell receptors. (12)

spirillum   Gram-negative bacteria having a helical body shape. (10)

spirochete   A bacterium with a tight, flexible spiral shape; a species of the phylum Spirochetes (Spirochaeta). (3, 10)

splenomegaly   Enlargement of the spleen. (21)

spontaneous generation   The theory, much debated in the nineteenth century, that under current Earth conditions life can arise spontaneously from nonliving matter. (1)

spore stain   A type of differential stain that is specific for the endospore coat of various bacteria, typically a firmicute species. (3)

spread plate   A method to grow separate bacterial colonies by plating serial dilutions of a liquid culture. (6)

staining   The process of treating microscopic specimens with a stain to enhance their detection or to visualize specific cell components. (3)

stalk   An extension of the cytoplasm and envelope that attaches a microbe to a substrate. (5)

staphylococcus   A member of the Gram-positive bacterial genus *Staphylococcus*, which forms a hexagonal arrangement of round cells resulting from septation in random orientations. (5)

static agent   An antimicrobial chemical that inhibits or controls growth. (13)

stationary phase   A period of cell culture, following exponential phase, during which the cell population appears to stop growing. (6)

sterilization   The destruction of all cells, spores, and viruses on an object. (13)

strain   A genetic variant of a microbial species. (10)

streptococcal pharyngitis   Strep throat; inflammation of the pharynx caused by a *Streptococcus pyogenes* infection. (20)

streptococcal pyogenic exotoxin (SPE)   An exotoxin produced by some strains of *Streptococcus pyogenes*; multiple forms; all are superantigens; associated with streptococcal toxic shock syndrome and scarlet fever. (19)

strict aerobe   An organism that performs aerobic respiration and can grow only in the presence of oxygen. (6)

strict anaerobe   An organism that cannot grow in the presence of oxygen. (6)

strongyloidiasis   Disease caused by the parasitic nematode *Strongyloides stercoralis*. (22)

structural formula   A representation of molecular structure in which each covalent bond is shown as a line between atoms. (4)

structural gene   A sequence of nucleotides that encodes a functional RNA molecule. (8)

subacute bacterial endocarditis (SBE)   Bacterial infection of a heart valve characterized by slow onset and vague symptoms; usually caused by a viridans streptococcus from the normal oral flora; compare with *acute bacterial endocarditis*. (21)

subacute sclerosing panencephalitis (SSPE)   A sequela of measles, possibly due to a latent measles infection, that manifests as a slow, progressive degenerative disease of the central nervous system. (19)

subcellular fractionation   Separation of the components of a cell by using chemical, physical, and biochemical processes. (5)

substrate-binding protein   A membrane protein that binds specific substrates and delivers them to their cognate ABC transporter. (5)

subunit vaccine   A vaccine containing only isolated antigens of a pathogen or fragments of a pathogen containing the antigens. (17)

superantigen   A molecule that directly stimulates T cells without undergoing antigen-presenting cell processing and surface presentation. (16)

supercoil   An extra twist or turn found in DNA, either positive (increases DNA winding) or negative (decreases DNA winding). (8)

suprapubic puncture   A method for collecting sterile urine by inserting a needle through the abdomen and directly into the bladder. (25)

surface receptor   A host cell membrane protein that is recognized and bound by a microbe having a complementary binding molecule on its surface. (12)

switch region   A repeating DNA sequence interspersed between antibody constant-region genes that serves as a recombination site during isotype switching. (16)

symbiont   An organism that lives in close association with another organism. (6)

symbiosis *pl.* **symbioses**   The intimate association of two species. (27)

symport   Coupled transport in which the molecules being transported move in the same direction across the membrane. (5)

symptom   A sensation related to disease that only a patient can feel, such as a headache, numbness, or fatigue; compare with *sign*. (2)

symptomatic surveillance   See **syndromic surveillance**. (26)

syncytium   A large multinucleated cell formed by the fusion of many uninuclear cells. (12, 20)

syndrome   Collection of signs and symptoms that occur together and signify a particular disease. (2)

syndromic surveillance   Also called *symptomatic surveillance*. Methods to identify clusters of patients with an unusual combination of symptoms that may represent a new emerging infectious disease. (26)

synergism   Cooperation between species in which both species benefit but can grow independently. The cooperation is less intimate than symbiosis. (27)

synergistic   Referring to antibiotics used in combinations whose mechanisms of action complement one another and hence intensify their effectiveness; compare with *antagonistic*. (13)

synthetic medium   A bacterial growth solution that contains defined, known components. (6)

syphilis   A sexually transmitted infection caused by the spirochete *Treponema pallidum*. (23)

systemic circulation   The system of vessels that moves blood between the heart and the rest of the body. (21)

systemic infection   An infection that spreads from its site of origin to many other parts of the body. (21)

systemic inflammatory response syndrome (SIRS)   A life-threatening condition whose symptoms include abnormal body temperature, rapid heart and breathing rates, and abnormal white cell counts. Patients with SIRS and a known infection are classified as having sepsis. (21)

tabes dorsalis   Loss of coordinated movement, especially as a result of syphilis. (23)

tachyzoite   One of the many cell types in the life cycle of *Toxoplasma gondii*; this motile cell grows and divides rapidly, spreading throughout the host. (21)

tapeworm   Also called *cestode*. A parasitic flatworm that infects many animal species via the fecal-oral route of transmission. (11, 22)

taxonomy   The description of distinct life-forms and their organization into categories. (9)

T cell   An adaptive immune cell that develops in the thymus and can give rise to antigen-specific helper cells and cytotoxic T cells. (15)

tegument   The contents of a virion between the capsid and the envelope. (12)

teichoic acid   A chain of phosphodiester-linked glycerol or ribitol that threads through and reinforces the cell wall in Gram-positive bacteria. (5)

temperate phage   A phage capable of lysogeny. (12)

template strand   A DNA strand (or an RNA strand in some viruses) that is used as a template for the synthesis of mRNA. (8)

termination   The final phase of DNA synthesis, in which DNA replication stops because DNA polymerase lands on a termination site. (8)

termination (*ter*) site   Also called *terminus*. A sequence of DNA that halts DNA replication by DNA polymerase. (8)

terminus   See **termination (*ter*) site**. (8)

tertiary (advanced) treatment   Late stage of wastewater treatment designed to remove remaining contaminants such as heavy metals, dissolved solids, phosphorus, and pathogens. (27)

tertiary structure   The third level of organization of proteins; the unique three-dimensional shape of a protein. (4)

tertiary syphilis   A final stage of syphilis, manifested by cardiovascular and nervous system symptoms. (23)

tetanospasmin   The tetanus-causing potent exotoxin produced by *Clostridium tetani*. (24)

tetanus   Deadly disease caused by the *Clostridium tetani* exotoxin, tetanospasmin, which interferes with nerve transmission and forces uncontrollable muscle contraction. (24)

therapeutic dose   The minimum dose of an antibiotic or antimicrobial per kilogram of body weight that stops the growth of a pathogen in a patient. (13)

thermocline   A region of the ocean where temperature decreases steeply with depth, and water density increases. (27)

thermocycler   Laboratory instrument that rapidly and precisely cycles the temperature during PCR. (25)

thermophile   An organism adapted for optimal growth at high temperatures, usually 55°C (131°F) or higher. (6)

threshold dose   The concentration of antigen needed to elicit adequate antibody production. (16)

thrombocytopenia   Low platelet count. (22)

thrombotic thrombocytopenic purpura (TTP)   A life-threatening sequela of Shiga toxin–producing pathogens such as *E. coli* O157:H7 in which microclots occur throughout the circulation owing to a drop in platelets. (22)

thrush   A disease of neonates, infants, and immunocompromised adults, caused by *Candida albicans* and characterized by a white coating in the mouth. (22)

tick   An arachnid ectoparasite that sucks host blood and in the process can deliver or receive a variety of microbial pathogens. (11)

tinea capitis   Dermatophytosis (ringworm) of the scalp. (19)

tinea corporis   Dermatophytosis (ringworm) of the body, primarily on the arms and legs. (19)

tinea cruris   Dermatophytosis (ringworm) of the groin. (19)

tinea pedis   Dermatophytosis (ringworm) of the feet. (19)

tinea unguium   Fungal infection of the nails, also called onychomycosis. (19)

tinea versicolor   Chronic infection of the skin caused by the yeast *Malassezia*. (19)

tissue tropism   The range of host tissue that a pathogen can infect. (12)

titer   A measure of concentration used to express the effective amount of antibody or antigen; also used to measure viral load. (17)

Toll-like receptor (TLR)   A member of a eukaryotic transmembrane glycoprotein family that recognizes a particular microbe-associated molecular pattern (MAMP) present on pathogenic microorganisms. (15)

tonsillitis   Inflammation of the tonsils. (20)

TORCH complex   Collection of microbes able to cross the placenta and infect the fetus; they include *Toxoplasma*, rubella, cytomegalovirus, herpes simplex 2 virus, and others. (21)

total magnification   The magnification of the ocular lens multiplied by the magnification of the objective lens. (3)

toxic dose   The maximum dose of an antibiotic or antimicrobial that is tolerated by a patient. (13)

toxoid vaccine   A vaccine composed of an inactivated microbial toxin. (17)

toxoplasmosis   Disease caused by the protozoan *Toxoplasma gondii*, which is typically asymptomatic but can be severe in the immunocompromised and in fetuses. (21, 24)

trachoma   A serious disease of the eye caused by *Chlamydia trachomatis*; damage to the cornea can lead to irreversible blindness. (19)

transcription   The synthesis of RNA complementary to a DNA template. (8)

transcytosis   Transport of vesicle-bound macromolecules across a cell; common in epithelial cells. The process is co-opted by some pathogens to aid in invasion of host tissue. (24)

transduction   The transfer of host genes between bacterial cells via a bacteriophage. (9, 12)

transformasome   A bacterial cell membrane protein complex that imports external DNA during transformation. (9)

transformation   1. The internalization of free DNA from the environment into bacterial cells. (8, 9) 2. The viral conversion of a normal cell to a cancer cell. (12)

transmembrane protein   A protein with one or more membrane-spanning regions. (4)

transmissible spongiform encephalopathy (TSE)   General term for a prion disease. (24)

transmission   The movement of a pathogen from one host to another. (12)

transmission electron microscopy (TEM)   Electron microscopy in which electron beams are transmitted through a thin specimen to reveal internal structure. (3)

transovarial transmission   The transfer of a pathogen from parent to offspring by infection of the egg cell. Typically seen in insects. (2)

transplacental transmission   Movement of a pathogen from an infected woman to her fetus across the placenta. (2)

transposable element   A segment of DNA that can move from one DNA region to another. (9)

transposase   An enzyme that catalyzes the transfer of a transposable element from one DNA region to another; often encoded by the transposable element. (9)

transposition   The process of moving a transposable element from one DNA region to another. (9)

transposon   A transposable DNA element that contains genes in addition to those required for transposition. (9)

trematode   See **fluke**. (11, 22)

trench mouth   Also called *Vincent's angina*. A severe form of gingivitis; also called necrotizing ulcerative gingivitis. (22)

tricarboxylic acid (TCA) cycle   Also called *citric acid cycle* or *Krebs cycle*. A metabolic cycle that catabolizes the acetyl group from acetyl-CoA into two CO_2 molecules with the concomitant production of NADH, $FADH_2$, and ATP. (7)

trichiasis   The condition in which eyelashes grow back toward the eye and touch the cornea or conjunctiva, as in trachoma. (19)

trichinosis   A helminthic caused by ingestion of *Trichinella spiralis* cysts embedded in pork or other meat; the disease affects muscles. (22)

trichomoniasis   Sexually transmitted infection of the protozoan *Trichomonas vaginalis*; infects the urogenital tract, often causing vaginitis. (23)

trophic level   A level of the food web representing the consumption of biomass of organisms from a previous level. (27)

trophozoite   Active, motile feeding stage of pathogenic protozoa. (22)

trypanosome   A parasitic excavate protist that has a cortical skeleton of microtubules culminating in a long flagellum. (11)

tubercle   A nodule seen in various organs that is characteristic of tuberculosis. (20)

tuberculin skin test   Diagnostic test for assessing whether a person is or has been infected with *Mycobacterium tuberculosis*. Positive readout is a delayed-type hypersensitivity reaction to tuberculin purified protein derivative. See also *Mantoux tuberculin skin test*. (20)

tuberculosis   Infectious disease of the lungs caused by *Mycobacterium tuberculosis*. (20)

tularemia   Also called *rabbit fever*. Disease caused by infection with the bacterium *Francisella tularensis*; symptoms and severity of the disease depend on the virulence of the infecting strain and how the bacteria enter the host. (21)

tumor   A mass of abnormal cells that forms within a tissue. (17)

tumor necrosis factor alpha (TNF-α)   A cytokine involved in systemic inflammation. (15)

turbidity   The cloudiness or opacity of a solution due to suspended solids, such as microbes. (6)

type I (immediate) hypersensitivity   An IgE-mediated allergic reaction that causes degranulation of mast cells within minutes of exposure to the antigen. The severe reaction known as anaphylaxis is triggered by type I hypersensitivity. (17)

type I pilus   A pilus that adheres to mannose residues on host cell surfaces. (18)

type II hypersensitivity   An immune response in which antibodies bind to the patient's own cell-surface antigens or to foreign antigens adsorbed onto the patient's cells. Antibody binding triggers cell-mediated cytotoxicity or activation of the complement cascade. (17)

type II secretion   A bacterial protein secretion system that uses a type IV pilus–like extension/retraction mechanism to push proteins out of the cell. (18)

type III hypersensitivity   Also called *immune complex disease*. An immune reaction triggered when IgG antibody binds to an excess of soluble foreign antigen in the blood. The immune complexes deposit in small blood vessels, where they interact with complement to initiate an inflammatory response. (17)

type III secretion   A bacterial protein secretion system that uses a molecular syringe to inject bacterial proteins into the host cytoplasm. (18)

type IV hypersensitivity   Also called *delayed-type hypersensitivity (DTH)*. An immune response that develops 24–72 hours after exposure to an antigen that the immune system recognizes as foreign. The response is triggered by antigen-specific T cells. It is delayed because the T cells need time to proliferate after being activated by the allergen. (17)

type IV pilus   A dynamic pilus that can repeatedly assemble and disassemble; it mediates twitching motility. (18)

type IV secretion   A bacterial protein secretion system that is homologous to the bacterial conjugation apparatus. (18)

typhoid fever   Also called *enteric fever*. Life-threatening systemic disease of humans caused by *Salmonella* Typhi; transmitted by fecal-oral route. (21, 22)

ubiquitination   Posttranslational attachment of the protein ubiquitin to another protein, which alters protein function or location or signals that the protein is to be degraded. (18)

uncoating   The release of a viral genome from its capsid, following entry of the virion into a host cell. (12)

upper respiratory tract   Upper airways of the body, which include the nasal passages, oral cavity, pharynx, and larynx. (20)

urethritis   Inflammation of the urethra. (23)

urinary tract infection (UTI)   Infection of the urinary tract; includes cystitis and pyelonephritis. (23)

urogenital route   Entry of a pathogen via the urethra or vagina. (2)

urosepsis   Life-threatening infection of the blood secondary to a urinary tract infection. (23)

use-dilution test   A method for measuring the efficacy of disinfectants against microbes dried onto a surface. (13)

vaccination   Exposure of an individual to a weakened version of a microbe or a microbial antigen to provoke immunity and prevent development of disease upon reexposure. (1, 16)

vaccine   The material used to vaccinate someone. (17)

vaccine-associated paralytic poliomyelitis (VAPP)   Polio acquired as the result of vaccination with a Sabin vaccine whose virus reverted to a virulent form. (24)

vaginitis   Inflammation of the vagina. (23)

vaginosis   Infection of the vagina. (23)

vancomycin   A glycopeptide antibiotic that inhibits bacterial cell wall synthesis by binding the D-alanine-D-alanine dipeptide; a mechanism distinct from penicillin inhibition. (13)

variable region   The amino-terminal portions of antibody light and heavy chains that confer specificity to antigen binding and define the antibody idiotype. (16)

variant Creutzfeldt-Jakob disease (vCJD)   A human form of mad cow disease (bovine spongiform encephalopathy), characterized by a slow deterioration of brain functions. It is invariably fatal. (24)

vasoactive factor   A cell signaling molecule that increases capillary permeability. (15)

vasodilation   Widening of blood vessels. (15)

vector   1. An organism (e.g., insect) that can carry infectious agents from one animal to another. (2) 2. In molecular biology, a molecule of DNA into which exogenous DNA can be inserted to be cloned. (8)

vegetative reproduction   See **asexual reproduction**. (11)

vehicle transmission   Movement of infectious agents between hosts via contaminated fomites, food, water, or air. (2)

Venereal Disease Research Laboratory (VDRL) test   A nonspecific test for syphilis that looks for the presence of anti-cardiolipin antibodies in blood or cerebrospinal fluid. (23)

venipuncture   The puncture of a vein with a needle to either withdraw blood or administer an intravenous injection. (25)

vertical gene transfer   Gene transfer from parent to offspring through reproduction. (9)

vertical transmission   In disease, the transfer of a pathogen from parent to offspring. See also *transovarial transmission*. (2)

vesicle   1. A small, membrane-enclosed sphere found within a cell. (4) 2. A small blister on the skin. (19)

viable   Capable of replicating—for instance, by forming a colony on an agar plate. (6)

Vi Ag (virulence antigen)   A polysaccharide capsule on the surface of *Salmonella* Typhi that renders the bacteria resistant to phagocytosis and serum complement. (22)

vibriosis   A disease caused by a *Vibrio* species. (22)

Vincent's angina   See **trench mouth**. (22)

viral load   The concentration of virus in the blood. (23)

viral sinusitis   Inflammation and congestion of the sinuses due to a viral infection of the upper respiratory tract. (20)

viridans streptococcus   Streptococci that are alpha-hemolytic on blood agar; most of the members are commensal bacteria of the oral cavity. (21)

virion   A virus particle. (12)

viroid   An infectious naked nucleic acid. (12)

virome   All the viruses that inhabit a particular organism or environment. (14)

virulence   A measure of the severity of a disease caused by a pathogenic agent. (2)

virulence factor   A trait of a pathogen that enhances the pathogen's disease-producing capability. (18)

virulent phage   A bacteriophage that reproduces entirely by a lytic cycle. (12)

virus   A noncellular particle containing a genome that can replicate only inside a host cell. (1, 12)

vulvovaginal candidiasis   Overgrowth of the vaginal commensal fungus *Candida*; commonly called a yeast infection. (23)

wart   Also called *papilloma*. A small growth on the skin caused by human papillomavirus. (19)

wastewater treatment plant   A series of wastewater transformations designed to lower biochemical oxygen demand and eliminate human pathogens before water is returned to local rivers. (27)

water activity   A measure of the water that is not bound to solutes and is available for use by organisms. (6)

wavelength   The distance between two adjacent peaks in the electromagnetic spectrum. (3)

western blot   A technique to detect specific proteins. Proteins are subjected to gel electrophoresis, transferred to a blot, and probed with enzyme-linked or fluorescently tagged antibodies that specifically bind the protein of interest. (17)

West Nile virus (WNV)   An RNA virus that is endemic to birds and is transmitted to the dead-end hosts, humans and horses, by mosquitoes, to cause West Nile encephalitis. (24)

wetland   A region of land that undergoes seasonal fluctuations in water level and aeration. (27)

wetland filtration   A wastewater treatment method that uses wetland microbial ecosystems to remove organic carbon and nitrogen from the wastewater. (27)

wetland restoration   The repair and reconstruction of wetlands and their ecosystems that were previously degraded by pollutants and urban/suburban development. (27)

wet mount   A technique to view living microbes with a microscope by placing the microbes in water on a slide under a coverslip. (3)

whey   The liquid portion of milk after proteins have precipitated out of solution, usually during cheese production. (27)

whipworm   *Trichiuris trichiuria*, a soil-dwelling helminthic parasite that burrows into the wall of the intestines, causing frequent bloody stools and rectal prolapse. (22)

white blood cell (WBC) differential   A laboratory test that counts the types of white blood cells in a patient's blood. (15)

whooping cough   Also called *pertussis*. A highly contagious respiratory infection, caused by *Bordetella pertussis*, whose defining symptom is a paroxysmal cough that lasts for weeks to months. (20)

X-linked agammaglobulinemia (Bruton's disease)   A disease in which the body fails to make B cells owing to mutation in the X-linked *BTK* gene; the gene product, Btk, is a protein tyrosine kinase essential for normal B-cell development. (17)

X-ray crystallography   A technique to determine the positions of atoms (atomic coordinates) within a molecule or molecular complex, based on the diffraction of X-rays by the molecule. (3)

yeast   A unicellular fungus. (11)

yersiniosis   A disease caused by species of *Yersinia* bacteria. (22)

yogurt   A semisolid food produced through acidification of milk by lactic acid–producing bacteria. (27)

zone of inhibition   A region of no bacterial growth on an agar plate due to the diffusion of a test antibiotic. Correlates with the minimal inhibitory concentration. (13)

zoonosis   See **zoonotic disease**. (10, 27)

zoonotic disease   Also called *zoonosis*. An infection that normally affects animals but can be transmitted to humans. (2)

zooxanthella *pl.* **zooxanthellae**   A phototrophic coral endosymbiont, most commonly a dinoflagellate of the genus *Symbiodinium*. (11)

zygote   Diploid cell produced by fusion of a male gamete and a female gamete. (11)

Chapter 1

p. 2 (composition book): courtesy of Anne DeMarinis; p. 2 (microbe): BSIP/UIG/Getty Images; p. 3 (left): courtesy of Dr. Charles L. Daley, National Jewish Health; pp. 3 & 4 (silhouette): sattva78/Shutterstock; p. 3 (right): CDC/Dr. Ray Butler; p. 5 (far left): Dr. Ralf Wagner/dr-ralf-wagner.de; p. 5 (center left): Scimat/ Photo Researchers, Inc.; p. 5 (center right): Electron Microscope Lab, University of California, Berkeley; p. 5 (far right): Linda Stannard, UCT/Photo Researchers, Inc; p. 6 (left): Bettmann/Corbis; p. 6 (right): Johner Images/Getty Images; p. 10 (top left): Bridgeman Art Library; p. 10 (top right): Reuters/Corbis; p. 10 (bottom left): © Arte & Immagini srl/CORBIS; p. 11 (top): Milton S. Eisenhower Library; p. 11 (center): Bettmann/ Corbis; p. 11 (bottom right): Brian J. Ford; p. 12: Institut Pasteur; p. 13 (bottom left): Corbis; p. 13 (bottom right): The Florence Nightingale Museum Trust, London; p. 14 (Figure 1.9): modified from CDC. 2010. *Flu-View: 2009–2010 Influenza Season Summary* (cdc. gov/flu/weekly/weeklyarchives2009-2010/09-10sum-mary.htm); p. 14 (bottom): Edith60/Shutterstock.com; p. 15 (top left): CDC; p. 15 (top right): © Dennis Kunkel Microscopy, Inc./Visuals Unlimited/Corbis; p. 15 (bottom left): Museum in the Robert Koch-Institut Berlin; p. 15 (bottom center): Museum in the Robert Koch-Institut Berlin; p. 15 (bottom right): Museum in the Robert Koch-Institut Berlin; p. 16: © Dennis Kunkel Microscopy, Inc./Visuals Unlimited/Corbis; p. 17 (top left): Max Planck Institute for Infection Biology/ Kaufmann; p. 17 (center left): © Donald Weber/VII/ Corbis; p. 17 (bottom left): The National Library of Medicine; p. 17 (bottom right): Cytographics/Visuals Unlimited; p. 19 (top): CDC/Douglas E. Jordan; p. 19 (bottom left): Hulton-Deutsch Collection/Corbis; p. 19 (bottom center): Bettmann/Corbis; p. 19 (bottom right): Corbis; p. 20: Institut Pasteur; p. 21 (top left): Corbis; p. 21 (top right): Mediscan/Visuals Unlimited; p. 21 (bottom right): From Grunewald et al. 21 November 2003. Three-Dimensional Structure of Herpes Simplex Virus from Cryo-Electron Tomography. *Science* 302(5649):1396–1398. Reprinted with permission from AAAS; p. 22: courtesy of Everglades National Park; p. 23 (top): American Society for Microbiology; p. 23 (bottom): Lynn Margulis, UMass, Amherst; p. 24 (top left): Tony Craddock/Science Source; p. 24 (top right): Carl Woese; p. 25 (bottom left): Photo Researchers, Inc.; p. 25 (bottom center): Omikron/Photo Researchers; p. 25 (bottom right): Barington Brown/ Photo Researchers; p. 26 (bottom left): Fred Sanger/ MRC Laboratory of Molecular Biology; p. 26 (bottom right): The Institute for Genomic Research; p. 27: Joan Slonczewski.

Chapter 2

p. 32: Pando Hall/Digital Vision/Getty Images; p. 33 (top): Science Photo Library/Custom Medical; pp. 33 & 34 (silhouette): Anne BA?? Pedersen/Getty Images; p. 33 (bottom): Getty Images/Kallista Images; p. 35 (top): Rajalakshmi Nambiar/Tyska Lab; p. 35 (bottom left): Jane Shemilt/Science Source; p. 35 (bottom center): Arthur Siegelman/Visuals Unlimited; p. 35 (bottom right): © 2007 doctorfungus.org; p. 36 (top left): John Greim/Science Source; p. 36 (top right): CDC; p. 36 (bottom left): CDC/Lois Norman; p. 36 (bottom right): CDC/Hermann; p. 37 (left): CDC/Lois Norman; p. 37 (right): CDC; p. 38 (left): R. L. Santos, R. M. Tsolis, A. J. Bumler, and L. G. Adams. 2003. Pathogenesis of Salmonella-induced enteritis. *Brazilian Journal of Medical and Biological Research* 36(1):3–12; p. 38: (right): Matthew Welch, University of California Berkeley; p. 40: Jamie Wilson/Shutterstock.com; p. 41 (top): Melinda Fawver/Shutterstock.com; p. 41 (bottom): CDC; p. 42 & 43: Science Source; p. 42 (bottom): Blend Images/ Shutterstock.com; p. 43 (top): Karabay, O. 26 January 2007. Tularaemic cervical lymphadenopathy. *Journal of the New Zealand Medical Association* 120(1248); p. 44 (cat): © GlobalP/iStockphoto.com; p. 44 (cheeseburger): © dehooks/iStockphoto.com; p. 46 (top): CDC/Brian Judd; p. 46 (bottom left): courtesy of Historical Collection & Services, Claude Moores Health Services Library, U. of Virginia; p. 46 (bottom center): CDC; p. 46 (bottom right): CDC/Science Source; p. 47 (left): GK Hart/Vikki Hart/Photodisc/Getty Images; p. 47 (right): Thomas H. Hahn docu-images; p. 49: DVARG/Shutterstock.com; p. 50 (top): FogStock/Vico Images/Alin Dragulin/Getty Images; p. 50 (bottom): Time & Life Pictures/Getty Images; p. 52 (top left): LABCONCO Corporation; p. 52 (top right): Image courtesy of J. Craig Venter Institute; p. 52 (bottom): Kevin Karem/CDC; p. 54: Copyright 2014 Elsevier Inc. All rights reserved. elsevierimages.com; p. 56 (top): Science Source; p. 56 (center left): Alain Pol, ISM/Science Source Images; p. 56 (bottom left): NASA Land-stat/University of Maryland Global Land Cover Facility; p. 56 (bottom right): NASA image by Jeff Schmaltz, MODIS Rapid Response Team at NASA GSFC; p. 57: © George Kourounis.

Chapter 3

p. 62 (can): Shutterstock; p. 62 (spoon): Shutterstock; p. 62 (chili): Shutterstock; p. 62 (bacteria): Science Source; pp. 63 & 64 (silhouette): Getty Images; p. 63: CDC/ Dr. George Lombard; p. 64 (top): Shutterstock; p. 64 (Bottom): Biophoto Associates/Science Source; p. 66 (left): Andrew Syred/Science Source; p. 66 (right): CDC/DPDx; p. 67: Michael Abbey/Science Source; p. 67: Wanderley de Souza (2013). Herbicides as Potential Chemotherapeutic Agents Against Parasitic Protozoa, Herbicides—Advances in Research, Dr. Andrew Price (Ed.), ISBN: 978-953-51-1122-1, InTech, DOI: 10.5772/56007. Available from: intechopen.com; p. 67: reprinted by permission from Macmillan Publishers Ltd: *Nature* 456, 750–754 (11 December 2008); p. 67: CDC/ Janice Haney Carr; p. 67: Dr. Stan Erlandsen/Science Faction/Getty Images; p. 68 (A): Science VU/Visuals Unlimited; p. 68 (B): Dennis Kunkel/Visuals Unlimited; p. 68 (C): Mediscan/Visuals Unlimited; p. 68 (D): Dennis Kunkel/Visuals Unlimited; p. 68 (E): Eye of Science/Photo Researchers, Inc.; p. 68 (F): Dennis Kunkel Microscopy; p. 69 (A): Ed Reschke/Peter Arnold/Getty Images; p. 69 (B): © BIODISC/Visuals Unlimited; p. 69 (C): © T.J. Beveridge/Visuals Unlimited; p. 69 (D): Kiseleve & Donald Fawcett/Visuals Unlimited; p. 69 (E): Steffen Klamt/U. Magdeburg; p. 72 (A): Michael W. Davidson; p. 72 (B): Michael W. Davidson; p. 75 (all): Steffen Klamt/U. Magdeburg; p. 76 (left): Henrik Larsson/Shutterstock; p. 76 (right): Biophoto Associates/ Science Source; p. 78 (both): courtesy of Cellscope; p. 79 (left): CDC/Dr. Mike Miller; p. 79 (right): Dr. Sandra Richter; p. 80 (left): CDC/Dr. George C. Kubica; p. 80 (center): Dr. Gladden Willis/Visuals Unlimited/Corbis; p. 80 (right): Jason C. Baker, Ph.D/Missouri Western State University; p. 81 (left): Arthure Siegelman/ Visuals Unlimited; p. 81 (right): Wim van Egmond/ Visuals Unlimited, Inc.; p. 82 (top): James R. Martin/ Shutterstock.com; p. 82 (bottom): reprinted by permission from P. L. Graumann and R. Losick. 2001. *J Bacteriol.* 183(13): 4052–4060; p. 83 (top left): CDC/Dr. William Cherry; p. 83 (top center): Michelle Clark and Joan Slonczewski; p. 83 (top right): Michelle Clark and Joan Slonczewski; p. 83 (bottom): Wah Chiu/Baylor College of Medicine; p. 84 (top): CDC; p. 84 (bottom): Veronica Burmeister/Visuals Unlimited; p. 87 (A): © Dennis Kunkel Microscopy, Inc./Visuals Unlimited/ Corbis; p. 87 (B): R. M. Macnab. 1976. *J. Clin. Microbiology* 4(3):258; p. 87 (C): CDC/Richard Eacklam; p. 87 (D): Mesnage et al. *J. Bacteriol.*

Chapter 4

p. 90: © MORTEZA NIKOUBAZL/Reuters/Corbis; p. 91 (left): © Ton Koene/Visuals Unlimited/Corbis; pp. 91 & 92 (silhouette): DRB Images, LLC/Getty Images; p. 91 (right): Juergen Berger/Science Source; p. 111 (Figure 4.29B): modified from Yin et al. 2006. *Science* 312:741; p. 114 (top): © Scientifica/RMF/Visuals Unlimited/Corbis; p. 114 (center): © ADAM; p. 114 (bottom): Brent Stirton/Getty Images; p. 116 (top): ERIC GRAVE/PHOTO RESEARCHERS; p. 116 (center): © Donkeyru | Dreamstime.com; p. 116 (bottom): photo by Dr. George Palade, courtesy of Dr. Marilyn Farquhar; p. 117: courtesy of Dr. Nicholas Arpaia/University of California, Berkeley.

Chapter 5

p. 122 (sandals): Shutterstock; p. 122 (leaves): Shutterstock; p. 122 (wood): Shutterstock; p. 123 (top): Paul Whitten/Science Source; pp. 123 & 124 (silhouette): Getty Images; p. 123 (bottom): Juergen Berger/Science Source; p. 124: MLA Izard, Jacques et al. 2009. Cryo-Electron Tomography Elucidates the Molecular Architecture of Treponema Pallidum, the Syphilis Spirochete. *Journal of Bacteriology* 191(24):7566, 7580. PMC. Web. 21 Nov. 2014; p. 126 (top left): Dennis

Kunkel Microscopy; p. 128 (Table 5.2): modified from F. Neidhardt and H. E. Umbarger. 1996. Chemical composition of *Escherichia coli*, p. 14. In *Escherichia coli* and *Salmonella: Cellular and Molecular Biology*, 2nd ed., ASM Press; p. 128 (Figure 5.3): modified from Protein Data Bank. *Molecule of the Month: Multidrug Resistance Transporters* (rcsb.org/pdb/education_discussion/molecule_of_the_month/download/MultidrugResistanceTransporters.pdf); p. 132: reprinted with permission from Holtje, J.V. Growth of the stress-bearing and shape maintaining murein sacculus of *Escherichia coli*; p. 134 (top): © David Scharf/Corbis; p. 134 (bottom left): Jenn Huls/Shutterstock.com; p. 134 (bottom right): David Scharf Photo; p. 136 (top left and right): courtesy of Benoit Zuber/MRC Laboratory of Molecular Biology; p. 137: De Agostini Picture Library/Science Source; p. 138 (left): © SPL/Custom Medical Stock Photo—All rights reserved; p. 138 (right): Ted Kinsman/Science Source; p. 139: reprinted from Rabinowitz, R. P., Lai, L. C., Jarvis, K., McDaniel, T. K., Kaper, J. B., Stone, K. D., Donnenberg, M. S. Attaching and effacing of host cells by enteropathogenic; p. 140: Kellenberger, E., Arnold-Schulz-Gahmen, B. 1992. Chromatins of low protein content. *FEMS Microbiol. Lett.* 79:361–370; p. 141 (center left): CDC/Janice Haney Carr/Jeff Hageman, M.H.S.; p. 141 (top right): © Dr. Ken Greer/Visuals Unlimited/Corbis; p. 141 (center right): CDC/Jeff Hageman, M.H.S.; p. 141 (bottom): All photos: Ahmed Touhami et al. 2004. *J. Bacteriol.* 186; p. 142: Tsute Chen and Margaret J. Duncan, The Forsyth Institute; p. 143 (center left): CAMR/A. Barry Dowsett, Robert Macnab. 1976. *J. Clinical Microbiology* 4:258; p. 143 (bottom left): David DeRosler/Brandeis University; p. 144 (left): W. Killi & F. Partensky, Bedford Institute of Oceanography; p. 144 (center): International Union of Microbiological Societies. From P. Anil Kumar et al. 2007. *Int. J. Syst. Evol. Microbiol.* 57:2110–2113; p. 144 (right): courtesy of the Sustainable Energy Research Group, University of Southampton; p. 145: EnCor Biotechnology, 1. Shaw, G., Yang, C., Ellis, R., Anderson, K., Parker, Mickle J., Scheff, S., Pike, B., Anderson, D. K., and Howland, D. R. 2005. Hyperphosphorylated neurofilament NF-H is a serum biomarker of axonal injury. *Biochem Biophys Res Commun.* 336:1268–1277; p. 147: © Visuals Unlimited/Corbis; p.147 (bottom right): © Dennis Kunkel Microscopy, Inc./Visuals Unlimited/Corbis; p. 148 (center): Don W. Fawcett/Science Source; p. 148 (right): Dr. George Chapman/Visuals Unlimited, Inc.; p. 150 (top and center): Dartmouth Electron Microscope Facility, Biomedia Museum U. of Paisley; p. 151: Michael Abbey/Visuals Unlimited.

Chapter 6

p. 156 (gloved hand and petri dish): Corbis; p. 156 (fingerprint): Shutterstock; p. 156 (sample in petri dish): Getty Images; p. 156: Medicimage/Science Source; pp. 157 & 158 (silhouette): Corbis; p. 157 (top): Medicimage/Science Source; p. 157 (center): © Nathan Reading/Flickr; p. 157 (bottom): CDC; p. 160 (left): D.J. Silverman, C.L. Wisseman, Jr. and A. Wadell; p. 160 (right): David Wood, PhD. University of South Alabama College of Medicine; p. 161: Wim van Egmond/Visuals Unlimited/Getty Images; p. 162 (left): Frank Dazzo, Michigan State University; p. 162 (right): Inga Spence/Photo Researchers, Inc.; p. 163 (top): Parm Randhawa, California Seed & Plant Lab, Inc.; p. 163 (bottom): Uni-

versity of Manitoba; p. 164 (top): John Foster; p. 164 (bottom): all photos by John Foster; p. 165 (left): Sfocato/Shutterstock.com; p. 165 (right): John Foster; p. 166: Filatov Alexey/Shutterstock.com; p. 167: Molecular Probes, Inc.; p. 168 (left): © Dennis Kunkel Microscopy, Inc.; p. 168 (right): Ellen Quardokus; p. 171 (right): courtesy of Global Medical Instrumentation, Inc.; p. 173 (top): Thomas D. Brock, U. of Wisconsin, Madison; p. 173 (center): B. Boonyaratanakornkit et al., University of California, Berkeley; p. 174 (top left): © Momatiuk-Eastcott/Corbis; p. 174 (top right): photo by Dudley Foster from RISE expedition, courtesy of William R. Normark, USGS; p. 174 (bottom left): © PASIEKA/Science Photo Library/Corbis; p. 174 (bottom right): Douglas G. Capone, U. of Southern California; p. 175: J. W. Deming & R. Colwell. *Appl. Environ. Microbial* 44:1222–1230; p. 176 (A): Wayne P. Armstrong; p. 176 (B): courtesy of S. DasSarma, U. of Maryland Biotechnology Institute; p. 177 (A): Kazuyoshi Nomachi/Corbis; p. 177 (B): © Dr. Peter Siver/Visuals Unlimited/Corbis; p. 177 (C): Wolfgang Kaehler/Corbis; p. 178 (top): Kwangshin Kim/Science Source; p. 178 (bottom): Rathman, M., Michael D. Sjaastad, and Stan Falkow. 1996. Acidification of phagosomes containing Salmonella typhimurium in murine macrophages. *Infection and Immunity* 64:2765–2773. Photo courtesy of Brett Finlay, U. of British Columbia; p. 179: Sebastian Kaulitzki/Shutterstock.com; p. 181 (top): Dr. Jack Bostrack/Visuals Unlimited, Inc.; p. 181 (bottom): courtesy of Joan Slonczewski; p. 182: eAlisa/Shutterstock.com; p. 183 (A): © J. D. Ruby, K. F. Gerencser; p. 183 (B): SCIMAT/Science Source; p. 184 (Figure 6.27): modified from H. C. Flemming and J. Wingender. 2010. *Nat. Rev. Microbiol.* 8:623–633; p. 186 (A): John Foster; p. 186 (B): John Foster; p. 186 (C): John Foster.

Chapter 7

p. 192 (umbrellas): Getty Images; p. 192 (sand): Shutterstock; p. 192 (thermometer): Shutterstock; pp. 193 & 194 (silhouette): Shutterstock; p. 193 (left): © Becton, Dickinson and Company; p. 193 (right): © Mediscan/Corbis; p. 203 (left): Dennis Strete/Fundamental Photographs, NYC; p. 203 (right): Troy Biological BD Worldwide; p. 205: © Dennis Kunkel Microscopy, Inc./Visuals Unlimited/Corbis; p. 206 (top): AP Photo/Gerald Herbert, File; p. 206 (bottom); Win McNamee/Getty Images; p. 207 (left): Gert Vrey/Shutterstock.com; p. 207 (right): Jason C. Baker, Ph.D./Missouri Western State University; p. 211 (top): Deyan Georgiev/Shutterstock.com; p. 211 (bottom left): Shun'ichi Ishii et al. 2008. BMC Microbiol, 8.6; p. 211 (bottom right): © Science VU/S. Watson/Visuals Unlimited/Corbis; p. 213: Wim van Egmond/Visuals Unlimited, Inc.; p. 214: © Jiang Kehong/Xinhua Press/Corbis; p. 215 (top left): Bill Bachman/Science Source; p. 215 (top right): © Dennis Kunkel Microscopy, Inc./Visuals Unlimited/Corbis.

Chapter 8

p. 222 (dental instruments): Shutterstock; p. 222 (smoke): Shutterstock; p. 223 (top): E+/Getty Images; pp. 223 & 224 (silhouette): Fotolia; p. 223 (bottom): Wallace Ambrose/Chapel Hill Analytical and Nanofabrication Lab, University of North Carolina; p. 224: Leonid Andronov/Shutterstock.com; p. 225: © Dr. Gopal Murti/Visuals Unlimited/Corbis; p. 234: Durund et al. 2007. *BMC Cell Biol.* 8:13; p. 236: Marligen Bio-

Sciences, Inc; p. 237: © Steve Gschmeissner/Science Photo Library/Corbis; p. 238 (Figure 8.21): modified from Alison J. Cody et al. 2003. *Infect. Genet. Evol.* 3:57; p. 239: Steve Greenberg, Ventura County Star, Calif., 2007; p. 240: © Dr. David Phillips/Visuals Unlimited/Corbis; p. 242: SPL/Science Source; p. 242 (inSight Figure 1): modified from Catharina C. Boehme et al. 2010. *N. Engl. J. Med.* 363:1005; p. 243 (inSight Figure 2): modified from Teruyuki Takahashi and Tomohiro Nakayama. 2006. *J. Clin. Microbiol.* 44(3):1029; p. 247 (B): O. L. Miller, Jr. 1982. *EMBO J.* 1:59.

Chapter 9

p. 258: Corbis; p. 259 (left): Dileepunnikri/Wikipedia Commons; pp. 259 & 260 (silhouette): iStock; p. 259 (right): Dr. Gary Gaugler/Science Source; p. 262: Dennis Kunkel Microscopy, Inc.; p. 269 (Gabon): El Albani et al. 2010. *Nature* 466:100; p. 269 (Bil'yakh): Andrew Knoll; p. 269 (Doushantuo): Xiao et al. 2000. Eumetazoan fossils in terminal Proterozoic phosphorites. *PNAS* 97:13684–13689; p. 269 (trilobite): Sinclair Stammers/SPL/Photo Researchers, Inc.; p. 271 (Figure 9.16): modified from Morgan Price et al. 2008. *Genome Biol.* 9:R4 and from Fabia Battistuzzi et al. 2004. *BMC Evol. Biol.* 4:44; p. 272 (top): © Dennis Kunkel Microscopy, Inc./Visuals Unlimited/Corbis; p. 272 (center): courtesy of Andrew Allen; p. 272 (bottom): Kelly et al. July 2004. ...Biofilm Microbes. *Applied and Environmental Microbiology* 70(7):4187–4192; p. 273: courtesy of Norman Pace; p. 275 (left): © Picsfive/Shutterstock.com; p. 275 (center): Wei Gao et al. 2010. PLoS Pathogens 6: N e1000944; p. 275 (right): Wei Gao et al. 2010. PLoS Pathogens 6: N e1000944; p. 276: Pfarr, K., Hoerauf, A. 2005. The Annotated Genome of Wolbachia. *PLoS Med* 2(4): e110 doi:10.1371/journal.pmed.0020110; p. 277 (top): © Tbcphotography | Dreamstime.com; p. 277 (Figure 9.20A): PRNewsFoto/Copan Diagnostics, Inc.; p. 277 (Figure 9.20B and C): modified from Ahmed Hiller et al. 2010. *PLOS Pathogens* 6:e1001108; p. 282 (Figure 9.23A): modified from Victor Santa Cruz. Database Central, Wastewater Organisms Database; p. 282 (Figure 9.23B): Victor Santa Cruz Productions; p. 282 (Figure 9.23C): Victor Santa Cruz Productions; p. 283: figure from Biolog, Inc. product video: "GEN III Microbial ID Overview," Biolog.com. Reprinted with permission from Biolog, Inc.

Chapter 10

p. 288 (surgical instruments): © Radius Images/Alamy; p. 288 (microbe): © Dennis Kunkel Microscopy, Inc./Visuals Unlimited/Corbis; p. 289 (center): © 4x6/iStockphoto; pp. 289 & 290 (top): Figure 10 from Onur Gonul, Sertac Aktop, Tulin Satilmis, Hasan Garip and Kamil Goker. 2013. Odontogenic Infections, A Textbook of Advanced Oral and Maxillofacial Surgery, Prof. Mohammad Hosein Kalantar Motamedi (Ed.), ISBN: 978-953-51-1146-7, InTech, DOI: 10.5772/54645. Available from: intechopen.com/books/a-textbook-of-advanced-oral-and-maxillofacial-surgery/odontogenic-infections; p. 289 (center): courtesy of Dr. A. T. Warfield, Dept. of Cellular Pathology, University Hospital Birmingham NHS Foundation Trust; p. 289 (bottom): © Dennis Kunkel Microscopy, Inc./Visuals Unlimited/Corbis; p. 293 (left): George Chapman/Visuals Unlimited; p. 293 (right): courtesy of Benjamin de Bivort/Harvard University—Wikipedia Commons; p. 295 (top): PNAS Cover, Dec. 1999/Marc

Chapter 27

O

O antigen, 137, 582. *See also* lipopolysaccharides
objective lens, **73–74**
observational study, **881**
occupation and disease susceptibility, 54
ocean habitats, 911–12
ocular lens, **74–75**
Okazaki, Reiji and Tsuneko, 229
Okazaki fragments, **229**, 230
oliguria, **778**
omalizumab (Xolair), 535
Onchocerca volvulus, 341
oncogene, **372**
oncogenic virus, 372
One Health Initiative, **897**
Onesimus, 19
oophoritis, **734**
open reading frames (ORFs), **278**
operons
    definition, **225**, 245
    gene organization in, 225f, 226
    lactose (*lac*) operon, 251, 252f
    transcription, 244f, 245
ophthalmia neonatorum, 621t, **633–34**, 792
opisthorchiasis, **760**
Opisthorchis, 342, 753t, 756f, 757, 760
O polysaccharide, 137. *See also* lipopolysaccharides
opportunistic pathogens, definition, **36**, **442**
opsonization
    antibodies, 470–71, 498, 589
    complement, 479, 529, 623, 631
    definition, **470**, **498**
    phagocytosis, 470–71
optical density, **168**
optics. *See also* light
    focal plane, 71, 72f
    focal point, 71, 72f
    interference, 72, 73, 75, 81, 82f
    magnification, 65
    magnification and refraction, 71, 72f, 73–74
    reflection, 71
    refraction, 71, 74
    refractive index, 71, 73–74, 81
optochin susceptibility disk test, 858f, **859**
oral and nasal cavities
    anatomy, 646, 647f, 723
    infections of oral cavity, 725–27
    microbiota, 431f, 433t, 434–35
oral-fecal route, 45t, **49, 50**, 53
oral rehydration therapy (ORT), 91, 92, 113, 114, 746
orbitals, **94**, 96
orchitis, **734, 795**
ORF Finder, 278
organ donation. *See* organ transplants
organelles. *See also specific types*
    dynamic transport between, 145–46
    endomembrane system, 117, 145f, 146–47
    endosymbiosis theory of eukaryotic evolution, 23–24, 148–49
    light microscopy, 66
    membranes, 115, 145–47
    overview, **115**, 145–47
organic acids, ionization, 109
organic bases, ionization, 109
organic molecules, **97**
organotrophy, **161, 195–96**

organ transplants
    bone marrow transplants, 512
    liver transplants for hepatitis C, 351–52
    thymus, 525
    transplant rejection, 544, 545f
    transplant-related rabies infections, 818
origin of replication (ori), on cell envelope, **140**
origin (oriC), in DNA replication, **228–30**
origins of life
    carbon isotope ratios, 269
    geological evidence, 269f
    microbes, 4
    microfossils, 269
    Miller experiments, 12
    Oró experiment, 12
    overview, 268–69
Oró, Juan, 12
oropharynx, **434**, 435
orthologs, **278**, 281
Orwell, George, 16
Oscillatoria, 177f, 311
oseltamivir (Tamiflu), 372, 374, 375f, 376, 416t, 417
osmolarity, **175**
osmolytes, 115
osmosis, **113**, 115
osmotic diarrhea, **727**, 730
osmotic shock, **113**
Osterholm, Michael T., 898, 901
Ostreococcus tauri, 145, 320, 321f
otitis media. *See* acute otitis media
otoscope, 658f
outbreak, **882**, 883
outer leaflet, 111
outehr membrane
    definition, **125**
    Gram stain, 79
    lipopolysaccharides, 125–26, 127f, 136f, 137
    pathogenic factors, 79
    subunits, 135–36
    toxic lipids, 112
oxacillin, 405f, 406, 409t, 619
oxazolidinones, 246t, 403f, 409t, 410f, 411
oxidation, **110**
oxidation–reduction reactions, 109–10, 199
oxidative burst, **471**
oxidative phosphorylation, 125, 148, **207**
oxidized, **199**
oxidizing agents, **110**
oxidoreductases, **208**, 209f, 210, 211
2-oxoglutarate (alpha-ketoglutarate), 197, 204, 217
oxygen as environmental limiting factor, 910–11
oxygenic photosynthesis, 195, **213**

P

Pace, Norman, 9t, 272, 273f
palmitic acid, 98f
pancarditis, 661, **685**
pancreas, 543, 630, 723f, 725, 745
pancreatitis, **745**
pandemics. *See also* influenza pandemics
    bubonic plague, 47, 882
    definition, 45t, **47, 882–83**
    examples, 48t
    HIV/AIDS, 47, 418, 784, 882
Pantoea agglomerans (formerly *Enterobacter agglomerans*), 897
papilloma, **616**

papillomavirus, 5f, 6, 357, 370. *See also* human papillomavirus
papular rash, 608
papules, **608**, 783
Paragonimus, 672
paralogs, **278**
Paramecium. *See also* ciliates
    conjugation, 335
    consumption by larger protists, 116, 334, 335f
    contractile vacuoles, 115, 151f, 334, 335f
    overview, 334, 335f
    size, 66f, 69f
paramyxovirus, **733**
parasitemia, 596
parasites, **35**. *See also specific types*
parasitic infections. *See also specific diseases and organisms*
    eye infections, 634–35
    gastrointestinal tract, 723f, 751–60, 762f
    notifiable infectious diseases, 890t
    protozoan central nervous system infections, 809f, 830–32, 834f
    protozoan digestive system infections, 751, 752t, 753–54
    protozoan pathogenesis, 595–97
    protozoan reproductive system infections, 789t, 794, 800f
    respiratory tract infections, 647f, 672
    systemic infections, 683f, 705–13, 714f
parasitism, **917**
parental strand, **228**
parenteral route, 45t, **49**, 50
paresthesia, **812**
parfocal lens systems, **74**
paromomycin, 331, 333
parotitis, **733–34**
parthenogenesis, **758**
parvovirus, 363
passive immunization, **553**
passive transport, **130**
Pasteur, Louis
    anaerobic bacteria, 181
    career, 12, 15
    fermentation, 7t, 12
    immunization with attenuated microbes, 19–20
    photograph, 12f
    rabies vaccine, 20
    spontaneous generation, 12
    swan-necked flask, 12f
pasteurization, **391**, 929. *See also* sterilization
pathogen-associated molecular pattern (PAMP), **464**, 693, 696
pathogenesis. *See also* endotoxins; exotoxins
    definitions and terminology, 5, 35–37, **568**
    microbial attachment, 570–75
    protozoan pathogenesis, 595–97
    Salmonella as bacterial model, 590
    tools of bacterial pathogenesis, 568–69
    viral pathogenesis, 590–95
pathogenicity, definition, **35**, 36
pathogenicity islands, **279**, 292, **568–69**, 590, 750
pathogen identification. *See also specific methods and organisms*
    acid-fast bacteria, 859
    API 20E (analytical profile index) strip, 854, 855t
    automated identification systems, 854, 856f
    bacitracin disk susceptibility test, 858–59